Tables for the hydraulic design of pipes, sewers and channels

Eighth edition – Volume I

Tables for the hydraulic design of pipes, sewers and channels

Eighth edition – Volume I

HR Wallingford and D. I. H. Barr

Thomas Telford, London

Published by Thomas Telford Publishing, Thomas Telford Ltd, 1 Heron Quay, London, E14 4JD. www.thomastelford.com

Distributors for Thomas Telford books are
USA: ASCE Press, 1801 Alexander Bell Drive, Reston, VA20191-4400, USA
Japan: Maruzen Co Ltd, Book Department, 3-10 Nihonbashi 2-chome, Chuo-ku, Tokyo 103
Australia: DA Books and Journals, 648 Whitehorse Road, Mitcham 3132, Victoria

First published 1963
Eighth edition 2006

A catalogue record for this book is available from the British Library

ISBN: 0 7277 3355 9

Set in Helvetica by D.I.H. Barr
Printed and bound in Great Britain by Cromwell Press, Trowbridge

Foreword to First Edition

Hydraulics Research Papers Nos 1 and 2 were published in 1958 under the titles *Resistance of fluids flowing in channels and pipes* and *Charts for the hydraulic design of channels and pipes*[1]. These dealt with the application of the Colebrook-White equation for turbulent-transitional flow in determining the discharge capacity of channels and pipes. The Wallingford Charts have achieved wide circulation, but there have been requests for the design data to be made available in tabular form.

With the collaboration of the Road Research Laboratory of the Department of Scientific and Industrial research, the present publication[2] has been prepared, as part of the programme of the Hydraulics Research Board, to meet this demand. It is hoped that it will be of particular value to civil engineers engaged on the design of urban drainage systems.

F H ALLEN
Director of Hydraulics Research

Hydraulics Research Station
Wallingford, Berks

March 1963

Foreword to Eighth Edition

The *Tables for the hydraulic design of pipes, sewers and channels* continue to provide a valuable reference for civil engineers working in the field of hydraulics.

The seventh edition[3] included the results of recent research on the hydraulic roughness of different materials and in particular on pipes with smooth internal coatings which had been undertaken by HR Wallingford with support from the Department of the Environment, Transport and the Regions. Extra material was also included to aid the calculation of temperature effects. There were modifications, now extended to Tables A, to make interpolation between entries easier, and to reduce the increments between entries in the Tables.

In this eighth edition there is extension of the treatment of part-full flows in pipes, leading on to a more direct approach to allowing for the effect of cross-sectional shape on flow in general. Also, the opportunity has been taken to revise and rearrange the introductory material on flow resistance.

It is hoped that by incorporating these changes into this new edition, the usefulness of these Tables to the industry will be enhanced.

Dr Colin Fenn
Managing Director
HR Wallingford
Wallingford, Oxfordshire

March 2005

Preface

This Eighth edition of the Wallingford Tables continues the two volume arrangement of the Sixth and Seventh[3] editions. The two volumes are designed both to be mutually supportive and to be individually free-standing in use.

The arrangements of the Sixth and Seventh editions provided a significant increase in the number of diameters treated by the established form of solution table (Tables A) for the Colebrook-White equation. This allowed coverage of sizes already associated with newer materials and planned for most pipes in the future. For this edition, the system of increments of gradient has been modified to reduce the need for interpolation, now matching that for Tables D of Volume II. Continued from the Seventh edition are the results of new work on the assessment of roughness size in commercial pipes manufactured from materials currently utilised to give a comparatively smooth finish and on the assessment of additional losses at bends in such pipes.

Volume II uses a newer, alternative, route to support the application of the unit size method. For this route, Manning equation tables (Tables D) act also as a carrier for obtaining solution of the Colebrook-White equation when combined with Tables E. For Volume II of the Seventh edition, the Manning equation tables were redone reducing the increment in gradient between entries to ease interpolation. This is continued here. As before, the coverage of discharges continues well into the order of scale of continental rivers.

In Volume II, a wide range of conduit and channel shapes is covered by tables of properties based on unit size, with key examples of these tables also included in Volume I. This gives illustration of solutions supported by the established form of Colebrook-White tables, as is possible for most conduits and smaller channels when the two volumes are used in conjunction.

In both volumes, the tables of unit properties provide aid for both gradually varied and rapidly varied flow problems. Also, there is more detailed coverage of the possible effects of variation in water temperature within the normal water resources and drainage range of temperatures.

Both volumes of this edition include a revised treatment of part-full flow in circular pipes and of the assessment of the effect of conduit shape on free-surface flow in general. The opportunity has been taken to revise and re-arrange much of the explanatory material.

The authors acknowledge the contribution of Ronald Baron, Computer Officer, Department of Civil Engineering, University of Strathclyde, to the production of the various forms of table.

Users of these Tables are invited to provide comments or corrections, particularly on conduit or channel shapes which are in common use but which are not covered. The authors are grateful for various comments which have been received already, many of which have influenced the content of this eighth edition.

Contents

(continued)

Contents (continued)

(continued)

Contents (continued)

Tables A - *Tables of Colebrook-White solutions*

Diameters 20 mm to 150 mm

Diameters 150 mm to 630 mm

Diameters 630 mm to 1250 mm

Diameters 1250 mm to 2400 mm

(continued)

Contents (continued)

Tables A (continued) *Tables of Colebrook-White solutions*

Diameters 2400 mm to 4500 mm

Annexure 1 to Tables A

Tables B

Values of proportioning exponents in equations (8), (9) and (10)

Tables C

Tables of properties of unit sections (and full proportional flow details for circular pipes only)

Annexure 2

Introductory survey

Early in the 1930's Nikuradse[4] completed his extensive experimental work on flow phenomena in smooth surfaced circular tubes. His paper[4] included an evaluation of his resistance data in logarithmic form, intended for practical use in piping design, although this had not been a prime objective of his programme. He went on to investigate flow in tubes which had been artificially roughened in a systematic manner. This led to a matching, logarithmic form, evaluation of his resistance data for rough turbulent flow[5].

At the end of the same decade Colebrook[6] published a combination of Nikuradse's[4,5] two evaluations of resistance data for pipes. The combination gave a medial evaluation of the resistance data obtained in typical pipes as installed in practice. For such pipes, this evaluation covered the full range of circumstances from smooth turbulent flow to rough turbulent flow. Thus, by its origins, it is physically compliant but also is more complex in form than other formulae for pipe flow, which are of a less physically compliant nature. The combined evaluation is known by different names, but that adopted most often is *Colebrook-White equation* which indicates the influence of C. M. White, Colebrook's collaborator and former research supervisor. Over the six decades since its inception, the Colebrook-White equation has been adopted increasingly in design; this process is still ongoing.

Both volumes of this treatment of resistance in circular pipes and other prismatic containments are centred on the use of the Colebrook-White equation. Tabular presentations are given which facilitate the numerical aspects of the design process. This volume (I) provides a comprehensive display of solutions over the range of standard pipe sizes. For Volume II there is adopted an alternative, condensed, basis for table aided solutions of the Colebrook-White equation which is carried on a tabulation of solutions of the earlier Manning[7] formula. This condensation allows coverage of hydraulic size up to that of continental rivers. It also allows inclusion of much geometric and hydraulic data for a considerable range of containment shapes. Naturally it provides also a convenient tabular approach to Manning formula solutions. In addition, it allows tabulation of solutions for other earlier resistance equations.

The two volumes are intended to be both individually free-standing and complimentary to each other. Explanatory and supporting material necessary for the effective independent use of either volume is included in that volume, and this necessitates a degree of duplication between volumes. Table 1 provides a summary of the main capacities of the table based approaches made available in the two volumes.

Review of hydraulic resistance

Nikuradse's logarithmic evaluations of smooth and rough turbulent resistance data

It is convenient to present the Nikuradse[4,5] evaluations in the re-arranged forms as follow.

TABLE 1 : Overall solution paths for uniform flow problems

Applications	**Resistance equations**	**Table sequences** (Volume shown I or II) A (I)	B (I)	C (I&II)	D (II)	E (II)	**Numerical** (Fig. 2; Table 2)
Water or sewage at normal temperature# in circular pipe, full bore flow.	C-W equation:-	●	o				
	or:-				♦	✶	
	Manning eq:-				♦		
As above but any fluid.	C-W equation:-						▲
Water or sewage in circular pipe flowing part-full. Solution for discharge or for proportional depth.	C-W equation:-	●	o	C1(a-e)			
	or:-			C1(a-e)	♦	✶	
Water or sewage in non-circular flow sections, including part-full circular pipes. All solutions covered.	C-W equation:-	●	o	■			
	or:-			■	♦	✶	
	Manning eq:-			■	♦		
As above but any fluid.	C-W equation:-			■			▲

● Tables A in Volume I give Colebrook-White solutions for circular pipes flowing full. The diameter range covered is 20 mm to 4000 mm plus 4500 mm. Use for water at standard temperature (15°C)#, with coverage of range of water temperature from 0°C to 35°C using the Annexure to Tables A.

o Tables B in Volume I give a full range of proportioning exponents for use in interpolations of Tables A, if and as required.

■ Tables C give details of geometric and hydraulic properties of cross-sections on a unit size basis. Key examples are included in Volume I and the full coverage is given in Volume II. Figure 3 outlines the combination of these tables with table aided, graphical or numerical solution of the Colebrook-White equation or the Manning equation.

♦ Tables D in Volume II allow of table-aided solution of the Manning equation, for circular pipes flowing full, in a parallel mode to that of the Colebrook-White eq. with Tables A. Use for water or sewage at normal temperature: the unqualified Manning equation may be less appropriate for smoother surfaces, but see Tables E below. The range of diameters covered is 15 mm to 65 m, with emphasis on the equivalent diameter function of the larger values.

✶ Tables E in Volume II allow selection of value of the Manning coefficient so as to give Colebrook-White solutions for 15°C from Tables D. An Annexure to Tables E gives adjustments through the range 0°C to 35°C.

▲ Figure 2 with Table 2 provides for non-tabular solution of the Colebrook-White equation for full bore flow of fluid of any viscosity. Diameter, D, and roughness size, k_s, are entered into the equations in metres. Thereby, the general method for non-circular flow sections, using Tables C, can be applied widely. Annexure 2 of Vol. II treats the Hazen-Williams formula.

Note:- The use of these tables for both gradually and rapidly varied flow assessments, and in treatment of different roughnesses around the wetted perimeter, is covered in the text.

These forms lead on directly to what is now the standard form of their combination. This involves also minor rounding of certain of the constants. As re-arranged, the Nikuradse equations describing, firstly, smooth turbulent flow[4] and, secondly, rough turbulent flow[5], as achieved in pipes roughened with uniformly graded sand grains, are

$$\frac{1}{\sqrt{\lambda}} = 2\log\left\{\frac{\mathbf{R}\sqrt{\lambda}}{2{\cdot}51}\right\} \qquad \text{(Smooth turbulent flow)} \qquad (1)$$

$$\frac{1}{\sqrt{\lambda}} = 2\log\left\{\frac{3{\cdot}71D}{k_s}\right\} \qquad \text{(Rough turbulent flow)} \qquad (2)$$

In which :

Friction factor, $\lambda = 2(Sg)D/V^2 = 1{\cdot}2337(Sg)D^5/Q^2$ (3)

Reynolds number, $\mathbf{R} = VD/\nu = 4Q/\pi\nu D$ (4)

and consequently, $\mathbf{R}\sqrt{\lambda} = \sqrt{(2Sg)}\, D^{3/2}/\nu$ (5)

Mean velocity, $V = Q/(\pi/4)D^2$ (6)

where D, k_s, Q, (Sg) and ν are diameter, roughness size, discharge, product of piezometric gradient and acceleration due to gravity and kinematic viscosity respectively.

Nikuradse's[5] tests in roughened pipes involved application of the sand grains to the semi-dry lacquered surface of the originally smooth pipes followed by overall re-lacquering. The original grading size of the grains was adopted as the sand roughness size. It was asserted by Nikuradse that the critical size of roughness was in the diametric direction and that this was not altered by the re-lacquer thickness which affected equally the grains and the boundary between them. Sufficiently close control of diameters appears to have been obtained by selection of the original pipe sizes that coincidence of relative roughness D/k_s of the test pipes in the sequence 1014, 504, 252, 120, 61·2 and 30 was normally obtained for two, and in one case three, of nominal sand grain sizes prepared by Nikuradse in the sequence 0·1, 0·2, 0·4, 0·8, and 1·6 mm, this using originally smooth pipes of 25, 50 and 100 mm nominal diameters.

The parameter in the RHS of Eq. (1), i.e. $\mathbf{R}\sqrt{\lambda}$, or $D/[\nu/\sqrt{(2SgD)}]$ is directly proportional to the parameter $D/[\nu/\sqrt{(\tau/\rho)}]$, where τ and ρ are boundary shear stress and mass density respectively. Also, the denominator of the transposed form is considered to provide a measure of the thickness of the laminar sub-layer existing between the turbulent flow and the smooth boundary. Thus Eqs (1) and (2) are fully compatible, with the change between then being substitution of a measure of roughness size, k_s, when rough turbulent flow obtains, for a measure of boundary layer thickness, $\nu/\sqrt{(\tau/\rho)}$, when smooth turbulent flow obtains. Such structure of equations for resistance in pipes and the potential for matching had been predicted by Nikuradse's supervisor Prandtl and by von Kármán. For the form of roughness adopted by Nikuradse, the value of the empirically determined proportionality quotient $k_s/[\nu/\sqrt{(2SgD)}]$ is seen to be $3{\cdot}71 \times 2{\cdot}51 = 9{\cdot}31$. Tests by Cope[8] and by Harris[9] support the validity of the concept of matching as between the smooth and the rough turbulent resistance laws (Eqs (1) and (2)).

The Colebrook-White equation

Colebrook[6] adopted a double inversion transitioning device to merge the influences of the bracketed portions of the RHS's of Eqs (1) and (2). This gave an overall value of the LHS friction coefficient $1/\sqrt{\lambda}$.

$$\frac{1}{\sqrt{\lambda}} = -2 \log \left\{ \frac{k_s}{3{\cdot}71D} + \frac{2{\cdot}51}{\boldsymbol{R}\sqrt{\lambda}} \right\} \qquad (7)$$

Then were a pipe assessed as completely smooth, i.e. $k_s = 0$, Eq. (7) reverts to Eq. (1). As the Reynolds number $\boldsymbol{R}$ becomes larger in a pipe with assessable roughness size, Eq. (7) tends in effect to Eq. (2), remaining implicit in respect of solution for $1/\sqrt{\lambda}$. With his uniform sand grain roughness, Nikuradse had found the stage of rough turbulent flow to be well defined and had used the mean values from the sets of data for pipes of different relative roughnesses k_s/D to confirm Eq. (2). Colebrook[6] demonstrated that the surfaces of commercially available pipes could be allocated an equivalent sand grain roughness size on the basis of resistance data obtained at sufficiently high values of Reynolds number that friction factor λ was independent of Reynolds number, i.e. using Eq. (2). Further that when so obtained k_s values were adopted in Eq. (7), it specified patterns of variation of value of friction factor λ with decreasing value of Reynolds number which were sufficiently close in aggregate to those obtained by experiments with a range of commercially available pipes that Eq. (7) could be adopted for general design guidance.

The most commonly reproduced graphical representation of Eq. (7) is the 1944 log-log plot by Moody[10], although the logic of the 1942 semi-log plot by Rouse[11] is compelling. Both show how Eq. (7) generates simple transitions running from individual rough turbulent evaluations to the common smooth turbulent evaluation. The individual graphical representations for selected values of relative roughness are geometrically similar to each other in the Rouse plot. This follows from its semi-log nature being in accord with the logarithmic form of Eq. (7). It follows that the Rouse plot corresponds more directly with the coalesced presentation of data and of turbulent transition routes as adopted by Nikuradse[5], Colebrook[6] and by Rouse[11], and as shown in most comprehensive texts.

The general sequence of Colebrook-White type transitions

The Colebrook-White equation (Eq. (7)) is equivalent to the following more generalised transition equation when y is made $1{\cdot}0$.

$$\frac{1}{\sqrt{\lambda}} = -2y \log \left\{ \left(\frac{k_s}{3{\cdot}71D} \right)^{1/y} + \left(\frac{2{\cdot}51}{\boldsymbol{R}\sqrt{\lambda}} \right)^{1/y} \right\} \qquad (7a)$$

In introducing Eq. (7) Colebrook[6] separated the data he had gathered from various sources into pipe material groupings. His empirically obtained trends of medial transitions for *Tar-coated cast-iron* and for *Wrought iron* correspond closely with the numerically generated transitions obtained by using Eq. (7a) for $y = 0{\cdot}9$ and for $y = 1{\cdot}2$ respectively. His *Galvanised iron* medial transition corresponds broadly with $y \approx 1{\cdot}3$. In his follow up to the Colebrook paper, Rouse[11] concentrated rather on assembling data which corresponded closely with $y = 1{\cdot}0$.

With hindsight we see that these first two verifications were complimentary in supporting Colebrook's contention that the basic transition equation (i.e. Eq. (7) or Eq. (7a) with $y = 1{\cdot}0$) gave good guidance for design procedures. The overall position has so remained following many further verifications.

Some special cases have been isolated. Included are notable examples which have followed from important and well conducted investigations. Straub, Bowers and Pilch [12] reported on extensive testing of significantly sized concrete pipes which were drawn randomly from normal stocks held by the State of Florida, and transported to the St Anthony Falls Hydraulic Laboratory in Minneapolis for testing. The tests on machine-tamped pipes were considered particularly significant and the data for these tests are well fitted by adopting $y = 0{\cdot}42$ in Eq. (7a).

There are circumstances where the transition from the smooth to the rough turbulent alignment is almost abrupt. Eq. (7a) with $y = 0{\cdot}2$ or less covers such cases. These include as diverse studies as those of Harris [9] and of Smith, Miller and Ferguson [13]. The former involved low values of Reynolds number and high values of relative roughness, and the latter involved the converse.

The foregoing and other similar examples do not modify the overall position that the use of the Colebrook-White equation is the best design option for estimation of resistance to flow in water supply and drainage generally.

Application of the Colebrook-White equation in design and tabular displays of solutions

Colebrook-White equation in design variable form

The design situation usually involves the use of the individual variables. If Eqs (3), (5) and (6) are substituted in Eq. (7), one obtains

$$\frac{V}{\sqrt{(2SgD)}} = \frac{0{\cdot}9003\,Q}{\sqrt{(Sg)}D^{2{\cdot}5}} = -2\log\left\{\frac{k_s}{3{\cdot}71D} + \frac{1{\cdot}775\,\nu}{\sqrt{(Sg)}D^{1{\cdot}5}}\right\} \quad (7b)$$

Eq. (7b) remains non-dimensional and is seen to be explicit in solution for mean velocity V or discharge Q. Solution for gradient S or for diameter D is implicit, this following from the implicit nature of the basic equation.

Direct solution equations and simplified forms

Fig. 1 (page 6) shows a selection of explicit approximations to Eqs (7b), and Eq. (7), which approximations remain fully non-dimensional. Fig. 2 (page 7) shows simplified versions which only apply for the SI units as specified and for water at 15°C and hence of kinematic viscosity $1{\cdot}141 \times 10^{-6}$ m^2s^{-1}.

However Table 2 gives correction factors for variations in viscosity as appropriate to other common fluids. Essentially Tables A, as follow, are tabulations of the simplified usage solution for discharge given in Fig. 2 (page 7), and with the complimentary velocities also listed.

Figure 1 : Colebrook-White equation and direct solution approximations

Solution for Q (or V) (i.e. Colebrook-White equation)

$$\frac{V}{\sqrt{(2SgD)}} = \frac{0{\cdot}9003Q}{\sqrt{(Sg)}D^{2\cdot5}} = -2\log\left\{\frac{k_s}{3{\cdot}71D} + \frac{1{\cdot}775\,\nu}{\sqrt{(Sg)}D^{1\cdot5}}\right\} \quad (7b)$$

Solution for S (Q as input variable for flow) (Barr[A] approximation)

$$\frac{0{\cdot}9003Q}{\sqrt{(Sg)}D^{2\cdot5}} = -1{\cdot}9\log\left\{\left(\frac{k_s}{3{\cdot}71D}\right)^{1\cdot053} + \left(\frac{4{\cdot}932\,\nu\,D}{Q}\right)^{0\cdot937}\right\} \quad (A)$$

Solution for D (Q as input variable for flow) (Pham[B] approximation)

$$\frac{0{\cdot}9003Q}{\sqrt{(Sg)}D^{2\cdot5}} = -1{\cdot}8844\log\left\{\frac{0{\cdot}365(Sg)^{0\cdot2}k_s}{Q^{0\cdot4}} + \frac{3{\cdot}55\,\nu}{Q^{0\cdot6}(Sg)^{0\cdot2}}\right\} \quad (B)$$

Alternative solution for S (Q as input variable for flow) (Marques and Sousa[C] approximation)

In their 1997 approximation, Marques and Sousa[C] adopted a rounded version of Eq. (13) to insert as an internal iteration in Eq. (7), i.e.

$$\frac{1}{\sqrt{\lambda}} = -2\log\left\{\frac{k_s}{3{\cdot}71D} - \frac{2{\cdot}51\times2}{\boldsymbol{R}}\log\left[\frac{k_s}{3{\cdot}71D} + \frac{5}{\boldsymbol{R}^{0\cdot89}}\right]\right\} \quad (7/13)$$

With the individual variables entered in the manner of Eq. (7b), this gives

$$\frac{V}{\sqrt{(2SgD)}} = \frac{0{\cdot}9003Q}{\sqrt{(Sg)}D^{2\cdot5}} = -2\log\left\{\frac{k_s}{3{\cdot}71D} - \frac{3{\cdot}943\nu D}{Q}\log\left[\frac{k_s}{3{\cdot}71D} + \left(\frac{4{\cdot}8\nu D}{Q}\right)^{0\cdot89}\right]\right\} \quad (C)$$

D is diameter of circular pipe flowing full
V is mean velocity
Q is discharge
S is piezometric gradient
g is acceleration due to gravity
ν is kinematic viscosity
k_s is equivalent sand roughness size
$\boldsymbol{R}$ is Reynolds number, VD/ν

The foregoing variables are in any coherent system of units such as pure SI (kg-m-s units)

A. BARR, D.I.H. Explicit Colebrook-White solutions. *Civil Engineering,* Sept. 1986, pp 19-31.

B. PHAM, Q.T. Explicit equations for the solution of turbulent pipe-flow problems. *Trans. Instn Chem. Engrs,* 1979, Vol. 57, pp 281-283.

C. MARQUES, J.A.A.de S. and SOUSA, J.J.O. Fórmula de Colebrook-White: velha mas actual. Soluções explícitas. *Proc. III Simpósio de hidráulica e recursos hídricos dos países de língua oficial Portuguesa,* Vol. I, Paper 37, Maputo, Moçambique, 1997.

Figure 2 : Colebrook-White equations in simplified usage mode (SU)

For SI units (D and k_s in m, Q in m^3s^{-1}); the fluid flowing in the first instance is water at 15°C (kinematic viscosity $1{\cdot}141 \times 10^{-6}$ m^2s^{-1}).

Solution for Q (based on Eq. 7b)

$$\frac{Q}{\sqrt{S}D^{2\cdot5}} = -\,6{\cdot}957 \log \left\{ \frac{k_s}{3{\cdot}71D} + \frac{(0{\cdot}647 \times 10^{-6})^{*}}{\sqrt{S}D^{1\cdot5}} \right\}$$

Solution for S (based on Eq. A of Fig. 1)

$$\frac{Q}{\sqrt{S}D^{2\cdot5}} = -\,6{\cdot}61 \log \left\{ \left(\frac{k_s}{3{\cdot}71D}\right)^{1\cdot053} + \left(\frac{(5{\cdot}63 \times 10^{-6})^{*} D}{Q}\right)^{0\cdot937} \right\}$$

Solution for D (Q given as flow variable) (based on Eq. B of Fig. 1)

$$\frac{Q}{\sqrt{S}D^{2\cdot5}} = -\,6{\cdot}555 \log \left\{ \frac{0{\cdot}576\, S^{0\cdot2}\, k_s}{Q^{0\cdot4}} + \frac{(2{\cdot}566 \times 10^{-6})^{*}}{Q^{0\cdot6}\, S^{0\cdot2}} \right\}$$

Alternative solution for S (based on Eq. C of Fig. 1)

$$\frac{Q}{\sqrt{S}D^{2\cdot5}} = -\,6{\cdot}957 \log \left\{ \frac{k_s}{3{\cdot}71D} - \frac{(4{\cdot}5\times10^{-6})^{*}D}{Q} \log\left[\frac{k_s}{3{\cdot}71D} + \left(\frac{(5{\cdot}5\times10^{-6})^{*}D}{Q}\right)^{0\cdot89} \right] \right\}$$

For other viscosities, multiply at * by the factor :-

$$\frac{\text{actual kinematic viscosity in } m^2s^{-1}}{1{\cdot}141 \times 10^{-6}\ m^2s^{-1}}$$

For other than circular pipes flowing full, D and Q are the values found for the equivalent pipe.

TABLE 2 : Values of multiplying factor for SU Colebrook-White equations

Fluid	Density, ρ (kgm^{-3})	Kin. visc., ν ($m^2s^{-1}\times10^6$)	Factor
Water (0°C)	999·9	1·787	1·566
Water (5°C)	1000·0	1·519	1·331
Water (10°C)	999·6	1·307	1·145
Water (15°C)	999·1	1·141	**1·00**
Water (20°C)	998·2	1·004	0·880
Water (25°C)	997·1	0·897	0·786
Water (30°C)	995·7	0·801	0·702
Water (35°C)	994·1	0·727	0·637
Water (40°C)	992·9	0·658	0·579
Water (50°C)	988·8	0·553	0·485
Water (60°C)	983·2	0·475	0·416
Water (70°C)	977·8	0·415	0·364
Water (80°C)	974·5	0·364	0·319
Water (90°C)	965·3	0·328	0·287
Water (100°C)	958·4	0·294	0·258
Salt water (0°C)	c 1027	c 1·923	c 1·69
Salt water (20°C)	c 1025	c 1·085	c 0·95
Salt water (40°C)	c 1019	c 0·881	c 0·77
Jet fuel (JR4)(15·6°C)	773	1·125	0·986
Paraffin oil (20°C)	800	2·375	2·08
Turpentine (26°C)	868	1·58	1·39
Kerosene (26°C)	820	2·00	1·75
Ethyl alcohol	785	1·40	1·23
Methyl alcohol	787	0·71	0·62
Propyl alcohol	800	2·40	2·10
Ether (26°C)	714	0·31	0·27
Linseed oil (26°C)	929	35·6	31·2
Crude oil (−10°C)	925	2000	1753
Crude oil (20°C)	855	74	64·9
Petrol (0°C)	c 716	0·8	0·70
Petrol (20°C)	c 716	0·59	0·52
Petrol (60°C)	c 716	0·4	0·35
Fuel oil (20°C)	940	1200	1052
Glycerin (20°C)	1258	1188	1041
SAE 10 oil (20°C)	918	89·3	78·3
SAE 30 oil (20°C)	918	479	420
Mercury (20°C)	13546	0·114	0·10
Air (Atmos. 20°C)	1·205	14·9	13·1
Hydrogen (Atmos. 20°C)	0·0839	107	93·8

Note :- SU Colebrook-White equations are given in Fig. 2. Then the above values are used as multiplier at * in these equations, for fluids other than water at standard temperature (15°C). For water, Tables A, when used in conjunction with Annexure 1, cover the temperature range 0°C-35°C.

A primary tabulation of solutions as provided in this volume
There is initial restriction to water at a standard temperature. Then it is possible within a single volume to tabulate the appropriate velocity and discharge values for a full range of standard pipe diameters. This is done for each of a wide range of roughness sizes and with listings of gradient with sufficiently small intervals for most practical applications. Tables A are the result.

An Annexure to Tables A provides tables of adjustments to solutions for the water temperature range 0°C - 35°C. However the Annexure demonstrates also that the need for significant adjustment is restricted to circumstances which are not typical of water supply and drainage practice.

Recommended roughness values
Appendix 1 provides roughness size k_s for many of the surfaces likely to be met in water supply and drainage practice. The earlier values of the set were based on findings by Colebrook[6,14], Rouse[11], King[15], Lamont[16] and others[17]. Ackers[1] showed the process of assessment of sources[11, 15-17] to give the original values. Subsequent additions include those due to Perkins and Gardiner[18] followed by Forty *et alia*[19], and those deriving from Escarameia and May[20], Escarameia[21] and others[22-25]. In some cases new to this edition it has seemed appropriate to enter values which are intermediate to those standard in the Tables A range. Then users can elect to adopt the closest standard value or to interpolate as suits their purposes.

In Appendix 1, the symbol # indicates entries which have been made, modified or confirmed following the review and experimental work undertaken at H R Wallingford during 1995 and 1996[20,21]. Also bold print indicates some values highlighted as particularly recommended for general design purposes.

Regarding the values for the roughness of a slimed sewer, this may vary considerably during any year. The normal value is that roughness which is exceeded for approximately half of the time. The poor value is that which is exceeded, generally on a continuous basis, for one month of the year. The value of k_s should be interpolated for velocities between 0·75 and 1·2 ms^{-1}.

The recommended roughness values for sewer rising mains follow from measurements on a range of pipe materials and sizes, as described in Forty *et alia*[19]. Note that the hydraulic roughness of sewer rising mains varies principally with the amount of slime that builds up inside the pipe and is normally not significantly affected by factors such as the jointing or the construction. *Primarily, the increasing roughness values are intended to cover for the loss of flow area.* In Appendix 1 the 'Normal' value represents the mean value of the measured hydraulic roughness while the 'Good' and 'Poor' values represent the values which are two standard deviations on each side of the 'Normal' value. This implies that 95% of the data lay between 'Good' and 'Poor' values.

The lists of roughness values in Appendix 1 should not be taken as absolving the engineer of the responsibility for checking the actual surface roughness achieved in particular projects by precise hydraulic tests whenever possible. Where such direct evidence is available from comparable projects it should clearly take precedence over the values quoted here. However, and as explained in following sections of this text, care must be exercised as to whether roughness sizes for closed conduits have been assessed as absolute, i.e. in round pipe flowing full if appropriate to the roughness form, or have been obtained in circumstances leading to configuration modified values. This applies in particular to proportionally larger roughnesses including those as obtain in the various forms of corrugated pipes and the like, whether annular or helical [22]. Quoted evaluations of roughness size should always be accompanied by details of the test circumstances.

Where direct evidence from tests or from past experience is not available, users may consider whether underestimation or overestimation of discharge, or of velocity, would be the more likely to be disadvantageous in their particular circumstances and choose accordingly from the alternative roughness values.

A condensed system of solutions as provided in Volume II
For this there is required much greater coverage of prismatic containment sizes and hence of discharges, but this occupying many fewer pages. This is achieved with Tables D based on the Manning [7] formula combined with Tables E to give factors for the adjustment of values obtained from Tables D so as to move from a Manning solution to a Colebrook-White solution.

Then it is appropriate also to provide comprehensive guidance on the use of Tables D to obtain free standing assessments using the Manning formula. In addition, the same basic approach as for Colebrook-White solutions can be used for additional resistance formulae. The Hazen-Williams formula is so covered in Volume II.

Further details of the tables in this volume

Tables of Colebrook-White solutions (Tables A1-A58)
There are five groupings, each showing 14 diameters. The ranges are 20 - 150 mm, 150 - 630 mm, 630 - 1250 mm, 1250 - 2400 mm and 2400 - 4000 mm plus 4500 mm, so avoiding the need for interpolation between pages in respect of diameter increments. For these five groupings in turn, each Table A relates to a particular roughness value. Experience has shown that continuous selections from the following sequence of sizes in millimetres - 0·003, 0·006, 0·015, 0.030.....15·0 allows a suitable compromise between discrimination and coverage for the diameters treated in Tables A. Larger roughnesses are included in the treatment in Volume II.

Within each table, solutions are given for the range of diameters and for ranges of gradient using 50 increments per order of magnitude change in gradient. For each of the combinations of roughness size, diameter and gradient, the mean velocity and the discharge for water at 15°C may be obtained directly.

In this volume, both standard pipe diameters and roughness sizes are stated in millimetres. Ratio values, D/k_s and the like, must be obtained with consistent units. Clearly, the foregoing ratio provides no difficulty. But, in general, diameter is used, and is operated upon, in terms of metres. This applies to its presence in non-dimensional equations, in other equations, and to its use in forming products and areas. It applies also to unit equivalent diameters as are provided in Tables C.

The Colebrook-White equation applies to turbulent flows. Velocity and discharge values are omitted from Tables A where Reynolds numbers (Eq. (4), page 3) are less than 2000. Also excluded are solutions corresponding to values of D/k_s less than about 5.

Tables of proportioning exponents (Tables B)

As a more accurate alternative to linear interpolation between table values, if this is thought needed, Tables B give proportioning exponents. The interpolation is based on the values of D/k_s and of DV for the table point which is selected as the base for the interpolation. The value of the product DV is the convenient substitute for Reynolds number where the value of kinematic viscosity is fixed. Reynolds number values are given also in Tables B.

Tables of properties of unit sections (Tables C)

The main sequence of these tables is contained in Volume II. This volume contains sample Tables C, including those for circular pipes. Examples demonstrate the use of Tables A with Tables C so that this combination can be adopted where preferred and where the size being treated, as quantified in terms of hydraulic diameter, is within the range covered by Tables A.

Circular section pipelines and sewers flowing full

Use of the Tables A

For a given roughness, the Tables relate values of pipe diameter, gradient and discharge or, alternatively, velocity. If two are specified, the Tables provide the appropriate value of the third variable.

What is the mean velocity and the discharge in a 1000 mm diameter pipe, with roughness size 0·06 mm, flowing full under a piezometric gradient of 0·000750? (Numerical solution of the Colebrook-White equation gives 1·0674 ms^{-1} and 0·8383 m^3s^{-1})

From Table A26, read the entries for the stated combination of diameter and gradient, i.e. 1·067 ms^{-1} and 0·8383 m^3s^{-1}.

Adjustment for effect of variation of temperature from standard

What are the discharges for the same basic circumstances but where the water temperature is, firstly, 0°C and, secondly, 35°C ? (Numerical solution of the Colebrook-White equation gives 0·8142 and 0·8584 m^3s^{-1} respectively)

The original solution of 0·8383 m^3s^{-1} from Table A26 was for water at the standard temperature of 15°C. Interpolating quite cursorily for gradient and diameter in page 4 of the Annexure to Tables A, one obtains multiplying factors of 0·970 and 1·025 for temperatures 0°C and 35°C respectively, giving corresponding estimates of discharge as 0·8132 and 0·8593 m^3s^{-1}.

These cases demonstrate extremes of variation in the context of water resources and drainage. The Annexure is intended both to allow such estimates to be made and to display clearly the patterns of the possible effects of variation of temperature from the standard 15°C, as these effects are influenced by pipe size, gradient and roughness size. Often these temperature effects are minor.

For estimates beyond the water resources range of temperatures one may adopt numerical solution using the equations in Fig. 2 with factors from Table 2. Alternatively, one may use *Tables for the calculation of friction in internal flows*[26]. This volume of tables has been prepared in the manner of the tables of Volume II to deal with fluid flows over a wide range of viscosities.

Interpolation between entries

If intermediate values between the entries in the Table are required then interpolation should be used. Simple linear interpolation may be sufficient in many cases.

> *In the original example, the pipe size is increased to 1025 mm and the gradient to 0·000775, roughness size remaining the same. Obtain the new flow values. (Numerical solution of the Colebrook-White equation now gives 1·1030 ms^{-1} and 0·9102 m^3s^{-1})*

On Table A26, linear proportioning between the values for 1000 mm diameter with 0·000750 gradient and 1050 mm diameter with a gradient of 0·000800 gives a velocity of 1·113 ms^{-1} and a discharge of 0·9122 m^3s^{-1}).

Multiplying factors on tabulated discharges for standard but non-tabulated diameters

Close estimates of discharges can be made quite simply, using the multiplying factors provided in Appendix 3. These were obtained using values of exponent x, as shown in the table, which relate to typical circumstances of use, of the size and type of pipe.

> *Estimate the discharge in a 1275 mm pipe with roughness size 0·30 mm, under a gradient of 0·0012. (Numerical solution gives 1·8211 m^3s^{-1})*

From Table A40 (or A28) for k_s of 0·30 mm, read 1·7292 m^3s^{-1} for 1250 mm diameter and gradient 0·0012. From Appendix 3, read factor 1·05 to adjust to 1275 mm (D_G) from 1250 mm diameter. The estimate is then 1·7292 x 1·05 = 1·816 m^3s^{-1} (cf. 1·821 m^3s^{-1}).

In cases where these approaches not sufficiently accurate, or do not cover some aspect of the variation from standard conditions, more sophisticated methods of interpolation can be adopted.

Proportioning equations

Tables B give the proportioning exponents which arise from the

Colebrook-White equation. Their use in interpolation is based on the values of D/k_s and of DV for the table point which is selected as the base for the interpolation. The value of the product DV is the convenient substitute for Reynolds number where the value of kinematic viscosity is fixed. Reynolds number values are given also.

In the proportioning equations (Eqs (8)-(10)), given values, table values and required values are indicated by subscripts G, T and R respectively.

$$Q_R = Q_T \left[\frac{D_G}{D_T} \right]^x \left[\frac{S_G}{S_T} \right]^{1/y} \left[\frac{(k_s)_T}{(k_s)_G} \right]^u \quad (8)$$

For V_R, Q_T is replaced by V_T and the exponent x is replaced by $(x-2)$

$$S_R = S_T \left[\frac{Q_G}{Q_T} \right]^y \left[\frac{D_T}{D_G} \right]^z \left[\frac{(k_s)_G}{(k_s)_T} \right]^v \quad (9)$$

$$D_R = D_T \left[\frac{Q_G}{Q_T} \right]^{1/x} \left[\frac{S_T}{S_G} \right]^{1/z} \left[\frac{(k_s)_G}{(k_s)_T} \right]^w \quad (10)$$

Intermediate values solution using proportioning exponents

For water at normal temperature, determine the discharge (Q_R) in a 560 mm diameter pipe (D_G) of 1 mm roughness size ($(k_s)_G$) under a piezometric gradient of 0·00775 (S_G) (Numerical solution gives 0·4746 m^3s^{-1}).

From Table A18, the discharge for a 525 mm diameter pipe (D_T) with 0·60 mm roughness size ($(k_s)_T$) under 0·00750 gradient (S_T) is read as 419·25 litres/sec or 0·4193 m^3s^{-1}, (i.e. Q_T), and V_T is noted as 1·937 ms^{-1}. At this table point

$$D/k_s = 0{\cdot}525 \,/\, 0{\cdot}0006 \text{ (or } 525 \,/\, 0{\cdot}60) = 875$$

$$DV = 0{\cdot}525 \times 1{\cdot}937 = 1{\cdot}017$$

Thus from Tables B1, B2 and B4, x, y and u values are read as 2·625, 1·98 and 0·115 respectively. Then substituting in Eq. (8)

$$Q_R = 0{\cdot}4193 \left[\frac{0{\cdot}560}{0{\cdot}525} \right]^{2{\cdot}625} \left[\frac{0{\cdot}00775}{0{\cdot}00750} \right]^{1/1{\cdot}98} \left[\frac{0{\cdot}6}{1{\cdot}0} \right]^{0{\cdot}11}$$

$$= 0{\cdot}4193 \times 1{\cdot}1846 \times 1{\cdot}0167 \times 0{\cdot}9429 = 0{\cdot}4762 \ m^3s^{-1}$$

For the reasons already given, including the use of Appendix 3, the need for the foregoing type of assessment in full is unlikely in practice. However, the more likely problem of the effect of a roughness size intermediate to the table values can be so assessed in isolation. Tables B display clearly the changing interactions of the design variables with changes in values of Reynolds number and of relative roughness when the physically compliant Colebrook-White equation is adopted.

Perimeters involving dissimilar roughness

The wetted perimeter of a channel or pipe may be composed of surfaces of dissimilar roughness. This occurs, for example, where sliming has taken place only below the usual water level or where the invert is of concrete and the walls of brick. Then there are two ways of computing discharges using these Tables.

The first method is based on the concept of an equivalent grain roughness for the whole perimeter. It utilises the expression

$$k_s = p_1\,k_{s1} + p_2\,k_{s2} \tag{11}$$

where p_1, p_2 denote the proportions of the total perimeter occupied by surfaces 1 and 2, and k_{s1}, k_{s2} denote the equivalent sand grain roughness of surfaces 1 and 2.

This method may be used in situations where the difference in roughness values is not excessive and where the two surfaces occupy similar proportions of the total wetted perimeter. It will also give approximate answers outside these ranges. These ranges can be defined (somewhat arbitrarily) as

$$0{\cdot}05 < k_{s1}/k_{s2} < 20 \quad \text{and} \quad 0{\cdot}33 < p_1/p_2 < 3{\cdot}0$$

This method provides a direct solution to the problem of composite roughness. The tabular system can be used without the need to resort to a successive approximation technique.

The second method is based on the concept of an equivalent friction factor for the whole wetted perimeter. It utilises the expression

$$\lambda_s = p_1\,\lambda_1 + p_2\,\lambda_2 \tag{12}$$

In Eq. (12) the suffices denote surfaces 1 and 2 as before. The method uses the following approximation[27] to the Colebrook-White equation

$$\frac{1}{\sqrt{\lambda}} = -2\log\left\{\frac{k_s}{3{\cdot}71D} + \frac{5{\cdot}1286}{\boldsymbol{R}^{0{\cdot}89}}\right\} \tag{13}$$

Later the use of an equivalent pipe diameter D_{ep} is explained, and on this basis the bracketed part of the RHS of Eq. (13) can be written

$$\left\{\frac{k_s}{3{\cdot}71D_{ep}} + \frac{5{\cdot}1286\,\nu^{0{\cdot}89}}{(D_{ep}V)^{0{\cdot}89}}\right\} \tag{13a}$$

Or, for water at normal temperature

$$\left\{\frac{k_s}{3{\cdot}71D_{ep}} + \frac{2{\cdot}63\times10^{-5}}{(D_{ep}V)^{0{\cdot}89}}\right\} \tag{13b}$$

This second method is not so restricted as the first method, but errors will occur, as follows, where there are large differences in the grain roughness of the two surfaces, say $k_{s1}/k_{s2} < 0{\cdot}01$ or $k_{s1}/k_{s2} > 100$. The method requires a successive approximation approach as follows

(i) Approximate to the composite value of k_s using $k_s = p_1 k_{s1} + p_2 k_{s2}$.

(ii) Use the appropriate tables to determine the flow, Q, and also the mean velocity, V, for the actual pipe flowing full, or for the equivalent pipe.

(iii) Determine values C_1 and C_2 of the expression (13b) by adopting k_{s1} and k_{s2} respectively.

(iv) Determine C from
$$\log C = \frac{\log C_1}{\sqrt{\{p_1 + (\log C_1 / \log C_2)^2 p_2\}}}$$

(v) Calculate the composite k_s value from
$$k_s = 3{\cdot}71\, D_{ep} \left\{ C - \frac{2{\cdot}63 \times 10^{-5}}{(D_{ep} V)^{0{\cdot}89}} \right\}$$

(vi) Use the appropriate tables to determine the flow, Q.

(vii) Repeat (iii) to (vi) until Q changes by less than the required tolerance.

These methods provide 'basic' estimates of discharge, as defined in the next section and applying there to pipes flowing part-full. By analogy, discontinuities in the flow boundary of a pipe flowing full, although lesser than between a solid boundary and the free surface, must also induce additional circulations. This suggests that a decrease from the originally calculated estimate of discharge is a prudent 'adjustment' in cases of significantly dissimilar roughness. In the absence of firm, experimental, information, some guidance can be taken from the treatment of part-full flows in pipes which follows.

Non-circular cross-sections of flow - general principles

The most common occurrences of non-circular cross-section of flow in artificial containments arise from part-full conditions in circular pipes. Then there are both full and part-full conditions in non-circular closed conduits. Conditions in artificial open conduits and canals lead on to those in natural channels. The approach presented in these Tables deals with all these circumstances from the reference base of tables of related values of variables for flow in the analogous circular containments.

The hydraulic quotient as primary measure of hydraulic size of a flow section

The hydraulic quotient is the area of the cross-section of flow, A, divided by the wetted perimeter, P, and seeks to establish the hydraulic size of a uniform flow circumstance within the active part of its containment. This linear quantity, R, is commonly known as hydraulic mean depth or as hydraulic radius, or variations on these. Hydraulic quotient is a long established and neutral terminology. For a circular pipe flowing full the value of R is $D/4$ from $(\pi/4)\,D^2/\pi D$. This value is unchanged for half depth flow because both numerator and denominator of the foregoing are halved.

The basic approach to correlation and prediction is to assume that the same mean velocity will exist in all cross-sections with the same hydraulic quotient regardless of the presence or absence of a free-surface, and with the other variables the same. With this assumption, the mean velocity V for a particular non-circular flow cross-section can be assessed as that, V_{ep}, in a circular pipe flowing full with the same hydraulic quotient R and with all other properties and constraints the same. The diameter of this conceptual equivalent pipe is $D_{ep} = 4R$.

However in practice V can differ from the 'basic' assessment, V_{ep}. We can obtain values of V_{ep} from Tables A or by other means, and may then require to adjust as appropriate. Defining the velocity ratio $V_r = V/V_{ep}$, the 'adjusted' assessment of mean velocity in the non-circular flow section is then

$$V = V_{ep} \times V_r \qquad (14)$$

It follows that the assessment of 'adjusted', i.e. actual, discharge Q through the cross-section is the product of adjusted mean velocity V and the flow cross-sectional area A, i.e.

$$Q = V_{ep} \times V_r \times A \qquad (14a)$$

Part-full circular pipes - proportional flow approach

Convenience of the use of proportional flow factors

As already indicated, circular pipes are by far the commonest artificial containments for both flowing full and flowing part-full applications. For part-full flows it is convenient to proceed from the solution for full pipe flow at the gradient to which the pipe is laid (the just full condition). For the approach provided in Table C1(a), the proportional factors are for discharge and are inclusive of the velocity ratio, as follows. Mean velocities can be evaluated from discharge and the cross-sectional area of flow.

Velocity ratio for half full flow, $V_{r(0\cdot5)}$, and extension to other proportional depths

$$V_{r(0\cdot5)} = 1\cdot0 - 70\cdot37\,[\,(\theta)^{0\cdot768} + 246\,]^{-1} \qquad (15)$$

$$\text{where}\quad \theta = \left\{ \frac{k_s}{D} + \frac{1}{416814\sqrt{S}D^{3/2}} \right\}^{-1} \qquad (15a)$$

Here θ is the transitioning element of the Colebrook-White equation. The foregoing combined evaluation is for water at 15°C, D and k_s in metres.

Eq. (15) is a solution for three selected points. These points are based, firstly, on the data analysis of Camp[28], who drew mainly from the data of Yarnell and Woodward[29] and of Wilcox[30]. This point is close to rough turbulent. Secondly and thirdly there are points representing smooth turbulent conditions from the more recent experimental findings of Sterling and Knight[31], these for 0·009 and 0·001 gradients. Primary checks came from the transition turbulent data of Ackers[32] and from fully rough turbulent data for large corrugated pipes[23]. Thus the evaluation applies to normal engineering and laboratory conditions. More detail is given in Annexure 3 of Volume II.

The parametric discharge ratio or quotient, Q_r, is defined as the basic assessment of discharge in a non-circular flow section divided by the discharge in the circular tube with the same wetted perimeter, other factors the same. The value is dependant mainly on the shape of the section, and is invariant with this if, for example, the Manning equation is adopted. With the Colebrook-White equation the value varies somewhat with the particular combination of size, gradient and roughness size. Always, the greatest value for a given wetted perimeter and given other variables is obtained for the open semi-circle. With Manning this value is $2\times 2^{2/3} = 3{\cdot}1748$.

The following equation gives an assessment of velocity ratio at proportional depth Y in a circular pipe.

$$V_{r(Y)} = 1 - \{1 - V_{r(0\cdot5)}\}\left\{\frac{(Q_{r(Y)} - 1)}{(Q_{r(0\cdot5)} - 1)}\right\} - B\sin(360Y)^\circ \qquad (16)$$

$$\text{where } B = 0{\cdot}0468 + 0{\cdot}0241\log(1 - V_{r(0\cdot5)}) \qquad (16a)$$

Apart from the last term, Eq. (16) proportions the defect in velocity ratio from one for half full by the ratio of values of $(Q_r - 1)$. Obviously Eq. (16) must give the value one for full pipe flow, as it does. There is asymmetry between the portions of the curve formed above and below half full. The value increases above one at very small proportional depths, as predicted by Pomeroy[33]. However, experimental findings suggest greater asymmetry in practice. For this, the last term of Eq. (16) provides a sinusoidal adjustment to the basic curve of variation of velocity ratio with depth to provide agreement with the findings of Camp[28] and of Sterling and Knight[31]. The sinusoidal adjustment follows indirectly from the approach of Hughes[34] to the evaluation of velocity ratio.

Eqs (15) and (16) together fit reasonably the data analysis of Camp[28] for conditions close to rough turbulent, except at smaller proportional depths, and the Colebrook-White transition data of Ackers[32]. There is good fit with the smooth turbulent data of Sterling and Knight[31].

Tabulation of proportional discharge

Eqs (14 - 16) allow the construction of Table C1(a) for proportional discharge against relative depth for a range of values of velocity ratio at the half full condition. Then to lead into this table, each column of Tables A includes the medial velocity ratio for the column.

Illustration of the proportional flow approach

Estimate the discharge at 0·66 proportional depth in a 3400 mm diameter conduit at a gradient of 0·0008. Roughness size is 1·5 mm.

It is assumed that the roughness size has been assessed for pipe full conditions. The full conduit discharge read from the 3400 mm diameter column at 0·0008 gradient on Table A53 is 16·406 m^3s^{-1}, with half-full velocity ratio $V_{r(0·5)}$ of 0·89 as medial for the column. In Table C1(a), the proportional discharge is 0·7066 giving the part-full discharge as 11·59 m^3s^{-1}. Table C1 gives unit sectional areas, i.e. 0·5499 m^2 at 0·66 proportional depth. If an estimate of mean velocity is required as well, the discharge of 11·59 m^3s^{-1} should be divided by $(0·5499 \times 3·40^2)$ m^2.

Calculation of depth of flow in part-full circular pipes

The converse problem of estimating depth, given discharge, also can be solved using Table C1(a). In theory, there may be alternative solutions above proportional depths from about 0·82 upwards, depending on the pattern of variation of velocity ratio. Conditions are likely to be unstable near and above the proportional depth for maximum discharge of about 0·92 upwards. Christensen[35] gave the following example.

Estimate the uniform depth for a discharge of 0·25 m^3s^{-1} in a 750 mm diameter pipe at a gradient of 0·00125. Roughness size is 5 mm and temperature is 10°C.

The Annexure to Tables A indicates the minimal effect of change of temperature from the standard 15°C with the stipulated conditions. Pipe full flow is estimated as follows, starting from Table A32 for roughness size k_s = 6 mm and then applying the appropriate element of Eq. (8). As 0·3188 m^3s^{-1} is read for 750 mm diameter and 0·00125 gradient, $V_{r(0·5)}$ is noted as 0·75. In this case D/k_s and DV are found to be 116 and 0·52 respectively giving u as 0·167.

$$Q_{full} = 0·3188 \left[\frac{6}{5} \right]^{0·167}$$

$$= 0·3287 \ m^3s^{-1}$$

With a proportional flow of 0·25/0·3287 = 0·761 and $V_{r(0·5)}$ as 0·75, Table C1(a) gives the proportional depth as about 0·743. Then 0·743 × 0·75 m = 0·557 m which compares with 0·49 m which is given by an iterative solution, and making the 'basic' assumption of the velocity ratio value as one.

Estimation of absolute roughness size with part-full data

In the context of these Tables, absolute resistance characteristics are those determined for the surfaces of circular pipes flowing full. In the case of part-full flow the uniform flow discharge may be known for a certain proportional depth. Then the discharge is divided by the corresponding proportional discharge from Table C1(a), using an estimated value of $V_{r(0·5)}$. This gives an estimation of pipe full discharge which is used together with diameter and gradient to search Tables A for a first estimate of roughness size, and a corresponding new estimate of $V_{r(0{\cdot}5)}$. If there is discrepancy between original and new estimates of $V_{r(0{\cdot}5)}$, the latter is used to re-estimate pipe full discharge, and so on until agreement is obtained and a final estimate of roughness size is given. Using the data of the previous example, even if the first $V_{r(0{\cdot}5)}$ estimate is as far out as 0·92, compared with 0·75, only one iteration, with approximate interpolations, is needed to get close to 5 mm for absolute roughness size.

Use of factors for temperature variation as given in Annexure 1
The estimates of $V_{r(0{\cdot}5)}$ for part-full flows, given in Tables A, assume the standard temperature of 15°C. This enables speedy predictions of part-full conditions in circular pipes at standard temperature using Table C1(a). The factors for temperature variations which are given in Annexure 1 to Tables A are for application to pipe full flow or to the equivalent pipe flow. In principle the diameter to be used in ascertaining the factor value for other than pipe full flow is the hydraulic diameter, $4R$, or diameter of the equivalent pipe, as is described in the next section. Nevertheless, the factors obtained from the Annexure 1 for the full pipe diameter can be used in conjunction with Table C1(a) to good effect. This is because the hydraulic diameter remains reasonably close to the pipe diameter down to about one third depth of flow.

To illustrate, consider the case of a 500 mm diameter pipe with roughness size 0·015 mm and a gradient of 0·00017. This is a quite smooth pipe of not insignificant size but with a relatively low value of pipe full Reynolds number, and with $V_{r(0{\cdot}5)}$ shown as 0·88 in Table A13. For these conditions, Annexure 1 shows the effect on discharge of a change of temperature from the standard temperature of 15°C to be about 5 % decrease and increase for change to 0°C and 35°C, respectively. If the pipe full factors are then applied to part-full results obtained from Table C1(a), the resulting deviations in assessment of part-full discharges with these temperatures remain within ± 0·3 % for the range down to one third proportional depth.

Where a part-full discharge at non-standard temperature is to be estimated, the proportional discharge for standard temperature, as obtained using Tables A and C1(a), is multiplied by the factor obtained from Annexure 1. If proportional depth is to be estimated with a non-standard temperature, the pipe full discharge at standard temperature is read from Tables A and is multiplied by the temperature factor from Annexure 1. The column medial value of $V_{r(0{\cdot}5)}$ is noted also. The temperature adjusted discharge value is used in forming the proportional flow quotient to apply to Table C1(a), where it is quite sufficient to use the value of $V_{r(0{\cdot}5)}$ for standard temperature as obtained from Tables A. Within the stipulated proportional depth range of 0·33 to about 0·88, the so predicted depths of flow are only slightly reduced for warmer water and slightly increased for colder water, compared with the value obtained for standard temperature.

Non-circular cross-sections of flow - general approach

Variety of shapes of containment and additional problems
Although part-full flow in circular pipes is much the most common free-surface case in man-made conduits, engineers and scientists

have concern with many other shapes of containment. The general approach organises succinctly the necessary information to deal with many conduit shapes, including part-full circular pipes. It treats also artificial and natural waterway circumstances. As with the proportional approach, it depends firstly on the core process of reference to the conceptual flow in a circular tube of the same hydraulic quotient as the cross-section in hand and with the same gradient and the same containment surface. It is re-emphasised that this process is embedded in the calculations for the 'adjusted' values in Table C1(a).

Apart from dealing with a wide range of shapes including circular pipes, the general approach extends on the scope of the proportional approach firstly by also dealing directly with assessment of gradient to achieve a stipulated depth in a given conduit or channel. Secondly it deals directly with determination of size of a given cross-sectional geometry for a stipulated duty.

First we establish the procedures for obtaining 'basic' solutions. These procedures are involved whether the roughness measure input is considered (i) to be absolute for the roughness form of solid boundary involved or (ii) to be configuration modified so as to be correct for the flow configuration of the solution. In either case the general approach utilises the concepts of hydraulic equivalence and 'unit size', as follow. This allows succinct presentation of the geometric and hydraulic properties of a wide range of conduit and channel shapes in Tables C. Then it is demonstrated, for case (i), how to move to 'adjusted' solutions which allow for the flow configuration of the solution. Alternatively in case (ii), one has considered that one has available from the start the correct value of the configuration modified roughness measure for the particular configuration of the solution. Then the procedure for obtaining the 'basic' solution gives the final result.

Hydraulic equivalence

For any steady flow in a non-circular section one can find an equivalent circular cross-section where the 'basic' flow characteristics are the same, in terms of hydraulic gradient and mean velocity, because the hydraulic quotient (hydraulic mean depth or hydraulic radius) is the same. This includes part-full flow in a circular pipe. Following Johnson[36] and Ackers[1], one can make the following comparison between a non-circular flow section and its equivalent pipe. Since

$$V = V_{ep} \tag{17}$$

$$Q/A = Q_{ep} / [(\pi/4)(4A/P)^2] \tag{17a}$$

Thus

$$Q_{ep} = Q \times 4\pi A/P^2 = Q \times J \tag{18}$$

Here A and P are cross-sectional flow area and wetted perimeter of the non-circular flow section, V and Q represent mean velocity and discharge and subscript ep indicates equivalent pipe in the sense of there being the same velocity, though not, of course, the same discharge. The discharge conversion factor $J = 4\pi A/P^2$ must apply also to the relation between flow areas, because discharge is mean velocity times flow area. This discharge conversion factor J is non-dimensional, depending on shape of flow cross-section only, and for a given overall conduit or channel shape can be listed against selected values of proportional or relative depths.

'Unit size' measures for shapes of conduits and channels
For any geometrically similar series of conduit or channel shapes, a 'unit size' can be defined in terms of a key dimension. Then, a multiplying factor *M* specifies the size of an actual example of the shape within the geometrically similar series. Also for each proportional or relative depth value, there is a 'unit' case wetted perimeter value P_u and a 'unit' case cross-sectional area of flow A_u. Hence for an actual example, MP_u and M^2A_u are the size of the wetted perimeter and of the area, respectively, at the specified proportional or relative depth. It follows that there is a 'unit' case value of hydraulic quotient (hydraulic mean depth or hydraulic radius), i.e. $R_u = A_u/P_u$ and hence a 'unit' case value of diameter of equivalent pipe, i.e. $D_{ep(u)} = 4R_u$.

Then with values of $D_{ep(u)}$ also listed against proportional or relative depths

$$D_{ep} = MD_{ep(u)} \tag{19}$$

and the third relationship for 'basic' calculations is

$$S_{ep} = S \tag{20}$$

For a 'basic' solution Eqs (18), (19) and (20) apply whatever the resistance equation to be used.

In Tables C values of linear variables $D_{ep(u)}$ and P_u etc. are quoted in metres, while the values of the other variables are quoted in metre-second units as appropriate, or are non-dimensional.

Illustrations of the 'basic' procedures

Finding discharge in a rectangular open channel
For a 'basic' solution for discharge using the general, 'unit size', method, the steps are as follows.

1) From Tables C determine, for the given shape, $D_{ep(u)}$ and *J*.
2) Determine multiplying factor *M* between 'unit' size used in Table C and real size.
3) Determine equivalent diameter of circular pipe, D_{ep} using Eq. (19).
4) Use the main Tables to determine discharge for pipe with diameter D_{ep} and given slope and roughness size, i.e. Q_{ep}.
5) Determine required discharge $Q = Q_{ep}/J$.

Estimate the uniform flow discharge at 1·20 m depth in a 2·40 m wide rectangular channel where the gradient is 0·0020 and where by inspection of the surface the roughness size is estimated as 1·50 mm. ('Basic' numerical solution 6·651 m^3s^{-1})

1) The relative depth is 0·50 and, from Table C14, $D_{ep(u)}$ and *J* are 1·00 m and 1·5708 respectively.
2) From the diagram on Table C14, the value of *M* is 2·40.

3) Then the diameter of the equivalent circular pipe, D_{ep}, is $2{\cdot}40 \times 1{\cdot}00 = 2{\cdot}40$ m, and table values for a 2400 mm pipe are appropriate.
4) From Table A42 (or A53) for k_s of 1·50 mm, the discharge for 2400 mm dia., 0·0020 gradient is 10·448 m^3s^{-1}, i.e. Q_{ep}.
5) The 'basic' estimated discharge in the channel is Q_{ep}/J, i.e. $10{\cdot}448 / 1{\cdot}5708 = 6{\cdot}651$ m^3s^{-1}, or in design terms, 6·7 m^3s^{-1}.

This solution route is shown schematically in Fig. 3 (page 23), together with the solutions for gradient and for the size of conduit or channel (of stipulated proportional depth of flow where there is a free surface). Again the demonstration of accuracy relates solely to the use of a table based procedure as opposed to a wholly computational approach. Also note that Fig. 3 includes the modification of the procedures from 'basic' to 'adjusted'. Naturally these coalesce if V_r is taken as one, i.e. the fundamental assumption of the simple application of the hydraulic quotient hypothesis.

Finding 'basic' (i) discharge, or (ii) gradient, or (iii) size where proportional depth is stipulated for an egg-shape sewer

In the case of the rectangular channel just treated, both the gradient and the diameter of the equivalent pipe corresponded to table values. Typically this is often true for gradient but rarely so when equivalent diameter is the unknown. To illustrate solution procedures more generally, consider a Form 1 egg-shape (see Table C2). The overall height is 1·8 m, so M is 1·8. The depth of flow is 0·540 m (i.e. a proportional depth (Y_N) of 0·30) and from inspection of the surface of the old brick construction, still well aligned, the absolute roughness size is estimated as 3 mm (k_s). At gradient 0·0010 (S), the 'basic' numerical solution for discharge is 0·2875 m^3s^{-1}.

For 0·30 proportional depth, Table C2 gives the values of $D_{ep(u)}$ and J as 0·5123 m and 1·9042 respectively. Fig. 3 shows that both values are needed for each of the three examples as follows.

(i) To estimate the 'basic' discharge Q, given gradient, size of conduit, roughness size and proportional depth

Substituting in Eq. (19)

$$D_{ep} = 1{\cdot}8 \times 0{\cdot}5123 = 0{\cdot}9221 \text{ m}$$

$$S_{ep} = 0{\cdot}0010 \text{ (given)}$$

Here, the stipulated gradient corresponds to a table value, on the Colebrook-White solution tables, as does the roughness size. Thus the nearest table value on Table A31 for k_s of 3 mm is the discharge for 900 mm diameter and 0·0010 gradient. This is 0·5131 m^3s^{-1}, and at this table point $D/k_s = 300$ and $DV = 0{\cdot}743$, giving (on Table B1) the exponent x as 2·647.

$$Q_{ep} = 0{\cdot}5131 \times [0{\cdot}9221 / 0{\cdot}900]^{2{\cdot}647} \quad \text{(i.e. from Eq. (8))}$$

$$= 0{\cdot}5471 \text{ m}^3\text{s}^{-1} = QJ \quad \text{(i.e. from Eq. (18))}$$

Then

$$Q = 0{\cdot}5471 / 1{\cdot}9042 = 0{\cdot}2873 \text{ m}^3\text{s}^{-1}$$

Figure 3 : Solution routes for uniform flow in non-circular cross-sections

Either we calculate in terms of an absolute measure of the roughness of the containing surface, as we believe would be obtained in tests in a circular tube flowing full, or we assume from past experience that a configuration modified measure can be estimated for the point of solution. The following solution route diagrams are based on the use of an equivalent sand roughness size k_s being an input. For the three explicit problems of finding (i) discharge (Q), (ii) friction gradient (S_f) or (iii) size (factor M), the proportional or the relative depth (Y_N) is known. Hence the values of equivalent diameter for the unit case ($D_{ep(u)}$) and of the equivalent discharge factor (J) can be read from the appropriate Table C. Similarly, and assuming an absolute measure of roughness size is to be used, an assessment of velocity ratio, i.e. $V_r = V_{adjusted}/V_{basic}$, is given, but we must vary this last figure if we consider the circumstances warrant a change. At a given proportional or relative depth and hence fixed cross-sectional area, the velocity ratio must also apply to discharge. It is emphasised that in the circumstance of knowledge of the value of velocity ration, 'basic' relates to the idealised condition of full compliance with the hydraulic mean depth or hydraulic radius hypothesis, while 'adjusted' indicates the actual flow condition, which often differs from the 'basic'. Chow[24] has used the terms 'mathematical' and 'practical' for 'basic' and 'adjusted' respectively.

In either approach, the first step is to find the basic value for the quantity to be estimated, as shown in the diagrams. This completes the process if the roughness measure is configuration modified.

Thus (i) *Find Q*:

$M \times D_{ep(u)} = D_{ep}$ ↘ ; k_s (either absolute or configuration modified) ↓

$S_f = S_{ep}$ ↗

[C-W] → $Q_{ep} = Q_{basic} \times J$; Hence Q_{basic}

and then $Q_{adjusted} = Q_{basic} \times V_r$

(ii) *Find S_f* :

$M \times D_{ep(u)} = D_{ep}$ ↘ ; k_s (either absolute or configuration modified) ↓

$Q \times J = Q_{ep}$ ↗

[C-W] → $S_{ep} = S_{f(basic)}$

and then $S_{f(adjusted)} \approx S_{f(basic)} \div V_r^2$

(iii) *Find size (i.e. factor M)* :

$Q \times J = Q_{ep}$ ↘ ; k_s (either absolute or configuration modified) ↓

$S_f = S_{ep}$ ↗

[C-W] → $D_{ep} = M_{basic} \times D_{ep(u)}$; Hence M_{basic}

and then $M_{adjusted} \approx M_{basic} \div V_r^{0.38}$

Where the roughness measure is applied as configuration modified, the'basic' solution is the result and there is no need for the operations involving the velocity ratio. This will apply in many if not most cases of open conduits and natural channels.

For the inherently implicit problem of finding (iv) normal depth (y_N) in a channel of known shape and size, find Q_s by route (i) above, where Q_s corresponds to that proportional, or that relative, depth which is specified for the medial condition on the appropriate Table C, and using the gradient and roughness size stipulated for the problem, either as an absolute or as a configuration modified value as appropriate. Evaluate Q/Q_s where Q is the stipulated discharge.

Then (iv) *Find y_N* : Q/Q_s → value of Y_N on the table of sequence C, using the 'adjusted' column if the roughness size is considered absolute, and the 'basic' column for the simpler procedure where the roughness measure is configuration modified. Hence $y_N = M \times Y_N$

Note : The Colebrook-White element of a solution, is indicated by the block containing 'C-W'. This may be aided by Tables, by Charts or be accomplished by solution of an equation.

(ii) To estimate gradient S, now given that Q = 0·2875 m^3s^{-1}

$$D_{ep} = 0{\cdot}9221 \text{ m (as before)}$$

$$Q_{ep} = 0{\cdot}2875 \times J = 0{\cdot}5475 \text{ m}^3\text{s}^{-1}$$

The discharge value on Table A31, for 900 mm diameter, 3 mm roughness size and gradient 0·00115, is 0·5503 m^3s^{-1}.

The discharge on the table chosen is that closest to 0·5475 m^3s^{-1} but this is arbitrary and the final result would be effectively the same with another choice. At the chosen table point D/k_s remains as 300 and *DV* is 0·798 giving *y* and *z* as 1·992 and 5·28 from Tables B2 and B3 respectively. Then substituting in Eq. (9)

$$S_{ep} = 0{\cdot}00115 \left[\frac{0{\cdot}5475}{0{\cdot}5503} \right]^{1{\cdot}992} \left[\frac{0{\cdot}900}{0{\cdot}9221} \right]^{5{\cdot}28}$$

$$= 0{\cdot}00100 = S$$

(iii) To find size of Form 1 egg-shape, k_s = 3 mm, which runs at 0·30 proportional depth with Q = 0·2875 m^3s^{-1} and S = 0·001

$$Q_{ep} = 0{\cdot}5475 \text{ m}^3\text{s}^{-1} \text{ (as before)}$$

$$S_{ep} = S = 0{\cdot}0010$$

The table point is that for 900 mm diameter (and gradient 0·0010) in Table A31 for roughness size of 3 mm. This shows a discharge of 0·5131 m^3s^{-1}. Then D/k_s and *DV* are 300 and 0·743 respectively, with the *x* value remaining as 2·647.

Substituting as relevant in Eq. (10)

$$D_{ep} = 0{\cdot}900 \times [0{\cdot}5475 / 0{\cdot}5131]^{1/2{\cdot}647} = 0{\cdot}9223 \text{ m}$$

However

$$D_{ep} = M\, D_{ep(u)} = M \times 0{\cdot}5123$$

This gives scale factor *M* as 1·800 on the basis of 1·000 m unit height of egg-shape. Thus the required egg-shape is estimated as 1·800 m (1800 mm) in overall height.

Finding 'basic' depth of flow in a conduit of specified boundary shape and size, given discharge, gradient and roughness size. Consider again the egg-shape sewer example. We wish to find the depth corresponding to a discharge of 0·2875 m^3s^{-1}.

Included in each Table C are 'basic' discharge ratios for representative conditions, these based on the just full discharge for closed conduits and on a selected proportional or relative depth for open conduits and channels. The conditions on which these ratios are based are given in the diagrams. In the case of Tables C2 and C14, the values are those of the foregoing examples.

Using the unit size method for estimation of discharge, as already illustrated, the 'basic' just full discharge in this particular conduit is estimated as 1·7625 m^3s^{-1}. Thus the discharge of 0·2875 m^3s^{-1} as stipulated gives a $Q/Q_{1\cdot00}$ ratio of 0·1631. In the 'basic' discharge ratio column of Table C2, this corresponds almost exactly to a proportional depth of 0·30 and hence to a depth of flow of 1·80 × 0·30 = 0·540 m. Fig. 3 (page 23) includes an abstract of the foregoing route for determination of normal depth.

Note that for changes in conduit size, gradient or roughness size, or for changes in temperature within the normal range encountered in drainage practice, the corresponding changes in 'basic' discharge ratios are small. Also that the invariant discharge ratios which could be calculated using the Manning formula are very similar. In contrast, the 'adjusted' discharge ratios are for the particular case, and must be taken as for guidance only for other cases, together with examination of Table C1(a). Also, the foregoing examples demonstrate the minimal nature of the discrepancies which arise solely from the adoption of table based procedures.

At this stage and as already emphasised the 'basic' solution obtained may be considered final if the roughness measure adopted is considered to be sufficiently appropriate for the configuration of the solution. Specifically here this would be with 3·0 mm taken as a suitable configuration modified roughness size for a better conditioned egg-shape than earlier posed. In some cases discrepancies may arise from the use of a single configuration modified roughness measure over the main zone of free-surface conditions to allow the 'basic' results so obtained be considered final. The potential occurrence of such discrepancies is common to all formulae and procedures where the roughness measures adopted are not absolute. This is demonstrated further in a following section where the 'adjustment' of 'basic' results is illustrated, following on from the adoption of absolute roughness measures.

Use of factors for temperature variation as given in Annexure 1
For flows with non-circular cross-sections, the multiplying factors on discharge (and velocity) given in the Annexure to Tables A apply directly to the flows in the equivalent pipe, and the values of the factors are defined by the diameter of the equivalent pipe D_{ep}, and the other parameters of the Annexure tables. Thus to estimate a discharge with a temperature other than the standard 15°C, the discharge for 15°C is found and the appropriate multiplying factor on discharge is then applied, using the relevant D_{ep} and gradient in evaluating the factor. It can be seen from Annexure 1 that the factors are comparatively insensitive to changes in gradient or (equivalent) diameter. To estimate either gradient or size, there should be a trial solution at standard temperature to allow the factor to be estimated from the Annexure, and the discharge *divided* by this factor. Then the required solution for gradient or for size is obtained using the thereby adjusted discharge.

Obtaining 'adjusted' solutions as appropriate

In using the Colebrook-White equation, one may have pre-determined the value of absolute roughness size k_s by inspection of the surface or, more probably, by knowledge of the testing of a sufficiently similar surface as the boundary of a circular pipe flowing full, and of generally similar value of hydraulic quotient. First we assume that this applies to the roughnesses adopted in the preceding examples.

In the case of the rectangular channel the velocity ratio V_r at relative depth Y of 0·50 is read from Table C14 as 0·946. The 'adjusted' solution is the product of the previously obtained 'basic' solution and this value of the ratio, i.e. $6{\cdot}651 \times 0{\cdot}961 = 6{\cdot}392$ m^3s^{-1}, or in design terms, 6·4 m^3s^{-1}.

The methods of generation of values of velocity ratio for Tables C are detailed in Volume II. The main purpose here is to demonstrate the use of Tables A in combination with Tables C.

For the assessment of 'adjusted' discharge in the egg-shape sewer the velocity ratio given in Table C2 for the stipulated size, gradient and roughness size and for a proportional depth Y of 0·30 is 0·803. The 'adjusted' solution is again the product of the 'basic' solution and the ratio value, i.e. $0{\cdot}2875 \times 0{\cdot}803 = 0{\cdot}2308$ m^3s^{-1}.

Discharge is an input for the two other explicit problems where proportional depth is stipulated. From Tables B1 and B2 it follows that 'adjusted' values of gradient and of size are obtained from the 'basic' values by division by V_r^2 and $V_r^{1/2{\cdot}65}$ respectively. In the egg-shape sewer example this gives 'adjusted' gradient as 0·00156 and size as 1·955 m, compared with the 'basic' assessments of 0·0010 and 1·80 m respectively. The exponents here applied to V_r are estimated from Tables B1 and B2 as typical for likely design conditions, and more or less coincide with the inbuilt exponents from the Manning formula.

The 'adjusted' just full discharge is estimated as 1·7731 m^3s^{-1} (the 'basic' estimate multiplied by the flowing full value of V_r from Table C2, i.e. 1·006), giving the required 'adjusted' $Q/Q_{1{\cdot}00}$ ratio as 0·162. Then the column for the 'adjusted' $Q/Q_{1{\cdot}00}$ ratio gives a proportional depth 0·335, compared with 0·30 for the 'basic' solution.

Configuration modified measures of surface resistance

There can be calculated the input value of roughness size k_s which is needed to obtain by 'basic' solutions the 'adjusted' results at 0·30 proportional depth. This is 12·5 mm compared with the assumed absolute size of 3·0 mm. While the same final results would be obtained for each of the four specific problems treated, the intermediate stage in the fourth problem gives only 1·4277 m^3s^{-1} as the evaluation of just full flow, compared with 1·7625 m^3s^{-1} with 3 mm roughness. Calculation for any pressurised pipe flow will be affected also. Clearly some discrepancy in estimates of part-full conditions may occur at any proportional depth other than that for which the configuration modified roughness size is intended. This is discussed further in Volume II under Review and in Annexure 3.

Other sources of resistance

When designing a pipe system on the basis of its full-bore capacity, allowances must be made for head losses which will occur at bends, manholes or other appurtenances involving changes in cross-section. Similar allowances should also be made under conditions of free-surface flow. Hydraulic handbooks, manuals and other technical publications provide two approaches to assessment of additional losses. Firstly, the length of pipe used in the flow calculations can be increased simply according to the nature and number of the deviations from uniform flow in a straight pipe. Details are given in Appendix 2, list (a). In principle, this method would be at its best when the flow is rough turbulent, but the values given are as thought medial for smoother pipes where applications more often arise.

Alternatively, there can be adopted head-loss coefficients, $\varepsilon_1, \varepsilon_2, \ldots$ for use in the expressions $h_1 = \varepsilon_1 V^2/2g$ etc., with $h_1, h_2, \ldots$ contributing cumulatively to the total head difference Σh necessary to drive the system. Here $V^2/2g$ is the kinetic head based simply on the mean velocity. Standard values for ε for pipe full flows are given also in Appendix 2, list (b).

Head losses at straight through open-channel manholes are generally small, the head loss coefficients being of the following approximate magnitudes.

	Part-full	*Full-bore*
Open-channel manhole	< 0·1	0·05 - 0·25
Open-channel manhole with bend	~ 0·3	~ 1·5
Open-channel manhole with pipe bend beyond manhole	~ 0·3	~ 0·3

If a manhole incorporates a junction, losses are much higher and depend on the relative magnitudes of the branches and the geometry of the junction. More detailed guidance is given by Chow[24], in Manuals[37-39], especially the last of these, and by Hager[40].

There is some loss of head at the entry to a pipe or conduit, which will depend upon the sharpness of the arris. Also, the kinetic energy of flow $\alpha(V^2/2g)$, where α is the Coriolis coefficient, is generally not recoverable at the exit. These factors may well prove important if the pipe or conduit is relatively short.

As indicated in Appendix 2, list (b), the head loss with a sharp-edged re-entrant inlet is approximately $V^2/2g$. With a flush headwall the loss coefficient drops to about 0·4, whilst a rounding as little as a sixth or seventh of the pipe diameter will almost eliminate the entrance loss, but not the need to allow for a surface elevation including kinetic head. Much information on losses at features and bends is given by Miller[41], by White[42] and by Fried and Idelchik[43].

Calculating with additional head losses present

In calculation, the equivalent additional length of pipe can come either from summation of diameter factors from such as Appendix 2, list (a), or from such as Appendix 2, list (b), as follows. From the original definition of the Darcy-Weisbach friction factor λ, the value of its reciprocal, i.e. $1/\lambda$, is the number of diameters to give the length of pipe in which accrue losses equivalent to the kinetic head $V^2/2g$. Consequently, if an additional loss is expressed as $\varepsilon\,(V^2/2g)$, ε/λ is the number of pipe diameters to give the equivalent length for the feature. This is constant in rough turbulent flow but otherwise varies with Reynolds number.

For selected values of D, S and k_s, the value of $1/\lambda$ can be found using the Tables as $0{\cdot}051(V^2/SD)$ for metre-second units, or the length in metres is found as $0{\cdot}051(V^2/S)$. This allows direct estimation of the required head for a selected discharge through a pipe system, by adding the length $\Sigma\varepsilon \times [0{\cdot}051(V^2/S)]$ to the actual pipe length before multiplying by the assessed gradient to obtain the required total head difference. It is re-iterated that information provided directly in terms of equivalent length of pipe may not sufficiently allow for the effect of variation in Reynolds number in smooth pipe flows.

Alternatively, values of the coefficient ε_f, to allow for the pipe flow resistance element in the following equation, are assessed as $SL/(V^2/2g)$, where L is the pipe length. This approach is preferable where discharge is required, especially in smooth pipe systems.

$$\Sigma\, h = (\,\varepsilon_1 + \varepsilon_2 \ldots + \varepsilon_f\,)\,(V^2/2g) \qquad (21)$$

Then using the Tables to establish successive trial values of V, closure is very rapid.

A 200 mm pipe, roughness size 0·03 mm, runs for 100 m between two tanks where the surface levels differs by 7·5 m. Allowing for entry, exit, valve and bend effects, the combined additional loss coefficient ε_c totals as 4·0. Estimate the discharge (Numerical solution using the Colebrook-White equation gives 0·1133 m^3s^{-1} with the pipe resistance gradient S_f at 0·04848).

The assessments can be tabulated as follows, using Table A14.

S_f (trial)	V (by Tables)	$\varepsilon_f = \dfrac{(2gL)S}{V^2}$	$\Sigma\,\varepsilon$ $(\varepsilon_c+\varepsilon_f)$	$V = \sqrt{\left\{\dfrac{2g\times\Sigma h}{\Sigma\,\varepsilon}\right\}}$ (check)
0·07500	4·529	7·171	11·171	3·628
[adjust gradient - $0{\cdot}07500 \times (3{\cdot}628\,/\,4{\cdot}529)^2 = 0{\cdot}0481$]				
0·04800	3·588	7·312	11·312	3·606
[adjust gradient - $0{\cdot}04800 \times (3{\cdot}606\,/\,3{\cdot}588)^2 = 0{\cdot}04848$]				
0·04848	3·606	7·313	11·313	3·606

The third assessment requires the use of an interpolating exponent from Table B2, but would be unnecessary in practice. From the second assessment the average of the velocities is 3·596ms^{-1}, which gives a discharge of 0·1130 m^3s^{-1} (cf. 0·1133 m^3s^{-1}).

In short culverts and conduits at hydraulically steep slopes, separation may occur at the inlet if it is square-edged. Then the inlet acts as a controlling section and precludes full-bore operation. It cannot be overemphasised that for very short culverts and conduits, friction loss may not be significant. Of primary concern is the relationship of inlet and outlet water levels to surrounding features, with the conduit tending to act as an orifice. For intermediate cases, these Tables may aid assessments of both gradually varied and rapidly varied flow conditions, as is shown in following sections.

Checks on mean velocity, Reynolds number and Froude number

With the tabular method now introduced (i.e. for solution for discharge, for gradient or for size), mean velocity in a section is the same as that for the flow in the equivalent pipe. Thus mean velocities may be read directly from the Colebrook-White solution tables. Reynolds number is DV/ν where the variables D, V and ν are in consistent units. The value is then given by $DV/(1{\cdot}141\times 10^{-6})$ for water at normal temperature (15°C) with kinematic viscosity value $1{\cdot}141\times 10^{-6}$ m^{2}s^{-1}, where D is the diameter for pipe-full flows and is the diameter of the equivalent pipe for cases other than pipe full. The Froude number is $V/\sqrt{(gy_{mean})}$ where mean depth y_{mean} is the cross-sectional area of flow divided by the free surface width. This can be evaluated using the Tables C.

Viscosities other than that of water at 15°C

The Annexure to Tables A provides a direct and comprehensive guide to the factors to be applied to velocity and discharge to allow for the effect of change in temperature of water. This treatment extends over the normal range of variation of water temperature in the civil engineering context. Outside this range, or for other fluids, the equations of Fig. 2 can be used, or the *Tables for the calculation of friction in internal flows*[26]. Then one is no longer using these Tables to determine the Colebrook-White element of the solution. However, it does allow Tables C be applied to the solution of flow problems involving fluids of any viscosity, provided that checks are made that the Reynolds number, in terms of D_{ep}, remains above 2000.

Critical depth and critical discharge

Tables C include values of critical discharge for the unit case, $Q_{c(u)}$, this being the product of unit area and the critical velocity for unit mean depth. Division of discharge in a conduit or channel by $M^{2{\cdot}5}$ gives the corresponding value of unit critical discharge. Hence, the corresponding proportional or relative depth is defined on the table and then multiplication of this by M gives the critical depth.

Alternatively, in a given shape of channel for which the scale factor M is known, division of a selected depth by M gives the proportional or relative depth. Then the corresponding unit critical discharge can be read from the appropriate table and multiplied by $M^{2{\cdot}5}$ to obtain the critical discharge for the chosen depth.

Gradually varied flow in prismatic channels

For gradually varied flows, the relative values of normal and of critical depths control the occurrence of flow profiles. This is described in appropriate texts. Then profile details can be estimated using the direct step method, adapted for use with Tables C.

This level of application fits well with the use of a brief routine on a programable hand calculator, but even this is not essential. With or without the use of such a routine, profile calculations to be undertaken immediately for any shape of conduit or channel for which a Table C is available, using Eq. (22) as follows. The increments of unit proportional or relative depth, as adopted for Tables C, are likely to be suitable as the basis for a series of steps.

Then for each successive step of that series, the only new information needed is the new end-of-step values of unit cross-sectional area A_u (subscript 2) and of unit wetted perimeter P_u. These are operated on by M^2 and M respectively, as is unit Δy. Here it is assumed that the brief routine includes transfer of the end-of-step values for one step to be start of step values (subscript 1) for the next.

$$\Delta x = \left[\Delta y + \frac{\alpha}{2g} (V_2^{\,2} - V_1^{\,2}) \right] \Big/ (S_o - S_f) \qquad (22)$$

Here Δx is the resulting distance along the conduit or channel, positive indicating the direction of flow; Δy is the direct step value, M times Δy_u, or $M \times (Y_2 - Y_1)$; α is the Coriolis coefficient; g is acceleration due to gravity; V is mean velocity, $Q/M^2 A_u$, with subscripts 1 and 2 corresponding to successive end of step positions; S_o is channel gradient, positive when sustaining flow; S_f is the friction gradient, which can be assessed using the second equation of Fig. 2. This equation re-arranges as follows.

$$S_f = \left\{ -\,6{\cdot}61\,(D^{2{\cdot}5}/Q)\,\log\left\{ \left(\frac{k_s}{3{\cdot}71D}\right)^{1{\cdot}053} + \left(\frac{(5.63\times10^{-6})\,D}{Q}\right)^{0{\cdot}937} \right\} \right\}^{-2} \qquad (23)$$

In comparing the procedure outlined here with existing solutions, the average value of unit area and wetted perimeter at the beginning and end of each step has been used before applying M^2 and M, respectively.

Solution for gradually varied flow in a circular pipe

Consider the application of the system to the following adaption of a problem treated by Chow[24].

> *An 1800 mm reinforced-concrete pipe culvert, 100 m long, is laid on a bed slope of 0·02 with a free outlet. Compute the flow profile if the culvert discharges 6·90 m^3s^{-1}. Take the roughness size k_s as 0·60 mm (as considered appropriate for the zone of proportional depth of specific relevance) and the Coriolis coefficient α as 1·0.*

Unit critical discharge is $6{\cdot}9/1{\cdot}8^{2{\cdot}5} = 1{\cdot}5873$ m^3s^{-1}. On Table C1, this corresponds to a proportional depth of 0·727, giving critical depth as 1·309 m. To find normal depth, $Q_{1{\cdot}00}$ is required. From Table A41 for roughness size k_s = 0.60 mm, an 1800 mm pipe at gradient 0·020 carries a discharge of 17·266 m^3s^{-1}.

The configuration modified roughness size has been selected as representative for the zone of proportional depth of specific interest. Then a sufficiently accurate estimate of normal depth for a discharge of 6·9 m^3s^{-1} can be obtained by applying the so calculated discharge ratio

$$6{\cdot}9/17{\cdot}266 = 0{\cdot}399$$

to the $V_{r(0{\cdot}5)} = 1{\cdot}00$ column in Table C1(a). This gives a proportional depth value just under 0·44 and hence an estimate of normal depth as

$$0{\cdot}44 \times 1{\cdot}8 \text{ m} = 0{\cdot}792 \text{ m}.$$

Thus, with critical depth established as 1·309 m, flow passes through critical depth at the inlet to the pipe and is then supercritical through the pipe with a profile of the S2 form.

A calculator aided solution is given in Table 3, (page 32) showing a proportional depth of about 0·462 to be reached at the end (100 m) of the pipe. For this illustration, there is given a full extract of the intermediate values of variables. Eq. (23) was included in the program. Such a basic routine for use with Tables C, once prepared, equally can be applied to any of the shapes treated, as provided extensively in Volume II.

Note that the surface levels for the varied flow assessed remained clearly within the main interest zone for part-full flow in a circular pipe. Then use of a single configuration modified roughness size for medial proportional depths is adequate for the demonstration in hand.

Rapidly varied flow

Tables C provide values of depth to centroid for unit size $y_{d(u)}$. This allows calculations for conditions involving the possible formation of a hydraulic jump.

There is a discharge of 3.38 m^3s^{-1} at a depth of 0·63 m in a horizontal 1800 mm pipe. Establish the conjugate depth.

The value of $M^{2{\cdot}5}$ is 4·347, giving Q_u as 0·778 m^3s^{-1}. A depth of 0·63 m gives a proportional depth of 0·35, and on Table C1 this shows $Q_{c(u)}$ as 0·3888 m^3s^{-1}, and thus flow to be supercritical (with a Froude number of 2·0).

Following Daugherty and Franzini[44], $(Q^2/Ag) + Ay_d$ is constant between conjugate depths, where Q is discharge, A is cross-sectional area of flow and y_d is depth to centroid of the cross-section. But under free surface conditions, $Q = M^{2{\cdot}5}Q_u$, $A = M^2A_u$ and $y_d = My_{d(u)}$. Thus

$$M^3\,[(Q_u{}^2/A_ug) + A_uy_{d(u)}\,] \qquad (24)$$

is also constant, and the operation of finding conjugate depth can be done in terms of table values for unit size. In this case, the value within the square bracket of Eq. (24) for the supercritical conditions is 0·2873 and this value is also found at a proportional depth just over 0·70.

The upper conjugate depth is then about 1·26 m. Thus a hydraulic jump could be formed in the pipe, with free-surface subcritical flow

TABLE 3 : Computation of S2 flow profile in a circular pipe

Given : Pipe diameter 1800 mm, i.e. multiplying factor $M = 1{\cdot}80$; Discharge $Q = 6{\cdot}9\,m^3s^{-1}$; Gradient $S_o = 0{\cdot}02000$; Roughness size $k_s = 0{\cdot}60$ mm ; Coriolis coefficient $\alpha = 1{\cdot}0$; Normal depth $y_N = 0{\cdot}792$ m and Critical depth $y_c = 1{\cdot}309$ m.

Y	Δy_u (m)	P_u (m)	A_u (m^2)	Δy (m)	ΔK (m)	J (calc.)	D_{ep} (m)	Q_{ep} (m^3s^{-1})	S_f	S_o-S_f	Δx (m)	x (m)	y (m)
0·727	--	2·0422	0·6116	--	--	--	--	--	--	--	--	--	1·3086
0·72	– 0·007	2·0264	0·6054	– 0·0126	0·01273	1·8477	2·1537	12·749	$4{\cdot}30\times10^{-3}$	0·01570	0·0081	0·0081	1·296
0·70	– 0·02	1·9823	0·5872	– 0·036	0·03972	1·8652	2·1420	12·870	$4{\cdot}51\times10^{-3}$	0·01550	0·2398	0·2479	1·260
0·68	– 0·02	1·9391	0·5687	– 0·036	0·04434	1·8892	2·1223	13·035	$4{\cdot}85\times10^{-3}$	0·01515	0·5506	0·7985	1·224
0·66	– 0·02	1·8965	0·5499	– 0·036	0·04972	1·9109	2·0998	13·185	$5{\cdot}24\times10^{-3}$	0·01476	0·9299	1·7284	1·188
0·64	– 0·02	1·8546	0·5308	– 0·036	0·05602	1·9303	2·0743	13·319	$5{\cdot}70\times10^{-3}$	0·01430	1·4002	3·1286	1·152
0·62	– 0·02	1·8132	0·5115	– 0·036	0·06310	1·9472	2·0461	13·436	$6{\cdot}23\times10^{-3}$	0·01377	1·9681	5·0966	1·116
0·60	– 0·02	1·7722	0·4920	– 0·036	0·07145	1·9619	2·0152	13·537	$6{\cdot}84\times10^{-3}$	0·01316	2·6940	7·7906	1·080
0·58	– 0·02	1·7315	0·4724	– 0·036	0·08091	1·9744	1·9818	13·642	$7{\cdot}56\times10^{-3}$	0·01244	3·6096	11·4003	1·044
0·56	– 0·02	1·6911	0·4526	– 0·036	0·09264	1·9846	1·9459	13·694	$8{\cdot}40\times10^{-3}$	0·01160	4·8817	16·2819	1·008
0·54	– 0·02	1·6509	0·4327	– 0·036	0·10622	1·9921	1·9073	13·746	$9{\cdot}39\times10^{-3}$	0·01061	6·6176	22·8996	0·972
0·52	– 0·02	1·6108	0·4127	– 0·036	0·12261	1·9972	1·8662	13·780	$10{\cdot}57\times10^{-3}$	0·00943	9·1832	32·0827	0·936
0·50	– 0·02	1·5708	0·3927	– 0·036	0·14181	1·9997	1·8226	13·798	$11{\cdot}98\times10^{-3}$	0·00802	13·1933	45·2761	0·900
0·49	– 0·01	1·5508	0·3827	– 0·018	0·07939	1·9999	1·7885	13·799	$13{\cdot}22\times10^{-3}$	0·00678	9·0751	54·3332	0·882
0·48	– 0·01	1·5308	0·3727	– 0·018	0·08586	1·9992	1·7650	13·795	$14{\cdot}16\times10^{-3}$	0·00585	11·6110	65·9442	0·864
0·47	– 0·01	1·5108	0·3627	– 0·018	0·09306	1·9978	1·7408	13·785	$15{\cdot}19\times10^{-3}$	0·00481	15·5905	81·5346	0·846
0·46	– 0·01	1·4907	0·3527	– 0·018	0·10109	1·9958	1·7161	13·771	$16{\cdot}33\times10^{-3}$	0·00367	22·6105	104·1451	0·828

Notes : (a) For assessment of change in kinetic head, ΔK, velocities are Q/M^2A_u.

(b) Assessments of J, D_{ep}, Q_{ep} and S_f are for mid-step averaged values of P_u and A_u. The latter values are read directly from Table C1, except for the starting values. Here the linear interpolation values have been assessed from the adjoining table values.

downstream, this starting from a proportional depth of 0·35 at a Froude number of about 2. The same does not hold for the case already treated for gradually varied flow, where towards the downstream end of the pipe there obtains the combination of proportional depth of about 0·46 and Froude number of 2·42. Then the within bracket value of 0·7969 for the supercritical flow is not re-attainable under subcritical conditions within the confines of the circular pipe.

The provision of values of depth to centroid for unit size $y_{d(u)}$, in Tables C, overcomes the problems arising from the section geometry. Methods for allowance for channel slope are treated in hydraulics texts, including that by Chow[30], but where illustration of solutions is usually confined to rectangular channel circumstances.

Review

Illustrations are to a greater accuracy than is physically significant. This has been done so that the user can appreciate the small degree of numerical variability that is intrinsic to the methods provided. In fact, the intention is to enable also that first estimates can be made as readily as possible. With experience, the user often may find it expedient to abstract and work with values to less accuracy than that of the significant figures given in the individual tables.

The possible effects of variation in water temperature from standard has been covered in some detail. Within the normal water resources and drainage range and for turbulent flows, such effects are potentially greatest in smooth pipes at low values of Reynolds number. In part-full flows in drainage conduits and sewers the effects are likely to be small. The tables provide a quantification of these effects. Often in the design process possible temperature effects will be overshadowed by variations resulting from the choice of roughness size.

As detailed in Volume II, there has been effort to incorporate explicitly the rather sparse findings which exist regarding friction head losses in free-surface flows in conduits in general. Many, if not most, design packages for such circumstances are not explicit in this matter but can often give fully satisfactory predictions, especially for part-full conditions in circular pipes. This results from an appropriate choice of configuration modified roughness size, or other measure of boundary characteristics, for that range of part-full flows that is of most concern. The illustrations provided in this volume demonstrate that it may not always be straightforward to arrive at such a value. The matter is discussed further under Review in Volume II.

It is hoped that the outlines herein will aid practitioners where more detailed considerations are relevant, as well as the Tables being useful as a standard office aid for the simpler of design situations.

References

1. ACKERS, P. *Resistance of fluids flowing in channels and pipes* and *Charts for the hydraulic design of channels and pipes* Hydraulics Research Papers Nos 1 & 2, H.M.S.O., London, 1958.

2. ACKERS, P. *Tables for the hydraulic design of storm-drains, sewers and pipes-lines* Hydraulics Research Paper No. 4, H.M.S.O., London, 1963.

3. HR WALLINGFORD and BARR, D. I. H. *Tables for the hydraulic design of pipes, sewers and channels*, 7th edition in two volumes. Thomas Telford Ltd, London, 1998.

4. NIKURADSE, J. Gesetzmäßigkeit der turbulenten Strömung in glatten Rohren. *Forsch. Arb. Ing.-Wes.* No.356 (1932).

5. NIKURADSE, J. Strömungsgesetze in rauhen Rohren. *Forsch. Arb. Ing.-Wes.* No.361 (1933).

6. COLEBROOK, C. F. Turbulent flow in pipes, with particular reference to the transition region between the smooth and the rough pipe laws. *J. Instn. Civ. Engrs*, 1939, Vol. 11, pp 133-156.

7. MANNING, R. On the flow of water in open channels and pipes. *Proc. Instn Civ. Engrs*, Ire., 1891, Vol. 20, p 161; 1895, Vol. 24, p 179.

8. COPE, W. F. The friction and heat transmission coefficients of rough pipes. *Proc. Instn Mech. Engnrs*, 1941, Vol. 145, pp 99-105.

9. HARRIS, C. W. *The influence of random roughness on flow in pipes.* University of Washington Engineering Experiment Station, Bull. No. 115, University of Washington Press, 1949.

10. MOODY, L.F. Friction factors for pipe flow. *Trans. Am. Soc. Mech. Engrs*, 1944, Vol. 66, pp 671-684.

11. ROUSE, H. Evaluation of boundary roughness. *Proc. 2nd Hydraulics Conf.*, University of Iowa, 1943, Bulletin 27.

12. STRAUB, L. G., Bowers C. E. and PILCH, M. *Resistance to flow in two types of concrete pipe.* University of Minnesota, St Anthony Falls Hyd. Lab., Tech. Paper No. 22, Series B, 1960.

13. SMITH, R. V., MILLER, J. S. and FERGUSON, J. W. *Flow of natural gas through experimental pipe lines and transmission lines.* Monograph 9, U. S. Bureau of Mines, American Gas Association, N. Y., 1956.

14. COLEBROOK, C. F. The flow of water in unlined, lined and partly lined rock tunnels. *J. Instn. Civ. Engrs*, 1958, Vol. 11, pp 103-132.

15. KING, H. W. *Handbook of Hydraulics.* McGraw-Hill, N.York, 1954, Sect. 6.

16. LAMONT, P. A. A review of pipe friction data and formulae, with a proposed set of exponential formulae based on the theory of roughness. *Proc.Instn Civ.Engrs*, Vol. 3, Part 3, 1954, pp. 248-274.

17. INSTITUTION of WATER ENGINEERS *Manual of British water supply practice.* Heffer, Cambridge, 1950.

18. PERKINS, J. A. and GARDINER, I. M. *The effect of sewage slime on the hydraulic roughness of pipes.* Report IT 218, Hydraulics Research Station, Wallingford , 1982.

19. FORTY, E. J., LAUCHLAN, C. and MAY, R. W. P. *Flow resistance of wastewater pumping mains.* Report SR641, Rev 1.0, H R Wallingford, Wallingford, March, 2004.

20. ESCARAMEIA, M. and MAY, R. W. P. *Literature review on roughness of pipes and sewers.* Report SR 432, HR Wallingford, Wallingford, 1995.

21. ESCARAMEIA, M. *Experimental determination of the hydraulic roughness of pipes.* Report SR 453, HR Wallingford, Wallingford, 1996.

22. AMERICAN IRON AND STEEL INSTITUTE *Handbook of steel drainage and highway construction products,* 1971.

23. YEN, B. C. Open channel flow resistance. *Journal of Hydraulic Engineering,* American Society of Civil Engineers, Vol. 128, No. 1, 2002, pp 20-39.

24. CHOW, V-T. *Open channel hydraulics.* McGraw-Hill, New York, 1959.

25. LAMONT, P. A. Metrication: hydraulic data and formulae *Water Services*, March, 1977.

26. H R WALLINGFORD and BARR, D. I. H. *Tables for the calculation of friction in internal flows.* Thomas Telford Services Ltd, London, 1995.

27. BARR, D. I. H. Two additional methods of direct solution of the Colebrook-White function. TN 128, *Proc. Instn Civ. Engrs*, Part 2, Dec. 1975, 3, p 827.

28. CAMP, T. R. Design of sewers to facilitate flow. *Sewage Works Journal*, Jan. 1946, pp 3-16. (also *Trans. Amer. Soc. Civ. Engrs*, Vol. 109, 1944, pp 240-243.)

29. YARNELL, D. L. and WOODWARD, S. H. *The flow of water in drain tile.* U. S. Department of Agriculture, Bull. No. 854, 1920.

30. WILCOX, E. R. *A comparative test of the flow of water in 8-inch concrete and vitrified clay sewer pipe.* University of Washington Engineering Experiment Station, Bull. No. 27, 1924.

31. STERLING, M. and KNIGHT, D. W. Resistance and boundary shear in circular conduits with flat beds running part full. *Proc. Instn Civ. Engrs, Water and Maritime Engng,* Vol. 142, 2000, pp. 229-240.

32. ACKERS, P. The hydraulic resistance of drainage conduits. *Proc. Instn Civ. Engrs,* Vol. 19, 1961, pp. 307-336.

33. POMEROY, R. D. Flow velocities in small sewers. *Journ. Water Poll. Control Fed.* 1967, Vol. 39, pp. 1525 - 1548. (also Flow velocities in pipelines. *Proc. Amer. Soc. Civ. Engrs,* Jrn. Hyd. Eng., Vol. 109, No 8, 1983, pp 1108-1117.)

34. HUGHES, W. C. *Relative velocity in a partially full pipe.* Proc. Int. Conf. for the Centennial of Manning's Formula and Kuichling's Rational Formula, Vol. *Catchment Runoff and Rational Formula,* Ed B. C. Yen, University of Virginia, 1989, Distributed by Water Resources Publications, Colorado, U. S.A.

35. CHRISTENSEN, B. A. Design of partially filled circular pipes: a rational approach to storm sewer analysis. *Proc. Intern. Syposium on Urban Hydrology, Hydraulics and Sediment Control,* 1984, University of Kentucky, Lexington, KY.

36. JOHNSON, S. P. A survey of flow calculation methods. *Pre-printed programme for June 19-21, 1934, meeting of the Amer. Soc. Mech. Engrs.,* University of California, Berkeley.

37. JOINT COMMITTEE OF ASCE AND THE WATER POLLUTION CONTROL FEDERATION *Design and construction of sanitary and storm sewers.* American Society of Civil Engineers Manual No. 37, 1969.

38. JOINT COMMITTEE OF ASCE AND THE WATER POLLUTION CONTROL FEDERATION *Gravity sanitary sewer design and construction.* American Society of Civil Engineers Manual No. 60, 1982.

39. JOINT COMMITTEE OF ASCE AND THE WATER ENVIRONMENT FEDERATION *Design and construction of urban stormwater management systems.* American Society of Civil Engineers Manual No. 77, 1992.

40. HAGER, W. H. *Wastewater hydraulics : theory and practice.* Springer-Verlag, Berlin, Heidelberg, 1999. (being updated translation of *Abwasserhydraulik,* Springer-Verlag, Berlin, Heidelberg, 1994

41. MILLER, D. S. *Internal flow systems* - 2nd Edition. BHRA (Information Services), Bedford, 1990.

42. WHITE, F. M. (Gen. Ed.) *General Electric - Fluid flow data book.* General Electric, Schenectady, New York, 1943 to 1983 and Genium Publishing, New York, 1984 onwards.

43. FRIED, E. and IDELCHIK, I. E. *Flow resistance - A design guide for engineers.* Hemisphere Publishing, New York, 1989.

44. DAUGHERTY, R. L. and FRANZINI, J. B. *Fluid mechanics with engineering applications,* 6th Edn. McGraw-Hill, New York, 1965.

Nomenclature

A Area of cross-section of flow

B Width of channel; horizontal size used in specifying relative depth.

D Diameter. When used in resistance formula there is the implication that flow is pipe-full, or that the equivalent diameter measure is being utilised. The diameter of a full pipe, or that of the equivalent pipe to a specified non-circular section of flow, is used in ratio D/k_s and in product *DV*. Values of diameter of standard pipe sizes are quoted in millimetres, and must be used consistently in respect of ratio D/k_s and used in metres in product *DV*, in equations, and in non-dimensional groups generally.

g Acceleration due to gravity, 9.80665 ms^{-2} in SI.

h Head loss in uniform flow

J Discharge conversion factor, $4\pi A/P^2$ where *A* and *P* relate to a particular flow circumstance.

l Length over which head loss is assessed.

L Length of pipe to which coefficient ε_f applies.

k_s Nikuradse equivalent sand roughness size; the linear measure of roughness size. Stated in millimetres, and to be used with consistency in evaluating D/k_s and in metres in equations generally.

log Common logarithm (base 10).

M Multiplying factor on unit measure of conduit or channel size to specify size of given example.

p A proportion of total wetted perimeter.

P Wetted perimeter of flow section.

Q Discharge (volume per unit time)

Q_r Parametric discharge ratio, i.e 'basic' discharge in a section divided by 'basic' discharge in the circular section running just full which has the same wetted perimeter, with the other variables unchanged.

R Hydraulic quotient of flow section, A/P, i.e. hydraulic mean depth or hydraulic radius, for full circle $\pi/4)D^2/\pi D$ or $(1/4)D$

$\mathbf{R}$ Reynolds number, VD/ν or $4Q/\pi\nu D$

S Hydraulic (piezometric) gradient; head loss per unit length of uniform flow, h/l.

V Mean velocity of flow through cross-section.

V_r Velocity ratio, being the actual mean velocity of flow through cross-section divided by the value predicted by the 'basic' calculation of mean velocity.

Δx Distance along channel in gradually varied flow, corresponding to direct step Δy.

y a depth of flow; vertical measure of size.

Y a non-dimensional depth of flow in terms of either vertical or horizontal measure of conduit or channel size.

Δy Direct step value in gradually varied flow.

α Coriolis coefficient; $\alpha V^2/2g$ is kinetic energy head.

ε Head loss coefficient, normally applied to $V^2/2g$.

θ Transitioning element of Colebrook-White equation evaluated for standard kinematic viscosity and standard gravity in SI units, see Eq (15a) in page 16.

λ Darcy-Weisbach friction factor, $2(Sg)D/V^2$.

ν Kinematic viscosity, i.e. dynamic viscosity divided by mass density. This is $1{\cdot}141 \times 10^{-6}$ m^{2}s^{-1} for water at 15°C.

Subscripts

d Depth to centroid.

ep Relating to equivalent pipe - i.e. circular pipe flowing full at same gradient and with same hydraulic quotient (hydraulic mean depth or hydraulic radius).

ep(u) Relating to equivalent pipe for unit case.

f Friction gradient in gradually varied flow; uniform flow friction gradient generally.

G Given value between table values in Colebrook-White solutions.

m Medial: typical case without necessarily implying specific numerical basis.

N Relating to normal depth.

o Relating to overall height measure in closed conduit, or to depth corresponding to upper limit of tabulated values in certain open channel flows i.e. y_N / y_o is (normal) proportional depth Y_N for flow which has y_N as normal depth. Also, channel gradient in gradually varied flow.

r Indicating velocity ratio as already defined.

R Required value between table values in Colebrook-White solutions

T Table value in Colebrook-White solutions.

u,(u) Relating to unit section.

c(u) Critical flow in unit section

d(u) Depth to centroid in unit section

m(u) Mean depth in unit section

0.50 (for example) - A proportional or a relative depth so defining a flow condition in a given conduit or channel.

Exponents

u,v,w,x,y,z exponents in proportioning equations 8, 9 and 10, as evaluated in Tables B.

APPENDIX 1 : Recommended roughness values

Classification ('Good' and 'Normal' assumed new and clean unless otherwise stated)	**Suitable values of k_s (mm)**		
	Good	**Normal**	**Poor**
Smooth materials (pipes)			
Drawn non-ferrous pipes of aluminium, brass, copper, lead etc, and non-metallic pipes of Alkathene, glass, perspex, PVC-U# etc	--	0·003	--
Asbestos-cement	0·015	**0·03#**	0·06#
Metal			
Spun bitumen or concrete lined	--	0·03	--
# Cast iron, epoxy lining, coupling joints	--	**0·015**	0·03
# Ductile iron, Polyethylene lining, push-fit joints	--	**0·06**	0·15
# Ductile iron, Polyurethane lining, push-fit joints	0·015	**0·03**	0·06
Wrought iron	0·03	0·06	0·15
Rusty wrought iron	0·15	0·6	3·0
Uncoated steel	0·015	0·03	0·06
# Rusty steel	--	0·15	0·3
# Steel, epoxy lining, push-fit joints	0·03	**0·06**	0·15
Galvanised iron, coated cast iron generally#	0·06	0·15	0·3
Uncoated cast iron	0·15	0·3	0·6
Tate relined pipes	0·15	0·3	0·6
152mm×51mm corrug'd plate, unpaved circular pipe, running full[22]	50	60	--
76mm×25mm corrug'd plate, unpaved circular pipe, running full[22]	25	30	--
Old tuberculated water mains as follows:			
Slight degree of attack	0·6	1·5	3·0
Moderate degree of attack	1·5	3·0	6·0
Appreciable degree of attack	6·0	15	30
Severe degree of attack	15	30	60
(Good: up to 20 years use; Normal: 40 to 50 years use; Poor: 80 to 100 years use)			
Wood			
Wood stave pipes, planed plank conduits	0·3	0·6	1·5
Concrete			
# Prestressed	0·03	0·06	0·15
Precast concrete pipes with "O" ring joints	0·06	0·15	0·6
Spun precast concrete pipes with "O" ring joints	0·06	0·15	0·3
Monolithic construction against steel forms	0·3	0·6	1·5
Monolithic construction against rough forms	0·6	1·5	--
Clayware			
Glazed or unglazed pipe:			
With sleeve joints	0·03	0·06	0·15
With spigot and socket joints and "O" ring seals			
-- dia < 0·150 m	--	0·03	--
-- dia > 0·150 m	--	0·06	--
Pitch fibre			
(lower value refers to full bore flow)	0·003	0·03	
# Glass reinforced plastic (GRP)	**0·03#**	0·06	--
uPVC			
# Twin-walled, with coupling joints	0·003	0·006	
Standard, with chemically cemented joints	--	0·03	--
Standard, with spigot and socket joints, "O" ring seals at 6 to 9 metre intervals	--	0·06	--

APPENDIX 1 : Recommended roughness values (continued)

Classification ('Good' and 'Normal' assumed new and clean unless otherwise stated)	Suitable values of k_s (mm)		
	Good	Normal	Poor
New relining of sewers			
Factory manufactured GRP	0·03	--	--
Brickwork			
Glazed	0·6	1·5	3·0
Well pointed	1·5	3·0	6·0
Old, in need of pointing	--	15	30
Slimed sewers (see specific coverage in text in page 9)			
Sewers slimed to about half depth; velocity, when flowing half full, approximately 0·75 ms^{-1}:			
Concrete, spun or vertically cast	--	3·0	6·0
Asbestos cement	--	3·0	6·0
Clayware	--	1·5	3·0
uPVC	--	0·6	1·5
Sewers slimed to about half depth; velocity, when flowing half full, approximately 1·2 ms^{-1}:			
Concrete, spun or vertically cast	--	1·5	3·0
Asbestos cement	--	0·6	1·5
Clayware	--	0·3	0·6
uPVC	--	0·15	0·3
Sewer rising mains (see specific coverage in text in page 9)			
All materials, operating as follows			
Mean velocity 0·5 ms^{-1}	0·3	3·0	30
Mean velocity 0·75 ms^{-1}	0·15	1·5	15
Mean velocity 1 ms^{-1}	0·06	0·6	6·0
Mean velocity 1·5 ms^{-1}	0·03	0·3	1·5
Mean velocity 2 ms^{-1}	0·015	0·15	1·5
Concrete channels (after Yen[23] following from Chow[24])			
Trowel finish	0·5	1·5	3·3
Float finish	1·5	3·3	5·0
Finished with gravel on bottom	3·3	7·0	18
Unfinished	2·0	7·0	18
Shortcrete, or Gunite, good section	5·0	14	43
Shortcrete, or Gunite, wavy section	10	33	70
Unlined rock tunnels			
Granite and other homogeneous rocks	60	150	300
Diagonally bedded slates	--	300	600
(values to be used with *design* diameter)			
Earth channels			
Straight uniform artificial channels	15	60	150
Straight natural channels, free from shoals, boulders and weeds	150	300	600

The background to the values given here is in the sub-section of the text entitled *'Recommended roughness sizes'* in pages 9 and 10. **This sub-section should be consulted before adoption of any of the values in this table.**

APPENDIX 2 : Allowances for additional head losses in turbulent flow

As explained in the section *Other sources of resistance*, the most basic method of allowing for additional head losses at bends and at appurtenances involving change in cross-section is to add pipe length. If the adjustment rule is not conditioned by the uniform flow conditions obtaining, this approach becomes more and more fallible the greater is any change in cross-sectional area of flow, temporary or permanent, that is involved. Given below (a) are indicative values for smooth pipes in terms of number of pipe diameters to be added to the pipe length. Where two diameters (D and d) are involved, the additional length is given in terms of the smaller diameter (d) and is to be added to the length of the smaller pipe. List (a) is based on material given in the annual Reference Handbook published by *Water Engineering and Management.*

Also given below (b) are indicative values, for smooth pipe systems, of the additional loss coefficient ε (as defined in text before Eq. (21)). Where two diameters (D and d) are involved, the velocity to be used is that in the smaller diameter (d). List (b) is based mainly on material compiled in *Mark's Standard handbook for mechanical engineers*, Editors E.A.Avallone and T.Baumuster, III, McGraw-Hill, New York, 1987. More extensive, slightly varying, lists are included in the various editions of the *Manual of British water engineering practice* which were published for the Institution of Water Engineers, now incorporated in the Chartered Institution of Water and Environmental Management.

The values for given items in two lists ((a) and (b)) for smooth pipes may not be fully compatible; where additional losses are a significant proportion of the total, references 41, 42 and 43 can provide extension of list (b) as appropriate for individual requirements. Also Escarameia[21] has provided new data on losses at bends which indicates that the standard values in list (b) are overestimates in many cases. The new assessments are for 90° bends and are indicated by #.

(a) Indicative numbers of pipe diameters to be added to flow length to allow for additional losses

Abrupt 90° elbow (mitre)	67	Sudden enlarge't, d/D=1/4	32
Standard 90° bend fitting	32	Sudden contract'n, d/D=3/4	7
Medium sweep 90° bend	25	Sudden contract'n, d/D=1/2	12
Long sweep 90° bend	20	Sudden contract'n, d/D=1/4	15
Standard 45° bend fitting	15	Borda, re-entrant, entrance	30
180° close return bend	72	Ordinary abrupt entrance	17
Standard tee, 90° turn	67	Globe valve, fully open	330
Standard tee, through flow	20	Angle valve, fully open	165
d/D=3/4 tee, through flow	25	Gate valve, fully open	7
d/D=1/2 tee, through flow	32	Gate valve, 1/4 closed	40
Sudden enlarge't, d/D=3/4	7	Gate valve, 1/2 closed	200
Sudden enlarge't, d/D=1/2	18	Gate valve, 3/4 closed	800

(b) Indicative values of additional loss coefficient ε, for summation in Eq. (21)

Standard 90° bend fitting	0·75	Gradual contraction	0·05
Long sweep 90° bend	0·45	Sudden cont'n, d/D=3/4	0·26
# Ductile iron with p'ethylene lining	0·23	Sudden cont'n, d/D=1/2	0·36
# Ductile iron with epoxy lining	0·23	Sudden cont'n, d/D=1/4	0·45
# Mitred glass reinforced plastic	0·10	Borda, re-entrant, entrance	1·0
# Mitred steel with epoxy lining	0·17	Sharp edged entry	0·4
# Short rad. C.I. with epoxy lining	0·60	Rounded entrance,	0·1
# Long rad. C.I. with epoxy lining	0·18	Well rounded entrance	0·05
# PVC-u externally ribbed	0·41	'Exit loss', without venturi	1·0
Standard 45° bend fitting	0·35	Globe valve, fully open	6·4
Standard tee, 90° turn	1·5	Globe valve, 1/2 closed	9·5
Standard tee, through flow	0·4	Gate valve, fully open	0·2
Sudden enlarge't, d/D=3/4	0·2	Gate valve, 1/4 closed	0·9
Sudden enlarge't, d/D=1/2	0·6	Gate valve, 1/2 closed	4·5
Sudden enlarge't, d/D=1/4	0·9	Gate valve, 3/4 closed	24

APPENDIX 3: Multiplying factors for discharges in pipes and lined tunnels

D_G (mm)	D_T (mm)	x	**Fact.**	Notes
12·7	20	2·72	0·29	0·5in
15·875	20	2·72	0·53	0·625in
19·05	20	2·72	0·88	0·75in
22·225	20	2·71	1·33	0·875in
25·4	25	2·71	1·04	1·0in
31·75	30	2·70	1·17	1·25in
38·10	40	2·70	0·88	1·5in
44·45	40	2·70	1·33	1·75in
50·8	50	2·69	1·04	2·0in
63·5	65	2·69	0·94	2·5in
67·0	65	2·69	1·08	2·64in
76·2	75	2·68	1·04	3·0in
88·9	90	2·68	0·97	3·5in
101·6	100	2·67	1·04	4·0in
108·0	100	2·67	1·23	4·25in
110	100	2·67	1·29	
127·0	125	2·66	1·04	5·0in
133·4	125	2·66	1·19	5·25in
152·4	150	2·65	1·04	6·0in
159·0	150	2·65	1·17	6·26in
160	150	2·65	1·19	
177·8	150	2·65	1·57	7·0in
203·2	200	2·64	1·04	8·0in
219·2	225	2·64	0·93	8·63in
228·6	225	2·64	1·04	9·0in
254·0	250	2·63	1·04	10·0in
267	275	2·63	0·93	10·51in
304·8	300	2·63	1·04	12·0in
315	300	2·63	1·14	
323·9	300	2·63	1·22	12·75in
355·6	350	2·63	1·04	14·0in
368	375	2·63	0·95	14·49in
381·0	375	2·63	1·04	15·0in
406·4	400	2·63	1·04	16·0in
419	400	2·63	1·13	16·5in
457·2	450	2·63	1·04	18·0in
508	500	2·63	1·04	20·0in
533·4	525	2·62	1·04	21·0in
560	525	2·62	1·18	
609·6	600	2·62	1·04	24in
685·8	675	2·62	1·04	27in
762	750	2·62	1·04	30in
838·2	825	2·61	1·04	33in
914	900	2·61	1·04	36in
990·6	1000	2·61	0·97	39in
1066·8	1050	2·61	1·04	42in
1143	1125	2·61	1·04	45in
1219·2	1200	2·61	1·04	48in, *
1275	1250	2·61	1·05	
1295·4	1300	2·61	0·99	51in
1371·6	1350	2·61	1·04	54in
1380	1400	2·61	0·96	*
1425	1400	2·61	1·05	
1447·8	1400	2·61	1·09	57in
1450	1400	2·61	1·10	*
1520	1500	2·61	1·03	
1524	1500	2·61	1·04	60in
1530	1500	2·61	1·05	*
1575	1600	2·60	0·96	
1660	1650	2·60	1·02	*
1676·4	1650	2·60	1·04	66in
1680	1650	2·60	1·05	
1700	1650	2·60	1·08	
1725	1650	2·60	1·12	
1828·8	1800	2·60	1·04	72in, *
1875	1800	2·60	1·11	
1900	1950	2·60	0·93	
1910	1950	2·60	0·95	*
1981·2	2000	2·60	0·98	78in
1990	2000	2·60	0·99	*
2133·6	2100	2·60	1·04	84in
2140	2100	2·60	1·05	*
2280	2250	2·60	1·03	*
2286·0	2250	2·60	1·04	90in
2300	2250	2·60	1·06	
2438·4	2400	2·60	1·04	96in, *
2590·8	2550	2·60	1·04	102in
2743·2	2700	2·60	1·04	108in
2750	2700	2·60	1·05	*
2895·6	2850	2·60	1·04	114in
2900	2850	2·60	1·05	*
3048·0	3000	2·60	1·04	120in, *
3200·4	3200	2·60	1·00	126in
3352·8	3400	2·60	0·96	132in, *
3505·2	3500	2·60	1·00	138in
3657·6	3600	2·60	1·04	144in
3670	3600	2·60	1·05	*
3800	4000	2·60	0·87	
3810	4000	2·60	0·88	*
3962·4	4000	2·60	0·97	156in
4120	4000	2·60	1·08	*
4267·2	4000	2·60	1·18	168in, *
4572·0	4500	2·60	1·04	180in
5000	4500	2·60	1·32	

x is exponent used in **Fact.** = $(D_G/D_T)^x$
* indicates tunnel size; one-pass
or granolithic lined.

To estimate pipe full discharge for diameter D_G, read discharge for diameter D_T, for same gradient and roughness size, from the appropriate table and multiply by the factor given above.

k_s = 0·003 mm
S = 0·00030 to 0·00100

ie hydraulic gradient =
1 in 3333 to 1 in 1000

Water (or sewage) at 15°C;
full bore conditions.

velocities in ms^{-1}
discharges in litres/sec

Gradient	**20**	**25**	**30**	**40**	**50**	**60**	**65**	**70**	**75**	**80**	**90**	**100**	**125**	**150**
0·000300				0·071	0·085	0·097	0·103	0·109	0·115	0·121	0·131	0·142	0·167	0·189
1/ 3333				0·0898	0·1667	0·2755	0·3433	0·4207	0·5082	0·6063	0·8363	1·1145	2·0434	3·3478
0·000320				0·074	0·088	0·101	0·107	0·113	0·119	0·125	0·136	0·147	0·173	0·197
1/ 3125				0·0933	0·1731	0·2861	0·3564	0·4367	0·5275	0·6293	0·8680	1·1565	2·1200	3·4726
0·000340				0·077	0·091	0·105	0·111	0·118	0·124	0·130	0·141	0·152	0·179	0·203
1/ 2941				0·0967	0·1794	0·2963	0·3692	0·4523	0·5463	0·6518	0·8988	1·1974	2·1945	3·5941
0·000360				0·080	0·094	0·108	0·115	0·121	0·128	0·134	0·146	0·158	0·185	0·210
1/ 2778				0·1000	0·1855	0·3064	0·3816	0·4676	0·5647	0·6736	0·9288	1·2372	2·2671	3·7124
0·000380				0·082	0·097	0·112	0·119	0·125	0·132	0·138	0·151	0·162	0·191	0·217
1/ 2632				0·1033	0·1914	0·3162	0·3938	0·4824	0·5826	0·6949	0·9581	1·2761	2·3379	3·8279
0·000400				0·085	0·100	0·115	0·122	0·129	0·136	0·142	0·155	0·167	0·196	0·223
1/ 2500				0·1065	0·1973	0·3257	0·4057	0·4969	0·6001	0·7158	0·9867	1·3142	2·4072	3·9407
0·000420				0·087	0·103	0·119	0·126	0·133	0·140	0·146	0·160	0·172	0·202	0·229
1/ 2381				0·1096	0·2030	0·3351	0·4173	0·5112	0·6172	0·7361	1·0147	1·3514	2·4749	4·0511
0·000440				0·090	0·106	0·122	0·129	0·136	0·144	0·150	0·164	0·177	0·207	0·235
1/ 2273				0·1126	0·2086	0·3443	0·4287	0·5251	0·6340	0·7561	1·0421	1·3878	2·5412	4·1591
0·000460				0·092	0·109	0·125	0·133	0·140	0·147	0·154	0·168	0·181	0·212	0·241
1/ 2174				0·1156	0·2141	0·3533	0·4399	0·5387	0·6505	0·7757	1·0690	1·4235	2·6062	4·2650
0·000480				0·094	0·112	0·128	0·136	0·143	0·151	0·158	0·172	0·186	0·218	0·247
1/ 2083				0·1185	0·2195	0·3621	0·4509	0·5521	0·6666	0·7949	1·0954	1·4585	2·6699	4·3689
0·000500			0·077	0·097	0·114	0·131	0·139	0·147	0·154	0·162	0·176	0·190	0·223	0·253
1/ 2000			0·0546	0·1214	0·2247	0·3708	0·4616	0·5653	0·6825	0·8138	1·1213	1·4929	2·7325	4·4709
0·000525			0·080	0·099	0·118	0·135	0·143	0·151	0·159	0·166	0·181	0·195	0·229	0·260
1/ 1905			0·0562	0·1250	0·2312	0·3814	0·4748	0·5814	0·7019	0·8369	1·1531	1·5351	2·8093	4·5958
0·000550			0·082	0·102	0·121	0·139	0·147	0·155	0·163	0·171	0·186	0·201	0·235	0·267
1/ 1818			0·0578	0·1284	0·2376	0·3918	0·4878	0·5972	0·7209	0·8596	1·1842	1·5764	2·8844	4·7182
0·000575			0·084	0·105	0·124	0·142	0·151	0·159	0·167	0·175	0·191	0·206	0·241	0·274
1/ 1739			0·0594	0·1318	0·2438	0·4020	0·5005	0·6127	0·7396	0·8818	1·2147	1·6168	2·9580	4·8381
0·000600			0·086	0·108	0·127	0·146	0·155	0·163	0·172	0·180	0·196	0·211	0·247	0·280
1/ 1667			0·0609	0·1352	0·2499	0·4121	0·5129	0·6279	0·7579	0·9036	1·2446	1·6565	3·0302	4·9557
0·000625			0·088	0·110	0·130	0·149	0·158	0·167	0·176	0·184	0·200	0·216	0·253	0·287
1/ 1600			0·0624	0·1384	0·2560	0·4219	0·5251	0·6429	0·7759	0·9250	1·2740	1·6955	3·1012	5·0712
0·000650			0·090	0·113	0·133	0·153	0·162	0·171	0·180	0·188	0·205	0·221	0·258	0·293
1/ 1538			0·0638	0·1417	0·2619	0·4316	0·5371	0·6575	0·7936	0·9460	1·3029	1·7338	3·1708	5·1846
0·000675			0·092	0·115	0·136	0·156	0·165	0·175	0·184	0·192	0·209	0·226	0·264	0·300
1/ 1481			0·0653	0·1448	0·2677	0·4411	0·5490	0·6720	0·8110	0·9667	1·3312	1·7715	3·2393	5·2961
0·000700			0·094	0·118	0·139	0·159	0·169	0·178	0·187	0·196	0·214	0·230	0·269	0·306
1/ 1429			0·0667	0·1480	0·2734	0·4505	0·5606	0·6862	0·8281	0·9870	1·3592	1·8086	3·3067	5·4058
0·000725			0·096	0·120	0·142	0·163	0·172	0·182	0·191	0·200	0·218	0·235	0·275	0·312
1/ 1379			0·0681	0·1510	0·2790	0·4597	0·5720	0·7002	0·8449	1·0070	1·3867	1·8450	3·3730	5·5138
0·000750			0·098	0·123	0·145	0·166	0·176	0·186	0·195	0·204	0·222	0·239	0·280	0·318
1/ 1333			0·0695	0·1541	0·2846	0·4688	0·5833	0·7139	0·8615	1·0268	1·4137	1·8810	3·4383	5·6201
0·000800			0·102	0·127	0·150	0·172	0·182	0·193	0·202	0·212	0·231	0·248	0·291	0·330
1/ 1250			0·0722	0·1600	0·2954	0·4866	0·6054	0·7409	0·8939	1·0654	1·4667	1·9513	3·5661	5·8282
0·000850		0·092	0·106	0·132	0·156	0·178	0·189	0·199	0·209	0·219	0·239	0·257	0·301	0·341
1/ 1176		0·0451	0·0748	0·1658	0·3060	0·5039	0·6269	0·7671	0·9255	1·1030	1·5183	2·0197	3·6905	6·0305
0·000900		0·095	0·110	0·136	0·161	0·184	0·195	0·206	0·216	0·227	0·247	0·266	0·311	0·352
1/ 1111		0·0466	0·0774	0·1714	0·3163	0·5208	0·6478	0·7927	0·9563	1·1396	1·5685	2·0863	3·8116	6·2276
0·000950		0·098	0·113	0·141	0·166	0·190	0·201	0·212	0·223	0·234	0·254	0·274	0·320	0·363
1/ 1053		0·0482	0·0799	0·1769	0·3264	0·5372	0·6682	0·8176	0·9863	1·1753	1·6176	2·1514	3·9297	6·4198
0·001000		0·101	0·117	0·145	0·171	0·196	0·207	0·219	0·230	0·241	0·262	0·282	0·330	0·374
1/ 1000		0·0496	0·0824	0·1823	0·3362	0·5533	0·6882	0·8420	1·0157	1·2102	1·6655	2·2149	4·0452	6·6076
		–	0·74	0·74	0·75	0·76	0·76	0·77	0·77	0·77	0·78	0·78	0·80	0·81

Column headings: (Equivalent) Pipe diameters in mm

$V_{r(0\cdot5)medial}$ for half-full circular pipes.

Gradient	(Equivalent) Pipe diameters in mm 20	25	30	40	50	60	65	70	75	80	90	100	125	150
0·00100		0·101	0·117	0·145	0·171	0·196	0·207	0·219	0·230	0·241	0·262	0·282	0·330	0·374
1/ 1000		0·0496	0·0824	0·1823	0·3362	0·5533	0·6882	0·8420	1·0157	1·2102	1·6655	2·2149	4·0452	6·6076
0·00105		0·104	0·120	0·149	0·176	0·201	0·213	0·225	0·236	0·248	0·269	0·290	0·339	0·384
1/ 952		0·0511	0·0848	0·1875	0·3458	0·5691	0·7078	0·8659	1·0444	1·2444	1·7123	2·2771	4·1581	6·7913
0·00110		0·107	0·123	0·153	0·181	0·207	0·219	0·231	0·243	0·254	0·276	0·298	0·348	0·394
1/ 909		0·0525	0·0871	0·1927	0·3553	0·5845	0·7269	0·8892	1·0726	1·2778	1·7582	2·3379	4·2686	6·9711
0·00115		0·110	0·127	0·157	0·186	0·212	0·225	0·237	0·249	0·261	0·283	0·305	0·357	0·404
1/ 870		0·0539	0·0894	0·1977	0·3645	0·5996	0·7457	0·9121	1·1001	1·3106	1·8032	2·3976	4·3770	7·1473
0·00120		0·113	0·130	0·161	0·190	0·217	0·230	0·243	0·255	0·267	0·290	0·313	0·365	0·414
1/ 833		0·0553	0·0917	0·2027	0·3736	0·6145	0·7641	0·9346	1·1272	1·3428	1·8473	2·4561	4·4832	7·3201
0·00125		0·115	0·133	0·165	0·195	0·222	0·236	0·249	0·261	0·273	0·297	0·320	0·374	0·424
1/ 800		0·0567	0·0939	0·2075	0·3825	0·6290	0·7822	0·9567	1·1537	1·3744	1·8907	2·5135	4·5875	7·4898
0·00130		0·118	0·136	0·169	0·199	0·228	0·241	0·254	0·267	0·280	0·304	0·327	0·382	0·433
1/ 769		0·0580	0·0961	0·2123	0·3913	0·6434	0·7999	0·9784	1·1798	1·4054	1·9332	2·5700	4·6900	7·6564
0·00135		0·121	0·139	0·173	0·204	0·233	0·246	0·260	0·273	0·286	0·310	0·334	0·390	0·443
1/ 741		0·0593	0·0983	0·2170	0·3999	0·6574	0·8174	0·9997	1·2055	1·4360	1·9751	2·6255	4·7907	7·8202
0·00140		0·123	0·142	0·176	0·208	0·237	0·252	0·265	0·279	0·292	0·317	0·341	0·398	0·452
1/ 714		0·0606	0·1004	0·2217	0·4083	0·6713	0·8346	1·0207	1·2308	1·4660	2·0162	2·6801	4·8898	7·9813
0·00145		0·126	0·145	0·180	0·212	0·242	0·257	0·271	0·284	0·298	0·323	0·348	0·406	0·461
1/ 690		0·0619	0·1025	0·2262	0·4167	0·6849	0·8515	1·0413	1·2556	1·4955	2·0568	2·7338	4·9873	8·1398
0·00150		0·129	0·148	0·184	0·216	0·247	0·262	0·276	0·290	0·303	0·330	0·355	0·414	0·469
1/ 667		0·0631	0·1046	0·2307	0·4249	0·6984	0·8682	1·0617	1·2801	1·5246	2·0967	2·7867	5·0834	8·2959
0·00160		0·134	0·154	0·191	0·225	0·256	0·271	0·286	0·301	0·315	0·342	0·368	0·430	0·487
1/ 625		0·0656	0·1086	0·2395	0·4410	0·7247	0·9008	1·1015	1·3280	1·5816	2·1748	2·8902	5·2713	8·6014
0·00170	0·116	0·138	0·159	0·197	0·233	0·265	0·281	0·296	0·311	0·326	0·354	0·381	0·444	0·504
1/ 588	0·0366	0·0680	0·1125	0·2481	0·4567	0·7503	0·9325	1·1402	1·3746	1·6370	2·2507	2·9909	5·4540	8·8983
0·00180	0·120	0·143	0·165	0·204	0·240	0·274	0·290	0·306	0·321	0·336	0·365	0·393	0·459	0·520
1/ 556	0·0378	0·0703	0·1163	0·2565	0·4719	0·7752	0·9634	1·1779	1·4200	1·6910	2·3248	3·0890	5·6320	9·1876
0·00190	0·124	0·148	0·170	0·211	0·248	0·283	0·299	0·316	0·331	0·347	0·377	0·405	0·473	0·536
1/ 526	0·0391	0·0726	0·1201	0·2646	0·4868	0·7995	0·9936	1·2148	1·4643	1·7437	2·3970	3·1848	5·8056	9·4698
0·00200	0·128	0·152	0·175	0·217	0·255	0·291	0·308	0·325	0·341	0·357	0·388	0·417	0·487	0·551
1/ 500	0·0403	0·0748	0·1237	0·2726	0·5014	0·8233	1·0231	1·2508	1·5076	1·7952	2·4675	3·2783	5·9752	9·7453
0·00210	0·132	0·157	0·180	0·223	0·263	0·299	0·317	0·334	0·351	0·367	0·399	0·429	0·500	0·567
1/ 476	0·0414	0·0770	0·1273	0·2804	0·5156	0·8466	1·0520	1·2860	1·5500	1·8455	2·5365	3·3697	6·1410	10·015
0·00220	0·136	0·161	0·185	0·229	0·270	0·307	0·326	0·343	0·360	0·377	0·409	0·440	0·514	0·582
1/ 455	0·0426	0·0791	0·1308	0·2880	0·5296	0·8694	1·0802	1·3204	1·5915	1·8948	2·6041	3·4592	6·3034	10·279
0·00230	0·139	0·165	0·190	0·235	0·277	0·315	0·334	0·352	0·369	0·387	0·420	0·452	0·527	0·596
1/ 435	0·0437	0·0812	0·1342	0·2955	0·5433	0·8917	1·1079	1·3542	1·6321	1·9431	2·6703	3·5470	6·4625	10·537
0·00240	0·143	0·170	0·195	0·241	0·284	0·323	0·342	0·360	0·378	0·396	0·430	0·463	0·539	0·611
1/ 417	0·0449	0·0833	0·1376	0·3029	0·5567	0·9136	1·1351	1·3874	1·6720	1·9905	2·7353	3·6330	6·6185	10·790
0·00250	0·146	0·174	0·199	0·247	0·290	0·331	0·350	0·369	0·387	0·405	0·440	0·473	0·552	0·625
1/ 400	0·0460	0·0853	0·1409	0·3101	0·5699	0·9352	1·1618	1·4199	1·7111	2·0371	2·7990	3·7175	6·7717	11·039
0·00260	0·150	0·178	0·204	0·252	0·297	0·338	0·358	0·377	0·396	0·414	0·450	0·484	0·564	0·639
1/ 385	0·0470	0·0873	0·1442	0·3172	0·5828	0·9563	1·1880	1·4519	1·7496	2·0828	2·8617	3·8005	6·9221	11·284
0·00270	0·153	0·182	0·209	0·258	0·303	0·346	0·366	0·385	0·405	0·423	0·460	0·494	0·576	0·652
1/ 370	0·0481	0·0892	0·1474	0·3242	0·5955	0·9771	1·2138	1·4833	1·7874	2·1277	2·9232	3·8821	7·0700	11·524
0·00280	0·156	0·186	0·213	0·263	0·310	0·353	0·373	0·393	0·413	0·432	0·469	0·505	0·588	0·665
1/ 357	0·0492	0·0911	0·1505	0·3310	0·6081	0·9975	1·2391	1·5142	1·8246	2·1719	2·9838	3·9623	7·2154	11·760
0·00290	0·160	0·190	0·217	0·269	0·316	0·360	0·381	0·401	0·421	0·441	0·478	0·515	0·600	0·679
1/ 345	0·0502	0·0930	0·1537	0·3378	0·6204	1·0177	1·2641	1·5447	1·8612	2·2155	3·0434	4·0413	7·3585	11·992
0·00300	0·163	0·193	0·222	0·274	0·322	0·367	0·388	0·409	0·429	0·449	0·488	0·524	0·611	0·692
1/ 333	0·0512	0·0949	0·1567	0·3445	0·6326	1·0375	1·2886	1·5746	1·8973	2·2583	3·1021	4·1191	7·4994	12·221
	0·74	0·74	0·75	0·76	0·77	0·78	0·78	0·79	0·79	0·79	0·80	0·81	0·83	0·84

$V_{r(0·5)medial}$ for half-full circular pipes.

A1

(p.3 of 6)

k_s = 0·003 mm
S = 0·00300 to 0·01000

ie hydraulic gradient =
1 in 333 to 1 in 100

Water (or sewage) at 15°C;
full bore conditions.

velocities in ms^{-1}
discharges in litres/sec

Gradient **(Equivalent) Pipe diameters in mm**

Gradient	20	25	30	40	50	60	65	70	75	80	90	100	125	150
0·00300	0163	0·193	0·222	0·274	0·322	0·367	0·388	0·409	0·429	0·449	0·488	0·524	0·611	0·692
1/ 333	0·0512	0·0949	0·1567	0·3445	0·6326	1·0375	1·2886	1·5746	1·8973	2·2583	3·1021	4·1191	7·4994	12·221
0·00320	0·169	0·201	0·230	0·284	0·334	0·381	0·403	0·424	0·445	0·466	0·506	0·544	0·634	0·717
1/ 313	0·0532	0·0985	0·1627	0·3575	0·6563	1·0763	1·3367	1·6333	1·9679	2·3422	3·2170	4·2712	7·7751	12·669
0·00340	0·175	0·208	0·238	0·295	0·346	0·394	0·417	0·439	0·461	0·482	0·523	0·563	0·655	0·742
1/ 294	0·0551	0·1021	0·1685	0·3702	0·6795	1·1141	1·3835	1·6904	2·0365	2·4237	3·3287	4·4192	8·0432	13·104
0·00360	0·181	0·215	0·246	0·304	0·358	0·407	0·431	0·454	0·476	0·498	0·540	0·581	0·677	0·766
1/ 278	0·0570	0·1056	0·1742	0·3825	0·7020	1·1509	1·4291	1·7460	2·1033	2·5032	3·4375	4·5633	8·3042	13·528
0·00380	0·187	0·222	0·254	0·314	0·369	0·420	0·444	0·468	0·491	0·513	0·557	0·599	0·697	0·789
1/ 263	0·0589	0·1089	0·1798	0·3946	0·7240	1·1868	1·4736	1·8002	2·1686	2·5807	3·5437	4·7039	8·5589	13·941
0·00400	0·193	0·229	0·262	0·323	0·380	0·432	0·457	0·482	0·505	0·528	0·573	0·616	0·718	0·812
1/ 250	0·0607	0·1123	0·1852	0·4064	0·7455	1·2218	1·5171	1·8532	2·2323	2·6564	3·6474	4·8412	8·8076	14·344
0·00420	0·199	0·235	0·269	0·333	0·390	0·444	0·470	0·495	0·519	0·543	0·589	0·634	0·738	0·834
1/ 238	0·0624	0·1155	0·1905	0·4180	0·7666	1·2561	1·5596	1·9050	2·2946	2·7305	3·7488	4·9755	9·0507	14·739
0·00440	0·204	0·242	0·277	0·342	0·401	0·456	0·483	0·508	0·533	0·558	0·605	0·650	0·757	0·856
1/ 227	0·0641	0·1187	0·1957	0·4293	0·7872	1·2897	1·6012	1·9558	2·3557	2·8030	3·8481	5·1070	9·2888	15·125
0·00460	0·210	0·248	0·284	0·350	0·411	0·468	0·495	0·521	0·547	0·572	0·620	0·667	0·776	0·877
1/ 217	0·0658	0·1218	0·2008	0·4403	0·8073	1·3226	1·6420	2·0055	2·4155	2·8740	3·9453	5·2358	9·5220	15·504
0·00480	0·215	0·254	0·291	0·359	0·421	0·479	0·507	0·534	0·560	0·586	0·635	0·683	0·795	0·898
1/ 208	0·0675	0·1248	0·2058	0·4512	0·8272	1·3549	1·6820	2·0543	2·4741	2·9437	4·0407	5·3621	9·7506	15·875
0·00500	0·220	0·260	0·298	0·368	0·431	0·490	0·519	0·546	0·573	0·599	0·650	0·699	0·813	0·919
1/ 200	0·0691	0·1278	0·2107	0·4619	0·8466	1·3866	1·7213	2·1022	2·5317	3·0121	4·1344	5·4861	9·9751	16·239
0·00525	0·226	0·268	0·307	0·378	0·443	0·504	0·533	0·561	0·589	0·616	0·668	0·718	0·835	0·944
1/ 190	0·0712	0·1315	0·2167	0·4750	0·8704	1·4255	1·7694	2·1608	2·6022	3·0959	4·2491	5·6380	10·250	16·685
0·00550	0·233	0·275	0·315	0·388	0·455	0·518	0·547	0·576	0·605	0·632	0·686	0·737	0·857	0·969
1/ 182	0·0731	0·1351	0·2226	0·4878	0·8938	1·4635	1·8165	2·2183	2·6713	3·1780	4·3614	5·7867	10·519	17·121
0·00575	0·239	0·282	0·323	0·398	0·467	0·531	0·561	0·591	0·620	0·648	0·703	0·755	0·879	0·993
1/ 174	0·0750	0·1386	0·2284	0·5003	0·9167	1·5008	1·8627	2·2746	2·7390	3·2584	4·4715	5·9324	10·783	17·549
0·00600	0·245	0·290	0·331	0·408	0·478	0·544	0·575	0·605	0·635	0·664	0·720	0·774	0·900	1·017
1/ 167	0·0769	0·1421	0·2341	0·5127	0·9391	1·5374	1·9080	2·3298	2·8054	3·3372	4·5794	6·0753	11·041	17·968
0·00625	0·251	0·296	0·339	0·418	0·489	0·556	0·588	0·619	0·650	0·679	0·736	0·791	0·920	1·040
1/ 160	0·0788	0·1455	0·2397	0·5248	0·9611	1·5733	1·9525	2·3840	2·8705	3·4146	4·6854	6·2155	11·295	18·380
0·00650	0·257	0·303	0·347	0·427	0·501	0·569	0·602	0·633	0·664	0·694	0·753	0·809	0·941	1·063
1/ 154	0·0806	0·1489	0·2451	0·5367	0·9828	1·6086	1·9961	2·4372	2·9345	3·4907	4·7894	6·3532	11·544	18·784
0·00675	0·262	0·310	0·354	0·436	0·511	0·581	0·614	0·647	0·678	0·709	0·769	0·826	0·961	1·085
1/ 148	0·0824	0·1522	0·2505	0·5484	1·0041	1·6432	2·0391	2·4896	2·9975	3·5654	4·8917	6·4886	11·789	19·181
0·00700	0·268	0·317	0·362	0·446	0·522	0·593	0·627	0·660	0·692	0·724	0·785	0·843	0·980	1·108
1/ 143	0·0842	0·1554	0·2558	0·5599	1·0250	1·6773	2·0813	2·5411	3·0594	3·6389	4·9923	6·6217	12·030	19·571
0·00725	0·274	0·323	0·369	0·455	0·533	0·605	0·640	0·673	0·706	0·738	0·800	0·860	1·000	1·129
1/ 138	0·0859	0·1586	0·2611	0·5712	1·0456	1·7109	2·1229	2·5917	3·1203	3·7113	5·0913	6·7528	12·267	19·955
0·00750	0·279	0·329	0·377	0·463	0·543	0·617	0·652	0·686	0·720	0·753	0·816	0·876	1·019	1·151
1/ 133	0·0876	0·1617	0·2662	0·5823	1·0659	1·7440	2·1638	2·6416	3·1802	3·7825	5·1888	6·8818	12·500	20·334
0·00800	0·290	0·342	0·391	0·481	0·563	0·640	0·676	0·712	0·746	0·780	0·846	0·908	1·056	1·193
1/ 125	0·0910	0·1679	0·2763	0·6042	1·1056	1·8086	2·2439	2·7393	3·2976	3·9218	5·3795	7·1341	12·956	21·073
0·00850	0·300	0·354	0·405	0·498	0·583	0·662	0·700	0·736	0·772	0·807	0·875	0·940	1·092	1·233
1/ 118	0·0943	0·1739	0·2861	0·6254	1·1442	1·8715	2·3218	2·8342	3·4117	4·0574	5·5649	7·3795	13·400	21·793
0·00900	0·310	0·366	0·418	0·514	0·602	0·684	0·723	0·760	0·797	0·833	0·903	0·970	1·127	1·273
1/ 111	0·0975	0·1797	0·2956	0·6461	1·1819	1·9328	2·3977	2·9267	3·5228	4·1893	5·7454	7·6184	13·832	22·493
0·00950	0·320	0·378	0·431	0·530	0·621	0·705	0·745	0·784	0·822	0·859	0·931	1·000	1·161	1·311
1/ 105	0·1006	0·1854	0·3049	0·6663	1·2186	1·9926	2·4717	3·0169	3·6312	4·3181	5·9215	7·8514	14·253	23·175
0·01000	0·330	0·389	0·444	0·546	0·639	0·725	0·767	0·807	0·846	0·884	0·958	1·029	1·195	1·349
1/ 100	0·1036	0·1910	0·3140	0·6860	1·2545	2·0510	2·5441	3·1050	3·7371	4·4438	6·0935	8·0789	14·664	23·842
	0·75	0·75	0·76	0·78	0·79	0·80	0·81	0·81	0·82	0·82	0·83	0·84	0·86	0·87

$V_{r(0\cdot5)medial}$ for half-full circular pipes.

k_s = 0·003 mm S = 0·00300 to 0·01000

Gradient	**(Equivalent) Pipe diameters in mm**													
	20	**25**	**30**	**40**	**50**	**60**	**65**	**70**	**75**	**80**	**90**	**100**	**125**	**150**
0·01000	0·330	0·389	0·444	0·546	0·639	0·725	0·767	0·807	0·846	0·884	0·958	1·029	1·195	1·349
1/ 100	0·1036	0·1910	0·3140	0·6860	1·2545	2·0510	2·5441	3·1050	3·7371	4·4438	6·0935	8·0789	14·664	23·842
0·01050	0·339	0·400	0·457	0·561	0·657	0·746	0·788	0·829	0·869	0·909	0·984	1·057	1·228	1·386
1/ 95	0·1066	0·1964	0·3229	0·7053	1·2895	2·1081	2·6148	3·1911	3·8406	4·5667	6·2617	8·3014	15·067	24·493
0·01100	0·349	0·411	0·469	0·576	0·674	0·765	0·809	0·851	0·892	0·932	1·010	1·085	1·260	1·422
1/ 91	0·1095	0·2017	0·3316	0·7242	1·3239	2·1640	2·6840	3·2754	3·9420	4·6871	6·4263	8·5191	15·460	25·131
0·01150	0·358	0·422	0·481	0·591	0·691	0·785	0·829	0·873	0·915	0·956	1·035	1·112	1·291	1·457
1/ 87	0·1124	0·2070	0·3402	0·7427	1·3576	2·2188	2·7518	3·3581	4·0413	4·8050	6·5875	8·7325	15·845	25·755
0·01200	0·367	0·432	0·493	0·606	0·708	0·804	0·849	0·894	0·937	0·979	1·060	1·138	1·322	1·492
1/ 83	0·1152	0·2121	0·3486	0·7609	1·3906	2·2725	2·8183	3·4391	4·1387	4·9206	6·7457	8·9416	16·223	26·368
0·01250	0·375	0·442	0·505	0·620	0·725	0·822	0·869	0·914	0·958	1·001	1·085	1·165	1·352	1·526
1/ 80	0·1179	0·2171	0·3568	0·7787	1·4230	2·3253	2·8836	3·5187	4·2342	5·0341	6·9009	9·1469	16·594	26·968
0·01300	0·384	0·452	0·516	0·634	0·741	0·841	0·888	0·935	0·980	1·024	1·109	1·190	1·382	1·559
1/ 77	0·1207	0·2221	0·3649	0·7963	1·4548	2·3771	2·9477	3·5968	4·3281	5·1455	7·0533	9·3485	16·958	27·558
0·01350	0·393	0·462	0·528	0·647	0·757	0·859	0·907	0·955	1·001	1·045	1·132	1·216	1·411	1·592
1/ 74	0·1233	0·2270	0·3729	0·8135	1·4861	2·4280	3·0108	3·6736	4·4204	5·2551	7·2031	9·5466	17·316	28·138
0·01400	0·401	0·472	0·539	0·661	0·773	0·876	0·926	0·974	1·021	1·067	1·155	1·240	1·440	1·625
1/ 71	0·1259	0·2318	0·3807	0·8304	1·5169	2·4781	3·0728	3·7491	4·5111	5·3628	7·3504	9·7414	17·668	28·708
0·01450	0·409	0·482	0·549	0·674	0·788	0·894	0·944	0·993	1·041	1·088	1·178	1·265	1·468	1·656
1/ 69	0·1285	0·2365	0·3884	0·8471	1·5472	2·5274	3·1338	3·8234	4·6004	5·4688	7·4953	9·9332	18·014	29·268
0·01500	0·417	0·491	0·560	0·687	0·803	0·911	0·962	1·013	1·061	1·109	1·201	1·289	1·496	1·687
1/ 67	0·1311	0·2411	0·3960	0·8635	1·5771	2·5759	3·1939	3·8966	4·6883	5·5732	7·6380	10·122	18·355	29·820
0·01600	0·433	0·510	0·581	0·713	0·833	0·945	0·998	1·050	1·100	1·149	1·244	1·336	1·550	1·749
1/ 63	0·1361	0·2502	0·4109	0·8957	1·6355	2·6709	3·3114	4·0397	4·8603	5·7773	7·9172	10·491	19·022	30·900
0·01700	0·449	0·528	0·602	0·738	0·862	0·977	1·032	1·086	1·138	1·189	1·287	1·381	1·603	1·808
1/ 59	0·1409	0·2591	0·4253	0·9270	1·6922	2·7632	3·4256	4·1789	5·0275	5·9758	8·1885	10·850	19·670	31·949
0·01800	0·464	0·545	0·622	0·762	0·890	1·009	1·066	1·121	1·175	1·227	1·329	1·426	1·654	1·866
1/ 56	0·1456	0·2677	0·4394	0·9574	1·7475	2·8531	3·5369	4·3145	5·1903	6·1691	8·4527	11·199	20·300	32·970
0·01900	0·478	0·563	0·641	0·786	0·917	1·040	1·099	1·155	1·211	1·265	1·369	1·469	1·704	1·922
1/ 53	0·1502	0·2761	0·4532	0·9871	1·8015	2·9409	3·6455	4·4466	5·3491	6·3576	8·7104	11·540	20·915	33·965
0·02000	0·493	0·579	0·660	0·809	0·944	1·070	1·131	1·189	1·246	1·301	1·409	1·512	1·753	1·977
1/ 50	0·1548	0·2843	0·4666	1·0162	1·8542	3·0265	3·7515	4·5757	5·5042	6·5417	8·9619	11·872	21·516	34·937
0·02100	0·507	0·596	0·679	0·831	0·971	1·100	1·162	1·222	1·280	1·337	1·447	1·553	1·801	2·031
1/ 47·6	0·1592	0·2924	0·4797	1·0446	1·9057	3·1103	3·8551	4·7020	5·6558	6·7216	9·2079	12·198	22·103	35·887
0·02200	0·520	0·612	0·697	0·853	0·996	1·129	1·192	1·254	1·314	1·372	1·485	1·594	1·848	2·083
1/ 45·5	0·1635	0·3003	0·4926	1·0724	1·9562	3·1923	3·9566	4·8255	5·8042	6·8977	9·4486	12·516	22·677	36·816
0·02300	0·534	0·627	0·715	0·875	1·021	1·157	1·222	1·285	1·347	1·407	1·522	1·633	1·894	2·135
1/ 43·5	0·1677	0·3080	0·5052	1·0996	2·0056	3·2726	4·0560	4·9465	5·9496	7·0703	9·6844	12·828	23·239	37·726
0·02400	0·547	0·643	0·732	0·896	1·046	1·185	1·252	1·316	1·379	1·440	1·559	1·672	1·939	2·185
1/ 41·7	0·1719	0·3156	0·5176	1·1263	2·0541	3·3514	4·1534	5·0652	6·0921	7·2394	9·9155	13·133	23·790	38·619
0·02500	0·560	0·658	0·749	0·917	1·070	1·213	1·280	1·346	1·411	1·473	1·594	1·710	1·983	2·235
1/ 40·0	0·1760	0·3230	0·5297	1·1526	2·1016	3·4287	4·2491	5·1817	6·2320	7·4054	10·142	13·433	24·331	39·494
0·02600	0·573	0·673	0·766	0·938	1·094	1·240	1·309	1·376	1·442	1·506	1·629	1·748	2·026	2·284
1/ 38·5	0·1800	0·3303	0·5416	1·1783	2·1484	3·5046	4·3430	5·2961	6·3694	7·5685	10·365	13·727	24·863	40·353
0·02700	0·585	0·688	0·783	0·958	1·118	1·266	1·337	1·405	1·472	1·538	1·664	1·785	2·069	2·331
1/ 37·0	0·1839	0·3375	0·5534	1·2037	2·1943	3·5793	4·4353	5·4085	6·5044	7·7287	10·584	14·017	25·384	41·198
0·02800	0·598	0·702	0·799	0·978	1·141	1·292	1·364	1·434	1·502	1·569	1·698	1·821	2·110	2·378
1/ 35·7	0·1878	0·3446	0·5649	1·2286	2·2395	3·6527	4·5261	5·5190	6·6371	7·8862	10·799	14·301	25·898	42·028
0·02900	0·610	0·716	0·815	0·997	1·163	1·317	1·391	1·462	1·532	1·600	1·731	1·857	2·151	2·425
1/ 34·5	0·1916	0·3516	0·5763	1·2531	2·2839	3·7249	4·6155	5·6278	6·7678	8·0412	11·011	14·581	26·402	42·844
0·03000	0·622	0·730	0·831	1·016	1·185	1·343	1·417	1·490	1·561	1·630	1·764	1·892	2·192	2·470
1/ 33·3	0·1954	0·3584	0·5874	1·2772	2·3277	3·7960	4·7034	5·7349	6·8964	8·1938	11·220	14·857	26·899	43·648
	0·76	0·77	0·78	0·80	0·81	0·83	0·83	0·84	0·85	0·85	0·86	0·87	0·89	0·90

$V_{r(0{\cdot}5)medial}$ for half-full circular pipes.

A1
(p.5 of 6)

k_s = 0·003 mm
S = 0·03000 to 0·10000

ie hydraulic gradient =
1 in 33·3 to 1 in 10·0

Water (or sewage) at 15°C;
full bore conditions.

velocities in ms^{-1}
discharges in litres/sec

Gradient **(Equivalent) Pipe diameters in mm**

Gradient	20	25	30	40	50	60	65	70	75	80	90	100	125	150
0·03000	0·622	0·730	0·831	1·016	1·185	1·343	1·417	1·490	1·561	1·630	1·764	1·892	2·192	2·470
1/ 33·3	0·1954	0·3584	0·5874	1·2772	2·3277	3·7960	4·7034	5·7349	6·8964	8·1938	11·220	14·857	26·899	43·648
0·03200	0·645	0·758	0·862	1·054	1·229	1·392	1·469	1·545	1·618	1·689	1·828	1·960	2·271	2·559
1/ 31·3	0·2027	0·3718	0·6093	1·3245	2·4133	3·9351	4·8755	5·9443	7·1479	8·4923	11·627	15·395	27·871	45·220
0·03400	0·668	0·784	0·892	1·091	1·272	1·440	1·520	1·598	1·673	1·747	1·890	2·027	2·348	2·645
1/ 29·4	0·2099	0·3849	0·6306	1·3704	2·4966	4·0703	5·0427	6·1479	7·3923	8·7823	12·024	15·919	28·815	46·747
0·03600	0·690	0·810	0·921	1·126	1·313	1·486	1·569	1·649	1·727	1·803	1·951	2·092	2·423	2·729
1/ 27·8	0·2169	0·3976	0·6513	1·4151	2·5777	4·2020	5·2056	6·3461	7·6304	9·0648	12·409	16·429	29·734	48·233
0·03800	0·712	0·835	0·950	1·161	1·353	1·532	1·617	1·699	1·780	1·858	2·010	2·155	2·496	2·811
1/ 26·3	0·2237	0·4100	0·6716	1·4588	2·6568	4·3304	5·3644	6·5395	7·8625	9·3401	12·785	16·926	30·630	49·682
0·04000	0·733	0·860	0·978	1·195	1·392	1·576	1·663	1·748	1·831	1·912	2·067	2·217	2·567	2·891
1/ 25·0	0·2304	0·4222	0·6913	1·5014	2·7340	4·4557	5·5194	6·7282	8·0891	9·6090	13·153	17·411	31·504	51·095
0·04200	0·754	0·884	1·005	1·228	1·431	1·619	1·709	1·796	1·881	1·964	2·124	2·277	2·637	2·970
1/ 23·8	0·2369	0·4340	0·7107	1·5431	2·8095	4·5783	5·6710	6·9127	8·3106	9·8718	13·512	17·885	32·359	52·477
0·04400	0·774	0·908	1·032	1·260	1·469	1·662	1·754	1·843	1·930	2·015	2·179	2·336	2·705	3·046
1/ 22·7	0·2432	0·4456	0·7296	1·5839	2·8834	4·6983	5·8194	7·0933	8·5274	10·129	13·863	18·349	33·195	53·829
0·04600	0·794	0·931	1·058	1·292	1·505	1·703	1·798	1·889	1·978	2·065	2·233	2·394	2·772	3·121
1/ 21·7	0·2495	0·4570	0·7481	1·6238	2·9558	4·8158	5·9647	7·2701	8·7397	10·381	14·207	18·804	34·014	55·152
0·04800	0·814	0·954	1·084	1·323	1·542	1·744	1·840	1·934	2·025	2·114	2·286	2·451	2·837	3·194
1/ 20·8	0·2556	0·4682	0·7663	1·6630	3·0268	4·9310	6·1072	7·4435	8·9479	10·628	14·544	19·249	34·816	56·450
0·05000	0·833	0·976	1·109	1·354	1·577	1·784	1·883	1·978	2·072	2·163	2·338	2·507	2·901	3·266
1/ 20·0	0·2617	0·4791	0·7842	1·7015	3·0965	5·0441	6·2470	7·6137	9·1521	10·870	14·875	19·686	35·604	57·722
0·05250	0·856	1·003	1·140	1·392	1·620	1·833	1·934	2·033	2·128	2·222	2·402	2·575	2·980	3·355
1/ 19·0	0·2690	0·4926	0·8061	1·7486	3·1818	5·1826	6·4182	7·8221	9·4023	11·167	15·280	20·221	36·568	59·280
0·05500	0·879	1·030	1·171	1·428	1·663	1·881	1·985	2·086	2·184	2·279	2·464	2·641	3·057	3·441
1/ 18·2	0·2763	0·5057	0·8275	1·7948	3·2653	5·3181	6·5858	8·0260	9·6471	11·457	15·676	20·745	37·511	60·804
0·05750	0·902	1·056	1·200	1·464	1·705	1·928	2·034	2·137	2·238	2·336	2·525	2·707	3·132	3·525
1/ 17·4	0·2833	0·5186	0·8485	1·8400	3·3472	5·4509	6·7499	8·2257	9·8868	11·742	16·065	21·258	38·434	62·297
0·06000	0·924	1·082	1·229	1·499	1·746	1·974	2·083	2·188	2·291	2·391	2·585	2·771	3·206	3·608
1/ 16·7	0·2903	0·5312	0·8690	1·8843	3·4274	5·5810	6·9109	8·4215	10·122	12·020	16·445	21·760	39·340	63·760
0·06250	0·946	1·108	1·258	1·534	1·786	2·019	2·130	2·238	2·343	2·446	2·644	2·833	3·278	3·689
1/ 16·0	0·2971	0·5436	0·8892	1·9278	3·5061	5·7087	7·0687	8·6137	10·352	12·294	16·819	22·253	40·227	65·194
0·06500	0·967	1·132	1·286	1·568	1·825	2·063	2·177	2·287	2·395	2·499	2·701	2·895	3·349	3·769
1/ 15·4	0·3038	0·5558	0·9091	1·9705	3·5835	5·8342	7·2238	8·8023	10·579	12·563	17·185	22·737	41·099	66·603
0·06750	0·988	1·157	1·314	1·602	1·864	2·107	2·223	2·335	2·445	2·552	2·758	2·956	3·419	3·847
1/ 14·8	0·3104	0·5678	0·9286	2·0125	3·6595	5·9574	7·3762	8·9877	10·801	12·826	17·545	23·213	41·956	67·986
0·07000	1·009	1·181	1·341	1·634	1·902	2·150	2·268	2·383	2·494	2·603	2·814	3·015	3·487	3·924
1/ 14·3	0·3169	0·5796	0·9478	2·0538	3·7342	6·0786	7·5260	9·1700	11·020	13·086	17·900	23·681	42·797	69·346
0·07250	1·029	1·204	1·368	1·667	1·939	2·192	2·312	2·429	2·543	2·654	2·868	3·074	3·555	4·000
1/ 13·8	0·3233	0·5912	0·9666	2·0945	3·8077	6·1978	7·6733	9·3493	11·235	13·341	18·248	24·141	43·625	70·684
0·07500	1·049	1·228	1·394	1·699	1·976	2·234	2·356	2·475	2·591	2·704	2·922	3·131	3·621	4·074
1/ 13·3	0·3295	0·6026	0·9852	2·1345	3·8801	6·3152	7·8184	9·5258	11·447	13·592	18·591	24·593	44·440	72·000
0·08000	1·088	1·273	1·445	1·761	2·048	2·315	2·442	2·565	2·685	2·802	3·028	3·244	3·751	4·220
1/ 12·5	0·3418	0·6250	1·0216	2·2127	4·0216	6·5447	8·1021	9·8709	11·861	14·084	19·261	25·479	46·034	74·573
0·08500	1·126	1·317	1·495	1·821	2·118	2·394	2·525	2·652	2·776	2·897	3·130	3·354	3·877	4·361
1/ 11·8	0·3538	0·6467	1·0570	2·2888	4·1592	6·7678	8·3778	10·206	12·264	14·561	19·913	26·339	47·581	77·072
0·09000	1·163	1·361	1·544	1·880	2·186	2·470	2·606	2·737	2·865	2·989	3·230	3·460	4·000	4·499
1/ 11·1	0·3654	0·6679	1·0914	2·3629	4·2932	6·9849	8·6462	10·533	12·656	15·025	20·546	27·175	49·087	79·504
0·09500	1·199	1·403	1·592	1·938	2·253	2·545	2·684	2·820	2·951	3·079	3·327	3·564	4·120	4·633
1/ 10·5	0·3768	0·6885	1·1250	2·4351	4·4238	7·1966	8·9078	10·851	13·038	15·478	21·164	27·991	50·555	81·874
0·10000	1·235	1·444	1·638	1·994	2·318	2·618	2·761	2·900	3·035	3·167	3·422	3·665	4·236	4·764
1/ 10·0	0·3879	0·7087	1·1578	2·5056	4·5513	7·4033	9·1633	11·162	13·410	15·920	21·768	28·787	51·987	84·187
	0·78	0·79	0·81	0·83	0·84	0·86	0·87	0·87	0·88	0·88	0·89	0·90	0·91	0·93

$V_{r(0{\cdot}5)medial}$ for half-full circular pipes.

k_s = **0·003 mm**
S = 0·10000 to 0·30000

ie hydraulic gradient =
1 in 10·0 to 1 in 3·3

Water (or sewage) at 15°C;
full bore conditions.

velocities in ms^{-1}
discharges in litres/sec

Gradient **(Equivalent) Pipe diameters in mm**

Gradient	20	25	30	40	50	60	65	70	75	80	90	100	125	150
0·10000	1·235	1·444	1·638	1·994	2·318	2·618	2·761	2·900	3·035	3·167	3·422	3·665	4·236	4·764
1/ 10·0	0·3879	0·7087	1·1578	2·5056	4·5513	7·4033	9·1633	11·162	13·410	15·920	21·768	28·787	51·987	84·187
0·10500	1·269	1·484	1·683	2·049	2·381	2·690	2·837	2·979	3·118	3·253	3·514	3·764	4·350	4·892
1/ 9·5	0·3988	0·7284	1·1899	2·5746	4·6760	7·6053	9·4129	11·465	13·775	16·352	22·357	29·565	53·387	86·446
0·11000	1·303	1·523	1·728	2·102	2·444	2·760	2·910	3·056	3·199	3·337	3·605	3·861	4·462	5·017
1/ 9·1	0·4094	0·7477	1·2213	2·6421	4·7979	7·8030	9·6571	11·762	14·131	16·775	22·933	30·326	54·755	88·655
0·11500	1·336	1·562	1·771	2·155	2·504	2·828	2·982	3·132	3·278	3·420	3·694	3·956	4·571	5·139
1/ 8·7	0·4198	0·7666	1·2520	2·7082	4·9174	7·9966	9·8963	12·053	14·480	17·189	23·498	31·071	56·095	90·819
0·12000	1·369	1·600	1·814	2·207	2·564	2·895	3·053	3·206	3·355	3·500	3·781	4·049	4·678	5·259
1/ 8·3	0·4300	0·7852	1·2822	2·7730	5·0345	8·1863	10·131	12·339	14·822	17·594	24·051	31·802	57·409	92·938
0·12500	1·401	1·637	1·856	2·257	2·623	2·961	3·122	3·279	3·431	3·579	3·866	4·140	4·783	5·377
1/ 8·0	0·4401	0·8034	1·3118	2·8366	5·1494	8·3725	10·361	12·618	15·158	17·992	24·594	32·518	58·697	95·017
0·13000	1·432	1·673	1·897	2·307	2·680	3·026	3·190	3·350	3·506	3·657	3·950	4·230	4·886	5·492
1/ 7·7	0·4499	0·8213	1·3408	2·8990	5·2623	8·5553	10·587	12·893	15·488	18·383	25·127	33·221	59·961	97·058
0·13500	1·463	1·709	1·937	2·356	2·737	3·089	3·257	3·420	3·579	3·734	4·032	4·318	4·987	5·606
1/ 7·4	0·4596	0·8388	1·3694	2·9604	5·3731	8·7349	10·809	13·163	15·811	18·767	25·650	33·912	61·203	99·062
0·14000	1·493	1·744	1·977	2·404	2·792	3·152	3·323	3·489	3·651	3·809	4·113	4·404	5·087	5·717
1/ 7·1	0·4691	0·8561	1·3975	3·0207	5·4822	8·9114	11·027	13·428	16·129	19·144	26·165	34·591	62·424	101·03
0·14500	1·523	1·779	2·016	2·451	2·847	3·213	3·388	3·557	3·722	3·882	4·192	4·489	5·185	5·827
1/ 6·9	0·4785	0·8731	1·4252	3·0801	5·5894	9·0851	11·241	13·689	16·443	19·515	26·671	35·259	63·624	102·97
0·15000	1·553	1·813	2·055	2·498	2·900	3·274	3·451	3·624	3·792	3·955	4·271	4·573	5·281	5·935
1/ 6·7	0·4877	0·8899	1·4524	3·1385	5·6950	9·2561	11·452	13·946	16·751	19·880	27·169	35·916	64·806	104·87
0·16000	1·610	1·880	2·130	2·589	3·006	3·392	3·576	3·754	3·928	4·097	4·424	4·737	5·469	6·145
1/ 6·3	0·5058	0·9226	1·5056	3·2528	5·9014	9·5903	11·865	14·448	17·353	20·594	28·143	37·201	67·115	108·60
0·17000	1·666	1·944	2·203	2·677	3·108	3·507	3·697	3·881	4·060	4·235	4·572	4·895	5·652	6·350
1/ 5·9	0·5234	0·9545	1·5573	3·3639	6·1020	9·9151	12·267	14·936	17·938	21·288	29·088	38·448	69·357	112·22
0·18000	1·720	2·008	2·274	2·763	3·207	3·619	3·814	4·004	4·189	4·369	4·717	5·050	5·829	6·549
1/ 5·6	0·5405	0·9855	1·6077	3·4721	6·2972	10·231	12·657	15·411	18·507	21·963	30·009	39·662	71·538	115·74
0·19000	1·773	2·069	2·344	2·847	3·304	3·728	3·929	4·125	4·315	4·500	4·858	5·201	6·003	6·743
1/ 5·3	0·5571	1·0157	1·6569	3·5775	6·4875	10·539	13·038	15·873	19·062	22·621	30·906	40·845	73·664	119·16
0·20000	1·825	2·129	2·412	2·929	3·399	3·834	4·041	4·242	4·437	4·628	4·996	5·348	6·172	6·933
1/ 5·0	0·5734	1·0453	1·7048	3·6804	6·6733	10·840	13·409	16·325	19·604	23·263	31·781	42·000	75·738	122·51
0·21000	1·876	2·188	2·478	3·009	3·491	3·938	4·150	4·357	4·557	4·753	5·130	5·491	6·337	7·117
1/ 4·8	0·5894	1·0742	1·7517	3·7810	6·8549	11·134	13·772	16·766	20·133	23·890	32·636	43·128	77·764	125·77
0·22000	1·926	2·246	2·543	3·087	3·582	4·039	4·257	4·469	4·674	4·875	5·261	5·632	6·498	7·298
1/ 4·5	0·6050	1·1024	1·7976	3·8794	7·0325	11·421	14·127	17·198	20·651	24·503	33·472	44·231	79·745	128·97
0·23000	1·974	2·302	2·607	3·164	3·670	4·139	4·362	4·579	4·789	4·994	5·390	5·769	6·656	7·475
1/ 4·3	0·6202	1·1301	1·8426	3·9758	7·2064	11·703	14·475	17·620	21·158	25·104	34·291	45·310	81·684	132·10
0·24000	2·022	2·358	2·669	3·239	3·757	4·237	4·465	4·686	4·902	5·111	5·516	5·904	6·811	7·648
1/ 4·2	0·6352	1·1572	1·8867	4·0703	7·3769	11·979	14·815	18·034	21·654	25·692	35·093	46·369	83·584	135·16
0·25000	2·069	2·412	2·730	3·313	3·842	4·332	4·565	4·792	5·012	5·226	5·640	6·036	6·963	7·818
1/ 4·0	0·6499	1·1839	1·9299	4·1630	7·5442	12·249	15·149	18·441	22·141	26·270	35·880	47·406	85·448	138·16
0·26000	2·115	2·465	2·790	3·385	3·926	4·426	4·664	4·895	5·120	5·339	5·761	6·166	7·112	7·985
1/ 3·8	0·6643	1·2100	1·9724	4·2541	7·7084	12·515	15·478	18·839	22·619	26·836	36·652	48·425	87·276	141·11
0·27000	2·160	2·517	2·849	3·456	4·008	4·519	4·761	4·997	5·226	5·450	5·881	6·293	7·258	8·149
1/ 3·7	0·6785	1·2358	2·0141	4·3435	7·8697	12·776	15·800	19·231	23·089	27·393	37·410	49·425	89·072	144·01
0·28000	2·204	2·569	2·907	3·526	4·089	4·609	4·857	5·097	5·331	5·558	5·998	6·418	7·402	8·310
1/ 3·6	0·6925	1·2610	2·0551	4·4314	8·0283	13·032	16·117	19·616	23·551	27·940	38·156	50·408	90·837	146·85
0·29000	2·248	2·620	2·965	3·595	4·168	4·699	4·951	5·196	5·434	5·666	6·113	6·541	7·543	8·468
1/ 3·4	0·7062	1·2859	2·0955	4·5179	8·1843	13·285	16·428	19·995	24·005	28·478	38·889	51·375	92·572	149·65
0·30000	2·291	2·670	3·021	3·663	4·246	4·786	5·043	5·292	5·535	5·771	6·226	6·662	7·683	8·624
1/ 3·3	0·7198	1·3104	2·1352	4·6031	8·3378	13·533	16·735	20·367	24·452	29·007	39·611	52·326	94·280	152·40
	0·80	0·82	0·83	0·85	0·87	0·89	0·89	0·90	0·90	0·91	0·92	0·92	0·94	0·94

$V_{r(0\cdot5)medial}$ **for half-full circular pipes.**

S = 0·10000 to 0·30000 k_s = **0·003 mm**

A2

(p.1 of 6)

k_s = 0·006 mm
S = 0·00030 to 0·00100

ie hydraulic gradient =
1 in 3333 to 1 in 1000

Water (or sewage) at 15°C;
full bore conditions.

velocities in ms^{-1}
discharges in litres/sec

Gradient	**20**	**25**	**30**	**40**	**50**	**60**	**65**	**70**	**75**	**80**	**90**	**100**	**125**	**150**
	(Equivalent) Pipe diameters in mm													
0·000300				0·071	0·085	0·097	0·103	0·109	0·115	0·121	0·131	0·142	0·166	0·189
1/ 3333				0·0897	0·1665	0·2752	0·3430	0·4203	0·5077	0·6057	0·8355	1·1133	2·0412	3·3440
0·000320				0·074	0·088	0·101	0·107	0·113	0·119	0·125	0·136	0·147	0·173	0·196
1/ 3125				0·0932	0·1729	0·2858	0·3561	0·4363	0·5270	0·6287	0·8671	1·1553	2·1177	3·4686
0·000340				0·077	0·091	0·105	0·111	0·117	0·124	0·130	0·141	0·152	0·179	0·203
1/ 2941				0·0966	0·1792	0·2961	0·3688	0·4519	0·5458	0·6511	0·8978	1·1961	2·1920	3·5898
0·000360				0·080	0·094	0·108	0·115	0·121	0·128	0·134	0·146	0·157	0·185	0·210
1/ 2778				0·1000	0·1853	0·3061	0·3813	0·4671	0·5641	0·6729	0·9278	1·2359	2·2645	3·7079
0·000380				0·082	0·097	0·112	0·119	0·125	0·132	0·138	0·150	0·162	0·190	0·216
1/ 2632				0·1032	0·1913	0·3158	0·3934	0·4819	0·5820	0·6942	0·9570	1·2747	2·3352	3·8231
0·000400				0·085	0·100	0·115	0·122	0·129	0·136	0·142	0·155	0·167	0·196	0·223
1/ 2500				0·1064	0·1971	0·3254	0·4053	0·4964	0·5995	0·7150	0·9856	1·3127	2·4043	3·9357
0·000420				0·087	0·103	0·118	0·126	0·133	0·140	0·146	0·159	0·172	0·201	0·229
1/ 2381				0·1095	0·2028	0·3348	0·4169	0·5106	0·6166	0·7353	1·0136	1·3498	2·4718	4·0458
0·000440				0·090	0·106	0·122	0·129	0·136	0·143	0·150	0·164	0·176	0·207	0·235
1/ 2273				0·1125	0·2084	0·3439	0·4283	0·5245	0·6333	0·7553	1·0409	1·3861	2·5380	4·1536
0·000460				0·092	0·109	0·125	0·132	0·140	0·147	0·154	0·168	0·181	0·212	0·241
1/ 2174				0·1155	0·2139	0·3529	0·4394	0·5382	0·6497	0·7748	1·0678	1·4218	2·6028	4·2593
0·000480				0·094	0·112	0·128	0·136	0·143	0·151	0·158	0·172	0·185	0·217	0·247
1/ 2083				0·1184	0·2192	0·3617	0·4504	0·5515	0·6659	0·7940	1·0941	1·4567	2·6665	4·3629
0·000500			0·077	0·097	0·114	0·131	0·139	0·147	0·154	0·162	0·176	0·190	0·222	0·253
1/ 2000			0·0546	0·1213	0·2245	0·3704	0·4611	0·5647	0·6817	0·8128	1·1200	1·4911	2·7289	4·4646
0·000525			0·079	0·099	0·118	0·135	0·143	0·151	0·159	0·166	0·181	0·195	0·229	0·260
1/ 1905			0·0562	0·1248	0·2310	0·3810	0·4743	0·5808	0·7011	0·8359	1·1517	1·5331	2·8054	4·5893
0·000550			0·082	0·102	0·121	0·138	0·147	0·155	0·163	0·171	0·186	0·200	0·235	0·267
1/ 1818			0·0578	0·1283	0·2373	0·3914	0·4872	0·5965	0·7201	0·8585	1·1827	1·5743	2·8804	4·7113
0·000575			0·084	0·105	0·124	0·142	0·151	0·159	0·167	0·175	0·191	0·206	0·241	0·273
1/ 1739			0·0593	0·1317	0·2436	0·4016	0·4999	0·6120	0·7387	0·8807	1·2131	1·6147	2·9538	4·8309
0·000600			0·086	0·107	0·127	0·146	0·154	0·163	0·171	0·180	0·195	0·211	0·247	0·280
1/ 1667			0·0608	0·1350	0·2497	0·4116	0·5123	0·6272	0·7570	0·9024	1·2430	1·6543	3·0259	4·9482
0·000625			0·088	0·110	0·130	0·149	0·158	0·167	0·175	0·184	0·200	0·216	0·252	0·287
1/ 1600			0·0623	0·1383	0·2557	0·4214	0·5245	0·6421	0·7749	0·9238	1·2723	1·6932	3·0966	5·0634
0·000650			0·090	0·113	0·133	0·152	0·162	0·171	0·179	0·188	0·205	0·220	0·258	0·293
1/ 1538			0·0638	0·1415	0·2616	0·4311	0·5365	0·6567	0·7926	0·9447	1·3011	1·7314	3·1661	5·1765
0·000675			0·092	0·115	0·136	0·156	0·165	0·174	0·183	0·192	0·209	0·225	0·264	0·299
1/ 1481			0·0652	0·1447	0·2674	0·4406	0·5483	0·6711	0·8099	0·9654	1·3294	1·7690	3·2344	5·2877
0·000700			0·094	0·118	0·139	0·159	0·169	0·178	0·187	0·196	0·213	0·230	0·269	0·305
1/ 1429			0·0666	0·1478	0·2731	0·4499	0·5599	0·6853	0·8270	0·9857	1·3573	1·8060	3·3016	5·3971
0·000725			0·096	0·120	0·142	0·162	0·172	0·182	0·191	0·200	0·218	0·235	0·274	0·312
1/ 1379			0·0680	0·1509	0·2787	0·4591	0·5713	0·6992	0·8438	1·0057	1·3847	1·8423	3·3678	5·5047
0·000750			0·098	0·122	0·145	0·166	0·176	0·185	0·195	0·204	0·222	0·239	0·280	0·318
1/ 1333			0·0694	0·1539	0·2842	0·4682	0·5825	0·7130	0·8603	1·0253	1·4117	1·8782	3·4329	5·6108
0·000800			0·102	0·127	0·150	0·172	0·182	0·192	0·202	0·212	0·230	0·248	0·290	0·329
1/ 1250			0·0721	0·1598	0·2951	0·4859	0·6046	0·7399	0·8927	1·0639	1·4646	1·9483	3·5603	5·8182
0·000850		0·092	0·106	0·132	0·156	0·178	0·189	0·199	0·209	0·219	0·238	0·257	0·300	0·341
1/ 1176		0·0450	0·0747	0·1656	0·3056	0·5032	0·6260	0·7660	0·9242	1·1013	1·5160	2·0165	3·6843	6·0199
0·000900		0·095	0·109	0·136	0·161	0·184	0·195	0·206	0·216	0·226	0·246	0·265	0·310	0·352
1/ 1111		0·0466	0·0773	0·1712	0·3159	0·5200	0·6469	0·7915	0·9549	1·1379	1·5661	2·0830	3·8051	6·2163
0·000950		0·098	0·113	0·141	0·166	0·190	0·201	0·212	0·223	0·233	0·254	0·273	0·320	0·363
1/ 1053		0·0481	0·0798	0·1767	0·3259	0·5365	0·6673	0·8164	0·9849	1·1735	1·6150	2·1478	3·9229	6·4080
0·001000		0·101	0·116	0·145	0·171	0·195	0·207	0·218	0·230	0·240	0·261	0·282	0·329	0·373
1/ 1000		0·0496	0·0823	0·1820	0·3357	0·5525	0·6872	0·8407	1·0141	1·2083	1·6628	2·2112	4·0380	6·5952
		–	0·74	0·74	0·75	0·76	0·76	0·77	0·77	0·77	0·78	0·78	0·80	0·81

$V_{r(0·5)medial}$ for half-full circular pipes.

k_s = **0·006 mm**
S = 0·00100 to 0·00300

ie hydraulic gradient =
1 in 1000 to 1 in 333

Water (or sewage) at 15°C;
full bore conditions.

velocities in ms^{-1}
discharges in litres/sec

A2

(p.2 of 6)

Gradient **(Equivalent) Pipe diameters in mm**

	20	25	30	40	50	60	65	70	75	80	90	100	125	150
0·00100		0·101	0·116	0·145	0·171	0·195	0·207	0·218	0·230	0·240	0·261	0·282	0·329	0·373
1/ 1000		0·0496	0·0823	0·1820	0·3357	0·5525	0·6872	0·8407	1·0141	1·2083	1·6628	2·2112	4·0380	6·5952
0·00105		0·104	0·120	0·149	0·176	0·201	0·213	0·225	0·236	0·247	0·269	0·289	0·338	0·384
1/ 952		0·0510	0·0847	0·1873	0·3453	0·5682	0·7067	0·8645	1·0428	1·2424	1·7095	2·2732	4·1505	6·7782
0·00110		0·107	0·123	0·153	0·181	0·206	0·219	0·231	0·242	0·254	0·276	0·297	0·347	0·394
1/ 909		0·0525	0·0870	0·1924	0·3548	0·5836	0·7258	0·8878	1·0708	1·2758	1·7553	2·3339	4·2607	6·9574
0·00115		0·110	0·126	0·157	0·185	0·212	0·224	0·237	0·249	0·260	0·283	0·305	0·356	0·404
1/ 870		0·0539	0·0893	0·1974	0·3640	0·5987	0·7445	0·9107	1·0983	1·3085	1·8001	2·3933	4·3687	7·1330
0·00120		0·113	0·130	0·161	0·190	0·217	0·230	0·242	0·255	0·267	0·290	0·312	0·365	0·413
1/ 833		0·0552	0·0916	0·2024	0·3730	0·6135	0·7628	0·9331	1·1253	1·3405	1·8441	2·4517	4·4746	7·3052
0·00125		0·115	0·133	0·165	0·195	0·222	0·235	0·248	0·261	0·273	0·297	0·319	0·373	0·423
1/ 800		0·0566	0·0938	0·2072	0·3819	0·6280	0·7809	0·9551	1·1518	1·3720	1·8873	2·5089	4·5785	7·4742
0·00130		0·118	0·136	0·169	0·199	0·227	0·241	0·254	0·267	0·279	0·303	0·327	0·381	0·432
1/ 769		0·0579	0·0960	0·2120	0·3906	0·6423	0·7986	0·9767	1·1778	1·4030	1·9297	2·5652	4·6806	7·6402
0·00135		0·121	0·139	0·172	0·203	0·232	0·246	0·259	0·272	0·285	0·310	0·334	0·390	0·442
1/ 741		0·0592	0·0981	0·2167	0·3992	0·6564	0·8160	0·9980	1·2034	1·4334	1·9714	2·6205	4·7810	7·8034
0·00140		0·123	0·142	0·176	0·208	0·237	0·251	0·265	0·278	0·291	0·316	0·341	0·398	0·451
1/ 714		0·0605	0·1003	0·2213	0·4077	0·6702	0·8331	1·0189	1·2286	1·4633	2·0125	2·6749	4·8797	7·9639
0·00145		0·126	0·145	0·180	0·212	0·242	0·256	0·270	0·284	0·297	0·323	0·347	0·406	0·460
1/ 690		0·0618	0·1023	0·2259	0·4160	0·6838	0·8500	1·0395	1·2533	1·4928	2·0529	2·7284	4·9769	8·1218
0·00150		0·128	0·148	0·183	0·216	0·247	0·261	0·275	0·289	0·303	0·329	0·354	0·413	0·468
1/ 667		0·0630	0·1044	0·2304	0·4242	0·6972	0·8666	1·0598	1·2777	1·5218	2·0926	2·7812	5·0726	8·2773
0·00160		0·133	0·153	0·190	0·224	0·256	0·271	0·286	0·300	0·314	0·341	0·367	0·429	0·486
1/ 625		0·0655	0·1084	0·2392	0·4403	0·7234	0·8992	1·0994	1·3255	1·5786	2·1705	2·8843	5·2598	8·5815
0·00170	0·116	0·138	0·159	0·197	0·232	0·265	0·281	0·296	0·311	0·325	0·353	0·380	0·443	0·502
1/ 588	0·0365	0·0679	0·1123	0·2477	0·4559	0·7489	0·9308	1·1380	1·3720	1·6338	2·2462	2·9847	5·4418	8·8773
0·00180	0·120	0·143	0·164	0·204	0·240	0·274	0·290	0·305	0·321	0·336	0·365	0·392	0·458	0·519
1/ 556	0·0378	0·0702	0·1162	0·2560	0·4711	0·7738	0·9616	1·1757	1·4172	1·6876	2·3199	3·0824	5·6190	9·1653
0·00190	0·124	0·148	0·170	0·210	0·247	0·282	0·299	0·315	0·331	0·346	0·376	0·405	0·472	0·535
1/ 526	0·0390	0·0725	0·1199	0·2642	0·4859	0·7980	0·9917	1·2124	1·4614	1·7401	2·3919	3·1778	5·7920	9·4463
0·00200	0·128	0·152	0·175	0·217	0·255	0·291	0·308	0·324	0·341	0·356	0·387	0·416	0·486	0·550
1/ 500	0·0402	0·0747	0·1235	0·2721	0·5004	0·8217	1·0211	1·2482	1·5045	1·7914	2·4622	3·2709	5·9609	9·7206
0·00210	0·132	0·157	0·180	0·223	0·262	0·299	0·316	0·333	0·350	0·366	0·398	0·428	0·499	0·565
1/ 476	0·0414	0·0769	0·1271	0·2799	0·5146	0·8449	1·0498	1·2833	1·5467	1·8416	2·5309	3·3620	6·1260	9·9889
0·00220	0·135	0·161	0·185	0·229	0·269	0·307	0·325	0·342	0·359	0·376	0·408	0·439	0·512	0·580
1/ 455	0·0425	0·0790	0·1306	0·2875	0·5286	0·8676	1·0780	1·3176	1·5880	1·8907	2·5982	3·4512	6·2877	10·251
0·00230	0·139	0·165	0·190	0·235	0·276	0·315	0·333	0·351	0·369	0·386	0·419	0·451	0·525	0·595
1/ 435	0·0437	0·0811	0·1340	0·2950	0·5422	0·8899	1·1056	1·3513	1·6285	1·9388	2·6641	3·5385	6·4460	10·509
0·00240	0·143	0·169	0·194	0·241	0·283	0·322	0·341	0·360	0·378	0·395	0·429	0·461	0·538	0·609
1/ 417	0·0448	0·0831	0·1374	0·3023	0·5556	0·9117	1·1327	1·3843	1·6683	1·9860	2·7288	3·6242	6·6014	10·761
0·00250	0·146	0·173	0·199	0·246	0·290	0·330	0·349	0·368	0·386	0·404	0·439	0·472	0·550	0·623
1/ 400	0·0459	0·0851	0·1407	0·3095	0·5687	0·9332	1·1592	1·4167	1·7073	2·0324	2·7923	3·7084	6·7538	11·008
0·00260	0·149	0·177	0·204	0·252	0·296	0·337	0·357	0·376	0·395	0·413	0·449	0·483	0·563	0·637
1/ 385	0·0470	0·0871	0·1439	0·3166	0·5816	0·9542	1·1853	1·4486	1·7456	2·0779	2·8547	3·7910	6·9035	11·252
0·00270	0·153	0·181	0·208	0·257	0·303	0·345	0·365	0·385	0·404	0·422	0·458	0·493	0·575	0·650
1/ 370	0·0480	0·0891	0·1471	0·3235	0·5943	0·9749	1·2110	1·4799	1·7832	2·1226	2·9160	3·8722	7·0507	11·490
0·00280	0·156	0·185	0·213	0·263	0·309	0·352	0·373	0·393	0·412	0·431	0·468	0·503	0·586	0·664
1/ 357	0·0491	0·0910	0·1503	0·3304	0·6068	0·9953	1·2363	1·5107	1·8203	2·1667	2·9763	3·9521	7·1954	11·725
0·00290	0·159	0·189	0·217	0·268	0·315	0·359	0·380	0·400	0·420	0·440	0·477	0·513	0·598	0·677
1/ 345	0·0501	0·0929	0·1534	0·3371	0·6191	1·0154	1·2611	1·5410	1·8567	2·2100	3·0357	4·0307	7·3378	11·957
0·00300	0·163	0·193	0·221	0·274	0·321	0·366	0·387	0·408	0·428	0·448	0·486	0·523	0·609	0·689
1/ 333	0·0511	0·0947	0·1564	0·3437	0·6312	1·0351	1·2856	1·5709	1·8926	2·2527	3·0941	4·1081	7·4781	12·184
	–	0·74	0·75	0·76	0·77	0·78	0·78	0·79	0·79	0·79	0·80	0·81	0·83	0·84

$V_{r(0\cdot5)medial}$ **for half-full circular pipes.**

k_s = 0·006 mm
S = 0·00300 to 0·01000

ie hydraulic gradient =
1 in 333 to 1 in 100

Water (or sewage) at 15°C;
full bore conditions.

velocities in ms^{-1}
discharges in litres/sec

Gradient **(Equivalent) Pipe diameters in mm**

Gradient	20	25	30	40	50	60	65	70	75	80	90	100	125	150
0·00300	0·163	0·193	0·221	0·274	0·321	0·366	0·387	0·408	0·428	0·448	0·486	0·523	0·609	0·689
1/ 333	0·0511	0·0947	0·1564	0·3437	0·6312	1·0351	1·2856	1·5709	1·8926	2·2527	3·0941	4·1081	7·4781	12·184
0·00320	0·169	0·200	0·230	0·284	0·334	0·380	0·402	0·423	0·444	0·465	0·504	0·542	0·632	0·715
1/ 313	0·0531	0·0984	0·1624	0·3567	0·6549	1·0738	1·3335	1·6293	1·9629	2·3362	3·2085	4·2595	7·7524	12·629
0·00340	0·175	0·208	0·238	0·294	0·345	0·393	0·416	0·438	0·460	0·481	0·522	0·561	0·653	0·739
1/ 294	0·0550	0·1019	0·1682	0·3694	0·6779	1·1114	1·3801	1·6861	2·0312	2·4173	3·3196	4·4068	8·0190	13·062
0·00360	0·181	0·215	0·246	0·304	0·357	0·406	0·430	0·453	0·475	0·497	0·539	0·579	0·675	0·763
1/ 278	0·0569	0·1054	0·1739	0·3817	0·7004	1·1480	1·4255	1·7414	2·0978	2·4964	3·4279	4·5502	8·2787	13·484
0·00380	0·187	0·221	0·254	0·313	0·368	0·419	0·443	0·467	0·490	0·512	0·555	0·597	0·695	0·786
1/ 263	0·0587	0·1087	0·1794	0·3937	0·7223	1·1837	1·4698	1·7954	2·1627	2·5736	3·5336	4·6901	8·5320	13·895
0·00400	0·193	0·228	0·261	0·323	0·379	0·431	0·456	0·480	0·504	0·527	0·572	0·615	0·715	0·809
1/ 250	0·0605	0·1120	0·1848	0·4055	0·7437	1·2187	1·5130	1·8482	2·2261	2·6489	3·6367	4·8267	8·7793	14·296
0·00420	0·198	0·235	0·269	0·332	0·389	0·443	0·469	0·494	0·518	0·542	0·588	0·632	0·735	0·831
1/ 238	0·0623	0·1153	0·1901	0·4170	0·7646	1·2528	1·5553	1·8997	2·2882	2·7226	3·7376	4·9602	9·0210	14·688
0·00440	0·204	0·241	0·276	0·341	0·400	0·455	0·481	0·507	0·532	0·556	0·603	0·648	0·754	0·853
1/ 227	0·0640	0·1184	0·1953	0·4282	0·7851	1·2862	1·5968	1·9502	2·3489	2·7947	3·8364	5·0910	9·2577	15·072
0·00460	0·209	0·248	0·283	0·350	0·410	0·467	0·493	0·520	0·545	0·570	0·618	0·665	0·773	0·874
1/ 217	0·0657	0·1215	0·2003	0·4393	0·8052	1·3190	1·6373	1·9997	2·4084	2·8654	3·9331	5·2191	9·4895	15·448
0·00480	0·214	0·254	0·290	0·358	0·420	0·478	0·505	0·532	0·558	0·584	0·633	0·681	0·792	0·895
1/ 208	0·0674	0·1246	0·2053	0·4501	0·8250	1·3511	1·6772	2·0483	2·4667	2·9347	4·0280	5·3447	9·7168	15·816
0·00500	0·220	0·260	0·297	0·367	0·430	0·489	0·517	0·545	0·571	0·597	0·648	0·696	0·810	0·915
1/ 200	0·0690	0·1275	0·2102	0·4607	0·8443	1·3827	1·7162	2·0959	2·5240	3·0028	4·1211	5·4680	9·9398	16·178
0·00525	0·226	0·267	0·306	0·377	0·442	0·503	0·532	0·560	0·587	0·614	0·666	0·715	0·832	0·941
1/ 190	0·0710	0·1312	0·2162	0·4737	0·8680	1·4213	1·7641	2·1543	2·5942	3·0861	4·2352	5·6190	10·213	16·621
0·00550	0·232	0·275	0·314	0·387	0·454	0·516	0·546	0·575	0·603	0·630	0·683	0·734	0·854	0·965
1/ 182	0·0730	0·1348	0·2221	0·4865	0·8913	1·4592	1·8110	2·2114	2·6629	3·1677	4·3469	5·7668	10·480	17·055
0·00575	0·238	0·282	0·322	0·397	0·466	0·529	0·560	0·589	0·618	0·646	0·700	0·753	0·875	0·989
1/ 174	0·0749	0·1383	0·2278	0·4990	0·9140	1·4963	1·8569	2·2674	2·7302	3·2477	4·4563	5·9116	10·742	17·479
0·00600	0·244	0·289	0·330	0·407	0·477	0·542	0·573	0·603	0·633	0·662	0·717	0·771	0·896	1·013
1/ 167	0·0768	0·1418	0·2335	0·5113	0·9363	1·5327	1·9020	2·3223	2·7962	3·3261	4·5636	6·0536	10·999	17·896
0·00625	0·250	0·296	0·338	0·416	0·488	0·555	0·587	0·617	0·648	0·677	0·734	0·789	0·917	1·036
1/ 160	0·0786	0·1452	0·2390	0·5233	0·9583	1·5684	1·9462	2·3762	2·8610	3·4030	4·6689	6·1929	11·251	18·304
0·00650	0·256	0·302	0·346	0·426	0·499	0·567	0·600	0·631	0·662	0·692	0·750	0·806	0·937	1·059
1/ 154	0·0804	0·1485	0·2445	0·5351	0·9798	1·6034	1·9896	2·4291	2·9246	3·4786	4·7723	6·3298	11·499	18·705
0·00675	0·262	0·309	0·353	0·435	0·510	0·579	0·612	0·645	0·676	0·707	0·766	0·823	0·957	1·081
1/ 148	0·0822	0·1518	0·2499	0·5468	1·0010	1·6379	2·0323	2·4812	2·9871	3·5529	4·8739	6·4643	11·742	19·100
0·00700	0·267	0·316	0·361	0·444	0·520	0·591	0·625	0·658	0·690	0·721	0·782	0·840	0·976	1·103
1/ 143	0·0840	0·1550	0·2551	0·5582	1·0218	1·6718	2·0743	2·5324	3·0487	3·6259	4·9739	6·5966	11·981	19·487
0·00725	0·273	0·322	0·368	0·453	0·531	0·603	0·638	0·671	0·704	0·736	0·797	0·856	0·995	1·124
1/ 138	0·0857	0·1582	0·2603	0·5695	1·0423	1·7052	2·1157	2·5827	3·1092	3·6978	5·0723	6·7267	12·216	19·868
0·00750	0·278	0·329	0·376	0·462	0·541	0·615	0·650	0·684	0·717	0·750	0·813	0·873	1·014	1·146
1/ 133	0·0874	0·1613	0·2655	0·5806	1·0625	1·7381	2·1564	2·6323	3·1688	3·7686	5·1691	6·8548	12·448	20·244
0·00800	0·289	0·341	0·390	0·479	0·561	0·637	0·674	0·709	0·744	0·777	0·842	0·905	1·051	1·187
1/ 125	0·0908	0·1674	0·2755	0·6023	1·1020	1·8024	2·2360	2·7293	3·2854	3·9071	5·3585	7·1054	12·901	20·978
0·00850	0·299	0·353	0·403	0·496	0·581	0·660	0·697	0·734	0·769	0·804	0·871	0·936	1·087	1·227
1/ 118	0·0940	0·1734	0·2852	0·6234	1·1404	1·8649	2·3134	2·8237	3·3988	4·0417	5·5426	7·3490	13·341	21·691
0·00900	0·309	0·365	0·417	0·512	0·600	0·681	0·720	0·758	0·794	0·830	0·899	0·966	1·122	1·267
1/ 111	0·0972	0·1792	0·2947	0·6440	1·1778	1·9258	2·3888	2·9155	3·5091	4·1728	5·7219	7·5862	13·769	22·385
0·00950	0·319	0·377	0·430	0·528	0·618	0·702	0·742	0·781	0·819	0·856	0·927	0·995	1·156	1·305
1/ 105	0·1003	0·1848	0·3040	0·6641	1·2143	1·9852	2·4623	3·0051	3·6168	4·3006	5·8967	7·8175	14·187	23·062
0·01000	0·329	0·388	0·443	0·544	0·637	0·723	0·764	0·804	0·842	0·880	0·954	1·024	1·189	1·342
1/ 100	0·1033	0·1904	0·3130	0·6837	1·2499	2·0432	2·5341	3·0926	3·7219	4·4254	6·0675	8·0433	14·595	23·723
	0·75	0·75	0·76	0·78	0·79	0·80	0·81	0·81	0·82	0·82	0·83	0·84	0·86	0·87

$V_{r(0\cdot5)medial}$ for half-full circular pipes.

k_s = 0·006 mm S = 0·00300 to 0·01000

k_s = 0·006 mm
S = 0·01000 to 0·03000

ie hydraulic gradient =
1 in 100 to 1 in 33·3

Water (or sewage) at 15°C;
full bore conditions.

velocities in ms^{-1}
discharges in litres/sec

Gradient	**(Equivalent) Pipe diameters in mm**													
	20	**25**	**30**	**40**	**50**	**60**	**65**	**70**	**75**	**80**	**90**	**100**	**125**	**150**
0·01000	0·329	0·388	0·443	0·544	0·637	0·723	0·764	0·804	0·842	0·880	0·954	1·024	1·189	1·342
1/ 100	0·1033	0·1904	0·3130	0·6837	1·2499	2·0432	2·5341	3·0926	3·7219	4·4254	6·0675	8·0433	14·595	23·723
0·01050	0·338	0·399	0·455	0·559	0·654	0·743	0·785	0·826	0·866	0·905	0·980	1·052	1·222	1·379
1/ 95	0·1063	0·1958	0·3219	0·7029	1·2848	2·0999	2·6044	3·1782	3·8247	4·5475	6·2343	8·2640	14·994	24·369
0·01100	0·348	0·410	0·468	0·574	0·672	0·762	0·806	0·848	0·889	0·928	1·006	1·080	1·254	1·415
1/ 91	0·1092	0·2011	0·3306	0·7216	1·3189	2·1555	2·6731	3·2619	3·9253	4·6669	6·3977	8·4800	15·384	25·000
0·01150	0·357	0·420	0·480	0·589	0·689	0·782	0·826	0·869	0·911	0·952	1·031	1·107	1·285	1·450
1/ 87	0·1120	0·2063	0·3391	0·7401	1·3524	2·2099	2·7404	3·3439	4·0239	4·7839	6·5577	8·6916	15·766	25·619
0·01200	0·366	0·431	0·491	0·603	0·705	0·800	0·846	0·890	0·933	0·975	1·055	1·133	1·315	1·484
1/ 83	0·1148	0·2114	0·3474	0·7581	1·3852	2·2632	2·8065	3·4244	4·1206	4·8987	6·7145	8·8991	16·141	26·226
0·01250	0·374	0·441	0·503	0·617	0·722	0·819	0·865	0·910	0·954	0·997	1·080	1·159	1·345	1·518
1/ 80	0·1176	0·2164	0·3556	0·7758	1·4174	2·3156	2·8713	3·5033	4·2154	5·0112	6·8685	9·1026	16·508	26·821
0·01300	0·383	0·451	0·514	0·631	0·738	0·837	0·884	0·930	0·975	1·019	1·103	1·184	1·375	1·551
1/ 77	0·1203	0·2214	0·3636	0·7932	1·4490	2·3670	2·9349	3·5808	4·3085	5·1218	7·0196	9·3025	16·869	27·405
0·01350	0·391	0·461	0·526	0·645	0·754	0·855	0·903	0·950	0·996	1·041	1·127	1·209	1·404	1·583
1/ 74	0·1229	0·2262	0·3715	0·8104	1·4801	2·4176	2·9975	3·6570	4·4001	5·2305	7·1682	9·4989	17·224	27·979
0·01400	0·400	0·471	0·537	0·658	0·769	0·873	0·922	0·970	1·016	1·062	1·150	1·234	1·432	1·615
1/ 71	0·1255	0·2310	0·3793	0·8272	1·5106	2·4673	3·0590	3·7320	4·4901	5·3373	7·3142	9·6920	17·572	28·543
0·01450	0·408	0·480	0·547	0·671	0·785	0·890	0·940	0·989	1·036	1·083	1·172	1·258	1·460	1·647
1/ 69	0·1281	0·2357	0·3870	0·8438	1·5407	2·5162	3·1196	3·8057	4·5786	5·4424	7·4579	9·8820	17·915	29·098
0·01500	0·416	0·489	0·558	0·684	0·800	0·907	0·958	1·008	1·056	1·103	1·195	1·282	1·487	1·678
1/ 67	0·1306	0·2403	0·3945	0·8601	1·5703	2·5643	3·1792	3·8783	4·6658	5·5459	7·5994	10·069	18·253	29·645
0·01600	0·432	0·508	0·579	0·710	0·829	0·940	0·993	1·045	1·095	1·144	1·238	1·329	1·541	1·738
1/ 63	0·1356	0·2493	0·4093	0·8920	1·6283	2·6586	3·2957	4·0202	4·8364	5·7483	7·8760	10·435	18·913	30·713
0·01700	0·447	0·526	0·599	0·735	0·858	0·973	1·027	1·080	1·132	1·183	1·280	1·374	1·593	1·797
1/ 59	0·1404	0·2581	0·4237	0·9230	1·6846	2·7501	3·4090	4·1582	5·0021	5·9451	8·1449	10·790	19·554	31·751
0·01800	0·462	0·543	0·619	0·759	0·886	1·004	1·061	1·115	1·169	1·221	1·321	1·418	1·644	1·854
1/ 56	0·1451	0·2667	0·4377	0·9533	1·7395	2·8393	3·5194	4·2926	5·1635	6·1366	8·4066	11·136	20·178	32·761
0·01900	0·476	0·560	0·638	0·782	0·913	1·035	1·093	1·149	1·204	1·258	1·362	1·461	1·694	1·910
1/ 53	0·1497	0·2750	0·4513	0·9828	1·7930	2·9263	3·6270	4·4236	5·3209	6·3233	8·6618	11·473	20·787	33·745
0·02000	0·491	0·577	0·657	0·805	0·940	1·065	1·125	1·183	1·239	1·294	1·401	1·503	1·742	1·964
1/ 50	0·1542	0·2832	0·4646	1·0116	1·8453	3·0112	3·7320	4·5515	5·4745	6·5057	8·9109	11·803	21·381	34·706
0·02100	0·505	0·593	0·676	0·827	0·966	1·094	1·156	1·215	1·273	1·330	1·439	1·544	1·790	2·017
1/ 47·6	0·1585	0·2912	0·4777	1·0398	1·8964	3·0942	3·8347	4·6766	5·6247	6·6839	9·1544	12·125	21·961	35·645
0·02200	0·518	0·609	0·694	0·849	0·991	1·123	1·186	1·247	1·306	1·364	1·476	1·584	1·836	2·069
1/ 45·5	0·1628	0·2990	0·4904	1·0673	1·9464	3·1755	3·9353	4·7989	5·7716	6·8583	9·3927	12·439	22·529	36·563
0·02300	0·532	0·625	0·712	0·871	1·016	1·151	1·216	1·278	1·339	1·398	1·513	1·623	1·881	2·120
1/ 43·5	0·1670	0·3067	0·5030	1·0944	1·9954	3·2551	4·0337	4·9188	5·9156	7·0291	9·6260	12·748	23·085	37·462
0·02400	0·545	0·640	0·729	0·892	1·041	1·179	1·245	1·309	1·371	1·432	1·549	1·662	1·926	2·170
1/ 41·7	0·1712	0·3142	0·5152	1·1208	2·0434	3·3331	4·1302	5·0363	6·0567	7·1965	9·8547	13·050	23·630	38·343
0·02500	0·558	0·655	0·746	0·913	1·065	1·206	1·273	1·339	1·402	1·464	1·584	1·699	1·969	2·219
1/ 40·0	0·1752	0·3216	0·5273	1·1469	2·0906	3·4097	4·2250	5·1516	6·1951	7·3608	10·079	13·347	24·164	39·208
0·02600	0·570	0·670	0·763	0·933	1·088	1·233	1·301	1·368	1·433	1·496	1·619	1·736	2·012	2·267
1/ 38·5	0·1792	0·3289	0·5391	1·1724	2·1369	3·4849	4·3180	5·2649	6·3311	7·5221	10·299	13·638	24·689	40·056
0·02700	0·583	0·685	0·779	0·953	1·111	1·259	1·329	1·397	1·463	1·528	1·653	1·773	2·054	2·314
1/ 37·0	0·1831	0·3360	0·5507	1·1975	2·1824	3·5588	4·4094	5·3761	6·4647	7·6806	10·516	13·924	25·205	40·890
0·02800	0·595	0·699	0·795	0·973	1·134	1·284	1·356	1·425	1·493	1·559	1·686	1·809	2·095	2·360
1/ 35·7	0·1870	0·3430	0·5622	1·2222	2·2271	3·6314	4·4992	5·4855	6·5960	7·8364	10·729	14·205	25·712	41·709
0·02900	0·607	0·713	0·811	0·992	1·157	1·310	1·383	1·453	1·522	1·590	1·719	1·844	2·136	2·406
1/ 34·5	0·1908	0·3499	0·5735	1·2465	2·2711	3·7029	4·5877	5·5931	6·7252	7·9898	10·938	14·482	26·210	42·515
0·03000	0·619	0·727	0·827	1·011	1·179	1·335	1·409	1·481	1·551	1·620	1·752	1·879	2·176	2·451
1/ 33·3	0·1945	0·3567	0·5845	1·2704	2·3145	3·7733	4·6747	5·6991	6·8524	8·1407	11·144	14·754	26·701	43·308
	0·76	0·77	0·78	0·80	0·81	0·83	0·83	0·84	0·84	0·85	0·86	0·87	0·88	0·90

$V_{r(0·5)medial}$ for half-full circular pipes.

$k_s = 0{\cdot}006$ **mm**
S = 0·03000 to 0·10000

ie hydraulic gradient =
1 in 33·3 to 1 in 10·0

Water (or sewage) at 15°C;
full bore conditions.

velocities in ms^{-1}
discharges in litres/sec

Gradient **(Equivalent) Pipe diameters in mm**

Gradient	20	25	30	40	50	60	65	70	75	80	90	100	125	150
0·03000	0·619	0·727	0·827	1·011	1·179	1·335	1·409	1·481	1·551	1·620	1·752	1·879	2·176	2·451
1/ 33·3	0·1945	0·3567	0·5845	1·2704	2·3145	3·7733	4·6747	5·6991	6·8524	8·1407	11·144	14·754	26·701	43·308
0·03200	0·642	0·754	0·858	1·048	1·222	1·383	1·460	1·535	1·607	1·678	1·815	1·946	2·254	2·538
1/ 31·3	0·2018	0·3700	0·6062	1·3172	2·3993	3·9109	4·8449	5·9062	7·1011	8·4357	11·547	15·286	27·660	44·859
0·03400	0·665	0·780	0·888	1·084	1·264	1·431	1·510	1·587	1·662	1·735	1·877	2·012	2·330	2·624
1/ 29·4	0·2089	0·3830	0·6273	1·3627	2·4817	4·0447	5·0103	6·1075	7·3428	8·7224	11·939	15·803	28·591	46·364
0·03600	0·687	0·806	0·917	1·120	1·305	1·477	1·558	1·638	1·715	1·791	1·937	2·076	2·404	2·707
1/ 27·8	0·2158	0·3956	0·6479	1·4070	2·5619	4·1749	5·1713	6·3035	7·5781	9·0015	12·320	16·307	29·498	47·829
0·03800	0·709	0·831	0·945	1·154	1·345	1·521	1·606	1·688	1·767	1·845	1·995	2·139	2·476	2·787
1/ 26·3	0·2226	0·4079	0·6679	1·4502	2·6402	4·3019	5·3283	6·4945	7·8074	9·2735	12·691	16·797	30·381	49·257
0·04000	0·730	0·855	0·973	1·188	1·384	1·565	1·652	1·736	1·818	1·898	2·052	2·200	2·546	2·866
1/ 25·0	0·2292	0·4199	0·6875	1·4924	2·7166	4·4258	5·4816	6·6810	8·0313	9·5390	13·054	17·276	31·244	50·650
0·04200	0·750	0·879	1·000	1·220	1·422	1·608	1·697	1·783	1·867	1·949	2·108	2·259	2·615	2·943
1/ 23·8	0·2357	0·4317	0·7067	1·5336	2·7912	4·5470	5·6313	6·8633	8·2500	9·7985	13·408	17·744	32·086	52·011
0·04400	0·770	0·903	1·026	1·253	1·459	1·650	1·741	1·830	1·916	2·000	2·162	2·317	2·682	3·019
1/ 22·7	0·2420	0·4432	0·7254	1·5740	2·8643	4·6655	5·7779	7·0416	8·4641	10·052	13·754	18·201	32·910	53·342
0·04600	0·790	0·926	1·052	1·284	1·495	1·691	1·784	1·875	1·963	2·049	2·215	2·375	2·747	3·092
1/ 21·7	0·2482	0·4545	0·7438	1·6135	2·9359	4·7816	5·9214	7·2162	8·6737	10·301	14·094	18·649	33·716	54·645
0·04800	0·809	0·948	1·078	1·315	1·531	1·731	1·827	1·920	2·010	2·098	2·268	2·430	2·812	3·164
1/ 20·8	0·2543	0·4655	0·7618	1·6523	3·0060	4·8954	6·0621	7·3874	8·8791	10·545	14·426	19·088	34·507	55·921
0·05000	0·828	0·970	1·103	1·345	1·566	1·771	1·868	1·963	2·055	2·145	2·319	2·485	2·875	3·235
1/ 20·0	0·2602	0·4764	0·7794	1·6904	3·0749	5·0071	6·2001	7·5553	9·0806	10·784	14·752	19·519	35·282	57·173
0·05250	0·852	0·998	1·133	1·382	1·609	1·819	1·919	2·017	2·111	2·204	2·382	2·552	2·952	3·322
1/ 19·0	0·2675	0·4897	0·8011	1·7370	3·1592	5·1438	6·3691	7·7610	9·3274	11·076	15·152	20·046	36·231	58·706
0·05500	0·874	1·024	1·163	1·419	1·651	1·867	1·969	2·069	2·166	2·261	2·443	2·618	3·028	3·407
1/ 18·2	0·2747	0·5027	0·8223	1·7826	3·2417	5·2776	6·5345	7·9621	9·5688	11·363	15·543	20·562	37·159	60·205
0·05750	0·897	1·050	1·193	1·454	1·692	1·913	2·018	2·120	2·219	2·316	2·503	2·682	3·102	3·490
1/ 17·4	0·2817	0·5154	0·8430	1·8272	3·3225	5·4086	6·6964	8·1591	9·8052	11·643	15·925	21·067	38·068	61·672
0·06000	0·919	1·075	1·221	1·489	1·732	1·958	2·066	2·170	2·272	2·371	2·562	2·745	3·175	3·571
1/ 16·7	0·2886	0·5279	0·8634	1·8710	3·4017	5·5370	6·8551	8·3522	10·037	11·918	16·300	21·562	38·958	63·110
0·06250	0·940	1·101	1·250	1·523	1·772	2·003	2·113	2·219	2·323	2·425	2·620	2·807	3·246	3·651
1/ 16·0	0·2953	0·5402	0·8833	1·9140	3·4794	5·6630	7·0108	8·5416	10·264	12·187	16·668	22·048	39·832	64·519
0·06500	0·961	1·125	1·277	1·557	1·811	2·047	2·159	2·268	2·374	2·477	2·677	2·868	3·316	3·729
1/ 15·4	0·3020	0·5523	0·9030	1·9562	3·5557	5·7867	7·1637	8·7275	10·487	12·452	17·029	22·524	40·689	65·903
0·06750	0·982	1·149	1·305	1·590	1·849	2·090	2·204	2·315	2·423	2·529	2·732	2·927	3·384	3·806
1/ 14·8	0·3085	0·5641	0·9222	1·9976	3·6307	5·9082	7·3138	8·9102	10·706	12·712	17·383	22·992	41·530	67·261
0·07000	1·002	1·173	1·332	1·622	1·887	2·132	2·249	2·362	2·472	2·580	2·787	2·986	3·452	3·882
1/ 14·3	0·3149	0·5758	0·9412	2·0384	3·7044	6·0276	7·4615	9·0897	10·922	12·967	17·732	23·452	42·358	68·597
0·07250	1·022	1·196	1·358	1·654	1·924	2·173	2·292	2·408	2·520	2·630	2·841	3·044	3·518	3·956
1/ 13·8	0·3212	0·5872	0·9598	2·0785	3·7769	6·1452	7·6067	9·2664	11·134	13·218	18·074	23·904	43·171	69·910
0·07500	1·042	1·219	1·384	1·685	1·960	2·214	2·335	2·453	2·567	2·679	2·894	3·100	3·583	4·029
1/ 13·3	0·3274	0·5985	0·9782	2·1180	3·8483	6·2608	7·7496	9·4402	11·342	13·466	18·412	24·350	43·972	71·202
0·08000	1·081	1·264	1·435	1·747	2·031	2·294	2·420	2·541	2·660	2·775	2·998	3·211	3·711	4·172
1/ 12·5	0·3396	0·6206	1·0141	2·1952	3·9878	6·4869	8·0290	9·7800	11·750	13·949	19·071	25·220	45·536	73·726
0·08500	1·119	1·308	1·484	1·807	2·100	2·372	2·501	2·627	2·749	2·868	3·098	3·319	3·834	4·311
1/ 11·8	0·3514	0·6421	1·0490	2·2703	4·1234	6·7066	8·3004	10·110	12·146	14·418	19·711	26·065	47·055	76·177
0·09000	1·155	1·351	1·532	1·865	2·167	2·448	2·581	2·711	2·837	2·959	3·196	3·423	3·955	4·446
1/ 11·1	0·3629	0·6630	1·0830	2·3433	4·2554	6·9203	8·5645	10·431	12·531	14·875	20·334	26·887	48·533	78·561
0·09500	1·191	1·392	1·579	1·921	2·233	2·521	2·659	2·792	2·922	3·048	3·292	3·525	4·072	4·577
1/ 10·5	0·3742	0·6834	1·1162	2·4145	4·3840	7·1287	8·8219	10·744	12·907	15·320	20·941	27·687	49·972	80·883
0·10000	1·226	1·433	1·625	1·977	2·297	2·593	2·734	2·871	3·004	3·134	3·385	3·625	4·187	4·705
1/ 10·0	0·3851	0·7033	1·1485	2·4840	4·5096	7·3320	9·0731	11·050	13·273	15·755	21·534	28·469	51·377	83·148
	0·78	0·79	0·80	0·82	0·84	0·86	0·86	0·87	0·87	0·88	0·89	0·89	0·91	0·92

$V_{r(0{\cdot}5)medial}$ **for half-full circular pipes.**

$k_s = 0{\cdot}006$ **mm** S = 0·03000 to 0·10000

Gradient	(Equivalent) Pipe diameters in mm													
	20	25	30	40	50	60	65	70	75	80	90	100	125	150
0·10000	1·226	1·433	1·625	1·977	2·297	2·593	2·734	2·871	3·004	3·134	3·385	3·625	4·187	4·705
1/ 10·0	0·3851	0·7033	1·1485	2·4840	4·5096	7·3320	9·0731	11·050	13·273	15·755	21·534	28·469	51·377	83·148
0·10500	1·260	1·472	1·670	2·031	2·359	2·663	2·808	2·949	3·085	3·219	3·476	3·722	4·298	4·830
1/ 9·5	0·3959	0·7227	1·1802	2·5519	4·6323	7·5307	9·3185	11·348	13·631	16·179	22·112	29·232	52·748	85·361
0·11000	1·293	1·511	1·713	2·084	2·420	2·732	2·881	3·025	3·165	3·301	3·565	3·817	4·408	4·953
1/ 9·1	0·4063	0·7418	1·2111	2·6184	4·7523	7·7250	9·5586	11·640	13·981	16·594	22·678	29·979	54·089	87·524
0·11500	1·326	1·549	1·756	2·135	2·480	2·799	2·951	3·099	3·242	3·382	3·652	3·910	4·515	5·073
1/ 8·7	0·4166	0·7604	1·2414	2·6835	4·8698	7·9153	9·7936	11·926	14·324	17·000	23·232	30·709	55·402	89·640
0·12000	1·358	1·586	1·798	2·186	2·539	2·865	3·021	3·172	3·318	3·461	3·737	4·001	4·619	5·190
1/ 8·3	0·4267	0·7787	1·2712	2·7473	4·9850	8·1018	10·024	12·206	14·660	17·398	23·774	31·425	56·688	91·714
0·12500	1·390	1·623	1·840	2·236	2·596	2·930	3·089	3·243	3·393	3·539	3·821	4·091	4·722	5·305
1/ 8·0	0·4366	0·7967	1·3003	2·8099	5·0980	8·2847	10·250	12·480	14·989	17·789	24·307	32·127	57·949	93·747
0·13000	1·421	1·659	1·880	2·285	2·653	2·994	3·156	3·313	3·466	3·615	3·903	4·178	4·823	5·418
1/ 7·7	0·4463	0·8143	1·3290	2·8713	5·2089	8·4642	10·472	12·750	15·313	18·172	24·829	32·816	59·186	95·742
0·13500	1·451	1·694	1·920	2·333	2·708	3·056	3·221	3·382	3·538	3·690	3·984	4·264	4·922	5·529
1/ 7·4	0·4559	0·8316	1·3571	2·9317	5·3179	8·6405	10·689	13·015	15·630	18·548	25·342	33·493	60·401	97·701
0·14000	1·481	1·729	1·959	2·380	2·763	3·117	3·286	3·449	3·609	3·764	4·063	4·349	5·019	5·638
1/ 7·1	0·4653	0·8487	1·3848	2·9910	5·4250	8·8139	10·903	13·275	15·942	18·918	25·846	34·158	61·595	99·625
0·14500	1·510	1·763	1·998	2·427	2·817	3·178	3·349	3·516	3·678	3·836	4·141	4·432	5·115	5·745
1/ 6·9	0·4745	0·8654	1·4120	3·0494	5·5303	8·9843	11·114	13·531	16·249	19·282	26·342	34·811	62·769	101·52
0·15000	1·539	1·797	2·035	2·472	2·869	3·237	3·412	3·581	3·746	3·907	4·217	4·514	5·209	5·850
1/ 6·7	0·4836	0·8819	1·4388	3·1069	5·6340	9·1521	11·321	13·783	16·551	19·639	26·830	35·455	63·924	103·38
0·16000	1·596	1·862	2·110	2·562	2·973	3·353	3·534	3·709	3·880	4·046	4·367	4·674	5·393	6·056
1/ 6·3	0·5014	0·9142	1·4912	3·2192	5·8366	9·4799	11·726	14·275	17·141	20·339	27·783	36·711	66·180	107·02
0·17000	1·651	1·926	2·182	2·649	3·073	3·465	3·652	3·833	4·010	4·181	4·513	4·830	5·571	6·256
1/ 5·9	0·5187	0·9455	1·5420	3·3283	6·0335	9·7984	12·119	14·753	17·714	21·018	28·708	37·932	68·370	110·55
0·18000	1·705	1·988	2·252	2·733	3·170	3·575	3·767	3·954	4·136	4·313	4·654	4·981	5·745	6·450
1/ 5·6	0·5356	0·9761	1·5916	3·4345	6·2250	10·108	12·502	15·218	18·272	21·678	29·608	39·118	70·500	113·98
0·19000	1·757	2·049	2·320	2·815	3·265	3·682	3·880	4·072	4·259	4·441	4·792	5·128	5·914	6·639
1/ 5·3	0·5520	1·0058	1·6399	3·5380	6·4116	10·410	12·874	15·671	18·815	22·322	30·485	40·274	72·574	117·32
0·20000	1·808	2·108	2·387	2·896	3·358	3·786	3·989	4·187	4·379	4·566	4·926	5·271	6·079	6·823
1/ 5·0	0·5680	1·0349	1·6870	3·6390	6·5937	10·704	13·238	16·112	19·344	22·950	31·340	41·402	74·597	120·58
0·21000	1·858	2·166	2·452	2·974	3·449	3·888	4·096	4·299	4·496	4·688	5·058	5·412	6·240	7·003
1/ 4·8	0·5837	1·0632	1·7331	3·7376	6·7716	10·992	13·593	16·544	19·862	23·563	32·175	42·503	76·573	123·76
0·22000	1·907	2·223	2·516	3·051	3·537	3·987	4·201	4·409	4·610	4·807	5·186	5·549	6·397	7·179
1/ 4·5	0·5990	1·0910	1·7781	3·8342	6·9456	11·273	13·940	16·966	20·368	24·162	32·992	43·579	78·504	126·87
0·23000	1·955	2·278	2·578	3·126	3·624	4·085	4·303	4·516	4·722	4·924	5·312	5·683	6·551	7·352
1/ 4·3	0·6141	1·1182	1·8223	3·9287	7·1159	11·549	14·280	17·379	20·863	24·749	33·791	44·633	80·394	129·92
0·24000	2·001	2·332	2·639	3·200	3·709	4·180	4·404	4·621	4·832	5·038	5·435	5·814	6·702	7·520
1/ 4·2	0·6288	1·1449	1·8655	4·0213	7·2828	11·819	14·613	17·784	21·348	25·323	34·574	45·665	82·245	132·90
0·25000	2·047	2·386	2·699	3·272	3·792	4·274	4·502	4·724	4·940	5·150	5·555	5·943	6·850	7·686
1/ 4·0	0·6432	1·1710	1·9079	4·1121	7·4465	12·083	14·940	18·181	21·824	25·887	35·341	46·676	84·059	135·82
0·26000	2·093	2·438	2·758	3·343	3·874	4·365	4·599	4·825	5·046	5·260	5·674	6·069	6·995	7·848
1/ 3·8	0·6574	1·1967	1·9496	4·2012	7·6071	12·343	15·260	18·570	22·291	26·440	36·095	47·669	85·839	138·69
0·27000	2·137	2·489	2·816	3·413	3·955	4·456	4·694	4·925	5·149	5·368	5·790	6·194	7·137	8·007
1/ 3·7	0·6714	1·2219	1·9905	4·2888	7·7649	12·598	15·575	18·952	22·749	26·983	36·834	48·644	87·587	141·50
0·28000	2·181	2·540	2·873	3·481	4·034	4·544	4·787	5·022	5·251	5·474	5·904	6·315	7·277	8·164
1/ 3·6	0·6851	1·2467	2·0307	4·3749	7·9199	12·848	15·884	19·328	23·199	27·516	37·561	49·601	89·304	144·26
0·29000	2·224	2·590	2·929	3·549	4·111	4·631	4·879	5·118	5·351	5·579	6·016	6·435	7·415	8·317
1/ 3·4	0·6986	1·2711	2·0703	4·4595	8·0724	13·095	16·188	19·698	23·642	28·041	38·275	50·543	90·992	146·98
0·30000	2·266	2·638	2·984	3·615	4·188	4·717	4·969	5·213	5·450	5·681	6·127	6·553	7·550	8·469
1/ 3·3	0·7118	1·2951	2·1092	4·5428	8·2224	13·337	16·488	20·061	24·078	28·557	38·978	51·469	92·652	149·65
	0·80	0·81	0·83	0·85	0·87	0·88	0·89	0·89	0·90	0·90	0·91	0·92	0·93	0·94

$V_{r(0·5)medial}$ for half-full circular pipes.

A3
(p.1 of 6)

k_s = 0·015 mm
S = 0·00030 to 0·00100

ie hydraulic gradient =
1 in 3333 to 1 in 1000

Water (or sewage) at 15°C;
full bore conditions.

velocities in ms^{-1}
discharges in litres/sec

Gradient	(Equivalent) Pipe diameters in mm													
	20	25	30	40	50	60	65	70	75	80	90	100	125	150
0·000300				0·071	0·085	0·097	0·103	0·109	0·115	0·120	0·131	0·141	0·166	0·189
1/ 3333				0·0895	0·1661	0·2745	0·3420	0·4191	0·5063	0·6040	0·8331	1·1100	2·0348	3·3328
0·000320				0·074	0·088	0·101	0·107	0·113	0·119	0·125	0·136	0·147	0·172	0·196
1/ 3125				0·0930	0·1725	0·2850	0·3551	0·4351	0·5255	0·6269	0·8645	1·1518	2·1108	3·4567
0·000340				0·077	0·091	0·104	0·111	0·117	0·123	0·129	0·141	0·152	0·178	0·202
1/ 2941				0·0964	0·1787	0·2952	0·3678	0·4506	0·5442	0·6492	0·8951	1·1924	2·1847	3·5772
0·000360				0·079	0·094	0·108	0·115	0·121	0·127	0·133	0·145	0·157	0·184	0·209
1/ 2778				0·0997	0·1848	0·3052	0·3802	0·4657	0·5624	0·6709	0·9249	1·2319	2·2567	3·6946
0·000380				0·082	0·097	0·111	0·118	0·125	0·131	0·138	0·150	0·162	0·190	0·216
1/ 2632				0·1029	0·1907	0·3149	0·3922	0·4805	0·5802	0·6920	0·9540	1·2705	2·3270	3·8091
0·000400				0·084	0·100	0·115	0·122	0·129	0·135	0·142	0·154	0·167	0·195	0·222
1/ 2500				0·1061	0·1965	0·3244	0·4041	0·4949	0·5976	0·7127	0·9824	1·3083	2·3957	3·9209
0·000420				0·087	0·103	0·118	0·125	0·132	0·139	0·146	0·159	0·171	0·201	0·228
1/ 2381				0·1092	0·2022	0·3337	0·4156	0·5090	0·6146	0·7330	1·0102	1·3452	2·4628	4·0303
0·000440				0·089	0·106	0·121	0·129	0·136	0·143	0·150	0·163	0·176	0·206	0·234
1/ 2273				0·1122	0·2078	0·3429	0·4269	0·5229	0·6313	0·7528	1·0374	1·3813	2·5286	4·1373
0·000460				0·092	0·109	0·124	0·132	0·139	0·147	0·154	0·167	0·180	0·211	0·240
1/ 2174				0·1152	0·2132	0·3518	0·4380	0·5364	0·6476	0·7722	1·0641	1·4167	2·5930	4·2423
0·000480				0·094	0·111	0·128	0·135	0·143	0·150	0·157	0·171	0·185	0·216	0·246
1/ 2083				0·1181	0·2186	0·3606	0·4489	0·5497	0·6636	0·7913	1·0903	1·4514	2·6562	4·3452
0·000500			0·077	0·096	0·114	0·131	0·139	0·146	0·154	0·161	0·175	0·189	0·222	0·252
1/ 2000			0·0544	0·1210	0·2238	0·3692	0·4596	0·5628	0·6794	0·8100	1·1160	1·4856	2·7182	4·4462
0·000525			0·079	0·099	0·117	0·134	0·142	0·150	0·158	0·166	0·180	0·194	0·228	0·259
1/ 1905			0·0560	0·1245	0·2303	0·3797	0·4727	0·5788	0·6986	0·8329	1·1475	1·5274	2·7942	4·5699
0·000550			0·081	0·102	0·120	0·138	0·146	0·154	0·162	0·170	0·185	0·200	0·234	0·265
1/ 1818			0·0576	0·1279	0·2366	0·3901	0·4855	0·5945	0·7175	0·8554	1·1783	1·5683	2·8686	4·6911
0·000575			0·084	0·104	0·124	0·142	0·150	0·158	0·167	0·175	0·190	0·205	0·240	0·272
1/ 1739			0·0591	0·1313	0·2428	0·4002	0·4981	0·6098	0·7360	0·8774	1·2085	1·6084	2·9416	4·8098
0·000600			0·086	0·107	0·127	0·145	0·154	0·162	0·171	0·179	0·195	0·210	0·246	0·279
1/ 1667			0·0606	0·1346	0·2488	0·4102	0·5105	0·6249	0·7542	0·8990	1·2382	1·6477	3·0131	4·9262
0·000625			0·088	0·110	0·130	0·149	0·157	0·166	0·175	0·183	0·199	0·215	0·251	0·285
1/ 1600			0·0621	0·1379	0·2548	0·4199	0·5226	0·6397	0·7720	0·9202	1·2673	1·6864	3·0833	5·0404
0·000650			0·090	0·112	0·133	0·152	0·161	0·170	0·179	0·187	0·204	0·220	0·257	0·292
1/ 1538			0·0636	0·1410	0·2607	0·4295	0·5345	0·6543	0·7895	0·9411	1·2959	1·7243	3·1523	5·1527
0·000675			0·092	0·115	0·136	0·155	0·165	0·174	0·183	0·191	0·208	0·224	0·262	0·298
1/ 1481			0·0650	0·1442	0·2664	0·4390	0·5462	0·6686	0·8068	0·9616	1·3240	1·7616	3·2201	5·2630
0·000700			0·094	0·117	0·139	0·159	0·168	0·177	0·186	0·195	0·212	0·229	0·268	0·304
1/ 1429			0·0664	0·1473	0·2721	0·4483	0·5578	0·6826	0·8237	0·9817	1·3517	1·7983	3·2867	5·3715
0·000725			0·096	0·120	0·141	0·162	0·172	0·181	0·190	0·199	0·217	0·234	0·273	0·310
1/ 1379			0·0678	0·1503	0·2777	0·4574	0·5691	0·6965	0·8404	1·0016	1·3789	1·8344	3·3524	5·4782
0·000750			0·098	0·122	0·144	0·165	0·175	0·185	0·194	0·203	0·221	0·238	0·278	0·316
1/ 1333			0·0692	0·1533	0·2832	0·4664	0·5803	0·7101	0·8568	1·0211	1·4057	1·8700	3·4170	5·5833
0·000800			0·102	0·127	0·150	0·171	0·181	0·191	0·201	0·211	0·229	0·247	0·289	0·328
1/ 1250			0·0719	0·1592	0·2940	0·4840	0·6022	0·7369	0·8890	1·0594	1·4582	1·9396	3·5434	5·7890
0·000850		0·091	0·105	0·131	0·155	0·177	0·188	0·198	0·208	0·218	0·237	0·256	0·299	0·339
1/ 1176		0·0449	0·0745	0·1650	0·3044	0·5012	0·6234	0·7628	0·9203	1·0966	1·5092	2·0073	3·6663	5·9889
0·000900		0·095	0·109	0·136	0·160	0·183	0·194	0·205	0·215	0·225	0·245	0·264	0·309	0·350
1/ 1111		0·0464	0·0770	0·1705	0·3147	0·5179	0·6442	0·7882	0·9507	1·1328	1·5589	2·0732	3·7860	6·1836
0·000950		0·098	0·113	0·140	0·165	0·189	0·200	0·211	0·222	0·232	0·253	0·272	0·318	0·361
1/ 1053		0·0479	0·0795	0·1760	0·3246	0·5342	0·6644	0·8129	0·9805	1·1682	1·6074	2·1375	3·9028	6·3734
0·001000		0·101	0·116	0·144	0·170	0·195	0·206	0·217	0·229	0·239	0·260	0·280	0·327	0·371
1/ 1000		0·0494	0·0820	0·1813	0·3344	0·5501	0·6842	0·8370	1·0095	1·2027	1·6548	2·2003	4·0169	6·5589
		–	0·74	0·74	0·75	0·76	0·76	0·76	0·77	0·77	0·78	0·78	0·80	0·81

$V_{r(0{\cdot}5)medial}$ for half-full circular pipes.

k_s = 0·015 mm S = 0·00030 to 0·00100

k_s = 0·015 mm
S = 0·00100 to 0·00300

ie hydraulic gradient =
1 in 1000 to 1 in 333

Water (or sewage) at 15°C;
full bore conditions.

velocities in ms^{-1}
discharges in litres/sec

Gradient	(Equivalent) Pipe diameters in mm													
	20	**25**	**30**	**40**	**50**	**60**	**65**	**70**	**75**	**80**	**90**	**100**	**125**	**150**
0·00100		0·101	0·116	0·144	0·170	0·195	0·206	0·217	0·229	0·239	0·260	0·280	0·327	0·371
1/ 1000		0·0494	0·0820	0·1813	0·3344	0·5501	0·6842	0·8370	1·0095	1·2027	1·6548	2·2003	4·0169	6·5589
0·00105		0·104	0·119	0·148	0·175	0·200	0·212	0·224	0·235	0·246	0·267	0·288	0·336	0·381
1/ 952		0·0509	0·0844	0·1865	0·3439	0·5657	0·7035	0·8606	1·0379	1·2365	1·7012	2·2618	4·1284	6·7401
0·00110		0·107	0·123	0·152	0·180	0·205	0·218	0·230	0·241	0·253	0·275	0·296	0·345	0·391
1/ 909		0·0523	0·0867	0·1916	0·3532	0·5810	0·7225	0·8837	1·0658	1·2696	1·7465	2·3219	4·2375	6·9176
0·00115		0·109	0·126	0·156	0·185	0·211	0·223	0·236	0·247	0·259	0·282	0·303	0·354	0·401
1/ 870		0·0537	0·0890	0·1966	0·3624	0·5960	0·7410	0·9064	1·0930	1·3020	1·7910	2·3809	4·3445	7·0914
0·00120		0·112	0·129	0·160	0·189	0·216	0·229	0·241	0·253	0·265	0·288	0·311	0·363	0·411
1/ 833		0·0550	0·0912	0·2015	0·3714	0·6107	0·7593	0·9286	1·1198	1·3339	1·8346	2·4387	4·4494	7·2618
0·00125		0·115	0·132	0·164	0·194	0·221	0·234	0·247	0·259	0·272	0·295	0·318	0·371	0·420
1/ 800		0·0564	0·0934	0·2064	0·3802	0·6251	0·7771	0·9504	1·1461	1·3651	1·8774	2·4954	4·5523	7·4291
0·00130		0·118	0·135	0·168	0·198	0·226	0·239	0·253	0·265	0·278	0·302	0·325	0·379	0·430
1/ 769		0·0577	0·0956	0·2111	0·3889	0·6392	0·7947	0·9719	1·1719	1·3957	1·9195	2·5511	4·6534	7·5934
0·00135		0·120	0·138	0·172	0·202	0·231	0·245	0·258	0·271	0·284	0·308	0·332	0·387	0·439
1/ 741		0·0590	0·0977	0·2158	0·3974	0·6532	0·8120	0·9929	1·1972	1·4259	1·9608	2·6059	4·7527	7·7548
0·00140		0·123	0·141	0·175	0·207	0·236	0·250	0·263	0·277	0·290	0·315	0·339	0·395	0·448
1/ 714		0·0603	0·0998	0·2204	0·4058	0·6669	0·8290	1·0137	1·2222	1·4556	2·0014	2·6598	4·8504	7·9135
0·00145		0·125	0·144	0·179	0·211	0·241	0·255	0·269	0·282	0·295	0·321	0·345	0·403	0·457
1/ 690		0·0615	0·1019	0·2249	0·4140	0·6804	0·8457	1·0341	1·2467	1·4847	2·0414	2·7128	4·9466	8·0697
0·00150		0·128	0·147	0·182	0·215	0·245	0·260	0·274	0·288	0·301	0·327	0·352	0·411	0·465
1/ 667		0·0628	0·1040	0·2293	0·4222	0·6936	0·8622	1·0542	1·2709	1·5135	2·0808	2·7650	5·0412	8·2235
0·00160		0·133	0·153	0·189	0·223	0·255	0·270	0·284	0·298	0·312	0·339	0·365	0·426	0·482
1/ 625		0·0652	0·1079	0·2380	0·4381	0·7196	0·8944	1·0935	1·3182	1·5697	2·1579	2·8671	5·2264	8·5242
0·00170	0·116	0·138	0·158	0·196	0·231	0·263	0·279	0·294	0·309	0·323	0·351	0·378	0·441	0·499
1/ 588	0·0364	0·0676	0·1118	0·2465	0·4535	0·7449	0·9257	1·1317	1·3642	1·6244	2·2328	2·9664	5·4064	8·8165
0·00180	0·120	0·142	0·164	0·203	0·239	0·272	0·288	0·304	0·319	0·334	0·362	0·390	0·455	0·515
1/ 556	0·0376	0·0699	0·1156	0·2548	0·4686	0·7696	0·9563	1·1690	1·4090	1·6777	2·3058	3·0631	5·5816	9·1011
0·00190	0·124	0·147	0·169	0·209	0·246	0·281	0·297	0·313	0·329	0·344	0·374	0·402	0·469	0·531
1/ 526	0·0388	0·0721	0·1193	0·2628	0·4834	0·7936	0·9861	1·2053	1·4527	1·7296	2·3770	3·1574	5·7526	9·3786
0·00200	0·127	0·151	0·174	0·215	0·253	0·289	0·306	0·322	0·338	0·354	0·385	0·414	0·482	0·546
1/ 500	0·0400	0·0743	0·1229	0·2707	0·4977	0·8171	1·0152	1·2408	1·4954	1·7804	2·4465	3·2496	5·9194	9·6495
0·00210	0·131	0·156	0·179	0·222	0·261	0·297	0·315	0·331	0·348	0·364	0·395	0·425	0·496	0·561
1/ 476	0·0412	0·0765	0·1265	0·2784	0·5118	0·8400	1·0436	1·2755	1·5372	1·8300	2·5145	3·3396	6·0826	9·9143
0·00220	0·135	0·160	0·184	0·228	0·268	0·305	0·323	0·340	0·357	0·374	0·406	0·436	0·509	0·576
1/ 455	0·0424	0·0786	0·1299	0·2860	0·5256	0·8625	1·0715	1·3095	1·5781	1·8786	2·5811	3·4277	6·2422	10·173
0·00230	0·138	0·164	0·189	0·233	0·275	0·313	0·331	0·349	0·366	0·383	0·416	0·447	0·521	0·590
1/ 435	0·0435	0·0807	0·1333	0·2934	0·5391	0·8845	1·0988	1·3428	1·6181	1·9262	2·6463	3·5141	6·3986	10·427
0·00240	0·142	0·168	0·193	0·239	0·281	0·320	0·339	0·357	0·375	0·392	0·426	0·458	0·534	0·604
1/ 417	0·0446	0·0827	0·1367	0·3006	0·5523	0·9061	1·1256	1·3755	1·6574	1·9729	2·7102	3·5988	6·5520	10·676
0·00250	0·145	0·173	0·198	0·245	0·288	0·328	0·347	0·366	0·384	0·402	0·436	0·469	0·546	0·618
1/ 400	0·0457	0·0847	0·1399	0·3078	0·5653	0·9274	1·1519	1·4075	1·6960	2·0187	2·7729	3·6818	6·7024	10·920
0·00260	0·149	0·177	0·203	0·250	0·294	0·335	0·355	0·374	0·392	0·411	0·446	0·479	0·558	0·632
1/ 385	0·0467	0·0867	0·1432	0·3148	0·5781	0·9482	1·1777	1·4390	1·7338	2·0637	2·8346	3·7634	6·8502	11·160
0·00270	0·152	0·180	0·207	0·256	0·301	0·343	0·363	0·382	0·401	0·419	0·455	0·489	0·570	0·645
1/ 370	0·0478	0·0886	0·1463	0·3217	0·5907	0·9687	1·2031	1·4700	1·7711	2·1079	2·8951	3·8436	6·9954	11·396
0·00280	0·155	0·184	0·211	0·261	0·307	0·350	0·370	0·390	0·409	0·428	0·464	0·499	0·582	0·658
1/ 357	0·0488	0·0905	0·1494	0·3284	0·6030	0·9888	1·2280	1·5004	1·8077	2·1514	2·9547	3·9225	7·1381	11·627
0·00290	0·159	0·188	0·216	0·267	0·313	0·357	0·377	0·398	0·417	0·437	0·474	0·509	0·593	0·671
1/ 345	0·0498	0·0924	0·1525	0·3351	0·6152	1·0087	1·2526	1·5304	1·8437	2·1942	3·0133	4·0001	7·2786	11·855
0·00300	0·162	0·192	0·220	0·272	0·319	0·364	0·385	0·405	0·425	0·445	0·483	0·519	0·604	0·684
1/ 333	0·0508	0·0942	0·1555	0·3417	0·6271	1·0282	1·2768	1·5599	1·8792	2·2364	3·0710	4·0765	7·4168	12·079
	–	0·74	0·75	0·76	0·77	0·78	0·78	0·78	0·79	0·79	0·80	0·81	0·82	0·84

$V_{r(0 \cdot 5)medial}$ for half-full circular pipes.

k_s = 0·015 mm
S = 0·00300 to 0·01000

ie hydraulic gradient =
1 in 333 to 1 in 100

Water (or sewage) at 15°C;
full bore conditions.

velocities in ms^{-1}
discharges in litres/sec

Gradient **(Equivalent) Pipe diameters in mm**

Gradient	20	25	30	40	50	60	65	70	75	80	90	100	125	150
0·00300	0·162	0·192	0·220	0·272	0·319	0·364	0·385	0·405	0·425	0·445	0·483	0·519	0·604	0·684
1/ 333	0·0508	0·0942	0·1555	0·3417	0·6271	1·0282	1·2768	1·5599	1·8792	2·2364	3·0710	4·0765	7·4168	12·079
0·00320	0·168	0·199	0·228	0·282	0·331	0·377	0·399	0·420	0·441	0·461	0·500	0·538	0·626	0·708
1/ 313	0·0528	0·0978	0·1614	0·3545	0·6506	1·0664	1·3241	1·6176	1·9486	2·3188	3·1838	4·2259	7·6872	12·518
0·00340	0·174	0·206	0·237	0·292	0·343	0·390	0·413	0·435	0·456	0·477	0·518	0·557	0·648	0·732
1/ 294	0·0547	0·1013	0·1672	0·3670	0·6733	1·1035	1·3702	1·6737	2·0160	2·3989	3·2935	4·3711	7·9500	12·944
0·00360	0·180	0·213	0·244	0·302	0·354	0·403	0·426	0·449	0·471	0·493	0·535	0·575	0·669	0·756
1/ 278	0·0566	0·1047	0·1728	0·3792	0·6955	1·1397	1·4150	1·7283	2·0817	2·4770	3·4003	4·5125	8·2058	13·359
0·00380	0·186	0·220	0·252	0·311	0·365	0·416	0·440	0·463	0·486	0·508	0·551	0·592	0·689	0·779
1/ 263	0·0584	0·1081	0·1783	0·3911	0·7172	1·1750	1·4587	1·7816	2·1458	2·5531	3·5045	4·6504	8·4551	13·763
0·00400	0·192	0·227	0·260	0·320	0·376	0·428	0·452	0·476	0·500	0·523	0·567	0·609	0·709	0·801
1/ 250	0·0602	0·1113	0·1836	0·4027	0·7383	1·2095	1·5014	1·8337	2·2083	2·6274	3·6062	4·7850	8·6986	14·157
0·00420	0·197	0·233	0·267	0·330	0·387	0·440	0·465	0·490	0·514	0·537	0·582	0·626	0·728	0·823
1/ 238	0·0619	0·1145	0·1889	0·4141	0·7590	1·2432	1·5431	1·8846	2·2695	2·7000	3·7056	4·9165	8·9365	14·543
0·00440	0·203	0·240	0·274	0·338	0·397	0·451	0·477	0·503	0·527	0·551	0·598	0·642	0·747	0·844
1/ 227	0·0636	0·1177	0·1940	0·4252	0·7793	1·2762	1·5840	1·9344	2·3294	2·7711	3·8029	5·0453	9·1693	14·920
0·00460	0·208	0·246	0·282	0·347	0·407	0·463	0·489	0·515	0·541	0·565	0·613	0·658	0·766	0·865
1/ 217	0·0653	0·1207	0·1990	0·4361	0·7991	1·3085	1·6240	1·9831	2·3880	2·8408	3·8982	5·1714	9·3973	15·290
0·00480	0·213	0·252	0·288	0·356	0·417	0·474	0·501	0·528	0·554	0·579	0·627	0·674	0·784	0·886
1/ 208	0·0670	0·1237	0·2039	0·4468	0·8186	1·3402	1·6633	2·0310	2·4455	2·9091	3·9916	5·2950	9·6208	15·652
0·00500	0·218	0·258	0·295	0·364	0·427	0·485	0·513	0·540	0·566	0·592	0·642	0·690	0·802	0·906
1/ 200	0·0686	0·1267	0·2088	0·4573	0·8377	1·3713	1·7018	2·0779	2·5020	2·9761	4·0833	5·4163	9·8400	16·007
0·00525	0·225	0·265	0·304	0·374	0·439	0·498	0·527	0·555	0·582	0·608	0·660	0·709	0·824	0·930
1/ 190	0·0705	0·1303	0·2147	0·4701	0·8611	1·4094	1·7490	2·1354	2·5711	3·0582	4·1956	5·5649	10·108	16·442
0·00550	0·231	0·273	0·312	0·384	0·450	0·512	0·541	0·570	0·597	0·624	0·677	0·727	0·845	0·955
1/ 182	0·0725	0·1339	0·2205	0·4827	0·8840	1·4467	1·7952	2·1917	2·6387	3·1385	4·3055	5·7102	10·371	16·868
0·00575	0·237	0·280	0·320	0·394	0·462	0·525	0·555	0·584	0·612	0·640	0·694	0·745	0·866	0·978
1/ 174	0·0744	0·1373	0·2262	0·4951	0·9064	1·4833	1·8404	2·2468	2·7050	3·2172	4·4131	5·8526	10·628	17·284
0·00600	0·243	0·287	0·328	0·404	0·473	0·537	0·568	0·598	0·627	0·655	0·710	0·763	0·887	1·001
1/ 167	0·0762	0·1408	0·2318	0·5072	0·9284	1·5191	1·8848	2·3009	2·7699	3·2943	4·5186	5·9921	10·880	17·693
0·00625	0·249	0·294	0·336	0·413	0·484	0·550	0·581	0·612	0·641	0·670	0·727	0·780	0·907	1·024
1/ 160	0·0781	0·1441	0·2372	0·5191	0·9500	1·5543	1·9283	2·3539	2·8337	3·3700	4·6221	6·1291	11·128	18·093
0·00650	0·254	0·300	0·343	0·422	0·495	0·562	0·594	0·625	0·656	0·685	0·743	0·797	0·927	1·046
1/ 154	0·0799	0·1474	0·2426	0·5307	0·9713	1·5888	1·9711	2·4060	2·8962	3·4443	4·7237	6·2635	11·371	18·486
0·00675	0·260	0·307	0·351	0·431	0·505	0·574	0·607	0·638	0·669	0·700	0·758	0·814	0·946	1·068
1/ 148	0·0816	0·1506	0·2479	0·5422	0·9921	1·6227	2·0131	2·4572	2·9578	3·5173	4·8236	6·3956	11·609	18·873
0·00700	0·265	0·313	0·358	0·440	0·516	0·586	0·619	0·652	0·683	0·714	0·774	0·831	0·965	1·089
1/ 143	0·0834	0·1538	0·2531	0·5535	1·0126	1·6561	2·0544	2·5075	3·0182	3·5891	4·9218	6·5254	11·844	19·252
0·00725	0·271	0·320	0·365	0·449	0·526	0·597	0·631	0·664	0·697	0·728	0·789	0·847	0·984	1·111
1/ 138	0·0851	0·1569	0·2582	0·5646	1·0328	1·6889	2·0951	2·5571	3·0777	3·6598	5·0183	6·6531	12·074	19·626
0·00750	0·276	0·326	0·372	0·458	0·536	0·609	0·643	0·677	0·710	0·742	0·804	0·863	1·002	1·131
1/ 133	0·0868	0·1600	0·2633	0·5755	1·0527	1·7213	2·1351	2·6058	3·1363	3·7293	5·1134	6·7789	12·301	19·993
0·00800	0·287	0·338	0·386	0·475	0·556	0·631	0·667	0·702	0·736	0·769	0·833	0·894	1·039	1·172
1/ 125	0·0901	0·1661	0·2732	0·5969	1·0915	1·7845	2·2133	2·7011	3·2508	3·8652	5·2993	7·0247	12·745	20·712
0·00850	0·297	0·350	0·400	0·492	0·575	0·653	0·690	0·726	0·761	0·795	0·861	0·925	1·074	1·212
1/ 118	0·0933	0·1720	0·2828	0·6177	1·1293	1·8459	2·2894	2·7938	3·3621	3·9974	5·4799	7·2635	13·176	21·409
0·00900	0·307	0·362	0·413	0·508	0·594	0·674	0·712	0·749	0·786	0·821	0·889	0·954	1·108	1·250
1/ 111	0·0964	0·1777	0·2921	0·6380	1·1661	1·9058	2·3634	2·8840	3·4705	4·1260	5·6557	7·4959	13·596	22·088
0·00950	0·317	0·373	0·426	0·523	0·612	0·695	0·734	0·772	0·809	0·846	0·916	0·983	1·141	1·287
1/ 105	0·0995	0·1833	0·3013	0·6577	1·2020	1·9641	2·4356	2·9719	3·5761	4·2514	5·8271	7·7225	14·004	22·750
0·01000	0·326	0·384	0·439	0·539	0·630	0·715	0·755	0·795	0·833	0·870	0·942	1·011	1·174	1·324
1/ 100	0·1025	0·1887	0·3102	0·6770	1·2370	2·0211	2·5061	3·0577	3·6792	4·3738	5·9944	7·9436	14·403	23·395
	0·74	0·75	0·76	0·77	0·79	0·80	0·80	0·81	0·81	0·82	0·83	0·83	0·85	0·87

$V_{r(0{\cdot}5)medial}$ for half-full circular pipes.

k_s = 0·015 mm
S = 0·01000 to 0·03000

ie hydraulic gradient =
1 in 100 to 1 in 33·3

Water (or sewage) at 15°C;
full bore conditions.

velocities in ms^{-1}
discharges in litres/sec

Gradient **(Equivalent) Pipe diameters in mm**

Gradient	20	25	30	40	50	60	65	70	75	80	90	100	125	150
0·01000	0·326	0·384	0·439	0·539	0·630	0·715	0·755	0·795	0·833	0·870	0·942	1·011	1·174	1·324
1/ 100	0·1025	0·1887	0·3102	0·6770	1·2370	2·0211	2·5061	3·0577	3·6792	4·3738	5·9944	7·9436	14·403	23·395
0·01050	0·335	0·395	0·451	0·554	0·647	0·735	0·776	0·816	0·856	0·894	0·968	1·039	1·205	1·360
1/ 95	0·1054	0·1941	0·3189	0·6959	1·2713	2·0768	2·5750	3·1416	3·7800	4·4934	6·1578	8·1597	14·793	24·026
0·01100	0·345	0·406	0·463	0·568	0·665	0·754	0·796	0·838	0·878	0·917	0·993	1·066	1·237	1·394
1/ 91	0·1083	0·1993	0·3274	0·7143	1·3048	2·1312	2·6424	3·2237	3·8786	4·6104	6·3177	8·3710	15·174	24·643
0·01150	0·354	0·416	0·475	0·583	0·681	0·773	0·816	0·859	0·900	0·940	1·018	1·092	1·267	1·429
1/ 87	0·1111	0·2044	0·3358	0·7324	1·3376	2·1846	2·7085	3·3041	3·9752	4·7250	6·4743	8·5780	15·548	25·246
0·01200	0·362	0·427	0·487	0·597	0·698	0·791	0·836	0·879	0·921	0·962	1·042	1·118	1·297	1·462
1/ 83	0·1138	0·2094	0·3440	0·7502	1·3698	2·2369	2·7732	3·3829	4·0698	4·8373	6·6278	8·7808	15·913	25·838
0·01250	0·371	0·437	0·498	0·611	0·714	0·809	0·855	0·899	0·942	0·984	1·065	1·143	1·326	1·495
1/ 80	0·1165	0·2144	0·3521	0·7676	1·4014	2·2882	2·8367	3·4602	4·1626	4·9475	6·7783	8·9798	16·272	26·418
0·01300	0·379	0·447	0·509	0·624	0·730	0·827	0·874	0·919	0·963	1·006	1·089	1·168	1·355	1·527
1/ 77	0·1192	0·2192	0·3600	0·7847	1·4324	2·3386	2·8990	3·5362	4·2538	5·0556	6·9261	9·1751	16·624	26·988
0·01350	0·388	0·456	0·520	0·638	0·745	0·845	0·892	0·938	0·983	1·027	1·112	1·193	1·383	1·559
1/ 74	0·1218	0·2240	0·3678	0·8015	1·4629	2·3881	2·9603	3·6107	4·3434	5·1619	7·0713	9·3669	16·970	27·547
0·01400	0·396	0·466	0·531	0·651	0·760	0·862	0·910	0·957	1·003	1·048	1·134	1·217	1·411	1·590
1/ 71	0·1244	0·2287	0·3754	0·8180	1·4928	2·4368	3·0205	3·6840	4·4314	5·2664	7·2140	9·5555	17·310	28·097
0·01450	0·404	0·475	0·542	0·664	0·775	0·879	0·928	0·976	1·023	1·068	1·156	1·240	1·438	1·621
1/ 69	0·1269	0·2333	0·3829	0·8342	1·5223	2·4847	3·0797	3·7562	4·5180	5·3691	7·3543	9·7410	17·644	28·637
0·01500	0·412	0·484	0·552	0·677	0·790	0·895	0·946	0·994	1·042	1·088	1·178	1·263	1·465	1·651
1/ 67	0·1294	0·2378	0·3903	0·8502	1·5513	2·5319	3·1380	3·8271	4·6032	5·4703	7·4925	9·9235	17·973	29·169
0·01600	0·427	0·503	0·573	0·702	0·819	0·928	0·980	1·031	1·080	1·128	1·220	1·309	1·517	1·709
1/ 63	0·1342	0·2467	0·4048	0·8815	1·6081	2·6240	3·2520	3·9659	4·7699	5·6680	7·7625	10·280	18·616	30·208
0·01700	0·442	0·520	0·593	0·726	0·847	0·960	1·013	1·066	1·116	1·166	1·261	1·353	1·568	1·767
1/ 59	0·1390	0·2553	0·4189	0·9120	1·6632	2·7136	3·3628	4·1007	4·9317	5·8600	8·0248	10·627	19·241	31·217
0·01800	0·457	0·537	0·612	0·749	0·874	0·991	1·046	1·100	1·152	1·203	1·302	1·396	1·617	1·822
1/ 56	0·1436	0·2638	0·4326	0·9416	1·7169	2·8008	3·4706	4·2320	5·0893	6·0469	8·2800	10·964	19·848	32·199
0·01900	0·471	0·554	0·631	0·772	0·901	1·021	1·078	1·133	1·187	1·239	1·341	1·438	1·666	1·876
1/ 53	0·1481	0·2720	0·4460	0·9705	1·7693	2·8857	3·5757	4·3598	5·2428	6·2290	8·5287	11·292	20·440	33·155
0·02000	0·485	0·570	0·649	0·795	0·927	1·050	1·108	1·165	1·221	1·275	1·379	1·479	1·713	1·929
1/ 50	0·1525	0·2800	0·4591	0·9987	1·8203	2·9686	3·6782	4·4846	5·3926	6·4068	8·7713	11·613	21·017	34·087
0·02100	0·499	0·586	0·668	0·817	0·953	1·079	1·139	1·197	1·254	1·309	1·416	1·518	1·759	1·980
1/ 47·6	0·1568	0·2878	0·4719	1·0262	1·8703	3·0497	3·7784	4·6066	5·5390	6·5804	9·0084	11·926	21·581	34·998
0·02200	0·513	0·602	0·685	0·838	0·977	1·107	1·168	1·228	1·286	1·343	1·452	1·557	1·803	2·031
1/ 45·5	0·1610	0·2955	0·4844	1·0532	1·9191	3·1289	3·8764	4·7259	5·6822	6·7503	9·2403	12·232	22·132	35·889
0·02300	0·526	0·617	0·703	0·859	1·002	1·134	1·197	1·258	1·318	1·376	1·488	1·596	1·847	2·080
1/ 43·5	0·1651	0·3030	0·4966	1·0796	1·9670	3·2066	3·9724	4·8426	5·8224	6·9165	9·4673	12·532	22·672	36·761
0·02400	0·539	0·632	0·720	0·880	1·026	1·161	1·225	1·288	1·349	1·408	1·523	1·633	1·891	2·129
1/ 41·7	0·1692	0·3104	0·5087	1·1055	2·0139	3·2827	4·0665	4·9571	5·9598	7·0795	9·6897	12·826	23·200	37·614
0·02500	0·551	0·647	0·736	0·900	1·049	1·187	1·253	1·317	1·380	1·440	1·557	1·670	1·933	2·176
1/ 40·0	0·1732	0·3176	0·5204	1·1309	2·0599	3·3573	4·1587	5·0694	6·0945	7·2393	9·9078	13·114	23·719	38·452
0·02600	0·564	0·662	0·753	0·920	1·072	1·213	1·281	1·346	1·409	1·471	1·591	1·706	1·974	2·222
1/ 38·5	0·1771	0·3247	0·5320	1·1559	2·1050	3·4305	4·2493	5·1796	6·2267	7·3961	10·122	13·396	24·227	39·273
0·02700	0·576	0·676	0·769	0·939	1·095	1·239	1·307	1·374	1·439	1·502	1·624	1·741	2·015	2·268
1/ 37·0	0·1809	0·3317	0·5434	1·1804	2·1494	3·5025	4·3382	5·2878	6·3566	7·5502	10·332	13·674	24·727	40·080
0·02800	0·588	0·690	0·785	0·958	1·117	1·264	1·334	1·402	1·468	1·532	1·657	1·776	2·055	2·313
1/ 35·7	0·1847	0·3386	0·5546	1·2045	2·1930	3·5732	4·4257	5·3942	6·4843	7·7016	10·539	13·947	25·218	40·873
0·02900	0·600	0·704	0·800	0·977	1·139	1·288	1·360	1·429	1·496	1·562	1·689	1·810	2·094	2·357
1/ 34·5	0·1884	0·3453	0·5656	1·2282	2·2359	3·6428	4·5117	5·4988	6·6099	7·8505	10·742	14·215	25·701	41·652
0·03000	0·611	0·717	0·815	0·996	1·160	1·313	1·385	1·456	1·524	1·591	1·720	1·844	2·133	2·400
1/ 33·3	0·1920	0·3520	0·5764	1·2515	2·2781	3·7113	4·5963	5·6018	6·7334	7·9970	10·942	14·479	26·176	42·418
	0·76	0·77	0·78	0·79	0·81	0·82	0·83	0·83	0·84	0·84	0·85	0·86	0·88	0·89

$V_{r(0\cdot5)medial}$ for half-full circular pipes.

$k_s = 0{\cdot}015$ mm
S = 0·03000 to 0·10000

ie hydraulic gradient =
1 in 33·3 to 1 in 10·0

Water (or sewage) at 15°C;
full bore conditions.

velocities in ms^{-1}
discharges in litres/sec

Gradient **(Equivalent) Pipe diameters in mm**

Gradient	20	25	30	40	50	60	65	70	75	80	90	100	125	150
0·03000	0·611	0·717	0·815	0·996	1·160	1·313	1·385	1·456	1·524	1·591	1·720	1·844	2·133	2·400
1/ 33·3	0·1920	0·3520	0·5764	1·2515	2·2781	3·7113	4·5963	5·6018	6·7334	7·9970	10·942	14·479	26·176	42·418
0·03200	0·634	0·744	0·845	1·032	1·202	1·360	1·435	1·508	1·579	1·648	1·781	1·909	2·209	2·485
1/ 31·3	0·1992	0·3650	0·5976	1·2971	2·3606	3·8451	4·7617	5·8030	6·9749	8·2834	11·333	14·995	27·103	43·916
0·03400	0·656	0·769	0·875	1·067	1·243	1·406	1·483	1·559	1·632	1·703	1·841	1·973	2·282	2·567
1/ 29·4	0·2061	0·3777	0·6182	1·3414	2·4408	3·9751	4·9224	5·9984	7·2094	8·5614	11·712	15·495	28·004	45·370
0·03600	0·678	0·794	0·903	1·102	1·283	1·451	1·531	1·608	1·684	1·757	1·899	2·035	2·353	2·647
1/ 27·8	0·2129	0·3900	0·6382	1·3846	2·5188	4·1016	5·0787	6·1885	7·4375	8·8319	12·081	15·983	28·880	46·784
0·03800	0·699	0·819	0·931	1·135	1·322	1·494	1·576	1·656	1·734	1·809	1·956	2·095	2·423	2·725
1/ 26·3	0·2195	0·4020	0·6578	1·4266	2·5949	4·2248	5·2310	6·3738	7·6598	9·0955	12·440	16·457	29·733	48·160
0·04000	0·719	0·843	0·958	1·168	1·359	1·537	1·621	1·703	1·783	1·861	2·011	2·154	2·491	2·801
1/ 25·0	0·2260	0·4137	0·6769	1·4677	2·6691	4·3450	5·3796	6·5545	7·8766	9·3525	12·791	16·920	30·565	49·503
0·04200	0·739	0·866	0·984	1·200	1·396	1·578	1·665	1·749	1·831	1·911	2·064	2·212	2·557	2·875
1/ 23·8	0·2323	0·4252	0·6955	1·5078	2·7416	4·4625	5·5247	6·7310	8·0884	9·6036	13·134	17·372	31·377	50·813
0·04400	0·759	0·889	1·010	1·231	1·432	1·619	1·708	1·794	1·878	1·959	2·117	2·268	2·622	2·948
1/ 22·7	0·2385	0·4364	0·7138	1·5470	2·8125	4·5774	5·6667	6·9037	8·2955	9·8492	13·468	17·814	32·172	52·095
0·04600	0·778	0·911	1·035	1·262	1·468	1·659	1·750	1·838	1·924	2·007	2·169	2·323	2·685	3·019
1/ 21·7	0·2445	0·4474	0·7316	1·5854	2·8819	4·6899	5·8056	7·0726	8·4982	10·089	13·796	18·246	32·949	53·349
0·04800	0·797	0·933	1·060	1·292	1·502	1·698	1·791	1·881	1·969	2·054	2·219	2·377	2·747	3·088
1/ 20·8	0·2504	0·4581	0·7491	1·6231	2·9499	4·8001	5·9417	7·2382	8·6968	10·325	14·117	18·670	33·710	54·576
0·05000	0·816	0·955	1·084	1·321	1·536	1·736	1·831	1·923	2·013	2·100	2·269	2·430	2·808	3·157
1/ 20·0	0·2562	0·4687	0·7663	1·6600	3·0166	4·9081	6·0753	7·4005	8·8916	10·556	14·432	19·085	34·456	55·780
0·05250	0·838	0·981	1·114	1·357	1·578	1·783	1·880	1·975	2·067	2·156	2·329	2·495	2·882	3·240
1/ 19·0	0·2634	0·4816	0·7874	1·7052	3·0983	5·0404	6·2386	7·5992	9·1299	10·838	14·817	19·593	35·370	57·253
0·05500	0·861	1·007	1·143	1·392	1·619	1·828	1·928	2·025	2·119	2·211	2·388	2·558	2·955	3·321
1/ 18·2	0·2703	0·4943	0·8079	1·7494	3·1782	5·1697	6·3984	7·7935	9·3629	11·114	15·194	20·090	36·262	58·692
0·05750	0·882	1·032	1·172	1·427	1·658	1·873	1·975	2·075	2·171	2·265	2·446	2·620	3·026	3·401
1/ 17·4	0·2772	0·5067	0·8281	1·7927	3·2564	5·2963	6·5548	7·9836	9·5910	11·385	15·562	20·576	37·135	60·100
0·06000	0·904	1·057	1·199	1·460	1·697	1·917	2·022	2·123	2·222	2·318	2·503	2·680	3·096	3·479
1/ 16·7	0·2839	0·5188	0·8479	1·8351	3·3330	5·4204	6·7080	8·1699	9·8144	11·650	15·923	21·052	37·991	61·479
0·06250	0·924	1·081	1·227	1·493	1·736	1·960	2·067	2·170	2·271	2·369	2·559	2·740	3·164	3·556
1/ 16·0	0·2904	0·5308	0·8672	1·8767	3·4081	5·5420	6·8583	8·3526	10·033	11·909	16·277	21·519	38·829	62·831
0·06500	0·945	1·105	1·254	1·526	1·773	2·002	2·111	2·217	2·320	2·420	2·613	2·798	3·231	3·631
1/ 15·4	0·2969	0·5425	0·8863	1·9176	3·4819	5·6613	7·0057	8·5318	10·248	12·164	16·625	21·977	39·652	64·157
0·06750	0·965	1·129	1·280	1·558	1·810	2·044	2·155	2·263	2·368	2·470	2·667	2·855	3·297	3·704
1/ 14·8	0·3032	0·5540	0·9049	1·9577	3·5543	5·7786	7·1504	8·7078	10·460	12·414	16·966	22·427	40·459	65·459
0·07000	0·985	1·152	1·306	1·589	1·846	2·084	2·198	2·308	2·414	2·519	2·719	2·912	3·362	3·777
1/ 14·3	0·3094	0·5653	0·9233	1·9971	3·6254	5·8937	7·2927	8·8807	10·667	12·660	17·301	22·869	41·253	66·737
0·07250	1·005	1·174	1·332	1·620	1·882	2·125	2·240	2·352	2·461	2·567	2·771	2·967	3·425	3·848
1/ 13·8	0·3156	0·5764	0·9414	2·0359	3·6954	6·0070	7·4326	9·0508	10·871	12·902	17·630	23·303	42·032	67·994
0·07500	1·024	1·197	1·357	1·650	1·917	2·164	2·281	2·395	2·506	2·614	2·822	3·021	3·488	3·918
1/ 13·3	0·3216	0·5873	0·9592	2·0740	3·7643	6·1184	7·5702	9·2180	11·071	13·139	17·954	23·730	42·799	69·230
0·08000	1·061	1·240	1·406	1·710	1·986	2·241	2·362	2·480	2·595	2·706	2·922	3·128	3·610	4·054
1/ 12·5	0·3334	0·6087	0·9939	2·1486	3·8988	6·3361	7·8390	9·5448	11·463	13·604	18·587	24·565	44·297	71·644
0·08500	1·098	1·282	1·454	1·767	2·052	2·316	2·441	2·563	2·681	2·796	3·018	3·231	3·728	4·187
1/ 11·8	0·3449	0·6295	1·0277	2·2210	4·0295	6·5475	8·1000	9·8621	11·844	14·054	19·201	25·375	45·751	73·986
0·09000	1·133	1·324	1·500	1·823	2·117	2·388	2·517	2·643	2·765	2·883	3·112	3·331	3·843	4·316
1/ 11·1	0·3560	0·6498	1·0605	2·2914	4·1565	6·7530	8·3538	10·171	12·213	14·493	19·798	26·162	47·164	76·263
0·09500	1·168	1·364	1·546	1·878	2·180	2·459	2·592	2·721	2·846	2·968	3·203	3·429	3·955	4·441
1/ 10·5	0·3669	0·6695	1·0926	2·3601	4·2803	6·9532	8·6010	10·471	12·574	14·919	20·380	26·929	48·539	78·479
0·10000	1·202	1·403	1·590	1·931	2·241	2·528	2·665	2·797	2·926	3·051	3·293	3·524	4·065	4·563
1/ 10·0	0·3775	0·6887	1·1238	2·4270	4·4010	7·1485	8·8420	10·764	12·925	15·336	20·947	27·676	49·880	80·639
	0·77	0·79	0·80	0·82	0·83	0·85	0·85	0·86	0·86	0·87	0·88	0·88	0·90	0·91

$V_{r(0{\cdot}5)medial}$ **for half-full circular pipes.**

k_s = **0·015 mm**
S = 0·10000 to 0·30000

ie hydraulic gradient =
1 in 10·0 to 1 in 3·3

Water (or sewage) at 15°C;
full bore conditions.

velocities in ms^{-1}
discharges in litres/sec

Gradient	**(Equivalent) Pipe diameters in mm**													
	20	**25**	**30**	**40**	**50**	**60**	**65**	**70**	**75**	**80**	**90**	**100**	**125**	**150**
0·10000	1·202	1·403	1·590	1·931	2·241	2·528	2·665	2·797	2·926	3·051	3·293	3·524	4·065	4·563
1/ 10·0	0·3775	0·6887	1·1238	2·4270	4·4010	7·1485	8·8420	10·764	12·925	15·336	20·947	27·676	49·880	80·639
0·10500	1·235	1·441	1·633	1·983	2·301	2·596	2·736	2·871	3·003	3·132	3·380	3·617	4·171	4·683
1/ 9·5	0·3879	0·7075	1·1543	2·4924	4·5189	7·3391	9·0774	11·050	13·268	15·742	21·500	28·406	51·189	82·748
0·11000	1·267	1·479	1·675	2·034	2·360	2·662	2·805	2·944	3·079	3·211	3·465	3·708	4·276	4·799
1/ 9·1	0·3981	0·7259	1·1842	2·5564	4·6342	7·5255	9·3075	11·330	13·603	16·139	22·041	29·119	52·469	84·808
0·11500	1·299	1·515	1·717	2·084	2·418	2·726	2·873	3·015	3·153	3·288	3·548	3·796	4·378	4·913
1/ 8·7	0·4080	0·7439	1·2134	2·6190	4·7470	7·7079	9·5327	11·603	13·931	16·528	22·570	29·817	53·720	86·824
0·12000	1·330	1·551	1·757	2·133	2·474	2·789	2·939	3·085	3·226	3·364	3·629	3·883	4·477	5·025
1/ 8·3	0·4177	0·7616	1·2420	2·6803	4·8576	7·8866	9·7532	11·871	14·252	16·908	23·089	30·500	54·945	88·797
0·12500	1·360	1·587	1·797	2·181	2·529	2·851	3·004	3·153	3·297	3·438	3·709	3·969	4·575	5·134
1/ 8·0	0·4273	0·7789	1·2701	2·7404	4·9659	8·0617	9·9694	12·134	14·567	17·281	23·597	31·170	56·146	90·731
0·13000	1·390	1·621	1·836	2·228	2·583	2·912	3·068	3·220	3·367	3·511	3·788	4·052	4·671	5·242
1/ 7·7	0·4367	0·7958	1·2976	2·7994	5·0722	8·2336	10·182	12·392	14·876	17·647	24·095	31·827	57·324	92·628
0·13500	1·419	1·655	1·874	2·274	2·636	2·972	3·131	3·286	3·436	3·582	3·864	4·134	4·765	5·347
1/ 7·4	0·4459	0·8125	1·3247	2·8573	5·1766	8·4023	10·390	12·645	15·179	18·006	24·585	32·471	58·480	94·490
0·14000	1·448	1·689	1·912	2·319	2·689	3·030	3·193	3·350	3·503	3·652	3·940	4·215	4·858	5·450
1/ 7·1	0·4550	0·8289	1·3513	2·9142	5·2792	8·5680	10·594	12·893	15·477	18·359	25·065	33·105	59·616	96·318
0·14500	1·477	1·722	1·949	2·364	2·740	3·088	3·253	3·414	3·570	3·721	4·014	4·294	4·949	5·552
1/ 6·9	0·4639	0·8451	1·3774	2·9702	5·3800	8·7310	10·795	13·137	15·770	18·706	25·538	33·727	60·732	98·115
0·15000	1·504	1·754	1·985	2·407	2·791	3·145	3·313	3·476	3·635	3·789	4·087	4·372	5·038	5·652
1/ 6·7	0·4726	0·8609	1·4032	3·0253	5·4792	8·8912	10·993	13·378	16·058	19·047	26·002	34·339	61·829	99·882
0·16000	1·559	1·817	2·056	2·493	2·889	3·255	3·429	3·598	3·762	3·922	4·230	4·524	5·213	5·847
1/ 6·3	0·4898	0·8919	1·4535	3·1329	5·6729	9·2043	11·380	13·847	16·621	19·714	26·909	35·535	63·972	103·33
0·17000	1·612	1·878	2·125	2·576	2·985	3·363	3·542	3·716	3·886	4·051	4·368	4·672	5·382	6·037
1/ 5·9	0·5064	0·9220	1·5023	3·2373	5·8610	9·5081	11·754	14·303	17·167	20·360	27·790	36·695	66·051	106·68
0·18000	1·663	1·938	2·192	2·657	3·078	3·467	3·652	3·831	4·006	4·176	4·503	4·816	5·547	6·221
1/ 5·6	0·5226	0·9513	1·5498	3·3388	6·0438	9·8034	12·119	14·745	17·697	20·988	28·645	37·822	68·071	109·93
0·19000	1·714	1·996	2·258	2·736	3·169	3·569	3·759	3·943	4·123	4·297	4·634	4·955	5·707	6·400
1/ 5·3	0·5384	0·9799	1·5960	3·4377	6·2219	10·091	12·474	15·176	18·214	21·600	29·478	38·919	70·037	113·09
0·20000	1·763	2·053	2·322	2·812	3·257	3·668	3·863	4·053	4·237	4·416	4·761	5·092	5·863	6·574
1/ 5·0	0·5538	1·0077	1·6411	3·5342	6·3955	10·371	12·820	15·597	18·717	22·197	30·290	39·989	71·954	116·18
0·21000	1·811	2·108	2·384	2·887	3·344	3·765	3·965	4·159	4·348	4·532	4·886	5·224	6·016	6·745
1/ 4·8	0·5688	1·0349	1·6852	3·6284	6·5650	10·645	13·157	16·007	19·209	22·779	31·082	41·033	73·824	119·19
0·22000	1·857	2·162	2·445	2·961	3·428	3·860	4·065	4·263	4·457	4·645	5·008	5·354	6·165	6·911
1/ 4·5	0·5835	1·0615	1·7283	3·7204	6·7306	10·913	13·488	16·408	19·689	23·348	31·857	42·053	75·651	122·13
0·23000	1·903	2·215	2·505	3·032	3·510	3·952	4·162	4·365	4·563	4·756	5·127	5·481	6·310	7·074
1/ 4·3	0·5979	1·0875	1·7704	3·8105	6·8927	11·174	13·810	16·800	20·159	23·904	32·614	43·050	77·438	125·00
0·24000	1·948	2·267	2·563	3·103	3·591	4·043	4·257	4·465	4·667	4·864	5·243	5·606	6·453	7·233
1/ 4·2	0·6120	1·1130	1·8117	3·8988	7·0515	11·431	14·127	17·184	20·619	24·449	33·356	44·027	79·188	127·82
0·25000	1·992	2·318	2·620	3·171	3·671	4·132	4·351	4·563	4·769	4·970	5·357	5·728	6·592	7·389
1/ 4·0	0·6259	1·1380	1·8522	3·9852	7·2071	11·682	14·436	17·560	21·070	24·983	34·082	44·984	80·902	130·57
0·26000	2·035	2·368	2·677	3·239	3·748	4·219	4·442	4·659	4·869	5·074	5·469	5·847	6·729	7·542
1/ 3·8	0·6394	1·1625	1·8919	4·0701	7·3597	11·928	14·740	17·929	21·512	25·506	34·795	45·923	82·583	133·28
0·27000	2·078	2·417	2·732	3·305	3·825	4·304	4·532	4·753	4·968	5·177	5·579	5·964	6·864	7·692
1/ 3·7	0·6528	1·1866	1·9309	4·1534	7·5095	12·170	15·039	18·291	21·946	26·020	35·494	46·844	84·233	135·93
0·28000	2·120	2·466	2·786	3·370	3·900	4·388	4·620	4·845	5·064	5·277	5·687	6·080	6·996	7·840
1/ 3·6	0·6659	1·2103	1·9693	4·2352	7·6567	12·407	15·332	18·647	22·372	26·525	36·181	47·749	85·853	138·54
0·29000	2·161	2·513	2·839	3·434	3·973	4·471	4·707	4·936	5·159	5·376	5·793	6·193	7·126	7·985
1/ 3·4	0·6788	1·2336	2·0069	4·3157	7·8014	12·641	15·620	18·997	22·791	27·021	36·856	48·638	87·445	141·10
0·30000	2·201	2·560	2·892	3·497	4·046	4·552	4·792	5·026	5·252	5·473	5·898	6·304	7·253	8·127
1/ 3·3	0·6914	1·2564	2·0440	4·3948	7·9437	12·871	15·903	19·341	23·203	27·509	37·520	49·512	89·010	143·62
	0·79	0·81	0·82	0·84	0·85	0·87	0·87	0·88	0·88	0·89	0·89	0·90	0·91	0·93

$V_{r(0\cdot5)medial}$ **for half-full circular pipes.**

A4
(p.1 of 6)

k_s = 0·030 mm
S = 0·00030 to 0·00100

ie hydraulic gradient =
1 in 3333 to 1 in 1000

Water (or sewage) at 15°C;
full bore conditions.

velocities in ms^{-1}
discharges in litres/sec

Gradient	**(Equivalent) Pipe diameters in mm**													
	20	**25**	**30**	**40**	**50**	**60**	**65**	**70**	**75**	**80**	**90**	**100**	**125**	**150**
0·000300				0·071	0·084	0·097	0·103	0·108	0·114	0·120	0·130	0·141	0·165	0·188
1/ 3333				0·0892	0·1654	0·2733	0·3406	0·4173	0·5040	0·6013	0·8292	1·1046	2·0243	3·3148
0·000320				0·074	0·087	0·100	0·107	0·113	0·118	0·124	0·135	0·146	0·171	0·195
1/ 3125				0·0926	0·1718	0·2838	0·3535	0·4331	0·5231	0·6240	0·8603	1·1460	2·0996	3·4376
0·000340				0·076	0·091	0·104	0·110	0·117	0·123	0·129	0·140	0·151	0·177	0·201
1/ 2941				0·0960	0·1780	0·2939	0·3661	0·4485	0·5416	0·6460	0·8907	1·1863	2·1729	3·5569
0·000360				0·079	0·094	0·107	0·114	0·120	0·127	0·133	0·145	0·156	0·183	0·208
1/ 2778				0·0993	0·1840	0·3038	0·3784	0·4635	0·5597	0·6675	0·9202	1·2255	2·2442	3·6731
0·000380				0·082	0·097	0·111	0·118	0·124	0·131	0·137	0·149	0·161	0·189	0·214
1/ 2632				0·1025	0·1899	0·3135	0·3904	0·4781	0·5773	0·6885	0·9490	1·2637	2·3138	3·7864
0·000400				0·084	0·100	0·114	0·121	0·128	0·135	0·141	0·154	0·166	0·194	0·221
1/ 2500				0·1056	0·1956	0·3229	0·4021	0·4924	0·5946	0·7090	0·9772	1·3011	2·3818	3·8971
0·000420				0·086	0·102	0·117	0·125	0·132	0·138	0·145	0·158	0·170	0·200	0·227
1/ 2381				0·1087	0·2013	0·3321	0·4135	0·5064	0·6114	0·7291	1·0047	1·3377	2·4483	4·0053
0·000440				0·089	0·105	0·121	0·128	0·135	0·142	0·149	0·162	0·175	0·205	0·233
1/ 2273				0·1117	0·2068	0·3412	0·4248	0·5201	0·6279	0·7488	1·0317	1·3735	2·5134	4·1113
0·000460				0·091	0·108	0·124	0·131	0·139	0·146	0·153	0·166	0·179	0·210	0·239
1/ 2174				0·1146	0·2122	0·3500	0·4358	0·5336	0·6441	0·7680	1·0581	1·4085	2·5771	4·2151
0·000480				0·094	0·111	0·127	0·135	0·142	0·149	0·157	0·170	0·184	0·215	0·244
1/ 2083				0·1175	0·2175	0·3587	0·4465	0·5468	0·6600	0·7869	1·0840	1·4429	2·6397	4·3168
0·000500			0·077	0·096	0·113	0·130	0·138	0·145	0·153	0·160	0·174	0·188	0·220	0·250
1/ 2000			0·0542	0·1204	0·2227	0·3672	0·4571	0·5597	0·6756	0·8054	1·1095	1·4767	2·7010	4·4167
0·000525			0·079	0·099	0·117	0·134	0·142	0·150	0·157	0·165	0·179	0·193	0·226	0·257
1/ 1905			0·0558	0·1239	0·2291	0·3777	0·4701	0·5756	0·6947	0·8282	1·1407	1·5181	2·7762	4·5390
0·000550			0·081	0·101	0·120	0·137	0·146	0·154	0·161	0·169	0·184	0·198	0·232	0·264
1/ 1818			0·0573	0·1273	0·2353	0·3879	0·4828	0·5911	0·7134	0·8504	1·1712	1·5586	2·8498	4·6588
0·000575			0·083	0·104	0·123	0·141	0·149	0·158	0·166	0·174	0·189	0·203	0·238	0·270
1/ 1739			0·0588	0·1306	0·2415	0·3980	0·4953	0·6063	0·7317	0·8722	1·2011	1·5982	2·9219	4·7760
0·000600			0·085	0·107	0·126	0·144	0·153	0·161	0·170	0·178	0·193	0·208	0·244	0·277
1/ 1667			0·0603	0·1339	0·2475	0·4078	0·5075	0·6212	0·7497	0·8936	1·2305	1·6372	2·9926	4·8910
0·000625			0·087	0·109	0·129	0·148	0·157	0·165	0·174	0·182	0·198	0·213	0·250	0·283
1/ 1600			0·0618	0·1371	0·2534	0·4175	0·5195	0·6359	0·7673	0·9146	1·2593	1·6754	3·0620	5·0039
0·000650			0·089	0·112	0·132	0·151	0·160	0·169	0·178	0·186	0·202	0·218	0·255	0·289
1/ 1538			0·0633	0·1403	0·2592	0·4270	0·5313	0·6503	0·7847	0·9352	1·2875	1·7129	3·1301	5·1147
0·000675			0·092	0·114	0·135	0·154	0·164	0·173	0·181	0·190	0·207	0·223	0·261	0·296
1/ 1481			0·0647	0·1434	0·2649	0·4364	0·5429	0·6645	0·8017	0·9555	1·3154	1·7498	3·1971	5·2236
0·000700			0·093	0·117	0·138	0·158	0·167	0·176	0·185	0·194	0·211	0·227	0·266	0·302
1/ 1429			0·0661	0·1465	0·2705	0·4456	0·5543	0·6784	0·8185	0·9754	1·3427	1·7860	3·2630	5·3307
0·000725			0·095	0·119	0·141	0·161	0·170	0·180	0·189	0·198	0·215	0·232	0·271	0·308
1/ 1379			0·0675	0·1495	0·2761	0·4546	0·5656	0·6921	0·8350	0·9950	1·3696	1·8217	3·3278	5·4361
0·000750			0·097	0·121	0·143	0·164	0·174	0·183	0·193	0·202	0·219	0·236	0·276	0·313
1/ 1333			0·0688	0·1525	0·2815	0·4635	0·5766	0·7056	0·8512	1·0144	1·3961	1·8569	3·3916	5·5398
0·000800			0·101	0·126	0·149	0·170	0·180	0·190	0·200	0·209	0·228	0·245	0·287	0·325
1/ 1250			0·0715	0·1583	0·2922	0·4810	0·5983	0·7320	0·8830	1·0522	1·4480	1·9256	3·5164	5·7427
0·000850			0·105	0·130	0·154	0·176	0·187	0·197	0·207	0·217	0·236	0·254	0·296	0·336
1/ 1176			0·0741	0·1640	0·3025	0·4979	0·6193	0·7577	0·9140	1·0889	1·4984	1·9925	3·6377	5·9399
0·000900		0·094	0·108	0·135	0·159	0·182	0·193	0·203	0·214	0·224	0·243	0·262	0·306	0·347
1/ 1111		0·0462	0·0766	0·1695	0·3127	0·5145	0·6398	0·7827	0·9441	1·1248	1·5475	2·0576	3·7558	6·1318
0·000950		0·097	0·112	0·139	0·164	0·188	0·199	0·210	0·220	0·231	0·251	0·270	0·315	0·358
1/ 1053		0·0477	0·0791	0·1749	0·3225	0·5306	0·6598	0·8071	0·9735	1·1597	1·5954	2·1211	3·8710	6·3190
0·001000		0·100	0·115	0·143	0·169	0·193	0·205	0·216	0·227	0·238	0·258	0·278	0·325	0·368
1/ 1000		0·0491	0·0815	0·1802	0·3322	0·5463	0·6794	0·8310	1·0022	1·1938	1·6422	2·1831	3·9835	6·5017
		–	0·74	0·74	0·75	0·76	0·76	0·76	0·77	0·77	0·78	0·78	0·80	0·81

$V_{r(0\cdot5)medial}$ for half-full circular pipes.

Gradient **(Equivalent) Pipe diameters in mm**

Gradient	20	25	30	40	50	60	65	70	75	80	90	100	125	150
0·00100		0·100	0·115	0·143	0·169	0·193	0·205	0·216	0·227	0·238	0·258	0·278	0·325	0·368
1/ 1000		0·0491	0·0815	0·1802	0·3322	0·5463	0·6794	0·8310	1·0022	1·1938	1·6422	2·1831	3·9835	6·5017
0·00105		0·103	0·119	0·147	0·174	0·199	0·210	0·222	0·233	0·244	0·265	0·286	0·334	0·378
1/ 952		0·0506	0·0838	0·1853	0·3416	0·5618	0·6985	0·8543	1·0302	1·2272	1·6879	2·2437	4·0934	6·6802
0·00110		0·106	0·122	0·151	0·179	0·204	0·216	0·228	0·239	0·251	0·272	0·293	0·342	0·388
1/ 909		0·0520	0·0862	0·1904	0·3508	0·5768	0·7172	0·8771	1·0577	1·2598	1·7327	2·3030	4·2010	6·8550
0·00115		0·109	0·125	0·155	0·183	0·209	0·222	0·234	0·246	0·257	0·279	0·301	0·351	0·398
1/ 870		0·0533	0·0884	0·1953	0·3599	0·5916	0·7355	0·8995	1·0846	1·2918	1·7765	2·3611	4·3063	7·0261
0·00120		0·111	0·128	0·159	0·188	0·214	0·227	0·239	0·251	0·263	0·286	0·308	0·359	0·407
1/ 833		0·0547	0·0907	0·2002	0·3687	0·6061	0·7535	0·9214	1·1110	1·3232	1·8195	2·4181	4·4096	7·1939
0·00125		0·114	0·131	0·163	0·192	0·219	0·232	0·245	0·257	0·269	0·293	0·315	0·368	0·416
1/ 800		0·0560	0·0928	0·2050	0·3775	0·6204	0·7712	0·9430	1·1369	1·3540	1·8617	2·4740	4·5110	7·3584
0·00130		0·117	0·134	0·167	0·197	0·224	0·238	0·251	0·263	0·275	0·299	0·322	0·376	0·426
1/ 769		0·0573	0·0950	0·2096	0·3860	0·6343	0·7885	0·9641	1·1624	1·3843	1·9032	2·5290	4·6105	7·5200
0·00135		0·119	0·137	0·170	0·201	0·229	0·243	0·256	0·269	0·281	0·306	0·329	0·384	0·435
1/ 741		0·0586	0·0971	0·2142	0·3944	0·6481	0·8055	0·9849	1·1874	1·4140	1·9439	2·5829	4·7083	7·6788
0·00140		0·122	0·140	0·174	0·205	0·234	0·248	0·261	0·274	0·287	0·312	0·336	0·392	0·443
1/ 714		0·0599	0·0992	0·2188	0·4027	0·6616	0·8223	1·0054	1·2120	1·4432	1·9840	2·6360	4·8045	7·8349
0·00145		0·125	0·143	0·178	0·209	0·239	0·253	0·266	0·280	0·293	0·318	0·342	0·399	0·452
1/ 690		0·0611	0·1012	0·2232	0·4109	0·6749	0·8388	1·0255	1·2362	1·4720	2·0234	2·6882	4·8991	7·9885
0·00150		0·127	0·146	0·181	0·213	0·243	0·258	0·272	0·285	0·298	0·324	0·349	0·407	0·461
1/ 667		0·0624	0·1032	0·2276	0·4189	0·6880	0·8550	1·0453	1·2600	1·5003	2·0622	2·7396	4·9922	8·1396
0·00160		0·132	0·152	0·188	0·221	0·252	0·267	0·282	0·296	0·310	0·336	0·362	0·422	0·477
1/ 625		0·0648	0·1072	0·2362	0·4346	0·7137	0·8868	1·0841	1·3066	1·5557	2·1381	2·8401	5·1743	8·4351
0·00170	0·115	0·137	0·157	0·195	0·229	0·261	0·277	0·291	0·306	0·320	0·348	0·374	0·436	0·494
1/ 588	0·0361	0·0671	0·1110	0·2446	0·4499	0·7386	0·9177	1·1217	1·3520	1·6096	2·2119	2·9378	5·3512	8·7222
0·00180	0·119	0·141	0·162	0·201	0·237	0·270	0·286	0·301	0·316	0·331	0·359	0·386	0·450	0·509
1/ 556	0·0373	0·0694	0·1148	0·2528	0·4647	0·7629	0·9478	1·1584	1·3961	1·6620	2·2837	3·0329	5·5234	9·0016
0·00190	0·123	0·146	0·168	0·207	0·244	0·278	0·294	0·310	0·326	0·341	0·370	0·398	0·464	0·525
1/ 526	0·0386	0·0716	0·1184	0·2607	0·4792	0·7865	0·9771	1·1942	1·4391	1·7132	2·3537	3·1257	5·6913	9·2739
0·00200	0·127	0·150	0·173	0·214	0·251	0·286	0·303	0·319	0·335	0·351	0·381	0·409	0·477	0·540
1/ 500	0·0397	0·0738	0·1220	0·2685	0·4934	0·8097	1·0058	1·2292	1·4811	1·7631	2·4221	3·2162	5·8551	9·5397
0·00210	0·130	0·155	0·178	0·220	0·258	0·294	0·312	0·328	0·345	0·360	0·391	0·421	0·490	0·555
1/ 476	0·0409	0·0759	0·1255	0·2761	0·5073	0·8323	1·0338	1·2633	1·5222	1·8119	2·4889	3·3047	6·0153	9·7995
0·00220	0·134	0·159	0·182	0·226	0·265	0·302	0·320	0·337	0·354	0·370	0·402	0·432	0·503	0·569
1/ 455	0·0420	0·0780	0·1289	0·2835	0·5208	0·8544	1·0612	1·2968	1·5624	1·8597	2·5543	3·3912	6·1719	10·054
0·00230	0·137	0·163	0·187	0·231	0·272	0·310	0·328	0·345	0·363	0·379	0·412	0·443	0·515	0·583
1/ 435	0·0432	0·0800	0·1323	0·2908	0·5341	0·8761	1·0881	1·3295	1·6018	1·9065	2·6183	3·4760	6·3254	10·302
0·00240	0·141	0·167	0·192	0·237	0·279	0·317	0·336	0·354	0·371	0·388	0·421	0·453	0·528	0·597
1/ 417	0·0442	0·0820	0·1355	0·2980	0·5472	0·8973	1·1144	1·3616	1·6404	1·9523	2·6811	3·5592	6·4757	10·546
0·00250	0·144	0·171	0·196	0·243	0·285	0·325	0·344	0·362	0·380	0·397	0·431	0·464	0·540	0·610
1/ 400	0·0453	0·0840	0·1388	0·3050	0·5600	0·9182	1·1403	1·3931	1·6783	1·9973	2·7427	3·6407	6·6233	10·785
0·00260	0·148	0·175	0·201	0·248	0·292	0·332	0·351	0·370	0·388	0·406	0·441	0·474	0·552	0·624
1/ 385	0·0464	0·0860	0·1419	0·3119	0·5725	0·9387	1·1657	1·4241	1·7155	2·0415	2·8032	3·7208	6·7681	11·020
0·00270	0·151	0·179	0·205	0·254	0·298	0·339	0·359	0·378	0·397	0·415	0·450	0·484	0·563	0·637
1/ 370	0·0474	0·0879	0·1451	0·3187	0·5849	0·9588	1·1906	1·4545	1·7521	2·0850	2·8627	3·7994	6·9103	11·251
0·00280	0·154	0·183	0·210	0·259	0·304	0·346	0·366	0·386	0·405	0·423	0·459	0·494	0·574	0·649
1/ 357	0·0484	0·0897	0·1481	0·3253	0·5970	0·9786	1·2151	1·4844	1·7880	2·1277	2·9211	3·8768	7·0502	11·477
0·00290	0·157	0·187	0·214	0·264	0·310	0·353	0·373	0·393	0·413	0·432	0·468	0·503	0·586	0·662
1/ 345	0·0494	0·0916	0·1512	0·3319	0·6090	0·9981	1·2393	1·5138	1·8234	2·1697	2·9786	3·9528	7·1877	11·700
0·00300	0·160	0·190	0·218	0·269	0·316	0·360	0·381	0·401	0·421	0·440	0·477	0·513	0·597	0·675
1/ 333	0·0504	0·0934	0·1541	0·3384	0·6208	1·0173	1·2630	1·5428	1·8582	2·2110	3·0352	4·0277	7·3231	11·920
	–	0·74	0·74	0·75	0·76	0·77	0·78	0·78	0·79	0·79	0·80	0·80	0·82	0·83

$V_{r(0·5)medial}$ for half-full circular pipes.

k_s = 0·030 mm
S = 0·00300 to 0·01000

ie hydraulic gradient =
1 in 333 to 1 in 100

Water (or sewage) at 15°C;
full bore conditions.

velocities in ms^{-1}
discharges in litres/sec

Gradient	**(Equivalent) Pipe diameters in mm**													
	20	**25**	**30**	**40**	**50**	**60**	**65**	**70**	**75**	**80**	**90**	**100**	**125**	**150**
0·00300	0·160	0·190	0·218	0·269	0·316	0·360	0·381	0·401	0·421	0·440	0·477	0·513	0·597	0·675
1/ 333	0·0504	0·0934	0·1541	0·3384	0·6208	1·0173	1·2630	1·5428	1·8582	2·2110	3·0352	4·0277	7·3231	11·920
0·00320	0·167	0·197	0·226	0·279	0·328	0·373	0·395	0·416	0·436	0·456	0·494	0·531	0·618	0·699
1/ 313	0·0524	0·0969	0·1599	0·3510	0·6438	1·0548	1·3095	1·5994	1·9263	2·2919	3·1458	4·1741	7·5877	12·348
0·00340	0·173	0·205	0·234	0·289	0·339	0·386	0·408	0·430	0·451	0·472	0·511	0·550	0·639	0·722
1/ 294	0·0542	0·1004	0·1656	0·3633	0·6662	1·0913	1·3547	1·6544	1·9924	2·3704	3·2533	4·3163	7·8447	12·765
0·00360	0·179	0·211	0·242	0·299	0·350	0·399	0·421	0·444	0·466	0·487	0·528	0·567	0·660	0·745
1/ 278	0·0561	0·1038	0·1711	0·3753	0·6880	1·1268	1·3986	1·7080	2·0568	2·4469	3·3579	4·4547	8·0949	13·170
0·00380	0·184	0·218	0·250	0·308	0·361	0·411	0·434	0·457	0·480	0·502	0·544	0·584	0·679	0·768
1/ 263	0·0579	0·1071	0·1765	0·3870	0·7092	1·1614	1·4415	1·7603	2·1196	2·5215	3·4599	4·5896	8·3386	13·565
0·00400	0·190	0·225	0·257	0·317	0·372	0·423	0·447	0·471	0·494	0·516	0·560	0·601	0·699	0·789
1/ 250	0·0596	0·1103	0·1818	0·3984	0·7300	1·1952	1·4834	1·8113	2·1809	2·5942	3·5594	4·7213	8·5764	13·950
0·00420	0·195	0·231	0·264	0·326	0·382	0·434	0·459	0·484	0·507	0·530	0·575	0·618	0·718	0·811
1/ 238	0·0613	0·1134	0·1869	0·4095	0·7503	1·2283	1·5243	1·8611	2·2408	2·6654	3·6567	4·8499	8·8088	14·326
0·00440	0·201	0·237	0·272	0·335	0·392	0·446	0·471	0·496	0·520	0·544	0·590	0·634	0·736	0·832
1/ 227	0·0630	0·1165	0·1919	0·4205	0·7702	1·2606	1·5643	1·9099	2·2994	2·7349	3·7518	4·9758	9·0361	14·694
0·00460	0·206	0·243	0·279	0·343	0·402	0·457	0·483	0·509	0·533	0·558	0·604	0·649	0·754	0·852
1/ 217	0·0647	0·1195	0·1969	0·4312	0·7896	1·2923	1·6035	1·9576	2·3568	2·8031	3·8450	5·0990	9·2585	15·054
0·00480	0·211	0·249	0·285	0·351	0·412	0·468	0·495	0·521	0·546	0·571	0·619	0·665	0·772	0·872
1/ 208	0·0663	0·1225	0·2017	0·4417	0·8087	1·3233	1·6419	2·0044	2·4130	2·8698	3·9363	5·2197	9·4766	15·407
0·00500	0·216	0·255	0·292	0·360	0·421	0·479	0·506	0·533	0·559	0·584	0·633	0·680	0·790	0·891
1/ 200	0·0679	0·1254	0·2065	0·4519	0·8274	1·3537	1·6796	2·0503	2·4682	2·9353	4·0258	5·3382	9·6904	15·753
0·00525	0·222	0·263	0·300	0·370	0·433	0·492	0·520	0·547	0·574	0·600	0·650	0·698	0·811	0·915
1/ 190	0·0698	0·1289	0·2123	0·4646	0·8503	1·3910	1·7258	2·1066	2·5358	3·0155	4·1355	5·4832	9·9521	16·177
0·00550	0·228	0·270	0·308	0·380	0·444	0·505	0·534	0·562	0·589	0·616	0·667	0·716	0·832	0·939
1/ 182	0·0717	0·1324	0·2180	0·4769	0·8727	1·4275	1·7709	2·1616	2·6019	3·0940	4·2427	5·6250	10·208	16·591
0·00575	0·234	0·277	0·316	0·389	0·456	0·518	0·547	0·576	0·604	0·631	0·683	0·734	0·852	0·962
1/ 174	0·0736	0·1358	0·2236	0·4890	0·8947	1·4633	1·8152	2·2155	2·6666	3·1708	4·3478	5·7638	10·459	16·996
0·00600	0·240	0·284	0·324	0·399	0·467	0·530	0·560	0·589	0·618	0·646	0·700	0·751	0·872	0·984
1/ 167	0·0754	0·1392	0·2290	0·5008	0·9162	1·4983	1·8585	2·2683	2·7300	3·2461	4·4507	5·8999	10·704	17·394
0·00625	0·246	0·290	0·332	0·408	0·477	0·542	0·573	0·603	0·632	0·660	0·715	0·768	0·892	1·006
1/ 160	0·0772	0·1425	0·2344	0·5125	0·9374	1·5327	1·9010	2·3201	2·7922	3·3200	4·5516	6·0333	10·945	17·783
0·00650	0·251	0·297	0·339	0·417	0·488	0·554	0·585	0·616	0·646	0·675	0·731	0·785	0·911	1·028
1/ 154	0·0790	0·1457	0·2397	0·5239	0·9581	1·5664	1·9428	2·3709	2·8533	3·3924	4·6507	6·1643	11·181	18·165
0·00675	0·257	0·303	0·346	0·426	0·498	0·566	0·598	0·629	0·659	0·689	0·746	0·801	0·930	1·049
1/ 148	0·0807	0·1489	0·2449	0·5351	0·9785	1·5995	1·9838	2·4208	2·9133	3·4637	4·7480	6·2929	11·413	18·541
0·00700	0·262	0·310	0·354	0·435	0·509	0·577	0·610	0·642	0·673	0·703	0·761	0·817	0·949	1·070
1/ 143	0·0824	0·1520	0·2500	0·5462	0·9986	1·6321	2·0241	2·4699	2·9723	3·5336	4·8436	6·4193	11·641	18·909
0·00725	0·268	0·316	0·361	0·443	0·519	0·589	0·622	0·654	0·686	0·717	0·776	0·833	0·967	1·091
1/ 138	0·0841	0·1550	0·2550	0·5570	1·0183	1·6642	2·0638	2·5182	3·0303	3·6025	4·9377	6·5436	11·865	19·272
0·00750	0·273	0·322	0·368	0·452	0·528	0·600	0·634	0·667	0·699	0·730	0·791	0·849	0·985	1·111
1/ 133	0·0858	0·1581	0·2599	0·5677	1·0377	1·6957	2·1028	2·5658	3·0873	3·6702	5·0303	6·6660	12·086	19·629
0·00800	0·283	0·334	0·381	0·468	0·548	0·622	0·657	0·691	0·724	0·757	0·819	0·879	1·020	1·150
1/ 125	0·0890	0·1640	0·2696	0·5886	1·0756	1·7574	2·1791	2·6587	3·1989	3·8026	5·2112	6·9050	12·517	20·326
0·00850	0·293	0·346	0·395	0·485	0·567	0·643	0·679	0·714	0·749	0·782	0·847	0·909	1·054	1·189
1/ 118	0·0922	0·1697	0·2790	0·6089	1·1125	1·8173	2·2532	2·7489	3·3073	3·9312	5·3868	7·1372	12·935	21·003
0·00900	0·303	0·357	0·408	0·500	0·585	0·663	0·701	0·737	0·772	0·807	0·874	0·937	1·087	1·226
1/ 111	0·0952	0·1754	0·2882	0·6287	1·1484	1·8756	2·3253	2·8367	3·4127	4·0563	5·5577	7·3630	13·342	21·661
0·00950	0·313	0·368	0·420	0·516	0·603	0·683	0·722	0·759	0·796	0·831	0·900	0·966	1·120	1·262
1/ 105	0·0982	0·1808	0·2971	0·6480	1·1833	1·9324	2·3956	2·9223	3·5155	4·1783	5·7243	7·5830	13·739	22·301
0·01000	0·322	0·379	0·433	0·531	0·620	0·703	0·743	0·781	0·818	0·855	0·925	0·993	1·151	1·297
1/ 100	0·1011	0·1862	0·3058	0·6668	1·2175	1·9879	2·4642	3·0058	3·6157	4·2972	5·8867	7·7976	14·126	22·926
	0·74	0·75	0·76	0·77	0·78	0·79	0·80	0·80	0·81	0·81	0·82	0·83	0·84	0·86

$V_{r(0{\cdot}5)medial}$ for half-full circular pipes.

k_s = 0·030 mm S = 0·00300 to 0·01000

k_s = 0·030 mm
S = 0·01000 to 0·03000

ie hydraulic gradient =
1 in 100 to 1 in 33·3

Water (or sewage) at 15°C;
full bore conditions.

velocities in ms^{-1}
discharges in litres/sec

Gradient	**(Equivalent) Pipe diameters in mm**													
	20	**25**	**30**	**40**	**50**	**60**	**65**	**70**	**75**	**80**	**90**	**100**	**125**	**150**
0·01000	0·322	0·379	0·433	0·531	0·620	0·703	0·743	0·781	0·818	0·855	0·925	0·993	1·151	1·297
1/ 100	0·1011	0·1862	0·3058	0·6668	1·2175	1·9879	2·4642	3·0058	3·6157	4·2972	5·8867	7·7976	14·126	22·926
0·01050	0·331	0·390	0·445	0·545	0·637	0·722	0·763	0·802	0·841	0·878	0·950	1·020	1·182	1·332
1/ 95	0·1040	0·1914	0·3143	0·6852	1·2508	2·0420	2·5312	3·0873	3·7137	4·4134	6·0454	8·0073	14·503	23·537
0·01100	0·340	0·400	0·456	0·560	0·654	0·741	0·783	0·823	0·862	0·901	0·975	1·046	1·212	1·366
1/ 91	0·1068	0·1965	0·3226	0·7032	1·2835	2·0950	2·5967	3·1671	3·8094	4·5270	6·2005	8·2122	14·873	24·133
0·01150	0·349	0·410	0·468	0·574	0·670	0·759	0·802	0·843	0·883	0·923	0·999	1·071	1·241	1·399
1/ 87	0·1095	0·2015	0·3308	0·7208	1·3154	2·1469	2·6609	3·2452	3·9032	4·6382	6·3524	8·4128	15·234	24·717
0·01200	0·357	0·420	0·479	0·587	0·686	0·777	0·821	0·863	0·904	0·944	1·022	1·096	1·270	1·431
1/ 83	0·1122	0·2064	0·3388	0·7381	1·3467	2·1977	2·7238	3·3217	3·9950	4·7472	6·5012	8·6094	15·588	25·289
0·01250	0·366	0·430	0·490	0·601	0·702	0·795	0·839	0·883	0·925	0·966	1·045	1·121	1·298	1·463
1/ 80	0·1149	0·2112	0·3467	0·7551	1·3774	2·2476	2·7854	3·3967	4·0851	4·8541	6·6471	8·8021	15·935	25·850
0·01300	0·374	0·440	0·501	0·614	0·717	0·812	0·858	0·902	0·945	0·987	1·067	1·145	1·326	1·494
1/ 77	0·1175	0·2159	0·3544	0·7717	1·4076	2·2965	2·8459	3·4704	4·1735	4·9589	6·7903	8·9912	16·275	26·400
0·01350	0·382	0·449	0·512	0·627	0·732	0·829	0·876	0·921	0·964	1·007	1·089	1·168	1·353	1·524
1/ 74	0·1200	0·2206	0·3620	0·7881	1·4372	2·3446	2·9054	3·5427	4·2603	5·0619	6·9309	9·1768	16·610	26·940
0·01400	0·390	0·459	0·523	0·640	0·747	0·846	0·893	0·939	0·984	1·027	1·111	1·192	1·380	1·554
1/ 71	0·1225	0·2252	0·3694	0·8041	1·4663	2·3918	2·9638	3·6138	4·3457	5·1631	7·0690	9·3593	16·938	27·470
0·01450	0·398	0·468	0·533	0·652	0·761	0·862	0·910	0·957	1·003	1·047	1·133	1·214	1·407	1·584
1/ 69	0·1250	0·2296	0·3767	0·8199	1·4949	2·4383	3·0212	3·6837	4·4296	5·2626	7·2049	9·5386	17·261	27·992
0·01500	0·406	0·477	0·543	0·665	0·776	0·879	0·928	0·975	1·021	1·066	1·154	1·237	1·432	1·613
1/ 67	0·1274	0·2341	0·3839	0·8355	1·5231	2·4840	3·0777	3·7525	4·5121	5·3605	7·3385	9·7151	17·579	28·505
0·01600	0·421	0·494	0·563	0·689	0·804	0·910	0·961	1·010	1·058	1·105	1·195	1·281	1·483	1·670
1/ 63	0·1322	0·2427	0·3981	0·8659	1·5782	2·5734	3·1882	3·8869	4·6734	5·5518	7·5996	10·060	18·199	29·506
0·01700	0·435	0·512	0·583	0·713	0·831	0·941	0·993	1·044	1·093	1·141	1·234	1·323	1·532	1·725
1/ 59	0·1368	0·2511	0·4118	0·8954	1·6317	2·6601	3·2954	4·0173	4·8300	5·7376	7·8531	10·395	18·801	30·479
0·01800	0·450	0·528	0·601	0·735	0·857	0·971	1·025	1·077	1·128	1·177	1·273	1·365	1·580	1·778
1/ 56	0·1413	0·2593	0·4251	0·9242	1·6837	2·7445	3·3998	4·1443	4·9823	5·9182	8·0995	10·720	19·387	31·424
0·01900	0·464	0·545	0·620	0·758	0·883	1·000	1·055	1·109	1·161	1·212	1·311	1·405	1·626	1·830
1/ 53	0·1457	0·2673	0·4381	0·9522	1·7344	2·8267	3·5014	4·2679	5·1307	6·0941	8·3396	11·037	19·957	32·344
0·02000	0·477	0·560	0·638	0·779	0·909	1·028	1·085	1·140	1·194	1·247	1·348	1·445	1·672	1·881
1/ 50	0·1500	0·2751	0·4508	0·9795	1·7839	2·9069	3·6005	4·3885	5·2754	6·2657	8·5737	11·346	20·513	33·241
0·02100	0·491	0·576	0·655	0·801	0·933	1·056	1·114	1·171	1·226	1·280	1·384	1·483	1·716	1·931
1/ 47·6	0·1541	0·2827	0·4632	1·0062	1·8322	2·9852	3·6973	4·5062	5·4167	6·4333	8·8023	11·648	21·055	34·117
0·02200	0·504	0·591	0·672	0·822	0·957	1·083	1·143	1·201	1·257	1·312	1·419	1·521	1·759	1·979
1/ 45·5	0·1582	0·2902	0·4753	1·0324	1·8794	3·0618	3·7920	4·6214	5·5549	6·5971	9·0258	11·943	21·586	34·973
0·02300	0·516	0·606	0·689	0·842	0·981	1·109	1·171	1·230	1·288	1·344	1·453	1·557	1·801	2·026
1/ 43·5	0·1622	0·2975	0·4872	1·0579	1·9257	3·1368	3·8846	4·7341	5·6901	6·7574	9·2445	12·231	22·105	35·811
0·02400	0·529	0·621	0·706	0·862	1·004	1·135	1·198	1·259	1·318	1·376	1·487	1·593	1·843	2·073
1/ 41·7	0·1662	0·3046	0·4989	1·0830	1·9710	3·2102	3·9754	4·8445	5·8225	6·9144	9·4587	12·514	22·613	36·631
0·02500	0·541	0·635	0·722	0·881	1·026	1·161	1·225	1·287	1·347	1·406	1·520	1·629	1·883	2·118
1/ 40·0	0·1700	0·3116	0·5103	1·1076	2·0154	3·2823	4·0644	4·9527	5·9524	7·0684	9·6687	12·791	23·111	37·434
0·02600	0·553	0·649	0·738	0·901	1·049	1·186	1·251	1·315	1·376	1·436	1·552	1·663	1·923	2·163
1/ 38·5	0·1738	0·3185	0·5215	1·1317	2·0590	3·3529	4·1517	5·0589	6·0798	7·2194	9·8747	13·063	23·600	38·223
0·02700	0·565	0·663	0·753	0·919	1·070	1·210	1·277	1·342	1·404	1·466	1·584	1·697	1·962	2·207
1/ 37·0	0·1775	0·3253	0·5325	1·1554	2·1018	3·4223	4·2374	5·1631	6·2048	7·3677	10·077	13·330	24·079	38·996
0·02800	0·577	0·676	0·769	0·938	1·092	1·235	1·302	1·368	1·432	1·495	1·615	1·731	2·001	2·250
1/ 35·7	0·1812	0·3319	0·5433	1·1786	2·1439	3·4905	4·3216	5·2656	6·3277	7·5134	10·276	13·592	24·550	39·756
0·02900	0·588	0·690	0·784	0·956	1·113	1·258	1·327	1·394	1·460	1·523	1·646	1·763	2·038	2·292
1/ 34·5	0·1848	0·3385	0·5540	1·2015	2·1853	3·5575	4·4044	5·3663	6·4486	7·6566	10·471	13·850	25·014	40·503
0·03000	0·600	0·703	0·799	0·974	1·134	1·282	1·352	1·420	1·487	1·551	1·676	1·796	2·075	2·334
1/ 33·3	0·1884	0·3449	0·5645	1·2241	2·2260	3·6234	4·4859	5·4654	6·5674	7·7975	10·663	14·103	25·469	41·238
	0·75	0·76	0·77	0·79	0·80	0·81	0·82	0·82	0·83	0·83	0·84	0·85	0·87	0·88

$V_{r(0·5)medial}$ for half-full circular pipes.

A4
(p.5 of 6)

k_s = 0·030 mm
S = 0·03000 to 0·10000

ie hydraulic gradient =
1 in 33·3 to 1 in 10·0

Water (or sewage) at 15°C;
full bore conditions.

velocities in ms^{-1}
discharges in litres/sec

Gradient	**(Equivalent) Pipe diameters in mm** 20	25	30	40	50	60	65	70	75	80	90	100	125	150
0·03000	0·600	0·703	0·799	0·974	1·134	1·282	1·352	1·420	1·487	1·551	1·676	1·796	2·075	2·334
1/ 33·3	0·1884	0·3449	0·5645	1·2241	2·2260	3·6234	4·4859	5·4654	6·5674	7·7975	10·663	14·103	25·469	41·238
0·03200	0·622	0·728	0·828	1·009	1·174	1·327	1·400	1·470	1·539	1·606	1·735	1·859	2·148	2·415
1/ 31·3	0·1953	0·3575	0·5849	1·2681	2·3055	3·7523	4·6451	5·6589	6·7995	8·0727	11·038	14·598	26·358	42·672
0·03400	0·643	0·753	0·856	1·043	1·214	1·371	1·446	1·519	1·590	1·659	1·792	1·920	2·218	2·494
1/ 29·4	0·2020	0·3698	0·6048	1·3108	2·3828	3·8773	4·7995	5·8467	7·0248	8·3397	11·402	15·078	27·221	44·064
0·03600	0·664	0·778	0·883	1·076	1·252	1·414	1·492	1·567	1·640	1·711	1·848	1·979	2·287	2·570
1/ 27·8	0·2086	0·3817	0·6242	1·3524	2·4579	3·9989	4·9497	6·0293	7·2439	8·5994	11·756	15·545	28·060	45·416
0·03800	0·684	0·801	0·910	1·108	1·289	1·456	1·536	1·613	1·688	1·761	1·902	2·037	2·353	2·645
1/ 26·3	0·2150	0·3933	0·6430	1·3929	2·5310	4·1173	5·0960	6·2072	7·4572	8·8522	12·101	16·000	28·877	46·733
0·04000	0·704	0·824	0·936	1·140	1·325	1·497	1·579	1·658	1·735	1·810	1·955	2·094	2·418	2·717
1/ 25·0	0·2212	0·4046	0·6614	1·4324	2·6023	4·2328	5·2386	6·3806	7·6651	9·0986	12·437	16·443	29·672	48·016
0·04200	0·723	0·847	0·961	1·171	1·361	1·537	1·621	1·702	1·781	1·858	2·007	2·149	2·481	2·788
1/ 23·8	0·2273	0·4156	0·6794	1·4710	2·6720	4·3456	5·3779	6·5499	7·8682	9·3393	12·765	16·876	30·449	49·268
0·04400	0·742	0·869	0·986	1·201	1·396	1·576	1·662	1·745	1·826	1·905	2·057	2·202	2·543	2·857
1/ 22·7	0·2332	0·4264	0·6970	1·5087	2·7401	4·4558	5·5140	6·7154	8·0666	9·5744	13·086	17·298	31·208	50·492
0·04600	0·761	0·890	1·010	1·230	1·429	1·614	1·702	1·787	1·870	1·951	2·106	2·255	2·604	2·925
1/ 21·7	0·2391	0·4370	0·7142	1·5456	2·8067	4·5636	5·6472	6·8772	8·2608	9·8045	13·399	17·712	31·951	51·688
0·04800	0·779	0·911	1·034	1·259	1·463	1·651	1·741	1·828	1·913	1·995	2·154	2·307	2·663	2·991
1/ 20·8	0·2448	0·4474	0·7310	1·5818	2·8720	4·6692	5·7776	7·0358	8·4509	10·030	13·706	18·117	32·678	52·860
0·05000	0·797	0·932	1·058	1·287	1·495	1·688	1·780	1·869	1·955	2·039	2·202	2·357	2·721	3·056
1/ 20·0	0·2504	0·4576	0·7476	1·6173	2·9360	4·7728	5·9055	7·1912	8·6372	10·251	14·007	18·514	33·390	54·008
0·05250	0·819	0·958	1·086	1·321	1·535	1·733	1·827	1·918	2·007	2·093	2·260	2·419	2·792	3·136
1/ 19·0	0·2573	0·4700	0·7678	1·6606	3·0143	4·8994	6·0619	7·3813	8·8652	10·521	14·375	18·999	34·261	55·412
0·05500	0·840	0·982	1·114	1·355	1·574	1·777	1·873	1·966	2·057	2·146	2·316	2·479	2·861	3·213
1/ 18·2	0·2640	0·4822	0·7876	1·7031	3·0908	5·0232	6·2147	7·5671	9·0880	10·785	14·735	19·473	35·112	56·783
0·05750	0·861	1·007	1·142	1·388	1·612	1·819	1·918	2·014	2·106	2·197	2·371	2·538	2·929	3·289
1/ 17·4	0·2706	0·4941	0·8069	1·7446	3·1657	5·1443	6·3643	7·7489	9·3059	11·043	15·087	19·937	35·945	58·125
0·06000	0·882	1·030	1·168	1·421	1·650	1·861	1·962	2·060	2·155	2·247	2·426	2·596	2·995	3·364
1/ 16·7	0·2770	0·5058	0·8259	1·7852	3·2390	5·2629	6·5107	7·9269	9·5193	11·296	15·431	20·391	36·760	59·438
0·06250	0·902	1·054	1·195	1·452	1·686	1·903	2·005	2·105	2·202	2·297	2·479	2·653	3·061	3·436
1/ 16·0	0·2833	0·5173	0·8445	1·8251	3·3109	5·3792	6·6543	8·1013	9·7284	11·544	15·769	20·837	37·559	60·725
0·06500	0·922	1·077	1·221	1·483	1·722	1·943	2·048	2·150	2·249	2·345	2·531	2·709	3·124	3·508
1/ 15·4	0·2895	0·5285	0·8627	1·8642	3·3814	5·4932	6·7951	8·2724	9·9336	11·787	16·100	21·273	38·342	61·987
0·06750	0·941	1·099	1·246	1·514	1·757	1·982	2·089	2·193	2·294	2·392	2·582	2·763	3·187	3·578
1/ 14·8	0·2956	0·5395	0·8807	1·9026	3·4506	5·6052	6·9333	8·4404	10·135	12·025	16·425	21·702	39·111	63·225
0·07000	0·960	1·121	1·271	1·544	1·792	2·021	2·130	2·236	2·339	2·439	2·632	2·817	3·249	3·647
1/ 14·3	0·3016	0·5504	0·8983	1·9403	3·5186	5·7152	7·0690	8·6053	10·333	12·260	16·744	22·122	39·866	64·441
0·07250	0·979	1·143	1·295	1·574	1·826	2·060	2·171	2·278	2·383	2·485	2·681	2·869	3·309	3·714
1/ 13·8	0·3075	0·5611	0·9156	1·9774	3·5855	5·8232	7·2025	8·7675	10·527	12·490	17·058	22·536	40·607	65·636
0·07500	0·997	1·164	1·319	1·603	1·860	2·097	2·210	2·320	2·426	2·530	2·730	2·921	3·368	3·781
1/ 13·3	0·3133	0·5715	0·9326	2·0139	3·6512	5·9295	7·3337	8·9269	10·718	12·716	17·366	22·943	41·337	66·811
0·08000	1·033	1·206	1·366	1·659	1·925	2·171	2·287	2·401	2·511	2·618	2·825	3·022	3·484	3·910
1/ 12·5	0·3246	0·5920	0·9659	2·0851	3·7796	6·1371	7·5899	9·2383	11·092	13·159	17·969	23·737	42·761	69·104
0·08500	1·068	1·247	1·412	1·714	1·988	2·242	2·362	2·479	2·593	2·703	2·916	3·120	3·597	4·036
1/ 11·8	0·3356	0·6120	0·9982	2·1543	3·9042	6·3385	7·8385	9·5404	11·454	13·587	18·553	24·507	44·142	71·328
0·09000	1·102	1·286	1·457	1·768	2·050	2·311	2·435	2·555	2·672	2·786	3·006	3·216	3·706	4·159
1/ 11·1	0·3463	0·6313	1·0296	2·2215	4·0254	6·5343	8·0802	9·8340	11·805	14·004	19·121	25·255	45·483	73·489
0·09500	1·135	1·325	1·500	1·820	2·110	2·378	2·506	2·630	2·750	2·867	3·092	3·308	3·813	4·278
1/ 10·5	0·3567	0·6502	1·0601	2·2869	4·1433	6·7248	8·3153	10·120	12·148	14·410	19·673	25·983	46·789	75·592
0·10000	1·168	1·362	1·542	1·871	2·169	2·444	2·575	2·702	2·825	2·945	3·177	3·399	3·916	4·394
1/ 10·0	0·3669	0·6686	1·0900	2·3508	4·2582	6·9106	8·5446	10·398	12·482	14·805	20·212	26·693	48·062	77·640
	0·77	0·78	0·79	0·81	0·82	0·83	0·84	0·84	0·85	0·85	0·86	0·87	0·88	0·90

$V_{r(0·5)medial}$ for half-full circular pipes.

Gradient	(Equivalent) Pipe diameters in mm													
	20	25	30	40	50	60	65	70	75	80	90	100	125	150
0·10000	1·168	1·362	1·542	1·871	2·169	2·444	2·575	2·702	2·825	2·945	3·177	3·399	3·916	4·394
1/ 10·0	0·3669	0·6686	1·0900	2·3508	4·2582	6·9106	8·5446	10·398	12·482	14·805	20·212	26·693	48·062	77·640
0·10500	1·199	1·399	1·583	1·920	2·226	2·508	2·642	2·773	2·899	3·022	3·260	3·487	4·018	4·507
1/ 9·5	0·3768	0·6865	1·1191	2·4130	4·3704	7·0919	8·7683	10·670	12·808	15·191	20·737	27·385	49·303	79·639
0·11000	1·230	1·434	1·623	1·969	2·282	2·571	2·708	2·842	2·971	3·097	3·340	3·573	4·116	4·617
1/ 9·1	0·3865	0·7041	1·1475	2·4739	4·4801	7·2690	8·9870	10·936	13·126	15·568	21·251	28·062	50·516	81·592
0·11500	1·260	1·469	1·663	2·016	2·336	2·632	2·773	2·909	3·042	3·171	3·419	3·657	4·213	4·725
1/ 8·7	0·3960	0·7212	1·1754	2·5335	4·5873	7·4423	9·2008	11·195	13·437	15·937	21·753	28·723	51·702	83·501
0·12000	1·290	1·504	1·701	2·063	2·390	2·692	2·836	2·975	3·111	3·242	3·497	3·740	4·308	4·831
1/ 8·3	0·4053	0·7381	1·2027	2·5918	4·6923	7·6119	9·4101	11·450	13·742	16·298	22·244	29·371	52·863	85·370
0·12500	1·319	1·537	1·739	2·108	2·442	2·751	2·898	3·040	3·178	3·313	3·572	3·820	4·400	4·935
1/ 8·0	0·4144	0·7546	1·2294	2·6490	4·7953	7·7782	9·6153	11·699	14·041	16·652	22·726	30·006	54·000	87·201
0·13000	1·348	1·570	1·776	2·153	2·494	2·809	2·958	3·103	3·244	3·382	3·647	3·900	4·491	5·036
1/ 7·7	0·4233	0·7707	1·2556	2·7051	4·8962	7·9412	9·8165	11·943	14·334	16·999	23·198	30·628	55·115	88·996
0·13500	1·375	1·602	1·813	2·196	2·544	2·865	3·018	3·166	3·310	3·450	3·719	3·977	4·580	5·136
1/ 7·4	0·4321	0·7866	1·2814	2·7601	4·9953	8·1012	10·014	12·183	14·621	17·339	23·662	31·238	56·210	90·757
0·14000	1·403	1·634	1·849	2·239	2·594	2·921	3·076	3·227	3·373	3·516	3·791	4·054	4·668	5·234
1/ 7·1	0·4408	0·8022	1·3066	2·8142	5·0926	8·2584	10·208	12·419	14·903	17·673	24·117	31·838	57·284	92·487
0·14500	1·430	1·666	1·884	2·282	2·642	2·975	3·134	3·287	3·436	3·581	3·861	4·129	4·754	5·330
1/ 6·9	0·4492	0·8176	1·3315	2·8673	5·1882	8·4128	10·398	12·650	15·181	18·002	24·564	32·427	58·340	94·186
0·15000	1·457	1·696	1·918	2·323	2·690	3·029	3·190	3·346	3·498	3·646	3·930	4·203	4·839	5·424
1/ 6·7	0·4576	0·8327	1·3560	2·9195	5·2823	8·5647	10·586	12·878	15·453	18·325	25·004	33·006	59·378	95·857
0·16000	1·508	1·756	1·986	2·405	2·784	3·134	3·300	3·462	3·618	3·771	4·065	4·346	5·004	5·609
1/ 6·3	0·4739	0·8621	1·4037	3·0216	5·4659	8·8611	10·952	13·322	15·986	18·955	25·862	34·137	61·404	99·117
0·17000	1·559	1·815	2·051	2·483	2·874	3·236	3·407	3·574	3·735	3·893	4·196	4·486	5·164	5·788
1/ 5·9	0·4897	0·8907	1·4500	3·1206	5·6440	9·1487	11·306	13·753	16·502	19·567	26·694	35·234	63·368	102·28
0·18000	1·608	1·871	2·115	2·560	2·963	3·335	3·511	3·682	3·849	4·011	4·323	4·622	5·319	5·962
1/ 5·6	0·5051	0·9185	1·4951	3·2168	5·8170	9·4281	11·651	14·172	17·004	20·161	27·503	36·299	65·276	105·35
0·19000	1·656	1·926	2·177	2·634	3·048	3·431	3·612	3·788	3·959	4·126	4·447	4·754	5·470	6·131
1/ 5·3	0·5201	0·9456	1·5389	3·3104	5·9855	9·7000	11·986	14·579	17·492	20·739	28·289	37·335	67·133	108·34
0·20000	1·702	1·980	2·238	2·707	3·132	3·524	3·711	3·891	4·067	4·238	4·567	4·882	5·618	6·295
1/ 5·0	0·5347	0·9720	1·5817	3·4017	6·1496	9·9650	12·313	14·976	17·967	21·302	29·056	38·345	68·942	111·25
0·21000	1·747	2·033	2·297	2·778	3·214	3·616	3·807	3·992	4·172	4·347	4·685	5·008	5·762	6·456
1/ 4·8	0·5490	0·9978	1·6234	3·4908	6·3098	10·224	12·632	15·363	18·431	21·852	29·804	39·330	70·706	114·09
0·22000	1·792	2·084	2·354	2·847	3·293	3·705	3·901	4·090	4·275	4·454	4·800	5·130	5·902	6·613
1/ 4·5	0·5629	1·0230	1·6641	3·5778	6·4664	10·476	12·944	15·742	18·885	22·388	30·535	40·292	72·430	116·86
0·23000	1·835	2·134	2·411	2·915	3·371	3·793	3·993	4·187	4·375	4·558	4·912	5·250	6·039	6·766
1/ 4·3	0·5766	1·0476	1·7040	3·6629	6·6194	10·723	13·249	16·112	19·328	22·913	31·249	41·233	74·115	119·57
0·24000	1·878	2·183	2·466	2·981	3·448	3·878	4·082	4·281	4·473	4·661	5·022	5·367	6·174	6·917
1/ 4·2	0·5899	1·0717	1·7431	3·7462	6·7693	10·965	13·547	16·474	19·762	23·427	31·948	42·154	75·765	122·23
0·25000	1·919	2·231	2·520	3·046	3·522	3·962	4·171	4·373	4·569	4·761	5·130	5·482	6·306	7·064
1/ 4·0	0·6030	1·0953	1·7813	3·8279	6·9161	11·202	13·839	16·829	20·187	23·931	32·633	43·057	77·380	124·83
0·26000	1·960	2·279	2·573	3·110	3·596	4·044	4·257	4·463	4·664	4·859	5·235	5·595	6·435	7·208
1/ 3·8	0·6159	1·1185	1·8189	3·9080	7·0600	11·434	14·126	17·177	20·604	24·424	33·305	43·941	78·964	127·37
0·27000	2·000	2·325	2·625	3·172	3·668	4·125	4·342	4·552	4·756	4·955	5·339	5·705	6·561	7·349
1/ 3·7	0·6285	1·1413	1·8557	3·9866	7·2013	11·662	14·407	17·518	21·013	24·908	33·964	44·809	80·519	129·87
0·28000	2·040	2·371	2·676	3·234	3·738	4·204	4·425	4·639	4·847	5·050	5·441	5·814	6·686	7·488
1/ 3·6	0·6409	1·1636	1·8919	4·0638	7·3401	11·886	14·683	17·853	21·415	25·384	34·611	45·661	82·044	132·33
0·29000	2·079	2·415	2·727	3·294	3·808	4·282	4·507	4·725	4·937	5·143	5·540	5·920	6·808	7·625
1/ 3·4	0·6530	1·1856	1·9274	4·1396	7·4764	12·106	14·954	18·183	21·809	25·851	35·247	46·499	83·544	134·74
0·30000	2·117	2·459	2·776	3·354	3·876	4·358	4·587	4·809	5·024	5·234	5·639	6·025	6·928	7·759
1/ 3·3	0·6650	1·2072	1·9624	4·2142	7·6104	12·322	15·221	18·507	22·197	26·310	35·872	47·322	85·017	137·11
	0·78	0·79	0·80	0·82	0·84	0·85	0·85	0·86	0·86	0·87	0·88	0·88	0·90	0·91

$V_{r(0{\cdot}5)\text{medial}}$ for half-full circular pipes.

A5

(p.1 of 6)

k_s = 0·060 mm
S = 0·00030 to 0·00100

ie hydraulic gradient =
1 in 3333 to 1 in 1000

Water (or sewage) at 15°C;
full bore conditions.

velocities in ms^{-1}
discharges in litres/sec

Gradient	(Equivalent) Pipe diameters in mm													
	20	25	30	40	50	60	65	70	75	80	90	100	125	150
0·000300				0·070	0·084	0·096	0·102	0·107	0·113	0·119	0·129	0·139	0·163	0·186
1/ 3333				0·0885	0·1641	0·2711	0·3377	0·4137	0·4996	0·5960	0·8216	1·0944	2·0044	3·2809
0·000320				0·073	0·087	0·100	0·106	0·112	0·117	0·123	0·134	0·145	0·169	0·192
1/ 3125				0·0919	0·1704	0·2814	0·3504	0·4293	0·5184	0·6183	0·8523	1·1351	2·0785	3·4014
0·000340				0·076	0·090	0·103	0·109	0·115	0·121	0·127	0·139	0·150	0·175	0·199
1/ 2941				0·0952	0·1765	0·2914	0·3629	0·4445	0·5367	0·6400	0·8822	1·1747	2·1505	3·5186
0·000360				0·078	0·093	0·106	0·113	0·119	0·126	0·132	0·143	0·154	0·181	0·206
1/ 2778				0·0985	0·1824	0·3011	0·3750	0·4592	0·5545	0·6612	0·9112	1·2133	2·2206	3·6327
0·000380				0·081	0·096	0·110	0·117	0·123	0·129	0·136	0·148	0·159	0·187	0·212
1/ 2632				0·1016	0·1882	0·3106	0·3868	0·4736	0·5718	0·6819	0·9396	1·2509	2·2890	3·7440
0·000400				0·083	0·099	0·113	0·120	0·127	0·133	0·140	0·152	0·164	0·192	0·218
1/ 2500				0·1047	0·1939	0·3199	0·3983	0·4877	0·5888	0·7020	0·9673	1·2876	2·3558	3·8526
0·000420				0·086	0·102	0·116	0·123	0·130	0·137	0·144	0·156	0·169	0·197	0·224
1/ 2381				0·1077	0·1994	0·3290	0·4096	0·5015	0·6054	0·7218	0·9943	1·3235	2·4210	3·9587
0·000440				0·088	0·104	0·119	0·127	0·134	0·141	0·147	0·160	0·173	0·202	0·230
1/ 2273				0·1107	0·2049	0·3379	0·4206	0·5150	0·6216	0·7411	1·0208	1·3587	2·4849	4·0626
0·000460				0·090	0·107	0·123	0·130	0·137	0·144	0·151	0·165	0·177	0·208	0·236
1/ 2174				0·1136	0·2102	0·3466	0·4314	0·5282	0·6375	0·7600	1·0468	1·3931	2·5474	4·1643
0·000480				0·093	0·110	0·126	0·133	0·141	0·148	0·155	0·169	0·182	0·213	0·241
1/ 2083				0·1165	0·2154	0·3551	0·4420	0·5412	0·6531	0·7786	1·0723	1·4269	2·6087	4·2640
0·000500				0·095	0·112	0·129	0·136	0·144	0·151	0·159	0·172	0·186	0·217	0·247
1/ 2000				0·1193	0·2205	0·3635	0·4525	0·5539	0·6684	0·7968	1·0972	1·4600	2·6689	4·3618
0·000525			0·078	0·098	0·116	0·132	0·140	0·148	0·156	0·163	0·177	0·191	0·223	0·254
1/ 1905			0·0553	0·1227	0·2268	0·3738	0·4652	0·5695	0·6872	0·8191	1·1279	1·5006	2·7426	4·4816
0·000550			0·080	0·100	0·119	0·136	0·144	0·152	0·160	0·167	0·182	0·196	0·229	0·260
1/ 1818			0·0568	0·1260	0·2330	0·3839	0·4777	0·5847	0·7056	0·8409	1·1578	1·5403	2·8147	4·5988
0·000575			0·082	0·103	0·122	0·139	0·148	0·156	0·164	0·172	0·187	0·201	0·235	0·267
1/ 1739			0·0583	0·1293	0·2390	0·3938	0·4900	0·5997	0·7236	0·8623	1·1871	1·5792	2·8853	4·7136
0·000600			0·085	0·105	0·125	0·143	0·151	0·160	0·168	0·176	0·191	0·206	0·241	0·273
1/ 1667			0·0598	0·1326	0·2449	0·4034	0·5020	0·6143	0·7412	0·8833	1·2159	1·6174	2·9545	4·8260
0·000625			0·087	0·108	0·128	0·146	0·155	0·163	0·172	0·180	0·196	0·211	0·246	0·279
1/ 1600			0·0612	0·1357	0·2507	0·4129	0·5137	0·6287	0·7585	0·9039	1·2442	1·6548	3·0224	4·9364
0·000650			0·089	0·111	0·131	0·149	0·158	0·167	0·176	0·184	0·200	0·215	0·252	0·285
1/ 1538			0·0626	0·1389	0·2564	0·4223	0·5253	0·6428	0·7755	0·9241	1·2719	1·6915	3·0891	5·0447
0·000675			0·091	0·113	0·133	0·153	0·162	0·171	0·179	0·188	0·204	0·220	0·257	0·291
1/ 1481			0·0640	0·1419	0·2620	0·4314	0·5367	0·6567	0·7922	0·9440	1·2991	1·7277	3·1546	5·1512
0·000700			0·093	0·115	0·136	0·156	0·165	0·174	0·183	0·192	0·208	0·224	0·262	0·297
1/ 1429			0·0654	0·1449	0·2676	0·4405	0·5479	0·6704	0·8087	0·9635	1·3259	1·7632	3·2190	5·2558
0·000725			0·094	0·118	0·139	0·159	0·168	0·178	0·187	0·196	0·213	0·229	0·267	0·303
1/ 1379			0·0668	0·1479	0·2730	0·4493	0·5589	0·6838	0·8248	0·9828	1·3523	1·7981	3·2824	5·3587
0·000750			0·096	0·120	0·142	0·162	0·172	0·181	0·190	0·199	0·217	0·233	0·273	0·309
1/ 1333			0·0681	0·1508	0·2783	0·4581	0·5697	0·6970	0·8408	1·0017	1·3782	1·8325	3·3447	5·4600
0·000800			0·100	0·125	0·147	0·168	0·178	0·188	0·197	0·207	0·225	0·242	0·282	0·320
1/ 1250			0·0707	0·1566	0·2888	0·4752	0·5910	0·7229	0·8719	1·0387	1·4290	1·8997	3·4666	5·6580
0·000850			0·104	0·129	0·152	0·174	0·184	0·194	0·204	0·214	0·232	0·250	0·292	0·331
1/ 1176			0·0733	0·1621	0·2990	0·4918	0·6116	0·7481	0·9022	1·0747	1·4783	1·9651	3·5851	5·8503
0·000900		0·093	0·107	0·133	0·157	0·180	0·190	0·201	0·211	0·221	0·240	0·258	0·302	0·342
1/ 1111		0·0457	0·0758	0·1675	0·3089	0·5080	0·6316	0·7726	0·9316	1·1097	1·5263	2·0287	3·7004	6·0375
0·000950		0·096	0·111	0·138	0·162	0·185	0·196	0·207	0·217	0·228	0·247	0·266	0·311	0·352
1/ 1053		0·0471	0·0782	0·1728	0·3185	0·5238	0·6512	0·7965	0·9604	1·1439	1·5731	2·0907	3·8127	6·2199
0·001000		0·099	0·114	0·142	0·167	0·191	0·202	0·213	0·224	0·234	0·254	0·274	0·320	0·362
1/ 1000		0·0486	0·0806	0·1780	0·3280	0·5392	0·6703	0·8198	0·9884	1·1772	1·6188	2·1512	3·9224	6·3978
		–	–	0·74	0·75	0·75	0·76	0·76	0·76	0·77	0·77	0·78	0·79	0·80

$V_{r(0\cdot5)medial}$ for half-full circular pipes.

k_s = 0·060 mm S = 0·00030 to 0·00100

k_s = 0·060 mm
S = 0·00100 to 0·00300

ie hydraulic gradient =
1 in 1000 to 1 in 333

Water (or sewage) at 15°C;
full bore conditions.

velocities in ms^{-1}
discharges in litres/sec

Gradient	**20**	**25**	**30**	**40**	**50**	**60**	**65**	**70**	**75**	**80**	**90**	**100**	**125**	**150**
		(Equivalent) Pipe diameters in mm												
0·00100		0·099	0·114	0·142	0·167	0·191	0·202	0·213	0·224	0·234	0·254	0·274	0·320	0·362
1/ 1000		0·0486	0·0806	0·1780	0·3280	0·5392	0·6703	0·8198	0·9884	1·1772	1·6188	2·1512	3·9224	6·3978
0·00105		0·102	0·117	0·146	0·172	0·196	0·208	0·219	0·230	0·241	0·261	0·281	0·328	0·372
1/ 952		0·0500	0·0829	0·1831	0·3372	0·5543	0·6890	0·8426	1·0159	1·2098	1·6634	2·2104	4·0295	6·5717
0·00110		0·105	0·120	0·150	0·176	0·201	0·213	0·225	0·236	0·247	0·268	0·289	0·337	0·382
1/ 909		0·0514	0·0851	0·1880	0·3463	0·5690	0·7073	0·8649	1·0427	1·2417	1·7071	2·2683	4·1343	6·7418
0·00115		0·107	0·124	0·153	0·181	0·206	0·219	0·230	0·242	0·253	0·275	0·296	0·345	0·391
1/ 870		0·0527	0·0874	0·1929	0·3551	0·5835	0·7252	0·8867	1·0690	1·2730	1·7499	2·3249	4·2370	6·9083
0·00120		0·110	0·127	0·157	0·185	0·211	0·224	0·236	0·248	0·259	0·282	0·303	0·353	0·400
1/ 833		0·0541	0·0896	0·1976	0·3638	0·5977	0·7428	0·9082	1·0948	1·3036	1·7919	2·3805	4·3375	7·0714
0·00125		0·113	0·130	0·161	0·190	0·216	0·229	0·241	0·254	0·265	0·288	0·310	0·361	0·409
1/ 800		0·0554	0·0917	0·2023	0·3723	0·6116	0·7600	0·9292	1·1200	1·3336	1·8330	2·4350	4·4362	7·2314
0·00130		0·115	0·133	0·165	0·194	0·221	0·234	0·247	0·259	0·271	0·294	0·317	0·369	0·418
1/ 769		0·0566	0·0938	0·2069	0·3807	0·6252	0·7770	0·9498	1·1449	1·3631	1·8734	2·4884	4·5330	7·3885
0·00135		0·118	0·136	0·168	0·198	0·226	0·239	0·252	0·265	0·277	0·301	0·324	0·377	0·427
1/ 741		0·0579	0·0959	0·2114	0·3889	0·6387	0·7936	0·9701	1·1693	1·3921	1·9131	2·5410	4·6281	7·5427
0·00140		0·120	0·139	0·172	0·202	0·231	0·244	0·257	0·270	0·283	0·307	0·330	0·385	0·435
1/ 714		0·0591	0·0979	0·2158	0·3970	0·6519	0·8100	0·9901	1·1932	1·4206	1·9521	2·5926	4·7216	7·6943
0·00145		0·123	0·141	0·175	0·206	0·235	0·249	0·262	0·275	0·288	0·313	0·337	0·392	0·444
1/ 690		0·0604	0·0999	0·2202	0·4049	0·6648	0·8261	1·0097	1·2168	1·4486	1·9905	2·6435	4·8135	7·8434
0·00150		0·125	0·144	0·179	0·210	0·240	0·254	0·267	0·281	0·294	0·319	0·343	0·400	0·452
1/ 667		0·0616	0·1019	0·2245	0·4128	0·6776	0·8419	1·0290	1·2401	1·4762	2·0282	2·6935	4·9040	7·9901
0·00160		0·130	0·150	0·185	0·218	0·248	0·263	0·277	0·291	0·304	0·330	0·355	0·414	0·468
1/ 625		0·0639	0·1057	0·2329	0·4281	0·7026	0·8729	1·0667	1·2855	1·5302	2·1021	2·7912	5·0808	8·2767
0·00170		0·135	0·155	0·192	0·226	0·257	0·272	0·287	0·301	0·315	0·342	0·367	0·428	0·484
1/ 588		0·0662	0·1095	0·2410	0·4430	0·7269	0·9030	1·1034	1·3296	1·5826	2·1738	2·8861	5·2525	8·5551
0·00180	0·117	0·139	0·160	0·198	0·233	0·265	0·281	0·296	0·311	0·325	0·353	0·379	0·442	0·499
1/ 556	0·0369	0·0684	0·1132	0·2490	0·4575	0·7506	0·9322	1·1391	1·3725	1·6336	2·2436	2·9785	5·4196	8·8258
0·00190	0·121	0·144	0·165	0·204	0·240	0·274	0·290	0·305	0·320	0·335	0·363	0·391	0·455	0·514
1/ 526	0·0380	0·0706	0·1167	0·2568	0·4716	0·7736	0·9608	1·1739	1·4143	1·6833	2·3116	3·0685	5·5823	9·0895
0·00200	0·125	0·148	0·170	0·210	0·247	0·282	0·298	0·314	0·329	0·345	0·374	0·402	0·468	0·529
1/ 500	0·0392	0·0727	0·1202	0·2643	0·4855	0·7961	0·9887	1·2079	1·4552	1·7318	2·3780	3·1564	5·7411	9·3468
0·00210	0·128	0·152	0·175	0·216	0·254	0·289	0·306	0·322	0·338	0·354	0·384	0·413	0·480	0·543
1/ 476	0·0403	0·0748	0·1236	0·2718	0·4989	0·8181	1·0159	1·2411	1·4951	1·7792	2·4429	3·2422	5·8962	9·5981
0·00220	0·132	0·157	0·180	0·222	0·261	0·297	0·314	0·331	0·347	0·363	0·394	0·423	0·493	0·557
1/ 455	0·0414	0·0769	0·1270	0·2790	0·5122	0·8396	1·0426	1·2736	1·5341	1·8255	2·5063	3·3261	6·0479	9·8438
0·00230	0·135	0·161	0·184	0·228	0·267	0·304	0·322	0·339	0·356	0·372	0·404	0·434	0·505	0·571
1/ 435	0·0425	0·0789	0·1302	0·2861	0·5251	0·8607	1·0687	1·3054	1·5724	1·8709	2·5684	3·4083	6·1963	10·084
0·00240	0·139	0·165	0·189	0·233	0·274	0·312	0·330	0·347	0·364	0·381	0·413	0·444	0·517	0·584
1/ 417	0·0436	0·0808	0·1334	0·2931	0·5378	0·8813	1·0942	1·3366	1·6098	1·9154	2·6292	3·4888	6·3418	10·320
0·00250	0·142	0·169	0·193	0·239	0·280	0·319	0·337	0·355	0·373	0·390	0·423	0·454	0·528	0·597
1/ 400	0·0446	0·0827	0·1366	0·2999	0·5502	0·9016	1·1193	1·3671	1·6466	1·9591	2·6889	3·5677	6·4844	10·551
0·00260	0·145	0·172	0·198	0·244	0·286	0·326	0·345	0·363	0·381	0·398	0·432	0·464	0·540	0·610
1/ 385	0·0457	0·0846	0·1397	0·3066	0·5624	0·9215	1·1439	1·3971	1·6826	2·0019	2·7475	3·6452	6·6244	10·777
0·00270	0·149	0·176	0·202	0·249	0·293	0·333	0·352	0·371	0·389	0·407	0·441	0·474	0·551	0·622
1/ 370	0·0467	0·0865	0·1427	0·3132	0·5744	0·9410	1·1681	1·4266	1·7181	2·0439	2·8050	3·7213	6·7618	11·000
0·00280	0·152	0·180	0·206	0·254	0·299	0·340	0·359	0·378	0·397	0·415	0·450	0·483	0·562	0·635
1/ 357	0·0477	0·0883	0·1457	0·3197	0·5862	0·9602	1·1919	1·4556	1·7529	2·0853	2·8616	3·7961	6·8968	11·219
0·00290	0·155	0·184	0·210	0·259	0·304	0·346	0·366	0·386	0·405	0·423	0·459	0·493	0·573	0·647
1/ 345	0·0487	0·0901	0·1486	0·3261	0·5978	0·9791	1·2153	1·4841	1·7871	2·1260	2·9172	3·8696	7·0296	11·434
0·00300	0·158	0·187	0·214	0·264	0·310	0·353	0·373	0·393	0·412	0·431	0·467	0·502	0·583	0·659
1/ 333	0·0496	0·0919	0·1515	0·3324	0·6092	0·9977	1·2383	1·5121	1·8208	2·1660	2·9719	3·9420	7·1603	11·645
	–	0·74	0·74	0·75	0·76	0·77	0·77	0·78	0·78	0·78	0·79	0·80	0·81	0·82

$V_{r(0{\cdot}5)medial}$ for half-full circular pipes.

k_s = 0·060 mm
S = 0·00300 to 0·01000

ie hydraulic gradient =
1 in 333 to 1 in 100

Water (or sewage) at 15°C;
full bore conditions.

velocities in ms^{-1}
discharges in litres/sec

Gradient **(Equivalent) Pipe diameters in mm**

Gradient	20	25	30	40	50	60	65	70	75	80	90	100	125	150
0·00300	0·158	0·187	0·214	0·264	0·310	0·353	0·373	0·393	0·412	0·431	0·467	0·502	0·583	0·659
1/ 333	0·0496	0·0919	0·1515	0·3324	0·6092	0·9977	1·2383	1·5121	1·8208	2·1660	2·9719	3·9420	7·1603	11·645
0·00320	0·164	0·194	0·222	0·274	0·322	0·366	0·387	0·407	0·427	0·446	0·484	0·520	0·604	0·682
1/ 313	0·0515	0·0953	0·1572	0·3446	0·6316	1·0341	1·2833	1·5670	1·8867	2·2442	3·0788	4·0834	7·4155	12·058
0·00340	0·170	0·201	0·230	0·284	0·333	0·378	0·400	0·421	0·442	0·462	0·500	0·537	0·624	0·705
1/ 294	0·0534	0·0987	0·1627	0·3566	0·6533	1·0694	1·3270	1·6202	1·9507	2·3201	3·1826	4·2206	7·6633	12·459
0·00360	0·176	0·208	0·238	0·293	0·343	0·390	0·413	0·434	0·456	0·476	0·516	0·554	0·644	0·727
1/ 278	0·0551	0·1020	0·1681	0·3682	0·6744	1·1037	1·3696	1·6720	2·0129	2·3940	3·2836	4·3542	7·9043	12·849
0·00380	0·181	0·214	0·245	0·302	0·354	0·402	0·425	0·448	0·469	0·491	0·532	0·571	0·663	0·749
1/ 263	0·0569	0·1052	0·1733	0·3795	0·6950	1·1372	1·4110	1·7225	2·0735	2·4660	3·3820	4·4842	8·1389	13·229
0·00400	0·187	0·221	0·252	0·311	0·364	0·414	0·437	0·460	0·483	0·505	0·547	0·587	0·682	0·770
1/ 250	0·0586	0·1083	0·1784	0·3906	0·7151	1·1699	1·4514	1·7717	2·1327	2·5362	3·4780	4·6111	8·3678	13·599
0·00420	0·192	0·227	0·259	0·319	0·374	0·425	0·449	0·473	0·496	0·518	0·561	0·603	0·700	0·790
1/ 238	0·0603	0·1113	0·1834	0·4014	0·7347	1·2018	1·4909	1·8198	2·1905	2·6047	3·5717	4·7350	8·5913	13·961
0·00440	0·197	0·233	0·266	0·328	0·384	0·436	0·461	0·485	0·509	0·532	0·576	0·618	0·718	0·810
1/ 227	0·0619	0·1143	0·1882	0·4119	0·7539	1·2330	1·5296	1·8669	2·2470	2·6718	3·6633	4·8561	8·8098	14·314
0·00460	0·202	0·239	0·273	0·336	0·394	0·447	0·472	0·497	0·521	0·545	0·590	0·633	0·735	0·830
1/ 217	0·0635	0·1172	0·1930	0·4223	0·7727	1·2636	1·5674	1·9129	2·3023	2·7375	3·7530	4·9747	9·0236	14·660
0·00480	0·207	0·245	0·280	0·344	0·403	0·457	0·484	0·509	0·533	0·557	0·604	0·648	0·752	0·849
1/ 208	0·0651	0·1201	0·1977	0·4324	0·7911	1·2935	1·6044	1·9580	2·3565	2·8018	3·8409	5·0908	9·2330	14·998
0·00500	0·212	0·250	0·286	0·352	0·412	0·468	0·494	0·520	0·545	0·570	0·617	0·663	0·769	0·868
1/ 200	0·0666	0·1229	0·2023	0·4424	0·8092	1·3229	1·6407	2·0023	2·4096	2·8648	3·9270	5·2047	9·4383	15·330
0·00525	0·218	0·257	0·294	0·362	0·423	0·481	0·508	0·534	0·560	0·585	0·634	0·680	0·790	0·890
1/ 190	0·0685	0·1264	0·2079	0·4546	0·8313	1·3588	1·6852	2·0564	2·4746	2·9420	4·0325	5·3440	9·6894	15·736
0·00550	0·224	0·264	0·302	0·371	0·434	0·493	0·521	0·548	0·575	0·600	0·650	0·698	0·810	0·913
1/ 182	0·0703	0·1297	0·2135	0·4665	0·8529	1·3940	1·7287	2·1094	2·5382	3·0174	4·1355	5·4802	9·9349	16·133
0·00575	0·230	0·271	0·310	0·380	0·445	0·505	0·534	0·562	0·589	0·615	0·666	0·715	0·829	0·935
1/ 174	0·0721	0·1330	0·2189	0·4781	0·8741	1·4284	1·7713	2·1612	2·6004	3·0913	4·2364	5·6134	10·175	16·522
0·00600	0·235	0·278	0·317	0·390	0·456	0·517	0·546	0·575	0·602	0·629	0·681	0·731	0·848	0·956
1/ 167	0·0739	0·1363	0·2241	0·4896	0·8948	1·4621	1·8130	2·2119	2·6614	3·1636	4·3352	5·7440	10·410	16·902
0·00625	0·241	0·284	0·324	0·399	0·466	0·529	0·559	0·588	0·616	0·643	0·697	0·748	0·867	0·978
1/ 160	0·0757	0·1395	0·2293	0·5008	0·9152	1·4951	1·8538	2·2617	2·7212	3·2345	4·4320	5·8719	10·641	17·274
0·00650	0·246	0·290	0·332	0·407	0·476	0·540	0·571	0·600	0·629	0·657	0·712	0·764	0·886	0·998
1/ 154	0·0774	0·1426	0·2344	0·5118	0·9351	1·5276	1·8940	2·3106	2·7798	3·3041	4·5271	5·9975	10·867	17·640
0·00675	0·252	0·297	0·339	0·416	0·486	0·552	0·583	0·613	0·642	0·671	0·726	0·779	0·904	1·019
1/ 148	0·0790	0·1456	0·2394	0·5226	0·9547	1·5594	1·9334	2·3585	2·8374	3·3724	4·6204	6·1207	11·089	17·999
0·00700	0·257	0·303	0·346	0·424	0·496	0·563	0·594	0·625	0·655	0·684	0·741	0·795	0·921	1·038
1/ 143	0·0807	0·1487	0·2444	0·5333	0·9740	1·5907	1·9721	2·4056	2·8940	3·4395	4·7121	6·2418	11·307	18·351
0·00725	0·262	0·309	0·353	0·433	0·506	0·573	0·606	0·637	0·668	0·697	0·755	0·810	0·939	1·058
1/ 138	0·0823	0·1516	0·2492	0·5437	0·9930	1·6215	2·0102	2·4520	2·9496	3·5055	4·8022	6·3609	11·522	18·698
0·00750	0·267	0·315	0·359	0·441	0·515	0·584	0·617	0·649	0·680	0·710	0·769	0·825	0·956	1·077
1/ 133	0·0839	0·1545	0·2540	0·5540	1·0116	1·6518	2·0476	2·4976	3·0043	3·5705	4·8908	6·4780	11·733	19·039
0·00800	0·277	0·326	0·372	0·457	0·534	0·605	0·639	0·672	0·704	0·736	0·796	0·854	0·990	1·115
1/ 125	0·0870	0·1603	0·2633	0·5741	1·0481	1·7110	2·1208	2·5866	3·1112	3·6973	5·0640	6·7067	12·144	19·704
0·00850	0·287	0·338	0·385	0·472	0·552	0·625	0·661	0·695	0·728	0·760	0·822	0·882	1·022	1·152
1/ 118	0·0901	0·1658	0·2723	0·5936	1·0835	1·7684	2·1918	2·6731	3·2150	3·8204	5·2320	6·9286	12·544	20·350
0·00900	0·296	0·349	0·398	0·488	0·569	0·645	0·681	0·716	0·751	0·784	0·848	0·910	1·054	1·187
1/ 111	0·0930	0·1712	0·2811	0·6126	1·1179	1·8243	2·2609	2·7571	3·3159	3·9401	5·3954	7·1444	12·933	20·977
0·00950	0·305	0·359	0·410	0·502	0·586	0·664	0·702	0·738	0·773	0·807	0·873	0·936	1·085	1·222
1/ 105	0·0959	0·1765	0·2897	0·6311	1·1514	1·8787	2·3281	2·8390	3·4142	4·0566	5·5545	7·3545	13·311	21·588
0·01000	0·314	0·370	0·422	0·517	0·603	0·683	0·721	0·758	0·794	0·830	0·897	0·962	1·115	1·255
1/ 100	0·0987	0·1816	0·2981	0·6492	1·1841	1·9317	2·3937	2·9188	3·5100	4·1702	5·7096	7·5593	13·679	22·183
	0·74	0·75	0·75	0·77	0·78	0·79	0·79	0·79	0·80	0·80	0·81	0·82	0·83	0·85

$V_{r(0{\cdot}5)medial}$ for half-full circular pipes.

$k_s = 0{\cdot}060$ **mm**
S = 0·01000 to 0·03000

ie hydraulic gradient =
1 in 100 to 1 in 33·3

Water (or sewage) at 15°C;
full bore conditions.

velocities in ms^{-1}
discharges in litres/sec

Gradient **(Equivalent) Pipe diameters in mm**

Gradient	20	25	30	40	50	60	65	70	75	80	90	100	125	150
0·01000	0·314	0·370	0·422	0·517	0·603	0·683	0·721	0·758	0·794	0·830	0·897	0·962	1·115	1·255
1/ 100	0·0987	0·1816	0·2981	0·6492	1·1841	1·9317	2·3937	2·9188	3·5100	4·1702	5·7096	7·5593	13·679	22·183
0·01050	0·323	0·380	0·433	0·531	0·619	0·702	0·741	0·779	0·816	0·852	0·921	0·988	1·144	1·288
1/ 95	0·1015	0·1866	0·3063	0·6668	1·2160	1·9836	2·4578	2·9967	3·6035	4·2812	5·8610	7·7592	14·039	22·764
0·01100	0·332	0·390	0·445	0·544	0·635	0·719	0·760	0·798	0·836	0·873	0·945	1·013	1·173	1·320
1/ 91	0·1042	0·1915	0·3143	0·6841	1·2473	2·0342	2·5204	3·0729	3·6950	4·3896	6·0090	7·9546	14·391	23·332
0·01150	0·340	0·400	0·456	0·558	0·651	0·737	0·778	0·818	0·857	0·894	0·967	1·037	1·201	1·352
1/ 87	0·1068	0·1963	0·3221	0·7010	1·2778	2·0838	2·5817	3·1475	3·7844	4·4957	6·1538	8·1458	14·735	23·887
0·01200	0·348	0·410	0·467	0·571	0·666	0·754	0·796	0·837	0·876	0·915	0·990	1·061	1·228	1·383
1/ 83	0·1094	0·2010	0·3298	0·7175	1·3078	2·1323	2·6417	3·2205	3·8721	4·5997	6·2956	8·3330	15·072	24·431
0·01250	0·356	0·419	0·477	0·584	0·681	0·771	0·814	0·855	0·896	0·935	1·011	1·084	1·255	1·413
1/ 80	0·1120	0·2057	0·3373	0·7337	1·3371	2·1799	2·7005	3·2921	3·9579	4·7015	6·4346	8·5164	15·402	24·964
0·01300	0·364	0·428	0·488	0·597	0·696	0·788	0·831	0·874	0·915	0·955	1·033	1·107	1·281	1·442
1/ 77	0·1144	0·2102	0·3447	0·7496	1·3659	2·2266	2·7583	3·3623	4·0422	4·8014	6·5709	8·6964	15·725	25·486
0·01350	0·372	0·437	0·498	0·609	0·710	0·804	0·848	0·892	0·934	0·975	1·054	1·130	1·307	1·471
1/ 74	0·1169	0·2146	0·3519	0·7652	1·3942	2·2725	2·8149	3·4312	4·1249	4·8995	6·7047	8·8730	16·043	25·999
0·01400	0·380	0·446	0·508	0·621	0·724	0·820	0·865	0·909	0·952	0·994	1·075	1·152	1·333	1·500
1/ 71	0·1193	0·2190	0·3591	0·7806	1·4219	2·3175	2·8706	3·4989	4·2061	4·9958	6·8362	9·0465	16·355	26·503
0·01450	0·387	0·455	0·518	0·633	0·738	0·835	0·882	0·926	0·970	1·013	1·095	1·174	1·358	1·528
1/ 69	0·1217	0·2233	0·3661	0·7957	1·4493	2·3618	2·9253	3·5655	4·2860	5·0905	6·9654	9·2170	16·662	26·997
0·01500	0·395	0·464	0·528	0·645	0·752	0·851	0·898	0·943	0·988	1·031	1·115	1·195	1·382	1·555
1/ 67	0·1240	0·2276	0·3730	0·8105	1·4761	2·4053	2·9791	3·6309	4·3646	5·1836	7·0924	9·3847	16·963	27·484
0·01600	0·409	0·480	0·547	0·668	0·779	0·881	0·929	0·977	1·023	1·067	1·154	1·237	1·430	1·609
1/ 63	0·1285	0·2358	0·3864	0·8396	1·5286	2·4904	3·0842	3·7588	4·5179	5·3655	7·3405	9·7121	17·552	28·434
0·01700	0·423	0·497	0·565	0·691	0·804	0·910	0·960	1·009	1·056	1·103	1·192	1·277	1·477	1·661
1/ 59	0·1330	0·2439	0·3995	0·8677	1·5795	2·5729	3·1862	3·8828	4·6668	5·5420	7·5812	10·030	18·123	29·356
0·01800	0·437	0·513	0·583	0·712	0·830	0·938	0·990	1·040	1·089	1·137	1·228	1·316	1·522	1·712
1/ 56	0·1373	0·2517	0·4123	0·8951	1·6290	2·6531	3·2853	4·0034	4·8114	5·7135	7·8151	10·338	18·678	30·251
0·01900	0·450	0·528	0·601	0·734	0·854	0·966	1·019	1·071	1·121	1·170	1·264	1·355	1·566	1·761
1/ 53	0·1415	0·2593	0·4247	0·9218	1·6773	2·7313	3·3819	4·1208	4·9523	5·8804	8·0428	10·639	19·218	31·122
0·02000	0·463	0·543	0·618	0·754	0·878	0·993	1·048	1·100	1·152	1·202	1·299	1·392	1·609	1·809
1/ 50	0·1455	0·2668	0·4368	0·9478	1·7243	2·8074	3·4760	4·2352	5·0895	6·0432	8·2647	10·932	19·744	31·971
0·02100	0·476	0·558	0·635	0·774	0·902	1·019	1·075	1·130	1·182	1·234	1·333	1·428	1·651	1·856
1/ 47·6	0·1495	0·2740	0·4486	0·9732	1·7702	2·8818	3·5678	4·3469	5·2235	6·2020	8·4814	11·218	20·258	32·800
0·02200	0·488	0·573	0·651	0·794	0·924	1·045	1·102	1·158	1·212	1·265	1·366	1·464	1·692	1·902
1/ 45·5	0·1534	0·2811	0·4601	0·9980	1·8150	2·9544	3·6576	4·4561	5·3545	6·3572	8·6930	11·497	20·760	33·609
0·02300	0·501	0·587	0·667	0·814	0·947	1·070	1·129	1·186	1·241	1·295	1·399	1·499	1·732	1·947
1/ 43·5	0·1573	0·2881	0·4714	1·0223	1·8589	3·0255	3·7454	4·5628	5·4826	6·5090	8·9000	11·770	21·250	34·401
0·02400	0·513	0·601	0·683	0·832	0·969	1·095	1·155	1·213	1·269	1·325	1·431	1·533	1·771	1·991
1/ 41·7	0·1610	0·2949	0·4825	1·0461	1·9019	3·0951	3·8314	4·6674	5·6080	6·6577	9·1027	12·037	21·731	35·176
0·02500	0·524	0·614	0·698	0·851	0·990	1·119	1·180	1·239	1·297	1·353	1·462	1·566	1·809	2·033
1/ 40·0	0·1647	0·3016	0·4934	1·0694	1·9440	3·1633	3·9156	4·7698	5·7308	6·8034	9·3013	12·299	22·202	35·935
0·02600	0·536	0·628	0·713	0·869	1·011	1·142	1·205	1·266	1·324	1·382	1·493	1·599	1·847	2·076
1/ 38·5	0·1683	0·3081	0·5040	1·0923	1·9853	3·2302	3·9983	4·8703	5·8514	6·9462	9·4961	12·556	22·663	36·679
0·02700	0·547	0·641	0·728	0·887	1·032	1·166	1·229	1·291	1·351	1·410	1·523	1·631	1·884	2·117
1/ 37·0	0·1718	0·3145	0·5145	1·1148	2·0259	3·2959	4·0794	4·9690	5·9697	7·0864	9·6872	12·808	23·116	37·409
0·02800	0·558	0·654	0·742	0·905	1·052	1·188	1·253	1·316	1·378	1·437	1·552	1·662	1·920	2·158
1/ 35·7	0·1753	0·3209	0·5247	1·1368	2·0657	3·3604	4·1591	5·0658	6·0858	7·2241	9·8749	13·056	23·561	38·127
0·02900	0·569	0·666	0·757	0·922	1·072	1·211	1·277	1·341	1·403	1·464	1·581	1·693	1·956	2·197
1/ 34·5	0·1787	0·3271	0·5348	1·1585	2·1049	3·4238	4·2374	5·1610	6·2000	7·3594	10·059	13·299	23·998	38·831
0·03000	0·580	0·679	0·771	0·939	1·092	1·233	1·300	1·365	1·429	1·491	1·610	1·724	1·991	2·237
1/ 33·3	0·1821	0·3332	0·5448	1·1798	2·1434	3·4861	4·3144	5·2546	6·3123	7·4925	10·241	13·539	24·428	39·524
	0·75	0·76	0·77	0·78	0·79	0·80	0·81	0·81	0·81	0·82	0·83	0·83	0·85	0·86

$V_{r(0{\cdot}5)medial}$ **for half-full circular pipes.**

k_s = 0·060 mm
S = 0·03000 to 0·10000

ie hydraulic gradient =
1 in 33·3 to 1 in 10·0

Water (or sewage) at 15°C;
full bore conditions.

velocities in ms^{-1}
discharges in litres/sec

Gradient **(Equivalent) Pipe diameters in mm**

Gradient	20	25	30	40	50	60	65	70	75	80	90	100	125	150
0·03000	0·580	0·679	0·771	0·939	1·092	1·233	1·300	1·365	1·429	1·491	1·610	1·724	1·991	2·237
1/ 33·3	0·1821	0·3332	0·5448	1·1798	2·1434	3·4861	4·3144	5·2546	6·3123	7·4925	10·241	13·539	24·428	39·524
0·03200	0·601	0·703	0·798	0·972	1·130	1·276	1·345	1·413	1·478	1·542	1·665	1·783	2·059	2·313
1/ 31·3	0·1887	0·3451	0·5642	1·2215	2·2186	3·6079	4·4647	5·4374	6·5315	7·7523	10·595	14·006	25·266	40·877
0·03400	0·621	0·727	0·825	1·004	1·167	1·318	1·389	1·459	1·527	1·592	1·719	1·841	2·125	2·387
1/ 29·4	0·1951	0·3567	0·5830	1·2619	2·2916	3·7259	4·6105	5·6147	6·7441	8·0042	10·938	14·459	26·080	42·188
0·03600	0·641	0·750	0·851	1·035	1·203	1·358	1·432	1·504	1·573	1·641	1·772	1·897	2·190	2·459
1/ 27·8	0·2013	0·3680	0·6013	1·3012	2·3625	3·8407	4·7523	5·7870	6·9507	8·2491	11·272	14·899	26·870	43·462
0·03800	0·660	0·772	0·876	1·066	1·238	1·398	1·474	1·547	1·619	1·688	1·823	1·951	2·252	2·530
1/ 26·3	0·2074	0·3790	0·6192	1·3395	2·4315	3·9524	4·8902	5·9546	7·1517	8·4873	11·597	15·327	27·639	44·701
0·04000	0·679	0·794	0·901	1·096	1·273	1·436	1·514	1·590	1·663	1·735	1·873	2·005	2·313	2·598
1/ 25·0	0·2133	0·3897	0·6366	1·3768	2·4988	4·0612	5·0246	6·1180	7·3476	8·7195	11·913	15·744	28·388	45·909
0·04200	0·697	0·815	0·925	1·125	1·306	1·474	1·554	1·631	1·706	1·780	1·921	2·056	2·373	2·665
1/ 23·8	0·2190	0·4001	0·6535	1·4132	2·5645	4·1675	5·1558	6·2775	7·5388	8·9461	12·222	16·151	29·119	47·087
0·04400	0·715	0·836	0·948	1·153	1·339	1·511	1·592	1·672	1·749	1·824	1·969	2·107	2·431	2·730
1/ 22·7	0·2247	0·4103	0·6701	1·4487	2·6286	4·2713	5·2840	6·4333	7·7256	9·1674	12·524	16·549	29·833	48·238
0·04600	0·733	0·856	0·971	1·181	1·371	1·547	1·630	1·711	1·790	1·867	2·015	2·157	2·488	2·793
1/ 21·7	0·2302	0·4203	0·6864	1·4835	2·6914	4·3728	5·4093	6·5856	7·9083	9·3839	12·819	16·938	30·531	49·363
0·04800	0·750	0·876	0·993	1·208	1·402	1·582	1·667	1·750	1·831	1·909	2·060	2·205	2·544	2·856
1/ 20·8	0·2356	0·4301	0·7023	1·5176	2·7528	4·4722	5·5320	6·7348	8·0871	9·5958	13·107	17·319	31·215	50·464
0·05000	0·767	0·896	1·016	1·234	1·433	1·616	1·703	1·788	1·870	1·950	2·105	2·253	2·598	2·917
1/ 20·0	0·2409	0·4397	0·7178	1·5510	2·8131	4·5696	5·6523	6·8809	8·2623	9·8034	13·390	17·692	31·884	51·543
0·05250	0·787	0·920	1·042	1·267	1·470	1·658	1·748	1·834	1·919	2·001	2·159	2·311	2·665	2·991
1/ 19·0	0·2473	0·4515	0·7369	1·5918	2·8867	4·6886	5·7993	7·0596	8·4766	10·057	13·736	18·148	32·702	52·862
0·05500	0·807	0·943	1·069	1·298	1·507	1·699	1·791	1·880	1·966	2·050	2·212	2·367	2·730	3·064
1/ 18·2	0·2537	0·4629	0·7555	1·6317	2·9586	4·8050	5·9429	7·2341	8·6859	10·305	14·074	18·593	33·502	54·150
0·05750	0·827	0·966	1·095	1·330	1·543	1·740	1·833	1·924	2·012	2·098	2·264	2·423	2·794	3·136
1/ 17·4	0·2599	0·4742	0·7737	1·6708	3·0290	4·9188	6·0834	7·4048	8·8905	10·548	14·404	19·029	34·284	55·410
0·06000	0·847	0·988	1·120	1·360	1·578	1·779	1·875	1·968	2·058	2·146	2·315	2·477	2·856	3·205
1/ 16·7	0·2659	0·4852	0·7916	1·7090	3·0979	5·0301	6·2209	7·5719	9·0909	10·785	14·728	19·455	35·049	56·643
0·06250	0·865	1·010	1·145	1·390	1·612	1·818	1·915	2·010	2·102	2·192	2·365	2·530	2·917	3·274
1/ 16·0	0·2719	0·4959	0·8091	1·7464	3·1654	5·1393	6·3556	7·7357	9·2871	11·018	15·044	19·873	35·798	57·850
0·06500	0·884	1·032	1·169	1·419	1·646	1·855	1·955	2·052	2·146	2·237	2·414	2·582	2·977	3·341
1/ 15·4	0·2777	0·5065	0·8262	1·7831	3·2316	5·2463	6·4877	7·8962	9·4796	11·246	15·355	20·282	36·533	59·034
0·06750	0·902	1·053	1·193	1·448	1·679	1·893	1·994	2·093	2·188	2·282	2·462	2·634	3·036	3·406
1/ 14·8	0·2835	0·5169	0·8430	1·8192	3·2965	5·3513	6·6174	8·0537	9·6684	11·469	15·660	20·684	37·254	60·196
0·07000	0·920	1·074	1·216	1·476	1·711	1·929	2·033	2·133	2·230	2·325	2·509	2·684	3·093	3·471
1/ 14·3	0·2891	0·5271	0·8596	1·8546	3·3603	5·4544	6·7446	8·2084	9·8539	11·689	15·959	21·079	37·961	61·336
0·07250	0·938	1·094	1·239	1·504	1·743	1·965	2·070	2·172	2·272	2·368	2·555	2·733	3·150	3·534
1/ 13·8	0·2946	0·5371	0·8758	1·8894	3·4230	5·5557	6·8697	8·3604	10·036	11·905	16·253	21·466	38·657	62·457
0·07500	0·955	1·114	1·262	1·531	1·775	2·000	2·107	2·211	2·312	2·411	2·600	2·782	3·206	3·597
1/ 13·3	0·3001	0·5469	0·8918	1·9236	3·4846	5·6554	6·9927	8·5098	10·215	12·117	16·542	21·847	39·341	63·558
0·08000	0·989	1·153	1·306	1·584	1·836	2·069	2·180	2·287	2·391	2·493	2·689	2·876	3·314	3·718
1/ 12·5	0·3107	0·5662	0·9230	1·9903	3·6049	5·8498	7·2327	8·8015	10·565	12·531	17·106	22·591	40·675	65·707
0·08500	1·022	1·191	1·349	1·635	1·895	2·136	2·250	2·361	2·468	2·573	2·775	2·968	3·420	3·836
1/ 11·8	0·3210	0·5848	0·9532	2·0551	3·7216	6·0384	7·4655	9·0843	10·904	12·933	17·654	23·312	41·968	67·791
0·09000	1·054	1·228	1·390	1·685	1·953	2·200	2·318	2·432	2·543	2·651	2·858	3·057	3·522	3·951
1/ 11·1	0·3310	0·6030	0·9826	2·1180	3·8349	6·2216	7·6916	9·3590	11·233	13·323	18·185	24·012	43·225	69·815
0·09500	1·085	1·264	1·431	1·734	2·009	2·263	2·384	2·501	2·615	2·726	2·940	3·144	3·622	4·062
1/ 10·5	0·3408	0·6206	1·0113	2·1793	3·9452	6·3998	7·9116	9·6263	11·554	13·703	18·702	24·694	44·447	71·784
0·10000	1·115	1·299	1·470	1·782	2·064	2·325	2·449	2·569	2·686	2·800	3·019	3·229	3·719	4·171
1/ 10·0	0·3503	0·6378	1·0392	2·2389	4·0527	6·5735	8·1259	9·8867	11·866	14·072	19·205	25·358	45·637	73·701
	0·76	0·77	0·78	0·79	0·80	0·81	0·82	0·82	0·83	0·83	0·84	0·85	0·86	0·87

$V_{r(0{\cdot}5)medial}$ for half-full circular pipes.

k_s = 0·060 mm S = 0·03000 to 0·10000

Gradient	**(Equivalent) Pipe diameters in mm** 20	25	30	40	50	60	65	70	75	80	90	100	125	150
0·10000	1·115	1·299	1·470	1·782	2·064	2·325	2·449	2·569	2·686	2·800	3·019	3·229	3·719	4·171
1/ 10·0	0·3503	0·6378	1·0392	2·2389	4·0527	6·5735	8·1259	9·8867	11·866	14·072	19·205	25·358	45·637	73·701
0·10500	1·145	1·334	1·509	1·828	2·117	2·385	2·512	2·635	2·755	2·871	3·096	3·311	3·814	4·277
1/ 9·5	0·3596	0·6546	1·0664	2·2972	4·1576	6·7429	8·3351	10·141	12·170	14·433	19·697	26·005	46·799	75·572
0·11000	1·173	1·367	1·546	1·873	2·170	2·443	2·573	2·700	2·822	2·941	3·172	3·392	3·906	4·380
1/ 9·1	0·3686	0·6710	1·0930	2·3540	4·2600	6·9084	8·5393	10·389	12·468	14·785	20·177	26·638	47·933	77·399
0·11500	1·202	1·400	1·583	1·918	2·221	2·501	2·634	2·763	2·888	3·010	3·245	3·470	3·996	4·481
1/ 8·7	0·3775	0·6871	1·1190	2·4097	4·3602	7·0703	8·7391	10·632	12·759	15·130	20·646	27·256	49·042	79·186
0·12000	1·229	1·432	1·619	1·961	2·271	2·557	2·693	2·824	2·952	3·077	3·318	3·547	4·085	4·580
1/ 8·3	0·3862	0·7028	1·1445	2·4641	4·4582	7·2287	8·9346	10·869	13·043	15·467	21·105	27·861	50·127	80·934
0·12500	1·256	1·463	1·654	2·003	2·319	2·612	2·750	2·885	3·015	3·143	3·388	3·623	4·171	4·677
1/ 8·0	0·3947	0·7182	1·1694	2·5175	4·5543	7·3839	9·1261	11·102	13·322	15·797	21·555	28·454	51·191	82·646
0·13000	1·283	1·494	1·689	2·045	2·367	2·665	2·807	2·944	3·077	3·207	3·458	3·697	4·256	4·772
1/ 7·7	0·4031	0·7333	1·1939	2·5698	4·6485	7·5360	9·3138	11·330	13·595	16·121	21·996	29·035	52·233	84·324
0·13500	1·309	1·524	1·723	2·086	2·415	2·718	2·862	3·002	3·138	3·270	3·526	3·769	4·340	4·865
1/ 7·4	0·4113	0·7481	1·2179	2·6211	4·7409	7·6853	9·4980	11·554	13·864	16·439	22·429	29·606	53·255	85·971
0·14000	1·335	1·554	1·756	2·126	2·461	2·770	2·917	3·059	3·198	3·332	3·592	3·841	4·421	4·956
1/ 7·1	0·4193	0·7626	1·2415	2·6715	4·8316	7·8318	9·6789	11·773	14·127	16·751	22·853	30·165	54·259	87·587
0·14500	1·360	1·583	1·789	2·165	2·506	2·821	2·970	3·115	3·256	3·393	3·658	3·911	4·502	5·046
1/ 6·9	0·4272	0·7769	1·2646	2·7210	4·9208	7·9758	9·8566	11·989	14·386	17·057	23·271	30·715	55·245	89·176
0·15000	1·385	1·611	1·821	2·204	2·551	2·871	3·023	3·171	3·314	3·453	3·722	3·980	4·581	5·135
1/ 6·7	0·4350	0·7910	1·2874	2·7697	5·0084	8·1174	10·031	12·202	14·640	17·359	23·681	31·256	56·214	90·737
0·16000	1·433	1·667	1·884	2·280	2·638	2·969	3·126	3·278	3·426	3·570	3·848	4·114	4·735	5·307
1/ 6·3	0·4502	0·8185	1·3319	2·8648	5·1795	8·3937	10·372	12·616	15·136	17·947	24·481	32·311	58·106	93·783
0·17000	1·480	1·722	1·945	2·353	2·722	3·063	3·225	3·382	3·535	3·684	3·970	4·244	4·884	5·474
1/ 5·9	0·4649	0·8451	1·3750	2·9569	5·3453	8·6616	10·703	13·017	15·618	18·517	25·258	33·334	59·940	96·736
0·18000	1·525	1·774	2·004	2·424	2·804	3·155	3·322	3·484	3·641	3·794	4·089	4·371	5·029	5·636
1/ 5·6	0·4792	0·8709	1·4169	3·0464	5·5064	8·9218	11·024	13·407	16·085	19·070	26·011	34·327	61·721	99·604
0·19000	1·570	1·826	2·062	2·494	2·884	3·245	3·416	3·582	3·744	3·901	4·204	4·494	5·171	5·794
1/ 5·3	0·4932	0·8961	1·4577	3·1335	5·6632	9·1749	11·336	13·787	16·540	19·609	26·745	35·293	63·453	102·39
0·20000	1·613	1·875	2·118	2·561	2·962	3·332	3·508	3·678	3·844	4·005	4·316	4·614	5·308	5·948
1/ 5·0	0·5068	0·9206	1·4974	3·2184	5·8159	9·4215	11·640	14·156	16·983	20·133	27·459	36·234	65·140	105·11
0·21000	1·655	1·924	2·173	2·627	3·038	3·417	3·597	3·772	3·942	4·107	4·426	4·730	5·442	6·098
1/ 4·8	0·5200	0·9446	1·5361	3·3012	5·9649	9·6620	11·937	14·517	17·415	20·645	28·156	37·153	66·786	107·76
0·22000	1·696	1·972	2·227	2·691	3·112	3·500	3·685	3·864	4·037	4·207	4·533	4·845	5·573	6·244
1/ 4·5	0·5329	0·9679	1·5740	3·3820	6·1103	9·8970	12·227	14·869	17·837	21·145	28·836	38·049	68·394	110·35
0·23000	1·737	2·018	2·279	2·754	3·184	3·582	3·770	3·953	4·131	4·304	4·637	4·956	5·701	6·388
1/ 4·3	0·5456	0·9908	1·6110	3·4611	6·2526	10·127	12·511	15·213	18·249	21·633	29·501	38·926	69·965	112·88
0·24000	1·776	2·064	2·330	2·816	3·255	3·661	3·854	4·041	4·222	4·399	4·740	5·065	5·827	6·528
1/ 4·2	0·5580	1·0132	1·6472	3·5385	6·3918	10·351	12·788	15·550	18·653	22·112	30·152	39·784	71·503	115·36
0·25000	1·815	2·109	2·381	2·876	3·325	3·739	3·936	4·126	4·312	4·492	4·840	5·172	5·949	6·665
1/ 4·0	0·5701	1·0351	1·6827	3·6143	6·5282	10·572	13·060	15·880	19·048	22·580	30·790	40·624	73·009	117·78
0·26000	1·853	2·152	2·430	2·935	3·393	3·815	4·016	4·210	4·399	4·583	4·938	5·277	6·070	6·800
1/ 3·8	0·5820	1·0566	1·7175	3·6886	6·6619	10·788	13·326	16·204	19·436	23·039	31·415	41·447	74·485	120·16
0·27000	1·890	2·195	2·478	2·993	3·460	3·890	4·095	4·293	4·486	4·673	5·035	5·380	6·188	6·931
1/ 3·7	0·5937	1·0777	1·7517	3·7615	6·7931	10·999	13·587	16·521	19·816	23·489	32·028	42·255	75·934	122·49
0·28000	1·926	2·238	2·526	3·050	3·525	3·964	4·172	4·374	4·570	4·761	5·129	5·481	6·303	7·061
1/ 3·6	0·6052	1·0984	1·7852	3·8331	6·9219	11·207	13·844	16·833	20·190	23·932	32·630	43·049	77·355	124·78
0·29000	1·962	2·279	2·572	3·106	3·590	4·036	4·248	4·453	4·653	4·847	5·222	5·580	6·417	7·188
1/ 3·4	0·6165	1·1187	1·8182	3·9035	7·0484	11·412	14·096	17·139	20·557	24·366	33·222	43·828	78·752	127·03
0·30000	1·998	2·320	2·618	3·161	3·653	4·107	4·323	4·532	4·735	4·932	5·314	5·678	6·529	7·313
1/ 3·3	0·6275	1·1388	1·8505	3·9726	7·1728	11·612	14·344	17·440	20·917	24·793	33·803	44·594	80·125	129·24
	0·77	0·78	0·79	0·80	0·81	0·82	0·83	0·83	0·84	0·84	0·85	0·86	0·87	0·88

$V_{r(0\cdot5)medial}$ for half-full circular pipes.

A6

(p.1 of 6)

k_s = 0·150 mm
S = 0·00030 to 0·00100

ie hydraulic gradient =
1 in 3333 to 1 in 1000

Water (or sewage) at 15°C;
full bore conditions.

velocities in ms^{-1}
discharges in litres/sec

Gradient **(Equivalent) Pipe diameters in mm**

Gradient	20	25	30	40	50	60	65	70	75	80	90	100	125	150
0·000300				0·069	0·082	0·094	0·099	0·105	0·110	0·116	0·126	0·136	0·159	0·181
1/ 3333				0·0866	0·1605	0·2649	0·3298	0·4040	0·4877	0·5816	0·8014	1·0669	1·9519	3·1919
0·000320				0·072	0·085	0·097	0·103	0·109	0·114	0·120	0·131	0·141	0·165	0·187
1/ 3125				0·0899	0·1665	0·2748	0·3421	0·4190	0·5058	0·6031	0·8309	1·1059	2·0229	3·3072
0·000340				0·074	0·088	0·101	0·107	0·113	0·118	0·124	0·135	0·146	0·170	0·193
1/ 2941				0·0931	0·1724	0·2844	0·3541	0·4335	0·5233	0·6239	0·8595	1·1439	2·0918	3·4192
0·000360				0·077	0·091	0·104	0·110	0·116	0·122	0·128	0·139	0·150	0·176	0·200
1/ 2778				0·0962	0·1781	0·2938	0·3657	0·4477	0·5404	0·6443	0·8874	1·1809	2·1588	3·5282
0·000380				0·079	0·094	0·107	0·114	0·120	0·126	0·132	0·144	0·155	0·181	0·206
1/ 2632				0·0993	0·1837	0·3029	0·3770	0·4616	0·5571	0·6641	0·9145	1·2169	2·2242	3·6343
0·000400				0·081	0·096	0·110	0·117	0·123	0·130	0·136	0·148	0·159	0·186	0·212
1/ 2500				0·1022	0·1891	0·3118	0·3881	0·4751	0·5733	0·6834	0·9410	1·2520	2·2879	3·7378
0·000420				0·084	0·099	0·113	0·120	0·127	0·133	0·140	0·152	0·164	0·192	0·217
1/ 2381				0·1052	0·1945	0·3205	0·3989	0·4883	0·5892	0·7023	0·9669	1·2863	2·3502	3·8390
0·000440				0·086	0·102	0·116	0·123	0·130	0·137	0·143	0·156	0·168	0·196	0·223
1/ 2273				0·1080	0·1997	0·3291	0·4095	0·5012	0·6048	0·7208	0·9923	1·3199	2·4111	3·9379
0·000460				0·088	0·104	0·119	0·127	0·134	0·140	0·147	0·160	0·172	0·201	0·228
1/ 2174				0·1108	0·2048	0·3374	0·4198	0·5138	0·6200	0·7389	1·0171	1·3528	2·4707	4·0347
0·000480				0·090	0·107	0·122	0·130	0·137	0·144	0·151	0·164	0·176	0·206	0·234
1/ 2083				0·1135	0·2098	0·3456	0·4300	0·5262	0·6349	0·7566	1·0414	1·3850	2·5291	4·1295
0·000500				0·092	0·109	0·125	0·133	0·140	0·147	0·154	0·167	0·180	0·211	0·239
1/ 2000				0·1162	0·2147	0·3536	0·4400	0·5384	0·6495	0·7740	1·0652	1·4166	2·5863	4·2225
0·000525			0·076	0·095	0·112	0·129	0·136	0·144	0·151	0·158	0·172	0·185	0·216	0·245
1/ 1905			0·0539	0·1195	0·2207	0·3635	0·4522	0·5533	0·6674	0·7953	1·0944	1·4553	2·6564	4·3363
0·000550			0·078	0·098	0·115	0·132	0·140	0·148	0·155	0·162	0·177	0·190	0·222	0·252
1/ 1818			0·0554	0·1227	0·2266	0·3731	0·4641	0·5678	0·6850	0·8161	1·1229	1·4931	2·7249	4·4475
0·000575			0·080	0·100	0·118	0·135	0·143	0·151	0·159	0·166	0·181	0·195	0·228	0·258
1/ 1739			0·0568	0·1259	0·2324	0·3825	0·4758	0·5821	0·7021	0·8365	1·1509	1·5301	2·7920	4·5564
0·000600			0·082	0·103	0·121	0·139	0·147	0·155	0·163	0·170	0·185	0·199	0·233	0·264
1/ 1667			0·0582	0·1290	0·2380	0·3918	0·4873	0·5961	0·7189	0·8565	1·1783	1·5664	2·8577	4·6630
0·000625			0·084	0·105	0·124	0·142	0·150	0·158	0·166	0·174	0·189	0·204	0·238	0·270
1/ 1600			0·0596	0·1320	0·2436	0·4008	0·4985	0·6098	0·7354	0·8761	1·2051	1·6020	2·9222	4·7676
0·000650			0·086	0·107	0·127	0·145	0·154	0·162	0·170	0·178	0·194	0·208	0·243	0·276
1/ 1538			0·0610	0·1350	0·2490	0·4097	0·5095	0·6233	0·7516	0·8953	1·2315	1·6369	2·9854	4·8702
0·000675			0·088	0·110	0·130	0·148	0·157	0·165	0·174	0·182	0·198	0·213	0·248	0·281
1/ 1481			0·0623	0·1379	0·2544	0·4185	0·5204	0·6365	0·7675	0·9142	1·2574	1·6711	3·0475	4·9709
0·000700			0·090	0·112	0·132	0·151	0·160	0·169	0·177	0·186	0·202	0·217	0·253	0·287
1/ 1429			0·0636	0·1408	0·2597	0·4271	0·5310	0·6495	0·7832	0·9328	1·2828	1·7048	3·1085	5·0698
0·000725			0·092	0·114	0·135	0·154	0·163	0·172	0·181	0·189	0·206	0·221	0·258	0·292
1/ 1379			0·0649	0·1436	0·2648	0·4355	0·5415	0·6622	0·7985	0·9511	1·3078	1·7380	3·1685	5·1671
0·000750			0·094	0·117	0·137	0·157	0·166	0·175	0·184	0·193	0·209	0·225	0·263	0·298
1/ 1333			0·0662	0·1464	0·2699	0·4438	0·5518	0·6748	0·8137	0·9691	1·3324	1·7706	3·2275	5·2628
0·000800			0·097	0·121	0·143	0·163	0·172	0·182	0·191	0·200	0·217	0·234	0·272	0·308
1/ 1250			0·0687	0·1519	0·2799	0·4601	0·5720	0·6994	0·8432	1·0042	1·3805	1·8343	3·3428	5·4498
0·000850			0·101	0·125	0·147	0·168	0·178	0·188	0·197	0·207	0·224	0·241	0·282	0·319
1/ 1176			0·0712	0·1572	0·2896	0·4759	0·5915	0·7233	0·8719	1·0383	1·4273	1·8961	3·4547	5·6312
0·000900			0·104	0·129	0·152	0·174	0·184	0·194	0·204	0·213	0·231	0·249	0·290	0·329
1/ 1111			0·0735	0·1624	0·2990	0·4913	0·6106	0·7465	0·8999	1·0715	1·4727	1·9562	3·5635	5·8077
0·000950		0·093	0·107	0·133	0·157	0·179	0·190	0·200	0·210	0·220	0·238	0·257	0·299	0·338
1/ 1053		0·0458	0·0758	0·1674	0·3082	0·5062	0·6291	0·7691	0·9271	1·1038	1·5169	2·0148	3·6694	5·9794
0·001000		0·096	0·110	0·137	0·162	0·184	0·195	0·206	0·216	0·226	0·245	0·264	0·307	0·348
1/ 1000		0·0471	0·0781	0·1723	0·3171	0·5209	0·6472	0·7912	0·9536	1·1353	1·5601	2·0719	3·7728	6·1469
		–	–	0·74	0·74	0·75	0·75	0·76	0·76	0·76	0·77	0·77	0·78	0·79

$V_{r(0·5)medial}$ for half-full circular pipes.

k_s = 0·150 mm S = 0·00030 to 0·00100

Gradient	(Equivalent) Pipe diameters in mm 20	25	30	40	50	60	65	70	75	80	90	100	125	150
0·00100		0·096	0·110	0·137	0·162	0·184	0·195	0·206	0·216	0·226	0·245	0·264	0·307	0·348
1/ 1000		0·0471	0·0781	0·1723	0·3171	0·5209	0·6472	0·7912	0·9536	1·1353	1·5601	2·0719	3·7728	6·1469
0·00105		0·099	0·114	0·141	0·166	0·189	0·200	0·211	0·222	0·232	0·252	0·271	0·316	0·357
1/ 952		0·0485	0·0803	0·1771	0·3259	0·5351	0·6649	0·8128	0·9795	1·1661	1·6022	2·1277	3·8737	6·3105
0·00110		0·101	0·117	0·145	0·170	0·194	0·206	0·217	0·227	0·238	0·258	0·278	0·324	0·366
1/ 909		0·0498	0·0824	0·1818	0·3344	0·5491	0·6822	0·8338	1·0049	1·1962	1·6435	2·1823	3·9723	6·4703
0·00115		0·104	0·120	0·148	0·175	0·199	0·211	0·222	0·233	0·244	0·265	0·285	0·332	0·375
1/ 870		0·0511	0·0846	0·1864	0·3428	0·5628	0·6992	0·8545	1·0297	1·2257	1·6838	2·2357	4·0688	6·6266
0·00120		0·107	0·123	0·152	0·179	0·204	0·216	0·227	0·239	0·250	0·271	0·291	0·339	0·384
1/ 833		0·0523	0·0866	0·1909	0·3510	0·5762	0·7157	0·8747	1·0540	1·2546	1·7233	2·2879	4·1633	6·7798
0·00125		0·109	0·125	0·155	0·183	0·208	0·221	0·232	0·244	0·255	0·277	0·298	0·347	0·392
1/ 800		0·0536	0·0887	0·1953	0·3591	0·5893	0·7320	0·8945	1·0778	1·2829	1·7620	2·3392	4·2559	6·9299
0·00130		0·112	0·128	0·159	0·187	0·213	0·225	0·237	0·249	0·261	0·283	0·304	0·354	0·400
1/ 769		0·0548	0·0907	0·1997	0·3670	0·6022	0·7480	0·9140	1·1012	1·3107	1·8000	2·3895	4·3468	7·0771
0·00135		0·114	0·131	0·162	0·191	0·217	0·230	0·242	0·254	0·266	0·289	0·311	0·361	0·409
1/ 741		0·0560	0·0926	0·2039	0·3748	0·6148	0·7637	0·9331	1·1242	1·3380	1·8374	2·4389	4·4360	7·2216
0·00140		0·116	0·134	0·166	0·195	0·222	0·235	0·247	0·260	0·272	0·295	0·317	0·369	0·417
1/ 714		0·0572	0·0946	0·2081	0·3824	0·6273	0·7791	0·9519	1·1468	1·3648	1·8740	2·4874	4·5237	7·3636
0·00145		0·119	0·136	0·169	0·199	0·226	0·239	0·252	0·265	0·277	0·300	0·323	0·376	0·425
1/ 690		0·0583	0·0965	0·2123	0·3899	0·6395	0·7942	0·9703	1·1689	1·3911	1·9101	2·5350	4·6098	7·5031
0·00150		0·121	0·139	0·172	0·202	0·230	0·244	0·257	0·270	0·282	0·306	0·329	0·383	0·432
1/ 667		0·0595	0·0983	0·2163	0·3973	0·6515	0·8091	0·9885	1·1908	1·4170	1·9455	2·5819	4·6946	7·6403
0·00160		0·126	0·144	0·178	0·210	0·239	0·253	0·266	0·279	0·292	0·317	0·340	0·396	0·448
1/ 625		0·0617	0·1020	0·2242	0·4117	0·6751	0·8382	1·0240	1·2334	1·4677	2·0148	2·6736	4·8600	7·9083
0·00170		0·130	0·149	0·185	0·217	0·247	0·261	0·275	0·289	0·302	0·327	0·352	0·409	0·462
1/ 588		0·0639	0·1055	0·2319	0·4258	0·6979	0·8665	1·0584	1·2748	1·5168	2·0820	2·7625	5·0206	8·1683
0·00180		0·134	0·154	0·191	0·224	0·255	0·269	0·284	0·298	0·311	0·338	0·363	0·422	0·477
1/ 556		0·0660	0·1090	0·2394	0·4394	0·7201	0·8940	1·0919	1·3150	1·5646	2·1473	2·8489	5·1767	8·4209
0·00190	0·117	0·139	0·159	0·196	0·231	0·262	0·277	0·292	0·307	0·321	0·348	0·373	0·434	0·490
1/ 526	0·0367	0·0680	0·1123	0·2467	0·4527	0·7417	0·9207	1·1245	1·3542	1·6111	2·2110	2·9330	5·3286	8·6669
0·00200	0·120	0·143	0·164	0·202	0·237	0·270	0·285	0·300	0·315	0·330	0·357	0·384	0·446	0·504
1/ 500	0·0378	0·0700	0·1156	0·2539	0·4656	0·7628	0·9469	1·1563	1·3925	1·6565	2·2730	3·0151	5·4767	8·9066
0·00210	0·124	0·147	0·168	0·208	0·244	0·277	0·293	0·309	0·324	0·338	0·367	0·394	0·458	0·517
1/ 476	0·0388	0·0720	0·1188	0·2608	0·4783	0·7834	0·9724	1·1874	1·4298	1·7008	2·3336	3·0952	5·6213	9·1406
0·00220	0·127	0·151	0·173	0·213	0·250	0·284	0·301	0·316	0·332	0·347	0·376	0·404	0·470	0·530
1/ 455	0·0399	0·0739	0·1220	0·2676	0·4907	0·8035	0·9973	1·2177	1·4663	1·7441	2·3928	3·1734	5·7626	9·3692
0·00230	0·130	0·154	0·177	0·218	0·256	0·291	0·308	0·324	0·340	0·355	0·385	0·414	0·481	0·543
1/ 435	0·0409	0·0758	0·1250	0·2743	0·5028	0·8232	1·0217	1·2475	1·5020	1·7865	2·4507	3·2500	5·9008	9·5929
0·00240	0·133	0·158	0·181	0·223	0·262	0·298	0·315	0·332	0·348	0·364	0·394	0·423	0·492	0·555
1/ 417	0·0419	0·0776	0·1280	0·2808	0·5146	0·8425	1·0456	1·2766	1·5369	1·8280	2·5074	3·3250	6·0361	9·8118
0·00250	0·137	0·162	0·185	0·229	0·268	0·305	0·322	0·339	0·356	0·372	0·403	0·433	0·503	0·567
1/ 400	0·0429	0·0794	0·1310	0·2872	0·5263	0·8615	1·0690	1·3051	1·5712	1·8686	2·5630	3·3985	6·1687	10·026
0·00260	0·140	0·165	0·189	0·234	0·274	0·311	0·329	0·346	0·363	0·380	0·411	0·442	0·513	0·579
1/ 385	0·0439	0·0812	0·1339	0·2935	0·5377	0·8800	1·0919	1·3331	1·6048	1·9085	2·6175	3·4706	6·2987	10·237
0·00270	0·143	0·169	0·193	0·238	0·280	0·318	0·336	0·354	0·371	0·387	0·420	0·451	0·524	0·591
1/ 370	0·0448	0·0829	0·1367	0·2997	0·5489	0·8982	1·1145	1·3605	1·6378	1·9477	2·6711	3·5413	6·4264	10·443
0·00280	0·146	0·172	0·197	0·243	0·285	0·324	0·343	0·361	0·378	0·395	0·428	0·460	0·534	0·602
1/ 357	0·0458	0·0847	0·1395	0·3057	0·5599	0·9161	1·1367	1·3875	1·6702	1·9862	2·7236	3·6108	6·5517	10·646
0·00290	0·149	0·176	0·201	0·248	0·291	0·330	0·349	0·367	0·385	0·403	0·436	0·468	0·544	0·614
1/ 345	0·0467	0·0863	0·1423	0·3117	0·5707	0·9337	1·1584	1·4140	1·7020	2·0240	2·7753	3·6792	6·6749	10·845
0·00300	0·151	0·179	0·205	0·253	0·296	0·336	0·356	0·374	0·392	0·410	0·444	0·477	0·554	0·625
1/ 333	0·0476	0·0880	0·1450	0·3176	0·5814	0·9511	1·1799	1·4401	1·7334	2·0612	2·8261	3·7463	6·7961	11·041
	–	0·74	0·74	0·75	0·75	0·76	0·76	0·77	0·77	0·77	0·78	0·78	0·80	0·81

$V_{r(0·5)medial}$ for half-full circular pipes.

A6
(p.3 of 6)

$k_s = 0{\cdot}150$ mm
S = 0·00300 to 0·01000

ie hydraulic gradient =
1 in 333 to 1 in 100

Water (or sewage) at 15°C;
full bore conditions.

velocities in ms^{-1}
discharges in litres/sec

Gradient	**(Equivalent) Pipe diameters in mm**													
	20	**25**	**30**	**40**	**50**	**60**	**65**	**70**	**75**	**80**	**90**	**100**	**125**	**150**
0·00300	0·151	0·179	0·205	0·253	0·296	0·336	0·356	0·374	0·392	0·410	0·444	0·477	0·554	0·625
1/ 333	0·0476	0·0880	0·1450	0·3176	0·5814	0·9511	1·1799	1·4401	1·7334	2·0612	2·8261	3·7463	6·7961	11·041
0·00320	0·157	0·186	0·213	0·262	0·307	0·348	0·368	0·387	0·406	0·425	0·460	0·494	0·573	0·646
1/ 313	0·0494	0·0912	0·1503	0·3290	0·6022	0·9849	1·2217	1·4911	1·7946	2·1338	2·9254	3·8775	7·0326	11·424
0·00340	0·163	0·192	0·220	0·271	0·317	0·360	0·380	0·400	0·420	0·439	0·475	0·510	0·592	0·667
1/ 294	0·0511	0·0944	0·1554	0·3401	0·6224	1·0177	1·2623	1·5405	1·8540	2·2043	3·0216	4·0047	7·2621	11·795
0·00360	0·168	0·198	0·227	0·279	0·327	0·371	0·392	0·413	0·433	0·452	0·490	0·526	0·610	0·688
1/ 278	0·0528	0·0974	0·1604	0·3509	0·6420	1·0496	1·3018	1·5886	1·9117	2·2728	3·1152	4·1284	7·4850	12·155
0·00380	0·173	0·205	0·234	0·288	0·337	0·382	0·404	0·425	0·445	0·465	0·504	0·541	0·628	0·708
1/ 263	0·0544	0·1004	0·1653	0·3615	0·6611	1·0807	1·3402	1·6354	1·9679	2·3394	3·2063	4·2487	7·7020	12·506
0·00400	0·178	0·210	0·241	0·296	0·346	0·393	0·415	0·437	0·458	0·478	0·518	0·556	0·645	0·727
1/ 250	0·0560	0·1033	0·1701	0·3717	0·6797	1·1109	1·3777	1·6810	2·0226	2·4044	3·2951	4·3661	7·9134	12·848
0·00420	0·183	0·216	0·247	0·304	0·355	0·403	0·426	0·448	0·470	0·491	0·532	0·570	0·662	0·746
1/ 238	0·0575	0·1062	0·1747	0·3818	0·6979	1·1405	1·4142	1·7255	2·0761	2·4678	3·3817	4·4805	8·1198	13·182
0·00440	0·188	0·222	0·254	0·312	0·365	0·414	0·437	0·460	0·482	0·503	0·545	0·585	0·678	0·764
1/ 227	0·0590	0·1089	0·1792	0·3916	0·7157	1·1694	1·4500	1·7690	2·1283	2·5298	3·4664	4·5924	8·3214	13·508
0·00460	0·193	0·227	0·260	0·319	0·373	0·424	0·447	0·471	0·493	0·515	0·558	0·599	0·694	0·782
1/ 217	0·0605	0·1117	0·1837	0·4012	0·7331	1·1976	1·4849	1·8115	2·1794	2·5905	3·5492	4·7018	8·5185	13·826
0·00480	0·197	0·233	0·266	0·327	0·382	0·433	0·458	0·482	0·505	0·527	0·571	0·612	0·710	0·800
1/ 208	0·0620	0·1143	0·1880	0·4106	0·7501	1·2253	1·5192	1·8532	2·2295	2·6498	3·6303	4·8089	8·7115	14·138
0·00500	0·202	0·238	0·272	0·334	0·391	0·443	0·468	0·492	0·516	0·539	0·583	0·626	0·725	0·817
1/ 200	0·0634	0·1169	0·1923	0·4198	0·7668	1·2524	1·5527	1·8940	2·2785	2·7080	3·7097	4·9139	8·9006	14·444
0·00525	0·207	0·245	0·279	0·343	0·401	0·455	0·480	0·505	0·529	0·553	0·598	0·642	0·744	0·839
1/ 190	0·0652	0·1201	0·1975	0·4310	0·7873	1·2856	1·5937	1·9440	2·3384	2·7791	3·8068	5·0422	9·1319	14·818
0·00550	0·213	0·251	0·287	0·352	0·411	0·466	0·492	0·518	0·543	0·567	0·613	0·658	0·763	0·859
1/ 182	0·0669	0·1233	0·2026	0·4421	0·8072	1·3180	1·6338	1·9928	2·3970	2·8486	3·9018	5·1676	9·3577	15·183
0·00575	0·218	0·257	0·294	0·360	0·421	0·477	0·504	0·530	0·556	0·580	0·628	0·674	0·781	0·879
1/ 174	0·0686	0·1263	0·2076	0·4528	0·8268	1·3498	1·6730	2·0405	2·4543	2·9166	3·9946	5·2902	9·5786	15·540
0·00600	0·223	0·263	0·301	0·369	0·431	0·488	0·516	0·542	0·568	0·593	0·642	0·689	0·798	0·899
1/ 167	0·0702	0·1293	0·2125	0·4634	0·8459	1·3808	1·7114	2·0873	2·5105	2·9832	4·0855	5·4102	9·7949	15·889
0·00625	0·229	0·269	0·307	0·377	0·440	0·499	0·527	0·554	0·581	0·606	0·656	0·704	0·815	0·919
1/ 160	0·0718	0·1322	0·2173	0·4737	0·8646	1·4112	1·7491	2·1330	2·5654	3·0484	4·1745	5·5279	10·007	16·232
0·00650	0·234	0·275	0·314	0·385	0·450	0·510	0·538	0·566	0·593	0·619	0·670	0·719	0·832	0·938
1/ 154	0·0734	0·1351	0·2220	0·4839	0·8830	1·4411	1·7860	2·1780	2·6193	3·1123	4·2618	5·6432	10·214	16·567
0·00675	0·239	0·281	0·321	0·393	0·459	0·520	0·549	0·577	0·605	0·632	0·683	0·733	0·849	0·956
1/ 148	0·0750	0·1380	0·2266	0·4938	0·9011	1·4704	1·8222	2·2220	2·6723	3·1751	4·3475	5·7564	10·418	16·897
0·00700	0·243	0·287	0·327	0·401	0·468	0·530	0·560	0·589	0·617	0·644	0·697	0·747	0·865	0·974
1/ 143	0·0765	0·1407	0·2311	0·5036	0·9188	1·4991	1·8578	2·2653	2·7242	3·2367	4·4317	5·8675	10·618	17·220
0·00725	0·248	0·292	0·333	0·408	0·477	0·540	0·570	0·600	0·628	0·656	0·710	0·761	0·881	0·992
1/ 138	0·0780	0·1435	0·2356	0·5132	0·9362	1·5274	1·8927	2·3079	2·7753	3·2973	4·5144	5·9767	10·815	17·538
0·00750	0·253	0·298	0·339	0·416	0·486	0·550	0·581	0·611	0·640	0·668	0·722	0·775	0·897	1·010
1/ 133	0·0795	0·1462	0·2400	0·5227	0·9534	1·5552	1·9271	2·3497	2·8255	3·3569	4·5957	6·0841	11·008	17·850
0·00800	0·262	0·308	0·352	0·431	0·503	0·569	0·601	0·632	0·662	0·691	0·747	0·801	0·928	1·045
1/ 125	0·0823	0·1514	0·2485	0·5412	0·9868	1·6095	1·9942	2·4314	2·9235	3·4731	4·7544	6·2937	11·386	18·460
0·00850	0·271	0·319	0·363	0·445	0·519	0·588	0·621	0·652	0·683	0·713	0·772	0·827	0·958	1·078
1/ 118	0·0851	0·1565	0·2568	0·5591	1·0193	1·6621	2·0593	2·5106	3·0186	3·5859	4·9083	6·4969	11·752	19·051
0·00900	0·280	0·329	0·375	0·459	0·535	0·606	0·640	0·672	0·704	0·735	0·795	0·852	0·987	1·111
1/ 111	0·0879	0·1615	0·2649	0·5765	1·0508	1·7133	2·1225	2·5875	3·1109	3·6954	5·0578	6·6944	12·107	19·625
0·00950	0·288	0·339	0·386	0·472	0·551	0·624	0·658	0·692	0·725	0·756	0·818	0·877	1·015	1·142
1/ 105	0·0905	0·1663	0·2728	0·5935	1·0815	1·7631	2·1841	2·6624	3·2008	3·8020	5·2033	6·8865	12·453	20·184
0·01000	0·296	0·348	0·397	0·485	0·566	0·641	0·676	0·711	0·744	0·777	0·840	0·901	1·042	1·173
1/ 100	0·0931	0·1710	0·2805	0·6100	1·1114	1·8116	2·2441	2·7354	3·2884	3·9059	5·3451	7·0737	12·790	20·728
	0·74	0·74	0·75	0·76	0·76	0·77	0·77	0·78	0·78	0·78	0·79	0·80	0·81	0·82

$V_{r(0{\cdot}5)medial}$ for half-full circular pipes.

$k_s = 0{\cdot}150$ mm S = 0·00300 to 0·01000

k_s = 0·150 mm
S = 0·01000 to 0·03000

ie hydraulic gradient =
1 in 100 to 1 in 33·3

Water (or sewage) at 15°C;
full bore conditions.

velocities in ms^{-1}
discharges in litres/sec

Gradient	(Equivalent) Pipe diameters in mm													
	20	**25**	**30**	**40**	**50**	**60**	**65**	**70**	**75**	**80**	**90**	**100**	**125**	**150**
0·01000	0·296	0·348	0·397	0·485	0·566	0·641	0·676	0·711	0·744	0·777	0·840	0·901	1·042	1·173
1/ 100	0·0931	0·1710	0·2805	0·6100	1·1114	1·8116	2·2441	2·7354	3·2884	3·9059	5·3451	7·0737	12·790	20·728
0·01050	0·304	0·358	0·407	0·498	0·581	0·657	0·694	0·729	0·764	0·797	0·862	0·924	1·069	1·203
1/ 95	0·0956	0·1756	0·2880	0·6261	1·1406	1·8589	2·3026	2·8066	3·3739	4·0073	5·4834	7·2563	13·119	21·259
0·01100	0·312	0·367	0·418	0·511	0·595	0·674	0·711	0·747	0·783	0·817	0·883	0·947	1·095	1·232
1/ 91	0·0981	0·1801	0·2953	0·6419	1·1691	1·9052	2·3598	2·8762	3·4574	4·1063	5·6185	7·4347	13·440	21·777
0·01150	0·320	0·376	0·428	0·523	0·610	0·690	0·728	0·765	0·801	0·836	0·904	0·969	1·121	1·261
1/ 87	0·1005	0·1845	0·3024	0·6573	1·1970	1·9504	2·4157	2·9442	3·5390	4·2031	5·7507	7·6092	13·754	22·284
0·01200	0·327	0·385	0·438	0·535	0·624	0·705	0·744	0·782	0·819	0·855	0·924	0·991	1·146	1·289
1/ 83	0·1029	0·1888	0·3094	0·6724	1·2243	1·9947	2·4704	3·0108	3·6189	4·2979	5·8800	7·7799	14·061	22·781
0·01250	0·335	0·393	0·447	0·547	0·637	0·721	0·761	0·799	0·837	0·874	0·944	1·012	1·170	1·317
1/ 80	0·1052	0·1930	0·3163	0·6872	1·2511	2·0381	2·5241	3·0760	3·6972	4·3907	6·0067	7·9472	14·362	23·267
0·01300	0·342	0·402	0·457	0·558	0·651	0·736	0·776	0·816	0·854	0·892	0·964	1·033	1·194	1·344
1/ 77	0·1075	0·1972	0·3230	0·7017	1·2773	2·0807	2·5766	3·1400	3·7740	4·4818	6·1309	8·1112	14·657	23·743
0·01350	0·349	0·410	0·466	0·570	0·664	0·751	0·792	0·832	0·871	0·909	0·983	1·053	1·218	1·370
1/ 74	0·1097	0·2012	0·3297	0·7159	1·3031	2·1224	2·6282	3·2028	3·8493	4·5711	6·2528	8·2721	14·947	24·211
0·01400	0·356	0·418	0·476	0·581	0·677	0·765	0·807	0·848	0·888	0·927	1·002	1·073	1·241	1·396
1/ 71	0·1119	0·2052	0·3362	0·7299	1·3284	2·1634	2·6789	3·2644	3·9233	4·6588	6·3725	8·4301	15·231	24·670
0·01450	0·363	0·426	0·485	0·592	0·689	0·779	0·822	0·864	0·904	0·944	1·020	1·093	1·264	1·422
1/ 69	0·1140	0·2091	0·3425	0·7436	1·3532	2·2037	2·7287	3·3250	3·9959	4·7450	6·4901	8·5853	15·510	25·121
0·01500	0·370	0·434	0·493	0·603	0·702	0·793	0·837	0·879	0·921	0·961	1·038	1·113	1·286	1·447
1/ 67	0·1162	0·2130	0·3488	0·7571	1·3776	2·2433	2·7776	3·3845	4·0674	4·8297	6·6057	8·7379	15·785	25·564
0·01600	0·383	0·449	0·511	0·624	0·726	0·821	0·866	0·910	0·952	0·994	1·074	1·150	1·330	1·496
1/ 63	0·1203	0·2205	0·3611	0·7835	1·4253	2·3206	2·8732	3·5007	4·2069	4·9951	6·8313	9·0357	16·321	26·429
0·01700	0·396	0·464	0·528	0·644	0·749	0·847	0·894	0·939	0·983	1·026	1·108	1·187	1·372	1·543
1/ 59	0·1243	0·2278	0·3730	0·8091	1·4716	2·3956	2·9658	3·6134	4·3421	5·1554	7·0500	9·3245	16·840	27·268
0·01800	0·408	0·479	0·544	0·664	0·772	0·873	0·921	0·967	1·013	1·057	1·142	1·223	1·413	1·589
1/ 56	0·1282	0·2349	0·3845	0·8340	1·5165	2·4684	3·0558	3·7229	4·4734	5·3111	7·2625	9·6050	17·345	28·082
0·01900	0·420	0·493	0·560	0·683	0·795	0·898	0·947	0·995	1·042	1·087	1·174	1·258	1·453	1·634
1/ 53	0·1320	0·2418	0·3958	0·8581	1·5603	2·5393	3·1434	3·8294	4·6012	5·4627	7·4692	9·8778	17·835	28·875
0·02000	0·432	0·506	0·575	0·702	0·816	0·923	0·973	1·022	1·070	1·116	1·206	1·292	1·492	1·678
1/ 50	0·1357	0·2486	0·4067	0·8817	1·6029	2·6083	3·2287	3·9332	4·7257	5·6103	7·6706	10·144	18·313	29·647
0·02100	0·444	0·520	0·591	0·720	0·838	0·946	0·998	1·048	1·097	1·145	1·237	1·325	1·530	1·720
1/ 47·6	0·1394	0·2552	0·4174	0·9047	1·6444	2·6757	3·3120	4·0344	4·8472	5·7543	7·8671	10·403	18·780	30·400
0·02200	0·455	0·533	0·605	0·738	0·858	0·970	1·023	1·074	1·124	1·173	1·267	1·357	1·567	1·762
1/ 45·5	0·1429	0·2616	0·4279	0·9272	1·6850	2·7415	3·3933	4·1333	4·9658	5·8950	8·0590	10·656	19·235	31·135
0·02300	0·466	0·546	0·620	0·755	0·878	0·992	1·047	1·099	1·150	1·200	1·296	1·388	1·604	1·803
1/ 43·5	0·1464	0·2679	0·4381	0·9491	1·7248	2·8058	3·4728	4·2300	5·0818	6·0325	8·2466	10·904	19·681	31·854
0·02400	0·477	0·558	0·634	0·772	0·898	1·015	1·070	1·124	1·176	1·227	1·325	1·419	1·639	1·842
1/ 41·7	0·1497	0·2740	0·4481	0·9707	1·7636	2·8688	3·5506	4·3246	5·1954	6·1671	8·4302	11·146	20·117	32·557
0·02500	0·487	0·571	0·648	0·789	0·918	1·036	1·093	1·148	1·201	1·253	1·353	1·449	1·674	1·881
1/ 40·0	0·1531	0·2800	0·4579	0·9917	1·8017	2·9305	3·6268	4·4173	5·3066	6·2990	8·6101	11·384	20·543	33·247
0·02600	0·498	0·583	0·661	0·806	0·937	1·058	1·115	1·171	1·226	1·279	1·381	1·479	1·708	1·920
1/ 38·5	0·1563	0·2860	0·4675	1·0124	1·8390	2·9909	3·7015	4·5082	5·4156	6·4282	8·7864	11·616	20·962	33·922
0·02700	0·508	0·594	0·675	0·822	0·955	1·079	1·138	1·195	1·250	1·304	1·408	1·508	1·742	1·957
1/ 37·0	0·1595	0·2918	0·4770	1·0327	1·8757	3·0503	3·7748	4·5974	5·5226	6·5551	8·9594	11·845	21·373	34·585
0·02800	0·518	0·606	0·688	0·838	0·974	1·099	1·159	1·217	1·274	1·329	1·435	1·537	1·774	1·994
1/ 35·7	0·1627	0·2975	0·4862	1·0526	1·9116	3·1086	3·8468	4·6849	5·6276	6·6796	9·1292	12·069	21·776	35·235
0·02900	0·528	0·617	0·701	0·853	0·992	1·120	1·181	1·240	1·297	1·353	1·461	1·565	1·807	2·030
1/ 34·5	0·1657	0·3031	0·4953	1·0721	1·9470	3·1658	3·9175	4·7709	5·7308	6·8019	9·2961	12·289	22·172	35·875
0·03000	0·537	0·629	0·713	0·868	1·009	1·140	1·202	1·262	1·320	1·377	1·487	1·592	1·838	2·066
1/ 33·3	0·1688	0·3086	0·5043	1·0914	1·9817	3·2221	3·9871	4·8555	5·8322	6·9222	9·4601	12·505	22·561	36·503
	0·74	0·75	0·75	0·76	0·77	0·78	0·78	0·79	0·79	0·79	0·80	0·80	0·82	0·83

$V_{r(0\cdot5)medial}$ for half-full circular pipes.

$k_s = 0{\cdot}150$ mm
S = 0·03000 to 0·10000

ie hydraulic gradient =
1 in 33·3 to 1 in 10·0

Water (or sewage) at 15°C;
full bore conditions.

velocities in ms^{-1}
discharges in litres/sec

Gradient	**(Equivalent) Pipe diameters in mm**													
	20	**25**	**30**	**40**	**50**	**60**	**65**	**70**	**75**	**80**	**90**	**100**	**125**	**150**
0·03000	0·537	0·629	0·713	0·868	1·009	1·140	1·202	1·262	1·320	1·377	1·487	1·592	1·838	2·066
1/ 33·3	0·1688	0·3086	0·5043	1·0914	1·9817	3·2221	3·9871	4·8555	5·8322	6·9222	9·4601	12·505	22·561	36·503
0·03200	0·556	0·651	0·738	0·898	1·044	1·178	1·242	1·305	1·365	1·424	1·537	1·646	1·900	2·135
1/ 31·3	0·1747	0·3193	0·5218	1·1289	2·0495	3·3319	4·1228	5·0205	6·0302	7·1569	9·7802	12·928	23·320	37·729
0·03400	0·574	0·672	0·762	0·927	1·077	1·216	1·282	1·346	1·408	1·469	1·586	1·698	1·960	2·202
1/ 29·4	0·1804	0·3297	0·5387	1·1653	2·1153	3·4384	4·2543	5·1805	6·2221	7·3844	10·091	13·337	24·057	38·917
0·03600	0·592	0·692	0·785	0·955	1·110	1·253	1·321	1·386	1·451	1·513	1·634	1·749	2·019	2·268
1/ 27·8	0·1860	0·3399	0·5552	1·2006	2·1791	3·5418	4·3821	5·3359	6·4085	7·6054	10·392	13·735	24·772	40·071
0·03800	0·609	0·712	0·808	0·983	1·141	1·288	1·358	1·426	1·492	1·556	1·680	1·798	2·075	2·331
1/ 26·3	0·1914	0·3497	0·5712	1·2350	2·2413	3·6424	4·5064	5·4870	6·5898	7·8203	10·685	14·122	25·467	41·194
0·04000	0·626	0·732	0·830	1·009	1·172	1·323	1·395	1·464	1·532	1·597	1·724	1·846	2·130	2·393
1/ 25·0	0·1967	0·3593	0·5868	1·2685	2·3018	3·7404	4·6275	5·6343	6·7665	8·0297	10·971	14·499	26·145	42·287
0·04200	0·643	0·751	0·852	1·035	1·202	1·357	1·430	1·501	1·571	1·638	1·768	1·893	2·184	2·453
1/ 23·8	0·2019	0·3687	0·6020	1·3012	2·3608	3·8361	4·7456	5·7779	6·9388	8·2340	11·249	14·866	26·806	43·354
0·04400	0·659	0·770	0·873	1·061	1·232	1·390	1·465	1·538	1·609	1·678	1·811	1·939	2·237	2·512
1/ 22·7	0·2069	0·3779	0·6169	1·3332	2·4185	3·9294	4·8610	5·9182	7·1071	8·4335	11·521	15·225	27·451	44·396
0·04600	0·674	0·788	0·893	1·086	1·260	1·422	1·499	1·573	1·646	1·717	1·853	1·983	2·288	2·570
1/ 21·7	0·2119	0·3868	0·6314	1·3644	2·4749	4·0207	4·9738	6·0553	7·2716	8·6285	11·787	15·576	28·082	45·414
0·04800	0·690	0·806	0·913	1·110	1·289	1·454	1·532	1·608	1·682	1·755	1·894	2·027	2·339	2·626
1/ 20·8	0·2167	0·3956	0·6457	1·3950	2·5301	4·1101	5·0841	6·1896	7·4326	8·8193	12·048	15·920	28·700	46·410
0·05000	0·705	0·823	0·933	1·134	1·316	1·485	1·565	1·642	1·718	1·792	1·934	2·070	2·388	2·682
1/ 20·0	0·2214	0·4042	0·6596	1·4249	2·5841	4·1976	5·1923	6·3210	7·5903	9·0063	12·302	16·256	29·304	47·386
0·05250	0·723	0·845	0·957	1·163	1·350	1·522	1·605	1·684	1·762	1·837	1·983	2·122	2·448	2·749
1/ 19·0	0·2272	0·4146	0·6767	1·4615	2·6502	4·3046	5·3244	6·4817	7·7831	9·2348	12·614	16·667	30·044	48·579
0·05500	0·741	0·866	0·981	1·191	1·383	1·559	1·643	1·725	1·804	1·882	2·031	2·173	2·507	2·815
1/ 18·2	0·2329	0·4249	0·6933	1·4972	2·7147	4·4091	5·4535	6·6387	7·9713	9·4580	12·918	17·069	30·765	49·744
0·05750	0·759	0·886	1·004	1·219	1·415	1·596	1·681	1·765	1·846	1·925	2·077	2·223	2·564	2·879
1/ 17·4	0·2384	0·4349	0·7096	1·5322	2·7778	4·5113	5·5797	6·7921	8·1554	9·6762	13·216	17·461	31·471	50·883
0·06000	0·776	0·906	1·026	1·247	1·446	1·631	1·719	1·804	1·887	1·967	2·123	2·272	2·621	2·942
1/ 16·7	0·2438	0·4447	0·7256	1·5664	2·8396	4·6112	5·7032	6·9423	8·3355	9·8897	13·507	17·846	32·162	51·997
0·06250	0·793	0·926	1·049	1·273	1·477	1·666	1·755	1·842	1·927	2·009	2·168	2·320	2·676	3·004
1/ 16·0	0·2491	0·4544	0·7412	1·5999	2·9001	4·7092	5·8242	7·0894	8·5119	10·099	13·792	18·222	32·838	53·089
0·06500	0·810	0·945	1·070	1·299	1·507	1·699	1·791	1·880	1·966	2·050	2·212	2·367	2·730	3·065
1/ 15·4	0·2543	0·4638	0·7565	1·6328	2·9594	4·8052	5·9427	7·2336	8·6849	10·304	14·072	18·591	33·501	54·158
0·06750	0·826	0·964	1·091	1·325	1·537	1·733	1·826	1·916	2·004	2·090	2·255	2·413	2·783	3·124
1/ 14·8	0·2594	0·4730	0·7715	1·6650	3·0175	4·8993	6·0591	7·3750	8·8545	10·505	14·346	18·952	34·151	55·208
0·07000	0·842	0·982	1·112	1·350	1·566	1·765	1·860	1·952	2·042	2·129	2·297	2·458	2·835	3·182
1/ 14·3	0·2644	0·4821	0·7862	1·6966	3·0747	4·9918	6·1733	7·5139	9·0211	10·702	14·615	19·308	34·790	56·238
0·07250	0·857	1·000	1·133	1·375	1·594	1·798	1·894	1·988	2·079	2·168	2·339	2·503	2·886	3·240
1/ 13·8	0·2694	0·4910	0·8007	1·7277	3·1308	5·0826	6·2855	7·6503	9·1847	10·896	14·880	19·657	35·417	57·250
0·07500	0·873	1·018	1·153	1·399	1·623	1·829	1·927	2·023	2·115	2·206	2·380	2·546	2·936	3·296
1/ 13·3	0·2742	0·4998	0·8150	1·7583	3·1859	5·1719	6·3958	7·7844	9·3456	11·087	15·140	20·000	36·034	58·245
0·08000	0·903	1·053	1·192	1·447	1·677	1·891	1·992	2·091	2·186	2·280	2·460	2·632	3·034	3·406
1/ 12·5	0·2836	0·5169	0·8428	1·8179	3·2936	5·3462	6·6110	8·0460	9·6594	11·459	15·647	20·669	37·237	60·186
0·08500	0·932	1·087	1·230	1·493	1·731	1·951	2·055	2·157	2·255	2·351	2·537	2·714	3·129	3·512
1/ 11·8	0·2928	0·5335	0·8697	1·8757	3·3979	5·5151	6·8196	8·2997	9·9637	11·820	16·139	21·318	38·403	62·068
0·09000	0·960	1·120	1·267	1·537	1·782	2·009	2·116	2·221	2·322	2·421	2·612	2·794	3·222	3·616
1/ 11·1	0·3017	0·5496	0·8959	1·9318	3·4992	5·6791	7·0222	8·5460	10·259	12·170	16·616	21·948	39·535	63·895
0·09500	0·988	1·152	1·303	1·581	1·832	2·065	2·176	2·283	2·387	2·489	2·685	2·872	3·311	3·716
1/ 10·5	0·3103	0·5653	0·9214	1·9865	3·5978	5·8386	7·2193	8·7856	10·546	12·510	17·081	22·560	40·637	65·672
0·10000	1·015	1·183	1·339	1·623	1·881	2·120	2·233	2·344	2·451	2·555	2·756	2·948	3·399	3·814
1/ 10·0	0·3188	0·5806	0·9462	2·0397	3·6938	5·9940	7·4112	9·0190	10·826	12·842	17·533	23·157	41·709	67·403
	0·75	0·75	0·76	0·77	0·78	0·79	0·79	0·79	0·80	0·80	0·81	0·81	0·82	0·83

$V_{r(0{\cdot}5)medial}$ **for half-full circular pipes.**

Gradient	**(Equivalent) Pipe diameters in mm**													
	20	**25**	**30**	**40**	**50**	**60**	**65**	**70**	**75**	**80**	**90**	**100**	**125**	**150**
0·10000	1·015	1·183	1·339	1·623	1·881	2·120	2·233	2·344	2·451	2·555	2·756	2·948	3·399	3·814
1/ 10·0	0·3188	0·5806	0·9462	2·0397	3·6938	5·9940	7·4112	9·0190	10·826	12·842	17·533	23·157	41·709	67·403
0·10500	1·041	1·213	1·373	1·664	1·929	2·174	2·290	2·403	2·512	2·619	2·825	3·023	3·484	3·910
1/ 9·5	0·3270	0·5955	0·9704	2·0916	3·7875	6·1456	7·5985	9·2466	11·099	13·166	17·974	23·740	42·756	69·092
0·11000	1·066	1·243	1·406	1·705	1·976	2·226	2·345	2·460	2·573	2·682	2·893	3·095	3·567	4·003
1/ 9·1	0·3350	0·6100	0·9940	2·1422	3·8789	6·2937	7·7813	9·4689	11·366	13·482	18·405	24·308	43·777	70·740
0·11500	1·091	1·272	1·439	1·744	2·021	2·277	2·399	2·517	2·632	2·744	2·959	3·166	3·649	4·094
1/ 8·7	0·3429	0·6243	1·0171	2·1918	3·9684	6·4384	7·9600	9·6863	11·627	13·791	18·826	24·864	44·776	72·352
0·12000	1·116	1·300	1·471	1·783	2·066	2·327	2·452	2·572	2·689	2·804	3·024	3·235	3·728	4·184
1/ 8·3	0·3506	0·6382	1·0397	2·2403	4·0559	6·5800	8·1350	9·8989	11·882	14·093	19·239	25·408	45·754	73·929
0·12500	1·140	1·328	1·502	1·821	2·109	2·376	2·503	2·626	2·746	2·863	3·088	3·303	3·806	4·271
1/ 8·0	0·3581	0·6519	1·0619	2·2878	4·1416	6·7187	8·3063	10·107	12·132	14·389	19·642	25·940	46·711	75·473
0·13000	1·163	1·355	1·533	1·858	2·152	2·424	2·554	2·679	2·801	2·920	3·150	3·369	3·883	4·357
1/ 7·7	0·3655	0·6652	1·0836	2·3343	4·2256	6·8547	8·4742	10·311	12·376	14·679	20·038	26·462	47·649	76·987
0·13500	1·186	1·382	1·563	1·894	2·194	2·472	2·603	2·731	2·856	2·977	3·211	3·435	3·958	4·441
1/ 7·4	0·3727	0·6784	1·1049	2·3800	4·3080	6·9881	8·6390	10·512	12·617	14·964	20·426	26·975	48·570	78·472
0·14000	1·209	1·408	1·593	1·930	2·235	2·518	2·652	2·782	2·909	3·033	3·271	3·499	4·031	4·523
1/ 7·1	0·3798	0·6912	1·1258	2·4248	4·3889	7·1190	8·8007	10·708	12·853	15·244	20·807	27·477	49·473	79·930
0·14500	1·231	1·434	1·622	1·965	2·276	2·563	2·700	2·833	2·962	3·087	3·330	3·561	4·104	4·604
1/ 6·9	0·3868	0·7039	1·1464	2·4689	4·4684	7·2476	8·9596	10·901	13·084	15·518	21·182	27·971	50·361	81·362
0·15000	1·253	1·459	1·650	1·999	2·316	2·608	2·747	2·882	3·013	3·141	3·387	3·623	4·175	4·684
1/ 6·7	0·3937	0·7163	1·1665	2·5122	4·5465	7·3741	9·1157	11·091	13·312	15·788	21·550	28·457	51·233	82·770
0·16000	1·296	1·509	1·706	2·066	2·393	2·695	2·839	2·978	3·114	3·246	3·500	3·744	4·314	4·839
1/ 6·3	0·4071	0·7406	1·2060	2·5967	4·6990	7·6208	9·4205	11·462	13·756	16·315	22·268	29·404	52·936	85·517
0·17000	1·337	1·557	1·760	2·132	2·468	2·780	2·928	3·072	3·211	3·347	3·610	3·861	4·448	4·990
1/ 5·9	0·4201	0·7642	1·2442	2·6786	4·8468	7·8600	9·7158	11·821	14·187	16·825	22·964	30·322	54·586	88·179
0·18000	1·377	1·603	1·813	2·195	2·542	2·862	3·014	3·162	3·306	3·446	3·716	3·974	4·579	5·136
1/ 5·6	0·4327	0·7870	1·2813	2·7581	4·9903	8·0922	10·003	12·170	14·605	17·321	23·640	31·214	56·188	90·764
0·19000	1·416	1·649	1·864	2·256	2·613	2·942	3·098	3·250	3·398	3·542	3·819	4·085	4·706	5·278
1/ 5·3	0·4450	0·8092	1·3174	2·8355	5·1298	8·3181	10·282	12·509	15·012	17·803	24·297	32·081	57·747	93·279
0·20000	1·454	1·693	1·913	2·316	2·682	3·020	3·180	3·336	3·488	3·635	3·920	4·192	4·829	5·417
1/ 5·0	0·4569	0·8309	1·3525	2·9109	5·2658	8·5381	10·553	12·839	15·408	18·273	24·937	32·926	59·265	95·728
0·21000	1·492	1·736	1·962	2·375	2·749	3·096	3·260	3·420	3·575	3·726	4·018	4·297	4·950	5·552
1/ 4·8	0·4686	0·8520	1·3868	2·9844	5·3984	8·7527	10·818	13·161	15·795	18·731	25·562	33·749	60·745	98·116
0·22000	1·528	1·778	2·009	2·432	2·815	3·170	3·338	3·502	3·661	3·815	4·114	4·400	5·068	5·684
1/ 4·5	0·4800	0·8727	1·4203	3·0562	5·5279	8·9623	11·077	13·476	16·172	19·178	26·171	34·554	62·191	100·45
0·23000	1·563	1·819	2·056	2·488	2·880	3·242	3·414	3·582	3·744	3·902	4·208	4·500	5·183	5·813
1/ 4·3	0·4911	0·8928	1·4530	3·1263	5·6545	9·1671	11·330	13·784	16·541	19·615	26·768	35·340	63·604	102·73
0·24000	1·598	1·859	2·101	2·542	2·943	3·313	3·489	3·660	3·826	3·987	4·299	4·598	5·296	5·940
1/ 4·2	0·5020	0·9126	1·4851	3·1950	5·7784	9·3676	11·578	14·085	16·902	20·043	27·351	36·110	64·987	104·96
0·25000	1·632	1·898	2·145	2·596	3·005	3·383	3·562	3·736	3·906	4·071	4·389	4·694	5·406	6·063
1/ 4·0	0·5127	0·9319	1·5165	3·2622	5·8997	9·5639	11·820	14·379	17·255	20·462	27·922	36·863	66·341	107·14
0·26000	1·665	1·937	2·189	2·648	3·065	3·451	3·634	3·811	3·984	4·153	4·477	4·788	5·514	6·184
1/ 3·8	0·5232	0·9509	1·5472	3·3282	6·0187	9·7564	12·058	14·668	17·602	20·873	28·482	37·602	67·669	109·29
0·27000	1·698	1·975	2·232	2·700	3·125	3·517	3·704	3·885	4·061	4·233	4·563	4·880	5·620	6·303
1/ 3·7	0·5334	0·9695	1·5774	3·3929	6·1353	9·9452	12·291	14·952	17·942	21·276	29·031	38·327	68·971	111·39
0·28000	1·730	2·012	2·273	2·750	3·183	3·583	3·773	3·957	4·137	4·311	4·648	4·971	5·724	6·420
1/ 3·6	0·5435	0·9877	1·6070	3·4563	6·2499	10·131	12·520	15·230	18·276	21·671	29·571	39·038	70·249	113·45
0·29000	1·762	2·049	2·315	2·800	3·240	3·647	3·841	4·028	4·211	4·389	4·731	5·059	5·827	6·535
1/ 3·4	0·5534	1·0057	1·6361	3·5187	6·3624	10·313	12·745	15·503	18·603	22·060	30·100	39·737	71·505	115·48
0·30000	1·793	2·085	2·355	2·849	3·297	3·711	3·907	4·098	4·284	4·465	4·813	5·147	5·927	6·647
1/ 3·3	0·5632	1·0233	1·6647	3·5800	6·4730	10·492	12·966	15·772	18·926	22·442	30·621	40·424	72·739	117·47
	0·75	0·76	0·76	0·77	0·78	0·79	0·79	0·80	0·80	0·80	0·81	0·82	0·83	0·84

$V_{r(0\cdot5)medial}$ for half-full circular pipes.

A7

(p.1 of 6)

k_s = 0·30 mm
S = 0·00030 to 0·00100

ie hydraulic gradient =
1 in 3333 to 1 in 1000

Water (or sewage) at 15°C;
full bore conditions.

velocities in ms^{-1}
discharges in litres/sec

Gradient	(Equivalent) Pipe diameters in mm 20	25	30	40	50	60	65	70	75	80	90	100	125	150
0·000300				0·067	0·079	0·091	0·096	0·101	0·107	0·112	0·122	0·131	0·153	0·174
1/ 3333				0·0838	0·1552	0·2561	0·3188	0·3903	0·4711	0·5616	0·7735	1·0293	1·8813	3·0738
0·000320				0·069	0·082	0·094	0·100	0·105	0·111	0·116	0·126	0·136	0·159	0·180
1/ 3125				0·0870	0·1610	0·2655	0·3305	0·4046	0·4883	0·5820	0·8014	1·0663	1·9484	3·1828
0·000340				0·072	0·085	0·097	0·103	0·109	0·114	0·120	0·130	0·140	0·164	0·186
1/ 2941				0·0900	0·1666	0·2746	0·3418	0·4184	0·5049	0·6018	0·8285	1·1022	2·0135	3·2886
0·000360				0·074	0·088	0·100	0·106	0·112	0·118	0·124	0·134	0·145	0·169	0·192
1/ 2778				0·0930	0·1720	0·2835	0·3528	0·4318	0·5211	0·6210	0·8549	1·1371	2·0769	3·3914
0·000380				0·076	0·090	0·103	0·110	0·116	0·122	0·127	0·138	0·149	0·174	0·198
1/ 2632				0·0959	0·1773	0·2922	0·3635	0·4449	0·5368	0·6397	0·8806	1·1712	2·1385	3·4916
0·000400				0·079	0·093	0·106	0·113	0·119	0·125	0·131	0·142	0·153	0·179	0·203
1/ 2500				0·0987	0·1825	0·3006	0·3740	0·4577	0·5522	0·6580	0·9056	1·2043	2·1986	3·5892
0·000420				0·081	0·095	0·109	0·116	0·122	0·128	0·134	0·146	0·157	0·184	0·208
1/ 2381				0·1015	0·1875	0·3088	0·3842	0·4702	0·5672	0·6759	0·9301	1·2367	2·2573	3·6844
0·000440				0·083	0·098	0·112	0·119	0·125	0·132	0·138	0·150	0·161	0·189	0·214
1/ 2273				0·1042	0·1924	0·3169	0·3942	0·4824	0·5819	0·6933	0·9540	1·2684	2·3147	3·7776
0·000460				0·085	0·100	0·115	0·122	0·128	0·135	0·141	0·154	0·165	0·193	0·219
1/ 2174				0·1068	0·1973	0·3248	0·4040	0·4943	0·5962	0·7104	0·9773	1·2994	2·3708	3·8686
0·000480				0·087	0·103	0·118	0·125	0·131	0·138	0·145	0·157	0·169	0·198	0·224
1/ 2083				0·1094	0·2020	0·3325	0·4136	0·5060	0·6103	0·7271	1·0003	1·3297	2·4258	3·9578
0·000500				0·089	0·105	0·120	0·127	0·134	0·141	0·148	0·161	0·173	0·202	0·229
1/ 2000				0·1119	0·2066	0·3401	0·4230	0·5175	0·6241	0·7435	1·0227	1·3595	2·4797	4·0453
0·000525				0·092	0·108	0·124	0·131	0·138	0·145	0·152	0·165	0·178	0·207	0·235
1/ 1905				0·1151	0·2123	0·3494	0·4345	0·5315	0·6410	0·7635	1·0502	1·3959	2·5456	4·1522
0·000550				0·094	0·111	0·127	0·134	0·142	0·149	0·156	0·169	0·182	0·213	0·241
1/ 1818				0·1181	0·2179	0·3585	0·4457	0·5452	0·6575	0·7831	1·0770	1·4314	2·6100	4·2567
0·000575			0·077	0·096	0·114	0·130	0·138	0·145	0·152	0·160	0·173	0·187	0·218	0·247
1/ 1739			0·0547	0·1211	0·2233	0·3673	0·4567	0·5586	0·6736	0·8023	1·1033	1·4662	2·6730	4·3588
0·000600			0·079	0·099	0·116	0·133	0·141	0·149	0·156	0·163	0·177	0·191	0·223	0·252
1/ 1667			0·0560	0·1240	0·2286	0·3760	0·4675	0·5718	0·6894	0·8211	1·1291	1·5003	2·7347	4·4589
0·000625			0·081	0·101	0·119	0·136	0·144	0·152	0·160	0·167	0·181	0·195	0·228	0·258
1/ 1600			0·0573	0·1269	0·2339	0·3846	0·4781	0·5847	0·7049	0·8395	1·1543	1·5337	2·7951	4·5570
0·000650			0·083	0·103	0·122	0·139	0·147	0·155	0·163	0·171	0·185	0·199	0·233	0·263
1/ 1538			0·0586	0·1297	0·2390	0·3929	0·4885	0·5973	0·7201	0·8576	1·1790	1·5665	2·8544	4·6531
0·000675			0·085	0·105	0·124	0·142	0·150	0·158	0·166	0·174	0·189	0·204	0·237	0·269
1/ 1481			0·0599	0·1324	0·2440	0·4011	0·4986	0·6097	0·7351	0·8754	1·2033	1·5986	2·9126	4·7475
0·000700			0·086	0·108	0·127	0·145	0·153	0·162	0·170	0·178	0·193	0·208	0·242	0·274
1/ 1429			0·0611	0·1351	0·2490	0·4092	0·5086	0·6219	0·7497	0·8928	1·2272	1·6302	2·9698	4·8402
0·000725			0·088	0·110	0·129	0·148	0·156	0·165	0·173	0·181	0·197	0·212	0·247	0·279
1/ 1379			0·0624	0·1378	0·2538	0·4172	0·5185	0·6339	0·7642	0·9099	1·2506	1·6613	3·0260	4·9313
0·000750			0·090	0·112	0·132	0·150	0·159	0·168	0·176	0·184	0·200	0·215	0·251	0·284
1/ 1333			0·0636	0·1404	0·2586	0·4250	0·5282	0·6457	0·7783	0·9268	1·2737	1·6918	3·0812	5·0209
0·000800			0·093	0·116	0·136	0·156	0·165	0·174	0·182	0·191	0·207	0·223	0·260	0·294
1/ 1250			0·0659	0·1455	0·2680	0·4402	0·5470	0·6687	0·8060	0·9597	1·3188	1·7514	3·1891	5·1958
0·000850			0·096	0·120	0·141	0·161	0·170	0·180	0·189	0·197	0·214	0·230	0·268	0·304
1/ 1176			0·0682	0·1505	0·2770	0·4550	0·5654	0·6911	0·8329	0·9916	1·3625	1·8093	3·2938	5·3654
0·000900			0·100	0·124	0·146	0·166	0·176	0·185	0·194	0·203	0·221	0·238	0·277	0·313
1/ 1111			0·0704	0·1554	0·2859	0·4694	0·5832	0·7128	0·8591	1·0227	1·4049	1·8655	3·3954	5·5303
0·000950			0·103	0·127	0·150	0·171	0·181	0·191	0·200	0·209	0·227	0·244	0·285	0·322
1/ 1053			0·0726	0·1601	0·2945	0·4834	0·6006	0·7340	0·8845	1·0529	1·4463	1·9203	3·4944	5·6907
0·001000		0·092	0·106	0·131	0·154	0·176	0·186	0·196	0·206	0·215	0·234	0·251	0·293	0·331
1/ 1000		0·0451	0·0747	0·1647	0·3028	0·4971	0·6175	0·7546	0·9093	1·0823	1·4866	1·9736	3·5909	5·8470
		–	–	0·74	0·74	0·74	0·75	0·75	0·75	0·75	0·76	0·76	0·77	0·78

$V_{r(0\cdot5)medial}$ for half-full circular pipes.

k_s = 0·30 mm S = 0·00030 to 0·00100

k_s = 0·30 mm
S = 0·00100 to 0·00300

ie hydraulic gradient =
1 in 1000 to 1 in 333

Water (or sewage) at 15°C;
full bore conditions.

velocities in ms^{-1}
discharges in litres/sec

Gradient	(Equivalent) Pipe diameters in mm													
	20	25	30	40	50	60	65	70	75	80	90	100	125	150
0·00100		0·092	0·106	0·131	0·154	0·176	0·186	0·196	0·206	0·215	0·234	0·251	0·293	0·331
1/ 1000		0·0451	0·0747	0·1647	0·3028	0·4971	0·6175	0·7546	0·9093	1·0823	1·4866	1·9736	3·5909	5·8470
0·00105		0·094	0·109	0·135	0·158	0·181	0·191	0·201	0·211	0·221	0·240	0·258	0·300	0·340
1/ 952		0·0464	0·0768	0·1692	0·3110	0·5104	0·6340	0·7748	0·9335	1·1111	1·5260	2·0257	3·6850	5·9996
0·00110		0·097	0·111	0·138	0·162	0·185	0·196	0·206	0·217	0·227	0·246	0·264	0·308	0·348
1/ 909		0·0476	0·0788	0·1736	0·3190	0·5234	0·6502	0·7944	0·9572	1·1392	1·5644	2·0766	3·7770	6·1486
0·00115		0·099	0·114	0·142	0·166	0·190	0·201	0·211	0·222	0·232	0·252	0·271	0·315	0·356
1/ 870		0·0488	0·0808	0·1779	0·3269	0·5362	0·6660	0·8137	0·9803	1·1667	1·6020	2·1263	3·8670	6·2944
0·00120		0·102	0·117	0·145	0·170	0·194	0·205	0·216	0·227	0·237	0·258	0·277	0·322	0·364
1/ 833		0·0500	0·0827	0·1821	0·3345	0·5487	0·6814	0·8326	1·0030	1·1936	1·6388	2·1751	3·9550	6·4370
0·00125		0·104	0·120	0·148	0·174	0·198	0·210	0·221	0·232	0·243	0·263	0·283	0·329	0·372
1/ 800		0·0512	0·0846	0·1862	0·3420	0·5609	0·6966	0·8510	1·0252	1·2200	1·6749	2·2228	4·0413	6·5768
0·00130		0·107	0·122	0·151	0·178	0·203	0·214	0·226	0·237	0·248	0·269	0·289	0·336	0·380
1/ 769		0·0523	0·0865	0·1902	0·3494	0·5729	0·7115	0·8692	1·0470	1·2458	1·7103	2·2696	4·1259	6·7139
0·00135		0·109	0·125	0·155	0·182	0·207	0·219	0·230	0·242	0·253	0·274	0·295	0·343	0·388
1/ 741		0·0534	0·0883	0·1942	0·3566	0·5847	0·7261	0·8869	1·0684	1·2712	1·7451	2·3156	4·2089	6·8484
0·00140		0·111	0·127	0·158	0·185	0·211	0·223	0·235	0·247	0·258	0·280	0·301	0·350	0·395
1/ 714		0·0545	0·0901	0·1981	0·3638	0·5963	0·7404	0·9044	1·0894	1·2962	1·7792	2·3607	4·2905	6·9805
0·00145		0·113	0·130	0·161	0·189	0·215	0·227	0·239	0·251	0·263	0·285	0·306	0·356	0·402
1/ 690		0·0556	0·0919	0·2020	0·3707	0·6077	0·7545	0·9216	1·1100	1·3207	1·8127	2·4050	4·3706	7·1102
0·00150		0·115	0·132	0·164	0·192	0·219	0·232	0·244	0·256	0·268	0·290	0·312	0·363	0·410
1/ 667		0·0566	0·0936	0·2057	0·3776	0·6189	0·7684	0·9385	1·1303	1·3448	1·8457	2·4486	4·4494	7·2378
0·00160		0·120	0·137	0·170	0·199	0·227	0·240	0·252	0·265	0·277	0·300	0·323	0·375	0·424
1/ 625		0·0587	0·0970	0·2131	0·3910	0·6407	0·7954	0·9715	1·1699	1·3918	1·9100	2·5338	4·6032	7·4869
0·00170		0·124	0·142	0·175	0·206	0·234	0·248	0·261	0·274	0·286	0·310	0·333	0·387	0·437
1/ 588		0·0607	0·1003	0·2203	0·4041	0·6619	0·8217	1·0034	1·2084	1·4375	1·9724	2·6163	4·7523	7·7284
0·00180		0·128	0·146	0·181	0·212	0·241	0·255	0·269	0·282	0·295	0·320	0·343	0·399	0·451
1/ 556		0·0627	0·1035	0·2272	0·4167	0·6825	0·8472	1·0345	1·2457	1·4818	2·0331	2·6966	4·8972	7·9630
0·00190		0·132	0·151	0·186	0·219	0·248	0·263	0·277	0·290	0·303	0·329	0·353	0·411	0·464
1/ 526		0·0646	0·1066	0·2340	0·4290	0·7026	0·8720	1·0648	1·2821	1·5250	2·0921	2·7746	5·0381	8·1913
0·00200	0·114	0·135	0·155	0·191	0·225	0·255	0·270	0·284	0·298	0·312	0·338	0·363	0·422	0·476
1/ 500	0·0359	0·0665	0·1097	0·2406	0·4410	0·7222	0·8962	1·0943	1·3175	1·5670	2·1496	2·8507	5·1755	8·4137
0·00210	0·117	0·139	0·159	0·197	0·231	0·262	0·277	0·292	0·306	0·320	0·347	0·372	0·433	0·488
1/ 476	0·0369	0·0683	0·1126	0·2471	0·4528	0·7412	0·9198	1·1230	1·3521	1·6081	2·2057	2·9249	5·3095	8·6307
0·00220	0·120	0·143	0·163	0·202	0·236	0·269	0·284	0·299	0·314	0·328	0·355	0·382	0·443	0·500
1/ 455	0·0378	0·0701	0·1155	0·2534	0·4642	0·7599	0·9429	1·1512	1·3858	1·6482	2·2606	2·9974	5·4405	8·8427
0·00230	0·123	0·146	0·167	0·207	0·242	0·275	0·291	0·306	0·321	0·336	0·364	0·391	0·454	0·512
1/ 435	0·0388	0·0718	0·1184	0·2595	0·4754	0·7781	0·9655	1·1787	1·4189	1·6874	2·3142	3·0683	5·5685	9·0499
0·00240	0·126	0·150	0·171	0·211	0·248	0·282	0·298	0·313	0·328	0·343	0·372	0·400	0·464	0·524
1/ 417	0·0397	0·0735	0·1212	0·2656	0·4864	0·7960	0·9876	1·2056	1·4512	1·7258	2·3667	3·1377	5·6938	9·2527
0·00250	0·129	0·153	0·175	0·216	0·253	0·288	0·304	0·320	0·336	0·351	0·380	0·408	0·474	0·535
1/ 400	0·0406	0·0752	0·1239	0·2715	0·4972	0·8135	1·0092	1·2319	1·4829	1·7634	2·4181	3·2057	5·8165	9·4515
0·00260	0·132	0·156	0·179	0·221	0·259	0·294	0·311	0·327	0·343	0·358	0·388	0·417	0·484	0·546
1/ 385	0·0415	0·0768	0·1266	0·2773	0·5077	0·8306	1·0305	1·2578	1·5140	1·8003	2·4685	3·2724	5·9369	9·6463
0·00270	0·135	0·160	0·183	0·225	0·264	0·300	0·317	0·333	0·350	0·365	0·396	0·425	0·493	0·557
1/ 370	0·0424	0·0784	0·1292	0·2830	0·5181	0·8475	1·0513	1·2832	1·5445	1·8365	2·5180	3·3378	6·0550	9·8374
0·00280	0·138	0·163	0·186	0·230	0·269	0·306	0·323	0·340	0·356	0·372	0·403	0·433	0·503	0·567
1/ 357	0·0433	0·0800	0·1318	0·2886	0·5282	0·8640	1·0718	1·3081	1·5744	1·8720	2·5666	3·4021	6·1709	10·025
0·00290	0·140	0·166	0·190	0·234	0·274	0·311	0·329	0·346	0·363	0·379	0·411	0·441	0·512	0·578
1/ 345	0·0441	0·0815	0·1343	0·2941	0·5382	0·8802	1·0919	1·3326	1·6038	1·9070	2·6143	3·4652	6·2848	10·209
0·00300	0·143	0·169	0·194	0·238	0·279	0·317	0·335	0·353	0·370	0·386	0·418	0·449	0·521	0·588
1/ 333	0·0449	0·0831	0·1368	0·2995	0·5480	0·8962	1·1117	1·3567	1·6328	1·9413	2·6613	3·5273	6·3968	10·391
	–	–	–	0·74	0·75	0·75	0·75	0·76	0·76	0·76	0·77	0·77	0·78	0·79

$V_{r(0·5)medial}$ for half-full circular pipes.

A7
(p.3 of 6)

k_s = 0·30 mm
S = 0·00300 to 0·01000

ie hydraulic gradient =
1 in 333 to 1 in 100

Water (or sewage) at 15°C;
full bore conditions.

velocities in ms^{-1}
discharges in litres/sec

Gradient **(Equivalent) Pipe diameters in mm**

Gradient	20	25	30	40	50	60	65	70	75	80	90	100	125	150
0·00300	0·143	0·169	0·194	0·238	0·279	0·317	0·335	0·353	0·370	0·386	0·418	0·449	0·521	0·588
1/ 333	0·0449	0·0831	0·1368	0·2995	0·5480	0·8962	1·1117	1·3567	1·6328	1·9413	2·6613	3·5273	6·3968	10·391
0·00320	0·148	0·175	0·200	0·247	0·289	0·328	0·347	0·365	0·382	0·400	0·433	0·465	0·539	0·608
1/ 313	0·0466	0·0861	0·1417	0·3100	0·5672	0·9274	1·1503	1·4037	1·6892	2·0083	2·7529	3·6484	6·6154	10·745
0·00340	0·153	0·181	0·207	0·255	0·298	0·339	0·358	0·377	0·395	0·412	0·447	0·479	0·556	0·627
1/ 294	0·0482	0·0889	0·1465	0·3203	0·5858	0·9577	1·1877	1·4493	1·7440	2·0733	2·8417	3·7658	6·8274	11·088
0·00360	0·158	0·187	0·214	0·263	0·308	0·349	0·369	0·388	0·407	0·425	0·460	0·494	0·573	0·646
1/ 278	0·0497	0·0918	0·1511	0·3302	0·6039	0·9870	1·2241	1·4936	1·7972	2·1365	2·9280	3·8799	7·0332	11·421
0·00380	0·163	0·193	0·220	0·271	0·317	0·359	0·380	0·399	0·419	0·437	0·473	0·508	0·589	0·665
1/ 263	0·0512	0·0945	0·1555	0·3399	0·6215	1·0156	1·2594	1·5367	1·8490	2·1979	3·0120	3·9909	7·2335	11·745
0·00400	0·168	0·198	0·226	0·278	0·325	0·369	0·390	0·410	0·430	0·449	0·486	0·522	0·605	0·682
1/ 250	0·0527	0·0972	0·1599	0·3494	0·6386	1·0435	1·2939	1·5787	1·8994	2·2578	3·0938	4·0990	7·4286	12·060
0·00420	0·172	0·203	0·232	0·285	0·334	0·379	0·400	0·421	0·441	0·461	0·499	0·535	0·621	0·700
1/ 238	0·0541	0·0998	0·1642	0·3586	0·6554	1·0707	1·3276	1·6196	1·9486	2·3162	3·1736	4·2045	7·6189	12·368
0·00440	0·177	0·208	0·238	0·293	0·342	0·388	0·410	0·431	0·452	0·472	0·511	0·548	0·636	0·717
1/ 227	0·0555	0·1023	0·1683	0·3676	0·6717	1·0973	1·3604	1·6597	1·9967	2·3732	3·2516	4·3076	7·8048	12·669
0·00460	0·181	0·214	0·244	0·300	0·350	0·397	0·420	0·441	0·463	0·483	0·523	0·561	0·651	0·734
1/ 217	0·0568	0·1048	0·1724	0·3764	0·6877	1·1232	1·3926	1·6988	2·0437	2·4290	3·3278	4·4084	7·9866	12·963
0·00480	0·185	0·219	0·250	0·306	0·358	0·406	0·429	0·451	0·473	0·494	0·535	0·574	0·665	0·750
1/ 208	0·0582	0·1073	0·1764	0·3850	0·7033	1·1487	1·4240	1·7371	2·0897	2·4836	3·4024	4·5070	8·1645	13·251
0·00500	0·189	0·223	0·255	0·313	0·366	0·415	0·438	0·461	0·483	0·505	0·546	0·586	0·679	0·766
1/ 200	0·0595	0·1097	0·1803	0·3935	0·7187	1·1736	1·4549	1·7746	2·1348	2·5371	3·4755	4·6036	8·3387	13·533
0·00525	0·194	0·229	0·262	0·321	0·376	0·426	0·450	0·473	0·496	0·518	0·560	0·601	0·697	0·785
1/ 190	0·0611	0·1126	0·1851	0·4038	0·7374	1·2040	1·4925	1·8205	2·1899	2·6025	3·5648	4·7217	8·5517	13·878
0·00550	0·199	0·235	0·268	0·329	0·385	0·436	0·461	0·485	0·508	0·530	0·574	0·616	0·714	0·804
1/ 182	0·0627	0·1154	0·1897	0·4139	0·7557	1·2338	1·5293	1·8653	2·2437	2·6664	3·6521	4·8370	8·7598	14·214
0·00575	0·204	0·241	0·275	0·337	0·394	0·447	0·472	0·496	0·520	0·543	0·587	0·630	0·730	0·823
1/ 174	0·0642	0·1182	0·1943	0·4238	0·7736	1·2629	1·5653	1·9091	2·2963	2·7288	3·7374	4·9498	8·9632	14·543
0·00600	0·209	0·246	0·281	0·345	0·403	0·457	0·482	0·507	0·531	0·555	0·601	0·644	0·747	0·841
1/ 167	0·0657	0·1210	0·1988	0·4334	0·7911	1·2914	1·6006	1·9520	2·3478	2·7899	3·8209	5·0601	9·1622	14·865
0·00625	0·214	0·252	0·287	0·352	0·412	0·467	0·493	0·518	0·543	0·567	0·613	0·658	0·762	0·859
1/ 160	0·0672	0·1237	0·2031	0·4429	0·8083	1·3192	1·6351	1·9940	2·3983	2·8498	3·9027	5·1682	9·3572	15·181
0·00650	0·218	0·257	0·293	0·360	0·420	0·476	0·503	0·529	0·554	0·579	0·626	0·672	0·778	0·877
1/ 154	0·0686	0·1263	0·2074	0·4522	0·8251	1·3466	1·6689	2·0352	2·4477	2·9085	3·9829	5·2742	9·5483	15·490
0·00675	0·223	0·263	0·299	0·367	0·429	0·486	0·513	0·539	0·565	0·590	0·638	0·685	0·793	0·894
1/ 148	0·0700	0·1289	0·2117	0·4613	0·8417	1·3734	1·7021	2·0756	2·4963	2·9661	4·0616	5·3782	9·7358	15·793
0·00700	0·227	0·268	0·305	0·374	0·437	0·495	0·523	0·550	0·576	0·601	0·651	0·698	0·808	0·911
1/ 143	0·0714	0·1314	0·2158	0·4702	0·8579	1·3998	1·7347	2·1153	2·5439	3·0226	4·1389	5·4803	9·9199	16·091
0·00725	0·232	0·273	0·311	0·381	0·445	0·504	0·532	0·560	0·586	0·612	0·663	0·711	0·823	0·927
1/ 138	0·0728	0·1339	0·2199	0·4790	0·8738	1·4257	1·7667	2·1543	2·5907	3·0782	4·2148	5·5806	10·101	16·383
0·00750	0·236	0·278	0·317	0·388	0·453	0·513	0·542	0·570	0·597	0·623	0·674	0·723	0·838	0·943
1/ 133	0·0741	0·1364	0·2239	0·4876	0·8895	1·4511	1·7982	2·1926	2·6367	3·1328	4·2894	5·6792	10·279	16·671
0·00800	0·244	0·288	0·328	0·401	0·469	0·531	0·560	0·589	0·617	0·644	0·697	0·748	0·866	0·975
1/ 125	0·0767	0·1411	0·2317	0·5045	0·9201	1·5008	1·8596	2·2674	2·7266	3·2394	4·4350	5·8716	10·625	17·232
0·00850	0·252	0·297	0·338	0·414	0·484	0·548	0·578	0·608	0·637	0·665	0·719	0·771	0·893	1·006
1/ 118	0·0793	0·1458	0·2393	0·5209	0·9497	1·5489	1·9192	2·3399	2·8136	3·3427	4·5761	6·0581	10·962	17·776
0·00900	0·260	0·306	0·349	0·427	0·498	0·564	0·596	0·626	0·656	0·685	0·741	0·794	0·920	1·036
1/ 111	0·0818	0·1503	0·2466	0·5367	0·9785	1·5957	1·9770	2·4104	2·8982	3·4431	4·7132	6·2393	11·288	18·304
0·00950	0·268	0·315	0·359	0·439	0·513	0·580	0·613	0·644	0·675	0·704	0·762	0·817	0·946	1·065
1/ 105	0·0842	0·1547	0·2538	0·5522	1·0065	1·6412	2·0333	2·4789	2·9805	3·5407	4·8465	6·4155	11·606	18·818
0·01000	0·275	0·324	0·369	0·451	0·527	0·596	0·629	0·661	0·693	0·723	0·782	0·839	0·971	1·093
1/ 100	0·0865	0·1590	0·2608	0·5673	1·0338	1·6856	2·0882	2·5456	3·0606	3·6358	4·9764	6·5871	11·916	19·319
	–	0·74	0·74	0·75	0·75	0·76	0·76	0·76	0·77	0·77	0·77	0·78	0·79	0·80

$V_{r(0·5)medial}$ for half-full circular pipes.

k_s = 0·30 mm S = 0·00300 to 0·01000

$k_s = 0{\cdot}30$ **mm**
S = 0·01000 to 0·03000

ie hydraulic gradient =
1 in 100 to 1 in 33·3

Water (or sewage) at 15°C;
full bore conditions.

velocities in ms^{-1}
discharges in litres/sec

Gradient **(Equivalent) Pipe diameters in mm**

Gradient	20	25	30	40	50	60	65	70	75	80	90	100	125	150
0·01000	0·275	0·324	0·369	0·451	0·527	0·596	0·629	0·661	0·693	0·723	0·782	0·839	0·971	1·093
1/ 100	0·0865	0·1590	0·2608	0·5673	1·0338	1·6856	2·0882	2·5456	3·0606	3·6358	4·9764	6·5871	11·916	19·319
0·01050	0·283	0·332	0·379	0·463	0·540	0·611	0·645	0·678	0·710	0·742	0·802	0·860	0·996	1·121
1/ 95	0·0888	0·1631	0·2676	0·5820	1·0605	1·7288	2·1417	2·6108	3·1388	3·7286	5·1032	6·7546	12·217	19·807
0·01100	0·290	0·341	0·388	0·475	0·553	0·626	0·661	0·695	0·728	0·760	0·822	0·881	1·020	1·148
1/ 91	0·0910	0·1672	0·2742	0·5963	1·0865	1·7710	2·1939	2·6744	3·2152	3·8192	5·2269	6·9181	12·512	20·284
0·01150	0·297	0·349	0·397	0·486	0·566	0·641	0·677	0·711	0·745	0·777	0·841	0·901	1·043	1·174
1/ 87	0·0932	0·1712	0·2807	0·6103	1·1119	1·8123	2·2450	2·7365	3·2899	3·9077	5·3479	7·0780	12·800	20·750
0·01200	0·304	0·357	0·406	0·497	0·579	0·655	0·692	0·727	0·761	0·795	0·859	0·921	1·066	1·200
1/ 83	0·0954	0·1751	0·2871	0·6241	1·1368	1·8528	2·2950	2·7974	3·3629	3·9944	5·4663	7·2344	13·083	21·206
0·01250	0·310	0·365	0·415	0·507	0·591	0·669	0·706	0·742	0·777	0·812	0·877	0·941	1·089	1·225
1/ 80	0·0975	0·1789	0·2933	0·6375	1·1612	1·8924	2·3439	2·8570	3·4345	4·0793	5·5823	7·3876	13·359	21·653
0·01300	0·317	0·372	0·424	0·518	0·604	0·683	0·721	0·758	0·793	0·828	0·895	0·960	1·111	1·250
1/ 77	0·0995	0·1827	0·2994	0·6507	1·1851	1·9312	2·3919	2·9154	3·5046	4·1626	5·6960	7·5379	13·630	22·091
0·01350	0·323	0·380	0·432	0·528	0·616	0·696	0·735	0·772	0·809	0·844	0·913	0·979	1·132	1·274
1/ 74	0·1016	0·1864	0·3054	0·6637	1·2085	1·9692	2·4390	2·9727	3·5735	4·2442	5·8075	7·6852	13·895	22·520
0·01400	0·330	0·387	0·440	0·538	0·627	0·710	0·749	0·787	0·824	0·860	0·930	0·997	1·154	1·298
1/ 71	0·1035	0·1900	0·3113	0·6764	1·2316	2·0066	2·4853	3·0290	3·6410	4·3244	5·9170	7·8299	14·156	22·942
0·01450	0·336	0·394	0·449	0·548	0·639	0·723	0·763	0·801	0·839	0·876	0·947	1·015	1·174	1·322
1/ 69	0·1055	0·1935	0·3171	0·6889	1·2542	2·0434	2·5307	3·0843	3·7074	4·4031	6·0246	7·9720	14·412	23·356
0·01500	0·342	0·401	0·457	0·558	0·650	0·735	0·776	0·816	0·854	0·891	0·964	1·033	1·195	1·345
1/ 67	0·1074	0·1970	0·3228	0·7011	1·2764	2·0795	2·5753	3·1386	3·7727	4·4806	6·1303	8·1117	14·664	23·764
0·01600	0·354	0·415	0·472	0·577	0·672	0·760	0·802	0·843	0·883	0·921	0·996	1·068	1·235	1·390
1/ 63	0·1111	0·2038	0·3339	0·7251	1·3198	2·1499	2·6625	3·2447	3·9000	4·6316	6·3366	8·3843	15·155	24·558
0·01700	0·365	0·429	0·488	0·595	0·694	0·785	0·828	0·870	0·911	0·951	1·027	1·101	1·274	1·433
1/ 59	0·1148	0·2104	0·3447	0·7483	1·3619	2·2182	2·7469	3·3475	4·0234	4·7781	6·5366	8·6486	15·632	25·328
0·01800	0·377	0·442	0·502	0·613	0·714	0·808	0·853	0·896	0·938	0·979	1·058	1·134	1·311	1·476
1/ 56	0·1183	0·2169	0·3551	0·7709	1·4028	2·2845	2·8290	3·4474	4·1433	4·9203	6·7309	8·9052	16·094	26·076
0·01900	0·387	0·455	0·517	0·631	0·735	0·831	0·877	0·921	0·964	1·006	1·088	1·166	1·348	1·517
1/ 53	0·1217	0·2231	0·3653	0·7928	1·4425	2·3491	2·9088	3·5445	4·2599	5·0586	6·9198	9·1548	16·544	26·804
0·02000	0·398	0·467	0·531	0·648	0·754	0·853	0·900	0·946	0·990	1·033	1·117	1·197	1·384	1·557
1/ 50	0·1251	0·2292	0·3752	0·8142	1·4812	2·4119	2·9865	3·6391	4·3735	5·1934	7·1038	9·3979	16·982	27·512
0·02100	0·408	0·479	0·545	0·664	0·774	0·875	0·923	0·970	1·015	1·059	1·145	1·227	1·419	1·596
1/ 47·6	0·1283	0·2351	0·3849	0·8350	1·5190	2·4732	3·0623	3·7314	4·4843	5·3248	7·2833	9·6351	17·410	28·204
0·02200	0·419	0·491	0·558	0·681	0·792	0·896	0·945	0·993	1·040	1·085	1·172	1·256	1·453	1·634
1/ 45·5	0·1315	0·2409	0·3943	0·8554	1·5559	2·5331	3·1363	3·8215	4·5925	5·4532	7·4586	9·8667	17·827	28·879
0·02300	0·429	0·502	0·571	0·697	0·811	0·917	0·967	1·016	1·063	1·110	1·199	1·285	1·486	1·672
1/ 43·5	0·1346	0·2466	0·4036	0·8753	1·5920	2·5916	3·2087	3·9096	4·6982	5·5787	7·6300	10·093	18·235	29·538
0·02400	0·438	0·514	0·584	0·712	0·829	0·937	0·988	1·038	1·087	1·134	1·226	1·313	1·518	1·708
1/ 41·7	0·1377	0·2521	0·4126	0·8948	1·6272	2·6489	3·2796	3·9958	4·8018	5·7015	7·7977	10·315	18·634	30·184
0·02500	0·448	0·525	0·596	0·727	0·846	0·957	1·009	1·060	1·110	1·158	1·252	1·341	1·550	1·744
1/ 40·0	0·1407	0·2576	0·4215	0·9139	1·6618	2·7050	3·3489	4·0802	4·9031	5·8217	7·9619	10·532	19·026	30·816
0·02600	0·457	0·536	0·609	0·742	0·864	0·976	1·030	1·082	1·132	1·182	1·277	1·368	1·582	1·779
1/ 38·5	0·1436	0·2629	0·4302	0·9326	1·6957	2·7600	3·4170	4·1630	5·0025	5·9396	8·1229	10·744	19·409	31·436
0·02700	0·466	0·546	0·621	0·757	0·881	0·995	1·050	1·103	1·154	1·205	1·302	1·395	1·612	1·813
1/ 37·0	0·1464	0·2681	0·4387	0·9509	1·7289	2·8140	3·4837	4·2442	5·1000	6·0553	8·2808	10·953	19·785	32·044
0·02800	0·475	0·557	0·632	0·771	0·897	1·014	1·070	1·124	1·176	1·227	1·326	1·421	1·642	1·847
1/ 35·7	0·1493	0·2733	0·4470	0·9690	1·7616	2·8670	3·5492	4·3239	5·1957	6·1688	8·4359	11·158	20·154	32·641
0·02900	0·484	0·567	0·644	0·785	0·913	1·032	1·089	1·144	1·197	1·249	1·350	1·446	1·672	1·880
1/ 34·5	0·1520	0·2783	0·4552	0·9867	1·7936	2·9190	3·6135	4·4023	5·2897	6·2804	8·5882	11·359	20·517	33·227
0·03000	0·493	0·577	0·655	0·799	0·930	1·050	1·108	1·164	1·218	1·271	1·374	1·471	1·701	1·913
1/ 33·3	0·1548	0·2833	0·4633	1·0041	1·8252	2·9702	3·6768	4·4792	5·3821	6·3900	8·7379	11·557	20·873	33·804
	–	0·74	0·74	0·75	0·76	0·76	0·76	0·77	0·77	0·77	0·78	0·78	0·79	0·80

$V_{r(0{\cdot}5)medial}$ **for half-full circular pipes.**

A7
(p.5 of 6)

k_s = 0·30 mm
S = 0·03000 to 0·10000

ie hydraulic gradient =
1 in 33·3 to 1 in 10·0

Water (or sewage) at 15°C;
full bore conditions.

velocities in ms^{-1}
discharges in litres/sec

Gradient	**(Equivalent) Pipe diameters in mm**													
	20	**25**	**30**	**40**	**50**	**60**	**65**	**70**	**75**	**80**	**90**	**100**	**125**	**150**
0·03000	0·493	0·577	0·655	0·799	0·930	1·050	1·108	1·164	1·218	1·271	1·374	1·471	1·701	1·913
1/ 33·3	0·1548	0·2833	0·4633	1·0041	1·8252	2·9702	3·6768	4·4792	5·3821	6·3900	8·7379	11·557	20·873	33·804
0·03200	0·510	0·597	0·678	0·826	0·961	1·086	1·145	1·203	1·259	1·314	1·419	1·521	1·758	1·977
1/ 31·3	0·1601	0·2929	0·4791	1·0380	1·8867	3·0700	3·8002	4·6294	5·5625	6·6039	9·0301	11·943	21·569	34·928
0·03400	0·526	0·616	0·699	0·852	0·991	1·120	1·181	1·241	1·299	1·355	1·464	1·568	1·813	2·038
1/ 29·4	0·1652	0·3023	0·4944	1·0710	1·9463	3·1667	3·9198	4·7750	5·7373	6·8113	9·3132	12·317	22·243	36·018
0·03600	0·542	0·634	0·720	0·878	1·021	1·153	1·216	1·278	1·337	1·395	1·507	1·614	1·866	2·098
1/ 27·8	0·1702	0·3114	0·5092	1·1029	2·0042	3·2607	4·0360	4·9164	5·9070	7·0126	9·5882	12·680	22·897	37·077
0·03800	0·557	0·652	0·741	0·902	1·049	1·186	1·250	1·313	1·374	1·434	1·549	1·659	1·918	2·156
1/ 26·3	0·1751	0·3203	0·5236	1·1340	2·0605	3·3521	4·1490	5·0540	6·0721	7·2085	9·8556	13·033	23·534	38·106
0·04000	0·572	0·670	0·761	0·927	1·077	1·217	1·284	1·348	1·411	1·472	1·590	1·703	1·968	2·213
1/ 25·0	0·1799	0·3289	0·5377	1·1643	2·1154	3·4411	4·2591	5·1879	6·2329	7·3993	10·116	13·378	24·154	39·109
0·04200	0·587	0·687	0·780	0·950	1·105	1·248	1·316	1·382	1·446	1·509	1·630	1·746	2·018	2·268
1/ 23·8	0·1845	0·3373	0·5514	1·1939	2·1689	3·5279	4·3664	5·3186	6·3898	7·5854	10·370	13·713	24·759	40·087
0·04400	0·602	0·704	0·799	0·973	1·131	1·278	1·347	1·415	1·481	1·545	1·669	1·788	2·066	2·322
1/ 22·7	0·1890	0·3456	0·5648	1·2227	2·2211	3·6127	4·4713	5·4462	6·5429	7·7671	10·618	14·041	25·350	41·042
0·04600	0·616	0·720	0·818	0·995	1·157	1·307	1·378	1·448	1·515	1·581	1·707	1·829	2·113	2·375
1/ 21·7	0·1934	0·3536	0·5779	1·2509	2·2722	3·6956	4·5738	5·5709	6·6927	7·9447	10·861	14·361	25·927	41·975
0·04800	0·629	0·736	0·836	1·017	1·183	1·336	1·409	1·479	1·548	1·615	1·745	1·868	2·159	2·427
1/ 20·8	0·1977	0·3615	0·5907	1·2785	2·3222	3·7767	4·6741	5·6930	6·8392	8·1185	11·098	14·675	26·492	42·889
0·05000	0·643	0·752	0·853	1·039	1·208	1·364	1·438	1·510	1·581	1·649	1·781	1·908	2·204	2·478
1/ 20·0	0·2020	0·3692	0·6033	1·3056	2·3711	3·8561	4·7723	5·8125	6·9827	8·2887	11·331	14·982	27·046	43·783
0·05250	0·659	0·771	0·875	1·065	1·238	1·398	1·474	1·548	1·620	1·690	1·826	1·955	2·259	2·540
1/ 19·0	0·2072	0·3786	0·6186	1·3386	2·4310	3·9532	4·8924	5·9586	7·1581	8·4968	11·615	15·357	27·722	44·877
0·05500	0·676	0·790	0·896	1·091	1·268	1·432	1·510	1·585	1·659	1·731	1·869	2·002	2·313	2·600
1/ 18·2	0·2122	0·3878	0·6336	1·3709	2·4894	4·0481	5·0096	6·1013	7·3294	8·7000	11·892	15·724	28·382	45·945
0·05750	0·691	0·808	0·917	1·116	1·297	1·464	1·544	1·622	1·697	1·770	1·912	2·048	2·365	2·659
1/ 17·4	0·2172	0·3968	0·6482	1·4025	2·5465	4·1408	5·1242	6·2408	7·4969	8·8987	12·164	16·082	29·028	46·989
0·06000	0·707	0·826	0·937	1·141	1·325	1·497	1·578	1·657	1·734	1·809	1·954	2·092	2·417	2·717
1/ 16·7	0·2220	0·4056	0·6626	1·4334	2·6024	4·2315	5·2364	6·3773	7·6607	9·0930	12·429	16·433	29·660	48·010
0·06250	0·722	0·844	0·957	1·165	1·353	1·528	1·611	1·692	1·770	1·847	1·995	2·136	2·467	2·773
1/ 16·0	0·2267	0·4142	0·6766	1·4636	2·6572	4·3203	5·3462	6·5110	7·8212	9·2834	12·689	16·776	30·278	49·010
0·06500	0·737	0·861	0·977	1·188	1·381	1·559	1·644	1·726	1·806	1·884	2·035	2·179	2·517	2·829
1/ 15·4	0·2314	0·4227	0·6904	1·4932	2·7109	4·4074	5·4539	6·6420	7·9785	9·4700	12·944	17·113	30·885	49·991
0·06750	0·751	0·878	0·996	1·211	1·407	1·589	1·675	1·759	1·841	1·920	2·074	2·221	2·565	2·883
1/ 14·8	0·2359	0·4310	0·7039	1·5223	2·7635	4·4928	5·5595	6·7705	8·1328	9·6530	13·194	17·443	31·480	50·953
0·07000	0·765	0·895	1·015	1·234	1·434	1·619	1·707	1·792	1·875	1·956	2·112	2·262	2·613	2·937
1/ 14·3	0·2404	0·4391	0·7171	1·5509	2·8152	4·5767	5·6632	6·8967	8·2843	9·8327	13·439	17·767	32·064	51·897
0·07250	0·779	0·911	1·033	1·256	1·460	1·648	1·737	1·824	1·909	1·991	2·150	2·303	2·660	2·989
1/ 13·8	0·2448	0·4471	0·7302	1·5789	2·8660	4·6591	5·7651	7·0207	8·4331	10·009	13·680	18·085	32·638	52·824
0·07500	0·793	0·927	1·051	1·278	1·485	1·676	1·768	1·856	1·942	2·026	2·188	2·343	2·706	3·041
1/ 13·3	0·2491	0·4550	0·7430	1·6065	2·9159	4·7400	5·8652	7·1425	8·5793	10·183	13·917	18·398	33·202	53·736
0·08000	0·820	0·958	1·086	1·321	1·535	1·732	1·826	1·918	2·007	2·093	2·260	2·420	2·795	3·142
1/ 12·5	0·2576	0·4703	0·7679	1·6603	3·0133	4·8981	6·0606	7·3803	8·8647	10·521	14·379	19·009	34·302	55·515
0·08500	0·846	0·988	1·121	1·363	1·583	1·786	1·883	1·978	2·069	2·158	2·331	2·496	2·882	3·239
1/ 11·8	0·2657	0·4852	0·7921	1·7124	3·1077	5·0512	6·2499	7·6107	9·1414	10·849	14·827	19·601	35·368	57·239
0·09000	0·871	1·018	1·154	1·403	1·629	1·839	1·939	2·036	2·130	2·222	2·399	2·569	2·966	3·334
1/ 11·1	0·2737	0·4996	0·8157	1·7631	3·1993	5·1999	6·4338	7·8345	9·4100	11·168	15·262	20·176	36·404	58·913
0·09500	0·896	1·046	1·186	1·442	1·675	1·890	1·993	2·092	2·189	2·283	2·466	2·640	3·049	3·426
1/ 10·5	0·2814	0·5137	0·8385	1·8123	3·2885	5·3446	6·6126	8·0521	9·6712	11·478	15·685	20·735	37·411	60·542
0·10000	0·920	1·074	1·218	1·480	1·719	1·940	2·045	2·147	2·247	2·344	2·530	2·709	3·128	3·516
1/ 10·0	0·2889	0·5273	0·8608	1·8603	3·3753	5·4854	6·7868	8·2641	9·9257	11·780	16·098	21·279	38·392	62·128
	0·74	0·74	0·74	0·75	0·76	0·76	0·77	0·77	0·77	0·78	0·78	0·78	0·79	0·80

$V_{r(0\cdot5)medial}$ for half-full circular pipes.

k_s = 0·30 mm S = 0·03000 to 0·10000

k_s = 0·30 mm
S = 0·10000 to 0·30000

ie hydraulic gradient =
1 in 10·0 to 1 in 3·3

Water (or sewage) at 15°C;
full bore conditions.

velocities in ms^{-1}
discharges in litres/sec

Gradient **(Equivalent) Pipe diameters in mm**

Gradient	20	25	30	40	50	60	65	70	75	80	90	100	125	150
0·10000	0·920	1·074	1·218	1·480	1·719	1·940	2·045	2·147	2·247	2·344	2·530	2·709	3·128	3·516
1/ 10·0	0·2889	0·5273	0·8608	1·8603	3·3753	5·4854	6·7868	8·2641	9·9257	11·780	16·098	21·279	38·392	62·128
0·10500	0·943	1·101	1·249	1·518	1·762	1·989	2·096	2·201	2·303	2·402	2·594	2·777	3·206	3·603
1/ 9·5	0·2963	0·5407	0·8825	1·9071	3·4600	5·6229	6·9567	8·4708	10·174	12·074	16·500	21·810	39·349	63·675
0·11000	0·966	1·128	1·279	1·554	1·804	2·036	2·146	2·254	2·358	2·459	2·655	2·843	3·283	3·689
1/ 9·1	0·3034	0·5537	0·9037	1·9528	3·5427	5·7570	7·1226	8·6727	10·416	12·362	16·892	22·329	40·283	65·185
0·11500	0·988	1·154	1·308	1·589	1·845	2·083	2·195	2·305	2·411	2·515	2·716	2·908	3·357	3·772
1/ 8·7	0·3104	0·5665	0·9245	1·9974	3·6236	5·8882	7·2848	8·8701	10·653	12·643	17·276	22·835	41·196	66·662
0·12000	1·010	1·179	1·337	1·624	1·886	2·128	2·243	2·355	2·464	2·570	2·775	2·971	3·430	3·854
1/ 8·3	0·3173	0·5789	0·9448	2·0411	3·7027	6·0166	7·4435	9·0632	10·885	12·918	17·651	23·331	42·090	68·106
0·12500	1·031	1·204	1·365	1·658	1·925	2·172	2·290	2·404	2·515	2·623	2·832	3·032	3·501	3·934
1/ 8·0	0·3240	0·5911	0·9647	2·0839	3·7801	6·1423	7·5989	9·2523	11·112	13·187	18·019	23·817	42·965	69·521
0·13000	1·052	1·229	1·392	1·692	1·964	2·216	2·336	2·452	2·566	2·676	2·889	3·093	3·571	4·013
1/ 7·7	0·3306	0·6031	0·9841	2·1259	3·8561	6·2655	7·7512	9·4377	11·335	13·451	18·380	24·293	43·823	70·908
0·13500	1·073	1·253	1·419	1·724	2·002	2·259	2·381	2·500	2·615	2·728	2·945	3·153	3·640	4·090
1/ 7·4	0·3371	0·6148	1·0033	2·1670	3·9306	6·3864	7·9007	9·6196	11·553	13·710	18·733	24·760	44·665	72·269
0·14000	1·093	1·276	1·446	1·757	2·039	2·301	2·425	2·546	2·664	2·778	2·999	3·211	3·707	4·165
1/ 7·1	0·3434	0·6264	1·0220	2·2074	4·0037	6·5050	8·0474	9·7981	11·767	13·964	19·080	25·219	45·491	73·604
0·14500	1·113	1·299	1·472	1·788	2·076	2·342	2·469	2·592	2·711	2·828	3·053	3·268	3·773	4·239
1/ 6·9	0·3496	0·6377	1·0405	2·2471	4·0756	6·6216	8·1915	9·9735	11·978	14·214	19·421	25·669	46·303	74·916
0·15000	1·132	1·322	1·498	1·819	2·112	2·382	2·511	2·636	2·758	2·877	3·106	3·325	3·838	4·312
1/ 6·7	0·3557	0·6488	1·0586	2·2862	4·1462	6·7361	8·3332	10·146	12·185	14·459	19·756	26·112	47·100	76·206
0·16000	1·170	1·366	1·548	1·880	2·182	2·461	2·595	2·724	2·849	2·972	3·208	3·435	3·965	4·455
1/ 6·3	0·3677	0·6705	1·0939	2·3623	4·2840	6·9597	8·6095	10·482	12·588	14·938	20·410	26·976	48·657	78·722
0·17000	1·207	1·409	1·596	1·939	2·250	2·538	2·675	2·808	2·938	3·064	3·308	3·541	4·088	4·593
1/ 5·9	0·3793	0·6916	1·1282	2·4361	4·4176	7·1764	8·8774	10·808	12·980	15·403	21·044	27·813	50·165	81·161
0·18000	1·243	1·451	1·643	1·996	2·316	2·613	2·754	2·891	3·024	3·154	3·405	3·645	4·207	4·727
1/ 5·6	0·3905	0·7120	1·1615	2·5077	4·5472	7·3868	9·1376	11·125	13·360	15·853	21·660	28·626	51·630	83·529
0·19000	1·278	1·491	1·689	2·051	2·380	2·685	2·830	2·971	3·108	3·241	3·499	3·745	4·323	4·857
1/ 5·3	0·4014	0·7319	1·1938	2·5774	4·6734	7·5914	9·3906	11·433	13·729	16·292	22·258	29·417	53·054	85·832
0·20000	1·312	1·530	1·734	2·105	2·443	2·755	2·904	3·049	3·189	3·326	3·590	3·844	4·436	4·984
1/ 5·0	0·4121	0·7513	1·2254	2·6453	4·7962	7·7907	9·6370	11·733	14·089	16·719	22·841	30·187	54·442	88·075
0·21000	1·345	1·569	1·777	2·158	2·504	2·824	2·977	3·125	3·269	3·409	3·680	3·939	4·547	5·108
1/ 4·8	0·4225	0·7702	1·2561	2·7115	4·9161	7·9851	9·8773	12·025	14·440	17·135	23·410	30·938	55·795	90·263
0·22000	1·377	1·607	1·820	2·209	2·563	2·891	3·047	3·199	3·346	3·490	3·767	4·033	4·654	5·229
1/ 4·5	0·4326	0·7886	1·2861	2·7761	5·0331	8·1749	10·112	12·311	14·783	17·542	23·965	31·671	57·117	92·399
0·23000	1·409	1·643	1·861	2·259	2·622	2·957	3·116	3·271	3·422	3·569	3·852	4·124	4·760	5·347
1/ 4·3	0·4425	0·8066	1·3155	2·8393	5·1474	8·3605	10·341	12·590	15·118	17·939	24·508	32·388	58·409	94·488
0·24000	1·440	1·679	1·902	2·309	2·679	3·021	3·184	3·342	3·496	3·646	3·936	4·213	4·863	5·463
1/ 4·2	0·4522	0·8243	1·3442	2·9012	5·2594	8·5420	10·566	12·863	15·446	18·328	25·039	33·090	59·673	96·531
0·25000	1·470	1·714	1·941	2·357	2·734	3·084	3·250	3·412	3·569	3·722	4·018	4·301	4·963	5·576
1/ 4·0	0·4618	0·8416	1·3723	2·9617	5·3690	8·7199	10·786	13·130	15·767	18·709	25·559	33·777	60·911	98·533
0·26000	1·499	1·749	1·980	2·404	2·789	3·146	3·315	3·480	3·640	3·796	4·098	4·386	5·062	5·687
1/ 3·8	0·4711	0·8585	1·3999	3·0211	5·4764	8·8942	11·001	13·393	16·082	19·083	26·069	34·450	62·124	100·49
0·27000	1·529	1·783	2·019	2·450	2·843	3·206	3·379	3·547	3·710	3·869	4·176	4·470	5·159	5·796
1/ 3·7	0·4802	0·8751	1·4270	3·0793	5·5818	9·0652	11·213	13·650	16·391	19·449	26·569	35·111	63·314	102·42
0·28000	1·557	1·816	2·056	2·496	2·895	3·266	3·442	3·612	3·779	3·941	4·254	4·553	5·255	5·903
1/ 3·6	0·4892	0·8914	1·4535	3·1365	5·6853	9·2330	11·420	13·903	16·694	19·809	27·060	35·760	64·483	104·31
0·29000	1·585	1·849	2·093	2·541	2·947	3·324	3·503	3·677	3·846	4·011	4·329	4·634	5·348	6·008
1/ 3·4	0·4980	0·9075	1·4796	3·1926	5·7869	9·3979	11·624	14·151	16·992	20·162	27·542	36·397	65·631	106·16
0·30000	1·613	1·881	2·129	2·585	2·998	3·381	3·563	3·740	3·912	4·080	4·404	4·714	5·440	6·111
1/ 3·3	0·5067	0·9232	1·5052	3·2478	5·8868	9·5600	11·824	14·394	17·285	20·509	28·016	37·023	66·759	107·99
	0·74	0·74	0·75	0·75	0·76	0·77	0·77	0·77	0·77	0·78	0·78	0·79	0·80	0·80

$V_{r(0\cdot5)medial}$ for half-full circular pipes.

k_s = 0·60 mm
S = 0·00030 to 0·00100

ie hydraulic gradient =
1 in 3333 to 1 in 1000

Water (or sewage) at 15°C;
full bore conditions.

velocities in ms^{-1}
discharges in litres/sec

Gradient	(Equivalent) Pipe diameters in mm													
	20	**25**	**30**	**40**	**50**	**60**	**65**	**70**	**75**	**80**	**90**	**100**	**125**	**150**
0·000300				0·063	0·075	0·086	0·091	0·096	0·101	0·106	0·115	0·124	0·145	0·164
1/ 3333				0·0794	0·1470	0·2424	0·3016	0·3693	0·4456	0·5311	0·7313	0·9728	1·7771	2·9024
0·000320				0·066	0·078	0·089	0·094	0·099	0·104	0·109	0·119	0·128	0·150	0·170
1/ 3125				0·0823	0·1523	0·2510	0·3124	0·3824	0·4614	0·5500	0·7571	1·0070	1·8392	3·0032
0·000340				0·068	0·080	0·092	0·097	0·103	0·108	0·113	0·123	0·132	0·155	0·175
1/ 2941				0·0851	0·1575	0·2595	0·3229	0·3952	0·4768	0·5682	0·7821	1·0402	1·8994	3·1010
0·000360				0·070	0·083	0·095	0·100	0·106	0·111	0·117	0·127	0·137	0·160	0·181
1/ 2778				0·0879	0·1625	0·2677	0·3330	0·4076	0·4917	0·5860	0·8065	1·0725	1·9578	3·1960
0·000380				0·072	0·085	0·097	0·103	0·109	0·115	0·120	0·130	0·141	0·164	0·186
1/ 2632				0·0906	0·1673	0·2757	0·3429	0·4197	0·5063	0·6033	0·8301	1·1039	2·0148	3·2884
0·000400				0·074	0·088	0·100	0·106	0·112	0·118	0·123	0·134	0·144	0·169	0·191
1/ 2500				0·0932	0·1721	0·2834	0·3526	0·4314	0·5204	0·6201	0·8532	1·1344	2·0702	3·3785
0·000420				0·076	0·090	0·103	0·109	0·115	0·121	0·127	0·138	0·148	0·173	0·196
1/ 2381				0·0957	0·1767	0·2910	0·3620	0·4429	0·5343	0·6365	0·8758	1·1643	2·1243	3·4664
0·000440				0·078	0·092	0·106	0·112	0·118	0·124	0·130	0·141	0·152	0·177	0·201
1/ 2273				0·0982	0·1813	0·2985	0·3712	0·4541	0·5478	0·6526	0·8978	1·1935	2·1772	3·5522
0·000460				0·080	0·095	0·108	0·115	0·121	0·127	0·133	0·145	0·156	0·182	0·206
1/ 2174				0·1006	0·1857	0·3057	0·3802	0·4651	0·5610	0·6683	0·9193	1·2220	2·2289	3·6361
0·000480				0·082	0·097	0·111	0·117	0·124	0·130	0·136	0·148	0·159	0·186	0·210
1/ 2083				0·1030	0·1901	0·3128	0·3890	0·4759	0·5739	0·6837	0·9404	1·2499	2·2795	3·7183
0·000500				0·084	0·099	0·113	0·120	0·126	0·133	0·139	0·151	0·163	0·190	0·215
1/ 2000				0·1053	0·1943	0·3198	0·3977	0·4864	0·5866	0·6988	0·9610	1·2773	2·3291	3·7988
0·000525				0·086	0·102	0·116	0·123	0·130	0·136	0·143	0·155	0·167	0·195	0·221
1/ 1905				0·1082	0·1996	0·3283	0·4082	0·4993	0·6021	0·7172	0·9863	1·3107	2·3897	3·8972
0·000550				0·088	0·104	0·119	0·126	0·133	0·140	0·146	0·159	0·171	0·200	0·226
1/ 1818				0·1110	0·2047	0·3366	0·4186	0·5119	0·6173	0·7352	1·0109	1·3434	2·4489	3·9934
0·000575				0·090	0·107	0·122	0·129	0·136	0·143	0·150	0·163	0·175	0·204	0·231
1/ 1739				0·1137	0·2096	0·3448	0·4287	0·5242	0·6321	0·7528	1·0351	1·3754	2·5068	4·0874
0·000600				0·093	0·109	0·125	0·132	0·139	0·146	0·153	0·166	0·179	0·209	0·237
1/ 1667				0·1164	0·2145	0·3528	0·4385	0·5363	0·6466	0·7700	1·0587	1·4067	2·5635	4·1794
0·000625			0·076	0·095	0·112	0·128	0·135	0·142	0·150	0·157	0·170	0·183	0·213	0·242
1/ 1600			0·0538	0·1190	0·2193	0·3606	0·4482	0·5481	0·6608	0·7869	1·0818	1·4373	2·6191	4·2695
0·000650			0·078	0·097	0·114	0·130	0·138	0·145	0·153	0·160	0·174	0·187	0·218	0·247
1/ 1538			0·0550	0·1216	0·2240	0·3682	0·4577	0·5597	0·6747	0·8035	1·1045	1·4674	2·6735	4·3579
0·000675			0·079	0·099	0·116	0·133	0·141	0·148	0·156	0·163	0·177	0·191	0·222	0·252
1/ 1481			0·0561	0·1241	0·2286	0·3758	0·4671	0·5711	0·6884	0·8198	1·1268	1·4969	2·7269	4·4446
0·000700			0·081	0·101	0·119	0·136	0·144	0·151	0·159	0·166	0·181	0·194	0·226	0·256
1/ 1429			0·0573	0·1266	0·2331	0·3831	0·4762	0·5822	0·7019	0·8357	1·1487	1·5258	2·7794	4·5297
0·000725			0·083	0·103	0·121	0·138	0·146	0·154	0·162	0·169	0·184	0·198	0·231	0·261
1/ 1379			0·0584	0·1290	0·2376	0·3904	0·4852	0·5932	0·7151	0·8514	1·1702	1·5543	2·8309	4·6134
0·000750			0·084	0·105	0·123	0·141	0·149	0·157	0·165	0·172	0·187	0·201	0·235	0·266
1/ 1333			0·0595	0·1314	0·2420	0·3975	0·4941	0·6040	0·7280	0·8668	1·1913	1·5823	2·8816	4·6956
0·000800			0·087	0·108	0·128	0·146	0·154	0·162	0·171	0·178	0·194	0·208	0·243	0·275
1/ 1250			0·0616	0·1361	0·2505	0·4115	0·5113	0·6251	0·7534	0·8969	1·2325	1·6369	2·9805	4·8561
0·000850			0·090	0·112	0·132	0·150	0·159	0·168	0·176	0·184	0·200	0·215	0·251	0·284
1/ 1176			0·0637	0·1406	0·2588	0·4250	0·5281	0·6455	0·7779	0·9261	1·2725	1·6898	3·0764	5·0117
0·000900			0·093	0·115	0·136	0·155	0·164	0·173	0·181	0·190	0·206	0·222	0·258	0·292
1/ 1111			0·0657	0·1450	0·2668	0·4381	0·5444	0·6653	0·8018	0·9545	1·3114	1·7413	3·1696	5·1629
0·000950			0·096	0·119	0·140	0·159	0·169	0·178	0·187	0·195	0·212	0·228	0·266	0·300
1/ 1053			0·0677	0·1493	0·2747	0·4509	0·5602	0·6847	0·8250	0·9821	1·3492	1·7913	3·2602	5·3099
0·001000			0·098	0·122	0·144	0·164	0·173	0·183	0·192	0·201	0·218	0·234	0·273	0·309
1/ 1000			0·0696	0·1535	0·2823	0·4634	0·5756	0·7035	0·8477	1·0090	1·3860	1·8401	3·3485	5·4531
			–	–	–	0·74	0·74	0·74	0·74	0·74	0·75	0·75	0·76	0·76

$V_{r(0{\cdot}5)medial}$ for half-full circular pipes.

k_s = 0·60 mm S = 0·00030 to 0·00100

Gradient **(Equivalent) Pipe diameters in mm**

Gradient	20	25	30	40	50	60	65	70	75	80	90	100	125	150
0·00100			0·098	0·122	0·144	0·164	0·173	0·183	0·192	0·201	0·218	0·234	0·273	0·309
1/ 1000			0·0696	0·1535	0·2823	0·4634	0·5756	0·7035	0·8477	1·0090	1·3860	1·8401	3·3485	5·4531
0·00105			0·101	0·125	0·148	0·168	0·178	0·188	0·197	0·206	0·224	0·240	0·280	0·316
1/ 952			0·0715	0·1576	0·2897	0·4755	0·5907	0·7218	0·8698	1·0353	1·4219	1·8877	3·4346	5·5929
0·00110			0·104	0·129	0·151	0·172	0·182	0·192	0·202	0·211	0·229	0·246	0·287	0·324
1/ 909			0·0733	0·1616	0·2970	0·4874	0·6054	0·7398	0·8913	1·0609	1·4570	1·9342	3·5187	5·7294
0·00115		0·092	0·106	0·132	0·155	0·176	0·187	0·197	0·207	0·216	0·234	0·252	0·293	0·332
1/ 870		0·0454	0·0751	0·1655	0·3041	0·4990	0·6198	0·7573	0·9124	1·0860	1·4913	1·9796	3·6010	5·8628
0·00120		0·095	0·109	0·135	0·158	0·181	0·191	0·201	0·211	0·221	0·240	0·258	0·300	0·339
1/ 833		0·0465	0·0769	0·1693	0·3111	0·5104	0·6339	0·7745	0·9331	1·1105	1·5249	2·0241	3·6815	5·9934
0·00125		0·097	0·111	0·138	0·162	0·184	0·195	0·206	0·216	0·226	0·245	0·263	0·306	0·346
1/ 800		0·0475	0·0786	0·1730	0·3179	0·5215	0·6477	0·7913	0·9533	1·1345	1·5578	2·0676	3·7603	6·1213
0·00130		0·099	0·114	0·141	0·165	0·188	0·199	0·210	0·220	0·230	0·250	0·269	0·313	0·353
1/ 769		0·0485	0·0803	0·1767	0·3246	0·5324	0·6612	0·8078	0·9732	1·1581	1·5901	2·1104	3·8377	6·2467
0·00135		0·101	0·116	0·143	0·169	0·192	0·203	0·214	0·225	0·235	0·255	0·274	0·319	0·360
1/ 741		0·0496	0·0819	0·1803	0·3312	0·5431	0·6745	0·8240	0·9926	1·1812	1·6217	2·1523	3·9135	6·3697
0·00140		0·103	0·118	0·146	0·172	0·196	0·207	0·218	0·229	0·240	0·260	0·279	0·325	0·367
1/ 714		0·0505	0·0836	0·1839	0·3377	0·5537	0·6875	0·8399	1·0117	1·2039	1·6528	2·1934	3·9880	6·4905
0·00145		0·105	0·120	0·149	0·175	0·199	0·211	0·222	0·233	0·244	0·265	0·284	0·331	0·374
1/ 690		0·0515	0·0852	0·1873	0·3440	0·5640	0·7003	0·8555	1·0305	1·2262	1·6834	2·2338	4·0611	6·6092
0·00150		0·107	0·123	0·152	0·178	0·203	0·215	0·226	0·237	0·248	0·269	0·289	0·337	0·381
1/ 667		0·0525	0·0867	0·1908	0·3502	0·5742	0·7129	0·8709	1·0490	1·2481	1·7134	2·2736	4·1330	6·7258
0·00160		0·111	0·127	0·157	0·185	0·210	0·222	0·234	0·246	0·257	0·279	0·299	0·348	0·393
1/ 625		0·0543	0·0898	0·1974	0·3624	0·5940	0·7375	0·9008	1·0850	1·2909	1·7720	2·3511	4·2734	6·9535
0·00170		0·114	0·131	0·162	0·191	0·217	0·229	0·242	0·253	0·265	0·287	0·309	0·359	0·406
1/ 588		0·0562	0·0928	0·2039	0·3742	0·6132	0·7613	0·9299	1·1199	1·3324	1·8288	2·4263	4·4095	7·1741
0·00180		0·118	0·135	0·167	0·196	0·223	0·236	0·249	0·261	0·273	0·296	0·318	0·370	0·418
1/ 556		0·0579	0·0957	0·2102	0·3857	0·6319	0·7845	0·9581	1·1538	1·3727	1·8839	2·4994	4·5416	7·3884
0·00190		0·122	0·139	0·172	0·202	0·230	0·243	0·256	0·269	0·281	0·305	0·327	0·381	0·430
1/ 526		0·0596	0·0985	0·2163	0·3968	0·6501	0·8070	0·9856	1·1869	1·4120	1·9376	2·5704	4·6702	7·5969
0·00200		0·125	0·143	0·177	0·208	0·236	0·250	0·263	0·276	0·288	0·313	0·336	0·391	0·441
1/ 500		0·0613	0·1012	0·2223	0·4077	0·6678	0·8289	1·0123	1·2190	1·4502	1·9899	2·6396	4·7954	7·8000
0·00210		0·128	0·147	0·182	0·213	0·242	0·256	0·270	0·283	0·296	0·321	0·345	0·401	0·453
1/ 476		0·0630	0·1039	0·2281	0·4183	0·6851	0·8503	1·0384	1·2504	1·4874	2·0409	2·7072	4·9175	7·9981
0·00220		0·131	0·151	0·186	0·218	0·248	0·263	0·276	0·290	0·303	0·329	0·353	0·410	0·464
1/ 455		0·0645	0·1065	0·2338	0·4286	0·7020	0·8712	1·0639	1·2810	1·5238	2·0907	2·7731	5·0368	8·1915
0·00230		0·135	0·154	0·190	0·223	0·254	0·269	0·283	0·297	0·310	0·336	0·361	0·420	0·474
1/ 435		0·0661	0·1091	0·2394	0·4388	0·7185	0·8917	1·0888	1·3110	1·5594	2·1394	2·8376	5·1535	8·3806
0·00240	0·116	0·138	0·158	0·195	0·229	0·260	0·275	0·289	0·303	0·317	0·344	0·369	0·429	0·485
1/ 417	0·0365	0·0676	0·1116	0·2448	0·4487	0·7346	0·9117	1·1132	1·3403	1·5942	2·1871	2·9007	5·2676	8·5657
0·00250	0·119	0·141	0·161	0·199	0·233	0·265	0·281	0·295	0·310	0·324	0·351	0·377	0·438	0·495
1/ 400	0·0373	0·0691	0·1141	0·2501	0·4584	0·7504	0·9313	1·1371	1·3690	1·6284	2·2338	2·9625	5·3794	8·7470
0·00260	0·121	0·144	0·165	0·203	0·238	0·271	0·286	0·302	0·316	0·331	0·358	0·385	0·447	0·505
1/ 385	0·0381	0·0706	0·1165	0·2554	0·4679	0·7660	0·9505	1·1605	1·3972	1·6618	2·2796	3·0230	5·4890	8·9247
0·00270	0·124	0·147	0·168	0·207	0·243	0·276	0·292	0·308	0·323	0·337	0·365	0·392	0·456	0·515
1/ 370	0·0389	0·0720	0·1188	0·2605	0·4772	0·7812	0·9694	1·1835	1·4248	1·6946	2·3245	3·0825	5·5965	9·0990
0·00280	0·126	0·150	0·171	0·211	0·248	0·282	0·298	0·313	0·329	0·344	0·372	0·400	0·465	0·525
1/ 357	0·0397	0·0735	0·1211	0·2655	0·4864	0·7961	0·9879	1·2060	1·4519	1·7268	2·3685	3·1408	5·7020	9·2701
0·00290	0·129	0·152	0·175	0·215	0·252	0·287	0·303	0·319	0·335	0·350	0·379	0·407	0·473	0·534
1/ 345	0·0405	0·0749	0·1234	0·2705	0·4954	0·8108	1·0061	1·2282	1·4786	1·7585	2·4119	3·1981	5·8057	9·4383
0·00300	0·131	0·155	0·178	0·219	0·257	0·292	0·309	0·325	0·341	0·356	0·386	0·414	0·481	0·543
1/ 333	0·0412	0·0762	0·1257	0·2753	0·5043	0·8252	1·0239	1·2500	1·5048	1·7896	2·4544	3·2545	5·9076	9·6035
	–	–	–	–	0·74	0·74	0·74	0·75	0·75	0·75	0·75	0·76	0·76	0·77

$V_{r(0·5)medial}$ for half-full circular pipes.

k_s = 0·60 mm
S = 0·00300 to 0·01000

ie hydraulic gradient =
1 in 333 to 1 in 100

Water (or sewage) at 15°C;
full bore conditions.

velocities in ms^{-1}
discharges in litres/sec

Gradient	**(Equivalent) Pipe diameters in mm**													
	20	**25**	**30**	**40**	**50**	**60**	**65**	**70**	**75**	**80**	**90**	**100**	**125**	**150**
0·00300	0·131	0·155	0·178	0·219	0·257	0·292	0·309	0·325	0·341	0·356	0·386	0·414	0·481	0·543
1/ 333	0·0412	0·0762	0·1257	0·2753	0·5043	0·8252	1·0239	1·2500	1·5048	1·7896	2·4544	3·2545	5·9076	9·6035
0·00320	0·136	0·161	0·184	0·227	0·266	0·302	0·319	0·336	0·352	0·368	0·399	0·428	0·498	0·562
1/ 313	0·0427	0·0789	0·1300	0·2848	0·5216	0·8534	1·0588	1·2925	1·5559	1·8503	2·5375	3·3644	6·1065	9·9260
0·00340	0·140	0·166	0·190	0·234	0·274	0·311	0·329	0·347	0·363	0·380	0·412	0·442	0·513	0·579
1/ 294	0·0441	0·0815	0·1343	0·2940	0·5383	0·8807	1·0926	1·3337	1·6054	1·9091	2·6181	3·4710	6·2993	10·239
0·00360	0·145	0·171	0·196	0·241	0·282	0·321	0·339	0·357	0·374	0·391	0·424	0·455	0·529	0·597
1/ 278	0·0454	0·0840	0·1384	0·3030	0·5546	0·9072	1·1255	1·3738	1·6536	1·9663	2·6963	3·5746	6·4865	10·542
0·00380	0·149	0·176	0·201	0·248	0·291	0·330	0·349	0·367	0·385	0·402	0·436	0·468	0·543	0·613
1/ 263	0·0468	0·0864	0·1424	0·3117	0·5704	0·9330	1·1574	1·4127	1·7004	2·0219	2·7724	3·6753	6·6686	10·837
0·00400	0·153	0·181	0·207	0·255	0·298	0·339	0·358	0·377	0·395	0·413	0·447	0·480	0·558	0·630
1/ 250	0·0481	0·0888	0·1463	0·3202	0·5859	0·9582	1·1886	1·4506	1·7460	2·0761	2·8465	3·7733	6·8460	11·125
0·00420	0·157	0·186	0·212	0·261	0·306	0·348	0·367	0·387	0·405	0·424	0·459	0·493	0·572	0·645
1/ 238	0·0493	0·0912	0·1501	0·3284	0·6009	0·9827	1·2189	1·4877	1·7905	2·1289	2·9188	3·8690	7·0190	11·405
0·00440	0·161	0·190	0·218	0·268	0·314	0·356	0·376	0·396	0·415	0·434	0·470	0·505	0·586	0·661
1/ 227	0·0506	0·0934	0·1538	0·3365	0·6156	1·0066	1·2486	1·5238	1·8339	2·1805	2·9894	3·9624	7·1880	11·679
0·00460	0·165	0·195	0·223	0·274	0·321	0·364	0·385	0·405	0·425	0·444	0·481	0·516	0·599	0·676
1/ 217	0·0518	0·0957	0·1575	0·3444	0·6300	1·0300	1·2776	1·5591	1·8764	2·2309	3·0584	4·0538	7·3532	11·947
0·00480	0·169	0·199	0·228	0·280	0·328	0·372	0·394	0·414	0·434	0·454	0·491	0·528	0·612	0·691
1/ 208	0·0530	0·0978	0·1610	0·3521	0·6441	1·0529	1·3059	1·5937	1·9179	2·2803	3·1259	4·1432	7·5148	12·209
0·00500	0·172	0·204	0·233	0·286	0·335	0·380	0·402	0·423	0·443	0·463	0·502	0·539	0·625	0·705
1/ 200	0·0542	0·1000	0·1645	0·3597	0·6578	1·0754	1·3337	1·6276	1·9586	2·3286	3·1921	4·2307	7·6731	12·466
0·00525	0·177	0·209	0·239	0·294	0·344	0·390	0·412	0·434	0·455	0·475	0·514	0·552	0·641	0·723
1/ 190	0·0556	0·1026	0·1688	0·3690	0·6747	1·1028	1·3677	1·6690	2·0084	2·3877	3·2729	4·3377	7·8667	12·780
0·00550	0·181	0·214	0·245	0·301	0·352	0·400	0·422	0·444	0·466	0·487	0·527	0·566	0·656	0·741
1/ 182	0·0570	0·1051	0·1730	0·3780	0·6911	1·1296	1·4009	1·7094	2·0570	2·4455	3·3519	4·4422	8·0556	13·086
0·00575	0·186	0·219	0·250	0·308	0·360	0·409	0·432	0·454	0·476	0·498	0·539	0·579	0·671	0·757
1/ 174	0·0583	0·1076	0·1770	0·3868	0·7072	1·1558	1·4333	1·7489	2·1045	2·5019	3·4291	4·5444	8·2404	13·385
0·00600	0·190	0·224	0·256	0·315	0·368	0·418	0·442	0·464	0·487	0·509	0·551	0·591	0·686	0·774
1/ 167	0·0597	0·1100	0·1810	0·3955	0·7229	1·1814	1·4651	1·7876	2·1510	2·5571	3·5047	4·6444	8·4211	13·678
0·00625	0·194	0·229	0·262	0·321	0·376	0·427	0·451	0·474	0·497	0·519	0·563	0·604	0·701	0·790
1/ 160	0·0610	0·1124	0·1849	0·4040	0·7384	1·2065	1·4961	1·8255	2·1965	2·6112	3·5787	4·7423	8·5982	13·965
0·00650	0·198	0·234	0·267	0·328	0·384	0·435	0·460	0·484	0·507	0·530	0·574	0·616	0·715	0·806
1/ 154	0·0622	0·1148	0·1888	0·4123	0·7535	1·2311	1·5266	1·8626	2·2412	2·6642	3·6512	4·8383	8·7717	14·247
0·00675	0·202	0·238	0·272	0·335	0·391	0·444	0·469	0·493	0·517	0·540	0·585	0·628	0·729	0·822
1/ 148	0·0635	0·1171	0·1925	0·4204	0·7683	1·2553	1·5565	1·8991	2·2850	2·7162	3·7223	4·9324	8·9420	14·523
0·00700	0·206	0·243	0·278	0·341	0·399	0·452	0·478	0·503	0·527	0·551	0·596	0·640	0·742	0·837
1/ 143	0·0647	0·1193	0·1962	0·4284	0·7828	1·2790	1·5859	1·9348	2·3279	2·7672	3·7922	5·0248	9·1091	14·794
0·00725	0·210	0·248	0·283	0·347	0·406	0·461	0·487	0·512	0·536	0·561	0·607	0·651	0·756	0·852
1/ 138	0·0659	0·1215	0·1998	0·4363	0·7971	1·3023	1·6147	1·9700	2·3702	2·8174	3·8608	5·1156	9·2733	15·060
0·00750	0·214	0·252	0·288	0·353	0·413	0·469	0·495	0·521	0·546	0·570	0·617	0·663	0·769	0·867
1/ 133	0·0671	0·1237	0·2034	0·4440	0·8112	1·3252	1·6430	2·0045	2·4117	2·8667	3·9283	5·2049	9·4347	15·321
0·00800	0·221	0·261	0·298	0·365	0·427	0·484	0·512	0·538	0·564	0·589	0·638	0·685	0·794	0·896
1/ 125	0·0694	0·1280	0·2103	0·4591	0·8386	1·3698	1·6983	2·0719	2·4927	2·9629	4·0599	5·3791	9·7496	15·832
0·00850	0·228	0·269	0·307	0·377	0·441	0·500	0·528	0·555	0·582	0·608	0·658	0·706	0·819	0·924
1/ 118	0·0717	0·1321	0·2171	0·4737	0·8652	1·4131	1·7519	2·1372	2·5712	3·0561	4·1875	5·5479	10·055	16·327
0·00900	0·235	0·277	0·316	0·388	0·454	0·515	0·544	0·572	0·599	0·626	0·678	0·727	0·843	0·951
1/ 111	0·0739	0·1361	0·2236	0·4879	0·8910	1·4551	1·8040	2·2007	2·6475	3·1467	4·3113	5·7118	10·351	16·807
0·00950	0·242	0·285	0·325	0·399	0·467	0·529	0·559	0·588	0·616	0·644	0·697	0·748	0·867	0·978
1/ 105	0·0760	0·1400	0·2300	0·5017	0·9161	1·4960	1·8546	2·2624	2·7216	3·2348	4·4318	5·8712	10·639	17·274
0·01000	0·249	0·293	0·334	0·410	0·479	0·543	0·574	0·603	0·632	0·661	0·715	0·767	0·890	1·003
1/ 100	0·0781	0·1438	0·2362	0·5151	0·9406	1·5359	1·9040	2·3225	2·7939	3·3206	4·5492	6·0265	10·920	17·729
	–	–	–	0·74	0·74	0·74	0·75	0·75	0·75	0·75	0·76	0·76	0·77	0·77

$V_{r(0·5)medial}$ for half-full circular pipes.

Gradient	20	25	30	40	50	60	65	70	75	80	90	100	125	150
	(Equivalent) Pipe diameters in mm													
0·01000	0·249	0·293	0·334	0·410	0·479	0·543	0·574	0·603	0·632	0·661	0·715	0·767	0·890	1·003
1/ 100	0·0781	0·1438	0·2362	0·5151	0·9406	1·5359	1·9040	2·3225	2·7939	3·3206	4·5492	6·0265	10·920	17·729
0·01050	0·255	0·300	0·343	0·420	0·491	0·557	0·588	0·619	0·648	0·677	0·733	0·787	0·912	1·028
1/ 95	0·0801	0·1475	0·2423	0·5283	0·9645	1·5747	1·9521	2·3811	2·8643	3·4042	4·6637	6·1780	11·194	18·173
0·01100	0·261	0·308	0·351	0·431	0·503	0·570	0·602	0·634	0·664	0·694	0·751	0·805	0·934	1·053
1/ 91	0·0821	0·1511	0·2482	0·5411	0·9878	1·6127	1·9991	2·4383	2·9331	3·4859	4·7755	6·3259	11·462	18·607
0·01150	0·267	0·315	0·359	0·441	0·515	0·583	0·616	0·648	0·679	0·709	0·768	0·824	0·955	1·077
1/ 87	0·0840	0·1546	0·2539	0·5536	1·0106	1·6498	2·0450	2·4943	3·0004	3·5658	4·8848	6·4705	11·723	19·031
0·01200	0·273	0·322	0·367	0·450	0·526	0·596	0·630	0·662	0·694	0·725	0·785	0·842	0·976	1·100
1/ 83	0·0859	0·1581	0·2596	0·5659	1·0328	1·6861	2·0899	2·5491	3·0662	3·6440	4·9917	6·6120	11·979	19·445
0·01250	0·279	0·329	0·375	0·460	0·537	0·609	0·643	0·676	0·709	0·740	0·801	0·860	0·997	1·123
1/ 80	0·0877	0·1615	0·2651	0·5779	1·0547	1·7216	2·1339	2·6027	3·1307	3·7205	5·0964	6·7506	12·229	19·851
0·01300	0·285	0·336	0·383	0·469	0·548	0·621	0·656	0·690	0·723	0·755	0·817	0·877	1·017	1·146
1/ 77	0·0896	0·1648	0·2706	0·5896	1·0761	1·7565	2·1771	2·6553	3·1939	3·7956	5·1991	6·8865	12·475	20·249
0·01350	0·291	0·342	0·390	0·478	0·559	0·633	0·669	0·703	0·737	0·770	0·833	0·894	1·036	1·168
1/ 74	0·0913	0·1680	0·2759	0·6012	1·0971	1·7906	2·2194	2·7069	3·2559	3·8692	5·2998	7·0197	12·716	20·640
0·01400	0·296	0·349	0·398	0·487	0·569	0·645	0·681	0·717	0·751	0·784	0·849	0·910	1·055	1·190
1/ 71	0·0931	0·1712	0·2811	0·6125	1·1177	1·8242	2·2610	2·7575	3·3167	3·9415	5·3987	7·1505	12·952	21·023
0·01450	0·302	0·355	0·405	0·496	0·580	0·657	0·694	0·729	0·764	0·798	0·864	0·927	1·074	1·211
1/ 69	0·0948	0·1744	0·2863	0·6237	1·1379	1·8572	2·3018	2·8072	3·3765	4·0125	5·4958	7·2790	13·185	21·400
0·01500	0·307	0·362	0·412	0·505	0·590	0·668	0·706	0·742	0·778	0·812	0·879	0·943	1·093	1·232
1/ 67	0·0965	0·1775	0·2913	0·6346	1·1578	1·8896	2·3419	2·8561	3·4353	4·0822	5·5912	7·4053	13·413	21·770
0·01600	0·318	0·374	0·426	0·522	0·609	0·691	0·729	0·767	0·804	0·839	0·908	0·974	1·129	1·273
1/ 63	0·0998	0·1835	0·3012	0·6560	1·1967	1·9528	2·4202	2·9515	3·5499	4·2184	5·7775	7·6518	13·859	22·492
0·01700	0·328	0·386	0·440	0·538	0·629	0·712	0·752	0·791	0·829	0·865	0·937	1·005	1·164	1·312
1/ 59	0·1030	0·1893	0·3107	0·6767	1·2343	2·0141	2·4961	3·0440	3·6610	4·3504	5·9580	7·8907	14·290	23·192
0·01800	0·338	0·397	0·453	0·554	0·647	0·733	0·774	0·814	0·853	0·891	0·964	1·034	1·199	1·351
1/ 56	0·1061	0·1950	0·3200	0·6968	1·2709	2·0736	2·5698	3·1338	3·7690	4·4785	6·1333	8·1226	14·710	23·872
0·01900	0·347	0·408	0·465	0·570	0·665	0·754	0·796	0·837	0·877	0·916	0·991	1·063	1·232	1·388
1/ 53	0·1091	0·2005	0·3290	0·7163	1·3064	2·1315	2·6415	3·2211	3·8739	4·6032	6·3039	8·3482	15·118	24·533
0·02000	0·357	0·419	0·478	0·585	0·683	0·774	0·817	0·859	0·900	0·940	1·017	1·091	1·264	1·425
1/ 50	0·1120	0·2059	0·3378	0·7354	1·3411	2·1879	2·7113	3·3062	3·9762	4·7246	6·4699	8·5680	15·515	25·177
0·02100	0·366	0·430	0·490	0·600	0·700	0·793	0·838	0·881	0·923	0·963	1·042	1·118	1·296	1·460
1/ 47·6	0·1149	0·2111	0·3464	0·7539	1·3749	2·2429	2·7794	3·3892	4·0759	4·8430	6·6319	8·7823	15·902	25·805
0·02200	0·375	0·441	0·502	0·614	0·717	0·812	0·858	0·902	0·945	0·986	1·067	1·145	1·327	1·495
1/ 45·5	0·1177	0·2162	0·3547	0·7721	1·4078	2·2966	2·8458	3·4702	4·1732	4·9586	6·7901	8·9916	16·281	26·418
0·02300	0·383	0·451	0·513	0·629	0·733	0·831	0·877	0·922	0·966	1·009	1·092	1·171	1·357	1·529
1/ 43·5	0·1204	0·2213	0·3629	0·7898	1·4401	2·3491	2·9108	3·5494	4·2684	5·0716	6·9447	9·1962	16·651	27·017
0·02400	0·392	0·461	0·525	0·642	0·750	0·849	0·896	0·942	0·987	1·031	1·115	1·196	1·386	1·562
1/ 41·7	0·1231	0·2262	0·3709	0·8072	1·4717	2·4004	2·9744	3·6269	4·3616	5·1822	7·0960	9·3964	17·012	27·604
0·02500	0·400	0·470	0·536	0·656	0·765	0·867	0·915	0·962	1·008	1·053	1·139	1·221	1·415	1·595
1/ 40·0	0·1257	0·2310	0·3788	0·8242	1·5026	2·4507	3·0367	3·7028	4·4528	5·2906	7·2442	9·5925	17·367	28·178
0·02600	0·408	0·480	0·547	0·669	0·781	0·884	0·934	0·981	1·028	1·074	1·162	1·246	1·443	1·626
1/ 38·5	0·1283	0·2357	0·3865	0·8408	1·5329	2·5001	3·0978	3·7772	4·5422	5·3967	7·3895	9·7846	17·714	28·742
0·02700	0·416	0·489	0·557	0·682	0·796	0·901	0·952	1·000	1·048	1·094	1·184	1·270	1·471	1·658
1/ 37·0	0·1308	0·2403	0·3940	0·8572	1·5626	2·5484	3·1577	3·8502	4·6299	5·5009	7·5320	9·9732	18·055	29·294
0·02800	0·424	0·499	0·568	0·695	0·811	0·918	0·969	1·019	1·067	1·115	1·206	1·293	1·499	1·688
1/ 35·7	0·1333	0·2448	0·4014	0·8732	1·5917	2·5959	3·2165	3·9218	4·7160	5·6031	7·6718	10·158	18·390	29·836
0·02900	0·432	0·508	0·578	0·707	0·825	0·935	0·987	1·037	1·087	1·135	1·228	1·317	1·525	1·719
1/ 34·5	0·1357	0·2492	0·4087	0·8890	1·6204	2·6426	3·2743	3·9922	4·8006	5·7036	7·8092	10·340	18·718	30·369
0·03000	0·440	0·517	0·588	0·720	0·840	0·951	1·004	1·055	1·105	1·154	1·249	1·339	1·552	1·748
1/ 33·3	0·1381	0·2536	0·4159	0·9045	1·6486	2·6884	3·3310	4·0614	4·8837	5·8023	7·9443	10·519	19·041	30·892
$V_{r(0{\cdot}5)medial}$	–	–	–	0·74	0·74	0·75	0·75	0·75	0·75	0·75	0·76	0·76	0·77	0·77

$V_{r(0{\cdot}5)medial}$ for half-full circular pipes.

k_s = 0·60 mm
S = 0·03000 to 0·10000

ie hydraulic gradient =
1 in 33·3 to 1 in 10·0

Water (or sewage) at 15°C;
full bore conditions.

velocities in ms^{-1}
discharges in litres/sec

Gradient **(Equivalent) Pipe diameters in mm**

Gradient	20	25	30	40	50	60	65	70	75	80	90	100	125	150
0·03000	0·440	0·517	0·588	0·720	0·840	0·951	1·004	1·055	1·105	1·154	1·249	1·339	1·552	1·748
1/ 33·3	0·1381	0·2536	0·4159	0·9045	1·6486	2·6884	3·3310	4·0614	4·8837	5·8023	7·9443	10·519	19·041	30·892
0·03200	0·455	0·534	0·608	0·744	0·868	0·982	1·037	1·090	1·142	1·193	1·290	1·384	1·603	1·806
1/ 31·3	0·1428	0·2622	0·4298	0·9347	1·7035	2·7779	3·4418	4·1964	5·0460	5·9950	8·2078	10·867	19·672	31·914
0·03400	0·469	0·551	0·627	0·767	0·895	1·013	1·070	1·124	1·178	1·230	1·330	1·427	1·653	1·862
1/ 29·4	0·1473	0·2704	0·4433	0·9640	1·7568	2·8646	3·5492	4·3272	5·2032	6·1817	8·4632	11·205	20·283	32·904
0·03600	0·483	0·567	0·646	0·790	0·921	1·043	1·101	1·157	1·212	1·266	1·369	1·468	1·701	1·916
1/ 27·8	0·1517	0·2785	0·4565	0·9925	1·8085	2·9488	3·6534	4·4542	5·3559	6·3630	8·7112	11·533	20·876	33·865
0·03800	0·497	0·583	0·664	0·812	0·947	1·072	1·132	1·190	1·246	1·301	1·407	1·509	1·748	1·969
1/ 26·3	0·1560	0·2863	0·4692	1·0202	1·8589	3·0307	3·7549	4·5778	5·5044	6·5393	8·9524	11·853	21·453	34·801
0·04000	0·510	0·599	0·681	0·833	0·972	1·100	1·161	1·221	1·279	1·335	1·444	1·549	1·794	2·021
1/ 25·0	0·1602	0·2939	0·4817	1·0471	1·9079	3·1105	3·8536	4·6981	5·6490	6·7110	9·1874	12·163	22·015	35·711
0·04200	0·523	0·614	0·699	0·854	0·996	1·128	1·190	1·251	1·311	1·368	1·480	1·587	1·839	2·071
1/ 23·8	0·1642	0·3013	0·4938	1·0734	1·9557	3·1883	3·9500	4·8155	5·7901	6·8786	9·4165	12·467	22·563	36·600
0·04400	0·535	0·629	0·715	0·875	1·020	1·155	1·219	1·281	1·342	1·401	1·515	1·625	1·882	2·120
1/ 22·7	0·1682	0·3086	0·5057	1·0991	2·0023	3·2643	4·0440	4·9301	5·9278	7·0421	9·6403	12·763	23·098	37·467
0·04600	0·548	0·643	0·732	0·895	1·043	1·181	1·246	1·310	1·372	1·433	1·550	1·662	1·925	2·168
1/ 21·7	0·1720	0·3156	0·5173	1·1242	2·0480	3·3386	4·1360	5·0422	6·0625	7·2020	9·8590	13·052	23·621	38·315
0·04800	0·560	0·657	0·748	0·914	1·066	1·206	1·274	1·339	1·402	1·464	1·583	1·698	1·967	2·215
1/ 20·8	0·1758	0·3226	0·5286	1·1487	2·0926	3·4113	4·2260	5·1518	6·1942	7·3585	10·073	13·335	24·133	39·144
0·05000	0·572	0·671	0·764	0·933	1·088	1·232	1·300	1·367	1·431	1·494	1·616	1·733	2·007	2·261
1/ 20·0	0·1796	0·3294	0·5397	1·1728	2·1363	3·4824	4·3141	5·2592	6·3233	7·5117	10·283	13·613	24·634	39·957
0·05250	0·586	0·688	0·783	0·957	1·115	1·262	1·333	1·401	1·467	1·532	1·657	1·776	2·057	2·317
1/ 19·0	0·1841	0·3377	0·5533	1·2022	2·1898	3·5694	4·4218	5·3904	6·4810	7·6990	10·539	13·952	25·247	40·950
0·05500	0·600	0·704	0·801	0·980	1·142	1·292	1·364	1·434	1·502	1·568	1·696	1·819	2·106	2·372
1/ 18·2	0·1885	0·3458	0·5665	1·2309	2·2420	3·6544	4·5270	5·5186	6·6350	7·8819	10·789	14·283	25·845	41·920
0·05750	0·614	0·721	0·820	1·002	1·168	1·322	1·395	1·467	1·536	1·604	1·734	1·860	2·154	2·426
1/ 17·4	0·1929	0·3537	0·5795	1·2590	2·2930	3·7374	4·6298	5·6439	6·7856	8·0607	11·034	14·606	26·430	42·868
0·06000	0·627	0·736	0·838	1·024	1·193	1·351	1·426	1·498	1·569	1·638	1·772	1·900	2·200	2·478
1/ 16·7	0·1971	0·3615	0·5921	1·2864	2·3429	3·8187	4·7304	5·7665	6·9329	8·2357	11·273	14·923	27·003	43·796
0·06250	0·641	0·752	0·855	1·045	1·218	1·379	1·455	1·530	1·602	1·673	1·809	1·940	2·246	2·530
1/ 16·0	0·2013	0·3690	0·6046	1·3133	2·3918	3·8983	4·8290	5·8865	7·0772	8·4070	11·507	15·233	27·563	44·704
0·06500	0·654	0·767	0·872	1·066	1·243	1·406	1·484	1·560	1·634	1·706	1·845	1·978	2·291	2·580
1/ 15·4	0·2053	0·3765	0·6167	1·3397	2·4397	3·9763	4·9255	6·0042	7·2186	8·5749	11·737	15·537	28·113	45·595
0·06750	0·666	0·782	0·889	1·087	1·266	1·433	1·513	1·590	1·665	1·739	1·880	2·016	2·335	2·630
1/ 14·8	0·2093	0·3838	0·6287	1·3656	2·4868	4·0528	5·0203	6·1196	7·3574	8·7397	11·962	15·835	28·652	46·468
0·07000	0·679	0·796	0·906	1·107	1·290	1·460	1·541	1·620	1·696	1·771	1·915	2·053	2·378	2·678
1/ 14·3	0·2132	0·3910	0·6404	1·3909	2·5329	4·1279	5·1133	6·2330	7·4935	8·9014	12·184	16·128	29·181	47·326
0·07250	0·691	0·811	0·922	1·127	1·313	1·486	1·568	1·649	1·726	1·802	1·949	2·090	2·420	2·726
1/ 13·8	0·2171	0·3980	0·6519	1·4159	2·5782	4·2017	5·2046	6·3443	7·6273	9·0602	12·401	16·415	29·700	48·168
0·07500	0·703	0·825	0·938	1·146	1·336	1·512	1·596	1·677	1·756	1·834	1·983	2·126	2·462	2·773
1/ 13·3	0·2209	0·4049	0·6632	1·4404	2·6228	4·2743	5·2944	6·4537	7·7588	9·2164	12·615	16·698	30·211	48·996
0·08000	0·727	0·852	0·969	1·184	1·380	1·562	1·648	1·732	1·814	1·894	2·048	2·196	2·543	2·864
1/ 12·5	0·2283	0·4184	0·6853	1·4882	2·7098	4·4158	5·4697	6·6672	8·0154	9·5211	13·031	17·249	31·208	50·611
0·08500	0·749	0·879	1·000	1·221	1·423	1·610	1·700	1·786	1·871	1·953	2·112	2·264	2·622	2·953
1/ 11·8	0·2354	0·4315	0·7067	1·5346	2·7940	4·5529	5·6395	6·8741	8·2641	9·8164	13·435	17·784	32·174	52·177
0·09000	0·772	0·905	1·029	1·257	1·465	1·657	1·749	1·838	1·925	2·010	2·174	2·330	2·698	3·039
1/ 11·1	0·2424	0·4442	0·7275	1·5796	2·8759	4·6861	5·8044	7·0750	8·5055	10·103	13·827	18·303	33·112	53·697
0·09500	0·793	0·930	1·058	1·292	1·505	1·703	1·798	1·889	1·978	2·065	2·234	2·395	2·773	3·122
1/ 10·5	0·2492	0·4566	0·7477	1·6234	2·9554	4·8157	5·9648	7·2705	8·7404	10·382	14·209	18·807	34·024	55·176
0·10000	0·814	0·955	1·086	1·326	1·545	1·748	1·845	1·939	2·030	2·119	2·292	2·457	2·845	3·204
1/ 10·0	0·2557	0·4686	0·7674	1·6660	3·0330	4·9419	6·1210	7·4608	8·9691	10·654	14·580	19·299	34·913	56·616
	–	–	–	0·74	0·74	0·75	0·75	0·75	0·75	0·75	0·76	0·76	0·77	0·78

$V_{r(0\cdot5)medial}$ for half-full circular pipes.

Gradient	**(Equivalent) Pipe diameters in mm**													
	20	**25**	**30**	**40**	**50**	**60**	**65**	**70**	**75**	**80**	**90**	**100**	**125**	**150**
0·10000	0·814	0·955	1·086	1·326	1·545	1·748	1·845	1·939	2·030	2·119	2·292	2·457	2·845	3·204
1/ 10·0	0·2557	0·4686	0·7674	1·6660	3·0330	4·9419	6·1210	7·4608	8·9691	10·654	14·580	19·299	34·913	56·616
0·10500	0·834	0·979	1·113	1·359	1·583	1·791	1·891	1·987	2·081	2·172	2·349	2·518	2·916	3·283
1/ 9·5	0·2622	0·4804	0·7866	1·7076	3·1086	5·0649	6·2734	7·6464	9·1922	10·919	14·943	19·778	35·780	58·021
0·11000	0·854	1·002	1·139	1·391	1·621	1·834	1·935	2·034	2·130	2·224	2·404	2·578	2·985	3·361
1/ 9·1	0·2684	0·4919	0·8053	1·7482	3·1824	5·1851	6·4221	7·8277	9·4101	11·177	15·297	20·247	36·626	59·392
0·11500	0·874	1·025	1·165	1·423	1·658	1·875	1·979	2·080	2·178	2·274	2·459	2·636	3·052	3·437
1/ 8·7	0·2746	0·5031	0·8237	1·7879	3·2546	5·3025	6·5676	8·0049	9·6230	11·430	15·643	20·704	37·453	60·733
0·12000	0·893	1·047	1·191	1·454	1·693	1·916	2·022	2·125	2·225	2·323	2·512	2·693	3·118	3·511
1/ 8·3	0·2806	0·5140	0·8416	1·8268	3·3252	5·4175	6·7099	8·1783	9·8314	11·677	15·981	21·152	38·263	62·045
0·12500	0·912	1·069	1·215	1·484	1·729	1·956	2·064	2·169	2·272	2·371	2·564	2·749	3·183	3·584
1/ 8·0	0·2864	0·5248	0·8592	1·8648	3·3943	5·5301	6·8492	8·3481	10·035	11·920	16·313	21·591	39·055	63·330
0·13000	0·930	1·091	1·240	1·514	1·763	1·995	2·105	2·212	2·317	2·419	2·615	2·804	3·246	3·655
1/ 7·7	0·2922	0·5353	0·8764	1·9021	3·4621	5·6404	6·9858	8·5146	10·236	12·157	16·638	22·020	39·832	64·589
0·13500	0·948	1·112	1·264	1·543	1·797	2·033	2·146	2·255	2·361	2·465	2·665	2·857	3·308	3·725
1/ 7·4	0·2979	0·5457	0·8933	1·9387	3·5286	5·7486	7·1199	8·6778	10·432	12·390	16·956	22·442	40·595	65·825
0·14000	0·966	1·132	1·287	1·571	1·830	2·071	2·185	2·297	2·405	2·510	2·715	2·910	3·369	3·794
1/ 7·1	0·3034	0·5558	0·9098	1·9746	3·5939	5·8549	7·2514	8·8381	10·624	12·619	17·269	22·856	41·343	67·037
0·14500	0·983	1·153	1·310	1·599	1·863	2·108	2·224	2·337	2·448	2·555	2·763	2·962	3·429	3·861
1/ 6·9	0·3089	0·5658	0·9261	2·0099	3·6580	5·9593	7·3806	8·9956	10·814	12·844	17·576	23·263	42·078	68·229
0·15000	1·000	1·173	1·333	1·627	1·895	2·144	2·262	2·378	2·490	2·599	2·810	3·013	3·488	3·927
1/ 6·7	0·3142	0·5756	0·9421	2·0446	3·7211	6·0618	7·5076	9·1503	11·000	13·065	17·879	23·662	42·801	69·400
0·16000	1·033	1·211	1·377	1·681	1·958	2·215	2·337	2·456	2·572	2·685	2·903	3·112	3·603	4·057
1/ 6·3	0·3247	0·5947	0·9734	2·1122	3·8440	6·2620	7·7555	9·4523	11·362	13·496	18·468	24·442	44·210	71·684
0·17000	1·066	1·249	1·420	1·733	2·018	2·283	2·410	2·532	2·652	2·768	2·993	3·208	3·714	4·182
1/ 5·9	0·3348	0·6132	1·0036	2·1778	3·9632	6·4560	7·9957	9·7450	11·714	13·913	19·039	25·198	45·577	73·899
0·18000	1·097	1·286	1·461	1·784	2·077	2·350	2·480	2·606	2·729	2·849	3·080	3·302	3·822	4·303
1/ 5·6	0·3446	0·6312	1·0330	2·2415	4·0790	6·6444	8·2289	10·029	12·056	14·319	19·594	25·932	46·903	76·049
0·19000	1·127	1·321	1·502	1·833	2·135	2·415	2·548	2·678	2·804	2·927	3·165	3·393	3·927	4·422
1/ 5·3	0·3542	0·6487	1·0616	2·3034	4·1915	6·8276	8·4557	10·306	12·388	14·713	20·134	26·646	48·193	78·140
0·20000	1·157	1·356	1·541	1·881	2·191	2·478	2·615	2·748	2·877	3·004	3·247	3·481	4·030	4·537
1/ 5·0	0·3635	0·6657	1·0895	2·3637	4·3012	7·0061	8·6767	10·575	12·711	15·097	20·659	27·341	49·450	80·177
0·21000	1·186	1·390	1·580	1·928	2·245	2·539	2·680	2·816	2·949	3·078	3·328	3·568	4·129	4·649
1/ 4·8	0·3726	0·6823	1·1166	2·4225	4·4081	7·1801	8·8922	10·837	13·027	15·472	21·172	28·019	50·676	82·163
0·22000	1·214	1·423	1·617	1·974	2·298	2·600	2·743	2·883	3·018	3·151	3·407	3·652	4·227	4·759
1/ 4·5	0·3815	0·6985	1·1432	2·4800	4·5125	7·3501	9·1026	11·094	13·335	15·838	21·672	28·681	51·873	84·103
0·23000	1·242	1·455	1·654	2·018	2·350	2·658	2·805	2·948	3·087	3·222	3·484	3·734	4·322	4·867
1/ 4·3	0·3902	0·7144	1·1691	2·5361	4·6146	7·5162	9·3083	11·344	13·636	16·195	22·161	29·329	53·043	85·999
0·24000	1·269	1·487	1·690	2·062	2·401	2·716	2·866	3·011	3·153	3·292	3·559	3·815	4·416	4·972
1/ 4·2	0·3987	0·7299	1·1944	2·5911	4·7145	7·6788	9·5095	11·590	13·931	16·545	22·640	29·962	54·188	87·855
0·25000	1·295	1·518	1·725	2·105	2·451	2·772	2·925	3·074	3·219	3·360	3·632	3·894	4·507	5·074
1/ 4·0	0·4070	0·7451	1·2193	2·6449	4·8123	7·8380	9·7067	11·830	14·219	16·888	23·109	30·582	55·309	89·672
0·26000	1·321	1·548	1·759	2·147	2·500	2·827	2·983	3·135	3·283	3·427	3·705	3·971	4·597	5·175
1/ 3·8	0·4151	0·7600	1·2436	2·6976	4·9082	7·9941	9·8999	12·065	14·502	17·224	23·568	31·190	56·408	91·453
0·27000	1·347	1·578	1·793	2·188	2·548	2·881	3·041	3·195	3·345	3·492	3·776	4·047	4·684	5·274
1/ 3·7	0·4231	0·7746	1·2675	2·7494	5·0022	8·1472	10·089	12·296	14·780	17·554	24·019	31·787	57·486	93·200
0·28000	1·372	1·607	1·826	2·228	2·595	2·935	3·097	3·254	3·407	3·557	3·845	4·122	4·771	5·371
1/ 3·6	0·4310	0·7890	1·2910	2·8002	5·0946	8·2974	10·275	12·523	15·052	17·877	24·462	32·372	58·544	94·915
0·29000	1·396	1·636	1·859	2·268	2·641	2·987	3·152	3·312	3·468	3·620	3·913	4·195	4·855	5·466
1/ 3·4	0·4387	0·8031	1·3140	2·8501	5·1853	8·4450	10·458	12·745	15·320	18·195	24·896	32·947	59·584	96·600
0·30000	1·420	1·664	1·891	2·307	2·686	3·038	3·206	3·369	3·527	3·682	3·981	4·267	4·939	5·560
1/ 3·3	0·4463	0·8169	1·3367	2·8991	5·2744	8·5901	10·638	12·964	15·583	18·507	25·323	33·512	60·605	98·256
	–	–	–	0·74	0·74	0·75	0·75	0·75	0·75	0·76	0·76	0·76	0·77	0·78

$V_{r(0{\cdot}5)medial}$ for half-full circular pipes.

A9

(p.1 of 6)

k_s = **1·50 mm**
S = 0·00030 to 0·00100

ie hydraulic gradient =
1 in 3333 to 1 in 1000

Water (or sewage) at 15°C;
full bore conditions.

velocities in ms^{-1}
discharges in litres/sec

Gradient	(Equivalent) Pipe diameters in mm													
	20	25	30	40	50	60	65	70	75	80	90	100	125	150
0·000300					0·067	0·076	0·081	0·086	0·090	0·094	0·103	0·111	0·130	0·147
1/ 3333					0·1308	0·2160	0·2689	0·3293	0·3976	0·4741	0·6532	0·8695	1·5905	2·6003
0·000320				0·058	0·069	0·079	0·084	0·089	0·093	0·098	0·106	0·114	0·134	0·152
1/ 3125				0·0731	0·1354	0·2235	0·2782	0·3407	0·4113	0·4904	0·6756	0·8993	1·6446	2·6885
0·000340				0·060	0·071	0·082	0·087	0·091	0·096	0·101	0·110	0·118	0·138	0·157
1/ 2941				0·0755	0·1398	0·2307	0·2873	0·3518	0·4246	0·5062	0·6973	0·9281	1·6971	2·7739
0·000360				0·062	0·073	0·084	0·089	0·094	0·099	0·104	0·113	0·122	0·142	0·162
1/ 2778				0·0778	0·1441	0·2378	0·2960	0·3625	0·4375	0·5216	0·7185	0·9561	1·7480	2·8569
0·000380				0·064	0·076	0·087	0·092	0·097	0·102	0·107	0·116	0·125	0·146	0·166
1/ 2632				0·0801	0·1483	0·2447	0·3046	0·3729	0·4501	0·5366	0·7390	0·9834	1·7976	2·9377
0·000400				0·066	0·078	0·089	0·094	0·100	0·105	0·110	0·119	0·129	0·150	0·171
1/ 2500				0·0823	0·1524	0·2514	0·3129	0·3831	0·4623	0·5511	0·7590	1·0099	1·8459	3·0163
0·000420				0·067	0·080	0·091	0·097	0·102	0·107	0·112	0·122	0·132	0·154	0·175
1/ 2381				0·0845	0·1564	0·2579	0·3210	0·3930	0·4743	0·5653	0·7785	1·0358	1·8930	3·0930
0·000440				0·069	0·082	0·093	0·099	0·105	0·110	0·115	0·125	0·135	0·158	0·179
1/ 2273				0·0866	0·1603	0·2643	0·3289	0·4026	0·4859	0·5792	0·7975	1·0611	1·9390	3·1680
0·000460				0·071	0·084	0·096	0·101	0·107	0·113	0·118	0·128	0·138	0·162	0·183
1/ 2174				0·0887	0·1641	0·2705	0·3367	0·4121	0·4973	0·5928	0·8162	1·0858	1·9840	3·2412
0·000480				0·072	0·085	0·098	0·104	0·109	0·115	0·121	0·131	0·141	0·165	0·187
1/ 2083				0·0907	0·1678	0·2766	0·3442	0·4214	0·5085	0·6060	0·8344	1·1100	2·0280	3·3129
0·000500				0·074	0·087	0·100	0·106	0·112	0·118	0·123	0·134	0·144	0·169	0·191
1/ 2000				0·0927	0·1715	0·2826	0·3517	0·4304	0·5194	0·6190	0·8522	1·1337	2·0711	3·3831
0·000525				0·076	0·090	0·103	0·109	0·115	0·121	0·126	0·137	0·148	0·173	0·196
1/ 1905				0·0951	0·1759	0·2899	0·3607	0·4415	0·5328	0·6349	0·8741	1·1627	2·1238	3·4689
0·000550				0·078	0·092	0·105	0·111	0·118	0·124	0·129	0·141	0·152	0·177	0·201
1/ 1818				0·0975	0·1803	0·2970	0·3696	0·4524	0·5458	0·6505	0·8954	1·1910	2·1753	3·5527
0·000575				0·079	0·094	0·108	0·114	0·120	0·126	0·132	0·144	0·155	0·181	0·206
1/ 1739				0·0998	0·1845	0·3040	0·3783	0·4630	0·5586	0·6657	0·9162	1·2186	2·2256	3·6346
0·000600				0·081	0·096	0·110	0·117	0·123	0·129	0·135	0·147	0·159	0·185	0·210
1/ 1667				0·1021	0·1887	0·3109	0·3868	0·4733	0·5711	0·6805	0·9366	1·2457	2·2748	3·7148
0·000625				0·083	0·098	0·112	0·119	0·126	0·132	0·138	0·150	0·162	0·189	0·215
1/ 1600				0·1043	0·1928	0·3176	0·3951	0·4835	0·5833	0·6951	0·9566	1·2722	2·3231	3·7933
0·000650				0·085	0·100	0·115	0·122	0·128	0·135	0·141	0·153	0·165	0·193	0·219
1/ 1538				0·1065	0·1968	0·3241	0·4032	0·4934	0·5953	0·7093	0·9762	1·2982	2·3703	3·8703
0·000675				0·086	0·102	0·117	0·124	0·131	0·137	0·144	0·156	0·169	0·197	0·223
1/ 1481				0·1087	0·2007	0·3306	0·4112	0·5032	0·6070	0·7233	0·9954	1·3237	2·4167	3·9458
0·000700				0·088	0·104	0·119	0·126	0·133	0·140	0·147	0·159	0·172	0·201	0·227
1/ 1429				0·1108	0·2046	0·3369	0·4191	0·5128	0·6186	0·7371	1·0143	1·3487	2·4622	4·0199
0·000725				0·090	0·106	0·121	0·129	0·136	0·143	0·149	0·162	0·175	0·204	0·232
1/ 1379				0·1128	0·2084	0·3431	0·4268	0·5222	0·6299	0·7506	1·0328	1·3733	2·5070	4·0928
0·000750				0·091	0·108	0·124	0·131	0·138	0·145	0·152	0·165	0·178	0·208	0·236
1/ 1333				0·1149	0·2121	0·3492	0·4344	0·5315	0·6411	0·7638	1·0510	1·3975	2·5509	4·1643
0·000800				0·095	0·112	0·128	0·135	0·143	0·150	0·157	0·171	0·184	0·215	0·244
1/ 1250				0·1188	0·2194	0·3611	0·4492	0·5495	0·6628	0·7897	1·0866	1·4447	2·6367	4·3040
0·000850			0·078	0·098	0·115	0·132	0·140	0·147	0·155	0·162	0·176	0·190	0·222	0·251
1/ 1176			0·0554	0·1227	0·2264	0·3727	0·4635	0·5670	0·6839	0·8148	1·1210	1·4904	2·7199	4·4394
0·000900			0·081	0·101	0·119	0·136	0·144	0·152	0·159	0·167	0·181	0·195	0·228	0·259
1/ 1111			0·0571	0·1264	0·2333	0·3839	0·4774	0·5840	0·7044	0·8392	1·1545	1·5348	2·8006	4·5709
0·000950			0·083	0·103	0·122	0·140	0·148	0·156	0·164	0·172	0·187	0·201	0·235	0·266
1/ 1053			0·0587	0·1300	0·2399	0·3948	0·4910	0·6006	0·7243	0·8629	1·1870	1·5780	2·8792	4·6988
0·001000			0·085	0·106	0·125	0·143	0·152	0·160	0·168	0·176	0·192	0·206	0·241	0·273
1/ 1000			0·0603	0·1336	0·2464	0·4054	0·5042	0·6167	0·7438	0·8860	1·2187	1·6201	2·9557	4·8233
			–	–	–	–	–	–	–	–	0·74	0·74	0·74	0·75

$V_{r(0·5)medial}$ for half-full circular pipes.

k_s = 1·50 mm S = 0·00030 to 0·00100

Gradient **(Equivalent) Pipe diameters in mm**

Gradient	20	25	30	40	50	60	65	70	75	80	90	100	125	150
0·00100			0·085	0·106	0·125	0·143	0·152	0·160	0·168	0·176	0·192	0·206	0·241	0·273
1/ 1000			0·0603	0·1336	0·2464	0·4054	0·5042	0·6167	0·7438	0·8860	1·2187	1·6201	2·9557	4·8233
0·00105			0·088	0·109	0·129	0·147	0·156	0·164	0·173	0·181	0·196	0·211	0·247	0·280
1/ 952			0·0619	0·1370	0·2527	0·4158	0·5170	0·6324	0·7627	0·9085	1·2497	1·6611	3·0303	4·9448
0·00110			0·090	0·112	0·132	0·151	0·160	0·168	0·177	0·185	0·201	0·217	0·253	0·287
1/ 909			0·0634	0·1404	0·2589	0·4259	0·5296	0·6478	0·7812	0·9305	1·2799	1·7012	3·1032	5·0634
0·00115			0·092	0·114	0·135	0·154	0·163	0·172	0·181	0·189	0·206	0·222	0·259	0·293
1/ 870			0·0649	0·1437	0·2650	0·4358	0·5419	0·6628	0·7993	0·9520	1·3094	1·7404	3·1744	5·1794
0·00120			0·094	0·117	0·138	0·158	0·167	0·176	0·185	0·194	0·210	0·226	0·264	0·300
1/ 833			0·0664	0·1469	0·2709	0·4455	0·5539	0·6775	0·8170	0·9731	1·3383	1·7787	3·2441	5·2929
0·00125			0·096	0·119	0·141	0·161	0·170	0·180	0·189	0·198	0·215	0·231	0·270	0·306
1/ 800			0·0678	0·1501	0·2767	0·4550	0·5657	0·6919	0·8343	0·9937	1·3666	1·8163	3·3124	5·4041
0·00130			0·098	0·122	0·144	0·164	0·174	0·183	0·193	0·202	0·219	0·236	0·275	0·312
1/ 769			0·0693	0·1532	0·2823	0·4643	0·5772	0·7060	0·8513	1·0139	1·3943	1·8531	3·3793	5·5130
0·00135			0·100	0·124	0·147	0·167	0·177	0·187	0·196	0·206	0·223	0·241	0·281	0·318
1/ 741			0·0706	0·1562	0·2879	0·4734	0·5886	0·7198	0·8679	1·0337	1·4215	1·8892	3·4450	5·6199
0·00140			0·102	0·127	0·149	0·171	0·181	0·191	0·200	0·210	0·228	0·245	0·286	0·324
1/ 714			0·0720	0·1592	0·2934	0·4824	0·5997	0·7334	0·8843	1·0532	1·4482	1·9246	3·5094	5·7248
0·00145			0·104	0·129	0·152	0·174	0·184	0·194	0·204	0·213	0·232	0·249	0·291	0·330
1/ 690			0·0733	0·1621	0·2988	0·4912	0·6106	0·7467	0·9004	1·0723	1·4745	1·9594	3·5727	5·8279
0·00150		0·092	0·106	0·131	0·155	0·177	0·187	0·197	0·207	0·217	0·236	0·254	0·296	0·336
1/ 667		0·0450	0·0747	0·1650	0·3040	0·4998	0·6214	0·7598	0·9161	1·0911	1·5003	1·9937	3·6350	5·9292
0·00160		0·095	0·109	0·136	0·160	0·183	0·194	0·204	0·214	0·224	0·244	0·262	0·306	0·347
1/ 625		0·0466	0·0772	0·1706	0·3143	0·5167	0·6423	0·7854	0·9470	1·1278	1·5506	2·0605	3·7564	6·1269
0·00170		0·098	0·113	0·140	0·165	0·189	0·200	0·211	0·221	0·231	0·251	0·271	0·316	0·358
1/ 588		0·0481	0·0797	0·1761	0·3243	0·5331	0·6626	0·8102	0·9768	1·1633	1·5994	2·1252	3·8741	6·3185
0·00180		0·101	0·116	0·144	0·170	0·194	0·206	0·217	0·228	0·238	0·259	0·279	0·325	0·368
1/ 556		0·0495	0·0821	0·1814	0·3340	0·5490	0·6823	0·8343	1·0059	1·1978	1·6468	2·1881	3·9884	6·5045
0·00190		0·104	0·119	0·148	0·175	0·200	0·211	0·223	0·234	0·245	0·266	0·286	0·334	0·378
1/ 526		0·0510	0·0845	0·1865	0·3435	0·5644	0·7015	0·8578	1·0341	1·2314	1·6928	2·2492	4·0995	6·6855
0·00200		0·107	0·123	0·152	0·180	0·205	0·217	0·229	0·240	0·251	0·273	0·294	0·343	0·388
1/ 500		0·0523	0·0868	0·1915	0·3527	0·5795	0·7202	0·8806	1·0616	1·2641	1·7377	2·3088	4·2078	6·8617
0·00210		0·109	0·126	0·156	0·184	0·210	0·223	0·235	0·246	0·258	0·280	0·301	0·351	0·398
1/ 476		0·0537	0·0890	0·1964	0·3616	0·5942	0·7385	0·9028	1·0884	1·2960	1·7815	2·3668	4·3134	7·0336
0·00220		0·112	0·129	0·160	0·189	0·215	0·228	0·240	0·252	0·264	0·287	0·309	0·360	0·408
1/ 455		0·0550	0·0912	0·2012	0·3704	0·6085	0·7563	0·9246	1·1145	1·3272	1·8243	2·4236	4·4166	7·2015
0·00230		0·115	0·132	0·164	0·193	0·220	0·233	0·246	0·258	0·270	0·293	0·316	0·368	0·417
1/ 435		0·0563	0·0933	0·2059	0·3789	0·6225	0·7737	0·9458	1·1401	1·3576	1·8660	2·4790	4·5174	7·3656
0·00240		0·117	0·135	0·167	0·197	0·225	0·238	0·251	0·264	0·276	0·300	0·323	0·376	0·426
1/ 417		0·0576	0·0954	0·2104	0·3873	0·6362	0·7907	0·9666	1·1652	1·3874	1·9069	2·5333	4·6160	7·5262
0·00250		0·120	0·138	0·171	0·201	0·230	0·243	0·256	0·269	0·282	0·306	0·329	0·384	0·435
1/ 400		0·0588	0·0974	0·2149	0·3955	0·6497	0·8074	0·9870	1·1897	1·4166	1·9470	2·5864	4·7126	7·6834
0·00260		0·122	0·141	0·175	0·206	0·234	0·248	0·262	0·275	0·288	0·312	0·336	0·392	0·444
1/ 385		0·0600	0·0994	0·2193	0·4036	0·6628	0·8237	1·0070	1·2138	1·4452	1·9863	2·6385	4·8073	7·8376
0·00270		0·125	0·143	0·178	0·210	0·239	0·253	0·267	0·280	0·293	0·318	0·342	0·399	0·452
1/ 370		0·0612	0·1014	0·2236	0·4115	0·6758	0·8397	1·0266	1·2373	1·4732	2·0248	2·6896	4·9002	7·9888
0·00280		0·127	0·146	0·181	0·214	0·243	0·258	0·272	0·285	0·299	0·324	0·349	0·407	0·460
1/ 357		0·0624	0·1033	0·2278	0·4192	0·6885	0·8555	1·0458	1·2605	1·5008	2·0626	2·7398	4·9914	8·1373
0·00290		0·129	0·149	0·185	0·217	0·248	0·262	0·277	0·290	0·304	0·330	0·355	0·414	0·469
1/ 345		0·0635	0·1052	0·2320	0·4268	0·7009	0·8710	1·0647	1·2832	1·5279	2·0997	2·7890	5·0810	8·2831
0·00300		0·132	0·151	0·188	0·221	0·252	0·267	0·281	0·296	0·309	0·336	0·361	0·421	0·477
1/ 333		0·0647	0·1071	0·2361	0·4343	0·7132	0·8862	1·0832	1·3056	1·5545	2·1362	2·8375	5·1691	8·4264
		–	–	–	–	–	–	–	–	–	0·74	0·74	0·74	0·75

$V_{r(0{\cdot}5)medial}$ **for half-full circular pipes.**

A9
(p.3 of 6)

k_s = 1·50 mm
S = 0·00300 to 0·01000

ie hydraulic gradient =
1 in 333 to 1 in 100

Water (or sewage) at 15°C;
full bore conditions.

velocities in ms^{-1}
discharges in litres/sec

Gradient	20	25	30	40	50	60	65	70	75	80	90	100	125	150
	(Equivalent) Pipe diameters in mm													
0·00300		0·132	0·151	0·188	0·221	0·252	0·267	0·281	0·296	0·309	0·336	0·361	0·421	0·477
1/ 333		0·0647	0·1071	0·2361	0·4343	0·7132	0·8862	1·0832	1·3056	1·5545	2·1362	2·8375	5·1691	8·4264
0·00320	0·114	0·136	0·157	0·194	0·229	0·261	0·276	0·291	0·305	0·320	0·347	0·373	0·435	0·493
1/ 313	0·0360	0·0669	0·1107	0·2440	0·4489	0·7371	0·9158	1·1195	1·3492	1·6064	2·2074	2·9320	5·3409	8·7061
0·00340	0·118	0·141	0·162	0·200	0·236	0·269	0·285	0·300	0·315	0·330	0·358	0·385	0·449	0·508
1/ 294	0·0371	0·0690	0·1142	0·2518	0·4630	0·7602	0·9446	1·1546	1·3915	1·6567	2·2765	3·0236	5·5074	8·9771
0·00360	0·122	0·145	0·166	0·206	0·243	0·277	0·293	0·309	0·324	0·339	0·368	0·396	0·462	0·523
1/ 278	0·0382	0·0711	0·1177	0·2593	0·4768	0·7827	0·9725	1·1887	1·4326	1·7055	2·3435	3·1125	5·6691	9·2403
0·00380	0·125	0·149	0·171	0·212	0·250	0·285	0·301	0·317	0·333	0·349	0·379	0·407	0·475	0·537
1/ 263	0·0393	0·0731	0·1210	0·2665	0·4901	0·8046	0·9997	1·2218	1·4725	1·7530	2·4087	3·1991	5·8264	9·4963
0·00400	0·129	0·153	0·176	0·218	0·256	0·292	0·309	0·326	0·342	0·358	0·389	0·418	0·487	0·551
1/ 250	0·0404	0·0751	0·1242	0·2736	0·5031	0·8259	1·0261	1·2541	1·5114	1·7993	2·4722	3·2833	5·9796	9·7457
0·00420	0·132	0·157	0·180	0·223	0·263	0·299	0·317	0·334	0·351	0·367	0·398	0·429	0·499	0·565
1/ 238	0·0414	0·0770	0·1274	0·2806	0·5158	0·8467	1·0519	1·2856	1·5493	1·8444	2·5342	3·3655	6·1290	9·9889
0·00440	0·135	0·161	0·185	0·229	0·269	0·307	0·325	0·342	0·359	0·376	0·408	0·439	0·511	0·579
1/ 227	0·0424	0·0789	0·1305	0·2873	0·5282	0·8670	1·0771	1·3164	1·5864	1·8885	2·5947	3·4458	6·2749	10·226
0·00460	0·138	0·164	0·189	0·234	0·275	0·314	0·332	0·350	0·367	0·384	0·417	0·449	0·523	0·592
1/ 217	0·0434	0·0807	0·1335	0·2940	0·5404	0·8868	1·1017	1·3465	1·6226	1·9316	2·6538	3·5242	6·4175	10·458
0·00480	0·141	0·168	0·193	0·239	0·281	0·321	0·339	0·358	0·375	0·393	0·426	0·458	0·534	0·605
1/ 208	0·0444	0·0825	0·1364	0·3004	0·5522	0·9062	1·1258	1·3759	1·6580	1·9737	2·7117	3·6010	6·5570	10·686
0·00500	0·144	0·172	0·197	0·244	0·287	0·327	0·346	0·365	0·383	0·401	0·435	0·468	0·545	0·617
1/ 200	0·0454	0·0842	0·1393	0·3068	0·5638	0·9253	1·1494	1·4047	1·6928	2·0150	2·7683	3·6762	6·6937	10·908
0·00525	0·148	0·176	0·202	0·250	0·294	0·335	0·355	0·374	0·393	0·411	0·446	0·480	0·559	0·633
1/ 190	0·0465	0·0864	0·1429	0·3145	0·5780	0·9485	1·1783	1·4400	1·7352	2·0655	2·8376	3·7681	6·8607	11·180
0·00550	0·152	0·180	0·207	0·256	0·301	0·343	0·364	0·383	0·402	0·421	0·457	0·491	0·572	0·648
1/ 182	0·0476	0·0885	0·1463	0·3221	0·5919	0·9712	1·2064	1·4744	1·7766	2·1148	2·9052	3·8578	7·0239	11·446
0·00575	0·155	0·184	0·212	0·262	0·308	0·351	0·372	0·392	0·411	0·430	0·467	0·502	0·585	0·662
1/ 174	0·0488	0·0905	0·1497	0·3295	0·6054	0·9934	1·2340	1·5080	1·8171	2·1630	2·9713	3·9455	7·1833	11·705
0·00600	0·159	0·189	0·216	0·268	0·315	0·359	0·380	0·400	0·420	0·440	0·477	0·513	0·598	0·677
1/ 167	0·0498	0·0925	0·1530	0·3367	0·6187	1·0151	1·2609	1·5409	1·8567	2·2101	3·0360	4·0314	7·3394	11·959
0·00625	0·162	0·193	0·221	0·274	0·322	0·367	0·388	0·409	0·429	0·449	0·487	0·524	0·611	0·691
1/ 160	0·0509	0·0945	0·1562	0·3438	0·6317	1·0364	1·2873	1·5731	1·8955	2·2563	3·0994	4·1154	7·4922	12·208
0·00650	0·165	0·196	0·226	0·279	0·328	0·374	0·396	0·417	0·438	0·458	0·497	0·534	0·623	0·705
1/ 154	0·0519	0·0964	0·1594	0·3507	0·6444	1·0572	1·3132	1·6047	1·9336	2·3015	3·1615	4·1978	7·6420	12·452
0·00675	0·169	0·200	0·230	0·285	0·335	0·381	0·403	0·425	0·446	0·467	0·507	0·545	0·635	0·718
1/ 148	0·0530	0·0983	0·1625	0·3576	0·6569	1·0777	1·3386	1·6357	1·9709	2·3459	3·2224	4·2787	7·7889	12·691
0·00700	0·172	0·204	0·234	0·290	0·341	0·388	0·411	0·433	0·454	0·475	0·516	0·555	0·646	0·731
1/ 143	0·0540	0·1002	0·1656	0·3642	0·6692	1·0977	1·3635	1·6661	2·0075	2·3895	3·2822	4·3580	7·9331	12·926
0·00725	0·175	0·208	0·238	0·295	0·347	0·395	0·418	0·441	0·463	0·484	0·525	0·565	0·658	0·744
1/ 138	0·0550	0·1020	0·1686	0·3708	0·6812	1·1174	1·3879	1·6960	2·0435	2·4323	3·3410	4·4360	8·0748	13·156
0·00750	0·178	0·211	0·243	0·300	0·353	0·402	0·426	0·448	0·471	0·492	0·534	0·575	0·669	0·757
1/ 133	0·0559	0·1038	0·1715	0·3773	0·6930	1·1368	1·4120	1·7254	2·0789	2·4744	3·3987	4·5126	8·2141	13·383
0·00800	0·184	0·219	0·251	0·310	0·365	0·415	0·440	0·463	0·486	0·509	0·552	0·594	0·691	0·782
1/ 125	0·0578	0·1073	0·1773	0·3899	0·7161	1·1746	1·4589	1·7827	2·1479	2·5565	3·5114	4·6621	8·4858	13·825
0·00850	0·190	0·225	0·259	0·320	0·376	0·428	0·453	0·478	0·501	0·524	0·569	0·612	0·713	0·807
1/ 118	0·0596	0·1107	0·1829	0·4021	0·7385	1·2113	1·5044	1·8382	2·2148	2·6361	3·6206	4·8069	8·7492	14·254
0·00900	0·196	0·232	0·266	0·329	0·387	0·441	0·467	0·492	0·516	0·540	0·586	0·630	0·734	0·830
1/ 111	0·0614	0·1139	0·1883	0·4140	0·7602	1·2469	1·5486	1·8922	2·2797	2·7133	3·7266	4·9476	9·0050	14·670
0·00950	0·201	0·239	0·274	0·339	0·398	0·453	0·480	0·505	0·530	0·555	0·602	0·647	0·754	0·853
1/ 105	0·0632	0·1171	0·1935	0·4255	0·7814	1·2815	1·5915	1·9446	2·3429	2·7885	3·8298	5·0845	9·2537	15·075
0·01000	0·206	0·245	0·281	0·348	0·408	0·465	0·492	0·519	0·544	0·569	0·618	0·664	0·774	0·875
1/ 100	0·0648	0·1203	0·1987	0·4367	0·8020	1·3152	1·6334	1·9957	2·4044	2·8617	3·9302	5·2177	9·4960	15·469
	–	–	–	–	–	–	–	–	–	0·74	0·74	0·74	0·74	0·75

$V_{r(0\cdot5)medial}$ for half-full circular pipes.

k_s = 1·50 mm S = 0·00300 to 0·01000

k_s = 1·50 mm
S = 0·01000 to 0·03000

ie hydraulic gradient =
1 in 100 to 1 in 33·3

Water (or sewage) at 15°C;
full bore conditions.

velocities in ms^{-1}
discharges in litres/sec

Gradient	20	25	30	40	50	60	65	70	75	80	90	100	125	150
	(Equivalent) Pipe diameters in mm													
0·01000	0·206	0·245	0·281	0·348	0·408	0·465	0·492	0·519	0·544	0·569	0·618	0·664	0·774	0·875
1/ 100	0·0648	0·1203	0·1987	0·4367	0·8020	1·3152	1·6334	1·9957	2·4044	2·8617	3·9302	5·2177	9·4960	15·469
0·01050	0·212	0·251	0·288	0·356	0·419	0·477	0·505	0·532	0·558	0·584	0·633	0·681	0·793	0·897
1/ 95	0·0665	0·1233	0·2037	0·4477	0·8221	1·3481	1·6742	2·0456	2·4644	2·9331	4·0282	5·3477	9·7323	15·854
0·01100	0·217	0·257	0·295	0·365	0·429	0·488	0·517	0·544	0·571	0·597	0·648	0·697	0·812	0·918
1/ 91	0·0681	0·1263	0·2086	0·4584	0·8417	1·3802	1·7141	2·0942	2·5230	3·0028	4·1239	5·4747	9·9630	16·229
0·01150	0·222	0·263	0·302	0·373	0·438	0·499	0·528	0·557	0·584	0·611	0·663	0·713	0·830	0·939
1/ 87	0·0697	0·1292	0·2133	0·4689	0·8608	1·4116	1·7530	2·1418	2·5803	3·0709	4·2174	5·5987	10·189	16·597
0·01200	0·227	0·269	0·308	0·381	0·448	0·510	0·540	0·569	0·597	0·624	0·677	0·728	0·848	0·959
1/ 83	0·0712	0·1320	0·2180	0·4791	0·8796	1·4423	1·7911	2·1883	2·6364	3·1376	4·3089	5·7202	10·409	16·956
0·01250	0·231	0·275	0·315	0·389	0·457	0·521	0·551	0·580	0·609	0·637	0·691	0·743	0·866	0·979
1/ 80	0·0727	0·1348	0·2226	0·4891	0·8980	1·4724	1·8285	2·2339	2·6913	3·2029	4·3985	5·8391	10·625	17·308
0·01300	0·236	0·280	0·321	0·397	0·467	0·531	0·562	0·592	0·621	0·650	0·705	0·758	0·883	0·999
1/ 77	0·0742	0·1375	0·2271	0·4990	0·9160	1·5019	1·8651	2·2786	2·7451	3·2669	4·4864	5·9556	10·837	17·652
0·01350	0·241	0·286	0·327	0·405	0·476	0·541	0·573	0·603	0·633	0·662	0·719	0·773	0·900	1·018
1/ 74	0·0756	0·1402	0·2315	0·5086	0·9336	1·5308	1·9009	2·3224	2·7979	3·3297	4·5726	6·0699	11·045	17·991
0·01400	0·245	0·291	0·334	0·412	0·484	0·551	0·583	0·615	0·645	0·675	0·732	0·787	0·917	1·037
1/ 71	0·0770	0·1428	0·2358	0·5181	0·9510	1·5592	1·9362	2·3655	2·8497	3·3914	4·6571	6·1822	11·249	18·323
0·01450	0·250	0·296	0·340	0·420	0·493	0·561	0·594	0·626	0·657	0·687	0·745	0·801	0·933	1·055
1/ 69	0·0784	0·1454	0·2401	0·5274	0·9680	1·5871	1·9708	2·4077	2·9006	3·4519	4·7402	6·2924	11·449	18·649
0·01500	0·254	0·301	0·346	0·427	0·502	0·571	0·604	0·636	0·668	0·699	0·758	0·815	0·949	1·073
1/ 67	0·0798	0·1479	0·2442	0·5365	0·9848	1·6145	2·0048	2·4493	2·9506	3·5114	4·8219	6·4008	11·646	18·969
0·01600	0·263	0·311	0·357	0·441	0·518	0·590	0·624	0·658	0·690	0·722	0·783	0·842	0·980	1·109
1/ 63	0·0825	0·1529	0·2524	0·5544	1·0174	1·6680	2·0712	2·5304	3·0482	3·6276	4·9813	6·6122	12·031	19·595
0·01700	0·271	0·321	0·368	0·455	0·534	0·608	0·644	0·678	0·711	0·744	0·807	0·868	1·011	1·143
1/ 59	0·0851	0·1576	0·2603	0·5716	1·0491	1·7198	2·1356	2·6089	3·1428	3·7401	5·1358	6·8172	12·403	20·201
0·01800	0·279	0·331	0·379	0·468	0·550	0·626	0·662	0·698	0·732	0·766	0·831	0·893	1·040	1·176
1/ 56	0·0876	0·1623	0·2679	0·5884	1·0798	1·7701	2·1980	2·6852	3·2347	3·8494	5·2857	7·0162	12·765	20·790
0·01900	0·287	0·340	0·390	0·481	0·565	0·643	0·681	0·717	0·752	0·787	0·854	0·918	1·069	1·209
1/ 53	0·0900	0·1668	0·2754	0·6047	1·1098	1·8191	2·2588	2·7594	3·3241	3·9557	5·4316	7·2097	13·117	21·363
0·02000	0·294	0·349	0·400	0·494	0·580	0·660	0·699	0·736	0·772	0·808	0·876	0·942	1·097	1·240
1/ 50	0·0924	0·1712	0·2826	0·6206	1·1389	1·8668	2·3180	2·8317	3·4111	4·0592	5·5737	7·3982	13·459	21·920
0·02100	0·302	0·358	0·410	0·506	0·594	0·677	0·716	0·754	0·791	0·828	0·898	0·965	1·124	1·271
1/ 47·6	0·0948	0·1755	0·2897	0·6362	1·1673	1·9133	2·3757	2·9022	3·4960	4·1602	5·7123	7·5821	13·794	22·464
0·02200	0·309	0·366	0·420	0·518	0·609	0·693	0·733	0·772	0·810	0·847	0·919	0·988	1·151	1·301
1/ 45·5	0·0970	0·1797	0·2966	0·6513	1·1950	1·9587	2·4321	2·9710	3·5789	4·2588	5·8476	7·7616	14·120	22·995
0·02300	0·316	0·374	0·429	0·530	0·622	0·708	0·750	0·789	0·828	0·866	0·940	1·011	1·177	1·331
1/ 43·5	0·0993	0·1838	0·3034	0·6661	1·2221	2·0031	2·4872	3·0383	3·6599	4·3552	5·9799	7·9371	14·439	23·515
0·02400	0·323	0·383	0·439	0·542	0·636	0·724	0·766	0·807	0·846	0·885	0·960	1·032	1·202	1·359
1/ 41·7	0·1014	0·1878	0·3100	0·6806	1·2487	2·0466	2·5411	3·1041	3·7392	4·4495	6·1093	8·1088	14·751	24·023
0·02500	0·330	0·391	0·448	0·553	0·649	0·739	0·782	0·823	0·864	0·904	0·980	1·054	1·227	1·388
1/ 40·0	0·1036	0·1918	0·3165	0·6948	1·2747	2·0891	2·5939	3·1686	3·8168	4·5419	6·2360	8·2770	15·057	24·520
0·02600	0·336	0·399	0·457	0·564	0·662	0·754	0·797	0·840	0·881	0·922	1·000	1·075	1·251	1·415
1/ 38·5	0·1056	0·1956	0·3228	0·7087	1·3001	2·1308	2·6457	3·2318	3·8929	4·6324	6·3603	8·4418	15·356	25·008
0·02700	0·343	0·406	0·466	0·575	0·675	0·768	0·813	0·856	0·898	0·939	1·019	1·095	1·275	1·442
1/ 37·0	0·1077	0·1994	0·3291	0·7223	1·3251	2·1717	2·6964	3·2938	3·9676	4·7212	6·4822	8·6035	15·650	25·486
0·02800	0·349	0·414	0·474	0·585	0·687	0·782	0·828	0·872	0·915	0·957	1·038	1·116	1·299	1·469
1/ 35·7	0·1097	0·2031	0·3352	0·7357	1·3497	2·2119	2·7463	3·3547	4·0409	4·8084	6·6018	8·7622	15·939	25·956
0·02900	0·355	0·421	0·483	0·596	0·700	0·796	0·842	0·887	0·931	0·974	1·056	1·135	1·322	1·495
1/ 34·5	0·1117	0·2068	0·3412	0·7489	1·3738	2·2513	2·7952	3·4145	4·1129	4·8941	6·7193	8·9182	16·222	26·417
0·03000	0·362	0·429	0·491	0·606	0·712	0·810	0·857	0·903	0·947	0·990	1·074	1·155	1·345	1·521
1/ 33·3	0·1136	0·2103	0·3471	0·7618	1·3974	2·2901	2·8434	3·4732	4·1836	4·9782	6·8349	9·0714	16·501	26·871
	–	–	–	–	–	–	–	–	–	0·74	0·74	0·74	0·74	0·75

$V_{r(0·5)medial}$ **for half-full circular pipes.**

k_s = 1·50 mm
S = 0·03000 to 0·10000

ie hydraulic gradient =
1 in 33·3 to 1 in 10·0

Water (or sewage) at 15°C;
full bore conditions.

velocities in ms^{-1}
discharges in litres/sec

Gradient	**(Equivalent) Pipe diameters in mm**													
	20	**25**	**30**	**40**	**50**	**60**	**65**	**70**	**75**	**80**	**90**	**100**	**125**	**150**
0·03000	0·362	0·429	0·491	0·606	0·712	0·810	0·857	0·903	0·947	0·990	1·074	1·155	1·345	1·521
1/ 33·3	0·1136	0·2103	0·3471	0·7618	1·3974	2·2901	2·8434	3·4732	4·1836	4·9782	6·8349	9·0714	16·501	26·871
0·03200	0·374	0·443	0·507	0·626	0·735	0·837	0·885	0·932	0·978	1·023	1·110	1·193	1·389	1·571
1/ 31·3	0·1174	0·2173	0·3586	0·7870	1·4436	2·3658	2·9372	3·5879	4·3217	5·1425	7·0603	9·3705	17·044	27·755
0·03400	0·385	0·457	0·523	0·646	0·758	0·863	0·913	0·961	1·009	1·055	1·144	1·230	1·432	1·619
1/ 29·4	0·1211	0·2241	0·3698	0·8115	1·4884	2·4391	3·0283	3·6990	4·4555	5·3017	7·2787	9·6603	17·571	28·613
0·03600	0·397	0·470	0·538	0·665	0·780	0·888	0·939	0·989	1·038	1·085	1·177	1·266	1·474	1·666
1/ 27·8	0·1246	0·2307	0·3806	0·8352	1·5319	2·5103	3·1166	3·8069	4·5854	5·4563	7·4908	9·9417	18·083	29·445
0·03800	0·408	0·483	0·553	0·683	0·802	0·912	0·965	1·016	1·067	1·115	1·210	1·301	1·514	1·712
1/ 26·3	0·1281	0·2371	0·3911	0·8583	1·5742	2·5795	3·2026	3·9119	4·7118	5·6066	7·6972	10·215	18·580	30·255
0·04000	0·418	0·496	0·568	0·701	0·823	0·936	0·990	1·043	1·094	1·145	1·242	1·335	1·554	1·757
1/ 25·0	0·1315	0·2433	0·4014	0·8808	1·6154	2·6470	3·2863	4·0141	4·8349	5·7530	7·8981	10·482	19·065	31·044
0·04200	0·429	0·508	0·582	0·718	0·843	0·959	1·015	1·069	1·122	1·173	1·272	1·368	1·592	1·800
1/ 23·8	0·1348	0·2494	0·4114	0·9027	1·6556	2·7128	3·3679	4·1138	4·9549	5·8958	8·0941	10·742	19·538	31·813
0·04400	0·439	0·520	0·596	0·735	0·863	0·982	1·039	1·094	1·148	1·201	1·302	1·400	1·630	1·843
1/ 22·7	0·1380	0·2553	0·4212	0·9241	1·6948	2·7770	3·4476	4·2111	5·0722	6·0353	8·2855	10·996	19·999	32·564
0·04600	0·449	0·532	0·609	0·752	0·883	1·004	1·062	1·119	1·174	1·228	1·332	1·432	1·666	1·884
1/ 21·7	0·1411	0·2612	0·4308	0·9451	1·7332	2·8398	3·5256	4·3063	5·1868	6·1716	8·4725	11·244	20·450	33·299
0·04800	0·459	0·544	0·623	0·768	0·902	1·026	1·085	1·143	1·199	1·254	1·361	1·463	1·702	1·925
1/ 20·8	0·1442	0·2668	0·4401	0·9656	1·7707	2·9012	3·6018	4·3994	5·2989	6·3050	8·6556	11·487	20·892	34·017
0·05000	0·469	0·555	0·636	0·784	0·921	1·047	1·108	1·167	1·224	1·280	1·389	1·493	1·738	1·965
1/ 20·0	0·1472	0·2724	0·4493	0·9856	1·8075	2·9614	3·6765	4·4906	5·4087	6·4356	8·8349	11·725	21·324	34·721
0·05250	0·480	0·569	0·651	0·804	0·943	1·073	1·135	1·196	1·255	1·312	1·423	1·530	1·781	2·013
1/ 19·0	0·1509	0·2792	0·4605	1·0102	1·8524	3·0350	3·7678	4·6021	5·5429	6·5953	9·0540	12·016	21·852	35·581
0·05500	0·492	0·582	0·667	0·823	0·966	1·099	1·162	1·224	1·284	1·343	1·457	1·566	1·823	2·061
1/ 18·2	0·1545	0·2858	0·4714	1·0341	1·8963	3·1068	3·8569	4·7109	5·6740	6·7512	9·2679	12·299	22·368	36·421
0·05750	0·503	0·596	0·682	0·842	0·988	1·124	1·189	1·252	1·313	1·373	1·490	1·601	1·864	2·107
1/ 17·4	0·1580	0·2923	0·4821	1·0575	1·9391	3·1770	3·9441	4·8173	5·8021	6·9036	9·4771	12·577	22·873	37·242
0·06000	0·514	0·608	0·697	0·860	1·009	1·148	1·214	1·279	1·342	1·403	1·522	1·636	1·904	2·153
1/ 16·7	0·1614	0·2987	0·4926	1·0804	1·9811	3·2457	4·0293	4·9214	5·9275	7·0528	9·6818	12·848	23·366	38·045
0·06250	0·525	0·621	0·711	0·878	1·030	1·172	1·239	1·305	1·369	1·432	1·553	1·670	1·943	2·197
1/ 16·0	0·1648	0·3049	0·5028	1·1028	2·0222	3·3129	4·1128	5·0234	6·0502	7·1988	9·8822	13·114	23·850	38·832
0·06500	0·535	0·634	0·726	0·895	1·050	1·195	1·264	1·331	1·397	1·461	1·584	1·703	1·982	2·241
1/ 15·4	0·1681	0·3110	0·5128	1·1248	2·0625	3·3789	4·1947	5·1233	6·1706	7·3420	10·079	13·375	24·323	39·603
0·06750	0·545	0·646	0·739	0·912	1·071	1·218	1·288	1·357	1·423	1·489	1·615	1·736	2·020	2·284
1/ 14·8	0·1713	0·3170	0·5227	1·1464	2·1020	3·4436	4·2749	5·2214	6·2886	7·4824	10·271	13·631	24·788	40·359
0·07000	0·555	0·658	0·753	0·929	1·090	1·240	1·312	1·382	1·450	1·516	1·644	1·767	2·057	2·326
1/ 14·3	0·1745	0·3228	0·5324	1·1676	2·1408	3·5071	4·3538	5·3176	6·4045	7·6203	10·461	13·882	25·244	41·102
0·07250	0·565	0·669	0·767	0·946	1·110	1·262	1·335	1·406	1·475	1·543	1·674	1·799	2·094	2·367
1/ 13·8	0·1776	0·3286	0·5419	1·1884	2·1789	3·5695	4·4312	5·4121	6·5184	7·7557	10·646	14·128	25·693	41·831
0·07500	0·575	0·681	0·780	0·962	1·129	1·284	1·358	1·430	1·501	1·569	1·702	1·830	2·130	2·408
1/ 13·3	0·1807	0·3343	0·5512	1·2088	2·2163	3·6308	4·5073	5·5050	6·6302	7·8888	10·829	14·371	26·133	42·548
0·08000	0·594	0·703	0·806	0·994	1·166	1·326	1·403	1·478	1·550	1·621	1·758	1·890	2·200	2·487
1/ 12·5	0·1867	0·3453	0·5694	1·2487	2·2894	3·7504	4·6557	5·6864	6·8486	8·1485	11·185	14·843	26·993	43·947
0·08500	0·613	0·725	0·831	1·024	1·202	1·367	1·446	1·523	1·598	1·671	1·813	1·948	2·267	2·564
1/ 11·8	0·1925	0·3560	0·5871	1·2874	2·3602	3·8664	4·7996	5·8621	7·0602	8·4002	11·531	15·302	27·826	45·303
0·09000	0·631	0·747	0·855	1·054	1·237	1·407	1·489	1·568	1·645	1·720	1·865	2·005	2·333	2·638
1/ 11·1	0·1981	0·3664	0·6042	1·3249	2·4290	3·9789	4·9394	6·0327	7·2656	8·6446	11·866	15·747	28·634	46·619
0·09500	0·648	0·767	0·878	1·083	1·271	1·446	1·529	1·611	1·690	1·767	1·917	2·060	2·397	2·711
1/ 10·5	0·2036	0·3766	0·6209	1·3614	2·4959	4·0884	5·0753	6·1986	7·4654	8·8824	12·193	16·180	29·421	47·900
0·10000	0·665	0·787	0·901	1·112	1·304	1·484	1·569	1·653	1·734	1·813	1·966	2·114	2·460	2·781
1/ 10·0	0·2089	0·3864	0·6371	1·3970	2·5610	4·1951	5·2076	6·3603	7·6601	9·1139	12·510	16·601	30·187	49·147
	–	–	–	–	–	–	–	–	0·74	0·74	0·74	0·74	0·74	0·75

$V_{r(0\cdot5)medial}$ for half-full circular pipes.

k_s = 1·50 mm S = 0·03000 to 0·10000

Gradient	**(Equivalent) Pipe diameters in mm**													
	20	**25**	**30**	**40**	**50**	**60**	**65**	**70**	**75**	**80**	**90**	**100**	**125**	**150**
0·10000	0·665	0·787	0·901	1·112	1·304	1·484	1·569	1·653	1·734	1·813	1·966	2·114	2·460	2·781
1/ 10·0	0·2089	0·3864	0·6371	1·3970	2·5610	4·1951	5·2076	6·3603	7·6601	9·1139	12·510	16·601	30·187	49·147
0·10500	0·682	0·807	0·924	1·139	1·337	1·520	1·608	1·694	1·777	1·858	2·015	2·166	2·521	2·850
1/ 9·5	0·2142	0·3960	0·6530	1·4317	2·6246	4·2991	5·3367	6·5179	7·8499	9·3397	12·820	17·012	30·935	50·363
0·11000	0·698	0·826	0·946	1·166	1·368	1·556	1·646	1·734	1·819	1·902	2·063	2·217	2·580	2·917
1/ 9·1	0·2192	0·4054	0·6684	1·4656	2·6866	4·4007	5·4628	6·6718	8·0353	9·5602	13·123	17·414	31·664	51·551
0·11500	0·714	0·845	0·967	1·193	1·399	1·592	1·683	1·773	1·860	1·945	2·109	2·267	2·638	2·983
1/ 8·7	0·2242	0·4146	0·6835	1·4987	2·7472	4·5000	5·5860	6·8223	8·2164	9·7758	13·419	17·806	32·378	52·712
0·12000	0·729	0·863	0·988	1·218	1·429	1·626	1·720	1·811	1·900	1·987	2·155	2·316	2·695	3·047
1/ 8·3	0·2291	0·4236	0·6983	1·5311	2·8066	4·5971	5·7066	6·9695	8·3937	9·9867	13·708	18·190	33·076	53·848
0·12500	0·744	0·881	1·008	1·244	1·459	1·660	1·755	1·848	1·939	2·028	2·199	2·364	2·751	3·110
1/ 8·0	0·2338	0·4324	0·7128	1·5628	2·8647	4·6923	5·8247	7·1137	8·5674	10·193	13·991	18·566	33·759	54·961
0·13000	0·759	0·898	1·029	1·268	1·488	1·693	1·790	1·885	1·978	2·068	2·243	2·411	2·806	3·172
1/ 7·7	0·2385	0·4410	0·7270	1·5939	2·9217	4·7855	5·9404	7·2551	8·7376	10·396	14·269	18·935	34·429	56·051
0·13500	0·774	0·916	1·048	1·293	1·516	1·725	1·824	1·921	2·016	2·108	2·286	2·457	2·859	3·232
1/ 7·4	0·2431	0·4495	0·7410	1·6244	2·9776	4·8770	6·0540	7·3937	8·9045	10·594	14·542	19·296	35·087	57·121
0·14000	0·788	0·933	1·068	1·316	1·544	1·757	1·858	1·957	2·053	2·146	2·328	2·502	2·912	3·292
1/ 7·1	0·2476	0·4578	0·7546	1·6543	3·0324	4·9668	6·1654	7·5298	9·0684	10·789	14·809	19·651	35·732	58·171
0·14500	0·802	0·949	1·087	1·340	1·572	1·788	1·891	1·991	2·089	2·185	2·369	2·546	2·963	3·350
1/ 6·9	0·2520	0·4659	0·7681	1·6837	3·0863	5·0550	6·2749	7·6635	9·2294	10·981	15·072	20·000	36·366	59·203
0·15000	0·816	0·965	1·105	1·363	1·599	1·819	1·923	2·025	2·125	2·222	2·410	2·590	3·014	3·408
1/ 6·7	0·2563	0·4739	0·7813	1·7127	3·1393	5·1417	6·3825	7·7949	9·3876	11·169	15·331	20·343	36·989	60·217
0·16000	0·843	0·997	1·142	1·408	1·651	1·878	1·987	2·092	2·195	2·295	2·489	2·675	3·113	3·520
1/ 6·3	0·2648	0·4896	0·8070	1·7691	3·2426	5·3109	6·5925	8·0513	9·6964	11·536	15·835	21·012	38·204	62·195
0·17000	0·869	1·028	1·177	1·451	1·702	1·936	2·048	2·157	2·263	2·366	2·566	2·758	3·209	3·628
1/ 5·9	0·2730	0·5047	0·8320	1·8238	3·3428	5·4749	6·7960	8·2999	9·9956	11·892	16·323	21·660	39·382	64·113
0·18000	0·894	1·058	1·211	1·494	1·752	1·993	2·108	2·219	2·328	2·435	2·640	2·838	3·302	3·733
1/ 5·6	0·2810	0·5194	0·8562	1·8769	3·4400	5·6341	6·9936	8·5412	10·286	12·238	16·797	22·289	40·526	65·974
0·19000	0·919	1·087	1·245	1·535	1·800	2·047	2·166	2·280	2·392	2·502	2·713	2·916	3·393	3·836
1/ 5·3	0·2887	0·5338	0·8798	1·9285	3·5346	5·7890	7·1858	8·7758	10·569	12·574	17·259	22·901	41·639	67·785
0·20000	0·943	1·116	1·277	1·575	1·847	2·101	2·222	2·340	2·455	2·567	2·784	2·992	3·481	3·936
1/ 5·0	0·2963	0·5477	0·9028	1·9788	3·6267	5·9398	7·3730	9·0044	10·844	12·902	17·708	23·497	42·722	69·549
0·21000	0·966	1·143	1·309	1·614	1·893	2·153	2·277	2·398	2·515	2·630	2·852	3·066	3·567	4·033
1/ 4·8	0·3036	0·5613	0·9252	2·0278	3·7166	6·0869	7·5556	9·2274	11·113	13·221	18·146	24·079	43·779	71·269
0·22000	0·989	1·171	1·340	1·652	1·938	2·204	2·331	2·454	2·575	2·692	2·920	3·138	3·652	4·128
1/ 4·5	0·3108	0·5746	0·9471	2·0757	3·8043	6·2306	7·7339	9·4451	11·375	13·533	18·574	24·646	44·811	72·949
0·23000	1·012	1·197	1·370	1·689	1·981	2·253	2·383	2·510	2·633	2·753	2·985	3·209	3·734	4·221
1/ 4·3	0·3179	0·5876	0·9684	2·1226	3·8901	6·3710	7·9081	9·6579	11·631	13·838	18·993	25·201	45·820	74·591
0·24000	1·034	1·223	1·400	1·726	2·024	2·302	2·435	2·564	2·689	2·812	3·050	3·278	3·814	4·312
1/ 4·2	0·3247	0·6003	0·9894	2·1684	3·9740	6·5084	8·0787	9·8661	11·882	14·136	19·402	25·745	46·807	76·197
0·25000	1·055	1·248	1·429	1·761	2·066	2·349	2·485	2·617	2·745	2·870	3·113	3·346	3·893	4·401
1/ 4·0	0·3315	0·6127	1·0098	2·2132	4·0562	6·6430	8·2457	10·070	12·127	14·428	19·803	26·276	47·774	77·771
0·26000	1·076	1·273	1·457	1·796	2·107	2·396	2·534	2·669	2·800	2·927	3·175	3·412	3·970	4·488
1/ 3·8	0·3381	0·6249	1·0299	2·2572	4·1368	6·7749	8·4094	10·270	12·368	14·714	20·196	26·798	48·721	79·313
0·27000	1·097	1·297	1·485	1·831	2·147	2·442	2·583	2·720	2·853	2·983	3·235	3·477	4·046	4·574
1/ 3·7	0·3445	0·6369	1·0496	2·3004	4·2158	6·9042	8·5699	10·466	12·604	14·995	20·581	27·309	49·651	80·826
0·28000	1·117	1·321	1·512	1·864	2·187	2·487	2·630	2·770	2·905	3·038	3·295	3·541	4·120	4·658
1/ 3·6	0·3509	0·6486	1·0690	2·3427	4·2934	7·0312	8·7276	10·658	12·836	15·271	20·960	27·811	50·563	82·311
0·29000	1·137	1·345	1·539	1·897	2·225	2·531	2·677	2·819	2·957	3·092	3·353	3·604	4·193	4·740
1/ 3·4	0·3571	0·6601	1·0880	2·3843	4·3696	7·1560	8·8824	10·848	13·063	15·542	21·331	28·304	51·460	83·770
0·30000	1·156	1·368	1·566	1·930	2·264	2·574	2·723	2·867	3·008	3·145	3·411	3·666	4·265	4·822
1/ 3·3	0·3633	0·6715	1·1066	2·4252	4·4445	7·2786	9·0346	11·033	13·287	15·808	21·697	28·789	52·341	85·204
	–	–	–	–	–	–	–	–	0·74	0·74	0·74	0·74	0·74	0·75

$V_{r(0\cdot5)medial}$ for half-full circular pipes.

k_s = 3·0 mm
S = 0·00030 to 0·00100

ie hydraulic gradient =
1 in 3333 to 1 in 1000

Water (or sewage) at 15°C;
full bore conditions.

velocities in ms^{-1}
discharges in litres/sec

Gradient **(Equivalent) Pipe diameters in mm**

Gradient	20	25	30	40	50	60	65	70	75	80	90	100	125	150
0·000300					0·059	0·067	0·072	0·076	0·080	0·084	0·091	0·099	0·116	0·132
1/ 3333					0·1152	0·1908	0·2379	0·2917	0·3526	0·4209	0·5810	0·7747	1·4217	2·3304
0·000320					0·061	0·070	0·074	0·078	0·083	0·087	0·094	0·102	0·120	0·136
1/ 3125					0·1191	0·1973	0·2460	0·3016	0·3645	0·4351	0·6006	0·8008	1·4694	2·4084
0·000340					0·063	0·072	0·076	0·081	0·085	0·089	0·097	0·105	0·124	0·141
1/ 2941					0·1229	0·2036	0·2538	0·3112	0·3761	0·4489	0·6196	0·8260	1·5157	2·4840
0·000360					0·064	0·074	0·079	0·083	0·088	0·092	0·100	0·108	0·127	0·145
1/ 2778					0·1266	0·2097	0·2614	0·3205	0·3874	0·4623	0·6381	0·8506	1·5606	2·5575
0·000380					0·066	0·076	0·081	0·086	0·090	0·095	0·103	0·111	0·131	0·149
1/ 2632					0·1302	0·2156	0·2688	0·3296	0·3983	0·4754	0·6560	0·8745	1·6043	2·6289
0·000400				0·057	0·068	0·078	0·083	0·088	0·093	0·097	0·106	0·114	0·134	0·153
1/ 2500				0·0719	0·1338	0·2214	0·2760	0·3384	0·4090	0·4881	0·6735	0·8978	1·6468	2·6985
0·000420				0·059	0·070	0·080	0·085	0·090	0·095	0·100	0·109	0·117	0·138	0·157
1/ 2381				0·0738	0·1372	0·2271	0·2831	0·3470	0·4194	0·5005	0·6906	0·9205	1·6883	2·7663
0·000440				0·060	0·072	0·082	0·087	0·092	0·097	0·102	0·111	0·120	0·141	0·160
1/ 2273				0·0756	0·1405	0·2326	0·2899	0·3554	0·4295	0·5126	0·7072	0·9427	1·7289	2·8326
0·000460				0·062	0·073	0·084	0·089	0·094	0·099	0·104	0·114	0·123	0·144	0·164
1/ 2174				0·0773	0·1438	0·2380	0·2967	0·3637	0·4394	0·5244	0·7235	0·9643	1·7685	2·8973
0·000480				0·063	0·075	0·086	0·091	0·097	0·102	0·107	0·116	0·125	0·147	0·168
1/ 2083				0·0791	0·1470	0·2433	0·3032	0·3717	0·4491	0·5360	0·7395	0·9855	1·8073	2·9607
0·000500				0·064	0·076	0·088	0·093	0·099	0·104	0·109	0·119	0·128	0·150	0·171
1/ 2000				0·0808	0·1502	0·2484	0·3097	0·3796	0·4586	0·5473	0·7551	1·0063	1·8452	3·0228
0·000525				0·066	0·078	0·090	0·096	0·101	0·106	0·112	0·122	0·131	0·154	0·175
1/ 1905				0·0829	0·1540	0·2548	0·3175	0·3892	0·4703	0·5611	0·7742	1·0317	1·8916	3·0987
0·000550				0·068	0·080	0·092	0·098	0·104	0·109	0·114	0·125	0·135	0·158	0·180
1/ 1818				0·0849	0·1577	0·2609	0·3252	0·3986	0·4816	0·5747	0·7928	1·0565	1·9370	3·1728
0·000575				0·069	0·082	0·094	0·100	0·106	0·112	0·117	0·127	0·138	0·161	0·184
1/ 1739				0·0869	0·1614	0·2670	0·3327	0·4078	0·4927	0·5879	0·8110	1·0807	1·9813	3·2452
0·000600				0·071	0·084	0·097	0·102	0·108	0·114	0·120	0·130	0·141	0·165	0·188
1/ 1667				0·0888	0·1650	0·2729	0·3401	0·4168	0·5036	0·6008	0·8288	1·1044	2·0246	3·3161
0·000625				0·072	0·086	0·099	0·105	0·111	0·116	0·122	0·133	0·144	0·168	0·192
1/ 1600				0·0907	0·1685	0·2787	0·3473	0·4256	0·5142	0·6135	0·8463	1·1277	2·0671	3·3855
0·000650				0·074	0·088	0·101	0·107	0·113	0·119	0·125	0·136	0·146	0·172	0·195
1/ 1538				0·0926	0·1719	0·2843	0·3543	0·4343	0·5246	0·6259	0·8634	1·1504	2·1087	3·4535
0·000675				0·075	0·089	0·103	0·109	0·115	0·121	0·127	0·138	0·149	0·175	0·199
1/ 1481				0·0944	0·1753	0·2899	0·3613	0·4427	0·5349	0·6381	0·8802	1·1728	2·1496	3·5203
0·000700				0·077	0·091	0·104	0·111	0·117	0·123	0·129	0·141	0·152	0·178	0·203
1/ 1429				0·0962	0·1786	0·2954	0·3681	0·4511	0·5449	0·6501	0·8966	1·1947	2·1896	3·5858
0·000725				0·078	0·093	0·106	0·113	0·119	0·126	0·132	0·143	0·155	0·182	0·207
1/ 1379				0·0980	0·1819	0·3007	0·3747	0·4592	0·5548	0·6619	0·9128	1·2162	2·2290	3·6502
0·000750				0·079	0·094	0·108	0·115	0·121	0·128	0·134	0·146	0·158	0·185	0·210
1/ 1333				0·0997	0·1851	0·3060	0·3813	0·4673	0·5645	0·6734	0·9287	1·2374	2·2677	3·7134
0·000800				0·082	0·097	0·112	0·119	0·125	0·132	0·138	0·151	0·163	0·191	0·217
1/ 1250				0·1031	0·1913	0·3163	0·3941	0·4829	0·5834	0·6960	0·9598	1·2787	2·3432	3·8369
0·000850				0·085	0·101	0·115	0·123	0·129	0·136	0·143	0·156	0·168	0·197	0·224
1/ 1176				0·1063	0·1974	0·3263	0·4065	0·4981	0·6017	0·7178	0·9899	1·3187	2·4164	3·9565
0·000900				0·087	0·104	0·119	0·126	0·133	0·140	0·147	0·160	0·173	0·203	0·230
1/ 1111				0·1095	0·2033	0·3360	0·4186	0·5129	0·6195	0·7390	1·0191	1·3576	2·4875	4·0727
0·000950				0·090	0·106	0·122	0·130	0·137	0·144	0·151	0·165	0·178	0·208	0·237
1/ 1053				0·1126	0·2090	0·3454	0·4303	0·5272	0·6368	0·7597	1·0475	1·3954	2·5566	4·1857
0·001000				0·092	0·109	0·125	0·133	0·141	0·148	0·155	0·169	0·182	0·214	0·243
1/ 1000				0·1156	0·2146	0·3546	0·4417	0·5412	0·6537	0·7798	1·0752	1·4322	2·6239	4·2958
				–	–	–	–	–	–	–	–	–	–	–

$V_{r(0\cdot5)medial}$ for half-full circular pipes.

k_s = 3·0 mm S = 0·00030 to 0·00100

ie hydraulic gradient = 1 in 1000 to 1 in 333

velocities in ms^{-1}
discharges in litres/sec

Gradient **(Equivalent) Pipe diameters in mm**

Gradient	20	25	30	40	50	60	65	70	75	80	90	100	125	150
0·00100				0·092	0·109	0·125	0·133	0·141	0·148	0·155	0·169	0·182	0·214	0·243
1/ 1000				0·1156	0·2146	0·3546	0·4417	0·5412	0·6537	0·7798	1·0752	1·4322	2·6239	4·2958
0·00105				0·094	0·112	0·129	0·136	0·144	0·152	0·159	0·173	0·187	0·219	0·249
1/ 952				0·1186	0·2200	0·3635	0·4529	0·5548	0·6701	0·7994	1·1022	1·4682	2·6896	4·4031
0·00110			0·077	0·097	0·115	0·132	0·140	0·148	0·155	0·163	0·177	0·191	0·224	0·255
1/ 909			0·0544	0·1214	0·2253	0·3722	0·4637	0·5682	0·6862	0·8185	1·1286	1·5032	2·7537	4·5079
0·00115			0·079	0·099	0·117	0·135	0·143	0·151	0·159	0·167	0·181	0·196	0·230	0·261
1/ 870			0·0557	0·1243	0·2305	0·3808	0·4744	0·5812	0·7019	0·8372	1·1543	1·5375	2·8164	4·6104
0·00120			0·081	0·101	0·120	0·138	0·146	0·154	0·162	0·170	0·185	0·200	0·235	0·267
1/ 833			0·0569	0·1270	0·2356	0·3892	0·4848	0·5939	0·7173	0·8556	1·1795	1·5711	2·8778	4·7107
0·00125			0·082	0·103	0·122	0·141	0·149	0·158	0·166	0·174	0·189	0·204	0·239	0·272
1/ 800			0·0581	0·1297	0·2405	0·3973	0·4950	0·6064	0·7323	0·8735	1·2042	1·6040	2·9378	4·8089
0·00130			0·084	0·105	0·125	0·143	0·152	0·161	0·169	0·177	0·193	0·208	0·244	0·278
1/ 769			0·0593	0·1323	0·2454	0·4054	0·5049	0·6186	0·7471	0·8911	1·2285	1·6362	2·9967	4·9051
0·00135			0·086	0·107	0·127	0·146	0·155	0·164	0·172	0·181	0·197	0·212	0·249	0·283
1/ 741			0·0605	0·1349	0·2502	0·4132	0·5147	0·6306	0·7615	0·9083	1·2522	1·6678	3·0545	4·9995
0·00140			0·087	0·109	0·130	0·149	0·158	0·167	0·176	0·184	0·200	0·216	0·254	0·288
1/ 714			0·0616	0·1375	0·2549	0·4210	0·5244	0·6424	0·7757	0·9253	1·2755	1·6988	3·1112	5·0922
0·00145			0·089	0·111	0·132	0·152	0·161	0·170	0·179	0·187	0·204	0·220	0·258	0·293
1/ 690			0·0628	0·1399	0·2595	0·4286	0·5338	0·6539	0·7897	0·9419	1·2984	1·7292	3·1669	5·1833
0·00150			0·090	0·113	0·134	0·154	0·164	0·173	0·182	0·191	0·208	0·224	0·263	0·298
1/ 667			0·0639	0·1424	0·2640	0·4360	0·5431	0·6653	0·8034	0·9582	1·3209	1·7592	3·2216	5·2728
0·00160			0·093	0·117	0·139	0·159	0·169	0·179	0·188	0·197	0·215	0·231	0·271	0·308
1/ 625			0·0660	0·1472	0·2728	0·4506	0·5612	0·6875	0·8302	0·9901	1·3649	1·8176	3·3285	5·4474
0·00170			0·096	0·121	0·143	0·164	0·174	0·184	0·194	0·203	0·221	0·239	0·280	0·318
1/ 588			0·0681	0·1518	0·2814	0·4647	0·5788	0·7090	0·8561	1·0210	1·4074	1·8743	3·4320	5·6166
0·00180			0·099	0·124	0·148	0·169	0·180	0·190	0·199	0·209	0·228	0·246	0·288	0·327
1/ 556			0·0701	0·1563	0·2897	0·4784	0·5958	0·7298	0·8813	1·0511	1·4488	1·9293	3·5325	5·7809
0·00190			0·102	0·128	0·152	0·174	0·185	0·195	0·205	0·215	0·234	0·252	0·296	0·336
1/ 526			0·0721	0·1607	0·2978	0·4917	0·6124	0·7502	0·9058	1·0803	1·4890	1·9828	3·6303	5·9408
0·00200			0·105	0·131	0·156	0·179	0·189	0·200	0·210	0·221	0·240	0·259	0·304	0·345
1/ 500			0·0740	0·1650	0·3057	0·5047	0·6286	0·7699	0·9297	1·1087	1·5281	2·0349	3·7256	6·0965
0·00210		0·093	0·107	0·135	0·160	0·183	0·194	0·205	0·216	0·226	0·246	0·266	0·311	0·354
1/ 476		0·0455	0·0759	0·1691	0·3134	0·5174	0·6443	0·7892	0·9529	1·1365	1·5663	2·0857	3·8184	6·2483
0·00220		0·095	0·110	0·138	0·163	0·187	0·199	0·210	0·221	0·231	0·252	0·272	0·319	0·362
1/ 455		0·0466	0·0778	0·1732	0·3209	0·5298	0·6597	0·8080	0·9757	1·1635	1·6036	2·1353	3·9092	6·3966
0·00230		0·097	0·113	0·141	0·167	0·192	0·203	0·215	0·226	0·237	0·258	0·278	0·326	0·370
1/ 435		0·0477	0·0795	0·1772	0·3282	0·5419	0·6748	0·8265	0·9979	1·1900	1·6401	2·1838	3·9978	6·5415
0·00240		0·099	0·115	0·144	0·171	0·196	0·208	0·219	0·231	0·242	0·263	0·284	0·333	0·378
1/ 417		0·0488	0·0813	0·1811	0·3354	0·5537	0·6895	0·8445	1·0196	1·2159	1·6757	2·2313	4·0846	6·6833
0·00250		0·101	0·117	0·147	0·174	0·200	0·212	0·224	0·236	0·247	0·269	0·290	0·340	0·386
1/ 400		0·0498	0·0830	0·1849	0·3425	0·5653	0·7039	0·8621	1·0409	1·2413	1·7107	2·2778	4·1696	6·8222
0·00260		0·103	0·120	0·150	0·178	0·204	0·216	0·229	0·240	0·252	0·274	0·296	0·347	0·394
1/ 385		0·0508	0·0847	0·1886	0·3493	0·5766	0·7180	0·8794	1·0618	1·2662	1·7449	2·3233	4·2528	6·9583
0·00270		0·106	0·122	0·153	0·181	0·208	0·221	0·233	0·245	0·257	0·280	0·302	0·353	0·401
1/ 370		0·0518	0·0863	0·1923	0·3561	0·5878	0·7319	0·8964	1·0823	1·2906	1·7785	2·3680	4·3345	7·0919
0·00280		0·107	0·124	0·156	0·185	0·212	0·225	0·237	0·250	0·262	0·285	0·307	0·360	0·409
1/ 357		0·0528	0·0880	0·1959	0·3627	0·5987	0·7455	0·9130	1·1024	1·3145	1·8115	2·4119	4·4147	7·2230
0·00290		0·109	0·127	0·159	0·188	0·216	0·229	0·241	0·254	0·266	0·290	0·313	0·366	0·416
1/ 345		0·0537	0·0896	0·1994	0·3693	0·6094	0·7589	0·9294	1·1221	1·3381	1·8439	2·4550	4·4935	7·3518
0·00300		0·111	0·129	0·161	0·191	0·219	0·233	0·246	0·258	0·271	0·295	0·318	0·372	0·423
1/ 333		0·0547	0·0911	0·2029	0·3757	0·6200	0·7720	0·9455	1·1415	1·3612	1·8757	2·4974	4·5710	7·4783
		–	–	–	–	–	–	–	–	–	–	–	–	–

$V_{r(0\cdot5)medial}$ **for half-full circular pipes.**

A10
(p.3 of 6)

k_s = 3·0 mm
S = 0·00300 to 0·01000

ie hydraulic gradient =
1 in 333 to 1 in 100

Water (or sewage) at 15°C;
full bore conditions.

velocities in ms^{-1}
discharges in litres/sec

Gradient	**(Equivalent) Pipe diameters in mm**													
	20	**25**	**30**	**40**	**50**	**60**	**65**	**70**	**75**	**80**	**90**	**100**	**125**	**150**
0·00300		0·111	0·129	0·161	0·191	0·219	0·233	0·246	0·258	0·271	0·295	0·318	0·372	0·423
1/ 333		0·0547	0·0911	0·2029	0·3757	0·6200	0·7720	0·9455	1·1415	1·3612	1·8757	2·4974	4·5710	7·4783
0·00320		0·115	0·133	0·167	0·198	0·227	0·240	0·254	0·267	0·280	0·305	0·328	0·385	0·437
1/ 313		0·0565	0·0942	0·2096	0·3882	0·6406	0·7976	0·9768	1·1794	1·4063	1·9379	2·5800	4·7221	7·7253
0·00340		0·119	0·137	0·172	0·204	0·234	0·248	0·262	0·275	0·288	0·314	0·339	0·397	0·451
1/ 294		0·0583	0·0971	0·2162	0·4003	0·6606	0·8225	1·0073	1·2160	1·4500	1·9981	2·6601	4·8685	7·9647
0·00360		0·122	0·141	0·177	0·210	0·240	0·255	0·269	0·283	0·297	0·323	0·349	0·408	0·464
1/ 278		0·0600	0·1000	0·2226	0·4121	0·6800	0·8466	1·0368	1·2517	1·4925	2·0565	2·7379	5·0107	8·1971
0·00380		0·126	0·145	0·182	0·216	0·247	0·262	0·277	0·291	0·305	0·332	0·358	0·420	0·477
1/ 263		0·0617	0·1028	0·2288	0·4235	0·6988	0·8701	1·0655	1·2863	1·5338	2·1134	2·8136	5·1490	8·4232
0·00400		0·129	0·149	0·187	0·221	0·254	0·269	0·284	0·299	0·313	0·341	0·368	0·431	0·489
1/ 250		0·0633	0·1055	0·2348	0·4347	0·7172	0·8929	1·0935	1·3201	1·5740	2·1688	2·8873	5·2837	8·6434
0·00420		0·132	0·153	0·192	0·227	0·260	0·276	0·291	0·306	0·321	0·349	0·377	0·441	0·501
1/ 238		0·0649	0·1082	0·2407	0·4456	0·7351	0·9152	1·1208	1·3530	1·6133	2·2228	2·9592	5·4151	8·8581
0·00440		0·135	0·157	0·196	0·232	0·266	0·282	0·298	0·314	0·329	0·358	0·386	0·452	0·513
1/ 227		0·0665	0·1108	0·2464	0·4562	0·7526	0·9370	1·1474	1·3852	1·6516	2·2756	3·0294	5·5434	9·0678
0·00460	0·115	0·139	0·160	0·201	0·238	0·272	0·289	0·305	0·321	0·336	0·366	0·394	0·462	0·525
1/ 217	0·0363	0·0680	0·1133	0·2521	0·4666	0·7697	0·9583	1·1734	1·4166	1·6890	2·3271	3·0980	5·6688	9·2728
0·00480	0·118	0·142	0·164	0·205	0·243	0·278	0·295	0·312	0·328	0·343	0·374	0·403	0·472	0·536
1/ 208	0·0371	0·0695	0·1158	0·2576	0·4767	0·7864	0·9791	1·1989	1·4473	1·7257	2·3776	3·1651	5·7915	9·4734
0·00500	0·120	0·145	0·167	0·209	0·248	0·284	0·301	0·318	0·334	0·350	0·382	0·411	0·482	0·547
1/ 200	0·0378	0·0710	0·1182	0·2629	0·4867	0·8028	0·9995	1·2239	1·4774	1·7616	2·4270	3·2308	5·9117	9·6698
0·00525	0·123	0·148	0·171	0·214	0·254	0·291	0·309	0·326	0·343	0·359	0·391	0·422	0·494	0·561
1/ 190	0·0388	0·0728	0·1212	0·2695	0·4988	0·8229	1·0244	1·2544	1·5142	1·8054	2·4874	3·3112	6·0586	9·9099
0·00550	0·126	0·152	0·176	0·220	0·260	0·298	0·316	0·334	0·351	0·368	0·400	0·432	0·505	0·574
1/ 182	0·0397	0·0745	0·1241	0·2760	0·5107	0·8424	1·0488	1·2842	1·5502	1·8483	2·5464	3·3897	6·2020	10·144
0·00575	0·129	0·155	0·180	0·225	0·266	0·305	0·323	0·341	0·359	0·376	0·409	0·441	0·517	0·587
1/ 174	0·0406	0·0762	0·1269	0·2822	0·5223	0·8615	1·0725	1·3133	1·5853	1·8902	2·6041	3·4664	6·3422	10·374
0·00600	0·132	0·159	0·184	0·229	0·272	0·311	0·330	0·349	0·367	0·384	0·418	0·451	0·528	0·600
1/ 167	0·0415	0·0779	0·1297	0·2884	0·5337	0·8802	1·0958	1·3418	1·6197	1·9311	2·6605	3·5414	6·4794	10·598
0·00625	0·135	0·162	0·187	0·234	0·277	0·318	0·337	0·356	0·374	0·392	0·427	0·460	0·539	0·612
1/ 160	0·0424	0·0795	0·1324	0·2944	0·5448	0·8986	1·1186	1·3697	1·6534	1·9713	2·7157	3·6149	6·6138	10·817
0·00650	0·138	0·165	0·191	0·239	0·283	0·324	0·344	0·363	0·382	0·400	0·435	0·469	0·550	0·624
1/ 154	0·0433	0·0811	0·1351	0·3003	0·5557	0·9165	1·1410	1·3970	1·6864	2·0106	2·7699	3·6870	6·7455	11·033
0·00675	0·140	0·168	0·195	0·244	0·288	0·330	0·350	0·370	0·389	0·408	0·444	0·478	0·560	0·636
1/ 148	0·0441	0·0827	0·1377	0·3061	0·5664	0·9341	1·1629	1·4239	1·7187	2·0492	2·8230	3·7577	6·8747	11·244
0·00700	0·143	0·172	0·198	0·248	0·294	0·337	0·357	0·377	0·396	0·415	0·452	0·487	0·571	0·648
1/ 143	0·0449	0·0842	0·1403	0·3118	0·5769	0·9514	1·1844	1·4502	1·7505	2·0871	2·8752	3·8271	7·0015	11·451
0·00725	0·146	0·175	0·202	0·253	0·299	0·343	0·363	0·384	0·403	0·423	0·460	0·496	0·581	0·660
1/ 138	0·0457	0·0858	0·1428	0·3174	0·5872	0·9684	1·2055	1·4761	1·7817	2·1243	2·9264	3·8952	7·1261	11·655
0·00750	0·148	0·178	0·206	0·257	0·304	0·348	0·370	0·390	0·410	0·430	0·468	0·504	0·591	0·671
1/ 133	0·0465	0·0873	0·1453	0·3229	0·5973	0·9851	1·2263	1·5015	1·8124	2·1608	2·9767	3·9622	7·2486	11·855
0·00800	0·153	0·184	0·212	0·265	0·314	0·360	0·382	0·403	0·424	0·444	0·483	0·521	0·610	0·693
1/ 125	0·0481	0·0902	0·1501	0·3336	0·6171	1·0177	1·2669	1·5511	1·8723	2·2322	3·0750	4·0929	7·4875	12·245
0·00850	0·158	0·189	0·219	0·274	0·324	0·371	0·394	0·416	0·437	0·458	0·498	0·537	0·629	0·714
1/ 118	0·0496	0·0930	0·1548	0·3440	0·6363	1·0493	1·3062	1·5992	1·9303	2·3013	3·1702	4·2196	7·7191	12·624
0·00900	0·163	0·195	0·225	0·282	0·334	0·382	0·405	0·428	0·450	0·471	0·513	0·553	0·647	0·735
1/ 111	0·0511	0·0957	0·1593	0·3540	0·6549	1·0800	1·3443	1·6459	1·9867	2·3685	3·2627	4·3426	7·9439	12·992
0·00950	0·167	0·200	0·232	0·290	0·343	0·393	0·416	0·439	0·462	0·484	0·527	0·568	0·665	0·755
1/ 105	0·0525	0·0984	0·1638	0·3638	0·6730	1·1098	1·3814	1·6913	2·0415	2·4338	3·3526	4·4623	8·1626	13·349
0·01000	0·172	0·206	0·238	0·297	0·352	0·403	0·427	0·451	0·474	0·497	0·541	0·583	0·683	0·775
1/ 100	0·0539	0·1010	0·1681	0·3734	0·6906	1·1388	1·4176	1·7356	2·0949	2·4974	3·4402	4·5788	8·3756	13·697
	–	–	–	–	–	–	–	–	–	–	–	–	–	0·74

$V_{r(0·5)medial}$ for half-full circular pipes.

k_s = 3·0 mm S = 0·00300 to 0·01000

Gradient	**(Equivalent) Pipe diameters in mm**													
	20	**25**	**30**	**40**	**50**	**60**	**65**	**70**	**75**	**80**	**90**	**100**	**125**	**150**
0·01000	0·172	0·206	0·238	0·297	0·352	0·403	0·427	0·451	0·474	0·497	0·541	0·583	0·683	0·775
1/ 100	0·0539	0·1010	0·1681	0·3734	0·6906	1·1388	1·4176	1·7356	2·0949	2·4974	3·4402	4·5788	8·3756	13·697
0·01050	0·176	0·211	0·244	0·305	0·361	0·413	0·438	0·462	0·486	0·509	0·554	0·597	0·699	0·794
1/ 95	0·0552	0·1035	0·1723	0·3827	0·7078	1·1672	1·4528	1·7787	2·1469	2·5595	3·5256	4·6925	8·5834	14·037
0·01100	0·180	0·216	0·250	0·312	0·369	0·423	0·448	0·473	0·497	0·521	0·567	0·612	0·716	0·813
1/ 91	0·0566	0·1060	0·1764	0·3918	0·7246	1·1948	1·4872	1·8208	2·1977	2·6201	3·6090	4·8034	8·7862	14·368
0·01150	0·184	0·221	0·255	0·319	0·377	0·432	0·458	0·484	0·509	0·533	0·580	0·625	0·732	0·831
1/ 87	0·0579	0·1084	0·1804	0·4007	0·7411	1·2219	1·5209	1·8620	2·2474	2·6793	3·6906	4·9119	8·9845	14·692
0·01200	0·188	0·226	0·261	0·326	0·386	0·442	0·468	0·494	0·520	0·545	0·593	0·639	0·748	0·849
1/ 83	0·0591	0·1107	0·1843	0·4094	0·7571	1·2483	1·5538	1·9023	2·2961	2·7372	3·7703	5·0181	9·1786	15·010
0·01250	0·192	0·230	0·266	0·333	0·394	0·451	0·478	0·505	0·531	0·556	0·605	0·652	0·763	0·867
1/ 80	0·0604	0·1131	0·1881	0·4179	0·7729	1·2742	1·5861	1·9418	2·3437	2·7940	3·8485	5·1220	9·3686	15·320
0·01300	0·196	0·235	0·271	0·339	0·401	0·460	0·487	0·515	0·541	0·567	0·617	0·665	0·779	0·884
1/ 77	0·0616	0·1153	0·1919	0·4262	0·7883	1·2996	1·6177	1·9805	2·3904	2·8496	3·9251	5·2239	9·5549	15·625
0·01350	0·200	0·239	0·277	0·346	0·409	0·468	0·497	0·524	0·551	0·578	0·629	0·678	0·793	0·901
1/ 74	0·0628	0·1176	0·1956	0·4344	0·8034	1·3246	1·6487	2·0184	2·4361	2·9042	4·0002	5·3239	9·7376	15·923
0·01400	0·203	0·244	0·282	0·352	0·417	0·477	0·506	0·534	0·562	0·588	0·640	0·690	0·808	0·918
1/ 71	0·0639	0·1197	0·1992	0·4425	0·8183	1·3490	1·6791	2·0557	2·4811	2·9578	4·0740	5·4220	9·9169	16·216
0·01450	0·207	0·248	0·287	0·358	0·424	0·486	0·515	0·544	0·572	0·599	0·652	0·703	0·822	0·934
1/ 69	0·0651	0·1219	0·2028	0·4504	0·8328	1·3731	1·7090	2·0922	2·5252	3·0104	4·1464	5·5184	10·093	16·504
0·01500	0·211	0·253	0·292	0·365	0·431	0·494	0·524	0·553	0·581	0·609	0·663	0·715	0·837	0·950
1/ 67	0·0662	0·1240	0·2063	0·4581	0·8472	1·3967	1·7384	2·1282	2·5686	3·0621	4·2176	5·6132	10·266	16·787
0·01600	0·218	0·261	0·302	0·377	0·446	0·510	0·541	0·571	0·601	0·629	0·685	0·738	0·864	0·981
1/ 63	0·0684	0·1281	0·2131	0·4733	0·8752	1·4428	1·7957	2·1984	2·6533	3·1630	4·3566	5·7980	10·604	17·340
0·01700	0·225	0·269	0·311	0·388	0·460	0·526	0·558	0·589	0·619	0·649	0·706	0·761	0·891	1·012
1/ 59	0·0705	0·1321	0·2198	0·4880	0·9023	1·4874	1·8513	2·2664	2·7354	3·2608	4·4913	5·9772	10·932	17·875
0·01800	0·231	0·277	0·320	0·400	0·473	0·541	0·574	0·606	0·637	0·668	0·727	0·783	0·917	1·041
1/ 56	0·0726	0·1360	0·2262	0·5022	0·9286	1·5308	1·9053	2·3324	2·8151	3·3558	4·6220	6·1512	11·250	18·395
0·01900	0·238	0·285	0·329	0·411	0·486	0·556	0·590	0·623	0·655	0·686	0·747	0·805	0·942	1·070
1/ 53	0·0746	0·1397	0·2324	0·5161	0·9542	1·5730	1·9577	2·3967	2·8926	3·4482	4·7492	6·3204	11·559	18·900
0·02000	0·244	0·292	0·337	0·421	0·499	0·571	0·605	0·639	0·672	0·704	0·766	0·826	0·966	1·097
1/ 50	0·0766	0·1434	0·2385	0·5296	0·9791	1·6140	2·0089	2·4592	2·9681	3·5382	4·8731	6·4852	11·860	19·393
0·02100	0·250	0·299	0·346	0·432	0·511	0·585	0·620	0·655	0·688	0·721	0·785	0·846	0·990	1·125
1/ 47·6	0·0785	0·1470	0·2445	0·5428	1·0035	1·6541	2·0587	2·5203	3·0417	3·6259	4·9939	6·6459	12·154	19·873
0·02200	0·256	0·307	0·354	0·442	0·523	0·599	0·635	0·670	0·705	0·738	0·804	0·866	1·014	1·151
1/ 45·5	0·0804	0·1505	0·2503	0·5556	1·0272	1·6932	2·1074	2·5798	3·1136	3·7116	5·1119	6·8029	12·441	20·342
0·02300	0·262	0·313	0·362	0·452	0·535	0·612	0·649	0·685	0·721	0·755	0·822	0·886	1·037	1·177
1/ 43·5	0·0822	0·1539	0·2560	0·5682	1·0504	1·7315	2·1550	2·6381	3·1839	3·7954	5·2272	6·9563	12·721	20·800
0·02400	0·267	0·320	0·370	0·462	0·547	0·626	0·663	0·700	0·736	0·771	0·839	0·905	1·059	1·202
1/ 41·7	0·0840	0·1572	0·2615	0·5805	1·0732	1·7689	2·2015	2·6951	3·2526	3·8773	5·3400	7·1065	12·996	21·249
0·02500	0·273	0·327	0·378	0·472	0·558	0·639	0·677	0·715	0·751	0·787	0·857	0·924	1·081	1·227
1/ 40·0	0·0857	0·1605	0·2669	0·5925	1·0954	1·8056	2·2472	2·7509	3·3200	3·9576	5·4505	7·2535	13·265	21·688
0·02600	0·278	0·333	0·385	0·481	0·569	0·651	0·691	0·729	0·766	0·803	0·874	0·942	1·102	1·252
1/ 38·5	0·0875	0·1637	0·2723	0·6044	1·1172	1·8415	2·2919	2·8056	3·3860	4·0363	5·5589	7·3976	13·528	22·119
0·02700	0·284	0·340	0·393	0·490	0·580	0·664	0·704	0·743	0·781	0·818	0·890	0·960	1·123	1·276
1/ 37·0	0·0891	0·1668	0·2775	0·6159	1·1386	1·8767	2·3357	2·8593	3·4507	4·1134	5·6651	7·5390	13·786	22·541
0·02800	0·289	0·346	0·400	0·499	0·591	0·676	0·717	0·757	0·795	0·833	0·907	0·978	1·144	1·299
1/ 35·7	0·0908	0·1699	0·2826	0·6273	1·1596	1·9113	2·3787	2·9119	3·5143	4·1892	5·7694	7·6778	14·040	22·956
0·02900	0·294	0·352	0·407	0·508	0·601	0·688	0·730	0·770	0·810	0·848	0·923	0·995	1·164	1·322
1/ 34·5	0·0924	0·1730	0·2877	0·6385	1·1803	1·9453	2·4210	2·9637	3·5767	4·2636	5·8719	7·8141	14·289	23·363
0·03000	0·299	0·358	0·414	0·517	0·611	0·700	0·742	0·783	0·824	0·863	0·939	1·012	1·184	1·345
1/ 33·3	0·0940	0·1760	0·2926	0·6495	1·2005	1·9787	2·4626	3·0145	3·6381	4·3368	5·9726	7·9481	14·534	23·763
	–	–	–	–	–	–	–	–	–	–	–	–	–	0·74

$V_{r(0·5)medial}$ for half-full circular pipes.

A10
(p.5 of 6)

$k_s = 3{\cdot}0$ **mm**
S = 0·03000 to 0·10000

ie hydraulic gradient =
1 in 33·3 to 1 in 10·0

Water (or sewage) at 15°C;
full bore conditions.

velocities in ms^{-1}
discharges in litres/sec

Gradient	20	25	30	40	50	60	65	70	75	80	90	100	125	150
	(Equivalent) Pipe diameters in mm													
0·03000	0·299	0·358	0·414	0·517	0·611	0·700	0·742	0·783	0·824	0·863	0·939	1·012	1·184	1·345
1/ 33·3	0·0940	0·1760	0·2926	0·6495	1·2005	1·9787	2·4626	3·0145	3·6381	4·3368	5·9726	7·9481	14·534	23·763
0·03200	0·309	0·370	0·428	0·534	0·632	0·723	0·767	0·809	0·851	0·891	0·970	1·045	1·223	1·389
1/ 31·3	0·0971	0·1818	0·3023	0·6709	1·2401	2·0439	2·5437	3·1138	3·7579	4·4795	6·1691	8·2095	15·012	24·544
0·03400	0·319	0·382	0·441	0·550	0·651	0·745	0·790	0·834	0·877	0·919	1·000	1·078	1·261	1·432
1/ 29·4	0·1001	0·1874	0·3117	0·6916	1·2784	2·1070	2·6223	3·2100	3·8739	4·6178	6·3596	8·4629	15·475	25·301
0·03600	0·328	0·393	0·454	0·566	0·670	0·767	0·813	0·858	0·902	0·945	1·029	1·109	1·298	1·473
1/ 27·8	0·1031	0·1929	0·3208	0·7118	1·3157	2·1684	2·6986	3·3034	3·9866	4·7521	6·5445	8·7089	15·925	26·036
0·03800	0·337	0·404	0·466	0·582	0·689	0·788	0·836	0·882	0·927	0·971	1·057	1·139	1·333	1·514
1/ 26·3	0·1059	0·1982	0·3296	0·7314	1·3519	2·2280	2·7728	3·3942	4·0962	4·8828	6·7244	8·9482	16·362	26·751
0·04000	0·346	0·414	0·478	0·597	0·706	0·809	0·857	0·905	0·951	0·997	1·085	1·169	1·368	1·553
1/ 25·0	0·1087	0·2034	0·3382	0·7505	1·3872	2·2861	2·8451	3·4827	4·2030	5·0100	6·8995	9·1813	16·788	27·448
0·04200	0·355	0·425	0·490	0·612	0·724	0·829	0·879	0·927	0·975	1·021	1·111	1·198	1·402	1·592
1/ 23·8	0·1114	0·2085	0·3466	0·7691	1·4216	2·3428	2·9156	3·5690	4·3071	5·1341	7·0704	9·4086	17·204	28·127
0·04400	0·363	0·435	0·502	0·627	0·741	0·848	0·899	0·949	0·998	1·046	1·138	1·226	1·435	1·629
1/ 22·7	0·1141	0·2134	0·3548	0·7873	1·4552	2·3981	2·9845	3·6533	4·4088	5·2553	7·2372	9·6306	17·610	28·790
0·04600	0·371	0·445	0·513	0·641	0·758	0·867	0·920	0·971	1·020	1·069	1·163	1·254	1·467	1·666
1/ 21·7	0·1166	0·2182	0·3629	0·8051	1·4880	2·4522	3·0518	3·7356	4·5082	5·3737	7·4003	9·8476	18·006	29·438
0·04800	0·379	0·454	0·524	0·655	0·774	0·886	0·940	0·992	1·042	1·092	1·188	1·281	1·499	1·702
1/ 20·8	0·1192	0·2230	0·3707	0·8225	1·5201	2·5052	3·1176	3·8162	4·6054	5·4896	7·5599	10·060	18·394	30·073
0·05000	0·387	0·464	0·535	0·668	0·790	0·904	0·959	1·012	1·064	1·115	1·213	1·307	1·530	1·737
1/ 20·0	0·1216	0·2276	0·3784	0·8395	1·5516	2·5570	3·1821	3·8952	4·7007	5·6032	7·7162	10·268	18·774	30·694
0·05250	0·397	0·475	0·549	0·685	0·810	0·927	0·983	1·037	1·090	1·142	1·243	1·340	1·568	1·780
1/ 19·0	0·1247	0·2332	0·3878	0·8604	1·5901	2·6204	3·2609	3·9916	4·8171	5·7419	7·9072	10·522	19·239	31·453
0·05500	0·406	0·486	0·562	0·701	0·829	0·949	1·006	1·062	1·116	1·169	1·272	1·371	1·605	1·822
1/ 18·2	0·1276	0·2388	0·3970	0·8807	1·6276	2·6822	3·3379	4·0858	4·9308	5·8774	8·0938	10·770	19·693	32·195
0·05750	0·415	0·497	0·574	0·717	0·848	0·970	1·029	1·086	1·141	1·196	1·301	1·402	1·641	1·863
1/ 17·4	0·1305	0·2442	0·4059	0·9006	1·6643	2·7427	3·4132	4·1779	5·0419	6·0098	8·2761	11·013	20·136	32·919
0·06000	0·424	0·508	0·587	0·732	0·866	0·991	1·051	1·109	1·166	1·221	1·329	1·432	1·676	1·903
1/ 16·7	0·1333	0·2494	0·4147	0·9200	1·7003	2·8019	3·4868	4·2680	5·1506	6·1394	8·4545	11·250	20·570	33·629
0·06250	0·433	0·519	0·599	0·747	0·884	1·011	1·073	1·132	1·190	1·247	1·356	1·462	1·711	1·942
1/ 16·0	0·1361	0·2546	0·4233	0·9391	1·7354	2·8598	3·5589	4·3563	5·2571	6·2663	8·6293	11·483	20·995	34·323
0·06500	0·442	0·529	0·611	0·762	0·901	1·032	1·094	1·154	1·214	1·271	1·383	1·491	1·745	1·981
1/ 15·4	0·1388	0·2597	0·4317	0·9578	1·7699	2·9166	3·6296	4·4428	5·3615	6·3907	8·8005	11·710	21·411	35·004
0·06750	0·450	0·539	0·622	0·777	0·919	1·051	1·115	1·176	1·237	1·296	1·410	1·519	1·778	2·019
1/ 14·8	0·1415	0·2647	0·4400	0·9761	1·8038	2·9723	3·6989	4·5277	5·4639	6·5128	8·9686	11·934	21·820	35·672
0·07000	0·459	0·549	0·634	0·791	0·936	1·071	1·135	1·198	1·260	1·320	1·436	1·547	1·811	2·056
1/ 14·3	0·1441	0·2695	0·4481	0·9941	1·8370	3·0270	3·7670	4·6109	5·5644	6·6326	9·1335	12·153	22·221	36·327
0·07250	0·467	0·559	0·645	0·805	0·952	1·090	1·155	1·219	1·282	1·343	1·461	1·575	1·843	2·092
1/ 13·8	0·1467	0·2743	0·4561	1·0117	1·8696	3·0808	3·8338	4·6928	5·6631	6·7502	9·2955	12·369	22·615	36·971
0·07500	0·475	0·568	0·656	0·819	0·969	1·108	1·175	1·240	1·304	1·366	1·486	1·602	1·874	2·128
1/ 13·3	0·1492	0·2791	0·4639	1·0291	1·9017	3·1336	3·8995	4·7732	5·7602	6·8659	9·4547	12·581	23·002	37·604
0·08000	0·491	0·587	0·678	0·846	1·000	1·145	1·214	1·281	1·347	1·411	1·535	1·654	1·936	2·198
1/ 12·5	0·1541	0·2883	0·4792	1·0630	1·9642	3·2366	4·0278	4·9301	5·9495	7·0916	9·7654	12·994	23·758	38·839
0·08500	0·506	0·605	0·699	0·872	1·031	1·180	1·251	1·321	1·388	1·454	1·582	1·705	1·996	2·266
1/ 11·8	0·1589	0·2972	0·4940	1·0958	2·0249	3·3365	4·1520	5·0822	6·1330	7·3103	10·067	13·395	24·490	40·036
0·09000	0·521	0·623	0·719	0·897	1·061	1·214	1·288	1·359	1·429	1·497	1·628	1·755	2·054	2·331
1/ 11·1	0·1635	0·3058	0·5084	1·1277	2·0837	3·4335	4·2727	5·2299	6·3112	7·5227	10·359	13·784	25·201	41·199
0·09500	0·535	0·640	0·739	0·922	1·090	1·248	1·323	1·396	1·468	1·538	1·673	1·803	2·110	2·395
1/ 10·5	0·1680	0·3143	0·5224	1·1587	2·1410	3·5278	4·3901	5·3735	6·4845	7·7292	10·643	14·162	25·893	42·329
0·10000	0·549	0·657	0·758	0·946	1·119	1·280	1·357	1·433	1·506	1·578	1·717	1·850	2·165	2·458
1/ 10·0	0·1724	0·3225	0·5360	1·1889	2·1968	3·6197	4·5044	5·5134	6·6533	7·9304	10·920	14·531	26·566	43·430
	–	–	–	–	–	–	–	–	–	–	–	–	–	0·74

$V_{r(0{\cdot}5)medial}$ for half-full circular pipes.

$k_s = 3{\cdot}0$ **mm** S = 0·03000 to 0·10000

Gradient	**(Equivalent) Pipe diameters in mm**													
	20	**25**	**30**	**40**	**50**	**60**	**65**	**70**	**75**	**80**	**90**	**100**	**125**	**150**
0·10000	0·549	0·657	0·758	0·946	1·119	1·280	1·357	1·433	1·506	1·578	1·717	1·850	2·165	2·458
1/ 10·0	0·1724	0·3225	0·5360	1·1889	2·1968	3·6197	4·5044	5·5134	6·6533	7·9304	10·920	14·531	26·566	43·430
0·10500	0·562	0·673	0·777	0·970	1·147	1·312	1·391	1·468	1·543	1·617	1·759	1·896	2·218	2·518
1/ 9·5	0·1767	0·3305	0·5493	1·2183	2·2512	3·7093	4·6158	5·6499	6·8180	8·1266	11·191	14·890	27·223	44·504
0·11000	0·576	0·689	0·795	0·992	1·174	1·343	1·424	1·503	1·580	1·655	1·801	1·941	2·271	2·578
1/ 9·1	0·1809	0·3383	0·5623	1·2471	2·3043	3·7968	4·7247	5·7831	6·9787	8·3182	11·454	15·241	27·865	45·553
0·11500	0·589	0·705	0·813	1·015	1·200	1·373	1·456	1·537	1·615	1·692	1·841	1·984	2·322	2·636
1/ 8·7	0·1850	0·3459	0·5750	1·2752	2·3562	3·8823	4·8311	5·9133	7·1359	8·5055	11·712	15·584	28·492	46·577
0·12000	0·602	0·720	0·831	1·037	1·226	1·403	1·487	1·570	1·650	1·729	1·881	2·027	2·372	2·693
1/ 8·3	0·1890	0·3534	0·5874	1·3027	2·4070	3·9660	4·9352	6·0408	7·2897	8·6888	11·965	15·920	29·106	47·580
0·12500	0·614	0·735	0·848	1·058	1·251	1·432	1·518	1·602	1·684	1·764	1·920	2·069	2·421	2·748
1/ 8·0	0·1929	0·3607	0·5995	1·3297	2·4568	4·0480	5·0372	6·1656	7·4403	8·8683	12·212	16·249	29·707	48·563
0·13000	0·626	0·749	0·865	1·079	1·276	1·460	1·548	1·634	1·718	1·799	1·958	2·110	2·469	2·803
1/ 7·7	0·1967	0·3679	0·6115	1·3561	2·5055	4·1283	5·1372	6·2879	7·5879	9·0442	12·454	16·571	30·296	49·525
0·13500	0·638	0·764	0·882	1·100	1·300	1·488	1·578	1·665	1·750	1·834	1·995	2·150	2·516	2·856
1/ 7·4	0·2005	0·3749	0·6231	1·3820	2·5534	4·2071	5·2352	6·4079	7·7327	9·2168	12·691	16·887	30·874	50·470
0·14000	0·650	0·778	0·898	1·120	1·324	1·515	1·607	1·696	1·782	1·867	2·032	2·190	2·562	2·908
1/ 7·1	0·2042	0·3818	0·6346	1·4074	2·6003	4·2845	5·3315	6·5257	7·8748	9·3862	12·925	17·198	31·441	51·397
0·14500	0·662	0·792	0·914	1·140	1·348	1·542	1·635	1·726	1·814	1·900	2·068	2·228	2·607	2·960
1/ 6·9	0·2078	0·3886	0·6459	1·4324	2·6465	4·3604	5·4260	6·6414	8·0144	9·5526	13·154	17·502	31·998	52·308
0·15000	0·673	0·805	0·929	1·159	1·371	1·569	1·663	1·755	1·845	1·933	2·103	2·267	2·652	3·011
1/ 6·7	0·2114	0·3953	0·6570	1·4569	2·6918	4·4351	5·5190	6·7552	8·1517	9·7162	13·379	17·802	32·546	53·203
0·16000	0·695	0·832	0·960	1·198	1·416	1·620	1·718	1·813	1·906	1·996	2·172	2·341	2·739	3·110
1/ 6·3	0·2184	0·4083	0·6786	1·5048	2·7803	4·5809	5·7003	6·9771	8·4195	10·035	13·818	18·387	33·614	54·950
0·17000	0·717	0·857	0·990	1·234	1·460	1·670	1·771	1·869	1·965	2·058	2·239	2·413	2·824	3·205
1/ 5·9	0·2251	0·4209	0·6995	1·5513	2·8661	4·7221	5·8760	7·1922	8·6790	10·345	14·244	18·953	34·650	56·642
0·18000	0·737	0·882	1·018	1·270	1·502	1·719	1·822	1·923	2·022	2·118	2·304	2·483	2·905	3·298
1/ 5·6	0·2317	0·4331	0·7199	1·5964	2·9493	4·8593	6·0467	7·4010	8·9310	10·645	14·658	19·503	35·655	58·286
0·19000	0·758	0·907	1·046	1·305	1·543	1·766	1·872	1·976	2·077	2·176	2·367	2·551	2·985	3·389
1/ 5·3	0·2380	0·4450	0·7397	1·6402	3·0303	4·9927	6·2127	7·6042	9·1761	10·937	15·060	20·038	36·634	59·885
0·20000	0·777	0·930	1·074	1·339	1·583	1·812	1·921	2·027	2·131	2·232	2·429	2·618	3·063	3·477
1/ 5·0	0·2442	0·4566	0·7589	1·6829	3·1092	5·1226	6·3743	7·8020	9·4148	11·222	15·452	20·560	37·586	61·442
0·21000	0·797	0·953	1·100	1·372	1·623	1·857	1·968	2·077	2·184	2·288	2·489	2·682	3·139	3·563
1/ 4·8	0·2503	0·4680	0·7777	1·7246	3·1861	5·2493	6·5320	7·9950	9·6477	11·499	15·834	21·068	38·515	62·960
0·22000	0·816	0·976	1·126	1·405	1·661	1·900	2·015	2·126	2·235	2·342	2·548	2·746	3·212	3·647
1/ 4·5	0·2562	0·4790	0·7961	1·7652	3·2612	5·3730	6·6859	8·1834	9·8750	11·770	16·207	21·564	39·423	64·443
0·23000	0·834	0·998	1·152	1·436	1·698	1·943	2·060	2·174	2·286	2·394	2·605	2·807	3·285	3·729
1/ 4·3	0·2620	0·4898	0·8140	1·8050	3·3347	5·4940	6·8364	8·3676	10·097	12·035	16·572	22·050	40·309	65·893
0·24000	0·852	1·019	1·176	1·467	1·735	1·985	2·105	2·221	2·335	2·446	2·661	2·868	3·355	3·809
1/ 4·2	0·2676	0·5004	0·8316	1·8439	3·4065	5·6123	6·9837	8·5478	10·315	12·294	16·928	22·524	41·177	67·311
0·25000	0·870	1·040	1·201	1·498	1·771	2·026	2·148	2·267	2·383	2·496	2·716	2·927	3·425	3·888
1/ 4·0	0·2732	0·5107	0·8488	1·8820	3·4769	5·7282	7·1279	8·7243	10·528	12·548	17·278	22·989	42·027	68·700
0·26000	0·887	1·061	1·225	1·527	1·806	2·066	2·191	2·312	2·430	2·546	2·770	2·985	3·493	3·965
1/ 3·8	0·2786	0·5209	0·8656	1·9194	3·5458	5·8419	7·2693	8·8973	10·736	12·797	17·620	23·445	42·860	70·062
0·27000	0·904	1·081	1·248	1·557	1·840	2·106	2·232	2·356	2·477	2·594	2·823	3·042	3·559	4·040
1/ 3·7	0·2839	0·5308	0·8821	1·9560	3·6135	5·9533	7·4079	9·0670	10·941	13·041	17·957	23·892	43·677	71·398
0·28000	0·920	1·101	1·271	1·585	1·874	2·144	2·273	2·399	2·522	2·642	2·874	3·098	3·625	4·114
1/ 3·6	0·2892	0·5406	0·8984	1·9920	3·6799	6·0627	7·5441	9·2336	11·142	13·280	18·286	24·331	44·479	72·709
0·29000	0·937	1·121	1·293	1·613	1·907	2·182	2·314	2·442	2·567	2·689	2·925	3·153	3·689	4·187
1/ 3·4	0·2943	0·5502	0·9143	2·0273	3·7452	6·1702	7·6778	9·3973	11·340	13·516	18·610	24·762	45·267	73·997
0·30000	0·953	1·140	1·316	1·641	1·940	2·220	2·353	2·484	2·611	2·735	2·975	3·207	3·752	4·259
1/ 3·3	0·2993	0·5596	0·9300	2·0620	3·8093	6·2758	7·8092	9·5581	11·534	13·747	18·929	25·186	46·042	75·263
	–	–	–	–	–	–	–	–	–	–	–	–	–	0·74

$V_{r(0{\cdot}5)medial}$ for half-full circular pipes.

A11
(p.1 of 6)

$k_s = 6{\cdot}0$ mm
S = 0·00030 to 0·00100

ie hydraulic gradient =
1 in 3333 to 1 in 1000

Water (or sewage) at 15°C;
full bore conditions.

velocities in ms^{-1}
discharges in litres/sec

Gradient **(Equivalent) Pipe diameters in mm**

Gradient	20	25	30	40	50	60	65	70	75	80	90	100	125	150
0·000300					0·050	0·058	0·061	0·065	0·069	0·072	0·079	0·085	0·101	0·115
1/ 3333					0·0975	0·1626	0·2033	0·2499	0·3028	0·3621	0·5016	0·6708	1·2385	2·0393
0·000320					0·051	0·059	0·063	0·067	0·071	0·074	0·081	0·088	0·104	0·119
1/ 3125					0·1008	0·1681	0·2101	0·2583	0·3129	0·3742	0·5184	0·6932	1·2797	2·1070
0·000340					0·053	0·061	0·065	0·069	0·073	0·077	0·084	0·091	0·108	0·123
1/ 2941					0·1040	0·1733	0·2167	0·2664	0·3227	0·3860	0·5346	0·7149	1·3196	2·1726
0·000360					0·055	0·063	0·067	0·071	0·075	0·079	0·087	0·094	0·111	0·127
1/ 2778					0·1071	0·1785	0·2232	0·2743	0·3323	0·3974	0·5504	0·7360	1·3584	2·2363
0·000380					0·056	0·065	0·069	0·073	0·077	0·081	0·089	0·096	0·114	0·130
1/ 2632					0·1101	0·1835	0·2294	0·2820	0·3415	0·4085	0·5657	0·7564	1·3961	2·2983
0·000400					0·058	0·067	0·071	0·075	0·079	0·083	0·091	0·099	0·117	0·133
1/ 2500					0·1130	0·1884	0·2355	0·2894	0·3506	0·4193	0·5807	0·7764	1·4329	2·3587
0·000420					0·059	0·068	0·073	0·077	0·081	0·086	0·094	0·101	0·120	0·137
1/ 2381					0·1159	0·1931	0·2414	0·2967	0·3594	0·4298	0·5952	0·7958	1·4687	2·4176
0·000440					0·060	0·070	0·075	0·079	0·083	0·088	0·096	0·104	0·123	0·140
1/ 2273					0·1187	0·1978	0·2472	0·3038	0·3680	0·4401	0·6094	0·8148	1·5037	2·4751
0·000460					0·062	0·072	0·076	0·081	0·085	0·090	0·098	0·106	0·125	0·143
1/ 2174					0·1214	0·2023	0·2529	0·3108	0·3764	0·4501	0·6233	0·8334	1·5379	2·5313
0·000480					0·063	0·073	0·078	0·083	0·087	0·092	0·100	0·108	0·128	0·146
1/ 2083					0·1241	0·2067	0·2584	0·3176	0·3846	0·4600	0·6369	0·8516	1·5713	2·5863
0·000500					0·065	0·075	0·080	0·084	0·089	0·093	0·102	0·111	0·131	0·149
1/ 2000					0·1267	0·2111	0·2639	0·3243	0·3927	0·4696	0·6503	0·8694	1·6041	2·6402
0·000525					0·066	0·077	0·082	0·086	0·091	0·096	0·105	0·113	0·134	0·153
1/ 1905					0·1299	0·2164	0·2705	0·3324	0·4026	0·4814	0·6666	0·8911	1·6442	2·7060
0·000550					0·068	0·078	0·083	0·088	0·093	0·098	0·107	0·116	0·137	0·157
1/ 1818					0·1330	0·2216	0·2770	0·3404	0·4122	0·4929	0·6825	0·9123	1·6833	2·7703
0·000575				0·058	0·069	0·080	0·085	0·090	0·095	0·100	0·110	0·119	0·140	0·160
1/ 1739				0·0725	0·1360	0·2266	0·2833	0·3481	0·4216	0·5041	0·6980	0·9331	1·7215	2·8332
0·000600				0·059	0·071	0·082	0·087	0·092	0·098	0·102	0·112	0·121	0·143	0·164
1/ 1667				0·0741	0·1390	0·2316	0·2895	0·3557	0·4308	0·5151	0·7132	0·9534	1·7589	2·8947
0·000625				0·060	0·072	0·084	0·089	0·094	0·100	0·105	0·114	0·124	0·146	0·167
1/ 1600				0·0757	0·1420	0·2365	0·2956	0·3632	0·4398	0·5259	0·7281	0·9733	1·7956	2·9549
0·000650				0·061	0·074	0·085	0·091	0·096	0·102	0·107	0·117	0·126	0·149	0·171
1/ 1538				0·0772	0·1448	0·2412	0·3015	0·3705	0·4486	0·5364	0·7427	0·9928	1·8315	3·0139
0·000675				0·063	0·075	0·087	0·093	0·098	0·104	0·109	0·119	0·129	0·152	0·174
1/ 1481				0·0787	0·1476	0·2459	0·3073	0·3777	0·4573	0·5468	0·7571	1·0120	1·8667	3·0718
0·000700				0·064	0·077	0·089	0·094	0·100	0·105	0·111	0·121	0·131	0·155	0·177
1/ 1429				0·0802	0·1504	0·2505	0·3131	0·3847	0·4658	0·5570	0·7711	1·0307	1·9013	3·1287
0·000725				0·065	0·078	0·090	0·096	0·102	0·107	0·113	0·123	0·134	0·158	0·180
1/ 1379				0·0816	0·1531	0·2550	0·3187	0·3916	0·4742	0·5670	0·7849	1·0492	1·9353	3·1845
0·000750				0·066	0·079	0·092	0·098	0·104	0·109	0·115	0·126	0·136	0·160	0·183
1/ 1333				0·0831	0·1558	0·2594	0·3242	0·3984	0·4824	0·5768	0·7985	1·0673	1·9687	3·2394
0·000800				0·068	0·082	0·095	0·101	0·107	0·113	0·119	0·130	0·140	0·166	0·189
1/ 1250				0·0859	0·1610	0·2681	0·3350	0·4116	0·4984	0·5959	0·8250	1·1027	2·0338	3·3465
0·000850				0·070	0·085	0·098	0·104	0·110	0·116	0·122	0·134	0·145	0·171	0·195
1/ 1176				0·0886	0·1660	0·2764	0·3455	0·4245	0·5140	0·6145	0·8507	1·1370	2·0970	3·4503
0·000900				0·073	0·087	0·101	0·107	0·114	0·120	0·126	0·138	0·149	0·176	0·201
1/ 1111				0·0912	0·1709	0·2846	0·3556	0·4369	0·5291	0·6325	0·8756	1·1703	2·1583	3·5511
0·000950				0·075	0·089	0·103	0·110	0·117	0·123	0·129	0·141	0·153	0·181	0·206
1/ 1053				0·0937	0·1757	0·2925	0·3655	0·4491	0·5437	0·6501	0·8999	1·2027	2·2179	3·6491
0·001000				0·077	0·092	0·106	0·113	0·120	0·126	0·133	0·145	0·157	0·185	0·212
1/ 1000				0·0962	0·1803	0·3002	0·3751	0·4609	0·5580	0·6671	0·9235	1·2342	2·2760	3·7446
				–	–	–	–	–	–	–	–	–	–	–

$V_{r(0{\cdot}5)medial}$ for half-full circular pipes.

$k_s = 6{\cdot}0$ mm S = 0·00030 to 0·00100

k_s = 6·0 mm
S = 0·00100 to 0·00300

ie hydraulic gradient =
1 in 1000 to 1 in 333

Water (or sewage) at 15°C;
full bore conditions.

velocities in ms^{-1}
discharges in litres/sec

Gradient	20	25	30	40	50	60	65	70	75	80	90	100	125	150
	(Equivalent) Pipe diameters in mm													
0·00100				0·077	0·092	0·106	0·113	0·120	0·126	0·133	0·145	0·157	0·185	0·212
1/ 1000				0·0962	0·1803	0·3002	0·3751	0·4609	0·5580	0·6671	0·9235	1·2342	2·2760	3·7446
0·00105				0·078	0·094	0·109	0·116	0·123	0·129	0·136	0·149	0·161	0·190	0·217
1/ 952				0·0986	0·1848	0·3077	0·3845	0·4724	0·5720	0·6838	0·9465	1·2650	2·3327	3·8377
0·00110				0·080	0·096	0·111	0·119	0·126	0·133	0·139	0·152	0·165	0·195	0·222
1/ 909				0·1010	0·1893	0·3151	0·3937	0·4837	0·5856	0·7001	0·9690	1·2950	2·3880	3·9287
0·00115				0·082	0·099	0·114	0·121	0·129	0·136	0·142	0·156	0·169	0·199	0·227
1/ 870				0·1033	0·1936	0·3222	0·4026	0·4947	0·5989	0·7160	0·9910	1·3244	2·4421	4·0175
0·00120				0·084	0·101	0·116	0·124	0·131	0·139	0·146	0·159	0·172	0·203	0·232
1/ 833				0·1056	0·1978	0·3292	0·4114	0·5054	0·6119	0·7315	1·0125	1·3531	2·4950	4·1045
0·00125				0·086	0·103	0·119	0·127	0·134	0·141	0·149	0·162	0·176	0·208	0·237
1/ 800				0·1078	0·2019	0·3361	0·4200	0·5160	0·6247	0·7468	1·0336	1·3813	2·5468	4·1897
0·00130				0·087	0·105	0·121	0·129	0·137	0·144	0·152	0·166	0·179	0·212	0·242
1/ 769				0·1099	0·2060	0·3429	0·4284	0·5263	0·6372	0·7617	1·0543	1·4089	2·5976	4·2732
0·00135				0·089	0·107	0·124	0·132	0·139	0·147	0·154	0·169	0·183	0·216	0·246
1/ 741				0·1121	0·2100	0·3495	0·4367	0·5364	0·6494	0·7763	1·0745	1·4359	2·6475	4·3551
0·00140				0·091	0·109	0·126	0·134	0·142	0·150	0·157	0·172	0·186	0·220	0·251
1/ 714				0·1142	0·2139	0·3560	0·4448	0·5464	0·6615	0·7907	1·0944	1·4625	2·6964	4·4355
0·00145				0·092	0·111	0·128	0·136	0·145	0·152	0·160	0·175	0·190	0·224	0·255
1/ 690				0·1162	0·2177	0·3623	0·4527	0·5561	0·6733	0·8049	1·1139	1·4886	2·7444	4·5145
0·00150				0·094	0·113	0·130	0·139	0·147	0·155	0·163	0·178	0·193	0·227	0·260
1/ 667				0·1182	0·2215	0·3686	0·4605	0·5657	0·6849	0·8187	1·1331	1·5142	2·7917	4·5921
0·00160			0·076	0·097	0·117	0·135	0·143	0·152	0·160	0·168	0·184	0·199	0·235	0·268
1/ 625			0·0540	0·1222	0·2289	0·3808	0·4758	0·5845	0·7076	0·8458	1·1706	1·5643	2·8838	4·7436
0·00170			0·079	0·100	0·120	0·139	0·148	0·157	0·165	0·173	0·190	0·205	0·242	0·277
1/ 588			0·0557	0·1260	0·2360	0·3927	0·4906	0·6026	0·7296	0·8721	1·2069	1·6128	2·9731	4·8904
0·00180			0·081	0·103	0·124	0·143	0·152	0·161	0·170	0·179	0·195	0·211	0·249	0·285
1/ 556			0·0573	0·1297	0·2429	0·4042	0·5050	0·6203	0·7509	0·8976	1·2422	1·6599	3·0599	5·0329
0·00190			0·083	0·106	0·127	0·147	0·156	0·166	0·175	0·184	0·201	0·217	0·256	0·293
1/ 526			0·0589	0·1333	0·2496	0·4154	0·5189	0·6374	0·7717	0·9224	1·2765	1·7057	3·1442	5·1716
0·00200			0·086	0·109	0·130	0·151	0·160	0·170	0·179	0·188	0·206	0·223	0·263	0·300
1/ 500			0·0605	0·1368	0·2562	0·4263	0·5326	0·6541	0·7919	0·9466	1·3099	1·7503	3·2264	5·3066
0·00210			0·088	0·112	0·134	0·155	0·164	0·174	0·184	0·193	0·211	0·228	0·269	0·308
1/ 476			0·0620	0·1402	0·2626	0·4369	0·5458	0·6704	0·8116	0·9701	1·3425	1·7938	3·3065	5·4383
0·00220			0·090	0·114	0·137	0·158	0·168	0·178	0·188	0·198	0·216	0·234	0·276	0·315
1/ 455			0·0635	0·1436	0·2689	0·4473	0·5588	0·6864	0·8309	0·9931	1·3743	1·8363	3·3848	5·5669
0·00230			0·092	0·117	0·140	0·162	0·172	0·182	0·192	0·202	0·221	0·239	0·282	0·322
1/ 435			0·0649	0·1469	0·2750	0·4574	0·5715	0·7019	0·8497	1·0156	1·4054	1·8778	3·4612	5·6926
0·00240			0·094	0·119	0·143	0·165	0·176	0·186	0·196	0·206	0·226	0·244	0·288	0·329
1/ 417			0·0663	0·1501	0·2809	0·4674	0·5839	0·7171	0·8681	1·0376	1·4358	1·9185	3·5361	5·8157
0·00250			0·096	0·122	0·146	0·169	0·180	0·190	0·201	0·211	0·230	0·249	0·294	0·336
1/ 400			0·0677	0·1532	0·2868	0·4771	0·5960	0·7320	0·8861	1·0592	1·4657	1·9583	3·6094	5·9361
0·00260			0·098	0·124	0·149	0·172	0·183	0·194	0·205	0·215	0·235	0·254	0·300	0·343
1/ 385			0·0691	0·1563	0·2925	0·4866	0·6079	0·7466	0·9038	1·0803	1·4949	1·9973	3·6812	6·0542
0·00270			0·100	0·127	0·152	0·175	0·187	0·198	0·209	0·219	0·239	0·259	0·306	0·349
1/ 370			0·0704	0·1593	0·2982	0·4960	0·6196	0·7610	0·9212	1·1010	1·5235	2·0356	3·7517	6·1700
0·00280			0·101	0·129	0·155	0·179	0·190	0·201	0·212	0·223	0·244	0·264	0·311	0·356
1/ 357			0·0717	0·1622	0·3037	0·5051	0·6310	0·7750	0·9382	1·1214	1·5517	2·0731	3·8209	6·2838
0·00290			0·103	0·131	0·157	0·182	0·194	0·205	0·216	0·227	0·248	0·269	0·317	0·362
1/ 345			0·0730	0·1651	0·3091	0·5142	0·6423	0·7889	0·9549	1·1413	1·5793	2·1101	3·8888	6·3955
0·00300			0·105	0·134	0·160	0·185	0·197	0·209	0·220	0·231	0·253	0·273	0·322	0·368
1/ 333			0·0743	0·1680	0·3144	0·5230	0·6534	0·8025	0·9713	1·1610	1·6065	2·1463	3·9556	6·5052
			–	–	–	–	–	–	–	–	–	–	–	–

$V_{r(0\cdot5)medial}$ for half-full circular pipes.

A11
(p.3 of 6)

$k_s = 6{\cdot}0$ mm
S = 0·00300 to 0·01000

ie hydraulic gradient =
1 in 333 to 1 in 100

Water (or sewage) at 15°C;
full bore conditions.

velocities in ms^{-1}
discharges in litres/sec

(Equivalent) Pipe diameters in mm

Gradient	**20**	**25**	**30**	**40**	**50**	**60**	**65**	**70**	**75**	**80**	**90**	**100**	**125**	**150**
0·00300			0·105	0·134	0·160	0·185	0·197	0·209	0·220	0·231	0·253	0·273	0·322	0·368
1/ 333			0·0743	0·1680	0·3144	0·5230	0·6534	0·8025	0·9713	1·1610	1·6065	2·1463	3·9556	6·5052
0·00320			0·109	0·138	0·165	0·191	0·203	0·215	0·227	0·239	0·261	0·282	0·333	0·380
1/ 313			0·0767	0·1736	0·3248	0·5403	0·6749	0·8290	1·0034	1·1993	1·6594	2·2171	4·0860	6·7195
0·00340			0·112	0·142	0·171	0·197	0·210	0·222	0·234	0·246	0·269	0·291	0·343	0·392
1/ 294			0·0791	0·1790	0·3349	0·5571	0·6959	0·8547	1·0345	1·2365	1·7108	2·2857	4·2123	6·9271
0·00360			0·115	0·147	0·176	0·203	0·216	0·229	0·241	0·253	0·277	0·300	0·353	0·403
1/ 278			0·0815	0·1842	0·3447	0·5733	0·7162	0·8796	1·0647	1·2725	1·7607	2·3523	4·3350	7·1287
0·00380			0·118	0·151	0·180	0·208	0·222	0·235	0·248	0·260	0·284	0·308	0·363	0·414
1/ 263			0·0837	0·1893	0·3543	0·5892	0·7359	0·9039	1·0940	1·3076	1·8092	2·4171	4·4543	7·3248
0·00400			0·122	0·155	0·185	0·214	0·228	0·241	0·254	0·267	0·292	0·316	0·372	0·425
1/ 250			0·0859	0·1943	0·3635	0·6046	0·7552	0·9275	1·1226	1·3418	1·8564	2·4802	4·5704	7·5157
0·00420			0·125	0·158	0·190	0·219	0·233	0·247	0·260	0·274	0·299	0·324	0·382	0·436
1/ 238			0·0881	0·1991	0·3726	0·6196	0·7740	0·9505	1·1505	1·3751	1·9025	2·5417	4·6838	7·7020
0·00440			0·128	0·162	0·194	0·224	0·239	0·253	0·267	0·280	0·306	0·331	0·391	0·446
1/ 227			0·0902	0·2038	0·3814	0·6343	0·7923	0·9730	1·1777	1·4076	1·9475	2·6018	4·7944	7·8839
0·00460			0·130	0·166	0·199	0·229	0·244	0·259	0·273	0·286	0·313	0·339	0·399	0·456
1/ 217			0·0922	0·2085	0·3901	0·6487	0·8102	0·9950	1·2044	1·4394	1·9915	2·6606	4·9026	8·0617
0·00480			0·133	0·169	0·203	0·234	0·249	0·264	0·279	0·293	0·320	0·346	0·408	0·466
1/ 208			0·0942	0·2130	0·3985	0·6627	0·8278	1·0166	1·2304	1·4706	2·0346	2·7180	5·0084	8·2356
0·00500			0·136	0·173	0·207	0·239	0·255	0·270	0·284	0·299	0·326	0·353	0·417	0·476
1/ 200			0·0962	0·2174	0·4068	0·6764	0·8449	1·0376	1·2559	1·5010	2·0767	2·7743	5·1121	8·4060
0·00525			0·139	0·177	0·212	0·245	0·261	0·276	0·291	0·306	0·335	0·362	0·427	0·487
1/ 190			0·0986	0·2228	0·4169	0·6933	0·8659	1·0634	1·2871	1·5383	2·1282	2·8431	5·2388	8·6143
0·00550			0·143	0·182	0·217	0·251	0·267	0·283	0·298	0·313	0·342	0·371	0·437	0·499
1/ 182			0·1009	0·2281	0·4268	0·7097	0·8864	1·0886	1·3175	1·5747	2·1785	2·9103	5·3625	8·8176
0·00575			0·146	0·186	0·222	0·257	0·273	0·289	0·305	0·320	0·350	0·379	0·447	0·510
1/ 174			0·1032	0·2333	0·4365	0·7257	0·9064	1·1132	1·3473	1·6102	2·2277	2·9760	5·4835	9·0164
0·00600			0·149	0·190	0·227	0·262	0·279	0·296	0·312	0·327	0·358	0·387	0·456	0·521
1/ 167			0·1055	0·2383	0·4459	0·7414	0·9260	1·1372	1·3764	1·6450	2·2758	3·0403	5·6018	9·2109
0·00625			0·152	0·194	0·232	0·268	0·285	0·302	0·318	0·334	0·365	0·395	0·466	0·532
1/ 160			0·1077	0·2433	0·4552	0·7568	0·9452	1·1608	1·4050	1·6791	2·3230	3·1032	5·7177	9·4014
0·00650			0·155	0·197	0·236	0·273	0·291	0·308	0·324	0·341	0·372	0·403	0·475	0·543
1/ 154			0·1098	0·2481	0·4642	0·7719	0·9641	1·1839	1·4329	1·7125	2·3692	3·1649	5·8313	9·5881
0·00675			0·158	0·201	0·241	0·278	0·296	0·314	0·331	0·347	0·380	0·411	0·484	0·553
1/ 148			0·1119	0·2529	0·4731	0·7866	0·9825	1·2066	1·4603	1·7453	2·4145	3·2254	5·9427	9·7712
0·00700			0·161	0·205	0·245	0·283	0·302	0·319	0·337	0·354	0·387	0·418	0·493	0·563
1/ 143			0·1140	0·2576	0·4819	0·8012	1·0006	1·2288	1·4872	1·7774	2·4590	3·2848	6·0521	9·9510
0·00725			0·164	0·209	0·250	0·288	0·307	0·325	0·343	0·360	0·393	0·426	0·502	0·573
1/ 138			0·1160	0·2622	0·4904	0·8154	1·0184	1·2507	1·5137	1·8090	2·5026	3·3432	6·1596	10·128
0·00750			0·167	0·212	0·254	0·293	0·312	0·331	0·349	0·366	0·400	0·433	0·511	0·583
1/ 133			0·1180	0·2667	0·4989	0·8294	1·0359	1·2722	1·5397	1·8401	2·5456	3·4005	6·2652	10·301
0·00800			0·173	0·219	0·262	0·303	0·322	0·341	0·360	0·378	0·413	0·447	0·527	0·602
1/ 125			0·1219	0·2755	0·5153	0·8568	1·0701	1·3141	1·5904	1·9007	2·6294	3·5124	6·4713	10·640
0·00850			0·178	0·226	0·271	0·312	0·332	0·352	0·371	0·390	0·426	0·461	0·544	0·621
1/ 118			0·1257	0·2840	0·5313	0·8833	1·1032	1·3547	1·6395	1·9594	2·7106	3·6209	6·6710	10·968
0·00900			0·183	0·233	0·278	0·321	0·342	0·362	0·382	0·401	0·438	0·474	0·559	0·639
1/ 111			0·1294	0·2923	0·5468	0·9090	1·1353	1·3941	1·6873	2·0164	2·7895	3·7262	6·8650	11·287
0·00950			0·188	0·239	0·286	0·330	0·352	0·372	0·392	0·412	0·451	0·487	0·575	0·656
1/ 105			0·1330	0·3004	0·5618	0·9340	1·1665	1·4325	1·7337	2·0719	2·8662	3·8287	7·0536	11·597
0·01000			0·193	0·245	0·294	0·339	0·361	0·382	0·403	0·423	0·462	0·500	0·590	0·673
1/ 100			0·1365	0·3082	0·5765	0·9584	1·1970	1·4699	1·7789	2·1259	2·9409	3·9284	7·2373	11·899
			–	–	–	–	–	–	–	–	–	–	–	–

$V_{r(0{\cdot}5)medial}$ for half-full circular pipes.

$k_s = 6{\cdot}0$ mm S = 0·00300 to 0·01000

k_s = **6·0 mm**
S = 0·01000 to 0·03000

ie hydraulic gradient =
1 in 100 to 1 in 33·3

Water (or sewage) at 15°C;
full bore conditions.

velocities in ms^{-1}
discharges in litres/sec

A11

(p.4 of 6)

Gradient	(Equivalent) Pipe diameters in mm 20	25	30	40	50	60	65	70	75	80	90	100	125	150
0·01000			0·193	0·245	0·294	0·339	0·361	0·382	0·403	0·423	0·462	0·500	0·590	0·673
1/ 100			0·1365	0·3082	0·5765	0·9584	1·1970	1·4699	1·7789	2·1259	2·9409	3·9284	7·2373	11·899
0·01050			0·198	0·251	0·301	0·347	0·370	0·391	0·413	0·433	0·474	0·513	0·604	0·690
1/ 95			0·1399	0·3159	0·5908	0·9822	1·2267	1·5063	1·8230	2·1786	3·0138	4·0257	7·4165	12·194
0·01100			0·203	0·257	0·308	0·356	0·378	0·401	0·422	0·444	0·485	0·525	0·619	0·706
1/ 91			0·1432	0·3234	0·6048	1·0054	1·2556	1·5419	1·8661	2·2301	3·0849	4·1208	7·5915	12·481
0·01150			0·207	0·263	0·315	0·364	0·387	0·410	0·432	0·454	0·496	0·536	0·633	0·722
1/ 87			0·1464	0·3307	0·6185	1·0281	1·2840	1·5767	1·9082	2·2804	3·1545	4·2136	7·7625	12·762
0·01200			0·212	0·269	0·322	0·371	0·395	0·419	0·441	0·463	0·507	0·548	0·646	0·738
1/ 83			0·1496	0·3378	0·6318	1·0503	1·3117	1·6107	1·9493	2·3296	3·2225	4·3045	7·9299	13·037
0·01250			0·216	0·274	0·328	0·379	0·403	0·427	0·450	0·473	0·517	0·559	0·660	0·753
1/ 80			0·1527	0·3449	0·6449	1·0720	1·3389	1·6440	1·9897	2·3778	3·2892	4·3935	8·0938	13·307
0·01300			0·220	0·280	0·335	0·387	0·411	0·436	0·459	0·482	0·527	0·571	0·673	0·768
1/ 77			0·1557	0·3517	0·6577	1·0934	1·3655	1·6767	2·0292	2·4250	3·3545	4·4808	8·2544	13·571
0·01350			0·225	0·285	0·341	0·394	0·419	0·444	0·468	0·492	0·537	0·581	0·685	0·783
1/ 74			0·1587	0·3585	0·6703	1·1143	1·3916	1·7088	2·0680	2·4713	3·4186	4·5664	8·4120	13·830
0·01400			0·229	0·291	0·348	0·401	0·427	0·452	0·477	0·501	0·547	0·592	0·698	0·797
1/ 71			0·1617	0·3651	0·6827	1·1348	1·4172	1·7402	2·1061	2·5168	3·4815	4·6504	8·5667	14·084
0·01450			0·233	0·296	0·354	0·408	0·435	0·460	0·485	0·510	0·557	0·603	0·710	0·811
1/ 69			0·1645	0·3716	0·6948	1·1549	1·4424	1·7711	2·1435	2·5615	3·5433	4·7329	8·7187	14·334
0·01500			0·237	0·301	0·360	0·415	0·442	0·468	0·493	0·518	0·567	0·613	0·723	0·825
1/ 67			0·1674	0·3779	0·7068	1·1748	1·4671	1·8015	2·1802	2·6054	3·6040	4·8140	8·8681	14·579
0·01600			0·245	0·311	0·372	0·429	0·457	0·484	0·510	0·535	0·585	0·633	0·746	0·852
1/ 63			0·1729	0·3904	0·7300	1·2134	1·5154	1·8608	2·2519	2·6911	3·7225	4·9723	9·1595	15·058
0·01700			0·252	0·320	0·383	0·442	0·471	0·498	0·525	0·552	0·603	0·653	0·769	0·878
1/ 59			0·1783	0·4025	0·7526	1·2509	1·5622	1·9182	2·3215	2·7742	3·8374	5·1257	9·4420	15·523
0·01800			0·260	0·330	0·394	0·455	0·484	0·513	0·541	0·568	0·621	0·672	0·792	0·904
1/ 56			0·1835	0·4142	0·7745	1·2873	1·6076	1·9740	2·3889	2·8548	3·9489	5·2746	9·7163	15·973
0·01900			0·267	0·339	0·405	0·468	0·498	0·527	0·556	0·584	0·638	0·690	0·813	0·929
1/ 53			0·1885	0·4256	0·7958	1·3227	1·6518	2·0283	2·4546	2·9333	4·0574	5·4195	9·9830	16·412
0·02000			0·274	0·348	0·416	0·480	0·511	0·541	0·570	0·599	0·654	0·708	0·835	0·953
1/ 50			0·1934	0·4367	0·8166	1·3572	1·6949	2·0811	2·5185	3·0097	4·1631	5·5606	10·243	16·839
0·02100			0·280	0·356	0·426	0·492	0·523	0·554	0·584	0·614	0·671	0·726	0·855	0·976
1/ 47·6			0·1982	0·4475	0·8368	1·3908	1·7368	2·1327	2·5809	3·0842	4·2661	5·6982	10·496	17·255
0·02200			0·287	0·365	0·436	0·504	0·536	0·567	0·598	0·628	0·686	0·743	0·875	0·999
1/ 45·5			0·2029	0·4581	0·8566	1·4236	1·7778	2·1830	2·6418	3·1570	4·3667	5·8326	10·744	17·662
0·02300			0·294	0·373	0·446	0·515	0·548	0·580	0·611	0·642	0·702	0·759	0·895	1·022
1/ 43·5			0·2075	0·4685	0·8759	1·4557	1·8179	2·2322	2·7013	3·2281	4·4651	5·9639	10·986	18·060
0·02400			0·300	0·381	0·456	0·526	0·560	0·593	0·625	0·656	0·717	0·776	0·914	1·044
1/ 41·7			0·2120	0·4786	0·8948	1·4871	1·8571	2·2803	2·7596	3·2977	4·5613	6·0925	11·222	18·449
0·02500			0·306	0·389	0·465	0·537	0·571	0·605	0·638	0·670	0·732	0·792	0·933	1·066
1/ 40·0			0·2164	0·4885	0·9133	1·5179	1·8955	2·3275	2·8166	3·3658	4·6556	6·2184	11·454	18·830
0·02600			0·312	0·396	0·474	0·547	0·583	0·617	0·650	0·683	0·746	0·807	0·952	1·087
1/ 38·5			0·2207	0·4982	0·9314	1·5480	1·9332	2·3737	2·8725	3·4326	4·7480	6·3417	11·681	19·203
0·02700			0·318	0·404	0·483	0·558	0·594	0·629	0·663	0·696	0·761	0·823	0·970	1·107
1/ 37·0			0·2249	0·5077	0·9492	1·5776	1·9701	2·4190	2·9274	3·4982	4·8386	6·4628	11·904	19·569
0·02800			0·324	0·411	0·492	0·568	0·605	0·640	0·675	0·709	0·775	0·838	0·988	1·128
1/ 35·7			0·2291	0·5171	0·9667	1·6066	2·0063	2·4635	2·9812	3·5625	4·9276	6·5816	12·123	19·929
0·02900			0·330	0·419	0·501	0·578	0·615	0·651	0·687	0·721	0·788	0·853	1·005	1·148
1/ 34·5			0·2331	0·5263	0·9839	1·6351	2·0419	2·5072	3·0341	3·6257	5·0150	6·6983	12·338	20·282
0·03000			0·335	0·426	0·510	0·588	0·626	0·663	0·699	0·734	0·802	0·867	1·023	1·167
1/ 33·3			0·2371	0·5353	1·0007	1·6632	2·0769	2·5502	3·0861	3·6878	5·1009	6·8130	12·549	20·629
			–	–	–	–	–	–	–	–	–	–	–	–

$V_{r(0·5)medial}$ **for half-full circular pipes.**

S = 0·01000 to 0·03000 k_s = **6·0 mm**

A11
(p.5 of 6)

$k_s = 6{\cdot}0$ mm
S = 0·03000 to 0·10000

ie hydraulic gradient =
1 in 33·3 to 1 in 10·0

Water (or sewage) at 15°C;
full bore conditions.

velocities in ms^{-1}
discharges in litres/sec

Gradient	**20**	**25**	**30**	**40**	**50**	**60**	**65**	**70**	**75**	**80**	**90**	**100**	**125**	**150**
			(Equivalent) Pipe diameters in mm											
0·03000			0·335	0·426	0·510	0·588	0·626	0·663	0·699	0·734	0·802	0·867	1·023	1·167
1/ 33·3			0·2371	0·5353	1·0007	1·6632	2·0769	2·5502	3·0861	3·6878	5·1009	6·8130	12·549	20·629
0·03200			0·347	0·440	0·526	0·608	0·646	0·684	0·722	0·758	0·828	0·896	1·056	1·206
1/ 31·3			0·2450	0·5529	1·0337	1·7178	2·1452	2·6340	3·1875	3·8090	5·2685	7·0368	12·961	21·307
0·03400			0·357	0·454	0·543	0·626	0·666	0·706	0·744	0·781	0·854	0·924	1·089	1·243
1/ 29·4			0·2525	0·5700	1·0656	1·7708	2·2114	2·7152	3·2858	3·9265	5·4309	7·2538	13·361	21·963
0·03600			0·368	0·467	0·558	0·645	0·686	0·726	0·765	0·804	0·878	0·950	1·120	1·279
1/ 27·8			0·2599	0·5866	1·0965	1·8223	2·2756	2·7941	3·3813	4·0405	5·5886	7·4644	13·749	22·601
0·03800			0·378	0·480	0·574	0·662	0·705	0·746	0·786	0·826	0·903	0·976	1·151	1·314
1/ 26·3			0·2670	0·6027	1·1267	1·8724	2·3381	2·8708	3·4741	4·1515	5·7421	7·6693	14·126	23·221
0·04000			0·388	0·492	0·589	0·679	0·723	0·765	0·807	0·847	0·926	1·002	1·181	1·348
1/ 25·0			0·2740	0·6184	1·1560	1·9211	2·3990	2·9456	3·5645	4·2595	5·8915	7·8688	14·493	23·825
0·04200			0·397	0·504	0·603	0·696	0·741	0·784	0·827	0·868	0·949	1·027	1·210	1·382
1/ 23·8			0·2808	0·6337	1·1846	1·9687	2·4584	3·0185	3·6527	4·3649	6·0372	8·0635	14·852	24·414
0·04400			0·407	0·516	0·618	0·713	0·758	0·803	0·846	0·889	0·971	1·051	1·239	1·414
1/ 22·7			0·2874	0·6487	1·2126	2·0151	2·5163	3·0896	3·7388	4·4678	6·1795	8·2535	15·202	24·989
0·04600			0·416	0·528	0·631	0·729	0·775	0·821	0·865	0·909	0·993	1·075	1·267	1·446
1/ 21·7			0·2939	0·6633	1·2399	2·0605	2·5730	3·1592	3·8230	4·5684	6·3186	8·4393	15·544	25·551
0·04800			0·425	0·539	0·645	0·744	0·792	0·839	0·884	0·928	1·015	1·098	1·294	1·477
1/ 20·8			0·3003	0·6776	1·2666	2·1049	2·6285	3·2273	3·9054	4·6668	6·4547	8·6210	15·878	26·102
0·05000			0·434	0·550	0·658	0·760	0·808	0·856	0·902	0·948	1·036	1·120	1·321	1·508
1/ 20·0			0·3065	0·6916	1·2928	2·1484	2·6828	3·2939	3·9861	4·7632	6·5880	8·7990	16·206	26·640
0·05250			0·444	0·564	0·675	0·779	0·828	0·877	0·925	0·971	1·061	1·148	1·353	1·545
1/ 19·0			0·3141	0·7087	1·3248	2·2015	2·7491	3·3754	4·0847	4·8810	6·7510	9·0166	16·607	27·299
0·05500			0·455	0·577	0·691	0·797	0·848	0·898	0·946	0·994	1·086	1·175	1·385	1·581
1/ 18·2			0·3215	0·7255	1·3561	2·2534	2·8140	3·4550	4·1810	4·9960	6·9101	9·2291	16·998	27·942
0·05750			0·465	0·590	0·706	0·815	0·867	0·918	0·968	1·016	1·111	1·202	1·416	1·617
1/ 17·4			0·3287	0·7418	1·3866	2·3042	2·8773	3·5328	4·2751	5·1085	7·0656	9·4368	17·381	28·571
0·06000			0·475	0·603	0·721	0·832	0·886	0·938	0·989	1·038	1·135	1·227	1·447	1·652
1/ 16·7			0·3358	0·7578	1·4165	2·3538	2·9393	3·6089	4·3672	5·2185	7·2177	9·6400	17·755	29·186
0·06250			0·485	0·616	0·736	0·850	0·904	0·957	1·009	1·060	1·158	1·253	1·477	1·686
1/ 16·0			0·3428	0·7735	1·4458	2·4025	3·0000	3·6834	4·4574	5·3263	7·3668	9·8391	18·121	29·788
0·06500			0·495	0·628	0·751	0·867	0·922	0·976	1·029	1·081	1·181	1·278	1·506	1·719
1/ 15·4			0·3496	0·7888	1·4745	2·4501	3·0595	3·7565	4·5458	5·4319	7·5129	10·034	18·481	30·378
0·06750			0·504	0·640	0·765	0·883	0·940	0·995	1·049	1·101	1·203	1·302	1·535	1·752
1/ 14·8			0·3563	0·8039	1·5026	2·4969	3·1179	3·8282	4·6325	5·5356	7·6562	10·226	18·833	30·958
0·07000			0·513	0·651	0·779	0·899	0·957	1·013	1·068	1·121	1·226	1·326	1·563	1·784
1/ 14·3			0·3628	0·8187	1·5302	2·5428	3·1752	3·8985	4·7176	5·6373	7·7968	10·413	19·179	31·526
0·07250			0·522	0·663	0·793	0·915	0·974	1·031	1·087	1·141	1·247	1·349	1·591	1·816
1/ 13·8			0·3693	0·8332	1·5574	2·5879	3·2315	3·9676	4·8012	5·7372	7·9350	10·598	19·519	32·085
0·07500			0·531	0·674	0·807	0·931	0·991	1·049	1·105	1·161	1·269	1·372	1·618	1·847
1/ 13·3			0·3756	0·8475	1·5840	2·6322	3·2868	4·0356	4·8834	5·8354	8·0708	10·779	19·853	32·634
0·08000			0·549	0·697	0·833	0·962	1·023	1·083	1·142	1·199	1·310	1·418	1·671	1·907
1/ 12·5			0·3879	0·8753	1·6361	2·7186	3·3948	4·1681	5·0438	6·0270	8·3358	11·133	20·504	33·705
0·08500			0·566	0·718	0·859	0·991	1·055	1·116	1·177	1·236	1·351	1·461	1·722	1·966
1/ 11·8			0·3999	0·9023	1·6865	2·8024	3·4994	4·2966	5·1993	6·2128	8·5927	11·476	21·136	34·743
0·09000			0·582	0·739	0·884	1·020	1·085	1·149	1·211	1·272	1·390	1·504	1·772	2·023
1/ 11·1			0·4115	0·9286	1·7355	2·8838	3·6010	4·4213	5·3502	6·3931	8·8421	11·809	21·749	35·751
0·09500			0·598	0·759	0·908	1·048	1·115	1·180	1·244	1·307	1·428	1·545	1·821	2·079
1/ 10·5			0·4228	0·9540	1·7832	2·9630	3·6999	4·5426	5·4970	6·5685	9·0846	12·133	22·346	36·731
0·10000			0·614	0·779	0·932	1·075	1·144	1·211	1·277	1·341	1·465	1·585	1·868	2·133
1/ 10·0			0·4339	0·9789	1·8296	3·0400	3·7961	4·6608	5·6400	6·7393	9·3209	12·449	22·927	37·686
			–	–	–	–	–	–	–	–	–	–	–	–

$V_{r(0{\cdot}5)medial}$ for half-full circular pipes.

$k_s = 6{\cdot}0$ mm S = 0·03000 to 0·10000

k_s = 6·0 mm
S = 0·10000 to 0·30000

ie hydraulic gradient =
1 in 10·0 to 1 in 3·3

Water (or sewage) at 15°C;
full bore conditions.

velocities in ms^{-1}
discharges in litres/sec

Gradient	**20**	**25**	**30**	**40**	**50**	**60**	**65**	**70**	**75**	**80**	**90**	**100**	**125**	**150**
			(Equivalent) Pipe diameters in mm											
0·10000			0·614	0·779	0·932	1·075	1·144	1·211	1·277	1·341	1·465	1·585	1·868	2·133
1/ 10·0			0·4339	0·9789	1·8296	3·0400	3·7961	4·6608	5·6400	6·7393	9·3209	12·449	22·927	37·686
0·10500			0·629	0·798	0·955	1·102	1·172	1·241	1·308	1·374	1·501	1·624	1·914	2·185
1/ 9·5			0·4446	1·0031	1·8748	3·1152	3·8900	4·7760	5·7794	6·9060	9·5513	12·756	23·494	38·618
0·11000			0·644	0·817	0·977	1·128	1·200	1·270	1·339	1·406	1·537	1·662	1·960	2·237
1/ 9·1			0·4551	1·0268	1·9190	3·1886	3·9816	4·8886	5·9156	7·0686	9·7763	13·057	24·047	39·527
0·11500			0·658	0·835	0·999	1·153	1·227	1·299	1·369	1·438	1·571	1·700	2·004	2·287
1/ 8·7			0·4653	1·0499	1·9622	3·2604	4·0712	4·9986	6·0487	7·2277	9·9963	13·351	24·588	40·416
0·12000			0·673	0·853	1·021	1·178	1·253	1·327	1·399	1·469	1·605	1·736	2·047	2·336
1/ 8·3			0·4754	1·0725	2·0045	3·3306	4·1589	5·1062	6·1789	7·3833	10·211	13·638	25·117	41·286
0·12500			0·686	0·871	1·042	1·202	1·279	1·354	1·427	1·499	1·638	1·772	2·089	2·385
1/ 8·0			0·4852	1·0947	2·0459	3·3994	4·2448	5·2116	6·3065	7·5357	10·422	13·920	25·635	42·138
0·13000			0·700	0·888	1·063	1·226	1·305	1·381	1·456	1·529	1·671	1·807	2·130	2·432
1/ 7·7			0·4948	1·1164	2·0864	3·4668	4·3289	5·3149	6·4315	7·6851	10·629	14·195	26·143	42·973
0·13500			0·713	0·905	1·083	1·250	1·329	1·407	1·484	1·558	1·703	1·842	2·171	2·478
1/ 7·4			0·5043	1·1377	2·1263	3·5329	4·4115	5·4163	6·5541	7·8316	10·831	14·466	26·642	43·792
0·14000			0·727	0·922	1·103	1·272	1·354	1·433	1·511	1·587	1·734	1·876	2·211	2·524
1/ 7·1			0·5136	1·1586	2·1653	3·5978	4·4925	5·5158	6·6745	7·9755	11·030	14·732	27·131	44·596
0·14500			0·739	0·938	1·122	1·295	1·378	1·459	1·538	1·615	1·765	1·909	2·250	2·568
1/ 6·9			0·5227	1·1791	2·2037	3·6616	4·5721	5·6135	6·7928	8·1168	11·226	14·993	27·612	45·386
0·15000			0·752	0·954	1·142	1·317	1·401	1·484	1·564	1·642	1·795	1·942	2·288	2·612
1/ 6·7			0·5316	1·1993	2·2414	3·7243	4·6504	5·7096	6·9090	8·2557	11·418	15·249	28·084	46·162
0·16000			0·777	0·986	1·179	1·360	1·447	1·532	1·615	1·696	1·854	2·005	2·364	2·698
1/ 6·3			0·5491	1·2387	2·3150	3·8465	4·8031	5·8970	7·1358	8·5267	11·793	15·750	29·006	47·677
0·17000			0·801	1·016	1·215	1·402	1·492	1·580	1·665	1·749	1·911	2·067	2·436	2·781
1/ 5·9			0·5660	1·2769	2·3864	3·9651	4·9511	6·0787	7·3557	8·7894	12·156	16·235	29·899	49·145
0·18000			0·824	1·046	1·251	1·443	1·535	1·625	1·713	1·799	1·966	2·127	2·507	2·862
1/ 5·6			0·5825	1·3140	2·4556	4·0801	5·0947	6·2551	7·5691	9·0444	12·509	16·706	30·766	50·571
0·19000			0·847	1·074	1·285	1·483	1·577	1·670	1·760	1·849	2·020	2·185	2·576	2·940
1/ 5·3			0·5984	1·3500	2·5230	4·1921	5·2345	6·4267	7·7767	9·2924	12·852	17·164	31·610	51·957
0·20000			0·869	1·102	1·318	1·521	1·618	1·713	1·806	1·897	2·073	2·242	2·643	3·017
1/ 5·0			0·6140	1·3851	2·5886	4·3011	5·3706	6·5938	7·9789	9·5340	13·186	17·610	32·431	53·308
0·21000			0·890	1·130	1·351	1·559	1·658	1·756	1·851	1·944	2·124	2·298	2·708	3·091
1/ 4·8			0·6292	1·4194	2·6526	4·4074	5·5034	6·7568	8·1761	9·7697	13·512	18·045	33·233	54·625
0·22000			0·911	1·156	1·383	1·596	1·698	1·797	1·894	1·989	2·174	2·352	2·772	3·164
1/ 4·5			0·6440	1·4528	2·7151	4·5112	5·6330	6·9159	8·3687	9·9998	13·830	18·470	34·015	55·911
0·23000			0·932	1·182	1·414	1·631	1·736	1·837	1·937	2·034	2·223	2·405	2·834	3·235
1/ 4·3			0·6585	1·4855	2·7762	4·6127	5·7597	7·0715	8·5569	10·225	14·141	18·886	34·780	57·168
0·24000			0·952	1·208	1·444	1·667	1·773	1·877	1·979	2·078	2·271	2·456	2·895	3·305
1/ 4·2			0·6727	1·5175	2·8360	4·7120	5·8837	7·2237	8·7411	10·445	14·445	19·292	35·529	58·398
0·25000			0·971	1·233	1·474	1·701	1·810	1·916	2·019	2·121	2·318	2·507	2·955	3·373
1/ 4·0			0·6866	1·5489	2·8945	4·8093	6·0051	7·3728	8·9215	10·660	14·743	19·690	36·262	59·603
0·26000			0·991	1·257	1·503	1·735	1·846	1·954	2·059	2·163	2·363	2·557	3·013	3·440
1/ 3·8			0·7002	1·5796	2·9519	4·9046	6·1241	7·5189	9·0983	10·872	15·035	20·080	36·980	60·784
0·27000			1·010	1·281	1·532	1·768	1·881	1·991	2·099	2·204	2·408	2·605	3·071	3·505
1/ 3·7			0·7136	1·6097	3·0082	4·9981	6·2409	7·6622	9·2717	11·079	15·322	20·463	37·685	61·942
0·28000			1·028	1·304	1·560	1·800	1·915	2·028	2·137	2·245	2·453	2·653	3·127	3·570
1/ 3·6			0·7267	1·6393	3·0635	5·0899	6·3555	7·8030	9·4420	11·282	15·603	20·839	38·377	63·080
0·29000			1·046	1·328	1·588	1·832	1·949	2·063	2·175	2·284	2·496	2·700	3·183	3·633
1/ 3·4			0·7396	1·6683	3·1178	5·1801	6·4681	7·9412	9·6093	11·482	15·880	21·208	39·056	64·197
0·30000			1·064	1·350	1·615	1·863	1·983	2·099	2·212	2·323	2·539	2·746	3·237	3·695
1/ 3·3			0·7523	1·6969	3·1711	5·2687	6·5788	8·0770	9·7736	11·679	16·151	21·571	39·724	65·295
			–	–	–	–	–	–	–	–	–	–	–	–

$V_{r(0·5)medial}$ for half-full circular pipes.

S = 0·10000 to 0·30000 **k_s = 6·0 mm**

A12

(p.1 of 6)

k_s = 0·006 mm
S = 0·00010 to 0·00030

ie hydraulic gradient =
1 in 10000 to 1 in 3333

Water (or sewage) at 15°C;
full bore conditions.

velocities in ms^{-1}
discharges in litres/sec

Gradient **(Equivalent) Pipe diameters in mm**

Gradient	150	200	225	250	275	300	350	375	400	450	500	525	600	630
0·000100	0·101	0·124	0·135	0·145	0·155	0·165	0·183	0·192	0·201	0·218	0·234	0·241	0·264	0·273
1/ 10000	1·7879	3·9026	5·3665	7·1323	9·2219	11·656	17·641	21·229	25·241	34·605	45·872	52·262	74·650	85·027
0·000105	0·104	0·128	0·139	0·149	0·160	0·169	0·188	0·198	0·206	0·224	0·240	0·248	0·271	0·280
1/ 9524	1·8385	4·0123	5·5169	7·3317	9·4792	11·981	18·130	21·818	25·940	35·560	47·136	53·700	76·700	87·360
0·000110	0·107	0·131	0·142	0·153	0·164	0·174	0·193	0·203	0·212	0·229	0·246	0·255	0·278	0·288
1/ 9091	1·8882	4·1198	5·6642	7·5270	9·7311	12·299	18·609	22·394	26·623	36·495	48·373	55·109	78·707	89·643
0·000115	0·110	0·134	0·146	0·157	0·168	0·178	0·198	0·208	0·217	0·235	0·253	0·261	0·285	0·295
1/ 8696	1·9368	4·2251	5·8086	7·7184	9·9779	12·610	19·079	22·958	27·293	37·412	49·585	56·488	80·673	91·880
0·000120	0·112	0·138	0·150	0·161	0·172	0·183	0·203	0·213	0·222	0·241	0·259	0·267	0·292	0·302
1/ 8333	1·9845	4·3284	5·9503	7·9061	10·220	12·915	19·540	23·511	27·950	38·310	50·774	57·841	82·600	94·073
0·000125	0·115	0·141	0·153	0·165	0·176	0·187	0·208	0·218	0·228	0·246	0·265	0·273	0·299	0·309
1/ 8000	2·0314	4·4298	6·0893	8·0904	10·458	13·215	19·992	24·055	28·595	39·192	51·940	59·169	84·491	96·225
0·000130	0·118	0·144	0·157	0·169	0·180	0·191	0·212	0·223	0·233	0·252	0·270	0·279	0·305	0·315
1/ 7692	2·0775	4·5295	6·2259	8·2714	10·691	13·510	20·436	24·588	29·229	40·058	53·086	60·473	86·349	98·339
0·000135	0·120	0·147	0·160	0·172	0·184	0·195	0·217	0·227	0·238	0·257	0·276	0·285	0·312	0·322
1/ 7407	2·1228	4·6275	6·3602	8·4494	10·921	13·799	20·872	25·113	29·851	40·910	54·212	61·754	88·175	100·42
0·000140	0·123	0·150	0·163	0·176	0·188	0·199	0·221	0·232	0·242	0·262	0·282	0·291	0·318	0·329
1/ 7143	2·1673	4·7239	6·4923	8·6244	11·147	14·084	21·302	25·628	30·464	41·747	55·320	63·015	89·970	102·46
0·000145	0·125	0·153	0·167	0·179	0·191	0·203	0·226	0·237	0·247	0·268	0·287	0·297	0·324	0·335
1/ 6897	2·2112	4·8188	6·6223	8·7968	11·369	14·364	21·724	26·136	31·066	42·571	56·409	64·255	91·736	104·47
0·000150	0·128	0·156	0·170	0·183	0·195	0·207	0·230	0·241	0·252	0·273	0·293	0·302	0·331	0·341
1/ 6667	2·2544	4·9122	6·7504	8·9665	11·588	14·640	22·140	26·636	31·660	43·382	57·482	65·476	93·476	106·45
0·000160	0·132	0·162	0·176	0·189	0·202	0·215	0·239	0·250	0·261	0·283	0·303	0·313	0·343	0·354
1/ 6250	2·3389	5·0951	7·0009	9·2985	12·016	15·180	22·954	27·614	32·820	44·969	59·581	67·864	96·877	110·32
0·000170	0·137	0·168	0·182	0·196	0·209	0·222	0·247	0·259	0·270	0·292	0·314	0·324	0·354	0·366
1/ 5882	2·4212	5·2729	7·2447	9·6214	12·432	15·705	23·745	28·564	33·949	46·512	61·621	70·186	100·18	114·08
0·000180	0·142	0·173	0·188	0·202	0·216	0·229	0·255	0·267	0·279	0·302	0·324	0·335	0·366	0·378
1/ 5556	2·5013	5·4462	7·4821	9·9359	12·838	16·217	24·516	29·490	35·048	48·014	63·607	72·446	103·40	117·74
0·000190	0·146	0·179	0·194	0·209	0·223	0·236	0·263	0·275	0·287	0·311	0·334	0·345	0·377	0·389
1/ 5263	2·5796	5·6153	7·7137	10·243	13·233	16·715	25·268	30·393	36·120	49·479	65·544	74·651	106·54	121·32
0·000200	0·150	0·184	0·200	0·215	0·229	0·243	0·270	0·283	0·296	0·320	0·343	0·355	0·388	0·400
1/ 5000	2·6560	5·7805	7·9400	10·543	13·620	17·203	26·002	31·275	37·167	50·910	67·436	76·804	109·61	124·80
0·000210	0·155	0·189	0·205	0·221	0·236	0·250	0·278	0·291	0·304	0·329	0·353	0·365	0·398	0·411
1/ 4762	2·7308	5·9420	8·1613	10·836	13·998	17·679	26·720	32·138	38·190	52·309	69·286	78·908	112·60	128·21
0·000220	0·159	0·194	0·211	0·227	0·242	0·257	0·285	0·299	0·312	0·338	0·362	0·374	0·409	0·422
1/ 4545	2·8040	6·1002	8·3779	11·123	14·368	18·146	27·423	32·982	39·192	53·678	71·096	80·968	115·54	131·55
0·000230	0·163	0·199	0·216	0·232	0·248	0·263	0·292	0·306	0·320	0·346	0·371	0·383	0·419	0·432
1/ 4348	2·8758	6·2552	8·5903	11·404	14·730	18·603	28·111	33·809	40·174	55·020	72·869	82·986	118·41	134·82
0·000240	0·167	0·204	0·221	0·238	0·254	0·270	0·299	0·313	0·327	0·354	0·380	0·392	0·429	0·443
1/ 4167	2·9462	6·4073	8·7985	11·680	15·086	19·051	28·786	34·620	41·136	56·335	74·608	84·964	121·23	138·02
0·000250	0·171	0·209	0·226	0·243	0·260	0·276	0·306	0·321	0·335	0·362	0·389	0·401	0·439	0·453
1/ 4000	3·0153	6·5565	9·0029	11·950	15·435	19·491	29·449	35·415	42·081	57·626	76·314	86·906	123·99	141·16
0·000260	0·174	0·213	0·231	0·249	0·266	0·282	0·313	0·328	0·342	0·370	0·397	0·410	0·448	0·463
1/ 3846	3·0832	6·7032	9·2037	12·216	15·778	19·923	30·100	36·197	43·008	58·893	77·989	88·812	126·70	144·25
0·000270	0·178	0·218	0·236	0·254	0·271	0·288	0·320	0·335	0·350	0·378	0·406	0·419	0·458	0·472
1/ 3704	3·1500	6·8473	9·4011	12·478	16·114	20·347	30·740	36·965	43·920	60·139	79·636	90·686	129·37	147·29
0·000280	0·182	0·222	0·241	0·259	0·277	0·294	0·326	0·342	0·357	0·386	0·414	0·427	0·467	0·482
1/ 3571	3·2156	6·9891	9·5953	12·735	16·446	20·765	31·369	37·721	44·816	61·364	81·255	92·528	131·99	150·27
0·000290	0·186	0·227	0·246	0·265	0·282	0·300	0·332	0·348	0·364	0·393	0·422	0·436	0·476	0·491
1/ 3448	3·2803	7·1287	9·7864	12·988	16·772	21·176	31·988	38·464	45·699	62·570	82·849	94·341	134·57	153·20
0·000300	0·189	0·231	0·251	0·270	0·288	0·305	0·339	0·355	0·371	0·401	0·430	0·444	0·485	0·501
1/ 3333	3·3440	7·2661	9·9746	13·237	17·093	21·581	32·597	39·196	46·567	63·756	84·417	96·125	137·11	156·09
	0·79	0·80	0·81	0·82	0·83	0·84	0·85	0·85	0·86	0·87	0·88	0·88	0·89	0·90

$V_{r(0{\cdot}5)medial}$ for half-full circular pipes.

k_s = 0·006 mm S = 0·00010 to 0·00030

k_s = 0·006 mm
S = 0·00030 to 0·00100

ie hydraulic gradient =
1 in 3333 to 1 in 1000

Water (or sewage) at 15°C;
full bore conditions.

velocities in ms^{-1}
discharges in litres/sec

Gradient	(Equivalent) Pipe diameters in mm													
	150	200	225	250	275	300	350	375	400	450	500	525	600	630
0·000300	0·189	0·231	0·251	0·270	0·288	0·305	0·339	0·355	0·371	0·401	0·430	0·444	0·485	0·501
1/ 3333	3·3440	7·2661	9·9746	13·237	17·093	21·581	32·597	39·196	46·567	63·756	84·417	96·125	137·11	156·09
0·000320	0·196	0·240	0·260	0·280	0·298	0·317	0·351	0·368	0·384	0·415	0·446	0·460	0·503	0·519
1/ 3125	3·4686	7·5350	10·343	13·724	17·721	22·373	33·790	40·628	48·266	66·077	87·485	99·615	142·08	161·74
0·000340	0·203	0·248	0·269	0·289	0·309	0·327	0·363	0·380	0·397	0·430	0·461	0·476	0·520	0·536
1/ 2941	3·5898	7·7965	10·701	14·198	18·332	23·142	34·949	42·020	49·918	68·334	90·466	103·01	146·91	167·23
0·000360	0·210	0·256	0·278	0·299	0·319	0·338	0·375	0·393	0·410	0·443	0·476	0·491	0·536	0·554
1/ 2778	3·7079	8·0512	11·049	14·660	18·927	23·892	36·077	43·375	51·526	70·530	93·369	106·31	151·61	172·58
0·000380	0·216	0·264	0·286	0·308	0·328	0·348	0·386	0·405	0·423	0·457	0·490	0·506	0·552	0·570
1/ 2632	3·8231	8·2997	11·390	15·110	19·507	24·623	37·178	44·697	53·094	72·673	96·200	109·53	156·19	177·79
0·000400	0·223	0·272	0·295	0·317	0·338	0·358	0·398	0·416	0·435	0·470	0·504	0·520	0·568	0·587
1/ 2500	3·9357	8·5425	11·722	15·550	20·074	25·337	38·253	45·988	54·626	74·764	98·964	112·68	160·66	182·88
0·000420	0·229	0·279	0·303	0·326	0·347	0·368	0·409	0·428	0·447	0·483	0·518	0·535	0·584	0·603
1/ 2381	4·0458	8·7798	12·047	15·980	20·628	26·035	39·304	47·249	56·123	76·809	101·67	115·75	165·04	187·85
0·000440	0·235	0·287	0·311	0·334	0·356	0·378	0·419	0·439	0·458	0·496	0·531	0·549	0·599	0·618
1/ 2273	4·1536	9·0122	12·365	16·401	21·170	26·719	40·333	48·484	57·588	78·810	104·31	118·76	169·31	192·72
0·000460	0·241	0·294	0·319	0·343	0·365	0·387	0·430	0·450	0·470	0·508	0·544	0·562	0·614	0·634
1/ 2174	4·2593	9·2399	12·676	16·814	21·702	27·388	41·340	49·694	59·023	80·770	106·90	121·70	173·51	197·49
0·000480	0·247	0·301	0·327	0·351	0·374	0·397	0·440	0·461	0·481	0·520	0·557	0·576	0·628	0·649
1/ 2083	4·3629	9·4633	12·982	17·218	22·223	28·045	42·328	50·880	60·430	82·692	109·44	124·59	177·61	202·16
0·000500	0·253	0·308	0·334	0·359	0·383	0·406	0·450	0·471	0·492	0·532	0·570	0·589	0·642	0·663
1/ 2000	4·4646	9·6825	13·282	17·615	22·734	28·689	43·298	52·044	61·811	84·578	111·93	127·42	181·65	206·75
0·000525	0·260	0·317	0·343	0·369	0·393	0·417	0·462	0·484	0·505	0·546	0·586	0·605	0·660	0·681
1/ 1905	4·5893	9·9511	13·649	18·101	23·361	29·478	44·486	53·470	63·503	86·887	114·98	130·89	186·58	212·36
0·000550	0·267	0·325	0·352	0·378	0·404	0·428	0·474	0·497	0·519	0·561	0·601	0·620	0·677	0·699
1/ 1818	4·7113	10·214	14·009	18·578	23·974	30·251	45·648	54·866	65·158	89·148	117·97	134·29	191·41	217·85
0·000575	0·273	0·333	0·361	0·388	0·414	0·439	0·486	0·509	0·531	0·574	0·616	0·636	0·694	0·716
1/ 1739	4·8309	10·472	14·362	19·044	24·575	31·007	46·787	56·232	66·780	91·362	120·89	137·62	196·14	223·23
0·000600	0·280	0·341	0·370	0·397	0·424	0·449	0·498	0·521	0·544	0·588	0·630	0·651	0·710	0·733
1/ 1667	4·9482	10·724	14·707	19·501	25·164	31·749	47·903	57·573	68·369	93·532	123·76	140·88	200·78	228·51
0·000625	0·287	0·349	0·378	0·406	0·433	0·459	0·509	0·533	0·556	0·601	0·645	0·666	0·726	0·750
1/ 1600	5·0634	10·972	15·047	19·950	25·742	32·478	48·999	58·888	69·929	95·662	126·57	144·08	205·33	233·68
0·000650	0·293	0·357	0·387	0·415	0·443	0·470	0·520	0·545	0·569	0·615	0·659	0·680	0·742	0·766
1/ 1538	5·1765	11·216	15·380	20·391	26·309	33·193	50·075	60·179	71·461	97·753	129·33	147·22	209·80	238·77
0·000675	0·299	0·365	0·395	0·424	0·452	0·480	0·531	0·556	0·581	0·628	0·673	0·694	0·758	0·782
1/ 1481	5·2877	11·455	15·707	20·824	26·867	33·896	51·132	61·448	72·966	99·808	132·05	150·31	214·19	243·76
0·000700	0·305	0·372	0·403	0·433	0·462	0·489	0·542	0·568	0·592	0·640	0·686	0·708	0·773	0·798
1/ 1429	5·3971	11·691	16·029	21·250	27·416	34·587	52·172	62·696	74·447	101·83	134·72	153·34	218·51	248·67
0·000725	0·312	0·380	0·411	0·441	0·471	0·499	0·553	0·579	0·604	0·653	0·699	0·722	0·788	0·813
1/ 1379	5·5047	11·922	16·346	21·670	27·956	35·267	53·195	63·924	75·903	103·82	137·34	156·33	222·75	253·49
0·000750	0·318	0·387	0·419	0·450	0·480	0·508	0·563	0·590	0·615	0·665	0·713	0·736	0·803	0·828
1/ 1333	5·6108	12·151	16·658	22·083	28·488	35·936	54·202	65·133	77·337	105·77	139·93	159·27	226·93	258·25
0·000800	0·329	0·401	0·434	0·466	0·497	0·527	0·584	0·611	0·638	0·689	0·738	0·762	0·832	0·858
1/ 1250	5·8182	12·597	17·269	22·890	29·528	37·246	56·171	67·497	80·140	109·60	144·98	165·02	235·10	267·54
0·000850	0·341	0·415	0·449	0·482	0·514	0·545	0·604	0·632	0·659	0·713	0·763	0·788	0·860	0·887
1/ 1176	6·0199	13·031	17·862	23·675	30·538	38·519	58·086	69·794	82·865	113·32	149·89	170·60	243·05	276·57
0·000900	0·352	0·428	0·464	0·498	0·531	0·562	0·623	0·652	0·681	0·735	0·788	0·813	0·887	0·915
1/ 1111	6·2163	13·453	18·440	24·439	31·523	39·758	59·949	72·031	85·518	116·94	154·67	176·04	250·78	285·36
0·000950	0·363	0·441	0·478	0·513	0·547	0·580	0·642	0·672	0·701	0·757	0·811	0·838	0·914	0·943
1/ 1053	6·4080	13·866	19·004	25·185	32·482	40·967	61·767	74·212	88·104	120·47	159·33	181·34	258·31	293·93
0·001000	0·373	0·454	0·492	0·528	0·563	0·596	0·660	0·691	0·721	0·779	0·835	0·862	0·940	0·970
1/ 1000	6·5952	14·268	19·554	25·913	33·420	42·147	63·541	76·341	90·629	123·92	163·88	186·51	265·66	302·29
	0·81	0·83	0·84	0·85	0·86	0·87	0·88	0·88	0·89	0·90	0·91	0·91	0·92	0·92

$V_{r(0\cdot5)medial}$ for half-full circular pipes.

k_s = 0·006 mm
S = 0·00100 to 0·00300

ie hydraulic gradient =
1 in 1000 to 1 in 333

Water (or sewage) at 15°C;
full bore conditions.

velocities in ms^{-1}
discharges in litres/sec

Gradient **(Equivalent) Pipe diameters in mm**

Gradient	150	200	225	250	275	300	350	375	400	450	500	525	600	630
0·00100	0·373	0·454	0·492	0·528	0·563	0·596	0·660	0·691	0·721	0·779	0·835	0·862	0·940	0·970
1/ 1000	6·5952	14·268	19·554	25·913	33·420	42·147	63·541	76·341	90·629	123·92	163·88	186·51	265·66	302·29
0·00105	0·384	0·467	0·505	0·542	0·578	0·613	0·678	0·710	0·741	0·800	0·857	0·885	0·965	0·996
1/ 952	6·7782	14·662	20·092	26·625	34·336	43·301	65·275	78·423	93·097	127·28	168·33	191·57	272·85	310·46
0·00110	0·394	0·479	0·519	0·557	0·593	0·629	0·696	0·728	0·760	0·821	0·879	0·908	0·990	1·022
1/ 909	6·9574	15·047	20·619	27·321	35·232	44·429	66·973	80·459	95·512	130·58	172·68	196·52	279·88	318·46
0·00115	0·404	0·491	0·532	0·570	0·608	0·644	0·713	0·747	0·779	0·841	0·901	0·930	1·014	1·047
1/ 870	7·1330	15·425	21·135	28·003	36·111	45·535	68·635	82·454	97·878	133·81	176·94	201·36	286·77	326·29
0·00120	0·413	0·503	0·544	0·584	0·622	0·660	0·730	0·764	0·797	0·861	0·922	0·952	1·038	1·071
1/ 833	7·3052	15·795	21·641	28·672	36·972	46·620	70·265	84·410	100·20	136·97	181·12	206·11	293·52	333·96
0·00125	0·423	0·514	0·557	0·597	0·637	0·675	0·747	0·782	0·815	0·881	0·943	0·974	1·062	1·095
1/ 800	7·4742	16·158	22·138	29·329	37·817	47·683	71·864	86·328	102·47	140·08	185·21	210·77	300·14	341·49
0·00130	0·432	0·526	0·569	0·611	0·651	0·689	0·763	0·799	0·833	0·900	0·964	0·995	1·085	1·119
1/ 769	7·6402	16·515	22·625	29·974	38·647	48·728	73·434	88·212	104·71	143·12	189·24	215·35	306·64	348·88
0·00135	0·442	0·537	0·581	0·624	0·664	0·704	0·779	0·815	0·851	0·919	0·984	1·016	1·107	1·142
1/ 741	7·8034	16·865	23·104	30·607	39·462	49·755	74·977	90·064	106·90	146·12	193·19	219·84	313·03	356·14
0·00140	0·451	0·548	0·593	0·636	0·678	0·718	0·795	0·832	0·868	0·937	1·004	1·036	1·129	1·165
1/ 714	7·9639	17·210	23·576	31·230	40·264	50·764	76·494	91·884	109·06	149·06	197·08	224·26	319·31	363·28
0·00145	0·460	0·559	0·605	0·649	0·691	0·732	0·811	0·848	0·885	0·955	1·023	1·056	1·151	1·188
1/ 690	8·1218	17·549	24·039	31·843	41·053	51·757	77·987	93·675	111·18	151·96	200·90	228·61	325·49	370·30
0·00150	0·468	0·569	0·616	0·661	0·704	0·746	0·826	0·864	0·901	0·973	1·042	1·076	1·173	1·210
1/ 667	8·2773	17·883	24·496	32·447	41·830	52·735	79·456	95·437	113·27	154·81	204·66	232·88	331·57	377·21
0·00160	0·486	0·590	0·639	0·685	0·730	0·773	0·856	0·895	0·934	1·008	1·080	1·114	1·215	1·253
1/ 625	8·5815	18·537	25·389	33·627	43·349	54·647	82·329	98·884	117·36	160·38	212·02	241·25	343·45	390·72
0·00170	0·502	0·610	0·660	0·708	0·755	0·799	0·885	0·926	0·966	1·042	1·116	1·152	1·256	1·296
1/ 588	8·8773	19·172	26·257	34·774	44·825	56·505	85·121	102·23	121·33	165·80	219·17	249·38	355·00	403·85
0·00180	0·519	0·630	0·682	0·731	0·779	0·825	0·913	0·955	0·996	1·076	1·152	1·189	1·295	1·337
1/ 556	9·1653	19·790	27·102	35·891	46·262	58·314	87·839	105·49	125·19	171·07	226·12	257·29	366·23	416·62
0·00190	0·535	0·649	0·702	0·753	0·802	0·850	0·941	0·984	1·026	1·108	1·186	1·224	1·334	1·376
1/ 526	9·4463	20·393	27·926	36·980	47·664	60·078	90·488	108·67	128·96	176·21	232·90	265·00	377·19	429·07
0·00200	0·550	0·668	0·723	0·775	0·826	0·874	0·967	1·012	1·056	1·139	1·220	1·259	1·372	1·415
1/ 500	9·7206	20·982	28·730	38·044	49·032	61·800	93·075	111·77	132·64	181·23	239·52	272·52	387·87	441·22
0·00210	0·565	0·686	0·742	0·796	0·848	0·898	0·994	1·039	1·084	1·170	1·253	1·293	1·409	1·454
1/ 476	9·9889	21·558	29·517	39·083	50·369	63·482	95·603	114·81	136·24	186·13	245·99	279·87	398·32	453·10
0·00220	0·580	0·704	0·762	0·817	0·870	0·921	1·019	1·066	1·112	1·200	1·285	1·326	1·445	1·491
1/ 455	10·251	22·121	30·286	40·100	51·678	65·129	98·076	117·77	139·75	190·92	252·32	287·07	408·54	464·71
0·00230	0·595	0·722	0·781	0·837	0·892	0·944	1·045	1·093	1·140	1·230	1·317	1·359	1·480	1·527
1/ 435	10·509	22·673	31·040	41·096	52·959	66·742	100·50	120·68	143·20	195·62	258·51	294·11	418·54	476·08
0·00240	0·609	0·739	0·799	0·857	0·913	0·967	1·069	1·118	1·166	1·259	1·348	1·391	1·515	1·563
1/ 417	10·761	23·214	31·779	42·073	54·216	68·323	102·87	123·53	146·57	200·22	264·59	301·02	428·34	487·22
0·00250	0·623	0·756	0·817	0·877	0·934	0·989	1·093	1·144	1·193	1·287	1·378	1·422	1·549	1·598
1/ 400	11·008	23·745	32·504	43·031	55·449	69·874	105·20	126·32	149·88	204·74	270·54	307·79	437·96	498·16
0·00260	0·637	0·772	0·835	0·896	0·954	1·010	1·117	1·169	1·219	1·315	1·408	1·453	1·582	1·632
1/ 385	11·252	24·266	33·217	43·972	56·659	71·397	107·49	129·06	153·13	209·17	276·39	314·44	447·40	508·89
0·00270	0·650	0·789	0·853	0·915	0·974	1·031	1·141	1·193	1·244	1·343	1·437	1·483	1·615	1·666
1/ 370	11·490	24·779	33·916	44·897	57·848	72·894	109·74	131·76	156·33	213·53	282·14	320·97	456·68	519·43
0·00280	0·664	0·805	0·870	0·933	0·994	1·052	1·164	1·217	1·269	1·369	1·466	1·512	1·647	1·700
1/ 357	11·725	25·283	34·604	45·806	59·018	74·365	111·95	134·41	159·47	217·81	287·78	327·39	465·79	529·79
0·00290	0·677	0·821	0·887	0·951	1·013	1·073	1·186	1·241	1·294	1·396	1·494	1·542	1·679	1·732
1/ 345	11·957	25·778	35·281	46·701	60·168	75·813	114·12	137·02	162·56	222·02	293·34	333·70	474·76	539·98
0·00300	0·689	0·836	0·904	0·969	1·032	1·093	1·208	1·264	1·318	1·422	1·522	1·570	1·710	1·764
1/ 333	12·184	26·266	35·948	47·581	61·301	77·237	116·26	139·58	165·60	226·16	298·81	339·92	483·58	550·01
	0·84	0·86	0·87	0·88	0·89	0·90	0·91	0·91	0·92	0·92	0·93	0·93	0·94	0·94

$V_{r(0·5)medial}$ for half-full circular pipes.

k_s = 0·006 mm S = 0·00100 to 0·00300

k_s = 0·006 mm
S = 0·00300 to 0·01000

ie hydraulic gradient =
1 in 333 to 1 in 100

Water (or sewage) at 15°C;
full bore conditions.

velocities in ms^{-1}
discharges in litres/sec

Gradient	150	200	225	250	275	300	350	375	400	450	500	525	600	630
	(Equivalent) Pipe diameters in mm													
0·00300	0·689	0·836	0·904	0·969	1·032	1·093	1·208	1·264	1·318	1·422	1·522	1·570	1·710	1·764
1/ 333	12·184	26·266	35·948	47·581	61·301	77·237	116·26	139·58	165·60	226·16	298·81	339·92	483·58	550·01
0·00320	0·715	0·866	0·937	1·004	1·069	1·132	1·252	1·309	1·365	1·473	1·576	1·626	1·771	1·827
1/ 313	12·629	27·221	37·251	49·303	63·515	80·023	120·44	144·60	171·54	234·26	309·49	352·06	500·83	569·61
0·00340	0·739	0·896	0·969	1·038	1·106	1·170	1·294	1·353	1·411	1·522	1·629	1·681	1·831	1·888
1/ 294	13·062	28·148	38·518	50·976	65·667	82·730	124·50	149·47	177·32	242·14	319·88	363·87	517·58	588·65
0·00360	0·763	0·925	1·000	1·072	1·141	1·208	1·335	1·396	1·456	1·571	1·681	1·734	1·888	1·948
1/ 278	13·484	29·051	39·750	52·604	67·761	85·364	128·46	154·21	182·94	249·80	329·98	375·35	533·88	607·18
0·00380	0·786	0·953	1·030	1·104	1·175	1·244	1·375	1·438	1·499	1·618	1·731	1·786	1·944	2·006
1/ 263	13·895	29·931	40·952	54·192	69·803	87·933	132·31	158·83	188·42	257·26	339·82	386·54	549·77	625·23
0·00400	0·809	0·980	1·059	1·136	1·209	1·279	1·414	1·479	1·542	1·663	1·780	1·836	1·999	2·062
1/ 250	14·296	30·791	42·126	55·742	71·796	90·440	136·08	163·34	193·76	264·55	349·43	397·46	565·27	642·85
0·00420	0·831	1·007	1·088	1·166	1·242	1·314	1·453	1·519	1·583	1·708	1·827	1·885	2·053	2·117
1/ 238	14·688	31·631	43·273	57·256	73·744	92·889	139·75	167·75	198·99	271·67	358·82	408·13	580·42	660·06
0·00440	0·853	1·033	1·117	1·197	1·274	1·348	1·490	1·558	1·624	1·752	1·874	1·934	2·105	2·171
1/ 227	15·072	32·453	44·395	58·739	75·649	95·286	143·35	172·07	204·10	278·63	368·00	418·57	595·23	676·89
0·00460	0·874	1·059	1·144	1·226	1·305	1·381	1·527	1·596	1·664	1·795	1·920	1·981	2·156	2·224
1/ 217	15·448	33·258	45·494	60·190	77·515	97·633	146·87	176·29	209·10	285·45	376·99	428·79	609·73	693·37
0·00480	0·895	1·084	1·171	1·255	1·336	1·414	1·562	1·634	1·703	1·837	1·965	2·027	2·207	2·276
1/ 208	15·816	34·048	46·571	61·613	79·345	99·934	150·32	180·43	214·00	292·13	385·80	438·80	623·94	709·52
0·00500	0·915	1·108	1·198	1·284	1·366	1·446	1·598	1·670	1·741	1·878	2·009	2·072	2·256	2·327
1/ 200	16·178	34·822	47·629	63·008	81·139	102·19	153·71	184·49	218·81	298·68	394·45	448·62	637·88	725·35
0·00525	0·941	1·139	1·230	1·318	1·403	1·485	1·641	1·715	1·788	1·928	2·063	2·128	2·316	2·389
1/ 190	16·621	35·770	48·923	64·718	83·337	104·95	157·85	189·46	224·70	306·71	405·02	460·64	654·94	744·74
0·00550	0·965	1·168	1·262	1·352	1·439	1·523	1·683	1·759	1·834	1·978	2·115	2·182	2·375	2·450
1/ 182	17·055	36·698	50·189	66·390	85·486	107·66	161·91	194·32	230·46	314·56	415·37	472·40	671·62	763·70
0·00575	0·989	1·197	1·293	1·386	1·475	1·560	1·724	1·802	1·879	2·026	2·167	2·235	2·433	2·509
1/ 174	17·479	37·607	51·430	68·027	87·591	110·30	165·88	199·08	236·10	322·24	425·50	483·91	687·96	782·26
0·00600	1·013	1·225	1·324	1·419	1·509	1·597	1·765	1·845	1·923	2·073	2·218	2·288	2·490	2·568
1/ 167	17·896	38·498	52·645	69·633	89·655	112·90	169·77	203·75	241·63	329·77	435·43	495·20	703·96	800·44
0·00625	1·036	1·253	1·354	1·451	1·544	1·633	1·804	1·886	1·966	2·120	2·267	2·339	2·545	2·625
1/ 160	18·304	39·372	53·838	71·207	91·679	115·44	173·59	208·32	247·05	337·16	445·16	506·26	719·66	818·28
0·00650	1·059	1·281	1·384	1·482	1·577	1·669	1·843	1·927	2·008	2·165	2·316	2·389	2·600	2·681
1/ 154	18·705	40·230	55·009	72·753	93·666	117·94	177·34	212·81	252·37	344·40	454·72	517·12	735·06	835·78
0·00675	1·081	1·307	1·412	1·513	1·610	1·703	1·881	1·967	2·050	2·210	2·364	2·438	2·653	2·736
1/ 148	19·100	41·073	56·160	74·272	95·618	120·40	181·02	217·23	257·60	351·53	464·11	527·78	750·20	852·97
0·00700	1·103	1·334	1·441	1·543	1·642	1·737	1·919	2·006	2·091	2·254	2·411	2·487	2·706	2·791
1/ 143	19·487	41·902	57·291	75·765	97·537	122·81	184·64	221·56	262·74	358·52	473·33	538·27	765·07	869·87
0·00725	1·124	1·360	1·469	1·573	1·674	1·771	1·956	2·045	2·131	2·298	2·457	2·534	2·758	2·844
1/ 138	19·868	42·718	58·403	77·234	99·425	125·18	188·20	225·83	267·79	365·41	482·40	548·58	779·69	886·48
0·00750	1·146	1·385	1·496	1·603	1·705	1·804	1·992	2·083	2·171	2·340	2·502	2·581	2·808	2·896
1/ 133	20·244	43·520	59·498	78·679	101·28	127·52	191·70	230·03	272·77	372·18	491·33	558·72	794·08	902·83
0·00800	1·187	1·435	1·550	1·660	1·766	1·869	2·064	2·157	2·248	2·423	2·591	2·673	2·908	2·999
1/ 125	20·978	45·090	61·639	81·505	104·91	132·08	198·54	238·23	282·49	385·42	508·78	578·55	822·20	934·78
0·00850	1·227	1·484	1·603	1·716	1·826	1·931	2·133	2·229	2·323	2·504	2·677	2·762	3·005	3·098
1/ 118	21·691	46·615	63·719	84·250	108·44	136·52	205·19	246·20	291·93	398·28	525·72	597·80	849·51	965·80
0·00900	1·267	1·531	1·653	1·771	1·884	1·992	2·200	2·299	2·396	2·583	2·761	2·848	3·098	3·195
1/ 111	22·385	48·099	65·744	86·922	111·88	140·84	211·66	253·96	301·11	410·78	542·21	616·53	876·07	995·98
0·00950	1·305	1·577	1·703	1·824	1·940	2·052	2·266	2·368	2·467	2·659	2·843	2·932	3·190	3·289
1/ 105	23·062	49·545	67·717	89·526	115·22	145·04	217·97	261·52	310·06	422·97	558·27	634·78	901·95	1025·4
0·01000	1·342	1·622	1·752	1·876	1·995	2·110	2·329	2·435	2·537	2·734	2·923	3·015	3·279	3·381
1/ 100	23·723	50·957	69·642	92·067	118·49	149·15	224·12	268·89	318·80	434·86	573·94	652·58	927·19	1054·1
	0·87	0·89	0·90	0·91	0·92	0·92	0·93	0·93	0·94	0·94	0·95	0·95	0·96	0·96

$V_{r(0·5)medial}$ for half-full circular pipes.

k_s = 0·006 mm
S = 0·01000 to 0·03000

ie hydraulic gradient =
1 in 100 to 1 in 33·3

Water (or sewage) at 15°C;
full bore conditions.

velocities in ms^{-1}
discharges in litres/sec

Gradient **(Equivalent) Pipe diameters in mm**

Gradient	150	200	225	250	275	300	350	375	400	450	500	525	600	630
0·01000	1·342	1·622	1·752	1·876	1·995	2·110	2·329	2·435	2·537	2·734	2·923	3·015	3·279	3·381
1/ 100	23·723	50·957	69·642	92·067	118·49	149·15	224·12	268·89	318·80	434·86	573·94	652·58	927·19	1054·1
0·01050	1·379	1·666	1·799	1·926	2·049	2·167	2·392	2·500	2·605	2·807	3·001	3·095	3·366	3·471
1/ 95	24·369	52·337	71·524	94·550	121·68	153·16	230·13	276·09	327·33	446·48	589·24	669·97	951·85	1082·1
0·01100	1·415	1·709	1·845	1·976	2·101	2·222	2·453	2·564	2·671	2·879	3·077	3·173	3·452	3·559
1/ 91	25·000	53·687	73·365	96·979	124·80	157·08	236·01	283·14	335·67	457·84	604·21	686·98	975·96	1109·5
0·01150	1·450	1·751	1·890	2·024	2·153	2·277	2·513	2·626	2·736	2·949	3·152	3·250	3·535	3·645
1/ 87	25·619	55·008	75·167	99·357	127·85	160·92	241·77	290·03	343·84	468·96	618·86	703·62	999·56	1136·3
0·01200	1·484	1·792	1·935	2·072	2·203	2·330	2·571	2·687	2·800	3·017	3·225	3·326	3·617	3·729
1/ 83	26·226	56·304	76·934	101·69	130·85	164·68	247·41	296·79	351·84	479·85	633·22	719·93	1022·7	1162·5
0·01250	1·518	1·833	1·978	2·118	2·252	2·382	2·629	2·747	2·862	3·084	3·297	3·400	3·697	3·812
1/ 80	26·821	57·575	78·667	103·97	133·78	168·37	252·94	303·42	359·69	490·54	647·29	735·92	1045·4	1188·3
0·01300	1·551	1·872	2·021	2·164	2·301	2·433	2·685	2·806	2·924	3·150	3·367	3·472	3·776	3·893
1/ 77	27·405	58·822	80·368	106·22	136·67	172·00	258·37	309·92	367·39	501·03	661·11	751·62	1067·6	1213·6
0·01350	1·583	1·911	2·063	2·209	2·349	2·484	2·741	2·864	2·984	3·215	3·436	3·543	3·853	3·973
1/ 74	27·979	60·048	82·038	108·42	139·50	175·55	263·70	316·31	374·96	511·32	674·67	767·03	1089·4	1238·4
0·01400	1·615	1·950	2·105	2·253	2·395	2·533	2·795	2·921	3·043	3·279	3·504	3·613	3·929	4·051
1/ 71	28·543	61·252	83·681	110·59	142·28	179·05	268·94	322·59	382·40	521·44	688·00	782·18	1110·9	1262·8
0·01450	1·647	1·987	2·145	2·296	2·442	2·582	2·849	2·977	3·101	3·341	3·571	3·682	4·004	4·128
1/ 69	29·098	62·437	85·296	112·72	145·02	182·49	274·09	328·77	389·71	531·40	701·11	797·07	1132·0	1286·7
0·01500	1·678	2·025	2·185	2·339	2·487	2·630	2·902	3·032	3·158	3·403	3·636	3·750	4·077	4·203
1/ 67	29·645	63·603	86·886	114·82	147·71	185·88	279·16	334·84	396·90	541·19	714·01	811·72	1152·8	1310·3
0·01600	1·738	2·097	2·263	2·423	2·576	2·723	3·005	3·139	3·270	3·523	3·765	3·882	4·221	4·351
1/ 63	30·713	65·883	89·994	118·92	152·98	192·49	289·07	346·72	410·96	560·32	739·22	840·35	1193·4	1356·4
0·01700	1·797	2·168	2·339	2·504	2·662	2·814	3·105	3·244	3·379	3·640	3·889	4·010	4·360	4·495
1/ 59	31·751	68·097	93·012	122·90	158·09	198·92	298·70	358·25	424·61	578·90	763·69	868·15	1232·8	1401·2
0·01800	1·854	2·236	2·413	2·583	2·745	2·903	3·202	3·345	3·485	3·754	4·011	4·135	4·496	4·634
1/ 56	32·761	70·252	95·949	126·77	163·07	205·17	308·06	369·46	437·89	596·97	787·49	895·19	1271·1	1444·7
0·01900	1·910	2·303	2·485	2·659	2·827	2·989	3·297	3·444	3·588	3·864	4·129	4·257	4·627	4·770
1/ 53	33·745	72·352	98·811	130·55	167·91	211·26	317·18	380·39	450·83	614·58	810·67	921·52	1308·4	1487·0
0·02000	1·964	2·368	2·555	2·734	2·907	3·073	3·389	3·541	3·688	3·972	4·244	4·376	4·756	4·903
1/ 50	34·706	74·401	101·60	134·23	172·64	217·20	326·08	391·05	463·45	631·75	833·28	947·21	1344·8	1528·4
0·02100	2·017	2·432	2·624	2·808	2·984	3·155	3·480	3·635	3·786	4·078	4·356	4·491	4·882	5·032
1/ 47·6	35·645	76·403	104·33	137·83	177·26	223·00	334·77	401·46	475·78	648·52	855·37	972·29	1380·3	1568·7
0·02200	2·069	2·494	2·691	2·879	3·061	3·235	3·568	3·727	3·882	4·181	4·466	4·605	5·005	5·159
1/ 45·5	36·563	78·361	107·00	141·35	181·78	228·68	343·27	411·64	487·83	664·92	876·96	996·82	1415·1	1608·2
0·02300	2·120	2·555	2·757	2·950	3·135	3·314	3·654	3·817	3·976	4·282	4·574	4·716	5·125	5·283
1/ 43·5	37·462	80·278	109·61	144·79	186·20	234·23	351·59	421·60	499·63	680·97	898·09	1020·8	1449·1	1646·8
0·02400	2·170	2·615	2·821	3·018	3·208	3·391	3·739	3·906	4·068	4·381	4·679	4·824	5·243	5·404
1/ 41·7	38·343	82·157	112·17	148·17	190·54	239·68	359·74	431·37	511·18	696·69	918·79	1044·3	1482·4	1684·6
0·02500	2·219	2·674	2·884	3·086	3·279	3·466	3·822	3·992	4·158	4·477	4·783	4·931	5·358	5·523
1/ 40·0	39·208	84·000	114·68	151·48	194·79	245·02	367·73	440·94	522·51	712·10	939·09	1067·4	1515·0	1721·7
0·02600	2·267	2·731	2·946	3·152	3·350	3·540	3·904	4·077	4·247	4·573	4·884	5·035	5·472	5·640
1/ 38·5	40·056	85·808	117·15	154·72	198·96	250·25	375·57	450·33	533·63	727·23	959·00	1090·0	1547·1	1758·1
0·02700	2·314	2·788	3·007	3·217	3·419	3·613	3·984	4·161	4·333	4·666	4·984	5·138	5·583	5·755
1/ 37·0	40·890	87·585	119·57	157·91	203·05	255·40	383·27	459·56	544·55	742·08	978·56	1112·2	1578·5	1793·8
0·02800	2·360	2·843	3·067	3·281	3·486	3·685	4·062	4·243	4·419	4·758	5·082	5·239	5·692	5·867
1/ 35·7	41·709	89·331	121·95	161·05	207·08	260·46	390·84	468·62	555·28	756·68	997·77	1134·0	1609·4	1828·9
0·02900	2·406	2·898	3·126	3·344	3·553	3·755	4·140	4·324	4·503	4·848	5·178	5·338	5·800	5·978
1/ 34·5	42·515	91·048	124·29	164·13	211·04	265·43	398·28	477·53	565·83	771·03	1016·7	1155·5	1639·8	1863·4
0·03000	2·451	2·952	3·184	3·406	3·619	3·824	4·216	4·403	4·585	4·937	5·272	5·435	5·905	6·087
1/ 33·3	43·308	92·737	126·59	167·17	214·93	270·32	405·60	486·30	576·21	785·14	1035·2	1176·6	1669·7	1897·4
	0·90	0·92	0·93	0·93	0·94	0·94	0·95	0·95	0·95	0·96	0·96	0·96	0·97	0·97

$V_{r(0{\cdot}5)medial}$ for half-full circular pipes.

k_s = 0·006 mm S = 0·01000 to 0·03000

Gradient	**(Equivalent) Pipe diameters in mm**													
	150	**200**	**225**	**250**	**275**	**300**	**350**	**375**	**400**	**450**	**500**	**525**	**600**	**630**
0·03000	2·451	2·952	3·184	3·406	3·619	3·824	4·216	4·403	4·585	4·937	5·272	5·435	5·905	6·087
1/ 33·3	43·308	92·737	126·59	167·17	214·93	270·32	405·60	486·30	576·21	785·14	1035·2	1176·6	1669·7	1897·4
0·03200	2·538	3·057	3·297	3·526	3·747	3·959	4·364	4·558	4·747	5·110	5·457	5·626	6·112	6·299
1/ 31·3	44·859	96·039	131·09	173·10	222·54	279·88	419·91	503·43	596·49	812·72	1071·5	1217·8	1728·1	1963·7
0·03400	2·624	3·159	3·407	3·644	3·871	4·091	4·509	4·709	4·903	5·278	5·637	5·811	6·312	6·506
1/ 29·4	46·364	99·246	135·45	178·85	229·93	289·16	433·80	520·06	616·17	839·49	1106·8	1257·8	1784·8	2028·0
0·03600	2·707	3·258	3·514	3·758	3·992	4·218	4·649	4·855	5·056	5·442	5·811	5·990	6·507	6·706
1/ 27·8	47·829	102·37	139·70	184·45	237·12	298·18	447·30	536·24	635·32	865·52	1141·0	1296·7	1839·9	2090·6
0·03800	2·787	3·355	3·618	3·869	4·110	4·343	4·786	4·998	5·204	5·601	5·981	6·165	6·697	6·902
1/ 26·3	49·257	105·41	143·84	189·91	244·12	306·97	460·46	551·99	653·96	890·88	1174·4	1334·6	1893·5	2151·5
0·04000	2·866	3·450	3·719	3·977	4·225	4·464	4·919	5·137	5·349	5·757	6·147	6·336	6·882	7·092
1/ 25·0	50·650	108·37	147·88	195·23	250·95	315·55	473·29	567·36	672·15	915·60	1207·0	1371·6	1945·8	2210·9
0·04200	2·943	3·542	3·818	4·083	4·337	4·583	5·050	5·273	5·490	5·909	6·309	6·503	7·063	7·278
1/ 23·8	52·011	111·27	151·82	200·43	257·63	323·93	485·82	582·36	689·90	939·74	1238·7	1407·7	1996·9	2268·9
0·04400	3·019	3·632	3·915	4·187	4·447	4·698	5·177	5·406	5·628	6·057	6·467	6·666	7·239	7·460
1/ 22·7	53·342	114·10	155·68	205·51	264·15	332·11	498·07	597·03	707·26	963·34	1269·8	1442·9	2046·8	2325·6
0·04600	3·092	3·720	4·010	4·288	4·555	4·812	5·301	5·536	5·763	6·202	6·622	6·825	7·412	7·638
1/ 21·7	54·645	116·87	159·46	210·49	270·53	340·13	510·06	611·38	724·25	986·44	1300·2	1477·4	2095·7	2381·0
0·04800	3·164	3·807	4·103	4·387	4·660	4·923	5·423	5·663	5·896	6·345	6·773	6·981	7·581	7·812
1/ 20·8	55·921	119·59	163·15	215·36	276·78	347·98	521·80	625·44	740·89	1009·1	1329·9	1511·2	2143·5	2435·3
0·05000	3·235	3·891	4·195	4·485	4·763	5·032	5·543	5·788	6·026	6·484	6·922	7·134	7·747	7·983
1/ 20·0	57·173	122·25	166·78	220·14	282·92	355·68	533·31	639·23	757·20	1031·2	1359·1	1544·4	2190·4	2488·6
0·05250	3·322	3·995	4·306	4·604	4·890	5·165	5·690	5·940	6·184	6·655	7·104	7·321	7·950	8·192
1/ 19·0	58·706	125·51	171·22	225·99	290·42	365·10	547·40	656·09	777·15	1058·4	1394·8	1584·9	2247·8	2553·7
0·05500	3·407	4·097	4·415	4·720	5·013	5·295	5·833	6·090	6·340	6·821	7·281	7·504	8·148	8·396
1/ 18·2	60·205	128·70	175·56	231·70	297·75	374·30	561·17	672·58	796·66	1084·9	1429·7	1624·5	2303·9	2617·3
0·05750	3·490	4·196	4·522	4·834	5·134	5·423	5·973	6·236	6·492	6·984	7·455	7·683	8·342	8·596
1/ 17·4	61·672	131·82	179·81	237·30	304·93	383·32	574·65	688·71	815·75	1110·8	1463·8	1663·3	2358·7	2679·6
0·06000	3·571	4·293	4·627	4·946	5·252	5·548	6·110	6·379	6·640	7·144	7·626	7·859	8·532	8·792
1/ 16·7	63·110	134·88	183·97	242·78	311·97	392·14	587·85	704·51	834·45	1136·2	1497·3	1701·2	2412·4	2740·6
0·06250	3·651	4·389	4·729	5·055	5·368	5·670	6·244	6·519	6·786	7·301	7·792	8·031	8·718	8·983
1/ 16·0	64·519	137·87	188·05	248·16	318·86	400·80	600·79	720·00	852·78	1161·1	1530·0	1738·4	2465·1	2800·4
0·06500	3·729	4·482	4·830	5·163	5·482	5·790	6·376	6·657	6·929	7·454	7·956	8·199	8·901	9·172
1/ 15·4	65·903	140·82	192·05	253·43	325·62	409·29	613·48	735·20	870·76	1185·6	1562·2	1774·9	2516·7	2859·0
0·06750	3·806	4·574	4·929	5·268	5·594	5·908	6·506	6·792	7·070	7·605	8·117	8·365	9·080	9·356
1/ 14·8	67·261	143·70	195·98	258·61	332·27	417·62	625·95	750·12	888·41	1209·6	1593·7	1810·8	2567·4	2916·5
0·07000	3·882	4·665	5·026	5·372	5·704	6·024	6·633	6·924	7·208	7·753	8·275	8·527	9·257	9·538
1/ 14·3	68·597	146·54	199·84	263·69	338·79	425·81	638·19	764·78	905·75	1233·1	1624·7	1846·0	2617·2	2973·1
0·07250	3·956	4·753	5·122	5·474	5·812	6·138	6·758	7·055	7·343	7·899	8·430	8·687	9·430	9·716
1/ 13·8	69·910	149·33	203·64	268·70	345·21	433·87	650·23	779·19	922·80	1256·3	1655·2	1880·5	2666·2	3028·6
0·07500	4·029	4·841	5·216	5·574	5·918	6·250	6·881	7·183	7·477	8·042	8·583	8·844	9·600	9·891
1/ 13·3	71·202	152·08	207·38	273·62	351·52	441·79	662·07	793·36	939·57	1279·1	1685·2	1914·6	2714·3	3083·3
0·08000	4·172	5·012	5·399	5·770	6·126	6·469	7·122	7·434	7·737	8·322	8·881	9·151	9·932	10·23
1/ 12·5	73·726	157·44	214·68	283·23	363·85	457·26	685·20	821·05	972·32	1323·6	1743·7	1981·0	2808·3	3190·0
0·08500	4·311	5·177	5·577	5·960	6·327	6·681	7·355	7·677	7·990	8·594	9·170	9·449	10·26	10·57
1/ 11·8	76·177	162·65	221·76	292·56	375·82	472·28	707·64	847·91	1004·1	1366·8	1800·5	2045·5	2899·5	3293·5
0·09000	4·446	5·338	5·751	6·145	6·523	6·888	7·582	7·914	8·236	8·858	9·451	9·739	10·57	10·89
1/ 11·1	78·561	167·71	228·65	301·63	387·45	486·88	729·47	874·03	1035·0	1408·8	1855·7	2108·2	2988·2	3394·2
0·09500	4·577	5·495	5·919	6·325	6·714	7·089	7·803	8·144	8·476	9·115	9·725	10·02	10·87	11·20
1/ 10·5	80·883	172·64	235·36	310·47	398·78	501·10	750·71	899·46	1065·1	1449·6	1909·5	2169·2	3074·5	3492·1
0·10000	4·705	5·649	6·084	6·500	6·900	7·285	8·018	8·368	8·709	9·365	9·992	10·30	11·17	11·51
1/ 10·0	83·148	177·45	241·91	319·08	409·83	514·96	771·43	924·25	1094·4	1489·5	1961·9	2228·7	3158·6	3587·6
	0·92	0·94	0·94	0·95	0·95	0·96	0·96	0·96	0·97	0·97	0·97	0·97	0·98	0·98

$V_{r(0·5)medial}$ for half-full circular pipes.

A13

(p.1 of 6)

k_s = **0·015 mm**
S = 0·00010 to 0·00030

ie hydraulic gradient =
1 in 10000 to 1 in 3333

Water (or sewage) at 15°C;
full bore conditions.

velocities in ms^{-1}
discharges in litres/sec

Gradient	(Equivalent) Pipe diameters in mm													
	150	200	225	250	275	300	350	375	400	450	500	525	600	630
0·000100	0·101	0·124	0·135	0·145	0·155	0·164	0·183	0·192	0·200	0·217	0·233	0·241	0·263	0·272
1/ 10000	1·7841	3·8937	5·3538	7·1150	9·1988	11·626	17·593	21·171	25·170	34·504	45·735	52·103	74·414	84·753
0·000105	0·104	0·127	0·138	0·149	0·159	0·169	0·188	0·197	0·206	0·223	0·239	0·247	0·270	0·279
1/ 9524	1·8346	4·0030	5·5037	7·3135	9·4550	11·950	18·080	21·757	25·865	35·455	46·992	53·534	76·452	87·072
0·000110	0·107	0·131	0·142	0·153	0·163	0·174	0·193	0·202	0·211	0·229	0·246	0·254	0·277	0·287
1/ 9091	1·8840	4·1100	5·6504	7·5080	9·7057	12·266	18·557	22·330	26·546	36·385	48·222	54·934	78·447	89·342
0·000115	0·109	0·134	0·146	0·157	0·168	0·178	0·198	0·207	0·217	0·235	0·252	0·260	0·284	0·294
1/ 8696	1·9325	4·2149	5·7941	7·6985	9·9515	12·576	19·025	22·891	27·212	37·296	49·428	56·306	80·401	91·566
0·000120	0·112	0·137	0·149	0·161	0·172	0·182	0·203	0·212	0·222	0·240	0·258	0·266	0·291	0·301
1/ 8333	1·9800	4·3177	5·9351	7·8853	10·192	12·880	19·483	23·442	27·866	38·190	50·609	57·651	82·317	93·745
0·000125	0·115	0·141	0·153	0·164	0·176	0·186	0·207	0·217	0·227	0·246	0·264	0·272	0·298	0·308
1/ 8000	2·0267	4·4187	6·0735	8·0688	10·429	13·178	19·933	23·982	28·507	39·067	51·769	58·971	84·197	95·884
0·000130	0·117	0·144	0·156	0·168	0·179	0·191	0·212	0·222	0·232	0·251	0·269	0·278	0·304	0·314
1/ 7692	2·0726	4·5180	6·2095	8·2489	10·661	13·471	20·374	24·513	29·137	39·928	52·908	60·267	86·043	97·985
0·000135	0·120	0·147	0·160	0·172	0·183	0·195	0·216	0·227	0·237	0·256	0·275	0·284	0·311	0·321
1/ 7407	2·1177	4·6155	6·3432	8·4261	10·890	13·759	20·809	25·034	29·757	40·775	54·028	61·541	87·857	100·05
0·000140	0·122	0·150	0·163	0·175	0·187	0·199	0·221	0·231	0·242	0·262	0·281	0·290	0·317	0·327
1/ 7143	2·1621	4·7115	6·4747	8·6003	11·114	14·042	21·236	25·547	30·365	41·607	55·128	62·793	89·640	102·08
0·000145	0·125	0·153	0·166	0·179	0·191	0·203	0·225	0·236	0·246	0·267	0·286	0·296	0·323	0·334
1/ 6897	2·2057	4·8059	6·6041	8·7717	11·336	14·321	21·656	26·052	30·965	42·426	56·211	64·026	91·395	104·07
0·000150	0·127	0·156	0·169	0·182	0·195	0·206	0·229	0·240	0·251	0·272	0·292	0·301	0·329	0·340
1/ 6667	2·2487	4·8989	6·7315	8·9406	11·553	14·596	22·070	26·549	31·555	43·233	57·277	65·239	93·123	106·04
0·000160	0·132	0·162	0·176	0·189	0·202	0·214	0·238	0·249	0·260	0·282	0·302	0·312	0·341	0·353
1/ 6250	2·3329	5·0809	6·9808	9·2709	11·979	15·133	22·879	27·521	32·708	44·810	59·362	67·612	96·501	109·88
0·000170	0·137	0·167	0·182	0·195	0·209	0·221	0·246	0·258	0·269	0·291	0·313	0·323	0·353	0·364
1/ 5882	2·4148	5·2579	7·2233	9·5921	12·393	15·655	23·665	28·466	33·830	46·343	61·389	69·918	99·785	113·62
0·000180	0·141	0·173	0·188	0·202	0·215	0·229	0·254	0·266	0·278	0·301	0·323	0·333	0·364	0·376
1/ 5556	2·4946	5·4303	7·4595	9·9049	12·797	16·163	24·432	29·386	34·922	47·835	63·362	72·163	102·98	117·26
0·000190	0·146	0·178	0·193	0·208	0·222	0·236	0·262	0·274	0·286	0·310	0·333	0·343	0·375	0·388
1/ 5263	2·5725	5·5985	7·6899	10·210	13·190	16·659	25·179	30·284	35·987	49·291	65·286	74·353	106·10	120·80
0·000200	0·150	0·183	0·199	0·214	0·229	0·243	0·269	0·282	0·295	0·319	0·342	0·353	0·386	0·399
1/ 5000	2·6485	5·7628	7·9149	10·508	13·574	17·144	25·908	31·160	37·027	50·712	67·165	76·490	109·14	124·26
0·000210	0·154	0·189	0·205	0·220	0·235	0·249	0·277	0·290	0·303	0·328	0·351	0·363	0·397	0·409
1/ 4762	2·7229	5·9235	8·1350	10·800	13·950	17·617	26·622	32·017	38·044	52·101	69·001	78·579	112·11	127·65
0·000220	0·158	0·194	0·210	0·226	0·241	0·256	0·284	0·297	0·311	0·336	0·361	0·372	0·407	0·420
1/ 4545	2·7958	6·0808	8·3504	11·085	14·318	18·081	27·320	32·855	39·039	53·461	70·798	80·624	115·02	130·96
0·000230	0·162	0·198	0·215	0·232	0·247	0·262	0·291	0·305	0·318	0·345	0·370	0·382	0·417	0·430
1/ 4348	2·8672	6·2350	8·5615	11·364	14·678	18·535	28·004	33·676	40·013	54·792	72·558	82·626	117·87	134·20
0·000240	0·166	0·203	0·221	0·237	0·253	0·269	0·298	0·312	0·326	0·353	0·378	0·391	0·427	0·441
1/ 4167	2·9372	6·3861	8·7685	11·639	15·031	18·980	28·674	34·482	40·969	56·098	74·283	84·589	120·67	137·37
0·000250	0·170	0·208	0·226	0·243	0·259	0·275	0·305	0·319	0·333	0·361	0·387	0·400	0·436	0·451
1/ 4000	3·0060	6·5345	8·9717	11·908	15·378	19·417	29·332	35·272	41·907	57·379	75·976	86·515	123·41	140·49
0·000260	0·174	0·213	0·231	0·248	0·265	0·281	0·312	0·326	0·341	0·369	0·395	0·408	0·446	0·461
1/ 3846	3·0735	6·6803	9·1712	12·172	15·718	19·846	29·979	36·048	42·828	58·637	77·639	88·406	126·10	143·55
0·000270	0·178	0·217	0·236	0·253	0·270	0·287	0·318	0·333	0·348	0·376	0·404	0·417	0·455	0·470
1/ 3704	3·1399	6·8236	9·3674	12·432	16·053	20·268	30·614	36·811	43·733	59·873	79·272	90·265	128·74	146·56
0·000280	0·181	0·222	0·240	0·258	0·276	0·293	0·325	0·340	0·355	0·384	0·412	0·425	0·465	0·480
1/ 3571	3·2052	6·9645	9·5603	12·687	16·382	20·683	31·238	37·561	44·622	61·088	80·878	92·092	131·34	149·52
0·000290	0·185	0·226	0·245	0·264	0·281	0·298	0·331	0·347	0·362	0·392	0·420	0·434	0·474	0·489
1/ 3448	3·2695	7·1032	9·7502	12·938	16·706	21·091	31·853	38·299	45·498	62·284	82·458	93·889	133·90	152·43
0·000300	0·189	0·230	0·250	0·269	0·287	0·304	0·337	0·353	0·369	0·399	0·428	0·442	0·482	0·498
1/ 3333	3·3328	7·2398	9·9372	13·186	17·025	21·493	32·458	39·025	46·360	63·461	84·014	95·659	136·42	155·29
	0·78	0·80	0·81	0·82	0·83	0·83	0·85	0·85	0·86	0·87	0·88	0·88	0·89	0·89

$V_{r(0{\cdot}5)medial}$ for half-full circular pipes.

k_s = 0·015 mm S = 0·00010 to 0·00030

k_s = **0·015 mm**
S = 0·00030 to 0·00100

ie hydraulic gradient =
1 in 3333 to 1 in 1000

Water (or sewage) at 15°C;
full bore conditions.

velocities in ms^{-1}
discharges in litres/sec

Gradient **(Equivalent) Pipe diameters in mm**

Gradient	150	200	225	250	275	300	350	375	400	450	500	525	600	630
0·000300	0·189	0·230	0·250	0·269	0·287	0·304	0·337	0·353	0·369	0·399	0·428	0·442	0·482	0·498
1/ 3333	3·3328	7·2398	9·9372	13·186	17·025	21·493	32·458	39·025	46·360	63·461	84·014	95·659	136·42	155·29
0·000320	0·196	0·239	0·259	0·278	0·297	0·315	0·350	0·366	0·382	0·413	0·443	0·458	0·500	0·516
1/ 3125	3·4567	7·5070	10·303	13·670	17·649	22·279	33·641	40·445	48·045	65·763	87·055	99·119	141·34	160·89
0·000340	0·202	0·247	0·268	0·288	0·307	0·326	0·362	0·379	0·395	0·428	0·458	0·473	0·517	0·534
1/ 2941	3·5772	7·7667	10·658	14·141	18·255	23·043	34·791	41·826	49·683	68·000	90·010	102·48	146·12	166·33
0·000360	0·209	0·255	0·277	0·297	0·317	0·337	0·373	0·391	0·408	0·441	0·473	0·489	0·533	0·551
1/ 2778	3·6946	8·0198	11·005	14·599	18·846	23·787	35·911	43·170	51·278	70·178	92·887	105·75	150·78	171·62
0·000380	0·216	0·263	0·285	0·307	0·327	0·347	0·385	0·403	0·420	0·455	0·487	0·503	0·549	0·567
1/ 2632	3·8091	8·2665	11·342	15·046	19·421	24·512	37·002	44·481	52·833	72·301	95·692	108·94	155·32	176·78
0·000400	0·222	0·271	0·294	0·315	0·336	0·357	0·396	0·414	0·433	0·468	0·501	0·518	0·565	0·583
1/ 2500	3·9209	8·5076	11·672	15·482	19·984	25·220	38·068	45·761	54·351	74·374	98·430	112·06	159·75	181·82
0·000420	0·228	0·278	0·302	0·324	0·346	0·367	0·407	0·426	0·444	0·480	0·515	0·532	0·580	0·599
1/ 2381	4·0303	8·7432	11·995	15·909	20·533	25·913	39·110	47·011	55·834	76·400	101·11	115·10	164·07	186·74
0·000440	0·234	0·286	0·310	0·333	0·355	0·376	0·417	0·437	0·456	0·493	0·528	0·545	0·595	0·615
1/ 2273	4·1373	8·9739	12·310	16·327	21·071	26·591	40·130	48·235	57·286	78·382	103·72	118·08	168·31	191·56
0·000460	0·240	0·293	0·317	0·341	0·364	0·386	0·427	0·448	0·467	0·505	0·541	0·559	0·610	0·630
1/ 2174	4·2423	9·1999	12·620	16·736	21·598	27·254	41·129	49·434	58·708	80·323	106·29	121·00	172·46	196·27
0·000480	0·246	0·300	0·325	0·349	0·372	0·395	0·438	0·458	0·478	0·517	0·554	0·572	0·624	0·644
1/ 2083	4·3452	9·4215	12·923	17·137	22·115	27·905	42·108	50·609	60·102	82·226	108·80	123·85	176·52	200·90
0·000500	0·252	0·307	0·332	0·357	0·381	0·404	0·448	0·469	0·489	0·529	0·567	0·585	0·638	0·659
1/ 2000	4·4462	9·6390	13·220	17·531	22·622	28·544	43·068	51·762	61·470	84·093	111·27	126·66	180·51	205·43
0·000525	0·259	0·315	0·342	0·367	0·391	0·415	0·460	0·481	0·502	0·543	0·582	0·601	0·656	0·677
1/ 1905	4·5699	9·9055	13·585	18·013	23·243	29·326	44·245	53·175	63·145	86·379	114·29	130·09	185·39	210·98
0·000550	0·265	0·324	0·351	0·377	0·402	0·426	0·472	0·494	0·516	0·557	0·597	0·616	0·673	0·694
1/ 1818	4·6911	10·166	13·941	18·485	23·851	30·091	45·396	54·556	64·783	88·616	117·24	133·45	190·17	216·41
0·000575	0·272	0·332	0·359	0·386	0·412	0·436	0·484	0·506	0·528	0·571	0·612	0·632	0·689	0·711
1/ 1739	4·8098	10·422	14·291	18·947	24·446	30·841	46·524	55·910	66·388	90·807	120·13	136·74	194·84	221·73
0·000600	0·279	0·340	0·368	0·395	0·421	0·447	0·495	0·518	0·541	0·584	0·626	0·647	0·705	0·728
1/ 1667	4·9262	10·672	14·634	19·400	25·030	31·576	47·629	57·236	67·962	92·954	122·97	139·97	199·43	226·94
0·000625	0·285	0·348	0·376	0·404	0·431	0·457	0·506	0·530	0·553	0·598	0·640	0·661	0·721	0·744
1/ 1600	5·0404	10·918	14·970	19·845	25·602	32·297	48·713	58·538	69·505	95·060	125·75	143·13	203·92	232·05
0·000650	0·292	0·355	0·385	0·413	0·441	0·467	0·517	0·542	0·565	0·611	0·654	0·676	0·737	0·761
1/ 1538	5·1527	11·160	15·300	20·282	26·165	33·005	49·778	59·816	71·021	97·129	128·48	146·23	208·34	237·07
0·000675	0·298	0·363	0·393	0·422	0·450	0·477	0·528	0·553	0·577	0·623	0·668	0·690	0·752	0·776
1/ 1481	5·2630	11·397	15·625	20·711	26·717	33·701	50·825	61·071	72·510	99·160	131·16	149·28	212·67	242·00
0·000700	0·304	0·370	0·401	0·431	0·459	0·486	0·539	0·564	0·589	0·636	0·681	0·703	0·767	0·792
1/ 1429	5·3715	11·630	15·944	21·133	27·261	34·385	51·853	62·305	73·973	101·16	133·80	152·28	216·93	246·85
0·000725	0·310	0·378	0·409	0·439	0·468	0·496	0·549	0·575	0·600	0·648	0·695	0·717	0·782	0·807
1/ 1379	5·4782	11·860	16·258	21·548	27·795	35·058	52·865	63·520	75·413	103·12	136·39	155·23	221·13	251·61
0·000750	0·316	0·385	0·417	0·447	0·477	0·505	0·560	0·586	0·611	0·661	0·708	0·731	0·797	0·822
1/ 1333	5·5833	12·086	16·567	21·957	28·321	35·721	53·861	64·715	76·831	105·06	138·95	158·14	225·25	256·31
0·000800	0·328	0·399	0·432	0·464	0·494	0·524	0·580	0·607	0·633	0·684	0·733	0·757	0·825	0·852
1/ 1250	5·7890	12·528	17·171	22·757	29·350	37·017	55·809	67·052	79·602	108·84	143·94	163·81	233·32	265·47
0·000850	0·339	0·412	0·447	0·479	0·511	0·541	0·600	0·628	0·655	0·707	0·758	0·782	0·853	0·880
1/ 1176	5·9889	12·958	17·759	23·533	30·351	38·276	57·701	69·323	82·294	112·51	148·79	169·32	241·15	274·38
0·000900	0·350	0·426	0·461	0·495	0·527	0·559	0·619	0·648	0·676	0·730	0·782	0·807	0·880	0·908
1/ 1111	6·1836	13·376	18·331	24·290	31·324	39·501	59·543	71·533	84·915	116·09	153·50	174·69	248·77	283·05
0·000950	0·361	0·439	0·475	0·510	0·543	0·576	0·638	0·667	0·696	0·752	0·805	0·831	0·906	0·935
1/ 1053	6·3734	13·784	18·889	25·027	32·273	40·696	61·339	73·687	87·469	119·57	158·10	179·92	256·20	291·49
0·001000	0·371	0·451	0·489	0·525	0·559	0·592	0·656	0·686	0·716	0·773	0·828	0·855	0·932	0·962
1/ 1000	6·5589	14·183	19·433	25·747	33·200	41·862	63·092	75·790	89·962	122·97	162·59	185·02	263·45	299·73
	0·81	0·83	0·84	0·85	0·86	0·86	0·88	0·88	0·89	0·90	0·90	0·91	0·92	0·92

$V_{r(0\cdot5)medial}$ **for half-full circular pipes.**

A13
(p.3 of 6)

$k_s = 0{\cdot}015$ **mm**
S = 0·00100 to 0·00300

ie hydraulic gradient =
1 in 1000 to 1 in 333

Water (or sewage) at 15°C;
full bore conditions.

velocities in ms^{-1}
discharges in litres/sec

Gradient **(Equivalent) Pipe diameters in mm**

Gradient	150	200	225	250	275	300	350	375	400	450	500	525	600	630
0·00100	0·371	0·451	0·489	0·525	0·559	0·592	0·656	0·686	0·716	0·773	0·828	0·855	0·932	0·962
1/ 1000	6·5589	14·183	19·433	25·747	33·200	41·862	63·092	75·790	89·962	122·97	162·59	185·02	263·45	299·73
0·00105	0·381	0·464	0·502	0·539	0·574	0·608	0·674	0·705	0·735	0·794	0·850	0·878	0·957	0·987
1/ 952	6·7401	14·572	19·965	26·451	34·105	43·002	64·805	77·845	92·398	126·29	166·98	190·01	270·53	307·78
0·00110	0·391	0·476	0·515	0·553	0·589	0·624	0·691	0·723	0·754	0·815	0·872	0·900	0·981	1·013
1/ 909	6·9176	14·953	20·486	27·139	34·991	44·118	66·480	79·855	94·781	129·54	171·26	194·88	277·46	315·66
0·00115	0·401	0·488	0·528	0·567	0·604	0·640	0·708	0·741	0·773	0·835	0·894	0·922	1·005	1·037
1/ 870	7·0914	15·327	20·996	27·814	35·859	45·210	68·121	81·824	97·115	132·73	175·46	199·66	284·24	323·37
0·00120	0·411	0·500	0·541	0·580	0·618	0·655	0·725	0·758	0·791	0·854	0·915	0·944	1·029	1·062
1/ 833	7·2618	15·693	21·496	28·475	36·710	46·280	69·730	83·753	99·402	135·85	179·58	204·34	290·89	330·92
0·00125	0·420	0·511	0·553	0·593	0·632	0·670	0·741	0·775	0·809	0·873	0·935	0·965	1·052	1·085
1/ 800	7·4291	16·052	21·987	29·123	37·544	47·331	71·307	85·646	101·65	138·91	183·62	208·93	297·41	338·33
0·00130	0·430	0·522	0·565	0·606	0·646	0·684	0·757	0·792	0·826	0·892	0·955	0·986	1·074	1·109
1/ 769	7·5934	16·404	22·469	29·760	38·364	48·362	72·856	87·504	103·85	141·91	187·58	213·43	303·81	345·60
0·00135	0·439	0·533	0·577	0·619	0·659	0·699	0·773	0·809	0·844	0·911	0·975	1·006	1·097	1·132
1/ 741	7·7548	16·751	22·942	30·386	39·169	49·375	74·378	89·329	106·01	144·86	191·47	217·86	310·09	352·75
0·00140	0·448	0·544	0·589	0·632	0·673	0·713	0·789	0·825	0·861	0·929	0·995	1·026	1·119	1·154
1/ 714	7·9135	17·092	23·408	31·001	39·960	50·371	75·874	91·123	108·14	147·76	195·30	222·21	316·27	359·77
0·00145	0·457	0·555	0·600	0·644	0·686	0·726	0·804	0·841	0·877	0·947	1·014	1·046	1·140	1·176
1/ 690	8·0697	17·427	23·866	31·606	40·739	51·351	77·345	92·888	110·23	150·61	199·06	226·48	322·34	366·67
0·00150	0·465	0·565	0·612	0·656	0·699	0·740	0·819	0·857	0·894	0·965	1·033	1·066	1·161	1·198
1/ 667	8·2235	17·757	24·317	32·202	41·505	52·315	78·793	94·625	112·29	153·42	202·76	230·69	328·32	373·46
0·00160	0·482	0·586	0·634	0·680	0·724	0·767	0·848	0·887	0·926	0·999	1·070	1·104	1·202	1·241
1/ 625	8·5242	18·402	25·198	33·366	43·003	54·200	81·624	98·020	116·31	158·90	210·00	238·92	340·00	386·74
0·00170	0·499	0·606	0·655	0·703	0·749	0·793	0·877	0·917	0·957	1·033	1·105	1·141	1·243	1·282
1/ 588	8·8165	19·029	26·054	34·498	44·459	56·031	84·374	101·32	120·22	164·23	217·03	246·91	351·34	399·63
0·00180	0·515	0·625	0·676	0·725	0·772	0·818	0·905	0·946	0·987	1·065	1·140	1·176	1·282	1·322
1/ 556	9·1011	19·639	26·888	35·599	45·876	57·814	87·050	104·53	124·03	169·42	223·87	254·68	362·38	412·17
0·00190	0·531	0·644	0·697	0·747	0·796	0·842	0·932	0·975	1·016	1·097	1·174	1·211	1·320	1·361
1/ 526	9·3786	20·234	27·701	36·673	47·256	59·551	89·658	107·65	127·73	174·47	230·53	262·26	373·13	424·39
0·00200	0·546	0·663	0·717	0·768	0·818	0·866	0·958	1·002	1·045	1·128	1·207	1·246	1·357	1·400
1/ 500	9·6495	20·815	28·494	37·721	48·604	61·246	92·203	110·71	131·35	179·40	237·03	269·65	383·62	436·31
0·00210	0·561	0·681	0·736	0·789	0·840	0·890	0·984	1·029	1·073	1·158	1·240	1·279	1·393	1·437
1/ 476	9·9143	21·383	29·269	38·745	49·921	62·903	94·690	113·69	134·88	184·21	243·38	276·86	393·86	447·96
0·00220	0·576	0·698	0·755	0·810	0·862	0·913	1·009	1·056	1·101	1·188	1·271	1·312	1·428	1·474
1/ 455	10·173	21·938	30·027	39·747	51·209	64·524	97·122	116·60	138·34	188·92	249·59	283·92	403·88	459·34
0·00230	0·590	0·716	0·774	0·830	0·883	0·935	1·034	1·082	1·128	1·217	1·302	1·343	1·463	1·509
1/ 435	10·427	22·482	30·770	40·728	52·471	66·110	99·503	119·46	141·72	193·53	255·67	290·83	413·69	470·49
0·00240	0·604	0·733	0·792	0·849	0·904	0·957	1·058	1·107	1·154	1·245	1·332	1·375	1·497	1·544
1/ 417	10·676	23·015	31·498	41·689	53·707	67·665	101·84	122·26	145·04	198·05	261·63	297·61	423·30	481·40
0·00250	0·618	0·749	0·810	0·868	0·925	0·979	1·082	1·132	1·180	1·273	1·362	1·405	1·530	1·579
1/ 400	10·920	23·538	32·212	42·632	54·920	69·191	104·13	125·00	148·29	202·48	267·47	304·24	432·72	492·11
0·00260	0·632	0·766	0·828	0·887	0·945	1·000	1·106	1·156	1·205	1·300	1·391	1·436	1·563	1·612
1/ 385	11·160	24·052	32·913	43·558	56·110	70·688	106·37	127·70	151·48	206·83	273·21	310·76	441·97	502·62
0·00270	0·645	0·782	0·845	0·906	0·964	1·021	1·129	1·180	1·230	1·327	1·420	1·465	1·595	1·645
1/ 370	11·396	24·556	33·602	44·467	57·279	72·159	108·58	130·34	154·62	211·10	278·84	317·16	451·05	512·94
0·00280	0·658	0·797	0·862	0·924	0·984	1·041	1·151	1·204	1·255	1·354	1·448	1·494	1·627	1·678
1/ 357	11·627	25·052	34·279	45·361	58·429	73·604	110·75	132·94	157·70	215·30	284·37	323·45	459·97	523·08
0·00290	0·671	0·813	0·879	0·942	1·003	1·061	1·173	1·227	1·279	1·380	1·476	1·523	1·658	1·710
1/ 345	11·855	25·540	34·945	46·241	59·560	75·026	112·88	135·50	160·73	219·43	289·81	329·63	468·75	533·04
0·00300	0·684	0·828	0·895	0·960	1·021	1·081	1·195	1·250	1·303	1·405	1·503	1·551	1·688	1·741
1/ 333	12·079	26·020	35·600	47·106	60·672	76·425	114·98	138·02	163·71	223·49	295·17	335·72	477·38	542·85
	0·84	0·86	0·87	0·88	0·89	0·89	0·90	0·91	0·91	0·92	0·93	0·93	0·94	0·94

$V_{r(0{\cdot}5)medial}$ **for half-full circular pipes.**

$k_s = 0{\cdot}015$ **mm** S = 0·00100 to 0·00300

Gradient **(Equivalent) Pipe diameters in mm**

Gradient	150	200	225	250	275	300	350	375	400	450	500	525	600	630
0·00300	0·684	0·828	0·895	0·960	1·021	1·081	1·195	1·250	1·303	1·405	1·503	1·551	1·688	1·741
1/ 333	12·079	26·020	35·600	47·106	60·672	76·425	114·98	138·02	163·71	223·49	295·17	335·72	477·38	542·85
0·00320	0·708	0·858	0·928	0·994	1·058	1·120	1·238	1·294	1·349	1·455	1·557	1·606	1·748	1·803
1/ 313	12·518	26·959	36·881	48·798	62·847	79·160	119·08	142·93	169·53	231·43	305·63	347·61	494·25	562·02
0·00340	0·732	0·887	0·959	1·028	1·094	1·157	1·279	1·337	1·394	1·504	1·608	1·659	1·806	1·863
1/ 294	12·944	27·871	38·126	50·441	64·959	81·816	123·07	147·71	175·19	239·13	315·79	359·15	510·62	580·62
0·00360	0·756	0·915	0·989	1·060	1·128	1·194	1·319	1·379	1·438	1·551	1·659	1·711	1·862	1·921
1/ 278	13·359	28·759	39·337	52·040	67·015	84·400	126·94	152·36	180·70	246·63	325·67	370·38	526·55	598·72
0·00380	0·779	0·943	1·019	1·092	1·162	1·230	1·359	1·420	1·481	1·597	1·708	1·761	1·917	1·977
1/ 263	13·763	29·624	40·517	53·598	69·017	86·919	130·72	156·88	186·06	253·93	335·30	381·32	542·06	616·34
0·00400	0·801	0·970	1·048	1·123	1·195	1·264	1·397	1·460	1·522	1·641	1·755	1·811	1·971	2·032
1/ 250	14·157	30·468	41·669	55·119	70·972	89·376	134·40	161·30	191·29	261·06	344·69	391·99	557·19	633·53
0·00420	0·823	0·996	1·076	1·153	1·227	1·298	1·434	1·499	1·563	1·685	1·802	1·859	2·023	2·086
1/ 238	14·543	31·293	42·795	56·604	72·881	91·776	138·00	165·61	196·40	268·02	353·86	402·41	571·97	650·32
0·00440	0·844	1·022	1·104	1·183	1·258	1·332	1·471	1·538	1·603	1·728	1·848	1·906	2·074	2·139
1/ 227	14·920	32·100	43·895	58·057	74·748	94·123	141·52	169·83	201·39	274·82	362·82	412·59	586·42	666·73
0·00460	0·865	1·047	1·131	1·212	1·289	1·364	1·507	1·575	1·642	1·770	1·893	1·952	2·124	2·190
1/ 217	15·290	32·890	44·973	59·480	76·576	96·421	144·97	173·96	206·29	281·48	371·60	422·57	600·55	682·79
0·00480	0·886	1·072	1·158	1·240	1·319	1·396	1·542	1·612	1·680	1·811	1·936	1·997	2·173	2·241
1/ 208	15·652	33·664	46·029	60·874	78·367	98·673	148·34	178·00	211·08	288·00	380·19	432·33	614·40	698·53
0·00500	0·906	1·096	1·184	1·268	1·349	1·427	1·576	1·648	1·717	1·851	1·979	2·041	2·221	2·290
1/ 200	16·007	34·423	47·065	62·241	80·124	100·88	151·65	181·97	215·77	294·40	388·62	441·91	627·98	713·95
0·00525	0·930	1·125	1·216	1·302	1·385	1·465	1·618	1·692	1·763	1·900	2·032	2·096	2·280	2·351
1/ 190	16·442	35·353	48·333	63·914	82·274	103·58	155·70	186·83	221·52	302·23	398·94	453·63	644·59	732·82
0·00550	0·955	1·154	1·247	1·335	1·421	1·503	1·660	1·735	1·808	1·948	2·083	2·148	2·337	2·410
1/ 182	16·868	36·262	49·573	65·551	84·377	106·23	159·66	191·57	227·15	309·88	409·02	465·08	660·84	751·27
0·00575	0·978	1·183	1·277	1·368	1·455	1·539	1·700	1·777	1·851	1·996	2·133	2·200	2·393	2·468
1/ 174	17·284	37·153	50·788	67·153	86·435	108·81	163·54	196·22	232·65	317·37	418·89	476·29	676·73	769·32
0·00600	1·001	1·210	1·307	1·400	1·489	1·575	1·739	1·818	1·894	2·042	2·183	2·251	2·449	2·525
1/ 167	17·693	38·025	51·977	68·723	88·452	111·35	167·34	200·77	238·04	324·71	428·56	487·28	692·30	787·01
0·00625	1·024	1·238	1·337	1·431	1·523	1·610	1·778	1·858	1·936	2·087	2·231	2·301	2·502	2·580
1/ 160	18·093	38·881	53·144	70·262	90·430	113·83	171·06	205·23	243·32	331·91	438·03	498·04	707·56	804·34
0·00650	1·046	1·264	1·365	1·462	1·555	1·645	1·816	1·898	1·978	2·131	2·278	2·349	2·555	2·635
1/ 154	18·486	39·721	54·289	71·773	92·371	116·27	174·72	209·61	248·51	338·96	447·33	508·61	722·53	821·35
0·00675	1·068	1·291	1·394	1·492	1·587	1·679	1·853	1·937	2·018	2·175	2·325	2·397	2·607	2·688
1/ 148	18·873	40·546	55·414	73·257	94·277	118·67	178·31	213·91	253·60	345·90	456·46	518·98	737·23	838·05
0·00700	1·089	1·316	1·421	1·522	1·619	1·712	1·890	1·975	2·058	2·218	2·370	2·444	2·659	2·741
1/ 143	19·252	41·356	56·520	74·716	96·151	121·02	181·84	218·14	258·61	352·71	465·43	529·17	751·68	854·45
0·00725	1·111	1·342	1·449	1·551	1·650	1·745	1·926	2·013	2·097	2·260	2·415	2·491	2·709	2·793
1/ 138	19·626	42·154	57·607	76·150	97·993	123·34	185·30	222·29	263·53	359·40	474·25	539·19	765·87	870·58
0·00750	1·131	1·367	1·476	1·580	1·680	1·777	1·961	2·050	2·136	2·301	2·460	2·536	2·758	2·844
1/ 133	19·993	42·938	58·676	77·561	99·805	125·62	188·72	226·38	268·37	365·99	482·93	549·04	779·84	886·44
0·00800	1·172	1·416	1·528	1·636	1·740	1·840	2·031	2·122	2·211	2·382	2·546	2·625	2·855	2·943
1/ 125	20·712	44·472	60·767	80·318	103·35	130·07	195·38	234·37	277·82	378·86	499·88	568·30	807·12	917·42
0·00850	1·212	1·463	1·579	1·691	1·798	1·901	2·098	2·192	2·284	2·461	2·630	2·712	2·948	3·040
1/ 118	21·409	45·961	62·797	82·995	106·79	134·39	201·85	242·12	287·00	391·35	516·32	586·98	833·60	947·49
0·00900	1·250	1·509	1·629	1·744	1·854	1·961	2·163	2·260	2·355	2·537	2·711	2·795	3·039	3·133
1/ 111	22·088	47·409	64·771	85·600	110·13	138·59	208·15	249·66	295·93	403·49	532·32	605·15	859·34	976·72
0·00950	1·287	1·554	1·677	1·796	1·909	2·019	2·227	2·327	2·424	2·611	2·790	2·877	3·128	3·225
1/ 105	22·750	48·821	66·695	88·137	113·39	142·68	214·27	257·00	304·62	415·32	547·89	622·84	884·40	1005·2
0·01000	1·324	1·598	1·725	1·846	1·963	2·075	2·289	2·392	2·492	2·684	2·868	2·957	3·214	3·314
1/ 100	23·395	50·198	68·571	90·611	116·57	146·67	220·25	264·16	313·10	426·85	563·08	640·09	908·84	1032·9
	0·87	0·89	0·90	0·90	0·91	0·92	0·93	0·93	0·93	0·94	0·94	0·95	0·95	0·95

$V_{r(0·5)medial}$ **for half-full circular pipes.**

$k_s = 0{\cdot}015$ **mm**
S = 0·01000 to 0·03000

ie hydraulic gradient =
1 in 100 to 1 in 33·3

Water (or sewage) at 15°C;
full bore conditions.

velocities in ms^{-1}
discharges in litres/sec

Gradient **(Equivalent) Pipe diameters in mm**

Gradient	150	200	225	250	275	300	350	375	400	450	500	525	600	630
0·01000	1·324	1·598	1·725	1·846	1·963	2·075	2·289	2·392	2·492	2·684	2·868	2·957	3·214	3·314
1/ 100	23·395	50·198	68·571	90·611	116·57	146·67	220·25	264·16	313·10	426·85	563·08	640·09	908·84	1032·9
0·01050	1·360	1·641	1·771	1·895	2·015	2·130	2·350	2·455	2·557	2·755	2·943	3·035	3·299	3·401
1/ 95	24·026	51·543	70·404	93·028	119·67	150·57	226·09	271·15	321·38	438·12	577·91	656·93	932·69	1060·0
0·01100	1·394	1·683	1·816	1·943	2·066	2·184	2·409	2·517	2·622	2·824	3·017	3·111	3·381	3·485
1/ 91	24·643	52·858	72·196	95·391	122·70	154·38	231·80	277·99	329·47	449·12	592·40	673·39	956·01	1086·5
0·01150	1·429	1·723	1·860	1·990	2·116	2·237	2·467	2·578	2·685	2·892	3·089	3·185	3·462	3·569
1/ 87	25·246	54·145	73·951	97·704	125·67	158·11	237·38	284·68	337·39	459·89	606·58	689·49	978·82	1112·4
0·01200	1·462	1·764	1·903	2·037	2·165	2·289	2·524	2·637	2·747	2·958	3·160	3·258	3·541	3·650
1/ 83	25·838	55·406	75·669	99·970	128·58	161·77	242·85	291·23	345·14	470·44	620·47	705·27	1001·2	1137·8
0·01250	1·495	1·803	1·945	2·082	2·213	2·339	2·580	2·695	2·807	3·023	3·229	3·329	3·618	3·730
1/ 80	26·418	56·643	77·354	102·19	131·43	165·35	248·21	297·65	352·74	480·78	634·08	720·73	1023·1	1162·6
0·01300	1·527	1·842	1·987	2·126	2·260	2·389	2·635	2·752	2·866	3·087	3·297	3·399	3·694	3·808
1/ 77	26·988	57·857	79·007	104·37	134·23	168·86	253·47	303·95	360·20	490·93	647·44	735·89	1044·5	1187·0
0·01350	1·559	1·880	2·028	2·170	2·306	2·438	2·688	2·808	2·925	3·149	3·364	3·468	3·769	3·885
1/ 74	27·547	59·048	80·631	106·51	136·98	172·31	258·64	310·14	367·52	500·89	660·54	750·78	1065·6	1211·0
0·01400	1·590	1·917	2·068	2·213	2·352	2·486	2·741	2·863	2·982	3·211	3·430	3·536	3·842	3·960
1/ 71	28·097	60·219	82·227	108·61	139·68	175·70	263·71	316·22	374·72	510·67	673·42	765·41	1086·3	1234·5
0·01450	1·621	1·953	2·107	2·255	2·396	2·533	2·793	2·917	3·038	3·271	3·494	3·602	3·914	4·034
1/ 69	28·637	61·371	83·795	110·68	142·33	179·04	268·70	322·19	381·79	520·29	686·08	779·78	1106·7	1257·6
0·01500	1·651	1·990	2·146	2·296	2·440	2·579	2·844	2·970	3·094	3·331	3·558	3·667	3·985	4·107
1/ 67	29·169	62·504	85·338	112·72	144·95	182·32	273·61	328·07	388·74	529·75	698·53	793·92	1126·7	1280·3
0·01600	1·709	2·060	2·222	2·377	2·526	2·670	2·943	3·074	3·202	3·447	3·681	3·795	4·123	4·250
1/ 63	30·208	64·717	88·353	116·69	150·05	188·72	283·19	339·55	402·33	548·22	722·85	821·53	1165·8	1324·7
0·01700	1·767	2·128	2·296	2·456	2·610	2·758	3·040	3·175	3·307	3·560	3·802	3·919	4·257	4·388
1/ 59	31·217	66·866	91·280	120·55	155·00	194·94	292·50	350·69	415·51	566·15	746·44	848·32	1203·7	1367·8
0·01800	1·822	2·195	2·367	2·532	2·691	2·843	3·134	3·273	3·409	3·669	3·918	4·039	4·388	4·522
1/ 56	32·199	68·956	94·126	124·30	159·81	200·98	301·54	361·51	428·33	583·57	769·37	874·36	1240·6	1409·7
0·01900	1·876	2·260	2·437	2·607	2·770	2·927	3·226	3·369	3·508	3·776	4·032	4·156	4·515	4·653
1/ 53	33·155	70·992	96·898	127·95	164·50	206·87	310·35	372·06	440·81	600·54	791·70	899·71	1276·5	1450·4
0·02000	1·929	2·323	2·505	2·679	2·847	3·008	3·315	3·462	3·605	3·880	4·143	4·270	4·638	4·780
1/ 50	34·087	72·977	99·602	131·51	169·07	212·61	318·93	382·34	452·97	617·08	813·46	924·43	1311·5	1490·1
0·02100	1·980	2·385	2·571	2·750	2·922	3·087	3·402	3·553	3·699	3·981	4·251	4·382	4·759	4·905
1/ 47·6	34·998	74·915	102·24	134·99	173·53	218·21	327·31	392·37	464·85	633·22	834·71	948·55	1345·6	1528·9
0·02200	2·031	2·445	2·636	2·819	2·995	3·165	3·487	3·641	3·792	4·081	4·357	4·491	4·877	5·026
1/ 45·5	35·889	76·811	104·82	138·39	177·90	223·69	335·51	402·18	476·46	649·00	855·47	972·12	1379·0	1566·8
0·02300	2·080	2·504	2·700	2·887	3·067	3·240	3·571	3·728	3·882	4·178	4·460	4·597	4·993	5·145
1/ 43·5	36·761	78·665	107·35	141·72	182·17	229·05	343·52	411·78	487·81	664·43	875·78	995·18	1411·6	1603·8
0·02400	2·129	2·562	2·762	2·953	3·137	3·315	3·652	3·813	3·970	4·273	4·562	4·702	5·106	5·261
1/ 41·7	37·614	80·482	109·82	144·98	186·35	234·30	351·37	421·18	498·93	679·55	895·66	1017·8	1443·6	1640·1
0·02500	2·176	2·619	2·823	3·019	3·206	3·387	3·732	3·897	4·057	4·366	4·661	4·804	5·216	5·375
1/ 40·0	38·452	82·263	112·24	148·17	190·45	239·44	359·06	430·39	509·83	694·36	915·15	1039·9	1474·9	1675·6
0·02600	2·222	2·674	2·883	3·082	3·274	3·459	3·810	3·979	4·142	4·457	4·758	4·904	5·325	5·487
1/ 38·5	39·273	84·010	114·62	151·31	194·47	244·49	366·61	439·42	520·52	708·89	934·26	1061·6	1505·6	1710·5
0·02700	2·268	2·729	2·942	3·145	3·341	3·529	3·887	4·059	4·226	4·547	4·854	5·002	5·432	5·597
1/ 37·0	40·080	85·726	116·96	154·38	198·41	249·44	374·02	448·29	531·01	723·15	953·02	1082·9	1535·7	1744·7
0·02800	2·313	2·782	2·999	3·207	3·406	3·598	3·963	4·138	4·308	4·635	4·948	5·099	5·536	5·705
1/ 35·7	40·873	87·411	119·25	157·41	202·29	254·31	381·29	457·00	541·32	737·16	971·45	1103·8	1565·3	1778·3
0·02900	2·357	2·835	3·056	3·267	3·470	3·665	4·037	4·215	4·388	4·721	5·040	5·194	5·639	5·811
1/ 34·5	41·652	89·068	121·51	160·38	206·10	259·09	388·45	465·56	551·45	750·92	989·56	1124·4	1594·4	1811·3
0·03000	2·400	2·887	3·112	3·327	3·533	3·732	4·111	4·291	4·468	4·807	5·130	5·287	5·740	5·915
1/ 33·3	42·418	90·698	123·73	163·30	209·85	263·80	395·48	473·98	561·41	764·46	1007·4	1144·6	1623·0	1843·8
	0·89	0·91	0·92	0·92	0·93	0·93	0·94	0·94	0·95	0·95	0·96	0·96	0·96	0·96

$V_{r(0{\cdot}5)medial}$ **for half-full circular pipes.**

$k_s = 0{\cdot}015$ **mm** S = 0·01000 to 0·03000

k_s = 0·015 mm
S = 0·03000 to 0·10000

ie hydraulic gradient =
1 in 33·3 to 1 in 10·0

Water (or sewage) at 15°C;
full bore conditions.

velocities in ms^{-1}
discharges in litres/sec

Gradient	**(Equivalent) Pipe diameters in mm**													
	150	**200**	**225**	**250**	**275**	**300**	**350**	**375**	**400**	**450**	**500**	**525**	**600**	**630**
0·03000	2·400	2·887	3·112	3·327	3·533	3·732	4·111	4·291	4·468	4·807	5·130	5·287	5·740	5·915
1/ 33·3	42·418	90·698	123·73	163·30	209·85	263·80	395·48	473·98	561·41	764·46	1007·4	1144·6	1623·0	1843·8
0·03200	2·485	2·988	3·221	3·443	3·656	3·862	4·253	4·440	4·622	4·973	5·308	5·470	5·938	6·118
1/ 31·3	43·916	93·881	128·06	169·00	217·17	272·98	409·22	490·42	580·86	790·89	1042·1	1184·1	1678·9	1907·2
0·03400	2·567	3·087	3·326	3·556	3·776	3·988	4·392	4·585	4·773	5·134	5·479	5·647	6·130	6·316
1/ 29·4	45·370	96·971	132·26	174·54	224·27	281·90	422·54	506·37	599·73	816·54	1075·9	1222·3	1733·1	1968·7
0·03600	2·647	3·182	3·429	3·665	3·892	4·111	4·526	4·725	4·918	5·291	5·646	5·818	6·316	6·507
1/ 27·8	46·784	99·975	136·35	179·93	231·18	290·56	435·49	521·87	618·07	841·46	1108·6	1259·6	1785·7	2028·5
0·03800	2·725	3·275	3·529	3·772	4·005	4·230	4·657	4·862	5·061	5·443	5·809	5·986	6·497	6·694
1/ 26·3	48·160	102·90	140·33	185·17	237·90	299·00	448·10	536·96	635·92	865·71	1140·5	1295·8	1837·0	2086·6
0·04000	2·801	3·366	3·627	3·876	4·116	4·346	4·785	4·995	5·199	5·592	5·967	6·149	6·673	6·876
1/ 25·0	49·503	105·75	144·21	190·28	244·45	307·22	460·39	551·67	653·32	889·35	1171·6	1331·1	1886·9	2143·3
0·04200	2·875	3·455	3·722	3·978	4·223	4·460	4·910	5·125	5·334	5·737	6·122	6·308	6·846	7·053
1/ 23·8	50·813	108·54	148·00	195·27	250·85	315·25	472·39	566·03	670·31	912·42	1202·0	1365·5	1935·6	2198·6
0·04400	2·948	3·541	3·815	4·077	4·329	4·571	5·032	5·252	5·466	5·879	6·273	6·463	7·014	7·226
1/ 22·7	52·095	111·26	151·70	200·14	257·10	323·09	484·11	580·06	686·90	934·97	1231·6	1399·2	1983·2	2252·6
0·04600	3·019	3·626	3·906	4·174	4·432	4·679	5·151	5·376	5·595	6·017	6·420	6·615	7·179	7·396
1/ 21·7	53·349	113·92	155·32	204·91	263·22	330·77	495·57	593·78	703·13	957·02	1260·6	1432·1	2029·8	2305·5
0·04800	3·088	3·709	3·996	4·269	4·532	4·786	5·268	5·498	5·722	6·153	6·565	6·764	7·340	7·562
1/ 20·8	54·576	116·53	158·87	209·58	269·20	338·28	506·80	607·21	719·02	978·60	1289·0	1464·3	2075·3	2357·2
0·05000	3·157	3·790	4·083	4·363	4·631	4·890	5·382	5·617	5·846	6·286	6·706	6·910	7·498	7·724
1/ 20·0	55·780	119·08	162·35	214·15	275·07	345·64	517·80	620·38	734·59	999·75	1316·8	1495·9	2120·0	2407·9
0·05250	3·240	3·890	4·190	4·477	4·752	5·017	5·522	5·763	5·997	6·449	6·880	7·088	7·691	7·923
1/ 19·0	57·253	122·21	166·60	219·75	282·25	354·65	531·25	636·48	753·63	1025·6	1350·8	1534·5	2174·6	2469·8
0·05500	3·321	3·987	4·294	4·588	4·870	5·142	5·658	5·905	6·145	6·608	7·049	7·263	7·880	8·117
1/ 18·2	58·692	125·26	170·75	225·22	289·26	363·44	544·40	652·21	772·24	1050·9	1384·1	1572·2	2227·9	2530·4
0·05750	3·401	4·082	4·397	4·697	4·986	5·263	5·792	6·044	6·290	6·763	7·214	7·433	8·064	8·307
1/ 17·4	60·100	128·25	174·82	230·57	296·12	372·05	557·25	667·59	790·44	1075·6	1416·5	1609·1	2280·1	2589·6
0·06000	3·479	4·175	4·497	4·804	5·099	5·383	5·923	6·181	6·432	6·915	7·376	7·600	8·245	8·493
1/ 16·7	61·479	131·18	178·80	235·81	302·84	380·48	569·84	682·65	808·25	1099·8	1448·4	1645·2	2331·1	2647·5
0·06250	3·556	4·267	4·595	4·909	5·210	5·499	6·051	6·314	6·571	7·064	7·535	7·763	8·422	8·675
1/ 16·0	62·831	134·04	182·70	240·95	309·42	388·73	582·17	697·41	825·71	1123·5	1479·5	1680·6	2381·2	2704·3
0·06500	3·631	4·356	4·691	5·011	5·318	5·614	6·177	6·446	6·707	7·210	7·691	7·924	8·595	8·854
1/ 15·4	64·157	136·86	186·52	245·98	315·88	396·83	594·27	711·88	842·82	1146·8	1510·1	1715·2	2430·2	2760·0
0·06750	3·704	4·444	4·786	5·112	5·425	5·726	6·300	6·574	6·841	7·354	7·844	8·081	8·765	9·029
1/ 14·8	65·459	139·62	190·28	250·92	322·21	404·78	606·14	726·09	859·62	1169·6	1540·1	1749·3	2478·4	2814·6
0·07000	3·777	4·531	4·878	5·211	5·530	5·837	6·421	6·700	6·972	7·495	7·994	8·235	8·933	9·201
1/ 14·3	66·737	142·33	193·97	255·78	328·44	412·58	617·80	740·04	876·12	1192·0	1569·5	1782·7	2525·6	2868·2
0·07250	3·848	4·615	4·969	5·308	5·633	5·945	6·540	6·825	7·101	7·633	8·141	8·387	9·097	9·370
1/ 13·8	67·994	145·00	197·59	260·55	334·55	420·25	629·25	753·74	892·33	1214·0	1598·5	1815·6	2572·1	2920·9
0·07500	3·918	4·699	5·059	5·403	5·734	6·052	6·657	6·947	7·228	7·769	8·286	8·536	9·258	9·536
1/ 13·3	69·230	147·62	201·15	265·24	340·57	427·80	640·52	767·22	908·27	1235·6	1626·9	1847·8	2617·7	2972·7
0·08000	4·054	4·862	5·234	5·590	5·932	6·260	6·886	7·185	7·475	8·035	8·569	8·827	9·573	9·861
1/ 12·5	71·644	152·74	208·11	274·40	352·31	442·52	662·51	793·53	939·38	1277·9	1682·5	1910·9	2706·8	3073·9
0·08500	4·187	5·020	5·404	5·771	6·123	6·462	7·108	7·416	7·715	8·292	8·843	9·110	9·879	10·18
1/ 11·8	73·986	157·70	214·87	283·28	363·69	456·81	683·84	819·05	969·56	1318·8	1736·3	1972·0	2793·3	3171·9
0·09000	4·316	5·173	5·569	5·947	6·310	6·659	7·323	7·640	7·949	8·543	9·110	9·384	10·18	10·48
1/ 11·1	76·263	162·53	221·43	291·92	374·76	470·68	704·56	843·84	998·88	1358·7	1788·6	2031·4	2877·2	3267·2
0·09500	4·441	5·323	5·730	6·118	6·491	6·850	7·533	7·859	8·176	8·786	9·369	9·651	10·46	10·78
1/ 10·5	78·479	167·22	227·81	300·32	385·53	484·19	724·72	867·97	1027·4	1397·4	1839·6	2089·2	2958·9	3359·9
0·10000	4·563	5·469	5·886	6·285	6·668	7·036	7·737	8·072	8·397	9·024	9·621	9·911	10·75	11·07
1/ 10·0	80·639	171·80	234·04	308·51	396·02	497·35	744·37	891·48	1055·2	1435·1	1889·2	2145·5	3038·5	3450·2
	0·91	0·93	0·93	0·94	0·94	0·95	0·95	0·96	0·96	0·96	0·97	0·97	0·97	0·97

$V_{r(0·5)medial}$ for half-full circular pipes.

k_s = 0·030 mm
S = 0·00010 to 0·00030

ie hydraulic gradient =
1 in 10000 to 1 in 3333

Water (or sewage) at 15°C;
full bore conditions.

velocities in ms^{-1}
discharges in litres/sec

Gradient **(Equivalent) Pipe diameters in mm**

Gradient	150	200	225	250	275	300	350	375	400	450	500	525	600	630
0·000100	0·101	0·123	0·134	0·144	0·154	0·164	0·182	0·191	0·199	0·216	0·232	0·240	0·262	0·270
1/ 10000	1·7780	3·8792	5·3332	7·0868	9·1614	11·578	17·517	21·077	25·056	34·342	45·513	51·847	74·033	84·313
0·000105	0·103	0·127	0·138	0·148	0·159	0·168	0·187	0·196	0·205	0·222	0·238	0·246	0·269	0·278
1/ 9524	1·8281	3·9878	5·4821	7·2840	9·4157	11·899	18·000	21·658	25·746	35·285	46·760	53·265	76·053	86·611
0·000110	0·106	0·130	0·142	0·152	0·163	0·173	0·192	0·201	0·210	0·228	0·244	0·252	0·276	0·285
1/ 9091	1·8773	4·0941	5·6277	7·4770	9·6646	12·213	18·473	22·226	26·421	36·207	47·979	54·653	78·030	88·860
0·000115	0·109	0·134	0·145	0·156	0·167	0·177	0·197	0·206	0·216	0·233	0·250	0·259	0·283	0·292
1/ 8696	1·9254	4·1982	5·7705	7·6662	9·9086	12·520	18·937	22·783	27·082	37·111	49·174	56·013	79·966	91·062
0·000120	0·112	0·137	0·149	0·160	0·171	0·181	0·202	0·211	0·221	0·239	0·256	0·265	0·290	0·299
1/ 8333	1·9727	4·3004	5·9105	7·8516	10·148	12·822	19·391	23·329	27·730	37·997	50·345	57·346	81·863	93·221
0·000125	0·114	0·140	0·152	0·164	0·175	0·186	0·206	0·216	0·226	0·244	0·262	0·271	0·296	0·306
1/ 8000	2·0191	4·4007	6·0479	8·0337	10·382	13·118	19·837	23·865	28·366	38·866	51·494	58·653	83·725	95·339
0·000130	0·117	0·143	0·156	0·167	0·179	0·190	0·211	0·221	0·231	0·250	0·268	0·277	0·303	0·313
1/ 7692	2·0646	4·4992	6·1829	8·2125	10·613	13·408	20·275	24·391	28·990	39·720	52·623	59·937	85·553	97·419
0·000135	0·119	0·146	0·159	0·171	0·182	0·194	0·215	0·226	0·236	0·255	0·274	0·283	0·309	0·319
1/ 7407	2·1094	4·5961	6·3155	8·3883	10·840	13·694	20·706	24·908	29·604	40·558	53·731	61·199	87·349	99·462
0·000140	0·122	0·149	0·162	0·174	0·186	0·198	0·220	0·230	0·240	0·260	0·279	0·288	0·315	0·326
1/ 7143	2·1535	4·6913	6·4460	8·5611	11·062	13·975	21·129	25·417	30·207	41·383	54·822	62·439	89·114	101·47
0·000145	0·124	0·152	0·165	0·178	0·190	0·202	0·224	0·235	0·245	0·265	0·285	0·294	0·321	0·332
1/ 6897	2·1969	4·7850	6·5744	8·7312	11·282	14·252	21·546	25·917	30·801	42·194	55·894	63·659	90·851	103·45
0·000150	0·127	0·155	0·169	0·181	0·194	0·205	0·228	0·239	0·250	0·270	0·290	0·300	0·327	0·338
1/ 6667	2·2396	4·8774	6·7009	8·8987	11·498	14·524	21·956	26·409	31·385	42·993	56·950	64·861	92·561	105·39
0·000160	0·131	0·161	0·175	0·188	0·201	0·213	0·237	0·248	0·259	0·280	0·301	0·310	0·339	0·350
1/ 6250	2·3232	5·0579	6·9482	9·2263	11·920	15·056	22·758	27·372	32·528	44·555	59·014	67·209	95·903	109·19
0·000170	0·136	0·167	0·181	0·194	0·208	0·220	0·245	0·256	0·268	0·290	0·311	0·321	0·351	0·362
1/ 5882	2·4045	5·2335	7·1887	9·5448	12·331	15·573	23·537	28·309	33·639	46·072	61·020	69·491	99·151	112·89
0·000180	0·141	0·172	0·187	0·201	0·214	0·227	0·253	0·265	0·276	0·299	0·321	0·331	0·362	0·374
1/ 5556	2·4837	5·4046	7·4229	9·8549	12·730	16·077	24·296	29·220	34·721	47·550	62·972	71·712	102·31	116·48
0·000190	0·145	0·177	0·192	0·207	0·221	0·234	0·260	0·273	0·285	0·308	0·330	0·341	0·373	0·385
1/ 5263	2·5609	5·5714	7·6513	10·157	13·120	16·569	25·036	30·108	35·775	48·990	64·875	73·878	105·39	119·99
0·000200	0·149	0·183	0·198	0·213	0·227	0·241	0·268	0·280	0·293	0·317	0·340	0·351	0·383	0·396
1/ 5000	2·6364	5·7343	7·8744	10·453	13·501	17·048	25·758	30·976	36·804	50·396	66·733	75·991	108·40	123·41
0·000210	0·153	0·188	0·204	0·219	0·234	0·248	0·275	0·288	0·301	0·326	0·349	0·361	0·394	0·407
1/ 4762	2·7102	5·8936	8·0925	10·742	13·873	17·517	26·464	31·824	37·810	51·770	68·549	78·056	111·34	126·75
0·000220	0·157	0·193	0·209	0·225	0·240	0·254	0·282	0·296	0·309	0·334	0·358	0·370	0·404	0·417
1/ 4545	2·7825	6·0495	8·3060	11·024	14·237	17·976	27·155	32·653	38·794	53·114	70·325	80·077	114·21	130·02
0·000230	0·161	0·197	0·214	0·230	0·246	0·261	0·289	0·303	0·316	0·342	0·367	0·379	0·414	0·427
1/ 4348	2·8533	6·2023	8·5151	11·301	14·594	18·426	27·832	33·466	39·758	54·431	72·064	82·056	117·03	133·22
0·000240	0·165	0·202	0·219	0·236	0·252	0·267	0·296	0·310	0·324	0·350	0·376	0·388	0·424	0·437
1/ 4167	2·9227	6·3521	8·7202	11·572	14·944	18·867	28·495	34·262	40·703	55·721	73·769	83·995	119·79	136·36
0·000250	0·169	0·207	0·224	0·241	0·257	0·273	0·303	0·317	0·331	0·358	0·384	0·397	0·433	0·447
1/ 4000	2·9909	6·4991	8·9214	11·839	15·287	19·299	29·146	35·044	41·630	56·987	75·442	85·898	122·49	139·44
0·000260	0·173	0·211	0·229	0·247	0·263	0·279	0·310	0·324	0·339	0·366	0·393	0·405	0·443	0·457
1/ 3846	3·0579	6·6435	9·1190	12·100	15·624	19·724	29·785	35·811	42·541	58·230	77·084	87·766	125·15	142·46
0·000270	0·177	0·216	0·234	0·252	0·269	0·285	0·316	0·331	0·346	0·374	0·401	0·414	0·452	0·467
1/ 3704	3·1237	6·7854	9·3133	12·358	15·955	20·141	30·413	36·565	43·435	59·452	78·697	89·601	127·76	145·42
0·000280	0·180	0·220	0·239	0·257	0·274	0·291	0·323	0·338	0·353	0·381	0·409	0·422	0·461	0·476
1/ 3571	3·1884	6·9250	9·5043	12·610	16·281	20·551	31·031	37·306	44·314	60·652	80·283	91·404	130·32	148·34
0·000290	0·184	0·225	0·244	0·262	0·279	0·296	0·329	0·344	0·360	0·389	0·417	0·430	0·470	0·485
1/ 3448	3·2521	7·0623	9·6922	12·859	16·601	20·955	31·638	38·035	45·179	61·833	81·843	93·179	132·85	151·21
0·000300	0·188	0·229	0·248	0·267	0·285	0·302	0·335	0·351	0·366	0·396	0·425	0·439	0·479	0·494
1/ 3333	3·3148	7·1975	9·8772	13·104	16·916	21·352	32·236	38·753	46·031	62·996	83·378	94·925	135·33	154·03
	0·78	0·80	0·81	0·82	0·83	0·83	0·84	0·85	0·85	0·86	0·87	0·88	0·89	0·89

$V_{r(0·5)medial}$ for half-full circular pipes.

k_s = 0·030 mm
S = 0·00030 to 0·00100

ie hydraulic gradient =
1 in 3333 to 1 in 1000

Water (or sewage) at 15°C;
full bore conditions.

velocities in ms^{-1}
discharges in litres/sec

Gradient **(Equivalent) Pipe diameters in mm**

Gradient	150	200	225	250	275	300	350	375	400	450	500	525	600	630
0·000300	0·188	0·229	0·248	0·267	0·285	0·302	0·335	0·351	0·366	0·396	0·425	0·439	0·479	0·494
1/ 3333	3·3148	7·1975	9·8772	13·104	16·916	21·352	32·236	38·753	46·031	62·996	83·378	94·925	135·33	154·03
0·000320	0·195	0·238	0·258	0·277	0·295	0·313	0·347	0·364	0·380	0·410	0·440	0·454	0·496	0·512
1/ 3125	3·4376	7·4620	10·239	13·583	17·533	22·129	33·405	40·156	47·695	65·268	86·379	98·338	140·18	159·55
0·000340	0·201	0·246	0·266	0·286	0·305	0·324	0·359	0·376	0·392	0·424	0·455	0·470	0·512	0·529
1/ 2941	3·5569	7·7190	10·591	14·048	18·133	22·884	34·541	41·520	49·313	67·476	89·295	101·65	144·90	164·92
0·000360	0·208	0·254	0·275	0·295	0·315	0·334	0·371	0·388	0·405	0·438	0·469	0·484	0·529	0·546
1/ 2778	3·6731	7·9694	10·933	14·501	18·716	23·619	35·647	42·847	50·887	69·624	92·132	104·88	149·49	170·13
0·000380	0·214	0·261	0·283	0·304	0·325	0·344	0·382	0·400	0·417	0·451	0·483	0·499	0·545	0·562
1/ 2632	3·7864	8·2134	11·267	14·943	19·285	24·336	36·725	44·141	52·421	71·718	94·897	108·03	153·96	175·21
0·000400	0·221	0·269	0·292	0·313	0·334	0·354	0·393	0·411	0·429	0·464	0·497	0·513	0·560	0·578
1/ 2500	3·8971	8·4518	11·593	15·374	19·841	25·035	37·777	45·403	53·918	73·762	97·596	111·09	158·32	180·17
0·000420	0·227	0·276	0·300	0·322	0·343	0·364	0·403	0·422	0·441	0·476	0·510	0·527	0·575	0·594
1/ 2381	4·0053	8·6848	11·912	15·796	20·383	25·719	38·805	46·637	55·382	75·759	100·23	114·09	162·58	185·02
0·000440	0·233	0·284	0·307	0·330	0·352	0·373	0·414	0·433	0·452	0·489	0·524	0·541	0·590	0·609
1/ 2273	4·1113	8·9128	12·224	16·208	20·915	26·388	39·811	47·844	56·813	77·713	102·81	117·03	166·75	189·76
0·000460	0·239	0·291	0·315	0·338	0·361	0·383	0·424	0·444	0·463	0·501	0·536	0·554	0·604	0·624
1/ 2174	4·2151	9·1361	12·529	16·612	21·435	27·043	40·796	49·026	58·215	79·625	105·34	119·90	170·83	194·40
0·000480	0·244	0·298	0·323	0·346	0·369	0·392	0·434	0·454	0·474	0·512	0·549	0·567	0·618	0·638
1/ 2083	4·3168	9·3551	12·829	17·009	21·945	27·685	41·761	50·185	59·588	81·500	107·81	122·71	174·83	198·95
0·000500	0·250	0·305	0·330	0·354	0·378	0·401	0·444	0·465	0·485	0·524	0·561	0·580	0·632	0·653
1/ 2000	4·4167	9·5700	13·122	17·397	22·445	28·315	42·708	51·321	60·936	83·338	110·24	125·47	178·75	203·40
0·000525	0·257	0·313	0·339	0·364	0·388	0·411	0·456	0·477	0·498	0·538	0·577	0·595	0·649	0·670
1/ 1905	4·5390	9·8331	13·482	17·873	23·058	29·086	43·867	52·712	62·586	85·589	113·21	128·85	183·55	208·86
0·000550	0·264	0·321	0·348	0·374	0·398	0·422	0·468	0·490	0·511	0·552	0·591	0·610	0·666	0·687
1/ 1818	4·6588	10·091	13·834	18·339	23·657	29·841	45·002	54·073	64·199	87·791	116·11	132·15	188·25	214·20
0·000575	0·270	0·329	0·357	0·383	0·408	0·433	0·479	0·502	0·523	0·566	0·606	0·625	0·682	0·704
1/ 1739	4·7760	10·343	14·179	18·795	24·244	30·580	46·112	55·406	65·779	89·946	118·96	135·39	192·84	219·42
0·000600	0·277	0·337	0·365	0·392	0·418	0·443	0·491	0·513	0·536	0·579	0·620	0·640	0·698	0·720
1/ 1667	4·8910	10·590	14·517	19·242	24·819	31·304	47·201	56·712	67·328	92·058	121·75	138·56	197·35	224·54
0·000625	0·283	0·345	0·373	0·401	0·427	0·453	0·502	0·525	0·548	0·592	0·634	0·654	0·714	0·736
1/ 1600	5·0039	10·833	14·849	19·680	25·384	32·015	48·269	57·993	68·847	94·130	124·48	141·67	201·76	229·56
0·000650	0·289	0·352	0·382	0·410	0·437	0·463	0·513	0·536	0·560	0·605	0·648	0·669	0·729	0·752
1/ 1538	5·1147	11·071	15·174	20·111	25·938	32·712	49·317	59·250	70·338	96·164	127·17	144·72	206·10	234·49
0·000675	0·296	0·360	0·390	0·418	0·446	0·472	0·523	0·548	0·571	0·617	0·661	0·682	0·744	0·768
1/ 1481	5·2236	11·305	15·494	20·533	26·482	33·397	50·346	60·486	71·802	98·161	129·80	147·71	210·35	239·33
0·000700	0·302	0·367	0·398	0·427	0·455	0·482	0·534	0·559	0·583	0·630	0·674	0·696	0·759	0·783
1/ 1429	5·3307	11·535	15·809	20·949	27·017	34·071	51·359	61·700	73·241	100·12	132·39	150·66	214·54	244·08
0·000725	0·308	0·374	0·405	0·435	0·464	0·491	0·544	0·569	0·594	0·642	0·687	0·709	0·773	0·798
1/ 1379	5·4361	11·761	16·118	21·358	27·544	34·733	52·354	62·894	74·657	102·05	134·94	153·56	218·65	248·76
0·000750	0·313	0·381	0·413	0·443	0·472	0·501	0·554	0·580	0·605	0·654	0·700	0·723	0·788	0·813
1/ 1333	5·5398	11·984	16·423	21·761	28·062	35·385	53·334	64·069	76·050	103·95	137·45	156·41	222·70	253·36
0·000800	0·325	0·395	0·428	0·459	0·490	0·519	0·574	0·601	0·627	0·677	0·725	0·748	0·816	0·842
1/ 1250	5·7427	12·420	17·018	22·548	29·074	36·660	55·249	66·366	78·773	107·67	142·34	161·98	230·61	262·35
0·000850	0·336	0·409	0·443	0·475	0·506	0·536	0·594	0·621	0·648	0·700	0·749	0·773	0·843	0·870
1/ 1176	5·9399	12·843	17·597	23·313	30·058	37·898	57·108	68·597	81·418	111·27	147·10	167·38	238·29	271·08
0·000900	0·347	0·422	0·457	0·490	0·522	0·553	0·612	0·641	0·668	0·722	0·773	0·798	0·869	0·897
1/ 1111	6·1318	13·256	18·160	24·057	31·016	39·103	58·918	70·768	83·990	114·78	151·73	172·64	245·76	279·57
0·000950	0·358	0·435	0·471	0·505	0·538	0·570	0·631	0·660	0·688	0·743	0·796	0·821	0·895	0·923
1/ 1053	6·3190	13·657	18·709	24·782	31·949	40·278	60·682	72·883	86·497	118·20	156·24	177·77	253·03	287·84
0·001000	0·368	0·447	0·484	0·519	0·553	0·586	0·649	0·679	0·708	0·764	0·818	0·844	0·920	0·949
1/ 1000	6·5017	14·049	19·244	25·490	32·860	41·423	62·402	74·946	88·943	121·53	160·64	182·77	260·13	295·90
	0·81	0·83	0·84	0·85	0·85	0·86	0·87	0·88	0·88	0·89	0·90	0·90	0·91	0·91

$V_{r(0·5)medial}$ for half-full circular pipes.

A14

(p.3 of 6)

k_s = 0·030 mm
S = 0·00100 to 0·00300

ie hydraulic gradient =
1 in 1000 to 1 in 333

Water (or sewage) at 15°C;
full bore conditions.

velocities in ms^{-1}
discharges in litres/sec

Gradient	**(Equivalent) Pipe diameters in mm**													
	150	**200**	**225**	**250**	**275**	**300**	**350**	**375**	**400**	**450**	**500**	**525**	**600**	**630**
0·00100	0·368	0·447	0·484	0·519	0·553	0·586	0·649	0·679	0·708	0·764	0·818	0·844	0·920	0·949
1/ 1000	6·5017	14·049	19·244	25·490	32·860	41·423	62·402	74·946	88·943	121·53	160·64	182·77	260·13	295·90
0·00105	0·378	0·459	0·497	0·533	0·568	0·602	0·666	0·697	0·727	0·785	0·840	0·867	0·945	0·975
1/ 952	6·6802	14·433	19·768	26·182	33·749	42·543	64·083	76·962	91·332	124·79	164·93	187·65	267·06	303·78
0·00110	0·388	0·471	0·510	0·547	0·583	0·617	0·683	0·715	0·745	0·805	0·861	0·889	0·969	0·999
1/ 909	6·8550	14·807	20·280	26·858	34·620	43·638	65·727	78·934	93·669	127·98	169·13	192·43	273·84	311·49
0·00115	0·398	0·483	0·523	0·561	0·597	0·633	0·700	0·732	0·764	0·824	0·882	0·910	0·992	1·023
1/ 870	7·0261	15·174	20·781	27·521	35·472	44·710	67·337	80·864	95·956	131·09	173·24	197·10	280·47	319·02
0·00120	0·407	0·494	0·535	0·574	0·611	0·647	0·716	0·749	0·781	0·843	0·903	0·932	1·015	1·047
1/ 833	7·1939	15·534	21·273	28·170	36·307	45·760	68·914	82·755	98·197	134·15	177·27	201·68	286·97	326·41
0·00125	0·416	0·506	0·547	0·587	0·625	0·662	0·732	0·766	0·799	0·862	0·923	0·952	1·037	1·070
1/ 800	7·3584	15·887	21·755	28·807	37·126	46·791	70·460	84·610	100·40	137·14	181·22	206·17	293·34	333·65
0·00130	0·426	0·517	0·559	0·600	0·639	0·676	0·748	0·783	0·816	0·881	0·943	0·973	1·060	1·093
1/ 769	7·5200	16·234	22·228	29·432	37·930	47·802	71·978	86·430	102·55	140·08	185·10	210·57	299·60	340·75
0·00135	0·435	0·528	0·571	0·612	0·652	0·690	0·764	0·799	0·833	0·899	0·962	0·993	1·081	1·116
1/ 741	7·6788	16·574	22·693	30·046	38·719	48·795	73·469	88·217	104·67	142·97	188·90	214·90	305·74	347·73
0·00140	0·443	0·538	0·582	0·624	0·665	0·704	0·779	0·815	0·850	0·917	0·981	1·012	1·103	1·137
1/ 714	7·8349	16·909	23·149	30·649	39·495	49·771	74·934	89·974	106·75	145·80	192·64	219·15	311·77	354·58
0·00145	0·452	0·549	0·594	0·636	0·678	0·718	0·794	0·830	0·866	0·934	1·000	1·032	1·124	1·159
1/ 690	7·9885	17·238	23·599	31·243	40·259	50·731	76·375	91·701	108·80	148·59	196·32	223·33	317·70	361·32
0·00150	0·461	0·559	0·605	0·648	0·690	0·731	0·809	0·846	0·882	0·952	1·018	1·051	1·144	1·180
1/ 667	8·1396	17·562	24·041	31·827	41·010	51·676	77·792	93·401	110·81	151·34	199·94	227·44	323·53	367·95
0·00160	0·477	0·579	0·626	0·672	0·715	0·757	0·837	0·876	0·913	0·985	1·054	1·088	1·185	1·222
1/ 625	8·4351	18·195	24·905	32·968	42·478	53·522	80·562	96·722	114·75	156·70	207·01	235·47	334·93	380·90
0·00170	0·494	0·599	0·647	0·694	0·739	0·783	0·865	0·905	0·944	1·018	1·089	1·124	1·224	1·262
1/ 588	8·7222	18·810	25·745	34·077	43·903	55·315	83·252	99·947	118·57	161·90	213·87	243·27	345·99	393·47
0·00180	0·509	0·618	0·668	0·716	0·763	0·807	0·892	0·933	0·973	1·050	1·123	1·159	1·262	1·301
1/ 556	9·0016	19·408	26·562	35·156	45·290	57·059	85·868	103·08	122·28	166·97	220·54	250·86	356·75	405·70
0·00190	0·525	0·636	0·688	0·738	0·785	0·831	0·919	0·961	1·002	1·081	1·156	1·193	1·299	1·340
1/ 526	9·2739	19·991	27·358	36·207	46·641	58·758	88·417	106·14	125·90	171·90	227·04	258·24	367·23	417·60
0·00200	0·540	0·654	0·708	0·758	0·807	0·855	0·945	0·988	1·030	1·111	1·189	1·226	1·335	1·377
1/ 500	9·5397	20·560	28·134	37·232	47·959	60·416	90·903	109·12	129·43	176·71	233·38	265·44	377·44	429·21
0·00210	0·555	0·672	0·727	0·779	0·829	0·878	0·970	1·014	1·057	1·141	1·220	1·259	1·370	1·413
1/ 476	9·7995	21·116	28·893	38·234	49·247	62·035	93·332	112·03	132·88	181·40	239·57	272·48	387·42	440·54
0·00220	0·569	0·689	0·745	0·799	0·850	0·900	0·995	1·040	1·084	1·169	1·251	1·290	1·405	1·449
1/ 455	10·054	21·660	29·635	39·214	50·506	63·618	95·706	114·88	136·25	185·99	245·62	279·35	397·17	451·62
0·00230	0·583	0·706	0·764	0·818	0·871	0·922	1·019	1·065	1·111	1·198	1·281	1·322	1·438	1·484
1/ 435	10·302	22·193	30·362	40·173	51·739	65·168	98·030	117·66	139·55	190·49	251·54	286·08	406·71	462·46
0·00240	0·597	0·723	0·782	0·838	0·891	0·943	1·043	1·090	1·136	1·225	1·311	1·352	1·472	1·518
1/ 417	10·546	22·714	31·073	41·112	52·946	66·686	100·31	120·39	142·78	194·89	257·34	292·67	416·06	473·08
0·00250	0·610	0·739	0·799	0·856	0·911	0·964	1·066	1·114	1·161	1·253	1·340	1·382	1·504	1·551
1/ 400	10·785	23·226	31·771	42·034	54·130	68·175	102·54	123·06	145·95	199·20	263·03	299·13	425·22	483·48
0·00260	0·624	0·755	0·816	0·875	0·931	0·985	1·089	1·138	1·186	1·279	1·368	1·411	1·536	1·584
1/ 385	11·020	23·728	32·456	42·938	55·292	69·635	104·73	125·69	149·06	203·44	268·61	305·47	434·21	493·69
0·00270	0·637	0·771	0·833	0·893	0·950	1·005	1·111	1·161	1·210	1·305	1·396	1·440	1·567	1·616
1/ 370	11·251	24·221	33·129	43·826	56·433	71·070	106·88	128·27	152·12	207·59	274·08	311·69	443·03	503·71
0·00280	0·649	0·786	0·850	0·911	0·969	1·025	1·133	1·184	1·234	1·331	1·423	1·468	1·598	1·647
1/ 357	11·477	24·706	33·790	44·698	57·555	72·479	108·99	130·80	155·11	211·68	279·47	317·81	451·70	513·56
0·00290	0·662	0·802	0·866	0·928	0·988	1·045	1·154	1·207	1·258	1·356	1·450	1·496	1·628	1·679
1/ 345	11·700	25·182	34·440	45·556	58·657	73·865	111·07	133·29	158·06	215·69	284·76	323·82	460·22	523·24
0·00300	0·675	0·817	0·882	0·945	1·006	1·064	1·176	1·229	1·281	1·381	1·477	1·523	1·657	1·709
1/ 333	11·920	25·651	35·080	46·401	59·742	75·229	113·11	135·74	160·96	219·64	289·96	329·73	468·60	532·76
	0·83	0·85	0·86	0·87	0·88	0·89	0·90	0·90	0·91	0·91	0·92	0·92	0·93	0·93

$V_{r(0\cdot5)medial}$ for half-full circular pipes.

k_s = 0·030 mm S = 0·00100 to 0·00300

k_s = **0·030 mm**
S = 0·00300 to 0·01000

ie hydraulic gradient =
1 in 333 to 1 in 100

Water (or sewage) at 15°C;
full bore conditions.

velocities in ms^{-1}
discharges in litres/sec

Gradient	(Equivalent) Pipe diameters in mm													
	150	200	225	250	275	300	350	375	400	450	500	525	600	630
0·00300	0·675	0·817	0·882	0·945	1·006	1·064	1·176	1·229	1·281	1·381	1·477	1·523	1·657	1·709
1/ 333	11·920	25·651	35·080	46·401	59·742	75·229	113·11	135·74	160·96	219·64	289·96	329·73	468·60	532·76
0·00320	0·699	0·846	0·914	0·979	1·042	1·102	1·217	1·272	1·326	1·430	1·529	1·576	1·715	1·769
1/ 313	12·348	26·568	36·330	48·050	61·862	77·893	117·11	140·52	166·63	227·35	300·12	341·27	484·96	551·35
0·00340	0·722	0·874	0·944	1·011	1·076	1·139	1·257	1·314	1·370	1·477	1·579	1·628	1·771	1·827
1/ 294	12·765	27·458	37·544	49·652	63·919	80·479	120·98	145·17	172·13	234·84	309·98	352·48	500·85	569·39
0·00360	0·745	0·902	0·974	1·043	1·110	1·174	1·297	1·355	1·412	1·522	1·628	1·679	1·826	1·883
1/ 278	13·170	28·323	38·725	51·209	65·921	82·994	124·75	149·68	177·48	242·12	319·57	363·37	516·28	586·92
0·00380	0·768	0·928	1·003	1·074	1·143	1·209	1·335	1·395	1·454	1·567	1·675	1·728	1·879	1·938
1/ 263	13·565	29·167	39·874	52·727	67·870	85·443	128·42	154·08	182·68	249·21	328·91	373·97	531·32	604·00
0·00400	0·789	0·955	1·031	1·104	1·175	1·243	1·372	1·434	1·494	1·610	1·721	1·775	1·931	1·991
1/ 250	13·950	29·989	40·996	54·207	69·771	87·832	132·00	158·37	187·76	256·12	338·01	384·31	545·97	620·64
0·00420	0·811	0·980	1·059	1·134	1·206	1·276	1·408	1·472	1·534	1·653	1·767	1·822	1·982	2·043
1/ 238	14·326	30·793	42·092	55·652	71·627	90·165	135·49	162·55	192·72	262·87	346·90	394·41	560·27	636·89
0·00440	0·832	1·005	1·086	1·163	1·236	1·308	1·444	1·509	1·572	1·694	1·811	1·868	2·031	2·094
1/ 227	14·694	31·579	43·163	57·065	73·442	92·445	138·91	166·65	197·57	269·46	355·58	404·27	574·25	652·76
0·00460	0·852	1·030	1·112	1·191	1·266	1·339	1·479	1·545	1·610	1·735	1·854	1·912	2·079	2·144
1/ 217	15·054	32·348	44·211	58·448	75·218	94·677	142·25	170·65	202·31	275·91	364·08	413·93	587·93	668·30
0·00480	0·872	1·054	1·138	1·218	1·296	1·370	1·513	1·581	1·647	1·775	1·897	1·956	2·127	2·193
1/ 208	15·407	33·101	45·238	59·802	76·958	96·863	145·53	174·57	206·95	282·23	372·40	423·38	601·32	683·50
0·00500	0·891	1·077	1·163	1·245	1·324	1·401	1·546	1·615	1·683	1·813	1·938	1·999	2·173	2·240
1/ 200	15·753	33·839	46·245	61·130	78·663	99·005	148·73	178·42	211·50	288·42	380·55	432·64	614·45	698·41
0·00525	0·915	1·106	1·194	1·278	1·360	1·438	1·587	1·658	1·727	1·861	1·989	2·051	2·230	2·299
1/ 190	16·177	34·743	47·477	62·755	80·750	101·63	152·66	183·12	217·07	296·00	390·53	443·97	630·50	716·64
0·00550	0·939	1·134	1·224	1·311	1·394	1·474	1·627	1·700	1·771	1·908	2·039	2·102	2·285	2·356
1/ 182	16·591	35·627	48·682	64·343	82·790	104·19	156·50	187·72	222·51	303·40	400·28	455·05	646·19	734·45
0·00575	0·962	1·162	1·254	1·342	1·427	1·509	1·666	1·740	1·813	1·953	2·087	2·152	2·340	2·412
1/ 174	16·996	36·492	49·860	65·898	84·786	106·70	160·25	192·22	227·84	310·65	409·82	465·88	661·53	751·88
0·00600	0·984	1·189	1·283	1·373	1·460	1·544	1·704	1·780	1·855	1·998	2·135	2·201	2·393	2·467
1/ 167	17·394	37·339	51·015	67·420	86·741	109·15	163·93	196·62	233·05	317·74	419·16	476·49	676·56	768·94
0·00625	1·006	1·215	1·312	1·404	1·493	1·578	1·741	1·819	1·895	2·042	2·181	2·249	2·445	2·520
1/ 160	17·783	38·170	52·147	68·913	88·657	111·56	167·53	200·94	238·16	324·69	428·31	486·88	691·28	785·66
0·00650	1·028	1·241	1·339	1·434	1·524	1·612	1·778	1·858	1·935	2·084	2·227	2·296	2·496	2·573
1/ 154	18·165	38·985	53·257	70·377	90·537	113·92	171·07	205·17	243·17	331·51	437·29	497·08	705·72	802·06
0·00675	1·049	1·266	1·367	1·463	1·555	1·644	1·814	1·895	1·974	2·126	2·272	2·342	2·546	2·625
1/ 148	18·541	39·785	54·348	71·815	92·383	116·24	174·54	209·33	248·09	338·20	446·10	507·08	719·90	818·16
0·00700	1·070	1·291	1·394	1·492	1·586	1·677	1·850	1·932	2·013	2·168	2·316	2·388	2·595	2·675
1/ 143	18·909	40·571	55·419	73·227	94·196	118·52	177·95	213·41	252·93	344·78	454·75	516·91	733·82	833·96
0·00725	1·091	1·316	1·420	1·520	1·616	1·708	1·884	1·969	2·051	2·208	2·359	2·432	2·644	2·725
1/ 138	19·272	41·344	56·472	74·616	95·979	120·76	181·30	217·42	257·68	351·24	463·26	526·57	747·50	849·50
0·00750	1·111	1·340	1·446	1·548	1·645	1·740	1·919	2·004	2·088	2·248	2·402	2·476	2·691	2·774
1/ 133	19·629	42·105	57·508	75·982	97·732	122·96	184·60	221·37	262·35	357·59	471·62	536·07	760·95	864·78
0·00800	1·150	1·388	1·497	1·602	1·703	1·800	1·986	2·074	2·160	2·326	2·485	2·562	2·784	2·870
1/ 125	20·326	43·590	59·532	78·649	101·16	127·26	191·03	229·08	271·48	370·00	487·96	554·62	787·22	894·60
0·00850	1·189	1·433	1·547	1·655	1·759	1·859	2·050	2·142	2·231	2·402	2·566	2·645	2·874	2·963
1/ 118	21·003	45·032	61·496	81·238	104·48	131·43	197·28	236·56	280·33	382·04	503·80	572·62	812·70	923·54
0·00900	1·226	1·478	1·595	1·706	1·813	1·917	2·114	2·208	2·299	2·476	2·644	2·726	2·962	3·053
1/ 111	21·661	46·433	63·405	83·755	107·71	135·49	203·35	243·83	288·93	393·74	519·20	590·10	837·46	951·65
0·00950	1·262	1·521	1·641	1·756	1·866	1·973	2·175	2·272	2·366	2·547	2·721	2·805	3·047	3·141
1/ 105	22·301	47·798	65·264	86·205	110·86	139·44	209·26	250·91	297·31	405·13	534·19	607·12	861·56	979·01
0·01000	1·297	1·564	1·687	1·805	1·918	2·027	2·235	2·334	2·431	2·617	2·795	2·881	3·130	3·226
1/ 100	22·926	49·130	67·077	88·594	113·92	143·29	215·02	257·80	305·47	416·23	548·80	623·71	885·05	1005·7
	0·86	0·88	0·89	0·90	0·90	0·91	0·92	0·92	0·92	0·93	0·94	0·94	0·94	0·95

$V_{r(0·5)medial}$ for half-full circular pipes.

k_s = 0·030 mm
S = 0·01000 to 0·03000

ie hydraulic gradient =
1 in 100 to 1 in 33·3

Water (or sewage) at 15°C;
full bore conditions.

velocities in ms^{-1}
discharges in litres/sec

Gradient	150	200	225	250	275	300	350	375	400	450	500	525	600	630
	(Equivalent) Pipe diameters in mm													
0·01000	1·297	1·564	1·687	1·805	1·918	2·027	2·235	2·334	2·431	2·617	2·795	2·881	3·130	3·226
1/ 100	22·926	49·130	67·077	88·594	113·92	143·29	215·02	257·80	305·47	416·23	548·80	623·71	885·05	1005·7
0·01050	1·332	1·605	1·732	1·852	1·968	2·080	2·293	2·395	2·494	2·685	2·868	2·956	3·211	3·310
1/ 95	23·537	50·429	68·847	90·927	116·91	147·05	220·64	264·54	313·44	427·06	563·06	639·90	907·97	1031·7
0·01100	1·366	1·646	1·775	1·899	2·018	2·132	2·350	2·455	2·556	2·752	2·939	3·029	3·290	3·391
1/ 91	24·133	51·699	70·577	93·206	119·84	150·72	226·14	271·12	321·22	437·65	576·99	655·72	930·36	1057·1
0·01150	1·399	1·685	1·818	1·944	2·066	2·183	2·406	2·513	2·617	2·817	3·008	3·101	3·368	3·471
1/ 87	24·717	52·942	72·269	95·436	122·70	154·31	231·51	277·55	328·84	448·00	590·61	671·19	952·26	1082·0
0·01200	1·431	1·724	1·859	1·989	2·113	2·233	2·461	2·570	2·676	2·881	3·076	3·170	3·444	3·549
1/ 83	25·289	54·159	73·926	97·620	125·50	157·83	236·77	283·85	336·30	458·14	603·95	686·34	973·70	1106·3
0·01250	1·463	1·762	1·900	2·032	2·159	2·282	2·515	2·626	2·734	2·943	3·142	3·239	3·518	3·626
1/ 80	25·850	55·352	75·550	99·761	128·25	161·28	241·93	290·03	343·60	468·07	617·02	701·18	994·70	1130·2
0·01300	1·494	1·799	1·940	2·075	2·205	2·329	2·567	2·681	2·791	3·004	3·208	3·306	3·591	3·701
1/ 77	26·400	56·522	77·144	101·86	130·94	164·66	246·99	296·08	350·77	477·81	629·83	715·73	1015·3	1153·6
0·01350	1·524	1·836	1·980	2·117	2·249	2·376	2·619	2·735	2·847	3·064	3·272	3·372	3·662	3·774
1/ 74	26·940	57·671	78·708	103·92	133·59	167·98	251·95	302·03	357·80	487·37	642·41	730·01	1035·5	1176·5
0·01400	1·554	1·872	2·018	2·158	2·293	2·423	2·669	2·787	2·902	3·123	3·335	3·437	3·733	3·846
1/ 71	27·470	58·800	80·244	105·94	136·19	171·24	256·83	307·86	364·71	496·76	654·77	744·04	1055·4	1199·0
0·01450	1·584	1·907	2·056	2·199	2·336	2·468	2·719	2·839	2·956	3·181	3·397	3·501	3·802	3·918
1/ 69	27·992	59·909	81·754	107·93	138·74	174·44	261·62	313·60	371·50	505·99	666·90	757·82	1074·9	1221·2
0·01500	1·613	1·942	2·094	2·239	2·378	2·512	2·768	2·890	3·009	3·238	3·457	3·563	3·869	3·987
1/ 67	28·505	61·000	83·240	109·89	141·25	177·59	266·33	319·24	378·17	515·06	678·84	771·37	1094·1	1243·0
0·01600	1·670	2·010	2·166	2·317	2·461	2·599	2·864	2·990	3·113	3·350	3·576	3·686	4·002	4·124
1/ 63	29·506	63·130	86·140	113·71	146·15	183·75	275·53	330·25	391·20	532·77	702·14	797·82	1131·5	1285·5
0·01700	1·725	2·075	2·237	2·392	2·541	2·684	2·957	3·087	3·214	3·458	3·691	3·804	4·130	4·256
1/ 59	30·479	65·198	88·954	117·42	150·91	189·72	284·46	340·94	403·84	549·95	724·74	823·48	1167·8	1326·7
0·01800	1·778	2·139	2·306	2·465	2·618	2·766	3·047	3·181	3·311	3·563	3·803	3·919	4·255	4·384
1/ 56	31·424	67·207	91·688	121·02	155·53	195·52	293·13	351·32	416·12	566·64	746·69	848·41	1203·1	1366·7
0·01900	1·830	2·202	2·373	2·537	2·694	2·846	3·134	3·272	3·407	3·665	3·912	4·031	4·376	4·509
1/ 53	32·344	69·163	94·351	124·53	160·02	201·16	301·57	361·42	428·07	582·88	768·06	872·66	1237·4	1405·7
0·02000	1·881	2·262	2·438	2·606	2·768	2·924	3·220	3·361	3·499	3·764	4·018	4·140	4·495	4·631
1/ 50	33·241	71·070	96·946	127·94	164·41	206·66	309·79	371·26	439·72	598·71	788·88	896·30	1270·8	1443·6
0·02100	1·931	2·321	2·502	2·674	2·840	3·000	3·303	3·448	3·590	3·862	4·121	4·247	4·610	4·750
1/ 47·6	34·117	72·931	99·478	131·28	168·69	212·03	317·82	380·87	451·09	614·15	809·19	919·36	1303·5	1480·7
0·02200	1·979	2·379	2·564	2·741	2·910	3·074	3·385	3·533	3·678	3·956	4·222	4·351	4·723	4·866
1/ 45·5	34·973	74·750	101·95	134·54	172·87	217·28	325·66	390·26	462·19	629·24	829·04	941·89	1335·3	1516·9
0·02300	2·026	2·436	2·625	2·806	2·979	3·146	3·465	3·617	3·764	4·049	4·321	4·453	4·833	4·980
1/ 43·5	35·811	76·529	104·37	137·73	176·96	222·41	333·33	399·43	473·05	643·99	848·44	963·92	1366·5	1552·2
0·02400	2·073	2·491	2·685	2·869	3·047	3·218	3·543	3·698	3·849	4·140	4·418	4·552	4·941	5·091
1/ 41·7	36·631	78·271	106·74	140·85	180·96	227·43	340·83	408·42	483·68	658·44	867·44	985·49	1397·0	1586·9
0·02500	2·118	2·546	2·743	2·932	3·113	3·287	3·619	3·778	3·932	4·229	4·513	4·650	5·047	5·199
1/ 40·0	37·434	79·978	109·07	143·91	184·88	232·35	348·19	417·22	494·10	672·59	886·05	1006·6	1426·9	1620·8
0·02600	2·163	2·599	2·800	2·993	3·177	3·355	3·694	3·856	4·013	4·316	4·606	4·746	5·150	5·306
1/ 38·5	38·223	81·652	111·34	146·91	188·73	237·18	355·40	425·86	504·31	686·46	904·30	1027·3	1456·2	1654·1
0·02700	2·207	2·651	2·857	3·053	3·241	3·422	3·768	3·932	4·093	4·402	4·697	4·840	5·252	5·411
1/ 37·0	38·996	83·295	113·58	149·85	192·50	241·92	362·48	434·33	514·33	700·08	922·20	1047·7	1485·0	1686·7
0·02800	2·250	2·703	2·912	3·112	3·303	3·488	3·840	4·008	4·171	4·486	4·786	4·932	5·352	5·514
1/ 35·7	39·756	84·909	115·78	152·74	196·21	246·57	369·43	442·65	524·17	713·44	939·78	1067·6	1513·2	1718·7
0·02900	2·292	2·753	2·966	3·169	3·365	3·553	3·911	4·082	4·248	4·568	4·874	5·022	5·450	5·615
1/ 34·5	40·503	86·495	117·93	155·58	199·85	251·14	376·26	450·82	533·84	726·58	957·06	1087·2	1540·9	1750·2
0·03000	2·334	2·803	3·019	3·226	3·425	3·616	3·981	4·155	4·324	4·650	4·961	5·111	5·546	5·714
1/ 33·3	41·238	88·054	120·05	158·38	203·43	255·63	382·98	458·86	543·35	739·49	974·04	1106·5	1568·2	1781·2
	0·88	0·90	0·91	0·91	0·92	0·92	0·93	0·93	0·94	0·94	0·95	0·95	0·95	0·96

$V_{r(0·5)medial}$ for half-full circular pipes.

k_s = **0·030 mm**
S = 0·03000 to 0·10000

ie hydraulic gradient =
1 in 33·3 to 1 in 10·0

Water (or sewage) at 15°C;
full bore conditions.

velocities in ms^{-1}
discharges in litres/sec

Gradient	(Equivalent) Pipe diameters in mm 150	200	225	250	275	300	350	375	400	450	500	525	600	630
0·03000	2·334	2·803	3·019	3·226	3·425	3·616	3·981	4·155	4·324	4·650	4·961	5·111	5·546	5·714
1/ 33·3	41·238	88·054	120·05	158·38	203·43	255·63	382·98	458·86	543·35	739·49	974·04	1106·5	1568·2	1781·2
0·03200	2·415	2·900	3·124	3·337	3·543	3·741	4·117	4·297	4·471	4·808	5·130	5·285	5·735	5·908
1/ 31·3	42·672	91·099	124·20	163·83	210·43	264·41	396·09	474·55	561·90	764·70	1007·2	1144·1	1621·4	1841·6
0·03400	2·494	2·994	3·225	3·445	3·657	3·861	4·249	4·434	4·615	4·962	5·293	5·454	5·917	6·096
1/ 29·4	44·064	94·053	128·21	169·12	217·21	272·91	408·80	489·76	579·90	789·15	1039·3	1180·6	1673·0	1900·2
0·03600	2·570	3·085	3·323	3·550	3·768	3·978	4·377	4·568	4·754	5·111	5·452	5·618	6·094	6·278
1/ 27·8	45·416	96·923	132·12	174·26	223·80	281·18	421·15	504·54	597·38	812·89	1070·6	1216·1	1723·2	1957·1
0·03800	2·645	3·174	3·418	3·652	3·876	4·092	4·502	4·698	4·889	5·256	5·607	5·777	6·267	6·456
1/ 26·3	46·733	99·716	135·92	179·26	230·21	289·22	433·17	518·92	614·39	836·00	1100·9	1250·5	1771·9	2012·4
0·04000	2·717	3·261	3·511	3·751	3·981	4·203	4·624	4·825	5·021	5·398	5·758	5·932	6·435	6·629
1/ 25·0	48·016	102·44	139·62	184·13	236·46	297·06	444·88	532·93	630·96	858·51	1130·5	1284·1	1819·5	2066·4
0·04200	2·788	3·345	3·602	3·848	4·084	4·311	4·743	4·949	5·150	5·536	5·905	6·083	6·599	6·798
1/ 23·8	49·268	105·09	143·23	188·89	242·55	304·71	456·30	546·60	647·13	880·47	1159·4	1316·9	1865·8	2119·0
0·04400	2·857	3·428	3·691	3·943	4·184	4·416	4·859	5·070	5·275	5·671	6·048	6·231	6·759	6·962
1/ 22·7	50·492	107·69	146·76	193·53	248·51	312·18	467·46	559·95	662·92	901·92	1187·6	1348·9	1911·1	2170·3
0·04600	2·925	3·509	3·778	4·035	4·282	4·520	4·972	5·188	5·398	5·803	6·189	6·376	6·916	7·124
1/ 21·7	51·688	110·23	150·21	198·08	254·33	319·49	478·37	573·01	678·37	922·89	1215·2	1380·2	1955·4	2220·6
0·04800	2·991	3·588	3·863	4·126	4·378	4·621	5·083	5·304	5·519	5·932	6·326	6·517	7·069	7·281
1/ 20·8	52·860	112·71	153·59	202·52	260·04	326·64	489·05	585·79	693·48	943·42	1242·2	1410·9	1998·7	2269·8
0·05000	3·056	3·665	3·946	4·215	4·472	4·720	5·192	5·417	5·636	6·058	6·461	6·656	7·219	7·436
1/ 20·0	54·008	115·15	156·90	206·88	265·62	333·64	499·52	598·31	708·29	963·53	1268·6	1440·9	2041·1	2317·9
0·05250	3·136	3·760	4·048	4·323	4·587	4·841	5·325	5·556	5·780	6·213	6·626	6·826	7·402	7·625
1/ 19·0	55·412	118·12	160·95	212·21	272·45	342·21	512·31	613·62	726·39	988·12	1300·9	1477·6	2093·0	2376·8
0·05500	3·213	3·853	4·147	4·429	4·699	4·960	5·455	5·691	5·921	6·364	6·786	6·991	7·582	7·809
1/ 18·2	56·783	121·03	164·90	217·41	279·12	350·58	524·81	628·57	744·07	1012·1	1332·5	1513·4	2143·7	2434·3
0·05750	3·289	3·943	4·245	4·533	4·809	5·075	5·582	5·824	6·059	6·511	6·944	7·153	7·757	7·989
1/ 17·4	58·125	123·88	168·77	222·50	285·64	358·76	537·02	643·19	761·36	1035·6	1363·4	1548·4	2193·2	2490·5
0·06000	3·364	4·032	4·340	4·634	4·917	5·189	5·706	5·953	6·193	6·656	7·097	7·311	7·928	8·166
1/ 16·7	59·438	126·66	172·56	227·48	292·03	366·77	548·98	657·50	778·29	1058·6	1393·6	1582·7	2241·7	2545·5
0·06250	3·436	4·119	4·433	4·734	5·022	5·300	5·828	6·080	6·325	6·798	7·248	7·466	8·096	8·339
1/ 16·0	60·725	129·39	176·26	232·36	298·28	374·61	560·70	671·52	794·86	1081·1	1423·2	1616·3	2289·2	2599·4
0·06500	3·508	4·204	4·525	4·831	5·125	5·408	5·947	6·204	6·455	6·936	7·396	7·619	8·261	8·508
1/ 15·4	61·987	132·06	179·90	237·15	304·42	382·30	572·18	685·26	811·11	1103·2	1452·2	1649·2	2335·7	2652·3
0·06750	3·578	4·287	4·614	4·927	5·227	5·515	6·064	6·327	6·582	7·072	7·541	7·768	8·423	8·675
1/ 14·8	63·225	134·69	183·47	241·84	310·43	389·85	583·45	698·74	827·06	1124·8	1480·7	1681·5	2381·4	2704·1
0·07000	3·647	4·369	4·702	5·021	5·326	5·620	6·179	6·446	6·706	7·206	7·683	7·914	8·581	8·838
1/ 14·3	64·441	137·26	186·97	246·45	316·34	397·26	594·52	711·98	842·71	1146·1	1508·6	1713·3	2426·3	2755·0
0·07250	3·714	4·450	4·789	5·113	5·424	5·723	6·292	6·564	6·829	7·337	7·823	8·058	8·737	8·998
1/ 13·8	65·636	139·80	190·41	250·98	322·15	404·54	605·39	724·98	858·09	1167·0	1536·1	1744·4	2470·3	2805·0
0·07500	3·781	4·529	4·874	5·204	5·520	5·824	6·403	6·680	6·949	7·466	7·961	8·200	8·890	9·156
1/ 13·3	66·811	142·29	193·80	255·43	327·85	411·70	616·08	737·77	873·22	1187·5	1563·0	1775·0	2513·6	2854·2
0·08000	3·910	4·684	5·040	5·381	5·707	6·022	6·620	6·906	7·184	7·718	8·229	8·476	9·189	9·464
1/ 12·5	69·104	147·14	200·40	264·12	338·99	425·67	636·93	762·73	902·72	1227·6	1615·7	1834·8	2598·1	2950·1
0·08500	4·036	4·834	5·201	5·552	5·889	6·214	6·830	7·125	7·411	7·963	8·489	8·744	9·479	9·762
1/ 11·8	71·328	151·86	206·81	272·55	349·79	439·21	657·16	786·92	931·33	1266·4	1666·8	1892·8	2680·1	3043·0
0·09000	4·159	4·979	5·358	5·719	6·066	6·400	7·034	7·338	7·632	8·200	8·741	9·004	9·760	10·05
1/ 11·1	73·489	156·43	213·03	280·74	360·28	452·37	676·80	810·41	959·12	1304·1	1716·4	1949·0	2759·6	3133·3
0·09500	4·278	5·121	5·510	5·881	6·238	6·581	7·233	7·545	7·848	8·431	8·987	9·256	10·03	10·33
1/ 10·5	75·592	160·89	219·08	288·70	370·49	465·17	695·90	833·27	986·15	1340·8	1764·6	2003·8	2837·0	3221·2
0·10000	4·394	5·259	5·658	6·039	6·405	6·757	7·427	7·746	8·057	8·655	9·226	9·503	10·30	10·61
1/ 10·0	77·640	165·23	224·98	296·46	380·43	477·63	714·52	855·54	1012·5	1376·6	1811·6	2057·1	2912·4	3306·7
	0·90	0·91	0·92	0·92	0·93	0·93	0·94	0·94	0·95	0·95	0·95	0·96	0·96	0·96

$V_{r(0\cdot5)medial}$ **for half-full circular pipes.**

A15
(p.1 of 6)

k_s = 0·060 mm
S = 0·00010 to 0·00030

ie hydraulic gradient =
1 in 10000 to 1 in 3333

Water (or sewage) at 15°C;
full bore conditions.

velocities in ms^{-1}
discharges in litres/sec

Gradient **(Equivalent) Pipe diameters in mm**

Gradient	150	200	225	250	275	300	350	375	400	450	500	525	600	630
0·000100	0·100	0·123	0·133	0·143	0·153	0·162	0·181	0·189	0·198	0·214	0·230	0·237	0·259	0·268
1/ 10000	1·7661	3·8513	5·2937	7·0328	9·0898	11·485	17·371	20·898	24·840	34·036	45·096	51·365	73·320	83·490
0·000105	0·103	0·126	0·137	0·147	0·157	0·167	0·186	0·194	0·203	0·220	0·236	0·244	0·266	0·275
1/ 9524	1·8157	3·9585	5·4406	7·2274	9·3407	11·802	17·847	21·471	25·519	34·965	46·323	52·761	75·307	85·750
0·000110	0·105	0·129	0·140	0·151	0·161	0·171	0·190	0·199	0·208	0·226	0·242	0·250	0·273	0·282
1/ 9091	1·8642	4·0635	5·5844	7·4179	9·5863	12·111	18·314	22·031	26·184	35·873	47·523	54·127	77·250	87·961
0·000115	0·108	0·133	0·144	0·155	0·165	0·176	0·195	0·204	0·214	0·231	0·248	0·256	0·280	0·289
1/ 8696	1·9118	4·1663	5·7253	7·6045	9·8268	12·414	18·770	22·579	26·835	36·762	48·698	55·464	79·153	90·125
0·000120	0·111	0·136	0·147	0·159	0·169	0·180	0·200	0·209	0·219	0·237	0·254	0·262	0·287	0·296
1/ 8333	1·9585	4·2672	5·8634	7·7874	10·063	12·712	19·218	23·117	27·473	37·634	49·850	56·774	81·018	92·246
0·000125	0·113	0·139	0·151	0·162	0·173	0·184	0·204	0·214	0·224	0·242	0·260	0·268	0·293	0·303
1/ 8000	2·0043	4·3661	5·9990	7·9669	10·294	13·003	19·657	23·644	28·099	38·489	50·980	58·060	82·847	94·326
0·000130	0·116	0·142	0·154	0·166	0·177	0·188	0·209	0·219	0·228	0·247	0·265	0·274	0·299	0·309
1/ 7692	2·0493	4·4633	6·1321	8·1432	10·521	13·290	20·088	24·162	28·713	39·328	52·089	59·322	84·642	96·368
0·000135	0·118	0·145	0·158	0·169	0·181	0·192	0·213	0·223	0·233	0·252	0·271	0·280	0·306	0·316
1/ 7407	2·0936	4·5589	6·2629	8·3164	10·744	13·571	20·512	24·671	29·317	40·153	53·179	60·561	86·405	98·374
0·000140	0·121	0·148	0·161	0·173	0·185	0·196	0·218	0·228	0·238	0·258	0·276	0·285	0·312	0·322
1/ 7143	2·1371	4·6528	6·3915	8·4867	10·964	13·848	20·929	25·171	29·910	40·963	54·250	61·779	88·139	100·34
0·000145	0·123	0·151	0·164	0·176	0·188	0·200	0·222	0·232	0·243	0·263	0·282	0·291	0·318	0·328
1/ 6897	2·1799	4·7452	6·5181	8·6543	11·180	14·120	21·338	25·663	30·494	41·761	55·303	62·978	89·843	102·28
0·000150	0·126	0·154	0·167	0·180	0·192	0·204	0·226	0·237	0·247	0·268	0·287	0·296	0·324	0·334
1/ 6667	2·2220	4·8362	6·6427	8·8193	11·392	14·388	21·742	26·147	31·069	42·545	56·340	64·157	91·521	104·19
0·000160	0·130	0·160	0·173	0·186	0·199	0·211	0·234	0·245	0·256	0·277	0·297	0·307	0·335	0·346
1/ 6250	2·3045	5·0142	6·8863	9·1418	11·808	14·912	22·530	27·094	32·192	44·079	58·366	66·462	94·799	107·92
0·000170	0·135	0·165	0·179	0·193	0·206	0·218	0·242	0·254	0·265	0·287	0·307	0·317	0·347	0·358
1/ 5882	2·3847	5·1872	7·1231	9·4553	12·212	15·420	23·296	28·014	33·283	45·569	60·335	68·701	97·984	111·54
0·000180	0·139	0·170	0·185	0·199	0·212	0·225	0·250	0·262	0·273	0·296	0·317	0·327	0·358	0·369
1/ 5556	2·4627	5·3556	7·3537	9·7605	12·605	15·916	24·042	28·909	34·345	47·019	62·250	70·879	101·08	115·06
0·000190	0·144	0·176	0·191	0·205	0·219	0·232	0·257	0·270	0·282	0·305	0·327	0·337	0·368	0·380
1/ 5263	2·5389	5·5199	7·5785	10·058	12·989	16·399	24·769	29·781	35·380	48·432	64·116	73·002	104·10	118·50
0·000200	0·148	0·181	0·196	0·211	0·225	0·239	0·265	0·277	0·290	0·313	0·336	0·347	0·379	0·391
1/ 5000	2·6133	5·6802	7·7979	10·349	13·363	16·870	25·478	30·633	36·390	49·811	65·937	75·073	107·04	121·84
0·000210	0·152	0·186	0·202	0·217	0·231	0·245	0·272	0·285	0·297	0·322	0·345	0·356	0·389	0·401
1/ 4762	2·6860	5·8370	8·0124	10·632	13·728	17·331	26·171	31·465	37·376	51·158	67·716	77·096	109·92	125·12
0·000220	0·156	0·191	0·207	0·222	0·237	0·252	0·279	0·292	0·305	0·330	0·354	0·365	0·399	0·412
1/ 4545	2·7571	5·9903	8·2223	10·910	14·086	17·781	26·849	32·278	38·342	52·475	69·456	79·075	112·73	128·31
0·000230	0·160	0·195	0·212	0·228	0·243	0·258	0·286	0·299	0·313	0·338	0·362	0·374	0·408	0·422
1/ 4348	2·8269	6·1406	8·4278	11·182	14·436	18·223	27·513	33·075	39·287	53·765	71·159	81·012	115·49	131·45
0·000240	0·164	0·200	0·217	0·233	0·249	0·264	0·293	0·307	0·320	0·346	0·371	0·383	0·418	0·432
1/ 4167	2·8952	6·2879	8·6294	11·449	14·780	18·655	28·163	33·856	40·213	55·029	72·828	82·910	118·19	134·51
0·000250	0·168	0·205	0·222	0·239	0·255	0·270	0·299	0·313	0·327	0·354	0·379	0·392	0·427	0·441
1/ 4000	2·9623	6·4324	8·8270	11·710	15·117	19·079	28·801	34·622	41·121	56·269	74·465	84·772	120·83	137·52
0·000260	0·171	0·209	0·227	0·244	0·260	0·276	0·306	0·320	0·334	0·361	0·387	0·400	0·437	0·451
1/ 3846	3·0282	6·5742	9·0211	11·967	15·447	19·496	29·428	35·374	42·012	57·486	76·071	86·598	123·43	140·47
0·000270	0·175	0·214	0·232	0·249	0·266	0·282	0·312	0·327	0·341	0·369	0·395	0·408	0·446	0·460
1/ 3704	3·0929	6·7137	9·2118	12·219	15·772	19·905	30·043	36·112	42·888	58·681	77·649	88·393	125·98	143·37
0·000280	0·179	0·218	0·236	0·254	0·271	0·287	0·319	0·334	0·348	0·376	0·403	0·416	0·454	0·469
1/ 3571	3·1566	6·8507	9·3993	12·467	16·092	20·307	30·648	36·838	43·749	59·855	79·199	90·156	128·48	146·22
0·000290	0·182	0·222	0·241	0·259	0·276	0·293	0·325	0·340	0·355	0·384	0·411	0·424	0·463	0·478
1/ 3448	3·2192	6·9856	9·5838	12·711	16·406	20·703	31·243	37·552	44·595	61·010	80·724	91·890	130·95	149·02
0·000300	0·186	0·227	0·246	0·264	0·281	0·298	0·331	0·346	0·362	0·391	0·419	0·432	0·472	0·487
1/ 3333	3·2809	7·1183	9·7653	12·951	16·715	21·092	31·828	38·254	45·428	62·147	82·224	93·596	133·37	151·78
	0·78	0·80	0·81	0·81	0·82	0·83	0·84	0·84	0·85	0·86	0·87	0·87	0·88	0·88

$V_{r(0\cdot5)medial}$ for half-full circular pipes.

k_s = 0·060 mm S = 0·00010 to 0·00030

k_s = **0·060 mm**
S = 0·00030 to 0·00100

ie hydraulic gradient =
1 in 3333 to 1 in 1000

Water (or sewage) at 15°C;
full bore conditions.

velocities in ms^{-1}
discharges in litres/sec

Gradient **(Equivalent) Pipe diameters in mm**

Gradient	150	200	225	250	275	300	350	375	400	450	500	525	600	630
0·000300	0·186	0·227	0·246	0·264	0·281	0·298	0·331	0·346	0·362	0·391	0·419	0·432	0·472	0·487
1/ 3333	3·2809	7·1183	9·7653	12·951	16·715	21·092	31·828	38·254	45·428	62·147	82·224	93·596	133·37	151·78
0·000320	0·192	0·235	0·255	0·273	0·292	0·309	0·343	0·359	0·374	0·405	0·434	0·448	0·488	0·504
1/ 3125	3·4014	7·3778	10·120	13·421	17·319	21·853	32·972	39·627	47·056	64·368	85·156	96·929	138·11	157·16
0·000340	0·199	0·243	0·263	0·283	0·301	0·320	0·354	0·371	0·387	0·418	0·448	0·463	0·505	0·521
1/ 2941	3·5186	7·6300	10·465	13·877	17·906	22·592	34·083	40·960	48·637	66·524	88·003	100·17	142·71	162·39
0·000360	0·206	0·251	0·272	0·292	0·311	0·330	0·365	0·383	0·399	0·431	0·462	0·477	0·521	0·537
1/ 2778	3·6327	7·8754	10·801	14·321	18·478	23·312	35·165	42·257	50·175	68·623	90·772	103·31	147·18	167·48
0·000380	0·212	0·258	0·280	0·301	0·320	0·340	0·376	0·394	0·411	0·444	0·476	0·491	0·536	0·553
1/ 2632	3·7440	8·1147	11·128	14·753	19·034	24·013	36·218	43·521	51·674	70·666	93·469	106·38	151·54	172·43
0·000400	0·218	0·266	0·288	0·309	0·330	0·349	0·387	0·405	0·423	0·457	0·489	0·505	0·551	0·569
1/ 2500	3·8526	8·3482	11·447	15·175	19·578	24·697	37·246	44·754	53·136	72·660	96·100	109·37	155·79	177·26
0·000420	0·224	0·273	0·296	0·318	0·339	0·359	0·398	0·416	0·434	0·469	0·503	0·519	0·566	0·584
1/ 2381	3·9587	8·5764	11·759	15·588	20·109	25·365	38·250	45·959	54·563	74·608	98·670	112·29	159·94	181·97
0·000440	0·230	0·280	0·303	0·326	0·347	0·368	0·408	0·427	0·445	0·481	0·515	0·532	0·580	0·599
1/ 2273	4·0626	8·7996	12·064	15·991	20·628	26·018	39·232	47·137	55·960	76·512	101·18	115·15	163·99	186·59
0·000460	0·236	0·287	0·311	0·334	0·356	0·377	0·418	0·437	0·456	0·493	0·528	0·545	0·594	0·613
1/ 2174	4·1643	9·0182	12·363	16·386	21·136	26·658	40·193	48·290	57·327	78·376	103·64	117·95	167·96	191·10
0·000480	0·241	0·294	0·318	0·342	0·364	0·386	0·428	0·447	0·467	0·504	0·540	0·558	0·608	0·627
1/ 2083	4·2640	9·2324	12·656	16·773	21·634	27·285	41·135	49·419	58·666	80·202	106·05	120·69	171·85	195·52
0·000500	0·247	0·301	0·326	0·349	0·372	0·395	0·437	0·457	0·477	0·516	0·552	0·570	0·621	0·641
1/ 2000	4·3618	9·4426	12·943	17·153	22·123	27·900	42·058	50·527	59·979	81·992	108·41	123·37	175·66	199·85
0·000525	0·254	0·309	0·334	0·359	0·383	0·405	0·449	0·470	0·490	0·529	0·567	0·585	0·638	0·658
1/ 1905	4·4816	9·6999	13·294	17·618	22·721	28·652	43·188	51·883	61·586	84·184	111·30	126·66	180·33	205·15
0·000550	0·260	0·317	0·343	0·368	0·392	0·416	0·460	0·482	0·503	0·543	0·581	0·600	0·654	0·675
1/ 1818	4·5988	9·9516	13·638	18·072	23·305	29·388	44·294	53·209	63·157	86·326	114·13	129·87	184·89	210·34
0·000575	0·267	0·325	0·351	0·377	0·402	0·426	0·472	0·494	0·515	0·556	0·595	0·614	0·670	0·691
1/ 1739	4·7136	10·198	13·975	18·517	23·878	30·109	45·376	54·506	64·695	88·422	116·90	133·01	189·35	215·41
0·000600	0·273	0·332	0·360	0·386	0·411	0·436	0·483	0·505	0·527	0·569	0·609	0·629	0·685	0·707
1/ 1667	4·8260	10·439	14·305	18·953	24·439	30·814	46·436	55·777	66·202	90·476	119·60	136·09	193·73	220·38
0·000625	0·279	0·340	0·368	0·395	0·421	0·446	0·493	0·516	0·539	0·582	0·623	0·643	0·700	0·723
1/ 1600	4·9364	10·676	14·628	19·381	24·989	31·507	47·475	57·024	67·679	92·490	122·26	139·11	198·01	225·25
0·000650	0·285	0·347	0·376	0·403	0·430	0·455	0·504	0·527	0·550	0·594	0·636	0·656	0·715	0·738
1/ 1538	5·0447	10·909	14·946	19·800	25·529	32·186	48·495	58·247	69·129	94·466	124·87	142·07	202·22	230·03
0·000675	0·291	0·355	0·384	0·412	0·439	0·465	0·514	0·538	0·561	0·606	0·649	0·670	0·730	0·753
1/ 1481	5·1512	11·137	15·258	20·213	26·059	32·853	49·496	59·448	70·552	96·406	127·42	144·98	206·34	234·72
0·000700	0·297	0·362	0·391	0·420	0·448	0·474	0·525	0·549	0·573	0·618	0·662	0·683	0·744	0·768
1/ 1429	5·2558	11·362	15·565	20·618	26·580	33·508	50·480	60·628	71·951	98·312	129·94	147·84	210·40	239·33
0·000725	0·303	0·369	0·399	0·428	0·456	0·483	0·535	0·559	0·584	0·630	0·674	0·696	0·758	0·782
1/ 1379	5·3587	11·582	15·866	21·016	27·093	34·153	51·448	61·788	73·326	100·19	132·41	150·65	214·38	243·86
0·000750	0·309	0·376	0·407	0·436	0·465	0·492	0·545	0·570	0·594	0·642	0·687	0·709	0·772	0·797
1/ 1333	5·4600	11·800	16·163	21·408	27·597	34·787	52·400	62·930	74·678	102·03	134·84	153·41	218·31	248·31
0·000800	0·320	0·389	0·421	0·452	0·481	0·510	0·564	0·590	0·615	0·664	0·711	0·734	0·799	0·824
1/ 1250	5·6580	12·224	16·743	22·174	28·582	36·027	54·260	65·160	77·321	105·63	139·59	158·81	225·96	257·01
0·000850	0·331	0·402	0·435	0·467	0·497	0·527	0·583	0·610	0·636	0·686	0·734	0·758	0·825	0·852
1/ 1176	5·8503	12·637	17·306	22·918	29·538	37·230	56·066	67·326	79·887	109·13	144·20	164·05	233·40	265·46
0·000900	0·342	0·415	0·449	0·482	0·513	0·543	0·601	0·629	0·656	0·708	0·757	0·781	0·851	0·878
1/ 1111	6·0375	13·038	17·854	23·641	30·469	38·400	57·822	69·431	82·381	112·53	148·68	169·14	240·62	273·67
0·000950	0·352	0·427	0·462	0·496	0·528	0·559	0·619	0·647	0·675	0·728	0·779	0·804	0·876	0·904
1/ 1053	6·2199	13·429	18·387	24·346	31·375	39·540	59·532	71·481	84·811	115·84	153·04	174·10	247·66	281·66
0·001000	0·362	0·440	0·476	0·510	0·543	0·575	0·636	0·665	0·694	0·749	0·801	0·827	0·900	0·929
1/ 1000	6·3978	13·810	18·908	25·034	32·259	40·652	61·200	73·481	87·180	119·06	157·30	178·93	254·52	289·46
	0·80	0·82	0·83	0·84	0·85	0·85	0·87	0·87	0·87	0·88	0·89	0·89	0·90	0·91

$V_{r(0{\cdot}5)medial}$ for half-full circular pipes.

k_s = 0·060 mm
S = 0·00100 to 0·00300

ie hydraulic gradient =
1 in 1000 to 1 in 333

Water (or sewage) at 15°C;
full bore conditions.

velocities in ms^{-1}
discharges in litres/sec

Gradient **(Equivalent) Pipe diameters in mm**

Gradient	150	200	225	250	275	300	350	375	400	450	500	525	600	630
0·00100	0·362	0·440	0·476	0·510	0·543	0·575	0·636	0·665	0·694	0·749	0·801	0·827	0·900	0·929
1/ 1000	6·3978	13·810	18·908	25·034	32·259	40·652	61·200	73·481	87·180	119·06	157·30	178·93	254·52	289·46
0·00105	0·372	0·451	0·488	0·524	0·558	0·590	0·653	0·683	0·712	0·768	0·822	0·848	0·924	0·953
1/ 952	6·5717	14·183	19·416	25·705	33·122	41·737	62·829	75·434	89·493	122·21	161·45	183·65	261·21	297·07
0·00110	0·382	0·463	0·501	0·537	0·572	0·605	0·670	0·700	0·730	0·788	0·843	0·870	0·947	0·977
1/ 909	6·7418	14·547	19·914	26·362	33·967	42·799	64·421	77·342	91·755	125·30	165·51	188·27	267·76	304·50
0·00115	0·391	0·474	0·513	0·550	0·586	0·620	0·686	0·717	0·748	0·807	0·863	0·891	0·970	1·000
1/ 870	6·9083	14·903	20·400	27·005	34·793	43·838	65·979	79·210	93·967	128·31	169·49	192·78	274·16	311·78
0·00120	0·400	0·486	0·525	0·563	0·599	0·635	0·702	0·734	0·765	0·825	0·883	0·911	0·992	1·023
1/ 833	7·0714	15·253	20·877	27·634	35·602	44·855	67·506	81·040	96·135	131·26	173·38	197·20	280·43	318·90
0·00125	0·409	0·496	0·537	0·576	0·613	0·649	0·717	0·750	0·782	0·844	0·902	0·931	1·014	1·045
1/ 800	7·2314	15·595	21·345	28·251	36·395	45·853	69·002	82·833	98·259	134·15	177·19	201·53	286·58	325·88
0·00130	0·418	0·507	0·548	0·588	0·626	0·663	0·732	0·766	0·799	0·861	0·921	0·951	1·035	1·067
1/ 769	7·3885	15·931	21·803	28·857	37·174	46·832	70·470	84·592	100·34	136·99	180·93	205·78	292·60	332·73
0·00135	0·427	0·518	0·560	0·600	0·639	0·676	0·747	0·782	0·815	0·879	0·940	0·970	1·056	1·089
1/ 741	7·5427	16·262	22·254	29·452	37·938	47·793	71·911	86·320	102·39	139·78	184·60	209·96	298·52	339·45
0·00140	0·435	0·528	0·571	0·612	0·651	0·689	0·762	0·797	0·831	0·896	0·959	0·989	1·076	1·110
1/ 714	7·6943	16·586	22·696	30·036	38·689	48·737	73·326	88·017	104·40	142·52	188·21	214·05	304·33	346·05
0·00145	0·444	0·538	0·582	0·624	0·664	0·703	0·777	0·812	0·847	0·913	0·977	1·007	1·097	1·131
1/ 690	7·8434	16·905	23·132	30·610	39·427	49·665	74·718	89·685	106·37	145·21	191·75	218·08	310·04	352·54
0·00150	0·452	0·548	0·593	0·635	0·676	0·716	0·791	0·827	0·862	0·930	0·994	1·026	1·116	1·151
1/ 667	7·9901	17·219	23·560	31·176	40·154	50·578	76·087	91·325	108·32	147·85	195·24	222·04	315·66	358·92
0·00160	0·468	0·568	0·614	0·658	0·700	0·741	0·819	0·856	0·892	0·962	1·029	1·061	1·155	1·191
1/ 625	8·2767	17·832	24·396	32·280	41·573	52·362	78·761	94·530	112·11	153·02	202·05	229·78	326·63	371·38
0·00170	0·484	0·587	0·634	0·679	0·723	0·765	0·846	0·884	0·921	0·994	1·063	1·096	1·193	1·230
1/ 588	8·5551	18·427	25·208	33·351	42·950	54·093	81·356	97·640	115·80	158·03	208·66	237·29	337·27	383·47
0·00180	0·499	0·605	0·654	0·701	0·746	0·789	0·872	0·911	0·950	1·024	1·095	1·130	1·229	1·268
1/ 556	8·8258	19·006	25·998	34·394	44·289	55·776	83·879	100·66	119·38	162·91	215·08	244·58	347·62	395·22
0·00190	0·514	0·623	0·673	0·721	0·768	0·812	0·897	0·938	0·978	1·054	1·127	1·163	1·265	1·305
1/ 526	9·0895	19·570	26·767	35·408	45·593	57·415	86·335	103·61	122·86	167·65	221·33	251·69	357·68	406·65
0·00200	0·529	0·640	0·692	0·741	0·789	0·835	0·922	0·964	1·005	1·083	1·158	1·195	1·300	1·340
1/ 500	9·3468	20·120	27·517	36·398	46·864	59·013	88·729	106·48	126·26	172·28	227·42	258·61	367·49	417·80
0·00210	0·543	0·658	0·710	0·761	0·810	0·857	0·947	0·989	1·031	1·112	1·189	1·226	1·334	1·375
1/ 476	9·5981	20·657	28·249	37·364	48·106	60·573	91·067	109·28	129·58	176·79	233·37	265·36	377·07	428·68
0·00220	0·557	0·674	0·728	0·780	0·830	0·879	0·970	1·014	1·057	1·139	1·218	1·256	1·367	1·409
1/ 455	9·8438	21·182	28·965	38·309	49·319	62·098	93·352	112·01	132·82	181·21	239·18	271·97	386·43	439·31
0·00230	0·571	0·691	0·746	0·799	0·850	0·900	0·994	1·038	1·082	1·166	1·247	1·286	1·399	1·443
1/ 435	10·084	21·695	29·666	39·233	50·506	63·590	95·587	114·69	135·99	185·52	244·87	278·43	395·58	449·70
0·00240	0·584	0·707	0·763	0·818	0·870	0·920	1·016	1·062	1·107	1·193	1·275	1·315	1·431	1·475
1/ 417	10·320	22·198	30·352	40·138	51·669	65·051	97·776	117·31	139·10	189·75	250·43	284·75	404·54	459·88
0·00250	0·597	0·722	0·780	0·836	0·889	0·941	1·039	1·085	1·131	1·219	1·303	1·344	1·462	1·507
1/ 400	10·551	22·692	31·024	41·025	52·808	66·483	99·921	119·88	142·14	193·89	255·89	290·95	413·32	469·85
0·00260	0·610	0·738	0·797	0·853	0·908	0·960	1·060	1·108	1·155	1·245	1·330	1·372	1·492	1·539
1/ 385	10·777	23·176	31·684	41·895	53·926	67·887	102·03	122·41	145·12	197·95	261·24	297·02	421·94	479·63
0·00270	0·622	0·753	0·813	0·871	0·926	0·980	1·082	1·131	1·178	1·270	1·357	1·400	1·522	1·569
1/ 370	11·000	23·651	32·332	42·750	55·024	69·266	104·09	124·88	148·05	201·94	266·49	302·99	430·39	489·23
0·00280	0·635	0·768	0·829	0·888	0·945	0·999	1·103	1·153	1·201	1·294	1·383	1·427	1·552	1·600
1/ 357	11·219	24·118	32·968	43·589	56·102	70·621	106·12	127·31	150·93	205·86	271·65	308·84	438·69	498·66
0·00290	0·647	0·782	0·845	0·905	0·962	1·018	1·124	1·174	1·224	1·319	1·409	1·453	1·580	1·629
1/ 345	11·434	24·576	33·593	44·414	57·161	71·952	108·11	129·70	153·76	209·70	276·71	314·60	446·84	507·92
0·00300	0·659	0·797	0·860	0·921	0·980	1·036	1·144	1·196	1·246	1·342	1·435	1·479	1·609	1·659
1/ 333	11·645	25·028	34·209	45·226	58·203	73·261	110·07	132·05	156·54	213·49	281·70	320·26	454·86	517·03
	0·82	0·85	0·85	0·86	0·87	0·87	0·89	0·89	0·89	0·90	0·91	0·91	0·92	0·92

$V_{r(0·5)medial}$ for half-full circular pipes.

k_s = 0·060 mm S = 0·00100 to 0·00300

Gradient **(Equivalent) Pipe diameters in mm**

Gradient	150	200	225	250	275	300	350	375	400	450	500	525	600	630
0·00300	0·659	0·797	0·860	0·921	0·980	1·036	1·144	1·196	1·246	1·342	1·435	1·479	1·609	1·659
1/ 333	11·645	25·028	34·209	45·226	58·203	73·261	110·07	132·05	156·54	213·49	281·70	320·26	454·86	517·03
0·00320	0·682	0·825	0·891	0·954	1·014	1·073	1·184	1·237	1·289	1·389	1·484	1·530	1·664	1·716
1/ 313	12·058	25·909	35·410	46·811	60·239	75·818	113·90	136·64	161·97	220·88	291·43	331·31	470·52	534·81
0·00340	0·705	0·852	0·920	0·985	1·047	1·108	1·222	1·277	1·331	1·434	1·532	1·580	1·718	1·771
1/ 294	12·459	26·765	36·576	48·348	62·213	78·299	117·62	141·09	167·24	228·04	300·86	342·03	485·71	552·06
0·00360	0·727	0·878	0·948	1·015	1·080	1·142	1·260	1·317	1·372	1·478	1·579	1·628	1·770	1·825
1/ 278	12·849	27·597	37·710	49·843	64·133	80·710	121·23	145·41	172·36	235·01	310·03	352·45	500·46	568·81
0·00380	0·749	0·904	0·976	1·045	1·111	1·175	1·297	1·355	1·411	1·520	1·624	1·675	1·821	1·877
1/ 263	13·229	28·406	38·814	51·298	66·001	83·057	124·74	149·62	177·34	241·79	318·96	362·58	514·82	585·12
0·00400	0·770	0·929	1·003	1·074	1·142	1·207	1·332	1·392	1·450	1·562	1·669	1·721	1·870	1·928
1/ 250	13·599	29·196	39·889	52·717	67·823	85·345	128·17	153·72	182·20	248·39	327·66	372·46	528·81	601·01
0·00420	0·790	0·954	1·030	1·102	1·172	1·239	1·367	1·428	1·488	1·602	1·712	1·765	1·919	1·978
1/ 238	13·961	29·967	40·940	54·102	69·600	87·578	131·51	157·73	186·94	254·84	336·14	382·10	542·47	616·51
0·00440	0·810	0·978	1·055	1·130	1·201	1·270	1·401	1·463	1·524	1·642	1·754	1·809	1·966	2·026
1/ 227	14·314	30·720	41·966	55·455	71·338	89·760	134·78	161·64	191·57	261·14	344·44	391·52	555·80	631·66
0·00460	0·830	1·001	1·081	1·157	1·230	1·300	1·434	1·498	1·561	1·681	1·796	1·851	2·012	2·074
1/ 217	14·660	31·457	42·970	56·779	73·037	91·894	137·97	165·46	196·10	267·30	352·55	400·73	568·85	646·47
0·00480	0·849	1·024	1·105	1·183	1·258	1·330	1·467	1·532	1·596	1·719	1·836	1·893	2·057	2·120
1/ 208	14·998	32·179	43·953	58·075	74·701	93·984	141·10	169·21	200·53	273·33	360·48	409·74	581·62	660·97
0·00500	0·868	1·047	1·130	1·209	1·285	1·359	1·498	1·565	1·630	1·756	1·876	1·934	2·101	2·166
1/ 200	15·330	32·886	44·917	59·345	76·332	96·032	144·16	172·88	204·88	279·24	368·26	418·58	594·13	675·17
0·00525	0·890	1·074	1·159	1·241	1·319	1·394	1·537	1·606	1·673	1·801	1·924	1·983	2·155	2·222
1/ 190	15·736	33·751	46·095	60·898	78·326	98·536	147·91	177·37	210·19	286·46	377·77	429·38	609·42	692·54
0·00550	0·913	1·101	1·188	1·272	1·352	1·429	1·575	1·646	1·714	1·846	1·971	2·032	2·208	2·276
1/ 182	16·133	34·597	47·247	62·417	80·274	100·98	151·57	181·75	215·38	293·52	387·06	439·93	624·36	709·51
0·00575	0·935	1·128	1·217	1·302	1·384	1·462	1·613	1·684	1·754	1·889	2·018	2·080	2·260	2·329
1/ 174	16·522	35·424	48·374	63·901	82·180	103·38	155·16	186·04	220·46	300·43	396·15	450·25	638·97	726·10
0·00600	0·956	1·153	1·244	1·331	1·415	1·496	1·649	1·722	1·794	1·931	2·063	2·127	2·310	2·381
1/ 167	16·902	36·234	49·477	65·355	84·046	105·72	158·66	190·24	225·43	307·18	405·04	460·35	653·28	742·33
0·00625	0·978	1·179	1·272	1·360	1·446	1·528	1·685	1·760	1·833	1·973	2·107	2·172	2·360	2·432
1/ 160	17·274	37·027	50·558	66·780	85·875	108·02	162·10	194·36	230·29	313·80	413·76	470·24	667·29	758·24
0·00650	0·998	1·203	1·298	1·389	1·476	1·560	1·720	1·796	1·871	2·014	2·151	2·217	2·409	2·482
1/ 154	17·640	37·806	51·618	68·177	87·668	110·27	165·47	198·39	235·07	320·30	422·30	479·95	681·03	773·84
0·00675	1·019	1·228	1·324	1·417	1·506	1·591	1·754	1·832	1·908	2·054	2·193	2·261	2·456	2·532
1/ 148	17·999	38·570	52·658	69·548	89·428	112·48	168·77	202·35	239·75	326·67	430·69	489·47	694·51	789·15
0·00700	1·038	1·252	1·350	1·444	1·535	1·622	1·788	1·867	1·945	2·093	2·235	2·304	2·503	2·580
1/ 143	18·351	39·320	53·680	70·895	91·156	114·65	172·02	206·24	244·35	332·92	438·92	498·81	707·74	804·18
0·00725	1·058	1·275	1·375	1·471	1·563	1·652	1·821	1·902	1·980	2·132	2·277	2·347	2·549	2·627
1/ 138	18·698	40·058	54·685	72·219	92·855	116·78	175·21	210·06	248·88	339·07	447·01	508·00	720·75	818·94
0·00750	1·077	1·298	1·400	1·498	1·591	1·682	1·854	1·936	2·016	2·170	2·317	2·388	2·594	2·674
1/ 133	19·039	40·783	55·672	73·520	94·525	118·88	178·35	213·81	253·32	345·11	454·96	517·03	733·53	833·46
0·00800	1·115	1·343	1·449	1·550	1·646	1·740	1·917	2·002	2·085	2·244	2·396	2·470	2·683	2·765
1/ 125	19·704	42·199	57·601	76·061	97·785	122·97	184·47	221·14	262·00	356·91	470·49	534·66	758·49	861·79
0·00850	1·152	1·387	1·496	1·600	1·700	1·796	1·979	2·067	2·152	2·316	2·473	2·549	2·768	2·853
1/ 118	20·350	43·573	59·471	78·525	100·95	126·94	190·41	228·25	270·41	368·35	485·54	551·75	782·69	889·26
0·00900	1·187	1·429	1·541	1·648	1·751	1·850	2·039	2·129	2·217	2·386	2·547	2·625	2·851	2·938
1/ 111	20·977	44·908	61·288	80·920	104·02	130·80	196·18	235·16	278·59	379·46	500·16	568·35	806·19	915·95
0·00950	1·222	1·471	1·586	1·696	1·802	1·904	2·097	2·190	2·280	2·454	2·620	2·700	2·932	3·022
1/ 105	21·588	46·207	63·057	83·250	107·01	134·55	201·79	241·88	286·54	390·27	514·39	584·51	829·06	941·91
0·01000	1·255	1·511	1·629	1·742	1·851	1·955	2·154	2·249	2·342	2·520	2·690	2·773	3·011	3·103
1/ 100	22·183	47·473	64·781	85·521	109·92	138·21	207·26	248·43	294·29	400·81	528·25	600·25	851·34	967·20
	0·85	0·87	0·87	0·88	0·89	0·89	0·90	0·91	0·91	0·92	0·92	0·93	0·93	0·93

$V_{r(0{\cdot}5)medial}$ for half-full circular pipes.

A15
(p.5 of 6)

k_s = 0·060 mm
S = 0·01000 to 0·03000

ie hydraulic gradient =
1 in 100 to 1 in 33·3

Water (or sewage) at 15°C;
full bore conditions.

velocities in ms^{-1}
discharges in litres/sec

Gradient **(Equivalent) Pipe diameters in mm**

Gradient	150	200	225	250	275	300	350	375	400	450	500	525	600	630
0·01000	1·255	1·511	1·629	1·742	1·851	1·955	2·154	2·249	2·342	2·520	2·690	2·773	3·011	3·103
1/ 100	22·183	47·473	64·781	85·521	109·92	138·21	207·26	248·43	294·29	400·81	528·25	600·25	851·34	967·20
0·01050	1·288	1·550	1·672	1·787	1·899	2·006	2·210	2·307	2·402	2·585	2·759	2·844	3·088	3·182
1/ 95	22·764	48·709	66·463	87·738	112·77	141·78	212·60	254·82	301·85	411·08	541·77	615·60	873·07	991·88
0·01100	1·320	1·589	1·713	1·831	1·945	2·055	2·264	2·364	2·461	2·648	2·826	2·913	3·163	3·259
1/ 91	23·332	49·916	68·106	89·903	115·54	145·26	217·81	261·06	309·24	421·12	554·98	630·60	894·30	1016·0
0·01150	1·352	1·626	1·753	1·875	1·991	2·103	2·317	2·419	2·518	2·710	2·892	2·981	3·236	3·335
1/ 87	23·887	51·097	69·714	92·020	118·26	148·67	222·91	267·17	316·46	430·94	567·90	645·27	915·06	1039·5
0·01200	1·383	1·663	1·793	1·917	2·036	2·151	2·369	2·473	2·575	2·770	2·957	3·047	3·308	3·409
1/ 83	24·431	52·253	71·287	94·092	120·92	152·01	227·90	273·14	323·53	440·55	580·54	659·62	935·37	1062·6
0·01250	1·413	1·699	1·832	1·958	2·080	2·197	2·420	2·526	2·630	2·829	3·020	3·112	3·379	3·481
1/ 80	24·964	53·385	72·829	96·123	123·53	155·28	232·79	278·99	330·45	449·96	592·92	673·68	955·27	1085·2
0·01300	1·442	1·735	1·870	1·999	2·123	2·242	2·469	2·578	2·684	2·887	3·082	3·176	3·448	3·552
1/ 77	25·486	54·496	74·340	98·114	126·08	158·49	237·58	284·73	337·24	459·19	605·06	687·46	974·78	1107·3
0·01350	1·471	1·769	1·907	2·039	2·165	2·287	2·518	2·629	2·737	2·944	3·142	3·238	3·515	3·622
1/ 74	25·999	55·586	75·824	100·07	128·59	161·63	242·29	290·36	343·90	468·24	616·97	700·99	993·92	1129·1
0·01400	1·500	1·803	1·944	2·078	2·206	2·330	2·566	2·679	2·789	3·000	3·202	3·300	3·582	3·690
1/ 71	26·503	56·656	77·280	101·99	131·05	164·72	246·91	295·89	350·44	477·13	628·67	714·27	1012·7	1150·4
0·01450	1·528	1·837	1·980	2·116	2·247	2·373	2·613	2·728	2·840	3·055	3·260	3·360	3·647	3·758
1/ 69	26·997	57·708	78·712	103·87	133·47	167·76	251·44	301·32	356·87	485·87	640·16	727·31	1031·2	1171·4
0·01500	1·555	1·870	2·015	2·154	2·287	2·415	2·660	2·777	2·890	3·109	3·318	3·419	3·711	3·824
1/ 67	27·484	58·742	80·119	105·73	135·84	170·74	255·90	306·66	363·19	494·45	651·46	740·14	1049·3	1192·0
0·01600	1·609	1·934	2·084	2·228	2·365	2·498	2·750	2·871	2·988	3·214	3·430	3·535	3·837	3·953
1/ 63	28·434	60·761	82·866	109·34	140·48	176·56	264·61	317·09	375·52	511·21	673·50	765·17	1084·7	1232·2
0·01700	1·661	1·996	2·151	2·299	2·441	2·578	2·838	2·962	3·083	3·316	3·539	3·647	3·958	4·078
1/ 59	29·356	62·719	85·530	112·85	144·98	182·21	273·05	327·19	387·48	527·46	694·88	789·43	1119·1	1271·2
0·01800	1·712	2·057	2·216	2·368	2·515	2·655	2·923	3·051	3·176	3·416	3·645	3·756	4·076	4·199
1/ 56	30·251	64·620	88·118	116·26	149·35	187·70	281·25	337·01	399·09	543·24	715·63	813·00	1152·4	1309·0
0·01900	1·761	2·116	2·280	2·436	2·586	2·731	3·006	3·138	3·266	3·512	3·748	3·862	4·191	4·317
1/ 53	31·122	66·471	90·636	119·58	153·61	193·03	289·23	346·56	410·39	558·59	735·83	835·93	1184·9	1345·9
0·02000	1·809	2·173	2·341	2·502	2·656	2·804	3·087	3·222	3·353	3·606	3·848	3·965	4·302	4·433
1/ 50	31·971	68·274	93·090	122·81	157·75	198·23	297·00	355·86	421·39	573·55	755·50	858·26	1216·5	1381·7
0·02100	1·856	2·229	2·401	2·566	2·724	2·876	3·166	3·304	3·439	3·698	3·945	4·065	4·411	4·545
1/ 47·6	32·800	70·034	95·484	125·96	161·79	203·31	304·59	364·94	432·13	588·14	774·69	880·05	1247·3	1416·7
0·02200	1·902	2·284	2·460	2·629	2·790	2·946	3·243	3·384	3·522	3·788	4·041	4·164	4·518	4·654
1/ 45·5	33·609	71·753	97·822	129·04	165·74	208·26	311·99	373·80	442·62	602·39	793·44	901·33	1277·4	1450·9
0·02300	1·947	2·337	2·518	2·690	2·856	3·015	3·318	3·463	3·604	3·875	4·134	4·260	4·622	4·762
1/ 43·5	34·401	73·433	100·11	132·05	169·60	213·11	319·23	382·47	452·87	616·32	811·76	922·14	1306·9	1484·3
0·02400	1·991	2·390	2·574	2·750	2·919	3·082	3·392	3·540	3·684	3·961	4·226	4·354	4·724	4·867
1/ 41·7	35·176	75·078	102·35	135·00	173·38	217·85	326·32	390·95	462·91	629·95	829·70	942·50	1335·7	1517·0
0·02500	2·033	2·441	2·629	2·809	2·981	3·148	3·464	3·615	3·762	4·045	4·315	4·446	4·824	4·969
1/ 40·0	35·935	76·690	104·54	137·88	177·09	222·50	333·26	399·26	472·74	643·31	847·26	962·44	1363·9	1549·1
0·02600	2·076	2·491	2·683	2·867	3·043	3·212	3·535	3·689	3·839	4·127	4·403	4·536	4·922	5·070
1/ 38·5	36·679	78·270	106·69	140·71	180·72	227·05	340·07	407·40	482·37	656·40	864·48	981·99	1391·5	1580·5
0·02700	2·117	2·541	2·736	2·923	3·103	3·275	3·604	3·761	3·914	4·208	4·489	4·625	5·018	5·169
1/ 37·0	37·409	79·820	108·80	143·49	184·28	231·52	346·75	415·40	491·83	669·25	881·38	1001·2	1418·7	1611·3
0·02800	2·158	2·589	2·788	2·979	3·161	3·337	3·672	3·832	3·988	4·287	4·573	4·712	5·112	5·266
1/ 35·7	38·127	81·343	110·87	146·22	187·77	235·90	353·30	423·24	501·11	681·86	897·97	1020·0	1445·3	1641·5
0·02900	2·197	2·637	2·840	3·033	3·219	3·398	3·739	3·902	4·060	4·365	4·656	4·797	5·204	5·361
1/ 34·5	38·831	82·838	112·90	148·90	191·21	240·21	359·74	430·95	510·23	694·25	914·26	1038·5	1471·5	1671·2
0·03000	2·237	2·684	2·890	3·087	3·276	3·458	3·805	3·970	4·132	4·442	4·738	4·881	5·295	5·455
1/ 33·3	39·524	84·309	114·90	151·53	194·58	244·45	366·07	438·53	519·19	706·43	930·28	1056·7	1497·2	1700·4
	0·86	0·88	0·89	0·89	0·90	0·90	0·91	0·92	0·92	0·93	0·93	0·93	0·94	0·94

$V_{r(0{\cdot}5)medial}$ for half-full circular pipes.

k_s = **0·060 mm**
S = 0·03000 to 0·10000

ie hydraulic gradient =
1 in 33·3 to 1 in 10·0

Water (or sewage) at 15°C;
full bore conditions.

velocities in ms^{-1}
discharges in litres/sec

Gradient **(Equivalent) Pipe diameters in mm**

Gradient	150	200	225	250	275	300	350	375	400	450	500	525	600	630
0·03000	2·237	2·684	2·890	3·087	3·276	3·458	3·805	3·970	4·132	4·442	4·738	4·881	5·295	5·455
1/ 33·3	39·524	84·309	114·90	151·53	194·58	244·45	366·07	438·53	519·19	706·43	930·28	1056·7	1497·2	1700·4
0·03200	2·313	2·775	2·988	3·192	3·387	3·575	3·933	4·104	4·271	4·591	4·897	5·045	5·473	5·638
1/ 31·3	40·877	87·179	118·81	156·67	201·18	252·72	378·43	453·32	536·68	730·19	961·54	1092·2	1547·4	1757·4
0·03400	2·387	2·864	3·083	3·293	3·495	3·689	4·058	4·234	4·406	4·736	5·051	5·204	5·645	5·815
1/ 29·4	42·188	89·962	122·59	161·65	207·56	260·74	390·41	467·65	553·64	753·23	991·84	1126·6	1596·1	1812·6
0·03600	2·459	2·950	3·176	3·392	3·599	3·799	4·179	4·360	4·537	4·877	5·201	5·358	5·812	5·987
1/ 27·8	43·462	92·665	126·27	166·49	213·77	268·52	402·04	481·57	570·11	775·61	1021·3	1160·0	1643·4	1866·3
0·03800	2·530	3·033	3·266	3·488	3·701	3·906	4·296	4·483	4·664	5·014	5·347	5·509	5·975	6·154
1/ 26·3	44·701	95·294	129·85	171·20	219·81	276·10	413·36	495·12	586·13	797·37	1049·9	1192·5	1689·3	1918·5
0·04000	2·598	3·115	3·353	3·581	3·800	4·010	4·411	4·602	4·788	5·147	5·489	5·655	6·133	6·317
1/ 25·0	45·909	97·856	133·33	175·79	225·69	283·48	424·38	508·31	601·74	818·58	1077·8	1224·1	1734·1	1969·3
0·04200	2·665	3·194	3·439	3·672	3·896	4·112	4·523	4·719	4·910	5·277	5·628	5·797	6·288	6·477
1/ 23·8	47·087	100·36	136·73	180·26	231·43	290·67	435·14	521·18	616·96	839·26	1105·0	1255·0	1777·8	2018·9
0·04400	2·730	3·272	3·522	3·761	3·991	4·212	4·632	4·833	5·028	5·404	5·763	5·937	6·439	6·632
1/ 22·7	48·238	102·80	140·05	184·63	237·03	297·70	445·64	533·75	631·83	859·46	1131·5	1285·1	1820·4	2067·3
0·04600	2·793	3·348	3·604	3·848	4·083	4·309	4·739	4·944	5·144	5·528	5·895	6·073	6·586	6·784
1/ 21·7	49·363	105·18	143·29	188·91	242·51	304·58	455·91	546·04	646·37	879·20	1157·5	1314·6	1862·1	2114·6
0·04800	2·856	3·422	3·684	3·934	4·173	4·404	4·843	5·053	5·257	5·650	6·025	6·206	6·730	6·932
1/ 20·8	50·464	107·52	146·47	193·09	247·87	311·30	465·96	558·06	660·59	898·52	1182·9	1343·5	1903·0	2161·0
0·05000	2·917	3·495	3·762	4·017	4·262	4·497	4·945	5·159	5·368	5·769	6·151	6·337	6·872	7·078
1/ 20·0	51·543	109·81	149·58	197·18	253·12	317·89	475·80	569·84	674·52	917·45	1207·8	1371·7	1942·9	2206·3
0·05250	2·991	3·584	3·858	4·119	4·370	4·611	5·070	5·290	5·503	5·914	6·306	6·496	7·044	7·256
1/ 19·0	52·862	112·60	153·39	202·19	259·54	325·94	487·83	584·24	691·55	940·58	1238·2	1406·3	1991·8	2261·8
0·05500	3·064	3·671	3·951	4·219	4·475	4·722	5·193	5·417	5·636	6·056	6·458	6·652	7·213	7·429
1/ 18·2	54·150	115·34	157·10	207·08	265·81	333·81	499·58	598·30	708·18	963·18	1267·9	1440·0	2039·5	2315·9
0·05750	3·136	3·756	4·043	4·316	4·578	4·831	5·312	5·542	5·765	6·195	6·605	6·804	7·378	7·599
1/ 17·4	55·410	118·01	160·73	211·86	271·94	341·50	511·07	612·05	724·44	985·26	1297·0	1473·0	2086·1	2368·9
0·06000	3·205	3·839	4·132	4·411	4·679	4·938	5·429	5·663	5·892	6·331	6·750	6·953	7·540	7·765
1/ 16·7	56·643	120·62	164·29	216·54	277·94	349·02	522·31	625·50	740·35	1006·9	1325·4	1505·2	2131·8	2420·7
0·06250	3·274	3·921	4·220	4·505	4·778	5·042	5·543	5·783	6·016	6·464	6·892	7·099	7·698	7·928
1/ 16·0	57·850	123·18	167·77	221·12	283·81	356·39	533·32	638·67	755·94	1028·0	1353·2	1536·8	2176·5	2471·4
0·06500	3·341	4·001	4·305	4·596	4·875	5·144	5·655	5·900	6·137	6·594	7·031	7·242	7·853	8·088
1/ 15·4	59·034	125·69	171·18	225·62	289·57	363·62	544·11	651·59	771·21	1048·8	1380·5	1567·8	2220·3	2521·1
0·06750	3·406	4·079	4·390	4·686	4·970	5·244	5·765	6·014	6·256	6·722	7·167	7·383	8·005	8·244
1/ 14·8	60·196	128·15	174·53	230·02	295·22	370·70	554·70	664·26	786·20	1069·2	1407·3	1598·2	2263·3	2569·9
0·07000	3·471	4·156	4·472	4·774	5·064	5·343	5·873	6·127	6·373	6·848	7·301	7·521	8·154	8·398
1/ 14·3	61·336	130·57	177·82	234·35	300·77	377·66	565·09	676·70	800·91	1089·1	1433·6	1628·0	2305·5	2617·8
0·07250	3·534	4·232	4·553	4·861	5·156	5·440	5·980	6·238	6·488	6·972	7·433	7·656	8·301	8·549
1/ 13·8	62·457	132·95	181·05	238·60	306·22	384·50	575·30	688·92	815·37	1108·8	1459·4	1657·3	2346·9	2664·9
0·07500	3·597	4·306	4·633	4·946	5·246	5·535	6·084	6·346	6·602	7·093	7·562	7·789	8·445	8·697
1/ 13·3	63·558	135·28	184·23	242·78	311·58	391·22	585·34	700·93	829·57	1128·1	1484·8	1686·1	2387·7	2711·1
0·08000	3·718	4·451	4·789	5·112	5·422	5·720	6·288	6·559	6·822	7·330	7·814	8·049	8·726	8·987
1/ 12·5	65·707	139·84	190·42	250·93	322·03	404·33	604·93	724·37	857·30	1165·7	1534·3	1742·3	2467·2	2801·4
0·08500	3·836	4·592	4·940	5·273	5·592	5·900	6·485	6·764	7·036	7·559	8·058	8·300	8·998	9·267
1/ 11·8	67·791	144·26	196·43	258·84	332·16	417·04	623·92	747·09	884·17	1202·2	1582·3	1796·8	2544·2	2888·8
0·09000	3·951	4·728	5·087	5·429	5·758	6·075	6·676	6·964	7·244	7·782	8·296	8·545	9·263	9·540
1/ 11·1	69·815	148·55	202·26	266·51	342·00	429·38	642·35	769·15	910·26	1237·7	1628·9	1849·7	2619·1	2973·8
0·09500	4·062	4·861	5·230	5·582	5·919	6·244	6·863	7·158	7·446	7·999	8·527	8·782	9·521	9·805
1/ 10·5	71·784	152·72	207·93	273·98	351·57	441·39	660·29	790·61	935·64	1272·1	1674·2	1901·2	2691·9	3056·4
0·10000	4·171	4·991	5·369	5·730	6·076	6·410	7·044	7·348	7·642	8·210	8·752	9·014	9·771	10·06
1/ 10·0	73·701	156·78	213·46	281·26	360·90	453·09	677·76	811·51	960·37	1305·7	1718·4	1951·3	2762·8	3136·9
	0·87	0·89	0·90	0·90	0·91	0·91	0·92	0·92	0·93	0·93	0·94	0·94	0·94	0·95

$V_{r(0·5)medial}$ for half-full circular pipes.

k_s = 0·150 mm
S = 0·00010 to 0·00030

ie hydraulic gradient =
1 in 10000 to 1 in 3333

Water (or sewage) at 15°C;
full bore conditions.

velocities in ms^{-1}
discharges in litres/sec

Gradient **(Equivalent) Pipe diameters in mm**

Gradient	150	200	225	250	275	300	350	375	400	450	500	525	600	630
0·000100	0·098	0·120	0·130	0·140	0·150	0·159	0·177	0·185	0·193	0·209	0·224	0·232	0·253	0·261
1/ 10000	1·7333	3·7753	5·1866	6·8872	8·8977	11·238	16·984	20·426	24·270	33·235	44·010	50·116	71·485	81·378
0·000105	0·101	0·123	0·134	0·144	0·154	0·163	0·181	0·190	0·198	0·215	0·230	0·238	0·260	0·268
1/ 9524	1·7814	3·8791	5·3287	7·0752	9·1399	11·543	17·443	20·977	24·924	34·128	45·189	51·457	73·391	83·546
0·000110	0·103	0·127	0·138	0·148	0·158	0·168	0·186	0·195	0·203	0·220	0·236	0·244	0·266	0·275
1/ 9091	1·8285	3·9806	5·4676	7·2591	9·3768	11·842	17·892	21·516	25·564	35·001	46·342	52·768	75·255	85·666
0·000115	0·106	0·130	0·141	0·152	0·162	0·172	0·191	0·200	0·208	0·225	0·242	0·250	0·273	0·281
1/ 8696	1·8746	4·0800	5·6036	7·4391	9·6088	12·134	18·332	22·044	26·190	35·856	47·470	54·051	77·079	87·740
0·000120	0·109	0·133	0·144	0·155	0·166	0·176	0·195	0·204	0·213	0·231	0·247	0·255	0·279	0·288
1/ 8333	1·9197	4·1774	5·7369	7·6156	9·8360	12·420	18·762	22·561	26·803	36·692	48·576	55·308	78·866	89·771
0·000125	0·111	0·136	0·148	0·159	0·169	0·180	0·199	0·209	0·218	0·236	0·253	0·261	0·285	0·294
1/ 8000	1·9641	4·2730	5·8677	7·7886	10·059	12·701	19·185	23·068	27·404	37·513	49·659	56·540	80·617	91·762
0·000130	0·114	0·139	0·151	0·162	0·173	0·184	0·204	0·213	0·223	0·241	0·258	0·267	0·291	0·301
1/ 7692	2·0076	4·3668	5·9961	7·9585	10·278	12·976	19·599	23·565	27·994	38·318	50·722	57·749	82·336	93·716
0·000135	0·116	0·142	0·154	0·166	0·177	0·187	0·208	0·218	0·227	0·246	0·264	0·272	0·297	0·307
1/ 7407	2·0503	4·4589	6·1221	8·1253	10·493	13·247	20·006	24·053	28·573	39·109	51·766	58·936	84·023	95·634
0·000140	0·118	0·145	0·157	0·169	0·180	0·191	0·212	0·222	0·232	0·251	0·269	0·278	0·303	0·313
1/ 7143	2·0924	4·5495	6·2461	8·2893	10·704	13·513	20·406	24·533	29·142	39·886	52·792	60·102	85·680	97·518
0·000145	0·121	0·148	0·160	0·172	0·184	0·195	0·216	0·226	0·236	0·256	0·274	0·283	0·309	0·319
1/ 6897	2·1337	4·6386	6·3679	8·4505	10·911	13·774	20·799	25·005	29·702	40·649	53·800	61·249	87·310	99·371
0·000150	0·123	0·150	0·163	0·175	0·187	0·199	0·220	0·231	0·241	0·260	0·279	0·288	0·314	0·325
1/ 6667	2·1744	4·7263	6·4879	8·6092	11·116	14·032	21·186	25·470	30·253	41·401	54·792	62·377	88·912	101·19
0·000160	0·128	0·156	0·169	0·182	0·194	0·206	0·228	0·239	0·249	0·270	0·289	0·298	0·326	0·336
1/ 6250	2·2540	4·8976	6·7223	8·9193	11·515	14·535	21·942	26·377	31·329	42·869	56·729	64·580	92·043	104·75
0·000170	0·132	0·161	0·175	0·188	0·200	0·213	0·236	0·247	0·258	0·278	0·298	0·308	0·336	0·347
1/ 5882	2·3313	5·0641	6·9499	9·2204	11·903	15·023	22·676	27·258	32·373	44·293	58·610	66·718	95·081	108·21
0·000180	0·136	0·166	0·180	0·194	0·207	0·219	0·243	0·255	0·266	0·287	0·308	0·318	0·347	0·358
1/ 5556	2·4065	5·2260	7·1714	9·5133	12·280	15·498	23·390	28·114	33·388	45·679	60·438	68·797	98·034	111·56
0·000190	0·140	0·171	0·186	0·200	0·213	0·226	0·250	0·262	0·274	0·296	0·317	0·327	0·357	0·368
1/ 5263	2·4799	5·3838	7·3872	9·7987	12·647	15·960	24·085	28·948	34·377	47·028	62·219	70·821	100·91	114·83
0·000200	0·144	0·176	0·191	0·205	0·219	0·232	0·257	0·269	0·281	0·304	0·326	0·336	0·367	0·379
1/ 5000	2·5514	5·5378	7·5977	10·077	13·006	16·411	24·763	29·762	35·342	48·343	63·955	72·795	103·71	118·02
0·000210	0·148	0·181	0·196	0·211	0·225	0·238	0·264	0·277	0·289	0·312	0·334	0·345	0·376	0·389
1/ 4762	2·6213	5·6882	7·8033	10·349	13·356	16·852	25·426	30·556	36·284	49·628	65·650	74·722	106·45	121·13
0·000220	0·152	0·186	0·201	0·216	0·231	0·244	0·271	0·284	0·296	0·320	0·343	0·354	0·386	0·398
1/ 4545	2·6897	5·8353	8·0044	10·615	13·698	17·283	26·073	31·333	37·204	50·884	67·306	76·606	109·13	124·17
0·000230	0·156	0·190	0·206	0·222	0·236	0·250	0·278	0·291	0·303	0·328	0·351	0·362	0·395	0·408
1/ 4348	2·7567	5·9793	8·2012	10·875	14·033	17·704	26·706	32·093	38·105	52·112	68·927	78·449	111·74	127·15
0·000240	0·160	0·195	0·211	0·227	0·242	0·256	0·284	0·297	0·310	0·335	0·359	0·371	0·404	0·417
1/ 4167	2·8223	6·1203	8·3941	11·130	14·361	18·117	27·327	32·837	38·988	53·315	70·515	80·253	114·31	130·06
0·000250	0·163	0·199	0·216	0·232	0·247	0·262	0·290	0·304	0·317	0·343	0·367	0·379	0·413	0·426
1/ 4000	2·8867	6·2587	8·5831	11·380	14·682	18·522	27·935	33·567	39·852	54·495	72·071	82·022	116·82	132·91
0·000260	0·167	0·204	0·221	0·237	0·253	0·268	0·297	0·310	0·324	0·350	0·375	0·387	0·422	0·435
1/ 3846	2·9498	6·3944	8·7687	11·625	14·998	18·919	28·532	34·282	40·701	55·651	73·597	83·757	119·28	135·71
0·000270	0·170	0·208	0·225	0·242	0·258	0·273	0·303	0·317	0·331	0·357	0·382	0·395	0·430	0·444
1/ 3704	3·0119	6·5277	8·9509	11·866	15·308	19·309	29·118	34·985	41·534	56·787	75·095	85·460	121·70	138·46
0·000280	0·174	0·212	0·230	0·247	0·263	0·279	0·309	0·323	0·337	0·364	0·390	0·403	0·439	0·453
1/ 3571	3·0729	6·6588	9·1300	12·103	15·613	19·693	29·693	35·676	42·352	57·903	76·567	87·133	124·07	141·16
0·000290	0·177	0·216	0·234	0·251	0·268	0·284	0·315	0·329	0·343	0·371	0·397	0·410	0·447	0·461
1/ 3448	3·1328	6·7876	9·3060	12·336	15·912	20·069	30·259	36·354	43·157	59·000	78·014	88·778	126·41	143·81
0·000300	0·181	0·220	0·238	0·256	0·273	0·289	0·320	0·335	0·350	0·378	0·405	0·418	0·455	0·470
1/ 3333	3·1919	6·9144	9·4792	12·564	16·206	20·440	30·816	37·022	43·948	60·078	79·437	90·395	128·71	146·42
	0·78	0·79	0·80	0·81	0·81	0·82	0·83	0·83	0·84	0·85	0·85	0·86	0·87	0·87

$V_{r(0·5)medial}$ for half-full circular pipes.

k_s = 0·150 mm S = 0·00010 to 0·00030

k_s = **0·150 mm**
S = 0·00030 to 0·00100

ie hydraulic gradient =
1 in 3333 to 1 in 1000

Water (or sewage) at 15°C;
full bore conditions.

velocities in ms^{-1}
discharges in litres/sec

Gradient	(Equivalent) Pipe diameters in mm 150	200	225	250	275	300	350	375	400	450	500	525	600	630
0·000300	0·181	0·220	0·238	0·256	0·273	0·289	0·320	0·335	0·350	0·378	0·405	0·418	0·455	0·470
1/ 3333	3·1919	6·9144	9·4792	12·564	16·206	20·440	30·816	37·022	43·948	60·078	79·437	90·395	128·71	146·42
0·000320	0·187	0·228	0·247	0·265	0·283	0·299	0·332	0·347	0·362	0·391	0·419	0·432	0·471	0·486
1/ 3125	3·3072	7·1620	9·8176	13·012	16·782	21·164	31·903	38·326	45·493	62·185	82·215	93·553	133·19	151·52
0·000340	0·193	0·236	0·255	0·274	0·292	0·309	0·343	0·358	0·374	0·404	0·432	0·446	0·486	0·502
1/ 2941	3·4192	7·4025	10·146	13·446	17·340	21·867	32·958	39·591	46·993	64·229	84·911	96·618	137·54	156·46
0·000360	0·200	0·243	0·263	0·283	0·301	0·319	0·353	0·370	0·386	0·416	0·446	0·460	0·501	0·517
1/ 2778	3·5282	7·6363	10·466	13·868	17·883	22·550	33·984	40·821	48·451	66·216	87·531	99·596	141·77	161·27
0·000380	0·206	0·250	0·271	0·291	0·310	0·328	0·364	0·380	0·397	0·428	0·459	0·473	0·516	0·532
1/ 2632	3·6343	7·8641	10·777	14·279	18·412	23·215	34·982	42·019	49·870	68·150	90·082	102·49	145·88	165·94
0·000400	0·212	0·257	0·279	0·299	0·319	0·338	0·374	0·391	0·408	0·440	0·471	0·487	0·530	0·547
1/ 2500	3·7378	8·0863	11·080	14·680	18·928	23·864	35·956	43·186	51·253	70·035	92·568	105·32	149·89	170·50
0·000420	0·217	0·264	0·286	0·307	0·327	0·347	0·384	0·401	0·419	0·452	0·484	0·499	0·544	0·561
1/ 2381	3·8390	8·3032	11·376	15·071	19·431	24·497	36·907	44·326	52·604	71·876	94·995	108·08	153·81	174·95
0·000440	0·223	0·271	0·293	0·315	0·335	0·355	0·393	0·411	0·429	0·463	0·496	0·512	0·558	0·575
1/ 2273	3·9379	8·5153	11·666	15·454	19·923	25·116	37·835	45·440	53·924	73·674	97·367	110·77	157·63	179·29
0·000460	0·228	0·278	0·301	0·322	0·344	0·364	0·403	0·421	0·439	0·474	0·508	0·524	0·571	0·589
1/ 2174	4·0347	8·7228	11·949	15·828	20·405	25·722	38·744	46·529	55·215	75·434	99·686	113·41	161·37	183·54
0·000480	0·234	0·284	0·308	0·330	0·351	0·372	0·412	0·431	0·449	0·485	0·519	0·536	0·584	0·602
1/ 2083	4·1295	8·9261	12·227	16·195	20·876	26·315	39·634	47·596	56·479	77·156	101·96	115·99	165·03	187·70
0·000500	0·239	0·290	0·314	0·337	0·359	0·381	0·421	0·440	0·459	0·496	0·531	0·548	0·596	0·615
1/ 2000	4·2225	9·1254	12·499	16·554	21·338	26·896	40·507	48·642	57·718	78·845	104·18	118·52	168·62	191·78
0·000525	0·245	0·298	0·323	0·346	0·369	0·391	0·432	0·452	0·471	0·509	0·544	0·562	0·612	0·631
1/ 1905	4·3363	9·3693	12·832	16·994	21·904	27·607	41·573	49·921	59·234	80·909	106·91	121·61	173·01	196·77
0·000550	0·252	0·306	0·331	0·355	0·378	0·400	0·443	0·463	0·483	0·521	0·558	0·576	0·627	0·647
1/ 1818	4·4475	9·6077	13·158	17·424	22·456	28·302	42·616	51·171	60·715	82·927	109·56	124·64	177·30	201·64
0·000575	0·258	0·313	0·339	0·364	0·387	0·410	0·454	0·474	0·495	0·534	0·571	0·589	0·642	0·662
1/ 1739	4·5564	9·8409	13·476	17·844	22·997	28·982	43·636	52·394	62·163	84·900	112·17	127·59	181·49	206·41
0·000600	0·264	0·321	0·347	0·372	0·396	0·419	0·464	0·485	0·506	0·546	0·584	0·603	0·656	0·677
1/ 1667	4·6630	10·069	13·788	18·256	23·526	29·648	44·635	53·591	63·582	86·832	114·71	130·49	185·60	211·07
0·000625	0·270	0·328	0·354	0·380	0·405	0·429	0·474	0·496	0·517	0·558	0·597	0·616	0·671	0·692
1/ 1600	4·7676	10·293	14·093	18·660	24·045	30·300	45·613	54·765	64·972	88·726	117·21	133·32	189·62	215·64
0·000650	0·276	0·335	0·362	0·388	0·413	0·438	0·484	0·506	0·528	0·570	0·609	0·629	0·685	0·706
1/ 1538	4·8702	10·513	14·393	19·056	24·554	30·940	46·573	55·915	66·335	90·583	119·66	136·10	193·56	220·12
0·000675	0·281	0·342	0·369	0·396	0·422	0·447	0·494	0·516	0·539	0·581	0·622	0·641	0·698	0·720
1/ 1481	4·9709	10·729	14·688	19·444	25·054	31·568	47·516	57·045	67·673	92·405	122·06	138·83	197·44	224·52
0·000700	0·287	0·348	0·377	0·404	0·430	0·455	0·503	0·527	0·549	0·592	0·634	0·654	0·712	0·734
1/ 1429	5·0698	10·941	14·977	19·826	25·544	32·185	48·441	58·154	68·987	94·195	124·42	141·51	201·24	228·84
0·000725	0·292	0·355	0·384	0·412	0·438	0·464	0·513	0·536	0·559	0·603	0·645	0·666	0·725	0·748
1/ 1379	5·1671	11·149	15·261	20·201	26·027	32·792	49·350	59·244	70·278	95·953	126·73	144·15	204·97	233·08
0·000750	0·298	0·361	0·391	0·419	0·446	0·472	0·522	0·546	0·569	0·614	0·657	0·678	0·738	0·761
1/ 1333	5·2628	11·354	15·541	20·570	26·501	33·388	50·245	60·316	71·548	97·683	129·01	146·74	208·65	237·26
0·000800	0·308	0·374	0·405	0·434	0·462	0·489	0·540	0·565	0·589	0·635	0·680	0·701	0·763	0·787
1/ 1250	5·4498	11·754	16·087	21·291	27·427	34·553	51·991	62·409	74·027	101·06	133·46	151·79	215·82	245·40
0·000850	0·319	0·386	0·418	0·448	0·477	0·505	0·558	0·583	0·608	0·656	0·702	0·724	0·788	0·813
1/ 1176	5·6312	12·142	16·616	21·990	28·326	35·683	53·685	64·439	76·432	104·33	137·78	156·70	222·77	253·30
0·000900	0·329	0·399	0·431	0·462	0·492	0·520	0·575	0·601	0·627	0·676	0·723	0·746	0·812	0·837
1/ 1111	5·8077	12·519	17·131	22·670	29·199	36·781	55·331	66·412	78·769	107·52	141·97	161·46	229·53	260·98
0·000950	0·338	0·410	0·443	0·475	0·506	0·535	0·592	0·619	0·645	0·695	0·744	0·767	0·835	0·861
1/ 1053	5·9794	12·887	17·632	23·331	30·050	37·849	56·933	68·332	81·043	110·61	146·05	166·10	236·10	268·44
0·001000	0·348	0·422	0·456	0·488	0·520	0·550	0·608	0·636	0·663	0·714	0·764	0·788	0·858	0·884
1/ 1000	6·1469	13·245	18·121	23·976	30·878	38·891	58·495	70·203	83·259	113·63	150·03	170·61	242·50	275·72
	0·79	0·81	0·82	0·82	0·83	0·84	0·85	0·85	0·86	0·86	0·87	0·87	0·88	0·89

$V_{r(0\cdot5)medial}$ for half-full circular pipes.

k_s = 0·150 mm
S = 0·00100 to 0·00300

ie hydraulic gradient =
1 in 1000 to 1 in 333

Water (or sewage) at 15°C;
full bore conditions.

velocities in ms^{-1}
discharges in litres/sec

Gradient	(Equivalent) Pipe diameters in mm													
	150	**200**	**225**	**250**	**275**	**300**	**350**	**375**	**400**	**450**	**500**	**525**	**600**	**630**
0·00100	0·348	0·422	0·456	0·488	0·520	0·550	0·608	0·636	0·663	0·714	0·764	0·788	0·858	0·884
1/ 1000	6·1469	13·245	18·121	23·976	30·878	38·891	58·495	70·203	83·259	113·63	150·03	170·61	242·50	275·72
0·00105	0·357	0·433	0·468	0·501	0·533	0·565	0·624	0·652	0·680	0·733	0·784	0·809	0·880	0·907
1/ 952	6·3105	13·595	18·598	24·606	31·687	39·907	60·018	72·030	85·422	116·57	153·91	175·02	248·75	282·82
0·00110	0·366	0·444	0·479	0·514	0·547	0·579	0·639	0·668	0·697	0·751	0·803	0·828	0·901	0·930
1/ 909	6·4703	13·936	19·064	25·221	32·477	40·901	61·507	73·814	87·535	119·45	157·70	179·33	254·86	289·75
0·00115	0·375	0·454	0·491	0·526	0·560	0·592	0·654	0·684	0·713	0·769	0·822	0·848	0·922	0·951
1/ 870	6·6266	14·270	19·520	25·822	33·250	41·872	62·963	75·558	89·601	122·26	161·40	183·54	260·83	296·54
0·00120	0·384	0·465	0·502	0·538	0·573	0·606	0·669	0·700	0·729	0·786	0·841	0·867	0·943	0·973
1/ 833	6·7798	14·598	19·966	26·411	34·007	42·823	64·388	77·266	91·624	125·02	165·03	187·66	266·67	303·17
0·00125	0·392	0·475	0·513	0·550	0·585	0·619	0·684	0·715	0·745	0·803	0·859	0·886	0·963	0·993
1/ 800	6·9299	14·918	20·404	26·988	34·748	43·755	65·785	78·940	93·606	127·72	168·59	191·70	272·40	309·68
0·00130	0·400	0·485	0·524	0·561	0·597	0·632	0·698	0·730	0·760	0·820	0·876	0·904	0·983	1·014
1/ 769	7·0771	15·233	20·832	27·554	35·476	44·669	67·155	80·581	95·549	130·36	172·07	195·66	278·01	316·05
0·00135	0·409	0·495	0·535	0·573	0·609	0·645	0·712	0·744	0·776	0·836	0·894	0·922	1·003	1·034
1/ 741	7·2216	15·542	21·254	28·110	36·189	45·566	68·499	82·191	97·457	132·96	175·49	199·54	283·52	322·31
0·00140	0·417	0·504	0·545	0·584	0·621	0·657	0·726	0·758	0·790	0·852	0·911	0·939	1·022	1·054
1/ 714	7·3636	15·845	21·667	28·656	36·890	46·447	69·818	83·773	99·329	135·51	178·85	203·36	288·92	328·45
0·00145	0·425	0·514	0·555	0·595	0·633	0·669	0·739	0·773	0·805	0·868	0·928	0·957	1·041	1·073
1/ 690	7·5031	16·143	22·073	29·192	37·579	47·312	71·115	85·327	101·17	138·01	182·15	207·11	294·24	334·48
0·00150	0·432	0·523	0·565	0·605	0·644	0·681	0·752	0·786	0·819	0·883	0·944	0·974	1·059	1·092
1/ 667	7·6403	16·436	22·473	29·719	38·256	48·163	72·390	86·854	102·98	140·47	185·39	210·79	299·46	340·42
0·00160	0·448	0·541	0·585	0·626	0·666	0·705	0·778	0·813	0·848	0·913	0·976	1·007	1·095	1·129
1/ 625	7·9083	17·008	23·253	30·748	39·578	49·825	74·879	89·837	106·51	145·28	191·72	217·98	309·66	352·00
0·00170	0·462	0·559	0·604	0·647	0·688	0·728	0·803	0·840	0·875	0·943	1·008	1·039	1·130	1·165
1/ 588	8·1683	17·563	24·010	31·747	40·861	51·436	77·293	92·729	109·93	149·94	197·86	224·96	319·54	363·23
0·00180	0·477	0·576	0·622	0·666	0·709	0·750	0·828	0·865	0·901	0·971	1·038	1·070	1·164	1·200
1/ 556	8·4209	18·103	24·745	32·717	42·106	53·002	79·638	95·539	113·26	154·47	203·83	231·73	329·14	374·13
0·00190	0·490	0·593	0·640	0·686	0·729	0·771	0·851	0·890	0·927	0·999	1·068	1·101	1·197	1·234
1/ 526	8·6669	18·628	25·461	33·660	43·319	54·525	81·920	98·272	116·50	158·87	209·63	238·33	338·48	384·74
0·00200	0·504	0·609	0·658	0·704	0·749	0·792	0·875	0·914	0·952	1·026	1·096	1·131	1·229	1·267
1/ 500	8·9066	19·140	26·159	34·580	44·500	56·009	84·144	100·94	119·65	163·17	215·28	244·75	347·59	395·08
0·00210	0·517	0·625	0·675	0·723	0·769	0·813	0·897	0·937	0·977	1·052	1·125	1·160	1·261	1·300
1/ 476	9·1406	19·639	26·839	35·478	45·653	57·458	86·313	103·54	122·73	167·35	220·80	251·01	356·46	405·16
0·00220	0·530	0·641	0·692	0·741	0·788	0·833	0·919	0·960	1·001	1·078	1·152	1·188	1·291	1·331
1/ 455	9·3692	20·127	27·504	36·355	46·779	58·873	88·432	106·07	125·74	171·44	226·19	257·13	365·14	415·01
0·00230	0·543	0·656	0·708	0·758	0·806	0·852	0·941	0·983	1·024	1·103	1·179	1·215	1·321	1·362
1/ 435	9·5929	20·604	28·154	37·213	47·881	60·256	90·505	108·56	128·68	175·44	231·45	263·11	373·61	424·64
0·00240	0·555	0·671	0·724	0·775	0·824	0·872	0·962	1·005	1·047	1·128	1·205	1·242	1·351	1·392
1/ 417	9·8118	21·071	28·791	38·052	48·959	61·611	92·533	110·99	131·55	179·36	236·61	268·97	381·91	434·07
0·00250	0·567	0·685	0·740	0·792	0·842	0·890	0·982	1·026	1·069	1·152	1·231	1·269	1·379	1·422
1/ 400	10·026	21·528	29·415	38·875	50·015	62·938	94·520	113·37	134·37	183·19	241·66	274·71	390·04	443·30
0·00260	0·579	0·700	0·755	0·808	0·859	0·909	1·003	1·048	1·091	1·175	1·256	1·295	1·408	1·451
1/ 385	10·237	21·977	30·026	39·681	51·051	64·239	96·468	115·70	137·13	186·95	246·61	280·33	398·01	452·35
0·00270	0·591	0·714	0·770	0·824	0·877	0·927	1·023	1·068	1·113	1·199	1·281	1·320	1·435	1·480
1/ 370	10·443	22·418	30·626	40·472	52·067	65·515	98·380	117·99	139·84	190·64	251·47	285·85	405·83	461·23
0·00280	0·602	0·727	0·785	0·840	0·893	0·945	1·042	1·089	1·134	1·221	1·305	1·345	1·462	1·508
1/ 357	10·646	22·850	31·216	41·250	53·065	66·769	100·26	120·24	142·51	194·26	256·24	291·27	413·50	469·95
0·00290	0·614	0·741	0·800	0·856	0·910	0·962	1·061	1·109	1·155	1·244	1·329	1·370	1·489	1·535
1/ 345	10·845	23·275	31·795	42·013	54·045	68·000	102·10	122·45	145·12	197·82	260·92	296·59	421·05	478·51
0·00300	0·625	0·754	0·814	0·871	0·926	0·979	1·080	1·128	1·175	1·266	1·352	1·394	1·515	1·562
1/ 333	11·041	23·693	32·364	42·764	55·009	69·211	103·91	124·62	147·69	201·32	265·53	301·82	428·46	486·93
	0·81	0·82	0·83	0·84	0·85	0·85	0·86	0·87	0·87	0·88	0·88	0·89	0·90	0·90

$V_{r(0{\cdot}5)medial}$ for half-full circular pipes.

Gradient **(Equivalent) Pipe diameters in mm**

Gradient	150	200	225	250	275	300	350	375	400	450	500	525	600	630
0·00300	0·625	0·754	0·814	0·871	0·926	0·979	1·080	1·128	1·175	1·266	1·352	1·394	1·515	1·562
1/ 333	11·041	23·693	32·364	42·764	55·009	69·211	103·91	124·62	147·69	201·32	265·53	301·82	428·46	486·93
0·00320	0·646	0·780	0·842	0·901	0·958	1·013	1·117	1·167	1·215	1·309	1·398	1·441	1·567	1·615
1/ 313	11·424	24·508	33·475	44·229	56·890	71·574	107·45	128·86	152·71	208·15	274·52	312·03	442·93	503·36
0·00340	0·667	0·805	0·869	0·930	0·989	1·045	1·152	1·204	1·254	1·350	1·442	1·487	1·616	1·666
1/ 294	11·795	25·299	34·553	45·650	58·715	73·866	110·88	132·97	157·58	214·77	283·23	321·93	456·95	519·29
0·00360	0·688	0·830	0·895	0·958	1·018	1·076	1·187	1·240	1·292	1·391	1·486	1·532	1·664	1·715
1/ 278	12·155	26·068	35·600	47·030	60·487	76·092	114·22	136·96	162·30	221·19	291·70	331·54	470·57	534·76
0·00380	0·708	0·854	0·921	0·985	1·047	1·107	1·221	1·275	1·328	1·430	1·528	1·575	1·711	1·764
1/ 263	12·506	26·815	36·619	48·373	62·211	78·257	117·46	140·84	166·90	227·45	299·93	340·90	483·82	549·81
0·00400	0·727	0·877	0·946	1·012	1·076	1·137	1·254	1·309	1·364	1·468	1·568	1·617	1·757	1·811
1/ 250	12·848	27·544	37·611	49·682	63·891	80·367	120·62	144·63	171·38	233·54	307·95	350·01	496·73	564·47
0·00420	0·746	0·899	0·970	1·038	1·103	1·166	1·286	1·343	1·399	1·506	1·608	1·658	1·801	1·857
1/ 238	13·182	28·255	38·580	50·959	65·530	82·426	123·70	148·32	175·75	239·48	315·78	358·90	509·32	578·77
0·00440	0·764	0·921	0·994	1·064	1·130	1·195	1·317	1·376	1·432	1·542	1·647	1·698	1·845	1·901
1/ 227	13·508	28·949	39·525	52·206	67·131	84·436	126·71	151·92	180·01	245·29	323·42	367·58	521·62	592·73
0·00460	0·782	0·943	1·017	1·088	1·157	1·222	1·348	1·407	1·466	1·578	1·685	1·737	1·887	1·945
1/ 217	13·826	29·628	40·450	53·425	68·696	86·402	129·65	155·45	184·18	250·96	330·89	376·06	533·64	606·38
0·00480	0·800	0·964	1·040	1·113	1·182	1·250	1·377	1·439	1·498	1·613	1·722	1·776	1·929	1·988
1/ 208	14·138	30·293	41·356	54·618	70·228	88·326	132·53	158·89	188·27	256·52	338·21	384·37	545·41	619·75
0·00500	0·817	0·985	1·062	1·136	1·208	1·276	1·407	1·469	1·530	1·647	1·759	1·813	1·970	2·030
1/ 200	14·444	30·944	42·242	55·787	71·728	90·211	135·35	162·27	192·27	261·96	345·37	392·50	556·93	632·83
0·00525	0·839	1·010	1·090	1·166	1·239	1·309	1·443	1·507	1·569	1·689	1·804	1·859	2·020	2·081
1/ 190	14·818	31·740	43·327	57·216	73·563	92·515	138·80	166·40	197·16	268·61	354·12	402·45	571·02	648·83
0·00550	0·859	1·035	1·116	1·194	1·269	1·341	1·478	1·543	1·607	1·730	1·847	1·904	2·068	2·132
1/ 182	15·183	32·517	44·386	58·612	75·355	94·765	142·17	170·44	201·93	275·10	362·68	412·16	584·78	664·45
0·00575	0·879	1·059	1·142	1·222	1·298	1·372	1·512	1·579	1·644	1·770	1·890	1·948	2·116	2·181
1/ 174	15·540	33·277	45·421	59·977	77·107	96·965	145·46	174·38	206·60	281·45	371·04	421·65	598·23	679·73
0·00600	0·899	1·083	1·168	1·249	1·327	1·402	1·545	1·614	1·680	1·809	1·931	1·991	2·162	2·228
1/ 167	15·889	34·021	46·435	61·313	78·821	99·118	148·68	178·24	211·17	287·67	379·22	430·95	611·39	694·67
0·00625	0·919	1·106	1·193	1·276	1·355	1·432	1·578	1·648	1·716	1·847	1·972	2·033	2·208	2·275
1/ 160	16·232	34·750	47·427	62·621	80·501	101·23	151·84	182·02	215·64	293·75	387·23	440·05	624·28	709·31
0·00650	0·938	1·129	1·217	1·302	1·383	1·461	1·610	1·682	1·751	1·885	2·012	2·074	2·253	2·321
1/ 154	16·567	35·465	48·401	63·904	82·148	103·30	154·93	185·73	220·03	299·72	395·08	448·97	636·91	723·66
0·00675	0·956	1·151	1·241	1·327	1·410	1·490	1·642	1·715	1·785	1·921	2·051	2·114	2·296	2·367
1/ 148	16·897	36·166	49·356	65·163	83·763	105·32	157·97	189·36	224·33	305·57	402·79	457·72	649·31	737·74
0·00700	0·974	1·173	1·265	1·353	1·437	1·518	1·673	1·747	1·819	1·957	2·090	2·154	2·340	2·411
1/ 143	17·220	36·855	50·293	66·399	85·349	107·32	160·95	192·93	228·56	311·32	410·36	466·31	661·48	751·56
0·00725	0·992	1·195	1·288	1·377	1·463	1·546	1·703	1·779	1·852	1·993	2·128	2·193	2·382	2·455
1/ 138	17·538	37·531	51·214	67·613	86·907	109·27	163·88	196·44	232·71	316·97	417·79	474·75	673·44	765·14
0·00750	1·010	1·216	1·311	1·402	1·489	1·573	1·733	1·810	1·884	2·028	2·165	2·231	2·423	2·497
1/ 133	17·850	38·196	52·120	68·806	88·439	111·20	166·76	199·89	236·79	322·52	425·09	483·05	685·19	778·48
0·00800	1·045	1·257	1·355	1·449	1·539	1·626	1·792	1·871	1·948	2·096	2·238	2·306	2·504	2·581
1/ 125	18·460	39·494	53·887	71·135	91·429	114·95	172·38	206·61	244·75	333·34	439·35	499·24	708·12	804·52
0·00850	1·078	1·297	1·398	1·495	1·588	1·678	1·848	1·930	2·009	2·162	2·308	2·379	2·583	2·662
1/ 118	19·051	40·752	55·601	73·393	94·327	118·59	177·82	213·14	252·47	343·84	453·17	514·94	730·35	829·77
0·00900	1·111	1·336	1·440	1·540	1·635	1·728	1·903	1·987	2·069	2·226	2·376	2·449	2·659	2·741
1/ 111	19·625	41·974	57·265	75·586	97·141	122·12	183·11	219·47	259·97	354·04	466·59	530·18	751·94	854·28
0·00950	1·142	1·374	1·481	1·583	1·682	1·776	1·957	2·043	2·127	2·288	2·443	2·518	2·734	2·817
1/ 105	20·184	43·163	58·884	77·720	99·879	125·56	188·25	225·63	267·26	363·95	479·64	545·00	772·94	878·13
0·01000	1·173	1·411	1·521	1·626	1·726	1·824	2·009	2·097	2·183	2·349	2·508	2·584	2·806	2·892
1/ 100	20·728	44·321	60·461	79·798	102·55	128·91	193·26	231·63	274·36	373·61	492·36	559·44	793·39	901·35
	0·82	0·84	0·84	0·85	0·86	0·86	0·87	0·88	0·88	0·89	0·89	0·90	0·90	0·91

$V_{r(0·5)medial}$ **for half-full circular pipes.**

A16
(p.5 of 6)

k_s = 0·150 mm
S = 0·01000 to 0·03000

ie hydraulic gradient =
1 in 100 to 1 in 33·3

Water (or sewage) at 15°C;
full bore conditions.

velocities in ms^{-1}
discharges in litres/sec

Gradient **(Equivalent) Pipe diameters in mm**

Gradient	150	200	225	250	275	300	350	375	400	450	500	525	600	630
0·01000	1·173	1·411	1·521	1·626	1·726	1·824	2·009	2·097	2·183	2·349	2·508	2·584	2·806	2·892
1/ 100	20·728	44·321	60·461	79·798	102·55	128·91	193·26	231·63	274·36	373·61	492·36	559·44	793·39	901·35
0·01050	1·203	1·447	1·559	1·667	1·770	1·870	2·060	2·150	2·238	2·408	2·571	2·649	2·877	2·964
1/ 95	21·259	45·450	61·999	81·825	105·15	132·18	198·15	237·48	281·29	383·03	504·76	573·53	813·34	924·01
0·01100	1·232	1·482	1·597	1·707	1·813	1·915	2·109	2·202	2·292	2·466	2·632	2·713	2·946	3·035
1/ 91	21·777	46·554	63·501	83·805	107·69	135·37	202·93	243·20	288·06	392·24	516·87	587·28	832·82	946·13
0·01150	1·261	1·516	1·634	1·747	1·855	1·959	2·158	2·253	2·345	2·523	2·693	2·775	3·013	3·105
1/ 87	22·284	47·633	64·970	85·741	110·17	138·49	207·59	248·79	294·67	401·23	528·71	600·73	851·87	967·75
0·01200	1·289	1·550	1·670	1·785	1·896	2·002	2·205	2·302	2·396	2·578	2·752	2·836	3·079	3·172
1/ 83	22·781	48·688	66·408	87·635	112·60	141·54	212·16	254·26	301·14	410·03	540·30	613·89	870·50	988·92
0·01250	1·317	1·583	1·706	1·823	1·936	2·045	2·252	2·351	2·447	2·632	2·810	2·895	3·143	3·239
1/ 80	23·267	49·722	67·816	89·491	114·98	144·53	216·63	259·61	307·48	418·65	551·65	626·78	888·76	1009·6
0·01300	1·344	1·615	1·740	1·860	1·975	2·086	2·297	2·398	2·496	2·685	2·866	2·954	3·207	3·304
1/ 77	23·743	50·736	69·196	91·309	117·32	147·46	221·02	264·86	313·70	427·11	562·77	639·41	906·65	1030·0
0·01350	1·370	1·647	1·774	1·896	2·014	2·127	2·342	2·445	2·545	2·738	2·922	3·011	3·269	3·368
1/ 74	24·211	51·731	70·550	93·094	119·61	150·33	225·32	270·02	319·79	435·40	573·68	651·80	924·20	1049·9
0·01400	1·396	1·678	1·808	1·932	2·052	2·167	2·386	2·491	2·592	2·789	2·976	3·067	3·330	3·431
1/ 71	24·670	52·707	71·880	94·846	121·86	153·16	229·54	275·07	325·78	443·53	584·40	663·97	941·43	1069·5
0·01450	1·422	1·708	1·841	1·967	2·089	2·206	2·429	2·536	2·639	2·839	3·030	3·122	3·389	3·492
1/ 69	25·121	53·667	73·186	96·567	124·07	155·93	233·69	280·04	331·66	451·53	594·92	675·92	958·36	1088·7
0·01500	1·447	1·738	1·873	2·002	2·125	2·245	2·471	2·580	2·685	2·888	3·083	3·177	3·448	3·553
1/ 67	25·564	54·609	74·470	98·258	126·24	158·65	237·77	284·92	337·44	459·39	605·27	687·67	974·99	1107·6
0·01600	1·496	1·797	1·936	2·069	2·197	2·320	2·554	2·666	2·775	2·985	3·185	3·283	3·563	3·671
1/ 63	26·429	56·450	76·975	101·56	130·47	163·97	245·73	294·45	348·71	474·72	625·45	710·59	1007·5	1144·4
0·01700	1·543	1·854	1·997	2·134	2·266	2·393	2·634	2·750	2·862	3·078	3·285	3·385	3·674	3·786
1/ 59	27·268	58·234	79·404	104·76	134·58	169·13	253·44	303·69	359·64	489·59	645·02	732·81	1038·9	1180·2
0·01800	1·589	1·909	2·056	2·197	2·333	2·464	2·712	2·831	2·946	3·169	3·382	3·485	3·783	3·897
1/ 56	28·082	59·966	81·762	107·87	138·57	174·14	260·93	312·66	370·26	504·02	664·02	754·39	1069·5	1214·9
0·01900	1·634	1·962	2·114	2·259	2·398	2·532	2·788	2·910	3·029	3·257	3·476	3·582	3·888	4·005
1/ 53	28·875	61·651	84·057	110·89	142·45	179·01	268·21	321·38	380·58	518·06	682·49	775·37	1099·2	1248·6
0·02000	1·678	2·015	2·170	2·319	2·462	2·600	2·862	2·987	3·109	3·343	3·568	3·676	3·990	4·111
1/ 50	29·647	63·293	86·291	113·83	146·22	183·75	275·31	329·88	390·64	531·74	700·49	795·81	1128·2	1281·5
0·02100	1·720	2·066	2·225	2·378	2·524	2·665	2·933	3·062	3·187	3·427	3·657	3·768	4·090	4·214
1/ 47·6	30·400	64·894	88·471	116·71	149·91	188·38	282·23	338·17	400·45	545·08	718·05	815·75	1156·4	1313·5
0·02200	1·762	2·115	2·279	2·435	2·584	2·729	3·004	3·135	3·263	3·509	3·744	3·858	4·187	4·314
1/ 45·5	31·135	66·458	90·600	119·51	153·51	192·89	288·99	346·26	410·03	558·10	735·19	835·22	1184·0	1344·9
0·02300	1·803	2·164	2·331	2·491	2·644	2·791	3·072	3·207	3·337	3·589	3·830	3·946	4·283	4·412
1/ 43·5	31·854	67·986	92·681	122·25	157·03	197·31	295·60	354·17	419·39	570·83	751·95	854·25	1210·9	1375·5
0·02400	1·842	2·212	2·382	2·545	2·702	2·853	3·140	3·277	3·410	3·668	3·913	4·032	4·376	4·508
1/ 41·7	32·557	69·482	94·718	124·94	160·47	201·63	302·07	361·91	428·56	583·29	768·35	872·88	1237·3	1405·4
0·02500	1·881	2·258	2·432	2·599	2·758	2·912	3·205	3·345	3·482	3·744	3·995	4·116	4·467	4·603
1/ 40·0	33·247	70·947	96·712	127·56	163·84	205·87	308·40	369·50	437·53	595·50	784·42	891·12	1263·1	1434·7
0·02600	1·920	2·304	2·482	2·651	2·814	2·971	3·270	3·413	3·552	3·819	4·075	4·199	4·557	4·695
1/ 38·5	33·922	72·384	98·668	130·14	167·15	210·02	314·61	376·93	446·33	607·46	800·16	909·00	1288·4	1463·5
0·02700	1·957	2·349	2·530	2·703	2·869	3·029	3·333	3·479	3·620	3·893	4·154	4·280	4·645	4·785
1/ 37·0	34·585	73·793	100·59	132·67	170·39	214·09	320·70	384·22	454·96	619·19	815·60	926·54	1313·3	1491·7
0·02800	1·994	2·393	2·577	2·753	2·922	3·085	3·395	3·544	3·688	3·966	4·231	4·360	4·731	4·874
1/ 35·7	35·235	75·176	102·47	135·15	173·57	218·08	326·67	391·38	463·43	630·71	830·77	943·75	1337·7	1519·4
0·02900	2·030	2·436	2·624	2·803	2·975	3·141	3·456	3·607	3·754	4·037	4·307	4·438	4·816	4·961
1/ 34·5	35·875	76·535	104·32	137·59	176·70	222·01	332·55	398·41	471·75	642·03	845·66	960·67	1361·6	1546·6
0·03000	2·066	2·479	2·669	2·852	3·027	3·195	3·516	3·670	3·819	4·107	4·381	4·515	4·899	5·047
1/ 33·3	36·503	77·871	106·14	139·98	179·77	225·87	338·32	405·32	479·93	653·15	860·30	977·29	1385·1	1573·3
	0·83	0·85	0·85	0·86	0·86	0·87	0·88	0·88	0·89	0·89	0·90	0·90	0·91	0·91

$V_{r(0\cdot5)medial}$ for half-full circular pipes.

k_s = 0·150 mm S = 0·01000 to 0·03000

Gradient	(Equivalent) Pipe diameters in mm													
	150	**200**	**225**	**250**	**275**	**300**	**350**	**375**	**400**	**450**	**500**	**525**	**600**	**630**
0·03000	2·066	2·479	2·669	2·852	3·027	3·195	3·516	3·670	3·819	4·107	4·381	4·515	4·899	5·047
1/ 33·3	36·503	77·871	106·14	139·98	179·77	225·87	338·32	405·32	479·93	653·15	860·30	977·29	1385·1	1573·3
0·03200	2·135	2·562	2·759	2·947	3·128	3·302	3·633	3·792	3·946	4·243	4·527	4·664	5·061	5·214
1/ 31·3	37·729	80·477	109·69	144·66	185·77	233·40	349·58	418·81	495·89	674·85	888·86	1009·7	1431·1	1625·5
0·03400	2·202	2·642	2·845	3·039	3·226	3·405	3·747	3·910	4·069	4·375	4·668	4·810	5·219	5·377
1/ 29·4	38·917	83·004	113·12	149·19	191·59	240·70	360·50	431·88	511·36	695·89	916·55	1041·2	1475·6	1676·0
0·03600	2·268	2·720	2·929	3·129	3·321	3·505	3·857	4·025	4·189	4·504	4·805	4·951	5·372	5·534
1/ 27·8	40·071	85·457	116·46	153·59	197·23	247·78	371·10	444·58	526·38	716·31	943·43	1071·7	1518·8	1725·1
0·03800	2·331	2·796	3·011	3·216	3·413	3·603	3·964	4·137	4·305	4·629	4·938	5·088	5·521	5·687
1/ 26·3	41·194	87·843	119·71	157·87	202·72	254·68	381·41	456·92	541·00	736·18	969·58	1101·4	1560·9	1772·9
0·04000	2·393	2·870	3·090	3·301	3·503	3·698	4·069	4·246	4·418	4·750	5·068	5·221	5·665	5·836
1/ 25·0	42·287	90·168	122·88	162·04	208·07	261·39	391·46	468·95	555·23	755·53	995·05	1130·3	1601·9	1819·4
0·04200	2·453	2·942	3·168	3·384	3·591	3·791	4·171	4·352	4·529	4·869	5·194	5·352	5·807	5·982
1/ 23·8	43·354	92·435	125·96	166·10	213·29	267·94	401·25	480·68	569·11	774·41	1019·9	1158·5	1641·8	1864·7
0·04400	2·512	3·013	3·244	3·465	3·677	3·881	4·270	4·456	4·637	4·985	5·318	5·479	5·945	6·124
1/ 22·7	44·396	94·649	128·98	170·07	218·38	274·34	410·82	492·13	582·66	792·84	1044·2	1186·1	1680·8	1909·0
0·04600	2·570	3·082	3·318	3·544	3·761	3·969	4·367	4·557	4·742	5·098	5·439	5·603	6·080	6·263
1/ 21·7	45·414	96·814	131·92	173·95	223·36	280·59	420·17	503·33	595·91	810·85	1067·9	1213·0	1718·9	1952·3
0·04800	2·626	3·149	3·390	3·621	3·843	4·056	4·462	4·656	4·845	5·209	5·557	5·725	6·211	6·399
1/ 20·8	46·410	98·932	134·81	177·75	228·23	286·70	429·32	514·28	608·88	828·48	1091·1	1239·4	1756·3	1994·7
0·05000	2·682	3·215	3·461	3·697	3·923	4·141	4·555	4·754	4·946	5·318	5·672	5·844	6·341	6·532
1/ 20·0	47·386	101·01	137·63	181·47	233·01	292·70	438·28	525·01	621·58	845·75	1113·8	1265·2	1792·8	2036·2
0·05250	2·749	3·296	3·548	3·790	4·021	4·244	4·669	4·872	5·070	5·450	5·814	5·990	6·499	6·695
1/ 19·0	48·579	103·54	141·08	186·02	238·84	300·02	449·23	538·13	637·10	866·85	1141·6	1296·7	1837·5	2086·9
0·05500	2·815	3·375	3·633	3·880	4·117	4·346	4·780	4·988	5·190	5·580	5·952	6·132	6·653	6·854
1/ 18·2	49·744	106·02	144·45	190·46	244·54	307·17	459·93	550·94	652·26	887·46	1168·7	1327·5	1881·1	2136·4
0·05750	2·879	3·452	3·716	3·968	4·211	4·444	4·889	5·102	5·308	5·707	6·087	6·271	6·804	7·009
1/ 17·4	50·883	108·44	147·75	194·80	250·11	314·16	470·39	563·46	667·07	907·61	1195·2	1357·6	1923·7	2184·8
0·06000	2·942	3·527	3·797	4·055	4·303	4·541	4·995	5·213	5·424	5·831	6·219	6·408	6·951	7·161
1/ 16·7	51·997	110·81	150·97	199·04	255·55	321·00	480·62	575·71	681·57	927·32	1221·1	1387·1	1965·4	2232·2
0·06250	3·004	3·601	3·876	4·140	4·392	4·636	5·100	5·321	5·537	5·952	6·349	6·541	7·096	7·310
1/ 16·0	53·089	113·13	154·13	203·20	260·89	327·70	490·64	587·71	695·77	946·63	1246·6	1415·9	2006·3	2278·6
0·06500	3·065	3·673	3·954	4·223	4·481	4·729	5·202	5·428	5·648	6·071	6·476	6·671	7·237	7·456
1/ 15·4	54·158	115·40	157·22	207·28	266·12	334·27	500·46	599·47	709·69	965·55	1271·5	1444·2	2046·3	2324·1
0·06750	3·124	3·744	4·031	4·304	4·567	4·820	5·302	5·532	5·756	6·188	6·600	6·800	7·376	7·599
1/ 14·8	55·208	117·63	160·26	211·28	271·25	340·71	510·10	611·00	723·34	984·11	1295·9	1472·0	2085·6	2368·7
0·07000	3·182	3·814	4·106	4·384	4·652	4·910	5·400	5·635	5·863	6·302	6·722	6·925	7·513	7·739
1/ 14·3	56·238	119·82	163·24	215·21	276·29	347·03	519·56	622·33	736·75	1002·3	1319·9	1499·2	2124·2	2412·5
0·07250	3·240	3·882	4·179	4·463	4·735	4·997	5·497	5·735	5·968	6·415	6·842	7·049	7·647	7·877
1/ 13·8	57·250	121·97	166·17	219·06	281·24	353·25	528·85	633·45	749·91	1020·2	1343·4	1525·9	2162·1	2455·5
0·07500	3·296	3·950	4·252	4·540	4·817	5·084	5·592	5·834	6·071	6·525	6·960	7·171	7·779	8·013
1/ 13·3	58·245	124·09	169·04	222·86	286·10	359·35	537·98	644·39	762·85	1037·8	1366·6	1552·2	2199·3	2497·8
0·08000	3·406	4·081	4·393	4·691	4·977	5·252	5·777	6·028	6·272	6·741	7·190	7·408	8·036	8·278
1/ 12·5	60·186	128·21	174·66	230·25	295·60	371·27	555·80	665·73	788·10	1072·2	1411·8	1603·6	2272·0	2580·3
0·08500	3·512	4·208	4·530	4·837	5·132	5·416	5·956	6·215	6·466	6·951	7·413	7·637	8·285	8·534
1/ 11·8	62·068	132·21	180·10	237·42	304·79	382·81	573·07	686·41	812·58	1105·4	1455·6	1653·3	2342·4	2660·3
0·09000	3·616	4·332	4·663	4·979	5·282	5·574	6·131	6·397	6·655	7·154	7·630	7·860	8·526	8·783
1/ 11·1	63·895	136·09	185·39	244·39	313·73	394·03	589·84	706·49	836·34	1137·8	1498·1	1701·6	2410·8	2737·9
0·09500	3·716	4·452	4·792	5·117	5·428	5·729	6·300	6·573	6·839	7·351	7·840	8·077	8·762	9·025
1/ 10·5	65·672	139·87	190·53	251·16	322·42	404·93	606·16	726·02	859·45	1169·2	1539·5	1748·5	2477·3	2813·4
0·10000	3·814	4·569	4·918	5·251	5·571	5·879	6·465	6·746	7·018	7·544	8·046	8·289	8·991	9·261
1/ 10·0	67·403	143·55	195·54	257·75	330·88	415·56	622·05	745·04	881·97	1199·8	1579·7	1794·3	2542·1	2887·0
	0·83	0·85	0·86	0·86	0·87	0·87	0·88	0·89	0·89	0·90	0·90	0·90	0·91	0·91

$V_{r(0\cdot5)medial}$ for half-full circular pipes.

A17
(p.1 of 6)

k_s = 0·30 mm
S = 0·00010 to 0·00030

ie hydraulic gradient =
1 in 10000 to 1 in 3333

Water (or sewage) at 15°C;
full bore conditions.

velocities in ms^{-1}
discharges in litres/sec

Gradient **(Equivalent) Pipe diameters in mm**

Gradient	150	200	225	250	275	300	350	375	400	450	500	525	600	630
0·000100	0·095	0·117	0·127	0·136	0·145	0·154	0·171	0·179	0·187	0·202	0·217	0·224	0·244	0·252
1/ 10000	1·6864	3·6685	5·0372	6·6855	8·6333	10·900	16·460	19·790	23·508	32·173	42·582	48·478	69·106	78·652
0·000105	0·098	0·120	0·130	0·140	0·149	0·158	0·176	0·184	0·192	0·208	0·223	0·230	0·251	0·259
1/ 9524	1·7325	3·7677	5·1729	6·8651	8·8645	11·191	16·898	20·315	24·130	33·023	43·704	49·754	70·919	80·713
0·000110	0·101	0·123	0·133	0·143	0·153	0·162	0·180	0·189	0·197	0·213	0·228	0·236	0·257	0·265
1/ 9091	1·7775	3·8648	5·3056	7·0406	9·0905	11·475	17·326	20·829	24·739	33·854	44·801	51·001	72·690	82·727
0·000115	0·103	0·126	0·137	0·147	0·157	0·166	0·184	0·193	0·202	0·218	0·234	0·241	0·263	0·272
1/ 8696	1·8216	3·9597	5·4355	7·2124	9·3117	11·754	17·745	21·331	25·335	34·666	45·873	52·220	74·423	84·697
0·000120	0·106	0·129	0·140	0·150	0·160	0·170	0·189	0·198	0·206	0·223	0·239	0·247	0·269	0·278
1/ 8333	1·8648	4·0527	5·5627	7·3806	9·5283	12·027	18·155	21·823	25·918	35·462	46·923	53·415	76·119	86·625
0·000125	0·108	0·132	0·143	0·154	0·164	0·174	0·193	0·202	0·211	0·228	0·244	0·252	0·275	0·284
1/ 8000	1·9072	4·1439	5·6874	7·5455	9·7406	12·294	18·556	22·305	26·490	36·242	47·953	54·585	77·782	88·515
0·000130	0·110	0·135	0·146	0·157	0·168	0·178	0·197	0·206	0·215	0·233	0·249	0·257	0·281	0·290
1/ 7692	1·9488	4·2334	5·8097	7·7073	9·9489	12·556	18·951	22·778	27·051	37·007	48·962	55·732	79·412	90·369
0·000135	0·113	0·138	0·149	0·160	0·171	0·181	0·201	0·210	0·220	0·237	0·254	0·263	0·287	0·296
1/ 7407	1·9896	4·3212	5·9298	7·8662	10·153	12·814	19·338	23·242	27·601	37·757	49·953	56·859	81·012	92·187
0·000140	0·115	0·140	0·152	0·163	0·174	0·185	0·205	0·215	0·224	0·242	0·259	0·268	0·292	0·301
1/ 7143	2·0297	4·4075	6·0478	8·0223	10·354	13·066	19·718	23·698	28·142	38·495	50·926	57·965	82·584	93·974
0·000145	0·117	0·143	0·155	0·167	0·178	0·188	0·209	0·219	0·228	0·247	0·264	0·273	0·298	0·307
1/ 6897	2·0692	4·4924	6·1639	8·1757	10·552	13·315	20·091	24·146	28·673	39·220	51·882	59·052	84·128	95·729
0·000150	0·119	0·146	0·158	0·170	0·181	0·192	0·213	0·223	0·232	0·251	0·269	0·278	0·303	0·313
1/ 6667	2·1080	4·5759	6·2780	8·3266	10·746	13·560	20·459	24·587	29·195	39·932	52·823	60·122	85·647	97·455
0·000160	0·124	0·151	0·164	0·176	0·187	0·199	0·220	0·230	0·240	0·260	0·278	0·287	0·313	0·323
1/ 6250	2·1838	4·7390	6·5009	8·6213	11·125	14·037	21·176	25·448	30·216	41·324	54·659	62·209	88·611	100·83
0·000170	0·128	0·156	0·169	0·181	0·194	0·205	0·227	0·238	0·248	0·268	0·287	0·297	0·324	0·334
1/ 5882	2·2575	4·8972	6·7173	8·9074	11·493	14·500	21·872	26·283	31·206	42·674	56·440	64·234	91·487	104·09
0·000180	0·132	0·161	0·174	0·187	0·200	0·212	0·234	0·245	0·256	0·277	0·296	0·306	0·333	0·344
1/ 5556	2·3291	5·0511	6·9276	9·1854	11·851	14·951	22·549	27·094	32·168	43·985	58·170	66·201	94·281	107·27
0·000190	0·136	0·166	0·179	0·193	0·205	0·218	0·241	0·252	0·263	0·285	0·305	0·315	0·343	0·354
1/ 5263	2·3988	5·2010	7·1324	9·4561	12·200	15·389	23·207	27·884	33·104	45·262	59·855	68·116	97·000	110·36
0·000200	0·140	0·170	0·184	0·198	0·211	0·224	0·248	0·259	0·271	0·292	0·313	0·323	0·352	0·364
1/ 5000	2·4668	5·3471	7·3321	9·7201	12·539	15·817	23·849	28·654	34·017	46·507	61·496	69·982	99·650	113·37
0·000210	0·143	0·175	0·189	0·203	0·217	0·230	0·254	0·266	0·278	0·300	0·321	0·332	0·362	0·373
1/ 4762	2·5332	5·4898	7·5270	9·9777	12·871	16·234	24·476	29·406	34·908	47·721	63·098	71·803	102·24	116·31
0·000220	0·147	0·179	0·194	0·208	0·222	0·235	0·261	0·273	0·285	0·308	0·329	0·340	0·371	0·382
1/ 4545	2·5982	5·6292	7·7175	10·230	13·195	16·641	25·088	30·140	35·778	48·907	64·664	73·583	104·76	119·18
0·000230	0·151	0·184	0·199	0·213	0·227	0·241	0·267	0·279	0·291	0·315	0·337	0·348	0·379	0·391
1/ 4348	2·6617	5·7656	7·9039	10·476	13·512	17·040	25·687	30·858	36·629	50·068	66·194	75·322	107·23	121·99
0·000240	0·154	0·188	0·203	0·218	0·233	0·247	0·273	0·286	0·298	0·322	0·345	0·356	0·388	0·400
1/ 4167	2·7239	5·8992	8·0865	10·717	13·822	17·431	26·273	31·561	37·462	51·204	67·693	77·026	109·65	124·74
0·000250	0·158	0·192	0·208	0·223	0·238	0·252	0·279	0·292	0·305	0·329	0·352	0·364	0·396	0·409
1/ 4000	2·7849	6·0302	8·2653	10·953	14·126	17·813	26·848	32·250	38·279	52·317	69·161	78·695	112·02	127·43
0·000260	0·161	0·196	0·212	0·228	0·243	0·257	0·285	0·298	0·311	0·336	0·360	0·371	0·404	0·417
1/ 3846	2·8448	6·1586	8·4408	11·185	14·424	18·189	27·411	32·926	39·080	53·409	70·600	80·331	114·34	130·07
0·000270	0·164	0·200	0·217	0·233	0·248	0·263	0·291	0·304	0·317	0·343	0·367	0·379	0·412	0·426
1/ 3704	2·9035	6·2847	8·6130	11·413	14·717	18·557	27·964	33·589	39·866	54·480	72·013	81·937	116·62	132·66
0·000280	0·168	0·204	0·221	0·237	0·253	0·268	0·296	0·310	0·323	0·349	0·374	0·386	0·420	0·434
1/ 3571	2·9613	6·4086	8·7822	11·636	15·005	18·919	28·507	34·240	40·638	55·532	73·401	83·514	118·86	135·20
0·000290	0·171	0·208	0·225	0·242	0·257	0·273	0·302	0·316	0·329	0·356	0·381	0·393	0·428	0·442
1/ 3448	3·0180	6·5303	8·9485	11·856	15·287	19·274	29·041	34·880	41·396	56·565	74·764	85·063	121·06	137·70
0·000300	0·174	0·212	0·229	0·246	0·262	0·278	0·307	0·322	0·335	0·362	0·388	0·400	0·436	0·450
1/ 3333	3·0738	6·6500	9·1120	12·072	15·565	19·624	29·566	35·510	42·142	57·582	76·104	86·587	123·22	140·16
	0·77	0·78	0·79	0·79	0·80	0·80	0·81	0·82	0·82	0·83	0·84	0·84	0·85	0·85

$V_{r(0{\cdot}5)medial}$ for half-full circular pipes.

k_s = 0·30 mm S = 0·00010 to 0·00030

$k_s = 0{\cdot}30$ **mm**
S = 0·00030 to 0·00100

ie hydraulic gradient =
1 in 3333 to 1 in 1000

Water (or sewage) at 15°C;
full bore conditions.

velocities in ms^{-1}
discharges in litres/sec

Gradient	(Equivalent) Pipe diameters in mm													
	150	**200**	**225**	**250**	**275**	**300**	**350**	**375**	**400**	**450**	**500**	**525**	**600**	**630**
0·000300	0·174	0·212	0·229	0·246	0·262	0·278	0·307	0·322	0·335	0·362	0·388	0·400	0·436	0·450
1/ 3333	3·0738	6·6500	9·1120	12·072	15·565	19·624	29·566	35·510	42·142	57·582	76·104	86·587	123·22	140·16
0·000320	0·180	0·219	0·237	0·255	0·271	0·287	0·318	0·333	0·347	0·375	0·401	0·414	0·451	0·465
1/ 3125	3·1828	6·8838	9·4314	12·494	16·108	20·306	30·590	36·738	43·598	59·566	78·721	89·561	127·44	144·96
0·000340	0·186	0·226	0·245	0·263	0·280	0·297	0·328	0·343	0·358	0·387	0·414	0·427	0·465	0·480
1/ 2941	3·2886	7·1107	9·7411	12·903	16·634	20·968	31·584	37·930	45·009	61·490	81·258	92·445	131·54	149·61
0·000360	0·192	0·233	0·253	0·271	0·289	0·306	0·338	0·354	0·369	0·398	0·426	0·440	0·479	0·494
1/ 2778	3·3914	7·3312	10·042	13·301	17·145	21·612	32·549	39·087	46·381	63·359	83·723	95·246	135·51	154·13
0·000380	0·198	0·240	0·260	0·279	0·297	0·315	0·348	0·364	0·380	0·410	0·439	0·453	0·493	0·509
1/ 2632	3·4916	7·5458	10·335	13·688	17·643	22·238	33·489	40·214	47·716	65·178	86·121	97·972	139·38	158·52
0·000400	0·203	0·247	0·267	0·287	0·305	0·323	0·358	0·374	0·390	0·421	0·451	0·465	0·506	0·522
1/ 2500	3·5892	7·7550	10·621	14·065	18·128	22·848	34·404	41·312	49·017	66·951	88·458	100·63	143·15	162·80
0·000420	0·208	0·253	0·274	0·294	0·313	0·332	0·367	0·384	0·400	0·432	0·462	0·477	0·519	0·536
1/ 2381	3·6844	7·9592	10·900	14·433	18·602	23·443	35·298	42·383	50·286	68·680	90·739	103·22	146·83	166·98
0·000440	0·214	0·260	0·281	0·301	0·321	0·340	0·376	0·393	0·410	0·442	0·473	0·489	0·532	0·549
1/ 2273	3·7776	8·1587	11·172	14·793	19·064	24·025	36·171	43·429	51·526	70·369	92·966	105·75	150·42	171·06
0·000460	0·219	0·266	0·288	0·309	0·329	0·348	0·385	0·402	0·420	0·453	0·485	0·500	0·544	0·562
1/ 2174	3·8686	8·3538	11·438	15·145	19·516	24·594	37·024	44·453	52·738	72·021	95·144	108·23	153·93	175·06
0·000480	0·224	0·272	0·294	0·316	0·336	0·356	0·394	0·412	0·429	0·463	0·495	0·511	0·557	0·574
1/ 2083	3·9578	8·5449	11·699	15·489	19·959	25·151	37·860	45·454	53·925	73·638	97·276	110·65	157·37	178·96
0·000500	0·229	0·278	0·301	0·322	0·343	0·364	0·402	0·420	0·438	0·473	0·506	0·522	0·568	0·586
1/ 2000	4·0453	8·7321	11·955	15·827	20·393	25·696	38·678	46·436	55·088	75·222	99·364	113·02	160·73	182·79
0·000525	0·235	0·285	0·309	0·331	0·352	0·373	0·412	0·431	0·450	0·485	0·519	0·536	0·583	0·601
1/ 1905	4·1522	8·9611	12·267	16·240	20·924	26·364	39·679	47·636	56·510	77·159	101·92	115·92	164·85	187·47
0·000550	0·241	0·292	0·316	0·339	0·361	0·382	0·423	0·442	0·461	0·497	0·532	0·549	0·597	0·616
1/ 1818	4·2567	9·1848	12·573	16·643	21·442	27·015	40·657	48·808	57·898	79·051	104·41	118·76	168·87	192·03
0·000575	0·247	0·299	0·324	0·347	0·370	0·391	0·433	0·452	0·472	0·509	0·544	0·561	0·611	0·630
1/ 1739	4·3588	9·4036	12·871	17·037	21·949	27·653	41·613	49·954	59·256	80·901	106·85	121·53	172·80	196·50
0·000600	0·252	0·306	0·331	0·355	0·378	0·400	0·442	0·462	0·482	0·520	0·556	0·574	0·625	0·644
1/ 1667	4·4589	9·6178	13·164	17·423	22·445	28·277	42·549	51·076	60·585	82·711	109·24	124·24	176·65	200·87
0·000625	0·258	0·313	0·338	0·363	0·386	0·409	0·452	0·472	0·492	0·531	0·568	0·586	0·638	0·658
1/ 1600	4·5570	9·8277	13·450	17·801	22·931	28·888	43·466	52·175	61·887	84·485	111·57	126·90	180·42	205·16
0·000650	0·263	0·319	0·345	0·370	0·394	0·417	0·461	0·482	0·503	0·542	0·580	0·598	0·651	0·672
1/ 1538	4·6531	10·034	13·731	18·172	23·408	29·487	44·365	53·253	63·164	86·224	113·87	129·50	184·12	209·35
0·000675	0·269	0·326	0·352	0·378	0·402	0·425	0·470	0·492	0·513	0·553	0·591	0·610	0·664	0·685
1/ 1481	4·7475	10·236	14·007	18·536	23·876	30·076	45·247	54·310	64·417	87·930	116·12	132·06	187·74	213·47
0·000700	0·274	0·332	0·359	0·385	0·410	0·434	0·479	0·501	0·522	0·563	0·603	0·622	0·677	0·698
1/ 1429	4·8402	10·434	14·277	18·893	24·335	30·653	46·113	55·348	65·646	89·605	118·32	134·57	191·30	217·52
0·000725	0·279	0·338	0·366	0·392	0·417	0·442	0·488	0·510	0·532	0·574	0·614	0·633	0·689	0·711
1/ 1379	4·9313	10·629	14·543	19·244	24·786	31·221	46·964	56·368	66·855	91·251	120·49	137·03	194·80	221·49
0·000750	0·284	0·344	0·372	0·399	0·425	0·450	0·497	0·519	0·541	0·584	0·625	0·644	0·701	0·723
1/ 1333	5·0209	10·820	14·805	19·590	25·230	31·778	47·801	57·371	68·043	92·869	122·63	139·46	198·24	225·40
0·000800	0·294	0·356	0·385	0·413	0·439	0·465	0·514	0·537	0·560	0·604	0·646	0·666	0·725	0·748
1/ 1250	5·1958	11·195	15·315	20·264	26·096	32·867	49·434	59·328	70·362	96·027	126·79	144·19	204·95	233·02
0·000850	0·304	0·368	0·398	0·426	0·454	0·480	0·530	0·554	0·578	0·623	0·666	0·687	0·748	0·771
1/ 1176	5·3654	11·558	15·810	20·917	26·936	33·923	51·017	61·226	72·610	99·089	130·82	148·77	211·45	240·41
0·000900	0·313	0·379	0·410	0·439	0·467	0·494	0·546	0·571	0·595	0·642	0·686	0·708	0·770	0·794
1/ 1111	5·5303	11·910	16·291	21·552	27·752	34·949	52·555	63·070	74·793	102·06	134·74	153·23	217·77	247·59
0·000950	0·322	0·390	0·422	0·452	0·481	0·509	0·562	0·587	0·612	0·660	0·706	0·728	0·792	0·817
1/ 1053	5·6907	12·253	16·759	22·170	28·546	35·947	54·051	64·863	76·918	104·96	138·56	157·56	223·91	254·57
0·001000	0·331	0·401	0·433	0·464	0·494	0·522	0·577	0·603	0·629	0·678	0·725	0·747	0·813	0·838
1/ 1000	5·8470	12·587	17·215	22·772	29·319	36·919	55·509	66·610	78·987	107·77	142·27	161·78	229·90	261·37
	0·78	0·79	0·80	0·81	0·81	0·82	0·83	0·83	0·83	0·84	0·85	0·85	0·86	0·86

$V_{r(0{\cdot}5)medial}$ **for half-full circular pipes.**

A17
(p.3 of 6)

k_s = 0·30 mm
S = 0·00100 to 0·00300

ie hydraulic gradient =
1 in 1000 to 1 in 333

Water (or sewage) at 15°C;
full bore conditions.

velocities in ms^{-1}
discharges in litres/sec

Gradient	(Equivalent) Pipe diameters in mm													
	150	**200**	**225**	**250**	**275**	**300**	**350**	**375**	**400**	**450**	**500**	**525**	**600**	**630**
0·00100	0·331	0·401	0·433	0·464	0·494	0·522	0·577	0·603	0·629	0·678	0·725	0·747	0·813	0·838
1/ 1000	5·8470	12·587	17·215	22·772	29·319	36·919	55·509	66·610	78·987	107·77	142·27	161·78	229·90	261·37
0·00105	0·340	0·411	0·444	0·476	0·506	0·536	0·592	0·619	0·645	0·695	0·743	0·766	0·834	0·860
1/ 952	5·9996	12·914	17·660	23·359	30·074	37·868	56·931	68·315	81·006	110·52	145·89	165·90	235·74	268·01
0·00110	0·348	0·421	0·455	0·488	0·519	0·549	0·606	0·634	0·660	0·712	0·761	0·785	0·854	0·881
1/ 909	6·1486	13·232	18·095	23·932	30·811	38·794	58·320	69·980	82·978	113·21	149·43	169·92	241·44	274·49
0·00115	0·356	0·431	0·466	0·499	0·531	0·562	0·620	0·648	0·676	0·728	0·779	0·803	0·874	0·901
1/ 870	6·2944	13·544	18·520	24·493	31·532	39·700	59·678	71·607	84·906	115·84	152·89	173·85	247·02	280·82
0·00120	0·364	0·441	0·476	0·510	0·543	0·574	0·634	0·663	0·691	0·744	0·796	0·821	0·893	0·921
1/ 833	6·4370	13·849	18·936	25·042	32·237	40·587	61·008	73·200	86·793	118·40	156·28	177·70	252·48	287·02
0·00125	0·372	0·450	0·486	0·521	0·554	0·586	0·648	0·677	0·705	0·760	0·813	0·838	0·912	0·940
1/ 800	6·5768	14·147	19·343	25·580	32·928	41·456	62·310	74·761	88·641	120·92	159·60	181·47	257·82	293·09
0·00130	0·380	0·460	0·497	0·532	0·566	0·599	0·661	0·691	0·720	0·776	0·829	0·855	0·930	0·959
1/ 769	6·7139	14·440	19·743	26·107	33·605	42·307	63·586	76·291	90·453	123·39	162·85	185·16	263·06	299·04
0·00135	0·388	0·469	0·506	0·542	0·577	0·610	0·674	0·704	0·734	0·791	0·846	0·872	0·949	0·978
1/ 741	6·8484	14·728	20·135	26·624	34·270	43·143	64·839	77·792	92·231	125·81	166·04	188·78	268·20	304·88
0·00140	0·395	0·478	0·516	0·553	0·588	0·622	0·687	0·718	0·748	0·806	0·862	0·889	0·966	0·996
1/ 714	6·9805	15·010	20·520	27·132	34·923	43·963	66·068	79·265	93·976	128·19	169·17	192·34	273·25	310·62
0·00145	0·402	0·487	0·526	0·563	0·599	0·633	0·699	0·731	0·761	0·821	0·877	0·905	0·984	1·015
1/ 690	7·1102	15·287	20·898	27·631	35·564	44·769	67·276	80·713	95·691	130·52	172·25	195·84	278·20	316·25
0·00150	0·410	0·495	0·535	0·573	0·609	0·645	0·712	0·744	0·775	0·835	0·893	0·921	1·001	1·032
1/ 667	7·2378	15·560	21·270	28·122	36·194	45·561	68·464	82·136	97·377	132·82	175·27	199·28	283·08	321·78
0·00160	0·424	0·512	0·553	0·592	0·630	0·666	0·736	0·769	0·801	0·863	0·923	0·952	1·035	1·067
1/ 625	7·4869	16·092	21·995	29·079	37·424	47·107	70·781	84·913	100·67	137·29	181·17	205·98	292·59	332·59
0·00170	0·437	0·529	0·571	0·611	0·650	0·688	0·759	0·793	0·826	0·891	0·952	0·982	1·067	1·101
1/ 588	7·7284	16·608	22·699	30·008	38·617	48·606	73·028	87·606	103·86	141·64	186·89	212·48	301·80	343·06
0·00180	0·451	0·545	0·588	0·630	0·670	0·708	0·782	0·817	0·851	0·917	0·980	1·011	1·099	1·133
1/ 556	7·9630	17·109	23·382	30·909	39·775	50·062	75·210	90·221	106·95	145·85	192·45	218·79	310·76	353·23
0·00190	0·464	0·560	0·605	0·648	0·689	0·728	0·804	0·840	0·875	0·943	1·008	1·039	1·130	1·165
1/ 526	8·1913	17·596	24·047	31·786	40·902	51·479	77·333	92·764	109·96	149·95	197·85	224·94	319·46	363·12
0·00200	0·476	0·575	0·621	0·665	0·707	0·748	0·825	0·862	0·898	0·968	1·034	1·067	1·160	1·196
1/ 500	8·4137	18·071	24·695	32·641	42·000	52·858	79·400	95·242	112·90	153·95	203·12	230·92	327·95	372·76
0·00210	0·488	0·590	0·637	0·682	0·725	0·767	0·846	0·884	0·921	0·992	1·061	1·094	1·189	1·226
1/ 476	8·6307	18·535	25·327	33·475	43·071	54·204	81·418	97·659	115·76	157·85	208·25	236·75	336·22	382·16
0·00220	0·500	0·604	0·652	0·699	0·743	0·785	0·867	0·906	0·943	1·016	1·086	1·120	1·218	1·255
1/ 455	8·8427	18·987	25·944	34·289	44·117	55·519	83·387	100·02	118·56	161·65	213·27	242·45	344·30	391·34
0·00230	0·512	0·618	0·668	0·715	0·760	0·804	0·887	0·926	0·965	1·040	1·111	1·146	1·246	1·284
1/ 435	9·0499	19·430	26·547	35·085	45·140	56·804	85·313	102·33	121·29	165·38	218·17	248·02	352·20	400·31
0·00240	0·524	0·632	0·683	0·731	0·777	0·821	0·906	0·947	0·986	1·063	1·136	1·171	1·273	1·312
1/ 417	9·2527	19·863	27·138	35·864	46·140	58·062	87·198	104·59	123·97	169·02	222·97	253·47	359·93	409·09
0·00250	0·535	0·646	0·697	0·746	0·793	0·839	0·926	0·967	1·007	1·085	1·160	1·196	1·300	1·340
1/ 400	9·4515	20·287	27·716	36·627	47·121	59·293	89·044	106·80	126·59	172·58	227·67	258·81	367·50	417·69
0·00260	0·546	0·659	0·711	0·761	0·810	0·856	0·944	0·987	1·028	1·107	1·183	1·220	1·326	1·367
1/ 385	9·6463	20·703	28·283	37·375	48·082	60·501	90·854	108·97	129·15	176·08	232·28	264·04	374·92	426·12
0·00270	0·557	0·672	0·725	0·776	0·825	0·873	0·963	1·006	1·048	1·129	1·206	1·243	1·352	1·394
1/ 370	9·8374	21·111	28·839	38·109	49·024	61·686	92·629	111·09	131·67	179·51	236·79	269·18	382·20	434·39
0·00280	0·567	0·685	0·739	0·791	0·841	0·889	0·981	1·025	1·068	1·150	1·229	1·267	1·377	1·420
1/ 357	10·025	21·512	29·385	38·829	49·950	62·849	94·372	113·18	134·15	182·88	241·23	274·22	389·35	442·51
0·00290	0·578	0·697	0·753	0·805	0·856	0·905	0·999	1·043	1·087	1·171	1·251	1·290	1·402	1·445
1/ 345	10·209	21·905	29·922	39·537	50·859	63·992	96·084	115·23	136·58	186·18	245·59	279·17	396·37	450·48
0·00300	0·588	0·710	0·766	0·820	0·871	0·921	1·016	1·062	1·106	1·191	1·273	1·312	1·426	1·470
1/ 333	10·391	22·292	30·450	40·233	51·753	65·115	97·767	117·25	138·96	189·44	249·87	284·04	403·27	458·32
	0·79	0·80	0·81	0·82	0·82	0·83	0·84	0·84	0·84	0·85	0·86	0·86	0·87	0·87

$V_{r(0{\cdot}5)medial}$ for half-full circular pipes.

k_s = 0·30 mm S = 0·00100 to 0·00300

Gradient **(Equivalent) Pipe diameters in mm**

Gradient	150	200	225	250	275	300	350	375	400	450	500	525	600	630
0·00300	0·588	0·710	0·766	0·820	0·871	0·921	1·016	1·062	1·106	1·191	1·273	1·312	1·426	1·470
1/ 333	10·391	22·292	30·450	40·233	51·753	65·115	97·767	117·25	138·96	189·44	249·87	284·04	403·27	458·32
0·00320	0·608	0·734	0·792	0·847	0·901	0·952	1·050	1·097	1·143	1·231	1·315	1·356	1·474	1·519
1/ 313	10·745	23·047	31·479	41·591	53·497	67·307	101·05	121·19	143·63	195·78	258·23	293·53	416·73	473·62
0·00340	0·627	0·757	0·817	0·874	0·929	0·982	1·083	1·132	1·179	1·270	1·356	1·398	1·520	1·567
1/ 294	11·088	23·779	32·477	42·907	55·188	69·432	104·24	125·00	148·14	201·93	266·34	302·74	429·79	488·45
0·00360	0·646	0·780	0·841	0·900	0·957	1·011	1·116	1·165	1·214	1·307	1·397	1·440	1·565	1·613
1/ 278	11·421	24·490	33·446	44·186	56·830	71·495	107·33	128·71	152·53	207·90	274·20	311·68	442·46	502·85
0·00380	0·665	0·802	0·865	0·925	0·984	1·040	1·147	1·198	1·248	1·344	1·435	1·480	1·608	1·658
1/ 263	11·745	25·182	34·389	45·429	58·428	73·502	110·33	132·31	156·80	213·71	281·86	320·38	454·79	516·85
0·00400	0·682	0·823	0·888	0·950	1·010	1·068	1·177	1·230	1·281	1·379	1·473	1·519	1·651	1·702
1/ 250	12·060	25·856	35·307	46·641	59·984	75·458	113·26	135·82	160·95	219·37	289·31	328·85	466·80	530·49
0·00420	0·700	0·844	0·911	0·974	1·035	1·094	1·207	1·261	1·313	1·414	1·510	1·557	1·692	1·744
1/ 238	12·368	26·513	36·203	47·823	61·502	77·365	116·12	139·24	165·01	224·89	296·58	337·11	478·51	543·80
0·00440	0·717	0·864	0·933	0·998	1·060	1·121	1·236	1·291	1·345	1·448	1·547	1·595	1·733	1·786
1/ 227	12·669	27·155	37·078	48·976	62·984	79·227	118·91	142·59	168·97	230·28	303·68	345·17	489·95	556·79
0·00460	0·734	0·884	0·954	1·021	1·085	1·147	1·264	1·321	1·375	1·481	1·582	1·631	1·772	1·827
1/ 217	12·963	27·782	37·933	50·105	64·433	81·048	121·64	145·86	172·84	235·55	310·63	353·06	501·13	569·49
0·00480	0·750	0·904	0·975	1·043	1·109	1·172	1·292	1·350	1·406	1·513	1·617	1·667	1·811	1·867
1/ 208	13·251	28·396	38·770	51·209	65·851	82·830	124·31	149·05	176·63	240·70	317·42	360·78	512·07	581·92
0·00500	0·766	0·923	0·996	1·065	1·132	1·196	1·319	1·378	1·435	1·545	1·650	1·702	1·849	1·906
1/ 200	13·533	28·997	39·590	52·290	67·239	84·575	126·92	152·19	180·34	245·75	324·07	368·33	522·78	594·09
0·00525	0·785	0·946	1·021	1·092	1·161	1·227	1·352	1·413	1·471	1·584	1·692	1·744	1·895	1·954
1/ 190	13·878	29·733	40·592	53·612	68·937	86·708	130·12	156·02	184·87	251·93	332·20	377·57	535·88	608·97
0·00550	0·804	0·969	1·046	1·118	1·189	1·256	1·385	1·446	1·506	1·622	1·732	1·786	1·941	2·000
1/ 182	14·214	30·451	41·571	54·902	70·595	88·791	133·24	159·76	189·30	257·95	340·14	386·59	548·67	623·50
0·00575	0·823	0·992	1·070	1·144	1·216	1·285	1·417	1·480	1·541	1·659	1·772	1·827	1·985	2·046
1/ 174	14·543	31·152	42·527	56·164	72·216	90·828	136·29	163·41	193·63	263·85	347·91	395·41	561·17	637·70
0·00600	0·841	1·013	1·093	1·169	1·243	1·313	1·448	1·512	1·575	1·695	1·811	1·866	2·028	2·090
1/ 167	14·865	31·839	43·464	57·399	73·802	92·820	139·28	166·99	197·86	269·61	355·50	404·04	573·41	651·60
0·00625	0·859	1·035	1·116	1·194	1·269	1·341	1·478	1·544	1·608	1·731	1·848	1·906	2·070	2·134
1/ 160	15·181	32·512	44·381	58·609	75·355	94·772	142·20	170·49	202·01	275·26	362·94	412·50	585·39	665·21
0·00650	0·877	1·056	1·139	1·218	1·294	1·368	1·508	1·575	1·640	1·766	1·886	1·944	2·112	2·177
1/ 154	15·490	33·172	45·280	59·794	76·878	96·686	145·07	173·93	206·08	280·79	370·23	420·78	597·14	678·56
0·00675	0·894	1·076	1·161	1·242	1·319	1·394	1·537	1·605	1·672	1·800	1·922	1·981	2·153	2·219
1/ 148	15·793	33·819	46·161	60·957	78·372	98·562	147·88	177·29	210·07	286·22	377·39	428·91	608·66	691·64
0·00700	0·911	1·097	1·183	1·265	1·344	1·420	1·566	1·635	1·703	1·833	1·958	2·018	2·193	2·260
1/ 143	16·091	34·454	47·027	62·099	79·838	100·41	150·64	180·60	213·98	291·56	384·41	436·89	619·97	704·49
0·00725	0·927	1·117	1·204	1·288	1·368	1·446	1·594	1·665	1·733	1·866	1·993	2·054	2·232	2·300
1/ 138	16·383	35·078	47·878	63·221	81·279	102·22	153·35	183·85	217·83	296·79	391·31	444·72	631·08	717·11
0·00750	0·943	1·136	1·225	1·310	1·392	1·471	1·622	1·694	1·764	1·898	2·027	2·090	2·271	2·340
1/ 133	16·671	35·691	48·713	64·323	82·695	103·99	156·02	187·04	221·61	301·94	398·09	452·43	642·00	729·52
0·00800	0·975	1·174	1·266	1·354	1·439	1·520	1·676	1·750	1·822	1·962	2·095	2·159	2·346	2·418
1/ 125	17·232	36·888	50·345	66·475	85·458	107·47	161·22	193·28	228·99	311·98	411·32	467·46	663·31	753·73
0·00850	1·006	1·211	1·306	1·397	1·484	1·568	1·728	1·805	1·879	2·023	2·160	2·227	2·419	2·493
1/ 118	17·776	38·048	51·926	68·560	88·136	110·83	166·26	199·31	236·14	321·72	424·14	482·03	683·97	777·19
0·00900	1·036	1·247	1·345	1·438	1·528	1·614	1·779	1·858	1·934	2·082	2·224	2·292	2·490	2·566
1/ 111	18·304	39·175	53·461	70·585	90·737	114·10	171·15	205·18	243·09	331·17	436·60	496·18	704·02	799·97
0·00950	1·065	1·282	1·382	1·478	1·570	1·659	1·828	1·909	1·988	2·140	2·285	2·356	2·559	2·637
1/ 105	18·818	40·270	54·955	72·555	93·266	117·28	175·91	210·88	249·84	340·36	448·71	509·94	723·53	822·13
0·01000	1·093	1·316	1·419	1·517	1·612	1·703	1·877	1·960	2·041	2·196	2·345	2·418	2·626	2·707
1/ 100	19·319	41·338	56·410	74·473	95·730	120·37	180·55	216·44	256·42	349·32	460·51	523·34	742·53	843·72
	0·80	0·81	0·82	0·82	0·83	0·83	0·84	0·85	0·85	0·86	0·86	0·87	0·87	0·88

$V_{r(0\cdot5)medial}$ for half-full circular pipes.

$k_s = 0{\cdot}30$ **mm**
S = 0·01000 to 0·03000

ie hydraulic gradient =
1 in 100 to 1 in 33·3

Water (or sewage) at 15°C;
full bore conditions.

velocities in ms^{-1}
discharges in litres/sec

Gradient **(Equivalent) Pipe diameters in mm**

Gradient	150	200	225	250	275	300	350	375	400	450	500	525	600	630
0·01000	1·093	1·316	1·419	1·517	1·612	1·703	1·877	1·960	2·041	2·196	2·345	2·418	2·626	2·707
1/ 100	19·319	41·338	56·410	74·473	95·730	120·37	180·55	216·44	256·42	349·32	460·51	523·34	742·53	843·72
0·01050	1·121	1·349	1·454	1·555	1·652	1·746	1·924	2·009	2·092	2·251	2·404	2·478	2·692	2·774
1/ 95	19·807	42·379	57·829	76·345	98·134	123·39	185·07	221·86	262·84	358·05	472·01	536·41	761·06	864·77
0·01100	1·148	1·381	1·489	1·593	1·692	1·787	1·970	2·057	2·141	2·305	2·461	2·537	2·756	2·840
1/ 91	20·284	43·396	59·214	78·172	100·48	126·34	189·49	227·15	269·11	366·58	483·25	549·18	779·16	885·32
0·01150	1·174	1·413	1·523	1·629	1·730	1·828	2·014	2·104	2·190	2·357	2·517	2·595	2·818	2·905
1/ 87	20·750	44·390	60·569	79·959	102·77	129·22	193·81	232·33	275·23	374·92	494·23	561·65	796·85	905·42
0·01200	1·200	1·444	1·557	1·665	1·768	1·868	2·058	2·149	2·238	2·409	2·572	2·651	2·879	2·968
1/ 83	21·206	45·362	61·894	81·707	105·02	132·04	198·03	237·39	281·23	383·08	504·98	573·86	814·16	925·08
0·01250	1·225	1·474	1·589	1·699	1·805	1·907	2·101	2·194	2·285	2·459	2·625	2·706	2·939	3·029
1/ 80	21·653	46·315	63·192	83·419	107·22	134·81	202·17	242·34	287·10	391·07	515·51	585·82	831·11	944·34
0·01300	1·250	1·504	1·621	1·734	1·841	1·945	2·143	2·238	2·330	2·508	2·678	2·760	2·998	3·090
1/ 77	22·091	47·248	64·465	85·097	109·37	137·51	206·23	247·20	292·85	398·90	525·82	597·54	847·73	963·21
0·01350	1·274	1·533	1·653	1·767	1·877	1·983	2·185	2·281	2·375	2·556	2·730	2·813	3·056	3·149
1/ 74	22·520	48·164	65·713	86·743	111·49	140·17	210·21	251·97	298·50	406·59	535·95	609·04	864·03	981·73
0·01400	1·298	1·562	1·684	1·800	1·912	2·020	2·225	2·324	2·419	2·604	2·780	2·866	3·112	3·208
1/ 71	22·942	49·064	66·939	88·359	113·56	142·78	214·11	256·65	304·04	414·13	545·88	620·33	880·03	999·91
0·01450	1·322	1·590	1·714	1·832	1·946	2·056	2·265	2·365	2·463	2·650	2·830	2·917	3·168	3·265
1/ 69	23·356	49·947	68·143	89·947	115·60	145·34	217·95	261·25	309·48	421·54	555·64	631·42	895·75	1017·8
0·01500	1·345	1·617	1·744	1·864	1·980	2·092	2·305	2·406	2·505	2·696	2·879	2·967	3·223	3·321
1/ 67	23·764	50·815	69·326	91·507	117·61	147·86	221·72	265·77	314·83	428·82	565·23	642·32	911·20	1035·3
0·01600	1·390	1·671	1·802	1·926	2·046	2·161	2·381	2·486	2·588	2·786	2·974	3·065	3·329	3·431
1/ 63	24·558	52·509	71·635	94·552	121·52	152·77	229·08	274·58	325·27	443·03	583·95	663·58	941·35	1069·6
0·01700	1·433	1·724	1·858	1·986	2·110	2·229	2·455	2·563	2·669	2·872	3·066	3·161	3·433	3·538
1/ 59	25·328	54·152	73·873	97·504	125·31	157·53	236·21	283·13	335·39	456·80	602·10	684·19	970·56	1102·7
0·01800	1·476	1·774	1·913	2·045	2·172	2·294	2·527	2·639	2·747	2·956	3·156	3·253	3·533	3·641
1/ 56	26·076	55·746	76·046	100·37	128·99	162·16	243·14	291·43	345·22	470·18	619·71	704·21	998·94	1135·0
0·01900	1·517	1·824	1·966	2·101	2·232	2·358	2·597	2·712	2·823	3·038	3·243	3·343	3·631	3·742
1/ 53	26·804	57·297	78·159	103·16	132·57	166·65	249·88	299·50	354·77	483·18	636·85	723·67	1026·5	1166·3
0·02000	1·557	1·872	2·018	2·157	2·291	2·420	2·665	2·783	2·897	3·118	3·328	3·431	3·726	3·839
1/ 50	27·512	58·808	80·218	105·87	136·05	171·03	256·44	307·36	364·08	495·85	653·54	742·63	1053·4	1196·9
0·02100	1·596	1·919	2·068	2·211	2·348	2·480	2·732	2·852	2·970	3·195	3·411	3·516	3·818	3·935
1/ 47·6	28·204	60·281	82·226	108·52	139·45	175·31	262·84	315·03	373·16	508·21	669·81	761·13	1079·6	1226·6
0·02200	1·634	1·965	2·117	2·263	2·404	2·539	2·797	2·920	3·040	3·271	3·492	3·599	3·909	4·028
1/ 45·5	28·879	61·720	84·187	111·11	142·77	179·48	269·09	322·51	382·03	520·28	685·71	779·18	1105·2	1255·7
0·02300	1·672	2·009	2·166	2·315	2·458	2·597	2·860	2·986	3·109	3·345	3·571	3·681	3·997	4·119
1/ 43·5	29·538	63·127	86·104	113·63	146·02	183·56	275·19	329·83	390·69	532·07	701·25	796·83	1130·2	1284·1
0·02400	1·708	2·053	2·213	2·365	2·512	2·653	2·922	3·051	3·177	3·418	3·649	3·761	4·084	4·209
1/ 41·7	30·184	64·503	87·979	116·11	149·20	187·55	281·17	336·99	399·17	543·61	716·45	814·11	1154·7	1311·9
0·02500	1·744	2·096	2·259	2·415	2·564	2·709	2·983	3·115	3·243	3·489	3·725	3·839	4·169	4·296
1/ 40·0	30·816	65·851	89·816	118·53	152·31	191·46	287·03	344·01	407·48	554·91	731·34	831·02	1178·7	1339·2
0·02600	1·779	2·138	2·304	2·463	2·616	2·763	3·043	3·177	3·307	3·559	3·799	3·915	4·252	4·382
1/ 38·5	31·436	67·172	91·617	120·90	155·36	195·29	292·76	350·88	415·62	565·99	745·93	847·60	1202·2	1365·9
0·02700	1·813	2·179	2·349	2·510	2·666	2·816	3·101	3·238	3·371	3·627	3·872	3·991	4·334	4·466
1/ 37·0	32·044	68·469	93·383	123·23	158·35	199·04	298·39	357·63	423·60	576·86	760·25	863·87	1225·3	1392·1
0·02800	1·847	2·220	2·392	2·557	2·715	2·868	3·159	3·298	3·433	3·694	3·944	4·064	4·414	4·548
1/ 35·7	32·641	69·741	95·117	125·52	161·28	202·73	303·92	364·25	431·44	587·53	774·31	879·83	1247·9	1417·8
0·02900	1·880	2·260	2·435	2·603	2·764	2·919	3·215	3·357	3·495	3·760	4·014	4·137	4·492	4·629
1/ 34·5	33·227	70·991	96·821	127·77	164·17	206·36	309·35	370·75	439·14	598·01	788·11	895·52	1270·1	1443·0
0·03000	1·913	2·299	2·477	2·648	2·812	2·970	3·271	3·415	3·555	3·825	4·083	4·208	4·569	4·709
1/ 33·3	33·804	72·220	98·495	129·97	167·00	209·92	314·68	377·14	446·71	608·31	801·68	910·93	1292·0	1467·9
	0·80	0·81	0·82	0·83	0·83	0·84	0·85	0·85	0·85	0·86	0·87	0·87	0·88	0·88

$V_{r(0{\cdot}5)medial}$ **for half-full circular pipes.**

$k_s = 0{\cdot}30$ **mm** S = 0·01000 to 0·03000

$k_s = 0{\cdot}30$ mm
$S = 0{\cdot}03000$ to $0{\cdot}10000$

ie hydraulic gradient = 1 in 33·3 to 1 in 10·0

Water (or sewage) at 15°C; full bore conditions.

velocities in ms^{-1}
discharges in litres/sec

Gradient	**150**	**200**	**225**	**250**	**275**	**300**	**350**	**375**	**400**	**450**	**500**	**525**	**600**	**630**
0·03000	1·913	2·299	2·477	2·648	2·812	2·970	3·271	3·415	3·555	3·825	4·083	4·208	4·569	4·709
1/ 33·3	33·804	72·220	98·495	129·97	167·00	209·92	314·68	377·14	446·71	608·31	801·68	910·93	1292·0	1467·9
0·03200	1·977	2·375	2·559	2·736	2·905	3·068	3·379	3·528	3·672	3·951	4·218	4·347	4·720	4·864
1/ 31·3	34·928	74·617	101·76	134·28	172·54	216·87	325·09	389·61	461·48	628·41	828·16	941·01	1334·6	1516·3
0·03400	2·038	2·449	2·639	2·821	2·995	3·163	3·484	3·637	3·786	4·074	4·348	4·482	4·866	5·015
1/ 29·4	36·018	76·940	104·93	138·46	177·90	223·61	335·18	401·70	475·79	647·90	853·82	970·17	1376·0	1563·2
0·03600	2·098	2·521	2·716	2·903	3·083	3·256	3·586	3·743	3·897	4·193	4·475	4·612	5·008	5·161
1/ 27·8	37·077	79·197	108·00	142·51	183·10	230·15	344·98	413·44	489·69	666·81	878·74	998·48	1416·1	1608·8
0·03800	2·156	2·591	2·792	2·984	3·168	3·346	3·685	3·847	4·004	4·308	4·599	4·740	5·146	5·303
1/ 26·3	38·106	81·391	110·99	146·45	188·17	236·51	354·51	424·86	503·21	685·21	902·98	1026·0	1455·1	1653·2
0·04000	2·213	2·659	2·865	3·062	3·251	3·434	3·781	3·947	4·109	4·421	4·719	4·864	5·281	5·442
1/ 25·0	39·109	83·528	113·91	150·30	193·10	242·71	363·79	435·98	516·38	703·13	926·58	1052·8	1493·1	1696·3
0·04200	2·268	2·725	2·936	3·138	3·332	3·519	3·875	4·046	4·211	4·531	4·836	4·984	5·412	5·577
1/ 23·8	40·087	85·613	116·75	154·04	197·91	248·75	372·84	446·82	529·22	720·61	949·61	1079·0	1530·2	1738·5
0·04400	2·322	2·790	3·006	3·213	3·411	3·603	3·967	4·141	4·311	4·638	4·951	5·102	5·540	5·709
1/ 22·7	41·042	87·649	119·52	157·70	202·61	254·65	381·68	457·41	541·76	737·68	972·09	1104·5	1566·4	1779·6
0·04600	2·375	2·853	3·074	3·285	3·488	3·684	4·057	4·235	4·409	4·743	5·063	5·218	5·665	5·838
1/ 21·7	41·975	89·639	122·23	161·28	207·20	260·42	390·32	467·76	554·02	754·36	994·06	1129·5	1601·8	1819·8
0·04800	2·427	2·915	3·141	3·357	3·564	3·764	4·145	4·327	4·504	4·846	5·172	5·330	5·788	5·964
1/ 20·8	42·889	91·586	124·89	164·77	211·69	266·07	398·78	477·89	566·01	770·68	1015·6	1153·9	1636·4	1859·1
0·05000	2·478	2·976	3·206	3·427	3·638	3·842	4·231	4·417	4·598	4·946	5·280	5·441	5·908	6·088
1/ 20·0	43·783	93·493	127·48	168·20	216·09	271·59	407·06	487·81	577·76	786·67	1036·6	1177·8	1670·4	1897·7
0·05250	2·540	3·050	3·286	3·512	3·729	3·938	4·336	4·527	4·712	5·069	5·411	5·576	6·054	6·239
1/ 19·0	44·877	95·824	130·66	172·39	221·47	278·35	417·18	499·94	592·12	806·21	1062·4	1207·1	1711·8	1944·7
0·05500	2·600	3·123	3·364	3·595	3·817	4·031	4·439	4·634	4·823	5·189	5·539	5·708	6·197	6·386
1/ 18·2	45·945	98·100	133·76	176·48	226·73	284·95	427·06	511·78	606·14	825·30	1087·5	1235·6	1752·3	1990·7
0·05750	2·659	3·193	3·440	3·677	3·904	4·122	4·539	4·739	4·933	5·306	5·664	5·837	6·337	6·530
1/ 17·4	46·989	100·32	136·79	180·48	231·86	291·40	436·72	523·36	619·84	843·95	1112·1	1263·6	1791·9	2035·7
0·06000	2·717	3·263	3·515	3·756	3·988	4·212	4·637	4·841	5·039	5·421	5·786	5·963	6·474	6·671
1/ 16·7	48·010	102·50	139·76	184·39	236·88	297·71	446·18	534·68	633·25	862·20	1136·1	1290·9	1830·6	2079·7
0·06250	2·773	3·331	3·588	3·834	4·071	4·299	4·734	4·942	5·144	5·534	5·906	6·087	6·608	6·810
1/ 16·0	49·010	104·63	142·67	188·22	241·80	303·89	455·44	545·77	646·39	880·08	1159·7	1317·6	1868·5	2122·7
0·06500	2·829	3·397	3·660	3·911	4·152	4·385	4·828	5·040	5·246	5·644	6·024	6·208	6·740	6·945
1/ 15·4	49·991	106·72	145·51	191·98	246·63	309·95	464·51	556·65	659·26	897·60	1182·7	1343·8	1905·7	2165·0
0·06750	2·883	3·462	3·730	3·986	4·232	4·469	4·921	5·137	5·347	5·752	6·139	6·327	6·869	7·078
1/ 14·8	50·953	108·77	148·31	195·66	251·36	315·90	473·41	567·31	671·89	914·78	1205·4	1369·6	1942·1	2206·4
0·07000	2·937	3·526	3·799	4·060	4·310	4·552	5·011	5·231	5·445	5·858	6·252	6·443	6·995	7·208
1/ 14·3	51·897	110·79	151·05	199·28	256·00	321·73	482·15	577·78	684·29	931·66	1227·6	1394·8	1977·9	2247·0
0·07250	2·989	3·589	3·867	4·132	4·387	4·633	5·101	5·324	5·542	5·962	6·363	6·558	7·120	7·336
1/ 13·8	52·824	112·76	153·74	202·83	260·56	327·46	490·74	588·06	696·46	948·23	1249·4	1419·6	2013·1	2287·0
0·07500	3·041	3·651	3·933	4·203	4·462	4·712	5·188	5·416	5·638	6·065	6·473	6·670	7·242	7·462
1/ 13·3	53·736	114·71	156·39	206·32	265·05	333·09	499·17	598·17	708·43	964·52	1270·9	1444·0	2047·6	2326·2
0·08000	3·142	3·772	4·063	4·342	4·610	4·868	5·359	5·595	5·823	6·264	6·686	6·890	7·481	7·708
1/ 12·5	55·515	118·50	161·56	213·14	273·80	344·09	515·64	617·90	731·79	996·31	1312·8	1491·6	2115·1	2402·8
0·08500	3·239	3·889	4·189	4·476	4·752	5·019	5·525	5·768	6·004	6·458	6·893	7·103	7·712	7·946
1/ 11·8	57·239	122·17	166·57	219·74	282·28	354·74	531·59	637·01	754·42	1027·1	1353·4	1537·7	2180·4	2477·0
0·09000	3·334	4·003	4·312	4·607	4·891	5·165	5·686	5·936	6·178	6·646	7·093	7·310	7·936	8·177
1/ 11·1	58·913	125·74	171·43	226·15	290·51	365·08	547·09	655·58	776·40	1057·0	1392·8	1582·4	2243·9	2549·1
0·09500	3·426	4·113	4·430	4·734	5·026	5·307	5·843	6·099	6·349	6·829	7·288	7·511	8·154	8·402
1/ 10·5	60·542	129·21	176·16	232·39	298·52	375·15	562·16	673·63	797·78	1086·1	1431·1	1626·0	2305·6	2619·2
0·10000	3·516	4·221	4·546	4·858	5·157	5·446	5·995	6·258	6·514	7·007	7·479	7·707	8·367	8·621
1/ 10·0	62·128	132·59	180·77	238·46	306·32	384·95	576·83	691·21	818·60	1114·5	1468·4	1668·4	2365·7	2687·5
	0·80	0·82	0·82	0·83	0·83	0·84	0·85	0·85	0·86	0·86	0·87	0·87	0·88	0·88

$V_{r(0{\cdot}5)medial}$ for half-full circular pipes.

A18
(p.1 of 6)

k_s = 0·60 mm
S = 0·00010 to 0·00030

ie hydraulic gradient =
1 in 10000 to 1 in 3333

Water (or sewage) at 15°C;
full bore conditions.

velocities in ms^{-1}
discharges in litres/sec

Gradient	(Equivalent) Pipe diameters in mm													
	150	**200**	**225**	**250**	**275**	**300**	**350**	**375**	**400**	**450**	**500**	**525**	**600**	**630**
0·000100	0·091	0·112	0·121	0·130	0·139	0·147	0·163	0·171	0·178	0·193	0·206	0·213	0·232	0·240
1/ 10000	1·6121	3·5029	4·8076	6·3782	8·2336	10·392	15·685	18·852	22·389	30·631	40·527	46·132	65·735	74·805
0·000105	0·094	0·114	0·124	0·133	0·142	0·151	0·167	0·175	0·183	0·198	0·212	0·219	0·238	0·246
1/ 9524	1·6552	3·5957	4·9345	6·5461	8·4498	10·664	16·094	19·344	22·972	31·425	41·576	47·324	67·429	76·731
0·000110	0·096	0·117	0·127	0·137	0·146	0·155	0·171	0·179	0·187	0·202	0·217	0·224	0·244	0·252
1/ 9091	1·6974	3·6864	5·0586	6·7102	8·6610	10·930	16·494	19·823	23·540	32·201	42·600	48·489	69·084	78·613
0·000115	0·098	0·120	0·130	0·140	0·149	0·158	0·175	0·184	0·192	0·207	0·222	0·229	0·250	0·258
1/ 8696	1·7386	3·7752	5·1799	6·8707	8·8676	11·190	16·885	20·292	24·097	32·960	43·602	49·628	70·703	80·453
0·000120	0·101	0·123	0·133	0·143	0·153	0·162	0·179	0·188	0·196	0·212	0·227	0·234	0·256	0·264
1/ 8333	1·7790	3·8621	5·2987	7·0278	9·0698	11·445	17·267	20·752	24·641	33·703	44·582	50·743	72·287	82·254
0·000125	0·103	0·126	0·136	0·146	0·156	0·165	0·183	0·192	0·200	0·216	0·232	0·239	0·261	0·270
1/ 8000	1·8186	3·9472	5·4152	7·1818	9·2680	11·694	17·642	21·202	25·175	34·431	45·543	51·835	73·839	84·018
0·000130	0·105	0·128	0·139	0·149	0·159	0·169	0·187	0·196	0·204	0·221	0·237	0·244	0·267	0·275
1/ 7692	1·8574	4·0307	5·5293	7·3328	9·4624	11·939	18·010	21·643	25·698	35·144	46·485	52·906	75·360	85·747
0·000135	0·107	0·131	0·142	0·152	0·163	0·172	0·191	0·200	0·209	0·225	0·241	0·249	0·272	0·281
1/ 7407	1·8955	4·1127	5·6414	7·4809	9·6531	12·179	18·371	22·076	26·211	35·844	47·409	53·957	76·853	87·444
0·000140	0·109	0·133	0·145	0·155	0·166	0·176	0·195	0·204	0·213	0·230	0·246	0·254	0·277	0·286
1/ 7143	1·9330	4·1932	5·7514	7·6264	9·8404	12·415	18·725	22·501	26·715	36·532	48·316	54·988	78·318	89·110
0·000145	0·111	0·136	0·147	0·158	0·169	0·179	0·198	0·208	0·217	0·234	0·251	0·259	0·282	0·291
1/ 6897	1·9698	4·2723	5·8596	7·7694	10·024	12·646	19·073	22·919	27·210	37·207	49·208	56·002	79·758	90·747
0·000150	0·114	0·138	0·150	0·161	0·172	0·182	0·202	0·211	0·220	0·238	0·255	0·263	0·287	0·296
1/ 6667	2·0060	4·3501	5·9659	7·9100	10·205	12·874	19·416	23·329	27·697	37·872	50·084	56·998	81·174	92·357
0·000160	0·118	0·143	0·155	0·167	0·178	0·188	0·209	0·218	0·228	0·246	0·264	0·272	0·297	0·306
1/ 6250	2·0766	4·5020	6·1735	8·1845	10·559	13·319	20·084	24·131	28·648	39·168	51·795	58·943	83·937	95·498
0·000170	0·121	0·148	0·160	0·172	0·184	0·195	0·215	0·226	0·235	0·254	0·272	0·281	0·306	0·316
1/ 5882	2·1452	4·6493	6·3749	8·4507	10·901	13·750	20·732	24·908	29·569	40·425	53·453	60·829	86·616	98·543
0·000180	0·125	0·153	0·165	0·177	0·189	0·200	0·222	0·232	0·242	0·262	0·280	0·289	0·316	0·326
1/ 5556	2·2118	4·7925	6·5706	8·7094	11·234	14·169	21·361	25·663	30·464	41·646	55·064	62·661	89·218	101·50
0·000190	0·129	0·157	0·170	0·183	0·195	0·206	0·228	0·239	0·249	0·269	0·288	0·298	0·324	0·335
1/ 5263	2·2766	4·9318	6·7610	8·9611	11·558	14·577	21·974	26·398	31·335	42·833	56·632	64·443	91·749	104·38
0·000200	0·132	0·161	0·175	0·188	0·200	0·212	0·235	0·245	0·256	0·277	0·296	0·306	0·333	0·344
1/ 5000	2·3398	5·0675	6·9465	9·2063	11·874	14·974	22·571	27·114	32·184	43·991	58·159	66·179	94·215	107·18
0·000210	0·136	0·166	0·179	0·192	0·205	0·217	0·241	0·252	0·263	0·284	0·304	0·314	0·342	0·353
1/ 4762	2·4015	5·2000	7·1275	9·4456	12·182	15·361	23·153	27·812	33·012	45·120	59·649	67·872	96·620	109·91
0·000220	0·139	0·170	0·184	0·197	0·210	0·223	0·247	0·258	0·269	0·291	0·311	0·321	0·350	0·361
1/ 4545	2·4618	5·3295	7·3044	9·6794	12·482	15·740	23·721	28·494	33·820	46·222	61·103	69·527	98·970	112·59
0·000230	0·143	0·174	0·188	0·202	0·215	0·228	0·252	0·264	0·275	0·297	0·318	0·329	0·358	0·370
1/ 4348	2·5207	5·4561	7·4774	9·9080	12·776	16·110	24·277	29·161	34·611	47·301	62·526	71·144	101·27	115·20
0·000240	0·146	0·178	0·192	0·206	0·220	0·233	0·258	0·270	0·282	0·304	0·326	0·336	0·366	0·378
1/ 4167	2·5784	5·5800	7·6467	10·132	13·064	16·472	24·822	29·814	35·385	48·356	63·918	72·727	103·51	117·75
0·000250	0·149	0·181	0·196	0·211	0·225	0·238	0·264	0·276	0·288	0·311	0·332	0·343	0·374	0·386
1/ 4000	2·6350	5·7014	7·8126	10·351	13·346	16·827	25·355	30·454	36·143	49·389	65·282	74·277	105·72	120·25
0·000260	0·152	0·185	0·201	0·215	0·229	0·243	0·269	0·281	0·294	0·317	0·339	0·350	0·382	0·394
1/ 3846	2·6904	5·8204	7·9752	10·566	13·623	17·175	25·878	31·081	36·886	50·403	66·619	75·797	107·88	122·71
0·000270	0·155	0·189	0·205	0·220	0·234	0·248	0·274	0·287	0·299	0·323	0·346	0·357	0·389	0·401
1/ 3704	2·7448	5·9372	8·1348	10·777	13·894	17·517	26·390	31·696	37·615	51·397	67·931	77·288	109·99	125·12
0·000280	0·158	0·193	0·209	0·224	0·238	0·253	0·280	0·292	0·305	0·329	0·353	0·364	0·396	0·409
1/ 3571	2·7983	6·0519	8·2915	10·984	14·161	17·852	26·894	32·300	38·331	52·373	69·219	78·752	112·07	127·48
0·000290	0·161	0·196	0·212	0·228	0·243	0·257	0·285	0·298	0·311	0·335	0·359	0·370	0·404	0·416
1/ 3448	2·8508	6·1647	8·4455	11·187	14·423	18·182	27·389	32·893	39·035	53·332	70·484	80·191	114·12	129·80
0·000300	0·164	0·200	0·216	0·232	0·247	0·262	0·290	0·303	0·316	0·341	0·365	0·377	0·411	0·424
1/ 3333	2·9024	6·2755	8·5969	11·388	14·680	18·505	27·875	33·477	39·726	54·275	71·728	81·605	116·12	132·08
	0·76	0·77	0·77	0·78	0·78	0·79	0·79	0·80	0·80	0·81	0·81	0·82	0·82	0·83

$V_{r(0·5)medial}$ for half-full circular pipes.

k_s = 0·60 mm S = 0·00010 to 0·00030

k_s = 0·60 mm
S = 0·00030 to 0·00100

ie hydraulic gradient =
1 in 3333 to 1 in 1000

Water (or sewage) at 15°C;
full bore conditions.

velocities in ms^{-1}
discharges in litres/sec

Gradient	**(Equivalent) Pipe diameters in mm**													
	150	**200**	**225**	**250**	**275**	**300**	**350**	**375**	**400**	**450**	**500**	**525**	**600**	**630**
0·000300	0·164	0·200	0·216	0·232	0·247	0·262	0·290	0·303	0·316	0·341	0·365	0·377	0·411	0·424
1/ 3333	2·9024	6·2755	8·5969	11·388	14·680	18·505	27·875	33·477	39·726	54·275	71·728	81·605	116·12	132·08
0·000320	0·170	0·207	0·224	0·240	0·256	0·271	0·300	0·313	0·327	0·353	0·378	0·390	0·425	0·438
1/ 3125	3·0032	6·4918	8·8924	11·778	15·182	19·138	28·824	34·615	41·076	56·115	74·155	84·364	120·04	136·54
0·000340	0·175	0·213	0·231	0·248	0·264	0·279	0·309	0·323	0·337	0·364	0·390	0·402	0·438	0·452
1/ 2941	3·1010	6·7015	9·1789	12·157	15·669	19·750	29·745	35·719	42·384	57·899	76·508	87·039	123·84	140·85
0·000360	0·181	0·220	0·238	0·255	0·272	0·288	0·318	0·333	0·347	0·375	0·401	0·414	0·451	0·465
1/ 2778	3·1960	6·9053	9·4573	12·524	16·143	20·346	30·639	36·791	43·655	59·631	78·794	89·637	127·53	145·05
0·000380	0·186	0·226	0·245	0·262	0·280	0·296	0·327	0·343	0·357	0·386	0·413	0·426	0·464	0·478
1/ 2632	3·2884	7·1036	9·7281	12·882	16·603	20·925	31·508	37·834	44·891	61·316	81·017	92·164	131·12	149·13
0·000400	0·191	0·232	0·251	0·270	0·287	0·304	0·336	0·352	0·367	0·396	0·424	0·437	0·476	0·491
1/ 2500	3·3785	7·2968	9·9919	13·231	17·051	21·489	32·355	38·850	46·095	62·958	83·182	94·626	134·61	153·10
0·000420	0·196	0·238	0·258	0·276	0·294	0·312	0·345	0·361	0·376	0·406	0·434	0·448	0·488	0·504
1/ 2381	3·4664	7·4852	10·249	13·571	17·489	22·039	33·182	39·841	47·270	64·559	85·295	97·028	138·02	156·98
0·000440	0·201	0·244	0·264	0·283	0·302	0·319	0·353	0·369	0·385	0·416	0·445	0·459	0·500	0·516
1/ 2273	3·5522	7·6693	10·501	13·903	17·916	22·577	33·989	40·809	48·417	66·123	87·358	99·373	141·35	160·76
0·000460	0·206	0·250	0·270	0·290	0·309	0·327	0·361	0·378	0·394	0·425	0·455	0·470	0·511	0·528
1/ 2174	3·6361	7·8493	10·747	14·228	18·334	23·103	34·778	41·755	49·538	67·652	89·375	101·67	144·61	164·46
0·000480	0·210	0·255	0·276	0·296	0·316	0·334	0·370	0·386	0·403	0·435	0·465	0·480	0·523	0·539
1/ 2083	3·7183	8·0255	10·987	14·546	18·743	23·617	35·550	42·682	50·636	69·149	91·348	103·91	147·80	168·08
0·000500	0·215	0·261	0·282	0·303	0·322	0·341	0·377	0·395	0·412	0·444	0·475	0·490	0·534	0·551
1/ 2000	3·7988	8·1981	11·223	14·857	19·143	24·121	36·307	43·589	51·711	70·614	93·282	106·11	150·92	171·63
0·000525	0·221	0·268	0·290	0·310	0·331	0·350	0·387	0·405	0·422	0·455	0·487	0·503	0·547	0·564
1/ 1905	3·8972	8·4091	11·511	15·238	19·633	24·737	37·231	44·698	53·026	72·406	95·645	108·79	154·73	175·96
0·000550	0·226	0·274	0·297	0·318	0·339	0·358	0·396	0·415	0·432	0·466	0·499	0·515	0·560	0·578
1/ 1818	3·9934	8·6152	11·792	15·610	20·111	25·338	38·134	45·781	54·309	74·156	97·953	111·42	158·45	180·20
0·000575	0·231	0·281	0·304	0·325	0·346	0·367	0·406	0·424	0·442	0·477	0·510	0·527	0·573	0·591
1/ 1739	4·0874	8·8167	12·068	15·973	20·578	25·927	39·017	46·840	55·564	75·866	100·21	113·98	162·10	184·34
0·000600	0·237	0·287	0·310	0·333	0·354	0·375	0·415	0·433	0·452	0·488	0·522	0·538	0·586	0·604
1/ 1667	4·1794	9·0139	12·337	16·329	21·036	26·502	39·881	47·876	56·792	77·540	102·42	116·49	165·66	188·39
0·000625	0·242	0·293	0·317	0·340	0·362	0·383	0·423	0·443	0·462	0·498	0·533	0·549	0·598	0·617
1/ 1600	4·2695	9·2071	12·601	16·677	21·484	27·066	40·728	48·891	57·995	79·180	104·58	118·95	169·15	192·35
0·000650	0·247	0·299	0·323	0·347	0·369	0·391	0·432	0·452	0·471	0·508	0·543	0·561	0·610	0·630
1/ 1538	4·3579	9·3965	12·859	17·019	21·923	27·619	41·558	49·886	59·174	80·787	106·70	121·36	172·57	196·24
0·000675	0·252	0·305	0·330	0·354	0·376	0·398	0·440	0·461	0·480	0·518	0·554	0·572	0·622	0·642
1/ 1481	4·4446	9·5823	13·113	17·354	22·355	28·161	42·372	50·862	60·331	82·364	108·78	123·72	175·93	200·06
0·000700	0·256	0·311	0·336	0·360	0·383	0·406	0·449	0·469	0·489	0·528	0·564	0·582	0·634	0·654
1/ 1429	4·5297	9·7648	13·362	17·683	22·778	28·693	43·171	51·820	61·466	83·912	110·82	126·04	179·22	203·80
0·000725	0·261	0·317	0·342	0·367	0·390	0·413	0·457	0·478	0·498	0·537	0·575	0·593	0·645	0·666
1/ 1379	4·6134	9·9440	13·607	18·007	23·193	29·216	43·956	52·761	62·582	85·432	112·83	128·32	182·46	207·48
0·000750	0·266	0·322	0·348	0·373	0·397	0·421	0·465	0·486	0·507	0·547	0·585	0·603	0·657	0·677
1/ 1333	4·6956	10·120	13·848	18·324	23·602	29·730	44·727	53·687	63·679	86·927	114·80	130·57	185·64	211·09
0·000800	0·275	0·333	0·360	0·386	0·411	0·435	0·481	0·502	0·524	0·565	0·604	0·623	0·679	0·700
1/ 1250	4·8561	10·464	14·317	18·945	24·400	30·734	46·233	55·493	65·819	89·844	118·65	134·94	191·85	218·15
0·000850	0·284	0·344	0·372	0·398	0·424	0·449	0·496	0·518	0·540	0·583	0·623	0·643	0·700	0·722
1/ 1176	5·0117	10·798	14·772	19·546	25·173	31·706	47·693	57·243	67·893	92·672	122·37	139·18	197·87	224·99
0·000900	0·292	0·354	0·383	0·410	0·436	0·462	0·510	0·534	0·556	0·600	0·642	0·662	0·720	0·743
1/ 1111	5·1629	11·121	15·214	20·130	25·924	32·651	49·111	58·943	69·908	95·417	126·00	143·29	203·71	231·63
0·000950	0·300	0·364	0·393	0·422	0·449	0·475	0·525	0·549	0·572	0·617	0·660	0·680	0·741	0·764
1/ 1053	5·3099	11·436	15·644	20·698	26·654	33·569	50·490	60·597	71·867	98·088	129·52	147·30	209·40	238·09
0·001000	0·309	0·374	0·404	0·433	0·461	0·488	0·539	0·563	0·587	0·633	0·677	0·698	0·760	0·784
1/ 1000	5·4531	11·743	16·063	21·251	27·366	34·464	51·833	62·207	73·776	100·69	132·95	151·20	214·93	244·38
	0·76	0·78	0·78	0·79	0·79	0·79	0·80	0·81	0·81	0·82	0·82	0·82	0·83	0·83

$V_{r(0·5)medial}$ for half-full circular pipes.

A18
(p.3 of 6)

k_s = 0·60 mm
S = 0·00100 to 0·00300

ie hydraulic gradient =
1 in 1000 to 1 in 333

Water (or sewage) at 15°C;
full bore conditions.

velocities in ms^{-1}
discharges in litres/sec

Gradient **(Equivalent) Pipe diameters in mm**

Gradient	150	200	225	250	275	300	350	375	400	450	500	525	600	630
0·00100	0·309	0·374	0·404	0·433	0·461	0·488	0·539	0·563	0·587	0·633	0·677	0·698	0·760	0·784
1/ 1000	5·4531	11·743	16·063	21·251	27·366	34·464	51·833	62·207	73·776	100·69	132·95	151·20	214·93	244·38
0·00105	0·316	0·383	0·414	0·444	0·472	0·500	0·552	0·577	0·602	0·649	0·694	0·716	0·779	0·804
1/ 952	5·5929	12·043	16·472	21·791	28·060	35·337	53·143	63·778	75·638	103·23	136·30	155·00	220·33	250·52
0·00110	0·324	0·393	0·424	0·455	0·484	0·512	0·566	0·591	0·616	0·665	0·711	0·733	0·798	0·823
1/ 909	5·7294	12·335	16·871	22·318	28·737	36·190	54·422	65·313	77·456	105·70	139·56	158·72	225·61	256·52
0·00115	0·332	0·402	0·434	0·465	0·495	0·524	0·579	0·605	0·631	0·680	0·727	0·750	0·816	0·842
1/ 870	5·8628	12·621	17·261	22·833	29·400	37·023	55·673	66·812	79·233	108·13	142·76	162·35	230·76	262·37
0·00120	0·339	0·411	0·444	0·475	0·506	0·535	0·591	0·618	0·644	0·695	0·743	0·766	0·834	0·860
1/ 833	5·9934	12·900	17·643	23·337	30·048	37·839	56·897	68·280	80·972	110·50	145·89	165·90	235·81	268·11
0·00125	0·346	0·419	0·453	0·485	0·517	0·547	0·604	0·631	0·658	0·709	0·759	0·782	0·851	0·878
1/ 800	6·1213	13·174	18·017	23·831	30·683	38·638	58·096	69·718	82·676	112·82	148·95	169·38	240·75	273·72
0·00130	0·353	0·428	0·462	0·495	0·527	0·558	0·616	0·644	0·671	0·724	0·774	0·798	0·869	0·896
1/ 769	6·2467	13·443	18·384	24·315	31·306	39·421	59·271	71·127	84·346	115·10	151·95	172·79	245·59	279·23
0·00135	0·360	0·436	0·471	0·505	0·537	0·569	0·628	0·657	0·684	0·738	0·789	0·814	0·885	0·913
1/ 741	6·3697	13·706	18·743	24·790	31·917	40·189	60·424	72·509	85·984	117·33	154·89	176·14	250·34	284·63
0·00140	0·367	0·445	0·480	0·515	0·547	0·579	0·640	0·669	0·697	0·751	0·804	0·829	0·902	0·930
1/ 714	6·4905	13·965	19·096	25·257	32·516	40·943	61·556	73·866	87·592	119·52	157·78	179·42	255·00	289·93
0·00145	0·374	0·453	0·489	0·524	0·557	0·590	0·651	0·681	0·710	0·765	0·818	0·844	0·918	0·947
1/ 690	6·6092	14·219	19·443	25·715	33·105	41·684	62·668	75·199	89·172	121·67	160·62	182·65	259·58	295·13
0·00150	0·381	0·461	0·498	0·533	0·567	0·600	0·663	0·693	0·722	0·778	0·832	0·858	0·934	0·963
1/ 667	6·7258	14·469	19·784	26·165	33·684	42·412	63·761	76·510	90·724	123·79	163·41	185·82	264·09	300·25
0·00160	0·393	0·476	0·514	0·551	0·586	0·620	0·685	0·716	0·746	0·804	0·860	0·887	0·965	0·995
1/ 625	6·9535	14·956	20·450	27·044	34·814	43·833	65·893	79·067	93·754	127·92	168·86	192·01	272·87	310·24
0·00170	0·406	0·491	0·531	0·568	0·605	0·640	0·706	0·738	0·769	0·829	0·887	0·915	0·995	1·026
1/ 588	7·1741	15·429	21·094	27·895	35·909	45·211	67·960	81·545	96·691	131·92	174·14	198·01	281·39	319·92
0·00180	0·418	0·506	0·546	0·585	0·622	0·659	0·727	0·760	0·792	0·854	0·913	0·942	1·024	1·056
1/ 556	7·3884	15·888	21·721	28·723	36·973	46·548	69·967	83·952	99·543	135·81	179·26	203·84	289·66	329·31
0·00190	0·430	0·520	0·562	0·602	0·640	0·677	0·748	0·781	0·814	0·878	0·938	0·968	1·053	1·086
1/ 526	7·5969	16·334	22·330	29·527	38·007	47·849	71·919	86·293	102·32	139·59	184·25	209·51	297·71	338·46
0·00200	0·441	0·534	0·577	0·617	0·657	0·695	0·767	0·802	0·836	0·901	0·963	0·993	1·081	1·114
1/ 500	7·8000	16·769	22·923	30·311	39·015	49·116	73·821	88·573	105·02	143·27	189·10	215·03	305·54	347·36
0·00210	0·453	0·547	0·591	0·633	0·673	0·712	0·787	0·822	0·857	0·923	0·987	1·018	1·108	1·142
1/ 476	7·9981	17·193	23·502	31·075	39·998	50·352	75·676	90·797	107·65	146·86	193·84	220·41	313·18	356·05
0·00220	0·464	0·560	0·605	0·648	0·690	0·729	0·805	0·842	0·877	0·945	1·011	1·042	1·134	1·169
1/ 455	8·1915	17·607	24·068	31·822	40·957	51·559	77·487	92·968	110·23	150·37	198·47	225·67	320·64	364·53
0·00230	0·474	0·573	0·619	0·663	0·705	0·746	0·824	0·861	0·897	0·967	1·034	1·066	1·160	1·196
1/ 435	8·3806	18·012	24·620	32·551	41·895	52·739	79·258	95·091	112·74	153·80	202·99	230·81	327·94	372·82
0·00240	0·485	0·586	0·633	0·678	0·721	0·762	0·842	0·880	0·917	0·988	1·056	1·089	1·185	1·222
1/ 417	8·5657	18·408	25·161	33·265	42·814	53·894	80·990	97·169	115·20	157·15	207·41	235·83	335·08	380·93
0·00250	0·495	0·598	0·646	0·692	0·736	0·778	0·859	0·898	0·936	1·009	1·078	1·112	1·210	1·247
1/ 400	8·7470	18·796	25·690	33·965	43·713	55·025	82·687	99·203	117·62	160·44	211·74	240·76	342·07	388·87
0·00260	0·505	0·610	0·659	0·706	0·751	0·794	0·877	0·916	0·955	1·029	1·100	1·134	1·234	1·272
1/ 385	8·9247	19·176	26·210	34·650	44·594	56·133	84·351	101·20	119·98	163·66	215·99	245·59	348·92	396·66
0·00270	0·515	0·622	0·672	0·720	0·765	0·810	0·894	0·934	0·973	1·049	1·121	1·156	1·258	1·297
1/ 370	9·0990	19·549	26·719	35·323	45·459	57·221	85·983	103·15	122·30	166·82	220·16	250·32	355·64	404·30
0·00280	0·525	0·634	0·685	0·733	0·780	0·825	0·910	0·951	0·991	1·068	1·142	1·178	1·281	1·321
1/ 357	9·2701	19·916	27·219	35·983	46·308	58·288	87·584	105·07	124·57	169·92	224·25	254·97	362·24	411·80
0·00290	0·534	0·645	0·697	0·746	0·794	0·839	0·927	0·968	1·009	1·088	1·163	1·199	1·304	1·345
1/ 345	9·4383	20·276	27·710	36·631	47·141	59·337	89·158	106·96	126·81	172·96	228·26	259·54	368·72	419·16
0·00300	0·543	0·657	0·709	0·759	0·807	0·854	0·943	0·985	1·027	1·106	1·183	1·220	1·327	1·368
1/ 333	9·6035	20·629	28·193	37·269	47·961	60·368	90·705	108·82	129·01	175·96	232·21	264·02	375·09	426·41
	0·77	0·78	0·79	0·79	0·79	0·80	0·81	0·81	0·81	0·82	0·83	0·83	0·84	0·84

$V_{r(0·5)medial}$ for half-full circular pipes.

k_s = 0·60 mm S = 0·00100 to 0·00300

k_s = 0·60 mm
S = 0·00300 to 0·01000

ie hydraulic gradient =
1 in 333 to 1 in 100

Water (or sewage) at 15°C;
full bore conditions.

velocities in ms^{-1}
discharges in litres/sec

Gradient	**(Equivalent) Pipe diameters in mm**													
	150	**200**	**225**	**250**	**275**	**300**	**350**	**375**	**400**	**450**	**500**	**525**	**600**	**630**
0·00300	0·543	0·657	0·709	0·759	0·807	0·854	0·943	0·985	1·027	1·106	1·183	1·220	1·327	1·368
1/ 333	9·6035	20·629	28·193	37·269	47·961	60·368	90·705	108·82	129·01	175·96	232·21	264·02	375·09	426·41
0·00320	0·562	0·679	0·733	0·785	0·834	0·882	0·974	1·018	1·061	1·143	1·222	1·260	1·371	1·413
1/ 313	9·9260	21·319	29·135	38·513	49·560	62·379	93·723	112·43	133·29	181·80	239·92	272·78	387·53	440·53
0·00340	0·579	0·700	0·756	0·809	0·861	0·910	1·005	1·050	1·094	1·179	1·260	1·299	1·413	1·457
1/ 294	10·239	21·988	30·048	39·719	51·110	64·329	96·648	115·94	137·45	187·47	247·39	281·27	399·58	454·23
0·00360	0·597	0·721	0·778	0·833	0·886	0·937	1·034	1·081	1·126	1·213	1·297	1·337	1·455	1·500
1/ 278	10·542	22·638	30·934	40·889	52·616	66·222	99·488	119·35	141·49	192·96	254·64	289·51	411·28	467·52
0·00380	0·613	0·741	0·800	0·856	0·910	0·963	1·063	1·111	1·157	1·247	1·333	1·374	1·495	1·541
1/ 263	10·837	23·270	31·797	42·028	54·080	68·063	102·25	122·66	145·41	198·31	261·69	297·53	422·66	480·46
0·00400	0·630	0·760	0·821	0·879	0·935	0·988	1·091	1·140	1·188	1·280	1·368	1·411	1·534	1·582
1/ 250	11·125	23·885	32·637	43·137	55·506	69·856	104·94	125·89	149·23	203·52	268·56	305·34	433·74	493·05
0·00420	0·645	0·779	0·841	0·901	0·958	1·013	1·118	1·168	1·217	1·312	1·402	1·446	1·572	1·621
1/ 238	11·405	24·486	33·456	44·219	56·896	71·606	107·57	129·03	152·96	208·60	275·26	312·96	444·55	505·34
0·00440	0·661	0·798	0·862	0·922	0·981	1·037	1·145	1·196	1·246	1·343	1·435	1·480	1·610	1·660
1/ 227	11·679	25·072	34·256	45·276	58·255	73·314	110·13	132·11	156·60	213·57	281·81	320·39	455·11	517·34
0·00460	0·676	0·816	0·881	0·943	1·003	1·061	1·171	1·223	1·275	1·373	1·468	1·514	1·646	1·697
1/ 217	11·947	25·645	35·038	46·308	59·582	74·983	112·63	135·11	160·16	218·42	288·20	327·66	465·43	529·06
0·00480	0·691	0·834	0·900	0·964	1·025	1·084	1·196	1·250	1·302	1·403	1·500	1·546	1·682	1·734
1/ 208	12·209	26·206	35·804	47·319	60·882	76·617	115·09	138·05	163·65	223·16	294·46	334·78	475·52	540·54
0·00500	0·705	0·852	0·919	0·984	1·046	1·107	1·221	1·276	1·329	1·432	1·531	1·579	1·717	1·770
1/ 200	12·466	26·755	36·553	48·308	62·154	78·217	117·49	140·93	167·06	227·81	300·59	341·74	485·41	551·78
0·00525	0·723	0·873	0·942	1·009	1·073	1·134	1·252	1·308	1·363	1·468	1·569	1·618	1·760	1·814
1/ 190	12·780	27·426	37·469	49·518	63·709	80·173	120·42	144·45	171·23	233·49	308·08	350·26	497·50	565·51
0·00550	0·741	0·894	0·965	1·033	1·098	1·161	1·281	1·339	1·395	1·503	1·606	1·656	1·801	1·857
1/ 182	13·086	28·082	38·364	50·700	65·228	82·083	123·29	147·88	175·30	239·04	315·40	358·58	509·30	578·93
0·00575	0·757	0·914	0·987	1·056	1·123	1·188	1·311	1·369	1·427	1·537	1·643	1·694	1·842	1·899
1/ 174	13·385	28·723	39·239	51·854	66·713	83·950	126·09	151·24	179·28	244·46	322·55	366·70	520·84	592·04
0·00600	0·774	0·934	1·008	1·079	1·148	1·214	1·339	1·399	1·458	1·570	1·678	1·731	1·882	1·940
1/ 167	13·678	29·350	40·094	52·984	68·165	85·777	128·83	154·53	183·17	249·77	329·55	374·66	532·13	604·87
0·00625	0·790	0·954	1·029	1·102	1·172	1·239	1·367	1·428	1·488	1·603	1·713	1·767	1·921	1·981
1/ 160	13·965	29·964	40·933	54·091	69·588	87·567	131·51	157·75	186·99	254·97	336·40	382·45	543·19	617·44
0·00650	0·806	0·973	1·050	1·124	1·195	1·264	1·394	1·457	1·518	1·635	1·748	1·802	1·959	2·020
1/ 154	14·247	30·566	41·754	55·176	70·983	89·321	134·15	160·90	190·72	260·06	343·12	390·08	554·03	629·75
0·00675	0·822	0·992	1·070	1·146	1·218	1·288	1·421	1·485	1·547	1·667	1·781	1·837	1·997	2·059
1/ 148	14·523	31·157	42·560	56·240	72·351	91·041	136·73	164·00	194·39	265·06	349·71	397·57	564·66	641·84
0·00700	0·837	1·010	1·090	1·167	1·241	1·312	1·447	1·512	1·576	1·697	1·814	1·871	2·034	2·097
1/ 143	14·794	31·736	43·351	57·285	73·694	92·730	139·26	167·04	197·99	269·97	356·18	404·93	575·09	653·70
0·00725	0·852	1·028	1·110	1·188	1·263	1·335	1·473	1·539	1·604	1·728	1·846	1·904	2·070	2·134
1/ 138	15·060	32·306	44·129	58·311	75·013	94·389	141·75	170·02	201·53	274·78	362·53	412·15	585·35	665·35
0·00750	0·867	1·046	1·129	1·208	1·285	1·358	1·499	1·566	1·631	1·758	1·878	1·937	2·106	2·171
1/ 133	15·321	32·866	44·892	59·319	76·310	96·020	144·20	172·95	205·00	279·52	368·78	419·25	595·42	676·80
0·00800	0·896	1·081	1·167	1·249	1·327	1·403	1·548	1·618	1·685	1·816	1·940	2·001	2·175	2·243
1/ 125	15·832	33·958	46·383	61·288	78·840	99·202	148·97	178·68	211·79	288·76	380·97	433·10	615·09	699·14
0·00850	0·924	1·115	1·203	1·287	1·369	1·447	1·596	1·668	1·738	1·872	2·000	2·063	2·243	2·312
1/ 118	16·327	35·016	47·828	63·195	81·292	102·29	153·60	184·22	218·36	297·72	392·78	446·53	634·14	720·80
0·00900	0·951	1·147	1·238	1·325	1·409	1·489	1·643	1·717	1·788	1·927	2·059	2·123	2·308	2·380
1/ 111	16·807	36·044	49·231	65·047	83·674	105·28	158·09	189·61	224·74	306·42	404·25	459·56	652·65	741·83
0·00950	0·978	1·179	1·272	1·362	1·448	1·531	1·689	1·764	1·838	1·980	2·116	2·182	2·372	2·445
1/ 105	17·274	37·044	50·595	66·849	85·990	108·19	162·46	194·85	230·95	314·88	415·41	472·24	670·64	762·28
0·01000	1·003	1·210	1·306	1·398	1·486	1·571	1·733	1·810	1·886	2·032	2·171	2·239	2·434	2·509
1/ 100	17·729	38·018	51·924	68·604	88·246	111·03	166·72	199·95	237·00	323·12	426·27	484·59	688·17	782·20
	0·77	0·78	0·79	0·79	0·80	0·80	0·81	0·81	0·82	0·82	0·83	0·83	0·84	0·84

$V_{r(0·5)medial}$ for half-full circular pipes.

A18

(p.5 of 6)

$k_s = 0{\cdot}60$ **mm**
S = 0·01000 to 0·03000

ie hydraulic gradient =
1 in 100 to 1 in 33·3

Water (or sewage) at 15°C;
full bore conditions.

velocities in ms^{-1}
discharges in litres/sec

Gradient **(Equivalent) Pipe diameters in mm**

Gradient	150	200	225	250	275	300	350	375	400	450	500	525	600	630
0·01000	1·003	1·210	1·306	1·398	1·486	1·571	1·733	1·810	1·886	2·032	2·171	2·239	2·434	2·509
1/ 100	17·729	38·018	51·924	68·604	88·246	111·03	166·72	199·95	237·00	323·12	426·27	484·59	688·17	782·20
0·01050	1·028	1·240	1·339	1·432	1·523	1·610	1·776	1·855	1·933	2·082	2·225	2·294	2·494	2·572
1/ 95	18·173	38·968	53·220	70·315	90·446	113·80	170·87	204·93	242·90	331·16	436·87	496·64	705·27	801·63
0·01100	1·053	1·270	1·370	1·466	1·559	1·648	1·818	1·899	1·979	2·132	2·278	2·349	2·553	2·632
1/ 91	18·607	39·895	54·486	71·986	92·594	116·50	174·92	209·79	248·65	339·00	447·22	508·40	721·96	820·60
0·01150	1·077	1·299	1·401	1·500	1·594	1·685	1·859	1·943	2·024	2·180	2·329	2·402	2·611	2·692
1/ 87	19·031	40·802	55·723	73·620	94·694	119·14	178·89	214·54	254·28	346·67	457·33	519·90	738·28	839·15
0·01200	1·100	1·327	1·432	1·532	1·629	1·722	1·900	1·985	2·067	2·227	2·380	2·454	2·668	2·750
1/ 83	19·445	41·689	56·934	75·219	96·749	121·72	182·76	219·19	259·79	354·18	467·23	531·15	754·25	857·30
0·01250	1·123	1·355	1·462	1·564	1·663	1·758	1·939	2·026	2·110	2·273	2·429	2·505	2·723	2·807
1/ 80	19·851	42·558	58·120	76·784	98·762	124·26	186·56	223·74	265·19	361·53	476·92	542·16	769·89	875·07
0·01300	1·146	1·382	1·491	1·595	1·696	1·793	1·978	2·066	2·152	2·318	2·477	2·554	2·777	2·863
1/ 77	20·249	43·410	59·282	78·319	100·73	126·74	190·28	228·21	270·47	368·74	486·43	552·96	785·22	892·49
0·01350	1·168	1·408	1·520	1·626	1·729	1·827	2·016	2·106	2·194	2·363	2·525	2·603	2·830	2·918
1/ 74	20·640	44·245	60·422	79·824	102·67	129·17	193·94	232·59	275·66	375·81	495·75	563·56	800·26	909·58
0·01400	1·190	1·434	1·548	1·656	1·761	1·861	2·053	2·145	2·234	2·407	2·571	2·651	2·883	2·972
1/ 71	21·023	45·065	61·541	81·302	104·57	131·56	197·52	236·88	280·75	382·74	504·90	573·96	815·02	926·35
0·01450	1·211	1·460	1·575	1·686	1·792	1·894	2·090	2·183	2·274	2·449	2·617	2·699	2·934	3·025
1/ 69	21·400	45·871	62·641	82·753	106·44	133·91	201·04	241·10	285·76	389·56	513·89	584·17	829·52	942·83
0·01500	1·232	1·485	1·603	1·715	1·823	1·927	2·126	2·221	2·313	2·492	2·662	2·745	2·984	3·077
1/ 67	21·770	46·663	63·721	84·180	108·27	136·21	204·50	245·25	290·67	396·26	522·72	594·21	843·77	959·03
0·01600	1·273	1·535	1·656	1·772	1·883	1·991	2·196	2·294	2·389	2·574	2·750	2·835	3·083	3·178
1/ 63	22·492	48·208	65·830	86·964	111·85	140·71	211·25	253·35	300·27	409·33	539·96	613·81	871·58	990·63
0·01700	1·312	1·582	1·707	1·827	1·942	2·052	2·264	2·365	2·463	2·653	2·835	2·923	3·178	3·276
1/ 59	23·192	49·706	67·874	89·663	115·32	145·08	217·80	261·20	309·56	422·00	556·66	632·80	898·53	1021·3
0·01800	1·351	1·628	1·757	1·880	1·998	2·112	2·330	2·434	2·535	2·731	2·918	3·008	3·270	3·372
1/ 56	23·872	51·160	69·858	92·283	118·69	149·31	224·15	268·82	318·59	434·30	572·89	651·23	924·70	1051·0
0·01900	1·388	1·673	1·805	1·932	2·053	2·171	2·394	2·501	2·605	2·806	2·998	3·091	3·360	3·464
1/ 53	24·533	52·574	71·788	94·831	121·96	153·43	230·34	276·23	327·37	446·27	588·67	669·17	950·16	1079·9
0·02000	1·425	1·717	1·853	1·982	2·107	2·227	2·457	2·566	2·673	2·879	3·076	3·172	3·448	3·555
1/ 50	25·177	53·952	73·668	97·313	125·15	157·44	236·36	283·45	335·93	457·92	604·03	686·63	974·95	1108·1
0·02100	1·460	1·760	1·899	2·032	2·160	2·283	2·518	2·630	2·740	2·951	3·153	3·251	3·534	3·643
1/ 47·6	25·805	55·295	75·501	99·734	128·27	161·36	242·23	290·48	344·27	469·29	619·02	703·67	999·13	1135·6
0·02200	1·495	1·802	1·944	2·080	2·211	2·337	2·577	2·692	2·804	3·020	3·227	3·327	3·617	3·729
1/ 45·5	26·418	56·607	77·292	102·10	131·30	165·18	247·96	297·36	352·41	480·39	633·66	720·30	1022·7	1162·4
0·02300	1·529	1·843	1·988	2·127	2·261	2·390	2·636	2·753	2·868	3·089	3·300	3·403	3·699	3·813
1/ 43·5	27·017	57·890	79·042	104·41	134·27	168·92	253·57	304·08	360·37	491·24	647·96	736·57	1045·8	1188·6
0·02400	1·562	1·883	2·031	2·173	2·310	2·441	2·693	2·813	2·930	3·155	3·371	3·476	3·779	3·895
1/ 41·7	27·604	59·145	80·754	106·67	137·18	172·57	259·05	310·65	368·16	501·85	661·96	752·48	1068·4	1214·3
0·02500	1·595	1·922	2·073	2·218	2·358	2·492	2·748	2·871	2·990	3·221	3·441	3·548	3·857	3·976
1/ 40·0	28·178	60·374	82·432	108·88	140·03	176·15	264·42	317·09	375·79	512·25	675·67	768·06	1090·5	1239·4
0·02600	1·626	1·960	2·115	2·262	2·405	2·542	2·803	2·928	3·050	3·285	3·510	3·619	3·934	4·055
1/ 38·5	28·742	61·578	84·076	111·05	142·82	179·66	269·69	323·41	383·27	522·44	689·11	783·33	1112·2	1264·1
0·02700	1·658	1·998	2·155	2·306	2·451	2·590	2·857	2·984	3·108	3·348	3·577	3·688	4·009	4·133
1/ 37·0	29·294	62·760	85·688	113·18	145·56	183·10	274·85	329·60	390·61	532·44	702·29	798·32	1133·5	1288·3
0·02800	1·688	2·035	2·195	2·348	2·496	2·638	2·909	3·039	3·166	3·409	3·643	3·756	4·083	4·209
1/ 35·7	29·836	63·920	87·271	115·27	148·24	186·48	279·92	335·68	397·81	542·25	715·24	813·03	1154·3	1312·0
0·02900	1·719	2·071	2·234	2·390	2·540	2·685	2·961	3·093	3·222	3·470	3·707	3·822	4·155	4·284
1/ 34·5	30·369	65·060	88·826	117·33	150·88	189·80	284·90	341·65	404·89	551·89	727·95	827·47	1174·9	1335·3
0·03000	1·748	2·107	2·272	2·431	2·584	2·731	3·012	3·146	3·277	3·530	3·771	3·888	4·226	4·357
1/ 33·3	30·892	66·180	90·355	119·35	153·48	193·06	289·79	347·51	411·84	561·37	740·44	841·67	1195·0	1358·2
	0·77	0·79	0·79	0·80	0·80	0·80	0·81	0·82	0·82	0·82	0·83	0·83	0·84	0·84

$V_{r(0{\cdot}5)medial}$ **for half-full circular pipes.**

$k_s = 0{\cdot}60$ **mm** S = 0·01000 to 0·03000

$k_s = 0{\cdot}60$ mm
S = 0·03000 to 0·10000

ie hydraulic gradient =
1 in 33·3 to 1 in 10·0

Water (or sewage) at 15°C;
full bore conditions.

velocities in ms^{-1}
discharges in litres/sec

Gradient **(Equivalent) Pipe diameters in mm**

Gradient	150	200	225	250	275	300	350	375	400	450	500	525	600	630
0·03000	1·748	2·107	2·272	2·431	2·584	2·731	3·012	3·146	3·277	3·530	3·771	3·888	4·226	4·357
1/ 33·3	30·892	66·180	90·355	119·35	153·48	193·06	289·79	347·51	411·84	561·37	740·44	841·67	1195·0	1358·2
0·03200	1·806	2·176	2·347	2·511	2·669	2·821	3·111	3·250	3·385	3·646	3·895	4·016	4·366	4·500
1/ 31·3	31·914	68·365	93·337	123·28	158·54	199·43	299·34	358·97	425·41	579·86	764·82	869·38	1234·3	1402·9
0·03400	1·862	2·244	2·420	2·589	2·752	2·909	3·208	3·351	3·490	3·759	4·016	4·140	4·500	4·639
1/ 29·4	32·904	70·483	96·228	127·10	163·44	205·60	308·60	370·06	438·56	597·78	788·45	896·24	1272·5	1446·2
0·03600	1·916	2·309	2·491	2·665	2·832	2·993	3·301	3·448	3·592	3·868	4·132	4·261	4·631	4·774
1/ 27·8	33·865	72·540	99·035	130·81	168·21	211·59	317·59	380·84	451·33	615·17	811·39	922·32	1309·5	1488·3
0·03800	1·969	2·373	2·559	2·738	2·910	3·076	3·392	3·543	3·690	3·974	4·246	4·378	4·759	4·906
1/ 26·3	34·801	74·541	101·76	134·41	172·84	217·42	326·33	391·32	463·75	632·10	833·71	947·68	1345·5	1529·2
0·04000	2·021	2·435	2·626	2·810	2·986	3·156	3·480	3·636	3·787	4·078	4·357	4·492	4·883	5·033
1/ 25·0	35·711	76·489	104·42	137·92	177·36	223·09	334·85	401·53	475·84	648·58	855·44	972·38	1380·5	1569·0
0·04200	2·071	2·495	2·692	2·879	3·060	3·234	3·567	3·726	3·881	4·179	4·465	4·603	5·004	5·158
1/ 23·8	36·600	78·389	107·02	141·34	181·76	228·63	343·15	411·49	487·64	664·66	876·64	996·48	1414·7	1607·9
0·04400	2·120	2·554	2·755	2·948	3·132	3·311	3·651	3·814	3·972	4·278	4·570	4·712	5·122	5·280
1/ 22·7	37·467	80·245	109·55	144·69	186·05	234·03	351·26	421·21	499·16	680·35	897·34	1020·0	1448·1	1645·8
0·04600	2·168	2·612	2·817	3·014	3·203	3·386	3·733	3·900	4·062	4·374	4·673	4·818	5·237	5·399
1/ 21·7	38·315	82·059	112·02	147·96	190·26	239·31	359·19	430·71	510·42	695·70	917·57	1043·0	1480·8	1682·9
0·04800	2·215	2·669	2·878	3·079	3·272	3·459	3·814	3·984	4·149	4·469	4·774	4·922	5·350	5·515
1/ 20·8	39·144	83·834	114·45	151·15	194·37	244·48	366·94	440·01	521·44	710·71	937·37	1065·5	1512·7	1719·2
0·05000	2·261	2·724	2·938	3·143	3·340	3·530	3·893	4·066	4·235	4·561	4·873	5·024	5·461	5·629
1/ 20·0	39·957	85·572	116·82	154·29	198·39	249·55	374·54	449·12	532·23	725·42	956·76	1087·5	1544·0	1754·8
0·05250	2·317	2·791	3·011	3·221	3·423	3·618	3·989	4·167	4·340	4·674	4·993	5·148	5·596	5·769
1/ 19·0	40·950	87·697	119·72	158·11	203·31	255·74	383·82	460·25	545·42	743·39	980·46	1114·5	1582·2	1798·2
0·05500	2·372	2·858	3·082	3·297	3·504	3·703	4·084	4·266	4·443	4·784	5·111	5·270	5·728	5·905
1/ 18·2	41·920	89·771	122·55	161·85	208·12	261·78	392·89	471·12	558·30	760·94	1003·6	1140·8	1619·6	1840·7
0·05750	2·426	2·922	3·152	3·372	3·583	3·787	4·176	4·362	4·543	4·892	5·227	5·389	5·857	6·038
1/ 17·4	42·868	91·800	125·32	165·51	212·82	267·69	401·75	481·74	570·89	778·09	1026·2	1166·5	1656·0	1882·1
0·06000	2·478	2·985	3·220	3·444	3·660	3·869	4·266	4·456	4·641	4·998	5·339	5·505	5·983	6·168
1/ 16·7	43·796	93·784	128·02	169·08	217·41	273·47	410·42	492·14	583·21	794·88	1048·4	1191·7	1691·8	1922·7
0·06250	2·530	3·047	3·287	3·516	3·736	3·949	4·354	4·548	4·737	5·101	5·450	5·619	6·107	6·295
1/ 16·0	44·704	95·727	130·68	172·58	221·91	279·13	418·92	502·32	595·27	811·32	1070·0	1216·3	1726·7	1962·5
0·06500	2·580	3·108	3·352	3·586	3·810	4·027	4·441	4·638	4·831	5·203	5·558	5·730	6·228	6·420
1/ 15·4	45·595	97·632	133·28	176·01	226·33	284·68	427·24	512·30	607·10	827·43	1091·3	1240·4	1761·0	2001·4
0·06750	2·630	3·167	3·416	3·654	3·883	4·104	4·526	4·727	4·923	5·302	5·664	5·840	6·347	6·543
1/ 14·8	46·468	99·501	135·83	179·38	230·66	290·12	435·41	522·10	618·70	843·24	1112·1	1264·1	1794·6	2039·6
0·07000	2·678	3·226	3·479	3·722	3·955	4·180	4·609	4·814	5·014	5·400	5·768	5·947	6·464	6·663
1/ 14·3	47·326	101·34	138·33	182·69	234·90	295·46	443·42	531·71	630·09	858·76	1132·6	1287·4	1827·6	2077·1
0·07250	2·726	3·283	3·541	3·788	4·025	4·254	4·691	4·900	5·103	5·495	5·871	6·053	6·579	6·782
1/ 13·8	48·168	103·14	140·79	185·93	239·08	300·71	451·30	541·15	641·27	874·00	1152·7	1310·2	1860·1	2114·0
0·07500	2·773	3·339	3·602	3·853	4·094	4·327	4·771	4·984	5·191	5·590	5·971	6·156	6·691	6·898
1/ 13·3	48·996	104·91	143·21	189·12	243·18	305·87	459·04	550·43	652·27	888·99	1172·5	1332·7	1891·9	2150·2
0·08000	2·864	3·449	3·720	3·980	4·229	4·470	4·928	5·148	5·361	5·773	6·168	6·359	6·911	7·125
1/ 12·5	50·611	108·36	147·92	195·35	251·18	315·94	474·14	568·53	673·72	918·22	1211·0	1376·5	1954·1	2220·9
0·08500	2·953	3·556	3·835	4·103	4·360	4·608	5·080	5·306	5·527	5·952	6·358	6·555	7·124	7·344
1/ 11·8	52·177	111·71	152·49	201·39	258·94	325·69	488·78	586·08	694·51	946·55	1248·4	1419·0	2014·4	2289·4
0·09000	3·039	3·659	3·947	4·222	4·486	4·742	5·228	5·461	5·687	6·125	6·543	6·745	7·331	7·558
1/ 11·1	53·697	114·97	156·93	207·25	266·48	335·16	502·99	603·12	714·70	974·06	1284·6	1460·2	2072·9	2355·9
0·09500	3·122	3·760	4·055	4·338	4·610	4·872	5·372	5·611	5·844	6·293	6·722	6·931	7·533	7·765
1/ 10·5	55·176	118·13	161·25	212·94	273·80	344·38	516·81	619·69	734·34	1000·8	1319·9	1500·3	2129·8	2420·6
0·10000	3·204	3·858	4·161	4·451	4·730	4·999	5·512	5·757	5·996	6·457	6·897	7·111	7·729	7·967
1/ 10·0	56·616	121·21	165·45	218·50	280·94	353·35	530·28	635·83	753·47	1026·9	1354·3	1539·4	2185·3	2483·6
	0·78	0·79	0·79	0·80	0·80	0·81	0·81	0·82	0·82	0·83	0·83	0·83	0·84	0·84

$V_{r(0{\cdot}5)medial}$ for half-full circular pipes.

A19
(p.1 of 6)

k_s = 1·50 mm
S = 0·00010 to 0·00030

ie hydraulic gradient =
1 in 10000 to 1 in 3333

Water (or sewage) at 15°C;
full bore conditions.

velocities in ms^{-1}
discharges in litres/sec

Gradient **(Equivalent) Pipe diameters in mm**

Gradient	150	200	225	250	275	300	350	375	400	450	500	525	600	630
0·000100	0·083	0·101	0·110	0·118	0·126	0·134	0·148	0·155	0·162	0·175	0·188	0·194	0·212	0·219
1/ 10000	1·4666	3·1876	4·3755	5·8059	7·4958	9·4618	14·285	17·172	20·397	27·911	36·938	42·051	59·940	68·220
0·000105	0·085	0·104	0·113	0·121	0·129	0·137	0·152	0·159	0·166	0·180	0·193	0·199	0·217	0·224
1/ 9524	1·5047	3·2699	4·4881	5·9549	7·6878	9·7037	14·649	17·610	20·916	28·620	37·875	43·117	61·456	69·944
0·000110	0·087	0·107	0·116	0·124	0·133	0·141	0·156	0·163	0·170	0·184	0·198	0·204	0·223	0·230
1/ 9091	1·5419	3·3502	4·5981	6·1005	7·8754	9·9401	15·005	18·037	21·422	29·312	38·789	44·157	62·937	71·628
0·000115	0·089	0·109	0·118	0·127	0·136	0·144	0·160	0·167	0·174	0·189	0·202	0·209	0·228	0·235
1/ 8696	1·5783	3·4288	4·7056	6·2428	8·0588	10·171	15·353	18·455	21·918	29·989	39·684	45·174	64·384	73·274
0·000120	0·091	0·112	0·121	0·130	0·139	0·147	0·163	0·171	0·178	0·193	0·207	0·213	0·233	0·240
1/ 8333	1·6140	3·5056	4·8108	6·3821	8·2382	10·397	15·693	18·863	22·403	30·651	40·559	46·170	65·800	74·885
0·000125	0·093	0·114	0·124	0·133	0·142	0·150	0·167	0·174	0·182	0·197	0·211	0·218	0·238	0·245
1/ 8000	1·6489	3·5809	4·9139	6·5185	8·4140	10·619	16·027	19·264	22·878	31·300	41·416	47·145	67·187	76·462
0·000130	0·095	0·116	0·126	0·136	0·145	0·153	0·170	0·178	0·186	0·201	0·215	0·222	0·242	0·250
1/ 7692	1·6831	3·6548	5·0149	6·6523	8·5863	10·836	16·353	19·656	23·343	31·936	42·256	48·101	68·547	78·009
0·000135	0·097	0·119	0·129	0·138	0·147	0·156	0·173	0·181	0·189	0·205	0·219	0·227	0·247	0·255
1/ 7407	1·7167	3·7272	5·1141	6·7835	8·7554	11·049	16·674	20·041	23·800	32·560	43·080	49·038	69·881	79·526
0·000140	0·099	0·121	0·131	0·141	0·150	0·159	0·177	0·185	0·193	0·209	0·224	0·231	0·252	0·260
1/ 7143	1·7497	3·7983	5·2114	6·9123	8·9214	11·258	16·989	20·419	24·248	33·172	43·889	49·959	71·190	81·015
0·000145	0·101	0·123	0·133	0·143	0·153	0·162	0·180	0·188	0·196	0·212	0·228	0·235	0·256	0·265
1/ 6897	1·7821	3·8682	5·3070	7·0389	9·0845	11·464	17·298	20·790	24·689	33·774	44·684	50·863	72·476	82·478
0·000150	0·103	0·125	0·136	0·146	0·156	0·165	0·183	0·192	0·200	0·216	0·232	0·239	0·261	0·269
1/ 6667	1·8139	3·9368	5·4010	7·1633	9·2448	11·665	17·602	21·155	25·122	34·365	45·465	51·752	73·741	83·916
0·000160	0·106	0·130	0·140	0·151	0·161	0·171	0·189	0·198	0·207	0·223	0·239	0·247	0·270	0·278
1/ 6250	1·8761	4·0709	5·5845	7·4061	9·5576	12·060	18·195	21·867	25·967	35·519	46·990	53·486	76·208	86·723
0·000170	0·110	0·134	0·145	0·156	0·166	0·176	0·195	0·204	0·213	0·230	0·247	0·255	0·278	0·287
1/ 5882	1·9364	4·2008	5·7623	7·6415	9·8608	12·442	18·770	22·558	26·786	36·638	48·468	55·167	78·600	89·443
0·000180	0·113	0·138	0·149	0·160	0·171	0·181	0·201	0·210	0·219	0·237	0·254	0·262	0·286	0·295
1/ 5556	1·9949	4·3270	5·9350	7·8701	10·155	12·813	19·329	23·228	27·581	37·724	49·903	56·799	80·922	92·084
0·000190	0·116	0·142	0·153	0·165	0·176	0·186	0·207	0·216	0·226	0·244	0·261	0·270	0·294	0·304
1/ 5263	2·0519	4·4498	6·1030	8·0925	10·442	13·174	19·872	23·880	28·355	38·780	51·298	58·387	83·181	94·653
0·000200	0·119	0·145	0·158	0·169	0·180	0·191	0·212	0·222	0·232	0·250	0·268	0·277	0·302	0·312
1/ 5000	2·1074	4·5694	6·2667	8·3091	10·721	13·525	20·401	24·516	29·109	39·810	52·658	59·934	85·381	97·156
0·000210	0·122	0·149	0·162	0·174	0·185	0·196	0·217	0·228	0·237	0·257	0·275	0·284	0·310	0·320
1/ 4762	2·1615	4·6860	6·4263	8·5204	10·993	13·868	20·917	25·135	29·844	40·813	53·984	61·442	87·527	99·596
0·000220	0·125	0·153	0·166	0·178	0·190	0·201	0·223	0·233	0·243	0·263	0·282	0·291	0·317	0·327
1/ 4545	2·2143	4·8000	6·5822	8·7267	11·259	14·203	21·421	25·740	30·562	41·794	55·279	62·916	89·623	101·98
0·000230	0·128	0·156	0·169	0·182	0·194	0·206	0·228	0·238	0·249	0·269	0·288	0·297	0·324	0·335
1/ 4348	2·2660	4·9113	6·7347	8·9284	11·519	14·530	21·914	26·332	31·263	42·752	56·545	64·356	91·671	104·31
0·000240	0·131	0·160	0·173	0·186	0·198	0·210	0·233	0·244	0·254	0·275	0·294	0·304	0·331	0·342
1/ 4167	2·3166	5·0204	6·8838	9·1258	11·773	14·851	22·396	26·911	31·950	43·690	57·784	65·765	93·676	106·59
0·000250	0·134	0·163	0·177	0·190	0·202	0·215	0·238	0·249	0·260	0·280	0·300	0·310	0·338	0·349
1/ 4000	2·3662	5·1271	7·0299	9·3192	12·022	15·165	22·868	27·478	32·622	44·608	58·997	67·145	95·639	108·82
0·000260	0·137	0·167	0·180	0·194	0·207	0·219	0·243	0·254	0·265	0·286	0·307	0·316	0·345	0·356
1/ 3846	2·4147	5·2318	7·1731	9·5087	12·266	15·472	23·331	28·033	33·282	45·508	60·187	68·498	97·564	111·01
0·000270	0·139	0·170	0·184	0·197	0·211	0·223	0·247	0·259	0·270	0·292	0·312	0·323	0·352	0·363
1/ 3704	2·4624	5·3345	7·3136	9·6947	12·506	15·774	23·785	28·578	33·929	46·391	61·353	69·825	99·452	113·16
0·000280	0·142	0·173	0·187	0·201	0·215	0·227	0·252	0·264	0·275	0·297	0·318	0·329	0·358	0·370
1/ 3571	2·5092	5·4353	7·4516	9·8772	12·741	16·070	24·231	29·114	34·563	47·258	62·499	71·128	101·31	115·27
0·000290	0·145	0·176	0·191	0·205	0·218	0·231	0·256	0·268	0·280	0·302	0·324	0·334	0·365	0·376
1/ 3448	2·5551	5·5343	7·5871	10·057	12·972	16·361	24·669	29·639	35·187	48·110	63·624	72·408	103·13	117·34
0·000300	0·147	0·179	0·194	0·208	0·222	0·236	0·261	0·273	0·285	0·308	0·330	0·340	0·371	0·383
1/ 3333	2·6003	5·6316	7·7203	10·233	13·199	16·647	25·100	30·156	35·800	48·947	64·730	73·666	104·92	119·37
	0·74	0·75	0·75	0·76	0·76	0·76	0·77	0·77	0·77	0·78	0·78	0·78	0·79	0·79

$V_{r(0·5)medial}$ for half-full circular pipes.

k_s = 1·50 mm S = 0·00010 to 0·00030

k_s = **1·50 mm**
S = 0·00030 to 0·00100

ie hydraulic gradient =
1 in 3333 to 1 in 1000

Water (or sewage) at 15°C;
full bore conditions.

velocities in ms^{-1}
discharges in litres/sec

Gradient **(Equivalent) Pipe diameters in mm**

Gradient	150	200	225	250	275	300	350	375	400	450	500	525	600	630
0·000300	0·147	0·179	0·194	0·208	0·222	0·236	0·261	0·273	0·285	0·308	0·330	0·340	0·371	0·383
1/ 3333	2·6003	5·6316	7·7203	10·233	13·199	16·647	25·100	30·156	35·800	48·947	64·730	73·666	104·92	119·37
0·000320	0·152	0·185	0·201	0·215	0·230	0·243	0·270	0·282	0·294	0·318	0·341	0·352	0·383	0·396
1/ 3125	2·6885	5·8216	7·9802	10·577	13·642	17·205	25·940	31·165	36·996	50·581	66·888	76·121	108·41	123·34
0·000340	0·157	0·191	0·207	0·222	0·237	0·251	0·278	0·291	0·304	0·328	0·351	0·363	0·395	0·408
1/ 2941	2·7739	6·0057	8·2322	10·910	14·071	17·746	26·754	32·142	38·156	52·164	68·979	78·500	111·79	127·19
0·000360	0·162	0·197	0·213	0·229	0·244	0·258	0·286	0·300	0·313	0·338	0·362	0·373	0·407	0·420
1/ 2778	2·8569	6·1846	8·4768	11·234	14·489	18·272	27·544	33·091	39·282	53·702	71·010	80·810	115·08	130·93
0·000380	0·166	0·202	0·219	0·235	0·251	0·266	0·294	0·308	0·321	0·347	0·372	0·384	0·418	0·432
1/ 2632	2·9377	6·3585	8·7148	11·549	14·894	18·783	28·313	34·014	40·377	55·197	72·986	83·057	118·27	134·57
0·000400	0·171	0·208	0·225	0·242	0·257	0·273	0·302	0·316	0·330	0·356	0·382	0·394	0·429	0·443
1/ 2500	3·0163	6·5279	8·9466	11·856	15·289	19·281	29·062	34·914	41·444	56·654	74·910	85·246	121·39	138·11
0·000420	0·175	0·213	0·231	0·248	0·264	0·280	0·310	0·324	0·338	0·365	0·391	0·404	0·440	0·454
1/ 2381	3·0930	6·6932	9·1728	12·155	15·675	19·766	29·793	35·791	42·484	58·074	76·787	87·381	124·42	141·56
0·000440	0·179	0·218	0·236	0·254	0·270	0·286	0·317	0·332	0·346	0·374	0·400	0·413	0·451	0·465
1/ 2273	3·1680	6·8546	9·3936	12·447	16·051	20·240	30·507	36·647	43·500	59·462	78·619	89·466	127·39	144·93
0·000460	0·183	0·223	0·242	0·259	0·276	0·293	0·324	0·339	0·354	0·382	0·410	0·423	0·461	0·476
1/ 2174	3·2412	7·0124	9·6094	12·733	16·419	20·704	31·204	37·484	44·493	60·818	80·411	91·503	130·29	148·23
0·000480	0·187	0·228	0·247	0·265	0·282	0·299	0·331	0·347	0·362	0·391	0·418	0·432	0·471	0·486
1/ 2083	3·3129	7·1668	9·8207	13·012	16·779	21·157	31·886	38·304	45·465	62·145	82·164	93·498	133·13	151·46
0·000500	0·191	0·233	0·252	0·271	0·288	0·306	0·338	0·354	0·369	0·399	0·427	0·441	0·481	0·496
1/ 2000	3·3831	7·3180	10·028	13·286	17·132	21·602	32·555	39·106	46·417	63·444	83·881	95·450	135·90	154·61
0·000525	0·196	0·239	0·259	0·277	0·296	0·313	0·347	0·363	0·379	0·409	0·438	0·452	0·493	0·508
1/ 1905	3·4689	7·5028	10·280	13·621	17·563	22·145	33·372	40·087	47·580	65·033	85·979	97·838	139·30	158·48
0·000550	0·201	0·245	0·265	0·284	0·303	0·321	0·355	0·372	0·388	0·419	0·448	0·463	0·504	0·520
1/ 1818	3·5527	7·6833	10·527	13·947	17·984	22·675	34·170	41·044	48·717	66·584	88·029	100·17	142·61	162·25
0·000575	0·206	0·250	0·271	0·291	0·310	0·328	0·363	0·380	0·397	0·428	0·459	0·473	0·516	0·532
1/ 1739	3·6346	7·8598	10·769	14·267	18·395	23·193	34·949	41·981	49·827	68·101	90·032	102·45	145·86	165·93
0·000600	0·210	0·256	0·277	0·297	0·316	0·335	0·371	0·388	0·405	0·438	0·469	0·484	0·527	0·544
1/ 1667	3·7148	8·0324	11·005	14·579	18·798	23·700	35·713	42·897	50·914	69·585	91·992	104·68	149·03	169·54
0·000625	0·215	0·261	0·283	0·303	0·323	0·342	0·379	0·397	0·414	0·447	0·478	0·494	0·538	0·555
1/ 1600	3·7933	8·2015	11·236	14·886	19·192	24·197	36·460	43·794	51·978	71·038	93·912	106·86	152·13	173·07
0·000650	0·219	0·266	0·288	0·309	0·330	0·349	0·387	0·404	0·422	0·456	0·488	0·504	0·549	0·566
1/ 1538	3·8703	8·3673	11·463	15·186	19·579	24·684	37·193	44·673	53·021	72·462	95·793	109·00	155·18	176·54
0·000675	0·223	0·272	0·294	0·315	0·336	0·356	0·394	0·412	0·430	0·464	0·497	0·513	0·559	0·577
1/ 1481	3·9458	8·5299	11·686	15·480	19·958	25·162	37·911	45·536	54·045	73·860	97·639	111·10	158·16	179·93
0·000700	0·227	0·277	0·299	0·321	0·342	0·363	0·401	0·420	0·438	0·473	0·507	0·523	0·570	0·588
1/ 1429	4·0199	8·6896	11·904	15·769	20·330	25·630	38·617	46·383	55·050	75·232	99·452	113·16	161·10	183·27
0·000725	0·232	0·282	0·305	0·327	0·348	0·369	0·409	0·427	0·446	0·481	0·516	0·532	0·580	0·598
1/ 1379	4·0928	8·8464	12·119	16·053	20·696	26·091	39·310	47·215	56·036	76·579	101·23	115·19	163·98	186·54
0·000750	0·236	0·286	0·310	0·333	0·354	0·376	0·416	0·435	0·454	0·490	0·524	0·541	0·590	0·609
1/ 1333	4·1643	9·0006	12·330	16·332	21·055	26·544	39·991	48·033	57·007	77·904	102·98	117·18	166·81	189·76
0·000800	0·244	0·296	0·320	0·344	0·366	0·388	0·429	0·449	0·469	0·506	0·542	0·559	0·609	0·629
1/ 1250	4·3040	9·3014	12·741	16·877	21·756	27·427	41·320	49·628	58·899	80·488	106·40	121·06	172·33	196·04
0·000850	0·251	0·305	0·330	0·355	0·378	0·400	0·443	0·463	0·483	0·522	0·559	0·577	0·628	0·648
1/ 1176	4·4394	9·5929	13·140	17·404	22·436	28·283	42·609	51·175	60·734	82·993	109·70	124·82	177·68	202·13
0·000900	0·259	0·314	0·340	0·365	0·389	0·412	0·456	0·477	0·497	0·537	0·575	0·594	0·647	0·667
1/ 1111	4·5709	9·8760	13·527	17·917	23·096	29·115	43·860	52·677	62·515	85·426	112·92	128·48	182·88	208·05
0·000950	0·266	0·323	0·350	0·375	0·400	0·423	0·469	0·490	0·511	0·552	0·591	0·610	0·665	0·686
1/ 1053	4·6988	10·151	13·904	18·415	23·738	29·924	45·077	54·137	64·248	87·791	116·04	132·03	187·94	213·80
0·001000	0·273	0·332	0·359	0·385	0·410	0·434	0·481	0·503	0·525	0·566	0·606	0·626	0·682	0·704
1/ 1000	4·8233	10·420	14·271	18·901	24·364	30·711	46·262	55·560	65·936	90·096	119·09	135·50	192·86	219·40
	0·75	0·75	0·76	0·76	0·76	0·77	0·77	0·77	0·78	0·78	0·78	0·79	0·79	0·79

$V_{r(0·5)medial}$ for half-full circular pipes.

k_s = 1·50 mm
S = 0·00100 to 0·00300

ie hydraulic gradient =
1 in 1000 to 1 in 333

Water (or sewage) at 15°C;
full bore conditions.

velocities in ms^{-1}
discharges in litres/sec

Gradient	**(Equivalent) Pipe diameters in mm**													
	150	**200**	**225**	**250**	**275**	**300**	**350**	**375**	**400**	**450**	**500**	**525**	**600**	**630**
0·00100	0·273	0·332	0·359	0·385	0·410	0·434	0·481	0·503	0·525	0·566	0·606	0·626	0·682	0·704
1/ 1000	4·8233	10·420	14·271	18·901	24·364	30·711	46·262	55·560	65·936	90·096	119·09	135·50	192·86	219·40
0·00105	0·280	0·340	0·368	0·395	0·420	0·445	0·493	0·516	0·538	0·581	0·622	0·642	0·699	0·721
1/ 952	4·9448	10·681	14·629	19·374	24·973	31·480	47·418	56·948	67·582	92·343	122·05	138·87	197·67	224·86
0·00110	0·287	0·348	0·377	0·404	0·430	0·456	0·505	0·528	0·551	0·594	0·636	0·657	0·716	0·738
1/ 909	5·0634	10·937	14·978	19·837	25·569	32·230	48·546	58·303	69·189	94·537	124·95	142·17	202·36	230·19
0·00115	0·293	0·356	0·385	0·413	0·440	0·466	0·516	0·540	0·563	0·608	0·651	0·672	0·732	0·755
1/ 870	5·1794	11·186	15·320	20·289	26·151	32·963	49·650	59·627	70·760	96·683	127·79	145·39	206·94	235·41
0·00120	0·300	0·364	0·394	0·422	0·450	0·476	0·527	0·552	0·575	0·621	0·665	0·686	0·748	0·772
1/ 833	5·2929	11·431	15·654	20·731	26·721	33·681	50·729	60·923	72·297	98·782	130·56	148·55	211·43	240·51
0·00125	0·306	0·371	0·402	0·431	0·459	0·486	0·538	0·563	0·587	0·634	0·679	0·700	0·763	0·788
1/ 800	5·4041	11·670	15·982	21·164	27·279	34·384	51·787	62·192	73·803	100·84	133·28	151·64	215·82	245·51
0·00130	0·312	0·379	0·410	0·440	0·468	0·496	0·549	0·574	0·599	0·647	0·692	0·714	0·779	0·803
1/ 769	5·5130	11·905	16·302	21·589	27·826	35·072	52·823	63·437	75·279	102·85	135·94	154·66	220·13	250·40
0·00135	0·318	0·386	0·418	0·448	0·478	0·506	0·560	0·585	0·611	0·659	0·706	0·728	0·793	0·819
1/ 741	5·6199	12·135	16·617	22·005	28·362	35·748	53·840	64·657	76·727	104·83	138·55	157·63	224·35	255·21
0·00140	0·324	0·393	0·426	0·457	0·486	0·515	0·570	0·596	0·622	0·671	0·719	0·742	0·808	0·834
1/ 714	5·7248	12·361	16·926	22·414	28·889	36·412	54·838	65·855	78·148	106·77	141·11	160·55	228·50	259·93
0·00145	0·330	0·401	0·433	0·465	0·495	0·524	0·580	0·607	0·633	0·683	0·732	0·755	0·823	0·849
1/ 690	5·8279	12·583	17·230	22·816	29·406	37·063	55·819	67·032	79·544	108·68	143·63	163·41	232·57	264·56
0·00150	0·336	0·407	0·441	0·473	0·504	0·533	0·590	0·617	0·644	0·695	0·744	0·768	0·837	0·863
1/ 667	5·9292	12·801	17·528	23·211	29·915	37·704	56·782	68·189	80·916	110·55	146·10	166·23	236·58	269·11
0·00160	0·347	0·421	0·455	0·489	0·520	0·551	0·610	0·638	0·665	0·718	0·769	0·793	0·864	0·892
1/ 625	6·1269	13·226	18·111	23·981	30·907	38·954	58·663	70·446	83·594	114·21	150·93	171·72	244·39	278·00
0·00170	0·358	0·434	0·470	0·504	0·537	0·568	0·629	0·658	0·686	0·740	0·793	0·818	0·891	0·919
1/ 588	6·3185	13·639	18·675	24·728	31·868	40·165	60·485	72·634	86·189	117·75	155·61	177·05	251·96	286·61
0·00180	0·368	0·447	0·483	0·519	0·552	0·585	0·647	0·677	0·706	0·762	0·816	0·842	0·917	0·946
1/ 556	6·5045	14·039	19·223	25·453	32·802	41·341	62·255	74·758	88·709	121·19	160·16	182·22	259·32	294·97
0·00190	0·378	0·459	0·497	0·533	0·568	0·601	0·665	0·696	0·725	0·783	0·838	0·865	0·942	0·972
1/ 526	6·6855	14·429	19·756	26·158	33·710	42·485	63·976	76·824	91·160	124·54	164·58	187·24	266·47	303·10
0·00200	0·388	0·471	0·510	0·547	0·582	0·617	0·682	0·714	0·744	0·804	0·860	0·888	0·967	0·998
1/ 500	6·8617	14·809	20·275	26·845	34·595	43·599	65·652	78·837	93·547	127·80	168·88	192·14	273·43	311·03
0·00210	0·398	0·483	0·523	0·561	0·597	0·632	0·699	0·732	0·763	0·824	0·882	0·910	0·991	1·023
1/ 476	7·0336	15·179	20·781	27·515	35·458	44·686	67·288	80·800	95·875	130·97	173·08	196·92	280·23	318·75
0·00220	0·408	0·495	0·535	0·574	0·611	0·647	0·716	0·749	0·781	0·843	0·902	0·931	1·015	1·047
1/ 455	7·2015	15·540	21·276	28·169	36·300	45·747	68·884	82·716	98·149	134·08	177·18	201·58	286·86	326·30
0·00230	0·417	0·506	0·547	0·587	0·625	0·662	0·732	0·766	0·799	0·862	0·923	0·952	1·037	1·070
1/ 435	7·3656	15·893	21·759	28·808	37·124	46·785	70·445	84·590	100·37	137·11	181·19	206·14	293·34	333·67
0·00240	0·426	0·517	0·559	0·600	0·639	0·676	0·748	0·782	0·816	0·881	0·943	0·973	1·060	1·094
1/ 417	7·5262	16·239	22·232	29·434	37·930	47·800	71·972	86·423	102·55	140·08	185·11	210·60	299·69	340·89
0·00250	0·435	0·528	0·571	0·612	0·652	0·690	0·764	0·799	0·833	0·899	0·962	0·993	1·082	1·116
1/ 400	7·6834	16·578	22·695	30·047	38·719	48·794	73·468	88·218	104·68	142·99	188·95	214·97	305·90	347·96
0·00260	0·444	0·538	0·582	0·624	0·665	0·704	0·779	0·815	0·850	0·917	0·982	1·013	1·103	1·138
1/ 385	7·8376	16·910	23·149	30·648	39·493	49·768	74·934	89·978	106·76	145·84	192·72	219·25	312·00	354·88
0·00270	0·452	0·549	0·593	0·636	0·678	0·718	0·794	0·830	0·866	0·935	1·000	1·032	1·125	1·160
1/ 370	7·9888	17·235	23·594	31·237	40·251	50·724	76·372	91·704	108·81	148·64	196·41	223·45	317·97	361·68
0·00280	0·460	0·559	0·604	0·648	0·690	0·731	0·808	0·846	0·882	0·952	1·019	1·051	1·145	1·182
1/ 357	8·1373	17·555	24·032	31·815	40·996	51·663	77·784	93·399	110·82	151·38	200·04	227·58	323·84	368·35
0·00290	0·469	0·569	0·615	0·660	0·703	0·744	0·823	0·861	0·898	0·969	1·037	1·070	1·166	1·203
1/ 345	8·2831	17·869	24·461	32·383	41·728	52·584	79·170	95·064	112·80	154·08	203·60	231·63	329·60	374·90
0·00300	0·477	0·579	0·626	0·671	0·715	0·757	0·837	0·876	0·913	0·985	1·055	1·088	1·186	1·223
1/ 333	8·4264	18·177	24·883	32·942	42·447	53·490	80·534	96·700	114·74	156·73	207·10	235·61	335·26	381·34
	0·75	0·75	0·76	0·76	0·76	0·77	0·77	0·77	0·78	0·78	0·79	0·79	0·79	0·80

$V_{r(0\cdot5)medial}$ for half-full circular pipes.

$k_s = 1{\cdot}50$ mm
S = 0·00300 to 0·01000

ie hydraulic gradient =
1 in 333 to 1 in 100

Water (or sewage) at 15°C;
full bore conditions.

velocities in ms^{-1}
discharges in litres/sec

Gradient **(Equivalent) Pipe diameters in mm**

Gradient	150	200	225	250	275	300	350	375	400	450	500	525	600	630
0·00300	0·477	0·579	0·626	0·671	0·715	0·757	0·837	0·876	0·913	0·985	1·055	1·088	1·186	1·223
1/ 333	8·4264	18·177	24·883	32·942	42·447	53·490	80·534	96·700	114·74	156·73	207·10	235·61	335·26	381·34
0·00320	0·493	0·598	0·647	0·693	0·738	0·782	0·865	0·904	0·943	1·018	1·090	1·124	1·225	1·264
1/ 313	8·7061	18·779	25·707	34·032	43·851	55·258	83·193	99·893	118·52	161·90	213·93	243·38	346·31	393·91
0·00340	0·508	0·616	0·667	0·715	0·761	0·806	0·891	0·932	0·972	1·049	1·123	1·159	1·263	1·303
1/ 294	8·9771	19·363	26·505	35·088	45·211	56·972	85·771	102·99	122·19	166·91	220·55	250·91	357·02	406·09
0·00360	0·523	0·634	0·686	0·736	0·783	0·830	0·918	0·960	1·001	1·080	1·156	1·193	1·299	1·341
1/ 278	9·2403	19·930	27·280	36·113	46·532	58·635	88·274	105·99	125·76	171·77	226·98	258·22	367·42	417·92
0·00380	0·537	0·652	0·705	0·756	0·805	0·852	0·943	0·986	1·028	1·110	1·188	1·226	1·335	1·378
1/ 263	9·4963	20·481	28·034	37·111	47·817	60·253	90·709	108·91	129·22	176·51	233·23	265·33	377·53	429·42
0·00400	0·551	0·669	0·724	0·776	0·826	0·875	0·967	1·012	1·055	1·139	1·219	1·258	1·370	1·414
1/ 250	9·7457	21·017	28·768	38·082	49·068	61·829	93·080	111·76	132·60	181·12	239·32	272·26	387·39	440·62
0·00420	0·565	0·686	0·742	0·795	0·847	0·896	0·991	1·037	1·081	1·167	1·249	1·289	1·404	1·449
1/ 238	9·9889	21·541	29·484	39·030	50·288	63·367	95·392	114·54	135·89	185·61	245·26	279·01	396·99	451·55
0·00440	0·579	0·702	0·759	0·814	0·867	0·918	1·015	1·062	1·107	1·195	1·279	1·319	1·437	1·483
1/ 227	10·226	22·052	30·184	39·955	51·480	64·868	97·651	117·25	139·11	190·00	251·06	285·61	406·38	462·22
0·00460	0·592	0·718	0·776	0·832	0·886	0·938	1·038	1·086	1·132	1·222	1·307	1·349	1·470	1·516
1/ 217	10·458	22·552	30·867	40·860	52·645	66·335	99·858	119·90	142·25	194·29	256·73	292·06	415·55	472·65
0·00480	0·605	0·733	0·793	0·850	0·906	0·959	1·060	1·109	1·156	1·248	1·336	1·378	1·501	1·549
1/ 208	10·686	23·041	31·536	41·745	53·784	67·771	102·02	122·49	145·33	198·49	262·27	298·37	424·52	482·86
0·00500	0·617	0·749	0·810	0·868	0·924	0·979	1·082	1·132	1·180	1·274	1·363	1·407	1·533	1·581
1/ 200	10·908	23·520	32·191	42·611	54·901	69·177	104·13	125·03	148·34	202·61	267·70	304·55	433·31	492·85
0·00525	0·633	0·767	0·830	0·890	0·947	1·003	1·109	1·160	1·210	1·306	1·397	1·442	1·571	1·620
1/ 190	11·180	24·105	32·992	43·671	56·265	70·895	106·72	128·13	152·02	207·63	274·34	312·10	444·05	505·07
0·00550	0·648	0·785	0·849	0·911	0·970	1·027	1·135	1·188	1·238	1·336	1·430	1·476	1·608	1·659
1/ 182	11·446	24·677	33·774	44·705	57·598	72·573	109·24	131·16	155·62	212·54	280·83	319·47	454·54	517·00
0·00575	0·662	0·803	0·869	0·931	0·992	1·050	1·161	1·214	1·266	1·367	1·463	1·509	1·644	1·696
1/ 174	11·705	25·236	34·538	45·716	58·900	74·214	111·71	134·13	159·13	217·34	287·17	326·68	464·80	528·66
0·00600	0·677	0·821	0·887	0·951	1·013	1·073	1·186	1·241	1·294	1·396	1·494	1·542	1·679	1·733
1/ 167	11·959	25·782	35·286	46·706	60·174	75·819	114·13	137·02	162·57	222·04	293·37	333·74	474·83	540·07
0·00625	0·691	0·838	0·906	0·971	1·034	1·095	1·211	1·266	1·320	1·425	1·525	1·574	1·714	1·768
1/ 160	12·208	26·318	36·019	47·675	61·422	77·391	116·49	139·86	165·94	226·63	299·44	340·65	484·66	551·25
0·00650	0·705	0·854	0·924	0·991	1·055	1·117	1·235	1·292	1·347	1·453	1·555	1·605	1·748	1·804
1/ 154	12·452	26·843	36·737	48·625	62·646	78·932	118·81	142·65	169·24	231·14	305·39	347·42	494·29	562·20
0·00675	0·718	0·871	0·942	1·010	1·075	1·138	1·259	1·316	1·373	1·481	1·585	1·636	1·782	1·838
1/ 148	12·691	27·358	37·441	49·557	63·846	80·444	121·08	145·38	172·48	235·56	311·23	354·06	503·74	572·94
0·00700	0·731	0·887	0·959	1·028	1·095	1·159	1·282	1·341	1·398	1·508	1·614	1·666	1·814	1·872
1/ 143	12·926	27·863	38·132	50·471	65·024	81·928	123·32	148·06	175·66	239·90	316·97	360·58	513·01	583·49
0·00725	0·744	0·903	0·976	1·047	1·114	1·180	1·305	1·364	1·423	1·535	1·643	1·695	1·847	1·905
1/ 138	13·156	28·359	38·811	51·370	66·181	83·385	125·51	150·69	178·78	244·16	322·60	366·99	522·12	593·85
0·00750	0·757	0·918	0·993	1·064	1·133	1·200	1·327	1·388	1·447	1·562	1·671	1·724	1·878	1·938
1/ 133	13·383	28·847	39·479	52·253	67·318	84·818	127·67	153·28	181·85	248·36	328·13	373·28	531·08	604·04
0·00800	0·782	0·949	1·026	1·100	1·171	1·239	1·371	1·433	1·495	1·613	1·726	1·781	1·940	2·001
1/ 125	13·825	29·800	40·781	53·976	69·538	87·613	131·87	158·32	187·84	256·53	338·93	385·57	548·55	623·91
0·00850	0·807	0·978	1·057	1·134	1·207	1·278	1·413	1·478	1·541	1·663	1·779	1·836	2·000	2·063
1/ 118	14·254	30·722	42·044	55·647	71·689	90·323	135·95	163·22	193·64	264·46	349·40	397·47	565·48	643·17
0·00900	0·830	1·006	1·088	1·167	1·242	1·315	1·454	1·521	1·586	1·711	1·831	1·890	2·058	2·123
1/ 111	14·670	31·618	43·270	57·268	73·777	92·954	139·91	167·97	199·28	272·15	359·56	409·03	581·93	661·87
0·00950	0·853	1·034	1·118	1·199	1·276	1·351	1·494	1·563	1·629	1·758	1·882	1·941	2·115	2·182
1/ 105	15·075	32·490	44·462	58·846	75·809	95·512	143·75	172·59	204·76	279·63	369·45	420·28	597·92	680·06
0·01000	0·875	1·061	1·147	1·230	1·310	1·386	1·533	1·603	1·672	1·804	1·931	1·992	2·170	2·238
1/ 100	15·469	33·339	45·623	60·382	77·787	98·004	147·50	177·09	210·10	286·92	379·08	431·23	613·50	697·77
	0·75	0·76	0·76	0·76	0·76	0·77	0·77	0·78	0·78	0·78	0·79	0·79	0·80	0·80

$V_{r(0{\cdot}5)medial}$ for half-full circular pipes.

k_s = 1·50 mm
S = 0·01000 to 0·03000

ie hydraulic gradient =
1 in 100 to 1 in 33·3

Water (or sewage) at 15°C;
full bore conditions.

velocities in ms^{-1}
discharges in litres/sec

Gradient **(Equivalent) Pipe diameters in mm**

Gradient	150	200	225	250	275	300	350	375	400	450	500	525	600	630
0·01000	0·875	1·061	1·147	1·230	1·310	1·386	1·533	1·603	1·672	1·804	1·931	1·992	2·170	2·238
1/ 100	15·469	33·339	45·623	60·382	77·787	98·004	147·50	177·09	210·10	286·92	379·08	431·23	613·50	697·77
0·01050	0·897	1·088	1·176	1·261	1·342	1·421	1·571	1·643	1·713	1·849	1·978	2·041	2·224	2·294
1/ 95	15·854	34·167	46·755	61·880	79·717	100·44	151·16	181·48	215·30	294·03	388·47	441·91	628·69	715·05
0·01100	0·918	1·113	1·204	1·290	1·374	1·454	1·608	1·682	1·754	1·892	2·025	2·090	2·276	2·348
1/ 91	16·229	34·975	47·861	63·343	81·601	102·81	154·73	185·77	220·39	300·97	397·64	452·34	643·52	731·92
0·01150	0·939	1·138	1·231	1·320	1·405	1·487	1·645	1·720	1·793	1·935	2·071	2·137	2·327	2·401
1/ 87	16·597	35·766	48·942	64·774	83·443	105·13	158·22	189·96	225·36	307·76	406·60	462·54	658·02	748·41
0·01200	0·959	1·163	1·258	1·348	1·435	1·519	1·680	1·757	1·832	1·977	2·115	2·183	2·377	2·453
1/ 83	16·956	36·539	50·000	66·173	85·246	107·40	161·64	194·06	230·22	314·40	415·37	472·51	672·21	764·55
0·01250	0·979	1·187	1·284	1·376	1·465	1·551	1·715	1·793	1·870	2·018	2·159	2·228	2·427	2·503
1/ 80	17·308	37·296	51·036	67·544	87·011	109·62	164·98	198·07	234·98	320·90	423·96	482·28	686·11	780·36
0·01300	0·999	1·211	1·309	1·403	1·494	1·582	1·749	1·829	1·907	2·058	2·202	2·272	2·475	2·553
1/ 77	17·652	38·038	52·051	68·887	88·741	111·80	168·26	202·01	239·65	327·28	432·38	491·86	699·73	795·85
0·01350	1·018	1·234	1·334	1·430	1·523	1·612	1·782	1·864	1·944	2·097	2·244	2·316	2·522	2·602
1/ 74	17·991	38·767	53·047	70·205	90·438	113·94	171·48	205·87	244·23	333·53	440·64	501·26	713·09	811·04
0·01400	1·037	1·257	1·359	1·457	1·551	1·642	1·815	1·898	1·979	2·136	2·285	2·358	2·568	2·650
1/ 71	18·323	39·481	54·025	71·499	92·104	116·04	174·64	209·66	248·73	339·67	448·75	510·48	726·21	825·96
0·01450	1·055	1·279	1·383	1·482	1·578	1·671	1·847	1·932	2·014	2·174	2·326	2·400	2·614	2·697
1/ 69	18·649	40·184	54·986	72·769	93·741	118·10	177·74	213·38	253·14	345·70	456·71	519·54	739·10	840·61
0·01500	1·073	1·301	1·407	1·508	1·605	1·699	1·879	1·965	2·049	2·211	2·366	2·441	2·659	2·743
1/ 67	18·969	40·874	55·930	74·018	95·350	120·13	180·79	217·04	257·48	351·62	464·54	528·44	751·76	855·02
0·01600	1·109	1·344	1·453	1·558	1·658	1·755	1·941	2·030	2·116	2·284	2·444	2·521	2·746	2·833
1/ 63	19·595	42·220	57·772	76·456	98·488	124·08	186·73	224·18	265·95	363·19	479·81	545·81	776·47	883·12
0·01700	1·143	1·385	1·498	1·606	1·709	1·810	2·001	2·092	2·182	2·354	2·519	2·599	2·831	2·920
1/ 59	20·201	43·526	59·557	78·818	101·53	127·91	192·50	231·10	274·16	374·39	494·62	562·65	800·42	910·36
0·01800	1·176	1·426	1·541	1·652	1·759	1·862	2·059	2·153	2·245	2·422	2·592	2·675	2·913	3·005
1/ 56	20·790	44·793	61·291	81·111	104·48	131·63	198·10	237·82	282·13	385·28	508·99	579·00	823·68	936·80
0·01900	1·209	1·465	1·584	1·698	1·807	1·913	2·116	2·212	2·307	2·489	2·663	2·748	2·993	3·088
1/ 53	21·363	46·025	62·977	83·342	107·36	135·25	203·54	244·35	289·88	395·86	522·97	594·90	846·29	962·53
0·02000	1·240	1·503	1·625	1·742	1·855	1·963	2·171	2·270	2·367	2·554	2·733	2·820	3·071	3·168
1/ 50	21·920	47·226	64·619	85·515	110·16	138·78	208·84	250·72	297·43	406·17	536·59	610·39	868·32	987·58
0·02100	1·271	1·541	1·665	1·785	1·901	2·012	2·224	2·326	2·425	2·617	2·800	2·889	3·147	3·247
1/ 47·6	22·464	48·397	66·221	87·634	112·88	142·21	214·01	256·93	304·80	416·22	549·87	625·50	889·81	1012·0
0·02200	1·301	1·577	1·705	1·827	1·945	2·059	2·277	2·381	2·483	2·679	2·867	2·958	3·221	3·323
1/ 45·5	22·995	49·540	67·785	89·703	115·55	145·57	219·06	262·99	311·99	426·04	562·84	640·25	910·79	1035·9
0·02300	1·331	1·612	1·743	1·869	1·989	2·106	2·328	2·435	2·539	2·739	2·931	3·024	3·294	3·398
1/ 43·5	23·515	50·658	69·314	91·726	118·15	148·85	224·00	268·91	319·02	435·64	575·51	654·67	931·30	1059·2
0·02400	1·359	1·647	1·781	1·909	2·032	2·151	2·378	2·487	2·593	2·798	2·994	3·089	3·365	3·471
1/ 41·7	24·023	51·752	70·809	93·705	120·70	152·06	228·83	274·71	325·89	445·03	587·92	668·78	951·36	1082·0
0·02500	1·388	1·681	1·818	1·948	2·074	2·196	2·428	2·539	2·647	2·856	3·056	3·153	3·434	3·543
1/ 40·0	24·520	52·823	72·275	95·643	123·20	155·21	233·56	280·39	332·63	454·22	600·06	682·60	971·02	1104·4
0·02600	1·415	1·715	1·854	1·987	2·115	2·239	2·476	2·589	2·700	2·913	3·117	3·216	3·502	3·613
1/ 38·5	25·008	53·872	73·711	97·543	125·65	158·29	238·20	285·96	339·23	463·24	611·97	696·14	990·28	1126·3
0·02700	1·442	1·748	1·889	2·025	2·156	2·282	2·523	2·639	2·751	2·968	3·176	3·277	3·569	3·682
1/ 37·0	25·486	54·902	75·119	99·407	128·05	161·31	242·75	291·42	345·71	472·08	623·65	709·43	1009·2	1147·8
0·02800	1·469	1·780	1·924	2·062	2·196	2·324	2·569	2·687	2·802	3·023	3·235	3·337	3·635	3·750
1/ 35·7	25·956	55·913	76·502	101·24	130·40	164·28	247·21	296·78	352·07	480·76	635·12	722·47	1027·7	1168·9
0·02900	1·495	1·811	1·958	2·099	2·234	2·365	2·615	2·735	2·851	3·076	3·292	3·397	3·699	3·816
1/ 34·5	26·417	56·906	77·861	103·03	132·72	167·19	251·60	302·04	358·31	489·29	646·38	735·28	1045·9	1189·6
0·03000	1·521	1·842	1·992	2·135	2·273	2·406	2·660	2·782	2·900	3·129	3·348	3·455	3·763	3·882
1/ 33·3	26·871	57·882	79·196	104·80	134·99	170·06	255·91	307·22	364·45	497·67	657·45	747·87	1063·9	1210·0
	0·75	0·76	0·76	0·76	0·77	0·77	0·77	0·78	0·78	0·78	0·79	0·79	0·80	0·80

$V_{r(0·5)medial}$ for half-full circular pipes.

k_s = 1·50 mm S = 0·01000 to 0·03000

Gradient	(Equivalent) Pipe diameters in mm													
	150	200	225	250	275	300	350	375	400	450	500	525	600	630
0·03000	1·521	1·842	1·992	2·135	2·273	2·406	2·660	2·782	2·900	3·129	3·348	3·455	3·763	3·882
1/ 33·3	26·871	57·882	79·196	104·80	134·99	170·06	255·91	307·22	364·45	497·67	657·45	747·87	1063·9	1210·0
0·03200	1·571	1·903	2·057	2·205	2·348	2·485	2·747	2·873	2·996	3·232	3·458	3·568	3·886	4·009
1/ 31·3	27·755	59·787	81·801	108·25	139·43	175·65	264·32	317·31	376·43	514·02	679·05	772·44	1098·8	1249·7
0·03400	1·619	1·962	2·121	2·273	2·420	2·562	2·832	2·962	3·088	3·332	3·565	3·678	4·006	4·133
1/ 29·4	28·613	61·633	84·326	111·59	143·73	181·07	272·47	327·10	388·04	529·87	699·99	796·26	1132·7	1288·2
0·03600	1·666	2·019	2·182	2·339	2·490	2·636	2·914	3·048	3·178	3·428	3·669	3·785	4·122	4·253
1/ 27·8	29·445	63·425	86·777	114·83	147·91	186·33	280·39	336·60	399·31	545·26	720·32	819·38	1165·6	1325·6
0·03800	1·712	2·074	2·242	2·404	2·559	2·708	2·994	3·131	3·265	3·522	3·769	3·889	4·235	4·369
1/ 26·3	30·255	65·168	89·162	117·99	151·97	191·45	288·09	345·85	410·27	560·23	740·09	841·87	1197·6	1362·0
0·04000	1·757	2·128	2·301	2·466	2·625	2·779	3·072	3·213	3·350	3·614	3·867	3·990	4·346	4·483
1/ 25·0	31·044	66·866	91·484	121·06	155·93	196·43	295·59	354·85	420·95	574·81	759·34	863·77	1228·7	1397·4
0·04200	1·800	2·181	2·358	2·527	2·690	2·848	3·148	3·292	3·433	3·704	3·963	4·089	4·453	4·594
1/ 23·8	31·813	68·522	93·750	124·06	159·79	201·30	302·90	363·63	431·36	589·03	778·13	885·13	1259·1	1432·0
0·04400	1·843	2·233	2·413	2·587	2·754	2·915	3·223	3·370	3·514	3·791	4·056	4·185	4·558	4·702
1/ 22·7	32·564	70·139	95·961	126·98	163·56	206·04	310·04	372·20	441·53	602·91	796·47	906·00	1288·8	1465·7
0·04600	1·884	2·283	2·468	2·645	2·816	2·981	3·295	3·446	3·593	3·876	4·148	4·279	4·661	4·808
1/ 21·7	33·299	71·720	98·124	129·84	167·24	210·68	317·03	380·58	451·47	616·48	814·39	926·39	1317·8	1498·7
0·04800	1·925	2·332	2·521	2·702	2·876	3·045	3·366	3·520	3·670	3·960	4·237	4·372	4·761	4·911
1/ 20·8	34·017	73·266	100·24	132·64	170·85	215·22	323·86	388·78	461·20	629·76	831·93	946·34	1346·1	1531·0
0·05000	1·965	2·380	2·573	2·758	2·936	3·108	3·436	3·593	3·746	4·041	4·325	4·462	4·859	5·013
1/ 20·0	34·721	74·781	102·31	135·38	174·38	219·67	330·55	396·81	470·73	642·77	849·11	965·88	1373·9	1562·6
0·05250	2·013	2·439	2·637	2·826	3·009	3·185	3·521	3·682	3·839	4·141	4·431	4·572	4·979	5·137
1/ 19·0	35·581	76·632	104·84	138·73	178·70	225·11	338·72	406·63	482·37	658·67	870·11	989·77	1407·9	1601·2
0·05500	2·061	2·497	2·699	2·893	3·079	3·260	3·604	3·768	3·929	4·239	4·536	4·680	5·097	5·258
1/ 18·2	36·421	78·440	107·32	142·01	182·91	230·41	346·71	416·21	493·74	674·19	890·62	1013·1	1441·1	1639·0
0·05750	2·107	2·553	2·760	2·958	3·149	3·333	3·685	3·853	4·017	4·334	4·638	4·785	5·211	5·376
1/ 17·4	37·242	80·207	109·73	145·20	187·03	235·60	354·51	425·58	504·85	689·36	910·66	1035·9	1473·5	1675·8
0·06000	2·153	2·608	2·819	3·022	3·217	3·405	3·764	3·936	4·104	4·428	4·738	4·888	5·324	5·492
1/ 16·7	38·045	81·937	112·10	148·33	191·06	240·68	362·15	434·75	515·73	704·21	930·27	1058·2	1505·2	1711·9
0·06250	2·197	2·662	2·878	3·084	3·283	3·475	3·842	4·018	4·189	4·519	4·836	4·989	5·434	5·605
1/ 16·0	38·832	83·630	114·42	151·40	195·01	245·65	369·63	443·73	526·38	718·75	949·48	1080·0	1536·3	1747·3
0·06500	2·241	2·715	2·935	3·145	3·348	3·544	3·918	4·097	4·272	4·609	4·932	5·088	5·541	5·716
1/ 15·4	39·603	85·290	116·69	154·40	198·87	250·52	376·96	452·53	536·82	733·01	968·31	1101·5	1566·8	1781·9
0·06750	2·284	2·767	2·991	3·205	3·412	3·612	3·993	4·175	4·353	4·697	5·026	5·185	5·647	5·825
1/ 14·8	40·359	86·918	118·91	157·35	202·67	255·30	384·16	461·16	547·06	746·99	986·77	1122·5	1596·7	1815·9
0·07000	2·326	2·818	3·046	3·264	3·475	3·678	4·066	4·252	4·433	4·783	5·118	5·280	5·751	5·932
1/ 14·3	41·102	88·517	121·10	160·24	206·40	260·00	391·22	469·64	557·11	760·71	1004·9	1143·1	1626·0	1849·2
0·07250	2·367	2·868	3·100	3·322	3·537	3·743	4·138	4·328	4·512	4·868	5·209	5·374	5·853	6·037
1/ 13·8	41·831	90·087	123·25	163·08	210·05	264·61	398·15	477·96	566·99	774·19	1022·7	1163·3	1654·8	1882·0
0·07500	2·408	2·917	3·153	3·379	3·597	3·808	4·209	4·402	4·589	4·951	5·298	5·466	5·953	6·141
1/ 13·3	42·548	91·630	125·36	165·88	213·65	269·14	404·97	486·14	576·69	787·45	1040·2	1183·3	1683·1	1914·2
0·08000	2·487	3·013	3·256	3·490	3·715	3·933	4·347	4·546	4·740	5·114	5·472	5·645	6·148	6·342
1/ 12·5	43·947	94·642	129·48	171·33	220·67	277·98	418·27	502·11	595·63	813·30	1074·4	1222·1	1738·4	1977·1
0·08500	2·564	3·105	3·357	3·598	3·830	4·054	4·481	4·686	4·886	5·271	5·640	5·819	6·338	6·538
1/ 11·8	45·303	97·560	133·47	176·61	227·47	286·55	431·16	517·58	613·98	838·36	1107·5	1259·8	1791·9	2038·0
0·09000	2·638	3·196	3·454	3·702	3·941	4·172	4·611	4·822	5·028	5·424	5·804	5·988	6·522	6·727
1/ 11·1	46·619	100·39	137·35	181·74	234·08	294·87	443·68	532·61	631·81	862·70	1139·6	1296·3	1843·9	2097·1
0·09500	2·711	3·283	3·549	3·804	4·049	4·286	4·738	4·955	5·166	5·573	5·963	6·153	6·700	6·912
1/ 10·5	47·900	103·15	141·12	186·73	240·50	302·96	455·85	547·22	649·14	886·36	1170·9	1331·9	1894·5	2154·6
0·10000	2·781	3·369	3·642	3·903	4·155	4·397	4·861	5·083	5·300	5·718	6·118	6·313	6·875	7·092
1/ 10·0	49·147	105·84	144·79	191·58	246·76	310·84	467·71	561·45	666·02	909·41	1201·3	1366·5	1943·8	2210·6
	0·75	0·76	0·76	0·76	0·77	0·77	0·77	0·78	0·78	0·78	0·79	0·79	0·80	0·80

$V_{r(0{\cdot}5)medial}$ for half-full circular pipes.

A20
(p.1 of 6)

k_s = 3·0 mm
S = 0·00010 to 0·00030

ie hydraulic gradient =
1 in 10000 to 1 in 3333

Water (or sewage) at 15°C;
full bore conditions.

velocities in ms^{-1}
discharges in litres/sec

Gradient	(Equivalent) Pipe diameters in mm													
	150	200	225	250	275	300	350	375	400	450	500	525	600	630
0·000100	0·075	0·092	0·100	0·107	0·115	0·122	0·135	0·142	0·148	0·160	0·172	0·178	0·194	0·200
1/ 10000	1·3257	2·8894	3·9704	5·2734	6·8142	8·6080	13·014	15·653	18·603	25·482	33·753	38·441	54·854	62·456
0·000105	0·077	0·094	0·102	0·110	0·118	0·125	0·139	0·145	0·152	0·164	0·176	0·182	0·199	0·205
1/ 9524	1·3595	2·9627	4·0711	5·4069	6·9864	8·8253	13·341	16·047	19·071	26·122	34·600	39·405	56·229	64·020
0·000110	0·079	0·097	0·105	0·113	0·120	0·128	0·142	0·149	0·155	0·168	0·180	0·186	0·204	0·210
1/ 9091	1·3926	3·0344	4·1694	5·5372	7·1545	9·0375	13·662	16·432	19·528	26·747	35·427	40·347	57·570	65·547
0·000115	0·081	0·099	0·107	0·115	0·123	0·131	0·145	0·152	0·159	0·172	0·185	0·191	0·208	0·215
1/ 8696	1·4249	3·1045	4·2654	5·6646	7·3189	9·2449	13·974	16·808	19·974	27·358	36·236	41·267	58·882	67·040
0·000120	0·082	0·101	0·110	0·118	0·126	0·134	0·148	0·156	0·162	0·176	0·189	0·195	0·213	0·220
1/ 8333	1·4565	3·1730	4·3595	5·7892	7·4798	9·4478	14·281	17·176	20·411	27·956	37·027	42·167	60·166	68·501
0·000125	0·084	0·103	0·112	0·120	0·129	0·136	0·152	0·159	0·166	0·179	0·193	0·199	0·217	0·224
1/ 8000	1·4875	3·2402	4·4515	5·9113	7·6373	9·6466	14·581	17·537	20·839	28·542	37·801	43·049	61·423	69·932
0·000130	0·086	0·105	0·114	0·123	0·131	0·139	0·155	0·162	0·169	0·183	0·196	0·203	0·222	0·229
1/ 7692	1·5179	3·3060	4·5418	6·0310	7·7918	9·8415	14·875	17·890	21·259	29·116	38·561	43·914	62·655	71·334
0·000135	0·088	0·107	0·116	0·125	0·134	0·142	0·158	0·165	0·172	0·187	0·200	0·207	0·226	0·233
1/ 7407	1·5477	3·3705	4·6304	6·1484	7·9433	10·033	15·163	18·237	21·670	29·679	39·306	44·762	63·864	72·710
0·000140	0·089	0·109	0·119	0·128	0·136	0·145	0·161	0·168	0·176	0·190	0·204	0·211	0·230	0·238
1/ 7143	1·5769	3·4339	4·7173	6·2637	8·0920	10·220	15·446	18·577	22·074	30·232	40·037	45·595	65·050	74·060
0·000145	0·091	0·111	0·121	0·130	0·139	0·147	0·163	0·171	0·179	0·193	0·208	0·214	0·234	0·242
1/ 6897	1·6057	3·4962	4·8027	6·3769	8·2381	10·405	15·724	18·911	22·471	30·775	40·756	46·412	66·216	75·387
0·000150	0·092	0·113	0·123	0·132	0·141	0·150	0·166	0·174	0·182	0·197	0·211	0·218	0·238	0·246
1/ 6667	1·6339	3·5574	4·8866	6·4882	8·3817	10·586	15·998	19·240	22·862	31·308	41·462	47·216	67·361	76·691
0·000160	0·096	0·117	0·127	0·137	0·146	0·155	0·172	0·180	0·188	0·203	0·218	0·225	0·246	0·254
1/ 6250	1·6890	3·6769	5·0504	6·7054	8·6619	10·939	16·531	19·881	23·623	32·350	42·840	48·785	69·597	79·235
0·000170	0·099	0·121	0·131	0·141	0·150	0·160	0·177	0·186	0·194	0·210	0·225	0·232	0·254	0·262
1/ 5882	1·7424	3·7926	5·2092	6·9159	8·9335	11·282	17·048	20·502	24·361	33·359	44·176	50·305	71·764	81·701
0·000180	0·102	0·124	0·135	0·145	0·155	0·164	0·182	0·191	0·200	0·216	0·232	0·239	0·261	0·270
1/ 5556	1·7943	3·9050	5·3633	7·1203	9·1973	11·615	17·550	21·105	25·077	34·339	45·472	51·781	73·868	84·095
0·000190	0·104	0·128	0·139	0·149	0·159	0·169	0·187	0·196	0·205	0·222	0·238	0·246	0·268	0·277
1/ 5263	1·8448	4·0144	5·5133	7·3191	9·4538	11·938	18·038	21·692	25·774	35·293	46·734	53·217	75·914	86·424
0·000200	0·107	0·131	0·142	0·153	0·163	0·173	0·192	0·202	0·211	0·228	0·244	0·252	0·276	0·285
1/ 5000	1·8939	4·1209	5·6594	7·5128	9·7037	12·254	18·514	22·264	26·453	36·221	47·962	54·616	77·907	88·692
0·000210	0·110	0·134	0·146	0·157	0·167	0·178	0·197	0·207	0·216	0·233	0·250	0·259	0·282	0·292
1/ 4762	1·9419	4·2248	5·8018	7·7017	9·9474	12·561	18·978	22·821	27·115	37·127	49·161	55·980	79·851	90·905
0·000220	0·113	0·138	0·149	0·161	0·171	0·182	0·202	0·212	0·221	0·239	0·256	0·265	0·289	0·299
1/ 4545	1·9887	4·3262	5·9409	7·8861	10·185	12·861	19·431	23·366	27·761	38·011	50·331	57·312	81·749	93·065
0·000230	0·115	0·141	0·153	0·164	0·175	0·186	0·207	0·216	0·226	0·244	0·262	0·271	0·296	0·305
1/ 4348	2·0345	4·4254	6·0769	8·0664	10·418	13·155	19·874	23·898	28·393	38·876	51·474	58·614	83·605	95·177
0·000240	0·118	0·144	0·156	0·168	0·179	0·190	0·211	0·221	0·231	0·250	0·268	0·277	0·302	0·312
1/ 4167	2·0793	4·5225	6·2100	8·2428	10·646	13·442	20·307	24·419	29·012	39·721	52·594	59·888	85·420	97·243
0·000250	0·120	0·147	0·159	0·171	0·183	0·194	0·215	0·226	0·236	0·255	0·273	0·282	0·308	0·318
1/ 4000	2·1231	4·6175	6·3403	8·4156	10·869	13·723	20·731	24·929	29·617	40·550	53·690	61·135	87·199	99·267
0·000260	0·123	0·150	0·163	0·175	0·187	0·198	0·220	0·230	0·240	0·260	0·279	0·288	0·315	0·325
1/ 3846	2·1661	4·7107	6·4681	8·5850	11·087	13·999	21·147	25·429	30·211	41·362	54·764	62·359	88·942	101·25
0·000270	0·125	0·153	0·166	0·178	0·190	0·202	0·224	0·235	0·245	0·265	0·284	0·294	0·321	0·331
1/ 3704	2·2083	4·8021	6·5934	8·7512	11·302	14·270	21·555	25·919	30·793	42·159	55·818	63·558	90·652	103·20
0·000280	0·127	0·156	0·169	0·182	0·194	0·206	0·228	0·239	0·250	0·270	0·290	0·299	0·327	0·337
1/ 3571	2·2497	4·8918	6·7164	8·9143	11·512	14·535	21·956	26·400	31·365	42·941	56·853	64·736	92·330	105·11
0·000290	0·130	0·159	0·172	0·185	0·197	0·209	0·232	0·243	0·254	0·275	0·295	0·304	0·332	0·343
1/ 3448	2·2904	4·9799	6·8373	9·0745	11·719	14·796	22·349	26·873	31·926	43·709	57·869	65·893	93·979	106·98
0·000300	0·132	0·161	0·175	0·188	0·201	0·213	0·236	0·248	0·258	0·280	0·300	0·310	0·338	0·349
1/ 3333	2·3304	5·0665	6·9561	9·2320	11·922	15·052	22·736	27·338	32·478	44·464	58·868	67·030	95·600	108·83
	–	0·74	0·74	0·74	0·74	0·75	0·75	0·75	0·75	0·76	0·76	0·76	0·77	0·77

$V_{r(0·5)medial}$ for half-full circular pipes.

k_s = 3·0 mm S = 0·00010 to 0·00030

Gradient	(Equivalent) Pipe diameters in mm													
	150	200	225	250	275	300	350	375	400	450	500	525	600	630
0·000300	0·132	0·161	0·175	0·188	0·201	0·213	0·236	0·248	0·258	0·280	0·300	0·310	0·338	0·349
1/ 3333	2·3304	5·0665	6·9561	9·2320	11·922	15·052	22·736	27·338	32·478	44·464	58·868	67·030	95·600	108·83
0·000320	0·136	0·167	0·181	0·194	0·207	0·220	0·244	0·256	0·267	0·289	0·310	0·320	0·349	0·361
1/ 3125	2·4084	5·2356	7·1879	9·5393	12·318	15·553	23·491	28·245	33·555	45·937	60·817	69·249	98·762	112·43
0·000340	0·141	0·172	0·186	0·200	0·214	0·227	0·252	0·264	0·275	0·298	0·319	0·330	0·360	0·372
1/ 2941	2·4840	5·3994	7·4125	9·8372	12·703	16·037	24·222	29·124	34·599	47·365	62·707	71·400	101·83	115·91
0·000360	0·145	0·177	0·192	0·206	0·220	0·234	0·259	0·271	0·283	0·307	0·329	0·339	0·371	0·383
1/ 2778	2·5575	5·5585	7·6307	10·126	13·076	16·508	24·932	29·978	35·613	48·751	64·541	73·488	104·80	119·30
0·000380	0·149	0·182	0·197	0·212	0·226	0·240	0·266	0·279	0·291	0·315	0·338	0·349	0·381	0·393
1/ 2632	2·6289	5·7133	7·8428	10·408	13·439	16·966	25·623	30·808	36·598	50·100	66·325	75·519	107·70	122·60
0·000400	0·153	0·187	0·202	0·218	0·232	0·246	0·273	0·286	0·299	0·323	0·347	0·358	0·391	0·404
1/ 2500	2·6985	5·8640	8·0495	10·682	13·792	17·412	26·296	31·616	37·559	51·413	68·063	77·497	110·52	125·80
0·000420	0·157	0·191	0·208	0·223	0·238	0·252	0·280	0·293	0·306	0·331	0·355	0·367	0·401	0·414
1/ 2381	2·7663	6·0110	8·2511	10·949	14·137	17·847	26·952	32·405	38·495	52·694	69·758	79·426	113·27	128·93
0·000440	0·160	0·196	0·212	0·228	0·244	0·258	0·287	0·300	0·314	0·339	0·364	0·376	0·410	0·423
1/ 2273	2·8326	6·1545	8·4479	11·210	14·474	18·272	27·593	33·175	39·410	53·945	71·413	81·310	115·95	131·99
0·000460	0·164	0·200	0·217	0·234	0·249	0·264	0·293	0·307	0·321	0·347	0·372	0·384	0·419	0·433
1/ 2174	2·8973	6·2949	8·6403	11·465	14·803	18·687	28·219	33·928	40·304	55·168	73·031	83·152	118·58	134·98
0·000480	0·168	0·205	0·222	0·239	0·255	0·270	0·300	0·314	0·328	0·354	0·380	0·392	0·428	0·442
1/ 2083	2·9607	6·4322	8·8286	11·715	15·125	19·093	28·832	34·665	41·178	56·365	74·614	84·954	121·14	137·90
0·000500	0·171	0·209	0·227	0·244	0·260	0·276	0·306	0·320	0·335	0·362	0·388	0·401	0·437	0·452
1/ 2000	3·0228	6·5666	9·0130	11·959	15·440	19·491	29·433	35·386	42·035	57·537	76·165	86·719	123·66	140·76
0·000525	0·175	0·214	0·232	0·250	0·266	0·283	0·314	0·328	0·343	0·371	0·398	0·411	0·448	0·463
1/ 1905	3·0987	6·7310	9·2384	12·258	15·826	19·978	30·166	36·268	43·082	58·969	78·060	88·876	126·73	144·26
0·000550	0·180	0·219	0·238	0·256	0·273	0·289	0·321	0·336	0·351	0·380	0·407	0·420	0·459	0·474
1/ 1818	3·1728	6·8915	9·4585	12·550	16·202	20·453	30·883	37·129	44·104	60·368	79·910	90·983	129·74	147·68
0·000575	0·184	0·224	0·243	0·261	0·279	0·296	0·328	0·344	0·359	0·388	0·416	0·430	0·469	0·484
1/ 1739	3·2452	7·0484	9·6736	12·835	16·570	20·917	31·583	37·970	45·104	61·735	81·719	93·042	132·67	151·02
0·000600	0·188	0·229	0·249	0·267	0·285	0·302	0·335	0·351	0·367	0·397	0·425	0·439	0·479	0·495
1/ 1667	3·3161	7·2020	9·8841	13·114	16·931	21·371	32·269	38·794	46·082	63·073	83·489	95·057	135·54	154·28
0·000625	0·192	0·234	0·254	0·273	0·291	0·309	0·342	0·359	0·374	0·405	0·434	0·448	0·489	0·505
1/ 1600	3·3855	7·3523	10·090	13·387	17·283	21·816	32·940	39·601	47·040	64·383	85·223	97·030	138·35	157·48
0·000650	0·195	0·239	0·259	0·278	0·297	0·315	0·349	0·366	0·382	0·413	0·443	0·457	0·499	0·515
1/ 1538	3·4535	7·4997	10·292	13·655	17·629	22·252	33·598	40·391	47·979	65·667	86·922	98·964	141·11	160·62
0·000675	0·199	0·243	0·264	0·284	0·303	0·321	0·356	0·373	0·389	0·421	0·451	0·466	0·509	0·525
1/ 1481	3·5203	7·6443	10·491	13·918	17·968	22·680	34·243	41·167	48·900	66·927	88·589	100·86	143·81	163·70
0·000700	0·203	0·248	0·269	0·289	0·308	0·327	0·362	0·380	0·396	0·429	0·460	0·475	0·518	0·535
1/ 1429	3·5858	7·7862	10·685	14·176	18·301	23·100	34·877	41·928	49·804	68·164	90·225	102·72	146·47	166·72
0·000725	0·207	0·252	0·274	0·294	0·314	0·333	0·369	0·386	0·403	0·436	0·468	0·483	0·527	0·544
1/ 1379	3·6502	7·9257	10·876	14·430	18·628	23·513	35·499	42·676	50·692	69·379	91·832	104·55	149·08	169·69
0·000750	0·210	0·257	0·278	0·299	0·319	0·338	0·375	0·393	0·410	0·444	0·476	0·491	0·536	0·554
1/ 1333	3·7134	8·0627	11·064	14·679	18·949	23·918	36·110	43·411	51·565	70·573	93·412	106·35	151·64	172·60
0·000800	0·217	0·265	0·287	0·309	0·330	0·350	0·388	0·406	0·424	0·458	0·491	0·507	0·554	0·572
1/ 1250	3·8369	8·3301	11·431	15·165	19·576	24·709	37·304	44·846	53·268	72·902	96·494	109·86	156·64	178·30
0·000850	0·224	0·273	0·296	0·319	0·340	0·360	0·400	0·419	0·437	0·473	0·507	0·523	0·571	0·590
1/ 1176	3·9565	8·5893	11·786	15·636	20·184	25·476	38·461	46·236	54·919	75·161	99·482	113·26	161·49	183·81
0·000900	0·230	0·281	0·305	0·328	0·350	0·371	0·411	0·431	0·450	0·486	0·521	0·538	0·588	0·607
1/ 1111	4·0727	8·8409	12·131	16·093	20·775	26·221	39·584	47·586	56·522	77·353	102·38	116·56	166·19	189·17
0·000950	0·237	0·289	0·314	0·337	0·359	0·381	0·423	0·443	0·462	0·500	0·536	0·553	0·604	0·624
1/ 1053	4·1857	9·0857	12·467	16·538	21·349	26·945	40·677	48·899	58·081	79·486	105·20	119·78	170·77	194·38
0·001000	0·243	0·297	0·322	0·346	0·369	0·391	0·434	0·454	0·474	0·513	0·550	0·568	0·620	0·640
1/ 1000	4·2958	9·3241	12·794	16·972	21·908	27·650	41·741	50·177	59·599	81·563	107·95	122·90	175·23	199·45
	–	0·74	0·74	0·74	0·75	0·75	0·75	0·75	0·75	0·76	0·76	0·76	0·77	0·77

$V_{r(0\cdot5)medial}$ for half-full circular pipes.

A20
(p.3 of 6)

k_s = 3·0 mm
S = 0·00100 to 0·00300

ie hydraulic gradient =
1 in 1000 to 1 in 333

Water (or sewage) at 15°C;
full bore conditions.

velocities in ms^{-1}
discharges in litres/sec

Gradient	**(Equivalent) Pipe diameters in mm**													
	150	**200**	**225**	**250**	**275**	**300**	**350**	**375**	**400**	**450**	**500**	**525**	**600**	**630**
0·00100	0·243	0·297	0·322	0·346	0·369	0·391	0·434	0·454	0·474	0·513	0·550	0·568	0·620	0·640
1/ 1000	4·2958	9·3241	12·794	16·972	21·908	27·650	41·741	50·177	59·599	81·563	107·95	122·90	175·23	199·45
0·00105	0·249	0·304	0·330	0·354	0·378	0·401	0·445	0·466	0·486	0·526	0·563	0·582	0·635	0·656
1/ 952	4·4031	9·5566	13·113	17·394	22·453	28·338	42·778	51·425	61·080	83·589	110·63	125·95	179·57	204·40
0·00110	0·255	0·311	0·338	0·363	0·387	0·410	0·455	0·477	0·498	0·538	0·577	0·596	0·650	0·671
1/ 909	4·5079	9·7837	13·424	17·807	22·985	29·010	43·792	52·642	62·526	85·567	113·25	128·93	183·82	209·23
0·00115	0·261	0·318	0·345	0·371	0·396	0·420	0·465	0·487	0·509	0·550	0·590	0·609	0·665	0·686
1/ 870	4·6104	10·006	13·728	18·211	23·506	29·667	44·782	53·833	63·940	87·501	115·81	131·85	187·97	213·95
0·00120	0·267	0·325	0·353	0·379	0·404	0·429	0·476	0·498	0·520	0·562	0·603	0·622	0·679	0·701
1/ 833	4·7107	10·223	14·026	18·605	24·015	30·309	45·752	54·998	65·323	89·393	118·31	134·70	192·03	218·57
0·00125	0·272	0·332	0·360	0·387	0·413	0·438	0·485	0·508	0·531	0·574	0·615	0·635	0·693	0·716
1/ 800	4·8089	10·435	14·318	18·992	24·514	30·939	46·701	56·139	66·678	91·246	120·76	137·49	196·01	223·10
0·00130	0·278	0·339	0·367	0·395	0·421	0·446	0·495	0·518	0·541	0·585	0·627	0·648	0·707	0·730
1/ 769	4·9051	10·644	14·604	19·371	25·003	31·555	47·631	57·257	68·006	93·062	123·17	140·22	199·91	227·54
0·00135	0·283	0·345	0·374	0·402	0·429	0·455	0·505	0·528	0·552	0·596	0·639	0·660	0·721	0·744
1/ 741	4·9995	10·849	14·884	19·743	25·483	32·160	48·544	58·354	69·308	94·844	125·52	142·91	203·73	231·89
0·00140	0·288	0·352	0·381	0·410	0·437	0·463	0·514	0·538	0·562	0·607	0·651	0·672	0·734	0·758
1/ 714	5·0922	11·049	15·159	20·108	25·954	32·754	49·440	59·431	70·587	96·593	127·84	145·54	207·49	236·16
0·00145	0·293	0·358	0·388	0·417	0·445	0·472	0·523	0·548	0·572	0·618	0·663	0·684	0·747	0·771
1/ 690	5·1833	11·246	15·430	20·466	26·416	33·338	50·320	60·488	71·843	98·311	130·11	148·13	211·17	240·36
0·00150	0·298	0·364	0·395	0·424	0·452	0·480	0·532	0·557	0·582	0·629	0·674	0·696	0·760	0·784
1/ 667	5·2728	11·440	15·696	20·818	26·871	33·911	51·186	61·528	73·078	100·00	132·35	150·67	214·80	244·48
0·00160	0·308	0·376	0·408	0·438	0·467	0·496	0·550	0·575	0·601	0·649	0·696	0·719	0·785	0·810
1/ 625	5·4474	11·819	16·214	21·506	27·758	35·030	52·874	63·557	75·486	103·30	136·71	155·63	221·87	252·53
0·00170	0·318	0·388	0·420	0·452	0·482	0·511	0·567	0·593	0·619	0·670	0·718	0·741	0·809	0·835
1/ 588	5·6166	12·185	16·717	22·172	28·617	36·115	54·510	65·523	77·821	106·49	140·93	160·44	228·72	260·33
0·00180	0·327	0·399	0·433	0·465	0·496	0·526	0·583	0·611	0·637	0·689	0·739	0·763	0·832	0·859
1/ 556	5·7809	12·541	17·205	22·819	29·452	37·168	56·098	67·432	80·088	109·59	145·03	165·11	235·38	267·91
0·00190	0·336	0·410	0·445	0·478	0·510	0·540	0·599	0·627	0·655	0·708	0·759	0·784	0·855	0·883
1/ 526	5·9408	12·887	17·679	23·449	30·264	38·192	57·643	69·289	82·293	112·61	149·02	169·66	241·85	275·28
0·00200	0·345	0·421	0·456	0·490	0·523	0·554	0·615	0·644	0·672	0·726	0·779	0·804	0·878	0·906
1/ 500	6·0965	13·224	18·142	24·062	31·055	39·190	59·148	71·097	84·441	115·54	152·91	174·08	248·16	282·45
0·00210	0·354	0·431	0·468	0·502	0·536	0·568	0·630	0·660	0·689	0·745	0·798	0·824	0·899	0·929
1/ 476	6·2483	13·553	18·593	24·659	31·826	40·163	60·616	72·861	86·535	118·41	156·70	178·40	254·31	289·45
0·00220	0·362	0·442	0·479	0·514	0·549	0·582	0·645	0·675	0·705	0·762	0·817	0·844	0·921	0·950
1/ 455	6·3966	13·874	19·033	25·243	32·579	41·113	62·049	74·584	88·581	121·21	160·40	182·61	260·31	296·28
0·00230	0·370	0·452	0·490	0·526	0·561	0·595	0·659	0·691	0·721	0·779	0·835	0·863	0·941	0·972
1/ 435	6·5415	14·188	19·463	25·814	33·315	42·042	63·450	76·267	90·580	123·94	164·02	186·73	266·18	302·96
0·00240	0·378	0·461	0·500	0·537	0·573	0·608	0·674	0·705	0·736	0·796	0·853	0·881	0·962	0·993
1/ 417	6·6833	14·496	19·885	26·372	34·036	42·951	64·821	77·915	92·536	126·62	167·56	190·76	271·93	309·50
0·00250	0·386	0·471	0·510	0·548	0·585	0·620	0·688	0·720	0·752	0·813	0·871	0·899	0·982	1·013
1/ 400	6·8222	14·796	20·297	26·919	34·741	43·841	66·164	79·528	94·452	129·24	171·03	194·70	277·55	315·90
0·00260	0·394	0·480	0·521	0·559	0·597	0·633	0·701	0·734	0·767	0·829	0·888	0·917	1·001	1·034
1/ 385	6·9583	15·091	20·701	27·455	35·433	44·713	67·480	81·110	96·330	131·81	174·43	198·57	283·06	322·18
0·00270	0·401	0·490	0·531	0·570	0·608	0·645	0·715	0·748	0·781	0·845	0·905	0·935	1·020	1·053
1/ 370	7·0919	15·381	21·098	27·981	36·111	45·569	68·771	82·661	98·172	134·33	177·76	202·37	288·47	328·33
0·00280	0·409	0·499	0·540	0·581	0·619	0·657	0·728	0·762	0·796	0·860	0·922	0·952	1·039	1·073
1/ 357	7·2230	15·664	21·487	28·497	36·777	46·409	70·038	84·184	99·981	136·80	181·03	206·09	293·78	334·37
0·00290	0·416	0·507	0·550	0·591	0·630	0·668	0·741	0·776	0·810	0·875	0·938	0·969	1·057	1·092
1/ 345	7·3518	15·943	21·870	29·004	37·431	47·234	71·283	85·680	101·76	139·23	184·25	209·75	299·00	340·31
0·00300	0·423	0·516	0·559	0·601	0·641	0·680	0·754	0·789	0·824	0·890	0·954	0·986	1·076	1·110
1/ 333	7·4783	16·218	22·246	29·502	38·074	48·045	72·506	87·151	103·50	141·62	187·41	213·35	304·12	346·14
	–	0·74	0·74	0·74	0·75	0·75	0·75	0·75	0·76	0·76	0·76	0·76	0·77	0·77

$V_{r(0·5)medial}$ for half-full circular pipes.

k_s = 3·0 mm
S = 0·00300 to 0·01000

ie hydraulic gradient =
1 in 333 to 1 in 100

Water (or sewage) at 15°C;
full bore conditions.

velocities in ms^{-1}
discharges in litres/sec

Gradient **(Equivalent) Pipe diameters in mm**

Gradient	**150**	**200**	**225**	**250**	**275**	**300**	**350**	**375**	**400**	**450**	**500**	**525**	**600**	**630**
0·00300	0·423	0·516	0·559	0·601	0·641	0·680	0·754	0·789	0·824	0·890	0·954	0·986	1·076	1·110
1/ 333	7·4783	16·218	22·246	29·502	38·074	48·045	72·506	87·151	103·50	141·62	187·41	213·35	304·12	346·14
0·00320	0·437	0·533	0·578	0·621	0·662	0·702	0·778	0·815	0·851	0·920	0·986	1·018	1·111	1·147
1/ 313	7·7253	16·752	22·979	30·474	39·329	49·628	74·894	90·020	106·91	146·28	193·58	220·37	314·12	357·52
0·00340	0·451	0·550	0·596	0·640	0·683	0·724	0·802	0·840	0·877	0·948	1·016	1·049	1·145	1·182
1/ 294	7·9647	17·271	23·690	31·417	40·544	51·162	77·208	92·800	110·21	150·80	199·55	227·17	323·82	368·56
0·00360	0·464	0·566	0·613	0·659	0·702	0·745	0·826	0·865	0·903	0·976	1·046	1·080	1·179	1·217
1/ 278	8·1971	17·774	24·380	32·332	41·725	52·651	79·454	95·500	113·42	155·18	205·35	233·78	333·23	379·27
0·00380	0·477	0·581	0·630	0·677	0·722	0·765	0·849	0·888	0·927	1·003	1·075	1·110	1·211	1·250
1/ 263	8·4232	18·264	25·052	33·222	42·873	54·100	81·640	98·126	116·54	159·45	211·00	240·20	342·39	389·69
0·00400	0·489	0·597	0·647	0·694	0·741	0·785	0·871	0·912	0·952	1·029	1·103	1·138	1·242	1·283
1/ 250	8·6434	18·741	25·705	34·089	43·992	55·511	83·768	100·68	119·57	163·60	216·49	246·46	351·30	399·84
0·00420	0·501	0·611	0·663	0·712	0·759	0·805	0·892	0·934	0·975	1·054	1·130	1·167	1·273	1·314
1/ 238	8·8581	19·206	26·343	34·934	45·082	56·887	85·844	103·18	122·54	167·65	221·85	252·56	360·00	409·73
0·00440	0·513	0·626	0·678	0·728	0·777	0·824	0·913	0·956	0·998	1·079	1·157	1·194	1·303	1·345
1/ 227	9·0678	19·660	26·966	35·760	46·148	58·231	87·871	105·61	125·43	171·61	227·09	258·52	368·49	419·40
0·00460	0·525	0·640	0·694	0·745	0·794	0·842	0·934	0·978	1·021	1·103	1·183	1·221	1·333	1·376
1/ 217	9·2728	20·104	27·575	36·567	47·189	59·544	89·852	108·00	128·26	175·48	232·21	264·34	376·79	428·84
0·00480	0·536	0·654	0·708	0·761	0·812	0·861	0·954	0·999	1·043	1·127	1·208	1·247	1·361	1·405
1/ 208	9·4734	20·539	28·170	37·357	48·208	60·829	91·791	110·33	131·02	179·26	237·21	270·04	384·91	438·09
0·00500	0·547	0·667	0·723	0·777	0·828	0·878	0·974	1·020	1·064	1·150	1·233	1·273	1·389	1·434
1/ 200	9·6698	20·964	28·754	38·130	49·205	62·088	93·690	112·61	133·73	182·97	242·12	275·62	392·87	447·14
0·00525	0·561	0·684	0·741	0·796	0·849	0·900	0·998	1·045	1·091	1·179	1·264	1·305	1·424	1·470
1/ 190	9·9099	21·484	29·467	39·075	50·425	63·627	96·011	115·40	137·04	187·50	248·11	282·45	402·59	458·21
0·00550	0·574	0·700	0·759	0·815	0·869	0·921	1·021	1·069	1·116	1·207	1·293	1·336	1·457	1·505
1/ 182	10·144	21·992	30·163	39·998	51·616	65·129	98·277	118·12	140·28	191·92	253·96	289·11	412·09	469·01
0·00575	0·587	0·716	0·776	0·833	0·889	0·942	1·044	1·094	1·141	1·234	1·323	1·366	1·490	1·538
1/ 174	10·374	22·488	30·844	40·900	52·780	66·597	100·49	120·78	143·44	196·25	259·69	295·62	421·37	479·57
0·00600	0·600	0·731	0·792	0·851	0·908	0·962	1·067	1·117	1·166	1·261	1·351	1·395	1·522	1·572
1/ 167	10·598	22·974	31·509	41·783	53·919	68·034	102·66	123·39	146·53	200·48	265·28	301·99	430·45	489·91
0·00625	0·612	0·746	0·809	0·869	0·927	0·982	1·089	1·140	1·190	1·287	1·379	1·424	1·554	1·604
1/ 160	10·817	23·450	32·162	42·648	55·034	69·442	104·78	125·94	149·56	204·62	270·77	308·23	439·34	500·03
0·00650	0·624	0·761	0·825	0·886	0·945	1·002	1·111	1·163	1·214	1·312	1·406	1·452	1·585	1·636
1/ 154	11·033	23·916	32·801	43·496	56·128	70·821	106·86	128·44	152·53	208·68	276·14	314·35	448·06	509·95
0·00675	0·636	0·776	0·841	0·903	0·963	1·021	1·132	1·185	1·237	1·337	1·433	1·480	1·615	1·667
1/ 148	11·244	24·373	33·428	44·327	57·200	72·174	108·90	130·89	155·44	212·67	281·41	320·35	456·61	519·68
0·00700	0·648	0·790	0·856	0·920	0·981	1·040	1·153	1·207	1·260	1·362	1·460	1·507	1·645	1·698
1/ 143	11·451	24·822	34·044	45·143	58·253	73·503	110·91	133·30	158·30	216·58	286·59	326·24	465·01	529·24
0·00725	0·660	0·804	0·871	0·936	0·998	1·058	1·173	1·228	1·282	1·386	1·485	1·534	1·674	1·728
1/ 138	11·655	25·263	34·648	45·945	59·288	74·807	112·88	135·67	161·11	220·42	291·67	332·03	473·25	538·62
0·00750	0·671	0·818	0·886	0·952	1·015	1·076	1·193	1·249	1·304	1·410	1·511	1·560	1·702	1·757
1/ 133	11·855	25·697	35·243	46·733	60·304	76·090	114·81	137·99	163·87	224·20	296·67	337·72	481·36	547·85
0·00800	0·693	0·845	0·916	0·983	1·049	1·112	1·233	1·290	1·347	1·456	1·561	1·611	1·758	1·815
1/ 125	12·245	26·543	36·403	48·270	62·288	78·592	118·59	142·53	169·26	231·57	306·41	348·81	497·17	565·85
0·00850	0·714	0·871	0·944	1·014	1·081	1·146	1·271	1·330	1·388	1·501	1·609	1·661	1·813	1·871
1/ 118	12·624	27·363	37·527	49·760	64·210	81·018	122·24	146·92	174·48	238·71	315·86	359·57	512·50	583·29
0·00900	0·735	0·896	0·971	1·043	1·112	1·179	1·308	1·369	1·429	1·545	1·655	1·709	1·865	1·926
1/ 111	12·992	28·159	38·618	51·207	66·077	83·373	125·80	151·19	179·55	245·64	325·04	370·01	527·38	600·23
0·00950	0·755	0·921	0·998	1·072	1·143	1·212	1·343	1·407	1·468	1·587	1·701	1·756	1·916	1·978
1/ 105	13·349	28·933	39·680	52·614	67·893	85·663	129·25	155·35	184·48	252·39	333·96	380·17	541·86	616·70
0·01000	0·775	0·945	1·024	1·100	1·173	1·243	1·378	1·443	1·506	1·628	1·745	1·802	1·966	2·030
1/ 100	13·697	29·687	40·713	53·985	69·661	87·894	132·62	159·39	189·28	258·96	342·65	390·06	555·96	632·75
	0·74	0·74	0·74	0·74	0·75	0·75	0·75	0·75	0·76	0·76	0·76	0·76	0·77	0·77

$V_{r(0{\cdot}5)medial}$ for half-full circular pipes.

A20
(p.5 of 6)

k_s = 3·0 mm
S = 0·01000 to 0·03000

ie hydraulic gradient =
1 in 100 to 1 in 33·3

Water (or sewage) at 15°C;
full bore conditions.

velocities in ms^{-1}
discharges in litres/sec

Gradient	**(Equivalent) Pipe diameters in mm**													
	150	**200**	**225**	**250**	**275**	**300**	**350**	**375**	**400**	**450**	**500**	**525**	**600**	**630**
0·01000	0·775	0·945	1·024	1·100	1·173	1·243	1·378	1·443	1·506	1·628	1·745	1·802	1·966	2·030
1/ 100	13·697	29·687	40·713	53·985	69·661	87·894	132·62	159·39	189·28	258·96	342·65	390·06	555·96	632·75
0·01050	0·794	0·968	1·049	1·127	1·202	1·274	1·413	1·479	1·544	1·669	1·788	1·846	2·015	2·080
1/ 95	14·037	30·422	41·722	55·322	71·386	90·070	135·90	163·33	193·97	265·36	351·13	399·71	569·71	648·40
0·01100	0·813	0·991	1·074	1·154	1·230	1·304	1·446	1·514	1·580	1·708	1·830	1·890	2·062	2·129
1/ 91	14·368	31·141	42·707	56·627	73·070	92·195	139·11	167·19	198·54	271·62	359·41	409·13	583·13	663·68
0·01150	0·831	1·014	1·098	1·180	1·258	1·334	1·478	1·548	1·616	1·746	1·872	1·933	2·109	2·177
1/ 87	14·692	31·843	43·669	57·903	74·716	94·272	142·24	170·95	203·01	277·74	367·50	418·34	596·26	678·61
0·01200	0·849	1·035	1·122	1·205	1·285	1·362	1·510	1·581	1·650	1·784	1·912	1·974	2·154	2·224
1/ 83	15·010	32·529	44·611	59·152	76·327	96·304	145·30	174·63	207·39	283·72	375·41	427·36	609·10	693·23
0·01250	0·867	1·057	1·145	1·230	1·312	1·391	1·541	1·614	1·684	1·821	1·951	2·015	2·199	2·270
1/ 80	15·320	33·202	45·533	60·375	77·905	98·294	148·31	178·24	211·67	289·58	383·17	436·18	621·68	707·54
0·01300	0·884	1·078	1·168	1·254	1·338	1·418	1·572	1·646	1·718	1·857	1·990	2·055	2·242	2·315
1/ 77	15·625	33·862	46·437	61·573	79·451	100·24	151·25	181·78	215·87	295·32	390·77	444·83	634·01	721·57
0·01350	0·901	1·098	1·190	1·278	1·363	1·445	1·602	1·677	1·751	1·892	2·028	2·094	2·285	2·359
1/ 74	15·923	34·508	47·324	62·749	80·968	102·16	154·14	185·25	219·99	300·96	398·22	453·32	646·10	735·34
0·01400	0·918	1·119	1·212	1·302	1·388	1·472	1·632	1·708	1·783	1·927	2·065	2·133	2·327	2·402
1/ 71	16·216	35·143	48·195	63·903	82·457	104·04	156·97	188·65	224·03	306·49	405·54	461·65	657·97	748·85
0·01450	0·934	1·139	1·234	1·325	1·413	1·498	1·660	1·738	1·814	1·961	2·102	2·170	2·368	2·445
1/ 69	16·504	35·767	49·050	65·037	83·920	105·88	159·75	192·00	228·01	311·92	412·73	469·83	669·64	762·12
0·01500	0·950	1·158	1·255	1·348	1·437	1·524	1·689	1·768	1·845	1·995	2·138	2·208	2·409	2·487
1/ 67	16·787	36·380	49·891	66·152	85·358	107·70	162·49	195·29	231·91	317·26	419·80	477·88	681·10	775·17
0·01600	0·981	1·196	1·296	1·392	1·484	1·574	1·744	1·826	1·906	2·060	2·208	2·280	2·488	2·568
1/ 63	17·340	37·577	51·531	68·326	88·163	111·24	167·83	201·70	239·53	327·69	433·58	493·57	703·46	800·62
0·01700	1·012	1·233	1·336	1·435	1·530	1·622	1·798	1·883	1·965	2·124	2·276	2·350	2·565	2·647
1/ 59	17·875	38·736	53·120	70·433	90·882	114·67	173·00	207·92	246·91	337·79	446·95	508·78	725·14	825·29
0·01800	1·041	1·269	1·375	1·477	1·575	1·669	1·850	1·937	2·022	2·186	2·342	2·419	2·639	2·724
1/ 56	18·395	39·862	54·664	72·480	93·522	118·00	178·03	213·96	254·08	347·59	459·92	523·55	746·19	849·24
0·01900	1·070	1·304	1·413	1·517	1·618	1·715	1·901	1·990	2·077	2·245	2·407	2·485	2·712	2·799
1/ 53	18·900	40·957	56·165	74·470	96·089	121·24	182·91	219·83	261·05	357·13	472·54	537·92	766·66	872·54
0·02000	1·097	1·338	1·449	1·557	1·660	1·760	1·951	2·042	2·131	2·304	2·469	2·550	2·782	2·872
1/ 50	19·393	42·023	57·627	76·408	98·590	124·39	187·67	225·55	267·85	366·42	484·83	551·91	786·60	895·23
0·02100	1·125	1·371	1·485	1·595	1·701	1·803	1·999	2·093	2·184	2·361	2·530	2·613	2·851	2·943
1/ 47·6	19·873	43·063	59·054	78·299	101·03	127·47	192·31	231·13	274·47	375·48	496·82	565·55	806·04	917·36
0·02200	1·151	1·403	1·520	1·633	1·741	1·846	2·046	2·142	2·236	2·417	2·590	2·674	2·918	3·012
1/ 45·5	20·342	44·079	60·446	80·145	103·41	130·47	196·85	236·58	280·94	384·33	508·53	578·88	825·03	938·97
0·02300	1·177	1·435	1·554	1·669	1·780	1·887	2·092	2·190	2·286	2·471	2·648	2·734	2·984	3·080
1/ 43·5	20·800	45·072	61·807	81·950	105·74	133·41	201·28	241·90	287·26	392·98	519·97	591·90	843·60	960·10
0·02400	1·202	1·466	1·588	1·705	1·819	1·928	2·137	2·237	2·335	2·524	2·705	2·793	3·048	3·146
1/ 41·7	21·249	46·043	63·139	83·715	108·02	136·28	205·61	247·11	293·45	401·44	531·16	604·65	861·76	980·77
0·02500	1·227	1·496	1·621	1·741	1·856	1·968	2·181	2·284	2·383	2·576	2·761	2·851	3·111	3·211
1/ 40·0	21·688	46·994	64·444	85·445	110·25	139·10	209·86	252·21	299·51	409·73	542·13	617·13	879·55	1001·0
0·02600	1·252	1·526	1·653	1·775	1·893	2·007	2·224	2·329	2·431	2·627	2·816	2·907	3·172	3·275
1/ 38·5	22·119	47·927	65·722	87·140	112·44	141·86	214·02	257·21	305·44	417·85	552·88	629·36	896·98	1020·9
0·02700	1·276	1·555	1·684	1·809	1·929	2·045	2·267	2·373	2·477	2·677	2·869	2·963	3·233	3·337
1/ 37·0	22·541	48·842	66·977	88·803	114·58	144·56	218·10	262·12	311·27	425·82	563·42	641·36	914·08	1040·3
0·02800	1·299	1·583	1·715	1·842	1·965	2·083	2·309	2·417	2·523	2·727	2·922	3·017	3·292	3·399
1/ 35·7	22·956	49·740	68·208	90·435	116·69	147·22	222·11	266·94	316·99	433·64	573·77	653·15	930·87	1059·4
0·02900	1·322	1·611	1·746	1·875	1·999	2·120	2·349	2·460	2·567	2·775	2·974	3·071	3·351	3·459
1/ 34·5	23·363	50·622	69·417	92·038	118·76	149·83	226·05	271·67	322·61	441·33	583·94	664·72	947·37	1078·2
0·03000	1·345	1·639	1·776	1·907	2·034	2·156	2·390	2·502	2·611	2·822	3·025	3·123	3·408	3·518
1/ 33·3	23·763	51·489	70·606	93·614	120·79	152·39	229·92	276·32	328·13	448·88	593·93	676·09	963·58	1096·6
	0·74	0·74	0·74	0·74	0·75	0·75	0·75	0·75	0·76	0·76	0·76	0·76	0·77	0·77

$V_{r(0·5)medial}$ for half-full circular pipes.

k_s = 3·0 mm S = 0·01000 to 0·03000

Gradient	150	200	225	250	275	300	350	375	400	450	500	525	600	630
	(Equivalent) Pipe diameters in mm													
0·03000	1·345	1·639	1·776	1·907	2·034	2·156	2·390	2·502	2·611	2·822	3·025	3·123	3·408	3·518
1/ 33·3	23·763	51·489	70·606	93·614	120·79	152·39	229·92	276·32	328·13	448·88	593·93	676·09	963·58	1096·6
0·03200	1·389	1·693	1·834	1·970	2·100	2·227	2·468	2·584	2·697	2·915	3·124	3·226	3·520	3·633
1/ 31·3	24·544	53·181	72·926	96·689	124·76	157·40	237·47	285·39	338·90	463·62	613·43	698·29	995·21	1132·6
0·03400	1·432	1·745	1·891	2·030	2·165	2·295	2·544	2·664	2·780	3·005	3·220	3·325	3·628	3·745
1/ 29·4	25·301	54·820	75·174	99·670	128·60	162·25	244·78	294·18	349·34	477·90	632·33	719·80	1025·9	1167·5
0·03600	1·473	1·796	1·946	2·089	2·228	2·362	2·618	2·741	2·861	3·092	3·314	3·422	3·734	3·854
1/ 27·8	26·036	56·413	77·357	102·56	132·33	166·96	251·89	302·72	359·48	491·77	650·68	740·69	1055·6	1201·4
0·03800	1·514	1·845	1·999	2·147	2·289	2·427	2·690	2·816	2·939	3·177	3·405	3·515	3·836	3·960
1/ 26·3	26·751	57·961	79·480	105·38	135·97	171·54	258·80	311·03	369·34	505·26	668·52	761·00	1084·6	1234·4
0·04000	1·553	1·893	2·051	2·203	2·349	2·490	2·760	2·889	3·016	3·259	3·493	3·607	3·936	4·063
1/ 25·0	27·448	59·469	81·548	108·12	139·50	176·00	265·53	319·12	378·95	518·40	685·90	780·79	1112·8	1266·4
0·04200	1·592	1·940	2·102	2·257	2·407	2·552	2·828	2·961	3·090	3·340	3·580	3·696	4·033	4·163
1/ 23·8	28·127	60·940	83·565	110·79	142·95	180·36	272·09	327·01	388·32	531·21	702·86	800·09	1140·3	1297·7
0·04400	1·629	1·986	2·151	2·310	2·463	2·612	2·895	3·031	3·163	3·419	3·664	3·783	4·128	4·261
1/ 22·7	28·790	62·377	85·534	113·40	146·32	184·60	278·50	334·71	397·46	543·72	719·41	818·93	1167·1	1328·3
0·04600	1·666	2·030	2·200	2·362	2·519	2·670	2·960	3·099	3·234	3·496	3·746	3·868	4·221	4·357
1/ 21·7	29·438	63·781	87·459	115·96	149·61	188·76	284·77	342·24	406·41	555·96	735·59	837·35	1193·4	1358·2
0·04800	1·702	2·074	2·247	2·413	2·573	2·728	3·024	3·165	3·304	3·571	3·827	3·951	4·312	4·451
1/ 20·8	30·073	65·154	89·343	118·45	152·84	192·82	290·90	349·61	415·16	567·92	751·43	855·37	1219·1	1387·4
0·05000	1·737	2·117	2·293	2·463	2·626	2·784	3·086	3·231	3·372	3·645	3·906	4·033	4·401	4·543
1/ 20·0	30·694	66·500	91·187	120·90	155·99	196·80	296·90	356·82	423·72	579·64	766·94	873·03	1244·2	1416·0
0·05250	1·780	2·169	2·350	2·524	2·691	2·853	3·162	3·311	3·455	3·735	4·003	4·133	4·509	4·655
1/ 19·0	31·453	68·144	93·442	123·89	159·85	201·67	304·24	365·64	434·20	593·97	785·89	894·60	1275·0	1451·0
0·05500	1·822	2·220	2·405	2·583	2·755	2·920	3·237	3·389	3·537	3·823	4·097	4·230	4·615	4·764
1/ 18·2	32·195	69·750	95·644	126·81	163·61	206·42	311·41	374·26	444·42	607·96	804·40	915·67	1305·0	1485·2
0·05750	1·863	2·270	2·460	2·641	2·817	2·986	3·310	3·465	3·616	3·909	4·189	4·325	4·719	4·872
1/ 17·4	32·919	71·320	97·796	129·66	167·29	211·06	318·42	382·67	454·42	621·64	822·49	936·26	1334·3	1518·6
0·06000	1·903	2·319	2·513	2·698	2·877	3·050	3·381	3·539	3·694	3·993	4·279	4·418	4·821	4·976
1/ 16·7	33·629	72·856	99·902	132·45	170·90	215·61	325·27	390·91	464·20	635·02	840·19	956·42	1363·1	1551·3
0·06250	1·942	2·367	2·564	2·754	2·937	3·113	3·451	3·612	3·770	4·075	4·367	4·509	4·920	5·079
1/ 16·0	34·323	74·360	101·97	135·19	174·42	220·06	331·98	398·98	473·78	648·12	857·53	976·15	1391·2	1583·3
0·06500	1·981	2·414	2·615	2·809	2·995	3·175	3·519	3·684	3·845	4·156	4·454	4·599	5·018	5·180
1/ 15·4	35·004	75·835	103·99	137·87	177·88	224·42	338·56	406·89	483·17	660·97	874·52	995·50	1418·8	1614·7
0·06750	2·019	2·460	2·665	2·862	3·052	3·235	3·586	3·754	3·918	4·235	4·539	4·686	5·113	5·279
1/ 14·8	35·672	77·281	105·97	140·50	181·27	228·70	345·02	414·65	492·39	673·57	891·20	1014·5	1445·8	1645·4
0·07000	2·056	2·505	2·714	2·915	3·108	3·295	3·652	3·823	3·990	4·313	4·622	4·772	5·207	5·375
1/ 14·3	36·327	78·701	107·92	143·08	184·60	232·90	351·36	422·26	501·43	685·93	907·56	1033·1	1472·3	1675·7
0·07250	2·092	2·550	2·762	2·966	3·163	3·353	3·717	3·891	4·061	4·389	4·704	4·857	5·300	5·471
1/ 13·8	36·971	80·096	109·83	145·61	187·87	237·03	357·58	429·74	510·31	698·08	923·64	1051·4	1498·4	1705·3
0·07500	2·128	2·593	2·810	3·017	3·217	3·411	3·780	3·958	4·130	4·464	4·785	4·940	5·390	5·564
1/ 13·3	37·604	81·467	111·71	148·10	191·09	241·08	363·70	437·09	519·04	710·03	939·44	1069·4	1524·1	1734·5
0·08000	2·198	2·678	2·902	3·116	3·323	3·523	3·904	4·087	4·266	4·611	4·942	5·102	5·567	5·747
1/ 12·5	38·839	84·142	115·38	152·97	197·36	249·00	375·64	451·44	536·08	733·33	970·26	1104·5	1574·1	1791·4
0·08500	2·266	2·761	2·991	3·212	3·425	3·631	4·025	4·213	4·397	4·753	5·094	5·259	5·739	5·924
1/ 11·8	40·036	86·734	118·93	157·68	203·44	256·67	387·21	465·34	552·59	755·91	1000·1	1138·5	1622·5	1846·6
0·09000	2·331	2·841	3·078	3·305	3·525	3·736	4·141	4·336	4·525	4·891	5·241	5·412	5·905	6·096
1/ 11·1	41·199	89·252	122·38	162·25	209·35	264·11	398·44	478·84	568·62	777·84	1029·2	1171·5	1669·6	1900·1
0·09500	2·395	2·919	3·162	3·396	3·621	3·839	4·255	4·454	4·649	5·025	5·385	5·560	6·067	6·263
1/ 10·5	42·329	91·700	125·74	166·70	215·09	271·36	409·37	491·97	584·21	799·17	1057·4	1203·6	1715·4	1952·2
0·10000	2·458	2·995	3·245	3·484	3·715	3·939	4·365	4·570	4·770	5·155	5·525	5·705	6·225	6·425
1/ 10·0	43·430	94·085	129·01	171·04	220·68	278·41	420·01	504·76	599·40	819·94	1084·9	1234·9	1760·0	2003·0
	0·74	0·74	0·74	0·74	0·75	0·75	0·75	0·75	0·76	0·76	0·76	0·76	0·77	0·77

$V_{r(0{\cdot}5)medial}$ for half-full circular pipes.

A21

(p.1 of 6)

$k_s = 6{\cdot}0$ **mm**
S = 0·00010 to 0·00030

ie hydraulic gradient =
1 in 10000 to 1 in 3333

Water (or sewage) at 15°C;
full bore conditions.

velocities in ms^{-1}
discharges in litres/sec

Gradient **(Equivalent) Pipe diameters in mm**

Gradient	150	200	225	250	275	300	350	375	400	450	500	525	600	630
0·000100	0·066	0·081	0·089	0·095	0·102	0·109	0·121	0·127	0·133	0·144	0·154	0·160	0·175	0·181
1/ 10000	1·1667	2·5574	3·5219	4·6864	6·0657	7·6738	11·630	14·005	16·660	22·862	30·330	34·566	49·419	56·305
0·000105	0·068	0·083	0·091	0·098	0·105	0·111	0·124	0·130	0·136	0·147	0·158	0·164	0·179	0·185
1/ 9524	1·1962	2·6216	3·6103	4·8039	6·2176	7·8658	11·921	14·355	17·076	23·432	31·086	35·428	50·649	57·707
0·000110	0·069	0·085	0·093	0·100	0·107	0·114	0·127	0·133	0·139	0·151	0·162	0·168	0·183	0·190
1/ 9091	1·2249	2·6844	3·6965	4·9186	6·3659	8·0533	12·205	14·696	17·482	23·989	31·824	36·269	51·851	59·076
0·000115	0·071	0·087	0·095	0·102	0·110	0·117	0·130	0·136	0·142	0·154	0·166	0·171	0·188	0·194
1/ 8696	1·2530	2·7457	3·7809	5·0307	6·5109	8·2365	12·482	15·030	17·879	24·534	32·546	37·091	53·025	60·414
0·000120	0·072	0·089	0·097	0·105	0·112	0·119	0·133	0·139	0·145	0·158	0·169	0·175	0·192	0·198
1/ 8333	1·2805	2·8057	3·8634	5·1404	6·6528	8·4159	12·754	15·357	18·268	25·066	33·252	37·896	54·175	61·723
0·000125	0·074	0·091	0·099	0·107	0·114	0·122	0·135	0·142	0·148	0·161	0·173	0·179	0·196	0·202
1/ 8000	1·3074	2·8645	3·9443	5·2479	6·7917	8·5915	13·020	15·677	18·648	25·588	33·943	38·684	55·301	63·005
0·000130	0·075	0·093	0·101	0·109	0·117	0·124	0·138	0·145	0·151	0·164	0·176	0·182	0·199	0·206
1/ 7692	1·3338	2·9221	4·0235	5·3532	6·9279	8·7637	13·280	15·990	19·021	26·099	34·621	39·456	56·404	64·262
0·000135	0·077	0·095	0·103	0·111	0·119	0·126	0·141	0·148	0·154	0·167	0·180	0·186	0·203	0·210
1/ 7407	1·3596	2·9786	4·1012	5·4565	7·0615	8·9326	13·536	16·298	19·387	26·601	35·286	40·214	57·486	65·495
0·000140	0·078	0·097	0·105	0·113	0·121	0·129	0·143	0·150	0·157	0·170	0·183	0·189	0·207	0·214
1/ 7143	1·3851	3·0341	4·1775	5·5580	7·1927	9·0984	13·787	16·600	19·746	27·093	35·939	40·957	58·549	66·705
0·000145	0·080	0·098	0·107	0·115	0·123	0·131	0·146	0·153	0·160	0·173	0·186	0·193	0·211	0·218
1/ 6897	1·4100	3·0886	4·2525	5·6576	7·3215	9·2612	14·033	16·897	20·099	27·577	36·580	41·688	59·593	67·894
0·000150	0·081	0·100	0·109	0·117	0·125	0·133	0·148	0·156	0·163	0·176	0·190	0·196	0·214	0·222
1/ 6667	1·4345	3·1422	4·3262	5·7555	7·4481	9·4213	14·276	17·188	20·446	28·052	37·211	42·406	60·618	69·062
0·000160	0·084	0·103	0·112	0·121	0·130	0·138	0·153	0·161	0·168	0·182	0·196	0·202	0·221	0·229
1/ 6250	1·4824	3·2467	4·4699	5·9466	7·6952	9·7337	14·748	17·757	21·122	28·980	38·440	43·807	62·620	71·342
0·000170	0·087	0·107	0·116	0·125	0·134	0·142	0·158	0·166	0·173	0·188	0·202	0·209	0·228	0·236
1/ 5882	1·5288	3·3480	4·6093	6·1318	7·9347	10·036	15·207	18·309	21·778	29·879	39·632	45·165	64·560	73·552
0·000180	0·089	0·110	0·119	0·129	0·138	0·146	0·163	0·171	0·178	0·193	0·208	0·215	0·235	0·243
1/ 5556	1·5739	3·4464	4·7446	6·3116	8·1673	10·330	15·651	18·844	22·415	30·752	40·790	46·484	66·444	75·698
0·000190	0·092	0·113	0·123	0·132	0·141	0·150	0·167	0·175	0·183	0·199	0·213	0·221	0·241	0·250
1/ 5263	1·6177	3·5421	4·8762	6·4865	8·3934	10·616	16·084	19·365	23·034	31·601	41·915	47·767	68·276	77·785
0·000200	0·094	0·116	0·126	0·136	0·145	0·154	0·172	0·180	0·188	0·204	0·219	0·226	0·248	0·256
1/ 5000	1·6604	3·6353	5·0043	6·6569	8·6137	10·895	16·506	19·872	23·637	32·428	43·012	49·016	70·061	79·818
0·000210	0·096	0·119	0·129	0·139	0·149	0·158	0·176	0·184	0·193	0·209	0·225	0·232	0·254	0·262
1/ 4762	1·7020	3·7262	5·1293	6·8231	8·8286	11·166	16·917	20·367	24·225	33·235	44·081	50·234	71·801	81·800
0·000220	0·099	0·121	0·132	0·142	0·152	0·162	0·180	0·189	0·197	0·214	0·230	0·238	0·260	0·269
1/ 4545	1·7426	3·8149	5·2514	6·9853	9·0384	11·432	17·318	20·850	24·800	34·022	45·125	51·424	73·501	83·736
0·000230	0·101	0·124	0·135	0·146	0·156	0·165	0·184	0·193	0·202	0·219	0·235	0·243	0·266	0·275
1/ 4348	1·7824	3·9017	5·3707	7·1439	9·2435	11·691	17·711	21·323	25·361	34·792	46·146	52·587	75·162	85·629
0·000240	0·103	0·127	0·138	0·149	0·159	0·169	0·188	0·197	0·206	0·223	0·240	0·248	0·272	0·281
1/ 4167	1·8213	3·9866	5·4875	7·2991	9·4442	11·944	18·095	21·785	25·911	35·546	47·145	53·725	76·788	87·480
0·000250	0·105	0·130	0·141	0·152	0·162	0·172	0·192	0·201	0·210	0·228	0·245	0·253	0·277	0·286
1/ 4000	1·8593	4·0697	5·6018	7·4511	9·6407	12·193	18·471	22·237	26·449	36·284	48·123	54·839	78·380	89·294
0·000260	0·107	0·132	0·144	0·155	0·166	0·176	0·196	0·205	0·215	0·233	0·250	0·258	0·283	0·292
1/ 3846	1·8967	4·1512	5·7139	7·6001	9·8334	12·436	18·839	22·681	26·976	37·007	49·082	55·932	79·941	91·071
0·000270	0·109	0·135	0·146	0·158	0·169	0·179	0·200	0·209	0·219	0·237	0·255	0·263	0·288	0·298
1/ 3704	1·9333	4·2312	5·8239	7·7462	10·022	12·675	19·201	23·116	27·494	37·716	50·022	57·003	81·472	92·815
0·000280	0·111	0·137	0·149	0·161	0·172	0·183	0·203	0·213	0·223	0·242	0·259	0·268	0·293	0·303
1/ 3571	1·9692	4·3097	5·9318	7·8897	10·208	12·910	19·556	23·543	28·002	38·413	50·946	58·055	82·974	94·527
0·000290	0·113	0·140	0·152	0·164	0·175	0·186	0·207	0·217	0·227	0·246	0·264	0·273	0·299	0·309
1/ 3448	2·0045	4·3868	6·0378	8·0306	10·390	13·140	19·904	23·963	28·501	39·097	51·853	59·089	84·450	96·208
0·000300	0·115	0·142	0·154	0·166	0·178	0·189	0·210	0·221	0·231	0·250	0·269	0·278	0·304	0·314
1/ 3333	2·0393	4·4626	6·1421	8·1692	10·569	13·367	20·247	24·375	28·991	39·769	52·744	60·104	85·901	97·861
	–	–	–	–	–	–	0·74	0·74	0·74	0·74	0·74	0·75	0·75	0·75

$V_{r(0{\cdot}5)medial}$ **for half-full circular pipes.**

$k_s = 6{\cdot}0$ mm
S = 0·00030 to 0·00100

ie hydraulic gradient =
1 in 3333 to 1 in 1000

Water (or sewage) at 15°C;
full bore conditions.

velocities in ms^{-1}
discharges in litres/sec

Gradient	(Equivalent) Pipe diameters in mm													
	150	200	225	250	275	300	350	375	400	450	500	525	600	630
0·000300	0·115	0·142	0·154	0·166	0·178	0·189	0·210	0·221	0·231	0·250	0·269	0·278	0·304	0·314
1/ 3333	2·0393	4·4626	6·1421	8·1692	10·569	13·367	20·247	24·375	28·991	39·769	52·744	60·104	85·901	97·861
0·000320	0·119	0·147	0·160	0·172	0·184	0·195	0·217	0·228	0·238	0·258	0·277	0·287	0·314	0·324
1/ 3125	2·1070	4·6104	6·3454	8·4394	10·919	13·808	20·916	25·180	29·948	41·081	54·483	62·086	88·732	101·09
0·000340	0·123	0·151	0·165	0·177	0·190	0·201	0·224	0·235	0·246	0·266	0·286	0·296	0·324	0·334
1/ 2941	2·1726	4·7537	6·5425	8·7014	11·257	14·237	21·564	25·960	30·875	42·353	56·169	64·007	91·476	104·21
0·000360	0·127	0·156	0·169	0·182	0·195	0·207	0·231	0·242	0·253	0·274	0·294	0·304	0·333	0·344
1/ 2778	2·2363	4·8929	6·7339	8·9558	11·586	14·652	22·193	26·718	31·776	43·588	57·806	65·872	94·141	107·25
0·000380	0·130	0·160	0·174	0·187	0·200	0·213	0·237	0·249	0·260	0·282	0·303	0·313	0·342	0·354
1/ 2632	2·2983	5·0282	6·9200	9·2032	11·906	15·057	22·805	27·454	32·652	44·789	59·398	67·686	96·732	110·20
0·000400	0·133	0·164	0·179	0·192	0·206	0·219	0·243	0·255	0·267	0·289	0·310	0·321	0·351	0·363
1/ 2500	2·3587	5·1600	7·1013	9·4441	12·218	15·451	23·401	28·172	33·505	45·958	60·949	69·452	99·256	113·07
0·000420	0·137	0·168	0·183	0·197	0·211	0·224	0·249	0·261	0·273	0·296	0·318	0·329	0·360	0·372
1/ 2381	2·4176	5·2886	7·2781	9·6792	12·522	15·835	23·983	28·871	34·337	47·099	62·461	71·176	101·72	115·88
0·000440	0·140	0·172	0·187	0·202	0·216	0·229	0·255	0·268	0·280	0·303	0·326	0·337	0·368	0·381
1/ 2273	2·4751	5·4142	7·4508	9·9086	12·818	16·210	24·551	29·555	35·149	48·213	63·938	72·858	104·12	118·61
0·000460	0·143	0·176	0·192	0·206	0·221	0·235	0·261	0·274	0·286	0·310	0·333	0·344	0·377	0·389
1/ 2174	2·5313	5·5369	7·6195	10·133	13·109	16·577	25·106	30·223	35·944	49·302	65·382	74·503	106·47	121·29
0·000480	0·146	0·180	0·196	0·211	0·225	0·240	0·267	0·280	0·292	0·317	0·340	0·352	0·385	0·397
1/ 2083	2·5863	5·6570	7·7847	10·352	13·392	16·935	25·649	30·876	36·721	50·367	66·794	76·112	108·77	123·91
0·000500	0·149	0·184	0·200	0·215	0·230	0·245	0·272	0·285	0·298	0·323	0·347	0·359	0·393	0·406
1/ 2000	2·6402	5·7746	7·9464	10·567	13·670	17·287	26·181	31·516	37·482	51·411	68·178	77·688	111·02	126·47
0·000525	0·153	0·188	0·205	0·221	0·236	0·251	0·279	0·292	0·306	0·331	0·356	0·368	0·402	0·416
1/ 1905	2·7060	5·9183	8·1441	10·830	14·010	17·716	26·831	32·299	38·412	52·686	69·869	79·615	113·77	129·61
0·000550	0·157	0·193	0·210	0·226	0·241	0·257	0·285	0·299	0·313	0·339	0·364	0·376	0·412	0·426
1/ 1818	2·7703	6·0587	8·3372	11·087	14·342	18·136	27·465	33·063	39·321	53·932	71·520	81·496	116·46	132·67
0·000575	0·160	0·197	0·214	0·231	0·247	0·262	0·292	0·306	0·320	0·347	0·372	0·385	0·421	0·435
1/ 1739	2·8332	6·1959	8·5259	11·338	14·666	18·546	28·086	33·809	40·208	55·149	73·134	83·335	119·09	135·66
0·000600	0·164	0·201	0·219	0·236	0·252	0·268	0·298	0·313	0·327	0·354	0·381	0·393	0·430	0·445
1/ 1667	2·8947	6·3302	8·7105	11·583	14·984	18·947	28·693	34·540	41·077	56·341	74·713	85·134	121·66	138·59
0·000625	0·167	0·206	0·224	0·241	0·257	0·274	0·304	0·319	0·334	0·362	0·388	0·401	0·439	0·454
1/ 1600	2·9549	6·4617	8·8914	11·823	15·294	19·340	29·288	35·256	41·928	57·507	76·260	86·897	124·18	141·46
0·000650	0·171	0·210	0·228	0·246	0·263	0·279	0·310	0·326	0·340	0·369	0·396	0·409	0·448	0·463
1/ 1538	3·0139	6·5906	9·0686	12·059	15·599	19·725	29·871	35·957	42·762	58·651	77·776	88·624	126·64	144·27
0·000675	0·174	0·214	0·232	0·250	0·268	0·284	0·316	0·332	0·347	0·376	0·404	0·417	0·456	0·472
1/ 1481	3·0718	6·7170	9·2425	12·290	15·898	20·102	30·442	36·645	43·580	59·773	79·263	90·319	129·06	147·03
0·000700	0·177	0·218	0·237	0·255	0·273	0·290	0·322	0·338	0·353	0·383	0·411	0·425	0·465	0·480
1/ 1429	3·1287	6·8411	9·4132	12·517	16·191	20·473	31·004	37·321	44·384	60·874	80·723	91·982	131·44	149·73
0·000725	0·180	0·222	0·241	0·260	0·277	0·295	0·328	0·344	0·359	0·390	0·418	0·432	0·473	0·489
1/ 1379	3·1845	6·9630	9·5808	12·740	16·479	20·838	31·555	37·984	45·173	61·956	82·157	93·616	133·77	152·39
0·000750	0·183	0·225	0·245	0·264	0·282	0·300	0·334	0·350	0·366	0·396	0·426	0·440	0·481	0·497
1/ 1333	3·2394	7·0829	9·7456	12·959	16·763	21·196	32·097	38·637	45·948	63·019	83·567	95·222	136·07	155·00
0·000800	0·189	0·233	0·253	0·273	0·292	0·310	0·345	0·361	0·378	0·409	0·440	0·454	0·497	0·514
1/ 1250	3·3465	7·3167	10·067	13·386	17·315	21·894	33·154	39·909	47·461	65·094	86·317	98·355	140·55	160·10
0·000850	0·195	0·240	0·261	0·281	0·301	0·319	0·355	0·373	0·389	0·422	0·453	0·468	0·512	0·529
1/ 1176	3·4503	7·5433	10·379	13·801	17·851	22·571	34·179	41·143	48·928	67·104	88·983	101·39	144·88	165·04
0·000900	0·201	0·247	0·269	0·289	0·309	0·329	0·366	0·383	0·401	0·434	0·466	0·482	0·527	0·545
1/ 1111	3·5511	7·7634	10·681	14·203	18·371	23·229	35·174	42·340	50·352	69·057	91·571	104·34	149·10	169·84
0·000950	0·206	0·254	0·276	0·297	0·318	0·338	0·376	0·394	0·412	0·446	0·479	0·495	0·542	0·560
1/ 1053	3·6491	7·9774	10·976	14·594	18·877	23·868	36·142	43·505	51·737	70·956	94·088	107·21	153·19	174·51
0·001000	0·212	0·261	0·283	0·305	0·326	0·346	0·385	0·404	0·422	0·458	0·492	0·508	0·556	0·574
1/ 1000	3·7446	8·1858	11·262	14·975	19·370	24·491	37·085	44·639	53·086	72·805	96·540	110·00	157·18	179·06
	–	–	–	–	–	0·74	0·74	0·74	0·74	0·74	0·74	0·75	0·75	0·75

$V_{r(0{\cdot}5)medial}$ for half-full circular pipes.

k_s = 6·0 mm
S = 0·00100 to 0·00300

ie hydraulic gradient =
1 in 1000 to 1 in 333

Water (or sewage) at 15°C;
full bore conditions.

velocities in ms^{-1}
discharges in litres/sec

Gradient **(Equivalent) Pipe diameters in mm**

Gradient	150	200	225	250	275	300	350	375	400	450	500	525	600	630
0·00100	0·212	0·261	0·283	0·305	0·326	0·346	0·385	0·404	0·422	0·458	0·492	0·508	0·556	0·574
1/ 1000	3·7446	8·1858	11·262	14·975	19·370	24·491	37·085	44·639	53·086	72·805	96·540	110·00	157·18	179·06
0·00105	0·217	0·267	0·290	0·313	0·334	0·355	0·395	0·414	0·433	0·469	0·504	0·521	0·570	0·589
1/ 952	3·8377	8·3892	11·542	15·347	19·850	25·098	38·004	45·746	54·401	74·609	98·932	112·73	161·08	183·49
0·00110	0·222	0·273	0·297	0·320	0·342	0·363	0·404	0·424	0·443	0·480	0·516	0·533	0·583	0·603
1/ 909	3·9287	8·5877	11·815	15·709	20·319	25·691	38·902	46·826	55·686	76·370	101·27	115·39	164·88	187·82
0·00115	0·227	0·280	0·304	0·327	0·350	0·372	0·413	0·434	0·453	0·491	0·527	0·545	0·596	0·616
1/ 870	4·0175	8·7817	12·082	16·064	20·778	26·271	39·779	47·882	56·942	78·092	103·55	117·99	168·59	192·05
0·00120	0·232	0·286	0·310	0·334	0·357	0·380	0·422	0·443	0·463	0·502	0·539	0·557	0·609	0·629
1/ 833	4·1045	8·9716	12·343	16·411	21·227	26·839	40·638	48·916	58·170	79·777	105·78	120·53	172·23	196·19
0·00125	0·237	0·291	0·317	0·341	0·365	0·388	0·431	0·452	0·472	0·512	0·550	0·568	0·622	0·642
1/ 800	4·1897	9·1576	12·599	16·751	21·666	27·394	41·479	49·928	59·374	81·427	107·97	123·03	175·79	200·25
0·00130	0·242	0·297	0·323	0·348	0·372	0·395	0·440	0·461	0·482	0·522	0·561	0·580	0·634	0·655
1/ 769	4·2732	9·3399	12·850	17·084	22·097	27·939	42·303	50·920	60·553	83·044	110·11	125·47	179·28	204·22
0·00135	0·246	0·303	0·329	0·355	0·379	0·403	0·448	0·470	0·491	0·532	0·572	0·591	0·646	0·668
1/ 741	4·3551	9·5187	13·095	17·411	22·520	28·473	43·112	51·893	61·711	84·631	112·22	127·87	182·70	208·12
0·00140	0·251	0·309	0·335	0·361	0·386	0·410	0·456	0·478	0·500	0·542	0·582	0·602	0·658	0·680
1/ 714	4·4355	9·6943	13·337	17·732	22·935	28·998	43·906	52·849	62·846	86·188	114·28	130·22	186·06	211·95
0·00145	0·255	0·314	0·341	0·368	0·393	0·418	0·464	0·487	0·509	0·552	0·592	0·612	0·670	0·692
1/ 690	4·5145	9·8667	13·574	18·047	23·342	29·513	44·686	53·787	63·962	87·718	116·31	132·53	189·36	215·71
0·00150	0·260	0·319	0·347	0·374	0·400	0·425	0·472	0·495	0·518	0·561	0·603	0·623	0·681	0·704
1/ 667	4·5921	10·036	13·807	18·357	23·743	30·019	45·452	54·709	65·059	89·222	118·30	134·80	192·61	219·40
0·00160	0·268	0·330	0·359	0·386	0·413	0·439	0·488	0·512	0·535	0·579	0·622	0·643	0·704	0·727
1/ 625	4·7436	10·367	14·262	18·962	24·525	31·007	46·947	56·509	67·199	92·156	122·19	139·23	198·94	226·61
0·00170	0·277	0·340	0·370	0·398	0·426	0·452	0·503	0·527	0·551	0·597	0·642	0·663	0·725	0·749
1/ 588	4·8904	10·687	14·703	19·547	25·282	31·965	48·397	58·253	69·273	94·999	125·96	143·53	205·07	233·60
0·00180	0·285	0·350	0·381	0·410	0·438	0·465	0·518	0·543	0·567	0·615	0·660	0·682	0·746	0·771
1/ 556	5·0329	10·999	15·131	20·116	26·018	32·894	49·804	59·947	71·287	97·761	129·62	147·70	211·03	240·39
0·00190	0·293	0·360	0·391	0·421	0·450	0·478	0·532	0·558	0·583	0·632	0·678	0·701	0·767	0·792
1/ 526	5·1716	11·301	15·547	20·670	26·733	33·799	51·173	61·594	73·245	100·45	133·18	151·75	216·83	246·99
0·00200	0·300	0·369	0·401	0·432	0·462	0·491	0·546	0·572	0·598	0·648	0·696	0·719	0·787	0·813
1/ 500	5·3066	11·596	15·952	21·208	27·430	34·680	52·506	63·199	75·153	103·06	136·65	155·70	222·47	253·42
0·00210	0·308	0·378	0·411	0·443	0·473	0·503	0·559	0·586	0·613	0·664	0·713	0·737	0·806	0·833
1/ 476	5·4383	11·884	16·348	21·734	28·109	35·539	53·806	64·764	77·014	105·61	140·03	159·56	227·98	259·69
0·00220	0·315	0·387	0·421	0·453	0·484	0·515	0·572	0·600	0·627	0·680	0·730	0·754	0·825	0·853
1/ 455	5·5669	12·164	16·734	22·247	28·773	36·377	55·076	66·292	78·830	108·10	143·34	163·32	233·35	265·81
0·00230	0·322	0·396	0·430	0·463	0·495	0·526	0·585	0·614	0·641	0·695	0·746	0·771	0·844	0·872
1/ 435	5·6926	12·439	17·111	22·749	29·422	37·197	56·317	67·785	80·606	110·54	146·56	167·00	238·60	271·80
0·00240	0·329	0·404	0·440	0·473	0·506	0·538	0·598	0·627	0·655	0·710	0·763	0·788	0·862	0·891
1/ 417	5·8157	12·707	17·481	23·240	30·056	38·000	57·531	69·247	82·344	112·92	149·72	170·60	243·75	277·65
0·00250	0·336	0·413	0·449	0·483	0·517	0·549	0·610	0·640	0·669	0·725	0·778	0·804	0·880	0·909
1/ 400	5·9361	12·970	17·842	23·721	30·678	38·786	58·721	70·678	84·046	115·25	152·82	174·12	248·78	283·39
0·00260	0·343	0·421	0·458	0·493	0·527	0·560	0·622	0·653	0·682	0·739	0·794	0·820	0·897	0·927
1/ 385	6·0542	13·228	18·197	24·192	31·287	39·556	59·886	72·081	85·714	117·54	155·85	177·58	253·72	289·01
0·00270	0·349	0·429	0·466	0·502	0·537	0·570	0·634	0·665	0·695	0·753	0·809	0·836	0·914	0·945
1/ 370	6·1700	13·481	18·545	24·654	31·885	40·311	61·030	73·457	87·351	119·79	158·82	180·96	258·56	294·52
0·00280	0·356	0·437	0·475	0·511	0·547	0·581	0·646	0·677	0·708	0·767	0·824	0·851	0·931	0·962
1/ 357	6·2838	13·729	18·886	25·108	32·472	41·053	62·153	74·809	88·957	121·99	161·74	184·29	263·31	299·94
0·00290	0·362	0·445	0·483	0·521	0·556	0·591	0·657	0·689	0·720	0·781	0·838	0·866	0·948	0·979
1/ 345	6·3955	13·973	19·221	25·553	33·048	41·782	63·255	76·136	90·535	124·15	164·61	187·56	267·98	305·25
0·00300	0·368	0·452	0·492	0·529	0·566	0·601	0·669	0·701	0·733	0·794	0·853	0·881	0·964	0·996
1/ 333	6·5052	14·213	19·551	25·992	33·615	42·498	64·339	77·440	92·086	126·28	167·43	190·77	272·57	310·48
	–	–	–	–	–	0·74	0·74	0·74	0·74	0·74	0·74	0·75	0·75	0·75

$V_{r(0·5)medial}$ for half-full circular pipes.

k_s = 6·0 mm
S = 0·00300 to 0·01000

ie hydraulic gradient =
1 in 333 to 1 in 100

Water (or sewage) at 15°C;
full bore conditions.

velocities in ms^{-1}
discharges in litres/sec

Gradient **(Equivalent) Pipe diameters in mm**

Gradient	150	200	225	250	275	300	350	375	400	450	500	525	600	630
0·00300	0·368	0·452	0·492	0·529	0·566	0·601	0·669	0·701	0·733	0·794	0·853	0·881	0·964	0·996
1/ 333	6·5052	14·213	19·551	25·992	33·615	42·498	64·339	77·440	92·086	126·28	167·43	190·77	272·57	310·48
0·00320	0·380	0·467	0·508	0·547	0·585	0·621	0·691	0·724	0·757	0·820	0·881	0·910	0·996	1·029
1/ 313	6·7195	14·681	20·194	26·846	34·720	43·895	66·454	79·985	95·112	130·43	172·93	197·04	281·52	320·68
0·00340	0·392	0·482	0·524	0·564	0·603	0·640	0·712	0·747	0·780	0·845	0·908	0·938	1·026	1·060
1/ 294	6·9271	15·134	20·818	27·675	35·791	45·249	68·504	82·452	98·046	134·45	178·26	203·11	290·20	330·56
0·00360	0·403	0·496	0·539	0·580	0·620	0·659	0·733	0·768	0·803	0·870	0·934	0·966	1·056	1·091
1/ 278	7·1287	15·574	21·423	28·479	36·832	46·564	70·494	84·847	100·89	138·35	183·44	209·01	298·62	340·16
0·00380	0·414	0·509	0·554	0·596	0·637	0·677	0·753	0·789	0·825	0·894	0·960	0·992	1·085	1·121
1/ 263	7·3248	16·002	22·012	29·262	37·843	47·843	72·430	87·177	103·66	142·15	188·47	214·75	306·82	349·49
0·00400	0·425	0·523	0·568	0·612	0·654	0·694	0·772	0·810	0·846	0·917	0·985	1·018	1·113	1·150
1/ 250	7·5157	16·419	22·585	30·024	38·829	49·089	74·315	89·446	106·36	145·85	193·38	220·33	314·80	358·59
0·00420	0·436	0·536	0·582	0·627	0·670	0·712	0·792	0·830	0·867	0·940	1·009	1·043	1·141	1·179
1/ 238	7·7020	16·826	23·144	30·767	39·790	50·304	76·154	91·659	108·99	149·46	198·16	225·78	322·58	367·45
0·00440	0·446	0·548	0·596	0·642	0·686	0·728	0·810	0·849	0·888	0·962	1·033	1·068	1·168	1·207
1/ 227	7·8839	17·223	23·690	31·493	40·729	51·490	77·949	93·820	111·56	152·98	202·83	231·11	330·19	376·11
0·00460	0·456	0·561	0·609	0·656	0·701	0·745	0·828	0·869	0·908	0·984	1·056	1·092	1·194	1·234
1/ 217	8·0617	17·611	24·224	32·203	41·646	52·650	79·705	95·932	114·07	156·43	207·40	236·31	337·62	384·57
0·00480	0·466	0·573	0·622	0·670	0·716	0·761	0·846	0·887	0·927	1·005	1·079	1·115	1·220	1·260
1/ 208	8·2356	17·991	24·746	32·897	42·544	53·784	81·422	97·999	116·53	159·80	211·86	241·40	344·89	392·86
0·00500	0·476	0·585	0·635	0·684	0·731	0·777	0·864	0·906	0·946	1·025	1·101	1·138	1·245	1·286
1/ 200	8·4060	18·363	25·258	33·577	43·423	54·896	83·104	100·02	118·94	163·10	216·24	246·38	352·01	400·97
0·00525	0·487	0·599	0·651	0·701	0·749	0·796	0·885	0·928	0·970	1·051	1·129	1·166	1·276	1·318
1/ 190	8·6143	18·817	25·883	34·408	44·497	56·254	85·160	102·50	121·88	167·13	221·59	252·47	360·71	410·88
0·00550	0·499	0·613	0·666	0·717	0·767	0·815	0·906	0·950	0·993	1·076	1·155	1·194	1·306	1·349
1/ 182	8·8176	19·261	26·494	35·219	45·547	57·580	87·167	104·91	124·75	171·07	226·81	258·42	369·21	420·56
0·00575	0·510	0·627	0·681	0·734	0·784	0·833	0·926	0·971	1·015	1·100	1·181	1·221	1·335	1·379
1/ 174	9·0164	19·695	27·090	36·012	46·572	58·877	89·130	107·28	127·56	174·92	231·91	264·24	377·52	430·02
0·00600	0·521	0·640	0·696	0·749	0·801	0·851	0·946	0·992	1·037	1·124	1·207	1·247	1·364	1·409
1/ 167	9·2109	20·120	27·674	36·789	47·576	60·145	91·050	109·59	130·31	178·69	236·91	269·93	385·65	439·28
0·00625	0·532	0·654	0·710	0·765	0·818	0·868	0·966	1·013	1·058	1·147	1·231	1·273	1·392	1·438
1/ 160	9·4014	20·536	28·246	37·549	48·559	61·388	92·930	111·85	133·00	182·37	241·80	275·50	393·61	448·35
0·00650	0·543	0·667	0·725	0·780	0·834	0·886	0·985	1·033	1·079	1·169	1·256	1·298	1·420	1·467
1/ 154	9·5881	20·943	28·807	38·294	49·522	62·606	94·774	114·07	135·64	185·99	246·59	280·96	401·41	457·24
0·00675	0·553	0·679	0·738	0·795	0·850	0·903	1·004	1·052	1·100	1·192	1·280	1·323	1·447	1·495
1/ 148	9·7712	21·343	29·357	39·025	50·467	63·800	96·582	116·24	138·22	189·54	251·30	286·32	409·07	465·96
0·00700	0·563	0·692	0·752	0·810	0·865	0·919	1·022	1·072	1·120	1·214	1·303	1·347	1·473	1·522
1/ 143	9·9510	21·736	29·897	39·742	51·395	64·973	98·357	118·38	140·76	193·02	255·91	291·58	416·58	474·52
0·00725	0·573	0·704	0·765	0·824	0·881	0·935	1·040	1·091	1·140	1·235	1·326	1·371	1·499	1·549
1/ 138	10·128	22·121	30·427	40·447	52·306	66·125	100·10	120·48	143·26	196·44	260·45	296·75	423·96	482·93
0·00750	0·583	0·716	0·778	0·838	0·896	0·951	1·058	1·110	1·160	1·256	1·349	1·394	1·525	1·576
1/ 133	10·301	22·500	30·948	41·140	53·202	67·257	101·81	122·54	145·71	199·80	264·91	301·83	431·22	491·19
0·00800	0·602	0·740	0·804	0·866	0·925	0·983	1·093	1·146	1·198	1·298	1·393	1·440	1·575	1·627
1/ 125	10·640	23·240	31·965	42·491	54·950	69·467	105·16	126·57	150·50	206·37	273·60	311·74	445·37	507·31
0·00850	0·621	0·763	0·829	0·892	0·954	1·013	1·127	1·181	1·235	1·338	1·436	1·484	1·624	1·678
1/ 118	10·968	23·956	32·951	43·801	56·644	71·608	108·40	130·47	155·13	212·72	282·03	321·34	459·09	522·94
0·00900	0·639	0·785	0·853	0·918	0·981	1·042	1·159	1·216	1·270	1·376	1·478	1·528	1·671	1·726
1/ 111	11·287	24·652	33·908	45·073	58·289	73·687	111·55	134·25	159·64	218·90	290·22	330·67	472·41	538·12
0·00950	0·656	0·806	0·876	0·943	1·008	1·071	1·191	1·249	1·305	1·414	1·519	1·569	1·717	1·774
1/ 105	11·597	25·329	34·838	46·311	59·888	75·709	114·61	137·94	164·02	224·90	298·18	339·74	485·37	552·88
0·01000	0·673	0·827	0·899	0·968	1·035	1·099	1·222	1·281	1·339	1·451	1·558	1·610	1·761	1·820
1/ 100	11·899	25·988	35·745	47·516	61·447	77·679	117·59	141·52	168·28	230·75	305·93	348·57	497·99	567·25
	–	–	–	–	–	0·74	0·74	0·74	0·74	0·74	0·74	0·75	0·75	0·75

$V_{r(0·5)medial}$ for half-full circular pipes.

A21

(p.5 of 6)

$k_s = 6{\cdot}0$ **mm**
S = 0·01000 to 0·03000

ie hydraulic gradient =
1 in 100 to 1 in 33·3

Water (or sewage) at 15°C;
full bore conditions.

velocities in ms^{-1}
discharges in litres/sec

Gradient **(Equivalent) Pipe diameters in mm**

Gradient	150	200	225	250	275	300	350	375	400	450	500	525	600	630
0·01000	0·673	0·827	0·899	0·968	1·035	1·099	1·222	1·281	1·339	1·451	1·558	1·610	1·761	1·820
1/ 100	11·899	25·988	35·745	47·516	61·447	77·679	117·59	141·52	168·28	230·75	305·93	348·57	497·99	567·25
0·01050	0·690	0·848	0·921	0·992	1·060	1·126	1·252	1·313	1·372	1·487	1·597	1·650	1·805	1·865
1/ 95	12·194	26·631	36·629	48·691	62·966	79·600	120·50	145·02	172·44	236·46	313·50	357·19	510·30	581·27
0·01100	0·706	0·868	0·943	1·015	1·085	1·153	1·282	1·344	1·405	1·522	1·634	1·689	1·847	1·909
1/ 91	12·481	27·259	37·493	49·838	64·450	81·475	123·33	148·44	176·51	242·03	320·88	365·60	522·32	594·96
0·01150	0·722	0·887	0·964	1·038	1·110	1·179	1·311	1·374	1·436	1·556	1·671	1·727	1·889	1·952
1/ 87	12·762	27·873	38·337	50·960	65·901	83·309	126·11	151·78	180·48	247·47	328·10	373·83	534·07	608·34
0·01200	0·738	0·906	0·985	1·061	1·133	1·204	1·339	1·404	1·467	1·590	1·707	1·764	1·930	1·994
1/ 83	13·037	28·474	39·163	52·058	67·320	85·103	128·82	155·05	184·36	252·80	335·16	381·87	545·56	621·44
0·01250	0·753	0·925	1·005	1·082	1·157	1·229	1·367	1·433	1·497	1·622	1·742	1·800	1·969	2·035
1/ 80	13·307	29·062	39·971	53·133	68·710	86·860	131·48	158·25	188·17	258·02	342·08	389·75	556·82	634·26
0·01300	0·768	0·943	1·025	1·104	1·180	1·253	1·394	1·461	1·527	1·654	1·777	1·836	2·008	2·075
1/ 77	13·571	29·638	40·764	54·187	70·073	88·582	134·09	161·39	191·90	263·13	348·86	397·48	567·86	646·83
0·01350	0·783	0·961	1·045	1·125	1·202	1·277	1·420	1·489	1·556	1·686	1·811	1·871	2·047	2·115
1/ 74	13·830	30·204	41·542	55·220	71·409	90·272	136·65	164·46	195·56	268·15	355·51	405·06	578·68	659·16
0·01400	0·797	0·979	1·064	1·146	1·224	1·301	1·446	1·516	1·585	1·717	1·844	1·906	2·084	2·153
1/ 71	14·084	30·759	42·305	56·235	72·721	91·930	139·16	167·49	199·15	273·07	362·04	412·50	589·31	671·27
0·01450	0·811	0·996	1·083	1·166	1·246	1·324	1·472	1·543	1·613	1·747	1·877	1·939	2·121	2·192
1/ 69	14·334	31·304	43·055	57·232	74·010	93·560	141·62	170·45	202·68	277·91	368·45	419·80	599·75	683·16
0·01500	0·825	1·014	1·101	1·186	1·267	1·346	1·497	1·570	1·640	1·777	1·909	1·972	2·157	2·229
1/ 67	14·579	31·840	43·792	58·211	75·277	95·161	144·05	173·37	206·15	282·67	374·76	426·99	610·01	694·84
0·01600	0·852	1·047	1·138	1·225	1·309	1·390	1·546	1·621	1·694	1·836	1·971	2·037	2·228	2·302
1/ 63	15·058	32·886	45·230	60·123	77·748	98·285	148·78	179·06	212·91	291·95	387·06	441·00	630·03	717·65
0·01700	0·878	1·079	1·173	1·263	1·349	1·433	1·594	1·671	1·747	1·892	2·032	2·100	2·297	2·373
1/ 59	15·523	33·899	46·624	61·976	80·144	101·31	153·36	184·58	219·47	300·94	398·98	454·58	649·43	739·75
0·01800	0·904	1·110	1·207	1·299	1·388	1·475	1·640	1·720	1·797	1·947	2·091	2·161	2·364	2·442
1/ 56	15·973	34·884	47·978	63·775	82·470	104·25	157·81	189·93	225·84	309·67	410·55	467·77	668·27	761·21
0·01900	0·929	1·141	1·240	1·335	1·427	1·515	1·685	1·767	1·846	2·000	2·148	2·220	2·428	2·509
1/ 53	16·412	35·841	49·294	65·524	84·733	107·11	162·14	195·14	232·04	318·16	421·81	480·60	686·60	782·08
0·02000	0·953	1·171	1·272	1·370	1·464	1·555	1·729	1·813	1·894	2·052	2·204	2·278	2·491	2·574
1/ 50	16·839	36·773	50·576	67·228	86·936	109·90	166·36	200·22	238·07	326·43	432·78	493·09	704·45	802·41
0·02100	0·976	1·199	1·303	1·403	1·500	1·593	1·772	1·858	1·941	2·103	2·259	2·334	2·553	2·638
1/ 47·6	17·255	37·682	51·827	68·890	89·085	112·62	170·47	205·16	243·95	334·50	443·47	505·28	721·85	822·24
0·02200	0·999	1·228	1·334	1·436	1·535	1·631	1·814	1·901	1·987	2·153	2·312	2·389	2·613	2·700
1/ 45·5	17·662	38·570	53·048	70·513	91·184	115·27	174·48	210·00	249·70	342·38	453·92	517·18	738·85	841·60
0·02300	1·022	1·255	1·364	1·469	1·570	1·667	1·854	1·944	2·032	2·201	2·364	2·443	2·672	2·761
1/ 43·5	18·060	39·438	54·241	72·100	93·235	117·86	178·41	214·72	255·31	350·08	464·13	528·81	755·47	860·52
0·02400	1·044	1·282	1·394	1·500	1·604	1·703	1·894	1·986	2·075	2·249	2·415	2·495	2·729	2·820
1/ 41·7	18·449	40·288	55·409	73·652	95·242	120·40	182·25	219·34	260·81	357·61	474·11	540·19	771·72	879·04
0·02500	1·066	1·309	1·422	1·531	1·637	1·738	1·933	2·027	2·118	2·295	2·464	2·547	2·786	2·878
1/ 40·0	18·830	41·119	56·553	75·172	97·208	122·88	186·01	223·87	266·19	364·99	483·90	551·33	787·65	897·18
0·02600	1·087	1·335	1·451	1·562	1·669	1·773	1·972	2·067	2·160	2·340	2·513	2·597	2·841	2·935
1/ 38·5	19·203	41·935	57·674	76·663	99·135	125·32	189·69	228·31	271·47	372·23	493·49	562·26	803·25	914·96
0·02700	1·107	1·360	1·478	1·592	1·701	1·807	2·009	2·107	2·201	2·385	2·561	2·647	2·895	2·991
1/ 37·0	19·569	42·734	58·774	78·124	101·03	127·71	193·31	232·66	276·64	379·32	502·89	572·98	818·56	932·39
0·02800	1·128	1·385	1·505	1·621	1·732	1·840	2·046	2·145	2·242	2·429	2·608	2·695	2·948	3·046
1/ 35·7	19·929	43·519	59·854	79·559	102·88	130·05	196·86	236·93	281·72	386·29	512·13	583·50	833·59	949·51
0·02900	1·148	1·410	1·532	1·649	1·763	1·872	2·082	2·183	2·282	2·472	2·654	2·743	3·000	3·100
1/ 34·5	20·282	44·291	60·914	80·969	104·70	132·36	200·35	241·13	286·71	393·13	521·20	593·83	848·35	966·33
0·03000	1·167	1·434	1·558	1·678	1·793	1·905	2·118	2·221	2·321	2·514	2·700	2·790	3·052	3·153
1/ 33·3	20·629	45·049	61·957	82·354	106·49	134·62	203·78	245·25	291·62	399·85	530·11	603·99	862·86	982·86
	–	–	–	–	–	0·74	0·74	0·74	0·74	0·74	0·74	0·75	0·75	0·75

$V_{r(0{\cdot}5)medial}$ **for half-full circular pipes.**

$k_s = 6{\cdot}0$ **mm** S = 0·01000 to 0·03000

$k_s = 6{\cdot}0$ **mm**
S = 0·03000 to 0·10000

ie hydraulic gradient =
1 in 33·3 to 1 in 10·0

Water (or sewage) at 15°C;
full bore conditions.

velocities in ms^{-1}
discharges in litres/sec

Gradient **(Equivalent) Pipe diameters in mm**

Gradient	150	200	225	250	275	300	350	375	400	450	500	525	600	630
0·03000	1·167	1·434	1·558	1·678	1·793	1·905	2·118	2·221	2·321	2·514	2·700	2·790	3·052	3·153
1/ 33·3	20·629	45·049	61·957	82·354	106·49	134·62	203·78	245·25	291·62	399·85	530·11	603·99	862·86	982·86
0·03200	1·206	1·481	1·609	1·733	1·852	1·967	2·188	2·293	2·397	2·597	2·788	2·882	3·152	3·256
1/ 31·3	21·307	46·527	63·991	85·058	109·99	139·04	210·46	253·30	301·19	412·97	547·51	623·81	891·18	1015·1
0·03400	1·243	1·527	1·659	1·786	1·909	2·028	2·255	2·364	2·471	2·677	2·874	2·970	3·249	3·357
1/ 29·4	21·963	47·961	65·962	87·678	113·38	143·32	216·95	261·10	310·46	425·69	564·37	643·02	918·62	1046·4
0·03600	1·279	1·571	1·707	1·838	1·964	2·086	2·320	2·433	2·542	2·754	2·958	3·057	3·343	3·454
1/ 27·8	22·601	49·353	67·876	90·222	116·67	147·48	223·24	268·68	319·47	438·04	580·74	661·67	945·26	1076·7
0·03800	1·314	1·614	1·754	1·888	2·018	2·144	2·384	2·499	2·612	2·830	3·039	3·140	3·435	3·549
1/ 26·3	23·221	50·706	69·738	92·696	119·87	151·53	229·36	276·04	328·23	450·05	596·66	679·81	971·18	1106·2
0·04000	1·348	1·656	1·800	1·937	2·071	2·199	2·446	2·564	2·680	2·903	3·118	3·222	3·524	3·641
1/ 25·0	23·825	52·025	71·551	95·106	122·98	155·47	235·32	283·22	336·76	461·75	612·17	697·48	996·42	1135·0
0·04200	1·382	1·697	1·844	1·985	2·122	2·254	2·506	2·628	2·746	2·975	3·195	3·302	3·611	3·731
1/ 23·8	24·414	53·311	73·319	97·457	126·02	159·31	241·14	290·22	345·08	473·16	627·29	714·71	1021·0	1163·0
0·04400	1·414	1·737	1·887	2·032	2·172	2·307	2·565	2·690	2·811	3·045	3·270	3·379	3·696	3·819
1/ 22·7	24·989	54·567	75·046	99·752	128·99	163·06	246·82	297·05	353·21	484·30	642·06	731·54	1045·1	1190·4
0·04600	1·446	1·776	1·930	2·078	2·221	2·359	2·623	2·750	2·874	3·114	3·344	3·455	3·779	3·905
1/ 21·7	25·551	55·794	76·734	102·00	131·89	166·73	252·37	303·73	361·15	495·19	656·50	747·98	1068·6	1217·2
0·04800	1·477	1·814	1·971	2·123	2·268	2·409	2·679	2·809	2·936	3·181	3·415	3·530	3·861	3·989
1/ 20·8	26·102	56·995	78·386	104·19	134·73	170·31	257·80	310·27	368·92	505·84	670·62	764·08	1091·6	1243·4
0·05000	1·508	1·852	2·012	2·166	2·315	2·459	2·735	2·867	2·996	3·246	3·486	3·602	3·940	4·071
1/ 20·0	26·640	58·171	80·004	106·34	137·51	173·83	263·12	316·67	376·53	516·28	684·46	779·84	1114·1	1269·0
0·05250	1·545	1·897	2·062	2·220	2·372	2·520	2·802	2·938	3·070	3·326	3·572	3·691	4·038	4·171
1/ 19·0	27·299	59·609	81·981	108·97	140·91	178·12	269·62	324·49	385·84	529·03	701·37	799·11	1141·6	1300·4
0·05500	1·581	1·942	2·110	2·272	2·428	2·579	2·868	3·007	3·143	3·405	3·656	3·778	4·133	4·270
1/ 18·2	27·942	61·013	83·911	111·54	144·23	182·32	275·97	332·13	394·92	541·49	717·88	817·92	1168·5	1331·0
0·05750	1·617	1·986	2·158	2·323	2·483	2·637	2·933	3·075	3·213	3·481	3·738	3·863	4·226	4·366
1/ 17·4	28·571	62·385	85·799	114·04	147·47	186·42	282·17	339·60	403·80	553·67	734·02	836·31	1194·7	1360·9
0·06000	1·652	2·029	2·204	2·373	2·536	2·694	2·996	3·141	3·282	3·556	3·819	3·946	4·316	4·460
1/ 16·7	29·186	63·728	87·645	116·50	150·64	190·43	288·24	346·91	412·49	565·58	749·82	854·30	1220·5	1390·2
0·06250	1·686	2·070	2·250	2·422	2·589	2·750	3·058	3·206	3·350	3·629	3·898	4·028	4·406	4·552
1/ 16·0	29·788	65·043	89·454	118·90	153·75	194·36	294·19	354·07	421·00	577·25	765·28	871·93	1245·6	1418·8
0·06500	1·719	2·111	2·294	2·470	2·640	2·804	3·118	3·269	3·417	3·701	3·975	4·108	4·493	4·642
1/ 15·4	30·378	66·332	91·227	121·26	156·80	198·21	300·02	361·08	429·34	588·68	780·44	889·20	1270·3	1447·0
0·06750	1·752	2·152	2·338	2·517	2·690	2·858	3·178	3·332	3·482	3·772	4·051	4·186	4·578	4·730
1/ 14·8	30·958	67·597	92·966	123·57	159·79	201·99	305·74	367·96	437·52	599·90	795·32	906·15	1294·5	1474·5
0·07000	1·784	2·191	2·381	2·564	2·740	2·910	3·236	3·393	3·546	3·841	4·125	4·263	4·662	4·817
1/ 14·3	31·526	68·838	94·673	125·84	162·72	205·70	311·35	374·72	445·55	610·91	809·92	922·78	1318·3	1501·6
0·07250	1·816	2·230	2·423	2·609	2·788	2·962	3·293	3·453	3·608	3·909	4·198	4·338	4·745	4·902
1/ 13·8	32·085	70·057	96·350	128·07	165·60	209·34	316·86	381·35	453·44	621·73	824·26	939·12	1341·6	1528·2
0·07500	1·847	2·268	2·465	2·654	2·836	3·012	3·350	3·512	3·670	3·976	4·270	4·412	4·826	4·986
1/ 13·3	32·634	71·256	97·998	130·26	168·44	212·92	322·28	387·88	461·20	632·37	838·35	955·18	1364·6	1554·3
0·08000	1·907	2·343	2·546	2·741	2·929	3·111	3·460	3·627	3·791	4·107	4·410	4·557	4·984	5·150
1/ 12·5	33·705	73·594	101·21	134·53	173·96	219·91	332·86	400·60	476·33	653·11	865·86	986·52	1409·3	1605·3
0·08500	1·966	2·415	2·624	2·825	3·019	3·207	3·566	3·739	3·907	4·233	4·546	4·697	5·138	5·308
1/ 11·8	34·743	75·861	104·33	138·67	179·32	226·68	343·11	412·94	491·00	673·22	892·52	1016·9	1452·7	1654·7
0·09000	2·023	2·485	2·700	2·907	3·107	3·300	3·670	3·847	4·021	4·356	4·677	4·834	5·287	5·462
1/ 11·1	35·751	78·062	107·36	142·70	184·52	233·25	353·06	424·91	505·24	692·74	918·40	1046·4	1494·8	1702·7
0·09500	2·079	2·553	2·774	2·987	3·192	3·390	3·770	3·953	4·131	4·475	4·806	4·966	5·432	5·612
1/ 10·5	36·731	80·202	110·30	146·61	189·58	239·65	362·74	436·56	519·09	711·73	943·57	1075·1	1535·8	1749·4
0·10000	2·133	2·619	2·846	3·064	3·275	3·478	3·868	4·055	4·238	4·591	4·930	5·095	5·573	5·758
1/ 10·0	37·686	82·287	113·17	150·42	194·51	245·88	372·16	447·91	532·58	730·23	968·09	1103·0	1575·7	1794·8
	–	–	–	–	–	0·74	0·74	0·74	0·74	0·74	0·74	0·75	0·75	0·75

$V_{r(0{\cdot}5)medial}$ **for half-full circular pipes.**

A22

(p.1 of 6)

$k_s = 15{\cdot}0$ mm
S = 0·00010 to 0·00030

ie hydraulic gradient =
1 in 10000 to 1 in 3333

Water (or sewage) at 15°C;
full bore conditions.

velocities in ms^{-1}
discharges in litres/sec

Gradient **(Equivalent) Pipe diameters in mm**

Gradient	150	200	225	250	275	300	350	375	400	450	500	525	600	630
0·000100	0·053	0·067	0·073	0·079	0·085	0·090	0·101	0·106	0·111	0·121	0·130	0·135	0·148	0·154
1/ 10000	0·9407	2·0895	2·8918	3·8643	5·0199	6·3712	9·7089	11·718	13·970	19·242	25·610	29·230	41·951	47·862
0·000105	0·055	0·068	0·075	0·081	0·087	0·092	0·103	0·109	0·114	0·124	0·134	0·138	0·152	0·157
1/ 9524	0·9642	2·1415	2·9638	3·9604	5·1447	6·5296	9·9501	12·009	14·317	19·720	26·246	29·955	42·991	49·049
0·000110	0·056	0·070	0·076	0·083	0·089	0·095	0·106	0·111	0·117	0·127	0·137	0·142	0·156	0·161
1/ 9091	0·9871	2·1924	3·0341	4·0543	5·2666	6·6842	10·186	12·294	14·656	20·186	26·866	30·663	44·007	50·207
0·000115	0·057	0·071	0·078	0·084	0·091	0·097	0·108	0·114	0·119	0·130	0·140	0·145	0·159	0·165
1/ 8696	1·0096	2·2420	3·1028	4·1461	5·3858	6·8354	10·416	12·571	14·987	20·642	27·472	31·355	45·000	51·340
0·000120	0·058	0·073	0·080	0·086	0·093	0·099	0·111	0·116	0·122	0·133	0·143	0·148	0·163	0·168
1/ 8333	1·0315	2·2907	3·1700	4·2359	5·5024	6·9833	10·641	12·843	15·311	21·088	28·066	32·032	45·971	52·448
0·000125	0·060	0·074	0·081	0·088	0·095	0·101	0·113	0·119	0·124	0·135	0·146	0·151	0·166	0·172
1/ 8000	1·0530	2·3383	3·2359	4·3238	5·6166	7·1282	10·862	13·109	15·628	21·525	28·647	32·696	46·923	53·534
0·000130	0·061	0·076	0·083	0·090	0·096	0·103	0·115	0·121	0·127	0·138	0·149	0·154	0·169	0·175
1/ 7692	1·0740	2·3850	3·3005	4·4100	5·7285	7·2702	11·078	13·370	15·939	21·953	29·217	33·346	47·855	54·597
0·000135	0·062	0·077	0·085	0·092	0·098	0·105	0·117	0·123	0·129	0·141	0·152	0·157	0·172	0·178
1/ 7407	1·0947	2·4308	3·3638	4·4946	5·8384	7·4095	11·290	13·626	16·244	22·373	29·776	33·983	48·770	55·641
0·000140	0·063	0·079	0·086	0·093	0·100	0·107	0·120	0·126	0·132	0·143	0·154	0·160	0·176	0·182
1/ 7143	1·1150	2·4757	3·4259	4·5776	5·9461	7·5463	11·498	13·878	16·543	22·786	30·324	34·609	49·668	56·666
0·000145	0·064	0·080	0·088	0·095	0·102	0·109	0·122	0·128	0·134	0·146	0·157	0·163	0·179	0·185
1/ 6897	1·1349	2·5199	3·4870	4·6592	6·0520	7·6806	11·703	14·124	16·837	23·191	30·863	35·224	50·550	57·672
0·000150	0·065	0·082	0·089	0·097	0·104	0·111	0·124	0·130	0·136	0·148	0·160	0·166	0·182	0·188
1/ 6667	1·1545	2·5633	3·5470	4·7393	6·1561	7·8126	11·904	14·367	17·127	23·589	31·393	35·829	51·417	58·661
0·000160	0·067	0·084	0·092	0·100	0·107	0·114	0·128	0·134	0·141	0·153	0·165	0·171	0·188	0·194
1/ 6250	1·1927	2·6479	3·6641	4·8957	6·3592	8·0702	12·296	14·840	17·691	24·365	32·426	37·008	53·109	60·591
0·000170	0·070	0·087	0·095	0·103	0·110	0·118	0·132	0·139	0·145	0·158	0·170	0·176	0·194	0·200
1/ 5882	1·2297	2·7300	3·7776	5·0473	6·5560	8·3199	12·676	15·299	18·238	25·118	33·428	38·151	54·749	62·462
0·000180	0·072	0·089	0·098	0·106	0·114	0·121	0·136	0·143	0·149	0·163	0·175	0·181	0·199	0·206
1/ 5556	1·2657	2·8097	3·8878	5·1945	6·7471	8·5624	13·046	15·745	18·769	25·849	34·400	39·261	56·341	64·278
0·000190	0·074	0·092	0·100	0·109	0·117	0·124	0·139	0·146	0·153	0·167	0·180	0·186	0·205	0·212
1/ 5263	1·3007	2·8872	3·9950	5·3376	6·9329	8·7982	13·405	16·178	19·285	26·560	35·346	40·340	57·890	66·045
0·000200	0·076	0·094	0·103	0·112	0·120	0·128	0·143	0·150	0·157	0·171	0·185	0·191	0·210	0·217
1/ 5000	1·3347	2·9627	4·0994	5·4771	7·1140	9·0278	13·754	16·600	19·788	27·253	36·267	41·392	59·398	67·765
0·000210	0·077	0·097	0·106	0·114	0·123	0·131	0·147	0·154	0·161	0·176	0·189	0·196	0·215	0·223
1/ 4762	1·3679	3·0363	4·2013	5·6131	7·2905	9·2518	14·095	17·011	20·278	27·928	37·166	42·417	60·869	69·443
0·000220	0·079	0·099	0·108	0·117	0·126	0·134	0·150	0·158	0·165	0·180	0·194	0·201	0·220	0·228
1/ 4545	1·4004	3·1082	4·3007	5·7458	7·4629	9·4705	14·429	17·413	20·757	28·588	38·043	43·418	62·306	71·082
0·000230	0·081	0·101	0·111	0·120	0·128	0·137	0·153	0·161	0·169	0·184	0·198	0·205	0·225	0·233
1/ 4348	1·4321	3·1785	4·3979	5·8756	7·6315	9·6843	14·754	17·806	21·226	29·232	38·901	44·397	63·710	72·684
0·000240	0·083	0·103	0·113	0·122	0·131	0·140	0·157	0·165	0·173	0·188	0·202	0·210	0·230	0·238
1/ 4167	1·4631	3·2473	4·4930	6·0026	7·7964	9·8936	15·073	18·191	21·684	29·863	39·740	45·355	65·084	74·251
0·000250	0·085	0·106	0·115	0·125	0·134	0·143	0·160	0·168	0·176	0·192	0·207	0·214	0·235	0·243
1/ 4000	1·4935	3·3146	4·5861	6·1270	7·9579	10·098	15·385	18·567	22·132	30·481	40·562	46·293	66·430	75·786
0·000260	0·086	0·108	0·118	0·127	0·137	0·146	0·163	0·171	0·180	0·195	0·211	0·218	0·240	0·248
1/ 3846	1·5233	3·3807	4·6774	6·2490	8·1162	10·299	15·691	18·936	22·572	31·086	41·368	47·212	67·749	77·291
0·000270	0·088	0·110	0·120	0·130	0·139	0·148	0·166	0·175	0·183	0·199	0·215	0·222	0·244	0·253
1/ 3704	1·5525	3·4454	4·7670	6·3686	8·2715	10·496	15·991	19·298	23·004	31·680	42·158	48·114	69·042	78·767
0·000280	0·089	0·112	0·122	0·132	0·142	0·151	0·169	0·178	0·186	0·203	0·219	0·226	0·249	0·257
1/ 3571	1·5812	3·5090	4·8549	6·4860	8·4239	10·690	16·285	19·653	23·427	32·263	42·934	49·000	70·312	80·216
0·000290	0·091	0·114	0·124	0·134	0·144	0·154	0·172	0·181	0·190	0·206	0·223	0·230	0·253	0·262
1/ 3448	1·6094	3·5714	4·9413	6·6013	8·5737	10·880	16·574	20·003	23·843	32·836	43·696	49·869	71·560	81·639
0·000300	0·093	0·116	0·126	0·137	0·147	0·157	0·175	0·184	0·193	0·210	0·226	0·234	0·257	0·266
1/ 3333	1·6371	3·6328	5·0261	6·7147	8·7208	11·066	16·859	20·346	24·252	33·399	44·445	50·724	72·786	83·038
	–	–	–	–	–	–	–	–	–	–	–	–	–	–

$V_{r(0{\cdot}5)medial}$ for half-full circular pipes.

$k_s = 15{\cdot}0$ mm S = 0·00010 to 0·00030

k_s = 15·0 mm
S = 0·00030 to 0·00100

ie hydraulic gradient =
1 in 3333 to 1 in 1000

Water (or sewage) at 15°C;
full bore conditions.

velocities in ms^{-1}
discharges in litres/sec

Gradient	**(Equivalent) Pipe diameters in mm**													
	150	**200**	**225**	**250**	**275**	**300**	**350**	**375**	**400**	**450**	**500**	**525**	**600**	**630**
0·000300	0·093	0·116	0·126	0·137	0·147	0·157	0·175	0·184	0·193	0·210	0·226	0·234	0·257	0·266
1/ 3333	1·6371	3·6328	5·0261	6·7147	8·7208	11·066	16·859	20·346	24·252	33·399	44·445	50·724	72·786	83·038
0·000320	0·096	0·119	0·131	0·141	0·152	0·162	0·181	0·190	0·199	0·217	0·234	0·242	0·266	0·275
1/ 3125	1·6911	3·7526	5·1918	6·9358	9·0080	11·431	17·413	21·015	25·050	34·498	45·907	52·392	75·179	85·767
0·000340	0·099	0·123	0·135	0·146	0·156	0·167	0·187	0·196	0·205	0·224	0·241	0·249	0·274	0·284
1/ 2941	1·7435	3·8687	5·3523	7·1502	9·2864	11·784	17·951	21·664	25·823	35·562	47·323	54·008	77·498	88·413
0·000360	0·102	0·127	0·139	0·150	0·161	0·172	0·192	0·202	0·211	0·230	0·248	0·257	0·282	0·292
1/ 2778	1·7943	3·9814	5·5082	7·3584	9·5566	12·127	18·473	22·294	26·574	36·596	48·699	55·578	79·750	90·981
0·000380	0·104	0·130	0·142	0·154	0·165	0·176	0·197	0·207	0·217	0·236	0·255	0·264	0·290	0·300
1/ 2632	1·8438	4·0910	5·6598	7·5608	9·8195	12·460	18·981	22·907	27·305	37·602	50·036	57·104	81·940	93·480
0·000400	0·107	0·134	0·146	0·158	0·170	0·181	0·202	0·213	0·223	0·243	0·261	0·271	0·297	0·308
1/ 2500	1·8920	4·1977	5·8074	7·7580	10·076	12·785	19·476	23·503	28·016	38·581	51·339	58·591	84·073	95·913
0·000420	0·110	0·137	0·150	0·162	0·174	0·185	0·207	0·218	0·228	0·249	0·268	0·277	0·305	0·315
1/ 2381	1·9390	4·3019	5·9514	7·9503	10·325	13·102	19·958	24·086	28·710	39·536	52·610	60·041	86·154	98·286
0·000440	0·112	0·140	0·153	0·166	0·178	0·190	0·212	0·223	0·234	0·254	0·274	0·284	0·312	0·323
1/ 2273	1·9848	4·4036	6·0920	8·1381	10·569	13·411	20·429	24·654	29·387	40·469	53·851	61·458	88·185	100·60
0·000460	0·115	0·143	0·157	0·170	0·182	0·194	0·217	0·228	0·239	0·260	0·280	0·290	0·319	0·330
1/ 2174	2·0297	4·5030	6·2295	8·3217	10·807	13·713	20·890	25·209	30·049	41·381	55·064	62·842	90·171	102·87
0·000480	0·117	0·146	0·160	0·173	0·186	0·198	0·222	0·233	0·244	0·266	0·286	0·297	0·326	0·337
1/ 2083	2·0736	4·6002	6·3640	8·5013	11·041	14·009	21·340	25·753	30·697	42·273	56·251	64·196	92·114	105·09
0·000500	0·120	0·149	0·163	0·177	0·190	0·202	0·226	0·238	0·249	0·271	0·292	0·303	0·333	0·344
1/ 2000	2·1166	4·6955	6·4957	8·6772	11·269	14·299	21·781	26·286	31·332	43·146	57·413	65·523	94·017	107·26
0·000525	0·123	0·153	0·167	0·181	0·194	0·207	0·232	0·244	0·256	0·278	0·300	0·310	0·341	0·353
1/ 1905	2·1691	4·8119	6·6567	8·8923	11·548	14·653	22·321	26·936	32·107	44·214	58·834	67·144	96·343	109·91
0·000550	0·126	0·157	0·171	0·185	0·199	0·212	0·237	0·250	0·262	0·285	0·307	0·317	0·349	0·361
1/ 1818	2·2204	4·9256	6·8139	9·1022	11·821	14·999	22·847	27·572	32·865	45·257	60·221	68·727	98·614	112·50
0·000575	0·128	0·160	0·175	0·190	0·204	0·217	0·243	0·255	0·267	0·291	0·314	0·325	0·357	0·369
1/ 1739	2·2705	5·0367	6·9676	9·3075	12·087	15·337	23·362	28·193	33·605	46·276	61·577	70·275	100·83	115·03
0·000600	0·131	0·164	0·179	0·194	0·208	0·222	0·248	0·261	0·273	0·297	0·320	0·332	0·364	0·377
1/ 1667	2·3196	5·1454	7·1180	9·5083	12·348	15·668	23·866	28·801	34·329	47·274	62·904	71·789	103·01	117·51
0·000625	0·134	0·167	0·183	0·198	0·212	0·226	0·253	0·266	0·279	0·303	0·327	0·338	0·372	0·385
1/ 1600	2·3676	5·2519	7·2653	9·7050	12·603	15·992	24·359	29·396	35·039	48·250	64·204	73·272	105·13	119·94
0·000650	0·137	0·170	0·186	0·202	0·216	0·231	0·258	0·271	0·284	0·309	0·333	0·345	0·379	0·392
1/ 1538	2·4147	5·3563	7·4096	9·8978	12·854	16·309	24·843	29·980	35·734	49·208	65·478	74·726	107·22	122·32
0·000675	0·139	0·174	0·190	0·205	0·221	0·235	0·263	0·277	0·290	0·315	0·340	0·352	0·386	0·400
1/ 1481	2·4609	5·4587	7·5513	10·087	13·099	16·621	25·317	30·552	36·416	50·147	66·727	76·152	109·27	124·65
0·000700	0·142	0·177	0·193	0·209	0·225	0·239	0·268	0·282	0·295	0·321	0·346	0·358	0·394	0·407
1/ 1429	2·5063	5·5592	7·6903	10·273	13·340	16·927	25·783	31·114	37·086	51·069	67·954	77·552	111·27	126·94
0·000725	0·144	0·180	0·197	0·213	0·229	0·244	0·273	0·287	0·300	0·327	0·352	0·365	0·401	0·414
1/ 1379	2·5509	5·6580	7·8268	10·455	13·577	17·227	26·240	31·666	37·744	51·975	69·159	78·927	113·25	129·19
0·000750	0·147	0·183	0·200	0·217	0·233	0·248	0·277	0·292	0·306	0·332	0·358	0·371	0·407	0·422
1/ 1333	2·5946	5·7550	7·9610	10·634	13·810	17·522	26·690	32·208	38·390	52·865	70·343	80·278	115·19	131·40
0·000800	0·152	0·189	0·207	0·224	0·240	0·256	0·287	0·301	0·316	0·343	0·370	0·383	0·421	0·435
1/ 1250	2·6801	5·9444	8·2229	10·984	14·264	18·098	27·567	33·267	39·652	54·602	72·654	82·915	118·97	135·72
0·000850	0·156	0·195	0·213	0·231	0·248	0·264	0·295	0·310	0·325	0·354	0·381	0·395	0·434	0·449
1/ 1176	2·7629	6·1279	8·4767	11·323	14·704	18·657	28·417	34·293	40·875	56·285	74·894	85·471	122·64	139·90
0·000900	0·161	0·201	0·219	0·237	0·255	0·272	0·304	0·320	0·335	0·364	0·393	0·406	0·446	0·462
1/ 1111	2·8433	6·3061	8·7232	11·652	15·131	19·199	29·243	35·289	42·062	57·920	77·069	87·953	126·20	143·96
0·000950	0·165	0·206	0·225	0·244	0·262	0·279	0·312	0·328	0·344	0·374	0·403	0·417	0·459	0·475
1/ 1053	2·9215	6·4795	8·9629	11·972	15·547	19·726	30·046	36·258	43·217	59·510	79·184	90·367	129·66	147·91
0·001000	0·170	0·212	0·231	0·250	0·269	0·286	0·320	0·337	0·353	0·384	0·414	0·428	0·471	0·487
1/ 1000	2·9977	6·6483	9·1963	12·284	15·952	20·240	30·828	37·201	44·341	61·058	81·244	92·718	133·03	151·76
	–	–	–	–	–	–	–	–	–	–	–	–	–	–

$V_{r(0\cdot5)medial}$ **for half-full circular pipes.**

A22
(p.3 of 6)

k_s = 15·0 mm
S = 0·00100 to 0·00300

ie hydraulic gradient =
1 in 1000 to 1 in 333

Water (or sewage) at 15°C;
full bore conditions.

velocities in ms^{-1}
discharges in litres/sec

Gradient	150	200	225	250	275	300	350	375	400	450	500	525	600	630
	(Equivalent) Pipe diameters in mm													
0·00100	0·170	0·212	0·231	0·250	0·269	0·286	0·320	0·337	0·353	0·384	0·414	0·428	0·471	0·487
1/ 1000	2·9977	6·6483	9·1963	12·284	15·952	20·240	30·828	37·201	44·341	61·058	81·244	92·718	133·03	151·76
0·00105	0·174	0·217	0·237	0·256	0·275	0·293	0·328	0·345	0·362	0·393	0·424	0·439	0·482	0·499
1/ 952	3·0720	6·8129	9·4240	12·588	16·347	20·741	31·591	38·122	45·438	62·568	83·253	95·011	136·32	155·51
0·00110	0·178	0·222	0·243	0·262	0·282	0·300	0·336	0·353	0·370	0·403	0·434	0·449	0·493	0·511
1/ 909	3·1445	6·9737	9·6464	12·885	16·732	21·230	32·335	39·020	46·509	64·043	85·215	97·250	139·53	159·18
0·00115	0·182	0·227	0·248	0·268	0·288	0·307	0·344	0·361	0·378	0·412	0·444	0·459	0·505	0·522
1/ 870	3·2154	7·1309	9·8637	13·175	17·109	21·708	33·063	39·899	47·556	65·485	87·133	99·438	142·67	162·76
0·00120	0·186	0·232	0·253	0·274	0·294	0·314	0·351	0·369	0·387	0·421	0·453	0·469	0·515	0·533
1/ 833	3·2848	7·2847	10·076	13·459	17·478	22·176	33·776	40·758	48·581	66·895	89·010	101·58	145·75	166·27
0·00125	0·190	0·237	0·259	0·280	0·300	0·320	0·358	0·377	0·395	0·429	0·463	0·479	0·526	0·544
1/ 800	3·3528	7·4353	10·285	13·737	17·839	22·634	34·474	41·600	49·584	68·277	90·848	103·68	148·75	169·70
0·00130	0·193	0·241	0·264	0·285	0·306	0·327	0·365	0·384	0·402	0·438	0·472	0·488	0·537	0·555
1/ 769	3·4194	7·5829	10·489	14·010	18·193	23·083	35·157	42·425	50·568	69·631	92·649	105·73	151·70	173·06
0·00135	0·197	0·246	0·269	0·291	0·312	0·333	0·372	0·391	0·410	0·446	0·481	0·498	0·547	0·566
1/ 741	3·4847	7·7277	10·689	14·277	18·540	23·523	35·828	43·235	51·532	70·959	94·416	107·75	154·60	176·36
0·00140	0·201	0·251	0·274	0·296	0·318	0·339	0·379	0·399	0·418	0·454	0·490	0·507	0·557	0·576
1/ 714	3·5489	7·8699	10·886	14·540	18·881	23·956	36·487	44·029	52·479	72·263	96·151	109·73	157·44	179·60
0·00145	0·204	0·255	0·279	0·301	0·324	0·345	0·386	0·406	0·425	0·462	0·498	0·516	0·567	0·586
1/ 690	3·6119	8·0095	11·079	14·798	19·216	24·381	37·134	44·810	53·410	73·544	97·855	111·67	160·23	182·78
0·00150	0·208	0·259	0·283	0·307	0·329	0·351	0·393	0·413	0·432	0·470	0·507	0·525	0·576	0·596
1/ 667	3·6738	8·1468	11·269	15·051	19·545	24·798	37·770	45·577	54·324	74·803	99·530	113·59	162·97	185·91
0·00160	0·215	0·268	0·293	0·317	0·340	0·362	0·405	0·426	0·446	0·486	0·524	0·542	0·595	0·616
1/ 625	3·7947	8·4146	11·639	15·546	20·187	25·613	39·010	47·074	56·108	77·259	102·80	117·31	168·32	192·02
0·00170	0·221	0·276	0·302	0·326	0·350	0·374	0·418	0·439	0·460	0·501	0·540	0·559	0·614	0·635
1/ 588	3·9118	8·6741	11·998	16·025	20·810	26·403	40·213	48·525	57·837	79·640	105·97	120·93	173·50	197·93
0·00180	0·228	0·284	0·311	0·336	0·361	0·384	0·430	0·452	0·474	0·515	0·555	0·575	0·631	0·653
1/ 556	4·0255	8·9262	12·346	16·491	21·414	27·169	41·380	49·934	59·516	81·951	109·04	124·44	178·54	203·68
0·00190	0·234	0·292	0·319	0·345	0·370	0·395	0·442	0·465	0·487	0·529	0·571	0·591	0·649	0·671
1/ 526	4·1361	9·1713	12·685	16·943	22·002	27·915	42·516	51·304	61·149	84·200	112·03	127·85	183·44	209·26
0·00200	0·240	0·300	0·327	0·354	0·380	0·405	0·453	0·477	0·499	0·543	0·585	0·606	0·666	0·689
1/ 500	4·2438	9·4100	13·016	17·384	22·574	28·641	43·622	52·638	62·740	86·390	114·95	131·18	188·21	214·70
0·00210	0·246	0·307	0·335	0·363	0·389	0·415	0·465	0·488	0·512	0·557	0·600	0·621	0·682	0·706
1/ 476	4·3489	9·6429	13·338	17·814	23·133	29·350	44·700	53·940	64·291	88·525	117·79	134·42	192·86	220·01
0·00220	0·252	0·314	0·343	0·371	0·399	0·425	0·476	0·500	0·524	0·570	0·614	0·636	0·698	0·722
1/ 455	4·4515	9·8703	13·652	18·234	23·678	30·041	45·754	55·211	65·806	90·611	120·56	137·59	197·40	225·19
0·00230	0·258	0·321	0·351	0·380	0·408	0·435	0·486	0·511	0·535	0·583	0·628	0·650	0·714	0·739
1/ 435	4·5518	10·093	13·959	18·645	24·211	30·717	46·783	56·453	67·287	92·650	123·27	140·68	201·84	230·26
0·00240	0·263	0·328	0·359	0·388	0·416	0·444	0·497	0·522	0·547	0·595	0·641	0·664	0·729	0·755
1/ 417	4·6499	10·310	14·260	19·046	24·732	31·379	47·791	57·669	68·736	94·644	125·93	143·71	206·19	235·21
0·00250	0·269	0·335	0·366	0·396	0·425	0·453	0·507	0·533	0·558	0·607	0·655	0·678	0·744	0·770
1/ 400	4·7460	10·523	14·555	19·440	25·243	32·027	48·777	58·860	70·155	96·598	128·53	146·68	210·44	240·07
0·00260	0·274	0·342	0·373	0·404	0·433	0·462	0·517	0·543	0·569	0·619	0·668	0·691	0·759	0·785
1/ 385	4·8402	10·732	14·844	19·825	25·744	32·662	49·745	60·027	71·545	98·513	131·08	149·58	214·61	244·83
0·00270	0·279	0·348	0·380	0·412	0·442	0·471	0·527	0·554	0·580	0·631	0·680	0·704	0·774	0·800
1/ 370	4·9326	10·937	15·127	20·204	26·235	33·285	50·693	61·171	72·910	100·39	133·57	152·44	218·70	249·49
0·00280	0·284	0·355	0·387	0·419	0·450	0·480	0·537	0·564	0·591	0·643	0·693	0·717	0·788	0·815
1/ 357	5·0234	11·138	15·405	20·575	26·717	33·897	51·625	62·295	74·249	102·24	136·03	155·24	222·72	254·08
0·00290	0·289	0·361	0·394	0·427	0·458	0·488	0·546	0·574	0·601	0·654	0·705	0·730	0·802	0·830
1/ 345	5·1125	11·335	15·678	20·940	27·190	34·497	52·539	63·399	75·565	104·05	138·44	157·99	226·67	258·58
0·00300	0·294	0·367	0·401	0·434	0·466	0·496	0·555	0·584	0·612	0·665	0·717	0·742	0·815	0·844
1/ 333	5·2000	11·529	15·946	21·298	27·656	35·088	53·439	64·484	76·858	105·83	140·81	160·69	230·54	263·00
	–	–	–	–	–	–	–	–	–	–	–	–	–	–

$V_{r(0·5)medial}$ for half-full circular pipes.

k_s = 15·0 mm S = 0·00100 to 0·00300

k_s = 15·0 mm
S = 0·00300 to 0·01000

ie hydraulic gradient =
1 in 333 to 1 in 100

Water (or sewage) at 15°C;
full bore conditions.

velocities in ms^{-1}
discharges in litres/sec

Gradient	**(Equivalent) Pipe diameters in mm**													
	150	**200**	**225**	**250**	**275**	**300**	**350**	**375**	**400**	**450**	**500**	**525**	**600**	**630**
0·00300	0·294	0·367	0·401	0·434	0·466	0·496	0·555	0·584	0·612	0·665	0·717	0·742	0·815	0·844
1/ 333	5·2000	11·529	15·946	21·298	27·656	35·088	53·439	64·484	76·858	105·83	140·81	160·69	230·54	263·00
0·00320	0·304	0·379	0·414	0·448	0·481	0·513	0·574	0·603	0·632	0·687	0·741	0·767	0·842	0·871
1/ 313	5·3709	11·908	16·470	21·998	28·564	36·240	55·193	66·601	79·381	109·30	145·43	165·96	238·11	271·63
0·00340	0·313	0·391	0·427	0·462	0·496	0·528	0·591	0·622	0·651	0·708	0·763	0·790	0·868	0·898
1/ 294	5·5366	12·275	16·978	22·676	29·444	37·357	56·893	68·653	81·826	112·67	149·91	171·07	245·44	280·00
0·00360	0·322	0·402	0·439	0·475	0·510	0·544	0·609	0·640	0·670	0·729	0·786	0·813	0·893	0·924
1/ 278	5·6974	12·631	17·471	23·334	30·299	38·441	58·545	70·645	84·201	115·94	154·26	176·04	252·56	288·12
0·00380	0·331	0·413	0·451	0·488	0·524	0·559	0·625	0·657	0·688	0·749	0·807	0·836	0·918	0·950
1/ 263	5·8538	12·978	17·950	23·974	31·130	39·496	60·150	72·583	86·510	119·12	158·49	180·87	259·49	296·02
0·00400	0·340	0·424	0·463	0·501	0·538	0·573	0·641	0·674	0·706	0·768	0·828	0·857	0·942	0·974
1/ 250	6·0062	13·316	18·417	24·598	31·940	40·523	61·715	74·470	88·760	122·21	162·61	185·57	266·24	303·72
0·00420	0·348	0·434	0·475	0·513	0·551	0·587	0·657	0·691	0·724	0·787	0·849	0·878	0·965	0·998
1/ 238	6·1547	13·645	18·872	25·206	32·730	41·525	63·240	76·311	90·953	125·23	166·63	190·15	272·81	311·22
0·00440	0·356	0·445	0·486	0·526	0·564	0·601	0·673	0·707	0·741	0·806	0·869	0·899	0·988	1·022
1/ 227	6·2998	13·967	19·317	25·800	33·501	42·503	64·730	78·108	93·096	128·18	170·55	194·63	279·24	318·55
0·00460	0·365	0·455	0·497	0·537	0·577	0·615	0·688	0·723	0·757	0·824	0·888	0·919	1·010	1·045
1/ 217	6·4417	14·281	19·752	26·380	34·254	43·459	66·186	79·865	95·190	131·07	174·39	199·01	285·52	325·71
0·00480	0·372	0·464	0·507	0·549	0·589	0·628	0·703	0·739	0·774	0·842	0·907	0·939	1·032	1·067
1/ 208	6·5804	14·589	20·177	26·948	34·992	44·395	67·611	81·584	97·239	133·89	178·14	203·29	291·66	332·72
0·00500	0·380	0·474	0·518	0·560	0·601	0·641	0·717	0·754	0·790	0·859	0·926	0·958	1·053	1·089
1/ 200	6·7164	14·890	20·594	27·505	35·714	45·311	69·006	83·268	99·245	136·65	181·82	207·49	297·68	339·59
0·00525	0·389	0·486	0·531	0·574	0·616	0·657	0·735	0·773	0·809	0·880	0·949	0·982	1·079	1·116
1/ 190	6·8825	15·258	21·103	28·185	36·597	46·431	70·712	85·326	101·70	140·03	186·31	212·61	305·04	347·98
0·00550	0·399	0·497	0·543	0·588	0·631	0·672	0·752	0·791	0·828	0·901	0·971	1·005	1·104	1·143
1/ 182	7·0447	15·617	21·600	28·848	37·459	47·525	72·377	87·336	104·09	143·33	190·70	217·62	312·22	356·17
0·00575	0·408	0·508	0·555	0·601	0·645	0·687	0·769	0·809	0·847	0·921	0·993	1·028	1·129	1·168
1/ 174	7·2033	15·969	22·086	29·498	38·302	48·594	74·005	89·300	106·43	146·55	194·98	222·51	319·24	364·18
0·00600	0·416	0·519	0·567	0·614	0·659	0·702	0·786	0·826	0·865	0·941	1·014	1·050	1·153	1·193
1/ 167	7·3584	16·313	22·562	30·133	39·126	49·640	75·598	91·222	108·73	149·70	199·18	227·30	326·11	372·02
0·00625	0·425	0·530	0·579	0·627	0·672	0·717	0·802	0·843	0·883	0·961	1·035	1·072	1·177	1·218
1/ 160	7·5104	16·649	23·027	30·755	39·934	50·664	77·158	93·105	110·97	152·79	203·29	231·99	332·84	379·69
0·00650	0·433	0·540	0·591	0·639	0·686	0·731	0·818	0·860	0·901	0·980	1·056	1·093	1·200	1·242
1/ 154	7·6594	16·980	23·484	31·364	40·725	51·668	78·687	94·950	113·17	155·82	207·32	236·59	339·43	387·22
0·00675	0·442	0·551	0·602	0·651	0·699	0·745	0·833	0·876	0·918	0·998	1·076	1·114	1·223	1·266
1/ 148	7·8055	17·303	23·932	31·962	41·502	52·654	80·187	96·760	115·33	158·79	211·27	241·10	345·90	394·60
0·00700	0·450	0·561	0·613	0·663	0·712	0·759	0·849	0·892	0·935	1·017	1·096	1·134	1·246	1·289
1/ 143	7·9489	17·621	24·371	32·549	42·264	53·621	81·660	98·537	117·44	161·71	215·15	245·53	352·25	401·84
0·00725	0·458	0·571	0·624	0·675	0·724	0·772	0·864	0·908	0·951	1·035	1·115	1·154	1·268	1·312
1/ 138	8·0898	17·934	24·803	33·126	43·013	54·570	83·106	100·28	119·52	164·57	218·96	249·87	358·49	408·96
0·00750	0·466	0·581	0·634	0·686	0·737	0·785	0·879	0·923	0·967	1·052	1·134	1·174	1·290	1·334
1/ 133	8·2283	18·240	25·227	33·693	43·749	55·504	84·528	102·00	121·57	167·38	222·70	254·15	364·62	415·95
0·00800	0·481	0·600	0·655	0·709	0·761	0·811	0·907	0·954	0·999	1·087	1·171	1·213	1·332	1·378
1/ 125	8·4985	18·839	26·056	34·799	45·185	57·326	87·302	105·34	125·56	172·88	230·01	262·49	376·59	429·60
0·00850	0·496	0·618	0·675	0·731	0·784	0·836	0·935	0·983	1·030	1·120	1·208	1·250	1·373	1·421
1/ 118	8·7604	19·420	26·858	35·871	46·577	59·091	89·991	108·59	129·42	178·20	237·10	270·57	388·18	442·83
0·00900	0·510	0·636	0·695	0·752	0·807	0·860	0·962	1·012	1·060	1·153	1·243	1·286	1·413	1·462
1/ 111	9·0146	19·983	27·638	36·911	47·928	60·806	92·601	111·74	133·18	183·37	243·97	278·42	399·44	455·67
0·00950	0·524	0·654	0·714	0·773	0·829	0·884	0·989	1·039	1·089	1·185	1·277	1·321	1·451	1·502
1/ 105	9·2620	20·531	28·396	37·924	49·242	62·473	95·141	114·80	136·83	188·40	250·66	286·05	410·39	468·16
0·01000	0·538	0·671	0·733	0·793	0·851	0·907	1·015	1·066	1·117	1·215	1·310	1·356	1·489	1·541
1/ 100	9·5029	21·065	29·134	38·910	50·522	64·097	97·614	117·79	140·39	193·29	257·18	293·49	421·06	480·33
	–	–	–	–	–	–	–	–	–	–	–	–	–	–

$V_{r(0·5)medial}$ for half-full circular pipes.

A22
(p.5 of 6)

k_s = 15·0 mm
S = 0·01000 to 0·03000

ie hydraulic gradient =
1 in 100 to 1 in 33·3

Water (or sewage) at 15°C;
full bore conditions.

velocities in ms^{-1}
discharges in litres/sec

Gradient **(Equivalent) Pipe diameters in mm**

	150	200	225	250	275	300	350	375	400	450	500	525	600	630
0·01000	0·538	0·671	0·733	0·793	0·851	0·907	1·015	1·066	1·117	1·215	1·310	1·356	1·489	1·541
1/ 100	9·5029	21·065	29·134	38·910	50·522	64·097	97·614	117·79	140·39	193·29	257·18	293·49	421·06	480·33
0·01050	0·551	0·687	0·751	0·812	0·872	0·929	1·040	1·093	1·145	1·245	1·342	1·389	1·526	1·579
1/ 95	9·7378	21·586	29·854	39·871	51·771	65·681	100·03	120·70	143·85	198·07	263·53	300·74	431·46	492·19
0·01100	0·564	0·703	0·769	0·831	0·892	0·951	1·064	1·119	1·172	1·275	1·374	1·422	1·562	1·616
1/ 91	9·9672	22·094	30·557	40·810	52·990	67·228	102·38	123·54	147·24	202·73	269·73	307·82	441·62	503·78
0·01150	0·577	0·719	0·786	0·850	0·912	0·972	1·088	1·144	1·198	1·303	1·405	1·454	1·597	1·652
1/ 87	10·191	22·591	31·244	41·728	54·182	68·740	104·68	126·32	150·55	207·29	275·80	314·74	451·55	515·11
0·01200	0·589	0·735	0·803	0·868	0·932	0·993	1·111	1·168	1·224	1·331	1·435	1·485	1·631	1·688
1/ 83	10·411	23·078	31·917	42·626	55·348	70·219	106·94	129·04	153·79	211·75	281·73	321·51	461·26	526·19
0·01250	0·601	0·750	0·819	0·886	0·951	1·014	1·134	1·192	1·249	1·359	1·464	1·516	1·665	1·723
1/ 80	10·626	23·554	32·576	43·506	56·490	71·668	109·14	131·70	156·97	216·12	287·55	328·14	470·78	537·05
0·01300	0·613	0·765	0·836	0·904	0·970	1·034	1·157	1·216	1·274	1·386	1·493	1·546	1·698	1·757
1/ 77	10·836	24·021	33·221	44·368	57·610	73·088	111·30	134·31	160·08	220·40	293·24	334·64	480·10	547·68
0·01350	0·625	0·779	0·851	0·921	0·988	1·054	1·179	1·239	1·298	1·412	1·522	1·575	1·730	1·790
1/ 74	11·043	24·479	33·854	45·214	58·708	74·481	113·43	136·87	163·13	224·60	298·83	341·02	489·25	558·12
0·01400	0·636	0·793	0·867	0·938	1·007	1·073	1·201	1·262	1·322	1·438	1·550	1·604	1·762	1·823
1/ 71	11·246	24·928	34·476	46·044	59·786	75·849	115·51	139·38	166·12	228·73	304·32	347·28	498·23	568·37
0·01450	0·648	0·808	0·882	0·955	1·024	1·092	1·222	1·284	1·345	1·464	1·577	1·633	1·793	1·856
1/ 69	11·445	25·370	35·087	46·860	60·845	77·192	117·55	141·85	169·06	232·78	309·71	353·43	507·06	578·43
0·01500	0·659	0·821	0·898	0·971	1·042	1·111	1·243	1·306	1·368	1·489	1·604	1·661	1·824	1·887
1/ 67	11·641	25·804	35·687	47·661	61·885	78·513	119·57	144·27	171·95	236·76	315·00	359·48	515·73	588·32
0·01600	0·680	0·848	0·927	1·003	1·076	1·147	1·284	1·349	1·413	1·537	1·657	1·715	1·884	1·949
1/ 63	12·023	26·651	36·858	49·225	63·916	81·089	123·49	149·01	177·60	244·53	325·34	371·27	532·65	607·62
0·01700	0·701	0·874	0·956	1·034	1·109	1·182	1·323	1·391	1·457	1·585	1·708	1·768	1·942	2·009
1/ 59	12·394	27·472	37·993	50·741	65·884	83·586	127·29	153·60	183·06	252·05	335·35	382·70	549·04	626·33
0·01800	0·722	0·900	0·983	1·064	1·141	1·217	1·361	1·431	1·499	1·631	1·757	1·819	1·998	2·068
1/ 56	12·753	28·269	39·096	52·213	67·795	86·010	130·98	158·05	188·37	259·36	345·08	393·80	564·97	644·49
0·01900	0·741	0·924	1·010	1·093	1·173	1·250	1·399	1·470	1·540	1·675	1·806	1·869	2·053	2·124
1/ 53	13·103	29·044	40·168	53·645	69·654	88·368	134·57	162·38	193·54	266·47	354·54	404·59	580·45	662·16
0·02000	0·761	0·949	1·036	1·121	1·203	1·283	1·435	1·508	1·580	1·719	1·853	1·918	2·106	2·179
1/ 50	13·444	29·799	41·212	55·039	71·465	90·665	138·07	166·60	198·57	273·40	363·75	415·11	595·54	679·36
0·02100	0·780	0·972	1·062	1·149	1·233	1·314	1·471	1·546	1·619	1·761	1·898	1·965	2·158	2·233
1/ 47·6	13·776	30·535	42·230	56·399	73·230	92·905	141·48	170·72	203·47	280·15	372·74	425·36	610·25	696·15
0·02200	0·798	0·995	1·087	1·176	1·262	1·345	1·505	1·582	1·657	1·803	1·943	2·011	2·209	2·286
1/ 45·5	14·100	31·254	43·224	57·727	74·955	95·093	144·81	174·74	208·26	286·75	381·51	435·37	624·61	712·53
0·02300	0·816	1·017	1·112	1·202	1·290	1·376	1·539	1·618	1·695	1·843	1·987	2·056	2·259	2·337
1/ 43·5	14·417	31·957	44·196	59·025	76·640	97·231	148·07	178·67	212·95	293·19	390·09	445·16	638·65	728·55
0·02400	0·833	1·039	1·135	1·228	1·318	1·405	1·572	1·652	1·731	1·883	2·029	2·101	2·307	2·387
1/ 41·7	14·728	32·645	45·147	60·295	78·289	99·323	151·25	182·51	217·53	299·50	398·48	454·74	652·39	744·23
0·02500	0·851	1·061	1·159	1·254	1·345	1·434	1·605	1·687	1·767	1·922	2·071	2·144	2·355	2·437
1/ 40·0	15·032	33·318	46·079	61·539	79·904	101·37	154·38	186·28	222·01	305·68	406·70	464·12	665·85	759·58
0·02600	0·867	1·082	1·182	1·279	1·372	1·463	1·636	1·720	1·802	1·960	2·112	2·186	2·402	2·485
1/ 38·5	15·330	33·978	46·992	62·759	81·487	103·38	157·43	189·97	226·41	311·74	414·76	473·31	679·04	774·62
0·02700	0·884	1·102	1·204	1·303	1·398	1·490	1·668	1·753	1·836	1·997	2·153	2·228	2·447	2·532
1/ 37·0	15·622	34·626	47·888	63·955	83·040	105·35	160·43	193·59	230·73	317·68	422·66	482·33	691·98	789·38
0·02800	0·900	1·122	1·227	1·327	1·424	1·518	1·698	1·785	1·870	2·034	2·192	2·269	2·492	2·579
1/ 35·7	15·909	35·262	48·767	65·129	84·565	107·28	163·38	197·14	234·96	323·51	430·42	491·19	704·68	803·87
0·02900	0·916	1·142	1·248	1·350	1·449	1·545	1·728	1·817	1·903	2·070	2·231	2·309	2·536	2·624
1/ 34·5	16·190	35·886	49·630	66·282	86·062	109·18	166·27	200·63	239·12	329·24	438·04	499·88	717·16	818·10
0·03000	0·932	1·162	1·270	1·373	1·474	1·571	1·758	1·848	1·935	2·106	2·269	2·349	2·580	2·669
1/ 33·3	16·467	36·500	50·479	67·416	87·534	111·05	169·12	204·06	243·21	334·87	445·53	508·43	729·42	832·09
	–	–	–	–	–	–	–	–	–	–	–	–	–	–

$V_{r(0·5)medial}$ for half-full circular pipes.

k_s = 15·0 mm S = 0·01000 to 0·03000

$k_s = 15{\cdot}0$ **mm**
S = 0·03000 to 0·10000

ie hydraulic gradient =
1 in 33·3 to 1 in 10·0

Water (or sewage) at 15°C;
full bore conditions.

velocities in ms^{-1}
discharges in litres/sec

Gradient	**(Equivalent) Pipe diameters in mm**													
	150	**200**	**225**	**250**	**275**	**300**	**350**	**375**	**400**	**450**	**500**	**525**	**600**	**630**
0·03000	0·932	1·162	1·270	1·373	1·474	1·571	1·758	1·848	1·935	2·106	2·269	2·349	2·580	2·669
1/ 33·3	16·467	36·500	50·479	67·416	87·534	111·05	169·12	204·06	243·21	334·87	445·53	508·43	729·42	832·09
0·03200	0·962	1·200	1·311	1·418	1·522	1·623	1·815	1·908	1·999	2·175	2·344	2·426	2·664	2·757
1/ 31·3	17·008	37·698	52·136	69·628	90·406	114·69	174·66	210·76	251·19	345·85	460·15	525·11	753·35	859·39
0·03400	0·992	1·237	1·352	1·462	1·569	1·673	1·871	1·967	2·060	2·242	2·416	2·500	2·746	2·842
1/ 29·4	17·532	38·859	53·741	71·771	93·190	118·23	180·04	217·25	258·92	356·50	474·31	541·27	776·54	885·84
0·03600	1·021	1·273	1·391	1·505	1·614	1·721	1·926	2·024	2·120	2·307	2·486	2·573	2·826	2·924
1/ 27·8	18·040	39·986	55·300	73·853	95·892	121·65	185·26	223·55	266·43	366·84	488·07	556·97	799·06	911·53
0·03800	1·049	1·308	1·429	1·546	1·659	1·768	1·978	2·079	2·178	2·370	2·554	2·643	2·904	3·004
1/ 26·3	18·535	41·082	56·816	75·878	98·521	124·99	190·34	229·67	273·74	376·89	501·44	572·24	820·96	936·51
0·04000	1·076	1·342	1·466	1·586	1·702	1·814	2·030	2·134	2·235	2·431	2·620	2·712	2·979	3·082
1/ 25·0	19·017	42·150	58·292	77·850	101·08	128·24	195·29	235·64	280·85	386·69	514·47	587·10	842·29	960·85
0·04200	1·103	1·375	1·502	1·625	1·744	1·859	2·080	2·186	2·290	2·491	2·685	2·779	3·053	3·158
1/ 23·8	19·486	43·191	59·732	79·773	103·58	131·41	200·11	241·46	287·79	396·24	527·18	601·61	863·09	984·58
0·04400	1·129	1·407	1·538	1·663	1·785	1·903	2·129	2·238	2·344	2·550	2·748	2·845	3·124	3·233
1/ 22·7	19·945	44·208	61·139	81·651	106·02	134·50	204·82	247·15	294·56	405·56	539·59	615·77	883·41	1007·8
0·04600	1·154	1·439	1·572	1·701	1·825	1·946	2·177	2·288	2·397	2·607	2·810	2·908	3·195	3·306
1/ 21·7	20·394	45·202	62·513	83·487	108·40	137·52	209·43	252·70	301·18	414·68	551·72	629·61	903·27	1030·4
0·04800	1·179	1·470	1·606	1·737	1·864	1·987	2·224	2·337	2·448	2·663	2·870	2·971	3·263	3·377
1/ 20·8	20·833	46·175	63·858	85·283	110·73	140·48	213·93	258·14	307·66	423·60	563·59	643·15	922·70	1052·6
0·05000	1·203	1·500	1·639	1·773	1·903	2·028	2·269	2·385	2·499	2·718	2·930	3·032	3·331	3·446
1/ 20·0	21·262	47·127	65·176	87·042	113·02	143·38	218·34	263·46	314·01	432·34	575·21	656·42	941·73	1074·3
0·05250	1·233	1·537	1·680	1·817	1·950	2·079	2·325	2·444	2·561	2·786	3·002	3·107	3·413	3·531
1/ 19·0	21·788	48·291	66·786	89·192	115·81	146·92	223·74	269·97	321·76	443·02	589·42	672·63	964·99	1100·8
0·05500	1·262	1·573	1·719	1·860	1·996	2·127	2·380	2·502	2·621	2·851	3·073	3·180	3·493	3·614
1/ 18·2	22·301	49·428	68·358	91·292	118·53	150·38	229·00	276·33	329·34	453·45	603·29	688·46	987·70	1126·7
0·05750	1·290	1·609	1·758	1·902	2·041	2·175	2·434	2·558	2·680	2·915	3·142	3·252	3·572	3·696
1/ 17·4	22·802	50·540	69·895	93·344	121·20	153·76	234·15	282·54	336·74	463·64	616·86	703·94	1009·9	1152·1
0·06000	1·318	1·643	1·796	1·943	2·084	2·222	2·486	2·613	2·737	2·978	3·209	3·322	3·649	3·775
1/ 16·7	23·293	51·627	71·399	95·353	123·81	157·07	239·19	288·62	343·99	473·61	630·13	719·08	1031·6	1176·8
0·06250	1·345	1·677	1·833	1·983	2·127	2·268	2·537	2·667	2·794	3·039	3·275	3·390	3·724	3·853
1/ 16·0	23·773	52·692	72·871	97·320	126·36	160·31	244·12	294·57	351·08	483·38	643·12	733·91	1052·9	1201·1
0·06500	1·372	1·710	1·869	2·022	2·170	2·313	2·588	2·720	2·849	3·100	3·340	3·457	3·798	3·929
1/ 15·4	24·244	53·736	74·315	99·247	128·86	163·48	248·96	300·40	358·04	492·96	655·86	748·45	1073·8	1224·9
0·06750	1·398	1·743	1·905	2·060	2·211	2·357	2·637	2·772	2·903	3·159	3·404	3·523	3·870	4·004
1/ 14·8	24·706	54·760	75·731	101·14	131·32	166·60	253·70	306·13	364·86	502·35	668·36	762·71	1094·2	1248·2
0·07000	1·424	1·775	1·940	2·098	2·252	2·400	2·685	2·823	2·957	3·217	3·466	3·588	3·941	4·078
1/ 14·3	25·160	55·765	77·121	103·00	133·73	169·66	258·36	311·75	371·55	511·57	680·62	776·71	1114·3	1271·1
0·07250	1·449	1·806	1·974	2·135	2·291	2·443	2·733	2·873	3·009	3·273	3·528	3·651	4·011	4·150
1/ 13·8	25·606	56·752	78·487	104·82	136·10	172·66	262·93	317·27	378·13	520·62	692·67	790·46	1134·0	1293·6
0·07500	1·474	1·837	2·008	2·172	2·331	2·484	2·780	2·922	3·061	3·329	3·588	3·714	4·079	4·221
1/ 13·3	26·043	57·723	79·829	106·61	138·42	175·61	267·43	322·69	384·60	529·53	704·52	803·97	1153·4	1315·8
0·08000	1·522	1·898	2·074	2·243	2·407	2·566	2·871	3·018	3·161	3·439	3·706	3·836	4·213	4·359
1/ 12·5	26·898	59·617	82·448	110·11	142·97	181·37	276·20	333·28	397·21	546·90	727·62	830·35	1191·2	1358·9
0·08500	1·569	1·956	2·137	2·312	2·481	2·645	2·959	3·110	3·258	3·545	3·820	3·954	4·343	4·494
1/ 11·8	27·726	61·452	84·986	113·50	147·37	186·96	284·70	343·53	409·44	563·73	750·02	855·91	1227·9	1400·8
0·09000	1·614	2·013	2·199	2·379	2·553	2·722	3·045	3·201	3·353	3·647	3·931	4·068	4·469	4·624
1/ 11·1	28·530	63·234	87·451	116·79	151·64	192·38	292·96	353·50	421·31	580·08	771·77	880·72	1263·5	1441·4
0·09500	1·659	2·068	2·260	2·444	2·623	2·796	3·128	3·288	3·445	3·747	4·038	4·180	4·591	4·751
1/ 10·5	29·312	64·968	89·848	119·99	155·80	197·65	300·99	363·18	432·86	595·97	792·92	904·86	1298·1	1480·9
0·10000	1·702	2·122	2·318	2·508	2·691	2·869	3·210	3·374	3·534	3·845	4·143	4·289	4·711	4·874
1/ 10·0	30·074	66·656	92·182	123·11	159·85	202·79	308·81	372·62	444·11	611·46	813·52	928·37	1331·9	1519·3
	–	–	–	–	–	–	–	–	–	–	–	–	–	–

$V_{r(0{\cdot}5)medial}$ **for half-full circular pipes.**

Tables A1 to A22 before this point show

mean velocities V in ms^{-1} in black

and discharges Q in litres/sec in green

(i.e. Q in $m^3s^{-1}/1000$)

Tables A23 to A58 before this point show

mean velocities V in ms^{-1} in black

and discharges Q in m^3s^{-1} in blue

A23

(p.1 of 6)

k_s = 0·006 mm
S = 0·00010 to 0·00030

ie hydraulic gradient =
1 in 10000 to 1 in 3333

Water (or sewage) at 15°C;
full bore conditions.

velocities in ms^{-1}
discharges in m^3s^{-1}

Gradient **(Equivalent) Pipe diameters in mm**

Gradient	630	675	700	750	800	825	900	975	1000	1050	1100	1125	1200	1250
0·000100	0·273	0·286	0·293	0·306	0·320	0·326	0·345	0·364	0·370	0·382	0·394	0·400	0·417	0·428
1/ 10000	0·0850	0·1022	0·1126	0·1353	0·1607	0·1744	0·2197	0·2718	0·2907	0·3309	0·3743	0·3973	0·4715	0·5253
0·000105	0·280	0·293	0·301	0·315	0·328	0·335	0·355	0·374	0·380	0·393	0·405	0·411	0·428	0·440
1/ 9524	0·0874	0·1050	0·1157	0·1390	0·1650	0·1791	0·2257	0·2792	0·2986	0·3399	0·3845	0·4081	0·4843	0·5396
0·000110	0·288	0·301	0·308	0·323	0·337	0·344	0·364	0·384	0·390	0·403	0·415	0·421	0·439	0·451
1/ 9091	0·0896	0·1077	0·1187	0·1426	0·1693	0·1838	0·2316	0·2864	0·3063	0·3487	0·3945	0·4187	0·4968	0·5535
0·000115	0·295	0·309	0·316	0·331	0·345	0·352	0·373	0·393	0·400	0·413	0·425	0·432	0·450	0·462
1/ 8696	0·0919	0·1104	0·1217	0·1462	0·1735	0·1883	0·2373	0·2935	0·3139	0·3573	0·4042	0·4290	0·5091	0·5672
0·000120	0·302	0·316	0·324	0·339	0·353	0·361	0·382	0·402	0·409	0·422	0·435	0·442	0·461	0·473
1/ 8333	0·0941	0·1131	0·1246	0·1497	0·1777	0·1928	0·2430	0·3005	0·3214	0·3658	0·4138	0·4392	0·5211	0·5806
0·000125	0·309	0·323	0·331	0·346	0·362	0·369	0·391	0·412	0·418	0·432	0·445	0·452	0·471	0·484
1/ 8000	0·0962	0·1156	0·1274	0·1531	0·1817	0·1972	0·2485	0·3073	0·3287	0·3741	0·4232	0·4491	0·5329	0·5937
0·000130	0·315	0·330	0·338	0·354	0·369	0·377	0·399	0·421	0·428	0·441	0·455	0·462	0·481	0·494
1/ 7692	0·0983	0·1182	0·1302	0·1564	0·1857	0·2015	0·2539	0·3140	0·3358	0·3822	0·4324	0·4589	0·5445	0·6066
0·000135	0·322	0·337	0·345	0·362	0·377	0·385	0·408	0·429	0·437	0·451	0·465	0·471	0·491	0·505
1/ 7407	0·1004	0·1207	0·1329	0·1597	0·1896	0·2058	0·2593	0·3206	0·3429	0·3902	0·4414	0·4685	0·5559	0·6193
0·000140	0·329	0·344	0·352	0·369	0·385	0·393	0·416	0·438	0·445	0·460	0·474	0·481	0·501	0·515
1/ 7143	0·1025	0·1231	0·1356	0·1630	0·1934	0·2099	0·2645	0·3271	0·3498	0·3981	0·4503	0·4780	0·5671	0·6318
0·000145	0·335	0·351	0·359	0·376	0·392	0·400	0·424	0·447	0·454	0·469	0·483	0·490	0·511	0·525
1/ 6897	0·1045	0·1255	0·1383	0·1661	0·1972	0·2140	0·2697	0·3335	0·3566	0·4059	0·4591	0·4873	0·5781	0·6440
0·000150	0·341	0·357	0·366	0·383	0·400	0·408	0·432	0·455	0·463	0·478	0·492	0·499	0·521	0·535
1/ 6667	0·1064	0·1279	0·1409	0·1693	0·2010	0·2181	0·2747	0·3397	0·3633	0·4135	0·4677	0·4964	0·5889	0·6561
0·000160	0·354	0·370	0·379	0·397	0·414	0·423	0·447	0·471	0·479	0·495	0·510	0·517	0·539	0·554
1/ 6250	0·1103	0·1326	0·1460	0·1754	0·2082	0·2260	0·2847	0·3520	0·3764	0·4284	0·4846	0·5143	0·6101	0·6797
0·000170	0·366	0·383	0·392	0·411	0·428	0·437	0·463	0·487	0·496	0·511	0·527	0·535	0·558	0·573
1/ 5882	0·1141	0·1371	0·1510	0·1814	0·2153	0·2336	0·2943	0·3639	0·3892	0·4429	0·5009	0·5317	0·6307	0·7026
0·000180	0·378	0·395	0·405	0·424	0·442	0·451	0·477	0·503	0·511	0·528	0·544	0·552	0·575	0·591
1/ 5556	0·1177	0·1415	0·1558	0·1872	0·2222	0·2411	0·3037	0·3755	0·4016	0·4570	0·5169	0·5486	0·6507	0·7249
0·000190	0·389	0·407	0·417	0·437	0·455	0·465	0·492	0·518	0·527	0·544	0·560	0·568	0·593	0·608
1/ 5263	0·1213	0·1458	0·1605	0·1928	0·2289	0·2484	0·3129	0·3868	0·4137	0·4707	0·5324	0·5651	0·6703	0·7467
0·000200	0·400	0·419	0·429	0·449	0·468	0·478	0·506	0·533	0·542	0·559	0·576	0·585	0·610	0·626
1/ 5000	0·1248	0·1499	0·1652	0·1984	0·2355	0·2555	0·3218	0·3979	0·4255	0·4842	0·5476	0·5812	0·6894	0·7679
0·000210	0·411	0·430	0·441	0·461	0·481	0·491	0·520	0·547	0·556	0·574	0·592	0·600	0·626	0·643
1/ 4762	0·1282	0·1540	0·1697	0·2038	0·2419	0·2624	0·3306	0·4087	0·4370	0·4973	0·5624	0·5969	0·7080	0·7887
0·000220	0·422	0·442	0·452	0·473	0·494	0·504	0·533	0·561	0·571	0·589	0·607	0·616	0·642	0·659
1/ 4545	0·1315	0·1580	0·1741	0·2091	0·2481	0·2692	0·3391	0·4192	0·4483	0·5101	0·5769	0·6123	0·7263	0·8090
0·000230	0·432	0·453	0·464	0·485	0·506	0·516	0·546	0·575	0·585	0·604	0·622	0·631	0·658	0·675
1/ 4348	0·1348	0·1620	0·1784	0·2142	0·2543	0·2759	0·3475	0·4296	0·4594	0·5227	0·5912	0·6274	0·7441	0·8289
0·000240	0·443	0·463	0·475	0·496	0·518	0·528	0·559	0·589	0·599	0·618	0·637	0·646	0·673	0·691
1/ 4167	0·1380	0·1658	0·1826	0·2193	0·2603	0·2824	0·3557	0·4397	0·4702	0·5350	0·6051	0·6421	0·7616	0·8484
0·000250	0·453	0·474	0·485	0·508	0·530	0·540	0·572	0·602	0·612	0·632	0·651	0·661	0·689	0·707
1/ 4000	0·1412	0·1696	0·1868	0·2243	0·2662	0·2888	0·3638	0·4497	0·4808	0·5471	0·6188	0·6566	0·7788	0·8675
0·000260	0·463	0·484	0·496	0·519	0·541	0·552	0·584	0·615	0·626	0·646	0·665	0·675	0·704	0·722
1/ 3846	0·1443	0·1733	0·1908	0·2292	0·2720	0·2951	0·3717	0·4594	0·4913	0·5590	0·6322	0·6709	0·7957	0·8863
0·000270	0·472	0·494	0·506	0·530	0·552	0·564	0·596	0·628	0·639	0·659	0·679	0·689	0·718	0·737
1/ 3704	0·1473	0·1769	0·1949	0·2340	0·2777	0·3013	0·3794	0·4690	0·5015	0·5706	0·6454	0·6849	0·8123	0·9048
0·000280	0·482	0·504	0·517	0·540	0·564	0·575	0·608	0·641	0·651	0·672	0·693	0·703	0·733	0·752
1/ 3571	0·1503	0·1805	0·1988	0·2387	0·2833	0·3074	0·3871	0·4785	0·5116	0·5821	0·6583	0·6986	0·8285	0·9229
0·000290	0·491	0·514	0·527	0·551	0·575	0·586	0·620	0·653	0·664	0·685	0·706	0·716	0·747	0·767
1/ 3448	0·1532	0·1840	0·2027	0·2434	0·2888	0·3134	0·3946	0·4877	0·5215	0·5934	0·6711	0·7121	0·8446	0·9407
0·000300	0·501	0·524	0·537	0·561	0·585	0·597	0·632	0·666	0·676	0·698	0·719	0·730	0·761	0·781
1/ 3333	0·1561	0·1875	0·2065	0·2480	0·2942	0·3192	0·4020	0·4969	0·5313	0·6045	0·6836	0·7255	0·8603	0·9583
	0·90	0·90	0·90	0·91	0·91	0·91	0·92	0·93	0·93	0·93	0·93	0·93	0·94	0·94

$V_{r(0\cdot5)medial}$ for half-full circular pipes.

k_s = 0·006 mm S = 0·00010 to 0·00030

$k_s = 0{\cdot}006$ **mm**
S = 0·00030 to 0·00100

ie hydraulic gradient =
1 in 3333 to 1 in 1000

Water (or sewage) at 15°C;
full bore conditions.

velocities in ms^{-1}
discharges in m^3s^{-1}

Gradient **(Equivalent) Pipe diameters in mm**

Gradient	630	675	700	750	800	825	900	975	1000	1050	1100	1125	1200	1250
0·000300	0·501	0·524	0·537	0·561	0·585	0·597	0·632	0·666	0·676	0·698	0·719	0·730	0·761	0·781
1/ 3333	0·1561	0·1875	0·2065	0·2480	0·2942	0·3192	0·4020	0·4969	0·5313	0·6045	0·6836	0·7255	0·8603	0·9583
0·000320	0·519	0·543	0·556	0·582	0·606	0·619	0·655	0·689	0·701	0·723	0·745	0·756	0·788	0·809
1/ 3125	0·1617	0·1943	0·2139	0·2569	0·3049	0·3308	0·4165	0·5147	0·5504	0·6262	0·7081	0·7515	0·8912	0·9926
0·000340	0·536	0·561	0·575	0·601	0·627	0·640	0·677	0·713	0·724	0·748	0·770	0·781	0·814	0·836
1/ 2941	0·1672	0·2008	0·2212	0·2656	0·3152	0·3419	0·4305	0·5321	0·5690	0·6473	0·7320	0·7768	0·9212	1·0260
0·000360	0·554	0·579	0·593	0·620	0·647	0·660	0·698	0·735	0·747	0·771	0·795	0·806	0·840	0·863
1/ 2778	0·1726	0·2073	0·2283	0·2741	0·3252	0·3528	0·4442	0·5490	0·5870	0·6678	0·7552	0·8014	0·9503	1·0585
0·000380	0·570	0·597	0·611	0·639	0·666	0·680	0·719	0·757	0·770	0·794	0·818	0·830	0·865	0·888
1/ 2632	0·1778	0·2135	0·2351	0·2823	0·3350	0·3634	0·4576	0·5655	0·6046	0·6878	0·7778	0·8254	0·9787	1·0901
0·000400	0·587	0·614	0·628	0·657	0·685	0·699	0·740	0·779	0·792	0·817	0·842	0·854	0·890	0·913
1/ 2500	0·1829	0·2196	0·2419	0·2904	0·3445	0·3738	0·4706	0·5815	0·6218	0·7074	0·7999	0·8488	1·0065	1·1210
0·000420	0·603	0·630	0·646	0·675	0·704	0·718	0·760	0·800	0·813	0·839	0·864	0·877	0·914	0·938
1/ 2381	0·1879	0·2256	0·2484	0·2983	0·3539	0·3839	0·4833	0·5973	0·6386	0·7265	0·8215	0·8717	1·0336	1·1512
0·000440	0·618	0·647	0·662	0·693	0·722	0·737	0·779	0·821	0·834	0·861	0·887	0·899	0·937	0·962
1/ 2273	0·1927	0·2314	0·2548	0·3060	0·3630	0·3938	0·4958	0·6126	0·6550	0·7451	0·8426	0·8941	1·0601	1·1807
0·000460	0·634	0·663	0·679	0·710	0·740	0·755	0·798	0·841	0·854	0·882	0·908	0·922	0·960	0·986
1/ 2174	0·1975	0·2371	0·2611	0·3135	0·3720	0·4035	0·5080	0·6277	0·6711	0·7634	0·8632	0·9160	1·0861	1·2096
0·000480	0·649	0·678	0·695	0·726	0·757	0·773	0·817	0·860	0·875	0·902	0·930	0·943	0·983	1·009
1/ 2083	0·2022	0·2427	0·2673	0·3209	0·3807	0·4130	0·5199	0·6424	0·6869	0·7814	0·8835	0·9375	1·1116	1·2379
0·000500	0·663	0·694	0·710	0·743	0·775	0·790	0·836	0·880	0·894	0·923	0·951	0·964	1·005	1·031
1/ 2000	0·2067	0·2482	0·2734	0·3282	0·3893	0·4224	0·5316	0·6569	0·7024	0·7990	0·9034	0·9586	1·1365	1·2657
0·000525	0·681	0·713	0·730	0·763	0·795	0·811	0·858	0·904	0·918	0·948	0·976	0·990	1·032	1·059
1/ 1905	0·2124	0·2550	0·2808	0·3371	0·3999	0·4338	0·5460	0·6746	0·7213	0·8205	0·9277	0·9844	1·1671	1·2998
0·000550	0·699	0·731	0·748	0·783	0·816	0·832	0·880	0·927	0·942	0·972	1·001	1·016	1·058	1·086
1/ 1818	0·2179	0·2616	0·2880	0·3458	0·4102	0·4450	0·5601	0·6920	0·7398	0·8415	0·9515	1·0096	1·1970	1·3330
0·000575	0·716	0·749	0·767	0·802	0·836	0·853	0·902	0·950	0·965	0·996	1·026	1·041	1·084	1·113
1/ 1739	0·2232	0·2680	0·2951	0·3543	0·4203	0·4559	0·5738	0·7090	0·7580	0·8622	0·9748	1·0344	1·2263	1·3656
0·000600	0·733	0·767	0·785	0·821	0·856	0·873	0·923	0·972	0·988	1·019	1·050	1·065	1·110	1·139
1/ 1667	0·2285	0·2743	0·3021	0·3626	0·4302	0·4666	0·5873	0·7256	0·7758	0·8824	0·9976	1·0586	1·2550	1·3976
0·000625	0·750	0·784	0·803	0·839	0·875	0·893	0·944	0·994	1·010	1·042	1·073	1·089	1·135	1·164
1/ 1600	0·2337	0·2806	0·3089	0·3708	0·4399	0·4771	0·6005	0·7419	0·7932	0·9022	1·0201	1·0824	1·2832	1·4289
0·000650	0·766	0·801	0·820	0·858	0·894	0·912	0·964	1·015	1·032	1·064	1·097	1·112	1·159	1·189
1/ 1538	0·2388	0·2866	0·3156	0·3789	0·4494	0·4875	0·6135	0·7579	0·8103	0·9217	1·0421	1·1057	1·3108	1·4597
0·000675	0·782	0·818	0·837	0·875	0·913	0·931	0·984	1·036	1·053	1·087	1·119	1·135	1·183	1·214
1/ 1481	0·2438	0·2926	0·3222	0·3868	0·4587	0·4976	0·6263	0·7737	0·8272	0·9408	1·0637	1·1286	1·3379	1·4899
0·000700	0·798	0·834	0·854	0·893	0·931	0·950	1·004	1·057	1·074	1·108	1·142	1·158	1·207	1·238
1/ 1429	0·2487	0·2985	0·3287	0·3945	0·4680	0·5076	0·6388	0·7892	0·8437	0·9596	1·0849	1·1511	1·3646	1·5196
0·000725	0·813	0·850	0·871	0·910	0·949	0·968	1·024	1·077	1·095	1·130	1·164	1·180	1·230	1·262
1/ 1379	0·2535	0·3043	0·3351	0·4022	0·4770	0·5174	0·6512	0·8044	0·8600	0·9781	1·1058	1·1733	1·3909	1·5488
0·000750	0·828	0·866	0·887	0·927	0·967	0·986	1·043	1·097	1·115	1·151	1·185	1·202	1·253	1·286
1/ 1333	0·2582	0·3100	0·3413	0·4097	0·4859	0·5271	0·6633	0·8194	0·8760	0·9963	1·1264	1·1951	1·4167	1·5776
0·000800	0·858	0·897	0·919	0·961	1·001	1·021	1·080	1·137	1·155	1·192	1·228	1·245	1·297	1·331
1/ 1250	0·2675	0·3212	0·3536	0·4244	0·5033	0·5460	0·6870	0·8487	0·9073	1·0319	1·1666	1·2378	1·4672	1·6338
0·000850	0·887	0·928	0·950	0·993	1·035	1·056	1·116	1·175	1·194	1·232	1·269	1·287	1·341	1·376
1/ 1176	0·2766	0·3320	0·3655	0·4387	0·5203	0·5643	0·7101	0·8771	0·9377	1·0665	1·2056	1·2792	1·5163	1·6884
0·000900	0·915	0·957	0·980	1·024	1·068	1·089	1·152	1·212	1·232	1·270	1·309	1·327	1·383	1·419
1/ 1111	0·2854	0·3425	0·3771	0·4526	0·5367	0·5822	0·7326	0·9048	0·9673	1·1001	1·2436	1·3195	1·5640	1·7415
0·000950	0·943	0·986	1·009	1·055	1·100	1·122	1·186	1·248	1·268	1·308	1·348	1·367	1·424	1·461
1/ 1053	0·2939	0·3528	0·3884	0·4661	0·5528	0·5996	0·7544	0·9318	0·9962	1·1329	1·2807	1·3588	1·6106	1·7933
0·001000	0·970	1·014	1·038	1·085	1·131	1·153	1·219	1·283	1·304	1·345	1·386	1·406	1·464	1·502
1/ 1000	0·3023	0·3628	0·3994	0·4794	0·5685	0·6166	0·7758	0·9582	1·0243	1·1649	1·3168	1·3971	1·6560	1·8438
	0·92	0·93	0·93	0·93	0·94	0·94	0·94	0·95	0·95	0·95	0·95	0·95	0·96	0·96

$V_{r(0{\cdot}5)medial}$ **for half-full circular pipes.**

S = 0·00030 to 0·00100 $k_s = 0{\cdot}006$ **mm**

A23
(p.3 of 6)

k_s = 0·006 mm
S = 0·00100 to 0·00300

ie hydraulic gradient =
1 in 1000 to 1 in 333

Water (or sewage) at 15°C;
full bore conditions.

velocities in ms^{-1}
discharges in m^3s^{-1}

Gradient	**(Equivalent) Pipe diameters in mm**													
	630	**675**	**700**	**750**	**800**	**825**	**900**	**975**	**1000**	**1050**	**1100**	**1125**	**1200**	**1250**
0·00100	0·970	1·014	1·038	1·085	1·131	1·153	1·219	1·283	1·304	1·345	1·386	1·406	1·464	1·502
1/ 1000	0·3023	0·3628	0·3994	0·4794	0·5685	0·6166	0·7758	0·9582	1·0243	1·1649	1·3168	1·3971	1·6560	1·8438
0·00105	0·996	1·041	1·066	1·114	1·161	1·185	1·252	1·318	1·339	1·381	1·423	1·443	1·503	1·543
1/ 952	0·3105	0·3726	0·4102	0·4923	0·5838	0·6332	0·7967	0·9839	1·0518	1·1962	1·3521	1·4346	1·7003	1·8932
0·00110	1·022	1·068	1·093	1·143	1·191	1·215	1·284	1·352	1·373	1·417	1·459	1·480	1·542	1·582
1/ 909	0·3185	0·3822	0·4208	0·5049	0·5988	0·6494	0·8171	1·0091	1·0787	1·2267	1·3867	1·4712	1·7437	1·9415
0·00115	1·047	1·094	1·120	1·171	1·220	1·245	1·316	1·385	1·407	1·451	1·495	1·516	1·579	1·621
1/ 870	0·3263	0·3916	0·4311	0·5173	0·6134	0·6653	0·8370	1·0337	1·1051	1·2567	1·4205	1·5071	1·7862	1·9887
0·00120	1·071	1·120	1·147	1·198	1·249	1·274	1·347	1·417	1·440	1·485	1·530	1·552	1·616	1·658
1/ 833	0·3340	0·4008	0·4412	0·5294	0·6278	0·6809	0·8566	1·0579	1·1309	1·2860	1·4536	1·5422	1·8278	2·0351
0·00125	1·095	1·145	1·172	1·225	1·277	1·302	1·377	1·449	1·472	1·518	1·564	1·586	1·652	1·695
1/ 800	0·3415	0·4098	0·4512	0·5414	0·6419	0·6962	0·8758	1·0816	1·1562	1·3148	1·4861	1·5767	1·8686	2·0805
0·00130	1·119	1·170	1·198	1·252	1·305	1·330	1·406	1·480	1·504	1·551	1·597	1·620	1·688	1·732
1/ 769	0·3489	0·4187	0·4609	0·5530	0·6558	0·7112	0·8947	1·1049	1·1811	1·3430	1·5181	1·6106	1·9087	2·1251
0·00135	1·142	1·194	1·223	1·278	1·332	1·358	1·436	1·510	1·535	1·583	1·630	1·654	1·722	1·767
1/ 741	0·3561	0·4274	0·4705	0·5645	0·6694	0·7260	0·9132	1·1277	1·2055	1·3708	1·5494	1·6438	1·9481	2·1689
0·00140	1·165	1·218	1·247	1·303	1·358	1·385	1·464	1·541	1·565	1·615	1·663	1·687	1·757	1·802
1/ 714	0·3633	0·4359	0·4799	0·5758	0·6827	0·7405	0·9314	1·1502	1·2295	1·3981	1·5802	1·6765	1·9868	2·2120
0·00145	1·188	1·242	1·271	1·328	1·384	1·412	1·492	1·570	1·595	1·646	1·695	1·719	1·790	1·837
1/ 690	0·3703	0·4444	0·4892	0·5869	0·6959	0·7547	0·9493	1·1723	1·2531	1·4249	1·6105	1·7086	2·0248	2·2543
0·00150	1·210	1·265	1·295	1·353	1·410	1·438	1·520	1·599	1·625	1·676	1·726	1·751	1·823	1·871
1/ 667	0·3772	0·4526	0·4983	0·5978	0·7088	0·7688	0·9670	1·1940	1·2763	1·4513	1·6403	1·7403	2·0623	2·2960
0·00160	1·253	1·310	1·341	1·402	1·460	1·489	1·574	1·656	1·683	1·736	1·787	1·813	1·888	1·937
1/ 625	0·3907	0·4688	0·5161	0·6192	0·7341	0·7962	1·0014	1·2365	1·3217	1·5029	1·6986	1·8021	2·1355	2·3774
0·00170	1·296	1·354	1·386	1·449	1·509	1·539	1·627	1·711	1·739	1·794	1·847	1·873	1·951	2·002
1/ 588	0·4038	0·4846	0·5334	0·6399	0·7587	0·8228	1·0349	1·2778	1·3658	1·5530	1·7552	1·8621	2·2065	2·4565
0·00180	1·337	1·397	1·430	1·494	1·557	1·588	1·678	1·765	1·794	1·850	1·905	1·932	2·012	2·064
1/ 556	0·4166	0·4999	0·5503	0·6601	0·7826	0·8487	1·0675	1·3179	1·4088	1·6018	1·8103	1·9205	2·2757	2·5334
0·00190	1·376	1·439	1·472	1·539	1·603	1·635	1·728	1·818	1·847	1·905	1·961	1·989	2·072	2·126
1/ 526	0·4291	0·5148	0·5667	0·6798	0·8059	0·8740	1·0992	1·3571	1·4506	1·6493	1·8640	1·9774	2·3431	2·6084
0·00200	1·415	1·479	1·514	1·582	1·649	1·681	1·776	1·869	1·899	1·958	2·016	2·045	2·130	2·185
1/ 500	0·4412	0·5294	0·5827	0·6990	0·8287	0·8987	1·1302	1·3953	1·4914	1·6956	1·9163	2·0330	2·4088	2·6816
0·00210	1·454	1·519	1·555	1·625	1·693	1·726	1·824	1·919	1·950	2·011	2·070	2·100	2·187	2·243
1/ 476	0·4531	0·5436	0·5983	0·7178	0·8509	0·9227	1·1604	1·4326	1·5312	1·7409	1·9675	2·0872	2·4730	2·7530
0·00220	1·491	1·558	1·595	1·666	1·736	1·770	1·871	1·968	1·999	2·062	2·123	2·153	2·242	2·300
1/ 455	0·4647	0·5575	0·6137	0·7361	0·8726	0·9463	1·1900	1·4691	1·5702	1·7852	2·0175	2·1403	2·5359	2·8229
0·00230	1·527	1·596	1·634	1·707	1·778	1·813	1·916	2·015	2·048	2·112	2·174	2·205	2·297	2·356
1/ 435	0·4761	0·5711	0·6287	0·7541	0·8939	0·9694	1·2190	1·5048	1·6084	1·8286	2·0665	2·1922	2·5973	2·8913
0·00240	1·563	1·633	1·672	1·747	1·820	1·856	1·961	2·062	2·096	2·161	2·225	2·257	2·350	2·411
1/ 417	0·4872	0·5845	0·6433	0·7717	0·9147	0·9920	1·2474	1·5398	1·6458	1·8711	2·1145	2·2431	2·6576	2·9583
0·00250	1·598	1·670	1·709	1·786	1·860	1·897	2·004	2·108	2·142	2·209	2·275	2·307	2·402	2·464
1/ 400	0·4982	0·5976	0·6578	0·7890	0·9352	1·0141	1·2752	1·5741	1·6825	1·9128	2·1615	2·2930	2·7167	3·0241
0·00260	1·632	1·706	1·746	1·824	1·900	1·938	2·047	2·153	2·188	2·256	2·323	2·356	2·453	2·517
1/ 385	0·5089	0·6105	0·6719	0·8059	0·9553	1·0359	1·3025	1·6078	1·7185	1·9537	2·2077	2·3420	2·7747	3·0886
0·00270	1·666	1·741	1·782	1·862	1·940	1·978	2·090	2·198	2·233	2·303	2·371	2·405	2·504	2·568
1/ 370	0·5194	0·6231	0·6858	0·8226	0·9750	1·0573	1·3294	1·6409	1·7538	1·9938	2·2531	2·3901	2·8316	3·1519
0·00280	1·700	1·776	1·818	1·899	1·978	2·017	2·131	2·241	2·277	2·348	2·418	2·452	2·553	2·619
1/ 357	0·5298	0·6355	0·6995	0·8389	0·9944	1·0783	1·3557	1·6734	1·7886	2·0333	2·2977	2·4374	2·8876	3·2142
0·00290	1·732	1·810	1·852	1·935	2·016	2·056	2·172	2·284	2·321	2·393	2·464	2·499	2·602	2·669
1/ 345	0·5400	0·6477	0·7129	0·8550	1·0134	1·0989	1·3817	1·7054	1·8228	2·0722	2·3415	2·4839	2·9426	3·2754
0·00300	1·764	1·844	1·887	1·971	2·053	2·094	2·212	2·326	2·364	2·437	2·509	2·545	2·650	2·718
1/ 333	0·5500	0·6597	0·7261	0·8709	1·0322	1·1193	1·4072	1·7369	1·8564	2·1104	2·3847	2·5297	2·9968	3·3357
	0·94	0·95	0·95	0·95	0·95	0·96	0·96	0·96	0·96	0·96	0·97	0·97	0·97	0·97

$V_{r(0·5)medial}$ for half-full circular pipes.

k_s = 0·006 mm S = 0·00100 to 0·00300

k_s = **0·006 mm**
S = 0·00300 to 0·01000

ie hydraulic gradient =
1 in 333 to 1 in 100

Water (or sewage) at 15°C;
full bore conditions.

velocities in ms^{-1}
discharges in m^3s^{-1}

Gradient	**(Equivalent) Pipe diameters in mm**													
	630	**675**	**700**	**750**	**800**	**825**	**900**	**975**	**1000**	**1050**	**1100**	**1125**	**1200**	**1250**
0·00300	1·764	1·844	1·887	1·971	2·053	2·094	2·212	2·326	2·364	2·437	2·509	2·545	2·650	2·718
1/ 333	0·5500	0·6597	0·7261	0·8709	1·0322	1·1193	1·4072	1·7369	1·8564	2·1104	2·3847	2·5297	2·9968	3·3357
0·00320	1·827	1·909	1·954	2·041	2·126	2·168	2·290	2·409	2·447	2·523	2·598	2·635	2·743	2·814
1/ 313	0·5696	0·6832	0·7520	0·9018	1·0688	1·1590	1·4571	1·7984	1·9221	2·1850	2·4690	2·6191	3·1026	3·4534
0·00340	1·888	1·973	2·019	2·109	2·197	2·240	2·367	2·489	2·529	2·607	2·684	2·722	2·834	2·907
1/ 294	0·5887	0·7060	0·7770	0·9319	1·1044	1·1976	1·5056	1·8581	1·9860	2·2575	2·5509	2·7059	3·2054	3·5677
0·00360	1·948	2·035	2·083	2·176	2·266	2·311	2·441	2·567	2·608	2·689	2·768	2·807	2·923	2·998
1/ 278	0·6072	0·7282	0·8015	0·9611	1·1391	1·2351	1·5527	1·9163	2·0480	2·3281	2·6305	2·7904	3·3053	3·6789
0·00380	2·006	2·096	2·144	2·240	2·333	2·379	2·513	2·642	2·685	2·768	2·850	2·890	3·009	3·086
1/ 263	0·6252	0·7499	0·8253	0·9896	1·1728	1·2717	1·5986	1·9729	2·1085	2·3968	2·7081	2·8727	3·4027	3·7873
0·00400	2·062	2·155	2·205	2·303	2·399	2·446	2·583	2·716	2·760	2·845	2·929	2·971	3·093	3·172
1/ 250	0·6428	0·7710	0·8485	1·0175	1·2058	1·3074	1·6434	2·0281	2·1676	2·4638	2·7838	2·9529	3·4977	3·8930
0·00420	2·117	2·212	2·264	2·365	2·463	2·511	2·652	2·789	2·833	2·921	3·007	3·050	3·175	3·256
1/ 238	0·6601	0·7916	0·8712	1·0446	1·2379	1·3423	1·6872	2·0821	2·2252	2·5293	2·8577	3·0313	3·5905	3·9962
0·00440	2·171	2·268	2·321	2·425	2·525	2·575	2·719	2·859	2·905	2·995	3·083	3·127	3·255	3·339
1/ 227	0·6769	0·8118	0·8933	1·0712	1·2694	1·3764	1·7300	2·1348	2·2816	2·5933	2·9300	3·1080	3·6812	4·0971
0·00460	2·224	2·324	2·378	2·484	2·587	2·637	2·785	2·929	2·975	3·067	3·158	3·202	3·333	3·419
1/ 217	0·6934	0·8315	0·9151	1·0972	1·3002	1·4097	1·7719	2·1865	2·3367	2·6560	3·0008	3·1830	3·7700	4·1959
0·00480	2·276	2·378	2·433	2·541	2·647	2·698	2·850	2·996	3·044	3·138	3·231	3·276	3·410	3·498
1/ 208	0·7095	0·8509	0·9363	1·1227	1·3304	1·4424	1·8130	2·2371	2·3908	2·7174	3·0701	3·2566	3·8570	4·2926
0·00500	2·327	2·431	2·487	2·598	2·706	2·758	2·913	3·063	3·112	3·208	3·302	3·349	3·486	3·575
1/ 200	0·7254	0·8698	0·9572	1·1477	1·3599	1·4745	1·8532	2·2867	2·4438	2·7776	3·1381	3·3286	3·9423	4·3875
0·00525	2·389	2·496	2·554	2·667	2·778	2·832	2·991	3·144	3·194	3·293	3·390	3·437	3·578	3·670
1/ 190	0·7447	0·8931	0·9827	1·1783	1·3962	1·5138	1·9025	2·3474	2·5087	2·8513	3·2213	3·4169	4·0467	4·5037
0·00550	2·450	2·559	2·619	2·735	2·848	2·904	3·066	3·224	3·275	3·376	3·475	3·524	3·668	3·762
1/ 182	0·7637	0·9158	1·0077	1·2082	1·4316	1·5522	1·9507	2·4068	2·5721	2·9234	3·3027	3·5031	4·1488	4·6172
0·00575	2·509	2·621	2·682	2·801	2·917	2·974	3·140	3·301	3·354	3·458	3·559	3·609	3·757	3·853
1/ 174	0·7823	0·9380	1·0322	1·2375	1·4663	1·5897	1·9979	2·4649	2·6342	2·9939	3·3823	3·5876	4·2487	4·7283
0·00600	2·568	2·682	2·744	2·866	2·985	3·043	3·213	3·378	3·431	3·537	3·641	3·692	3·843	3·942
1/ 167	0·8004	0·9598	1·0561	1·2662	1·5002	1·6265	2·0441	2·5219	2·6951	3·0630	3·4603	3·6703	4·3466	4·8372
0·00625	2·625	2·742	2·805	2·930	3·051	3·110	3·284	3·453	3·507	3·616	3·722	3·774	3·928	4·029
1/ 160	0·8183	0·9811	1·0796	1·2943	1·5335	1·6626	2·0894	2·5777	2·7547	3·1307	3·5368	3·7514	4·4426	4·9439
0·00650	2·681	2·800	2·865	2·992	3·116	3·177	3·354	3·526	3·582	3·692	3·801	3·854	4·011	4·114
1/ 154	0·8358	1·0021	1·1027	1·3219	1·5662	1·6981	2·1338	2·6325	2·8133	3·1972	3·6119	3·8310	4·5367	5·0486
0·00675	2·736	2·858	2·924	3·054	3·180	3·242	3·423	3·598	3·655	3·768	3·878	3·933	4·093	4·198
1/ 148	0·8530	1·0227	1·1253	1·3491	1·5983	1·7329	2·1775	2·6863	2·8707	3·2625	3·6856	3·9091	4·6292	5·1515
0·00700	2·791	2·914	2·982	3·114	3·243	3·306	3·490	3·669	3·727	3·842	3·954	4·010	4·173	4·280
1/ 143	0·8699	1·0429	1·1476	1·3757	1·6299	1·7670	2·2204	2·7392	2·9272	3·3266	3·7580	3·9859	4·7200	5·2525
0·00725	2·844	2·970	3·039	3·173	3·304	3·368	3·557	3·738	3·798	3·915	4·029	4·086	4·252	4·361
1/ 138	0·8865	1·0628	1·1695	1·4019	1·6609	1·8007	2·2626	2·7912	2·9828	3·3897	3·8292	4·0614	4·8094	5·3519
0·00750	2·896	3·025	3·095	3·232	3·365	3·430	3·622	3·807	3·867	3·986	4·103	4·161	4·330	4·441
1/ 133	0·9028	1·0824	1·1910	1·4277	1·6914	1·8337	2·3041	2·8423	3·0374	3·4518	3·8992	4·1357	4·8973	5·4496
0·00800	2·999	3·132	3·204	3·346	3·484	3·551	3·749	3·941	4·003	4·126	4·247	4·307	4·482	4·596
1/ 125	0·9348	1·1207	1·2331	1·4781	1·7511	1·8984	2·3852	2·9423	3·1442	3·5730	4·0361	4·2808	5·0689	5·6406
0·00850	3·098	3·236	3·310	3·457	3·599	3·669	3·873	4·071	4·135	4·262	4·387	4·448	4·629	4·747
1/ 118	0·9658	1·1578	1·2739	1·5270	1·8090	1·9611	2·4640	3·0393	3·2478	3·6907	4·1690	4·4217	5·2356	5·8259
0·00900	3·195	3·337	3·414	3·564	3·711	3·783	3·994	4·197	4·264	4·394	4·523	4·586	4·773	4·894
1/ 111	0·9960	1·1940	1·3137	1·5746	1·8653	2·0222	2·5406	3·1337	3·3486	3·8052	4·2982	4·5587	5·3977	6·0062
0·00950	3·289	3·435	3·514	3·669	3·820	3·894	4·111	4·320	4·389	4·523	4·655	4·720	4·912	5·037
1/ 105	1·0254	1·2292	1·3524	1·6210	1·9202	2·0816	2·6152	3·2256	3·4468	3·9167	4·4240	4·6921	5·5555	6·1817
0·01000	3·381	3·531	3·612	3·772	3·927	4·003	4·225	4·440	4·511	4·649	4·784	4·851	5·048	5·177
1/ 100	1·0541	1·2635	1·3902	1·6662	1·9737	2·1396	2·6879	3·3152	3·5425	4·0254	4·5468	4·8223	5·7095	6·3529
	0·96	0·96	0·96	0·97	0·97	0·97	0·97	0·97	0·97	0·98	0·98	0·98	0·98	0·98

$V_{r(0·5)medial}$ for half-full circular pipes.

k_s = **0·006 mm**
S = 0·01000 to 0·03000

ie hydraulic gradient =
1 in 100 to 1 in 33·3

Water (or sewage) at 15°C;
full bore conditions.

velocities in ms^{-1}
discharges in m^3s^{-1}

Gradient **(Equivalent) Pipe diameters in mm**

Gradient	630	675	700	750	800	825	900	975	1000	1050	1100	1125	1200	1250
0·01000	3·381	3·531	3·612	3·772	3·927	4·003	4·225	4·440	4·511	4·649	4·784	4·851	5·048	5·177
1/ 100	1·0541	1·2635	1·3902	1·6662	1·9737	2·1396	2·6879	3·3152	3·5425	4·0254	4·5468	4·8223	5·7095	6·3529
0·01050	3·471	3·625	3·708	3·871	4·030	4·108	4·337	4·558	4·630	4·771	4·911	4·979	5·181	5·313
1/ 95	1·0821	1·2971	1·4270	1·7104	2·0259	2·1962	2·7590	3·4028	3·6361	4·1316	4·6666	4·9493	5·8598	6·5201
0·01100	3·559	3·716	3·802	3·969	4·132	4·212	4·446	4·672	4·746	4·891	5·034	5·104	5·311	5·446
1/ 91	1·1095	1·3299	1·4631	1·7535	2·0770	2·2516	2·8284	3·4883	3·7275	4·2354	4·7838	5·0736	6·0068	6·6835
0·01150	3·645	3·806	3·894	4·065	4·232	4·313	4·553	4·784	4·860	5·009	5·154	5·226	5·438	5·577
1/ 87	1·1363	1·3620	1·4984	1·7958	2·1271	2·3058	2·8964	3·5721	3·8169	4·3370	4·8985	5·1952	6·1506	6·8434
0·01200	3·729	3·894	3·983	4·159	4·329	4·413	4·658	4·894	4·971	5·124	5·273	5·346	5·563	5·704
1/ 83	1·1625	1·3934	1·5330	1·8372	2·1760	2·3589	2·9630	3·6541	3·9046	4·4365	5·0108	5·3142	6·2914	7·0000
0·01250	3·812	3·980	4·072	4·250	4·425	4·510	4·760	5·002	5·081	5·236	5·389	5·464	5·685	5·829
1/ 80	1·1883	1·4243	1·5669	1·8778	2·2241	2·4109	3·0283	3·7346	3·9905	4·5340	5·1209	5·4310	6·4295	7·1536
0·01300	3·893	4·065	4·158	4·341	4·518	4·606	4·861	5·108	5·188	5·347	5·502	5·579	5·805	5·952
1/ 77	1·2136	1·4545	1·6002	1·9176	2·2712	2·4620	3·0924	3·8135	4·0748	4·6297	5·2289	5·5455	6·5650	7·3042
0·01350	3·973	4·148	4·243	4·429	4·611	4·699	4·960	5·211	5·294	5·455	5·614	5·692	5·922	6·073
1/ 74	1·2384	1·4842	1·6328	1·9568	2·3175	2·5121	3·1553	3·8910	4·1576	4·7237	5·3350	5·6580	6·6980	7·4521
0·01400	4·051	4·229	4·326	4·516	4·701	4·792	5·057	5·313	5·397	5·562	5·723	5·803	6·038	6·191
1/ 71	1·2628	1·5134	1·6649	1·9952	2·3630	2·5614	3·2171	3·9671	4·2389	4·8160	5·4392	5·7684	6·8286	7·5974
0·01450	4·128	4·309	4·408	4·602	4·790	4·882	5·153	5·414	5·499	5·667	5·831	5·912	6·151	6·307
1/ 69	1·2867	1·5421	1·6965	2·0330	2·4077	2·6099	3·2779	4·0420	4·3188	4·9068	5·5416	5·8770	6·9571	7·7403
0·01500	4·203	4·388	4·489	4·686	4·877	4·971	5·246	5·512	5·599	5·770	5·937	6·020	6·263	6·422
1/ 67	1·3103	1·5704	1·7276	2·0701	2·4517	2·6575	3·3377	4·1156	4·3975	4·9961	5·6424	5·9839	7·0834	7·8808
0·01600	4·351	4·543	4·647	4·850	5·048	5·146	5·430	5·705	5·795	5·971	6·144	6·230	6·481	6·645
1/ 63	1·3564	1·6256	1·7882	2·1428	2·5376	2·7506	3·4544	4·2594	4·5511	5·1705	5·8393	6·1926	7·3303	8·1552
0·01700	4·495	4·692	4·800	5·010	5·214	5·315	5·608	5·892	5·984	6·167	6·345	6·434	6·693	6·862
1/ 59	1·4012	1·6791	1·8471	2·2133	2·6210	2·8410	3·5678	4·3990	4·7002	5·3398	6·0303	6·3951	7·5698	8·4215
0·01800	4·634	4·838	4·948	5·165	5·376	5·479	5·781	6·074	6·169	6·357	6·541	6·632	6·899	7·073
1/ 56	1·4447	1·7312	1·9044	2·2818	2·7021	2·9288	3·6780	4·5347	4·8452	5·5044	6·2161	6·5920	7·8027	8·6805
0·01900	4·770	4·980	5·093	5·316	5·533	5·639	5·950	6·251	6·349	6·542	6·731	6·825	7·100	7·279
1/ 53	1·4870	1·7819	1·9602	2·3486	2·7811	3·0144	3·7853	4·6669	4·9863	5·6646	6·3969	6·7838	8·0294	8·9326
0·02000	4·903	5·118	5·235	5·463	5·686	5·795	6·115	6·423	6·524	6·722	6·917	7·013	7·295	7·479
1/ 50	1·5284	1·8314	2·0146	2·4137	2·8581	3·0979	3·8900	4·7958	5·1240	5·8209	6·5733	6·9707	8·2505	9·1784
0·02100	5·032	5·253	5·373	5·607	5·836	5·948	6·275	6·592	6·695	6·899	7·098	7·196	7·486	7·675
1/ 47·6	1·5687	1·8797	2·0677	2·4773	2·9333	3·1794	3·9922	4·9216	5·2584	5·9735	6·7455	7·1533	8·4663	9·4184
0·02200	5·159	5·385	5·508	5·748	5·982	6·097	6·432	6·757	6·862	7·071	7·275	7·376	7·672	7·866
1/ 45·5	1·6082	1·9270	2·1196	2·5394	3·0068	3·2590	4·0921	5·0446	5·3898	6·1226	6·9138	7·3317	8·6773	9·6529
0·02300	5·283	5·514	5·640	5·886	6·125	6·242	6·586	6·918	7·026	7·239	7·448	7·551	7·855	8·053
1/ 43·5	1·6468	1·9732	2·1704	2·6002	3·0788	3·3370	4·1898	5·1650	5·5183	6·2685	7·0784	7·5062	8·8837	9·8824
0·02400	5·404	5·641	5·769	6·021	6·265	6·385	6·736	7·076	7·186	7·404	7·618	7·723	8·034	8·236
1/ 41·7	1·6846	2·0185	2·2202	2·6598	3·1493	3·4133	4·2855	5·2829	5·6442	6·4114	7·2397	7·6772	9·0859	10·107
0·02500	5·523	5·765	5·896	6·153	6·403	6·525	6·884	7·230	7·344	7·566	7·784	7·892	8·209	8·416
1/ 40·0	1·7217	2·0629	2·2690	2·7182	3·2183	3·4881	4·3794	5·3984	5·7676	6·5515	7·3977	7·8447	9·2840	10·327
0·02600	5·640	5·886	6·020	6·282	6·537	6·663	7·029	7·382	7·498	7·725	7·948	8·057	8·381	8·592
1/ 38·5	1·7581	2·1064	2·3169	2·7755	3·2861	3·5615	4·4714	5·5117	5·8886	6·6889	7·5528	8·0091	9·4783	10·543
0·02700	5·755	6·006	6·142	6·410	6·670	6·797	7·171	7·531	7·649	7·881	8·108	8·220	8·549	8·764
1/ 37·0	1·7938	2·1492	2·3639	2·8317	3·3526	3·6336	4·5618	5·6230	6·0075	6·8238	7·7050	8·1705	9·6691	10·756
0·02800	5·867	6·123	6·262	6·535	6·800	6·930	7·310	7·678	7·798	8·034	8·265	8·379	8·715	8·934
1/ 35·7	1·8289	2·1912	2·4100	2·8870	3·4180	3·7044	4·6506	5·7323	6·1242	6·9563	7·8545	8·3290	9·8565	10·964
0·02900	5·978	6·239	6·380	6·658	6·928	7·060	7·447	7·822	7·944	8·184	8·420	8·536	8·878	9·101
1/ 34·5	1·8634	2·2324	2·4554	2·9413	3·4822	3·7740	4·7378	5·8398	6·2389	7·0866	8·0015	8·4848	10·041	11·169
0·03000	6·087	6·352	6·496	6·779	7·053	7·188	7·582	7·963	8·087	8·332	8·572	8·690	9·038	9·265
1/ 33·3	1·8974	2·2731	2·5001	2·9947	3·5454	3·8425	4·8237	5·9454	6·3518	7·2147	8·1460	8·6380	10·222	11·370
	0·97	0·97	0·97	0·97	0·98	0·98	0·98	0·98	0·98	0·98	0·98	0·98	0·98	0·98

$V_{r(0\cdot5)medial}$ for half-full circular pipes.

k_s = 0·006 mm S = 0·01000 to 0·03000

Gradient	(Equivalent) Pipe diameters in mm													
	630	**675**	**700**	**750**	**800**	**825**	**900**	**975**	**1000**	**1050**	**1100**	**1125**	**1200**	**1250**
0·03000	6·087	6·352	6·496	6·779	7·053	7·188	7·582	7·963	8·087	8·332	8·572	8·690	9·038	9·265
1/ 33·3	1·8974	2·2731	2·5001	2·9947	3·5454	3·8425	4·8237	5·9454	6·3518	7·2147	8·1460	8·6380	10·222	11·370
0·03200	6·299	6·574	6·723	7·015	7·299	7·438	7·846	8·240	8·368	8·621	8·869	8·991	9·351	9·586
1/ 31·3	1·9637	2·3524	2·5873	3·0991	3·6689	3·9762	4·9913	6·1518	6·5722	7·4648	8·4283	8·9372	10·576	11·763
0·03400	6·506	6·789	6·943	7·244	7·537	7·681	8·102	8·508	8·640	8·901	9·157	9·283	9·654	9·897
1/ 29·4	2·0280	2·4294	2·6720	3·2004	3·7887	4·1060	5·1540	6·3521	6·7861	7·7076	8·7022	9·2276	10·919	12·145
0·03600	6·706	6·998	7·157	7·467	7·769	7·917	8·350	8·769	8·905	9·174	9·437	9·567	9·949	10·20
1/ 27·8	2·0906	2·5043	2·7542	3·2989	3·9051	4·2321	5·3121	6·5468	6·9940	7·9436	8·9685	9·5098	11·253	12·516
0·03800	6·902	7·202	7·365	7·684	7·995	8·147	8·592	9·022	9·163	9·439	9·710	9·843	10·24	10·49
1/ 26·3	2·1515	2·5772	2·8344	3·3947	4·0185	4·3549	5·4661	6·7363	7·1964	8·1733	9·2276	9·7846	11·577	12·877
0·04000	7·092	7·400	7·568	7·896	8·214	8·371	8·828	9·270	9·414	9·698	9·976	10·11	10·52	10·78
1/ 25·0	2·2109	2·6482	2·9125	3·4882	4·1290	4·4747	5·6162	6·9211	7·3937	8·3972	9·4803	10·052	11·894	13·229
0·04200	7·278	7·594	7·766	8·102	8·429	8·589	9·058	9·511	9·659	9·950	10·24	10·38	10·79	11·06
1/ 23·8	2·2689	2·7176	2·9888	3·5795	4·2370	4·5916	5·7628	7·1015	7·5863	8·6159	9·7270	10·314	12·203	13·572
0·04400	7·460	7·784	7·960	8·304	8·639	8·803	9·284	9·748	9·899	10·20	10·49	10·63	11·06	11·33
1/ 22·7	2·3256	2·7855	3·0633	3·6687	4·3424	4·7058	5·9060	7·2777	7·7746	8·8295	9·9680	10·569	12·505	13·908
0·04600	7·638	7·969	8·149	8·502	8·844	9·012	9·504	9·979	10·13	10·44	10·74	10·88	11·32	11·60
1/ 21·7	2·3810	2·8518	3·1363	3·7559	4·4456	4·8176	6·0461	7·4502	7·9587	9·0385	10·204	10·819	12·800	14·236
0·04800	7·812	8·151	8·335	8·695	9·045	9·217	9·719	10·20	10·36	10·67	10·98	11·13	11·57	11·86
1/ 20·8	2·4353	2·9168	3·2077	3·8414	4·5467	4·9270	6·1832	7·6190	8·1390	9·2431	10·435	11·064	13·090	14·558
0·05000	7·983	8·329	8·517	8·885	9·242	9·418	9·931	10·43	10·59	10·91	11·22	11·37	11·82	12·12
1/ 20·0	2·4886	2·9805	3·2777	3·9251	4·6457	5·0343	6·3177	7·7845	8·3157	9·4436	10·661	11·304	13·373	14·873
0·05250	8·192	8·547	8·739	9·117	9·483	9·663	10·19	10·70	10·86	11·19	11·51	11·67	12·13	12·43
1/ 19·0	2·5537	3·0584	3·3633	4·0276	4·7668	5·1655	6·4821	7·9869	8·5319	9·6889	10·938	11·597	13·720	15·258
0·05500	8·396	8·760	8·957	9·343	9·719	9·903	10·44	10·96	11·13	11·47	11·79	11·96	12·43	12·74
1/ 18·2	2·6173	3·1346	3·4470	4·1277	4·8852	5·2938	6·6429	8·1847	8·7431	9·9286	11·208	11·884	14·059	15·635
0·05750	8·596	8·968	9·170	9·565	9·949	10·14	10·69	11·22	11·40	11·74	12·07	12·24	12·72	13·04
1/ 17·4	2·6796	3·2091	3·5289	4·2256	5·0010	5·4192	6·8001	8·3782	8·9497	10·163	11·472	12·164	14·390	16·003
0·06000	8·792	9·172	9·378	9·782	10·17	10·37	10·93	11·48	11·65	12·00	12·34	12·51	13·01	13·33
1/ 16·7	2·7406	3·2820	3·6091	4·3216	5·1144	5·5420	6·9540	8·5676	9·1520	10·393	11·731	12·438	14·714	16·364
0·06250	8·983	9·371	9·582	9·995	10·40	10·59	11·17	11·72	11·91	12·26	12·61	12·78	13·29	13·62
1/ 16·0	2·8004	3·3535	3·6877	4·4156	5·2256	5·6624	7·1049	8·7533	9·3502	10·618	11·985	12·707	15·032	16·717
0·06500	9·172	9·567	9·783	10·20	10·61	10·81	11·40	11·97	12·15	12·52	12·87	13·05	13·57	13·90
1/ 15·4	2·8590	3·4237	3·7648	4·5078	5·3346	5·7805	7·2528	8·9353	9·5447	10·838	12·234	12·971	15·344	17·064
0·06750	9·356	9·760	9·979	10·41	10·83	11·03	11·63	12·21	12·40	12·77	13·13	13·31	13·84	14·18
1/ 14·8	2·9165	3·4925	3·8404	4·5983	5·4416	5·8964	7·3981	9·1141	9·7355	11·055	12·478	13·230	15·650	17·404
0·07000	9·538	9·949	10·17	10·61	11·03	11·24	11·85	12·44	12·63	13·01	13·38	13·57	14·10	14·45
1/ 14·3	2·9731	3·5601	3·9147	4·6872	5·5467	6·0102	7·5407	9·2896	9·9229	11·267	12·718	13·484	15·951	17·738
0·07250	9·716	10·13	10·36	10·81	11·24	11·45	12·07	12·67	12·87	13·25	13·63	13·82	14·36	14·72
1/ 13·8	3·0286	3·6266	3·9878	4·7745	5·6500	6·1221	7·6809	9·4621	10·107	11·476	12·954	13·734	16·246	18·066
0·07500	9·891	10·32	10·55	11·00	11·44	11·66	12·29	12·90	13·10	13·49	13·88	14·06	14·62	14·98
1/ 13·3	3·0833	3·6920	4·0596	4·8605	5·7516	6·2321	7·8188	9·6317	10·288	11·682	13·186	13·980	16·536	18·389
0·08000	10·23	10·67	10·91	11·38	11·84	12·06	12·71	13·34	13·55	13·95	14·35	14·55	15·12	15·50
1/ 12·5	3·1900	3·8196	4·1999	5·0282	5·9499	6·4469	8·0879	9·9629	10·642	12·083	13·639	14·460	17·103	19·019
0·08500	10·57	11·02	11·27	11·75	12·22	12·45	13·12	13·77	13·99	14·40	14·81	15·02	15·61	16·00
1/ 11·8	3·2935	3·9434	4·3360	5·1910	6·1424	6·6554	8·3491	10·284	10·985	12·473	14·078	14·925	17·653	19·630
0·09000	10·89	11·36	11·61	12·11	12·59	12·83	13·52	14·19	14·41	14·84	15·26	15·47	16·08	16·48
1/ 11·1	3·3942	4·0638	4·4683	5·3492	6·3294	6·8579	8·6028	10·596	11·318	12·851	14·504	15·377	18·188	20·224
0·09500	11·20	11·68	11·95	12·46	12·95	13·20	13·91	14·60	14·82	15·27	15·70	15·91	16·54	16·95
1/ 10·5	3·4921	4·1810	4·5970	5·5032	6·5114	7·0550	8·8498	10·900	11·643	13·219	14·919	15·818	18·708	20·802
0·10000	11·51	12·00	12·27	12·80	13·31	13·56	14·29	15·00	15·23	15·68	16·13	16·34	16·99	17·41
1/ 10·0	3·5876	4·2952	4·7225	5·6533	6·6888	7·2471	9·0905	11·196	11·959	13·578	15·324	16·246	19·215	21·365
	0·98	0·98	0·98	0·98	0·98	0·98	0·98	0·99	0·99	0·99	0·99	0·99	0·99	0·99

$V_{r(0{\cdot}5)medial}$ for half-full circular pipes.

A24

(p.1 of 6)

$k_s = 0{\cdot}015$ **mm**
S = 0·00010 to 0·00030

ie hydraulic gradient =
1 in 10000 to 1 in 3333

Water (or sewage) at 15°C;
full bore conditions.

velocities in ms^{-1}
discharges in m^3s^{-1}

Gradient **(Equivalent) Pipe diameters in mm**

Gradient	630	675	700	750	800	825	900	975	1000	1050	1100	1125	1200	1250
0·000100	0·272	0·285	0·292	0·305	0·319	0·325	0·344	0·363	0·369	0·381	0·392	0·398	0·415	0·426
1/ 10000	0·0848	0·1019	0·1122	0·1348	0·1601	0·1737	0·2189	0·2708	0·2896	0·3296	0·3729	0·3958	0·4696	0·5232
0·000105	0·279	0·292	0·300	0·314	0·327	0·334	0·354	0·373	0·379	0·391	0·403	0·409	0·426	0·438
1/ 9524	0·0871	0·1046	0·1153	0·1385	0·1645	0·1785	0·2249	0·2781	0·2975	0·3386	0·3830	0·4065	0·4823	0·5374
0·000110	0·287	0·300	0·307	0·322	0·336	0·343	0·363	0·382	0·389	0·401	0·413	0·420	0·437	0·449
1/ 9091	0·0893	0·1074	0·1183	0·1421	0·1687	0·1831	0·2307	0·2853	0·3052	0·3473	0·3929	0·4170	0·4947	0·5512
0·000115	0·294	0·307	0·315	0·330	0·344	0·351	0·372	0·392	0·398	0·411	0·424	0·430	0·448	0·460
1/ 8696	0·0916	0·1100	0·1212	0·1456	0·1729	0·1876	0·2364	0·2924	0·3127	0·3559	0·4026	0·4273	0·5069	0·5648
0·000120	0·301	0·315	0·322	0·337	0·352	0·359	0·380	0·401	0·408	0·421	0·434	0·440	0·459	0·471
1/ 8333	0·0937	0·1127	0·1241	0·1491	0·1770	0·1921	0·2420	0·2993	0·3201	0·3643	0·4121	0·4373	0·5189	0·5781
0·000125	0·308	0·322	0·330	0·345	0·360	0·367	0·389	0·410	0·417	0·430	0·443	0·450	0·469	0·482
1/ 8000	0·0959	0·1152	0·1269	0·1525	0·1810	0·1965	0·2475	0·3061	0·3273	0·3725	0·4214	0·4472	0·5306	0·5911
0·000130	0·314	0·329	0·337	0·353	0·368	0·376	0·398	0·419	0·426	0·440	0·453	0·460	0·479	0·492
1/ 7692	0·0980	0·1177	0·1297	0·1558	0·1850	0·2007	0·2529	0·3127	0·3344	0·3806	0·4305	0·4569	0·5421	0·6039
0·000135	0·321	0·336	0·344	0·360	0·376	0·383	0·406	0·428	0·435	0·449	0·462	0·469	0·489	0·502
1/ 7407	0·1000	0·1202	0·1324	0·1591	0·1889	0·2049	0·2582	0·3193	0·3414	0·3885	0·4395	0·4665	0·5534	0·6165
0·000140	0·327	0·343	0·351	0·367	0·383	0·391	0·414	0·436	0·443	0·458	0·472	0·479	0·499	0·512
1/ 7143	0·1021	0·1227	0·1351	0·1623	0·1927	0·2091	0·2634	0·3257	0·3483	0·3964	0·4483	0·4758	0·5645	0·6288
0·000145	0·334	0·349	0·358	0·375	0·391	0·399	0·422	0·445	0·452	0·467	0·481	0·488	0·509	0·522
1/ 6897	0·1041	0·1250	0·1377	0·1655	0·1964	0·2131	0·2685	0·3320	0·3550	0·4040	0·4570	0·4850	0·5754	0·6410
0·000150	0·340	0·356	0·365	0·382	0·398	0·406	0·430	0·453	0·461	0·475	0·490	0·497	0·518	0·532
1/ 6667	0·1060	0·1274	0·1403	0·1686	0·2001	0·2172	0·2736	0·3382	0·3617	0·4116	0·4656	0·4941	0·5861	0·6530
0·000160	0·353	0·369	0·378	0·395	0·412	0·421	0·445	0·469	0·477	0·492	0·507	0·515	0·537	0·551
1/ 6250	0·1099	0·1320	0·1454	0·1747	0·2073	0·2250	0·2834	0·3504	0·3747	0·4264	0·4823	0·5118	0·6071	0·6764
0·000170	0·364	0·381	0·391	0·409	0·426	0·435	0·461	0·485	0·493	0·509	0·525	0·532	0·555	0·570
1/ 5882	0·1136	0·1365	0·1504	0·1806	0·2144	0·2326	0·2930	0·3622	0·3873	0·4408	0·4985	0·5291	0·6276	0·6991
0·000180	0·376	0·394	0·403	0·422	0·440	0·449	0·475	0·501	0·509	0·525	0·541	0·549	0·572	0·588
1/ 5556	0·1173	0·1409	0·1552	0·1864	0·2212	0·2400	0·3023	0·3737	0·3996	0·4548	0·5143	0·5458	0·6474	0·7212
0·000190	0·388	0·406	0·415	0·435	0·453	0·463	0·489	0·516	0·524	0·541	0·557	0·566	0·590	0·605
1/ 5263	0·1208	0·1451	0·1598	0·1920	0·2279	0·2472	0·3114	0·3849	0·4116	0·4684	0·5297	0·5622	0·6668	0·7428
0·000200	0·399	0·417	0·427	0·447	0·466	0·476	0·503	0·530	0·539	0·556	0·573	0·582	0·606	0·622
1/ 5000	0·1243	0·1493	0·1644	0·1975	0·2344	0·2543	0·3203	0·3959	0·4233	0·4817	0·5448	0·5781	0·6857	0·7638
0·000210	0·409	0·428	0·439	0·459	0·479	0·489	0·517	0·545	0·554	0·571	0·589	0·597	0·623	0·639
1/ 4762	0·1276	0·1533	0·1689	0·2028	0·2407	0·2612	0·3289	0·4066	0·4348	0·4947	0·5595	0·5937	0·7042	0·7844
0·000220	0·420	0·440	0·450	0·471	0·491	0·501	0·530	0·559	0·568	0·586	0·604	0·613	0·639	0·656
1/ 4545	0·1310	0·1573	0·1733	0·2081	0·2469	0·2679	0·3374	0·4171	0·4460	0·5074	0·5738	0·6090	0·7222	0·8045
0·000230	0·430	0·450	0·461	0·483	0·503	0·514	0·543	0·572	0·582	0·600	0·619	0·628	0·654	0·672
1/ 4348	0·1342	0·1612	0·1775	0·2132	0·2530	0·2745	0·3457	0·4273	0·4569	0·5199	0·5879	0·6239	0·7399	0·8242
0·000240	0·441	0·461	0·472	0·494	0·515	0·526	0·556	0·586	0·595	0·614	0·633	0·642	0·670	0·687
1/ 4167	0·1374	0·1650	0·1817	0·2182	0·2590	0·2810	0·3538	0·4373	0·4676	0·5321	0·6017	0·6385	0·7573	0·8435
0·000250	0·451	0·472	0·483	0·505	0·527	0·538	0·569	0·599	0·609	0·628	0·647	0·657	0·685	0·703
1/ 4000	0·1405	0·1687	0·1858	0·2232	0·2648	0·2873	0·3618	0·4472	0·4782	0·5440	0·6152	0·6529	0·7743	0·8624
0·000260	0·461	0·482	0·493	0·516	0·538	0·549	0·581	0·612	0·622	0·642	0·661	0·671	0·699	0·718
1/ 3846	0·1436	0·1724	0·1899	0·2280	0·2706	0·2936	0·3697	0·4569	0·4885	0·5558	0·6285	0·6670	0·7910	0·8810
0·000270	0·470	0·492	0·504	0·527	0·550	0·561	0·593	0·625	0·635	0·655	0·675	0·685	0·714	0·733
1/ 3704	0·1466	0·1760	0·1939	0·2328	0·2762	0·2997	0·3774	0·4664	0·4987	0·5673	0·6416	0·6808	0·8073	0·8992
0·000280	0·480	0·502	0·514	0·538	0·561	0·572	0·605	0·637	0·648	0·668	0·689	0·699	0·728	0·747
1/ 3571	0·1495	0·1796	0·1978	0·2375	0·2818	0·3057	0·3849	0·4757	0·5087	0·5787	0·6544	0·6944	0·8235	0·9172
0·000290	0·489	0·512	0·524	0·548	0·571	0·583	0·617	0·649	0·660	0·681	0·702	0·712	0·742	0·762
1/ 3448	0·1524	0·1831	0·2016	0·2421	0·2872	0·3116	0·3924	0·4849	0·5185	0·5899	0·6670	0·7078	0·8393	0·9348
0·000300	0·498	0·521	0·534	0·558	0·582	0·594	0·628	0·662	0·672	0·694	0·715	0·725	0·756	0·776
1/ 3333	0·1553	0·1865	0·2054	0·2466	0·2926	0·3175	0·3997	0·4939	0·5281	0·6008	0·6794	0·7210	0·8549	0·9522
	0·89	0·90	0·90	0·91	0·91	0·91	0·92	0·92	0·93	0·93	0·93	0·93	0·94	0·94

$V_{r(0\cdot5)medial}$ **for half-full circular pipes.**

k_s = **0·015 mm**
S = 0·00030 to 0·00100

ie hydraulic gradient =
1 in 3333 to 1 in 1000

Water (or sewage) at 15°C;
full bore conditions.

velocities in ms^{-1}
discharges in m^3s^{-1}

Gradient	**(Equivalent) Pipe diameters in mm**													
	630	**675**	**700**	**750**	**800**	**825**	**900**	**975**	**1000**	**1050**	**1100**	**1125**	**1200**	**1250**
0·000300	0·498	0·521	0·534	0·558	0·582	0·594	0·628	0·662	0·672	0·694	0·715	0·725	0·756	0·776
1/ 3333	0·1553	0·1865	0·2054	0·2466	0·2926	0·3175	0·3997	0·4939	0·5281	0·6008	0·6794	0·7210	0·8549	0·9522
0·000320	0·516	0·540	0·553	0·578	0·603	0·615	0·651	0·685	0·697	0·719	0·740	0·751	0·783	0·804
1/ 3125	0·1609	0·1932	0·2128	0·2555	0·3031	0·3288	0·4140	0·5116	0·5470	0·6223	0·7037	0·7467	0·8854	0·9861
0·000340	0·534	0·558	0·572	0·598	0·623	0·636	0·673	0·708	0·720	0·743	0·765	0·776	0·809	0·830
1/ 2941	0·1663	0·1997	0·2200	0·2641	0·3133	0·3399	0·4279	0·5288	0·5654	0·6432	0·7273	0·7717	0·9150	1·0191
0·000360	0·551	0·576	0·590	0·617	0·643	0·656	0·694	0·731	0·743	0·766	0·789	0·801	0·835	0·857
1/ 2778	0·1716	0·2061	0·2269	0·2725	0·3233	0·3507	0·4415	0·5455	0·5832	0·6635	0·7502	0·7960	0·9438	1·0512
0·000380	0·567	0·593	0·607	0·635	0·662	0·676	0·715	0·752	0·765	0·789	0·813	0·825	0·859	0·882
1/ 2632	0·1768	0·2123	0·2338	0·2806	0·3329	0·3612	0·4547	0·5618	0·6006	0·6833	0·7725	0·8198	0·9719	1·0824
0·000400	0·583	0·610	0·625	0·653	0·681	0·695	0·735	0·774	0·786	0·811	0·836	0·848	0·884	0·907
1/ 2500	0·1818	0·2183	0·2404	0·2886	0·3424	0·3714	0·4675	0·5777	0·6176	0·7026	0·7944	0·8429	0·9993	1·1129
0·000420	0·599	0·627	0·642	0·671	0·700	0·714	0·755	0·795	0·808	0·833	0·858	0·871	0·907	0·931
1/ 2381	0·1867	0·2242	0·2469	0·2964	0·3516	0·3814	0·4801	0·5932	0·6342	0·7214	0·8157	0·8655	1·0261	1·1427
0·000440	0·615	0·643	0·658	0·688	0·717	0·732	0·774	0·815	0·828	0·854	0·880	0·893	0·930	0·955
1/ 2273	0·1916	0·2300	0·2533	0·3040	0·3606	0·3912	0·4924	0·6084	0·6505	0·7399	0·8365	0·8876	1·0523	1·1718
0·000460	0·630	0·659	0·674	0·705	0·735	0·750	0·793	0·835	0·848	0·875	0·902	0·915	0·953	0·978
1/ 2174	0·1963	0·2357	0·2595	0·3115	0·3695	0·4008	0·5045	0·6233	0·6663	0·7579	0·8569	0·9092	1·0779	1·2004
0·000480	0·644	0·674	0·690	0·722	0·752	0·767	0·812	0·854	0·868	0·896	0·923	0·936	0·975	1·001
1/ 2083	0·2009	0·2412	0·2656	0·3188	0·3782	0·4102	0·5163	0·6378	0·6819	0·7756	0·8769	0·9305	1·1030	1·2283
0·000500	0·659	0·689	0·706	0·738	0·769	0·785	0·830	0·873	0·888	0·916	0·943	0·957	0·997	1·023
1/ 2000	0·2054	0·2466	0·2716	0·3260	0·3867	0·4194	0·5279	0·6521	0·6972	0·7930	0·8965	0·9513	1·1277	1·2557
0·000525	0·677	0·708	0·725	0·758	0·790	0·806	0·852	0·897	0·911	0·940	0·969	0·983	1·024	1·051
1/ 1905	0·2110	0·2533	0·2789	0·3348	0·3971	0·4307	0·5420	0·6696	0·7159	0·8142	0·9205	0·9767	1·1578	1·2893
0·000550	0·694	0·726	0·743	0·777	0·810	0·826	0·874	0·920	0·935	0·964	0·993	1·008	1·050	1·077
1/ 1818	0·2164	0·2598	0·2861	0·3433	0·4072	0·4417	0·5559	0·6867	0·7342	0·8350	0·9440	1·0016	1·1873	1·3221
0·000575	0·711	0·744	0·762	0·796	0·830	0·847	0·895	0·942	0·958	0·988	1·018	1·032	1·075	1·104
1/ 1739	0·2217	0·2662	0·2931	0·3518	0·4172	0·4526	0·5695	0·7035	0·7521	0·8554	0·9670	1·0260	1·2162	1·3542
0·000600	0·728	0·761	0·779	0·815	0·849	0·866	0·916	0·964	0·980	1·011	1·041	1·056	1·100	1·129
1/ 1667	0·2269	0·2724	0·2999	0·3600	0·4270	0·4631	0·5828	0·7199	0·7696	0·8753	0·9895	1·0499	1·2445	1·3857
0·000625	0·744	0·778	0·797	0·833	0·868	0·886	0·937	0·986	1·002	1·033	1·064	1·080	1·125	1·154
1/ 1600	0·2321	0·2786	0·3067	0·3681	0·4366	0·4735	0·5959	0·7360	0·7868	0·8949	1·0116	1·0733	1·2722	1·4166
0·000650	0·761	0·795	0·814	0·851	0·887	0·905	0·957	1·007	1·023	1·056	1·087	1·103	1·149	1·179
1/ 1538	0·2371	0·2846	0·3133	0·3760	0·4460	0·4837	0·6087	0·7518	0·8037	0·9140	1·0333	1·0963	1·2994	1·4469
0·000675	0·776	0·812	0·831	0·869	0·906	0·924	0·977	1·028	1·044	1·077	1·110	1·126	1·173	1·203
1/ 1481	0·2420	0·2905	0·3198	0·3838	0·4552	0·4937	0·6212	0·7673	0·8203	0·9329	1·0546	1·1189	1·3262	1·4766
0·000700	0·792	0·828	0·848	0·886	0·924	0·942	0·996	1·048	1·065	1·099	1·132	1·148	1·196	1·227
1/ 1429	0·2468	0·2963	0·3262	0·3915	0·4643	0·5036	0·6336	0·7826	0·8366	0·9514	1·0755	1·1411	1·3525	1·5059
0·000725	0·807	0·844	0·864	0·903	0·941	0·960	1·015	1·068	1·086	1·120	1·153	1·170	1·219	1·251
1/ 1379	0·2516	0·3020	0·3325	0·3990	0·4732	0·5133	0·6458	0·7976	0·8526	0·9696	1·0961	1·1629	1·3783	1·5346
0·000750	0·822	0·860	0·880	0·920	0·959	0·978	1·034	1·088	1·106	1·140	1·175	1·192	1·241	1·274
1/ 1333	0·2563	0·3076	0·3387	0·4064	0·4820	0·5228	0·6577	0·8123	0·8684	0·9876	1·1163	1·1844	1·4037	1·5629
0·000800	0·852	0·890	0·911	0·953	0·993	1·013	1·071	1·127	1·145	1·181	1·216	1·234	1·285	1·319
1/ 1250	0·2655	0·3186	0·3508	0·4209	0·4992	0·5414	0·6811	0·8412	0·8992	1·0226	1·1559	1·2263	1·4534	1·6182
0·000850	0·880	0·920	0·942	0·985	1·026	1·047	1·106	1·164	1·183	1·220	1·257	1·275	1·328	1·362
1/ 1176	0·2744	0·3293	0·3625	0·4350	0·5158	0·5595	0·7038	0·8692	0·9292	1·0566	1·1943	1·2671	1·5017	1·6719
0·000900	0·908	0·949	0·972	1·016	1·059	1·080	1·141	1·201	1·220	1·258	1·296	1·315	1·369	1·405
1/ 1111	0·2830	0·3397	0·3740	0·4487	0·5321	0·5771	0·7259	0·8965	0·9583	1·0897	1·2317	1·3067	1·5486	1·7241
0·000950	0·935	0·978	1·001	1·046	1·090	1·112	1·175	1·236	1·256	1·296	1·334	1·353	1·410	1·446
1/ 1053	0·2915	0·3498	0·3851	0·4621	0·5479	0·5942	0·7475	0·9230	0·9867	1·1219	1·2681	1·3454	1·5943	1·7750
0·001000	0·962	1·005	1·029	1·075	1·121	1·143	1·208	1·271	1·292	1·332	1·372	1·391	1·449	1·487
1/ 1000	0·2997	0·3597	0·3960	0·4751	0·5633	0·6109	0·7685	0·9489	1·0143	1·1534	1·3036	1·3830	1·6389	1·8246
	0·92	0·92	0·93	0·93	0·93	0·94	0·94	0·94	0·95	0·95	0·95	0·95	0·95	0·96

$V_{r(0·5)medial}$ for half-full circular pipes.

A24
(p.3 of 6)

$k_s = 0{\cdot}015$ mm
S = 0·00100 to 0·00300

ie hydraulic gradient =
1 in 1000 to 1 in 333

Water (or sewage) at 15°C;
full bore conditions.

velocities in ms^{-1}
discharges in m^3s^{-1}

Gradient **(Equivalent) Pipe diameters in mm**

Gradient	630	675	700	750	800	825	900	975	1000	1050	1100	1125	1200	1250
0·00100	0·962	1·005	1·029	1·075	1·121	1·143	1·208	1·271	1·292	1·332	1·372	1·391	1·449	1·487
1/ 1000	0·2997	0·3597	0·3960	0·4751	0·5633	0·6109	0·7685	0·9489	1·0143	1·1534	1·3036	1·3830	1·6389	1·8246
0·00105	0·987	1·032	1·056	1·104	1·151	1·173	1·240	1·305	1·326	1·367	1·408	1·428	1·488	1·526
1/ 952	0·3078	0·3693	0·4066	0·4878	0·5784	0·6273	0·7890	0·9742	1·0414	1·1841	1·3383	1·4198	1·6825	1·8731
0·00110	1·013	1·058	1·083	1·132	1·180	1·203	1·272	1·338	1·360	1·402	1·444	1·465	1·525	1·565
1/ 909	0·3157	0·3788	0·4170	0·5003	0·5931	0·6432	0·8091	0·9990	1·0678	1·2142	1·3723	1·4558	1·7251	1·9205
0·00115	1·037	1·084	1·110	1·160	1·209	1·233	1·303	1·370	1·393	1·436	1·479	1·500	1·562	1·603
1/ 870	0·3234	0·3880	0·4271	0·5124	0·6075	0·6589	0·8287	1·0232	1·0937	1·2436	1·4055	1·4910	1·7668	1·9669
0·00120	1·062	1·110	1·136	1·187	1·237	1·261	1·333	1·402	1·425	1·469	1·513	1·535	1·598	1·640
1/ 833	0·3309	0·3971	0·4371	0·5244	0·6217	0·6742	0·8480	1·0469	1·1191	1·2724	1·4380	1·5255	1·8076	2·0123
0·00125	1·085	1·134	1·161	1·213	1·264	1·289	1·363	1·433	1·456	1·502	1·547	1·569	1·634	1·676
1/ 800	0·3383	0·4060	0·4469	0·5361	0·6355	0·6892	0·8668	1·0702	1·1439	1·3006	1·4699	1·5594	1·8477	2·0569
0·00130	1·109	1·159	1·186	1·239	1·291	1·317	1·392	1·464	1·488	1·534	1·580	1·602	1·668	1·712
1/ 769	0·3456	0·4147	0·4564	0·5476	0·6491	0·7040	0·8854	1·0930	1·1683	1·3283	1·5012	1·5926	1·8870	2·1006
0·00135	1·132	1·183	1·211	1·265	1·318	1·344	1·420	1·494	1·518	1·565	1·612	1·635	1·703	1·747
1/ 741	0·3527	0·4232	0·4659	0·5588	0·6625	0·7185	0·9035	1·1155	1·1923	1·3556	1·5320	1·6252	1·9256	2·1435
0·00140	1·154	1·206	1·235	1·290	1·344	1·371	1·448	1·524	1·548	1·596	1·644	1·667	1·736	1·781
1/ 714	0·3598	0·4316	0·4751	0·5699	0·6756	0·7327	0·9214	1·1375	1·2158	1·3823	1·5622	1·6572	1·9635	2·1857
0·00145	1·176	1·229	1·258	1·315	1·370	1·397	1·476	1·553	1·578	1·627	1·675	1·699	1·769	1·815
1/ 690	0·3667	0·4399	0·4842	0·5808	0·6886	0·7467	0·9390	1·1592	1·2390	1·4086	1·5919	1·6887	2·0008	2·2272
0·00150	1·198	1·252	1·281	1·339	1·395	1·423	1·503	1·581	1·607	1·657	1·706	1·730	1·802	1·848
1/ 667	0·3735	0·4481	0·4932	0·5916	0·7013	0·7605	0·9563	1·1805	1·2618	1·4345	1·6211	1·7197	2·0375	2·2680
0·00160	1·241	1·297	1·327	1·387	1·444	1·473	1·556	1·637	1·663	1·715	1·766	1·791	1·865	1·913
1/ 625	0·3867	0·4640	0·5107	0·6125	0·7261	0·7874	0·9901	1·2221	1·3063	1·4851	1·6782	1·7803	2·1091	2·3477
0·00170	1·282	1·340	1·371	1·433	1·492	1·522	1·608	1·691	1·718	1·772	1·824	1·850	1·926	1·976
1/ 588	0·3996	0·4794	0·5277	0·6329	0·7502	0·8135	1·0229	1·2626	1·3495	1·5342	1·7336	1·8390	2·1787	2·4251
0·00180	1·322	1·382	1·414	1·477	1·539	1·569	1·658	1·744	1·772	1·827	1·881	1·908	1·986	2·037
1/ 556	0·4122	0·4944	0·5442	0·6527	0·7736	0·8389	1·0548	1·3019	1·3915	1·5819	1·7875	1·8962	2·2463	2·5004
0·00190	1·361	1·423	1·456	1·521	1·585	1·616	1·707	1·795	1·824	1·881	1·936	1·964	2·044	2·097
1/ 526	0·4244	0·5091	0·5603	0·6720	0·7965	0·8637	1·0859	1·3402	1·4325	1·6284	1·8400	1·9519	2·3122	2·5737
0·00200	1·400	1·463	1·497	1·564	1·629	1·661	1·755	1·845	1·875	1·933	1·990	2·018	2·101	2·155
1/ 500	0·4363	0·5234	0·5760	0·6908	0·8188	0·8878	1·1162	1·3776	1·4724	1·6738	1·8912	2·0062	2·3765	2·6452
0·00210	1·437	1·502	1·537	1·605	1·672	1·705	1·801	1·894	1·924	1·984	2·043	2·072	2·157	2·212
1/ 476	0·4480	0·5373	0·5914	0·7092	0·8405	0·9114	1·1458	1·4141	1·5114	1·7180	1·9413	2·0592	2·4393	2·7150
0·00220	1·474	1·540	1·576	1·646	1·715	1·748	1·847	1·942	1·973	2·034	2·094	2·124	2·211	2·268
1/ 455	0·4593	0·5510	0·6064	0·7272	0·8618	0·9345	1·1748	1·4498	1·5495	1·7613	1·9901	2·1111	2·5006	2·7832
0·00230	1·509	1·577	1·614	1·686	1·756	1·790	1·891	1·989	2·020	2·083	2·145	2·175	2·264	2·322
1/ 435	0·4705	0·5643	0·6210	0·7448	0·8826	0·9571	1·2031	1·4847	1·5868	1·8037	2·0380	2·1618	2·5606	2·8500
0·00240	1·544	1·613	1·651	1·725	1·797	1·832	1·935	2·034	2·067	2·131	2·194	2·225	2·316	2·376
1/ 417	0·4814	0·5774	0·6354	0·7620	0·9030	0·9792	1·2309	1·5189	1·6234	1·8452	2·0849	2·2115	2·6194	2·9154
0·00250	1·579	1·649	1·688	1·763	1·836	1·872	1·978	2·079	2·113	2·178	2·242	2·274	2·367	2·428
1/ 400	0·4921	0·5902	0·6495	0·7789	0·9230	1·0008	1·2581	1·5525	1·6592	1·8859	2·1308	2·2602	2·6771	2·9795
0·00260	1·612	1·685	1·724	1·801	1·875	1·912	2·020	2·123	2·157	2·224	2·290	2·322	2·417	2·479
1/ 385	0·5026	0·6028	0·6634	0·7955	0·9427	1·0221	1·2848	1·5854	1·6943	1·9259	2·1759	2·3080	2·7337	3·0424
0·00270	1·645	1·719	1·759	1·837	1·914	1·951	2·061	2·167	2·201	2·269	2·336	2·369	2·466	2·530
1/ 370	0·5129	0·6152	0·6770	0·8118	0·9619	1·0430	1·3110	1·6177	1·7289	1·9651	2·2201	2·3549	2·7892	3·1042
0·00280	1·678	1·753	1·794	1·874	1·951	1·990	2·101	2·209	2·244	2·314	2·382	2·416	2·514	2·579
1/ 357	0·5231	0·6273	0·6903	0·8278	0·9809	1·0635	1·3367	1·6494	1·7628	2·0036	2·2636	2·4011	2·8437	3·1649
0·00290	1·710	1·786	1·828	1·909	1·988	2·027	2·141	2·251	2·287	2·358	2·427	2·461	2·562	2·628
1/ 345	0·5330	0·6392	0·7035	0·8435	0·9995	1·0837	1·3621	1·6806	1·7961	2·0414	2·3064	2·4464	2·8974	3·2245
0·00300	1·741	1·819	1·861	1·944	2·025	2·064	2·180	2·292	2·329	2·401	2·471	2·506	2·608	2·675
1/ 333	0·5429	0·6510	0·7164	0·8590	1·0178	1·1036	1·3870	1·7113	1·8289	2·0787	2·3484	2·4910	2·9501	3·2832
	0·94	0·94	0·94	0·95	0·95	0·95	0·96	0·96	0·96	0·96	0·96	0·96	0·97	0·97

$V_{r(0{\cdot}5)medial}$ for half-full circular pipes.

$k_s = 0{\cdot}015$ mm S = 0·00100 to 0·00300

$k_s = 0{\cdot}015$ **mm**
S = 0·00300 to 0·01000

ie hydraulic gradient =
1 in 333 to 1 in 100

Water (or sewage) at 15°C;
full bore conditions.

velocities in ms^{-1}
discharges in m^3s^{-1}

Gradient **(Equivalent) Pipe diameters in mm**

Gradient	630	675	700	750	800	825	900	975	1000	1050	1100	1125	1200	1250
0·00300	1·741	1·819	1·861	1·944	2·025	2·064	2·180	2·292	2·329	2·401	2·471	2·506	2·608	2·675
1/ 333	0·5429	0·6510	0·7164	0·8590	1·0178	1·1036	1·3870	1·7113	1·8289	2·0787	2·3484	2·4910	2·9501	3·2832
0·00320	1·803	1·883	1·927	2·013	2·096	2·137	2·257	2·372	2·410	2·485	2·558	2·594	2·700	2·769
1/ 313	0·5620	0·6739	0·7416	0·8892	1·0536	1·1423	1·4357	1·7713	1·8930	2·1515	2·4306	2·5781	3·0532	3·3978
0·00340	1·863	1·946	1·991	2·079	2·165	2·207	2·331	2·450	2·489	2·566	2·642	2·679	2·788	2·859
1/ 294	0·5806	0·6962	0·7661	0·9185	1·0883	1·1800	1·4829	1·8295	1·9552	2·2221	2·5103	2·6626	3·1532	3·5090
0·00360	1·921	2·006	2·053	2·144	2·232	2·276	2·403	2·526	2·566	2·646	2·723	2·761	2·874	2·948
1/ 278	0·5987	0·7179	0·7900	0·9471	1·1221	1·2166	1·5288	1·8861	2·0156	2·2907	2·5878	2·7448	3·2504	3·6172
0·00380	1·977	2·065	2·113	2·207	2·298	2·343	2·474	2·600	2·641	2·723	2·802	2·842	2·958	3·033
1/ 263	0·6163	0·7390	0·8132	0·9749	1·1550	1·2522	1·5736	1·9413	2·0745	2·3576	2·6633	2·8248	3·3451	3·7224
0·00400	2·032	2·123	2·172	2·268	2·362	2·408	2·542	2·672	2·714	2·798	2·880	2·920	3·039	3·117
1/ 250	0·6335	0·7596	0·8358	1·0020	1·1871	1·2870	1·6172	1·9950	2·1319	2·4228	2·7369	2·9029	3·4374	3·8251
0·00420	2·086	2·179	2·229	2·328	2·424	2·471	2·609	2·742	2·786	2·872	2·956	2·997	3·119	3·199
1/ 238	0·6503	0·7797	0·8579	1·0285	1·2184	1·3210	1·6598	2·0475	2·1880	2·4865	2·8088	2·9791	3·5275	3·9253
0·00440	2·139	2·234	2·285	2·387	2·485	2·533	2·674	2·811	2·856	2·943	3·029	3·072	3·197	3·278
1/ 227	0·6667	0·7994	0·8795	1·0543	1·2491	1·3541	1·7014	2·0988	2·2428	2·5487	2·8790	3·0535	3·6156	4·0233
0·00460	2·190	2·288	2·340	2·444	2·545	2·594	2·739	2·878	2·924	3·014	3·102	3·145	3·273	3·357
1/ 217	0·6828	0·8186	0·9007	1·0797	1·2790	1·3866	1·7422	2·1490	2·2964	2·6096	2·9477	3·1264	3·7018	4·1191
0·00480	2·241	2·340	2·394	2·500	2·603	2·653	2·801	2·944	2·991	3·083	3·173	3·217	3·348	3·433
1/ 208	0·6985	0·8374	0·9214	1·1045	1·3084	1·4184	1·7821	2·1981	2·3489	2·6692	3·0150	3·1977	3·7862	4·2130
0·00500	2·290	2·392	2·447	2·555	2·660	2·712	2·863	3·009	3·056	3·150	3·242	3·287	3·421	3·508
1/ 200	0·7139	0·8559	0·9417	1·1288	1·3372	1·4496	1·8212	2·2463	2·4004	2·7276	3·0810	3·2677	3·8689	4·3050
0·00525	2·351	2·455	2·512	2·622	2·730	2·783	2·938	3·088	3·136	3·233	3·327	3·373	3·510	3·600
1/ 190	0·7328	0·8785	0·9666	1·1585	1·3724	1·4877	1·8691	2·3053	2·4634	2·7991	3·1617	3·3532	3·9701	4·4175
0·00550	2·410	2·517	2·575	2·688	2·799	2·853	3·012	3·165	3·215	3·313	3·410	3·458	3·598	3·689
1/ 182	0·7513	0·9006	0·9909	1·1876	1·4068	1·5250	1·9158	2·3629	2·5249	2·8690	3·2405	3·4368	4·0690	4·5275
0·00575	2·468	2·577	2·636	2·753	2·866	2·921	3·083	3·240	3·291	3·392	3·491	3·540	3·683	3·777
1/ 174	0·7693	0·9222	1·0146	1·2161	1·4404	1·5615	1·9616	2·4193	2·5851	2·9374	3·3177	3·5187	4·1658	4·6351
0·00600	2·525	2·636	2·697	2·816	2·931	2·988	3·154	3·314	3·367	3·470	3·571	3·620	3·767	3·863
1/ 167	0·7870	0·9434	1·0379	1·2439	1·4734	1·5972	2·0064	2·4745	2·6441	3·0043	3·3933	3·5988	4·2605	4·7404
0·00625	2·580	2·694	2·756	2·878	2·996	3·053	3·223	3·387	3·440	3·545	3·649	3·699	3·849	3·947
1/ 160	0·8043	0·9641	1·0607	1·2713	1·5057	1·6323	2·0504	2·5286	2·7019	3·0699	3·4673	3·6773	4·3534	4·8437
0·00650	2·635	2·751	2·814	2·938	3·059	3·118	3·291	3·458	3·512	3·620	3·725	3·777	3·930	4·030
1/ 154	0·8213	0·9845	1·0831	1·2981	1·5375	1·6666	2·0934	2·5817	2·7586	3·1343	3·5400	3·7543	4·4445	4·9450
0·00675	2·688	2·807	2·872	2·998	3·121	3·181	3·357	3·528	3·583	3·693	3·800	3·853	4·009	4·111
1/ 148	0·8380	1·0045	1·1051	1·3244	1·5686	1·7004	2·1358	2·6338	2·8142	3·1975	3·6113	3·8299	4·5339	5·0444
0·00700	2·741	2·862	2·928	3·056	3·181	3·243	3·423	3·596	3·653	3·764	3·874	3·928	4·086	4·190
1/ 143	0·8545	1·0241	1·1267	1·3502	1·5992	1·7335	2·1773	2·6850	2·8689	3·2596	3·6813	3·9042	4·6217	5·1420
0·00725	2·793	2·916	2·983	3·114	3·241	3·304	3·487	3·664	3·721	3·835	3·946	4·001	4·163	4·268
1/ 138	0·8706	1·0434	1·1479	1·3756	1·6292	1·7661	2·2182	2·7353	2·9226	3·3206	3·7502	3·9771	4·7080	5·2380
0·00750	2·844	2·969	3·037	3·170	3·300	3·364	3·550	3·730	3·789	3·904	4·017	4·073	4·238	4·345
1/ 133	0·8864	1·0624	1·1688	1·4006	1·6588	1·7981	2·2583	2·7847	2·9755	3·3806	3·8179	4·0489	4·7929	5·3324
0·00800	2·943	3·073	3·143	3·281	3·415	3·481	3·673	3·859	3·920	4·039	4·157	4·214	4·384	4·495
1/ 125	0·9174	1·0995	1·2096	1·4494	1·7165	1·8607	2·3368	2·8814	3·0787	3·4977	3·9501	4·1891	4·9587	5·5167
0·00850	3·040	3·173	3·246	3·388	3·526	3·594	3·793	3·985	4·047	4·171	4·292	4·351	4·527	4·641
1/ 118	0·9475	1·1355	1·2491	1·4968	1·7726	1·9214	2·4129	2·9751	3·1788	3·6114	4·0784	4·3251	5·1195	5·6955
0·00900	3·133	3·271	3·346	3·492	3·635	3·705	3·909	4·107	4·171	4·298	4·423	4·484	4·665	4·783
1/ 111	0·9767	1·1705	1·2876	1·5428	1·8270	1·9804	2·4869	3·0663	3·2762	3·7219	4·2031	4·4573	5·2758	5·8693
0·00950	3·225	3·366	3·443	3·594	3·740	3·812	4·022	4·226	4·292	4·422	4·550	4·614	4·799	4·921
1/ 105	1·0052	1·2046	1·3251	1·5877	1·8801	2·0378	2·5590	3·1550	3·3709	3·8294	4·3245	4·5860	5·4279	6·0384
0·01000	3·314	3·459	3·538	3·693	3·843	3·917	4·133	4·342	4·410	4·544	4·675	4·740	4·930	5·055
1/ 100	1·0329	1·2378	1·3616	1·6314	1·9318	2·0938	2·6292	3·2415	3·4633	3·9343	4·4428	4·7114	5·5762	6·2033
	0·95	0·96	0·96	0·96	0·96	0·96	0·97	0·97	0·97	0·97	0·97	0·97	0·97	0·98

$V_{r(0\cdot5)medial}$ **for half-full circular pipes.**

k_s = 0·015 mm
S = 0·01000 to 0·03000

ie hydraulic gradient =
1 in 100 to 1 in 33·3

Water (or sewage) at 15°C;
full bore conditions.

velocities in ms^{-1}
discharges in m^3s^{-1}

Gradient **(Equivalent) Pipe diameters in mm**

Gradient	630	675	700	750	800	825	900	975	1000	1050	1100	1125	1200	1250
0·01000	3·314	3·459	3·538	3·693	3·843	3·917	4·133	4·342	4·410	4·544	4·675	4·740	4·930	5·055
1/ 100	1·0329	1·2378	1·3616	1·6314	1·9318	2·0938	2·6292	3·2415	3·4633	3·9343	4·4428	4·7114	5·5762	6·2033
0·01050	3·401	3·550	3·631	3·789	3·944	4·019	4·241	4·455	4·524	4·662	4·796	4·863	5·058	5·186
1/ 95	1·0600	1·2702	1·3972	1·6740	1·9822	2·1485	2·6978	3·3259	3·5534	4·0366	4·5582	4·8338	5·7210	6·3642
0·01100	3·485	3·638	3·721	3·884	4·042	4·119	4·346	4·565	4·637	4·777	4·915	4·983	5·184	5·314
1/ 91	1·0865	1·3019	1·4321	1·7157	2·0316	2·2019	2·7648	3·4084	3·6415	4·1366	4·6711	4·9534	5·8624	6·5214
0·01150	3·569	3·725	3·810	3·976	4·138	4·217	4·449	4·673	4·746	4·890	5·031	5·101	5·306	5·439
1/ 87	1·1124	1·3329	1·4662	1·7565	2·0798	2·2542	2·8303	3·4890	3·7277	4·2344	4·7814	5·0704	6·0007	6·6752
0·01200	3·650	3·810	3·896	4·066	4·232	4·313	4·550	4·779	4·854	5·001	5·145	5·216	5·426	5·562
1/ 83	1·1378	1·3633	1·4995	1·7965	2·1270	2·3054	2·8945	3·5680	3·8120	4·3302	4·8895	5·1850	6·1362	6·8258
0·01250	3·730	3·893	3·981	4·155	4·324	4·406	4·649	4·883	4·959	5·109	5·256	5·329	5·543	5·682
1/ 80	1·1626	1·3930	1·5322	1·8356	2·1733	2·3555	2·9574	3·6455	3·8947	4·4241	4·9954	5·2972	6·2689	6·9733
0·01300	3·808	3·974	4·065	4·242	4·414	4·498	4·746	4·984	5·062	5·215	5·366	5·440	5·658	5·800
1/ 77	1·1870	1·4222	1·5643	1·8740	2·2188	2·4047	3·0190	3·7214	3·9758	4·5161	5·0993	5·4073	6·3991	7·1180
0·01350	3·885	4·054	4·147	4·327	4·503	4·589	4·841	5·084	5·164	5·320	5·473	5·549	5·771	5·916
1/ 74	1·2110	1·4509	1·5958	1·9117	2·2633	2·4530	3·0796	3·7959	4·0554	4·6064	5·2012	5·5154	6·5268	7·2600
0·01400	3·960	4·133	4·227	4·411	4·590	4·678	4·934	5·182	5·263	5·422	5·578	5·655	5·882	6·030
1/ 71	1·2345	1·4790	1·6268	1·9487	2·3071	2·5004	3·1390	3·8691	4·1336	4·6952	5·3013	5·6215	6·6523	7·3995
0·01450	4·034	4·210	4·306	4·493	4·675	4·765	5·026	5·279	5·361	5·523	5·682	5·760	5·991	6·141
1/ 69	1·2576	1·5067	1·6572	1·9851	2·3501	2·5470	3·1975	3·9410	4·2104	4·7823	5·3997	5·7258	6·7755	7·5366
0·01500	4·107	4·286	4·384	4·574	4·760	4·850	5·116	5·373	5·457	5·622	5·784	5·863	6·098	6·251
1/ 67	1·2803	1·5339	1·6871	2·0209	2·3924	2·5929	3·2549	4·0118	4·2859	4·8681	5·4964	5·8283	6·8968	7·6713
0·01600	4·250	4·435	4·535	4·732	4·924	5·018	5·293	5·558	5·645	5·815	5·982	6·065	6·307	6·466
1/ 63	1·3247	1·5870	1·7455	2·0907	2·4751	2·6824	3·3671	4·1499	4·4334	5·0355	5·6853	6·0285	7·1335	7·9344
0·01700	4·388	4·579	4·683	4·886	5·083	5·180	5·464	5·738	5·827	6·003	6·175	6·260	6·510	6·674
1/ 59	1·3678	1·6385	1·8021	2·1585	2·5552	2·7692	3·4760	4·2839	4·5765	5·1979	5·8685	6·2227	7·3630	8·1896
0·01800	4·522	4·719	4·826	5·035	5·238	5·338	5·630	5·912	6·004	6·185	6·363	6·450	6·708	6·876
1/ 56	1·4097	1·6886	1·8572	2·2244	2·6331	2·8536	3·5817	4·4141	4·7155	5·3556	6·0465	6·4114	7·5861	8·4376
0·01900	4·653	4·855	4·965	5·180	5·389	5·492	5·792	6·082	6·176	6·362	6·545	6·635	6·899	7·072
1/ 53	1·4504	1·7374	1·9107	2·2885	2·7089	2·9357	3·6847	4·5408	4·8508	5·5092	6·2197	6·5950	7·8032	8·6789
0·02000	4·780	4·988	5·101	5·321	5·536	5·642	5·950	6·247	6·344	6·535	6·722	6·815	7·087	7·264
1/ 50	1·4901	1·7849	1·9630	2·3510	2·7828	3·0157	3·7850	4·6642	4·9827	5·6588	6·3886	6·7740	8·0147	8·9140
0·02100	4·905	5·117	5·233	5·460	5·680	5·788	6·104	6·409	6·508	6·704	6·896	6·990	7·269	7·451
1/ 47·6	1·5289	1·8313	2·0140	2·4120	2·8549	3·0939	3·8829	4·7847	5·1114	5·8049	6·5533	6·9486	8·2211	9·1435
0·02200	5·026	5·244	5·363	5·594	5·820	5·930	6·254	6·566	6·668	6·869	7·065	7·162	7·447	7·633
1/ 45·5	1·5668	1·8766	2·0638	2·4715	2·9254	3·1702	3·9786	4·9025	5·2371	5·9476	6·7143	7·1192	8·4228	9·3676
0·02300	5·145	5·368	5·489	5·726	5·957	6·070	6·401	6·720	6·825	7·030	7·231	7·330	7·622	7·812
1/ 43·5	1·6038	1·9209	2·1125	2·5298	2·9943	3·2448	4·0721	5·0176	5·3601	6·0871	6·8717	7·2861	8·6200	9·5868
0·02400	5·261	5·489	5·613	5·856	6·091	6·207	6·545	6·871	6·978	7·188	7·393	7·494	7·792	7·987
1/ 41·7	1·6401	1·9643	2·1602	2·5869	3·0618	3·3179	4·1637	5·1303	5·4804	6·2237	7·0258	7·4494	8·8131	9·8014
0·02500	5·375	5·608	5·735	5·982	6·223	6·341	6·686	7·019	7·128	7·342	7·552	7·655	7·960	8·158
1/ 40·0	1·6756	2·0068	2·2069	2·6428	3·1279	3·3895	4·2534	5·2408	5·5984	6·3575	7·1768	7·6095	9·0022	10·012
0·02600	5·487	5·725	5·854	6·106	6·352	6·472	6·824	7·164	7·275	7·494	7·708	7·813	8·124	8·326
1/ 38·5	1·7105	2·0486	2·2528	2·6976	3·1927	3·4597	4·3415	5·3491	5·7140	6·4888	7·3248	7·7664	9·1877	10·218
0·02700	5·597	5·839	5·971	6·228	6·478	6·601	6·960	7·307	7·420	7·642	7·861	7·968	8·285	8·491
1/ 37·0	1·7447	2·0895	2·2978	2·7515	3·2564	3·5287	4·4278	5·4554	5·8275	6·6176	7·4701	7·9204	9·3697	10·420
0·02800	5·705	5·951	6·085	6·348	6·603	6·728	7·093	7·447	7·562	7·788	8·011	8·120	8·443	8·653
1/ 35·7	1·7783	2·1297	2·3419	2·8043	3·3189	3·5963	4·5126	5·5597	5·9390	6·7440	7·6128	8·0716	9·5485	10·619
0·02900	5·811	6·062	6·198	6·465	6·725	6·852	7·224	7·584	7·701	7·932	8·158	8·270	8·598	8·812
1/ 34·5	1·8113	2·1692	2·3854	2·8562	3·3803	3·6628	4·5960	5·6623	6·0485	6·8683	7·7530	8·2202	9·7241	10·814
0·03000	5·915	6·170	6·309	6·581	6·845	6·974	7·353	7·719	7·838	8·073	8·303	8·417	8·751	8·968
1/ 33·3	1·8438	2·2080	2·4281	2·9073	3·4406	3·7282	4·6779	5·7631	6·1562	6·9905	7·8908	8·3663	9·8967	11·006
	0·96	0·97	0·97	0·97	0·97	0·97	0·97	0·98	0·98	0·98	0·98	0·98	0·98	0·98

$V_{r(0\cdot5)medial}$ for half-full circular pipes.

k_s = 0·015 mm S = 0·01000 to 0·03000

Gradient	**(Equivalent) Pipe diameters in mm**													
	630	**675**	**700**	**750**	**800**	**825**	**900**	**975**	**1000**	**1050**	**1100**	**1125**	**1200**	**1250**
0·03000	5·915	6·170	6·309	6·581	6·845	6·974	7·353	7·719	7·838	8·073	8·303	8·417	8·751	8·968
1/ 33·3	1·8438	2·2080	2·4281	2·9073	3·4406	3·7282	4·6779	5·7631	6·1562	6·9905	7·8908	8·3663	9·8967	11·006
0·03200	6·118	6·382	6·526	6·806	7·079	7·213	7·605	7·983	8·106	8·349	8·586	8·704	9·049	9·274
1/ 31·3	1·9072	2·2839	2·5114	3·0070	3·5585	3·8559	4·8379	5·9600	6·3663	7·2290	8·1599	8·6515	10·234	11·380
0·03400	6·316	6·588	6·736	7·025	7·307	7·445	7·849	8·238	8·365	8·616	8·861	8·982	9·338	9·570
1/ 29·4	1·9687	2·3574	2·5923	3·1037	3·6728	3·9797	4·9930	6·1509	6·5702	7·4603	8·4208	8·9280	10·561	11·744
0·03600	6·507	6·788	6·940	7·238	7·528	7·670	8·085	8·487	8·618	8·875	9·128	9·252	9·618	9·857
1/ 27·8	2·0285	2·4289	2·6708	3·1976	3·7839	4·1000	5·1437	6·3364	6·7682	7·6850	8·6743	9·1967	10·878	12·097
0·03800	6·694	6·982	7·139	7·445	7·743	7·889	8·316	8·728	8·863	9·128	9·387	9·515	9·892	10·14
1/ 26·3	2·0866	2·4985	2·7472	3·2891	3·8920	4·2171	5·2904	6·5168	6·9609	7·9037	8·9209	9·4581	11·187	12·440
0·04000	6·876	7·171	7·332	7·647	7·952	8·102	8·541	8·964	9·102	9·374	9·640	9·771	10·16	10·41
1/ 25·0	2·1433	2·5663	2·8217	3·3782	3·9973	4·3311	5·4333	6·6927	7·1487	8·1167	9·1612	9·7128	11·488	12·774
0·04200	7·053	7·356	7·521	7·843	8·157	8·310	8·760	9·194	9·335	9·614	9·887	10·02	10·42	10·68
1/ 23·8	2·1986	2·6324	2·8944	3·4651	4·1000	4·4424	5·5728	6·8642	7·3319	8·3246	9·3957	9·9613	11·782	13·101
0·04400	7·226	7·537	7·706	8·036	8·357	8·514	8·974	9·418	9·563	9·848	10·13	10·27	10·67	10·94
1/ 22·7	2·2526	2·6970	2·9654	3·5500	4·2004	4·5511	5·7090	7·0318	7·5108	8·5276	9·6247	10·204	12·069	13·420
0·04600	7·396	7·713	7·886	8·224	8·552	8·713	9·183	9·638	9·786	10·08	10·36	10·50	10·92	11·19
1/ 21·7	2·3055	2·7602	3·0349	3·6331	4·2986	4·6575	5·8422	7·1957	7·6858	8·7262	9·8487	10·441	12·349	13·731
0·04800	7·562	7·886	8·063	8·408	8·743	8·907	9·388	9·853	10·00	10·30	10·59	10·74	11·16	11·44
1/ 20·8	2·3572	2·8221	3·1029	3·7144	4·3947	4·7615	5·9726	7·3562	7·8571	8·9205	10·068	10·674	12·624	14·037
0·05000	7·724	8·056	8·236	8·588	8·930	9·098	9·589	10·06	10·22	10·52	10·82	10·97	11·40	11·68
1/ 20·0	2·4079	2·8827	3·1695	3·7940	4·4889	4·8635	6·1003	7·5133	8·0249	9·1109	10·283	10·901	12·893	14·335
0·05250	7·923	8·263	8·447	8·808	9·159	9·331	9·835	10·32	10·48	10·79	11·10	11·25	11·69	11·98
1/ 19·0	2·4698	2·9568	3·2509	3·8914	4·6040	4·9882	6·2565	7·7055	8·2301	9·3437	10·545	11·180	13·221	14·701
0·05500	8·117	8·465	8·654	9·024	9·383	9·559	10·07	10·57	10·73	11·05	11·37	11·52	11·97	12·27
1/ 18·2	2·5304	3·0292	3·3305	3·9865	4·7164	5·1100	6·4091	7·8932	8·4305	9·5711	10·802	11·451	13·543	15·058
0·05750	8·307	8·663	8·856	9·234	9·602	9·782	10·31	10·82	10·98	11·31	11·63	11·79	12·25	12·55
1/ 17·4	2·5896	3·1000	3·4083	4·0796	4·8264	5·2291	6·5583	8·0767	8·6265	9·7934	11·052	11·717	13·857	15·407
0·06000	8·493	8·857	9·054	9·440	9·816	10·00	10·54	11·06	11·23	11·56	11·89	12·05	12·52	12·83
1/ 16·7	2·6475	3·1693	3·4844	4·1706	4·9340	5·3456	6·7043	8·2563	8·8183	10·011	11·298	11·977	14·164	15·748
0·06250	8·675	9·046	9·248	9·642	10·03	10·21	10·76	11·29	11·47	11·81	12·14	12·31	12·79	13·11
1/ 16·0	2·7043	3·2372	3·5590	4·2598	5·0395	5·4598	6·8474	8·4323	9·0061	10·224	11·538	12·232	14·465	16·083
0·06500	8·854	9·232	9·438	9·840	10·23	10·42	10·98	11·53	11·70	12·05	12·39	12·56	13·05	13·37
1/ 15·4	2·7600	3·3038	3·6322	4·3473	5·1429	5·5718	6·9876	8·6048	9·1904	10·433	11·774	12·482	14·760	16·411
0·06750	9·029	9·415	9·625	10·03	10·43	10·63	11·20	11·75	11·93	12·29	12·63	12·80	13·31	13·64
1/ 14·8	2·8146	3·3691	3·7039	4·4331	5·2443	5·6816	7·1252	8·7741	9·3711	10·638	12·005	12·727	15·050	16·733
0·07000	9·201	9·594	9·808	10·23	10·63	10·83	11·41	11·97	12·16	12·52	12·87	13·05	13·56	13·89
1/ 14·3	2·8682	3·4332	3·7744	4·5174	5·3439	5·7895	7·2603	8·9403	9·5486	10·840	12·232	12·968	15·334	17·049
0·07250	9·370	9·770	9·988	10·41	10·83	11·03	11·62	12·19	12·38	12·75	13·11	13·28	13·81	14·15
1/ 13·8	2·9209	3·4962	3·8437	4·6002	5·4418	5·8955	7·3931	9·1037	9·7229	11·037	12·455	13·204	15·614	17·359
0·07500	9·536	9·943	10·16	10·60	11·02	11·22	11·83	12·41	12·60	12·97	13·34	13·52	14·05	14·39
1/ 13·3	2·9727	3·5582	3·9118	4·6816	5·5380	5·9997	7·5236	9·2642	9·8944	11·232	12·675	13·437	15·888	17·665
0·08000	9·861	10·28	10·51	10·96	11·39	11·60	12·23	12·83	13·02	13·41	13·79	13·97	14·52	14·88
1/ 12·5	3·0739	3·6791	4·0447	4·8405	5·7258	6·2031	7·7783	9·5776	10·229	11·611	13·103	13·890	16·424	18·260
0·08500	10·18	10·61	10·84	11·31	11·75	11·97	12·62	13·23	13·44	13·83	14·22	14·42	14·98	15·35
1/ 11·8	3·1719	3·7964	4·1735	4·9946	5·9079	6·4002	8·0253	9·8814	10·553	11·979	13·518	14·330	16·944	18·838
0·09000	10·48	10·93	11·17	11·64	12·11	12·33	12·99	13·63	13·84	14·25	14·65	14·85	15·43	15·81
1/ 11·1	3·2672	3·9103	4·2987	5·1442	6·0847	6·5918	8·2652	10·176	10·868	12·337	13·921	14·757	17·449	19·399
0·09500	10·78	11·24	11·49	11·97	12·45	12·68	13·36	14·01	14·23	14·65	15·06	15·26	15·86	16·25
1/ 10·5	3·3599	4·0212	4·4205	5·2898	6·2568	6·7781	8·4986	10·464	11·175	12·685	14·313	15·173	17·940	19·944
0·10000	11·07	11·54	11·79	12·29	12·78	13·02	13·72	14·39	14·61	15·04	15·46	15·67	16·29	16·69
1/ 10·0	3·4502	4·1291	4·5391	5·4317	6·4245	6·9596	8·7260	10·743	11·473	13·023	14·695	15·578	18·418	20·476
	0·97	0·97	0·97	0·98	0·98	0·98	0·98	0·98	0·98	0·98	0·98	0·98	0·98	0·98

$V_{r(0{\cdot}5)medial}$ **for half-full circular pipes.**

A25

(p.1 of 6)

$k_s = 0{\cdot}030$ **mm**
S = 0·00010 to 0·00030

ie hydraulic gradient =
1 in 10000 to 1 in 3333

Water (or sewage) at 15°C;
full bore conditions.

velocities in ms^{-1}
discharges in m^3s^{-1}

Gradient **(Equivalent) Pipe diameters in mm**

Gradient	630	675	700	750	800	825	900	975	1000	1050	1100	1125	1200	1250
0·000100	0·270	0·283	0·290	0·304	0·317	0·323	0·342	0·361	0·367	0·378	0·390	0·396	0·413	0·424
1/ 10000	0·0843	0·1013	0·1116	0·1341	0·1592	0·1728	0·2177	0·2692	0·2879	0·3276	0·3706	0·3933	0·4666	0·5199
0·000105	0·278	0·291	0·298	0·312	0·325	0·332	0·351	0·370	0·376	0·389	0·400	0·406	0·424	0·435
1/ 9524	0·0866	0·1041	0·1147	0·1377	0·1635	0·1775	0·2236	0·2765	0·2957	0·3365	0·3806	0·4039	0·4792	0·5339
0·000110	0·285	0·298	0·306	0·320	0·334	0·341	0·360	0·380	0·386	0·399	0·411	0·417	0·435	0·446
1/ 9091	0·0889	0·1068	0·1176	0·1413	0·1677	0·1820	0·2293	0·2836	0·3033	0·3451	0·3904	0·4143	0·4915	0·5476
0·000115	0·292	0·306	0·313	0·328	0·342	0·349	0·369	0·389	0·396	0·408	0·421	0·427	0·445	0·457
1/ 8696	0·0911	0·1094	0·1205	0·1448	0·1719	0·1865	0·2350	0·2905	0·3107	0·3536	0·4000	0·4245	0·5035	0·5610
0·000120	0·299	0·313	0·321	0·336	0·350	0·357	0·378	0·398	0·405	0·418	0·431	0·437	0·456	0·468
1/ 8333	0·0932	0·1120	0·1234	0·1482	0·1759	0·1909	0·2405	0·2974	0·3180	0·3619	0·4093	0·4344	0·5153	0·5741
0·000125	0·306	0·320	0·328	0·343	0·358	0·365	0·387	0·407	0·414	0·427	0·440	0·447	0·466	0·478
1/ 8000	0·0953	0·1146	0·1262	0·1516	0·1799	0·1952	0·2459	0·3041	0·3252	0·3700	0·4185	0·4442	0·5269	0·5870
0·000130	0·313	0·327	0·335	0·351	0·366	0·373	0·395	0·416	0·423	0·437	0·450	0·457	0·476	0·489
1/ 7692	0·0974	0·1170	0·1289	0·1549	0·1838	0·1995	0·2513	0·3107	0·3322	0·3780	0·4276	0·4538	0·5383	0·5996
0·000135	0·319	0·334	0·342	0·358	0·373	0·381	0·403	0·425	0·432	0·446	0·459	0·466	0·486	0·499
1/ 7407	0·0995	0·1195	0·1316	0·1581	0·1877	0·2036	0·2565	0·3171	0·3391	0·3859	0·4365	0·4632	0·5494	0·6120
0·000140	0·326	0·341	0·349	0·365	0·381	0·389	0·411	0·433	0·440	0·455	0·468	0·475	0·495	0·509
1/ 7143	0·1015	0·1219	0·1343	0·1613	0·1914	0·2077	0·2616	0·3235	0·3459	0·3936	0·4452	0·4725	0·5604	0·6242
0·000145	0·332	0·347	0·356	0·372	0·388	0·396	0·419	0·442	0·449	0·463	0·477	0·484	0·505	0·518
1/ 6897	0·1034	0·1243	0·1369	0·1644	0·1952	0·2118	0·2667	0·3297	0·3526	0·4012	0·4538	0·4816	0·5712	0·6362
0·000150	0·338	0·354	0·362	0·379	0·396	0·404	0·427	0·450	0·457	0·472	0·486	0·493	0·514	0·528
1/ 6667	0·1054	0·1266	0·1395	0·1675	0·1988	0·2157	0·2717	0·3359	0·3592	0·4087	0·4622	0·4905	0·5818	0·6481
0·000160	0·350	0·367	0·375	0·393	0·410	0·418	0·442	0·466	0·474	0·489	0·504	0·511	0·533	0·547
1/ 6250	0·1092	0·1312	0·1445	0·1735	0·2059	0·2235	0·2814	0·3479	0·3720	0·4233	0·4787	0·5080	0·6025	0·6711
0·000170	0·362	0·379	0·388	0·406	0·424	0·432	0·457	0·482	0·490	0·505	0·521	0·528	0·551	0·565
1/ 5882	0·1129	0·1356	0·1494	0·1794	0·2129	0·2310	0·2909	0·3596	0·3845	0·4375	0·4947	0·5250	0·6227	0·6936
0·000180	0·374	0·391	0·400	0·419	0·437	0·446	0·472	0·497	0·505	0·521	0·537	0·545	0·568	0·583
1/ 5556	0·1165	0·1399	0·1541	0·1851	0·2196	0·2383	0·3001	0·3709	0·3966	0·4513	0·5103	0·5416	0·6423	0·7154
0·000190	0·385	0·403	0·412	0·431	0·450	0·459	0·486	0·512	0·520	0·537	0·553	0·561	0·585	0·600
1/ 5263	0·1200	0·1441	0·1587	0·1906	0·2262	0·2454	0·3091	0·3820	0·4085	0·4647	0·5255	0·5577	0·6614	0·7366
0·000200	0·396	0·414	0·424	0·444	0·463	0·472	0·500	0·526	0·535	0·552	0·569	0·577	0·601	0·617
1/ 5000	0·1234	0·1482	0·1632	0·1960	0·2326	0·2524	0·3178	0·3928	0·4200	0·4778	0·5404	0·5734	0·6800	0·7574
0·000210	0·407	0·425	0·436	0·456	0·475	0·485	0·513	0·540	0·549	0·567	0·584	0·592	0·617	0·634
1/ 4762	0·1267	0·1522	0·1677	0·2013	0·2389	0·2592	0·3264	0·4033	0·4313	0·4907	0·5548	0·5888	0·6982	0·7776
0·000220	0·417	0·436	0·447	0·467	0·487	0·497	0·526	0·554	0·563	0·581	0·599	0·607	0·633	0·650
1/ 4545	0·1300	0·1562	0·1720	0·2065	0·2450	0·2658	0·3347	0·4137	0·4423	0·5032	0·5690	0·6038	0·7160	0·7974
0·000230	0·427	0·447	0·458	0·479	0·499	0·509	0·539	0·568	0·577	0·595	0·613	0·622	0·648	0·666
1/ 4348	0·1332	0·1600	0·1762	0·2116	0·2510	0·2724	0·3429	0·4238	0·4531	0·5155	0·5829	0·6185	0·7334	0·8168
0·000240	0·437	0·458	0·469	0·490	0·511	0·521	0·552	0·581	0·590	0·609	0·628	0·637	0·664	0·681
1/ 4167	0·1364	0·1638	0·1803	0·2165	0·2569	0·2787	0·3509	0·4337	0·4637	0·5275	0·5965	0·6329	0·7505	0·8358
0·000250	0·447	0·468	0·479	0·501	0·523	0·533	0·564	0·594	0·604	0·623	0·642	0·651	0·678	0·696
1/ 4000	0·1394	0·1675	0·1844	0·2214	0·2627	0·2850	0·3588	0·4434	0·4741	0·5393	0·6098	0·6471	0·7672	0·8545
0·000260	0·457	0·478	0·490	0·512	0·534	0·545	0·576	0·607	0·617	0·636	0·655	0·665	0·693	0·711
1/ 3846	0·1425	0·1711	0·1884	0·2262	0·2684	0·2911	0·3665	0·4529	0·4842	0·5509	0·6229	0·6609	0·7837	0·8728
0·000270	0·467	0·488	0·500	0·523	0·545	0·556	0·588	0·619	0·629	0·649	0·669	0·679	0·707	0·726
1/ 3704	0·1454	0·1746	0·1923	0·2309	0·2739	0·2972	0·3741	0·4623	0·4943	0·5622	0·6357	0·6746	0·7998	0·8907
0·000280	0·476	0·498	0·510	0·533	0·556	0·567	0·600	0·631	0·642	0·662	0·682	0·692	0·721	0·740
1/ 3571	0·1483	0·1781	0·1962	0·2355	0·2794	0·3031	0·3816	0·4715	0·5041	0·5734	0·6483	0·6880	0·8157	0·9084
0·000290	0·485	0·507	0·520	0·543	0·567	0·578	0·611	0·644	0·654	0·675	0·695	0·705	0·735	0·754
1/ 3448	0·1512	0·1816	0·1999	0·2400	0·2848	0·3089	0·3889	0·4805	0·5137	0·5844	0·6608	0·7011	0·8313	0·9257
0·000300	0·494	0·517	0·529	0·553	0·577	0·589	0·623	0·656	0·666	0·687	0·708	0·718	0·749	0·768
1/ 3333	0·1540	0·1850	0·2037	0·2445	0·2901	0·3147	0·3961	0·4894	0·5233	0·5952	0·6730	0·7141	0·8466	0·9428
	0·89	0·90	0·90	0·90	0·91	0·91	0·92	0·92	0·92	0·93	0·93	0·93	0·93	0·94

$V_{r(0{\cdot}5)medial}$ **for half-full circular pipes.**

$k_s = 0{\cdot}030$ **mm** S = 0·00010 to 0·00030

$k_s = 0{\cdot}030$ mm
S = 0·00030 to 0·00100

ie hydraulic gradient =
1 in 3333 to 1 in 1000

Water (or sewage) at 15°C;
full bore conditions.

velocities in ms^{-1}
discharges in m^3s^{-1}

Gradient	(Equivalent) Pipe diameters in mm													
	630	675	700	750	800	825	900	975	1000	1050	1100	1125	1200	1250
0·000300	0·494	0·517	0·529	0·553	0·577	0·589	0·623	0·656	0·666	0·687	0·708	0·718	0·749	0·768
1/ 3333	0·1540	0·1850	0·2037	0·2445	0·2901	0·3147	0·3961	0·4894	0·5233	0·5952	0·6730	0·7141	0·8466	0·9428
0·000320	0·512	0·535	0·548	0·573	0·598	0·610	0·645	0·679	0·690	0·712	0·733	0·744	0·775	0·795
1/ 3125	0·1596	0·1916	0·2110	0·2532	0·3004	0·3259	0·4102	0·5068	0·5419	0·6163	0·6968	0·7394	0·8766	0·9762
0·000340	0·529	0·553	0·567	0·592	0·618	0·630	0·666	0·701	0·713	0·735	0·758	0·769	0·801	0·822
1/ 2941	0·1649	0·1980	0·2180	0·2617	0·3105	0·3368	0·4239	0·5237	0·5599	0·6369	0·7200	0·7640	0·9057	1·0086
0·000360	0·546	0·571	0·584	0·611	0·637	0·650	0·687	0·723	0·735	0·759	0·781	0·793	0·826	0·848
1/ 2778	0·1701	0·2043	0·2249	0·2700	0·3202	0·3474	0·4372	0·5401	0·5775	0·6568	0·7426	0·7879	0·9340	1·0401
0·000380	0·562	0·588	0·602	0·629	0·656	0·669	0·708	0·745	0·757	0·781	0·804	0·816	0·850	0·873
1/ 2632	0·1752	0·2104	0·2316	0·2780	0·3298	0·3577	0·4502	0·5561	0·5946	0·6763	0·7645	0·8112	0·9616	1·0708
0·000400	0·578	0·604	0·619	0·647	0·675	0·688	0·728	0·766	0·778	0·803	0·827	0·839	0·874	0·897
1/ 2500	0·1802	0·2163	0·2382	0·2859	0·3391	0·3678	0·4629	0·5718	0·6113	0·6952	0·7859	0·8339	0·9885	1·1007
0·000420	0·594	0·621	0·635	0·664	0·693	0·706	0·747	0·786	0·799	0·824	0·849	0·861	0·897	0·921
1/ 2381	0·1850	0·2221	0·2446	0·2935	0·3481	0·3776	0·4752	0·5870	0·6276	0·7137	0·8069	0·8561	1·0148	1·1299
0·000440	0·609	0·637	0·652	0·681	0·710	0·724	0·766	0·806	0·819	0·845	0·871	0·883	0·920	0·944
1/ 2273	0·1898	0·2278	0·2508	0·3010	0·3570	0·3873	0·4873	0·6019	0·6435	0·7319	0·8273	0·8778	1·0405	1·1585
0·000460	0·624	0·652	0·668	0·698	0·728	0·742	0·785	0·826	0·839	0·866	0·892	0·904	0·942	0·967
1/ 2174	0·1944	0·2334	0·2569	0·3084	0·3657	0·3967	0·4991	0·6165	0·6591	0·7496	0·8473	0·8990	1·0656	1·1865
0·000480	0·638	0·667	0·683	0·714	0·745	0·759	0·803	0·845	0·859	0·886	0·912	0·925	0·964	0·989
1/ 2083	0·1989	0·2388	0·2629	0·3155	0·3742	0·4059	0·5107	0·6308	0·6744	0·7669	0·8670	0·9198	1·0902	1·2139
0·000500	0·653	0·682	0·698	0·730	0·761	0·776	0·821	0·864	0·878	0·905	0·933	0·946	0·985	1·011
1/ 2000	0·2034	0·2442	0·2688	0·3226	0·3826	0·4150	0·5221	0·6449	0·6894	0·7840	0·8862	0·9402	1·1144	1·2408
0·000525	0·670	0·701	0·717	0·750	0·781	0·797	0·843	0·887	0·901	0·929	0·957	0·971	1·011	1·038
1/ 1905	0·2089	0·2507	0·2760	0·3312	0·3928	0·4260	0·5360	0·6620	0·7077	0·8048	0·9097	0·9652	1·1439	1·2737
0·000550	0·687	0·718	0·735	0·769	0·801	0·817	0·864	0·909	0·924	0·953	0·981	0·996	1·037	1·064
1/ 1818	0·2142	0·2571	0·2830	0·3397	0·4028	0·4369	0·5496	0·6788	0·7256	0·8252	0·9327	0·9896	1·1728	1·3058
0·000575	0·704	0·736	0·753	0·788	0·821	0·837	0·885	0·931	0·946	0·976	1·005	1·020	1·062	1·090
1/ 1739	0·2194	0·2634	0·2899	0·3479	0·4126	0·4475	0·5630	0·6952	0·7432	0·8451	0·9553	1·0135	1·2011	1·3373
0·000600	0·720	0·753	0·771	0·806	0·840	0·857	0·905	0·953	0·968	0·999	1·028	1·043	1·087	1·115
1/ 1667	0·2245	0·2695	0·2967	0·3560	0·4222	0·4579	0·5760	0·7113	0·7604	0·8647	0·9773	1·0369	1·2288	1·3681
0·000625	0·736	0·770	0·788	0·824	0·859	0·876	0·926	0·974	0·990	1·021	1·051	1·066	1·111	1·139
1/ 1600	0·2296	0·2755	0·3033	0·3639	0·4315	0·4681	0·5888	0·7271	0·7772	0·8838	0·9990	1·0598	1·2560	1·3983
0·000650	0·752	0·786	0·805	0·841	0·877	0·894	0·945	0·995	1·011	1·042	1·074	1·089	1·134	1·164
1/ 1538	0·2345	0·2814	0·3098	0·3717	0·4408	0·4781	0·6014	0·7426	0·7938	0·9026	1·0202	1·0823	1·2826	1·4279
0·000675	0·768	0·803	0·822	0·859	0·895	0·913	0·965	1·015	1·031	1·064	1·095	1·111	1·157	1·187
1/ 1481	0·2393	0·2872	0·3162	0·3794	0·4498	0·4879	0·6137	0·7578	0·8100	0·9211	1·0410	1·1044	1·3088	1·4570
0·000700	0·783	0·819	0·838	0·876	0·913	0·931	0·984	1·035	1·052	1·085	1·117	1·133	1·180	1·211
1/ 1429	0·2441	0·2929	0·3224	0·3869	0·4587	0·4975	0·6258	0·7727	0·8260	0·9392	1·0615	1·1262	1·3345	1·4856
0·000725	0·798	0·834	0·854	0·892	0·930	0·948	1·002	1·055	1·072	1·105	1·138	1·154	1·202	1·234
1/ 1379	0·2488	0·2985	0·3286	0·3943	0·4675	0·5070	0·6377	0·7874	0·8417	0·9570	1·0816	1·1475	1·3597	1·5137
0·000750	0·813	0·850	0·870	0·909	0·947	0·966	1·021	1·074	1·091	1·125	1·159	1·176	1·224	1·256
1/ 1333	0·2534	0·3040	0·3347	0·4015	0·4761	0·5163	0·6494	0·8018	0·8571	0·9746	1·1014	1·1685	1·3846	1·5414
0·000800	0·842	0·880	0·900	0·941	0·981	1·000	1·057	1·112	1·130	1·165	1·200	1·217	1·267	1·300
1/ 1250	0·2623	0·3148	0·3465	0·4157	0·4929	0·5346	0·6723	0·8301	0·8873	1·0088	1·1401	1·2095	1·4332	1·5954
0·000850	0·870	0·909	0·930	0·972	1·013	1·033	1·092	1·148	1·167	1·203	1·239	1·257	1·309	1·343
1/ 1176	0·2711	0·3253	0·3580	0·4295	0·5092	0·5522	0·6945	0·8575	0·9165	1·0421	1·1777	1·2493	1·4803	1·6479
0·000900	0·897	0·937	0·959	1·003	1·045	1·065	1·126	1·184	1·203	1·241	1·278	1·296	1·349	1·384
1/ 1111	0·2796	0·3354	0·3692	0·4429	0·5251	0·5695	0·7161	0·8841	0·9450	1·0744	1·2142	1·2881	1·5261	1·6988
0·000950	0·923	0·965	0·988	1·032	1·075	1·097	1·159	1·219	1·239	1·277	1·315	1·334	1·389	1·425
1/ 1053	0·2878	0·3453	0·3801	0·4560	0·5406	0·5862	0·7372	0·9100	0·9727	1·1059	1·2497	1·3258	1·5707	1·7485
0·001000	0·949	0·992	1·015	1·061	1·105	1·127	1·191	1·253	1·273	1·313	1·352	1·371	1·427	1·464
1/ 1000	0·2959	0·3550	0·3908	0·4687	0·5556	0·6026	0·7577	0·9353	0·9998	1·1366	1·2844	1·3625	1·6142	1·7969
	0·91	0·92	0·92	0·93	0·93	0·93	0·94	0·94	0·94	0·94	0·95	0·95	0·95	0·95

$V_{r(0{\cdot}5)medial}$ for half-full circular pipes.

A25
(p.3 of 6)

k_s = 0·030 mm
S = 0·00100 to 0·00300

ie hydraulic gradient =
1 in 1000 to 1 in 333

Water (or sewage) at 15°C;
full bore conditions.

velocities in ms^{-1}
discharges in m^3s^{-1}

Gradient **(Equivalent) Pipe diameters in mm**

Gradient	630	675	700	750	800	825	900	975	1000	1050	1100	1125	1200	1250
0·00100	0·949	0·992	1·015	1·061	1·105	1·127	1·191	1·253	1·273	1·313	1·352	1·371	1·427	1·464
1/ 1000	0·2959	0·3550	0·3908	0·4687	0·5556	0·6026	0·7577	0·9353	0·9998	1·1366	1·2844	1·3625	1·6142	1·7969
0·00105	0·975	1·018	1·042	1·089	1·135	1·157	1·223	1·286	1·307	1·347	1·387	1·407	1·465	1·503
1/ 952	0·3038	0·3645	0·4011	0·4812	0·5704	0·6185	0·7778	0·9601	1·0261	1·1666	1·3183	1·3984	1·6567	1·8441
0·00110	0·999	1·044	1·069	1·117	1·163	1·186	1·253	1·318	1·339	1·381	1·422	1·442	1·502	1·540
1/ 909	0·3115	0·3737	0·4113	0·4933	0·5848	0·6341	0·7973	0·9842	1·0519	1·1959	1·3514	1·4335	1·6982	1·8903
0·00115	1·023	1·069	1·095	1·144	1·191	1·215	1·283	1·350	1·372	1·414	1·456	1·477	1·538	1·577
1/ 870	0·3190	0·3827	0·4212	0·5052	0·5989	0·6494	0·8165	1·0078	1·0772	1·2246	1·3837	1·4679	1·7389	1·9355
0·00120	1·047	1·094	1·120	1·170	1·219	1·243	1·313	1·381	1·403	1·447	1·489	1·511	1·573	1·613
1/ 833	0·3264	0·3916	0·4310	0·5169	0·6127	0·6644	0·8353	1·0310	1·1019	1·2526	1·4154	1·5015	1·7787	1·9798
0·00125	1·070	1·118	1·145	1·196	1·246	1·270	1·342	1·411	1·434	1·478	1·522	1·544	1·607	1·649
1/ 800	0·3337	0·4002	0·4405	0·5283	0·6262	0·6790	0·8537	1·0537	1·1262	1·2802	1·4465	1·5344	1·8177	2·0231
0·00130	1·093	1·142	1·169	1·221	1·272	1·297	1·370	1·441	1·464	1·510	1·554	1·576	1·641	1·683
1/ 769	0·3408	0·4087	0·4499	0·5395	0·6395	0·6934	0·8718	1·0759	1·1499	1·3072	1·4770	1·5668	1·8559	2·0657
0·00135	1·116	1·166	1·193	1·246	1·298	1·324	1·398	1·470	1·494	1·540	1·586	1·608	1·674	1·717
1/ 741	0·3477	0·4171	0·4591	0·5505	0·6525	0·7075	0·8895	1·0978	1·1733	1·3337	1·5070	1·5985	1·8935	2·1075
0·00140	1·137	1·189	1·216	1·271	1·324	1·350	1·426	1·499	1·523	1·570	1·617	1·640	1·707	1·751
1/ 714	0·3546	0·4253	0·4681	0·5614	0·6653	0·7214	0·9069	1·1192	1·1962	1·3597	1·5364	1·6297	1·9304	2·1485
0·00145	1·159	1·211	1·239	1·295	1·349	1·375	1·453	1·527	1·552	1·600	1·647	1·670	1·739	1·784
1/ 690	0·3613	0·4334	0·4770	0·5720	0·6779	0·7351	0·9240	1·1404	1·2188	1·3853	1·5653	1·6604	1·9667	2·1888
0·00150	1·180	1·233	1·262	1·318	1·373	1·400	1·479	1·555	1·580	1·629	1·677	1·701	1·770	1·816
1/ 667	0·3680	0·4413	0·4857	0·5825	0·6903	0·7485	0·9409	1·1611	1·2409	1·4105	1·5937	1·6905	2·0023	2·2285
0·00160	1·222	1·277	1·306	1·365	1·421	1·449	1·531	1·609	1·635	1·686	1·735	1·760	1·832	1·879
1/ 625	0·3809	0·4568	0·5028	0·6029	0·7145	0·7747	0·9738	1·2017	1·2842	1·4597	1·6492	1·7494	2·0720	2·3060
0·00170	1·262	1·319	1·349	1·410	1·468	1·497	1·581	1·662	1·689	1·741	1·792	1·817	1·892	1·940
1/ 588	0·3935	0·4719	0·5193	0·6227	0·7379	0·8001	1·0057	1·2410	1·3263	1·5075	1·7031	1·8065	2·1396	2·3812
0·00180	1·301	1·360	1·391	1·453	1·514	1·543	1·630	1·713	1·741	1·794	1·847	1·873	1·950	2·000
1/ 556	0·4057	0·4865	0·5354	0·6420	0·7608	0·8249	1·0368	1·2792	1·3671	1·5538	1·7555	1·8621	2·2053	2·4543
0·00190	1·340	1·399	1·432	1·496	1·558	1·588	1·677	1·763	1·791	1·847	1·901	1·928	2·006	2·058
1/ 526	0·4176	0·5008	0·5511	0·6608	0·7830	0·8489	1·0670	1·3165	1·4069	1·5990	1·8065	1·9161	2·2693	2·5254
0·00200	1·377	1·438	1·472	1·537	1·601	1·632	1·723	1·812	1·841	1·898	1·953	1·981	2·062	2·114
1/ 500	0·4292	0·5147	0·5664	0·6791	0·8047	0·8724	1·0964	1·3528	1·4457	1·6431	1·8562	1·9688	2·3316	2·5947
0·00210	1·413	1·476	1·511	1·578	1·643	1·675	1·769	1·859	1·889	1·947	2·004	2·032	2·115	2·170
1/ 476	0·4405	0·5283	0·5813	0·6970	0·8258	0·8954	1·1252	1·3882	1·4836	1·6861	1·9047	2·0203	2·3925	2·6624
0·00220	1·449	1·513	1·548	1·617	1·684	1·717	1·813	1·906	1·936	1·996	2·054	2·083	2·168	2·223
1/ 455	0·4516	0·5415	0·5959	0·7144	0·8465	0·9178	1·1533	1·4229	1·5205	1·7281	1·9522	2·0706	2·4520	2·7286
0·00230	1·484	1·550	1·586	1·656	1·724	1·758	1·856	1·951	1·982	2·043	2·103	2·133	2·219	2·276
1/ 435	0·4625	0·5545	0·6102	0·7315	0·8667	0·9397	1·1808	1·4567	1·5567	1·7692	1·9986	2·1198	2·5101	2·7933
0·00240	1·518	1·585	1·622	1·694	1·764	1·798	1·898	1·996	2·027	2·090	2·151	2·181	2·270	2·328
1/ 417	0·4731	0·5672	0·6242	0·7483	0·8865	0·9611	1·2078	1·4899	1·5922	1·8094	2·0440	2·1679	2·5671	2·8566
0·00250	1·551	1·620	1·657	1·731	1·802	1·837	1·940	2·039	2·071	2·135	2·198	2·228	2·319	2·378
1/ 400	0·4835	0·5797	0·6379	0·7647	0·9059	0·9822	1·2342	1·5224	1·6269	1·8489	2·0885	2·2151	2·6230	2·9187
0·00260	1·584	1·654	1·692	1·767	1·840	1·876	1·981	2·082	2·115	2·180	2·244	2·275	2·368	2·428
1/ 385	0·4937	0·5919	0·6513	0·7808	0·9250	1·0028	1·2600	1·5543	1·6610	1·8875	2·1322	2·2614	2·6777	2·9796
0·00270	1·616	1·688	1·727	1·803	1·877	1·914	2·021	2·124	2·157	2·224	2·289	2·321	2·415	2·477
1/ 370	0·5037	0·6039	0·6645	0·7966	0·9437	1·0231	1·2855	1·5856	1·6944	1·9255	2·1750	2·3069	2·7315	3·0394
0·00280	1·647	1·721	1·760	1·838	1·914	1·951	2·060	2·165	2·199	2·267	2·333	2·366	2·462	2·525
1/ 357	0·5136	0·6157	0·6775	0·8121	0·9621	1·0430	1·3104	1·6164	1·7273	1·9628	2·2171	2·3515	2·7843	3·0981
0·00290	1·679	1·753	1·793	1·873	1·950	1·988	2·098	2·205	2·240	2·309	2·377	2·410	2·508	2·572
1/ 345	0·5232	0·6273	0·6902	0·8273	0·9801	1·0625	1·3350	1·6466	1·7596	1·9995	2·2585	2·3954	2·8361	3·1558
0·00300	1·709	1·785	1·826	1·907	1·985	2·024	2·136	2·245	2·281	2·351	2·419	2·453	2·553	2·618
1/ 333	0·5328	0·6387	0·7027	0·8424	0·9979	1·0818	1·3591	1·6764	1·7913	2·0355	2·2992	2·4385	2·8872	3·2125
	0·93	0·94	0·94	0·94	0·95	0·95	0·95	0·95	0·95	0·96	0·96	0·96	0·96	0·96

$V_{r(0\cdot5)medial}$ for half-full circular pipes.

k_s = 0·030 mm S = 0·00100 to 0·00300

Gradient	**(Equivalent) Pipe diameters in mm** 630	675	700	750	800	825	900	975	1000	1050	1100	1125	1200	1250
0·00300	1·709	1·785	1·826	1·907	1·985	2·024	2·136	2·245	2·281	2·351	2·419	2·453	2·553	2·618
1/ 333	0·5328	0·6387	0·7027	0·8424	0·9979	1·0818	1·3591	1·6764	1·7913	2·0355	2·2992	2·4385	2·8872	3·2125
0·00320	1·769	1·847	1·890	1·973	2·054	2·094	2·210	2·323	2·360	2·432	2·503	2·538	2·641	2·708
1/ 313	0·5513	0·6610	0·7272	0·8717	1·0325	1·1193	1·4062	1·7344	1·8533	2·1059	2·3786	2·5227	2·9868	3·3233
0·00340	1·827	1·907	1·951	2·037	2·121	2·162	2·282	2·398	2·436	2·511	2·584	2·620	2·726	2·796
1/ 294	0·5694	0·6826	0·7510	0·9001	1·0662	1·1558	1·4519	1·7907	1·9135	2·1742	2·4557	2·6044	3·0834	3·4308
0·00360	1·883	1·966	2·011	2·100	2·186	2·228	2·352	2·472	2·511	2·588	2·663	2·700	2·809	2·881
1/ 278	0·5869	0·7036	0·7741	0·9277	1·0989	1·1912	1·4964	1·8454	1·9719	2·2406	2·5306	2·6838	3·1773	3·5352
0·00380	1·938	2·023	2·070	2·161	2·249	2·293	2·420	2·543	2·583	2·662	2·740	2·778	2·890	2·964
1/ 263	0·6040	0·7240	0·7965	0·9546	1·1307	1·2257	1·5396	1·8987	2·0288	2·3052	2·6035	2·7611	3·2687	3·6368
0·00400	1·991	2·079	2·127	2·220	2·311	2·356	2·486	2·613	2·654	2·735	2·814	2·854	2·969	3·044
1/ 250	0·6206	0·7439	0·8184	0·9808	1·1617	1·2593	1·5818	1·9506	2·0843	2·3681	2·6746	2·8365	3·3578	3·7358
0·00420	2·043	2·133	2·182	2·278	2·371	2·417	2·551	2·680	2·723	2·806	2·887	2·927	3·046	3·123
1/ 238	0·6369	0·7634	0·8398	1·0064	1·1920	1·2921	1·6229	2·0013	2·1384	2·4296	2·7439	2·9100	3·4447	3·8325
0·00440	2·094	2·186	2·237	2·335	2·430	2·477	2·614	2·747	2·790	2·875	2·959	3·000	3·121	3·200
1/ 227	0·6528	0·7824	0·8607	1·0315	1·2216	1·3242	1·6631	2·0508	2·1913	2·4896	2·8117	2·9818	3·5297	3·9269
0·00460	2·144	2·238	2·290	2·390	2·488	2·536	2·676	2·812	2·856	2·943	3·028	3·070	3·194	3·275
1/ 217	0·6683	0·8010	0·8812	1·0559	1·2505	1·3555	1·7025	2·0992	2·2430	2·5483	2·8779	3·0521	3·6128	4·0193
0·00480	2·193	2·289	2·342	2·444	2·544	2·593	2·737	2·875	2·920	3·009	3·097	3·140	3·266	3·349
1/ 208	0·6835	0·8192	0·9012	1·0799	1·2789	1·3862	1·7410	2·1467	2·2936	2·6058	2·9428	3·1208	3·6941	4·1097
0·00500	2·240	2·339	2·393	2·497	2·599	2·650	2·796	2·937	2·984	3·074	3·163	3·207	3·337	3·421
1/ 200	0·6984	0·8370	0·9208	1·1033	1·3066	1·4163	1·7787	2·1931	2·3433	2·6621	3·0064	3·1882	3·7738	4·1983
0·00525	2·299	2·400	2·455	2·562	2·667	2·718	2·868	3·014	3·061	3·154	3·245	3·290	3·423	3·509
1/ 190	0·7166	0·8588	0·9448	1·1321	1·3406	1·4531	1·8248	2·2499	2·4040	2·7310	3·0841	3·2706	3·8712	4·3067
0·00550	2·356	2·460	2·516	2·626	2·733	2·786	2·939	3·088	3·136	3·232	3·325	3·371	3·507	3·596
1/ 182	0·7345	0·8802	0·9682	1·1601	1·3738	1·4891	1·8699	2·3055	2·4632	2·7983	3·1601	3·3511	3·9664	4·4125
0·00575	2·412	2·518	2·575	2·688	2·798	2·851	3·009	3·161	3·210	3·308	3·403	3·451	3·589	3·680
1/ 174	0·7519	0·9010	0·9911	1·1876	1·4063	1·5242	1·9140	2·3598	2·5212	2·8642	3·2343	3·4299	4·0595	4·5160
0·00600	2·467	2·575	2·634	2·749	2·861	2·916	3·077	3·232	3·282	3·382	3·480	3·528	3·670	3·763
1/ 167	0·7689	0·9214	1·0136	1·2144	1·4380	1·5587	1·9572	2·4129	2·5780	2·9286	3·3070	3·5070	4·1507	4·6173
0·00625	2·520	2·631	2·691	2·808	2·923	2·979	3·143	3·302	3·353	3·455	3·555	3·604	3·749	3·843
1/ 160	0·7857	0·9414	1·0356	1·2408	1·4692	1·5924	1·9995	2·4650	2·6336	2·9917	3·3783	3·5825	4·2400	4·7166
0·00650	2·573	2·686	2·747	2·867	2·984	3·041	3·208	3·370	3·423	3·527	3·628	3·679	3·826	3·923
1/ 154	0·8021	0·9611	1·0572	1·2666	1·4997	1·6255	2·0410	2·5161	2·6882	3·0536	3·4482	3·6566	4·3275	4·8140
0·00675	2·625	2·740	2·802	2·924	3·043	3·102	3·272	3·437	3·491	3·597	3·701	3·752	3·902	4·001
1/ 148	0·8182	0·9803	1·0783	1·2919	1·5297	1·6580	2·0817	2·5662	2·7417	3·1144	3·5167	3·7292	4·4135	4·9095
0·00700	2·675	2·792	2·856	2·981	3·102	3·161	3·335	3·503	3·558	3·666	3·771	3·823	3·977	4·077
1/ 143	0·8340	0·9993	1·0991	1·3168	1·5591	1·6899	2·1217	2·6154	2·7943	3·1741	3·5840	3·8006	4·4978	5·0033
0·00725	2·725	2·844	2·909	3·036	3·159	3·220	3·397	3·568	3·624	3·733	3·841	3·894	4·050	4·152
1/ 138	0·8495	1·0178	1·1196	1·3413	1·5881	1·7212	2·1610	2·6638	2·8459	3·2327	3·6502	3·8707	4·5808	5·0955
0·00750	2·774	2·895	2·961	3·090	3·216	3·277	3·458	3·631	3·688	3·800	3·909	3·963	4·122	4·226
1/ 133	0·8648	1·0361	1·1397	1·3653	1·6165	1·7520	2·1996	2·7113	2·8967	3·2904	3·7153	3·9397	4·6623	5·1861
0·00800	2·870	2·995	3·063	3·197	3·326	3·390	3·576	3·756	3·814	3·930	4·043	4·099	4·263	4·370
1/ 125	0·8946	1·0718	1·1789	1·4123	1·6720	1·8122	2·2750	2·8042	2·9959	3·4029	3·8422	4·0743	4·8214	5·3630
0·00850	2·963	3·092	3·162	3·300	3·434	3·499	3·691	3·876	3·937	4·056	4·173	4·230	4·400	4·510
1/ 118	0·9235	1·1064	1·2170	1·4578	1·7259	1·8705	2·3482	2·8942	3·0920	3·5120	3·9654	4·2048	4·9757	5·5346
0·00900	3·053	3·186	3·258	3·400	3·538	3·605	3·803	3·994	4·056	4·178	4·299	4·358	4·532	4·646
1/ 111	0·9516	1·1401	1·2540	1·5020	1·7782	1·9272	2·4192	2·9817	3·1854	3·6181	4·0850	4·3316	5·1257	5·7012
0·00950	3·141	3·277	3·352	3·497	3·639	3·708	3·911	4·108	4·172	4·298	4·421	4·482	4·661	4·778
1/ 105	0·9790	1·1728	1·2899	1·5451	1·8291	1·9823	2·4884	3·0668	3·2763	3·7212	4·2014	4·4550	5·2715	5·8634
0·01000	3·226	3·367	3·443	3·592	3·738	3·809	4·017	4·219	4·284	4·414	4·540	4·603	4·787	4·907
1/ 100	1·0057	1·2047	1·3250	1·5871	1·8788	2·0361	2·5557	3·1498	3·3649	3·8218	4·3148	4·5753	5·4137	6·0214
	0·95	0·95	0·95	0·95	0·96	0·96	0·96	0·96	0·96	0·97	0·97	0·97	0·97	0·97

$V_{r(0 \cdot 5)medial}$ for half-full circular pipes.

k_s = 0·030 mm
S = 0·01000 to 0·03000

ie hydraulic gradient =
1 in 100 to 1 in 33·3

Water (or sewage) at 15°C;
full bore conditions.

velocities in ms^{-1}
discharges in m^3s^{-1}

Gradient	**(Equivalent) Pipe diameters in mm**													
	630	**675**	**700**	**750**	**800**	**825**	**900**	**975**	**1000**	**1050**	**1100**	**1125**	**1200**	**1250**
0·01000	3·226	3·367	3·443	3·592	3·738	3·809	4·017	4·219	4·284	4·414	4·540	4·603	4·787	4·907
1/ 100	1·0057	1·2047	1·3250	1·5871	1·8788	2·0361	2·5557	3·1498	3·3649	3·8218	4·3148	4·5753	5·4137	6·0214
0·01050	3·310	3·454	3·532	3·685	3·834	3·907	4·121	4·327	4·394	4·527	4·657	4·721	4·909	5·032
1/ 95	1·0317	1·2359	1·3592	1·6280	1·9272	2·0886	2·6215	3·2307	3·4514	3·9199	4·4255	4·6926	5·5524	6·1756
0·01100	3·391	3·539	3·619	3·776	3·928	4·003	4·222	4·433	4·502	4·638	4·771	4·836	5·029	5·155
1/ 91	1·0571	1·2663	1·3927	1·6680	1·9745	2·1398	2·6857	3·3098	3·5358	4·0157	4·5336	4·8072	5·6878	6·3262
0·01150	3·471	3·622	3·704	3·864	4·020	4·097	4·320	4·537	4·607	4·746	4·882	4·949	5·146	5·275
1/ 87	1·0820	1·2960	1·4254	1·7071	2·0208	2·1899	2·7485	3·3871	3·6183	4·1094	4·6393	4·9192	5·8203	6·4734
0·01200	3·549	3·703	3·787	3·951	4·110	4·188	4·417	4·638	4·710	4·852	4·991	5·059	5·261	5·392
1/ 83	1·1063	1·3252	1·4574	1·7454	2·0660	2·2390	2·8100	3·4627	3·6992	4·2011	4·7428	5·0289	5·9499	6·6175
0·01250	3·626	3·783	3·868	4·036	4·199	4·278	4·512	4·737	4·811	4·955	5·097	5·167	5·373	5·507
1/ 80	1·1302	1·3537	1·4887	1·7830	2·1104	2·2870	2·8703	3·5369	3·7783	4·2910	4·8442	5·1364	6·0770	6·7587
0·01300	3·701	3·861	3·948	4·119	4·285	4·366	4·605	4·835	4·910	5·057	5·202	5·273	5·483	5·620
1/ 77	1·1536	1·3817	1·5195	1·8197	2·1539	2·3341	2·9293	3·6096	3·8560	4·3791	4·9436	5·2417	6·2015	6·8971
0·01350	3·774	3·938	4·027	4·201	4·370	4·453	4·696	4·930	5·007	5·157	5·305	5·377	5·591	5·731
1/ 74	1·1765	1·4091	1·5497	1·8558	2·1966	2·3804	2·9873	3·6809	3·9321	4·4655	5·0411	5·3451	6·3237	7·0330
0·01400	3·846	4·013	4·104	4·281	4·453	4·538	4·785	5·024	5·102	5·255	5·405	5·479	5·697	5·840
1/ 71	1·1990	1·4361	1·5793	1·8913	2·2385	2·4258	3·0442	3·7509	4·0069	4·5504	5·1368	5·4466	6·4437	7·1663
0·01450	3·918	4·087	4·179	4·360	4·535	4·621	4·873	5·116	5·195	5·351	5·504	5·580	5·802	5·946
1/ 69	1·2212	1·4626	1·6084	1·9261	2·2797	2·4704	3·1001	3·8198	4·0804	4·6338	5·2309	5·5463	6·5615	7·2973
0·01500	3·987	4·160	4·254	4·437	4·616	4·703	4·959	5·207	5·287	5·446	5·602	5·678	5·904	6·051
1/ 67	1·2430	1·4886	1·6371	1·9604	2·3202	2·5143	3·1551	3·8874	4·1527	4·7158	5·3234	5·6444	6·6774	7·4262
0·01600	4·124	4·302	4·399	4·589	4·773	4·864	5·128	5·384	5·467	5·631	5·792	5·871	6·104	6·256
1/ 63	1·2855	1·5395	1·6929	2·0272	2·3992	2·5999	3·2624	4·0195	4·2937	4·8758	5·5040	5·8357	6·9036	7·6776
0·01700	4·256	4·440	4·540	4·735	4·926	5·019	5·292	5·555	5·641	5·810	5·976	6·057	6·298	6·455
1/ 59	1·3267	1·5888	1·7471	2·0920	2·4759	2·6829	3·3664	4·1475	4·4304	5·0309	5·6790	6·0213	7·1229	7·9213
0·01800	4·384	4·574	4·677	4·878	5·074	5·170	5·451	5·722	5·810	5·984	6·155	6·239	6·486	6·648
1/ 56	1·3667	1·6367	1·7998	2·1550	2·5503	2·7635	3·4675	4·2719	4·5632	5·1816	5·8490	6·2015	7·3359	8·1581
0·01900	4·509	4·704	4·810	5·017	5·218	5·316	5·605	5·884	5·975	6·153	6·329	6·415	6·670	6·835
1/ 53	1·4057	1·6833	1·8510	2·2163	2·6228	2·8420	3·5658	4·3929	4·6924	5·3283	6·0144	6·3768	7·5431	8·3884
0·02000	4·631	4·831	4·939	5·152	5·358	5·459	5·756	6·042	6·135	6·318	6·498	6·587	6·848	7·018
1/ 50	1·4436	1·7287	1·9009	2·2760	2·6934	2·9184	3·6616	4·5108	4·8183	5·4711	6·1756	6·5476	7·7450	8·6128
0·02100	4·750	4·955	5·066	5·284	5·495	5·599	5·903	6·196	6·291	6·479	6·664	6·755	7·022	7·197
1/ 47·6	1·4807	1·7730	1·9496	2·3342	2·7622	2·9930	3·7551	4·6258	4·9411	5·6105	6·3328	6·7142	7·9420	8·8317
0·02200	4·866	5·076	5·190	5·412	5·629	5·735	6·046	6·346	6·444	6·637	6·825	6·918	7·192	7·371
1/ 45·5	1·5169	1·8163	1·9972	2·3911	2·8295	3·0659	3·8464	4·7381	5·0611	5·7466	6·4863	6·8770	8·1343	9·0455
0·02300	4·980	5·194	5·310	5·538	5·760	5·869	6·186	6·493	6·593	6·790	6·983	7·078	7·359	7·541
1/ 43·5	1·5522	1·8586	2·0437	2·4468	2·8953	3·1371	3·9356	4·8480	5·1784	5·8797	6·6365	7·0361	8·3224	9·2546
0·02400	5·091	5·310	5·429	5·662	5·888	5·999	6·324	6·637	6·739	6·941	7·138	7·235	7·521	7·708
1/ 41·7	1·5869	1·9000	2·0892	2·5012	2·9596	3·2068	4·0230	4·9555	5·2932	6·0100	6·7834	7·1919	8·5065	9·4591
0·02500	5·199	5·423	5·545	5·782	6·013	6·127	6·458	6·778	6·883	7·088	7·289	7·389	7·681	7·871
1/ 40·0	1·6208	1·9406	2·1338	2·5545	3·0227	3·2751	4·1086	5·0608	5·4056	6·1376	6·9273	7·3444	8·6868	9·6595
0·02600	5·306	5·534	5·658	5·901	6·136	6·252	6·590	6·917	7·023	7·233	7·438	7·539	7·837	8·031
1/ 38·5	1·6541	1·9804	2·1775	2·6068	3·0845	3·3421	4·1925	5·1640	5·5158	6·2627	7·0685	7·4940	8·8635	9·8560
0·02700	5·411	5·643	5·770	6·017	6·257	6·375	6·720	7·052	7·161	7·374	7·584	7·687	7·990	8·188
1/ 37·0	1·6867	2·0195	2·2204	2·6582	3·1452	3·4078	4·2748	5·2653	5·6240	6·3854	7·2069	7·6407	9·0370	10·049
0·02800	5·514	5·750	5·879	6·131	6·376	6·496	6·847	7·185	7·296	7·513	7·727	7·832	8·141	8·343
1/ 35·7	1·7187	2·0578	2·2626	2·7085	3·2047	3·4723	4·3556	5·3648	5·7302	6·5059	7·3428	7·7848	9·2072	10·238
0·02900	5·615	5·856	5·987	6·243	6·492	6·614	6·971	7·316	7·429	7·650	7·867	7·974	8·289	8·494
1/ 34·5	1·7502	2·0954	2·3039	2·7580	3·2633	3·5357	4·4350	5·4625	5·8345	6·6243	7·4764	7·9263	9·3745	10·424
0·03000	5·714	5·959	6·092	6·353	6·606	6·731	7·094	7·445	7·559	7·785	8·005	8·114	8·434	8·643
1/ 33·3	1·7812	2·1325	2·3446	2·8067	3·3208	3·5980	4·5131	5·5585	5·9371	6·7406	7·6076	8·0655	9·5389	10·607
	0·96	0·96	0·96	0·96	0·96	0·96	0·97	0·97	0·97	0·97	0·97	0·97	0·97	0·97

$V_{r(0·5)medial}$ for half-full circular pipes.

k_s = 0·030 mm S = 0·01000 to 0·03000

Gradient	(Equivalent) Pipe diameters in mm													
	630	**675**	**700**	**750**	**800**	**825**	**900**	**975**	**1000**	**1050**	**1100**	**1125**	**1200**	**1250**
0·03000	5·714	5·959	6·092	6·353	6·606	6·731	7·094	7·445	7·559	7·785	8·005	8·114	8·434	8·643
1/ 33·3	1·7812	2·1325	2·3446	2·8067	3·3208	3·5980	4·5131	5·5585	5·9371	6·7406	7·6076	8·0655	9·5389	10·607
0·03200	5·908	6·161	6·299	6·568	6·830	6·958	7·334	7·696	7·814	8·047	8·275	8·387	8·718	8·934
1/ 31·3	1·8416	2·2047	2·4240	2·9017	3·4330	3·7196	4·6654	5·7460	6·1372	6·9677	7·8638	8·3370	9·8598	10·963
0·03400	6·096	6·357	6·499	6·776	7·046	7·179	7·566	7·939	8·061	8·301	8·536	8·652	8·993	9·215
1/ 29·4	1·9002	2·2748	2·5010	2·9937	3·5419	3·8374	4·8131	5·9277	6·3313	7·1879	8·1122	8·6002	10·171	11·309
0·03600	6·278	6·547	6·693	6·979	7·257	7·393	7·791	8·176	8·301	8·548	8·790	8·909	9·260	9·489
1/ 27·8	1·9571	2·3428	2·5758	3·0832	3·6476	3·9519	4·9566	6·1042	6·5197	7·4018	8·3534	8·8559	10·473	11·645
0·03800	6·456	6·732	6·882	7·176	7·461	7·601	8·011	8·406	8·535	8·788	9·037	9·159	9·520	9·755
1/ 26·3	2·0124	2·4090	2·6486	3·1702	3·7505	4·0633	5·0961	6·2759	6·7031	7·6098	8·5880	9·1046	10·767	11·971
0·04000	6·629	6·912	7·066	7·368	7·661	7·804	8·224	8·630	8·762	9·022	9·277	9·403	9·773	10·01
1/ 25·0	2·0664	2·4735	2·7194	3·2549	3·8506	4·1718	5·2321	6·4432	6·8817	7·8124	8·8166	9·3468	11·053	12·290
0·04200	6·798	7·088	7·246	7·555	7·855	8·002	8·433	8·848	8·984	9·251	9·512	9·641	10·02	10·27
1/ 23·8	2·1190	2·5365	2·7886	3·3376	3·9484	4·2777	5·3647	6·6063	7·0559	8·0101	9·0396	9·5832	11·333	12·600
0·04400	6·962	7·260	7·421	7·738	8·045	8·196	8·636	9·062	9·200	9·474	9·741	9·873	10·26	10·51
1/ 22·7	2·1703	2·5979	2·8561	3·4184	4·0438	4·3811	5·4942	6·7657	7·2260	8·2032	9·2573	9·8140	11·605	12·903
0·04600	7·124	7·428	7·593	7·916	8·231	8·385	8·835	9·270	9·412	9·692	9·965	10·10	10·50	10·76
1/ 21·7	2·2206	2·6580	2·9221	3·4973	4·1372	4·4821	5·6208	6·9215	7·3924	8·3919	9·4702	10·040	11·872	13·199
0·04800	7·281	7·592	7·761	8·091	8·412	8·570	9·030	9·475	9·620	9·905	10·18	10·32	10·73	10·99
1/ 20·8	2·2698	2·7168	2·9868	3·5746	4·2285	4·5811	5·7448	7·0740	7·5552	8·5766	9·6785	10·260	12·133	13·489
0·05000	7·436	7·753	7·925	8·263	8·590	8·751	9·221	9·675	9·823	10·11	10·40	10·54	10·95	11·22
1/ 20·0	2·3179	2·7744	3·0500	3·6503	4·3180	4·6779	5·8661	7·2233	7·7146	8·7575	9·8826	10·477	12·388	13·774
0·05250	7·625	7·950	8·126	8·472	8·808	8·973	9·454	9·919	10·07	10·37	10·66	10·81	11·23	11·51
1/ 19·0	2·3768	2·8448	3·1274	3·7428	4·4273	4·7964	6·0145	7·4058	7·9096	8·9787	10·132	10·741	12·701	14·121
0·05500	7·809	8·142	8·323	8·677	9·020	9·189	9·682	10·16	10·31	10·62	10·92	11·07	11·50	11·78
1/ 18·2	2·4343	2·9136	3·2030	3·8332	4·5342	4·9121	6·1594	7·5841	8·0999	9·1947	10·376	10·999	13·006	14·460
0·05750	7·989	8·330	8·515	8·877	9·228	9·401	9·905	10·39	10·55	10·86	11·17	11·32	11·76	12·05
1/ 17·4	2·4905	2·9808	3·2769	3·9215	4·6386	5·0252	6·3011	7·7584	8·2860	9·4058	10·614	11·252	13·304	14·791
0·06000	8·166	8·514	8·703	9·072	9·431	9·608	10·12	10·62	10·78	11·10	11·41	11·57	12·02	12·32
1/ 16·7	2·5455	3·0466	3·3492	4·0080	4·7408	5·1358	6·4398	7·9290	8·4681	9·6124	10·847	11·499	13·596	15·116
0·06250	8·339	8·694	8·887	9·264	9·631	9·810	10·34	10·84	11·01	11·33	11·65	11·81	12·27	12·58
1/ 16·0	2·5994	3·1111	3·4200	4·0927	4·8409	5·2442	6·5756	8·0961	8·6465	9·8148	11·075	11·741	13·882	15·433
0·06500	8·508	8·870	9·067	9·452	9·826	10·01	10·55	11·06	11·23	11·56	11·89	12·05	12·52	12·83
1/ 15·4	2·6523	3·1742	3·4894	4·1757	4·9390	5·3505	6·7087	8·2598	8·8214	10·013	11·299	11·978	14·162	15·745
0·06750	8·675	9·044	9·244	9·636	10·02	10·20	10·75	11·28	11·45	11·79	12·12	12·28	12·77	13·08
1/ 14·8	2·7041	3·2362	3·5575	4·2571	5·0353	5·4548	6·8393	8·4205	8·9929	10·208	11·518	12·210	14·437	16·051
0·07000	8·838	9·214	9·418	9·817	10·21	10·40	10·95	11·49	11·66	12·01	12·35	12·51	13·00	13·32
1/ 14·3	2·7550	3·2971	3·6244	4·3371	5·1298	5·5571	6·9675	8·5783	9·1614	10·399	11·734	12·439	14·707	16·350
0·07250	8·998	9·381	9·589	9·995	10·39	10·58	11·15	11·70	11·88	12·23	12·57	12·74	13·24	13·56
1/ 13·8	2·8050	3·3569	3·6901	4·4157	5·2227	5·6577	7·0935	8·7332	9·3268	10·587	11·946	12·663	14·972	16·645
0·07500	9·156	9·545	9·756	10·17	10·57	10·77	11·34	11·90	12·08	12·44	12·79	12·96	13·47	13·80
1/ 13·3	2·8542	3·4157	3·7547	4·4929	5·3139	5·7565	7·2173	8·8855	9·4895	10·771	12·154	12·884	15·233	16·935
0·08000	9·464	9·866	10·08	10·51	10·93	11·13	11·72	12·30	12·49	12·86	13·22	13·39	13·92	14·26
1/ 12·5	2·9501	3·5303	3·8807	4·6436	5·4920	5·9494	7·4589	9·1828	9·8068	11·131	12·560	13·314	15·742	17·500
0·08500	9·762	10·18	10·40	10·84	11·27	11·48	12·09	12·69	12·88	13·26	13·63	13·81	14·35	14·71
1/ 11·8	3·0430	3·6415	4·0029	4·7897	5·6647	6·1364	7·6931	9·4709	10·114	11·480	12·954	13·732	16·235	18·048
0·09000	10·05	10·48	10·71	11·16	11·60	11·82	12·45	13·06	13·26	13·65	14·03	14·22	14·78	15·14
1/ 11·1	3·1333	3·7495	4·1215	4·9315	5·8324	6·3180	7·9206	9·7507	10·413	11·819	13·336	14·137	16·713	18·580
0·09500	10·33	10·77	11·01	11·48	11·93	12·15	12·80	13·42	13·63	14·03	14·42	14·62	15·19	15·56
1/ 10·5	3·2212	3·8545	4·2369	5·0695	5·9955	6·4946	8·1418	10·023	10·704	12·149	13·708	14·531	17·179	19·098
0·10000	10·61	11·06	11·30	11·78	12·24	12·47	13·14	13·78	13·99	14·40	14·81	15·00	15·59	15·97
1/ 10·0	3·3067	3·9568	4·3494	5·2039	6·1543	6·6666	8·3573	10·288	10·987	12·470	14·070	14·915	17·633	19·602
	0·96	0·96	0·96	0·97	0·97	0·97	0·97	0·97	0·97	0·97	0·98	0·98	0·98	0·98

$\mathbf{V_{r(0{\cdot}5)medial}}$ **for half-full circular pipes.**

A26

(p.1 of 6)

k_s = **0·060 mm**
S = 0·00010 to 0·00030

ie hydraulic gradient =
1 in 10000 to 1 in 3333

Water (or sewage) at 15°C;
full bore conditions.

velocities in ms^{-1}
discharges in m^3s^{-1}

Gradient **(Equivalent) Pipe diameters in mm**

Gradient	630	675	700	750	800	825	900	975	1000	1050	1100	1125	1200	1250
0·000100	0·268	0·280	0·287	0·300	0·313	0·320	0·338	0·357	0·362	0·374	0·386	0·391	0·408	0·419
1/ 10000	0·0835	0·1003	0·1105	0·1327	0·1575	0·1710	0·2153	0·2662	0·2847	0·3240	0·3664	0·3888	0·4612	0·5138
0·000105	0·275	0·288	0·295	0·309	0·322	0·328	0·348	0·366	0·372	0·384	0·396	0·402	0·419	0·430
1/ 9524	0·0858	0·1030	0·1135	0·1363	0·1618	0·1756	0·2211	0·2734	0·2923	0·3326	0·3762	0·3992	0·4735	0·5275
0·000110	0·282	0·295	0·302	0·316	0·330	0·337	0·356	0·375	0·382	0·394	0·406	0·412	0·429	0·441
1/ 9091	0·0880	0·1057	0·1164	0·1398	0·1659	0·1801	0·2268	0·2803	0·2998	0·3411	0·3858	0·4094	0·4856	0·5409
0·000115	0·289	0·303	0·310	0·324	0·338	0·345	0·365	0·385	0·391	0·404	0·416	0·422	0·440	0·451
1/ 8696	0·0901	0·1083	0·1193	0·1432	0·1700	0·1845	0·2323	0·2872	0·3071	0·3494	0·3952	0·4194	0·4974	0·5540
0·000120	0·296	0·310	0·317	0·332	0·346	0·353	0·374	0·394	0·400	0·413	0·426	0·432	0·450	0·462
1/ 8333	0·0922	0·1108	0·1221	0·1466	0·1740	0·1888	0·2377	0·2939	0·3143	0·3576	0·4044	0·4291	0·5089	0·5669
0·000125	0·303	0·317	0·324	0·339	0·354	0·361	0·382	0·402	0·409	0·422	0·435	0·441	0·460	0·472
1/ 8000	0·0943	0·1133	0·1248	0·1499	0·1779	0·1930	0·2431	0·3005	0·3213	0·3655	0·4134	0·4387	0·5203	0·5795
0·000130	0·309	0·323	0·331	0·347	0·362	0·369	0·390	0·411	0·418	0·431	0·444	0·451	0·470	0·482
1/ 7692	0·0964	0·1158	0·1275	0·1531	0·1817	0·1972	0·2483	0·3069	0·3282	0·3734	0·4222	0·4481	0·5314	0·5919
0·000135	0·316	0·330	0·338	0·354	0·369	0·376	0·398	0·420	0·426	0·440	0·453	0·460	0·479	0·492
1/ 7407	0·0984	0·1182	0·1301	0·1563	0·1855	0·2012	0·2534	0·3132	0·3349	0·3811	0·4309	0·4573	0·5423	0·6040
0·000140	0·322	0·337	0·345	0·361	0·376	0·384	0·406	0·428	0·435	0·449	0·462	0·469	0·489	0·502
1/ 7143	0·1003	0·1205	0·1327	0·1594	0·1892	0·2053	0·2585	0·3194	0·3416	0·3886	0·4395	0·4663	0·5530	0·6159
0·000145	0·328	0·343	0·352	0·368	0·384	0·391	0·414	0·436	0·443	0·457	0·471	0·478	0·498	0·511
1/ 6897	0·1023	0·1229	0·1353	0·1625	0·1928	0·2092	0·2634	0·3256	0·3481	0·3960	0·4479	0·4753	0·5636	0·6277
0·000150	0·334	0·350	0·358	0·375	0·391	0·399	0·422	0·444	0·451	0·466	0·480	0·487	0·507	0·521
1/ 6667	0·1042	0·1251	0·1378	0·1655	0·1964	0·2131	0·2683	0·3316	0·3545	0·4034	0·4561	0·4840	0·5739	0·6392
0·000160	0·346	0·362	0·371	0·388	0·405	0·413	0·437	0·460	0·467	0·482	0·497	0·504	0·525	0·539
1/ 6250	0·1079	0·1296	0·1427	0·1714	0·2034	0·2207	0·2778	0·3433	0·3671	0·4176	0·4723	0·5011	0·5942	0·6618
0·000170	0·358	0·374	0·383	0·401	0·418	0·427	0·451	0·475	0·483	0·498	0·513	0·521	0·543	0·557
1/ 5882	0·1115	0·1340	0·1475	0·1771	0·2102	0·2280	0·2871	0·3548	0·3793	0·4315	0·4879	0·5178	0·6139	0·6837
0·000180	0·369	0·386	0·395	0·414	0·431	0·440	0·465	0·490	0·498	0·514	0·529	0·537	0·560	0·574
1/ 5556	0·1151	0·1382	0·1522	0·1827	0·2168	0·2352	0·2961	0·3659	0·3912	0·4450	0·5032	0·5339	0·6330	0·7050
0·000190	0·380	0·398	0·407	0·426	0·444	0·453	0·479	0·505	0·513	0·529	0·545	0·553	0·576	0·591
1/ 5263	0·1185	0·1423	0·1567	0·1881	0·2232	0·2422	0·3048	0·3767	0·4027	0·4581	0·5180	0·5497	0·6517	0·7257
0·000200	0·391	0·409	0·419	0·438	0·457	0·466	0·493	0·519	0·527	0·544	0·560	0·568	0·592	0·608
1/ 5000	0·1218	0·1463	0·1611	0·1934	0·2295	0·2490	0·3134	0·3872	0·4140	0·4709	0·5325	0·5650	0·6699	0·7460
0·000210	0·401	0·420	0·430	0·450	0·469	0·478	0·506	0·532	0·541	0·558	0·575	0·583	0·608	0·624
1/ 4762	0·1251	0·1502	0·1654	0·1986	0·2356	0·2556	0·3218	0·3975	0·4250	0·4835	0·5466	0·5800	0·6876	0·7657
0·000220	0·412	0·431	0·441	0·461	0·481	0·490	0·519	0·546	0·555	0·572	0·590	0·598	0·623	0·640
1/ 4545	0·1283	0·1541	0·1697	0·2037	0·2416	0·2621	0·3299	0·4076	0·4358	0·4957	0·5604	0·5947	0·7050	0·7850
0·000230	0·422	0·441	0·452	0·472	0·492	0·502	0·531	0·559	0·568	0·586	0·604	0·613	0·638	0·655
1/ 4348	0·1314	0·1578	0·1738	0·2086	0·2475	0·2685	0·3379	0·4175	0·4463	0·5077	0·5740	0·6090	0·7219	0·8039
0·000240	0·432	0·451	0·462	0·483	0·504	0·514	0·543	0·572	0·581	0·600	0·618	0·627	0·653	0·670
1/ 4167	0·1345	0·1615	0·1778	0·2135	0·2532	0·2747	0·3457	0·4271	0·4566	0·5194	0·5872	0·6230	0·7386	0·8224
0·000250	0·441	0·461	0·472	0·494	0·515	0·525	0·556	0·585	0·594	0·613	0·632	0·641	0·667	0·685
1/ 4000	0·1375	0·1651	0·1818	0·2182	0·2589	0·2808	0·3534	0·4366	0·4668	0·5309	0·6002	0·6368	0·7549	0·8406
0·000260	0·451	0·471	0·483	0·505	0·526	0·537	0·567	0·597	0·607	0·626	0·645	0·654	0·682	0·699
1/ 3846	0·1405	0·1687	0·1857	0·2229	0·2644	0·2868	0·3609	0·4459	0·4767	0·5422	0·6129	0·6503	0·7709	0·8584
0·000270	0·460	0·481	0·492	0·515	0·537	0·548	0·579	0·609	0·619	0·639	0·658	0·668	0·696	0·714
1/ 3704	0·1434	0·1721	0·1895	0·2275	0·2698	0·2927	0·3683	0·4550	0·4864	0·5532	0·6254	0·6636	0·7866	0·8759
0·000280	0·469	0·491	0·502	0·525	0·547	0·558	0·590	0·621	0·632	0·651	0·671	0·681	0·709	0·728
1/ 3571	0·1462	0·1755	0·1933	0·2320	0·2752	0·2985	0·3756	0·4640	0·4960	0·5641	0·6377	0·6766	0·8020	0·8930
0·000290	0·478	0·500	0·512	0·535	0·558	0·569	0·602	0·633	0·644	0·664	0·684	0·694	0·723	0·741
1/ 3448	0·1490	0·1789	0·1970	0·2364	0·2804	0·3042	0·3827	0·4728	0·5054	0·5748	0·6498	0·6894	0·8172	0·9099
0·000300	0·487	0·509	0·521	0·545	0·568	0·579	0·613	0·645	0·655	0·676	0·696	0·706	0·736	0·755
1/ 3333	0·1518	0·1822	0·2006	0·2408	0·2856	0·3098	0·3898	0·4814	0·5147	0·5853	0·6617	0·7021	0·8321	0·9265
	0·88	0·89	0·89	0·90	0·90	0·90	0·91	0·91	0·92	0·92	0·92	0·92	0·93	0·93

$V_{r(0·5)medial}$ for half-full circular pipes.

k_s = 0·060 mm S = 0·00010 to 0·00030

Gradient	**(Equivalent) Pipe diameters in mm**													
	630	**675**	**700**	**750**	**800**	**825**	**900**	**975**	**1000**	**1050**	**1100**	**1125**	**1200**	**1250**
0·000300	0·487	0·509	0·521	0·545	0·568	0·579	0·613	0·645	0·655	0·676	0·696	0·706	0·736	0·755
1/ 3333	0·1518	0·1822	0·2006	0·2408	0·2856	0·3098	0·3898	0·4814	0·5147	0·5853	0·6617	0·7021	0·8321	0·9265
0·000320	0·504	0·527	0·540	0·564	0·588	0·600	0·634	0·668	0·678	0·700	0·721	0·731	0·762	0·781
1/ 3125	0·1572	0·1887	0·2077	0·2493	0·2957	0·3207	0·4035	0·4984	0·5328	0·6059	0·6849	0·7267	0·8613	0·9589
0·000340	0·521	0·545	0·558	0·583	0·608	0·620	0·655	0·690	0·701	0·723	0·744	0·755	0·787	0·807
1/ 2941	0·1624	0·1949	0·2146	0·2576	0·3054	0·3313	0·4168	0·5148	0·5503	0·6258	0·7074	0·7506	0·8895	0·9904
0·000360	0·537	0·562	0·575	0·601	0·627	0·639	0·676	0·711	0·722	0·745	0·767	0·778	0·811	0·832
1/ 2778	0·1675	0·2010	0·2213	0·2656	0·3150	0·3416	0·4298	0·5308	0·5674	0·6452	0·7293	0·7738	0·9170	1·0210
0·000380	0·553	0·578	0·592	0·619	0·645	0·658	0·695	0·732	0·744	0·767	0·790	0·801	0·835	0·856
1/ 2632	0·1724	0·2070	0·2278	0·2734	0·3242	0·3517	0·4424	0·5463	0·5840	0·6641	0·7507	0·7964	0·9438	1·0508
0·000400	0·569	0·595	0·609	0·636	0·663	0·676	0·715	0·752	0·764	0·788	0·812	0·823	0·858	0·880
1/ 2500	0·1773	0·2127	0·2342	0·2810	0·3333	0·3615	0·4547	0·5615	0·6002	0·6825	0·7715	0·8185	0·9699	1·0798
0·000420	0·584	0·610	0·625	0·653	0·681	0·694	0·734	0·772	0·784	0·809	0·833	0·845	0·880	0·903
1/ 2381	0·1820	0·2184	0·2404	0·2885	0·3421	0·3710	0·4667	0·5763	0·6161	0·7005	0·7918	0·8400	0·9954	1·1082
0·000440	0·599	0·626	0·641	0·670	0·698	0·712	0·752	0·791	0·804	0·829	0·854	0·866	0·902	0·926
1/ 2273	0·1866	0·2239	0·2465	0·2958	0·3507	0·3804	0·4785	0·5908	0·6315	0·7181	0·8116	0·8610	1·0203	1·1359
0·000460	0·613	0·641	0·656	0·686	0·715	0·729	0·770	0·810	0·823	0·849	0·874	0·887	0·924	0·948
1/ 2174	0·1911	0·2293	0·2525	0·3029	0·3592	0·3895	0·4900	0·6050	0·6467	0·7353	0·8310	0·8816	1·0447	1·1630
0·000480	0·627	0·656	0·671	0·701	0·731	0·745	0·788	0·829	0·842	0·869	0·894	0·907	0·945	0·969
1/ 2083	0·1955	0·2346	0·2583	0·3099	0·3674	0·3985	0·5012	0·6188	0·6615	0·7521	0·8500	0·9018	1·0686	1·1895
0·000500	0·641	0·670	0·686	0·717	0·747	0·762	0·805	0·847	0·861	0·888	0·914	0·927	0·965	0·991
1/ 2000	0·1999	0·2398	0·2640	0·3167	0·3755	0·4073	0·5122	0·6324	0·6760	0·7686	0·8687	0·9216	1·0919	1·2156
0·000525	0·658	0·688	0·704	0·736	0·767	0·782	0·826	0·869	0·883	0·911	0·938	0·951	0·991	1·016
1/ 1905	0·2052	0·2462	0·2710	0·3251	0·3854	0·4180	0·5257	0·6491	0·6938	0·7888	0·8915	0·9457	1·1205	1·2474
0·000550	0·675	0·705	0·722	0·754	0·786	0·802	0·847	0·891	0·905	0·934	0·962	0·975	1·016	1·042
1/ 1818	0·2103	0·2524	0·2778	0·3333	0·3951	0·4285	0·5389	0·6653	0·7112	0·8085	0·9138	0·9694	1·1485	1·2785
0·000575	0·691	0·722	0·739	0·773	0·805	0·821	0·867	0·912	0·927	0·956	0·984	0·998	1·040	1·067
1/ 1739	0·2154	0·2585	0·2845	0·3413	0·4046	0·4388	0·5518	0·6812	0·7282	0·8279	0·9356	0·9925	1·1759	1·3089
0·000600	0·707	0·739	0·756	0·790	0·823	0·840	0·887	0·933	0·948	0·978	1·007	1·021	1·063	1·091
1/ 1667	0·2204	0·2644	0·2911	0·3491	0·4139	0·4489	0·5645	0·6968	0·7448	0·8468	0·9569	1·0151	1·2026	1·3387
0·000625	0·723	0·755	0·773	0·808	0·842	0·858	0·907	0·954	0·969	0·999	1·029	1·044	1·087	1·115
1/ 1600	0·2252	0·2703	0·2975	0·3568	0·4230	0·4587	0·5768	0·7121	0·7611	0·8653	0·9778	1·0373	1·2289	1·3679
0·000650	0·738	0·771	0·789	0·825	0·859	0·876	0·926	0·974	0·989	1·020	1·051	1·065	1·109	1·138
1/ 1538	0·2300	0·2760	0·3038	0·3644	0·4319	0·4684	0·5890	0·7271	0·7771	0·8835	0·9983	1·0590	1·2546	1·3965
0·000675	0·753	0·787	0·805	0·842	0·877	0·894	0·945	0·993	1·009	1·041	1·072	1·087	1·132	1·161
1/ 1481	0·2347	0·2816	0·3099	0·3718	0·4407	0·4779	0·6009	0·7418	0·7928	0·9013	1·0185	1·0804	1·2799	1·4247
0·000700	0·768	0·802	0·821	0·858	0·894	0·911	0·963	1·013	1·029	1·061	1·093	1·108	1·154	1·183
1/ 1429	0·2393	0·2871	0·3160	0·3791	0·4493	0·4872	0·6126	0·7562	0·8082	0·9188	1·0383	1·1014	1·3047	1·4523
0·000725	0·782	0·818	0·837	0·874	0·911	0·929	0·981	1·032	1·048	1·081	1·113	1·129	1·175	1·206
1/ 1379	0·2439	0·2925	0·3220	0·3862	0·4578	0·4964	0·6242	0·7704	0·8234	0·9360	1·0577	1·1220	1·3291	1·4794
0·000750	0·797	0·832	0·852	0·890	0·927	0·946	0·999	1·051	1·067	1·101	1·133	1·149	1·196	1·227
1/ 1333	0·2483	0·2979	0·3279	0·3932	0·4661	0·5054	0·6355	0·7843	0·8383	0·9530	1·0768	1·1423	1·3531	1·5061
0·000800	0·824	0·862	0·882	0·921	0·960	0·978	1·034	1·087	1·104	1·139	1·172	1·189	1·238	1·270
1/ 1250	0·2570	0·3083	0·3393	0·4070	0·4824	0·5231	0·6576	0·8116	0·8674	0·9860	1·1142	1·1819	1·4000	1·5582
0·000850	0·852	0·890	0·911	0·951	0·991	1·010	1·067	1·122	1·140	1·176	1·210	1·228	1·278	1·311
1/ 1176	0·2655	0·3184	0·3505	0·4203	0·4981	0·5402	0·6791	0·8380	0·8957	1·0181	1·1504	1·2203	1·4454	1·6087
0·000900	0·878	0·917	0·939	0·981	1·022	1·042	1·100	1·157	1·175	1·212	1·248	1·265	1·317	1·351
1/ 1111	0·2737	0·3283	0·3613	0·4333	0·5135	0·5568	0·6999	0·8637	0·9231	1·0493	1·1856	1·2576	1·4896	1·6578
0·000950	0·904	0·944	0·966	1·009	1·051	1·072	1·132	1·190	1·209	1·247	1·284	1·302	1·355	1·390
1/ 1053	0·2817	0·3378	0·3718	0·4459	0·5284	0·5729	0·7202	0·8888	0·9499	1·0797	1·2198	1·2939	1·5325	1·7056
0·001000	0·929	0·970	0·993	1·037	1·080	1·101	1·163	1·223	1·243	1·281	1·319	1·337	1·392	1·428
1/ 1000	0·2895	0·3472	0·3821	0·4582	0·5430	0·5887	0·7400	0·9132	0·9759	1·1092	1·2532	1·3293	1·5744	1·7522
	0·91	0·91	0·91	0·92	0·92	0·92	0·93	0·93	0·93	0·94	0·94	0·94	0·94	0·94

$V_{r(0{\cdot}5)\mathrm{medial}}$ **for half-full circular pipes.**

$k_s = 0{\cdot}060$ mm
S = 0·00100 to 0·00300

ie hydraulic gradient =
1 in 1000 to 1 in 333

Water (or sewage) at 15°C;
full bore conditions.

velocities in ms^{-1}
discharges in m^3s^{-1}

Gradient **(Equivalent) Pipe diameters in mm**

	630	675	700	750	800	825	900	975	1000	1050	1100	1125	1200	1250
0·00100	0·929	0·970	0·993	1·037	1·080	1·101	1·163	1·223	1·243	1·281	1·319	1·337	1·392	1·428
1/ 1000	0·2895	0·3472	0·3821	0·4582	0·5430	0·5887	0·7400	0·9132	0·9759	1·1092	1·2532	1·3293	1·5744	1·7522
0·00105	0·953	0·996	1·019	1·064	1·108	1·130	1·194	1·255	1·275	1·314	1·353	1·372	1·428	1·465
1/ 952	0·2971	0·3563	0·3921	0·4702	0·5572	0·6041	0·7593	0·9370	1·0013	1·1381	1·2858	1·3639	1·6153	1·7977
0·00110	0·977	1·021	1·044	1·091	1·136	1·158	1·223	1·286	1·307	1·347	1·387	1·406	1·464	1·501
1/ 909	0·3045	0·3652	0·4019	0·4819	0·5710	0·6192	0·7782	0·9602	1·0262	1·1663	1·3177	1·3977	1·6553	1·8421
0·00115	1·000	1·045	1·069	1·117	1·163	1·186	1·252	1·317	1·338	1·379	1·419	1·439	1·498	1·537
1/ 870	0·3118	0·3739	0·4115	0·4934	0·5846	0·6339	0·7967	0·9830	1·0505	1·1939	1·3488	1·4307	1·6944	1·8856
0·00120	1·023	1·069	1·094	1·142	1·190	1·213	1·281	1·346	1·368	1·410	1·451	1·472	1·532	1·571
1/ 833	0·3189	0·3824	0·4208	0·5046	0·5979	0·6483	0·8147	1·0052	1·0743	1·2209	1·3793	1·4630	1·7326	1·9281
0·00125	1·045	1·092	1·117	1·167	1·215	1·239	1·309	1·376	1·398	1·441	1·483	1·504	1·565	1·605
1/ 800	0·3259	0·3908	0·4300	0·5156	0·6109	0·6624	0·8325	1·0271	1·0976	1·2474	1·4092	1·4947	1·7701	1·9698
0·00130	1·067	1·115	1·141	1·192	1·241	1·265	1·336	1·404	1·427	1·471	1·514	1·535	1·598	1·638
1/ 769	0·3327	0·3990	0·4391	0·5264	0·6237	0·6762	0·8498	1·0485	1·1205	1·2734	1·4385	1·5258	1·8068	2·0107
0·00135	1·089	1·137	1·164	1·216	1·266	1·290	1·363	1·432	1·455	1·500	1·544	1·566	1·630	1·671
1/ 741	0·3394	0·4070	0·4479	0·5370	0·6363	0·6898	0·8669	1·0695	1·1429	1·2989	1·4673	1·5563	1·8429	2·0508
0·00140	1·110	1·160	1·186	1·239	1·290	1·315	1·389	1·460	1·483	1·529	1·574	1·596	1·661	1·703
1/ 714	0·3461	0·4149	0·4566	0·5474	0·6486	0·7032	0·8836	1·0901	1·1649	1·3239	1·4955	1·5862	1·8783	2·0902
0·00145	1·131	1·181	1·209	1·262	1·314	1·340	1·415	1·487	1·511	1·557	1·603	1·625	1·692	1·735
1/ 690	0·3525	0·4227	0·4651	0·5576	0·6607	0·7163	0·9001	1·1104	1·1866	1·3485	1·5233	1·6157	1·9132	2·1289
0·00150	1·151	1·203	1·230	1·285	1·338	1·364	1·440	1·514	1·538	1·585	1·632	1·655	1·722	1·766
1/ 667	0·3589	0·4304	0·4735	0·5677	0·6726	0·7292	0·9163	1·1303	1·2079	1·3727	1·5506	1·6446	1·9474	2·1670
0·00160	1·191	1·244	1·273	1·329	1·384	1·411	1·490	1·566	1·591	1·640	1·688	1·711	1·781	1·826
1/ 625	0·3714	0·4453	0·4900	0·5873	0·6958	0·7544	0·9479	1·1692	1·2495	1·4199	1·6039	1·7011	2·0142	2·2413
0·00170	1·230	1·285	1·314	1·373	1·429	1·457	1·538	1·617	1·642	1·693	1·742	1·766	1·838	1·885
1/ 588	0·3835	0·4598	0·5059	0·6064	0·7184	0·7788	0·9785	1·2070	1·2898	1·4657	1·6556	1·7559	2·0790	2·3134
0·00180	1·268	1·324	1·355	1·415	1·473	1·501	1·585	1·666	1·692	1·744	1·795	1·820	1·894	1·942
1/ 556	0·3952	0·4738	0·5213	0·6249	0·7403	0·8026	1·0083	1·2437	1·3290	1·5102	1·7058	1·8091	2·1420	2·3834
0·00190	1·305	1·362	1·394	1·455	1·515	1·545	1·631	1·714	1·741	1·794	1·846	1·872	1·948	1·998
1/ 526	0·4067	0·4875	0·5364	0·6429	0·7616	0·8257	1·0373	1·2794	1·3671	1·5535	1·7546	1·8609	2·2033	2·4515
0·00200	1·340	1·400	1·432	1·495	1·557	1·587	1·675	1·760	1·788	1·843	1·896	1·923	2·001	2·052
1/ 500	0·4178	0·5009	0·5511	0·6605	0·7824	0·8482	1·0655	1·3142	1·4043	1·5956	1·8023	1·9114	2·2630	2·5179
0·00210	1·375	1·436	1·469	1·534	1·597	1·628	1·718	1·806	1·834	1·890	1·945	1·972	2·052	2·105
1/ 476	0·4287	0·5139	0·5654	0·6776	0·8027	0·8702	1·0931	1·3481	1·4405	1·6368	1·8487	1·9607	2·3212	2·5827
0·00220	1·409	1·472	1·505	1·572	1·636	1·668	1·761	1·850	1·879	1·937	1·993	2·021	2·103	2·156
1/ 455	0·4393	0·5266	0·5794	0·6944	0·8225	0·8916	1·1200	1·3813	1·4760	1·6770	1·8941	2·0088	2·3782	2·6460
0·00230	1·443	1·506	1·541	1·609	1·675	1·707	1·802	1·894	1·923	1·982	2·040	2·068	2·152	2·207
1/ 435	0·4497	0·5391	0·5931	0·7108	0·8419	0·9126	1·1464	1·4137	1·5106	1·7164	1·9385	2·0559	2·4338	2·7079
0·00240	1·475	1·540	1·576	1·645	1·713	1·746	1·843	1·936	1·967	2·027	2·086	2·115	2·200	2·256
1/ 417	0·4599	0·5512	0·6065	0·7268	0·8609	0·9332	1·1722	1·4455	1·5445	1·7549	1·9820	2·1019	2·4883	2·7684
0·00250	1·507	1·574	1·610	1·681	1·750	1·783	1·882	1·978	2·009	2·070	2·130	2·160	2·247	2·304
1/ 400	0·4699	0·5632	0·6196	0·7425	0·8794	0·9533	1·1974	1·4766	1·5778	1·7926	2·0245	2·1471	2·5417	2·8278
0·00260	1·539	1·607	1·643	1·716	1·786	1·820	1·921	2·019	2·050	2·113	2·174	2·205	2·294	2·352
1/ 385	0·4796	0·5749	0·6325	0·7579	0·8977	0·9731	1·2222	1·5071	1·6103	1·8296	2·0663	2·1913	2·5940	2·8860
0·00270	1·569	1·639	1·676	1·750	1·821	1·857	1·959	2·059	2·091	2·155	2·217	2·248	2·339	2·398
1/ 370	0·4892	0·5864	0·6451	0·7731	0·9156	0·9924	1·2465	1·5370	1·6423	1·8659	2·1072	2·2348	2·6454	2·9431
0·00280	1·600	1·670	1·708	1·783	1·856	1·892	1·997	2·098	2·131	2·196	2·260	2·291	2·384	2·444
1/ 357	0·4987	0·5977	0·6575	0·7879	0·9331	1·0115	1·2704	1·5664	1·6737	1·9015	2·1475	2·2774	2·6958	2·9991
0·00290	1·629	1·701	1·740	1·816	1·891	1·927	2·034	2·137	2·170	2·236	2·301	2·333	2·427	2·489
1/ 345	0·5079	0·6087	0·6697	0·8025	0·9504	1·0302	1·2938	1·5953	1·7046	1·9365	2·1870	2·3193	2·7453	3·0542
0·00300	1·659	1·732	1·771	1·849	1·925	1·962	2·070	2·175	2·209	2·276	2·342	2·375	2·470	2·533
1/ 333	0·5170	0·6196	0·6817	0·8168	0·9674	1·0486	1·3169	1·6237	1·7349	1·9710	2·2258	2·3605	2·7940	3·1084
	0·92	0·93	0·93	0·93	0·94	0·94	0·94	0·94	0·95	0·95	0·95	0·95	0·95	0·95

$V_{r(0\cdot5)medial}$ for half-full circular pipes.

$k_s = 0{\cdot}060$ mm S = 0·00100 to 0·00300

$k_s = 0{\cdot}060$ **mm**
S = 0·00300 to 0·01000

ie hydraulic gradient =
1 in 333 to 1 in 100

Water (or sewage) at 15°C;
full bore conditions.

velocities in ms^{-1}
discharges in m^3s^{-1}

Gradient **(Equivalent) Pipe diameters in mm**

Gradient	630	675	700	750	800	825	900	975	1000	1050	1100	1125	1200	1250
0·00300	1·659	1·732	1·771	1·849	1·925	1·962	2·070	2·175	2·209	2·276	2·342	2·375	2·470	2·533
1/ 333	0·5170	0·6196	0·6817	0·8168	0·9674	1·0486	1·3169	1·6237	1·7349	1·9710	2·2258	2·3605	2·7940	3·1084
0·00320	1·716	1·791	1·832	1·912	1·990	2·029	2·141	2·249	2·284	2·354	2·422	2·456	2·555	2·619
1/ 313	0·5348	0·6409	0·7051	0·8448	1·0005	1·0845	1·3619	1·6791	1·7941	2·0382	2·3017	2·4409	2·8891	3·2141
0·00340	1·771	1·849	1·891	1·974	2·054	2·094	2·209	2·321	2·357	2·429	2·499	2·534	2·636	2·703
1/ 294	0·5521	0·6616	0·7278	0·8720	1·0326	1·1193	1·4055	1·7329	1·8515	2·1033	2·3752	2·5189	2·9813	3·3166
0·00360	1·825	1·905	1·948	2·034	2·116	2·157	2·276	2·391	2·428	2·502	2·575	2·610	2·715	2·784
1/ 278	0·5688	0·6816	0·7498	0·8984	1·0638	1·1531	1·4479	1·7851	1·9073	2·1667	2·4467	2·5946	3·0709	3·4162
0·00380	1·877	1·959	2·004	2·092	2·177	2·219	2·341	2·459	2·497	2·573	2·648	2·684	2·792	2·863
1/ 263	0·5851	0·7011	0·7713	0·9241	1·0942	1·1860	1·4892	1·8359	1·9615	2·2282	2·5162	2·6683	3·1580	3·5130
0·00400	1·928	2·012	2·058	2·148	2·236	2·279	2·404	2·525	2·565	2·643	2·719	2·757	2·867	2·940
1/ 250	0·6010	0·7202	0·7922	0·9491	1·1238	1·2181	1·5294	1·8854	2·0144	2·2883	2·5839	2·7401	3·2429	3·6074
0·00420	1·978	2·064	2·111	2·204	2·293	2·337	2·466	2·590	2·630	2·710	2·788	2·827	2·941	3·015
1/ 238	0·6165	0·7387	0·8126	0·9735	1·1527	1·2493	1·5686	1·9337	2·0660	2·3468	2·6500	2·8101	3·3257	3·6995
0·00440	2·026	2·115	2·163	2·258	2·349	2·394	2·526	2·653	2·695	2·776	2·856	2·896	3·012	3·088
1/ 227	0·6317	0·7568	0·8325	0·9974	1·1809	1·2799	1·6069	1·9809	2·1164	2·4040	2·7145	2·8785	3·4066	3·7894
0·00460	2·074	2·165	2·214	2·310	2·404	2·450	2·585	2·715	2·757	2·841	2·923	2·963	3·082	3·160
1/ 217	0·6465	0·7746	0·8520	1·0207	1·2085	1·3098	1·6444	2·0270	2·1656	2·4599	2·7776	2·9454	3·4857	3·8774
0·00480	2·120	2·213	2·263	2·362	2·458	2·505	2·643	2·775	2·819	2·904	2·988	3·029	3·150	3·230
1/ 208	0·6610	0·7919	0·8711	1·0435	1·2355	1·3390	1·6811	2·0722	2·2138	2·5147	2·8394	3·0109	3·5631	3·9634
0·00500	2·166	2·261	2·312	2·413	2·511	2·559	2·699	2·835	2·879	2·966	3·051	3·094	3·218	3·298
1/ 200	0·6752	0·8089	0·8898	1·0659	1·2619	1·3677	1·7170	2·1164	2·2611	2·5683	2·8999	3·0750	3·6389	4·0477
0·00525	2·222	2·319	2·371	2·475	2·575	2·624	2·768	2·907	2·952	3·042	3·129	3·172	3·299	3·382
1/ 190	0·6925	0·8297	0·9126	1·0932	1·2943	1·4027	1·7609	2·1705	2·3188	2·6338	2·9738	3·1534	3·7316	4·1508
0·00550	2·276	2·375	2·429	2·535	2·638	2·688	2·835	2·978	3·024	3·116	3·205	3·249	3·380	3·464
1/ 182	0·7095	0·8500	0·9349	1·1199	1·3258	1·4369	1·8038	2·2233	2·3752	2·6979	3·0461	3·2300	3·8222	4·2514
0·00575	2·329	2·431	2·486	2·594	2·699	2·751	2·901	3·047	3·094	3·188	3·280	3·325	3·458	3·545
1/ 174	0·7261	0·8699	0·9567	1·1460	1·3567	1·4704	1·8458	2·2749	2·4304	2·7605	3·1167	3·3049	3·9107	4·3499
0·00600	2·381	2·485	2·542	2·652	2·759	2·812	2·966	3·115	3·163	3·259	3·352	3·399	3·534	3·623
1/ 167	0·7423	0·8893	0·9781	1·1716	1·3870	1·5031	1·8868	2·3255	2·4844	2·8217	3·1859	3·3782	3·9974	4·4462
0·00625	2·432	2·538	2·596	2·709	2·818	2·872	3·029	3·181	3·231	3·328	3·424	3·471	3·609	3·700
1/ 160	0·7582	0·9083	0·9990	1·1966	1·4166	1·5352	1·9271	2·3750	2·5372	2·8817	3·2536	3·4500	4·0822	4·5406
0·00650	2·482	2·591	2·649	2·764	2·876	2·931	3·091	3·246	3·297	3·396	3·493	3·542	3·683	3·775
1/ 154	0·7738	0·9270	1·0196	1·2212	1·4456	1·5667	1·9665	2·4235	2·5891	2·9406	3·3200	3·5204	4·1654	4·6331
0·00675	2·532	2·642	2·702	2·819	2·933	2·988	3·152	3·310	3·361	3·463	3·562	3·611	3·755	3·849
1/ 148	0·7891	0·9453	1·0397	1·2453	1·4741	1·5975	2·0052	2·4712	2·6399	2·9983	3·3851	3·5894	4·2471	4·7238
0·00700	2·580	2·692	2·753	2·872	2·988	3·045	3·212	3·372	3·425	3·528	3·629	3·679	3·826	3·922
1/ 143	0·8042	0·9633	1·0594	1·2689	1·5021	1·6278	2·0432	2·5179	2·6899	3·0550	3·4491	3·6572	4·3272	4·8129
0·00725	2·627	2·741	2·803	2·925	3·043	3·101	3·270	3·434	3·487	3·592	3·695	3·746	3·896	3·993
1/ 138	0·8189	0·9810	1·0789	1·2922	1·5296	1·6576	2·0805	2·5639	2·7389	3·1107	3·5119	3·7238	4·4060	4·9004
0·00750	2·674	2·790	2·853	2·977	3·097	3·156	3·328	3·494	3·549	3·656	3·760	3·812	3·964	4·063
1/ 133	0·8335	0·9983	1·0980	1·3150	1·5566	1·6869	2·1172	2·6090	2·7872	3·1654	3·5737	3·7893	4·4834	4·9865
0·00800	2·765	2·885	2·950	3·077	3·202	3·262	3·441	3·613	3·669	3·779	3·887	3·941	4·098	4·200
1/ 125	0·8618	1·0322	1·1352	1·3596	1·6093	1·7440	2·1888	2·6972	2·8813	3·2723	3·6942	3·9171	4·6344	5·1544
0·00850	2·853	2·976	3·044	3·175	3·303	3·366	3·550	3·727	3·785	3·899	4·010	4·065	4·227	4·333
1/ 118	0·8893	1·0651	1·1714	1·4028	1·6604	1·7994	2·2582	2·7826	2·9726	3·3758	3·8111	4·0409	4·7809	5·3172
0·00900	2·938	3·066	3·135	3·270	3·402	3·467	3·656	3·838	3·898	4·015	4·130	4·186	4·353	4·462
1/ 111	0·9159	1·0970	1·2065	1·4448	1·7101	1·8531	2·3256	2·8656	3·0612	3·4764	3·9246	4·1613	4·9231	5·4754
0·00950	3·022	3·152	3·224	3·363	3·498	3·565	3·759	3·946	4·007	4·128	4·246	4·304	4·475	4·587
1/ 105	0·9419	1·1281	1·2406	1·4857	1·7584	1·9055	2·3912	2·9464	3·1474	3·5743	4·0350	4·2783	5·0615	5·6292
0·01000	3·103	3·237	3·310	3·453	3·592	3·660	3·859	4·052	4·114	4·238	4·359	4·419	4·595	4·709
1/ 100	0·9672	1·1584	1·2739	1·5255	1·8054	1·9565	2·4551	3·0250	3·2314	3·6696	4·1425	4·3923	5·1963	5·7790
	0·93	0·94	0·94	0·94	0·94	0·95	0·95	0·95	0·95	0·95	0·96	0·96	0·96	0·96

$V_{r(0{\cdot}5)medial}$ **for half-full circular pipes.**

A26
(p.5 of 6)

k_s = 0·060 mm
S = 0·01000 to 0·03000

ie hydraulic gradient =
1 in 100 to 1 in 33·3

Water (or sewage) at 15°C;
full bore conditions.

velocities in ms^{-1}
discharges in m^3s^{-1}

Gradient **(Equivalent) Pipe diameters in mm**

Gradient	630	675	700	750	800	825	900	975	1000	1050	1100	1125	1200	1250
0·01000	3·103	3·237	3·310	3·453	3·592	3·660	3·859	4·052	4·114	4·238	4·359	4·419	4·595	4·709
1/ 100	0·9672	1·1584	1·2739	1·5255	1·8054	1·9565	2·4551	3·0250	3·2314	3·6696	4·1425	4·3923	5·1963	5·7790
0·01050	3·182	3·320	3·394	3·541	3·683	3·753	3·957	4·154	4·219	4·345	4·469	4·531	4·711	4·828
1/ 95	0·9919	1·1879	1·3063	1·5643	1·8514	2·0062	2·5174	3·1017	3·3134	3·7626	4·2475	4·5035	5·3278	5·9252
0·01100	3·259	3·400	3·477	3·627	3·772	3·844	4·053	4·255	4·321	4·450	4·577	4·640	4·824	4·945
1/ 91	1·0160	1·2167	1·3380	1·6022	1·8962	2·0548	2·5783	3·1767	3·3934	3·8535	4·3499	4·6121	5·4562	6·0679
0·01150	3·335	3·479	3·557	3·711	3·860	3·933	4·146	4·353	4·420	4·553	4·683	4·747	4·935	5·058
1/ 87	1·0395	1·2449	1·3690	1·6393	1·9400	2·1022	2·6378	3·2499	3·4716	3·9422	4·4501	4·7183	5·5817	6·2074
0·01200	3·409	3·556	3·636	3·793	3·945	4·020	4·238	4·449	4·518	4·653	4·786	4·851	5·044	5·170
1/ 83	1·0626	1·2725	1·3993	1·6755	1·9829	2·1487	2·6961	3·3216	3·5482	4·0291	4·5481	4·8223	5·7045	6·3440
0·01250	3·481	3·632	3·713	3·873	4·029	4·105	4·328	4·543	4·613	4·751	4·887	4·954	5·150	5·279
1/ 80	1·0852	1·2996	1·4290	1·7111	2·0250	2·1942	2·7532	3·3919	3·6232	4·1143	4·6442	4·9240	5·8249	6·4778
0·01300	3·552	3·706	3·789	3·952	4·111	4·188	4·416	4·635	4·707	4·848	4·986	5·054	5·255	5·385
1/ 77	1·1073	1·3260	1·4582	1·7459	2·0662	2·2389	2·8091	3·4607	3·6967	4·1977	4·7383	5·0238	5·9428	6·6089
0·01350	3·622	3·778	3·863	4·029	4·191	4·270	4·502	4·726	4·799	4·942	5·083	5·153	5·357	5·490
1/ 74	1·1291	1·3520	1·4867	1·7801	2·1066	2·2826	2·8640	3·5283	3·7688	4·2796	4·8307	5·1217	6·0586	6·7375
0·01400	3·690	3·850	3·936	4·105	4·270	4·351	4·587	4·814	4·889	5·035	5·179	5·249	5·457	5·593
1/ 71	1·1504	1·3776	1·5148	1·8137	2·1463	2·3256	2·9178	3·5946	3·8396	4·3599	4·9213	5·2178	6·1722	6·8638
0·01450	3·758	3·920	4·008	4·180	4·347	4·430	4·670	4·902	4·977	5·126	5·272	5·344	5·556	5·694
1/ 69	1·1714	1·4026	1·5423	1·8466	2·1853	2·3678	2·9708	3·6597	3·9092	4·4389	5·0104	5·3122	6·2838	6·9879
0·01500	3·824	3·989	4·078	4·253	4·424	4·507	4·752	4·987	5·064	5·216	5·364	5·438	5·653	5·794
1/ 67	1·1920	1·4273	1·5694	1·8791	2·2236	2·4094	3·0228	3·7238	3·9776	4·5165	5·0980	5·4051	6·3935	7·1098
0·01600	3·953	4·123	4·216	4·396	4·572	4·659	4·911	5·155	5·234	5·391	5·544	5·620	5·842	5·988
1/ 63	1·2322	1·4754	1·6223	1·9423	2·2984	2·4904	3·1243	3·8487	4·1110	4·6679	5·2688	5·5862	6·6076	7·3478
0·01700	4·078	4·253	4·349	4·535	4·717	4·806	5·066	5·317	5·399	5·560	5·718	5·796	6·026	6·176
1/ 59	1·2712	1·5220	1·6736	2·0036	2·3709	2·5689	3·2228	3·9699	4·2404	4·8147	5·4344	5·7617	6·8151	7·5785
0·01800	4·199	4·380	4·478	4·670	4·857	4·948	5·216	5·475	5·559	5·725	5·888	5·968	6·204	6·358
1/ 56	1·3090	1·5673	1·7234	2·0632	2·4413	2·6452	3·3183	4·0875	4·3660	4·9573	5·5953	5·9322	7·0166	7·8025
0·01900	4·317	4·503	4·604	4·801	4·993	5·087	5·362	5·628	5·715	5·885	6·052	6·135	6·377	6·536
1/ 53	1·3459	1·6114	1·7718	2·1211	2·5098	2·7193	3·4113	4·2019	4·4882	5·0960	5·7517	6·0981	7·2127	8·0204
0·02000	4·433	4·623	4·726	4·929	5·126	5·222	5·505	5·777	5·866	6·041	6·213	6·297	6·546	6·709
1/ 50	1·3817	1·6543	1·8189	2·1775	2·5765	2·7916	3·5019	4·3134	4·6073	5·2311	5·9041	6·2596	7·4036	8·2327
0·02100	4·545	4·740	4·846	5·053	5·255	5·354	5·643	5·923	6·014	6·193	6·369	6·456	6·711	6·877
1/ 47·6	1·4167	1·6962	1·8650	2·2326	2·6416	2·8621	3·5902	4·4221	4·7234	5·3628	6·0528	6·4172	7·5899	8·4397
0·02200	4·654	4·854	4·963	5·175	5·382	5·483	5·779	6·065	6·158	6·342	6·522	6·611	6·872	7·042
1/ 45·5	1·4509	1·7371	1·9099	2·2863	2·7051	2·9309	3·6765	4·5283	4·8367	5·4915	6·1980	6·5710	7·7718	8·6419
0·02300	4·762	4·966	5·077	5·294	5·505	5·609	5·912	6·204	6·299	6·487	6·671	6·762	7·029	7·203
1/ 43·5	1·4843	1·7770	1·9538	2·3389	2·7673	2·9983	3·7609	4·6321	4·9476	5·6173	6·3399	6·7215	7·9496	8·8395
0·02400	4·867	5·075	5·189	5·411	5·626	5·732	6·041	6·340	6·438	6·629	6·817	6·910	7·183	7·361
1/ 41·7	1·5170	1·8162	1·9968	2·3903	2·8281	3·0641	3·8434	4·7337	5·0561	5·7404	6·4788	6·8687	8·1236	9·0329
0·02500	4·969	5·182	5·298	5·525	5·745	5·853	6·169	6·473	6·573	6·769	6·960	7·055	7·334	7·515
1/ 40·0	1·5491	1·8545	2·0389	2·4407	2·8876	3·1286	3·9242	4·8332	5·1623	5·8610	6·6148	7·0129	8·2940	9·2224
0·02600	5·070	5·287	5·405	5·636	5·861	5·971	6·293	6·604	6·705	6·905	7·101	7·197	7·481	7·666
1/ 38·5	1·5805	1·8920	2·0802	2·4900	2·9460	3·1919	4·0035	4·9307	5·2665	5·9792	6·7481	7·1542	8·4611	9·4081
0·02700	5·169	5·390	5·511	5·746	5·975	6·087	6·415	6·732	6·836	7·039	7·238	7·337	7·626	7·815
1/ 37·0	1·6113	1·9289	2·1207	2·5385	3·0033	3·2539	4·0812	5·0264	5·3686	6·0951	6·8789	7·2928	8·6249	9·5902
0·02800	5·266	5·491	5·614	5·854	6·087	6·201	6·535	6·858	6·963	7·170	7·374	7·474	7·768	7·961
1/ 35·7	1·6415	1·9651	2·1605	2·5860	3·0595	3·3148	4·1575	5·1203	5·4689	6·2089	7·0073	7·4289	8·7858	9·7691
0·02900	5·361	5·591	5·715	5·959	6·197	6·313	6·653	6·982	7·089	7·300	7·506	7·608	7·908	8·104
1/ 34·5	1·6712	2·0006	2·1995	2·6328	3·1147	3·3746	4·2325	5·2126	5·5675	6·3207	7·1335	7·5627	8·9439	9·9447
0·03000	5·455	5·688	5·815	6·063	6·305	6·423	6·769	7·103	7·212	7·427	7·637	7·740	8·045	8·244
1/ 33·3	1·7004	2·0355	2·2379	2·6787	3·1690	3·4334	4·3062	5·3033	5·6643	6·4306	7·2575	7·6941	9·0992	10·117
	0·94	0·94	0·95	0·95	0·95	0·95	0·95	0·96	0·96	0·96	0·96	0·96	0·96	0·96

$V_{r(0\cdot5)medial}$ for half-full circular pipes.

k_s = 0·060 mm S = 0·01000 to 0·03000

k_s = 0·060 mm
S = 0·03000 to 0·10000

ie hydraulic gradient =
1 in 33·3 to 1 in 10·0

Water (or sewage) at 15°C;
full bore conditions.

velocities in ms^{-1}
discharges in m^3s^{-1}

Gradient **(Equivalent) Pipe diameters in mm**

Gradient	630	675	700	750	800	825	900	975	1000	1050	1100	1125	1200	1250
0·03000	5·455	5·688	5·815	6·063	6·305	6·423	6·769	7·103	7·212	7·427	7·637	7·740	8·045	8·244
1/ 33·3	1·7004	2·0355	2·2379	2·6787	3·1690	3·4334	4·3062	5·3033	5·6643	6·4306	7·2575	7·6941	9·0992	10·117
0·03200	5·638	5·879	6·010	6·266	6·515	6·638	6·995	7·340	7·453	7·674	7·891	7·998	8·314	8·519
1/ 31·3	1·7574	2·1037	2·3129	2·7683	3·2750	3·5482	4·4500	5·4803	5·8533	6·6451	7·4994	7·9505	9·4024	10·454
0·03400	5·815	6·063	6·199	6·463	6·720	6·846	7·214	7·570	7·686	7·914	8·138	8·248	8·573	8·785
1/ 29·4	1·8126	2·1698	2·3855	2·8552	3·3777	3·6594	4·5894	5·6518	6·0365	6·8530	7·7339	8·1991	9·6962	10·781
0·03600	5·987	6·243	6·382	6·654	6·918	7·048	7·427	7·793	7·912	8·147	8·378	8·491	8·826	9·044
1/ 27·8	1·8663	2·2340	2·4560	2·9395	3·4774	3·7674	4·7248	5·8184	6·2144	7·0549	7·9617	8·4405	9·9815	11·098
0·03800	6·154	6·417	6·560	6·839	7·111	7·244	7·634	8·010	8·133	8·374	8·611	8·728	9·071	9·295
1/ 26·3	1·9185	2·2964	2·5246	3·0216	3·5744	3·8725	4·8565	5·9805	6·3874	7·2512	8·1832	8·6753	10·259	11·407
0·04000	6·317	6·587	6·734	7·020	7·299	7·436	7·836	8·221	8·347	8·595	8·838	8·958	9·310	9·540
1/ 25·0	1·9693	2·3572	2·5914	3·1015	3·6689	3·9749	4·9847	6·1383	6·5560	7·4425	8·3990	8·9040	10·529	11·707
0·04200	6·477	6·753	6·903	7·197	7·482	7·622	8·032	8·428	8·557	8·811	9·059	9·182	9·543	9·779
1/ 23·8	2·0189	2·4165	2·6566	3·1795	3·7611	4·0747	5·1098	6·2923	6·7204	7·6290	8·6094	9·1271	10·793	12·000
0·04400	6·632	6·915	7·068	7·369	7·662	7·805	8·224	8·629	8·761	9·021	9·276	9·401	9·771	10·01
1/ 22·7	2·0673	2·4744	2·7202	3·2556	3·8511	4·1722	5·2320	6·4426	6·8809	7·8112	8·8149	9·3449	11·051	12·286
0·04600	6·784	7·073	7·230	7·538	7·837	7·983	8·412	8·826	8·961	9·227	9·487	9·615	9·993	10·24
1/ 21·7	2·1146	2·5310	2·7825	3·3300	3·9391	4·2675	5·3514	6·5896	7·0378	7·9893	9·0158	9·5579	11·302	12·566
0·04800	6·932	7·228	7·388	7·702	8·008	8·158	8·596	9·018	9·156	9·428	9·694	9·825	10·21	10·46
1/ 20·8	2·1610	2·5864	2·8434	3·4028	4·0252	4·3607	5·4683	6·7334	7·1914	8·1636	9·2124	9·7662	11·548	12·840
0·05000	7·078	7·379	7·543	7·864	8·176	8·328	8·776	9·207	9·348	9·625	9·896	10·03	10·42	10·68
1/ 20·0	2·2063	2·6407	2·9030	3·4742	4·1095	4·4521	5·5827	6·8742	7·3418	8·3342	9·4049	9·9703	11·790	13·108
0·05250	7·256	7·565	7·733	8·061	8·381	8·537	8·995	9·438	9·582	9·866	10·14	10·28	10·69	10·95
1/ 19·0	2·2618	2·7070	2·9759	3·5614	4·2126	4·5637	5·7226	7·0464	7·5256	8·5429	9·6403	10·220	12·085	13·436
0·05500	7·429	7·746	7·918	8·254	8·581	8·741	9·210	9·663	9·810	10·10	10·39	10·53	10·94	11·21
1/ 18·2	2·3159	2·7718	3·0471	3·6465	4·3133	4·6727	5·8593	7·2145	7·7051	8·7466	9·8701	10·463	12·372	13·756
0·05750	7·599	7·923	8·098	8·442	8·777	8·941	9·420	9·883	10·03	10·33	10·62	10·77	11·19	11·46
1/ 17·4	2·3689	2·8351	3·1167	3·7297	4·4117	4·7793	5·9928	7·3788	7·8806	8·9457	10·095	10·701	12·654	14·069
0·06000	7·765	8·096	8·275	8·627	8·968	9·136	9·626	10·10	10·25	10·56	10·85	11·00	11·43	11·71
1/ 16·7	2·4207	2·8971	3·1848	3·8112	4·5080	4·8836	6·1235	7·5397	8·0524	9·1406	10·315	10·934	12·929	14·375
0·06250	7·928	8·265	8·449	8·807	9·156	9·327	9·827	10·31	10·47	10·78	11·08	11·23	11·67	11·96
1/ 16·0	2·4714	2·9578	3·2515	3·8909	4·6023	4·9857	6·2515	7·6972	8·2206	9·3314	10·530	11·163	13·199	14·674
0·06500	8·088	8·432	8·619	8·984	9·340	9·514	10·02	10·52	10·68	10·99	11·30	11·45	11·90	12·20
1/ 15·4	2·5211	3·0173	3·3168	3·9691	4·6947	5·0859	6·3770	7·8516	8·3854	9·5185	10·741	11·386	13·463	14·968
0·06750	8·244	8·595	8·785	9·158	9·520	9·698	10·22	10·72	10·88	11·20	11·52	11·68	12·13	12·43
1/ 14·8	2·5699	3·0756	3·3810	4·0458	4·7854	5·1841	6·5000	8·0030	8·5471	9·7021	10·948	11·606	13·723	15·257
0·07000	8·398	8·755	8·949	9·328	9·697	9·878	10·41	10·92	11·08	11·41	11·73	11·89	12·36	12·66
1/ 14·3	2·6178	3·1329	3·4439	4·1211	4·8744	5·2805	6·6209	8·1517	8·7059	9·8822	11·151	11·821	13·977	15·540
0·07250	8·549	8·912	9·110	9·496	9·871	10·06	10·59	11·11	11·28	11·62	11·94	12·11	12·58	12·89
1/ 13·8	2·6649	3·1892	3·5058	4·1951	4·9619	5·3752	6·7395	8·2978	8·8619	10·059	11·351	12·033	14·228	15·818
0·07500	8·697	9·067	9·268	9·660	10·04	10·23	10·78	11·31	11·48	11·82	12·15	12·31	12·80	13·11
1/ 13·3	2·7111	3·2445	3·5666	4·2678	5·0478	5·4683	6·8562	8·4413	9·0152	10·233	11·547	12·241	14·473	16·091
0·08000	8·987	9·368	9·576	9·982	10·38	10·57	11·14	11·68	11·86	12·21	12·55	12·72	13·22	13·55
1/ 12·5	2·8014	3·3524	3·6852	4·4097	5·2156	5·6500	7·0838	8·7214	9·3143	10·573	11·930	12·647	14·953	16·624
0·08500	9·267	9·661	9·875	10·29	10·70	10·90	11·48	12·04	12·23	12·59	12·94	13·12	13·63	13·97
1/ 11·8	2·8888	3·4571	3·8002	4·5472	5·3781	5·8261	7·3045	8·9930	9·6042	10·902	12·301	13·040	15·418	17·141
0·09000	9·540	9·945	10·16	10·60	11·01	11·22	11·82	12·40	12·59	12·96	13·32	13·50	14·03	14·38
1/ 11·1	2·9738	3·5587	3·9118	4·6808	5·5360	5·9970	7·5187	9·2566	9·8857	11·221	12·662	13·422	15·870	17·643
0·09500	9·805	10·22	10·45	10·89	11·32	11·53	12·15	12·74	12·94	13·32	13·69	13·88	14·42	14·77
1/ 10·5	3·0564	3·6575	4·0204	4·8106	5·6895	6·1633	7·7271	9·5130	10·160	11·532	13·012	13·794	16·309	18·131
0·10000	10·06	10·49	10·72	11·18	11·62	11·83	12·47	13·08	13·28	13·67	14·05	14·24	14·80	15·16
1/ 10·0	3·1369	3·7537	4·1262	4·9371	5·8391	6·3253	7·9301	9·7628	10·426	11·834	13·353	14·156	16·736	18·606
	0·95	0·95	0·95	0·95	0·95	0·96	0·96	0·96	0·96	0·96	0·96	0·96	0·97	0·97

$V_{r(0·5)medial}$ for half-full circular pipes.

A27

(p.1 of 6)

k_s = 0·150 mm
S = 0·00010 to 0·00030

ie hydraulic gradient =
1 in 10000 to 1 in 3333

Water (or sewage) at 15°C;
full bore conditions.

velocities in ms^{-1}
discharges in m^3s^{-1}

Gradient	**(Equivalent) Pipe diameters in mm**													
	630	**675**	**700**	**750**	**800**	**825**	**900**	**975**	**1000**	**1050**	**1100**	**1125**	**1200**	**1250**
0·000100	0·261	0·273	0·280	0·293	0·305	0·311	0·329	0·347	0·352	0·364	0·375	0·380	0·396	0·406
1/ 10000	0·0814	0·0977	0·1076	0·1292	0·1534	0·1664	0·2095	0·2588	0·2768	0·3148	0·3560	0·3778	0·4479	0·4988
0·000105	0·268	0·280	0·287	0·300	0·313	0·319	0·338	0·356	0·362	0·373	0·384	0·390	0·406	0·417
1/ 9524	0·0835	0·1003	0·1105	0·1327	0·1574	0·1708	0·2150	0·2657	0·2841	0·3231	0·3654	0·3877	0·4597	0·5119
0·000110	0·275	0·287	0·294	0·308	0·321	0·328	0·346	0·365	0·371	0·383	0·394	0·400	0·417	0·428
1/ 9091	0·0857	0·1029	0·1133	0·1360	0·1614	0·1751	0·2204	0·2723	0·2912	0·3312	0·3745	0·3974	0·4711	0·5247
0·000115	0·281	0·294	0·302	0·315	0·329	0·335	0·355	0·374	0·380	0·392	0·404	0·409	0·427	0·438
1/ 8696	0·0877	0·1054	0·1160	0·1393	0·1653	0·1793	0·2257	0·2789	0·2982	0·3392	0·3835	0·4069	0·4824	0·5372
0·000120	0·288	0·301	0·308	0·323	0·336	0·343	0·363	0·382	0·388	0·401	0·413	0·419	0·436	0·448
1/ 8333	0·0898	0·1078	0·1187	0·1425	0·1691	0·1834	0·2309	0·2853	0·3050	0·3469	0·3922	0·4162	0·4934	0·5495
0·000125	0·294	0·308	0·315	0·330	0·344	0·351	0·371	0·390	0·397	0·409	0·422	0·428	0·446	0·458
1/ 8000	0·0918	0·1102	0·1213	0·1457	0·1728	0·1875	0·2360	0·2915	0·3117	0·3545	0·4008	0·4253	0·5042	0·5615
0·000130	0·301	0·314	0·322	0·337	0·351	0·358	0·379	0·399	0·405	0·418	0·431	0·437	0·455	0·467
1/ 7692	0·0937	0·1125	0·1239	0·1488	0·1765	0·1914	0·2410	0·2977	0·3183	0·3620	0·4093	0·4343	0·5148	0·5732
0·000135	0·307	0·321	0·329	0·344	0·358	0·365	0·386	0·407	0·413	0·427	0·439	0·446	0·464	0·477
1/ 7407	0·0956	0·1148	0·1264	0·1518	0·1801	0·1953	0·2458	0·3037	0·3247	0·3693	0·4175	0·4430	0·5252	0·5848
0·000140	0·313	0·327	0·335	0·350	0·365	0·373	0·394	0·415	0·422	0·435	0·448	0·454	0·473	0·486
1/ 7143	0·0975	0·1171	0·1289	0·1548	0·1836	0·1992	0·2507	0·3096	0·3310	0·3765	0·4257	0·4517	0·5354	0·5962
0·000145	0·319	0·333	0·341	0·357	0·372	0·380	0·401	0·423	0·429	0·443	0·456	0·463	0·482	0·495
1/ 6897	0·0994	0·1193	0·1314	0·1577	0·1871	0·2029	0·2554	0·3155	0·3373	0·3836	0·4337	0·4601	0·5454	0·6073
0·000150	0·325	0·340	0·348	0·363	0·379	0·387	0·409	0·430	0·437	0·451	0·465	0·471	0·491	0·504
1/ 6667	0·1012	0·1215	0·1338	0·1606	0·1905	0·2066	0·2600	0·3212	0·3434	0·3906	0·4415	0·4685	0·5553	0·6183
0·000160	0·336	0·351	0·360	0·376	0·392	0·400	0·423	0·445	0·452	0·467	0·481	0·488	0·508	0·521
1/ 6250	0·1048	0·1258	0·1385	0·1662	0·1971	0·2138	0·2691	0·3324	0·3553	0·4041	0·4569	0·4847	0·5745	0·6397
0·000170	0·347	0·363	0·372	0·389	0·405	0·413	0·437	0·460	0·467	0·482	0·496	0·504	0·525	0·538
1/ 5882	0·1082	0·1299	0·1430	0·1717	0·2036	0·2208	0·2779	0·3432	0·3669	0·4173	0·4717	0·5005	0·5932	0·6605
0·000180	0·358	0·374	0·383	0·401	0·418	0·426	0·450	0·474	0·482	0·497	0·512	0·519	0·541	0·555
1/ 5556	0·1116	0·1339	0·1475	0·1770	0·2099	0·2277	0·2865	0·3538	0·3782	0·4301	0·4862	0·5159	0·6114	0·6807
0·000190	0·368	0·385	0·394	0·412	0·430	0·438	0·463	0·488	0·496	0·511	0·526	0·534	0·556	0·571
1/ 5263	0·1148	0·1378	0·1518	0·1821	0·2160	0·2343	0·2948	0·3641	0·3892	0·4426	0·5003	0·5308	0·6291	0·7004
0·000200	0·379	0·396	0·405	0·424	0·442	0·450	0·476	0·501	0·509	0·525	0·541	0·549	0·571	0·586
1/ 5000	0·1180	0·1417	0·1560	0·1872	0·2220	0·2408	0·3029	0·3741	0·3999	0·4547	0·5140	0·5453	0·6463	0·7195
0·000210	0·389	0·406	0·416	0·435	0·453	0·462	0·489	0·514	0·522	0·539	0·555	0·563	0·586	0·602
1/ 4762	0·1211	0·1454	0·1601	0·1921	0·2278	0·2471	0·3108	0·3838	0·4103	0·4666	0·5274	0·5595	0·6631	0·7382
0·000220	0·398	0·416	0·426	0·446	0·464	0·474	0·501	0·527	0·535	0·552	0·569	0·577	0·601	0·616
1/ 4545	0·1242	0·1490	0·1641	0·1969	0·2335	0·2532	0·3186	0·3934	0·4205	0·4782	0·5405	0·5734	0·6795	0·7565
0·000230	0·408	0·426	0·437	0·456	0·476	0·485	0·513	0·539	0·548	0·565	0·582	0·591	0·615	0·631
1/ 4348	0·1271	0·1526	0·1680	0·2016	0·2390	0·2593	0·3261	0·4027	0·4305	0·4895	0·5533	0·5870	0·6956	0·7744
0·000240	0·417	0·436	0·447	0·467	0·486	0·496	0·524	0·552	0·561	0·578	0·595	0·604	0·629	0·645
1/ 4167	0·1301	0·1561	0·1718	0·2062	0·2445	0·2652	0·3336	0·4119	0·4403	0·5006	0·5658	0·6003	0·7113	0·7919
0·000250	0·426	0·446	0·456	0·477	0·497	0·507	0·536	0·564	0·573	0·591	0·608	0·617	0·643	0·659
1/ 4000	0·1329	0·1595	0·1756	0·2107	0·2498	0·2710	0·3408	0·4208	0·4498	0·5115	0·5781	0·6133	0·7267	0·8090
0·000260	0·435	0·455	0·466	0·487	0·507	0·517	0·547	0·575	0·585	0·603	0·621	0·630	0·656	0·673
1/ 3846	0·1357	0·1629	0·1793	0·2151	0·2551	0·2766	0·3479	0·4296	0·4592	0·5222	0·5901	0·6261	0·7418	0·8258
0·000270	0·444	0·464	0·475	0·497	0·518	0·528	0·558	0·587	0·596	0·615	0·633	0·642	0·669	0·686
1/ 3704	0·1385	0·1662	0·1829	0·2195	0·2602	0·2822	0·3549	0·4382	0·4684	0·5326	0·6019	0·6386	0·7567	0·8423
0·000280	0·453	0·473	0·485	0·506	0·528	0·538	0·569	0·598	0·608	0·627	0·646	0·655	0·682	0·700
1/ 3571	0·1412	0·1694	0·1865	0·2237	0·2653	0·2877	0·3618	0·4467	0·4775	0·5429	0·6136	0·6509	0·7712	0·8585
0·000290	0·461	0·482	0·494	0·516	0·538	0·548	0·579	0·609	0·619	0·639	0·658	0·667	0·695	0·713
1/ 3448	0·1438	0·1726	0·1900	0·2279	0·2702	0·2930	0·3686	0·4550	0·4864	0·5530	0·6250	0·6630	0·7855	0·8744
0·000300	0·470	0·491	0·503	0·525	0·547	0·558	0·590	0·620	0·630	0·650	0·669	0·679	0·707	0·725
1/ 3333	0·1464	0·1757	0·1934	0·2320	0·2751	0·2983	0·3752	0·4632	0·4951	0·5629	0·6362	0·6749	0·7996	0·8901
	0·87	0·87	0·88	0·88	0·89	0·89	0·89	0·90	0·90	0·90	0·91	0·91	0·91	0·91

$V_{r(0·5)medial}$ for half-full circular pipes.

k_s = 0·150 mm
S = 0·00030 to 0·00100

ie hydraulic gradient =
1 in 3333 to 1 in 1000

Water (or sewage) at 15°C;
full bore conditions.

velocities in ms^{-1}
discharges in m^3s^{-1}

Gradient **(Equivalent) Pipe diameters in mm**

Gradient	630	675	700	750	800	825	900	975	1000	1050	1100	1125	1200	1250
0·000300	0·470	0·491	0·503	0·525	0·547	0·558	0·590	0·620	0·630	0·650	0·669	0·679	0·707	0·725
1/ 3333	0·1464	0·1757	0·1934	0·2320	0·2751	0·2983	0·3752	0·4632	0·4951	0·5629	0·6362	0·6749	0·7996	0·8901
0·000320	0·486	0·508	0·520	0·543	0·566	0·577	0·610	0·642	0·652	0·672	0·692	0·702	0·731	0·750
1/ 3125	0·1515	0·1818	0·2001	0·2401	0·2846	0·3087	0·3882	0·4792	0·5122	0·5823	0·6581	0·6981	0·8271	0·9207
0·000340	0·502	0·525	0·537	0·561	0·585	0·596	0·630	0·663	0·673	0·694	0·715	0·725	0·755	0·774
1/ 2941	0·1565	0·1877	0·2066	0·2479	0·2939	0·3187	0·4007	0·4947	0·5287	0·6011	0·6793	0·7206	0·8537	0·9503
0·000360	0·517	0·541	0·553	0·578	0·602	0·614	0·649	0·683	0·694	0·715	0·737	0·747	0·778	0·798
1/ 2778	0·1613	0·1935	0·2130	0·2555	0·3028	0·3284	0·4129	0·5097	0·5448	0·6194	0·6999	0·7425	0·8796	0·9791
0·000380	0·532	0·556	0·569	0·595	0·620	0·632	0·668	0·702	0·714	0·736	0·758	0·768	0·800	0·821
1/ 2632	0·1659	0·1991	0·2191	0·2628	0·3116	0·3379	0·4248	0·5244	0·5605	0·6372	0·7200	0·7638	0·9048	1·0071
0·000400	0·547	0·572	0·585	0·611	0·637	0·649	0·686	0·721	0·733	0·756	0·778	0·789	0·822	0·843
1/ 2500	0·1705	0·2045	0·2251	0·2700	0·3201	0·3471	0·4364	0·5387	0·5757	0·6545	0·7396	0·7845	0·9293	1·0344
0·000420	0·561	0·586	0·600	0·627	0·653	0·666	0·704	0·740	0·752	0·775	0·798	0·810	0·843	0·865
1/ 2381	0·1749	0·2099	0·2310	0·2771	0·3284	0·3561	0·4477	0·5526	0·5906	0·6714	0·7586	0·8047	0·9533	1·0610
0·000440	0·575	0·601	0·615	0·643	0·669	0·683	0·721	0·758	0·771	0·794	0·818	0·829	0·864	0·886
1/ 2273	0·1793	0·2151	0·2367	0·2839	0·3365	0·3649	0·4588	0·5662	0·6052	0·6879	0·7773	0·8245	0·9767	1·0871
0·000460	0·589	0·615	0·630	0·658	0·685	0·699	0·738	0·776	0·789	0·813	0·837	0·849	0·884	0·907
1/ 2174	0·1835	0·2202	0·2423	0·2906	0·3445	0·3735	0·4696	0·5795	0·6194	0·7041	0·7955	0·8439	0·9996	1·1125
0·000480	0·602	0·629	0·644	0·673	0·701	0·714	0·755	0·794	0·806	0·831	0·856	0·868	0·904	0·927
1/ 2083	0·1877	0·2252	0·2478	0·2972	0·3522	0·3819	0·4801	0·5926	0·6333	0·7199	0·8134	0·8628	1·0219	1·1374
0·000500	0·615	0·643	0·658	0·687	0·716	0·730	0·771	0·811	0·824	0·849	0·874	0·887	0·923	0·947
1/ 2000	0·1918	0·2300	0·2532	0·3036	0·3598	0·3902	0·4905	0·6053	0·6469	0·7354	0·8309	0·8813	1·0439	1·1618
0·000525	0·631	0·660	0·675	0·705	0·734	0·749	0·791	0·832	0·845	0·871	0·897	0·909	0·947	0·971
1/ 1905	0·1968	0·2360	0·2597	0·3115	0·3692	0·4003	0·5032	0·6209	0·6636	0·7543	0·8522	0·9040	1·0707	1·1916
0·000550	0·647	0·676	0·692	0·722	0·753	0·767	0·810	0·852	0·866	0·892	0·919	0·932	0·970	0·995
1/ 1818	0·2016	0·2418	0·2662	0·3192	0·3783	0·4101	0·5155	0·6362	0·6799	0·7728	0·8731	0·9261	1·0969	1·2208
0·000575	0·662	0·692	0·708	0·739	0·770	0·785	0·829	0·872	0·886	0·913	0·940	0·953	0·993	1·018
1/ 1739	0·2064	0·2476	0·2724	0·3267	0·3872	0·4198	0·5276	0·6511	0·6958	0·7909	0·8935	0·9478	1·1225	1·2493
0·000600	0·677	0·707	0·724	0·756	0·788	0·803	0·848	0·892	0·906	0·934	0·961	0·975	1·015	1·041
1/ 1667	0·2111	0·2531	0·2786	0·3341	0·3959	0·4292	0·5395	0·6657	0·7114	0·8086	0·9135	0·9690	1·1476	1·2772
0·000625	0·692	0·723	0·740	0·772	0·805	0·820	0·866	0·911	0·925	0·954	0·982	0·996	1·036	1·063
1/ 1600	0·2156	0·2586	0·2846	0·3413	0·4044	0·4385	0·5511	0·6800	0·7267	0·8259	0·9331	0·9898	1·1722	1·3045
0·000650	0·706	0·738	0·755	0·788	0·821	0·837	0·884	0·930	0·944	0·973	1·002	1·016	1·058	1·085
1/ 1538	0·2201	0·2640	0·2905	0·3483	0·4128	0·4475	0·5625	0·6940	0·7417	0·8430	0·9523	1·0101	1·1963	1·3313
0·000675	0·720	0·752	0·770	0·804	0·837	0·854	0·902	0·948	0·963	0·993	1·022	1·036	1·079	1·106
1/ 1481	0·2245	0·2693	0·2963	0·3553	0·4210	0·4564	0·5736	0·7078	0·7564	0·8596	0·9712	1·0301	1·2199	1·3576
0·000700	0·734	0·767	0·785	0·820	0·854	0·870	0·919	0·966	0·981	1·012	1·041	1·056	1·099	1·127
1/ 1429	0·2288	0·2744	0·3020	0·3621	0·4290	0·4652	0·5846	0·7213	0·7708	0·8760	0·9897	1·0497	1·2431	1·3834
0·000725	0·748	0·781	0·799	0·835	0·869	0·886	0·936	0·984	0·999	1·030	1·061	1·075	1·119	1·148
1/ 1379	0·2331	0·2795	0·3076	0·3688	0·4370	0·4738	0·5954	0·7345	0·7850	0·8921	1·0078	1·0690	1·2659	1·4087
0·000750	0·761	0·795	0·814	0·850	0·885	0·902	0·953	1·001	1·017	1·049	1·079	1·094	1·139	1·168
1/ 1333	0·2373	0·2845	0·3131	0·3754	0·4447	0·4822	0·6060	0·7476	0·7989	0·9080	1·0257	1·0879	1·2883	1·4337
0·000800	0·787	0·822	0·841	0·879	0·915	0·933	0·985	1·035	1·052	1·084	1·116	1·132	1·178	1·208
1/ 1250	0·2454	0·2943	0·3238	0·3882	0·4599	0·4987	0·6266	0·7731	0·8261	0·9389	1·0606	1·1249	1·3321	1·4823
0·000850	0·813	0·849	0·868	0·907	0·944	0·963	1·017	1·068	1·085	1·119	1·152	1·168	1·215	1·246
1/ 1176	0·2533	0·3037	0·3342	0·4007	0·4747	0·5146	0·6467	0·7978	0·8525	0·9688	1·0944	1·1608	1·3745	1·5295
0·000900	0·837	0·874	0·895	0·934	0·973	0·992	1·047	1·101	1·118	1·152	1·186	1·203	1·252	1·284
1/ 1111	0·2610	0·3129	0·3443	0·4128	0·4890	0·5302	0·6662	0·8217	0·8781	0·9979	1·1272	1·1956	1·4157	1·5753
0·000950	0·861	0·899	0·920	0·961	1·001	1·020	1·077	1·132	1·150	1·185	1·220	1·237	1·287	1·320
1/ 1053	0·2684	0·3219	0·3541	0·4245	0·5029	0·5453	0·6851	0·8451	0·9031	1·0262	1·1592	1·2295	1·4558	1·6199
0·001000	0·884	0·924	0·945	0·987	1·028	1·048	1·106	1·162	1·181	1·217	1·253	1·270	1·322	1·355
1/ 1000	0·2757	0·3306	0·3637	0·4360	0·5165	0·5600	0·7036	0·8678	0·9274	1·0538	1·1903	1·2625	1·4948	1·6633
	0·89	0·89	0·89	0·90	0·90	0·90	0·91	0·91	0·91	0·92	0·92	0·92	0·92	0·93

$V_{r(0\cdot5)medial}$ for half-full circular pipes.

$k_s = 0{\cdot}150$ **mm**
S = 0·00100 to 0·00300

ie hydraulic gradient =
1 in 1000 to 1 in 333

Water (or sewage) at 15°C;
full bore conditions.

velocities in ms^{-1}
discharges in m^3s^{-1}

Gradient	**(Equivalent) Pipe diameters in mm**													
	630	**675**	**700**	**750**	**800**	**825**	**900**	**975**	**1000**	**1050**	**1100**	**1125**	**1200**	**1250**
0·00100	0·884	0·924	0·945	0·987	1·028	1·048	1·106	1·162	1·181	1·217	1·253	1·270	1·322	1·355
1/ 1000	0·2757	0·3306	0·3637	0·4360	0·5165	0·5600	0·7036	0·8678	0·9274	1·0538	1·1903	1·2625	1·4948	1·6633
0·00105	0·907	0·948	0·969	1·012	1·054	1·074	1·134	1·192	1·211	1·248	1·285	1·302	1·355	1·390
1/ 952	0·2828	0·3391	0·3731	0·4472	0·5298	0·5743	0·7216	0·8900	0·9511	1·0807	1·2207	1·2947	1·5329	1·7057
0·00110	0·930	0·971	0·993	1·037	1·080	1·101	1·162	1·221	1·240	1·278	1·316	1·334	1·388	1·424
1/ 909	0·2898	0·3474	0·3822	0·4581	0·5427	0·5883	0·7392	0·9117	0·9742	1·1070	1·2504	1·3261	1·5701	1·7470
0·00115	0·951	0·993	1·016	1·061	1·105	1·126	1·189	1·249	1·269	1·308	1·346	1·365	1·420	1·457
1/ 870	0·2965	0·3555	0·3911	0·4688	0·5554	0·6020	0·7563	0·9329	0·9968	1·1327	1·2794	1·3569	1·6065	1·7875
0·00120	0·973	1·016	1·039	1·085	1·129	1·151	1·215	1·277	1·297	1·337	1·376	1·395	1·452	1·489
1/ 833	0·3032	0·3634	0·3999	0·4793	0·5677	0·6155	0·7732	0·9536	1·0190	1·1578	1·3078	1·3870	1·6421	1·8271
0·00125	0·993	1·037	1·061	1·108	1·154	1·176	1·241	1·304	1·325	1·366	1·405	1·425	1·483	1·520
1/ 800	0·3097	0·3712	0·4084	0·4895	0·5799	0·6286	0·7897	0·9739	1·0407	1·1825	1·3356	1·4165	1·6770	1·8659
0·00130	1·014	1·059	1·083	1·131	1·177	1·200	1·267	1·331	1·352	1·393	1·434	1·454	1·513	1·551
1/ 769	0·3161	0·3789	0·4168	0·4996	0·5918	0·6415	0·8058	0·9938	1·0619	1·2066	1·3628	1·4454	1·7111	1·9039
0·00135	1·034	1·080	1·105	1·153	1·200	1·224	1·292	1·357	1·379	1·421	1·462	1·483	1·543	1·582
1/ 741	0·3223	0·3864	0·4251	0·5094	0·6034	0·6541	0·8217	1·0134	1·0828	1·2303	1·3896	1·4737	1·7447	1·9412
0·00140	1·054	1·100	1·126	1·175	1·223	1·247	1·316	1·383	1·405	1·448	1·490	1·511	1·572	1·612
1/ 714	0·3284	0·3937	0·4331	0·5191	0·6149	0·6665	0·8373	1·0325	1·1033	1·2536	1·4158	1·5016	1·7776	1·9778
0·00145	1·073	1·120	1·146	1·197	1·246	1·270	1·340	1·408	1·430	1·474	1·517	1·538	1·600	1·641
1/ 690	0·3345	0·4009	0·4411	0·5286	0·6261	0·6787	0·8526	1·0514	1·1234	1·2764	1·4416	1·5289	1·8100	2·0138
0·00150	1·092	1·140	1·166	1·218	1·268	1·292	1·364	1·433	1·456	1·500	1·544	1·565	1·628	1·670
1/ 667	0·3404	0·4080	0·4489	0·5380	0·6372	0·6907	0·8676	1·0699	1·1432	1·2989	1·4670	1·5558	1·8418	2·0491
0·00160	1·129	1·179	1·206	1·259	1·311	1·336	1·410	1·481	1·505	1·551	1·596	1·618	1·683	1·726
1/ 625	0·3520	0·4219	0·4642	0·5562	0·6588	0·7141	0·8970	1·1061	1·1818	1·3428	1·5165	1·6083	1·9038	2·1182
0·00170	1·165	1·217	1·244	1·299	1·352	1·378	1·455	1·528	1·552	1·600	1·646	1·669	1·737	1·781
1/ 588	0·3632	0·4353	0·4789	0·5739	0·6797	0·7368	0·9254	1·1411	1·2193	1·3853	1·5645	1·6591	1·9640	2·1851
0·00180	1·200	1·253	1·282	1·338	1·393	1·420	1·498	1·574	1·599	1·647	1·695	1·719	1·788	1·834
1/ 556	0·3741	0·4484	0·4933	0·5911	0·7001	0·7588	0·9530	1·1752	1·2556	1·4266	1·6111	1·7085	2·0224	2·2500
0·00190	1·234	1·289	1·318	1·376	1·432	1·460	1·540	1·618	1·644	1·694	1·743	1·767	1·838	1·885
1/ 526	0·3847	0·4611	0·5073	0·6078	0·7198	0·7803	0·9799	1·2083	1·2910	1·4667	1·6564	1·7566	2·0793	2·3133
0·00200	1·267	1·323	1·353	1·413	1·470	1·499	1·582	1·662	1·688	1·739	1·789	1·814	1·887	1·935
1/ 500	0·3951	0·4735	0·5209	0·6241	0·7391	0·8012	1·0061	1·2405	1·3255	1·5058	1·7006	1·8034	2·1347	2·3748
0·00210	1·300	1·357	1·388	1·449	1·508	1·537	1·622	1·704	1·730	1·783	1·835	1·860	1·935	1·984
1/ 476	0·4052	0·4855	0·5341	0·6400	0·7579	0·8215	1·0317	1·2720	1·3591	1·5440	1·7436	1·8491	2·1887	2·4349
0·00220	1·331	1·390	1·422	1·484	1·544	1·574	1·661	1·745	1·772	1·826	1·879	1·905	1·982	2·032
1/ 455	0·4150	0·4973	0·5471	0·6555	0·7763	0·8414	1·0566	1·3027	1·3919	1·5813	1·7857	1·8937	2·2414	2·4936
0·00230	1·362	1·422	1·455	1·518	1·580	1·610	1·699	1·785	1·813	1·868	1·922	1·949	2·027	2·079
1/ 435	0·4246	0·5089	0·5598	0·6707	0·7942	0·8609	1·0810	1·3328	1·4240	1·6177	1·8268	1·9373	2·2930	2·5509
0·00240	1·392	1·454	1·487	1·552	1·615	1·646	1·737	1·824	1·853	1·909	1·965	1·992	2·072	2·124
1/ 417	0·4341	0·5202	0·5722	0·6856	0·8118	0·8799	1·1049	1·3622	1·4554	1·6534	1·8671	1·9800	2·3435	2·6070
0·00250	1·422	1·484	1·518	1·585	1·649	1·681	1·774	1·863	1·892	1·950	2·006	2·034	2·116	2·169
1/ 400	0·4433	0·5312	0·5843	0·7001	0·8290	0·8985	1·1283	1·3910	1·4862	1·6883	1·9065	2·0218	2·3929	2·6620
0·00260	1·451	1·515	1·549	1·617	1·683	1·715	1·810	1·901	1·931	1·989	2·047	2·075	2·159	2·213
1/ 385	0·4523	0·5420	0·5962	0·7144	0·8459	0·9168	1·1512	1·4192	1·5163	1·7225	1·9451	2·0627	2·4414	2·7159
0·00270	1·480	1·544	1·580	1·649	1·716	1·749	1·845	1·938	1·968	2·028	2·087	2·116	2·201	2·256
1/ 370	0·4612	0·5527	0·6079	0·7283	0·8624	0·9348	1·1737	1·4469	1·5459	1·7561	1·9831	2·1029	2·4889	2·7687
0·00280	1·508	1·574	1·609	1·680	1·748	1·782	1·880	1·974	2·005	2·066	2·126	2·155	2·242	2·298
1/ 357	0·4699	0·5631	0·6194	0·7421	0·8787	0·9524	1·1958	1·4741	1·5750	1·7891	2·0203	2·1424	2·5356	2·8206
0·00290	1·535	1·602	1·639	1·710	1·780	1·814	1·914	2·010	2·042	2·104	2·164	2·194	2·282	2·340
1/ 345	0·4785	0·5734	0·6307	0·7556	0·8946	0·9697	1·2175	1·5009	1·6035	1·8215	2·0568	2·1812	2·5814	2·8716
0·00300	1·562	1·630	1·668	1·740	1·811	1·846	1·947	2·045	2·077	2·140	2·202	2·233	2·322	2·381
1/ 333	0·4869	0·5834	0·6418	0·7688	0·9103	0·9867	1·2388	1·5271	1·6316	1·8534	2·0928	2·2193	2·6265	2·9217
	0·90	0·90	0·90	0·91	0·91	0·91	0·92	0·92	0·92	0·93	0·93	0·93	0·93	0·93

$V_{r(0{\cdot}5)\text{medial}}$ **for half-full circular pipes.**

$k_s = 0{\cdot}150$ **mm** S = 0·00100 to 0·00300

k_s = 0·150 mm
S = 0·00300 to 0·01000

ie hydraulic gradient =
1 in 333 to 1 in 100

Water (or sewage) at 15°C;
full bore conditions.

velocities in ms^{-1}
discharges in m^3s^{-1}

Gradient (Equivalent) Pipe diameters in mm

Gradient	630	675	700	750	800	825	900	975	1000	1050	1100	1125	1200	1250
0·00300	1·562	1·630	1·668	1·740	1·811	1·846	1·947	2·045	2·077	2·140	2·202	2·233	2·322	2·381
1/ 333	0·4869	0·5834	0·6418	0·7688	0·9103	0·9867	1·2388	1·5271	1·6316	1·8534	2·0928	2·2193	2·6265	2·9217
0·00320	1·615	1·685	1·724	1·799	1·872	1·908	2·013	2·114	2·147	2·212	2·276	2·307	2·400	2·461
1/ 313	0·5034	0·6031	0·6634	0·7947	0·9410	1·0198	1·2804	1·5784	1·6863	1·9155	2·1629	2·2936	2·7144	3·0195
0·00340	1·666	1·739	1·778	1·856	1·931	1·968	2·076	2·181	2·215	2·282	2·348	2·380	2·475	2·538
1/ 294	0·5193	0·6222	0·6843	0·8198	0·9706	1·0520	1·3208	1·6280	1·7394	1·9758	2·2309	2·3657	2·7997	3·1143
0·00360	1·715	1·790	1·831	1·911	1·988	2·026	2·138	2·245	2·280	2·349	2·417	2·450	2·549	2·613
1/ 278	0·5348	0·6407	0·7047	0·8442	0·9995	1·0832	1·3599	1·6763	1·7909	2·0343	2·2969	2·4357	2·8825	3·2063
0·00380	1·764	1·841	1·883	1·964	2·044	2·083	2·198	2·308	2·344	2·415	2·485	2·519	2·620	2·686
1/ 263	0·5498	0·6587	0·7245	0·8679	1·0275	1·1136	1·3980	1·7232	1·8410	2·0912	2·3612	2·5038	2·9630	3·2959
0·00400	1·811	1·890	1·933	2·017	2·099	2·139	2·256	2·369	2·406	2·479	2·550	2·586	2·689	2·757
1/ 250	0·5645	0·6763	0·7438	0·8910	1·0548	1·1432	1·4352	1·7689	1·8898	2·1466	2·4237	2·5701	3·0414	3·3831
0·00420	1·857	1·938	1·982	2·068	2·152	2·193	2·313	2·429	2·467	2·541	2·615	2·651	2·757	2·826
1/ 238	0·5788	0·6934	0·7626	0·9135	1·0815	1·1721	1·4714	1·8135	1·9375	2·2007	2·4847	2·6348	3·1179	3·4682
0·00440	1·901	1·984	2·029	2·118	2·203	2·245	2·368	2·487	2·526	2·602	2·677	2·714	2·823	2·894
1/ 227	0·5927	0·7101	0·7810	0·9355	1·1075	1·2003	1·5067	1·8571	1·9840	2·2535	2·5443	2·6980	3·1926	3·5512
0·00460	1·945	2·030	2·076	2·166	2·254	2·297	2·423	2·544	2·584	2·662	2·739	2·776	2·887	2·960
1/ 217	0·6064	0·7264	0·7990	0·9570	1·1329	1·2278	1·5413	1·8996	2·0294	2·3051	2·6026	2·7597	3·2657	3·6324
0·00480	1·988	2·075	2·122	2·214	2·303	2·347	2·476	2·600	2·641	2·720	2·799	2·837	2·951	3·025
1/ 208	0·6197	0·7424	0·8165	0·9780	1·1578	1·2548	1·5751	1·9413	2·0739	2·3556	2·6596	2·8202	3·3371	3·7119
0·00500	2·030	2·118	2·166	2·260	2·352	2·397	2·528	2·655	2·696	2·778	2·857	2·897	3·013	3·088
1/ 200	0·6328	0·7581	0·8338	0·9987	1·1822	1·2812	1·6082	1·9821	2·1175	2·4051	2·7154	2·8794	3·4071	3·7898
0·00525	2·081	2·172	2·221	2·318	2·411	2·457	2·592	2·722	2·764	2·847	2·929	2·969	3·088	3·166
1/ 190	0·6488	0·7772	0·8548	1·0239	1·2120	1·3135	1·6487	2·0320	2·1708	2·4655	2·7837	2·9517	3·4927	3·8849
0·00550	2·132	2·224	2·275	2·373	2·469	2·516	2·654	2·787	2·830	2·916	2·999	3·041	3·162	3·241
1/ 182	0·6645	0·7959	0·8754	1·0485	1·2411	1·3450	1·6883	2·0807	2·2228	2·5246	2·8503	3·0224	3·5763	3·9778
0·00575	2·181	2·275	2·327	2·428	2·526	2·574	2·715	2·851	2·895	2·982	3·068	3·110	3·234	3·315
1/ 174	0·6797	0·8142	0·8955	1·0725	1·2696	1·3759	1·7269	2·1283	2·2736	2·5823	2·9155	3·0914	3·6580	4·0686
0·00600	2·228	2·325	2·378	2·481	2·581	2·630	2·774	2·913	2·958	3·047	3·135	3·178	3·305	3·388
1/ 167	0·6947	0·8321	0·9151	1·0960	1·2974	1·4060	1·7648	2·1749	2·3234	2·6388	2·9792	3·1590	3·7379	4·1575
0·00625	2·275	2·374	2·428	2·533	2·635	2·686	2·832	2·974	3·020	3·111	3·201	3·245	3·374	3·459
1/ 160	0·7093	0·8496	0·9344	1·1191	1·3247	1·4356	1·8018	2·2205	2·3721	2·6941	3·0417	3·2252	3·8162	4·2446
0·00650	2·321	2·422	2·477	2·584	2·689	2·740	2·889	3·034	3·081	3·174	3·265	3·310	3·442	3·528
1/ 154	0·7237	0·8668	0·9533	1·1417	1·3514	1·4646	1·8382	2·2652	2·4199	2·7484	3·1029	3·2901	3·8929	4·3299
0·00675	2·367	2·469	2·525	2·634	2·741	2·793	2·945	3·093	3·141	3·235	3·328	3·374	3·509	3·597
1/ 148	0·7377	0·8837	0·9718	1·1639	1·3777	1·4930	1·8738	2·3091	2·4668	2·8016	3·1629	3·3538	3·9682	4·4136
0·00700	2·411	2·516	2·572	2·684	2·792	2·845	3·000	3·150	3·199	3·296	3·390	3·437	3·574	3·663
1/ 143	0·7516	0·9002	0·9900	1·1856	1·4034	1·5209	1·9088	2·3522	2·5128	2·8538	3·2219	3·4163	4·0421	4·4958
0·00725	2·455	2·561	2·619	2·732	2·842	2·896	3·054	3·207	3·257	3·355	3·451	3·499	3·638	3·729
1/ 138	0·7651	0·9164	1·0079	1·2070	1·4287	1·5483	1·9431	2·3945	2·5580	2·9051	3·2798	3·4777	4·1147	4·5765
0·00750	2·497	2·606	2·665	2·780	2·892	2·947	3·108	3·263	3·314	3·413	3·511	3·559	3·701	3·794
1/ 133	0·7785	0·9324	1·0254	1·2280	1·4536	1·5752	1·9769	2·4361	2·6024	2·9556	3·3367	3·5380	4·1861	4·6558
0·00800	2·581	2·693	2·754	2·873	2·988	3·045	3·211	3·372	3·424	3·527	3·628	3·678	3·824	3·920
1/ 125	0·8045	0·9636	1·0597	1·2690	1·5021	1·6278	2·0428	2·5173	2·6891	3·0540	3·4478	3·6557	4·3253	4·8106
0·00850	2·662	2·777	2·840	2·963	3·082	3·140	3·312	3·477	3·531	3·637	3·741	3·793	3·944	4·042
1/ 118	0·8298	0·9938	1·0929	1·3088	1·5491	1·6787	2·1067	2·5959	2·7731	3·1494	3·5554	3·7699	4·4603	4·9607
0·00900	2·741	2·859	2·924	3·050	3·173	3·233	3·409	3·579	3·635	3·744	3·851	3·904	4·060	4·161
1/ 111	0·8543	1·0231	1·1252	1·3474	1·5948	1·7282	2·1688	2·6723	2·8547	3·2420	3·6599	3·8807	4·5913	5·1064
0·00950	2·817	2·939	3·005	3·135	3·261	3·323	3·504	3·679	3·736	3·848	3·958	4·012	4·172	4·277
1/ 105	0·8781	1·0517	1·1565	1·3849	1·6392	1·7763	2·2291	2·7466	2·9341	3·3321	3·7616	3·9885	4·7188	5·2481
0·01000	2·892	3·017	3·085	3·218	3·347	3·411	3·596	3·776	3·834	3·949	4·062	4·118	4·282	4·389
1/ 100	0·9014	1·0795	1·1871	1·4215	1·6824	1·8232	2·2879	2·8190	3·0114	3·4198	3·8606	4·0935	4·8429	5·3862
	0·91	0·91	0·91	0·92	0·92	0·92	0·92	0·93	0·93	0·93	0·93	0·93	0·94	0·94

$V_{r(0·5)medial}$ for half-full circular pipes.

k_s = 0·150 mm
S = 0·01000 to 0·03000

ie hydraulic gradient =
1 in 100 to 1 in 33·3

Water (or sewage) at 15°C;
full bore conditions.

velocities in ms^{-1}
discharges in m^3s^{-1}

Gradient **(Equivalent) Pipe diameters in mm**

Gradient	630	675	700	750	800	825	900	975	1000	1050	1100	1125	1200	1250
0·01000	2·892	3·017	3·085	3·218	3·347	3·411	3·596	3·776	3·834	3·949	4·062	4·118	4·282	4·389
1/ 100	0·9014	1·0795	1·1871	1·4215	1·6824	1·8232	2·2879	2·8190	3·0114	3·4198	3·8606	4·0935	4·8429	5·3862
0·01050	2·964	3·092	3·162	3·298	3·431	3·496	3·686	3·870	3·930	4·048	4·164	4·221	4·389	4·499
1/ 95	0·9240	1·1066	1·2169	1·4572	1·7246	1·8689	2·3452	2·8896	3·0868	3·5054	3·9572	4·1959	4·9640	5·5208
0·01100	3·035	3·166	3·238	3·377	3·513	3·580	3·774	3·963	4·024	4·145	4·263	4·322	4·494	4·606
1/ 91	0·9461	1·1331	1·2460	1·4920	1·7658	1·9135	2·4012	2·9585	3·1604	3·5890	4·0516	4·2959	5·0823	5·6523
0·01150	3·105	3·239	3·312	3·454	3·593	3·661	3·860	4·053	4·116	4·239	4·360	4·420	4·596	4·711
1/ 87	0·9678	1·1590	1·2745	1·5261	1·8061	1·9572	2·4559	3·0259	3·2324	3·6707	4·1438	4·3936	5·1979	5·7809
0·01200	3·172	3·309	3·384	3·530	3·672	3·741	3·945	4·141	4·205	4·332	4·455	4·516	4·696	4·813
1/ 83	0·9889	1·1843	1·3023	1·5594	1·8455	1·9999	2·5095	3·0919	3·3028	3·7507	4·2340	4·4893	5·3110	5·9066
0·01250	3·239	3·379	3·455	3·604	3·748	3·819	4·027	4·228	4·293	4·422	4·548	4·611	4·794	4·914
1/ 80	1·0096	1·2091	1·3296	1·5920	1·8841	2·0417	2·5619	3·1565	3·3718	3·8290	4·3224	4·5830	5·4218	6·0298
0·01300	3·304	3·447	3·524	3·676	3·824	3·896	4·108	4·312	4·379	4·511	4·639	4·703	4·890	5·012
1/ 77	1·0300	1·2334	1·3563	1·6240	1·9220	2·0827	2·6133	3·2198	3·4394	3·9057	4·4090	4·6748	5·5304	6·1505
0·01350	3·368	3·513	3·592	3·747	3·898	3·971	4·187	4·396	4·464	4·598	4·729	4·794	4·984	5·108
1/ 74	1·0499	1·2573	1·3825	1·6554	1·9591	2·1229	2·6637	3·2819	3·5057	3·9810	4·4940	4·7649	5·6369	6·2690
0·01400	3·431	3·579	3·659	3·817	3·970	4·045	4·265	4·477	4·547	4·683	4·817	4·882	5·077	5·203
1/ 71	1·0695	1·2807	1·4083	1·6862	1·9956	2·1624	2·7132	3·3428	3·5708	4·0549	4·5774	4·8533	5·7415	6·3853
0·01450	3·492	3·643	3·725	3·885	4·041	4·118	4·341	4·557	4·628	4·767	4·903	4·970	5·167	5·296
1/ 69	1·0887	1·3037	1·4336	1·7165	2·0314	2·2012	2·7619	3·4027	3·6348	4·1275	4·6593	4·9402	5·8442	6·4995
0·01500	3·553	3·706	3·790	3·953	4·111	4·189	4·417	4·636	4·708	4·849	4·988	5·056	5·257	5·388
1/ 67	1·1076	1·3263	1·4584	1·7462	2·0665	2·2393	2·8097	3·4616	3·6977	4·1989	4·7398	5·0255	5·9452	6·6117
0·01600	3·671	3·830	3·916	4·084	4·248	4·328	4·563	4·790	4·864	5·010	5·153	5·223	5·431	5·566
1/ 63	1·1444	1·3704	1·5069	1·8043	2·1352	2·3137	2·9030	3·5764	3·8203	4·3382	4·8970	5·1921	6·1422	6·8308
0·01700	3·786	3·949	4·038	4·211	4·380	4·463	4·705	4·939	5·016	5·166	5·313	5·386	5·600	5·739
1/ 59	1·1802	1·4132	1·5539	1·8605	2·2017	2·3858	2·9934	3·6878	3·9392	4·4731	5·0493	5·3536	6·3332	7·0431
0·01800	3·897	4·065	4·156	4·335	4·509	4·594	4·843	5·084	5·163	5·317	5·469	5·544	5·764	5·907
1/ 56	1·2149	1·4547	1·5996	1·9152	2·2664	2·4558	3·0812	3·7959	4·0547	4·6042	5·1972	5·5104	6·5186	7·2493
0·01900	4·005	4·178	4·272	4·455	4·634	4·721	4·977	5·225	5·306	5·464	5·620	5·697	5·923	6·071
1/ 53	1·2486	1·4951	1·6440	1·9683	2·3292	2·5239	3·1665	3·9010	4·1670	4·7317	5·3411	5·6629	6·6989	7·4498
0·02000	4·111	4·288	4·384	4·572	4·756	4·845	5·108	5·362	5·445	5·608	5·768	5·846	6·078	6·230
1/ 50	1·2815	1·5344	1·6872	2·0201	2·3904	2·5902	3·2497	4·0034	4·2764	4·8559	5·4812	5·8115	6·8746	7·6451
0·02100	4·214	4·395	4·494	4·687	4·874	4·966	5·236	5·496	5·581	5·748	5·911	5·992	6·230	6·385
1/ 47·6	1·3135	1·5728	1·7294	2·0705	2·4502	2·6549	3·3308	4·1033	4·3830	4·9770	5·6179	5·9564	7·0459	7·8356
0·02200	4·314	4·500	4·601	4·798	4·990	5·085	5·360	5·626	5·713	5·884	6·052	6·135	6·378	6·537
1/ 45·5	1·3449	1·6103	1·7706	2·1198	2·5085	2·7180	3·4100	4·2008	4·4872	5·0952	5·7513	6·0978	7·2132	8·0216
0·02300	4·412	4·602	4·705	4·907	5·104	5·200	5·482	5·754	5·843	6·018	6·189	6·274	6·522	6·685
1/ 43·5	1·3755	1·6469	1·8109	2·1680	2·5655	2·7798	3·4875	4·2962	4·5890	5·2108	5·8818	6·2361	7·3768	8·2034
0·02400	4·508	4·702	4·808	5·014	5·215	5·313	5·601	5·879	5·970	6·148	6·323	6·410	6·664	6·830
1/ 41·7	1·4054	1·6827	1·8503	2·2152	2·6212	2·8402	3·5632	4·3895	4·6887	5·3239	6·0094	6·3715	7·5368	8·3814
0·02500	4·603	4·801	4·908	5·119	5·323	5·424	5·718	6·002	6·094	6·276	6·455	6·543	6·803	6·972
1/ 40·0	1·4347	1·7179	1·8889	2·2614	2·6759	2·8994	3·6374	4·4808	4·7863	5·4347	6·1344	6·5040	7·6935	8·5556
0·02600	4·695	4·897	5·006	5·221	5·430	5·532	5·832	6·121	6·216	6·402	6·584	6·674	6·938	7·111
1/ 38·5	1·4635	1·7523	1·9267	2·3066	2·7294	2·9574	3·7102	4·5704	4·8819	5·5433	6·2570	6·6339	7·8471	8·7264
0·02700	4·785	4·991	5·103	5·322	5·534	5·639	5·944	6·239	6·335	6·525	6·710	6·802	7·072	7·247
1/ 37·0	1·4917	1·7860	1·9638	2·3510	2·7819	3·0143	3·7815	4·6583	4·9758	5·6498	6·3772	6·7613	7·9978	8·8940
0·02800	4·874	5·084	5·197	5·420	5·637	5·743	6·054	6·355	6·453	6·646	6·835	6·928	7·202	7·382
1/ 35·7	1·5194	1·8191	2·0002	2·3946	2·8335	3·0701	3·8516	4·7445	5·0679	5·7544	6·4952	6·8864	8·1458	9·0585
0·02900	4·961	5·175	5·290	5·517	5·738	5·846	6·162	6·468	6·568	6·764	6·957	7·051	7·331	7·513
1/ 34·5	1·5466	1·8517	2·0360	2·4374	2·8841	3·1250	3·9204	4·8292	5·1584	5·8571	6·6111	7·0093	8·2911	9·2200
0·03000	5·047	5·264	5·382	5·612	5·837	5·947	6·269	6·580	6·681	6·881	7·077	7·173	7·457	7·643
1/ 33·3	1·5733	1·8837	2·0712	2·4795	2·9339	3·1789	3·9880	4·9125	5·2473	5·9581	6·7251	7·1301	8·4339	9·3788
	0·91	0·91	0·92	0·92	0·92	0·92	0·93	0·93	0·93	0·93	0·94	0·94	0·94	0·94

$V_{r(0\cdot5)medial}$ for half-full circular pipes.

Gradient **(Equivalent) Pipe diameters in mm**

Gradient	630	675	700	750	800	825	900	975	1000	1050	1100	1125	1200	1250
0·03000	5·047	5·264	5·382	5·612	5·837	5·947	6·269	6·580	6·681	6·881	7·077	7·173	7·457	7·643
1/ 33·3	1·5733	1·8837	2·0712	2·4795	2·9339	3·1789	3·9880	4·9125	5·2473	5·9581	6·7251	7·1301	8·4339	9·3788
0·03200	5·214	5·438	5·560	5·798	6·030	6·144	6·476	6·797	6·902	7·108	7·310	7·410	7·704	7·895
1/ 31·3	1·6255	1·9461	2·1398	2·5616	3·0311	3·2842	4·1200	5·0750	5·4208	6·1551	6·9474	7·3658	8·7126	9·6887
0·03400	5·377	5·607	5·733	5·979	6·217	6·334	6·677	7·008	7·116	7·329	7·537	7·640	7·942	8·140
1/ 29·4	1·6760	2·0066	2·2063	2·6412	3·1252	3·3862	4·2479	5·2325	5·5890	6·3460	7·1628	7·5942	8·9827	9·9891
0·03600	5·534	5·772	5·901	6·154	6·399	6·520	6·872	7·213	7·324	7·543	7·757	7·863	8·174	8·377
1/ 27·8	1·7251	2·0654	2·2709	2·7185	3·2166	3·4852	4·3721	5·3854	5·7524	6·5314	7·3721	7·8160	9·2450	10·281
0·03800	5·687	5·931	6·064	6·324	6·576	6·700	7·062	7·412	7·526	7·751	7·971	8·080	8·400	8·609
1/ 26·3	1·7729	2·1225	2·3337	2·7937	3·3056	3·5816	4·4928	5·5341	5·9112	6·7117	7·5756	8·0318	9·5001	10·564
0·04000	5·836	6·087	6·223	6·489	6·748	6·875	7·247	7·606	7·723	7·954	8·180	8·291	8·620	8·834
1/ 25·0	1·8194	2·1782	2·3949	2·8669	3·3922	3·6754	4·6105	5·6790	6·0659	6·8874	7·7738	8·2419	9·7486	10·841
0·04200	5·982	6·239	6·378	6·651	6·917	7·047	7·428	7·796	7·916	8·152	8·383	8·498	8·834	9·053
1/ 23·8	1·8647	2·2325	2·4546	2·9383	3·4766	3·7669	4·7253	5·8203	6·2169	7·0587	7·9671	8·4468	9·9910	11·110
0·04400	6·124	6·387	6·530	6·809	7·081	7·214	7·604	7·980	8·103	8·345	8·582	8·699	9·043	9·268
1/ 22·7	1·9090	2·2855	2·5129	3·0081	3·5591	3·8562	4·8373	5·9583	6·3642	7·2260	8·1559	8·6470	10·228	11·373
0·04600	6·263	6·532	6·678	6·963	7·241	7·377	7·776	8·161	8·287	8·534	8·776	8·896	9·248	9·477
1/ 21·7	1·9523	2·3373	2·5698	3·0762	3·6397	3·9436	4·9468	6·0931	6·5083	7·3895	8·3404	8·8426	10·459	11·630
0·04800	6·399	6·673	6·822	7·114	7·398	7·537	7·944	8·338	8·466	8·719	8·966	9·088	9·448	9·682
1/ 20·8	1·9947	2·3880	2·6256	3·1429	3·7186	4·0291	5·0540	6·2251	6·6492	7·5495	8·5209	9·0340	10·685	11·882
0·05000	6·532	6·812	6·964	7·262	7·552	7·694	8·109	8·511	8·642	8·900	9·152	9·277	9·644	9·883
1/ 20·0	2·0362	2·4377	2·6802	3·2083	3·7959	4·1128	5·1590	6·3544	6·7872	7·7062	8·6978	9·2215	10·907	12·129
0·05250	6·695	6·982	7·138	7·443	7·740	7·885	8·311	8·722	8·857	9·121	9·380	9·507	9·883	10·13
1/ 19·0	2·0869	2·4984	2·7469	3·2881	3·8904	4·2151	5·2873	6·5123	6·9560	7·8977	8·9139	9·4506	11·178	12·430
0·05500	6·854	7·147	7·307	7·619	7·923	8·072	8·508	8·929	9·066	9·337	9·602	9·733	10·12	10·37
1/ 18·2	2·1364	2·5576	2·8120	3·3661	3·9826	4·3150	5·4126	6·6666	7·1207	8·0848	9·1250	9·6743	11·443	12·724
0·05750	7·009	7·309	7·472	7·792	8·102	8·255	8·701	9·131	9·271	9·548	9·819	9·953	10·35	10·60
1/ 17·4	2·1848	2·6156	2·8757	3·4423	4·0727	4·4127	5·5350	6·8174	7·2818	8·2676	9·3313	9·8931	11·701	13·012
0·06000	7·161	7·468	7·634	7·961	8·278	8·434	8·889	9·329	9·472	9·755	10·03	10·17	10·57	10·83
1/ 16·7	2·2322	2·6723	2·9380	3·5169	4·1610	4·5082	5·6549	6·9650	7·4394	8·4465	9·5332	10·107	11·954	13·293
0·06250	7·310	7·623	7·793	8·126	8·450	8·609	9·073	9·522	9·669	9·957	10·24	10·38	10·79	11·06
1/ 16·0	2·2786	2·7278	2·9991	3·5899	4·2474	4·6018	5·7722	7·1095	7·5937	8·6217	9·7309	10·317	12·202	13·568
0·06500	7·456	7·775	7·948	8·288	8·618	8·780	9·254	9·712	9·861	10·16	10·44	10·59	11·00	11·28
1/ 15·4	2·3241	2·7822	3·0589	3·6615	4·3320	4·6936	5·8873	7·2511	7·7450	8·7934	9·9247	10·522	12·445	13·839
0·06750	7·599	7·924	8·101	8·447	8·784	8·949	9·432	9·898	10·05	10·35	10·64	10·79	11·21	11·49
1/ 14·8	2·3687	2·8356	3·1176	3·7318	4·4151	4·7836	6·0001	7·3901	7·8934	8·9619	10·115	10·724	12·683	14·104
0·07000	7·739	8·071	8·251	8·603	8·946	9·114	9·606	10·08	10·24	10·54	10·84	10·99	11·42	11·70
1/ 14·3	2·4125	2·8880	3·1752	3·8007	4·4967	4·8719	6·1109	7·5265	8·0391	9·1273	10·301	10·922	12·917	14·364
0·07250	7·877	8·214	8·398	8·756	9·105	9·276	9·777	10·26	10·42	10·73	11·03	11·18	11·62	11·91
1/ 13·8	2·4555	2·9395	3·2318	3·8684	4·5768	4·9587	6·2197	7·6605	8·1822	9·2897	10·485	11·116	13·147	14·619
0·07500	8·013	8·356	8·542	8·907	9·262	9·436	9·945	10·44	10·60	10·91	11·22	11·37	11·82	12·12
1/ 13·3	2·4978	2·9901	3·2874	3·9350	4·6555	5·0440	6·3267	7·7922	8·3228	9·4494	10·665	11·307	13·373	14·870
0·08000	8·278	8·632	8·824	9·201	9·567	9·747	10·27	10·78	10·95	11·27	11·59	11·75	12·21	12·52
1/ 12·5	2·5803	3·0888	3·3960	4·0649	4·8091	5·2104	6·5354	8·0491	8·5973	9·7609	11·017	11·680	13·814	15·360
0·08500	8·534	8·899	9·098	9·486	9·864	10·05	10·59	11·11	11·29	11·62	11·95	12·11	12·59	12·90
1/ 11·8	2·6603	3·1845	3·5012	4·1908	4·9580	5·3718	6·7376	8·2982	8·8633	10·063	11·357	12·041	14·241	15·835
0·09000	8·783	9·159	9·363	9·763	10·15	10·34	10·90	11·44	11·61	11·96	12·30	12·47	12·96	13·28
1/ 11·1	2·7379	3·2775	3·6033	4·3130	5·1026	5·5284	6·9340	8·5400	9·1216	10·356	11·688	12·392	14·656	16·297
0·09500	9·025	9·411	9·621	10·03	10·43	10·63	11·20	11·75	11·93	12·29	12·64	12·81	13·32	13·65
1/ 10·5	2·8134	3·3678	3·7027	4·4319	5·2433	5·6807	7·1251	8·7752	9·3728	10·641	12·010	12·733	15·059	16·745
0·10000	9·261	9·657	9·873	10·29	10·70	10·90	11·49	12·06	12·25	12·61	12·97	13·14	13·66	14·00
1/ 10·0	2·8870	3·4559	3·7994	4·5477	5·3802	5·8291	7·3111	9·0043	9·6175	10·919	12·323	13·065	15·452	17·182
	0·91	0·92	0·92	0·92	0·92	0·93	0·93	0·93	0·93	0·94	0·94	0·94	0·94	0·94

$V_{r(0\cdot5)medial}$ for half-full circular pipes.

A28
(p.1 of 6)

$k_s = 0{\cdot}30$ **mm**
S = 0·00010 to 0·00030

ie hydraulic gradient =
1 in 10000 to 1 in 3333

Water (or sewage) at 15°C;
full bore conditions.

velocities in ms^{-1}
discharges in m^3s^{-1}

Gradient	**(Equivalent) Pipe diameters in mm**													
	630	**675**	**700**	**750**	**800**	**825**	**900**	**975**	**1000**	**1050**	**1100**	**1125**	**1200**	**1250**
0·000100	0·252	0·264	0·270	0·283	0·295	0·300	0·318	0·334	0·340	0·351	0·361	0·366	0·382	0·392
1/ 10000	0·0787	0·0944	0·1040	0·1248	0·1481	0·1606	0·2021	0·2497	0·2669	0·3036	0·3432	0·3641	0·4316	0·4806
0·000105	0·259	0·271	0·277	0·290	0·302	0·308	0·326	0·343	0·349	0·360	0·370	0·376	0·392	0·402
1/ 9524	0·0807	0·0969	0·1067	0·1281	0·1519	0·1648	0·2074	0·2562	0·2739	0·3115	0·3521	0·3736	0·4428	0·4930
0·000110	0·265	0·278	0·284	0·297	0·310	0·316	0·334	0·352	0·357	0·369	0·380	0·385	0·401	0·412
1/ 9091	0·0827	0·0993	0·1094	0·1313	0·1557	0·1689	0·2125	0·2625	0·2806	0·3191	0·3608	0·3828	0·4537	0·5052
0·000115	0·272	0·284	0·291	0·304	0·317	0·323	0·342	0·360	0·366	0·377	0·389	0·394	0·411	0·421
1/ 8696	0·0847	0·1017	0·1120	0·1344	0·1594	0·1729	0·2175	0·2687	0·2872	0·3267	0·3693	0·3918	0·4644	0·5170
0·000120	0·278	0·291	0·298	0·311	0·324	0·331	0·350	0·368	0·374	0·386	0·397	0·403	0·420	0·431
1/ 8333	0·0866	0·1040	0·1145	0·1374	0·1630	0·1768	0·2224	0·2747	0·2937	0·3340	0·3776	0·4006	0·4748	0·5286
0·000125	0·284	0·297	0·304	0·318	0·331	0·338	0·357	0·376	0·382	0·394	0·406	0·412	0·429	0·440
1/ 8000	0·0885	0·1063	0·1170	0·1404	0·1665	0·1806	0·2273	0·2807	0·3001	0·3412	0·3857	0·4092	0·4850	0·5400
0·000130	0·290	0·303	0·310	0·324	0·338	0·345	0·365	0·384	0·390	0·402	0·414	0·420	0·438	0·449
1/ 7692	0·0904	0·1085	0·1194	0·1433	0·1700	0·1844	0·2320	0·2865	0·3063	0·3483	0·3937	0·4177	0·4950	0·5512
0·000135	0·296	0·309	0·317	0·331	0·345	0·352	0·372	0·391	0·398	0·410	0·423	0·429	0·446	0·458
1/ 7407	0·0922	0·1107	0·1218	0·1462	0·1734	0·1881	0·2366	0·2922	0·3124	0·3552	0·4015	0·4260	0·5049	0·5621
0·000140	0·301	0·315	0·323	0·337	0·352	0·359	0·379	0·399	0·405	0·418	0·431	0·437	0·455	0·467
1/ 7143	0·0940	0·1128	0·1242	0·1490	0·1767	0·1917	0·2412	0·2978	0·3184	0·3620	0·4092	0·4342	0·5145	0·5728
0·000145	0·307	0·321	0·329	0·344	0·358	0·365	0·386	0·406	0·413	0·426	0·439	0·445	0·463	0·475
1/ 6897	0·0957	0·1149	0·1265	0·1518	0·1800	0·1953	0·2456	0·3033	0·3243	0·3687	0·4168	0·4422	0·5240	0·5834
0·000150	0·313	0·327	0·335	0·350	0·365	0·372	0·393	0·414	0·420	0·433	0·446	0·453	0·472	0·484
1/ 6667	0·0975	0·1170	0·1288	0·1545	0·1832	0·1988	0·2500	0·3088	0·3301	0·3753	0·4242	0·4501	0·5333	0·5938
0·000160	0·323	0·338	0·346	0·362	0·377	0·385	0·407	0·428	0·435	0·448	0·462	0·468	0·488	0·500
1/ 6250	0·1008	0·1210	0·1332	0·1599	0·1896	0·2056	0·2586	0·3193	0·3414	0·3882	0·4387	0·4654	0·5516	0·6140
0·000170	0·334	0·349	0·357	0·374	0·389	0·397	0·420	0·441	0·449	0·463	0·476	0·483	0·503	0·516
1/ 5882	0·1041	0·1249	0·1375	0·1650	0·1957	0·2122	0·2669	0·3296	0·3523	0·4006	0·4528	0·4804	0·5692	0·6337
0·000180	0·344	0·360	0·368	0·385	0·401	0·409	0·432	0·455	0·462	0·477	0·491	0·498	0·518	0·532
1/ 5556	0·1073	0·1287	0·1417	0·1700	0·2016	0·2187	0·2750	0·3396	0·3630	0·4127	0·4665	0·4949	0·5864	0·6528
0·000190	0·354	0·370	0·379	0·396	0·413	0·421	0·445	0·468	0·475	0·490	0·505	0·512	0·533	0·547
1/ 5263	0·1104	0·1324	0·1458	0·1749	0·2074	0·2249	0·2829	0·3493	0·3734	0·4245	0·4798	0·5090	0·6031	0·6713
0·000200	0·364	0·380	0·389	0·407	0·424	0·432	0·457	0·480	0·488	0·503	0·518	0·526	0·548	0·562
1/ 5000	0·1134	0·1360	0·1498	0·1797	0·2130	0·2310	0·2906	0·3587	0·3834	0·4360	0·4927	0·5227	0·6193	0·6894
0·000210	0·373	0·390	0·399	0·417	0·435	0·443	0·468	0·493	0·501	0·516	0·532	0·539	0·562	0·576
1/ 4762	0·1163	0·1396	0·1536	0·1843	0·2185	0·2370	0·2980	0·3680	0·3933	0·4472	0·5053	0·5361	0·6352	0·7071
0·000220	0·382	0·400	0·409	0·427	0·445	0·454	0·480	0·505	0·513	0·529	0·545	0·552	0·575	0·590
1/ 4545	0·1192	0·1430	0·1574	0·1888	0·2239	0·2428	0·3053	0·3770	0·4029	0·4581	0·5177	0·5492	0·6507	0·7243
0·000230	0·391	0·409	0·419	0·438	0·456	0·465	0·491	0·517	0·525	0·541	0·557	0·565	0·589	0·604
1/ 4348	0·1220	0·1464	0·1611	0·1933	0·2291	0·2485	0·3125	0·3858	0·4123	0·4688	0·5298	0·5620	0·6658	0·7412
0·000240	0·400	0·418	0·428	0·447	0·466	0·475	0·502	0·528	0·537	0·553	0·570	0·578	0·602	0·617
1/ 4167	0·1247	0·1497	0·1647	0·1976	0·2343	0·2541	0·3195	0·3944	0·4215	0·4792	0·5416	0·5745	0·6806	0·7576
0·000250	0·409	0·427	0·437	0·457	0·476	0·485	0·513	0·540	0·548	0·565	0·582	0·590	0·615	0·631
1/ 4000	0·1274	0·1529	0·1683	0·2019	0·2393	0·2595	0·3263	0·4028	0·4306	0·4895	0·5531	0·5868	0·6952	0·7738
0·000260	0·417	0·436	0·446	0·466	0·486	0·495	0·524	0·551	0·559	0·577	0·594	0·602	0·627	0·643
1/ 3846	0·1301	0·1561	0·1718	0·2060	0·2442	0·2649	0·3330	0·4111	0·4394	0·4995	0·5645	0·5988	0·7094	0·7896
0·000270	0·426	0·445	0·455	0·476	0·496	0·505	0·534	0·562	0·571	0·588	0·606	0·614	0·640	0·656
1/ 3704	0·1327	0·1592	0·1752	0·2101	0·2491	0·2701	0·3396	0·4192	0·4481	0·5094	0·5756	0·6106	0·7234	0·8052
0·000280	0·434	0·453	0·464	0·485	0·505	0·515	0·544	0·572	0·581	0·599	0·617	0·626	0·652	0·669
1/ 3571	0·1352	0·1622	0·1785	0·2141	0·2538	0·2753	0·3461	0·4272	0·4566	0·5191	0·5866	0·6222	0·7371	0·8204
0·000290	0·442	0·462	0·472	0·494	0·514	0·524	0·554	0·583	0·592	0·610	0·629	0·637	0·664	0·681
1/ 3448	0·1377	0·1652	0·1818	0·2181	0·2585	0·2803	0·3525	0·4350	0·4650	0·5286	0·5973	0·6336	0·7506	0·8354
0·000300	0·450	0·470	0·481	0·502	0·523	0·534	0·564	0·593	0·603	0·621	0·640	0·649	0·675	0·693
1/ 3333	0·1402	0·1681	0·1851	0·2220	0·2631	0·2853	0·3587	0·4428	0·4732	0·5379	0·6079	0·6448	0·7638	0·8502
	0·85	0·86	0·86	0·86	0·87	0·87	0·88	0·88	0·88	0·89	0·89	0·89	0·89	0·90

$V_{r(0{\cdot}5)medial}$ **for half-full circular pipes.**

$k_s = 0{\cdot}30$ **mm** S = 0·00010 to 0·00030

$k_s = 0{\cdot}30$ **mm**
S = 0·00030 to 0·00100

ie hydraulic gradient =
1 in 3333 to 1 in 1000

Water (or sewage) at 15°C;
full bore conditions.

velocities in ms^{-1}
discharges in m^3s^{-1}

Gradient	**(Equivalent) Pipe diameters in mm**													
	630	**675**	**700**	**750**	**800**	**825**	**900**	**975**	**1000**	**1050**	**1100**	**1125**	**1200**	**1250**
0·000300	0·450	0·470	0·481	0·502	0·523	0·534	0·564	0·593	0·603	0·621	0·640	0·649	0·675	0·693
1/ 3333	0·1402	0·1681	0·1851	0·2220	0·2631	0·2853	0·3587	0·4428	0·4732	0·5379	0·6079	0·6448	0·7638	0·8502
0·000320	0·465	0·486	0·497	0·520	0·541	0·552	0·583	0·613	0·623	0·642	0·661	0·671	0·698	0·716
1/ 3125	0·1450	0·1739	0·1914	0·2295	0·2721	0·2950	0·3709	0·4578	0·4893	0·5562	0·6285	0·6666	0·7897	0·8789
0·000340	0·480	0·502	0·513	0·536	0·559	0·570	0·602	0·633	0·643	0·663	0·682	0·692	0·720	0·739
1/ 2941	0·1496	0·1795	0·1975	0·2369	0·2808	0·3044	0·3827	0·4724	0·5049	0·5739	0·6484	0·6878	0·8147	0·9068
0·000360	0·494	0·517	0·529	0·552	0·575	0·587	0·620	0·652	0·662	0·683	0·703	0·713	0·742	0·761
1/ 2778	0·1541	0·1849	0·2035	0·2440	0·2892	0·3136	0·3942	0·4865	0·5200	0·5911	0·6678	0·7084	0·8391	0·9339
0·000380	0·509	0·531	0·544	0·568	0·592	0·603	0·637	0·670	0·681	0·702	0·723	0·733	0·763	0·782
1/ 2632	0·1585	0·1901	0·2093	0·2510	0·2974	0·3225	0·4054	0·5003	0·5347	0·6078	0·6867	0·7284	0·8628	0·9602
0·000400	0·522	0·546	0·558	0·583	0·608	0·620	0·654	0·688	0·699	0·721	0·742	0·752	0·783	0·803
1/ 2500	0·1628	0·1953	0·2149	0·2577	0·3054	0·3312	0·4163	0·5137	0·5490	0·6241	0·7051	0·7479	0·8859	0·9859
0·000420	0·536	0·560	0·573	0·598	0·623	0·635	0·671	0·706	0·717	0·739	0·761	0·772	0·803	0·824
1/ 2381	0·1670	0·2003	0·2204	0·2643	0·3132	0·3396	0·4269	0·5268	0·5630	0·6399	0·7230	0·7669	0·9084	1·0109
0·000440	0·549	0·573	0·587	0·613	0·638	0·651	0·687	0·723	0·734	0·757	0·779	0·790	0·823	0·844
1/ 2273	0·1711	0·2052	0·2258	0·2708	0·3209	0·3479	0·4373	0·5396	0·5767	0·6555	0·7405	0·7855	0·9303	1·0354
0·000460	0·562	0·587	0·600	0·627	0·653	0·666	0·703	0·739	0·751	0·774	0·797	0·809	0·842	0·863
1/ 2174	0·1751	0·2099	0·2310	0·2771	0·3283	0·3560	0·4474	0·5521	0·5900	0·6706	0·7577	0·8037	0·9518	1·0593
0·000480	0·574	0·600	0·614	0·641	0·668	0·681	0·719	0·756	0·768	0·792	0·815	0·826	0·860	0·882
1/ 2083	0·1790	0·2146	0·2362	0·2832	0·3356	0·3639	0·4574	0·5643	0·6031	0·6855	0·7744	0·8214	0·9729	1·0827
0·000500	0·586	0·613	0·627	0·655	0·682	0·695	0·734	0·772	0·784	0·808	0·832	0·844	0·878	0·901
1/ 2000	0·1828	0·2192	0·2412	0·2893	0·3428	0·3716	0·4671	0·5763	0·6159	0·7000	0·7909	0·8389	0·9935	1·1056
0·000525	0·601	0·628	0·643	0·671	0·699	0·713	0·753	0·792	0·804	0·829	0·853	0·865	0·901	0·924
1/ 1905	0·1875	0·2248	0·2474	0·2966	0·3515	0·3811	0·4790	0·5910	0·6316	0·7178	0·8109	0·8601	1·0186	1·1336
0·000550	0·616	0·644	0·658	0·688	0·716	0·730	0·771	0·811	0·824	0·849	0·874	0·886	0·922	0·946
1/ 1818	0·1920	0·2303	0·2534	0·3038	0·3600	0·3903	0·4906	0·6053	0·6468	0·7352	0·8305	0·8809	1·0432	1·1610
0·000575	0·630	0·658	0·674	0·704	0·733	0·747	0·789	0·829	0·843	0·869	0·894	0·907	0·944	0·968
1/ 1739	0·1965	0·2356	0·2593	0·3109	0·3684	0·3994	0·5019	0·6193	0·6618	0·7521	0·8497	0·9012	1·0673	1·1877
0·000600	0·644	0·673	0·689	0·719	0·749	0·764	0·806	0·848	0·861	0·888	0·914	0·927	0·964	0·989
1/ 1667	0·2009	0·2409	0·2651	0·3178	0·3765	0·4082	0·5130	0·6330	0·6764	0·7687	0·8684	0·9211	1·0908	1·2139
0·000625	0·658	0·687	0·703	0·735	0·765	0·780	0·824	0·866	0·879	0·907	0·933	0·946	0·985	1·010
1/ 1600	0·2052	0·2460	0·2707	0·3245	0·3845	0·4169	0·5239	0·6464	0·6907	0·7850	0·8868	0·9406	1·1139	1·2395
0·000650	0·672	0·702	0·718	0·750	0·781	0·796	0·840	0·883	0·897	0·925	0·952	0·965	1·005	1·031
1/ 1538	0·2094	0·2510	0·2762	0·3312	0·3924	0·4254	0·5346	0·6595	0·7048	0·8010	0·9048	0·9597	1·1365	1·2647
0·000675	0·685	0·715	0·732	0·764	0·796	0·811	0·857	0·901	0·915	0·943	0·971	0·984	1·024	1·051
1/ 1481	0·2135	0·2560	0·2817	0·3377	0·4001	0·4337	0·5450	0·6724	0·7186	0·8166	0·9225	0·9784	1·1586	1·2893
0·000700	0·698	0·729	0·746	0·779	0·811	0·827	0·873	0·918	0·932	0·961	0·989	1·003	1·044	1·070
1/ 1429	0·2175	0·2608	0·2870	0·3441	0·4076	0·4419	0·5553	0·6851	0·7321	0·8320	0·9398	0·9968	1·1804	1·3135
0·000725	0·711	0·742	0·759	0·793	0·826	0·842	0·889	0·934	0·949	0·978	1·007	1·021	1·063	1·090
1/ 1379	0·2215	0·2656	0·2922	0·3503	0·4150	0·4500	0·5654	0·6975	0·7454	0·8471	0·9569	1·0149	1·2018	1·3373
0·000750	0·723	0·755	0·773	0·807	0·840	0·857	0·904	0·951	0·966	0·995	1·024	1·039	1·081	1·109
1/ 1333	0·2254	0·2702	0·2974	0·3565	0·4223	0·4579	0·5753	0·7097	0·7584	0·8619	0·9736	1·0326	1·2228	1·3607
0·000800	0·748	0·781	0·799	0·834	0·869	0·885	0·935	0·983	0·998	1·029	1·059	1·074	1·117	1·146
1/ 1250	0·2330	0·2794	0·3074	0·3685	0·4366	0·4733	0·5947	0·7336	0·7839	0·8908	1·0063	1·0673	1·2638	1·4063
0·000850	0·771	0·805	0·824	0·861	0·896	0·913	0·964	1·014	1·030	1·061	1·092	1·108	1·153	1·182
1/ 1176	0·2404	0·2882	0·3171	0·3802	0·4504	0·4883	0·6135	0·7567	0·8086	0·9189	1·0380	1·1009	1·3035	1·4505
0·000900	0·794	0·829	0·849	0·886	0·923	0·941	0·993	1·044	1·060	1·093	1·125	1·140	1·187	1·217
1/ 1111	0·2476	0·2968	0·3266	0·3915	0·4638	0·5028	0·6317	0·7792	0·8326	0·9462	1·0687	1·1335	1·3421	1·4934
0·000950	0·817	0·853	0·873	0·911	0·949	0·967	1·021	1·073	1·090	1·123	1·156	1·172	1·220	1·251
1/ 1053	0·2546	0·3052	0·3358	0·4025	0·4768	0·5169	0·6494	0·8010	0·8560	0·9727	1·0987	1·1652	1·3797	1·5352
0·001000	0·838	0·876	0·896	0·935	0·974	0·993	1·048	1·101	1·119	1·153	1·187	1·203	1·252	1·284
1/ 1000	0·2614	0·3133	0·3448	0·4132	0·4895	0·5307	0·6667	0·8223	0·8787	0·9985	1·1278	1·1962	1·4162	1·5759
	0·86	0·87	0·87	0·87	0·88	0·88	0·89	0·89	0·89	0·89	0·90	0·90	0·90	0·90

$V_{r(0{\cdot}5)medial}$ **for half-full circular pipes.**

$k_s = 0{\cdot}30$ **mm**
S = 0·00100 to 0·00300

ie hydraulic gradient =
1 in 1000 to 1 in 333

Water (or sewage) at 15°C;
full bore conditions.

velocities in ms^{-1}
discharges in m^3s^{-1}

Gradient	**(Equivalent) Pipe diameters in mm**													
	630	**675**	**700**	**750**	**800**	**825**	**900**	**975**	**1000**	**1050**	**1100**	**1125**	**1200**	**1250**
0·00100	0·838	0·876	0·896	0·935	0·974	0·993	1·048	1·101	1·119	1·153	1·187	1·203	1·252	1·284
1/ 1000	0·2614	0·3133	0·3448	0·4132	0·4895	0·5307	0·6667	0·8223	0·8787	0·9985	1·1278	1·1962	1·4162	1·5759
0·00105	0·860	0·898	0·919	0·959	0·998	1·018	1·074	1·129	1·147	1·182	1·217	1·234	1·284	1·316
1/ 952	0·2680	0·3213	0·3535	0·4237	0·5019	0·5441	0·6835	0·8431	0·9009	1·0237	1·1563	1·2263	1·4519	1·6155
0·00110	0·881	0·920	0·941	0·982	1·023	1·042	1·100	1·156	1·175	1·211	1·246	1·263	1·315	1·348
1/ 909	0·2745	0·3290	0·3620	0·4339	0·5140	0·5572	0·7000	0·8633	0·9225	1·0483	1·1840	1·2558	1·4868	1·6543
0·00115	0·901	0·941	0·962	1·005	1·046	1·066	1·126	1·183	1·202	1·238	1·274	1·292	1·345	1·379
1/ 870	0·2808	0·3366	0·3704	0·4439	0·5258	0·5700	0·7161	0·8832	0·9437	1·0723	1·2112	1·2845	1·5208	1·6922
0·00120	0·921	0·961	0·984	1·027	1·069	1·090	1·150	1·209	1·228	1·266	1·302	1·321	1·374	1·409
1/ 833	0·2870	0·3441	0·3785	0·4537	0·5374	0·5826	0·7318	0·9026	0·9644	1·0958	1·2377	1·3127	1·5541	1·7292
0·00125	0·940	0·982	1·004	1·049	1·092	1·113	1·175	1·234	1·254	1·292	1·330	1·348	1·403	1·439
1/ 800	0·2931	0·3513	0·3865	0·4633	0·5487	0·5948	0·7472	0·9215	0·9847	1·1189	1·2638	1·3403	1·5868	1·7655
0·00130	0·959	1·002	1·025	1·070	1·114	1·135	1·198	1·259	1·279	1·318	1·357	1·376	1·431	1·468
1/ 769	0·2990	0·3585	0·3944	0·4726	0·5598	0·6069	0·7623	0·9402	1·0046	1·1415	1·2893	1·3673	1·6188	1·8011
0·00135	0·978	1·021	1·045	1·091	1·135	1·157	1·222	1·284	1·304	1·344	1·383	1·402	1·459	1·496
1/ 741	0·3049	0·3654	0·4021	0·4819	0·5707	0·6187	0·7772	0·9584	1·0241	1·1636	1·3143	1·3939	1·6502	1·8360
0·00140	0·996	1·040	1·064	1·111	1·157	1·179	1·244	1·308	1·328	1·369	1·409	1·428	1·486	1·524
1/ 714	0·3106	0·3723	0·4096	0·4909	0·5814	0·6303	0·7917	0·9764	1·0433	1·1854	1·3388	1·4199	1·6810	1·8703
0·00145	1·015	1·059	1·084	1·131	1·178	1·200	1·267	1·331	1·352	1·394	1·434	1·454	1·513	1·551
1/ 690	0·3162	0·3791	0·4170	0·4998	0·5920	0·6417	0·8060	0·9940	1·0621	1·2068	1·3630	1·4455	1·7113	1·9040
0·00150	1·032	1·078	1·103	1·151	1·198	1·221	1·289	1·355	1·376	1·418	1·459	1·479	1·539	1·578
1/ 667	0·3218	0·3857	0·4243	0·5085	0·6023	0·6529	0·8201	1·0113	1·0806	1·2278	1·3867	1·4706	1·7410	1·9371
0·00160	1·067	1·114	1·140	1·190	1·238	1·262	1·332	1·400	1·422	1·465	1·508	1·529	1·591	1·631
1/ 625	0·3326	0·3986	0·4385	0·5256	0·6225	0·6747	0·8475	1·0451	1·1167	1·2688	1·4330	1·5197	1·7991	2·0016
0·00170	1·101	1·149	1·175	1·227	1·277	1·302	1·374	1·444	1·466	1·511	1·555	1·577	1·640	1·682
1/ 588	0·3431	0·4112	0·4523	0·5421	0·6420	0·6959	0·8741	1·0778	1·1517	1·3085	1·4778	1·5673	1·8553	2·0642
0·00180	1·133	1·183	1·210	1·263	1·315	1·340	1·415	1·486	1·510	1·556	1·601	1·623	1·689	1·732
1/ 556	0·3532	0·4234	0·4657	0·5581	0·6610	0·7165	0·8999	1·1097	1·1857	1·3471	1·5214	1·6135	1·9100	2·1250
0·00190	1·165	1·216	1·244	1·299	1·352	1·378	1·454	1·528	1·552	1·599	1·645	1·668	1·736	1·780
1/ 526	0·3631	0·4352	0·4788	0·5737	0·6794	0·7365	0·9250	1·1406	1·2187	1·3846	1·5638	1·6584	1·9631	2·1841
0·00200	1·196	1·248	1·277	1·333	1·387	1·414	1·492	1·568	1·593	1·641	1·689	1·712	1·782	1·827
1/ 500	0·3728	0·4467	0·4915	0·5889	0·6974	0·7560	0·9494	1·1707	1·2509	1·4212	1·6050	1·7022	2·0149	2·2417
0·00210	1·226	1·280	1·309	1·367	1·422	1·450	1·530	1·607	1·633	1·682	1·731	1·755	1·826	1·872
1/ 476	0·3822	0·4580	0·5038	0·6037	0·7150	0·7750	0·9733	1·2001	1·2823	1·4568	1·6453	1·7448	2·0654	2·2979
0·00220	1·255	1·311	1·341	1·399	1·456	1·484	1·567	1·646	1·672	1·723	1·773	1·797	1·870	1·917
1/ 455	0·3913	0·4690	0·5159	0·6182	0·7321	0·7936	0·9966	1·2288	1·3130	1·4917	1·6846	1·7865	2·1147	2·3527
0·00230	1·284	1·341	1·371	1·431	1·490	1·518	1·602	1·683	1·710	1·762	1·813	1·838	1·912	1·961
1/ 435	0·4003	0·4797	0·5277	0·6323	0·7488	0·8117	1·0194	1·2569	1·3429	1·5257	1·7230	1·8273	2·1629	2·4063
0·00240	1·312	1·370	1·401	1·463	1·522	1·552	1·637	1·720	1·747	1·800	1·853	1·878	1·954	2·004
1/ 417	0·4091	0·4902	0·5393	0·6462	0·7652	0·8295	1·0416	1·2843	1·3723	1·5590	1·7606	1·8671	2·2101	2·4588
0·00250	1·340	1·399	1·431	1·493	1·554	1·584	1·672	1·756	1·784	1·838	1·891	1·918	1·995	2·045
1/ 400	0·4177	0·5005	0·5506	0·6598	0·7813	0·8468	1·0635	1·3112	1·4010	1·5916	1·7974	1·9062	2·2563	2·5102
0·00260	1·367	1·427	1·460	1·523	1·586	1·616	1·705	1·792	1·820	1·875	1·929	1·956	2·035	2·087
1/ 385	0·4261	0·5106	0·5617	0·6730	0·7970	0·8639	1·0849	1·3376	1·4291	1·6236	1·8335	1·9445	2·3016	2·5605
0·00270	1·394	1·455	1·488	1·553	1·616	1·647	1·738	1·826	1·855	1·911	1·967	1·994	2·074	2·127
1/ 370	0·4344	0·5205	0·5726	0·6861	0·8124	0·8806	1·1059	1·3634	1·4568	1·6550	1·8690	1·9820	2·3460	2·6100
0·00280	1·420	1·482	1·516	1·582	1·646	1·678	1·771	1·860	1·889	1·947	2·003	2·031	2·113	2·166
1/ 357	0·4425	0·5303	0·5833	0·6989	0·8276	0·8970	1·1265	1·3888	1·4839	1·6858	1·9037	2·0189	2·3896	2·6585
0·00290	1·445	1·509	1·543	1·610	1·676	1·708	1·802	1·894	1·923	1·982	2·039	2·067	2·151	2·205
1/ 345	0·4505	0·5398	0·5938	0·7115	0·8425	0·9132	1·1467	1·4138	1·5105	1·7160	1·9379	2·0551	2·4325	2·7061
0·00300	1·470	1·535	1·570	1·638	1·705	1·738	1·834	1·926	1·957	2·016	2·074	2·103	2·188	2·243
1/ 333	0·4583	0·5492	0·6041	0·7238	0·8571	0·9290	1·1666	1·4383	1·5367	1·7458	1·9714	2·0907	2·4746	2·7529
	0·87	0·88	0·88	0·88	0·89	0·89	0·89	0·90	0·90	0·90	0·90	0·90	0·91	0·91

$V_{r(0{\cdot}5)medial}$ **for half-full circular pipes.**

$k_s = 0{\cdot}30$ **mm** S = 0·00100 to 0·00300

Gradient **(Equivalent) Pipe diameters in mm**

Gradient	630	675	700	750	800	825	900	975	1000	1050	1100	1125	1200	1250
0·00300	1·470	1·535	1·570	1·638	1·705	1·738	1·834	1·926	1·957	2·016	2·074	2·103	2·188	2·243
1/ 333	0·4583	0·5492	0·6041	0·7238	0·8571	0·9290	1·1666	1·4383	1·5367	1·7458	1·9714	2·0907	2·4746	2·7529
0·00320	1·519	1·586	1·622	1·693	1·762	1·796	1·895	1·990	2·022	2·083	2·143	2·173	2·261	2·318
1/ 313	0·4736	0·5675	0·6243	0·7479	0·8857	0·9600	1·2054	1·4861	1·5878	1·8038	2·0369	2·1601	2·5567	2·8443
0·00340	1·567	1·636	1·673	1·746	1·817	1·852	1·954	2·053	2·085	2·148	2·210	2·241	2·331	2·390
1/ 294	0·4884	0·5853	0·6438	0·7713	0·9133	0·9899	1·2430	1·5325	1·6373	1·8600	2·1004	2·2274	2·6364	2·9329
0·00360	1·613	1·684	1·722	1·797	1·870	1·906	2·011	2·113	2·146	2·211	2·275	2·307	2·399	2·460
1/ 278	0·5028	0·6025	0·6628	0·7940	0·9402	1·0190	1·2796	1·5775	1·6854	1·9146	2·1621	2·2928	2·7137	3·0189
0·00380	1·658	1·731	1·770	1·847	1·922	1·959	2·067	2·171	2·205	2·272	2·338	2·371	2·466	2·528
1/ 263	0·5169	0·6193	0·6812	0·8161	0·9663	1·0474	1·3151	1·6213	1·7322	1·9677	2·2220	2·3564	2·7889	3·1025
0·00400	1·702	1·776	1·817	1·896	1·973	2·011	2·122	2·229	2·263	2·332	2·400	2·433	2·531	2·595
1/ 250	0·5305	0·6356	0·6992	0·8376	0·9918	1·0749	1·3497	1·6639	1·7777	2·0195	2·2804	2·4183	2·8622	3·1840
0·00420	1·744	1·821	1·862	1·943	2·022	2·061	2·175	2·284	2·320	2·390	2·460	2·494	2·594	2·659
1/ 238	0·5438	0·6516	0·7167	0·8586	1·0166	1·1019	1·3835	1·7055	1·8222	2·0699	2·3374	2·4787	2·9336	3·2635
0·00440	1·786	1·864	1·907	1·990	2·071	2·110	2·226	2·339	2·375	2·447	2·518	2·553	2·656	2·723
1/ 227	0·5568	0·6671	0·7338	0·8791	1·0409	1·1281	1·4164	1·7461	1·8655	2·1192	2·3930	2·5377	3·0034	3·3411
0·00460	1·827	1·907	1·950	2·035	2·118	2·158	2·277	2·392	2·429	2·503	2·575	2·611	2·716	2·784
1/ 217	0·5695	0·6823	0·7505	0·8991	1·0646	1·1538	1·4487	1·7858	1·9080	2·1674	2·4474	2·5953	3·0716	3·4169
0·00480	1·867	1·948	1·993	2·080	2·164	2·205	2·327	2·444	2·482	2·557	2·631	2·668	2·775	2·845
1/ 208	0·5819	0·6972	0·7669	0·9187	1·0877	1·1789	1·4802	1·8247	1·9495	2·2145	2·5006	2·6517	3·1383	3·4912
0·00500	1·906	1·989	2·034	2·123	2·209	2·251	2·375	2·495	2·534	2·611	2·686	2·723	2·833	2·904
1/ 200	0·5941	0·7118	0·7829	0·9379	1·1105	1·2035	1·5111	1·8627	1·9901	2·2606	2·5527	2·7070	3·2037	3·5639
0·00525	1·954	2·039	2·085	2·176	2·264	2·308	2·435	2·557	2·597	2·676	2·753	2·791	2·903	2·977
1/ 190	0·6090	0·7296	0·8025	0·9614	1·1382	1·2336	1·5488	1·9092	2·0398	2·3171	2·6164	2·7745	3·2836	3·6527
0·00550	2·000	2·087	2·135	2·228	2·318	2·363	2·493	2·618	2·659	2·740	2·819	2·858	2·972	3·047
1/ 182	0·6235	0·7470	0·8216	0·9843	1·1653	1·2630	1·5857	1·9546	2·0883	2·3722	2·6786	2·8404	3·3616	3·7395
0·00575	2·046	2·135	2·184	2·279	2·371	2·416	2·549	2·677	2·719	2·802	2·883	2·922	3·040	3·116
1/ 174	0·6377	0·7640	0·8403	1·0067	1·1918	1·2917	1·6217	1·9990	2·1357	2·4260	2·7394	2·9049	3·4379	3·8243
0·00600	2·090	2·182	2·231	2·328	2·423	2·469	2·605	2·736	2·778	2·863	2·945	2·986	3·106	3·184
1/ 167	0·6516	0·7807	0·8586	1·0286	1·2178	1·3198	1·6570	2·0425	2·1821	2·4787	2·7989	2·9680	3·5125	3·9073
0·00625	2·134	2·227	2·278	2·377	2·473	2·520	2·659	2·793	2·836	2·922	3·006	3·048	3·170	3·250
1/ 160	0·6652	0·7970	0·8766	1·0500	1·2432	1·3473	1·6915	2·0850	2·2276	2·5303	2·8571	3·0298	3·5856	3·9886
0·00650	2·177	2·272	2·323	2·424	2·523	2·571	2·712	2·848	2·893	2·981	3·067	3·109	3·234	3·315
1/ 154	0·6786	0·8129	0·8941	1·0711	1·2681	1·3743	1·7253	2·1267	2·2721	2·5809	2·9142	3·0903	3·6572	4·0683
0·00675	2·219	2·316	2·368	2·471	2·571	2·620	2·764	2·903	2·949	3·038	3·125	3·169	3·296	3·379
1/ 148	0·6916	0·8286	0·9114	1·0917	1·2925	1·4008	1·7586	2·1676	2·3158	2·6305	2·9703	3·1497	3·7275	4·1464
0·00700	2·260	2·359	2·412	2·517	2·619	2·669	2·816	2·957	3·003	3·094	3·183	3·227	3·357	3·441
1/ 143	0·7045	0·8440	0·9283	1·1120	1·3164	1·4268	1·7911	2·2078	2·3587	2·6793	3·0252	3·2080	3·7965	4·2232
0·00725	2·300	2·401	2·455	2·562	2·666	2·717	2·866	3·010	3·057	3·149	3·240	3·285	3·417	3·503
1/ 138	0·7171	0·8591	0·9449	1·1319	1·3400	1·4523	1·8232	2·2472	2·4009	2·7271	3·0793	3·2653	3·8643	4·2985
0·00750	2·340	2·442	2·498	2·606	2·712	2·764	2·915	3·062	3·110	3·204	3·296	3·342	3·476	3·563
1/ 133	0·7295	0·8740	0·9612	1·1514	1·3631	1·4774	1·8546	2·2860	2·4423	2·7741	3·1324	3·3216	3·9309	4·3726
0·00800	2·418	2·523	2·581	2·693	2·802	2·855	3·012	3·163	3·213	3·310	3·405	3·452	3·591	3·681
1/ 125	0·7537	0·9029	0·9931	1·1896	1·4083	1·5263	1·9161	2·3617	2·5231	2·8659	3·2360	3·4315	4·0608	4·5172
0·00850	2·493	2·602	2·661	2·776	2·889	2·944	3·105	3·261	3·312	3·413	3·511	3·559	3·702	3·795
1/ 118	0·7772	0·9310	1·0240	1·2266	1·4521	1·5737	1·9756	2·4350	2·6015	2·9549	3·3364	3·5379	4·1868	4·6573
0·00900	2·566	2·678	2·739	2·858	2·973	3·030	3·196	3·357	3·409	3·512	3·613	3·663	3·810	3·906
1/ 111	0·8000	0·9583	1·0540	1·2625	1·4946	1·6198	2·0334	2·5062	2·6775	3·0413	3·4339	3·6413	4·3091	4·7933
0·00950	2·637	2·752	2·815	2·937	3·056	3·114	3·285	3·450	3·503	3·609	3·713	3·764	3·915	4·014
1/ 105	0·8221	0·9849	1·0832	1·2974	1·5359	1·6646	2·0896	2·5755	2·7515	3·1253	3·5288	3·7419	4·4281	4·9256
0·01000	2·707	2·824	2·888	3·014	3·136	3·196	3·371	3·540	3·595	3·704	3·810	3·863	4·018	4·119
1/ 100	0·8437	1·0107	1·1116	1·3315	1·5762	1·7082	2·1443	2·6429	2·8236	3·2071	3·6211	3·8398	4·5440	5·0545
	0·88	0·88	0·88	0·89	0·89	0·89	0·90	0·90	0·90	0·90	0·91	0·91	0·91	0·91

$V_{r(0\cdot5)medial}$ for half-full circular pipes.

A28
(p.5 of 6)

$k_s = 0{\cdot}30$ mm
S = 0·01000 to 0·03000

ie hydraulic gradient =
1 in 100 to 1 in 33·3

Water (or sewage) at 15°C;
full bore conditions.

velocities in ms^{-1}
discharges in m^3s^{-1}

Gradient **(Equivalent) Pipe diameters in mm**

Gradient	630	675	700	750	800	825	900	975	1000	1050	1100	1125	1200	1250
0·01000	2·707	2·824	2·888	3·014	3·136	3·196	3·371	3·540	3·595	3·704	3·810	3·863	4·018	4·119
1/ 100	0·8437	1·0107	1·1116	1·3315	1·5762	1·7082	2·1443	2·6429	2·8236	3·2071	3·6211	3·8398	4·5440	5·0545
0·01050	2·774	2·895	2·960	3·089	3·214	3·275	3·455	3·628	3·685	3·796	3·905	3·959	4·118	4·221
1/ 95	0·8648	1·0359	1·1393	1·3646	1·6155	1·7508	2·1977	2·7087	2·8938	3·2869	3·7112	3·9354	4·6570	5·1802
0·01100	2·840	2·964	3·031	3·162	3·290	3·353	3·537	3·714	3·772	3·886	3·998	4·053	4·215	4·321
1/ 91	0·8853	1·0605	1·1664	1·3970	1·6538	1·7923	2·2499	2·7730	2·9625	3·3649	3·7992	4·0286	4·7673	5·3029
0·01150	2·905	3·031	3·100	3·234	3·365	3·429	3·617	3·798	3·857	3·974	4·088	4·145	4·311	4·419
1/ 87	0·9054	1·0846	1·1928	1·4287	1·6913	1·8330	2·3008	2·8358	3·0296	3·4410	3·8852	4·1198	4·8752	5·4229
0·01200	2·968	3·097	3·167	3·304	3·438	3·503	3·695	3·880	3·941	4·060	4·177	4·234	4·404	4·515
1/ 83	0·9251	1·1081	1·2187	1·4597	1·7280	1·8727	2·3507	2·8972	3·0952	3·5156	3·9694	4·2090	4·9808	5·5403
0·01250	3·029	3·161	3·233	3·373	3·509	3·576	3·772	3·961	4·023	4·144	4·264	4·322	4·495	4·608
1/ 80	0·9443	1·1312	1·2441	1·4901	1·7639	1·9116	2·3996	2·9574	3·1595	3·5886	4·0518	4·2964	5·0842	5·6552
0·01300	3·090	3·224	3·297	3·440	3·579	3·647	3·847	4·040	4·103	4·227	4·349	4·408	4·585	4·700
1/ 77	0·9632	1·1538	1·2689	1·5198	1·7991	1·9498	2·4475	3·0164	3·2225	3·6602	4·1325	4·3821	5·1855	5·7679
0·01350	3·149	3·286	3·361	3·506	3·648	3·718	3·921	4·118	4·182	4·308	4·432	4·493	4·673	4·790
1/ 74	0·9817	1·1760	1·2933	1·5490	1·8337	1·9872	2·4944	3·0743	3·2843	3·7303	4·2118	4·4661	5·2849	5·8785
0·01400	3·208	3·347	3·423	3·571	3·715	3·786	3·993	4·194	4·259	4·388	4·514	4·576	4·759	4·879
1/ 71	0·9999	1·1977	1·3173	1·5777	1·8676	2·0240	2·5405	3·1311	3·3450	3·7993	4·2896	4·5486	5·3825	5·9870
0·01450	3·265	3·407	3·484	3·635	3·782	3·854	4·065	4·268	4·335	4·466	4·594	4·657	4·844	4·966
1/ 69	1·0178	1·2191	1·3408	1·6058	1·9009	2·0601	2·5858	3·1869	3·4046	3·8670	4·3660	4·6296	5·4783	6·0936
0·01500	3·321	3·466	3·544	3·698	3·847	3·920	4·135	4·342	4·409	4·543	4·673	4·738	4·927	5·051
1/ 67	1·0353	1·2401	1·3639	1·6335	1·9337	2·0956	2·6303	3·2417	3·4632	3·9335	4·4411	4·7092	5·5725	6·1984
0·01600	3·431	3·580	3·661	3·820	3·974	4·050	4·271	4·485	4·555	4·693	4·827	4·894	5·090	5·218
1/ 63	1·0696	1·2811	1·4090	1·6875	1·9976	2·1648	2·7172	3·3488	3·5775	4·0633	4·5877	4·8646	5·7564	6·4029
0·01700	3·538	3·691	3·775	3·938	4·097	4·175	4·404	4·624	4·696	4·838	4·977	5·045	5·247	5·379
1/ 59	1·1027	1·3209	1·4527	1·7398	2·0595	2·2319	2·8014	3·4525	3·6883	4·1891	4·7297	5·0152	5·9346	6·6010
0·01800	3·641	3·799	3·885	4·053	4·217	4·297	4·532	4·759	4·833	4·979	5·122	5·193	5·400	5·536
1/ 56	1·1350	1·3595	1·4951	1·7906	2·1196	2·2970	2·8831	3·5532	3·7959	4·3113	4·8676	5·1615	6·1076	6·7934
0·01900	3·742	3·904	3·992	4·165	4·333	4·416	4·657	4·890	4·966	5·116	5·263	5·336	5·549	5·688
1/ 53	1·1663	1·3970	1·5364	1·8400	2·1781	2·3604	2·9627	3·6512	3·9006	4·4301	5·0018	5·3037	6·2758	6·9806
0·02000	3·839	4·006	4·097	4·274	4·446	4·531	4·779	5·018	5·096	5·250	5·401	5·475	5·694	5·837
1/ 50	1·1969	1·4336	1·5766	1·8882	2·2350	2·4221	3·0401	3·7466	4·0025	4·5459	5·1324	5·4422	6·4397	7·1628
0·02100	3·935	4·106	4·199	4·380	4·557	4·644	4·897	5·143	5·223	5·380	5·535	5·611	5·835	5·982
1/ 47·6	1·2266	1·4692	1·6158	1·9351	2·2906	2·4823	3·1156	3·8396	4·1019	4·6588	5·2599	5·5774	6·5996	7·3406
0·02200	4·028	4·203	4·298	4·484	4·665	4·754	5·013	5·264	5·346	5·508	5·666	5·744	5·973	6·123
1/ 45·5	1·2557	1·5040	1·6541	1·9810	2·3448	2·5411	3·1894	3·9305	4·1990	4·7690	5·3843	5·7093	6·7557	7·5142
0·02300	4·119	4·298	4·395	4·585	4·770	4·861	5·127	5·383	5·467	5·632	5·794	5·873	6·108	6·261
1/ 43·5	1·2841	1·5381	1·6915	2·0258	2·3979	2·5986	3·2615	4·0193	4·2938	4·8768	5·5059	5·8383	6·9082	7·6839
0·02400	4·209	4·391	4·490	4·685	4·874	4·966	5·238	5·500	5·585	5·754	5·919	6·000	6·240	6·397
1/ 41·7	1·3119	1·5714	1·7281	2·0696	2·4498	2·6548	3·3320	4·1062	4·3867	4·9822	5·6249	5·9645	7·0575	7·8499
0·02500	4·296	4·482	4·584	4·782	4·975	5·069	5·346	5·614	5·701	5·873	6·042	6·125	6·370	6·529
1/ 40·0	1·3392	1·6040	1·7640	2·1126	2·5006	2·7099	3·4011	4·1914	4·4776	5·0854	5·7415	6·0881	7·2037	8·0126
0·02600	4·382	4·572	4·675	4·877	5·074	5·170	5·453	5·726	5·815	5·990	6·162	6·247	6·496	6·659
1/ 38·5	1·3659	1·6360	1·7991	2·1546	2·5504	2·7638	3·4688	4·2748	4·5668	5·1867	5·8558	6·2092	7·3471	8·1719
0·02700	4·466	4·659	4·765	4·971	5·171	5·269	5·557	5·835	5·926	6·105	6·280	6·366	6·621	6·787
1/ 37·0	1·3921	1·6673	1·8336	2·1959	2·5992	2·8168	3·5353	4·3567	4·6542	5·2859	5·9678	6·3280	7·4877	8·3283
0·02800	4·548	4·745	4·853	5·062	5·266	5·367	5·660	5·943	6·035	6·217	6·396	6·483	6·743	6·912
1/ 35·7	1·4178	1·6981	1·8675	2·2365	2·6472	2·8688	3·6005	4·4370	4·7400	5·3834	6·0779	6·4447	7·6257	8·4818
0·02900	4·629	4·830	4·939	5·152	5·360	5·462	5·760	6·049	6·143	6·328	6·509	6·599	6·862	7·034
1/ 34·5	1·4430	1·7283	1·9007	2·2763	2·6943	2·9198	3·6646	4·5159	4·8243	5·4792	6·1859	6·5593	7·7612	8·6326
0·03000	4·709	4·913	5·024	5·241	5·452	5·556	5·859	6·152	6·248	6·436	6·621	6·712	6·980	7·155
1/ 33·3	1·4679	1·7581	1·9334	2·3154	2·7406	2·9700	3·7275	4·5935	4·9072	5·5733	6·2922	6·6719	7·8945	8·7808
	0·88	0·88	0·88	0·89	0·89	0·89	0·90	0·90	0·90	0·91	0·91	0·91	0·91	0·91

$V_{r(0{\cdot}5)medial}$ for half-full circular pipes.

$k_s = 0{\cdot}30$ mm S = 0·01000 to 0·03000

$k_s = 0{\cdot}30$ mm
S = 0·03000 to 0·10000

ie hydraulic gradient =
1 in 33·3 to 1 in 10·0

Water (or sewage) at 15°C;
full bore conditions.

velocities in ms^{-1}
discharges in m^3s^{-1}

Gradient **(Equivalent) Pipe diameters in mm**

Gradient	630	675	700	750	800	825	900	975	1000	1050	1100	1125	1200	1250
0·03000	4·709	4·913	5·024	5·241	5·452	5·556	5·859	6·152	6·248	6·436	6·621	6·712	6·980	7·155
1/ 33·3	1·4679	1·7581	1·9334	2·3154	2·7406	2·9700	3·7275	4·5935	4·9072	5·5733	6·2922	6·6719	7·8945	8·7808
0·03200	4·864	5·075	5·190	5·414	5·632	5·739	6·052	6·355	6·454	6·648	6·839	6·933	7·210	7·391
1/ 31·3	1·5163	1·8161	1·9972	2·3918	2·8310	3·0679	3·8504	4·7449	5·0689	5·7569	6·4994	6·8917	8·1545	9·0699
0·03400	5·015	5·232	5·350	5·581	5·806	5·917	6·240	6·552	6·653	6·854	7·051	7·147	7·433	7·619
1/ 29·4	1·5632	1·8723	2·0590	2·4658	2·9186	3·1628	3·9695	4·8916	5·2256	5·9348	6·7003	7·1047	8·4065	9·3502
0·03600	5·161	5·385	5·506	5·744	5·976	6·089	6·421	6·742	6·847	7·053	7·256	7·356	7·649	7·841
1/ 27·8	1·6088	1·9269	2·1190	2·5376	3·0036	3·2550	4·0851	5·0340	5·3778	6·1076	6·8954	7·3115	8·6512	9·6223
0·03800	5·303	5·533	5·658	5·902	6·140	6·257	6·598	6·928	7·036	7·248	7·455	7·558	7·860	8·057
1/ 26·3	1·6532	1·9800	2·1774	2·6075	3·0863	3·3446	4·1975	5·1726	5·5258	6·2757	7·0851	7·5127	8·8891	9·8870
0·04000	5·442	5·677	5·806	6·056	6·300	6·420	6·770	7·109	7·219	7·437	7·650	7·755	8·065	8·267
1/ 25·0	1·6963	2·0317	2·2343	2·6756	3·1669	3·4319	4·3071	5·3075	5·6699	6·4394	7·2699	7·7086	9·1209	10·145
0·04200	5·577	5·818	5·950	6·207	6·457	6·579	6·938	7·285	7·398	7·621	7·840	7·947	8·265	8·472
1/ 23·8	1·7385	2·0821	2·2897	2·7420	3·2455	3·5170	4·4139	5·4391	5·8105	6·5990	7·4501	7·8997	9·3470	10·396
0·04400	5·709	5·956	6·090	6·353	6·609	6·735	7·102	7·457	7·573	7·801	8·025	8·135	8·460	8·672
1/ 22·7	1·7796	2·1313	2·3439	2·8068	3·3222	3·6002	4·5182	5·5677	5·9478	6·7549	7·6261	8·0863	9·5678	10·642
0·04600	5·838	6·091	6·228	6·497	6·759	6·887	7·262	7·625	7·744	7·977	8·206	8·318	8·651	8·867
1/ 21·7	1·8198	2·1795	2·3968	2·8702	3·3972	3·6814	4·6202	5·6933	6·0820	6·9073	7·7982	8·2687	9·7836	10·882
0·04800	5·964	6·222	6·363	6·637	6·905	7·036	7·419	7·790	7·911	8·149	8·383	8·498	8·837	9·059
1/ 20·8	1·8591	2·2266	2·4486	2·9322	3·4706	3·7609	4·7200	5·8162	6·2133	7·0565	7·9665	8·4472	9·9947	11·117
0·05000	6·088	6·351	6·494	6·775	7·047	7·181	7·573	7·951	8·075	8·318	8·556	8·674	9·020	9·246
1/ 20·0	1·8977	2·2727	2·4993	2·9930	3·5425	3·8388	4·8177	5·9366	6·3419	7·2025	8·1313	8·6220	10·201	11·347
0·05250	6·239	6·509	6·655	6·943	7·222	7·359	7·761	8·148	8·275	8·524	8·768	8·889	9·244	9·475
1/ 19·0	1·9447	2·3291	2·5613	3·0672	3·6303	3·9340	4·9371	6·0838	6·4991	7·3810	8·3328	8·8356	10·454	11·628
0·05500	6·386	6·662	6·813	7·107	7·393	7·533	7·944	8·341	8·470	8·725	8·975	9·099	9·462	9·699
1/ 18·2	1·9907	2·3842	2·6219	3·1397	3·7161	4·0270	5·0537	6·2274	6·6526	7·5553	8·5296	9·0442	10·701	11·902
0·05750	6·530	6·813	6·967	7·267	7·560	7·703	8·123	8·529	8·661	8·922	9·178	9·304	9·675	9·917
1/ 17·4	2·0357	2·4380	2·6810	3·2105	3·7999	4·1178	5·1677	6·3679	6·8026	7·7257	8·7219	9·2482	10·942	12·170
0·06000	6·671	6·960	7·117	7·424	7·723	7·870	8·299	8·713	8·848	9·115	9·376	9·505	9·884	10·13
1/ 16·7	2·0797	2·4906	2·7389	3·2799	3·8820	4·2067	5·2793	6·5053	6·9494	7·8924	8·9101	9·4477	11·178	12·433
0·06250	6·810	7·104	7·264	7·578	7·883	8·032	8·470	8·893	9·031	9·303	9·570	9·701	10·09	10·34
1/ 16·0	2·1227	2·5422	2·7957	3·3478	3·9623	4·2938	5·3886	6·6400	7·0932	8·0557	9·0945	9·6432	11·410	12·690
0·06500	6·945	7·245	7·409	7·728	8·040	8·192	8·639	9·070	9·211	9·488	9·760	9·894	10·29	10·55
1/ 15·4	2·1650	2·5928	2·8512	3·4143	4·0411	4·3792	5·4956	6·7719	7·2342	8·2157	9·2751	9·8347	11·636	12·942
0·06750	7·078	7·384	7·551	7·876	8·193	8·349	8·804	9·243	9·387	9·669	9·946	10·08	10·49	10·75
1/ 14·8	2·2064	2·6424	2·9058	3·4796	4·1184	4·4629	5·6007	6·9013	7·3724	8·3727	9·4524	10·023	11·859	13·189
0·07000	7·208	7·520	7·690	8·021	8·344	8·502	8·966	9·414	9·560	9·847	10·13	10·27	10·68	10·95
1/ 14·3	2·2470	2·6910	2·9593	3·5437	4·1942	4·5451	5·7038	7·0284	7·5081	8·5268	9·6263	10·207	12·077	13·432
0·07250	7·336	7·654	7·826	8·164	8·492	8·653	9·125	9·581	9·729	10·02	10·31	10·45	10·87	11·14
1/ 13·8	2·2870	2·7389	3·0119	3·6067	4·2687	4·6258	5·8051	7·1532	7·6414	8·6782	9·7972	10·388	12·291	13·671
0·07500	7·462	7·785	7·961	8·304	8·638	8·802	9·282	9·745	9·896	10·19	10·49	10·63	11·05	11·33
1/ 13·3	2·3262	2·7859	3·0636	3·6686	4·3420	4·7052	5·9047	7·2758	7·7725	8·8270	9·9652	10·566	12·502	13·905
0·08000	7·708	8·041	8·223	8·577	8·922	9·092	9·587	10·07	10·22	10·53	10·83	10·98	11·42	11·70
1/ 12·5	2·4028	2·8776	3·1644	3·7893	4·4848	4·8600	6·0989	7·5152	8·0282	9·1174	10·293	10·914	12·913	14·362
0·08500	7·946	8·290	8·477	8·842	9·198	9·372	9·883	10·38	10·54	10·85	11·17	11·32	11·77	12·06
1/ 11·8	2·4770	2·9665	3·2622	3·9063	4·6233	5·0100	6·2872	7·7472	8·2760	9·3988	10·611	11·251	13·311	14·805
0·09000	8·177	8·531	8·723	9·099	9·465	9·645	10·17	10·68	10·84	11·17	11·49	11·65	12·11	12·41
1/ 11·1	2·5491	3·0528	3·3571	4·0200	4·7578	5·1558	6·4701	7·9724	8·5166	9·6720	10·919	11·578	13·698	15·235
0·09500	8·402	8·766	8·963	9·350	9·726	9·910	10·45	10·97	11·14	11·48	11·81	11·97	12·44	12·76
1/ 10·5	2·6192	3·1367	3·4494	4·1305	4·8886	5·2975	6·6479	8·1915	8·7506	9·9377	11·219	11·896	14·075	15·654
0·10000	8·621	8·994	9·197	9·593	9·979	10·17	10·72	11·26	11·43	11·78	12·11	12·28	12·77	13·09
1/ 10·0	2·6875	3·2185	3·5393	4·2381	5·0160	5·4355	6·8211	8·4049	8·9785	10·197	11·511	12·206	14·441	16·062
	0·88	0·88	0·89	0·89	0·89	0·90	0·90	0·90	0·91	0·91	0·91	0·91	0·91	0·92

$V_{r(0{\cdot}5)medial}$ for half-full circular pipes.

A29
(p.1 of 6)

k_s = 0·60 mm
S = 0·00010 to 0·00030

ie hydraulic gradient =
1 in 10000 to 1 in 3333

Water (or sewage) at 15°C;
full bore conditions.

velocities in ms^{-1}
discharges in m^3s^{-1}

Gradient **(Equivalent) Pipe diameters in mm**

Gradient	630	675	700	750	800	825	900	975	1000	1050	1100	1125	1200	1250
0·000100	0·240	0·251	0·257	0·269	0·280	0·286	0·302	0·318	0·323	0·333	0·343	0·348	0·362	0·372
1/ 10000	0·0748	0·0898	0·0989	0·1187	0·1407	0·1526	0·1921	0·2372	0·2536	0·2884	0·3260	0·3458	0·4099	0·4564
0·000105	0·246	0·257	0·264	0·275	0·287	0·293	0·310	0·326	0·331	0·342	0·352	0·357	0·372	0·381
1/ 9524	0·0767	0·0921	0·1014	0·1217	0·1443	0·1566	0·1970	0·2433	0·2601	0·2957	0·3343	0·3547	0·4203	0·4680
0·000110	0·252	0·264	0·270	0·282	0·294	0·300	0·317	0·334	0·339	0·350	0·360	0·365	0·381	0·391
1/ 9091	0·0786	0·0944	0·1039	0·1247	0·1479	0·1604	0·2018	0·2492	0·2664	0·3029	0·3424	0·3633	0·4305	0·4793
0·000115	0·258	0·270	0·276	0·289	0·301	0·307	0·325	0·341	0·347	0·358	0·369	0·374	0·389	0·400
1/ 8696	0·0805	0·0966	0·1063	0·1276	0·1513	0·1641	0·2065	0·2550	0·2726	0·3099	0·3503	0·3717	0·4405	0·4904
0·000120	0·264	0·276	0·282	0·295	0·308	0·314	0·332	0·349	0·355	0·366	0·377	0·382	0·398	0·408
1/ 8333	0·0823	0·0987	0·1087	0·1304	0·1547	0·1678	0·2111	0·2606	0·2786	0·3168	0·3581	0·3799	0·4502	0·5013
0·000125	0·270	0·282	0·288	0·302	0·314	0·321	0·339	0·357	0·362	0·374	0·385	0·390	0·407	0·417
1/ 8000	0·0840	0·1008	0·1110	0·1332	0·1580	0·1713	0·2155	0·2662	0·2845	0·3236	0·3657	0·3880	0·4598	0·5119
0·000130	0·275	0·288	0·294	0·308	0·321	0·327	0·346	0·364	0·370	0·381	0·393	0·398	0·415	0·426
1/ 7692	0·0857	0·1029	0·1133	0·1360	0·1612	0·1749	0·2200	0·2716	0·2904	0·3302	0·3732	0·3959	0·4692	0·5223
0·000135	0·281	0·293	0·300	0·314	0·327	0·334	0·353	0·371	0·377	0·389	0·400	0·406	0·423	0·434
1/ 7407	0·0874	0·1049	0·1155	0·1386	0·1644	0·1783	0·2243	0·2770	0·2961	0·3366	0·3805	0·4037	0·4784	0·5326
0·000140	0·286	0·299	0·306	0·320	0·333	0·340	0·359	0·378	0·384	0·396	0·408	0·414	0·431	0·442
1/ 7143	0·0891	0·1069	0·1177	0·1413	0·1675	0·1817	0·2285	0·2822	0·3017	0·3430	0·3877	0·4113	0·4874	0·5426
0·000145	0·291	0·304	0·312	0·326	0·339	0·346	0·366	0·385	0·391	0·403	0·415	0·421	0·439	0·450
1/ 6897	0·0907	0·1089	0·1199	0·1439	0·1706	0·1850	0·2327	0·2873	0·3072	0·3493	0·3947	0·4188	0·4962	0·5525
0·000150	0·296	0·310	0·317	0·331	0·345	0·352	0·372	0·392	0·398	0·410	0·423	0·429	0·446	0·458
1/ 6667	0·0924	0·1108	0·1220	0·1464	0·1736	0·1883	0·2368	0·2924	0·3126	0·3554	0·4017	0·4261	0·5050	0·5621
0·000160	0·306	0·320	0·328	0·343	0·357	0·364	0·385	0·405	0·411	0·424	0·437	0·443	0·462	0·473
1/ 6250	0·0955	0·1146	0·1262	0·1514	0·1795	0·1946	0·2448	0·3023	0·3231	0·3674	0·4152	0·4405	0·5220	0·5811
0·000170	0·316	0·330	0·338	0·354	0·368	0·376	0·397	0·418	0·424	0·438	0·451	0·457	0·476	0·488
1/ 5882	0·0985	0·1183	0·1302	0·1562	0·1852	0·2008	0·2526	0·3118	0·3333	0·3790	0·4283	0·4544	0·5384	0·5994
0·000180	0·326	0·340	0·348	0·364	0·379	0·387	0·409	0·430	0·437	0·451	0·464	0·471	0·490	0·503
1/ 5556	0·1015	0·1218	0·1341	0·1609	0·1907	0·2068	0·2601	0·3211	0·3433	0·3903	0·4411	0·4679	0·5544	0·6172
0·000190	0·335	0·350	0·358	0·374	0·390	0·398	0·420	0·442	0·449	0·463	0·477	0·484	0·504	0·517
1/ 5263	0·1044	0·1252	0·1379	0·1654	0·1961	0·2127	0·2674	0·3302	0·3529	0·4013	0·4535	0·4811	0·5700	0·6345
0·000200	0·344	0·359	0·368	0·384	0·401	0·408	0·432	0·454	0·461	0·476	0·490	0·497	0·517	0·531
1/ 5000	0·1072	0·1286	0·1416	0·1698	0·2013	0·2183	0·2746	0·3390	0·3623	0·4120	0·4656	0·4939	0·5852	0·6514
0·000210	0·353	0·369	0·377	0·394	0·411	0·419	0·443	0·466	0·473	0·488	0·502	0·509	0·530	0·544
1/ 4762	0·1099	0·1319	0·1452	0·1741	0·2064	0·2239	0·2816	0·3476	0·3715	0·4224	0·4773	0·5064	0·6000	0·6679
0·000220	0·361	0·377	0·386	0·404	0·421	0·429	0·453	0·477	0·484	0·500	0·514	0·522	0·543	0·557
1/ 4545	0·1126	0·1351	0·1487	0·1784	0·2114	0·2293	0·2884	0·3560	0·3805	0·4326	0·4889	0·5186	0·6144	0·6839
0·000230	0·370	0·386	0·395	0·413	0·430	0·439	0·464	0·488	0·496	0·511	0·526	0·534	0·556	0·570
1/ 4348	0·1152	0·1382	0·1521	0·1825	0·2163	0·2346	0·2950	0·3642	0·3893	0·4425	0·5001	0·5305	0·6285	0·6996
0·000240	0·378	0·395	0·404	0·422	0·440	0·449	0·474	0·499	0·507	0·522	0·538	0·545	0·568	0·583
1/ 4167	0·1178	0·1413	0·1555	0·1865	0·2211	0·2398	0·3015	0·3722	0·3978	0·4523	0·5111	0·5422	0·6424	0·7150
0·000250	0·386	0·403	0·413	0·431	0·449	0·458	0·484	0·509	0·517	0·533	0·549	0·557	0·580	0·595
1/ 4000	0·1203	0·1443	0·1588	0·1905	0·2258	0·2449	0·3079	0·3801	0·4062	0·4618	0·5219	0·5536	0·6559	0·7301
0·000260	0·394	0·411	0·421	0·440	0·458	0·467	0·494	0·519	0·528	0·544	0·560	0·568	0·592	0·607
1/ 3846	0·1227	0·1472	0·1620	0·1944	0·2304	0·2498	0·3142	0·3878	0·4145	0·4712	0·5325	0·5648	0·6692	0·7449
0·000270	0·401	0·419	0·429	0·449	0·467	0·477	0·503	0·530	0·538	0·555	0·571	0·579	0·603	0·619
1/ 3704	0·1251	0·1501	0·1652	0·1982	0·2349	0·2547	0·3203	0·3953	0·4226	0·4804	0·5428	0·5758	0·6822	0·7593
0·000280	0·409	0·427	0·437	0·457	0·476	0·485	0·513	0·539	0·548	0·565	0·582	0·590	0·614	0·630
1/ 3571	0·1275	0·1529	0·1683	0·2019	0·2393	0·2595	0·3263	0·4028	0·4305	0·4894	0·5530	0·5866	0·6950	0·7736
0·000290	0·416	0·435	0·445	0·465	0·485	0·494	0·522	0·549	0·558	0·575	0·592	0·601	0·626	0·642
1/ 3448	0·1298	0·1557	0·1714	0·2056	0·2437	0·2642	0·3322	0·4101	0·4383	0·4982	0·5630	0·5972	0·7075	0·7875
0·000300	0·424	0·443	0·453	0·473	0·493	0·503	0·531	0·559	0·568	0·585	0·603	0·611	0·636	0·653
1/ 3333	0·1321	0·1584	0·1744	0·2092	0·2479	0·2689	0·3380	0·4172	0·4459	0·5069	0·5728	0·6077	0·7199	0·8013
	0·83	0·83	0·83	0·84	0·84	0·84	0·85	0·85	0·86	0·86	0·86	0·86	0·87	0·87

$V_{r(0·5)medial}$ for half-full circular pipes.

k_s = 0·60 mm S = 0·00010 to 0·00030

k_s = **0·60 mm**
S = 0·00030 to 0·00100

ie hydraulic gradient =
1 in 3333 to 1 in 1000

Water (or sewage) at 15°C;
full bore conditions.

velocities in ms^{-1}
discharges in m^3s^{-1}

Gradient **(Equivalent) Pipe diameters in mm**

Gradient	630	675	700	750	800	825	900	975	1000	1050	1100	1125	1200	1250
0·000300	0·424	0·443	0·453	0·473	0·493	0·503	0·531	0·559	0·568	0·585	0·603	0·611	0·636	0·653
1/ 3333	0·1321	0·1584	0·1744	0·2092	0·2479	0·2689	0·3380	0·4172	0·4459	0·5069	0·5728	0·6077	0·7199	0·8013
0·000320	0·438	0·458	0·468	0·489	0·510	0·520	0·549	0·578	0·587	0·605	0·623	0·632	0·658	0·675
1/ 3125	0·1365	0·1638	0·1803	0·2162	0·2563	0·2779	0·3494	0·4312	0·4609	0·5239	0·5920	0·6280	0·7439	0·8280
0·000340	0·452	0·472	0·483	0·505	0·526	0·536	0·566	0·596	0·605	0·624	0·643	0·652	0·678	0·696
1/ 2941	0·1409	0·1690	0·1860	0·2230	0·2643	0·2866	0·3604	0·4448	0·4754	0·5404	0·6106	0·6477	0·7673	0·8540
0·000360	0·465	0·486	0·498	0·520	0·541	0·552	0·583	0·613	0·623	0·643	0·662	0·671	0·698	0·716
1/ 2778	0·1450	0·1740	0·1915	0·2297	0·2722	0·2951	0·3711	0·4579	0·4894	0·5564	0·6287	0·6669	0·7899	0·8792
0·000380	0·478	0·500	0·512	0·534	0·557	0·568	0·600	0·631	0·641	0·660	0·680	0·690	0·718	0·736
1/ 2632	0·1491	0·1789	0·1969	0·2361	0·2798	0·3034	0·3814	0·4708	0·5031	0·5719	0·6462	0·6855	0·8120	0·9038
0·000400	0·491	0·513	0·525	0·549	0·571	0·583	0·615	0·647	0·658	0·678	0·698	0·708	0·737	0·756
1/ 2500	0·1531	0·1836	0·2021	0·2424	0·2873	0·3115	0·3916	0·4832	0·5165	0·5871	0·6633	0·7036	0·8335	0·9277
0·000420	0·504	0·526	0·538	0·562	0·586	0·597	0·631	0·664	0·674	0·695	0·716	0·726	0·755	0·775
1/ 2381	0·1570	0·1883	0·2072	0·2485	0·2945	0·3193	0·4014	0·4954	0·5295	0·6018	0·6800	0·7213	0·8544	0·9510
0·000440	0·516	0·539	0·551	0·576	0·600	0·612	0·646	0·679	0·690	0·712	0·733	0·743	0·774	0·793
1/ 2273	0·1608	0·1928	0·2122	0·2545	0·3016	0·3270	0·4111	0·5073	0·5422	0·6163	0·6963	0·7386	0·8749	0·9737
0·000460	0·528	0·551	0·564	0·589	0·614	0·626	0·661	0·695	0·706	0·728	0·749	0·760	0·791	0·812
1/ 2174	0·1645	0·1973	0·2171	0·2603	0·3085	0·3345	0·4205	0·5189	0·5546	0·6304	0·7122	0·7555	0·8949	0·9960
0·000480	0·539	0·563	0·576	0·602	0·627	0·639	0·675	0·710	0·722	0·744	0·766	0·777	0·809	0·829
1/ 2083	0·1681	0·2016	0·2219	0·2660	0·3153	0·3419	0·4297	0·5303	0·5667	0·6442	0·7278	0·7720	0·9144	1·0177
0·000500	0·551	0·575	0·589	0·615	0·640	0·653	0·690	0·725	0·737	0·760	0·782	0·793	0·825	0·847
1/ 2000	0·1716	0·2058	0·2265	0·2716	0·3219	0·3490	0·4387	0·5414	0·5786	0·6577	0·7431	0·7882	0·9336	1·0391
0·000525	0·564	0·590	0·603	0·630	0·657	0·669	0·707	0·743	0·755	0·779	0·802	0·813	0·846	0·868
1/ 1905	0·1760	0·2110	0·2322	0·2785	0·3300	0·3578	0·4498	0·5550	0·5932	0·6742	0·7617	0·8080	0·9570	1·0651
0·000550	0·578	0·604	0·618	0·646	0·672	0·685	0·724	0·761	0·773	0·797	0·821	0·832	0·866	0·889
1/ 1818	0·1802	0·2161	0·2378	0·2852	0·3379	0·3664	0·4606	0·5683	0·6074	0·6903	0·7800	0·8273	0·9799	1·0906
0·000575	0·591	0·618	0·632	0·660	0·688	0·701	0·741	0·779	0·791	0·815	0·839	0·851	0·886	0·909
1/ 1739	0·1843	0·2211	0·2433	0·2917	0·3457	0·3748	0·4711	0·5813	0·6213	0·7061	0·7978	0·8462	1·0023	1·1154
0·000600	0·604	0·631	0·646	0·675	0·703	0·717	0·757	0·796	0·808	0·833	0·858	0·870	0·906	0·929
1/ 1667	0·1884	0·2259	0·2486	0·2981	0·3533	0·3830	0·4814	0·5940	0·6348	0·7216	0·8152	0·8647	1·0241	1·1398
0·000625	0·617	0·645	0·660	0·689	0·718	0·732	0·773	0·812	0·825	0·851	0·876	0·888	0·924	0·948
1/ 1600	0·1924	0·2307	0·2538	0·3044	0·3607	0·3911	0·4915	0·6065	0·6481	0·7367	0·8323	0·8828	1·0456	1·1636
0·000650	0·630	0·658	0·673	0·703	0·732	0·746	0·788	0·829	0·842	0·868	0·893	0·906	0·943	0·967
1/ 1538	0·1962	0·2353	0·2590	0·3105	0·3679	0·3989	0·5014	0·6187	0·6612	0·7515	0·8490	0·9006	1·0666	1·1870
0·000675	0·642	0·670	0·686	0·716	0·746	0·761	0·803	0·845	0·858	0·885	0·911	0·923	0·961	0·986
1/ 1481	0·2001	0·2399	0·2640	0·3165	0·3751	0·4067	0·5111	0·6307	0·6740	0·7660	0·8654	0·9180	1·0872	1·2099
0·000700	0·654	0·683	0·699	0·730	0·760	0·775	0·818	0·860	0·874	0·901	0·928	0·941	0·979	1·004
1/ 1429	0·2038	0·2444	0·2689	0·3225	0·3821	0·4143	0·5207	0·6424	0·6865	0·7803	0·8815	0·9351	1·1074	1·2324
0·000725	0·666	0·695	0·711	0·743	0·774	0·789	0·833	0·876	0·890	0·917	0·944	0·958	0·997	1·022
1/ 1379	0·2075	0·2488	0·2738	0·3283	0·3890	0·4217	0·5300	0·6540	0·6989	0·7943	0·8974	0·9518	1·1273	1·2545
0·000750	0·677	0·707	0·724	0·756	0·787	0·803	0·848	0·891	0·905	0·933	0·961	0·974	1·014	1·040
1/ 1333	0·2111	0·2531	0·2786	0·3340	0·3957	0·4291	0·5392	0·6653	0·7110	0·8081	0·9129	0·9683	1·1468	1·2763
0·000800	0·700	0·731	0·748	0·781	0·814	0·829	0·876	0·921	0·935	0·964	0·993	1·007	1·048	1·075
1/ 1250	0·2182	0·2616	0·2879	0·3451	0·4089	0·4434	0·5572	0·6875	0·7347	0·8350	0·9433	1·0005	1·1849	1·3187
0·000850	0·722	0·754	0·771	0·806	0·839	0·855	0·903	0·950	0·965	0·994	1·024	1·038	1·080	1·108
1/ 1176	0·2250	0·2698	0·2969	0·3559	0·4217	0·4572	0·5746	0·7089	0·7576	0·8610	0·9727	1·0318	1·2219	1·3598
0·000900	0·743	0·776	0·794	0·829	0·864	0·881	0·930	0·977	0·993	1·024	1·054	1·068	1·112	1·141
1/ 1111	0·2316	0·2777	0·3056	0·3664	0·4341	0·4707	0·5915	0·7298	0·7799	0·8863	1·0013	1·0621	1·2577	1·3997
0·000950	0·764	0·798	0·816	0·852	0·888	0·905	0·956	1·005	1·021	1·052	1·083	1·098	1·143	1·172
1/ 1053	0·2381	0·2855	0·3141	0·3766	0·4462	0·4838	0·6079	0·7500	0·8015	0·9109	1·0291	1·0915	1·2926	1·4385
0·001000	0·784	0·819	0·838	0·875	0·911	0·929	0·981	1·031	1·047	1·080	1·111	1·127	1·173	1·203
1/ 1000	0·2444	0·2930	0·3224	0·3866	0·4580	0·4965	0·6240	0·7698	0·8226	0·9349	1·0562	1·1202	1·3266	1·4763
	0·83	0·84	0·84	0·85	0·85	0·85	0·86	0·86	0·86	0·86	0·87	0·87	0·87	0·88

$V_{r(0\cdot5)medial}$ for half-full circular pipes.

S = 0·00030 to 0·00100 **k_s = 0·60 mm**

A29
(p.3 of 6)

k_s = 0·60 mm
S = 0·00100 to 0·00300

ie hydraulic gradient =
1 in 1000 to 1 in 333

Water (or sewage) at 15°C;
full bore conditions.

velocities in ms^{-1}
discharges in m^3s^{-1}

Gradient	(Equivalent) Pipe diameters in mm													
	630	**675**	**700**	**750**	**800**	**825**	**900**	**975**	**1000**	**1050**	**1100**	**1125**	**1200**	**1250**
0·00100	0·784	0·819	0·838	0·875	0·911	0·929	0·981	1·031	1·047	1·080	1·111	1·127	1·173	1·203
1/ 1000	0·2444	0·2930	0·3224	0·3866	0·4580	0·4965	0·6240	0·7698	0·8226	0·9349	1·0562	1·1202	1·3266	1·4763
0·00105	0·804	0·839	0·859	0·897	0·934	0·952	1·005	1·057	1·074	1·107	1·139	1·155	1·202	1·233
1/ 952	0·2505	0·3004	0·3305	0·3962	0·4695	0·5090	0·6396	0·7890	0·8432	0·9583	1·0826	1·1482	1·3598	1·5132
0·00110	0·823	0·859	0·879	0·918	0·956	0·975	1·029	1·082	1·099	1·133	1·166	1·183	1·231	1·262
1/ 909	0·2565	0·3076	0·3384	0·4057	0·4807	0·5211	0·6548	0·8079	0·8633	0·9811	1·1084	1·1756	1·3921	1·5492
0·00115	0·842	0·879	0·899	0·939	0·978	0·997	1·053	1·107	1·124	1·159	1·193	1·210	1·259	1·291
1/ 870	0·2624	0·3146	0·3461	0·4150	0·4916	0·5330	0·6698	0·8262	0·8830	1·0034	1·1336	1·2023	1·4238	1·5844
0·00120	0·860	0·898	0·919	0·960	0·999	1·019	1·076	1·131	1·149	1·184	1·219	1·236	1·286	1·319
1/ 833	0·2681	0·3215	0·3537	0·4240	0·5023	0·5446	0·6843	0·8442	0·9022	1·0253	1·1582	1·2285	1·4547	1·6189
0·00125	0·878	0·917	0·938	0·980	1·020	1·040	1·098	1·154	1·173	1·209	1·244	1·262	1·313	1·347
1/ 800	0·2737	0·3282	0·3611	0·4329	0·5128	0·5560	0·6986	0·8619	0·9210	1·0467	1·1824	1·2541	1·4851	1·6526
0·00130	0·896	0·936	0·957	1·000	1·041	1·061	1·120	1·177	1·196	1·233	1·269	1·287	1·339	1·374
1/ 769	0·2792	0·3348	0·3684	0·4416	0·5231	0·5672	0·7126	0·8791	0·9395	1·0676	1·2061	1·2792	1·5148	1·6857
0·00135	0·913	0·954	0·976	1·019	1·061	1·081	1·142	1·200	1·219	1·257	1·294	1·312	1·365	1·400
1/ 741	0·2846	0·3412	0·3755	0·4501	0·5332	0·5781	0·7264	0·8961	0·9576	1·0882	1·2293	1·3038	1·5439	1·7181
0·00140	0·930	0·971	0·994	1·038	1·081	1·102	1·163	1·222	1·242	1·280	1·318	1·336	1·390	1·426
1/ 714	0·2899	0·3476	0·3825	0·4585	0·5431	0·5888	0·7399	0·9127	0·9753	1·1084	1·2521	1·3280	1·5726	1·7499
0·00145	0·947	0·989	1·012	1·056	1·100	1·121	1·184	1·244	1·264	1·303	1·341	1·360	1·415	1·451
1/ 690	0·2951	0·3538	0·3893	0·4667	0·5529	0·5994	0·7531	0·9290	0·9928	1·1282	1·2745	1·3518	1·6007	1·7812
0·00150	0·963	1·006	1·029	1·075	1·119	1·141	1·204	1·266	1·286	1·325	1·364	1·383	1·440	1·477
1/ 667	0·3003	0·3600	0·3961	0·4748	0·5625	0·6098	0·7662	0·9451	1·0100	1·1477	1·2965	1·3751	1·6283	1·8120
0·00160	0·995	1·039	1·063	1·110	1·156	1·179	1·244	1·308	1·329	1·369	1·409	1·429	1·487	1·525
1/ 625	0·3102	0·3719	0·4092	0·4905	0·5811	0·6300	0·7916	0·9764	1·0434	1·1858	1·3395	1·4207	1·6822	1·8720
0·00170	1·026	1·072	1·097	1·145	1·192	1·215	1·283	1·348	1·370	1·412	1·453	1·474	1·534	1·573
1/ 588	0·3199	0·3835	0·4220	0·5058	0·5992	0·6496	0·8162	1·0068	1·0759	1·2226	1·3811	1·4648	1·7345	1·9301
0·00180	1·056	1·103	1·129	1·179	1·227	1·251	1·321	1·388	1·410	1·453	1·496	1·517	1·579	1·619
1/ 556	0·3293	0·3948	0·4344	0·5207	0·6168	0·6687	0·8401	1·0363	1·1074	1·2584	1·4215	1·5077	1·7853	1·9866
0·00190	1·086	1·134	1·160	1·211	1·261	1·286	1·357	1·426	1·449	1·494	1·537	1·559	1·622	1·664
1/ 526	0·3385	0·4057	0·4464	0·5351	0·6339	0·6872	0·8634	1·0650	1·1381	1·2932	1·4608	1·5494	1·8346	2·0415
0·00200	1·114	1·164	1·191	1·243	1·294	1·319	1·393	1·464	1·487	1·533	1·577	1·600	1·665	1·707
1/ 500	0·3474	0·4164	0·4582	0·5492	0·6505	0·7053	0·8861	1·0929	1·1679	1·3272	1·4991	1·5900	1·8827	2·0950
0·00210	1·142	1·193	1·220	1·274	1·327	1·352	1·428	1·500	1·524	1·571	1·617	1·639	1·706	1·750
1/ 476	0·3560	0·4268	0·4696	0·5629	0·6668	0·7229	0·9082	1·1202	1·1970	1·3602	1·5365	1·6296	1·9296	2·1471
0·00220	1·169	1·221	1·249	1·304	1·358	1·384	1·461	1·536	1·560	1·608	1·655	1·678	1·747	1·791
1/ 455	0·3645	0·4370	0·4808	0·5763	0·6826	0·7400	0·9297	1·1468	1·2255	1·3925	1·5730	1·6683	1·9754	2·1981
0·00230	1·196	1·249	1·278	1·334	1·389	1·416	1·495	1·571	1·596	1·645	1·693	1·716	1·786	1·832
1/ 435	0·3728	0·4469	0·4917	0·5894	0·6981	0·7568	0·9508	1·1728	1·2532	1·4241	1·6086	1·7061	2·0201	2·2479
0·00240	1·222	1·276	1·305	1·363	1·419	1·447	1·527	1·605	1·630	1·680	1·729	1·754	1·825	1·871
1/ 417	0·3809	0·4566	0·5024	0·6022	0·7133	0·7733	0·9715	1·1983	1·2804	1·4550	1·6435	1·7431	2·0639	2·2966
0·00250	1·247	1·303	1·333	1·391	1·449	1·477	1·559	1·638	1·664	1·715	1·765	1·790	1·863	1·910
1/ 400	0·3889	0·4662	0·5129	0·6147	0·7282	0·7894	0·9917	1·2232	1·3071	1·4853	1·6777	1·7794	2·1068	2·3443
0·00260	1·272	1·329	1·359	1·419	1·478	1·506	1·590	1·671	1·697	1·750	1·801	1·826	1·900	1·948
1/ 385	0·3967	0·4755	0·5231	0·6270	0·7427	0·8052	1·0115	1·2476	1·3332	1·5149	1·7112	1·8149	2·1489	2·3911
0·00270	1·297	1·354	1·386	1·447	1·506	1·535	1·621	1·703	1·730	1·783	1·835	1·861	1·936	1·986
1/ 370	0·4043	0·4846	0·5332	0·6391	0·7570	0·8206	1·0310	1·2716	1·3588	1·5440	1·7440	1·8497	2·1901	2·4370
0·00280	1·321	1·379	1·411	1·473	1·534	1·564	1·651	1·735	1·762	1·816	1·869	1·895	1·972	2·023
1/ 357	0·4118	0·4936	0·5431	0·6509	0·7710	0·8358	1·0500	1·2951	1·3839	1·5726	1·7763	1·8839	2·2306	2·4820
0·00290	1·345	1·404	1·436	1·500	1·561	1·592	1·680	1·766	1·794	1·849	1·902	1·929	2·007	2·059
1/ 345	0·4192	0·5025	0·5528	0·6626	0·7848	0·8508	1·0688	1·3182	1·4086	1·6006	1·8080	1·9175	2·2704	2·5263
0·00300	1·368	1·428	1·461	1·526	1·588	1·619	1·709	1·796	1·824	1·880	1·935	1·962	2·042	2·094
1/ 333	0·4264	0·5111	0·5623	0·6740	0·7983	0·8654	1·0872	1·3409	1·4329	1·6282	1·8391	1·9506	2·3095	2·5698
	0·84	0·84	0·85	0·85	0·85	0·85	0·86	0·86	0·87	0·87	0·87	0·87	0·88	0·88

$V_{r(0·5)medial}$ for half-full circular pipes.

k_s = 0·60 mm S = 0·00100 to 0·00300

Gradient **(Equivalent) Pipe diameters in mm**

Gradient	630	675	700	750	800	825	900	975	1000	1050	1100	1125	1200	1250
0·00300	1·368	1·428	1·461	1·526	1·588	1·619	1·709	1·796	1·824	1·880	1·935	1·962	2·042	2·094
1/ 333	0·4264	0·5111	0·5623	0·6740	0·7983	0·8654	1·0872	1·3409	1·4329	1·6282	1·8391	1·9506	2·3095	2·5698
0·00320	1·413	1·476	1·510	1·576	1·641	1·673	1·766	1·855	1·885	1·943	1·999	2·027	2·109	2·163
1/ 313	0·4405	0·5281	0·5810	0·6963	0·8248	0·8941	1·1232	1·3853	1·4803	1·6820	1·8999	2·0150	2·3857	2·6546
0·00340	1·457	1·521	1·557	1·625	1·692	1·724	1·820	1·913	1·943	2·003	2·061	2·090	2·175	2·230
1/ 294	0·4542	0·5445	0·5990	0·7179	0·8504	0·9218	1·1580	1·4282	1·5262	1·7342	1·9588	2·0774	2·4597	2·7369
0·00360	1·500	1·566	1·602	1·673	1·741	1·775	1·873	1·969	2·000	2·061	2·121	2·151	2·238	2·295
1/ 278	0·4675	0·5604	0·6165	0·7389	0·8752	0·9488	1·1919	1·4700	1·5707	1·7848	2·0160	2·1381	2·5314	2·8167
0·00380	1·541	1·609	1·646	1·719	1·789	1·824	1·925	2·023	2·055	2·118	2·180	2·210	2·300	2·359
1/ 263	0·4805	0·5759	0·6336	0·7593	0·8994	0·9750	1·2248	1·5105	1·6141	1·8341	2·0716	2·1971	2·6013	2·8944
0·00400	1·582	1·652	1·689	1·764	1·836	1·872	1·976	2·076	2·109	2·173	2·237	2·268	2·360	2·420
1/ 250	0·4931	0·5910	0·6502	0·7792	0·9230	1·0005	1·2568	1·5501	1·6563	1·8820	2·1257	2·2545	2·6693	2·9701
0·00420	1·621	1·693	1·732	1·808	1·882	1·918	2·025	2·128	2·161	2·228	2·292	2·324	2·419	2·480
1/ 238	0·5053	0·6057	0·6664	0·7986	0·9459	1·0254	1·2881	1·5886	1·6975	1·9288	2·1786	2·3106	2·7356	3·0438
0·00440	1·660	1·733	1·773	1·851	1·926	1·964	2·073	2·178	2·213	2·280	2·347	2·380	2·476	2·539
1/ 227	0·5173	0·6201	0·6822	0·8176	0·9684	1·0497	1·3186	1·6262	1·7377	1·9745	2·2302	2·3653	2·8004	3·1159
0·00460	1·697	1·772	1·813	1·893	1·970	2·008	2·120	2·227	2·263	2·332	2·400	2·433	2·532	2·596
1/ 217	0·5291	0·6341	0·6976	0·8361	0·9903	1·0735	1·3485	1·6630	1·7770	2·0192	2·2806	2·4188	2·8637	3·1863
0·00480	1·734	1·810	1·852	1·934	2·013	2·052	2·166	2·276	2·312	2·382	2·452	2·486	2·587	2·653
1/ 208	0·5405	0·6479	0·7128	0·8542	1·0117	1·0967	1·3777	1·6990	1·8155	2·0629	2·3300	2·4711	2·9256	3·2552
0·00500	1·770	1·848	1·891	1·974	2·055	2·094	2·211	2·323	2·360	2·432	2·503	2·538	2·640	2·708
1/ 200	0·5518	0·6613	0·7276	0·8720	1·0328	1·1195	1·4063	1·7343	1·8532	2·1057	2·3783	2·5223	2·9863	3·3227
0·00525	1·814	1·894	1·938	2·023	2·106	2·146	2·265	2·381	2·418	2·492	2·565	2·601	2·706	2·775
1/ 190	0·5655	0·6778	0·7457	0·8936	1·0584	1·1474	1·4412	1·7774	1·8992	2·1580	2·4374	2·5850	3·0604	3·4052
0·00550	1·857	1·939	1·984	2·071	2·156	2·197	2·319	2·437	2·475	2·551	2·625	2·662	2·770	2·840
1/ 182	0·5789	0·6939	0·7634	0·9148	1·0835	1·1745	1·4754	1·8195	1·9442	2·2090	2·4951	2·6462	3·1328	3·4858
0·00575	1·899	1·983	2·028	2·118	2·204	2·247	2·372	2·492	2·531	2·609	2·685	2·722	2·833	2·905
1/ 174	0·5920	0·7096	0·7806	0·9355	1·1080	1·2011	1·5087	1·8606	1·9881	2·2590	2·5514	2·7060	3·2036	3·5645
0·00600	1·940	2·026	2·072	2·163	2·252	2·296	2·423	2·546	2·586	2·665	2·743	2·781	2·894	2·967
1/ 167	0·6049	0·7250	0·7976	0·9558	1·1320	1·2271	1·5414	1·9009	2·0311	2·3078	2·6066	2·7645	3·2729	3·6416
0·00625	1·981	2·068	2·115	2·208	2·299	2·343	2·473	2·599	2·640	2·721	2·800	2·839	2·954	3·029
1/ 160	0·6174	0·7400	0·8141	0·9756	1·1555	1·2526	1·5733	1·9403	2·0733	2·3557	2·6607	2·8218	3·3407	3·7171
0·00650	2·020	2·109	2·158	2·252	2·345	2·390	2·522	2·651	2·692	2·775	2·855	2·895	3·013	3·089
1/ 154	0·6298	0·7548	0·8304	0·9951	1·1785	1·2775	1·6047	1·9789	2·1145	2·4026	2·7136	2·8780	3·4072	3·7910
0·00675	2·059	2·150	2·199	2·296	2·390	2·436	2·571	2·701	2·744	2·828	2·910	2·951	3·070	3·148
1/ 148	0·6418	0·7693	0·8463	1·0142	1·2011	1·3020	1·6354	2·0168	2·1550	2·4486	2·7656	2·9331	3·4724	3·8636
0·00700	2·097	2·189	2·240	2·338	2·434	2·481	2·618	2·751	2·795	2·880	2·964	3·005	3·127	3·206
1/ 143	0·6537	0·7835	0·8619	1·0329	1·2233	1·3261	1·6656	2·0540	2·1948	2·4938	2·8166	2·9871	3·5365	3·9348
0·00725	2·134	2·228	2·280	2·380	2·477	2·525	2·665	2·800	2·844	2·931	3·017	3·059	3·183	3·263
1/ 138	0·6653	0·7974	0·8773	1·0513	1·2451	1·3497	1·6953	2·0906	2·2339	2·5381	2·8667	3·0403	3·5993	4·0048
0·00750	2·171	2·267	2·319	2·421	2·520	2·568	2·711	2·848	2·893	2·982	3·068	3·111	3·237	3·319
1/ 133	0·6768	0·8111	0·8924	1·0694	1·2665	1·3729	1·7244	2·1265	2·2722	2·5817	2·9159	3·0925	3·6612	4·0736
0·00800	2·243	2·342	2·395	2·500	2·603	2·653	2·800	2·942	2·988	3·080	3·169	3·214	3·344	3·429
1/ 125	0·6991	0·8379	0·9218	1·1046	1·3083	1·4182	1·7813	2·1966	2·3471	2·6668	3·0120	3·1944	3·7818	4·2078
0·00850	2·312	2·414	2·469	2·578	2·683	2·735	2·887	3·033	3·081	3·175	3·267	3·313	3·447	3·535
1/ 118	0·7208	0·8639	0·9503	1·1388	1·3488	1·4620	1·8364	2·2646	2·4197	2·7493	3·1052	3·2932	3·8987	4·3378
0·00900	2·380	2·484	2·541	2·653	2·762	2·815	2·971	3·121	3·171	3·268	3·363	3·409	3·548	3·638
1/ 111	0·7418	0·8891	0·9781	1·1720	1·3881	1·5047	1·8899	2·3305	2·4902	2·8294	3·1956	3·3891	4·0122	4·4641
0·00950	2·445	2·553	2·611	2·726	2·838	2·892	3·053	3·207	3·258	3·357	3·455	3·503	3·645	3·738
1/ 105	0·7623	0·9136	1·0050	1·2043	1·4263	1·5461	1·9419	2·3947	2·5588	2·9073	3·2835	3·4823	4·1226	4·5869
0·01000	2·509	2·620	2·680	2·797	2·912	2·968	3·132	3·291	3·343	3·445	3·545	3·595	3·740	3·835
1/ 100	0·7822	0·9374	1·0313	1·2358	1·4636	1·5865	1·9926	2·4572	2·6255	2·9831	3·3692	3·5732	4·2301	4·7066
	0·84	0·85	0·85	0·85	0·86	0·86	0·86	0·87	0·87	0·87	0·87	0·87	0·88	0·88

$V_{r(0·5)medial}$ for half-full circular pipes.

A29

(p.5 of 6)

$k_s = 0{\cdot}60$ **mm**
S = 0·01000 to 0·03000

ie hydraulic gradient =
1 in 100 to 1 in 33·3

Water (or sewage) at 15°C;
full bore conditions.

velocities in ms^{-1}
discharges in m^3s^{-1}

Gradient **(Equivalent) Pipe diameters in mm**

Gradient	630	675	700	750	800	825	900	975	1000	1050	1100	1125	1200	1250
0·01000	2·509	2·620	2·680	2·797	2·912	2·968	3·132	3·291	3·343	3·445	3·545	3·595	3·740	3·835
1/ 100	0·7822	0·9374	1·0313	1·2358	1·4636	1·5865	1·9926	2·4572	2·6255	2·9831	3·3692	3·5732	4·2301	4·7066
0·01050	2·572	2·685	2·746	2·867	2·984	3·041	3·210	3·373	3·426	3·531	3·633	3·684	3·833	3·930
1/ 95	0·8016	0·9607	1·0569	1·2665	1·4999	1·6259	2·0421	2·5181	2·6907	3·0571	3·4528	3·6618	4·3350	4·8232
0·01100	2·632	2·748	2·811	2·935	3·055	3·113	3·286	3·452	3·507	3·614	3·719	3·771	3·924	4·023
1/ 91	0·8206	0·9835	1·0819	1·2964	1·5354	1·6643	2·0903	2·5777	2·7543	3·1293	3·5343	3·7483	4·4374	4·9372
0·01150	2·692	2·810	2·875	3·001	3·124	3·184	3·360	3·530	3·586	3·696	3·803	3·856	4·012	4·114
1/ 87	0·8392	1·0057	1·1063	1·3257	1·5700	1·7019	2·1375	2·6358	2·8164	3·2000	3·6141	3·8329	4·5375	5·0485
0·01200	2·750	2·871	2·937	3·066	3·191	3·252	3·433	3·607	3·663	3·775	3·885	3·939	4·099	4·203
1/ 83	0·8573	1·0274	1·1302	1·3544	1·6040	1·7387	2·1837	2·6928	2·8773	3·2691	3·6921	3·9156	4·6355	5·1575
0·01250	2·807	2·931	2·998	3·129	3·257	3·320	3·504	3·681	3·739	3·853	3·966	4·021	4·183	4·290
1/ 80	0·8751	1·0487	1·1537	1·3824	1·6372	1·7747	2·2289	2·7485	2·9368	3·3367	3·7685	3·9967	4·7314	5·2642
0·01300	2·863	2·989	3·057	3·191	3·322	3·386	3·573	3·755	3·814	3·930	4·044	4·101	4·267	4·375
1/ 77	0·8925	1·0696	1·1766	1·4099	1·6698	1·8100	2·2732	2·8032	2·9952	3·4031	3·8435	4·0761	4·8255	5·3689
0·01350	2·918	3·046	3·116	3·253	3·385	3·451	3·642	3·826	3·887	4·005	4·122	4·179	4·348	4·459
1/ 74	0·9096	1·0901	1·1992	1·4369	1·7017	1·8446	2·3167	2·8568	3·0525	3·4681	3·9169	4·1541	4·9177	5·4715
0·01400	2·972	3·102	3·173	3·312	3·448	3·514	3·709	3·897	3·958	4·079	4·198	4·256	4·428	4·541
1/ 71	0·9264	1·1102	1·2213	1·4634	1·7331	1·8786	2·3594	2·9094	3·1087	3·5320	3·9891	4·2306	5·0082	5·5722
0·01450	3·025	3·158	3·230	3·371	3·509	3·577	3·775	3·966	4·028	4·151	4·272	4·332	4·507	4·621
1/ 69	0·9428	1·1299	1·2430	1·4894	1·7639	1·9120	2·4013	2·9611	3·1640	3·5948	4·0599	4·3057	5·0972	5·6712
0·01500	3·077	3·212	3·285	3·429	3·569	3·638	3·839	4·034	4·098	4·223	4·345	4·406	4·584	4·701
1/ 67	0·9590	1·1493	1·2643	1·5150	1·7942	1·9448	2·4426	3·0119	3·2182	3·6564	4·1296	4·3796	5·1846	5·7684
0·01600	3·178	3·318	3·394	3·542	3·687	3·758	3·966	4·167	4·232	4·362	4·488	4·551	4·735	4·855
1/ 63	0·9906	1·1872	1·3060	1·5649	1·8533	2·0089	2·5230	3·1111	3·3242	3·7768	4·2655	4·5237	5·3552	5·9582
0·01700	3·276	3·420	3·498	3·652	3·801	3·874	4·088	4·296	4·363	4·496	4·627	4·691	4·881	5·005
1/ 59	1·0213	1·2239	1·3463	1·6133	1·9105	2·0709	2·6009	3·2072	3·4268	3·8934	4·3972	4·6634	5·5205	6·1422
0·01800	3·372	3·520	3·600	3·758	3·912	3·987	4·207	4·421	4·490	4·627	4·762	4·828	5·023	5·151
1/ 56	1·0510	1·2595	1·3855	1·6602	1·9661	2·1312	2·6766	3·3005	3·5265	4·0067	4·5251	4·7990	5·6811	6·3208
0·01900	3·464	3·617	3·699	3·861	4·019	4·096	4·323	4·542	4·614	4·754	4·892	4·961	5·161	5·292
1/ 53	1·0799	1·2942	1·4237	1·7059	2·0202	2·1898	2·7502	3·3912	3·6235	4·1168	4·6495	4·9309	5·8372	6·4945
0·02000	3·555	3·711	3·796	3·962	4·124	4·203	4·436	4·660	4·734	4·878	5·020	5·090	5·296	5·430
1/ 50	1·1081	1·3279	1·4608	1·7504	2·0729	2·2469	2·8219	3·4796	3·7179	4·2241	4·7706	5·0594	5·9893	6·6637
0·02100	3·643	3·803	3·890	4·060	4·226	4·307	4·546	4·776	4·851	4·999	5·144	5·216	5·427	5·565
1/ 47·6	1·1356	1·3609	1·4970	1·7938	2·1243	2·3026	2·8918	3·5658	3·8100	4·3287	4·8888	5·1847	6·1376	6·8287
0·02200	3·729	3·893	3·982	4·156	4·326	4·409	4·653	4·889	4·966	5·117	5·266	5·339	5·555	5·696
1/ 45·5	1·1624	1·3930	1·5324	1·8361	2·1744	2·3570	2·9601	3·6500	3·9000	4·4309	5·0042	5·3071	6·2825	6·9898
0·02300	3·813	3·981	4·072	4·250	4·423	4·509	4·758	4·999	5·078	5·232	5·384	5·459	5·680	5·824
1/ 43·5	1·1886	1·4244	1·5669	1·8776	2·2235	2·4101	3·0268	3·7323	3·9879	4·5308	5·1170	5·4267	6·4240	7·1473
0·02400	3·895	4·066	4·160	4·342	4·519	4·606	4·861	5·107	5·187	5·345	5·501	5·577	5·803	5·950
1/ 41·7	1·2143	1·4552	1·6008	1·9181	2·2715	2·4621	3·0921	3·8128	4·0739	4·6285	5·2273	5·5437	6·5626	7·3014
0·02500	3·976	4·151	4·246	4·431	4·612	4·701	4·961	5·212	5·294	5·456	5·614	5·692	5·923	6·073
1/ 40·0	1·2394	1·4853	1·6339	1·9578	2·3185	2·5131	3·1561	3·8916	4·1582	4·7242	5·3354	5·6583	6·6982	7·4524
0·02600	4·055	4·233	4·330	4·520	4·704	4·795	5·060	5·316	5·399	5·564	5·726	5·805	6·040	6·193
1/ 38·5	1·2641	1·5148	1·6664	1·9967	2·3645	2·5630	3·2188	3·9689	4·2407	4·8180	5·4414	5·7707	6·8312	7·6003
0·02700	4·133	4·314	4·413	4·606	4·794	4·886	5·156	5·417	5·503	5·670	5·835	5·916	6·155	6·312
1/ 37·0	1·2883	1·5438	1·6982	2·0348	2·4097	2·6120	3·2803	4·0447	4·3217	4·9101	5·5453	5·8809	6·9617	7·7455
0·02800	4·209	4·394	4·494	4·691	4·882	4·976	5·251	5·517	5·604	5·775	5·942	6·025	6·269	6·428
1/ 35·7	1·3120	1·5722	1·7295	2·0723	2·4541	2·6600	3·3407	4·1192	4·4013	5·0004	5·6473	5·9891	7·0897	7·8879
0·02900	4·284	4·472	4·574	4·774	4·969	5·064	5·344	5·615	5·703	5·877	6·048	6·132	6·380	6·542
1/ 34·5	1·3353	1·6001	1·7602	2·1091	2·4976	2·7073	3·4000	4·1923	4·4794	5·0891	5·7475	6·0954	7·2155	8·0279
0·03000	4·357	4·548	4·652	4·856	5·054	5·151	5·436	5·711	5·801	5·978	6·152	6·237	6·489	6·654
1/ 33·3	1·3582	1·6276	1·7904	2·1453	2·5405	2·7537	3·4582	4·2641	4·5562	5·1764	5·8460	6·1998	7·3392	8·1654
	0·84	0·85	0·85	0·85	0·86	0·86	0·86	0·87	0·87	0·87	0·87	0·88	0·88	0·88

$V_{r(0\cdot5)medial}$ **for half-full circular pipes.**

$k_s = 0{\cdot}60$ **mm** S = 0·01000 to 0·03000

$k_s = 0{\cdot}60$ mm
S = 0·03000 to 0·10000

ie hydraulic gradient =
1 in 33·3 to 1 in 10·0

Water (or sewage) at 15°C;
full bore conditions.

velocities in ms^{-1}
discharges in m^3s^{-1}

Gradient **(Equivalent) Pipe diameters in mm**

Gradient	630	675	700	750	800	825	900	975	1000	1050	1100	1125	1200	1250
0·03000	4·357	4·548	4·652	4·856	5·054	5·151	5·436	5·711	5·801	5·978	6·152	6·237	6·489	6·654
1/ 33·3	1·3582	1·6276	1·7904	2·1453	2·5405	2·7537	3·4582	4·2641	4·5562	5·1764	5·8460	6·1998	7·3392	8·1654
0·03200	4·500	4·698	4·805	5·016	5·220	5·321	5·615	5·899	5·992	6·175	6·354	6·442	6·703	6·873
1/ 31·3	1·4029	1·6811	1·8493	2·2158	2·6240	2·8443	3·5720	4·4043	4·7060	5·3466	6·0382	6·4036	7·5804	8·4338
0·03400	4·639	4·843	4·954	5·170	5·381	5·485	5·788	6·081	6·177	6·365	6·550	6·641	6·909	7·084
1/ 29·4	1·4462	1·7330	1·9064	2·2842	2·7050	2·9320	3·6822	4·5402	4·8512	5·5115	6·2245	6·6012	7·8143	8·6940
0·03600	4·774	4·984	5·098	5·321	5·538	5·644	5·956	6·258	6·356	6·550	6·740	6·834	7·110	7·290
1/ 27·8	1·4883	1·7834	1·9618	2·3506	2·7836	3·0173	3·7892	4·6722	4·9921	5·6717	6·4054	6·7930	8·0413	8·9466
0·03800	4·906	5·121	5·238	5·467	5·690	5·799	6·120	6·430	6·531	6·730	6·925	7·022	7·305	7·490
1/ 26·3	1·5292	1·8324	2·0157	2·4152	2·8601	3·1002	3·8933	4·8005	5·1293	5·8274	6·5813	6·9795	8·2621	9·1922
0·04000	5·033	5·254	5·374	5·609	5·838	5·951	6·279	6·597	6·701	6·905	7·106	7·204	7·495	7·685
1/ 25·0	1·5690	1·8802	2·0683	2·4782	2·9346	3·1809	3·9947	4·9255	5·2628	5·9792	6·7526	7·1612	8·4772	9·4315
0·04200	5·158	5·384	5·507	5·748	5·983	6·098	6·435	6·760	6·867	7·076	7·281	7·383	7·681	7·876
1/ 23·8	1·6079	1·9268	2·1195	2·5395	3·0073	3·2597	4·0936	5·0474	5·3931	6·1272	6·9197	7·3385	8·6870	9·6649
0·04400	5·280	5·511	5·637	5·884	6·124	6·242	6·587	6·920	7·029	7·243	7·453	7·557	7·862	8·061
1/ 22·7	1·6458	1·9722	2·1695	2·5994	3·0782	3·3366	4·1902	5·1665	5·5203	6·2716	7·0829	7·5115	8·8918	9·8927
0·04600	5·399	5·636	5·764	6·016	6·262	6·382	6·735	7·076	7·187	7·406	7·621	7·727	8·039	8·243
1/ 21·7	1·6829	2·0167	2·2184	2·6580	3·1476	3·4117	4·2845	5·2829	5·6446	6·4129	7·2424	7·6807	9·0920	10·115
0·04800	5·515	5·757	5·889	6·146	6·397	6·520	6·880	7·228	7·342	7·566	7·785	7·893	8·212	8·420
1/ 20·8	1·7192	2·0602	2·2662	2·7153	3·2155	3·4853	4·3769	5·3967	5·7663	6·5511	7·3985	7·8462	9·2879	10·333
0·05000	5·629	5·876	6·010	6·273	6·529	6·655	7·022	7·378	7·494	7·722	7·946	8·056	8·382	8·594
1/ 20·0	1·7548	2·1028	2·3131	2·7714	3·2819	3·5573	4·4674	5·5082	5·8854	6·6865	7·5513	8·0083	9·4798	10·547
0·05250	5·769	6·022	6·159	6·429	6·691	6·819	7·196	7·560	7·679	7·913	8·143	8·256	8·589	8·807
1/ 19·0	1·7982	2·1548	2·3703	2·8401	3·3631	3·6454	4·5779	5·6446	6·0310	6·8519	7·7382	8·2064	9·7143	10·808
0·05500	5·905	6·164	6·304	6·580	6·849	6·980	7·366	7·738	7·860	8·100	8·335	8·450	8·792	9·015
1/ 18·2	1·8407	2·2056	2·4263	2·9070	3·4425	3·7314	4·6859	5·7776	6·1733	7·0135	7·9206	8·3999	9·9433	11·063
0·05750	6·038	6·303	6·447	6·728	7·003	7·137	7·532	7·913	8·037	8·282	8·522	8·641	8·990	9·218
1/ 17·4	1·8821	2·2553	2·4809	2·9725	3·5200	3·8154	4·7914	5·9078	6·3123	7·1714	8·0990	8·5890	10·167	11·312
0·06000	6·168	6·438	6·585	6·873	7·154	7·291	7·694	8·083	8·210	8·460	8·706	8·827	9·183	9·416
1/ 16·7	1·9227	2·3040	2·5344	3·0366	3·5959	3·8976	4·8947	6·0351	6·4483	7·3259	8·2735	8·7741	10·386	11·555
0·06250	6·295	6·571	6·722	7·016	7·302	7·442	7·853	8·250	8·380	8·635	8·886	9·009	9·373	9·611
1/ 16·0	1·9625	2·3516	2·5868	3·0994	3·6702	3·9782	4·9958	6·1597	6·5815	7·4772	8·4444	8·9553	10·601	11·794
0·06500	6·420	6·702	6·855	7·155	7·447	7·590	8·009	8·414	8·546	8·807	9·062	9·188	9·559	9·801
1/ 15·4	2·0014	2·3983	2·6381	3·1609	3·7430	4·0571	5·0949	6·2819	6·7121	7·6256	8·6119	9·1330	10·811	12·028
0·06750	6·543	6·830	6·986	7·291	7·589	7·734	8·162	8·574	8·709	8·975	9·235	9·363	9·741	9·988
1/ 14·8	2·0396	2·4441	2·6885	3·2212	3·8145	4·1345	5·1922	6·4018	6·8402	7·7711	8·7762	9·3072	11·017	12·257
0·07000	6·663	6·955	7·114	7·425	7·728	7·877	8·312	8·732	8·869	9·140	9·405	9·535	9·920	10·17
1/ 14·3	2·0771	2·4890	2·7379	3·2804	3·8846	4·2106	5·2876	6·5195	6·9659	7·9139	8·9375	9·4783	11·220	12·483
0·07250	6·782	7·079	7·241	7·557	7·865	8·016	8·459	8·887	9·027	9·302	9·571	9·704	10·10	10·35
1/ 13·8	2·1140	2·5332	2·7865	3·3386	3·9535	4·2852	5·3814	6·6351	7·0894	8·0542	9·0960	9·6463	11·419	12·704
0·07500	6·898	7·200	7·365	7·687	8·000	8·154	8·604	9·039	9·181	9·461	9·735	9·871	10·27	10·53
1/ 13·3	2·1502	2·5765	2·8342	3·3958	4·0212	4·3586	5·4735	6·7487	7·2108	8·1922	9·2517	9·8115	11·614	12·921
0·08000	7·125	7·437	7·607	7·939	8·263	8·422	8·887	9·336	9·483	9·772	10·06	10·19	10·61	10·88
1/ 12·5	2·2209	2·6612	2·9274	3·5074	4·1533	4·5018	5·6534	6·9704	7·4477	8·4613	9·5556	10·134	11·996	13·346
0·08500	7·344	7·666	7·841	8·184	8·518	8·681	9·160	9·624	9·775	10·07	10·36	10·51	10·93	11·21
1/ 11·8	2·2894	2·7433	3·0176	3·6156	4·2814	4·6406	5·8276	7·1853	7·6773	8·7221	9·8501	10·446	12·365	13·757
0·09000	7·558	7·889	8·069	8·422	8·765	8·933	9·426	9·903	10·06	10·37	10·67	10·81	11·25	11·54
1/ 11·1	2·3559	2·8230	3·1053	3·7206	4·4057	4·7754	5·9969	7·3939	7·9002	8·9753	10·136	10·749	12·724	14·157
0·09500	7·765	8·105	8·290	8·653	9·006	9·179	9·685	10·18	10·33	10·65	10·96	11·11	11·56	11·85
1/ 10·5	2·4206	2·9005	3·1905	3·8227	4·5267	4·9065	6·1615	7·5969	8·1170	9·2216	10·414	11·044	13·074	14·545
0·10000	7·967	8·316	8·506	8·878	9·240	9·417	9·937	10·44	10·60	10·93	11·24	11·40	11·86	12·16
1/ 10·0	2·4836	2·9760	3·2736	3·9222	4·6445	5·0342	6·3218	7·7945	8·3282	9·4615	10·685	11·332	13·414	14·923
	0·84	0·85	0·85	0·85	0·86	0·86	0·86	0·87	0·87	0·87	0·87	0·88	0·88	0·88

$V_{r(0{\cdot}5)medial}$ for half-full circular pipes.

A30

(p.1 of 6)

k_s = 1·50 mm
S = 0·00010 to 0·00030

ie hydraulic gradient =
1 in 10000 to 1 in 3333

Water (or sewage) at 15°C;
full bore conditions.

velocities in ms^{-1}
discharges in m^3s^{-1}

Gradient (Equivalent) Pipe diameters in mm

Gradient	630	675	700	750	800	825	900	975	1000	1050	1100	1125	1200	1250
0·000100	0·219	0·229	0·234	0·245	0·255	0·261	0·276	0·290	0·295	0·304	0·313	0·318	0·331	0·340
1/ 10000	0·0682	0·0819	0·0902	0·1083	0·1284	0·1393	0·1753	0·2166	0·2316	0·2634	0·2978	0·3160	0·3746	0·4171
0·000105	0·224	0·235	0·240	0·251	0·262	0·267	0·283	0·297	0·302	0·312	0·321	0·326	0·339	0·348
1/ 9524	0·0699	0·0840	0·0925	0·1110	0·1317	0·1428	0·1797	0·2220	0·2374	0·2700	0·3053	0·3239	0·3840	0·4276
0·000110	0·230	0·240	0·246	0·257	0·268	0·274	0·289	0·305	0·309	0·319	0·329	0·334	0·348	0·357
1/ 9091	0·0716	0·0860	0·0947	0·1137	0·1348	0·1462	0·1841	0·2274	0·2431	0·2765	0·3126	0·3316	0·3931	0·4378
0·000115	0·235	0·246	0·252	0·263	0·274	0·280	0·296	0·311	0·317	0·327	0·336	0·341	0·356	0·365
1/ 8696	0·0733	0·0880	0·0969	0·1163	0·1379	0·1496	0·1883	0·2326	0·2486	0·2828	0·3197	0·3392	0·4021	0·4478
0·000120	0·240	0·251	0·257	0·269	0·280	0·286	0·302	0·318	0·323	0·334	0·344	0·349	0·363	0·373
1/ 8333	0·0749	0·0899	0·0990	0·1188	0·1409	0·1529	0·1924	0·2377	0·2541	0·2890	0·3267	0·3466	0·4109	0·4575
0·000125	0·245	0·257	0·263	0·275	0·286	0·292	0·309	0·325	0·330	0·341	0·351	0·356	0·371	0·381
1/ 8000	0·0765	0·0918	0·1011	0·1213	0·1439	0·1561	0·1964	0·2426	0·2594	0·2950	0·3335	0·3539	0·4195	0·4671
0·000130	0·250	0·262	0·268	0·280	0·292	0·298	0·315	0·332	0·337	0·348	0·358	0·363	0·378	0·388
1/ 7692	0·0780	0·0936	0·1031	0·1238	0·1468	0·1592	0·2004	0·2475	0·2646	0·3010	0·3402	0·3610	0·4279	0·4765
0·000135	0·255	0·267	0·273	0·286	0·298	0·304	0·321	0·338	0·343	0·354	0·365	0·370	0·386	0·396
1/ 7407	0·0795	0·0955	0·1051	0·1262	0·1496	0·1623	0·2043	0·2523	0·2697	0·3068	0·3468	0·3680	0·4362	0·4857
0·000140	0·260	0·272	0·278	0·291	0·303	0·309	0·327	0·344	0·350	0·361	0·372	0·377	0·393	0·403
1/ 7143	0·0810	0·0973	0·1071	0·1285	0·1524	0·1654	0·2081	0·2570	0·2748	0·3125	0·3533	0·3748	0·4443	0·4947
0·000145	0·265	0·277	0·283	0·296	0·309	0·315	0·333	0·350	0·356	0·367	0·378	0·384	0·400	0·410
1/ 6897	0·0825	0·0990	0·1090	0·1308	0·1552	0·1683	0·2118	0·2616	0·2797	0·3181	0·3596	0·3816	0·4523	0·5036
0·000150	0·269	0·282	0·288	0·301	0·314	0·320	0·339	0·357	0·362	0·374	0·385	0·391	0·407	0·417
1/ 6667	0·0839	0·1007	0·1109	0·1331	0·1579	0·1713	0·2155	0·2662	0·2846	0·3236	0·3659	0·3882	0·4601	0·5123
0·000160	0·278	0·291	0·298	0·311	0·325	0·331	0·350	0·368	0·374	0·386	0·398	0·404	0·420	0·431
1/ 6250	0·0867	0·1041	0·1146	0·1376	0·1631	0·1770	0·2227	0·2750	0·2940	0·3344	0·3780	0·4011	0·4754	0·5294
0·000170	0·287	0·300	0·307	0·321	0·335	0·341	0·361	0·380	0·386	0·398	0·410	0·416	0·433	0·445
1/ 5882	0·0894	0·1074	0·1182	0·1419	0·1683	0·1825	0·2296	0·2836	0·3032	0·3448	0·3898	0·4136	0·4902	0·5459
0·000180	0·295	0·309	0·316	0·331	0·345	0·351	0·372	0·391	0·397	0·410	0·422	0·428	0·446	0·458
1/ 5556	0·0921	0·1105	0·1217	0·1461	0·1732	0·1879	0·2364	0·2920	0·3121	0·3550	0·4013	0·4258	0·5046	0·5619
0·000190	0·304	0·317	0·325	0·340	0·354	0·361	0·382	0·402	0·408	0·421	0·434	0·440	0·459	0·471
1/ 5263	0·0947	0·1136	0·1251	0·1501	0·1780	0·1931	0·2430	0·3001	0·3208	0·3648	0·4124	0·4376	0·5186	0·5775
0·000200	0·312	0·326	0·334	0·349	0·364	0·371	0·392	0·413	0·419	0·432	0·445	0·452	0·471	0·483
1/ 5000	0·0972	0·1166	0·1284	0·1541	0·1827	0·1982	0·2494	0·3080	0·3293	0·3744	0·4233	0·4491	0·5323	0·5926
0·000210	0·320	0·334	0·342	0·358	0·373	0·380	0·402	0·423	0·430	0·443	0·457	0·463	0·482	0·495
1/ 4762	0·0996	0·1195	0·1316	0·1579	0·1873	0·2032	0·2556	0·3157	0·3375	0·3838	0·4339	0·4603	0·5456	0·6075
0·000220	0·327	0·342	0·350	0·366	0·382	0·389	0·411	0·433	0·440	0·454	0·467	0·474	0·494	0·507
1/ 4545	0·1020	0·1224	0·1348	0·1617	0·1918	0·2080	0·2617	0·3232	0·3455	0·3930	0·4442	0·4713	0·5586	0·6219
0·000230	0·335	0·350	0·358	0·374	0·390	0·398	0·421	0·443	0·450	0·464	0·478	0·485	0·505	0·518
1/ 4348	0·1043	0·1252	0·1378	0·1654	0·1962	0·2128	0·2677	0·3306	0·3534	0·4019	0·4543	0·4820	0·5713	0·6360
0·000240	0·342	0·358	0·366	0·383	0·399	0·407	0·430	0·452	0·460	0·474	0·488	0·495	0·516	0·530
1/ 4167	0·1066	0·1279	0·1408	0·1690	0·2004	0·2174	0·2735	0·3378	0·3611	0·4106	0·4642	0·4925	0·5837	0·6499
0·000250	0·349	0·365	0·374	0·391	0·407	0·415	0·439	0·462	0·469	0·484	0·499	0·506	0·527	0·541
1/ 4000	0·1088	0·1306	0·1438	0·1726	0·2046	0·2219	0·2792	0·3448	0·3686	0·4192	0·4739	0·5027	0·5958	0·6634
0·000260	0·356	0·372	0·381	0·398	0·415	0·424	0·448	0·471	0·479	0·494	0·509	0·516	0·537	0·551
1/ 3846	0·1110	0·1332	0·1467	0·1760	0·2087	0·2264	0·2848	0·3517	0·3760	0·4276	0·4833	0·5128	0·6078	0·6767
0·000270	0·363	0·380	0·389	0·406	0·423	0·432	0·456	0·480	0·488	0·503	0·518	0·526	0·548	0·562
1/ 3704	0·1132	0·1358	0·1495	0·1794	0·2128	0·2308	0·2903	0·3585	0·3833	0·4358	0·4927	0·5227	0·6195	0·6897
0·000280	0·370	0·387	0·396	0·414	0·431	0·440	0·465	0·489	0·497	0·513	0·528	0·536	0·558	0·572
1/ 3571	0·1153	0·1383	0·1523	0·1828	0·2167	0·2351	0·2957	0·3652	0·3904	0·4439	0·5018	0·5324	0·6310	0·7025
0·000290	0·376	0·394	0·403	0·421	0·439	0·448	0·473	0·498	0·506	0·522	0·537	0·545	0·568	0·583
1/ 3448	0·1173	0·1408	0·1550	0·1860	0·2206	0·2393	0·3010	0·3717	0·3974	0·4519	0·5108	0·5419	0·6422	0·7151
0·000300	0·383	0·400	0·410	0·428	0·446	0·455	0·481	0·506	0·515	0·531	0·547	0·555	0·578	0·593
1/ 3333	0·1194	0·1433	0·1577	0·1893	0·2244	0·2434	0·3062	0·3781	0·4042	0·4597	0·5196	0·5513	0·6533	0·7274
	0·79	0·80	0·80	0·80	0·80	0·81	0·81	0·81	0·82	0·82	0·82	0·82	0·83	0·83

$V_{r(0{\cdot}5)medial}$ for half-full circular pipes.

k_s = 1·50 mm S = 0·00010 to 0·00030

$k_s = 1{\cdot}50$ **mm**
S = 0·00030 to 0·00100

ie hydraulic gradient =
1 in 3333 to 1 in 1000

Water (or sewage) at 15°C;
full bore conditions.

velocities in ms^{-1}
discharges in m^3s^{-1}

Gradient **(Equivalent) Pipe diameters in mm**

	630	675	700	750	800	825	900	975	1000	1050	1100	1125	1200	1250
0·000300	0·383	0·400	0·410	0·428	0·446	0·455	0·481	0·506	0·515	0·531	0·547	0·555	0·578	0·593
1/ 3333	0·1194	0·1433	0·1577	0·1893	0·2244	0·2434	0·3062	0·3781	0·4042	0·4597	0·5196	0·5513	0·6533	0·7274
0·000320	0·396	0·414	0·423	0·443	0·461	0·470	0·497	0·523	0·532	0·548	0·565	0·573	0·597	0·612
1/ 3125	0·1233	0·1480	0·1630	0·1955	0·2319	0·2515	0·3164	0·3907	0·4176	0·4749	0·5368	0·5695	0·6750	0·7515
0·000340	0·408	0·427	0·437	0·456	0·476	0·485	0·513	0·540	0·548	0·566	0·582	0·591	0·615	0·631
1/ 2941	0·1272	0·1526	0·1680	0·2016	0·2391	0·2593	0·3262	0·4028	0·4306	0·4897	0·5535	0·5872	0·6959	0·7748
0·000360	0·420	0·439	0·449	0·470	0·490	0·499	0·528	0·555	0·564	0·582	0·599	0·608	0·633	0·650
1/ 2778	0·1309	0·1571	0·1730	0·2076	0·2461	0·2669	0·3358	0·4146	0·4432	0·5040	0·5697	0·6044	0·7163	0·7975
0·000380	0·432	0·451	0·462	0·483	0·503	0·513	0·542	0·571	0·580	0·598	0·616	0·625	0·651	0·668
1/ 2632	0·1346	0·1615	0·1778	0·2133	0·2529	0·2743	0·3451	0·4261	0·4555	0·5180	0·5855	0·6211	0·7361	0·8195
0·000400	0·443	0·463	0·474	0·496	0·516	0·527	0·557	0·586	0·595	0·614	0·632	0·641	0·668	0·685
1/ 2500	0·1381	0·1657	0·1825	0·2189	0·2596	0·2815	0·3541	0·4373	0·4675	0·5316	0·6008	0·6374	0·7554	0·8410
0·000420	0·454	0·475	0·486	0·508	0·529	0·540	0·571	0·600	0·610	0·629	0·648	0·657	0·685	0·702
1/ 2381	0·1416	0·1699	0·1870	0·2244	0·2661	0·2886	0·3630	0·4482	0·4791	0·5448	0·6158	0·6533	0·7742	0·8620
0·000440	0·465	0·486	0·498	0·520	0·542	0·553	0·584	0·615	0·625	0·644	0·663	0·673	0·701	0·719
1/ 2273	0·1449	0·1739	0·1915	0·2297	0·2724	0·2954	0·3716	0·4589	0·4905	0·5577	0·6304	0·6688	0·7926	0·8824
0·000460	0·476	0·497	0·509	0·532	0·554	0·565	0·597	0·629	0·639	0·659	0·678	0·688	0·717	0·735
1/ 2174	0·1482	0·1779	0·1958	0·2349	0·2786	0·3021	0·3800	0·4693	0·5016	0·5704	0·6447	0·6840	0·8105	0·9024
0·000480	0·486	0·508	0·520	0·543	0·566	0·577	0·610	0·642	0·653	0·673	0·693	0·703	0·732	0·751
1/ 2083	0·1515	0·1817	0·2001	0·2400	0·2846	0·3087	0·3883	0·4794	0·5125	0·5828	0·6587	0·6988	0·8281	0·9220
0·000500	0·496	0·518	0·531	0·555	0·578	0·589	0·623	0·656	0·666	0·687	0·708	0·718	0·747	0·767
1/ 2000	0·1546	0·1855	0·2042	0·2451	0·2905	0·3151	0·3964	0·4894	0·5232	0·5949	0·6724	0·7133	0·8453	0·9411
0·000525	0·508	0·531	0·544	0·569	0·592	0·604	0·639	0·672	0·683	0·704	0·725	0·735	0·766	0·786
1/ 1905	0·1585	0·1902	0·2093	0·2512	0·2978	0·3230	0·4062	0·5016	0·5362	0·6097	0·6891	0·7311	0·8664	0·9645
0·000550	0·520	0·544	0·557	0·582	0·607	0·619	0·654	0·688	0·699	0·721	0·742	0·753	0·784	0·805
1/ 1818	0·1622	0·1947	0·2143	0·2571	0·3049	0·3306	0·4159	0·5135	0·5489	0·6242	0·7055	0·7484	0·8869	0·9874
0·000575	0·532	0·556	0·570	0·595	0·620	0·633	0·669	0·703	0·715	0·737	0·759	0·770	0·802	0·823
1/ 1739	0·1659	0·1991	0·2192	0·2630	0·3118	0·3381	0·4253	0·5252	0·5614	0·6383	0·7214	0·7654	0·9070	1·0097
0·000600	0·544	0·569	0·582	0·608	0·634	0·646	0·683	0·719	0·730	0·753	0·776	0·787	0·819	0·841
1/ 1667	0·1695	0·2034	0·2239	0·2687	0·3186	0·3455	0·4345	0·5365	0·5735	0·6521	0·7371	0·7820	0·9266	1·0316
0·000625	0·555	0·580	0·594	0·621	0·647	0·660	0·697	0·734	0·745	0·769	0·792	0·803	0·836	0·858
1/ 1600	0·1731	0·2077	0·2286	0·2743	0·3252	0·3527	0·4436	0·5477	0·5855	0·6657	0·7524	0·7982	0·9459	1·0530
0·000650	0·566	0·592	0·606	0·633	0·660	0·673	0·711	0·748	0·760	0·784	0·807	0·819	0·853	0·875
1/ 1538	0·1765	0·2118	0·2332	0·2798	0·3317	0·3597	0·4524	0·5586	0·5971	0·6790	0·7674	0·8141	0·9647	1·0740
0·000675	0·577	0·603	0·618	0·645	0·673	0·686	0·725	0·763	0·775	0·799	0·823	0·835	0·869	0·892
1/ 1481	0·1799	0·2159	0·2377	0·2851	0·3381	0·3666	0·4611	0·5694	0·6086	0·6920	0·7821	0·8297	0·9832	1·0946
0·000700	0·588	0·615	0·629	0·657	0·685	0·699	0·738	0·777	0·789	0·814	0·838	0·850	0·885	0·908
1/ 1429	0·1833	0·2199	0·2421	0·2904	0·3443	0·3734	0·4697	0·5799	0·6199	0·7048	0·7966	0·8451	1·0014	1·1148
0·000725	0·598	0·625	0·640	0·669	0·697	0·711	0·751	0·791	0·803	0·828	0·853	0·865	0·901	0·925
1/ 1379	0·1865	0·2238	0·2464	0·2956	0·3505	0·3801	0·4780	0·5902	0·6309	0·7174	0·8108	0·8601	1·0192	1·1347
0·000750	0·609	0·636	0·651	0·681	0·709	0·723	0·764	0·804	0·817	0·843	0·868	0·880	0·917	0·941
1/ 1333	0·1898	0·2277	0·2506	0·3007	0·3565	0·3866	0·4863	0·6004	0·6418	0·7297	0·8247	0·8749	1·0368	1·1542
0·000800	0·629	0·657	0·673	0·703	0·733	0·747	0·790	0·831	0·844	0·871	0·896	0·909	0·947	0·972
1/ 1250	0·1960	0·2352	0·2589	0·3106	0·3683	0·3994	0·5023	0·6202	0·6630	0·7538	0·8520	0·9038	1·0710	1·1923
0·000850	0·648	0·678	0·694	0·725	0·755	0·770	0·814	0·856	0·870	0·898	0·924	0·937	0·976	1·002
1/ 1176	0·2021	0·2425	0·2670	0·3203	0·3797	0·4118	0·5179	0·6394	0·6835	0·7772	0·8783	0·9318	1·1042	1·2292
0·000900	0·667	0·698	0·714	0·746	0·777	0·793	0·838	0·881	0·896	0·924	0·951	0·965	1·005	1·031
1/ 1111	0·2080	0·2496	0·2748	0·3296	0·3908	0·4238	0·5330	0·6581	0·7035	0·7998	0·9040	0·9590	1·1364	1·2651
0·000950	0·686	0·717	0·734	0·767	0·799	0·815	0·861	0·906	0·920	0·949	0·977	0·991	1·032	1·059
1/ 1053	0·2138	0·2565	0·2824	0·3387	0·4016	0·4355	0·5477	0·6763	0·7229	0·8219	0·9289	0·9854	1·1677	1·2999
0·001000	0·704	0·736	0·753	0·787	0·820	0·836	0·884	0·929	0·944	0·974	1·003	1·017	1·059	1·087
1/ 1000	0·2194	0·2632	0·2898	0·3476	0·4121	0·4469	0·5621	0·6939	0·7418	0·8434	0·9532	1·0112	1·1982	1·3339
	0·79	0·80	0·80	0·80	0·81	0·81	0·81	0·82	0·82	0·82	0·82	0·82	0·83	0·83

$V_{r(0{\cdot}5)medial}$ **for half-full circular pipes.**

A30

(p.3 of 6)

k_s = 1·50 mm
S = 0·00100 to 0·00300

ie hydraulic gradient =
1 in 1000 to 1 in 333

Water (or sewage) at 15°C;
full bore conditions.

velocities in ms^{-1}
discharges in m^3s^{-1}

Gradient	**(Equivalent) Pipe diameters in mm**													
	630	**675**	**700**	**750**	**800**	**825**	**900**	**975**	**1000**	**1050**	**1100**	**1125**	**1200**	**1250**
0·00100	0·704	0·736	0·753	0·787	0·820	0·836	0·884	0·929	0·944	0·974	1·003	1·017	1·059	1·087
1/ 1000	0·2194	0·2632	0·2898	0·3476	0·4121	0·4469	0·5621	0·6939	0·7418	0·8434	0·9532	1·0112	1·1982	1·3339
0·00105	0·721	0·754	0·772	0·806	0·840	0·857	0·905	0·953	0·968	0·998	1·028	1·043	1·086	1·114
1/ 952	0·2249	0·2698	0·2970	0·3563	0·4223	0·4580	0·5760	0·7112	0·7602	0·8643	0·9768	1·0363	1·2279	1·3670
0·00110	0·738	0·772	0·790	0·826	0·860	0·877	0·927	0·975	0·991	1·022	1·052	1·067	1·111	1·140
1/ 909	0·2302	0·2762	0·3040	0·3647	0·4324	0·4689	0·5897	0·7280	0·7782	0·8848	1·0000	1·0608	1·2570	1·3994
0·00115	0·755	0·789	0·808	0·844	0·880	0·897	0·948	0·997	1·013	1·045	1·076	1·091	1·137	1·166
1/ 870	0·2354	0·2824	0·3109	0·3730	0·4421	0·4795	0·6030	0·7445	0·7958	0·9048	1·0226	1·0848	1·2854	1·4310
0·00120	0·772	0·806	0·825	0·862	0·899	0·916	0·968	1·019	1·035	1·068	1·099	1·115	1·161	1·191
1/ 833	0·2405	0·2886	0·3176	0·3810	0·4517	0·4899	0·6161	0·7606	0·8130	0·9244	1·0447	1·1083	1·3132	1·4619
0·00125	0·788	0·823	0·842	0·880	0·917	0·935	0·988	1·040	1·057	1·090	1·122	1·138	1·185	1·216
1/ 800	0·2455	0·2946	0·3242	0·3889	0·4611	0·5000	0·6289	0·7764	0·8299	0·9435	1·0663	1·1312	1·3404	1·4922
0·00130	0·803	0·840	0·859	0·898	0·936	0·954	1·008	1·061	1·078	1·111	1·144	1·161	1·209	1·240
1/ 769	0·2504	0·3004	0·3307	0·3967	0·4703	0·5100	0·6414	0·7918	0·8464	0·9623	1·0876	1·1538	1·3671	1·5219
0·00135	0·819	0·856	0·876	0·915	0·954	0·972	1·028	1·081	1·098	1·133	1·166	1·183	1·232	1·264
1/ 741	0·2552	0·3062	0·3370	0·4043	0·4793	0·5198	0·6537	0·8070	0·8626	0·9807	1·1084	1·1759	1·3933	1·5510
0·00140	0·834	0·871	0·892	0·932	0·971	0·990	1·046	1·101	1·119	1·154	1·188	1·205	1·255	1·287
1/ 714	0·2599	0·3118	0·3433	0·4118	0·4881	0·5294	0·6657	0·8219	0·8785	0·9988	1·1288	1·1975	1·4189	1·5796
0·00145	0·849	0·887	0·908	0·949	0·988	1·008	1·065	1·120	1·138	1·174	1·209	1·226	1·277	1·310
1/ 690	0·2646	0·3174	0·3494	0·4191	0·4968	0·5388	0·6776	0·8365	0·8941	1·0166	1·1489	1·2188	1·4442	1·6077
0·00150	0·863	0·902	0·923	0·965	1·005	1·025	1·083	1·140	1·158	1·194	1·230	1·247	1·299	1·333
1/ 667	0·2691	0·3229	0·3554	0·4263	0·5054	0·5481	0·6892	0·8509	0·9095	1·0341	1·1686	1·2398	1·4690	1·6353
0·00160	0·892	0·932	0·954	0·997	1·039	1·059	1·119	1·177	1·196	1·234	1·270	1·288	1·342	1·376
1/ 625	0·2780	0·3335	0·3671	0·4404	0·5220	0·5661	0·7120	0·8789	0·9395	1·0682	1·2072	1·2806	1·5174	1·6892
0·00170	0·919	0·961	0·983	1·028	1·071	1·092	1·154	1·214	1·233	1·272	1·310	1·328	1·383	1·419
1/ 588	0·2866	0·3439	0·3785	0·4540	0·5382	0·5837	0·7340	0·9061	0·9685	1·1012	1·2445	1·3202	1·5643	1·7414
0·00180	0·946	0·989	1·012	1·058	1·102	1·124	1·187	1·249	1·269	1·309	1·348	1·367	1·423	1·460
1/ 556	0·2950	0·3539	0·3895	0·4672	0·5539	0·6007	0·7554	0·9325	0·9968	1·1333	1·2807	1·3587	1·6098	1·7921
0·00190	0·972	1·016	1·040	1·087	1·132	1·155	1·220	1·283	1·304	1·345	1·385	1·404	1·463	1·501
1/ 526	0·3031	0·3636	0·4003	0·4801	0·5691	0·6172	0·7762	0·9582	1·0242	1·1645	1·3160	1·3961	1·6541	1·8414
0·00200	0·998	1·043	1·067	1·115	1·162	1·185	1·252	1·317	1·338	1·380	1·421	1·441	1·501	1·540
1/ 500	0·3110	0·3731	0·4107	0·4927	0·5840	0·6333	0·7964	0·9832	1·0509	1·1948	1·3503	1·4325	1·6973	1·8894
0·00210	1·023	1·069	1·094	1·143	1·191	1·214	1·283	1·350	1·371	1·414	1·456	1·477	1·538	1·578
1/ 476	0·3188	0·3824	0·4209	0·5049	0·5985	0·6491	0·8162	1·0076	1·0770	1·2245	1·3838	1·4680	1·7394	1·9363
0·00220	1·047	1·094	1·120	1·170	1·219	1·243	1·313	1·381	1·404	1·448	1·491	1·512	1·574	1·615
1/ 455	0·3263	0·3915	0·4309	0·5168	0·6127	0·6644	0·8355	1·0314	1·1024	1·2534	1·4165	1·5027	1·7804	1·9820
0·00230	1·070	1·119	1·145	1·196	1·246	1·271	1·343	1·413	1·435	1·480	1·524	1·546	1·610	1·652
1/ 435	0·3337	0·4003	0·4406	0·5285	0·6265	0·6794	0·8544	1·0547	1·1273	1·2817	1·4485	1·5366	1·8206	2·0267
0·00240	1·094	1·143	1·170	1·222	1·273	1·298	1·372	1·443	1·466	1·512	1·557	1·579	1·645	1·687
1/ 417	0·3409	0·4090	0·4501	0·5399	0·6400	0·6941	0·8728	1·0775	1·1517	1·3094	1·4797	1·5698	1·8599	2·0705
0·00250	1·116	1·167	1·194	1·247	1·300	1·325	1·400	1·473	1·497	1·543	1·589	1·612	1·679	1·722
1/ 400	0·3480	0·4174	0·4595	0·5511	0·6533	0·7085	0·8909	1·0998	1·1755	1·3365	1·5104	1·6023	1·8984	2·1133
0·00260	1·138	1·190	1·218	1·272	1·326	1·352	1·428	1·502	1·526	1·574	1·621	1·644	1·712	1·756
1/ 385	0·3549	0·4257	0·4686	0·5621	0·6663	0·7226	0·9086	1·1216	1·1989	1·3630	1·5404	1·6341	1·9361	2·1553
0·00270	1·160	1·213	1·241	1·297	1·351	1·378	1·456	1·531	1·556	1·604	1·652	1·675	1·745	1·790
1/ 370	0·3617	0·4339	0·4776	0·5728	0·6790	0·7364	0·9260	1·1431	1·2218	1·3891	1·5698	1·6654	1·9732	2·1965
0·00280	1·182	1·235	1·264	1·321	1·376	1·403	1·482	1·559	1·584	1·634	1·682	1·706	1·777	1·823
1/ 357	0·3683	0·4419	0·4864	0·5834	0·6916	0·7500	0·9430	1·1642	1·2443	1·4147	1·5987	1·6960	2·0095	2·2369
0·00290	1·203	1·257	1·286	1·344	1·400	1·428	1·509	1·587	1·612	1·663	1·712	1·737	1·808	1·855
1/ 345	0·3749	0·4498	0·4950	0·5938	0·7039	0·7633	0·9598	1·1848	1·2664	1·4398	1·6271	1·7262	2·0452	2·2767
0·00300	1·223	1·278	1·308	1·367	1·424	1·452	1·535	1·614	1·640	1·691	1·742	1·766	1·839	1·887
1/ 333	0·3813	0·4575	0·5035	0·6040	0·7159	0·7764	0·9763	1·2052	1·2882	1·4645	1·6551	1·7558	2·0802	2·3157
	0·80	0·80	0·80	0·80	0·81	0·81	0·81	0·82	0·82	0·82	0·82	0·83	0·83	0·83

$V_{r(0·5)medial}$ for half-full circular pipes.

k_s = 1·50 mm S = 0·00100 to 0·00300

k_s = 1·50 mm
S = 0·00300 to 0·01000

ie hydraulic gradient =
1 in 333 to 1 in 100

Water (or sewage) at 15°C;
full bore conditions.

velocities in ms^{-1}
discharges in m^3s^{-1}

Gradient **(Equivalent) Pipe diameters in mm**

Gradient	**630**	**675**	**700**	**750**	**800**	**825**	**900**	**975**	**1000**	**1050**	**1100**	**1125**	**1200**	**1250**
0·00300	1·223	1·278	1·308	1·367	1·424	1·452	1·535	1·614	1·640	1·691	1·742	1·766	1·839	1·887
1/ 333	0·3813	0·4575	0·5035	0·6040	0·7159	0·7764	0·9763	1·2052	1·2882	1·4645	1·6551	1·7558	2·0802	2·3157
0·00320	1·264	1·321	1·352	1·412	1·471	1·500	1·585	1·667	1·694	1·747	1·799	1·824	1·900	1·949
1/ 313	0·3939	0·4726	0·5201	0·6239	0·7395	0·8020	1·0084	1·2448	1·3306	1·5127	1·7095	1·8135	2·1487	2·3919
0·00340	1·303	1·361	1·393	1·456	1·517	1·547	1·634	1·719	1·746	1·801	1·854	1·881	1·959	2·009
1/ 294	0·4061	0·4872	0·5362	0·6432	0·7624	0·8267	1·0396	1·2833	1·3717	1·5594	1·7623	1·8695	2·2150	2·4657
0·00360	1·341	1·401	1·434	1·498	1·561	1·592	1·682	1·769	1·797	1·853	1·908	1·935	2·015	2·068
1/ 278	0·4179	0·5013	0·5518	0·6619	0·7846	0·8508	1·0698	1·3206	1·4116	1·6048	1·8136	1·9239	2·2794	2·5374
0·00380	1·378	1·440	1·473	1·539	1·604	1·635	1·728	1·817	1·847	1·904	1·961	1·989	2·071	2·125
1/ 263	0·4294	0·5151	0·5670	0·6801	0·8061	0·8742	1·0992	1·3569	1·4504	1·6489	1·8634	1·9768	2·3421	2·6072
0·00400	1·414	1·477	1·512	1·580	1·646	1·678	1·773	1·865	1·895	1·954	2·012	2·041	2·125	2·180
1/ 250	0·4406	0·5286	0·5818	0·6978	0·8272	0·8970	1·1279	1·3923	1·4882	1·6919	1·9120	2·0283	2·4031	2·6751
0·00420	1·449	1·514	1·549	1·619	1·686	1·720	1·817	1·911	1·942	2·002	2·062	2·091	2·177	2·234
1/ 238	0·4516	0·5417	0·5962	0·7151	0·8477	0·9192	1·1558	1·4268	1·5250	1·7338	1·9593	2·0785	2·4626	2·7413
0·00440	1·483	1·550	1·586	1·657	1·726	1·760	1·860	1·956	1·988	2·050	2·110	2·140	2·229	2·287
1/ 227	0·4622	0·5545	0·6103	0·7320	0·8677	0·9409	1·1831	1·4605	1·5610	1·7747	2·0056	2·1276	2·5207	2·8060
0·00460	1·516	1·584	1·622	1·694	1·765	1·800	1·902	2·000	2·032	2·096	2·158	2·189	2·279	2·338
1/ 217	0·4727	0·5670	0·6241	0·7485	0·8872	0·9621	1·2098	1·4934	1·5962	1·8147	2·0508	2·1755	2·5775	2·8693
0·00480	1·549	1·619	1·657	1·731	1·803	1·839	1·943	2·043	2·076	2·141	2·205	2·236	2·328	2·388
1/ 208	0·4829	0·5792	0·6375	0·7647	0·9064	0·9829	1·2359	1·5256	1·6307	1·8539	2·0950	2·2225	2·6331	2·9311
0·00500	1·581	1·652	1·691	1·767	1·841	1·877	1·983	2·086	2·119	2·185	2·250	2·282	2·376	2·438
1/ 200	0·4929	0·5912	0·6507	0·7805	0·9251	1·0032	1·2615	1·5572	1·6644	1·8922	2·1383	2·2684	2·6876	2·9917
0·00525	1·620	1·693	1·733	1·810	1·886	1·923	2·032	2·137	2·172	2·239	2·306	2·339	2·435	2·498
1/ 190	0·5051	0·6059	0·6669	0·7998	0·9481	1·0281	1·2927	1·5957	1·7056	1·9391	2·1913	2·3246	2·7541	3·0658
0·00550	1·659	1·733	1·774	1·853	1·931	1·969	2·080	2·188	2·223	2·292	2·360	2·394	2·493	2·557
1/ 182	0·5170	0·6202	0·6826	0·8187	0·9704	1·0524	1·3232	1·6334	1·7459	1·9848	2·2430	2·3794	2·8191	3·1381
0·00575	1·696	1·772	1·814	1·895	1·974	2·013	2·127	2·237	2·273	2·344	2·413	2·448	2·549	2·615
1/ 174	0·5287	0·6342	0·6980	0·8372	0·9923	1·0761	1·3531	1·6702	1·7852	2·0295	2·2935	2·4330	2·8826	3·2088
0·00600	1·733	1·810	1·853	1·936	2·017	2·056	2·173	2·285	2·322	2·394	2·465	2·500	2·604	2·671
1/ 167	0·5401	0·6479	0·7131	0·8552	1·0137	1·0993	1·3822	1·7062	1·8237	2·0733	2·3430	2·4855	2·9447	3·2780
0·00625	1·768	1·848	1·891	1·976	2·058	2·099	2·218	2·333	2·370	2·444	2·516	2·552	2·658	2·726
1/ 160	0·5512	0·6613	0·7278	0·8729	1·0347	1·1220	1·4108	1·7415	1·8614	2·1162	2·3914	2·5369	3·0056	3·3457
0·00650	1·804	1·885	1·929	2·015	2·099	2·141	2·262	2·379	2·417	2·492	2·566	2·603	2·710	2·780
1/ 154	0·5622	0·6744	0·7423	0·8903	1·0553	1·1443	1·4388	1·7761	1·8984	2·1582	2·4389	2·5872	3·0653	3·4121
0·00675	1·838	1·921	1·966	2·054	2·139	2·182	2·305	2·424	2·463	2·540	2·615	2·652	2·762	2·834
1/ 148	0·5729	0·6873	0·7564	0·9073	1·0754	1·1662	1·4663	1·8100	1·9346	2·1994	2·4854	2·6366	3·1238	3·4773
0·00700	1·872	1·956	2·002	2·091	2·179	2·222	2·347	2·469	2·509	2·587	2·663	2·701	2·813	2·886
1/ 143	0·5835	0·6999	0·7704	0·9240	1·0952	1·1876	1·4933	1·8433	1·9702	2·2398	2·5312	2·6851	3·1812	3·5412
0·00725	1·905	1·991	2·037	2·129	2·218	2·261	2·389	2·513	2·553	2·633	2·711	2·749	2·863	2·937
1/ 138	0·5939	0·7124	0·7840	0·9404	1·1146	1·2087	1·5198	1·8760	2·0052	2·2796	2·5761	2·7328	3·2377	3·6040
0·00750	1·938	2·025	2·072	2·165	2·256	2·300	2·430	2·556	2·597	2·678	2·757	2·796	2·912	2·987
1/ 133	0·6040	0·7246	0·7975	0·9565	1·1337	1·2294	1·5459	1·9082	2·0395	2·3186	2·6202	2·7796	3·2931	3·6658
0·00800	2·001	2·091	2·140	2·236	2·330	2·376	2·510	2·640	2·682	2·766	2·848	2·888	3·007	3·085
1/ 125	0·6239	0·7484	0·8237	0·9880	1·1710	1·2699	1·5967	1·9709	2·1066	2·3949	2·7063	2·8709	3·4013	3·7862
0·00850	2·063	2·156	2·206	2·305	2·402	2·449	2·587	2·721	2·765	2·851	2·936	2·977	3·100	3·180
1/ 118	0·6432	0·7715	0·8491	1·0184	1·2072	1·3090	1·6459	2·0317	2·1715	2·4687	2·7898	2·9595	3·5062	3·9030
0·00900	2·123	2·219	2·271	2·372	2·471	2·520	2·662	2·800	2·845	2·934	3·021	3·064	3·190	3·273
1/ 111	0·6619	0·7939	0·8738	1·0481	1·2422	1·3471	1·6938	2·0907	2·2346	2·5405	2·8708	3·0454	3·6081	4·0163
0·00950	2·182	2·280	2·333	2·437	2·539	2·589	2·736	2·877	2·923	3·014	3·104	3·148	3·278	3·363
1/ 105	0·6801	0·8158	0·8978	1·0768	1·2764	1·3841	1·7403	2·1481	2·2960	2·6102	2·9497	3·1291	3·7071	4·1266
0·01000	2·238	2·339	2·394	2·501	2·605	2·657	2·807	2·952	2·999	3·093	3·185	3·230	3·363	3·450
1/ 100	0·6978	0·8370	0·9212	1·1049	1·3096	1·4201	1·7856	2·2040	2·3558	2·6782	3·0264	3·2105	3·8036	4·2340
	0·80	0·80	0·80	0·81	0·81	0·81	0·81	0·82	0·82	0·82	0·83	0·83	0·83	0·83

$V_{r(0·5)medial}$ for half-full circular pipes.

S = 0·00300 to 0·01000 **k_s = 1·50 mm**

A30
(p.5 of 6)

k_s = 1·50 mm
S = 0·01000 to 0·03000

ie hydraulic gradient = 1 in 100 to 1 in 33·3

Water (or sewage) at 15°C; full bore conditions.

velocities in ms^{-1}
discharges in m^3s^{-1}

Gradient	(Equivalent) Pipe diameters in mm													
	630	675	700	750	800	825	900	975	1000	1050	1100	1125	1200	1250
0·01000	2·238	2·339	2·394	2·501	2·605	2·657	2·807	2·952	2·999	3·093	3·185	3·230	3·363	3·450
1/ 100	0·6978	0·8370	0·9212	1·1049	1·3096	1·4201	1·7856	2·2040	2·3558	2·6782	3·0264	3·2105	3·8036	4·2340
0·01050	2·294	2·397	2·453	2·563	2·670	2·722	2·876	3·025	3·074	3·169	3·263	3·310	3·446	3·536
1/ 95	0·7151	0·8577	0·9440	1·1322	1·3420	1·4553	1·8298	2·2586	2·4141	2·7444	3·1013	3·2899	3·8977	4·3387
0·01100	2·348	2·453	2·511	2·623	2·733	2·787	2·944	3·096	3·146	3·244	3·340	3·388	3·528	3·619
1/ 91	0·7319	0·8780	0·9663	1·1589	1·3737	1·4896	1·8729	2·3118	2·4710	2·8091	3·1744	3·3675	3·9896	4·4410
0·01150	2·401	2·509	2·567	2·682	2·794	2·849	3·010	3·166	3·217	3·317	3·416	3·464	3·607	3·700
1/ 87	0·7484	0·8977	0·9881	1·1851	1·4046	1·5232	1·9151	2·3639	2·5266	2·8724	3·2459	3·4433	4·0794	4·5410
0·01200	2·453	2·563	2·623	2·740	2·855	2·911	3·075	3·234	3·286	3·389	3·489	3·539	3·685	3·780
1/ 83	0·7646	0·9171	1·0094	1·2106	1·4349	1·5560	1·9564	2·4148	2·5811	2·9343	3·3158	3·5175	4·1673	4·6388
0·01250	2·503	2·616	2·677	2·797	2·914	2·971	3·139	3·301	3·354	3·459	3·561	3·612	3·761	3·858
1/ 80	0·7804	0·9361	1·0302	1·2356	1·4645	1·5881	1·9968	2·4647	2·6344	2·9949	3·3843	3·5902	4·2534	4·7346
0·01300	2·553	2·668	2·730	2·852	2·971	3·030	3·201	3·367	3·421	3·527	3·632	3·683	3·835	3·935
1/ 77	0·7958	0·9546	1·0507	1·2601	1·4936	1·6196	2·0364	2·5136	2·6867	3·0543	3·4515	3·6614	4·3378	4·8286
0·01350	2·602	2·719	2·782	2·907	3·028	3·088	3·262	3·431	3·486	3·595	3·701	3·754	3·909	4·010
1/ 74	0·8110	0·9729	1·0707	1·2842	1·5221	1·6506	2·0753	2·5616	2·7379	3·1126	3·5173	3·7313	4·4205	4·9207
0·01400	2·650	2·769	2·833	2·960	3·084	3·144	3·322	3·494	3·550	3·661	3·769	3·823	3·980	4·083
1/ 71	0·8260	0·9908	1·0904	1·3078	1·5501	1·6809	2·1134	2·6087	2·7883	3·1698	3·5820	3·7998	4·5018	5·0111
0·01450	2·697	2·818	2·884	3·013	3·139	3·200	3·381	3·556	3·613	3·726	3·836	3·890	4·051	4·156
1/ 69	0·8406	1·0083	1·1098	1·3310	1·5776	1·7107	2·1509	2·6549	2·8377	3·2260	3·6455	3·8672	4·5816	5·0999
0·01500	2·743	2·866	2·933	3·064	3·192	3·255	3·439	3·617	3·675	3·789	3·902	3·957	4·120	4·227
1/ 67	0·8550	1·0256	1·1288	1·3538	1·6046	1·7400	2·1878	2·7004	2·8863	3·2812	3·7079	3·9334	4·6600	5·1873
0·01600	2·833	2·960	3·029	3·165	3·297	3·362	3·552	3·736	3·796	3·914	4·030	4·087	4·256	4·366
1/ 63	0·8831	1·0593	1·1659	1·3983	1·6573	1·7972	2·2596	2·7891	2·9811	3·3890	3·8297	4·0626	4·8131	5·3576
0·01700	2·920	3·052	3·123	3·263	3·399	3·466	3·661	3·851	3·913	4·035	4·154	4·213	4·387	4·500
1/ 59	0·9104	1·0920	1·2018	1·4414	1·7084	1·8526	2·3293	2·8751	3·0730	3·4935	3·9477	4·1878	4·9614	5·5227
0·01800	3·005	3·140	3·214	3·357	3·498	3·566	3·768	3·963	4·026	4·152	4·275	4·335	4·514	4·631
1/ 56	0·9368	1·1237	1·2367	1·4833	1·7581	1·9064	2·3969	2·9586	3·1622	3·5949	4·0624	4·3094	5·1054	5·6831
0·01900	3·088	3·226	3·302	3·450	3·594	3·664	3·871	4·071	4·137	4·266	4·392	4·454	4·638	4·758
1/ 53	0·9625	1·1546	1·2707	1·5240	1·8063	1·9587	2·4627	3·0398	3·2490	3·6936	4·1738	4·4277	5·2455	5·8390
0·02000	3·168	3·310	3·388	3·539	3·687	3·760	3·972	4·177	4·244	4·377	4·506	4·570	4·759	4·882
1/ 50	0·9876	1·1846	1·3038	1·5637	1·8533	2·0097	2·5268	3·1189	3·3335	3·7897	4·2824	4·5428	5·3820	5·9909
0·02100	3·247	3·392	3·472	3·627	3·778	3·853	4·070	4·281	4·349	4·485	4·618	4·683	4·876	5·003
1/ 47·6	1·0120	1·2139	1·3360	1·6023	1·8992	2·0594	2·5893	3·1960	3·4160	3·8834	4·3883	4·6552	5·5151	6·1390
0·02200	3·323	3·472	3·553	3·712	3·867	3·943	4·166	4·382	4·452	4·590	4·726	4·794	4·991	5·120
1/ 45·5	1·0359	1·2425	1·3675	1·6401	1·9439	2·1080	2·6503	3·2713	3·4965	3·9749	4·4917	4·7649	5·6450	6·2836
0·02300	3·398	3·550	3·633	3·796	3·954	4·032	4·260	4·480	4·552	4·694	4·833	4·901	5·104	5·236
1/ 43·5	1·0592	1·2705	1·3983	1·6770	1·9877	2·1554	2·7100	3·3449	3·5752	4·0643	4·5928	4·8721	5·7720	6·4250
0·02400	3·471	3·627	3·712	3·878	4·040	4·119	4·352	4·577	4·650	4·795	4·937	5·007	5·214	5·348
1/ 41·7	1·0820	1·2979	1·4284	1·7132	2·0305	2·2018	2·7683	3·4170	3·6522	4·1519	4·6917	4·9770	5·8963	6·5634
0·02500	3·543	3·702	3·788	3·958	4·123	4·204	4·441	4·671	4·746	4·894	5·039	5·110	5·321	5·459
1/ 40·0	1·1044	1·3247	1·4579	1·7485	2·0724	2·2473	2·8255	3·4875	3·7276	4·2376	4·7886	5·0798	6·0181	6·6989
0·02600	3·613	3·775	3·864	4·036	4·205	4·287	4·530	4·764	4·840	4·991	5·139	5·212	5·427	5·567
1/ 38·5	1·1263	1·3510	1·4869	1·7832	2·1135	2·2919	2·8815	3·5567	3·8015	4·3216	4·8835	5·1805	6·1374	6·8317
0·02700	3·682	3·847	3·937	4·113	4·285	4·369	4·616	4·855	4·933	5·086	5·237	5·311	5·530	5·673
1/ 37·0	1·1478	1·3767	1·5152	1·8172	2·1539	2·3356	2·9365	3·6245	3·8740	4·4040	4·9766	5·2793	6·2544	6·9620
0·02800	3·750	3·918	4·010	4·189	4·364	4·449	4·701	4·944	5·023	5·180	5·333	5·409	5·632	5·777
1/ 35·7	1·1689	1·4020	1·5431	1·8506	2·1934	2·3785	2·9905	3·6911	3·9452	4·4850	5·0681	5·3763	6·3693	7·0899
0·02900	3·816	3·987	4·081	4·263	4·441	4·528	4·784	5·031	5·112	5·271	5·427	5·504	5·732	5·880
1/ 34·5	1·1896	1·4269	1·5704	1·8834	2·2323	2·4207	3·0435	3·7565	4·0151	4·5644	5·1579	5·4716	6·4822	7·2155
0·03000	3·882	4·056	4·151	4·336	4·517	4·606	4·866	5·118	5·200	5·362	5·520	5·599	5·830	5·980
1/ 33·3	1·2100	1·4513	1·5973	1·9157	2·2705	2·4621	3·0956	3·8208	4·0838	4·6426	5·2462	5·5652	6·5931	7·3390
$V_{r(0·5)medial}$ for half-full circular pipes.	0·80	0·80	0·80	0·81	0·81	0·81	0·82	0·82	0·82	0·82	0·83	0·83	0·83	0·83

k_s = 1·50 mm S = 0·01000 to 0·03000

k_s = 1·50 mm
S = 0·03000 to 0·10000

ie hydraulic gradient =
1 in 33·3 to 1 in 10·0

Water (or sewage) at 15°C;
full bore conditions.

velocities in ms^{-1}
discharges in m^3s^{-1}

Gradient	**(Equivalent) Pipe diameters in mm**													
	630	**675**	**700**	**750**	**800**	**825**	**900**	**975**	**1000**	**1050**	**1100**	**1125**	**1200**	**1250**
0·03000	3·882	4·056	4·151	4·336	4·517	4·606	4·866	5·118	5·200	5·362	5·520	5·599	5·830	5·980
1/ 33·3	1·2100	1·4513	1·5973	1·9157	2·2705	2·4621	3·0956	3·8208	4·0838	4·6426	5·2462	5·5652	6·5931	7·3390
0·03200	4·009	4·189	4·287	4·479	4·665	4·757	5·026	5·286	5·370	5·538	5·702	5·782	6·021	6·177
1/ 31·3	1·2497	1·4990	1·6498	1·9786	2·3451	2·5430	3·1972	3·9463	4·2179	4·7950	5·4184	5·7479	6·8096	7·5799
0·03400	4·133	4·318	4·419	4·617	4·809	4·904	5·181	5·448	5·536	5·708	5·877	5·961	6·206	6·367
1/ 29·4	1·2882	1·5452	1·7006	2·0396	2·4174	2·6213	3·2957	4·0679	4·3478	4·9427	5·5853	5·9250	7·0194	7·8134
0·03600	4·253	4·443	4·547	4·751	4·949	5·046	5·331	5·607	5·697	5·874	6·048	6·134	6·387	6·552
1/ 27·8	1·3256	1·5901	1·7500	2·0988	2·4875	2·6974	3·3914	4·1860	4·4740	5·0862	5·7474	6·0969	7·2231	8·0402
0·03800	4·369	4·565	4·672	4·881	5·085	5·184	5·477	5·760	5·853	6·035	6·214	6·302	6·562	6·731
1/ 26·3	1·3620	1·6337	1·7980	2·1564	2·5558	2·7714	3·4844	4·3008	4·5968	5·2257	5·9051	6·2642	7·4212	8·2607
0·04000	4·483	4·684	4·794	5·008	5·217	5·319	5·620	5·910	6·005	6·192	6·375	6·466	6·732	6·906
1/ 25·0	1·3974	1·6762	1·8448	2·2125	2·6223	2·8435	3·5751	4·4126	4·7163	5·3616	6·0586	6·4271	7·6141	8·4755
0·04200	4·594	4·800	4·912	5·132	5·346	5·451	5·759	6·056	6·153	6·345	6·533	6·626	6·899	7·077
1/ 23·8	1·4320	1·7176	1·8904	2·2672	2·6871	2·9138	3·6634	4·5217	4·8329	5·4941	6·2084	6·5859	7·8023	8·6850
0·04400	4·702	4·913	5·028	5·253	5·472	5·579	5·894	6·199	6·298	6·494	6·687	6·782	7·061	7·244
1/ 22·7	1·4657	1·7581	1·9349	2·3206	2·7504	2·9825	3·7497	4·6282	4·9467	5·6235	6·3546	6·7411	7·9861	8·8895
0·04600	4·808	5·024	5·141	5·371	5·595	5·705	6·027	6·338	6·440	6·641	6·837	6·934	7·220	7·407
1/ 21·7	1·4987	1·7977	1·9785	2·3728	2·8123	3·0496	3·8341	4·7323	5·0580	5·7500	6·4976	6·8927	8·1658	9·0895
0·04800	4·911	5·132	5·252	5·487	5·715	5·828	6·157	6·475	6·579	6·783	6·984	7·083	7·376	7·566
1/ 20·8	1·5310	1·8364	2·0211	2·4239	2·8728	3·1152	3·9166	4·8342	5·1669	5·8738	6·6375	7·0411	8·3415	9·2851
0·05000	5·013	5·238	5·360	5·600	5·833	5·948	6·284	6·608	6·715	6·923	7·129	7·230	7·528	7·722
1/ 20·0	1·5626	1·8743	2·0628	2·4739	2·9321	3·1795	3·9975	4·9340	5·2736	5·9950	6·7745	7·1864	8·5137	9·4768
0·05250	5·137	5·367	5·493	5·738	5·978	6·095	6·439	6·772	6·880	7·095	7·305	7·408	7·714	7·913
1/ 19·0	1·6012	1·9206	2·1138	2·5351	3·0046	3·2581	4·0963	5·0560	5·4039	6·1432	6·9419	7·3640	8·7241	9·7110
0·05500	5·258	5·494	5·622	5·873	6·118	6·239	6·591	6·931	7·043	7·262	7·477	7·583	7·895	8·100
1/ 18·2	1·6390	1·9659	2·1636	2·5948	3·0754	3·3349	4·1928	5·1751	5·5312	6·2879	7·1054	7·5375	8·9296	9·9397
0·05750	5·376	5·617	5·748	6·006	6·256	6·379	6·739	7·087	7·201	7·425	7·645	7·753	8·073	8·282
1/ 17·4	1·6758	2·0101	2·2123	2·6532	3·1446	3·4099	4·2871	5·2915	5·6556	6·4294	7·2652	7·7070	9·1304	10·163
0·06000	5·492	5·738	5·872	6·135	6·391	6·516	6·884	7·240	7·356	7·585	7·809	7·920	8·247	8·460
1/ 16·7	1·7119	2·0534	2·2599	2·7103	3·2123	3·4833	4·3794	5·4054	5·7774	6·5678	7·4216	7·8729	9·3269	10·382
0·06250	5·605	5·857	5·993	6·262	6·523	6·651	7·026	7·389	7·508	7·741	7·971	8·084	8·417	8·635
1/ 16·0	1·7473	2·0958	2·3066	2·7662	3·2786	3·5552	4·4698	5·5169	5·8966	6·7033	7·5748	8·0354	9·5194	10·596
0·06500	5·716	5·973	6·112	6·386	6·652	6·783	7·165	7·536	7·657	7·895	8·129	8·244	8·584	8·806
1/ 15·4	1·7819	2·1373	2·3523	2·8211	3·3436	3·6257	4·5584	5·6263	6·0135	6·8361	7·7249	8·1946	9·7081	10·806
0·06750	5·825	6·087	6·229	6·507	6·779	6·912	7·302	7·679	7·803	8·045	8·284	8·401	8·747	8·974
1/ 14·8	1·8159	2·1781	2·3971	2·8749	3·4073	3·6948	4·6453	5·7335	6·1281	6·9665	7·8721	8·3508	9·8931	11·012
0·07000	5·932	6·198	6·343	6·627	6·903	7·039	7·436	7·820	7·946	8·193	8·436	8·555	8·908	9·138
1/ 14·3	1·8492	2·2181	2·4412	2·9277	3·4699	3·7627	4·7306	5·8388	6·2406	7·0944	8·0167	8·5042	10·075	11·214
0·07250	6·037	6·308	6·456	6·744	7·025	7·163	7·568	7·959	8·087	8·338	8·585	8·707	9·066	9·300
1/ 13·8	1·8820	2·2574	2·4844	2·9795	3·5314	3·8293	4·8144	5·9423	6·3512	7·2201	8·1587	8·6548	10·253	11·413
0·07500	6·141	6·416	6·566	6·860	7·146	7·286	7·697	8·095	8·225	8·481	8·732	8·856	9·221	9·459
1/ 13·3	1·9142	2·2960	2·5269	3·0305	3·5918	3·8948	4·8968	6·0439	6·4598	7·3436	8·2983	8·8029	10·429	11·608
0·08000	6·342	6·627	6·782	7·085	7·380	7·525	7·950	8·361	8·495	8·759	9·019	9·146	9·524	9·770
1/ 12·5	1·9771	2·3714	2·6099	3·1300	3·7097	4·0227	5·0575	6·2423	6·6719	7·5846	8·5706	9·0918	10·771	11·989
0·08500	6·538	6·831	6·991	7·303	7·608	7·757	8·195	8·618	8·757	9·029	9·296	9·428	9·817	10·07
1/ 11·8	2·0380	2·4444	2·6903	3·2264	3·8240	4·1466	5·2133	6·4345	6·8773	7·8182	8·8346	9·3718	11·103	12·358
0·09000	6·727	7·029	7·193	7·515	7·828	7·982	8·432	8·868	9·011	9·291	9·566	9·702	10·10	10·36
1/ 11·1	2·0971	2·5154	2·7683	3·3200	3·9349	4·2669	5·3645	6·6212	7·0769	8·0450	9·0909	9·6436	11·425	12·717
0·09500	6·912	7·222	7·391	7·721	8·043	8·201	8·664	9·111	9·258	9·546	9·828	9·968	10·38	10·65
1/ 10·5	2·1546	2·5844	2·8443	3·4111	4·0428	4·3839	5·5116	6·8028	7·2709	8·2656	9·3401	9·9081	11·738	13·066
0·10000	7·092	7·410	7·583	7·922	8·252	8·414	8·889	9·348	9·498	9·794	10·08	10·23	10·65	10·92
1/ 10·0	2·2106	2·6516	2·9182	3·4998	4·1480	4·4979	5·6549	6·9796	7·4599	8·4805	9·5829	10·166	12·043	13·405
	0·80	0·80	0·80	0·81	0·81	0·81	0·82	0·82	0·82	0·82	0·83	0·83	0·83	0·83

$V_{r(0·5)medial}$ for half-full circular pipes.

A31
(p.1 of 6)

$k_s = 3{\cdot}0$ **mm**
S = 0·00010 to 0·00030

ie hydraulic gradient =
1 in 10000 to 1 in 3333

Water (or sewage) at 15°C;
full bore conditions.

velocities in ms^{-1}
discharges in m^3s^{-1}

Gradient	(Equivalent) Pipe diameters in mm													
	630	**675**	**700**	**750**	**800**	**825**	**900**	**975**	**1000**	**1050**	**1100**	**1125**	**1200**	**1250**
0·000100	0·200	0·210	0·215	0·225	0·234	0·239	0·253	0·267	0·271	0·280	0·288	0·292	0·305	0·313
1/ 10000	0·0625	0·0750	0·0826	0·0993	0·1178	0·1278	0·1610	0·1990	0·2128	0·2421	0·2738	0·2906	0·3447	0·3839
0·000105	0·205	0·215	0·220	0·230	0·240	0·245	0·259	0·273	0·278	0·287	0·295	0·300	0·312	0·321
1/ 9524	0·0640	0·0769	0·0847	0·1017	0·1207	0·1310	0·1650	0·2040	0·2181	0·2482	0·2807	0·2978	0·3533	0·3935
0·000110	0·210	0·220	0·225	0·236	0·246	0·251	0·266	0·280	0·284	0·293	0·302	0·307	0·320	0·328
1/ 9091	0·0655	0·0787	0·0867	0·1042	0·1236	0·1341	0·1689	0·2088	0·2233	0·2541	0·2873	0·3049	0·3616	0·4028
0·000115	0·215	0·225	0·230	0·241	0·252	0·257	0·272	0·286	0·291	0·300	0·309	0·314	0·327	0·336
1/ 8696	0·0670	0·0805	0·0887	0·1065	0·1264	0·1372	0·1728	0·2136	0·2284	0·2598	0·2938	0·3118	0·3698	0·4120
0·000120	0·220	0·230	0·236	0·246	0·257	0·262	0·277	0·292	0·297	0·307	0·316	0·321	0·334	0·343
1/ 8333	0·0685	0·0823	0·0906	0·1088	0·1292	0·1402	0·1765	0·2182	0·2333	0·2655	0·3002	0·3186	0·3779	0·4209
0·000125	0·224	0·235	0·240	0·252	0·262	0·268	0·283	0·298	0·303	0·313	0·322	0·327	0·341	0·350
1/ 8000	0·0699	0·0840	0·0925	0·1111	0·1319	0·1431	0·1802	0·2227	0·2382	0·2710	0·3065	0·3252	0·3857	0·4297
0·000130	0·229	0·239	0·245	0·257	0·268	0·273	0·289	0·304	0·309	0·319	0·329	0·334	0·348	0·357
1/ 7692	0·0713	0·0857	0·0944	0·1133	0·1345	0·1459	0·1838	0·2272	0·2429	0·2764	0·3126	0·3317	0·3934	0·4382
0·000135	0·233	0·244	0·250	0·262	0·273	0·278	0·294	0·310	0·315	0·325	0·335	0·340	0·355	0·364
1/ 7407	0·0727	0·0873	0·0962	0·1155	0·1371	0·1488	0·1873	0·2316	0·2476	0·2817	0·3186	0·3381	0·4010	0·4466
0·000140	0·238	0·249	0·255	0·266	0·278	0·283	0·300	0·316	0·321	0·331	0·341	0·346	0·361	0·371
1/ 7143	0·0741	0·0890	0·0980	0·1177	0·1396	0·1515	0·1908	0·2358	0·2522	0·2869	0·3245	0·3444	0·4084	0·4549
0·000145	0·242	0·253	0·259	0·271	0·283	0·289	0·305	0·322	0·327	0·337	0·348	0·353	0·368	0·377
1/ 6897	0·0754	0·0906	0·0997	0·1198	0·1421	0·1542	0·1942	0·2401	0·2567	0·2921	0·3303	0·3505	0·4157	0·4630
0·000150	0·246	0·257	0·264	0·276	0·288	0·293	0·311	0·327	0·332	0·343	0·354	0·359	0·374	0·384
1/ 6667	0·0767	0·0921	0·1015	0·1218	0·1446	0·1569	0·1976	0·2442	0·2611	0·2971	0·3360	0·3566	0·4229	0·4710
0·000160	0·254	0·266	0·272	0·285	0·297	0·303	0·321	0·338	0·343	0·354	0·365	0·371	0·386	0·396
1/ 6250	0·0792	0·0952	0·1048	0·1259	0·1494	0·1621	0·2041	0·2523	0·2698	0·3069	0·3471	0·3683	0·4368	0·4866
0·000170	0·262	0·274	0·281	0·294	0·306	0·313	0·331	0·348	0·354	0·365	0·377	0·382	0·398	0·409
1/ 5882	0·0817	0·0981	0·1081	0·1298	0·1540	0·1671	0·2104	0·2601	0·2781	0·3164	0·3579	0·3798	0·4504	0·5017
0·000180	0·270	0·282	0·289	0·302	0·315	0·322	0·340	0·359	0·364	0·376	0·388	0·393	0·410	0·421
1/ 5556	0·0841	0·1010	0·1112	0·1336	0·1585	0·1720	0·2166	0·2677	0·2863	0·3257	0·3683	0·3909	0·4635	0·5163
0·000190	0·277	0·290	0·297	0·311	0·324	0·331	0·350	0·368	0·375	0·387	0·398	0·404	0·421	0·432
1/ 5263	0·0864	0·1038	0·1143	0·1373	0·1629	0·1768	0·2226	0·2751	0·2942	0·3347	0·3785	0·4016	0·4763	0·5305
0·000200	0·285	0·298	0·305	0·319	0·333	0·339	0·359	0·378	0·384	0·397	0·409	0·415	0·432	0·444
1/ 5000	0·0887	0·1065	0·1173	0·1409	0·1672	0·1814	0·2284	0·2823	0·3019	0·3434	0·3884	0·4122	0·4888	0·5444
0·000210	0·292	0·305	0·312	0·327	0·341	0·348	0·368	0·388	0·394	0·407	0·419	0·425	0·443	0·455
1/ 4762	0·0909	0·1092	0·1202	0·1444	0·1714	0·1859	0·2341	0·2893	0·3094	0·3520	0·3981	0·4224	0·5009	0·5580
0·000220	0·299	0·312	0·320	0·335	0·349	0·356	0·377	0·397	0·403	0·416	0·429	0·435	0·453	0·465
1/ 4545	0·0931	0·1118	0·1231	0·1478	0·1754	0·1903	0·2397	0·2962	0·3167	0·3603	0·4075	0·4324	0·5128	0·5712
0·000230	0·305	0·319	0·327	0·342	0·357	0·364	0·385	0·406	0·412	0·426	0·438	0·445	0·464	0·476
1/ 4348	0·0952	0·1143	0·1259	0·1512	0·1794	0·1946	0·2451	0·3029	0·3239	0·3685	0·4167	0·4422	0·5244	0·5841
0·000240	0·312	0·326	0·334	0·350	0·365	0·372	0·394	0·414	0·421	0·435	0·448	0·454	0·474	0·486
1/ 4167	0·0972	0·1168	0·1286	0·1545	0·1833	0·1989	0·2504	0·3095	0·3309	0·3765	0·4257	0·4518	0·5358	0·5967
0·000250	0·318	0·333	0·341	0·357	0·372	0·380	0·402	0·423	0·430	0·444	0·457	0·464	0·484	0·496
1/ 4000	0·0993	0·1192	0·1313	0·1577	0·1871	0·2030	0·2556	0·3159	0·3378	0·3843	0·4346	0·4612	0·5469	0·6091
0·000260	0·325	0·340	0·348	0·364	0·380	0·387	0·410	0·432	0·439	0·453	0·466	0·473	0·493	0·506
1/ 3846	0·1013	0·1216	0·1339	0·1608	0·1908	0·2070	0·2607	0·3222	0·3445	0·3919	0·4432	0·4703	0·5578	0·6212
0·000270	0·331	0·346	0·355	0·371	0·387	0·395	0·418	0·440	0·447	0·461	0·475	0·482	0·503	0·516
1/ 3704	0·1032	0·1239	0·1365	0·1639	0·1945	0·2110	0·2657	0·3284	0·3511	0·3995	0·4517	0·4794	0·5685	0·6331
0·000280	0·337	0·353	0·361	0·378	0·394	0·402	0·425	0·448	0·455	0·470	0·484	0·491	0·512	0·525
1/ 3571	0·1051	0·1262	0·1390	0·1669	0·1981	0·2149	0·2706	0·3344	0·3576	0·4068	0·4601	0·4882	0·5789	0·6448
0·000290	0·343	0·359	0·368	0·385	0·401	0·409	0·433	0·456	0·463	0·478	0·493	0·500	0·521	0·535
1/ 3448	0·1070	0·1285	0·1415	0·1699	0·2016	0·2187	0·2754	0·3404	0·3640	0·4141	0·4683	0·4969	0·5892	0·6563
0·000300	0·349	0·365	0·374	0·391	0·408	0·416	0·440	0·464	0·471	0·486	0·501	0·508	0·530	0·544
1/ 3333	0·1088	0·1307	0·1439	0·1728	0·2051	0·2225	0·2802	0·3462	0·3702	0·4212	0·4763	0·5054	0·5994	0·6676
	0·77	0·77	0·77	0·77	0·78	0·78	0·78	0·78	0·79	0·79	0·79	0·79	0·79	0·80

$V_{r(0{\cdot}5)medial}$ **for half-full circular pipes.**

$k_s = 3{\cdot}0$ **mm** S = 0·00010 to 0·00030

$k_s = 3{\cdot}0$ mm
S = 0·00030 to 0·00100

ie hydraulic gradient =
1 in 3333 to 1 in 1000

Water (or sewage) at 15°C;
full bore conditions.

velocities in ms^{-1}
discharges in m^3s^{-1}

Gradient **(Equivalent) Pipe diameters in mm**

Gradient	630	675	700	750	800	825	900	975	1000	1050	1100	1125	1200	1250
0·000300	0·349	0·365	0·374	0·391	0·408	0·416	0·440	0·464	0·471	0·486	0·501	0·508	0·530	0·544
1/ 3333	0·1088	0·1307	0·1439	0·1728	0·2051	0·2225	0·2802	0·3462	0·3702	0·4212	0·4763	0·5054	0·5994	0·6676
0·000320	0·361	0·377	0·386	0·404	0·421	0·430	0·455	0·479	0·487	0·502	0·518	0·525	0·547	0·562
1/ 3125	0·1124	0·1350	0·1487	0·1786	0·2119	0·2299	0·2894	0·3577	0·3824	0·4351	0·4920	0·5221	0·6191	0·6896
0·000340	0·372	0·389	0·398	0·417	0·435	0·443	0·469	0·494	0·502	0·518	0·534	0·542	0·564	0·579
1/ 2941	0·1159	0·1392	0·1533	0·1841	0·2184	0·2370	0·2984	0·3688	0·3943	0·4486	0·5072	0·5383	0·6383	0·7109
0·000360	0·383	0·400	0·410	0·429	0·447	0·456	0·483	0·508	0·517	0·533	0·549	0·557	0·581	0·596
1/ 2778	0·1193	0·1433	0·1578	0·1895	0·2248	0·2439	0·3071	0·3795	0·4058	0·4616	0·5220	0·5540	0·6569	0·7317
0·000380	0·393	0·411	0·421	0·441	0·460	0·469	0·496	0·522	0·531	0·548	0·564	0·573	0·597	0·613
1/ 2632	0·1226	0·1472	0·1621	0·1947	0·2310	0·2506	0·3155	0·3900	0·4170	0·4744	0·5364	0·5692	0·6750	0·7518
0·000400	0·404	0·422	0·432	0·452	0·472	0·481	0·509	0·536	0·545	0·562	0·579	0·588	0·612	0·629
1/ 2500	0·1258	0·1511	0·1664	0·1998	0·2370	0·2572	0·3238	0·4002	0·4279	0·4868	0·5504	0·5841	0·6926	0·7714
0·000420	0·414	0·433	0·443	0·463	0·483	0·493	0·522	0·549	0·558	0·576	0·594	0·602	0·628	0·644
1/ 2381	0·1289	0·1548	0·1705	0·2048	0·2429	0·2636	0·3318	0·4101	0·4385	0·4988	0·5641	0·5986	0·7098	0·7906
0·000440	0·423	0·443	0·454	0·474	0·495	0·505	0·534	0·562	0·572	0·590	0·608	0·616	0·642	0·659
1/ 2273	0·1320	0·1585	0·1746	0·2096	0·2487	0·2698	0·3397	0·4198	0·4489	0·5106	0·5774	0·6127	0·7266	0·8093
0·000460	0·433	0·453	0·464	0·485	0·506	0·516	0·546	0·575	0·584	0·603	0·621	0·630	0·657	0·674
1/ 2174	0·1350	0·1621	0·1785	0·2143	0·2543	0·2759	0·3474	0·4293	0·4590	0·5222	0·5905	0·6266	0·7430	0·8275
0·000480	0·442	0·463	0·474	0·496	0·517	0·527	0·558	0·587	0·597	0·616	0·635	0·644	0·671	0·689
1/ 2083	0·1379	0·1656	0·1824	0·2190	0·2598	0·2819	0·3549	0·4386	0·4689	0·5335	0·6032	0·6401	0·7591	0·8454
0·000500	0·452	0·472	0·484	0·506	0·528	0·538	0·569	0·600	0·609	0·629	0·648	0·657	0·685	0·703
1/ 2000	0·1408	0·1690	0·1862	0·2235	0·2652	0·2877	0·3622	0·4476	0·4786	0·5445	0·6157	0·6534	0·7748	0·8629
0·000525	0·463	0·484	0·496	0·519	0·541	0·552	0·584	0·614	0·625	0·644	0·664	0·674	0·702	0·721
1/ 1905	0·1443	0·1732	0·1908	0·2291	0·2718	0·2949	0·3712	0·4588	0·4905	0·5580	0·6310	0·6696	0·7940	0·8843
0·000550	0·474	0·496	0·507	0·531	0·553	0·565	0·597	0·629	0·639	0·660	0·680	0·690	0·719	0·738
1/ 1818	0·1477	0·1773	0·1953	0·2345	0·2782	0·3018	0·3800	0·4696	0·5021	0·5712	0·6459	0·6854	0·8128	0·9052
0·000575	0·484	0·507	0·519	0·543	0·566	0·577	0·611	0·643	0·654	0·675	0·695	0·705	0·735	0·754
1/ 1739	0·1510	0·1814	0·1997	0·2398	0·2845	0·3087	0·3886	0·4802	0·5135	0·5841	0·6605	0·7009	0·8311	0·9256
0·000600	0·495	0·518	0·530	0·555	0·578	0·590	0·624	0·657	0·668	0·689	0·710	0·720	0·751	0·771
1/ 1667	0·1543	0·1853	0·2040	0·2450	0·2907	0·3153	0·3970	0·4906	0·5245	0·5967	0·6748	0·7160	0·8490	0·9456
0·000625	0·505	0·528	0·541	0·566	0·590	0·602	0·637	0·671	0·682	0·703	0·725	0·735	0·766	0·787
1/ 1600	0·1575	0·1891	0·2083	0·2501	0·2967	0·3219	0·4052	0·5007	0·5354	0·6091	0·6887	0·7309	0·8666	0·9652
0·000650	0·515	0·539	0·552	0·577	0·602	0·614	0·650	0·684	0·695	0·717	0·739	0·750	0·781	0·802
1/ 1538	0·1606	0·1929	0·2124	0·2550	0·3026	0·3283	0·4133	0·5107	0·5461	0·6212	0·7024	0·7454	0·8838	0·9844
0·000675	0·525	0·549	0·563	0·588	0·613	0·626	0·662	0·697	0·709	0·731	0·753	0·764	0·796	0·817
1/ 1481	0·1637	0·1966	0·2165	0·2599	0·3084	0·3346	0·4212	0·5205	0·5565	0·6331	0·7159	0·7596	0·9008	1·0032
0·000700	0·535	0·559	0·573	0·599	0·625	0·637	0·674	0·710	0·722	0·745	0·767	0·778	0·811	0·833
1/ 1429	0·1667	0·2002	0·2205	0·2647	0·3141	0·3407	0·4289	0·5301	0·5668	0·6448	0·7291	0·7736	0·9173	1·0217
0·000725	0·544	0·569	0·583	0·610	0·636	0·649	0·686	0·723	0·734	0·758	0·781	0·792	0·826	0·847
1/ 1379	0·1697	0·2038	0·2244	0·2694	0·3196	0·3468	0·4366	0·5395	0·5768	0·6562	0·7420	0·7874	0·9336	1·0398
0·000750	0·554	0·579	0·593	0·620	0·647	0·660	0·698	0·735	0·747	0·771	0·794	0·806	0·840	0·862
1/ 1333	0·1726	0·2073	0·2283	0·2740	0·3251	0·3527	0·4441	0·5488	0·5867	0·6675	0·7547	0·8009	0·9497	1·0577
0·000800	0·572	0·598	0·613	0·641	0·668	0·682	0·721	0·759	0·772	0·796	0·820	0·832	0·867	0·890
1/ 1250	0·1783	0·2141	0·2358	0·2831	0·3359	0·3644	0·4587	0·5668	0·6061	0·6894	0·7796	0·8273	0·9809	1·0925
0·000850	0·590	0·617	0·632	0·661	0·689	0·703	0·743	0·783	0·795	0·821	0·846	0·858	0·894	0·918
1/ 1176	0·1838	0·2207	0·2431	0·2918	0·3462	0·3756	0·4729	0·5843	0·6248	0·7107	0·8037	0·8528	1·0112	1·1262
0·000900	0·607	0·635	0·650	0·680	0·709	0·723	0·765	0·805	0·819	0·845	0·870	0·883	0·920	0·944
1/ 1111	0·1892	0·2272	0·2501	0·3003	0·3563	0·3866	0·4866	0·6013	0·6430	0·7314	0·8271	0·8776	1·0406	1·1590
0·000950	0·624	0·652	0·668	0·699	0·728	0·743	0·786	0·828	0·841	0·868	0·894	0·907	0·945	0·970
1/ 1053	0·1944	0·2334	0·2570	0·3086	0·3661	0·3972	0·5000	0·6179	0·6606	0·7515	0·8498	0·9017	1·0692	1·1908
0·001000	0·640	0·669	0·685	0·717	0·747	0·762	0·806	0·849	0·863	0·891	0·918	0·931	0·970	0·996
1/ 1000	0·1994	0·2395	0·2637	0·3166	0·3757	0·4076	0·5131	0·6340	0·6779	0·7711	0·8719	0·9253	1·0971	1·2219
	0·77	0·77	0·77	0·78	0·78	0·78	0·78	0·79	0·79	0·79	0·79	0·79	0·80	0·80

$V_{r(0\cdot5)medial}$ for half-full circular pipes.

$k_s = 3{\cdot}0$ mm
S = 0·00100 to 0·00300

ie hydraulic gradient =
1 in 1000 to 1 in 333

Water (or sewage) at 15°C;
full bore conditions.

velocities in ms^{-1}
discharges in m^3s^{-1}

Gradient **(Equivalent) Pipe diameters in mm**

Gradient	630	675	700	750	800	825	900	975	1000	1050	1100	1125	1200	1250
0·00100	0·640	0·669	0·685	0·717	0·747	0·762	0·806	0·849	0·863	0·891	0·918	0·931	0·970	0·996
1/ 1000	0·1994	0·2395	0·2637	0·3166	0·3757	0·4076	0·5131	0·6340	0·6779	0·7711	0·8719	0·9253	1·0971	1·2219
0·00105	0·656	0·686	0·702	0·735	0·766	0·781	0·826	0·870	0·884	0·913	0·940	0·954	0·994	1·020
1/ 952	0·2044	0·2454	0·2703	0·3245	0·3850	0·4177	0·5258	0·6497	0·6947	0·7902	0·8936	0·9482	1·1243	1·2521
0·00110	0·671	0·702	0·719	0·752	0·784	0·800	0·846	0·891	0·905	0·934	0·962	0·976	1·018	1·044
1/ 909	0·2092	0·2512	0·2767	0·3322	0·3941	0·4275	0·5382	0·6650	0·7111	0·8089	0·9146	0·9706	1·1508	1·2817
0·00115	0·686	0·718	0·735	0·769	0·802	0·818	0·865	0·911	0·926	0·955	0·984	0·998	1·040	1·068
1/ 870	0·2140	0·2569	0·2829	0·3397	0·4030	0·4372	0·5503	0·6800	0·7271	0·8271	0·9353	0·9924	1·1768	1·3106
0·00120	0·701	0·733	0·751	0·785	0·819	0·835	0·884	0·930	0·946	0·976	1·005	1·020	1·063	1·091
1/ 833	0·2186	0·2625	0·2890	0·3470	0·4117	0·4466	0·5622	0·6947	0·7428	0·8450	0·9554	1·0138	1·2021	1·3388
0·00125	0·716	0·749	0·767	0·802	0·836	0·853	0·902	0·950	0·965	0·996	1·026	1·041	1·085	1·114
1/ 800	0·2231	0·2679	0·2950	0·3542	0·4202	0·4558	0·5738	0·7091	0·7582	0·8625	0·9752	1·0348	1·2270	1·3665
0·00130	0·730	0·764	0·782	0·818	0·853	0·870	0·920	0·969	0·984	1·016	1·047	1·062	1·106	1·136
1/ 769	0·2275	0·2732	0·3009	0·3612	0·4285	0·4649	0·5852	0·7232	0·7732	0·8796	0·9946	1·0554	1·2514	1·3937
0·00135	0·744	0·778	0·797	0·833	0·869	0·886	0·938	0·987	1·003	1·035	1·067	1·082	1·128	1·157
1/ 741	0·2319	0·2784	0·3066	0·3681	0·4367	0·4738	0·5964	0·7370	0·7880	0·8964	1·0136	1·0755	1·2753	1·4203
0·00140	0·758	0·792	0·811	0·849	0·885	0·903	0·955	1·005	1·022	1·054	1·086	1·102	1·148	1·179
1/ 714	0·2362	0·2836	0·3123	0·3749	0·4448	0·4825	0·6074	0·7506	0·8025	0·9129	1·0322	1·0953	1·2987	1·4464
0·00145	0·771	0·807	0·826	0·864	0·901	0·919	0·972	1·023	1·040	1·073	1·105	1·121	1·169	1·200
1/ 690	0·2404	0·2886	0·3178	0·3816	0·4527	0·4911	0·6182	0·7639	0·8167	0·9291	1·0505	1·1148	1·3218	1·4721
0·00150	0·784	0·820	0·840	0·879	0·916	0·934	0·988	1·041	1·058	1·091	1·124	1·141	1·189	1·220
1/ 667	0·2445	0·2936	0·3233	0·3881	0·4604	0·4995	0·6288	0·7770	0·8307	0·9450	1·0686	1·1339	1·3444	1·4973
0·00160	0·810	0·847	0·868	0·907	0·946	0·965	1·021	1·075	1·093	1·127	1·161	1·178	1·228	1·260
1/ 625	0·2525	0·3032	0·3339	0·4009	0·4756	0·5159	0·6495	0·8025	0·8581	0·9761	1·1037	1·1712	1·3886	1·5465
0·00170	0·835	0·874	0·894	0·935	0·975	0·995	1·052	1·108	1·126	1·162	1·197	1·215	1·266	1·299
1/ 588	0·2603	0·3126	0·3442	0·4133	0·4903	0·5319	0·6695	0·8273	0·8845	1·0062	1·1378	1·2073	1·4315	1·5942
0·00180	0·859	0·899	0·920	0·963	1·004	1·024	1·083	1·140	1·159	1·196	1·232	1·250	1·302	1·337
1/ 556	0·2679	0·3217	0·3542	0·4253	0·5045	0·5473	0·6890	0·8514	0·9103	1·0355	1·1708	1·2424	1·4731	1·6406
0·00190	0·883	0·924	0·946	0·989	1·031	1·052	1·113	1·172	1·191	1·229	1·266	1·284	1·338	1·374
1/ 526	0·2753	0·3305	0·3640	0·4370	0·5184	0·5624	0·7079	0·8747	0·9353	1·0639	1·2030	1·2765	1·5135	1·6856
0·00200	0·906	0·948	0·970	1·015	1·058	1·079	1·142	1·202	1·222	1·261	1·299	1·318	1·373	1·409
1/ 500	0·2825	0·3391	0·3735	0·4484	0·5319	0·5770	0·7264	0·8975	0·9596	1·0916	1·2343	1·3097	1·5529	1·7295
0·00210	0·929	0·971	0·994	1·040	1·084	1·106	1·170	1·232	1·252	1·292	1·331	1·350	1·407	1·444
1/ 476	0·2894	0·3476	0·3827	0·4595	0·5451	0·5913	0·7444	0·9198	0·9834	1·1187	1·2649	1·3422	1·5914	1·7723
0·00220	0·950	0·994	1·018	1·065	1·110	1·132	1·198	1·261	1·282	1·322	1·362	1·382	1·440	1·478
1/ 455	0·2963	0·3558	0·3918	0·4703	0·5579	0·6053	0·7619	0·9414	1·0066	1·1450	1·2947	1·3738	1·6289	1·8141
0·00230	0·972	1·017	1·041	1·089	1·135	1·158	1·225	1·289	1·310	1·352	1·393	1·413	1·473	1·512
1/ 435	0·3030	0·3638	0·4006	0·4809	0·5705	0·6189	0·7791	0·9627	1·0293	1·1708	1·3239	1·4048	1·6656	1·8550
0·00240	0·993	1·038	1·063	1·112	1·159	1·183	1·251	1·317	1·339	1·381	1·423	1·444	1·504	1·544
1/ 417	0·3095	0·3716	0·4092	0·4913	0·5828	0·6323	0·7959	0·9834	1·0514	1·1961	1·3524	1·4350	1·7015	1·8949
0·00250	1·013	1·060	1·085	1·135	1·183	1·207	1·277	1·344	1·366	1·410	1·452	1·474	1·536	1·576
1/ 400	0·3159	0·3793	0·4177	0·5014	0·5949	0·6453	0·8123	1·0037	1·0732	1·2208	1·3803	1·4647	1·7366	1·9341
0·00260	1·034	1·081	1·107	1·158	1·207	1·231	1·302	1·371	1·394	1·438	1·481	1·503	1·566	1·607
1/ 385	0·3222	0·3868	0·4260	0·5114	0·6067	0·6581	0·8285	1·0237	1·0945	1·2450	1·4077	1·4938	1·7711	1·9725
0·00270	1·053	1·102	1·128	1·180	1·230	1·255	1·327	1·397	1·420	1·465	1·510	1·531	1·596	1·638
1/ 370	0·3283	0·3942	0·4341	0·5212	0·6183	0·6707	0·8443	1·0432	1·1154	1·2688	1·4346	1·5223	1·8049	2·0101
0·00280	1·073	1·122	1·149	1·201	1·253	1·278	1·352	1·423	1·446	1·492	1·537	1·560	1·625	1·668
1/ 357	0·3344	0·4015	0·4421	0·5307	0·6296	0·6831	0·8598	1·0624	1·1359	1·2921	1·4610	1·5503	1·8381	2·0471
0·00290	1·092	1·142	1·169	1·223	1·275	1·300	1·376	1·448	1·472	1·519	1·565	1·587	1·654	1·698
1/ 345	0·3403	0·4086	0·4499	0·5402	0·6408	0·6952	0·8751	1·0812	1·1560	1·3150	1·4869	1·5778	1·8707	2·0834
0·00300	1·110	1·161	1·189	1·244	1·297	1·323	1·399	1·473	1·497	1·545	1·591	1·614	1·682	1·727
1/ 333	0·3461	0·4156	0·4577	0·5494	0·6518	0·7071	0·8901	1·0998	1·1758	1·3376	1·5124	1·6048	1·9027	2·1190
	0·77	0·77	0·77	0·78	0·78	0·78	0·78	0·79	0·79	0·79	0·79	0·79	0·80	0·80

$V_{r(0{\cdot}5)medial}$ for half-full circular pipes.

Gradient **(Equivalent) Pipe diameters in mm**

Gradient	630	675	700	750	800	825	900	975	1000	1050	1100	1125	1200	1250
0·00300	1·110	1·161	1·189	1·244	1·297	1·323	1·399	1·473	1·497	1·545	1·591	1·614	1·682	1·727
1/ 333	0·3461	0·4156	0·4577	0·5494	0·6518	0·7071	0·8901	1·0998	1·1758	1·3376	1·5124	1·6048	1·9027	2·1190
0·00320	1·147	1·200	1·228	1·285	1·339	1·366	1·445	1·521	1·546	1·595	1·644	1·667	1·738	1·783
1/ 313	0·3575	0·4293	0·4727	0·5675	0·6732	0·7303	0·9193	1·1359	1·2145	1·3815	1·5621	1·6575	1·9652	2·1887
0·00340	1·182	1·237	1·266	1·324	1·381	1·408	1·490	1·568	1·594	1·645	1·694	1·719	1·791	1·838
1/ 294	0·3686	0·4425	0·4873	0·5850	0·6940	0·7529	0·9477	1·1709	1·2519	1·4241	1·6102	1·7086	2·0258	2·2561
0·00360	1·217	1·273	1·303	1·363	1·421	1·449	1·533	1·614	1·640	1·692	1·744	1·769	1·843	1·892
1/ 278	0·3793	0·4554	0·5015	0·6020	0·7141	0·7747	0·9752	1·2049	1·2883	1·4655	1·6570	1·7582	2·0847	2·3216
0·00380	1·250	1·308	1·339	1·400	1·460	1·489	1·575	1·658	1·685	1·739	1·791	1·817	1·894	1·944
1/ 263	0·3897	0·4679	0·5152	0·6185	0·7338	0·7960	1·0020	1·2380	1·3237	1·5057	1·7025	1·8065	2·1419	2·3854
0·00400	1·283	1·342	1·374	1·437	1·498	1·528	1·616	1·701	1·729	1·784	1·838	1·865	1·943	1·994
1/ 250	0·3998	0·4801	0·5286	0·6346	0·7529	0·8167	1·0280	1·2702	1·3581	1·5449	1·7468	1·8535	2·1976	2·4474
0·00420	1·314	1·375	1·408	1·472	1·535	1·566	1·656	1·743	1·772	1·828	1·884	1·911	1·991	2·044
1/ 238	0·4097	0·4920	0·5417	0·6503	0·7715	0·8369	1·0535	1·3017	1·3917	1·5831	1·7900	1·8994	2·2520	2·5080
0·00440	1·345	1·407	1·441	1·507	1·571	1·603	1·695	1·785	1·814	1·871	1·928	1·956	2·038	2·092
1/ 227	0·4194	0·5036	0·5545	0·6657	0·7897	0·8567	1·0783	1·3323	1·4245	1·6204	1·8322	1·9441	2·3050	2·5671
0·00460	1·376	1·439	1·473	1·541	1·606	1·639	1·733	1·825	1·855	1·913	1·971	2·000	2·084	2·139
1/ 217	0·4288	0·5149	0·5670	0·6807	0·8075	0·8760	1·1026	1·3623	1·4566	1·6569	1·8734	1·9879	2·3569	2·6249
0·00480	1·405	1·470	1·505	1·574	1·641	1·674	1·771	1·864	1·895	1·955	2·014	2·043	2·129	2·185
1/ 208	0·4381	0·5260	0·5792	0·6953	0·8249	0·8948	1·1263	1·3917	1·4880	1·6926	1·9138	2·0307	2·4077	2·6814
0·00500	1·434	1·500	1·536	1·606	1·675	1·709	1·807	1·902	1·934	1·995	2·055	2·085	2·173	2·230
1/ 200	0·4471	0·5369	0·5912	0·7097	0·8419	0·9133	1·1496	1·4204	1·5187	1·7275	1·9533	2·0726	2·4574	2·7368
0·00525	1·470	1·537	1·574	1·646	1·716	1·751	1·852	1·950	1·981	2·044	2·106	2·137	2·227	2·285
1/ 190	0·4582	0·5502	0·6058	0·7273	0·8627	0·9359	1·1781	1·4556	1·5563	1·7703	2·0016	2·1239	2·5182	2·8044
0·00550	1·505	1·574	1·611	1·685	1·757	1·792	1·895	1·995	2·028	2·093	2·156	2·187	2·279	2·339
1/ 182	0·4690	0·5631	0·6201	0·7444	0·8831	0·9580	1·2058	1·4899	1·5929	1·8120	2·0488	2·1740	2·5775	2·8705
0·00575	1·538	1·609	1·648	1·723	1·796	1·832	1·938	2·040	2·074	2·140	2·204	2·236	2·330	2·392
1/ 174	0·4796	0·5758	0·6341	0·7612	0·9029	0·9795	1·2330	1·5234	1·6288	1·8528	2·0949	2·2229	2·6355	2·9351
0·00600	1·572	1·644	1·683	1·760	1·835	1·872	1·980	2·084	2·119	2·186	2·252	2·284	2·381	2·443
1/ 167	0·4899	0·5882	0·6477	0·7776	0·9224	1·0006	1·2595	1·5562	1·6639	1·8927	2·1400	2·2708	2·6923	2·9983
0·00625	1·604	1·678	1·718	1·796	1·873	1·911	2·021	2·127	2·162	2·231	2·298	2·332	2·430	2·494
1/ 160	0·5000	0·6004	0·6611	0·7936	0·9415	1·0213	1·2855	1·5884	1·6982	1·9318	2·1842	2·3176	2·7479	3·0602
0·00650	1·636	1·711	1·752	1·832	1·910	1·948	2·061	2·170	2·205	2·275	2·344	2·378	2·478	2·543
1/ 154	0·5100	0·6123	0·6742	0·8094	0·9601	1·0416	1·3110	1·6199	1·7319	1·9701	2·2275	2·3636	2·8024	3·1209
0·00675	1·667	1·744	1·785	1·867	1·947	1·986	2·100	2·211	2·247	2·319	2·389	2·423	2·525	2·592
1/ 148	0·5197	0·6240	0·6871	0·8248	0·9784	1·0614	1·3360	1·6508	1·7649	2·0077	2·2700	2·4087	2·8558	3·1804
0·00700	1·698	1·776	1·818	1·901	1·982	2·022	2·139	2·252	2·288	2·361	2·433	2·468	2·571	2·639
1/ 143	0·5292	0·6354	0·6997	0·8400	0·9964	1·0809	1·3606	1·6811	1·7974	2·0445	2·3117	2·4529	2·9083	3·2389
0·00725	1·728	1·807	1·850	1·935	2·017	2·058	2·177	2·292	2·329	2·403	2·476	2·511	2·617	2·686
1/ 138	0·5386	0·6467	0·7121	0·8549	1·0141	1·1001	1·3847	1·7109	1·8292	2·0808	2·3527	2·4964	2·9598	3·2963
0·00750	1·757	1·838	1·882	1·968	2·052	2·093	2·214	2·331	2·369	2·444	2·518	2·554	2·662	2·732
1/ 133	0·5478	0·6578	0·7243	0·8695	1·0315	1·1189	1·4084	1·7402	1·8605	2·1164	2·3929	2·5391	3·0105	3·3527
0·00800	1·815	1·899	1·944	2·033	2·119	2·162	2·287	2·407	2·447	2·524	2·601	2·638	2·749	2·822
1/ 125	0·5658	0·6794	0·7481	0·8981	1·0653	1·1557	1·4547	1·7973	1·9216	2·1859	2·4715	2·6225	3·1093	3·4627
0·00850	1·871	1·957	2·004	2·095	2·185	2·229	2·357	2·481	2·522	2·602	2·681	2·720	2·834	2·909
1/ 118	0·5833	0·7003	0·7712	0·9257	1·0982	1·1913	1·4995	1·8527	1·9809	2·2532	2·5477	2·7033	3·2051	3·5694
0·00900	1·926	2·014	2·062	2·156	2·248	2·293	2·425	2·554	2·595	2·678	2·759	2·799	2·916	2·993
1/ 111	0·6002	0·7207	0·7935	0·9526	1·1300	1·2259	1·5430	1·9065	2·0384	2·3186	2·6216	2·7818	3·2981	3·6730
0·00950	1·978	2·069	2·119	2·215	2·310	2·356	2·492	2·624	2·667	2·751	2·834	2·875	2·996	3·075
1/ 105	0·6167	0·7404	0·8153	0·9788	1·1611	1·2595	1·5854	1·9588	2·0943	2·3823	2·6935	2·8581	3·3886	3·7738
0·01000	2·030	2·123	2·174	2·273	2·370	2·417	2·557	2·692	2·736	2·823	2·908	2·950	3·074	3·155
1/ 100	0·6327	0·7597	0·8365	1·0042	1·1913	1·2923	1·6266	2·0097	2·1487	2·4442	2·7636	2·9324	3·4767	3·8719
	0·77	0·77	0·77	0·78	0·78	0·78	0·78	0·79	0·79	0·79	0·79	0·79	0·80	0·80

$V_{r(0{\cdot}5)medial}$ for half-full circular pipes.

A31

(p.5 of 6)

$k_s = 3{\cdot}0$ **mm**
S = 0·01000 to 0·03000

ie hydraulic gradient =
1 in 100 to 1 in 33·3

Water (or sewage) at 15°C;
full bore conditions.

velocities in ms^{-1}
discharges in m^3s^{-1}

Gradient **(Equivalent) Pipe diameters in mm**

Gradient	630	675	700	750	800	825	900	975	1000	1050	1100	1125	1200	1250
0·01000	2·030	2·123	2·174	2·273	2·370	2·417	2·557	2·692	2·736	2·823	2·908	2·950	3·074	3·155
1/ 100	0·6327	0·7597	0·8365	1·0042	1·1913	1·2923	1·6266	2·0097	2·1487	2·4442	2·7636	2·9324	3·4767	3·8719
0·01050	2·080	2·175	2·227	2·329	2·429	2·477	2·620	2·758	2·803	2·893	2·980	3·023	3·150	3·233
1/ 95	0·6484	0·7785	0·8572	1·0290	1·2207	1·3242	1·6668	2·0594	2·2019	2·5046	2·8319	3·0049	3·5627	3·9676
0·01100	2·129	2·227	2·280	2·384	2·486	2·536	2·682	2·823	2·870	2·961	3·050	3·094	3·224	3·309
1/ 91	0·6637	0·7968	0·8774	1·0533	1·2495	1·3554	1·7061	2·1079	2·2537	2·5636	2·8986	3·0757	3·6466	4·0611
0·01150	2·177	2·277	2·331	2·438	2·542	2·593	2·742	2·887	2·934	3·027	3·119	3·164	3·297	3·384
1/ 87	0·6786	0·8148	0·8972	1·0770	1·2776	1·3859	1·7445	2·1554	2·3044	2·6213	2·9638	3·1449	3·7286	4·1524
0·01200	2·224	2·326	2·381	2·490	2·596	2·649	2·801	2·949	2·997	3·092	3·186	3·232	3·368	3·457
1/ 83	0·6932	0·8323	0·9165	1·1002	1·3051	1·4158	1·7820	2·2018	2·3541	2·6777	3·0276	3·2126	3·8089	4·2418
0·01250	2·270	2·374	2·431	2·542	2·650	2·703	2·859	3·010	3·059	3·156	3·252	3·299	3·437	3·528
1/ 80	0·7075	0·8495	0·9354	1·1229	1·3320	1·4450	1·8188	2·2472	2·4026	2·7330	3·0901	3·2789	3·8875	4·3294
0·01300	2·315	2·421	2·479	2·592	2·703	2·757	2·916	3·070	3·120	3·219	3·316	3·364	3·505	3·598
1/ 77	0·7216	0·8663	0·9540	1·1452	1·3585	1·4737	1·8549	2·2918	2·4503	2·7872	3·1514	3·3439	3·9646	4·4152
0·01350	2·359	2·467	2·526	2·642	2·754	2·809	2·971	3·128	3·179	3·280	3·379	3·428	3·572	3·666
1/ 74	0·7353	0·8829	0·9722	1·1670	1·3844	1·5018	1·8903	2·3355	2·4970	2·8403	3·2114	3·4077	4·0402	4·4994
0·01400	2·402	2·513	2·573	2·690	2·805	2·861	3·026	3·186	3·238	3·340	3·441	3·491	3·638	3·734
1/ 71	0·7488	0·8991	0·9900	1·1885	1·4098	1·5294	1·9250	2·3784	2·5429	2·8925	3·2704	3·4702	4·1144	4·5820
0·01450	2·445	2·557	2·618	2·738	2·854	2·912	3·079	3·242	3·295	3·400	3·502	3·553	3·702	3·800
1/ 69	0·7621	0·9150	1·0076	1·2095	1·4348	1·5565	1·9591	2·4205	2·5879	2·9438	3·3284	3·5317	4·1872	4·6632
0·01500	2·487	2·601	2·663	2·785	2·903	2·961	3·132	3·297	3·351	3·458	3·562	3·614	3·766	3·865
1/ 67	0·7752	0·9307	1·0248	1·2302	1·4593	1·5831	1·9926	2·4619	2·6322	2·9941	3·3853	3·5921	4·2589	4·7429
0·01600	2·568	2·686	2·750	2·876	2·999	3·059	3·235	3·406	3·461	3·571	3·679	3·732	3·889	3·992
1/ 63	0·8006	0·9612	1·0585	1·2706	1·5072	1·6351	2·0580	2·5428	2·7186	3·0924	3·4964	3·7101	4·3987	4·8986
0·01700	2·647	2·769	2·835	2·965	3·091	3·153	3·335	3·511	3·568	3·681	3·792	3·847	4·009	4·115
1/ 59	0·8253	0·9909	1·0911	1·3097	1·5537	1·6854	2·1214	2·6211	2·8023	3·1877	3·6041	3·8243	4·5342	5·0495
0·01800	2·724	2·849	2·917	3·051	3·181	3·244	3·431	3·612	3·672	3·788	3·903	3·959	4·125	4·234
1/ 56	0·8492	1·0196	1·1227	1·3478	1·5988	1·7343	2·1830	2·6971	2·8837	3·2802	3·7087	3·9353	4·6657	5·1960
0·01900	2·799	2·927	2·997	3·134	3·268	3·333	3·526	3·712	3·772	3·892	4·010	4·068	4·239	4·350
1/ 53	0·8725	1·0476	1·1535	1·3847	1·6426	1·7819	2·2429	2·7711	2·9627	3·3701	3·8104	4·0432	4·7937	5·3385
0·02000	2·872	3·004	3·075	3·216	3·353	3·420	3·617	3·808	3·870	3·993	4·114	4·173	4·349	4·463
1/ 50	0·8952	1·0748	1·1835	1·4207	1·6853	1·8283	2·3012	2·8432	3·0398	3·4577	3·9095	4·1483	4·9183	5·4773
0·02100	2·943	3·078	3·151	3·295	3·436	3·505	3·707	3·902	3·966	4·092	4·215	4·276	4·456	4·574
1/ 47·6	0·9174	1·1014	1·2128	1·4559	1·7270	1·8734	2·3580	2·9134	3·1149	3·5432	4·0061	4·2508	5·0398	5·6126
0·02200	3·012	3·150	3·226	3·373	3·517	3·587	3·794	3·994	4·059	4·188	4·315	4·377	4·561	4·681
1/ 45·5	0·9390	1·1273	1·2414	1·4901	1·7677	1·9176	2·4136	2·9820	3·1883	3·6266	4·1004	4·3509	5·1585	5·7448
0·02300	3·080	3·221	3·298	3·449	3·596	3·668	3·879	4·084	4·151	4·282	4·412	4·476	4·664	4·787
1/ 43·5	0·9601	1·1527	1·2693	1·5237	1·8074	1·9607	2·4679	3·0491	3·2600	3·7082	4·1927	4·4488	5·2745	5·8740
0·02400	3·146	3·291	3·369	3·523	3·673	3·747	3·963	4·172	4·240	4·375	4·507	4·572	4·764	4·890
1/ 41·7	0·9808	1·1775	1·2966	1·5565	1·8463	2·0029	2·5210	3·1147	3·3301	3·7880	4·2829	4·5446	5·3880	6·0004
0·02500	3·211	3·358	3·439	3·596	3·749	3·824	4·045	4·258	4·328	4·465	4·600	4·666	4·862	4·990
1/ 40·0	1·0010	1·2018	1·3234	1·5886	1·8844	2·0442	2·5730	3·1790	3·3989	3·8662	4·3713	4·6383	5·4992	6·1242
0·02600	3·275	3·425	3·507	3·667	3·823	3·900	4·125	4·342	4·413	4·553	4·691	4·759	4·959	5·089
1/ 38·5	1·0209	1·2257	1·3496	1·6201	1·9218	2·0848	2·6240	3·2420	3·4662	3·9428	4·4579	4·7302	5·6082	6·2456
0·02700	3·337	3·490	3·574	3·737	3·896	3·974	4·203	4·425	4·497	4·640	4·780	4·849	5·053	5·186
1/ 37·0	1·0403	1·2490	1·3753	1·6510	1·9584	2·1245	2·6740	3·3038	3·5323	4·0179	4·5429	4·8204	5·7151	6·3646
0·02800	3·399	3·554	3·639	3·806	3·968	4·047	4·281	4·506	4·580	4·725	4·868	4·938	5·146	5·282
1/ 35·7	1·0594	1·2720	1·4006	1·6813	1·9944	2·1635	2·7231	3·3645	3·5971	4·0917	4·6263	4·9089	5·8200	6·4815
0·02900	3·459	3·617	3·704	3·873	4·038	4·119	4·356	4·586	4·661	4·809	4·954	5·026	5·237	5·375
1/ 34·5	1·0782	1·2945	1·4254	1·7111	2·0297	2·2018	2·7714	3·4241	3·6609	4·1642	4·7082	4·9959	5·9231	6·5963
0·03000	3·518	3·679	3·767	3·939	4·107	4·189	4·431	4·665	4·741	4·891	5·039	5·112	5·327	5·467
1/ 33·3	1·0966	1·3166	1·4498	1·7403	2·0644	2·2395	2·8188	3·4826	3·7235	4·2354	4·7888	5·0813	6·0244	6·7091
	0·77	0·77	0·77	0·78	0·78	0·78	0·78	0·79	0·79	0·79	0·79	0·79	0·80	0·80

$V_{r(0{\cdot}5)medial}$ **for half-full circular pipes.**

$k_s = 3{\cdot}0$ **mm** S = 0·01000 to 0·03000

Gradient **(Equivalent) Pipe diameters in mm**

Gradient	630	675	700	750	800	825	900	975	1000	1050	1100	1125	1200	1250
0·03000	3·518	3·679	3·767	3·939	4·107	4·189	4·431	4·665	4·741	4·891	5·039	5·112	5·327	5·467
1/ 33·3	1·0966	1·3166	1·4498	1·7403	2·0644	2·2395	2·8188	3·4826	3·7235	4·2354	4·7888	5·0813	6·0244	6·7091
0·03200	3·633	3·800	3·891	4·069	4·242	4·327	4·576	4·818	4·896	5·052	5·204	5·280	5·502	5·646
1/ 31·3	1·1326	1·3599	1·4974	1·7975	2·1322	2·3130	2·9113	3·5969	3·8457	4·3744	4·9459	5·2481	6·2221	6·9292
0·03400	3·745	3·917	4·011	4·194	4·373	4·460	4·717	4·966	5·047	5·207	5·365	5·442	5·671	5·820
1/ 29·4	1·1675	1·4017	1·5435	1·8528	2·1979	2·3842	3·0010	3·7077	3·9641	4·5091	5·0982	5·4097	6·4137	7·1426
0·03600	3·854	4·031	4·127	4·316	4·499	4·590	4·854	5·110	5·194	5·359	5·520	5·600	5·835	5·989
1/ 27·8	1·2014	1·4424	1·5883	1·9066	2·2616	2·4534	3·0880	3·8153	4·0791	4·6399	5·2461	5·5666	6·5997	7·3498
0·03800	3·960	4·141	4·240	4·434	4·623	4·715	4·987	5·250	5·336	5·505	5·672	5·754	5·995	6·153
1/ 26·3	1·2344	1·4820	1·6318	1·9589	2·3236	2·5207	3·1727	3·9199	4·1909	4·7672	5·3899	5·7192	6·7807	7·5513
0·04000	4·063	4·249	4·350	4·549	4·743	4·838	5·117	5·387	5·475	5·649	5·819	5·903	6·151	6·313
1/ 25·0	1·2664	1·5205	1·6742	2·0098	2·3841	2·5862	3·2552	4·0218	4·2999	4·8911	5·5300	5·8679	6·9569	7·7476
0·04200	4·163	4·354	4·458	4·662	4·860	4·958	5·243	5·520	5·610	5·788	5·963	6·049	6·303	6·469
1/ 23·8	1·2977	1·5581	1·7156	2·0594	2·4430	2·6501	3·3356	4·1211	4·4061	5·0119	5·6667	6·0129	7·1288	7·9390
0·04400	4·261	4·457	4·563	4·771	4·975	5·074	5·367	5·650	5·742	5·924	6·103	6·191	6·452	6·622
1/ 22·7	1·3283	1·5948	1·7560	2·1079	2·5005	2·7125	3·4141	4·2182	4·5099	5·1299	5·8001	6·1544	7·2967	8·1259
0·04600	4·357	4·557	4·666	4·879	5·086	5·188	5·487	5·777	5·871	6·058	6·240	6·331	6·597	6·770
1/ 21·7	1·3582	1·6306	1·7955	2·1553	2·5567	2·7735	3·4909	4·3130	4·6113	5·2453	5·9305	6·2928	7·4607	8·3086
0·04800	4·451	4·655	4·766	4·984	5·196	5·300	5·605	5·901	5·998	6·188	6·375	6·467	6·739	6·916
1/ 20·8	1·3874	1·6657	1·8342	2·2017	2·6117	2·8332	3·5660	4·4058	4·7105	5·3581	6·0581	6·4282	7·6213	8·4874
0·05000	4·543	4·751	4·864	5·087	5·303	5·409	5·721	6·023	6·121	6·316	6·506	6·600	6·878	7·059
1/ 20·0	1·4160	1·7001	1·8720	2·2472	2·6656	2·8917	3·6396	4·4967	4·8077	5·4687	6·1831	6·5608	7·7785	8·6625
0·05250	4·655	4·868	4·985	5·212	5·434	5·543	5·862	6·172	6·273	6·472	6·667	6·763	7·048	7·233
1/ 19·0	1·4510	1·7421	1·9183	2·3027	2·7315	2·9631	3·7295	4·6079	4·9265	5·6038	6·3359	6·7229	7·9707	8·8765
0·05500	4·764	4·983	5·102	5·335	5·562	5·674	6·000	6·317	6·420	6·624	6·824	6·923	7·214	7·404
1/ 18·2	1·4852	1·7831	1·9634	2·3569	2·7958	3·0329	3·8173	4·7163	5·0425	5·7357	6·4850	6·8812	8·1583	9·0855
0·05750	4·872	5·095	5·217	5·455	5·687	5·801	6·135	6·459	6·565	6·773	6·977	7·078	7·376	7·570
1/ 17·4	1·5186	1·8232	2·0076	2·4099	2·8587	3·1011	3·9032	4·8224	5·1559	5·8647	6·6309	7·0359	8·3417	9·2898
0·06000	4·976	5·205	5·329	5·572	5·810	5·926	6·267	6·598	6·706	6·919	7·128	7·231	7·534	7·733
1/ 16·7	1·5513	1·8625	2·0508	2·4618	2·9202	3·1678	3·9872	4·9262	5·2668	5·9909	6·7735	7·1873	8·5212	9·4896
0·06250	5·079	5·312	5·439	5·687	5·929	6·048	6·397	6·734	6·844	7·061	7·275	7·380	7·690	7·892
1/ 16·0	1·5833	1·9009	2·0931	2·5126	2·9805	3·2332	4·0694	5·0278	5·3755	6·1145	6·9133	7·3356	8·6970	9·6854
0·06500	5·180	5·417	5·547	5·800	6·047	6·168	6·523	6·867	6·980	7·201	7·419	7·526	7·842	8·049
1/ 15·4	1·6147	1·9386	2·1346	2·5624	3·0395	3·2972	4·1501	5·1274	5·4820	6·2356	7·0502	7·4809	8·8693	9·8773
0·06750	5·279	5·521	5·652	5·911	6·162	6·286	6·648	6·998	7·113	7·339	7·560	7·669	7·992	8·202
1/ 14·8	1·6454	1·9755	2·1753	2·6112	3·0974	3·3601	4·2292	5·2251	5·5864	6·3545	7·1846	7·6235	9·0383	10·066
0·07000	5·375	5·622	5·756	6·019	6·275	6·401	6·770	7·127	7·243	7·473	7·699	7·810	8·138	8·353
1/ 14·3	1·6757	2·0118	2·2152	2·6591	3·1543	3·4218	4·3068	5·3210	5·6890	6·4711	7·3165	7·7635	9·2043	10·250
0·07250	5·471	5·721	5·858	6·126	6·386	6·514	6·890	7·253	7·372	7·606	7·835	7·948	8·282	8·501
1/ 13·8	1·7053	2·0474	2·2544	2·7062	3·2102	3·4824	4·3831	5·4153	5·7897	6·5857	7·4460	7·9009	9·3672	10·432
0·07500	5·564	5·819	5·958	6·230	6·496	6·626	7·008	7·377	7·498	7·736	7·969	8·084	8·424	8·646
1/ 13·3	1·7345	2·0824	2·2930	2·7525	3·2651	3·5419	4·4580	5·5079	5·8887	6·6983	7·5734	8·0360	9·5274	10·610
0·08000	5·747	6·010	6·154	6·435	6·709	6·843	7·237	7·619	7·744	7·989	8·231	8·350	8·700	8·930
1/ 12·5	1·7914	2·1508	2·3682	2·8428	3·3722	3·6581	4·6043	5·6886	6·0820	6·9181	7·8219	8·2997	9·8400	10·958
0·08500	5·924	6·195	6·343	6·633	6·915	7·054	7·460	7·854	7·982	8·235	8·484	8·607	8·968	9·205
1/ 11·8	1·8466	2·2170	2·4412	2·9304	3·4760	3·7708	4·7461	5·8637	6·2692	7·1311	8·0627	8·5552	10·143	11·296
0·09000	6·096	6·375	6·527	6·825	7·116	7·259	7·677	8·081	8·214	8·474	8·730	8·856	9·228	9·471
1/ 11·1	1·9001	2·2813	2·5120	3·0154	3·5768	3·8801	4·8837	6·0338	6·4510	7·3379	8·2965	8·8033	10·437	11·623
0·09500	6·263	6·550	6·706	7·012	7·311	7·458	7·887	8·303	8·439	8·707	8·969	9·099	9·481	9·731
1/ 10·5	1·9522	2·3438	2·5808	3·0980	3·6749	3·9865	5·0176	6·1992	6·6279	7·5391	8·5239	9·0447	10·723	11·942
0·10000	6·425	6·720	6·880	7·195	7·501	7·651	8·092	8·519	8·658	8·933	9·203	9·336	9·728	9·984
1/ 10·0	2·0030	2·4048	2·6479	3·1785	3·7704	4·0901	5·1480	6·3603	6·8001	7·7350	8·7454	9·2797	11·002	12·252
	0·77	0·77	0·77	0·78	0·78	0·78	0·78	0·79	0·79	0·79	0·79	0·79	0·80	0·80

$V_{r(0{\cdot}5)medial}$ for half-full circular pipes.

A32
(p.1 of 6)

k_s = 6·0 mm
S = 0·00010 to 0·00030

ie hydraulic gradient =
1 in 10000 to 1 in 3333

Water (or sewage) at 15°C;
full bore conditions.

velocities in ms^{-1}
discharges in m^3s^{-1}

Gradient	**(Equivalent) Pipe diameters in mm**													
	630	**675**	**700**	**750**	**800**	**825**	**900**	**975**	**1000**	**1050**	**1100**	**1125**	**1200**	**1250**
0·000100	0·181	0·189	0·194	0·203	0·212	0·216	0·229	0·242	0·246	0·254	0·262	0·266	0·277	0·284
1/ 10000	0·0563	0·0677	0·0746	0·0897	0·1065	0·1156	0·1458	0·1804	0·1930	0·2197	0·2486	0·2639	0·3133	0·3491
0·000105	0·185	0·194	0·199	0·208	0·217	0·222	0·235	0·248	0·252	0·260	0·268	0·272	0·284	0·292
1/ 9524	0·0577	0·0694	0·0765	0·0919	0·1092	0·1185	0·1494	0·1849	0·1978	0·2252	0·2548	0·2705	0·3211	0·3578
0·000110	0·190	0·199	0·203	0·213	0·222	0·227	0·240	0·254	0·258	0·266	0·274	0·279	0·291	0·298
1/ 9091	0·0591	0·0710	0·0783	0·0941	0·1118	0·1213	0·1530	0·1893	0·2025	0·2305	0·2608	0·2769	0·3287	0·3663
0·000115	0·194	0·203	0·208	0·218	0·227	0·232	0·246	0·259	0·264	0·272	0·281	0·285	0·297	0·305
1/ 8696	0·0604	0·0726	0·0800	0·0962	0·1143	0·1241	0·1564	0·1936	0·2070	0·2357	0·2667	0·2831	0·3361	0·3745
0·000120	0·198	0·207	0·213	0·223	0·232	0·237	0·251	0·265	0·269	0·278	0·287	0·291	0·304	0·312
1/ 8333	0·0617	0·0742	0·0818	0·0983	0·1168	0·1268	0·1598	0·1977	0·2115	0·2408	0·2725	0·2893	0·3433	0·3826
0·000125	0·202	0·212	0·217	0·227	0·237	0·242	0·256	0·270	0·275	0·284	0·293	0·297	0·310	0·318
1/ 8000	0·0630	0·0758	0·0835	0·1004	0·1192	0·1294	0·1631	0·2018	0·2159	0·2458	0·2782	0·2953	0·3505	0·3906
0·000130	0·206	0·216	0·221	0·232	0·242	0·247	0·262	0·276	0·280	0·290	0·299	0·303	0·316	0·325
1/ 7692	0·0643	0·0773	0·0851	0·1024	0·1216	0·1320	0·1664	0·2059	0·2202	0·2507	0·2837	0·3011	0·3574	0·3983
0·000135	0·210	0·220	0·225	0·236	0·246	0·252	0·267	0·281	0·286	0·295	0·304	0·309	0·322	0·331
1/ 7407	0·0655	0·0787	0·0868	0·1043	0·1239	0·1345	0·1696	0·2098	0·2244	0·2555	0·2891	0·3069	0·3643	0·4059
0·000140	0·214	0·224	0·230	0·240	0·251	0·256	0·271	0·286	0·291	0·301	0·310	0·314	0·328	0·337
1/ 7143	0·0667	0·0802	0·0884	0·1062	0·1262	0·1370	0·1727	0·2137	0·2286	0·2602	0·2945	0·3126	0·3710	0·4134
0·000145	0·218	0·228	0·234	0·245	0·256	0·261	0·276	0·291	0·296	0·306	0·315	0·320	0·334	0·343
1/ 6897	0·0679	0·0816	0·0900	0·1081	0·1284	0·1394	0·1758	0·2175	0·2326	0·2648	0·2997	0·3181	0·3776	0·4208
0·000150	0·222	0·232	0·238	0·249	0·260	0·265	0·281	0·296	0·301	0·311	0·321	0·326	0·340	0·349
1/ 6667	0·0691	0·0830	0·0915	0·1100	0·1306	0·1418	0·1788	0·2212	0·2366	0·2694	0·3048	0·3236	0·3841	0·4280
0·000160	0·229	0·240	0·246	0·257	0·268	0·274	0·290	0·306	0·311	0·321	0·331	0·336	0·351	0·360
1/ 6250	0·0713	0·0858	0·0945	0·1136	0·1350	0·1465	0·1847	0·2285	0·2444	0·2783	0·3149	0·3342	0·3967	0·4421
0·000170	0·236	0·247	0·253	0·265	0·277	0·283	0·299	0·316	0·321	0·331	0·342	0·347	0·362	0·371
1/ 5882	0·0736	0·0884	0·0974	0·1171	0·1391	0·1510	0·1904	0·2356	0·2520	0·2869	0·3246	0·3446	0·4090	0·4558
0·000180	0·243	0·254	0·261	0·273	0·285	0·291	0·308	0·325	0·330	0·341	0·352	0·357	0·372	0·382
1/ 5556	0·0757	0·0910	0·1003	0·1206	0·1432	0·1554	0·1959	0·2424	0·2593	0·2952	0·3341	0·3546	0·4209	0·4690
0·000190	0·250	0·261	0·268	0·280	0·293	0·299	0·316	0·334	0·339	0·350	0·361	0·367	0·382	0·393
1/ 5263	0·0778	0·0935	0·1031	0·1239	0·1471	0·1597	0·2013	0·2491	0·2665	0·3034	0·3433	0·3644	0·4325	0·4819
0·000200	0·256	0·268	0·275	0·288	0·300	0·307	0·325	0·342	0·348	0·359	0·371	0·376	0·392	0·403
1/ 5000	0·0798	0·0960	0·1057	0·1271	0·1510	0·1639	0·2066	0·2556	0·2734	0·3113	0·3522	0·3739	0·4437	0·4945
0·000210	0·262	0·275	0·282	0·295	0·308	0·314	0·333	0·351	0·357	0·368	0·380	0·385	0·402	0·413
1/ 4762	0·0818	0·0983	0·1084	0·1303	0·1547	0·1679	0·2117	0·2620	0·2802	0·3190	0·3609	0·3831	0·4547	0·5068
0·000220	0·269	0·281	0·288	0·302	0·315	0·322	0·341	0·359	0·365	0·377	0·389	0·395	0·412	0·423
1/ 4545	0·0837	0·1007	0·1109	0·1333	0·1584	0·1719	0·2167	0·2681	0·2868	0·3265	0·3695	0·3922	0·4655	0·5187
0·000230	0·275	0·288	0·295	0·309	0·322	0·329	0·348	0·367	0·373	0·386	0·398	0·403	0·421	0·432
1/ 4348	0·0856	0·1030	0·1134	0·1364	0·1620	0·1758	0·2216	0·2742	0·2933	0·3339	0·3778	0·4010	0·4760	0·5304
0·000240	0·281	0·294	0·301	0·315	0·329	0·336	0·356	0·375	0·381	0·394	0·406	0·412	0·430	0·442
1/ 4167	0·0875	0·1052	0·1159	0·1393	0·1655	0·1796	0·2264	0·2801	0·2996	0·3411	0·3860	0·4097	0·4863	0·5419
0·000250	0·286	0·300	0·307	0·322	0·336	0·343	0·363	0·383	0·389	0·402	0·415	0·421	0·439	0·451
1/ 4000	0·0893	0·1074	0·1183	0·1422	0·1689	0·1833	0·2311	0·2859	0·3058	0·3482	0·3940	0·4182	0·4963	0·5531
0·000260	0·292	0·306	0·314	0·328	0·343	0·350	0·370	0·391	0·397	0·410	0·423	0·429	0·448	0·460
1/ 3846	0·0911	0·1095	0·1207	0·1450	0·1722	0·1870	0·2357	0·2916	0·3119	0·3551	0·4018	0·4265	0·5062	0·5641
0·000270	0·298	0·312	0·320	0·335	0·349	0·356	0·378	0·398	0·405	0·418	0·431	0·437	0·456	0·468
1/ 3704	0·0928	0·1116	0·1230	0·1478	0·1755	0·1905	0·2402	0·2972	0·3179	0·3619	0·4095	0·4346	0·5159	0·5749
0·000280	0·303	0·318	0·325	0·341	0·356	0·363	0·385	0·405	0·412	0·426	0·439	0·445	0·465	0·477
1/ 3571	0·0945	0·1136	0·1252	0·1505	0·1788	0·1940	0·2446	0·3027	0·3237	0·3685	0·4170	0·4426	0·5254	0·5854
0·000290	0·309	0·323	0·331	0·347	0·362	0·369	0·391	0·413	0·419	0·433	0·447	0·453	0·473	0·486
1/ 3448	0·0962	0·1157	0·1275	0·1532	0·1819	0·1975	0·2490	0·3080	0·3295	0·3751	0·4244	0·4505	0·5347	0·5958
0·000300	0·314	0·329	0·337	0·353	0·368	0·376	0·398	0·420	0·427	0·441	0·454	0·461	0·481	0·494
1/ 3333	0·0979	0·1177	0·1296	0·1558	0·1851	0·2009	0·2532	0·3133	0·3351	0·3815	0·4317	0·4582	0·5439	0·6061
	0·75	0·75	0·75	0·75	0·76	0·76	0·76	0·76	0·76	0·76	0·77	0·77	0·77	0·77

$V_{r(0\cdot5)medial}$ for half-full circular pipes.

Gradient	**(Equivalent) Pipe diameters in mm**													
	630	**675**	**700**	**750**	**800**	**825**	**900**	**975**	**1000**	**1050**	**1100**	**1125**	**1200**	**1250**
0·000300	0·314	0·329	0·337	0·353	0·368	0·376	0·398	0·420	0·427	0·441	0·454	0·461	0·481	0·494
1/ 3333	0·0979	0·1177	0·1296	0·1558	0·1851	0·2009	0·2532	0·3133	0·3351	0·3815	0·4317	0·4582	0·5439	0·6061
0·000320	0·324	0·340	0·348	0·364	0·380	0·388	0·411	0·433	0·441	0·455	0·469	0·476	0·497	0·510
1/ 3125	0·1011	0·1215	0·1339	0·1610	0·1912	0·2075	0·2616	0·3236	0·3462	0·3941	0·4459	0·4733	0·5617	0·6260
0·000340	0·334	0·350	0·359	0·376	0·392	0·400	0·424	0·447	0·454	0·469	0·484	0·491	0·512	0·526
1/ 2941	0·1042	0·1253	0·1381	0·1659	0·1971	0·2139	0·2697	0·3336	0·3568	0·4062	0·4597	0·4879	0·5791	0·6453
0·000360	0·344	0·360	0·369	0·387	0·403	0·412	0·436	0·460	0·468	0·483	0·498	0·505	0·527	0·541
1/ 2778	0·1072	0·1289	0·1421	0·1708	0·2028	0·2201	0·2775	0·3433	0·3672	0·4181	0·4730	0·5021	0·5959	0·6641
0·000380	0·354	0·370	0·379	0·397	0·415	0·423	0·448	0·472	0·480	0·496	0·511	0·519	0·541	0·556
1/ 2632	0·1102	0·1325	0·1460	0·1755	0·2084	0·2262	0·2851	0·3528	0·3773	0·4295	0·4860	0·5159	0·6123	0·6823
0·000400	0·363	0·380	0·389	0·408	0·425	0·434	0·460	0·485	0·493	0·509	0·525	0·533	0·555	0·570
1/ 2500	0·1131	0·1359	0·1498	0·1800	0·2138	0·2321	0·2926	0·3620	0·3872	0·4407	0·4987	0·5294	0·6282	0·7001
0·000420	0·372	0·389	0·399	0·418	0·436	0·445	0·471	0·497	0·505	0·522	0·538	0·546	0·569	0·585
1/ 2381	0·1159	0·1393	0·1535	0·1845	0·2191	0·2378	0·2998	0·3709	0·3967	0·4517	0·5110	0·5425	0·6438	0·7174
0·000440	0·381	0·398	0·408	0·427	0·446	0·455	0·482	0·509	0·517	0·534	0·550	0·559	0·583	0·598
1/ 2273	0·1186	0·1426	0·1571	0·1889	0·2243	0·2434	0·3069	0·3797	0·4061	0·4623	0·5231	0·5553	0·6590	0·7344
0·000460	0·389	0·407	0·417	0·437	0·456	0·466	0·493	0·520	0·529	0·546	0·563	0·571	0·596	0·612
1/ 2174	0·1213	0·1458	0·1607	0·1931	0·2294	0·2489	0·3138	0·3882	0·4153	0·4727	0·5349	0·5678	0·6738	0·7509
0·000480	0·397	0·416	0·427	0·447	0·466	0·476	0·504	0·531	0·540	0·558	0·575	0·584	0·609	0·625
1/ 2083	0·1239	0·1490	0·1641	0·1973	0·2343	0·2543	0·3206	0·3966	0·4242	0·4829	0·5464	0·5800	0·6884	0·7671
0·000500	0·406	0·425	0·435	0·456	0·476	0·486	0·514	0·542	0·551	0·569	0·587	0·596	0·621	0·638
1/ 2000	0·1265	0·1520	0·1675	0·2014	0·2391	0·2596	0·3272	0·4048	0·4330	0·4929	0·5577	0·5920	0·7026	0·7830
0·000525	0·416	0·435	0·446	0·467	0·488	0·498	0·527	0·556	0·565	0·583	0·601	0·610	0·637	0·654
1/ 1905	0·1296	0·1558	0·1717	0·2064	0·2451	0·2660	0·3353	0·4148	0·4437	0·5051	0·5715	0·6067	0·7200	0·8023
0·000550	0·426	0·446	0·457	0·478	0·499	0·509	0·540	0·569	0·578	0·597	0·616	0·625	0·652	0·669
1/ 1818	0·1327	0·1595	0·1757	0·2112	0·2509	0·2723	0·3432	0·4246	0·4542	0·5170	0·5850	0·6210	0·7370	0·8213
0·000575	0·435	0·456	0·467	0·489	0·510	0·521	0·552	0·582	0·591	0·611	0·629	0·639	0·666	0·684
1/ 1739	0·1357	0·1631	0·1797	0·2160	0·2565	0·2784	0·3510	0·4342	0·4644	0·5287	0·5982	0·6350	0·7536	0·8398
0·000600	0·445	0·466	0·477	0·499	0·521	0·532	0·564	0·594	0·604	0·624	0·643	0·653	0·681	0·699
1/ 1667	0·1386	0·1666	0·1836	0·2207	0·2620	0·2844	0·3585	0·4436	0·4744	0·5401	0·6111	0·6487	0·7698	0·8579
0·000625	0·454	0·475	0·487	0·510	0·532	0·543	0·575	0·606	0·617	0·637	0·656	0·666	0·695	0·713
1/ 1600	0·1415	0·1701	0·1874	0·2252	0·2675	0·2903	0·3659	0·4527	0·4842	0·5512	0·6237	0·6621	0·7857	0·8756
0·000650	0·463	0·485	0·497	0·520	0·543	0·554	0·587	0·618	0·629	0·649	0·669	0·679	0·709	0·728
1/ 1538	0·1443	0·1734	0·1911	0·2297	0·2728	0·2961	0·3732	0·4617	0·4938	0·5622	0·6361	0·6752	0·8013	0·8930
0·000675	0·472	0·494	0·506	0·530	0·553	0·564	0·598	0·630	0·641	0·662	0·682	0·692	0·722	0·742
1/ 1481	0·1470	0·1767	0·1948	0·2341	0·2780	0·3017	0·3803	0·4705	0·5033	0·5729	0·6483	0·6881	0·8166	0·9100
0·000700	0·480	0·503	0·515	0·540	0·563	0·575	0·609	0·642	0·653	0·674	0·695	0·705	0·735	0·755
1/ 1429	0·1497	0·1800	0·1983	0·2384	0·2831	0·3073	0·3873	0·4792	0·5125	0·5835	0·6602	0·7008	0·8316	0·9267
0·000725	0·489	0·512	0·525	0·549	0·573	0·585	0·620	0·653	0·664	0·686	0·707	0·717	0·748	0·769
1/ 1379	0·1524	0·1832	0·2019	0·2426	0·2881	0·3127	0·3942	0·4877	0·5216	0·5938	0·6719	0·7132	0·8464	0·9432
0·000750	0·497	0·521	0·534	0·559	0·583	0·595	0·630	0·664	0·676	0·698	0·719	0·730	0·761	0·782
1/ 1333	0·1550	0·1863	0·2053	0·2468	0·2931	0·3181	0·4010	0·4960	0·5306	0·6040	0·6834	0·7254	0·8609	0·9593
0·000800	0·514	0·538	0·551	0·577	0·602	0·615	0·651	0·686	0·698	0·720	0·743	0·754	0·786	0·807
1/ 1250	0·1601	0·1925	0·2121	0·2549	0·3027	0·3285	0·4141	0·5123	0·5480	0·6238	0·7059	0·7492	0·8892	0·9909
0·000850	0·529	0·554	0·568	0·595	0·621	0·634	0·671	0·707	0·719	0·743	0·766	0·777	0·810	0·832
1/ 1176	0·1650	0·1984	0·2186	0·2628	0·3120	0·3387	0·4269	0·5282	0·5649	0·6431	0·7276	0·7723	0·9166	1·0214
0·000900	0·545	0·571	0·585	0·612	0·639	0·652	0·691	0·728	0·740	0·764	0·788	0·800	0·834	0·856
1/ 1111	0·1698	0·2042	0·2250	0·2704	0·3211	0·3485	0·4393	0·5435	0·5813	0·6618	0·7488	0·7948	0·9432	1·0511
0·000950	0·560	0·586	0·601	0·629	0·656	0·670	0·710	0·748	0·760	0·785	0·810	0·822	0·857	0·880
1/ 1053	0·1745	0·2098	0·2312	0·2778	0·3299	0·3581	0·4514	0·5584	0·5973	0·6799	0·7693	0·8166	0·9691	1·0799
0·001000	0·574	0·602	0·616	0·645	0·673	0·687	0·728	0·767	0·780	0·806	0·831	0·843	0·879	0·903
1/ 1000	0·1791	0·2152	0·2372	0·2851	0·3385	0·3674	0·4631	0·5730	0·6128	0·6976	0·7893	0·8379	0·9943	1·1080
	0·75	0·75	0·75	0·75	0·76	0·76	0·76	0·76	0·76	0·76	0·77	0·77	0·77	0·77

$V_{r(0{\cdot}5)medial}$ **for half-full circular pipes.**

$k_s = 6{\cdot}0$ mm
S = 0·00100 to 0·00300

ie hydraulic gradient =
1 in 1000 to 1 in 333

Water (or sewage) at 15°C;
full bore conditions.

velocities in ms^{-1}
discharges in m^3s^{-1}

Gradient **(Equivalent) Pipe diameters in mm**

Gradient	630	675	700	750	800	825	900	975	1000	1050	1100	1125	1200	1250
0·00100	0·574	0·602	0·616	0·645	0·673	0·687	0·728	0·767	0·780	0·806	0·831	0·843	0·879	0·903
1/ 1000	0·1791	0·2152	0·2372	0·2851	0·3385	0·3674	0·4631	0·5730	0·6128	0·6976	0·7893	0·8379	0·9943	1·1080
0·00105	0·589	0·616	0·632	0·661	0·690	0·704	0·746	0·786	0·800	0·826	0·851	0·864	0·901	0·925
1/ 952	0·1835	0·2206	0·2430	0·2921	0·3469	0·3765	0·4746	0·5871	0·6280	0·7149	0·8089	0·8586	1·0189	1·1354
0·00110	0·603	0·631	0·646	0·677	0·706	0·721	0·764	0·805	0·818	0·845	0·871	0·884	0·922	0·947
1/ 909	0·1878	0·2258	0·2488	0·2990	0·3551	0·3854	0·4858	0·6010	0·6428	0·7317	0·8279	0·8788	1·0430	1·1622
0·00115	0·616	0·645	0·661	0·692	0·722	0·737	0·781	0·823	0·837	0·864	0·891	0·904	0·943	0·968
1/ 870	0·1921	0·2309	0·2544	0·3057	0·3631	0·3941	0·4967	0·6145	0·6573	0·7482	0·8466	0·8986	1·0664	1·1884
0·00120	0·629	0·659	0·675	0·707	0·738	0·753	0·798	0·841	0·855	0·883	0·910	0·923	0·963	0·989
1/ 833	0·1962	0·2358	0·2599	0·3123	0·3709	0·4026	0·5074	0·6278	0·6714	0·7643	0·8648	0·9180	1·0894	1·2140
0·00125	0·642	0·673	0·689	0·722	0·753	0·769	0·814	0·858	0·873	0·901	0·929	0·943	0·983	1·010
1/ 800	0·2002	0·2407	0·2652	0·3188	0·3786	0·4109	0·5179	0·6407	0·6853	0·7801	0·8827	0·9369	1·1119	1·2390
0·00130	0·655	0·686	0·703	0·736	0·768	0·784	0·830	0·875	0·890	0·919	0·947	0·961	1·003	1·030
1/ 769	0·2042	0·2455	0·2705	0·3251	0·3861	0·4190	0·5282	0·6534	0·6989	0·7956	0·9002	0·9555	1·1340	1·2636
0·00135	0·668	0·699	0·716	0·750	0·783	0·799	0·846	0·892	0·907	0·936	0·965	0·980	1·022	1·049
1/ 741	0·2081	0·2502	0·2757	0·3313	0·3934	0·4270	0·5383	0·6659	0·7122	0·8108	0·9174	0·9737	1·1556	1·2877
0·00140	0·680	0·712	0·729	0·764	0·797	0·814	0·862	0·908	0·924	0·954	0·983	0·998	1·041	1·069
1/ 714	0·2119	0·2548	0·2807	0·3374	0·4007	0·4349	0·5482	0·6781	0·7253	0·8257	0·9342	0·9916	1·1768	1·3114
0·00145	0·692	0·725	0·742	0·777	0·811	0·828	0·877	0·924	0·940	0·970	1·000	1·015	1·059	1·088
1/ 690	0·2157	0·2593	0·2857	0·3434	0·4078	0·4426	0·5579	0·6902	0·7382	0·8403	0·9508	1·0092	1·1977	1·3346
0·00150	0·704	0·737	0·755	0·791	0·825	0·842	0·892	0·940	0·956	0·987	1·018	1·033	1·077	1·106
1/ 667	0·2194	0·2637	0·2906	0·3493	0·4148	0·4502	0·5674	0·7020	0·7508	0·8547	0·9671	1·0265	1·2182	1·3575
0·00160	0·727	0·761	0·780	0·817	0·852	0·870	0·921	0·971	0·987	1·020	1·051	1·067	1·113	1·142
1/ 625	0·2266	0·2724	0·3002	0·3607	0·4284	0·4650	0·5861	0·7250	0·7755	0·8828	0·9988	1·0602	1·2582	1·4021
0·00170	0·749	0·785	0·804	0·842	0·879	0·897	0·950	1·001	1·018	1·051	1·083	1·099	1·147	1·178
1/ 588	0·2336	0·2808	0·3094	0·3719	0·4416	0·4793	0·6041	0·7474	0·7994	0·9100	1·0296	1·0929	1·2970	1·4453
0·00180	0·771	0·808	0·827	0·866	0·904	0·923	0·977	1·030	1·047	1·081	1·115	1·131	1·180	1·212
1/ 556	0·2404	0·2890	0·3184	0·3827	0·4544	0·4932	0·6217	0·7691	0·8226	0·9364	1·0595	1·1246	1·3346	1·4872
0·00190	0·792	0·830	0·850	0·890	0·929	0·948	1·004	1·058	1·076	1·111	1·145	1·162	1·212	1·245
1/ 526	0·2470	0·2969	0·3271	0·3932	0·4669	0·5068	0·6388	0·7902	0·8452	0·9621	1·0886	1·1555	1·3713	1·5280
0·00200	0·813	0·851	0·872	0·913	0·953	0·973	1·030	1·086	1·104	1·140	1·175	1·193	1·244	1·278
1/ 500	0·2534	0·3046	0·3357	0·4034	0·4791	0·5199	0·6554	0·8108	0·8672	0·9871	1·1169	1·1855	1·4069	1·5678
0·00210	0·833	0·872	0·894	0·936	0·977	0·997	1·056	1·113	1·131	1·168	1·204	1·222	1·275	1·309
1/ 476	0·2597	0·3122	0·3440	0·4134	0·4909	0·5328	0·6716	0·8308	0·8886	1·0116	1·1445	1·2149	1·4417	1·6065
0·00220	0·853	0·893	0·915	0·958	1·000	1·020	1·081	1·139	1·158	1·196	1·233	1·251	1·305	1·340
1/ 455	0·2658	0·3195	0·3521	0·4231	0·5025	0·5454	0·6874	0·8504	0·9096	1·0354	1·1715	1·2435	1·4757	1·6444
0·00230	0·872	0·913	0·935	0·979	1·022	1·043	1·105	1·165	1·184	1·223	1·260	1·279	1·334	1·370
1/ 435	0·2718	0·3267	0·3600	0·4327	0·5138	0·5576	0·7029	0·8695	0·9300	1·0587	1·1978	1·2715	1·5089	1·6814
0·00240	0·891	0·933	0·956	1·000	1·044	1·066	1·129	1·190	1·210	1·249	1·288	1·307	1·363	1·400
1/ 417	0·2777	0·3338	0·3677	0·4420	0·5249	0·5697	0·7180	0·8882	0·9500	1·0815	1·2236	1·2988	1·5414	1·7176
0·00250	0·909	0·952	0·975	1·021	1·066	1·088	1·152	1·214	1·235	1·275	1·314	1·334	1·391	1·429
1/ 400	0·2834	0·3407	0·3753	0·4511	0·5357	0·5814	0·7328	0·9066	0·9697	1·1038	1·2489	1·3257	1·5732	1·7530
0·00260	0·927	0·971	0·995	1·041	1·087	1·109	1·175	1·238	1·259	1·300	1·340	1·360	1·419	1·457
1/ 385	0·2890	0·3474	0·3828	0·4600	0·5463	0·5929	0·7474	0·9246	0·9889	1·1257	1·2737	1·3519	1·6044	1·7878
0·00270	0·945	0·989	1·014	1·061	1·108	1·130	1·197	1·262	1·283	1·325	1·366	1·386	1·446	1·485
1/ 370	0·2945	0·3540	0·3901	0·4688	0·5567	0·6043	0·7616	0·9422	1·0078	1·1472	1·2980	1·3777	1·6350	1·8219
0·00280	0·962	1·008	1·032	1·081	1·128	1·151	1·219	1·285	1·307	1·349	1·391	1·411	1·472	1·512
1/ 357	0·2999	0·3605	0·3973	0·4774	0·5670	0·6154	0·7756	0·9595	1·0263	1·1682	1·3218	1·4030	1·6650	1·8554
0·00290	0·979	1·025	1·051	1·100	1·148	1·172	1·241	1·308	1·330	1·373	1·416	1·436	1·498	1·539
1/ 345	0·3053	0·3669	0·4043	0·4859	0·5770	0·6263	0·7894	0·9765	1·0445	1·1889	1·3452	1·4279	1·6945	1·8882
0·00300	0·996	1·043	1·069	1·119	1·168	1·192	1·262	1·330	1·353	1·397	1·440	1·461	1·524	1·565
1/ 333	0·3105	0·3732	0·4112	0·4942	0·5869	0·6370	0·8029	0·9932	1·0623	1·2093	1·3682	1·4523	1·7235	1·9205
	0·75	0·75	0·75	0·75	0·76	0·76	0·76	0·76	0·76	0·76	0·77	0·77	0·77	0·77

$V_{r(0{\cdot}5)medial}$ for half-full circular pipes.

Gradient **(Equivalent) Pipe diameters in mm**

Gradient	630	675	700	750	800	825	900	975	1000	1050	1100	1125	1200	1250
0·00300	0·996	1·043	1·069	1·119	1·168	1·192	1·262	1·330	1·353	1·397	1·440	1·461	1·524	1·565
1/ 333	0·3105	0·3732	0·4112	0·4942	0·5869	0·6370	0·8029	0·9932	1·0623	1·2093	1·3682	1·4523	1·7235	1·9205
0·00320	1·029	1·077	1·104	1·155	1·206	1·231	1·303	1·374	1·397	1·442	1·487	1·509	1·574	1·616
1/ 313	0·3207	0·3855	0·4247	0·5105	0·6062	0·6579	0·8292	1·0258	1·0972	1·2490	1·4132	1·5000	1·7801	1·9836
0·00340	1·060	1·110	1·138	1·191	1·243	1·269	1·344	1·416	1·440	1·487	1·533	1·556	1·622	1·666
1/ 294	0·3306	0·3974	0·4378	0·5262	0·6248	0·6782	0·8548	1·0574	1·1310	1·2875	1·4567	1·5462	1·8349	2·0447
0·00360	1·091	1·143	1·171	1·226	1·279	1·305	1·383	1·457	1·482	1·530	1·577	1·601	1·670	1·715
1/ 278	0·3402	0·4089	0·4505	0·5415	0·6430	0·6979	0·8796	1·0881	1·1638	1·3248	1·4990	1·5911	1·8882	2·1040
0·00380	1·121	1·174	1·203	1·259	1·314	1·341	1·421	1·497	1·522	1·572	1·621	1·645	1·715	1·762
1/ 263	0·3495	0·4201	0·4629	0·5563	0·6606	0·7170	0·9037	1·1180	1·1958	1·3612	1·5401	1·6347	1·9400	2·1617
0·00400	1·150	1·205	1·234	1·292	1·348	1·376	1·458	1·536	1·562	1·613	1·663	1·687	1·760	1·807
1/ 250	0·3586	0·4310	0·4749	0·5708	0·6778	0·7357	0·9272	1·1471	1·2269	1·3966	1·5801	1·6772	1·9904	2·2179
0·00420	1·179	1·234	1·265	1·324	1·382	1·410	1·494	1·574	1·601	1·653	1·704	1·729	1·803	1·852
1/ 238	0·3675	0·4417	0·4867	0·5849	0·6946	0·7538	0·9502	1·1754	1·2572	1·4311	1·6192	1·7187	2·0396	2·2728
0·00440	1·207	1·263	1·294	1·355	1·414	1·443	1·529	1·611	1·638	1·692	1·744	1·770	1·846	1·896
1/ 227	0·3761	0·4521	0·4981	0·5987	0·7109	0·7716	0·9725	1·2031	1·2868	1·4648	1·6573	1·7592	2·0876	2·3263
0·00460	1·234	1·292	1·324	1·386	1·446	1·476	1·563	1·648	1·675	1·730	1·783	1·810	1·887	1·938
1/ 217	0·3846	0·4623	0·5093	0·6121	0·7269	0·7890	0·9944	1·2302	1·3157	1·4978	1·6946	1·7987	2·1346	2·3786
0·00480	1·260	1·320	1·352	1·415	1·477	1·508	1·597	1·683	1·711	1·767	1·822	1·849	1·928	1·980
1/ 208	0·3929	0·4722	0·5203	0·6253	0·7426	0·8059	1·0158	1·2566	1·3441	1·5300	1·7311	1·8375	2·1806	2·4298
0·00500	1·286	1·347	1·380	1·445	1·508	1·539	1·630	1·718	1·747	1·803	1·859	1·887	1·968	2·021
1/ 200	0·4010	0·4820	0·5311	0·6382	0·7579	0·8226	1·0368	1·2826	1·3718	1·5616	1·7668	1·8754	2·2256	2·4799
0·00525	1·318	1·380	1·414	1·480	1·545	1·577	1·670	1·760	1·790	1·848	1·905	1·933	2·016	2·071
1/ 190	0·4109	0·4939	0·5442	0·6540	0·7766	0·8429	1·0624	1·3143	1·4057	1·6002	1·8105	1·9217	2·2806	2·5412
0·00550	1·349	1·413	1·447	1·515	1·581	1·614	1·709	1·802	1·832	1·892	1·950	1·979	2·064	2·120
1/ 182	0·4206	0·5055	0·5570	0·6694	0·7949	0·8628	1·0874	1·3452	1·4388	1·6379	1·8531	1·9670	2·3343	2·6011
0·00575	1·379	1·445	1·480	1·549	1·617	1·650	1·748	1·842	1·873	1·934	1·994	2·023	2·110	2·167
1/ 174	0·4300	0·5169	0·5695	0·6845	0·8128	0·8822	1·1119	1·3755	1·4712	1·6747	1·8948	2·0112	2·3868	2·6596
0·00600	1·409	1·476	1·512	1·583	1·652	1·686	1·785	1·882	1·913	1·976	2·037	2·067	2·156	2·214
1/ 167	0·4393	0·5280	0·5818	0·6992	0·8303	0·9012	1·1358	1·4051	1·5029	1·7107	1·9356	2·0545	2·4381	2·7168
0·00625	1·438	1·506	1·543	1·615	1·686	1·721	1·822	1·921	1·953	2·016	2·079	2·110	2·200	2·260
1/ 160	0·4484	0·5389	0·5938	0·7136	0·8474	0·9198	1·1593	1·4341	1·5339	1·7461	1·9755	2·0969	2·4884	2·7729
0·00650	1·467	1·536	1·574	1·647	1·719	1·755	1·858	1·959	1·992	2·056	2·120	2·151	2·244	2·304
1/ 154	0·4572	0·5496	0·6056	0·7278	0·8642	0·9380	1·1823	1·4625	1·5643	1·7807	2·0147	2·1385	2·5378	2·8278
0·00675	1·495	1·565	1·604	1·679	1·752	1·788	1·894	1·996	2·030	2·096	2·160	2·192	2·287	2·348
1/ 148	0·4660	0·5601	0·6171	0·7417	0·8807	0·9559	1·2048	1·4904	1·5941	1·8146	2·0531	2·1792	2·5861	2·8817
0·00700	1·522	1·594	1·633	1·710	1·784	1·821	1·929	2·033	2·067	2·134	2·200	2·233	2·329	2·391
1/ 143	0·4745	0·5704	0·6285	0·7553	0·8969	0·9734	1·2269	1·5178	1·6234	1·8479	2·0908	2·2193	2·6336	2·9347
0·00725	1·549	1·622	1·662	1·740	1·816	1·853	1·963	2·069	2·104	2·172	2·239	2·272	2·370	2·434
1/ 138	0·4829	0·5805	0·6396	0·7687	0·9128	0·9907	1·2487	1·5447	1·6521	1·8807	2·1278	2·2586	2·6803	2·9866
0·00750	1·576	1·650	1·690	1·770	1·847	1·885	1·996	2·104	2·140	2·209	2·277	2·311	2·410	2·475
1/ 133	0·4912	0·5904	0·6505	0·7818	0·9284	1·0076	1·2700	1·5711	1·6804	1·9128	2·1642	2·2972	2·7261	3·0377
0·00800	1·627	1·704	1·746	1·828	1·908	1·947	2·062	2·173	2·210	2·282	2·352	2·387	2·490	2·557
1/ 125	0·5073	0·6098	0·6719	0·8075	0·9589	1·0407	1·3117	1·6227	1·7355	1·9756	2·2353	2·3726	2·8156	3·1374
0·00850	1·678	1·757	1·800	1·884	1·966	2·007	2·125	2·240	2·278	2·352	2·425	2·460	2·566	2·635
1/ 118	0·5229	0·6286	0·6926	0·8324	0·9884	1·0728	1·3521	1·6726	1·7890	2·0364	2·3041	2·4457	2·9023	3·2340
0·00900	1·726	1·808	1·852	1·939	2·023	2·065	2·187	2·305	2·344	2·420	2·495	2·532	2·641	2·712
1/ 111	0·5381	0·6468	0·7127	0·8565	1·0171	1·1039	1·3913	1·7212	1·8409	2·0955	2·3709	2·5166	2·9865	3·3278
0·00950	1·774	1·857	1·903	1·992	2·079	2·122	2·247	2·368	2·408	2·486	2·563	2·601	2·713	2·786
1/ 105	0·5529	0·6646	0·7322	0·8800	1·0450	1·1342	1·4295	1·7683	1·8914	2·1530	2·4359	2·5856	3·0684	3·4191
0·01000	1·820	1·905	1·952	2·044	2·133	2·177	2·305	2·430	2·471	2·551	2·630	2·669	2·784	2·859
1/ 100	0·5673	0·6819	0·7513	0·9029	1·0722	1·1636	1·4667	1·8143	1·9405	2·2090	2·4993	2·6528	3·1481	3·5080
	0·75	0·75	0·75	0·75	0·76	0·76	0·76	0·76	0·76	0·76	0·77	0·77	0·77	0·77

$V_{r(0{\cdot}5)medial}$ for half-full circular pipes.

$k_s = 6{\cdot}0\,mm$
S = 0·01000 to 0·03000

ie hydraulic gradient =
1 in 100 to 1 in 33·3

Water (or sewage) at 15°C;
full bore conditions.

velocities in ms^{-1}
discharges in m^3s^{-1}

Gradient **(Equivalent) Pipe diameters in mm**

Gradient	630	675	700	750	800	825	900	975	1000	1050	1100	1125	1200	1250
0·01000	1·820	1·905	1·952	2·044	2·133	2·177	2·305	2·430	2·471	2·551	2·630	2·669	2·784	2·859
1/ 100	0·5673	0·6819	0·7513	0·9029	1·0722	1·1636	1·4667	1·8143	1·9405	2·2090	2·4993	2·6528	3·1481	3·5080
0·01050	1·865	1·953	2·000	2·094	2·186	2·231	2·362	2·490	2·532	2·614	2·695	2·735	2·852	2·929
1/ 95	0·5813	0·6987	0·7698	0·9252	1·0986	1·1924	1·5029	1·8591	1·9885	2·2635	2·5610	2·7184	3·2259	3·5946
0·01100	1·909	1·999	2·047	2·144	2·237	2·283	2·418	2·549	2·591	2·676	2·758	2·799	2·919	2·998
1/ 91	0·5950	0·7152	0·7880	0·9470	1·1245	1·2205	1·5383	1·9029	2·0353	2·3168	2·6213	2·7824	3·3019	3·6793
0·01150	1·952	2·043	2·094	2·192	2·287	2·334	2·472	2·606	2·650	2·736	2·820	2·862	2·985	3·066
1/ 87	0·6083	0·7313	0·8057	0·9683	1·1498	1·2479	1·5729	1·9457	2·0811	2·3689	2·6803	2·8449	3·3761	3·7620
0·01200	1·994	2·087	2·139	2·239	2·337	2·385	2·526	2·662	2·707	2·795	2·881	2·924	3·049	3·132
1/ 83	0·6214	0·7470	0·8230	0·9891	1·1746	1·2748	1·6067	1·9876	2·1259	2·4199	2·7379	2·9062	3·4488	3·8430
0·01250	2·035	2·131	2·183	2·285	2·385	2·434	2·578	2·717	2·763	2·852	2·940	2·984	3·112	3·196
1/ 80	0·6343	0·7624	0·8400	1·0095	1·1988	1·3011	1·6399	2·0286	2·1697	2·4698	2·7944	2·9661	3·5199	3·9222
0·01300	2·075	2·173	2·226	2·330	2·432	2·482	2·629	2·771	2·817	2·909	2·999	3·043	3·174	3·259
1/ 77	0·6468	0·7775	0·8567	1·0295	1·2225	1·3269	1·6724	2·0688	2·2127	2·5188	2·8498	3·0249	3·5897	3·9999
0·01350	2·115	2·214	2·268	2·375	2·479	2·529	2·679	2·824	2·871	2·964	3·056	3·101	3·234	3·322
1/ 74	0·6592	0·7923	0·8730	1·0492	1·2459	1·3522	1·7043	2·1082	2·2549	2·5668	2·9041	3·0825	3·6581	4·0762
0·01400	2·153	2·255	2·310	2·418	2·524	2·576	2·728	2·876	2·924	3·019	3·112	3·158	3·294	3·383
1/ 71	0·6713	0·8069	0·8890	1·0684	1·2687	1·3770	1·7355	2·1469	2·2963	2·6139	2·9574	3·1391	3·7252	4·1510
0·01450	2·192	2·295	2·351	2·461	2·569	2·622	2·776	2·926	2·975	3·072	3·167	3·214	3·352	3·442
1/ 69	0·6832	0·8212	0·9048	1·0873	1·2912	1·4014	1·7663	2·1850	2·3370	2·6602	3·0098	3·1947	3·7912	4·2245
0·01500	2·229	2·334	2·391	2·503	2·613	2·666	2·824	2·977	3·026	3·125	3·221	3·269	3·409	3·501
1/ 67	0·6948	0·8352	0·9202	1·1059	1·3133	1·4254	1·7965	2·2223	2·3769	2·7057	3·0613	3·2494	3·8560	4·2968
0·01600	2·302	2·411	2·470	2·585	2·698	2·754	2·917	3·074	3·126	3·227	3·327	3·376	3·521	3·616
1/ 63	0·7176	0·8626	0·9504	1·1422	1·3564	1·4721	1·8554	2·2953	2·4549	2·7945	3·1617	3·3560	3·9826	4·4378
0·01700	2·373	2·485	2·546	2·665	2·782	2·839	3·006	3·169	3·222	3·327	3·429	3·480	3·630	3·728
1/ 59	0·7397	0·8892	0·9797	1·1774	1·3981	1·5175	1·9126	2·3659	2·5305	2·8805	3·2591	3·4593	4·1052	4·5744
0·01800	2·442	2·557	2·620	2·742	2·862	2·921	3·094	3·261	3·315	3·423	3·529	3·581	3·735	3·836
1/ 56	0·7612	0·9150	1·0081	1·2116	1·4387	1·5615	1·9681	2·4346	2·6039	2·9641	3·3536	3·5597	4·2242	4·7071
0·01900	2·509	2·627	2·691	2·818	2·941	3·001	3·178	3·350	3·406	3·517	3·626	3·679	3·837	3·941
1/ 53	0·7821	0·9401	1·0358	1·2448	1·4781	1·6043	2·0220	2·5013	2·6753	3·0453	3·4455	3·6572	4·3400	4·8361
0·02000	2·574	2·695	2·761	2·891	3·017	3·079	3·261	3·437	3·495	3·608	3·720	3·775	3·937	4·043
1/ 50	0·8024	0·9645	1·0627	1·2771	1·5166	1·6460	2·0746	2·5663	2·7448	3·1245	3·5351	3·7523	4·4528	4·9618
0·02100	2·638	2·762	2·830	2·962	3·092	3·155	3·342	3·522	3·581	3·698	3·812	3·868	4·034	4·143
1/ 47·6	0·8222	0·9884	1·0890	1·3087	1·5540	1·6867	2·1258	2·6297	2·8126	3·2017	3·6224	3·8450	4·5628	5·0844
0·02200	2·700	2·827	2·896	3·032	3·164	3·229	3·420	3·605	3·665	3·785	3·901	3·959	4·129	4·241
1/ 45·5	0·8416	1·0116	1·1146	1·3395	1·5906	1·7264	2·1759	2·6916	2·8789	3·2770	3·7077	3·9355	4·6703	5·2040
0·02300	2·761	2·891	2·961	3·100	3·236	3·302	3·497	3·686	3·748	3·870	3·989	4·048	4·222	4·336
1/ 43·5	0·8605	1·0344	1·1397	1·3696	1·6264	1·7652	2·2248	2·7521	2·9436	3·3507	3·7911	4·0240	4·7753	5·3211
0·02400	2·820	2·953	3·025	3·167	3·305	3·373	3·572	3·765	3·829	3·953	4·075	4·135	4·313	4·429
1/ 41·7	0·8790	1·0566	1·1642	1·3991	1·6614	1·8032	2·2727	2·8114	3·0069	3·4228	3·8726	4·1106	4·8780	5·4355
0·02500	2·878	3·014	3·087	3·232	3·373	3·443	3·646	3·843	3·908	4·034	4·159	4·221	4·402	4·521
1/ 40·0	0·8972	1·0784	1·1882	1·4280	1·6957	1·8404	2·3196	2·8693	3·0689	3·4934	3·9525	4·1954	4·9786	5·5477
0·02600	2·935	3·073	3·149	3·296	3·440	3·511	3·718	3·919	3·985	4·114	4·241	4·304	4·489	4·610
1/ 38·5	0·9150	1·0998	1·2118	1·4563	1·7293	1·8768	2·3655	2·9262	3·1297	3·5626	4·0308	4·2785	5·0773	5·6576
0·02700	2·991	3·132	3·209	3·359	3·506	3·578	3·789	3·994	4·061	4·193	4·322	4·386	4·575	4·698
1/ 37·0	0·9324	1·1208	1·2348	1·4840	1·7622	1·9126	2·4106	2·9820	3·1894	3·6305	4·1076	4·3600	5·1740	5·7654
0·02800	3·046	3·189	3·268	3·421	3·570	3·644	3·859	4·067	4·135	4·270	4·402	4·467	4·659	4·784
1/ 35·7	0·9495	1·1413	1·2575	1·5113	1·7946	1·9477	2·4548	3·0367	3·2479	3·6972	4·1830	4·4400	5·2690	5·8712
0·02900	3·100	3·246	3·325	3·481	3·633	3·708	3·927	4·139	4·209	4·345	4·480	4·546	4·741	4·869
1/ 34·5	0·9663	1·1615	1·2798	1·5380	1·8263	1·9822	2·4983	3·0905	3·3054	3·7626	4·2571	4·5187	5·3623	5·9752
0·03000	3·153	3·301	3·382	3·541	3·696	3·771	3·994	4·210	4·281	4·420	4·556	4·624	4·822	4·952
1/ 33·3	0·9829	1·1814	1·3017	1·5643	1·8576	2·0161	2·5410	3·1433	3·3620	3·8270	4·3299	4·5959	5·4540	6·0773
	0·75	0·75	0·75	0·75	0·76	0·76	0·76	0·76	0·76	0·76	0·77	0·77	0·77	0·77

$V_{r(0{\cdot}5)medial}$ for half-full circular pipes.

Gradient	(Equivalent) Pipe diameters in mm													
	630	**675**	**700**	**750**	**800**	**825**	**900**	**975**	**1000**	**1050**	**1100**	**1125**	**1200**	**1250**
0·03000	3·153	3·301	3·382	3·541	3·696	3·771	3·994	4·210	4·281	4·420	4·556	4·624	4·822	4·952
1/ 33·3	0·9829	1·1814	1·3017	1·5643	1·8576	2·0161	2·5410	3·1433	3·3620	3·8270	4·3299	4·5959	5·4540	6·0773
0·03200	3·256	3·410	3·493	3·657	3·817	3·895	4·125	4·348	4·421	4·565	4·706	4·775	4·981	5·115
1/ 31·3	1·0151	1·2202	1·3444	1·6157	1·9185	2·0822	2·6244	3·2464	3·4723	3·9525	4·4720	4·7467	5·6329	6·2767
0·03400	3·357	3·515	3·601	3·770	3·934	4·015	4·252	4·482	4·557	4·705	4·851	4·922	5·134	5·272
1/ 29·4	1·0464	1·2577	1·3858	1·6654	1·9776	2·1463	2·7052	3·3464	3·5792	4·0742	4·6096	4·8928	5·8063	6·4699
0·03600	3·454	3·617	3·705	3·879	4·048	4·132	4·376	4·612	4·689	4·842	4·991	5·065	5·283	5·425
1/ 27·8	1·0767	1·2942	1·4260	1·7137	2·0350	2·2086	2·7837	3·4434	3·6830	4·1924	4·7433	5·0347	5·9747	6·6576
0·03800	3·549	3·716	3·807	3·985	4·159	4·245	4·496	4·738	4·818	4·974	5·128	5·204	5·428	5·574
1/ 26·3	1·1062	1·3297	1·4651	1·7607	2·0907	2·2691	2·8600	3·5378	3·7839	4·3073	4·8733	5·1728	6·1385	6·8401
0·04000	3·641	3·812	3·906	4·089	4·267	4·355	4·612	4·862	4·943	5·104	5·261	5·339	5·569	5·719
1/ 25·0	1·1350	1·3643	1·5031	1·8064	2·1451	2·3281	2·9343	3·6298	3·8823	4·4192	5·0000	5·3072	6·2980	7·0178
0·04200	3·731	3·907	4·002	4·190	4·373	4·463	4·726	4·982	5·065	5·230	5·391	5·471	5·706	5·860
1/ 23·8	1·1630	1·3980	1·5403	1·8511	2·1981	2·3856	3·0068	3·7194	3·9782	4·5284	5·1235	5·4383	6·4536	7·1911
0·04400	3·819	3·999	4·097	4·289	4·476	4·568	4·838	5·099	5·184	5·353	5·518	5·600	5·841	5·998
1/ 22·7	1·1904	1·4309	1·5765	1·8946	2·2498	2·4418	3·0776	3·8070	4·0718	4·6350	5·2441	5·5663	6·6055	7·3604
0·04600	3·905	4·088	4·189	4·385	4·576	4·671	4·946	5·214	5·301	5·473	5·642	5·726	5·972	6·133
1/ 21·7	1·2172	1·4630	1·6120	1·9372	2·3004	2·4967	3·1467	3·8926	4·1633	4·7392	5·3620	5·6914	6·7540	7·5259
0·04800	3·989	4·176	4·279	4·479	4·675	4·771	5·053	5·326	5·415	5·591	5·764	5·849	6·100	6·265
1/ 20·8	1·2434	1·4945	1·6467	1·9789	2·3499	2·5504	3·2144	3·9763	4·2529	4·8412	5·4773	5·8139	6·8993	7·6878
0·05000	4·071	4·263	4·367	4·572	4·771	4·869	5·157	5·436	5·527	5·706	5·882	5·969	6·226	6·394
1/ 20·0	1·2690	1·5254	1·6806	2·0197	2·3984	2·6030	3·2807	4·0583	4·3406	4·9410	5·5903	5·9338	7·0416	7·8464
0·05250	4·171	4·368	4·475	4·685	4·889	4·990	5·284	5·570	5·663	5·847	6·028	6·117	6·380	6·552
1/ 19·0	1·3004	1·5630	1·7221	2·0696	2·4576	2·6673	3·3618	4·1586	4·4479	5·0631	5·7284	6·0803	7·2155	8·0402
0·05500	4·270	4·471	4·580	4·795	5·004	5·107	5·409	5·701	5·797	5·985	6·170	6·261	6·530	6·706
1/ 18·2	1·3310	1·5998	1·7627	2·1183	2·5155	2·7301	3·4409	4·2565	4·5526	5·1822	5·8632	6·2235	7·3853	8·2294
0·05750	4·366	4·571	4·683	4·903	5·117	5·222	5·530	5·829	5·927	6·119	6·308	6·402	6·677	6·857
1/ 17·4	1·3609	1·6358	1·8023	2·1660	2·5720	2·7915	3·5183	4·3522	4·6549	5·2987	5·9950	6·3634	7·5514	8·4144
0·06000	4·460	4·670	4·784	5·008	5·227	5·334	5·649	5·955	6·054	6·251	6·444	6·539	6·821	7·004
1/ 16·7	1·3902	1·6710	1·8411	2·2126	2·6273	2·8515	3·5940	4·4458	4·7551	5·4127	6·1240	6·5003	7·7138	8·5954
0·06250	4·552	4·766	4·883	5·112	5·335	5·444	5·766	6·077	6·179	6·380	6·577	6·674	6·961	7·149
1/ 16·0	1·4188	1·7055	1·8791	2·2582	2·6815	2·9103	3·6681	4·5375	4·8531	5·5244	6·2503	6·6343	7·8729	8·7727
0·06500	4·642	4·860	4·979	5·213	5·440	5·552	5·880	6·198	6·302	6·506	6·707	6·806	7·099	7·290
1/ 15·4	1·4470	1·7393	1·9163	2·3029	2·7347	2·9680	3·7408	4·6274	4·9493	5·6338	6·3741	6·7658	8·0289	8·9465
0·06750	4·730	4·953	5·074	5·312	5·544	5·658	5·992	6·316	6·422	6·630	6·835	6·936	7·234	7·429
1/ 14·8	1·4745	1·7724	1·9528	2·3468	2·7868	3·0246	3·8120	4·7156	5·0436	5·7411	6·4956	6·8947	8·1818	9·1169
0·07000	4·817	5·044	5·167	5·410	5·646	5·762	6·102	6·432	6·540	6·752	6·961	7·063	7·367	7·566
1/ 14·3	1·5016	1·8049	1·9887	2·3899	2·8379	3·0801	3·8820	4·8021	5·1361	5·8465	6·6148	7·0212	8·3320	9·2843
0·07250	4·902	5·133	5·259	5·505	5·746	5·864	6·210	6·546	6·655	6·871	7·084	7·189	7·498	7·699
1/ 13·8	1·5282	1·8369	2·0239	2·4322	2·8882	3·1346	3·9507	4·8871	5·2271	5·9500	6·7319	7·1455	8·4795	9·4486
0·07500	4·986	5·221	5·349	5·600	5·844	5·964	6·316	6·658	6·769	6·989	7·205	7·311	7·626	7·831
1/ 13·3	1·5543	1·8683	2·0585	2·4738	2·9375	3·1882	4·0183	4·9707	5·3165	6·0518	6·8470	7·2677	8·6245	9·6102
0·08000	5·150	5·392	5·524	5·783	6·036	6·160	6·524	6·876	6·991	7·218	7·441	7·551	7·876	8·088
1/ 12·5	1·6053	1·9296	2·1260	2·5550	3·0339	3·2928	4·1501	5·1338	5·4908	6·2503	7·0716	7·5061	8·9074	9·9254
0·08500	5·308	5·558	5·694	5·961	6·222	6·349	6·724	7·088	7·206	7·440	7·670	7·784	8·118	8·337
1/ 11·8	1·6547	1·9890	2·1914	2·6336	3·1273	3·3942	4·2779	5·2918	5·6599	6·4427	7·2893	7·7372	9·1816	10·231
0·09000	5·462	5·719	5·859	6·134	6·402	6·534	6·919	7·293	7·415	7·656	7·893	8·009	8·354	8·579
1/ 11·1	1·7027	2·0467	2·2550	2·7100	3·2180	3·4926	4·4019	5·4452	5·8240	6·6295	7·5007	7·9615	9·4478	10·528
0·09500	5·612	5·876	6·020	6·302	6·577	6·713	7·109	7·493	7·619	7·866	8·109	8·229	8·583	8·814
1/ 10·5	1·7494	2·1028	2·3168	2·7843	3·3062	3·5883	4·5226	5·5945	5·9836	6·8112	7·7063	8·1797	9·7068	10·816
0·10000	5·758	6·029	6·177	6·466	6·748	6·887	7·294	7·688	7·817	8·070	8·320	8·443	8·806	9·043
1/ 10·0	1·7948	2·1574	2·3770	2·8566	3·3921	3·6815	4·6401	5·7398	6·1391	6·9882	7·9065	8·3923	9·9590	11·097
	0·75	0·75	0·75	0·75	0·76	0·76	0·76	0·76	0·76	0·76	0·77	0·77	0·77	0·77

$V_{r(0{\cdot}5)medial}$ for half-full circular pipes.

A33

(p.1 of 6)

k_s = 15·0 mm
S = 0·00010 to 0·00030

ie hydraulic gradient =
1 in 10000 to 1 in 3333

Water (or sewage) at 15°C;
full bore conditions.

velocities in ms^{-1}
discharges in m^3s^{-1}

Gradient **(Equivalent) Pipe diameters in mm**

Gradient	630	675	700	750	800	825	900	975	1000	1050	1100	1125	1200	1250
0·000100	0·154	0·161	0·165	0·173	0·181	0·185	0·197	0·208	0·211	0·219	0·226	0·229	0·239	0·246
1/ 10000	0·0479	0·0577	0·0636	0·0766	0·0911	0·0990	0·1251	0·1551	0·1660	0·1892	0·2144	0·2277	0·2706	0·3019
0·000105	0·157	0·165	0·169	0·178	0·186	0·190	0·202	0·213	0·217	0·224	0·231	0·235	0·245	0·252
1/ 9524	0·0490	0·0591	0·0652	0·0785	0·0934	0·1015	0·1282	0·1589	0·1701	0·1939	0·2197	0·2333	0·2773	0·3093
0·000110	0·161	0·169	0·173	0·182	0·190	0·194	0·206	0·218	0·222	0·229	0·237	0·240	0·251	0·258
1/ 9091	0·0502	0·0605	0·0667	0·0804	0·0956	0·1039	0·1312	0·1627	0·1741	0·1985	0·2249	0·2388	0·2839	0·3166
0·000115	0·165	0·173	0·177	0·186	0·194	0·199	0·211	0·223	0·227	0·234	0·242	0·246	0·257	0·264
1/ 8696	0·0513	0·0618	0·0682	0·0822	0·0978	0·1062	0·1342	0·1664	0·1781	0·2030	0·2299	0·2442	0·2903	0·3238
0·000120	0·168	0·177	0·181	0·190	0·199	0·203	0·215	0·228	0·232	0·239	0·247	0·251	0·262	0·270
1/ 8333	0·0524	0·0632	0·0697	0·0839	0·0999	0·1085	0·1371	0·1700	0·1819	0·2073	0·2349	0·2495	0·2965	0·3308
0·000125	0·172	0·180	0·185	0·194	0·203	0·207	0·220	0·232	0·236	0·244	0·252	0·256	0·268	0·275
1/ 8000	0·0535	0·0645	0·0711	0·0857	0·1019	0·1107	0·1399	0·1735	0·1857	0·2116	0·2397	0·2546	0·3027	0·3376
0·000130	0·175	0·184	0·189	0·198	0·207	0·211	0·224	0·237	0·241	0·249	0·257	0·261	0·273	0·281
1/ 7692	0·0546	0·0658	0·0726	0·0874	0·1040	0·1129	0·1427	0·1769	0·1893	0·2158	0·2445	0·2597	0·3087	0·3443
0·000135	0·178	0·187	0·192	0·202	0·211	0·215	0·229	0·241	0·246	0·254	0·262	0·266	0·278	0·286
1/ 7407	0·0556	0·0670	0·0739	0·0890	0·1059	0·1151	0·1454	0·1803	0·1930	0·2199	0·2492	0·2646	0·3146	0·3509
0·000140	0·182	0·191	0·196	0·205	0·215	0·219	0·233	0·246	0·250	0·259	0·267	0·271	0·283	0·291
1/ 7143	0·0567	0·0683	0·0753	0·0907	0·1079	0·1172	0·1481	0·1836	0·1965	0·2240	0·2537	0·2695	0·3203	0·3573
0·000145	0·185	0·194	0·199	0·209	0·218	0·223	0·237	0·250	0·255	0·263	0·272	0·276	0·288	0·296
1/ 6897	0·0577	0·0695	0·0766	0·0923	0·1098	0·1193	0·1507	0·1869	0·2000	0·2280	0·2582	0·2743	0·3260	0·3637
0·000150	0·188	0·197	0·203	0·212	0·222	0·227	0·241	0·255	0·259	0·268	0·276	0·281	0·293	0·301
1/ 6667	0·0587	0·0707	0·0779	0·0939	0·1117	0·1213	0·1533	0·1901	0·2034	0·2319	0·2627	0·2790	0·3316	0·3699
0·000160	0·194	0·204	0·209	0·219	0·230	0·234	0·249	0·263	0·268	0·277	0·285	0·290	0·303	0·311
1/ 6250	0·0606	0·0730	0·0805	0·0970	0·1154	0·1253	0·1583	0·1963	0·2101	0·2395	0·2713	0·2881	0·3425	0·3820
0·000170	0·200	0·210	0·216	0·226	0·237	0·242	0·257	0·271	0·276	0·285	0·294	0·299	0·312	0·321
1/ 5882	0·0625	0·0752	0·0830	0·1000	0·1189	0·1292	0·1632	0·2024	0·2166	0·2469	0·2797	0·2970	0·3531	0·3938
0·000180	0·206	0·216	0·222	0·233	0·243	0·249	0·264	0·279	0·284	0·293	0·303	0·307	0·321	0·330
1/ 5556	0·0643	0·0774	0·0854	0·1029	0·1224	0·1329	0·1680	0·2083	0·2229	0·2541	0·2878	0·3057	0·3633	0·4053
0·000190	0·212	0·222	0·228	0·239	0·250	0·256	0·271	0·287	0·292	0·301	0·311	0·316	0·330	0·339
1/ 5263	0·0660	0·0796	0·0878	0·1057	0·1257	0·1366	0·1726	0·2140	0·2290	0·2610	0·2957	0·3140	0·3733	0·4164
0·000200	0·217	0·228	0·234	0·245	0·257	0·262	0·278	0·294	0·299	0·309	0·319	0·324	0·339	0·348
1/ 5000	0·0678	0·0816	0·0900	0·1084	0·1290	0·1402	0·1771	0·2195	0·2350	0·2678	0·3034	0·3222	0·3830	0·4272
0·000210	0·223	0·234	0·240	0·252	0·263	0·269	0·285	0·301	0·307	0·317	0·327	0·332	0·347	0·357
1/ 4762	0·0694	0·0837	0·0923	0·1111	0·1322	0·1436	0·1815	0·2250	0·2408	0·2744	0·3109	0·3302	0·3925	0·4378
0·000220	0·228	0·239	0·245	0·257	0·269	0·275	0·292	0·308	0·314	0·324	0·335	0·340	0·355	0·365
1/ 4545	0·0711	0·0856	0·0945	0·1137	0·1353	0·1470	0·1857	0·2303	0·2465	0·2809	0·3182	0·3380	0·4017	0·4481
0·000230	0·233	0·245	0·251	0·263	0·275	0·281	0·299	0·315	0·321	0·332	0·342	0·348	0·363	0·373
1/ 4348	0·0727	0·0876	0·0966	0·1163	0·1384	0·1503	0·1899	0·2355	0·2520	0·2872	0·3254	0·3456	0·4108	0·4582
0·000240	0·238	0·250	0·256	0·269	0·281	0·287	0·305	0·322	0·328	0·339	0·350	0·355	0·371	0·381
1/ 4167	0·0743	0·0894	0·0987	0·1188	0·1414	0·1536	0·1940	0·2405	0·2574	0·2934	0·3324	0·3530	0·4196	0·4681
0·000250	0·243	0·255	0·262	0·275	0·287	0·293	0·311	0·329	0·335	0·346	0·357	0·362	0·379	0·389
1/ 4000	0·0758	0·0913	0·1007	0·1213	0·1443	0·1567	0·1980	0·2455	0·2628	0·2995	0·3393	0·3603	0·4283	0·4777
0·000260	0·248	0·260	0·267	0·280	0·293	0·299	0·317	0·335	0·341	0·353	0·364	0·370	0·386	0·397
1/ 3846	0·0773	0·0931	0·1027	0·1237	0·1471	0·1598	0·2020	0·2504	0·2680	0·3054	0·3460	0·3675	0·4368	0·4872
0·000270	0·253	0·265	0·272	0·285	0·298	0·305	0·324	0·342	0·348	0·359	0·371	0·377	0·394	0·405
1/ 3704	0·0788	0·0949	0·1047	0·1260	0·1500	0·1629	0·2058	0·2552	0·2731	0·3113	0·3526	0·3745	0·4451	0·4965
0·000280	0·257	0·270	0·277	0·291	0·304	0·310	0·329	0·348	0·354	0·366	0·378	0·384	0·401	0·412
1/ 3571	0·0802	0·0966	0·1066	0·1284	0·1527	0·1659	0·2096	0·2598	0·2781	0·3170	0·3591	0·3814	0·4533	0·5056
0·000290	0·262	0·275	0·282	0·296	0·309	0·316	0·335	0·354	0·360	0·373	0·385	0·390	0·408	0·419
1/ 3448	0·0816	0·0983	0·1085	0·1306	0·1554	0·1688	0·2133	0·2645	0·2830	0·3226	0·3654	0·3881	0·4613	0·5146
0·000300	0·266	0·280	0·287	0·301	0·314	0·321	0·341	0·360	0·367	0·379	0·391	0·397	0·415	0·427
1/ 3333	0·0830	0·1000	0·1103	0·1329	0·1581	0·1717	0·2170	0·2690	0·2879	0·3281	0·3717	0·3948	0·4692	0·5234
	–	–	–	0·74	0·74	0·74	0·74	0·74	0·74	0·74	0·74	0·74	0·74	0·74

$V_{r(0\cdot5)medial}$ for half-full circular pipes.

k_s = 15·0 mm S = 0·00010 to 0·00030

$k_s = 15{\cdot}0$ **mm**
S = 0·00030 to 0·00100

ie hydraulic gradient =
1 in 3333 to 1 in 1000

Water (or sewage) at 15°C;
full bore conditions.

velocities in ms^{-1}
discharges in m^3s^{-1}

Gradient **(Equivalent) Pipe diameters in mm**

Gradient	630	675	700	750	800	825	900	975	1000	1050	1100	1125	1200	1250
0·000300	0·266	0·280	0·287	0·301	0·314	0·321	0·341	0·360	0·367	0·379	0·391	0·397	0·415	0·427
1/ 3333	0·0830	0·1000	0·1103	0·1329	0·1581	0·1717	0·2170	0·2690	0·2879	0·3281	0·3717	0·3948	0·4692	0·5234
0·000320	0·275	0·289	0·296	0·311	0·325	0·332	0·352	0·372	0·379	0·391	0·404	0·410	0·429	0·441
1/ 3125	0·0858	0·1033	0·1140	0·1372	0·1633	0·1774	0·2241	0·2778	0·2973	0·3389	0·3839	0·4077	0·4847	0·5406
0·000340	0·284	0·298	0·305	0·320	0·335	0·342	0·363	0·384	0·390	0·403	0·416	0·423	0·442	0·454
1/ 2941	0·0884	0·1065	0·1175	0·1415	0·1683	0·1828	0·2310	0·2864	0·3065	0·3494	0·3957	0·4203	0·4996	0·5573
0·000360	0·292	0·306	0·314	0·330	0·345	0·352	0·374	0·395	0·402	0·415	0·429	0·435	0·455	0·467
1/ 2778	0·0910	0·1096	0·1209	0·1456	0·1732	0·1881	0·2377	0·2947	0·3154	0·3595	0·4072	0·4325	0·5141	0·5734
0·000380	0·300	0·315	0·323	0·339	0·354	0·362	0·384	0·406	0·413	0·427	0·440	0·447	0·467	0·480
1/ 2632	0·0935	0·1126	0·1242	0·1496	0·1780	0·1933	0·2442	0·3028	0·3241	0·3694	0·4184	0·4444	0·5282	0·5892
0·000400	0·308	0·323	0·331	0·347	0·363	0·371	0·394	0·416	0·423	0·438	0·452	0·459	0·479	0·493
1/ 2500	0·0959	0·1155	0·1274	0·1535	0·1826	0·1983	0·2506	0·3107	0·3325	0·3790	0·4293	0·4559	0·5419	0·6045
0·000420	0·315	0·331	0·339	0·356	0·372	0·380	0·404	0·426	0·434	0·448	0·463	0·470	0·491	0·505
1/ 2381	0·0983	0·1184	0·1306	0·1573	0·1871	0·2032	0·2568	0·3183	0·3407	0·3883	0·4399	0·4672	0·5553	0·6194
0·000440	0·323	0·339	0·347	0·364	0·381	0·389	0·413	0·436	0·444	0·459	0·474	0·481	0·503	0·517
1/ 2273	0·1006	0·1212	0·1337	0·1610	0·1915	0·2080	0·2628	0·3259	0·3488	0·3975	0·4503	0·4782	0·5684	0·6340
0·000460	0·330	0·346	0·355	0·373	0·390	0·398	0·422	0·446	0·454	0·469	0·484	0·492	0·514	0·528
1/ 2174	0·1029	0·1239	0·1367	0·1646	0·1958	0·2127	0·2688	0·3332	0·3566	0·4064	0·4604	0·4890	0·5812	0·6483
0·000480	0·337	0·354	0·363	0·381	0·398	0·407	0·432	0·456	0·464	0·479	0·495	0·503	0·525	0·540
1/ 2083	0·1051	0·1266	0·1396	0·1681	0·2000	0·2173	0·2745	0·3404	0·3643	0·4152	0·4703	0·4995	0·5937	0·6623
0·000500	0·344	0·361	0·370	0·388	0·406	0·415	0·440	0·465	0·473	0·489	0·505	0·513	0·536	0·551
1/ 2000	0·1073	0·1292	0·1425	0·1716	0·2042	0·2218	0·2802	0·3474	0·3718	0·4238	0·4800	0·5098	0·6060	0·6759
0·000525	0·353	0·370	0·379	0·398	0·416	0·425	0·451	0·477	0·485	0·502	0·518	0·526	0·549	0·564
1/ 1905	0·1099	0·1324	0·1460	0·1759	0·2092	0·2273	0·2871	0·3560	0·3810	0·4342	0·4919	0·5224	0·6210	0·6926
0·000550	0·361	0·379	0·388	0·407	0·426	0·435	0·462	0·488	0·497	0·513	0·530	0·538	0·562	0·578
1/ 1818	0·1125	0·1355	0·1495	0·1800	0·2142	0·2326	0·2939	0·3644	0·3900	0·4445	0·5035	0·5347	0·6356	0·7090
0·000575	0·369	0·387	0·397	0·417	0·436	0·445	0·472	0·499	0·508	0·525	0·542	0·550	0·575	0·591
1/ 1739	0·1150	0·1386	0·1528	0·1841	0·2190	0·2379	0·3005	0·3726	0·3987	0·4545	0·5148	0·5468	0·6499	0·7249
0·000600	0·377	0·396	0·406	0·426	0·445	0·455	0·483	0·510	0·519	0·536	0·553	0·562	0·587	0·603
1/ 1667	0·1175	0·1416	0·1561	0·1880	0·2237	0·2430	0·3070	0·3806	0·4073	0·4643	0·5259	0·5585	0·6639	0·7405
0·000625	0·385	0·404	0·414	0·434	0·454	0·464	0·493	0·520	0·529	0·547	0·565	0·574	0·599	0·616
1/ 1600	0·1199	0·1445	0·1594	0·1919	0·2283	0·2480	0·3133	0·3884	0·4157	0·4739	0·5368	0·5701	0·6776	0·7558
0·000650	0·392	0·412	0·422	0·443	0·463	0·473	0·502	0·531	0·540	0·558	0·576	0·585	0·611	0·628
1/ 1538	0·1223	0·1473	0·1625	0·1957	0·2328	0·2529	0·3195	0·3961	0·4240	0·4832	0·5474	0·5814	0·6910	0·7708
0·000675	0·400	0·420	0·430	0·451	0·472	0·482	0·512	0·541	0·550	0·569	0·587	0·596	0·623	0·640
1/ 1481	0·1247	0·1502	0·1656	0·1994	0·2373	0·2577	0·3256	0·4037	0·4321	0·4925	0·5578	0·5925	0·7042	0·7855
0·000700	0·407	0·427	0·438	0·460	0·481	0·491	0·521	0·551	0·560	0·579	0·598	0·607	0·634	0·652
1/ 1429	0·1269	0·1529	0·1687	0·2031	0·2416	0·2625	0·3316	0·4111	0·4400	0·5015	0·5681	0·6033	0·7171	0·7999
0·000725	0·414	0·435	0·446	0·468	0·489	0·500	0·531	0·560	0·570	0·589	0·608	0·618	0·645	0·663
1/ 1379	0·1292	0·1556	0·1717	0·2067	0·2459	0·2671	0·3375	0·4184	0·4478	0·5104	0·5782	0·6140	0·7299	0·8141
0·000750	0·422	0·442	0·454	0·476	0·498	0·508	0·540	0·570	0·580	0·600	0·619	0·628	0·656	0·675
1/ 1333	0·1314	0·1583	0·1746	0·2102	0·2501	0·2717	0·3433	0·4256	0·4555	0·5191	0·5881	0·6245	0·7423	0·8280
0·000800	0·435	0·457	0·469	0·492	0·514	0·525	0·557	0·589	0·599	0·619	0·639	0·649	0·678	0·697
1/ 1250	0·1357	0·1635	0·1803	0·2172	0·2583	0·2806	0·3546	0·4395	0·4704	0·5362	0·6074	0·6450	0·7667	0·8552
0·000850	0·449	0·471	0·483	0·507	0·530	0·541	0·574	0·607	0·617	0·638	0·659	0·669	0·699	0·718
1/ 1176	0·1399	0·1685	0·1859	0·2238	0·2663	0·2893	0·3655	0·4531	0·4849	0·5527	0·6261	0·6649	0·7903	0·8815
0·000900	0·462	0·485	0·497	0·521	0·545	0·557	0·591	0·624	0·635	0·657	0·678	0·688	0·719	0·739
1/ 1111	0·1440	0·1734	0·1913	0·2303	0·2740	0·2977	0·3761	0·4662	0·4990	0·5687	0·6442	0·6842	0·8133	0·9071
0·000950	0·475	0·498	0·511	0·536	0·560	0·572	0·607	0·642	0·653	0·675	0·697	0·707	0·739	0·759
1/ 1053	0·1479	0·1782	0·1965	0·2367	0·2815	0·3058	0·3864	0·4790	0·5127	0·5843	0·6619	0·7030	0·8356	0·9320
0·001000	0·487	0·511	0·524	0·550	0·575	0·587	0·623	0·658	0·670	0·692	0·715	0·726	0·758	0·779
1/ 1000	0·1518	0·1828	0·2016	0·2428	0·2889	0·3138	0·3964	0·4915	0·5260	0·5995	0·6791	0·7213	0·8573	0·9562
	–	–	–	0·74	0·74	0·74	0·74	0·74	0·74	0·74	0·74	0·74	0·74	0·74

$V_{r(0{\cdot}5)medial}$ **for half-full circular pipes.**

A33

(p.3 of 6)

k_s = **15·0 mm**
S = 0·00100 to 0·00300

ie hydraulic gradient =
1 in 1000 to 1 in 333

Water (or sewage) at 15°C;
full bore conditions.

velocities in ms^{-1}
discharges in m^3s^{-1}

Gradient	**(Equivalent) Pipe diameters in mm**													
	630	**675**	**700**	**750**	**800**	**825**	**900**	**975**	**1000**	**1050**	**1100**	**1125**	**1200**	**1250**
0·00100	0·487	0·511	0·524	0·550	0·575	0·587	0·623	0·658	0·670	0·692	0·715	0·726	0·758	0·779
1/ 1000	0·1518	0·1828	0·2016	0·2428	0·2889	0·3138	0·3964	0·4915	0·5260	0·5995	0·6791	0·7213	0·8573	0·9562
0·00105	0·499	0·523	0·537	0·563	0·589	0·602	0·639	0·675	0·686	0·709	0·732	0·744	0·777	0·798
1/ 952	0·1555	0·1873	0·2066	0·2488	0·2960	0·3215	0·4062	0·5036	0·5390	0·6143	0·6959	0·7391	0·8785	0·9799
0·00110	0·511	0·536	0·550	0·576	0·603	0·616	0·654	0·690	0·702	0·726	0·750	0·761	0·795	0·817
1/ 909	0·1592	0·1917	0·2115	0·2547	0·3030	0·3291	0·4158	0·5155	0·5517	0·6288	0·7123	0·7565	0·8992	1·0029
0·00115	0·522	0·548	0·562	0·589	0·616	0·630	0·668	0·706	0·718	0·743	0·766	0·778	0·813	0·836
1/ 870	0·1628	0·1961	0·2162	0·2604	0·3098	0·3365	0·4252	0·5271	0·5641	0·6430	0·7283	0·7735	0·9194	1·0255
0·00120	0·533	0·560	0·574	0·602	0·630	0·643	0·683	0·721	0·734	0·759	0·783	0·795	0·830	0·854
1/ 833	0·1663	0·2003	0·2209	0·2660	0·3165	0·3438	0·4343	0·5384	0·5763	0·6568	0·7440	0·7902	0·9392	1·0476
0·00125	0·544	0·571	0·586	0·615	0·643	0·656	0·697	0·736	0·749	0·774	0·799	0·811	0·848	0·871
1/ 800	0·1697	0·2044	0·2255	0·2715	0·3230	0·3509	0·4433	0·5495	0·5882	0·6703	0·7593	0·8065	0·9586	1·0692
0·00130	0·555	0·583	0·597	0·627	0·655	0·669	0·711	0·751	0·764	0·790	0·815	0·827	0·864	0·889
1/ 769	0·1731	0·2085	0·2299	0·2769	0·3294	0·3578	0·4521	0·5604	0·5998	0·6836	0·7744	0·8224	0·9776	1·0904
0·00135	0·566	0·594	0·609	0·639	0·668	0·682	0·724	0·765	0·778	0·805	0·830	0·843	0·881	0·905
1/ 741	0·1764	0·2124	0·2343	0·2822	0·3357	0·3646	0·4607	0·5711	0·6112	0·6967	0·7892	0·8381	0·9962	1·1112
0·00140	0·576	0·605	0·620	0·650	0·680	0·695	0·737	0·779	0·793	0·819	0·846	0·859	0·897	0·922
1/ 714	0·1796	0·2163	0·2386	0·2874	0·3418	0·3713	0·4692	0·5816	0·6225	0·7095	0·8036	0·8535	1·0145	1·1316
0·00145	0·586	0·615	0·631	0·662	0·692	0·707	0·751	0·793	0·807	0·834	0·861	0·874	0·913	0·938
1/ 690	0·1828	0·2202	0·2428	0·2924	0·3479	0·3779	0·4775	0·5919	0·6335	0·7220	0·8179	0·8686	1·0325	1·1516
0·00150	0·596	0·626	0·642	0·673	0·704	0·719	0·763	0·806	0·820	0·848	0·875	0·889	0·929	0·954
1/ 667	0·1859	0·2239	0·2470	0·2974	0·3539	0·3844	0·4856	0·6020	0·6443	0·7344	0·8319	0·8835	1·0501	1·1713
0·00160	0·616	0·646	0·663	0·695	0·727	0·743	0·788	0·833	0·847	0·876	0·904	0·918	0·959	0·986
1/ 625	0·1920	0·2313	0·2551	0·3072	0·3655	0·3970	0·5016	0·6218	0·6655	0·7585	0·8592	0·9125	1·0846	1·2097
0·00170	0·635	0·666	0·683	0·717	0·749	0·766	0·813	0·858	0·873	0·903	0·932	0·946	0·989	1·016
1/ 588	0·1979	0·2384	0·2630	0·3167	0·3767	0·4092	0·5170	0·6409	0·6860	0·7818	0·8856	0·9406	1·1180	1·2470
0·00180	0·653	0·686	0·703	0·738	0·771	0·788	0·836	0·883	0·899	0·929	0·959	0·974	1·017	1·046
1/ 556	0·2037	0·2453	0·2706	0·3259	0·3877	0·4211	0·5320	0·6595	0·7059	0·8045	0·9113	0·9679	1·1504	1·2832
0·00190	0·671	0·704	0·722	0·758	0·792	0·809	0·859	0·908	0·923	0·955	0·985	1·000	1·045	1·074
1/ 526	0·2093	0·2521	0·2780	0·3348	0·3983	0·4327	0·5466	0·6776	0·7252	0·8266	0·9363	0·9944	1·1820	1·3184
0·00200	0·689	0·723	0·741	0·778	0·813	0·830	0·882	0·931	0·947	0·979	1·011	1·026	1·072	1·102
1/ 500	0·2147	0·2586	0·2853	0·3435	0·4086	0·4439	0·5608	0·6952	0·7441	0·8481	0·9607	1·0203	1·2127	1·3526
0·00210	0·706	0·741	0·760	0·797	0·833	0·851	0·903	0·954	0·971	1·004	1·036	1·052	1·099	1·129
1/ 476	0·2200	0·2650	0·2923	0·3520	0·4187	0·4549	0·5747	0·7124	0·7625	0·8690	0·9844	1·0455	1·2426	1·3860
0·00220	0·722	0·758	0·777	0·816	0·853	0·871	0·925	0·977	0·994	1·027	1·060	1·077	1·125	1·156
1/ 455	0·2252	0·2713	0·2992	0·3603	0·4286	0·4656	0·5882	0·7292	0·7804	0·8895	1·0076	1·0701	1·2719	1·4187
0·00230	0·739	0·775	0·795	0·834	0·872	0·891	0·945	0·999	1·016	1·050	1·084	1·101	1·150	1·182
1/ 435	0·2303	0·2774	0·3059	0·3684	0·4382	0·4761	0·6014	0·7456	0·7980	0·9095	1·0302	1·0941	1·3005	1·4506
0·00240	0·755	0·792	0·812	0·852	0·891	0·910	0·966	1·020	1·038	1·073	1·107	1·124	1·175	1·207
1/ 417	0·2352	0·2833	0·3125	0·3763	0·4477	0·4863	0·6144	0·7616	0·8152	0·9291	1·0524	1·1177	1·3285	1·4818
0·00250	0·770	0·808	0·829	0·869	0·909	0·928	0·986	1·041	1·059	1·095	1·130	1·148	1·199	1·232
1/ 400	0·2401	0·2892	0·3190	0·3841	0·4569	0·4963	0·6271	0·7773	0·8320	0·9482	1·0741	1·1408	1·3559	1·5124
0·00260	0·785	0·824	0·845	0·887	0·927	0·947	1·005	1·062	1·080	1·117	1·153	1·170	1·223	1·257
1/ 385	0·2448	0·2949	0·3253	0·3917	0·4660	0·5062	0·6395	0·7928	0·8485	0·9670	1·0954	1·1634	1·3828	1·5423
0·00270	0·800	0·840	0·861	0·904	0·945	0·965	1·024	1·082	1·101	1·138	1·175	1·193	1·246	1·281
1/ 370	0·2495	0·3005	0·3315	0·3992	0·4749	0·5158	0·6517	0·8079	0·8646	0·9855	1·1163	1·1855	1·4091	1·5717
0·00280	0·815	0·855	0·877	0·920	0·962	0·983	1·043	1·102	1·121	1·159	1·196	1·215	1·269	1·304
1/ 357	0·2541	0·3060	0·3376	0·4065	0·4836	0·5253	0·6636	0·8227	0·8805	1·0036	1·1368	1·2073	1·4350	1·6006
0·00290	0·830	0·870	0·893	0·936	0·979	1·000	1·062	1·121	1·141	1·179	1·217	1·236	1·291	1·327
1/ 345	0·2586	0·3115	0·3435	0·4137	0·4921	0·5346	0·6754	0·8373	0·8961	1·0213	1·1569	1·2287	1·4604	1·6289
0·00300	0·844	0·885	0·908	0·952	0·996	1·017	1·080	1·141	1·160	1·200	1·238	1·257	1·313	1·350
1/ 333	0·2630	0·3168	0·3494	0·4208	0·5006	0·5437	0·6870	0·8516	0·9114	1·0388	1·1767	1·2497	1·4854	1·6568
	–	–	–	0·74	0·74	0·74	0·74	0·74	0·74	0·74	0·74	0·74	0·74	0·74

$V_{r(0\cdot5)medial}$ for half-full circular pipes.

k_s = 15·0 mm S = 0·00100 to 0·00300

k_s = **15·0 mm**
S = 0·00300 to 0·01000

ie hydraulic gradient =
1 in 333 to 1 in 100

Water (or sewage) at 15°C;
full bore conditions.

velocities in ms^{-1}
discharges in m^3s^{-1}

Gradient	**(Equivalent) Pipe diameters in mm**													
	630	**675**	**700**	**750**	**800**	**825**	**900**	**975**	**1000**	**1050**	**1100**	**1125**	**1200**	**1250**
0·00300	0·844	0·885	0·908	0·952	0·996	1·017	1·080	1·141	1·160	1·200	1·238	1·257	1·313	1·350
1/ 333	0·2630	0·3168	0·3494	0·4208	0·5006	0·5437	0·6870	0·8516	0·9114	1·0388	1·1767	1·2497	1·4854	1·6568
0·00320	0·871	0·914	0·938	0·984	1·029	1·051	1·115	1·178	1·199	1·239	1·279	1·298	1·356	1·394
1/ 313	0·2716	0·3272	0·3609	0·4346	0·5170	0·5616	0·7095	0·8795	0·9413	1·0729	1·2153	1·2907	1·5341	1·7111
0·00340	0·898	0·942	0·967	1·014	1·060	1·083	1·150	1·214	1·235	1·277	1·318	1·338	1·398	1·437
1/ 294	0·2800	0·3373	0·3720	0·4480	0·5329	0·5789	0·7313	0·9066	0·9703	1·1059	1·2527	1·3304	1·5813	1·7638
0·00360	0·924	0·970	0·995	1·043	1·091	1·114	1·183	1·250	1·271	1·314	1·356	1·377	1·439	1·479
1/ 278	0·2881	0·3470	0·3828	0·4609	0·5484	0·5957	0·7526	0·9329	0·9985	1·1380	1·2891	1·3690	1·6272	1·8150
0·00380	0·950	0·996	1·022	1·072	1·121	1·145	1·215	1·284	1·306	1·350	1·394	1·415	1·478	1·520
1/ 263	0·2960	0·3566	0·3933	0·4736	0·5634	0·6120	0·7732	0·9585	1·0258	1·1692	1·3244	1·4066	1·6718	1·8647
0·00400	0·974	1·022	1·048	1·100	1·150	1·175	1·247	1·317	1·340	1·385	1·430	1·452	1·517	1·559
1/ 250	0·3037	0·3658	0·4035	0·4859	0·5780	0·6279	0·7933	0·9834	1·0525	1·1996	1·3588	1·4431	1·7153	1·9132
0·00420	0·998	1·048	1·074	1·127	1·178	1·204	1·278	1·350	1·373	1·420	1·465	1·488	1·554	1·598
1/ 238	0·3112	0·3749	0·4135	0·4979	0·5923	0·6434	0·8129	1·0077	1·0785	1·2292	1·3924	1·4788	1·7576	1·9605
0·00440	1·022	1·072	1·100	1·154	1·206	1·232	1·308	1·381	1·406	1·453	1·500	1·523	1·591	1·635
1/ 227	0·3185	0·3837	0·4232	0·5096	0·6063	0·6586	0·8320	1·0314	1·1039	1·2582	1·4252	1·5136	1·7990	2·0066
0·00460	1·045	1·096	1·124	1·179	1·233	1·260	1·337	1·413	1·437	1·486	1·533	1·557	1·626	1·672
1/ 217	0·3257	0·3923	0·4327	0·5211	0·6199	0·6734	0·8507	1·0546	1·1287	1·2864	1·4572	1·5476	1·8395	2·0517
0·00480	1·067	1·120	1·149	1·205	1·260	1·287	1·366	1·443	1·468	1·518	1·566	1·590	1·661	1·708
1/ 208	0·3327	0·4008	0·4420	0·5323	0·6332	0·6879	0·8690	1·0773	1·1530	1·3141	1·4886	1·5809	1·8790	2·0959
0·00500	1·089	1·143	1·172	1·230	1·286	1·313	1·394	1·473	1·498	1·549	1·599	1·623	1·696	1·743
1/ 200	0·3396	0·4090	0·4512	0·5433	0·6463	0·7021	0·8870	1·0995	1·1768	1·3412	1·5193	1·6135	1·9178	2·1391
0·00525	1·116	1·171	1·201	1·260	1·318	1·346	1·429	1·509	1·535	1·587	1·638	1·663	1·738	1·786
1/ 190	0·3480	0·4191	0·4623	0·5567	0·6623	0·7194	0·9089	1·1267	1·2059	1·3744	1·5568	1·6534	1·9652	2·1920
0·00550	1·143	1·199	1·230	1·290	1·349	1·377	1·462	1·545	1·571	1·625	1·677	1·702	1·779	1·828
1/ 182	0·3562	0·4290	0·4732	0·5698	0·6779	0·7363	0·9303	1·1532	1·2342	1·4067	1·5935	1·6923	2·0115	2·2436
0·00575	1·168	1·226	1·257	1·319	1·379	1·408	1·495	1·579	1·607	1·661	1·714	1·741	1·818	1·869
1/ 174	0·3642	0·4387	0·4838	0·5826	0·6931	0·7529	0·9512	1·1791	1·2620	1·4383	1·6293	1·7304	2·0567	2·2940
0·00600	1·193	1·252	1·284	1·347	1·409	1·439	1·527	1·613	1·641	1·697	1·751	1·778	1·858	1·910
1/ 167	0·3720	0·4481	0·4942	0·5952	0·7080	0·7691	0·9717	1·2045	1·2891	1·4693	1·6643	1·7676	2·1009	2·3434
0·00625	1·218	1·278	1·311	1·375	1·438	1·468	1·559	1·647	1·675	1·732	1·787	1·815	1·896	1·949
1/ 160	0·3797	0·4573	0·5044	0·6074	0·7226	0·7850	0·9917	1·2294	1·3157	1·4996	1·6987	1·8040	2·1443	2·3917
0·00650	1·242	1·303	1·337	1·402	1·466	1·498	1·590	1·679	1·708	1·766	1·823	1·851	1·934	1·988
1/ 154	0·3872	0·4664	0·5144	0·6195	0·7369	0·8005	1·0113	1·2537	1·3418	1·5293	1·7323	1·8398	2·1867	2·4391
0·00675	1·266	1·328	1·362	1·429	1·494	1·526	1·620	1·711	1·741	1·800	1·858	1·886	1·970	2·025
1/ 148	0·3946	0·4753	0·5242	0·6313	0·7510	0·8158	1·0306	1·2776	1·3674	1·5585	1·7653	1·8748	2·2284	2·4856
0·00700	1·289	1·353	1·387	1·455	1·521	1·554	1·650	1·743	1·773	1·833	1·892	1·921	2·007	2·063
1/ 143	0·4018	0·4840	0·5339	0·6429	0·7648	0·8307	1·0495	1·3011	1·3925	1·5871	1·7977	1·9093	2·2693	2·5312
0·00725	1·312	1·377	1·412	1·481	1·548	1·582	1·679	1·773	1·804	1·865	1·925	1·955	2·042	2·099
1/ 138	0·4090	0·4926	0·5433	0·6543	0·7783	0·8455	1·0681	1·3241	1·4171	1·6152	1·8296	1·9431	2·3095	2·5760
0·00750	1·334	1·400	1·436	1·506	1·575	1·609	1·708	1·804	1·835	1·897	1·958	1·988	2·077	2·135
1/ 133	0·4160	0·5010	0·5526	0·6654	0·7916	0·8599	1·0864	1·3467	1·4414	1·6428	1·8609	1·9763	2·3490	2·6200
0·00800	1·378	1·446	1·483	1·556	1·627	1·661	1·764	1·863	1·895	1·959	2·022	2·053	2·145	2·205
1/ 125	0·4296	0·5175	0·5707	0·6873	0·8176	0·8881	1·1220	1·3909	1·4887	1·6967	1·9219	2·0411	2·4260	2·7060
0·00850	1·421	1·491	1·529	1·604	1·677	1·713	1·818	1·920	1·954	2·020	2·085	2·117	2·211	2·273
1/ 118	0·4428	0·5334	0·5883	0·7084	0·8428	0·9155	1·1566	1·4337	1·5345	1·7489	1·9811	2·1040	2·5007	2·7893
0·00900	1·462	1·534	1·573	1·650	1·725	1·762	1·871	1·976	2·010	2·078	2·145	2·178	2·275	2·339
1/ 111	0·4557	0·5489	0·6054	0·7290	0·8672	0·9420	1·1901	1·4753	1·5790	1·7996	2·0385	2·1650	2·5733	2·8702
0·00950	1·502	1·576	1·616	1·695	1·773	1·811	1·922	2·030	2·066	2·135	2·204	2·238	2·338	2·403
1/ 105	0·4682	0·5639	0·6220	0·7490	0·8910	0·9678	1·2227	1·5158	1·6223	1·8490	2·0944	2·2243	2·6438	2·9489
0·01000	1·541	1·617	1·658	1·739	1·819	1·858	1·972	2·083	2·119	2·191	2·261	2·296	2·398	2·465
1/ 100	0·4803	0·5786	0·6381	0·7684	0·9141	0·9930	1·2545	1·5552	1·6644	1·8970	2·1488	2·2821	2·7125	3·0255
	–	–	–	0·74	0·74	0·74	0·74	0·74	0·74	0·74	0·74	0·74	0·74	0·74

$V_{r(0·5)medial}$ **for half-full circular pipes.**

A33
(p.5 of 6)

$k_s = 15{\cdot}0$ **mm**
S = 0·01000 to 0·03000

ie hydraulic gradient =
1 in 100 to 1 in 33·3

Water (or sewage) at 15°C;
full bore conditions.

velocities in ms^{-1}
discharges in m^3s^{-1}

Gradient **(Equivalent) Pipe diameters in mm**

Gradient	630	675	700	750	800	825	900	975	1000	1050	1100	1125	1200	1250
0·01000	1·541	1·617	1·658	1·739	1·819	1·858	1·972	2·083	2·119	2·191	2·261	2·296	2·398	2·465
1/ 100	0·4803	0·5786	0·6381	0·7684	0·9141	0·9930	1·2545	1·5552	1·6644	1·8970	2·1488	2·2821	2·7125	3·0255
0·01050	1·579	1·657	1·699	1·782	1·864	1·903	2·021	2·134	2·172	2·245	2·317	2·353	2·458	2·526
1/ 95	0·4922	0·5929	0·6539	0·7874	0·9367	1·0175	1·2855	1·5936	1·7055	1·9439	2·2019	2·3385	2·7795	3·1002
0·01100	1·616	1·696	1·739	1·824	1·907	1·948	2·068	2·185	2·223	2·298	2·372	2·408	2·515	2·586
1/ 91	0·5038	0·6068	0·6693	0·8059	0·9588	1·0415	1·3158	1·6311	1·7457	1·9896	2·2537	2·3935	2·8449	3·1732
0·01150	1·652	1·734	1·778	1·865	1·950	1·992	2·115	2·234	2·273	2·349	2·425	2·462	2·572	2·644
1/ 87	0·5151	0·6205	0·6843	0·8241	0·9803	1·0649	1·3453	1·6677	1·7849	2·0344	2·3044	2·4473	2·9089	3·2445
0·01200	1·688	1·771	1·816	1·905	1·992	2·035	2·160	2·282	2·322	2·400	2·477	2·515	2·627	2·701
1/ 83	0·5262	0·6338	0·6991	0·8418	1·0014	1·0878	1·3743	1·7036	1·8233	2·0781	2·3540	2·5000	2·9714	3·3143
0·01250	1·723	1·808	1·854	1·945	2·033	2·077	2·205	2·329	2·369	2·449	2·528	2·567	2·682	2·756
1/ 80	0·5370	0·6469	0·7135	0·8592	1·0221	1·1102	1·4026	1·7388	1·8609	2·1210	2·4025	2·5516	3·0327	3·3827
0·01300	1·757	1·843	1·891	1·983	2·074	2·118	2·248	2·375	2·416	2·498	2·578	2·618	2·735	2·811
1/ 77	0·5477	0·6597	0·7276	0·8762	1·0423	1·1322	1·4304	1·7732	1·8978	2·1630	2·4501	2·6021	3·0928	3·4497
0·01350	1·790	1·879	1·927	2·021	2·113	2·158	2·291	2·420	2·462	2·546	2·627	2·668	2·787	2·865
1/ 74	0·5581	0·6723	0·7415	0·8929	1·0622	1·1538	1·4577	1·8070	1·9340	2·2042	2·4968	2·6517	3·1517	3·5154
0·01400	1·823	1·913	1·962	2·058	2·152	2·198	2·333	2·465	2·508	2·592	2·676	2·717	2·838	2·917
1/ 71	0·5684	0·6846	0·7551	0·9093	1·0817	1·1750	1·4844	1·8402	1·9695	2·2447	2·5426	2·7004	3·2096	3·5799
0·01450	1·856	1·947	1·997	2·095	2·190	2·237	2·375	2·508	2·552	2·638	2·723	2·765	2·888	2·969
1/ 69	0·5784	0·6967	0·7685	0·9254	1·1008	1·1958	1·5107	1·8727	2·0043	2·2844	2·5876	2·7482	3·2664	3·6433
0·01500	1·887	1·980	2·031	2·130	2·228	2·275	2·415	2·551	2·596	2·683	2·769	2·812	2·938	3·020
1/ 67	0·5883	0·7086	0·7816	0·9412	1·1197	1·2162	1·5366	1·9048	2·0386	2·3235	2·6319	2·7952	3·3223	3·7056
0·01600	1·949	2·045	2·098	2·200	2·301	2·350	2·495	2·635	2·681	2·771	2·860	2·904	3·034	3·119
1/ 63	0·6076	0·7319	0·8072	0·9721	1·1564	1·2561	1·5870	1·9673	2·1055	2·3997	2·7182	2·8868	3·4312	3·8272
0·01700	2·009	2·108	2·162	2·268	2·371	2·422	2·571	2·716	2·763	2·857	2·948	2·994	3·127	3·215
1/ 59	0·6263	0·7544	0·8321	1·0020	1·1920	1·2948	1·6358	2·0278	2·1703	2·4736	2·8019	2·9757	3·5369	3·9450
0·01800	2·068	2·169	2·225	2·334	2·440	2·492	2·646	2·795	2·843	2·939	3·034	3·080	3·218	3·308
1/ 56	0·6445	0·7763	0·8562	1·0310	1·2266	1·3324	1·6832	2·0866	2·2332	2·5453	2·8831	3·0620	3·6394	4·0594
0·01900	2·124	2·229	2·286	2·398	2·507	2·561	2·718	2·871	2·921	3·020	3·117	3·165	3·306	3·399
1/ 53	0·6622	0·7976	0·8797	1·0593	1·2602	1·3689	1·7294	2·1438	2·2944	2·6150	2·9622	3·1459	3·7392	4·1706
0·02000	2·179	2·287	2·345	2·460	2·572	2·627	2·789	2·946	2·997	3·098	3·198	3·247	3·392	3·487
1/ 50	0·6794	0·8183	0·9026	1·0868	1·2929	1·4044	1·7743	2·1995	2·3540	2·6830	3·0391	3·2277	3·8363	4·2790
0·02100	2·233	2·343	2·403	2·521	2·636	2·692	2·858	3·019	3·071	3·175	3·277	3·327	3·476	3·573
1/ 47·6	0·6961	0·8385	0·9249	1·1137	1·3249	1·4391	1·8181	2·2538	2·4122	2·7493	3·1142	3·3074	3·9311	4·3847
0·02200	2·286	2·398	2·460	2·580	2·698	2·756	2·925	3·090	3·144	3·250	3·354	3·406	3·558	3·657
1/ 45·5	0·7125	0·8582	0·9466	1·1399	1·3560	1·4730	1·8609	2·3069	2·4690	2·8140	3·1875	3·3852	4·0236	4·4879
0·02300	2·337	2·452	2·515	2·638	2·758	2·817	2·991	3·159	3·214	3·323	3·429	3·482	3·638	3·739
1/ 43·5	0·7286	0·8775	0·9679	1·1655	1·3865	1·5061	1·9028	2·3587	2·5245	2·8772	3·2591	3·4613	4·1140	4·5888
0·02400	2·387	2·505	2·569	2·695	2·818	2·878	3·055	3·227	3·283	3·394	3·503	3·557	3·716	3·820
1/ 41·7	0·7442	0·8964	0·9887	1·1906	1·4164	1·5385	1·9437	2·4095	2·5788	2·9391	3·3293	3·5358	4·2025	4·6875
0·02500	2·437	2·557	2·622	2·751	2·876	2·937	3·118	3·294	3·351	3·464	3·576	3·630	3·792	3·898
1/ 40·0	0·7596	0·9149	1·0091	1·2152	1·4456	1·5703	1·9838	2·4592	2·6320	2·9997	3·3979	3·6087	4·2892	4·7841
0·02600	2·485	2·607	2·674	2·805	2·933	2·996	3·180	3·359	3·417	3·533	3·646	3·702	3·868	3·976
1/ 38·5	0·7746	0·9330	1·0291	1·2392	1·4742	1·6014	2·0231	2·5079	2·6841	3·0592	3·4652	3·6802	4·3742	4·8789
0·02700	2·532	2·657	2·725	2·858	2·989	3·053	3·241	3·423	3·483	3·600	3·716	3·773	3·941	4·051
1/ 37·0	0·7894	0·9508	1·0487	1·2628	1·5023	1·6319	2·0616	2·5557	2·7352	3·1174	3·5312	3·7503	4·4575	4·9719
0·02800	2·579	2·706	2·775	2·911	3·044	3·109	3·300	3·486	3·547	3·666	3·784	3·842	4·014	4·126
1/ 35·7	0·8039	0·9683	1·0680	1·2860	1·5299	1·6618	2·0995	2·6026	2·7854	3·1747	3·5961	3·8191	4·5393	5·0631
0·02900	2·624	2·754	2·824	2·962	3·097	3·164	3·359	3·548	3·609	3·731	3·851	3·910	4·085	4·199
1/ 34·5	0·8181	0·9854	1·0869	1·3088	1·5570	1·6912	2·1366	2·6487	2·8347	3·2309	3·6597	3·8867	4·6197	5·1527
0·03000	2·669	2·801	2·872	3·013	3·150	3·218	3·416	3·608	3·671	3·795	3·917	3·977	4·155	4·271
1/ 33·3	0·8321	1·0023	1·1055	1·3312	1·5836	1·7202	2·1732	2·6939	2·8832	3·2861	3·7223	3·9532	4·6987	5·2408
	–	–	–	0·74	0·74	0·74	0·74	0·74	0·74	0·74	0·74	0·74	0·74	0·74

$V_{r(0{\cdot}5)medial}$ **for half-full circular pipes.**

$k_s = 15{\cdot}0$ **mm** S = 0·01000 to 0·03000

Gradient **(Equivalent) Pipe diameters in mm**

Gradient	630	675	700	750	800	825	900	975	1000	1050	1100	1125	1200	1250
0·03000	2·669	2·801	2·872	3·013	3·150	3·218	3·416	3·608	3·671	3·795	3·917	3·977	4·155	4·271
1/ 33·3	0·8321	1·0023	1·1055	1·3312	1·5836	1·7202	2·1732	2·6939	2·8832	3·2861	3·7223	3·9532	4·6987	5·2408
0·03200	2·757	2·893	2·967	3·112	3·254	3·323	3·528	3·727	3·791	3·919	4·045	4·107	4·291	4·411
1/ 31·3	0·8594	1·0351	1·1417	1·3748	1·6355	1·7766	2·2445	2·7823	2·9778	3·3939	3·8444	4·0829	4·8528	5·4127
0·03400	2·842	2·982	3·058	3·208	3·354	3·426	3·637	3·841	3·908	4·040	4·170	4·234	4·423	4·546
1/ 29·4	0·8858	1·0670	1·1769	1·4171	1·6859	1·8313	2·3135	2·8679	3·0694	3·4984	3·9627	4·2085	5·0022	5·5794
0·03600	2·924	3·068	3·147	3·301	3·451	3·525	3·742	3·953	4·021	4·157	4·291	4·357	4·551	4·678
1/ 27·8	0·9115	1·0979	1·2110	1·4582	1·7347	1·8844	2·3806	2·9511	3·1585	3·5998	4·0776	4·3306	5·1472	5·7411
0·03800	3·004	3·152	3·233	3·391	3·546	3·622	3·845	4·061	4·132	4·271	4·408	4·476	4·676	4·807
1/ 26·3	0·9365	1·1280	1·2442	1·4982	1·7823	1·9360	2·4459	3·0320	3·2450	3·6985	4·1894	4·4493	5·2883	5·8985
0·04000	3·082	3·234	3·317	3·479	3·638	3·716	3·945	4·166	4·239	4·382	4·523	4·592	4·797	4·931
1/ 25·0	0·9608	1·1573	1·2765	1·5371	1·8286	1·9863	2·5094	3·1108	3·3293	3·7945	4·2982	4·5649	5·4257	6·0517
0·04200	3·158	3·314	3·399	3·565	3·728	3·808	4·042	4·269	4·344	4·490	4·635	4·706	4·916	5·053
1/ 23·8	0·9846	1·1859	1·3080	1·5751	1·8738	2·0354	2·5714	3·1876	3·4116	3·8883	4·4044	4·6776	5·5597	6·2012
0·04400	3·233	3·392	3·479	3·649	3·815	3·897	4·137	4·370	4·446	4·596	4·744	4·816	5·032	5·172
1/ 22·7	1·0078	1·2138	1·3388	1·6122	1·9179	2·0833	2·6319	3·2626	3·4918	3·9798	4·5080	4·7877	5·6905	6·3471
0·04600	3·306	3·468	3·557	3·731	3·901	3·985	4·230	4·468	4·546	4·699	4·850	4·925	5·145	5·288
1/ 21·7	1·0304	1·2411	1·3689	1·6484	1·9610	2·1301	2·6911	3·3360	3·5703	4·0692	4·6094	4·8953	5·8184	6·4898
0·04800	3·377	3·543	3·634	3·811	3·985	4·071	4·321	4·564	4·644	4·801	4·955	5·031	5·255	5·402
1/ 20·8	1·0526	1·2678	1·3984	1·6839	2·0032	2·1759	2·7490	3·4077	3·6471	4·1568	4·7085	5·0006	5·9436	6·6294
0·05000	3·446	3·616	3·709	3·890	4·067	4·154	4·410	4·658	4·739	4·900	5·057	5·134	5·364	5·514
1/ 20·0	1·0743	1·2940	1·4272	1·7186	2·0445	2·2208	2·8057	3·4780	3·7224	4·2425	4·8056	5·1037	6·0662	6·7661
0·05250	3·531	3·705	3·800	3·986	4·168	4·257	4·519	4·773	4·857	5·021	5·182	5·261	5·496	5·650
1/ 19·0	1·1008	1·3259	1·4625	1·7610	2·0950	2·2757	2·8750	3·5639	3·8143	4·3473	4·9243	5·2298	6·2160	6·9332
0·05500	3·614	3·793	3·890	4·080	4·266	4·357	4·626	4·886	4·971	5·139	5·304	5·385	5·625	5·783
1/ 18·2	1·1267	1·3571	1·4969	1·8025	2·1443	2·3292	2·9426	3·6478	3·9041	4·4496	5·0402	5·3529	6·3623	7·0964
0·05750	3·696	3·878	3·977	4·172	4·362	4·455	4·729	4·996	5·083	5·254	5·423	5·506	5·752	5·913
1/ 17·4	1·1521	1·3876	1·5305	1·8430	2·1925	2·3816	3·0088	3·7298	3·9918	4·5496	5·1535	5·4732	6·5053	7·2559
0·06000	3·775	3·961	4·063	4·261	4·456	4·551	4·831	5·103	5·192	5·367	5·539	5·625	5·876	6·040
1/ 16·7	1·1768	1·4175	1·5635	1·8827	2·2396	2·4328	3·0735	3·8100	4·0777	4·6475	5·2644	5·5909	6·6452	7·4120
0·06250	3·853	4·043	4·146	4·349	4·548	4·645	4·931	5·208	5·299	5·478	5·654	5·741	5·997	6·164
1/ 16·0	1·2011	1·4467	1·5957	1·9215	2·2858	2·4830	3·1369	3·8886	4·1618	4·7433	5·3729	5·7062	6·7822	7·5648
0·06500	3·929	4·123	4·228	4·435	4·638	4·737	5·029	5·311	5·404	5·586	5·766	5·854	6·116	6·286
1/ 15·4	1·2249	1·4754	1·6273	1·9595	2·3311	2·5322	3·1990	3·9656	4·2442	4·8373	5·4793	5·8192	6·9166	7·7147
0·06750	4·004	4·201	4·309	4·520	4·726	4·827	5·124	5·413	5·507	5·693	5·876	5·966	6·232	6·406
1/ 14·8	1·2482	1·5035	1·6583	1·9969	2·3755	2·5804	3·2600	4·0411	4·3251	4·9294	5·5837	5·9301	7·0483	7·8616
0·07000	4·078	4·279	4·388	4·603	4·813	4·916	5·218	5·512	5·608	5·797	5·983	6·075	6·346	6·524
1/ 14·3	1·2711	1·5311	1·6887	2·0335	2·4191	2·6278	3·3198	4·1153	4·4044	5·0199	5·6862	6·0389	7·1777	8·0059
0·07250	4·150	4·354	4·466	4·684	4·898	5·003	5·311	5·609	5·707	5·900	6·089	6·183	6·459	6·639
1/ 13·8	1·2936	1·5582	1·7186	2·0695	2·4619	2·6743	3·3786	4·1882	4·4824	5·1087	5·7869	6·1458	7·3048	8·1476
0·07500	4·221	4·429	4·542	4·765	4·982	5·088	5·402	5·705	5·805	6·001	6·193	6·289	6·569	6·753
1/ 13·3	1·3158	1·5848	1·7480	2·1049	2·5040	2·7200	3·4363	4·2598	4·5590	5·1961	5·8858	6·2509	7·4296	8·2869
0·08000	4·359	4·574	4·691	4·921	5·145	5·255	5·579	5·893	5·995	6·198	6·397	6·495	6·785	6·974
1/ 12·5	1·3589	1·6368	1·8054	2·1739	2·5862	2·8092	3·5490	4·3995	4·7086	5·3665	6·0788	6·4559	7·6733	8·5587
0·08500	4·494	4·715	4·836	5·072	5·303	5·417	5·750	6·074	6·180	6·388	6·593	6·695	6·994	7·189
1/ 11·8	1·4008	1·6872	1·8609	2·2409	2·6658	2·8957	3·6583	4·5349	4·8535	5·5317	6·2660	6·6546	7·9095	8·8222
0·09000	4·624	4·852	4·976	5·219	5·457	5·574	5·917	6·250	6·359	6·574	6·785	6·889	7·196	7·397
1/ 11·1	1·4414	1·7361	1·9149	2·3058	2·7431	2·9797	3·7643	4·6664	4·9942	5·6921	6·4476	6·8476	8·1388	9·0780
0·09500	4·751	4·985	5·112	5·362	5·607	5·727	6·079	6·421	6·533	6·754	6·971	7·078	7·394	7·600
1/ 10·5	1·4809	1·7837	1·9674	2·3690	2·8182	3·0613	3·8675	4·7943	5·1311	5·8481	6·6243	7·0352	8·3619	9·3267
0·10000	4·874	5·114	5·245	5·502	5·752	5·876	6·237	6·588	6·703	6·929	7·152	7·261	7·586	7·798
1/ 10·0	1·5193	1·8300	2·0185	2·4306	2·8915	3·1408	3·9680	4·9188	5·2644	6·0000	6·7964	7·2180	8·5791	9·5691
	–	–	–	0·74	0·74	0·74	0·74	0·74	0·74	0·74	0·74	0·74	0·74	0·74

$V_{r(0{\cdot}5)medial}$ **for half-full circular pipes.**

A34
(p.1 of 6)

k_s = 30·0 mm
S = 0·00010 to 0·00030

ie hydraulic gradient =
1 in 10000 to 1 in 3333

Water (or sewage) at 15°C;
full bore conditions.

velocities in ms^{-1}
discharges in m^3s^{-1}

Gradient	630	675	700	750	800	825	900	975	1000	1050	1100	1125	1200	1250
	(Equivalent) Pipe diameters in mm													
0·000100	0·133	0·140	0·143	0·151	0·158	0·161	0·172	0·182	0·185	0·192	0·198	0·201	0·210	0·216
1/ 10000	0·0414	0·0499	0·0551	0·0665	0·0793	0·0862	0·1092	0·1357	0·1453	0·1658	0·1881	0·1999	0·2379	0·2657
0·000105	0·136	0·143	0·147	0·154	0·162	0·165	0·176	0·186	0·190	0·196	0·203	0·206	0·216	0·222
1/ 9524	0·0424	0·0512	0·0565	0·0682	0·0813	0·0884	0·1119	0·1390	0·1489	0·1699	0·1927	0·2048	0·2438	0·2722
0·000110	0·139	0·146	0·150	0·158	0·165	0·169	0·180	0·191	0·194	0·201	0·208	0·211	0·221	0·227
1/ 9091	0·0434	0·0524	0·0578	0·0698	0·0832	0·0904	0·1145	0·1423	0·1524	0·1739	0·1973	0·2096	0·2496	0·2786
0·000115	0·142	0·150	0·154	0·162	0·169	0·173	0·184	0·195	0·198	0·205	0·212	0·216	0·226	0·232
1/ 8696	0·0444	0·0536	0·0591	0·0714	0·0851	0·0925	0·1171	0·1455	0·1558	0·1779	0·2017	0·2143	0·2552	0·2849
0·000120	0·145	0·153	0·157	0·165	0·173	0·177	0·188	0·199	0·203	0·210	0·217	0·220	0·230	0·237
1/ 8333	0·0453	0·0547	0·0604	0·0729	0·0869	0·0945	0·1196	0·1486	0·1592	0·1817	0·2061	0·2190	0·2607	0·2911
0·000125	0·148	0·156	0·160	0·168	0·176	0·180	0·192	0·203	0·207	0·214	0·221	0·225	0·235	0·242
1/ 8000	0·0463	0·0558	0·0617	0·0744	0·0887	0·0964	0·1221	0·1517	0·1625	0·1854	0·2103	0·2235	0·2661	0·2971
0·000130	0·151	0·159	0·163	0·172	0·180	0·184	0·196	0·207	0·211	0·218	0·226	0·229	0·240	0·247
1/ 7692	0·0472	0·0569	0·0629	0·0759	0·0904	0·0983	0·1245	0·1547	0·1657	0·1891	0·2145	0·2279	0·2713	0·3030
0·000135	0·154	0·162	0·167	0·175	0·183	0·187	0·199	0·211	0·215	0·223	0·230	0·234	0·244	0·252
1/ 7407	0·0481	0·0580	0·0641	0·0773	0·0922	0·1002	0·1269	0·1577	0·1689	0·1927	0·2186	0·2323	0·2765	0·3087
0·000140	0·157	0·165	0·170	0·178	0·187	0·191	0·203	0·215	0·219	0·227	0·234	0·238	0·249	0·256
1/ 7143	0·0490	0·0591	0·0653	0·0787	0·0939	0·1021	0·1292	0·1606	0·1720	0·1963	0·2226	0·2365	0·2816	0·3144
0·000145	0·160	0·168	0·173	0·181	0·190	0·194	0·207	0·219	0·223	0·231	0·238	0·242	0·253	0·261
1/ 6897	0·0498	0·0601	0·0664	0·0801	0·0955	0·1039	0·1315	0·1634	0·1750	0·1997	0·2265	0·2407	0·2866	0·3200
0·000150	0·163	0·171	0·176	0·185	0·193	0·198	0·210	0·223	0·227	0·235	0·242	0·246	0·258	0·265
1/ 6667	0·0507	0·0612	0·0676	0·0815	0·0972	0·1056	0·1338	0·1662	0·1780	0·2032	0·2304	0·2448	0·2915	0·3255
0·000160	0·168	0·177	0·181	0·191	0·200	0·204	0·217	0·230	0·234	0·242	0·250	0·254	0·266	0·274
1/ 6250	0·0523	0·0632	0·0698	0·0842	0·1004	0·1091	0·1382	0·1717	0·1839	0·2098	0·2380	0·2529	0·3011	0·3361
0·000170	0·173	0·182	0·187	0·196	0·206	0·210	0·224	0·237	0·241	0·250	0·258	0·262	0·274	0·282
1/ 5882	0·0540	0·0651	0·0719	0·0868	0·1034	0·1125	0·1424	0·1770	0·1895	0·2163	0·2453	0·2607	0·3103	0·3465
0·000180	0·178	0·187	0·192	0·202	0·212	0·217	0·230	0·244	0·248	0·257	0·266	0·270	0·282	0·291
1/ 5556	0·0555	0·0670	0·0740	0·0893	0·1064	0·1157	0·1466	0·1821	0·1950	0·2226	0·2524	0·2682	0·3193	0·3566
0·000190	0·183	0·192	0·198	0·208	0·218	0·222	0·237	0·251	0·255	0·264	0·273	0·277	0·290	0·299
1/ 5263	0·0570	0·0689	0·0760	0·0918	0·1094	0·1189	0·1506	0·1871	0·2004	0·2287	0·2594	0·2756	0·3281	0·3663
0·000200	0·188	0·197	0·203	0·213	0·223	0·228	0·243	0·257	0·262	0·271	0·280	0·284	0·298	0·306
1/ 5000	0·0585	0·0707	0·0780	0·0941	0·1122	0·1220	0·1545	0·1920	0·2056	0·2346	0·2661	0·2828	0·3366	0·3759
0·000210	0·192	0·202	0·208	0·218	0·229	0·234	0·249	0·263	0·268	0·278	0·287	0·292	0·305	0·314
1/ 4762	0·0600	0·0724	0·0799	0·0965	0·1150	0·1250	0·1583	0·1967	0·2107	0·2404	0·2727	0·2898	0·3450	0·3852
0·000220	0·197	0·207	0·213	0·224	0·234	0·239	0·255	0·270	0·275	0·284	0·294	0·298	0·312	0·321
1/ 4545	0·0614	0·0741	0·0818	0·0988	0·1177	0·1280	0·1621	0·2013	0·2156	0·2461	0·2791	0·2966	0·3531	0·3942
0·000230	0·201	0·212	0·217	0·229	0·239	0·245	0·260	0·276	0·281	0·291	0·300	0·305	0·319	0·328
1/ 4348	0·0628	0·0758	0·0837	0·1010	0·1203	0·1308	0·1657	0·2059	0·2205	0·2516	0·2854	0·3033	0·3610	0·4031
0·000240	0·206	0·216	0·222	0·233	0·245	0·250	0·266	0·282	0·287	0·297	0·307	0·312	0·326	0·336
1/ 4167	0·0641	0·0774	0·0855	0·1031	0·1229	0·1337	0·1693	0·2103	0·2252	0·2570	0·2915	0·3098	0·3688	0·4118
0·000250	0·210	0·221	0·227	0·238	0·250	0·255	0·272	0·288	0·293	0·303	0·313	0·318	0·333	0·342
1/ 4000	0·0655	0·0790	0·0872	0·1053	0·1255	0·1364	0·1728	0·2147	0·2299	0·2624	0·2975	0·3162	0·3764	0·4203
0·000260	0·214	0·225	0·231	0·243	0·255	0·260	0·277	0·293	0·299	0·309	0·319	0·324	0·339	0·349
1/ 3846	0·0667	0·0806	0·0890	0·1074	0·1280	0·1391	0·1762	0·2189	0·2344	0·2676	0·3034	0·3224	0·3839	0·4286
0·000270	0·218	0·229	0·236	0·248	0·259	0·265	0·282	0·299	0·304	0·315	0·325	0·331	0·346	0·356
1/ 3704	0·0680	0·0821	0·0907	0·1094	0·1304	0·1418	0·1796	0·2231	0·2389	0·2727	0·3092	0·3286	0·3912	0·4368
0·000280	0·222	0·234	0·240	0·252	0·264	0·270	0·287	0·304	0·310	0·321	0·331	0·337	0·352	0·362
1/ 3571	0·0693	0·0836	0·0923	0·1114	0·1328	0·1444	0·1829	0·2272	0·2433	0·2777	0·3149	0·3346	0·3984	0·4448
0·000290	0·226	0·238	0·244	0·257	0·269	0·275	0·293	0·310	0·315	0·326	0·337	0·343	0·358	0·369
1/ 3448	0·0705	0·0851	0·0940	0·1134	0·1352	0·1469	0·1861	0·2312	0·2476	0·2826	0·3205	0·3406	0·4054	0·4527
0·000300	0·230	0·242	0·248	0·261	0·273	0·280	0·298	0·315	0·321	0·332	0·343	0·348	0·365	0·375
1/ 3333	0·0717	0·0866	0·0956	0·1153	0·1375	0·1495	0·1893	0·2352	0·2519	0·2874	0·3260	0·3464	0·4124	0·4604
	–	–	–	–	–	–	–	–	–	–	–	–	–	–

$V_{r(0\cdot5)medial}$ for half-full circular pipes.

k_s = 30·0 mm S = 0·00010 to 0·00030

k_s = **30·0 mm**
S = 0·00030 to 0·00100

ie hydraulic gradient =
1 in 3333 to 1 in 1000

Water (or sewage) at 15°C;
full bore conditions.

velocities in ms^{-1}
discharges in m^3s^{-1}

Gradient	**(Equivalent) Pipe diameters in mm**													
	630	**675**	**700**	**750**	**800**	**825**	**900**	**975**	**1000**	**1050**	**1100**	**1125**	**1200**	**1250**
0·000300	0·230	0·242	0·248	0·261	0·273	0·280	0·298	0·315	0·321	0·332	0·343	0·348	0·365	0·375
1/ 3333	0·0717	0·0866	0·0956	0·1153	0·1375	0·1495	0·1893	0·2352	0·2519	0·2874	0·3260	0·3464	0·4124	0·4604
0·000320	0·238	0·250	0·257	0·270	0·282	0·289	0·307	0·325	0·331	0·343	0·354	0·360	0·377	0·387
1/ 3125	0·0741	0·0894	0·0987	0·1191	0·1420	0·1544	0·1955	0·2429	0·2601	0·2969	0·3367	0·3578	0·4259	0·4755
0·000340	0·245	0·258	0·264	0·278	0·291	0·298	0·317	0·335	0·341	0·353	0·365	0·371	0·388	0·399
1/ 2941	0·0763	0·0922	0·1018	0·1228	0·1464	0·1591	0·2015	0·2504	0·2681	0·3060	0·3470	0·3688	0·4390	0·4902
0·000360	0·252	0·265	0·272	0·286	0·300	0·306	0·326	0·345	0·351	0·364	0·376	0·382	0·399	0·411
1/ 2778	0·0786	0·0948	0·1047	0·1264	0·1506	0·1637	0·2074	0·2576	0·2759	0·3149	0·3571	0·3795	0·4518	0·5044
0·000380	0·259	0·272	0·280	0·294	0·308	0·315	0·335	0·355	0·361	0·374	0·386	0·392	0·410	0·422
1/ 2632	0·0807	0·0974	0·1076	0·1298	0·1547	0·1682	0·2131	0·2647	0·2835	0·3235	0·3669	0·3899	0·4642	0·5182
0·000400	0·266	0·279	0·287	0·302	0·316	0·323	0·344	0·364	0·370	0·383	0·396	0·402	0·421	0·433
1/ 2500	0·0828	0·1000	0·1104	0·1332	0·1588	0·1726	0·2186	0·2716	0·2909	0·3319	0·3764	0·4000	0·4762	0·5317
0·000420	0·272	0·286	0·294	0·309	0·324	0·331	0·352	0·373	0·379	0·393	0·406	0·412	0·431	0·444
1/ 2381	0·0849	0·1024	0·1131	0·1365	0·1627	0·1769	0·2240	0·2783	0·2980	0·3401	0·3857	0·4099	0·4880	0·5448
0·000440	0·279	0·293	0·301	0·316	0·331	0·339	0·360	0·382	0·388	0·402	0·415	0·422	0·442	0·454
1/ 2273	0·0869	0·1049	0·1158	0·1397	0·1665	0·1810	0·2293	0·2848	0·3051	0·3481	0·3948	0·4196	0·4995	0·5577
0·000460	0·285	0·300	0·308	0·323	0·339	0·346	0·369	0·390	0·397	0·411	0·425	0·432	0·452	0·465
1/ 2174	0·0888	0·1072	0·1184	0·1429	0·1703	0·1851	0·2344	0·2913	0·3119	0·3560	0·4037	0·4290	0·5107	0·5702
0·000480	0·291	0·306	0·314	0·330	0·346	0·354	0·376	0·398	0·406	0·420	0·434	0·441	0·461	0·475
1/ 2083	0·0907	0·1095	0·1209	0·1459	0·1739	0·1891	0·2395	0·2975	0·3186	0·3636	0·4124	0·4382	0·5217	0·5825
0·000500	0·297	0·312	0·321	0·337	0·353	0·361	0·384	0·407	0·414	0·429	0·443	0·450	0·471	0·484
1/ 2000	0·0926	0·1118	0·1234	0·1489	0·1775	0·1930	0·2444	0·3037	0·3252	0·3711	0·4209	0·4473	0·5325	0·5945
0·000525	0·304	0·320	0·329	0·345	0·362	0·370	0·394	0·417	0·424	0·439	0·454	0·461	0·482	0·496
1/ 1905	0·0949	0·1145	0·1265	0·1526	0·1819	0·1978	0·2505	0·3112	0·3333	0·3803	0·4313	0·4583	0·5456	0·6092
0·000550	0·312	0·328	0·336	0·354	0·370	0·379	0·403	0·427	0·434	0·450	0·465	0·472	0·494	0·508
1/ 1818	0·0971	0·1172	0·1295	0·1562	0·1862	0·2024	0·2564	0·3185	0·3411	0·3893	0·4415	0·4691	0·5585	0·6235
0·000575	0·319	0·335	0·344	0·362	0·379	0·387	0·412	0·436	0·444	0·460	0·475	0·483	0·505	0·520
1/ 1739	0·0993	0·1199	0·1324	0·1597	0·1904	0·2070	0·2621	0·3257	0·3488	0·3980	0·4514	0·4797	0·5710	0·6376
0·000600	0·325	0·342	0·351	0·369	0·387	0·396	0·421	0·446	0·454	0·470	0·485	0·493	0·516	0·531
1/ 1667	0·1014	0·1225	0·1352	0·1632	0·1945	0·2114	0·2678	0·3327	0·3563	0·4066	0·4611	0·4900	0·5833	0·6513
0·000625	0·332	0·349	0·359	0·377	0·395	0·404	0·430	0·455	0·463	0·479	0·495	0·503	0·526	0·542
1/ 1600	0·1035	0·1250	0·1380	0·1665	0·1985	0·2158	0·2733	0·3395	0·3636	0·4150	0·4706	0·5001	0·5954	0·6647
0·000650	0·339	0·356	0·366	0·384	0·403	0·412	0·438	0·464	0·472	0·489	0·505	0·513	0·537	0·552
1/ 1538	0·1056	0·1275	0·1407	0·1698	0·2024	0·2201	0·2787	0·3463	0·3708	0·4232	0·4799	0·5100	0·6072	0·6779
0·000675	0·345	0·363	0·373	0·392	0·410	0·420	0·446	0·473	0·481	0·498	0·515	0·523	0·547	0·563
1/ 1481	0·1076	0·1299	0·1434	0·1731	0·2063	0·2243	0·2840	0·3529	0·3779	0·4313	0·4891	0·5197	0·6187	0·6908
0·000700	0·352	0·370	0·380	0·399	0·418	0·427	0·455	0·481	0·490	0·507	0·524	0·532	0·557	0·573
1/ 1429	0·1096	0·1323	0·1461	0·1763	0·2101	0·2284	0·2892	0·3593	0·3848	0·4392	0·4981	0·5293	0·6301	0·7035
0·000725	0·358	0·376	0·386	0·406	0·425	0·435	0·463	0·490	0·499	0·516	0·533	0·542	0·567	0·583
1/ 1379	0·1115	0·1346	0·1487	0·1794	0·2138	0·2324	0·2944	0·3657	0·3917	0·4470	0·5069	0·5386	0·6413	0·7160
0·000750	0·364	0·383	0·393	0·413	0·433	0·442	0·471	0·498	0·507	0·525	0·543	0·551	0·577	0·593
1/ 1333	0·1134	0·1369	0·1512	0·1824	0·2175	0·2364	0·2994	0·3720	0·3984	0·4546	0·5156	0·5479	0·6522	0·7282
0·000800	0·376	0·395	0·406	0·427	0·447	0·457	0·486	0·515	0·524	0·542	0·560	0·569	0·596	0·613
1/ 1250	0·1172	0·1414	0·1562	0·1884	0·2246	0·2442	0·3092	0·3842	0·4114	0·4695	0·5325	0·5658	0·6736	0·7521
0·000850	0·387	0·407	0·418	0·440	0·461	0·471	0·501	0·530	0·540	0·559	0·578	0·587	0·614	0·632
1/ 1176	0·1208	0·1458	0·1610	0·1942	0·2315	0·2517	0·3188	0·3960	0·4241	0·4840	0·5489	0·5833	0·6944	0·7753
0·000900	0·399	0·419	0·430	0·452	0·474	0·485	0·516	0·546	0·556	0·575	0·594	0·604	0·632	0·650
1/ 1111	0·1243	0·1500	0·1656	0·1999	0·2382	0·2590	0·3280	0·4075	0·4364	0·4980	0·5648	0·6002	0·7145	0·7977
0·000950	0·410	0·431	0·442	0·465	0·487	0·498	0·530	0·561	0·571	0·591	0·611	0·620	0·649	0·668
1/ 1053	0·1277	0·1541	0·1702	0·2054	0·2448	0·2661	0·3370	0·4187	0·4484	0·5117	0·5803	0·6166	0·7341	0·8196
0·001000	0·420	0·442	0·454	0·477	0·500	0·511	0·543	0·575	0·586	0·606	0·626	0·636	0·666	0·685
1/ 1000	0·1310	0·1581	0·1746	0·2107	0·2511	0·2730	0·3458	0·4295	0·4600	0·5250	0·5954	0·6327	0·7532	0·8409
	–	–	–	–	–	–	–	–	–	–	–	–	–	–

$V_{r(0{\cdot}5)medial}$ for half-full circular pipes.

A34
(p.3 of 6)

k_s = 30·0 mm
S = 0·00100 to 0·00300

ie hydraulic gradient =
1 in 1000 to 1 in 333

Water (or sewage) at 15°C;
full bore conditions.

velocities in ms^{-1}
discharges in m^3s^{-1}

Gradient **(Equivalent) Pipe diameters in mm**

Gradient	630	675	700	750	800	825	900	975	1000	1050	1100	1125	1200	1250
0·00100	0·420	0·442	0·454	0·477	0·500	0·511	0·543	0·575	0·586	0·606	0·626	0·636	0·666	0·685
1/ 1000	0·1310	0·1581	0·1746	0·2107	0·2511	0·2730	0·3458	0·4295	0·4600	0·5250	0·5954	0·6327	0·7532	0·8409
0·00105	0·431	0·453	0·465	0·489	0·512	0·523	0·557	0·590	0·600	0·621	0·642	0·652	0·682	0·702
1/ 952	0·1342	0·1620	0·1789	0·2159	0·2573	0·2798	0·3543	0·4402	0·4714	0·5379	0·6101	0·6483	0·7718	0·8617
0·00110	0·441	0·463	0·476	0·500	0·524	0·536	0·570	0·603	0·614	0·636	0·657	0·668	0·698	0·719
1/ 909	0·1374	0·1659	0·1831	0·2210	0·2634	0·2863	0·3626	0·4505	0·4825	0·5506	0·6244	0·6636	0·7900	0·8820
0·00115	0·451	0·474	0·487	0·511	0·536	0·548	0·583	0·617	0·628	0·650	0·672	0·683	0·714	0·735
1/ 870	0·1405	0·1696	0·1873	0·2259	0·2693	0·2928	0·3708	0·4606	0·4933	0·5630	0·6385	0·6785	0·8077	0·9018
0·00120	0·460	0·484	0·497	0·522	0·547	0·560	0·595	0·630	0·642	0·664	0·686	0·697	0·730	0·751
1/ 833	0·1435	0·1732	0·1913	0·2308	0·2751	0·2991	0·3788	0·4706	0·5040	0·5751	0·6522	0·6931	0·8251	0·9212
0·00125	0·470	0·494	0·507	0·533	0·559	0·571	0·608	0·643	0·655	0·678	0·700	0·712	0·745	0·766
1/ 800	0·1465	0·1768	0·1952	0·2356	0·2808	0·3053	0·3866	0·4803	0·5144	0·5870	0·6657	0·7074	0·8421	0·9402
0·00130	0·479	0·504	0·517	0·544	0·570	0·582	0·620	0·656	0·668	0·691	0·714	0·726	0·759	0·781
1/ 769	0·1494	0·1803	0·1991	0·2402	0·2863	0·3113	0·3942	0·4898	0·5245	0·5986	0·6789	0·7214	0·8588	0·9588
0·00135	0·488	0·513	0·527	0·554	0·581	0·593	0·632	0·668	0·681	0·704	0·728	0·740	0·774	0·796
1/ 741	0·1522	0·1837	0·2029	0·2448	0·2918	0·3172	0·4018	0·4991	0·5345	0·6100	0·6918	0·7351	0·8752	0·9771
0·00140	0·497	0·523	0·537	0·564	0·591	0·604	0·643	0·681	0·693	0·717	0·741	0·753	0·788	0·811
1/ 714	0·1550	0·1871	0·2066	0·2493	0·2972	0·3231	0·4091	0·5083	0·5444	0·6212	0·7045	0·7486	0·8912	0·9951
0·00145	0·506	0·532	0·546	0·574	0·602	0·615	0·655	0·693	0·705	0·730	0·754	0·766	0·802	0·825
1/ 690	0·1578	0·1904	0·2103	0·2537	0·3024	0·3288	0·4164	0·5173	0·5540	0·6322	0·7170	0·7619	0·9070	1·0127
0·00150	0·515	0·541	0·556	0·584	0·612	0·626	0·666	0·705	0·717	0·743	0·767	0·780	0·816	0·839
1/ 667	0·1605	0·1937	0·2139	0·2581	0·3076	0·3344	0·4235	0·5261	0·5635	0·6430	0·7292	0·7749	0·9225	1·0300
0·00160	0·532	0·559	0·574	0·603	0·632	0·646	0·688	0·728	0·741	0·767	0·793	0·805	0·842	0·867
1/ 625	0·1657	0·2000	0·2209	0·2665	0·3177	0·3454	0·4374	0·5434	0·5820	0·6641	0·7532	0·8003	0·9528	1·0638
0·00170	0·548	0·576	0·592	0·622	0·651	0·666	0·709	0·750	0·764	0·791	0·817	0·830	0·868	0·894
1/ 588	0·1708	0·2062	0·2277	0·2747	0·3275	0·3560	0·4509	0·5601	0·5999	0·6846	0·7764	0·8250	0·9821	1·0965
0·00180	0·564	0·593	0·609	0·640	0·670	0·685	0·729	0·772	0·786	0·814	0·841	0·854	0·894	0·919
1/ 556	0·1758	0·2122	0·2343	0·2827	0·3370	0·3663	0·4639	0·5764	0·6173	0·7044	0·7989	0·8489	1·0106	1·1283
0·00190	0·579	0·609	0·625	0·657	0·689	0·704	0·749	0·793	0·807	0·836	0·864	0·877	0·918	0·945
1/ 526	0·1806	0·2180	0·2407	0·2905	0·3462	0·3764	0·4767	0·5922	0·6342	0·7237	0·8208	0·8722	1·0383	1·1593
0·00200	0·594	0·625	0·642	0·675	0·707	0·722	0·769	0·814	0·828	0·858	0·886	0·900	0·942	0·969
1/ 500	0·1853	0·2237	0·2470	0·2980	0·3552	0·3862	0·4891	0·6076	0·6507	0·7425	0·8421	0·8948	1·0653	1·1894
0·00210	0·609	0·640	0·658	0·691	0·724	0·740	0·788	0·834	0·849	0·879	0·908	0·922	0·965	0·993
1/ 476	0·1899	0·2292	0·2531	0·3054	0·3640	0·3957	0·5011	0·6226	0·6667	0·7609	0·8629	0·9170	1·0916	1·2188
0·00220	0·623	0·656	0·673	0·708	0·741	0·758	0·806	0·853	0·869	0·899	0·929	0·944	0·988	1·017
1/ 455	0·1943	0·2346	0·2590	0·3126	0·3725	0·4050	0·5129	0·6372	0·6824	0·7788	0·8832	0·9385	1·1173	1·2475
0·00230	0·637	0·670	0·688	0·723	0·758	0·775	0·824	0·873	0·888	0·920	0·950	0·965	1·010	1·039
1/ 435	0·1987	0·2399	0·2649	0·3196	0·3809	0·4141	0·5245	0·6515	0·6978	0·7963	0·9031	0·9596	1·1424	1·2755
0·00240	0·651	0·685	0·703	0·739	0·774	0·791	0·842	0·891	0·908	0·939	0·971	0·986	1·032	1·062
1/ 417	0·2030	0·2450	0·2706	0·3265	0·3891	0·4230	0·5357	0·6656	0·7128	0·8134	0·9225	0·9803	1·1670	1·3030
0·00250	0·665	0·699	0·718	0·754	0·790	0·808	0·860	0·910	0·926	0·959	0·991	1·007	1·053	1·084
1/ 400	0·2072	0·2501	0·2761	0·3332	0·3971	0·4318	0·5468	0·6793	0·7275	0·8302	0·9415	1·0005	1·1911	1·3298
0·00260	0·678	0·713	0·732	0·769	0·806	0·824	0·877	0·928	0·945	0·978	1·010	1·026	1·074	1·105
1/ 385	0·2113	0·2550	0·2816	0·3398	0·4050	0·4403	0·5576	0·6927	0·7419	0·8467	0·9602	1·0203	1·2147	1·3562
0·00270	0·691	0·726	0·746	0·784	0·821	0·839	0·893	0·946	0·963	0·996	1·030	1·046	1·094	1·126
1/ 370	0·2153	0·2599	0·2870	0·3463	0·4127	0·4487	0·5683	0·7059	0·7561	0·8628	0·9785	1·0398	1·2378	1·3820
0·00280	0·703	0·740	0·759	0·798	0·836	0·855	0·910	0·963	0·980	1·015	1·049	1·065	1·115	1·147
1/ 357	0·2193	0·2647	0·2922	0·3526	0·4203	0·4570	0·5787	0·7189	0·7699	0·8786	0·9964	1·0589	1·2605	1·4074
0·00290	0·716	0·753	0·773	0·812	0·851	0·870	0·926	0·980	0·998	1·033	1·067	1·084	1·134	1·167
1/ 345	0·2231	0·2694	0·2974	0·3589	0·4277	0·4650	0·5889	0·7316	0·7836	0·8942	1·0141	1·0776	1·2829	1·4323
0·00300	0·728	0·766	0·786	0·826	0·866	0·885	0·942	0·997	1·015	1·050	1·085	1·103	1·154	1·187
1/ 333	0·2270	0·2740	0·3025	0·3650	0·4351	0·4730	0·5990	0·7441	0·7970	0·9095	1·0314	1·0960	1·3048	1·4568
	–	–	–	–	–	–	–	–	–	–	–	–	–	–

$V_{r(0·5)medial}$ for half-full circular pipes.

k_s = 30·0 mm S = 0·00100 to 0·00300

Gradient	**(Equivalent) Pipe diameters in mm**													
	630	**675**	**700**	**750**	**800**	**825**	**900**	**975**	**1000**	**1050**	**1100**	**1125**	**1200**	**1250**
0·00300	0·728	0·766	0·786	0·826	0·866	0·885	0·942	0·997	1·015	1·050	1·085	1·103	1·154	1·187
1/ 333	0·2270	0·2740	0·3025	0·3650	0·4351	0·4730	0·5990	0·7441	0·7970	0·9095	1·0314	1·0960	1·3048	1·4568
0·00320	0·752	0·791	0·812	0·853	0·894	0·914	0·972	1·029	1·048	1·085	1·121	1·139	1·192	1·226
1/ 313	0·2344	0·2830	0·3124	0·3770	0·4493	0·4885	0·6187	0·7686	0·8231	0·9393	1·0653	1·1320	1·3476	1·5046
0·00340	0·775	0·815	0·837	0·880	0·921	0·942	1·002	1·061	1·080	1·118	1·155	1·174	1·228	1·264
1/ 294	0·2416	0·2917	0·3221	0·3886	0·4632	0·5036	0·6377	0·7922	0·8484	0·9682	1·0981	1·1668	1·3891	1·5509
0·00360	0·798	0·839	0·861	0·905	0·948	0·969	1·031	1·092	1·112	1·151	1·189	1·208	1·264	1·300
1/ 278	0·2486	0·3001	0·3314	0·3999	0·4766	0·5182	0·6562	0·8152	0·8731	0·9963	1·1299	1·2007	1·4294	1·5959
0·00380	0·819	0·862	0·885	0·930	0·974	0·996	1·060	1·122	1·142	1·182	1·222	1·241	1·298	1·336
1/ 263	0·2554	0·3084	0·3405	0·4108	0·4897	0·5324	0·6742	0·8375	0·8970	1·0236	1·1609	1·2336	1·4685	1·6396
0·00400	0·841	0·884	0·908	0·954	0·999	1·022	1·087	1·151	1·172	1·213	1·253	1·273	1·332	1·371
1/ 250	0·2621	0·3164	0·3493	0·4215	0·5024	0·5462	0·6917	0·8593	0·9203	1·0502	1·1910	1·2656	1·5067	1·6822
0·00420	0·862	0·906	0·930	0·978	1·024	1·047	1·114	1·179	1·201	1·243	1·284	1·305	1·365	1·405
1/ 238	0·2686	0·3242	0·3580	0·4319	0·5148	0·5597	0·7088	0·8805	0·9430	1·0762	1·2204	1·2969	1·5439	1·7238
0·00440	0·882	0·927	0·952	1·001	1·048	1·072	1·140	1·207	1·229	1·272	1·314	1·335	1·397	1·438
1/ 227	0·2749	0·3318	0·3664	0·4421	0·5269	0·5729	0·7255	0·9013	0·9652	1·1015	1·2492	1·3274	1·5802	1·7643
0·00460	0·902	0·948	0·973	1·023	1·072	1·096	1·166	1·234	1·257	1·301	1·344	1·365	1·429	1·470
1/ 217	0·2811	0·3393	0·3746	0·4520	0·5388	0·5857	0·7418	0·9215	0·9869	1·1263	1·2773	1·3573	1·6158	1·8040
0·00480	0·921	0·968	0·994	1·045	1·095	1·119	1·191	1·261	1·284	1·329	1·373	1·395	1·459	1·502
1/ 208	0·2871	0·3466	0·3827	0·4618	0·5503	0·5983	0·7577	0·9413	1·0082	1·1505	1·3047	1·3865	1·6505	1·8428
0·00500	0·940	0·988	1·015	1·067	1·117	1·142	1·216	1·287	1·310	1·356	1·401	1·424	1·489	1·533
1/ 200	0·2930	0·3537	0·3906	0·4713	0·5617	0·6107	0·7734	0·9608	1·0289	1·1742	1·3316	1·4151	1·6846	1·8808
0·00525	0·963	1·013	1·040	1·093	1·145	1·171	1·246	1·319	1·342	1·390	1·436	1·459	1·526	1·570
1/ 190	0·3003	0·3625	0·4002	0·4829	0·5756	0·6258	0·7925	0·9845	1·0544	1·2032	1·3645	1·4500	1·7262	1·9273
0·00550	0·986	1·037	1·064	1·119	1·172	1·198	1·275	1·350	1·374	1·422	1·470	1·493	1·562	1·607
1/ 182	0·3073	0·3710	0·4096	0·4943	0·5891	0·6405	0·8111	1·0077	1·0792	1·2315	1·3967	1·4841	1·7668	1·9726
0·00575	1·008	1·060	1·088	1·144	1·198	1·225	1·304	1·380	1·405	1·454	1·503	1·527	1·597	1·644
1/ 174	0·3143	0·3793	0·4189	0·5054	0·6024	0·6549	0·8294	1·0303	1·1034	1·2592	1·4281	1·5175	1·8065	2·0170
0·00600	1·030	1·083	1·112	1·169	1·224	1·251	1·332	1·410	1·435	1·486	1·535	1·559	1·632	1·679
1/ 167	0·3210	0·3875	0·4279	0·5163	0·6153	0·6690	0·8472	1·0525	1·1272	1·2863	1·4588	1·5501	1·8454	2·0604
0·00625	1·051	1·105	1·135	1·193	1·249	1·277	1·359	1·439	1·465	1·516	1·567	1·592	1·665	1·714
1/ 160	0·3276	0·3955	0·4367	0·5269	0·6280	0·6828	0·8647	1·0742	1·1504	1·3128	1·4889	1·5821	1·8835	2·1029
0·00650	1·072	1·127	1·157	1·216	1·274	1·303	1·386	1·467	1·494	1·546	1·598	1·623	1·698	1·748
1/ 154	0·3341	0·4033	0·4453	0·5374	0·6405	0·6963	0·8818	1·0955	1·1732	1·3388	1·5184	1·6134	1·9208	2·1445
0·00675	1·092	1·149	1·179	1·240	1·298	1·327	1·413	1·495	1·522	1·576	1·628	1·654	1·731	1·781
1/ 148	0·3405	0·4110	0·4538	0·5476	0·6527	0·7096	0·8986	1·1163	1·1956	1·3644	1·5473	1·6442	1·9574	2·1854
0·00700	1·112	1·170	1·201	1·262	1·322	1·352	1·438	1·523	1·550	1·605	1·658	1·684	1·762	1·813
1/ 143	0·3467	0·4185	0·4622	0·5577	0·6646	0·7226	0·9151	1·1368	1·2175	1·3894	1·5757	1·6744	1·9933	2·2255
0·00725	1·132	1·190	1·222	1·285	1·346	1·376	1·464	1·550	1·578	1·633	1·687	1·714	1·794	1·846
1/ 138	0·3529	0·4260	0·4703	0·5675	0·6764	0·7354	0·9313	1·1570	1·2391	1·4140	1·6036	1·7040	2·0286	2·2649
0·00750	1·151	1·211	1·243	1·307	1·369	1·399	1·489	1·576	1·605	1·661	1·716	1·744	1·824	1·877
1/ 133	0·3589	0·4332	0·4784	0·5772	0·6880	0·7480	0·9472	1·1767	1·2603	1·4382	1·6310	1·7331	2·0633	2·3036
0·00800	1·189	1·250	1·284	1·349	1·414	1·445	1·538	1·628	1·657	1·715	1·773	1·801	1·884	1·939
1/ 125	0·3707	0·4475	0·4941	0·5962	0·7105	0·7725	0·9783	1·2153	1·3016	1·4853	1·6845	1·7900	2·1309	2·3792
0·00850	1·226	1·289	1·323	1·391	1·457	1·490	1·585	1·678	1·708	1·768	1·827	1·856	1·942	1·998
1/ 118	0·3821	0·4612	0·5093	0·6145	0·7324	0·7963	1·0084	1·2527	1·3417	1·5311	1·7363	1·8451	2·1965	2·4524
0·00900	1·261	1·326	1·362	1·431	1·499	1·533	1·631	1·727	1·758	1·819	1·880	1·910	1·998	2·056
1/ 111	0·3932	0·4746	0·5240	0·6323	0·7537	0·8194	1·0377	1·2891	1·3806	1·5755	1·7867	1·8986	2·2602	2·5235
0·00950	1·296	1·363	1·399	1·471	1·540	1·575	1·676	1·774	1·806	1·869	1·932	1·962	2·053	2·113
1/ 105	0·4040	0·4876	0·5384	0·6497	0·7743	0·8418	1·0661	1·3244	1·4184	1·6186	1·8357	1·9506	2·3222	2·5927
0·01000	1·330	1·398	1·435	1·509	1·580	1·616	1·719	1·820	1·853	1·918	1·982	2·013	2·107	2·168
1/ 100	0·4145	0·5003	0·5524	0·6666	0·7944	0·8637	1·0938	1·3588	1·4552	1·6607	1·8834	2·0013	2·3825	2·6600
	–	–	–	–	–	–	–	–	–	–	–	–	–	–

$V_{r(0·5)medial}$ for half-full circular pipes.

A34
(p.5 of 6)

k_s = 30·0 mm
S = 0·01000 to 0·03000

ie hydraulic gradient =
1 in 100 to 1 in 33·3

Water (or sewage) at 15°C;
full bore conditions.

velocities in ms^{-1}
discharges in m^3s^{-1}

Gradient	**(Equivalent) Pipe diameters in mm**													
	630	**675**	**700**	**750**	**800**	**825**	**900**	**975**	**1000**	**1050**	**1100**	**1125**	**1200**	**1250**
0·01000	1·330	1·398	1·435	1·509	1·580	1·616	1·719	1·820	1·853	1·918	1·982	2·013	2·107	2·168
1/ 100	0·4145	0·5003	0·5524	0·6666	0·7944	0·8637	1·0938	1·3588	1·4552	1·6607	1·8834	2·0013	2·3825	2·6600
0·01050	1·362	1·433	1·471	1·546	1·619	1·656	1·762	1·865	1·899	1·965	2·031	2·063	2·159	2·221
1/ 95	0·4247	0·5126	0·5660	0·6830	0·8140	0·8850	1·1208	1·3924	1·4912	1·7017	1·9299	2·0507	2·4413	2·7257
0·01100	1·394	1·466	1·505	1·582	1·658	1·695	1·803	1·909	1·943	2·012	2·079	2·112	2·209	2·273
1/ 91	0·4347	0·5247	0·5794	0·6991	0·8332	0·9059	1·1472	1·4251	1·5263	1·7418	1·9753	2·0990	2·4988	2·7899
0·01150	1·426	1·499	1·539	1·618	1·695	1·733	1·844	1·952	1·987	2·057	2·125	2·159	2·259	2·325
1/ 87	0·4445	0·5365	0·5924	0·7148	0·8519	0·9262	1·1730	1·4572	1·5606	1·7809	2·0197	2·1462	2·5550	2·8526
0·01200	1·456	1·531	1·572	1·653	1·731	1·770	1·883	1·994	2·030	2·101	2·171	2·206	2·308	2·375
1/ 83	0·4540	0·5480	0·6051	0·7302	0·8703	0·9462	1·1982	1·4885	1·5942	1·8192	2·0631	2·1924	2·6099	2·9140
0·01250	1·487	1·563	1·605	1·687	1·767	1·806	1·922	2·035	2·072	2·144	2·216	2·251	2·355	2·423
1/ 80	0·4634	0·5593	0·6176	0·7452	0·8882	0·9657	1·2229	1·5192	1·6270	1·8567	2·1057	2·2376	2·6637	2·9740
0·01300	1·516	1·594	1·637	1·720	1·802	1·842	1·960	2·075	2·113	2·187	2·260	2·296	2·402	2·471
1/ 77	0·4726	0·5704	0·6299	0·7600	0·9058	0·9848	1·2471	1·5493	1·6593	1·8935	2·1474	2·2819	2·7165	3·0330
0·01350	1·545	1·624	1·668	1·753	1·836	1·877	1·998	2·115	2·153	2·228	2·303	2·339	2·448	2·519
1/ 74	0·4816	0·5813	0·6419	0·7745	0·9231	1·0036	1·2709	1·5788	1·6909	1·9296	2·1883	2·3254	2·7683	3·0907
0·01400	1·573	1·654	1·698	1·785	1·870	1·912	2·034	2·153	2·192	2·269	2·345	2·382	2·493	2·565
1/ 71	0·4904	0·5920	0·6536	0·7887	0·9400	1·0220	1·2942	1·6078	1·7219	1·9650	2·2285	2·3680	2·8191	3·1475
0·01450	1·601	1·684	1·728	1·817	1·903	1·946	2·070	2·192	2·231	2·309	2·386	2·424	2·537	2·610
1/ 69	0·4991	0·6024	0·6652	0·8027	0·9566	1·0401	1·3171	1·6363	1·7524	1·9998	2·2679	2·4100	2·8690	3·2032
0·01500	1·628	1·712	1·758	1·848	1·936	1·979	2·106	2·229	2·269	2·349	2·427	2·466	2·580	2·655
1/ 67	0·5076	0·6127	0·6766	0·8164	0·9730	1·0579	1·3397	1·6642	1·7824	2·0340	2·3067	2·4512	2·9180	3·2579
0·01600	1·682	1·768	1·816	1·909	1·999	2·044	2·175	2·302	2·344	2·426	2·507	2·547	2·665	2·742
1/ 63	0·5243	0·6328	0·6988	0·8432	1·0049	1·0925	1·3836	1·7188	1·8408	2·1007	2·3824	2·5316	3·0137	3·3648
0·01700	1·734	1·823	1·872	1·967	2·061	2·107	2·242	2·373	2·416	2·501	2·584	2·625	2·747	2·826
1/ 59	0·5404	0·6523	0·7203	0·8691	1·0359	1·1262	1·4262	1·7717	1·8975	2·1654	2·4557	2·6095	3·1065	3·4684
0·01800	1·784	1·876	1·926	2·024	2·121	2·168	2·307	2·442	2·486	2·573	2·659	2·701	2·826	2·908
1/ 56	0·5561	0·6712	0·7412	0·8943	1·0659	1·1588	1·4675	1·8231	1·9525	2·2281	2·5269	2·6851	3·1966	3·5689
0·01900	1·833	1·927	1·979	2·080	2·179	2·227	2·370	2·509	2·554	2·644	2·732	2·775	2·904	2·988
1/ 53	0·5713	0·6896	0·7615	0·9188	1·0951	1·1906	1·5078	1·8731	2·0060	2·2892	2·5961	2·7587	3·2842	3·6667
0·02000	1·880	1·977	2·030	2·134	2·235	2·285	2·432	2·574	2·620	2·712	2·803	2·847	2·979	3·066
1/ 50	0·5862	0·7075	0·7813	0·9427	1·1236	1·2215	1·5469	1·9217	2·0581	2·3487	2·6636	2·8304	3·3695	3·7620
0·02100	1·927	2·026	2·080	2·187	2·290	2·342	2·492	2·637	2·685	2·779	2·872	2·918	3·053	3·141
1/ 47·6	0·6006	0·7250	0·8006	0·9660	1·1513	1·2517	1·5851	1·9692	2·1090	2·4067	2·7294	2·9003	3·4527	3·8549
0·02200	1·972	2·074	2·129	2·238	2·344	2·397	2·550	2·700	2·748	2·845	2·940	2·986	3·125	3·215
1/ 45·5	0·6148	0·7421	0·8194	0·9887	1·1784	1·2812	1·6224	2·0156	2·1586	2·4633	2·7936	2·9686	3·5340	3·9456
0·02300	2·017	2·120	2·177	2·288	2·397	2·451	2·608	2·760	2·810	2·909	3·006	3·054	3·195	3·287
1/ 43·5	0·6286	0·7588	0·8378	1·0109	1·2049	1·3100	1·6589	2·0609	2·2071	2·5187	2·8564	3·0353	3·6134	4·0343
0·02400	2·060	2·166	2·224	2·338	2·449	2·503	2·664	2·820	2·871	2·971	3·070	3·119	3·264	3·358
1/ 41·7	0·6421	0·7751	0·8558	1·0327	1·2308	1·3381	1·6946	2·1052	2·2546	2·5729	2·9178	3·1006	3·6911	4·1211
0·02500	2·102	2·211	2·270	2·386	2·499	2·555	2·719	2·878	2·930	3·033	3·134	3·184	3·331	3·427
1/ 40·0	0·6554	0·7911	0·8735	1·0540	1·2562	1·3657	1·7295	2·1486	2·3011	2·6259	2·9780	3·1645	3·7673	4·2061
0·02600	2·144	2·254	2·315	2·433	2·549	2·605	2·773	2·935	2·988	3·093	3·196	3·247	3·397	3·495
1/ 38·5	0·6683	0·8067	0·8908	1·0749	1·2811	1·3928	1·7638	2·1912	2·3467	2·6779	3·0370	3·2272	3·8419	4·2894
0·02700	2·185	2·297	2·359	2·479	2·597	2·655	2·825	2·991	3·045	3·152	3·257	3·308	3·462	3·562
1/ 37·0	0·6811	0·8221	0·9078	1·0953	1·3055	1·4193	1·7974	2·2329	2·3914	2·7290	3·0949	3·2887	3·9151	4·3711
0·02800	2·225	2·340	2·402	2·525	2·645	2·704	2·877	3·046	3·101	3·209	3·316	3·369	3·525	3·627
1/ 35·7	0·6936	0·8372	0·9244	1·1154	1·3294	1·4454	1·8304	2·2739	2·4353	2·7790	3·1516	3·3490	3·9869	4·4513
0·02900	2·264	2·381	2·445	2·570	2·692	2·752	2·928	3·099	3·156	3·266	3·375	3·429	3·588	3·691
1/ 34·5	0·7059	0·8520	0·9408	1·1352	1·3530	1·4709	1·8628	2·3141	2·4784	2·8282	3·2074	3·4083	4·0575	4·5301
0·03000	2·303	2·422	2·486	2·613	2·738	2·799	2·978	3·152	3·210	3·322	3·433	3·487	3·649	3·755
1/ 33·3	0·7179	0·8666	0·9569	1·1546	1·3761	1·4961	1·8946	2·3537	2·5207	2·8766	3·2623	3·4666	4·1269	4·6076
	–	–	–	–	–	–	–	–	–	–	–	–	–	–

$V_{r(0\cdot5)medial}$ for half-full circular pipes.

k_s = 30·0 mm S = 0·01000 to 0·03000

k_s = **30·0 mm**
S = 0·03000 to 0·10000

ie hydraulic gradient =
1 in 33·3 to 1 in 10·0

Water (or sewage) at 15°C;
full bore conditions.

velocities in ms^{-1}
discharges in m^3s^{-1}

Gradient **(Equivalent) Pipe diameters in mm**

Gradient	630	675	700	750	800	825	900	975	1000	1050	1100	1125	1200	1250
0·03000	2·303	2·422	2·486	2·613	2·738	2·799	2·978	3·152	3·210	3·322	3·433	3·487	3·649	3·755
1/ 33·3	0·7179	0·8666	0·9569	1·1546	1·3761	1·4961	1·8946	2·3537	2·5207	2·8766	3·2623	3·4666	4·1269	4·6076
0·03200	2·379	2·501	2·568	2·699	2·827	2·891	3·076	3·256	3·315	3·431	3·545	3·602	3·769	3·878
1/ 31·3	0·7415	0·8950	0·9883	1·1925	1·4212	1·5452	1·9568	2·4309	2·6034	2·9709	3·3693	3·5803	4·2622	4·7587
0·03400	2·452	2·578	2·647	2·782	2·914	2·980	3·171	3·356	3·417	3·537	3·654	3·713	3·885	3·997
1/ 29·4	0·7643	0·9226	1·0187	1·2292	1·4650	1·5927	2·0170	2·5057	2·6836	3·0624	3·4730	3·6905	4·3934	4·9052
0·03600	2·523	2·653	2·724	2·863	2·999	3·066	3·262	3·453	3·516	3·639	3·760	3·820	3·997	4·113
1/ 27·8	0·7865	0·9493	1·0482	1·2648	1·5075	1·6389	2·0755	2·5784	2·7614	3·1512	3·5737	3·7975	4·5208	5·0474
0·03800	2·592	2·726	2·798	2·941	3·081	3·150	3·352	3·548	3·612	3·739	3·864	3·925	4·107	4·226
1/ 26·3	0·8080	0·9753	1·0769	1·2995	1·5488	1·6838	2·1324	2·6490	2·8370	3·2375	3·6716	3·9016	4·6447	5·1857
0·04000	2·659	2·796	2·871	3·018	3·161	3·232	3·439	3·640	3·706	3·836	3·964	4·027	4·213	4·335
1/ 25·0	0·8290	1·0007	1·1049	1·3332	1·5890	1·7276	2·1878	2·7179	2·9107	3·3217	3·7670	4·0029	4·7653	5·3204
0·04200	2·725	2·865	2·942	3·092	3·239	3·312	3·524	3·730	3·798	3·931	4·062	4·126	4·318	4·443
1/ 23·8	0·8495	1·0254	1·1322	1·3662	1·6283	1·7702	2·2418	2·7850	2·9826	3·4037	3·8600	4·1018	4·8830	5·4518
0·04400	2·789	2·933	3·011	3·165	3·316	3·390	3·607	3·818	3·887	4·023	4·157	4·224	4·419	4·547
1/ 22·7	0·8695	1·0495	1·1589	1·3983	1·6666	1·8119	2·2946	2·8505	3·0528	3·4838	3·9509	4·1983	4·9979	5·5801
0·04600	2·852	2·999	3·079	3·236	3·390	3·466	3·688	3·904	3·974	4·114	4·251	4·319	4·518	4·649
1/ 21·7	0·8890	1·0731	1·1849	1·4297	1·7040	1·8526	2·3461	2·9146	3·1214	3·5621	4·0397	4·2927	5·1103	5·7056
0·04800	2·913	3·063	3·145	3·306	3·463	3·540	3·767	3·988	4·060	4·202	4·342	4·411	4·616	4·749
1/ 20·8	0·9081	1·0962	1·2104	1·4605	1·7407	1·8925	2·3966	2·9773	3·1886	3·6387	4·1266	4·3850	5·2202	5·8283
0·05000	2·973	3·126	3·210	3·374	3·534	3·613	3·845	4·070	4·144	4·289	4·432	4·502	4·711	4·847
1/ 20·0	0·9269	1·1188	1·2354	1·4906	1·7766	1·9315	2·4460	3·0387	3·2543	3·7137	4·2117	4·4754	5·3278	5·9485
0·05250	3·047	3·204	3·289	3·457	3·622	3·702	3·940	4·170	4·246	4·395	4·541	4·614	4·827	4·967
1/ 19·0	0·9498	1·1464	1·2659	1·5274	1·8205	1·9792	2·5064	3·1137	3·3347	3·8055	4·3157	4·5860	5·4594	6·0954
0·05500	3·118	3·279	3·367	3·539	3·707	3·790	4·033	4·269	4·346	4·498	4·648	4·722	4·941	5·084
1/ 18·2	0·9721	1·1734	1·2957	1·5634	1·8633	2·0258	2·5654	3·1870	3·4132	3·8950	4·4172	4·6939	5·5879	6·2388
0·05750	3·189	3·353	3·442	3·618	3·790	3·875	4·123	4·365	4·443	4·599	4·753	4·828	5·052	5·198
1/ 17·4	0·9940	1·1998	1·3248	1·5985	1·9052	2·0713	2·6231	3·2586	3·4899	3·9826	4·5165	4·7994	5·7135	6·3791
0·06000	3·257	3·425	3·516	3·696	3·872	3·958	4·212	4·458	4·539	4·698	4·855	4·932	5·161	5·310
1/ 16·7	1·0153	1·2256	1·3533	1·6329	1·9462	2·1159	2·6795	3·3287	3·5650	4·0682	4·6137	4·9026	5·8364	6·5163
0·06250	3·324	3·496	3·589	3·772	3·952	4·040	4·299	4·550	4·633	4·795	4·955	5·034	5·267	5·419
1/ 16·0	1·0363	1·2509	1·3812	1·6666	1·9863	2·1595	2·7348	3·3974	3·6385	4·1521	4·7088	5·0037	5·9568	6·6506
0·06500	3·390	3·565	3·660	3·847	4·030	4·120	4·384	4·640	4·724	4·890	5·053	5·134	5·371	5·527
1/ 15·4	1·0568	1·2756	1·4085	1·6996	2·0256	2·2023	2·7889	3·4647	3·7105	4·2344	4·8021	5·1028	6·0747	6·7824
0·06750	3·455	3·633	3·730	3·920	4·107	4·198	4·467	4·729	4·814	4·983	5·149	5·231	5·474	5·632
1/ 14·8	1·0769	1·2999	1·4354	1·7320	2·0642	2·2442	2·8421	3·5307	3·7812	4·3150	4·8936	5·2000	6·1905	6·9116
0·07000	3·518	3·699	3·798	3·992	4·182	4·275	4·549	4·816	4·903	5·075	5·244	5·327	5·574	5·735
1/ 14·3	1·0967	1·3238	1·4617	1·7638	2·1021	2·2854	2·8942	3·5955	3·8506	4·3942	4·9834	5·2955	6·3041	7·0384
0·07250	3·580	3·765	3·865	4·063	4·256	4·351	4·630	4·901	4·990	5·165	5·337	5·422	5·673	5·837
1/ 13·8	1·1161	1·3472	1·4876	1·7950	2·1393	2·3259	2·9455	3·6591	3·9188	4·4720	5·0716	5·3892	6·4156	7·1630
0·07500	3·642	3·829	3·932	4·132	4·329	4·425	4·709	4·985	5·075	5·253	5·428	5·514	5·770	5·937
1/ 13·3	1·1352	1·3703	1·5130	1·8257	2·1759	2·3656	2·9958	3·7217	3·9858	4·5485	5·1583	5·4813	6·5253	7·2855
0·08000	3·761	3·955	4·060	4·268	4·471	4·571	4·864	5·148	5·241	5·425	5·606	5·695	5·959	6·131
1/ 12·5	1·1724	1·4152	1·5626	1·8855	2·2473	2·4432	3·0941	3·8437	4·1165	4·6976	5·3275	5·6611	6·7393	7·5244
0·08500	3·877	4·076	4·185	4·399	4·608	4·711	5·013	5·307	5·403	5·592	5·778	5·870	6·142	6·320
1/ 11·8	1·2085	1·4588	1·6107	1·9436	2·3164	2·5184	3·1893	3·9620	4·2432	4·8422	5·4914	5·8354	6·9468	7·7560
0·09000	3·989	4·195	4·307	4·527	4·742	4·848	5·159	5·460	5·559	5·754	5·946	6·041	6·320	6·503
1/ 11·1	1·2435	1·5011	1·6574	1·9999	2·3836	2·5914	3·2818	4·0769	4·3662	4·9826	5·6506	6·0045	7·1482	7·9808
0·09500	4·099	4·310	4·425	4·651	4·872	4·981	5·300	5·610	5·712	5·912	6·109	6·206	6·494	6·682
1/ 10·5	1·2776	1·5422	1·7029	2·0547	2·4489	2·6625	3·3717	4·1886	4·4859	5·1192	5·8055	6·1691	7·3441	8·1996
0·10000	4·205	4·422	4·540	4·772	4·999	5·110	5·438	5·756	5·860	6·066	6·268	6·367	6·662	6·855
1/ 10·0	1·3108	1·5823	1·7471	2·1081	2·5125	2·7316	3·4593	4·2974	4·6024	5·2521	5·9563	6·3294	7·5349	8·4126
	–	–	–	–	–	–	–	–	–	–	–	–	–	–

$V_{r(0·5)medial}$ for half-full circular pipes.

A35
(p.1 of 6)

k_s = 0·006 mm
S = 0·00010 to 0·00030

ie hydraulic gradient =
1 in 10000 to 1 in 3333

Water (or sewage) at 15°C;
full bore conditions.

velocities in ms^{-1}
discharges in m^3s^{-1}

Gradient	(Equivalent) Pipe diameters in mm													
	1250	**1300**	**1350**	**1400**	**1500**	**1600**	**1650**	**1800**	**1950**	**2000**	**2100**	**2200**	**2250**	**2400**
0·000100	0·428	0·439	0·450	0·461	0·482	0·502	0·512	0·542	0·570	0·579	0·598	0·616	0·625	0·651
1/ 10000	0·5253	0·5828	0·6441	0·7092	0·8512	1·0096	1·0952	1·3783	1·7027	1·8203	2·0705	2·3408	2·4837	2·9443
0·000105	0·440	0·451	0·462	0·473	0·495	0·516	0·526	0·556	0·586	0·595	0·614	0·632	0·641	0·668
1/ 9524	0·5396	0·5986	0·6616	0·7284	0·8743	1·0369	1·1248	1·4155	1·7486	1·8694	2·1262	2·4038	2·5505	3·0235
0·000110	0·451	0·463	0·474	0·485	0·507	0·529	0·540	0·571	0·601	0·610	0·630	0·649	0·658	0·685
1/ 9091	0·5535	0·6141	0·6786	0·7472	0·8968	1·0636	1·1538	1·4519	1·7935	1·9174	2·1808	2·4655	2·6160	3·1010
0·000115	0·462	0·474	0·486	0·497	0·520	0·542	0·553	0·585	0·615	0·625	0·645	0·664	0·674	0·702
1/ 8696	0·5672	0·6293	0·6954	0·7656	0·9189	1·0898	1·1821	1·4876	1·8375	1·9645	2·2343	2·5259	2·6800	3·1769
0·000120	0·473	0·485	0·497	0·509	0·532	0·555	0·566	0·598	0·630	0·640	0·660	0·680	0·690	0·719
1/ 8333	0·5806	0·6441	0·7118	0·7837	0·9405	1·1155	1·2099	1·5225	1·8806	2·0105	2·2867	2·5850	2·7428	3·2512
0·000125	0·484	0·496	0·508	0·521	0·544	0·567	0·579	0·612	0·644	0·654	0·675	0·695	0·705	0·735
1/ 8000	0·5937	0·6587	0·7279	0·8014	0·9617	1·1406	1·2372	1·5568	1·9229	2·0558	2·3380	2·6431	2·8044	3·3242
0·000130	0·494	0·507	0·520	0·532	0·556	0·580	0·591	0·625	0·658	0·668	0·690	0·710	0·721	0·751
1/ 7692	0·6066	0·6730	0·7437	0·8188	0·9826	1·1653	1·2640	1·5905	1·9645	2·1001	2·3885	2·7001	2·8648	3·3958
0·000135	0·505	0·518	0·530	0·543	0·568	0·592	0·603	0·638	0·671	0·682	0·704	0·725	0·735	0·766
1/ 7407	0·6193	0·6870	0·7592	0·8358	1·0031	1·1896	1·2903	1·6235	2·0053	2·1438	2·4381	2·7561	2·9242	3·4661
0·000140	0·515	0·528	0·541	0·554	0·579	0·604	0·616	0·651	0·685	0·696	0·718	0·740	0·750	0·781
1/ 7143	0·6318	0·7009	0·7744	0·8526	1·0232	1·2134	1·3162	1·6560	2·0454	2·1866	2·4868	2·8111	2·9826	3·5353
0·000145	0·525	0·538	0·552	0·565	0·590	0·615	0·627	0·663	0·698	0·709	0·732	0·754	0·765	0·797
1/ 6897	0·6440	0·7145	0·7895	0·8691	1·0430	1·2369	1·3416	1·6880	2·0849	2·2288	2·5347	2·8653	3·0400	3·6033
0·000150	0·535	0·548	0·562	0·575	0·601	0·627	0·639	0·676	0·711	0·723	0·745	0·768	0·779	0·811
1/ 6667	0·6561	0·7278	0·8042	0·8854	1·0625	1·2600	1·3667	1·7195	2·1237	2·2703	2·5819	2·9186	3·0966	3·6702
0·000160	0·554	0·568	0·582	0·596	0·623	0·649	0·662	0·700	0·737	0·748	0·772	0·795	0·807	0·840
1/ 6250	0·6797	0·7540	0·8331	0·9172	1·1006	1·3052	1·4156	1·7810	2·1996	2·3514	2·6741	3·0228	3·2071	3·8011
0·000170	0·573	0·587	0·602	0·616	0·644	0·671	0·684	0·723	0·761	0·774	0·798	0·822	0·834	0·868
1/ 5882	0·7026	0·7794	0·8612	0·9481	1·1377	1·3491	1·4632	1·8408	2·2734	2·4303	2·7637	3·1240	3·3144	3·9282
0·000180	0·591	0·606	0·621	0·635	0·664	0·692	0·706	0·746	0·785	0·798	0·823	0·848	0·860	0·896
1/ 5556	0·7249	0·8042	0·8885	0·9782	1·1737	1·3918	1·5095	1·8990	2·3452	2·5070	2·8509	3·2225	3·4189	4·0520
0·000190	0·608	0·624	0·639	0·654	0·684	0·713	0·727	0·769	0·809	0·822	0·848	0·873	0·885	0·922
1/ 5263	0·7467	0·8283	0·9152	1·0075	1·2089	1·4334	1·5547	1·9557	2·4151	2·5818	2·9358	3·3185	3·5207	4·1725
0·000200	0·626	0·642	0·658	0·673	0·703	0·733	0·748	0·790	0·832	0·845	0·872	0·898	0·910	0·948
1/ 5000	0·7679	0·8518	0·9412	1·0361	1·2432	1·4740	1·5987	2·0111	2·4834	2·6547	3·0188	3·4121	3·6201	4·2902
0·000210	0·643	0·659	0·675	0·691	0·722	0·753	0·768	0·812	0·854	0·868	0·895	0·922	0·935	0·974
1/ 4762	0·7887	0·8749	0·9666	1·0641	1·2767	1·5138	1·6418	2·0652	2·5502	2·7260	3·0998	3·5037	3·7172	4·4052
0·000220	0·659	0·676	0·693	0·709	0·741	0·772	0·788	0·832	0·876	0·890	0·918	0·945	0·959	0·999
1/ 4545	0·8090	0·8974	0·9915	1·0914	1·3095	1·5526	1·6839	2·1181	2·6154	2·7958	3·1791	3·5932	3·8122	4·5177
0·000230	0·675	0·693	0·710	0·726	0·759	0·791	0·807	0·853	0·897	0·912	0·940	0·968	0·982	1·023
1/ 4348	0·8289	0·9194	1·0158	1·1182	1·3416	1·5906	1·7251	2·1699	2·6794	2·8641	3·2567	3·6809	3·9052	4·6278
0·000240	0·691	0·709	0·726	0·743	0·777	0·810	0·826	0·873	0·918	0·933	0·962	0·991	1·005	1·047
1/ 4167	0·8484	0·9410	1·0397	1·1445	1·3731	1·6279	1·7655	2·2207	2·7420	2·9310	3·3328	3·7669	3·9963	4·7357
0·000250	0·707	0·725	0·743	0·760	0·794	0·828	0·844	0·892	0·939	0·954	0·984	1·013	1·028	1·070
1/ 4000	0·8675	0·9622	1·0631	1·1702	1·4039	1·6645	1·8052	2·2705	2·8034	2·9967	3·4074	3·8511	4·0857	4·8416
0·000260	0·722	0·741	0·759	0·777	0·812	0·846	0·862	0·911	0·959	0·974	1·005	1·035	1·050	1·093
1/ 3846	0·8863	0·9831	1·0861	1·1955	1·4343	1·7004	1·8441	2·3194	2·8638	3·0612	3·4807	3·9339	4·1735	4·9455
0·000270	0·737	0·756	0·775	0·793	0·828	0·863	0·880	0·930	0·979	0·995	1·026	1·056	1·071	1·116
1/ 3704	0·9048	1·0035	1·1087	1·2204	1·4640	1·7357	1·8824	2·3675	2·9230	3·1245	3·5526	4·0152	4·2597	5·0475
0·000280	0·752	0·771	0·790	0·809	0·845	0·881	0·898	0·949	0·998	1·014	1·046	1·077	1·093	1·138
1/ 3571	0·9229	1·0236	1·1309	1·2448	1·4933	1·7704	1·9200	2·4147	2·9813	3·1867	3·6233	4·0951	4·3444	5·1479
0·000290	0·767	0·786	0·805	0·824	0·861	0·897	0·915	0·967	1·017	1·034	1·066	1·098	1·114	1·160
1/ 3448	0·9407	1·0434	1·1527	1·2688	1·5221	1·8045	1·9569	2·4612	3·0386	3·2480	3·6929	4·1736	4·4277	5·2466
0·000300	0·781	0·801	0·820	0·840	0·877	0·914	0·932	0·985	1·036	1·053	1·086	1·118	1·134	1·181
1/ 3333	0·9583	1·0629	1·1742	1·2925	1·5505	1·8381	1·9933	2·5069	3·0950	3·3082	3·7614	4·2510	4·5098	5·3437
	0·94	0·94	0·94	0·95	0·95	0·95	0·95	0·96	0·96	0·96	0·96	0·97	0·97	0·97

$V_{r(0{\cdot}5)medial}$ for half-full circular pipes.

Gradient	**(Equivalent) Pipe diameters in mm**													
	1250	**1300**	**1350**	**1400**	**1500**	**1600**	**1650**	**1800**	**1950**	**2000**	**2100**	**2200**	**2250**	**2400**
0·000300	0·781	0·801	0·820	0·840	0·877	0·914	0·932	0·985	1·036	1·053	1·086	1·118	1·134	1·181
1/ 3333	0·9583	1·0629	1·1742	1·2925	1·5505	1·8381	1·9933	2·5069	3·0950	3·3082	3·7614	4·2510	4·5098	5·3437
0·000320	0·809	0·829	0·850	0·870	0·909	0·947	0·966	1·020	1·073	1·091	1·125	1·158	1·175	1·223
1/ 3125	0·9926	1·1009	1·2162	1·3387	1·6059	1·9037	2·0645	2·5962	3·2052	3·4260	3·8952	4·4021	4·6701	5·5335
0·000340	0·836	0·857	0·878	0·899	0·939	0·979	0·998	1·054	1·109	1·127	1·162	1·197	1·214	1·264
1/ 2941	1·0260	1·1379	1·2570	1·3836	1·6597	1·9674	2·1336	2·6831	3·3123	3·5404	4·0252	4·5490	4·8258	5·7179
0·000360	0·863	0·884	0·906	0·927	0·969	1·009	1·029	1·088	1·144	1·162	1·199	1·234	1·252	1·304
1/ 2778	1·0585	1·1739	1·2968	1·4273	1·7121	2·0295	2·2009	2·7676	3·4165	3·6518	4·1517	4·6919	4·9774	5·8973
0·000380	0·888	0·911	0·933	0·955	0·998	1·039	1·060	1·120	1·178	1·197	1·234	1·271	1·289	1·342
1/ 2632	1·0901	1·2090	1·3355	1·4699	1·7631	2·0900	2·2664	2·8499	3·5180	3·7603	4·2750	4·8311	5·1250	6·0721
0·000400	0·913	0·937	0·959	0·982	1·026	1·069	1·090	1·152	1·211	1·231	1·269	1·307	1·325	1·380
1/ 2500	1·1210	1·2432	1·3733	1·5115	1·8130	2·1490	2·3304	2·9302	3·6171	3·8661	4·3953	4·9670	5·2691	6·2427
0·000420	0·938	0·962	0·985	1·008	1·053	1·097	1·119	1·182	1·244	1·264	1·303	1·342	1·361	1·417
1/ 2381	1·1512	1·2766	1·4102	1·5522	1·8617	2·2066	2·3929	3·0088	3·7139	3·9696	4·5128	5·0997	5·4099	6·4093
0·000440	0·962	0·986	1·010	1·034	1·080	1·126	1·148	1·213	1·275	1·296	1·336	1·376	1·395	1·453
1/ 2273	1·1807	1·3094	1·4464	1·5919	1·9093	2·2631	2·4540	3·0856	3·8086	4·0708	4·6278	5·2296	5·5476	6·5723
0·000460	0·986	1·011	1·035	1·059	1·107	1·153	1·176	1·242	1·306	1·327	1·369	1·409	1·429	1·488
1/ 2174	1·2096	1·3414	1·4818	1·6308	1·9559	2·3183	2·5139	3·1607	3·9013	4·1698	4·7404	5·3567	5·6824	6·7319
0·000480	1·009	1·034	1·059	1·084	1·133	1·180	1·203	1·271	1·337	1·358	1·400	1·442	1·462	1·523
1/ 2083	1·2379	1·3728	1·5164	1·6690	2·0016	2·3724	2·5726	3·2344	3·9922	4·2669	4·8507	5·4812	5·8145	6·8883
0·000500	1·031	1·057	1·083	1·108	1·158	1·206	1·230	1·299	1·367	1·389	1·432	1·474	1·495	1·557
1/ 2000	1·2657	1·4036	1·5504	1·7064	2·0465	2·4255	2·6301	3·3067	4·0813	4·3621	4·9588	5·6034	5·9441	7·0417
0·000525	1·059	1·086	1·112	1·138	1·189	1·239	1·263	1·334	1·403	1·426	1·470	1·513	1·535	1·598
1/ 1905	1·2998	1·4413	1·5921	1·7522	2·1014	2·4905	2·7006	3·3952	4·1904	4·4787	5·0913	5·7530	6·1027	7·2294
0·000550	1·086	1·114	1·141	1·167	1·220	1·270	1·295	1·368	1·439	1·462	1·507	1·552	1·574	1·639
1/ 1818	1·3330	1·4782	1·6328	1·7970	2·1551	2·5541	2·7695	3·4817	4·2971	4·5927	5·2208	5·8993	6·2578	7·4130
0·000575	1·113	1·141	1·169	1·196	1·249	1·301	1·327	1·402	1·474	1·497	1·544	1·590	1·612	1·678
1/ 1739	1·3656	1·5144	1·6727	1·8409	2·2076	2·6163	2·8370	3·5665	4·4016	4·7043	5·3476	6·0425	6·4097	7·5928
0·000600	1·139	1·168	1·196	1·224	1·278	1·332	1·358	1·434	1·508	1·532	1·580	1·626	1·649	1·717
1/ 1667	1·3976	1·5498	1·7118	1·8839	2·2591	2·6774	2·9031	3·6495	4·5040	4·8137	5·4719	6·1828	6·5585	7·7690
0·000625	1·164	1·194	1·223	1·251	1·307	1·361	1·388	1·466	1·542	1·566	1·615	1·663	1·686	1·756
1/ 1600	1·4289	1·5845	1·7501	1·9261	2·3097	2·7372	2·9680	3·7310	4·6044	4·9210	5·5938	6·3204	6·7045	7·9417
0·000650	1·189	1·219	1·249	1·278	1·335	1·391	1·418	1·498	1·575	1·600	1·650	1·698	1·722	1·793
1/ 1538	1·4597	1·6186	1·7878	1·9675	2·3593	2·7959	3·0316	3·8109	4·7029	5·0263	5·7134	6·4555	6·8477	8·1113
0·000675	1·214	1·245	1·275	1·305	1·363	1·419	1·447	1·528	1·607	1·633	1·683	1·733	1·758	1·830
1/ 1481	1·4899	1·6521	1·8248	2·0081	2·4080	2·8536	3·0942	3·8894	4·7997	5·1297	5·8309	6·5882	6·9884	8·2778
0·000700	1·238	1·269	1·300	1·330	1·390	1·447	1·476	1·559	1·639	1·665	1·717	1·767	1·792	1·866
1/ 1429	1·5196	1·6850	1·8611	2·0481	2·4559	2·9103	3·1556	3·9666	4·8949	5·2314	5·9464	6·7186	7·1267	8·4415
0·000725	1·262	1·294	1·325	1·356	1·416	1·475	1·504	1·589	1·670	1·697	1·750	1·801	1·827	1·902
1/ 1379	1·5488	1·7174	1·8968	2·0874	2·5030	2·9661	3·2161	4·0425	4·9884	5·3314	6·0599	6·8468	7·2627	8·6024
0·000750	1·286	1·318	1·350	1·381	1·443	1·503	1·532	1·618	1·701	1·728	1·782	1·834	1·860	1·937
1/ 1333	1·5776	1·7493	1·9320	2·1261	2·5494	3·0210	3·2756	4·1172	5·0805	5·4297	6·1717	6·9730	7·3965	8·7608
0·000800	1·331	1·365	1·398	1·430	1·494	1·556	1·586	1·675	1·761	1·790	1·845	1·899	1·926	2·005
1/ 1250	1·6338	1·8116	2·0008	2·2018	2·6400	3·1283	3·3919	4·2632	5·2605	5·6220	6·3901	7·2197	7·6581	9·0704
0·000850	1·376	1·410	1·444	1·478	1·544	1·608	1·639	1·731	1·820	1·849	1·906	1·962	1·990	2·071
1/ 1176	1·6884	1·8721	2·0676	2·2752	2·7280	3·2325	3·5048	4·4050	5·4353	5·8088	6·6022	7·4592	7·9121	9·3709
0·000900	1·419	1·455	1·490	1·524	1·592	1·658	1·690	1·785	1·877	1·907	1·966	2·024	2·052	2·136
1/ 1111	1·7415	1·9309	2·1326	2·3467	2·8136	3·3339	3·6147	4·5429	5·6053	5·9904	6·8086	7·6922	8·1591	9·6634
0·000950	1·461	1·498	1·534	1·570	1·639	1·707	1·741	1·838	1·932	1·963	2·024	2·083	2·113	2·199
1/ 1053	1·7933	1·9883	2·1959	2·4164	2·8971	3·4327	3·7217	4·6773	5·7710	6·1674	7·0096	7·9192	8·3999	9·9483
0·001000	1·502	1·540	1·577	1·614	1·685	1·755	1·789	1·890	1·987	2·018	2·080	2·142	2·172	2·260
1/ 1000	1·8438	2·0443	2·2577	2·4844	2·9785	3·5291	3·8262	4·8085	5·9327	6·3401	7·2058	8·1407	8·6348	10·226
	0·96	0·96	0·96	0·96	0·96	0·97	0·97	0·97	0·97	0·97	0·97	0·98	0·98	0·98

$V_{r(0·5)medial}$ for half-full circular pipes.

$k_s = 0{\cdot}006$ **mm**
S = 0·00100 to 0·00300

ie hydraulic gradient =
1 in 1000 to 1 in 333

Water (or sewage) at 15°C;
full bore conditions.

velocities in ms^{-1}
discharges in m^3s^{-1}

Gradient **(Equivalent) Pipe diameters in mm**

Gradient	1250	1300	1350	1400	1500	1600	1650	1800	1950	2000	2100	2200	2250	2400
0·00100	1·502	1·540	1·577	1·614	1·685	1·755	1·789	1·890	1·987	2·018	2·080	2·142	2·172	2·260
1/ 1000	1·8438	2·0443	2·2577	2·4844	2·9785	3·5291	3·8262	4·8085	5·9327	6·3401	7·2058	8·1407	8·6348	10·226
0·00105	1·543	1·581	1·619	1·657	1·730	1·802	1·837	1·940	2·039	2·072	2·136	2·198	2·229	2·321
1/ 952	1·8932	2·0990	2·3181	2·5508	3·0580	3·6232	3·9283	4·9366	6·0906	6·5088	7·3974	8·3571	8·8642	10·498
0·00110	1·582	1·622	1·661	1·699	1·775	1·848	1·884	1·989	2·091	2·124	2·190	2·254	2·286	2·379
1/ 909	1·9415	2·1525	2·3772	2·6157	3·1358	3·7153	4·0281	5·0619	6·2450	6·6738	7·5848	8·5686	9·0885	10·763
0·00115	1·621	1·661	1·701	1·741	1·818	1·893	1·930	2·037	2·142	2·176	2·243	2·309	2·341	2·437
1/ 870	1·9887	2·2049	2·4350	2·6793	3·2120	3·8055	4·1258	5·1846	6·3962	6·8353	7·7683	8·7758	9·3082	11·023
0·00120	1·658	1·700	1·741	1·781	1·860	1·937	1·974	2·085	2·191	2·226	2·295	2·362	2·395	2·493
1/ 833	2·0351	2·2562	2·4916	2·7416	3·2866	3·8939	4·2216	5·3048	6·5443	6·9936	7·9480	8·9787	9·5233	11·278
0·00125	1·695	1·738	1·780	1·821	1·901	1·980	2·018	2·131	2·240	2·276	2·346	2·414	2·448	2·548
1/ 800	2·0805	2·3066	2·5472	2·8027	3·3598	3·9805	4·3155	5·4226	6·6896	7·1488	8·1242	9·1777	9·7344	11·527
0·00130	1·732	1·775	1·818	1·860	1·942	2·022	2·061	2·176	2·288	2·324	2·396	2·466	2·500	2·602
1/ 769	2·1251	2·3560	2·6018	2·8627	3·4317	4·0656	4·4076	5·5383	6·8321	7·3011	8·2972	9·3730	9·9414	11·772
0·00135	1·767	1·812	1·855	1·898	1·982	2·064	2·104	2·221	2·335	2·372	2·445	2·516	2·551	2·655
1/ 741	2·1689	2·4045	2·6553	2·9216	3·5023	4·1491	4·4982	5·6519	6·9722	7·4507	8·4671	9·5648	10·145	12·013
0·00140	1·802	1·848	1·892	1·936	2·021	2·104	2·145	2·265	2·381	2·418	2·493	2·566	2·602	2·708
1/ 714	2·2120	2·4522	2·7080	2·9796	3·5716	4·2312	4·5871	5·7636	7·1098	7·5977	8·6341	9·7533	10·345	12·250
0·00145	1·837	1·883	1·928	1·973	2·060	2·145	2·186	2·308	2·426	2·464	2·540	2·615	2·651	2·759
1/ 690	2·2543	2·4991	2·7598	3·0365	3·6398	4·3119	4·6746	5·8734	7·2451	7·7422	8·7982	9·9386	10·541	12·482
0·00150	1·871	1·918	1·964	2·009	2·098	2·184	2·226	2·351	2·471	2·510	2·587	2·662	2·700	2·810
1/ 667	2·2960	2·5453	2·8107	3·0925	3·7069	4·3913	4·7607	5·9814	7·3782	7·8844	8·9598	10·121	10·735	12·711
0·00160	1·937	1·986	2·033	2·080	2·172	2·261	2·305	2·434	2·558	2·598	2·678	2·756	2·795	2·909
1/ 625	2·3774	2·6355	2·9103	3·2021	3·8381	4·5466	4·9289	6·1926	7·6384	8·1624	9·2754	10·477	11·112	13·158
0·00170	2·002	2·052	2·101	2·149	2·244	2·336	2·382	2·514	2·642	2·684	2·766	2·847	2·887	3·004
1/ 588	2·4565	2·7232	3·0070	3·3084	3·9655	4·6974	5·0923	6·3976	7·8910	8·4323	9·5819	10·823	11·479	13·592
0·00180	2·064	2·116	2·167	2·216	2·314	2·409	2·456	2·592	2·725	2·768	2·853	2·936	2·977	3·098
1/ 556	2·5334	2·8084	3·1011	3·4119	4·0894	4·8440	5·2512	6·5971	8·1368	8·6948	9·8800	11·160	11·836	14·014
0·00190	2·126	2·178	2·231	2·282	2·382	2·480	2·528	2·669	2·805	2·849	2·936	3·022	3·064	3·189
1/ 526	2·6084	2·8915	3·1928	3·5127	4·2101	4·9869	5·4061	6·7914	8·3762	8·9505	10·170	11·488	12·184	14·425
0·00200	2·185	2·240	2·293	2·346	2·449	2·550	2·599	2·743	2·883	2·928	3·018	3·106	3·150	3·277
1/ 500	2·6816	2·9726	3·2823	3·6111	4·3279	5·1263	5·5571	6·9809	8·6097	9·2000	10·454	11·808	12·523	14·827
0·00210	2·243	2·299	2·354	2·408	2·514	2·617	2·668	2·816	2·959	3·006	3·098	3·188	3·233	3·364
1/ 476	2·7530	3·0517	3·3696	3·7072	4·4429	5·2625	5·7047	7·1660	8·8378	9·4437	10·730	12·120	12·854	15·218
0·00220	2·300	2·357	2·414	2·469	2·578	2·684	2·735	2·887	3·034	3·082	3·176	3·269	3·314	3·449
1/ 455	2·8229	3·1291	3·4551	3·8011	4·5554	5·3956	5·8490	7·3471	9·0609	9·6819	11·001	12·425	13·178	15·601
0·00230	2·356	2·415	2·472	2·529	2·640	2·748	2·801	2·957	3·107	3·156	3·253	3·347	3·394	3·532
1/ 435	2·8913	3·2049	3·5387	3·8931	4·6656	5·5259	5·9902	7·5243	9·2792	9·9151	11·266	12·724	13·495	15·976
0·00240	2·411	2·471	2·529	2·588	2·701	2·812	2·866	3·025	3·179	3·229	3·328	3·424	3·472	3·613
1/ 417	2·9583	3·2792	3·6207	3·9832	4·7735	5·6536	6·1285	7·6979	9·4931	10·144	11·525	13·017	13·805	16·344
0·00250	2·464	2·525	2·586	2·645	2·761	2·874	2·930	3·092	3·249	3·300	3·401	3·500	3·549	3·692
1/ 400	3·0241	3·3520	3·7010	4·0716	4·8793	5·7788	6·2642	7·8681	9·7028	10·368	11·780	13·304	14·110	16·704
0·00260	2·517	2·579	2·641	2·701	2·820	2·935	2·992	3·158	3·318	3·370	3·473	3·574	3·624	3·770
1/ 385	3·0886	3·4235	3·7799	4·1583	4·9831	5·9017	6·3974	8·0352	9·9086	10·587	12·029	13·586	14·408	17·057
0·00270	2·568	2·632	2·695	2·757	2·878	2·995	3·053	3·222	3·385	3·439	3·544	3·647	3·698	3·847
1/ 370	3·1519	3·4937	3·8574	4·2435	5·0851	6·0224	6·5281	8·1992	10·111	10·803	12·274	13·863	14·702	17·404
0·00280	2·619	2·684	2·748	2·811	2·934	3·054	3·113	3·285	3·452	3·506	3·613	3·718	3·770	3·922
1/ 357	3·2142	3·5627	3·9335	4·3272	5·1853	6·1410	6·6566	8·3604	10·309	11·015	12·515	14·134	14·990	17·745
0·00290	2·669	2·735	2·800	2·864	2·990	3·112	3·172	3·348	3·517	3·573	3·682	3·789	3·841	3·997
1/ 345	3·2754	3·6305	4·0084	4·4095	5·2839	6·2576	6·7830	8·5189	10·504	11·224	12·752	14·402	15·273	18·080
0·00300	2·718	2·786	2·852	2·917	3·045	3·169	3·230	3·409	3·582	3·638	3·749	3·858	3·911	4·069
1/ 333	3·3357	3·6973	4·0820	4·4905	5·3808	6·3723	6·9073	8·6749	10·697	11·429	12·985	14·665	15·552	18·410
	0·97	0·97	0·97	0·97	0·98	0·98	0·98	0·98	0·98	0·98	0·98	0·98	0·98	0·98

$V_{r(0{\cdot}5)medial}$ **for half-full circular pipes.**

$k_s = 0{\cdot}006$ **mm** S = 0·00100 to 0·00300

Gradient **(Equivalent) Pipe diameters in mm**

Gradient	1250	1300	1350	1400	1500	1600	1650	1800	1950	2000	2100	2200	2250	2400
0·00300	2·718	2·786	2·852	2·917	3·045	3·169	3·230	3·409	3·582	3·638	3·749	3·858	3·911	4·069
1/ 333	3·3357	3·6973	4·0820	4·4905	5·3808	6·3723	6·9073	8·6749	10·697	11·429	12·985	14·665	15·552	18·410
0·00320	2·814	2·884	2·952	3·020	3·152	3·281	3·344	3·529	3·707	3·766	3·880	3·993	4·048	4·212
1/ 313	3·4534	3·8277	4·2259	4·6488	5·5703	6·5965	7·1502	8·9796	11·072	11·830	13·440	15·178	16·097	19·054
0·00340	2·907	2·979	3·050	3·120	3·256	3·389	3·454	3·645	3·829	3·889	4·008	4·124	4·181	4·350
1/ 294	3·5677	3·9543	4·3657	4·8024	5·7542	6·8141	7·3860	9·2754	11·436	12·219	13·882	15·677	16·626	19·680
0·00360	2·998	3·072	3·145	3·217	3·357	3·494	3·561	3·758	3·948	4·010	4·132	4·252	4·311	4·485
1/ 278	3·6789	4·0775	4·5016	4·9519	5·9331	7·0258	7·6153	9·5631	11·791	12·598	14·312	16·162	17·140	20·288
0·00380	3·086	3·162	3·237	3·311	3·456	3·597	3·666	3·868	4·064	4·127	4·253	4·376	4·437	4·615
1/ 263	3·7873	4·1975	4·6340	5·0975	6·1074	7·2320	7·8387	9·8433	12·136	12·966	14·730	16·635	17·641	20·880
0·00400	3·172	3·251	3·328	3·404	3·552	3·697	3·768	3·976	4·176	4·242	4·371	4·497	4·560	4·743
1/ 250	3·8930	4·3146	4·7632	5·2395	6·2774	7·4331	8·0567	10·117	12·472	13·326	15·138	17·095	18·129	21·458
0·00420	3·256	3·337	3·416	3·494	3·646	3·795	3·867	4·080	4·286	4·354	4·486	4·616	4·679	4·868
1/ 238	3·9962	4·4289	4·8894	5·3782	6·4434	7·6296	8·2695	10·384	12·801	13·677	15·537	17·545	18·606	22·022
0·00440	3·339	3·421	3·502	3·582	3·738	3·890	3·965	4·183	4·394	4·463	4·598	4·731	4·797	4·990
1/ 227	4·0971	4·5407	5·0127	5·5138	6·6058	7·8216	8·4775	10·645	13·123	14·020	15·927	17·985	19·072	22·573
0·00460	3·419	3·503	3·586	3·668	3·828	3·984	4·060	4·283	4·499	4·570	4·708	4·844	4·911	5·109
1/ 217	4·1959	4·6501	5·1334	5·6465	6·7646	8·0096	8·6812	10·900	13·437	14·356	16·308	18·415	19·528	23·113
0·00480	3·498	3·584	3·669	3·753	3·916	4·075	4·153	4·382	4·602	4·674	4·816	4·955	5·024	5·226
1/ 208	4·2926	4·7573	5·2517	5·7766	6·9203	8·1937	8·8807	11·150	13·745	14·685	16·682	18·837	19·975	23·641
0·00500	3·575	3·663	3·750	3·835	4·002	4·165	4·245	4·478	4·704	4·777	4·922	5·064	5·134	5·340
1/ 200	4·3875	4·8624	5·3677	5·9041	7·0729	8·3742	9·0762	11·395	14·047	15·008	17·048	19·250	20·414	24·159
0·00525	3·670	3·760	3·849	3·937	4·108	4·275	4·357	4·596	4·827	4·903	5·051	5·197	5·269	5·480
1/ 190	4·5037	4·9910	5·5096	6·0601	7·2596	8·5951	9·3156	11·695	14·417	15·403	17·496	19·756	20·950	24·793
0·00550	3·762	3·855	3·946	4·036	4·211	4·382	4·466	4·711	4·948	5·026	5·178	5·327	5·401	5·617
1/ 182	4·6172	5·1168	5·6484	6·2126	7·4422	8·8111	9·5495	11·989	14·778	15·788	17·934	20·250	21·474	25·413
0·00575	3·853	3·948	4·041	4·133	4·313	4·487	4·573	4·824	5·067	5·146	5·302	5·454	5·530	5·752
1/ 174	4·7283	5·2399	5·7842	6·3619	7·6208	9·0224	9·7785	12·276	15·132	16·166	18·363	20·734	21·986	26·019
0·00600	3·942	4·039	4·134	4·228	4·412	4·590	4·678	4·935	5·183	5·264	5·423	5·579	5·656	5·883
1/ 167	4·8372	5·3604	5·9172	6·5082	7·7959	9·2295	10·003	12·557	15·478	16·536	18·783	21·208	22·489	26·613
0·00625	4·029	4·128	4·225	4·321	4·509	4·691	4·781	5·043	5·296	5·379	5·542	5·701	5·780	6·011
1/ 160	4·9439	5·4787	6·0476	6·6515	7·9675	9·4324	10·223	12·833	15·817	16·898	19·194	21·672	22·981	27·195
0·00650	4·114	4·215	4·314	4·412	4·604	4·790	4·882	5·149	5·408	5·492	5·658	5·821	5·901	6·138
1/ 154	5·0486	5·5947	6·1756	6·7922	8·1358	9·6316	10·438	13·104	16·150	17·254	19·598	22·128	23·464	27·766
0·00675	4·198	4·301	4·402	4·502	4·697	4·888	4·981	5·254	5·517	5·603	5·773	5·939	6·021	6·262
1/ 148	5·1515	5·7086	6·3012	6·9304	8·3011	9·8271	10·650	13·369	16·477	17·603	19·994	22·575	23·938	28·327
0·00700	4·280	4·385	4·488	4·590	4·789	4·983	5·078	5·356	5·625	5·713	5·885	6·054	6·138	6·383
1/ 143	5·2525	5·8205	6·4247	7·0661	8·4636	10·019	10·858	13·630	16·799	17·947	20·384	23·015	24·404	28·878
0·00725	4·361	4·468	4·573	4·677	4·880	5·077	5·174	5·457	5·731	5·820	5·996	6·168	6·253	6·503
1/ 138	5·3519	5·9305	6·5461	7·1996	8·6233	10·208	11·063	13·887	17·115	18·284	20·767	23·447	24·862	29·419
0·00750	4·441	4·550	4·657	4·762	4·969	5·170	5·268	5·556	5·835	5·926	6·104	6·280	6·366	6·621
1/ 133	5·4496	6·0388	6·6655	7·3308	8·7803	10·394	11·264	14·139	17·425	18·616	21·143	23·872	25·313	29·952
0·00800	4·596	4·709	4·820	4·929	5·142	5·350	5·452	5·750	6·038	6·132	6·317	6·498	6·588	6·851
1/ 125	5·6406	6·2502	6·8988	7·5873	9·0872	10·757	11·657	14·632	18·032	19·264	21·879	24·702	26·193	30·992
0·00850	4·747	4·864	4·978	5·091	5·311	5·525	5·630	5·938	6·235	6·332	6·523	6·710	6·802	7·074
1/ 118	5·8259	6·4555	7·1253	7·8362	9·3851	11·109	12·039	15·110	18·621	19·893	22·593	25·507	27·047	32·002
0·00900	4·894	5·014	5·132	5·248	5·475	5·696	5·804	6·121	6·427	6·527	6·723	6·916	7·011	7·291
1/ 111	6·0062	6·6551	7·3455	8·0783	9·6747	11·452	12·410	15·576	19·194	20·504	23·287	26·290	27·877	32·983
0·00950	5·037	5·160	5·282	5·401	5·634	5·862	5·973	6·299	6·614	6·716	6·919	7·117	7·214	7·502
1/ 105	6·1817	6·8495	7·5600	8·3140	9·9568	11·785	12·771	16·029	19·751	21·100	23·963	27·053	28·685	33·939
0·01000	5·177	5·303	5·428	5·550	5·790	6·023	6·138	6·472	6·796	6·901	7·109	7·312	7·413	7·708
1/ 100	6·3529	7·0391	7·7691	8·5439	10·232	12·111	13·124	16·470	20·295	21·681	24·622	27·796	29·473	34·871
	0·98	0·98	0·98	0·98	0·98	0·98	0·98	0·99	0·99	0·99	0·99	0·99	0·99	0·99

$V_{r(0·5)medial}$ for half-full circular pipes.

A35
(p.5 of 6)

$k_s = 0{\cdot}006$ **mm**
S = 0·01000 to 0·03000

ie hydraulic gradient =
1 in 100 to 1 in 33·3

Water (or sewage) at 15°C;
full bore conditions.

velocities in ms^{-1}
discharges in m^3s^{-1}

Gradient **(Equivalent) Pipe diameters in mm**

	1250	1300	1350	1400	1500	1600	1650	1800	1950	2000	2100	2200	2250	2400
0·01000	5·177	5·303	5·428	5·550	5·790	6·023	6·138	6·472	6·796	6·901	7·109	7·312	7·413	7·708
1/ 100	6·3529	7·0391	7·7691	8·5439	10·232	12·111	13·124	16·470	20·295	21·681	24·622	27·796	29·473	34·871
0·01050	5·313	5·443	5·570	5·696	5·942	6·181	6·299	6·642	6·973	7·082	7·295	7·503	7·606	7·909
1/ 95	6·5201	7·2243	7·9733	8·7684	10·500	12·428	13·468	16·901	20·826	22·247	25·265	28·522	30·243	35·780
0·01100	5·446	5·579	5·710	5·839	6·091	6·336	6·456	6·807	7·147	7·258	7·476	7·690	7·795	8·106
1/ 91	6·6835	7·4052	8·1730	8·9878	10·763	12·739	13·804	17·323	21·345	22·801	25·894	29·232	30·995	36·669
0·01150	5·577	5·712	5·846	5·978	6·236	6·487	6·610	6·970	7·317	7·430	7·654	7·872	7·980	8·298
1/ 87	6·8434	7·5823	8·3683	9·2025	11·020	13·042	14·133	17·735	21·852	23·344	26·509	29·926	31·730	37·539
0·01200	5·704	5·843	5·980	6·115	6·378	6·635	6·760	7·128	7·484	7·599	7·828	8·051	8·161	8·486
1/ 83	7·0000	7·7557	8·5596	9·4128	11·271	13·340	14·455	18·139	22·350	23·874	27·112	30·606	32·451	38·390
0·01250	5·829	5·971	6·111	6·249	6·518	6·780	6·908	7·284	7·647	7·765	7·998	8·227	8·339	8·671
1/ 80	7·1536	7·9258	8·7472	9·6190	11·518	13·631	14·771	18·535	22·837	24·395	27·702	31·272	33·157	39·225
0·01300	5·952	6·097	6·240	6·380	6·655	6·922	7·053	7·436	7·807	7·928	8·165	8·398	8·513	8·852
1/ 77	7·3042	8·0926	8·9312	9·8212	11·760	13·918	15·081	18·924	23·315	24·905	28·282	31·925	33·850	40·044
0·01350	6·073	6·220	6·366	6·509	6·789	7·062	7·195	7·586	7·964	8·087	8·330	8·567	8·684	9·029
1/ 74	7·4521	8·2563	9·1118	10·020	11·997	14·198	15·385	19·305	23·784	25·406	28·850	32·567	34·530	40·848
0·01400	6·191	6·342	6·490	6·636	6·921	7·199	7·335	7·733	8·118	8·244	8·491	8·733	8·852	9·204
1/ 71	7·5974	8·4172	9·2893	10·215	12·231	14·474	15·684	19·679	24·245	25·898	29·409	33·197	35·197	41·637
0·01450	6·307	6·461	6·612	6·760	7·051	7·334	7·472	7·878	8·270	8·398	8·649	8·896	9·017	9·375
1/ 69	7·7403	8·5754	9·4638	10·407	12·460	14·745	15·978	20·047	24·698	26·382	29·958	33·816	35·854	42·413
0·01500	6·422	6·578	6·731	6·883	7·179	7·466	7·608	8·020	8·419	8·549	8·805	9·056	9·180	9·544
1/ 67	7·8808	8·7310	9·6354	10·595	12·686	15·012	16·267	20·410	25·144	26·858	30·498	34·425	36·500	43·176
0·01600	6·645	6·807	6·966	7·122	7·428	7·726	7·871	8·298	8·711	8·845	9·110	9·369	9·497	9·873
1/ 63	8·1552	9·0349	9·9705	10·964	13·126	15·533	16·831	21·117	26·014	27·787	31·552	35·615	37·760	44·666
0·01700	6·862	7·029	7·193	7·354	7·670	7·977	8·128	8·568	8·993	9·132	9·405	9·673	9·805	10·19
1/ 59	8·4215	9·3298	10·296	11·321	13·554	16·039	17·379	21·803	26·858	28·689	32·575	36·769	38·984	46·112
0·01800	7·073	7·245	7·414	7·580	7·905	8·221	8·376	8·830	9·268	9·411	9·692	9·968	10·10	10·50
1/ 56	8·6805	9·6165	10·612	11·669	13·970	16·530	17·911	22·470	27·679	29·565	33·570	37·891	40·173	47·517
0·01900	7·279	7·455	7·629	7·800	8·134	8·459	8·619	9·085	9·536	9·683	9·972	10·26	10·39	10·81
1/ 53	8·9326	9·8956	10·920	12·007	14·374	17·009	18·429	23·119	28·478	30·418	34·538	38·983	41·330	48·885
0·02000	7·479	7·660	7·839	8·014	8·358	8·691	8·855	9·334	9·796	9·947	10·24	10·54	10·68	11·10
1/ 50	9·1784	10·168	11·220	12·337	14·769	17·475	18·934	23·752	29·257	31·250	35·482	40·047	42·458	50·219
0·02100	7·675	7·861	8·043	8·223	8·575	8·918	9·086	9·577	10·05	10·21	10·51	10·81	10·96	11·39
1/ 47·6	9·4184	10·433	11·513	12·659	15·154	17·931	19·428	24·370	30·017	32·062	36·403	41·086	43·560	51·520
0·02200	7·866	8·056	8·243	8·428	8·788	9·139	9·311	9·814	10·30	10·46	10·77	11·08	11·23	11·67
1/ 45·5	9·6529	10·693	11·800	12·974	15·531	18·376	19·910	24·974	30·760	32·855	37·303	42·102	44·636	52·792
0·02300	8·053	8·248	8·439	8·628	8·997	9·356	9·532	10·05	10·54	10·71	11·02	11·34	11·49	11·94
1/ 43·5	9·8824	10·947	12·080	13·282	15·899	18·811	20·381	25·565	31·487	33·631	38·183	43·095	45·689	54·036
0·02400	8·236	8·435	8·631	8·824	9·201	9·568	9·748	10·27	10·78	10·95	11·27	11·59	11·75	12·21
1/ 41·7	10·107	11·196	12·354	13·583	16·259	19·237	20·843	26·143	32·199	34·391	39·046	44·067	46·719	55·254
0·02500	8·416	8·619	8·819	9·016	9·401	9·776	9·959	10·50	11·02	11·18	11·52	11·84	12·00	12·48
1/ 40·0	10·327	11·440	12·623	13·879	16·613	19·655	21·295	26·710	32·896	35·136	39·891	45·020	47·729	56·447
0·02600	8·592	8·799	9·003	9·204	9·597	9·979	10·17	10·71	11·24	11·42	11·76	12·09	12·25	12·74
1/ 38·5	10·543	11·679	12·887	14·169	16·960	20·065	21·739	27·266	33·580	35·866	40·719	45·955	48·720	57·618
0·02700	8·764	8·976	9·184	9·389	9·790	10·18	10·37	10·93	11·47	11·64	11·99	12·33	12·50	12·99
1/ 37·0	10·756	11·914	13·146	14·453	17·300	20·467	22·175	27·812	34·252	36·583	41·533	46·872	49·692	58·767
0·02800	8·934	9·150	9·362	9·570	9·979	10·38	10·57	11·14	11·69	11·87	12·22	12·57	12·74	13·24
1/ 35·7	10·964	12·145	13·400	14·733	17·634	20·862	22·603	28·348	34·911	37·287	42·331	47·773	50·647	59·895
0·02900	9·101	9·320	9·536	9·749	10·16	10·57	10·77	11·35	11·91	12·09	12·45	12·80	12·97	13·48
1/ 34·5	11·169	12·371	13·650	15·007	17·963	21·250	23·023	28·875	35·559	37·979	43·116	48·659	51·585	61·004
0·03000	9·265	9·488	9·708	9·924	10·35	10·76	10·96	11·55	12·12	12·31	12·67	13·03	13·21	13·73
1/ 33·3	11·370	12·594	13·896	15·278	18·286	21·632	23·437	29·393	36·196	38·660	43·888	49·530	52·508	62·094
	0·98	0·99	0·99	0·99	0·99	0·99	0·99	0·99	0·99	0·99	0·99	0·99	0·99	0·99

$V_{r(0{\cdot}5)medial}$ **for half-full circular pipes.**

$k_s = 0{\cdot}006$ **mm** S = 0·01000 to 0·03000

Gradient	**(Equivalent) Pipe diameters in mm**													
	1250	**1300**	**1350**	**1400**	**1500**	**1600**	**1650**	**1800**	**1950**	**2000**	**2100**	**2200**	**2250**	**2400**
0·03000	9·265	9·488	9·708	9·924	10·35	10·76	10·96	11·55	12·12	12·31	12·67	13·03	13·21	13·73
1/ 33·3	11·370	12·594	13·896	15·278	18·286	21·632	23·437	29·393	36·196	38·660	43·888	49·530	52·508	62·094
0·03200	9·586	9·816	10·04	10·27	10·70	11·13	11·34	11·95	12·54	12·73	13·11	13·48	13·66	14·20
1/ 31·3	11·763	13·030	14·376	15·805	18·917	22·378	24·244	30·404	37·441	39·988	45·396	51·230	54·310	64·223
0·03400	9·897	10·13	10·37	10·60	11·05	11·49	11·70	12·33	12·94	13·14	13·53	13·91	14·10	14·65
1/ 29·4	12·145	13·452	14·842	16·317	19·529	23·101	25·028	31·386	38·648	41·277	46·858	52·879	56·058	66·288
0·03600	10·20	10·44	10·69	10·92	11·39	11·84	12·06	12·71	13·33	13·54	13·94	14·33	14·53	15·10
1/ 27·8	12·516	13·863	15·295	16·815	20·124	23·805	25·789	32·339	39·821	42·530	48·279	54·481	57·756	68·295
0·03800	10·49	10·75	10·99	11·24	11·72	12·18	12·41	13·07	13·72	13·93	14·34	14·74	14·94	15·53
1/ 26·3	12·877	14·262	15·736	17·299	20·703	24·489	26·530	33·268	40·963	43·749	49·662	56·041	59·409	70·248
0·04000	10·78	11·04	11·29	11·54	12·03	12·51	12·75	13·43	14·09	14·30	14·73	15·14	15·35	15·95
1/ 25·0	13·229	14·652	16·165	17·771	21·267	25·156	27·252	34·172	42·076	44·937	51·010	57·561	61·020	72·151
0·04200	11·06	11·33	11·59	11·84	12·35	12·84	13·07	13·78	14·45	14·67	15·11	15·53	15·74	16·36
1/ 23·8	13·572	15·032	16·585	18·232	21·818	25·807	27·957	35·055	43·162	46·096	52·325	59·044	62·592	74·008
0·04400	11·33	11·61	11·87	12·14	12·65	13·15	13·40	14·11	14·81	15·03	15·48	15·91	16·13	16·76
1/ 22·7	13·908	15·404	16·994	18·682	22·356	26·443	28·646	35·918	44·223	47·229	53·610	60·494	64·128	75·823
0·04600	11·60	11·88	12·15	12·42	12·95	13·46	13·71	14·45	15·16	15·39	15·84	16·29	16·51	17·15
1/ 21·7	14·236	15·767	17·395	19·122	22·883	27·065	29·320	36·762	45·261	48·337	54·867	61·911	65·630	77·597
0·04800	11·86	12·15	12·43	12·70	13·24	13·76	14·02	14·77	15·50	15·73	16·20	16·65	16·88	17·54
1/ 20·8	14·558	16·123	17·788	19·554	23·398	27·675	29·980	37·588	46·277	49·422	56·098	63·299	67·101	79·335
0·05000	12·12	12·41	12·70	12·98	13·53	14·06	14·32	15·09	15·83	16·07	16·54	17·01	17·24	17·91
1/ 20·0	14·873	16·472	18·172	19·976	23·903	28·271	30·626	38·398	47·272	50·485	57·303	64·658	68·542	81·037
0·05250	12·43	12·73	13·02	13·31	13·88	14·42	14·69	15·48	16·24	16·48	16·97	17·45	17·68	18·37
1/ 19·0	15·258	16·899	18·643	20·493	24·521	29·002	31·417	39·387	48·490	51·785	58·778	66·321	70·304	83·118
0·05500	12·74	13·05	13·35	13·64	14·22	14·78	15·05	15·86	16·63	16·89	17·39	17·87	18·11	18·82
1/ 18·2	15·635	17·315	19·102	20·998	25·125	29·715	32·189	40·355	49·679	53·055	60·218	67·945	72·025	85·152
0·05750	13·04	13·35	13·66	13·96	14·55	15·13	15·41	16·23	17·02	17·28	17·79	18·29	18·54	19·26
1/ 17·4	16·003	17·723	19·552	21·492	25·715	30·412	32·944	41·301	50·843	54·297	61·627	69·534	73·708	87·140
0·06000	13·33	13·65	13·97	14·28	14·88	15·47	15·75	16·59	17·41	17·67	18·19	18·70	18·95	19·69
1/ 16·7	16·364	18·122	19·992	21·975	26·293	31·095	33·684	42·227	51·981	55·512	63·006	71·089	75·356	89·087
0·06250	13·62	13·95	14·27	14·58	15·20	15·80	16·09	16·95	17·78	18·05	18·58	19·10	19·36	20·11
1/ 16·0	16·717	18·513	20·423	22·449	26·859	31·765	34·409	43·134	53·097	56·703	64·357	72·612	76·971	90·994
0·06500	13·90	14·24	14·56	14·88	15·51	16·12	16·42	17·30	18·15	18·42	18·96	19·49	19·76	20·53
1/ 15·4	17·064	18·897	20·846	22·914	27·415	32·421	35·119	44·024	54·191	57·872	65·682	74·106	78·554	92·864
0·06750	14·18	14·52	14·85	15·18	15·82	16·45	16·75	17·64	18·51	18·79	19·34	19·88	20·15	20·93
1/ 14·8	17·404	19·273	21·261	23·370	27·960	33·065	35·817	44·897	55·265	59·018	66·982	75·573	80·108	94·700
0·07000	14·45	14·80	15·14	15·47	16·12	16·76	17·07	17·98	18·86	19·14	19·71	20·26	20·53	21·33
1/ 14·3	17·738	19·643	21·669	23·817	28·495	33·697	36·501	45·754	56·319	60·143	68·259	77·012	81·633	96·502
0·07250	14·72	15·07	15·42	15·76	16·42	17·07	17·39	18·31	19·21	19·50	20·07	20·63	20·91	21·72
1/ 13·8	18·066	20·006	22·069	24·257	29·021	34·319	37·174	46·597	57·356	61·250	69·514	78·427	83·133	98·273
0·07500	14·98	15·34	15·69	16·04	16·72	17·37	17·69	18·64	19·55	19·84	20·43	21·00	21·28	22·11
1/ 13·3	18·389	20·363	22·463	24·690	29·538	34·930	37·836	47·426	58·374	62·337	70·747	79·818	84·607	100·01
0·08000	15·50	15·87	16·23	16·59	17·29	17·97	18·30	19·27	20·21	20·52	21·12	21·71	22·00	22·86
1/ 12·5	19·019	21·061	23·232	25·535	30·548	36·123	39·128	49·043	60·363	64·460	73·155	82·533	87·484	103·41
0·08500	16·00	16·38	16·75	17·12	17·84	18·54	18·88	19·89	20·86	21·17	21·80	22·40	22·70	23·59
1/ 11·8	19·630	21·737	23·978	26·354	31·528	37·280	40·381	50·611	62·291	66·519	75·490	85·166	90·274	106·71
0·09000	16·48	16·87	17·26	17·64	18·38	19·10	19·45	20·49	21·49	21·81	22·45	23·08	23·39	24·30
1/ 11·1	20·224	22·395	24·702	27·150	32·479	38·404	41·598	52·135	64·165	68·519	77·758	87·723	92·984	109·91
0·09500	16·95	17·35	17·75	18·14	18·90	19·64	20·01	21·07	22·10	22·43	23·09	23·73	24·05	24·98
1/ 10·5	20·802	23·034	25·408	27·925	33·405	39·498	42·782	53·618	65·988	70·465	79·966	90·212	95·621	113·02
0·10000	17·41	17·82	18·23	18·63	19·41	20·17	20·55	21·64	22·69	23·03	23·71	24·37	24·70	25·65
1/ 10·0	21·365	23·658	26·095	28·680	34·307	40·564	43·936	55·063	67·765	72·362	82·117	92·637	98·191	116·06
	0·99	0·99	0·99	0·99	0·99	0·99	0·99	0·99	0·99	0·99	0·99	0·99	0·99	0·99

$V_{r(0\cdot5)medial}$ for half-full circular pipes.

A36
(p.1 of 6)

k_s = **0·015 mm**
S = 0·00010 to 0·00030

ie hydraulic gradient =
1 in 10000 to 1 in 3333

Water (or sewage) at 15°C;
full bore conditions.

velocities in ms^{-1}
discharges in m^3s^{-1}

Gradient **(Equivalent) Pipe diameters in mm**

Gradient	1250	1300	1350	1400	1500	1600	1650	1800	1950	2000	2100	2200	2250	2400
0·000100	0·426	0·437	0·448	0·459	0·480	0·500	0·510	0·539	0·567	0·577	0·595	0·613	0·622	0·648
1/ 10000	0·5232	0·5805	0·6414	0·7062	0·8476	1·0052	1·0904	1·3720	1·6948	1·8118	2·0606	2·3294	2·4716	2·9297
0·000105	0·438	0·449	0·460	0·471	0·493	0·513	0·524	0·554	0·583	0·592	0·611	0·629	0·638	0·665
1/ 9524	0·5374	0·5962	0·6588	0·7253	0·8704	1·0323	1·1197	1·4090	1·7403	1·8605	2·1159	2·3919	2·5378	3·0081
0·000110	0·449	0·461	0·472	0·483	0·505	0·527	0·537	0·568	0·598	0·607	0·627	0·645	0·655	0·682
1/ 9091	0·5512	0·6115	0·6757	0·7439	0·8928	1·0588	1·1485	1·4451	1·7848	1·9081	2·1700	2·4530	2·6027	3·0849
0·000115	0·460	0·472	0·484	0·495	0·518	0·540	0·550	0·582	0·612	0·622	0·642	0·661	0·671	0·699
1/ 8696	0·5648	0·6265	0·6923	0·7622	0·9147	1·0848	1·1766	1·4804	1·8284	1·9547	2·2230	2·5129	2·6661	3·1601
0·000120	0·471	0·483	0·495	0·507	0·530	0·552	0·563	0·595	0·627	0·637	0·657	0·676	0·686	0·715
1/ 8333	0·5781	0·6413	0·7086	0·7801	0·9362	1·1102	1·2042	1·5151	1·8712	2·0004	2·2749	2·5715	2·7283	3·2337
0·000125	0·482	0·494	0·506	0·518	0·542	0·565	0·576	0·609	0·641	0·651	0·671	0·692	0·702	0·731
1/ 8000	0·5911	0·6557	0·7246	0·7977	0·9572	1·1351	1·2312	1·5490	1·9131	2·0452	2·3258	2·6290	2·7893	3·3060
0·000130	0·492	0·505	0·517	0·529	0·553	0·577	0·588	0·622	0·654	0·665	0·686	0·706	0·717	0·746
1/ 7692	0·6039	0·6699	0·7402	0·8149	0·9779	1·1596	1·2578	1·5824	1·9543	2·0891	2·3758	2·6855	2·8492	3·3769
0·000135	0·502	0·515	0·528	0·540	0·565	0·589	0·600	0·635	0·668	0·679	0·700	0·721	0·731	0·762
1/ 7407	0·6165	0·6839	0·7556	0·8319	0·9982	1·1837	1·2839	1·6152	1·9947	2·1323	2·4249	2·7409	2·9080	3·4465
0·000140	0·512	0·526	0·538	0·551	0·576	0·600	0·612	0·647	0·681	0·692	0·714	0·735	0·746	0·777
1/ 7143	0·6288	0·6976	0·7708	0·8485	1·0182	1·2073	1·3095	1·6474	2·0344	2·1748	2·4731	2·7954	2·9658	3·5150
0·000145	0·522	0·536	0·549	0·562	0·587	0·612	0·624	0·660	0·694	0·706	0·728	0·749	0·760	0·792
1/ 6897	0·6410	0·7110	0·7856	0·8649	1·0378	1·2306	1·3347	1·6791	2·0735	2·2166	2·5206	2·8490	3·0227	3·5823
0·000150	0·532	0·546	0·559	0·572	0·598	0·623	0·636	0·672	0·707	0·719	0·741	0·763	0·774	0·807
1/ 6667	0·6530	0·7243	0·8003	0·8810	1·0571	1·2535	1·3595	1·7102	2·1120	2·2576	2·5673	2·9018	3·0786	3·6485
0·000160	0·551	0·565	0·579	0·593	0·620	0·646	0·658	0·696	0·732	0·744	0·768	0·790	0·802	0·835
1/ 6250	0·6764	0·7503	0·8289	0·9125	1·0949	1·2982	1·4080	1·7712	2·1871	2·3380	2·6585	3·0049	3·1880	3·7780
0·000170	0·570	0·584	0·599	0·613	0·640	0·667	0·681	0·719	0·757	0·769	0·793	0·817	0·829	0·863
1/ 5882	0·6991	0·7754	0·8568	0·9431	1·1316	1·3417	1·4551	1·8304	2·2601	2·4160	2·7472	3·1050	3·2942	3·9038
0·000180	0·588	0·603	0·617	0·632	0·661	0·688	0·702	0·742	0·781	0·793	0·818	0·842	0·854	0·890
1/ 5556	0·7212	0·8000	0·8838	0·9729	1·1673	1·3840	1·5010	1·8880	2·3312	2·4919	2·8334	3·2025	3·3975	4·0261
0·000190	0·605	0·621	0·636	0·651	0·680	0·709	0·723	0·764	0·804	0·817	0·842	0·867	0·880	0·916
1/ 5263	0·7428	0·8239	0·9102	1·0020	1·2021	1·4252	1·5457	1·9441	2·4004	2·5659	2·9175	3·2974	3·4982	4·1453
0·000200	0·622	0·638	0·654	0·669	0·699	0·729	0·743	0·786	0·826	0·840	0·866	0·892	0·905	0·942
1/ 5000	0·7638	0·8472	0·9360	1·0303	1·2361	1·4654	1·5893	1·9989	2·4679	2·6380	2·9995	3·3900	3·5965	4·2616
0·000210	0·639	0·655	0·671	0·687	0·718	0·748	0·763	0·807	0·848	0·862	0·889	0·916	0·929	0·967
1/ 4762	0·7844	0·8700	0·9611	1·0580	1·2692	1·5047	1·6319	2·0524	2·5339	2·7086	3·0796	3·4805	3·6924	4·3752
0·000220	0·656	0·672	0·689	0·705	0·737	0·768	0·783	0·827	0·870	0·884	0·912	0·939	0·952	0·992
1/ 4545	0·8045	0·8923	0·9858	1·0851	1·3017	1·5432	1·6735	2·1047	2·5985	2·7775	3·1580	3·5690	3·7863	4·4863
0·000230	0·672	0·689	0·706	0·722	0·755	0·786	0·802	0·847	0·891	0·906	0·934	0·962	0·975	1·016
1/ 4348	0·8242	0·9141	1·0099	1·1116	1·3335	1·5808	1·7143	2·1559	2·6617	2·8450	3·2347	3·6556	3·8781	4·5951
0·000240	0·687	0·705	0·722	0·739	0·772	0·805	0·820	0·867	0·912	0·927	0·956	0·984	0·998	1·039
1/ 4167	0·8435	0·9355	1·0335	1·1376	1·3646	1·6177	1·7543	2·2062	2·7236	2·9112	3·3098	3·7405	3·9682	4·7017
0·000250	0·703	0·721	0·738	0·756	0·789	0·823	0·839	0·886	0·932	0·947	0·977	1·006	1·020	1·062
1/ 4000	0·8624	0·9565	1·0566	1·1630	1·3951	1·6538	1·7935	2·2554	2·7843	2·9761	3·3835	3·8238	4·0564	4·8062
0·000260	0·718	0·736	0·754	0·772	0·806	0·840	0·857	0·905	0·952	0·968	0·998	1·027	1·042	1·085
1/ 3846	0·8810	0·9771	1·0794	1·1881	1·4251	1·6893	1·8320	2·3037	2·8439	3·0397	3·4559	3·9055	4·1431	4·9087
0·000270	0·733	0·751	0·770	0·788	0·823	0·858	0·874	0·924	0·972	0·987	1·018	1·049	1·063	1·107
1/ 3704	0·8992	0·9973	1·1017	1·2126	1·4545	1·7242	1·8698	2·3512	2·9024	3·1023	3·5269	3·9857	4·2282	5·0095
0·000280	0·747	0·766	0·785	0·803	0·839	0·875	0·892	0·942	0·991	1·007	1·038	1·069	1·084	1·129
1/ 3571	0·9172	1·0172	1·1237	1·2368	1·4835	1·7585	1·9069	2·3978	2·9599	3·1637	3·5968	4·0646	4·3118	5·1084
0·000290	0·762	0·781	0·800	0·819	0·856	0·891	0·909	0·960	1·010	1·026	1·058	1·090	1·105	1·151
1/ 3448	0·9348	1·0367	1·1453	1·2605	1·5119	1·7922	1·9435	2·4437	3·0165	3·2242	3·6654	4·1421	4·3941	5·2058
0·000300	0·776	0·796	0·815	0·834	0·871	0·908	0·926	0·978	1·029	1·045	1·078	1·110	1·125	1·172
1/ 3333	0·9522	1·0560	1·1665	1·2839	1·5400	1·8253	1·9794	2·4889	3·0722	3·2836	3·7330	4·2184	4·4750	5·3016
	0·94	0·94	0·94	0·94	0·95	0·95	0·95	0·96	0·96	0·96	0·96	0·96	0·96	0·97

$V_{r(0·5)medial}$ **for half-full circular pipes.**

k_s = **0·015 mm** S = 0·00010 to 0·00030

$k_s = 0{\cdot}015$ **mm**
S = 0·00030 to 0·00100

ie hydraulic gradient =
1 in 3333 to 1 in 1000

Water (or sewage) at 15°C;
full bore conditions.

velocities in ms^{-1}
discharges in m^3s^{-1}

Gradient	**(Equivalent) Pipe diameters in mm**													
	1250	**1300**	**1350**	**1400**	**1500**	**1600**	**1650**	**1800**	**1950**	**2000**	**2100**	**2200**	**2250**	**2400**
0·000300	0·776	0·796	0·815	0·834	0·871	0·908	0·926	0·978	1·029	1·045	1·078	1·110	1·125	1·172
1/ 3333	0·9522	1·0560	1·1665	1·2839	1·5400	1·8253	1·9794	2·4889	3·0722	3·2836	3·7330	4·2184	4·4750	5·3016
0·000320	0·804	0·824	0·844	0·864	0·902	0·940	0·959	1·013	1·065	1·082	1·116	1·149	1·165	1·213
1/ 3125	0·9861	1·0936	1·2080	1·3296	1·5947	1·8902	2·0497	2·5771	3·1809	3·3999	3·8650	4·3675	4·6331	5·4887
0·000340	0·830	0·851	0·872	0·893	0·932	0·971	0·990	1·046	1·100	1·118	1·153	1·187	1·204	1·253
1/ 2941	1·0191	1·1301	1·2484	1·3740	1·6479	1·9531	2·1179	2·6628	3·2866	3·5127	3·9933	4·5123	4·7866	5·6705
0·000360	0·857	0·878	0·900	0·921	0·962	1·002	1·022	1·079	1·135	1·153	1·189	1·224	1·241	1·293
1/ 2778	1·0512	1·1657	1·2876	1·4171	1·6996	2·0144	2·1843	2·7461	3·3893	3·6225	4·1180	4·6532	4·9360	5·8472
0·000380	0·882	0·904	0·926	0·948	0·990	1·032	1·052	1·111	1·168	1·187	1·224	1·260	1·278	1·331
1/ 2632	1·0824	1·2003	1·3259	1·4592	1·7500	2·0740	2·2490	2·8273	3·4895	3·7295	4·2395	4·7904	5·0815	6·0195
0·000400	0·907	0·930	0·952	0·975	1·018	1·060	1·081	1·142	1·201	1·220	1·258	1·295	1·314	1·368
1/ 2500	1·1129	1·2341	1·3632	1·5002	1·7991	2·1323	2·3121	2·9066	3·5871	3·8338	4·3580	4·9242	5·2234	6·1874
0·000420	0·931	0·955	0·978	1·001	1·045	1·089	1·110	1·173	1·233	1·253	1·292	1·330	1·349	1·404
1/ 2381	1·1427	1·2671	1·3996	1·5403	1·8472	2·1891	2·3737	2·9840	3·6825	3·9358	4·4738	5·0549	5·3621	6·3515
0·000440	0·955	0·979	1·003	1·026	1·072	1·116	1·138	1·202	1·264	1·285	1·324	1·363	1·383	1·439
1/ 2273	1·1718	1·2994	1·4353	1·5796	1·8942	2·2447	2·4340	3·0596	3·7758	4·0354	4·5870	5·1828	5·4976	6·5119
0·000460	0·978	1·003	1·027	1·051	1·098	1·144	1·166	1·231	1·295	1·316	1·356	1·396	1·416	1·474
1/ 2174	1·2004	1·3311	1·4702	1·6179	1·9401	2·2992	2·4930	3·1337	3·8671	4·1330	4·6978	5·3079	5·6303	6·6689
0·000480	1·001	1·026	1·051	1·075	1·123	1·170	1·193	1·260	1·325	1·346	1·388	1·429	1·449	1·508
1/ 2083	1·2283	1·3620	1·5044	1·6556	1·9852	2·3525	2·5508	3·2063	3·9566	4·2286	4·8064	5·4305	5·7603	6·8228
0·000500	1·023	1·049	1·074	1·099	1·148	1·196	1·219	1·288	1·354	1·376	1·418	1·460	1·481	1·542
1/ 2000	1·2557	1·3924	1·5379	1·6924	2·0294	2·4048	2·6075	3·2775	4·0443	4·3223	4·9128	5·5507	5·8877	6·9736
0·000525	1·051	1·077	1·103	1·129	1·179	1·228	1·252	1·322	1·390	1·412	1·456	1·499	1·520	1·582
1/ 1905	1·2893	1·4296	1·5790	1·7376	2·0835	2·4689	2·6769	3·3646	4·1517	4·4370	5·0431	5·6978	6·0437	7·1582
0·000550	1·077	1·104	1·131	1·157	1·209	1·259	1·284	1·356	1·425	1·448	1·493	1·537	1·558	1·622
1/ 1818	1·3221	1·4660	1·6191	1·7817	2·1364	2·5315	2·7447	3·4498	4·2566	4·5491	5·1705	5·8416	6·1962	7·3386
0·000575	1·104	1·131	1·159	1·186	1·238	1·290	1·315	1·388	1·460	1·483	1·529	1·574	1·596	1·661
1/ 1739	1·3542	1·5016	1·6584	1·8250	2·1881	2·5928	2·8112	3·5331	4·3594	4·6589	5·2952	5·9824	6·3455	7·5152
0·000600	1·129	1·158	1·185	1·213	1·267	1·319	1·345	1·421	1·493	1·517	1·564	1·610	1·633	1·699
1/ 1667	1·3857	1·5365	1·6969	1·8673	2·2388	2·6528	2·8762	3·6148	4·4601	4·7664	5·4173	6·1202	6·4917	7·6882
0·000625	1·154	1·183	1·212	1·240	1·295	1·349	1·375	1·452	1·526	1·551	1·599	1·646	1·669	1·737
1/ 1600	1·4166	1·5707	1·7347	1·9088	2·2886	2·7117	2·9400	3·6949	4·5588	4·8719	5·5371	6·2555	6·6351	7·8579
0·000650	1·179	1·209	1·238	1·266	1·323	1·377	1·404	1·483	1·559	1·584	1·633	1·680	1·704	1·774
1/ 1538	1·4469	1·6042	1·7717	1·9496	2·3374	2·7695	3·0027	3·7735	4·6556	4·9754	5·6546	6·3881	6·7757	8·0243
0·000675	1·203	1·233	1·263	1·292	1·350	1·406	1·433	1·513	1·591	1·616	1·666	1·715	1·739	1·810
1/ 1481	1·4766	1·6372	1·8081	1·9896	2·3853	2·8262	3·0641	3·8507	4·7507	5·0770	5·7700	6·5184	6·9139	8·1878
0·000700	1·227	1·258	1·288	1·318	1·376	1·433	1·461	1·543	1·622	1·648	1·699	1·748	1·773	1·845
1/ 1429	1·5059	1·6696	1·8439	2·0289	2·4324	2·8820	3·1246	3·9265	4·8442	5·1768	5·8834	6·6464	7·0496	8·3484
0·000725	1·251	1·282	1·313	1·343	1·403	1·461	1·489	1·572	1·653	1·679	1·731	1·782	1·807	1·880
1/ 1379	1·5346	1·7015	1·8791	2·0676	2·4788	2·9368	3·1840	4·0011	4·9361	5·2750	5·9948	6·7723	7·1831	8·5062
0·000750	1·274	1·306	1·337	1·368	1·428	1·487	1·516	1·601	1·683	1·710	1·762	1·814	1·840	1·915
1/ 1333	1·5629	1·7328	1·9137	2·1057	2·5243	2·9907	3·2425	4·0744	5·0265	5·3715	6·1045	6·8961	7·3143	8·6616
0·000800	1·319	1·352	1·384	1·416	1·479	1·540	1·570	1·657	1·742	1·770	1·824	1·878	1·904	1·982
1/ 1250	1·6182	1·7941	1·9813	2·1800	2·6134	3·0961	3·3567	4·2178	5·2031	5·5602	6·3188	7·1380	7·5708	8·9650
0·000850	1·362	1·396	1·430	1·463	1·528	1·591	1·622	1·712	1·800	1·828	1·884	1·940	1·967	2·047
1/ 1176	1·6719	1·8536	2·0469	2·2522	2·6998	3·1985	3·4676	4·3569	5·3746	5·7434	6·5268	7·3728	7·8198	9·2596
0·000900	1·405	1·440	1·475	1·509	1·575	1·640	1·672	1·765	1·855	1·885	1·943	2·000	2·028	2·110
1/ 1111	1·7241	1·9114	2·1108	2·3225	2·7839	3·2980	3·5754	4·4923	5·5413	5·9215	6·7291	7·6011	8·0619	9·5460
0·000950	1·446	1·483	1·518	1·553	1·622	1·689	1·721	1·817	1·910	1·940	2·000	2·058	2·087	2·172
1/ 1053	1·7750	1·9678	2·1730	2·3909	2·8658	3·3949	3·6805	4·6241	5·7037	6·0950	6·9261	7·8235	8·2977	9·8249
0·001000	1·487	1·524	1·560	1·596	1·667	1·736	1·769	1·868	1·963	1·994	2·055	2·115	2·145	2·232
1/ 1000	1·8246	2·0228	2·2337	2·4576	2·9457	3·4895	3·7829	4·7527	5·8621	6·2642	7·1182	8·0404	8·5277	10·097
	0·96	0·96	0·96	0·96	0·96	0·96	0·97	0·97	0·97	0·97	0·97	0·97	0·97	0·98

$V_{r(0{\cdot}5)medial}$ **for half-full circular pipes.**

$k_s = 0{\cdot}015$ mm
S = 0·00100 to 0·00300

ie hydraulic gradient =
1 in 1000 to 1 in 333

Water (or sewage) at 15°C;
full bore conditions.

velocities in ms^{-1}
discharges in m^3s^{-1}

Gradient **(Equivalent) Pipe diameters in mm**

Gradient	1250	1300	1350	1400	1500	1600	1650	1800	1950	2000	2100	2200	2250	2400
0·00100	1·487	1·524	1·560	1·596	1·667	1·736	1·769	1·868	1·963	1·994	2·055	2·115	2·145	2·232
1/ 1000	1·8246	2·0228	2·2337	2·4576	2·9457	3·4895	3·7829	4·7527	5·8621	6·2642	7·1182	8·0404	8·5277	10·097
0·00105	1·526	1·564	1·602	1·639	1·711	1·781	1·816	1·917	2·015	2·047	2·109	2·171	2·201	2·291
1/ 952	1·8731	2·0765	2·2929	2·5228	3·0238	3·5818	3·8830	4·8782	6·0168	6·4294	7·3058	8·2522	8·7522	10·363
0·00110	1·565	1·604	1·642	1·680	1·754	1·826	1·862	1·965	2·065	2·098	2·162	2·225	2·256	2·348
1/ 909	1·9205	2·1290	2·3509	2·5865	3·1001	3·6721	3·9808	5·0010	6·1680	6·5909	7·4892	8·4592	8·9717	10·622
0·00115	1·603	1·643	1·682	1·721	1·797	1·870	1·906	2·012	2·115	2·148	2·214	2·279	2·310	2·404
1/ 870	1·9669	2·1804	2·4076	2·6488	3·1747	3·7605	4·0766	5·1211	6·3160	6·7490	7·6687	8·6618	9·1865	10·876
0·00120	1·640	1·681	1·721	1·760	1·838	1·913	1·950	2·059	2·163	2·198	2·265	2·331	2·363	2·459
1/ 833	2·0123	2·2307	2·4632	2·7099	3·2479	3·8470	4·1703	5·2387	6·4609	6·9038	7·8445	8·8602	9·3969	11·125
0·00125	1·676	1·718	1·759	1·799	1·879	1·956	1·993	2·104	2·211	2·246	2·315	2·382	2·415	2·513
1/ 800	2·0569	2·2801	2·5176	2·7698	3·3196	3·9319	4·2623	5·3541	6·6030	7·0556	8·0169	9·0547	9·6031	11·369
0·00130	1·712	1·754	1·796	1·837	1·918	1·997	2·036	2·149	2·258	2·293	2·363	2·432	2·466	2·566
1/ 769	2·1006	2·3285	2·5711	2·8286	3·3900	4·0152	4·3525	5·4673	6·7424	7·2045	8·1859	9·2456	9·8054	11·608
0·00135	1·747	1·790	1·833	1·875	1·957	2·038	2·077	2·192	2·303	2·340	2·411	2·481	2·516	2·618
1/ 741	2·1435	2·3761	2·6236	2·8863	3·4590	4·0969	4·4411	5·5784	6·8793	7·3507	8·3519	9·4329	10·004	11·843
0·00140	1·781	1·825	1·869	1·912	1·996	2·078	2·118	2·235	2·349	2·386	2·458	2·530	2·565	2·669
1/ 714	2·1857	2·4228	2·6751	2·9430	3·5269	4·1772	4·5281	5·6876	7·0138	7·4944	8·5150	9·6170	10·199	12·074
0·00145	1·815	1·860	1·904	1·948	2·034	2·117	2·158	2·277	2·393	2·430	2·505	2·578	2·613	2·719
1/ 690	2·2272	2·4688	2·7258	2·9988	3·5937	4·2562	4·6137	5·7949	7·1460	7·6356	8·6754	9·7980	10·391	12·301
0·00150	1·848	1·894	1·939	1·984	2·071	2·156	2·197	2·319	2·436	2·475	2·550	2·624	2·661	2·768
1/ 667	2·2680	2·5140	2·7757	3·0536	3·6593	4·3339	4·6979	5·9005	7·2761	7·7745	8·8331	9·9760	10·580	12·524
0·00160	1·913	1·961	2·007	2·053	2·143	2·231	2·274	2·400	2·521	2·561	2·639	2·716	2·754	2·865
1/ 625	2·3477	2·6023	2·8732	3·1608	3·7876	4·4857	4·8623	6·1068	7·5301	8·0458	9·1411	10·324	10·948	12·960
0·00170	1·976	2·025	2·073	2·121	2·214	2·304	2·349	2·478	2·604	2·645	2·726	2·805	2·844	2·958
1/ 588	2·4251	2·6880	2·9678	3·2648	3·9121	4·6330	5·0219	6·3069	7·7766	8·3092	9·4401	10·661	11·306	13·383
0·00180	2·037	2·088	2·138	2·187	2·282	2·375	2·421	2·555	2·684	2·726	2·809	2·891	2·931	3·049
1/ 556	2·5004	2·7714	3·0598	3·3659	4·0332	4·7762	5·1771	6·5016	8·0163	8·5652	9·7308	10·989	11·654	13·794
0·00190	2·097	2·149	2·200	2·251	2·349	2·445	2·492	2·629	2·762	2·806	2·891	2·975	3·016	3·138
1/ 526	2·5737	2·8526	3·1494	3·4644	4·1511	4·9157	5·3282	6·6911	8·2498	8·8145	10·014	11·309	11·992	14·195
0·00200	2·155	2·209	2·261	2·313	2·414	2·513	2·561	2·702	2·839	2·883	2·971	3·057	3·099	3·224
1/ 500	2·6452	2·9318	3·2367	3·5605	4·2660	5·0517	5·4756	6·8759	8·4774	9·0576	10·290	11·620	12·323	14·585
0·00210	2·212	2·267	2·321	2·374	2·478	2·579	2·628	2·773	2·913	2·959	3·049	3·137	3·180	3·308
1/ 476	2·7150	3·0091	3·3220	3·6543	4·3783	5·1845	5·6194	7·0564	8·6996	9·2949	10·559	11·924	12·645	14·966
0·00220	2·268	2·324	2·379	2·433	2·540	2·643	2·694	2·842	2·986	3·032	3·125	3·215	3·259	3·391
1/ 455	2·7832	3·0847	3·4054	3·7460	4·4880	5·3143	5·7601	7·2327	8·9167	9·5268	10·822	12·221	12·960	15·339
0·00230	2·322	2·380	2·436	2·492	2·600	2·706	2·758	2·910	3·057	3·105	3·199	3·291	3·337	3·471
1/ 435	2·8500	3·1586	3·4870	3·8357	4·5954	5·4413	5·8977	7·4053	9·1292	9·7538	11·080	12·512	13·268	15·703
0·00240	2·376	2·434	2·492	2·549	2·660	2·768	2·821	2·977	3·127	3·175	3·272	3·366	3·413	3·550
1/ 417	2·9154	3·2311	3·5670	3·9236	4·7006	5·5657	6·0325	7·5743	9·3373	9·9760	11·332	12·797	13·570	16·060
0·00250	2·428	2·488	2·547	2·605	2·718	2·829	2·883	3·042	3·195	3·245	3·343	3·440	3·487	3·627
1/ 400	2·9795	3·3021	3·6453	4·0097	4·8037	5·6877	6·1646	7·7400	9·5413	10·194	11·580	13·076	13·866	16·410
0·00260	2·479	2·540	2·600	2·660	2·776	2·888	2·944	3·105	3·262	3·313	3·413	3·512	3·560	3·703
1/ 385	3·0424	3·3718	3·7222	4·0942	4·9048	5·8073	6·2942	7·9025	9·7414	10·408	11·822	13·349	14·156	16·753
0·00270	2·530	2·592	2·653	2·714	2·832	2·947	3·003	3·168	3·328	3·380	3·482	3·582	3·632	3·778
1/ 370	3·1042	3·4402	3·7977	4·1772	5·0041	5·9248	6·4214	8·0620	9·9378	10·617	12·060	13·618	14·441	17·090
0·00280	2·579	2·642	2·705	2·766	2·887	3·004	3·062	3·230	3·392	3·445	3·550	3·652	3·702	3·851
1/ 357	3·1649	3·5074	3·8718	4·2587	5·1016	6·0401	6·5464	8·2187	10·131	10·823	12·294	13·882	14·721	17·421
0·00290	2·628	2·692	2·756	2·819	2·941	3·061	3·119	3·290	3·456	3·510	3·616	3·720	3·771	3·923
1/ 345	3·2245	3·5734	3·9447	4·3388	5·1975	6·1535	6·6693	8·3728	10·320	11·026	12·524	14·141	14·996	17·746
0·00300	2·675	2·741	2·806	2·870	2·995	3·116	3·176	3·350	3·518	3·573	3·681	3·787	3·839	3·993
1/ 333	3·2832	3·6384	4·0164	4·4176	5·2918	6·2651	6·7901	8·5243	10·507	11·225	12·750	14·396	15·266	18·066
	0·97	0·97	0·97	0·97	0·97	0·97	0·97	0·98	0·98	0·98	0·98	0·98	0·98	0·98

$V_{r(0{\cdot}5)medial}$ for half-full circular pipes.

$k_s = 0{\cdot}015$ mm S = 0·00100 to 0·00300

k_s = **0·015 mm**
S = 0·00300 to 0·01000

ie hydraulic gradient =
1 in 333 to 1 in 100

Water (or sewage) at 15°C;
full bore conditions.

velocities in ms^{-1}
discharges in m^3s^{-1}

Gradient **(Equivalent) Pipe diameters in mm**

Gradient	1250	1300	1350	1400	1500	1600	1650	1800	1950	2000	2100	2200	2250	2400
0·00300	2·675	2·741	2·806	2·870	2·995	3·116	3·176	3·350	3·518	3·573	3·681	3·787	3·839	3·993
1/ 333	3·2832	3·6384	4·0164	4·4176	5·2918	6·2651	6·7901	8·5243	10·507	11·225	12·750	14·396	15·266	18·066
0·00320	2·769	2·837	2·904	2·970	3·099	3·224	3·286	3·466	3·640	3·697	3·809	3·918	3·972	4·131
1/ 313	3·3978	3·7654	4·1564	4·5716	5·4760	6·4830	7·0261	8·8202	10·871	11·614	13·192	14·895	15·794	18·690
0·00340	2·859	2·930	2·999	3·067	3·200	3·330	3·393	3·579	3·759	3·817	3·932	4·045	4·101	4·265
1/ 294	3·5090	3·8885	4·2923	4·7209	5·6548	6·6944	7·2551	9·1073	11·225	11·992	13·620	15·378	16·307	19·296
0·00360	2·948	3·020	3·091	3·161	3·298	3·432	3·497	3·689	3·874	3·934	4·053	4·169	4·226	4·395
1/ 278	3·6172	4·0083	4·4244	4·8662	5·8286	6·8999	7·4777	9·3864	11·568	12·359	14·037	15·848	16·805	19·885
0·00380	3·033	3·108	3·181	3·253	3·394	3·531	3·599	3·795	3·986	4·048	4·170	4·289	4·348	4·522
1/ 263	3·7224	4·1249	4·5531	5·0076	5·9977	7·1000	7·6945	9·6581	11·903	12·716	14·442	16·305	17·290	20·458
0·00400	3·117	3·193	3·269	3·343	3·487	3·628	3·697	3·899	4·095	4·158	4·284	4·407	4·467	4·646
1/ 250	3·8251	4·2386	4·6785	5·1454	6·1627	7·2950	7·9058	9·9230	12·229	13·064	14·837	16·751	17·762	21·017
0·00420	3·199	3·277	3·354	3·430	3·578	3·723	3·794	4·001	4·201	4·267	4·395	4·521	4·583	4·766
1/ 238	3·9253	4·3496	4·8009	5·2800	6·3237	7·4855	8·1121	10·182	12·547	13·404	15·223	17·186	18·223	21·562
0·00440	3·278	3·359	3·438	3·515	3·668	3·816	3·888	4·100	4·305	4·372	4·504	4·633	4·697	4·884
1/ 227	4·0233	4·4581	4·9206	5·4115	6·4811	7·6716	8·3136	10·434	12·858	13·736	15·600	17·612	18·674	22·095
0·00460	3·357	3·439	3·519	3·599	3·755	3·906	3·980	4·198	4·407	4·476	4·610	4·742	4·808	4·999
1/ 217	4·1191	4·5642	5·0377	5·5402	6·6350	7·8536	8·5108	10·681	13·162	14·061	15·969	18·028	19·115	22·616
0·00480	3·433	3·517	3·600	3·681	3·840	3·995	4·071	4·293	4·507	4·577	4·715	4·850	4·916	5·112
1/ 208	4·2130	4·6681	5·1523	5·6662	6·7858	8·0319	8·7039	10·923	13·460	14·379	16·330	18·435	19·547	23·126
0·00500	3·508	3·594	3·678	3·761	3·924	4·082	4·159	4·386	4·605	4·676	4·817	4·955	5·023	5·223
1/ 200	4·3050	4·7700	5·2647	5·7897	6·9335	8·2066	8·8932	11·161	13·752	14·691	16·683	18·834	19·970	23·626
0·00525	3·600	3·688	3·774	3·859	4·026	4·188	4·267	4·500	4·724	4·798	4·942	5·083	5·153	5·358
1/ 190	4·4175	4·8946	5·4021	5·9408	7·1143	8·4203	9·1246	11·451	14·109	15·072	17·116	19·322	20·487	24·238
0·00550	3·689	3·779	3·868	3·955	4·126	4·292	4·373	4·611	4·841	4·916	5·064	5·209	5·280	5·490
1/ 182	4·5275	5·0164	5·5365	6·0885	7·2909	8·6291	9·3508	11·734	14·458	15·445	17·539	19·799	20·993	24·835
0·00575	3·777	3·869	3·960	4·049	4·224	4·393	4·477	4·720	4·955	5·032	5·183	5·331	5·404	5·619
1/ 174	4·6351	5·1355	5·6679	6·2329	7·4637	8·8334	9·5721	12·012	14·799	15·809	17·953	20·266	21·487	25·420
0·00600	3·863	3·957	4·050	4·141	4·319	4·493	4·578	4·827	5·067	5·146	5·300	5·451	5·526	5·746
1/ 167	4·7404	5·2522	5·7966	6·3743	7·6328	9·0334	9·7888	12·283	15·133	16·166	18·357	20·723	21·972	25·992
0·00625	3·947	4·043	4·138	4·231	4·413	4·590	4·677	4·931	5·177	5·257	5·415	5·569	5·645	5·869
1/ 160	4·8437	5·3665	5·9227	6·5129	7·7986	9·2295	10·001	12·549	15·461	16·515	18·754	21·170	22·446	26·553
0·00650	4·030	4·128	4·224	4·319	4·505	4·686	4·775	5·034	5·285	5·366	5·527	5·685	5·762	5·991
1/ 154	4·9450	5·4787	6·0464	6·6489	7·9612	9·4217	10·209	12·810	15·782	16·858	19·143	21·609	22·911	27·103
0·00675	4·111	4·211	4·309	4·406	4·595	4·780	4·870	5·135	5·390	5·473	5·637	5·798	5·877	6·110
1/ 148	5·0444	5·5887	6·1678	6·7823	8·1208	9·6104	10·414	13·066	16·097	17·195	19·525	22·040	23·368	27·642
0·00700	4·190	4·292	4·392	4·491	4·684	4·872	4·964	5·234	5·494	5·579	5·746	5·909	5·990	6·227
1/ 143	5·1420	5·6969	6·2870	6·9134	8·2776	9·7958	10·614	13·318	16·407	17·525	19·900	22·463	23·816	28·172
0·00725	4·268	4·372	4·474	4·575	4·771	4·963	5·056	5·331	5·596	5·682	5·852	6·019	6·101	6·343
1/ 138	5·2380	5·8031	6·4042	7·0422	8·4317	9·9779	10·812	13·565	16·711	17·850	20·269	22·879	24·257	28·693
0·00750	4·345	4·451	4·555	4·657	4·857	5·052	5·147	5·426	5·696	5·784	5·957	6·126	6·210	6·456
1/ 133	5·3324	5·9076	6·5195	7·1689	8·5832	10·157	11·006	13·808	17·010	18·170	20·631	23·288	24·690	29·205
0·00800	4·495	4·605	4·712	4·818	5·025	5·226	5·324	5·613	5·891	5·982	6·161	6·336	6·423	6·677
1/ 125	5·5167	6·1117	6·7446	7·4162	8·8790	10·507	11·385	14·283	17·594	18·793	21·339	24·086	25·536	30·205
0·00850	4·641	4·754	4·864	4·974	5·187	5·394	5·496	5·794	6·081	6·175	6·359	6·540	6·629	6·891
1/ 118	5·6955	6·3097	6·9629	7·6562	9·1660	10·846	11·752	14·743	18·161	19·398	22·025	24·860	26·357	31·175
0·00900	4·783	4·899	5·013	5·125	5·345	5·558	5·663	5·969	6·265	6·362	6·552	6·738	6·829	7·099
1/ 111	5·8693	6·5021	7·1751	7·8894	9·4449	11·176	12·109	15·190	18·711	19·986	22·692	25·612	27·154	32·117
0·00950	4·921	5·040	5·157	5·272	5·498	5·718	5·826	6·141	6·445	6·544	6·739	6·930	7·025	7·302
1/ 105	6·0384	6·6893	7·3817	8·1163	9·7164	11·497	12·457	15·626	19·247	20·558	23·341	26·344	27·930	33·034
0·01000	5·055	5·177	5·298	5·416	5·648	5·874	5·984	6·307	6·619	6·721	6·922	7·118	7·215	7·500
1/ 100	6·2033	6·8719	7·5830	8·3376	9·9809	11·809	12·795	16·050	19·769	21·115	23·974	27·058	28·686	33·927
	0·98	0·98	0·98	0·98	0·98	0·98	0·98	0·98	0·98	0·98	0·99	0·99	0·99	0·99

$V_{r(0·5)medial}$ for half-full circular pipes.

k_s = 0·015 mm
S = 0·01000 to 0·03000

ie hydraulic gradient =
1 in 100 to 1 in 33·3

Water (or sewage) at 15°C;
full bore conditions.

velocities in ms^{-1}
discharges in m^3s^{-1}

Gradient **(Equivalent) Pipe diameters in mm**

Gradient	1250	1300	1350	1400	1500	1600	1650	1800	1950	2000	2100	2200	2250	2400
0·01000	5·055	5·177	5·298	5·416	5·648	5·874	5·984	6·307	6·619	6·721	6·922	7·118	7·215	7·500
1/ 100	6·2033	6·8719	7·5830	8·3376	9·9809	11·809	12·795	16·050	19·769	21·115	23·974	27·058	28·686	33·927
0·01050	5·186	5·311	5·435	5·556	5·794	6·025	6·139	6·470	6·790	6·894	7·100	7·301	7·400	7·692
1/ 95	6·3642	7·0500	7·7795	8·5535	10·239	12·115	13·126	16·464	20·278	21·659	24·591	27·754	29·424	34·799
0·01100	5·314	5·443	5·569	5·693	5·937	6·174	6·290	6·629	6·957	7·064	7·274	7·480	7·582	7·881
1/ 91	6·5214	7·2241	7·9714	8·7644	10·491	12·413	13·449	16·869	20·776	22·191	25·194	28·434	30·145	35·651
0·01150	5·439	5·571	5·700	5·828	6·077	6·319	6·437	6·785	7·120	7·229	7·444	7·655	7·759	8·065
1/ 87	6·6752	7·3944	8·1592	8·9708	10·738	12·704	13·765	17·265	21·263	22·711	25·784	29·099	30·850	36·484
0·01200	5·562	5·696	5·829	5·959	6·213	6·461	6·582	6·937	7·279	7·391	7·611	7·826	7·932	8·245
1/ 83	6·8258	7·5611	8·3430	9·1728	10·980	12·990	14·074	17·652	21·739	23·219	26·361	29·750	31·540	37·299
0·01250	5·682	5·820	5·954	6·087	6·347	6·600	6·724	7·086	7·436	7·550	7·774	7·994	8·102	8·422
1/ 80	6·9733	7·7244	8·5232	9·3707	11·216	13·270	14·377	18·032	22·206	23·718	26·926	30·388	32·216	38·098
0·01300	5·800	5·940	6·078	6·213	6·479	6·736	6·863	7·232	7·589	7·705	7·934	8·159	8·269	8·595
1/ 77	7·1180	7·8846	8·6998	9·5648	11·449	13·544	14·674	18·404	22·664	24·207	27·481	31·013	32·879	38·881
0·01350	5·916	6·059	6·199	6·337	6·607	6·870	6·999	7·376	7·739	7·858	8·091	8·320	8·433	8·765
1/ 74	7·2600	8·0418	8·8732	9·7553	11·676	13·813	14·966	18·769	23·113	24·686	28·025	31·627	33·529	39·650
0·01400	6·030	6·175	6·318	6·459	6·734	7·002	7·133	7·517	7·887	8·008	8·246	8·479	8·593	8·931
1/ 71	7·3995	8·1962	9·0434	9·9424	11·900	14·078	15·252	19·128	23·555	25·157	28·560	32·230	34·168	40·405
0·01450	6·141	6·289	6·435	6·578	6·858	7·131	7·264	7·655	8·032	8·155	8·397	8·634	8·751	9·095
1/ 69	7·5366	8·3479	9·2107	10·126	12·120	14·338	15·533	19·480	23·988	25·620	29·085	32·822	34·796	41·146
0·01500	6·251	6·402	6·550	6·696	6·981	7·258	7·394	7·791	8·175	8·300	8·546	8·788	8·906	9·257
1/ 67	7·6713	8·4971	9·3753	10·307	12·336	14·593	15·810	19·827	24·414	26·075	29·601	33·404	35·413	41·876
0·01600	6·466	6·621	6·774	6·925	7·220	7·506	7·646	8·057	8·454	8·583	8·837	9·087	9·209	9·571
1/ 63	7·9344	8·7883	9·6964	10·660	12·758	15·092	16·350	20·503	25·247	26·964	30·609	34·541	36·617	43·299
0·01700	6·674	6·834	6·992	7·147	7·451	7·747	7·891	8·315	8·724	8·857	9·119	9·377	9·503	9·876
1/ 59	8·1896	9·0708	10·008	11·002	13·167	15·575	16·874	21·159	26·053	27·825	31·586	35·643	37·786	44·679
0·01800	6·876	7·041	7·203	7·363	7·676	7·980	8·129	8·565	8·986	9·123	9·394	9·658	9·789	10·17
1/ 56	8·4376	9·3453	10·311	11·335	13·565	16·045	17·383	21·796	26·837	28·662	32·536	36·714	38·920	46·019
0·01900	7·072	7·242	7·409	7·573	7·895	8·208	8·361	8·809	9·242	9·383	9·660	9·932	10·07	10·46
1/ 53	8·6789	9·6124	10·605	11·658	13·952	16·503	17·878	22·416	27·600	29·477	33·459	37·756	40·024	47·324
0·02000	7·264	7·438	7·610	7·778	8·108	8·429	8·587	9·047	9·491	9·635	9·920	10·20	10·34	10·74
1/ 50	8·9140	9·8727	10·892	11·974	14·329	16·948	18·360	23·021	28·343	30·270	34·359	38·771	41·100	48·595
0·02100	7·451	7·629	7·805	7·978	8·317	8·646	8·807	9·278	9·733	9·882	10·17	10·46	10·60	11·02
1/ 47·6	9·1435	10·127	11·172	12·281	14·697	17·383	18·831	23·610	29·068	31·044	35·238	39·761	42·150	49·835
0·02200	7·633	7·816	7·996	8·173	8·520	8·857	9·022	9·505	9·971	10·12	10·42	10·71	10·86	11·28
1/ 45·5	9·3676	10·375	11·446	12·582	15·056	17·808	19·291	24·186	29·777	31·800	36·096	40·729	43·175	51·046
0·02300	7·812	7·999	8·183	8·364	8·719	9·063	9·232	9·726	10·20	10·36	10·66	10·96	11·11	11·55
1/ 43·5	9·5868	10·617	11·713	12·876	15·407	18·223	19·741	24·749	30·469	32·540	36·934	41·675	44·177	52·230
0·02400	7·987	8·178	8·366	8·551	8·913	9·265	9·438	9·943	10·43	10·59	10·90	11·21	11·36	11·80
1/ 41·7	9·8014	10·855	11·975	13·164	15·751	18·629	20·181	25·301	31·147	33·264	37·755	42·600	45·159	53·389
0·02500	8·158	8·353	8·545	8·735	9·104	9·464	9·640	10·15	10·65	10·81	11·13	11·45	11·60	12·05
1/ 40·0	10·012	11·088	12·232	13·446	16·088	19·028	20·612	25·841	31·811	33·973	38·560	43·507	46·120	54·525
0·02600	8·326	8·525	8·721	8·914	9·291	9·658	9·837	10·36	10·87	11·04	11·36	11·68	11·84	12·30
1/ 38·5	10·218	11·316	12·484	13·722	16·419	19·418	21·035	26·370	32·463	34·668	39·348	44·397	47·062	55·638
0·02700	8·491	8·694	8·894	9·090	9·475	9·848	10·03	10·57	11·08	11·25	11·58	11·91	12·07	12·54
1/ 37·0	10·420	11·540	12·730	13·993	16·743	19·801	21·450	26·890	33·102	35·350	40·122	45·269	47·987	56·731
0·02800	8·653	8·860	9·063	9·263	9·655	10·04	10·22	10·77	11·29	11·47	11·80	12·13	12·30	12·78
1/ 35·7	10·619	11·760	12·973	14·260	17·062	20·178	21·857	27·400	33·729	36·020	40·882	46·126	48·895	57·803
0·02900	8·812	9·022	9·230	9·433	9·832	10·22	10·41	10·96	11·50	11·67	12·02	12·36	12·52	13·01
1/ 34·5	10·814	11·976	13·211	14·522	17·374	20·547	22·257	27·901	34·345	36·678	41·628	46·968	49·787	58·857
0·03000	8·968	9·183	9·393	9·601	10·01	10·40	10·59	11·16	11·70	11·88	12·23	12·57	12·74	13·24
1/ 33·3	11·006	12·188	13·445	14·779	17·682	20·911	22·651	28·394	34·951	37·325	42·362	47·795	50·664	59·893
	0·98	0·98	0·98	0·98	0·98	0·98	0·99	0·99	0·99	0·99	0·99	0·99	0·99	0·99

$V_{r(0\cdot5)medial}$ for half-full circular pipes.

k_s = 0·015 mm S = 0·01000 to 0·03000

k_s = **0·015 mm**
S = 0·03000 to 0·10000

ie hydraulic gradient =
1 in 33·3 to 1 in 10·0

Water (or sewage) at 15°C;
full bore conditions.

velocities in ms^{-1}
discharges in m^3s^{-1}

Gradient	**(Equivalent) Pipe diameters in mm**													
	1250	**1300**	**1350**	**1400**	**1500**	**1600**	**1650**	**1800**	**1950**	**2000**	**2100**	**2200**	**2250**	**2400**
0·03000	8·968	9·183	9·393	9·601	10·01	10·40	10·59	11·16	11·70	11·88	12·23	12·57	12·74	13·24
1/ 33·3	11·006	12·188	13·445	14·779	17·682	20·911	22·651	28·394	34·951	37·325	42·362	47·795	50·664	59·893
0·03200	9·274	9·495	9·713	9·927	10·35	10·75	10·95	11·54	12·10	12·28	12·64	13·00	13·17	13·69
1/ 31·3	11·380	12·603	13·902	15·281	18·282	21·620	23·419	29·356	36·134	38·588	43·794	49·410	52·375	61·914
0·03400	9·570	9·798	10·02	10·24	10·68	11·10	11·30	11·90	12·48	12·67	13·05	13·41	13·59	14·12
1/ 29·4	11·744	13·005	14·346	15·768	18·865	22·308	24·164	30·288	37·281	39·812	45·183	50·976	54·035	63·875
0·03600	9·857	10·09	10·32	10·55	11·00	11·43	11·64	12·26	12·86	13·05	13·43	13·81	14·00	14·54
1/ 27·8	12·097	13·395	14·776	16·241	19·430	22·976	24·887	31·194	38·395	41·001	46·532	52·497	55·646	65·778
0·03800	10·14	10·38	10·62	10·85	11·31	11·75	11·97	12·60	13·22	13·42	13·81	14·20	14·39	14·95
1/ 26·3	12·440	13·775	15·195	16·701	19·980	23·626	25·591	32·075	39·478	42·158	47·844	53·977	57·214	67·630
0·04000	10·41	10·66	10·90	11·14	11·61	12·07	12·29	12·94	13·57	13·78	14·18	14·58	14·77	15·35
1/ 25·0	12·774	14·146	15·603	17·150	20·516	24·259	26·277	32·934	40·534	43·285	49·122	55·418	58·742	69·435
0·04200	10·68	10·93	11·18	11·43	11·91	12·37	12·60	13·27	13·92	14·13	14·54	14·95	15·15	15·74
1/ 23·8	13·101	14·507	16·002	17·587	21·039	24·877	26·946	33·772	41·564	44·384	50·369	56·824	60·232	71·195
0·04400	10·94	11·20	11·45	11·70	12·19	12·67	12·91	13·59	14·25	14·47	14·89	15·31	15·51	16·12
1/ 22·7	13·420	14·860	16·391	18·015	21·550	25·481	27·599	34·590	42·570	45·458	51·587	58·198	61·687	72·914
0·04600	11·19	11·46	11·72	11·97	12·48	12·97	13·21	13·91	14·58	14·80	15·24	15·66	15·87	16·49
1/ 21·7	13·731	15·205	16·771	18·433	22·049	26·071	28·238	35·390	43·554	46·508	52·778	59·541	63·110	74·595
0·04800	11·44	11·71	11·98	12·24	12·75	13·25	13·50	14·21	14·91	15·13	15·57	16·01	16·22	16·85
1/ 20·8	14·037	15·542	17·143	18·842	22·538	26·648	28·864	36·173	44·516	47·536	53·944	60·855	64·503	76·240
0·05000	11·68	11·96	12·23	12·50	13·03	13·54	13·79	14·52	15·22	15·45	15·90	16·35	16·57	17·21
1/ 20·0	14·335	15·873	17·508	19·242	23·017	27·214	29·476	36·939	45·459	48·543	55·086	62·142	65·867	77·851
0·05250	11·98	12·26	12·54	12·82	13·36	13·88	14·14	14·88	15·61	15·84	16·31	16·76	16·99	17·64
1/ 19·0	14·701	16·278	17·954	19·732	23·603	27·906	30·225	37·877	46·612	49·773	56·481	63·716	67·535	79·821
0·05500	12·27	12·56	12·85	13·13	13·68	14·22	14·48	15·24	15·98	16·23	16·70	17·17	17·39	18·07
1/ 18·2	15·058	16·673	18·390	20·211	24·174	28·581	30·956	38·793	47·738	50·975	57·844	65·253	69·164	81·745
0·05750	12·55	12·85	13·14	13·43	14·00	14·54	14·81	15·60	16·35	16·60	17·09	17·56	17·80	18·49
1/ 17·4	15·407	17·059	18·815	20·678	24·733	29·242	31·672	39·688	48·838	52·150	59·177	66·755	70·756	83·625
0·06000	12·83	13·14	13·44	13·73	14·31	14·87	15·14	15·94	16·71	16·97	17·46	17·95	18·19	18·89
1/ 16·7	15·748	17·437	19·232	21·136	25·281	29·888	32·371	40·564	49·916	53·300	60·481	68·226	72·314	85·466
0·06250	13·11	13·42	13·72	14·02	14·61	15·18	15·46	16·28	17·07	17·32	17·83	18·33	18·57	19·29
1/ 16·0	16·083	17·807	19·640	21·585	25·817	30·521	33·057	41·422	50·971	54·427	61·759	69·666	73·841	87·269
0·06500	13·37	13·69	14·00	14·31	14·91	15·49	15·77	16·61	17·41	17·68	18·19	18·70	18·95	19·68
1/ 15·4	16·411	18·170	20·041	22·024	26·342	31·142	33·729	42·264	52·005	55·531	63·011	71·079	75·337	89·036
0·06750	13·64	13·96	14·28	14·59	15·20	15·79	16·08	16·93	17·75	18·02	18·55	19·06	19·32	20·06
1/ 14·8	16·733	18·527	20·433	22·456	26·858	31·751	34·389	43·089	53·020	56·614	64·240	72·464	76·805	90·770
0·07000	13·89	14·22	14·54	14·86	15·48	16·09	16·39	17·25	18·09	18·36	18·90	19·42	19·68	20·44
1/ 14·3	17·049	18·876	20·819	22·879	27·364	32·349	35·036	43·900	54·017	57·678	65·446	73·824	78·246	92·472
0·07250	14·15	14·48	14·81	15·13	15·77	16·38	16·68	17·56	18·41	18·69	19·24	19·77	20·04	20·81
1/ 13·8	17·359	19·220	21·198	23·295	27·861	32·937	35·672	44·696	54·996	58·723	66·632	75·160	79·662	94·145
0·07500	14·39	14·73	15·07	15·40	16·04	16·67	16·98	17·87	18·74	19·02	19·57	20·12	20·39	21·17
1/ 13·3	17·665	19·558	21·570	23·704	28·350	33·514	36·297	45·479	55·958	59·750	67·797	76·474	81·054	95·789
0·08000	14·88	15·23	15·58	15·92	16·58	17·23	17·55	18·47	19·37	19·66	20·23	20·79	21·07	21·88
1/ 12·5	18·260	20·217	22·297	24·503	29·304	34·641	37·517	47·006	57·836	61·755	70·070	79·037	83·770	98·997
0·08500	15·35	15·71	16·07	16·42	17·11	17·77	18·10	19·05	19·98	20·28	20·87	21·45	21·73	22·57
1/ 11·8	18·838	20·856	23·001	25·277	30·229	35·734	38·700	48·487	59·656	63·698	72·274	81·522	86·404	102·11
0·09000	15·81	16·18	16·55	16·91	17·61	18·30	18·64	19·62	20·57	20·88	21·48	22·08	22·37	23·24
1/ 11·1	19·399	21·477	23·686	26·028	31·127	36·795	39·849	49·925	61·424	65·585	74·414	83·935	88·961	105·13
0·09500	16·25	16·64	17·01	17·38	18·11	18·81	19·16	20·17	21·14	21·46	22·09	22·70	23·00	23·89
1/ 10·5	19·944	22·080	24·351	26·760	32·001	37·827	40·967	51·324	63·144	67·421	76·496	86·282	91·448	108·06
0·10000	16·69	17·08	17·47	17·85	18·59	19·31	19·67	20·70	21·70	22·03	22·67	23·30	23·61	24·52
1/ 10·0	20·476	22·669	25·000	27·472	32·852	38·833	42·055	52·686	64·819	69·209	78·524	88·569	93·871	110·93
	0·98	0·98	0·99	0·99	0·99	0·99	0·99	0·99	0·99	0·99	0·99	0·99	0·99	0·99

$V_{r(0{\cdot}5)medial}$ for half-full circular pipes.

S = 0·03000 to 0·10000 **k_s = 0·015 mm**

$k_s = 0{\cdot}030$ mm
S = 0·00010 to 0·00030

ie hydraulic gradient =
1 in 10000 to 1 in 3333

Water (or sewage) at 15°C;
full bore conditions.

velocities in ms^{-1}
discharges in m^3s^{-1}

Gradient	**(Equivalent) Pipe diameters in mm**													
	1250	**1300**	**1350**	**1400**	**1500**	**1600**	**1650**	**1800**	**1950**	**2000**	**2100**	**2200**	**2250**	**2400**
0·000100	0·424	0·434	0·445	0·456	0·476	0·497	0·506	0·535	0·563	0·572	0·590	0·608	0·617	0·643
1/ 10000	0·5199	0·5767	0·6372	0·7016	0·8419	0·9983	1·0828	1·3623	1·6824	1·7985	2·0452	2·3118	2·4527	2·9068
0·000105	0·435	0·446	0·457	0·468	0·489	0·510	0·520	0·550	0·578	0·588	0·606	0·624	0·633	0·660
1/ 9524	0·5339	0·5922	0·6544	0·7204	0·8645	1·0251	1·1118	1·3987	1·7273	1·8465	2·0998	2·3734	2·5181	2·9843
0·000110	0·446	0·458	0·469	0·480	0·502	0·523	0·533	0·564	0·593	0·603	0·622	0·640	0·649	0·676
1/ 9091	0·5476	0·6074	0·6711	0·7388	0·8866	1·0513	1·1402	1·4343	1·7713	1·8935	2·1531	2·4337	2·5820	3·0600
0·000115	0·457	0·469	0·480	0·492	0·514	0·536	0·546	0·577	0·608	0·617	0·637	0·656	0·665	0·693
1/ 8696	0·5610	0·6223	0·6875	0·7569	0·9082	1·0769	1·1680	1·4692	1·8143	1·9395	2·2054	2·4927	2·6446	3·1341
0·000120	0·468	0·480	0·492	0·503	0·526	0·548	0·559	0·591	0·622	0·632	0·652	0·671	0·681	0·709
1/ 8333	0·5741	0·6368	0·7036	0·7746	0·9294	1·1020	1·1952	1·5034	1·8565	1·9845	2·2566	2·5506	2·7059	3·2067
0·000125	0·478	0·491	0·503	0·514	0·538	0·560	0·571	0·604	0·635	0·646	0·666	0·686	0·696	0·725
1/ 8000	0·5870	0·6511	0·7194	0·7919	0·9502	1·1266	1·2219	1·5370	1·8978	2·0287	2·3068	2·6073	2·7661	3·2779
0·000130	0·489	0·501	0·513	0·526	0·549	0·572	0·584	0·617	0·649	0·660	0·680	0·701	0·711	0·740
1/ 7692	0·5996	0·6651	0·7349	0·8089	0·9706	1·1508	1·2481	1·5699	1·9384	2·0721	2·3561	2·6629	2·8251	3·3477
0·000135	0·499	0·511	0·524	0·536	0·561	0·584	0·596	0·630	0·662	0·673	0·694	0·715	0·725	0·755
1/ 7407	0·6120	0·6789	0·7501	0·8257	0·9906	1·1745	1·2738	1·6022	1·9783	2·1146	2·4044	2·7175	2·8830	3·4163
0·000140	0·509	0·522	0·534	0·547	0·572	0·596	0·608	0·642	0·676	0·686	0·708	0·729	0·739	0·770
1/ 7143	0·6242	0·6924	0·7650	0·8421	1·0103	1·1978	1·2991	1·6339	2·0174	2·1565	2·4520	2·7712	2·9400	3·4837
0·000145	0·518	0·532	0·545	0·558	0·583	0·607	0·619	0·654	0·688	0·700	0·721	0·743	0·754	0·785
1/ 6897	0·6362	0·7057	0·7797	0·8583	1·0297	1·2208	1·3240	1·6652	2·0559	2·1976	2·4987	2·8240	2·9960	3·5500
0·000150	0·528	0·542	0·555	0·568	0·593	0·618	0·631	0·666	0·701	0·712	0·735	0·757	0·767	0·799
1/ 6667	0·6481	0·7188	0·7941	0·8742	1·0487	1·2434	1·3484	1·6959	2·0938	2·2381	2·5447	2·8760	3·0511	3·6153
0·000160	0·547	0·561	0·575	0·588	0·615	0·640	0·653	0·690	0·726	0·738	0·761	0·783	0·794	0·827
1/ 6250	0·6711	0·7444	0·8224	0·9053	1·0860	1·2875	1·3962	1·7560	2·1679	2·3173	2·6346	2·9775	3·1587	3·7427
0·000170	0·565	0·580	0·594	0·608	0·635	0·662	0·675	0·713	0·750	0·762	0·786	0·809	0·821	0·855
1/ 5882	0·6936	0·7693	0·8498	0·9354	1·1222	1·3303	1·4427	1·8143	2·2398	2·3941	2·7219	3·0761	3·2633	3·8664
0·000180	0·583	0·598	0·612	0·627	0·655	0·682	0·696	0·735	0·773	0·786	0·810	0·834	0·846	0·881
1/ 5556	0·7154	0·7934	0·8765	0·9648	1·1574	1·3720	1·4878	1·8710	2·3097	2·4688	2·8068	3·1719	3·3649	3·9867
0·000190	0·600	0·616	0·631	0·645	0·674	0·703	0·716	0·757	0·796	0·809	0·834	0·859	0·871	0·907
1/ 5263	0·7366	0·8170	0·9025	0·9934	1·1916	1·4126	1·5318	1·9263	2·3778	2·5416	2·8895	3·2653	3·4639	4·1039
0·000200	0·617	0·633	0·648	0·663	0·693	0·722	0·736	0·778	0·818	0·832	0·858	0·883	0·895	0·932
1/ 5000	0·7574	0·8400	0·9279	1·0213	1·2251	1·4522	1·5748	1·9802	2·4443	2·6126	2·9701	3·3564	3·5605	4·2182
0·000210	0·634	0·650	0·666	0·681	0·712	0·741	0·756	0·799	0·840	0·854	0·880	0·906	0·919	0·957
1/ 4762	0·7776	0·8624	0·9527	1·0486	1·2577	1·4908	1·6167	2·0328	2·5092	2·6819	3·0489	3·4453	3·6548	4·3299
0·000220	0·650	0·666	0·683	0·699	0·730	0·760	0·775	0·819	0·861	0·875	0·903	0·929	0·942	0·981
1/ 4545	0·7974	0·8844	0·9769	1·0753	1·2897	1·5287	1·6577	2·0843	2·5726	2·7497	3·1259	3·5323	3·7471	4·4390
0·000230	0·666	0·682	0·699	0·715	0·748	0·779	0·794	0·839	0·882	0·896	0·924	0·952	0·965	1·005
1/ 4348	0·8168	0·9059	1·0007	1·1014	1·3210	1·5657	1·6978	2·1347	2·6348	2·8161	3·2013	3·6174	3·8373	4·5458
0·000240	0·681	0·698	0·715	0·732	0·765	0·797	0·812	0·858	0·903	0·917	0·946	0·974	0·987	1·028
1/ 4167	0·8358	0·9269	1·0239	1·1269	1·3516	1·6020	1·7371	2·1840	2·6956	2·8811	3·2751	3·7008	3·9257	4·6504
0·000250	0·696	0·714	0·731	0·748	0·782	0·814	0·830	0·877	0·923	0·937	0·966	0·995	1·009	1·051
1/ 4000	0·8545	0·9476	1·0467	1·1520	1·3816	1·6375	1·7757	2·2324	2·7552	2·9448	3·3475	3·7825	4·0124	4·7530
0·000260	0·711	0·729	0·747	0·764	0·799	0·832	0·848	0·896	0·942	0·957	0·987	1·016	1·031	1·073
1/ 3846	0·8728	0·9679	1·0691	1·1766	1·4111	1·6724	1·8135	2·2799	2·8138	3·0073	3·4185	3·8627	4·0974	4·8536
0·000270	0·726	0·744	0·762	0·780	0·815	0·849	0·865	0·914	0·961	0·977	1·007	1·037	1·052	1·095
1/ 3704	0·8907	0·9878	1·0911	1·2008	1·4401	1·7067	1·8506	2·3265	2·8712	3·0687	3·4883	3·9414	4·1809	4·9525
0·000280	0·740	0·759	0·777	0·795	0·831	0·866	0·883	0·932	0·980	0·996	1·027	1·057	1·072	1·116
1/ 3571	0·9084	1·0073	1·1127	1·2245	1·4685	1·7404	1·8871	2·3723	2·9277	3·1291	3·5568	4·0188	4·2630	5·0495
0·000290	0·754	0·773	0·792	0·811	0·847	0·882	0·899	0·950	0·999	1·015	1·046	1·077	1·092	1·137
1/ 3448	0·9257	1·0265	1·1339	1·2479	1·4965	1·7735	1·9230	2·4174	2·9832	3·1884	3·6242	4·0949	4·3436	5·1450
0·000300	0·768	0·788	0·807	0·826	0·862	0·898	0·916	0·967	1·017	1·033	1·065	1·097	1·112	1·158
1/ 3333	0·9428	1·0455	1·1548	1·2709	1·5240	1·8061	1·9583	2·4617	3·0379	3·2467	3·6905	4·1697	4·4230	5·2389
	0·94	0·94	0·94	0·94	0·94	0·95	0·95	0·95	0·96	0·96	0·96	0·96	0·96	0·96

$V_{r(0\cdot5)medial}$ for half-full circular pipes.

Gradient **(Equivalent) Pipe diameters in mm**

Gradient	1250	1300	1350	1400	1500	1600	1650	1800	1950	2000	2100	2200	2250	2400
0·000300	0·768	0·788	0·807	0·826	0·862	0·898	0·916	0·967	1·017	1·033	1·065	1·097	1·112	1·158
1/ 3333	0·9428	1·0455	1·1548	1·2709	1·5240	1·8061	1·9583	2·4617	3·0379	3·2467	3·6905	4·1697	4·4230	5·2389
0·000320	0·795	0·816	0·835	0·855	0·893	0·930	0·948	1·001	1·053	1·070	1·103	1·135	1·151	1·199
1/ 3125	0·9762	1·0824	1·1956	1·3157	1·5778	1·8697	2·0273	2·5483	3·1446	3·3607	3·8199	4·3159	4·5780	5·4223
0·000340	0·822	0·843	0·863	0·883	0·922	0·961	0·979	1·034	1·088	1·105	1·139	1·173	1·189	1·238
1/ 2941	1·0086	1·1183	1·2352	1·3593	1·6299	1·9315	2·0943	2·6323	3·2482	3·4714	3·9456	4·4578	4·7285	5·6003
0·000360	0·848	0·869	0·890	0·911	0·951	0·991	1·010	1·067	1·121	1·139	1·174	1·209	1·226	1·276
1/ 2778	1·0401	1·1533	1·2737	1·4017	1·6807	1·9916	2·1594	2·7141	3·3489	3·5790	4·0678	4·5958	4·8748	5·7734
0·000380	0·873	0·894	0·916	0·937	0·979	1·020	1·040	1·098	1·154	1·173	1·209	1·244	1·262	1·313
1/ 2632	1·0708	1·1873	1·3113	1·4430	1·7302	2·0501	2·2228	2·7937	3·4470	3·6838	4·1868	4·7301	5·0172	5·9420
0·000400	0·897	0·919	0·942	0·964	1·006	1·048	1·068	1·128	1·186	1·205	1·242	1·279	1·297	1·350
1/ 2500	1·1007	1·2204	1·3479	1·4833	1·7784	2·1072	2·2847	2·8713	3·5427	3·7860	4·3029	4·8612	5·1562	6·1063
0·000420	0·921	0·944	0·967	0·989	1·033	1·076	1·097	1·158	1·218	1·237	1·275	1·312	1·331	1·385
1/ 2381	1·1299	1·2528	1·3837	1·5226	1·8255	2·1629	2·3451	2·9471	3·6361	3·8858	4·4162	4·9891	5·2918	6·2668
0·000440	0·944	0·968	0·991	1·014	1·059	1·103	1·124	1·187	1·248	1·268	1·307	1·345	1·364	1·420
1/ 2273	1·1585	1·2845	1·4186	1·5610	1·8715	2·2174	2·4041	3·0212	3·7274	3·9833	4·5270	5·1141	5·4244	6·4237
0·000460	0·967	0·991	1·015	1·039	1·085	1·129	1·151	1·216	1·278	1·298	1·338	1·378	1·397	1·454
1/ 2174	1·1865	1·3155	1·4528	1·5987	1·9166	2·2708	2·4619	3·0937	3·8167	4·0788	4·6354	5·2365	5·5541	6·5772
0·000480	0·989	1·014	1·038	1·062	1·110	1·155	1·178	1·244	1·307	1·328	1·369	1·409	1·429	1·487
1/ 2083	1·2139	1·3459	1·4863	1·6355	1·9607	2·3230	2·5185	3·1648	3·9042	4·1723	4·7416	5·3563	5·6812	6·7275
0·000500	1·011	1·036	1·061	1·086	1·134	1·181	1·204	1·271	1·336	1·357	1·399	1·440	1·460	1·520
1/ 2000	1·2408	1·3756	1·5192	1·6716	2·0040	2·3742	2·5740	3·2344	3·9900	4·2639	4·8456	5·4738	5·8057	6·8748
0·000525	1·038	1·064	1·089	1·115	1·164	1·212	1·236	1·305	1·371	1·393	1·436	1·478	1·498	1·560
1/ 1905	1·2737	1·4121	1·5594	1·7159	2·0569	2·4369	2·6419	3·3196	4·0950	4·3760	4·9729	5·6175	5·9581	7·0551
0·000550	1·064	1·091	1·117	1·143	1·193	1·242	1·267	1·337	1·406	1·428	1·472	1·515	1·536	1·598
1/ 1818	1·3058	1·4477	1·5987	1·7591	2·1087	2·4981	2·7083	3·4029	4·1976	4·4856	5·0974	5·7580	6·1070	7·2312
0·000575	1·090	1·117	1·144	1·170	1·222	1·272	1·297	1·369	1·439	1·462	1·507	1·551	1·573	1·637
1/ 1739	1·3373	1·4826	1·6372	1·8014	2·1593	2·5581	2·7732	3·4844	4·2980	4·5928	5·2191	5·8954	6·2528	7·4036
0·000600	1·115	1·143	1·170	1·197	1·250	1·301	1·327	1·401	1·472	1·495	1·541	1·586	1·608	1·674
1/ 1667	1·3681	1·5167	1·6749	1·8428	2·2089	2·6168	2·8368	3·5642	4·3963	4·6979	5·3384	6·0300	6·3955	7·5724
0·000625	1·139	1·168	1·196	1·224	1·278	1·330	1·356	1·431	1·504	1·528	1·575	1·621	1·644	1·710
1/ 1600	1·3983	1·5502	1·7118	1·8834	2·2576	2·6743	2·8992	3·6424	4·4927	4·8008	5·4553	6·1620	6·5353	7·7378
0·000650	1·164	1·193	1·221	1·249	1·305	1·358	1·384	1·462	1·536	1·560	1·608	1·655	1·678	1·746
1/ 1538	1·4279	1·5830	1·7480	1·9233	2·3053	2·7308	2·9604	3·7192	4·5872	4·9018	5·5699	6·2914	6·6725	7·9001
0·000675	1·187	1·217	1·246	1·275	1·331	1·386	1·413	1·491	1·567	1·592	1·641	1·688	1·712	1·782
1/ 1481	1·4570	1·6152	1·7836	1·9624	2·3521	2·7862	3·0204	3·7945	4·6800	5·0009	5·6825	6·4184	6·8072	8·0595
0·000700	1·211	1·241	1·271	1·300	1·357	1·413	1·440	1·520	1·598	1·623	1·673	1·721	1·745	1·816
1/ 1429	1·4856	1·6469	1·8186	2·0008	2·3981	2·8407	3·0794	3·8685	4·7712	5·0983	5·7931	6·5432	6·9396	8·2159
0·000725	1·234	1·264	1·295	1·324	1·383	1·439	1·467	1·549	1·628	1·653	1·704	1·754	1·778	1·850
1/ 1379	1·5137	1·6781	1·8530	2·0386	2·4434	2·8942	3·1374	3·9413	4·8608	5·1940	5·9017	6·6659	7·0696	8·3698
0·000750	1·256	1·287	1·318	1·348	1·408	1·466	1·494	1·577	1·657	1·683	1·735	1·785	1·810	1·884
1/ 1333	1·5414	1·7087	1·8868	2·0758	2·4879	2·9468	3·1945	4·0128	4·9490	5·2882	6·0086	6·7865	7·1975	8·5211
0·000800	1·300	1·332	1·364	1·396	1·457	1·517	1·546	1·632	1·715	1·742	1·795	1·847	1·873	1·949
1/ 1250	1·5954	1·7685	1·9528	2·1484	2·5748	3·0497	3·3059	4·1526	5·1211	5·4721	6·2174	7·0222	7·4473	8·8165
0·000850	1·343	1·376	1·409	1·441	1·505	1·566	1·597	1·685	1·771	1·799	1·854	1·907	1·934	2·012
1/ 1176	1·6479	1·8266	2·0169	2·2189	2·6592	3·1495	3·4141	4·2883	5·2882	5·6505	6·4200	7·2508	7·6897	9·1031
0·000900	1·384	1·419	1·453	1·486	1·551	1·615	1·646	1·737	1·825	1·854	1·910	1·966	1·993	2·074
1/ 1111	1·6988	1·8831	2·0792	2·2874	2·7412	3·2465	3·5192	4·4201	5·4505	5·8239	6·6169	7·4730	7·9253	9·3817
0·000950	1·425	1·460	1·495	1·529	1·596	1·662	1·694	1·787	1·878	1·908	1·966	2·023	2·051	2·134
1/ 1053	1·7485	1·9381	2·1399	2·3541	2·8210	3·3410	3·6215	4·5485	5·6086	5·9928	6·8086	7·6893	8·1546	9·6530
0·001000	1·464	1·501	1·536	1·572	1·640	1·707	1·740	1·837	1·930	1·960	2·020	2·078	2·107	2·192
1/ 1000	1·7969	1·9917	2·1990	2·4191	2·8989	3·4331	3·7213	4·6736	5·7628	6·1574	6·9955	7·9002	8·3782	9·9174
	0·95	0·95	0·95	0·96	0·96	0·96	0·96	0·97	0·97	0·97	0·97	0·97	0·97	0·97

$V_{r(0\cdot5)medial}$ for half-full circular pipes.

k_s = 0·030 mm
S = 0·00100 to 0·00300

ie hydraulic gradient =
1 in 1000 to 1 in 333

Water (or sewage) at 15°C;
full bore conditions.

velocities in ms^{-1}
discharges in m^3s^{-1}

Gradient **(Equivalent) Pipe diameters in mm**

Gradient	1250	1300	1350	1400	1500	1600	1650	1800	1950	2000	2100	2200	2250	2400
0·00100	1·464	1·501	1·536	1·572	1·640	1·707	1·740	1·837	1·930	1·960	2·020	2·078	2·107	2·192
1/ 1000	1·7969	1·9917	2·1990	2·4191	2·8989	3·4331	3·7213	4·6736	5·7628	6·1574	6·9955	7·9002	8·3782	9·9174
0·00105	1·503	1·540	1·577	1·613	1·683	1·752	1·786	1·885	1·980	2·011	2·072	2·132	2·162	2·249
1/ 952	1·8441	2·0440	2·2568	2·4826	2·9749	3·5230	3·8187	4·7958	5·9132	6·3181	7·1779	8·1061	8·5965	10·175
0·00110	1·540	1·579	1·616	1·653	1·725	1·796	1·830	1·932	2·029	2·061	2·124	2·185	2·216	2·305
1/ 909	1·8903	2·0952	2·3132	2·5447	3·0491	3·6109	3·9139	4·9152	6·0602	6·4751	7·3561	8·3072	8·8097	10·428
0·00115	1·577	1·616	1·655	1·693	1·767	1·839	1·874	1·977	2·077	2·110	2·174	2·237	2·268	2·360
1/ 870	1·9355	2·1453	2·3685	2·6054	3·1218	3·6968	4·0071	5·0320	6·2041	6·6287	7·5305	8·5040	9·0183	10·674
0·00120	1·613	1·653	1·692	1·731	1·807	1·881	1·917	2·022	2·125	2·158	2·223	2·288	2·320	2·413
1/ 833	1·9798	2·1943	2·4226	2·6649	3·1930	3·7810	4·0983	5·1464	6·3449	6·7791	7·7012	8·6967	9·2226	10·916
0·00125	1·649	1·689	1·729	1·769	1·846	1·922	1·958	2·066	2·171	2·205	2·272	2·337	2·370	2·465
1/ 800	2·0231	2·2423	2·4756	2·7232	3·2627	3·8635	4·1877	5·2584	6·4829	6·9265	7·8686	8·8855	9·4227	11·152
0·00130	1·683	1·725	1·766	1·806	1·885	1·962	1·999	2·110	2·216	2·251	2·319	2·386	2·419	2·517
1/ 769	2·0657	2·2895	2·5276	2·7803	3·3312	3·9445	4·2753	5·3684	6·6183	7·0711	8·0327	9·0707	9·6191	11·385
0·00135	1·717	1·760	1·801	1·843	1·923	2·001	2·040	2·152	2·261	2·296	2·366	2·434	2·468	2·567
1/ 741	2·1075	2·3357	2·5786	2·8365	3·3983	4·0239	4·3614	5·4763	6·7512	7·2130	8·1938	9·2525	9·8118	11·612
0·00140	1·751	1·794	1·837	1·878	1·960	2·040	2·079	2·194	2·304	2·340	2·411	2·481	2·515	2·616
1/ 714	2·1485	2·3812	2·6288	2·8916	3·4643	4·1020	4·4459	5·5823	6·8817	7·3524	8·3520	9·4310	10·001	11·836
0·00145	1·784	1·828	1·871	1·914	1·997	2·078	2·118	2·235	2·347	2·384	2·456	2·527	2·562	2·665
1/ 690	2·1888	2·4259	2·6781	2·9458	3·5291	4·1787	4·5290	5·6865	7·0099	7·4894	8·5075	9·6065	10·187	12·056
0·00150	1·816	1·861	1·905	1·948	2·033	2·116	2·156	2·275	2·389	2·427	2·500	2·573	2·608	2·713
1/ 667	2·2285	2·4698	2·7265	2·9991	3·5929	4·2541	4·6107	5·7889	7·1361	7·6241	8·6604	9·7790	10·370	12·272
0·00160	1·879	1·925	1·971	2·016	2·104	2·189	2·231	2·354	2·472	2·511	2·587	2·661	2·698	2·806
1/ 625	2·3060	2·5556	2·8212	3·1031	3·7174	4·4014	4·7703	5·9890	7·3824	7·8871	8·9590	10·116	10·727	12·695
0·00170	1·940	1·988	2·035	2·081	2·172	2·260	2·303	2·430	2·552	2·592	2·670	2·747	2·785	2·897
1/ 588	2·3812	2·6389	2·9131	3·2041	3·8383	4·5443	4·9251	6·1831	7·6213	8·1423	9·2486	10·443	11·073	13·104
0·00180	2·000	2·049	2·098	2·145	2·238	2·329	2·374	2·504	2·630	2·671	2·751	2·831	2·870	2·985
1/ 556	2·4543	2·7198	3·0023	3·3022	3·9557	4·6832	5·0756	6·3717	7·8535	8·3903	9·5300	10·760	11·410	13·502
0·00190	2·058	2·108	2·158	2·207	2·303	2·396	2·442	2·576	2·705	2·748	2·831	2·912	2·952	3·070
1/ 526	2·5254	2·7986	3·0893	3·3978	4·0700	4·8184	5·2220	6·5553	8·0795	8·6316	9·8040	11·069	11·738	13·889
0·00200	2·114	2·166	2·217	2·268	2·366	2·462	2·509	2·646	2·779	2·822	2·908	2·991	3·032	3·154
1/ 500	2·5947	2·8754	3·1740	3·4909	4·1814	4·9501	5·3648	6·7342	8·2998	8·8669	10·071	11·371	12·057	14·267
0·00210	2·170	2·223	2·275	2·327	2·428	2·526	2·574	2·715	2·851	2·895	2·983	3·069	3·111	3·235
1/ 476	2·6624	2·9504	3·2567	3·5818	4·2902	5·0788	5·5041	6·9089	8·5148	9·0965	10·332	11·665	12·369	14·635
0·00220	2·223	2·278	2·332	2·384	2·488	2·588	2·638	2·782	2·921	2·967	3·056	3·144	3·187	3·315
1/ 455	2·7286	3·0236	3·3375	3·6706	4·3965	5·2044	5·6402	7·0795	8·7248	9·3208	10·586	11·952	12·673	14·995
0·00230	2·276	2·332	2·387	2·441	2·547	2·650	2·700	2·848	2·990	3·037	3·128	3·218	3·262	3·392
1/ 435	2·7933	3·0953	3·4165	3·7575	4·5004	5·3274	5·7734	7·2464	8·9303	9·5402	10·835	12·233	12·971	15·347
0·00240	2·328	2·385	2·441	2·496	2·604	2·709	2·761	2·912	3·058	3·105	3·199	3·290	3·336	3·469
1/ 417	2·8566	3·1654	3·4939	3·8426	4·6022	5·4477	5·9037	7·4099	9·1314	9·7549	11·079	12·508	13·262	15·692
0·00250	2·378	2·437	2·494	2·550	2·661	2·768	2·821	2·975	3·124	3·172	3·268	3·361	3·407	3·543
1/ 400	2·9187	3·2342	3·5698	3·9260	4·7019	5·5656	6·0315	7·5700	9·3285	9·9654	11·318	12·777	13·548	16·029
0·00260	2·428	2·487	2·546	2·603	2·716	2·826	2·879	3·037	3·188	3·238	3·335	3·431	3·478	3·616
1/ 385	2·9796	3·3016	3·6442	4·0077	4·7997	5·6813	6·1568	7·7270	9·5217	10·172	11·552	13·041	13·828	16·360
0·00270	2·477	2·537	2·597	2·656	2·770	2·882	2·937	3·097	3·252	3·302	3·402	3·499	3·547	3·688
1/ 370	3·0394	3·3678	3·7172	4·0880	4·8957	5·7948	6·2797	7·8811	9·7114	10·374	11·782	13·301	14·103	16·685
0·00280	2·525	2·586	2·647	2·707	2·824	2·938	2·993	3·157	3·314	3·366	3·467	3·566	3·615	3·759
1/ 357	3·0981	3·4328	3·7889	4·1668	4·9900	5·9063	6·4005	8·0324	9·8977	10·573	12·007	13·555	14·373	17·004
0·00290	2·572	2·634	2·696	2·757	2·876	2·992	3·049	3·215	3·375	3·428	3·531	3·632	3·682	3·828
1/ 345	3·1558	3·4967	3·8593	4·2442	5·0827	6·0159	6·5191	8·1812	10·081	10·769	12·229	13·806	14·638	17·318
0·00300	2·618	2·682	2·745	2·807	2·928	3·046	3·103	3·272	3·436	3·489	3·594	3·697	3·747	3·896
1/ 333	3·2125	3·5595	3·9286	4·3204	5·1738	6·1236	6·6358	8·3274	10·261	10·961	12·447	14·052	14·899	17·626
	0·96	0·96	0·96	0·97	0·97	0·97	0·97	0·97	0·97	0·98	0·98	0·98	0·98	0·98

$V_{r(0\cdot5)medial}$ for half-full circular pipes.

k_s = 0·030 mm S = 0·00100 to 0·00300

ie hydraulic gradient = 1 in 333 to 1 in 100

velocities in ms^{-1}
discharges in m^3s^{-1}

Gradient **(Equivalent) Pipe diameters in mm**

Gradient	1250	1300	1350	1400	1500	1600	1650	1800	1950	2000	2100	2200	2250	2400
0·00300	2·618	2·682	2·745	2·807	2·928	3·046	3·103	3·272	3·436	3·489	3·594	3·697	3·747	3·896
1/ 333	3·2125	3·5595	3·9286	4·3204	5·1738	6·1236	6·6358	8·3274	10·261	10·961	12·447	14·052	14·899	17·626
0·00320	2·708	2·774	2·839	2·903	3·028	3·150	3·210	3·385	3·553	3·608	3·717	3·823	3·875	4·029
1/ 313	3·3233	3·6822	4·0639	4·4691	5·3517	6·3339	6·8636	8·6129	10·612	11·336	12·873	14·532	15·408	18·228
0·00340	2·796	2·864	2·931	2·997	3·126	3·252	3·313	3·493	3·667	3·724	3·836	3·945	3·999	4·158
1/ 294	3·4308	3·8012	4·1952	4·6133	5·5242	6·5379	7·0846	8·8898	10·953	11·700	13·286	14·998	15·902	18·811
0·00360	2·881	2·951	3·020	3·088	3·221	3·350	3·414	3·599	3·778	3·837	3·952	4·064	4·120	4·284
1/ 278	3·5352	3·9168	4·3227	4·7535	5·6918	6·7361	7·2993	9·1588	11·284	12·053	13·687	15·450	16·381	19·378
0·00380	2·964	3·036	3·107	3·177	3·313	3·446	3·511	3·702	3·886	3·946	4·064	4·180	4·237	4·405
1/ 263	3·6368	4·0293	4·4468	4·8899	5·8550	6·9290	7·5082	9·4206	11·606	12·397	14·078	15·891	16·848	19·930
0·00400	3·044	3·118	3·191	3·263	3·403	3·540	3·607	3·802	3·991	4·053	4·174	4·293	4·352	4·524
1/ 250	3·7358	4·1390	4·5678	5·0228	6·0140	7·1170	7·7118	9·6758	11·920	12·733	14·458	16·320	17·303	20·468
0·00420	3·123	3·199	3·274	3·347	3·491	3·631	3·700	3·900	4·094	4·157	4·281	4·403	4·463	4·640
1/ 238	3·8325	4·2460	4·6858	5·1525	6·1692	7·3004	7·9105	9·9247	12·226	13·060	14·829	16·738	17·747	20·992
0·00440	3·200	3·278	3·354	3·429	3·577	3·720	3·790	3·996	4·194	4·259	4·386	4·511	4·572	4·754
1/ 227	3·9269	4·3506	4·8011	5·2793	6·3208	7·4797	8·1046	10·168	12·526	13·379	15·192	17·147	18·180	21·504
0·00460	3·275	3·355	3·433	3·510	3·661	3·807	3·879	4·089	4·292	4·358	4·488	4·616	4·679	4·864
1/ 217	4·0193	4·4528	4·9139	5·4032	6·4690	7·6549	8·2944	10·406	12·818	13·692	15·546	17·547	18·604	22·005
0·00480	3·349	3·430	3·510	3·589	3·743	3·893	3·966	4·181	4·388	4·456	4·589	4·719	4·783	4·973
1/ 208	4·1097	4·5529	5·0243	5·5246	6·6141	7·8265	8·4802	10·639	13·105	13·998	15·893	17·939	19·019	22·496
0·00500	3·421	3·504	3·586	3·666	3·823	3·976	4·051	4·270	4·482	4·551	4·687	4·820	4·886	5·079
1/ 200	4·1983	4·6510	5·1325	5·6435	6·7563	7·9946	8·6622	10·867	13·385	14·297	16·233	18·322	19·426	22·976
0·00525	3·509	3·594	3·678	3·761	3·922	4·078	4·155	4·380	4·597	4·668	4·807	4·943	5·011	5·209
1/ 190	4·3067	4·7710	5·2648	5·7889	6·9302	8·2001	8·8848	11·146	13·728	14·664	16·649	18·791	19·923	23·563
0·00550	3·596	3·683	3·768	3·853	4·018	4·178	4·257	4·487	4·709	4·782	4·924	5·064	5·133	5·335
1/ 182	4·4125	4·8881	5·3940	5·9309	7·1000	8·4009	9·1023	11·418	14·064	15·022	17·055	19·249	20·408	24·137
0·00575	3·680	3·769	3·857	3·943	4·112	4·276	4·356	4·592	4·819	4·893	5·039	5·182	5·252	5·459
1/ 174	4·5160	5·0028	5·5204	6·0698	7·2661	8·5972	9·3149	11·684	14·391	15·372	17·452	19·697	20·883	24·698
0·00600	3·763	3·854	3·943	4·031	4·204	4·372	4·454	4·694	4·926	5·002	5·151	5·297	5·369	5·581
1/ 167	4·6173	5·1149	5·6442	6·2057	7·4287	8·7894	9·5231	11·945	14·712	15·714	17·841	20·136	21·348	25·247
0·00625	3·843	3·936	4·028	4·118	4·294	4·465	4·549	4·795	5·032	5·109	5·261	5·410	5·483	5·700
1/ 160	4·7166	5·2248	5·7654	6·3389	7·5880	8·9777	9·7270	12·201	15·027	16·050	18·222	20·565	21·803	25·785
0·00650	3·923	4·018	4·111	4·203	4·382	4·557	4·643	4·893	5·135	5·214	5·369	5·521	5·596	5·816
1/ 154	4·8140	5·3326	5·8842	6·4695	7·7442	9·1624	9·9270	12·451	15·335	16·379	18·595	20·986	22·249	26·313
0·00675	4·001	4·097	4·192	4·286	4·469	4·647	4·734	4·990	5·236	5·316	5·474	5·629	5·706	5·931
1/ 148	4·9095	5·4384	6·0009	6·5977	7·8975	9·3435	10·123	12·697	15·637	16·702	18·961	21·400	22·687	26·830
0·00700	4·077	4·176	4·272	4·368	4·554	4·736	4·824	5·084	5·335	5·417	5·578	5·736	5·814	6·043
1/ 143	5·0033	5·5422	6·1154	6·7236	8·0480	9·5214	10·316	12·938	15·934	17·019	19·321	21·805	23·117	27·338
0·00725	4·152	4·252	4·351	4·448	4·638	4·823	4·913	5·178	5·433	5·516	5·680	5·841	5·920	6·153
1/ 138	5·0955	5·6443	6·2279	6·8472	8·1959	9·6962	10·505	13·175	16·226	17·330	19·675	22·204	23·539	27·837
0·00750	4·226	4·328	4·428	4·527	4·720	4·908	5·000	5·269	5·529	5·614	5·781	5·944	6·025	6·262
1/ 133	5·1861	5·7446	6·3386	6·9688	8·3413	9·8681	10·691	13·409	16·513	17·637	20·022	22·596	23·955	28·328
0·00800	4·370	4·476	4·579	4·681	4·881	5·075	5·170	5·448	5·717	5·804	5·976	6·145	6·229	6·474
1/ 125	5·3630	5·9405	6·5546	7·2062	8·6250	10·203	11·054	13·864	17·072	18·234	20·700	23·360	24·765	29·286
0·00850	4·510	4·619	4·725	4·831	5·036	5·237	5·335	5·621	5·898	5·989	6·166	6·340	6·426	6·679
1/ 118	5·5346	6·1304	6·7640	7·4363	8·9002	10·529	11·407	14·305	17·615	18·814	21·358	24·102	25·551	30·214
0·00900	4·646	4·758	4·868	4·976	5·188	5·394	5·495	5·790	6·075	6·168	6·351	6·530	6·618	6·878
1/ 111	5·7012	6·3149	6·9675	7·6598	9·1675	10·845	11·749	14·733	18·142	19·377	21·996	24·822	26·314	31·116
0·00950	4·778	4·893	5·006	5·117	5·335	5·547	5·650	5·954	6·247	6·342	6·530	6·714	6·805	7·072
1/ 105	5·8634	6·4944	7·1654	7·8774	9·4276	11·152	12·082	15·150	18·655	19·924	22·618	25·523	27·057	31·994
0·01000	4·907	5·025	5·141	5·255	5·478	5·696	5·802	6·113	6·414	6·512	6·705	6·894	6·987	7·261
1/ 100	6·0214	6·6693	7·3583	8·0893	9·6810	11·452	12·406	15·557	19·155	20·458	23·223	26·206	27·781	32·849
	0·97	0·97	0·97	0·97	0·97	0·98	0·98	0·98	0·98	0·98	0·98	0·98	0·98	0·98

$V_{r(0{\cdot}5)medial}$ **for half-full circular pipes.**

A37
(p.5 of 6)

k_s = 0·030 mm
S = 0·01000 to 0·03000

ie hydraulic gradient =
1 in 100 to 1 in 33·3

Water (or sewage) at 15°C;
full bore conditions.

velocities in ms^{-1}
discharges in m^3s^{-1}

Gradient **(Equivalent) Pipe diameters in mm**

Gradient	1250	1300	1350	1400	1500	1600	1650	1800	1950	2000	2100	2200	2250	2400
0·01000	4·907	5·025	5·141	5·255	5·478	5·696	5·802	6·113	6·414	6·512	6·705	6·894	6·987	7·261
1/ 100	6·0214	6·6693	7·3583	8·0893	9·6810	11·452	12·406	15·557	19·155	20·458	23·223	26·206	27·781	32·849
0·01050	5·032	5·153	5·272	5·389	5·618	5·841	5·950	6·269	6·577	6·678	6·875	7·069	7·164	7·446
1/ 95	6·1756	6·8400	7·5465	8·2961	9·9283	11·744	12·723	15·953	19·643	20·978	23·813	26·872	28·486	33·683
0·01100	5·155	5·279	5·401	5·520	5·755	5·983	6·095	6·421	6·737	6·839	7·042	7·240	7·338	7·626
1/ 91	6·3262	7·0067	7·7303	8·4981	10·170	12·029	13·032	16·340	20·119	21·487	24·390	27·522	29·176	34·497
0·01150	5·275	5·402	5·526	5·649	5·889	6·122	6·236	6·570	6·893	6·998	7·205	7·407	7·507	7·802
1/ 87	6·4734	7·1697	7·9100	8·6956	10·406	12·308	13·334	16·719	20·584	21·984	24·954	28·158	29·850	35·294
0·01200	5·392	5·522	5·649	5·774	6·019	6·257	6·374	6·716	7·045	7·153	7·364	7·571	7·673	7·974
1/ 83	6·6175	7·3292	8·0860	8·8889	10·637	12·581	13·630	17·089	21·040	22·470	25·506	28·781	30·510	36·073
0·01250	5·507	5·640	5·769	5·897	6·147	6·391	6·510	6·858	7·195	7·304	7·520	7·732	7·836	8·143
1/ 80	6·7587	7·4855	8·2583	9·0782	10·863	12·849	13·919	17·452	21·486	22·947	26·047	29·390	31·156	36·836
0·01300	5·620	5·755	5·887	6·018	6·273	6·521	6·643	6·998	7·341	7·453	7·673	7·889	7·995	8·308
1/ 77	6·8971	7·6387	8·4273	9·2639	11·085	13·111	14·203	17·808	21·924	23·414	26·577	29·988	31·789	37·585
0·01350	5·731	5·868	6·003	6·136	6·396	6·649	6·773	7·135	7·485	7·599	7·823	8·043	8·151	8·470
1/ 74	7·0330	7·7890	8·5930	9·4460	11·303	13·369	14·482	18·157	22·353	23·872	27·096	30·574	32·411	38·319
0·01400	5·840	5·979	6·117	6·252	6·517	6·775	6·901	7·270	7·626	7·742	7·971	8·195	8·305	8·630
1/ 71	7·1663	7·9367	8·7558	9·6249	11·517	13·621	14·756	18·499	22·775	24·322	27·607	31·150	33·021	39·040
0·01450	5·946	6·089	6·229	6·367	6·636	6·898	7·027	7·402	7·765	7·883	8·115	8·343	8·456	8·786
1/ 69	7·2973	8·0817	8·9157	9·8006	11·727	13·870	15·025	18·836	23·189	24·765	28·109	31·716	33·620	39·748
0·01500	6·051	6·196	6·339	6·479	6·753	7·020	7·150	7·532	7·901	8·021	8·258	8·490	8·604	8·940
1/ 67	7·4262	8·2243	9·0730	9·9733	11·934	14·114	15·289	19·167	23·596	25·199	28·602	32·272	34·209	40·444
0·01600	6·256	6·406	6·553	6·698	6·981	7·256	7·391	7·786	8·167	8·291	8·536	8·775	8·893	9·240
1/ 63	7·6776	8·5026	9·3798	10·310	12·337	14·590	15·805	19·813	24·390	26·047	29·564	33·357	35·359	41·803
0·01700	6·455	6·609	6·761	6·910	7·202	7·486	7·625	8·032	8·425	8·553	8·805	9·052	9·173	9·532
1/ 59	7·9213	8·7724	9·6773	10·637	12·727	15·052	16·305	20·439	25·161	26·870	30·497	34·409	36·474	43·120
0·01800	6·648	6·807	6·963	7·116	7·417	7·709	7·852	8·271	8·675	8·807	9·066	9·321	9·446	9·814
1/ 56	8·1581	9·0344	9·9663	10·955	13·107	15·500	16·790	21·048	25·909	27·668	31·403	35·430	37·557	44·399
0·01900	6·835	6·999	7·159	7·317	7·626	7·926	8·073	8·504	8·919	9·054	9·321	9·582	9·711	10·09
1/ 53	8·3884	9·2894	10·247	11·264	13·476	15·937	17·263	21·639	26·636	28·445	32·284	36·424	38·610	45·644
0·02000	7·018	7·186	7·350	7·512	7·829	8·138	8·289	8·730	9·156	9·295	9·569	9·837	9·969	10·36
1/ 50	8·6128	9·5378	10·521	11·565	13·836	16·362	17·723	22·216	27·345	29·202	33·142	37·392	39·636	46·856
0·02100	7·197	7·368	7·537	7·703	8·028	8·344	8·499	8·951	9·388	9·530	9·811	10·09	10·22	10·62
1/ 47·6	8·8317	9·7801	10·788	11·858	14·187	16·776	18·172	22·778	28·037	29·940	33·980	38·337	40·637	48·038
0·02200	7·371	7·547	7·719	7·889	8·222	8·545	8·704	9·167	9·614	9·760	10·05	10·33	10·47	10·87
1/ 45·5	9·0455	10·017	11·049	12·145	14·529	17·181	18·610	23·327	28·712	30·661	34·798	39·259	41·615	49·193
0·02300	7·541	7·721	7·898	8·071	8·412	8·742	8·904	9·378	9·835	9·984	10·28	10·56	10·71	11·12
1/ 43·5	9·2546	10·248	11·304	12·425	14·864	17·577	19·039	23·864	29·372	31·366	35·597	40·161	42·570	50·322
0·02400	7·708	7·891	8·072	8·250	8·597	8·935	9·100	9·584	10·05	10·20	10·50	10·80	10·94	11·37
1/ 41·7	9·4591	10·475	11·554	12·699	15·192	17·965	19·459	24·389	30·018	32·056	36·380	41·043	43·505	51·426
0·02500	7·871	8·059	8·243	8·424	8·779	9·124	9·293	9·787	10·26	10·42	10·72	11·02	11·17	11·61
1/ 40·0	9·6595	10·696	11·799	12·968	15·514	18·344	19·870	24·904	30·651	32·732	37·146	41·907	44·421	52·508
0·02600	8·031	8·222	8·410	8·595	8·957	9·309	9·481	9·985	10·47	10·63	10·94	11·25	11·40	11·84
1/ 38·5	9·8560	10·914	12·038	13·231	15·828	18·716	20·273	25·408	31·272	33·394	37·897	42·755	45·319	53·569
0·02700	8·188	8·383	8·575	8·763	9·132	9·490	9·666	10·18	10·67	10·84	11·15	11·47	11·62	12·07
1/ 37·0	10·049	11·127	12·274	13·490	16·137	19·081	20·668	25·903	31·880	34·044	38·635	43·586	46·200	54·610
0·02800	8·343	8·541	8·736	8·928	9·303	9·668	9·847	10·37	10·88	11·04	11·36	11·68	11·84	12·30
1/ 35·7	10·238	11·337	12·505	13·744	16·441	19·440	21·056	26·389	32·478	34·682	39·358	44·402	47·065	55·631
0·02900	8·494	8·696	8·894	9·090	9·472	9·844	10·03	10·56	11·07	11·24	11·57	11·89	12·05	12·52
1/ 34·5	10·424	11·542	12·731	13·993	16·739	19·792	21·437	26·867	33·065	35·308	40·069	45·203	47·914	56·635
0·03000	8·643	8·848	9·050	9·249	9·638	10·02	10·20	10·74	11·26	11·44	11·77	12·10	12·26	12·74
1/ 33·3	10·607	11·745	12·954	14·238	17·031	20·138	21·812	27·336	33·642	35·924	40·768	45·991	48·749	57·621
	0·97	0·98	0·98	0·98	0·98	0·98	0·98	0·98	0·98	0·98	0·98	0·98	0·98	0·99

$V_{r(0·5)medial}$ for half-full circular pipes.

k_s = 0·030 mm S = 0·01000 to 0·03000

k_s = 0·030 mm
S = 0·03000 to 0·10000

ie hydraulic gradient =
1 in 33·3 to 1 in 10·0

Water (or sewage) at 15°C;
full bore conditions.

velocities in ms^{-1}
discharges in m^3s^{-1}

Gradient	**(Equivalent) Pipe diameters in mm**													
	1250	**1300**	**1350**	**1400**	**1500**	**1600**	**1650**	**1800**	**1950**	**2000**	**2100**	**2200**	**2250**	**2400**
0·03000	8·643	8·848	9·050	9·249	9·638	10·02	10·20	10·74	11·26	11·44	11·77	12·10	12·26	12·74
1/ 33·3	10·607	11·745	12·954	14·238	17·031	20·138	21·812	27·336	33·642	35·924	40·768	45·991	48·749	57·621
0·03200	8·934	9·146	9·354	9·560	9·961	10·35	10·54	11·10	11·64	11·82	12·16	12·50	12·67	13·16
1/ 31·3	10·963	12·139	13·390	14·716	17·603	20·813	22·543	28·251	34·768	37·127	42·131	47·529	50·379	59·546
0·03400	9·215	9·434	9·649	9·861	10·27	10·68	10·87	11·45	12·01	12·19	12·55	12·90	13·07	13·58
1/ 29·4	11·309	12·522	13·812	15·180	18·157	21·468	23·252	29·139	35·860	38·292	43·453	49·020	51·958	61·412
0·03600	9·489	9·714	9·935	10·15	10·58	10·99	11·20	11·79	12·36	12·55	12·92	13·28	13·45	13·98
1/ 27·8	11·645	12·894	14·221	15·630	18·695	22·103	23·940	30·001	36·920	39·423	44·737	50·467	53·492	63·224
0·03800	9·755	9·986	10·21	10·44	10·88	11·30	11·51	12·12	12·71	12·90	13·28	13·65	13·83	14·37
1/ 26·3	11·971	13·255	14·620	16·068	19·219	22·722	24·610	30·840	37·951	40·524	45·985	51·875	54·984	64·987
0·04000	10·01	10·25	10·49	10·71	11·16	11·60	11·81	12·44	13·04	13·24	13·63	14·01	14·19	14·74
1/ 25·0	12·290	13·607	15·008	16·494	19·728	23·324	25·262	31·656	38·955	41·597	47·202	53·247	56·438	66·704
0·04200	10·27	10·51	10·75	10·99	11·45	11·89	12·11	12·75	13·37	13·57	13·97	14·36	14·55	15·11
1/ 23·8	12·600	13·951	15·387	16·910	20·226	23·912	25·899	32·453	39·935	42·642	48·388	54·584	57·855	68·378
0·04400	10·51	10·76	11·01	11·25	11·72	12·18	12·40	13·06	13·69	13·90	14·30	14·70	14·90	15·48
1/ 22·7	12·903	14·286	15·757	17·317	20·711	24·486	26·520	33·231	40·891	43·664	49·546	55·891	59·240	70·014
0·04600	10·76	11·01	11·26	11·51	11·99	12·46	12·69	13·36	14·01	14·22	14·63	15·04	15·24	15·83
1/ 21·7	13·199	14·614	16·118	17·714	21·186	25·047	27·127	33·992	41·827	44·662	50·679	57·168	60·593	71·612
0·04800	10·99	11·25	11·51	11·76	12·25	12·73	12·96	13·65	14·31	14·53	14·95	15·37	15·57	16·18
1/ 20·8	13·489	14·935	16·472	18·103	21·651	25·596	27·722	34·736	42·742	45·639	51·787	58·417	61·917	73·177
0·05000	11·22	11·49	11·75	12·01	12·51	13·00	13·24	13·94	14·61	14·83	15·27	15·69	15·90	16·51
1/ 20·0	13·774	15·250	16·819	18·483	22·106	26·134	28·304	35·465	43·638	46·596	52·873	59·641	63·215	74·709
0·05250	11·51	11·78	12·05	12·31	12·82	13·32	13·57	14·29	14·98	15·20	15·65	16·08	16·30	16·93
1/ 19·0	14·121	15·634	17·243	18·949	22·662	26·791	29·016	36·356	44·734	47·766	54·200	61·138	64·800	76·582
0·05500	11·78	12·06	12·34	12·60	13·13	13·64	13·89	14·63	15·34	15·57	16·02	16·47	16·69	17·33
1/ 18·2	14·460	16·009	17·656	19·403	23·206	27·433	29·711	37·226	45·804	48·908	55·495	62·599	66·348	78·411
0·05750	12·05	12·34	12·62	12·89	13·43	13·96	14·21	14·96	15·69	15·92	16·39	16·84	17·07	17·73
1/ 17·4	14·791	16·376	18·061	19·848	23·737	28·060	30·390	38·077	46·850	50·025	56·762	64·027	67·862	80·199
0·06000	12·32	12·61	12·89	13·18	13·73	14·26	14·52	15·29	16·03	16·27	16·75	17·21	17·44	18·11
1/ 16·7	15·116	16·735	18·457	20·283	24·257	28·674	31·055	38·909	47·874	51·118	58·001	65·425	69·343	81·948
0·06250	12·58	12·87	13·17	13·45	14·01	14·56	14·83	15·61	16·37	16·61	17·10	17·57	17·80	18·49
1/ 16·0	15·433	17·087	18·844	20·709	24·766	29·276	31·706	39·725	48·876	52·188	59·215	66·794	70·794	83·662
0·06500	12·83	13·13	13·43	13·72	14·30	14·85	15·13	15·92	16·70	16·95	17·44	17·92	18·16	18·86
1/ 15·4	15·745	17·432	19·224	21·126	25·265	29·865	32·345	40·524	49·859	53·237	60·405	68·135	72·216	85·342
0·06750	13·08	13·39	13·69	13·99	14·57	15·14	15·42	16·23	17·02	17·27	17·78	18·27	18·51	19·23
1/ 14·8	16·051	17·770	19·597	21·536	25·754	30·444	32·971	41·308	50·823	54·267	61·573	69·452	73·611	86·989
0·07000	13·32	13·64	13·95	14·25	14·85	15·42	15·71	16·54	17·33	17·60	18·11	18·61	18·86	19·59
1/ 14·3	16·350	18·102	19·963	21·938	26·235	31·012	33·586	42·078	51·770	55·277	62·719	70·744	74·980	88·607
0·07250	13·56	13·88	14·20	14·51	15·11	15·70	15·99	16·83	17·65	17·91	18·43	18·94	19·20	19·94
1/ 13·8	16·645	18·428	20·323	22·333	26·707	31·569	34·190	42·834	52·700	56·270	63·845	72·013	76·325	90·196
0·07500	13·80	14·12	14·44	14·76	15·38	15·97	16·27	17·12	17·95	18·22	18·75	19·27	19·53	20·28
1/ 13·3	16·935	18·748	20·676	22·721	27·171	32·118	34·783	43·577	53·613	57·245	64·951	73·261	77·648	91·758
0·08000	14·26	14·60	14·93	15·25	15·89	16·51	16·81	17·69	18·55	18·83	19·38	19·91	20·18	20·96
1/ 12·5	17·500	19·374	21·366	23·479	28·076	33·187	35·942	45·027	55·397	59·149	67·111	75·696	80·228	94·806
0·08500	14·71	15·05	15·39	15·73	16·38	17·02	17·33	18·25	19·13	19·42	19·98	20·53	20·81	21·61
1/ 11·8	18·048	19·980	22·034	24·213	28·954	34·224	37·064	46·433	57·125	60·994	69·203	78·056	82·729	97·760
0·09000	15·14	15·50	15·85	16·19	16·87	17·52	17·84	18·78	19·69	19·99	20·57	21·14	21·42	22·24
1/ 11·1	18·580	20·569	22·684	24·926	29·806	35·231	38·155	47·798	58·804	62·786	71·236	80·348	85·157	100·63
0·09500	15·56	15·93	16·29	16·64	17·34	18·01	18·34	19·31	20·24	20·54	21·14	21·72	22·01	22·86
1/ 10·5	19·098	21·142	23·315	25·620	30·636	36·211	39·215	49·126	60·436	64·529	73·213	82·576	87·519	103·42
0·10000	15·97	16·35	16·72	17·08	17·79	18·48	18·82	19·81	20·77	21·08	21·69	22·29	22·59	23·46
1/ 10·0	19·602	21·700	23·930	26·295	31·443	37·165	40·248	50·419	62·026	66·226	75·138	84·748	89·820	106·14
	0·98	0·98	0·98	0·98	0·98	0·98	0·98	0·98	0·98	0·98	0·99	0·99	0·99	0·99

$V_{r(0·5)medial}$ for half-full circular pipes.

A38
(p.1 of 6)

k_s = 0·060 mm
S = 0·00010 to 0·00030

ie hydraulic gradient =
1 in 10000 to 1 in 3333

Water (or sewage) at 15°C;
full bore conditions.

velocities in ms^{-1}
discharges in m^3s^{-1}

Gradient **(Equivalent) Pipe diameters in mm**

Gradient	1250	1300	1350	1400	1500	1600	1650	1800	1950	2000	2100	2200	2250	2400
0·000100	0·419	0·429	0·440	0·450	0·471	0·490	0·500	0·528	0·556	0·565	0·583	0·600	0·608	0·634
1/ 10000	0·5138	0·5699	0·6296	0·6931	0·8315	0·9858	1·0691	1·3446	1·6601	1·7745	2·0176	2·2802	2·4191	2·8663
0·000105	0·430	0·441	0·452	0·462	0·483	0·503	0·513	0·542	0·571	0·580	0·598	0·616	0·624	0·650
1/ 9524	0·5275	0·5851	0·6464	0·7115	0·8536	1·0120	1·0975	1·3803	1·7041	1·8215	2·0710	2·3405	2·4829	2·9419
0·000110	0·441	0·452	0·463	0·474	0·495	0·516	0·526	0·556	0·585	0·594	0·613	0·631	0·640	0·667
1/ 9091	0·5409	0·5999	0·6628	0·7296	0·8752	1·0376	1·1253	1·4151	1·7471	1·8674	2·1231	2·3994	2·5454	3·0159
0·000115	0·451	0·463	0·474	0·485	0·507	0·529	0·539	0·570	0·599	0·609	0·628	0·646	0·656	0·683
1/ 8696	0·5540	0·6145	0·6789	0·7472	0·8964	1·0627	1·1524	1·4493	1·7891	1·9123	2·1742	2·4570	2·6065	3·0882
0·000120	0·462	0·474	0·485	0·497	0·519	0·541	0·551	0·583	0·613	0·623	0·642	0·661	0·671	0·698
1/ 8333	0·5669	0·6287	0·6946	0·7646	0·9171	1·0872	1·1790	1·4827	1·8303	1·9564	2·2242	2·5135	2·6664	3·1590
0·000125	0·472	0·484	0·496	0·508	0·531	0·553	0·564	0·596	0·626	0·636	0·656	0·676	0·685	0·714
1/ 8000	0·5795	0·6427	0·7100	0·7815	0·9375	1·1113	1·2051	1·5154	1·8707	1·9995	2·2732	2·5688	2·7251	3·2285
0·000130	0·482	0·495	0·507	0·519	0·542	0·564	0·576	0·608	0·640	0·650	0·670	0·690	0·700	0·729
1/ 7692	0·5919	0·6564	0·7252	0·7982	0·9574	1·1349	1·2307	1·5476	1·9103	2·0418	2·3213	2·6231	2·7826	3·2966
0·000135	0·492	0·505	0·517	0·529	0·553	0·576	0·587	0·621	0·653	0·663	0·684	0·704	0·714	0·743
1/ 7407	0·6040	0·6699	0·7400	0·8145	0·9770	1·1581	1·2559	1·5791	1·9492	2·0834	2·3685	2·6764	2·8392	3·3635
0·000140	0·502	0·515	0·527	0·540	0·564	0·587	0·599	0·633	0·665	0·676	0·697	0·718	0·728	0·758
1/ 7143	0·6159	0·6831	0·7546	0·8306	0·9963	1·1809	1·2806	1·6101	1·9874	2·1242	2·4148	2·7288	2·8947	3·4292
0·000145	0·511	0·524	0·537	0·550	0·574	0·598	0·610	0·645	0·678	0·689	0·710	0·731	0·742	0·772
1/ 6897	0·6277	0·6961	0·7690	0·8464	1·0152	1·2033	1·3048	1·6406	2·0250	2·1644	2·4604	2·7803	2·9493	3·4938
0·000150	0·521	0·534	0·547	0·560	0·585	0·609	0·621	0·657	0·690	0·702	0·723	0·745	0·755	0·786
1/ 6667	0·6392	0·7089	0·7831	0·8619	1·0338	1·2253	1·3287	1·6706	2·0620	2·2038	2·5053	2·8309	3·0030	3·5573
0·000160	0·539	0·553	0·566	0·580	0·606	0·631	0·643	0·680	0·715	0·726	0·749	0·771	0·782	0·814
1/ 6250	0·6618	0·7339	0·8107	0·8923	1·0702	1·2684	1·3754	1·7292	2·1341	2·2810	2·5929	2·9298	3·1078	3·6814
0·000170	0·557	0·571	0·585	0·599	0·626	0·652	0·664	0·702	0·738	0·750	0·773	0·796	0·807	0·840
1/ 5882	0·6837	0·7582	0·8375	0·9218	1·1055	1·3102	1·4207	1·7860	2·2042	2·3558	2·6779	3·0258	3·2096	3·8018
0·000180	0·574	0·589	0·603	0·617	0·645	0·672	0·685	0·724	0·761	0·773	0·797	0·821	0·832	0·866
1/ 5556	0·7050	0·7818	0·8636	0·9504	1·1398	1·3508	1·4647	1·8413	2·2723	2·4286	2·7605	3·1191	3·3085	3·9189
0·000190	0·591	0·606	0·621	0·636	0·664	0·692	0·705	0·745	0·783	0·796	0·820	0·844	0·856	0·891
1/ 5263	0·7257	0·8048	0·8889	0·9783	1·1732	1·3904	1·5076	1·8951	2·3386	2·4994	2·8410	3·2099	3·4049	4·0329
0·000200	0·608	0·623	0·638	0·653	0·682	0·711	0·725	0·765	0·805	0·818	0·843	0·868	0·880	0·916
1/ 5000	0·7460	0·8272	0·9137	1·0055	1·2058	1·4289	1·5494	1·9476	2·4033	2·5685	2·9194	3·2985	3·4988	4·1440
0·000210	0·624	0·640	0·655	0·670	0·700	0·729	0·744	0·785	0·826	0·839	0·865	0·890	0·903	0·940
1/ 4762	0·7657	0·8491	0·9378	1·0321	1·2376	1·4666	1·5902	1·9988	2·4664	2·6359	2·9960	3·3849	3·5904	4·2524
0·000220	0·640	0·656	0·672	0·687	0·718	0·748	0·762	0·805	0·847	0·860	0·887	0·913	0·926	0·963
1/ 4545	0·7850	0·8705	0·9615	1·0581	1·2687	1·5034	1·6301	2·0489	2·5281	2·7018	3·0709	3·4694	3·6800	4·3584
0·000230	0·655	0·672	0·688	0·704	0·735	0·766	0·781	0·824	0·867	0·881	0·908	0·934	0·948	0·986
1/ 4348	0·8039	0·8914	0·9846	1·0835	1·2992	1·5395	1·6691	2·0979	2·5885	2·7663	3·1441	3·5521	3·7677	4·4621
0·000240	0·670	0·687	0·704	0·720	0·752	0·783	0·798	0·843	0·887	0·901	0·928	0·956	0·969	1·009
1/ 4167	0·8224	0·9119	1·0072	1·1084	1·3290	1·5747	1·7074	2·1458	2·6476	2·8295	3·2158	3·6331	3·8535	4·5637
0·000250	0·685	0·702	0·719	0·736	0·769	0·800	0·816	0·862	0·906	0·920	0·949	0·977	0·990	1·031
1/ 4000	0·8406	0·9320	1·0294	1·1328	1·3582	1·6093	1·7449	2·1929	2·7056	2·8914	3·2861	3·7124	3·9377	4·6632
0·000260	0·699	0·717	0·734	0·751	0·785	0·817	0·833	0·880	0·925	0·940	0·969	0·997	1·011	1·052
1/ 3846	0·8584	0·9518	1·0512	1·1567	1·3869	1·6432	1·7816	2·2390	2·7624	2·9521	3·3551	3·7902	4·0202	4·7608
0·000270	0·714	0·732	0·749	0·767	0·801	0·834	0·850	0·898	0·944	0·959	0·988	1·017	1·031	1·074
1/ 3704	0·8759	0·9711	1·0725	1·1802	1·4150	1·6766	1·8177	2·2843	2·8182	3·0117	3·4227	3·8666	4·1012	4·8567
0·000280	0·728	0·746	0·764	0·782	0·816	0·850	0·867	0·915	0·962	0·977	1·007	1·037	1·051	1·094
1/ 3571	0·8930	0·9902	1·0935	1·2033	1·4426	1·7093	1·8532	2·3288	2·8730	3·0702	3·4892	3·9417	4·1807	4·9508
0·000290	0·741	0·760	0·778	0·796	0·832	0·866	0·883	0·932	0·980	0·996	1·026	1·056	1·071	1·115
1/ 3448	0·9099	1·0089	1·1142	1·2260	1·4698	1·7414	1·8880	2·3725	2·9269	3·1278	3·5546	4·0154	4·2589	5·0433
0·000300	0·755	0·774	0·793	0·811	0·847	0·882	0·899	0·949	0·998	1·014	1·045	1·075	1·090	1·135
1/ 3333	0·9265	1·0272	1·1345	1·2483	1·4965	1·7731	1·9223	2·4155	2·9798	3·1844	3·6188	4·0880	4·3359	5·1342
	0·93	0·93	0·93	0·94	0·94	0·94	0·94	0·95	0·95	0·95	0·95	0·96	0·96	0·96

$V_{r(0·5)medial}$ for half-full circular pipes.

k_s = 0·060 mm S = 0·00010 to 0·00030

k_s = 0·060 mm
S = 0·00030 to 0·00100

ie hydraulic gradient =
1 in 3333 to 1 in 1000

Water (or sewage) at 15°C;
full bore conditions.

velocities in ms^{-1}
discharges in m^3s^{-1}

Gradient	**(Equivalent) Pipe diameters in mm**													
	1250	**1300**	**1350**	**1400**	**1500**	**1600**	**1650**	**1800**	**1950**	**2000**	**2100**	**2200**	**2250**	**2400**
0·000300	0·755	0·774	0·793	0·811	0·847	0·882	0·899	0·949	0·998	1·014	1·045	1·075	1·090	1·135
1/ 3333	0·9265	1·0272	1·1345	1·2483	1·4965	1·7731	1·9223	2·4155	2·9798	3·1844	3·6188	4·0880	4·3359	5·1342
0·000320	0·781	0·801	0·820	0·839	0·876	0·913	0·930	0·982	1·032	1·049	1·081	1·113	1·128	1·174
1/ 3125	0·9589	1·0632	1·1741	1·2919	1·5487	1·8348	1·9892	2·4995	3·0833	3·2949	3·7443	4·2296	4·4860	5·3119
0·000340	0·807	0·827	0·847	0·867	0·905	0·942	0·961	1·014	1·066	1·083	1·116	1·149	1·165	1·212
1/ 2941	0·9904	1·0980	1·2126	1·3342	1·5994	1·8948	2·0542	2·5810	3·1837	3·4021	3·8661	4·3670	4·6317	5·4842
0·000360	0·832	0·853	0·873	0·893	0·933	0·971	0·990	1·045	1·099	1·116	1·150	1·184	1·201	1·249
1/ 2778	1·0210	1·1319	1·2500	1·3754	1·6486	1·9530	2·1173	2·6602	3·2812	3·5063	3·9844	4·5006	4·7733	5·6517
0·000380	0·856	0·878	0·899	0·919	0·960	1·000	1·019	1·076	1·131	1·148	1·184	1·218	1·235	1·285
1/ 2632	1·0508	1·1649	1·2864	1·4154	1·6966	2·0098	2·1788	2·7373	3·3762	3·6078	4·0996	4·6306	4·9111	5·8147
0·000400	0·880	0·902	0·924	0·945	0·987	1·027	1·047	1·105	1·162	1·180	1·216	1·251	1·269	1·320
1/ 2500	1·0798	1·1971	1·3219	1·4544	1·7433	2·0651	2·2387	2·8124	3·4688	3·7066	4·2118	4·7573	5·0455	5·9736
0·000420	0·903	0·926	0·948	0·970	1·012	1·054	1·074	1·134	1·192	1·211	1·248	1·284	1·302	1·355
1/ 2381	1·1082	1·2285	1·3566	1·4925	1·7889	2·1190	2·2972	2·8858	3·5592	3·8032	4·3214	4·8810	5·1766	6·1286
0·000440	0·926	0·949	0·971	0·994	1·038	1·080	1·101	1·162	1·221	1·241	1·279	1·316	1·334	1·388
1/ 2273	1·1359	1·2592	1·3904	1·5298	1·8335	2·1718	2·3543	2·9575	3·6475	3·8975	4·4285	5·0018	5·3047	6·2802
0·000460	0·948	0·971	0·995	1·017	1·062	1·106	1·127	1·190	1·250	1·270	1·309	1·347	1·366	1·421
1/ 2174	1·1630	1·2892	1·4236	1·5662	1·8771	2·2234	2·4102	3·0276	3·7338	3·9897	4·5332	5·1200	5·4300	6·4284
0·000480	0·969	0·993	1·017	1·041	1·086	1·131	1·153	1·217	1·279	1·299	1·338	1·377	1·397	1·453
1/ 2083	1·1895	1·3186	1·4560	1·6019	1·9198	2·2739	2·4650	3·0963	3·8184	4·0800	4·6358	5·2357	5·5527	6·5734
0·000500	0·991	1·015	1·039	1·063	1·110	1·156	1·178	1·243	1·306	1·327	1·367	1·407	1·427	1·484
1/ 2000	1·2156	1·3475	1·4878	1·6369	1·9617	2·3234	2·5186	3·1636	3·9013	4·1685	4·7362	5·3491	5·6729	6·7156
0·000525	1·016	1·042	1·067	1·091	1·139	1·186	1·209	1·276	1·340	1·361	1·403	1·444	1·464	1·523
1/ 1905	1·2474	1·3827	1·5267	1·6796	2·0129	2·3840	2·5842	3·2459	4·0026	4·2768	4·8591	5·4878	5·8200	6·8895
0·000550	1·042	1·068	1·093	1·118	1·167	1·215	1·239	1·307	1·373	1·395	1·438	1·479	1·500	1·560
1/ 1818	1·2785	1·4172	1·5647	1·7214	2·0629	2·4432	2·6484	3·3263	4·1016	4·3826	4·9792	5·6233	5·9636	7·0594
0·000575	1·067	1·093	1·119	1·145	1·195	1·244	1·268	1·338	1·406	1·428	1·471	1·514	1·535	1·597
1/ 1739	1·3089	1·4509	1·6019	1·7623	2·1118	2·5011	2·7111	3·4049	4·1985	4·4860	5·0966	5·7558	6·1041	7·2255
0·000600	1·091	1·118	1·145	1·171	1·222	1·272	1·297	1·368	1·438	1·460	1·505	1·548	1·570	1·633
1/ 1667	1·3387	1·4839	1·6383	1·8023	2·1597	2·5578	2·7725	3·4819	4·2933	4·5873	5·2116	5·8856	6·2417	7·3882
0·000625	1·115	1·142	1·170	1·196	1·249	1·300	1·325	1·398	1·469	1·492	1·537	1·582	1·604	1·668
1/ 1600	1·3679	1·5162	1·6740	1·8416	2·2067	2·6133	2·8327	3·5574	4·3862	4·6865	5·3243	6·0127	6·3764	7·5475
0·000650	1·138	1·166	1·194	1·221	1·275	1·327	1·352	1·427	1·499	1·523	1·569	1·615	1·637	1·703
1/ 1538	1·3965	1·5479	1·7090	1·8800	2·2527	2·6678	2·8917	3·6314	4·4774	4·7839	5·4348	6·1374	6·5086	7·7038
0·000675	1·161	1·190	1·218	1·246	1·300	1·353	1·379	1·456	1·529	1·553	1·600	1·647	1·670	1·737
1/ 1481	1·4247	1·5791	1·7434	1·9178	2·2979	2·7212	2·9496	3·7040	4·5668	4·8794	5·5432	6·2598	6·6383	7·8572
0·000700	1·183	1·213	1·242	1·270	1·325	1·380	1·406	1·484	1·559	1·583	1·631	1·678	1·702	1·770
1/ 1429	1·4523	1·6096	1·7771	1·9549	2·3423	2·7737	3·0065	3·7754	4·6546	4·9732	5·6497	6·3799	6·7657	8·0078
0·000725	1·206	1·235	1·265	1·294	1·350	1·405	1·432	1·511	1·587	1·612	1·661	1·709	1·733	1·803
1/ 1379	1·4794	1·6397	1·8103	1·9913	2·3860	2·8253	3·0624	3·8455	4·7409	5·0654	5·7543	6·4980	6·8909	8·1558
0·000750	1·227	1·258	1·287	1·317	1·374	1·430	1·458	1·538	1·616	1·641	1·691	1·740	1·764	1·835
1/ 1333	1·5061	1·6692	1·8429	2·0272	2·4289	2·8761	3·1174	3·9144	4·8258	5·1560	5·8572	6·6141	7·0139	8·3013
0·000800	1·270	1·301	1·332	1·362	1·422	1·480	1·508	1·591	1·671	1·698	1·749	1·800	1·824	1·898
1/ 1250	1·5582	1·7270	1·9065	2·0972	2·5126	2·9752	3·2247	4·0490	4·9915	5·3330	6·0581	6·8408	7·2542	8·5854
0·000850	1·311	1·343	1·375	1·406	1·468	1·528	1·557	1·642	1·725	1·752	1·805	1·857	1·883	1·959
1/ 1176	1·6087	1·7829	1·9683	2·1651	2·5939	3·0713	3·3288	4·1795	5·1522	5·5046	6·2529	7·0606	7·4872	8·8609
0·000900	1·351	1·384	1·417	1·449	1·512	1·574	1·604	1·692	1·777	1·805	1·860	1·914	1·940	2·018
1/ 1111	1·6578	1·8373	2·0283	2·2310	2·6728	3·1646	3·4299	4·3063	5·3083	5·6713	6·4421	7·2741	7·7136	9·1286
0·000950	1·390	1·424	1·458	1·491	1·556	1·619	1·650	1·741	1·828	1·857	1·913	1·968	1·995	2·075
1/ 1053	1·7056	1·8903	2·0867	2·2953	2·7496	3·2555	3·5284	4·4297	5·4603	5·8336	6·6263	7·4820	7·9339	9·3890
0·001000	1·428	1·463	1·498	1·532	1·598	1·663	1·695	1·788	1·878	1·907	1·965	2·022	2·049	2·132
1/ 1000	1·7522	1·9419	2·1437	2·3578	2·8245	3·3441	3·6243	4·5500	5·6084	5·9917	6·8058	7·6845	8·1486	9·6428
	0·94	0·95	0·95	0·95	0·95	0·95	0·96	0·96	0·96	0·96	0·96	0·97	0·97	0·97

$V_{r(0·5)medial}$ for half-full circular pipes.

S = 0·00030 to 0·00100 **k_s = 0·060 mm**

A38
(p.3 of 6)

k_s = 0·060 mm
S = 0·00100 to 0·00300

ie hydraulic gradient =
1 in 1000 to 1 in 333

Water (or sewage) at 15°C;
full bore conditions.

velocities in ms^{-1}
discharges in m^3s^{-1}

Gradient **(Equivalent) Pipe diameters in mm**

Gradient	1250	1300	1350	1400	1500	1600	1650	1800	1950	2000	2100	2200	2250	2400
0·00100	1·428	1·463	1·498	1·532	1·598	1·663	1·695	1·788	1·878	1·907	1·965	2·022	2·049	2·132
1/ 1000	1·7522	1·9419	2·1437	2·3578	2·8245	3·3441	3·6243	4·5500	5·6084	5·9917	6·8058	7·6845	8·1486	9·6428
0·00105	1·465	1·501	1·536	1·571	1·640	1·706	1·739	1·834	1·926	1·956	2·016	2·074	2·102	2·186
1/ 952	1·7977	1·9922	2·1992	2·4189	2·8976	3·4305	3·7179	4·6674	5·7528	6·1460	6·9809	7·8821	8·3581	9·8905
0·00110	1·501	1·538	1·574	1·610	1·680	1·748	1·782	1·879	1·974	2·004	2·065	2·124	2·154	2·240
1/ 909	1·8421	2·0414	2·2535	2·4786	2·9690	3·5149	3·8094	4·7821	5·8940	6·2968	7·1520	8·0751	8·5627	10·132
0·00115	1·537	1·574	1·611	1·648	1·720	1·789	1·823	1·923	2·020	2·051	2·113	2·174	2·204	2·292
1/ 870	1·8856	2·0896	2·3066	2·5369	3·0388	3·5975	3·8988	4·8942	6·0320	6·4442	7·3193	8·2639	8·7628	10·369
0·00120	1·571	1·610	1·648	1·685	1·758	1·829	1·864	1·966	2·065	2·097	2·160	2·223	2·253	2·343
1/ 833	1·9281	2·1367	2·3586	2·5941	3·1072	3·6783	3·9864	5·0040	6·1671	6·5885	7·4831	8·4487	8·9587	10·601
0·00125	1·605	1·645	1·683	1·721	1·796	1·869	1·904	2·009	2·109	2·142	2·207	2·270	2·301	2·393
1/ 800	1·9698	2·1828	2·4095	2·6500	3·1741	3·7575	4·0722	5·1115	6·2995	6·7299	7·6435	8·6297	9·1506	10·827
0·00130	1·638	1·679	1·718	1·757	1·833	1·907	1·944	2·050	2·153	2·186	2·252	2·317	2·349	2·443
1/ 769	2·0107	2·2281	2·4594	2·7049	3·2398	3·8352	4·1563	5·2169	6·4293	6·8685	7·8009	8·8072	9·3387	11·050
0·00135	1·671	1·712	1·752	1·792	1·870	1·945	1·982	2·091	2·195	2·230	2·297	2·363	2·395	2·491
1/ 741	2·0508	2·2725	2·5084	2·7588	3·3042	3·9114	4·2389	5·3204	6·5567	7·0045	7·9552	8·9814	9·5233	11·268
0·00140	1·703	1·745	1·786	1·827	1·906	1·983	2·020	2·131	2·237	2·272	2·341	2·408	2·441	2·538
1/ 714	2·0902	2·3162	2·5565	2·8117	3·3675	3·9862	4·3199	5·4220	6·6818	7·1381	8·1068	9·1524	9·7046	11·482
0·00145	1·735	1·777	1·819	1·860	1·941	2·019	2·058	2·170	2·278	2·314	2·384	2·452	2·486	2·585
1/ 690	2·1289	2·3590	2·6038	2·8637	3·4297	4·0598	4·3996	5·5219	6·8047	7·2693	8·2558	9·3204	9·8827	11·693
0·00150	1·766	1·809	1·852	1·893	1·975	2·055	2·094	2·209	2·319	2·355	2·426	2·495	2·530	2·630
1/ 667	2·1670	2·4012	2·6503	2·9148	3·4909	4·1321	4·4779	5·6200	6·9255	7·3983	8·4022	9·4856	10·058	11·900
0·00160	1·826	1·871	1·915	1·958	2·043	2·125	2·166	2·284	2·398	2·435	2·508	2·580	2·616	2·720
1/ 625	2·2413	2·4835	2·7411	3·0145	3·6102	4·2732	4·6308	5·8117	7·1613	7·6502	8·6880	9·8081	10·400	12·304
0·00170	1·885	1·931	1·976	2·021	2·108	2·193	2·235	2·357	2·475	2·513	2·588	2·662	2·699	2·806
1/ 588	2·3134	2·5633	2·8291	3·1113	3·7260	4·4101	4·7790	5·9975	7·3900	7·8944	8·9652	10·121	10·731	12·696
0·00180	1·942	1·990	2·036	2·082	2·172	2·260	2·302	2·428	2·549	2·588	2·666	2·742	2·780	2·890
1/ 556	2·3834	2·6408	2·9147	3·2053	3·8384	4·5430	4·9231	6·1780	7·6122	8·1316	9·2344	10·424	11·053	13·076
0·00190	1·998	2·046	2·094	2·142	2·234	2·324	2·368	2·497	2·621	2·662	2·742	2·820	2·859	2·972
1/ 526	2·4515	2·7163	2·9979	3·2968	3·9478	4·6724	5·0632	6·3536	7·8283	8·3624	9·4963	10·720	11·366	13·446
0·00200	2·052	2·102	2·151	2·200	2·294	2·387	2·432	2·564	2·692	2·733	2·815	2·896	2·935	3·052
1/ 500	2·5179	2·7898	3·0790	3·3859	4·0544	4·7985	5·1997	6·5247	8·0389	8·5873	9·7516	11·008	11·671	13·807
0·00210	2·105	2·156	2·206	2·256	2·353	2·448	2·494	2·630	2·761	2·803	2·887	2·970	3·010	3·130
1/ 476	2·5827	2·8615	3·1581	3·4729	4·1585	4·9215	5·3330	6·6917	8·2444	8·8067	10·001	11·289	11·969	14·159
0·00220	2·156	2·209	2·260	2·311	2·411	2·507	2·555	2·694	2·828	2·871	2·958	3·042	3·083	3·206
1/ 455	2·6460	2·9316	3·2354	3·5578	4·2601	5·0416	5·4631	6·8548	8·4451	9·0210	10·244	11·563	12·260	14·503
0·00230	2·207	2·260	2·313	2·365	2·467	2·566	2·614	2·756	2·893	2·938	3·026	3·112	3·155	3·280
1/ 435	2·7079	3·0001	3·3110	3·6409	4·3594	5·1591	5·5903	7·0142	8·6413	9·2306	10·482	11·831	12·544	14·839
0·00240	2·256	2·311	2·365	2·418	2·522	2·623	2·673	2·818	2·958	3·003	3·093	3·182	3·225	3·353
1/ 417	2·7684	3·0672	3·3849	3·7222	4·4567	5·2741	5·7149	7·1703	8·8334	9·4357	10·714	12·094	12·823	15·168
0·00250	2·304	2·360	2·415	2·470	2·576	2·679	2·730	2·878	3·021	3·067	3·159	3·249	3·293	3·424
1/ 400	2·8278	3·1329	3·4574	3·8018	4·5520	5·3867	5·8369	7·3232	9·0216	9·6366	10·942	12·351	13·095	15·490
0·00260	2·352	2·409	2·465	2·520	2·629	2·734	2·786	2·937	3·083	3·130	3·224	3·316	3·361	3·494
1/ 385	2·8860	3·1973	3·5285	3·8799	4·6454	5·4972	5·9565	7·4731	9·2060	9·8336	11·166	12·603	13·362	15·806
0·00270	2·398	2·456	2·514	2·570	2·681	2·788	2·841	2·995	3·143	3·192	3·287	3·381	3·427	3·562
1/ 370	2·9431	3·2605	3·5982	3·9566	4·7371	5·6055	6·0739	7·6202	9·3870	10·027	11·385	12·851	13·625	16·116
0·00280	2·444	2·503	2·562	2·619	2·732	2·841	2·894	3·051	3·203	3·252	3·349	3·444	3·491	3·630
1/ 357	2·9991	3·3226	3·6667	4·0318	4·8271	5·7120	6·1891	7·7646	9·5647	10·217	11·600	13·094	13·882	16·420
0·00290	2·489	2·549	2·609	2·667	2·782	2·893	2·947	3·107	3·261	3·311	3·410	3·507	3·555	3·696
1/ 345	3·0542	3·3836	3·7340	4·1057	4·9155	5·8165	6·3023	7·9065	9·7393	10·403	11·812	13·332	14·135	16·719
0·00300	2·533	2·594	2·655	2·714	2·831	2·944	2·999	3·162	3·319	3·370	3·470	3·569	3·618	3·761
1/ 333	3·1084	3·4436	3·8001	4·1784	5·0024	5·9193	6·4137	8·0460	9·9110	10·586	12·020	13·567	14·384	17·013
	0·95	0·96	0·96	0·96	0·96	0·96	0·96	0·97	0·97	0·97	0·97	0·97	0·97	0·97

$V_{r(0\cdot5)medial}$ for half-full circular pipes.

k_s = 0·060 mm S = 0·00100 to 0·00300

k_s = 0·060 mm
S = 0·00300 to 0·01000

ie hydraulic gradient =
1 in 333 to 1 in 100

Water (or sewage) at 15°C;
full bore conditions.

velocities in ms^{-1}
discharges in m^3s^{-1}

Gradient	**(Equivalent) Pipe diameters in mm**													
	1250	**1300**	**1350**	**1400**	**1500**	**1600**	**1650**	**1800**	**1950**	**2000**	**2100**	**2200**	**2250**	**2400**
0·00300	2·533	2·594	2·655	2·714	2·831	2·944	2·999	3·162	3·319	3·370	3·470	3·569	3·618	3·761
1/ 333	3·1084	3·4436	3·8001	4·1784	5·0024	5·9193	6·4137	8·0460	9·9110	10·586	12·020	13·567	14·384	17·013
0·00320	2·619	2·683	2·745	2·807	2·927	3·044	3·101	3·269	3·431	3·484	3·588	3·689	3·740	3·887
1/ 313	3·2141	3·5606	3·9292	4·3203	5·1721	6·1199	6·6309	8·3182	10·246	10·944	12·426	14·025	14·869	17·586
0·00340	2·703	2·768	2·832	2·896	3·020	3·141	3·200	3·373	3·540	3·594	3·701	3·806	3·858	4·010
1/ 294	3·3166	3·6741	4·0543	4·4578	5·3366	6·3144	6·8416	8·5821	10·571	11·291	12·819	14·469	15·340	18·143
0·00360	2·784	2·851	2·917	2·983	3·110	3·234	3·295	3·473	3·645	3·701	3·811	3·920	3·973	4·130
1/ 278	3·4162	3·7843	4·1759	4·5914	5·4964	6·5033	7·0461	8·8385	10·886	11·628	13·201	14·900	15·796	18·683
0·00380	2·863	2·932	3·000	3·067	3·198	3·326	3·388	3·571	3·748	3·805	3·919	4·030	4·085	4·246
1/ 263	3·5130	3·8916	4·2942	4·7214	5·6519	6·6870	7·2452	9·0879	11·193	11·955	13·573	15·319	16·241	19·208
0·00400	2·940	3·011	3·081	3·149	3·284	3·415	3·479	3·667	3·848	3·907	4·023	4·137	4·194	4·359
1/ 250	3·6074	3·9961	4·4094	4·8481	5·8033	6·8661	7·4391	9·3308	11·492	12·274	13·935	15·728	16·674	19·720
0·00420	3·015	3·087	3·159	3·230	3·368	3·502	3·568	3·760	3·946	4·006	4·125	4·242	4·300	4·469
1/ 238	3·6995	4·0980	4·5219	4·9716	5·9511	7·0408	7·6283	9·5679	11·784	12·586	14·289	16·126	17·096	20·219
0·00440	3·088	3·162	3·236	3·308	3·449	3·587	3·654	3·851	4·041	4·103	4·225	4·345	4·404	4·577
1/ 227	3·7894	4·1976	4·6317	5·0923	6·0954	7·2114	7·8130	9·7993	12·068	12·890	14·634	16·515	17·509	20·706
0·00460	3·160	3·236	3·311	3·385	3·529	3·670	3·738	3·940	4·134	4·198	4·322	4·445	4·505	4·682
1/ 217	3·8774	4·2949	4·7391	5·2103	6·2365	7·3782	7·9937	10·026	12·347	13·187	14·971	16·896	17·912	21·183
0·00480	3·230	3·308	3·384	3·460	3·607	3·751	3·821	4·027	4·226	4·290	4·418	4·543	4·604	4·786
1/ 208	3·9634	4·3902	4·8441	5·3258	6·3746	7·5414	8·1705	10·247	12·619	13·478	15·301	17·268	18·307	21·650
0·00500	3·298	3·378	3·456	3·533	3·684	3·830	3·902	4·112	4·315	4·381	4·511	4·639	4·702	4·887
1/ 200	4·0477	4·4836	4·9471	5·4389	6·5099	7·7013	8·3437	10·464	12·886	13·763	15·625	17·633	18·694	22·106
0·00525	3·382	3·464	3·544	3·623	3·777	3·928	4·001	4·216	4·424	4·492	4·625	4·756	4·820	5·010
1/ 190	4·1508	4·5977	5·0729	5·5772	6·6752	7·8968	8·5554	10·729	13·213	14·112	16·020	18·079	19·166	22·665
0·00550	3·464	3·548	3·630	3·711	3·869	4·022	4·098	4·318	4·531	4·600	4·737	4·871	4·937	5·131
1/ 182	4·2514	4·7091	5·1958	5·7122	6·8367	8·0877	8·7621	10·988	13·531	14·452	16·406	18·515	19·628	23·211
0·00575	3·545	3·630	3·714	3·797	3·958	4·115	4·192	4·418	4·635	4·706	4·846	4·983	5·050	5·249
1/ 174	4·3499	4·8181	5·3160	5·8443	6·9946	8·2744	8·9643	11·242	13·843	14·785	16·784	18·940	20·079	23·744
0·00600	3·623	3·710	3·796	3·880	4·046	4·206	4·285	4·515	4·737	4·810	4·952	5·092	5·161	5·364
1/ 167	4·4462	4·9247	5·4336	5·9735	7·1492	8·4570	9·1621	11·490	14·148	15·110	17·153	19·357	20·521	24·266
0·00625	3·700	3·789	3·876	3·963	4·131	4·295	4·375	4·610	4·837	4·911	5·057	5·200	5·270	5·477
1/ 160	4·5406	5·0292	5·5488	6·1001	7·3006	8·6360	9·3559	11·732	14·447	15·429	17·515	19·765	20·953	24·777
0·00650	3·775	3·866	3·955	4·043	4·215	4·382	4·464	4·704	4·935	5·011	5·159	5·305	5·377	5·588
1/ 154	4·6331	5·1316	5·6617	6·2242	7·4490	8·8114	9·5459	11·970	14·739	15·742	17·870	20·165	21·377	25·278
0·00675	3·849	3·942	4·033	4·122	4·298	4·468	4·552	4·796	5·032	5·108	5·260	5·408	5·481	5·696
1/ 148	4·7238	5·2320	5·7725	6·3459	7·5946	8·9835	9·7322	12·204	15·027	16·049	18·218	20·558	21·793	25·770
0·00700	3·922	4·016	4·109	4·200	4·379	4·552	4·637	4·886	5·126	5·204	5·358	5·509	5·584	5·803
1/ 143	4·8129	5·3307	5·8813	6·4655	7·7375	9·1524	9·9152	12·433	15·309	16·350	18·559	20·943	22·202	26·252
0·00725	3·993	4·089	4·183	4·276	4·458	4·635	4·721	4·974	5·219	5·298	5·455	5·609	5·685	5·908
1/ 138	4·9004	5·4276	5·9882	6·5829	7·8779	9·3184	10·095	12·658	15·586	16·646	18·895	21·322	22·603	26·726
0·00750	4·063	4·161	4·257	4·351	4·536	4·716	4·804	5·061	5·310	5·391	5·551	5·707	5·784	6·011
1/ 133	4·9865	5·5229	6·0932	6·6984	8·0160	9·4816	10·272	12·880	15·858	16·936	19·225	21·694	22·997	27·192
0·00800	4·200	4·301	4·400	4·498	4·689	4·874	4·965	5·231	5·488	5·572	5·736	5·898	5·978	6·212
1/ 125	5·1544	5·7088	6·2982	6·9236	8·2854	9·8000	10·617	13·312	16·389	17·504	19·869	22·420	23·767	28·102
0·00850	4·333	4·437	4·539	4·640	4·836	5·028	5·121	5·396	5·660	5·747	5·917	6·083	6·165	6·407
1/ 118	5·3172	5·8890	6·4970	7·1420	8·5465	10·109	10·951	13·730	16·905	18·054	20·493	23·124	24·513	28·983
0·00900	4·462	4·569	4·674	4·777	4·980	5·177	5·273	5·555	5·828	5·917	6·092	6·263	6·347	6·596
1/ 111	5·4754	6·0640	6·6900	7·3541	8·8001	10·408	11·275	14·137	17·405	18·588	21·099	23·807	25·237	29·839
0·00950	4·587	4·697	4·805	4·911	5·119	5·322	5·421	5·711	5·991	6·082	6·262	6·438	6·525	6·780
1/ 105	5·6292	6·2343	6·8778	7·5605	9·0468	10·700	11·591	14·532	17·892	19·107	21·688	24·472	25·942	30·672
0·01000	4·709	4·822	4·933	5·042	5·255	5·463	5·565	5·862	6·150	6·243	6·428	6·608	6·697	6·959
1/ 100	5·7790	6·4002	7·0607	7·7614	9·2871	10·984	11·899	14·918	18·366	19·614	22·263	25·120	26·629	31·483
	0·96	0·96	0·96	0·96	0·97	0·97	0·97	0·97	0·97	0·97	0·97	0·97	0·98	0·98

$V_{r(0·5)medial}$ for half-full circular pipes.

A38
(p.5 of 6)

k_s = **0·060 mm**
S = 0·01000 to 0·03000

ie hydraulic gradient =
1 in 100 to 1 in 33·3

Water (or sewage) at 15°C;
full bore conditions.

velocities in ms^{-1}
discharges in m^3s^{-1}

Gradient **(Equivalent) Pipe diameters in mm**

Gradient	1250	1300	1350	1400	1500	1600	1650	1800	1950	2000	2100	2200	2250	2400
0·01000	4·709	4·822	4·933	5·042	5·255	5·463	5·565	5·862	6·150	6·243	6·428	6·608	6·697	6·959
1/ 100	5·7790	6·4002	7·0607	7·7614	9·2871	10·984	11·899	14·918	18·366	19·614	22·263	25·120	26·629	31·483
0·01050	4·828	4·944	5·057	5·169	5·388	5·601	5·705	6·010	6·304	6·400	6·589	6·774	6·866	7·134
1/ 95	5·9252	6·5620	7·2391	7·9575	9·5216	11·261	12·199	15·294	18·828	20·107	22·823	25·752	27·298	32·274
0·01100	4·945	5·063	5·179	5·294	5·518	5·735	5·842	6·154	6·456	6·554	6·747	6·937	7·030	7·305
1/ 91	6·0679	6·7200	7·4134	8·1490	9·7505	11·532	12·492	15·661	19·279	20·589	23·370	26·369	27·952	33·047
0·01150	5·058	5·179	5·298	5·415	5·644	5·867	5·976	6·295	6·603	6·704	6·902	7·095	7·191	7·472
1/ 87	6·2074	6·8745	7·5837	8·3361	9·9743	11·796	12·778	16·020	19·721	21·061	23·904	26·972	28·591	33·802
0·01200	5·170	5·293	5·415	5·534	5·768	5·996	6·107	6·433	6·748	6·851	7·053	7·251	7·348	7·635
1/ 83	6·3440	7·0256	7·7504	8·5193	10·193	12·055	13·059	16·371	20·153	21·522	24·428	27·562	29·217	34·541
0·01250	5·279	5·405	5·529	5·651	5·890	6·122	6·236	6·568	6·890	6·994	7·201	7·403	7·502	7·795
1/ 80	6·4778	7·1737	7·9137	8·6987	10·408	12·309	13·333	16·715	20·576	21·974	24·940	28·140	29·829	35·265
0·01300	5·385	5·514	5·641	5·765	6·009	6·245	6·361	6·701	7·029	7·135	7·346	7·552	7·653	7·952
1/ 77	6·6089	7·3188	8·0738	8·8746	10·618	12·557	13·602	17·052	20·991	22·417	25·443	28·707	30·430	35·975
0·01350	5·490	5·621	5·750	5·877	6·125	6·367	6·485	6·831	7·165	7·274	7·488	7·698	7·801	8·106
1/ 74	6·7375	7·4612	8·2308	9·0472	10·824	12·801	13·866	17·383	21·398	22·851	25·935	29·262	31·019	36·671
0·01400	5·593	5·727	5·858	5·987	6·240	6·486	6·606	6·959	7·299	7·409	7·628	7·841	7·947	8·257
1/ 71	6·8638	7·6010	8·3849	9·2165	11·027	13·040	14·126	17·707	21·797	23·277	26·419	29·808	31·597	37·354
0·01450	5·694	5·830	5·964	6·095	6·353	6·603	6·725	7·084	7·430	7·543	7·765	7·983	8·090	8·406
1/ 69	6·9879	7·7384	8·5364	9·3830	11·226	13·276	14·380	18·026	22·189	23·696	26·894	30·344	32·166	38·026
0·01500	5·794	5·932	6·068	6·202	6·463	6·718	6·842	7·207	7·559	7·674	7·900	8·121	8·230	8·551
1/ 67	7·1098	7·8734	8·6852	9·5465	11·422	13·507	14·631	18·340	22·575	24·108	27·362	30·871	32·724	38·686
0·01600	5·988	6·130	6·271	6·409	6·679	6·942	7·071	7·448	7·811	7·930	8·163	8·392	8·504	8·836
1/ 63	7·3478	8·1368	8·9757	9·8657	11·803	13·958	15·119	18·952	23·328	24·912	28·273	31·899	33·814	39·973
0·01700	6·176	6·323	6·467	6·610	6·889	7·159	7·292	7·681	8·055	8·178	8·418	8·654	8·770	9·112
1/ 59	7·5785	8·3922	9·2573	10·175	12·173	14·395	15·593	19·545	24·057	25·691	29·157	32·896	34·870	41·221
0·01800	6·358	6·509	6·658	6·805	7·092	7·371	7·507	7·907	8·293	8·418	8·666	8·908	9·028	9·380
1/ 56	7·8025	8·6402	9·5308	10·476	12·532	14·820	16·052	20·121	24·766	26·447	30·015	33·864	35·896	42·433
0·01900	6·536	6·691	6·844	6·995	7·290	7·576	7·716	8·127	8·523	8·652	8·907	9·156	9·279	9·640
1/ 53	8·0204	8·8814	9·7968	10·768	12·882	15·232	16·500	20·681	25·455	27·183	30·850	34·805	36·893	43·612
0·02000	6·709	6·868	7·025	7·180	7·482	7·776	7·920	8·341	8·748	8·881	9·142	9·397	9·523	9·894
1/ 50	8·2327	9·1163	10·056	11·053	13·222	15·635	16·935	21·226	26·126	27·899	31·663	35·722	37·865	44·760
0·02100	6·877	7·041	7·202	7·360	7·670	7·971	8·119	8·551	8·967	9·103	9·370	9·632	9·762	10·14
1/ 47·6	8·4397	9·3455	10·309	11·330	13·554	16·027	17·360	21·758	26·780	28·598	32·456	36·616	38·813	45·880
0·02200	7·042	7·209	7·374	7·536	7·854	8·162	8·313	8·755	9·181	9·320	9·594	9·862	9·994	10·38
1/ 45·5	8·6419	9·5693	10·555	11·601	13·878	16·410	17·775	22·278	27·420	29·281	33·230	37·490	39·738	46·974
0·02300	7·203	7·374	7·543	7·709	8·033	8·348	8·503	8·954	9·391	9·533	9·813	10·09	10·22	10·62
1/ 43·5	8·8395	9·7881	10·797	11·866	14·195	16·785	18·181	22·786	28·045	29·948	33·987	38·343	40·643	48·043
0·02400	7·361	7·536	7·708	7·877	8·208	8·530	8·688	9·150	9·595	9·741	10·03	10·31	10·44	10·85
1/ 41·7	9·0329	10·002	11·033	12·126	14·505	17·151	18·577	23·283	28·656	30·601	34·728	39·179	41·528	49·089
0·02500	7·515	7·694	7·869	8·042	8·380	8·709	8·870	9·341	9·796	9·944	10·24	10·52	10·66	11·08
1/ 40·0	9·2224	10·212	11·264	12·380	14·809	17·510	18·966	23·770	29·255	31·240	35·453	39·997	42·396	50·113
0·02600	7·666	7·848	8·028	8·204	8·549	8·884	9·048	9·529	9·992	10·14	10·44	10·73	10·88	11·30
1/ 38·5	9·4081	10·417	11·491	12·629	15·107	17·862	19·347	24·248	29·842	31·867	36·164	40·799	43·246	51·118
0·02700	7·815	8·000	8·183	8·363	8·714	9·056	9·223	9·713	10·19	10·34	10·64	10·94	11·09	11·52
1/ 37·0	9·5902	10·619	11·713	12·873	15·399	18·207	19·721	24·716	30·418	32·482	36·862	41·586	44·079	52·103
0·02800	7·961	8·150	8·335	8·518	8·876	9·224	9·395	9·893	10·37	10·53	10·84	11·14	11·29	11·73
1/ 35·7	9·7691	10·817	11·931	13·113	15·686	18·546	20·088	25·175	30·983	33·085	37·546	42·358	44·898	53·070
0·02900	8·104	8·296	8·485	8·671	9·036	9·390	9·563	10·07	10·56	10·72	11·03	11·34	11·49	11·94
1/ 34·5	9·9447	11·011	12·146	13·349	15·967	18·879	20·448	25·627	31·539	33·678	38·219	43·117	45·702	54·020
0·03000	8·244	8·440	8·632	8·822	9·192	9·552	9·729	10·25	10·74	10·91	11·23	11·54	11·69	12·15
1/ 33·3	10·117	11·203	12·356	13·580	16·244	19·206	20·803	26·071	32·085	34·261	38·880	43·862	46·492	54·954
	0·96	0·97	0·97	0·97	0·97	0·97	0·97	0·97	0·97	0·97	0·98	0·98	0·98	0·98

$V_{r(0·5)medial}$ **for half-full circular pipes.**

Gradient	(Equivalent) Pipe diameters in mm 1250	1300	1350	1400	1500	1600	1650	1800	1950	2000	2100	2200	2250	2400
0·03000	8·244	8·440	8·632	8·822	9·192	9·552	9·729	10·25	10·74	10·91	11·23	11·54	11·69	12·15
1/ 33·3	10·117	11·203	12·356	13·580	16·244	19·206	20·803	26·071	32·085	34·261	38·880	43·862	46·492	54·954
0·03200	8·519	8·721	8·920	9·115	9·498	9·870	10·05	10·59	11·10	11·27	11·60	11·92	12·08	12·55
1/ 31·3	10·454	11·576	12·768	14·032	16·784	19·845	21·494	26·937	33·150	35·399	40·171	45·317	48·035	56·776
0·03400	8·785	8·993	9·198	9·400	9·794	10·18	10·37	10·92	11·45	11·62	11·96	12·29	12·46	12·94
1/ 29·4	10·781	11·937	13·166	14·470	17·308	20·464	22·164	27·776	34·182	36·501	41·421	46·728	49·529	58·542
0·03600	9·044	9·258	9·469	9·676	10·08	10·48	10·67	11·24	11·78	11·96	12·31	12·65	12·82	13·32
1/ 27·8	11·098	12·288	13·553	14·895	17·817	21·065	22·815	28·591	35·185	37·571	42·635	48·097	50·981	60·257
0·03800	9·295	9·515	9·732	9·945	10·36	10·77	10·97	11·55	12·11	12·29	12·65	13·00	13·18	13·69
1/ 26·3	11·407	12·630	13·930	15·309	18·311	21·649	23·448	29·384	36·160	38·612	43·817	49·429	52·392	61·925
0·04000	9·540	9·766	9·988	10·21	10·63	11·05	11·25	11·85	12·43	12·61	12·98	13·34	13·52	14·05
1/ 25·0	11·707	12·962	14·297	15·712	18·793	22·218	24·065	30·156	37·110	39·626	44·967	50·727	53·767	63·550
0·04200	9·779	10·01	10·24	10·46	10·90	11·33	11·54	12·15	12·74	12·93	13·31	13·68	13·86	14·40
1/ 23·8	12·000	13·287	14·654	16·105	19·263	22·774	24·666	30·909	38·036	40·615	46·089	51·992	55·109	65·134
0·04400	10·01	10·25	10·48	10·71	11·16	11·60	11·81	12·44	13·04	13·24	13·62	14·00	14·19	14·74
1/ 22·7	12·286	13·603	15·004	16·489	19·722	23·316	25·253	31·645	38·941	41·581	47·185	53·228	56·418	66·682
0·04600	10·24	10·48	10·72	10·96	11·41	11·86	12·08	12·72	13·34	13·54	13·93	14·32	14·51	15·07
1/ 21·7	12·566	13·913	15·345	16·864	20·170	23·846	25·827	32·363	39·825	42·525	48·256	54·436	57·698	68·194
0·04800	10·46	10·71	10·95	11·19	11·66	12·12	12·34	12·99	13·62	13·83	14·23	14·63	14·83	15·40
1/ 20·8	12·840	14·216	15·679	17·231	20·609	24·365	26·389	33·067	40·690	43·449	49·304	55·618	58·951	69·675
0·05000	10·68	10·93	11·18	11·43	11·91	12·37	12·60	13·27	13·91	14·12	14·53	14·94	15·14	15·72
1/ 20·0	13·108	14·513	16·006	17·591	21·039	24·873	26·939	33·756	41·538	44·354	50·330	56·775	60·178	71·124
0·05250	10·95	11·21	11·46	11·71	12·20	12·68	12·91	13·60	14·26	14·47	14·89	15·31	15·51	16·11
1/ 19·0	13·436	14·876	16·406	18·030	21·564	25·494	27·611	34·598	42·574	45·460	51·585	58·190	61·678	72·896
0·05500	11·21	11·47	11·73	11·99	12·49	12·98	13·22	13·92	14·59	14·81	15·25	15·67	15·88	16·50
1/ 18·2	13·756	15·230	16·797	18·459	22·077	26·100	28·268	35·420	43·585	46·540	52·810	59·572	63·142	74·626
0·05750	11·46	11·74	12·00	12·26	12·78	13·28	13·52	14·24	14·93	15·15	15·59	16·03	16·24	16·87
1/ 17·4	14·069	15·576	17·179	18·879	22·579	26·693	28·910	36·224	44·574	47·595	54·008	60·923	64·573	76·317
0·06000	11·71	11·99	12·26	12·53	13·05	13·56	13·81	14·54	15·25	15·48	15·93	16·37	16·59	17·24
1/ 16·7	14·375	15·915	17·553	19·289	23·070	27·273	29·538	37·011	45·541	48·628	55·179	62·244	65·974	77·972
0·06250	11·96	12·24	12·52	12·79	13·33	13·85	14·10	14·85	15·57	15·80	16·26	16·71	16·94	17·59
1/ 16·0	14·674	16·247	17·918	19·691	23·550	27·841	30·153	37·781	46·489	49·640	56·327	63·539	67·346	79·593
0·06500	12·20	12·49	12·77	13·05	13·59	14·12	14·38	15·14	15·88	16·12	16·59	17·05	17·28	17·95
1/ 15·4	14·968	16·572	18·277	20·086	24·022	28·398	30·756	38·536	47·418	50·632	57·452	64·808	68·691	81·182
0·06750	12·43	12·73	13·01	13·30	13·85	14·40	14·66	15·43	16·18	16·43	16·91	17·38	17·61	18·29
1/ 14·8	15·257	16·891	18·629	20·472	24·484	28·944	31·347	39·277	48·329	51·605	58·556	66·052	70·010	82·740
0·07000	12·66	12·96	13·26	13·55	14·11	14·66	14·93	15·72	16·48	16·73	17·22	17·70	17·93	18·63
1/ 14·3	15·540	17·205	18·974	20·852	24·938	29·480	31·928	40·004	49·223	52·560	59·639	67·274	71·304	84·270
0·07250	12·89	13·19	13·49	13·79	14·36	14·92	15·20	16·00	16·78	17·03	17·53	18·01	18·25	18·96
1/ 13·8	15·818	17·512	19·314	21·225	25·383	30·007	32·498	40·719	50·102	53·498	60·703	68·474	72·576	85·773
0·07500	13·11	13·42	13·73	14·03	14·61	15·18	15·46	16·28	17·07	17·32	17·83	18·32	18·57	19·29
1/ 13·3	16·091	17·815	19·647	21·591	25·821	30·524	33·059	41·421	50·966	54·420	61·749	69·654	73·827	87·250
0·08000	13·55	13·87	14·18	14·49	15·10	15·68	15·97	16·82	17·63	17·90	18·42	18·93	19·18	19·92
1/ 12·5	16·624	18·405	20·298	22·306	26·676	31·535	34·153	42·791	52·651	56·219	63·790	71·956	76·266	90·133
0·08500	13·97	14·30	14·62	14·94	15·56	16·17	16·47	17·34	18·18	18·45	18·99	19·52	19·78	20·54
1/ 11·8	17·141	18·977	20·929	22·999	27·505	32·514	35·213	44·119	54·284	57·963	65·768	74·187	78·631	92·926
0·09000	14·38	14·72	15·05	15·38	16·02	16·64	16·95	17·84	18·71	18·99	19·54	20·09	20·35	21·14
1/ 11·1	17·643	19·532	21·541	23·672	28·309	33·464	36·243	45·408	55·870	59·656	67·689	76·353	80·926	95·639
0·09500	14·77	15·12	15·47	15·80	16·46	17·10	17·42	18·34	19·22	19·51	20·08	20·64	20·91	21·72
1/ 10·5	18·131	20·073	22·137	24·326	29·092	34·389	37·244	46·662	57·412	61·303	69·558	78·460	83·159	98·277
0·10000	15·16	15·52	15·87	16·22	16·89	17·55	17·87	18·82	19·73	20·02	20·61	21·18	21·46	22·29
1/ 10·0	18·606	20·599	22·717	24·964	29·854	35·290	38·219	47·884	58·915	62·907	71·377	80·512	85·334	100·85
	0·97	0·97	0·97	0·97	0·97	0·97	0·97	0·97	0·98	0·98	0·98	0·98	0·98	0·98

$V_{r(0·5)medial}$ for half-full circular pipes.

k_s = 0·150 mm
S = 0·00010 to 0·00030

ie hydraulic gradient =
1 in 10000 to 1 in 3333

Water (or sewage) at 15°C;
full bore conditions.

velocities in ms^{-1}
discharges in m^3s^{-1}

Gradient **(Equivalent) Pipe diameters in mm**

Gradient	1250	1300	1350	1400	1500	1600	1650	1800	1950	2000	2100	2200	2250	2400
0·000100	0·406	0·417	0·427	0·437	0·456	0·475	0·485	0·512	0·538	0·547	0·564	0·580	0·589	0·613
1/ 10000	0·4988	0·5531	0·6110	0·6724	0·8064	0·9557	1·0363	1·3026	1·6075	1·7180	1·9528	2·2064	2·3404	2·7720
0·000105	0·417	0·428	0·438	0·448	0·468	0·488	0·497	0·525	0·552	0·561	0·578	0·595	0·604	0·629
1/ 9524	0·5119	0·5676	0·6270	0·6900	0·8275	0·9806	1·0633	1·3366	1·6493	1·7627	2·0035	2·2636	2·4011	2·8438
0·000110	0·428	0·438	0·449	0·459	0·480	0·500	0·510	0·538	0·566	0·575	0·593	0·610	0·619	0·644
1/ 9091	0·5247	0·5818	0·6426	0·7072	0·8481	1·0050	1·0897	1·3697	1·6902	1·8064	2·0531	2·3196	2·4604	2·9140
0·000115	0·438	0·449	0·460	0·470	0·491	0·512	0·522	0·551	0·579	0·589	0·607	0·625	0·633	0·659
1/ 8696	0·5372	0·5957	0·6579	0·7241	0·8682	1·0289	1·1156	1·4022	1·7302	1·8491	2·1016	2·3743	2·5185	2·9827
0·000120	0·448	0·459	0·470	0·481	0·502	0·523	0·534	0·564	0·592	0·602	0·620	0·639	0·648	0·674
1/ 8333	0·5495	0·6093	0·6729	0·7405	0·8880	1·0522	1·1409	1·4339	1·7693	1·8909	2·1491	2·4279	2·5753	3·0499
0·000125	0·458	0·469	0·480	0·492	0·513	0·535	0·545	0·576	0·605	0·615	0·634	0·653	0·662	0·689
1/ 8000	0·5615	0·6226	0·6876	0·7567	0·9073	1·0751	1·1657	1·4651	1·8076	1·9318	2·1956	2·4804	2·6310	3·1158
0·000130	0·467	0·479	0·490	0·502	0·524	0·546	0·557	0·588	0·618	0·628	0·647	0·666	0·675	0·703
1/ 7692	0·5732	0·6356	0·7020	0·7725	0·9263	1·0976	1·1900	1·4956	1·8453	1·9720	2·2412	2·5319	2·6856	3·1804
0·000135	0·477	0·489	0·500	0·512	0·535	0·557	0·568	0·599	0·630	0·640	0·660	0·679	0·689	0·717
1/ 7407	0·5848	0·6484	0·7162	0·7881	0·9449	1·1196	1·2139	1·5255	1·8822	2·0114	2·2860	2·5825	2·7391	3·2438
0·000140	0·486	0·498	0·510	0·522	0·545	0·568	0·579	0·611	0·642	0·653	0·673	0·692	0·702	0·731
1/ 7143	0·5962	0·6610	0·7300	0·8034	0·9632	1·1412	1·2373	1·5550	1·9184	2·0501	2·3299	2·6321	2·7918	3·3060
0·000145	0·495	0·507	0·520	0·532	0·555	0·578	0·589	0·622	0·654	0·665	0·685	0·705	0·715	0·744
1/ 6897	0·6073	0·6734	0·7437	0·8184	0·9811	1·1625	1·2604	1·5839	1·9540	2·0882	2·3731	2·6809	2·8435	3·3672
0·000150	0·504	0·516	0·529	0·541	0·565	0·589	0·600	0·634	0·666	0·677	0·697	0·718	0·728	0·758
1/ 6667	0·6183	0·6855	0·7571	0·8331	0·9988	1·1834	1·2830	1·6123	1·9890	2·1256	2·4156	2·7288	2·8943	3·4273
0·000160	0·521	0·534	0·547	0·560	0·585	0·609	0·621	0·655	0·689	0·700	0·721	0·742	0·753	0·784
1/ 6250	0·6397	0·7093	0·7833	0·8619	1·0333	1·2242	1·3273	1·6678	2·0574	2·1986	2·4985	2·8224	2·9935	3·5447
0·000170	0·538	0·552	0·565	0·578	0·604	0·629	0·641	0·677	0·711	0·722	0·745	0·766	0·777	0·809
1/ 5882	0·6605	0·7323	0·8087	0·8899	1·0668	1·2638	1·3702	1·7216	2·1237	2·2694	2·5790	2·9132	3·0898	3·6585
0·000180	0·555	0·569	0·582	0·596	0·622	0·648	0·660	0·697	0·733	0·744	0·767	0·790	0·801	0·833
1/ 5556	0·6807	0·7547	0·8334	0·9170	1·0993	1·3023	1·4119	1·7739	2·1881	2·3383	2·6571	3·0014	3·1833	3·7691
0·000190	0·571	0·585	0·599	0·613	0·640	0·666	0·679	0·717	0·754	0·766	0·789	0·812	0·823	0·857
1/ 5263	0·7004	0·7765	0·8575	0·9435	1·1309	1·3397	1·4524	1·8248	2·2508	2·4052	2·7331	3·0872	3·2743	3·8767
0·000200	0·586	0·601	0·615	0·630	0·657	0·684	0·698	0·737	0·774	0·786	0·810	0·834	0·846	0·880
1/ 5000	0·7195	0·7977	0·8809	0·9692	1·1618	1·3762	1·4920	1·8744	2·3119	2·4705	2·8072	3·1708	3·3629	3·9816
0·000210	0·602	0·617	0·631	0·646	0·674	0·702	0·716	0·756	0·794	0·807	0·831	0·856	0·868	0·903
1/ 4762	0·7382	0·8184	0·9038	0·9944	1·1919	1·4118	1·5306	1·9228	2·3715	2·5342	2·8795	3·2524	3·4494	4·0839
0·000220	0·616	0·632	0·647	0·662	0·691	0·719	0·733	0·774	0·814	0·826	0·852	0·877	0·889	0·925
1/ 4545	0·7565	0·8387	0·9261	1·0189	1·2213	1·4466	1·5682	1·9701	2·4298	2·5964	2·9501	3·3321	3·5339	4·1839
0·000230	0·631	0·647	0·662	0·678	0·707	0·736	0·751	0·792	0·833	0·846	0·872	0·897	0·910	0·946
1/ 4348	0·7744	0·8585	0·9479	1·0429	1·2500	1·4807	1·6051	2·0163	2·4867	2·6572	3·0192	3·4101	3·6166	4·2816
0·000240	0·645	0·661	0·677	0·693	0·723	0·753	0·768	0·810	0·851	0·865	0·891	0·917	0·930	0·968
1/ 4167	0·7919	0·8778	0·9693	1·0665	1·2782	1·5140	1·6412	2·0616	2·5425	2·7167	3·0868	3·4864	3·6975	4·3773
0·000250	0·659	0·676	0·692	0·708	0·739	0·769	0·784	0·828	0·870	0·883	0·910	0·937	0·950	0·988
1/ 4000	0·8090	0·8968	0·9903	1·0895	1·3057	1·5466	1·6765	2·1059	2·5971	2·7751	3·1530	3·5611	3·7767	4·4710
0·000260	0·673	0·690	0·706	0·722	0·754	0·785	0·800	0·845	0·888	0·902	0·929	0·956	0·969	1·009
1/ 3846	0·8258	0·9154	1·0108	1·1121	1·3328	1·5786	1·7112	2·1494	2·6506	2·8323	3·2180	3·6344	3·8544	4·5629
0·000270	0·686	0·703	0·720	0·737	0·769	0·801	0·816	0·861	0·905	0·919	0·947	0·975	0·989	1·029
1/ 3704	0·8423	0·9337	1·0310	1·1343	1·3593	1·6100	1·7452	2·1921	2·7032	2·8884	3·2817	3·7063	3·9307	4·6530
0·000280	0·700	0·717	0·734	0·751	0·784	0·816	0·832	0·878	0·922	0·937	0·966	0·994	1·007	1·048
1/ 3571	0·8585	0·9517	1·0508	1·1560	1·3854	1·6408	1·7786	2·2340	2·7548	2·9435	3·3443	3·7769	4·0055	4·7416
0·000290	0·713	0·730	0·748	0·765	0·798	0·831	0·847	0·894	0·939	0·954	0·983	1·012	1·026	1·067
1/ 3448	0·8744	0·9693	1·0702	1·1774	1·4110	1·6711	1·8114	2·2752	2·8055	2·9976	3·4057	3·8463	4·0791	4·8286
0·000300	0·725	0·743	0·761	0·779	0·813	0·846	0·862	0·910	0·956	0·971	1·001	1·030	1·044	1·086
1/ 3333	0·8901	0·9866	1·0894	1·1985	1·4362	1·7009	1·8437	2·3156	2·8553	3·0509	3·4662	3·9145	4·1514	4·9141
	0·91	0·92	0·92	0·92	0·93	0·93	0·93	0·93	0·94	0·94	0·94	0·94	0·94	0·95

$V_{r(0·5)medial}$ for half-full circular pipes.

k_s = 0·150 mm S = 0·00010 to 0·00030

Gradient	(Equivalent) Pipe diameters in mm													
	1250	**1300**	**1350**	**1400**	**1500**	**1600**	**1650**	**1800**	**1950**	**2000**	**2100**	**2200**	**2250**	**2400**
0·000300	0·725	0·743	0·761	0·779	0·813	0·846	0·862	0·910	0·956	0·971	1·001	1·030	1·044	1·086
1/ 3333	0·8901	0·9866	1·0894	1·1985	1·4362	1·7009	1·8437	2·3156	2·8553	3·0509	3·4662	3·9145	4·1514	4·9141
0·000320	0·750	0·769	0·787	0·805	0·841	0·875	0·892	0·941	0·989	1·004	1·035	1·065	1·080	1·123
1/ 3125	0·9207	1·0205	1·1267	1·2395	1·4853	1·7590	1·9067	2·3946	2·9526	3·1548	3·5841	4·0477	4·2925	5·0810
0·000340	0·774	0·794	0·812	0·831	0·867	0·903	0·920	0·971	1·020	1·036	1·068	1·099	1·114	1·159
1/ 2941	0·9503	1·0533	1·1629	1·2793	1·5330	1·8154	1·9678	2·4712	3·0470	3·2556	3·6985	4·1767	4·4294	5·2428
0·000360	0·798	0·818	0·837	0·856	0·894	0·930	0·948	1·000	1·051	1·067	1·100	1·132	1·147	1·194
1/ 2778	0·9791	1·0852	1·1981	1·3180	1·5793	1·8702	2·0271	2·5457	3·1386	3·3534	3·8096	4·3021	4·5623	5·4000
0·000380	0·821	0·841	0·861	0·881	0·919	0·957	0·975	1·029	1·081	1·098	1·131	1·164	1·180	1·227
1/ 2632	1·0071	1·1162	1·2324	1·3557	1·6243	1·9235	2·0849	2·6181	3·2278	3·4487	3·9178	4·4241	4·6916	5·5530
0·000400	0·843	0·864	0·884	0·904	0·944	0·982	1·001	1·057	1·110	1·127	1·162	1·195	1·212	1·260
1/ 2500	1·0344	1·1465	1·2657	1·3923	1·6682	1·9754	2·1411	2·6886	3·3146	3·5414	4·0231	4·5430	4·8176	5·7020
0·000420	0·865	0·886	0·907	0·928	0·968	1·008	1·027	1·084	1·138	1·156	1·191	1·226	1·243	1·293
1/ 2381	1·0610	1·1760	1·2983	1·4281	1·7110	2·0261	2·1960	2·7574	3·3994	3·6320	4·1258	4·6590	4·9406	5·8473
0·000440	0·886	0·908	0·929	0·950	0·992	1·032	1·052	1·110	1·166	1·184	1·220	1·255	1·273	1·324
1/ 2273	1·0871	1·2048	1·3301	1·4631	1·7529	2·0756	2·2496	2·8247	3·4822	3·7204	4·2262	4·7722	5·0606	5·9893
0·000460	0·907	0·929	0·951	0·973	1·015	1·056	1·077	1·136	1·193	1·212	1·249	1·285	1·302	1·355
1/ 2174	1·1125	1·2330	1·3612	1·4973	1·7938	2·1239	2·3021	2·8904	3·5631	3·8068	4·3243	4·8830	5·1780	6·1281
0·000480	0·927	0·950	0·972	0·994	1·038	1·080	1·101	1·161	1·220	1·239	1·276	1·313	1·331	1·385
1/ 2083	1·1374	1·2606	1·3916	1·5307	1·8338	2·1713	2·3534	2·9547	3·6423	3·8914	4·4204	4·9913	5·2929	6·2640
0·000500	0·947	0·970	0·993	1·016	1·060	1·103	1·124	1·186	1·246	1·265	1·303	1·341	1·360	1·414
1/ 2000	1·1618	1·2876	1·4214	1·5635	1·8731	2·2177	2·4036	3·0178	3·7199	3·9743	4·5145	5·0975	5·4055	6·3971
0·000525	0·971	0·995	1·019	1·042	1·087	1·131	1·153	1·216	1·277	1·297	1·337	1·375	1·394	1·450
1/ 1905	1·1916	1·3206	1·4579	1·6036	1·9210	2·2744	2·4651	3·0948	3·8148	4·0756	4·6295	5·2273	5·5431	6·5598
0·000550	0·995	1·019	1·043	1·067	1·114	1·159	1·181	1·246	1·308	1·329	1·369	1·408	1·428	1·485
1/ 1818	1·2208	1·3529	1·4935	1·6427	1·9679	2·3298	2·5251	3·1701	3·9075	4·1746	4·7418	5·3541	5·6775	6·7187
0·000575	1·018	1·043	1·068	1·092	1·139	1·186	1·208	1·275	1·339	1·360	1·401	1·441	1·461	1·519
1/ 1739	1·2493	1·3845	1·5283	1·6810	2·0137	2·3840	2·5838	3·2437	3·9981	4·2714	4·8517	5·4780	5·8089	6·8740
0·000600	1·041	1·066	1·092	1·116	1·165	1·212	1·235	1·303	1·368	1·390	1·432	1·473	1·493	1·553
1/ 1667	1·2772	1·4154	1·5624	1·7184	2·0585	2·4370	2·6412	3·3157	4·0868	4·3661	4·9592	5·5993	5·9374	7·0261
0·000625	1·063	1·089	1·115	1·140	1·190	1·238	1·262	1·331	1·398	1·419	1·462	1·504	1·525	1·586
1/ 1600	1·3045	1·4456	1·5958	1·7551	2·1024	2·4890	2·6975	3·3862	4·1736	4·4588	5·0645	5·7181	6·0634	7·1750
0·000650	1·085	1·111	1·138	1·164	1·214	1·263	1·287	1·358	1·426	1·448	1·492	1·535	1·556	1·618
1/ 1538	1·3313	1·4753	1·6285	1·7911	2·1455	2·5399	2·7527	3·4554	4·2588	4·5498	5·1677	5·8346	6·1869	7·3210
0·000675	1·106	1·133	1·160	1·186	1·238	1·288	1·313	1·385	1·454	1·477	1·521	1·565	1·586	1·650
1/ 1481	1·3576	1·5044	1·6606	1·8264	2·1877	2·5899	2·8068	3·5233	4·3423	4·6390	5·2690	5·9489	6·3080	7·4642
0·000700	1·127	1·155	1·182	1·209	1·261	1·312	1·338	1·411	1·481	1·505	1·550	1·594	1·616	1·681
1/ 1429	1·3834	1·5330	1·6921	1·8611	2·2292	2·6389	2·8599	3·5899	4·4244	4·7266	5·3684	6·0611	6·4270	7·6049
0·000725	1·148	1·176	1·204	1·231	1·285	1·336	1·362	1·436	1·508	1·532	1·578	1·623	1·646	1·712
1/ 1379	1·4087	1·5611	1·7231	1·8951	2·2699	2·6871	2·9121	3·6554	4·5050	4·8127	5·4661	6·1713	6·5438	7·7430
0·000750	1·168	1·197	1·225	1·253	1·307	1·360	1·386	1·462	1·535	1·559	1·606	1·652	1·675	1·742
1/ 1333	1·4337	1·5887	1·7536	1·9286	2·3100	2·7345	2·9635	3·7197	4·5842	4·8973	5·5621	6·2797	6·6587	7·8788
0·000800	1·208	1·237	1·267	1·295	1·351	1·406	1·433	1·511	1·587	1·611	1·660	1·708	1·731	1·800
1/ 1250	1·4823	1·6425	1·8130	1·9940	2·3882	2·8270	3·0636	3·8453	4·7388	5·0624	5·7496	6·4912	6·8829	8·1439
0·000850	1·246	1·277	1·307	1·336	1·394	1·451	1·478	1·559	1·637	1·662	1·712	1·762	1·786	1·857
1/ 1176	1·5295	1·6948	1·8707	2·0573	2·4640	2·9166	3·1608	3·9670	4·8887	5·2225	5·9312	6·6962	7·1002	8·4008
0·000900	1·284	1·315	1·346	1·376	1·436	1·494	1·522	1·605	1·686	1·712	1·763	1·814	1·839	1·912
1/ 1111	1·5753	1·7455	1·9266	2·1188	2·5376	3·0037	3·2551	4·0853	5·0342	5·3779	6·1077	6·8953	7·3113	8·6504
0·000950	1·320	1·352	1·384	1·415	1·477	1·536	1·565	1·651	1·733	1·760	1·813	1·865	1·890	1·966
1/ 1053	1·6199	1·7949	1·9811	2·1787	2·6092	3·0884	3·3468	4·2003	5·1758	5·5292	6·2793	7·0890	7·5166	8·8931
0·001000	1·355	1·388	1·421	1·453	1·516	1·577	1·607	1·695	1·779	1·807	1·861	1·914	1·941	2·018
1/ 1000	1·6633	1·8430	2·0341	2·2370	2·6790	3·1709	3·4362	4·3124	5·3138	5·6765	6·4466	7·2776	7·7166	9·1296
	0·93	0·93	0·93	0·93	0·94	0·94	0·94	0·94	0·95	0·95	0·95	0·95	0·95	0·95

$V_{r(0\cdot5)medial}$ for half-full circular pipes.

A39
(p.3 of 6)

$k_s = 0{\cdot}150$ **mm**
S = 0·00100 to 0·00300

ie hydraulic gradient =
1 in 1000 to 1 in 333

Water (or sewage) at 15°C;
full bore conditions.

velocities in ms^{-1}
discharges in m^3s^{-1}

Gradient **(Equivalent) Pipe diameters in mm**

Gradient	1250	1300	1350	1400	1500	1600	1650	1800	1950	2000	2100	2200	2250	2400
0·00100	1·355	1·388	1·421	1·453	1·516	1·577	1·607	1·695	1·779	1·807	1·861	1·914	1·941	2·018
1/ 1000	1·6633	1·8430	2·0341	2·2370	2·6790	3·1709	3·4362	4·3124	5·3138	5·6765	6·4466	7·2776	7·7166	9·1296
0·00105	1·390	1·424	1·457	1·490	1·555	1·617	1·648	1·738	1·824	1·853	1·908	1·963	1·990	2·069
1/ 952	1·7057	1·8899	2·0859	2·2939	2·7471	3·2514	3·5234	4·4217	5·4484	5·8202	6·6097	7·4617	7·9117	9·3602
0·00110	1·424	1·458	1·493	1·526	1·592	1·656	1·688	1·780	1·868	1·897	1·954	2·010	2·038	2·119
1/ 909	1·7470	1·9357	2·1364	2·3495	2·8135	3·3300	3·6086	4·5284	5·5798	5·9606	6·7690	7·6414	8·1022	9·5855
0·00115	1·457	1·492	1·527	1·562	1·629	1·694	1·727	1·821	1·911	1·941	1·999	2·056	2·085	2·168
1/ 870	1·7875	1·9805	2·1859	2·4038	2·8785	3·4069	3·6918	4·6328	5·7083	6·0978	6·9247	7·8171	8·2885	9·8057
0·00120	1·489	1·525	1·561	1·596	1·665	1·732	1·765	1·861	1·953	1·984	2·043	2·102	2·130	2·215
1/ 833	1·8271	2·0244	2·2342	2·4569	2·9421	3·4821	3·7733	4·7350	5·8340	6·2320	7·0771	7·9891	8·4707	10·021
0·00125	1·520	1·557	1·594	1·630	1·700	1·768	1·802	1·900	1·995	2·026	2·086	2·146	2·175	2·262
1/ 800	1·8659	2·0673	2·2816	2·5090	3·0044	3·5558	3·8531	4·8350	5·9572	6·3636	7·2264	8·1575	8·6493	10·232
0·00130	1·551	1·589	1·626	1·663	1·735	1·804	1·839	1·939	2·035	2·067	2·129	2·189	2·219	2·308
1/ 769	1·9039	2·1094	2·3280	2·5601	3·0655	3·6280	3·9314	4·9331	6·0779	6·4925	7·3727	8·3226	8·8243	10·439
0·00135	1·582	1·620	1·658	1·696	1·769	1·840	1·874	1·976	2·075	2·107	2·170	2·232	2·263	2·352
1/ 741	1·9412	2·1507	2·3736	2·6102	3·1254	3·6989	4·0081	5·0293	6·1963	6·6190	7·5163	8·4846	8·9960	10·642
0·00140	1·612	1·651	1·690	1·728	1·802	1·874	1·910	2·014	2·114	2·146	2·211	2·274	2·305	2·396
1/ 714	1·9778	2·1913	2·4183	2·6593	3·1842	3·7684	4·0835	5·1238	6·3126	6·7432	7·6572	8·6436	9·1646	10·841
0·00145	1·641	1·681	1·720	1·759	1·835	1·908	1·944	2·050	2·152	2·185	2·251	2·315	2·347	2·440
1/ 690	2·0138	2·2311	2·4623	2·7076	3·2420	3·8368	4·1575	5·2166	6·4268	6·8651	7·7957	8·7998	9·3302	11·037
0·00150	1·670	1·710	1·750	1·790	1·867	1·942	1·978	2·086	2·190	2·223	2·290	2·355	2·388	2·482
1/ 667	2·0491	2·2703	2·5055	2·7551	3·2989	3·9040	4·2303	5·3078	6·5391	6·9851	7·9318	8·9534	9·4929	11·230
0·00160	1·726	1·768	1·809	1·850	1·930	2·007	2·045	2·156	2·263	2·298	2·367	2·434	2·467	2·565
1/ 625	2·1182	2·3467	2·5898	2·8478	3·4097	4·0351	4·3723	5·4858	6·7582	7·2191	8·1974	9·2530	9·8106	11·605
0·00170	1·781	1·824	1·866	1·908	1·990	2·070	2·109	2·224	2·334	2·370	2·441	2·511	2·545	2·646
1/ 588	2·1851	2·4208	2·6715	2·9376	3·5172	4·1621	4·5099	5·6584	6·9707	7·4459	8·4548	9·5435	10·118	11·969
0·00180	1·834	1·878	1·922	1·965	2·049	2·131	2·172	2·289	2·403	2·440	2·513	2·585	2·620	2·724
1/ 556	2·2500	2·4927	2·7509	3·0249	3·6216	4·2856	4·6436	5·8259	7·1769	7·6662	8·7048	9·8256	10·417	12·323
0·00190	1·885	1·931	1·976	2·020	2·107	2·191	2·233	2·354	2·470	2·508	2·583	2·657	2·693	2·800
1/ 526	2·3133	2·5627	2·8281	3·1097	3·7231	4·4056	4·7737	5·9889	7·3776	7·8804	8·9480	10·100	10·708	12·666
0·00200	1·935	1·982	2·028	2·074	2·163	2·249	2·292	2·416	2·536	2·575	2·652	2·727	2·764	2·874
1/ 500	2·3748	2·6309	2·9033	3·1924	3·8220	4·5226	4·9004	6·1477	7·5730	8·0892	9·1849	10·367	10·992	13·001
0·00210	1·984	2·032	2·080	2·126	2·217	2·306	2·350	2·477	2·600	2·640	2·719	2·796	2·834	2·946
1/ 476	2·4349	2·6974	2·9767	3·2730	3·9185	4·6366	5·0239	6·3026	7·7636	8·2927	9·4159	10·628	11·268	13·328
0·00220	2·032	2·081	2·130	2·177	2·271	2·361	2·406	2·536	2·662	2·703	2·784	2·863	2·902	3·017
1/ 455	2·4936	2·7624	3·0483	3·3518	4·0127	4·7480	5·1446	6·4538	7·9498	8·4915	9·6415	10·882	11·538	13·647
0·00230	2·079	2·129	2·179	2·227	2·323	2·416	2·461	2·594	2·723	2·765	2·847	2·928	2·968	3·086
1/ 435	2·5509	2·8259	3·1184	3·4288	4·1048	4·8569	5·2625	6·6017	8·1318	8·6859	9·8621	11·131	11·802	13·959
0·00240	2·124	2·176	2·226	2·276	2·374	2·469	2·515	2·651	2·783	2·825	2·910	2·992	3·033	3·153
1/ 417	2·6070	2·8880	3·1869	3·5041	4·1949	4·9635	5·3780	6·7464	8·3099	8·8761	10·078	11·375	12·060	14·264
0·00250	2·169	2·222	2·273	2·324	2·424	2·521	2·568	2·707	2·841	2·885	2·971	3·055	3·097	3·219
1/ 400	2·6620	2·9489	3·2541	3·5779	4·2832	5·0679	5·4910	6·8881	8·4843	9·0624	10·289	11·613	12·313	14·563
0·00260	2·213	2·267	2·319	2·371	2·473	2·571	2·620	2·761	2·898	2·943	3·031	3·117	3·159	3·284
1/ 385	2·7159	3·0086	3·3199	3·6502	4·3697	5·1702	5·6019	7·0270	8·6553	9·2449	10·497	11·847	12·560	14·856
0·00270	2·256	2·311	2·364	2·417	2·521	2·621	2·671	2·815	2·954	3·000	3·089	3·177	3·220	3·347
1/ 370	2·7687	3·0671	3·3844	3·7212	4·4546	5·2706	5·7106	7·1633	8·8231	9·4241	10·700	12·077	12·804	15·143
0·00280	2·298	2·354	2·409	2·463	2·568	2·670	2·721	2·868	3·009	3·056	3·147	3·236	3·280	3·410
1/ 357	2·8206	3·1246	3·4478	3·7909	4·5379	5·3692	5·8174	7·2971	8·9877	9·5999	10·899	12·302	13·042	15·425
0·00290	2·340	2·397	2·452	2·507	2·614	2·719	2·770	2·919	3·064	3·111	3·203	3·294	3·339	3·471
1/ 345	2·8716	3·1810	3·5101	3·8593	4·6198	5·4660	5·9222	7·4285	9·1495	9·7727	11·096	12·523	13·277	15·703
0·00300	2·381	2·438	2·495	2·551	2·660	2·766	2·818	2·970	3·117	3·165	3·259	3·352	3·397	3·531
1/ 333	2·9217	3·2365	3·5713	3·9266	4·7003	5·5611	6·0253	7·5577	9·3085	9·9425	11·288	12·740	13·507	15·975
	0·93	0·94	0·94	0·94	0·94	0·94	0·95	0·95	0·95	0·95	0·95	0·96	0·96	0·96

$V_{r(0{\cdot}5)medial}$ **for half-full circular pipes.**

$k_s = 0{\cdot}150$ **mm** S = 0·00100 to 0·00300

Gradient **(Equivalent) Pipe diameters in mm**

Gradient	1250	1300	1350	1400	1500	1600	1650	1800	1950	2000	2100	2200	2250	2400
0·00300	2·381	2·438	2·495	2·551	2·660	2·766	2·818	2·970	3·117	3·165	3·259	3·352	3·397	3·531
1/ 333	2·9217	3·2365	3·5713	3·9266	4·7003	5·5611	6·0253	7·5577	9·3085	9·9425	11·288	12·740	13·507	15·975
0·00320	2·461	2·520	2·578	2·636	2·749	2·858	2·912	3·069	3·221	3·270	3·368	3·463	3·510	3·649
1/ 313	3·0195	3·3448	3·6907	4·0578	4·8573	5·7468	6·2264	7·8098	9·6188	10·274	11·664	13·165	13·957	16·507
0·00340	2·538	2·599	2·659	2·719	2·835	2·948	3·003	3·165	3·321	3·372	3·473	3·571	3·620	3·763
1/ 294	3·1143	3·4497	3·8065	4·1851	5·0095	5·9268	6·4214	8·0542	9·9195	10·595	12·029	13·576	14·393	17·022
0·00360	2·613	2·676	2·738	2·799	2·918	3·035	3·092	3·258	3·419	3·472	3·575	3·676	3·726	3·873
1/ 278	3·2063	3·5517	3·9189	4·3086	5·1573	6·1016	6·6107	8·2914	10·212	10·907	12·383	13·975	14·816	17·522
0·00380	2·686	2·751	2·814	2·877	3·000	3·119	3·178	3·349	3·514	3·568	3·674	3·779	3·830	3·981
1/ 263	3·2959	3·6508	4·0283	4·4288	5·3011	6·2716	6·7948	8·5222	10·496	11·210	12·727	14·363	15·228	18·009
0·00400	2·757	2·823	2·889	2·953	3·079	3·202	3·262	3·437	3·607	3·662	3·771	3·878	3·931	4·086
1/ 250	3·3831	3·7474	4·1348	4·5459	5·4412	6·4372	6·9742	8·7470	10·772	11·506	13·062	14·742	15·629	18·483
0·00420	2·826	2·894	2·961	3·027	3·156	3·282	3·343	3·524	3·697	3·754	3·866	3·975	4·029	4·188
1/ 238	3·4682	3·8416	4·2387	4·6601	5·5778	6·5987	7·1491	8·9663	11·042	11·794	13·389	15·111	16·020	18·945
0·00440	2·894	2·964	3·032	3·100	3·232	3·360	3·423	3·608	3·786	3·844	3·958	4·070	4·125	4·288
1/ 227	3·5512	3·9336	4·3402	4·7716	5·7112	6·7564	7·3200	9·1804	11·306	12·075	13·709	15·471	16·402	19·397
0·00460	2·960	3·031	3·101	3·171	3·306	3·437	3·501	3·690	3·872	3·931	4·048	4·163	4·219	4·385
1/ 217	3·6324	4·0235	4·4394	4·8806	5·8416	6·9106	7·4870	9·3897	11·563	12·350	14·021	15·823	16·775	19·838
0·00480	3·025	3·098	3·169	3·240	3·378	3·512	3·578	3·770	3·956	4·017	4·136	4·253	4·311	4·481
1/ 208	3·7119	4·1115	4·5364	4·9873	5·9692	7·0615	7·6504	9·5946	11·815	12·620	14·326	16·168	17·141	20·270
0·00500	3·088	3·163	3·236	3·308	3·449	3·586	3·653	3·849	4·039	4·101	4·223	4·342	4·401	4·574
1/ 200	3·7898	4·1977	4·6315	5·0918	6·0941	7·2093	7·8105	9·7952	12·062	12·883	14·626	16·506	17·498	20·693
0·00525	3·166	3·242	3·317	3·391	3·535	3·675	3·744	3·946	4·140	4·203	4·328	4·451	4·511	4·688
1/ 190	3·8849	4·3030	4·7477	5·2195	6·2469	7·3899	8·0061	10·040	12·364	13·205	14·991	16·918	17·936	21·210
0·00550	3·241	3·319	3·396	3·472	3·619	3·763	3·834	4·040	4·239	4·304	4·431	4·557	4·618	4·800
1/ 182	3·9778	4·4059	4·8611	5·3442	6·3961	7·5663	8·1972	10·280	12·659	13·520	15·349	17·321	18·363	21·715
0·00575	3·315	3·395	3·474	3·551	3·702	3·849	3·921	4·132	4·335	4·402	4·532	4·660	4·723	4·909
1/ 174	4·0686	4·5065	4·9721	5·4661	6·5419	7·7387	8·3839	10·514	12·947	13·828	15·698	17·715	18·780	22·209
0·00600	3·388	3·469	3·549	3·628	3·783	3·933	4·006	4·222	4·430	4·497	4·631	4·762	4·826	5·016
1/ 167	4·1575	4·6049	5·0806	5·5854	6·6846	7·9074	8·5667	10·743	13·229	14·129	16·039	18·100	19·189	22·691
0·00625	3·459	3·542	3·624	3·704	3·862	4·015	4·090	4·310	4·522	4·591	4·727	4·861	4·927	5·120
1/ 160	4·2446	4·7012	5·1869	5·7023	6·8244	8·0727	8·7457	10·967	13·505	14·424	16·374	18·478	19·589	23·164
0·00650	3·528	3·613	3·697	3·779	3·939	4·096	4·172	4·396	4·613	4·683	4·822	4·958	5·025	5·223
1/ 154	4·3299	4·7957	5·2911	5·8168	6·9614	8·2347	8·9211	11·187	13·776	14·713	16·702	18·848	19·981	23·628
0·00675	3·597	3·683	3·768	3·852	4·015	4·175	4·253	4·481	4·702	4·773	4·915	5·054	5·122	5·323
1/ 148	4·4136	4·8884	5·3934	5·9291	7·0958	8·3936	9·0932	11·403	14·041	14·996	17·024	19·211	20·366	24·083
0·00700	3·663	3·751	3·838	3·923	4·090	4·252	4·332	4·564	4·789	4·862	5·006	5·147	5·217	5·422
1/ 143	4·4958	4·9794	5·4937	6·0394	7·2277	8·5496	9·2622	11·615	14·302	15·274	17·339	19·567	20·744	24·529
0·00725	3·729	3·819	3·907	3·994	4·163	4·328	4·409	4·646	4·875	4·949	5·096	5·240	5·310	5·519
1/ 138	4·5765	5·0688	5·5923	6·1478	7·3573	8·7028	9·4282	11·823	14·558	15·548	17·650	19·917	21·115	24·968
0·00750	3·794	3·885	3·975	4·063	4·235	4·403	4·486	4·726	4·959	5·035	5·184	5·330	5·402	5·614
1/ 133	4·6558	5·1566	5·6892	6·2543	7·4847	8·8534	9·5913	12·027	14·809	15·817	17·955	20·261	21·479	25·399
0·00800	3·920	4·014	4·107	4·198	4·376	4·550	4·634	4·883	5·123	5·201	5·356	5·507	5·581	5·800
1/ 125	4·8106	5·3280	5·8783	6·4621	7·7333	9·1473	9·9097	12·426	15·300	16·341	18·550	20·933	22·191	26·240
0·00850	4·042	4·139	4·235	4·329	4·512	4·691	4·779	5·035	5·283	5·363	5·522	5·678	5·755	5·981
1/ 118	4·9607	5·4942	6·0615	6·6635	7·9742	9·4322	10·218	12·813	15·776	16·849	19·126	21·583	22·881	27·055
0·00900	4·161	4·261	4·359	4·456	4·645	4·829	4·919	5·183	5·437	5·520	5·684	5·844	5·923	6·156
1/ 111	5·1064	5·6555	6·2395	6·8591	8·2081	9·7088	10·518	13·188	16·238	17·343	19·686	22·215	23·550	27·847
0·00950	4·277	4·379	4·480	4·579	4·774	4·963	5·055	5·326	5·588	5·673	5·841	6·006	6·087	6·326
1/ 105	5·2481	5·8125	6·4126	7·0493	8·4357	9·9778	10·809	13·553	16·688	17·823	20·231	22·830	24·202	28·617
0·01000	4·389	4·494	4·598	4·700	4·899	5·093	5·188	5·466	5·734	5·822	5·994	6·163	6·246	6·492
1/ 100	5·3862	5·9653	6·5812	7·2346	8·6573	10·240	11·093	13·909	17·126	18·290	20·762	23·428	24·836	29·367
	0·94	0·94	0·94	0·94	0·95	0·95	0·95	0·95	0·95	0·96	0·96	0·96	0·96	0·96

$V_{r(0\cdot5)medial}$ **for half-full circular pipes.**

A39

(p.5 of 6)

$k_s = 0{\cdot}150$ **mm**
S = 0·01000 to 0·03000

ie hydraulic gradient =
1 in 100 to 1 in 33·3

Water (or sewage) at 15°C;
full bore conditions.

velocities in ms^{-1}
discharges in m^3s^{-1}

Gradient **(Equivalent) Pipe diameters in mm**

Gradient	1250	1300	1350	1400	1500	1600	1650	1800	1950	2000	2100	2200	2250	2400
0·01000	4·389	4·494	4·598	4·700	4·899	5·093	5·188	5·466	5·734	5·822	5·994	6·163	6·246	6·492
1/ 100	5·3862	5·9653	6·5812	7·2346	8·6573	10·240	11·093	13·909	17·126	18·290	20·762	23·428	24·836	29·367
0·01050	4·499	4·607	4·713	4·817	5·021	5·220	5·317	5·602	5·877	5·967	6·144	6·317	6·402	6·653
1/ 95	5·5208	6·1144	6·7456	7·4153	8·8735	10·495	11·370	14·256	17·553	18·746	21·279	24·012	25·455	30·098
0·01100	4·606	4·716	4·825	4·932	5·141	5·344	5·444	5·735	6·017	6·109	6·290	6·467	6·554	6·811
1/ 91	5·6523	6·2600	6·9062	7·5918	9·0846	10·745	11·640	14·595	17·970	19·191	21·784	24·582	26·059	30·813
0·01150	4·711	4·823	4·934	5·044	5·258	5·465	5·567	5·866	6·153	6·247	6·432	6·613	6·703	6·965
1/ 87	5·7809	6·4023	7·0632	7·7643	9·2909	10·989	11·904	14·926	18·377	19·627	22·278	25·139	26·650	31·511
0·01200	4·813	4·928	5·042	5·153	5·372	5·584	5·688	5·993	6·287	6·383	6·572	6·757	6·848	7·116
1/ 83	5·9066	6·5416	7·2168	7·9331	9·4929	11·228	12·163	15·250	18·776	20·052	22·762	25·685	27·228	32·194
0·01250	4·914	5·031	5·147	5·261	5·484	5·701	5·807	6·118	6·418	6·516	6·708	6·897	6·990	7·264
1/ 80	6·0298	6·6780	7·3672	8·0984	9·6906	11·462	12·416	15·568	19·167	20·470	23·235	26·219	27·794	32·863
0·01300	5·012	5·132	5·250	5·366	5·593	5·815	5·923	6·240	6·546	6·646	6·842	7·035	7·130	7·409
1/ 77	6·1505	6·8116	7·5146	8·2605	9·8844	11·691	12·665	15·879	19·549	20·878	23·699	26·742	28·349	33·519
0·01350	5·108	5·231	5·351	5·469	5·701	5·926	6·037	6·360	6·672	6·773	6·974	7·170	7·267	7·552
1/ 74	6·2690	6·9428	7·6593	8·4195	10·075	11·916	12·908	16·184	19·925	21·279	24·154	27·256	28·893	34·163
0·01400	5·203	5·328	5·450	5·571	5·807	6·036	6·149	6·478	6·795	6·899	7·103	7·303	7·401	7·691
1/ 71	6·3853	7·0715	7·8013	8·5755	10·261	12·136	13·147	16·483	20·294	21·673	24·601	27·760	29·427	34·794
0·01450	5·296	5·423	5·548	5·670	5·910	6·144	6·258	6·593	6·916	7·022	7·229	7·433	7·533	7·828
1/ 69	6·4995	7·1980	7·9408	8·7288	10·445	12·353	13·382	16·778	20·656	22·060	25·040	28·255	29·952	35·415
0·01500	5·388	5·517	5·643	5·768	6·012	6·250	6·366	6·707	7·036	7·143	7·354	7·561	7·663	7·963
1/ 67	6·6117	7·3223	8·0779	8·8795	10·625	12·566	13·613	17·067	21·012	22·440	25·471	28·741	30·468	36·024
0·01600	5·566	5·699	5·830	5·959	6·211	6·457	6·577	6·929	7·268	7·379	7·597	7·811	7·916	8·226
1/ 63	6·8308	7·5648	8·3454	9·1735	10·977	12·982	14·063	17·631	21·707	23·182	26·313	29·691	31·475	37·214
0·01700	5·739	5·876	6·011	6·144	6·404	6·657	6·781	7·144	7·494	7·608	7·833	8·053	8·161	8·481
1/ 59	7·0431	7·7999	8·6047	9·4585	11·317	13·385	14·500	18·178	22·380	23·901	27·129	30·612	32·450	38·367
0·01800	5·907	6·048	6·187	6·324	6·592	6·852	6·979	7·352	7·713	7·830	8·061	8·288	8·400	8·729
1/ 56	7·2493	8·0282	8·8565	9·7352	11·648	13·776	14·923	18·710	23·033	24·599	27·921	31·505	33·398	39·487
0·01900	6·071	6·216	6·358	6·499	6·774	7·041	7·172	7·555	7·925	8·046	8·284	8·517	8·631	8·969
1/ 53	7·4498	8·2502	9·1013	10·004	11·970	14·157	15·336	19·226	23·669	25·278	28·692	32·375	34·319	40·576
0·02000	6·230	6·379	6·525	6·669	6·951	7·225	7·360	7·753	8·133	8·257	8·500	8·739	8·857	9·204
1/ 50	7·6451	8·4665	9·3399	10·266	12·284	14·528	15·737	19·729	24·289	25·939	29·442	33·221	35·217	41·637
0·02100	6·385	6·538	6·688	6·835	7·124	7·405	7·543	7·946	8·335	8·462	8·712	8·957	9·077	9·433
1/ 47·6	7·8356	8·6774	9·5725	10·522	12·590	14·889	16·129	20·220	24·893	26·584	30·174	34·047	36·092	42·672
0·02200	6·537	6·693	6·846	6·997	7·293	7·581	7·722	8·134	8·533	8·662	8·918	9·169	9·292	9·656
1/ 45·5	8·0216	8·8833	9·7996	10·772	12·888	15·242	16·511	20·699	25·482	27·214	30·889	34·853	36·947	43·682
0·02300	6·685	6·844	7·001	7·156	7·458	7·753	7·897	8·318	8·726	8·858	9·120	9·376	9·502	9·874
1/ 43·5	8·2034	9·0847	10·022	11·016	13·180	15·587	16·885	21·168	26·059	27·830	31·588	35·642	37·782	44·670
0·02400	6·830	6·993	7·153	7·311	7·620	7·920	8·068	8·499	8·915	9·050	9·317	9·579	9·708	10·09
1/ 41·7	8·3814	9·2817	10·239	11·255	13·466	15·925	17·251	21·626	26·623	28·432	32·271	36·413	38·600	45·636
0·02500	6·972	7·138	7·302	7·463	7·778	8·085	8·235	8·675	9·100	9·238	9·511	9·778	9·909	10·30
1/ 40·0	8·5556	9·4746	10·452	11·488	13·746	16·256	17·609	22·075	27·176	29·022	32·941	37·168	39·400	46·583
0·02600	7·111	7·281	7·448	7·612	7·934	8·246	8·400	8·848	9·281	9·422	9·700	9·972	10·11	10·50
1/ 38·5	8·7264	9·6637	10·660	11·718	14·020	16·580	17·960	22·515	27·717	29·600	33·597	37·909	40·185	47·510
0·02700	7·247	7·420	7·591	7·758	8·086	8·404	8·561	9·018	9·459	9·603	9·886	10·16	10·30	10·70
1/ 37·0	8·8940	9·8492	10·865	11·943	14·289	16·898	18·305	22·947	28·248	30·168	34·241	38·635	40·955	48·420
0·02800	7·382	7·558	7·731	7·901	8·235	8·560	8·719	9·184	9·633	9·780	10·07	10·35	10·49	10·90
1/ 35·7	9·0585	10·031	11·066	12·163	14·553	17·210	18·643	23·371	28·770	30·725	34·873	39·348	41·711	49·314
0·02900	7·513	7·692	7·869	8·042	8·382	8·712	8·874	9·348	9·805	9·954	10·25	10·54	10·68	11·09
1/ 34·5	9·2200	10·210	11·263	12·380	14·812	17·517	18·975	23·787	29·282	31·272	35·494	40·048	42·453	50·191
0·03000	7·643	7·825	8·004	8·181	8·526	8·862	9·027	9·508	9·974	10·13	10·42	10·72	10·86	11·29
1/ 33·3	9·3788	10·386	11·457	12·593	15·067	17·818	19·301	24·196	29·786	31·809	36·104	40·737	43·183	51·054
	0·94	0·94	0·94	0·95	0·95	0·95	0·95	0·95	0·96	0·96	0·96	0·96	0·96	0·96

$V_{r(0{\cdot}5)medial}$ **for half-full circular pipes.**

$k_s = 0{\cdot}150$ **mm** S = 0·01000 to 0·03000

Gradient **(Equivalent) Pipe diameters in mm**

Gradient	1250	1300	1350	1400	1500	1600	1650	1800	1950	2000	2100	2200	2250	2400
0·03000	7·643	7·825	8·004	8·181	8·526	8·862	9·027	9·508	9·974	10·13	10·42	10·72	10·86	11·29
1/ 33·3	9·3788	10·386	11·457	12·593	15·067	17·818	19·301	24·196	29·786	31·809	36·104	40·737	43·183	51·054
0·03200	7·895	8·083	8·269	8·451	8·808	9·155	9·325	9·822	10·30	10·46	10·77	11·07	11·22	11·66
1/ 31·3	9·6887	10·729	11·836	13·009	15·565	18·407	19·938	24·994	30·768	32·858	37·295	42·080	44·606	52·737
0·03400	8·140	8·334	8·525	8·713	9·081	9·438	9·613	10·13	10·62	10·78	11·10	11·41	11·57	12·02
1/ 29·4	9·9891	11·062	12·202	13·412	16·047	18·977	20·556	25·768	31·721	33·875	38·448	43·382	45·986	54·368
0·03600	8·377	8·577	8·774	8·967	9·346	9·713	9·894	10·42	10·93	11·10	11·42	11·74	11·90	12·37
1/ 27·8	10·281	11·385	12·558	13·804	16·515	19·530	21·155	26·520	32·645	34·863	39·569	44·646	47·326	55·952
0·03800	8·609	8·814	9·016	9·214	9·603	9·981	10·17	10·71	11·23	11·40	11·74	12·07	12·23	12·71
1/ 26·3	10·564	11·699	12·905	14·184	16·970	20·068	21·739	27·250	33·545	35·823	40·659	45·875	48·630	57·492
0·04000	8·834	9·044	9·251	9·455	9·854	10·24	10·43	10·99	11·53	11·70	12·05	12·38	12·55	13·04
1/ 25·0	10·841	12·005	13·242	14·555	17·414	20·593	22·306	27·962	34·420	36·758	41·720	47·073	49·899	58·992
0·04200	9·053	9·269	9·481	9·690	10·10	10·50	10·69	11·26	11·81	11·99	12·34	12·69	12·86	13·36
1/ 23·8	11·110	12·303	13·571	14·917	17·846	21·104	22·860	28·656	35·275	37·671	42·756	48·241	51·137	60·456
0·04400	9·268	9·488	9·706	9·920	10·34	10·74	10·94	11·53	12·09	12·27	12·64	12·99	13·17	13·68
1/ 22·7	11·373	12·594	13·893	15·270	18·269	21·604	23·401	29·334	36·109	38·561	43·766	49·381	52·346	61·885
0·04600	9·477	9·703	9·925	10·14	10·57	10·99	11·19	11·79	12·36	12·55	12·92	13·28	13·46	13·99
1/ 21·7	11·630	12·879	14·207	15·615	18·682	22·092	23·930	29·997	36·924	39·432	44·755	50·496	53·527	63·282
0·04800	9·682	9·913	10·14	10·36	10·80	11·23	11·43	12·04	12·63	12·82	13·20	13·57	13·75	14·29
1/ 20·8	11·882	13·158	14·514	15·953	19·086	22·569	24·447	30·645	37·722	40·284	45·722	51·587	54·684	64·648
0·05000	9·883	10·12	10·35	10·58	11·02	11·46	11·67	12·29	12·89	13·09	13·47	13·85	14·04	14·59
1/ 20·0	12·129	13·431	14·815	16·284	19·481	23·037	24·954	31·280	38·504	41·119	46·669	52·655	55·816	65·987
0·05250	10·13	10·37	10·61	10·84	11·30	11·74	11·96	12·60	13·21	13·41	13·81	14·20	14·39	14·95
1/ 19·0	12·430	13·764	15·183	16·688	19·965	23·609	25·573	32·056	39·459	42·139	47·826	53·961	57·200	67·623
0·05500	10·37	10·62	10·86	11·10	11·57	12·02	12·24	12·90	13·52	13·73	14·13	14·53	14·73	15·30
1/ 18·2	12·724	14·090	15·542	17·083	20·437	24·167	26·178	32·814	40·392	43·135	48·956	55·236	58·552	69·221
0·05750	10·60	10·86	11·10	11·35	11·83	12·29	12·52	13·19	13·83	14·04	14·45	14·86	15·06	15·65
1/ 17·4	13·012	14·408	15·893	17·469	20·899	24·713	26·769	33·555	41·303	44·108	50·061	56·483	59·873	70·782
0·06000	10·83	11·09	11·34	11·59	12·08	12·56	12·79	13·47	14·13	14·34	14·77	15·18	15·38	15·98
1/ 16·7	13·293	14·720	16·237	17·846	21·350	25·247	27·348	34·280	42·196	45·061	51·142	57·702	61·166	72·310
0·06250	11·06	11·32	11·58	11·83	12·33	12·82	13·05	13·75	14·42	14·64	15·07	15·49	15·70	16·31
1/ 16·0	13·568	15·025	16·574	18·216	21·793	25·770	27·914	34·990	43·069	45·994	52·201	58·897	62·432	73·807
0·06500	11·28	11·55	11·81	12·07	12·58	13·07	13·31	14·02	14·71	14·93	15·37	15·80	16·01	16·64
1/ 15·4	13·839	15·324	16·903	18·579	22·226	26·283	28·470	35·686	43·926	46·908	53·239	60·068	63·673	75·274
0·06750	11·49	11·77	12·04	12·30	12·82	13·32	13·57	14·29	14·99	15·22	15·66	16·10	16·32	16·96
1/ 14·8	14·104	15·617	17·227	18·935	22·652	26·786	29·014	36·368	44·766	47·806	54·257	61·216	64·890	76·713
0·07000	11·70	11·98	12·26	12·53	13·05	13·57	13·82	14·56	15·27	15·50	15·95	16·40	16·62	17·27
1/ 14·3	14·364	15·905	17·545	19·284	23·069	27·280	29·549	37·038	45·590	48·686	55·256	62·344	66·085	78·126
0·07250	11·91	12·20	12·48	12·75	13·29	13·81	14·06	14·81	15·54	15·77	16·24	16·69	16·92	17·58
1/ 13·8	14·619	16·188	17·857	19·627	23·480	27·765	30·074	37·697	46·401	49·551	56·238	63·451	67·259	79·513
0·07500	12·12	12·41	12·69	12·97	13·51	14·05	14·31	15·07	15·80	16·04	16·52	16·98	17·21	17·88
1/ 13·3	14·870	16·466	18·163	19·964	23·883	28·241	30·590	38·344	47·197	50·402	57·203	64·540	68·413	80·877
0·08000	12·52	12·81	13·11	13·40	13·96	14·51	14·78	15·56	16·32	16·57	17·06	17·54	17·77	18·47
1/ 12·5	15·360	17·009	18·762	20·621	24·669	29·171	31·598	39·606	48·750	52·061	59·086	66·664	70·665	83·538
0·08500	12·90	13·21	13·51	13·81	14·39	14·96	15·23	16·05	16·83	17·08	17·59	18·08	18·32	19·04
1/ 11·8	15·835	17·535	19·342	21·259	25·432	30·073	32·574	40·830	50·256	53·669	60·910	68·722	72·847	86·118
0·09000	13·28	13·60	13·91	14·21	14·81	15·39	15·68	16·51	17·32	17·58	18·10	18·60	18·85	19·59
1/ 11·1	16·297	18·045	19·905	21·878	26·172	30·948	33·522	42·018	51·718	55·230	62·682	70·721	74·965	88·622
0·09500	13·65	13·97	14·29	14·60	15·22	15·82	16·11	16·97	17·79	18·06	18·59	19·12	19·37	20·13
1/ 10·5	16·745	18·542	20·453	22·480	26·892	31·799	34·444	43·173	53·140	56·748	64·406	72·665	77·026	91·058
0·10000	14·00	14·33	14·66	14·98	15·61	16·23	16·53	17·41	18·26	18·53	19·08	19·61	19·88	20·65
1/ 10·0	17·182	19·026	20·986	23·066	27·593	32·628	35·342	44·299	54·526	58·227	66·084	74·559	79·033	93·430
	0·94	0·94	0·95	0·95	0·95	0·95	0·95	0·95	0·96	0·96	0·96	0·96	0·96	0·96

$V_{r(0\cdot5)medial}$ for half-full circular pipes.

A40

(p.1 of 6)

k_s = 0·30 mm
S = 0·00010 to 0·00030

ie hydraulic gradient =
1 in 10000 to 1 in 3333

Water (or sewage) at 15°C;
full bore conditions.

velocities in ms^{-1}
discharges in m^3s^{-1}

Gradient	**(Equivalent) Pipe diameters in mm**													
	1250	**1300**	**1350**	**1400**	**1500**	**1600**	**1650**	**1800**	**1950**	**2000**	**2100**	**2200**	**2250**	**2400**
0·000100	0·392	0·401	0·411	0·421	0·439	0·458	0·466	0·493	0·518	0·526	0·542	0·558	0·566	0·589
1/ 10000	0·4806	0·5329	0·5885	0·6476	0·7764	0·9199	0·9973	1·2533	1·5461	1·6523	1·8777	2·1212	2·2498	2·6641
0·000105	0·402	0·412	0·422	0·432	0·451	0·469	0·478	0·505	0·531	0·539	0·556	0·572	0·580	0·604
1/ 9524	0·4930	0·5466	0·6037	0·6643	0·7964	0·9436	1·0230	1·2854	1·5858	1·6946	1·9258	2·1754	2·3073	2·7322
0·000110	0·412	0·422	0·432	0·442	0·462	0·481	0·490	0·517	0·544	0·553	0·570	0·586	0·594	0·619
1/ 9091	0·5052	0·5601	0·6185	0·6806	0·8159	0·9667	1·0480	1·3169	1·6245	1·7359	1·9727	2·2284	2·3635	2·7987
0·000115	0·421	0·432	0·442	0·452	0·473	0·492	0·502	0·530	0·557	0·565	0·583	0·600	0·608	0·633
1/ 8696	0·5170	0·5732	0·6330	0·6965	0·8350	0·9893	1·0725	1·3476	1·6623	1·7764	2·0186	2·2803	2·4185	2·8637
0·000120	0·431	0·442	0·452	0·463	0·483	0·503	0·513	0·541	0·569	0·578	0·596	0·613	0·622	0·647
1/ 8333	0·5286	0·5861	0·6472	0·7122	0·8537	1·0114	1·0965	1·3777	1·6994	1·8159	2·0636	2·3310	2·4723	2·9273
0·000125	0·440	0·451	0·462	0·473	0·493	0·514	0·524	0·553	0·581	0·590	0·608	0·626	0·635	0·661
1/ 8000	0·5400	0·5987	0·6611	0·7274	0·8720	1·0330	1·1199	1·4071	1·7356	1·8547	2·1076	2·3807	2·5249	2·9896
0·000130	0·449	0·460	0·471	0·482	0·504	0·524	0·535	0·564	0·593	0·602	0·621	0·639	0·648	0·674
1/ 7692	0·5512	0·6110	0·6748	0·7424	0·8900	1·0543	1·1430	1·4360	1·7712	1·8927	2·1508	2·4294	2·5766	3·0507
0·000135	0·458	0·469	0·481	0·492	0·514	0·535	0·545	0·575	0·605	0·614	0·633	0·652	0·661	0·688
1/ 7407	0·5621	0·6232	0·6881	0·7572	0·9076	1·0751	1·1655	1·4643	1·8061	1·9300	2·1931	2·4772	2·6273	3·1106
0·000140	0·467	0·478	0·490	0·501	0·523	0·545	0·555	0·586	0·616	0·626	0·645	0·664	0·673	0·701
1/ 7143	0·5728	0·6351	0·7013	0·7716	0·9249	1·0956	1·1877	1·4921	1·8404	1·9666	2·2347	2·5241	2·6770	3·1695
0·000145	0·475	0·487	0·499	0·510	0·533	0·555	0·566	0·597	0·628	0·637	0·657	0·676	0·686	0·713
1/ 6897	0·5834	0·6468	0·7142	0·7858	0·9419	1·1157	1·2095	1·5195	1·8741	2·0026	2·2755	2·5702	2·7259	3·2273
0·000150	0·484	0·496	0·508	0·520	0·542	0·565	0·576	0·608	0·639	0·649	0·669	0·688	0·698	0·726
1/ 6667	0·5938	0·6583	0·7269	0·7997	0·9586	1·1355	1·2309	1·5464	1·9072	2·0380	2·3157	2·6155	2·7740	3·2842
0·000160	0·500	0·513	0·525	0·537	0·561	0·584	0·595	0·628	0·660	0·671	0·691	0·711	0·721	0·750
1/ 6250	0·6140	0·6807	0·7516	0·8270	0·9912	1·1741	1·2727	1·5988	1·9718	2·1070	2·3940	2·7040	2·8677	3·3951
0·000170	0·516	0·529	0·542	0·554	0·579	0·603	0·614	0·648	0·681	0·692	0·713	0·734	0·744	0·774
1/ 5882	0·6337	0·7025	0·7757	0·8534	1·0228	1·2115	1·3133	1·6497	2·0344	2·1739	2·4700	2·7897	2·9587	3·5026
0·000180	0·532	0·545	0·558	0·571	0·596	0·621	0·633	0·668	0·702	0·713	0·734	0·756	0·766	0·797
1/ 5556	0·6528	0·7236	0·7990	0·8790	1·0535	1·2478	1·3526	1·6990	2·0953	2·2389	2·5438	2·8730	3·0470	3·6071
0·000190	0·547	0·561	0·574	0·587	0·613	0·638	0·651	0·687	0·721	0·733	0·755	0·777	0·788	0·820
1/ 5263	0·6713	0·7442	0·8217	0·9040	1·0834	1·2831	1·3909	1·7471	2·1545	2·3021	2·6156	2·9540	3·1329	3·7087
0·000200	0·562	0·576	0·589	0·603	0·630	0·655	0·668	0·705	0·741	0·752	0·775	0·798	0·809	0·842
1/ 5000	0·6894	0·7642	0·8438	0·9283	1·1125	1·3176	1·4283	1·7939	2·2121	2·3637	2·6855	3·0330	3·2166	3·8077
0·000210	0·576	0·590	0·605	0·618	0·646	0·672	0·685	0·723	0·760	0·772	0·795	0·818	0·830	0·863
1/ 4762	0·7071	0·7838	0·8654	0·9520	1·1409	1·3512	1·4647	1·8396	2·2684	2·4238	2·7537	3·1100	3·2982	3·9042
0·000220	0·590	0·605	0·619	0·633	0·661	0·688	0·702	0·740	0·778	0·790	0·814	0·838	0·850	0·884
1/ 4545	0·7243	0·8028	0·8864	0·9752	1·1686	1·3840	1·5002	1·8842	2·3233	2·4824	2·8204	3·1852	3·3779	3·9985
0·000230	0·604	0·619	0·634	0·648	0·677	0·704	0·718	0·758	0·796	0·808	0·833	0·857	0·869	0·904
1/ 4348	0·7412	0·8215	0·9070	0·9978	1·1957	1·4161	1·5350	1·9278	2·3770	2·5398	2·8855	3·2587	3·4559	4·0907
0·000240	0·617	0·633	0·648	0·663	0·692	0·720	0·734	0·774	0·814	0·826	0·851	0·876	0·888	0·924
1/ 4167	0·7576	0·8398	0·9272	1·0200	1·2223	1·4475	1·5690	1·9705	2·4296	2·5959	2·9492	3·3306	3·5321	4·1809
0·000250	0·631	0·646	0·662	0·677	0·706	0·735	0·749	0·791	0·831	0·844	0·870	0·895	0·907	0·944
1/ 4000	0·7738	0·8577	0·9470	1·0417	1·2483	1·4782	1·6023	2·0123	2·4811	2·6509	3·0117	3·4011	3·6069	4·2693
0·000260	0·643	0·659	0·675	0·691	0·721	0·750	0·765	0·807	0·848	0·861	0·887	0·913	0·926	0·963
1/ 3846	0·7896	0·8752	0·9663	1·0630	1·2737	1·5084	1·6350	2·0532	2·5316	2·7049	3·0729	3·4702	3·6801	4·3560
0·000270	0·656	0·672	0·688	0·704	0·735	0·765	0·780	0·823	0·864	0·878	0·905	0·931	0·944	0·982
1/ 3704	0·8052	0·8925	0·9853	1·0839	1·2987	1·5380	1·6670	2·0934	2·5811	2·7577	3·1329	3·5380	3·7520	4·4409
0·000280	0·669	0·685	0·701	0·717	0·749	0·779	0·794	0·838	0·881	0·894	0·922	0·948	0·961	1·000
1/ 3571	0·8204	0·9094	1·0040	1·1044	1·3233	1·5670	1·6985	2·1329	2·6297	2·8097	3·1919	3·6045	3·8225	4·5244
0·000290	0·681	0·698	0·714	0·731	0·762	0·794	0·809	0·853	0·897	0·911	0·938	0·965	0·979	1·018
1/ 3448	0·8354	0·9260	1·0223	1·1246	1·3474	1·5955	1·7294	2·1717	2·6775	2·8607	3·2498	3·6699	3·8918	4·6064
0·000300	0·693	0·710	0·727	0·743	0·776	0·808	0·823	0·868	0·912	0·927	0·955	0·982	0·996	1·036
1/ 3333	0·8502	0·9423	1·0403	1·1444	1·3711	1·6236	1·7598	2·2098	2·7244	2·9108	3·3068	3·7342	3·9599	4·6869
	0·90	0·90	0·90	0·90	0·91	0·91	0·91	0·92	0·92	0·92	0·92	0·93	0·93	0·93

$V_{r(0\cdot5)medial}$ for half-full circular pipes.

Gradient **(Equivalent) Pipe diameters in mm**

Gradient	1250	1300	1350	1400	1500	1600	1650	1800	1950	2000	2100	2200	2250	2400
0·000300	0·693	0·710	0·727	0·743	0·776	0·808	0·823	0·868	0·912	0·927	0·955	0·982	0·996	1·036
1/ 3333	0·8502	0·9423	1·0403	1·1444	1·3711	1·6236	1·7598	2·2098	2·7244	2·9108	3·3068	3·7342	3·9599	4·6869
0·000320	0·716	0·734	0·751	0·768	0·802	0·835	0·851	0·898	0·943	0·958	0·987	1·015	1·029	1·071
1/ 3125	0·8789	0·9741	1·0755	1·1830	1·4174	1·6783	1·8191	2·2842	2·8160	3·0087	3·4179	3·8596	4·0929	4·8442
0·000340	0·739	0·757	0·775	0·793	0·827	0·861	0·878	0·926	0·973	0·988	1·018	1·047	1·062	1·104
1/ 2941	0·9068	1·0050	1·1095	1·2205	1·4622	1·7314	1·8766	2·3563	2·9048	3·1036	3·5256	3·9811	4·2218	4·9966
0·000360	0·761	0·780	0·798	0·816	0·852	0·887	0·904	0·953	1·002	1·017	1·048	1·078	1·093	1·137
1/ 2778	0·9339	1·0350	1·1426	1·2569	1·5058	1·7829	1·9324	2·4263	2·9911	3·1957	3·6302	4·0992	4·3469	5·1446
0·000380	0·782	0·802	0·821	0·839	0·876	0·912	0·929	0·980	1·030	1·046	1·077	1·109	1·124	1·169
1/ 2632	0·9602	1·0642	1·1748	1·2923	1·5482	1·8331	1·9868	2·4945	3·0750	3·2853	3·7319	4·2140	4·4687	5·2886
0·000400	0·803	0·823	0·843	0·862	0·899	0·936	0·954	1·006	1·057	1·074	1·106	1·138	1·154	1·200
1/ 2500	0·9859	1·0926	1·2062	1·3268	1·5894	1·8819	2·0397	2·5608	3·1567	3·3726	3·8310	4·3259	4·5872	5·4288
0·000420	0·824	0·844	0·864	0·884	0·922	0·960	0·978	1·032	1·084	1·101	1·134	1·167	1·183	1·230
1/ 2381	1·0109	1·1204	1·2368	1·3604	1·6297	1·9295	2·0913	2·6256	3·2365	3·4578	3·9277	4·4350	4·7029	5·5657
0·000440	0·844	0·864	0·885	0·905	0·944	0·983	1·002	1·057	1·110	1·127	1·161	1·195	1·211	1·260
1/ 2273	1·0354	1·1475	1·2667	1·3933	1·6690	1·9761	2·1417	2·6888	3·3143	3·5409	4·0221	4·5415	4·8159	5·6992
0·000460	0·863	0·884	0·905	0·926	0·966	1·005	1·025	1·081	1·135	1·153	1·188	1·222	1·239	1·289
1/ 2174	1·0593	1·1739	1·2959	1·4254	1·7075	2·0215	2·1910	2·7506	3·3904	3·6222	4·1144	4·6457	4·9263	5·8299
0·000480	0·882	0·904	0·925	0·946	0·988	1·028	1·047	1·105	1·160	1·178	1·214	1·249	1·266	1·317
1/ 2083	1·0827	1·1999	1·3245	1·4568	1·7451	2·0661	2·2392	2·8111	3·4649	3·7018	4·2047	4·7476	5·0344	5·9577
0·000500	0·901	0·923	0·945	0·966	1·008	1·049	1·069	1·128	1·185	1·203	1·240	1·275	1·293	1·345
1/ 2000	1·1056	1·2252	1·3525	1·4876	1·7820	2·1097	2·2864	2·8703	3·5379	3·7797	4·2932	4·8474	5·1402	6·0828
0·000525	0·924	0·946	0·969	0·991	1·034	1·076	1·096	1·156	1·214	1·233	1·271	1·307	1·325	1·378
1/ 1905	1·1336	1·2563	1·3867	1·5252	1·8270	2·1629	2·3442	2·9427	3·6270	3·8749	4·4013	4·9695	5·2696	6·2358
0·000550	0·946	0·969	0·992	1·015	1·059	1·102	1·123	1·184	1·244	1·263	1·301	1·339	1·357	1·411
1/ 1818	1·1610	1·2866	1·4202	1·5620	1·8710	2·2150	2·4005	3·0134	3·7141	3·9680	4·5069	5·0887	5·3960	6·3853
0·000575	0·968	0·992	1·015	1·038	1·083	1·127	1·148	1·211	1·272	1·292	1·331	1·369	1·388	1·444
1/ 1739	1·1877	1·3162	1·4529	1·5979	1·9140	2·2659	2·4557	3·0826	3·7993	4·0589	4·6101	5·2052	5·5195	6·5314
0·000600	0·989	1·013	1·037	1·061	1·107	1·152	1·174	1·238	1·300	1·320	1·360	1·399	1·419	1·475
1/ 1667	1·2139	1·3452	1·4848	1·6331	1·9561	2·3157	2·5096	3·1502	3·8826	4·1479	4·7112	5·3192	5·6403	6·6743
0·000625	1·010	1·035	1·059	1·083	1·130	1·176	1·198	1·264	1·327	1·348	1·389	1·429	1·448	1·506
1/ 1600	1·2395	1·3736	1·5162	1·6675	1·9973	2·3644	2·5624	3·2165	3·9642	4·2350	4·8101	5·4308	5·7587	6·8143
0·000650	1·031	1·056	1·081	1·105	1·153	1·200	1·223	1·290	1·354	1·375	1·417	1·457	1·478	1·537
1/ 1538	1·2647	1·4014	1·5469	1·7013	2·0377	2·4122	2·6142	3·2814	4·0442	4·3205	4·9071	5·5403	5·8748	6·9515
0·000675	1·051	1·076	1·102	1·127	1·176	1·223	1·246	1·315	1·380	1·402	1·444	1·486	1·506	1·566
1/ 1481	1·2893	1·4287	1·5770	1·7344	2·0774	2·4591	2·6650	3·3452	4·1226	4·4043	5·0023	5·6477	5·9886	7·0862
0·000700	1·070	1·097	1·122	1·148	1·198	1·246	1·270	1·339	1·406	1·428	1·471	1·513	1·534	1·596
1/ 1429	1·3135	1·4555	1·6066	1·7669	2·1163	2·5052	2·7149	3·4077	4·1997	4·4865	5·0957	5·7531	6·1004	7·2183
0·000725	1·090	1·116	1·143	1·169	1·219	1·268	1·293	1·363	1·432	1·454	1·498	1·541	1·562	1·624
1/ 1379	1·3373	1·4819	1·6356	1·7989	2·1545	2·5504	2·7639	3·4692	4·2753	4·5674	5·1874	5·8567	6·2101	7·3482
0·000750	1·109	1·136	1·163	1·189	1·240	1·291	1·315	1·387	1·456	1·479	1·524	1·567	1·589	1·653
1/ 1333	1·3607	1·5077	1·6642	1·8303	2·1921	2·5948	2·8121	3·5296	4·3497	4·6468	5·2776	5·9584	6·3181	7·4758
0·000800	1·146	1·174	1·202	1·229	1·282	1·334	1·359	1·433	1·505	1·528	1·575	1·620	1·642	1·708
1/ 1250	1·4063	1·5583	1·7199	1·8916	2·2654	2·6816	2·9061	3·6474	4·4949	4·8018	5·4536	6·1571	6·5286	7·7248
0·000850	1·182	1·211	1·239	1·267	1·322	1·376	1·402	1·478	1·552	1·576	1·624	1·670	1·693	1·761
1/ 1176	1·4505	1·6072	1·7740	1·9509	2·3365	2·7657	2·9972	3·7617	4·6356	4·9521	5·6242	6·3496	6·7327	7·9662
0·000900	1·217	1·247	1·276	1·305	1·361	1·416	1·443	1·522	1·598	1·623	1·672	1·720	1·743	1·813
1/ 1111	1·4934	1·6548	1·8264	2·0086	2·4055	2·8474	3·0856	3·8726	4·7722	5·0980	5·7899	6·5365	6·9309	8·2006
0·000950	1·251	1·282	1·312	1·341	1·399	1·456	1·483	1·564	1·642	1·668	1·718	1·767	1·792	1·863
1/ 1053	1·5352	1·7010	1·8775	2·0647	2·4727	2·9268	3·1717	3·9805	4·9051	5·2400	5·9510	6·7184	7·1237	8·4285
0·001000	1·284	1·315	1·346	1·377	1·436	1·494	1·523	1·606	1·686	1·712	1·763	1·814	1·839	1·912
1/ 1000	1·5759	1·7461	1·9272	2·1194	2·5381	3·0041	3·2555	4·0857	5·0345	5·3782	6·1079	6·8955	7·3115	8·6506
	0·90	0·91	0·91	0·91	0·91	0·92	0·92	0·92	0·93	0·93	0·93	0·93	0·93	0·94

$V_{r(0\cdot5)medial}$ for half-full circular pipes.

A40
(p.3 of 6)

$k_s = 0{\cdot}30$ **mm**
S = 0·00100 to 0·00300

ie hydraulic gradient =
1 in 1000 to 1 in 333

Water (or sewage) at 15°C;
full bore conditions.

velocities in ms^{-1}
discharges in m^3s^{-1}

Gradient **(Equivalent) Pipe diameters in mm**

	1250	1300	1350	1400	1500	1600	1650	1800	1950	2000	2100	2200	2250	2400
0·00100	1·284	1·315	1·346	1·377	1·436	1·494	1·523	1·606	1·686	1·712	1·763	1·814	1·839	1·912
1/ 1000	1·5759	1·7461	1·9272	2·1194	2·5381	3·0041	3·2555	4·0857	5·0345	5·3782	6·1079	6·8955	7·3115	8·6506
0·00105	1·316	1·349	1·380	1·411	1·472	1·532	1·561	1·646	1·728	1·755	1·808	1·859	1·885	1·960
1/ 952	1·6155	1·7900	1·9757	2·1727	2·6019	3·0796	3·3373	4·1882	5·1608	5·5131	6·2610	7·0683	7·4946	8·8672
0·00110	1·348	1·381	1·413	1·445	1·508	1·568	1·598	1·685	1·769	1·797	1·851	1·904	1·930	2·007
1/ 909	1·6543	1·8330	2·0230	2·2247	2·6642	3·1533	3·4171	4·2883	5·2841	5·6447	6·4105	7·2370	7·6735	9·0787
0·00115	1·379	1·413	1·446	1·478	1·542	1·604	1·635	1·724	1·810	1·838	1·893	1·947	1·974	2·053
1/ 870	1·6922	1·8749	2·0693	2·2756	2·7251	3·2253	3·4951	4·3862	5·4046	5·7735	6·5566	7·4019	7·8483	9·2854
0·00120	1·409	1·443	1·477	1·511	1·576	1·639	1·670	1·761	1·849	1·878	1·934	1·990	2·017	2·097
1/ 833	1·7292	1·9160	2·1146	2·3254	2·7847	3·2958	3·5715	4·4819	5·5225	5·8994	6·6996	7·5632	8·0194	9·4877
0·00125	1·439	1·474	1·508	1·542	1·609	1·674	1·705	1·798	1·888	1·917	1·975	2·031	2·059	2·141
1/ 800	1·7655	1·9562	2·1590	2·3742	2·8430	3·3649	3·6463	4·5757	5·6380	6·0228	6·8397	7·7213	8·1869	9·6859
0·00130	1·468	1·503	1·539	1·573	1·641	1·707	1·740	1·834	1·926	1·956	2·014	2·072	2·100	2·184
1/ 769	1·8011	1·9956	2·2024	2·4220	2·9003	3·4326	3·7196	4·6677	5·7513	6·1437	6·9770	7·8762	8·3512	9·8801
0·00135	1·496	1·533	1·568	1·604	1·673	1·740	1·773	1·870	1·963	1·993	2·053	2·112	2·141	2·226
1/ 741	1·8360	2·0343	2·2451	2·4689	2·9564	3·4989	3·7916	4·7579	5·8623	6·2623	7·1116	8·0282	8·5123	10·071
0·00140	1·524	1·561	1·598	1·634	1·704	1·773	1·806	1·905	1·999	2·030	2·091	2·151	2·181	2·267
1/ 714	1·8703	2·0722	2·2870	2·5149	3·0115	3·5641	3·8622	4·8464	5·9714	6·3788	7·2438	8·1774	8·6705	10·258
0·00145	1·551	1·589	1·626	1·663	1·735	1·804	1·839	1·939	2·035	2·067	2·129	2·190	2·220	2·308
1/ 690	1·9040	2·1095	2·3281	2·5602	3·0656	3·6281	3·9315	4·9334	6·0785	6·4932	7·3737	8·3239	8·8258	10·441
0·00150	1·578	1·617	1·655	1·692	1·765	1·836	1·871	1·972	2·071	2·103	2·166	2·228	2·258	2·348
1/ 667	1·9371	2·1462	2·3686	2·6046	3·1188	3·6911	3·9997	5·0189	6·1837	6·6056	7·5013	8·4680	8·9785	10·622
0·00160	1·631	1·671	1·710	1·748	1·824	1·897	1·933	2·038	2·139	2·172	2·238	2·302	2·333	2·426
1/ 625	2·0016	2·2177	2·4475	2·6914	3·2226	3·8139	4·1327	5·1857	6·3892	6·8250	7·7504	8·7491	9·2765	10·974
0·00170	1·682	1·723	1·763	1·803	1·881	1·956	1·993	2·101	2·206	2·240	2·307	2·373	2·406	2·501
1/ 588	2·0642	2·2870	2·5240	2·7754	3·3232	3·9329	4·2617	5·3474	6·5883	7·0377	7·9918	9·0215	9·5654	11·316
0·00180	1·732	1·774	1·815	1·856	1·936	2·014	2·052	2·163	2·271	2·306	2·375	2·443	2·476	2·575
1/ 556	2·1250	2·3543	2·5982	2·8571	3·4210	4·0485	4·3869	5·5044	6·7816	7·2442	8·2263	9·2861	9·8458	11·648
0·00190	1·780	1·823	1·866	1·908	1·990	2·069	2·109	2·223	2·334	2·370	2·441	2·511	2·545	2·646
1/ 526	2·1841	2·4198	2·6705	2·9365	3·5160	4·1609	4·5087	5·6572	6·9697	7·4450	8·4543	9·5434	10·119	11·970
0·00200	1·827	1·871	1·915	1·958	2·042	2·124	2·164	2·282	2·395	2·432	2·505	2·576	2·612	2·715
1/ 500	2·2417	2·4836	2·7408	3·0138	3·6086	4·2704	4·6273	5·8059	7·1529	7·6407	8·6763	9·7940	10·384	12·284
0·00210	1·872	1·918	1·963	2·007	2·093	2·177	2·218	2·339	2·455	2·493	2·568	2·641	2·677	2·783
1/ 476	2·2979	2·5458	2·8095	3·0893	3·6988	4·3772	4·7430	5·9510	7·3315	7·8315	8·8930	10·038	10·643	12·591
0·00220	1·917	1·964	2·010	2·055	2·143	2·229	2·271	2·394	2·513	2·552	2·629	2·704	2·741	2·849
1/ 455	2·3527	2·6065	2·8765	3·1630	3·7870	4·4815	4·8560	6·0927	7·5060	8·0178	9·1045	10·277	10·896	12·890
0·00230	1·961	2·008	2·055	2·101	2·192	2·280	2·323	2·449	2·570	2·610	2·688	2·765	2·803	2·914
1/ 435	2·4063	2·6659	2·9420	3·2350	3·8732	4·5834	4·9664	6·2312	7·6765	8·2000	9·3112	10·510	11·144	13·183
0·00240	2·004	2·052	2·100	2·147	2·240	2·329	2·373	2·502	2·626	2·667	2·747	2·825	2·864	2·977
1/ 417	2·4588	2·7240	3·0061	3·3055	3·9575	4·6832	5·0745	6·3667	7·8434	8·3782	9·5136	10·739	11·386	13·469
0·00250	2·045	2·095	2·144	2·192	2·286	2·378	2·423	2·554	2·681	2·722	2·804	2·884	2·923	3·039
1/ 400	2·5102	2·7809	3·0689	3·3745	4·0401	4·7809	5·1804	6·4994	8·0068	8·5527	9·7117	10·962	11·623	13·749
0·00260	2·087	2·137	2·187	2·236	2·332	2·425	2·471	2·605	2·735	2·777	2·860	2·942	2·982	3·100
1/ 385	2·5605	2·8367	3·1304	3·4421	4·1211	4·8767	5·2841	6·6295	8·1670	8·7238	9·9059	11·182	11·855	14·024
0·00270	2·127	2·178	2·229	2·279	2·377	2·472	2·519	2·655	2·787	2·830	2·915	2·998	3·039	3·160
1/ 370	2·6100	2·8914	3·1908	3·5085	4·2005	4·9706	5·3859	6·7571	8·3242	8·8917	10·096	11·397	12·083	14·294
0·00280	2·166	2·219	2·271	2·321	2·421	2·518	2·566	2·705	2·839	2·883	2·969	3·054	3·095	3·218
1/ 357	2·6585	2·9451	3·2501	3·5736	4·2785	5·0628	5·4858	6·8824	8·4784	9·0564	10·284	11·608	12·307	14·558
0·00290	2·205	2·259	2·311	2·363	2·464	2·563	2·611	2·753	2·890	2·934	3·022	3·108	3·151	3·276
1/ 345	2·7061	2·9979	3·3083	3·6376	4·3551	5·1534	5·5839	7·0055	8·6300	9·2183	10·467	11·815	12·527	14·818
0·00300	2·243	2·298	2·351	2·404	2·507	2·607	2·657	2·801	2·940	2·985	3·074	3·162	3·205	3·332
1/ 333	2·7529	3·0498	3·3655	3·7006	4·4303	5·2425	5·6804	7·1265	8·7789	9·3774	10·648	12·019	12·743	15·074
	0·91	0·91	0·91	0·92	0·92	0·92	0·92	0·93	0·93	0·93	0·93	0·94	0·94	0·94

$V_{r(0{\cdot}5)medial}$ **for half-full circular pipes.**

$k_s = 0{\cdot}30$ **mm** S = 0·00100 to 0·00300

k_s = 0·30 mm
S = 0·00300 to 0·01000

ie hydraulic gradient =
1 in 333 to 1 in 100

Water (or sewage) at 15°C;
full bore conditions.

velocities in ms^{-1}
discharges in m^3s^{-1}

Gradient **(Equivalent) Pipe diameters in mm**

	1250	1300	1350	1400	1500	1600	1650	1800	1950	2000	2100	2200	2250	2400
0·00300	2·243	2·298	2·351	2·404	2·507	2·607	2·657	2·801	2·940	2·985	3·074	3·162	3·205	3·332
1/ 333	2·7529	3·0498	3·3655	3·7006	4·4303	5·2425	5·6804	7·1265	8·7789	9·3774	10·648	12·019	12·743	15·074
0·00320	2·318	2·374	2·429	2·484	2·590	2·694	2·745	2·893	3·037	3·084	3·176	3·266	3·311	3·442
1/ 313	2·8443	3·1510	3·4772	3·8233	4·5772	5·4162	5·8686	7·3625	9·0696	9·6878	11·000	12·417	13·164	15·572
0·00340	2·390	2·448	2·505	2·561	2·671	2·778	2·830	2·983	3·131	3·179	3·275	3·368	3·414	3·549
1/ 294	2·9329	3·2491	3·5854	3·9423	4·7196	5·5846	6·0511	7·5913	9·3513	9·9886	11·342	12·802	13·573	16·055
0·00360	2·460	2·520	2·578	2·636	2·749	2·859	2·913	3·070	3·223	3·272	3·370	3·466	3·513	3·653
1/ 278	3·0189	3·3443	3·6905	4·0578	4·8579	5·7481	6·2282	7·8134	9·6248	10·281	11·673	13·176	13·970	16·524
0·00380	2·528	2·589	2·650	2·709	2·825	2·938	2·993	3·155	3·312	3·363	3·463	3·562	3·611	3·754
1/ 263	3·1025	3·4370	3·7927	4·1702	4·9923	5·9072	6·4005	8·0295	9·8909	10·565	11·996	13·540	14·356	16·981
0·00400	2·595	2·657	2·719	2·780	2·899	3·015	3·072	3·238	3·399	3·451	3·554	3·655	3·705	3·852
1/ 250	3·1840	3·5272	3·8922	4·2796	5·1233	6·0621	6·5684	8·2399	10·150	10·842	12·310	13·895	14·732	17·425
0·00420	2·659	2·724	2·787	2·849	2·971	3·090	3·148	3·319	3·483	3·537	3·643	3·746	3·797	3·948
1/ 238	3·2635	3·6152	3·9893	4·3863	5·2510	6·2132	6·7321	8·4452	10·403	11·112	12·617	14·241	15·098	17·859
0·00440	2·723	2·788	2·853	2·917	3·042	3·164	3·223	3·398	3·566	3·621	3·729	3·835	3·887	4·041
1/ 227	3·3411	3·7012	4·0842	4·4906	5·3758	6·3607	6·8919	8·6456	10·650	11·375	12·916	14·578	15·456	18·282
0·00460	2·784	2·852	2·918	2·983	3·111	3·235	3·296	3·475	3·647	3·703	3·813	3·922	3·975	4·133
1/ 217	3·4169	3·7852	4·1768	4·5925	5·4977	6·5050	7·0481	8·8415	10·891	11·633	13·208	14·908	15·806	18·696
0·00480	2·845	2·914	2·981	3·048	3·179	3·305	3·368	3·550	3·726	3·783	3·896	4·007	4·061	4·222
1/ 208	3·4912	3·8674	4·2675	4·6922	5·6170	6·6461	7·2010	9·0332	11·127	11·885	13·494	15·231	16·148	19·101
0·00500	2·904	2·974	3·043	3·112	3·245	3·374	3·438	3·624	3·803	3·862	3·977	4·090	4·146	4·310
1/ 200	3·5639	3·9479	4·3564	4·7898	5·7339	6·7843	7·3508	9·2210	11·358	12·132	13·775	15·548	16·484	19·497
0·00525	2·977	3·048	3·119	3·189	3·326	3·458	3·523	3·714	3·898	3·958	4·076	4·192	4·249	4·417
1/ 190	3·6527	4·0463	4·4649	4·9092	5·8767	6·9533	7·5338	9·4505	11·641	12·434	14·117	15·934	16·894	19·982
0·00550	3·047	3·121	3·193	3·265	3·404	3·540	3·607	3·802	3·990	4·052	4·173	4·291	4·349	4·522
1/ 182	3·7395	4·1424	4·5710	5·0257	6·0162	7·1182	7·7125	9·6746	11·917	12·728	14·452	16·312	17·294	20·455
0·00575	3·116	3·192	3·266	3·339	3·482	3·621	3·689	3·888	4·080	4·143	4·267	4·388	4·448	4·624
1/ 174	3·8243	4·2364	4·6746	5·1396	6·1525	7·2795	7·8872	9·8937	12·186	13·017	14·779	16·681	17·685	20·918
0·00600	3·184	3·261	3·337	3·411	3·557	3·699	3·769	3·972	4·169	4·233	4·359	4·483	4·544	4·724
1/ 167	3·9073	4·3283	4·7760	5·2511	6·2859	7·4373	8·0581	10·108	12·450	13·298	15·099	17·042	18·068	21·371
0·00625	3·250	3·329	3·406	3·482	3·631	3·776	3·847	4·055	4·255	4·321	4·450	4·576	4·638	4·822
1/ 160	3·9886	4·4183	4·8753	5·3603	6·4166	7·5918	8·2256	10·318	12·709	13·574	15·412	17·396	18·443	21·814
0·00650	3·315	3·395	3·474	3·552	3·703	3·851	3·924	4·136	4·340	4·407	4·539	4·667	4·731	4·918
1/ 154	4·0683	4·5065	4·9727	5·4673	6·5446	7·7433	8·3897	10·524	12·962	13·845	15·720	17·742	18·810	22·249
0·00675	3·379	3·460	3·541	3·620	3·775	3·925	3·999	4·215	4·423	4·492	4·626	4·757	4·822	5·012
1/ 148	4·1464	4·5931	5·0682	5·5723	6·6702	7·8919	8·5507	10·726	13·211	14·111	16·021	18·082	19·171	22·675
0·00700	3·441	3·524	3·606	3·687	3·844	3·998	4·073	4·293	4·505	4·574	4·711	4·845	4·911	5·105
1/ 143	4·2232	4·6781	5·1619	5·6754	6·7936	8·0378	8·7087	10·924	13·455	14·371	16·317	18·416	19·525	23·093
0·00725	3·503	3·587	3·671	3·753	3·913	4·069	4·145	4·369	4·585	4·656	4·795	4·931	4·998	5·196
1/ 138	4·2985	4·7615	5·2540	5·7766	6·9147	8·1811	8·8639	11·118	13·694	14·627	16·607	18·744	19·872	23·504
0·00750	3·563	3·649	3·734	3·817	3·980	4·139	4·217	4·444	4·664	4·736	4·877	5·016	5·084	5·285
1/ 133	4·3726	4·8436	5·3445	5·8761	7·0338	8·3219	9·0165	11·310	13·930	14·879	16·893	19·066	20·214	23·908
0·00800	3·681	3·770	3·857	3·943	4·112	4·276	4·356	4·591	4·818	4·892	5·038	5·181	5·252	5·459
1/ 125	4·5172	5·0037	5·5211	6·0702	7·2661	8·5967	9·3142	11·683	14·390	15·370	17·450	19·695	20·881	24·697
0·00850	3·795	3·887	3·977	4·066	4·239	4·408	4·491	4·733	4·967	5·044	5·194	5·342	5·414	5·628
1/ 118	4·6573	5·1588	5·6923	6·2584	7·4913	8·8631	9·6028	12·045	14·835	15·845	17·990	20·305	21·527	25·461
0·00900	3·906	4·000	4·093	4·184	4·363	4·537	4·622	4·871	5·112	5·191	5·346	5·497	5·572	5·792
1/ 111	4·7933	5·3095	5·8585	6·4411	7·7100	9·1217	9·8829	12·396	15·268	16·308	18·515	20·897	22·155	26·203
0·00950	4·014	4·111	4·206	4·300	4·483	4·662	4·749	5·006	5·253	5·334	5·493	5·649	5·725	5·952
1/ 105	4·9256	5·4560	6·0202	6·6189	7·9226	9·3733	10·155	12·738	15·688	16·757	19·025	21·472	22·765	26·925
0·01000	4·119	4·218	4·316	4·412	4·601	4·784	4·874	5·136	5·390	5·473	5·636	5·796	5·875	6·107
1/ 100	5·0545	5·5988	6·1777	6·7920	8·1298	9·6183	10·421	13·071	16·098	17·195	19·522	22·033	23·359	27·628
	0·91	0·91	0·92	0·92	0·92	0·92	0·93	0·93	0·93	0·93	0·94	0·94	0·94	0·94

$V_{r(0·5)medial}$ for half-full circular pipes.

$k_s = 0{\cdot}30$ **mm**
S = 0·01000 to 0·03000

ie hydraulic gradient =
1 in 100 to 1 in 33·3

Water (or sewage) at 15°C;
full bore conditions.

velocities in ms^{-1}
discharges in m^3s^{-1}

Gradient **(Equivalent) Pipe diameters in mm**

Gradient	1250	1300	1350	1400	1500	1600	1650	1800	1950	2000	2100	2200	2250	2400
0·01000	4·119	4·218	4·316	4·412	4·601	4·784	4·874	5·136	5·390	5·473	5·636	5·796	5·875	6·107
1/ 100	5·0545	5·5988	6·1777	6·7920	8·1298	9·6183	10·421	13·071	16·098	17·195	19·522	22·033	23·359	27·628
0·01050	4·221	4·323	4·423	4·522	4·715	4·903	4·995	5·264	5·524	5·609	5·776	5·940	6·021	6·259
1/ 95	5·1802	5·7380	6·3312	6·9608	8·3318	9·8572	10·680	13·395	16·498	17·621	20·006	22·580	23·939	28·313
0·01100	4·321	4·425	4·528	4·629	4·826	5·019	5·113	5·389	5·655	5·742	5·913	6·080	6·163	6·407
1/ 91	5·3029	5·8739	6·4812	7·1256	8·5291	10·091	10·933	13·712	16·888	18·038	20·480	23·114	24·505	28·982
0·01150	4·419	4·525	4·630	4·734	4·936	5·132	5·228	5·510	5·783	5·871	6·046	6·218	6·302	6·551
1/ 87	5·4229	6·0068	6·6278	7·2868	8·7219	10·319	11·180	14·022	17·270	18·446	20·942	23·636	25·058	29·637
0·01200	4·515	4·623	4·731	4·836	5·042	5·243	5·341	5·629	5·908	5·998	6·177	6·352	6·438	6·693
1/ 83	5·5403	6·1368	6·7712	7·4445	8·9106	10·542	11·421	14·325	17·643	18·844	21·395	24·147	25·600	30·277
0·01250	4·608	4·719	4·829	4·936	5·147	5·352	5·452	5·746	6·030	6·123	6·305	6·484	6·572	6·831
1/ 80	5·6552	6·2641	6·9117	7·5989	9·0954	10·760	11·658	14·622	18·009	19·235	21·838	24·647	26·130	30·904
0·01300	4·700	4·813	4·925	5·035	5·249	5·458	5·561	5·861	6·150	6·245	6·430	6·613	6·703	6·967
1/ 77	5·7679	6·3890	7·0494	7·7503	9·2766	10·975	11·890	14·913	18·367	19·618	22·273	25·137	26·650	31·519
0·01350	4·790	4·906	5·019	5·131	5·350	5·563	5·667	5·973	6·268	6·364	6·554	6·739	6·831	7·100
1/ 74	5·8785	6·5114	7·1845	7·8988	9·4542	11·185	12·118	15·199	18·719	19·993	22·699	25·618	27·160	32·122
0·01400	4·879	4·996	5·112	5·226	5·449	5·666	5·772	6·083	6·383	6·481	6·674	6·863	6·957	7·231
1/ 71	5·9870	6·6316	7·3171	8·0445	9·6287	11·391	12·342	15·479	19·064	20·362	23·117	26·090	27·660	32·714
0·01450	4·966	5·085	5·203	5·319	5·546	5·766	5·874	6·191	6·497	6·597	6·793	6·985	7·080	7·360
1/ 69	6·0936	6·7496	7·4474	8·1877	9·8000	11·594	12·561	15·754	19·403	20·724	23·528	26·554	28·152	33·295
0·01500	5·051	5·173	5·292	5·410	5·641	5·865	5·975	6·297	6·608	6·710	6·910	7·105	7·202	7·486
1/ 67	6·1984	6·8657	7·5754	8·3284	9·9684	11·793	12·777	16·025	19·736	21·080	23·932	27·010	28·635	33·867
0·01600	5·218	5·343	5·467	5·589	5·827	6·059	6·172	6·505	6·826	6·931	7·137	7·339	7·439	7·733
1/ 63	6·4029	7·0921	7·8252	8·6031	10·297	12·182	13·198	16·553	20·386	21·774	24·720	27·899	29·578	34·982
0·01700	5·379	5·509	5·636	5·762	6·007	6·246	6·363	6·706	7·037	7·145	7·358	7·566	7·669	7·972
1/ 59	6·6010	7·3116	8·0673	8·8692	10·616	12·559	13·606	17·065	21·016	22·447	25·484	28·761	30·492	36·062
0·01800	5·536	5·669	5·800	5·929	6·182	6·428	6·549	6·901	7·242	7·353	7·572	7·786	7·892	8·203
1/ 56	6·7934	7·5247	8·3024	9·1276	10·925	12·924	14·002	17·562	21·628	23·100	26·226	29·599	31·380	37·112
0·01900	5·688	5·825	5·960	6·093	6·352	6·605	6·729	7·091	7·441	7·555	7·780	8·001	8·109	8·429
1/ 53	6·9806	7·7319	8·5311	9·3790	11·226	13·280	14·388	18·045	22·223	23·736	26·948	30·413	32·243	38·132
0·02000	5·837	5·977	6·116	6·252	6·518	6·777	6·904	7·276	7·635	7·752	7·983	8·209	8·321	8·649
1/ 50	7·1628	7·9338	8·7538	9·6238	11·519	13·627	14·763	18·516	22·803	24·355	27·650	31·206	33·083	39·126
0·02100	5·982	6·126	6·267	6·407	6·680	6·945	7·076	7·457	7·825	7·945	8·181	8·413	8·527	8·863
1/ 47·6	7·3406	8·1307	8·9710	9·8626	11·804	13·965	15·129	18·975	23·368	24·959	28·336	31·979	33·903	40·096
0·02200	6·123	6·270	6·416	6·558	6·838	7·110	7·243	7·633	8·009	8·132	8·374	8·611	8·728	9·072
1/ 45·5	7·5142	8·3230	9·1831	10·096	12·083	14·295	15·487	19·423	23·920	25·548	29·005	32·734	34·704	41·043
0·02300	6·261	6·412	6·560	6·706	6·992	7·270	7·406	7·805	8·190	8·316	8·563	8·806	8·925	9·277
1/ 43·5	7·6839	8·5109	9·3904	10·324	12·356	14·617	15·836	19·861	24·460	26·125	29·659	33·473	35·487	41·968
0·02400	6·397	6·551	6·702	6·851	7·143	7·427	7·566	7·973	8·367	8·495	8·748	8·996	9·118	9·477
1/ 41·7	7·8499	8·6948	9·5933	10·547	12·623	14·933	16·178	20·290	24·988	26·688	30·299	34·195	36·252	42·874
0·02500	6·529	6·686	6·841	6·993	7·291	7·581	7·723	8·139	8·540	8·671	8·929	9·182	9·306	9·673
1/ 40·0	8·0126	8·8749	9·7920	10·765	12·884	15·242	16·513	20·710	25·505	27·241	30·926	34·902	37·002	43·761
0·02600	6·659	6·819	6·977	7·132	7·436	7·731	7·876	8·300	8·710	8·843	9·106	9·364	9·491	9·865
1/ 38·5	8·1719	9·0514	9·9867	10·979	13·140	15·545	16·842	21·122	26·011	27·782	31·541	35·596	37·737	44·630
0·02700	6·787	6·950	7·110	7·269	7·578	7·879	8·027	8·459	8·876	9·012	9·280	9·543	9·672	10·05
1/ 37·0	8·3283	9·2246	10·178	11·189	13·392	15·842	17·164	21·525	26·509	28·313	32·143	36·276	38·459	45·483
0·02800	6·912	7·078	7·241	7·403	7·718	8·024	8·175	8·615	9·040	9·178	9·451	9·719	9·851	10·24
1/ 35·7	8·4818	9·3946	10·365	11·395	13·639	16·134	17·480	21·922	26·997	28·834	32·735	36·944	39·166	46·320
0·02900	7·034	7·204	7·370	7·534	7·855	8·167	8·320	8·768	9·200	9·341	9·619	9·891	10·03	10·42
1/ 34·5	8·6326	9·5616	10·550	11·598	13·881	16·421	17·790	22·311	27·476	29·346	33·316	37·600	39·862	47·142
0·03000	7·155	7·327	7·497	7·663	7·990	8·307	8·463	8·918	9·358	9·501	9·784	10·06	10·20	10·60
1/ 33·3	8·7808	9·7257	10·731	11·797	14·119	16·702	18·095	22·694	27·947	29·850	33·888	38·244	40·545	47·950
	0·91	0·92	0·92	0·92	0·92	0·93	0·93	0·93	0·93	0·94	0·94	0·94	0·94	0·94

$V_{r(0\cdot5)medial}$ **for half-full circular pipes.**

Gradient **(Equivalent) Pipe diameters in mm**

Gradient	1250	1300	1350	1400	1500	1600	1650	1800	1950	2000	2100	2200	2250	2400
0·03000	7·155	7·327	7·497	7·663	7·990	8·307	8·463	8·918	9·358	9·501	9·784	10·06	10·20	10·60
1/ 33·3	8·7808	9·7257	10·731	11·797	14·119	16·702	18·095	22·694	27·947	29·850	33·888	38·244	40·545	47·950
0·03200	7·391	7·569	7·743	7·916	8·253	8·581	8·741	9·212	9·666	9·814	10·11	10·39	10·53	10·95
1/ 31·3	9·0699	10·046	11·084	12·185	14·584	17·252	18·691	23·441	28·867	30·832	35·002	39·502	41·879	49·527
0·03400	7·619	7·802	7·983	8·160	8·508	8·846	9·011	9·496	9·964	10·12	10·42	10·71	10·86	11·29
1/ 29·4	9·3502	10·356	11·426	12·562	15·034	17·785	19·268	24·164	29·758	31·783	36·083	40·722	43·171	51·056
0·03600	7·841	8·029	8·215	8·398	8·755	9·103	9·273	9·772	10·25	10·41	10·72	11·02	11·17	11·61
1/ 27·8	9·6223	10·658	11·759	12·927	15·472	18·302	19·829	24·867	30·623	32·708	37·132	41·906	44·426	52·540
0·03800	8·057	8·250	8·441	8·629	8·996	9·353	9·528	10·04	10·54	10·70	11·02	11·33	11·48	11·93
1/ 26·3	9·8870	10·951	12·082	13·283	15·897	18·806	20·374	25·551	31·465	33·606	38·152	43·057	45·647	53·983
0·04000	8·267	8·465	8·661	8·854	9·230	9·597	9·777	10·30	10·81	10·98	11·30	11·62	11·78	12·24
1/ 25·0	10·145	11·236	12·397	13·629	16·311	19·296	20·905	26·216	32·285	34·482	39·146	44·178	46·836	55·389
0·04200	8·472	8·675	8·876	9·073	9·459	9·835	10·02	10·56	11·08	11·25	11·58	11·91	12·07	12·55
1/ 23·8	10·396	11·515	12·704	13·967	16·715	19·774	21·422	26·866	33·084	35·336	40·115	45·272	47·996	56·760
0·04400	8·672	8·880	9·085	9·287	9·682	10·07	10·26	10·81	11·34	11·51	11·86	12·19	12·36	12·84
1/ 22·7	10·642	11·787	13·004	14·296	17·110	20·240	21·928	27·500	33·865	36·169	41·062	46·340	49·128	58·099
0·04600	8·867	9·080	9·290	9·497	9·901	10·29	10·49	11·05	11·59	11·77	12·12	12·47	12·63	13·13
1/ 21·7	10·882	12·052	13·298	14·619	17·496	20·697	22·422	28·119	34·628	36·984	41·987	47·384	50·235	59·408
0·04800	9·059	9·276	9·490	9·701	10·11	10·52	10·71	11·29	11·84	12·03	12·38	12·73	12·91	13·42
1/ 20·8	11·117	12·313	13·585	14·934	17·873	21·143	22·906	28·726	35·375	37·782	42·892	48·406	51·318	60·689
0·05000	9·246	9·468	9·687	9·902	10·32	10·73	10·93	11·52	12·09	12·27	12·64	13·00	13·17	13·69
1/ 20·0	11·347	12·567	13·866	15·243	18·243	21·580	23·380	29·320	36·106	38·563	43·779	49·407	52·378	61·943
0·05250	9·475	9·703	9·927	10·15	10·58	11·00	11·20	11·81	12·39	12·58	12·95	13·32	13·50	14·03
1/ 19·0	11·628	12·879	14·209	15·621	18·695	22·115	23·959	30·046	37·000	39·518	44·863	50·629	53·675	63·476
0·05500	9·699	9·932	10·16	10·39	10·83	11·26	11·47	12·09	12·68	12·88	13·26	13·63	13·82	14·36
1/ 18·2	11·902	13·183	14·544	15·989	19·136	22·636	24·524	30·755	37·873	40·450	45·921	51·824	54·941	64·973
0·05750	9·917	10·16	10·39	10·62	11·07	11·51	11·73	12·36	12·97	13·17	13·56	13·94	14·13	14·69
1/ 17·4	12·170	13·480	14·872	16·350	19·567	23·147	25·076	31·447	38·726	41·361	46·955	52·991	56·178	66·436
0·06000	10·13	10·37	10·61	10·85	11·31	11·76	11·98	12·62	13·25	13·45	13·85	14·24	14·43	15·00
1/ 16·7	12·433	13·771	15·193	16·702	19·989	23·646	25·617	32·125	39·561	42·253	47·967	54·133	57·389	67·868
0·06250	10·34	10·59	10·83	11·07	11·55	12·00	12·23	12·89	13·52	13·73	14·14	14·53	14·73	15·31
1/ 16·0	12·690	14·055	15·507	17·048	20·402	24·135	26·147	32·790	40·378	43·126	48·959	55·252	58·575	69·270
0·06500	10·55	10·80	11·05	11·29	11·77	12·24	12·47	13·14	13·79	14·00	14·42	14·82	15·02	15·62
1/ 15·4	12·942	14·334	15·815	17·386	20·808	24·614	26·666	33·440	41·180	43·982	49·930	56·348	59·737	70·645
0·06750	10·75	11·01	11·26	11·51	12·00	12·48	12·71	13·39	14·05	14·27	14·69	15·11	15·31	15·91
1/ 14·8	13·189	14·608	16·117	17·718	21·205	25·084	27·175	34·079	41·966	44·821	50·883	57·424	60·878	71·993
0·07000	10·95	11·21	11·47	11·72	12·22	12·71	12·94	13·64	14·31	14·53	14·96	15·38	15·59	16·21
1/ 14·3	13·432	14·877	16·414	18·044	21·595	25·545	27·675	34·706	42·737	45·646	51·819	58·480	61·997	73·317
0·07250	11·14	11·41	11·67	11·93	12·44	12·93	13·17	13·88	14·56	14·79	15·23	15·66	15·87	16·49
1/ 13·8	13·671	15·141	16·705	18·365	21·978	25·998	28·166	35·321	43·496	46·455	52·738	59·517	63·096	74·617
0·07500	11·33	11·60	11·87	12·13	12·65	13·15	13·40	14·12	14·81	15·04	15·49	15·92	16·14	16·78
1/ 13·3	13·905	15·401	16·991	18·679	22·355	26·444	28·648	35·926	44·241	47·251	53·641	60·536	64·177	75·895
0·08000	11·70	11·98	12·26	12·53	13·07	13·58	13·84	14·58	15·30	15·53	16·00	16·45	16·67	17·33
1/ 12·5	14·362	15·907	17·550	19·293	23·090	27·313	29·590	37·107	45·695	48·804	55·404	62·525	66·286	78·388
0·08500	12·06	12·35	12·64	12·92	13·47	14·00	14·27	15·03	15·77	16·01	16·49	16·96	17·19	17·86
1/ 11·8	14·805	16·398	18·091	19·889	23·802	28·156	30·503	38·252	47·104	50·309	57·112	64·453	68·329	80·805
0·09000	12·41	12·71	13·01	13·30	13·86	14·41	14·68	15·47	16·23	16·48	16·97	17·45	17·68	18·38
1/ 11·1	15·235	16·874	18·617	20·467	24·494	28·974	31·389	39·363	48·472	51·770	58·771	66·325	70·314	83·152
0·09500	12·76	13·06	13·36	13·66	14·24	14·81	15·08	15·89	16·68	16·93	17·43	17·93	18·17	18·89
1/ 10·5	15·654	17·338	19·129	21·029	25·166	29·769	32·251	40·443	49·803	53·191	60·384	68·146	72·244	85·434
0·10000	13·09	13·40	13·71	14·02	14·61	15·19	15·48	16·31	17·11	17·37	17·89	18·39	18·64	19·38
1/ 10·0	16·062	17·789	19·627	21·576	25·821	30·544	33·090	41·496	51·099	54·576	61·956	69·919	74·124	87·657
	0·92	0·92	0·92	0·92	0·92	0·93	0·93	0·93	0·93	0·94	0·94	0·94	0·94	0·94

$V_{r(0{\cdot}5)medial}$ **for half-full circular pipes.**

A41

(p.1 of 6)

k_s = 0·60 mm
S = 0·00010 to 0·00030

ie hydraulic gradient =
1 in 10000 to 1 in 3333

Water (or sewage) at 15°C;
full bore conditions.

velocities in ms^{-1}
discharges in m^3s^{-1}

Gradient **(Equivalent) Pipe diameters in mm**

Gradient	1250	1300	1350	1400	1500	1600	1650	1800	1950	2000	2100	2200	2250	2400
0·000100	0·372	0·381	0·390	0·399	0·417	0·434	0·443	0·467	0·491	0·499	0·514	0·529	0·537	0·559
1/ 10000	0·4564	0·5060	0·5588	0·6148	0·7370	0·8731	0·9466	1·1894	1·4672	1·5678	1·7817	2·0126	2·1346	2·5276
0·000105	0·381	0·391	0·400	0·410	0·428	0·445	0·454	0·479	0·504	0·512	0·527	0·543	0·550	0·573
1/ 9524	0·4680	0·5188	0·5730	0·6304	0·7557	0·8953	0·9706	1·2195	1·5043	1·6075	1·8267	2·0634	2·1885	2·5913
0·000110	0·391	0·400	0·410	0·419	0·438	0·456	0·465	0·491	0·516	0·524	0·540	0·556	0·564	0·587
1/ 9091	0·4793	0·5314	0·5868	0·6457	0·7740	0·9169	0·9940	1·2489	1·5405	1·6462	1·8707	2·1131	2·2412	2·6536
0·000115	0·400	0·410	0·419	0·429	0·448	0·467	0·476	0·502	0·528	0·536	0·553	0·569	0·577	0·600
1/ 8696	0·4904	0·5437	0·6004	0·6606	0·7919	0·9381	1·0169	1·2777	1·5760	1·6841	1·9137	2·1616	2·2926	2·7145
0·000120	0·408	0·419	0·429	0·439	0·458	0·477	0·486	0·513	0·539	0·548	0·565	0·581	0·589	0·613
1/ 8333	0·5013	0·5557	0·6137	0·6752	0·8093	0·9587	1·0394	1·3058	1·6106	1·7211	1·9558	2·2091	2·3430	2·7741
0·000125	0·417	0·428	0·438	0·448	0·468	0·487	0·496	0·524	0·551	0·559	0·577	0·593	0·602	0·626
1/ 8000	0·5119	0·5675	0·6267	0·6895	0·8265	0·9790	1·0613	1·3334	1·6446	1·7574	1·9970	2·2556	2·3923	2·8325
0·000130	0·426	0·436	0·447	0·457	0·477	0·497	0·506	0·535	0·562	0·571	0·588	0·605	0·614	0·639
1/ 7692	0·5223	0·5790	0·6394	0·7035	0·8432	0·9989	1·0829	1·3604	1·6779	1·7930	2·0374	2·3012	2·4407	2·8897
0·000135	0·434	0·445	0·455	0·466	0·486	0·506	0·516	0·545	0·573	0·582	0·600	0·617	0·626	0·651
1/ 7407	0·5326	0·5904	0·6519	0·7173	0·8597	1·0184	1·1040	1·3869	1·7106	1·8279	2·0770	2·3460	2·4881	2·9458
0·000140	0·442	0·453	0·464	0·475	0·496	0·516	0·526	0·555	0·584	0·593	0·611	0·629	0·637	0·663
1/ 7143	0·5426	0·6015	0·6642	0·7308	0·8759	1·0375	1·1247	1·4129	1·7426	1·8621	2·1159	2·3899	2·5347	3·0009
0·000145	0·450	0·461	0·472	0·483	0·505	0·525	0·536	0·565	0·594	0·603	0·622	0·640	0·649	0·675
1/ 6897	0·5525	0·6124	0·6763	0·7440	0·8918	1·0563	1·1451	1·4385	1·7741	1·8958	2·1541	2·4331	2·5804	3·0551
0·000150	0·458	0·469	0·481	0·492	0·513	0·535	0·545	0·575	0·604	0·614	0·633	0·651	0·660	0·687
1/ 6667	0·5621	0·6232	0·6881	0·7571	0·9074	1·0748	1·1651	1·4636	1·8051	1·9288	2·1917	2·4755	2·6254	3·1083
0·000160	0·473	0·485	0·497	0·508	0·531	0·552	0·563	0·594	0·625	0·635	0·654	0·673	0·682	0·710
1/ 6250	0·5811	0·6441	0·7112	0·7825	0·9378	1·1108	1·2042	1·5127	1·8655	1·9934	2·2650	2·5582	2·7132	3·2121
0·000170	0·488	0·501	0·513	0·524	0·547	0·570	0·581	0·613	0·644	0·654	0·674	0·694	0·704	0·732
1/ 5882	0·5994	0·6644	0·7337	0·8072	0·9674	1·1458	1·2421	1·5602	1·9241	2·0560	2·3360	2·6384	2·7982	3·3128
0·000180	0·503	0·515	0·528	0·540	0·564	0·587	0·598	0·631	0·663	0·674	0·694	0·715	0·725	0·754
1/ 5556	0·6172	0·6842	0·7554	0·8311	0·9960	1·1797	1·2788	1·6063	1·9810	2·1167	2·4050	2·7163	2·8808	3·4105
0·000190	0·517	0·530	0·543	0·555	0·579	0·603	0·615	0·649	0·682	0·693	0·714	0·735	0·745	0·775
1/ 5263	0·6345	0·7034	0·7766	0·8544	1·0239	1·2127	1·3146	1·6512	2·0363	2·1758	2·4722	2·7921	2·9612	3·5055
0·000200	0·531	0·544	0·557	0·570	0·595	0·619	0·631	0·666	0·700	0·711	0·733	0·754	0·764	0·795
1/ 5000	0·6514	0·7221	0·7973	0·8771	1·0511	1·2449	1·3495	1·6950	2·0902	2·2334	2·5375	2·8659	3·0394	3·5981
0·000210	0·544	0·558	0·571	0·584	0·610	0·635	0·647	0·683	0·717	0·729	0·751	0·773	0·784	0·815
1/ 4762	0·6679	0·7403	0·8174	0·8992	1·0776	1·2763	1·3835	1·7376	2·1427	2·2895	2·6013	2·9379	3·1158	3·6884
0·000220	0·557	0·571	0·585	0·598	0·624	0·650	0·663	0·699	0·735	0·746	0·769	0·791	0·802	0·835
1/ 4545	0·6839	0·7581	0·8370	0·9208	1·1035	1·3069	1·4167	1·7793	2·1941	2·3444	2·6636	3·0082	3·1903	3·7766
0·000230	0·570	0·584	0·598	0·612	0·639	0·665	0·678	0·715	0·751	0·763	0·787	0·809	0·821	0·854
1/ 4348	0·6996	0·7755	0·8563	0·9420	1·1288	1·3368	1·4491	1·8200	2·2443	2·3980	2·7245	3·0769	3·2632	3·8628
0·000240	0·583	0·597	0·611	0·625	0·653	0·679	0·693	0·731	0·768	0·780	0·804	0·827	0·839	0·873
1/ 4167	0·7150	0·7926	0·8751	0·9626	1·1536	1·3662	1·4809	1·8599	2·2934	2·4504	2·7840	3·1442	3·3345	3·9472
0·000250	0·595	0·610	0·624	0·639	0·667	0·694	0·707	0·746	0·784	0·796	0·821	0·844	0·856	0·891
1/ 4000	0·7301	0·8092	0·8935	0·9829	1·1778	1·3949	1·5120	1·8989	2·3414	2·5018	2·8424	3·2100	3·4043	4·0298
0·000260	0·607	0·622	0·637	0·651	0·680	0·708	0·721	0·761	0·800	0·812	0·837	0·861	0·873	0·909
1/ 3846	0·7449	0·8256	0·9115	1·0028	1·2016	1·4230	1·5424	1·9372	2·3886	2·5522	2·8996	3·2746	3·4728	4·1108
0·000270	0·619	0·634	0·649	0·664	0·693	0·721	0·735	0·776	0·815	0·828	0·853	0·878	0·890	0·926
1/ 3704	0·7593	0·8416	0·9292	1·0222	1·2249	1·4506	1·5724	1·9747	2·4348	2·6016	2·9557	3·3379	3·5399	4·1903
0·000280	0·630	0·646	0·661	0·676	0·706	0·735	0·749	0·790	0·830	0·844	0·869	0·894	0·907	0·943
1/ 3571	0·7736	0·8574	0·9466	1·0413	1·2478	1·4777	1·6017	2·0115	2·4802	2·6501	3·0107	3·4001	3·6058	4·2683
0·000290	0·642	0·658	0·673	0·689	0·719	0·748	0·763	0·805	0·845	0·859	0·885	0·911	0·923	0·960
1/ 3448	0·7875	0·8729	0·9637	1·0601	1·2703	1·5043	1·6306	2·0477	2·5248	2·6977	3·0648	3·4612	3·6706	4·3449
0·000300	0·653	0·669	0·685	0·701	0·731	0·761	0·776	0·819	0·860	0·874	0·900	0·926	0·939	0·977
1/ 3333	0·8013	0·8881	0·9805	1·0786	1·2924	1·5305	1·6589	2·0833	2·5687	2·7445	3·1180	3·5212	3·7343	4·4202
	0·87	0·87	0·87	0·88	0·88	0·88	0·89	0·89	0·90	0·90	0·90	0·90	0·90	0·91

$V_{r(0\cdot5)medial}$ for half-full circular pipes.

$k_s = 0{\cdot}60$ **mm**
S = 0·00030 to 0·00100

ie hydraulic gradient =
1 in 3333 to 1 in 1000

Water (or sewage) at 15°C;
full bore conditions.

velocities in ms^{-1}
discharges in m^3s^{-1}

Gradient	**(Equivalent) Pipe diameters in mm**													
	1250	**1300**	**1350**	**1400**	**1500**	**1600**	**1650**	**1800**	**1950**	**2000**	**2100**	**2200**	**2250**	**2400**
0·000300	0·653	0·669	0·685	0·701	0·731	0·761	0·776	0·819	0·860	0·874	0·900	0·926	0·939	0·977
1/ 3333	0·8013	0·8881	0·9805	1·0786	1·2924	1·5305	1·6589	2·0833	2·5687	2·7445	3·1180	3·5212	3·7343	4·4202
0·000320	0·675	0·691	0·708	0·724	0·756	0·787	0·802	0·846	0·889	0·903	0·930	0·957	0·970	1·010
1/ 3125	0·8280	0·9178	1·0133	1·1146	1·3355	1·5815	1·7142	2·1527	2·6542	2·8359	3·2218	3·6384	3·8585	4·5672
0·000340	0·696	0·713	0·730	0·747	0·779	0·811	0·827	0·872	0·917	0·931	0·959	0·987	1·001	1·041
1/ 2941	0·8540	0·9465	1·0450	1·1495	1·3773	1·6310	1·7679	2·2200	2·7371	2·9245	3·3224	3·7519	3·9789	4·7096
0·000360	0·716	0·734	0·752	0·769	0·802	0·835	0·851	0·898	0·943	0·958	0·987	1·016	1·030	1·072
1/ 2778	0·8792	0·9745	1·0758	1·1834	1·4179	1·6791	1·8199	2·2853	2·8176	3·0105	3·4200	3·8622	4·0958	4·8479
0·000380	0·736	0·755	0·773	0·790	0·825	0·858	0·875	0·923	0·970	0·985	1·015	1·044	1·059	1·101
1/ 2632	0·9038	1·0017	1·1058	1·2164	1·4574	1·7258	1·8706	2·3489	2·8959	3·0941	3·5150	3·9694	4·2095	4·9824
0·000400	0·756	0·775	0·793	0·811	0·847	0·881	0·898	0·947	0·995	1·011	1·042	1·072	1·087	1·130
1/ 2500	0·9277	1·0281	1·1350	1·2485	1·4959	1·7713	1·9199	2·4108	2·9722	3·1756	3·6075	4·0739	4·3202	5·1135
0·000420	0·775	0·794	0·813	0·831	0·868	0·903	0·920	0·971	1·020	1·036	1·068	1·098	1·114	1·159
1/ 2381	0·9510	1·0540	1·1635	1·2799	1·5334	1·8157	1·9680	2·4712	3·0466	3·2551	3·6978	4·1757	4·4282	5·2413
0·000440	0·793	0·813	0·832	0·851	0·888	0·925	0·942	0·994	1·044	1·061	1·093	1·125	1·140	1·186
1/ 2273	0·9737	1·0792	1·1914	1·3105	1·5701	1·8591	2·0150	2·5301	3·1192	3·3327	3·7859	4·2752	4·5337	5·3661
0·000460	0·812	0·832	0·851	0·871	0·909	0·946	0·964	1·017	1·068	1·085	1·118	1·150	1·166	1·213
1/ 2174	0·9960	1·1038	1·2186	1·3404	1·6059	1·9015	2·0609	2·5878	3·1902	3·4085	3·8721	4·3725	4·6368	5·4881
0·000480	0·829	0·850	0·870	0·890	0·929	0·966	0·985	1·039	1·091	1·109	1·142	1·175	1·192	1·240
1/ 2083	1·0177	1·1279	1·2452	1·3697	1·6409	1·9429	2·1059	2·6442	3·2597	3·4828	3·9564	4·4676	4·7377	5·6074
0·000500	0·847	0·868	0·888	0·908	0·948	0·987	1·005	1·061	1·114	1·132	1·166	1·200	1·216	1·265
1/ 2000	1·0391	1·1516	1·2713	1·3983	1·6752	1·9836	2·1499	2·6994	3·3278	3·5555	4·0389	4·5608	4·8366	5·7243
0·000525	0·868	0·889	0·910	0·931	0·972	1·011	1·031	1·087	1·142	1·160	1·195	1·230	1·247	1·297
1/ 1905	1·0651	1·1804	1·3031	1·4334	1·7172	2·0332	2·2037	2·7669	3·4110	3·6443	4·1398	4·6748	4·9573	5·8673
0·000550	0·889	0·911	0·932	0·953	0·995	1·035	1·055	1·113	1·169	1·188	1·224	1·259	1·276	1·328
1/ 1818	1·0906	1·2086	1·3342	1·4676	1·7582	2·0817	2·2562	2·8329	3·4922	3·7311	4·2384	4·7860	5·0753	6·0068
0·000575	0·909	0·931	0·953	0·975	1·018	1·059	1·079	1·139	1·196	1·215	1·252	1·288	1·305	1·358
1/ 1739	1·1154	1·2362	1·3647	1·5010	1·7982	2·1291	2·3076	2·8973	3·5716	3·8159	4·3347	4·8947	5·1906	6·1432
0·000600	0·929	0·952	0·974	0·996	1·040	1·082	1·103	1·163	1·222	1·241	1·279	1·316	1·334	1·387
1/ 1667	1·1398	1·2632	1·3944	1·5338	1·8374	2·1755	2·3579	2·9604	3·6493	3·8989	4·4290	5·0012	5·3034	6·2767
0·000625	0·948	0·972	0·995	1·017	1·061	1·105	1·126	1·188	1·247	1·267	1·305	1·343	1·362	1·416
1/ 1600	1·1636	1·2896	1·4236	1·5658	1·8758	2·2209	2·4071	3·0221	3·7254	3·9802	4·5213	5·1054	5·4139	6·4074
0·000650	0·967	0·991	1·015	1·038	1·083	1·127	1·148	1·211	1·272	1·292	1·331	1·370	1·389	1·445
1/ 1538	1·1870	1·3155	1·4522	1·5972	1·9134	2·2654	2·4553	3·0827	3·8000	4·0599	4·6118	5·2075	5·5222	6·5356
0·000675	0·986	1·010	1·034	1·058	1·104	1·148	1·170	1·235	1·297	1·317	1·357	1·396	1·416	1·472
1/ 1481	1·2099	1·3409	1·4802	1·6281	1·9503	2·3091	2·5027	3·1421	3·8732	4·1381	4·7005	5·3077	5·6285	6·6613
0·000700	1·004	1·029	1·053	1·077	1·124	1·170	1·192	1·258	1·321	1·342	1·382	1·422	1·442	1·500
1/ 1429	1·2324	1·3658	1·5077	1·6583	1·9866	2·3520	2·5491	3·2004	3·9450	4·2148	4·7877	5·4061	5·7328	6·7847
0·000725	1·022	1·047	1·072	1·097	1·144	1·191	1·214	1·280	1·345	1·366	1·407	1·448	1·468	1·527
1/ 1379	1·2545	1·3903	1·5347	1·6881	2·0222	2·3941	2·5948	3·2576	4·0156	4·2902	4·8733	5·5027	5·8352	6·9059
0·000750	1·040	1·066	1·091	1·116	1·164	1·211	1·234	1·302	1·368	1·389	1·431	1·473	1·493	1·553
1/ 1333	1·2763	1·4144	1·5613	1·7173	2·0571	2·4355	2·6396	3·3139	4·0849	4·3643	4·9574	5·5977	5·9360	7·0250
0·000800	1·075	1·101	1·127	1·153	1·203	1·252	1·275	1·345	1·413	1·435	1·479	1·521	1·542	1·604
1/ 1250	1·3187	1·4614	1·6132	1·7743	2·1254	2·5163	2·7272	3·4238	4·2203	4·5088	5·1216	5·7830	6·1325	7·2575
0·000850	1·108	1·135	1·162	1·188	1·240	1·290	1·315	1·387	1·457	1·480	1·525	1·569	1·590	1·654
1/ 1176	1·3598	1·5069	1·6634	1·8295	2·1916	2·5946	2·8120	3·5302	4·3514	4·6490	5·2807	5·9627	6·3229	7·4828
0·000900	1·141	1·169	1·196	1·223	1·277	1·328	1·354	1·428	1·500	1·523	1·569	1·614	1·637	1·702
1/ 1111	1·3997	1·5511	1·7122	1·8832	2·2558	2·6706	2·8944	3·6336	4·4788	4·7850	5·4352	6·1371	6·5079	7·7016
0·000950	1·172	1·201	1·229	1·257	1·312	1·365	1·391	1·467	1·541	1·565	1·613	1·659	1·682	1·749
1/ 1053	1·4385	1·5941	1·7597	1·9354	2·3183	2·7446	2·9745	3·7342	4·6027	4·9173	5·5855	6·3067	6·6877	7·9144
0·001000	1·203	1·233	1·262	1·290	1·346	1·401	1·428	1·506	1·582	1·606	1·655	1·703	1·726	1·795
1/ 1000	1·4763	1·6360	1·8059	1·9862	2·3792	2·8166	3·0526	3·8321	4·7234	5·0462	5·7319	6·4720	6·8629	8·1217
	0·88	0·88	0·88	0·88	0·89	0·89	0·89	0·90	0·90	0·90	0·90	0·91	0·91	0·91

$V_{r(0{\cdot}5)medial}$ **for half-full circular pipes.**

$k_s = 0{\cdot}60$ **mm**
S = 0·00100 to 0·00300

ie hydraulic gradient =
1 in 1000 to 1 in 333

Water (or sewage) at 15°C;
full bore conditions.

velocities in ms^{-1}
discharges in m^3s^{-1}

Gradient	**(Equivalent) Pipe diameters in mm**													
	1250	**1300**	**1350**	**1400**	**1500**	**1600**	**1650**	**1800**	**1950**	**2000**	**2100**	**2200**	**2250**	**2400**
0·00100	1·203	1·233	1·262	1·290	1·346	1·401	1·428	1·506	1·582	1·606	1·655	1·703	1·726	1·795
1/ 1000	1·4763	1·6360	1·8059	1·9862	2·3792	2·8166	3·0526	3·8321	4·7234	5·0462	5·7319	6·4720	6·8629	8·1217
0·00105	1·233	1·263	1·293	1·322	1·380	1·436	1·463	1·543	1·621	1·646	1·696	1·745	1·769	1·840
1/ 952	1·5132	1·6769	1·8510	2·0358	2·4385	2·8869	3·1287	3·9276	4·8410	5·1720	5·8746	6·6331	7·0338	8·3238
0·00110	1·262	1·293	1·324	1·354	1·413	1·470	1·498	1·580	1·659	1·685	1·736	1·786	1·811	1·884
1/ 909	1·5492	1·7168	1·8950	2·0842	2·4965	2·9555	3·2031	4·0209	4·9560	5·2947	6·0140	6·7905	7·2007	8·5212
0·00115	1·291	1·323	1·354	1·385	1·445	1·503	1·532	1·616	1·697	1·724	1·776	1·827	1·852	1·926
1/ 870	1·5844	1·7558	1·9380	2·1315	2·5532	3·0225	3·2757	4·1121	5·0683	5·4147	6·1503	6·9443	7·3638	8·7142
0·00120	1·319	1·352	1·383	1·415	1·476	1·536	1·565	1·651	1·734	1·761	1·814	1·866	1·892	1·968
1/ 833	1·6189	1·7939	1·9802	2·1778	2·6086	3·0881	3·3468	4·2013	5·1782	5·5321	6·2836	7·0948	7·5234	8·9030
0·00125	1·347	1·380	1·412	1·444	1·507	1·568	1·598	1·685	1·770	1·798	1·852	1·905	1·931	2·009
1/ 800	1·6526	1·8313	2·0214	2·2232	2·6629	3·1524	3·4165	4·2887	5·2859	5·6471	6·4142	7·2423	7·6797	9·0880
0·00130	1·374	1·407	1·440	1·473	1·537	1·599	1·630	1·719	1·805	1·833	1·889	1·943	1·970	2·049
1/ 769	1·6857	1·8679	2·0618	2·2677	2·7161	3·2154	3·4847	4·3743	5·3914	5·7598	6·5422	7·3868	7·8329	9·2692
0·00135	1·400	1·434	1·468	1·501	1·567	1·630	1·661	1·752	1·840	1·869	1·925	1·980	2·008	2·088
1/ 741	1·7181	1·9039	2·1015	2·3113	2·7683	3·2772	3·5517	4·4583	5·4949	5·8704	6·6678	7·5285	7·9832	9·4470
0·00140	1·426	1·461	1·495	1·529	1·596	1·660	1·692	1·784	1·874	1·903	1·961	2·017	2·045	2·127
1/ 714	1·7499	1·9391	2·1404	2·3541	2·8196	3·3379	3·6174	4·5408	5·5965	5·9790	6·7911	7·6677	8·1307	9·6216
0·00145	1·451	1·487	1·522	1·557	1·624	1·690	1·722	1·816	1·907	1·937	1·996	2·053	2·081	2·165
1/ 690	1·7812	1·9738	2·1787	2·3961	2·8700	3·3974	3·6820	4·6218	5·6963	6·0856	6·9122	7·8044	8·2756	9·7930
0·00150	1·477	1·513	1·548	1·583	1·652	1·719	1·752	1·848	1·940	1·970	2·030	2·088	2·117	2·202
1/ 667	1·8120	2·0079	2·2163	2·4375	2·9194	3·4560	3·7454	4·7014	5·7945	6·1904	7·0312	7·9387	8·4181	9·9616
0·00160	1·525	1·563	1·600	1·636	1·707	1·776	1·810	1·909	2·004	2·036	2·097	2·157	2·187	2·275
1/ 625	1·8720	2·0743	2·2896	2·5181	3·0160	3·5703	3·8693	4·8568	5·9859	6·3949	7·2634	8·2009	8·6961	10·290
0·00170	1·573	1·611	1·649	1·687	1·760	1·831	1·866	1·968	2·066	2·099	2·162	2·224	2·255	2·345
1/ 588	1·9301	2·1387	2·3607	2·5963	3·1096	3·6811	3·9893	5·0074	6·1714	6·5931	7·4885	8·4549	8·9655	10·609
0·00180	1·619	1·658	1·697	1·736	1·811	1·884	1·920	2·025	2·127	2·160	2·225	2·289	2·321	2·414
1/ 556	1·9866	2·2013	2·4297	2·6722	3·2005	3·7886	4·1058	5·1537	6·3516	6·7856	7·7070	8·7016	9·2270	10·919
0·00190	1·664	1·704	1·744	1·784	1·861	1·936	1·973	2·081	2·185	2·219	2·287	2·352	2·385	2·480
1/ 526	2·0415	2·2621	2·4969	2·7460	3·2889	3·8932	4·2192	5·2959	6·5268	6·9727	7·9196	8·9416	9·4815	11·220
0·00200	1·707	1·749	1·790	1·831	1·910	1·987	2·025	2·136	2·243	2·278	2·346	2·414	2·447	2·545
1/ 500	2·0950	2·3214	2·5623	2·8179	3·3750	3·9951	4·3296	5·4344	6·6975	7·1551	8·1266	9·1753	9·7293	11·513
0·00210	1·750	1·792	1·835	1·876	1·957	2·036	2·075	2·189	2·298	2·334	2·405	2·474	2·508	2·608
1/ 476	2·1471	2·3792	2·6260	2·8881	3·4590	4·0945	4·4373	5·5695	6·8640	7·3329	8·3285	9·4033	9·9710	11·799
0·00220	1·791	1·835	1·878	1·921	2·004	2·085	2·124	2·241	2·353	2·389	2·462	2·532	2·567	2·670
1/ 455	2·1981	2·4356	2·6883	2·9565	3·5410	4·1915	4·5424	5·7015	7·0265	7·5065	8·5257	9·6258	10·207	12·078
0·00230	1·832	1·877	1·921	1·964	2·049	2·132	2·172	2·291	2·406	2·443	2·517	2·589	2·625	2·730
1/ 435	2·2479	2·4908	2·7492	3·0235	3·6211	4·2864	4·6452	5·8304	7·1854	7·6762	8·7185	9·8434	10·438	12·351
0·00240	1·871	1·917	1·962	2·007	2·094	2·178	2·219	2·341	2·458	2·496	2·572	2·645	2·682	2·789
1/ 417	2·2966	2·5448	2·8088	3·0890	3·6995	4·3792	4·7458	5·9566	7·3409	7·8423	8·9071	10·056	10·663	12·618
0·00250	1·910	1·957	2·003	2·048	2·137	2·223	2·266	2·389	2·509	2·548	2·625	2·700	2·737	2·847
1/ 400	2·3443	2·5976	2·8671	3·1532	3·7764	4·4701	4·8443	6·0802	7·4932	8·0050	9·0918	10·265	10·884	12·879
0·00260	1·948	1·996	2·043	2·089	2·180	2·268	2·311	2·437	2·559	2·599	2·677	2·754	2·792	2·904
1/ 385	2·3911	2·6494	2·9243	3·2160	3·8516	4·5592	4·9408	6·2014	7·6424	8·1644	9·2728	10·469	11·101	13·136
0·00270	1·986	2·034	2·082	2·129	2·221	2·311	2·355	2·484	2·608	2·649	2·728	2·807	2·845	2·959
1/ 370	2·4370	2·7003	2·9804	3·2777	3·9255	4·6466	5·0355	6·3202	7·7888	8·3208	9·4504	10·670	11·314	13·387
0·00280	2·023	2·072	2·121	2·169	2·262	2·354	2·398	2·530	2·656	2·697	2·779	2·859	2·898	3·014
1/ 357	2·4820	2·7502	3·0355	3·3383	3·9980	4·7324	5·1285	6·4368	7·9325	8·4743	9·6248	10·867	11·522	13·634
0·00290	2·059	2·109	2·158	2·207	2·303	2·396	2·441	2·575	2·703	2·745	2·828	2·909	2·949	3·067
1/ 345	2·5263	2·7992	3·0896	3·3978	4·0692	4·8167	5·2198	6·5514	8·0737	8·6252	9·7960	11·060	11·727	13·877
0·00300	2·094	2·145	2·196	2·245	2·342	2·437	2·483	2·619	2·750	2·793	2·877	2·959	3·000	3·120
1/ 333	2·5698	2·8474	3·1427	3·4562	4·1392	4·8995	5·3096	6·6641	8·2125	8·7734	9·9644	11·250	11·929	14·115
	0·88	0·88	0·88	0·88	0·89	0·89	0·89	0·90	0·90	0·90	0·91	0·91	0·91	0·91

$\mathbf{V_{r(0{\cdot}5)medial}}$ **for half-full circular pipes.**

$k_s = 0{\cdot}60$ **mm** S = 0·00100 to 0·00300

k_s = **0·60 mm**
S = 0·00300 to 0·01000

ie hydraulic gradient =
1 in 333 to 1 in 100

Water (or sewage) at 15°C;
full bore conditions.

velocities in ms^{-1}
discharges in m^3s^{-1}

Gradient **(Equivalent) Pipe diameters in mm**

Gradient	1250	1300	1350	1400	1500	1600	1650	1800	1950	2000	2100	2200	2250	2400
0·00300	2·094	2·145	2·196	2·245	2·342	2·437	2·483	2·619	2·750	2·793	2·877	2·959	3·000	3·120
1/ 333	2·5698	2·8474	3·1427	3·4562	4·1392	4·8995	5·3096	6·6641	8·2125	8·7734	9·9644	11·250	11·929	14·115
0·00320	2·163	2·216	2·268	2·319	2·420	2·517	2·565	2·705	2·841	2·885	2·972	3·057	3·099	3·223
1/ 313	2·6546	2·9414	3·2465	3·5703	4·2758	5·0612	5·4847	6·8838	8·4833	9·0626	10·293	11·621	12·322	14·580
0·00340	2·230	2·285	2·338	2·391	2·495	2·595	2·644	2·789	2·928	2·974	3·064	3·152	3·195	3·323
1/ 294	2·7369	3·0325	3·3470	3·6809	4·4082	5·2178	5·6545	7·0968	8·7457	9·3430	10·611	11·980	12·703	15·031
0·00360	2·295	2·351	2·407	2·461	2·567	2·671	2·722	2·870	3·014	3·061	3·153	3·243	3·288	3·419
1/ 278	2·8167	3·1210	3·4447	3·7882	4·5368	5·3699	5·8193	7·3037	9·0005	9·6152	10·920	12·329	13·073	15·469
0·00380	2·359	2·416	2·473	2·529	2·638	2·744	2·797	2·949	3·097	3·145	3·240	3·333	3·378	3·513
1/ 263	2·8944	3·2070	3·5396	3·8927	4·6618	5·5179	5·9797	7·5048	9·2483	9·8799	11·221	12·668	13·433	15·894
0·00400	2·420	2·479	2·537	2·595	2·707	2·816	2·870	3·026	3·178	3·227	3·324	3·420	3·467	3·605
1/ 250	2·9701	3·2909	3·6321	3·9944	4·7836	5·6620	6·1358	7·7008	9·4897	10·138	11·514	12·999	13·783	16·309
0·00420	2·480	2·541	2·601	2·659	2·774	2·886	2·941	3·101	3·256	3·307	3·407	3·504	3·553	3·694
1/ 238	3·0438	3·3726	3·7223	4·0936	4·9023	5·8026	6·2881	7·8919	9·7252	10·389	11·799	13·321	14·125	16·713
0·00440	2·539	2·601	2·662	2·722	2·840	2·954	3·010	3·175	3·333	3·385	3·487	3·587	3·636	3·782
1/ 227	3·1159	3·4524	3·8104	4·1904	5·0183	5·9398	6·4368	8·0785	9·9551	10·635	12·078	13·636	14·459	17·108
0·00460	2·596	2·660	2·722	2·784	2·904	3·021	3·078	3·246	3·409	3·462	3·566	3·668	3·719	3·867
1/ 217	3·1863	3·5305	3·8965	4·2851	5·1317	6·0740	6·5822	8·2609	10·180	10·875	12·351	13·944	14·785	17·494
0·00480	2·653	2·717	2·781	2·844	2·967	3·086	3·145	3·316	3·482	3·536	3·643	3·747	3·799	3·951
1/ 208	3·2552	3·6068	3·9808	4·3778	5·2426	6·2052	6·7245	8·4393	10·400	11·110	12·618	14·245	15·105	17·872
0·00500	2·708	2·774	2·839	2·903	3·028	3·150	3·210	3·385	3·554	3·610	3·718	3·825	3·877	4·032
1/ 200	3·3227	3·6816	4·0633	4·4685	5·3512	6·3338	6·8638	8·6141	10·615	11·340	12·879	14·540	15·417	18·242
0·00525	2·775	2·843	2·909	2·975	3·103	3·228	3·290	3·469	3·643	3·699	3·811	3·920	3·974	4·132
1/ 190	3·4052	3·7730	4·1642	4·5794	5·4840	6·4909	7·0340	8·8278	10·878	11·621	13·198	14·900	15·799	18·694
0·00550	2·840	2·910	2·978	3·045	3·177	3·305	3·367	3·551	3·729	3·786	3·901	4·012	4·067	4·230
1/ 182	3·4858	3·8622	4·2627	4·6877	5·6137	6·6444	7·2003	9·0364	11·135	11·896	13·510	15·252	16·173	19·136
0·00575	2·905	2·976	3·045	3·114	3·248	3·379	3·443	3·631	3·813	3·872	3·989	4·103	4·159	4·325
1/ 174	3·5645	3·9495	4·3590	4·7936	5·7404	6·7944	7·3629	9·2404	11·387	12·164	13·815	15·596	16·537	19·567
0·00600	2·967	3·040	3·111	3·181	3·319	3·452	3·518	3·710	3·895	3·956	4·075	4·191	4·249	4·419
1/ 167	3·6416	4·0348	4·4532	4·8972	5·8645	6·9412	7·5219	9·4399	11·632	12·427	14·113	15·933	16·894	19·989
0·00625	3·029	3·103	3·176	3·247	3·387	3·524	3·591	3·786	3·976	4·037	4·159	4·278	4·337	4·510
1/ 160	3·7171	4·1184	4·5454	4·9986	5·9859	7·0849	7·6777	9·6353	11·873	12·684	14·405	16·263	17·244	20·403
0·00650	3·089	3·165	3·239	3·312	3·455	3·594	3·662	3·862	4·055	4·118	4·242	4·363	4·423	4·600
1/ 154	3·7910	4·2004	4·6359	5·0981	6·1050	7·2258	7·8303	9·8269	12·109	12·936	14·691	16·586	17·587	20·808
0·00675	3·148	3·225	3·301	3·375	3·521	3·663	3·732	3·936	4·132	4·196	4·323	4·447	4·508	4·688
1/ 148	3·8636	4·2808	4·7246	5·1956	6·2218	7·3640	7·9801	10·015	12·341	13·183	14·972	16·903	17·923	21·206
0·00700	3·206	3·285	3·362	3·437	3·586	3·730	3·801	4·008	4·208	4·274	4·402	4·528	4·591	4·774
1/ 143	3·9348	4·3597	4·8117	5·2914	6·3364	7·4997	8·1271	10·199	12·568	13·426	15·248	17·214	18·253	21·596
0·00725	3·263	3·343	3·421	3·498	3·649	3·796	3·868	4·079	4·283	4·350	4·480	4·609	4·672	4·859
1/ 138	4·0048	4·4372	4·8972	5·3854	6·4491	7·6330	8·2715	10·380	12·791	13·664	15·519	17·520	18·577	21·980
0·00750	3·319	3·400	3·480	3·558	3·712	3·861	3·935	4·149	4·357	4·424	4·557	4·688	4·752	4·942
1/ 133	4·0736	4·5134	4·9813	5·4779	6·5598	7·7640	8·4135	10·559	13·011	13·899	15·785	17·820	18·895	22·357
0·00800	3·429	3·512	3·595	3·676	3·834	3·989	4·064	4·286	4·500	4·570	4·707	4·842	4·909	5·104
1/ 125	4·2078	4·6620	5·1453	5·6583	6·7757	8·0196	8·6904	10·906	13·439	14·356	16·304	18·407	19·517	23·092
0·00850	3·535	3·621	3·706	3·789	3·953	4·112	4·190	4·418	4·639	4·711	4·853	4·992	5·060	5·262
1/ 118	4·3378	4·8061	5·3043	5·8331	6·9851	8·2673	8·9589	11·243	13·854	14·800	16·808	18·975	20·120	23·805
0·00900	3·638	3·726	3·814	3·900	4·068	4·231	4·312	4·547	4·774	4·848	4·994	5·137	5·207	5·415
1/ 111	4·4641	4·9460	5·4587	6·0029	7·1884	8·5079	9·2195	11·570	14·257	15·230	17·296	19·527	20·705	24·497
0·00950	3·738	3·829	3·919	4·007	4·180	4·348	4·430	4·672	4·905	4·981	5·131	5·278	5·350	5·564
1/ 105	4·5869	5·0821	5·6089	6·1680	7·3861	8·7418	9·4730	11·888	14·649	15·649	17·772	20·063	21·274	25·170
0·01000	3·835	3·929	4·021	4·111	4·289	4·461	4·546	4·793	5·033	5·111	5·265	5·415	5·490	5·709
1/ 100	4·7066	5·2146	5·7552	6·3288	7·5786	8·9697	9·7200	12·198	15·030	16·056	18·235	20·586	21·828	25·826
	0·88	0·88	0·88	0·89	0·89	0·89	0·90	0·90	0·90	0·90	0·91	0·91	0·91	0·91

$V_{r(0\cdot5)medial}$ for half-full circular pipes.

$k_s = 0{\cdot}60$ mm
S = 0·01000 to 0·03000

ie hydraulic gradient =
1 in 100 to 1 in 33·3

Water (or sewage) at 15°C;
full bore conditions.

velocities in ms^{-1}
discharges in m^3s^{-1}

Gradient **(Equivalent) Pipe diameters in mm**

Gradient	1250	1300	1350	1400	1500	1600	1650	1800	1950	2000	2100	2200	2250	2400
0·01000	3·835	3·929	4·021	4·111	4·289	4·461	4·546	4·793	5·033	5·111	5·265	5·415	5·490	5·709
1/ 100	4·7066	5·2146	5·7552	6·3288	7·5786	8·9697	9·7200	12·198	15·030	16·056	18·235	20·586	21·828	25·826
0·01050	3·930	4·026	4·120	4·213	4·395	4·572	4·658	4·912	5·157	5·237	5·395	5·550	5·626	5·850
1/ 95	4·8232	5·3439	5·8978	6·4857	7·7664	9·1919	9·9608	12·500	15·403	16·454	18·686	21·096	22·368	26·465
0·01100	4·023	4·121	4·218	4·313	4·499	4·680	4·768	5·028	5·279	5·361	5·522	5·681	5·758	5·988
1/ 91	4·9372	5·4701	6·0371	6·6389	7·9498	9·4090	10·196	12·795	15·766	16·842	19·127	21·593	22·896	27·090
0·01150	4·114	4·214	4·313	4·410	4·600	4·785	4·876	5·142	5·398	5·482	5·647	5·809	5·888	6·123
1/ 87	5·0485	5·5935	6·1733	6·7886	8·1291	9·6211	10·426	13·084	16·121	17·222	19·558	22·080	23·412	27·700
0·01200	4·203	4·305	4·406	4·505	4·699	4·888	4·981	5·252	5·515	5·600	5·769	5·934	6·015	6·255
1/ 83	5·1575	5·7142	6·3065	6·9351	8·3045	9·8287	10·651	13·366	16·469	17·593	19·980	22·556	23·917	28·297
0·01250	4·290	4·394	4·497	4·598	4·797	4·989	5·084	5·361	5·629	5·716	5·888	6·056	6·140	6·384
1/ 80	5·2642	5·8325	6·4370	7·0786	8·4763	10·032	10·871	13·642	16·810	17·957	20·393	23·022	24·411	28·882
0·01300	4·375	4·481	4·586	4·690	4·892	5·089	5·185	5·467	5·740	5·829	6·005	6·177	6·261	6·511
1/ 77	5·3689	5·9484	6·5649	7·2192	8·6446	10·231	11·087	13·913	17·144	18·314	20·798	23·480	24·896	29·456
0·01350	4·459	4·567	4·674	4·779	4·985	5·186	5·284	5·572	5·850	5·941	6·119	6·295	6·381	6·635
1/ 74	5·4715	6·0621	6·6904	7·3572	8·8098	10·427	11·299	14·179	17·471	18·663	21·195	23·928	25·371	30·018
0·01400	4·541	4·651	4·760	4·867	5·077	5·281	5·381	5·674	5·958	6·050	6·232	6·410	6·498	6·757
1/ 71	5·5722	6·1737	6·8135	7·4926	8·9720	10·619	11·507	14·440	17·792	19·007	21·585	24·368	25·838	30·570
0·01450	4·621	4·734	4·845	4·954	5·167	5·375	5·477	5·775	6·063	6·157	6·343	6·524	6·614	6·877
1/ 69	5·6712	6·2833	6·9345	7·6256	9·1313	10·807	11·711	14·696	18·108	19·344	21·968	24·800	26·296	31·112
0·01500	4·701	4·815	4·928	5·039	5·256	5·467	5·571	5·874	6·167	6·263	6·451	6·636	6·727	6·995
1/ 67	5·7684	6·3911	7·0534	7·7564	9·2878	10·992	11·912	14·948	18·418	19·675	22·345	25·225	26·747	31·645
0·01600	4·855	4·973	5·090	5·204	5·429	5·647	5·754	6·067	6·370	6·469	6·663	6·854	6·948	7·225
1/ 63	5·9582	6·6013	7·2854	8·0115	9·5933	11·354	12·303	15·439	19·024	20·322	23·079	26·054	27·626	32·685
0·01700	5·005	5·127	5·247	5·365	5·596	5·821	5·931	6·254	6·567	6·668	6·869	7·065	7·162	7·448
1/ 59	6·1422	6·8051	7·5103	8·2588	9·8893	11·704	12·683	15·916	19·611	20·949	23·791	26·858	28·478	33·693
0·01800	5·151	5·276	5·399	5·521	5·759	5·990	6·104	6·436	6·757	6·862	7·068	7·271	7·370	7·664
1/ 56	6·3208	7·0030	7·7287	8·4989	10·177	12·044	13·052	16·378	20·181	21·558	24·482	27·638	29·305	34·672
0·01900	5·292	5·421	5·548	5·673	5·917	6·155	6·272	6·613	6·943	7·051	7·263	7·470	7·573	7·875
1/ 53	6·4945	7·1954	7·9410	8·7324	10·456	12·375	13·410	16·828	20·735	22·150	25·155	28·397	30·110	35·624
0·02000	5·430	5·562	5·692	5·820	6·071	6·315	6·435	6·785	7·124	7·234	7·452	7·665	7·770	8·080
1/ 50	6·6637	7·3828	8·1479	8·9599	10·729	12·698	13·759	17·266	21·275	22·727	25·809	29·137	30·894	36·551
0·02100	5·565	5·700	5·833	5·965	6·222	6·472	6·594	6·953	7·300	7·413	7·636	7·854	7·962	8·280
1/ 47·6	6·8287	7·5657	8·3496	9·1817	10·994	13·012	14·100	17·694	21·801	23·289	26·448	29·857	31·658	37·456
0·02200	5·696	5·834	5·971	6·105	6·368	6·624	6·750	7·117	7·472	7·588	7·816	8·040	8·150	8·475
1/ 45·5	6·9898	7·7442	8·5466	9·3983	11·254	13·319	14·433	18·111	22·315	23·838	27·072	30·561	32·405	38·339
0·02300	5·824	5·966	6·105	6·243	6·512	6·773	6·902	7·277	7·640	7·759	7·992	8·221	8·333	8·666
1/ 43·5	7·1473	7·9187	8·7392	9·6101	11·507	13·619	14·758	18·519	22·818	24·375	27·681	31·250	33·134	39·202
0·02400	5·950	6·095	6·237	6·377	6·652	6·919	7·051	7·434	7·805	7·926	8·164	8·398	8·513	8·852
1/ 41·7	7·3014	8·0894	8·9276	9·8173	11·755	13·912	15·076	18·918	23·310	24·900	28·278	31·923	33·848	40·047
0·02500	6·073	6·221	6·366	6·509	6·790	7·062	7·196	7·588	7·966	8·090	8·333	8·571	8·689	9·035
1/ 40·0	7·4524	8·2566	9·1122	10·020	11·998	14·200	15·387	19·309	23·791	25·415	28·862	32·582	34·548	40·874
0·02600	6·193	6·344	6·492	6·638	6·924	7·203	7·339	7·738	8·124	8·250	8·498	8·741	8·861	9·214
1/ 38·5	7·6003	8·4205	9·2930	10·219	12·236	14·482	15·693	19·692	24·263	25·919	29·435	33·229	35·233	41·685
0·02700	6·312	6·465	6·616	6·765	7·057	7·340	7·479	7·886	8·279	8·408	8·660	8·908	9·030	9·390
1/ 37·0	7·7455	8·5813	9·4705	10·414	12·470	14·758	15·992	20·068	24·726	26·414	29·996	33·863	35·905	42·480
0·02800	6·428	6·584	6·738	6·890	7·186	7·475	7·617	8·031	8·432	8·562	8·820	9·072	9·196	9·563
1/ 35·7	7·8879	8·7391	9·6447	10·606	12·699	15·029	16·286	20·437	25·181	26·899	30·548	34·485	36·565	43·261
0·02900	6·542	6·701	6·858	7·012	7·314	7·608	7·752	8·174	8·581	8·714	8·976	9·233	9·359	9·732
1/ 34·5	8·0279	8·8942	9·8158	10·794	12·925	15·296	16·575	20·799	25·627	27·376	31·089	35·097	37·213	44·028
0·03000	6·654	6·816	6·975	7·132	7·439	7·738	7·884	8·314	8·728	8·863	9·130	9·391	9·520	9·899
1/ 33·3	8·1654	9·0466	9·9839	10·979	13·146	15·558	16·859	21·155	26·066	27·845	31·622	35·698	37·851	44·782
	0·88	0·88	0·89	0·89	0·89	0·89	0·90	0·90	0·90	0·91	0·91	0·91	0·91	0·91

$V_{r(0{\cdot}5)medial}$ for half-full circular pipes.

$k_s = 0{\cdot}60$ mm S = 0·01000 to 0·03000

k_s = **0·60 mm**
S = 0·03000 to 0·10000

ie hydraulic gradient =
1 in 33·3 to 1 in 10·0

Water (or sewage) at 15°C;
full bore conditions.

velocities in ms^{-1}
discharges in m^3s^{-1}

Gradient **(Equivalent) Pipe diameters in mm**

Gradient	1250	1300	1350	1400	1500	1600	1650	1800	1950	2000	2100	2200	2250	2400
0·03000	6·654	6·816	6·975	7·132	7·439	7·738	7·884	8·314	8·728	8·863	9·130	9·391	9·520	9·899
1/ 33·3	8·1654	9·0466	9·9839	10·979	13·146	15·558	16·859	21·155	26·066	27·845	31·622	35·698	37·851	44·782
0·03200	6·873	7·040	7·204	7·366	7·684	7·992	8·144	8·587	9·015	9·154	9·430	9·699	9·832	10·22
1/ 31·3	8·4338	9·3439	10·312	11·340	13·578	16·069	17·413	21·850	26·923	28·760	32·660	36·870	39·094	46·253
0·03400	7·084	7·257	7·426	7·594	7·920	8·239	8·395	8·851	9·293	9·437	9·720	9·998	10·14	10·54
1/ 29·4	8·6940	9·6321	10·630	11·689	13·997	16·565	17·950	22·524	27·753	29·646	33·667	38·007	40·299	47·678
0·03600	7·290	7·468	7·642	7·814	8·151	8·478	8·638	9·108	9·563	9·711	10·00	10·29	10·43	10·85
1/ 27·8	8·9466	9·9119	10·939	12·029	14·403	17·046	18·471	23·178	28·558	30·507	34·645	39·110	41·469	49·062
0·03800	7·490	7·673	7·852	8·029	8·374	8·711	8·876	9·358	9·825	9·977	10·28	10·57	10·72	11·14
1/ 26·3	9·1922	10·184	11·239	12·359	14·799	17·514	18·978	23·814	29·342	31·344	35·596	40·184	42·607	50·409
0·04000	7·685	7·872	8·056	8·238	8·592	8·937	9·107	9·602	10·08	10·24	10·54	10·85	10·99	11·43
1/ 25·0	9·4315	10·449	11·532	12·681	15·184	17·970	19·472	24·434	30·106	32·160	36·522	41·229	43·716	51·720
0·04200	7·876	8·067	8·256	8·441	8·805	9·158	9·332	9·839	10·33	10·49	10·81	11·11	11·27	11·72
1/ 23·8	9·6649	10·708	11·817	12·995	15·559	18·414	19·954	25·038	30·850	32·955	37·425	42·249	44·797	52·999
0·04400	8·061	8·257	8·450	8·640	9·012	9·374	9·552	10·07	10·57	10·74	11·06	11·38	11·53	11·99
1/ 22·7	9·8927	10·960	12·096	13·301	15·926	18·848	20·424	25·629	31·577	33·732	38·307	43·244	45·852	54·248
0·04600	8·243	8·443	8·641	8·835	9·215	9·585	9·767	10·30	10·81	10·98	11·31	11·63	11·79	12·26
1/ 21·7	10·115	11·207	12·368	13·600	16·285	19·272	20·884	26·205	32·288	34·491	39·169	44·217	46·884	55·468
0·04800	8·420	8·625	8·827	9·025	9·414	9·792	9·977	10·52	11·04	11·22	11·55	11·88	12·05	12·53
1/ 20·8	10·333	11·448	12·635	13·893	16·636	19·688	21·334	26·770	32·983	35·234	40·012	45·170	47·894	56·663
0·05000	8·594	8·803	9·009	9·212	9·608	9·994	10·18	10·74	11·27	11·45	11·79	12·13	12·29	12·78
1/ 20·0	10·547	11·685	12·895	14·180	16·979	20·094	21·774	27·323	33·665	35·962	40·839	46·102	48·883	57·833
0·05250	8·807	9·021	9·232	9·440	9·846	10·24	10·44	11·00	11·55	11·73	12·08	12·43	12·60	13·10
1/ 19·0	10·808	11·974	13·214	14·531	17·399	20·591	22·313	27·998	34·497	36·851	41·848	47·242	50·091	59·263
0·05500	9·015	9·234	9·450	9·662	10·08	10·48	10·68	11·26	11·82	12·01	12·37	12·72	12·89	13·41
1/ 18·2	11·063	12·256	13·526	14·874	17·809	21·076	22·839	28·658	35·310	37·719	42·834	48·355	51·271	60·659
0·05750	9·218	9·442	9·662	9·880	10·30	10·72	10·92	11·52	12·09	12·28	12·65	13·01	13·19	13·71
1/ 17·4	11·312	12·532	13·830	15·208	18·210	21·551	23·353	29·303	36·104	38·568	43·798	49·443	52·425	62·024
0·06000	9·416	9·645	9·870	10·09	10·53	10·95	11·16	11·76	12·35	12·54	12·92	13·29	13·47	14·01
1/ 16·7	11·555	12·802	14·128	15·536	18·602	22·015	23·856	29·934	36·882	39·398	44·741	50·508	53·554	63·359
0·06250	9·611	9·844	10·07	10·30	10·74	11·18	11·39	12·01	12·60	12·80	13·18	13·56	13·75	14·29
1/ 16·0	11·794	13·067	14·420	15·857	18·986	22·470	24·348	30·552	37·643	40·212	45·665	51·551	54·660	64·667
0·06500	9·801	10·04	10·27	10·51	10·96	11·40	11·61	12·24	12·85	13·05	13·45	13·83	14·02	14·58
1/ 15·4	12·028	13·326	14·706	16·171	19·363	22·915	24·831	31·158	38·390	41·009	46·571	52·573	55·743	65·949
0·06750	9·988	10·23	10·47	10·71	11·17	11·61	11·83	12·48	13·10	13·30	13·70	14·09	14·29	14·86
1/ 14·8	12·257	13·580	14·987	16·480	19·732	23·352	25·305	31·752	39·122	41·791	47·459	53·575	56·806	67·207
0·07000	10·17	10·42	10·66	10·90	11·37	11·83	12·05	12·71	13·34	13·55	13·95	14·35	14·55	15·13
1/ 14·3	12·483	13·829	15·262	16·783	20·095	23·781	25·770	32·336	39·841	42·559	48·331	54·560	57·850	68·441
0·07250	10·35	10·60	10·85	11·10	11·57	12·04	12·27	12·93	13·58	13·79	14·20	14·61	14·81	15·40
1/ 13·8	12·704	14·075	15·533	17·080	20·451	24·203	26·226	32·909	40·547	43·313	49·187	55·526	58·875	69·654
0·07500	10·53	10·79	11·04	11·29	11·77	12·24	12·48	13·15	13·81	14·02	14·44	14·86	15·06	15·66
1/ 13·3	12·921	14·316	15·799	17·372	20·801	24·617	26·675	33·472	41·241	44·054	50·029	56·477	59·882	70·846
0·08000	10·88	11·14	11·40	11·66	12·16	12·65	12·88	13·59	14·26	14·48	14·92	15·34	15·56	16·17
1/ 12·5	13·346	14·786	16·317	17·943	21·484	25·425	27·551	34·571	42·595	45·501	51·671	58·331	61·848	73·172
0·08500	11·21	11·48	11·75	12·02	12·53	13·04	13·28	14·00	14·70	14·93	15·38	15·82	16·03	16·67
1/ 11·8	13·757	15·241	16·820	18·496	22·146	26·209	28·400	35·636	43·907	46·903	53·263	60·128	63·753	75·426
0·09000	11·54	11·82	12·09	12·36	12·90	13·41	13·67	14·41	15·13	15·36	15·82	16·28	16·50	17·16
1/ 11·1	14·157	15·684	17·309	19·033	22·789	26·970	29·224	36·670	45·181	48·264	54·809	61·873	65·604	77·614
0·09500	11·85	12·14	12·42	12·70	13·25	13·78	14·04	14·81	15·54	15·78	16·26	16·72	16·95	17·63
1/ 10·5	14·545	16·114	17·783	19·555	23·414	27·710	30·026	37·676	46·421	49·588	56·312	63·570	67·403	79·743
0·10000	12·16	12·46	12·75	13·03	13·59	14·14	14·41	15·19	15·95	16·19	16·68	17·16	17·39	18·09
1/ 10·0	14·923	16·533	18·246	20·064	24·023	28·430	30·807	38·656	47·628	50·877	57·776	65·223	69·155	81·817
	0·88	0·88	0·89	0·89	0·89	0·89	0·90	0·90	0·90	0·91	0·91	0·91	0·91	0·91

$V_{r(0\cdot5)medial}$ for half-full circular pipes.

A42

(p.1 of 6)

k_s = 1·50 mm
S = 0·00010 to 0·00030

ie hydraulic gradient =
1 in 10000 to 1 in 3333

Water (or sewage) at 15°C;
full bore conditions.

velocities in ms^{-1}
discharges in m^3s^{-1}

Gradient **(Equivalent) Pipe diameters in mm**

Gradient	1250	1300	1350	1400	1500	1600	1650	1800	1950	2000	2100	2200	2250	2400
0·000100	0·340	0·348	0·357	0·365	0·381	0·397	0·405	0·428	0·450	0·457	0·471	0·485	0·492	0·512
1/ 10000	0·4171	0·4625	0·5108	0·5621	0·6740	0·7987	0·8660	1·0885	1·3432	1·4355	1·6316	1·8434	1·9554	2·3159
0·000105	0·348	0·357	0·366	0·374	0·391	0·407	0·415	0·438	0·461	0·468	0·483	0·497	0·504	0·525
1/ 9524	0·4276	0·4741	0·5236	0·5762	0·6909	0·8187	0·8877	1·1157	1·3768	1·4714	1·6724	1·8895	2·0042	2·3738
0·000110	0·357	0·366	0·375	0·383	0·400	0·417	0·425	0·449	0·472	0·480	0·494	0·509	0·516	0·537
1/ 9091	0·4378	0·4854	0·5361	0·5900	0·7074	0·8382	0·9089	1·1423	1·4095	1·5064	1·7122	1·9344	2·0519	2·4302
0·000115	0·365	0·374	0·383	0·392	0·409	0·426	0·435	0·459	0·483	0·490	0·506	0·520	0·528	0·549
1/ 8696	0·4478	0·4965	0·5483	0·6034	0·7235	0·8573	0·9296	1·1683	1·4416	1·5406	1·7511	1·9784	2·0985	2·4854
0·000120	0·373	0·382	0·391	0·401	0·418	0·436	0·444	0·469	0·493	0·501	0·517	0·532	0·539	0·561
1/ 8333	0·4575	0·5073	0·5603	0·6166	0·7393	0·8760	0·9498	1·1938	1·4730	1·5742	1·7892	2·0214	2·1441	2·5394
0·000125	0·381	0·390	0·400	0·409	0·427	0·445	0·453	0·479	0·503	0·512	0·527	0·543	0·550	0·573
1/ 8000	0·4671	0·5179	0·5720	0·6295	0·7547	0·8943	0·9696	1·2187	1·5037	1·6070	1·8265	2·0635	2·1888	2·5923
0·000130	0·388	0·398	0·408	0·417	0·436	0·454	0·463	0·488	0·514	0·522	0·538	0·554	0·561	0·584
1/ 7692	0·4765	0·5283	0·5835	0·6421	0·7699	0·9122	0·9891	1·2431	1·5338	1·6392	1·8630	2·1048	2·2326	2·6441
0·000135	0·396	0·406	0·416	0·425	0·444	0·462	0·471	0·498	0·523	0·532	0·548	0·564	0·572	0·596
1/ 7407	0·4857	0·5385	0·5948	0·6545	0·7847	0·9298	1·0081	1·2670	1·5633	1·6707	1·8989	2·1453	2·2755	2·6950
0·000140	0·403	0·413	0·423	0·433	0·452	0·471	0·480	0·507	0·533	0·542	0·558	0·575	0·583	0·607
1/ 7143	0·4947	0·5485	0·6058	0·6667	0·7993	0·9471	1·0269	1·2905	1·5923	1·7017	1·9341	2·1851	2·3177	2·7449
0·000145	0·410	0·421	0·431	0·441	0·460	0·479	0·489	0·516	0·543	0·551	0·568	0·585	0·593	0·618
1/ 6897	0·5036	0·5584	0·6167	0·6786	0·8136	0·9641	1·0452	1·3136	1·6208	1·7321	1·9687	2·2241	2·3591	2·7940
0·000150	0·417	0·428	0·438	0·448	0·468	0·488	0·497	0·525	0·552	0·561	0·578	0·595	0·604	0·628
1/ 6667	0·5123	0·5681	0·6274	0·6904	0·8277	0·9807	1·0633	1·3363	1·6488	1·7620	2·0026	2·2625	2·3998	2·8422
0·000160	0·431	0·442	0·453	0·463	0·484	0·504	0·514	0·543	0·570	0·579	0·597	0·615	0·624	0·649
1/ 6250	0·5294	0·5869	0·6482	0·7133	0·8552	1·0133	1·0986	1·3806	1·7034	1·8204	2·0690	2·3374	2·4793	2·9362
0·000170	0·445	0·456	0·467	0·478	0·499	0·520	0·530	0·559	0·588	0·597	0·616	0·634	0·643	0·669
1/ 5882	0·5459	0·6052	0·6684	0·7355	0·8818	1·0448	1·1328	1·4236	1·7564	1·8770	2·1332	2·4100	2·5563	3·0274
0·000180	0·458	0·469	0·481	0·492	0·514	0·535	0·545	0·576	0·605	0·615	0·634	0·653	0·662	0·689
1/ 5556	0·5619	0·6230	0·6880	0·7571	0·9076	1·0754	1·1659	1·4652	1·8078	1·9319	2·1957	2·4805	2·6310	3·1159
0·000190	0·471	0·482	0·494	0·505	0·528	0·550	0·560	0·592	0·622	0·632	0·651	0·671	0·680	0·708
1/ 5263	0·5775	0·6403	0·7071	0·7781	0·9328	1·1052	1·1982	1·5058	1·8578	1·9853	2·2564	2·5491	2·7037	3·2020
0·000200	0·483	0·495	0·507	0·519	0·542	0·564	0·575	0·607	0·638	0·649	0·669	0·688	0·698	0·726
1/ 5000	0·5926	0·6571	0·7257	0·7985	0·9573	1·1342	1·2297	1·5453	1·9065	2·0374	2·3155	2·6159	2·7746	3·2858
0·000210	0·495	0·507	0·520	0·532	0·555	0·578	0·589	0·622	0·654	0·665	0·685	0·705	0·715	0·744
1/ 4762	0·6075	0·6735	0·7438	0·8184	0·9812	1·1625	1·2603	1·5838	1·9540	2·0881	2·3732	2·6810	2·8437	3·3676
0·000220	0·507	0·519	0·532	0·544	0·568	0·592	0·603	0·637	0·670	0·680	0·701	0·722	0·732	0·762
1/ 4545	0·6219	0·6895	0·7615	0·8379	1·0045	1·1901	1·2903	1·6214	2·0004	2·1377	2·4295	2·7446	2·9111	3·4475
0·000230	0·518	0·531	0·544	0·557	0·581	0·605	0·617	0·652	0·685	0·696	0·717	0·738	0·749	0·779
1/ 4348	0·6360	0·7052	0·7788	0·8569	1·0273	1·2171	1·3196	1·6582	2·0457	2·1862	2·4845	2·8068	2·9770	3·5255
0·000240	0·530	0·543	0·556	0·569	0·594	0·618	0·631	0·666	0·700	0·711	0·733	0·754	0·765	0·796
1/ 4167	0·6499	0·7205	0·7957	0·8755	1·0496	1·2435	1·3482	1·6942	2·0901	2·2336	2·5384	2·8676	3·0416	3·6019
0·000250	0·541	0·554	0·567	0·581	0·606	0·631	0·644	0·680	0·714	0·726	0·748	0·770	0·781	0·813
1/ 4000	0·6634	0·7355	0·8123	0·8938	1·0714	1·2694	1·3762	1·7294	2·1335	2·2800	2·5911	2·9272	3·1048	3·6767
0·000260	0·551	0·565	0·579	0·592	0·618	0·644	0·656	0·693	0·729	0·740	0·763	0·785	0·796	0·829
1/ 3846	0·6767	0·7502	0·8285	0·9116	1·0929	1·2948	1·4037	1·7639	2·1761	2·3255	2·6429	2·9856	3·1667	3·7501
0·000270	0·562	0·576	0·590	0·604	0·630	0·656	0·669	0·706	0·743	0·754	0·778	0·800	0·812	0·845
1/ 3704	0·6897	0·7647	0·8444	0·9292	1·1139	1·3197	1·4307	1·7978	2·2179	2·3701	2·6936	3·0429	3·2275	3·8220
0·000280	0·572	0·587	0·601	0·615	0·642	0·668	0·681	0·720	0·756	0·768	0·792	0·815	0·827	0·860
1/ 3571	0·7025	0·7788	0·8601	0·9464	1·1345	1·3441	1·4572	1·8311	2·2589	2·4140	2·7434	3·0991	3·2871	3·8926
0·000290	0·583	0·597	0·612	0·626	0·653	0·680	0·694	0·732	0·770	0·782	0·806	0·830	0·841	0·876
1/ 3448	0·7151	0·7928	0·8755	0·9633	1·1548	1·3681	1·4832	1·8637	2·2992	2·4570	2·7923	3·1544	3·3457	3·9620
0·000300	0·593	0·608	0·622	0·637	0·665	0·692	0·706	0·745	0·783	0·796	0·820	0·844	0·856	0·891
1/ 3333	0·7274	0·8064	0·8906	0·9799	1·1747	1·3917	1·5088	1·8958	2·3388	2·4993	2·8404	3·2087	3·4033	4·0302
	0·83	0·83	0·83	0·83	0·84	0·84	0·84	0·85	0·85	0·85	0·86	0·86	0·86	0·86

$V_{r(0·5)medial}$ for half-full circular pipes.

k_s = 1·50 mm S = 0·00010 to 0·00030

Gradient **(Equivalent) Pipe diameters in mm**

Gradient	1250	1300	1350	1400	1500	1600	1650	1800	1950	2000	2100	2200	2250	2400
0·000300	0·593	0·608	0·622	0·637	0·665	0·692	0·706	0·745	0·783	0·796	0·820	0·844	0·856	0·891
1/ 3333	0·7274	0·8064	0·8906	0·9799	1·1747	1·3917	1·5088	1·8958	2·3388	2·4993	2·8404	3·2087	3·4033	4·0302
0·000320	0·612	0·628	0·643	0·658	0·687	0·715	0·729	0·770	0·809	0·822	0·847	0·872	0·884	0·920
1/ 3125	0·7515	0·8331	0·9200	1·0123	1·2135	1·4377	1·5587	1·9585	2·4161	2·5819	2·9342	3·3146	3·5157	4·1632
0·000340	0·631	0·647	0·663	0·678	0·708	0·737	0·752	0·794	0·834	0·847	0·873	0·899	0·912	0·949
1/ 2941	0·7748	0·8590	0·9486	1·0438	1·2512	1·4823	1·6070	2·0192	2·4909	2·6619	3·0251	3·4173	3·6246	4·2921
0·000360	0·650	0·666	0·682	0·698	0·729	0·759	0·774	0·817	0·858	0·872	0·899	0·925	0·938	0·976
1/ 2778	0·7975	0·8841	0·9763	1·0743	1·2878	1·5256	1·6540	2·0782	2·5637	2·7396	3·1134	3·5170	3·7303	4·4173
0·000380	0·668	0·685	0·701	0·717	0·749	0·780	0·795	0·839	0·882	0·896	0·924	0·951	0·964	1·003
1/ 2632	0·8195	0·9086	1·0033	1·1040	1·3233	1·5677	1·6996	2·1356	2·6344	2·8152	3·1992	3·6140	3·8332	4·5391
0·000400	0·685	0·702	0·719	0·736	0·768	0·800	0·816	0·861	0·905	0·920	0·948	0·976	0·989	1·030
1/ 2500	0·8410	0·9324	1·0296	1·1329	1·3580	1·6087	1·7441	2·1914	2·7033	2·8888	3·2829	3·7085	3·9334	4·6577
0·000420	0·702	0·720	0·737	0·754	0·788	0·820	0·836	0·883	0·928	0·942	0·971	1·000	1·014	1·055
1/ 2381	0·8620	0·9556	1·0552	1·1611	1·3918	1·6488	1·7875	2·2459	2·7705	2·9606	3·3645	3·8006	4·0311	4·7734
0·000440	0·719	0·737	0·755	0·772	0·806	0·839	0·856	0·903	0·950	0·965	0·994	1·023	1·038	1·080
1/ 2273	0·8824	0·9782	1·0803	1·1886	1·4247	1·6878	1·8298	2·2991	2·8361	3·0307	3·4441	3·8906	4·1265	4·8864
0·000460	0·735	0·754	0·772	0·790	0·824	0·858	0·875	0·924	0·971	0·987	1·017	1·047	1·061	1·105
1/ 2174	0·9024	1·0004	1·1047	1·2155	1·4570	1·7260	1·8712	2·3511	2·9002	3·0992	3·5220	3·9785	4·2197	4·9968
0·000480	0·751	0·770	0·789	0·807	0·842	0·877	0·894	0·944	0·992	1·008	1·039	1·069	1·084	1·128
1/ 2083	0·9220	1·0221	1·1287	1·2418	1·4886	1·7634	1·9117	2·4020	2·9629	3·1663	3·5982	4·0646	4·3110	5·1048
0·000500	0·767	0·786	0·805	0·823	0·860	0·895	0·913	0·964	1·013	1·029	1·060	1·091	1·107	1·152
1/ 2000	0·9411	1·0433	1·1521	1·2676	1·5195	1·8000	1·9514	2·4518	3·0244	3·2319	3·6728	4·1488	4·4004	5·2106
0·000525	0·786	0·806	0·825	0·844	0·881	0·918	0·935	0·987	1·038	1·054	1·087	1·119	1·134	1·180
1/ 1905	0·9645	1·0693	1·1808	1·2992	1·5572	1·8448	1·9999	2·5127	3·0995	3·3122	3·7640	4·2518	4·5096	5·3400
0·000550	0·805	0·825	0·844	0·864	0·902	0·939	0·957	1·011	1·062	1·079	1·112	1·145	1·161	1·208
1/ 1818	0·9874	1·0946	1·2087	1·3299	1·5941	1·8884	2·0473	2·5722	3·1729	3·3906	3·8530	4·3524	4·6163	5·4663
0·000575	0·823	0·843	0·864	0·883	0·922	0·960	0·979	1·034	1·086	1·104	1·138	1·171	1·187	1·236
1/ 1739	1·0097	1·1194	1·2361	1·3600	1·6302	1·9311	2·0936	2·6304	3·2446	3·4672	3·9401	4·4507	4·7206	5·5897
0·000600	0·841	0·862	0·882	0·903	0·942	0·981	1·000	1·056	1·110	1·128	1·162	1·196	1·213	1·262
1/ 1667	1·0316	1·1436	1·2629	1·3895	1·6655	1·9729	2·1389	2·6873	3·3147	3·5422	4·0253	4·5469	4·8226	5·7105
0·000625	0·858	0·879	0·901	0·921	0·962	1·002	1·021	1·078	1·133	1·151	1·186	1·221	1·238	1·288
1/ 1600	1·0530	1·1674	1·2891	1·4183	1·7000	2·0139	2·1832	2·7430	3·3835	3·6156	4·1087	4·6412	4·9225	5·8288
0·000650	0·875	0·897	0·919	0·940	0·981	1·022	1·041	1·099	1·155	1·174	1·210	1·245	1·263	1·314
1/ 1538	1·0740	1·1906	1·3148	1·4466	1·7339	2·0540	2·2267	2·7976	3·4508	3·6876	4·1905	4·7335	5·0205	5·9448
0·000675	0·892	0·914	0·936	0·958	1·000	1·041	1·061	1·120	1·178	1·196	1·233	1·269	1·287	1·339
1/ 1481	1·0946	1·2135	1·3400	1·4743	1·7671	2·0933	2·2694	2·8512	3·5169	3·7582	4·2707	4·8241	5·1166	6·0585
0·000700	0·908	0·931	0·953	0·975	1·018	1·060	1·081	1·141	1·199	1·218	1·256	1·292	1·311	1·364
1/ 1429	1·1148	1·2359	1·3647	1·5015	1·7997	2·1320	2·3112	2·9038	3·5817	3·8275	4·3494	4·9131	5·2109	6·1702
0·000725	0·925	0·948	0·970	0·993	1·037	1·079	1·100	1·161	1·221	1·240	1·278	1·315	1·334	1·388
1/ 1379	1·1347	1·2579	1·3890	1·5283	1·8318	2·1699	2·3524	2·9554	3·6455	3·8955	4·4268	5·0004	5·3035	6·2799
0·000750	0·941	0·964	0·987	1·010	1·054	1·098	1·119	1·181	1·242	1·261	1·300	1·338	1·357	1·412
1/ 1333	1·1542	1·2795	1·4129	1·5545	1·8633	2·2072	2·3928	3·0062	3·7081	3·9624	4·5028	5·0863	5·3946	6·3877
0·000800	0·972	0·996	1·020	1·043	1·089	1·134	1·156	1·220	1·283	1·303	1·343	1·382	1·401	1·458
1/ 1250	1·1923	1·3218	1·4595	1·6058	1·9247	2·2800	2·4717	3·1053	3·8303	4·0930	4·6511	5·2538	5·5723	6·5981
0·000850	1·002	1·027	1·051	1·075	1·123	1·169	1·192	1·258	1·322	1·343	1·384	1·425	1·445	1·504
1/ 1176	1·2292	1·3627	1·5047	1·6555	1·9843	2·3505	2·5481	3·2013	3·9487	4·2195	4·7949	5·4162	5·7445	6·8020
0·000900	1·031	1·057	1·082	1·107	1·156	1·203	1·226	1·295	1·361	1·382	1·425	1·466	1·487	1·547
1/ 1111	1·2651	1·4024	1·5486	1·7038	2·0421	2·4190	2·6224	3·2946	4·0637	4·3424	4·9345	5·5739	5·9118	6·9999
0·000950	1·059	1·086	1·112	1·137	1·187	1·236	1·260	1·330	1·398	1·420	1·464	1·507	1·528	1·590
1/ 1053	1·2999	1·4410	1·5912	1·7507	2·0984	2·4856	2·6946	3·3853	4·1755	4·4619	5·0703	5·7273	6·0744	7·1925
0·001000	1·087	1·114	1·141	1·167	1·218	1·269	1·293	1·365	1·435	1·457	1·502	1·546	1·568	1·631
1/ 1000	1·3339	1·4787	1·6328	1·7964	2·1531	2·5505	2·7649	3·4736	4·2845	4·5783	5·2025	5·8766	6·2328	7·3800
	0·83	0·83	0·83	0·84	0·84	0·84	0·85	0·85	0·85	0·86	0·86	0·86	0·86	0·87

$V_{r(0·5)medial}$ for half-full circular pipes.

A42
(p.3 of 6)

k_s = **1·50 mm**
S = 0·00100 to 0·00300

ie hydraulic gradient =
1 in 1000 to 1 in 333

Water (or sewage) at 15°C;
full bore conditions.

velocities in ms^{-1}
discharges in m^3s^{-1}

Gradient **(Equivalent) Pipe diameters in mm**

Gradient	1250	1300	1350	1400	1500	1600	1650	1800	1950	2000	2100	2200	2250	2400
0·00100	1·087	1·114	1·141	1·167	1·218	1·269	1·293	1·365	1·435	1·457	1·502	1·546	1·568	1·631
1/ 1000	1·3339	1·4787	1·6328	1·7964	2·1531	2·5505	2·7649	3·4736	4·2845	4·5783	5·2025	5·8766	6·2328	7·3800
0·00105	1·114	1·142	1·169	1·196	1·249	1·300	1·325	1·399	1·470	1·493	1·539	1·584	1·606	1·672
1/ 952	1·3670	1·5154	1·6733	1·8410	2·2066	2·6138	2·8335	3·5598	4·3907	4·6918	5·3315	6·0223	6·3873	7·5629
0·00110	1·140	1·169	1·197	1·224	1·278	1·331	1·356	1·432	1·505	1·529	1·576	1·622	1·644	1·711
1/ 909	1·3994	1·5512	1·7129	1·8846	2·2587	2·6756	2·9005	3·6439	4·4944	4·8027	5·4575	6·1646	6·5382	7·7416
0·00115	1·166	1·195	1·224	1·252	1·307	1·361	1·387	1·464	1·539	1·563	1·611	1·658	1·681	1·750
1/ 870	1·4310	1·5863	1·7516	1·9271	2·3097	2·7360	2·9660	3·7261	4·5958	4·9111	5·5806	6·3036	6·6857	7·9162
0·00120	1·191	1·221	1·250	1·279	1·335	1·390	1·417	1·496	1·572	1·597	1·646	1·694	1·718	1·788
1/ 833	1·4619	1·6206	1·7894	1·9688	2·3596	2·7951	3·0300	3·8066	4·6951	5·0171	5·7011	6·4397	6·8300	8·0870
0·00125	1·216	1·246	1·276	1·305	1·363	1·419	1·446	1·527	1·605	1·630	1·680	1·729	1·753	1·825
1/ 800	1·4922	1·6541	1·8265	2·0095	2·4085	2·8529	3·0928	3·8854	4·7923	5·1209	5·8191	6·5729	6·9713	8·2543
0·00130	1·240	1·271	1·301	1·331	1·390	1·447	1·475	1·557	1·637	1·662	1·713	1·763	1·788	1·861
1/ 769	1·5219	1·6871	1·8629	2·0495	2·4564	2·9097	3·1543	3·9626	4·8875	5·2227	5·9347	6·7036	7·1098	8·4183
0·00135	1·264	1·295	1·326	1·357	1·417	1·475	1·503	1·587	1·668	1·694	1·746	1·797	1·822	1·896
1/ 741	1·5510	1·7194	1·8985	2·0887	2·5034	2·9653	3·2146	4·0384	4·9810	5·3226	6·0482	6·8317	7·2457	8·5792
0·00140	1·287	1·319	1·351	1·382	1·443	1·502	1·531	1·616	1·699	1·725	1·778	1·830	1·856	1·931
1/ 714	1·5796	1·7510	1·9335	2·1272	2·5495	3·0200	3·2738	4·1128	5·0727	5·4206	6·1595	6·9575	7·3791	8·7371
0·00145	1·310	1·343	1·375	1·406	1·468	1·529	1·558	1·645	1·729	1·756	1·810	1·863	1·889	1·966
1/ 690	1·6077	1·7822	1·9679	2·1650	2·5949	3·0736	3·3320	4·1859	5·1628	5·5168	6·2689	7·0810	7·5101	8·8923
0·00150	1·333	1·366	1·398	1·431	1·494	1·555	1·585	1·673	1·758	1·786	1·841	1·895	1·921	1·999
1/ 667	1·6353	1·8128	2·0017	2·2022	2·6394	3·1264	3·3892	4·2577	5·2514	5·6115	6·3764	7·2025	7·6389	9·0447
0·00160	1·376	1·411	1·444	1·478	1·543	1·606	1·637	1·728	1·816	1·845	1·902	1·957	1·984	2·065
1/ 625	1·6892	1·8725	2·0676	2·2747	2·7263	3·2293	3·5007	4·3978	5·4242	5·7961	6·5862	7·4394	7·8902	9·3422
0·00170	1·419	1·454	1·489	1·523	1·590	1·656	1·688	1·782	1·872	1·902	1·960	2·017	2·046	2·129
1/ 588	1·7414	1·9304	2·1315	2·3450	2·8105	3·3291	3·6089	4·5336	5·5916	5·9751	6·7896	7·6691	8·1338	9·6306
0·00180	1·460	1·497	1·532	1·568	1·637	1·704	1·737	1·833	1·927	1·957	2·017	2·076	2·105	2·191
1/ 556	1·7921	1·9866	2·1935	2·4133	2·8923	3·4259	3·7139	4·6655	5·7543	6·1489	6·9870	7·8921	8·3703	9·9106
0·00190	1·501	1·538	1·575	1·611	1·682	1·751	1·785	1·884	1·980	2·011	2·073	2·133	2·163	2·251
1/ 526	1·8414	2·0412	2·2539	2·4797	2·9719	3·5201	3·8160	4·7938	5·9124	6·3179	7·1790	8·1090	8·6003	10·183
0·00200	1·540	1·578	1·616	1·653	1·726	1·796	1·831	1·933	2·031	2·063	2·127	2·189	2·219	2·310
1/ 500	1·8894	2·0944	2·3126	2·5443	3·0493	3·6119	3·9155	4·9187	6·0665	6·4825	7·3661	8·3202	8·8244	10·448
0·00210	1·578	1·617	1·656	1·694	1·768	1·841	1·877	1·981	2·082	2·115	2·179	2·243	2·274	2·367
1/ 476	1·9363	2·1464	2·3700	2·6074	3·1249	3·7014	4·0125	5·0405	6·2168	6·6430	7·5485	8·5263	9·0429	10·707
0·00220	1·615	1·655	1·695	1·734	1·810	1·884	1·921	2·028	2·131	2·164	2·231	2·296	2·328	2·423
1/ 455	1·9820	2·1971	2·4259	2·6690	3·1987	3·7888	4·1072	5·1595	6·3635	6·7998	7·7266	8·7274	9·2563	10·959
0·00230	1·652	1·693	1·733	1·773	1·851	1·927	1·964	2·073	2·179	2·213	2·281	2·348	2·380	2·477
1/ 435	2·0267	2·2466	2·4806	2·7291	3·2708	3·8742	4·1998	5·2758	6·5069	6·9531	7·9007	8·9241	9·4648	11·206
0·00240	1·687	1·729	1·770	1·811	1·891	1·968	2·007	2·118	2·226	2·261	2·330	2·398	2·432	2·531
1/ 417	2·0705	2·2951	2·5342	2·7880	3·3414	3·9578	4·2904	5·3896	6·6472	7·1030	8·0711	9·1165	9·6689	11·448
0·00250	1·722	1·765	1·807	1·849	1·930	2·009	2·048	2·162	2·272	2·308	2·378	2·448	2·482	2·583
1/ 400	2·1133	2·3426	2·5866	2·8457	3·4105	4·0396	4·3791	5·5011	6·7847	7·2499	8·2380	9·3050	9·8688	11·685
0·00260	1·756	1·800	1·843	1·885	1·968	2·049	2·089	2·205	2·317	2·354	2·426	2·496	2·531	2·634
1/ 385	2·1553	2·3891	2·6380	2·9023	3·4783	4·1199	4·4661	5·6103	6·9194	7·3938	8·4015	9·4897	10·065	11·917
0·00270	1·790	1·834	1·878	1·921	2·006	2·088	2·129	2·247	2·361	2·398	2·472	2·544	2·580	2·684
1/ 370	2·1965	2·4348	2·6884	2·9577	3·5447	4·1986	4·5514	5·7175	7·0516	7·5350	8·5620	9·6709	10·257	12·144
0·00280	1·823	1·868	1·913	1·957	2·043	2·127	2·168	2·288	2·405	2·443	2·517	2·591	2·627	2·734
1/ 357	2·2369	2·4796	2·7379	3·0122	3·6100	4·2759	4·6352	5·8227	7·1813	7·6736	8·7195	9·8488	10·446	12·367
0·00290	1·855	1·901	1·947	1·991	2·079	2·164	2·206	2·329	2·447	2·486	2·562	2·637	2·674	2·782
1/ 345	2·2767	2·5237	2·7865	3·0656	3·6741	4·3518	4·7175	5·9260	7·3087	7·8098	8·8742	10·024	10·631	12·587
0·00300	1·887	1·934	1·980	2·026	2·115	2·201	2·244	2·369	2·489	2·529	2·606	2·682	2·720	2·830
1/ 333	2·3157	2·5669	2·8343	3·1182	3·7371	4·4264	4·7983	6·0276	7·4340	7·9436	9·0263	10·195	10·813	12·802
	0·83	0·83	0·84	0·84	0·84	0·84	0·85	0·85	0·86	0·86	0·86	0·86	0·86	0·87

$V_{r(0·5)medial}$ for half-full circular pipes.

k_s = 1·50 mm
S = 0·00300 to 0·01000

ie hydraulic gradient =
1 in 333 to 1 in 100

Water (or sewage) at 15°C;
full bore conditions.

velocities in ms^{-1}
discharges in m^3s^{-1}

Gradient	(Equivalent) Pipe diameters in mm													
	1250	**1300**	**1350**	**1400**	**1500**	**1600**	**1650**	**1800**	**1950**	**2000**	**2100**	**2200**	**2250**	**2400**
0·00300	1·887	1·934	1·980	2·026	2·115	2·201	2·244	2·369	2·489	2·529	2·606	2·682	2·720	2·830
1/ 333	2·3157	2·5669	2·8343	3·1182	3·7371	4·4264	4·7983	6·0276	7·4340	7·9436	9·0263	10·195	10·813	12·802
0·00320	1·949	1·998	2·045	2·092	2·184	2·274	2·318	2·447	2·571	2·612	2·692	2·770	2·809	2·923
1/ 313	2·3919	2·6514	2·9276	3·2208	3·8600	4·5719	4·9561	6·2258	7·6784	8·2048	9·3230	10·530	11·168	13·223
0·00340	2·009	2·059	2·108	2·157	2·252	2·344	2·389	2·522	2·650	2·692	2·775	2·856	2·896	3·013
1/ 294	2·4657	2·7332	3·0179	3·3202	3·9791	4·7130	5·1090	6·4178	7·9152	8·4579	9·6105	10·855	11·513	13·631
0·00360	2·068	2·119	2·170	2·220	2·317	2·412	2·459	2·595	2·727	2·770	2·855	2·939	2·980	3·101
1/ 278	2·5374	2·8127	3·1057	3·4167	4·0947	4·8500	5·2575	6·6044	8·1452	8·7036	9·8898	11·171	11·847	14·027
0·00380	2·125	2·177	2·229	2·281	2·381	2·478	2·526	2·667	2·802	2·847	2·934	3·019	3·061	3·186
1/ 263	2·6072	2·8900	3·1910	3·5106	4·2072	4·9832	5·4019	6·7858	8·3689	8·9427	10·161	11·477	12·173	14·412
0·00400	2·180	2·234	2·287	2·340	2·443	2·543	2·592	2·736	2·875	2·921	3·010	3·098	3·141	3·269
1/ 250	2·6751	2·9653	3·2741	3·6020	4·3168	5·1130	5·5426	6·9624	8·5868	9·1755	10·426	11·776	12·489	14·787
0·00420	2·234	2·289	2·344	2·398	2·503	2·606	2·656	2·804	2·946	2·993	3·085	3·175	3·219	3·350
1/ 238	2·7413	3·0387	3·3552	3·6912	4·4237	5·2396	5·6798	7·1347	8·7993	9·4025	10·684	12·067	12·799	15·153
0·00440	2·287	2·343	2·399	2·454	2·562	2·667	2·719	2·870	3·016	3·063	3·157	3·249	3·295	3·429
1/ 227	2·8060	3·1104	3·4343	3·7783	4·5280	5·3631	5·8138	7·3030	9·0068	9·6242	10·936	12·352	13·100	15·510
0·00460	2·338	2·396	2·453	2·510	2·620	2·727	2·780	2·935	3·084	3·132	3·228	3·323	3·369	3·506
1/ 217	2·8693	3·1805	3·5117	3·8634	4·6300	5·4840	5·9447	7·4675	9·2096	9·8410	11·182	12·630	13·395	15·860
0·00480	2·388	2·448	2·506	2·564	2·677	2·786	2·840	2·998	3·150	3·200	3·298	3·394	3·442	3·581
1/ 208	2·9311	3·2491	3·5874	3·9467	4·7299	5·6022	6·0729	7·6284	9·4081	10·053	11·423	12·902	13·684	16·201
0·00500	2·438	2·498	2·558	2·617	2·732	2·844	2·899	3·060	3·215	3·266	3·366	3·464	3·513	3·655
1/ 200	2·9917	3·3162	3·6616	4·0283	4·8276	5·7179	6·1984	7·7861	9·6025	10·261	11·659	13·169	13·967	16·536
0·00525	2·498	2·560	2·621	2·682	2·799	2·914	2·971	3·135	3·295	3·347	3·449	3·550	3·600	3·746
1/ 190	3·0658	3·3983	3·7522	4·1280	4·9471	5·8594	6·3517	7·9787	9·8400	10·515	11·947	13·495	14·312	16·945
0·00550	2·557	2·621	2·683	2·745	2·866	2·983	3·041	3·209	3·373	3·426	3·531	3·634	3·684	3·834
1/ 182	3·1381	3·4785	3·8407	4·2254	5·0638	5·9976	6·5015	8·1668	10·072	10·762	12·229	13·813	14·649	17·344
0·00575	2·615	2·680	2·744	2·807	2·930	3·050	3·109	3·282	3·448	3·503	3·610	3·715	3·767	3·920
1/ 174	3·2088	3·5568	3·9272	4·3205	5·1778	6·1327	6·6479	8·3507	10·299	11·005	12·504	14·124	14·979	17·735
0·00600	2·671	2·737	2·803	2·867	2·993	3·116	3·176	3·352	3·523	3·578	3·688	3·795	3·848	4·005
1/ 167	3·2780	3·6335	4·0119	4·4137	5·2894	6·2648	6·7912	8·5307	10·521	11·242	12·774	14·428	15·302	18·117
0·00625	2·726	2·794	2·861	2·926	3·055	3·180	3·242	3·422	3·596	3·652	3·764	3·874	3·928	4·087
1/ 160	3·3457	3·7086	4·0948	4·5049	5·3987	6·3943	6·9315	8·7069	10·738	11·474	13·038	14·726	15·618	18·491
0·00650	2·780	2·850	2·917	2·984	3·116	3·243	3·306	3·489	3·667	3·725	3·839	3·951	4·006	4·168
1/ 154	3·4121	3·7822	4·1761	4·5943	5·5058	6·5212	7·0690	8·8796	10·951	11·702	13·296	15·018	15·928	18·858
0·00675	2·834	2·904	2·973	3·041	3·175	3·305	3·369	3·556	3·737	3·796	3·912	4·026	4·082	4·248
1/ 148	3·4773	3·8544	4·2558	4·6820	5·6109	6·6456	7·2039	9·0491	11·160	11·925	13·550	15·304	16·232	19·217
0·00700	2·886	2·957	3·028	3·097	3·233	3·366	3·431	3·621	3·806	3·866	3·984	4·100	4·157	4·326
1/ 143	3·5412	3·9253	4·3340	4·7680	5·7141	6·7678	7·3364	9·2154	11·365	12·144	13·799	15·586	16·530	19·571
0·00725	2·937	3·010	3·082	3·152	3·291	3·426	3·492	3·686	3·873	3·934	4·055	4·173	4·231	4·403
1/ 138	3·6040	3·9949	4·4109	4·8526	5·8154	6·8878	7·4665	9·3788	11·567	12·359	14·044	15·862	16·823	19·917
0·00750	2·987	3·061	3·134	3·206	3·347	3·484	3·552	3·749	3·939	4·001	4·124	4·244	4·303	4·478
1/ 133	3·6658	4·0633	4·4865	4·9357	5·9150	7·0057	7·5943	9·5394	11·765	12·571	14·284	16·134	17·111	20·258
0·00800	3·085	3·162	3·237	3·312	3·457	3·599	3·668	3·872	4·069	4·133	4·259	4·384	4·445	4·625
1/ 125	3·7862	4·1969	4·6339	5·0979	6·1093	7·2359	7·8438	9·8527	12·151	12·984	14·753	16·663	17·673	20·924
0·00850	3·180	3·259	3·337	3·414	3·564	3·710	3·781	3·991	4·194	4·260	4·391	4·519	4·582	4·768
1/ 118	3·9030	4·3263	4·7768	5·2551	6·2977	7·4590	8·0856	10·156	12·526	13·384	15·208	17·177	18·217	21·568
0·00900	3·273	3·354	3·434	3·513	3·667	3·818	3·891	4·107	4·316	4·384	4·518	4·650	4·715	4·906
1/ 111	4·0163	4·4519	4·9155	5·4077	6·4806	7·6756	8·3204	10·451	12·889	13·773	15·649	17·676	18·746	22·195
0·00950	3·363	3·446	3·528	3·609	3·768	3·922	3·998	4·220	4·434	4·504	4·642	4·777	4·844	5·041
1/ 105	4·1266	4·5741	5·0504	5·5561	6·6585	7·8862	8·5488	10·738	13·243	14·151	16·079	18·161	19·261	22·804
0·01000	3·450	3·536	3·620	3·703	3·866	4·024	4·102	4·330	4·550	4·621	4·763	4·902	4·970	5·172
1/ 100	4·2340	4·6932	5·1819	5·7007	6·8317	8·0914	8·7712	11·018	13·587	14·519	16·497	18·633	19·762	23·397
	0·83	0·83	0·84	0·84	0·84	0·85	0·85	0·85	0·86	0·86	0·86	0·86	0·86	0·87

$V_{r(0·5)medial}$ for half-full circular pipes.

k_s = 1·50 mm
S = 0·01000 to 0·03000

ie hydraulic gradient =
1 in 100 to 1 in 33·3

Water (or sewage) at 15°C;
full bore conditions.

velocities in ms^{-1}
discharges in m^3s^{-1}

Gradient **(Equivalent) Pipe diameters in mm**

Gradient	1250	1300	1350	1400	1500	1600	1650	1800	1950	2000	2100	2200	2250	2400
0·01000	3·450	3·536	3·620	3·703	3·866	4·024	4·102	4·330	4·550	4·621	4·763	4·902	4·970	5·172
1/ 100	4·2340	4·6932	5·1819	5·7007	6·8317	8·0914	8·7712	11·018	13·587	14·519	16·497	18·633	19·762	23·397
0·01050	3·536	3·623	3·710	3·795	3·962	4·124	4·203	4·437	4·662	4·736	4·881	5·023	5·093	5·300
1/ 95	4·3387	4·8093	5·3100	5·8417	7·0007	8·2915	8·9881	11·290	13·923	14·878	16·905	19·094	20·250	23·975
0·01100	3·619	3·709	3·797	3·884	4·055	4·221	4·303	4·541	4·772	4·847	4·996	5·141	5·213	5·425
1/ 91	4·4410	4·9226	5·4352	5·9794	7·1657	8·4870	9·1999	11·556	14·252	15·228	17·303	19·544	20·727	24·540
0·01150	3·700	3·792	3·883	3·972	4·146	4·316	4·399	4·643	4·879	4·956	5·108	5·257	5·330	5·547
1/ 87	4·5410	5·0335	5·5576	6·1140	7·3270	8·6780	9·4070	11·816	14·572	15·571	17·693	19·983	21·194	25·092
0·01200	3·780	3·874	3·966	4·057	4·236	4·409	4·494	4·743	4·984	5·063	5·218	5·370	5·445	5·666
1/ 83	4·6388	5·1419	5·6773	6·2457	7·4848	8·8649	9·6096	12·071	14·886	15·906	18·074	20·414	21·650	25·632
0·01250	3·858	3·954	4·048	4·141	4·323	4·500	4·587	4·841	5·087	5·168	5·326	5·481	5·557	5·783
1/ 80	4·7346	5·2481	5·7945	6·3747	7·6393	9·0479	9·8080	12·320	15·193	16·235	18·447	20·835	22·097	26·161
0·01300	3·935	4·032	4·128	4·223	4·409	4·589	4·678	4·937	5·188	5·270	5·431	5·590	5·668	5·898
1/ 77	4·8286	5·3522	5·9095	6·5011	7·7909	9·2274	10·002	12·564	15·495	16·557	18·812	21·248	22·535	26·680
0·01350	4·010	4·109	4·207	4·304	4·493	4·677	4·767	5·032	5·287	5·371	5·535	5·696	5·776	6·010
1/ 74	4·9207	5·4543	6·0222	6·6252	7·9395	9·4034	10·193	12·804	15·790	16·872	19·171	21·653	22·965	27·189
0·01400	4·083	4·185	4·285	4·383	4·575	4·763	4·855	5·124	5·384	5·469	5·637	5·801	5·882	6·120
1/ 71	5·0111	5·5545	6·1329	6·7469	8·0854	9·5761	10·381	13·039	16·080	17·182	19·523	22·051	23·387	27·688
0·01450	4·156	4·259	4·360	4·461	4·656	4·847	4·941	5·215	5·480	5·566	5·737	5·904	5·986	6·229
1/ 69	5·0999	5·6530	6·2416	6·8665	8·2287	9·7459	10·565	13·270	16·365	17·487	19·869	22·442	23·801	28·179
0·01500	4·227	4·332	4·435	4·537	4·736	4·930	5·025	5·304	5·574	5·661	5·835	6·005	6·089	6·335
1/ 67	5·1873	5·7498	6·3484	6·9840	8·3695	9·9127	10·745	13·497	16·645	17·786	20·209	22·826	24·208	28·661
0·01600	4·366	4·474	4·581	4·686	4·892	5·092	5·190	5·478	5·757	5·847	6·026	6·202	6·288	6·543
1/ 63	5·3576	5·9386	6·5569	7·2134	8·6443	10·238	11·098	13·940	17·192	18·370	20·873	23·575	25·003	29·602
0·01700	4·500	4·612	4·722	4·830	5·042	5·249	5·350	5·647	5·934	6·028	6·212	6·393	6·482	6·745
1/ 59	5·5227	6·1216	6·7590	7·4357	8·9107	10·554	11·440	14·370	17·721	18·936	21·516	24·301	25·773	30·514
0·01800	4·631	4·746	4·859	4·971	5·189	5·401	5·506	5·811	6·106	6·202	6·392	6·578	6·670	6·941
1/ 56	5·6831	6·2993	6·9552	7·6515	9·1694	10·860	11·772	14·787	18·236	19·485	22·140	25·007	26·521	31·399
0·01900	4·758	4·876	4·992	5·107	5·331	5·549	5·657	5·970	6·274	6·373	6·568	6·759	6·853	7·131
1/ 53	5·8390	6·4722	7·1460	7·8615	9·4209	11·158	12·095	15·192	18·736	20·020	22·747	25·693	27·249	32·260
0·02000	4·882	5·003	5·122	5·240	5·470	5·694	5·804	6·126	6·437	6·538	6·738	6·935	7·031	7·317
1/ 50	5·9909	6·6405	7·3319	8·0659	9·6659	11·448	12·410	15·588	19·223	20·541	23·339	26·361	27·957	33·099
0·02100	5·003	5·127	5·249	5·369	5·605	5·835	5·947	6·277	6·596	6·700	6·905	7·106	7·205	7·497
1/ 47·6	6·1390	6·8047	7·5131	8·2653	9·9049	11·731	12·716	15·973	19·698	21·048	23·916	27·012	28·648	33·917
0·02200	5·120	5·247	5·373	5·496	5·737	5·972	6·087	6·425	6·751	6·858	7·068	7·273	7·375	7·674
1/ 45·5	6·2836	6·9650	7·6902	8·4601	10·138	12·007	13·016	16·349	20·162	21·544	24·479	27·648	29·323	34·716
0·02300	5·236	5·365	5·493	5·619	5·866	6·106	6·224	6·569	6·903	7·012	7·227	7·437	7·541	7·847
1/ 43·5	6·4250	7·1217	7·8632	8·6504	10·366	12·278	13·309	16·717	20·616	22·029	25·030	28·270	29·983	35·497
0·02400	5·348	5·481	5·612	5·740	5·992	6·238	6·358	6·711	7·052	7·163	7·382	7·597	7·703	8·015
1/ 41·7	6·5634	7·2751	8·0325	8·8367	10·590	12·542	13·595	17·077	21·060	22·503	25·569	28·879	30·628	36·261
0·02500	5·459	5·594	5·728	5·859	6·116	6·367	6·489	6·849	7·197	7·311	7·534	7·754	7·862	8·181
1/ 40·0	6·6989	7·4253	8·1983	9·0191	10·808	12·801	13·876	17·429	21·494	22·967	26·096	29·475	31·260	37·009
0·02600	5·567	5·705	5·841	5·975	6·237	6·493	6·618	6·985	7·340	7·456	7·684	7·908	8·018	8·343
1/ 38·5	6·8317	7·5725	8·3608	9·1979	11·022	13·055	14·151	17·775	21·920	23·423	26·613	30·059	31·880	37·743
0·02700	5·673	5·814	5·952	6·089	6·356	6·617	6·744	7·118	7·480	7·598	7·830	8·058	8·171	8·502
1/ 37·0	6·9620	7·7169	8·5203	9·3733	11·233	13·303	14·421	18·114	22·338	23·869	27·121	30·632	32·487	38·462
0·02800	5·777	5·921	6·062	6·201	6·473	6·738	6·868	7·249	7·617	7·737	7·974	8·206	8·321	8·658
1/ 35·7	7·0899	7·8586	8·6768	9·5454	11·439	13·548	14·686	18·446	22·748	24·307	27·619	31·195	33·084	39·169
0·02900	5·880	6·026	6·169	6·311	6·588	6·857	6·990	7·377	7·752	7·874	8·115	8·352	8·468	8·812
1/ 34·5	7·2155	7·9979	8·8305	9·7146	11·642	13·788	14·946	18·773	23·151	24·738	28·108	31·747	33·670	39·862
0·03000	5·980	6·129	6·275	6·419	6·700	6·975	7·109	7·504	7·885	8·009	8·254	8·494	8·613	8·962
1/ 33·3	7·3390	8·1347	8·9816	9·8808	11·841	14·024	15·202	19·094	23·547	25·161	28·589	32·290	34·246	40·544
	0·83	0·83	0·84	0·84	0·84	0·85	0·85	0·85	0·86	0·86	0·86	0·86	0·86	0·87

$V_{r(0·5)medial}$ for half-full circular pipes.

Gradient **(Equivalent) Pipe diameters in mm**

Gradient	1250	1300	1350	1400	1500	1600	1650	1800	1950	2000	2100	2200	2250	2400
0·03000	5·980	6·129	6·275	6·419	6·700	6·975	7·109	7·504	7·885	8·009	8·254	8·494	8·613	8·962
1/ 33·3	7·3390	8·1347	8·9816	9·8808	11·841	14·024	15·202	19·094	23·547	25·161	28·589	32·290	34·246	40·544
0·03200	6·177	6·330	6·481	6·629	6·920	7·204	7·343	7·750	8·143	8·272	8·525	8·773	8·896	9·256
1/ 31·3	7·5799	8·4018	9·2765	10·205	12·229	14·484	15·701	19·721	24·320	25·987	29·527	33·350	35·370	41·875
0·03400	6·367	6·525	6·680	6·834	7·134	7·426	7·569	7·989	8·394	8·527	8·788	9·043	9·170	9·541
1/ 29·4	7·8134	8·6606	9·5622	10·520	12·606	14·930	16·184	20·328	25·069	26·787	30·437	34·377	36·459	43·165
0·03600	6·552	6·714	6·874	7·032	7·341	7·641	7·789	8·220	8·638	8·774	9·043	9·306	9·436	9·818
1/ 27·8	8·0402	8·9119	9·8397	10·825	12·972	15·363	16·654	20·918	25·797	27·565	31·320	35·374	37·517	44·417
0·03800	6·731	6·898	7·063	7·225	7·542	7·851	8·002	8·446	8·875	9·015	9·290	9·561	9·694	10·09
1/ 26·3	8·2607	9·1564	10·110	11·122	13·328	15·785	17·111	21·492	26·504	28·320	32·179	36·344	38·546	45·635
0·04000	6·906	7·078	7·246	7·413	7·738	8·055	8·210	8·665	9·105	9·249	9·532	9·810	9·946	10·35
1/ 25·0	8·4755	9·3944	10·372	11·411	13·674	16·195	17·555	22·051	27·193	29·057	33·015	37·289	39·548	46·821
0·04200	7·077	7·253	7·426	7·596	7·929	8·254	8·413	8·879	9·330	9·478	9·768	10·05	10·19	10·61
1/ 23·8	8·6850	9·6266	10·629	11·693	14·012	16·595	17·989	22·596	27·865	29·775	33·831	38·211	40·525	47·978
0·04400	7·244	7·423	7·600	7·775	8·116	8·448	8·611	9·089	9·550	9·701	9·998	10·29	10·43	10·86
1/ 22·7	8·8895	9·8534	10·879	11·968	14·342	16·986	18·413	23·128	28·521	30·476	34·627	39·110	41·479	49·108
0·04600	7·407	7·590	7·771	7·950	8·299	8·638	8·805	9·293	9·765	9·919	10·22	10·52	10·67	11·10
1/ 21·7	9·0895	10·075	11·124	12·237	14·665	17·368	18·827	23·648	29·163	31·161	35·406	39·990	42·412	50·212
0·04800	7·566	7·754	7·939	8·121	8·477	8·824	8·994	9·493	9·975	10·13	10·44	10·75	10·90	11·34
1/ 20·8	9·2851	10·292	11·363	12·501	14·980	17·742	19·232	24·157	29·790	31·832	36·168	40·850	43·324	51·292
0·05000	7·722	7·914	8·102	8·288	8·652	9·006	9·180	9·689	10·18	10·34	10·66	10·97	11·12	11·57
1/ 20·0	9·4768	10·504	11·598	12·759	15·289	18·108	19·629	24·655	30·405	32·489	36·914	41·693	44·218	52·351
0·05250	7·913	8·109	8·303	8·493	8·866	9·229	9·407	9·928	10·43	10·60	10·92	11·24	11·40	11·86
1/ 19·0	9·7110	10·764	11·884	13·074	15·667	18·556	20·114	25·264	31·156	33·291	37·827	42·724	45·311	53·644
0·05500	8·100	8·300	8·498	8·693	9·075	9·446	9·628	10·16	10·68	10·85	11·18	11·50	11·66	12·14
1/ 18·2	9·9397	11·017	12·164	13·382	16·036	18·993	20·588	25·859	31·890	34·075	38·717	43·729	46·378	54·907
0·05750	8·282	8·487	8·689	8·889	9·279	9·659	9·845	10·39	10·92	11·09	11·43	11·76	11·93	12·41
1/ 17·4	10·163	11·265	12·438	13·683	16·397	19·420	21·051	26·441	32·607	34·842	39·588	44·713	47·421	56·142
0·06000	8·460	8·670	8·876	9·080	9·478	9·866	10·06	10·61	11·15	11·33	11·68	12·02	12·18	12·68
1/ 16·7	10·382	11·508	12·706	13·977	16·750	19·838	21·504	27·010	33·309	35·591	40·440	45·675	48·441	57·350
0·06250	8·635	8·849	9·060	9·267	9·674	10·07	10·26	10·83	11·38	11·56	11·92	12·26	12·43	12·94
1/ 16·0	10·596	11·745	12·968	14·266	17·095	20·247	21·947	27·567	33·996	36·326	41·274	46·617	49·441	58·533
0·06500	8·806	9·024	9·239	9·451	9·866	10·27	10·47	11·05	11·61	11·79	12·15	12·51	12·68	13·19
1/ 15·4	10·806	11·978	13·225	14·549	17·434	20·648	22·382	28·113	34·670	37·045	42·092	47·541	50·420	59·693
0·06750	8·974	9·196	9·415	9·631	10·05	10·47	10·67	11·26	11·83	12·02	12·38	12·74	12·92	13·45
1/ 14·8	11·012	12·206	13·477	14·826	17·767	21·042	22·809	28·649	35·330	37·751	42·894	48·447	51·381	60·830
0·07000	9·138	9·365	9·588	9·808	10·24	10·66	10·86	11·47	12·05	12·24	12·61	12·98	13·16	13·69
1/ 14·3	11·214	12·430	13·724	15·098	18·093	21·428	23·228	29·175	35·979	38·445	43·682	49·336	52·324	61·947
0·07250	9·300	9·531	9·758	9·982	10·42	10·85	11·06	11·67	12·26	12·45	12·83	13·21	13·39	13·94
1/ 13·8	11·413	12·650	13·967	15·366	18·413	21·808	23·639	29·692	36·616	39·125	44·455	50·210	53·251	63·044
0·07500	9·459	9·694	9·925	10·15	10·60	11·03	11·24	11·87	12·47	12·67	13·05	13·43	13·62	14·17
1/ 13·3	11·608	12·867	14·206	15·628	18·728	22·181	24·044	30·200	37·242	39·795	45·216	51·069	54·162	64·122
0·08000	9·770	10·01	10·25	10·49	10·95	11·39	11·61	12·26	12·88	13·08	13·48	13·88	14·07	14·64
1/ 12·5	11·989	13·289	14·672	16·141	19·343	22·908	24·832	31·191	38·464	41·100	46·699	52·745	55·939	66·226
0·08500	10·07	10·32	10·57	10·81	11·28	11·74	11·97	12·63	13·28	13·49	13·90	14·30	14·50	15·09
1/ 11·8	12·358	13·698	15·124	16·638	19·938	23·614	25·597	32·151	39·649	42·366	48·137	54·369	57·661	68·265
0·09000	10·36	10·62	10·87	11·12	11·61	12·09	12·32	13·00	13·66	13·88	14·30	14·72	14·92	15·53
1/ 11·1	12·717	14·096	15·563	17·121	20·517	24·299	26·340	33·084	40·799	43·595	49·533	55·945	59·333	70·245
0·09500	10·65	10·91	11·17	11·43	11·93	12·42	12·66	13·36	14·04	14·26	14·69	15·12	15·33	15·95
1/ 10·5	13·066	14·482	15·990	17·590	21·079	24·965	27·062	33·991	41·917	44·790	50·891	57·479	60·960	72·171
0·10000	10·92	11·19	11·46	11·72	12·24	12·74	12·98	13·70	14·40	14·63	15·07	15·51	15·73	16·37
1/ 10·0	13·405	14·859	16·405	18·048	21·627	25·614	27·765	34·874	43·007	45·954	52·214	58·973	62·544	74·046
	0·83	0·83	0·84	0·84	0·84	0·85	0·85	0·85	0·86	0·86	0·86	0·86	0·86	0·87

$V_{r(0·5)medial}$ for half-full circular pipes.

A43
(p.1 of 6)

$k_s = 3{\cdot}0$ mm
S = 0·00010 to 0·00030

ie hydraulic gradient =
1 in 10000 to 1 in 3333

Water (or sewage) at 15°C;
full bore conditions.

velocities in ms^{-1}
discharges in m^3s^{-1}

Gradient	(Equivalent) Pipe diameters in mm													
	1250	**1300**	**1350**	**1400**	**1500**	**1600**	**1650**	**1800**	**1950**	**2000**	**2100**	**2200**	**2250**	**2400**
0·000100	0·313	0·321	0·329	0·336	0·352	0·366	0·374	0·395	0·415	0·422	0·435	0·448	0·455	0·473
1/ 10000	0·3839	0·4258	0·4705	0·5179	0·6213	0·7366	0·7988	1·0047	1·2404	1·3259	1·5076	1·7039	1·8077	2·1420
0·000105	0·321	0·329	0·337	0·345	0·360	0·375	0·383	0·405	0·426	0·433	0·446	0·459	0·466	0·485
1/ 9524	0·3935	0·4364	0·4822	0·5308	0·6367	0·7549	0·8187	1·0297	1·2713	1·3589	1·5451	1·7463	1·8526	2·1952
0·000110	0·328	0·337	0·345	0·353	0·369	0·384	0·392	0·414	0·436	0·443	0·457	0·470	0·477	0·497
1/ 9091	0·4028	0·4468	0·4936	0·5434	0·6519	0·7728	0·8381	1·0541	1·3014	1·3911	1·5817	1·7876	1·8965	2·2472
0·000115	0·336	0·344	0·353	0·361	0·377	0·393	0·401	0·424	0·446	0·453	0·467	0·481	0·488	0·508
1/ 8696	0·4120	0·4569	0·5048	0·5557	0·6666	0·7903	0·8571	1·0779	1·3309	1·4226	1·6175	1·8280	1·9393	2·2980
0·000120	0·343	0·352	0·360	0·369	0·385	0·402	0·410	0·433	0·455	0·463	0·477	0·491	0·498	0·519
1/ 8333	0·4209	0·4668	0·5158	0·5677	0·6811	0·8074	0·8756	1·1013	1·3597	1·4534	1·6525	1·8676	1·9813	2·3477
0·000125	0·350	0·359	0·368	0·376	0·393	0·410	0·418	0·442	0·465	0·472	0·487	0·501	0·509	0·530
1/ 8000	0·4297	0·4765	0·5265	0·5795	0·6952	0·8242	0·8938	1·1241	1·3879	1·4835	1·6868	1·9063	2·0224	2·3964
0·000130	0·357	0·366	0·375	0·384	0·401	0·418	0·426	0·451	0·474	0·482	0·497	0·511	0·519	0·540
1/ 7692	0·4382	0·4861	0·5370	0·5911	0·7091	0·8406	0·9117	1·1465	1·4155	1·5131	1·7204	1·9443	2·0627	2·4441
0·000135	0·364	0·373	0·382	0·391	0·409	0·426	0·435	0·459	0·483	0·491	0·506	0·521	0·529	0·551
1/ 7407	0·4466	0·4954	0·5473	0·6024	0·7227	0·8568	0·9291	1·1685	1·4427	1·5421	1·7533	1·9816	2·1022	2·4909
0·000140	0·371	0·380	0·389	0·399	0·417	0·434	0·443	0·468	0·492	0·500	0·516	0·531	0·538	0·561
1/ 7143	0·4549	0·5046	0·5574	0·6136	0·7360	0·8726	0·9463	1·1901	1·4693	1·5705	1·7857	2·0181	2·1410	2·5369
0·000145	0·377	0·387	0·396	0·406	0·424	0·442	0·450	0·476	0·501	0·509	0·525	0·540	0·548	0·571
1/ 6897	0·4630	0·5136	0·5674	0·6245	0·7492	0·8881	0·9632	1·2113	1·4955	1·5985	1·8175	2·0540	2·1791	2·5820
0·000150	0·384	0·394	0·403	0·413	0·431	0·449	0·458	0·484	0·509	0·518	0·534	0·550	0·557	0·581
1/ 6667	0·4710	0·5224	0·5771	0·6353	0·7621	0·9034	0·9797	1·2321	1·5212	1·6260	1·8487	2·0893	2·2165	2·6264
0·000160	0·396	0·407	0·417	0·426	0·445	0·464	0·473	0·500	0·526	0·535	0·551	0·568	0·576	0·600
1/ 6250	0·4866	0·5397	0·5962	0·6562	0·7872	0·9332	1·0121	1·2728	1·5714	1·6796	1·9097	2·1582	2·2896	2·7130
0·000170	0·409	0·419	0·429	0·440	0·459	0·479	0·488	0·516	0·542	0·551	0·568	0·585	0·594	0·618
1/ 5882	0·5017	0·5564	0·6147	0·6766	0·8116	0·9621	1·0434	1·3122	1·6200	1·7316	1·9688	2·2250	2·3604	2·7969
0·000180	0·421	0·431	0·442	0·452	0·473	0·492	0·502	0·531	0·558	0·567	0·585	0·602	0·611	0·636
1/ 5556	0·5163	0·5726	0·6326	0·6963	0·8353	0·9902	1·0738	1·3505	1·6672	1·7821	2·0261	2·2898	2·4292	2·8783
0·000190	0·432	0·443	0·454	0·465	0·486	0·506	0·516	0·545	0·574	0·583	0·601	0·619	0·628	0·654
1/ 5263	0·5305	0·5884	0·6501	0·7155	0·8583	1·0175	1·1034	1·3877	1·7131	1·8312	2·0819	2·3529	2·4961	2·9576
0·000200	0·444	0·455	0·466	0·477	0·498	0·519	0·530	0·560	0·589	0·598	0·617	0·635	0·644	0·671
1/ 5000	0·5444	0·6038	0·6671	0·7342	0·8808	1·0441	1·1323	1·4239	1·7579	1·8790	2·1363	2·4143	2·5613	3·0348
0·000210	0·455	0·466	0·478	0·489	0·511	0·532	0·543	0·573	0·603	0·613	0·632	0·651	0·660	0·687
1/ 4762	0·5580	0·6188	0·6836	0·7525	0·9026	1·0700	1·1604	1·4593	1·8015	1·9256	2·1893	2·4742	2·6248	3·1100
0·000220	0·465	0·477	0·489	0·500	0·523	0·545	0·556	0·587	0·617	0·627	0·647	0·666	0·676	0·704
1/ 4545	0·5712	0·6335	0·6998	0·7703	0·9240	1·0953	1·1878	1·4938	1·8441	1·9711	2·2411	2·5327	2·6868	3·1835
0·000230	0·476	0·488	0·500	0·512	0·535	0·557	0·568	0·600	0·631	0·642	0·662	0·681	0·691	0·720
1/ 4348	0·5841	0·6478	0·7156	0·7877	0·9449	1·1201	1·2147	1·5275	1·8858	2·0156	2·2917	2·5899	2·7475	3·2554
0·000240	0·486	0·499	0·511	0·523	0·546	0·569	0·580	0·613	0·645	0·655	0·676	0·696	0·706	0·735
1/ 4167	0·5967	0·6618	0·7311	0·8047	0·9653	1·1443	1·2409	1·5605	1·9265	2·0592	2·3412	2·6458	2·8068	3·3257
0·000250	0·496	0·509	0·521	0·534	0·558	0·581	0·592	0·626	0·658	0·669	0·690	0·710	0·721	0·750
1/ 4000	0·6091	0·6755	0·7463	0·8214	0·9853	1·1680	1·2667	1·5929	1·9664	2·1019	2·3897	2·7006	2·8650	3·3946
0·000260	0·506	0·519	0·532	0·544	0·569	0·592	0·604	0·638	0·672	0·682	0·704	0·725	0·735	0·765
1/ 3846	0·6212	0·6890	0·7611	0·8378	1·0049	1·1913	1·2919	1·6246	2·0055	2·1437	2·4372	2·7543	2·9219	3·4620
0·000270	0·516	0·529	0·542	0·555	0·580	0·604	0·616	0·651	0·684	0·695	0·717	0·738	0·749	0·780
1/ 3704	0·6331	0·7022	0·7757	0·8538	1·0242	1·2141	1·3166	1·6557	2·0439	2·1847	2·4838	2·8070	2·9778	3·5283
0·000280	0·525	0·539	0·552	0·565	0·590	0·615	0·627	0·663	0·697	0·708	0·730	0·752	0·763	0·794
1/ 3571	0·6448	0·7152	0·7900	0·8696	1·0431	1·2365	1·3409	1·6862	2·0816	2·2249	2·5296	2·8587	3·0327	3·5932
0·000290	0·535	0·548	0·562	0·575	0·601	0·626	0·638	0·674	0·709	0·721	0·743	0·765	0·776	0·808
1/ 3448	0·6563	0·7279	0·8041	0·8851	1·0616	1·2585	1·3647	1·7162	2·1186	2·2645	2·5745	2·9095	3·0866	3·6571
0·000300	0·544	0·558	0·571	0·585	0·611	0·637	0·649	0·686	0·722	0·733	0·756	0·779	0·790	0·822
1/ 3333	0·6676	0·7404	0·8179	0·9003	1·0799	1·2801	1·3882	1·7456	2·1549	2·3034	2·6187	2·9595	3·1395	3·7198
	0·80	0·80	0·80	0·80	0·80	0·81	0·81	0·81	0·82	0·82	0·82	0·82	0·83	0·83

$V_{r(0{\cdot}5)medial}$ for half-full circular pipes.

$k_s = 3{\cdot}0$ mm S = 0·00010 to 0·00030

$k_s = 3{\cdot}0$ mm
S = 0·00030 to 0·00100

ie hydraulic gradient =
1 in 3333 to 1 in 1000

Water (or sewage) at 15°C;
full bore conditions.

velocities in ms^{-1}
discharges in m^3s^{-1}

Gradient	(Equivalent) Pipe diameters in mm													
	1250	1300	1350	1400	1500	1600	1650	1800	1950	2000	2100	2200	2250	2400
0·000300	0·544	0·558	0·571	0·585	0·611	0·637	0·649	0·686	0·722	0·733	0·756	0·779	0·790	0·822
1/ 3333	0·6676	0·7404	0·8179	0·9003	1·0799	1·2801	1·3882	1·7456	2·1549	2·3034	2·6187	2·9595	3·1395	3·7198
0·000320	0·562	0·576	0·590	0·604	0·631	0·658	0·671	0·709	0·745	0·757	0·781	0·804	0·816	0·849
1/ 3125	0·6896	0·7648	0·8449	0·9299	1·1155	1·3223	1·4339	1·8031	2·2259	2·3792	2·7050	3·0569	3·2429	3·8423
0·000340	0·579	0·594	0·609	0·623	0·651	0·678	0·691	0·730	0·768	0·781	0·805	0·829	0·841	0·876
1/ 2941	0·7109	0·7885	0·8710	0·9587	1·1499	1·3631	1·4782	1·8588	2·2947	2·4527	2·7885	3·1513	3·3431	3·9609
0·000360	0·596	0·611	0·626	0·641	0·670	0·698	0·711	0·752	0·791	0·803	0·829	0·853	0·865	0·901
1/ 2778	0·7317	0·8114	0·8964	0·9866	1·1834	1·4028	1·5213	1·9130	2·3615	2·5241	2·8697	3·2430	3·4403	4·0762
0·000380	0·613	0·628	0·643	0·659	0·688	0·717	0·731	0·772	0·812	0·826	0·851	0·877	0·889	0·926
1/ 2632	0·7518	0·8338	0·9211	1·0138	1·2160	1·4414	1·5631	1·9656	2·4264	2·5935	2·9486	3·3322	3·5349	4·1882
0·000400	0·629	0·645	0·660	0·676	0·706	0·736	0·750	0·793	0·834	0·847	0·873	0·899	0·912	0·950
1/ 2500	0·7714	0·8555	0·9451	1·0402	1·2477	1·4790	1·6039	2·0168	2·4897	2·6611	3·0255	3·4190	3·6271	4·2974
0·000420	0·644	0·661	0·677	0·693	0·724	0·754	0·769	0·812	0·854	0·868	0·895	0·922	0·935	0·973
1/ 2381	0·7906	0·8768	0·9685	1·0660	1·2787	1·5157	1·6437	2·0668	2·5514	2·7271	3·1004	3·5038	3·7169	4·4038
0·000440	0·659	0·676	0·693	0·709	0·741	0·772	0·787	0·831	0·874	0·889	0·916	0·943	0·957	0·996
1/ 2273	0·8093	0·8975	0·9914	1·0912	1·3089	1·5515	1·6825	2·1157	2·6116	2·7915	3·1736	3·5865	3·8047	4·5078
0·000460	0·674	0·691	0·708	0·725	0·757	0·789	0·805	0·850	0·894	0·909	0·937	0·965	0·978	1·019
1/ 2174	0·8275	0·9178	1·0138	1·1159	1·3384	1·5865	1·7205	2·1634	2·6705	2·8544	3·2452	3·6673	3·8905	4·6094
0·000480	0·689	0·706	0·724	0·741	0·774	0·806	0·822	0·869	0·914	0·928	0·957	0·986	1·000	1·041
1/ 2083	0·8454	0·9376	1·0357	1·1400	1·3673	1·6208	1·7576	2·2101	2·7282	2·9160	3·3152	3·7465	3·9744	4·7089
0·000500	0·703	0·721	0·739	0·756	0·790	0·823	0·839	0·886	0·932	0·947	0·977	1·006	1·020	1·062
1/ 2000	0·8629	0·9570	1·0572	1·1636	1·3956	1·6543	1·7940	2·2558	2·7846	2·9764	3·3838	3·8240	4·0566	4·8062
0·000525	0·721	0·739	0·757	0·775	0·809	0·843	0·860	0·908	0·956	0·971	1·001	1·031	1·046	1·089
1/ 1905	0·8843	0·9807	1·0834	1·1924	1·4302	1·6953	1·8384	2·3117	2·8536	3·0501	3·4676	3·9187	4·1571	4·9253
0·000550	0·738	0·756	0·775	0·793	0·828	0·863	0·880	0·930	0·978	0·994	1·025	1·055	1·070	1·114
1/ 1818	0·9052	1·0039	1·1090	1·2206	1·4640	1·7354	1·8818	2·3663	2·9210	3·1221	3·5495	4·0112	4·2552	5·0415
0·000575	0·754	0·773	0·792	0·811	0·847	0·883	0·900	0·951	1·000	1·016	1·048	1·079	1·094	1·140
1/ 1739	0·9256	1·0265	1·1340	1·2481	1·4970	1·7745	1·9243	2·4196	2·9868	3·1925	3·6295	4·1016	4·3511	5·1551
0·000600	0·771	0·790	0·809	0·828	0·865	0·902	0·919	0·971	1·022	1·038	1·070	1·102	1·118	1·164
1/ 1667	0·9456	1·0487	1·1585	1·2751	1·5293	1·8128	1·9658	2·4718	3·0512	3·2613	3·7077	4·1900	4·4449	5·2663
0·000625	0·787	0·806	0·826	0·845	0·883	0·920	0·938	0·991	1·043	1·060	1·093	1·125	1·141	1·188
1/ 1600	0·9652	1·0704	1·1824	1·3014	1·5610	1·8503	2·0065	2·5230	3·1143	3·3288	3·7844	4·2767	4·5368	5·3751
0·000650	0·802	0·822	0·843	0·862	0·901	0·939	0·957	1·011	1·064	1·081	1·114	1·147	1·164	1·212
1/ 1538	0·9844	1·0917	1·2059	1·3273	1·5920	1·8871	2·0463	2·5731	3·1762	3·3949	3·8596	4·3616	4·6269	5·4819
0·000675	0·817	0·838	0·859	0·879	0·918	0·956	0·975	1·030	1·084	1·101	1·136	1·169	1·186	1·235
1/ 1481	1·0032	1·1126	1·2290	1·3527	1·6224	1·9231	2·0854	2·6222	3·2369	3·4597	3·9333	4·4449	4·7153	5·5865
0·000700	0·833	0·854	0·874	0·895	0·935	0·974	0·993	1·049	1·104	1·122	1·157	1·191	1·208	1·258
1/ 1429	1·0217	1·1331	1·2516	1·3776	1·6523	1·9585	2·1238	2·6705	3·2964	3·5234	4·0057	4·5267	4·8020	5·6893
0·000725	0·847	0·869	0·890	0·911	0·952	0·991	1·011	1·068	1·123	1·141	1·177	1·212	1·229	1·280
1/ 1379	1·0398	1·1532	1·2739	1·4021	1·6816	1·9933	2·1615	2·7179	3·3549	3·5859	4·0767	4·6070	4·8872	5·7903
0·000750	0·862	0·884	0·905	0·926	0·968	1·008	1·028	1·086	1·143	1·161	1·197	1·233	1·250	1·302
1/ 1333	1·0577	1·1730	1·2957	1·4261	1·7105	2·0275	2·1986	2·7645	3·4125	3·6474	4·1466	4·6859	4·9710	5·8895
0·000800	0·890	0·913	0·935	0·957	1·000	1·042	1·062	1·122	1·180	1·199	1·237	1·273	1·291	1·345
1/ 1250	1·0925	1·2116	1·3384	1·4730	1·7668	2·0942	2·2709	2·8554	3·5247	3·7673	4·2830	4·8400	5·1344	6·0831
0·000850	0·918	0·941	0·964	0·986	1·031	1·074	1·095	1·157	1·217	1·236	1·275	1·313	1·331	1·386
1/ 1176	1·1262	1·2490	1·3797	1·5185	1·8213	2·1588	2·3410	2·9435	3·6334	3·8836	4·4151	4·9893	5·2928	6·2707
0·000900	0·944	0·968	0·992	1·015	1·061	1·105	1·127	1·190	1·252	1·272	1·312	1·351	1·370	1·426
1/ 1111	1·1590	1·2853	1·4198	1·5627	1·8743	2·2216	2·4091	3·0291	3·7390	3·9964	4·5434	5·1343	5·4466	6·4529
0·000950	0·970	0·995	1·019	1·043	1·090	1·135	1·158	1·223	1·286	1·307	1·348	1·388	1·407	1·466
1/ 1053	1·1908	1·3206	1·4588	1·6056	1·9258	2·2826	2·4753	3·1123	3·8417	4·1062	4·6682	5·2753	5·5962	6·6301
0·001000	0·996	1·021	1·046	1·070	1·118	1·165	1·188	1·255	1·320	1·341	1·383	1·424	1·444	1·504
1/ 1000	1·2219	1·3550	1·4968	1·6474	1·9759	2·3421	2·5397	3·1934	3·9418	4·2131	4·7897	5·4126	5·7419	6·8027
	0·80	0·80	0·80	0·80	0·81	0·81	0·81	0·81	0·82	0·82	0·82	0·82	0·83	0·83

$V_{r(0{\cdot}5)medial}$ for half-full circular pipes.

A43

(p.3 of 6)

k_s = 3·0 mm
S = 0·00100 to 0·00300

ie hydraulic gradient =
1 in 1000 to 1 in 333

Water (or sewage) at 15°C;
full bore conditions.

velocities in ms^{-1}
discharges in m^3s^{-1}

Gradient **(Equivalent) Pipe diameters in mm**

Gradient	1250	1300	1350	1400	1500	1600	1650	1800	1950	2000	2100	2200	2250	2400
0·00100	0·996	1·021	1·046	1·070	1·118	1·165	1·188	1·255	1·320	1·341	1·383	1·424	1·444	1·504
1/ 1000	1·2219	1·3550	1·4968	1·6474	1·9759	2·3421	2·5397	3·1934	3·9418	4·2131	4·7897	5·4126	5·7419	6·8027
0·00105	1·020	1·046	1·072	1·097	1·146	1·194	1·217	1·286	1·353	1·374	1·417	1·459	1·480	1·541
1/ 952	1·2521	1·3886	1·5339	1·6882	2·0249	2·4000	2·6026	3·2724	4·0393	4·3174	4·9083	5·5466	5·8840	6·9710
0·00110	1·044	1·071	1·097	1·123	1·173	1·222	1·246	1·316	1·384	1·407	1·451	1·494	1·515	1·577
1/ 909	1·2817	1·4214	1·5701	1·7281	2·0726	2·4567	2·6640	3·3496	4·1346	4·4192	5·0240	5·6774	6·0227	7·1354
0·00115	1·068	1·095	1·122	1·148	1·199	1·249	1·274	1·346	1·416	1·438	1·483	1·527	1·549	1·613
1/ 870	1·3106	1·4534	1·6055	1·7670	2·1193	2·5120	2·7240	3·4250	4·2277	4·5188	5·1372	5·8052	6·1583	7·2961
0·00120	1·091	1·119	1·146	1·173	1·225	1·276	1·301	1·375	1·446	1·469	1·515	1·560	1·582	1·648
1/ 833	1·3388	1·4848	1·6401	1·8051	2·1650	2·5662	2·7827	3·4989	4·3188	4·6162	5·2479	5·9303	6·2910	7·4533
0·00125	1·114	1·142	1·170	1·197	1·250	1·303	1·328	1·403	1·476	1·500	1·546	1·592	1·615	1·682
1/ 800	1·3665	1·5154	1·6740	1·8424	2·2098	2·6192	2·8403	3·5712	4·4081	4·7115	5·3563	6·0529	6·4210	7·6073
0·00130	1·136	1·164	1·193	1·221	1·275	1·329	1·355	1·431	1·505	1·529	1·577	1·624	1·647	1·715
1/ 769	1·3937	1·5455	1·7073	1·8790	2·2536	2·6712	2·8966	3·6421	4·4956	4·8050	5·4626	6·1730	6·5484	7·7582
0·00135	1·157	1·187	1·216	1·244	1·300	1·354	1·381	1·459	1·534	1·559	1·607	1·655	1·678	1·748
1/ 741	1·4203	1·5751	1·7399	1·9149	2·2967	2·7222	2·9519	3·7116	4·5814	4·8967	5·5669	6·2908	6·6734	7·9062
0·00140	1·179	1·208	1·238	1·267	1·324	1·379	1·406	1·485	1·562	1·587	1·637	1·685	1·709	1·780
1/ 714	1·4464	1·6040	1·7719	1·9501	2·3389	2·7723	3·0062	3·7798	4·6656	4·9868	5·6692	6·4064	6·7961	8·0516
0·00145	1·200	1·230	1·260	1·289	1·347	1·403	1·431	1·512	1·590	1·615	1·666	1·715	1·740	1·811
1/ 690	1·4721	1·6325	1·8033	1·9847	2·3804	2·8214	3·0595	3·8469	4·7484	5·0752	5·7697	6·5200	6·9166	8·1943
0·00150	1·220	1·251	1·281	1·311	1·370	1·427	1·455	1·538	1·617	1·643	1·694	1·745	1·769	1·842
1/ 667	1·4973	1·6605	1·8342	2·0187	2·4212	2·8698	3·1120	3·9128	4·8297	5·1621	5·8685	6·6317	7·0350	8·3346
0·00160	1·260	1·292	1·324	1·355	1·415	1·474	1·503	1·588	1·670	1·697	1·750	1·802	1·827	1·903
1/ 625	1·5465	1·7151	1·8945	2·0851	2·5008	2·9641	3·2142	4·0413	4·9884	5·3317	6·0614	6·8495	7·2661	8·6084
0·00170	1·299	1·332	1·364	1·396	1·459	1·520	1·550	1·637	1·722	1·749	1·804	1·857	1·884	1·962
1/ 588	1·5942	1·7680	1·9529	2·1494	2·5779	3·0555	3·3134	4·1660	5·1422	5·4961	6·2482	7·0607	7·4901	8·8738
0·00180	1·337	1·371	1·404	1·437	1·501	1·564	1·595	1·685	1·772	1·800	1·856	1·911	1·939	2·018
1/ 556	1·6406	1·8193	2·0097	2·2119	2·6528	3·1443	3·4096	4·2870	5·2915	5·6557	6·4297	7·2657	7·7076	9·1315
0·00190	1·374	1·408	1·443	1·476	1·542	1·607	1·638	1·731	1·820	1·850	1·907	1·964	1·992	2·074
1/ 526	1·6856	1·8693	2·0649	2·2726	2·7256	3·2306	3·5032	4·4046	5·4368	5·8110	6·6061	7·4652	7·9192	9·3821
0·00200	1·409	1·445	1·480	1·515	1·583	1·649	1·681	1·776	1·868	1·898	1·957	2·015	2·044	2·128
1/ 500	1·7295	1·9180	2·1186	2·3318	2·7966	3·3147	3·5944	4·5193	5·5782	5·9622	6·7780	7·6594	8·1252	9·6261
0·00210	1·444	1·481	1·517	1·552	1·622	1·689	1·723	1·820	1·914	1·945	2·005	2·065	2·094	2·180
1/ 476	1·7723	1·9654	2·1711	2·3895	2·8658	3·3967	3·6833	4·6311	5·7162	6·1097	6·9457	7·8488	8·3262	9·8642
0·00220	1·478	1·516	1·553	1·589	1·660	1·729	1·763	1·863	1·959	1·991	2·053	2·113	2·143	2·232
1/ 455	1·8141	2·0118	2·2222	2·4458	2·9334	3·4768	3·7701	4·7402	5·8509	6·2537	7·1094	8·0338	8·5224	10·097
0·00230	1·512	1·550	1·587	1·625	1·697	1·768	1·803	1·905	2·003	2·035	2·099	2·161	2·192	2·282
1/ 435	1·8550	2·0571	2·2723	2·5009	2·9994	3·5550	3·8550	4·8469	5·9827	6·3944	7·2694	8·2146	8·7142	10·324
0·00240	1·544	1·583	1·622	1·660	1·734	1·806	1·842	1·946	2·046	2·079	2·144	2·208	2·239	2·331
1/ 417	1·8949	2·1014	2·3213	2·5548	3·0640	3·6316	3·9381	4·9514	6·1115	6·5322	7·4260	8·3915	8·9019	10·546
0·00250	1·576	1·616	1·655	1·694	1·770	1·844	1·880	1·986	2·089	2·122	2·188	2·253	2·285	2·379
1/ 400	1·9341	2·1448	2·3692	2·6075	3·1273	3·7067	4·0194	5·0536	6·2377	6·6671	7·5793	8·5648	9·0857	10·764
0·00260	1·607	1·648	1·688	1·727	1·805	1·880	1·917	2·025	2·130	2·164	2·232	2·298	2·330	2·427
1/ 385	1·9725	2·1874	2·4162	2·6593	3·1894	3·7802	4·0991	5·1538	6·3614	6·7993	7·7296	8·7347	9·2659	10·977
0·00270	1·638	1·679	1·720	1·760	1·839	1·916	1·954	2·064	2·171	2·206	2·274	2·342	2·375	2·473
1/ 370	2·0101	2·2291	2·4623	2·7100	3·2502	3·8523	4·1774	5·2522	6·4828	6·9290	7·8771	8·9013	9·4426	11·187
0·00280	1·668	1·710	1·752	1·793	1·873	1·951	1·990	2·102	2·211	2·246	2·316	2·385	2·418	2·518
1/ 357	2·0471	2·2701	2·5076	2·7598	3·3100	3·9231	4·2541	5·3487	6·6019	7·0563	8·0218	9·0648	9·6161	11·392
0·00290	1·698	1·741	1·783	1·825	1·906	1·986	2·025	2·139	2·250	2·286	2·357	2·427	2·461	2·563
1/ 345	2·0834	2·3104	2·5521	2·8088	3·3686	3·9927	4·3295	5·4435	6·7189	7·1814	8·1640	9·2255	9·7865	11·594
0·00300	1·727	1·770	1·813	1·856	1·939	2·020	2·059	2·176	2·288	2·325	2·397	2·468	2·503	2·607
1/ 333	2·1190	2·3499	2·5958	2·8569	3·4263	4·0610	4·4037	5·5367	6·8340	7·3043	8·3037	9·3834	9·9540	11·793
	0·80	0·80	0·80	0·80	0·81	0·81	0·81	0·82	0·82	0·82	0·82	0·83	0·83	0·83

$V_{r(0·5)medial}$ for half-full circular pipes.

$k_s = 3{\cdot}0$ **mm**
S = 0·00300 to 0·01000

ie hydraulic gradient =
1 in 333 to 1 in 100

Water (or sewage) at 15°C;
full bore conditions.

velocities in ms^{-1}
discharges in m^3s^{-1}

Gradient	(Equivalent) Pipe diameters in mm													
	1250	**1300**	**1350**	**1400**	**1500**	**1600**	**1650**	**1800**	**1950**	**2000**	**2100**	**2200**	**2250**	**2400**
0·00300	1·727	1·770	1·813	1·856	1·939	2·020	2·059	2·176	2·288	2·325	2·397	2·468	2·503	2·607
1/ 333	2·1190	2·3499	2·5958	2·8569	3·4263	4·0610	4·4037	5·5367	6·8340	7·3043	8·3037	9·3834	9·9540	11·793
0·00320	1·783	1·829	1·873	1·917	2·003	2·086	2·127	2·247	2·363	2·401	2·476	2·550	2·586	2·692
1/ 313	2·1887	2·4271	2·6810	2·9507	3·5389	4·1944	4·5483	5·7185	7·0584	7·5442	8·5764	9·6915	10·281	12·180
0·00340	1·838	1·885	1·931	1·976	2·064	2·150	2·193	2·316	2·436	2·475	2·552	2·628	2·665	2·775
1/ 294	2·2561	2·5020	2·7637	3·0417	3·6479	4·3237	4·6885	5·8948	7·2759	7·7767	8·8407	9·9901	10·598	12·555
0·00360	1·892	1·940	1·987	2·033	2·124	2·213	2·256	2·384	2·507	2·547	2·627	2·704	2·743	2·856
1/ 278	2·3216	2·5746	2·8439	3·1300	3·7539	4·4492	4·8246	6·0659	7·4871	8·0024	9·0973	10·280	10·905	12·920
0·00380	1·944	1·993	2·041	2·089	2·183	2·274	2·318	2·449	2·576	2·617	2·699	2·779	2·818	2·934
1/ 263	2·3854	2·6453	2·9220	3·2159	3·8569	4·5713	4·9570	6·2323	7·6925	8·2219	9·3469	10·562	11·204	13·274
0·00400	1·994	2·045	2·094	2·143	2·239	2·333	2·379	2·513	2·643	2·685	2·769	2·851	2·891	3·010
1/ 250	2·4474	2·7141	2·9980	3·2995	3·9572	4·6902	5·0859	6·3944	7·8926	8·4358	9·5900	10·837	11·496	13·619
0·00420	2·044	2·095	2·146	2·196	2·295	2·390	2·437	2·575	2·708	2·752	2·837	2·921	2·963	3·085
1/ 238	2·5080	2·7812	3·0721	3·3811	4·0551	4·8062	5·2117	6·5525	8·0877	8·6443	9·8270	11·105	11·780	13·956
0·00440	2·092	2·145	2·197	2·248	2·349	2·447	2·495	2·636	2·772	2·816	2·904	2·990	3·033	3·158
1/ 227	2·5671	2·8468	3·1445	3·4608	4·1506	4·9194	5·3345	6·7069	8·2783	8·8480	10·059	11·366	12·058	14·285
0·00460	2·139	2·193	2·246	2·299	2·402	2·502	2·551	2·695	2·834	2·880	2·969	3·057	3·101	3·229
1/ 217	2·6249	2·9108	3·2153	3·5387	4·2440	5·0301	5·4545	6·8578	8·4645	9·0471	10·285	11·622	12·329	14·606
0·00480	2·185	2·240	2·295	2·348	2·453	2·556	2·606	2·753	2·895	2·942	3·033	3·123	3·167	3·298
1/ 208	2·6814	2·9735	3·2846	3·6149	4·3354	5·1384	5·5720	7·0055	8·6468	9·2418	10·506	11·872	12·594	14·920
0·00500	2·230	2·286	2·342	2·397	2·504	2·608	2·660	2·810	2·955	3·002	3·096	3·188	3·233	3·366
1/ 200	2·7368	3·0349	3·3524	3·6896	4·4249	5·2445	5·6870	7·1501	8·8253	9·4326	10·723	12·117	12·854	15·228
0·00525	2·285	2·343	2·400	2·456	2·566	2·673	2·725	2·879	3·028	3·077	3·172	3·266	3·313	3·449
1/ 190	2·8044	3·1100	3·4353	3·7808	4·5343	5·3742	5·8276	7·3269	9·0434	9·6658	10·988	12·417	13·172	15·605
0·00550	2·339	2·398	2·457	2·514	2·626	2·736	2·790	2·947	3·099	3·149	3·247	3·343	3·391	3·531
1/ 182	2·8705	3·1833	3·5162	3·8699	4·6412	5·5008	5·9649	7·4995	9·2564	9·8935	11·247	12·709	13·482	15·972
0·00575	2·392	2·452	2·512	2·570	2·685	2·797	2·852	3·013	3·169	3·220	3·320	3·419	3·467	3·610
1/ 174	2·9351	3·2549	3·5953	3·9569	4·7456	5·6246	6·0991	7·6682	9·4647	10·116	11·500	12·995	13·785	16·332
0·00600	2·443	2·505	2·566	2·626	2·743	2·858	2·914	3·078	3·237	3·289	3·392	3·492	3·542	3·688
1/ 167	2·9983	3·3250	3·6728	4·0422	4·8478	5·7457	6·2304	7·8333	9·6685	10·334	11·748	13·275	14·082	16·683
0·00625	2·494	2·557	2·619	2·680	2·800	2·917	2·974	3·142	3·304	3·357	3·462	3·564	3·615	3·764
1/ 160	3·0602	3·3936	3·7486	4·1256	4·9478	5·8643	6·3590	7·9950	9·8680	10·547	11·990	13·549	14·373	17·027
0·00650	2·543	2·607	2·671	2·733	2·855	2·974	3·033	3·204	3·370	3·424	3·530	3·635	3·686	3·838
1/ 154	3·1209	3·4609	3·8229	4·2074	5·0459	5·9805	6·4851	8·1535	10·064	10·756	12·228	13·817	14·658	17·365
0·00675	2·592	2·657	2·722	2·785	2·910	3·031	3·091	3·265	3·434	3·489	3·598	3·704	3·757	3·912
1/ 148	3·1804	3·5269	3·8958	4·2876	5·1422	6·0946	6·6088	8·3089	10·255	10·961	12·461	14·081	14·937	17·696
0·00700	2·639	2·706	2·772	2·836	2·963	3·087	3·148	3·325	3·497	3·553	3·664	3·772	3·826	3·983
1/ 143	3·2389	3·5917	3·9674	4·3664	5·2366	6·2065	6·7302	8·4615	10·444	11·163	12·690	14·339	15·211	18·021
0·00725	2·686	2·754	2·821	2·887	3·016	3·142	3·203	3·384	3·559	3·616	3·729	3·839	3·894	4·054
1/ 138	3·2963	3·6554	4·0377	4·4438	5·3294	6·3165	6·8494	8·6114	10·629	11·360	12·915	14·594	15·481	18·340
0·00750	2·732	2·801	2·869	2·936	3·067	3·195	3·258	3·442	3·620	3·678	3·792	3·905	3·960	4·123
1/ 133	3·3527	3·7179	4·1068	4·5198	5·4206	6·4246	6·9666	8·7588	10·811	11·555	13·136	14·843	15·746	18·654
0·00800	2·822	2·893	2·963	3·033	3·168	3·300	3·365	3·555	3·739	3·799	3·917	4·033	4·090	4·259
1/ 125	3·4627	3·8400	4·2416	4·6682	5·5986	6·6355	7·1953	9·0463	11·166	11·934	13·567	15·330	16·263	19·266
0·00850	2·909	2·982	3·055	3·126	3·266	3·402	3·469	3·664	3·854	3·916	4·038	4·157	4·216	4·390
1/ 118	3·5694	3·9583	4·3723	4·8120	5·7710	6·8399	7·4169	9·3250	11·509	12·302	13·984	15·803	16·763	19·859
0·00900	2·993	3·069	3·143	3·217	3·361	3·501	3·569	3·771	3·966	4·029	4·155	4·278	4·338	4·517
1/ 111	3·6730	4·0732	4·4992	4·9517	5·9385	7·0384	7·6321	9·5955	11·843	12·658	14·390	16·261	17·250	20·436
0·00950	3·075	3·153	3·229	3·305	3·453	3·597	3·667	3·874	4·074	4·140	4·269	4·395	4·457	4·641
1/ 105	3·7738	4·1849	4·6226	5·0875	6·1014	7·2314	7·8415	9·8587	12·168	13·006	14·785	16·707	17·723	20·996
0·01000	3·155	3·235	3·313	3·391	3·542	3·690	3·763	3·975	4·180	4·247	4·380	4·509	4·573	4·762
1/ 100	3·8719	4·2937	4·7428	5·2198	6·2600	7·4194	8·0453	10·115	12·485	13·344	15·169	17·141	18·184	21·542
	0·80	0·80	0·80	0·80	0·81	0·81	0·81	0·82	0·82	0·82	0·82	0·83	0·83	0·83

$V_{r(0{\cdot}5)medial}$ **for half-full circular pipes.**

k_s = **3·0 mm**
S = 0·01000 to 0·03000

ie hydraulic gradient =
1 in 100 to 1 in 33·3

Water (or sewage) at 15°C;
full bore conditions.

velocities in ms^{-1}
discharges in m^3s^{-1}

Gradient **(Equivalent) Pipe diameters in mm**

Gradient	1250	1300	1350	1400	1500	1600	1650	1800	1950	2000	2100	2200	2250	2400
0·01000	3·155	3·235	3·313	3·391	3·542	3·690	3·763	3·975	4·180	4·247	4·380	4·509	4·573	4·762
1/ 100	3·8719	4·2937	4·7428	5·2198	6·2600	7·4194	8·0453	10·115	12·485	13·344	15·169	17·141	18·184	21·542
0·01050	3·233	3·315	3·395	3·475	3·630	3·781	3·856	4·073	4·284	4·352	4·488	4·621	4·686	4·879
1/ 95	3·9676	4·3998	4·8600	5·3488	6·4147	7·6028	8·2442	10·365	12·793	13·673	15·544	17·565	18·633	22·074
0·01100	3·309	3·393	3·475	3·556	3·715	3·870	3·946	4·169	4·385	4·455	4·594	4·730	4·797	4·994
1/ 91	4·0611	4·5035	4·9745	5·4748	6·5658	7·7819	8·4383	10·609	13·094	13·995	15·910	17·978	19·072	22·594
0·01150	3·384	3·469	3·553	3·636	3·799	3·957	4·035	4·263	4·483	4·555	4·697	4·836	4·904	5·107
1/ 87	4·1524	4·6048	5·0864	5·5979	6·7135	7·9569	8·6281	10·848	13·389	14·310	16·268	18·383	19·501	23·102
0·01200	3·457	3·544	3·630	3·715	3·881	4·043	4·122	4·355	4·580	4·653	4·798	4·940	5·010	5·217
1/ 83	4·2418	4·7039	5·1959	5·7184	6·8580	8·1282	8·8139	11·081	13·677	14·618	16·618	18·778	19·920	23·599
0·01250	3·528	3·617	3·705	3·791	3·961	4·126	4·207	4·444	4·674	4·749	4·897	5·042	5·113	5·324
1/ 80	4·3294	4·8010	5·3031	5·8364	6·9995	8·2959	8·9957	11·310	13·959	14·920	16·961	19·166	20·331	24·086
0·01300	3·598	3·689	3·778	3·867	4·039	4·208	4·290	4·533	4·767	4·843	4·994	5·142	5·215	5·430
1/ 77	4·4152	4·8961	5·4082	5·9521	7·1383	8·4603	9·1740	11·534	14·236	15·216	17·297	19·546	20·734	24·563
0·01350	3·666	3·759	3·850	3·940	4·116	4·288	4·372	4·619	4·858	4·936	5·089	5·240	5·314	5·533
1/ 74	4·4994	4·9895	5·5113	6·0656	7·2744	8·6216	9·3489	11·754	14·507	15·506	17·627	19·918	21·129	25·032
0·01400	3·734	3·828	3·921	4·013	4·192	4·367	4·453	4·704	4·947	5·026	5·183	5·336	5·412	5·635
1/ 71	4·5820	5·0811	5·6125	6·1770	7·4079	8·7799	9·5206	11·970	14·774	15·790	17·950	20·284	21·517	25·491
0·01450	3·800	3·896	3·990	4·084	4·266	4·444	4·531	4·787	5·034	5·115	5·274	5·431	5·508	5·735
1/ 69	4·6632	5·1711	5·7120	6·2864	7·5392	8·9354	9·6892	12·182	15·035	16·070	18·268	20·643	21·898	25·943
0·01500	3·865	3·963	4·059	4·154	4·339	4·520	4·609	4·869	5·121	5·203	5·365	5·523	5·602	5·833
1/ 67	4·7429	5·2596	5·8097	6·3939	7·6681	9·0883	9·8550	12·390	15·292	16·345	18·581	20·996	22·273	26·386
0·01600	3·992	4·093	4·192	4·290	4·482	4·668	4·760	5·029	5·289	5·373	5·541	5·705	5·786	6·024
1/ 63	4·8986	5·4322	6·0003	6·6038	7·9198	9·3866	10·178	12·797	15·794	16·881	19·191	21·685	23·004	27·252
0·01700	4·115	4·219	4·321	4·422	4·620	4·812	4·907	5·184	5·451	5·539	5·711	5·880	5·964	6·210
1/ 59	5·0495	5·5995	6·1852	6·8072	8·1637	9·6756	10·492	13·191	16·281	17·401	19·781	22·353	23·712	28·091
0·01800	4·234	4·341	4·446	4·550	4·754	4·952	5·049	5·334	5·610	5·700	5·877	6·051	6·137	6·390
1/ 56	5·1960	5·7620	6·3646	7·0047	8·4005	9·9563	10·796	13·573	16·753	17·906	20·355	23·001	24·400	28·906
0·01900	4·350	4·460	4·568	4·675	4·884	5·088	5·188	5·480	5·763	5·856	6·038	6·217	6·305	6·565
1/ 53	5·3385	5·9200	6·5391	7·1967	8·6309	10·229	11·092	13·945	17·212	18·397	20·913	23·632	25·069	29·699
0·02000	4·463	4·576	4·687	4·797	5·011	5·220	5·322	5·623	5·913	6·008	6·195	6·378	6·469	6·735
1/ 50	5·4773	6·0739	6·7091	7·3838	8·8552	10·495	11·381	14·308	17·660	18·875	21·457	24·246	25·720	30·470
0·02100	4·574	4·689	4·803	4·915	5·135	5·349	5·454	5·762	6·059	6·156	6·348	6·536	6·629	6·902
1/ 47·6	5·6126	6·2240	6·8749	7·5663	9·0741	10·755	11·662	14·662	18·096	19·341	21·987	24·845	26·356	31·223
0·02200	4·681	4·800	4·916	5·031	5·256	5·475	5·582	5·897	6·202	6·301	6·497	6·690	6·785	7·064
1/ 45·5	5·7448	6·3706	7·0368	7·7444	9·2877	11·008	11·936	15·007	18·522	19·797	22·505	25·430	26·976	31·958
0·02300	4·787	4·907	5·027	5·144	5·374	5·598	5·708	6·030	6·341	6·443	6·644	6·840	6·937	7·223
1/ 43·5	5·8740	6·5138	7·1950	7·9186	9·4966	11·255	12·205	15·344	18·939	20·242	23·011	26·002	27·583	32·677
0·02400	4·890	5·013	5·135	5·255	5·490	5·718	5·831	6·160	6·478	6·582	6·787	6·987	7·087	7·379
1/ 41·7	6·0004	6·6540	7·3499	8·0890	9·7009	11·498	12·467	15·674	19·346	20·677	23·506	26·561	28·176	33·380
0·02500	4·990	5·117	5·241	5·363	5·603	5·836	5·951	6·287	6·612	6·718	6·927	7·132	7·233	7·531
1/ 40·0	6·1242	6·7913	7·5015	8·2559	9·9011	11·735	12·725	15·998	19·745	21·104	23·991	27·109	28·758	34·069
0·02600	5·089	5·218	5·345	5·469	5·714	5·952	6·069	6·411	6·743	6·851	7·064	7·273	7·376	7·680
1/ 38·5	6·2456	6·9259	7·6502	8·4195	10·097	11·967	12·977	16·315	20·136	21·522	24·466	27·647	29·328	34·744
0·02700	5·186	5·317	5·446	5·574	5·823	6·065	6·185	6·533	6·871	6·981	7·198	7·411	7·517	7·826
1/ 37·0	6·3646	7·0579	7·7960	8·5800	10·290	12·195	13·224	16·626	20·520	21·932	24·932	28·173	29·886	35·406
0·02800	5·282	5·415	5·546	5·676	5·930	6·177	6·298	6·653	6·997	7·109	7·331	7·548	7·655	7·970
1/ 35·7	6·4815	7·1875	7·9391	8·7375	10·479	12·419	13·467	16·931	20·897	22·335	25·390	28·691	30·435	36·056
0·02900	5·375	5·511	5·645	5·777	6·035	6·286	6·410	6·771	7·121	7·235	7·460	7·681	7·790	8·111
1/ 34·5	6·5963	7·3148	8·0797	8·8922	10·664	12·639	13·705	17·231	21·267	22·730	25·840	29·199	30·974	36·694
0·03000	5·467	5·605	5·741	5·875	6·138	6·394	6·519	6·887	7·243	7·359	7·588	7·813	7·923	8·250
1/ 33·3	6·7091	7·4399	8·2179	9·0443	10·847	12·855	13·940	17·525	21·631	23·119	26·282	29·698	31·504	37·322
	0·80	0·80	0·80	0·80	0·81	0·81	0·81	0·82	0·82	0·82	0·82	0·83	0·83	0·83

$V_{r(0\cdot5)medial}$ **for half-full circular pipes.**

k_s = **3·0 mm** S = 0·01000 to 0·03000

Gradient	(Equivalent) Pipe diameters in mm													
	1250	**1300**	**1350**	**1400**	**1500**	**1600**	**1650**	**1800**	**1950**	**2000**	**2100**	**2200**	**2250**	**2400**
0·03000	5·467	5·605	5·741	5·875	6·138	6·394	6·519	6·887	7·243	7·359	7·588	7·813	7·923	8·250
1/ 33·3	6·7091	7·4399	8·2179	9·0443	10·847	12·855	13·940	17·525	21·631	23·119	26·282	29·698	31·504	37·322
0·03200	5·646	5·789	5·930	6·068	6·339	6·604	6·733	7·113	7·480	7·600	7·837	8·069	8·183	8·521
1/ 31·3	6·9292	7·6840	8·4876	9·3411	11·203	13·277	14·397	18·100	22·340	23·878	27·144	30·672	32·537	38·546
0·03400	5·820	5·967	6·112	6·255	6·535	6·807	6·940	7·332	7·711	7·834	8·078	8·317	8·435	8·783
1/ 29·4	7·1426	7·9206	8·7489	9·6287	11·547	13·686	14·840	18·658	23·028	24·613	27·980	31·617	33·539	39·733
0·03600	5·989	6·140	6·289	6·436	6·724	7·004	7·142	7·545	7·934	8·062	8·312	8·559	8·680	9·038
1/ 27·8	7·3498	8·1504	9·0027	9·9080	11·882	14·083	15·271	19·199	23·696	25·327	28·791	32·534	34·512	40·885
0·03800	6·153	6·309	6·462	6·613	6·908	7·196	7·338	7·751	8·152	8·283	8·540	8·793	8·918	9·285
1/ 26·3	7·5513	8·3738	9·2495	10·180	12·208	14·469	15·689	19·725	24·346	26·021	29·580	33·425	35·458	42·006
0·04000	6·313	6·473	6·630	6·785	7·088	7·383	7·528	7·953	8·364	8·498	8·762	9·022	9·150	9·527
1/ 25·0	7·7476	8·5915	9·4899	10·444	12·525	14·845	16·097	20·238	24·978	26·697	30·349	34·294	36·379	43·097
0·04200	6·469	6·633	6·794	6·952	7·263	7·566	7·714	8·149	8·570	8·708	8·979	9·244	9·376	9·762
1/ 23·8	7·9390	8·8037	9·7244	10·702	12·835	15·212	16·495	20·738	25·595	27·357	31·099	35·141	37·278	44·162
0·04400	6·622	6·789	6·954	7·116	7·434	7·744	7·896	8·341	8·772	8·913	9·190	9·462	9·596	9·992
1/ 22·7	8·1259	9·0110	9·9533	10·954	13·137	15·570	16·883	21·226	26·198	28·001	31·831	35·969	38·155	45·202
0·04600	6·770	6·942	7·110	7·276	7·601	7·918	8·073	8·529	8·969	9·113	9·397	9·675	9·812	10·22
1/ 21·7	8·3086	9·2136	10·177	11·201	13·432	15·920	17·263	21·703	26·787	28·630	32·546	36·777	39·013	46·218
0·04800	6·916	7·091	7·263	7·433	7·765	8·088	8·247	8·712	9·162	9·309	9·599	9·883	10·02	10·44
1/ 20·8	8·4874	9·4119	10·396	11·442	13·721	16·262	17·634	22·170	27·363	29·246	33·247	37·568	39·853	47·212
0·05000	7·059	7·237	7·413	7·586	7·925	8·255	8·417	8·892	9·351	9·501	9·797	10·09	10·23	10·65
1/ 20·0	8·6625	9·6060	10·611	11·678	14·005	16·598	17·998	22·627	27·928	29·849	33·932	38·343	40·675	48·186
0·05250	7·233	7·416	7·596	7·773	8·121	8·459	8·625	9·112	9·582	9·736	10·04	10·34	10·48	10·91
1/ 19·0	8·8765	9·8434	10·873	11·966	14·350	17·008	18·443	23·186	28·618	30·587	34·771	39·291	41·679	49·376
0·05500	7·404	7·591	7·775	7·956	8·312	8·658	8·828	9·326	9·808	9·965	10·28	10·58	10·73	11·17
1/ 18·2	9·0855	10·075	11·129	12·248	14·688	17·408	18·877	23·732	29·291	31·307	35·589	40·215	42·661	50·539
0·05750	7·570	7·761	7·950	8·135	8·499	8·853	9·027	9·536	10·03	10·19	10·51	10·82	10·97	11·42
1/ 17·4	9·2898	10·302	11·379	12·523	15·019	17·800	19·301	24·266	29·950	32·011	36·389	41·120	43·620	51·675
0·06000	7·733	7·928	8·121	8·310	8·682	9·043	9·221	9·741	10·24	10·41	10·73	11·05	11·21	11·67
1/ 16·7	9·4896	10·523	11·624	12·793	15·342	18·183	19·716	24·788	30·594	32·699	37·172	42·004	44·558	52·786
0·06250	7·892	8·092	8·288	8·482	8·861	9·230	9·411	9·942	10·46	10·62	10·95	11·28	11·44	11·91
1/ 16·0	9·6854	10·740	11·864	13·056	15·658	18·558	20·123	25·299	31·225	33·374	37·939	42·871	45·477	53·875
0·06500	8·049	8·252	8·452	8·650	9·036	9·413	9·598	10·14	10·66	10·83	11·17	11·50	11·66	12·14
1/ 15·4	9·8773	10·953	12·099	13·315	15·968	18·925	20·522	25·800	31·844	34·035	38·690	43·720	46·378	54·942
0·06750	8·202	8·409	8·613	8·815	9·208	9·592	9·780	10·33	10·87	11·04	11·38	11·72	11·89	12·38
1/ 14·8	10·066	11·162	12·329	13·569	16·273	19·286	20·913	26·292	32·450	34·683	39·428	44·553	47·262	55·989
0·07000	8·353	8·564	8·771	8·976	9·377	9·768	9·960	10·52	11·07	11·24	11·59	11·94	12·10	12·60
1/ 14·3	10·250	11·367	12·555	13·818	16·571	19·640	21·297	26·775	33·046	35·320	40·151	45·371	48·129	57·017
0·07250	8·501	8·715	8·927	9·135	9·544	9·941	10·14	10·71	11·26	11·44	11·80	12·15	12·32	12·83
1/ 13·8	10·432	11·568	12·778	14·063	16·865	19·988	21·674	27·249	33·631	35·945	40·862	46·174	48·981	58·026
0·07500	8·646	8·864	9·079	9·291	9·707	10·11	10·31	10·89	11·45	11·64	12·00	12·35	12·53	13·05
1/ 13·3	10·610	11·766	12·996	14·303	17·153	20·330	22·044	27·715	34·206	36·560	41·561	46·963	49·819	59·019
0·08000	8·930	9·155	9·377	9·596	10·03	10·44	10·65	11·25	11·83	12·02	12·39	12·76	12·94	13·47
1/ 12·5	10·958	12·152	13·423	14·772	17·716	20·997	22·768	28·624	35·328	37·759	42·924	48·504	51·453	60·955
0·08500	9·205	9·437	9·666	9·892	10·33	10·76	10·98	11·59	12·19	12·39	12·77	13·15	13·34	13·89
1/ 11·8	11·296	12·526	13·836	15·227	18·261	21·643	23·469	29·505	36·416	38·922	44·246	49·997	53·037	62·831
0·09000	9·471	9·711	9·946	10·18	10·63	11·08	11·29	11·93	12·55	12·75	13·14	13·53	13·73	14·29
1/ 11·1	11·623	12·889	14·237	15·669	18·791	22·271	24·149	30·361	37·472	40·050	45·529	51·447	54·575	64·653
0·09500	9·731	9·977	10·22	10·46	10·92	11·38	11·60	12·26	12·89	13·10	13·51	13·90	14·10	14·68
1/ 10·5	11·942	13·243	14·627	16·098	19·306	22·881	24·811	31·193	38·499	41·148	46·777	52·857	56·071	66·425
0·10000	9·984	10·24	10·48	10·73	11·21	11·68	11·90	12·58	13·23	13·44	13·86	14·27	14·47	15·06
1/ 10·0	12·252	13·587	15·007	16·517	19·808	23·476	25·456	32·003	39·499	42·217	47·992	54·231	57·528	68·151
	0·80	0·80	0·80	0·80	0·81	0·81	0·81	0·82	0·82	0·82	0·82	0·83	0·83	0·83

$V_{r(0{\cdot}5)medial}$ for half-full circular pipes.

A44

(p.1 of 6)

k_s = 6·0 mm
S = 0·00010 to 0·00030

ie hydraulic gradient =
1 in 10000 to 1 in 3333

Water (or sewage) at 15°C;
full bore conditions.

velocities in ms^{-1}
discharges in m^3s^{-1}

Gradient **(Equivalent) Pipe diameters in mm**

Gradient	1250	1300	1350	1400	1500	1600	1650	1800	1950	2000	2100	2200	2250	2400
0·000100	0·284	0·292	0·299	0·306	0·320	0·334	0·341	0·361	0·380	0·386	0·398	0·410	0·416	0·434
1/ 10000	0·3491	0·3874	0·4282	0·4716	0·5662	0·6717	0·7287	0·9174	1·1336	1·2121	1·3789	1·5592	1·6545	1·9618
0·000105	0·292	0·299	0·307	0·314	0·328	0·342	0·349	0·369	0·389	0·395	0·408	0·420	0·426	0·444
1/ 9524	0·3578	0·3970	0·4388	0·4833	0·5802	0·6884	0·7468	0·9401	1·1618	1·2422	1·4131	1·5978	1·6955	2·0104
0·000110	0·298	0·306	0·314	0·321	0·336	0·350	0·358	0·378	0·398	0·405	0·418	0·430	0·437	0·455
1/ 9091	0·3663	0·4064	0·4492	0·4947	0·5939	0·7046	0·7644	0·9624	1·1892	1·2715	1·4465	1·6356	1·7356	2·0579
0·000115	0·305	0·313	0·321	0·329	0·344	0·358	0·366	0·387	0·407	0·414	0·427	0·440	0·446	0·465
1/ 8696	0·3745	0·4156	0·4594	0·5059	0·6073	0·7205	0·7817	0·9841	1·2160	1·3002	1·4791	1·6725	1·7747	2·1043
0·000120	0·312	0·320	0·328	0·336	0·351	0·366	0·373	0·395	0·416	0·423	0·436	0·449	0·456	0·475
1/ 8333	0·3826	0·4246	0·4693	0·5168	0·6204	0·7361	0·7986	1·0053	1·2423	1·3282	1·5110	1·7085	1·8130	2·1497
0·000125	0·318	0·327	0·335	0·343	0·358	0·374	0·381	0·403	0·425	0·432	0·445	0·459	0·465	0·485
1/ 8000	0·3906	0·4334	0·4790	0·5275	0·6333	0·7513	0·8151	1·0261	1·2680	1·3557	1·5423	1·7439	1·8505	2·1942
0·000130	0·325	0·333	0·341	0·349	0·366	0·381	0·389	0·411	0·433	0·440	0·454	0·468	0·475	0·495
1/ 7692	0·3983	0·4420	0·4885	0·5380	0·6459	0·7663	0·8313	1·0465	1·2932	1·3827	1·5729	1·7785	1·8873	2·2378
0·000135	0·331	0·339	0·348	0·356	0·372	0·388	0·396	0·419	0·441	0·449	0·463	0·477	0·484	0·504
1/ 7407	0·4059	0·4505	0·4979	0·5483	0·6582	0·7809	0·8472	1·0665	1·3179	1·4091	1·6030	1·8125	1·9233	2·2805
0·000140	0·337	0·346	0·354	0·363	0·379	0·396	0·404	0·427	0·449	0·457	0·471	0·486	0·493	0·513
1/ 7143	0·4134	0·4588	0·5071	0·5584	0·6704	0·7953	0·8628	1·0862	1·3422	1·4350	1·6325	1·8459	1·9587	2·3225
0·000145	0·343	0·352	0·361	0·369	0·386	0·403	0·411	0·434	0·457	0·465	0·480	0·494	0·501	0·523
1/ 6897	0·4208	0·4669	0·5161	0·5683	0·6823	0·8095	0·8781	1·1055	1·3660	1·4605	1·6615	1·8787	1·9935	2·3637
0·000150	0·349	0·358	0·367	0·376	0·393	0·409	0·418	0·442	0·465	0·473	0·488	0·503	0·510	0·531
1/ 6667	0·4280	0·4749	0·5249	0·5781	0·6940	0·8233	0·8932	1·1244	1·3894	1·4856	1·6900	1·9109	2·0277	2·4043
0·000160	0·360	0·370	0·379	0·388	0·406	0·423	0·431	0·456	0·481	0·488	0·504	0·519	0·527	0·549
1/ 6250	0·4421	0·4906	0·5422	0·5971	0·7168	0·8504	0·9226	1·1614	1·4352	1·5345	1·7456	1·9737	2·0944	2·4833
0·000170	0·371	0·381	0·391	0·400	0·418	0·436	0·445	0·471	0·495	0·504	0·520	0·535	0·543	0·566
1/ 5882	0·4558	0·5057	0·5590	0·6155	0·7390	0·8767	0·9511	1·1973	1·4795	1·5818	1·7994	2·0347	2·1590	2·5600
0·000180	0·382	0·392	0·402	0·411	0·430	0·449	0·458	0·484	0·510	0·518	0·535	0·551	0·559	0·582
1/ 5556	0·4690	0·5205	0·5752	0·6334	0·7605	0·9022	0·9788	1·2321	1·5225	1·6278	1·8518	2·0938	2·2218	2·6344
0·000190	0·393	0·403	0·413	0·423	0·442	0·461	0·470	0·497	0·524	0·532	0·549	0·566	0·574	0·598
1/ 5263	0·4819	0·5348	0·5911	0·6509	0·7814	0·9270	1·0057	1·2660	1·5643	1·6726	1·9026	2·1514	2·2829	2·7068
0·000200	0·403	0·413	0·424	0·434	0·454	0·473	0·483	0·510	0·537	0·546	0·564	0·581	0·589	0·614
1/ 5000	0·4945	0·5487	0·6065	0·6678	0·8018	0·9512	1·0319	1·2990	1·6051	1·7161	1·9522	2·2074	2·3423	2·7773
0·000210	0·413	0·424	0·434	0·445	0·465	0·485	0·495	0·523	0·551	0·560	0·578	0·595	0·604	0·629
1/ 4762	0·5068	0·5623	0·6215	0·6844	0·8216	0·9747	1·0575	1·3311	1·6448	1·7586	2·0006	2·2620	2·4003	2·8460
0·000220	0·423	0·434	0·444	0·455	0·476	0·496	0·506	0·535	0·564	0·573	0·591	0·609	0·618	0·644
1/ 4545	0·5187	0·5756	0·6362	0·7005	0·8410	0·9978	1·0824	1·3626	1·6836	1·8001	2·0478	2·3154	2·4569	2·9132
0·000230	0·432	0·443	0·454	0·465	0·487	0·507	0·518	0·548	0·576	0·586	0·605	0·623	0·632	0·658
1/ 4348	0·5304	0·5886	0·6505	0·7163	0·8600	1·0202	1·1068	1·3933	1·7216	1·8407	2·0939	2·3676	2·5123	2·9788
0·000240	0·442	0·453	0·464	0·475	0·497	0·518	0·529	0·559	0·589	0·599	0·618	0·636	0·645	0·673
1/ 4167	0·5419	0·6013	0·6646	0·7318	0·8785	1·0423	1·1307	1·4233	1·7587	1·8804	2·1390	2·4186	2·5664	3·0430
0·000250	0·451	0·462	0·474	0·485	0·507	0·529	0·540	0·571	0·601	0·611	0·630	0·649	0·659	0·687
1/ 4000	0·5531	0·6137	0·6783	0·7469	0·8967	1·0638	1·1541	1·4527	1·7951	1·9193	2·1833	2·4686	2·6195	3·1059
0·000260	0·460	0·472	0·483	0·495	0·518	0·540	0·550	0·582	0·613	0·623	0·643	0·662	0·672	0·700
1/ 3846	0·5641	0·6259	0·6918	0·7618	0·9145	1·0849	1·1770	1·4816	1·8307	1·9574	2·2266	2·5176	2·6715	3·1675
0·000270	0·468	0·481	0·493	0·504	0·527	0·550	0·561	0·593	0·625	0·635	0·655	0·675	0·685	0·714
1/ 3704	0·5749	0·6379	0·7050	0·7763	0·9320	1·1057	1·1995	1·5099	1·8657	1·9947	2·2691	2·5657	2·7225	3·2280
0·000280	0·477	0·489	0·502	0·514	0·537	0·560	0·571	0·604	0·636	0·647	0·667	0·687	0·697	0·727
1/ 3571	0·5854	0·6496	0·7180	0·7906	0·9492	1·1260	1·2215	1·5377	1·9000	2·0314	2·3108	2·6129	2·7726	3·2874
0·000290	0·486	0·498	0·510	0·523	0·547	0·570	0·581	0·615	0·647	0·658	0·679	0·700	0·710	0·740
1/ 3448	0·5958	0·6611	0·7307	0·8046	0·9660	1·1460	1·2432	1·5650	1·9337	2·0675	2·3518	2·6592	2·8217	3·3457
0·000300	0·494	0·507	0·519	0·532	0·556	0·580	0·591	0·626	0·659	0·669	0·691	0·712	0·722	0·752
1/ 3333	0·6061	0·6725	0·7432	0·8184	0·9826	1·1656	1·2645	1·5918	1·9668	2·1029	2·3921	2·7048	2·8701	3·4030
	0·77	0·77	0·77	0·77	0·78	0·78	0·78	0·78	0·79	0·79	0·79	0·79	0·79	0·80

$V_{r(0·5)medial}$ for half-full circular pipes.

k_s = 6·0 mm S = 0·00010 to 0·00030

k_s = 6·0 mm
S = 0·00030 to 0·00100

ie hydraulic gradient =
1 in 3333 to 1 in 1000

Water (or sewage) at 15°C;
full bore conditions.

velocities in ms^{-1}
discharges in m^3s^{-1}

Gradient **(Equivalent) Pipe diameters in mm**

Gradient	1250	1300	1350	1400	1500	1600	1650	1800	1950	2000	2100	2200	2250	2400
0·000300	0·494	0·507	0·519	0·532	0·556	0·580	0·591	0·626	0·659	0·669	0·691	0·712	0·722	0·752
1/ 3333	0·6061	0·6725	0·7432	0·8184	0·9826	1·1656	1·2645	1·5918	1·9668	2·1029	2·3921	2·7048	2·8701	3·4030
0·000320	0·510	0·523	0·536	0·549	0·574	0·599	0·611	0·646	0·680	0·691	0·713	0·735	0·746	0·777
1/ 3125	0·6260	0·6946	0·7677	0·8454	1·0149	1·2040	1·3061	1·6441	2·0315	2·1720	2·4708	2·7937	2·9644	3·5148
0·000340	0·526	0·539	0·553	0·566	0·592	0·617	0·630	0·666	0·701	0·713	0·735	0·758	0·769	0·801
1/ 2941	0·6453	0·7160	0·7914	0·8715	1·0462	1·2411	1·3464	1·6948	2·0941	2·2390	2·5470	2·8798	3·0558	3·6232
0·000360	0·541	0·555	0·569	0·583	0·609	0·635	0·648	0·685	0·722	0·733	0·757	0·780	0·791	0·824
1/ 2778	0·6641	0·7369	0·8144	0·8968	1·0766	1·2772	1·3855	1·7441	2·1550	2·3041	2·6210	2·9635	3·1446	3·7284
0·000380	0·556	0·570	0·585	0·599	0·626	0·653	0·666	0·704	0·741	0·754	0·777	0·801	0·813	0·847
1/ 2632	0·6823	0·7571	0·8368	0·9214	1·1062	1·3123	1·4236	1·7920	2·2142	2·3673	2·6929	3·0449	3·2309	3·8308
0·000400	0·570	0·585	0·600	0·614	0·642	0·670	0·683	0·723	0·761	0·773	0·798	0·822	0·834	0·869
1/ 2500	0·7001	0·7768	0·8586	0·9454	1·1350	1·3464	1·4606	1·8386	2·2718	2·4289	2·7630	3·1241	3·3150	3·9305
0·000420	0·585	0·600	0·615	0·629	0·658	0·686	0·700	0·740	0·780	0·792	0·817	0·842	0·854	0·890
1/ 2381	0·7174	0·7961	0·8798	0·9688	1·1631	1·3798	1·4968	1·8841	2·3280	2·4890	2·8314	3·2014	3·3970	4·0277
0·000440	0·598	0·614	0·629	0·644	0·674	0·702	0·717	0·758	0·798	0·811	0·837	0·862	0·875	0·911
1/ 2273	0·7344	0·8148	0·9006	0·9917	1·1905	1·4123	1·5321	1·9285	2·3829	2·5477	2·8981	3·2769	3·4771	4·1227
0·000460	0·612	0·628	0·643	0·659	0·689	0·718	0·733	0·775	0·816	0·829	0·856	0·881	0·894	0·932
1/ 2174	0·7509	0·8332	0·9209	1·0140	1·2173	1·4441	1·5666	1·9720	2·4366	2·6051	2·9634	3·3506	3·5554	4·2155
0·000480	0·625	0·641	0·657	0·673	0·704	0·734	0·748	0·792	0·833	0·847	0·874	0·900	0·913	0·952
1/ 2083	0·7671	0·8512	0·9407	1·0359	1·2436	1·4752	1·6004	2·0145	2·4891	2·6612	3·0272	3·4228	3·6320	4·3063
0·000500	0·638	0·655	0·671	0·687	0·718	0·749	0·764	0·808	0·851	0·865	0·892	0·919	0·932	0·972
1/ 2000	0·7830	0·8688	0·9602	1·0573	1·2693	1·5057	1·6334	2·0561	2·5405	2·7162	3·0898	3·4935	3·7070	4·3952
0·000525	0·654	0·671	0·687	0·704	0·736	0·767	0·783	0·828	0·872	0·886	0·914	0·942	0·955	0·996
1/ 1905	0·8023	0·8903	0·9839	1·0835	1·3007	1·5430	1·6739	2·1070	2·6033	2·7834	3·1662	3·5799	3·7987	4·5039
0·000550	0·669	0·687	0·704	0·720	0·753	0·786	0·801	0·848	0·892	0·907	0·936	0·964	0·978	1·019
1/ 1818	0·8213	0·9113	1·0071	1·1090	1·3313	1·5794	1·7133	2·1566	2·6647	2·8490	3·2408	3·6643	3·8883	4·6101
0·000575	0·684	0·702	0·719	0·737	0·770	0·803	0·819	0·867	0·912	0·927	0·957	0·986	1·000	1·042
1/ 1739	0·8398	0·9318	1·0298	1·1340	1·3613	1·6149	1·7519	2·2052	2·7247	2·9132	3·3138	3·7468	3·9758	4·7138
0·000600	0·699	0·717	0·735	0·753	0·787	0·821	0·837	0·885	0·932	0·947	0·977	1·007	1·021	1·064
1/ 1667	0·8579	0·9519	1·0520	1·1584	1·3907	1·6497	1·7897	2·2527	2·7834	2·9759	3·3852	3·8275	4·0614	4·8154
0·000625	0·713	0·732	0·750	0·768	0·803	0·837	0·854	0·904	0·951	0·967	0·998	1·028	1·043	1·086
1/ 1600	0·8756	0·9715	1·0737	1·1824	1·4194	1·6838	1·8266	2·2992	2·8409	3·0374	3·4551	3·9066	4·1453	4·9148
0·000650	0·728	0·746	0·765	0·783	0·819	0·854	0·871	0·921	0·970	0·986	1·017	1·048	1·063	1·108
1/ 1538	0·8930	0·9908	1·0951	1·2058	1·4476	1·7172	1·8629	2·3448	2·8972	3·0976	3·5236	3·9841	4·2275	5·0123
0·000675	0·742	0·761	0·780	0·798	0·835	0·870	0·888	0·939	0·989	1·005	1·037	1·068	1·084	1·129
1/ 1481	0·9100	1·0097	1·1160	1·2288	1·4752	1·7500	1·8984	2·3896	2·9525	3·1567	3·5908	4·0601	4·3082	5·1079
0·000700	0·755	0·775	0·794	0·813	0·850	0·886	0·904	0·956	1·007	1·023	1·056	1·088	1·103	1·150
1/ 1429	0·9267	1·0283	1·1365	1·2514	1·5023	1·7821	1·9333	2·4335	3·0068	3·2147	3·6568	4·1347	4·3873	5·2018
0·000725	0·769	0·788	0·808	0·827	0·865	0·902	0·920	0·973	1·025	1·041	1·075	1·107	1·123	1·170
1/ 1379	0·9432	1·0465	1·1566	1·2736	1·5289	1·8137	1·9676	2·4766	3·0601	3·2717	3·7216	4·2080	4·4651	5·2940
0·000750	0·782	0·802	0·822	0·842	0·880	0·918	0·936	0·990	1·042	1·059	1·093	1·126	1·142	1·190
1/ 1333	0·9593	1·0645	1·1764	1·2954	1·5551	1·8448	2·0013	2·5191	3·1125	3·3277	3·7854	4·2800	4·5415	5·3846
0·000800	0·807	0·828	0·849	0·869	0·909	0·948	0·967	1·022	1·076	1·094	1·129	1·163	1·180	1·229
1/ 1250	0·9909	1·0994	1·2151	1·3380	1·6062	1·9054	2·0670	2·6018	3·2147	3·4370	3·9097	4·4205	4·6907	5·5614
0·000850	0·832	0·854	0·875	0·896	0·937	0·977	0·996	1·054	1·110	1·128	1·164	1·199	1·216	1·267
1/ 1176	1·0214	1·1333	1·2526	1·3792	1·6557	1·9641	2·1307	2·6820	3·3138	3·5430	4·0302	4·5568	4·8352	5·7328
0·000900	0·856	0·879	0·900	0·922	0·964	1·005	1·025	1·085	1·142	1·160	1·197	1·234	1·251	1·304
1/ 1111	1·0511	1·1662	1·2889	1·4193	1·7038	2·0212	2·1926	2·7599	3·4100	3·6458	4·1472	4·6890	4·9756	5·8992
0·000950	0·880	0·903	0·925	0·947	0·991	1·033	1·054	1·114	1·173	1·192	1·230	1·267	1·286	1·340
1/ 1053	1·0799	1·1983	1·3243	1·4583	1·7506	2·0766	2·2528	2·8356	3·5035	3·7458	4·2609	4·8177	5·1121	6·0610
0·001000	0·903	0·926	0·949	0·972	1·016	1·060	1·081	1·143	1·204	1·223	1·262	1·300	1·319	1·375
1/ 1000	1·1080	1·2294	1·3588	1·4962	1·7961	2·1307	2·3114	2·9093	3·5947	3·8433	4·3718	4·9430	5·2450	6·2186
	0·77	0·77	0·77	0·77	0·78	0·78	0·78	0·78	0·79	0·79	0·79	0·79	0·79	0·80

$V_{r(0·5)medial}$ for half-full circular pipes.

S = 0·00030 to 0·00100 **k_s = 6·0 mm**

$k_s = 6{\cdot}0$ mm
S = 0·00100 to 0·00300

ie hydraulic gradient =
1 in 1000 to 1 in 333

Water (or sewage) at 15°C;
full bore conditions.

velocities in ms^{-1}
discharges in m^3s^{-1}

Gradient **(Equivalent) Pipe diameters in mm**

Gradient	1250	1300	1350	1400	1500	1600	1650	1800	1950	2000	2100	2200	2250	2400
0·00100	0·903	0·926	0·949	0·972	1·016	1·060	1·081	1·143	1·204	1·223	1·262	1·300	1·319	1·375
1/ 1000	1·1080	1·2294	1·3588	1·4962	1·7961	2·1307	2·3114	2·9093	3·5947	3·8433	4·3718	4·9430	5·2450	6·2186
0·00105	0·925	0·949	0·973	0·996	1·042	1·086	1·108	1·172	1·233	1·254	1·293	1·332	1·352	1·409
1/ 952	1·1354	1·2599	1·3924	1·5332	1·8405	2·1834	2·3685	2·9813	3·6835	3·9383	4·4799	5·0652	5·3747	6·3724
0·00110	0·947	0·972	0·996	1·019	1·066	1·112	1·134	1·199	1·262	1·283	1·324	1·364	1·384	1·442
1/ 909	1·1622	1·2895	1·4252	1·5693	1·8839	2·2348	2·4243	3·0515	3·7703	4·0311	4·5854	5·1845	5·5013	6·5225
0·00115	0·968	0·993	1·018	1·042	1·090	1·137	1·159	1·226	1·291	1·312	1·354	1·395	1·415	1·474
1/ 870	1·1884	1·3186	1·4573	1·6047	1·9263	2·2851	2·4789	3·1202	3·8552	4·1218	4·6886	5·3012	5·6251	6·6692
0·00120	0·989	1·015	1·040	1·065	1·114	1·161	1·184	1·253	1·319	1·340	1·383	1·425	1·445	1·506
1/ 833	1·2140	1·3470	1·4887	1·6392	1·9678	2·3343	2·5323	3·1874	3·9382	4·2106	4·7895	5·4153	5·7462	6·8128
0·00125	1·010	1·036	1·061	1·087	1·137	1·185	1·209	1·278	1·346	1·368	1·411	1·454	1·475	1·537
1/ 800	1·2390	1·3748	1·5194	1·6731	2·0084	2·3825	2·5846	3·2532	4·0195	4·2975	4·8884	5·5271	5·8648	6·9535
0·00130	1·030	1·056	1·083	1·108	1·159	1·208	1·233	1·304	1·373	1·395	1·439	1·483	1·504	1·568
1/ 769	1·2636	1·4021	1·5495	1·7063	2·0483	2·4298	2·6358	3·3177	4·0992	4·3827	4·9853	5·6367	5·9811	7·0913
0·00135	1·049	1·076	1·103	1·130	1·181	1·232	1·256	1·329	1·399	1·422	1·467	1·511	1·533	1·597
1/ 741	1·2877	1·4288	1·5791	1·7388	2·0873	2·4761	2·6861	3·3810	4·1774	4·4663	5·0804	5·7442	6·0952	7·2265
0·00140	1·069	1·096	1·123	1·150	1·203	1·254	1·279	1·353	1·424	1·448	1·494	1·539	1·561	1·627
1/ 714	1·3114	1·4551	1·6081	1·7708	2·1257	2·5216	2·7355	3·4431	4·2541	4·5483	5·1737	5·8497	6·2071	7·3593
0·00145	1·088	1·116	1·143	1·171	1·224	1·276	1·302	1·377	1·450	1·473	1·520	1·566	1·589	1·656
1/ 690	1·3346	1·4809	1·6366	1·8021	2·1634	2·5663	2·7839	3·5041	4·3295	4·6289	5·2654	5·9533	6·3171	7·4896
0·00150	1·106	1·135	1·163	1·191	1·245	1·298	1·324	1·401	1·475	1·499	1·546	1·593	1·616	1·684
1/ 667	1·3575	1·5062	1·6646	1·8330	2·2004	2·6102	2·8316	3·5641	4·4036	4·7081	5·3555	6·0552	6·4252	7·6178
0·00160	1·142	1·172	1·201	1·230	1·286	1·341	1·368	1·447	1·523	1·548	1·597	1·645	1·669	1·739
1/ 625	1·4021	1·5557	1·7193	1·8932	2·2726	2·6959	2·9245	3·6811	4·5481	4·8627	5·5313	6·2540	6·6361	7·8679
0·00170	1·178	1·208	1·238	1·268	1·326	1·382	1·410	1·491	1·570	1·596	1·646	1·696	1·720	1·793
1/ 588	1·4453	1·6036	1·7723	1·9515	2·3427	2·7790	3·0147	3·7945	4·6882	5·0125	5·7017	6·4466	6·8405	8·1102
0·00180	1·212	1·243	1·274	1·305	1·364	1·422	1·451	1·534	1·615	1·642	1·694	1·745	1·770	1·845
1/ 556	1·4872	1·6502	1·8237	2·0082	2·4107	2·8596	3·1021	3·9046	4·8243	5·1579	5·8672	6·6337	7·0390	8·3455
0·00190	1·245	1·277	1·309	1·340	1·402	1·461	1·491	1·577	1·660	1·687	1·740	1·793	1·819	1·895
1/ 526	1·5280	1·6954	1·8738	2·0632	2·4768	2·9381	3·1872	4·0117	4·9566	5·2994	6·0281	6·8157	7·2321	8·5744
0·00200	1·278	1·311	1·343	1·375	1·438	1·499	1·529	1·618	1·703	1·731	1·786	1·840	1·866	1·945
1/ 500	1·5678	1·7395	1·9225	2·1169	2·5412	3·0145	3·2701	4·1160	5·0855	5·4372	6·1848	6·9929	7·4201	8·7973
0·00210	1·309	1·343	1·376	1·409	1·474	1·536	1·567	1·657	1·745	1·773	1·830	1·885	1·912	1·993
1/ 476	1·6065	1·7825	1·9700	2·1692	2·6040	3·0890	3·3510	4·2178	5·2112	5·5716	6·3377	7·1657	7·6035	9·0148
0·00220	1·340	1·375	1·409	1·442	1·508	1·573	1·604	1·697	1·786	1·815	1·873	1·929	1·957	2·040
1/ 455	1·6444	1·8245	2·0164	2·2203	2·6654	3·1618	3·4299	4·3171	5·3339	5·7028	6·4869	7·3345	7·7826	9·2271
0·00230	1·370	1·406	1·440	1·475	1·542	1·608	1·640	1·735	1·826	1·856	1·915	1·973	2·001	2·086
1/ 435	1·6814	1·8656	2·0618	2·2703	2·7253	3·2329	3·5071	4·4143	5·4539	5·8311	6·6329	7·4994	7·9576	9·4346
0·00240	1·400	1·436	1·471	1·507	1·575	1·643	1·675	1·772	1·866	1·896	1·956	2·015	2·044	2·130
1/ 417	1·7176	1·9058	2·1062	2·3192	2·7840	3·3025	3·5826	4·5093	5·5713	5·9566	6·7756	7·6609	8·1289	9·6377
0·00250	1·429	1·465	1·502	1·538	1·608	1·676	1·710	1·809	1·904	1·935	1·997	2·057	2·087	2·174
1/ 400	1·7530	1·9451	2·1497	2·3671	2·8415	3·3707	3·6565	4·6023	5·6863	6·0795	6·9155	7·8189	8·2966	9·8365
0·00260	1·457	1·494	1·532	1·568	1·640	1·710	1·744	1·844	1·942	1·974	2·036	2·098	2·128	2·217
1/ 385	1·7878	1·9837	2·1923	2·4140	2·8978	3·4375	3·7290	4·6936	5·7990	6·2000	7·0525	7·9739	8·4611	10·031
0·00270	1·485	1·523	1·561	1·598	1·671	1·742	1·777	1·880	1·979	2·011	2·075	2·138	2·169	2·260
1/ 370	1·8219	2·0215	2·2341	2·4600	2·9530	3·5030	3·8001	4·7830	5·9096	6·3182	7·1870	8·1259	8·6224	10·223
0·00280	1·512	1·551	1·589	1·627	1·702	1·774	1·810	1·914	2·015	2·048	2·113	2·177	2·208	2·301
1/ 357	1·8554	2·0586	2·2751	2·5052	3·0073	3·5673	3·8699	4·8709	6·0181	6·4343	7·3189	8·2751	8·7807	10·410
0·00290	1·539	1·578	1·618	1·656	1·732	1·806	1·842	1·948	2·051	2·084	2·151	2·215	2·247	2·342
1/ 345	1·8882	2·0951	2·3154	2·5496	3·0606	3·6305	3·9384	4·9572	6·1247	6·5482	7·4486	8·4217	8·9362	10·595
0·00300	1·565	1·605	1·645	1·685	1·762	1·837	1·873	1·981	2·086	2·120	2·187	2·253	2·286	2·382
1/ 333	1·9205	2·1309	2·3550	2·5932	3·1129	3·6927	4·0058	5·0420	6·2295	6·6603	7·5760	8·5658	9·0891	10·776
	0·77	0·77	0·77	0·77	0·78	0·78	0·78	0·78	0·79	0·79	0·79	0·79	0·79	0·80

$V_{r(0{\cdot}5)medial}$ **for half-full circular pipes.**

$k_s = 6{\cdot}0$ mm S = 0·00100 to 0·00300

Gradient	(Equivalent) Pipe diameters in mm 1250	1300	1350	1400	1500	1600	1650	1800	1950	2000	2100	2200	2250	2400
0·00300	1·565	1·605	1·645	1·685	1·762	1·837	1·873	1·981	2·086	2·120	2·187	2·253	2·286	2·382
1/ 333	1·9205	2·1309	2·3550	2·5932	3·1129	3·6927	4·0058	5·0420	6·2295	6·6603	7·5760	8·5658	9·0891	10·776
0·00320	1·616	1·658	1·699	1·740	1·819	1·897	1·935	2·046	2·154	2·190	2·259	2·327	2·361	2·460
1/ 313	1·9836	2·2009	2·4324	2·6783	3·2151	3·8139	4·1373	5·2075	6·4339	6·8789	7·8246	8·8469	9·3874	11·130
0·00340	1·666	1·709	1·752	1·793	1·875	1·955	1·994	2·109	2·221	2·257	2·329	2·399	2·434	2·536
1/ 294	2·0447	2·2687	2·5073	2·7608	3·3141	3·9313	4·2647	5·3679	6·6321	7·0907	8·0656	9·1193	9·6765	11·472
0·00360	1·715	1·759	1·802	1·846	1·930	2·012	2·052	2·171	2·285	2·323	2·396	2·469	2·504	2·610
1/ 278	2·1040	2·3345	2·5800	2·8409	3·4103	4·0454	4·3885	5·5236	6·8245	7·2964	8·2996	9·3839	9·9572	11·805
0·00380	1·762	1·807	1·852	1·896	1·983	2·067	2·109	2·230	2·348	2·386	2·462	2·536	2·573	2·681
1/ 263	2·1617	2·3985	2·6508	2·9188	3·5038	4·1563	4·5088	5·6751	7·0116	7·4965	8·5272	9·6412	10·230	12·129
0·00400	1·807	1·854	1·900	1·945	2·034	2·121	2·163	2·288	2·409	2·448	2·526	2·602	2·640	2·751
1/ 250	2·2179	2·4609	2·7197	2·9947	3·5949	4·2644	4·6260	5·8226	7·1939	7·6914	8·7488	9·8918	10·496	12·444
0·00420	1·852	1·900	1·947	1·993	2·085	2·173	2·217	2·345	2·468	2·509	2·588	2·666	2·705	2·819
1/ 238	2·2728	2·5217	2·7869	3·0688	3·6838	4·3698	4·7403	5·9665	7·3716	7·8814	8·9650	10·136	10·755	12·752
0·00440	1·896	1·945	1·993	2·040	2·134	2·225	2·269	2·400	2·526	2·568	2·649	2·729	2·769	2·885
1/ 227	2·3263	2·5811	2·8526	3·1410	3·7705	4·4727	4·8520	6·1070	7·5452	8·0670	9·1761	10·375	11·009	13·052
0·00460	1·938	1·988	2·038	2·086	2·182	2·275	2·320	2·454	2·583	2·626	2·709	2·791	2·831	2·950
1/ 217	2·3786	2·6392	2·9167	3·2117	3·8553	4·5733	4·9611	6·2443	7·7149	8·2484	9·3825	10·608	11·256	13·345
0·00480	1·980	2·031	2·082	2·131	2·229	2·324	2·370	2·507	2·639	2·682	2·767	2·851	2·892	3·013
1/ 208	2·4298	2·6960	2·9795	3·2808	3·9383	4·6717	5·0679	6·3787	7·8809	8·4259	9·5844	10·836	11·499	13·633
0·00500	2·021	2·073	2·125	2·175	2·275	2·371	2·419	2·558	2·693	2·737	2·824	2·910	2·952	3·076
1/ 200	2·4799	2·7516	3·0410	3·3485	4·0196	4·7681	5·1724	6·5103	8·0435	8·5998	9·7821	11·060	11·736	13·914
0·00525	2·071	2·124	2·177	2·229	2·331	2·430	2·479	2·622	2·760	2·805	2·894	2·981	3·025	3·152
1/ 190	2·5412	2·8196	3·1161	3·4312	4·1189	4·8859	5·3002	6·6712	8·2423	8·8123	10·024	11·333	12·026	14·258
0·00550	2·120	2·174	2·228	2·281	2·386	2·487	2·537	2·683	2·825	2·871	2·962	3·052	3·096	3·226
1/ 182	2·6011	2·8860	3·1895	3·5120	4·2159	5·0010	5·4250	6·8283	8·4364	9·0197	10·260	11·600	12·309	14·593
0·00575	2·167	2·223	2·278	2·333	2·439	2·543	2·594	2·744	2·888	2·936	3·029	3·120	3·165	3·298
1/ 174	2·6596	2·9509	3·2613	3·5910	4·3107	5·1134	5·5470	6·9818	8·6261	9·2226	10·491	11·861	12·586	14·921
0·00600	2·214	2·271	2·327	2·383	2·492	2·598	2·650	2·803	2·951	2·999	3·094	3·187	3·233	3·369
1/ 167	2·7168	3·0144	3·3314	3·6683	4·4035	5·2235	5·6664	7·1321	8·8117	9·4210	10·716	12·116	12·856	15·243
0·00625	2·260	2·318	2·375	2·432	2·543	2·652	2·705	2·861	3·011	3·061	3·158	3·253	3·300	3·439
1/ 160	2·7729	3·0766	3·4002	3·7440	4·4943	5·3313	5·7833	7·2792	8·9935	9·6154	10·937	12·366	13·122	15·557
0·00650	2·304	2·364	2·423	2·480	2·594	2·704	2·758	2·917	3·071	3·121	3·220	3·318	3·366	3·507
1/ 154	2·8278	3·1376	3·4676	3·8182	4·5834	5·4369	5·8979	7·4234	9·1717	9·8059	11·154	12·611	13·382	15·865
0·00675	2·348	2·409	2·469	2·528	2·643	2·756	2·811	2·973	3·130	3·181	3·282	3·381	3·430	3·574
1/ 148	2·8817	3·1974	3·5337	3·8910	4·6707	5·5405	6·0103	7·5649	9·3465	9·9928	11·367	12·852	13·637	16·168
0·00700	2·391	2·453	2·514	2·574	2·692	2·806	2·862	3·027	3·187	3·239	3·342	3·443	3·493	3·639
1/ 143	2·9347	3·2561	3·5986	3·9624	4·7565	5·6423	6·1207	7·7038	9·5181	10·176	11·575	13·088	13·887	16·464
0·00725	2·434	2·497	2·559	2·620	2·739	2·856	2·913	3·081	3·244	3·297	3·401	3·504	3·554	3·704
1/ 138	2·9866	3·3138	3·6623	4·0326	4·8407	5·7422	6·2291	7·8402	9·6866	10·356	11·780	13·319	14·133	16·756
0·00750	2·475	2·539	2·602	2·664	2·786	2·905	2·963	3·134	3·299	3·353	3·459	3·564	3·615	3·767
1/ 133	3·0377	3·3705	3·7249	4·1016	4·9235	5·8404	6·3356	7·9743	9·8523	10·534	11·982	13·547	14·375	17·042
0·00800	2·557	2·623	2·688	2·752	2·878	3·000	3·060	3·237	3·407	3·463	3·573	3·681	3·734	3·891
1/ 125	3·1374	3·4811	3·8472	4·2362	5·0851	6·0320	6·5435	8·2360	10·176	10·879	12·375	13·992	14·846	17·602
0·00850	2·635	2·703	2·770	2·837	2·966	3·092	3·154	3·336	3·512	3·570	3·683	3·794	3·849	4·011
1/ 118	3·2340	3·5883	3·9656	4·3666	5·2417	6·2178	6·7450	8·4896	10·489	11·214	12·756	14·422	15·303	18·144
0·00900	2·712	2·782	2·851	2·919	3·052	3·182	3·246	3·433	3·614	3·673	3·790	3·904	3·960	4·127
1/ 111	3·3278	3·6924	4·0807	4·4933	5·3937	6·3981	6·9407	8·7358	10·793	11·539	13·126	14·841	15·747	18·670
0·00950	2·786	2·858	2·929	2·999	3·136	3·269	3·335	3·527	3·713	3·774	3·894	4·011	4·069	4·240
1/ 105	3·4191	3·7936	4·1925	4·6165	5·5416	6·5735	7·1309	8·9753	11·089	11·856	13·486	15·247	16·179	19·182
0·01000	2·859	2·932	3·005	3·077	3·217	3·354	3·422	3·619	3·810	3·872	3·995	4·115	4·175	4·350
1/ 100	3·5080	3·8922	4·3015	4·7365	5·6856	6·7444	7·3163	9·2086	11·377	12·164	13·836	15·644	16·599	19·680
	0·77	0·77	0·77	0·77	0·78	0·78	0·78	0·78	0·79	0·79	0·79	0·79	0·79	0·80

$V_{r(0{\cdot}5)medial}$ **for half-full circular pipes.**

k_s = 6·0 mm
S = 0·01000 to 0·03000

ie hydraulic gradient =
1 in 100 to 1 in 33·3

Water (or sewage) at 15°C;
full bore conditions.

velocities in ms^{-1}
discharges in m^3s^{-1}

Gradient **(Equivalent) Pipe diameters in mm**

Gradient	1250	1300	1350	1400	1500	1600	1650	1800	1950	2000	2100	2200	2250	2400
0·01000	2·859	2·932	3·005	3·077	3·217	3·354	3·422	3·619	3·810	3·872	3·995	4·115	4·175	4·350
1/ 100	3·5080	3·8922	4·3015	4·7365	5·6856	6·7444	7·3163	9·2086	11·377	12·164	13·836	15·644	16·599	19·680
0·01050	2·929	3·005	3·079	3·153	3·297	3·437	3·506	3·708	3·904	3·968	4·093	4·217	4·278	4·458
1/ 95	3·5946	3·9884	4·4078	4·8535	5·8261	6·9110	7·4970	9·4361	11·658	12·464	14·178	16·030	17·009	20·166
0·01100	2·998	3·076	3·152	3·227	3·375	3·518	3·589	3·795	3·996	4·061	4·190	4·316	4·379	4·563
1/ 91	3·6793	4·0823	4·5116	4·9678	5·9633	7·0737	7·6735	9·6582	11·933	12·758	14·512	16·408	17·410	20·641
0·01150	3·066	3·145	3·223	3·300	3·450	3·597	3·669	3·881	4·085	4·152	4·284	4·413	4·477	4·665
1/ 87	3·7620	4·1741	4·6130	5·0795	6·0974	7·2328	7·8461	9·8754	12·201	13·045	14·838	16·776	17·801	21·105
0·01200	3·132	3·212	3·292	3·371	3·525	3·675	3·748	3·964	4·173	4·242	4·376	4·508	4·573	4·766
1/ 83	3·8430	4·2639	4·7123	5·1888	6·2286	7·3884	8·0149	10·088	12·464	13·325	15·157	17·137	18·184	21·559
0·01250	3·196	3·279	3·360	3·440	3·597	3·750	3·826	4·046	4·259	4·329	4·466	4·601	4·668	4·864
1/ 80	3·9222	4·3519	4·8095	5·2958	6·3571	7·5408	8·1802	10·296	12·721	13·600	15·470	17·491	18·559	22·004
0·01300	3·259	3·344	3·427	3·508	3·669	3·825	3·901	4·126	4·344	4·415	4·555	4·692	4·760	4·960
1/ 77	3·9999	4·4381	4·9048	5·4007	6·4830	7·6902	8·3423	10·500	12·973	13·870	15·776	17·837	18·927	22·440
0·01350	3·322	3·407	3·492	3·575	3·739	3·898	3·976	4·205	4·427	4·499	4·642	4·782	4·851	5·055
1/ 74	4·0762	4·5227	4·9983	5·5037	6·6066	7·8368	8·5013	10·700	13·220	14·134	16·077	18·177	19·288	22·867
0·01400	3·383	3·470	3·556	3·641	3·807	3·969	4·049	4·282	4·508	4·582	4·727	4·870	4·940	5·148
1/ 71	4·1510	4·6057	5·0900	5·6047	6·7278	7·9806	8·6573	10·896	13·463	14·393	16·372	18·511	19·642	23·287
0·01450	3·442	3·531	3·619	3·705	3·875	4·040	4·120	4·358	4·588	4·663	4·811	4·956	5·027	5·239
1/ 69	4·2245	4·6873	5·1802	5·7039	6·8470	8·1220	8·8106	11·089	13·701	14·648	16·662	18·839	19·990	23·699
0·01500	3·501	3·592	3·681	3·769	3·941	4·109	4·191	4·432	4·666	4·742	4·893	5·041	5·113	5·328
1/ 67	4·2968	4·7674	5·2688	5·8015	6·9641	8·2609	8·9613	11·279	13·935	14·899	16·947	19·161	20·331	24·105
0·01600	3·616	3·710	3·802	3·892	4·070	4·243	4·328	4·578	4·819	4·898	5·053	5·206	5·281	5·503
1/ 63	4·4378	4·9238	5·4416	5·9918	7·1926	8·5319	9·2553	11·649	14·392	15·388	17·503	19·789	20·998	24·895
0·01700	3·728	3·824	3·919	4·012	4·195	4·374	4·462	4·719	4·968	5·049	5·209	5·366	5·444	5·672
1/ 59	4·5744	5·0755	5·6092	6·1763	7·4140	8·7946	9·5403	12·008	14·835	15·861	18·042	20·399	21·645	25·662
0·01800	3·836	3·935	4·032	4·129	4·317	4·501	4·591	4·856	5·112	5·195	5·360	5·522	5·602	5·837
1/ 56	4·7071	5·2227	5·7718	6·3554	7·6290	9·0496	9·8169	12·356	15·266	16·321	18·565	20·990	22·273	26·406
0·01900	3·941	4·043	4·143	4·242	4·435	4·624	4·717	4·989	5·252	5·338	5·507	5·673	5·755	5·997
1/ 53	4·8361	5·3658	5·9301	6·5297	7·8381	9·2977	10·086	12·695	15·684	16·769	19·074	21·566	22·883	27·130
0·02000	4·043	4·148	4·251	4·352	4·551	4·744	4·840	5·118	5·388	5·476	5·650	5·821	5·905	6·153
1/ 50	4·9618	5·5053	6·0842	6·6994	8·0418	9·5393	10·348	13·025	16·092	17·204	19·570	22·126	23·478	27·835
0·02100	4·143	4·250	4·356	4·459	4·663	4·862	4·959	5·245	5·521	5·612	5·790	5·964	6·051	6·305
1/ 47·6	5·0844	5·6413	6·2345	6·8649	8·2405	9·7749	10·604	13·346	16·489	17·629	20·053	22·673	24·058	28·522
0·02200	4·241	4·350	4·458	4·564	4·773	4·976	5·076	5·368	5·651	5·744	5·926	6·105	6·193	6·453
1/ 45·5	5·2040	5·7741	6·3812	7·0265	8·4345	10·005	10·853	13·660	16·877	18·044	20·525	23·206	24·624	29·194
0·02300	4·336	4·448	4·558	4·667	4·880	5·088	5·190	5·489	5·778	5·873	6·059	6·242	6·332	6·598
1/ 43·5	5·3211	5·9039	6·5247	7·1844	8·6241	10·230	11·097	13·968	17·257	18·450	20·986	23·728	25·177	29·850
0·02400	4·429	4·544	4·656	4·768	4·985	5·197	5·302	5·607	5·903	5·999	6·189	6·376	6·468	6·740
1/ 41·7	5·4355	6·0309	6·6651	7·3390	8·8097	10·450	11·336	14·268	17·628	18·847	21·438	24·238	25·719	30·492
0·02500	4·521	4·637	4·752	4·866	5·088	5·305	5·411	5·723	6·024	6·123	6·317	6·508	6·602	6·879
1/ 40·0	5·5477	6·1553	6·8026	7·4904	8·9914	10·666	11·570	14·562	17·992	19·236	21·880	24·738	26·249	31·121
0·02600	4·610	4·729	4·847	4·962	5·189	5·410	5·518	5·836	6·144	6·244	6·442	6·637	6·733	7·015
1/ 38·5	5·6576	6·2773	6·9373	7·6388	9·1695	10·877	11·799	14·851	18·348	19·617	22·313	25·228	26·769	31·737
0·02700	4·698	4·819	4·939	5·057	5·288	5·513	5·623	5·947	6·261	6·363	6·565	6·763	6·861	7·149
1/ 37·0	5·7654	6·3969	7·0695	7·7843	9·3442	11·084	12·024	15·134	18·698	19·990	22·739	25·709	27·280	32·342
0·02800	4·784	4·908	5·030	5·150	5·385	5·614	5·727	6·056	6·376	6·480	6·686	6·887	6·987	7·280
1/ 35·7	5·8712	6·5143	7·1993	7·9272	9·5157	11·288	12·245	15·412	19·041	20·357	23·156	26·181	27·780	32·936
0·02900	4·869	4·995	5·119	5·241	5·480	5·713	5·828	6·164	6·489	6·595	6·804	7·009	7·111	7·409
1/ 34·5	5·9752	6·6296	7·3268	8·0676	9·6842	11·487	12·461	15·684	19·378	20·718	23·566	26·644	28·272	33·519
0·03000	4·952	5·080	5·206	5·330	5·574	5·811	5·928	6·269	6·599	6·707	6·920	7·129	7·232	7·536
1/ 33·3	6·0773	6·7430	7·4521	8·2055	9·8498	11·684	12·675	15·953	19·709	21·072	23·969	27·100	28·755	34·092
	0·77	0·77	0·77	0·77	0·78	0·78	0·78	0·78	0·79	0·79	0·79	0·79	0·79	0·80

$V_{r(0{\cdot}5)medial}$ for half-full circular pipes.

k_s = 6·0 mm S = 0·01000 to 0·03000

$k_s = 6{\cdot}0$ **mm**
S = 0·03000 to 0·10000

ie hydraulic gradient =
1 in 33·3 to 1 in 10·0

Water (or sewage) at 15°C;
full bore conditions.

velocities in ms^{-1}
discharges in m^3s^{-1}

A44

(p.6 of 6)

Gradient **(Equivalent) Pipe diameters in mm**

Gradient	1250	1300	1350	1400	1500	1600	1650	1800	1950	2000	2100	2200	2250	2400
0·03000	4·952	5·080	5·206	5·330	5·574	5·811	5·928	6·269	6·599	6·707	6·920	7·129	7·232	7·536
1/ 33·3	6·0773	6·7430	7·4521	8·2055	9·8498	11·684	12·675	15·953	19·709	21·072	23·969	27·100	28·755	34·092
0·03200	5·115	5·247	5·377	5·505	5·757	6·002	6·122	6·475	6·816	6·927	7·147	7·363	7·469	7·783
1/ 31·3	6·2767	6·9642	7·6965	8·4747	10·173	12·067	13·090	16·476	20·356	21·763	24·755	27·989	29·699	35·210
0·03400	5·272	5·408	5·542	5·675	5·934	6·186	6·310	6·674	7·026	7·141	7·367	7·590	7·699	8·023
1/ 29·4	6·4699	7·1786	7·9335	8·7356	10·486	12·439	13·493	16·983	20·982	22·433	25·517	28·850	30·613	36·294
0·03600	5·425	5·565	5·703	5·839	6·106	6·366	6·493	6·867	7·230	7·348	7·581	7·810	7·923	8·255
1/ 27·8	6·6576	7·3868	8·1635	8·9889	10·790	12·799	13·885	17·476	21·591	23·084	26·257	29·687	31·501	37·346
0·03800	5·574	5·718	5·860	5·999	6·273	6·540	6·671	7·056	7·428	7·549	7·789	8·024	8·140	8·482
1/ 26·3	6·8401	7·5893	8·3873	9·2353	11·086	13·150	14·265	17·955	22·183	23·716	26·977	30·501	32·364	38·370
0·04000	5·719	5·866	6·012	6·155	6·436	6·710	6·845	7·239	7·621	7·745	7·991	8·232	8·351	8·702
1/ 25·0	7·0178	7·7865	8·6052	9·4753	11·374	13·492	14·636	18·421	22·759	24·333	27·678	31·293	33·205	39·367
0·04200	5·860	6·011	6·160	6·307	6·595	6·876	7·014	7·418	7·809	7·937	8·188	8·436	8·557	8·917
1/ 23·8	7·1911	7·9788	8·8178	9·7093	11·655	13·825	14·997	18·876	23·321	24·934	28·361	32·066	34·025	40·339
0·04400	5·998	6·153	6·305	6·456	6·751	7·038	7·179	7·592	7·993	8·123	8·381	8·634	8·759	9·127
1/ 22·7	7·3604	8·1666	9·0254	9·9379	11·929	14·151	15·350	19·320	23·870	25·521	29·029	32·821	34·826	41·289
0·04600	6·133	6·291	6·447	6·601	6·902	7·196	7·340	7·763	8·172	8·306	8·570	8·828	8·956	9·332
1/ 21·7	7·5259	8·3502	9·2282	10·161	12·197	14·469	15·695	19·755	24·407	26·094	29·681	33·559	35·609	42·217
0·04800	6·265	6·426	6·586	6·743	7·051	7·351	7·498	7·930	8·348	8·485	8·754	9·018	9·148	9·533
1/ 20·8	7·6878	8·5298	9·4268	10·380	12·460	14·780	16·033	20·180	24·932	26·656	30·320	34·281	36·375	43·125
0·05000	6·394	6·559	6·722	6·882	7·196	7·503	7·653	8·094	8·520	8·660	8·934	9·204	9·337	9·729
1/ 20·0	7·8464	8·7058	9·6212	10·594	12·717	15·085	16·364	20·596	25·446	27·205	30·945	34·988	37·125	44·014
0·05250	6·552	6·721	6·888	7·052	7·374	7·688	7·842	8·294	8·731	8·874	9·155	9·431	9·568	9·970
1/ 19·0	8·0402	8·9208	9·8589	10·856	13·031	15·457	16·768	21·105	26·074	27·877	31·710	35·852	38·042	45·102
0·05500	6·706	6·879	7·050	7·218	7·548	7·869	8·026	8·489	8·936	9·082	9·371	9·653	9·793	10·20
1/ 18·2	8·2294	9·1308	10·091	11·111	13·338	15·821	17·163	21·601	26·688	28·533	32·456	36·696	38·937	46·163
0·05750	6·857	7·034	7·208	7·380	7·717	8·046	8·207	8·680	9·137	9·287	9·581	9·870	10·01	10·43
1/ 17·4	8·4144	9·3361	10·318	11·361	13·637	16·177	17·548	22·087	27·288	29·175	33·186	37·520	39·812	47·201
0·06000	7·004	7·185	7·363	7·539	7·883	8·219	8·383	8·866	9·334	9·486	9·787	10·08	10·23	10·66
1/ 16·7	8·5954	9·5369	10·540	11·605	13·931	16·525	17·926	22·562	27·875	29·802	33·899	38·328	40·669	48·216
0·06250	7·149	7·333	7·515	7·694	8·046	8·388	8·556	9·049	9·526	9·682	9·989	10·29	10·44	10·88
1/ 16·0	8·7727	9·7336	10·757	11·845	14·218	16·866	18·296	23·027	28·450	30·417	34·599	39·118	41·508	49·210
0·06500	7·290	7·479	7·664	7·847	8·205	8·554	8·726	9·228	9·715	9·874	10·19	10·49	10·65	11·09
1/ 15·4	8·9465	9·9264	10·970	12·079	14·500	17·200	18·658	23·483	29·013	31·019	35·284	39·893	42·330	50·185
0·06750	7·429	7·621	7·810	7·996	8·362	8·717	8·892	9·404	9·900	10·06	10·38	10·69	10·85	11·30
1/ 14·8	9·1169	10·116	11·179	12·309	14·776	17·527	19·013	23·931	29·566	31·610	35·956	40·653	43·136	51·141
0·07000	7·566	7·761	7·953	8·143	8·515	8·877	9·055	9·577	10·08	10·25	10·57	10·89	11·05	11·51
1/ 14·3	9·2843	10·301	11·384	12·535	15·047	17·849	19·362	24·370	30·109	32·191	36·616	41·399	43·928	52·080
0·07250	7·699	7·898	8·094	8·287	8·666	9·035	9·216	9·746	10·26	10·43	10·76	11·08	11·24	11·72
1/ 13·8	9·4486	10·484	11·586	12·757	15·314	18·165	19·705	24·801	30·642	32·760	37·264	42·132	44·706	53·002
0·07500	7·831	8·033	8·233	8·429	8·814	9·189	9·373	9·913	10·44	10·61	10·94	11·27	11·44	11·92
1/ 13·3	9·6102	10·663	11·784	12·975	15·575	18·476	20·042	25·226	31·166	33·321	37·901	42·852	45·470	53·908
0·08000	8·088	8·297	8·503	8·705	9·103	9·490	9·681	10·24	10·78	10·95	11·30	11·64	11·81	12·31
1/ 12·5	9·9254	11·013	12·171	13·401	16·086	19·082	20·700	26·053	32·188	34·414	39·144	44·258	46·961	55·676
0·08500	8·337	8·552	8·764	8·973	9·383	9·783	9·979	10·55	11·11	11·29	11·65	12·00	12·17	12·69
1/ 11·8	10·231	11·352	12·545	13·814	16·582	19·669	21·337	26·855	33·179	35·473	40·349	45·620	48·407	57·390
0·09000	8·579	8·800	9·018	9·234	9·655	10·07	10·27	10·86	11·43	11·62	11·99	12·35	12·53	13·05
1/ 11·1	10·528	11·681	12·909	14·214	17·062	20·239	21·955	27·634	34·141	36·501	41·519	46·943	49·810	59·054
0·09500	8·814	9·041	9·266	9·487	9·920	10·34	10·55	11·16	11·75	11·94	12·32	12·69	12·87	13·41
1/ 10·5	10·816	12·001	13·263	14·604	17·530	20·794	22·557	28·391	35·076	37·502	42·657	48·229	51·175	60·672
0·10000	9·043	9·276	9·506	9·733	10·18	10·61	10·82	11·45	12·05	12·25	12·64	13·02	13·21	13·76
1/ 10·0	11·097	12·313	13·607	14·983	17·985	21·334	23·143	29·129	35·988	38·476	43·765	49·482	52·505	62·249
	0·77	0·77	0·77	0·77	0·78	0·78	0·78	0·78	0·79	0·79	0·79	0·79	0·79	0·80

$V_{r(0{\cdot}5)medial}$ **for half-full circular pipes.**

A45
(p.1 of 6)

k_s = **15·0 mm**
S = 0·00010 to 0·00030

ie hydraulic gradient =
1 in 10000 to 1 in 3333

Water (or sewage) at 15°C;
full bore conditions.

velocities in ms^{-1}
discharges in m^3s^{-1}

Gradient	**(Equivalent) Pipe diameters in mm**													
	1250	**1300**	**1350**	**1400**	**1500**	**1600**	**1650**	**1800**	**1950**	**2000**	**2100**	**2200**	**2250**	**2400**
0·000100	0·246	0·253	0·259	0·266	0·278	0·290	0·296	0·314	0·331	0·337	0·348	0·359	0·364	0·380
1/ 10000	0·3019	0·3353	0·3709	0·4087	0·4915	0·5839	0·6338	0·7994	0·9894	1·0584	1·2051	1·3639	1·4479	1·7189
0·000105	0·252	0·259	0·266	0·272	0·285	0·298	0·304	0·322	0·339	0·345	0·357	0·368	0·373	0·389
1/ 9524	0·3093	0·3436	0·3801	0·4189	0·5036	0·5983	0·6495	0·8191	1·0138	1·0845	1·2349	1·3976	1·4837	1·7614
0·000110	0·258	0·265	0·272	0·279	0·292	0·305	0·311	0·329	0·347	0·353	0·365	0·376	0·382	0·399
1/ 9091	0·3166	0·3517	0·3890	0·4287	0·5155	0·6124	0·6648	0·8384	1·0377	1·1101	1·2640	1·4306	1·5187	1·8029
0·000115	0·264	0·271	0·278	0·285	0·298	0·311	0·318	0·337	0·355	0·361	0·373	0·385	0·391	0·408
1/ 8696	0·3238	0·3596	0·3978	0·4384	0·5271	0·6262	0·6798	0·8573	1·0611	1·1351	1·2925	1·4628	1·5529	1·8435
0·000120	0·270	0·277	0·284	0·291	0·305	0·318	0·325	0·344	0·363	0·369	0·381	0·393	0·399	0·416
1/ 8333	0·3308	0·3673	0·4064	0·4478	0·5385	0·6397	0·6945	0·8758	1·0840	1·1595	1·3203	1·4943	1·5863	1·8832
0·000125	0·275	0·282	0·290	0·297	0·311	0·325	0·331	0·351	0·370	0·377	0·389	0·401	0·407	0·425
1/ 8000	0·3376	0·3749	0·4148	0·4571	0·5496	0·6529	0·7088	0·8939	1·1063	1·1835	1·3476	1·5252	1·6191	1·9221
0·000130	0·281	0·288	0·296	0·303	0·317	0·331	0·338	0·358	0·378	0·384	0·397	0·409	0·415	0·433
1/ 7692	0·3443	0·3824	0·4230	0·4662	0·5605	0·6659	0·7229	0·9116	1·1283	1·2070	1·3743	1·5554	1·6512	1·9602
0·000135	0·286	0·294	0·301	0·309	0·323	0·338	0·345	0·365	0·385	0·392	0·404	0·417	0·423	0·442
1/ 7407	0·3509	0·3897	0·4311	0·4751	0·5712	0·6786	0·7367	0·9290	1·1498	1·2300	1·4006	1·5851	1·6827	1·9976
0·000140	0·291	0·299	0·307	0·314	0·329	0·344	0·351	0·372	0·392	0·399	0·412	0·425	0·431	0·450
1/ 7143	0·3573	0·3968	0·4390	0·4838	0·5817	0·6911	0·7502	0·9461	1·1710	1·2526	1·4263	1·6142	1·7136	2·0343
0·000145	0·296	0·304	0·312	0·320	0·335	0·350	0·357	0·378	0·399	0·406	0·419	0·432	0·439	0·458
1/ 6897	0·3637	0·4039	0·4468	0·4924	0·5920	0·7033	0·7635	0·9629	1·1917	1·2748	1·4516	1·6428	1·7440	2·0704
0·000150	0·301	0·309	0·317	0·325	0·341	0·356	0·363	0·385	0·406	0·413	0·426	0·440	0·446	0·465
1/ 6667	0·3699	0·4108	0·4544	0·5008	0·6022	0·7154	0·7766	0·9794	1·2121	1·2966	1·4764	1·6709	1·7739	2·1058
0·000160	0·311	0·320	0·328	0·336	0·352	0·367	0·375	0·398	0·419	0·426	0·440	0·454	0·461	0·481
1/ 6250	0·3820	0·4243	0·4694	0·5173	0·6220	0·7389	0·8021	1·0115	1·2519	1·3392	1·5249	1·7258	1·8321	2·1750
0·000170	0·321	0·330	0·338	0·346	0·363	0·379	0·387	0·410	0·432	0·439	0·454	0·468	0·475	0·496
1/ 5882	0·3938	0·4374	0·4838	0·5332	0·6411	0·7617	0·8268	1·0427	1·2905	1·3805	1·5719	1·7790	1·8886	2·2420
0·000180	0·330	0·339	0·348	0·356	0·373	0·390	0·398	0·422	0·445	0·452	0·467	0·482	0·489	0·510
1/ 5556	0·4053	0·4501	0·4979	0·5487	0·6597	0·7838	0·8508	1·0730	1·3280	1·4206	1·6176	1·8306	1·9434	2·3071
0·000190	0·339	0·348	0·357	0·366	0·384	0·401	0·409	0·433	0·457	0·465	0·480	0·495	0·502	0·524
1/ 5263	0·4164	0·4624	0·5115	0·5638	0·6779	0·8053	0·8742	1·1024	1·3644	1·4596	1·6619	1·8809	1·9967	2·3704
0·000200	0·348	0·357	0·367	0·376	0·394	0·411	0·419	0·444	0·469	0·477	0·492	0·508	0·515	0·538
1/ 5000	0·4272	0·4745	0·5249	0·5784	0·6955	0·8262	0·8969	1·1311	1·3999	1·4975	1·7052	1·9298	2·0486	2·4320
0·000210	0·357	0·366	0·376	0·385	0·403	0·421	0·430	0·455	0·480	0·488	0·504	0·520	0·528	0·551
1/ 4762	0·4378	0·4862	0·5378	0·5927	0·7127	0·8467	0·9191	1·1591	1·4345	1·5346	1·7473	1·9775	2·0993	2·4922
0·000220	0·365	0·375	0·385	0·394	0·413	0·431	0·440	0·466	0·492	0·500	0·516	0·532	0·540	0·564
1/ 4545	0·4481	0·4977	0·5505	0·6067	0·7295	0·8666	0·9408	1·1864	1·4683	1·5707	1·7885	2·0241	2·1488	2·5509
0·000230	0·373	0·383	0·393	0·403	0·422	0·441	0·450	0·477	0·503	0·511	0·528	0·544	0·553	0·577
1/ 4348	0·4582	0·5089	0·5629	0·6204	0·7459	0·8861	0·9619	1·2131	1·5014	1·6061	1·8287	2·0696	2·1971	2·6083
0·000240	0·381	0·392	0·402	0·412	0·431	0·450	0·460	0·487	0·514	0·522	0·539	0·556	0·564	0·589
1/ 4167	0·4681	0·5198	0·5750	0·6337	0·7620	0·9052	0·9827	1·2392	1·5337	1·6406	1·8681	2·1142	2·2444	2·6644
0·000250	0·389	0·400	0·410	0·420	0·440	0·460	0·469	0·497	0·524	0·533	0·550	0·568	0·576	0·601
1/ 4000	0·4777	0·5306	0·5869	0·6468	0·7777	0·9239	1·0029	1·2648	1·5654	1·6745	1·9067	2·1579	2·2907	2·7194
0·000260	0·397	0·408	0·418	0·429	0·449	0·469	0·478	0·507	0·535	0·544	0·561	0·579	0·588	0·613
1/ 3846	0·4872	0·5411	0·5986	0·6597	0·7931	0·9422	1·0228	1·2899	1·5964	1·7077	1·9445	2·2006	2·3362	2·7733
0·000270	0·405	0·415	0·426	0·437	0·457	0·478	0·487	0·517	0·545	0·554	0·572	0·590	0·599	0·625
1/ 3704	0·4965	0·5514	0·6100	0·6722	0·8083	0·9602	1·0423	1·3145	1·6268	1·7403	1·9816	2·2426	2·3807	2·8262
0·000280	0·412	0·423	0·434	0·445	0·466	0·486	0·496	0·526	0·555	0·564	0·583	0·601	0·610	0·636
1/ 3571	0·5056	0·5616	0·6212	0·6846	0·8231	0·9778	1·0615	1·3386	1·6567	1·7723	2·0180	2·2838	2·4244	2·8781
0·000290	0·419	0·431	0·442	0·453	0·474	0·495	0·505	0·535	0·565	0·574	0·593	0·611	0·621	0·647
1/ 3448	0·5146	0·5715	0·6322	0·6967	0·8377	0·9952	1·0803	1·3623	1·6861	1·8037	2·0537	2·3243	2·4674	2·9291
0·000300	0·427	0·438	0·449	0·460	0·482	0·503	0·514	0·545	0·574	0·584	0·603	0·622	0·631	0·659
1/ 3333	0·5234	0·5813	0·6430	0·7086	0·8520	1·0122	1·0988	1·3857	1·7149	1·8345	2·0889	2·3640	2·5096	2·9792
	0·74	0·75	0·75	0·75	0·75	0·75	0·75	0·75	0·76	0·76	0·76	0·76	0·76	0·76

$V_{r(0·5)medial}$ for half-full circular pipes.

k_s = 15·0 mm S = 0·00010 to 0·00030

k_s = 15·0 mm
S = 0·00030 to 0·00100

ie hydraulic gradient =
1 in 3333 to 1 in 1000

Water (or sewage) at 15°C;
full bore conditions.

velocities in ms^{-1}
discharges in m^3s^{-1}

Gradient	**(Equivalent) Pipe diameters in mm**													
	1250	**1300**	**1350**	**1400**	**1500**	**1600**	**1650**	**1800**	**1950**	**2000**	**2100**	**2200**	**2250**	**2400**
0·000300	0·427	0·438	0·449	0·460	0·482	0·503	0·514	0·545	0·574	0·584	0·603	0·622	0·631	0·659
1/ 3333	0·5234	0·5813	0·6430	0·7086	0·8520	1·0122	1·0988	1·3857	1·7149	1·8345	2·0889	2·3640	2·5096	2·9792
0·000320	0·441	0·452	0·464	0·475	0·498	0·520	0·531	0·562	0·593	0·603	0·623	0·642	0·652	0·680
1/ 3125	0·5406	0·6004	0·6641	0·7319	0·8800	1·0454	1·1349	1·4312	1·7713	1·8947	2·1575	2·4416	2·5920	3·0770
0·000340	0·454	0·466	0·478	0·490	0·513	0·536	0·547	0·580	0·611	0·622	0·642	0·662	0·672	0·701
1/ 2941	0·5573	0·6189	0·6846	0·7545	0·9071	1·0776	1·1698	1·4753	1·8258	1·9531	2·2239	2·5169	2·6718	3·1718
0·000360	0·467	0·480	0·492	0·504	0·528	0·552	0·563	0·597	0·629	0·640	0·661	0·681	0·691	0·721
1/ 2778	0·5734	0·6369	0·7045	0·7764	0·9335	1·1089	1·2038	1·5181	1·8788	2·0098	2·2885	2·5899	2·7494	3·2639
0·000380	0·480	0·493	0·506	0·518	0·543	0·567	0·578	0·613	0·646	0·657	0·679	0·700	0·710	0·741
1/ 2632	0·5892	0·6543	0·7238	0·7977	0·9591	1·1393	1·2368	1·5597	1·9303	2·0649	2·3512	2·6609	2·8248	3·3534
0·000400	0·493	0·506	0·519	0·532	0·557	0·581	0·593	0·629	0·663	0·674	0·696	0·718	0·729	0·761
1/ 2500	0·6045	0·6714	0·7426	0·8184	0·9840	1·1690	1·2690	1·6003	1·9805	2·1186	2·4124	2·7301	2·8982	3·4406
0·000420	0·505	0·518	0·532	0·545	0·571	0·596	0·608	0·644	0·680	0·691	0·714	0·736	0·747	0·779
1/ 2381	0·6194	0·6880	0·7610	0·8387	1·0083	1·1979	1·3003	1·6398	2·0295	2·1710	2·4720	2·7976	2·9699	3·5256
0·000440	0·517	0·531	0·544	0·558	0·584	0·610	0·622	0·660	0·696	0·707	0·731	0·753	0·765	0·798
1/ 2273	0·6340	0·7042	0·7789	0·8584	1·0321	1·2261	1·3310	1·6784	2·0773	2·2221	2·5302	2·8635	3·0398	3·6086
0·000460	0·528	0·542	0·556	0·570	0·597	0·624	0·636	0·674	0·711	0·723	0·747	0·770	0·782	0·816
1/ 2174	0·6483	0·7200	0·7964	0·8777	1·0553	1·2537	1·3609	1·7162	2·1240	2·2721	2·5871	2·9279	3·1082	3·6898
0·000480	0·540	0·554	0·568	0·582	0·610	0·637	0·650	0·689	0·727	0·739	0·763	0·787	0·799	0·833
1/ 2083	0·6623	0·7355	0·8136	0·8966	1·0780	1·2807	1·3902	1·7532	2·1697	2·3210	2·6428	2·9909	3·1751	3·7692
0·000500	0·551	0·566	0·580	0·594	0·623	0·650	0·664	0·703	0·742	0·754	0·779	0·803	0·815	0·850
1/ 2000	0·6759	0·7507	0·8304	0·9151	1·1003	1·3071	1·4189	1·7893	2·2145	2·3689	2·6973	3·0526	3·2406	3·8470
0·000525	0·564	0·580	0·594	0·609	0·638	0·666	0·680	0·721	0·760	0·773	0·798	0·823	0·835	0·871
1/ 1905	0·6926	0·7692	0·8509	0·9378	1·1275	1·3394	1·4540	1·8336	2·2693	2·4275	2·7640	3·1281	3·3207	3·9421
0·000550	0·578	0·593	0·608	0·624	0·653	0·682	0·696	0·738	0·778	0·791	0·817	0·842	0·855	0·892
1/ 1818	0·7090	0·7874	0·8710	0·9598	1·1540	1·3710	1·4882	1·8767	2·3227	2·4846	2·8291	3·2017	3·3989	4·0349
0·000575	0·591	0·607	0·622	0·638	0·668	0·697	0·712	0·754	0·795	0·809	0·835	0·861	0·874	0·912
1/ 1739	0·7249	0·8051	0·8906	0·9814	1·1800	1·4018	1·5217	1·9190	2·3749	2·5405	2·8927	3·2738	3·4753	4·1256
0·000600	0·603	0·620	0·636	0·651	0·682	0·712	0·727	0·770	0·812	0·826	0·853	0·880	0·893	0·932
1/ 1667	0·7405	0·8224	0·9097	1·0026	1·2054	1·4320	1·5545	1·9603	2·4261	2·5952	2·9550	3·3442	3·5501	4·2144
0·000625	0·616	0·632	0·649	0·665	0·696	0·727	0·742	0·786	0·829	0·843	0·871	0·898	0·911	0·951
1/ 1600	0·7558	0·8394	0·9285	1·0233	1·2303	1·4615	1·5866	2·0007	2·4761	2·6488	3·0160	3·4132	3·6234	4·3014
0·000650	0·628	0·645	0·662	0·678	0·710	0·741	0·757	0·802	0·846	0·860	0·888	0·916	0·929	0·970
1/ 1538	0·7708	0·8560	0·9469	1·0435	1·2547	1·4905	1·6180	2·0404	2·5252	2·7013	3·0757	3·4809	3·6952	4·3866
0·000675	0·640	0·657	0·674	0·691	0·724	0·755	0·771	0·817	0·862	0·876	0·905	0·933	0·947	0·988
1/ 1481	0·7855	0·8724	0·9650	1·0634	1·2786	1·5189	1·6488	2·0793	2·5733	2·7527	3·1344	3·5472	3·7656	4·4703
0·000700	0·652	0·669	0·687	0·704	0·737	0·769	0·785	0·832	0·877	0·892	0·922	0·950	0·964	1·006
1/ 1429	0·7999	0·8884	0·9827	1·0830	1·3021	1·5468	1·6791	2·1175	2·6206	2·8033	3·1919	3·6123	3·8348	4·5523
0·000725	0·663	0·681	0·699	0·716	0·750	0·783	0·799	0·847	0·893	0·908	0·938	0·967	0·982	1·024
1/ 1379	0·8141	0·9041	1·0001	1·1022	1·3251	1·5742	1·7089	2·1550	2·6670	2·8529	3·2485	3·6763	3·9027	4·6330
0·000750	0·675	0·693	0·711	0·728	0·763	0·796	0·813	0·861	0·908	0·924	0·954	0·984	0·998	1·042
1/ 1333	0·8280	0·9196	1·0172	1·1210	1·3478	1·6011	1·7381	2·1918	2·7126	2·9018	3·3040	3·7392	3·9695	4·7122
0·000800	0·697	0·716	0·734	0·752	0·788	0·822	0·840	0·890	0·938	0·954	0·985	1·016	1·031	1·076
1/ 1250	0·8552	0·9498	1·0506	1·1578	1·3921	1·6537	1·7951	2·2638	2·8017	2·9970	3·4125	3·8619	4·0997	4·8668
0·000850	0·718	0·738	0·757	0·775	0·812	0·848	0·865	0·917	0·967	0·983	1·016	1·047	1·063	1·109
1/ 1176	0·8815	0·9790	1·0830	1·1935	1·4349	1·7046	1·8504	2·3335	2·8879	3·0893	3·5176	3·9809	4·2260	5·0167
0·000900	0·739	0·759	0·779	0·798	0·836	0·872	0·891	0·944	0·995	1·012	1·045	1·078	1·094	1·141
1/ 1111	0·9071	1·0074	1·1144	1·2281	1·4766	1·7541	1·9041	2·4012	2·9717	3·1789	3·6196	4·0963	4·3486	5·1622
0·000950	0·759	0·780	0·800	0·820	0·858	0·896	0·915	0·969	1·022	1·040	1·074	1·107	1·124	1·172
1/ 1053	0·9320	1·0351	1·1449	1·2618	1·5171	1·8022	1·9563	2·4670	3·0532	3·2661	3·7188	4·2086	4·4678	5·3038
0·001000	0·779	0·800	0·821	0·841	0·881	0·920	0·939	0·995	1·049	1·067	1·102	1·136	1·153	1·203
1/ 1000	0·9562	1·0620	1·1747	1·2946	1·5565	1·8490	2·0072	2·5311	3·1326	3·3510	3·8155	4·3180	4·5839	5·4416
	0·74	0·75	0·75	0·75	0·75	0·75	0·75	0·75	0·76	0·76	0·76	0·76	0·76	0·76

$V_{r(0·5)medial}$ for half-full circular pipes.

k_s = 15·0 mm
S = 0·00100 to 0·00300

ie hydraulic gradient =
1 in 1000 to 1 in 333

Water (or sewage) at 15°C;
full bore conditions.

velocities in ms^{-1}
discharges in m^3s^{-1}

Gradient **(Equivalent) Pipe diameters in mm**

Gradient	1250	1300	1350	1400	1500	1600	1650	1800	1950	2000	2100	2200	2250	2400
0·00100	0·779	0·800	0·821	0·841	0·881	0·920	0·939	0·995	1·049	1·067	1·102	1·136	1·153	1·203
1/ 1000	0·9562	1·0620	1·1747	1·2946	1·5565	1·8490	2·0072	2·5311	3·1326	3·3510	3·8155	4·3180	4·5839	5·4416
0·00105	0·798	0·820	0·841	0·862	0·903	0·942	0·962	1·019	1·075	1·093	1·129	1·164	1·181	1·233
1/ 952	0·9799	1·0882	1·2037	1·3266	1·5950	1·8947	2·0568	2·5937	3·2100	3·4338	3·9098	4·4247	4·6972	5·5761
0·00110	0·817	0·839	0·861	0·882	0·924	0·965	0·985	1·043	1·100	1·119	1·155	1·191	1·209	1·262
1/ 909	1·0029	1·1138	1·2321	1·3578	1·6325	1·9393	2·1052	2·6548	3·2855	3·5146	4·0018	4·5289	4·8078	5·7074
0·00115	0·836	0·858	0·880	0·902	0·945	0·986	1·007	1·067	1·125	1·144	1·181	1·218	1·236	1·290
1/ 870	1·0255	1·1389	1·2598	1·3884	1·6692	1·9830	2·1526	2·7145	3·3594	3·5936	4·0918	4·6307	4·9159	5·8357
0·00120	0·854	0·877	0·899	0·921	0·965	1·007	1·028	1·090	1·149	1·169	1·207	1·244	1·263	1·318
1/ 833	1·0476	1·1634	1·2869	1·4182	1·7052	2·0256	2·1989	2·7729	3·4317	3·6710	4·1799	4·7304	5·0217	5·9613
0·00125	0·871	0·895	0·918	0·940	0·985	1·028	1·050	1·112	1·173	1·193	1·232	1·270	1·289	1·345
1/ 800	1·0692	1·1874	1·3135	1·4475	1·7403	2·0674	2·2443	2·8301	3·5025	3·7467	4·2661	4·8280	5·1253	6·0843
0·00130	0·889	0·912	0·936	0·959	1·004	1·049	1·070	1·134	1·196	1·216	1·256	1·295	1·315	1·372
1/ 769	1·0904	1·2110	1·3395	1·4762	1·7748	2·1084	2·2887	2·8862	3·5719	3·8209	4·3506	4·9236	5·2268	6·2048
0·00135	0·905	0·930	0·954	0·977	1·023	1·069	1·091	1·156	1·219	1·239	1·280	1·320	1·340	1·398
1/ 741	1·1112	1·2340	1·3650	1·5043	1·8087	2·1486	2·3324	2·9412	3·6400	3·8938	4·4336	5·0175	5·3264	6·3230
0·00140	0·922	0·947	0·971	0·995	1·042	1·088	1·111	1·177	1·241	1·262	1·304	1·344	1·364	1·423
1/ 714	1·1316	1·2567	1·3901	1·5319	1·8419	2·1880	2·3752	2·9952	3·7068	3·9653	4·5150	5·1096	5·4242	6·4391
0·00145	0·938	0·964	0·988	1·013	1·061	1·108	1·130	1·198	1·263	1·285	1·327	1·368	1·388	1·449
1/ 690	1·1516	1·2790	1·4147	1·5591	1·8745	2·2268	2·4172	3·0482	3·7725	4·0355	4·5949	5·2001	5·5203	6·5532
0·00150	0·954	0·980	1·005	1·030	1·079	1·126	1·150	1·218	1·285	1·307	1·349	1·391	1·412	1·473
1/ 667	1·1713	1·3008	1·4389	1·5858	1·9066	2·2649	2·4586	3·1004	3·8370	4·1045	4·6735	5·2890	5·6147	6·6652
0·00160	0·986	1·012	1·038	1·064	1·114	1·163	1·188	1·258	1·327	1·349	1·394	1·437	1·458	1·522
1/ 625	1·2097	1·3435	1·4861	1·6378	1·9691	2·3392	2·5393	3·2021	3·9629	4·2392	4·8268	5·4626	5·7989	6·8839
0·00170	1·016	1·043	1·070	1·097	1·149	1·199	1·224	1·297	1·368	1·391	1·436	1·481	1·503	1·569
1/ 588	1·2470	1·3849	1·5319	1·6882	2·0298	2·4112	2·6175	3·3007	4·0849	4·3697	4·9754	5·6307	5·9774	7·0959
0·00180	1·046	1·074	1·101	1·129	1·182	1·234	1·260	1·335	1·407	1·431	1·478	1·524	1·547	1·614
1/ 556	1·2832	1·4251	1·5763	1·7372	2·0886	2·4812	2·6934	3·3964	4·2034	4·4964	5·1198	5·7941	6·1508	7·3017
0·00190	1·074	1·103	1·131	1·159	1·214	1·268	1·294	1·371	1·446	1·470	1·519	1·566	1·589	1·658
1/ 526	1·3184	1·4641	1·6195	1·7848	2·1459	2·5492	2·7672	3·4895	4·3186	4·6197	5·2601	5·9529	6·3194	7·5018
0·00200	1·102	1·132	1·161	1·190	1·246	1·301	1·328	1·407	1·484	1·509	1·558	1·607	1·631	1·701
1/ 500	1·3526	1·5022	1·6616	1·8312	2·2017	2·6154	2·8391	3·5802	4·4309	4·7398	5·3968	6·1076	6·4836	7·6968
0·00210	1·129	1·160	1·190	1·219	1·277	1·333	1·361	1·442	1·520	1·546	1·597	1·646	1·671	1·743
1/ 476	1·3860	1·5393	1·7027	1·8765	2·2561	2·6801	2·9093	3·6687	4·5403	4·8569	5·5301	6·2585	6·6438	7·8869
0·00220	1·156	1·187	1·218	1·248	1·307	1·364	1·393	1·476	1·556	1·582	1·634	1·685	1·710	1·784
1/ 455	1·4187	1·5756	1·7428	1·9206	2·3092	2·7432	2·9778	3·7551	4·6472	4·9712	5·6603	6·4058	6·8002	8·0726
0·00230	1·182	1·214	1·245	1·276	1·336	1·395	1·424	1·509	1·591	1·618	1·671	1·723	1·749	1·825
1/ 435	1·4506	1·6110	1·7820	1·9638	2·3611	2·8048	3·0447	3·8395	4·7517	5·0830	5·7876	6·5498	6·9531	8·2541
0·00240	1·207	1·240	1·272	1·303	1·365	1·425	1·455	1·541	1·625	1·653	1·707	1·760	1·786	1·864
1/ 417	1·4818	1·6457	1·8203	2·0061	2·4119	2·8652	3·1103	3·9221	4·8539	5·1923	5·9121	6·6908	7·1027	8·4317
0·00250	1·232	1·265	1·298	1·330	1·393	1·454	1·485	1·573	1·659	1·687	1·742	1·796	1·823	1·902
1/ 400	1·5124	1·6796	1·8579	2·0475	2·4617	2·9243	3·1744	4·0030	4·9541	5·2994	6·0341	6·8288	7·2492	8·6056
0·00260	1·257	1·290	1·324	1·356	1·421	1·483	1·514	1·604	1·692	1·720	1·777	1·832	1·859	1·940
1/ 385	1·5423	1·7129	1·8947	2·0880	2·5104	2·9822	3·2373	4·0823	5·0522	5·4044	6·1536	6·9641	7·3928	8·7761
0·00270	1·281	1·315	1·349	1·382	1·448	1·512	1·543	1·635	1·724	1·753	1·811	1·867	1·895	1·977
1/ 370	1·5717	1·7455	1·9308	2·1278	2·5583	3·0391	3·2990	4·1601	5·1485	5·5074	6·2709	7·0968	7·5337	8·9433
0·00280	1·304	1·339	1·374	1·408	1·474	1·539	1·571	1·665	1·756	1·785	1·844	1·901	1·930	2·013
1/ 357	1·6006	1·7776	1·9662	2·1669	2·6052	3·0949	3·3596	4·2365	5·2430	5·6085	6·3860	7·2270	7·6720	9·1075
0·00290	1·327	1·363	1·398	1·433	1·500	1·567	1·599	1·694	1·787	1·817	1·876	1·935	1·964	2·049
1/ 345	1·6289	1·8090	2·0011	2·2053	2·6514	3·1497	3·4190	4·3115	5·3358	5·7078	6·4991	7·3550	7·8078	9·2687
0·00300	1·350	1·386	1·422	1·457	1·526	1·593	1·626	1·723	1·817	1·848	1·908	1·968	1·997	2·084
1/ 333	1·6568	1·8400	2·0353	2·2430	2·6967	3·2035	3·4775	4·3852	5·4271	5·8054	6·6102	7·4808	7·9414	9·4272
	0·74	0·75	0·75	0·75	0·75	0·75	0·75	0·75	0·76	0·76	0·76	0·76	0·76	0·76

$V_{r(0\cdot5)medial}$ for half-full circular pipes.

k_s = 15·0 mm S = 0·00100 to 0·00300

$k_s = 15{\cdot}0$ mm
S = 0·00300 to 0·01000

ie hydraulic gradient =
1 in 333 to 1 in 100

Water (or sewage) at 15°C;
full bore conditions.

velocities in ms^{-1}
discharges in m^3s^{-1}

Gradient (Equivalent) Pipe diameters in mm

Gradient	1250	1300	1350	1400	1500	1600	1650	1800	1950	2000	2100	2200	2250	2400
0·00300	1·350	1·386	1·422	1·457	1·526	1·593	1·626	1·723	1·817	1·848	1·908	1·968	1·997	2·084
1/ 333	1·6568	1·8400	2·0353	2·2430	2·6967	3·2035	3·4775	4·3852	5·4271	5·8054	6·6102	7·4808	7·9414	9·4272
0·00320	1·394	1·432	1·469	1·505	1·576	1·646	1·680	1·780	1·877	1·909	1·971	2·032	2·063	2·152
1/ 313	1·7111	1·9004	2·1021	2·3166	2·7852	3·3086	3·5916	4·5291	5·6051	5·9959	6·8271	7·7262	8·2019	9·7365
0·00340	1·437	1·476	1·514	1·551	1·625	1·696	1·731	1·835	1·935	1·967	2·032	2·095	2·126	2·218
1/ 294	1·7638	1·9589	2·1668	2·3879	2·8710	3·4105	3·7022	4·6685	5·7777	6·1805	7·0372	7·9640	8·4544	10·036
0·00360	1·479	1·519	1·558	1·596	1·672	1·745	1·782	1·888	1·991	2·024	2·091	2·156	2·188	2·283
1/ 278	1·8150	2·0157	2·2296	2·4572	2·9542	3·5094	3·8096	4·8039	5·9453	6·3597	7·2413	8·1950	8·6995	10·327
0·00380	1·520	1·560	1·600	1·640	1·718	1·793	1·830	1·940	2·045	2·080	2·148	2·215	2·248	2·345
1/ 263	1·8647	2·0709	2·2908	2·5245	3·0352	3·6056	3·9140	4·9356	6·1082	6·5340	7·4398	8·4196	8·9380	10·610
0·00400	1·559	1·601	1·642	1·683	1·762	1·840	1·878	1·990	2·098	2·134	2·204	2·272	2·306	2·406
1/ 250	1·9132	2·1248	2·3503	2·5901	3·1141	3·6993	4·0157	5·0639	6·2670	6·7038	7·6331	8·6384	9·1703	10·886
0·00420	1·598	1·640	1·683	1·724	1·806	1·885	1·924	2·039	2·150	2·187	2·258	2·329	2·363	2·466
1/ 238	1·9605	2·1773	2·4083	2·6541	3·1910	3·7907	4·1149	5·1889	6·4218	6·8694	7·8217	8·8518	9·3968	11·155
0·00440	1·635	1·679	1·722	1·765	1·848	1·930	1·970	2·087	2·201	2·238	2·311	2·383	2·419	2·524
1/ 227	2·0066	2·2285	2·4650	2·7166	3·2661	3·8799	4·2118	5·3111	6·5729	7·0311	8·0058	9·0602	9·6180	11·418
0·00460	1·672	1·717	1·761	1·804	1·890	1·973	2·014	2·134	2·250	2·288	2·363	2·437	2·473	2·581
1/ 217	2·0517	2·2786	2·5205	2·7777	3·3396	3·9671	4·3065	5·4305	6·7207	7·1892	8·1858	9·2638	9·8342	11·674
0·00480	1·708	1·754	1·799	1·843	1·930	2·016	2·057	2·180	2·299	2·338	2·414	2·489	2·527	2·636
1/ 208	2·0959	2·3276	2·5747	2·8374	3·4114	4·0525	4·3991	5·5473	6·8653	7·3439	8·3619	9·4631	10·046	11·925
0·00500	1·743	1·790	1·836	1·881	1·970	2·057	2·100	2·225	2·346	2·386	2·464	2·541	2·579	2·690
1/ 200	2·1391	2·3757	2·6278	2·8960	3·4818	4·1361	4·4898	5·6617	7·0069	7·4953	8·5343	9·6583	10·253	12·171
0·00525	1·786	1·834	1·881	1·928	2·019	2·108	2·152	2·280	2·404	2·445	2·525	2·604	2·642	2·757
1/ 190	2·1920	2·4343	2·6927	2·9675	3·5678	4·2383	4·6008	5·8016	7·1800	7·6805	8·7452	9·8969	10·506	12·472
0·00550	1·828	1·877	1·925	1·973	2·066	2·158	2·202	2·334	2·461	2·502	2·584	2·665	2·705	2·822
1/ 182	2·2436	2·4916	2·7561	3·0373	3·6518	4·3380	4·7090	5·9382	7·3490	7·8613	8·9510	10·130	10·754	12·765
0·00575	1·869	1·919	1·969	2·017	2·113	2·206	2·252	2·386	2·516	2·559	2·642	2·725	2·765	2·885
1/ 174	2·2940	2·5477	2·8181	3·1056	3·7339	4·4355	4·8149	6·0717	7·5142	8·0380	9·1522	10·358	10·995	13·052
0·00600	1·910	1·961	2·011	2·061	2·158	2·254	2·300	2·437	2·570	2·614	2·699	2·783	2·825	2·947
1/ 167	2·3434	2·6025	2·8787	3·1724	3·8142	4·5310	4·9185	6·2023	7·6758	8·2109	9·3491	10·580	11·232	13·333
0·00625	1·949	2·001	2·053	2·103	2·203	2·300	2·348	2·488	2·623	2·668	2·755	2·841	2·883	3·008
1/ 160	2·3917	2·6562	2·9381	3·2379	3·8929	4·6244	5·0200	6·3302	7·8342	8·3803	9·5419	10·799	11·463	13·608
0·00650	1·988	2·041	2·093	2·145	2·247	2·346	2·394	2·537	2·675	2·720	2·809	2·897	2·940	3·068
1/ 154	2·4391	2·7088	2·9963	3·3020	3·9700	4·7160	5·1194	6·4556	7·9893	8·5463	9·7309	11·012	11·691	13·878
0·00675	2·025	2·080	2·133	2·186	2·289	2·390	2·440	2·585	2·726	2·772	2·863	2·952	2·996	3·126
1/ 148	2·4856	2·7604	3·0534	3·3649	4·0456	4·8059	5·2169	6·5786	8·1416	8·7091	9·9164	11·222	11·913	14·142
0·00700	2·063	2·118	2·172	2·226	2·331	2·434	2·485	2·633	2·776	2·823	2·916	3·006	3·051	3·183
1/ 143	2·5312	2·8111	3·1094	3·4267	4·1199	4·8941	5·3127	6·6994	8·2910	8·8690	10·098	11·428	12·132	14·402
0·00725	2·099	2·155	2·211	2·265	2·373	2·477	2·529	2·679	2·825	2·873	2·967	3·060	3·105	3·240
1/ 138	2·5760	2·8608	3·1645	3·4874	4·1928	4·9808	5·4068	6·8180	8·4378	9·0260	10·277	11·631	12·347	14·657
0·00750	2·135	2·192	2·249	2·304	2·413	2·520	2·572	2·725	2·874	2·922	3·018	3·112	3·158	3·295
1/ 133	2·6200	2·9098	3·2186	3·5470	4·2645	5·0659	5·4992	6·9345	8·5821	9·1803	10·453	11·829	12·558	14·907
0·00800	2·205	2·264	2·322	2·380	2·492	2·602	2·656	2·814	2·968	3·018	3·117	3·214	3·262	3·403
1/ 125	2·7060	3·0052	3·3242	3·6634	4·4044	5·2321	5·6796	7·1620	8·8636	9·4814	10·796	12·218	12·970	15·396
0·00850	2·273	2·334	2·394	2·453	2·569	2·682	2·738	2·901	3·059	3·111	3·213	3·313	3·362	3·508
1/ 118	2·7893	3·0977	3·4265	3·7761	4·5400	5·3932	5·8544	7·3825	9·1364	9·7733	11·128	12·594	13·369	15·870
0·00900	2·339	2·401	2·463	2·524	2·644	2·760	2·817	2·985	3·148	3·201	3·306	3·409	3·460	3·610
1/ 111	2·8702	3·1876	3·5259	3·8857	4·6717	5·5496	6·0242	7·5966	9·4013	10·057	11·451	12·959	13·757	16·330
0·00950	2·403	2·467	2·531	2·593	2·716	2·836	2·895	3·067	3·234	3·289	3·397	3·502	3·555	3·709
1/ 105	2·9489	3·2749	3·6225	3·9922	4·7997	5·7017	6·1893	7·8048	9·6590	10·332	11·765	13·314	14·134	16·778
0·01000	2·465	2·531	2·597	2·661	2·787	2·909	2·970	3·147	3·318	3·374	3·485	3·593	3·647	3·805
1/ 100	3·0255	3·3600	3·7166	4·0959	4·9244	5·8498	6·3501	8·0076	9·9100	10·601	12·070	13·660	14·501	17·214
	0·74	0·75	0·75	0·75	0·75	0·75	0·75	0·75	0·76	0·76	0·76	0·76	0·76	0·76

$V_{r(0{\cdot}5)medial}$ for half-full circular pipes.

A45
(p.5 of 6)

k_s = **15·0 mm**
S = 0·01000 to 0·03000

ie hydraulic gradient =
1 in 100 to 1 in 33·3

Water (or sewage) at 15°C;
full bore conditions.

velocities in ms^{-1}
discharges in m^3s^{-1}

Gradient **(Equivalent) Pipe diameters in mm**

Gradient	1250	1300	1350	1400	1500	1600	1650	1800	1950	2000	2100	2200	2250	2400
0·01000	2·465	2·531	2·597	2·661	2·787	2·909	2·970	3·147	3·318	3·374	3·485	3·593	3·647	3·805
1/ 100	3·0255	3·3600	3·7166	4·0959	4·9244	5·8498	6·3501	8·0076	9·9100	10·601	12·070	13·660	14·501	17·214
0·01050	2·526	2·594	2·661	2·726	2·855	2·981	3·043	3·225	3·400	3·458	3·571	3·682	3·737	3·899
1/ 95	3·1002	3·4430	3·8084	4·1971	5·0461	5·9943	6·5070	8·2054	10·155	10·863	12·368	13·997	14·859	17·639
0·01100	2·586	2·655	2·723	2·791	2·923	3·052	3·115	3·300	3·480	3·539	3·655	3·769	3·825	3·991
1/ 91	3·1732	3·5241	3·8981	4·2958	5·1648	6·1354	6·6601	8·3985	10·394	11·118	12·660	14·327	15·209	18·054
0·01150	2·644	2·715	2·785	2·853	2·988	3·120	3·185	3·375	3·559	3·619	3·737	3·854	3·911	4·081
1/ 87	3·2445	3·6033	3·9857	4·3924	5·2809	6·2733	6·8099	8·5873	10·627	11·368	12·944	14·649	15·551	18·460
0·01200	2·701	2·773	2·844	2·915	3·053	3·187	3·253	3·447	3·635	3·696	3·818	3·936	3·995	4·168
1/ 83	3·3143	3·6808	4·0715	4·4869	5·3945	6·4083	6·9564	8·7720	10·856	11·613	13·223	14·964	15·885	18·857
0·01250	2·756	2·830	2·903	2·975	3·116	3·253	3·320	3·518	3·710	3·773	3·896	4·018	4·078	4·254
1/ 80	3·3827	3·7567	4·1554	4·5795	5·5058	6·5404	7·0998	8·9529	11·080	11·852	13·495	15·272	16·213	19·246
0·01300	2·811	2·886	2·961	3·034	3·177	3·317	3·386	3·588	3·784	3·847	3·973	4·097	4·158	4·339
1/ 77	3·4497	3·8311	4·2377	4·6702	5·6149	6·6700	7·2405	9·1303	11·299	12·087	13·763	15·575	16·534	19·627
0·01350	2·865	2·941	3·017	3·092	3·238	3·381	3·451	3·656	3·856	3·921	4·049	4·175	4·238	4·421
1/ 74	3·5154	3·9041	4·3185	4·7591	5·7218	6·7971	7·3784	9·3042	11·515	12·317	14·025	15·872	16·849	20·001
0·01400	2·917	2·995	3·072	3·148	3·297	3·443	3·514	3·723	3·926	3·993	4·123	4·252	4·315	4·502
1/ 71	3·5799	3·9758	4·3977	4·8465	5·8268	6·9218	7·5138	9·4750	11·726	12·543	14·282	16·163	17·158	20·368
0·01450	2·969	3·048	3·127	3·204	3·356	3·504	3·576	3·789	3·996	4·063	4·196	4·327	4·392	4·582
1/ 69	3·6433	4·0462	4·4756	4·9323	5·9300	7·0444	7·6468	9·6427	11·934	12·765	14·535	16·449	17·462	20·729
0·01500	3·020	3·100	3·180	3·259	3·413	3·563	3·637	3·854	4·064	4·133	4·268	4·401	4·467	4·660
1/ 67	3·7056	4·1153	4·5521	5·0166	6·0314	7·1648	7·7776	9·8076	12·138	12·984	14·783	16·730	17·760	21·083
0·01600	3·119	3·202	3·285	3·366	3·525	3·680	3·757	3·981	4·198	4·268	4·408	4·546	4·613	4·813
1/ 63	3·8272	4·2503	4·7014	5·1812	6·2292	7·3998	8·0327	10·129	12·536	13·410	15·268	17·279	18·343	21·775
0·01700	3·215	3·301	3·386	3·469	3·634	3·794	3·872	4·103	4·327	4·400	4·544	4·685	4·755	4·961
1/ 59	3·9450	4·3812	4·8462	5·3407	6·4210	7·6276	8·2800	10·441	12·922	13·822	15·738	17·811	18·908	22·445
0·01800	3·308	3·396	3·484	3·570	3·739	3·904	3·985	4·222	4·452	4·527	4·676	4·821	4·893	5·105
1/ 56	4·0594	4·5082	4·9867	5·4955	6·6072	7·8488	8·5201	10·744	13·296	14·223	16·195	18·327	19·456	23·096
0·01900	3·399	3·490	3·579	3·668	3·841	4·011	4·094	4·338	4·574	4·651	4·804	4·953	5·027	5·245
1/ 53	4·1706	4·6318	5·1234	5·6461	6·7883	8·0639	8·7536	11·038	13·661	14·613	16·639	18·830	19·989	23·729
0·02000	3·487	3·580	3·672	3·763	3·941	4·115	4·200	4·450	4·693	4·772	4·929	5·082	5·158	5·382
1/ 50	4·2790	4·7521	5·2565	5·7929	6·9646	8·2734	8·9810	11·325	14·016	14·993	17·071	19·319	20·508	24·345
0·02100	3·573	3·669	3·763	3·856	4·039	4·216	4·304	4·560	4·809	4·890	5·050	5·208	5·285	5·514
1/ 47·6	4·3847	4·8695	5·3863	5·9359	7·1366	8·4778	9·2028	11·605	14·362	15·363	17·492	19·796	21·015	24·947
0·02200	3·657	3·755	3·852	3·947	4·134	4·316	4·405	4·668	4·922	5·005	5·169	5·330	5·410	5·644
1/ 45·5	4·4879	4·9841	5·5131	6·0756	7·3046	8·6773	9·4194	11·878	14·700	15·724	17·904	20·262	21·509	25·534
0·02300	3·739	3·839	3·938	4·036	4·226	4·413	4·504	4·773	5·033	5·118	5·285	5·450	5·531	5·771
1/ 43·5	4·5888	5·0961	5·6370	6·2122	7·4688	8·8723	9·6311	12·145	15·030	16·078	18·307	20·717	21·993	26·108
0·02400	3·820	3·922	4·023	4·122	4·317	4·508	4·601	4·875	5·141	5·228	5·399	5·567	5·650	5·895
1/ 41·7	4·6875	5·2058	5·7583	6·3458	7·6295	9·0632	9·8383	12·406	15·354	16·424	18·700	21·163	22·466	26·669
0·02500	3·898	4·003	4·106	4·207	4·406	4·601	4·696	4·976	5·247	5·336	5·510	5·682	5·767	6·017
1/ 40·0	4·7841	5·3131	5·8770	6·4767	7·7868	9·2501	10·041	12·662	15·670	16·762	19·086	21·600	22·929	27·219
0·02600	3·976	4·082	4·187	4·291	4·494	4·692	4·789	5·074	5·351	5·441	5·620	5·795	5·881	6·136
1/ 38·5	4·8789	5·4184	5·9934	6·6050	7·9411	9·4333	10·240	12·913	15·981	17·094	19·464	22·027	23·383	27·758
0·02700	4·051	4·160	4·267	4·372	4·579	4·781	4·880	5·171	5·453	5·545	5·727	5·905	5·993	6·253
1/ 37·0	4·9719	5·5216	6·1076	6·7308	8·0923	9·6130	10·435	13·159	16·285	17·420	19·835	22·447	23·829	28·287
0·02800	4·126	4·236	4·345	4·453	4·663	4·869	4·970	5·266	5·553	5·647	5·832	6·013	6·103	6·368
1/ 35·7	5·0631	5·6229	6·2197	6·8544	8·2409	9·7895	10·627	13·400	16·584	17·740	20·199	22·859	24·266	28·806
0·02900	4·199	4·311	4·422	4·532	4·746	4·955	5·058	5·359	5·651	5·747	5·935	6·120	6·211	6·480
1/ 34·5	5·1527	5·7225	6·3298	6·9757	8·3867	9·9628	10·815	13·637	16·877	18·054	20·556	23·264	24·696	29·316
0·03000	4·271	4·385	4·498	4·609	4·827	5·040	5·144	5·451	5·748	5·845	6·036	6·224	6·317	6·591
1/ 33·3	5·2408	5·8203	6·4380	7·0950	8·5301	10·133	11·000	13·871	17·166	18·363	20·908	23·661	25·118	29·817
	0·74	0·75	0·75	0·75	0·75	0·75	0·75	0·75	0·76	0·76	0·76	0·76	0·76	0·76

$V_{r(0\cdot5)medial}$ **for half-full circular pipes.**

k_s = **15·0 mm** S = 0·01000 to 0·03000

k_s = 15·0 mm
S = 0·03000 to 0·10000

ie hydraulic gradient =
1 in 33·3 to 1 in 10·0

Water (or sewage) at 15°C;
full bore conditions.

velocities in ms^{-1}
discharges in m^3s^{-1}

Gradient	**(Equivalent) Pipe diameters in mm**													
	1250	**1300**	**1350**	**1400**	**1500**	**1600**	**1650**	**1800**	**1950**	**2000**	**2100**	**2200**	**2250**	**2400**
0·03000	4·271	4·385	4·498	4·609	4·827	5·040	5·144	5·451	5·748	5·845	6·036	6·224	6·317	6·591
1/ 33·3	5·2408	5·8203	6·4380	7·0950	8·5301	10·133	11·000	13·871	17·166	18·363	20·908	23·661	25·118	29·817
0·03200	4·411	4·529	4·645	4·760	4·985	5·205	5·313	5·630	5·936	6·037	6·234	6·429	6·524	6·807
1/ 31·3	5·4127	6·0112	6·6492	7·3277	8·8099	10·465	11·361	14·326	17·729	18·965	21·594	24·437	25·942	30·795
0·03400	4·546	4·668	4·788	4·907	5·139	5·365	5·477	5·803	6·119	6·222	6·426	6·627	6·725	7·017
1/ 29·4	5·5794	6·1963	6·8539	7·5532	9·0811	10·788	11·710	14·767	18·275	19·549	22·258	25·190	26·740	31·743
0·03600	4·678	4·804	4·927	5·049	5·288	5·521	5·635	5·971	6·297	6·403	6·613	6·819	6·920	7·220
1/ 27·8	5·7411	6·3759	7·0526	7·7722	9·3444	11·100	12·050	15·195	18·805	20·115	22·904	25·920	27·516	32·664
0·03800	4·807	4·935	5·062	5·187	5·433	5·672	5·790	6·135	6·469	6·578	6·794	7·006	7·110	7·418
1/ 26·3	5·8985	6·5507	7·2459	7·9853	9·6005	11·405	12·380	15·611	19·320	20·667	23·531	26·630	28·270	33·559
0·04000	4·931	5·063	5·194	5·322	5·574	5·820	5·940	6·294	6·637	6·749	6·970	7·188	7·295	7·611
1/ 25·0	6·0517	6·7209	7·4342	8·1927	9·8499	11·701	12·702	16·017	19·822	21·204	24·143	27·322	29·004	34·431
0·04200	5·053	5·189	5·322	5·454	5·712	5·963	6·087	6·450	6·801	6·916	7·143	7·365	7·475	7·799
1/ 23·8	6·2012	6·8868	7·6178	8·3951	10·093	11·990	13·015	16·412	20·311	21·727	24·739	27·997	29·721	35·281
0·04400	5·172	5·311	5·447	5·582	5·846	6·104	6·230	6·601	6·961	7·079	7·311	7·538	7·651	7·982
1/ 22·7	6·3471	7·0489	7·7970	8·5926	10·331	12·272	13·322	16·799	20·789	22·239	25·321	28·656	30·420	36·111
0·04600	5·288	5·430	5·570	5·707	5·977	6·241	6·370	6·750	7·118	7·238	7·475	7·708	7·823	8·162
1/ 21·7	6·4898	7·2074	7·9723	8·7858	10·563	12·548	13·621	17·176	21·257	22·738	25·890	29·300	31·104	36·923
0·04800	5·402	5·547	5·689	5·830	6·106	6·375	6·507	6·895	7·271	7·394	7·636	7·874	7·991	8·337
1/ 20·8	6·6294	7·3624	8·1438	8·9748	10·790	12·818	13·914	17·546	21·714	23·228	26·447	29·930	31·773	37·717
0·05000	5·514	5·661	5·807	5·950	6·232	6·507	6·641	7·037	7·421	7·546	7·793	8·036	8·156	8·509
1/ 20·0	6·7661	7·5142	8·3117	9·1599	11·013	13·082	14·201	17·907	22·162	23·707	26·993	30·547	32·428	38·495
0·05250	5·650	5·801	5·950	6·097	6·386	6·667	6·805	7·211	7·604	7·732	7·986	8·234	8·357	8·719
1/ 19·0	6·9332	7·6998	8·5170	9·3861	11·285	13·405	14·552	18·350	22·709	24·292	27·659	31·302	33·229	39·446
0·05500	5·783	5·938	6·090	6·241	6·536	6·824	6·966	7·381	7·783	7·914	8·174	8·428	8·554	8·925
1/ 18·2	7·0964	7·8810	8·7175	9·6070	11·550	13·721	14·894	18·782	23·243	24·864	28·310	32·038	34·011	40·374
0·05750	5·913	6·071	6·227	6·381	6·683	6·978	7·122	7·547	7·958	8·092	8·357	8·618	8·746	9·125
1/ 17·4	7·2559	8·0582	8·9134	9·8229	11·810	14·029	15·229	19·204	23·766	25·423	28·946	32·759	34·775	41·281
0·06000	6·040	6·202	6·361	6·518	6·827	7·128	7·275	7·709	8·129	8·266	8·537	8·803	8·934	9·321
1/ 16·7	7·4120	8·2315	9·1051	10·034	12·064	14·331	15·557	19·617	24·277	25·969	29·569	33·463	35·523	42·169
0·06250	6·164	6·329	6·492	6·653	6·968	7·275	7·425	7·868	8·297	8·437	8·713	8·985	9·118	9·514
1/ 16·0	7·5648	8·4013	9·2929	10·241	12·313	14·626	15·877	20·021	24·778	26·505	30·179	34·153	36·256	43·039
0·06500	6·286	6·455	6·621	6·785	7·106	7·419	7·572	8·024	8·461	8·604	8·886	9·162	9·299	9·702
1/ 15·4	7·7147	8·5677	9·4770	10·444	12·557	14·916	16·192	20·418	25·268	27·030	30·777	34·830	36·974	43·891
0·06750	6·406	6·578	6·747	6·914	7·241	7·560	7·717	8·177	8·622	8·768	9·055	9·337	9·476	9·887
1/ 14·8	7·8616	8·7309	9·6575	10·643	12·796	15·200	16·500	20·807	25·750	27·545	31·363	35·493	37·678	44·728
0·07000	6·524	6·699	6·871	7·041	7·374	7·699	7·858	8·327	8·780	8·929	9·221	9·508	9·650	10·07
1/ 14·3	8·0059	8·8911	9·8347	10·838	13·031	15·479	16·803	21·189	26·222	28·050	31·938	36·144	38·370	45·548
0·07250	6·639	6·817	6·992	7·165	7·504	7·835	7·997	8·474	8·936	9·087	9·384	9·677	9·821	10·25
1/ 13·8	8·1476	9·0485	10·009	11·030	13·261	15·753	17·100	21·564	26·687	28·547	32·504	36·784	39·049	46·355
0·07500	6·753	6·934	7·112	7·288	7·633	7·969	8·134	8·619	9·089	9·242	9·545	9·842	9·989	10·42
1/ 13·3	8·2869	9·2032	10·180	11·219	13·488	16·023	17·393	21·932	27·143	29·035	33·059	37·413	39·717	47·147
0·08000	6·974	7·161	7·345	7·527	7·883	8·230	8·401	8·902	9·387	9·545	9·858	10·16	10·32	10·76
1/ 12·5	8·5587	9·5051	10·514	11·587	13·930	16·548	17·963	22·652	28·033	29·987	34·144	38·640	41·019	48·693
0·08500	7·189	7·381	7·571	7·758	8·126	8·484	8·660	9·176	9·676	9·839	10·16	10·48	10·63	11·09
1/ 11·8	8·8222	9·7976	10·837	11·943	14·359	17·057	18·516	23·349	28·896	30·910	35·195	39·830	42·282	50·192
0·09000	7·397	7·596	7·791	7·983	8·361	8·730	8·911	9·442	9·956	10·12	10·46	10·78	10·94	11·42
1/ 11·1	9·0780	10·082	11·152	12·290	14·775	17·552	19·053	24·026	29·734	31·806	36·215	40·984	43·508	51·647
0·09500	7·600	7·804	8·004	8·202	8·590	8·969	9·155	9·700	10·23	10·40	10·74	11·08	11·24	11·73
1/ 10·5	9·3267	10·358	11·457	12·626	15·180	18·033	19·575	24·684	30·548	32·678	37·208	42·107	44·700	53·063
0·10000	7·798	8·006	8·212	8·415	8·813	9·202	9·393	9·952	10·49	10·67	11·02	11·36	11·53	12·03
1/ 10·0	9·5691	10·627	11·755	12·954	15·575	18·501	20·084	25·326	31·342	33·527	38·174	43·201	45·861	54·441
	0·74	0·75	0·75	0·75	0·75	0·75	0·75	0·75	0·76	0·76	0·76	0·76	0·76	0·76

$V_{r(0 \cdot 5)medial}$ for half-full circular pipes.

A46

(p.1 of 6)

k_s = 30·0 mm
S = 0·00010 to 0·00030

ie hydraulic gradient =
1 in 10000 to 1 in 3333

Water (or sewage) at 15°C;
full bore conditions.

velocities in ms^{-1}
discharges in m^3s^{-1}

Gradient **(Equivalent) Pipe diameters in mm**

Gradient	1250	1300	1350	1400	1500	1600	1650	1800	1950	2000	2100	2200	2250	2400
0·000100	0·216	0·222	0·228	0·234	0·246	0·257	0·262	0·279	0·294	0·299	0·310	0·320	0·324	0·339
1/ 10000	0·2657	0·2953	0·3270	0·3606	0·4343	0·5167	0·5613	0·7091	0·8791	0·9408	1·0723	1·2147	1·2901	1·5334
0·000105	0·222	0·228	0·234	0·240	0·252	0·263	0·269	0·286	0·302	0·307	0·317	0·327	0·332	0·347
1/ 9524	0·2722	0·3026	0·3351	0·3696	0·4450	0·5295	0·5751	0·7266	0·9008	0·9641	1·0989	1·2447	1·3220	1·5713
0·000110	0·227	0·233	0·240	0·246	0·258	0·270	0·275	0·292	0·309	0·314	0·325	0·335	0·340	0·356
1/ 9091	0·2786	0·3098	0·3429	0·3783	0·4555	0·5419	0·5887	0·7438	0·9220	0·9868	1·1247	1·2741	1·3531	1·6083
0·000115	0·232	0·239	0·245	0·251	0·264	0·276	0·282	0·299	0·316	0·321	0·332	0·343	0·348	0·364
1/ 8696	0·2849	0·3167	0·3507	0·3868	0·4658	0·5541	0·6019	0·7605	0·9428	1·0090	1·1500	1·3027	1·3836	1·6445
0·000120	0·237	0·244	0·250	0·257	0·269	0·282	0·288	0·305	0·322	0·328	0·339	0·350	0·355	0·371
1/ 8333	0·2911	0·3236	0·3582	0·3951	0·4758	0·5661	0·6149	0·7769	0·9631	1·0307	1·1748	1·3308	1·4133	1·6799
0·000125	0·242	0·249	0·255	0·262	0·275	0·287	0·294	0·312	0·329	0·335	0·346	0·357	0·363	0·379
1/ 8000	0·2971	0·3302	0·3656	0·4033	0·4856	0·5777	0·6276	0·7929	0·9829	1·0520	1·1990	1·3582	1·4425	1·7146
0·000130	0·247	0·254	0·260	0·267	0·280	0·293	0·299	0·318	0·336	0·342	0·353	0·364	0·370	0·387
1/ 7692	0·3030	0·3368	0·3729	0·4113	0·4953	0·5892	0·6400	0·8086	1·0024	1·0729	1·2228	1·3852	1·4711	1·7486
0·000135	0·252	0·259	0·265	0·272	0·286	0·299	0·305	0·324	0·342	0·348	0·360	0·371	0·377	0·394
1/ 7407	0·3087	0·3432	0·3800	0·4191	0·5047	0·6004	0·6522	0·8240	1·0215	1·0933	1·2461	1·4116	1·4991	1·7819
0·000140	0·256	0·263	0·270	0·277	0·291	0·304	0·311	0·330	0·348	0·354	0·366	0·378	0·384	0·401
1/ 7143	0·3144	0·3495	0·3870	0·4268	0·5140	0·6115	0·6642	0·8392	1·0403	1·1134	1·2690	1·4375	1·5267	1·8146
0·000145	0·261	0·268	0·275	0·282	0·296	0·310	0·316	0·336	0·355	0·361	0·373	0·385	0·391	0·408
1/ 6897	0·3200	0·3557	0·3938	0·4344	0·5231	0·6223	0·6760	0·8540	1·0587	1·1331	1·2915	1·4630	1·5537	1·8468
0·000150	0·265	0·273	0·280	0·287	0·301	0·315	0·322	0·341	0·361	0·367	0·379	0·391	0·397	0·415
1/ 6667	0·3255	0·3618	0·4006	0·4418	0·5320	0·6329	0·6875	0·8686	1·0768	1·1525	1·3136	1·4880	1·5803	1·8784
0·000160	0·274	0·282	0·289	0·296	0·311	0·325	0·332	0·353	0·372	0·379	0·392	0·404	0·410	0·429
1/ 6250	0·3361	0·3737	0·4137	0·4563	0·5495	0·6537	0·7101	0·8972	1·1122	1·1903	1·3567	1·5368	1·6322	1·9400
0·000170	0·282	0·290	0·298	0·306	0·321	0·335	0·342	0·363	0·384	0·391	0·404	0·417	0·423	0·442
1/ 5882	0·3465	0·3852	0·4264	0·4704	0·5664	0·6739	0·7320	0·9248	1·1464	1·2270	1·3985	1·5842	1·6824	1·9998
0·000180	0·291	0·299	0·307	0·314	0·330	0·345	0·352	0·374	0·395	0·402	0·415	0·429	0·435	0·455
1/ 5556	0·3566	0·3964	0·4388	0·4840	0·5829	0·6934	0·7532	0·9516	1·1797	1·2626	1·4391	1·6301	1·7313	2·0578
0·000190	0·299	0·307	0·315	0·323	0·339	0·354	0·362	0·384	0·406	0·413	0·427	0·441	0·447	0·467
1/ 5263	0·3663	0·4072	0·4509	0·4973	0·5988	0·7124	0·7739	0·9777	1·2121	1·2972	1·4785	1·6748	1·7787	2·1142
0·000200	0·306	0·315	0·323	0·331	0·348	0·364	0·371	0·394	0·416	0·424	0·438	0·452	0·459	0·479
1/ 5000	0·3759	0·4178	0·4626	0·5102	0·6144	0·7309	0·7940	1·0031	1·2436	1·3310	1·5170	1·7184	1·8250	2·1692
0·000210	0·314	0·323	0·331	0·340	0·356	0·373	0·381	0·404	0·427	0·434	0·449	0·463	0·470	0·491
1/ 4762	0·3852	0·4281	0·4740	0·5228	0·6296	0·7490	0·8136	1·0279	1·2743	1·3639	1·5545	1·7608	1·8701	2·2228
0·000220	0·321	0·330	0·339	0·348	0·365	0·381	0·389	0·413	0·437	0·444	0·459	0·474	0·481	0·503
1/ 4545	0·3942	0·4382	0·4852	0·5352	0·6444	0·7667	0·8328	1·0521	1·3043	1·3960	1·5911	1·8023	1·9141	2·2751
0·000230	0·328	0·338	0·347	0·355	0·373	0·390	0·398	0·423	0·447	0·454	0·470	0·485	0·492	0·514
1/ 4348	0·4031	0·4481	0·4961	0·5472	0·6589	0·7839	0·8515	1·0758	1·3336	1·4274	1·6268	1·8428	1·9571	2·3263
0·000240	0·336	0·345	0·354	0·363	0·381	0·398	0·407	0·432	0·456	0·464	0·480	0·495	0·503	0·525
1/ 4167	0·4118	0·4577	0·5068	0·5590	0·6731	0·8008	0·8699	1·0990	1·3624	1·4581	1·6619	1·8825	1·9993	2·3763
0·000250	0·342	0·352	0·361	0·371	0·389	0·406	0·415	0·441	0·466	0·474	0·490	0·505	0·513	0·536
1/ 4000	0·4203	0·4672	0·5172	0·5705	0·6870	0·8173	0·8878	1·1216	1·3905	1·4882	1·6961	1·9213	2·0405	2·4254
0·000260	0·349	0·359	0·369	0·378	0·396	0·415	0·423	0·450	0·475	0·483	0·499	0·515	0·523	0·547
1/ 3846	0·4286	0·4764	0·5275	0·5818	0·7006	0·8335	0·9054	1·1439	1·4180	1·5177	1·7298	1·9594	2·0810	2·4734
0·000270	0·356	0·366	0·376	0·385	0·404	0·422	0·432	0·458	0·484	0·492	0·509	0·525	0·533	0·557
1/ 3704	0·4368	0·4855	0·5375	0·5929	0·7140	0·8494	0·9227	1·1657	1·4450	1·5466	1·7627	1·9967	2·1206	2·5206
0·000280	0·362	0·373	0·382	0·392	0·411	0·430	0·439	0·466	0·493	0·501	0·518	0·535	0·543	0·567
1/ 3571	0·4448	0·4944	0·5474	0·6038	0·7271	0·8650	0·9396	1·1871	1·4716	1·5750	1·7951	2·0334	2·1596	2·5668
0·000290	0·369	0·379	0·389	0·399	0·419	0·438	0·447	0·475	0·501	0·510	0·527	0·544	0·553	0·577
1/ 3448	0·4527	0·5032	0·5571	0·6145	0·7400	0·8803	0·9562	1·2081	1·4976	1·6029	1·8269	2·0694	2·1978	2·6123
0·000300	0·375	0·386	0·396	0·406	0·426	0·445	0·455	0·483	0·510	0·519	0·536	0·554	0·562	0·587
1/ 3333	0·4604	0·5118	0·5666	0·6250	0·7526	0·8954	0·9726	1·2288	1·5233	1·6303	1·8581	2·1048	2·2354	2·6570
	-	-	-	-	0·74	0·74	0·74	0·74	0·74	0·74	0·74	0·74	0·74	0·74

$V_{r(0·5)medial}$ for half-full circular pipes.

k_s = 30·0 mm S = 0·00010 to 0·00030

$k_s = 30{\cdot}0$ **mm**
S = 0·00030 to 0·00100

ie hydraulic gradient =
1 in 3333 to 1 in 1000

Water (or sewage) at 15°C;
full bore conditions.

velocities in ms^{-1}
discharges in m^3s^{-1}

Gradient **(Equivalent) Pipe diameters in mm**

Gradient	1250	1300	1350	1400	1500	1600	1650	1800	1950	2000	2100	2200	2250	2400
0·000300	0·375	0·386	0·396	0·406	0·426	0·445	0·455	0·483	0·510	0·519	0·536	0·554	0·562	0·587
1/ 3333	0·4604	0·5118	0·5666	0·6250	0·7526	0·8954	0·9726	1·2288	1·5233	1·6303	1·8581	2·1048	2·2354	2·6570
0·000320	0·387	0·398	0·409	0·419	0·440	0·460	0·470	0·499	0·527	0·536	0·554	0·572	0·581	0·607
1/ 3125	0·4755	0·5286	0·5852	0·6455	0·7773	0·9247	1·0045	1·2691	1·5733	1·6838	1·9191	2·1739	2·3087	2·7442
0·000340	0·399	0·411	0·421	0·432	0·453	0·474	0·484	0·514	0·543	0·552	0·571	0·589	0·599	0·625
1/ 2941	0·4902	0·5449	0·6033	0·6654	0·8013	0·9532	1·0355	1·3082	1·6217	1·7357	1·9782	2·2408	2·3798	2·8287
0·000360	0·411	0·422	0·434	0·445	0·467	0·488	0·498	0·529	0·559	0·569	0·588	0·607	0·616	0·643
1/ 2778	0·5044	0·5607	0·6208	0·6847	0·8245	0·9809	1·0655	1·3461	1·6687	1·7860	2·0356	2·3058	2·4489	2·9107
0·000380	0·422	0·434	0·446	0·457	0·479	0·501	0·512	0·543	0·574	0·584	0·604	0·623	0·633	0·661
1/ 2632	0·5182	0·5761	0·6378	0·7035	0·8471	1·0078	1·0947	1·3830	1·7145	1·8350	2·0914	2·3690	2·5160	2·9905
0·000400	0·433	0·445	0·457	0·469	0·492	0·514	0·525	0·558	0·589	0·599	0·620	0·639	0·649	0·678
1/ 2500	0·5317	0·5911	0·6544	0·7218	0·8691	1·0340	1·1232	1·4190	1·7591	1·8827	2·1458	2·4306	2·5814	3·0682
0·000420	0·444	0·456	0·468	0·480	0·504	0·527	0·538	0·571	0·604	0·614	0·635	0·655	0·665	0·695
1/ 2381	0·5448	0·6057	0·6705	0·7396	0·8906	1·0595	1·1509	1·4540	1·8025	1·9292	2·1988	2·4907	2·6452	3·1440
0·000440	0·454	0·467	0·479	0·492	0·516	0·539	0·551	0·585	0·618	0·629	0·650	0·671	0·681	0·711
1/ 2273	0·5577	0·6199	0·6863	0·7570	0·9116	1·0845	1·1780	1·4883	1·8450	1·9746	2·2505	2·5493	2·7075	3·2180
0·000460	0·465	0·478	0·490	0·503	0·527	0·551	0·563	0·598	0·632	0·643	0·664	0·686	0·696	0·727
1/ 2174	0·5702	0·6339	0·7018	0·7740	0·9321	1·1089	1·2045	1·5217	1·8864	2·0190	2·3011	2·6066	2·7683	3·2904
0·000480	0·475	0·488	0·501	0·514	0·539	0·563	0·575	0·611	0·645	0·656	0·679	0·700	0·711	0·743
1/ 2083	0·5825	0·6475	0·7169	0·7907	0·9522	1·1327	1·2304	1·5545	1·9270	2·0624	2·3507	2·6627	2·8279	3·3612
0·000500	0·484	0·498	0·511	0·524	0·550	0·575	0·587	0·623	0·659	0·670	0·693	0·715	0·726	0·758
1/ 2000	0·5945	0·6609	0·7317	0·8070	0·9718	1·1561	1·2558	1·5866	1·9668	2·1050	2·3992	2·7176	2·8862	3·4305
0·000525	0·496	0·510	0·524	0·537	0·564	0·589	0·602	0·639	0·675	0·687	0·710	0·733	0·744	0·777
1/ 1905	0·6092	0·6772	0·7497	0·8270	0·9958	1·1847	1·2868	1·6258	2·0154	2·1570	2·4584	2·7848	2·9575	3·5153
0·000550	0·508	0·522	0·536	0·550	0·577	0·603	0·616	0·654	0·691	0·703	0·726	0·750	0·761	0·795
1/ 1818	0·6235	0·6931	0·7674	0·8464	1·0193	1·2125	1·3171	1·6640	2·0628	2·2078	2·5163	2·8503	3·0272	3·5980
0·000575	0·520	0·534	0·548	0·562	0·590	0·617	0·630	0·669	0·706	0·719	0·743	0·767	0·778	0·813
1/ 1739	0·6376	0·7087	0·7847	0·8655	1·0422	1·2398	1·3468	1·7015	2·1092	2·2574	2·5729	2·9144	3·0952	3·6789
0·000600	0·531	0·545	0·560	0·574	0·602	0·630	0·643	0·683	0·721	0·734	0·759	0·783	0·795	0·831
1/ 1667	0·6513	0·7240	0·8015	0·8841	1·0646	1·2665	1·3757	1·7381	2·1546	2·3060	2·6282	2·9771	3·1618	3·7581
0·000625	0·542	0·557	0·572	0·586	0·615	0·643	0·657	0·697	0·736	0·749	0·774	0·799	0·812	0·848
1/ 1600	0·6647	0·7389	0·8181	0·9023	1·0866	1·2926	1·4041	1·7739	2·1990	2·3536	2·6825	3·0385	3·2270	3·8356
0·000650	0·552	0·568	0·583	0·598	0·627	0·656	0·670	0·711	0·751	0·764	0·790	0·815	0·828	0·865
1/ 1538	0·6779	0·7536	0·8343	0·9202	1·1081	1·3182	1·4319	1·8091	2·2426	2·4002	2·7356	3·0987	3·2910	3·9116
0·000675	0·563	0·579	0·594	0·609	0·639	0·668	0·682	0·724	0·765	0·779	0·805	0·831	0·843	0·881
1/ 1481	0·6908	0·7679	0·8502	0·9377	1·1292	1·3434	1·4592	1·8435	2·2854	2·4459	2·7877	3·1578	3·3537	3·9862
0·000700	0·573	0·589	0·605	0·620	0·651	0·680	0·695	0·738	0·779	0·793	0·820	0·846	0·859	0·897
1/ 1429	0·7035	0·7820	0·8658	0·9550	1·1499	1·3680	1·4860	1·8774	2·3273	2·4908	2·8389	3·2158	3·4153	4·0593
0·000725	0·583	0·600	0·616	0·631	0·662	0·692	0·707	0·751	0·793	0·807	0·834	0·861	0·874	0·913
1/ 1379	0·7160	0·7959	0·8811	0·9719	1·1703	1·3922	1·5123	1·9106	2·3685	2·5349	2·8892	3·2727	3·4757	4·1312
0·000750	0·593	0·610	0·626	0·642	0·674	0·704	0·719	0·764	0·807	0·821	0·848	0·876	0·889	0·929
1/ 1333	0·7282	0·8095	0·8962	0·9885	1·1903	1·4161	1·5382	1·9433	2·4090	2·5783	2·9386	3·3287	3·5352	4·2018
0·000800	0·613	0·630	0·647	0·663	0·696	0·727	0·743	0·789	0·833	0·848	0·876	0·904	0·918	0·959
1/ 1250	0·7521	0·8360	0·9256	1·0209	1·2294	1·4625	1·5887	2·0071	2·4881	2·6629	3·0350	3·4379	3·6512	4·3397
0·000850	0·632	0·649	0·667	0·684	0·717	0·750	0·766	0·813	0·859	0·874	0·903	0·932	0·947	0·989
1/ 1176	0·7753	0·8618	0·9541	1·0524	1·2672	1·5075	1·6376	2·0689	2·5647	2·7449	3·1284	3·5437	3·7636	4·4733
0·000900	0·650	0·668	0·686	0·703	0·738	0·772	0·788	0·837	0·884	0·899	0·929	0·959	0·974	1·017
1/ 1111	0·7977	0·8868	0·9818	1·0829	1·3040	1·5513	1·6851	2·1289	2·6390	2·8245	3·2192	3·6465	3·8727	4·6030
0·000950	0·668	0·686	0·705	0·723	0·758	0·793	0·810	0·860	0·908	0·924	0·955	0·986	1·001	1·045
1/ 1053	0·8196	0·9111	1·0087	1·1126	1·3397	1·5938	1·7313	2·1872	2·7114	2·9019	3·3074	3·7464	3·9789	4·7292
0·001000	0·685	0·704	0·723	0·742	0·778	0·813	0·831	0·882	0·931	0·948	0·980	1·011	1·027	1·073
1/ 1000	0·8409	0·9348	1·0349	1·1415	1·3746	1·6352	1·7763	2·2441	2·7818	2·9773	3·3934	3·8438	4·0823	4·8521
	–	–	–	–	0·74	0·74	0·74	0·74	0·74	0·74	0·74	0·74	0·74	0·74

$V_{r(0{\cdot}5)medial}$ **for half-full circular pipes.**

k_s = **30·0 mm**
S = 0·00100 to 0·00300

ie hydraulic gradient =
1 in 1000 to 1 in 333

Water (or sewage) at 15°C;
full bore conditions.

velocities in ms^{-1}
discharges in m^3s^{-1}

Gradient **(Equivalent) Pipe diameters in mm**

Gradient	1250	1300	1350	1400	1500	1600	1650	1800	1950	2000	2100	2200	2250	2400
0·00100	0·685	0·704	0·723	0·742	0·778	0·813	0·831	0·882	0·931	0·948	0·980	1·011	1·027	1·073
1/ 1000	0·8409	0·9348	1·0349	1·1415	1·3746	1·6352	1·7763	2·2441	2·7818	2·9773	3·3934	3·8438	4·0823	4·8521
0·00105	0·702	0·722	0·741	0·760	0·797	0·833	0·851	0·904	0·954	0·971	1·004	1·036	1·052	1·099
1/ 952	0·8617	0·9579	1·0605	1·1697	1·4085	1·6756	1·8201	2·2995	2·8506	3·0509	3·4772	3·9388	4·1831	4·9719
0·00110	0·719	0·739	0·758	0·778	0·816	0·853	0·871	0·925	0·977	0·994	1·028	1·061	1·077	1·125
1/ 909	0·8820	0·9804	1·0854	1·1972	1·4417	1·7151	1·8630	2·3536	2·9177	3·1227	3·5590	4·0315	4·2816	5·0890
0·00115	0·735	0·755	0·775	0·795	0·834	0·872	0·891	0·946	0·999	1·016	1·051	1·084	1·101	1·150
1/ 870	0·9018	1·0025	1·1099	1·2241	1·4741	1·7536	1·9049	2·4065	2·9833	3·1929	3·6390	4·1221	4·3778	5·2034
0·00120	0·751	0·772	0·792	0·812	0·852	0·891	0·910	0·966	1·020	1·038	1·073	1·108	1·125	1·175
1/ 833	0·9212	1·0240	1·1337	1·2505	1·5058	1·7913	1·9459	2·4583	3·0474	3·2616	3·7173	4·2108	4·4720	5·3153
0·00125	0·766	0·787	0·808	0·829	0·870	0·909	0·929	0·986	1·041	1·060	1·095	1·131	1·148	1·199
1/ 800	0·9402	1·0452	1·1571	1·2763	1·5369	1·8283	1·9860	2·5090	3·1103	3·3288	3·7940	4·2976	4·5642	5·4250
0·00130	0·781	0·803	0·824	0·846	0·887	0·927	0·947	1·006	1·062	1·081	1·117	1·153	1·171	1·223
1/ 769	0·9588	1·0659	1·1800	1·3016	1·5673	1·8645	2·0253	2·5587	3·1719	3·3948	3·8692	4·3828	4·6547	5·5324
0·00135	0·796	0·818	0·840	0·862	0·904	0·945	0·965	1·025	1·082	1·101	1·138	1·175	1·193	1·246
1/ 741	0·9771	1·0862	1·2025	1·3264	1·5972	1·9000	2·0639	2·6075	3·2324	3·4595	3·9429	4·4663	4·7433	5·6378
0·00140	0·811	0·833	0·856	0·877	0·920	0·962	0·983	1·043	1·102	1·121	1·159	1·196	1·215	1·269
1/ 714	0·9951	1·1061	1·2246	1·3507	1·6265	1·9349	2·1018	2·6554	3·2917	3·5230	4·0153	4·5483	4·8304	5·7413
0·00145	0·825	0·848	0·871	0·893	0·937	0·979	1·000	1·062	1·122	1·141	1·180	1·218	1·236	1·292
1/ 690	1·0127	1·1257	1·2463	1·3746	1·6553	1·9692	2·1390	2·7024	3·3500	3·5853	4·0864	4·6288	4·9159	5·8430
0·00150	0·839	0·863	0·886	0·908	0·953	0·996	1·017	1·080	1·141	1·161	1·200	1·238	1·258	1·314
1/ 667	1·0300	1·1450	1·2676	1·3981	1·6836	2·0029	2·1756	2·7486	3·4073	3·6467	4·1562	4·7079	5·0000	5·9429
0·00160	0·867	0·891	0·915	0·938	0·984	1·029	1·051	1·116	1·178	1·199	1·239	1·279	1·299	1·357
1/ 625	1·0638	1·1825	1·3092	1·4440	1·7388	2·0686	2·2470	2·8387	3·5190	3·7663	4·2926	4·8624	5·1640	6·1378
0·00170	0·894	0·918	0·943	0·967	1·014	1·060	1·083	1·150	1·215	1·236	1·277	1·319	1·339	1·399
1/ 588	1·0965	1·2189	1·3495	1·4885	1·7924	2·1322	2·3162	2·9261	3·6274	3·8822	4·4247	5·0121	5·3230	6·3268
0·00180	0·919	0·945	0·970	0·995	1·044	1·091	1·115	1·183	1·250	1·272	1·315	1·357	1·378	1·439
1/ 556	1·1283	1·2543	1·3886	1·5316	1·8443	2·1941	2·3833	3·0110	3·7325	3·9948	4·5530	5·1574	5·4773	6·5102
0·00190	0·945	0·971	0·997	1·022	1·072	1·121	1·145	1·216	1·284	1·306	1·351	1·394	1·415	1·479
1/ 526	1·1593	1·2887	1·4267	1·5736	1·8949	2·2542	2·4487	3·0935	3·8349	4·1043	4·6778	5·2988	5·6275	6·6887
0·00200	0·969	0·996	1·023	1·049	1·100	1·150	1·175	1·247	1·317	1·340	1·386	1·430	1·452	1·517
1/ 500	1·1894	1·3221	1·4638	1·6145	1·9441	2·3128	2·5123	3·1739	3·9345	4·2109	4·7994	5·4364	5·7737	6·8625
0·00210	0·993	1·021	1·048	1·075	1·127	1·179	1·204	1·278	1·350	1·373	1·420	1·465	1·488	1·554
1/ 476	1·2188	1·3548	1·4999	1·6544	1·9922	2·3699	2·5743	3·2523	4·0317	4·3150	4·9179	5·5707	5·9163	7·0320
0·00220	1·017	1·045	1·073	1·100	1·154	1·206	1·232	1·308	1·382	1·406	1·453	1·500	1·523	1·591
1/ 455	1·2475	1·3867	1·5352	1·6933	2·0391	2·4257	2·6349	3·3289	4·1266	4·4165	5·0337	5·7018	6·0556	7·1975
0·00230	1·039	1·068	1·097	1·125	1·180	1·234	1·260	1·338	1·413	1·437	1·486	1·534	1·557	1·627
1/ 435	1·2755	1·4179	1·5698	1·7314	2·0849	2·4802	2·6942	3·4037	4·2193	4·5158	5·1468	5·8300	6·1917	7·3593
0·00240	1·062	1·091	1·120	1·149	1·205	1·260	1·287	1·366	1·443	1·468	1·518	1·567	1·591	1·662
1/ 417	1·3030	1·4484	1·6035	1·7686	2·1297	2·5336	2·7521	3·4769	4·3101	4·6130	5·2575	5·9554	6·3249	7·5176
0·00250	1·084	1·114	1·143	1·173	1·230	1·286	1·314	1·395	1·473	1·499	1·549	1·599	1·624	1·696
1/ 400	1·3298	1·4782	1·6366	1·8051	2·1737	2·5859	2·8089	3·5486	4·3990	4·7081	5·3660	6·0783	6·4553	7·6726
0·00260	1·105	1·136	1·166	1·196	1·254	1·312	1·340	1·422	1·502	1·528	1·580	1·631	1·656	1·730
1/ 385	1·3562	1·5075	1·6690	1·8409	2·2167	2·6371	2·8645	3·6189	4·4861	4·8014	5·4723	6·1987	6·5832	7·8246
0·00270	1·126	1·157	1·188	1·219	1·278	1·337	1·365	1·449	1·531	1·557	1·610	1·662	1·687	1·763
1/ 370	1·3820	1·5363	1·7008	1·8760	2·2590	2·6873	2·9191	3·6879	4·5716	4·8928	5·5765	6·3168	6·7086	7·9737
0·00280	1·147	1·179	1·210	1·241	1·302	1·361	1·390	1·476	1·559	1·586	1·640	1·692	1·718	1·795
1/ 357	1·4074	1·5645	1·7320	1·9104	2·3004	2·7366	2·9727	3·7556	4·6555	4·9826	5·6789	6·4327	6·8317	8·1200
0·00290	1·167	1·200	1·231	1·263	1·325	1·385	1·415	1·502	1·586	1·614	1·669	1·722	1·749	1·827
1/ 345	1·4323	1·5922	1·7627	1·9442	2·3412	2·7851	3·0253	3·8221	4·7380	5·0708	5·7794	6·5466	6·9527	8·2638
0·00300	1·187	1·220	1·253	1·285	1·347	1·409	1·439	1·528	1·614	1·642	1·697	1·752	1·779	1·858
1/ 333	1·4568	1·6194	1·7928	1·9775	2·3812	2·8327	3·0771	3·8874	4·8190	5·1576	5·8782	6·6585	7·0716	8·4051
	–	–	–	–	0·74	0·74	0·74	0·74	0·74	0·74	0·74	0·74	0·74	0·74

$V_{r(0\cdot5)medial}$ **for half-full circular pipes.**

k_s = 30·0 mm S = 0·00100 to 0·00300

k_s = 30·0 mm
S = 0·00300 to 0·01000

ie hydraulic gradient =
1 in 333 to 1 in 100

Water (or sewage) at 15°C;
full bore conditions.

velocities in ms^{-1}
discharges in m^3s^{-1}

Gradient **(Equivalent) Pipe diameters in mm**

Gradient	1250	1300	1350	1400	1500	1600	1650	1800	1950	2000	2100	2200	2250	2400
0·00300	1·187	1·220	1·253	1·285	1·347	1·409	1·439	1·528	1·614	1·642	1·697	1·752	1·779	1·858
1/ 333	1·4568	1·6194	1·7928	1·9775	2·3812	2·8327	3·0771	3·8874	4·8190	5·1576	5·8782	6·6585	7·0716	8·4051
0·00320	1·226	1·260	1·294	1·327	1·392	1·455	1·486	1·578	1·667	1·696	1·753	1·809	1·837	1·919
1/ 313	1·5046	1·6725	1·8517	2·0423	2·4593	2·9256	3·1780	4·0149	4·9770	5·3267	6·0711	6·8769	7·3035	8·6808
0·00340	1·264	1·299	1·333	1·368	1·435	1·500	1·532	1·626	1·718	1·748	1·807	1·865	1·893	1·978
1/ 294	1·5509	1·7240	1·9087	2·1052	2·5350	3·0157	3·2758	4·1385	5·1302	5·4907	6·2579	7·0886	7·5283	8·9480
0·00360	1·300	1·337	1·372	1·407	1·476	1·543	1·576	1·673	1·768	1·798	1·859	1·919	1·948	2·035
1/ 278	1·5959	1·7740	1·9640	2·1662	2·6085	3·1031	3·3708	4·2585	5·2790	5·6499	6·4394	7·2941	7·7466	9·2074
0·00380	1·336	1·373	1·410	1·446	1·517	1·586	1·620	1·719	1·816	1·848	1·910	1·971	2·002	2·091
1/ 263	1·6396	1·8226	2·0178	2·2256	2·6800	3·1882	3·4632	4·3752	5·4237	5·8048	6·6159	7·4941	7·9589	9·4598
0·00400	1·371	1·409	1·446	1·483	1·556	1·627	1·662	1·764	1·863	1·896	1·960	2·023	2·054	2·145
1/ 250	1·6822	1·8700	2·0703	2·2834	2·7496	3·2710	3·5532	4·4889	5·5646	5·9556	6·7878	7·6888	8·1657	9·7056
0·00420	1·405	1·444	1·482	1·520	1·594	1·667	1·703	1·808	1·909	1·943	2·008	2·073	2·104	2·198
1/ 238	1·7238	1·9162	2·1214	2·3398	2·8176	3·3518	3·6409	4·5998	5·7020	6·1027	6·9554	7·8787	8·3674	9·9453
0·00440	1·438	1·478	1·517	1·556	1·632	1·706	1·743	1·850	1·954	1·988	2·055	2·121	2·154	2·250
1/ 227	1·7643	1·9613	2·1713	2·3949	2·8839	3·4307	3·7266	4·7080	5·8363	6·2463	7·1191	8·0641	8·5643	10·179
0·00460	1·470	1·511	1·551	1·591	1·669	1·745	1·782	1·892	1·998	2·033	2·102	2·169	2·202	2·301
1/ 217	1·8040	2·0053	2·2201	2·4487	2·9487	3·5078	3·8104	4·8139	5·9674	6·3867	7·2792	8·2454	8·7569	10·408
0·00480	1·502	1·543	1·584	1·625	1·705	1·782	1·820	1·932	2·041	2·077	2·147	2·216	2·250	2·350
1/ 208	1·8428	2·0485	2·2679	2·5014	3·0121	3·5833	3·8924	4·9174	6·0958	6·5241	7·4357	8·4227	8·9452	10·632
0·00500	1·533	1·575	1·617	1·658	1·740	1·819	1·858	1·972	2·083	2·120	2·191	2·261	2·296	2·399
1/ 200	1·8808	2·0907	2·3147	2·5530	3·0743	3·6572	3·9727	5·0188	6·2215	6·6587	7·5891	8·5964	9·1297	10·851
0·00525	1·570	1·614	1·657	1·699	1·783	1·864	1·904	2·021	2·135	2·172	2·245	2·317	2·353	2·458
1/ 190	1·9273	2·1424	2·3719	2·6161	3·1502	3·7475	4·0708	5·1428	6·3752	6·8231	7·7765	8·8088	9·3552	11·119
0·00550	1·607	1·652	1·696	1·739	1·825	1·908	1·949	2·069	2·185	2·223	2·298	2·372	2·408	2·516
1/ 182	1·9726	2·1928	2·4277	2·6777	3·2243	3·8357	4·1666	5·2639	6·5252	6·9837	7·9596	9·0161	9·5754	11·381
0·00575	1·644	1·689	1·734	1·779	1·866	1·951	1·992	2·115	2·234	2·273	2·350	2·425	2·462	2·572
1/ 174	2·0170	2·2421	2·4822	2·7378	3·2968	3·9220	4·2602	5·3822	6·6719	7·1407	8·1385	9·2187	9·7906	11·637
0·00600	1·679	1·726	1·771	1·817	1·906	1·993	2·035	2·161	2·282	2·322	2·400	2·477	2·515	2·628
1/ 167	2·0604	2·2903	2·5356	2·7967	3·3677	4·0063	4·3519	5·4979	6·8154	7·2943	8·3135	9·4170	10·001	11·887
0·00625	1·714	1·761	1·808	1·854	1·945	2·034	2·077	2·205	2·329	2·370	2·450	2·528	2·567	2·682
1/ 160	2·1029	2·3376	2·5879	2·8544	3·4372	4·0890	4·4416	5·6113	6·9560	7·4447	8·4850	9·6113	10·207	12·132
0·00650	1·748	1·796	1·844	1·891	1·984	2·074	2·118	2·249	2·375	2·417	2·498	2·578	2·618	2·735
1/ 154	2·1445	2·3839	2·6392	2·9110	3·5053	4·1699	4·5296	5·7225	7·0938	7·5922	8·6530	9·8016	10·410	12·373
0·00675	1·781	1·830	1·879	1·927	2·021	2·113	2·159	2·292	2·421	2·463	2·546	2·628	2·668	2·787
1/ 148	2·1854	2·4293	2·6895	2·9664	3·5721	4·2494	4·6159	5·8315	7·2289	7·7368	8·8179	9·9884	10·608	12·608
0·00700	1·813	1·864	1·913	1·962	2·058	2·152	2·198	2·334	2·465	2·508	2·593	2·676	2·717	2·838
1/ 143	2·2255	2·4739	2·7388	3·0209	3·6376	4·3274	4·7006	5·9385	7·3616	7·8788	8·9797	10·172	10·803	12·840
0·00725	1·846	1·897	1·947	1·997	2·095	2·190	2·237	2·375	2·509	2·552	2·638	2·723	2·765	2·888
1/ 138	2·2649	2·5177	2·7873	3·0743	3·7020	4·4040	4·7838	6·0437	7·4919	8·0183	9·1387	10·352	10·994	13·067
0·00750	1·877	1·929	1·981	2·031	2·131	2·228	2·276	2·416	2·551	2·596	2·684	2·770	2·812	2·938
1/ 133	2·3036	2·5607	2·8350	3·1269	3·7653	4·4793	4·8656	6·1470	7·6200	8·1554	9·2949	10·529	11·182	13·290
0·00800	1·939	1·993	2·046	2·098	2·201	2·301	2·350	2·495	2·635	2·681	2·772	2·861	2·905	3·034
1/ 125	2·3792	2·6447	2·9280	3·2295	3·8888	4·6262	5·0252	6·3486	7·8699	8·4229	9·5998	10·874	11·549	13·726
0·00850	1·998	2·054	2·109	2·162	2·268	2·372	2·423	2·572	2·716	2·764	2·857	2·949	2·994	3·128
1/ 118	2·4524	2·7261	3·0181	3·3289	4·0085	4·7686	5·1799	6·5440	8·1122	8·6821	9·8953	11·209	11·904	14·149
0·00900	2·056	2·113	2·170	2·225	2·334	2·440	2·493	2·646	2·795	2·844	2·940	3·034	3·081	3·218
1/ 111	2·5235	2·8051	3·1056	3·4254	4·1247	4·9069	5·3301	6·7338	8·3474	8·9338	10·182	11·534	12·249	14·559
0·00950	2·113	2·171	2·229	2·286	2·398	2·507	2·561	2·719	2·872	2·922	3·020	3·117	3·165	3·306
1/ 105	2·5927	2·8820	3·1907	3·5193	4·2378	5·0413	5·4762	6·9183	8·5761	9·1787	10·461	11·850	12·585	14·958
0·01000	2·168	2·228	2·287	2·346	2·460	2·573	2·628	2·789	2·946	2·998	3·099	3·198	3·247	3·392
1/ 100	2·6600	2·9569	3·2736	3·6107	4·3479	5·1723	5·6185	7·0980	8·7989	9·4172	10·733	12·158	12·912	15·347
	–	–	–	–	0·74	0·74	0·74	0·74	0·74	0·74	0·74	0·74	0·74	0·74

$V_{r(0{\cdot}5)medial}$ for half-full circular pipes.

A46
(p.5 of 6)

k_s = 30·0 mm
S = 0·01000 to 0·03000

ie hydraulic gradient =
1 in 100 to 1 in 33·3

Water (or sewage) at 15°C;
full bore conditions.

velocities in ms^{-1}
discharges in m^3s^{-1}

Gradient **(Equivalent) Pipe diameters in mm**

Gradient	1250	1300	1350	1400	1500	1600	1650	1800	1950	2000	2100	2200	2250	2400
0·01000	2·168	2·228	2·287	2·346	2·460	2·573	2·628	2·789	2·946	2·998	3·099	3·198	3·247	3·392
1/ 100	2·6600	2·9569	3·2736	3·6107	4·3479	5·1723	5·6185	7·0980	8·7989	9·4172	10·733	12·158	12·912	15·347
0·01050	2·221	2·283	2·344	2·403	2·521	2·636	2·692	2·858	3·019	3·072	3·175	3·277	3·328	3·476
1/ 95	2·7257	3·0299	3·3545	3·6999	4·4553	5·3001	5·7572	7·2734	9·0163	9·6497	10·998	12·458	13·231	15·726
0·01100	2·273	2·336	2·399	2·460	2·581	2·698	2·756	2·926	3·090	3·144	3·250	3·354	3·406	3·558
1/ 91	2·7899	3·1013	3·4334	3·7870	4·5601	5·4248	5·8927	7·4445	9·2285	9·8768	11·257	12·751	13·542	16·096
0·01150	2·325	2·389	2·453	2·515	2·639	2·759	2·818	2·991	3·160	3·215	3·323	3·430	3·482	3·638
1/ 87	2·8526	3·1710	3·5106	3·8721	4·6626	5·5467	6·0252	7·6119	9·4359	10·099	11·510	13·038	13·847	16·458
0·01200	2·375	2·440	2·505	2·569	2·695	2·818	2·878	3·056	3·227	3·284	3·395	3·504	3·557	3·716
1/ 83	2·9140	3·2392	3·5861	3·9554	4·7629	5·6661	6·1548	7·7756	9·6388	10·316	11·758	13·318	14·144	16·812
0·01250	2·423	2·491	2·557	2·622	2·751	2·876	2·938	3·119	3·294	3·351	3·465	3·576	3·631	3·793
1/ 80	2·9740	3·3060	3·6601	4·0369	4·8611	5·7829	6·2817	7·9360	9·8376	10·529	12·000	13·593	14·436	17·158
0·01300	2·471	2·540	2·608	2·674	2·805	2·933	2·996	3·180	3·359	3·418	3·533	3·647	3·703	3·868
1/ 77	3·0330	3·3715	3·7326	4·1169	4·9574	5·8974	6·4061	8·0931	10·032	10·737	12·238	13·862	14·722	17·498
0·01350	2·519	2·588	2·657	2·725	2·859	2·989	3·053	3·241	3·423	3·483	3·601	3·716	3·773	3·942
1/ 74	3·0907	3·4357	3·8037	4·1953	5·0519	6·0098	6·5282	8·2473	10·224	10·942	12·471	14·126	15·002	17·831
0·01400	2·565	2·636	2·706	2·775	2·911	3·044	3·109	3·300	3·486	3·547	3·667	3·784	3·842	4·014
1/ 71	3·1475	3·4987	3·8735	4·2723	5·1446	6·1201	6·6480	8·3987	10·411	11·143	12·700	14·385	15·278	18·159
0·01450	2·610	2·683	2·754	2·824	2·963	3·098	3·164	3·359	3·548	3·610	3·732	3·851	3·910	4·085
1/ 69	3·2032	3·5607	3·9421	4·3480	5·2357	6·2284	6·7656	8·5473	10·596	11·340	12·925	14·640	15·548	18·480
0·01500	2·655	2·728	2·801	2·873	3·013	3·151	3·218	3·416	3·608	3·671	3·795	3·917	3·977	4·155
1/ 67	3·2579	3·6215	4·0095	4·4223	5·3252	6·3349	6·8813	8·6935	10·777	11·534	13·145	14·890	15·814	18·796
0·01600	2·742	2·818	2·893	2·967	3·112	3·254	3·324	3·528	3·727	3·792	3·920	4·046	4·108	4·291
1/ 63	3·3648	3·7403	4·1410	4·5673	5·4998	6·5427	7·1070	8·9786	11·130	11·912	13·577	15·379	16·333	19·413
0·01700	2·826	2·905	2·982	3·058	3·208	3·354	3·426	3·637	3·842	3·908	4·040	4·170	4·234	4·423
1/ 59	3·4684	3·8555	4·2684	4·7079	5·6691	6·7441	7·3258	9·2550	11·473	12·279	13·994	15·852	16·835	20·010
0·01800	2·908	2·989	3·068	3·147	3·301	3·451	3·525	3·742	3·953	4·022	4·158	4·291	4·357	4·551
1/ 56	3·5689	3·9672	4·3922	4·8444	5·8335	6·9396	7·5382	9·5233	11·805	12·635	14·400	16·312	17·324	20·590
0·01900	2·988	3·071	3·153	3·233	3·392	3·546	3·622	3·845	4·061	4·132	4·272	4·409	4·476	4·676
1/ 53	3·6667	4·0760	4·5126	4·9772	5·9934	7·1298	7·7448	9·7843	12·129	12·981	14·795	16·759	17·798	21·154
0·02000	3·066	3·151	3·234	3·317	3·480	3·638	3·716	3·945	4·167	4·239	4·382	4·523	4·593	4·798
1/ 50	3·7620	4·1819	4·6298	5·1065	6·1491	7·3150	7·9460	10·038	12·444	13·318	15·179	17·194	18·261	21·704
0·02100	3·141	3·228	3·314	3·399	3·566	3·728	3·808	4·042	4·270	4·344	4·491	4·635	4·706	4·916
1/ 47·6	3·8549	4·2851	4·7441	5·2326	6·3009	7·4957	8·1422	10·286	12·751	13·647	15·554	17·619	18·712	22·240
0·02200	3·215	3·304	3·392	3·479	3·650	3·816	3·898	4·137	4·370	4·446	4·596	4·744	4·817	5·032
1/ 45·5	3·9456	4·3860	4·8558	5·3558	6·4492	7·6721	8·3338	10·528	13·051	13·968	15·920	18·033	19·152	22·763
0·02300	3·287	3·379	3·469	3·557	3·732	3·902	3·985	4·230	4·468	4·546	4·700	4·851	4·925	5·145
1/ 43·5	4·0343	4·4846	4·9649	5·4762	6·5942	7·8445	8·5211	10·765	13·345	14·282	16·278	18·439	19·582	23·275
0·02400	3·358	3·451	3·543	3·634	3·812	3·985	4·071	4·321	4·564	4·644	4·801	4·955	5·031	5·256
1/ 41·7	4·1211	4·5810	5·0717	5·5939	6·7360	8·0133	8·7044	10·997	13·632	14·590	16·628	18·835	20·004	23·776
0·02500	3·427	3·523	3·616	3·709	3·890	4·068	4·155	4·411	4·659	4·740	4·900	5·057	5·135	5·364
1/ 40·0	4·2061	4·6755	5·1763	5·7093	6·8749	8·1785	8·8839	11·223	13·913	14·890	16·971	19·224	20·416	24·266
0·02600	3·495	3·592	3·688	3·782	3·967	4·148	4·237	4·498	4·751	4·834	4·997	5·157	5·236	5·470
1/ 38·5	4·2894	4·7681	5·2788	5·8224	7·0111	8·3405	9·0599	11·446	14·188	15·185	17·307	19·604	20·821	24·747
0·02700	3·562	3·661	3·758	3·854	4·043	4·227	4·318	4·584	4·841	4·926	5·092	5·256	5·336	5·574
1/ 37·0	4·3711	4·8590	5·3794	5·9333	7·1447	8·4994	9·2325	11·664	14·459	15·475	17·637	19·978	21·217	25·218
0·02800	3·627	3·728	3·827	3·925	4·117	4·305	4·397	4·668	4·930	5·016	5·185	5·352	5·434	5·677
1/ 35·7	4·4513	4·9481	5·4781	6·0422	7·2758	8·6554	9·4019	11·878	14·724	15·759	17·961	20·345	21·607	25·681
0·02900	3·691	3·794	3·895	3·995	4·190	4·381	4·475	4·750	5·018	5·105	5·277	5·447	5·530	5·777
1/ 34·5	4·5301	5·0357	5·5751	6·1491	7·4046	8·8086	9·5683	12·088	14·985	16·038	18·278	20·705	21·989	26·135
0·03000	3·755	3·859	3·961	4·063	4·262	4·456	4·551	4·832	5·103	5·192	5·368	5·540	5·625	5·876
1/ 33·3	4·6076	5·1218	5·6704	6·2543	7·5312	8·9592	9·7319	12·295	15·241	16·312	18·591	21·059	22·365	26·582
	–	–	–	–	0·74	0·74	0·74	0·74	0·74	0·74	0·74	0·74	0·74	0·74

$V_{r(0·5)medial}$ for half-full circular pipes.

k_s = 30·0 mm S = 0·01000 to 0·03000

k_s = 30·0 mm
S = 0·03000 to 0·10000

ie hydraulic gradient =
1 in 33·3 to 1 in 10·0

Water (or sewage) at 15°C;
full bore conditions.

velocities in ms^{-1}
discharges in m^3s^{-1}

Gradient **(Equivalent) Pipe diameters in mm**

Gradient	1250	1300	1350	1400	1500	1600	1650	1800	1950	2000	2100	2200	2250	2400
0·03000	3·755	3·859	3·961	4·063	4·262	4·456	4·551	4·832	5·103	5·192	5·368	5·540	5·625	5·876
1/ 33·3	4·6076	5·1218	5·6704	6·2543	7·5312	8·9592	9·7319	12·295	15·241	16·312	18·591	21·059	22·365	26·582
0·03200	3·878	3·985	4·091	4·196	4·402	4·602	4·701	4·990	5·271	5·362	5·544	5·721	5·809	6·069
1/ 31·3	4·7587	5·2898	5·8564	6·4594	7·7782	9·2530	10·051	12·698	15·741	16·847	19·201	21·749	23·098	27·454
0·03400	3·997	4·108	4·217	4·325	4·537	4·744	4·845	5·144	5·433	5·528	5·714	5·898	5·988	6·255
1/ 29·4	4·9052	5·4526	6·0367	6·6582	8·0176	9·5378	10·360	13·089	16·225	17·365	19·792	22·419	23·809	28·299
0·03600	4·113	4·227	4·340	4·451	4·669	4·881	4·986	5·293	5·590	5·688	5·880	6·069	6·162	6·437
1/ 27·8	5·0474	5·6107	6·2117	6·8513	8·2500	9·8143	10·661	13·468	16·696	17·869	20·365	23·069	24·500	29·119
0·03800	4·226	4·343	4·459	4·573	4·796	5·015	5·122	5·438	5·744	5·844	6·041	6·235	6·331	6·613
1/ 26·3	5·1857	5·7645	6·3819	7·0390	8·4761	10·083	10·953	13·837	17·153	18·358	20·924	23·701	25·171	29·917
0·04000	4·335	4·456	4·574	4·691	4·921	5·145	5·256	5·579	5·893	5·995	6·198	6·397	6·495	6·785
1/ 25·0	5·3204	5·9142	6·5477	7·2219	8·6963	10·345	11·238	14·197	17·599	18·835	21·467	24·317	25·825	30·695
0·04200	4·443	4·566	4·687	4·807	5·043	5·272	5·385	5·717	6·038	6·144	6·351	6·555	6·655	6·953
1/ 23·8	5·4518	6·0603	6·7094	7·4002	8·9111	10·601	11·515	14·547	18·033	19·300	21·997	24·917	26·463	31·453
0·04400	4·547	4·673	4·798	4·920	5·161	5·396	5·512	5·851	6·180	6·288	6·500	6·709	6·812	7·116
1/ 22·7	5·5801	6·2029	6·8673	7·5744	9·1208	10·850	11·786	14·890	18·458	19·755	22·515	25·503	27·085	32·193
0·04600	4·649	4·778	4·905	5·031	5·277	5·518	5·636	5·983	6·319	6·429	6·647	6·860	6·965	7·276
1/ 21·7	5·7056	6·3423	7·0217	7·7446	9·3258	11·094	12·051	15·225	18·873	20·199	23·021	26·077	27·694	32·917
0·04800	4·749	4·881	5·011	5·139	5·391	5·636	5·757	6·112	6·455	6·568	6·790	7·007	7·115	7·433
1/ 20·8	5·8283	6·4787	7·1727	7·9112	9·5264	11·333	12·310	15·552	19·279	20·633	23·516	26·638	28·290	33·625
0·05000	4·847	4·982	5·114	5·245	5·502	5·753	5·876	6·238	6·588	6·703	6·930	7·152	7·262	7·586
1/ 20·0	5·9485	6·6124	7·3206	8·0744	9·7228	11·566	12·564	15·873	19·676	21·059	24·001	27·187	28·873	34·318
0·05250	4·967	5·105	5·241	5·375	5·638	5·895	6·021	6·392	6·751	6·869	7·101	7·329	7·441	7·773
1/ 19·0	6·0954	6·7757	7·5014	8·2738	9·9630	11·852	12·874	16·265	20·162	21·579	24·594	27·858	29·586	35·165
0·05500	5·084	5·225	5·364	5·501	5·771	6·033	6·163	6·542	6·910	7·030	7·268	7·501	7·616	7·956
1/ 18·2	6·2388	6·9351	7·6779	8·4685	10·197	12·131	13·177	16·647	20·637	22·086	25·173	28·514	30·283	35·993
0·05750	5·198	5·342	5·485	5·625	5·900	6·169	6·301	6·689	7·065	7·188	7·431	7·670	7·787	8·135
1/ 17·4	6·3791	7·0910	7·8505	8·6588	10·427	12·404	13·473	17·022	21·100	22·583	25·738	29·155	30·963	36·802
0·06000	5·310	5·457	5·603	5·746	6·027	6·302	6·437	6·833	7·217	7·343	7·591	7·835	7·955	8·310
1/ 16·7	6·5163	7·2435	8·0194	8·8451	10·651	12·670	13·763	17·388	21·554	23·069	26·292	29·782	31·629	37·594
0·06250	5·419	5·570	5·718	5·864	6·151	6·432	6·569	6·974	7·366	7·494	7·747	7·996	8·119	8·481
1/ 16·0	6·6506	7·3929	8·1847	9·0275	10·871	12·932	14·047	17·746	21·999	23·544	26·834	30·396	32·281	38·369
0·06500	5·527	5·680	5·831	5·981	6·273	6·559	6·700	7·112	7·512	7·643	7·901	8·155	8·280	8·649
1/ 15·4	6·7824	7·5393	8·3468	9·2063	11·086	13·188	14·325	18·098	22·434	24·011	27·366	30·998	32·921	39·129
0·06750	5·632	5·788	5·942	6·094	6·393	6·684	6·827	7·247	7·655	7·788	8·051	8·310	8·437	8·814
1/ 14·8	6·9116	7·6829	8·5058	9·3817	11·297	13·439	14·598	18·443	22·862	24·468	27·887	31·588	33·548	39·874
0·07000	5·735	5·895	6·051	6·206	6·510	6·807	6·952	7·380	7·796	7·931	8·199	8·462	8·592	8·976
1/ 14·3	7·0384	7·8239	8·6619	9·5538	11·504	13·686	14·866	18·781	23·281	24·917	28·399	32·168	34·164	40·606
0·07250	5·837	5·999	6·159	6·316	6·625	6·927	7·076	7·511	7·934	8·072	8·344	8·612	8·744	9·135
1/ 13·8	7·1630	7·9624	8·8153	9·7229	11·708	13·928	15·129	19·113	23·693	25·358	28·901	32·738	34·768	41·325
0·07500	5·937	6·101	6·264	6·424	6·739	7·046	7·197	7·640	8·069	8·210	8·487	8·759	8·894	9·291
1/ 13·3	7·2855	8·0985	8·9660	9·8892	11·908	14·166	15·388	19·440	24·098	25·792	29·395	33·297	35·363	42·031
0·08000	6·131	6·302	6·469	6·635	6·960	7·277	7·433	7·890	8·334	8·479	8·765	9·047	9·186	9·596
1/ 12·5	7·5244	8·3641	9·2600	10·213	12·299	14·631	15·893	20·078	24·889	26·637	30·360	34·389	36·523	43·409
0·08500	6·320	6·495	6·668	6·839	7·174	7·501	7·661	8·133	8·590	8·740	9·035	9·325	9·468	9·891
1/ 11·8	7·7560	8·6216	9·5450	10·528	12·677	15·081	16·382	20·696	25·655	27·457	31·294	35·448	37·647	44·745
0·09000	6·503	6·684	6·862	7·037	7·382	7·718	7·883	8·369	8·839	8·993	9·297	9·595	9·743	10·18
1/ 11·1	7·9808	8·8715	9·8218	10·833	13·045	15·518	16·857	21·296	26·399	28·253	32·201	36·475	38·738	46·043
0·09500	6·682	6·867	7·050	7·230	7·584	7·930	8·099	8·598	9·082	9·240	9·552	9·858	10·01	10·46
1/ 10·5	8·1996	9·1147	10·091	11·130	13·402	15·943	17·319	21·879	27·122	29·028	33·084	37·475	39·800	47·304
0·10000	6·855	7·045	7·233	7·418	7·781	8·136	8·310	8·821	9·318	9·480	9·800	10·11	10·27	10·73
1/ 10·0	8·4126	9·3514	10·353	11·419	13·750	16·358	17·769	22·448	27·827	29·782	33·943	38·449	40·834	48·533
	–	–	–	–	0·74	0·74	0·74	0·74	0·74	0·74	0·74	0·74	0·74	0·74

$V_{r(0·5)medial}$ for half-full circular pipes.

A47
(p.1 of 6)

k_s = 0·015 mm
S = 0·00003 to 0·00010

ie hydraulic gradient =
1 in 33333 to 1 in 10000

Water (or sewage) at 15°C;
full bore conditions.

velocities in ms^{-1}
discharges in m^3s^{-1}

Gradient	**(Equivalent) Pipe diameters in mm**													
	2400	**2500**	**2550**	**2600**	**2700**	**2800**	**2850**	**3000**	**3200**	**3400**	**3500**	**3600**	**4000**	**4500**
0·0000300	0·336	0·345	0·350	0·354	0·363	0·371	0·376	0·388	0·404	0·420	0·428	0·436	0·466	0·502
1/ 33333	1·5218	1·6951	1·7862	1·8802	2·0772	2·2866	2·3959	2·7432	3·2523	3·8160	4·1190	4·4364	5·8553	7·9829
0·0000320	0·348	0·358	0·362	0·367	0·376	0·385	0·389	0·402	0·419	0·435	0·443	0·451	0·483	0·520
1/ 31250	1·5763	1·7559	1·8502	1·9475	2·1516	2·3684	2·4817	2·8413	3·3685	3·9522	4·2660	4·5946	6·0638	8·2668
0·0000340	0·360	0·370	0·374	0·379	0·388	0·398	0·402	0·415	0·433	0·450	0·458	0·466	0·499	0·537
1/ 29412	1·6294	1·8149	1·9124	2·0130	2·2239	2·4479	2·5649	2·9366	3·4814	4·0845	4·4087	4·7483	6·2663	8·5425
0·0000360	0·372	0·381	0·386	0·391	0·401	0·410	0·415	0·429	0·447	0·464	0·473	0·481	0·514	0·554
1/ 27778	1·6810	1·8724	1·9729	2·0767	2·2942	2·5253	2·6460	3·0293	3·5912	4·2133	4·5476	4·8979	6·4635	8·8108
0·0000380	0·383	0·393	0·398	0·403	0·413	0·422	0·427	0·441	0·460	0·478	0·487	0·496	0·530	0·570
1/ 26316	1·7313	1·9284	2·0319	2·1387	2·3627	2·6007	2·7250	3·1197	3·6983	4·3388	4·6830	5·0436	6·6555	9·0723
0·0000400	0·394	0·404	0·409	0·414	0·424	0·434	0·439	0·454	0·473	0·491	0·500	0·509	0·545	0·586
1/ 25000	1·7804	1·9830	2·0894	2·1993	2·4296	2·6743	2·8021	3·2079	3·8027	4·4612	4·8152	5·1859	6·8430	9·3274
0·0000420	0·404	0·415	0·420	0·425	0·436	0·446	0·451	0·466	0·486	0·505	0·514	0·523	0·559	0·602
1/ 23810	1·8284	2·0364	2·1457	2·2585	2·4950	2·7462	2·8774	3·2941	3·9048	4·5809	4·9442	5·3249	7·0261	9·5766
0·0000440	0·415	0·426	0·431	0·436	0·447	0·457	0·463	0·478	0·498	0·517	0·527	0·536	0·573	0·617
1/ 22727	1·8753	2·0887	2·2007	2·3164	2·5589	2·8165	2·9511	3·3784	4·0046	4·6979	5·0705	5·4608	7·2053	9·8204
0·0000460	0·425	0·436	0·441	0·447	0·458	0·469	0·474	0·490	0·510	0·530	0·540	0·550	0·587	0·632
1/ 21739	1·9213	2·1398	2·2546	2·3732	2·6216	2·8854	3·0233	3·4609	4·1024	4·8125	5·1942	5·5939	7·3806	10·059
0·0000480	0·435	0·446	0·452	0·457	0·469	0·480	0·485	0·501	0·522	0·542	0·552	0·562	0·601	0·647
1/ 20833	1·9663	2·1900	2·3075	2·4287	2·6829	2·9529	3·0940	3·5418	4·1982	4·9248	5·3153	5·7244	7·5525	10·293
0·0000500	0·444	0·456	0·462	0·468	0·479	0·490	0·496	0·512	0·534	0·555	0·565	0·575	0·614	0·662
1/ 20000	2·0105	2·2392	2·3593	2·4833	2·7431	3·0192	3·1634	3·6212	4·2922	5·0350	5·4342	5·8523	7·7211	10·522
0·0000525	0·456	0·468	0·474	0·480	0·492	0·503	0·509	0·526	0·548	0·569	0·580	0·590	0·631	0·679
1/ 19048	2·0646	2·2994	2·4227	2·5500	2·8168	3·1002	3·2483	3·7183	4·4072	5·1698	5·5797	6·0089	7·9274	10·803
0·0000550	0·468	0·480	0·487	0·493	0·505	0·516	0·522	0·539	0·562	0·584	0·595	0·605	0·647	0·697
1/ 18182	2·1175	2·3583	2·4848	2·6153	2·8889	3·1795	3·3314	3·8134	4·5198	5·3017	5·7220	6·1621	8·1293	11·078
0·0000575	0·480	0·492	0·498	0·505	0·517	0·529	0·535	0·553	0·576	0·598	0·609	0·620	0·663	0·713
1/ 17391	2·1694	2·4160	2·5455	2·6792	2·9595	3·2572	3·4127	3·9064	4·6300	5·4309	5·8613	6·3121	8·3269	11·347
0·0000600	0·491	0·504	0·510	0·516	0·529	0·541	0·547	0·566	0·589	0·612	0·623	0·635	0·678	0·730
1/ 16667	2·2202	2·4725	2·6051	2·7419	3·0287	3·3333	3·4924	3·9976	4·7379	5·5574	5·9978	6·4591	8·5205	11·610
0·0000625	0·502	0·515	0·522	0·528	0·541	0·553	0·560	0·578	0·602	0·626	0·637	0·649	0·693	0·746
1/ 16000	2·2700	2·5280	2·6635	2·8034	3·0966	3·4079	3·5706	4·0870	4·8439	5·6816	6·1317	6·6033	8·7104	11·869
0·0000650	0·513	0·526	0·533	0·539	0·552	0·565	0·572	0·591	0·615	0·639	0·651	0·663	0·708	0·762
1/ 15385	2·3189	2·5824	2·7208	2·8637	3·1632	3·4812	3·6474	4·1748	4·9478	5·8034	6·2632	6·7448	8·8968	12·122
0·0000675	0·523	0·537	0·544	0·551	0·564	0·577	0·584	0·603	0·628	0·652	0·664	0·676	0·723	0·778
1/ 14815	2·3670	2·6359	2·7772	2·9230	3·2286	3·5532	3·7227	4·2611	5·0499	5·9231	6·3923	6·8837	9·0799	12·371
0·0000700	0·534	0·548	0·555	0·562	0·575	0·589	0·595	0·615	0·640	0·665	0·678	0·690	0·737	0·793
1/ 14286	2·4142	2·6885	2·8325	2·9812	3·2929	3·6239	3·7969	4·3458	5·1503	6·0407	6·5192	7·0204	9·2598	12·616
0·0000725	0·544	0·558	0·565	0·572	0·586	0·600	0·607	0·627	0·653	0·678	0·691	0·703	0·751	0·808
1/ 13793	2·4606	2·7402	2·8870	3·0385	3·3562	3·6935	3·8697	4·4292	5·2490	6·1564	6·6440	7·1547	9·4368	12·857
0·0000750	0·554	0·569	0·576	0·583	0·597	0·611	0·618	0·638	0·665	0·691	0·703	0·716	0·765	0·823
1/ 13333	2·5064	2·7911	2·9406	3·0949	3·4184	3·7620	3·9415	4·5112	5·3462	6·2703	6·7668	7·2869	9·6109	13·094
0·0000800	0·574	0·589	0·596	0·604	0·618	0·633	0·640	0·661	0·688	0·715	0·728	0·741	0·792	0·852
1/ 12500	2·5957	2·8905	3·0453	3·2052	3·5401	3·8959	4·0817	4·6716	5·5360	6·4928	7·0069	7·5453	9·9513	13·557
0·0000850	0·593	0·609	0·616	0·624	0·639	0·654	0·661	0·683	0·711	0·739	0·753	0·766	0·818	0·881
1/ 11765	2·6825	2·9872	3·1471	3·3122	3·6583	4·0259	4·2179	4·8274	5·7205	6·7089	7·2400	7·7963	10·282	14·006
0·0000900	0·612	0·628	0·636	0·643	0·659	0·674	0·682	0·704	0·734	0·762	0·776	0·790	0·844	0·908
1/ 11111	2·7670	3·0812	3·2461	3·4164	3·7733	4·1524	4·3504	4·9789	5·8999	6·9192	7·4668	8·0404	10·603	14·444
0·0000950	0·630	0·646	0·655	0·663	0·679	0·694	0·702	0·725	0·755	0·785	0·799	0·813	0·869	0·935
1/ 10526	2·8493	3·1728	3·3426	3·5180	3·8854	4·2757	4·4795	5·1266	6·0747	7·1240	7·6878	8·2783	10·917	14·870
0·0001000	0·648	0·665	0·673	0·681	0·698	0·714	0·722	0·746	0·777	0·807	0·821	0·836	0·893	0·961
1/ 10000	2·9297	3·2622	3·4368	3·6170	3·9948	4·3959	4·6055	5·2707	6·2453	7·3239	7·9035	8·5104	11·222	15·285
	0·95	0·95	0·96	0·96	0·96	0·96	0·96	0·96	0·96	0·97	0·97	0·97	0·97	0·97

$V_{r(0·5)medial}$ for half-full circular pipes.

k_s = 0·015 mm S = 0·00003 to 0·00010

k_s = 0·015 mm
S = 0·00010 to 0·00030

ie hydraulic gradient =
1 in 10000 to 1 in 3333

Water (or sewage) at 15°C;
full bore conditions.

velocities in ms^{-1}
discharges in m^3s^{-1}

Gradient	**(Equivalent) Pipe diameters in mm**													
	2400	**2500**	**2550**	**2600**	**2700**	**2800**	**2850**	**3000**	**3200**	**3400**	**3500**	**3600**	**4000**	**4500**
0·000100	0·648	0·665	0·673	0·681	0·698	0·714	0·722	0·746	0·777	0·807	0·821	0·836	0·893	0·961
1/ 10000	2·9297	3·2622	3·4368	3·6170	3·9948	4·3959	4·6055	5·2707	6·2453	7·3239	7·9035	8·5104	11·222	15·285
0·000105	0·665	0·682	0·691	0·699	0·716	0·733	0·741	0·766	0·797	0·828	0·843	0·858	0·917	0·987
1/ 9524	3·0081	3·3495	3·5288	3·7138	4·1016	4·5135	4·7286	5·4114	6·4120	7·5192	8·1141	8·7372	11·521	15·691
0·000110	0·682	0·700	0·709	0·717	0·735	0·752	0·760	0·785	0·818	0·849	0·865	0·880	0·940	1·012
1/ 9091	3·0849	3·4350	3·6187	3·8085	4·2061	4·6284	4·8490	5·5491	6·5750	7·7102	8·3201	8·9589	11·813	16·089
0·000115	0·699	0·717	0·726	0·735	0·752	0·770	0·779	0·804	0·837	0·870	0·886	0·901	0·963	1·036
1/ 8696	3·1601	3·5186	3·7068	3·9012	4·3084	4·7409	4·9668	5·6839	6·7345	7·8972	8·5218	9·1760	12·099	16·477
0·000120	0·715	0·733	0·743	0·752	0·770	0·788	0·797	0·823	0·857	0·890	0·906	0·922	0·985	1·060
1/ 8333	3·2337	3·6006	3·7932	3·9920	4·4087	4·8512	5·0823	5·8160	6·8909	8·0804	8·7194	9·3887	12·379	16·858
0·000125	0·731	0·750	0·759	0·769	0·787	0·805	0·814	0·841	0·876	0·910	0·926	0·943	1·007	1·083
1/ 8000	3·3060	3·6809	3·8778	4·0811	4·5070	4·9593	5·1956	5·9455	7·0442	8·2600	8·9132	9·5973	12·653	17·232
0·000130	0·746	0·766	0·776	0·785	0·804	0·823	0·832	0·859	0·895	0·929	0·946	0·963	1·028	1·106
1/ 7692	3·3769	3·7598	3·9609	4·1685	4·6035	5·0655	5·3067	6·0726	7·1947	8·4363	9·1034	9·8020	12·923	17·598
0·000135	0·762	0·782	0·792	0·801	0·821	0·840	0·849	0·877	0·913	0·948	0·966	0·983	1·049	1·129
1/ 7407	3·4465	3·8373	4·0426	4·2544	4·6983	5·1697	5·4159	6·1975	7·3425	8·6095	9·2902	10·003	13·188	17·958
0·000140	0·777	0·797	0·807	0·817	0·837	0·856	0·866	0·894	0·931	0·967	0·985	1·002	1·070	1·151
1/ 7143	3·5150	3·9135	4·1228	4·3388	4·7915	5·2722	5·5232	6·3202	7·4878	8·7797	9·4738	10·201	13·448	18·312
0·000145	0·792	0·813	0·823	0·833	0·853	0·873	0·882	0·911	0·949	0·985	1·003	1·021	1·090	1·173
1/ 6897	3·5823	3·9884	4·2017	4·4218	4·8831	5·3729	5·6288	6·4409	7·6306	8·9471	9·6543	10·395	13·704	18·659
0·000150	0·807	0·828	0·838	0·848	0·869	0·889	0·899	0·928	0·966	1·004	1·022	1·040	1·111	1·195
1/ 6667	3·6485	4·0621	4·2793	4·5035	4·9733	5·4721	5·7326	6·5596	7·7712	9·1117	9·8319	10·586	13·955	19·002
0·000160	0·835	0·857	0·868	0·878	0·899	0·920	0·930	0·961	1·000	1·039	1·058	1·077	1·150	1·237
1/ 6250	3·7780	4·2062	4·4310	4·6631	5·1495	5·6659	5·9356	6·7917	8·0459	9·4336	10·179	10·960	14·447	19·670
0·000170	0·863	0·885	0·896	0·907	0·929	0·951	0·961	0·993	1·034	1·073	1·093	1·112	1·188	1·278
1/ 5882	3·9038	4·3461	4·5784	4·8182	5·3206	5·8541	6·1327	7·0171	8·3126	9·7461	10·516	11·323	14·925	20·319
0·000180	0·890	0·913	0·925	0·936	0·958	0·980	0·991	1·024	1·066	1·107	1·127	1·147	1·225	1·317
1/ 5556	4·0261	4·4823	4·7218	4·9690	5·4870	6·0371	6·3244	7·2363	8·5721	10·050	10·844	11·675	15·389	20·951
0·000190	0·916	0·940	0·952	0·964	0·987	1·009	1·021	1·054	1·097	1·140	1·160	1·181	1·261	1·356
1/ 5263	4·1453	4·6149	4·8614	5·1160	5·6492	6·2155	6·5112	7·4499	8·8249	10·346	11·163	12·019	15·842	21·566
0·000200	0·942	0·967	0·979	0·991	1·014	1·038	1·049	1·083	1·128	1·171	1·193	1·214	1·296	1·394
1/ 5000	4·2616	4·7443	4·9977	5·2593	5·8075	6·3895	6·6935	7·6583	9·0715	10·635	11·475	12·355	16·283	22·166
0·000210	0·967	0·992	1·005	1·017	1·041	1·065	1·077	1·112	1·158	1·202	1·224	1·246	1·330	1·431
1/ 4762	4·3752	4·8707	5·1308	5·3994	5·9620	6·5595	6·8715	7·8618	9·3124	10·917	11·779	12·682	16·714	22·752
0·000220	0·992	1·017	1·030	1·043	1·068	1·092	1·104	1·140	1·187	1·233	1·255	1·277	1·364	1·467
1/ 4545	4·4863	4·9943	5·2610	5·5363	6·1132	6·7257	7·0456	8·0608	9·5479	11·193	12·077	13·002	17·136	23·325
0·000230	1·016	1·042	1·055	1·068	1·094	1·119	1·131	1·168	1·216	1·263	1·286	1·308	1·396	1·502
1/ 4348	4·5951	5·1153	5·3885	5·6704	6·2612	6·8884	7·2160	8·2556	9·7785	11·463	12·368	13·316	17·548	23·886
0·000240	1·039	1·066	1·080	1·093	1·119	1·145	1·157	1·195	1·244	1·292	1·315	1·338	1·429	1·536
1/ 4167	4·7017	5·2339	5·5133	5·8018	6·4061	7·0478	7·3829	8·4465	10·004	11·728	12·654	13·623	17·952	24·435
0·000250	1·062	1·090	1·104	1·117	1·144	1·170	1·183	1·221	1·271	1·320	1·344	1·368	1·460	1·570
1/ 4000	4·8062	5·3502	5·6358	5·9306	6·5483	7·2041	7·5466	8·6337	10·226	11·987	12·933	13·924	18·349	24·973
0·000260	1·085	1·113	1·127	1·141	1·168	1·195	1·208	1·247	1·299	1·348	1·373	1·397	1·491	1·603
1/ 3846	4·9087	5·4643	5·7559	6·0570	6·6878	7·3575	7·7073	8·8174	10·443	12·242	13·208	14·220	18·738	25·502
0·000270	1·107	1·136	1·150	1·164	1·192	1·219	1·233	1·273	1·325	1·376	1·401	1·425	1·521	1·636
1/ 3704	5·0095	5·5763	5·8740	6·1812	6·8248	7·5082	7·8651	8·9977	10·657	12·492	13·478	14·510	19·119	26·021
0·000280	1·129	1·158	1·173	1·187	1·216	1·243	1·257	1·298	1·351	1·403	1·428	1·454	1·551	1·668
1/ 3571	5·1084	5·6865	5·9899	6·3032	6·9595	7·6563	8·0201	9·1750	10·866	12·738	13·743	14·795	19·495	26·531
0·000290	1·151	1·181	1·195	1·210	1·239	1·267	1·281	1·323	1·377	1·430	1·455	1·481	1·581	1·700
1/ 3448	5·2058	5·7948	6·1040	6·4231	7·0919	7·8018	8·1726	9·3493	11·073	12·979	14·003	15·075	19·864	27·032
0·000300	1·172	1·202	1·217	1·232	1·261	1·290	1·305	1·347	1·402	1·456	1·482	1·508	1·610	1·731
1/ 3333	5·3016	5·9013	6·2162	6·5412	7·2221	7·9451	8·3226	9·5208	11·276	13·217	14·259	15·351	20·226	27·525
	0·97	0·97	0·97	0·97	0·97	0·97	0·97	0·97	0·97	0·98	0·98	0·98	0·98	0·98

$V_{r(0{\cdot}5)medial}$ for half-full circular pipes.

$k_s = 0{\cdot}015$ mm
S = 0·00030 to 0·00100

ie hydraulic gradient =
1 in 3333 to 1 in 1000

Water (or sewage) at 15°C;
full bore conditions.

velocities in ms^{-1}
discharges in m^3s^{-1}

Gradient	(Equivalent) Pipe diameters in mm													
	2400	2500	2550	2600	2700	2800	2850	3000	3200	3400	3500	3600	4000	4500
0·000300	1·172	1·202	1·217	1·232	1·261	1·290	1·305	1·347	1·402	1·456	1·482	1·508	1·610	1·731
1/ 3333	5·3016	5·9013	6·2162	6·5412	7·2221	7·9451	8·3226	9·5208	11·276	13·217	14·259	15·351	20·226	27·525
0·000320	1·213	1·245	1·260	1·275	1·306	1·336	1·351	1·394	1·451	1·507	1·534	1·561	1·666	1·791
1/ 3125	5·4887	6·1095	6·4355	6·7719	7·4767	8·2250	8·6157	9·8559	11·672	13·681	14·760	15·890	20·936	28·488
0·000340	1·253	1·286	1·302	1·318	1·349	1·380	1·395	1·440	1·499	1·557	1·585	1·613	1·721	1·850
1/ 2941	5·6705	6·3117	6·6484	6·9958	7·7238	8·4967	8·9003	10·181	12·057	14·132	15·246	16·413	21·624	29·424
0·000360	1·293	1·326	1·342	1·359	1·391	1·423	1·439	1·485	1·546	1·605	1·634	1·662	1·774	1·907
1/ 2778	5·8472	6·5084	6·8554	7·2137	7·9642	8·7610	9·1771	10·498	12·431	14·570	15·719	16·922	22·293	30·333
0·000380	1·331	1·365	1·382	1·399	1·432	1·465	1·481	1·529	1·591	1·652	1·682	1·711	1·826	1·963
1/ 2632	6·0195	6·6999	7·0572	7·4259	8·1984	9·0185	9·4467	10·806	12·796	14·997	16·180	17·418	22·945	31·219
0·000400	1·368	1·403	1·420	1·438	1·472	1·505	1·522	1·571	1·635	1·698	1·728	1·759	1·877	2·017
1/ 2500	6·1874	6·8868	7·2539	7·6329	8·4268	9·2696	9·7097	11·106	13·152	15·414	16·629	17·901	23·581	32·082
0·000420	1·404	1·440	1·458	1·476	1·511	1·545	1·562	1·613	1·678	1·743	1·774	1·805	1·926	2·070
1/ 2381	6·3515	7·0693	7·4461	7·8351	8·6499	9·5149	9·9666	11·400	13·499	15·821	17·068	18·373	24·202	32·926
0·000440	1·439	1·477	1·495	1·513	1·549	1·584	1·602	1·653	1·721	1·786	1·819	1·850	1·974	2·122
1/ 2273	6·5119	7·2478	7·6340	8·0327	8·8680	9·7547	10·218	11·687	13·839	16·218	17·496	18·834	24·809	33·750
0·000460	1·474	1·512	1·531	1·549	1·586	1·622	1·640	1·693	1·762	1·829	1·862	1·895	2·022	2·173
1/ 2174	6·6689	7·4224	7·8179	8·2262	9·0814	9·9894	10·463	11·968	14·171	16·608	17·916	19·286	25·403	34·557
0·000480	1·508	1·547	1·566	1·585	1·623	1·660	1·678	1·732	1·803	1·871	1·905	1·938	2·068	2·223
1/ 2083	6·8228	7·5935	7·9981	8·4157	9·2906	10·219	10·704	12·243	14·497	16·989	18·327	19·728	25·985	35·347
0·000500	1·542	1·581	1·601	1·620	1·658	1·696	1·715	1·770	1·842	1·912	1·947	1·981	2·113	2·271
1/ 2000	6·9736	7·7613	8·1748	8·6015	9·4956	10·445	10·940	12·513	14·816	17·362	18·730	20·162	26·555	36·122
0·000525	1·582	1·623	1·643	1·663	1·702	1·741	1·760	1·817	1·891	1·963	1·998	2·033	2·169	2·331
1/ 1905	7·1582	7·9666	8·3909	8·8289	9·7465	10·721	11·229	12·843	15·206	17·820	19·223	20·693	27·253	37·070
0·000550	1·622	1·664	1·684	1·705	1·745	1·785	1·804	1·863	1·938	2·012	2·048	2·084	2·223	2·389
1/ 1818	7·3386	8·1673	8·6023	9·0513	9·9918	10·990	11·512	13·166	15·588	18·267	19·705	21·211	27·935	37·996
0·000575	1·661	1·704	1·725	1·746	1·787	1·828	1·848	1·907	1·985	2·060	2·097	2·134	2·276	2·446
1/ 1739	7·5152	8·3638	8·8092	9·2689	10·232	11·254	11·788	13·482	15·962	18·704	20·177	21·719	28·603	38·903
0·000600	1·699	1·743	1·765	1·786	1·828	1·870	1·890	1·951	2·030	2·107	2·145	2·183	2·328	2·502
1/ 1667	7·6882	8·5562	9·0118	9·4820	10·467	11·513	12·059	13·791	16·328	19·133	20·639	22·216	29·257	39·791
0·000625	1·737	1·781	1·803	1·825	1·868	1·911	1·932	1·994	2·075	2·154	2·192	2·231	2·379	2·557
1/ 1600	7·8579	8·7449	9·2105	9·6910	10·698	11·766	12·324	14·094	16·687	19·553	21·092	22·704	29·898	40·661
0·000650	1·774	1·819	1·842	1·864	1·908	1·951	1·973	2·036	2·119	2·199	2·238	2·277	2·429	2·610
1/ 1538	8·0243	8·9300	9·4054	9·8960	10·924	12·015	12·585	14·392	17·039	19·965	21·537	23·182	30·527	41·515
0·000675	1·810	1·856	1·879	1·902	1·947	1·991	2·013	2·077	2·162	2·244	2·284	2·324	2·478	2·663
1/ 1481	8·1878	9·1118	9·5968	10·097	11·146	12·259	12·840	14·684	17·384	20·370	21·973	23·651	31·144	42·353
0·000700	1·845	1·893	1·916	1·939	1·985	2·030	2·052	2·118	2·204	2·287	2·328	2·369	2·527	2·715
1/ 1429	8·3484	9·2904	9·7848	10·295	11·364	12·499	13·091	14·971	17·724	20·767	22·402	24·113	31·750	43·177
0·000725	1·880	1·928	1·952	1·976	2·022	2·068	2·091	2·158	2·245	2·330	2·372	2·413	2·574	2·766
1/ 1379	8·5062	9·4660	9·9697	10·490	11·579	12·735	13·338	15·254	18·058	21·158	22·823	24·566	32·347	43·987
0·000750	1·915	1·964	1·988	2·012	2·059	2·106	2·129	2·197	2·286	2·373	2·415	2·457	2·621	2·816
1/ 1333	8·6616	9·6388	10·152	10·681	11·790	12·967	13·581	15·531	18·386	21·543	23·238	25·012	32·933	44·783
0·000800	1·982	2·032	2·057	2·082	2·131	2·179	2·203	2·274	2·366	2·455	2·499	2·543	2·712	2·914
1/ 1250	8·9650	9·9762	10·507	11·055	12·202	13·420	14·056	16·074	19·028	22·294	24·047	25·883	34·079	46·338
0·000850	2·047	2·099	2·125	2·150	2·201	2·251	2·276	2·348	2·443	2·536	2·581	2·626	2·800	3·008
1/ 1176	9·2596	10·304	10·852	11·418	12·602	13·860	14·517	16·600	19·650	23·023	24·833	26·729	35·190	47·847
0·000900	2·110	2·164	2·191	2·217	2·269	2·320	2·346	2·421	2·519	2·614	2·661	2·707	2·886	3·101
1/ 1111	9·5460	10·622	11·187	11·770	12·992	14·288	14·965	17·112	20·256	23·731	25·597	27·551	36·271	49·314
0·000950	2·172	2·227	2·255	2·282	2·335	2·388	2·414	2·491	2·592	2·690	2·738	2·785	2·970	3·190
1/ 1053	9·8249	10·933	11·514	12·114	13·371	14·704	15·401	17·610	20·845	24·421	26·341	28·352	37·323	50·743
0·001000	2·232	2·289	2·317	2·345	2·400	2·454	2·481	2·560	2·663	2·764	2·813	2·862	3·052	3·278
1/ 1000	10·097	11·235	11·832	12·449	13·740	15·111	15·826	18·096	21·420	25·094	27·067	29·132	38·349	52·135
	0·98	0·98	0·98	0·98	0·98	0·98	0·98	0·98	0·98	0·98	0·98	0·98	0·99	0·99

$V_{r(0{\cdot}5)medial}$ for half-full circular pipes.

$k_s = 0{\cdot}015$ mm S = 0·00030 to 0·00100

ie hydraulic gradient = 1 in 1000 to 1 in 333

velocities in ms^{-1}
discharges in m^3s^{-1}

Gradient	(Equivalent) Pipe diameters in mm													
	2400	**2500**	**2550**	**2600**	**2700**	**2800**	**2850**	**3000**	**3200**	**3400**	**3500**	**3600**	**4000**	**4500**
0·00100	2·232	2·289	2·317	2·345	2·400	2·454	2·481	2·560	2·663	2·764	2·813	2·862	3·052	3·278
1/ 1000	10·097	11·235	11·832	12·449	13·740	15·111	15·826	18·096	21·420	25·094	27·067	29·132	38·349	52·135
0·00105	2·291	2·349	2·378	2·406	2·463	2·518	2·546	2·627	2·733	2·836	2·887	2·937	3·131	3·364
1/ 952	10·363	11·530	12·143	12·776	14·101	15·507	16·242	18·571	21·981	25·751	27·775	29·894	39·351	53·495
0·00110	2·348	2·408	2·437	2·467	2·524	2·581	2·610	2·693	2·801	2·907	2·959	3·010	3·209	3·447
1/ 909	10·622	11·819	12·447	13·096	14·454	15·895	16·647	19·035	22·530	26·393	28·467	30·639	40·330	54·823
0·00115	2·404	2·465	2·496	2·525	2·585	2·643	2·672	2·757	2·868	2·976	3·029	3·082	3·286	3·529
1/ 870	10·876	12·102	12·745	13·409	14·799	16·274	17·045	19·489	23·066	27·021	29·145	31·367	41·288	56·123
0·00120	2·459	2·522	2·553	2·583	2·644	2·703	2·733	2·820	2·933	3·044	3·098	3·152	3·360	3·609
1/ 833	11·125	12·378	13·036	13·715	15·137	16·646	17·434	19·933	23·592	27·636	29·808	32·081	42·226	57·396
0·00125	2·513	2·577	2·609	2·640	2·702	2·762	2·793	2·882	2·997	3·110	3·166	3·220	3·433	3·687
1/ 800	11·369	12·650	13·322	14·015	15·468	17·010	17·815	20·369	24·107	28·239	30·458	32·780	43·145	58·644
0·00130	2·566	2·631	2·663	2·695	2·758	2·821	2·851	2·942	3·060	3·175	3·232	3·288	3·505	3·764
1/ 769	11·608	12·916	13·602	14·310	15·793	17·367	18·189	20·796	24·612	28·831	31·096	33·467	44·047	59·867
0·00135	2·618	2·684	2·717	2·750	2·814	2·877	2·909	3·001	3·122	3·239	3·297	3·354	3·576	3·840
1/ 741	11·843	13·177	13·877	14·599	16·112	17·718	18·556	21·216	25·108	29·412	31·722	34·140	44·932	61·068
0·00140	2·669	2·737	2·770	2·803	2·869	2·933	2·965	3·060	3·183	3·302	3·361	3·419	3·645	3·914
1/ 714	12·074	13·434	14·147	14·884	16·426	18·063	18·917	21·628	25·596	29·982	32·337	34·802	45·801	62·248
0·00145	2·719	2·788	2·822	2·856	2·923	2·988	3·021	3·117	3·242	3·364	3·424	3·483	3·713	3·987
1/ 690	12·301	13·686	14·413	15·163	16·734	18·401	19·272	22·033	26·075	30·543	32·941	35·452	46·656	63·408
0·00150	2·768	2·839	2·873	2·908	2·976	3·043	3·076	3·173	3·301	3·425	3·486	3·546	3·780	4·059
1/ 667	12·524	13·934	14·674	15·438	17·037	18·734	19·621	22·432	26·546	31·094	33·536	36·091	47·496	64·548
0·00160	2·865	2·937	2·973	3·009	3·079	3·148	3·182	3·284	3·415	3·543	3·606	3·668	3·910	4·199
1/ 625	12·960	14·419	15·184	15·974	17·629	19·385	20·302	23·210	27·466	32·171	34·697	37·340	49·137	66·775
0·00170	2·958	3·033	3·070	3·107	3·179	3·251	3·286	3·390	3·526	3·658	3·723	3·788	4·037	4·334
1/ 588	13·383	14·889	15·680	16·495	18·204	20·016	20·963	23·965	28·359	33·216	35·823	38·552	50·730	68·936
0·00180	3·049	3·126	3·164	3·202	3·277	3·350	3·387	3·494	3·634	3·770	3·837	3·903	4·160	4·466
1/ 556	13·794	15·347	16·161	17·002	18·762	20·630	21·605	24·699	29·227	34·232	36·918	39·730	52·278	71·035
0·00190	3·138	3·217	3·256	3·295	3·372	3·447	3·485	3·595	3·739	3·879	3·948	4·016	4·280	4·595
1/ 526	14·195	15·792	16·630	17·495	19·306	21·228	22·231	25·414	30·072	35·220	37·984	40·877	53·784	73·079
0·00200	3·224	3·306	3·346	3·386	3·464	3·542	3·580	3·694	3·842	3·985	4·056	4·126	4·397	4·720
1/ 500	14·585	16·226	17·087	17·975	19·836	21·810	22·841	26·111	30·896	36·184	39·023	41·995	55·253	75·072
0·00210	3·308	3·392	3·433	3·474	3·555	3·634	3·674	3·790	3·942	4·089	4·161	4·233	4·511	4·842
1/ 476	14·966	16·650	17·533	18·444	20·353	22·379	23·436	26·791	31·700	37·125	40·037	43·086	56·686	77·016
0·00220	3·391	3·476	3·518	3·560	3·643	3·725	3·765	3·884	4·039	4·190	4·264	4·338	4·622	4·962
1/ 455	15·339	17·064	17·969	18·903	20·859	22·935	24·018	27·455	32·485	38·045	41·029	44·152	58·087	78·916
0·00230	3·471	3·559	3·602	3·645	3·730	3·813	3·854	3·976	4·135	4·289	4·365	4·440	4·731	5·079
1/ 435	15·703	17·469	18·395	19·352	21·354	23·478	24·588	28·106	33·254	38·944	41·998	45·195	59·457	80·775
0·00240	3·550	3·640	3·684	3·728	3·814	3·899	3·942	4·066	4·228	4·386	4·464	4·540	4·838	5·193
1/ 417	16·060	17·866	18·813	19·791	21·838	24·011	25·145	28·742	34·007	39·825	42·948	46·216	60·799	82·595
0·00250	3·627	3·719	3·764	3·809	3·897	3·984	4·027	4·154	4·320	4·481	4·561	4·639	4·943	5·305
1/ 400	16·410	18·255	19·223	20·221	22·313	24·533	25·691	29·366	34·744	40·688	43·878	47·217	62·113	84·378
0·00260	3·703	3·796	3·843	3·888	3·978	4·067	4·111	4·241	4·410	4·575	4·655	4·735	5·045	5·415
1/ 385	16·753	18·636	19·624	20·644	22·779	25·044	26·227	29·978	35·468	41·534	44·790	48·199	63·403	86·126
0·00270	3·778	3·873	3·920	3·966	4·058	4·149	4·194	4·326	4·498	4·666	4·749	4·830	5·146	5·523
1/ 370	17·090	19·010	20·018	21·058	23·236	25·547	26·753	30·579	36·178	42·365	45·686	49·162	64·668	87·843
0·00280	3·851	3·948	3·996	4·043	4·137	4·229	4·275	4·410	4·585	4·756	4·840	4·923	5·245	5·629
1/ 357	17·421	19·378	20·405	21·465	23·685	26·040	27·270	31·169	36·875	43·181	46·566	50·108	65·911	89·528
0·00290	3·923	4·021	4·070	4·118	4·214	4·308	4·354	4·492	4·670	4·844	4·930	5·014	5·342	5·733
1/ 345	17·746	19·740	20·786	21·865	24·126	26·525	27·777	31·749	37·561	43·983	47·430	51·038	67·133	91·185
0·00300	3·993	4·094	4·143	4·192	4·290	4·385	4·432	4·572	4·754	4·931	5·018	5·104	5·438	5·836
1/ 333	18·066	20·095	21·160	22·259	24·560	27·002	28·277	32·319	38·235	44·772	48·280	51·953	68·334	92·814
	0·98	0·98	0·98	0·98	0·98	0·99	0·99	0·99	0·99	0·99	0·99	0·99	0·99	0·99

$V_{r(0{\cdot}5)medial}$ for half-full circular pipes.

A47
(p.5 of 6)

k_s = 0·015 mm
S = 0·00300 to 0·01000

ie hydraulic gradient =
1 in 333 to 1 in 100

Water (or sewage) at 15°C;
full bore conditions.

velocities in ms^{-1}
discharges in m^3s^{-1}

Gradient **(Equivalent) Pipe diameters in mm**

Gradient	2400	2500	2550	2600	2700	2800	2850	3000	3200	3400	3500	3600	4000	4500
0·00300	3·993	4·094	4·143	4·192	4·290	4·385	4·432	4·572	4·754	4·931	5·018	5·104	5·438	5·836
1/ 333	18·066	20·095	21·160	22·259	24·560	27·002	28·277	32·319	38·235	44·772	48·280	51·953	68·334	92·814
0·00320	4·131	4·235	4·286	4·337	4·438	4·536	4·585	4·730	4·918	5·101	5·191	5·279	5·624	6·036
1/ 313	18·690	20·789	21·891	23·027	25·408	27·933	29·252	33·433	39·551	46·312	49·940	53·738	70·679	95·994
0·00340	4·265	4·372	4·425	4·478	4·581	4·683	4·734	4·883	5·077	5·265	5·358	5·450	5·805	6·230
1/ 294	19·296	21·463	22·600	23·773	26·230	28·837	30·197	34·513	40·828	47·805	51·550	55·470	72·953	99·078
0·00360	4·395	4·506	4·560	4·614	4·721	4·826	4·878	5·031	5·231	5·425	5·521	5·615	5·981	6·418
1/ 278	19·885	22·117	23·289	24·498	27·029	29·715	31·117	35·563	42·069	49·257	53·115	57·153	75·164	102·08
0·00380	4·522	4·636	4·691	4·747	4·857	4·965	5·018	5·176	5·381	5·581	5·679	5·776	6·153	6·601
1/ 263	20·458	22·755	23·960	25·203	27·807	30·569	32·011	36·585	43·276	50·670	54·638	58·791	77·315	104·99
0·00400	4·646	4·762	4·819	4·876	4·989	5·100	5·155	5·317	5·527	5·733	5·833	5·933	6·319	6·780
1/ 250	21·017	23·376	24·613	25·890	28·565	31·402	32·883	37·581	44·453	52·047	56·122	60·388	79·411	107·84
0·00420	4·766	4·886	4·944	5·003	5·118	5·232	5·288	5·454	5·670	5·881	5·984	6·086	6·482	6·955
1/ 238	21·562	23·982	25·251	26·561	29·305	32·215	33·735	38·553	45·602	53·391	57·571	61·946	81·457	110·61
0·00440	4·884	5·006	5·066	5·126	5·245	5·361	5·418	5·588	5·810	6·025	6·131	6·235	6·641	7·125
1/ 227	22·095	24·574	25·875	27·217	30·028	33·010	34·566	39·503	46·725	54·704	58·986	63·468	83·456	113·32
0·00460	4·999	5·124	5·186	5·247	5·368	5·487	5·546	5·720	5·946	6·167	6·275	6·382	6·797	7·292
1/ 217	22·616	25·153	26·485	27·858	30·735	33·787	35·380	40·431	47·822	55·988	60·370	64·957	85·411	115·97
0·00480	5·112	5·240	5·303	5·365	5·489	5·611	5·671	5·849	6·080	6·305	6·416	6·525	6·949	7·455
1/ 208	23·126	25·721	27·082	28·486	31·427	34·547	36·176	41·341	48·897	57·245	61·725	66·414	87·325	118·56
0·00500	5·223	5·353	5·417	5·481	5·607	5·732	5·793	5·975	6·211	6·441	6·554	6·665	7·098	7·615
1/ 200	23·626	26·276	27·667	29·102	32·106	35·293	36·956	42·232	49·950	58·477	63·053	67·842	89·200	121·11
0·00525	5·358	5·491	5·558	5·623	5·752	5·880	5·943	6·129	6·371	6·607	6·722	6·837	7·281	7·810
1/ 190	24·238	26·956	28·382	29·854	32·935	36·204	37·911	43·322	51·238	59·984	64·677	69·589	91·493	124·21
0·00550	5·490	5·627	5·694	5·761	5·894	6·024	6·089	6·279	6·527	6·769	6·887	7·004	7·459	8·001
1/ 182	24·835	27·620	29·082	30·589	33·746	37·095	38·843	44·387	52·496	61·455	66·263	71·295	93·733	127·25
0·00575	5·619	5·759	5·828	5·897	6·032	6·166	6·232	6·427	6·680	6·927	7·048	7·168	7·633	8·188
1/ 174	25·420	28·270	29·765	31·308	34·539	37·966	39·755	45·428	53·727	62·895	67·815	72·963	95·924	130·22
0·00600	5·746	5·889	5·959	6·029	6·168	6·304	6·372	6·571	6·830	7·083	7·206	7·329	7·804	8·371
1/ 167	25·992	28·906	30·435	32·012	35·315	38·819	40·648	46·448	54·931	64·304	69·333	74·597	98·068	133·13
0·00625	5·869	6·016	6·088	6·159	6·301	6·440	6·509	6·712	6·977	7·235	7·361	7·486	7·971	8·550
1/ 160	26·553	29·529	31·091	32·702	36·076	39·655	41·523	47·446	56·112	65·684	70·821	76·197	100·17	135·98
0·00650	5·991	6·140	6·214	6·287	6·431	6·573	6·643	6·851	7·121	7·384	7·513	7·640	8·135	8·725
1/ 154	27·103	30·141	31·734	33·379	36·822	40·474	42·381	48·426	57·269	67·038	72·280	77·766	102·23	138·77
0·00675	6·110	6·262	6·337	6·412	6·559	6·704	6·775	6·987	7·262	7·530	7·662	7·791	8·296	8·897
1/ 148	27·642	30·740	32·366	34·043	37·554	41·278	43·222	49·387	58·405	68·367	73·712	79·306	104·25	141·51
0·00700	6·227	6·382	6·459	6·535	6·685	6·832	6·905	7·120	7·401	7·674	7·808	7·940	8·454	9·067
1/ 143	28·172	31·329	32·986	34·695	38·273	42·068	44·049	50·331	59·521	69·671	75·118	80·819	106·24	144·20
0·00725	6·343	6·500	6·578	6·655	6·808	6·958	7·032	7·252	7·537	7·815	7·951	8·086	8·609	9·233
1/ 138	28·693	31·908	33·595	35·335	38·979	42·844	44·862	51·259	60·617	70·953	76·500	82·305	108·19	146·84
0·00750	6·456	6·616	6·695	6·774	6·929	7·082	7·157	7·381	7·671	7·954	8·092	8·229	8·762	9·396
1/ 133	29·205	32·478	34·194	35·965	39·673	43·607	45·660	52·171	61·694	72·214	77·859	83·766	110·11	149·44
0·00800	6·677	6·843	6·924	7·006	7·166	7·324	7·402	7·633	7·933	8·225	8·368	8·510	9·060	9·715
1/ 125	30·205	33·589	35·364	37·195	41·029	45·097	47·220	53·952	63·798	74·674	80·510	86·618	113·85	154·52
0·00850	6·891	7·062	7·147	7·230	7·396	7·559	7·639	7·877	8·186	8·488	8·635	8·781	9·349	10·02
1/ 118	31·175	34·666	36·498	38·388	42·345	46·542	48·732	55·678	65·838	77·061	83·082	89·384	117·48	159·44
0·00900	7·099	7·275	7·362	7·449	7·619	7·786	7·869	8·114	8·433	8·743	8·895	9·045	9·629	10·33
1/ 111	32·117	35·714	37·600	39·547	43·622	47·945	50·202	57·356	67·820	79·379	85·580	92·070	121·01	164·22
0·00950	7·302	7·483	7·572	7·661	7·836	8·008	8·093	8·345	8·673	8·991	9·148	9·302	9·902	10·62
1/ 105	33·034	36·732	38·673	40·674	44·866	49·311	51·631	58·989	69·749	81·634	88·011	94·685	124·44	168·87
0·01000	7·500	7·685	7·777	7·868	8·048	8·224	8·312	8·570	8·906	9·233	9·394	9·552	10·17	10·90
1/ 100	33·927	37·725	39·718	41·773	46·077	50·642	53·025	60·579	71·628	83·832	90·380	97·232	127·78	173·40
	0·99	0·99	0·99	0·99	0·99	0·99	0·99	0·99	0·99	0·99	0·99	0·99	0·99	0·99

$V_{r(0·5)medial}$ for half-full circular pipes.

k_s = 0·015 mm S = 0·00300 to 0·01000

k_s = 0·015 mm
S = 0·01000 to 0·03000

ie hydraulic gradient =
1 in 100 to 1 in 33·3

Water (or sewage) at 15°C;
full bore conditions.

velocities in ms^{-1}
discharges in m^3s^{-1}

Gradient	**(Equivalent) Pipe diameters in mm**													
	2400	**2500**	**2550**	**2600**	**2700**	**2800**	**2850**	**3000**	**3200**	**3400**	**3500**	**3600**	**4000**	**4500**
0·01000	7·500	7·685	7·777	7·868	8·048	8·224	8·312	8·570	8·906	9·233	9·394	9·552	10·17	10·90
1/ 100	33·927	37·725	39·718	41·773	46·077	50·642	53·025	60·579	71·628	83·832	90·380	97·232	127·78	173·40
0·01050	7·692	7·883	7·977	8·070	8·254	8·435	8·525	8·790	9·134	9·470	9·634	9·797	10·43	11·18
1/ 95	34·799	38·694	40·738	42·846	47·259	51·941	54·384	62·132	73·462	85·977	92·691	99·718	131·04	177·82
0·01100	7·881	8·076	8·172	8·267	8·456	8·641	8·733	9·004	9·357	9·700	9·869	10·04	10·68	11·45
1/ 91	35·651	39·641	41·734	43·893	48·414	53·210	55·712	63·648	75·254	88·072	94·949	102·15	134·23	182·13
0·01150	8·065	8·264	8·363	8·460	8·653	8·843	8·937	9·214	9·575	9·926	10·10	10·27	10·93	11·72
1/ 87	36·484	40·567	42·708	44·918	49·544	54·450	57·011	65·130	77·005	90·120	97·157	104·52	137·35	186·36
0·01200	8·245	8·449	8·549	8·649	8·846	9·040	9·136	9·419	9·788	10·15	10·32	10·50	11·17	11·98
1/ 83	37·299	41·473	43·662	45·921	50·649	55·665	58·282	66·582	78·720	92·125	99·317	106·84	140·40	190·49
0·01250	8·422	8·630	8·732	8·834	9·035	9·233	9·331	9·621	9·997	10·36	10·54	10·72	11·41	12·23
1/ 80	38·098	42·360	44·596	46·903	51·732	56·855	59·528	68·004	80·399	94·089	101·43	109·12	143·38	194·54
0·01300	8·595	8·807	8·912	9·016	9·221	9·423	9·523	9·818	10·20	10·58	10·76	10·94	11·64	12·48
1/ 77	38·881	43·231	45·512	47·866	52·794	58·021	60·749	69·398	82·046	96·015	103·51	111·35	146·31	198·51
0·01350	8·765	8·981	9·088	9·194	9·403	9·609	9·710	10·01	10·40	10·78	10·97	11·15	11·87	12·73
1/ 74	39·650	44·085	46·411	48·811	53·836	59·166	61·947	70·766	83·662	97·905	105·55	113·54	149·19	202·40
0·01400	8·931	9·152	9·261	9·368	9·582	9·791	9·895	10·20	10·60	10·99	11·18	11·37	12·10	12·97
1/ 71	40·405	44·924	47·294	49·740	54·859	60·290	63·124	72·109	85·249	99·761	107·55	115·69	152·01	206·22
0·01450	9·095	9·320	9·430	9·540	9·757	9·971	10·08	10·39	10·79	11·19	11·38	11·57	12·32	13·20
1/ 69	41·146	45·748	48·162	50·652	55·865	61·394	64·280	73·429	86·808	101·58	109·51	117·81	154·78	209·98
0·01500	9·257	9·485	9·597	9·709	9·930	10·15	10·25	10·57	10·98	11·39	11·58	11·78	12·53	13·43
1/ 67	41·876	46·558	49·014	51·548	56·853	62·480	65·416	74·726	88·341	103·38	111·44	119·88	157·51	213·67
0·01600	9·571	9·807	9·923	10·04	10·27	10·49	10·60	10·93	11·36	11·77	11·97	12·18	12·96	13·89
1/ 63	43·299	48·140	50·679	53·298	58·783	64·599	67·635	77·259	91·332	106·87	115·21	123·94	162·83	220·88
0·01700	9·876	10·12	10·24	10·36	10·59	10·82	10·94	11·28	11·72	12·14	12·35	12·56	13·37	14·33
1/ 59	44·679	49·673	52·293	54·996	60·653	66·654	69·786	79·714	94·233	110·27	118·87	127·87	167·98	227·87
0·01800	10·17	10·42	10·55	10·67	10·91	11·15	11·27	11·61	12·07	12·51	12·72	12·94	13·77	14·75
1/ 56	46·019	51·163	53·861	56·644	62·471	68·650	71·875	82·100	97·051	113·56	122·42	131·69	172·99	234·65
0·01900	10·46	10·72	10·85	10·97	11·22	11·46	11·59	11·94	12·41	12·86	13·08	13·30	14·15	15·17
1/ 53	47·324	52·612	55·386	58·248	64·239	70·593	73·908	84·420	99·792	116·77	125·87	135·40	177·87	241·26
0·02000	10·74	11·01	11·14	11·27	11·52	11·77	11·90	12·26	12·74	13·20	13·43	13·66	14·53	15·57
1/ 50	48·595	54·025	56·873	59·811	65·961	72·485	75·889	86·681	102·46	119·89	129·24	139·02	182·62	247·69
0·02100	11·02	11·29	11·42	11·55	11·81	12·07	12·20	12·57	13·06	13·54	13·77	14·00	14·90	15·97
1/ 47·6	49·835	55·402	58·323	61·335	67·642	74·331	77·821	88·887	105·07	122·94	132·52	142·55	187·25	253·96
0·02200	11·28	11·56	11·70	11·83	12·10	12·36	12·49	12·88	13·38	13·87	14·11	14·34	15·26	16·35
1/ 45·5	51·046	56·748	59·739	62·825	69·284	76·134	79·709	91·042	107·61	125·91	135·73	146·00	191·77	260·09
0·02300	11·55	11·83	11·97	12·11	12·38	12·65	12·78	13·18	13·69	14·19	14·43	14·67	15·61	16·73
1/ 43·5	52·230	58·064	61·124	64·281	70·889	77·897	81·554	93·149	110·10	128·82	138·86	149·37	196·20	266·08
0·02400	11·80	12·09	12·23	12·38	12·66	12·93	13·07	13·47	13·99	14·50	14·75	15·00	15·96	17·10
1/ 41·7	53·389	59·352	62·480	65·706	72·460	79·623	83·360	95·210	112·54	131·67	141·93	152·67	200·53	271·95
0·02500	12·05	12·35	12·49	12·64	12·92	13·21	13·34	13·76	14·29	14·81	15·06	15·32	16·29	17·46
1/ 40·0	54·525	60·614	63·808	67·102	73·999	81·313	85·130	97·230	114·92	134·46	144·94	155·90	204·77	277·69
0·02600	12·30	12·60	12·75	12·90	13·19	13·47	13·62	14·04	14·58	15·11	15·37	15·63	16·63	17·81
1/ 38·5	55·638	61·851	65·109	68·471	75·508	82·971	86·864	99·210	117·26	137·19	147·88	159·07	208·92	283·32
0·02700	12·54	12·85	13·00	13·15	13·45	13·74	13·88	14·31	14·87	15·41	15·67	15·93	16·95	18·16
1/ 37·0	56·731	63·065	66·387	69·814	76·988	84·597	88·566	101·15	119·56	139·87	150·77	162·18	213·00	288·85
0·02800	12·78	13·09	13·24	13·40	13·70	14·00	14·15	14·58	15·15	15·70	15·97	16·23	17·27	18·50
1/ 35·7	57·803	64·256	67·641	71·133	78·442	86·193	90·237	103·06	121·81	142·51	153·61	165·23	217·01	294·27
0·02900	13·01	13·33	13·49	13·64	13·95	14·25	14·40	14·85	15·42	15·98	16·26	16·53	17·58	18·84
1/ 34·5	58·857	65·427	68·873	72·428	79·870	87·761	91·879	104·93	124·02	145·10	156·40	168·23	220·94	299·60
0·03000	13·24	13·56	13·72	13·88	14·19	14·50	14·66	15·11	15·69	16·26	16·54	16·82	17·89	19·17
1/ 33·3	59·893	66·578	70·084	73·702	81·273	89·303	93·493	106·78	126·20	147·64	159·14	171·18	224·81	304·84
	0·99	0·99	0·99	0·99	0·99	0·99	0·99	0·99	0·99	0·99	0·99	0·99	0·99	0·99

$V_{r(0·5)medial}$ for half-full circular pipes.

$k_s = 0{\cdot}030$ **mm**
S = 0·00003 to 0·00010

ie hydraulic gradient =
1 in 33333 to 1 in 10000

Water (or sewage) at 15°C;
full bore conditions.

velocities in ms^{-1}
discharges in m^3s^{-1}

Gradient	(Equivalent) Pipe diameters in mm													
	2400	**2500**	**2550**	**2600**	**2700**	**2800**	**2850**	**3000**	**3200**	**3400**	**3500**	**3600**	**4000**	**4500**
0·0000300	0·335	0·344	0·348	0·352	0·361	0·369	0·374	0·386	0·402	0·418	0·426	0·433	0·463	0·499
1/ 33333	1·5145	1·6869	1·7775	1·8710	2·0669	2·2751	2·3839	2·7292	3·2353	3·7957	4·0968	4·4123	5·8224	7·9363
0·0000320	0·347	0·356	0·360	0·365	0·374	0·383	0·387	0·400	0·417	0·433	0·441	0·449	0·480	0·517
1/ 31250	1·5686	1·7472	1·8409	1·9377	2·1407	2·3562	2·4688	2·8264	3·3504	3·9306	4·2424	4·5690	6·0288	8·2172
0·0000340	0·358	0·368	0·373	0·377	0·386	0·395	0·400	0·413	0·430	0·447	0·456	0·464	0·496	0·534
1/ 29412	1·6212	1·8057	1·9026	2·0026	2·2123	2·4350	2·5513	2·9207	3·4622	4·0616	4·3837	4·7212	6·2293	8·4901
0·0000360	0·370	0·379	0·384	0·389	0·399	0·408	0·413	0·426	0·444	0·461	0·470	0·478	0·511	0·551
1/ 27778	1·6723	1·8626	1·9625	2·0657	2·2819	2·5116	2·6316	3·0126	3·5710	4·1891	4·5213	4·8692	6·4243	8·7554
0·0000380	0·381	0·391	0·396	0·401	0·410	0·420	0·425	0·439	0·457	0·475	0·484	0·493	0·526	0·567
1/ 26316	1·7222	1·9181	2·0210	2·1272	2·3498	2·5863	2·7098	3·1021	3·6769	4·3133	4·6552	5·0135	6·6143	9·0140
0·0000400	0·391	0·402	0·407	0·412	0·422	0·432	0·437	0·451	0·470	0·488	0·497	0·506	0·541	0·583
1/ 25000	1·7708	1·9722	2·0780	2·1872	2·4161	2·6592	2·7862	3·1894	3·7803	4·4345	4·7860	5·1542	6·7997	9·2662
0·0000420	0·402	0·413	0·418	0·423	0·433	0·443	0·448	0·463	0·483	0·501	0·511	0·520	0·556	0·598
1/ 23810	1·8183	2·0251	2·1337	2·2458	2·4808	2·7304	2·8608	3·2747	3·8813	4·5528	4·9137	5·2917	6·9808	9·5126
0·0000440	0·412	0·423	0·428	0·434	0·444	0·455	0·460	0·475	0·495	0·514	0·524	0·533	0·570	0·613
1/ 22727	1·8648	2·0769	2·1882	2·3032	2·5441	2·8000	2·9337	3·3581	3·9801	4·6686	5·0386	5·4261	7·1579	9·7535
0·0000460	0·422	0·433	0·439	0·444	0·455	0·466	0·471	0·487	0·507	0·527	0·536	0·546	0·583	0·628
1/ 21739	1·9103	2·1275	2·2416	2·3593	2·6061	2·8682	3·0051	3·4398	4·0768	4·7819	5·1608	5·5577	7·3313	9·9893
0·0000480	0·432	0·444	0·449	0·455	0·466	0·477	0·482	0·498	0·519	0·539	0·549	0·559	0·597	0·643
1/ 20833	1·9549	2·1771	2·2938	2·4143	2·6668	2·9350	3·0751	3·5198	4·1716	4·8930	5·2806	5·6867	7·5011	10·220
0·0000500	0·442	0·453	0·459	0·465	0·476	0·487	0·493	0·509	0·530	0·551	0·561	0·571	0·610	0·657
1/ 20000	1·9987	2·2258	2·3451	2·4683	2·7264	3·0005	3·1437	3·5983	4·2645	5·0019	5·3981	5·8131	7·6677	10·447
0·0000525	0·454	0·466	0·471	0·477	0·489	0·500	0·506	0·523	0·544	0·566	0·576	0·586	0·626	0·674
1/ 19048	2·0522	2·2854	2·4079	2·5343	2·7993	3·0807	3·2277	3·6944	4·3783	5·1352	5·5419	5·9679	7·8715	10·724
0·0000550	0·465	0·477	0·483	0·489	0·501	0·513	0·519	0·536	0·558	0·580	0·591	0·601	0·642	0·691
1/ 18182	2·1046	2·3437	2·4693	2·5989	2·8706	3·1591	3·3098	3·7883	4·4895	5·2655	5·6825	6·1193	8·0709	10·995
0·0000575	0·477	0·489	0·495	0·501	0·514	0·526	0·531	0·549	0·572	0·594	0·605	0·616	0·658	0·708
1/ 17391	2·1558	2·4007	2·5293	2·6621	2·9404	3·2359	3·3902	3·8803	4·5984	5·3931	5·8202	6·2674	8·2660	11·261
0·0000600	0·488	0·500	0·507	0·513	0·525	0·538	0·544	0·562	0·585	0·608	0·619	0·630	0·673	0·724
1/ 16667	2·2061	2·4567	2·5882	2·7241	3·0088	3·3111	3·4690	3·9704	4·7051	5·5181	5·9550	6·4126	8·4571	11·521
0·0000625	0·499	0·512	0·518	0·525	0·537	0·550	0·556	0·574	0·598	0·621	0·633	0·644	0·688	0·740
1/ 16000	2·2553	2·5115	2·6460	2·7848	3·0758	3·3849	3·5463	4·0588	4·8097	5·6407	6·0873	6·5549	8·6446	11·775
0·0000650	0·509	0·523	0·529	0·536	0·549	0·561	0·568	0·586	0·611	0·635	0·646	0·658	0·703	0·756
1/ 15385	2·3037	2·5653	2·7027	2·8445	3·1417	3·4573	3·6221	4·1455	4·9124	5·7610	6·2170	6·6946	8·8285	12·026
0·0000675	0·520	0·533	0·540	0·547	0·560	0·573	0·579	0·599	0·623	0·648	0·659	0·671	0·717	0·772
1/ 14815	2·3512	2·6181	2·7583	2·9030	3·2063	3·5284	3·6966	4·2307	5·0132	5·8792	6·3445	6·8318	9·0091	12·271
0·0000700	0·530	0·544	0·551	0·558	0·571	0·584	0·591	0·610	0·636	0·660	0·672	0·684	0·731	0·787
1/ 14286	2·3979	2·6701	2·8130	2·9606	3·2699	3·5983	3·7698	4·3144	5·1123	5·9953	6·4697	6·9666	9·1866	12·513
0·0000725	0·540	0·554	0·561	0·568	0·582	0·596	0·602	0·622	0·648	0·673	0·685	0·697	0·745	0·802
1/ 13793	2·4438	2·7212	2·8668	3·0172	3·3323	3·6670	3·8418	4·3967	5·2097	6·1095	6·5929	7·0992	9·3612	12·750
0·0000750	0·550	0·565	0·572	0·579	0·593	0·607	0·613	0·633	0·660	0·685	0·698	0·710	0·759	0·816
1/ 13333	2·4890	2·7714	2·9198	3·0729	3·3938	3·7346	3·9126	4·4777	5·3056	6·2218	6·7141	7·2296	9·5329	12·983
0·0000800	0·570	0·585	0·592	0·599	0·614	0·628	0·635	0·656	0·683	0·709	0·722	0·735	0·785	0·845
1/ 12500	2·5772	2·8697	3·0232	3·1818	3·5140	3·8668	4·0510	4·6359	5·4930	6·4413	6·9508	7·4844	9·8684	13·440
0·0000850	0·589	0·604	0·612	0·619	0·634	0·649	0·656	0·678	0·706	0·733	0·746	0·760	0·811	0·873
1/ 11765	2·6630	2·9651	3·1237	3·2875	3·6306	3·9951	4·1854	4·7896	5·6749	6·6544	7·1807	7·7319	10·194	13·882
0·0000900	0·607	0·623	0·631	0·639	0·654	0·669	0·677	0·699	0·728	0·756	0·770	0·783	0·836	0·900
1/ 11111	2·7463	3·0579	3·2214	3·3903	3·7441	4·1199	4·3161	4·9391	5·8518	6·8617	7·4043	7·9725	10·511	14·313
0·0000950	0·625	0·641	0·649	0·657	0·673	0·689	0·697	0·719	0·749	0·778	0·792	0·806	0·861	0·926
1/ 10526	2·8276	3·1483	3·3166	3·4905	3·8547	4·2415	4·4435	5·0847	6·0242	7·0636	7·6221	8·2069	10·819	14·733
0·0001000	0·643	0·659	0·668	0·676	0·692	0·708	0·716	0·739	0·770	0·800	0·814	0·829	0·885	0·952
1/ 10000	2·9068	3·2365	3·4095	3·5882	3·9625	4·3601	4·5677	5·2268	6·1923	7·2606	7·8345	8·4356	11·120	15·142
	0·95	0·95	0·95	0·95	0·96	0·96	0·96	0·96	0·96	0·96	0·97	0·97	0·97	0·97

$V_{r(0{\cdot}5)medial}$ **for half-full circular pipes.**

$k_s = 0{\cdot}030$ **mm** S = 0·00003 to 0·00010

Gradient **(Equivalent) Pipe diameters in mm**

Gradient	2400	2500	2550	2600	2700	2800	2850	3000	3200	3400	3500	3600	4000	4500
0·000100	0·643	0·659	0·668	0·676	0·692	0·708	0·716	0·739	0·770	0·800	0·814	0·829	0·885	0·952
1/ 10000	2·9068	3·2365	3·4095	3·5882	3·9625	4·3601	4·5677	5·2268	6·1923	7·2606	7·8345	8·4356	11·120	15·142
0·000105	0·660	0·677	0·685	0·694	0·710	0·727	0·735	0·759	0·790	0·821	0·836	0·851	0·908	0·977
1/ 9524	2·9843	3·3226	3·5002	3·6836	4·0679	4·4759	4·6890	5·3655	6·3565	7·4529	8·0420	8·6588	11·414	15·541
0·000110	0·676	0·694	0·703	0·711	0·728	0·745	0·754	0·778	0·810	0·842	0·857	0·872	0·931	1·002
1/ 9091	3·0600	3·4068	3·5890	3·7770	4·1709	4·5892	4·8077	5·5011	6·5170	7·6410	8·2448	8·8772	11·702	15·932
0·000115	0·693	0·711	0·720	0·729	0·746	0·763	0·772	0·797	0·830	0·862	0·878	0·893	0·954	1·026
1/ 8696	3·1341	3·4893	3·6758	3·8683	4·2717	4·7001	4·9238	5·6339	6·6741	7·8251	8·4433	9·0908	11·983	16·314
0·000120	0·709	0·727	0·736	0·745	0·763	0·781	0·790	0·815	0·849	0·882	0·898	0·914	0·975	1·049
1/ 8333	3·2067	3·5701	3·7609	3·9578	4·3705	4·8087	5·0376	5·7640	6·8281	8·0054	8·6378	9·3001	12·258	16·688
0·000125	0·725	0·743	0·753	0·762	0·780	0·798	0·807	0·833	0·868	0·901	0·918	0·934	0·997	1·072
1/ 8000	3·2779	3·6493	3·8443	4·0456	4·4673	4·9152	5·1491	5·8915	6·9790	8·1822	8·8285	9·5053	12·528	17·055
0·000130	0·740	0·759	0·769	0·778	0·797	0·815	0·824	0·851	0·886	0·920	0·937	0·954	1·018	1·095
1/ 7692	3·3477	3·7270	3·9261	4·1317	4·5624	5·0197	5·2585	6·0166	7·1271	8·3556	9·0156	9·7067	12·793	17·415
0·000135	0·755	0·775	0·785	0·794	0·813	0·832	0·841	0·869	0·904	0·939	0·956	0·973	1·039	1·117
1/ 7407	3·4163	3·8033	4·0065	4·2162	4·6557	5·1223	5·3660	6·1395	7·2725	8·5260	9·1993	9·9044	13·054	17·769
0·000140	0·770	0·790	0·800	0·810	0·829	0·848	0·858	0·886	0·922	0·957	0·975	0·992	1·059	1·139
1/ 7143	3·4837	3·8783	4·0855	4·2993	4·7474	5·2232	5·4716	6·2602	7·4154	8·6933	9·3798	10·099	13·309	18·116
0·000145	0·785	0·805	0·815	0·825	0·845	0·864	0·874	0·902	0·939	0·976	0·993	1·011	1·079	1·161
1/ 6897	3·5500	3·9521	4·1632	4·3811	4·8376	5·3223	5·5755	6·3790	7·5559	8·8579	9·5572	10·290	13·560	18·458
0·000150	0·799	0·820	0·830	0·840	0·860	0·880	0·890	0·919	0·957	0·993	1·012	1·029	1·099	1·182
1/ 6667	3·6153	4·0246	4·2396	4·4614	4·9263	5·4199	5·6776	6·4958	7·6941	9·0197	9·7318	10·477	13·808	18·794
0·000160	0·827	0·849	0·859	0·870	0·891	0·911	0·921	0·951	0·990	1·028	1·047	1·065	1·137	1·223
1/ 6250	3·7427	4·1664	4·3889	4·6185	5·0996	5·6105	5·8772	6·7239	7·9641	9·3360	10·073	10·845	14·291	19·450
0·000170	0·855	0·877	0·888	0·899	0·920	0·941	0·952	0·983	1·023	1·062	1·081	1·100	1·175	1·263
1/ 5882	3·8664	4·3041	4·5338	4·7710	5·2679	5·7955	6·0710	6·9454	8·2262	9·6429	10·404	11·201	14·759	20·087
0·000180	0·881	0·904	0·915	0·927	0·949	0·970	0·981	1·013	1·055	1·095	1·115	1·134	1·211	1·302
1/ 5556	3·9867	4·4379	4·6748	4·9193	5·4315	5·9754	6·2594	7·1608	8·4810	9·9414	10·726	11·547	15·215	20·706
0·000190	0·907	0·931	0·942	0·954	0·976	0·999	1·010	1·043	1·085	1·127	1·147	1·168	1·246	1·340
1/ 5263	4·1039	4·5683	4·8121	5·0637	5·5909	6·1506	6·4429	7·3706	8·7293	10·232	11·039	11·885	15·659	21·309
0·000200	0·932	0·957	0·968	0·980	1·004	1·027	1·038	1·072	1·115	1·158	1·179	1·200	1·281	1·377
1/ 5000	4·2182	4·6954	4·9460	5·2046	5·7463	6·3215	6·6219	7·5752	8·9713	10·516	11·345	12·214	16·092	21·897
0·000210	0·957	0·982	0·994	1·006	1·030	1·054	1·065	1·100	1·145	1·189	1·210	1·231	1·314	1·413
1/ 4762	4·3299	4·8196	5·0767	5·3421	5·8981	6·4884	6·7967	7·7750	9·2077	10·792	11·644	12·535	16·514	22·471
0·000220	0·981	1·007	1·019	1·032	1·056	1·080	1·092	1·128	1·174	1·218	1·241	1·262	1·347	1·448
1/ 4545	4·4390	4·9410	5·2046	5·4766	6·0465	6·6516	6·9676	7·9703	9·4387	11·063	11·935	12·849	16·927	23·032
0·000230	1·005	1·031	1·044	1·056	1·081	1·106	1·118	1·155	1·202	1·248	1·270	1·292	1·379	1·483
1/ 4348	4·5458	5·0598	5·3297	5·6082	6·1917	6·8113	7·1348	8·1614	9·6648	11·328	12·221	13·156	17·331	23·581
0·000240	1·028	1·054	1·068	1·081	1·106	1·132	1·144	1·181	1·229	1·276	1·299	1·322	1·411	1·516
1/ 4167	4·6504	5·1762	5·4522	5·7371	6·3340	6·9676	7·2985	8·3486	9·8863	11·587	12·500	13·457	17·727	24·118
0·000250	1·051	1·078	1·091	1·104	1·131	1·156	1·169	1·207	1·256	1·304	1·328	1·351	1·442	1·550
1/ 4000	4·7530	5·2903	5·5724	5·8635	6·4734	7·1210	7·4591	8·5321	10·103	11·841	12·775	13·752	18·115	24·645
0·000260	1·073	1·101	1·114	1·128	1·155	1·181	1·194	1·233	1·283	1·332	1·356	1·379	1·472	1·582
1/ 3846	4·8536	5·4023	5·6903	5·9875	6·6103	7·2714	7·6166	8·7121	10·316	12·091	13·044	14·041	18·496	25·162
0·000270	1·095	1·123	1·137	1·151	1·178	1·205	1·218	1·258	1·309	1·359	1·383	1·407	1·502	1·614
1/ 3704	4·9525	5·5122	5·8060	6·1093	6·7446	7·4191	7·7712	8·8889	10·525	12·335	13·308	14·325	18·869	25·670
0·000280	1·116	1·145	1·159	1·173	1·201	1·228	1·242	1·282	1·334	1·385	1·410	1·435	1·531	1·645
1/ 3571	5·0495	5·6202	5·9197	6·2289	6·8766	7·5642	7·9232	9·0625	10·731	12·576	13·567	14·604	19·236	26·168
0·000290	1·137	1·167	1·181	1·195	1·224	1·252	1·265	1·306	1·359	1·411	1·437	1·462	1·559	1·676
1/ 3448	5·1450	5·7263	6·0315	6·3465	7·0063	7·7068	8·0726	9·2332	10·933	12·812	13·822	14·879	19·597	26·658
0·000300	1·158	1·188	1·203	1·217	1·246	1·274	1·288	1·330	1·384	1·437	1·463	1·488	1·588	1·706
1/ 3333	5·2389	5·8308	6·1415	6·4621	7·1339	7·8471	8·2195	9·4012	11·131	13·045	14·073	15·149	19·952	27·140
	0·96	0·97	0·97	0·97	0·97	0·97	0·97	0·97	0·97	0·97	0·97	0·98	0·98	0·98

$V_{r(0\cdot5)medial}$ for half-full circular pipes.

A48
(p.3 of 6)

k_s = **0·030 mm**
S = 0·00030 to 0·00100

ie hydraulic gradient =
1 in 3333 to 1 in 1000

Water (or sewage) at 15°C;
full bore conditions.

velocities in ms^{-1}
discharges in m^3s^{-1}

Gradient **(Equivalent) Pipe diameters in mm**

Gradient	2400	2500	2550	2600	2700	2800	2850	3000	3200	3400	3500	3600	4000	4500
0·000300	1·158	1·188	1·203	1·217	1·246	1·274	1·288	1·330	1·384	1·437	1·463	1·488	1·588	1·706
1/ 3333	5·2389	5·8308	6·1415	6·4621	7·1339	7·8471	8·2195	9·4012	11·131	13·045	14·073	15·149	19·952	27·140
0·000320	1·199	1·229	1·245	1·260	1·290	1·319	1·333	1·376	1·432	1·487	1·514	1·540	1·643	1·766
1/ 3125	5·4223	6·0348	6·3563	6·6881	7·3832	8·1212	8·5065	9·7292	11·519	13·499	14·562	15·676	20·645	28·081
0·000340	1·238	1·270	1·285	1·301	1·332	1·362	1·377	1·421	1·479	1·535	1·563	1·590	1·696	1·823
1/ 2941	5·6003	6·2328	6·5648	6·9074	7·6252	8·3872	8·7850	10·047	11·896	13·940	15·038	16·187	21·318	28·994
0·000360	1·276	1·309	1·325	1·341	1·373	1·404	1·420	1·465	1·525	1·583	1·611	1·639	1·748	1·879
1/ 2778	5·7734	6·4253	6·7675	7·1206	7·8604	8·6458	9·0558	10·357	12·262	14·369	15·500	16·684	21·971	29·882
0·000380	1·313	1·347	1·364	1·380	1·413	1·445	1·461	1·508	1·569	1·629	1·658	1·687	1·799	1·933
1/ 2632	5·9420	6·6128	6·9649	7·3283	8·0895	8·8976	9·3195	10·658	12·618	14·786	15·950	17·168	22·608	30·746
0·000400	1·350	1·384	1·401	1·418	1·452	1·485	1·501	1·549	1·612	1·673	1·703	1·733	1·848	1·986
1/ 2500	6·1063	6·7956	7·1573	7·5307	8·3129	9·1431	9·5766	10·952	12·966	15·193	16·388	17·640	23·228	31·588
0·000420	1·385	1·421	1·438	1·456	1·490	1·524	1·541	1·590	1·654	1·717	1·748	1·778	1·897	2·038
1/ 2381	6·2668	6·9741	7·3453	7·7284	8·5309	9·3828	9·8276	11·239	13·305	15·590	16·817	18·101	23·834	32·411
0·000440	1·420	1·456	1·474	1·492	1·527	1·562	1·579	1·630	1·696	1·760	1·791	1·823	1·944	2·088
1/ 2273	6·4237	7·1485	7·5289	7·9216	8·7441	9·6171	10·073	11·519	13·637	15·978	17·235	18·551	24·426	33·214
0·000460	1·454	1·491	1·509	1·528	1·564	1·599	1·617	1·668	1·736	1·802	1·834	1·866	1·990	2·138
1/ 2174	6·5772	7·3192	7·7087	8·1106	8·9526	9·8463	10·313	11·794	13·961	16·358	17·645	18·992	25·005	34·000
0·000480	1·487	1·525	1·544	1·562	1·599	1·636	1·653	1·706	1·775	1·843	1·876	1·908	2·035	2·186
1/ 2083	6·7275	7·4864	7·8847	8·2957	9·1568	10·071	10·548	12·062	14·279	16·730	18·045	19·423	25·572	34·770
0·000500	1·520	1·558	1·578	1·597	1·634	1·671	1·690	1·744	1·814	1·883	1·916	1·950	2·079	2·234
1/ 2000	6·8748	7·6502	8·0572	8·4772	9·3570	10·291	10·778	12·325	14·590	17·094	18·438	19·846	26·128	35·524
0·000525	1·560	1·599	1·619	1·638	1·677	1·715	1·734	1·789	1·861	1·932	1·966	2·001	2·133	2·292
1/ 1905	7·0551	7·8507	8·2682	8·6992	9·6019	10·560	11·060	12·647	14·971	17·540	18·919	20·363	26·808	36·447
0·000550	1·598	1·639	1·659	1·679	1·719	1·758	1·777	1·834	1·908	1·980	2·015	2·050	2·186	2·348
1/ 1818	7·2312	8·0466	8·4745	8·9161	9·8412	10·823	11·336	12·962	15·343	17·975	19·389	20·868	27·472	37·349
0·000575	1·637	1·678	1·699	1·719	1·760	1·799	1·819	1·877	1·953	2·027	2·063	2·099	2·238	2·404
1/ 1739	7·4036	8·2382	8·6763	9·1284	10·075	11·080	11·605	13·270	15·707	18·402	19·848	21·363	28·122	38·230
0·000600	1·674	1·717	1·738	1·758	1·800	1·840	1·861	1·920	1·997	2·073	2·110	2·146	2·288	2·458
1/ 1667	7·5724	8·4260	8·8739	9·3363	10·305	11·332	11·869	13·572	16·064	18·819	20·298	21·847	28·758	39·094
0·000625	1·710	1·754	1·776	1·797	1·839	1·881	1·901	1·962	2·041	2·118	2·156	2·193	2·338	2·511
1/ 1600	7·7378	8·6099	9·0676	9·5400	10·529	11·579	12·128	13·867	16·413	19·228	20·739	22·321	29·381	39·940
0·000650	1·746	1·791	1·813	1·834	1·878	1·920	1·941	2·003	2·083	2·162	2·201	2·239	2·387	2·563
1/ 1538	7·9001	8·7904	9·2576	9·7398	10·750	11·822	12·381	14·157	16·756	19·629	21·172	22·786	29·993	40·770
0·000675	1·782	1·827	1·849	1·871	1·915	1·959	1·980	2·043	2·125	2·205	2·245	2·283	2·435	2·615
1/ 1481	8·0595	8·9676	9·4441	9·9360	10·966	12·060	12·630	14·441	17·092	20·023	21·596	23·243	30·593	41·584
0·000700	1·816	1·862	1·885	1·908	1·952	1·996	2·018	2·083	2·166	2·248	2·288	2·328	2·481	2·665
1/ 1429	8·2159	9·1416	9·6274	10·129	11·179	12·293	12·875	14·721	17·423	20·410	22·013	23·692	31·183	42·384
0·000725	1·850	1·897	1·920	1·943	1·989	2·034	2·056	2·121	2·207	2·290	2·331	2·371	2·528	2·714
1/ 1379	8·3698	9·3126	9·8074	10·318	11·388	12·523	13·115	14·996	17·747	20·790	22·423	24·133	31·762	43·170
0·000750	1·884	1·931	1·955	1·978	2·025	2·070	2·093	2·160	2·246	2·331	2·372	2·413	2·573	2·763
1/ 1333	8·5211	9·4808	9·9845	10·504	11·593	12·749	13·352	15·266	18·067	21·163	22·826	24·566	32·332	43·943
0·000800	1·949	1·998	2·023	2·047	2·095	2·142	2·165	2·234	2·324	2·411	2·454	2·497	2·661	2·858
1/ 1250	8·8165	9·8093	10·330	10·868	11·994	13·190	13·814	15·793	18·691	21·893	23·613	25·413	33·444	45·453
0·000850	2·012	2·063	2·088	2·113	2·163	2·212	2·236	2·307	2·399	2·489	2·534	2·577	2·747	2·950
1/ 1176	9·1031	10·128	10·666	11·221	12·384	13·617	14·261	16·305	19·296	22·601	24·376	26·234	34·523	46·916
0·000900	2·074	2·126	2·152	2·178	2·229	2·279	2·304	2·377	2·472	2·565	2·611	2·656	2·831	3·039
1/ 1111	9·3817	10·438	10·992	11·564	12·762	14·033	14·697	16·802	19·883	23·289	25·118	27·032	35·571	48·339
0·000950	2·134	2·188	2·214	2·241	2·293	2·345	2·370	2·446	2·543	2·639	2·686	2·732	2·912	3·126
1/ 1053	9·6530	10·739	11·310	11·898	13·130	14·438	15·121	17·286	20·456	23·959	25·840	27·808	36·592	49·723
0·001000	2·192	2·248	2·275	2·302	2·356	2·409	2·435	2·512	2·613	2·711	2·759	2·806	2·991	3·211
1/ 1000	9·9174	11·033	11·619	12·223	13·489	14·832	15·534	17·758	21·014	24·612	26·543	28·565	37·587	51·073
	0·97	0·97	0·97	0·98	0·98	0·98	0·98	0·98	0·98	0·98	0·98	0·98	0·98	0·99

$V_{r(0·5)medial}$ for half-full circular pipes.

k_s = 0·030 mm S = 0·00030 to 0·00100

$k_s = 0{\cdot}030$ mm
S = 0·00100 to 0·00300

ie hydraulic gradient =
1 in 1000 to 1 in 333

Water (or sewage) at 15°C;
full bore conditions.

velocities in ms^{-1}
discharges in m^3s^{-1}

Gradient **(Equivalent) Pipe diameters in mm**

Gradient	2400	2500	2550	2600	2700	2800	2850	3000	3200	3400	3500	3600	4000	4500
0·00100	2·192	2·248	2·275	2·302	2·356	2·409	2·435	2·512	2·613	2·711	2·759	2·806	2·991	3·211
1/ 1000	9·9174	11·033	11·619	12·223	13·489	14·832	15·534	17·758	21·014	24·612	26·543	28·565	37·587	51·073
0·00105	2·249	2·306	2·334	2·362	2·417	2·471	2·498	2·577	2·681	2·781	2·830	2·879	3·068	3·294
1/ 952	10·175	11·320	11·921	12·541	13·840	15·217	15·937	18·219	21·558	25·249	27·230	29·304	38·557	52·390
0·00110	2·305	2·363	2·392	2·421	2·477	2·532	2·560	2·641	2·747	2·850	2·900	2·950	3·144	3·375
1/ 909	10·428	11·601	12·216	12·851	14·182	15·594	16·331	18·669	22·090	25·871	27·901	30·026	39·506	53·676
0·00115	2·360	2·419	2·449	2·478	2·535	2·592	2·620	2·703	2·811	2·917	2·968	3·019	3·218	3·454
1/ 870	10·674	11·875	12·505	13·155	14·517	15·962	16·716	19·109	22·610	26·480	28·558	30·732	40·433	54·934
0·00120	2·413	2·474	2·504	2·534	2·593	2·651	2·679	2·764	2·875	2·982	3·035	3·087	3·290	3·531
1/ 833	10·916	12·143	12·788	13·452	14·845	16·322	17·093	19·540	23·120	27·076	29·200	31·423	41·341	56·166
0·00125	2·465	2·527	2·558	2·589	2·649	2·708	2·737	2·824	2·937	3·047	3·100	3·154	3·361	3·607
1/ 800	11·152	12·407	13·065	13·744	15·166	16·675	17·463	19·962	23·619	27·660	29·830	32·100	42·231	57·372
0·00130	2·517	2·580	2·611	2·642	2·704	2·764	2·794	2·883	2·998	3·110	3·165	3·219	3·430	3·682
1/ 769	11·385	12·665	13·336	14·029	15·481	17·021	17·825	20·376	24·108	28·233	30·447	32·765	43·103	58·556
0·00135	2·567	2·632	2·664	2·695	2·758	2·820	2·850	2·940	3·057	3·172	3·228	3·283	3·498	3·755
1/ 741	11·612	12·918	13·603	14·310	15·790	17·361	18·181	20·782	24·589	28·795	31·053	33·416	43·959	59·717
0·00140	2·616	2·682	2·715	2·747	2·811	2·874	2·905	2·997	3·116	3·232	3·289	3·346	3·565	3·826
1/ 714	11·836	13·167	13·865	14·585	16·094	17·695	18·531	21·182	25·061	29·347	31·648	34·057	44·800	60·857
0·00145	2·665	2·732	2·765	2·798	2·863	2·927	2·959	3·052	3·174	3·292	3·350	3·408	3·631	3·897
1/ 690	12·056	13·411	14·122	14·856	16·393	18·023	18·874	21·574	25·524	29·890	32·233	34·686	45·626	61·978
0·00150	2·713	2·781	2·815	2·848	2·914	2·979	3·012	3·107	3·230	3·351	3·410	3·468	3·695	3·966
1/ 667	12·272	13·652	14·375	15·122	16·686	18·346	19·212	21·959	25·980	30·423	32·808	35·304	46·439	63·080
0·00160	2·806	2·877	2·912	2·946	3·014	3·082	3·115	3·213	3·341	3·466	3·527	3·587	3·822	4·101
1/ 625	12·695	14·121	14·869	15·642	17·259	18·975	19·871	22·712	26·870	31·464	33·930	36·511	48·025	65·230
0·00170	2·897	2·969	3·005	3·041	3·111	3·181	3·215	3·316	3·448	3·577	3·640	3·702	3·944	4·233
1/ 588	13·104	14·576	15·349	16·146	17·815	19·586	20·510	23·443	27·733	32·474	35·019	37·682	49·563	67·316
0·00180	2·985	3·060	3·097	3·133	3·206	3·277	3·313	3·417	3·553	3·685	3·750	3·814	4·063	4·360
1/ 556	13·502	15·019	15·814	16·636	18·355	20·179	21·132	24·152	28·572	33·456	36·077	38·820	51·057	69·342
0·00190	3·070	3·147	3·185	3·223	3·298	3·371	3·407	3·515	3·654	3·790	3·857	3·923	4·179	4·484
1/ 526	13·889	15·449	16·268	17·112	18·881	20·757	21·736	24·843	29·388	34·410	37·106	39·927	52·511	71·314
0·00200	3·154	3·233	3·272	3·311	3·387	3·462	3·500	3·610	3·753	3·893	3·961	4·029	4·291	4·605
1/ 500	14·267	15·869	16·709	17·577	19·393	21·320	22·326	25·516	30·184	35·341	38·109	41·006	53·928	73·235
0·00210	3·235	3·316	3·356	3·396	3·474	3·552	3·590	3·703	3·850	3·993	4·063	4·132	4·401	4·723
1/ 476	14·635	16·278	17·140	18·030	19·893	21·869	22·901	26·173	30·960	36·249	39·088	42·059	55·310	75·109
0·00220	3·315	3·398	3·439	3·479	3·560	3·639	3·678	3·793	3·944	4·090	4·162	4·233	4·509	4·838
1/ 455	14·995	16·678	17·562	18·473	20·381	22·406	23·462	26·814	31·718	37·136	40·044	43·087	56·661	76·940
0·00230	3·392	3·477	3·519	3·561	3·643	3·724	3·764	3·882	4·036	4·186	4·259	4·332	4·614	4·950
1/ 435	15·347	17·070	17·973	18·906	20·859	22·931	24·012	27·442	32·459	38·003	40·979	44·093	57·981	78·731
0·00240	3·469	3·555	3·598	3·641	3·725	3·807	3·848	3·969	4·126	4·279	4·354	4·429	4·717	5·060
1/ 417	15·692	17·453	18·377	19·330	21·326	23·444	24·550	28·056	33·185	38·852	41·894	45·077	59·274	80·483
0·00250	3·543	3·632	3·676	3·719	3·805	3·889	3·931	4·054	4·215	4·371	4·448	4·523	4·818	5·168
1/ 400	16·029	17·828	18·772	19·745	21·784	23·948	25·077	28·657	33·896	39·684	42·790	46·041	60·540	82·200
0·00260	3·616	3·707	3·751	3·796	3·883	3·969	4·012	4·138	4·301	4·461	4·539	4·616	4·916	5·274
1/ 385	16·360	18·196	19·159	20·153	22·233	24·441	25·593	29·247	34·594	40·500	43·670	46·987	61·782	83·884
0·00270	3·688	3·780	3·826	3·871	3·960	4·048	4·091	4·220	4·386	4·549	4·629	4·707	5·013	5·378
1/ 370	16·685	18·557	19·539	20·552	22·674	24·925	26·100	29·826	35·278	41·300	44·532	47·915	63·000	85·535
0·00280	3·759	3·853	3·899	3·945	4·036	4·125	4·169	4·300	4·470	4·635	4·717	4·797	5·109	5·480
1/ 357	17·004	18·912	19·912	20·945	23·107	25·401	26·598	30·395	35·950	42·086	45·379	48·826	64·196	87·157
0·00290	3·828	3·924	3·971	4·018	4·110	4·201	4·246	4·379	4·552	4·721	4·803	4·885	5·202	5·580
1/ 345	17·318	19·260	20·279	21·331	23·532	25·868	27·087	30·954	36·610	42·858	46·212	49·721	65·372	88·751
0·00300	3·896	3·993	4·041	4·089	4·183	4·276	4·321	4·457	4·633	4·804	4·888	4·971	5·294	5·679
1/ 333	17·626	19·603	20·640	21·710	23·951	26·328	27·568	31·503	37·259	43·618	47·030	50·601	66·527	90·317
	0·98	0·98	0·98	0·98	0·98	0·98	0·98	0·98	0·98	0·98	0·99	0·99	0·99	0·99

$V_{r(0\cdot5)medial}$ for half-full circular pipes.

A48
(p.5 of 6)

k_s = **0·030 mm**
S = 0·00300 to 0·01000

ie hydraulic gradient =
1 in 333 to 1 in 100

Water (or sewage) at 15°C;
full bore conditions.

velocities in ms^{-1}
discharges in m^3s^{-1}

Gradient **(Equivalent) Pipe diameters in mm**

Gradient	2400	2500	2550	2600	2700	2800	2850	3000	3200	3400	3500	3600	4000	4500
0·00300	3·896	3·993	4·041	4·089	4·183	4·276	4·321	4·457	4·633	4·804	4·888	4·971	5·294	5·679
1/ 333	17·626	19·603	20·640	21·710	23·951	26·328	27·568	31·503	37·259	43·618	47·030	50·601	66·527	90·317
0·00320	4·029	4·130	4·179	4·228	4·326	4·421	4·469	4·608	4·790	4·967	5·054	5·140	5·474	5·871
1/ 313	18·228	20·272	21·344	22·450	24·767	27·225	28·507	32·575	38·526	45·100	48·627	52·319	68·783	93·375
0·00340	4·158	4·262	4·313	4·364	4·464	4·563	4·611	4·756	4·943	5·126	5·215	5·304	5·648	6·057
1/ 294	18·811	20·920	22·027	23·168	25·558	28·094	29·418	33·615	39·754	46·537	50·176	53·985	70·970	96·339
0·00360	4·284	4·390	4·443	4·495	4·598	4·700	4·750	4·898	5·091	5·279	5·372	5·463	5·817	6·239
1/ 278	19·378	21·550	22·690	23·866	26·327	28·939	30·302	34·625	40·948	47·932	51·681	55·603	73·094	99·219
0·00380	4·405	4·515	4·569	4·623	4·729	4·833	4·885	5·037	5·236	5·429	5·524	5·617	5·981	6·415
1/ 263	19·930	22·164	23·335	24·544	27·076	29·761	31·163	35·607	42·109	49·291	53·144	57·177	75·161	102·02
0·00400	4·524	4·637	4·692	4·747	4·856	4·963	5·016	5·173	5·376	5·575	5·672	5·768	6·141	6·586
1/ 250	20·468	22·761	23·964	25·206	27·805	30·562	32·001	36·565	43·240	50·614	54·571	58·711	77·175	104·75
0·00420	4·640	4·756	4·813	4·869	4·981	5·090	5·145	5·305	5·514	5·717	5·817	5·915	6·298	6·754
1/ 238	20·992	23·344	24·578	25·851	28·516	31·344	32·819	37·499	44·344	51·905	55·962	60·208	79·140	107·41
0·00440	4·754	4·872	4·930	4·988	5·102	5·214	5·270	5·434	5·648	5·856	5·958	6·059	6·450	6·917
1/ 227	21·504	23·913	25·177	26·481	29·211	32·107	33·618	38·411	45·422	53·166	57·321	61·670	81·059	110·01
0·00460	4·864	4·985	5·045	5·104	5·221	5·336	5·392	5·560	5·779	5·992	6·096	6·199	6·600	7·077
1/ 217	22·005	24·470	25·763	27·097	29·890	32·853	34·400	39·304	46·476	54·399	58·650	63·099	82·935	112·56
0·00480	4·973	5·096	5·157	5·217	5·337	5·454	5·512	5·684	5·907	6·125	6·231	6·336	6·746	7·234
1/ 208	22·496	25·015	26·337	27·700	30·555	33·584	35·165	40·177	47·508	55·606	59·951	64·498	84·771	115·05
0·00500	5·079	5·205	5·267	5·329	5·450	5·570	5·630	5·805	6·033	6·255	6·364	6·471	6·889	7·387
1/ 200	22·976	25·549	26·899	28·291	31·207	34·300	35·914	41·032	48·519	56·788	61·225	65·868	86·570	117·48
0·00525	5·209	5·338	5·402	5·465	5·590	5·713	5·773	5·953	6·186	6·414	6·525	6·636	7·064	7·574
1/ 190	23·563	26·202	27·586	29·014	32·004	35·175	36·830	42·078	49·755	58·233	62·783	67·543	88·769	120·46
0·00550	5·335	5·468	5·533	5·598	5·725	5·851	5·914	6·097	6·337	6·569	6·684	6·796	7·235	7·757
1/ 182	24·137	26·839	28·257	29·719	32·782	36·030	37·725	43·100	50·962	59·645	64·304	69·180	90·917	123·37
0·00575	5·459	5·595	5·661	5·728	5·858	5·987	6·051	6·239	6·483	6·721	6·838	6·954	7·402	7·936
1/ 174	24·698	27·463	28·913	30·410	33·542	36·865	38·600	44·099	52·142	61·025	65·792	70·779	93·017	126·22
0·00600	5·581	5·719	5·787	5·855	5·988	6·120	6·185	6·377	6·627	6·870	6·990	7·107	7·566	8·111
1/ 167	25·247	28·073	29·555	31·085	34·287	37·683	39·456	45·077	53·297	62·376	67·248	72·345	95·072	129·01
0·00625	5·700	5·841	5·910	5·979	6·116	6·250	6·316	6·513	6·768	7·016	7·138	7·258	7·726	8·283
1/ 160	25·785	28·671	30·185	31·746	35·016	38·485	40·295	46·034	54·428	63·699	68·674	73·879	97·085	131·73
0·00650	5·816	5·960	6·031	6·102	6·241	6·378	6·445	6·645	6·906	7·159	7·283	7·406	7·883	8·451
1/ 154	26·313	29·257	30·802	32·395	35·731	39·270	41·117	46·973	55·538	64·997	70·072	75·383	99·059	134·41
0·00675	5·931	6·077	6·150	6·221	6·363	6·503	6·572	6·776	7·041	7·299	7·426	7·551	8·037	8·616
1/ 148	26·830	29·832	31·407	33·032	36·433	40·041	41·924	47·894	56·626	66·270	71·444	76·858	101·00	137·03
0·00700	6·043	6·192	6·266	6·339	6·484	6·626	6·696	6·904	7·174	7·437	7·566	7·693	8·188	8·778
1/ 143	27·338	30·397	32·001	33·656	37·122	40·798	42·716	48·799	57·695	67·519	72·790	78·306	102·90	139·61
0·00725	6·153	6·305	6·380	6·455	6·602	6·746	6·818	7·029	7·304	7·572	7·703	7·833	8·337	8·937
1/ 138	27·837	30·951	32·585	34·270	37·799	41·542	43·495	49·688	58·744	68·747	74·114	79·729	104·76	142·14
0·00750	6·262	6·416	6·493	6·568	6·718	6·865	6·938	7·153	7·433	7·705	7·838	7·970	8·483	9·094
1/ 133	28·328	31·497	33·159	34·874	38·464	42·272	44·260	50·561	59·777	69·954	75·414	81·128	106·60	144·63
0·00800	6·474	6·633	6·712	6·790	6·945	7·097	7·172	7·394	7·683	7·964	8·102	8·239	8·768	9·399
1/ 125	29·286	32·561	34·279	36·052	39·763	43·699	45·753	52·266	61·791	72·310	77·953	83·858	110·18	149·48
0·00850	6·679	6·844	6·925	7·005	7·165	7·321	7·399	7·628	7·926	8·216	8·358	8·499	9·045	9·695
1/ 118	30·214	33·593	35·366	37·194	41·022	45·082	47·201	53·919	63·743	74·593	80·414	86·505	113·66	154·19
0·00900	6·878	7·048	7·131	7·214	7·378	7·540	7·619	7·855	8·162	8·460	8·607	8·751	9·313	9·982
1/ 111	31·116	34·596	36·420	38·303	42·245	46·426	48·607	55·524	65·640	76·811	82·804	89·075	117·03	158·76
0·00950	7·072	7·246	7·332	7·418	7·586	7·752	7·834	8·076	8·391	8·698	8·848	8·997	9·574	10·26
1/ 105	31·994	35·571	37·447	39·382	43·435	47·732	49·975	57·086	67·485	78·969	85·129	91·576	120·31	163·21
0·01000	7·261	7·440	7·528	7·616	7·789	7·959	8·043	8·291	8·615	8·929	9·084	9·236	9·829	10·53
1/ 100	32·849	36·521	38·447	40·434	44·594	49·006	51·308	58·608	69·282	81·071	87·395	94·012	123·51	167·54
	0·98	0·98	0·98	0·98	0·98	0·99	0·99	0·99	0·99	0·99	0·99	0·99	0·99	0·99

$V_{r(0\cdot5)medial}$ **for half-full circular pipes.**

k_s = 0·030 mm S = 0·00300 to 0·01000

$k_s = 0{\cdot}030$ mm
S = 0·01000 to 0·03000

ie hydraulic gradient =
1 in 100 to 1 in 33·3

Water (or sewage) at 15°C;
full bore conditions.

velocities in ms^{-1}
discharges in m^3s^{-1}

Gradient **(Equivalent) Pipe diameters in mm**

Gradient	2400	2500	2550	2600	2700	2800	2850	3000	3200	3400	3500	3600	4000	4500
0·01000	7·261	7·440	7·528	7·616	7·789	7·959	8·043	8·291	8·615	8·929	9·084	9·236	9·829	10·53
1/ 100	32·849	36·521	38·447	40·434	44·594	49·006	51·308	58·608	69·282	81·071	87·395	94·012	123·51	167·54
0·01050	7·446	7·629	7·719	7·809	7·986	8·160	8·247	8·501	8·833	9·155	9·313	9·470	10·08	10·80
1/ 95	33·683	37·447	39·422	41·459	45·724	50·248	52·608	60·092	71·036	83·121	89·604	96·388	126·63	171·77
0·01100	7·626	7·813	7·906	7·997	8·179	8·357	8·446	8·706	9·045	9·376	9·538	9·698	10·32	11·06
1/ 91	34·497	38·353	40·375	42·461	46·828	51·460	53·878	61·541	72·748	85·123	91·762	98·709	129·67	175·89
0·01150	7·802	7·993	8·088	8·182	8·367	8·550	8·640	8·907	9·254	9·591	9·757	9·920	10·56	11·31
1/ 87	35·294	39·237	41·306	43·440	47·908	52·646	55·119	62·959	74·422	87·081	93·872	100·98	132·65	179·93
0·01200	7·974	8·170	8·267	8·362	8·552	8·738	8·831	9·103	9·457	9·802	9·971	10·14	10·79	11·56
1/ 83	36·073	40·104	42·218	44·399	48·964	53·807	56·334	64·346	76·060	88·997	95·937	103·20	135·56	183·88
0·01250	8·143	8·343	8·441	8·539	8·733	8·923	9·017	9·295	9·657	10·01	10·18	10·35	11·01	11·80
1/ 80	36·836	40·952	43·110	45·337	49·999	54·944	57·524	65·704	77·665	90·873	97·959	105·37	138·42	187·74
0·01300	8·308	8·512	8·613	8·713	8·910	9·104	9·200	9·484	9·852	10·21	10·39	10·56	11·24	12·04
1/ 77	37·585	41·783	43·986	46·258	51·014	56·058	58·690	67·036	79·238	92·713	99·941	107·50	141·22	191·53
0·01350	8·470	8·678	8·781	8·883	9·084	9·282	9·379	9·668	10·04	10·41	10·59	10·77	11·46	12·28
1/ 74	38·319	42·599	44·845	47·161	52·009	57·152	59·835	68·342	80·782	94·518	101·89	109·60	143·96	195·25
0·01400	8·630	8·841	8·946	9·050	9·254	9·456	9·556	9·850	10·23	10·61	10·79	10·97	11·67	12·51
1/ 71	39·040	43·400	45·688	48·047	52·987	58·225	60·958	69·625	82·297	96·290	103·80	111·65	146·66	198·90
0·01450	8·786	9·002	9·108	9·214	9·422	9·627	9·729	10·03	10·42	10·80	10·98	11·17	11·88	12·73
1/ 69	39·748	44·187	46·516	48·918	53·947	59·280	62·062	70·885	83·786	98·030	105·67	113·67	149·30	202·49
0·01500	8·940	9·159	9·268	9·375	9·587	9·796	9·899	10·20	10·60	10·99	11·17	11·36	12·09	12·95
1/ 67	40·444	44·961	47·330	49·774	54·890	60·316	63·147	72·124	85·249	99·741	107·52	115·65	151·90	206·01
0·01600	9·240	9·467	9·579	9·690	9·909	10·12	10·23	10·55	10·95	11·35	11·55	11·74	12·49	13·39
1/ 63	41·803	46·471	48·919	51·445	56·732	62·339	65·265	74·541	88·104	103·08	111·11	119·52	156·98	212·89
0·01700	9·532	9·765	9·880	9·995	10·22	10·44	10·55	10·88	11·30	11·71	11·91	12·11	12·88	13·81
1/ 59	43·120	47·934	50·459	53·064	58·517	64·300	67·318	76·885	90·872	106·32	114·60	123·27	161·90	219·56
0·01800	9·814	10·05	10·17	10·29	10·52	10·75	10·86	11·20	11·63	12·06	12·26	12·47	13·26	14·21
1/ 56	44·399	49·356	51·955	54·637	60·251	66·205	69·311	79·160	93·560	109·46	117·99	126·91	166·68	226·04
0·01900	10·09	10·34	10·46	10·58	10·82	11·05	11·17	11·51	11·96	12·39	12·61	12·82	13·63	14·61
1/ 53	45·644	50·738	53·411	56·167	61·938	68·057	71·250	81·374	96·175	112·52	121·28	130·46	171·33	232·34
0·02000	10·36	10·61	10·74	10·86	11·10	11·35	11·46	11·82	12·28	12·72	12·94	13·16	13·99	14·99
1/ 50	46·856	52·085	54·828	57·658	63·581	69·862	73·140	83·530	98·722	115·50	124·49	133·91	175·86	238·47
0·02100	10·62	10·88	11·01	11·13	11·38	11·63	11·75	12·11	12·58	13·04	13·26	13·49	14·35	15·37
1/ 47·6	48·038	53·399	56·211	59·111	65·183	71·622	74·982	85·634	101·21	118·40	127·62	137·27	180·28	244·46
0·02200	10·87	11·14	11·27	11·40	11·66	11·91	12·04	12·41	12·89	13·35	13·58	13·81	14·69	15·74
1/ 45·5	49·193	54·682	57·561	60·531	66·748	73·341	76·782	87·688	103·63	121·24	130·68	140·56	184·59	250·30
0·02300	11·12	11·40	11·53	11·66	11·93	12·18	12·31	12·69	13·18	13·66	13·89	14·13	15·03	16·10
1/ 43·5	50·322	55·936	58·881	61·919	68·278	75·022	78·541	89·696	106·01	124·01	133·67	143·78	188·81	256·02
0·02400	11·37	11·65	11·78	11·92	12·19	12·45	12·58	12·97	13·47	13·96	14·20	14·43	15·35	16·45
1/ 41·7	51·426	57·164	60·173	63·278	69·776	76·667	80·262	91·661	108·33	126·73	136·60	146·92	192·94	261·61
0·02500	11·61	11·89	12·03	12·17	12·44	12·71	12·85	13·24	13·75	14·25	14·50	14·74	15·68	16·79
1/ 40·0	52·508	58·366	61·439	64·608	71·242	78·278	81·948	93·586	110·60	129·38	139·46	150·00	196·98	267·08
0·02600	11·84	12·13	12·27	12·41	12·69	12·97	13·10	13·51	14·03	14·54	14·79	15·03	15·99	17·13
1/ 38·5	53·569	59·545	62·679	65·912	72·680	79·857	83·601	95·473	112·83	131·99	142·27	153·02	200·94	272·45
0·02700	12·07	12·37	12·51	12·66	12·94	13·22	13·36	13·77	14·30	14·82	15·07	15·32	16·30	17·46
1/ 37·0	54·610	60·701	63·896	67·192	74·090	81·406	85·223	97·324	115·01	134·55	145·02	155·98	204·83	277·72
0·02800	12·30	12·60	12·75	12·89	13·18	13·47	13·61	14·03	14·57	15·10	15·35	15·61	16·60	17·79
1/ 35·7	55·631	61·836	65·090	68·448	75·475	82·926	86·815	99·141	117·16	137·06	147·73	158·89	208·65	282·88
0·02900	12·52	12·82	12·98	13·12	13·42	13·71	13·85	14·28	14·83	15·37	15·63	15·89	16·90	18·11
1/ 34·5	56·635	62·951	66·264	69·681	76·835	84·420	88·378	100·93	119·27	139·52	150·38	161·75	212·39	287·96
0·03000	12·74	13·05	13·20	13·35	13·65	13·95	14·09	14·53	15·09	15·63	15·90	16·17	17·20	18·42
1/ 33·3	57·621	64·047	67·418	70·894	78·172	85·889	89·915	102·68	121·34	141·94	152·99	164·56	216·08	292·95
	0·99	0·99	0·99	0·99	0·99	0·99	0·99	0·99	0·99	0·99	0·99	0·99	0·99	0·99

$V_{r(0{\cdot}5)medial}$ for half-full circular pipes.

A49

(p.1 of 6)

k_s = 0·060 mm
S = 0·00003 to 0·00010

ie hydraulic gradient =
1 in 33333 to 1 in 10000

Water (or sewage) at 15°C;
full bore conditions.

velocities in ms^{-1}
discharges in m^3s^{-1}

Gradient	**(Equivalent) Pipe diameters in mm**													
	2400	**2500**	**2550**	**2600**	**2700**	**2800**	**2850**	**3000**	**3200**	**3400**	**3500**	**3600**	**4000**	**4500**
0·0000300	0·332	0·341	0·345	0·349	0·358	0·366	0·370	0·382	0·398	0·414	0·422	0·429	0·459	0·494
1/ 33333	1·5010	1·6717	1·7613	1·8539	2·0478	2·2539	2·3615	2·7032	3·2039	3·7582	4·0560	4·3680	5·7621	7·8514
0·0000320	0·344	0·353	0·357	0·362	0·370	0·379	0·383	0·396	0·412	0·429	0·436	0·444	0·475	0·511
1/ 31250	1·5543	1·7310	1·8238	1·9196	2·1204	2·3337	2·4451	2·7988	3·3171	3·8908	4·1991	4·5220	5·9649	8·1272
0·0000340	0·355	0·364	0·369	0·374	0·383	0·392	0·396	0·409	0·426	0·443	0·451	0·459	0·490	0·528
1/ 29412	1·6060	1·7885	1·8844	1·9834	2·1908	2·4111	2·5262	2·8916	3·4270	4·0196	4·3380	4·6715	6·1617	8·3949
0·0000360	0·366	0·376	0·381	0·385	0·395	0·404	0·408	0·422	0·439	0·457	0·465	0·473	0·506	0·544
1/ 27778	1·6563	1·8445	1·9434	2·0455	2·2593	2·4865	2·6051	2·9818	3·5338	4·1448	4·4730	4·8169	6·3532	8·6552
0·0000380	0·377	0·387	0·392	0·397	0·406	0·416	0·420	0·434	0·452	0·470	0·479	0·487	0·520	0·560
1/ 26316	1·7053	1·8991	2·0009	2·1059	2·3261	2·5599	2·6820	3·0697	3·6379	4·2667	4·6046	4·9585	6·5396	8·9088
0·0000400	0·388	0·398	0·403	0·408	0·418	0·427	0·432	0·446	0·465	0·483	0·492	0·501	0·535	0·576
1/ 25000	1·7532	1·9523	2·0569	2·1649	2·3912	2·6315	2·7570	3·1555	3·7394	4·3857	4·7329	5·0966	6·7215	9·1560
0·0000420	0·398	0·408	0·413	0·419	0·429	0·439	0·444	0·458	0·477	0·496	0·505	0·514	0·549	0·591
1/ 23810	1·7999	2·0043	2·1117	2·2225	2·4547	2·7014	2·8302	3·2393	3·8386	4·5018	4·8582	5·2314	6·8990	9·3974
0·0000440	0·408	0·419	0·424	0·429	0·440	0·450	0·455	0·470	0·489	0·508	0·518	0·527	0·563	0·606
1/ 22727	1·8455	2·0551	2·1652	2·2788	2·5169	2·7698	2·9019	3·3211	3·9355	4·6154	4·9807	5·3633	7·0726	9·6334
0·0000460	0·418	0·429	0·434	0·440	0·450	0·461	0·466	0·481	0·501	0·521	0·530	0·540	0·576	0·620
1/ 21739	1·8902	2·1049	2·2176	2·3339	2·5777	2·8367	2·9719	3·4013	4·0304	4·7265	5·1006	5·4923	7·2425	9·8644
0·0000480	0·428	0·439	0·444	0·450	0·461	0·471	0·477	0·492	0·513	0·533	0·542	0·552	0·590	0·634
1/ 20833	1·9340	2·1536	2·2689	2·3879	2·6374	2·9022	3·0406	3·4798	4·1233	4·8354	5·2180	5·6187	7·4089	10·091
0·0000500	0·437	0·448	0·454	0·460	0·471	0·482	0·487	0·503	0·524	0·544	0·554	0·564	0·603	0·648
1/ 20000	1·9770	2·2014	2·3192	2·4409	2·6958	2·9665	3·1079	3·5568	4·2144	4·9422	5·3331	5·7426	7·5720	10·312
0·0000525	0·449	0·460	0·466	0·472	0·483	0·495	0·500	0·517	0·538	0·559	0·569	0·579	0·618	0·665
1/ 19048	2·0295	2·2599	2·3808	2·5057	2·7673	3·0452	3·1903	3·6510	4·3259	5·0728	5·4740	5·8942	7·7715	10·584
0·0000550	0·460	0·472	0·478	0·484	0·496	0·507	0·513	0·530	0·551	0·573	0·583	0·594	0·634	0·682
1/ 18182	2·0809	2·3170	2·4410	2·5690	2·8372	3·1221	3·2708	3·7430	4·4349	5·2004	5·6117	6·0425	7·9666	10·849
0·0000575	0·471	0·483	0·490	0·496	0·507	0·519	0·525	0·542	0·565	0·587	0·597	0·608	0·649	0·698
1/ 17391	2·1312	2·3730	2·5000	2·6310	2·9057	3·1973	3·3496	3·8332	4·5415	5·3254	5·7465	6·1875	8·1575	11·108
0·0000600	0·482	0·495	0·501	0·507	0·519	0·531	0·537	0·555	0·578	0·600	0·611	0·622	0·664	0·714
1/ 16667	2·1805	2·4278	2·5577	2·6918	2·9727	3·2710	3·4268	3·9214	4·6460	5·4478	5·8785	6·3296	8·3445	11·362
0·0000625	0·493	0·506	0·512	0·518	0·531	0·543	0·549	0·567	0·590	0·613	0·624	0·636	0·679	0·730
1/ 16000	2·2288	2·4816	2·6143	2·7513	3·0384	3·3433	3·5025	4·0080	4·7484	5·5677	6·0079	6·4688	8·5278	11·611
0·0000650	0·503	0·516	0·523	0·529	0·542	0·554	0·561	0·579	0·603	0·626	0·638	0·649	0·693	0·745
1/ 15385	2·2762	2·5343	2·6698	2·8097	3·1029	3·4142	3·5768	4·0929	4·8489	5·6854	6·1348	6·6054	8·7076	11·856
0·0000675	0·513	0·527	0·533	0·540	0·553	0·566	0·572	0·591	0·615	0·639	0·651	0·662	0·707	0·761
1/ 14815	2·3227	2·5861	2·7243	2·8671	3·1662	3·4838	3·6497	4·1762	4·9476	5·8010	6·2595	6·7396	8·8841	12·096
0·0000700	0·524	0·537	0·544	0·551	0·564	0·577	0·583	0·602	0·627	0·651	0·663	0·675	0·721	0·775
1/ 14286	2·3684	2·6369	2·7779	2·9235	3·2284	3·5522	3·7213	4·2581	5·0445	5·9145	6·3819	6·8714	9·0575	12·331
0·0000725	0·533	0·547	0·554	0·561	0·575	0·588	0·594	0·614	0·639	0·664	0·676	0·688	0·734	0·790
1/ 13793	2·4134	2·6870	2·8306	2·9789	3·2896	3·6195	3·7918	4·3387	5·1398	6·0262	6·5023	7·0009	9·2280	12·563
0·0000750	0·543	0·557	0·564	0·571	0·585	0·599	0·605	0·625	0·651	0·676	0·688	0·700	0·748	0·804
1/ 13333	2·4576	2·7362	2·8824	3·0334	3·3497	3·6856	3·8611	4·4179	5·2335	6·1359	6·6207	7·1284	9·3957	12·791
0·0000800	0·562	0·577	0·584	0·591	0·606	0·620	0·626	0·647	0·674	0·699	0·712	0·725	0·774	0·832
1/ 12500	2·5440	2·8323	2·9836	3·1399	3·4673	3·8149	3·9964	4·5726	5·4166	6·3504	6·8520	7·3773	9·7232	13·236
0·0000850	0·581	0·596	0·603	0·611	0·625	0·640	0·647	0·668	0·696	0·722	0·736	0·749	0·799	0·859
1/ 11765	2·6279	2·9256	3·0819	3·2433	3·5813	3·9403	4·1278	4·7228	5·5943	6·5585	7·0764	7·6188	10·041	13·668
0·0000900	0·599	0·614	0·622	0·630	0·645	0·660	0·667	0·689	0·717	0·745	0·758	0·772	0·824	0·886
1/ 11111	2·7094	3·0163	3·1774	3·3438	3·6922	4·0623	4·2555	4·8688	5·7671	6·7609	7·2947	7·8536	10·350	14·088
0·0000950	0·616	0·632	0·640	0·648	0·664	0·679	0·687	0·709	0·738	0·766	0·780	0·794	0·848	0·911
1/ 10526	2·7889	3·1047	3·2705	3·4416	3·8003	4·1810	4·3799	5·0110	5·9353	6·9579	7·5071	8·0823	10·651	14·496
0·0001000	0·634	0·650	0·658	0·666	0·682	0·698	0·706	0·729	0·758	0·788	0·802	0·816	0·871	0·937
1/ 10000	2·8663	3·1908	3·3612	3·5371	3·9056	4·2968	4·5011	5·1496	6·0994	7·1500	7·7143	8·3052	10·944	14·895
	0·95	0·95	0·95	0·95	0·95	0·95	0·95	0·96	0·96	0·96	0·96	0·96	0·97	0·97

$V_{r(0\cdot5)medial}$ for half-full circular pipes.

k_s = 0·060 mm S = 0·00003 to 0·00010

$k_s = 0{\cdot}060$ **mm**
S = 0·00010 to 0·00030

ie hydraulic gradient =
1 in 10000 to 1 in 3333

Water (or sewage) at 15°C;
full bore conditions.

velocities in ms^{-1}
discharges in m^3s^{-1}

Gradient **(Equivalent) Pipe diameters in mm**

Gradient	2400	2500	2550	2600	2700	2800	2850	3000	3200	3400	3500	3600	4000	4500
0·000100	0·634	0·650	0·658	0·666	0·682	0·698	0·706	0·729	0·758	0·788	0·802	0·816	0·871	0·937
1/ 10000	2·8663	3·1908	3·3612	3·5371	3·9056	4·2968	4·5011	5·1496	6·0994	7·1500	7·7143	8·3052	10·944	14·895
0·000105	0·650	0·667	0·675	0·684	0·700	0·716	0·724	0·748	0·778	0·808	0·823	0·837	0·894	0·961
1/ 9524	2·9419	3·2750	3·4498	3·6303	4·0084	4·4099	4·6196	5·2850	6·2595	7·3375	7·9166	8·5229	11·230	15·284
0·000110	0·667	0·684	0·692	0·701	0·718	0·734	0·742	0·766	0·798	0·828	0·843	0·858	0·916	0·985
1/ 9091	3·0159	3·3572	3·5364	3·7214	4·1089	4·5204	4·7353	5·4172	6·4160	7·5208	8·1142	8·7356	11·510	15·664
0·000115	0·683	0·700	0·709	0·718	0·735	0·752	0·760	0·785	0·817	0·848	0·863	0·879	0·938	1·008
1/ 8696	3·0882	3·4377	3·6211	3·8105	4·2073	4·6286	4·8485	5·5467	6·5692	7·7002	8·3076	8·9437	11·784	16·036
0·000120	0·698	0·716	0·725	0·734	0·752	0·769	0·777	0·803	0·835	0·867	0·883	0·899	0·959	1·031
1/ 8333	3·1590	3·5165	3·7041	3·8978	4·3036	4·7345	4·9594	5·6735	6·7192	7·8758	8·4970	9·1475	12·052	16·400
0·000125	0·714	0·732	0·741	0·750	0·768	0·786	0·794	0·820	0·854	0·886	0·902	0·918	0·980	1·054
1/ 8000	3·2285	3·5937	3·7855	3·9834	4·3980	4·8383	5·0682	5·7977	6·8662	8·0479	8·6826	9·3472	12·315	16·756
0·000130	0·729	0·748	0·757	0·766	0·784	0·802	0·811	0·837	0·872	0·905	0·921	0·938	1·000	1·076
1/ 7692	3·2966	3·6695	3·8653	4·0673	4·4907	4·9401	5·1748	5·9196	7·0104	8·2168	8·8647	9·5432	12·572	17·106
0·000135	0·743	0·763	0·772	0·782	0·800	0·819	0·828	0·854	0·889	0·923	0·940	0·956	1·021	1·097
1/ 7407	3·3635	3·7439	3·9436	4·1497	4·5816	5·0401	5·2795	6·0393	7·1519	8·3825	9·0434	9·7355	12·825	17·450
0·000140	0·758	0·778	0·787	0·797	0·816	0·834	0·844	0·871	0·907	0·941	0·958	0·975	1·040	1·118
1/ 7143	3·4292	3·8170	4·0206	4·2307	4·6709	5·1383	5·3823	6·1568	7·2910	8·5453	9·2190	9·9244	13·074	17·787
0·000145	0·772	0·792	0·802	0·812	0·831	0·850	0·860	0·887	0·924	0·959	0·976	0·993	1·060	1·139
1/ 6897	3·4938	3·8889	4·0962	4·3103	4·7587	5·2348	5·4834	6·2723	7·4276	8·7054	9·3916	10·110	13·318	18·119
0·000150	0·786	0·807	0·817	0·827	0·846	0·866	0·875	0·903	0·940	0·976	0·994	1·011	1·079	1·160
1/ 6667	3·5573	3·9595	4·1706	4·3886	4·8451	5·3298	5·5828	6·3860	7·5621	8·8628	9·5613	10·293	13·558	18·445
0·000160	0·814	0·835	0·845	0·855	0·876	0·896	0·906	0·935	0·973	1·010	1·028	1·046	1·116	1·200
1/ 6250	3·6814	4·0975	4·3160	4·5414	5·0138	5·5152	5·7770	6·6079	7·8246	9·1701	9·8927	10·649	14·027	19·082
0·000170	0·840	0·862	0·873	0·883	0·904	0·925	0·935	0·965	1·005	1·043	1·062	1·080	1·152	1·239
1/ 5882	3·8018	4·2315	4·4570	4·6898	5·1774	5·6951	5·9654	6·8232	8·0793	9·4683	10·214	10·995	14·482	19·700
0·000180	0·866	0·889	0·900	0·910	0·932	0·953	0·964	0·995	1·035	1·075	1·094	1·113	1·188	1·276
1/ 5556	3·9189	4·3617	4·5941	4·8340	5·3365	5·8700	6·1486	7·0325	8·3268	9·7582	10·527	11·332	14·924	20·300
0·000190	0·891	0·914	0·926	0·937	0·959	0·981	0·992	1·024	1·065	1·106	1·126	1·145	1·222	1·313
1/ 5263	4·0329	4·4884	4·7276	4·9744	5·4914	6·0403	6·3269	7·2363	8·5678	10·040	10·831	11·659	15·355	20·884
0·000200	0·916	0·940	0·951	0·963	0·985	1·008	1·019	1·052	1·095	1·136	1·157	1·177	1·255	1·349
1/ 5000	4·1440	4·6120	4·8577	5·1112	5·6424	6·2063	6·5007	7·4349	8·8028	10·315	11·128	11·978	15·774	21·454
0·000210	0·940	0·964	0·976	0·988	1·011	1·034	1·046	1·079	1·123	1·166	1·187	1·207	1·288	1·384
1/ 4762	4·2524	4·7326	4·9847	5·2448	5·7898	6·3683	6·6703	7·6288	9·0321	10·584	11·417	12·289	16·184	22·010
0·000220	0·963	0·988	1·000	1·012	1·036	1·060	1·072	1·106	1·151	1·195	1·216	1·237	1·320	1·418
1/ 4545	4·3584	4·8505	5·1088	5·3754	5·9339	6·5266	6·8361	7·8182	9·2562	10·846	11·700	12·594	16·584	22·553
0·000230	0·986	1·012	1·024	1·037	1·061	1·085	1·097	1·132	1·178	1·223	1·245	1·267	1·351	1·451
1/ 4348	4·4621	4·9659	5·2303	5·5031	6·0748	6·6815	6·9983	8·0036	9·4753	11·103	11·977	12·891	16·975	23·084
0·000240	1·009	1·035	1·047	1·060	1·085	1·110	1·122	1·158	1·205	1·251	1·273	1·295	1·381	1·484
1/ 4167	4·5637	5·0788	5·3492	5·6282	6·2127	6·8332	7·1571	8·1850	9·6900	11·354	12·247	13·183	17·358	23·604
0·000250	1·031	1·057	1·070	1·083	1·109	1·134	1·146	1·183	1·231	1·278	1·301	1·323	1·411	1·516
1/ 4000	4·6632	5·1895	5·4657	5·7508	6·3480	6·9818	7·3128	8·3629	9·9003	11·600	12·513	13·468	17·734	24·114
0·000260	1·052	1·079	1·093	1·106	1·132	1·158	1·170	1·208	1·257	1·304	1·328	1·351	1·441	1·548
1/ 3846	4·7608	5·2980	5·5800	5·8710	6·4806	7·1276	7·4654	8·5373	10·107	11·842	12·773	13·749	18·102	24·614
0·000270	1·074	1·101	1·115	1·128	1·155	1·181	1·194	1·232	1·282	1·330	1·354	1·378	1·469	1·578
1/ 3704	4·8567	5·4046	5·6922	5·9890	6·6108	7·2707	7·6152	8·7085	10·309	12·079	13·029	14·023	18·463	25·104
0·000280	1·094	1·122	1·136	1·150	1·177	1·204	1·217	1·256	1·307	1·356	1·380	1·404	1·498	1·609
1/ 3571	4·9508	5·5093	5·8024	6·1049	6·7386	7·4112	7·7624	8·8766	10·508	12·311	13·280	14·293	18·818	25·586
0·000290	1·115	1·143	1·157	1·171	1·199	1·226	1·239	1·279	1·331	1·381	1·406	1·430	1·525	1·639
1/ 3448	5·0433	5·6121	5·9107	6·2188	6·8643	7·5494	7·9070	9·0419	10·703	12·540	13·526	14·559	19·167	26·059
0·000300	1·135	1·164	1·178	1·192	1·220	1·248	1·262	1·302	1·355	1·406	1·431	1·456	1·553	1·668
1/ 3333	5·1342	5·7133	6·0172	6·3308	6·9878	7·6852	8·0493	9·2044	10·895	12·765	13·769	14·820	19·510	26·525
	0·96	0·96	0·96	0·96	0·96	0·96	0·96	0·97	0·97	0·97	0·97	0·97	0·97	0·98

$V_{r(0{\cdot}5)medial}$ **for half-full circular pipes.**

k_s = **0·060 mm**
S = 0·00030 to 0·00100

ie hydraulic gradient =
1 in 3333 to 1 in 1000

Water (or sewage) at 15°C;
full bore conditions.

velocities in ms^{-1}
discharges in m^3s^{-1}

Gradient	**(Equivalent) Pipe diameters in mm**													
	2400	**2500**	**2550**	**2600**	**2700**	**2800**	**2850**	**3000**	**3200**	**3400**	**3500**	**3600**	**4000**	**4500**
0·000300	1·135	1·164	1·178	1·192	1·220	1·248	1·262	1·302	1·355	1·406	1·431	1·456	1·553	1·668
1/ 3333	5·1342	5·7133	6·0172	6·3308	6·9878	7·6852	8·0493	9·2044	10·895	12·765	13·769	14·820	19·510	26·525
0·000320	1·174	1·204	1·219	1·234	1·263	1·291	1·305	1·347	1·401	1·454	1·480	1·506	1·606	1·725
1/ 3125	5·3119	5·9108	6·2252	6·5496	7·2291	7·9504	8·3270	9·5217	11·271	13·204	14·242	15·329	20·179	27·433
0·000340	1·212	1·243	1·258	1·274	1·303	1·333	1·348	1·391	1·447	1·501	1·528	1·555	1·657	1·780
1/ 2941	5·4842	6·1024	6·4269	6·7618	7·4632	8·2077	8·5963	9·8295	11·635	13·630	14·702	15·823	20·829	28·315
0·000360	1·249	1·281	1·297	1·312	1·343	1·374	1·389	1·433	1·491	1·547	1·574	1·602	1·708	1·834
1/ 2778	5·6517	6·2887	6·6230	6·9680	7·6907	8·4577	8·8581	10·129	11·988	14·044	15·148	16·303	21·460	29·171
0·000380	1·285	1·318	1·334	1·350	1·382	1·413	1·428	1·474	1·533	1·591	1·620	1·648	1·757	1·887
1/ 2632	5·8147	6·4699	6·8138	7·1687	7·9121	8·7010	9·1129	10·420	12·332	14·447	15·582	16·770	22·074	30·004
0·000400	1·320	1·354	1·371	1·387	1·420	1·452	1·467	1·514	1·575	1·634	1·664	1·692	1·804	1·938
1/ 2500	5·9736	6·6466	6·9998	7·3643	8·1279	8·9382	9·3612	10·703	12·668	14·839	16·005	17·226	22·672	30·816
0·000420	1·355	1·389	1·406	1·423	1·456	1·489	1·505	1·553	1·616	1·677	1·706	1·736	1·851	1·987
1/ 2381	6·1286	6·8190	7·1814	7·5553	8·3385	9·1696	9·6036	10·980	12·995	15·223	16·418	17·670	23·256	31·608
0·000440	1·388	1·423	1·441	1·458	1·492	1·526	1·543	1·592	1·656	1·718	1·748	1·779	1·896	2·036
1/ 2273	6·2802	6·9875	7·3587	7·7418	8·5442	9·3958	9·8403	11·251	13·315	15·597	16·822	18·104	23·826	32·382
0·000460	1·421	1·457	1·475	1·493	1·527	1·562	1·579	1·629	1·694	1·758	1·789	1·820	1·940	2·084
1/ 2174	6·4284	7·1523	7·5322	7·9243	8·7455	9·6170	10·072	11·515	13·628	15·963	17·217	18·529	24·384	33·139
0·000480	1·453	1·490	1·508	1·526	1·562	1·597	1·614	1·666	1·733	1·798	1·830	1·861	1·984	2·130
1/ 2083	6·5734	7·3136	7·7020	8·1029	8·9425	9·8335	10·299	11·774	13·934	16·321	17·603	18·944	24·930	33·880
0·000500	1·484	1·522	1·541	1·559	1·596	1·631	1·649	1·702	1·770	1·836	1·869	1·901	2·026	2·176
1/ 2000	6·7156	7·4717	7·8685	8·2779	9·1355	10·046	10·521	12·028	14·234	16·672	17·981	19·352	25·465	34·606
0·000525	1·523	1·562	1·581	1·599	1·637	1·674	1·692	1·746	1·815	1·884	1·917	1·950	2·079	2·232
1/ 1905	6·8895	7·6650	8·0720	8·4920	9·3716	10·305	10·792	12·338	14·601	17·102	18·444	19·849	26·120	35·493
0·000550	1·560	1·600	1·619	1·639	1·677	1·715	1·733	1·788	1·860	1·930	1·964	1·998	2·129	2·286
1/ 1818	7·0594	7·8539	8·2708	8·7011	9·6023	10·559	11·058	12·642	14·959	17·521	18·896	20·336	26·759	36·360
0·000575	1·597	1·638	1·658	1·677	1·716	1·755	1·774	1·830	1·904	1·975	2·010	2·045	2·179	2·340
1/ 1739	7·2255	8·0386	8·4653	8·9056	9·8278	10·806	11·317	12·938	15·310	17·931	19·339	20·812	27·384	37·208
0·000600	1·633	1·674	1·695	1·715	1·755	1·794	1·814	1·871	1·946	2·019	2·055	2·090	2·228	2·392
1/ 1667	7·3882	8·2194	8·6557	9·1059	10·049	11·049	11·571	13·228	15·653	18·333	19·771	21·277	27·995	38·038
0·000625	1·668	1·711	1·731	1·752	1·793	1·833	1·853	1·912	1·988	2·063	2·099	2·135	2·275	2·443
1/ 1600	7·5475	8·3966	8·8422	9·3020	10·265	11·287	11·820	13·513	15·989	18·726	20·195	21·733	28·595	38·851
0·000650	1·703	1·746	1·767	1·788	1·830	1·871	1·891	1·951	2·029	2·105	2·142	2·179	2·322	2·493
1/ 1538	7·7038	8·5704	9·0251	9·4944	10·477	11·520	12·065	13·791	16·319	19·112	20·611	22·181	29·182	39·648
0·000675	1·737	1·781	1·802	1·824	1·866	1·908	1·929	1·990	2·069	2·147	2·185	2·222	2·368	2·542
1/ 1481	7·8572	8·7409	9·2046	9·6832	10·685	11·749	12·304	14·065	16·642	19·490	21·019	22·620	29·759	40·430
0·000700	1·770	1·815	1·837	1·859	1·902	1·945	1·966	2·028	2·109	2·188	2·226	2·265	2·413	2·590
1/ 1429	8·0078	8·9084	9·3809	9·8686	10·890	11·974	12·539	14·334	16·960	19·862	21·420	23·051	30·325	41·199
0·000725	1·803	1·848	1·871	1·893	1·937	1·980	2·002	2·065	2·148	2·228	2·267	2·306	2·457	2·638
1/ 1379	8·1558	9·0729	9·5541	10·051	11·091	12·194	12·770	14·598	17·272	20·227	21·814	23·474	30·882	41·953
0·000750	1·835	1·881	1·904	1·927	1·972	2·016	2·037	2·102	2·186	2·267	2·307	2·347	2·501	2·684
1/ 1333	8·3013	9·2347	9·7245	10·230	11·288	12·411	12·998	14·857	17·579	20·586	22·201	23·890	31·428	42·695
0·000800	1·898	1·946	1·969	1·993	2·039	2·084	2·107	2·174	2·260	2·345	2·386	2·427	2·586	2·776
1/ 1250	8·5854	9·5505	10·057	10·580	11·674	12·835	13·441	15·364	18·178	21·287	22·956	24·703	32·496	44·143
0·000850	1·959	2·008	2·032	2·057	2·104	2·151	2·174	2·243	2·332	2·419	2·462	2·504	2·668	2·864
1/ 1176	8·8609	9·8568	10·379	10·919	12·048	13·246	13·872	15·856	18·759	21·967	23·689	25·491	33·531	45·547
0·000900	2·018	2·069	2·094	2·119	2·168	2·216	2·240	2·311	2·403	2·492	2·536	2·580	2·748	2·950
1/ 1111	9·1286	10·154	10·693	11·248	12·411	13·645	14·290	16·333	19·323	22·627	24·401	26·257	34·537	46·911
0·000950	2·075	2·128	2·153	2·179	2·229	2·279	2·304	2·376	2·471	2·563	2·608	2·653	2·826	3·033
1/ 1053	9·3890	10·444	10·997	11·569	12·765	14·034	14·696	16·797	19·872	23·269	25·093	27·002	35·515	48·238
0·001000	2·132	2·185	2·212	2·238	2·290	2·341	2·366	2·440	2·537	2·632	2·678	2·724	2·902	3·114
1/ 1000	9·6428	10·726	11·294	11·881	13·109	14·412	15·093	17·250	20·407	23·895	25·768	27·727	36·468	49·530
	0·97	0·97	0·97	0·97	0·97	0·97	0·97	0·97	0·97	0·98	0·98	0·98	0·98	0·98

$V_{r(0\cdot5)medial}$ for half-full circular pipes.

k_s = 0·060 mm S = 0·00030 to 0·00100

Gradient	**(Equivalent) Pipe diameters in mm**													
	2400	**2500**	**2550**	**2600**	**2700**	**2800**	**2850**	**3000**	**3200**	**3400**	**3500**	**3600**	**4000**	**4500**
0·00100	2·132	2·185	2·212	2·238	2·290	2·341	2·366	2·440	2·537	2·632	2·678	2·724	2·902	3·114
1/ 1000	9·6428	10·726	11·294	11·881	13·109	14·412	15·093	17·250	20·407	23·895	25·768	27·727	36·468	49·530
0·00105	2·186	2·241	2·268	2·295	2·348	2·401	2·426	2·503	2·602	2·699	2·747	2·794	2·976	3·194
1/ 952	9·8905	11·001	11·584	12·186	13·445	14·782	15·479	17·692	20·929	24·506	26·426	28·435	37·398	50·792
0·00110	2·240	2·296	2·324	2·351	2·406	2·459	2·486	2·564	2·666	2·765	2·813	2·862	3·048	3·271
1/ 909	10·132	11·270	11·867	12·483	13·774	15·142	15·857	18·123	21·439	25·103	27·069	29·127	38·307	52·023
0·00115	2·292	2·350	2·378	2·406	2·462	2·516	2·544	2·624	2·728	2·829	2·879	2·928	3·119	3·347
1/ 870	10·369	11·533	12·144	12·775	14·095	15·495	16·226	18·545	21·937	25·686	27·698	29·803	39·195	53·228
0·00120	2·343	2·402	2·431	2·460	2·517	2·573	2·600	2·682	2·788	2·892	2·943	2·993	3·188	3·421
1/ 833	10·601	11·791	12·415	13·059	14·409	15·840	16·588	18·958	22·425	26·257	28·313	30·465	40·064	54·407
0·00125	2·393	2·453	2·483	2·512	2·570	2·627	2·656	2·739	2·848	2·954	3·005	3·057	3·256	3·493
1/ 800	10·827	12·043	12·681	13·339	14·717	16·179	16·942	19·362	22·903	26·816	28·916	31·113	40·916	55·561
0·00130	2·443	2·504	2·534	2·564	2·623	2·681	2·710	2·795	2·906	3·014	3·067	3·119	3·322	3·565
1/ 769	11·050	12·290	12·941	13·612	15·018	16·510	17·289	19·759	23·372	27·364	29·507	31·749	41·750	56·693
0·00135	2·491	2·553	2·584	2·614	2·675	2·734	2·764	2·850	2·963	3·073	3·127	3·180	3·388	3·634
1/ 741	11·268	12·533	13·196	13·881	15·315	16·836	17·630	20·148	23·832	27·902	30·086	32·373	42·570	57·804
0·00140	2·538	2·602	2·633	2·664	2·726	2·786	2·816	2·904	3·019	3·131	3·186	3·241	3·452	3·703
1/ 714	11·482	12·771	13·447	14·145	15·605	17·155	17·964	20·530	24·283	28·430	30·656	32·985	43·374	58·894
0·00145	2·585	2·649	2·681	2·713	2·775	2·837	2·867	2·957	3·075	3·189	3·244	3·300	3·514	3·770
1/ 690	11·693	13·005	13·693	14·404	15·891	17·469	18·293	20·905	24·727	28·949	31·215	33·587	44·164	59·966
0·00150	2·630	2·696	2·729	2·761	2·825	2·887	2·918	3·010	3·129	3·245	3·302	3·358	3·576	3·837
1/ 667	11·900	13·235	13·936	14·658	16·172	17·778	18·616	21·274	25·163	29·459	31·765	34·178	44·941	61·019
0·00160	2·720	2·788	2·821	2·855	2·920	2·985	3·017	3·112	3·235	3·354	3·413	3·471	3·697	3·966
1/ 625	12·304	13·684	14·408	15·156	16·720	18·380	19·246	21·994	26·014	30·455	32·838	35·333	46·456	63·074
0·00170	2·806	2·876	2·911	2·945	3·013	3·080	3·113	3·210	3·337	3·461	3·521	3·581	3·814	4·091
1/ 588	12·696	14·120	14·867	15·638	17·252	18·964	19·858	22·692	26·839	31·420	33·879	36·452	47·926	65·067
0·00180	2·890	2·963	2·998	3·034	3·103	3·172	3·206	3·306	3·437	3·564	3·626	3·688	3·927	4·213
1/ 556	13·076	14·543	15·312	16·106	17·768	19·531	20·452	23·371	27·641	32·358	34·889	37·539	49·353	67·002
0·00190	2·972	3·046	3·083	3·119	3·191	3·262	3·296	3·400	3·534	3·664	3·729	3·792	4·038	4·331
1/ 526	13·446	14·954	15·745	16·561	18·270	20·083	21·029	24·030	28·420	33·270	35·872	38·596	50·742	68·885
0·00200	3·052	3·128	3·166	3·203	3·276	3·349	3·385	3·491	3·628	3·762	3·828	3·893	4·146	4·447
1/ 500	13·807	15·355	16·167	17·005	18·760	20·621	21·592	24·673	29·180	34·159	36·830	39·626	52·095	70·719
0·00210	3·130	3·208	3·246	3·284	3·360	3·434	3·471	3·579	3·720	3·858	3·925	3·992	4·251	4·559
1/ 476	14·159	15·746	16·579	17·438	19·237	21·145	22·141	25·300	29·921	35·025	37·765	40·631	53·414	72·508
0·00220	3·206	3·286	3·325	3·364	3·441	3·517	3·555	3·666	3·810	3·951	4·020	4·088	4·353	4·669
1/ 455	14·503	16·128	16·981	17·861	19·703	21·658	22·678	25·913	30·645	35·872	38·677	41·613	54·703	74·255
0·00230	3·280	3·362	3·402	3·442	3·521	3·599	3·637	3·751	3·898	4·042	4·113	4·182	4·453	4·776
1/ 435	14·839	16·502	17·374	18·274	20·159	22·159	23·202	26·512	31·352	36·700	39·569	42·572	55·963	75·963
0·00240	3·353	3·436	3·477	3·518	3·599	3·678	3·718	3·834	3·984	4·131	4·204	4·275	4·552	4·881
1/ 417	15·168	16·867	17·759	18·679	20·605	22·649	23·715	27·098	32·045	37·510	40·442	43·511	57·196	77·635
0·00250	3·424	3·509	3·551	3·593	3·675	3·756	3·796	3·915	4·069	4·219	4·292	4·365	4·648	4·984
1/ 400	15·490	17·226	18·136	19·076	21·043	23·129	24·218	27·672	32·723	38·303	41·298	44·431	58·404	79·273
0·00260	3·494	3·581	3·624	3·666	3·750	3·833	3·874	3·994	4·151	4·304	4·380	4·454	4·742	5·085
1/ 385	15·806	17·577	18·506	19·464	21·471	23·600	24·711	28·234	33·388	39·081	42·136	45·333	59·588	80·878
0·00270	3·562	3·651	3·695	3·738	3·823	3·908	3·949	4·072	4·233	4·389	4·465	4·541	4·834	5·184
1/ 370	16·116	17·921	18·868	19·845	21·891	24·062	25·194	28·787	34·041	39·844	42·959	46·218	60·750	82·453
0·00280	3·630	3·720	3·764	3·808	3·896	3·981	4·024	4·149	4·312	4·471	4·549	4·626	4·925	5·281
1/ 357	16·420	18·259	19·224	20·220	22·304	24·515	25·669	29·329	34·681	40·594	43·766	47·086	61·890	83·999
0·00290	3·696	3·787	3·833	3·878	3·966	4·054	4·097	4·224	4·391	4·552	4·631	4·710	5·014	5·377
1/ 345	16·719	18·592	19·574	20·588	22·710	24·961	26·136	29·861	35·311	41·330	44·560	47·940	63·011	85·517
0·00300	3·761	3·854	3·900	3·946	4·036	4·125	4·169	4·299	4·467	4·632	4·713	4·792	5·102	5·471
1/ 333	17·013	18·918	19·918	20·949	23·108	25·399	26·594	30·385	35·929	42·053	45·340	48·779	64·112	87·010
	0·97	0·97	0·97	0·97	0·98	0·98	0·98	0·98	0·98	0·98	0·98	0·98	0·98	0·98

$V_{r(0{\cdot}5)medial}$ for half-full circular pipes.

A49
(p.5 of 6)

k_s = **0·060 mm**
S = 0·00300 to 0·01000

ie hydraulic gradient =
1 in 333 to 1 in 100

Water (or sewage) at 15°C;
full bore conditions.

velocities in ms^{-1}
discharges in m^3s^{-1}

Gradient **(Equivalent) Pipe diameters in mm**

Gradient	2400	2500	2550	2600	2700	2800	2850	3000	3200	3400	3500	3600	4000	4500
0·00300	3·761	3·854	3·900	3·946	4·036	4·125	4·169	4·299	4·467	4·632	4·713	4·792	5·102	5·471
1/ 333	17·013	18·918	19·918	20·949	23·108	25·399	26·594	30·385	35·929	42·053	45·340	48·779	64·112	87·010
0·00320	3·887	3·984	4·031	4·079	4·172	4·264	4·309	4·443	4·618	4·787	4·871	4·953	5·273	5·654
1/ 313	17·586	19·556	20·589	21·655	23·886	26·253	27·489	31·406	37·136	43·466	46·862	50·415	66·261	89·924
0·00340	4·010	4·110	4·159	4·207	4·304	4·398	4·445	4·583	4·763	4·938	5·024	5·109	5·439	5·832
1/ 294	18·143	20·174	21·239	22·339	24·640	27·082	28·356	32·397	38·307	44·835	48·337	52·002	68·344	92·748
0·00360	4·130	4·232	4·283	4·333	4·432	4·529	4·577	4·719	4·904	5·085	5·173	5·260	5·600	6·004
1/ 278	18·683	20·774	21·871	23·003	25·373	27·887	29·199	33·359	39·443	46·164	49·770	53·543	70·368	95·491
0·00380	4·246	4·351	4·403	4·454	4·556	4·656	4·705	4·852	5·042	5·227	5·318	5·408	5·756	6·172
1/ 263	19·208	21·358	22·486	23·649	26·085	28·670	30·018	34·294	40·549	47·457	51·164	55·043	72·336	98·159
0·00400	4·359	4·467	4·520	4·573	4·677	4·780	4·831	4·981	5·176	5·366	5·459	5·551	5·909	6·335
1/ 250	19·720	21·927	23·084	24·279	26·779	29·432	30·816	35·206	41·626	48·717	52·522	56·503	74·253	100·76
0·00420	4·469	4·580	4·634	4·689	4·795	4·901	4·953	5·106	5·306	5·501	5·597	5·691	6·058	6·495
1/ 238	20·219	22·481	23·668	24·893	27·457	30·176	31·595	36·095	42·677	49·946	53·846	57·927	76·123	103·29
0·00440	4·577	4·690	4·746	4·802	4·911	5·019	5·072	5·229	5·434	5·633	5·731	5·828	6·203	6·650
1/ 227	20·706	23·023	24·239	25·493	28·118	30·902	32·356	36·964	43·703	51·146	55·139	59·318	77·949	105·77
0·00460	4·682	4·798	4·855	4·912	5·024	5·134	5·188	5·349	5·559	5·763	5·862	5·961	6·345	6·802
1/ 217	21·183	23·553	24·796	26·079	28·764	31·613	33·099	37·812	44·706	52·319	56·404	60·678	79·735	108·19
0·00480	4·786	4·904	4·962	5·020	5·134	5·247	5·302	5·467	5·681	5·889	5·991	6·092	6·484	6·951
1/ 208	21·650	24·072	25·342	26·653	29·397	32·308	33·827	38·643	45·687	53·467	57·641	62·009	81·482	110·56
0·00500	4·887	5·007	5·067	5·126	5·243	5·357	5·414	5·582	5·800	6·013	6·117	6·220	6·620	7·097
1/ 200	22·106	24·579	25·876	27·215	30·017	32·988	34·539	39·457	46·649	54·592	58·853	63·313	83·193	112·88
0·00525	5·010	5·134	5·195	5·255	5·375	5·493	5·551	5·723	5·946	6·164	6·271	6·377	6·787	7·275
1/ 190	22·665	25·200	26·530	27·902	30·774	33·821	35·410	40·452	47·824	55·966	60·335	64·906	85·285	115·71
0·00550	5·131	5·257	5·320	5·382	5·504	5·625	5·684	5·860	6·089	6·312	6·421	6·529	6·949	7·450
1/ 182	23·211	25·806	27·168	28·573	31·514	34·633	36·261	41·423	48·972	57·309	61·782	66·462	87·328	118·48
0·00575	5·249	5·378	5·442	5·505	5·630	5·754	5·814	5·995	6·229	6·457	6·568	6·679	7·108	7·620
1/ 174	23·744	26·399	27·792	29·229	32·237	35·428	37·093	42·373	50·094	58·621	63·196	67·983	89·324	121·19
0·00600	5·364	5·496	5·561	5·626	5·754	5·880	5·942	6·126	6·365	6·598	6·712	6·825	7·264	7·786
1/ 167	24·266	26·979	28·402	29·871	32·945	36·205	37·907	43·302	51·192	59·905	64·580	69·472	91·279	123·83
0·00625	5·477	5·612	5·678	5·745	5·875	6·004	6·067	6·255	6·499	6·737	6·853	6·968	7·416	7·949
1/ 160	24·777	27·547	29·000	30·499	33·638	36·967	38·704	44·212	52·267	61·163	65·936	70·930	93·193	126·43
0·00650	5·588	5·725	5·793	5·861	5·994	6·125	6·190	6·381	6·630	6·872	6·991	7·109	7·565	8·109
1/ 154	25·278	28·104	29·586	31·116	34·317	37·713	39·485	45·104	53·321	62·396	67·265	72·359	95·069	128·97
0·00675	5·696	5·837	5·906	5·975	6·110	6·244	6·310	6·505	6·759	7·006	7·127	7·247	7·712	8·266
1/ 148	25·770	28·650	30·161	31·720	34·984	38·446	40·252	45·980	54·356	63·606	68·568	73·761	96·909	131·47
0·00700	5·803	5·946	6·016	6·086	6·224	6·360	6·428	6·626	6·885	7·136	7·260	7·382	7·856	8·420
1/ 143	26·252	29·187	30·726	32·314	35·639	39·165	41·005	46·839	55·371	64·793	69·848	75·137	98·716	133·92
0·00725	5·908	6·053	6·125	6·196	6·337	6·475	6·544	6·746	7·009	7·265	7·391	7·515	7·997	8·571
1/ 138	26·726	29·714	31·281	32·897	36·281	39·871	41·744	47·683	56·368	65·960	71·105	76·490	100·49	136·32
0·00750	6·011	6·159	6·232	6·304	6·447	6·588	6·658	6·863	7·131	7·391	7·519	7·645	8·136	8·720
1/ 133	27·192	30·231	31·826	33·470	36·913	40·565	42·471	48·513	57·349	67·106	72·341	77·819	102·24	138·69
0·00800	6·212	6·365	6·440	6·515	6·663	6·808	6·880	7·092	7·369	7·638	7·770	7·900	8·407	9·010
1/ 125	28·102	31·242	32·889	34·589	38·147	41·920	43·889	50·132	59·262	69·344	74·753	80·412	105·64	143·30
0·00850	6·407	6·564	6·642	6·719	6·871	7·021	7·095	7·314	7·599	7·877	8·013	8·147	8·669	9·291
1/ 118	28·983	32·222	33·921	35·673	39·342	43·233	45·264	51·702	61·116	71·513	77·090	82·926	108·94	147·77
0·00900	6·596	6·758	6·838	6·917	7·074	7·228	7·305	7·530	7·823	8·109	8·249	8·387	8·924	9·564
1/ 111	29·839	33·173	34·922	36·726	40·503	44·509	46·599	53·226	62·917	73·619	79·360	85·367	112·14	152·12
0·00950	6·780	6·947	7·029	7·110	7·271	7·430	7·508	7·740	8·041	8·334	8·478	8·620	9·172	9·830
1/ 105	30·672	34·099	35·896	37·750	41·632	45·749	47·897	54·709	64·669	75·667	81·568	87·742	115·26	156·34
0·01000	6·959	7·130	7·215	7·298	7·463	7·626	7·706	7·944	8·253	8·554	8·702	8·847	9·414	10·09
1/ 100	31·483	35·000	36·845	38·748	42·732	46·957	49·162	56·153	66·375	77·663	83·718	90·055	118·30	160·46
$V_{r(0·5)medial}$	0·98	0·98	0·98	0·98	0·98	0·98	0·98	0·98	0·98	0·98	0·98	0·98	0·98	0·99

$V_{r(0·5)medial}$ for half-full circular pipes.

k_s = 0·060 mm S = 0·00300 to 0·01000

Gradient **(Equivalent) Pipe diameters in mm**

Gradient	2400	2500	2550	2600	2700	2800	2850	3000	3200	3400	3500	3600	4000	4500
0·01000	6·959	7·130	7·215	7·298	7·463	7·626	7·706	7·944	8·253	8·554	8·702	8·847	9·414	10·09
1/ 100	31·483	35·000	36·845	38·748	42·732	46·957	49·162	56·153	66·375	77·663	83·718	90·055	118·30	160·46
0·01050	7·134	7·309	7·396	7·481	7·651	7·817	7·900	8·143	8·460	8·768	8·920	9·069	9·649	10·34
1/ 95	32·274	35·879	37·770	39·721	43·805	48·136	50·396	57·561	68·039	79·609	85·816	92·311	121·26	164·47
0·01100	7·305	7·484	7·573	7·660	7·834	8·004	8·089	8·338	8·662	8·978	9·132	9·285	9·880	10·59
1/ 91	33·047	36·738	38·674	40·671	44·852	49·287	51·601	58·937	69·664	81·510	87·864	94·514	124·15	168·39
0·01150	7·472	7·655	7·746	7·835	8·013	8·187	8·273	8·528	8·860	9·182	9·341	9·497	10·10	10·83
1/ 87	33·802	37·577	39·557	41·600	45·876	50·412	52·778	60·282	71·253	83·368	89·867	96·667	126·98	172·22
0·01200	7·635	7·823	7·915	8·007	8·188	8·366	8·454	8·714	9·053	9·383	9·544	9·704	10·32	11·06
1/ 83	34·541	38·399	40·422	42·509	46·879	51·513	53·931	61·597	72·807	85·186	91·826	98·774	129·74	175·97
0·01250	7·795	7·986	8·081	8·174	8·359	8·541	8·631	8·897	9·242	9·579	9·744	9·907	10·54	11·30
1/ 80	35·265	39·203	41·269	43·400	47·860	52·591	55·060	62·886	74·330	86·966	93·745	100·84	132·45	179·64
0·01300	7·952	8·147	8·243	8·339	8·527	8·713	8·804	9·075	9·428	9·771	9·939	10·11	10·75	11·52
1/ 77	35·975	39·992	42·099	44·272	48·822	53·648	56·166	64·149	75·822	88·712	95·627	102·86	135·11	183·24
0·01350	8·106	8·305	8·403	8·500	8·692	8·881	8·974	9·251	9·610	9·959	10·13	10·30	10·96	11·74
1/ 74	36·671	40·765	42·913	45·128	49·766	54·685	57·251	65·389	77·287	90·424	97·472	104·85	137·71	186·77
0·01400	8·257	8·459	8·559	8·658	8·854	9·046	9·141	9·423	9·789	10·14	10·32	10·49	11·16	11·96
1/ 71	37·354	41·525	43·712	45·969	50·693	55·703	58·317	66·605	78·724	92·105	99·284	106·79	140·27	190·24
0·01450	8·406	8·611	8·713	8·814	9·013	9·209	9·306	9·592	9·964	10·33	10·50	10·68	11·36	12·18
1/ 69	38·026	42·271	44·498	46·795	51·603	56·703	59·364	67·800	80·136	93·756	101·06	108·71	142·78	193·64
0·01500	8·551	8·761	8·864	8·967	9·169	9·368	9·467	9·758	10·14	10·51	10·69	10·86	11·56	12·39
1/ 67	38·686	43·004	45·269	47·606	52·498	57·686	60·393	68·975	81·524	95·379	102·81	110·59	145·25	196·99
0·01600	8·836	9·052	9·159	9·265	9·474	9·680	9·782	10·08	10·47	10·85	11·04	11·23	11·94	12·80
1/ 63	39·973	44·435	46·775	49·190	54·244	59·603	62·400	71·267	84·232	98·546	106·23	114·26	150·07	203·52
0·01700	9·112	9·335	9·445	9·554	9·770	9·982	10·09	10·40	10·80	11·19	11·38	11·57	12·31	13·19
1/ 59	41·221	45·822	48·235	50·725	55·936	61·462	64·346	73·489	86·856	101·62	109·53	117·82	154·74	209·85
0·01800	9·380	9·609	9·722	9·835	10·06	10·27	10·38	10·70	11·12	11·52	11·72	11·91	12·67	13·58
1/ 56	42·433	47·169	49·653	52·215	57·579	63·268	66·236	75·646	89·405	104·60	112·75	121·27	159·28	215·99
0·01900	9·640	9·876	9·992	10·11	10·34	10·56	10·67	11·00	11·42	11·84	12·04	12·24	13·03	13·96
1/ 53	43·612	48·479	51·032	53·665	59·177	65·023	68·074	77·745	91·884	107·50	115·87	124·63	163·69	221·97
0·02000	9·894	10·14	10·26	10·37	10·61	10·84	10·95	11·29	11·73	12·15	12·36	12·57	13·37	14·32
1/ 50	44·760	49·755	52·375	55·077	60·734	66·734	69·864	79·788	94·299	110·32	118·91	127·91	167·98	227·79
0·02100	10·14	10·39	10·51	10·63	10·87	11·11	11·23	11·57	12·02	12·45	12·67	12·88	13·70	14·68
1/ 47·6	45·880	50·999	53·685	56·454	62·253	68·402	71·610	81·782	96·654	113·07	121·88	131·10	172·17	233·47
0·02200	10·38	10·64	10·76	10·89	11·13	11·37	11·49	11·85	12·30	12·75	12·97	13·19	14·03	15·03
1/ 45·5	46·974	52·215	54·964	57·799	63·735	70·030	73·315	83·729	98·954	115·76	124·78	134·22	176·26	239·02
0·02300	10·62	10·88	11·01	11·13	11·38	11·63	11·75	12·11	12·58	13·04	13·26	13·49	14·34	15·37
1/ 43·5	48·043	53·403	56·214	59·114	65·185	71·623	74·982	85·632	101·20	118·39	127·61	137·26	180·26	244·44
0·02400	10·85	11·12	11·25	11·38	11·63	11·88	12·01	12·38	12·86	13·32	13·55	13·78	14·66	15·70
1/ 41·7	49·089	54·565	57·438	60·401	66·603	73·181	76·613	87·494	103·40	120·97	130·39	140·25	184·18	249·74
0·02500	11·08	11·35	11·48	11·61	11·88	12·13	12·26	12·64	13·12	13·60	13·83	14·07	14·96	16·03
1/ 40·0	50·113	55·704	58·636	61·661	67·992	74·707	78·210	89·317	105·56	123·49	133·10	143·17	188·01	254·94
0·02600	11·30	11·58	11·71	11·85	12·11	12·38	12·51	12·89	13·39	13·87	14·11	14·35	15·26	16·35
1/ 38·5	51·118	56·820	59·811	62·896	69·354	76·203	79·776	91·105	107·67	125·96	135·76	146·03	191·77	260·03
0·02700	11·52	11·80	11·94	12·07	12·35	12·61	12·75	13·14	13·65	14·14	14·38	14·62	15·55	16·66
1/ 37·0	52·103	57·915	60·963	64·107	70·690	77·670	81·312	92·859	109·74	128·38	138·38	148·84	195·45	265·02
0·02800	11·73	12·02	12·16	12·30	12·58	12·85	12·98	13·38	13·90	14·40	14·65	14·89	15·84	16·97
1/ 35·7	53·070	58·990	62·094	65·297	72·001	79·110	82·820	94·580	111·77	130·76	140·94	151·59	199·07	269·93
0·02900	11·94	12·23	12·38	12·52	12·80	13·08	13·21	13·62	14·15	14·66	14·91	15·16	16·12	17·27
1/ 34·5	54·020	60·045	63·205	66·465	73·289	80·525	84·301	96·271	113·77	133·09	143·46	154·30	202·63	274·74
0·03000	12·15	12·44	12·59	12·73	13·02	13·30	13·44	13·85	14·39	14·91	15·17	15·42	16·40	17·57
1/ 33·3	54·954	61·083	64·298	67·614	74·555	81·916	85·757	97·933	115·74	135·39	145·93	156·96	206·12	279·48
	0·98	0·98	0·98	0·98	0·98	0·98	0·98	0·98	0·98	0·98	0·98	0·98	0·99	0·99

$V_{r(0{\cdot}5)medial}$ for half-full circular pipes.

A50
(p.1 of 6)

k_s = 0·150 mm
S = 0·00003 to 0·00010

ie hydraulic gradient =
1 in 33333 to 1 in 10000

Water (or sewage) at 15°C;
full bore conditions.

velocities in ms^{-1}
discharges in m^3s^{-1}

Gradient	**(Equivalent) Pipe diameters in mm**													
	2400	**2500**	**2550**	**2600**	**2700**	**2800**	**2850**	**3000**	**3200**	**3400**	**3500**	**3600**	**4000**	**4500**
0·0000300	0·324	0·333	0·337	0·341	0·349	0·357	0·361	0·373	0·389	0·404	0·411	0·418	0·447	0·481
1/ 33333	1·4666	1·6330	1·7204	1·8106	1·9996	2·2003	2·3052	2·6380	3·1255	3·6650	3·9548	4·2583	5·6141	7·6446
0·0000320	0·336	0·344	0·349	0·353	0·361	0·370	0·374	0·386	0·402	0·418	0·425	0·433	0·462	0·497
1/ 31250	1·5179	1·6900	1·7805	1·8738	2·0694	2·2771	2·3855	2·7298	3·2343	3·7923	4·0922	4·4061	5·8086	7·9089
0·0000340	0·347	0·356	0·360	0·364	0·373	0·382	0·386	0·399	0·415	0·431	0·439	0·447	0·477	0·513
1/ 29412	1·5676	1·7454	1·8388	1·9352	2·1371	2·3515	2·4635	2·8190	3·3397	3·9159	4·2254	4·5496	5·9973	8·1654
0·0000360	0·357	0·367	0·371	0·376	0·385	0·394	0·398	0·411	0·428	0·445	0·453	0·461	0·492	0·529
1/ 27778	1·6160	1·7993	1·8955	1·9948	2·2029	2·4239	2·5393	2·9056	3·4423	4·0360	4·3549	4·6889	6·1807	8·4146
0·0000380	0·368	0·377	0·382	0·387	0·396	0·405	0·410	0·423	0·440	0·457	0·466	0·474	0·506	0·544
1/ 26316	1·6631	1·8517	1·9507	2·0529	2·2670	2·4944	2·6131	2·9900	3·5421	4·1529	4·4810	4·8246	6·3592	8·6571
0·0000400	0·378	0·388	0·392	0·397	0·407	0·416	0·421	0·435	0·453	0·470	0·479	0·487	0·520	0·559
1/ 25000	1·7091	1·9028	2·0045	2·1095	2·3295	2·5630	2·6850	3·0722	3·6394	4·2669	4·6039	4·9569	6·5332	8·8935
0·0000420	0·388	0·398	0·403	0·408	0·418	0·427	0·432	0·446	0·464	0·482	0·491	0·500	0·533	0·574
1/ 23810	1·7539	1·9527	2·0570	2·1647	2·3904	2·6301	2·7552	3·1525	3·7344	4·3781	4·7239	5·0860	6·7031	9·1242
0·0000440	0·397	0·408	0·413	0·418	0·428	0·438	0·443	0·457	0·476	0·494	0·503	0·512	0·547	0·588
1/ 22727	1·7977	2·0014	2·1083	2·2187	2·4500	2·6956	2·8239	3·2309	3·8272	4·4868	4·8411	5·2121	6·8690	9·3497
0·0000460	0·407	0·417	0·423	0·428	0·438	0·448	0·453	0·468	0·487	0·506	0·515	0·524	0·560	0·602
1/ 21739	1·8406	2·0491	2·1586	2·2716	2·5083	2·7597	2·8910	3·3077	3·9180	4·5931	4·9557	5·3355	7·0313	9·5702
0·0000480	0·416	0·427	0·432	0·438	0·448	0·458	0·463	0·479	0·498	0·517	0·527	0·536	0·572	0·615
1/ 20833	1·8825	2·0958	2·2077	2·3233	2·5654	2·8225	2·9567	3·3828	4·0068	4·6972	5·0680	5·4563	7·1902	9·7860
0·0000500	0·425	0·436	0·442	0·447	0·458	0·468	0·474	0·489	0·509	0·529	0·538	0·548	0·585	0·629
1/ 20000	1·9237	2·1415	2·2559	2·3740	2·6213	2·8840	3·0211	3·4564	4·0940	4·7992	5·1780	5·5746	7·3459	9·9975
0·0000525	0·436	0·448	0·453	0·459	0·470	0·481	0·486	0·502	0·522	0·542	0·552	0·562	0·600	0·645
1/ 19048	1·9740	2·1975	2·3149	2·4360	2·6898	2·9592	3·0999	3·5465	4·2005	4·9239	5·3125	5·7194	7·5363	10·256
0·0000550	0·447	0·459	0·465	0·470	0·481	0·493	0·498	0·514	0·535	0·556	0·566	0·576	0·615	0·661
1/ 18182	2·0232	2·2522	2·3725	2·4966	2·7566	3·0327	3·1769	3·6344	4·3046	5·0458	5·4440	5·8608	7·7224	10·509
0·0000575	0·458	0·470	0·476	0·481	0·493	0·504	0·510	0·526	0·548	0·569	0·579	0·589	0·629	0·676
1/ 17391	2·0713	2·3057	2·4288	2·5558	2·8220	3·1046	3·2522	3·7205	4·4064	5·1651	5·5725	5·9992	7·9044	10·756
0·0000600	0·468	0·480	0·486	0·492	0·504	0·516	0·521	0·538	0·560	0·582	0·592	0·603	0·643	0·692
1/ 16667	2·1184	2·3581	2·4840	2·6139	2·8860	3·1750	3·3259	3·8048	4·5061	5·2818	5·6984	6·1346	8·0825	10·998
0·0000625	0·478	0·491	0·497	0·503	0·515	0·527	0·533	0·550	0·572	0·594	0·605	0·616	0·657	0·706
1/ 16000	2·1645	2·4095	2·5380	2·6707	2·9488	3·2440	3·3981	3·8873	4·6038	5·3962	5·8217	6·2674	8·2571	11·235
0·0000650	0·488	0·501	0·507	0·514	0·526	0·538	0·544	0·561	0·584	0·607	0·618	0·629	0·671	0·721
1/ 15385	2·2098	2·4598	2·5911	2·7265	3·0103	3·3116	3·4690	3·9683	4·6996	5·5083	5·9427	6·3975	8·4282	11·468
0·0000675	0·498	0·511	0·518	0·524	0·536	0·549	0·555	0·573	0·596	0·619	0·630	0·641	0·684	0·735
1/ 14815	2·2542	2·5092	2·6431	2·7813	3·0707	3·3780	3·5385	4·0478	4·7936	5·6184	6·0614	6·5252	8·5962	11·696
0·0000700	0·508	0·521	0·528	0·534	0·547	0·559	0·565	0·584	0·608	0·631	0·642	0·653	0·697	0·749
1/ 14286	2·2979	2·5578	2·6942	2·8350	3·1301	3·4433	3·6068	4·1258	4·8859	5·7265	6·1780	6·6507	8·7612	11·920
0·0000725	0·517	0·531	0·537	0·544	0·557	0·570	0·576	0·595	0·619	0·642	0·654	0·665	0·710	0·763
1/ 13793	2·3408	2·6055	2·7444	2·8879	3·1884	3·5074	3·6739	4·2025	4·9766	5·8328	6·2925	6·7739	8·9233	12·140
0·0000750	0·527	0·540	0·547	0·554	0·567	0·580	0·586	0·605	0·630	0·654	0·666	0·677	0·723	0·777
1/ 13333	2·3829	2·6524	2·7938	2·9398	3·2457	3·5704	3·7399	4·2780	5·0659	5·9372	6·4052	6·8951	9·0827	12·356
0·0000800	0·545	0·559	0·566	0·573	0·586	0·600	0·606	0·626	0·652	0·676	0·689	0·701	0·748	0·803
1/ 12500	2·4653	2·7440	2·8903	3·0413	3·3576	3·6934	3·8688	4·4252	5·2400	6·1412	6·6251	7·1318	9·3938	12·779
0·0000850	0·563	0·577	0·584	0·591	0·605	0·619	0·626	0·646	0·673	0·698	0·711	0·723	0·772	0·829
1/ 11765	2·5452	2·8328	2·9838	3·1397	3·4662	3·8128	3·9938	4·5681	5·4090	6·3390	6·8384	7·3612	9·6956	13·188
0·0000900	0·580	0·595	0·602	0·609	0·624	0·638	0·645	0·666	0·693	0·719	0·732	0·745	0·795	0·854
1/ 11111	2·6228	2·9192	3·0747	3·2353	3·5717	3·9288	4·1152	4·7068	5·5731	6·5311	7·0456	7·5842	9·9888	13·587
0·0000950	0·596	0·612	0·619	0·627	0·642	0·656	0·664	0·685	0·713	0·740	0·753	0·766	0·818	0·879
1/ 10526	2·6984	3·0032	3·1632	3·3284	3·6743	4·0416	4·2334	4·8419	5·7329	6·7181	7·2472	7·8012	10·274	13·974
0·0001000	0·613	0·628	0·636	0·644	0·659	0·674	0·682	0·704	0·732	0·760	0·774	0·787	0·840	0·902
1/ 10000	2·7720	3·0851	3·2494	3·4191	3·7744	4·1516	4·3486	4·9735	5·8886	6·9004	7·4437	8·0126	10·552	14·351
	0·94	0·94	0·94	0·94	0·94	0·94	0·94	0·95	0·95	0·95	0·95	0·95	0·96	0·96

$V_{r(0\cdot5)medial}$ for half-full circular pipes.

k_s = 0·150 mm S = 0·00003 to 0·00010

k_s = 0·150 mm
S = 0·00010 to 0·00030

ie hydraulic gradient =
1 in 10000 to 1 in 3333

Water (or sewage) at 15°C;
full bore conditions.

velocities in ms^{-1}
discharges in m^3s^{-1}

Gradient	**(Equivalent) Pipe diameters in mm**													
	2400	**2500**	**2550**	**2600**	**2700**	**2800**	**2850**	**3000**	**3200**	**3400**	**3500**	**3600**	**4000**	**4500**
0·000100	0·613	0·628	0·636	0·644	0·659	0·674	0·682	0·704	0·732	0·760	0·774	0·787	0·840	0·902
1/ 10000	2·7720	3·0851	3·2494	3·4191	3·7744	4·1516	4·3486	4·9735	5·8886	6·9004	7·4437	8·0126	10·552	14·351
0·000105	0·629	0·645	0·653	0·661	0·676	0·692	0·699	0·722	0·751	0·780	0·794	0·807	0·861	0·926
1/ 9524	2·8438	3·1650	3·3336	3·5076	3·8721	4·2589	4·4610	5·1020	6·0405	7·0782	7·6355	8·2189	10·823	14·720
0·000110	0·644	0·661	0·669	0·677	0·693	0·709	0·716	0·740	0·770	0·799	0·813	0·827	0·882	0·948
1/ 9091	2·9140	3·2431	3·4158	3·5940	3·9674	4·3638	4·5708	5·2274	6·1889	7·2520	7·8228	8·4205	11·088	15·079
0·000115	0·659	0·676	0·685	0·693	0·709	0·725	0·733	0·757	0·788	0·817	0·832	0·847	0·903	0·970
1/ 8696	2·9827	3·3194	3·4962	3·6786	4·0607	4·4664	4·6781	5·3501	6·3340	7·4219	8·0060	8·6176	11·347	15·431
0·000120	0·674	0·691	0·700	0·708	0·725	0·742	0·750	0·774	0·805	0·836	0·851	0·866	0·923	0·992
1/ 8333	3·0499	3·3942	3·5749	3·7614	4·1521	4·5668	4·7833	5·4703	6·4760	7·5882	8·1853	8·8105	11·601	15·775
0·000125	0·689	0·706	0·715	0·724	0·741	0·758	0·766	0·791	0·823	0·854	0·869	0·884	0·943	1·013
1/ 8000	3·1158	3·4674	3·6520	3·8425	4·2416	4·6651	4·8863	5·5879	6·6152	7·7511	8·3610	8·9995	11·849	16·113
0·000130	0·703	0·721	0·730	0·739	0·756	0·773	0·782	0·807	0·840	0·871	0·887	0·902	0·962	1·034
1/ 7692	3·1804	3·5393	3·7276	3·9221	4·3293	4·7616	4·9873	5·7033	6·7517	7·9108	8·5332	9·1848	12·093	16·443
0·000135	0·717	0·735	0·744	0·753	0·771	0·789	0·797	0·823	0·856	0·889	0·904	0·920	0·981	1·054
1/ 7407	3·2438	3·6098	3·8018	4·0001	4·4154	4·8562	5·0864	5·8166	6·8856	8·0676	8·7022	9·3667	12·332	16·768
0·000140	0·731	0·749	0·759	0·768	0·786	0·804	0·813	0·839	0·873	0·906	0·922	0·938	1·000	1·074
1/ 7143	3·3060	3·6790	3·8747	4·0768	4·5000	4·9492	5·1837	5·9278	7·0171	8·2215	8·8682	9·5452	12·567	17·086
0·000145	0·744	0·763	0·773	0·782	0·800	0·819	0·828	0·854	0·889	0·922	0·939	0·955	1·018	1·094
1/ 6897	3·3672	3·7470	3·9463	4·1521	4·5830	5·0405	5·2793	6·0371	7·1463	8·3728	9·0313	9·7207	12·798	17·400
0·000150	0·758	0·777	0·787	0·796	0·815	0·833	0·842	0·869	0·904	0·939	0·955	0·972	1·036	1·113
1/ 6667	3·4273	3·8139	4·0167	4·2261	4·6647	5·1303	5·3733	6·1445	7·2734	8·5215	9·1916	9·8932	13·024	17·707
0·000160	0·784	0·804	0·813	0·823	0·843	0·862	0·871	0·899	0·935	0·971	0·988	1·005	1·072	1·151
1/ 6250	3·5447	3·9444	4·1542	4·3707	4·8242	5·3055	5·5569	6·3542	7·5214	8·8118	9·5046	10·230	13·467	18·308
0·000170	0·809	0·829	0·840	0·850	0·870	0·889	0·899	0·928	0·965	1·002	1·019	1·037	1·106	1·188
1/ 5882	3·6585	4·0710	4·2875	4·5109	4·9788	5·4755	5·7348	6·5575	7·7618	9·0932	9·8080	10·556	13·896	18·890
0·000180	0·833	0·854	0·865	0·875	0·896	0·916	0·926	0·956	0·994	1·032	1·050	1·068	1·139	1·223
1/ 5556	3·7691	4·1940	4·4169	4·6471	5·1291	5·6406	5·9077	6·7551	7·9954	9·3666	10·103	10·873	14·313	19·456
0·000190	0·857	0·879	0·890	0·900	0·921	0·942	0·952	0·983	1·022	1·061	1·080	1·099	1·171	1·258
1/ 5263	3·8767	4·3136	4·5429	4·7796	5·2752	5·8013	6·0759	6·9473	8·2227	9·6326	10·390	11·182	14·718	20·007
0·000200	0·880	0·903	0·914	0·925	0·946	0·968	0·978	1·009	1·050	1·090	1·109	1·128	1·203	1·292
1/ 5000	3·9816	4·4303	4·6657	4·9087	5·4177	5·9579	6·2399	7·1345	8·4441	9·8918	10·669	11·483	15·113	20·543
0·000210	0·903	0·926	0·937	0·948	0·970	0·992	1·003	1·035	1·077	1·117	1·137	1·157	1·233	1·325
1/ 4762	4·0839	4·5440	4·7855	5·0347	5·5566	6·1106	6·3998	7·3173	8·6602	10·145	10·942	11·776	15·499	21·066
0·000220	0·925	0·948	0·960	0·971	0·994	1·017	1·028	1·060	1·103	1·145	1·165	1·185	1·263	1·357
1/ 4545	4·1839	4·6552	4·9025	5·1578	5·6924	6·2598	6·5560	7·4957	8·8712	10·392	11·208	12·062	15·875	21·577
0·000230	0·946	0·970	0·982	0·994	1·017	1·040	1·052	1·085	1·129	1·171	1·192	1·213	1·293	1·388
1/ 4348	4·2816	4·7638	5·0169	5·2781	5·8251	6·4057	6·7087	7·6702	9·0776	10·633	11·468	12·343	16·244	22·076
0·000240	0·968	0·992	1·004	1·016	1·040	1·063	1·075	1·109	1·154	1·197	1·218	1·240	1·321	1·419
1/ 4167	4·3773	4·8702	5·1289	5·3959	5·9550	6·5485	6·8583	7·8411	9·2795	10·870	11·723	12·617	16·604	22·565
0·000250	0·988	1·013	1·026	1·038	1·062	1·086	1·098	1·133	1·178	1·223	1·244	1·266	1·349	1·449
1/ 4000	4·4710	4·9744	5·2386	5·5113	6·0823	6·6883	7·0047	8·0084	9·4774	11·101	11·973	12·885	16·957	23·044
0·000260	1·009	1·034	1·047	1·059	1·084	1·108	1·121	1·156	1·203	1·248	1·270	1·292	1·377	1·478
1/ 3846	4·5629	5·0766	5·3462	5·6244	6·2071	6·8255	7·1483	8·1724	9·6713	11·328	12·218	13·149	17·303	23·514
0·000270	1·029	1·055	1·067	1·080	1·105	1·130	1·143	1·179	1·226	1·272	1·295	1·317	1·404	1·507
1/ 3704	4·6530	5·1769	5·4517	5·7354	6·3295	6·9601	7·2892	8·3334	9·8616	11·551	12·458	13·407	17·642	23·974
0·000280	1·048	1·075	1·088	1·101	1·126	1·152	1·164	1·201	1·249	1·296	1·319	1·342	1·430	1·536
1/ 3571	4·7416	5·2753	5·5554	5·8444	6·4498	7·0922	7·4276	8·4914	10·049	11·770	12·693	13·661	17·976	24·427
0·000290	1·067	1·094	1·108	1·121	1·147	1·173	1·186	1·223	1·272	1·320	1·343	1·367	1·457	1·564
1/ 3448	4·8286	5·3720	5·6572	5·9515	6·5679	7·2220	7·5635	8·6467	10·232	11·984	12·925	13·910	18·303	24·871
0·000300	1·086	1·114	1·127	1·141	1·167	1·194	1·207	1·245	1·295	1·343	1·367	1·391	1·482	1·591
1/ 3333	4·9141	5·4671	5·7573	6·0568	6·6840	7·3496	7·6971	8·7994	10·413	12·196	13·153	14·155	18·625	25·308
	0·95	0·95	0·95	0·95	0·95	0·95	0·95	0·96	0·96	0·96	0·96	0·96	0·96	0·97

$V_{r(0\cdot5)medial}$ for half-full circular pipes.

k_s = **0·150 mm**
S = 0·00030 to 0·00100

ie hydraulic gradient =
1 in 3333 to 1 in 1000

Water (or sewage) at 15°C;
full bore conditions.

velocities in ms^{-1}
discharges in m^3s^{-1}

Gradient	(Equivalent) Pipe diameters in mm													
	2400	**2500**	**2550**	**2600**	**2700**	**2800**	**2850**	**3000**	**3200**	**3400**	**3500**	**3600**	**4000**	**4500**
0·000300	1·086	1·114	1·127	1·141	1·167	1·194	1·207	1·245	1·295	1·343	1·367	1·391	1·482	1·591
1/ 3333	4·9141	5·4671	5·7573	6·0568	6·6840	7·3496	7·6971	8·7994	10·413	12·196	13·153	14·155	18·625	25·308
0·000320	1·123	1·152	1·166	1·179	1·207	1·234	1·247	1·287	1·338	1·389	1·413	1·438	1·532	1·645
1/ 3125	5·0810	5·6527	5·9527	6·2623	6·9107	7·5987	7·9579	9·0973	10·765	12·608	13·597	14·633	19·253	26·160
0·000340	1·159	1·188	1·203	1·217	1·245	1·273	1·287	1·328	1·381	1·433	1·458	1·483	1·581	1·697
1/ 2941	5·2428	5·8326	6·1421	6·4615	7·1304	7·8402	8·2108	9·3862	11·106	13·008	14·028	15·096	19·862	26·987
0·000360	1·194	1·224	1·239	1·253	1·283	1·311	1·326	1·368	1·422	1·475	1·502	1·527	1·628	1·747
1/ 2778	5·4000	6·0074	6·3261	6·6550	7·3438	8·0748	8·4563	9·6667	11·438	13·396	14·447	15·546	20·454	27·789
0·000380	1·227	1·258	1·274	1·289	1·319	1·348	1·363	1·406	1·462	1·517	1·544	1·570	1·673	1·796
1/ 2632	5·5530	6·1774	6·5051	6·8433	7·5515	8·3030	8·6953	9·9397	11·761	13·773	14·854	15·984	21·029	28·570
0·000400	1·260	1·292	1·308	1·323	1·354	1·385	1·400	1·444	1·501	1·558	1·585	1·612	1·718	1·844
1/ 2500	5·7020	6·3431	6·6795	7·0267	7·7538	8·5253	8·9281	10·206	12·075	14·141	15·250	16·411	21·590	29·331
0·000420	1·293	1·325	1·341	1·357	1·389	1·420	1·435	1·480	1·540	1·597	1·625	1·653	1·762	1·891
1/ 2381	5·8473	6·5047	6·8497	7·2057	7·9511	8·7422	9·1551	10·465	12·382	14·500	15·637	16·827	22·137	30·073
0·000440	1·324	1·357	1·374	1·390	1·422	1·454	1·470	1·516	1·577	1·636	1·665	1·693	1·804	1·936
1/ 2273	5·9893	6·6626	7·0159	7·3804	8·1439	8·9540	9·3769	10·718	12·681	14·851	16·015	17·234	22·671	30·797
0·000460	1·355	1·389	1·406	1·422	1·455	1·488	1·504	1·551	1·613	1·673	1·703	1·732	1·846	1·981
1/ 2174	6·1281	6·8169	7·1783	7·5513	8·3324	9·1611	9·5938	10·966	12·974	15·193	16·384	17·631	23·193	31·506
0·000480	1·385	1·419	1·437	1·454	1·488	1·521	1·537	1·586	1·649	1·710	1·741	1·770	1·886	2·025
1/ 2083	6·2640	6·9680	7·3373	7·7185	8·5168	9·3638	9·8060	11·208	13·261	15·529	16·746	18·020	23·704	32·199
0·000500	1·414	1·450	1·467	1·485	1·519	1·553	1·570	1·619	1·684	1·747	1·777	1·808	1·926	2·067
1/ 2000	6·3971	7·1159	7·4931	7·8823	8·6975	9·5624	10·014	11·446	13·541	15·857	17·100	18·401	24·204	32·878
0·000525	1·450	1·486	1·504	1·522	1·558	1·592	1·610	1·660	1·726	1·791	1·822	1·854	1·975	2·119
1/ 1905	6·5598	7·2968	7·6835	8·0826	8·9183	9·8051	10·268	11·736	13·885	16·259	17·533	18·867	24·816	33·708
0·000550	1·485	1·522	1·541	1·559	1·595	1·631	1·648	1·700	1·768	1·834	1·866	1·898	2·022	2·170
1/ 1818	6·7187	7·4735	7·8695	8·2782	9·1340	10·042	10·516	12·020	14·220	16·651	17·956	19·321	25·414	34·518
0·000575	1·519	1·558	1·577	1·595	1·632	1·669	1·686	1·740	1·809	1·876	1·909	1·942	2·069	2·220
1/ 1739	6·8740	7·6462	8·0513	8·4694	9·3449	10·274	10·759	12·297	14·547	17·034	18·369	19·766	25·998	35·311
0·000600	1·553	1·592	1·611	1·630	1·668	1·705	1·724	1·778	1·849	1·918	1·951	1·985	2·114	2·269
1/ 1667	7·0261	7·8152	8·2293	8·6566	9·5513	10·501	10·996	12·568	14·868	17·409	18·773	20·201	26·570	36·086
0·000625	1·586	1·626	1·645	1·665	1·704	1·741	1·760	1·816	1·888	1·958	1·992	2·027	2·159	2·317
1/ 1600	7·1750	7·9808	8·4036	8·8399	9·7535	10·723	11·229	12·834	15·182	17·777	19·170	20·627	27·130	36·846
0·000650	1·618	1·659	1·679	1·699	1·738	1·777	1·796	1·852	1·926	1·998	2·033	2·068	2·203	2·364
1/ 1538	7·3210	8·1431	8·5745	9·0196	9·9517	10·941	11·457	13·094	15·490	18·137	19·558	21·045	27·678	37·591
0·000675	1·650	1·691	1·712	1·732	1·772	1·812	1·831	1·889	1·964	2·037	2·072	2·108	2·245	2·409
1/ 1481	7·4642	8·3024	8·7421	9·1959	10·146	11·154	11·681	13·350	15·792	18·491	19·939	21·455	28·217	38·321
0·000700	1·681	1·723	1·744	1·765	1·805	1·846	1·865	1·924	2·000	2·075	2·111	2·147	2·288	2·455
1/ 1429	7·6049	8·4587	8·9067	9·3690	10·337	11·364	11·900	13·601	16·089	18·838	20·313	21·857	28·746	39·038
0·000725	1·712	1·754	1·776	1·797	1·838	1·879	1·899	1·959	2·037	2·112	2·150	2·186	2·329	2·499
1/ 1379	7·7430	8·6123	9·0684	9·5390	10·525	11·570	12·116	13·847	16·380	19·179	20·681	22·253	29·265	39·743
0·000750	1·742	1·785	1·807	1·828	1·870	1·912	1·932	1·993	2·072	2·149	2·187	2·224	2·369	2·542
1/ 1333	7·8788	8·7633	9·2274	9·7062	10·709	11·773	12·328	14·089	16·666	19·514	21·042	22·641	29·775	40·435
0·000800	1·800	1·845	1·868	1·890	1·933	1·976	1·997	2·060	2·142	2·221	2·260	2·299	2·449	2·627
1/ 1250	8·1439	9·0580	9·5376	10·032	11·069	12·168	12·742	14·562	17·225	20·168	21·747	23·400	30·772	41·787
0·000850	1·857	1·903	1·926	1·949	1·994	2·038	2·060	2·125	2·209	2·291	2·331	2·371	2·526	2·710
1/ 1176	8·4008	9·3436	9·8383	10·349	11·417	12·551	13·143	15·020	17·767	20·802	22·430	24·135	31·737	43·097
0·000900	1·912	1·960	1·984	2·007	2·053	2·099	2·121	2·188	2·275	2·359	2·400	2·441	2·600	2·790
1/ 1111	8·6504	9·6210	10·130	10·656	11·756	12·924	13·533	15·466	18·293	21·417	23·094	24·849	32·675	44·369
0·000950	1·966	2·015	2·039	2·063	2·111	2·158	2·181	2·249	2·338	2·425	2·467	2·509	2·673	2·868
1/ 1053	8·8931	9·8909	10·414	10·955	12·086	13·286	13·912	15·899	18·805	22·016	23·739	25·543	33·587	45·606
0·001000	2·018	2·069	2·093	2·118	2·167	2·215	2·239	2·309	2·400	2·489	2·533	2·576	2·743	2·943
1/ 1000	9·1296	10·154	10·691	11·246	12·407	13·638	14·281	16·320	19·303	22·599	24·368	26·219	34·476	46·811
	0·95	0·96	0·96	0·96	0·96	0·96	0·96	0·96	0·96	0·96	0·97	0·97	0·97	0·97

$V_{r(0\cdot5)medial}$ for half-full circular pipes.

k_s = 0·150 mm S = 0·00030 to 0·00100

Gradient	**(Equivalent) Pipe diameters in mm**													
	2400	**2500**	**2550**	**2600**	**2700**	**2800**	**2850**	**3000**	**3200**	**3400**	**3500**	**3600**	**4000**	**4500**
0·00100	2·018	2·069	2·093	2·118	2·167	2·215	2·239	2·309	2·400	2·489	2·533	2·576	2·743	2·943
1/ 1000	9·1296	10·154	10·691	11·246	12·407	13·638	14·281	16·320	19·303	22·599	24·368	26·219	34·476	46·811
0·00105	2·069	2·121	2·146	2·172	2·222	2·271	2·295	2·367	2·461	2·552	2·597	2·641	2·812	3·017
1/ 952	9·3602	10·410	10·961	11·529	12·720	13·982	14·641	16·732	19·790	23·168	24·981	26·879	35·342	47·987
0·00110	2·119	2·172	2·198	2·224	2·275	2·325	2·350	2·424	2·520	2·613	2·659	2·704	2·880	3·089
1/ 909	9·5855	10·661	11·225	11·807	13·025	14·318	14·993	17·133	20·264	23·724	25·580	27·523	36·189	49·135
0·00115	2·168	2·222	2·248	2·275	2·327	2·379	2·404	2·479	2·577	2·673	2·720	2·766	2·946	3·160
1/ 870	9·8057	10·905	11·482	12·078	13·324	14·647	15·337	17·526	20·729	24·267	26·166	28·153	37·016	50·257
0·00120	2·215	2·270	2·298	2·325	2·378	2·431	2·457	2·534	2·634	2·731	2·779	2·826	3·010	3·229
1/ 833	10·021	11·145	11·735	12·343	13·617	14·968	15·673	17·910	21·183	24·798	26·739	28·769	37·826	51·355
0·00125	2·262	2·318	2·346	2·374	2·428	2·482	2·509	2·587	2·689	2·789	2·837	2·886	3·073	3·297
1/ 800	10·232	11·380	11·982	12·603	13·903	15·283	16·003	18·287	21·628	25·319	27·300	29·373	38·619	52·431
0·00130	2·308	2·365	2·393	2·422	2·477	2·532	2·559	2·639	2·743	2·845	2·895	2·944	3·135	3·363
1/ 769	10·439	11·610	12·224	12·857	14·184	15·591	16·326	18·656	22·064	25·829	27·850	29·965	39·396	53·485
0·00135	2·352	2·411	2·440	2·469	2·525	2·581	2·609	2·690	2·797	2·900	2·951	3·001	3·196	3·428
1/ 741	10·642	11·835	12·461	13·107	14·459	15·894	16·643	19·018	22·492	26·330	28·390	30·545	40·159	54·519
0·00140	2·396	2·456	2·486	2·515	2·573	2·630	2·658	2·741	2·849	2·954	3·006	3·057	3·255	3·492
1/ 714	10·841	12·057	12·694	13·352	14·730	16·191	16·954	19·373	22·912	26·821	28·919	31·115	40·907	55·535
0·00145	2·440	2·501	2·531	2·560	2·619	2·677	2·706	2·790	2·900	3·007	3·060	3·112	3·314	3·555
1/ 690	11·037	12·274	12·924	13·593	14·996	16·483	17·260	19·722	23·325	27·304	29·440	31·675	41·642	56·532
0·00150	2·482	2·544	2·575	2·605	2·665	2·724	2·753	2·839	2·951	3·060	3·113	3·166	3·371	3·616
1/ 667	11·230	12·488	13·149	13·830	15·257	16·770	17·560	20·065	23·730	27·779	29·951	32·225	42·365	57·513
0·00160	2·565	2·629	2·661	2·692	2·754	2·814	2·845	2·933	3·049	3·162	3·217	3·271	3·484	3·736
1/ 625	11·605	12·906	13·588	14·292	15·766	17·330	18·146	20·735	24·522	28·705	30·950	33·299	43·776	59·426
0·00170	2·646	2·712	2·744	2·776	2·840	2·903	2·934	3·025	3·144	3·260	3·317	3·374	3·592	3·853
1/ 588	11·969	13·310	14·014	14·740	16·260	17·873	18·715	21·384	25·289	29·602	31·917	34·340	45·143	61·280
0·00180	2·724	2·792	2·825	2·858	2·924	2·988	3·020	3·114	3·237	3·356	3·415	3·473	3·698	3·966
1/ 556	12·323	13·703	14·428	15·175	16·740	18·400	19·267	22·014	26·034	30·474	32·857	35·350	46·470	63·081
0·00190	2·800	2·869	2·904	2·938	3·005	3·072	3·104	3·201	3·327	3·450	3·510	3·570	3·801	4·076
1/ 526	12·666	14·086	14·830	15·598	17·207	18·913	19·803	22·627	26·758	31·322	33·771	36·333	47·762	64·832
0·00200	2·874	2·945	2·981	3·016	3·085	3·153	3·186	3·286	3·415	3·541	3·603	3·664	3·901	4·184
1/ 500	13·001	14·458	15·222	16·010	17·661	19·412	20·326	23·225	27·464	32·148	34·661	37·291	49·020	66·538
0·00210	2·946	3·019	3·055	3·091	3·162	3·232	3·266	3·368	3·500	3·630	3·693	3·755	3·998	4·288
1/ 476	13·328	14·821	15·604	16·412	18·104	19·899	20·836	23·807	28·153	32·953	35·529	38·225	50·246	68·202
0·00220	3·017	3·092	3·129	3·165	3·238	3·309	3·344	3·448	3·584	3·716	3·781	3·845	4·094	4·390
1/ 455	13·647	15·176	15·978	16·805	18·537	20·375	21·334	24·376	28·825	33·740	36·377	39·137	51·444	69·827
0·00230	3·086	3·162	3·200	3·237	3·312	3·384	3·421	3·527	3·666	3·801	3·867	3·933	4·187	4·490
1/ 435	13·959	15·522	16·342	17·189	18·960	20·840	21·821	24·932	29·482	34·509	37·206	40·029	52·615	71·415
0·00240	3·153	3·231	3·270	3·308	3·384	3·458	3·495	3·604	3·746	3·884	3·951	4·018	4·278	4·588
1/ 417	14·264	15·861	16·700	17·564	19·375	21·295	22·297	25·476	30·125	35·261	38·017	40·901	53·761	72·969
0·00250	3·219	3·299	3·338	3·377	3·455	3·531	3·568	3·679	3·824	3·965	4·034	4·102	4·367	4·684
1/ 400	14·563	16·194	17·049	17·932	19·780	21·741	22·764	26·009	30·755	35·998	38·811	41·755	54·884	74·491
0·00260	3·284	3·365	3·406	3·445	3·524	3·602	3·640	3·753	3·901	4·044	4·115	4·184	4·455	4·778
1/ 385	14·856	16·519	17·392	18·293	20·178	22·178	23·221	26·531	31·372	36·720	39·590	42·593	55·984	75·983
0·00270	3·347	3·430	3·471	3·512	3·592	3·671	3·710	3·826	3·976	4·122	4·194	4·265	4·541	4·870
1/ 370	15·143	16·839	17·728	18·646	20·568	22·606	23·670	27·044	31·978	37·429	40·354	43·415	57·063	77·447
0·00280	3·410	3·494	3·536	3·577	3·659	3·740	3·779	3·897	4·050	4·199	4·272	4·344	4·625	4·960
1/ 357	15·425	17·153	18·059	18·993	20·951	23·027	24·110	27·547	32·573	38·124	41·104	44·221	58·122	78·884
0·00290	3·471	3·557	3·600	3·642	3·725	3·807	3·847	3·967	4·123	4·274	4·349	4·422	4·708	5·049
1/ 345	15·703	17·461	18·383	19·334	21·327	23·440	24·543	28·041	33·157	38·808	41·840	45·013	59·163	80·295
0·00300	3·531	3·619	3·662	3·705	3·789	3·873	3·914	4·036	4·194	4·348	4·424	4·499	4·789	5·136
1/ 333	15·975	17·764	18·702	19·670	21·696	23·846	24·968	28·526	33·731	39·479	42·564	45·792	60·186	81·683
	0·96	0·96	0·96	0·96	0·96	0·96	0·96	0·96	0·97	0·97	0·97	0·97	0·97	0·97

$\mathbf{V_{r(0{\cdot}5)medial}}$ **for half-full circular pipes.**

k_s = 0·150 mm
S = 0·00300 to 0·01000

ie hydraulic gradient =
1 in 333 to 1 in 100

Water (or sewage) at 15°C;
full bore conditions.

velocities in ms^{-1}
discharges in m^3s^{-1}

Gradient **(Equivalent) Pipe diameters in mm**

Gradient	2400	2500	2550	2600	2700	2800	2850	3000	3200	3400	3500	3600	4000	4500
0·00300	3·531	3·619	3·662	3·705	3·789	3·873	3·914	4·036	4·194	4·348	4·424	4·499	4·789	5·136
1/ 333	15·975	17·764	18·702	19·670	21·696	23·846	24·968	28·526	33·731	39·479	42·564	45·792	60·186	81·683
0·00320	3·649	3·739	3·784	3·828	3·915	4·001	4·044	4·170	4·333	4·493	4·571	4·648	4·948	5·306
1/ 313	16·507	18·354	19·324	20·324	22·418	24·639	25·798	29·474	34·851	40·790	43·977	47·312	62·182	84·390
0·00340	3·763	3·856	3·902	3·947	4·038	4·126	4·170	4·300	4·468	4·633	4·713	4·793	5·102	5·471
1/ 294	17·022	18·927	19·927	20·958	23·117	25·407	26·603	30·393	35·937	42·060	45·346	48·785	64·117	87·014
0·00360	3·873	3·969	4·016	4·063	4·156	4·247	4·293	4·426	4·599	4·768	4·851	4·933	5·252	5·631
1/ 278	17·522	19·483	20·512	21·574	23·796	26·153	27·384	31·285	36·991	43·294	46·676	50·215	65·995	89·561
0·00380	3·981	4·079	4·128	4·176	4·271	4·365	4·412	4·549	4·727	4·901	4·986	5·070	5·397	5·787
1/ 263	18·009	20·024	21·082	22·173	24·456	26·879	28·143	32·153	38·017	44·494	47·969	51·606	67·822	92·039
0·00400	4·086	4·187	4·237	4·286	4·384	4·480	4·528	4·668	4·851	5·029	5·117	5·203	5·539	5·939
1/ 250	18·483	20·551	21·636	22·756	25·100	27·586	28·883	32·998	39·015	45·662	49·229	52·961	69·602	94·453
0·00420	4·188	4·291	4·343	4·393	4·493	4·592	4·641	4·785	4·972	5·155	5·244	5·333	5·677	6·087
1/ 238	18·945	21·065	22·177	23·325	25·727	28·275	29·605	33·822	39·990	46·802	50·457	54·283	71·338	96·807
0·00440	4·288	4·394	4·446	4·498	4·600	4·701	4·751	4·899	5·091	5·277	5·369	5·460	5·812	6·231
1/ 227	19·397	21·567	22·706	23·880	26·340	28·948	30·310	34·627	40·941	47·915	51·657	55·573	73·033	99·106
0·00460	4·385	4·494	4·547	4·600	4·705	4·808	4·859	5·010	5·206	5·397	5·491	5·584	5·944	6·373
1/ 217	19·838	22·058	23·222	24·423	26·939	29·606	30·999	35·414	41·871	49·003	52·830	56·834	74·690	101·35
0·00480	4·481	4·591	4·646	4·700	4·807	4·913	4·965	5·119	5·319	5·515	5·610	5·705	6·073	6·511
1/ 208	20·270	22·538	23·728	24·955	27·525	30·250	31·673	36·184	42·781	50·068	53·977	58·069	76·311	103·55
0·00500	4·574	4·687	4·743	4·798	4·908	5·015	5·068	5·226	5·430	5·629	5·727	5·824	6·199	6·646
1/ 200	20·693	23·008	24·222	25·475	28·099	30·881	32·333	36·938	43·672	51·110	55·101	59·278	77·899	105·71
0·00525	4·688	4·804	4·861	4·918	5·030	5·140	5·195	5·356	5·566	5·770	5·870	5·969	6·353	6·812
1/ 190	21·210	23·583	24·827	26·112	28·800	31·652	33·140	37·859	44·761	52·385	56·475	60·756	79·840	108·34
0·00550	4·800	4·919	4·977	5·035	5·150	5·263	5·318	5·483	5·698	5·907	6·009	6·111	6·504	6·973
1/ 182	21·715	24·144	25·418	26·733	29·485	32·405	33·928	38·759	45·825	53·629	57·817	62·198	81·735	110·91
0·00575	4·909	5·030	5·090	5·149	5·267	5·382	5·439	5·608	5·827	6·041	6·146	6·249	6·652	7·131
1/ 174	22·209	24·693	25·996	27·340	30·155	33·140	34·699	39·639	46·865	54·846	59·128	63·609	83·588	113·42
0·00600	5·016	5·140	5·201	5·261	5·381	5·499	5·557	5·730	5·954	6·172	6·279	6·385	6·796	7·286
1/ 167	22·691	25·229	26·561	27·934	30·810	33·860	35·452	40·500	47·882	56·036	60·411	64·989	85·400	115·88
0·00625	5·120	5·247	5·309	5·371	5·493	5·614	5·673	5·849	6·078	6·300	6·410	6·518	6·937	7·437
1/ 160	23·164	25·755	27·114	28·516	31·452	34·566	36·191	41·343	48·879	57·202	61·668	66·341	87·176	118·29
0·00650	5·223	5·352	5·415	5·478	5·603	5·726	5·786	5·966	6·199	6·426	6·538	6·648	7·076	7·586
1/ 154	23·628	26·271	27·657	29·087	32·081	35·257	36·914	42·170	49·856	58·345	62·900	67·666	88·916	120·65
0·00675	5·323	5·455	5·520	5·584	5·711	5·836	5·898	6·080	6·318	6·550	6·663	6·775	7·212	7·731
1/ 148	24·083	26·776	28·189	29·647	32·698	35·935	37·624	42·980	50·814	59·466	64·108	68·966	90·624	122·96
0·00700	5·422	5·556	5·622	5·687	5·817	5·944	6·007	6·193	6·435	6·671	6·787	6·901	7·345	7·874
1/ 143	24·529	27·272	28·711	30·196	33·304	36·601	38·321	43·776	51·755	60·566	65·294	70·242	92·299	125·24
0·00725	5·519	5·655	5·722	5·789	5·921	6·050	6·114	6·304	6·550	6·790	6·908	7·024	7·476	8·015
1/ 138	24·968	27·760	29·225	30·736	33·899	37·254	39·006	44·558	52·679	61·647	66·460	71·495	93·946	127·47
0·00750	5·614	5·753	5·821	5·889	6·023	6·155	6·220	6·412	6·663	6·907	7·027	7·145	7·605	8·153
1/ 133	25·399	28·239	29·729	31·266	34·484	37·897	39·678	45·327	53·587	62·710	67·605	72·727	95·564	129·66
0·00800	5·800	5·943	6·014	6·084	6·222	6·358	6·426	6·625	6·883	7·135	7·259	7·381	7·856	8·422
1/ 125	26·240	29·174	30·713	32·301	35·625	39·151	40·991	46·826	55·359	64·783	69·840	75·131	98·721	133·94
0·00850	5·981	6·128	6·201	6·273	6·415	6·556	6·625	6·830	7·097	7·357	7·484	7·610	8·099	8·683
1/ 118	27·055	30·080	31·667	33·304	36·731	40·367	42·264	48·279	57·077	66·793	72·006	77·461	101·78	138·09
0·00900	6·156	6·307	6·382	6·456	6·603	6·747	6·819	7·030	7·304	7·572	7·703	7·832	8·336	8·936
1/ 111	27·847	30·960	32·593	34·278	37·806	41·547	43·499	49·690	58·744	68·744	74·109	79·723	104·75	142·12
0·00950	6·326	6·482	6·558	6·635	6·785	6·934	7·007	7·224	7·506	7·781	7·915	8·048	8·566	9·183
1/ 105	28·617	31·816	33·494	35·226	38·850	42·695	44·701	51·063	60·367	70·642	76·155	81·924	107·64	146·04
0·01000	6·492	6·651	6·730	6·809	6·963	7·115	7·191	7·413	7·702	7·984	8·122	8·259	8·790	9·423
1/ 100	29·367	32·650	34·372	36·149	39·868	43·813	45·872	52·400	61·947	72·490	78·147	84·067	110·46	149·86
	0·96	0·96	0·96	0·96	0·96	0·96	0·96	0·97	0·97	0·97	0·97	0·97	0·97	0·97

$V_{r(0·5)medial}$ for half-full circular pipes.

k_s = 0·150 mm S = 0·00300 to 0·01000

Gradient	(Equivalent) Pipe diameters in mm													
	2400	**2500**	**2550**	**2600**	**2700**	**2800**	**2850**	**3000**	**3200**	**3400**	**3500**	**3600**	**4000**	**4500**
0·01000	6·492	6·651	6·730	6·809	6·963	7·115	7·191	7·413	7·702	7·984	8·122	8·259	8·790	9·423
1/ 100	29·367	32·650	34·372	36·149	39·868	43·813	45·872	52·400	61·947	72·490	78·147	84·067	110·46	149·86
0·01050	6·653	6·817	6·898	6·978	7·136	7·292	7·370	7·598	7·894	8·183	8·324	8·464	9·008	9·657
1/ 95	30·098	33·463	35·228	37·049	40·860	44·904	47·013	53·704	63·488	74·293	80·091	86·157	113·20	153·59
0·01100	6·811	6·979	7·062	7·144	7·306	7·465	7·544	7·778	8·081	8·377	8·522	8·665	9·222	9·885
1/ 91	30·813	34·257	36·064	37·927	41·830	45·968	48·128	54·977	64·992	76·054	81·988	88·198	115·88	157·22
0·01150	6·965	7·137	7·221	7·305	7·471	7·634	7·715	7·954	8·264	8·566	8·714	8·861	9·430	10·11
1/ 87	31·511	35·033	36·880	38·787	42·777	47·009	49·218	56·221	66·463	77·775	83·843	90·194	118·50	160·77
0·01200	7·116	7·292	7·378	7·464	7·633	7·800	7·882	8·126	8·443	8·752	8·903	9·053	9·634	10·33
1/ 83	32·194	35·792	37·680	39·627	43·704	48·028	50·284	57·439	67·903	79·458	85·658	92·146	121·07	164·25
0·01250	7·264	7·443	7·531	7·619	7·792	7·962	8·046	8·295	8·618	8·933	9·088	9·241	9·834	10·54
1/ 80	32·863	36·536	38·463	40·451	44·612	49·025	51·329	58·632	69·312	81·108	87·436	94·058	123·58	167·66
0·01300	7·409	7·592	7·682	7·771	7·947	8·121	8·206	8·460	8·790	9·111	9·269	9·425	10·03	10·75
1/ 77	33·519	37·265	39·230	41·258	45·502	50·003	52·352	59·801	70·694	82·724	89·178	95·932	126·04	171·00
0·01350	7·552	7·737	7·829	7·920	8·100	8·276	8·364	8·622	8·959	9·286	9·447	9·605	10·22	10·96
1/ 74	34·163	37·980	39·983	42·049	46·375	50·962	53·357	60·948	72·049	84·310	90·888	97·771	128·46	174·27
0·01400	7·691	7·880	7·974	8·066	8·249	8·429	8·518	8·782	9·124	9·457	9·621	9·783	10·41	11·16
1/ 71	34·794	38·682	40·722	42·826	47·232	51·904	54·342	62·074	73·380	85·866	92·566	99·576	130·83	177·49
0·01450	7·828	8·021	8·116	8·210	8·396	8·580	8·670	8·938	9·287	9·626	9·792	9·957	10·60	11·36
1/ 69	35·415	39·372	41·448	43·590	48·073	52·829	55·311	63·180	74·687	87·396	94·214	101·35	133·16	180·64
0·01500	7·963	8·159	8·256	8·351	8·541	8·727	8·819	9·092	9·446	9·791	9·961	10·13	10·78	11·55
1/ 67	36·024	40·050	42·162	44·340	48·901	53·738	56·262	64·267	75·972	88·899	95·834	103·09	135·44	183·75
0·01600	8·226	8·428	8·528	8·627	8·823	9·015	9·111	9·392	9·758	10·11	10·29	10·46	11·13	11·93
1/ 63	37·214	41·373	43·554	45·804	50·515	55·512	58·120	66·387	78·478	91·831	98·995	106·49	139·91	189·80
0·01700	8·481	8·689	8·792	8·894	9·096	9·295	9·393	9·683	10·06	10·43	10·61	10·79	11·48	12·30
1/ 59	38·367	42·654	44·903	47·223	52·080	57·231	59·920	68·443	80·908	94·673	102·06	109·79	144·24	195·67
0·01800	8·729	8·943	9·049	9·154	9·361	9·566	9·667	9·965	10·35	10·73	10·92	11·10	11·81	12·66
1/ 56	39·487	43·899	46·213	48·601	53·599	58·901	61·667	70·439	83·267	97·433	105·03	112·99	148·44	201·37
0·01900	8·969	9·190	9·298	9·406	9·620	9·829	9·933	10·24	10·64	11·03	11·22	11·41	12·14	13·01
1/ 53	40·576	45·110	47·488	49·941	55·077	60·525	63·367	72·381	85·561	100·12	107·93	116·10	152·53	206·92
0·02000	9·204	9·430	9·542	9·652	9·871	10·09	10·19	10·51	10·92	11·32	11·51	11·70	12·45	13·35
1/ 50	41·637	46·289	48·729	51·246	56·516	62·106	65·023	74·271	87·796	102·73	110·75	119·13	156·51	212·32
0·02100	9·433	9·664	9·779	9·892	10·12	10·34	10·45	10·77	11·19	11·60	11·80	11·99	12·76	13·68
1/ 47·6	42·672	47·439	49·940	52·520	57·920	63·649	66·638	76·116	89·976	105·28	113·49	122·09	160·39	217·58
0·02200	9·656	9·893	10·01	10·13	10·36	10·58	10·69	11·02	11·45	11·87	12·08	12·28	13·07	14·00
1/ 45·5	43·682	48·562	51·122	53·763	59·291	65·155	68·215	77·917	92·104	107·77	116·18	124·97	164·19	222·73
0·02300	9·874	10·12	10·24	10·36	10·59	10·82	10·93	11·27	11·71	12·14	12·35	12·56	13·36	14·32
1/ 43·5	44·670	49·660	52·277	54·978	60·631	66·627	69·756	79·677	94·185	110·21	118·80	127·80	167·89	227·75
0·02400	10·09	10·34	10·46	10·58	10·82	11·05	11·17	11·52	11·96	12·40	12·61	12·83	13·65	14·63
1/ 41·7	45·636	50·734	53·408	56·167	61·942	68·068	71·264	81·399	96·220	112·59	121·37	130·56	171·52	232·67
0·02500	10·30	10·55	10·67	10·80	11·04	11·28	11·40	11·75	12·21	12·66	12·88	13·09	13·93	14·93
1/ 40·0	46·583	51·786	54·516	57·331	63·226	69·479	72·741	83·086	98·214	114·92	123·88	133·26	175·07	237·49
0·02600	10·50	10·76	10·89	11·01	11·26	11·51	11·63	11·99	12·45	12·91	13·13	13·35	14·21	15·23
1/ 38·5	47·510	52·817	55·601	58·473	64·485	70·862	74·189	84·740	100·17	117·21	126·35	135·91	178·55	242·21
0·02700	10·70	10·97	11·10	11·22	11·48	11·73	11·85	12·22	12·69	13·16	13·38	13·61	14·48	15·52
1/ 37·0	48·420	53·829	56·666	59·593	65·720	72·218	75·610	86·362	102·09	119·45	128·77	138·51	181·97	246·84
0·02800	10·90	11·17	11·30	11·43	11·69	11·94	12·07	12·44	12·93	13·40	13·63	13·86	14·75	15·81
1/ 35·7	49·314	54·822	57·711	60·692	66·932	73·550	77·004	87·954	103·97	121·65	131·14	141·07	185·32	251·39
0·02900	11·09	11·37	11·50	11·63	11·90	12·16	12·29	12·66	13·16	13·64	13·87	14·11	15·01	16·09
1/ 34·5	50·191	55·797	58·738	61·772	68·123	74·858	78·373	89·518	105·82	123·81	133·47	143·57	188·61	255·85
0·03000	11·29	11·56	11·70	11·83	12·10	12·37	12·50	12·88	13·38	13·87	14·11	14·35	15·27	16·36
1/ 33·3	51·054	56·756	59·747	62·833	69·293	76·144	79·720	91·056	107·63	125·94	135·76	146·04	191·85	260·24
	0·96	0·96	0·96	0·96	0·96	0·97	0·97	0·97	0·97	0·97	0·97	0·97	0·97	0·98

$V_{r(0\cdot5)medial}$ for half-full circular pipes.

A51

(p.1 of 6)

k_s = 0·30 mm
S = 0·00003 to 0·00010

ie hydraulic gradient =
1 in 33333 to 1 in 10000

Water (or sewage) at 15°C;
full bore conditions.

velocities in ms^{-1}
discharges in m^3s^{-1}

Gradient **(Equivalent) Pipe diameters in mm**

Gradient	2400	2500	2550	2600	2700	2800	2850	3000	3200	3400	3500	3600	4000	4500
0·0000300	0·314	0·323	0·327	0·331	0·339	0·346	0·350	0·362	0·376	0·391	0·398	0·405	0·432	0·465
1/ 33333	1·4226	1·5838	1·6683	1·7557	1·9386	2·1329	2·2343	2·5563	3·0279	3·5496	3·8298	4·1232	5·4335	7·3949
0·0000320	0·325	0·334	0·338	0·342	0·350	0·358	0·362	0·374	0·389	0·404	0·412	0·419	0·447	0·481
1/ 31250	1·4716	1·6383	1·7258	1·8161	2·0053	2·2062	2·3111	2·6441	3·1317	3·6711	3·9609	4·2643	5·6190	7·6470
0·0000340	0·336	0·345	0·349	0·353	0·362	0·370	0·374	0·386	0·402	0·417	0·425	0·432	0·461	0·496
1/ 29412	1·5192	1·6912	1·7815	1·8747	2·0699	2·2773	2·3855	2·7291	3·2324	3·7890	4·0880	4·4011	5·7990	7·8915
0·0000360	0·346	0·355	0·359	0·364	0·372	0·381	0·385	0·398	0·414	0·430	0·438	0·445	0·475	0·511
1/ 27778	1·5654	1·7426	1·8356	1·9316	2·1328	2·3463	2·4579	2·8118	3·3302	3·9035	4·2115	4·5339	5·9738	8·1289
0·0000380	0·356	0·365	0·370	0·374	0·383	0·392	0·396	0·409	0·426	0·442	0·450	0·458	0·489	0·526
1/ 26316	1·6103	1·7926	1·8882	1·9870	2·1939	2·4135	2·5282	2·8922	3·4253	4·0150	4·3317	4·6633	6·1439	8·3600
0·0000400	0·366	0·375	0·380	0·384	0·394	0·403	0·407	0·420	0·437	0·454	0·462	0·471	0·502	0·540
1/ 25000	1·6542	1·8413	1·9396	2·0410	2·2534	2·4790	2·5968	2·9706	3·5181	4·1236	4·4488	4·7893	6·3096	8·5851
0·0000420	0·375	0·385	0·390	0·394	0·404	0·413	0·418	0·431	0·449	0·466	0·474	0·483	0·515	0·554
1/ 23810	1·6969	1·8889	1·9896	2·0937	2·3116	2·5429	2·6637	3·0471	3·6085	4·2295	4·5630	4·9122	6·4713	8·8047
0·0000440	0·384	0·394	0·399	0·404	0·414	0·423	0·428	0·442	0·460	0·477	0·486	0·494	0·528	0·567
1/ 22727	1·7387	1·9353	2·0386	2·1451	2·3683	2·6053	2·7291	3·1218	3·6969	4·3330	4·6746	5·0323	6·6292	9·0192
0·0000460	0·393	0·404	0·409	0·414	0·423	0·433	0·438	0·452	0·470	0·488	0·497	0·506	0·540	0·580
1/ 21739	1·7795	1·9808	2·0864	2·1954	2·4239	2·6664	2·7930	3·1949	3·7834	4·4342	4·7837	5·1497	6·7837	9·2289
0·0000480	0·402	0·413	0·418	0·423	0·433	0·443	0·448	0·462	0·481	0·499	0·508	0·517	0·552	0·593
1/ 20833	1·8195	2·0252	2·1332	2·2447	2·4782	2·7261	2·8556	3·2664	3·8679	4·5332	4·8905	5·2646	6·9348	9·4342
0·0000500	0·411	0·421	0·427	0·432	0·442	0·452	0·457	0·472	0·491	0·510	0·519	0·528	0·564	0·606
1/ 20000	1·8587	2·0688	2·1791	2·2929	2·5315	2·7847	2·9169	3·3364	3·9508	4·6303	4·9952	5·3772	7·0829	9·6353
0·0000525	0·421	0·432	0·438	0·443	0·454	0·464	0·469	0·484	0·504	0·523	0·532	0·542	0·578	0·621
1/ 19048	1·9066	2·1221	2·2352	2·3520	2·5966	2·8562	2·9918	3·4221	4·0521	4·7489	5·1231	5·5149	7·2639	9·8812
0·0000550	0·432	0·443	0·448	0·454	0·465	0·475	0·480	0·496	0·516	0·536	0·545	0·555	0·592	0·636
1/ 18182	1·9533	2·1741	2·2900	2·4096	2·6602	2·9261	3·0650	3·5058	4·1511	4·8648	5·2481	5·6493	7·4407	10·121
0·0000575	0·442	0·453	0·459	0·464	0·475	0·486	0·492	0·508	0·528	0·548	0·558	0·568	0·606	0·651
1/ 17391	1·9991	2·2250	2·3436	2·4660	2·7223	2·9945	3·1366	3·5876	4·2479	4·9781	5·3702	5·7808	7·6136	10·356
0·0000600	0·452	0·463	0·469	0·475	0·486	0·497	0·503	0·519	0·540	0·561	0·571	0·581	0·619	0·666
1/ 16667	2·0439	2·2748	2·3960	2·5211	2·7832	3·0614	3·2067	3·6677	4·3426	5·0890	5·4898	5·9095	7·7829	10·586
0·0000625	0·461	0·473	0·479	0·485	0·497	0·508	0·513	0·530	0·551	0·572	0·583	0·593	0·633	0·680
1/ 16000	2·0878	2·3236	2·4474	2·5752	2·8428	3·1270	3·2753	3·7461	4·4354	5·1977	5·6070	6·0355	7·9486	10·811
0·0000650	0·471	0·483	0·489	0·495	0·507	0·518	0·524	0·541	0·563	0·584	0·595	0·605	0·645	0·694
1/ 15385	2·1308	2·3715	2·4978	2·6282	2·9013	3·1913	3·3426	3·8230	4·5264	5·3042	5·7218	6·1591	8·1111	11·032
0·0000675	0·480	0·493	0·499	0·505	0·517	0·529	0·534	0·552	0·574	0·596	0·606	0·617	0·658	0·707
1/ 14815	2·1730	2·4184	2·5472	2·6802	2·9587	3·2543	3·4087	3·8985	4·6157	5·4087	5·8345	6·2804	8·2706	11·248
0·0000700	0·489	0·502	0·508	0·514	0·527	0·539	0·544	0·562	0·585	0·607	0·618	0·629	0·671	0·721
1/ 14286	2·2144	2·4645	2·5958	2·7312	3·0150	3·3162	3·4735	3·9726	4·7033	5·5113	5·9452	6·3994	8·4271	11·461
0·0000725	0·499	0·511	0·518	0·524	0·536	0·548	0·554	0·572	0·596	0·618	0·629	0·640	0·683	0·734
1/ 13793	2·2552	2·5098	2·6434	2·7814	3·0704	3·3771	3·5372	4·0454	4·7894	5·6121	6·0539	6·5164	8·5809	11·670
0·0000750	0·507	0·520	0·527	0·533	0·546	0·558	0·564	0·582	0·606	0·629	0·640	0·651	0·695	0·747
1/ 13333	2·2952	2·5543	2·6903	2·8307	3·1248	3·4369	3·5999	4·1170	4·8741	5·7112	6·1608	6·6314	8·7322	11·875
0·0000800	0·525	0·538	0·545	0·551	0·564	0·577	0·583	0·602	0·627	0·650	0·662	0·674	0·718	0·772
1/ 12500	2·3734	2·6412	2·7818	2·9270	3·2310	3·5536	3·7221	4·2567	5·0393	5·9047	6·3693	6·8558	9·0272	12·276
0·0000850	0·541	0·555	0·562	0·569	0·582	0·596	0·602	0·621	0·647	0·671	0·683	0·695	0·741	0·796
1/ 11765	2·4491	2·7255	2·8706	3·0203	3·3339	3·6668	3·8406	4·3921	5·1995	6·0922	6·5716	7·0734	9·3133	12·664
0·0000900	0·558	0·572	0·579	0·586	0·600	0·613	0·620	0·640	0·666	0·691	0·703	0·716	0·763	0·820
1/ 11111	2·5227	2·8074	2·9568	3·1110	3·4339	3·7768	3·9558	4·5237	5·3551	6·2744	6·7680	7·2847	9·5912	13·041
0·0000950	0·573	0·588	0·595	0·603	0·617	0·631	0·638	0·658	0·685	0·711	0·723	0·736	0·785	0·843
1/ 10526	2·5943	2·8870	3·0406	3·1992	3·5312	3·8837	4·0678	4·6517	5·5065	6·4516	6·9590	7·4903	9·8615	13·408
0·0001000	0·589	0·604	0·611	0·619	0·633	0·648	0·655	0·676	0·703	0·730	0·743	0·756	0·806	0·866
1/ 10000	2·6641	2·9646	3·1223	3·2851	3·6260	3·9879	4·1769	4·7763	5·6539	6·6242	7·1452	7·6906	10·125	13·766
	0·92	0·92	0·93	0·93	0·93	0·93	0·93	0·93	0·94	0·94	0·94	0·94	0·95	0·95

$V_{r(0{\cdot}5)medial}$ for half-full circular pipes.

k_s = 0·30 mm S = 0·00003 to 0·00010

k_s = 0·30 mm
S = 0·00010 to 0·00030

ie hydraulic gradient =
1 in 10000 to 1 in 3333

Water (or sewage) at 15°C;
full bore conditions.

velocities in ms^{-1}
discharges in m^3s^{-1}

Gradient **(Equivalent) Pipe diameters in mm**

Gradient	2400	2500	2550	2600	2700	2800	2850	3000	3200	3400	3500	3600	4000	4500
0·000100	0·589	0·604	0·611	0·619	0·633	0·648	0·655	0·676	0·703	0·730	0·743	0·756	0·806	0·866
1/ 10000	2·6641	2·9646	3·1223	3·2851	3·6260	3·9879	4·1769	4·7763	5·6539	6·6242	7·1452	7·6906	10·125	13·766
0·000105	0·604	0·619	0·627	0·635	0·649	0·664	0·671	0·693	0·721	0·748	0·762	0·775	0·826	0·887
1/ 9524	2·7322	3·0403	3·2020	3·3689	3·7185	4·0896	4·2833	4·8980	5·7978	6·7926	7·3268	7·8860	10·382	14·115
0·000110	0·619	0·634	0·642	0·650	0·665	0·680	0·688	0·710	0·738	0·766	0·780	0·793	0·846	0·909
1/ 9091	2·7987	3·1142	3·2799	3·4508	3·8089	4·1889	4·3873	5·0168	5·9383	6·9571	7·5041	8·0768	10·633	14·455
0·000115	0·633	0·649	0·657	0·665	0·681	0·696	0·704	0·726	0·755	0·784	0·798	0·812	0·866	0·930
1/ 8696	2·8637	3·1865	3·3560	3·5309	3·8972	4·2860	4·4889	5·1329	6·0757	7·1179	7·6775	8·2633	10·878	14·788
0·000120	0·647	0·664	0·672	0·680	0·696	0·711	0·719	0·742	0·772	0·801	0·816	0·830	0·885	0·950
1/ 8333	2·9273	3·2573	3·4305	3·6092	3·9836	4·3810	4·5884	5·2466	6·2101	7·2753	7·8472	8·4459	11·118	15·114
0·000125	0·661	0·678	0·686	0·694	0·711	0·727	0·735	0·758	0·789	0·818	0·833	0·847	0·903	0·970
1/ 8000	2·9896	3·3266	3·5034	3·6860	4·0683	4·4740	4·6859	5·3580	6·3418	7·4295	8·0134	8·6248	11·353	15·433
0·000130	0·674	0·692	0·700	0·708	0·725	0·741	0·750	0·773	0·805	0·835	0·850	0·865	0·922	0·990
1/ 7692	3·0507	3·3945	3·5750	3·7612	4·1513	4·5653	4·7814	5·4671	6·4709	7·5806	8·1764	8·8001	11·583	15·746
0·000135	0·688	0·705	0·714	0·722	0·739	0·756	0·764	0·789	0·820	0·851	0·866	0·881	0·940	1·009
1/ 7407	3·1106	3·4612	3·6452	3·8350	4·2327	4·6548	4·8751	5·5742	6·5976	7·7289	8·3362	8·9721	11·810	16·053
0·000140	0·701	0·718	0·727	0·736	0·753	0·770	0·779	0·803	0·836	0·867	0·883	0·898	0·957	1·028
1/ 7143	3·1695	3·5266	3·7141	3·9075	4·3127	4·7427	4·9672	5·6794	6·7219	7·8744	8·4932	9·1409	12·032	16·354
0·000145	0·713	0·732	0·741	0·749	0·767	0·784	0·793	0·818	0·851	0·883	0·899	0·914	0·975	1·047
1/ 6897	3·2273	3·5909	3·7818	3·9787	4·3912	4·8290	5·0576	5·7827	6·8441	8·0174	8·6474	9·3068	12·250	16·650
0·000150	0·726	0·744	0·754	0·763	0·780	0·798	0·807	0·832	0·866	0·899	0·915	0·930	0·992	1·065
1/ 6667	3·2842	3·6542	3·8483	4·0487	4·4684	4·9139	5·1464	5·8843	6·9642	8·1580	8·7990	9·4699	12·464	16·941
0·000160	0·750	0·770	0·779	0·788	0·807	0·825	0·834	0·860	0·895	0·929	0·945	0·962	1·025	1·101
1/ 6250	3·3951	3·7775	3·9782	4·1853	4·6191	5·0795	5·3199	6·0824	7·1986	8·4324	9·0947	9·7882	12·882	17·509
0·000170	0·774	0·794	0·804	0·813	0·832	0·851	0·860	0·888	0·923	0·958	0·975	0·992	1·057	1·136
1/ 5882	3·5026	3·8971	4·1041	4·3178	4·7652	5·2401	5·4880	6·2745	7·4258	8·6983	9·3815	10·097	13·288	18·060
0·000180	0·797	0·818	0·828	0·837	0·857	0·876	0·886	0·914	0·951	0·986	1·004	1·021	1·089	1·169
1/ 5556	3·6071	4·0133	4·2264	4·4464	4·9071	5·3961	5·6513	6·4611	7·6464	8·9566	9·6600	10·396	13·682	18·595
0·000190	0·820	0·841	0·851	0·861	0·881	0·901	0·911	0·940	0·977	1·014	1·032	1·050	1·119	1·202
1/ 5263	3·7087	4·1262	4·3454	4·5715	5·0451	5·5478	5·8102	6·6426	7·8610	9·2078	9·9308	10·688	14·065	19·115
0·000200	0·842	0·863	0·874	0·884	0·905	0·925	0·935	0·965	1·003	1·041	1·060	1·078	1·149	1·234
1/ 5000	3·8077	4·2363	4·4613	4·6934	5·1796	5·6956	5·9649	6·8194	8·0701	9·4526	10·195	10·972	14·438	19·621
0·000210	0·863	0·885	0·896	0·906	0·928	0·948	0·959	0·989	1·029	1·067	1·086	1·105	1·178	1·265
1/ 4762	3·9042	4·3437	4·5743	4·8123	5·3107	5·8397	6·1159	6·9919	8·2741	9·6913	10·452	11·249	14·802	20·115
0·000220	0·884	0·906	0·917	0·928	0·950	0·971	0·982	1·013	1·054	1·093	1·112	1·132	1·206	1·295
1/ 4545	3·9985	4·4485	4·6847	4·9284	5·4388	5·9805	6·2633	7·1603	8·4732	9·9244	10·703	11·519	15·158	20·598
0·000230	0·904	0·927	0·938	0·950	0·972	0·994	1·004	1·036	1·078	1·118	1·138	1·158	1·234	1·325
1/ 4348	4·0907	4·5511	4·7926	5·0419	5·5641	6·1181	6·4074	7·3250	8·6680	10·152	10·949	11·783	15·505	21·070
0·000240	0·924	0·948	0·959	0·971	0·993	1·015	1·026	1·059	1·101	1·143	1·163	1·183	1·261	1·354
1/ 4167	4·1809	4·6514	4·8983	5·1530	5·6866	6·2528	6·5484	7·4861	8·8585	10·375	11·190	12·042	15·845	21·531
0·000250	0·944	0·968	0·979	0·991	1·014	1·037	1·048	1·081	1·125	1·167	1·188	1·208	1·287	1·382
1/ 4000	4·2693	4·7497	5·0017	5·2618	5·8066	6·3848	6·6866	7·6440	9·0451	10·594	11·425	12·295	16·178	21·983
0·000260	0·963	0·987	0·999	1·011	1·035	1·058	1·069	1·103	1·147	1·190	1·212	1·232	1·313	1·410
1/ 3846	4·3560	4·8460	5·1031	5·3685	5·9243	6·5141	6·8220	7·7987	9·2281	10·808	11·656	12·544	16·505	22·426
0·000270	0·982	1·006	1·019	1·031	1·055	1·079	1·090	1·125	1·170	1·214	1·235	1·256	1·339	1·437
1/ 3704	4·4409	4·9405	5·2026	5·4732	6·0397	6·6410	6·9548	7·9505	9·4076	11·018	11·883	12·788	16·825	22·861
0·000280	1·000	1·025	1·038	1·050	1·075	1·099	1·111	1·146	1·192	1·236	1·258	1·280	1·364	1·464
1/ 3571	4·5244	5·0333	5·3003	5·5759	6·1531	6·7655	7·0853	8·0995	9·5838	11·224	12·105	13·027	17·140	23·288
0·000290	1·018	1·044	1·057	1·069	1·094	1·119	1·131	1·167	1·213	1·259	1·281	1·303	1·389	1·491
1/ 3448	4·6064	5·1244	5·3963	5·6768	6·2644	6·8879	7·2134	8·2459	9·7569	11·427	12·323	13·262	17·449	23·707
0·000300	1·036	1·062	1·075	1·088	1·113	1·138	1·150	1·187	1·234	1·281	1·303	1·326	1·413	1·517
1/ 3333	4·6869	5·2140	5·4906	5·7761	6·3738	7·0082	7·3393	8·3898	9·9270	11·626	12·538	13·493	17·752	24·119
	0·93	0·93	0·93	0·93	0·94	0·94	0·94	0·94	0·94	0·95	0·95	0·95	0·95	0·96

$V_{r(0·5)medial}$ for half-full circular pipes.

k_s = 0·30 mm
S = 0·00030 to 0·00100

ie hydraulic gradient =
1 in 3333 to 1 in 1000

Water (or sewage) at 15°C;
full bore conditions.

velocities in ms^{-1}
discharges in m^3s^{-1}

Gradient **(Equivalent) Pipe diameters in mm**

Gradient	2400	2500	2550	2600	2700	2800	2850	3000	3200	3400	3500	3600	4000	4500
0·000300	1·036	1·062	1·075	1·088	1·113	1·138	1·150	1·187	1·234	1·281	1·303	1·326	1·413	1·517
1/ 3333	4·6869	5·2140	5·4906	5·7761	6·3738	7·0082	7·3393	8·3898	9·9270	11·626	12·538	13·493	17·752	24·119
0·000320	1·071	1·098	1·111	1·124	1·151	1·176	1·189	1·227	1·276	1·323	1·347	1·370	1·460	1·567
1/ 3125	4·8442	5·3889	5·6747	5·9697	6·5874	7·2429	7·5851	8·6706	10·259	12·015	12·957	13·944	18·345	24·924
0·000340	1·104	1·132	1·146	1·160	1·187	1·213	1·226	1·265	1·316	1·365	1·389	1·413	1·506	1·616
1/ 2941	4·9966	5·5584	5·8531	6·1573	6·7944	7·4704	7·8233	8·9428	10·581	12·391	13·363	14·381	18·919	25·703
0·000360	1·137	1·166	1·180	1·194	1·222	1·249	1·263	1·303	1·354	1·405	1·430	1·454	1·550	1·664
1/ 2778	5·1446	5·7229	6·0264	6·3396	6·9954	7·6914	8·0547	9·2071	10·893	12·757	13·758	14·805	19·477	26·460
0·000380	1·169	1·198	1·213	1·227	1·256	1·284	1·298	1·339	1·392	1·444	1·470	1·495	1·593	1·710
1/ 2632	5·2886	5·8830	6·1950	6·5169	7·1910	7·9063	8·2797	9·4642	11·197	13·113	14·141	15·218	20·019	27·196
0·000400	1·200	1·230	1·245	1·260	1·289	1·318	1·332	1·374	1·429	1·482	1·509	1·535	1·635	1·755
1/ 2500	5·4288	6·0390	6·3592	6·6896	7·3815	8·1157	8·4989	9·7146	11·494	13·460	14·515	15·620	20·548	27·914
0·000420	1·230	1·261	1·277	1·292	1·322	1·351	1·366	1·409	1·465	1·520	1·547	1·573	1·676	1·799
1/ 2381	5·5657	6·1911	6·5193	6·8580	7·5673	8·3199	8·7127	9·9589	11·782	13·798	14·879	16·012	21·063	28·613
0·000440	1·260	1·292	1·307	1·323	1·353	1·384	1·398	1·443	1·500	1·556	1·584	1·611	1·716	1·842
1/ 2273	5·6992	6·3397	6·6757	7·0225	7·7487	8·5193	8·9215	10·197	12·065	14·128	15·235	16·395	21·566	29·296
0·000460	1·289	1·321	1·337	1·353	1·384	1·415	1·430	1·476	1·534	1·592	1·620	1·647	1·755	1·884
1/ 2174	5·8299	6·4849	6·8286	7·1833	7·9261	8·7142	9·1256	10·431	12·340	14·451	15·583	16·769	22·058	29·964
0·000480	1·317	1·350	1·366	1·383	1·415	1·446	1·462	1·508	1·568	1·626	1·655	1·683	1·794	1·925
1/ 2083	5·9577	6·6270	6·9782	7·3406	8·0996	8·9050	9·3254	10·659	12·610	14·766	15·924	17·135	22·540	30·617
0·000500	1·345	1·378	1·395	1·412	1·444	1·477	1·492	1·540	1·601	1·660	1·690	1·719	1·831	1·965
1/ 2000	6·0828	6·7662	7·1248	7·4948	8·2696	9·0918	9·5210	10·882	12·874	15·076	16·257	17·494	23·011	31·257
0·000525	1·378	1·413	1·430	1·447	1·481	1·514	1·530	1·578	1·641	1·702	1·732	1·762	1·877	2·015
1/ 1905	6·2358	6·9363	7·3039	7·6832	8·4774	9·3202	9·7601	11·156	13·197	15·454	16·665	17·933	23·588	32·040
0·000550	1·411	1·447	1·464	1·482	1·516	1·550	1·567	1·616	1·680	1·743	1·773	1·804	1·922	2·063
1/ 1818	6·3853	7·1025	7·4788	7·8671	8·6803	9·5432	9·9936	11·422	13·513	15·823	17·063	18·361	24·151	32·804
0·000575	1·444	1·480	1·498	1·516	1·551	1·585	1·602	1·653	1·719	1·783	1·814	1·845	1·966	2·110
1/ 1739	6·5314	7·2649	7·6498	8·0470	8·8787	9·7613	10·222	11·683	13·821	16·184	17·452	18·779	24·701	33·550
0·000600	1·475	1·512	1·531	1·549	1·585	1·620	1·637	1·689	1·756	1·821	1·854	1·885	2·008	2·155
1/ 1667	6·6743	7·4238	7·8171	8·2230	9·0728	9·9746	10·445	11·938	14·123	16·537	17·833	19·189	25·239	34·281
0·000625	1·506	1·544	1·563	1·581	1·618	1·654	1·672	1·724	1·793	1·860	1·892	1·925	2·050	2·200
1/ 1600	6·8143	7·5795	7·9810	8·3953	9·2630	10·184	10·664	12·188	14·419	16·883	18·206	19·590	25·766	34·997
0·000650	1·537	1·575	1·594	1·613	1·650	1·687	1·705	1·759	1·829	1·897	1·930	1·963	2·092	2·245
1/ 1538	6·9515	7·7321	8·1417	8·5643	9·4493	10·388	10·879	12·433	14·708	17·222	18·571	19·984	26·283	35·698
0·000675	1·566	1·606	1·625	1·644	1·682	1·720	1·738	1·793	1·864	1·934	1·968	2·001	2·132	2·288
1/ 1481	7·0862	7·8818	8·2993	8·7301	9·6322	10·589	11·089	12·674	14·992	17·555	18·930	20·369	26·790	36·386
0·000700	1·596	1·636	1·655	1·675	1·714	1·752	1·771	1·826	1·899	1·969	2·004	2·038	2·172	2·330
1/ 1429	7·2183	8·0288	8·4540	8·8928	9·8116	10·787	11·295	12·910	15·271	17·881	19·282	20·748	27·288	37·062
0·000725	1·624	1·665	1·685	1·705	1·744	1·783	1·802	1·859	1·933	2·005	2·040	2·075	2·210	2·372
1/ 1379	7·3482	8·1731	8·6060	9·0526	9·9879	10·980	11·498	13·141	15·545	18·202	19·628	21·120	27·777	37·725
0·000750	1·653	1·694	1·714	1·735	1·775	1·814	1·834	1·891	1·966	2·040	2·075	2·111	2·249	2·413
1/ 1333	7·4758	8·3150	8·7554	9·2097	10·161	11·171	11·698	13·369	15·815	18·517	19·968	21·486	28·258	38·378
0·000800	1·708	1·750	1·771	1·792	1·834	1·875	1·895	1·954	2·032	2·107	2·144	2·181	2·323	2·493
1/ 1250	7·7248	8·5919	9·0469	9·5163	10·499	11·542	12·087	13·814	16·340	19·132	20·631	22·199	29·196	39·650
0·000850	1·761	1·805	1·827	1·848	1·891	1·933	1·954	2·015	2·095	2·173	2·211	2·249	2·396	2·571
1/ 1176	7·9662	8·8603	9·3294	9·8135	10·827	11·903	12·464	14·245	16·850	19·729	21·274	22·891	30·105	40·884
0·000900	1·813	1·858	1·880	1·903	1·947	1·990	2·011	2·074	2·157	2·237	2·276	2·315	2·466	2·646
1/ 1111	8·2006	9·1209	9·6038	10·102	11·145	12·253	12·830	14·663	17·345	20·308	21·898	23·563	30·987	42·082
0·000950	1·863	1·910	1·933	1·956	2·001	2·045	2·067	2·132	2·216	2·299	2·339	2·379	2·534	2·719
1/ 1053	8·4285	9·3744	9·8706	10·383	11·455	12·593	13·187	15·070	17·826	20·871	22·505	24·216	31·846	43·247
0·001000	1·912	1·960	1·984	2·007	2·053	2·099	2·121	2·188	2·275	2·359	2·401	2·442	2·601	2·791
1/ 1000	8·6506	9·6213	10·131	10·656	11·757	12·924	13·534	15·467	18·295	21·420	23·097	24·852	32·682	44·382
	0·94	0·94	0·94	0·94	0·94	0·94	0·94	0·94	0·95	0·95	0·95	0·95	0·95	0·96

$V_{r(0{\cdot}5)medial}$ for half-full circular pipes.

Gradient	**(Equivalent) Pipe diameters in mm**													
	2400	**2500**	**2550**	**2600**	**2700**	**2800**	**2850**	**3000**	**3200**	**3400**	**3500**	**3600**	**4000**	**4500**
0·00100	1·912	1·960	1·984	2·007	2·053	2·099	2·121	2·188	2·275	2·359	2·401	2·442	2·601	2·791
1/ 1000	8·6506	9·6213	10·131	10·656	11·757	12·924	13·534	15·467	18·295	21·420	23·097	24·852	32·682	44·382
0·00105	1·960	2·009	2·033	2·057	2·105	2·151	2·175	2·243	2·332	2·418	2·461	2·503	2·666	2·860
1/ 952	8·8672	9·8621	10·384	10·923	12·051	13·247	13·872	15·853	18·752	21·955	23·674	25·473	33·497	45·489
0·00110	2·007	2·057	2·082	2·106	2·155	2·203	2·226	2·296	2·387	2·476	2·519	2·562	2·729	2·928
1/ 909	9·0787	10·097	10·632	11·183	12·338	13·563	14·202	16·231	19·198	22·477	24·237	26·079	34·294	46·569
0·00115	2·053	2·104	2·129	2·154	2·204	2·253	2·277	2·348	2·441	2·532	2·576	2·620	2·791	2·995
1/ 870	9·2854	10·327	10·874	11·438	12·619	13·872	14·526	16·600	19·635	22·988	24·787	26·671	35·072	47·626
0·00120	2·097	2·150	2·176	2·201	2·252	2·302	2·327	2·399	2·494	2·587	2·632	2·677	2·852	3·060
1/ 833	9·4877	10·552	11·111	11·687	12·893	14·173	14·842	16·961	20·062	23·488	25·326	27·251	35·834	48·660
0·00125	2·141	2·195	2·221	2·247	2·299	2·350	2·375	2·450	2·546	2·641	2·687	2·733	2·911	3·123
1/ 800	9·6859	10·772	11·342	11·931	13·162	14·469	15·151	17·315	20·480	23·977	25·854	27·818	36·580	49·672
0·00130	2·184	2·239	2·265	2·292	2·345	2·397	2·423	2·499	2·597	2·694	2·741	2·788	2·969	3·186
1/ 769	9·8801	10·988	11·570	12·170	13·426	14·759	15·455	17·661	20·890	24·457	26·371	28·375	37·311	50·665
0·00135	2·226	2·282	2·309	2·336	2·390	2·443	2·469	2·547	2·647	2·746	2·794	2·841	3·026	3·247
1/ 741	10·071	11·200	11·793	12·404	13·685	15·043	15·752	18·001	21·292	24·927	26·879	28·921	38·029	51·638
0·00140	2·267	2·324	2·352	2·380	2·434	2·488	2·515	2·594	2·697	2·796	2·845	2·894	3·082	3·307
1/ 714	10·258	11·408	12·012	12·635	13·939	15·323	16·045	18·335	21·687	25·389	27·377	29·456	38·733	52·594
0·00145	2·308	2·366	2·394	2·422	2·478	2·533	2·560	2·640	2·745	2·846	2·896	2·946	3·137	3·366
1/ 690	10·441	11·612	12·227	12·861	14·188	15·597	16·332	18·663	22·074	25·843	27·866	29·983	39·425	53·533
0·00150	2·348	2·407	2·436	2·464	2·521	2·577	2·604	2·686	2·792	2·896	2·946	2·996	3·191	3·424
1/ 667	10·622	11·813	12·438	13·083	14·434	15·866	16·614	18·986	22·456	26·289	28·347	30·500	40·105	54·456
0·00160	2·426	2·486	2·516	2·546	2·604	2·662	2·691	2·775	2·885	2·991	3·044	3·096	3·297	3·537
1/ 625	10·974	12·205	12·851	13·517	14·912	16·392	17·165	19·615	23·199	27·160	29·285	31·510	41·432	56·256
0·00170	2·501	2·564	2·595	2·625	2·685	2·745	2·774	2·861	2·974	3·084	3·138	3·192	3·399	3·647
1/ 588	11·316	12·585	13·251	13·937	15·376	16·902	17·698	20·224	23·920	28·003	30·195	32·488	42·718	58·001
0·00180	2·575	2·639	2·671	2·702	2·764	2·825	2·856	2·945	3·061	3·175	3·230	3·285	3·499	3·753
1/ 556	11·648	12·954	13·639	14·346	15·826	17·397	18·217	20·816	24·620	28·823	31·078	33·438	43·966	59·696
0·00190	2·646	2·712	2·745	2·777	2·841	2·903	2·935	3·026	3·146	3·262	3·319	3·376	3·595	3·857
1/ 526	11·970	13·312	14·016	14·743	16·264	17·878	18·721	21·392	25·301	29·619	31·937	34·363	45·181	61·344
0·00200	2·715	2·783	2·817	2·850	2·915	2·980	3·011	3·106	3·228	3·348	3·406	3·464	3·690	3·958
1/ 500	12·284	13·662	14·384	15·130	16·691	18·347	19·211	21·953	25·964	30·395	32·774	35·263	46·364	62·949
0·00210	2·783	2·853	2·887	2·921	2·988	3·054	3·087	3·183	3·309	3·431	3·491	3·551	3·781	4·056
1/ 476	12·591	14·002	14·743	15·507	17·107	18·804	19·690	22·500	26·611	31·152	33·590	36·141	47·517	64·515
0·00220	2·849	2·920	2·955	2·990	3·059	3·126	3·160	3·259	3·387	3·513	3·574	3·635	3·871	4·153
1/ 455	12·890	14·335	15·093	15·875	17·513	19·251	20·158	23·034	27·242	31·891	34·386	36·998	48·644	66·044
0·00230	2·914	2·987	3·022	3·058	3·128	3·197	3·231	3·333	3·464	3·592	3·655	3·717	3·959	4·247
1/ 435	13·183	14·660	15·436	16·235	17·910	19·687	20·615	23·556	27·860	32·614	35·165	37·836	49·745	67·538
0·00240	2·977	3·051	3·088	3·124	3·196	3·267	3·302	3·405	3·539	3·670	3·734	3·798	4·044	4·338
1/ 417	13·469	14·979	15·771	16·588	18·299	20·115	21·062	24·067	28·464	33·321	35·928	38·656	50·823	69·001
0·00250	3·039	3·115	3·152	3·189	3·263	3·335	3·370	3·476	3·613	3·746	3·812	3·877	4·128	4·429
1/ 400	13·749	15·290	16·099	16·933	18·680	20·533	21·500	24·568	29·055	34·013	36·674	39·459	51·878	70·433
0·00260	3·100	3·177	3·215	3·253	3·328	3·401	3·438	3·545	3·685	3·821	3·888	3·954	4·211	4·517
1/ 385	14·024	15·596	16·420	17·271	19·053	20·943	21·929	25·058	29·635	34·692	37·406	40·246	52·913	71·837
0·00270	3·160	3·238	3·277	3·316	3·392	3·467	3·504	3·613	3·756	3·894	3·962	4·030	4·291	4·603
1/ 370	14·294	15·895	16·736	17·603	19·419	21·345	22·351	25·539	30·204	35·358	38·123	41·018	53·927	73·214
0·00280	3·218	3·298	3·338	3·377	3·454	3·531	3·568	3·680	3·825	3·966	4·036	4·104	4·371	4·688
1/ 357	14·558	16·190	17·046	17·929	19·778	21·740	22·764	26·012	30·762	36·011	38·828	41·776	54·924	74·566
0·00290	3·276	3·357	3·397	3·437	3·516	3·594	3·632	3·746	3·893	4·037	4·108	4·177	4·449	4·772
1/ 345	14·818	16·479	17·350	18·249	20·131	22·128	23·170	26·476	31·311	36·653	39·520	42·521	55·902	75·894
0·00300	3·332	3·415	3·456	3·496	3·577	3·656	3·695	3·810	3·960	4·107	4·178	4·249	4·525	4·854
1/ 333	15·074	16·763	17·649	18·563	20·478	22·509	23·569	26·932	31·850	37·284	40·201	43·253	56·864	77·199
	0·94	0·94	0·94	0·94	0·94	0·94	0·95	0·95	0·95	0·95	0·95	0·95	0·96	0·96

$V_{r(0·5)medial}$ for half-full circular pipes.

$k_s = 0{\cdot}30$ **mm**
S = 0·00300 to 0·01000

ie hydraulic gradient =
1 in 333 to 1 in 100

Water (or sewage) at 15°C;
full bore conditions.

velocities in ms^{-1}
discharges in m^3s^{-1}

Gradient **(Equivalent) Pipe diameters in mm**

Gradient	2400	2500	2550	2600	2700	2800	2850	3000	3200	3400	3500	3600	4000	4500
0·00300	3·332	3·415	3·456	3·496	3·577	3·656	3·695	3·810	3·960	4·107	4·178	4·249	4·525	4·854
1/ 333	15·074	16·763	17·649	18·563	20·478	22·509	23·569	26·932	31·850	37·284	40·201	43·253	56·864	77·199
0·00320	3·442	3·528	3·570	3·612	3·695	3·776	3·817	3·936	4·091	4·242	4·316	4·390	4·674	5·014
1/ 313	15·572	17·317	18·232	19·177	21·155	23·253	24·348	27·821	32·902	38·516	41·528	44·681	58·741	79·746
0·00340	3·549	3·637	3·681	3·724	3·809	3·893	3·935	4·058	4·218	4·374	4·450	4·526	4·819	5·169
1/ 294	16·055	17·854	18·798	19·772	21·811	23·974	25·103	28·684	33·922	39·709	42·814	46·065	60·560	82·214
0·00360	3·653	3·744	3·788	3·833	3·921	4·007	4·050	4·176	4·341	4·501	4·580	4·658	4·960	5·320
1/ 278	16·524	18·376	19·347	20·349	22·448	24·674	25·836	29·521	34·912	40·868	44·064	47·409	62·326	84·611
0·00380	3·754	3·847	3·893	3·939	4·029	4·118	4·162	4·292	4·461	4·625	4·706	4·786	5·096	5·467
1/ 263	16·981	18·883	19·881	20·911	23·068	25·355	26·549	30·336	35·875	41·995	45·279	48·716	64·044	86·942
0·00400	3·852	3·948	3·995	4·042	4·134	4·225	4·271	4·404	4·577	4·746	4·829	4·911	5·230	5·609
1/ 250	17·425	19·377	20·402	21·458	23·671	26·019	27·244	31·129	36·813	43·092	46·462	49·989	65·717	89·213
0·00420	3·948	4·046	4·094	4·142	4·237	4·331	4·377	4·513	4·691	4·864	4·949	5·033	5·359	5·749
1/ 238	17·859	19·859	20·909	21·992	24·260	26·665	27·921	31·903	37·728	44·163	47·616	51·231	67·349	91·427
0·00440	4·041	4·142	4·191	4·240	4·337	4·433	4·480	4·620	4·802	4·979	5·066	5·152	5·486	5·885
1/ 227	18·282	20·330	21·405	22·513	24·834	27·297	28·582	32·658	38·621	45·209	48·744	52·443	68·942	93·590
0·00460	4·133	4·235	4·286	4·336	4·436	4·533	4·582	4·725	4·911	5·092	5·181	5·269	5·610	6·017
1/ 217	18·696	20·790	21·889	23·023	25·396	27·915	29·229	33·397	39·494	46·230	49·845	53·628	70·500	95·704
0·00480	4·222	4·327	4·379	4·430	4·532	4·632	4·681	4·827	5·017	5·202	5·293	5·383	5·731	6·148
1/ 208	19·101	21·240	22·363	23·521	25·946	28·519	29·861	34·120	40·349	47·230	50·923	54·788	72·024	97·772
0·00500	4·310	4·417	4·470	4·522	4·626	4·728	4·778	4·927	5·121	5·310	5·403	5·494	5·850	6·275
1/ 200	19·497	21·681	22·827	24·009	26·484	29·110	30·481	34·827	41·186	48·210	51·979	55·924	73·517	99·798
0·00525	4·417	4·527	4·581	4·634	4·741	4·845	4·897	5·049	5·248	5·442	5·537	5·631	5·995	6·431
1/ 190	19·982	22·220	23·394	24·606	27·143	29·834	31·238	35·693	42·208	49·407	53·270	57·312	75·341	102·27
0·00550	4·522	4·634	4·689	4·744	4·853	4·960	5·013	5·169	5·372	5·570	5·668	5·764	6·137	6·583
1/ 182	20·455	22·746	23·948	25·188	27·785	30·540	31·978	36·537	43·207	50·576	54·530	58·668	77·123	104·69
0·00575	4·624	4·739	4·795	4·851	4·963	5·072	5·126	5·286	5·494	5·696	5·796	5·894	6·276	6·731
1/ 174	20·918	23·261	24·490	25·758	28·413	31·230	32·701	37·363	44·184	51·718	55·762	59·993	78·864	107·05
0·00600	4·724	4·841	4·899	4·956	5·070	5·182	5·237	5·400	5·613	5·819	5·921	6·021	6·411	6·877
1/ 167	21·371	23·764	25·020	26·315	29·028	31·906	33·408	38·171	45·139	52·836	56·967	61·290	80·569	109·37
0·00625	4·822	4·942	5·001	5·059	5·175	5·289	5·345	5·512	5·729	5·940	6·044	6·146	6·544	7·019
1/ 160	21·814	24·257	25·539	26·861	29·630	32·567	34·101	38·962	46·074	53·931	58·148	62·560	82·238	111·63
0·00650	4·918	5·040	5·100	5·160	5·278	5·394	5·452	5·622	5·843	6·058	6·164	6·268	6·674	7·159
1/ 154	22·249	24·740	26·047	27·396	30·220	33·216	34·780	39·738	46·992	55·005	59·305	63·805	83·873	113·85
0·00675	5·012	5·137	5·198	5·259	5·379	5·498	5·556	5·729	5·955	6·174	6·282	6·388	6·802	7·296
1/ 148	22·675	25·214	26·546	27·921	30·799	33·852	35·446	40·499	47·891	56·057	60·440	65·026	85·478	116·03
0·00700	5·105	5·231	5·294	5·356	5·478	5·599	5·659	5·835	6·065	6·288	6·398	6·506	6·928	7·430
1/ 143	23·093	25·680	27·036	28·436	31·367	34·477	36·099	41·246	48·774	57·091	61·554	66·225	87·054	118·17
0·00725	5·196	5·324	5·388	5·451	5·576	5·699	5·759	5·939	6·172	6·400	6·512	6·622	7·051	7·562
1/ 138	23·504	26·137	27·517	28·942	31·925	35·090	36·742	41·980	49·642	58·106	62·648	67·402	88·601	120·27
0·00750	5·285	5·416	5·481	5·545	5·672	5·797	5·858	6·041	6·278	6·510	6·623	6·736	7·172	7·692
1/ 133	23·908	26·586	27·990	29·440	32·474	35·693	37·373	42·701	50·494	59·104	63·724	68·559	90·122	122·33
0·00800	5·459	5·595	5·661	5·728	5·859	5·988	6·051	6·240	6·485	6·724	6·842	6·957	7·408	7·945
1/ 125	24·697	27·462	28·913	30·410	33·544	36·869	38·605	44·108	52·158	61·051	65·823	70·818	93·090	126·36
0·00850	5·628	5·768	5·837	5·905	6·040	6·173	6·239	6·433	6·686	6·932	7·053	7·172	7·637	8·190
1/ 118	25·461	28·312	29·808	31·351	34·582	38·010	39·799	45·472	53·771	62·938	67·858	73·006	95·966	130·26
0·00900	5·792	5·936	6·007	6·077	6·216	6·353	6·420	6·620	6·881	7·134	7·258	7·381	7·859	8·429
1/ 111	26·203	29·137	30·676	32·264	35·589	39·117	40·958	46·796	55·336	64·771	69·833	75·132	98·759	134·05
0·00950	5·952	6·099	6·172	6·244	6·387	6·528	6·597	6·803	7·070	7·330	7·458	7·584	8·075	8·660
1/ 105	26·925	29·939	31·521	33·153	36·569	40·194	42·085	48·084	56·859	66·553	71·755	77·199	101·48	137·74
0·01000	6·107	6·258	6·333	6·407	6·554	6·698	6·769	6·980	7·254	7·521	7·653	7·782	8·286	8·886
1/ 100	27·628	30·721	32·344	34·018	37·524	41·243	43·183	49·339	58·342	68·289	73·626	79·212	104·12	141·33
	0·94	0·94	0·94	0·94	0·95	0·95	0·95	0·95	0·95	0·95	0·95	0·95	0·96	0·96

$V_{r(0{\cdot}5)medial}$ **for half-full circular pipes.**

$k_s = 0{\cdot}30$ **mm** S = 0·00300 to 0·01000

Gradient	**(Equivalent) Pipe diameters in mm**													
	2400	**2500**	**2550**	**2600**	**2700**	**2800**	**2850**	**3000**	**3200**	**3400**	**3500**	**3600**	**4000**	**4500**
0·01000	6·107	6·258	6·333	6·407	6·554	6·698	6·769	6·980	7·254	7·521	7·653	7·782	8·286	8·886
1/ 100	27·628	30·721	32·344	34·018	37·524	41·243	43·183	49·339	58·342	68·289	73·626	79·212	104·12	141·33
0·01050	6·259	6·414	6·490	6·566	6·716	6·864	6·937	7·153	7·434	7·708	7·842	7·975	8·491	9·106
1/ 95	28·313	31·483	33·146	34·862	38·454	42·266	44·255	50·562	59·789	69·982	75·451	81·176	106·70	144·83
0·01100	6·407	6·565	6·644	6·721	6·875	7·026	7·101	7·322	7·610	7·890	8·028	8·163	8·692	9·321
1/ 91	28·982	32·227	33·930	35·686	39·363	43·264	45·300	51·757	61·202	71·635	77·234	83·093	109·22	148·25
0·01150	6·551	6·714	6·794	6·873	7·030	7·185	7·261	7·487	7·781	8·068	8·209	8·348	8·888	9·532
1/ 87	29·637	32·955	34·695	36·491	40·252	44·241	46·323	52·925	62·582	73·251	78·976	84·967	111·68	151·59
0·01200	6·693	6·859	6·940	7·022	7·182	7·340	7·418	7·649	7·949	8·242	8·386	8·528	9·079	9·737
1/ 83	30·277	33·667	35·445	37·280	41·121	45·196	47·323	54·068	63·934	74·832	80·681	86·801	114·09	154·86
0·01250	6·831	7·001	7·084	7·167	7·331	7·492	7·572	7·807	8·114	8·413	8·559	8·704	9·267	9·939
1/ 80	30·904	34·364	36·179	38·052	41·973	46·132	48·303	55·187	65·257	76·381	82·350	88·597	116·45	158·07
0·01300	6·967	7·140	7·225	7·309	7·476	7·641	7·722	7·963	8·275	8·580	8·729	8·877	9·451	10·14
1/ 77	31·519	35·047	36·898	38·808	42·807	47·049	49·263	56·284	66·554	77·899	83·987	90·358	118·77	161·21
0·01350	7·100	7·276	7·363	7·449	7·619	7·787	7·870	8·115	8·434	8·744	8·896	9·047	9·632	10·33
1/ 74	32·122	35·718	37·604	39·551	43·626	47·949	50·205	57·360	67·826	79·388	85·592	92·085	121·04	164·29
0·01400	7·231	7·410	7·499	7·587	7·760	7·930	8·015	8·264	8·589	8·905	9·060	9·213	9·809	10·52
1/ 71	32·714	36·376	38·297	40·279	44·429	48·832	51·130	58·417	69·075	80·850	87·168	93·780	123·27	167·31
0·01450	7·360	7·542	7·632	7·721	7·898	8·071	8·157	8·411	8·741	9·063	9·221	9·377	9·983	10·71
1/ 69	33·295	37·022	38·977	40·995	45·219	49·700	52·038	59·454	70·302	82·286	88·716	95·445	125·45	170·28
0·01500	7·486	7·672	7·763	7·854	8·033	8·210	8·297	8·555	8·891	9·219	9·379	9·538	10·15	10·89
1/ 67	33·867	37·658	39·646	41·698	45·995	50·552	52·931	60·474	71·508	83·697	90·237	97·082	127·61	173·20
0·01600	7·733	7·924	8·019	8·112	8·298	8·480	8·570	8·837	9·184	9·522	9·688	9·852	10·49	11·25
1/ 63	34·982	38·897	40·952	43·071	47·509	52·216	54·673	62·464	73·861	86·451	93·206	100·28	131·80	178·89
0·01700	7·972	8·169	8·266	8·363	8·554	8·742	8·835	9·110	9·467	9·816	9·987	10·16	10·81	11·60
1/ 59	36·062	40·099	42·217	44·401	48·976	53·829	56·362	64·393	76·142	89·119	96·084	103·37	135·87	184·41
0·01800	8·203	8·407	8·507	8·606	8·803	8·996	9·092	9·375	9·743	10·10	10·28	10·45	11·13	11·93
1/ 56	37·112	41·266	43·445	45·693	50·401	55·395	58·001	66·266	78·356	91·711	98·877	106·38	139·82	189·77
0·01900	8·429	8·638	8·741	8·843	9·045	9·244	9·342	9·632	10·01	10·38	10·56	10·74	11·43	12·26
1/ 53	38·132	42·400	44·640	46·950	51·787	56·918	59·596	68·088	80·510	94·232	101·59	109·30	143·66	194·99
0·02000	8·649	8·863	8·969	9·073	9·281	9·485	9·585	9·883	10·27	10·65	10·83	11·02	11·73	12·58
1/ 50	39·126	43·506	45·803	48·173	53·136	58·401	61·149	69·862	82·607	96·687	104·24	112·15	147·40	200·07
0·02100	8·863	9·083	9·191	9·298	9·510	9·719	9·823	10·13	10·53	10·91	11·10	11·29	12·02	12·89
1/ 47·6	40·096	44·584	46·938	49·367	54·453	59·848	62·664	71·593	84·653	99·081	106·82	114·92	151·05	205·02
0·02200	9·072	9·297	9·408	9·518	9·735	9·949	10·05	10·37	10·77	11·17	11·36	11·56	12·30	13·19
1/ 45·5	41·043	45·636	48·046	50·532	55·738	61·261	64·143	73·282	86·651	101·42	109·34	117·64	154·62	209·85
0·02300	9·277	9·507	9·620	9·732	9·954	10·17	10·28	10·60	11·02	11·42	11·62	11·82	12·58	13·49
1/ 43·5	41·968	46·665	49·129	51·672	56·995	62·642	65·589	74·934	88·604	103·70	111·81	120·29	158·10	214·58
0·02400	9·477	9·712	9·827	9·942	10·17	10·39	10·50	10·83	11·25	11·67	11·87	12·07	12·85	13·78
1/ 41·7	42·874	47·672	50·189	52·786	58·224	63·993	67·003	76·550	90·515	105·94	114·22	122·88	161·51	219·21
0·02500	9·673	9·913	10·03	10·15	10·38	10·61	10·72	11·05	11·49	11·91	12·12	12·32	13·12	14·07
1/ 40·0	43·761	48·658	51·227	53·878	59·428	65·316	68·389	78·133	92·386	108·13	116·58	125·42	164·85	223·74
0·02600	9·865	10·11	10·23	10·35	10·59	10·82	10·93	11·27	11·72	12·15	12·36	12·57	13·38	14·35
1/ 38·5	44·630	49·625	52·245	54·948	60·609	66·613	69·747	79·685	94·221	110·28	118·89	127·91	168·12	228·18
0·02700	10·05	10·30	10·43	10·55	10·79	11·02	11·14	11·49	11·94	12·38	12·59	12·81	13·63	14·62
1/ 37·0	45·483	50·573	53·243	55·998	61·767	67·886	71·079	81·207	96·020	112·38	121·16	130·35	171·33	232·53
0·02800	10·24	10·49	10·62	10·74	10·99	11·23	11·35	11·70	12·16	12·61	12·83	13·04	13·88	14·89
1/ 35·7	46·320	51·503	54·223	57·029	62·903	69·135	72·387	82·701	97·786	114·45	123·39	132·75	174·48	236·81
0·02900	10·42	10·68	10·81	10·93	11·18	11·43	11·55	11·91	12·37	12·83	13·05	13·27	14·13	15·15
1/ 34·5	47·142	52·418	55·185	58·041	64·020	70·362	73·672	84·168	99·522	116·48	125·58	135·11	177·58	241·01
0·03000	10·60	10·86	10·99	11·12	11·37	11·62	11·75	12·11	12·59	13·05	13·28	13·50	14·37	15·41
1/ 33·3	47·950	53·316	56·131	59·036	65·117	71·568	74·934	85·611	101·23	118·48	127·73	137·42	180·62	245·14
	0·94	0·94	0·94	0·94	0·95	0·95	0·95	0·95	0·95	0·95	0·95	0·96	0·96	0·96

$V_{r(0·5)medial}$ for half-full circular pipes.

A52
(p.1 of 6)

k_s = 0·60 mm
S = 0·00003 to 0·00010

ie hydraulic gradient =
1 in 33333 to 1 in 10000

Water (or sewage) at 15°C;
full bore conditions.

velocities in ms^{-1}
discharges in m^3s^{-1}

Gradient **(Equivalent) Pipe diameters in mm**

Gradient	2400	2500	2550	2600	2700	2800	2850	3000	3200	3400	3500	3600	4000	4500
0·0000300	0·301	0·309	0·313	0·316	0·324	0·331	0·335	0·346	0·360	0·374	0·380	0·387	0·413	0·444
1/ 33333	1·3613	1·5153	1·5961	1·6796	1·8544	2·0400	2·1369	2·4445	2·8950	3·3932	3·6607	3·9409	5·1919	7·0643
0·0000320	0·311	0·319	0·323	0·327	0·335	0·343	0·346	0·358	0·372	0·386	0·393	0·400	0·427	0·459
1/ 31250	1·4075	1·5667	1·6503	1·7365	1·9172	2·1091	2·2093	2·5272	2·9928	3·5077	3·7843	4·0739	5·3668	7·3019
0·0000340	0·321	0·329	0·333	0·337	0·345	0·353	0·357	0·369	0·384	0·399	0·406	0·413	0·441	0·474
1/ 29412	1·4523	1·6165	1·7027	1·7917	1·9781	2·1761	2·2794	2·6074	3·0876	3·6188	3·9041	4·2028	5·5364	7·5323
0·0000360	0·331	0·339	0·343	0·348	0·356	0·364	0·368	0·380	0·395	0·410	0·418	0·425	0·454	0·488
1/ 27778	1·4958	1·6649	1·7537	1·8453	2·0373	2·2411	2·3475	2·6852	3·1798	3·7267	4·0204	4·3280	5·7011	7·7561
0·0000380	0·340	0·349	0·353	0·357	0·366	0·374	0·378	0·391	0·407	0·422	0·430	0·437	0·466	0·501
1/ 26316	1·5381	1·7120	1·8033	1·8975	2·0948	2·3044	2·4138	2·7610	3·2694	3·8316	4·1336	4·4498	5·8613	7·9737
0·0000400	0·349	0·358	0·363	0·367	0·376	0·384	0·388	0·401	0·417	0·433	0·441	0·449	0·479	0·515
1/ 25000	1·5794	1·7579	1·8516	1·9483	2·1509	2·3660	2·4783	2·8348	3·3567	3·9339	4·2438	4·5684	6·0174	8·1858
0·0000420	0·358	0·367	0·372	0·376	0·385	0·394	0·398	0·411	0·428	0·444	0·452	0·460	0·491	0·528
1/ 23810	1·6196	1·8026	1·8987	1·9979	2·2056	2·4261	2·5413	2·9067	3·4418	4·0336	4·3514	4·6841	6·1696	8·3926
0·0000440	0·367	0·376	0·381	0·385	0·395	0·404	0·408	0·421	0·438	0·455	0·463	0·471	0·503	0·540
1/ 22727	1·6589	1·8463	1·9447	2·0463	2·2590	2·4849	2·6028	2·9770	3·5250	4·1310	4·4564	4·7971	6·3183	8·5946
0·0000460	0·375	0·385	0·390	0·394	0·404	0·413	0·417	0·431	0·448	0·465	0·474	0·482	0·514	0·553
1/ 21739	1·6973	1·8891	1·9897	2·0936	2·3112	2·5423	2·6629	3·0458	3·6063	4·2262	4·5591	4·9076	6·4637	8·7920
0·0000480	0·383	0·393	0·398	0·403	0·413	0·422	0·427	0·440	0·458	0·476	0·484	0·493	0·526	0·565
1/ 20833	1·7349	1·9309	2·0337	2·1399	2·3624	2·5985	2·7218	3·1130	3·6859	4·3194	4·6596	5·0158	6·6059	8·9852
0·0000500	0·392	0·402	0·407	0·412	0·421	0·431	0·436	0·450	0·468	0·486	0·495	0·503	0·537	0·577
1/ 20000	1·7717	1·9718	2·0769	2·1853	2·4124	2·6535	2·7794	3·1789	3·7638	4·4106	4·7580	5·1217	6·7452	9·1745
0·0000525	0·402	0·412	0·417	0·422	0·432	0·442	0·447	0·461	0·480	0·498	0·507	0·516	0·550	0·591
1/ 19048	1·8167	2·0219	2·1296	2·2408	2·4736	2·7208	2·8499	3·2594	3·8591	4·5222	4·8783	5·2511	6·9155	9·4058
0·0000550	0·411	0·422	0·427	0·432	0·442	0·453	0·458	0·472	0·491	0·510	0·519	0·528	0·564	0·606
1/ 18182	1·8607	2·0708	2·1811	2·2949	2·5334	2·7865	2·9187	3·3381	3·9521	4·6312	4·9958	5·3776	7·0818	9·6318
0·0000575	0·421	0·432	0·437	0·442	0·453	0·463	0·468	0·483	0·503	0·522	0·531	0·540	0·576	0·619
1/ 17391	1·9037	2·1186	2·2314	2·3479	2·5918	2·8508	2·9860	3·4150	4·0431	4·7377	5·1107	5·5012	7·2444	9·8526
0·0000600	0·430	0·441	0·447	0·452	0·463	0·473	0·478	0·494	0·514	0·533	0·543	0·552	0·589	0·633
1/ 16667	1·9457	2·1654	2·2807	2·3997	2·6490	2·9136	3·0518	3·4902	4·1321	4·8419	5·2231	5·6221	7·4035	10·069
0·0000625	0·439	0·450	0·456	0·462	0·472	0·483	0·488	0·504	0·525	0·545	0·554	0·564	0·602	0·646
1/ 16000	1·9869	2·2112	2·3290	2·4505	2·7050	2·9752	3·1163	3·5639	4·2193	4·9440	5·3332	5·7406	7·5594	10·280
0·0000650	0·448	0·460	0·465	0·471	0·482	0·493	0·498	0·514	0·535	0·556	0·566	0·575	0·614	0·659
1/ 15385	2·0273	2·2562	2·3763	2·5002	2·7599	3·0356	3·1795	3·6362	4·3048	5·0441	5·4411	5·8568	7·7121	10·488
0·0000675	0·457	0·469	0·474	0·480	0·491	0·503	0·508	0·524	0·546	0·566	0·577	0·587	0·626	0·672
1/ 14815	2·0669	2·3002	2·4227	2·5490	2·8138	3·0948	3·2415	3·7070	4·3886	5·1423	5·5470	5·9707	7·8620	10·692
0·0000700	0·465	0·477	0·483	0·489	0·501	0·512	0·518	0·534	0·556	0·577	0·587	0·598	0·637	0·685
1/ 14286	2·1058	2·3435	2·4682	2·5970	2·8667	3·1529	3·3024	3·7766	4·4710	5·2387	5·6509	6·0825	8·0091	10·891
0·0000725	0·474	0·486	0·492	0·498	0·510	0·521	0·527	0·544	0·566	0·587	0·598	0·608	0·649	0·697
1/ 13793	2·1441	2·3860	2·5130	2·6441	2·9186	3·2100	3·3622	3·8450	4·5518	5·3334	5·7530	6·1924	8·1537	11·088
0·0000750	0·482	0·495	0·501	0·507	0·519	0·530	0·536	0·553	0·576	0·598	0·608	0·619	0·660	0·709
1/ 13333	2·1816	2·4278	2·5570	2·6903	2·9697	3·2662	3·4210	3·9122	4·6313	5·4265	5·8534	6·3004	8·2957	11·281
0·0000800	0·498	0·511	0·517	0·524	0·536	0·548	0·554	0·572	0·595	0·618	0·629	0·640	0·682	0·733
1/ 12500	2·2549	2·5093	2·6429	2·7807	3·0693	3·3757	3·5357	4·0433	4·7864	5·6081	6·0493	6·5112	8·5730	11·657
0·0000850	0·514	0·527	0·534	0·540	0·553	0·565	0·572	0·590	0·614	0·637	0·648	0·660	0·704	0·756
1/ 11765	2·3260	2·5884	2·7261	2·8682	3·1659	3·4819	3·6469	4·1704	4·9368	5·7842	6·2392	6·7155	8·8417	12·022
0·0000900	0·529	0·543	0·550	0·556	0·569	0·582	0·589	0·607	0·632	0·656	0·668	0·679	0·724	0·778
1/ 11111	2·3950	2·6651	2·8069	2·9532	3·2597	3·5851	3·7549	4·2938	5·0828	5·9551	6·4235	6·9139	9·1026	12·377
0·0000950	0·544	0·558	0·565	0·572	0·585	0·599	0·605	0·624	0·650	0·674	0·686	0·698	0·745	0·800
1/ 10526	2·4622	2·7398	2·8855	3·0359	3·3510	3·6854	3·8600	4·4139	5·2248	6·1215	6·6029	7·1069	9·3564	12·721
0·0001000	0·559	0·573	0·580	0·587	0·601	0·614	0·621	0·641	0·667	0·692	0·704	0·717	0·764	0·821
1/ 10000	2·5276	2·8126	2·9621	3·1165	3·4399	3·7831	3·9623	4·5309	5·3632	6·2835	6·7776	7·2949	9·6037	13·057
	0·90	0·90	0·90	0·91	0·91	0·91	0·91	0·91	0·92	0·92	0·92	0·92	0·93	0·93

$V_{r(0·5)medial}$ for half-full circular pipes.

Gradient	(Equivalent) Pipe diameters in mm													
	2400	**2500**	**2550**	**2600**	**2700**	**2800**	**2850**	**3000**	**3200**	**3400**	**3500**	**3600**	**4000**	**4500**
0·000100	0·559	0·573	0·580	0·587	0·601	0·614	0·621	0·641	0·667	0·692	0·704	0·717	0·764	0·821
1/ 10000	2·5276	2·8126	2·9621	3·1165	3·4399	3·7831	3·9623	4·5309	5·3632	6·2835	6·7776	7·2949	9·6037	13·057
0·000105	0·573	0·587	0·595	0·602	0·616	0·630	0·637	0·657	0·684	0·709	0·722	0·735	0·783	0·842
1/ 9524	2·5913	2·8835	3·0368	3·1951	3·5266	3·8784	4·0621	4·6449	5·4981	6·4415	6·9480	7·4782	9·8448	13·385
0·000110	0·587	0·602	0·609	0·616	0·631	0·645	0·652	0·673	0·700	0·726	0·739	0·752	0·802	0·862
1/ 9091	2·6536	2·9528	3·1098	3·2718	3·6113	3·9715	4·1596	4·7563	5·6299	6·5958	7·1143	7·6572	10·080	13·705
0·000115	0·600	0·615	0·623	0·630	0·645	0·660	0·667	0·688	0·716	0·743	0·756	0·769	0·820	0·881
1/ 8696	2·7145	3·0205	3·1811	3·3469	3·6940	4·0625	4·2549	4·8652	5·7588	6·7466	7·2770	7·8323	10·311	14·017
0·000120	0·613	0·629	0·637	0·644	0·659	0·674	0·682	0·703	0·732	0·759	0·773	0·786	0·838	0·901
1/ 8333	2·7741	3·0868	3·2509	3·4203	3·7750	4·1515	4·3481	4·9718	5·8848	6·8942	7·4362	8·0036	10·536	14·323
0·000125	0·626	0·642	0·650	0·658	0·673	0·688	0·696	0·718	0·747	0·775	0·789	0·803	0·856	0·919
1/ 8000	2·8325	3·1517	3·3193	3·4922	3·8544	4·2388	4·4394	5·0762	6·0083	7·0388	7·5921	8·1713	10·756	14·623
0·000130	0·639	0·655	0·663	0·671	0·687	0·702	0·710	0·733	0·762	0·791	0·805	0·819	0·873	0·938
1/ 7692	2·8897	3·2154	3·3863	3·5627	3·9321	4·3243	4·5290	5·1785	6·1293	7·1805	7·7449	8·3357	10·973	14·917
0·000135	0·651	0·668	0·676	0·684	0·700	0·716	0·724	0·747	0·777	0·806	0·821	0·835	0·890	0·956
1/ 7407	2·9458	3·2778	3·4520	3·6318	4·0084	4·4081	4·6168	5·2789	6·2481	7·3195	7·8948	8·4971	11·185	15·205
0·000140	0·663	0·680	0·689	0·697	0·713	0·729	0·737	0·761	0·791	0·821	0·836	0·850	0·907	0·974
1/ 7143	3·0009	3·3391	3·5165	3·6997	4·0833	4·4905	4·7030	5·3774	6·3646	7·4560	8·0420	8·6554	11·393	15·488
0·000145	0·675	0·692	0·701	0·709	0·726	0·742	0·750	0·774	0·806	0·836	0·851	0·866	0·923	0·991
1/ 6897	3·0551	3·3993	3·5799	3·7664	4·1569	4·5713	4·7877	5·4742	6·4791	7·5901	8·1865	8·8110	11·598	15·766
0·000150	0·687	0·705	0·713	0·722	0·739	0·755	0·764	0·788	0·820	0·850	0·866	0·881	0·939	1·008
1/ 6667	3·1083	3·4585	3·6422	3·8319	4·2292	4·6508	4·8710	5·5694	6·5917	7·7219	8·3286	8·9639	11·799	16·039
0·000160	0·710	0·728	0·737	0·746	0·763	0·781	0·789	0·814	0·847	0·879	0·894	0·910	0·970	1·042
1/ 6250	3·2121	3·5739	3·7638	3·9598	4·3703	4·8060	5·0334	5·7550	6·8113	7·9790	8·6059	9·2622	12·191	16·572
0·000170	0·732	0·751	0·760	0·769	0·787	0·805	0·814	0·840	0·873	0·906	0·922	0·938	1·000	1·074
1/ 5882	3·3128	3·6859	3·8817	4·0838	4·5071	4·9563	5·1909	5·9350	7·0242	8·2282	8·8747	9·5514	12·572	17·088
0·000180	0·754	0·773	0·782	0·792	0·810	0·829	0·838	0·864	0·899	0·933	0·950	0·966	1·030	1·106
1/ 5556	3·4105	3·7946	3·9962	4·2042	4·6399	5·1024	5·3438	6·1097	7·2309	8·4703	9·1357	9·8323	12·941	17·590
0·000190	0·775	0·795	0·804	0·814	0·833	0·852	0·861	0·888	0·924	0·959	0·976	0·993	1·058	1·137
1/ 5263	3·5055	3·9003	4·1075	4·3213	4·7691	5·2444	5·4925	6·2797	7·4319	8·7057	9·3895	10·105	13·300	18·078
0·000200	0·795	0·816	0·826	0·835	0·855	0·874	0·884	0·912	0·948	0·984	1·002	1·019	1·086	1·167
1/ 5000	3·5981	4·0033	4·2159	4·4353	4·8949	5·3827	5·6374	6·4452	7·6278	8·9350	9·6368	10·371	13·650	18·553
0·000210	0·815	0·836	0·846	0·856	0·876	0·896	0·906	0·935	0·972	1·009	1·027	1·044	1·113	1·196
1/ 4762	3·6884	4·1037	4·3217	4·5466	5·0177	5·5176	5·7787	6·6067	7·8188	9·1586	9·8779	10·631	13·991	19·017
0·000220	0·835	0·856	0·866	0·877	0·897	0·917	0·927	0·957	0·995	1·033	1·051	1·069	1·140	1·224
1/ 4545	3·7766	4·2018	4·4250	4·6552	5·1375	5·6494	5·9166	6·7644	8·0053	9·3770	10·113	10·884	14·325	19·469
0·000230	0·854	0·876	0·886	0·897	0·918	0·938	0·949	0·979	1·018	1·056	1·075	1·094	1·166	1·252
1/ 4348	3·8628	4·2977	4·5259	4·7614	5·2547	5·7782	6·0515	6·9185	8·1876	9·5905	10·344	11·132	14·650	19·912
0·000240	0·873	0·895	0·906	0·916	0·938	0·959	0·969	1·000	1·040	1·079	1·099	1·117	1·191	1·279
1/ 4167	3·9472	4·3915	4·6247	4·8653	5·3694	5·9043	6·1835	7·0694	8·3660	9·7994	10·569	11·374	14·969	20·345
0·000250	0·891	0·913	0·925	0·936	0·957	0·979	0·990	1·021	1·062	1·102	1·121	1·141	1·216	1·306
1/ 4000	4·0298	4·4834	4·7215	4·9671	5·4817	6·0277	6·3128	7·2171	8·5408	10·004	10·790	11·612	15·281	20·769
0·000260	0·909	0·932	0·943	0·954	0·977	0·999	1·009	1·041	1·083	1·124	1·144	1·164	1·240	1·332
1/ 3846	4·1108	4·5735	4·8163	5·0669	5·5917	6·1487	6·4395	7·3619	8·7121	10·205	11·006	11·845	15·587	21·184
0·000270	0·926	0·950	0·961	0·973	0·995	1·018	1·029	1·062	1·104	1·146	1·166	1·186	1·264	1·358
1/ 3704	4·1903	4·6619	4·9094	5·1648	5·6997	6·2674	6·5638	7·5040	8·8801	10·401	11·218	12·073	15·888	21·592
0·000280	0·943	0·967	0·979	0·991	1·014	1·037	1·048	1·081	1·125	1·167	1·188	1·208	1·288	1·383
1/ 3571	4·2683	4·7486	5·0007	5·2608	5·8057	6·3839	6·6858	7·6434	9·0451	10·594	11·426	12·297	16·182	21·992
0·000290	0·960	0·985	0·997	1·009	1·032	1·055	1·067	1·101	1·145	1·188	1·209	1·230	1·311	1·408
1/ 3448	4·3449	4·8338	5·0904	5·3552	5·9098	6·4984	6·8057	7·7804	9·2071	10·784	11·631	12·517	16·472	22·385
0·000300	0·977	1·002	1·014	1·026	1·050	1·074	1·085	1·120	1·165	1·208	1·230	1·251	1·333	1·432
1/ 3333	4·4202	4·9176	5·1786	5·4480	6·0122	6·6109	6·9235	7·9151	9·3664	10·971	11·832	12·733	16·756	22·772
	0·91	0·91	0·91	0·91	0·91	0·91	0·92	0·92	0·92	0·92	0·93	0·93	0·93	0·94

$V_{r(0{\cdot}5)medial}$ for half-full circular pipes.

A52
(p.3 of 6)

k_s = 0·60 mm
S = 0·00030 to 0·00100

ie hydraulic gradient =
1 in 3333 to 1 in 1000

Water (or sewage) at 15°C;
full bore conditions.

velocities in ms^{-1}
discharges in m^3s^{-1}

Gradient **(Equivalent) Pipe diameters in mm**

Gradient	2400	2500	2550	2600	2700	2800	2850	3000	3200	3400	3500	3600	4000	4500
0·000300	0·977	1·002	1·014	1·026	1·050	1·074	1·085	1·120	1·165	1·208	1·230	1·251	1·333	1·432
1/ 3333	4·4202	4·9176	5·1786	5·4480	6·0122	6·6109	6·9235	7·9151	9·3664	10·971	11·832	12·733	16·756	22·772
0·000320	1·010	1·035	1·048	1·060	1·085	1·109	1·121	1·157	1·203	1·248	1·271	1·292	1·378	1·479
1/ 3125	4·5672	5·0810	5·3507	5·6290	6·2119	6·8305	7·1534	8·1778	9·6771	11·334	12·224	13·156	17·312	23·526
0·000340	1·041	1·067	1·080	1·093	1·119	1·144	1·156	1·193	1·241	1·287	1·310	1·333	1·420	1·525
1/ 2941	4·7096	5·2395	5·5175	5·8045	6·4055	7·0432	7·3762	8·4324	9·9784	11·687	12·604	13·565	17·850	24·257
0·000360	1·072	1·099	1·112	1·125	1·152	1·177	1·190	1·228	1·277	1·325	1·348	1·372	1·462	1·570
1/ 2778	4·8479	5·3933	5·6795	5·9748	6·5934	7·2499	7·5926	8·6797	10·271	12·030	12·974	13·962	18·372	24·966
0·000380	1·101	1·129	1·143	1·157	1·184	1·210	1·223	1·262	1·312	1·362	1·386	1·410	1·502	1·613
1/ 2632	4·9824	5·5429	5·8370	6·1405	6·7762	7·4508	7·8030	8·9202	10·555	12·363	13·333	14·349	18·881	25·657
0·000400	1·130	1·159	1·173	1·187	1·215	1·242	1·255	1·295	1·347	1·397	1·422	1·447	1·542	1·655
1/ 2500	5·1135	5·6886	5·9905	6·3019	6·9543	7·6466	8·0080	9·1545	10·832	12·687	13·683	14·725	19·376	26·329
0·000420	1·159	1·188	1·202	1·217	1·245	1·273	1·287	1·327	1·381	1·432	1·458	1·483	1·580	1·697
1/ 2381	5·2413	5·8308	6·1401	6·4594	7·1280	7·8375	8·2079	9·3830	11·103	13·004	14·024	15·092	19·858	26·985
0·000440	1·186	1·216	1·231	1·246	1·275	1·303	1·317	1·359	1·413	1·466	1·492	1·518	1·618	1·737
1/ 2273	5·3661	5·9696	6·2863	6·6131	7·2976	8·0240	8·4032	9·6061	11·367	13·313	14·357	15·451	20·330	27·625
0·000460	1·213	1·244	1·259	1·274	1·304	1·333	1·347	1·390	1·445	1·500	1·526	1·552	1·654	1·776
1/ 2174	5·4881	6·1053	6·4291	6·7634	7·4634	8·2062	8·5940	9·8242	11·625	13·615	14·683	15·801	20·791	28·251
0·000480	1·240	1·271	1·286	1·302	1·332	1·362	1·376	1·420	1·477	1·532	1·559	1·586	1·690	1·815
1/ 2083	5·6074	6·2380	6·5689	6·9104	7·6256	8·3845	8·7807	10·038	11·877	13·910	15·002	16·144	21·242	28·863
0·000500	1·265	1·297	1·313	1·329	1·360	1·390	1·405	1·450	1·508	1·564	1·592	1·619	1·726	1·853
1/ 2000	5·7243	6·3680	6·7058	7·0544	7·7845	8·5592	8·9636	10·247	12·124	14·200	15·314	16·480	21·683	29·463
0·000525	1·297	1·330	1·346	1·362	1·394	1·425	1·440	1·486	1·545	1·603	1·631	1·659	1·768	1·899
1/ 1905	5·8673	6·5270	6·8732	7·2304	7·9787	8·7727	9·1872	10·502	12·426	14·553	15·695	16·890	22·223	30·196
0·000550	1·328	1·361	1·378	1·394	1·427	1·459	1·474	1·521	1·582	1·641	1·670	1·699	1·810	1·944
1/ 1818	6·0068	6·6822	7·0366	7·4023	8·1683	8·9812	9·4055	10·752	12·721	14·899	16·068	17·291	22·750	30·912
0·000575	1·358	1·392	1·409	1·426	1·459	1·492	1·508	1·556	1·618	1·678	1·708	1·737	1·851	1·988
1/ 1739	6·1432	6·8339	7·1963	7·5704	8·3537	9·1849	9·6189	10·995	13·010	15·237	16·432	17·683	23·266	31·612
0·000600	1·387	1·422	1·440	1·457	1·491	1·524	1·541	1·589	1·653	1·715	1·745	1·775	1·892	2·031
1/ 1667	6·2767	6·9824	7·3526	7·7348	8·5351	9·3843	9·8277	11·234	13·292	15·567	16·788	18·066	23·770	32·297
0·000625	1·416	1·452	1·470	1·487	1·522	1·556	1·573	1·622	1·687	1·750	1·781	1·812	1·931	2·073
1/ 1600	6·4074	7·1277	7·5057	7·8958	8·7127	9·5796	10·032	11·468	13·569	15·891	17·137	18·442	24·264	32·968
0·000650	1·445	1·481	1·499	1·517	1·552	1·587	1·604	1·655	1·721	1·785	1·817	1·848	1·969	2·114
1/ 1538	6·5356	7·2703	7·6558	8·0536	8·8869	9·7711	10·233	11·697	13·840	16·208	17·479	18·810	24·748	33·625
0·000675	1·472	1·510	1·528	1·546	1·582	1·617	1·635	1·687	1·754	1·819	1·852	1·883	2·007	2·155
1/ 1481	6·6613	7·4101	7·8030	8·2085	9·0577	9·9588	10·429	11·922	14·105	16·519	17·815	19·171	25·223	34·270
0·000700	1·500	1·538	1·556	1·575	1·611	1·647	1·665	1·718	1·786	1·853	1·886	1·918	2·044	2·195
1/ 1429	6·7847	7·5473	7·9475	8·3605	9·2254	10·143	10·622	12·142	14·366	16·825	18·144	19·526	25·689	34·903
0·000725	1·527	1·565	1·584	1·603	1·640	1·677	1·695	1·748	1·818	1·886	1·920	1·952	2·081	2·234
1/ 1379	6·9059	7·6821	8·0894	8·5098	9·3901	10·324	10·812	12·359	14·623	17·125	18·468	19·874	26·147	35·525
0·000750	1·553	1·592	1·611	1·630	1·668	1·706	1·724	1·779	1·850	1·919	1·953	1·986	2·117	2·272
1/ 1333	7·0250	7·8146	8·2290	8·6565	9·5520	10·502	10·998	12·572	14·875	17·420	18·786	20·216	26·597	36·136
0·000800	1·604	1·645	1·665	1·684	1·723	1·762	1·781	1·837	1·911	1·982	2·017	2·052	2·186	2·347
1/ 1250	7·2575	8·0732	8·5012	8·9429	9·8679	10·850	11·362	12·987	15·366	17·995	19·407	20·884	27·475	37·329
0·000850	1·654	1·696	1·716	1·737	1·777	1·817	1·836	1·894	1·970	2·043	2·080	2·115	2·254	2·420
1/ 1176	7·4828	8·3238	8·7651	9·2204	10·174	11·186	11·714	13·390	15·843	18·553	20·008	21·531	28·326	38·484
0·000900	1·702	1·745	1·766	1·787	1·829	1·870	1·890	1·950	2·027	2·103	2·140	2·177	2·320	2·490
1/ 1111	7·7016	8·5671	9·0213	9·4899	10·471	11·513	12·057	13·781	16·305	19·095	20·592	22·159	29·153	39·607
0·000950	1·749	1·793	1·815	1·837	1·879	1·921	1·942	2·003	2·083	2·161	2·199	2·237	2·384	2·559
1/ 1053	7·9144	8·8038	9·2705	9·7521	10·761	11·831	12·390	14·162	16·755	19·622	21·160	22·771	29·956	40·699
0·001000	1·795	1·840	1·863	1·885	1·929	1·972	1·993	2·056	2·138	2·218	2·257	2·296	2·446	2·626
1/ 1000	8·1217	9·0343	9·5132	10·007	11·042	12·140	12·714	14·532	17·193	20·135	21·713	23·366	30·739	41·762
	0·91	0·91	0·91	0·91	0·92	0·92	0·92	0·92	0·92	0·93	0·93	0·93	0·93	0·94

$V_{r(0·5)medial}$ for half-full circular pipes.

k_s = 0·60 mm S = 0·00030 to 0·00100

k_s = 0·60 mm
S = 0·00100 to 0·00300

ie hydraulic gradient =
1 in 1000 to 1 in 333

Water (or sewage) at 15°C;
full bore conditions.

velocities in ms^{-1}
discharges in m^3s^{-1}

Gradient	**(Equivalent) Pipe diameters in mm**													
	2400	**2500**	**2550**	**2600**	**2700**	**2800**	**2850**	**3000**	**3200**	**3400**	**3500**	**3600**	**4000**	**4500**
0·00100	1·795	1·840	1·863	1·885	1·929	1·972	1·993	2·056	2·138	2·218	2·257	2·296	2·446	2·626
1/ 1000	8·1217	9·0343	9·5132	10·007	11·042	12·140	12·714	14·532	17·193	20·135	21·713	23·366	30·739	41·762
0·00105	1·840	1·886	1·909	1·932	1·977	2·021	2·043	2·107	2·191	2·273	2·313	2·353	2·507	2·691
1/ 952	8·3238	9·2591	9·7499	10·256	11·317	12·442	13·030	14·894	17·621	20·635	22·253	23·947	31·503	42·799
0·00110	1·884	1·931	1·954	1·978	2·023	2·069	2·091	2·157	2·243	2·327	2·368	2·408	2·566	2·755
1/ 909	8·5212	9·4787	9·9811	10·500	11·585	12·737	13·339	15·247	18·038	21·124	22·780	24·514	32·249	43·811
0·00115	1·926	1·975	1·999	2·022	2·069	2·115	2·138	2·206	2·294	2·379	2·421	2·463	2·624	2·817
1/ 870	8·7142	9·6933	10·207	10·737	11·847	13·026	13·641	15·591	18·446	21·601	23·295	25·068	32·977	44·801
0·00120	1·968	2·017	2·042	2·066	2·114	2·161	2·185	2·253	2·343	2·431	2·474	2·516	2·681	2·878
1/ 833	8·9030	9·9033	10·428	10·970	12·104	13·308	13·936	15·929	18·846	22·069	23·799	25·610	33·691	45·770
0·00125	2·009	2·059	2·084	2·109	2·158	2·206	2·230	2·300	2·392	2·481	2·525	2·568	2·737	2·937
1/ 800	9·0880	10·109	10·645	11·198	12·355	13·584	14·225	16·260	19·237	22·527	24·293	26·141	34·389	46·718
0·00130	2·049	2·100	2·126	2·151	2·201	2·250	2·274	2·346	2·440	2·531	2·575	2·619	2·791	2·996
1/ 769	9·2692	10·311	10·857	11·421	12·602	13·855	14·509	16·584	19·620	22·975	24·777	26·662	35·074	47·648
0·00135	2·088	2·141	2·167	2·192	2·243	2·293	2·318	2·391	2·486	2·579	2·625	2·670	2·845	3·053
1/ 741	9·4470	10·508	11·065	11·640	12·843	14·120	14·787	16·901	19·996	23·416	25·251	27·173	35·745	48·560
0·00140	2·127	2·180	2·207	2·233	2·285	2·336	2·361	2·435	2·532	2·627	2·673	2·719	2·897	3·110
1/ 714	9·6216	10·702	11·270	11·855	13·080	14·381	15·060	17·213	20·365	23·848	25·717	27·674	36·405	49·456
0·00145	2·165	2·219	2·246	2·273	2·325	2·377	2·403	2·479	2·577	2·673	2·721	2·767	2·949	3·165
1/ 690	9·7930	10·893	11·470	12·066	13·313	14·637	15·328	17·520	20·727	24·272	26·175	28·167	37·052	50·335
0·00150	2·202	2·257	2·285	2·312	2·365	2·418	2·444	2·521	2·622	2·719	2·767	2·815	2·999	3·219
1/ 667	9·9616	11·081	11·668	12·274	13·542	14·889	15·592	17·821	21·084	24·689	26·625	28·651	37·689	51·200
0·00160	2·275	2·332	2·360	2·388	2·443	2·498	2·525	2·604	2·708	2·809	2·859	2·908	3·098	3·325
1/ 625	10·290	11·446	12·053	12·679	13·989	15·380	16·106	18·409	21·779	25·503	27·503	29·595	38·931	52·886
0·00170	2·345	2·404	2·433	2·462	2·519	2·575	2·603	2·685	2·792	2·896	2·947	2·997	3·194	3·428
1/ 588	10·609	11·801	12·426	13·071	14·422	15·856	16·605	18·979	22·453	26·292	28·353	30·511	40·135	54·521
0·00180	2·414	2·474	2·504	2·534	2·592	2·650	2·679	2·763	2·873	2·980	3·033	3·085	3·287	3·528
1/ 556	10·919	12·145	12·788	13·452	14·843	16·319	17·089	19·532	23·107	27·058	29·179	31·399	41·304	56·109
0·00190	2·480	2·542	2·573	2·604	2·664	2·723	2·753	2·839	2·952	3·062	3·116	3·170	3·377	3·625
1/ 526	11·220	12·480	13·141	13·823	15·252	16·768	17·560	20·070	23·744	27·804	29·983	32·264	42·441	57·653
0·00200	2·545	2·609	2·640	2·672	2·733	2·794	2·824	2·913	3·029	3·142	3·198	3·252	3·465	3·720
1/ 500	11·513	12·806	13·484	14·184	15·650	17·206	18·018	20·594	24·364	28·529	30·765	33·106	43·548	59·157
0·00210	2·608	2·674	2·706	2·738	2·801	2·864	2·895	2·986	3·105	3·220	3·277	3·333	3·551	3·812
1/ 476	11·799	13·124	13·819	14·536	16·039	17·633	18·465	21·105	24·968	29·237	31·529	33·927	44·628	60·623
0·00220	2·670	2·737	2·770	2·803	2·868	2·931	2·963	3·056	3·178	3·296	3·354	3·412	3·635	3·902
1/ 455	12·078	13·434	14·146	14·880	16·418	18·050	18·902	21·604	25·558	29·928	32·274	34·729	45·683	62·055
0·00230	2·730	2·799	2·832	2·866	2·932	2·998	3·030	3·125	3·250	3·371	3·430	3·489	3·717	3·990
1/ 435	12·351	13·738	14·466	15·217	16·789	18·458	19·329	22·092	26·136	30·604	33·002	35·513	46·713	63·455
0·00240	2·789	2·859	2·894	2·928	2·996	3·062	3·095	3·193	3·320	3·444	3·504	3·564	3·798	4·076
1/ 417	12·618	14·035	14·778	15·545	17·152	18·857	19·747	22·570	26·700	31·265	33·715	36·280	47·722	64·825
0·00250	2·847	2·918	2·954	2·989	3·058	3·126	3·160	3·259	3·389	3·515	3·577	3·638	3·876	4·160
1/ 400	12·879	14·326	15·085	15·868	17·508	19·248	20·156	23·037	27·253	31·913	34·413	37·031	48·710	66·167
0·00260	2·904	2·976	3·012	3·048	3·119	3·188	3·222	3·324	3·456	3·585	3·648	3·710	3·953	4·243
1/ 385	13·136	14·611	15·385	16·183	17·856	19·631	20·557	23·495	27·795	32·547	35·098	37·768	49·678	67·482
0·00270	2·959	3·033	3·070	3·106	3·178	3·249	3·284	3·388	3·522	3·653	3·718	3·781	4·029	4·324
1/ 370	13·387	14·891	15·679	16·493	18·198	20·006	20·950	23·945	28·327	33·170	35·769	38·490	50·628	68·772
0·00280	3·014	3·089	3·127	3·164	3·237	3·309	3·345	3·450	3·587	3·721	3·786	3·851	4·103	4·404
1/ 357	13·634	15·165	15·968	16·797	18·533	20·375	21·337	24·386	28·849	33·781	36·428	39·199	51·561	70·038
0·00290	3·067	3·144	3·182	3·220	3·294	3·368	3·404	3·511	3·651	3·787	3·854	3·919	4·176	4·482
1/ 345	13·877	15·435	16·252	17·096	18·863	20·737	21·716	24·820	29·362	34·381	37·075	39·896	52·477	71·282
0·00300	3·120	3·198	3·237	3·275	3·351	3·426	3·463	3·572	3·713	3·852	3·920	3·987	4·248	4·559
1/ 333	14·115	15·700	16·531	17·390	19·187	21·093	22·089	25·246	29·866	34·971	37·712	40·580	53·377	72·505
	0·91	0·91	0·92	0·92	0·92	0·92	0·92	0·92	0·93	0·93	0·93	0·93	0·93	0·94

$V_{r(0{\cdot}5)medial}$ for half-full circular pipes.

k_s = **0·60 mm**
S = 0·00300 to 0·01000

ie hydraulic gradient =
1 in 333 to 1 in 100

Water (or sewage) at 15°C;
full bore conditions.

velocities in ms^{-1}
discharges in m^3s^{-1}

Gradient **(Equivalent) Pipe diameters in mm**

Gradient	2400	2500	2550	2600	2700	2800	2850	3000	3200	3400	3500	3600	4000	4500
0·00300	3·120	3·198	3·237	3·275	3·351	3·426	3·463	3·572	3·713	3·852	3·920	3·987	4·248	4·559
1/ 333	14·115	15·700	16·531	17·390	19·187	21·093	22·089	25·246	29·866	34·971	37·712	40·580	53·377	72·505
0·00320	3·223	3·304	3·344	3·383	3·461	3·538	3·577	3·689	3·836	3·979	4·049	4·118	4·387	4·709
1/ 313	14·580	16·217	17·076	17·962	19·819	21·788	22·816	26·077	30·849	36·123	38·953	41·916	55·134	74·890
0·00340	3·323	3·406	3·447	3·488	3·568	3·648	3·687	3·803	3·954	4·102	4·174	4·245	4·523	4·854
1/ 294	15·031	16·718	17·604	18·518	20·431	22·461	23·521	26·883	31·802	37·238	40·156	43·211	56·836	77·202
0·00360	3·419	3·505	3·547	3·589	3·672	3·754	3·794	3·914	4·069	4·221	4·295	4·369	4·654	4·995
1/ 278	15·469	17·205	18·116	19·057	21·026	23·115	24·206	27·665	32·728	38·322	41·325	44·468	58·489	79·447
0·00380	3·513	3·601	3·645	3·688	3·773	3·857	3·899	4·022	4·181	4·337	4·413	4·489	4·782	5·133
1/ 263	15·894	17·679	18·615	19·581	21·604	23·751	24·872	28·426	33·628	39·376	42·461	45·690	60·097	81·631
0·00400	3·605	3·695	3·740	3·784	3·872	3·958	4·000	4·126	4·290	4·450	4·528	4·606	4·907	5·266
1/ 250	16·309	18·140	19·101	20·092	22·168	24·371	25·520	29·168	34·504	40·402	43·568	46·881	61·663	83·758
0·00420	3·694	3·787	3·833	3·878	3·968	4·056	4·100	4·229	4·397	4·560	4·641	4·720	5·029	5·397
1/ 238	16·713	18·590	19·574	20·590	22·717	24·975	26·153	29·890	35·359	41·403	44·647	48·043	63·191	85·832
0·00440	3·782	3·877	3·923	3·970	4·061	4·152	4·196	4·328	4·500	4·668	4·750	4·831	5·147	5·524
1/ 227	17·108	19·029	20·036	21·076	23·254	25·565	26·771	30·596	36·194	42·381	45·701	49·177	64·682	87·857
0·00460	3·867	3·964	4·012	4·059	4·153	4·245	4·291	4·426	4·602	4·773	4·857	4·940	5·263	5·649
1/ 217	17·494	19·458	20·489	21·552	23·779	26·141	27·374	31·286	37·011	43·336	46·732	50·286	66·140	89·837
0·00480	3·951	4·050	4·098	4·147	4·243	4·337	4·384	4·522	4·701	4·876	4·962	5·047	5·377	5·770
1/ 208	17·872	19·878	20·931	22·017	24·292	26·705	27·965	31·962	37·809	44·271	47·740	51·370	67·567	91·775
0·00500	4·032	4·133	4·183	4·233	4·330	4·427	4·474	4·615	4·798	4·977	5·065	5·151	5·488	5·890
1/ 200	18·242	20·290	21·364	22·473	24·794	27·258	28·544	32·623	38·591	45·187	48·727	52·433	68·964	93·672
0·00525	4·132	4·236	4·287	4·338	4·438	4·536	4·585	4·730	4·917	5·100	5·190	5·279	5·624	6·036
1/ 190	18·694	20·792	21·893	23·030	25·409	27·933	29·251	33·431	39·547	46·306	49·934	53·731	70·671	95·991
0·00550	4·230	4·336	4·388	4·440	4·543	4·644	4·694	4·841	5·033	5·221	5·313	5·403	5·757	6·178
1/ 182	19·136	21·283	22·410	23·573	26·009	28·593	29·942	34·220	40·481	47·399	51·112	54·999	72·339	98·256
0·00575	4·325	4·434	4·487	4·540	4·645	4·748	4·799	4·950	5·147	5·338	5·432	5·525	5·886	6·317
1/ 174	19·567	21·763	22·916	24·105	26·595	29·238	30·617	34·992	41·393	48·467	52·264	56·239	73·969	100·47
0·00600	4·419	4·529	4·584	4·638	4·745	4·851	4·903	5·057	5·258	5·453	5·549	5·644	6·013	6·453
1/ 167	19·989	22·233	23·410	24·625	27·169	29·868	31·277	35·747	42·286	49·513	53·392	57·452	75·564	102·64
0·00625	4·510	4·623	4·679	4·734	4·843	4·951	5·004	5·162	5·367	5·566	5·664	5·761	6·137	6·587
1/ 160	20·403	22·693	23·895	25·135	27·731	30·486	31·924	36·486	43·160	50·537	54·496	58·640	77·126	104·76
0·00650	4·600	4·715	4·772	4·828	4·940	5·049	5·104	5·264	5·473	5·677	5·777	5·875	6·259	6·717
1/ 154	20·808	23·144	24·369	25·634	28·282	31·092	32·558	37·210	44·018	51·540	55·578	59·804	78·657	106·84
0·00675	4·688	4·805	4·863	4·920	5·034	5·146	5·201	5·365	5·578	5·785	5·887	5·988	6·379	6·846
1/ 148	21·206	23·586	24·835	26·124	28·822	31·686	33·180	37·921	44·858	52·524	56·639	60·946	80·159	108·88
0·00700	4·774	4·893	4·952	5·011	5·127	5·241	5·297	5·463	5·680	5·892	5·995	6·098	6·496	6·972
1/ 143	21·596	24·020	25·292	26·605	29·353	32·269	33·791	38·619	45·684	53·491	57·681	62·067	81·633	110·88
0·00725	4·859	4·980	5·040	5·100	5·218	5·334	5·391	5·560	5·781	5·996	6·102	6·206	6·611	7·095
1/ 138	21·980	24·447	25·741	27·077	29·874	32·842	34·391	39·305	46·495	54·440	58·705	63·169	83·082	112·84
0·00750	4·942	5·066	5·127	5·187	5·307	5·425	5·483	5·656	5·880	6·099	6·206	6·312	6·725	7·217
1/ 133	22·357	24·866	26·183	27·541	30·386	33·405	34·980	39·978	47·291	55·373	59·711	64·251	84·505	114·78
0·00800	5·104	5·232	5·295	5·358	5·482	5·603	5·664	5·842	6·074	6·299	6·410	6·520	6·946	7·454
1/ 125	23·092	25·684	27·044	28·447	31·385	34·503	36·131	41·293	48·846	57·193	61·674	66·363	87·283	118·55
0·00850	5·262	5·394	5·459	5·523	5·651	5·776	5·838	6·022	6·261	6·494	6·608	6·721	7·160	7·684
1/ 118	23·805	26·477	27·878	29·325	32·354	35·568	37·246	42·567	50·353	58·958	63·576	68·410	89·975	122·21
0·00900	5·415	5·551	5·617	5·684	5·815	5·944	6·008	6·197	6·443	6·682	6·800	6·916	7·368	7·907
1/ 111	24·497	27·246	28·689	30·177	33·294	36·602	38·328	43·804	51·817	60·671	65·424	70·398	92·589	125·76
0·00950	5·564	5·703	5·772	5·840	5·975	6·108	6·173	6·367	6·620	6·866	6·987	7·106	7·570	8·124
1/ 105	25·170	27·995	29·477	31·006	34·209	37·607	39·381	45·007	53·240	62·337	67·220	72·331	95·131	129·21
0·01000	5·709	5·852	5·922	5·992	6·130	6·267	6·334	6·533	6·792	7·045	7·169	7·291	7·767	8·336
1/ 100	25·826	28·724	30·245	31·814	35·100	38·587	40·407	46·179	54·626	63·960	68·970	74·214	97·608	132·57
	0·91	0·92	0·92	0·92	0·92	0·92	0·92	0·92	0·93	0·93	0·93	0·93	0·94	0·94

$V_{r(0{\cdot}5)medial}$ for half-full circular pipes.

k_s = 0·60 mm S = 0·00300 to 0·01000

Gradient	**(Equivalent) Pipe diameters in mm**													
	2400	**2500**	**2550**	**2600**	**2700**	**2800**	**2850**	**3000**	**3200**	**3400**	**3500**	**3600**	**4000**	**4500**
0·01000	5·709	5·852	5·922	5·992	6·130	6·267	6·334	6·533	6·792	7·045	7·169	7·291	7·767	8·336
1/ 100	25·826	28·724	30·245	31·814	35·100	38·587	40·407	46·179	54·626	63·960	68·970	74·214	97·608	132·57
0·01050	5·850	5·997	6·069	6·140	6·282	6·422	6·491	6·695	6·960	7·219	7·346	7·472	7·960	8·542
1/ 95	26·465	29·435	30·994	32·602	35·969	39·542	41·407	47·322	55·978	65·543	70·677	76·051	100·02	135·85
0·01100	5·988	6·138	6·212	6·285	6·430	6·573	6·644	6·853	7·124	7·389	7·519	7·648	8·147	8·743
1/ 91	27·090	30·130	31·725	33·371	36·817	40·475	42·384	48·439	57·298	67·089	72·344	77·844	102·38	139·05
0·01150	6·123	6·276	6·352	6·427	6·575	6·721	6·793	7·007	7·285	7·556	7·689	7·820	8·331	8·940
1/ 87	27·700	30·809	32·440	34·122	37·647	41·386	43·338	49·530	58·589	68·600	73·973	79·597	104·69	142·18
0·01200	6·255	6·412	6·489	6·565	6·717	6·866	6·940	7·158	7·442	7·719	7·854	7·988	8·510	9·133
1/ 83	28·297	31·473	33·139	34·858	38·458	42·278	44·272	50·597	59·852	70·078	75·567	81·312	106·94	145·25
0·01250	6·384	6·544	6·623	6·701	6·856	7·008	7·083	7·306	7·596	7·878	8·017	8·153	8·686	9·321
1/ 80	28·882	32·123	33·824	35·579	39·253	43·152	45·187	51·643	61·088	71·526	77·128	82·992	109·15	148·25
0·01300	6·511	6·674	6·754	6·834	6·992	7·147	7·224	7·451	7·746	8·034	8·176	8·315	8·858	9·506
1/ 77	29·456	32·761	34·495	36·285	40·032	44·009	46·084	52·668	62·300	72·945	78·659	84·639	111·32	151·19
0·01350	6·635	6·801	6·883	6·965	7·125	7·284	7·362	7·593	7·894	8·188	8·332	8·474	9·027	9·687
1/ 74	30·018	33·386	35·154	36·977	40·797	44·849	46·964	53·673	63·489	74·337	80·160	86·254	113·44	154·07
0·01400	6·757	6·926	7·010	7·093	7·256	7·418	7·497	7·733	8·039	8·338	8·485	8·630	9·193	9·866
1/ 71	30·570	34·000	35·800	37·657	41·547	45·673	47·827	54·660	64·657	75·704	81·633	87·840	115·53	156·90
0·01450	6·877	7·049	7·134	7·219	7·385	7·549	7·630	7·870	8·182	8·486	8·635	8·783	9·356	10·04
1/ 69	31·112	34·603	36·435	38·325	42·284	46·484	48·676	55·629	65·803	77·046	83·081	89·397	117·57	159·69
0·01500	6·995	7·170	7·257	7·342	7·512	7·678	7·761	8·005	8·322	8·631	8·783	8·933	9·516	10·21
1/ 67	31·645	35·196	37·059	38·982	43·008	47·280	49·509	56·582	66·930	78·366	84·504	90·928	119·59	162·42
0·01600	7·225	7·406	7·495	7·584	7·758	7·931	8·016	8·268	8·596	8·915	9·072	9·227	9·829	10·55
1/ 63	32·685	36·353	38·277	40·263	44·421	48·833	51·136	58·441	69·129	80·940	87·280	93·915	123·52	167·75
0·01700	7·448	7·634	7·726	7·817	7·998	8·175	8·263	8·523	8·861	9·190	9·351	9·511	10·13	10·87
1/ 59	33·693	37·474	39·458	41·505	45·791	50·339	52·713	60·243	71·261	83·436	89·970	96·810	127·32	172·93
0·01800	7·664	7·856	7·951	8·044	8·230	8·413	8·503	8·770	9·118	9·457	9·623	9·787	10·43	11·19
1/ 56	34·672	38·563	40·604	42·710	47·121	51·801	54·244	61·993	73·330	85·859	92·583	99·622	131·02	177·95
0·01900	7·875	8·072	8·169	8·265	8·456	8·644	8·736	9·011	9·368	9·716	9·887	10·06	10·71	11·50
1/ 53	35·624	39·621	41·719	43·883	48·415	53·223	55·733	63·694	75·343	88·215	95·124	102·36	134·61	182·83
0·02000	8·080	8·282	8·381	8·480	8·676	8·869	8·964	9·245	9·612	9·969	10·14	10·32	10·99	11·79
1/ 50	36·551	40·653	42·804	45·025	49·675	54·608	57·184	65·352	77·303	90·510	97·599	105·02	138·12	187·58
0·02100	8·280	8·487	8·589	8·690	8·891	9·088	9·186	9·474	9·850	10·22	10·40	10·57	11·26	12·09
1/ 47·6	37·456	41·658	43·863	46·139	50·904	55·959	58·598	66·968	79·215	92·749	100·01	107·62	141·53	192·22
0·02200	8·475	8·687	8·791	8·895	9·100	9·302	9·402	9·697	10·08	10·46	10·64	10·82	11·53	12·37
1/ 45·5	38·339	42·640	44·897	47·226	52·104	57·278	59·979	68·547	81·082	94·935	102·37	110·15	144·87	196·75
0·02300	8·666	8·882	8·989	9·095	9·305	9·512	9·614	9·916	10·31	10·69	10·88	11·07	11·79	12·65
1/ 43·5	39·202	43·600	45·908	48·290	53·276	58·568	61·329	70·090	82·907	97·072	104·67	112·63	148·13	201·18
0·02400	8·852	9·074	9·183	9·291	9·505	9·716	9·821	10·13	10·53	10·92	11·11	11·30	12·04	12·92
1/ 41·7	40·047	44·540	46·897	49·330	54·424	59·829	62·651	71·599	84·693	99·162	106·93	115·06	151·32	205·51
0·02500	9·035	9·261	9·373	9·483	9·702	9·917	10·02	10·34	10·75	11·15	11·34	11·54	12·29	13·19
1/ 40·0	40·874	45·460	47·866	50·349	55·548	61·065	63·945	73·078	86·442	101·21	109·14	117·43	154·44	209·75
0·02600	9·214	9·445	9·558	9·671	9·894	10·11	10·22	10·54	10·96	11·37	11·57	11·77	12·53	13·45
1/ 38·5	41·685	46·362	48·815	51·348	56·650	62·276	65·213	74·527	88·156	103·22	111·30	119·76	157·50	213·91
0·02700	9·390	9·625	9·741	9·856	10·08	10·31	10·42	10·74	11·17	11·59	11·79	11·99	12·77	13·71
1/ 37·0	42·480	47·246	49·747	52·327	57·731	63·464	66·457	75·949	89·838	105·19	113·42	122·05	160·51	217·99
0·02800	9·563	9·802	9·920	10·04	10·27	10·50	10·61	10·94	11·38	11·80	12·01	12·21	13·01	13·96
1/ 35·7	43·261	48·114	50·661	53·289	58·792	64·631	67·678	77·345	91·489	107·12	115·51	124·29	163·46	222·00
0·02900	9·732	9·976	10·10	10·21	10·45	10·68	10·80	11·14	11·58	12·01	12·22	12·43	13·24	14·21
1/ 34·5	44·028	48·967	51·559	54·234	59·834	65·776	68·878	78·716	93·110	109·02	117·55	126·49	166·35	225·93
0·03000	9·899	10·15	10·27	10·39	10·63	10·87	10·98	11·33	11·78	12·21	12·43	12·64	13·46	14·45
1/ 33·3	44·782	49·806	52·442	55·162	60·858	66·902	70·057	80·063	94·704	110·88	119·57	128·66	169·20	229·80
	0·91	0·92	0·92	0·92	0·92	0·92	0·92	0·92	0·93	0·93	0·93	0·93	0·94	0·94

$V_{r(0{\cdot}5)medial}$ **for half-full circular pipes.**

A53

(p.1 of 6)

$k_s = 1{\cdot}50$ mm
S = 0·00003 to 0·00010

ie hydraulic gradient =
1 in 33333 to 1 in 10000

Water (or sewage) at 15°C;
full bore conditions.

velocities in ms^{-1}
discharges in m^3s^{-1}

Gradient **(Equivalent) Pipe diameters in mm**

Gradient	2400	2500	2550	2600	2700	2800	2850	3000	3200	3400	3500	3600	4000	4500
0·0000300	0·278	0·285	0·289	0·292	0·299	0·306	0·309	0·320	0·333	0·345	0·352	0·358	0·382	0·411
1/ 33333	1·2575	1·3998	1·4745	1·5516	1·7132	1·8848	1·9744	2·2587	2·6751	3·1358	3·3832	3·6423	4·7993	6·5315
0·0000320	0·287	0·295	0·298	0·302	0·309	0·316	0·320	0·330	0·344	0·357	0·363	0·370	0·395	0·424
1/ 31250	1·2995	1·4466	1·5238	1·6034	1·7704	1·9476	2·0402	2·3340	2·7642	3·2402	3·4958	3·7635	4·9589	6·7485
0·0000340	0·296	0·304	0·308	0·311	0·319	0·326	0·330	0·341	0·354	0·368	0·375	0·381	0·407	0·438
1/ 29412	1·3402	1·4919	1·5715	1·6536	1·8258	2·0086	2·1040	2·4070	2·8506	3·3414	3·6050	3·8810	5·1136	6·9589
0·0000360	0·305	0·313	0·317	0·321	0·328	0·336	0·340	0·351	0·365	0·379	0·386	0·392	0·419	0·450
1/ 27778	1·3798	1·5359	1·6178	1·7024	1·8796	2·0678	2·1660	2·4778	2·9345	3·4397	3·7110	3·9951	5·2638	7·1631
0·0000380	0·313	0·322	0·326	0·330	0·337	0·345	0·349	0·360	0·375	0·389	0·396	0·403	0·431	0·463
1/ 26316	1·4182	1·5787	1·6629	1·7498	1·9320	2·1253	2·2263	2·5468	3·0161	3·5353	3·8141	4·1061	5·4100	7·3618
0·0000400	0·322	0·330	0·334	0·338	0·346	0·354	0·358	0·370	0·385	0·400	0·407	0·414	0·442	0·475
1/ 25000	1·4557	1·6204	1·7068	1·7960	1·9829	2·1814	2·2850	2·6139	3·0956	3·6284	3·9146	4·2142	5·5523	7·5553
0·0000420	0·330	0·338	0·343	0·347	0·355	0·363	0·367	0·379	0·395	0·410	0·417	0·424	0·453	0·487
1/ 23810	1·4922	1·6610	1·7496	1·8411	2·0327	2·2361	2·3423	2·6794	3·1731	3·7192	4·0125	4·3196	5·6911	7·7441
0·0000440	0·338	0·346	0·351	0·355	0·363	0·372	0·376	0·388	0·404	0·419	0·427	0·434	0·464	0·499
1/ 22727	1·5279	1·7007	1·7914	1·8851	2·0812	2·2895	2·3982	2·7434	3·2488	3·8079	4·1082	4·4226	5·8266	7·9284
0·0000460	0·345	0·354	0·359	0·363	0·372	0·380	0·385	0·397	0·413	0·429	0·437	0·444	0·474	0·510
1/ 21739	1·5628	1·7395	1·8323	1·9281	2·1287	2·3417	2·4529	2·8059	3·3228	3·8946	4·2017	4·5233	5·9591	8·1085
0·0000480	0·353	0·362	0·367	0·371	0·380	0·389	0·393	0·406	0·422	0·438	0·446	0·454	0·485	0·521
1/ 20833	1·5969	1·7775	1·8723	1·9701	2·1751	2·3928	2·5064	2·8670	3·3952	3·9794	4·2932	4·6217	6·0888	8·2848
0·0000500	0·360	0·370	0·374	0·379	0·388	0·397	0·401	0·414	0·431	0·447	0·456	0·464	0·495	0·532
1/ 20000	1·6304	1·8147	1·9115	2·0114	2·2206	2·4428	2·5588	2·9270	3·4661	4·0625	4·3828	4·7182	6·2158	8·4575
0·0000525	0·369	0·379	0·384	0·388	0·398	0·407	0·411	0·424	0·442	0·459	0·467	0·475	0·507	0·545
1/ 19048	1·6712	1·8602	1·9594	2·0617	2·2762	2·5039	2·6228	3·0002	3·5528	4·1641	4·4923	4·8361	6·3710	8·6685
0·0000550	0·378	0·388	0·393	0·398	0·407	0·416	0·421	0·435	0·452	0·470	0·478	0·486	0·519	0·558
1/ 18182	1·7111	1·9046	2·0061	2·1109	2·3305	2·5636	2·6854	3·0717	3·6375	4·2632	4·5993	4·9512	6·5226	8·8746
0·0000575	0·387	0·397	0·402	0·407	0·416	0·426	0·431	0·444	0·463	0·480	0·489	0·497	0·531	0·571
1/ 17391	1·7501	1·9480	2·0518	2·1590	2·3836	2·6220	2·7465	3·1416	3·7202	4·3602	4·7039	5·0638	6·6707	9·0761
0·0000600	0·395	0·405	0·411	0·416	0·425	0·435	0·440	0·454	0·473	0·491	0·500	0·508	0·542	0·583
1/ 16667	1·7883	1·9904	2·0966	2·2061	2·4355	2·6791	2·8063	3·2100	3·8012	4·4550	4·8062	5·1739	6·8157	9·2732
0·0000625	0·404	0·414	0·419	0·424	0·434	0·444	0·449	0·464	0·482	0·501	0·510	0·519	0·554	0·595
1/ 16000	1·8257	2·0320	2·1404	2·2522	2·4864	2·7351	2·8649	3·2770	3·8805	4·5479	4·9064	5·2817	6·9577	9·4663
0·0000650	0·412	0·422	0·428	0·433	0·443	0·453	0·458	0·473	0·492	0·511	0·520	0·529	0·565	0·607
1/ 15385	1·8623	2·0728	2·1833	2·2973	2·5363	2·7899	2·9224	3·3427	3·9582	4·6390	5·0046	5·3875	7·0969	9·6555
0·0000675	0·420	0·430	0·436	0·441	0·452	0·462	0·467	0·482	0·502	0·521	0·530	0·539	0·576	0·619
1/ 14815	1·8983	2·1128	2·2254	2·3417	2·5852	2·8437	2·9787	3·4071	4·0345	4·7283	5·1010	5·4912	7·2334	9·8411
0·0000700	0·427	0·438	0·444	0·449	0·460	0·470	0·476	0·491	0·511	0·530	0·540	0·549	0·586	0·630
1/ 14286	1·9336	2·1521	2·2668	2·3852	2·6332	2·8965	3·0340	3·4703	4·1093	4·8160	5·1956	5·5930	7·3675	10·023
0·0000725	0·435	0·446	0·452	0·457	0·468	0·479	0·484	0·500	0·520	0·540	0·550	0·559	0·597	0·641
1/ 13793	1·9682	2·1906	2·3074	2·4279	2·6804	2·9484	3·0883	3·5325	4·1828	4·9022	5·2885	5·6930	7·4991	10·202
0·0000750	0·443	0·454	0·460	0·465	0·476	0·487	0·492	0·508	0·529	0·549	0·559	0·569	0·607	0·653
1/ 13333	2·0023	2·2286	2·3473	2·4699	2·7267	2·9994	3·1417	3·5935	4·2551	4·9868	5·3798	5·7913	7·6285	10·378
0·0000800	0·457	0·469	0·475	0·481	0·492	0·503	0·509	0·525	0·547	0·567	0·578	0·588	0·627	0·674
1/ 12500	2·0688	2·3025	2·4252	2·5519	2·8172	3·0989	3·2459	3·7127	4·3961	5·1520	5·5580	5·9831	7·8810	10·722
0·0000850	0·472	0·484	0·490	0·496	0·507	0·519	0·525	0·542	0·564	0·585	0·596	0·606	0·647	0·695
1/ 11765	2·1332	2·3742	2·5007	2·6313	2·9049	3·1953	3·3469	3·8281	4·5328	5·3122	5·7307	6·1690	8·1258	11·054
0·0000900	0·485	0·498	0·504	0·510	0·522	0·534	0·540	0·557	0·580	0·602	0·613	0·624	0·666	0·715
1/ 11111	2·1958	2·4439	2·5741	2·7085	2·9900	3·2889	3·4450	3·9403	4·6655	5·4677	5·8984	6·3495	8·3634	11·378
0·0000950	0·499	0·512	0·518	0·524	0·537	0·549	0·555	0·573	0·596	0·619	0·630	0·641	0·684	0·735
1/ 10526	2·2567	2·5116	2·6454	2·7835	3·0728	3·3800	3·5404	4·0493	4·7946	5·6189	6·0616	6·5251	8·5945	11·692
0·0001000	0·512	0·525	0·532	0·538	0·551	0·563	0·570	0·588	0·612	0·635	0·647	0·658	0·702	0·754
1/ 10000	2·3159	2·5775	2·7148	2·8566	3·1535	3·4687	3·6333	4·1556	4·9204	5·7662	6·2205	6·6961	8·8197	11·998
	0·86	0·86	0·86	0·87	0·87	0·87	0·87	0·87	0·88	0·88	0·88	0·88	0·89	0·90

$V_{r(0{\cdot}5)medial}$ for half-full circular pipes.

$k_s = 1{\cdot}50$ mm S = 0·00003 to 0·00010

k_s = **1·50 mm**
S = 0·00010 to 0·00030

ie hydraulic gradient =
1 in 10000 to 1 in 3333

Water (or sewage) at 15°C;
full bore conditions.

velocities in ms^{-1}
discharges in m^3s^{-1}

Gradient **(Equivalent) Pipe diameters in mm**

Gradient	2400	2500	2550	2600	2700	2800	2850	3000	3200	3400	3500	3600	4000	4500
0·000100	0·512	0·525	0·532	0·538	0·551	0·563	0·570	0·588	0·612	0·635	0·647	0·658	0·702	0·754
1/ 10000	2·3159	2·5775	2·7148	2·8566	3·1535	3·4687	3·6333	4·1556	4·9204	5·7662	6·2205	6·6961	8·8197	11·998
0·000105	0·525	0·538	0·545	0·551	0·565	0·577	0·584	0·603	0·627	0·651	0·663	0·674	0·719	0·773
1/ 9524	2·3738	2·6419	2·7826	2·9279	3·2322	3·5552	3·7239	4·2592	5·0430	5·9099	6·3754	6·8629	9·0392	12·296
0·000110	0·537	0·551	0·558	0·565	0·578	0·591	0·598	0·617	0·642	0·666	0·678	0·690	0·736	0·791
1/ 9091	2·4302	2·7047	2·8487	2·9975	3·3090	3·6397	3·8124	4·3603	5·1627	6·0502	6·5267	7·0257	9·2536	12·588
0·000115	0·549	0·564	0·570	0·577	0·591	0·605	0·611	0·631	0·656	0·681	0·694	0·706	0·753	0·809
1/ 8696	2·4854	2·7661	2·9134	3·0655	3·3841	3·7223	3·8988	4·4592	5·2798	6·1873	6·6746	7·1849	9·4632	12·873
0·000120	0·561	0·576	0·583	0·590	0·604	0·618	0·624	0·645	0·671	0·696	0·709	0·721	0·769	0·827
1/ 8333	2·5394	2·8262	2·9767	3·1321	3·4575	3·8031	3·9835	4·5560	5·3943	6·3215	6·8194	7·3407	9·6683	13·152
0·000125	0·573	0·588	0·595	0·602	0·616	0·630	0·637	0·658	0·685	0·711	0·724	0·736	0·785	0·844
1/ 8000	2·5923	2·8850	3·0387	3·1973	3·5295	3·8822	4·0664	4·6508	5·5065	6·4529	6·9611	7·4933	9·8692	13·425
0·000130	0·584	0·599	0·607	0·614	0·629	0·643	0·650	0·671	0·698	0·725	0·738	0·751	0·801	0·861
1/ 7692	2·6441	2·9427	3·0994	3·2612	3·6001	3·9598	4·1476	4·7437	5·6165	6·5817	7·1001	7·6428	10·066	13·692
0·000135	0·596	0·611	0·619	0·626	0·641	0·655	0·663	0·684	0·712	0·739	0·752	0·765	0·816	0·877
1/ 7407	2·6950	2·9993	3·1590	3·3239	3·6692	4·0359	4·2273	4·8348	5·7243	6·7081	7·2364	7·7895	10·259	13·955
0·000140	0·607	0·622	0·630	0·638	0·653	0·668	0·675	0·697	0·725	0·753	0·766	0·779	0·831	0·894
1/ 7143	2·7449	3·0548	3·2175	3·3854	3·7372	4·1106	4·3056	4·9243	5·8302	6·8321	7·3702	7·9336	10·449	14·213
0·000145	0·618	0·633	0·641	0·649	0·664	0·679	0·687	0·709	0·738	0·766	0·780	0·793	0·846	0·910
1/ 6897	2·7940	3·1094	3·2750	3·4459	3·8039	4·1840	4·3824	5·0121	5·9342	6·9540	7·5016	8·0750	10·635	14·466
0·000150	0·628	0·644	0·652	0·660	0·676	0·691	0·699	0·721	0·751	0·779	0·793	0·807	0·861	0·925
1/ 6667	2·8422	3·1630	3·3315	3·5053	3·8695	4·2561	4·4579	5·0985	6·0365	7·0738	7·6308	8·2141	10·818	14·715
0·000160	0·649	0·666	0·674	0·682	0·698	0·714	0·722	0·745	0·775	0·805	0·819	0·834	0·889	0·956
1/ 6250	2·9362	3·2677	3·4417	3·6213	3·9975	4·3968	4·6053	5·2670	6·2360	7·3075	7·8829	8·4854	11·175	15·201
0·000170	0·669	0·686	0·695	0·703	0·720	0·736	0·744	0·768	0·799	0·830	0·845	0·859	0·917	0·985
1/ 5882	3·0274	3·3691	3·5485	3·7336	4·1215	4·5332	4·7482	5·4304	6·4293	7·5340	8·1272	8·7483	11·521	15·671
0·000180	0·689	0·706	0·715	0·724	0·741	0·758	0·766	0·791	0·823	0·854	0·869	0·885	0·944	1·014
1/ 5556	3·1159	3·4676	3·6522	3·8428	4·2419	4·6657	4·8869	5·5890	6·6170	7·7539	8·3644	9·0037	11·858	16·128
0·000190	0·708	0·726	0·735	0·744	0·761	0·779	0·787	0·813	0·845	0·878	0·893	0·909	0·970	1·042
1/ 5263	3·2020	3·5634	3·7531	3·9489	4·3591	4·7945	5·0218	5·7433	6·7996	7·9678	8·5952	9·2520	12·184	16·573
0·000200	0·726	0·745	0·754	0·763	0·781	0·799	0·808	0·834	0·868	0·901	0·917	0·933	0·995	1·069
1/ 5000	3·2858	3·6567	3·8514	4·0523	4·4732	4·9200	5·1533	5·8935	6·9775	8·1762	8·8199	9·4939	12·503	17·006
0·000210	0·744	0·763	0·773	0·782	0·801	0·819	0·828	0·854	0·889	0·923	0·940	0·956	1·020	1·096
1/ 4762	3·3676	3·7477	3·9472	4·1531	4·5845	5·0423	5·2814	6·0401	7·1509	8·3794	9·0391	9·7298	12·813	17·428
0·000220	0·762	0·782	0·791	0·801	0·820	0·838	0·848	0·875	0·910	0·945	0·962	0·979	1·044	1·122
1/ 4545	3·4475	3·8366	4·0408	4·2516	4·6931	5·1618	5·4066	6·1832	7·3203	8·5778	9·2531	9·9602	13·117	17·840
0·000230	0·779	0·799	0·809	0·819	0·838	0·857	0·867	0·895	0·931	0·966	0·983	1·001	1·067	1·147
1/ 4348	3·5255	3·9234	4·1323	4·3478	4·7993	5·2787	5·5289	6·3231	7·4859	8·7718	9·4623	10·185	13·413	18·243
0·000240	0·796	0·817	0·827	0·837	0·856	0·876	0·885	0·914	0·951	0·987	1·005	1·022	1·090	1·172
1/ 4167	3·6019	4·0084	4·2218	4·4420	4·9033	5·3930	5·6486	6·4600	7·6479	8·9616	9·6671	10·406	13·703	18·638
0·000250	0·813	0·834	0·844	0·854	0·874	0·894	0·904	0·933	0·971	1·008	1·026	1·043	1·113	1·196
1/ 4000	3·6767	4·0916	4·3094	4·5342	5·0051	5·5049	5·7659	6·5940	7·8066	9·1475	9·8676	10·621	13·987	19·024
0·000260	0·829	0·850	0·861	0·871	0·892	0·912	0·922	0·951	0·990	1·028	1·046	1·064	1·135	1·220
1/ 3846	3·7501	4·1732	4·3954	4·6246	5·1048	5·6146	5·8808	6·7254	7·9621	9·3297	10·064	10·833	14·266	19·402
0·000270	0·845	0·866	0·877	0·888	0·909	0·929	0·940	0·970	1·009	1·047	1·066	1·085	1·157	1·243
1/ 3704	3·8220	4·2533	4·4796	4·7133	5·2027	5·7223	5·9935	6·8543	8·1147	9·5084	10·257	11·041	14·539	19·774
0·000280	0·860	0·882	0·893	0·904	0·925	0·946	0·957	0·988	1·028	1·067	1·086	1·105	1·178	1·266
1/ 3571	3·8926	4·3319	4·5624	4·8003	5·2988	5·8279	6·1042	6·9808	8·2644	9·6839	10·446	11·244	14·807	20·138
0·000290	0·876	0·898	0·909	0·920	0·942	0·963	0·974	1·005	1·046	1·086	1·105	1·124	1·199	1·289
1/ 3448	3·9620	4·4090	4·6437	4·8859	5·3932	5·9317	6·2129	7·1051	8·4116	9·8562	10·632	11·444	15·070	20·497
0·000300	0·891	0·914	0·925	0·936	0·958	0·980	0·991	1·022	1·064	1·104	1·124	1·144	1·220	1·311
1/ 3333	4·0302	4·4849	4·7236	4·9699	5·4859	6·0337	6·3197	7·2273	8·5562	10·026	10·815	11·641	15·329	20·849
	0·86	0·87	0·87	0·87	0·87	0·87	0·87	0·88	0·88	0·88	0·89	0·89	0·89	0·90

$V_{r(0·5)medial}$ **for half-full circular pipes.**

A53
(p.3 of 6)

k_s = 1·50 mm
S = 0·00030 to 0·00100

ie hydraulic gradient =
1 in 3333 to 1 in 1000

Water (or sewage) at 15°C;
full bore conditions.

velocities in ms^{-1}
discharges in m^3s^{-1}

Gradient **(Equivalent) Pipe diameters in mm**

Gradient	2400	2500	2550	2600	2700	2800	2850	3000	3200	3400	3500	3600	4000	4500
0·000300	0·891	0·914	0·925	0·936	0·958	0·980	0·991	1·022	1·064	1·104	1·124	1·144	1·220	1·311
1/ 3333	4·0302	4·4849	4·7236	4·9699	5·4859	6·0337	6·3197	7·2273	8·5562	10·026	10·815	11·641	15·329	20·849
0·000320	0·920	0·944	0·955	0·967	0·990	1·012	1·023	1·056	1·099	1·141	1·161	1·181	1·260	1·354
1/ 3125	4·1632	4·6329	4·8795	5·1339	5·6670	6·2328	6·5282	7·4657	8·8383	10·356	11·171	12·025	15·835	21·535
0·000340	0·949	0·973	0·985	0·997	1·020	1·044	1·055	1·089	1·133	1·176	1·197	1·218	1·299	1·396
1/ 2941	4·2921	4·7764	5·0305	5·2929	5·8424	6·4257	6·7303	7·6967	9·1118	10·677	11·517	12·397	16·324	22·201
0·000360	0·976	1·001	1·014	1·026	1·050	1·074	1·086	1·121	1·166	1·210	1·232	1·253	1·337	1·437
1/ 2778	4·4173	4·9157	5·1773	5·4472	6·0127	6·6131	6·9265	7·9211	9·3773	10·988	11·852	12·758	16·800	22·847
0·000380	1·003	1·029	1·042	1·054	1·079	1·104	1·116	1·151	1·198	1·244	1·266	1·288	1·374	1·476
1/ 2632	4·5391	5·0512	5·3199	5·5973	6·1784	6·7953	7·1173	8·1393	9·6356	11·290	12·179	13·109	17·262	23·476
0·000400	1·030	1·056	1·069	1·082	1·107	1·132	1·145	1·182	1·229	1·276	1·299	1·321	1·409	1·515
1/ 2500	4·6577	5·1831	5·4589	5·7436	6·3398	6·9727	7·3032	8·3518	9·8871	11·585	12·497	13·451	17·712	24·088
0·000420	1·055	1·082	1·095	1·109	1·135	1·161	1·173	1·211	1·260	1·308	1·331	1·354	1·444	1·552
1/ 2381	4·7734	5·3118	5·5945	5·8862	6·4972	7·1458	7·4845	8·5591	10·132	11·872	12·807	13·785	18·152	24·686
0·000440	1·080	1·108	1·121	1·135	1·162	1·188	1·201	1·239	1·290	1·339	1·363	1·386	1·479	1·589
1/ 2273	4·8864	5·4375	5·7268	6·0254	6·6509	7·3148	7·6615	8·7615	10·372	12·153	13·109	14·111	18·580	25·269
0·000460	1·105	1·133	1·147	1·161	1·188	1·215	1·228	1·267	1·319	1·369	1·393	1·418	1·512	1·625
1/ 2174	4·9968	5·5604	5·8562	6·1615	6·8011	7·4800	7·8345	8·9593	10·606	12·427	13·405	14·429	19·000	25·839
0·000480	1·128	1·157	1·171	1·186	1·214	1·241	1·255	1·295	1·347	1·398	1·423	1·448	1·545	1·660
1/ 2083	5·1048	5·6806	5·9828	6·2947	6·9481	7·6417	8·0038	9·1529	10·835	12·696	13·695	14·741	19·410	26·396
0·000500	1·152	1·181	1·196	1·210	1·239	1·267	1·281	1·322	1·375	1·427	1·453	1·478	1·577	1·694
1/ 2000	5·2106	5·7983	6·1068	6·4252	7·0921	7·8000	8·1696	9·3425	11·060	12·959	13·978	15·046	19·812	26·943
0·000525	1·180	1·211	1·225	1·240	1·269	1·298	1·312	1·354	1·409	1·463	1·489	1·515	1·616	1·736
1/ 1905	5·3400	5·9422	6·2583	6·5846	7·2681	7·9935	8·3723	9·5742	11·334	13·280	14·325	15·419	20·303	27·610
0·000550	1·208	1·239	1·254	1·270	1·299	1·329	1·343	1·386	1·443	1·497	1·524	1·551	1·654	1·777
1/ 1818	5·4663	6·0827	6·4063	6·7403	7·4399	8·1825	8·5702	9·8005	11·602	13·594	14·663	15·783	20·782	28·262
0·000575	1·236	1·267	1·283	1·298	1·329	1·359	1·374	1·418	1·475	1·531	1·558	1·586	1·691	1·817
1/ 1739	5·5897	6·2201	6·5510	6·8925	7·6079	8·3672	8·7637	10·022	11·864	13·900	14·994	16·139	21·251	28·900
0·000600	1·262	1·295	1·310	1·326	1·357	1·388	1·403	1·448	1·507	1·564	1·592	1·620	1·728	1·856
1/ 1667	5·7105	6·3545	6·6925	7·0414	7·7722	8·5480	8·9530	10·238	12·120	14·201	15·318	16·488	21·710	29·523
0·000625	1·288	1·321	1·338	1·354	1·386	1·417	1·432	1·478	1·538	1·596	1·625	1·653	1·763	1·895
1/ 1600	5·8288	6·4861	6·8312	7·1873	7·9332	8·7250	9·1384	10·450	12·371	14·495	15·635	16·829	22·159	30·134
0·000650	1·314	1·348	1·364	1·381	1·413	1·445	1·461	1·508	1·569	1·628	1·657	1·686	1·798	1·932
1/ 1538	5·9448	6·6152	6·9670	7·3302	8·0910	8·8985	9·3201	10·658	12·617	14·783	15·946	17·163	22·599	30·733
0·000675	1·339	1·373	1·390	1·407	1·440	1·473	1·489	1·537	1·599	1·659	1·689	1·718	1·833	1·969
1/ 1481	6·0585	6·7417	7·1003	7·4705	8·2458	9·0687	9·4984	10·862	12·858	15·065	16·251	17·492	23·031	31·320
0·000700	1·364	1·399	1·416	1·433	1·467	1·500	1·516	1·565	1·628	1·690	1·720	1·750	1·867	2·006
1/ 1429	6·1702	6·8660	7·2312	7·6081	8·3977	9·2358	9·6734	11·062	13·095	15·343	16·550	17·814	23·455	31·896
0·000725	1·388	1·424	1·441	1·458	1·493	1·527	1·543	1·593	1·657	1·720	1·751	1·781	1·900	2·041
1/ 1379	6·2799	6·9880	7·3597	7·7433	8·5469	9·3999	9·8453	11·258	13·327	15·615	16·844	18·130	23·872	32·463
0·000750	1·412	1·448	1·466	1·483	1·518	1·553	1·570	1·620	1·686	1·749	1·781	1·812	1·932	2·076
1/ 1333	6·3877	7·1080	7·4861	7·8763	8·6936	9·5612	10·014	11·452	13·556	15·883	17·133	18·441	24·281	33·019
0·000800	1·458	1·496	1·514	1·532	1·568	1·604	1·621	1·673	1·741	1·807	1·839	1·871	1·996	2·144
1/ 1250	6·5981	7·3421	7·7326	8·1356	8·9799	9·8760	10·344	11·829	14·002	16·406	17·696	19·048	25·080	34·105
0·000850	1·504	1·542	1·561	1·580	1·617	1·653	1·672	1·725	1·795	1·863	1·896	1·929	2·057	2·211
1/ 1176	6·8020	7·5689	7·9715	8·3869	9·2573	10·181	10·663	12·194	14·435	16·912	18·243	19·636	25·854	35·158
0·000900	1·547	1·587	1·606	1·626	1·664	1·702	1·720	1·775	1·847	1·917	1·951	1·985	2·117	2·275
1/ 1111	6·9999	7·7892	8·2035	8·6310	9·5266	10·477	10·974	12·549	14·855	17·404	18·773	20·207	26·606	36·180
0·000950	1·590	1·630	1·650	1·670	1·710	1·748	1·767	1·824	1·898	1·970	2·005	2·040	2·175	2·337
1/ 1053	7·1925	8·0034	8·4291	8·8684	9·7886	10·765	11·275	12·894	15·263	17·883	19·289	20·762	27·337	37·174
0·001000	1·631	1·673	1·694	1·714	1·754	1·794	1·814	1·872	1·947	2·021	2·057	2·093	2·232	2·398
1/ 1000	7·3800	8·2121	8·6489	9·0996	10·044	11·046	11·569	13·230	15·661	18·349	19·792	21·303	28·049	38·142
	0·87	0·87	0·87	0·87	0·87	0·87	0·88	0·88	0·88	0·89	0·89	0·89	0·89	0·90

$V_{r(0{\cdot}5)medial}$ for half-full circular pipes.

k_s = 1·50 mm S = 0·00030 to 0·00100

$k_s = 1{\cdot}50$ **mm**
S = 0·00100 to 0·00300

ie hydraulic gradient =
1 in 1000 to 1 in 333

Water (or sewage) at 15°C;
full bore conditions.

velocities in ms^{-1}
discharges in m^3s^{-1}

Gradient (Equivalent) Pipe diameters in mm

Gradient	2400	2500	2550	2600	2700	2800	2850	3000	3200	3400	3500	3600	4000	4500
0·00100	1·631	1·673	1·694	1·714	1·754	1·794	1·814	1·872	1·947	2·021	2·057	2·093	2·232	2·398
1/ 1000	7·3800	8·2121	8·6489	9·0996	10·044	11·046	11·569	13·230	15·661	18·349	19·792	21·303	28·049	38·142
0·00105	1·672	1·714	1·735	1·756	1·798	1·838	1·858	1·918	1·995	2·071	2·108	2·145	2·287	2·458
1/ 952	7·5629	8·4156	8·8632	9·3251	10·293	11·320	11·856	13·557	16·049	18·803	20·282	21·831	28·744	39·086
0·00110	1·711	1·755	1·776	1·798	1·840	1·882	1·902	1·963	2·043	2·120	2·158	2·195	2·341	2·516
1/ 909	7·7416	8·6144	9·0725	9·5453	10·536	11·587	12·136	13·877	16·427	19·247	20·761	22·346	29·422	40·009
0·00115	1·750	1·794	1·817	1·838	1·882	1·924	1·945	2·008	2·089	2·168	2·206	2·245	2·394	2·572
1/ 870	7·9162	8·8086	9·2771	9·7605	10·773	11·848	12·409	14·190	16·798	19·681	21·229	22·850	30·085	40·910
0·00120	1·788	1·833	1·856	1·878	1·922	1·966	1·987	2·051	2·134	2·214	2·254	2·293	2·446	2·628
1/ 833	8·0870	8·9987	9·4773	9·9711	11·006	12·104	12·677	14·496	17·160	20·105	21·687	23·342	30·733	41·792
0·00125	1·825	1·871	1·894	1·917	1·962	2·006	2·028	2·093	2·178	2·260	2·301	2·341	2·496	2·682
1/ 800	8·2543	9·1849	9·6733	10·177	11·233	12·354	12·939	14·796	17·515	20·521	22·135	23·825	31·369	42·656
0·00130	1·861	1·908	1·932	1·955	2·001	2·046	2·069	2·135	2·221	2·305	2·346	2·387	2·546	2·735
1/ 769	8·4183	9·3674	9·8655	10·380	11·456	12·600	13·196	15·090	17·863	20·928	22·574	24·298	31·991	43·502
0·00135	1·896	1·945	1·969	1·992	2·039	2·085	2·108	2·176	2·263	2·349	2·391	2·433	2·594	2·787
1/ 741	8·5792	9·5464	10·054	10·578	11·675	12·840	13·449	15·378	18·204	21·328	23·006	24·762	32·602	44·333
0·00140	1·931	1·981	2·005	2·029	2·077	2·124	2·147	2·216	2·305	2·392	2·435	2·477	2·642	2·839
1/ 714	8·7371	9·7221	10·239	10·773	11·890	13·077	13·696	15·661	18·539	21·720	23·429	25·217	33·202	45·148
0·00145	1·966	2·016	2·040	2·065	2·114	2·161	2·185	2·255	2·346	2·435	2·478	2·521	2·689	2·889
1/ 690	8·8923	9·8947	10·421	10·964	12·101	13·309	13·939	15·939	18·868	22·106	23·845	25·665	33·791	45·949
0·00150	1·999	2·050	2·075	2·100	2·150	2·198	2·222	2·294	2·386	2·476	2·521	2·565	2·735	2·939
1/ 667	9·0447	10·064	10·600	11·152	12·309	13·537	14·178	16·213	19·191	22·485	24·253	26·105	34·370	46·736
0·00160	2·065	2·118	2·144	2·170	2·221	2·271	2·296	2·369	2·465	2·558	2·604	2·649	2·825	3·035
1/ 625	9·3422	10·395	10·948	11·519	12·714	13·982	14·644	16·746	19·822	23·224	25·050	26·963	35·500	48·272
0·00170	2·129	2·183	2·210	2·236	2·289	2·341	2·366	2·442	2·541	2·637	2·684	2·731	2·912	3·129
1/ 588	9·6306	10·716	11·286	11·874	13·106	14·413	15·096	17·262	20·434	23·940	25·823	27·795	36·595	49·760
0·00180	2·191	2·247	2·274	2·301	2·356	2·409	2·435	2·513	2·615	2·713	2·762	2·810	2·997	3·220
1/ 556	9·9106	11·028	11·614	12·219	13·487	14·832	15·535	17·764	21·027	24·636	26·574	28·602	37·658	51·206
0·00190	2·251	2·308	2·337	2·365	2·420	2·475	2·502	2·582	2·686	2·788	2·838	2·887	3·079	3·308
1/ 526	10·183	11·331	11·933	12·555	13·857	15·240	15·962	18·252	21·605	25·312	27·303	29·388	38·692	52·612
0·00200	2·310	2·368	2·397	2·426	2·483	2·539	2·567	2·649	2·756	2·861	2·912	2·962	3·159	3·394
1/ 500	10·448	11·626	12·244	12·882	14·218	15·637	16·377	18·727	22·167	25·971	28·014	30·153	39·699	53·981
0·00210	2·367	2·427	2·457	2·486	2·545	2·602	2·631	2·715	2·825	2·931	2·984	3·036	3·237	3·478
1/ 476	10·707	11·914	12·547	13·201	14·570	16·024	16·783	19·191	22·716	26·614	28·707	30·899	40·681	55·316
0·00220	2·423	2·484	2·515	2·545	2·605	2·664	2·693	2·779	2·891	3·000	3·054	3·107	3·314	3·560
1/ 455	10·959	12·195	12·843	13·512	14·914	16·402	17·179	19·643	23·252	27·242	29·384	31·627	41·640	56·620
0·00230	2·477	2·540	2·571	2·602	2·663	2·724	2·753	2·842	2·956	3·068	3·123	3·177	3·388	3·640
1/ 435	11·206	12·469	13·132	13·817	15·250	16·771	17·566	20·086	23·775	27·855	30·046	32·340	42·578	57·895
0·00240	2·531	2·595	2·627	2·658	2·721	2·782	2·813	2·903	3·020	3·134	3·190	3·246	3·461	3·719
1/ 417	11·448	12·738	13·416	14·115	15·579	17·133	17·944	20·519	24·288	28·456	30·693	33·036	43·495	59·142
0·00250	2·583	2·649	2·681	2·713	2·777	2·840	2·871	2·963	3·082	3·199	3·256	3·313	3·533	3·795
1/ 400	11·685	13·002	13·693	14·406	15·901	17·487	18·315	20·943	24·790	29·043	31·328	33·719	44·394	60·364
0·00260	2·634	2·701	2·734	2·767	2·832	2·896	2·928	3·022	3·144	3·262	3·321	3·378	3·603	3·871
1/ 385	11·917	13·260	13·965	14·692	16·216	17·834	18·678	21·358	25·282	29·620	31·949	34·388	45·274	61·561
0·00270	2·684	2·753	2·787	2·820	2·886	2·952	2·984	3·079	3·204	3·325	3·384	3·443	3·672	3·945
1/ 370	12·144	13·513	14·231	14·973	16·526	18·174	19·035	21·766	25·764	30·185	32·559	35·044	46·138	62·736
0·00280	2·734	2·803	2·838	2·872	2·939	3·006	3·039	3·136	3·262	3·386	3·446	3·506	3·739	4·017
1/ 357	12·367	13·761	14·493	15·248	16·830	18·508	19·385	22·166	26·238	30·740	33·157	35·688	46·986	63·889
0·00290	2·782	2·853	2·888	2·923	2·992	3·059	3·093	3·191	3·320	3·446	3·507	3·568	3·805	4·088
1/ 345	12·587	14·005	14·750	15·518	17·128	18·837	19·729	22·559	26·703	31·285	33·745	36·321	47·819	65·021
0·00300	2·830	2·902	2·938	2·973	3·043	3·112	3·146	3·246	3·377	3·505	3·567	3·629	3·870	4·158
1/ 333	12·802	14·245	15·003	15·784	17·422	19·159	20·067	22·946	27·160	31·821	34·323	36·943	48·638	66·135
	0·87	0·87	0·87	0·87	0·87	0·88	0·88	0·88	0·88	0·89	0·89	0·89	0·89	0·90

$V_{r(0{\cdot}5)medial}$ **for half-full circular pipes.**

k_s = 1·50 mm
S = 0·00300 to 0·01000

ie hydraulic gradient =
1 in 333 to 1 in 100

Water (or sewage) at 15°C;
full bore conditions.

velocities in ms^{-1}
discharges in m^3s^{-1}

Gradient **(Equivalent) Pipe diameters in mm**

	2400	**2500**	**2550**	**2600**	**2700**	**2800**	**2850**	**3000**	**3200**	**3400**	**3500**	**3600**	**4000**	**4500**
0·00300	2·830	2·902	2·938	2·973	3·043	3·112	3·146	3·246	3·377	3·505	3·567	3·629	3·870	4·158
1/ 333	12·802	14·245	15·003	15·784	17·422	19·159	20·067	22·946	27·160	31·821	34·323	36·943	48·638	66·135
0·00320	2·923	2·997	3·034	3·071	3·143	3·214	3·249	3·353	3·488	3·620	3·685	3·749	3·998	4·295
1/ 313	13·223	14·714	15·496	16·303	17·994	19·789	20·726	23·700	28·053	32·866	35·451	38·157	50·236	68·307
0·00340	3·013	3·090	3·128	3·165	3·240	3·313	3·349	3·456	3·596	3·732	3·798	3·864	4·121	4·427
1/ 294	13·631	15·167	15·974	16·806	18·549	20·399	21·365	24·430	28·918	33·879	36·544	39·333	51·784	70·412
0·00360	3·101	3·180	3·219	3·257	3·334	3·409	3·446	3·557	3·700	3·840	3·909	3·976	4·240	4·556
1/ 278	14·027	15·608	16·438	17·294	19·088	20·992	21·986	25·140	29·757	34·863	37·605	40·475	53·287	72·456
0·00380	3·186	3·267	3·307	3·347	3·425	3·503	3·541	3·654	3·802	3·945	4·016	4·086	4·357	4·681
1/ 263	14·412	16·036	16·889	17·769	19·612	21·568	22·589	25·830	30·574	35·820	38·637	41·586	54·750	74·444
0·00400	3·269	3·352	3·393	3·434	3·514	3·594	3·633	3·749	3·901	4·048	4·120	4·192	4·470	4·803
1/ 250	14·787	16·454	17·328	18·231	20·122	22·129	23·177	26·502	31·370	36·752	39·642	42·668	56·174	76·381
0·00420	3·350	3·435	3·477	3·519	3·601	3·683	3·723	3·842	3·997	4·148	4·222	4·296	4·581	4·921
1/ 238	15·153	16·861	17·757	18·682	20·620	22·677	23·750	27·157	32·146	37·661	40·622	43·723	57·563	78·269
0·00440	3·429	3·516	3·559	3·602	3·686	3·770	3·811	3·933	4·091	4·246	4·322	4·397	4·689	5·037
1/ 227	15·510	17·258	18·176	19·123	21·106	23·211	24·310	27·798	32·903	38·548	41·580	44·753	58·920	80·113
0·00460	3·506	3·595	3·639	3·683	3·769	3·854	3·897	4·021	4·183	4·341	4·419	4·496	4·794	5·151
1/ 217	15·860	17·647	18·585	19·553	21·581	23·734	24·857	28·423	33·644	39·416	42·516	45·761	60·245	81·916
0·00480	3·581	3·672	3·717	3·762	3·850	3·937	3·980	4·108	4·273	4·435	4·514	4·593	4·897	5·261
1/ 208	16·201	18·027	18·985	19·974	22·046	24·245	25·393	29·036	34·368	40·265	43·431	46·746	61·543	83·680
0·00500	3·655	3·748	3·794	3·840	3·930	4·019	4·063	4·193	4·362	4·526	4·607	4·687	4·999	5·370
1/ 200	16·536	18·399	19·377	20·387	22·501	24·746	25·917	29·635	35·078	41·096	44·328	47·711	62·814	85·407
0·00525	3·746	3·841	3·888	3·935	4·027	4·118	4·163	4·296	4·469	4·638	4·721	4·803	5·122	5·503
1/ 190	16·945	18·854	19·857	20·891	23·058	25·358	26·558	30·368	35·946	42·113	45·424	48·891	64·367	87·519
0·00550	3·834	3·932	3·980	4·028	4·122	4·215	4·261	4·397	4·575	4·748	4·833	4·916	5·243	5·632
1/ 182	17·344	19·299	20·325	21·383	23·601	25·955	27·184	31·084	36·793	43·105	46·494	50·043	65·883	89·581
0·00575	3·920	4·020	4·069	4·118	4·215	4·310	4·357	4·496	4·678	4·855	4·941	5·027	5·361	5·759
1/ 174	17·735	19·733	20·782	21·865	24·132	26·539	27·796	31·783	37·621	44·075	47·541	51·169	67·366	91·596
0·00600	4·005	4·107	4·157	4·207	4·306	4·403	4·451	4·593	4·778	4·959	5·048	5·135	5·476	5·883
1/ 167	18·117	20·158	21·230	22·336	24·652	27·111	28·395	32·468	38·431	45·024	48·565	52·271	68·816	93·568
0·00625	4·087	4·191	4·243	4·294	4·395	4·494	4·543	4·688	4·877	5·061	5·152	5·241	5·589	6·005
1/ 160	18·491	20·575	21·668	22·797	25·161	27·671	28·981	33·138	39·224	45·954	49·567	53·350	70·237	95·500
0·00650	4·168	4·275	4·327	4·379	4·482	4·583	4·633	4·781	4·974	5·162	5·254	5·345	5·700	6·124
1/ 154	18·858	20·983	22·098	23·249	25·660	28·219	29·556	33·795	40·002	46·865	50·550	54·408	71·629	97·393
0·00675	4·248	4·356	4·409	4·462	4·567	4·670	4·721	4·872	5·069	5·260	5·354	5·447	5·809	6·240
1/ 148	19·217	21·383	22·520	23·693	26·150	28·758	30·119	34·440	40·765	47·759	51·514	55·446	72·995	99·250
0·00700	4·326	4·436	4·491	4·544	4·651	4·756	4·808	4·962	5·162	5·357	5·453	5·547	5·915	6·355
1/ 143	19·571	21·776	22·933	24·128	26·630	29·286	30·673	35·073	41·514	48·636	52·460	56·464	74·336	101·07
0·00725	4·403	4·515	4·570	4·625	4·734	4·840	4·893	5·050	5·253	5·452	5·549	5·646	6·020	6·468
1/ 138	19·917	22·162	23·340	24·556	27·102	29·805	31·216	35·694	42·250	49·498	53·390	57·465	75·653	102·86
0·00750	4·478	4·592	4·648	4·704	4·815	4·923	4·977	5·136	5·343	5·545	5·644	5·742	6·123	6·578
1/ 133	20·258	22·541	23·739	24·976	27·566	30·315	31·751	36·305	42·973	50·345	54·304	58·448	76·948	104·62
0·00800	4·625	4·743	4·801	4·859	4·973	5·085	5·141	5·305	5·519	5·727	5·830	5·931	6·324	6·794
1/ 125	20·924	23·281	24·519	25·796	28·471	31·311	32·793	37·497	44·384	51·998	56·087	60·367	79·474	108·06
0·00850	4·768	4·889	4·949	5·008	5·126	5·242	5·299	5·468	5·689	5·904	6·009	6·113	6·519	7·004
1/ 118	21·568	23·999	25·274	26·591	29·349	32·276	33·804	38·653	45·751	53·600	57·814	62·227	81·922	111·39
0·00900	4·906	5·031	5·093	5·154	5·275	5·394	5·453	5·627	5·854	6·075	6·184	6·291	6·708	7·207
1/ 111	22·195	24·695	26·008	27·363	30·200	33·212	34·785	39·774	47·079	55·156	59·492	64·033	84·299	114·62
0·00950	5·041	5·169	5·232	5·295	5·419	5·542	5·602	5·781	6·014	6·242	6·353	6·463	6·892	7·404
1/ 105	22·804	25·373	26·722	28·113	31·029	34·124	35·739	40·865	48·370	56·668	61·124	65·789	86·611	117·76
0·01000	5·172	5·303	5·368	5·433	5·560	5·686	5·748	5·932	6·171	6·404	6·518	6·631	7·072	7·597
1/ 100	23·397	26·033	27·417	28·845	31·836	35·011	36·669	41·928	49·628	58·142	62·714	67·500	88·863	120·82
	0·87	0·87	0·87	0·87	0·87	0·88	0·88	0·88	0·88	0·89	0·89	0·89	0·90	0·90

$V_{r(0·5)medial}$ for half-full circular pipes.

$k_s = 1{\cdot}50$ mm
S = 0·01000 to 0·03000

ie hydraulic gradient =
1 in 100 to 1 in 33·3

Water (or sewage) at 15°C;
full bore conditions.

velocities in ms^{-1}
discharges in m^3s^{-1}

Gradient **(Equivalent) Pipe diameters in mm**

Gradient	2400	2500	2550	2600	2700	2800	2850	3000	3200	3400	3500	3600	4000	4500
0·01000	5·172	5·303	5·368	5·433	5·560	5·686	5·748	5·932	6·171	6·404	6·518	6·631	7·072	7·597
1/ 100	23·397	26·033	27·417	28·845	31·836	35·011	36·669	41·928	49·628	58·142	62·714	67·500	88·863	120·82
0·01050	5·300	5·435	5·501	5·567	5·698	5·826	5·890	6·078	6·323	6·562	6·679	6·795	7·246	7·785
1/ 95	23·975	26·677	28·094	29·558	32·623	35·876	37·575	42·965	50·855	59·579	64·264	69·168	91·060	123·81
0·01100	5·425	5·563	5·631	5·698	5·832	5·964	6·029	6·221	6·472	6·717	6·837	6·955	7·417	7·968
1/ 91	24·540	27·305	28·756	30·254	33·392	36·722	38·460	43·977	52·053	60·983	65·777	70·797	93·204	126·73
0·01150	5·547	5·688	5·757	5·827	5·963	6·098	6·164	6·361	6·618	6·868	6·991	7·112	7·584	8·147
1/ 87	25·092	27·919	29·403	30·935	34·143	37·548	39·326	44·966	53·224	62·354	67·257	72·390	95·301	129·58
0·01200	5·666	5·810	5·881	5·952	6·092	6·229	6·297	6·498	6·760	7·016	7·141	7·265	7·747	8·323
1/ 83	25·632	28·521	30·036	31·601	34·878	38·356	40·172	45·934	54·370	63·697	68·705	73·948	97·352	132·37
0·01250	5·783	5·930	6·003	6·075	6·217	6·358	6·427	6·632	6·900	7·160	7·288	7·415	7·907	8·494
1/ 80	26·161	29·109	30·656	32·253	35·598	39·148	41·001	46·882	55·492	65·011	70·123	75·474	99·361	135·10
0·01300	5·898	6·048	6·122	6·195	6·341	6·484	6·555	6·764	7·037	7·302	7·433	7·562	8·064	8·663
1/ 77	26·680	29·686	31·264	32·893	36·303	39·924	41·814	47·812	56·592	66·300	71·513	76·970	101·33	137·77
0·01350	6·010	6·163	6·238	6·313	6·462	6·607	6·680	6·893	7·171	7·442	7·575	7·706	8·217	8·828
1/ 74	27·189	30·252	31·860	33·520	36·996	40·685	42·611	48·723	57·671	67·564	72·876	78·438	103·26	140·40
0·01400	6·120	6·276	6·353	6·429	6·580	6·729	6·802	7·020	7·302	7·578	7·714	7·848	8·368	8·990
1/ 71	27·688	30·808	32·445	34·135	37·675	41·432	43·394	49·618	58·730	68·805	74·214	79·878	105·16	142·98
0·01450	6·229	6·387	6·466	6·543	6·697	6·848	6·923	7·144	7·432	7·713	7·850	7·987	8·516	9·149
1/ 69	28·179	31·354	33·020	34·740	38·343	42·166	44·163	50·497	59·770	70·024	75·529	81·293	107·02	145·51
0·01500	6·335	6·497	6·576	6·655	6·811	6·965	7·041	7·266	7·559	7·844	7·985	8·123	8·662	9·306
1/ 67	28·661	31·890	33·585	35·335	38·999	42·888	44·918	51·361	60·793	71·222	76·821	82·684	108·85	148·00
0·01600	6·543	6·710	6·792	6·874	7·035	7·194	7·272	7·505	7·807	8·102	8·247	8·390	8·946	9·611
1/ 63	29·602	32·937	34·688	36·495	40·279	44·296	46·393	53·047	62·788	73·559	79·343	85·398	112·42	152·86
0·01700	6·745	6·917	7·001	7·085	7·252	7·415	7·496	7·736	8·048	8·351	8·501	8·648	9·222	9·907
1/ 59	30·514	33·952	35·756	37·619	41·520	45·660	47·822	54·681	64·722	75·825	81·786	88·028	115·89	157·57
0·01800	6·941	7·117	7·205	7·291	7·462	7·630	7·714	7·960	8·281	8·594	8·747	8·899	9·490	10·19
1/ 56	31·399	34·937	36·794	38·710	42·724	46·985	49·209	56·267	66·600	78·025	84·159	90·582	119·25	162·14
0·01900	7·131	7·313	7·402	7·491	7·667	7·840	7·925	8·178	8·508	8·829	8·987	9·143	9·750	10·47
1/ 53	32·260	35·895	37·803	39·772	43·896	48·273	50·559	57·810	68·426	80·164	86·467	93·065	122·52	166·58
0·02000	7·317	7·503	7·595	7·686	7·866	8·044	8·131	8·391	8·729	9·059	9·221	9·381	10·00	10·75
1/ 50	33·099	36·829	38·786	40·806	45·037	49·529	51·873	59·313	70·205	82·248	88·715	95·485	125·70	170·91
0·02100	7·497	7·688	7·782	7·876	8·060	8·242	8·332	8·598	8·945	9·283	9·449	9·613	10·25	11·01
1/ 47·6	33·917	37·739	39·744	41·814	46·150	50·753	53·155	60·779	71·940	84·281	90·907	97·844	128·81	175·14
0·02200	7·674	7·869	7·966	8·061	8·250	8·436	8·529	8·801	9·156	9·501	9·671	9·839	10·49	11·27
1/ 45·5	34·716	38·628	40·680	42·799	47·237	51·948	54·407	62·210	73·634	86·265	93·048	100·15	131·84	179·26
0·02300	7·847	8·046	8·145	8·243	8·436	8·626	8·720	8·999	9·362	9·715	9·889	10·06	10·73	11·52
1/ 43·5	35·497	39·496	41·595	43·762	48·300	53·116	55·631	63·610	75·290	88·205	95·140	102·40	134·81	183·29
0·02400	8·015	8·219	8·320	8·420	8·617	8·812	8·908	9·193	9·563	9·924	10·10	10·28	10·96	11·77
1/ 41·7	36·261	40·347	42·491	44·704	49·339	54·259	56·828	64·979	76·911	90·104	97·188	104·60	137·71	187·23
0·02500	8·181	8·389	8·492	8·594	8·795	8·994	9·092	9·382	9·760	10·13	10·31	10·49	11·18	12·02
1/ 40·0	37·009	41·179	43·368	45·626	50·357	55·379	58·001	66·319	78·498	91·963	99·193	106·76	140·55	191·10
0·02600	8·343	8·555	8·660	8·764	8·970	9·172	9·272	9·568	9·954	10·33	10·51	10·70	11·41	12·25
1/ 38·5	37·743	41·995	44·227	46·530	51·355	56·476	59·150	67·634	80·053	93·785	101·16	108·88	143·34	194·88
0·02700	8·502	8·718	8·825	8·931	9·140	9·347	9·449	9·751	10·14	10·53	10·71	10·90	11·62	12·49
1/ 37·0	38·462	42·796	45·070	47·417	52·334	57·553	60·278	68·923	81·579	95·573	103·09	110·95	146·07	198·60
0·02800	8·658	8·878	8·987	9·095	9·308	9·518	9·622	9·930	10·33	10·72	10·91	11·10	11·84	12·72
1/ 35·7	39·169	43·582	45·898	48·288	53·295	58·610	61·385	70·188	83·077	97·328	104·98	112·99	148·75	202·24
0·02900	8·812	9·036	9·146	9·256	9·473	9·687	9·793	10·11	10·51	10·91	11·10	11·30	12·05	12·94
1/ 34·5	39·862	44·354	46·711	49·143	54·239	59·648	62·472	71·432	84·548	99·051	106·84	114·99	151·38	205·83
0·03000	8·962	9·190	9·303	9·414	9·635	9·853	9·960	10·28	10·69	11·10	11·29	11·49	12·25	13·16
1/ 33·3	40·544	45·112	47·510	49·984	55·167	60·668	63·540	72·653	85·995	100·75	108·67	116·96	153·97	209·35
	0·87	0·87	0·87	0·87	0·87	0·88	0·88	0·88	0·88	0·89	0·89	0·89	0·90	0·90

$\mathbf{V_{r(0\cdot5)medial}}$ **for half-full circular pipes.**

k_s = 3·0 mm
S = 0·00003 to 0·00010

ie hydraulic gradient =
1 in 33333 to 1 in 10000

Water (or sewage) at 15°C;
full bore conditions.

velocities in ms^{-1}
discharges in m^3s^{-1}

Gradient **(Equivalent) Pipe diameters in mm**

Gradient	2400	2500	2550	2600	2700	2800	2850	3000	3200	3400	3500	3600	4000	4500
0·0000300	0·258	0·265	0·268	0·271	0·278	0·284	0·288	0·297	0·309	0·321	0·327	0·333	0·355	0·382
1/ 33333	1·1674	1·2997	1·3692	1·4410	1·5914	1·7511	1·8345	2·0992	2·4871	2·9162	3·1468	3·3883	4·4671	6·0831
0·0000320	0·267	0·274	0·277	0·280	0·287	0·294	0·297	0·307	0·319	0·332	0·338	0·344	0·367	0·395
1/ 31250	1·2061	1·3428	1·4146	1·4888	1·6441	1·8091	1·8952	2·1687	2·5694	3·0127	3·2509	3·5004	4·6148	6·2841
0·0000340	0·275	0·282	0·286	0·289	0·296	0·303	0·306	0·316	0·329	0·342	0·348	0·355	0·379	0·407
1/ 29412	1·2436	1·3846	1·4586	1·5350	1·6952	1·8653	1·9541	2·2361	2·6492	3·1063	3·3518	3·6090	4·7579	6·4790
0·0000360	0·283	0·290	0·294	0·298	0·305	0·312	0·315	0·326	0·339	0·352	0·359	0·365	0·390	0·419
1/ 27778	1·2800	1·4251	1·5013	1·5800	1·7448	1·9199	2·0113	2·3015	2·7266	3·1971	3·4498	3·7145	4·8969	6·6681
0·0000380	0·291	0·298	0·302	0·306	0·313	0·320	0·324	0·335	0·348	0·362	0·368	0·375	0·400	0·431
1/ 26316	1·3154	1·4645	1·5429	1·6237	1·7931	1·9729	2·0669	2·3651	2·8020	3·2854	3·5451	3·8171	5·0321	6·8521
0·0000400	0·298	0·306	0·310	0·314	0·321	0·329	0·332	0·343	0·358	0·371	0·378	0·385	0·411	0·442
1/ 25000	1·3499	1·5030	1·5833	1·6663	1·8401	2·0247	2·1211	2·4271	2·8754	3·3714	3·6379	3·9170	5·1638	7·0313
0·0000420	0·306	0·314	0·318	0·322	0·329	0·337	0·341	0·352	0·366	0·381	0·388	0·394	0·421	0·453
1/ 23810	1·3836	1·5404	1·6228	1·7078	1·8859	2·0751	2·1739	2·4875	2·9469	3·4553	3·7284	4·0145	5·2922	7·2061
0·0000440	0·313	0·321	0·325	0·329	0·337	0·345	0·349	0·360	0·375	0·390	0·397	0·404	0·431	0·464
1/ 22727	1·4164	1·5770	1·6613	1·7483	1·9307	2·1243	2·2255	2·5465	3·0168	3·5373	3·8168	4·1096	5·4176	7·3768
0·0000460	0·320	0·329	0·333	0·337	0·345	0·353	0·357	0·368	0·384	0·398	0·406	0·413	0·441	0·474
1/ 21739	1·4486	1·6128	1·6990	1·7880	1·9745	2·1725	2·2759	2·6042	3·0852	3·6173	3·9032	4·2027	5·5402	7·5436
0·0000480	0·327	0·336	0·340	0·344	0·352	0·360	0·364	0·376	0·392	0·407	0·414	0·422	0·450	0·485
1/ 20833	1·4800	1·6477	1·7358	1·8267	2·0173	2·2196	2·3253	2·6607	3·1520	3·6957	3·9878	4·2937	5·6601	7·7068
0·0000500	0·334	0·343	0·347	0·351	0·360	0·368	0·372	0·384	0·400	0·416	0·423	0·431	0·460	0·495
1/ 20000	1·5108	1·6820	1·7719	1·8647	2·0592	2·2657	2·3736	2·7159	3·2175	3·7725	4·0706	4·3828	5·7776	7·8667
0·0000525	0·342	0·351	0·356	0·360	0·369	0·377	0·381	0·394	0·410	0·426	0·434	0·441	0·471	0·507
1/ 19048	1·5484	1·7239	1·8160	1·9112	2·1105	2·3221	2·4326	2·7835	3·2975	3·8663	4·1718	4·4918	5·9212	8·0621
0·0000550	0·350	0·360	0·364	0·369	0·377	0·386	0·390	0·403	0·420	0·436	0·444	0·452	0·482	0·519
1/ 18182	1·5852	1·7648	1·8591	1·9565	2·1605	2·3772	2·4903	2·8495	3·3757	3·9579	4·2706	4·5982	6·0614	8·2530
0·0000575	0·358	0·368	0·372	0·377	0·386	0·395	0·399	0·412	0·429	0·446	0·454	0·462	0·493	0·531
1/ 17391	1·6211	1·8048	1·9012	2·0008	2·2095	2·4310	2·5467	2·9140	3·4521	4·0474	4·3672	4·7022	6·1984	8·4395
0·0000600	0·366	0·376	0·380	0·385	0·394	0·403	0·408	0·421	0·439	0·455	0·464	0·472	0·504	0·542
1/ 16667	1·6562	1·8439	1·9424	2·0442	2·2573	2·4837	2·6019	2·9771	3·5268	4·1350	4·4618	4·8040	6·3325	8·6220
0·0000625	0·374	0·383	0·388	0·393	0·402	0·412	0·416	0·430	0·448	0·465	0·473	0·482	0·514	0·553
1/ 16000	1·6906	1·8822	1·9828	2·0866	2·3042	2·5353	2·6559	3·0389	3·6000	4·2208	4·5544	4·9036	6·4639	8·8008
0·0000650	0·381	0·391	0·396	0·401	0·410	0·420	0·425	0·438	0·457	0·474	0·483	0·491	0·525	0·564
1/ 15385	1·7244	1·9198	2·0224	2·1283	2·3502	2·5858	2·7089	3·0995	3·6718	4·3050	4·6451	5·0013	6·5926	8·9760
0·0000675	0·388	0·399	0·404	0·409	0·418	0·428	0·433	0·447	0·465	0·483	0·492	0·501	0·535	0·575
1/ 14815	1·7575	1·9566	2·0612	2·1691	2·3953	2·6354	2·7608	3·1590	3·7422	4·3875	4·7341	5·0972	6·7189	9·1478
0·0000700	0·396	0·406	0·411	0·416	0·426	0·436	0·441	0·455	0·474	0·492	0·501	0·510	0·545	0·586
1/ 14286	1·7900	1·9928	2·0993	2·2092	2·4395	2·6841	2·8118	3·2173	3·8113	4·4685	4·8215	5·1913	6·8429	9·3166
0·0000725	0·403	0·413	0·418	0·424	0·434	0·444	0·449	0·463	0·482	0·501	0·510	0·519	0·554	0·596
1/ 13793	1·8219	2·0283	2·1367	2·2486	2·4830	2·7319	2·8619	3·2746	3·8792	4·5480	4·9074	5·2837	6·9646	9·4823
0·0000750	0·410	0·420	0·426	0·431	0·441	0·451	0·456	0·471	0·491	0·510	0·519	0·528	0·564	0·606
1/ 13333	1·8533	2·0632	2·1735	2·2873	2·5257	2·7789	2·9112	3·3309	3·9459	4·6262	4·9917	5·3745	7·0843	9·6452
0·0000800	0·423	0·434	0·440	0·445	0·456	0·466	0·471	0·487	0·507	0·526	0·536	0·545	0·582	0·626
1/ 12500	1·9145	2·1314	2·2452	2·3628	2·6091	2·8707	3·0073	3·4409	4·0760	4·7788	5·1564	5·5518	7·3179	9·9631
0·0000850	0·436	0·448	0·453	0·459	0·470	0·481	0·486	0·502	0·523	0·543	0·553	0·562	0·600	0·646
1/ 11765	1·9738	2·1974	2·3148	2·4360	2·6899	2·9596	3·1004	3·5474	4·2022	4·9267	5·3159	5·7235	7·5442	10·271
0·0000900	0·449	0·461	0·466	0·472	0·484	0·495	0·500	0·516	0·538	0·558	0·569	0·579	0·618	0·665
1/ 11111	2·0314	2·2615	2·3823	2·5071	2·7684	3·0459	3·1908	3·6508	4·3247	5·0703	5·4709	5·8903	7·7640	10·570
0·0000950	0·461	0·473	0·479	0·485	0·497	0·508	0·514	0·531	0·553	0·574	0·584	0·595	0·635	0·683
1/ 10526	2·0874	2·3239	2·4480	2·5762	2·8447	3·1299	3·2788	3·7515	4·4439	5·2100	5·6216	6·0526	7·9778	10·861
0·0001000	0·473	0·486	0·492	0·498	0·510	0·522	0·527	0·545	0·567	0·589	0·600	0·610	0·651	0·701
1/ 10000	2·1420	2·3846	2·5120	2·6435	2·9191	3·2116	3·3644	3·8495	4·5600	5·3461	5·7683	6·2106	8·1860	11·145
	0·83	0·83	0·83	0·83	0·83	0·84	0·84	0·84	0·84	0·85	0·85	0·85	0·86	0·86

$V_{r(0·5)medial}$ for half-full circular pipes.

k_s = 3·0 mm S = 0·00003 to 0·00010

$k_s = 3{\cdot}0$ **mm**
S = 0·00010 to 0·00030

ie hydraulic gradient =
1 in 10000 to 1 in 3333

Water (or sewage) at 15°C;
full bore conditions.

velocities in ms^{-1}
discharges in m^3s^{-1}

Gradient	**(Equivalent) Pipe diameters in mm**													
	2400	**2500**	**2550**	**2600**	**2700**	**2800**	**2850**	**3000**	**3200**	**3400**	**3500**	**3600**	**4000**	**4500**
0·000100	0·473	0·486	0·492	0·498	0·510	0·522	0·527	0·545	0·567	0·589	0·600	0·610	0·651	0·701
1/ 10000	2·1420	2·3846	2·5120	2·6435	2·9191	3·2116	3·3644	3·8495	4·5600	5·3461	5·7683	6·2106	8·1860	11·145
0·000105	0·485	0·498	0·504	0·510	0·522	0·535	0·540	0·558	0·581	0·603	0·614	0·625	0·668	0·718
1/ 9524	2·1952	2·4439	2·5744	2·7092	2·9916	3·2914	3·4480	3·9450	4·6731	5·4787	5·9115	6·3647	8·3891	11·421
0·000110	0·497	0·510	0·516	0·522	0·535	0·547	0·553	0·571	0·595	0·618	0·629	0·640	0·683	0·735
1/ 9091	2·2472	2·5017	2·6354	2·7733	3·0623	3·3693	3·5296	4·0384	4·7837	5·6083	6·0513	6·5152	8·5874	11·691
0·000115	0·508	0·521	0·528	0·534	0·547	0·560	0·566	0·584	0·608	0·632	0·643	0·655	0·699	0·752
1/ 8696	2·2980	2·5583	2·6949	2·8360	3·1316	3·4454	3·6093	4·1296	4·8917	5·7349	6·1879	6·6623	8·7812	11·955
0·000120	0·519	0·532	0·539	0·546	0·559	0·572	0·578	0·597	0·621	0·645	0·657	0·669	0·714	0·768
1/ 8333	2·3477	2·6136	2·7532	2·8973	3·1993	3·5199	3·6873	4·2188	4·9974	5·8589	6·3216	6·8062	8·9709	12·213
0·000125	0·530	0·543	0·550	0·557	0·570	0·583	0·590	0·609	0·634	0·659	0·671	0·683	0·729	0·784
1/ 8000	2·3964	2·6678	2·8103	2·9574	3·2656	3·5929	3·7638	4·3063	5·1010	5·9802	6·4526	6·9472	9·1566	12·466
0·000130	0·540	0·554	0·561	0·568	0·582	0·595	0·602	0·621	0·647	0·672	0·684	0·696	0·743	0·799
1/ 7692	2·4441	2·7209	2·8662	3·0163	3·3306	3·6644	3·8387	4·3920	5·2025	6·0992	6·5809	7·0854	9·3387	12·713
0·000135	0·551	0·565	0·572	0·579	0·593	0·606	0·613	0·633	0·659	0·685	0·697	0·709	0·757	0·815
1/ 7407	2·4909	2·7730	2·9211	3·0740	3·3944	3·7345	3·9122	4·4760	5·3020	6·2159	6·7068	7·2209	9·5173	12·957
0·000140	0·561	0·575	0·583	0·590	0·604	0·618	0·625	0·645	0·671	0·697	0·710	0·722	0·771	0·830
1/ 7143	2·5369	2·8242	2·9750	3·1307	3·4570	3·8034	3·9843	4·5586	5·3998	6·3305	6·8304	7·3540	9·6927	13·195
0·000145	0·571	0·586	0·593	0·600	0·615	0·629	0·636	0·656	0·683	0·710	0·723	0·735	0·785	0·844
1/ 6897	2·5820	2·8744	3·0279	3·1864	3·5184	3·8710	4·0552	4·6396	5·4958	6·4430	6·9518	7·4847	9·8649	13·430
0·000150	0·581	0·596	0·603	0·610	0·625	0·639	0·647	0·668	0·695	0·722	0·735	0·748	0·798	0·859
1/ 6667	2·6264	2·9238	3·0800	3·2411	3·5789	3·9375	4·1248	4·7193	5·5901	6·5536	7·0711	7·6132	10·034	13·660
0·000160	0·600	0·615	0·623	0·631	0·646	0·661	0·668	0·690	0·718	0·746	0·759	0·773	0·825	0·887
1/ 6250	2·7130	3·0202	3·1815	3·3480	3·6968	4·0672	4·2607	4·8747	5·7742	6·7694	7·3040	7·8638	10·364	14·109
0·000170	0·618	0·634	0·642	0·650	0·666	0·681	0·689	0·711	0·740	0·769	0·783	0·796	0·850	0·915
1/ 5882	2·7969	3·1136	3·2798	3·4515	3·8111	4·1930	4·3924	5·0254	5·9527	6·9786	7·5297	8·1068	10·685	14·545
0·000180	0·636	0·653	0·661	0·669	0·685	0·701	0·709	0·732	0·762	0·791	0·805	0·820	0·875	0·941
1/ 5556	2·8783	3·2043	3·3754	3·5520	3·9221	4·3151	4·5203	5·1717	6·1260	7·1817	7·7488	8·3427	10·995	14·968
0·000190	0·654	0·671	0·679	0·687	0·704	0·720	0·728	0·752	0·783	0·813	0·828	0·842	0·899	0·967
1/ 5263	2·9576	3·2925	3·4683	3·6498	4·0300	4·4338	4·6447	5·3140	6·2945	7·3792	7·9619	8·5721	11·298	15·380
0·000200	0·671	0·688	0·697	0·705	0·722	0·739	0·747	0·771	0·803	0·834	0·849	0·864	0·922	0·992
1/ 5000	3·0348	3·3784	3·5588	3·7450	4·1352	4·5495	4·7658	5·4526	6·4586	7·5716	8·1695	8·7956	11·592	15·780
0·000210	0·687	0·705	0·714	0·723	0·740	0·757	0·766	0·791	0·823	0·855	0·870	0·886	0·945	1·017
1/ 4762	3·1100	3·4622	3·6470	3·8379	4·2377	4·6623	4·8840	5·5878	6·6187	7·7593	8·3720	9·0136	11·879	16·171
0·000220	0·704	0·722	0·731	0·740	0·758	0·775	0·784	0·809	0·842	0·875	0·891	0·906	0·968	1·041
1/ 4545	3·1835	3·5440	3·7332	3·9285	4·3378	4·7724	4·9994	5·7198	6·7750	7·9425	8·5696	9·2264	12·160	16·553
0·000230	0·720	0·738	0·747	0·757	0·775	0·793	0·801	0·827	0·861	0·895	0·911	0·927	0·989	1·064
1/ 4348	3·2554	3·6239	3·8174	4·0172	4·4357	4·8801	5·1122	5·8488	6·9278	8·1216	8·7629	9·4345	12·434	16·926
0·000240	0·735	0·754	0·764	0·773	0·791	0·810	0·819	0·845	0·880	0·914	0·930	0·947	1·011	1·087
1/ 4167	3·3257	3·7022	3·8999	4·1039	4·5315	4·9855	5·2225	5·9750	7·0774	8·2969	8·9520	9·6380	12·702	17·291
0·000250	0·750	0·770	0·779	0·789	0·808	0·826	0·836	0·863	0·898	0·933	0·950	0·966	1·032	1·110
1/ 4000	3·3946	3·7789	3·9806	4·1889	4·6253	5·0886	5·3306	6·0987	7·2238	8·4685	9·1372	9·8374	12·965	17·649
0·000260	0·765	0·785	0·795	0·805	0·824	0·843	0·852	0·880	0·916	0·951	0·969	0·986	1·052	1·132
1/ 3846	3·4620	3·8540	4·0597	4·2722	4·7172	5·1898	5·4366	6·2199	7·3673	8·6368	9·3187	10·033	13·222	17·999
0·000270	0·780	0·800	0·810	0·820	0·840	0·859	0·868	0·897	0·934	0·969	0·987	1·004	1·072	1·153
1/ 3704	3·5283	3·9277	4·1374	4·3538	4·8074	5·2890	5·5405	6·3388	7·5081	8·8018	9·4968	10·225	13·475	18·343
0·000280	0·794	0·815	0·825	0·835	0·855	0·875	0·884	0·913	0·951	0·987	1·005	1·023	1·092	1·175
1/ 3571	3·5932	4·0000	4·2136	4·4340	4·8959	5·3864	5·6425	6·4555	7·6463	8·9638	9·6716	10·413	13·723	18·680
0·000290	0·808	0·829	0·840	0·850	0·870	0·890	0·900	0·929	0·968	1·005	1·023	1·041	1·111	1·195
1/ 3448	3·6571	4·0711	4·2884	4·5128	4·9829	5·4820	5·7427	6·5701	7·7821	9·1230	9·8433	10·598	13·967	19·012
0·000300	0·822	0·844	0·854	0·865	0·885	0·906	0·916	0·945	0·984	1·022	1·041	1·059	1·130	1·216
1/ 3333	3·7198	4·1409	4·3620	4·5902	5·0684	5·5761	5·8412	6·6828	7·9156	9·2794	10·012	10·779	14·206	19·338
	0·83	0·83	0·83	0·83	0·84	0·84	0·84	0·84	0·84	0·85	0·85	0·85	0·86	0·86

$V_{r(0\cdot5)medial}$ **for half-full circular pipes.**

S = 0·00010 to 0·00030 $k_s = 3{\cdot}0$ **mm**

k_s = 3·0 mm
S = 0·00030 to 0·00100

ie hydraulic gradient =
1 in 3333 to 1 in 1000

Water (or sewage) at 15°C;
full bore conditions.

velocities in ms^{-1}
discharges in m^3s^{-1}

Gradient	(Equivalent) Pipe diameters in mm													
	2400	**2500**	**2550**	**2600**	**2700**	**2800**	**2850**	**3000**	**3200**	**3400**	**3500**	**3600**	**4000**	**4500**
0·000300	0·822	0·844	0·854	0·865	0·885	0·906	0·916	0·945	0·984	1·022	1·041	1·059	1·130	1·216
1/ 3333	3·7198	4·1409	4·3620	4·5902	5·0684	5·5761	5·8412	6·6828	7·9156	9·2794	10·012	10·779	14·206	19·338
0·000320	0·849	0·871	0·882	0·893	0·914	0·935	0·946	0·977	1·017	1·056	1·075	1·094	1·168	1·256
1/ 3125	3·8423	4·2772	4·5056	4·7413	5·2351	5·7596	6·0334	6·9027	8·1759	9·5846	10·341	11·134	14·673	19·974
0·000340	0·876	0·898	0·909	0·921	0·943	0·964	0·975	1·007	1·048	1·088	1·108	1·128	1·204	1·295
1/ 2941	3·9609	4·4093	4·6447	4·8877	5·3968	5·9374	6·2197	7·1158	8·4283	9·8804	10·660	11·477	15·126	20·590
0·000360	0·901	0·924	0·936	0·947	0·970	0·992	1·003	1·036	1·078	1·120	1·140	1·160	1·239	1·332
1/ 2778	4·0762	4·5376	4·7798	5·0298	5·5538	6·1101	6·4006	7·3227	8·6734	10·168	10·970	11·811	15·566	21·188
0·000380	0·926	0·950	0·962	0·973	0·997	1·020	1·031	1·064	1·108	1·151	1·172	1·192	1·273	1·369
1/ 2632	4·1882	4·6623	4·9112	5·1681	5·7064	6·2780	6·5765	7·5239	8·9117	10·447	11·272	12·135	15·993	21·770
0·000400	0·950	0·975	0·987	0·999	1·023	1·046	1·058	1·092	1·137	1·181	1·202	1·223	1·306	1·404
1/ 2500	4·2974	4·7838	5·0392	5·3028	5·8551	6·4416	6·7478	7·7199	9·1438	10·719	11·565	12·452	16·410	22·337
0·000420	0·973	0·999	1·011	1·024	1·048	1·072	1·084	1·119	1·165	1·210	1·232	1·254	1·338	1·439
1/ 2381	4·4038	4·9023	5·1640	5·4341	6·0001	6·6011	6·9149	7·9111	9·3702	10·985	11·852	12·760	16·816	22·890
0·000440	0·996	1·022	1·035	1·048	1·073	1·097	1·110	1·146	1·193	1·238	1·261	1·283	1·370	1·473
1/ 2273	4·5078	5·0180	5·2859	5·5624	6·1417	6·7569	7·0781	8·0978	9·5913	11·244	12·131	13·061	17·212	23·429
0·000460	1·019	1·045	1·058	1·071	1·097	1·122	1·135	1·171	1·219	1·266	1·289	1·312	1·401	1·506
1/ 2174	4·6094	5·1311	5·4050	5·6877	6·2801	6·9091	7·2376	8·2802	9·8074	11·497	12·405	13·355	17·600	23·957
0·000480	1·041	1·068	1·081	1·094	1·121	1·146	1·159	1·197	1·246	1·294	1·317	1·340	1·431	1·539
1/ 2083	4·7089	5·2418	5·5216	5·8104	6·4156	7·0581	7·3937	8·4588	10·019	11·745	12·672	13·643	17·979	24·473
0·000500	1·062	1·090	1·104	1·117	1·144	1·170	1·183	1·221	1·271	1·320	1·344	1·368	1·460	1·571
1/ 2000	4·8062	5·3502	5·6358	5·9306	6·5482	7·2041	7·5466	8·6336	10·226	11·988	12·934	13·925	18·351	24·979
0·000525	1·089	1·117	1·131	1·145	1·172	1·199	1·212	1·252	1·303	1·353	1·378	1·402	1·496	1·609
1/ 1905	4·9253	5·4827	5·7753	6·0774	6·7104	7·3824	7·7334	8·8474	10·479	12·284	13·254	14·269	18·805	25·597
0·000550	1·114	1·143	1·158	1·172	1·200	1·227	1·241	1·281	1·334	1·385	1·410	1·435	1·532	1·647
1/ 1818	5·0415	5·6120	5·9116	6·2208	6·8687	7·5566	7·9158	9·0561	10·726	12·574	13·567	14·606	19·248	26·201
0·000575	1·140	1·169	1·184	1·198	1·227	1·255	1·269	1·310	1·364	1·416	1·442	1·467	1·566	1·684
1/ 1739	5·1551	5·7385	6·0448	6·3610	7·0234	7·7268	8·0942	9·2601	10·968	12·857	13·872	14·935	19·682	26·790
0·000600	1·164	1·194	1·209	1·224	1·253	1·282	1·296	1·338	1·393	1·447	1·473	1·499	1·600	1·721
1/ 1667	5·2663	5·8623	6·1751	6·4981	7·1749	7·8934	8·2687	9·4597	11·204	13·134	14·171	15·257	20·106	27·368
0·000625	1·188	1·219	1·234	1·249	1·279	1·308	1·323	1·366	1·422	1·477	1·503	1·530	1·633	1·756
1/ 1600	5·3751	5·9834	6·3028	6·6325	7·3232	8·0566	8·4396	9·6552	11·436	13·406	14·464	15·572	20·521	27·933
0·000650	1·212	1·243	1·259	1·274	1·304	1·334	1·349	1·393	1·450	1·506	1·533	1·560	1·665	1·791
1/ 1538	5·4819	6·1022	6·4279	6·7641	7·4686	8·2165	8·6071	9·8469	11·663	13·672	14·751	15·881	20·929	28·487
0·000675	1·235	1·267	1·283	1·298	1·329	1·360	1·375	1·420	1·478	1·535	1·562	1·590	1·697	1·825
1/ 1481	5·5865	6·2188	6·5507	6·8933	7·6112	8·3734	8·7714	10·035	11·885	13·933	15·033	16·184	21·328	29·031
0·000700	1·258	1·290	1·306	1·322	1·354	1·385	1·400	1·446	1·505	1·563	1·591	1·619	1·728	1·859
1/ 1429	5·6893	6·3331	6·6712	7·0201	7·7511	8·5274	8·9327	10·219	12·104	14·189	15·309	16·482	21·720	29·564
0·000725	1·280	1·313	1·329	1·346	1·378	1·409	1·425	1·471	1·532	1·591	1·619	1·648	1·759	1·892
1/ 1379	5·7903	6·4455	6·7895	7·1446	7·8886	8·6786	9·0912	10·401	12·319	14·441	15·580	16·774	22·105	30·089
0·000750	1·302	1·336	1·352	1·369	1·401	1·434	1·450	1·497	1·558	1·618	1·647	1·676	1·789	1·924
1/ 1333	5·8895	6·5559	6·9058	7·2670	8·0238	8·8273	9·2469	10·579	12·530	14·688	15·847	17·061	22·484	30·604
0·000800	1·345	1·379	1·397	1·414	1·447	1·481	1·497	1·546	1·609	1·671	1·701	1·731	1·848	1·987
1/ 1250	6·0831	6·7714	7·1328	7·5059	8·2875	9·1174	9·5508	10·926	12·941	15·171	16·368	17·622	23·222	31·609
0·000850	1·386	1·422	1·440	1·457	1·492	1·526	1·543	1·593	1·659	1·722	1·754	1·785	1·905	2·049
1/ 1176	6·2707	6·9803	7·3528	7·7374	8·5431	9·3986	9·8454	11·263	13·340	15·638	16·873	18·165	23·938	32·583
0·000900	1·426	1·463	1·482	1·500	1·535	1·571	1·588	1·640	1·707	1·772	1·805	1·836	1·960	2·108
1/ 1111	6·4529	7·1831	7·5664	7·9622	8·7913	9·6716	10·131	11·591	13·728	16·092	17·363	18·693	24·633	33·529
0·000950	1·466	1·504	1·522	1·541	1·578	1·614	1·632	1·685	1·754	1·821	1·854	1·887	2·014	2·166
1/ 1053	6·6301	7·3803	7·7742	8·1808	9·0327	9·9372	10·410	11·909	14·105	16·534	17·839	19·206	25·309	34·449
0·001000	1·504	1·543	1·562	1·581	1·619	1·656	1·674	1·729	1·799	1·868	1·902	1·936	2·066	2·222
1/ 1000	6·8027	7·5724	7·9766	8·3937	9·2678	10·196	10·680	12·219	14·472	16·964	18·303	19·705	25·968	35·346
	0·83	0·83	0·83	0·83	0·84	0·84	0·84	0·84	0·85	0·85	0·85	0·85	0·86	0·86

$V_{r(0·5)medial}$ for half-full circular pipes.

k_s = 3·0 mm S = 0·00030 to 0·00100

$k_s = 3{\cdot}0$ mm
S = 0·00100 to 0·00300

ie hydraulic gradient =
1 in 1000 to 1 in 333

Water (or sewage) at 15°C;
full bore conditions.

velocities in ms^{-1}
discharges in m^3s^{-1}

Gradient	(Equivalent) Pipe diameters in mm													
	2400	2500	2550	2600	2700	2800	2850	3000	3200	3400	3500	3600	4000	4500
0·00100	1·504	1·543	1·562	1·581	1·619	1·656	1·674	1·729	1·799	1·868	1·902	1·936	2·066	2·222
1/ 1000	6·8027	7·5724	7·9766	8·3937	9·2678	10·196	10·680	12·219	14·472	16·964	18·303	19·705	25·968	35·346
0·00105	1·541	1·581	1·601	1·620	1·659	1·697	1·716	1·771	1·844	1·915	1·949	1·984	2·118	2·277
1/ 952	6·9710	7·7598	8·1739	8·6014	9·4971	10·448	10·945	12·521	14·830	17·384	18·756	20·193	26·610	36·220
0·00110	1·577	1·618	1·638	1·658	1·698	1·737	1·756	1·813	1·887	1·960	1·995	2·031	2·167	2·331
1/ 909	7·1354	7·9428	8·3666	8·8042	9·7210	10·694	11·203	12·816	15·179	17·794	19·198	20·669	27·237	37·073
0·00115	1·613	1·655	1·675	1·696	1·736	1·776	1·796	1·854	1·930	2·004	2·040	2·076	2·216	2·383
1/ 870	7·2961	8·1216	8·5550	9·0024	9·9398	10·935	11·455	13·105	15·521	18·194	19·630	21·134	27·850	37·907
0·00120	1·648	1·690	1·711	1·732	1·773	1·814	1·834	1·894	1·971	2·047	2·084	2·121	2·264	2·435
1/ 833	7·4533	8·2966	8·7394	9·1964	10·154	11·171	11·702	13·387	15·855	18·586	20·053	21·589	28·450	38·724
0·00125	1·682	1·725	1·747	1·768	1·810	1·852	1·872	1·933	2·012	2·089	2·127	2·165	2·311	2·485
1/ 800	7·6073	8·4680	8·9199	9·3864	10·364	11·401	11·943	13·663	16·183	18·970	20·467	22·035	29·038	39·523
0·00130	1·715	1·759	1·781	1·803	1·846	1·888	1·909	1·971	2·052	2·131	2·170	2·208	2·357	2·534
1/ 769	7·7582	8·6360	9·0969	9·5726	10·569	11·628	12·180	13·934	16·504	19·346	20·873	22·472	29·613	40·307
0·00135	1·748	1·793	1·815	1·837	1·881	1·924	1·946	2·009	2·091	2·171	2·211	2·250	2·402	2·583
1/ 741	7·9062	8·8008	9·2704	9·7552	10·771	11·849	12·413	14·200	16·819	19·715	21·271	22·901	30·178	41·076
0·00140	1·780	1·826	1·849	1·871	1·916	1·960	1·982	2·046	2·130	2·211	2·252	2·291	2·446	2·630
1/ 714	8·0516	8·9626	9·4408	9·9345	10·969	12·067	12·641	14·461	17·128	20·078	21·662	23·322	30·733	41·830
0·00145	1·811	1·858	1·881	1·904	1·950	1·995	2·017	2·082	2·167	2·251	2·291	2·332	2·489	2·677
1/ 690	8·1943	9·1215	9·6082	10·111	11·163	12·281	12·865	14·718	17·431	20·433	22·046	23·735	31·277	42·572
0·00150	1·842	1·890	1·914	1·937	1·983	2·029	2·051	2·118	2·205	2·289	2·331	2·372	2·532	2·723
1/ 667	8·3346	9·2777	9·7727	10·284	11·355	12·491	13·085	14·970	17·730	20·783	22·424	24·141	31·813	43·300
0·00160	1·903	1·952	1·976	2·001	2·048	2·095	2·119	2·187	2·277	2·364	2·407	2·450	2·615	2·812
1/ 625	8·6084	9·5824	10·094	10·622	11·727	12·902	13·515	15·461	18·312	21·466	23·160	24·934	32·857	44·722
0·00170	1·962	2·012	2·037	2·062	2·111	2·160	2·184	2·255	2·347	2·437	2·481	2·525	2·695	2·899
1/ 588	8·8738	9·8778	10·405	10·949	12·089	13·299	13·931	15·938	18·876	22·127	23·874	25·702	33·870	46·100
0·00180	2·018	2·071	2·097	2·122	2·173	2·223	2·247	2·320	2·415	2·508	2·553	2·598	2·773	2·983
1/ 556	9·1315	10·165	10·707	11·267	12·440	13·685	14·336	16·400	19·424	22·770	24·567	26·448	34·853	47·438
0·00190	2·074	2·128	2·154	2·180	2·232	2·284	2·309	2·384	2·481	2·577	2·623	2·670	2·850	3·065
1/ 526	9·3821	10·444	11·001	11·576	12·781	14·061	14·729	16·850	19·957	23·394	25·240	27·174	35·809	48·739
0·00200	2·128	2·183	2·210	2·237	2·290	2·343	2·369	2·446	2·546	2·644	2·692	2·739	2·924	3·144
1/ 500	9·6261	10·715	11·287	11·877	13·114	14·427	15·112	17·289	20·476	24·003	25·897	27·881	36·740	50·006
0·00210	2·180	2·237	2·265	2·292	2·347	2·401	2·428	2·506	2·609	2·709	2·758	2·807	2·996	3·222
1/ 476	9·8642	10·980	11·566	12·171	13·438	14·784	15·486	17·716	20·983	24·596	26·537	28·570	37·648	51·242
0·00220	2·232	2·290	2·318	2·346	2·402	2·457	2·485	2·565	2·670	2·773	2·823	2·873	3·067	3·298
1/ 455	10·097	11·239	11·839	12·458	13·755	15·132	15·851	18·134	21·477	25·176	27·162	29·243	38·535	52·449
0·00230	2·282	2·341	2·370	2·399	2·456	2·513	2·541	2·623	2·731	2·835	2·887	2·938	3·136	3·372
1/ 435	10·324	11·492	12·105	12·738	14·064	15·472	16·208	18·542	21·960	25·742	27·774	29·901	39·402	53·629
0·00240	2·331	2·392	2·421	2·451	2·509	2·567	2·595	2·680	2·789	2·896	2·949	3·001	3·203	3·445
1/ 417	10·546	11·739	12·366	13·012	14·367	15·806	16·557	18·941	22·433	26·296	28·371	30·545	40·250	54·784
0·00250	2·379	2·441	2·471	2·501	2·561	2·620	2·649	2·735	2·847	2·956	3·010	3·063	3·269	3·516
1/ 400	10·764	11·982	12·621	13·281	14·664	16·132	16·899	19·332	22·896	26·839	28·957	31·175	41·081	55·914
0·00260	2·427	2·489	2·520	2·551	2·612	2·672	2·701	2·789	2·903	3·015	3·069	3·123	3·334	3·585
1/ 385	10·977	12·219	12·871	13·544	14·954	16·452	17·234	19·715	23·350	27·371	29·531	31·793	41·895	57·023
0·00270	2·473	2·537	2·568	2·600	2·662	2·723	2·753	2·842	2·959	3·072	3·128	3·183	3·397	3·654
1/ 370	11·187	12·452	13·117	13·803	15·240	16·765	17·562	20·091	23·795	27·893	30·094	32·399	42·694	58·110
0·00280	2·518	2·583	2·616	2·647	2·711	2·773	2·804	2·895	3·013	3·129	3·185	3·242	3·460	3·721
1/ 357	11·392	12·681	13·358	14·056	15·520	17·073	17·885	20·460	24·232	28·406	30·647	32·994	43·478	59·177
0·00290	2·563	2·629	2·662	2·694	2·759	2·822	2·853	2·946	3·066	3·184	3·242	3·299	3·521	3·787
1/ 345	11·594	12·906	13·595	14·305	15·795	17·376	18·202	20·823	24·662	28·909	31·190	33·579	44·249	60·225
0·00300	2·607	2·674	2·707	2·741	2·806	2·870	2·902	2·996	3·119	3·239	3·297	3·355	3·581	3·852
1/ 333	11·793	13·127	13·827	14·550	16·065	17·673	18·513	21·179	25·084	29·403	31·724	34·154	45·006	61·256
	0·83	0·83	0·83	0·83	0·84	0·84	0·84	0·84	0·85	0·85	0·85	0·85	0·86	0·86

$V_{r(0{\cdot}5)medial}$ for half-full circular pipes.

$k_s = 3{\cdot}0$ mm
S = 0·00300 to 0·01000

ie hydraulic gradient =
1 in 333 to 1 in 100

Water (or sewage) at 15°C;
full bore conditions.

velocities in ms^{-1}
discharges in m^3s^{-1}

Gradient	**(Equivalent) Pipe diameters in mm**													
	2400	**2500**	**2550**	**2600**	**2700**	**2800**	**2850**	**3000**	**3200**	**3400**	**3500**	**3600**	**4000**	**4500**
0·00300	2·607	2·674	2·707	2·741	2·806	2·870	2·902	2·996	3·119	3·239	3·297	3·355	3·581	3·852
1/ 333	11·793	13·127	13·827	14·550	16·065	17·673	18·513	21·179	25·084	29·403	31·724	34·154	45·006	61·256
0·00320	2·692	2·762	2·796	2·830	2·898	2·964	2·997	3·095	3·221	3·345	3·406	3·466	3·699	3·978
1/ 313	12·180	13·558	14·281	15·028	16·592	18·254	19·121	21·874	25·907	30·369	32·765	35·275	46·483	63·266
0·00340	2·775	2·847	2·883	2·918	2·987	3·056	3·090	3·190	3·321	3·448	3·510	3·572	3·813	4·100
1/ 294	12·555	13·975	14·721	15·491	17·104	18·816	19·710	22·548	26·705	31·304	33·774	36·361	47·915	65·215
0·00360	2·856	2·930	2·966	3·002	3·074	3·144	3·179	3·282	3·417	3·548	3·612	3·676	3·924	4·219
1/ 278	12·920	14·381	15·148	15·940	17·600	19·362	20·282	23·203	27·480	32·213	34·754	37·416	49·305	67·107
0·00380	2·934	3·010	3·048	3·085	3·158	3·231	3·267	3·373	3·511	3·645	3·711	3·777	4·031	4·335
1/ 263	13·274	14·776	15·564	16·378	18·083	19·893	20·838	23·839	28·234	33·096	35·708	38·443	50·657	68·947
0·00400	3·010	3·088	3·127	3·165	3·240	3·315	3·351	3·460	3·602	3·740	3·808	3·875	4·136	4·448
1/ 250	13·619	15·160	15·969	16·804	18·553	20·410	21·380	24·459	28·968	33·956	36·636	39·442	51·974	70·740
0·00420	3·085	3·165	3·204	3·243	3·320	3·397	3·434	3·546	3·691	3·832	3·902	3·971	4·238	4·558
1/ 238	13·956	15·535	16·363	17·219	19·012	20·915	21·909	25·063	29·684	34·796	37·541	40·417	53·259	72·488
0·00440	3·158	3·239	3·280	3·320	3·399	3·477	3·515	3·629	3·778	3·923	3·994	4·064	4·338	4·665
1/ 227	14·285	15·900	16·749	17·625	19·459	21·407	22·425	25·654	30·383	35·615	38·426	41·369	54·513	74·195
0·00460	3·229	3·312	3·353	3·394	3·475	3·555	3·594	3·711	3·863	4·011	4·084	4·156	4·436	4·770
1/ 217	14·606	16·258	17·126	18·021	19·897	21·889	22·929	26·231	31·067	36·416	39·290	42·299	55·739	75·863
0·00480	3·298	3·383	3·426	3·467	3·550	3·631	3·672	3·791	3·946	4·097	4·172	4·245	4·531	4·873
1/ 208	14·920	16·608	17·494	18·409	20·325	22·360	23·423	26·795	31·735	37·200	40·136	43·210	56·938	77·496
0·00500	3·366	3·453	3·496	3·539	3·623	3·706	3·747	3·869	4·027	4·182	4·258	4·333	4·625	4·973
1/ 200	15·228	16·951	17·855	18·789	20·745	22·822	23·906	27·348	32·390	37·968	40·964	44·101	58·113	79·095
0·00525	3·449	3·539	3·583	3·626	3·713	3·798	3·840	3·965	4·127	4·285	4·363	4·440	4·739	5·096
1/ 190	15·605	17·370	18·297	19·253	21·258	23·386	24·497	28·024	33·191	38·906	41·976	45·191	59·549	81·050
0·00550	3·531	3·622	3·667	3·712	3·800	3·887	3·930	4·058	4·224	4·386	4·466	4·544	4·850	5·216
1/ 182	15·972	17·779	18·728	19·707	21·758	23·936	25·074	28·684	33·972	39·822	42·965	46·255	60·952	82·958
0·00575	3·610	3·703	3·749	3·795	3·886	3·975	4·019	4·149	4·319	4·485	4·566	4·646	4·959	5·333
1/ 174	16·332	18·179	19·149	20·150	22·248	24·475	25·638	29·329	34·736	40·718	43·931	47·295	62·322	84·824
0·00600	3·688	3·783	3·830	3·877	3·969	4·060	4·105	4·239	4·412	4·581	4·664	4·746	5·066	5·448
1/ 167	16·683	18·570	19·561	20·584	22·727	25·002	26·190	29·961	35·484	41·594	44·876	48·313	63·664	86·649
0·00625	3·764	3·861	3·909	3·957	4·051	4·144	4·190	4·326	4·503	4·676	4·761	4·844	5·171	5·561
1/ 160	17·027	18·954	19·965	21·009	23·196	25·518	26·730	30·579	36·216	42·453	45·802	49·310	64·977	88·437
0·00650	3·838	3·938	3·987	4·035	4·132	4·226	4·273	4·412	4·592	4·768	4·855	4·940	5·273	5·671
1/ 154	17·365	19·329	20·360	21·425	23·655	26·023	27·260	31·185	36·934	43·294	46·710	50·287	66·265	90·189
0·00675	3·912	4·013	4·063	4·112	4·210	4·307	4·355	4·496	4·680	4·859	4·947	5·035	5·374	5·779
1/ 148	17·696	19·698	20·749	21·833	24·106	26·519	27·780	31·779	37·638	44·119	47·600	51·246	67·528	91·908
0·00700	3·983	4·086	4·137	4·188	4·288	4·386	4·435	4·578	4·766	4·949	5·038	5·127	5·472	5·885
1/ 143	18·021	20·059	21·130	22·234	24·549	27·006	28·290	32·363	38·329	44·929	48·474	52·187	68·768	93·595
0·00725	4·054	4·159	4·211	4·262	4·364	4·464	4·513	4·660	4·850	5·036	5·128	5·218	5·569	5·989
1/ 138	18·340	20·415	21·504	22·628	24·984	27·485	28·791	32·936	39·008	45·725	49·333	53·111	69·986	95·253
0·00750	4·123	4·230	4·283	4·335	4·438	4·540	4·590	4·739	4·933	5·122	5·215	5·307	5·665	6·092
1/ 133	18·654	20·764	21·872	23·015	25·411	27·955	29·283	33·500	39·675	46·507	50·177	54·020	71·183	96·882
0·00800	4·259	4·369	4·423	4·477	4·584	4·689	4·741	4·895	5·095	5·290	5·386	5·481	5·850	6·291
1/ 125	19·266	21·445	22·590	23·771	26·245	28·872	30·244	34·599	40·977	48·033	51·823	55·792	73·518	100·06
0·00850	4·390	4·503	4·559	4·615	4·725	4·833	4·887	5·045	5·252	5·453	5·552	5·650	6·031	6·485
1/ 118	19·859	22·106	23·285	24·503	27·053	29·762	31·176	35·664	42·239	49·512	53·419	57·510	75·782	103·14
0·00900	4·517	4·634	4·692	4·749	4·862	4·974	5·029	5·192	5·404	5·612	5·713	5·814	6·205	6·673
1/ 111	20·436	22·747	23·961	25·213	27·838	30·625	32·080	36·699	43·464	50·949	54·969	59·179	77·980	106·13
0·00950	4·641	4·761	4·820	4·879	4·995	5·110	5·167	5·334	5·553	5·765	5·870	5·973	6·376	6·856
1/ 105	20·996	23·371	24·618	25·905	28·601	31·465	32·960	37·705	44·656	52·345	56·476	60·801	80·118	109·04
0·01000	4·762	4·885	4·946	5·006	5·125	5·243	5·301	5·473	5·697	5·915	6·023	6·129	6·541	7·034
1/ 100	21·542	23·978	25·258	26·578	29·345	32·282	33·817	38·685	45·817	53·706	57·944	62·381	82·201	111·88
	0·83	0·83	0·83	0·83	0·84	0·84	0·84	0·84	0·85	0·85	0·85	0·85	0·86	0·86

$V_{r(0{\cdot}5)medial}$ for half-full circular pipes.

$k_s = 3{\cdot}0$ mm S = 0·00300 to 0·01000

$k_s = 3{\cdot}0$ **mm**
S = 0·01000 to 0·03000

ie hydraulic gradient =
1 in 100 to 1 in 33·3

Water (or sewage) at 15°C;
full bore conditions.

velocities in ms^{-1}
discharges in m^3s^{-1}

Gradient	**(Equivalent) Pipe diameters in mm**													
	2400	**2500**	**2550**	**2600**	**2700**	**2800**	**2850**	**3000**	**3200**	**3400**	**3500**	**3600**	**4000**	**4500**
0·01000	4·762	4·885	4·946	5·006	5·125	5·243	5·301	5·473	5·697	5·915	6·023	6·129	6·541	7·034
1/ 100	21·542	23·978	25·258	26·578	29·345	32·282	33·817	38·685	45·817	53·706	57·944	62·381	82·201	111·88
0·01050	4·879	5·006	5·068	5·130	5·252	5·372	5·432	5·608	5·838	6·061	6·171	6·280	6·703	7·208
1/ 95	22·074	24·571	25·882	27·235	30·070	33·080	34·652	39·641	46·949	55·033	59·375	63·923	84·232	114·64
0·01100	4·994	5·123	5·187	5·250	5·376	5·499	5·560	5·740	5·975	6·204	6·317	6·428	6·861	7·378
1/ 91	22·594	25·150	26·491	27·876	30·778	33·859	35·468	40·575	48·054	56·329	60·773	65·428	86·215	117·34
0·01150	5·107	5·239	5·304	5·369	5·496	5·622	5·685	5·869	6·109	6·344	6·459	6·572	7·015	7·544
1/ 87	23·102	25·715	27·087	28·503	31·470	34·620	36·266	41·487	49·135	57·595	62·140	66·899	88·153	119·98
0·01200	5·217	5·351	5·418	5·484	5·615	5·743	5·807	5·996	6·241	6·480	6·598	6·714	7·166	7·706
1/ 83	23·599	26·269	27·670	29·117	32·147	35·366	37·046	42·380	50·192	58·835	63·477	68·338	90·050	122·56
0·01250	5·324	5·462	5·530	5·597	5·731	5·862	5·927	6·119	6·370	6·614	6·734	6·852	7·314	7·865
1/ 80	24·086	26·811	28·241	29·717	32·811	36·095	37·810	43·254	51·228	60·048	64·787	69·748	91·908	125·09
0·01300	5·430	5·570	5·639	5·708	5·844	5·978	6·044	6·240	6·496	6·745	6·867	6·988	7·459	8·021
1/ 77	24·563	27·342	28·800	30·306	33·461	36·810	38·560	44·111	52·243	61·238	66·070	71·130	93·729	127·57
0·01350	5·533	5·676	5·747	5·817	5·956	6·092	6·160	6·359	6·620	6·873	6·998	7·121	7·601	8·174
1/ 74	25·032	27·863	29·349	30·884	34·099	37·512	39·294	44·952	53·238	62·405	67·329	72·486	95·515	130·00
0·01400	5·635	5·780	5·852	5·924	6·065	6·204	6·273	6·476	6·741	7·000	7·127	7·252	7·740	8·324
1/ 71	25·491	28·374	29·888	31·451	34·725	38·201	40·016	45·777	54·216	63·551	68·565	73·816	97·268	132·38
0·01450	5·735	5·883	5·956	6·029	6·172	6·314	6·384	6·591	6·861	7·124	7·253	7·380	7·877	8·471
1/ 69	25·943	28·877	30·418	32·008	35·340	38·877	40·724	46·588	55·176	64·676	69·779	75·123	98·991	134·73
0·01500	5·833	5·983	6·058	6·132	6·278	6·422	6·493	6·704	6·978	7·245	7·377	7·507	8·012	8·616
1/ 67	26·386	29·371	30·938	32·555	35·944	39·542	41·421	47·385	56·119	65·782	70·973	76·408	100·68	137·03
0·01600	6·024	6·180	6·257	6·333	6·484	6·632	6·706	6·923	7·207	7·483	7·619	7·753	8·275	8·899
1/ 63	27·252	30·335	31·953	33·623	37·123	40·839	42·780	48·939	57·961	67·941	73·301	78·915	103·99	141·53
0·01700	6·210	6·370	6·449	6·528	6·683	6·837	6·912	7·137	7·429	7·714	7·853	7·992	8·530	9·173
1/ 59	28·091	31·269	32·937	34·659	38·266	42·097	44·097	50·446	59·745	70·033	75·558	81·345	107·19	145·89
0·01800	6·390	6·555	6·636	6·717	6·877	7·035	7·113	7·344	7·644	7·937	8·081	8·223	8·777	9·439
1/ 56	28·906	32·176	33·892	35·664	39·376	43·318	45·376	51·909	61·478	72·064	77·750	83·704	110·30	150·12
0·01900	6·565	6·734	6·818	6·901	7·066	7·228	7·308	7·545	7·854	8·155	8·303	8·449	9·018	9·697
1/ 53	29·699	33·058	34·821	36·642	40·456	44·505	46·620	53·332	63·164	74·039	79·881	85·999	113·32	154·23
0·02000	6·735	6·910	6·995	7·081	7·249	7·416	7·498	7·741	8·058	8·367	8·518	8·668	9·252	9·949
1/ 50	30·470	33·917	35·726	37·594	41·507	45·662	47·832	54·718	64·805	75·963	81·957	88·234	116·27	158·24
0·02100	6·902	7·080	7·168	7·256	7·429	7·599	7·683	7·932	8·257	8·573	8·729	8·883	9·481	10·20
1/ 47·6	31·223	34·755	36·609	38·523	42·533	46·790	49·014	56·070	66·406	77·840	83·982	90·413	119·14	162·15
0·02200	7·064	7·247	7·337	7·427	7·603	7·778	7·864	8·119	8·451	8·775	8·934	9·092	9·704	10·44
1/ 45·5	31·958	35·573	37·471	39·430	43·534	47·892	50·167	57·390	67·969	79·672	85·959	92·542	121·94	165·96
0·02300	7·223	7·410	7·502	7·594	7·774	7·953	8·041	8·302	8·641	8·973	9·135	9·296	9·922	10·67
1/ 43·5	32·677	36·373	38·313	40·316	44·513	48·969	51·295	58·680	69·497	81·464	87·891	94·622	124·68	169·70
0·02400	7·379	7·569	7·663	7·757	7·942	8·124	8·214	8·480	8·827	9·166	9·332	9·496	10·14	10·90
1/ 41·7	33·380	37·156	39·138	41·184	45·471	50·022	52·399	59·943	70·993	83·216	89·782	96·658	127·37	173·35
0·02500	7·531	7·725	7·822	7·917	8·106	8·291	8·383	8·655	9·009	9·355	9·524	9·692	10·34	11·12
1/ 40·0	34·069	37·922	39·945	42·033	46·409	51·054	53·480	61·180	72·457	84·933	91·634	98·652	129·99	176·92
0·02600	7·680	7·878	7·977	8·074	8·266	8·456	8·549	8·827	9·188	9·540	9·713	9·884	10·55	11·34
1/ 38·5	34·744	38·673	40·736	42·866	47·328	52·066	54·540	62·392	73·893	86·615	93·449	100·61	132·57	180·43
0·02700	7·826	8·029	8·129	8·228	8·424	8·617	8·712	8·995	9·363	9·722	9·898	10·07	10·75	11·56
1/ 37·0	35·406	39·410	41·513	43·683	48·230	53·058	55·579	63·581	75·301	88·266	95·230	102·52	135·09	183·86
0·02800	7·970	8·176	8·278	8·379	8·578	8·775	8·872	9·160	9·535	9·900	10·08	10·26	10·95	11·77
1/ 35·7	36·056	40·134	42·275	44·485	49·115	54·032	56·599	64·748	76·683	89·886	96·978	104·41	137·57	187·24
0·02900	8·111	8·321	8·424	8·527	8·730	8·930	9·029	9·322	9·704	10·08	10·26	10·44	11·14	11·98
1/ 34·5	36·694	40·844	43·023	45·273	49·985	54·988	57·601	65·894	78·041	91·478	98·695	106·25	140·01	190·55
0·03000	8·250	8·463	8·568	8·673	8·879	9·083	9·184	9·482	9·869	10·25	10·43	10·62	11·33	12·19
1/ 33·3	37·322	41·543	43·759	46·047	50·840	55·929	58·586	67·021	79·375	93·042	100·38	108·07	142·40	193·81
	0·83	0·83	0·83	0·83	0·84	0·84	0·84	0·84	0·85	0·85	0·85	0·85	0·86	0·86

$V_{r(0{\cdot}5)medial}$ **for half-full circular pipes.**

A55

(p.1 of 6)

$k_s = 6{\cdot}0$ mm
S = 0·00003 to 0·00010

ie hydraulic gradient =
1 in 33333 to 1 in 10000

Water (or sewage) at 15°C;
full bore conditions.

velocities in ms^{-1}
discharges in m^3s^{-1}

Gradient **(Equivalent) Pipe diameters in mm**

Gradient	2400	2500	2550	2600	2700	2800	2850	3000	3200	3400	3500	3600	4000	4500
0·0000300	0·237	0·243	0·246	0·249	0·255	0·261	0·264	0·273	0·285	0·296	0·301	0·307	0·328	0·353
1/ 33333	1·0715	1·1935	1·2575	1·3236	1·4622	1·6095	1·6864	1·9306	2·2886	2·6849	2·8979	3·1211	4·1184	5·6137
0·0000320	0·245	0·251	0·254	0·258	0·264	0·270	0·273	0·282	0·294	0·305	0·311	0·317	0·339	0·365
1/ 31250	1·1069	1·2328	1·2990	1·3673	1·5105	1·6626	1·7420	1·9943	2·3641	2·7734	2·9934	3·2239	4·2541	5·7986
0·0000340	0·252	0·259	0·262	0·265	0·272	0·278	0·282	0·291	0·303	0·315	0·321	0·327	0·349	0·376
1/ 29412	1·1411	1·2710	1·3392	1·4096	1·5572	1·7140	1·7959	2·0560	2·4372	2·8592	3·0860	3·3236	4·3856	5·9778
0·0000360	0·260	0·266	0·270	0·273	0·280	0·286	0·290	0·299	0·312	0·324	0·330	0·336	0·359	0·387
1/ 27778	1·1744	1·3081	1·3782	1·4507	1·6026	1·7640	1·8482	2·1159	2·5082	2·9425	3·1759	3·4204	4·5133	6·1517
0·0000380	0·267	0·274	0·277	0·281	0·288	0·294	0·298	0·308	0·320	0·333	0·339	0·345	0·369	0·397
1/ 26316	1·2068	1·3441	1·4162	1·4907	1·6468	1·8125	1·8992	2·1742	2·5772	3·0235	3·2633	3·5145	4·6375	6·3210
0·0000400	0·274	0·281	0·285	0·288	0·295	0·302	0·305	0·316	0·329	0·342	0·348	0·354	0·379	0·408
1/ 25000	1·2383	1·3792	1·4532	1·5296	1·6898	1·8599	1·9487	2·2309	2·6445	3·1024	3·3485	3·6062	4·7584	6·4858
0·0000420	0·281	0·288	0·292	0·295	0·302	0·310	0·313	0·323	0·337	0·350	0·357	0·363	0·388	0·418
1/ 23810	1·2690	1·4134	1·4893	1·5676	1·7317	1·9060	1·9971	2·2863	2·7101	3·1793	3·4315	3·6957	4·8764	6·6466
0·0000440	0·287	0·295	0·299	0·302	0·310	0·317	0·320	0·331	0·345	0·358	0·365	0·372	0·397	0·428
1/ 22727	1·2991	1·4469	1·5245	1·6047	1·7726	1·9511	2·0443	2·3403	2·7742	3·2545	3·5126	3·7830	4·9916	6·8035
0·0000460	0·294	0·301	0·305	0·309	0·317	0·324	0·328	0·339	0·353	0·367	0·373	0·380	0·406	0·437
1/ 21739	1·3284	1·4796	1·5589	1·6409	1·8127	1·9952	2·0905	2·3931	2·8368	3·3279	3·5919	3·8684	5·1042	6·9570
0·0000480	0·300	0·308	0·312	0·316	0·323	0·331	0·335	0·346	0·360	0·374	0·381	0·388	0·415	0·447
1/ 20833	1·3571	1·5115	1·5926	1·6764	1·8518	2·0383	2·1356	2·4448	2·8981	3·3998	3·6694	3·9519	5·2144	7·1071
0·0000500	0·306	0·314	0·318	0·322	0·330	0·338	0·342	0·353	0·368	0·382	0·389	0·396	0·424	0·456
1/ 20000	1·3852	1·5429	1·6256	1·7111	1·8902	2·0805	2·1799	2·4955	2·9581	3·4702	3·7454	4·0337	5·3223	7·2542
0·0000525	0·314	0·322	0·326	0·330	0·338	0·346	0·350	0·362	0·377	0·392	0·399	0·406	0·434	0·467
1/ 19048	1·4196	1·5811	1·6660	1·7535	1·9371	2·1321	2·2339	2·5574	3·0314	3·5562	3·8382	4·1337	5·4542	7·4339
0·0000550	0·321	0·330	0·334	0·338	0·346	0·354	0·358	0·370	0·386	0·401	0·408	0·416	0·444	0·478
1/ 18182	1·4532	1·6185	1·7053	1·7950	1·9829	2·1825	2·2867	2·6178	3·1030	3·6402	3·9289	4·2313	5·5830	7·6094
0·0000575	0·328	0·337	0·341	0·346	0·354	0·362	0·367	0·379	0·395	0·410	0·418	0·425	0·454	0·489
1/ 17391	1·4860	1·6551	1·7438	1·8355	2·0276	2·2317	2·3383	2·6769	3·1730	3·7223	4·0175	4·3268	5·7089	7·7809
0·0000600	0·336	0·344	0·349	0·353	0·362	0·370	0·374	0·387	0·403	0·419	0·427	0·434	0·464	0·500
1/ 16667	1·5181	1·6908	1·7815	1·8752	2·0714	2·2799	2·3888	2·7347	3·2415	3·8027	4·1042	4·4201	5·8321	7·9488
0·0000625	0·343	0·352	0·356	0·360	0·369	0·378	0·382	0·395	0·411	0·428	0·435	0·443	0·474	0·510
1/ 16000	1·5495	1·7258	1·8184	1·9140	2·1143	2·3271	2·4383	2·7913	3·3086	3·8814	4·1892	4·5116	5·9527	8·1132
0·0000650	0·349	0·359	0·363	0·368	0·377	0·385	0·390	0·403	0·420	0·436	0·444	0·452	0·483	0·520
1/ 15385	1·5804	1·7601	1·8546	1·9520	2·1564	2·3734	2·4868	2·8468	3·3744	3·9585	4·2724	4·6012	6·0710	8·2743
0·0000675	0·356	0·365	0·370	0·375	0·384	0·393	0·397	0·410	0·428	0·444	0·453	0·461	0·492	0·530
1/ 14815	1·6106	1·7938	1·8900	1·9894	2·1976	2·4188	2·5343	2·9012	3·4389	4·0342	4·3541	4·6892	6·1870	8·4324
0·0000700	0·363	0·372	0·377	0·382	0·391	0·400	0·405	0·418	0·435	0·453	0·461	0·469	0·501	0·540
1/ 14286	1·6403	1·8269	1·9249	2·0260	2·2381	2·4633	2·5810	2·9546	3·5022	4·1084	4·4342	4·7755	6·3009	8·5876
0·0000725	0·369	0·379	0·384	0·388	0·398	0·407	0·412	0·425	0·443	0·461	0·469	0·477	0·510	0·550
1/ 13793	1·6694	1·8593	1·9591	2·0620	2·2779	2·5071	2·6269	3·0071	3·5644	4·1814	4·5130	4·8603	6·4127	8·7400
0·0000750	0·375	0·385	0·390	0·395	0·405	0·414	0·419	0·433	0·451	0·468	0·477	0·486	0·519	0·559
1/ 13333	1·6981	1·8912	1·9927	2·0974	2·3169	2·5501	2·6719	3·0587	3·6256	4·2531	4·5904	4·9437	6·5227	8·8899
0·0000800	0·388	0·398	0·403	0·408	0·418	0·428	0·433	0·447	0·466	0·484	0·493	0·502	0·536	0·577
1/ 12500	1·7540	1·9535	2·0583	2·1665	2·3932	2·6341	2·7599	3·1594	3·7449	4·3931	4·7414	5·1063	6·7372	9·1822
0·0000850	0·400	0·410	0·415	0·421	0·431	0·441	0·446	0·461	0·480	0·499	0·508	0·517	0·553	0·595
1/ 11765	1·8082	2·0139	2·1219	2·2334	2·4671	2·7154	2·8451	3·2569	3·8605	4·5287	4·8878	5·2639	6·9451	9·4655
0·0000900	0·411	0·422	0·428	0·433	0·443	0·454	0·459	0·474	0·494	0·513	0·523	0·532	0·569	0·612
1/ 11111	1·8608	2·0725	2·1836	2·2984	2·5389	2·7944	2·9279	3·3516	3·9728	4·6604	5·0299	5·4170	7·1470	9·7406
0·0000950	0·423	0·434	0·439	0·445	0·456	0·466	0·472	0·487	0·508	0·527	0·537	0·547	0·584	0·629
1/ 10526	1·9120	2·1294	2·2436	2·3616	2·6087	2·8712	3·0084	3·4438	4·0820	4·7884	5·1681	5·5658	7·3434	10·008
0·0001000	0·434	0·445	0·451	0·456	0·468	0·478	0·484	0·500	0·521	0·541	0·551	0·561	0·600	0·646
1/ 10000	1·9618	2·1850	2·3021	2·4231	2·6767	2·9461	3·0868	3·5335	4·1883	4·9132	5·3028	5·7108	7·5347	10·269
	0·79	0·80	0·80	0·80	0·80	0·80	0·80	0·81	0·81	0·81	0·81	0·81	0·82	0·83

$V_{r(0{\cdot}5)medial}$ for half-full circular pipes.

$k_s = 6{\cdot}0$ mm S = 0·00003 to 0·00010

Gradient **(Equivalent) Pipe diameters in mm**

Gradient	2400	2500	2550	2600	2700	2800	2850	3000	3200	3400	3500	3600	4000	4500
0·000100	0·434	0·445	0·451	0·456	0·468	0·478	0·484	0·500	0·521	0·541	0·551	0·561	0·600	0·646
1/ 10000	1·9618	2·1850	2·3021	2·4231	2·6767	2·9461	3·0868	3·5335	4·1883	4·9132	5·3028	5·7108	7·5347	10·269
0·000105	0·444	0·456	0·462	0·468	0·479	0·490	0·496	0·512	0·534	0·555	0·565	0·575	0·614	0·662
1/ 9524	2·0104	2·2391	2·3592	2·4832	2·7430	3·0190	3·1632	3·6210	4·2920	5·0349	5·4341	5·8522	7·7212	10·523
0·000110	0·455	0·467	0·473	0·479	0·490	0·502	0·508	0·524	0·546	0·568	0·578	0·589	0·629	0·677
1/ 9091	2·0579	2·2920	2·4149	2·5418	2·8078	3·0903	3·2379	3·7065	4·3933	5·1537	5·5623	5·9903	7·9033	10·771
0·000115	0·465	0·477	0·484	0·490	0·501	0·513	0·519	0·536	0·559	0·580	0·591	0·602	0·643	0·693
1/ 8696	2·1043	2·3436	2·4693	2·5991	2·8711	3·1600	3·3109	3·7900	4·4923	5·2698	5·6876	6·1253	8·0814	11·014
0·000120	0·475	0·488	0·494	0·500	0·512	0·524	0·530	0·548	0·571	0·593	0·604	0·615	0·657	0·707
1/ 8333	2·1497	2·3942	2·5226	2·6552	2·9330	3·2281	3·3823	3·8718	4·5892	5·3834	5·8103	6·2574	8·2556	11·251
0·000125	0·485	0·498	0·504	0·510	0·523	0·535	0·541	0·559	0·582	0·605	0·616	0·627	0·671	0·722
1/ 8000	2·1942	2·4437	2·5748	2·7101	2·9937	3·2949	3·4522	3·9518	4·6841	5·4947	5·9304	6·3867	8·4262	11·484
0·000130	0·495	0·508	0·514	0·521	0·533	0·546	0·552	0·570	0·594	0·617	0·629	0·640	0·684	0·736
1/ 7692	2·2378	2·4923	2·6259	2·7639	3·0531	3·3603	3·5208	4·0303	4·7771	5·6038	6·0481	6·5135	8·5935	11·712
0·000135	0·504	0·517	0·524	0·531	0·543	0·556	0·562	0·581	0·605	0·629	0·641	0·652	0·697	0·750
1/ 7407	2·2805	2·5399	2·6761	2·8167	3·1114	3·4245	3·5880	4·1073	4·8683	5·7108	6·1636	6·6379	8·7576	11·935
0·000140	0·513	0·527	0·534	0·540	0·553	0·566	0·573	0·592	0·616	0·641	0·652	0·664	0·710	0·764
1/ 7143	2·3225	2·5866	2·7253	2·8685	3·1687	3·4875	3·6541	4·1829	4·9579	5·8159	6·2770	6·7600	8·9186	12·155
0·000145	0·523	0·536	0·543	0·550	0·563	0·576	0·583	0·602	0·627	0·652	0·664	0·676	0·722	0·778
1/ 6897	2·3637	2·6325	2·7737	2·9195	3·2249	3·5494	3·7189	4·2571	5·0459	5·9191	6·3883	6·8799	9·0768	12·370
0·000150	0·531	0·545	0·552	0·559	0·573	0·586	0·593	0·613	0·638	0·663	0·675	0·687	0·735	0·791
1/ 6667	2·4043	2·6777	2·8213	2·9695	3·2802	3·6102	3·7827	4·3300	5·1323	6·0205	6·4978	6·9978	9·2323	12·582
0·000160	0·549	0·563	0·571	0·578	0·592	0·606	0·612	0·633	0·659	0·685	0·698	0·710	0·759	0·817
1/ 6250	2·4833	2·7657	2·9140	3·0672	3·3881	3·7290	3·9070	4·4724	5·3011	6·2184	6·7114	7·2278	9·5357	12·996
0·000170	0·566	0·581	0·588	0·596	0·610	0·624	0·631	0·652	0·679	0·706	0·719	0·732	0·782	0·842
1/ 5882	2·5600	2·8511	3·0040	3·1618	3·4926	3·8440	4·0276	4·6104	5·4646	6·4102	6·9184	7·4507	9·8298	13·396
0·000180	0·582	0·598	0·605	0·613	0·628	0·642	0·650	0·671	0·699	0·727	0·740	0·753	0·805	0·867
1/ 5556	2·6344	2·9339	3·0913	3·2537	3·5941	3·9557	4·1446	4·7443	5·6234	6·5964	7·1194	7·6671	10·115	13·785
0·000190	0·598	0·614	0·622	0·630	0·645	0·660	0·668	0·690	0·718	0·746	0·760	0·774	0·827	0·891
1/ 5263	2·7068	3·0145	3·1762	3·3431	3·6928	4·0644	4·2584	4·8746	5·7778	6·7776	7·3148	7·8776	10·393	14·164
0·000200	0·614	0·630	0·638	0·646	0·662	0·677	0·685	0·708	0·737	0·766	0·780	0·794	0·849	0·914
1/ 5000	2·7773	3·0931	3·2589	3·4301	3·7890	4·1702	4·3693	5·0016	5·9282	6·9540	7·5052	8·0827	10·663	14·532
0·000210	0·629	0·646	0·654	0·662	0·678	0·694	0·702	0·725	0·755	0·785	0·799	0·814	0·870	0·936
1/ 4762	2·8460	3·1696	3·3396	3·5150	3·8828	4·2734	4·4775	5·1253	6·0749	7·1261	7·6909	8·2827	10·927	14·892
0·000220	0·644	0·661	0·669	0·678	0·694	0·710	0·718	0·742	0·773	0·803	0·818	0·833	0·890	0·958
1/ 4545	2·9132	3·2444	3·4183	3·5979	3·9744	4·3742	4·5831	5·2462	6·2181	7·2941	7·8723	8·4780	11·185	15·243
0·000230	0·658	0·676	0·684	0·693	0·710	0·726	0·735	0·759	0·791	0·821	0·837	0·852	0·910	0·980
1/ 4348	2·9788	3·3175	3·4953	3·6790	4·0639	4·4727	4·6863	5·3644	6·3582	7·4583	8·0495	8·6688	11·437	15·586
0·000240	0·673	0·690	0·699	0·708	0·725	0·742	0·750	0·775	0·808	0·839	0·855	0·870	0·930	1·001
1/ 4167	3·0430	3·3890	3·5707	3·7583	4·1515	4·5691	4·7873	5·4800	6·4952	7·6190	8·2230	8·8556	11·683	15·922
0·000250	0·687	0·705	0·714	0·722	0·740	0·757	0·766	0·791	0·824	0·857	0·872	0·888	0·949	1·022
1/ 4000	3·1059	3·4590	3·6445	3·8359	4·2373	4·6635	4·8862	5·5932	6·6294	7·7764	8·3928	9·0385	11·924	16·250
0·000260	0·700	0·719	0·728	0·737	0·755	0·772	0·781	0·807	0·841	0·874	0·890	0·906	0·968	1·042
1/ 3846	3·1675	3·5277	3·7168	3·9121	4·3213	4·7561	4·9831	5·7042	6·7609	7·9307	8·5593	9·2179	12·161	16·573
0·000270	0·714	0·732	0·742	0·751	0·769	0·787	0·796	0·822	0·857	0·890	0·907	0·923	0·986	1·062
1/ 3704	3·2280	3·5950	3·7877	3·9867	4·4038	4·8468	5·0783	5·8130	6·8899	8·0820	8·7227	9·3937	12·393	16·889
0·000280	0·727	0·746	0·755	0·765	0·783	0·802	0·811	0·837	0·872	0·907	0·923	0·940	1·004	1·081
1/ 3571	3·2874	3·6611	3·8574	4·0601	4·4848	4·9359	5·1716	5·9199	7·0166	8·2306	8·8830	9·5664	12·621	17·199
0·000290	0·740	0·759	0·769	0·778	0·797	0·816	0·825	0·852	0·888	0·923	0·940	0·957	1·022	1·101
1/ 3448	3·3457	3·7260	3·9258	4·1321	4·5643	5·0235	5·2633	6·0249	7·1410	8·3765	9·0405	9·7360	12·844	17·504
0·000300	0·752	0·772	0·782	0·792	0·811	0·830	0·839	0·867	0·903	0·938	0·956	0·973	1·040	1·119
1/ 3333	3·4030	3·7899	3·9931	4·2028	4·6425	5·1095	5·3535	6·1280	7·2633	8·5200	9·1953	9·9027	13·064	17·804
	0·80	0·80	0·80	0·80	0·80	0·80	0·80	0·81	0·81	0·81	0·81	0·81	0·82	0·83

$V_{r(0{\cdot}5)medial}$ for half-full circular pipes.

A55
(p.3 of 6)

$k_s = 6{\cdot}0$ mm
S = 0·00030 to 0·00100

ie hydraulic gradient =
1 in 3333 to 1 in 1000

Water (or sewage) at 15°C;
full bore conditions.

velocities in ms^{-1}
discharges in m^3s^{-1}

Gradient **(Equivalent) Pipe diameters in mm**

Gradient	2400	2500	2550	2600	2700	2800	2850	3000	3200	3400	3500	3600	4000	4500
0·000300	0·752	0·772	0·782	0·792	0·811	0·830	0·839	0·867	0·903	0·938	0·956	0·973	1·040	1·119
1/ 3333	3·4030	3·7899	3·9931	4·2028	4·6425	5·1095	5·3535	6·1280	7·2633	8·5200	9·1953	9·9027	13·064	17·804
0·000320	0·777	0·797	0·808	0·818	0·837	0·857	0·867	0·895	0·933	0·969	0·987	1·005	1·074	1·156
1/ 3125	3·5148	3·9144	4·1243	4·3409	4·7950	5·2774	5·5294	6·3294	7·5019	8·7998	9·4973	10·228	13·493	18·388
0·000340	0·801	0·822	0·832	0·843	0·863	0·883	0·893	0·923	0·962	0·999	1·018	1·036	1·107	1·192
1/ 2941	3·6232	4·0351	4·2514	4·4748	4·9429	5·4401	5·6998	6·5245	7·7331	9·0711	9·7901	10·543	13·909	18·955
0·000360	0·824	0·846	0·857	0·867	0·888	0·909	0·919	0·950	0·989	1·028	1·047	1·066	1·139	1·226
1/ 2778	3·7284	4·1523	4·3749	4·6047	5·0864	5·5981	5·8654	6·7140	7·9577	9·3345	10·074	10·849	14·313	19·505
0·000380	0·847	0·869	0·880	0·891	0·913	0·934	0·945	0·976	1·017	1·056	1·076	1·095	1·170	1·260
1/ 2632	3·8308	4·2663	4·4950	4·7311	5·2261	5·7517	6·0264	6·8982	8·1761	9·5906	10·351	11·147	14·706	20·041
0·000400	0·869	0·892	0·903	0·914	0·937	0·958	0·969	1·001	1·043	1·084	1·104	1·124	1·201	1·293
1/ 2500	3·9305	4·3773	4·6120	4·8543	5·3620	5·9014	6·1832	7·0777	8·3888	9·8401	10·620	11·437	15·088	20·562
0·000420	0·890	0·914	0·925	0·937	0·960	0·982	0·993	1·026	1·069	1·111	1·131	1·151	1·230	1·325
1/ 2381	4·0277	4·4856	4·7261	4·9743	5·4947	6·0474	6·3361	7·2528	8·5963	10·083	10·883	11·720	15·461	21·070
0·000440	0·911	0·935	0·947	0·959	0·982	1·005	1·017	1·050	1·094	1·137	1·158	1·179	1·259	1·356
1/ 2273	4·1227	4·5913	4·8375	5·0916	5·6242	6·1899	6·4854	7·4237	8·7988	10·321	11·139	11·996	15·826	21·567
0·000460	0·932	0·956	0·969	0·981	1·004	1·028	1·040	1·074	1·119	1·162	1·184	1·205	1·288	1·387
1/ 2174	4·2155	4·6947	4·9464	5·2062	5·7508	6·3292	6·6314	7·5908	8·9969	10·553	11·390	12·266	16·182	22·052
0·000480	0·952	0·977	0·989	1·002	1·026	1·050	1·062	1·097	1·143	1·187	1·209	1·231	1·315	1·416
1/ 2083	4·3063	4·7958	5·0529	5·3183	5·8747	6·4656	6·7742	7·7543	9·1906	10·781	11·635	12·530	16·530	22·527
0·000500	0·972	0·997	1·010	1·022	1·047	1·072	1·084	1·120	1·166	1·212	1·234	1·256	1·343	1·446
1/ 2000	4·3952	4·8949	5·1573	5·4282	5·9960	6·5991	6·9141	7·9144	9·3804	11·003	11·875	12·789	16·872	22·992
0·000525	0·996	1·022	1·035	1·048	1·073	1·098	1·111	1·147	1·195	1·242	1·265	1·287	1·376	1·481
1/ 1905	4·5039	5·0159	5·2848	5·5624	6·1443	6·7623	7·0851	8·1101	9·6124	11·275	12·169	13·105	17·289	23·560
0·000550	1·019	1·046	1·059	1·072	1·098	1·124	1·137	1·174	1·223	1·271	1·295	1·318	1·408	1·516
1/ 1818	4·6101	5·1341	5·4094	5·6935	6·2891	6·9216	7·2521	8·3012	9·8389	11·541	12·456	13·414	17·696	24·115
0·000575	1·042	1·069	1·083	1·097	1·123	1·149	1·162	1·201	1·251	1·300	1·324	1·347	1·440	1·550
1/ 1739	4·7138	5·2497	5·5311	5·8217	6·4306	7·0774	7·4153	8·4880	10·060	11·801	12·736	13·716	18·094	24·658
0·000600	1·064	1·092	1·106	1·120	1·147	1·174	1·187	1·227	1·278	1·328	1·352	1·376	1·471	1·584
1/ 1667	4·8154	5·3628	5·6502	5·9470	6·5691	7·2298	7·5750	8·6708	10·277	12·055	13·010	14·011	18·484	25·188
0·000625	1·086	1·115	1·129	1·143	1·171	1·198	1·212	1·252	1·304	1·355	1·380	1·405	1·501	1·616
1/ 1600	4·9148	5·4735	5·7669	6·0698	6·7047	7·3791	7·7314	8·8498	10·489	12·304	13·279	14·300	18·865	25·708
0·000650	1·108	1·137	1·152	1·166	1·194	1·222	1·236	1·277	1·330	1·382	1·408	1·433	1·531	1·648
1/ 1538	5·0123	5·5821	5·8813	6·1902	6·8377	7·5254	7·8847	9·0253	10·697	12·548	13·542	14·584	19·239	26·218
0·000675	1·129	1·159	1·174	1·188	1·217	1·245	1·260	1·301	1·355	1·408	1·434	1·460	1·560	1·680
1/ 1481	5·1079	5·6885	5·9935	6·3083	6·9681	7·6690	8·0351	9·1974	10·901	12·787	13·800	14·862	19·606	26·718
0·000700	1·150	1·180	1·195	1·210	1·239	1·268	1·283	1·325	1·380	1·434	1·461	1·487	1·589	1·711
1/ 1429	5·2018	5·7931	6·1036	6·4242	7·0961	7·8099	8·1827	9·3664	11·101	13·022	14·054	15·135	19·966	27·209
0·000725	1·170	1·201	1·216	1·231	1·261	1·291	1·305	1·349	1·405	1·460	1·487	1·513	1·617	1·741
1/ 1379	5·2940	5·8957	6·2118	6·5380	7·2219	7·9483	8·3277	9·5324	11·298	13·252	14·303	15·403	20·320	27·691
0·000750	1·190	1·222	1·237	1·253	1·283	1·313	1·328	1·372	1·429	1·485	1·512	1·539	1·645	1·771
1/ 1333	5·3846	5·9966	6·3181	6·6500	7·3455	8·0843	8·4702	9·6955	11·491	13·479	14·548	15·667	20·668	28·164
0·000800	1·229	1·262	1·278	1·294	1·325	1·356	1·371	1·417	1·476	1·533	1·562	1·590	1·699	1·829
1/ 1250	5·5614	6·1936	6·5255	6·8683	7·5867	8·3497	8·7483	10·014	11·869	13·922	15·025	16·181	21·346	29·089
0·000850	1·267	1·301	1·317	1·333	1·366	1·398	1·414	1·460	1·521	1·581	1·610	1·639	1·751	1·885
1/ 1176	5·7328	6·3844	6·7266	7·0799	7·8205	8·6070	9·0179	10·322	12·234	14·351	15·488	16·679	22·004	29·985
0·000900	1·304	1·338	1·355	1·372	1·406	1·438	1·455	1·503	1·565	1·626	1·657	1·686	1·802	1·940
1/ 1111	5·8992	6·5697	6·9219	7·2854	8·0474	8·8568	9·2796	10·622	12·589	14·767	15·937	17·163	22·642	30·855
0·000950	1·340	1·375	1·393	1·410	1·444	1·478	1·495	1·544	1·608	1·671	1·702	1·732	1·851	1·993
1/ 1053	6·0610	6·7499	7·1117	7·4853	8·2682	9·0998	9·5341	10·913	12·935	15·172	16·375	17·634	23·263	31·701
0·001000	1·375	1·411	1·429	1·447	1·482	1·516	1·533	1·584	1·650	1·715	1·746	1·777	1·899	2·045
1/ 1000	6·2186	6·9255	7·2967	7·6800	8·4832	9·3364	9·7821	11·197	13·271	15·567	16·800	18·093	23·868	32·525
	0·80	0·80	0·80	0·80	0·80	0·80	0·80	0·81	0·81	0·81	0·81	0·82	0·82	0·83

$V_{r(0\cdot5)medial}$ for half-full circular pipes.

$k_s = 6{\cdot}0$ mm S = 0·00030 to 0·00100

k_s = 6·0 mm
S = 0·00100 to 0·00300

ie hydraulic gradient =
1 in 1000 to 1 in 333

Water (or sewage) at 15°C;
full bore conditions.

velocities in ms^{-1}
discharges in m^3s^{-1}

Gradient **(Equivalent) Pipe diameters in mm**

Gradient	2400	2500	2550	2600	2700	2800	2850	3000	3200	3400	3500	3600	4000	4500
0·00100	1·375	1·411	1·429	1·447	1·482	1·516	1·533	1·584	1·650	1·715	1·746	1·777	1·899	2·045
1/ 1000	6·2186	6·9255	7·2967	7·6800	8·4832	9·3364	9·7821	11·197	13·271	15·567	16·800	18·093	23·868	32·525
0·00105	1·409	1·446	1·464	1·482	1·518	1·554	1·571	1·623	1·691	1·757	1·789	1·821	1·946	2·096
1/ 952	6·3724	7·0967	7·4771	7·8698	8·6929	9·5672	10·024	11·474	13·599	15·951	17·216	18·540	24·458	33·329
0·00110	1·442	1·480	1·499	1·517	1·554	1·590	1·608	1·661	1·731	1·798	1·831	1·864	1·992	2·145
1/ 909	6·5225	7·2639	7·6532	8·0552	8·8977	9·7926	10·260	11·744	13·919	16·327	17·621	18·976	25·034	34·114
0·00115	1·474	1·513	1·532	1·551	1·589	1·626	1·644	1·699	1·770	1·839	1·873	1·906	2·037	2·193
1/ 870	6·6692	7·4273	7·8254	8·2364	9·0979	10·013	10·491	12·008	14·232	16·694	18·017	19·403	25·597	34·881
0·00120	1·506	1·546	1·565	1·585	1·623	1·661	1·680	1·735	1·808	1·878	1·913	1·947	2·081	2·240
1/ 833	6·8128	7·5872	7·9939	8·4137	9·2937	10·228	10·717	12·267	14·539	17·054	18·405	19·821	26·148	35·632
0·00125	1·537	1·578	1·598	1·617	1·657	1·695	1·715	1·771	1·845	1·917	1·952	1·987	2·124	2·287
1/ 800	6·9535	7·7438	8·1589	8·5874	9·4856	10·440	10·938	12·520	14·839	17·406	18·785	20·230	26·687	36·367
0·00130	1·568	1·609	1·629	1·649	1·690	1·729	1·749	1·806	1·882	1·955	1·991	2·027	2·166	2·332
1/ 769	7·0913	7·8973	8·3206	8·7576	9·6736	10·646	11·155	12·768	15·133	17·751	19·157	20·631	27·216	37·088
0·00135	1·597	1·640	1·660	1·681	1·722	1·762	1·782	1·841	1·918	1·992	2·029	2·065	2·207	2·376
1/ 741	7·2265	8·0479	8·4793	8·9246	9·8580	10·849	11·367	13·012	15·421	18·089	19·523	21·024	27·735	37·795
0·00140	1·627	1·670	1·691	1·712	1·753	1·794	1·815	1·875	1·953	2·029	2·066	2·103	2·248	2·420
1/ 714	7·3593	8·1957	8·6350	9·0885	10·039	11·049	11·576	13·251	15·705	18·421	19·881	21·410	28·244	38·489
0·00145	1·656	1·699	1·721	1·742	1·784	1·826	1·847	1·908	1·987	2·065	2·103	2·141	2·287	2·463
1/ 690	7·4896	8·3409	8·7880	9·2496	10·217	11·244	11·781	13·485	15·983	18·748	20·233	21·790	28·745	39·170
0·00150	1·684	1·728	1·750	1·772	1·815	1·857	1·878	1·940	2·021	2·100	2·139	2·177	2·327	2·505
1/ 667	7·6178	8·4836	8·9383	9·4078	10·392	11·437	11·983	13·716	16·256	19·068	20·579	22·162	29·236	39·840
0·00160	1·739	1·785	1·808	1·830	1·875	1·918	1·940	2·004	2·088	2·169	2·209	2·249	2·403	2·587
1/ 625	7·8679	8·7621	9·2317	9·7166	10·733	11·812	12·376	14·166	16·790	19·694	21·255	22·890	30·196	41·148
0·00170	1·793	1·840	1·863	1·886	1·932	1·977	2·000	2·066	2·152	2·236	2·277	2·318	2·477	2·667
1/ 588	8·1102	9·0320	9·5161	10·016	11·063	12·176	12·757	14·603	17·307	20·301	21·909	23·595	31·126	42·415
0·00180	1·845	1·893	1·917	1·941	1·988	2·035	2·058	2·126	2·214	2·301	2·343	2·385	2·549	2·744
1/ 556	8·3455	9·2941	9·7922	10·307	11·384	12·529	13·127	15·026	17·809	20·890	22·545	24·279	32·029	43·645
0·00190	1·895	1·945	1·970	1·994	2·043	2·091	2·114	2·184	2·275	2·364	2·408	2·451	2·619	2·819
1/ 526	8·5744	9·5490	10·061	10·589	11·697	12·873	13·487	15·438	18·297	21·462	23·163	24·945	32·907	44·842
0·00200	1·945	1·996	2·021	2·046	2·096	2·145	2·169	2·241	2·334	2·425	2·470	2·514	2·687	2·893
1/ 500	8·7973	9·7972	10·322	10·864	12·001	13·208	13·838	15·840	18·773	22·020	23·765	25·593	33·762	46·008
0·00210	1·993	2·045	2·071	2·097	2·148	2·198	2·223	2·296	2·392	2·485	2·531	2·577	2·753	2·964
1/ 476	9·0148	10·039	10·577	11·133	12·297	13·534	14·180	16·231	19·237	22·564	24·353	26·226	34·597	47·144
0·00220	2·040	2·093	2·120	2·146	2·198	2·250	2·275	2·350	2·448	2·544	2·591	2·637	2·818	3·034
1/ 455	9·2271	10·276	10·827	11·395	12·587	13·853	14·514	16·613	19·690	23·096	24·926	26·843	35·411	48·254
0·00230	2·086	2·140	2·168	2·195	2·248	2·300	2·326	2·403	2·503	2·601	2·649	2·696	2·881	3·102
1/ 435	9·4346	10·507	11·070	11·651	12·870	14·164	14·840	16·987	20·133	23·615	25·487	27·447	36·207	49·339
0·00240	2·130	2·187	2·214	2·242	2·296	2·350	2·376	2·455	2·557	2·657	2·706	2·755	2·943	3·169
1/ 417	9·6377	10·733	11·308	11·902	13·147	14·469	15·160	17·352	20·566	24·123	26·035	28·037	36·987	50·401
0·00250	2·174	2·232	2·260	2·288	2·344	2·398	2·425	2·506	2·610	2·712	2·762	2·811	3·004	3·234
1/ 400	9·8365	10·955	11·542	12·148	13·418	14·768	15·472	17·710	20·990	24·621	26·572	28·616	37·750	51·441
0·00260	2·217	2·276	2·305	2·333	2·390	2·446	2·473	2·555	2·662	2·766	2·817	2·867	3·064	3·298
1/ 385	10·031	11·172	11·770	12·389	13·684	15·060	15·779	18·061	21·406	25·109	27·099	29·183	38·498	52·460
0·00270	2·260	2·319	2·349	2·378	2·436	2·492	2·521	2·604	2·712	2·818	2·870	2·922	3·122	3·361
1/ 370	10·223	11·385	11·995	12·625	13·945	15·347	16·080	18·406	21·814	25·588	27·615	29·739	39·231	53·460
0·00280	2·301	2·362	2·392	2·422	2·480	2·538	2·567	2·652	2·762	2·870	2·923	2·975	3·179	3·423
1/ 357	10·410	11·594	12·215	12·856	14·201	15·629	16·375	18·744	22·215	26·057	28·122	30·285	39·952	54·441
0·00290	2·342	2·404	2·434	2·464	2·524	2·583	2·612	2·699	2·811	2·921	2·975	3·028	3·236	3·484
1/ 345	10·595	11·799	12·431	13·084	14·453	15·906	16·665	19·076	22·608	26·519	28·620	30·822	40·659	55·405
0·00300	2·382	2·445	2·476	2·507	2·567	2·627	2·657	2·745	2·859	2·971	3·026	3·080	3·291	3·543
1/ 333	10·776	12·001	12·644	13·308	14·700	16·178	16·950	19·402	22·995	26·972	29·110	31·349	41·355	56·353
	0·80	0·80	0·80	0·80	0·80	0·80	0·80	0·81	0·81	0·81	0·81	0·82	0·82	0·83

$V_{r(0{\cdot}5)medial}$ for half-full circular pipes.

k_s = 6·0 mm
S = 0·00300 to 0·01000

ie hydraulic gradient =
1 in 333 to 1 in 100

Water (or sewage) at 15°C;
full bore conditions.

velocities in ms^{-1}
discharges in m^3s^{-1}

Gradient **(Equivalent) Pipe diameters in mm**

Gradient	2400	2500	2550	2600	2700	2800	2850	3000	3200	3400	3500	3600	4000	4500
0·00300	2·382	2·445	2·476	2·507	2·567	2·627	2·657	2·745	2·859	2·971	3·026	3·080	3·291	3·543
1/ 333	10·776	12·001	12·644	13·308	14·700	16·178	16·950	19·402	22·995	26·972	29·110	31·349	41·355	56·353
0·00320	2·460	2·525	2·557	2·589	2·652	2·714	2·744	2·835	2·953	3·068	3·125	3·181	3·399	3·660
1/ 313	11·130	12·395	13·059	13·745	15·182	16·709	17·506	20·039	23·750	27·857	30·065	32·377	42·711	58·202
0·00340	2·536	2·603	2·636	2·669	2·733	2·797	2·829	2·922	3·044	3·163	3·221	3·279	3·504	3·772
1/ 294	11·472	12·776	13·461	14·168	15·650	17·223	18·045	20·656	24·481	28·715	30·991	33·374	44·027	59·994
0·00360	2·610	2·678	2·712	2·746	2·813	2·878	2·911	3·007	3·132	3·254	3·315	3·374	3·605	3·882
1/ 278	11·805	13·147	13·852	14·579	16·104	17·723	18·569	21·255	25·191	29·548	31·890	34·342	45·303	61·734
0·00380	2·681	2·752	2·787	2·821	2·890	2·957	2·991	3·089	3·218	3·344	3·405	3·466	3·704	3·988
1/ 263	12·129	13·507	14·231	14·979	16·545	18·209	19·078	21·837	25·882	30·358	32·764	35·284	46·545	63·426
0·00400	2·751	2·823	2·859	2·895	2·965	3·034	3·068	3·170	3·302	3·431	3·494	3·557	3·800	4·092
1/ 250	12·444	13·858	14·601	15·368	16·975	18·682	19·574	22·405	26·554	31·147	33·615	36·201	47·755	65·075
0·00420	2·819	2·893	2·930	2·966	3·038	3·109	3·144	3·248	3·383	3·515	3·580	3·644	3·894	4·193
1/ 238	12·752	14·201	14·962	15·748	17·395	19·144	20·058	22·959	27·210	31·917	34·446	37·095	48·935	66·682
0·00440	2·885	2·961	2·999	3·036	3·110	3·182	3·218	3·324	3·463	3·598	3·665	3·730	3·986	4·291
1/ 227	13·052	14·535	15·314	16·119	17·804	19·595	20·530	23·499	27·851	32·668	35·257	37·968	50·087	68·252
0·00460	2·950	3·028	3·066	3·104	3·180	3·254	3·290	3·399	3·541	3·679	3·747	3·814	4·075	4·388
1/ 217	13·345	14·862	15·659	16·481	18·204	20·035	20·991	24·028	28·477	33·403	36·050	38·822	51·213	69·787
0·00480	3·013	3·093	3·132	3·171	3·248	3·324	3·361	3·472	3·617	3·758	3·828	3·896	4·163	4·482
1/ 208	13·633	15·182	15·996	16·836	18·596	20·466	21·443	24·545	29·090	34·121	36·825	39·657	52·315	71·288
0·00500	3·076	3·157	3·197	3·236	3·315	3·392	3·431	3·544	3·692	3·836	3·906	3·976	4·249	4·575
1/ 200	13·914	15·495	16·326	17·183	18·980	20·888	21·885	25·051	29·690	34·825	37·585	40·476	53·394	72·758
0·00525	3·152	3·235	3·276	3·316	3·397	3·476	3·515	3·632	3·783	3·930	4·003	4·075	4·354	4·688
1/ 190	14·258	15·878	16·729	17·607	19·449	21·405	22·426	25·670	30·424	35·686	38·513	41·475	54·713	74·556
0·00550	3·226	3·311	3·353	3·394	3·477	3·558	3·598	3·717	3·872	4·023	4·097	4·171	4·456	4·798
1/ 182	14·593	16·252	17·123	18·022	19·907	21·908	22·954	26·274	31·140	36·526	39·420	42·452	56·001	76·311
0·00575	3·298	3·385	3·428	3·471	3·555	3·638	3·679	3·801	3·959	4·113	4·189	4·264	4·557	4·906
1/ 174	14·921	16·617	17·508	18·427	20·354	22·401	23·470	26·865	31·840	37·347	40·306	43·406	57·260	78·027
0·00600	3·369	3·458	3·502	3·545	3·631	3·716	3·758	3·882	4·044	4·202	4·279	4·356	4·655	5·012
1/ 167	15·243	16·975	17·885	18·824	20·792	22·883	23·975	27·443	32·525	38·151	41·174	44·340	58·492	79·705
0·00625	3·439	3·529	3·574	3·619	3·706	3·793	3·836	3·962	4·128	4·289	4·368	4·446	4·751	5·115
1/ 160	15·557	17·325	18·253	19·212	21·221	23·355	24·470	28·009	33·196	38·938	42·023	45·255	59·699	81·349
0·00650	3·507	3·599	3·645	3·690	3·780	3·868	3·912	4·041	4·209	4·374	4·454	4·534	4·845	5·216
1/ 154	15·865	17·668	18·615	19·593	21·642	23·818	24·955	28·564	33·854	39·709	42·855	46·151	60·881	82·961
0·00675	3·574	3·668	3·714	3·761	3·852	3·942	3·986	4·118	4·290	4·457	4·539	4·620	4·937	5·316
1/ 148	16·168	18·005	18·970	19·966	22·054	24·272	25·430	29·108	34·499	40·466	43·672	47·031	62·041	84·542
0·00700	3·639	3·735	3·783	3·830	3·923	4·014	4·059	4·194	4·368	4·539	4·623	4·705	5·028	5·413
1/ 143	16·464	18·335	19·318	20·333	22·459	24·717	25·897	29·643	35·132	41·208	44·474	47·894	63·180	86·093
0·00725	3·704	3·801	3·850	3·897	3·992	4·085	4·131	4·268	4·446	4·619	4·704	4·789	5·117	5·509
1/ 138	16·756	18·660	19·660	20·693	22·857	25·155	26·356	30·167	35·754	41·938	45·261	48·742	64·299	87·618
0·00750	3·767	3·866	3·915	3·964	4·060	4·155	4·202	4·341	4·522	4·698	4·785	4·871	5·204	5·603
1/ 133	17·042	18·979	19·996	21·047	23·247	25·585	26·806	30·683	36·366	42·655	46·035	49·576	65·398	89·116
0·00800	3·891	3·993	4·044	4·094	4·194	4·291	4·340	4·483	4·670	4·852	4·942	5·030	5·375	5·787
1/ 125	17·602	19·602	20·652	21·737	24·010	26·425	27·686	31·690	37·559	44·055	47·545	51·202	67·544	92·039
0·00850	4·011	4·116	4·168	4·220	4·323	4·424	4·473	4·621	4·814	5·002	5·094	5·185	5·540	5·965
1/ 118	18·144	20·205	21·288	22·406	24·749	27·238	28·538	32·666	38·715	45·411	49·009	52·778	69·623	94·873
0·00900	4·127	4·236	4·289	4·343	4·448	4·552	4·603	4·755	4·953	5·147	5·242	5·336	5·701	6·138
1/ 111	18·670	20·791	21·906	23·056	25·467	28·028	29·366	33·613	39·838	46·728	50·430	54·309	71·642	97·624
0·00950	4·240	4·352	4·407	4·462	4·570	4·677	4·729	4·886	5·089	5·288	5·385	5·482	5·857	6·306
1/ 105	19·182	21·361	22·506	23·688	26·165	28·796	30·171	34·534	40·930	48·009	51·813	55·798	73·606	100·30
0·01000	4·350	4·465	4·521	4·578	4·689	4·798	4·852	5·013	5·221	5·425	5·525	5·624	6·010	6·470
1/ 100	19·680	21·917	23·091	24·304	26·845	29·545	30·955	35·432	41·993	49·256	53·159	57·248	75·519	102·91
	0·80	0·80	0·80	0·80	0·80	0·80	0·80	0·81	0·81	0·81	0·81	0·82	0·82	0·83

$V_{r(0\cdot5)medial}$ for half-full circular pipes.

Gradient	**(Equivalent) Pipe diameters in mm**													
	2400	**2500**	**2550**	**2600**	**2700**	**2800**	**2850**	**3000**	**3200**	**3400**	**3500**	**3600**	**4000**	**4500**
0·01000	4·350	4·465	4·521	4·578	4·689	4·798	4·852	5·013	5·221	5·425	5·525	5·624	6·010	6·470
1/ 100	19·680	21·917	23·091	24·304	26·845	29·545	30·955	35·432	41·993	49·256	53·159	57·248	75·519	102·91
0·01050	4·458	4·575	4·633	4·691	4·804	4·917	4·972	5·136	5·350	5·559	5·662	5·763	6·158	6·630
1/ 95	20·166	22·458	23·662	24·904	27·508	30·275	31·719	36·307	43·031	50·473	54·472	58·662	77·384	105·45
0·01100	4·563	4·683	4·742	4·801	4·918	5·032	5·089	5·257	5·476	5·690	5·795	5·899	6·303	6·786
1/ 91	20·641	22·987	24·219	25·490	28·156	30·987	32·466	37·162	44·043	51·661	55·755	60·042	79·206	107·93
0·01150	4·665	4·788	4·849	4·909	5·028	5·146	5·204	5·375	5·599	5·818	5·925	6·031	6·445	6·939
1/ 87	21·105	23·503	24·763	26·063	28·789	31·684	33·196	37·997	45·034	52·822	57·008	61·392	80·986	110·36
0·01200	4·766	4·891	4·953	5·015	5·136	5·256	5·316	5·491	5·720	5·943	6·053	6·161	6·583	7·088
1/ 83	21·559	24·009	25·296	26·624	29·408	32·365	33·910	38·815	46·002	53·959	58·234	62·713	82·728	112·73
0·01250	4·864	4·992	5·055	5·118	5·242	5·365	5·425	5·604	5·838	6·066	6·178	6·288	6·719	7·234
1/ 80	22·004	24·504	25·818	27·173	30·015	33·033	34·610	39·615	46·951	55·072	59·436	64·007	84·435	115·06
0·01300	4·960	5·091	5·155	5·219	5·346	5·471	5·533	5·715	5·954	6·186	6·300	6·413	6·852	7·378
1/ 77	22·440	24·990	26·329	27·712	30·609	33·687	35·295	40·400	47·881	56·163	60·613	65·274	86·107	117·33
0·01350	5·055	5·188	5·254	5·319	5·448	5·575	5·638	5·824	6·067	6·304	6·420	6·535	6·983	7·518
1/ 74	22·867	25·466	26·831	28·240	31·193	34·329	35·968	41·170	48·794	57·233	61·768	66·518	87·748	119·57
0·01400	5·148	5·283	5·350	5·417	5·548	5·678	5·742	5·931	6·178	6·419	6·538	6·655	7·111	7·656
1/ 71	23·287	25·933	27·323	28·758	31·765	34·959	36·628	41·925	49·689	58·283	62·901	67·739	89·358	121·76
0·01450	5·239	5·377	5·445	5·512	5·646	5·778	5·843	6·036	6·288	6·533	6·654	6·773	7·237	7·792
1/ 69	23·699	26·393	27·807	29·267	32·328	35·578	37·276	42·668	50·569	59·315	64·015	68·938	90·940	123·92
0·01500	5·328	5·469	5·538	5·607	5·743	5·877	5·943	6·139	6·395	6·645	6·767	6·889	7·361	7·925
1/ 67	24·105	26·844	28·282	29·768	32·881	36·187	37·914	43·397	51·434	60·329	65·110	70·117	92·495	126·04
0·01600	5·503	5·648	5·720	5·791	5·931	6·070	6·138	6·341	6·605	6·863	6·989	7·115	7·602	8·185
1/ 63	24·895	27·725	29·210	30·744	33·959	37·374	39·158	44·821	53·121	62·308	67·246	72·417	95·529	130·17
0·01700	5·672	5·822	5·896	5·969	6·114	6·256	6·327	6·536	6·808	7·074	7·205	7·334	7·836	8·437
1/ 59	25·662	28·578	30·109	31·691	35·005	38·524	40·363	46·201	54·756	64·226	69·316	74·646	98·470	134·18
0·01800	5·837	5·991	6·067	6·142	6·291	6·438	6·511	6·726	7·006	7·279	7·413	7·546	8·063	8·681
1/ 56	26·406	29·407	30·983	32·610	36·020	39·642	41·534	47·540	56·344	66·089	71·326	76·811	101·33	138·07
0·01900	5·997	6·155	6·233	6·310	6·463	6·614	6·689	6·910	7·198	7·479	7·617	7·753	8·284	8·919
1/ 53	27·130	30·213	31·832	33·503	37·007	40·728	42·672	48·843	57·888	67·900	73·280	78·916	104·10	141·86
0·02000	6·153	6·315	6·395	6·474	6·631	6·786	6·863	7·089	7·385	7·673	7·815	7·954	8·499	9·151
1/ 50	27·835	30·998	32·659	34·374	37·969	41·786	43·781	50·113	59·392	69·665	75·185	80·967	106·81	145·54
0·02100	6·305	6·471	6·553	6·634	6·795	6·954	7·032	7·265	7·567	7·863	8·008	8·151	8·709	9·377
1/ 47·6	28·522	31·764	33·466	35·223	38·906	42·819	44·862	51·350	60·859	71·385	77·042	82·967	109·45	149·14
0·02200	6·453	6·623	6·707	6·790	6·955	7·118	7·198	7·436	7·745	8·048	8·196	8·343	8·914	9·598
1/ 45·5	29·194	32·511	34·253	36·052	39·822	43·826	45·918	52·559	62·292	73·065	78·855	84·919	112·02	152·65
0·02300	6·598	6·772	6·858	6·943	7·112	7·278	7·360	7·603	7·919	8·228	8·380	8·530	9·115	9·814
1/ 43·5	29·850	33·242	35·023	36·863	40·717	44·812	46·950	53·741	63·692	74·708	80·628	86·828	114·54	156·08
0·02400	6·740	6·918	7·005	7·092	7·264	7·434	7·518	7·766	8·090	8·405	8·561	8·714	9·311	10·02
1/ 41·7	30·492	33·957	35·777	37·656	41·593	45·776	47·960	54·897	65·062	76·315	82·362	88·696	117·00	159·43
0·02500	6·879	7·060	7·150	7·239	7·414	7·587	7·673	7·926	8·257	8·579	8·737	8·894	9·503	10·23
1/ 40·0	31·121	34·658	36·515	38·432	42·451	46·720	48·949	56·029	66·404	77·889	84·061	90·525	119·42	162·72
0·02600	7·015	7·200	7·291	7·382	7·561	7·738	7·825	8·083	8·420	8·749	8·910	9·070	9·691	10·43
1/ 38·5	31·737	35·344	37·238	39·194	43·292	47·645	49·919	57·139	67·720	79·432	85·726	92·318	121·78	165·95
0·02700	7·149	7·337	7·430	7·523	7·705	7·885	7·974	8·237	8·581	8·915	9·080	9·243	9·876	10·63
1/ 37·0	32·342	36·017	37·948	39·940	44·117	48·553	50·870	58·227	69·010	80·945	87·359	94·077	124·10	169·11
0·02800	7·280	7·472	7·567	7·661	7·847	8·030	8·120	8·389	8·738	9·079	9·247	9·412	10·06	10·83
1/ 35·7	32·936	36·679	38·644	40·673	44·927	49·444	51·804	59·296	70·276	82·431	88·962	95·804	126·38	172·21
0·02900	7·409	7·604	7·701	7·796	7·986	8·172	8·264	8·537	8·893	9·240	9·410	9·579	10·24	11·02
1/ 34·5	33·519	37·328	39·328	41·393	45·722	50·319	52·721	60·346	71·520	83·890	90·537	97·500	128·62	175·26
0·03000	7·536	7·734	7·832	7·930	8·122	8·312	8·406	8·683	9·045	9·398	9·571	9·743	10·41	11·21
1/ 33·3	34·092	37·966	40·001	42·101	46·504	51·180	53·622	61·378	72·743	85·324	92·085	99·167	130·82	178·26
	0·80	0·80	0·80	0·80	0·80	0·80	0·80	0·81	0·81	0·81	0·81	0·82	0·82	0·83

$V_{r(0{\cdot}5)medial}$ for half-full circular pipes.

k_s = 15·0 mm
S = 0·00003 to 0·00010

ie hydraulic gradient =
1 in 33333 to 1 in 10000

Water (or sewage) at 15°C;
full bore conditions.

velocities in ms^{-1}
discharges in m^3s^{-1}

Gradient **(Equivalent) Pipe diameters in mm**

Gradient	2400	2500	2550	2600	2700	2800	2850	3000	3200	3400	3500	3600	4000	4500
0·0000300	0·208	0·213	0·216	0·219	0·225	0·230	0·233	0·241	0·251	0·261	0·266	0·271	0·290	0·313
1/ 33333	0·9402	1·0480	1·1046	1·1631	1·2857	1·4161	1·4842	1·7006	2·0181	2·3699	2·5591	2·7574	3·6445	4·9767
0·0000320	0·215	0·221	0·223	0·226	0·232	0·238	0·240	0·248	0·259	0·270	0·275	0·280	0·300	0·323
1/ 31250	0·9712	1·0825	1·1410	1·2014	1·3280	1·4626	1·5330	1·7565	2·0844	2·4478	2·6432	2·8480	3·7643	5·1402
0·0000340	0·221	0·227	0·230	0·233	0·239	0·245	0·248	0·256	0·267	0·278	0·283	0·288	0·309	0·333
1/ 29412	1·0011	1·1159	1·1762	1·2384	1·3690	1·5078	1·5803	1·8107	2·1487	2·5233	2·7247	2·9358	3·8804	5·2987
0·0000360	0·228	0·234	0·237	0·240	0·246	0·252	0·255	0·264	0·275	0·286	0·291	0·297	0·318	0·343
1/ 27778	1·0302	1·1483	1·2103	1·2744	1·4088	1·5516	1·6262	1·8633	2·2112	2·5966	2·8039	3·0211	3·9931	5·4526
0·0000380	0·234	0·240	0·244	0·247	0·253	0·259	0·262	0·271	0·282	0·294	0·299	0·305	0·326	0·352
1/ 26316	1·0585	1·1799	1·2436	1·3094	1·4475	1·5942	1·6709	1·9145	2·2719	2·6679	2·8809	3·1041	4·1027	5·6023
0·0000400	0·240	0·247	0·250	0·253	0·259	0·266	0·269	0·278	0·290	0·301	0·307	0·313	0·335	0·361
1/ 25000	1·0861	1·2106	1·2760	1·3435	1·4852	1·6357	1·7144	1·9643	2·3310	2·7374	2·9559	3·1849	4·2095	5·7481
0·0000420	0·246	0·253	0·256	0·259	0·266	0·272	0·275	0·285	0·297	0·309	0·315	0·321	0·343	0·370
1/ 23810	1·1130	1·2406	1·3076	1·3768	1·5219	1·6762	1·7568	2·0130	2·3887	2·8051	3·0290	3·2637	4·3136	5·8903
0·0000440	0·252	0·259	0·262	0·265	0·272	0·279	0·282	0·291	0·304	0·316	0·322	0·328	0·351	0·379
1/ 22727	1·1393	1·2698	1·3384	1·4093	1·5578	1·7157	1·7983	2·0604	2·4450	2·8712	3·1004	3·3406	4·4153	6·0291
0·0000460	0·258	0·265	0·268	0·271	0·278	0·285	0·288	0·298	0·311	0·323	0·330	0·336	0·359	0·388
1/ 21739	1·1649	1·2984	1·3686	1·4410	1·5929	1·7544	1·8388	2·1068	2·5001	2·9359	3·1703	3·4159	4·5147	6·1648
0·0000480	0·263	0·270	0·274	0·277	0·284	0·291	0·294	0·304	0·318	0·330	0·337	0·343	0·367	0·396
1/ 20833	1·1900	1·3264	1·3981	1·4721	1·6273	1·7922	1·8784	2·1522	2·5540	2·9992	3·2386	3·4895	4·6120	6·2976
0·0000500	0·268	0·276	0·279	0·283	0·290	0·297	0·301	0·311	0·324	0·337	0·344	0·350	0·375	0·404
1/ 20000	1·2146	1·3538	1·4270	1·5025	1·6609	1·8292	1·9172	2·1967	2·6068	3·0611	3·3055	3·5615	4·7073	6·4277
0·0000525	0·275	0·283	0·286	0·290	0·297	0·304	0·308	0·318	0·332	0·345	0·352	0·359	0·384	0·414
1/ 19048	1·2447	1·3873	1·4623	1·5397	1·7020	1·8745	1·9647	2·2511	2·6712	3·1369	3·3872	3·6497	4·8237	6·5867
0·0000550	0·282	0·289	0·293	0·297	0·304	0·312	0·315	0·326	0·340	0·354	0·360	0·367	0·393	0·424
1/ 18182	1·2741	1·4201	1·4968	1·5760	1·7421	1·9187	2·0110	2·3041	2·7342	3·2108	3·4671	3·7357	4·9374	6·7419
0·0000575	0·288	0·296	0·300	0·304	0·311	0·319	0·322	0·333	0·348	0·362	0·368	0·375	0·402	0·433
1/ 17391	1·3028	1·4520	1·5305	1·6115	1·7814	1·9619	2·0563	2·3560	2·7958	3·2831	3·5451	3·8198	5·0485	6·8936
0·0000600	0·294	0·302	0·306	0·310	0·318	0·325	0·329	0·340	0·355	0·369	0·376	0·383	0·410	0·443
1/ 16667	1·3308	1·4833	1·5635	1·6462	1·8197	2·0042	2·1006	2·4068	2·8560	3·3538	3·6215	3·9021	5·1573	7·0421
0·0000625	0·300	0·308	0·312	0·316	0·324	0·332	0·336	0·348	0·362	0·377	0·384	0·391	0·419	0·452
1/ 16000	1·3583	1·5140	1·5958	1·6802	1·8573	2·0456	2·1440	2·4565	2·9150	3·4231	3·6963	3·9827	5·2638	7·1875
0·0000650	0·306	0·315	0·319	0·323	0·331	0·339	0·343	0·354	0·370	0·385	0·392	0·399	0·427	0·461
1/ 15385	1·3853	1·5440	1·6274	1·7136	1·8942	2·0862	2·1865	2·5052	2·9728	3·4910	3·7696	4·0617	5·3682	7·3301
0·0000675	0·312	0·321	0·325	0·329	0·337	0·345	0·349	0·361	0·377	0·392	0·399	0·407	0·435	0·470
1/ 14815	1·4117	1·5735	1·6585	1·7463	1·9303	2·1260	2·2282	2·5530	3·0296	3·5576	3·8416	4·1391	5·4706	7·4699
0·0000700	0·318	0·326	0·331	0·335	0·343	0·352	0·356	0·368	0·384	0·399	0·407	0·414	0·443	0·478
1/ 14286	1·4377	1·6024	1·6890	1·7784	1·9658	2·1651	2·2692	2·6000	3·0852	3·6230	3·9122	4·2152	5·5711	7·6071
0·0000725	0·323	0·332	0·337	0·341	0·349	0·358	0·362	0·374	0·390	0·406	0·414	0·421	0·451	0·487
1/ 13793	1·4632	1·6308	1·7189	1·8099	2·0007	2·2034	2·3094	2·6461	3·1399	3·6872	3·9815	4·2899	5·6699	7·7419
0·0000750	0·329	0·338	0·342	0·347	0·355	0·364	0·368	0·381	0·397	0·413	0·421	0·429	0·459	0·495
1/ 13333	1·4882	1·6588	1·7484	1·8409	2·0349	2·2412	2·3490	2·6914	3·1937	3·7503	4·0497	4·3634	5·7669	7·8745
0·0000800	0·340	0·349	0·354	0·358	0·367	0·376	0·380	0·393	0·410	0·427	0·435	0·443	0·474	0·511
1/ 12500	1·5371	1·7133	1·8058	1·9014	2·1018	2·3148	2·4261	2·7798	3·2986	3·8735	4·1827	4·5067	5·9563	8·1330
0·0000850	0·350	0·360	0·364	0·369	0·378	0·388	0·392	0·405	0·423	0·440	0·448	0·456	0·489	0·527
1/ 11765	1·5845	1·7661	1·8615	1·9600	2·1666	2·3862	2·5009	2·8655	3·4003	3·9929	4·3116	4·6456	6·1398	8·3836
0·0000900	0·360	0·370	0·375	0·380	0·389	0·399	0·403	0·417	0·435	0·453	0·461	0·470	0·503	0·542
1/ 11111	1·6306	1·8174	1·9155	2·0169	2·2295	2·4554	2·5735	2·9487	3·4990	4·1088	4·4368	4·7804	6·3181	8·6270
0·0000950	0·370	0·380	0·385	0·390	0·400	0·410	0·414	0·429	0·447	0·465	0·474	0·483	0·517	0·557
1/ 10526	1·6753	1·8673	1·9681	2·0723	2·2907	2·5228	2·6442	3·0296	3·5950	4·2216	4·5585	4·9116	6·4914	8·8636
0·0001000	0·380	0·390	0·395	0·400	0·410	0·420	0·425	0·440	0·459	0·477	0·486	0·495	0·530	0·572
1/ 10000	1·7189	1·9158	2·0193	2·1262	2·3503	2·5885	2·7130	3·1084	3·6885	4·3314	4·6771	5·0394	6·6602	9·0941
	0·76	0·76	0·76	0·76	0·76	0·77	0·77	0·77	0·77	0·77	0·77	0·77	0·78	0·78

$V_{r(0\cdot5)medial}$ for half-full circular pipes.

$k_s = 15{\cdot}0$ mm
S = 0·00010 to 0·00030

ie hydraulic gradient =
1 in 10000 to 1 in 3333

Water (or sewage) at 15°C;
full bore conditions.

velocities in ms^{-1}
discharges in m^3s^{-1}

Gradient **(Equivalent) Pipe diameters in mm**

	2400	**2500**	**2550**	**2600**	**2700**	**2800**	**2850**	**3000**	**3200**	**3400**	**3500**	**3600**	**4000**	**4500**
0·000100	0·380	0·390	0·395	0·400	0·410	0·420	0·425	0·440	0·459	0·477	0·486	0·495	0·530	0·572
1/ 10000	1·7189	1·9158	2·0193	2·1262	2·3503	2·5885	2·7130	3·1084	3·6885	4·3314	4·6771	5·0394	6·6602	9·0941
0·000105	0·389	0·400	0·405	0·410	0·421	0·431	0·436	0·451	0·470	0·489	0·498	0·507	0·543	0·586
1/ 9524	1·7614	1·9632	2·0693	2·1788	2·4084	2·6525	2·7800	3·1853	3·7797	4·4385	4·7927	5·1640	6·8249	9·3190
0·000110	0·399	0·409	0·415	0·420	0·431	0·441	0·446	0·461	0·481	0·500	0·510	0·519	0·556	0·600
1/ 9091	1·8029	2·0095	2·1180	2·2301	2·4652	2·7150	2·8456	3·2603	3·8688	4·5431	4·9056	5·2856	6·9857	9·5385
0·000115	0·408	0·419	0·424	0·429	0·440	0·451	0·456	0·472	0·492	0·512	0·521	0·531	0·568	0·613
1/ 8696	1·8435	2·0547	2·1657	2·2803	2·5207	2·7761	2·9096	3·3337	3·9559	4·6453	5·0160	5·4046	7·1429	9·7531
0·000120	0·416	0·428	0·433	0·439	0·450	0·461	0·466	0·482	0·502	0·523	0·533	0·542	0·581	0·626
1/ 8333	1·8832	2·0990	2·2123	2·3294	2·5749	2·8359	2·9723	3·4055	4·0411	4·7453	5·1240	5·5209	7·2967	9·9631
0·000125	0·425	0·436	0·442	0·448	0·459	0·470	0·476	0·492	0·513	0·533	0·544	0·554	0·593	0·639
1/ 8000	1·9221	2·1423	2·2580	2·3775	2·6281	2·8944	3·0336	3·4758	4·1245	4·8433	5·2298	5·6349	7·4473	10·169
0·000130	0·433	0·445	0·451	0·457	0·468	0·479	0·485	0·501	0·523	0·544	0·554	0·565	0·604	0·652
1/ 7692	1·9602	2·1848	2·3028	2·4247	2·6802	2·9518	3·0938	3·5447	4·2063	4·9393	5·3335	5·7466	7·5949	10·370
0·000135	0·442	0·454	0·460	0·465	0·477	0·489	0·494	0·511	0·533	0·554	0·565	0·575	0·616	0·664
1/ 7407	1·9976	2·2265	2·3467	2·4709	2·7314	3·0081	3·1528	3·6123	4·2865	5·0335	5·4352	5·8562	7·7397	10·568
0·000140	0·450	0·462	0·468	0·474	0·486	0·498	0·503	0·520	0·543	0·565	0·575	0·586	0·627	0·677
1/ 7143	2·0343	2·2674	2·3899	2·5163	2·7815	3·0634	3·2107	3·6787	4·3652	5·1260	5·5351	5·9638	7·8819	10·762
0·000145	0·458	0·470	0·476	0·482	0·494	0·506	0·512	0·530	0·552	0·575	0·585	0·596	0·638	0·689
1/ 6897	2·0704	2·3076	2·4322	2·5609	2·8308	3·1177	3·2676	3·7439	4·4426	5·2168	5·6331	6·0695	8·0216	10·953
0·000150	0·465	0·478	0·484	0·491	0·503	0·515	0·521	0·539	0·562	0·584	0·596	0·606	0·649	0·700
1/ 6667	2·1058	2·3471	2·4738	2·6048	2·8793	3·1711	3·3235	3·8080	4·5186	5·3061	5·7295	6·1733	8·1588	11·140
0·000160	0·481	0·494	0·500	0·507	0·519	0·532	0·538	0·556	0·580	0·604	0·615	0·626	0·671	0·723
1/ 6250	2·1750	2·4242	2·5551	2·6903	2·9738	3·2752	3·4327	3·9330	4·6670	5·4803	5·9176	6·3760	8·4266	11·506
0·000170	0·496	0·509	0·516	0·522	0·535	0·548	0·555	0·574	0·598	0·622	0·634	0·646	0·691	0·746
1/ 5882	2·2420	2·4989	2·6338	2·7732	3·0655	3·3761	3·5384	4·0542	4·8108	5·6491	6·0999	6·5724	8·6862	11·860
0·000180	0·510	0·524	0·531	0·537	0·551	0·564	0·571	0·590	0·616	0·640	0·652	0·664	0·711	0·767
1/ 5556	2·3071	2·5714	2·7103	2·8537	3·1544	3·4741	3·6411	4·1718	4·9504	5·8130	6·2770	6·7631	8·9383	12·204
0·000190	0·524	0·538	0·545	0·552	0·566	0·580	0·586	0·606	0·632	0·658	0·670	0·683	0·731	0·788
1/ 5263	2·3704	2·6419	2·7846	2·9320	3·2410	3·5694	3·7410	4·2863	5·0862	5·9725	6·4491	6·9486	9·1834	12·539
0·000200	0·538	0·552	0·559	0·567	0·581	0·595	0·602	0·622	0·649	0·675	0·688	0·700	0·750	0·809
1/ 5000	2·4320	2·7106	2·8570	3·0082	3·3253	3·6622	3·8383	4·3977	5·2184	6·1278	6·6168	7·1293	9·4222	12·865
0·000210	0·551	0·566	0·573	0·581	0·595	0·609	0·617	0·638	0·665	0·692	0·705	0·718	0·768	0·829
1/ 4762	2·4922	2·7777	2·9277	3·0826	3·4075	3·7527	3·9332	4·5064	5·3474	6·2792	6·7804	7·3055	9·6551	13·183
0·000220	0·564	0·579	0·587	0·594	0·609	0·624	0·631	0·653	0·681	0·708	0·721	0·735	0·786	0·848
1/ 4545	2·5509	2·8431	2·9966	3·1552	3·4877	3·8411	4·0258	4·6126	5·4734	6·4271	6·9401	7·4776	9·8824	13·493
0·000230	0·577	0·592	0·600	0·608	0·623	0·638	0·645	0·667	0·696	0·724	0·738	0·751	0·804	0·867
1/ 4348	2·6083	2·9071	3·0641	3·2262	3·5662	3·9276	4·1164	4·7164	5·5965	6·5717	7·0962	7·6458	10·105	13·797
0·000240	0·589	0·605	0·613	0·621	0·636	0·652	0·659	0·682	0·711	0·739	0·753	0·767	0·821	0·886
1/ 4167	2·6644	2·9697	3·1300	3·2957	3·6430	4·0121	4·2050	4·8179	5·7169	6·7132	7·2489	7·8104	10·322	14·094
0·000250	0·601	0·617	0·626	0·634	0·649	0·665	0·673	0·696	0·726	0·755	0·769	0·783	0·838	0·904
1/ 4000	2·7194	3·0309	3·1946	3·3637	3·7182	4·0949	4·2918	4·9173	5·8349	6·8517	7·3985	7·9715	10·535	14·385
0·000260	0·613	0·630	0·638	0·646	0·662	0·678	0·686	0·709	0·740	0·770	0·784	0·799	0·855	0·922
1/ 3846	2·7733	3·0910	3·2580	3·4304	3·7919	4·1761	4·3769	5·0148	5·9506	6·9875	7·5452	8·1295	10·744	14·670
0·000270	0·625	0·642	0·650	0·658	0·675	0·691	0·699	0·723	0·754	0·784	0·799	0·814	0·871	0·940
1/ 3704	2·8262	3·1500	3·3201	3·4958	3·8642	4·2557	4·4603	5·1104	6·0640	7·1207	7·6890	8·2845	10·949	14·949
0·000280	0·636	0·653	0·662	0·671	0·687	0·704	0·712	0·736	0·768	0·799	0·814	0·829	0·887	0·957
1/ 3571	2·8781	3·2078	3·3811	3·5600	3·9351	4·3339	4·5423	5·2043	6·1754	7·2515	7·8302	8·4366	11·150	15·224
0·000290	0·647	0·665	0·674	0·682	0·699	0·716	0·725	0·749	0·781	0·813	0·828	0·844	0·903	0·974
1/ 3448	2·9291	3·2647	3·4410	3·6230	4·0049	4·4107	4·6227	5·2965	6·2848	7·3800	7·9689	8·5861	11·347	15·494
0·000300	0·659	0·676	0·685	0·694	0·711	0·729	0·737	0·762	0·795	0·827	0·842	0·858	0·918	0·991
1/ 3333	2·9792	3·3205	3·4998	3·6850	4·0734	4·4861	4·7018	5·3871	6·3923	7·5062	8·1052	8·7330	11·541	15·759
	0·76	0·76	0·76	0·76	0·77	0·77	0·77	0·77	0·77	0·77	0·77	0·78	0·78	0·78

$V_{r(0{\cdot}5)medial}$ for half-full circular pipes.

k_s = 15·0 mm
S = 0·00030 to 0·00100

ie hydraulic gradient =
1 in 3333 to 1 in 1000

Water (or sewage) at 15°C;
full bore conditions.

velocities in ms^{-1}
discharges in m^3s^{-1}

Gradient **(Equivalent) Pipe diameters in mm**

Gradient	**2400**	**2500**	**2550**	**2600**	**2700**	**2800**	**2850**	**3000**	**3200**	**3400**	**3500**	**3600**	**4000**	**4500**
0·000300	0·659	0·676	0·685	0·694	0·711	0·729	0·737	0·762	0·795	0·827	0·842	0·858	0·918	0·991
1/ 3333	2·9792	3·3205	3·4998	3·6850	4·0734	4·4861	4·7018	5·3871	6·3923	7·5062	8·1052	8·7330	11·541	15·759
0·000320	0·680	0·699	0·708	0·717	0·735	0·752	0·761	0·787	0·821	0·854	0·870	0·886	0·949	1·023
1/ 3125	3·0770	3·4295	3·6147	3·8060	4·2071	4·6334	4·8561	5·5639	6·6021	7·7526	8·3712	9·0196	11·920	16·276
0·000340	0·701	0·720	0·730	0·739	0·757	0·776	0·785	0·811	0·846	0·880	0·897	0·913	0·978	1·055
1/ 2941	3·1718	3·5352	3·7261	3·9232	4·3367	4·7761	5·0057	5·7353	6·8055	7·9913	8·6291	9·2974	12·287	16·777
0·000360	0·721	0·741	0·751	0·760	0·779	0·798	0·807	0·835	0·871	0·906	0·923	0·940	1·006	1·085
1/ 2778	3·2639	3·6377	3·8342	4·0371	4·4625	4·9147	5·1510	5·9017	7·0029	8·2232	8·8794	9·5671	12·644	17·264
0·000380	0·741	0·761	0·771	0·781	0·801	0·820	0·830	0·858	0·895	0·931	0·948	0·966	1·034	1·115
1/ 2632	3·3534	3·7375	3·9393	4·1478	4·5849	5·0494	5·2922	6·0635	7·1949	8·4487	9·1229	9·8294	12·990	17·737
0·000400	0·761	0·781	0·791	0·802	0·822	0·841	0·851	0·880	0·918	0·955	0·973	0·991	1·061	1·144
1/ 2500	3·4406	3·8347	4·0418	4·2556	4·7041	5·1807	5·4298	6·2211	7·3820	8·6683	9·3600	10·085	13·328	18·198
0·000420	0·779	0·801	0·811	0·821	0·842	0·862	0·872	0·902	0·941	0·978	0·997	1·015	1·087	1·172
1/ 2381	3·5256	3·9294	4·1416	4·3608	4·8203	5·3087	5·5640	6·3749	7·5644	8·8825	9·5913	10·334	13·657	18·648
0·000440	0·798	0·819	0·830	0·841	0·862	0·882	0·893	0·923	0·963	1·001	1·020	1·039	1·112	1·200
1/ 2273	3·6086	4·0220	4·2392	4·4635	4·9338	5·4338	5·6950	6·5250	7·7425	9·0917	9·8172	10·577	13·979	19·087
0·000460	0·816	0·838	0·849	0·860	0·881	0·902	0·913	0·944	0·984	1·024	1·043	1·063	1·137	1·227
1/ 2174	3·6898	4·1124	4·3345	4·5639	5·0448	5·5560	5·8231	6·6717	7·9167	9·2961	10·038	10·815	14·293	19·516
0·000480	0·833	0·856	0·867	0·878	0·900	0·922	0·932	0·964	1·006	1·046	1·066	1·085	1·162	1·253
1/ 2083	3·7692	4·2010	4·4278	4·6621	5·1534	5·6755	5·9484	6·8153	8·0870	9·4962	10·254	11·048	14·601	19·936
0·000500	0·850	0·873	0·885	0·896	0·919	0·941	0·952	0·984	1·026	1·068	1·088	1·108	1·186	1·279
1/ 2000	3·8470	4·2877	4·5192	4·7583	5·2597	5·7927	6·0712	6·9559	8·2539	9·6921	10·466	11·276	14·902	20·347
0·000525	0·871	0·895	0·907	0·918	0·941	0·964	0·975	1·008	1·052	1·094	1·115	1·135	1·215	1·311
1/ 1905	3·9421	4·3936	4·6309	4·8759	5·3897	5·9358	6·2212	7·1278	8·4578	9·9316	10·724	11·555	15·270	20·850
0·000550	0·892	0·916	0·928	0·940	0·964	0·987	0·998	1·032	1·076	1·120	1·141	1·162	1·244	1·342
1/ 1818	4·0349	4·4971	4·7399	4·9907	5·5166	6·0756	6·3677	7·2957	8·6570	10·165	10·977	11·827	15·630	21·341
0·000575	0·912	0·937	0·949	0·961	0·985	1·009	1·021	1·055	1·101	1·145	1·167	1·188	1·272	1·372
1/ 1739	4·1256	4·5982	4·8465	5·1030	5·6407	6·2122	6·5109	7·4597	8·8517	10·394	11·223	12·093	15·981	21·821
0·000600	0·932	0·957	0·969	0·982	1·006	1·031	1·043	1·078	1·124	1·169	1·192	1·214	1·299	1·402
1/ 1667	4·2144	4·6972	4·9508	5·2128	5·7621	6·3459	6·6510	7·6203	9·0422	10·618	11·465	12·353	16·325	22·290
0·000625	0·951	0·977	0·989	1·002	1·027	1·052	1·064	1·100	1·147	1·194	1·216	1·239	1·326	1·430
1/ 1600	4·3014	4·7941	5·0530	5·3203	5·8810	6·4768	6·7882	7·7775	9·2287	10·837	11·702	12·608	16·662	22·750
0·000650	0·970	0·996	1·009	1·022	1·047	1·073	1·085	1·122	1·170	1·217	1·240	1·263	1·352	1·459
1/ 1538	4·3866	4·8891	5·1531	5·4258	5·9975	6·6052	6·9227	7·9316	9·4116	11·051	11·933	12·858	16·992	23·201
0·000675	0·988	1·015	1·028	1·041	1·067	1·093	1·106	1·143	1·193	1·240	1·264	1·287	1·378	1·487
1/ 1481	4·4703	4·9823	5·2513	5·5292	6·1118	6·7311	7·0547	8·0828	9·5910	11·262	12·161	13·103	17·316	23·643
0·000700	1·006	1·034	1·047	1·061	1·087	1·113	1·126	1·164	1·214	1·263	1·287	1·311	1·403	1·514
1/ 1429	4·5523	5·0738	5·3477	5·6307	6·2240	6·8547	7·1842	8·2312	9·7670	11·469	12·384	13·343	17·634	24·077
0·000725	1·024	1·052	1·066	1·079	1·106	1·133	1·146	1·185	1·236	1·286	1·310	1·334	1·428	1·541
1/ 1379	4·6330	5·1636	5·4425	5·7304	6·3343	6·9761	7·3114	8·3769	9·9400	11·672	12·603	13·579	17·946	24·503
0·000750	1·042	1·070	1·084	1·098	1·125	1·152	1·166	1·205	1·257	1·308	1·332	1·357	1·453	1·567
1/ 1333	4·7122	5·2520	5·5356	5·8284	6·4426	7·0954	7·4365	8·5202	10·110	11·872	12·819	13·812	18·253	24·922
0·000800	1·076	1·105	1·119	1·134	1·162	1·190	1·204	1·245	1·298	1·350	1·376	1·401	1·500	1·618
1/ 1250	4·8668	5·4243	5·7172	6·0197	6·6540	7·3282	7·6805	8·7998	10·442	12·261	13·240	14·265	18·852	25·740
0·000850	1·109	1·139	1·154	1·169	1·198	1·227	1·241	1·283	1·338	1·392	1·418	1·445	1·546	1·668
1/ 1176	5·0167	5·5913	5·8932	6·2051	6·8589	7·5538	7·9170	9·0707	10·763	12·639	13·647	14·704	19·432	26·532
0·000900	1·141	1·172	1·187	1·203	1·233	1·262	1·277	1·320	1·377	1·432	1·460	1·486	1·591	1·717
1/ 1111	5·1622	5·7535	6·0642	6·3851	7·0579	7·7730	8·1467	9·3338	11·075	13·005	14·043	15·131	19·996	27·302
0·000950	1·172	1·204	1·220	1·236	1·266	1·297	1·312	1·357	1·415	1·472	1·500	1·527	1·635	1·764
1/ 1053	5·3038	5·9113	6·2305	6·5601	7·2514	7·9861	8·3700	9·5897	11·379	13·362	14·428	15·545	20·544	28·050
0·001000	1·203	1·236	1·252	1·268	1·299	1·331	1·346	1·392	1·452	1·510	1·539	1·567	1·677	1·810
1/ 1000	5·4416	6·0649	6·3924	6·7306	7·4398	8·1936	8·5875	9·8390	11·675	13·709	14·803	15·949	21·078	28·779
	0·76	0·76	0·76	0·76	0·77	0·77	0·77	0·77	0·77	0·77	0·77	0·78	0·78	0·78

$V_{r(0·5)medial}$ for half-full circular pipes.

k_s = 15·0 mm S = 0·00030 to 0·00100

Gradient **(Equivalent) Pipe diameters in mm**

Gradient	2400	2500	2550	2600	2700	2800	2850	3000	3200	3400	3500	3600	4000	4500
0·00100	1·203	1·236	1·252	1·268	1·299	1·331	1·346	1·392	1·452	1·510	1·539	1·567	1·677	1·810
1/ 1000	5·4416	6·0649	6·3924	6·7306	7·4398	8·1936	8·5875	9·8390	11·675	13·709	14·803	15·949	21·078	28·779
0·00105	1·233	1·266	1·283	1·299	1·332	1·364	1·379	1·426	1·488	1·547	1·577	1·606	1·719	1·854
1/ 952	5·5761	6·2148	6·5503	6·8969	7·6237	8·3961	8·7997	10·082	11·963	14·048	15·169	16·343	21·599	29·490
0·00110	1·262	1·296	1·313	1·330	1·363	1·396	1·412	1·460	1·523	1·584	1·614	1·643	1·759	1·898
1/ 909	5·7074	6·3611	6·7045	7·0593	7·8031	8·5937	9·0069	10·319	12·245	14·378	15·526	16·728	22·107	30·184
0·00115	1·290	1·325	1·342	1·360	1·394	1·427	1·444	1·493	1·557	1·619	1·650	1·680	1·799	1·941
1/ 870	5·8357	6·5041	6·8553	7·2180	7·9786	8·7869	9·2094	10·551	12·520	14·702	15·875	17·104	22·604	30·863
0·00120	1·318	1·354	1·371	1·389	1·423	1·458	1·475	1·525	1·590	1·654	1·685	1·717	1·837	1·982
1/ 833	5·9613	6·6441	7·0028	7·3733	8·1503	8·9760	9·4075	10·778	12·790	15·018	16·216	17·472	23·091	31·527
0·00125	1·345	1·381	1·399	1·417	1·453	1·488	1·505	1·556	1·623	1·688	1·720	1·752	1·875	2·023
1/ 800	6·0843	6·7811	7·1473	7·5254	8·3184	9·1612	9·6016	11·001	13·053	15·328	16·551	17·833	23·567	32·177
0·00130	1·372	1·409	1·427	1·445	1·482	1·517	1·535	1·587	1·655	1·722	1·754	1·787	1·913	2·063
1/ 769	6·2048	6·9155	7·2889	7·6745	8·4832	9·3427	9·7918	11·219	13·312	15·631	16·879	18·186	24·034	32·815
0·00135	1·398	1·436	1·454	1·473	1·510	1·546	1·564	1·617	1·687	1·754	1·788	1·821	1·949	2·103
1/ 741	6·3230	7·0473	7·4278	7·8208	8·6449	9·5207	9·9785	11·433	13·566	15·929	17·200	18·532	24·492	33·440
0·00140	1·423	1·462	1·481	1·500	1·538	1·575	1·593	1·647	1·718	1·787	1·821	1·854	1·985	2·141
1/ 714	6·4391	7·1766	7·5641	7·9644	8·8036	9·6955	10·162	11·642	13·815	16·222	17·516	18·873	24·941	34·054
0·00145	1·449	1·488	1·507	1·527	1·565	1·602	1·621	1·676	1·748	1·818	1·853	1·887	2·020	2·179
1/ 690	6·5532	7·3037	7·6981	8·1054	8·9595	9·8672	10·342	11·849	14·059	16·509	17·826	19·207	25·383	34·657
0·00150	1·473	1·513	1·533	1·553	1·592	1·630	1·649	1·705	1·778	1·849	1·885	1·919	2·054	2·216
1/ 667	6·6652	7·4286	7·8297	8·2440	9·1127	10·036	10·518	12·051	14·300	16·791	18·131	19·535	25·817	35·249
0·00160	1·522	1·563	1·583	1·604	1·644	1·683	1·703	1·761	1·836	1·910	1·946	1·982	2·122	2·289
1/ 625	6·8839	7·6724	8·0866	8·5145	9·4117	10·365	10·864	12·447	14·769	17·342	18·726	20·176	26·664	36·406
0·00170	1·569	1·611	1·632	1·653	1·694	1·735	1·755	1·815	1·893	1·969	2·006	2·043	2·187	2·360
1/ 588	7·0959	7·9086	8·3356	8·7766	9·7014	10·684	11·198	12·830	15·224	17·876	19·302	20·797	27·485	37·527
0·00180	1·614	1·658	1·680	1·701	1·744	1·785	1·806	1·868	1·948	2·026	2·064	2·102	2·251	2·428
1/ 556	7·3017	8·1380	8·5774	9·0312	9·9828	10·994	11·523	13·202	15·665	18·394	19·862	21·400	28·282	38·615
0·00190	1·658	1·703	1·726	1·748	1·791	1·834	1·856	1·919	2·001	2·082	2·121	2·160	2·312	2·494
1/ 526	7·5018	8·3610	8·8125	9·2787	10·256	11·296	11·839	13·564	16·094	18·899	20·407	21·987	29·057	39·673
0·00200	1·701	1·748	1·770	1·793	1·838	1·882	1·904	1·969	2·053	2·136	2·176	2·216	2·372	2·559
1/ 500	7·6968	8·5783	9·0415	9·5199	10·523	11·589	12·146	13·916	16·513	19·390	20·937	22·558	29·812	40·704
0·00210	1·743	1·791	1·814	1·837	1·883	1·929	1·951	2·017	2·104	2·188	2·230	2·271	2·431	2·623
1/ 476	7·8869	8·7902	9·2649	9·7551	10·783	11·875	12·446	14·260	16·921	19·869	21·454	23·116	30·549	41·709
0·00220	1·784	1·833	1·857	1·881	1·928	1·974	1·997	2·065	2·153	2·240	2·282	2·324	2·488	2·684
1/ 455	8·0726	8·9972	9·4830	9·9847	11·037	12·155	12·739	14·596	17·319	20·336	21·959	23·660	31·268	42·691
0·00230	1·825	1·874	1·899	1·923	1·971	2·018	2·042	2·111	2·202	2·290	2·334	2·377	2·544	2·745
1/ 435	8·2541	9·1994	9·6962	10·209	11·285	12·428	13·026	14·924	17·708	20·794	22·453	24·192	31·971	43·651
0·00240	1·864	1·914	1·939	1·964	2·013	2·062	2·086	2·157	2·249	2·340	2·384	2·428	2·599	2·804
1/ 417	8·4317	9·3974	9·9048	10·429	11·528	12·696	13·306	15·245	18·089	21·241	22·936	24·712	32·658	44·590
0·00250	1·902	1·954	1·979	2·005	2·055	2·104	2·129	2·201	2·296	2·388	2·433	2·478	2·652	2·861
1/ 400	8·6056	9·5912	10·109	10·644	11·765	12·957	13·580	15·559	18·462	21·679	23·409	25·222	33·332	45·510
0·00260	1·940	1·993	2·019	2·044	2·096	2·146	2·171	2·245	2·341	2·435	2·481	2·527	2·705	2·918
1/ 385	8·7761	9·7812	10·309	10·855	11·999	13·214	13·849	15·868	18·828	22·109	23·873	25·721	33·992	46·411
0·00270	1·977	2·031	2·057	2·083	2·136	2·187	2·212	2·288	2·386	2·481	2·529	2·575	2·757	2·974
1/ 370	8·9433	9·9676	10·506	11·062	12·227	13·466	14·113	16·170	19·187	22·530	24·327	26·211	34·640	47·295
0·00280	2·013	2·068	2·095	2·122	2·175	2·227	2·253	2·330	2·429	2·527	2·575	2·622	2·807	3·028
1/ 357	9·1075	10·151	10·699	11·265	12·452	13·713	14·372	16·467	19·539	22·943	24·774	26·693	35·276	48·163
0·00290	2·049	2·104	2·132	2·159	2·213	2·266	2·293	2·371	2·472	2·572	2·621	2·669	2·857	3·082
1/ 345	9·2687	10·330	10·888	11·464	12·672	13·956	14·627	16·758	19·885	23·350	25·213	27·165	35·900	49·016
0·00300	2·084	2·140	2·168	2·196	2·251	2·305	2·332	2·411	2·515	2·616	2·665	2·714	2·906	3·135
1/ 333	9·4272	10·507	11·074	11·660	12·889	14·195	14·877	17·045	20·225	23·749	25·644	27·630	36·514	49·854
	0·76	0·76	0·76	0·76	0·77	0·77	0·77	0·77	0·77	0·77	0·77	0·78	0·78	0·78

$V_{r(0{\cdot}5)medial}$ for half-full circular pipes.

A56
(p.5 of 6)

k_s = **15·0 mm**
S = 0·00300 to 0·01000

ie hydraulic gradient =
1 in 333 to 1 in 100

Water (or sewage) at 15°C;
full bore conditions.

velocities in ms^{-1}
discharges in m^3s^{-1}

Gradient **(Equivalent) Pipe diameters in mm**

Gradient	2400	2500	2550	2600	2700	2800	2850	3000	3200	3400	3500	3600	4000	4500
0·00300	2·084	2·140	2·168	2·196	2·251	2·305	2·332	2·411	2·515	2·616	2·665	2·714	2·906	3·135
1/ 333	9·4272	10·507	11·074	11·660	12·889	14·195	14·877	17·045	20·225	23·749	25·644	27·630	36·514	49·854
0·00320	2·152	2·211	2·240	2·268	2·325	2·381	2·409	2·490	2·597	2·702	2·753	2·803	3·001	3·237
1/ 313	9·7365	10·852	11·438	12·043	13·312	14·660	15·365	17·604	20·888	24·528	26·485	28·536	37·712	51·490
0·00340	2·218	2·279	2·308	2·338	2·397	2·454	2·483	2·567	2·677	2·785	2·838	2·890	3·093	3·337
1/ 294	10·036	11·186	11·790	12·413	13·721	15·111	15·838	18·146	21·531	25·283	27·300	29·414	38·873	53·075
0·00360	2·283	2·345	2·375	2·406	2·466	2·525	2·555	2·642	2·755	2·865	2·920	2·974	3·183	3·434
1/ 278	10·327	11·510	12·132	12·773	14·119	15·550	16·297	18·672	22·156	26·016	28·092	30·267	40·000	54·614
0·00380	2·345	2·409	2·441	2·472	2·534	2·595	2·625	2·714	2·830	2·944	3·000	3·055	3·270	3·528
1/ 263	10·610	11·826	12·464	13·123	14·506	15·976	16·744	19·184	22·763	26·729	28·862	31·097	41·096	56·110
0·00400	2·406	2·472	2·504	2·536	2·599	2·662	2·693	2·784	2·904	3·020	3·078	3·134	3·355	3·620
1/ 250	10·886	12·133	12·788	13·464	14·883	16·391	17·179	19·682	23·355	27·424	29·612	31·905	42·164	57·568
0·00420	2·466	2·533	2·566	2·599	2·664	2·728	2·759	2·853	2·976	3·095	3·154	3·212	3·438	3·709
1/ 238	11·155	12·433	13·104	13·797	15·251	16·796	17·603	20·168	23·931	28·101	30·343	32·693	43·205	58·990
0·00440	2·524	2·592	2·626	2·660	2·726	2·792	2·824	2·920	3·046	3·168	3·228	3·287	3·519	3·796
1/ 227	11·418	12·725	13·412	14·122	15·610	17·191	18·018	20·643	24·495	28·762	31·057	33·462	44·222	60·379
0·00460	2·581	2·651	2·685	2·720	2·788	2·855	2·888	2·986	3·114	3·239	3·301	3·361	3·598	3·882
1/ 217	11·674	13·011	13·714	14·439	15·961	17·578	18·423	21·107	25·045	29·409	31·756	34·215	45·216	61·736
0·00480	2·636	2·708	2·743	2·778	2·848	2·916	2·950	3·050	3·181	3·309	3·372	3·434	3·676	3·965
1/ 208	11·925	13·291	14·009	14·750	16·304	17·956	18·819	21·561	25·584	30·042	32·439	34·951	46·189	63·064
0·00500	2·690	2·763	2·800	2·835	2·906	2·976	3·011	3·113	3·247	3·377	3·441	3·505	3·751	4·047
1/ 200	12·171	13·565	14·298	15·054	16·640	18·326	19·207	22·006	26·112	30·661	33·108	35·671	47·142	64·364
0·00525	2·757	2·832	2·869	2·905	2·978	3·050	3·085	3·190	3·327	3·460	3·526	3·591	3·844	4·147
1/ 190	12·472	13·900	14·651	15·426	17·051	18·779	19·682	22·550	26·757	31·419	33·925	36·553	48·306	65·954
0·00550	2·822	2·898	2·936	2·974	3·048	3·122	3·158	3·265	3·405	3·542	3·609	3·676	3·935	4·245
1/ 182	12·765	14·228	14·996	15·789	17·453	19·221	20·145	23·080	27·386	32·158	34·724	37·413	49·443	67·506
0·00575	2·885	2·964	3·002	3·041	3·117	3·192	3·229	3·339	3·482	3·622	3·690	3·758	4·023	4·340
1/ 174	13·052	14·547	15·333	16·144	17·845	19·653	20·598	23·599	28·002	32·881	35·504	38·254	50·554	69·024
0·00600	2·947	3·027	3·067	3·106	3·184	3·260	3·298	3·410	3·557	3·699	3·770	3·839	4·110	4·433
1/ 167	13·333	14·860	15·663	16·491	18·229	20·076	21·041	24·107	28·604	33·588	36·268	39·077	51·642	70·509
0·00625	3·008	3·090	3·130	3·170	3·249	3·328	3·366	3·481	3·630	3·776	3·847	3·918	4·194	4·525
1/ 160	13·608	15·167	15·986	16·831	18·605	20·490	21·475	24·604	29·194	34·281	37·016	39·883	52·707	71·963
0·00650	3·068	3·151	3·192	3·233	3·314	3·394	3·433	3·550	3·702	3·851	3·924	3·996	4·277	4·614
1/ 154	13·878	15·467	16·302	17·165	18·973	20·896	21·900	25·091	29·773	34·960	37·749	40·673	53·751	73·388
0·00675	3·126	3·211	3·253	3·295	3·377	3·458	3·498	3·617	3·772	3·924	3·998	4·072	4·359	4·702
1/ 148	14·142	15·762	16·613	17·492	19·335	21·294	22·317	25·569	30·340	35·626	38·469	41·448	54·775	74·786
0·00700	3·183	3·270	3·313	3·355	3·439	3·522	3·563	3·684	3·842	3·996	4·072	4·147	4·439	4·789
1/ 143	14·402	16·051	16·918	17·813	19·690	21·684	22·727	26·039	30·897	36·280	39·175	42·208	55·780	76·159
0·00725	3·240	3·328	3·371	3·414	3·500	3·584	3·626	3·749	3·910	4·067	4·144	4·220	4·517	4·873
1/ 138	14·657	16·335	17·217	18·128	20·038	22·068	23·129	26·500	31·444	36·922	39·868	42·955	56·768	77·507
0·00750	3·295	3·385	3·429	3·473	3·560	3·645	3·688	3·813	3·977	4·136	4·215	4·292	4·595	4·957
1/ 133	14·907	16·615	17·512	18·438	20·381	22·446	23·525	26·953	31·981	37·553	40·550	43·690	57·738	78·832
0·00800	3·403	3·496	3·541	3·587	3·676	3·765	3·809	3·938	4·107	4·272	4·353	4·433	4·745	5·119
1/ 125	15·396	17·160	18·086	19·043	21·049	23·182	24·296	27·837	33·030	38·785	41·880	45·123	59·632	81·418
0·00850	3·508	3·603	3·650	3·697	3·790	3·881	3·926	4·059	4·233	4·403	4·487	4·569	4·891	5·277
1/ 118	15·870	17·688	18·643	19·629	21·697	23·896	25·044	28·694	34·047	39·979	43·169	46·512	61·468	83·924
0·00900	3·610	3·708	3·756	3·804	3·899	3·993	4·040	4·177	4·356	4·531	4·617	4·702	5·033	5·430
1/ 111	16·330	18·201	19·183	20·198	22·327	24·588	25·770	29·526	35·034	41·138	44·421	47·860	63·250	86·357
0·00950	3·709	3·809	3·859	3·909	4·006	4·103	4·150	4·291	4·476	4·655	4·744	4·831	5·171	5·579
1/ 105	16·778	18·700	19·709	20·752	22·938	25·262	26·477	30·335	35·994	42·266	45·638	49·172	64·983	88·724
0·01000	3·805	3·908	3·959	4·010	4·110	4·209	4·258	4·403	4·592	4·776	4·867	4·956	5·306	5·724
1/ 100	17·214	19·185	20·221	21·291	23·534	25·919	27·165	31·123	36·930	43·364	46·824	50·450	66·672	91·029
	0·76	0·76	0·76	0·76	0·77	0·77	0·77	0·77	0·77	0·77	0·77	0·78	0·78	0·78

$V_{r(0\cdot5)medial}$ for half-full circular pipes.

k_s = 15·0 mm S = 0·00300 to 0·01000

k_s = 15·0 mm
S = 0·01000 to 0·03000

ie hydraulic gradient =
1 in 100 to 1 in 33·3

Water (or sewage) at 15°C;
full bore conditions.

velocities in ms^{-1}
discharges in m^3s^{-1}

Gradient	(Equivalent) Pipe diameters in mm													
	2400	**2500**	**2550**	**2600**	**2700**	**2800**	**2850**	**3000**	**3200**	**3400**	**3500**	**3600**	**4000**	**4500**
0·01000	3·805	3·908	3·959	4·010	4·110	4·209	4·258	4·403	4·592	4·776	4·867	4·956	5·306	5·724
1/ 100	17·214	19·185	20·221	21·291	23·534	25·919	27·165	31·123	36·930	43·364	46·824	50·450	66·672	91·029
0·01050	3·899	4·005	4·057	4·109	4·212	4·313	4·363	4·512	4·705	4·894	4·987	5·079	5·437	5·865
1/ 95	17·639	19·659	20·721	21·817	24·116	26·559	27·836	31·892	37·842	44·435	47·980	51·696	68·318	93·277
0·01100	3·991	4·099	4·153	4·206	4·311	4·415	4·466	4·618	4·816	5·009	5·104	5·198	5·565	6·003
1/ 91	18·054	20·122	21·208	22·330	24·683	27·184	28·491	32·642	38·732	45·481	49·109	52·912	69·926	95·473
0·01150	4·081	4·191	4·246	4·300	4·408	4·514	4·566	4·722	4·924	5·122	5·219	5·315	5·690	6·138
1/ 87	18·460	20·574	21·685	22·832	25·238	27·795	29·131	33·376	39·603	46·503	50·213	54·102	71·498	97·619
0·01200	4·168	4·282	4·337	4·393	4·503	4·611	4·665	4·823	5·030	5·232	5·331	5·429	5·812	6·270
1/ 83	18·857	21·017	22·152	23·324	25·781	28·393	29·758	34·094	40·455	47·503	51·293	55·265	73·036	99·718
0·01250	4·254	4·370	4·427	4·484	4·596	4·706	4·761	4·923	5·134	5·340	5·441	5·541	5·932	6·399
1/ 80	19·246	21·450	22·608	23·805	26·313	28·978	30·371	34·797	41·289	48·483	52·351	56·405	74·542	101·77
0·01300	4·339	4·456	4·515	4·572	4·687	4·799	4·855	5·020	5·236	5·446	5·549	5·651	6·049	6·526
1/ 77	19·627	21·875	23·056	24·276	26·834	29·552	30·973	35·486	42·107	49·443	53·388	57·522	76·018	103·79
0·01350	4·421	4·541	4·601	4·659	4·776	4·891	4·948	5·116	5·335	5·550	5·655	5·759	6·165	6·650
1/ 74	20·001	22·292	23·496	24·739	27·345	30·115	31·563	36·162	42·909	50·385	54·405	58·618	77·467	105·77
0·01400	4·502	4·625	4·685	4·745	4·864	4·981	5·038	5·210	5·433	5·651	5·759	5·865	6·278	6·772
1/ 71	20·368	22·701	23·927	25·193	27·847	30·668	32·142	36·826	43·697	51·310	55·404	59·694	78·888	107·71
0·01450	4·582	4·706	4·768	4·829	4·950	5·069	5·128	5·302	5·529	5·751	5·860	5·968	6·389	6·892
1/ 69	20·729	23·103	24·350	25·639	28·340	31·211	32·711	37·478	44·470	52·218	56·384	60·751	80·285	109·62
0·01500	4·660	4·787	4·849	4·912	5·034	5·155	5·215	5·393	5·624	5·850	5·961	6·070	6·498	7·010
1/ 67	21·083	23·498	24·767	26·077	28·824	31·745	33·271	38·119	45·231	53·111	57·348	61·789	81·658	111·49
0·01600	4·813	4·944	5·009	5·073	5·199	5·325	5·386	5·570	5·808	6·042	6·156	6·270	6·711	7·240
1/ 63	21·775	24·269	25·579	26·932	29·770	32·786	34·362	39·369	46·714	54·853	59·229	63·816	84·336	115·15
0·01700	4·961	5·096	5·163	5·229	5·360	5·488	5·552	5·741	5·987	6·228	6·346	6·463	6·918	7·463
1/ 59	22·445	25·016	26·366	27·761	30·686	33·795	35·420	40·581	48·152	56·541	61·053	65·780	86·932	118·69
0·01800	5·105	5·244	5·312	5·380	5·515	5·648	5·713	5·907	6·161	6·408	6·530	6·650	7·118	7·679
1/ 56	23·096	25·741	27·131	28·566	31·576	34·775	36·447	41·757	49·548	58·181	62·823	67·688	89·452	122·13
0·01900	5·245	5·388	5·458	5·528	5·666	5·802	5·870	6·069	6·330	6·584	6·709	6·832	7·313	7·890
1/ 53	23·729	26·446	27·874	29·349	32·441	35·728	37·445	42·902	50·906	59·775	64·544	69·542	91·903	125·48
0·02000	5·382	5·528	5·600	5·671	5·813	5·953	6·022	6·227	6·494	6·755	6·883	7·010	7·503	8·095
1/ 50	24·345	27·134	28·599	30·112	33·284	36·656	38·418	44·016	52·228	61·328	66·221	71·349	94·291	128·74
0·02100	5·514	5·664	5·738	5·812	5·957	6·100	6·171	6·381	6·654	6·922	7·053	7·183	7·689	8·294
1/ 47·6	24·947	27·804	29·305	30·855	34·106	37·561	39·367	45·103	53·518	62·843	67·857	73·111	96·620	131·92
0·02200	5·644	5·797	5·873	5·948	6·097	6·244	6·316	6·531	6·811	7·084	7·219	7·352	7·870	8·490
1/ 45·5	25·534	28·458	29·995	31·581	34·909	38·445	40·294	46·165	54·778	64·322	69·454	74·832	98·894	135·02
0·02300	5·771	5·928	6·005	6·082	6·234	6·384	6·458	6·678	6·964	7·244	7·381	7·517	8·047	8·681
1/ 43·5	26·108	29·098	30·669	32·291	35·694	39·310	41·199	47·203	56·009	65·767	71·015	76·514	101·12	138·06
0·02400	5·895	6·055	6·134	6·213	6·368	6·521	6·597	6·821	7·114	7·400	7·540	7·679	8·220	8·867
1/ 41·7	26·669	29·724	31·328	32·986	36·461	40·155	42·085	48·218	57·214	67·182	72·542	78·160	103·29	141·03
0·02500	6·017	6·180	6·261	6·341	6·500	6·656	6·733	6·962	7·261	7·552	7·695	7·837	8·389	9·050
1/ 40·0	27·219	30·337	31·974	33·666	37·213	40·983	42·953	49·212	58·394	68·567	74·038	79·772	105·42	143·94
0·02600	6·136	6·303	6·385	6·467	6·628	6·788	6·866	7·100	7·404	7·702	7·848	7·992	8·555	9·229
1/ 38·5	27·758	30·937	32·608	34·333	37·950	41·795	43·804	50·187	59·550	69·925	75·505	81·352	107·51	146·79
0·02700	6·253	6·423	6·506	6·590	6·755	6·917	6·997	7·235	7·546	7·848	7·997	8·145	8·718	9·405
1/ 37·0	28·287	31·527	33·229	34·987	38·673	42·591	44·639	51·143	60·685	71·258	76·943	82·901	109·56	149·58
0·02800	6·368	6·540	6·626	6·711	6·878	7·044	7·126	7·368	7·684	7·992	8·144	8·294	8·878	9·578
1/ 35·7	28·806	32·105	33·839	35·629	39·383	43·373	45·458	52·082	61·798	72·565	78·355	84·423	111·57	152·33
0·02900	6·480	6·656	6·743	6·829	7·000	7·169	7·252	7·498	7·820	8·134	8·288	8·441	9·035	9·747
1/ 34·5	29·316	32·674	34·438	36·260	40·080	44·141	46·262	53·004	62·892	73·850	79·742	85·917	113·54	155·02
0·03000	6·591	6·770	6·858	6·946	7·120	7·291	7·376	7·627	7·954	8·273	8·430	8·585	9·190	9·914
1/ 33·3	29·817	33·232	35·027	36·880	40·765	44·895	47·053	53·910	63·968	75·112	81·105	87·386	115·48	157·67
	0·76	0·76	0·76	0·76	0·77	0·77	0·77	0·77	0·77	0·77	0·77	0·78	0·78	0·78

$V_{r(0·5)medial}$ for half-full circular pipes.

A57
(p.1 of 6)

k_s = 30·0 mm
S = 0·00003 to 0·00010

ie hydraulic gradient =
1 in 33333 to 1 in 10000

Water (or sewage) at 15°C;
full bore conditions.

velocities in ms^{-1}
discharges in m^3s^{-1}

Gradient **(Equivalent) Pipe diameters in mm**

Gradient	2400	2500	2550	2600	2700	2800	2850	3000	3200	3400	3500	3600	4000	4500
0·0000300	0·186	0·191	0·193	0·196	0·201	0·206	0·208	0·216	0·225	0·234	0·239	0·243	0·261	0·282
1/ 33333	0·8393	0·9361	0·9871	1·0397	1·1501	1·2675	1·3288	1·5239	1·8104	2·1282	2·2992	2·4784	3·2813	4·4888
0·0000320	0·192	0·197	0·200	0·202	0·207	0·213	0·215	0·223	0·232	0·242	0·247	0·251	0·270	0·292
1/ 31250	0·8668	0·9669	1·0195	1·0738	1·1879	1·3091	1·3725	1·5740	1·8699	2·1980	2·3747	2·5598	3·3890	4·6362
0·0000340	0·198	0·203	0·206	0·208	0·214	0·219	0·222	0·230	0·240	0·250	0·254	0·259	0·278	0·300
1/ 29412	0·8936	0·9967	1·0509	1·1069	1·2245	1·3494	1·4148	1·6225	1·9275	2·2658	2·4478	2·6387	3·4934	4·7790
0·0000360	0·203	0·209	0·212	0·215	0·220	0·226	0·228	0·236	0·247	0·257	0·262	0·267	0·286	0·309
1/ 27778	0·9195	1·0256	1·0814	1·1391	1·2600	1·3886	1·4559	1·6696	1·9835	2·3316	2·5189	2·7153	3·5948	4·9177
0·0000380	0·209	0·215	0·218	0·220	0·226	0·232	0·234	0·243	0·253	0·264	0·269	0·274	0·294	0·318
1/ 26316	0·9447	1·0538	1·1111	1·1703	1·2946	1·4267	1·4958	1·7154	2·0379	2·3955	2·5880	2·7898	3·6934	5·0526
0·0000400	0·214	0·220	0·223	0·226	0·232	0·238	0·241	0·249	0·260	0·271	0·276	0·281	0·302	0·326
1/ 25000	0·9693	1·0812	1·1400	1·2008	1·3283	1·4638	1·5347	1·7600	2·0909	2·4578	2·6553	2·8623	3·7895	5·1840
0·0000420	0·220	0·226	0·229	0·232	0·238	0·244	0·247	0·255	0·266	0·277	0·283	0·288	0·309	0·334
1/ 23810	0·9933	1·1079	1·1682	1·2305	1·3611	1·5000	1·5727	1·8036	2·1426	2·5186	2·7209	2·9331	3·8832	5·3121
0·0000440	0·225	0·231	0·234	0·237	0·243	0·249	0·252	0·261	0·273	0·284	0·289	0·295	0·316	0·342
1/ 22727	1·0167	1·1340	1·1957	1·2595	1·3932	1·5354	1·6097	1·8460	2·1930	2·5779	2·7850	3·0022	3·9746	5·4372
0·0000460	0·230	0·236	0·239	0·243	0·249	0·255	0·258	0·267	0·279	0·290	0·296	0·302	0·323	0·350
1/ 21739	1·0396	1·1596	1·2226	1·2878	1·4245	1·5699	1·6459	1·8876	2·2424	2·6359	2·8477	3·0697	4·0641	5·5596
0·0000480	0·235	0·241	0·245	0·248	0·254	0·260	0·264	0·273	0·285	0·297	0·302	0·308	0·330	0·357
1/ 20833	1·0620	1·1845	1·2490	1·3155	1·4552	1·6037	1·6814	1·9282	2·2907	2·6927	2·9090	3·1358	4·1516	5·6792
0·0000500	0·240	0·246	0·250	0·253	0·259	0·266	0·269	0·278	0·291	0·303	0·309	0·314	0·337	0·364
1/ 20000	1·0839	1·2090	1·2748	1·3427	1·4853	1·6368	1·7161	1·9680	2·3380	2·7483	2·9690	3·2005	4·2372	5·7964
0·0000525	0·246	0·252	0·256	0·259	0·266	0·272	0·276	0·285	0·298	0·310	0·316	0·322	0·346	0·373
1/ 19048	1·1107	1·2389	1·3063	1·3759	1·5220	1·6773	1·7585	2·0167	2·3958	2·8162	3·0424	3·2796	4·3420	5·9397
0·0000550	0·251	0·258	0·262	0·265	0·272	0·279	0·282	0·292	0·305	0·317	0·324	0·330	0·354	0·382
1/ 18182	1·1369	1·2681	1·3371	1·4083	1·5578	1·7168	1·8000	2·0642	2·4522	2·8825	3·1141	3·3569	4·4442	6·0796
0·0000575	0·257	0·264	0·268	0·271	0·278	0·285	0·289	0·299	0·312	0·325	0·331	0·337	0·362	0·391
1/ 17391	1·1624	1·2966	1·3671	1·4400	1·5929	1·7555	1·8405	2·1106	2·5074	2·9474	3·1842	3·4324	4·5442	6·2164
0·0000600	0·262	0·270	0·273	0·277	0·284	0·291	0·295	0·305	0·318	0·332	0·338	0·344	0·369	0·399
1/ 16667	1·1875	1·3245	1·3966	1·4710	1·6272	1·7933	1·8801	2·1561	2·5613	3·0108	3·2527	3·5063	4·6420	6·3502
0·0000625	0·268	0·275	0·279	0·283	0·290	0·297	0·301	0·311	0·325	0·338	0·345	0·352	0·377	0·408
1/ 16000	1·2120	1·3519	1·4254	1·5014	1·6608	1·8303	1·9189	2·2006	2·6142	3·0730	3·3198	3·5787	4·7378	6·4812
0·0000650	0·273	0·281	0·285	0·288	0·296	0·303	0·307	0·317	0·331	0·345	0·352	0·359	0·384	0·416
1/ 15385	1·2360	1·3787	1·4537	1·5311	1·6937	1·8666	1·9569	2·2442	2·6660	3·1339	3·3856	3·6496	4·8317	6·6097
0·0000675	0·278	0·286	0·290	0·294	0·301	0·309	0·313	0·324	0·338	0·352	0·359	0·365	0·392	0·424
1/ 14815	1·2596	1·4050	1·4814	1·5603	1·7260	1·9021	1·9942	2·2870	2·7169	3·1936	3·4502	3·7192	4·9239	6·7357
0·0000700	0·284	0·291	0·295	0·299	0·307	0·315	0·318	0·329	0·344	0·358	0·365	0·372	0·399	0·431
1/ 14286	1·2827	1·4308	1·5086	1·5890	1·7577	1·9371	2·0309	2·3290	2·7668	3·2523	3·5136	3·7875	5·0143	6·8593
0·0000725	0·289	0·297	0·301	0·305	0·312	0·320	0·324	0·335	0·350	0·365	0·372	0·379	0·406	0·439
1/ 13793	1·3055	1·4561	1·5353	1·6172	1·7888	1·9714	2·0669	2·3703	2·8158	3·3099	3·5758	3·8546	5·1031	6·9808
0·0000750	0·294	0·302	0·306	0·310	0·318	0·326	0·330	0·341	0·356	0·371	0·378	0·385	0·413	0·446
1/ 13333	1·3278	1·4810	1·5616	1·6448	1·8194	2·0051	2·1022	2·4108	2·8640	3·3665	3·6370	3·9205	5·1904	7·1003
0·0000800	0·303	0·312	0·316	0·320	0·328	0·336	0·340	0·352	0·368	0·383	0·390	0·398	0·427	0·461
1/ 12500	1·3714	1·5297	1·6129	1·6988	1·8792	2·0710	2·1712	2·4900	2·9580	3·4770	3·7564	4·0492	5·3608	7·3333
0·0000850	0·312	0·321	0·326	0·330	0·338	0·347	0·351	0·363	0·379	0·395	0·402	0·410	0·440	0·475
1/ 11765	1·4136	1·5768	1·6626	1·7512	1·9371	2·1348	2·2381	2·5667	3·0491	3·5841	3·8721	4·1739	5·5259	7·5591
0·0000900	0·322	0·331	0·335	0·339	0·348	0·357	0·361	0·374	0·390	0·406	0·414	0·422	0·452	0·489
1/ 11111	1·4547	1·6225	1·7108	1·8020	1·9933	2·1967	2·3031	2·6411	3·1375	3·6881	3·9844	4·2950	5·6862	7·7784
0·0000950	0·330	0·340	0·344	0·349	0·358	0·367	0·371	0·384	0·401	0·417	0·425	0·434	0·465	0·502
1/ 10526	1·4946	1·6670	1·7577	1·8514	2·0479	2·2569	2·3662	2·7136	3·2236	3·7893	4·0937	4·4128	5·8421	7·9917
0·0001000	0·339	0·348	0·353	0·358	0·367	0·376	0·381	0·394	0·411	0·428	0·437	0·445	0·477	0·516
1/ 10000	1·5334	1·7104	1·8034	1·8995	2·1012	2·3156	2·4277	2·7841	3·3074	3·8878	4·2001	4·5275	5·9940	8·1994
	0·74	0·74	0·75	0·75	0·75	0·75	0·75	0·75	0·75	0·75	0·75	0·75	0·76	0·76

$V_{r(0{\cdot}5)medial}$ for half-full circular pipes.

k_s = 30·0 mm S = 0·00003 to 0·00010

$k_s = 30{\cdot}0$ **mm**
S = 0·00010 to 0·00030

ie hydraulic gradient =
1 in 10000 to 1 in 3333

Water (or sewage) at 15°C;
full bore conditions.

velocities in ms^{-1}
discharges in m^3s^{-1}

Gradient **(Equivalent) Pipe diameters in mm**

Gradient	2400	2500	2550	2600	2700	2800	2850	3000	3200	3400	3500	3600	4000	4500
0·000100	0·339	0·348	0·353	0·358	0·367	0·376	0·381	0·394	0·411	0·428	0·437	0·445	0·477	0·516
1/ 10000	1·5334	1·7104	1·8034	1·8995	2·1012	2·3156	2·4277	2·7841	3·3074	3·8878	4·2001	4·5275	5·9940	8·1994
0·000105	0·347	0·357	0·362	0·367	0·376	0·385	0·390	0·404	0·421	0·439	0·447	0·456	0·489	0·528
1/ 9524	1·5713	1·7527	1·8480	1·9465	2·1531	2·3729	2·4877	2·8529	3·3891	3·9838	4·3039	4·6394	6·1421	8·4020
0·000110	0·356	0·365	0·370	0·375	0·385	0·394	0·399	0·413	0·431	0·449	0·458	0·467	0·500	0·541
1/ 9091	1·6083	1·7939	1·8915	1·9923	2·2038	2·4287	2·5463	2·9201	3·4689	4·0777	4·4052	4·7487	6·2867	8·5999
0·000115	0·364	0·374	0·379	0·384	0·394	0·403	0·408	0·422	0·441	0·459	0·468	0·477	0·512	0·553
1/ 8696	1·6445	1·8343	1·9341	2·0372	2·2534	2·4834	2·6036	2·9858	3·5470	4·1694	4·5043	4·8555	6·4281	8·7933
0·000120	0·371	0·382	0·387	0·392	0·402	0·412	0·417	0·431	0·451	0·469	0·478	0·487	0·523	0·565
1/ 8333	1·6799	1·8738	1·9757	2·0810	2·3019	2·5368	2·6596	3·0501	3·6233	4·2591	4·6012	4·9600	6·5664	8·9825
0·000125	0·379	0·390	0·395	0·400	0·410	0·420	0·426	0·440	0·460	0·479	0·488	0·497	0·533	0·576
1/ 8000	1·7146	1·9125	2·0165	2·1239	2·3494	2·5892	2·7145	3·1130	3·6981	4·3470	4·6962	5·0623	6·7019	9·1678
0·000130	0·387	0·397	0·403	0·408	0·418	0·429	0·434	0·449	0·469	0·488	0·498	0·507	0·544	0·588
1/ 7692	1·7486	1·9504	2·0564	2·1660	2·3960	2·6405	2·7683	3·1747	3·7714	4·4331	4·7892	5·1626	6·8347	9·3495
0·000135	0·394	0·405	0·410	0·416	0·426	0·437	0·442	0·458	0·478	0·498	0·507	0·517	0·554	0·599
1/ 7407	1·7819	1·9875	2·0956	2·2073	2·4416	2·6908	2·8211	3·2352	3·8432	4·5176	4·8805	5·2610	6·9650	9·5277
0·000140	0·401	0·412	0·418	0·423	0·434	0·445	0·450	0·466	0·487	0·507	0·517	0·526	0·564	0·610
1/ 7143	1·8146	2·0240	2·1341	2·2479	2·4865	2·7402	2·8729	3·2946	3·9138	4·6006	4·9702	5·3576	7·0929	9·7026
0·000145	0·408	0·420	0·425	0·431	0·442	0·453	0·458	0·474	0·495	0·516	0·526	0·536	0·574	0·621
1/ 6897	1·8468	2·0599	2·1719	2·2877	2·5305	2·7888	2·9238	3·3530	3·9831	4·6821	5·0582	5·4525	7·2185	9·8744
0·000150	0·415	0·427	0·433	0·438	0·450	0·461	0·466	0·482	0·504	0·525	0·535	0·545	0·584	0·631
1/ 6667	1·8784	2·0951	2·2091	2·3268	2·5738	2·8365	2·9738	3·4103	4·0513	4·7622	5·1447	5·5458	7·3419	10·043
0·000160	0·429	0·441	0·447	0·453	0·464	0·476	0·481	0·498	0·520	0·542	0·552	0·563	0·603	0·652
1/ 6250	1·9400	2·1639	2·2816	2·4032	2·6583	2·9296	3·0714	3·5222	4·1842	4·9184	5·3135	5·7277	7·5828	10·373
0·000170	0·442	0·454	0·461	0·467	0·479	0·490	0·496	0·514	0·536	0·558	0·569	0·580	0·622	0·672
1/ 5882	1·9998	2·2305	2·3519	2·4772	2·7401	3·0198	3·1660	3·6307	4·3131	5·0699	5·4771	5·9041	7·8163	10·692
0·000180	0·455	0·468	0·474	0·480	0·492	0·505	0·511	0·529	0·552	0·575	0·586	0·597	0·640	0·692
1/ 5556	2·0578	2·2952	2·4201	2·5491	2·8196	3·1074	3·2578	3·7360	4·4382	5·2169	5·6360	6·0754	8·0431	11·002
0·000190	0·467	0·480	0·487	0·493	0·506	0·518	0·525	0·543	0·567	0·590	0·602	0·613	0·658	0·711
1/ 5263	2·1142	2·3582	2·4864	2·6190	2·8970	3·1926	3·3472	3·8385	4·5599	5·3600	5·7905	6·2419	8·2636	11·304
0·000200	0·479	0·493	0·500	0·506	0·519	0·532	0·538	0·557	0·582	0·606	0·617	0·629	0·675	0·729
1/ 5000	2·1692	2·4195	2·5511	2·6870	2·9723	3·2756	3·4342	3·9382	4·6784	5·4993	5·9410	6·4042	8·4783	11·598
0·000210	0·491	0·505	0·512	0·519	0·532	0·545	0·552	0·571	0·596	0·621	0·633	0·645	0·691	0·747
1/ 4762	2·2228	2·4793	2·6141	2·7534	3·0457	3·3565	3·5190	4·0355	4·7940	5·6352	6·0878	6·5624	8·6878	11·884
0·000220	0·503	0·517	0·524	0·531	0·544	0·558	0·565	0·584	0·610	0·635	0·648	0·660	0·708	0·765
1/ 4545	2·2751	2·5376	2·6757	2·8183	3·1174	3·4355	3·6019	4·1306	4·9068	5·7678	6·2312	6·7169	8·8923	12·164
0·000230	0·514	0·529	0·536	0·543	0·557	0·570	0·577	0·597	0·624	0·650	0·662	0·675	0·724	0·782
1/ 4348	2·3263	2·5947	2·7358	2·8816	3·1875	3·5128	3·6829	4·2234	5·0172	5·8975	6·3713	6·8679	9·0923	12·438
0·000240	0·525	0·540	0·547	0·554	0·569	0·583	0·590	0·610	0·637	0·664	0·676	0·689	0·739	0·799
1/ 4167	2·3763	2·6506	2·7947	2·9437	3·2561	3·5884	3·7621	4·3143	5·1251	6·0244	6·5084	7·0157	9·2879	12·705
0·000250	0·536	0·551	0·559	0·566	0·580	0·595	0·602	0·623	0·650	0·677	0·690	0·703	0·754	0·815
1/ 4000	2·4254	2·7052	2·8524	3·0044	3·3233	3·6624	3·8397	4·4033	5·2309	6·1487	6·6426	7·1605	9·4795	12·967
0·000260	0·547	0·562	0·570	0·577	0·592	0·607	0·614	0·635	0·663	0·691	0·704	0·717	0·769	0·831
1/ 3846	2·4734	2·7588	2·9089	3·0639	3·3891	3·7350	3·9158	4·4906	5·3345	6·2705	6·7742	7·3023	9·6673	13·224
0·000270	0·557	0·573	0·580	0·588	0·603	0·618	0·626	0·647	0·676	0·704	0·718	0·731	0·784	0·847
1/ 3704	2·5206	2·8114	2·9643	3·1223	3·4537	3·8062	3·9904	4·5762	5·4362	6·3901	6·9033	7·4415	9·8516	13·476
0·000280	0·567	0·583	0·591	0·599	0·614	0·629	0·637	0·659	0·688	0·717	0·731	0·745	0·798	0·863
1/ 3571	2·5668	2·8630	3·0188	3·1796	3·5171	3·8761	4·0637	4·6602	5·5360	6·5074	7·0301	7·5781	10·032	13·724
0·000290	0·577	0·594	0·602	0·609	0·625	0·641	0·648	0·671	0·701	0·729	0·744	0·758	0·812	0·878
1/ 3448	2·6123	2·9138	3·0722	3·2360	3·5794	3·9447	4·1357	4·7427	5·6340	6·6226	7·1546	7·7123	10·210	13·967
0·000300	0·587	0·604	0·612	0·620	0·636	0·652	0·659	0·682	0·713	0·742	0·756	0·771	0·826	0·893
1/ 3333	2·6570	2·9636	3·1248	3·2913	3·6406	4·0122	4·2064	4·8238	5·7304	6·7359	7·2769	7·8442	10·385	14·205
	0·74	0·74	0·75	0·75	0·75	0·75	0·75	0·75	0·75	0·75	0·75	0·75	0·76	0·76

$V_{r(0{\cdot}5)medial}$ for half-full circular pipes.

A57
(p.3 of 6)

k_s = 30·0 mm
S = 0·00030 to 0·00100

ie hydraulic gradient =
1 in 3333 to 1 in 1000

Water (or sewage) at 15°C;
full bore conditions.

velocities in ms^{-1}
discharges in m^3s^{-1}

Gradient **(Equivalent) Pipe diameters in mm**

Gradient	2400	2500	2550	2600	2700	2800	2850	3000	3200	3400	3500	3600	4000	4500
0·000300	0·587	0·604	0·612	0·620	0·636	0·652	0·659	0·682	0·713	0·742	0·756	0·771	0·826	0·893
1/ 3333	2·6570	2·9636	3·1248	3·2913	3·6406	4·0122	4·2064	4·8238	5·7304	6·7359	7·2769	7·8442	10·385	14·205
0·000320	0·607	0·624	0·632	0·640	0·657	0·673	0·681	0·705	0·736	0·766	0·781	0·796	0·853	0·922
1/ 3125	2·7442	3·0608	3·2273	3·3993	3·7601	4·1438	4·3444	4·9821	5·9184	6·9569	7·5157	8·1015	10·725	14·671
0·000340	0·625	0·643	0·651	0·660	0·677	0·694	0·702	0·727	0·759	0·790	0·805	0·820	0·880	0·951
1/ 2941	2·8287	3·1551	3·3267	3·5040	3·8759	4·2714	4·4782	5·1355	6·1006	7·1710	7·7471	8·3510	11·056	15·123
0·000360	0·643	0·661	0·670	0·679	0·697	0·714	0·722	0·748	0·781	0·813	0·829	0·844	0·905	0·978
1/ 2778	2·9107	3·2466	3·4232	3·6056	3·9883	4·3953	4·6081	5·2844	6·2775	7·3790	7·9718	8·5932	11·376	15·562
0·000380	0·661	0·680	0·689	0·698	0·716	0·733	0·742	0·768	0·802	0·835	0·851	0·867	0·930	1·005
1/ 2632	2·9905	3·3356	3·5170	3·7044	4·0976	4·5158	4·7344	5·4293	6·4496	7·5813	8·1903	8·8287	11·688	15·988
0·000400	0·678	0·697	0·707	0·716	0·734	0·752	0·761	0·788	0·823	0·857	0·873	0·890	0·954	1·031
1/ 2500	3·0682	3·4223	3·6084	3·8007	4·2041	4·6331	4·8574	5·5704	6·6172	7·7783	8·4031	9·0582	11·992	16·404
0·000420	0·695	0·714	0·724	0·734	0·752	0·771	0·780	0·808	0·843	0·878	0·895	0·912	0·978	1·057
1/ 2381	3·1440	3·5068	3·6976	3·8946	4·3080	4·7476	4·9774	5·7080	6·7807	7·9705	8·6107	9·2819	12·288	16·809
0·000440	0·711	0·731	0·741	0·751	0·770	0·789	0·799	0·827	0·863	0·899	0·916	0·933	1·001	1·082
1/ 2273	3·2180	3·5894	3·7846	3·9863	4·4094	4·8594	5·0946	5·8424	6·9403	8·1581	8·8134	9·5004	12·577	17·205
0·000460	0·727	0·748	0·758	0·768	0·787	0·807	0·817	0·845	0·882	0·919	0·937	0·954	1·023	1·106
1/ 2174	3·2904	3·6701	3·8697	4·0759	4·5086	4·9686	5·2092	5·9737	7·0964	8·3415	9·0116	9·7140	12·860	17·591
0·000480	0·743	0·764	0·774	0·784	0·804	0·824	0·834	0·863	0·901	0·939	0·957	0·975	1·045	1·130
1/ 2083	3·3612	3·7491	3·9530	4·1636	4·6056	5·0755	5·3212	6·1023	7·2491	8·5210	9·2054	9·9230	13·137	17·970
0·000500	0·758	0·780	0·790	0·800	0·821	0·841	0·851	0·881	0·920	0·958	0·977	0·995	1·067	1·153
1/ 2000	3·4305	3·8264	4·0345	4·2495	4·7006	5·1802	5·4310	6·2282	7·3986	8·6968	9·3953	10·128	13·408	18·341
0·000525	0·777	0·799	0·810	0·820	0·841	0·862	0·872	0·903	0·943	0·982	1·001	1·020	1·093	1·182
1/ 1905	3·5153	3·9209	4·1342	4·3545	4·8167	5·3082	5·5652	6·3820	7·5814	8·9116	9·6274	10·378	13·739	18·794
0·000550	0·795	0·818	0·829	0·839	0·861	0·882	0·893	0·924	0·965	1·005	1·024	1·044	1·119	1·209
1/ 1818	3·5980	4·0133	4·2315	4·4570	4·9301	5·4332	5·6962	6·5322	7·7598	9·1214	9·8540	10·622	14·062	19·236
0·000575	0·813	0·836	0·847	0·858	0·880	0·902	0·913	0·945	0·987	1·027	1·047	1·067	1·144	1·237
1/ 1739	3·6789	4·1035	4·3267	4·5572	5·0409	5·5553	5·8243	6·6791	7·9343	9·3264	10·076	10·861	14·378	19·668
0·000600	0·831	0·854	0·865	0·877	0·899	0·922	0·933	0·965	1·008	1·049	1·070	1·090	1·169	1·263
1/ 1667	3·7581	4·1918	4·4197	4·6553	5·1494	5·6748	5·9496	6·8228	8·1050	9·5271	10·292	11·095	14·688	20·092
0·000625	0·848	0·872	0·883	0·895	0·918	0·941	0·952	0·985	1·029	1·071	1·092	1·112	1·193	1·289
1/ 1600	3·8356	4·2782	4·5109	4·7513	5·2556	5·7919	6·0723	6·9635	8·2722	9·7236	10·505	11·323	14·991	20·506
0·000650	0·865	0·889	0·901	0·913	0·936	0·959	0·971	1·005	1·049	1·092	1·113	1·135	1·217	1·315
1/ 1538	3·9116	4·3630	4·6003	4·8454	5·3597	5·9066	6·1926	7·1015	8·4360	9·9162	10·713	11·548	15·288	20·912
0·000675	0·881	0·906	0·918	0·930	0·954	0·978	0·989	1·024	1·069	1·113	1·135	1·156	1·240	1·340
1/ 1481	3·9862	4·4461	4·6879	4·9378	5·4619	6·0192	6·3106	7·2368	8·5968	10·105	10·917	11·768	15·579	21·311
0·000700	0·897	0·922	0·935	0·947	0·971	0·995	1·007	1·043	1·089	1·133	1·156	1·177	1·262	1·365
1/ 1429	4·0593	4·5278	4·7740	5·0284	5·5621	6·1297	6·4264	7·3696	8·7546	10·291	11·117	11·984	15·865	21·702
0·000725	0·913	0·939	0·951	0·964	0·989	1·013	1·025	1·061	1·108	1·153	1·176	1·198	1·285	1·389
1/ 1379	4·1312	4·6079	4·8585	5·1174	5·6606	6·2382	6·5402	7·5001	8·9096	10·473	11·314	12·196	16·146	22·086
0·000750	0·929	0·955	0·968	0·980	1·006	1·030	1·043	1·079	1·127	1·173	1·196	1·219	1·307	1·412
1/ 1333	4·2018	4·6867	4·9416	5·2050	5·7574	6·3449	6·6521	7·6284	9·0619	10·652	11·508	12·405	16·422	22·464
0·000800	0·959	0·986	0·999	1·013	1·039	1·064	1·077	1·115	1·164	1·212	1·235	1·259	1·350	1·459
1/ 1250	4·3397	4·8405	5·1037	5·3757	5·9463	6·5530	6·8703	7·8786	9·3592	11·001	11·885	12·811	16·961	23·200
0·000850	0·989	1·016	1·030	1·044	1·071	1·097	1·110	1·149	1·200	1·249	1·273	1·297	1·391	1·504
1/ 1176	4·4733	4·9895	5·2608	5·5412	6·1293	6·7548	7·0818	8·1212	9·6473	11·340	12·251	13·206	17·483	23·915
0·000900	1·017	1·046	1·060	1·074	1·102	1·129	1·142	1·182	1·234	1·285	1·310	1·335	1·432	1·547
1/ 1111	4·6030	5·1342	5·4134	5·7019	6·3071	6·9507	7·2871	8·3567	9·9271	11·669	12·606	13·589	17·990	24·608
0·000950	1·045	1·075	1·089	1·103	1·132	1·160	1·174	1·215	1·268	1·320	1·346	1·372	1·471	1·590
1/ 1053	4·7292	5·2749	5·5618	5·8582	6·4800	7·1412	7·4869	8·5857	10·199	11·989	12·952	13·961	18·483	25·283
0·001000	1·073	1·103	1·117	1·132	1·161	1·190	1·204	1·246	1·301	1·355	1·381	1·407	1·509	1·631
1/ 1000	4·8521	5·4120	5·7063	6·0104	6·6483	7·3267	7·6814	8·8088	10·464	12·300	13·288	14·324	18·963	25·939
	0·74	0·74	0·75	0·75	0·75	0·75	0·75	0·75	0·75	0·75	0·75	0·75	0·76	0·76

$V_{r(0\cdot5)medial}$ for half-full circular pipes.

k_s = 30·0 mm S = 0·00030 to 0·00100

k_s = 30·0 mm
S = 0·00100 to 0·00300

ie hydraulic gradient =
1 in 1000 to 1 in 333

Water (or sewage) at 15°C;
full bore conditions.

velocities in ms^{-1}
discharges in m^3s^{-1}

Gradient	**(Equivalent) Pipe diameters in mm**													
	2400	**2500**	**2550**	**2600**	**2700**	**2800**	**2850**	**3000**	**3200**	**3400**	**3500**	**3600**	**4000**	**4500**
0·00100	1·073	1·103	1·117	1·132	1·161	1·190	1·204	1·246	1·301	1·355	1·381	1·407	1·509	1·631
1/ 1000	4·8521	5·4120	5·7063	6·0104	6·6483	7·3267	7·6814	8·8088	10·464	12·300	13·288	14·324	18·963	25·939
0·00105	1·099	1·130	1·145	1·160	1·190	1·219	1·234	1·277	1·333	1·388	1·415	1·442	1·546	1·671
1/ 952	4·9719	5·5457	5·8473	6·1589	6·8126	7·5077	7·8712	9·0264	10·723	12·604	13·616	14·678	19·431	26·580
0·00110	1·125	1·156	1·172	1·187	1·218	1·248	1·263	1·307	1·365	1·421	1·449	1·476	1·583	1·711
1/ 909	5·0890	5·6762	5·9849	6·3039	6·9729	7·6844	8·0564	9·2389	10·975	12·901	13·937	15·023	19·889	27·206
0·00115	1·150	1·182	1·198	1·214	1·245	1·276	1·291	1·336	1·395	1·453	1·481	1·509	1·618	1·749
1/ 870	5·2034	5·8038	6·1195	6·4456	7·1297	7·8572	8·2376	9·4466	11·222	13·191	14·250	15·361	20·336	27·817
0·00120	1·175	1·208	1·224	1·240	1·272	1·303	1·319	1·365	1·425	1·484	1·513	1·542	1·653	1·787
1/ 833	5·3153	5·9287	6·2511	6·5842	7·2831	8·0262	8·4148	9·6498	11·463	13·474	14·557	15·692	20·773	28·416
0·00125	1·199	1·233	1·249	1·266	1·298	1·330	1·346	1·393	1·455	1·515	1·544	1·573	1·687	1·824
1/ 800	5·4250	6·0510	6·3801	6·7200	7·4333	8·1918	8·5883	9·8488	11·700	13·752	14·857	16·015	21·202	29·002
0·00130	1·223	1·257	1·274	1·291	1·324	1·357	1·373	1·421	1·484	1·545	1·575	1·605	1·721	1·860
1/ 769	5·5324	6·1708	6·5064	6·8532	7·5805	8·3540	8·7585	10·044	11·931	14·025	15·151	16·332	21·622	29·576
0·00135	1·246	1·281	1·298	1·315	1·349	1·383	1·399	1·448	1·512	1·574	1·605	1·635	1·753	1·895
1/ 741	5·6378	6·2884	6·6304	6·9837	7·7250	8·5132	8·9253	10·235	12·159	14·292	15·440	16·644	22·034	30·140
0·00140	1·269	1·305	1·322	1·340	1·374	1·408	1·425	1·475	1·540	1·603	1·634	1·665	1·786	1·930
1/ 714	5·7413	6·4038	6·7521	7·1119	7·8667	8·6695	9·0892	10·423	12·382	14·554	15·723	16·949	22·438	30·693
0·00145	1·292	1·328	1·346	1·363	1·398	1·433	1·450	1·501	1·567	1·631	1·663	1·695	1·817	1·964
1/ 690	5·8430	6·5172	6·8716	7·2378	8·0060	8·8230	9·2501	10·608	12·601	14·812	16·002	17·249	22·835	31·236
0·00150	1·314	1·350	1·369	1·387	1·422	1·457	1·475	1·526	1·594	1·659	1·692	1·724	1·848	1·998
1/ 667	5·9429	6·6286	6·9891	7·3616	8·1429	8·9738	9·4082	10·789	12·817	15·065	16·275	17·544	23·226	31·770
0·00160	1·357	1·395	1·413	1·432	1·469	1·505	1·523	1·576	1·646	1·714	1·747	1·780	1·909	2·063
1/ 625	6·1378	6·8461	7·2184	7·6031	8·4100	9·2682	9·7168	11·143	13·237	15·559	16·809	18·119	23·987	32·812
0·00170	1·399	1·438	1·457	1·476	1·514	1·552	1·570	1·625	1·697	1·766	1·801	1·835	1·968	2·127
1/ 588	6·3268	7·0568	7·4406	7·8371	8·6689	9·5535	10·016	11·486	13·644	16·038	17·327	18·677	24·726	33·822
0·00180	1·439	1·479	1·499	1·519	1·558	1·597	1·616	1·672	1·746	1·818	1·853	1·888	2·025	2·188
1/ 556	6·5102	7·2615	7·6564	8·0644	8·9203	9·8305	10·306	11·819	14·040	16·503	17·829	19·219	25·443	34·803
0·00190	1·479	1·520	1·540	1·561	1·601	1·640	1·660	1·718	1·794	1·868	1·904	1·940	2·080	2·248
1/ 526	6·6887	7·4605	7·8662	8·2854	9·1648	10·100	10·589	12·143	14·425	16·956	18·318	19·746	26·140	35·757
0·00200	1·517	1·559	1·580	1·601	1·642	1·683	1·703	1·763	1·840	1·916	1·953	1·990	2·134	2·307
1/ 500	6·8625	7·6543	8·0706	8·5007	9·4029	10·362	10·864	12·458	14·800	17·396	18·794	20·259	26·819	36·686
0·00210	1·554	1·598	1·619	1·641	1·683	1·724	1·745	1·806	1·886	1·963	2·002	2·039	2·187	2·364
1/ 476	7·0320	7·8434	8·2700	8·7106	9·6351	10·618	11·132	12·766	15·165	17·826	19·258	20·759	27·482	37·592
0·00220	1·591	1·635	1·657	1·679	1·722	1·765	1·786	1·849	1·930	2·010	2·049	2·087	2·238	2·419
1/ 455	7·1975	8·0280	8·4646	8·9157	9·8619	10·868	11·394	13·067	15·522	18·246	19·711	21·248	28·128	38·477
0·00230	1·627	1·672	1·695	1·717	1·761	1·805	1·826	1·890	1·973	2·055	2·095	2·134	2·289	2·474
1/ 435	7·3593	8·2085	8·6549	9·1161	10·084	11·113	11·650	13·360	15·871	18·656	20·154	21·725	28·761	39·342
0·00240	1·662	1·708	1·731	1·754	1·799	1·844	1·866	1·931	2·016	2·099	2·140	2·180	2·338	2·527
1/ 417	7·5176	8·3851	8·8411	9·3122	10·301	11·352	11·901	13·648	16·212	19·057	20·588	22·192	29·379	40·188
0·00250	1·696	1·743	1·767	1·790	1·836	1·882	1·904	1·971	2·057	2·142	2·184	2·225	2·386	2·579
1/ 400	7·6726	8·5580	9·0234	9·5042	10·513	11·586	12·147	13·929	16·547	19·450	21·012	22·650	29·985	41·017
0·00260	1·730	1·778	1·802	1·826	1·873	1·919	1·942	2·010	2·098	2·185	2·227	2·269	2·433	2·630
1/ 385	7·8246	8·7275	9·2021	9·6925	10·721	11·815	12·387	14·205	16·875	19·835	21·428	23·099	30·579	41·829
0·00270	1·763	1·812	1·836	1·860	1·908	1·955	1·979	2·048	2·138	2·226	2·270	2·313	2·480	2·680
1/ 370	7·9737	8·8938	9·3775	9·8772	10·925	12·040	12·623	14·476	17·196	20·213	21·837	23·539	31·162	42·626
0·00280	1·795	1·845	1·870	1·894	1·943	1·991	2·015	2·085	2·177	2·267	2·311	2·355	2·525	2·729
1/ 357	8·1200	9·0570	9·5496	10·058	11·126	12·261	12·855	14·741	17·512	20·584	22·237	23·971	31·734	43·408
0·00290	1·827	1·878	1·903	1·928	1·978	2·027	2·051	2·122	2·216	2·307	2·352	2·397	2·570	2·778
1/ 345	8·2638	9·2174	9·7186	10·237	11·323	12·478	13·082	15·002	17·822	20·949	22·631	24·395	32·295	44·177
0·00300	1·858	1·910	1·936	1·961	2·011	2·061	2·086	2·159	2·254	2·347	2·392	2·438	2·614	2·825
1/ 333	8·4051	9·3749	9·8848	10·412	11·517	12·692	13·306	15·259	18·126	21·307	23·018	24·812	32·848	44·932
	0·74	0·74	0·75	0·75	0·75	0·75	0·75	0·75	0·75	0·75	0·75	0·75	0·76	0·76

$V_{r(0{\cdot}5)medial}$ for half-full circular pipes.

k_s = 30·0 mm
S = 0·00300 to 0·01000

ie hydraulic gradient =
1 in 333 to 1 in 100

Water (or sewage) at 15°C;
full bore conditions.

velocities in ms^{-1}
discharges in m^3s^{-1}

Gradient	**(Equivalent) Pipe diameters in mm**													
	2400	**2500**	**2550**	**2600**	**2700**	**2800**	**2850**	**3000**	**3200**	**3400**	**3500**	**3600**	**4000**	**4500**
0·00300	1·858	1·910	1·936	1·961	2·011	2·061	2·086	2·159	2·254	2·347	2·392	2·438	2·614	2·825
1/ 333	8·4051	9·3749	9·8848	10·412	11·517	12·692	13·306	15·259	18·126	21·307	23·018	24·812	32·848	44·932
0·00320	1·919	1·972	1·999	2·025	2·077	2·129	2·154	2·230	2·328	2·424	2·471	2·518	2·700	2·918
1/ 313	8·6808	9·6825	10·209	10·753	11·894	13·108	13·743	15·759	18·721	22·006	23·773	25·626	33·925	46·406
0·00340	1·978	2·033	2·061	2·088	2·141	2·194	2·221	2·298	2·399	2·498	2·547	2·595	2·783	3·008
1/ 294	8·9480	9·9805	10·523	11·084	12·260	13·511	14·166	16·245	19·297	22·683	24·505	26·415	34·969	47·834
0·00360	2·035	2·092	2·120	2·148	2·203	2·258	2·285	2·365	2·469	2·571	2·621	2·670	2·863	3·095
1/ 278	9·2074	10·270	10·828	11·405	12·616	13·903	14·576	16·716	19·857	23·341	25·215	27·181	35·983	49·221
0·00380	2·091	2·150	2·178	2·207	2·264	2·320	2·348	2·430	2·537	2·641	2·693	2·744	2·942	3·180
1/ 263	9·4598	10·551	11·125	11·718	12·962	14·284	14·976	17·174	20·401	23·980	25·906	27·926	36·969	50·570
0·00400	2·145	2·205	2·235	2·264	2·323	2·380	2·409	2·493	2·603	2·710	2·763	2·815	3·018	3·262
1/ 250	9·7056	10·825	11·414	12·022	13·298	14·655	15·365	17·620	20·931	24·603	26·579	28·651	37·930	51·884
0·00420	2·198	2·260	2·290	2·320	2·380	2·439	2·468	2·554	2·667	2·777	2·831	2·884	3·093	3·343
1/ 238	9·9453	11·093	11·696	12·319	13·627	15·017	15·744	18·055	21·448	25·211	27·236	29·359	38·866	53·165
0·00440	2·250	2·313	2·344	2·375	2·436	2·496	2·526	2·614	2·730	2·842	2·897	2·952	3·166	3·421
1/ 227	10·179	11·354	11·971	12·609	13·948	15·371	16·115	18·480	21·953	25·804	27·877	30·050	39·781	54·416
0·00460	2·301	2·365	2·397	2·428	2·491	2·552	2·583	2·673	2·791	2·906	2·963	3·019	3·237	3·498
1/ 217	10·408	11·609	12·240	12·893	14·261	15·716	16·477	18·895	22·446	26·384	28·503	30·725	40·675	55·639
0·00480	2·350	2·416	2·448	2·481	2·544	2·607	2·638	2·731	2·851	2·969	3·026	3·083	3·306	3·574
1/ 208	10·632	11·859	12·504	13·170	14·568	16·054	16·832	19·302	22·929	26·952	29·117	31·386	41·550	56·836
0·00500	2·399	2·466	2·499	2·532	2·597	2·661	2·693	2·787	2·910	3·030	3·089	3·147	3·375	3·647
1/ 200	10·851	12·103	12·762	13·442	14·868	16·385	17·179	19·700	23·402	27·508	29·717	32·033	42·407	58·008
0·00525	2·458	2·527	2·561	2·594	2·661	2·727	2·759	2·856	2·982	3·105	3·165	3·225	3·458	3·737
1/ 190	11·119	12·402	13·077	13·774	15·236	16·790	17·603	20·186	23·980	28·187	30·451	32·825	43·454	59·441
0·00550	2·516	2·586	2·621	2·655	2·724	2·791	2·824	2·923	3·052	3·178	3·239	3·301	3·539	3·825
1/ 182	11·381	12·694	13·385	14·098	15·594	17·185	18·017	20·661	24·544	28·850	31·168	33·597	44·477	60·840
0·00575	2·572	2·644	2·680	2·715	2·785	2·854	2·888	2·989	3·120	3·249	3·312	3·375	3·619	3·911
1/ 174	11·637	12·980	13·685	14·415	15·945	17·572	18·422	21·126	25·096	29·499	31·868	34·352	45·477	62·207
0·00600	2·628	2·701	2·737	2·773	2·845	2·915	2·950	3·053	3·188	3·319	3·384	3·447	3·697	3·995
1/ 167	11·887	13·259	13·980	14·725	16·288	17·950	18·818	21·580	25·636	30·133	32·554	35·091	46·455	63·545
0·00625	2·682	2·757	2·794	2·831	2·903	2·975	3·011	3·116	3·253	3·387	3·453	3·519	3·773	4·078
1/ 160	12·132	13·532	14·268	15·028	16·623	18·320	19·207	22·025	26·164	30·755	33·225	35·815	47·413	64·856
0·00650	2·735	2·811	2·849	2·887	2·961	3·034	3·070	3·178	3·318	3·454	3·522	3·588	3·848	4·159
1/ 154	12·373	13·800	14·551	15·326	16·953	18·683	19·587	22·462	26·682	31·364	33·883	36·524	48·352	66·140
0·00675	2·787	2·865	2·903	2·942	3·017	3·092	3·129	3·238	3·381	3·520	3·589	3·657	3·921	4·238
1/ 148	12·608	14·063	14·828	15·618	17·276	19·038	19·960	22·890	27·191	31·961	34·529	37·220	49·273	67·400
0·00700	2·838	2·918	2·957	2·996	3·073	3·149	3·186	3·298	3·443	3·585	3·655	3·724	3·993	4·316
1/ 143	12·840	14·321	15·100	15·905	17·593	19·388	20·326	23·310	27·690	32·548	35·162	37·903	50·177	68·637
0·00725	2·888	2·969	3·009	3·049	3·127	3·204	3·243	3·356	3·504	3·648	3·719	3·790	4·064	4·392
1/ 138	13·067	14·575	15·367	16·186	17·904	19·731	20·686	23·722	28·180	33·124	35·785	38·574	51·066	69·852
0·00750	2·938	3·020	3·061	3·101	3·181	3·259	3·298	3·413	3·564	3·711	3·783	3·854	4·133	4·467
1/ 133	13·290	14·824	15·630	16·463	18·210	20·068	21·040	24·128	28·662	33·690	36·396	39·233	51·939	71·047
0·00800	3·034	3·119	3·161	3·202	3·285	3·366	3·406	3·525	3·681	3·832	3·907	3·981	4·269	4·614
1/ 125	13·726	15·310	16·143	17·003	18·808	20·727	21·730	24·919	29·602	34·795	37·590	40·520	53·642	73·377
0·00850	3·128	3·215	3·258	3·301	3·386	3·470	3·511	3·634	3·794	3·950	4·027	4·103	4·400	4·756
1/ 118	14·149	15·781	16·640	17·526	19·386	21·365	22·399	25·686	30·513	35·866	38·747	41·767	55·293	75·635
0·00900	3·218	3·308	3·353	3·397	3·484	3·570	3·613	3·739	3·904	4·065	4·144	4·222	4·528	4·894
1/ 111	14·559	16·239	17·122	18·034	19·949	21·984	23·048	26·431	31·398	36·906	39·871	42·978	56·896	77·828
0·00950	3·306	3·399	3·445	3·490	3·580	3·668	3·712	3·842	4·011	4·176	4·258	4·338	4·652	5·028
1/ 105	14·958	16·684	17·591	18·529	20·495	22·586	23·680	27·155	32·258	37·918	40·963	44·156	58·456	79·961
0·01000	3·392	3·487	3·534	3·581	3·673	3·763	3·808	3·941	4·115	4·285	4·368	4·451	4·773	5·158
1/ 100	15·347	17·117	18·048	19·010	21·028	23·173	24·295	27·861	33·096	38·903	42·027	45·303	59·974	82·038
	0·74	0·74	0·75	0·75	0·75	0·75	0·75	0·75	0·75	0·75	0·75	0·75	0·76	0·76

$V_{r(0{\cdot}5)medial}$ for half-full circular pipes.

k_s = 30·0 mm S = 0·00300 to 0·01000

Gradient	**(Equivalent) Pipe diameters in mm**													
	2400	**2500**	**2550**	**2600**	**2700**	**2800**	**2850**	**3000**	**3200**	**3400**	**3500**	**3600**	**4000**	**4500**
0·01000	3·392	3·487	3·534	3·581	3·673	3·763	3·808	3·941	4·115	4·285	4·368	4·451	4·773	5·158
1/ 100	15·347	17·117	18·048	19·010	21·028	23·173	24·295	27·861	33·096	38·903	42·027	45·303	59·974	82·038
0·01050	3·476	3·573	3·621	3·669	3·763	3·856	3·902	4·039	4·217	4·391	4·476	4·561	4·890	5·286
1/ 95	15·726	17·540	18·494	19·480	21·547	23·746	24·895	28·549	33·913	39·864	43·065	46·422	61·455	84·064
0·01100	3·558	3·657	3·707	3·755	3·852	3·947	3·994	4·134	4·316	4·494	4·581	4·668	5·006	5·410
1/ 91	16·096	17·953	18·929	19·938	22·054	24·304	25·481	29·221	34·712	40·802	44·079	47·515	62·902	86·043
0·01150	3·638	3·740	3·790	3·840	3·938	4·036	4·084	4·227	4·413	4·595	4·684	4·773	5·118	5·532
1/ 87	16·458	18·357	19·355	20·386	22·550	24·851	26·054	29·877	35·492	41·719	45·070	48·583	64·316	87·977
0·01200	3·716	3·820	3·871	3·922	4·023	4·123	4·172	4·318	4·508	4·694	4·785	4·876	5·228	5·651
1/ 83	16·812	18·751	19·771	20·825	23·035	25·385	26·614	30·520	36·255	42·616	46·039	49·628	65·699	89·869
0·01250	3·793	3·899	3·951	4·003	4·106	4·208	4·258	4·407	4·601	4·791	4·884	4·976	5·336	5·767
1/ 80	17·158	19·138	20·179	21·254	23·510	25·909	27·163	31·149	37·003	43·495	46·988	50·651	67·054	91·722
0·01300	3·868	3·976	4·029	4·082	4·187	4·291	4·342	4·494	4·692	4·885	4·981	5·075	5·442	5·881
1/ 77	17·498	19·517	20·579	21·675	23·976	26·422	27·701	31·766	37·736	44·356	47·919	51·654	68·382	93·539
0·01350	3·942	4·052	4·106	4·160	4·267	4·373	4·425	4·580	4·781	4·979	5·075	5·171	5·545	5·993
1/ 74	17·831	19·889	20·971	22·088	24·432	26·925	28·229	32·372	38·455	45·201	48·832	52·638	69·684	95·321
0·01400	4·014	4·126	4·182	4·237	4·346	4·453	4·506	4·664	4·869	5·070	5·169	5·266	5·647	6·103
1/ 71	18·159	20·254	21·355	22·493	24·881	27·419	28·747	32·966	39·160	46·031	49·728	53·604	70·963	97·070
0·01450	4·085	4·199	4·256	4·312	4·422	4·532	4·586	4·746	4·955	5·160	5·260	5·360	5·747	6·211
1/ 69	18·480	20·612	21·733	22·892	25·321	27·905	29·256	33·549	39·854	46·846	50·608	54·553	72·219	98·788
0·01500	4·155	4·271	4·328	4·385	4·498	4·609	4·664	4·827	5·040	5·248	5·350	5·451	5·845	6·318
1/ 67	18·796	20·965	22·105	23·283	25·754	28·382	29·756	34·123	40·535	47·647	51·474	55·486	73·454	100·48
0·01600	4·291	4·411	4·470	4·529	4·646	4·760	4·817	4·986	5·205	5·420	5·526	5·630	6·037	6·525
1/ 63	19·413	21·652	22·830	24·047	26·599	29·313	30·732	35·242	41·864	49·209	53·162	57·306	75·863	103·77
0·01700	4·423	4·547	4·608	4·669	4·789	4·907	4·966	5·139	5·366	5·587	5·696	5·803	6·223	6·726
1/ 59	20·010	22·319	23·533	24·787	27·417	30·215	31·677	36·327	43·153	50·724	54·798	59·069	78·198	106·97
0·01800	4·551	4·679	4·741	4·804	4·927	5·049	5·110	5·288	5·521	5·749	5·861	5·971	6·403	6·921
1/ 56	20·590	22·966	24·215	25·505	28·212	31·091	32·596	37·380	44·404	52·195	56·387	60·782	80·465	110·07
0·01900	4·676	4·807	4·871	4·936	5·062	5·188	5·250	5·433	5·672	5·906	6·021	6·135	6·579	7·110
1/ 53	21·154	23·595	24·879	26·204	28·985	31·943	33·489	38·404	45·621	53·625	57·932	62·448	82·670	113·08
0·02000	4·798	4·932	4·998	5·064	5·194	5·322	5·386	5·574	5·820	6·060	6·178	6·294	6·750	7·295
1/ 50	21·704	24·208	25·525	26·885	29·738	32·773	34·359	39·402	46·806	55·018	59·437	64·070	84·818	116·02
0·02100	4·916	5·053	5·121	5·189	5·322	5·454	5·519	5·712	5·964	6·209	6·330	6·450	6·916	7·475
1/ 47·6	22·240	24·806	26·155	27·549	30·473	33·582	35·208	40·375	47·962	56·377	60·905	65·652	86·913	118·89
0·02200	5·032	5·172	5·242	5·311	5·448	5·582	5·649	5·846	6·104	6·356	6·479	6·602	7·079	7·651
1/ 45·5	22·763	25·390	26·771	28·197	31·190	34·372	36·036	41·325	49·091	57·704	62·338	67·197	88·958	121·68
0·02300	5·145	5·289	5·360	5·430	5·570	5·708	5·776	5·978	6·241	6·498	6·625	6·750	7·238	7·823
1/ 43·5	23·275	25·961	27·373	28·831	31·891	35·145	36·846	42·254	50·194	59·000	63·739	68·708	90·958	124·42
0·02400	5·256	5·402	5·475	5·547	5·690	5·830	5·900	6·106	6·375	6·638	6·767	6·895	7·394	7·991
1/ 41·7	23·776	26·519	27·961	29·451	32·577	35·901	37·639	43·163	51·274	60·269	65·110	70·185	92·914	127·10
0·02500	5·364	5·514	5·588	5·662	5·807	5·951	6·022	6·232	6·507	6·775	6·907	7·037	7·546	8·156
1/ 40·0	24·266	27·066	28·538	30·059	33·249	36·641	38·415	44·053	52·331	61·512	66·453	71·633	94·830	129·72
0·02600	5·470	5·623	5·699	5·774	5·922	6·069	6·141	6·356	6·636	6·909	7·044	7·177	7·696	8·318
1/ 38·5	24·747	27·602	29·103	30·654	33·907	37·367	39·176	44·925	53·367	62·731	67·769	73·051	96·708	132·29
0·02700	5·574	5·730	5·807	5·884	6·035	6·184	6·258	6·477	6·762	7·041	7·178	7·314	7·842	8·476
1/ 37·0	25·218	28·128	29·658	31·238	34·553	38·079	39·922	45·781	54·384	63·926	69·060	74·443	98·550	134·81
0·02800	5·677	5·835	5·914	5·992	6·146	6·298	6·373	6·596	6·886	7·170	7·310	7·448	7·986	8·632
1/ 35·7	25·681	28·644	30·202	31·811	35·187	38·778	40·655	46·621	55·382	65·099	70·327	75·809	100·36	137·28
0·02900	5·777	5·939	6·018	6·098	6·254	6·409	6·486	6·712	7·008	7·297	7·439	7·580	8·128	8·784
1/ 34·5	26·135	29·151	30·736	32·374	35·810	39·464	41·374	47·447	56·362	66·251	71·572	77·151	102·14	139·71
0·03000	5·876	6·040	6·121	6·202	6·361	6·519	6·597	6·827	7·128	7·422	7·566	7·709	8·267	8·935
1/ 33·3	26·582	29·649	31·262	32·928	36·422	40·139	42·082	48·258	57·326	67·384	72·796	78·470	103·88	142·10
	0·74	0·74	0·75	0·75	0·75	0·75	0·75	0·75	0·75	0·75	0·75	0·75	0·76	0·76

$V_{r(0{\cdot}5)medial}$ for half-full circular pipes.

k_s = 60·0 mm
S = 0·00003 to 0·00010

ie hydraulic gradient =
1 in 33333 to 1 in 10000

Water (or sewage) at 15°C;
full bore conditions.

velocities in ms^{-1}
discharges in m^3s^{-1}

Gradient	**(Equivalent) Pipe diameters in mm**													
	2400	**2500**	**2550**	**2600**	**2700**	**2800**	**2850**	**3000**	**3200**	**3400**	**3500**	**3600**	**4000**	**4500**
0·0000300	0·163	0·168	0·170	0·172	0·177	0·182	0·184	0·190	0·199	0·208	0·212	0·216	0·232	0·251
1/ 33333	0·7376	0·8235	0·8688	0·9155	1·0136	1·1179	1·1725	1·3462	1·6015	1·8851	2·0378	2·1979	2·9162	3·9985
0·0000320	0·168	0·173	0·176	0·178	0·183	0·188	0·190	0·197	0·206	0·214	0·219	0·223	0·240	0·260
1/ 31250	0·7618	0·8506	0·8973	0·9455	1·0468	1·1546	1·2110	1·3904	1·6541	1·9469	2·1046	2·2701	3·0119	4·1297
0·0000340	0·174	0·179	0·181	0·184	0·188	0·193	0·196	0·203	0·212	0·221	0·225	0·230	0·247	0·268
1/ 29412	0·7853	0·8768	0·9249	0·9747	1·0791	1·1902	1·2483	1·4332	1·7051	2·0069	2·1695	2·3400	3·1046	4·2569
0·0000360	0·179	0·184	0·186	0·189	0·194	0·199	0·201	0·209	0·218	0·227	0·232	0·237	0·254	0·275
1/ 27778	0·8081	0·9022	0·9517	1·0029	1·1104	1·2247	1·2845	1·4748	1·7545	2·0651	2·2324	2·4079	3·1947	4·3804
0·0000380	0·184	0·189	0·191	0·194	0·199	0·204	0·207	0·214	0·224	0·234	0·238	0·243	0·261	0·283
1/ 26316	0·8302	0·9270	0·9778	1·0304	1·1408	1·2583	1·3198	1·5153	1·8026	2·1217	2·2936	2·4739	3·2823	4·5005
0·0000400	0·188	0·194	0·196	0·199	0·204	0·210	0·212	0·220	0·230	0·240	0·245	0·249	0·268	0·290
1/ 25000	0·8518	0·9511	1·0033	1·0572	1·1705	1·2910	1·3541	1·5547	1·8495	2·1769	2·3532	2·5382	3·3676	4·6175
0·0000420	0·193	0·199	0·201	0·204	0·209	0·215	0·218	0·225	0·236	0·246	0·251	0·256	0·275	0·298
1/ 23810	0·8729	0·9746	1·0281	1·0834	1·1994	1·3229	1·3875	1·5931	1·8952	2·2307	2·4114	2·6009	3·4508	4·7316
0·0000440	0·197	0·203	0·206	0·209	0·214	0·220	0·223	0·231	0·241	0·251	0·257	0·262	0·281	0·305
1/ 22727	0·8934	0·9975	1·0523	1·1089	1·2277	1·3541	1·4202	1·6306	1·9398	2·2832	2·4682	2·6622	3·5321	4·8430
0·0000460	0·202	0·208	0·211	0·214	0·219	0·225	0·228	0·236	0·247	0·257	0·262	0·267	0·287	0·311
1/ 21739	0·9135	1·0200	1·0759	1·1338	1·2553	1·3845	1·4522	1·6673	1·9834	2·3346	2·5237	2·7220	3·6115	4·9519
0·0000480	0·206	0·212	0·215	0·218	0·224	0·230	0·233	0·241	0·252	0·263	0·268	0·273	0·294	0·318
1/ 20833	0·9332	1·0419	1·0991	1·1582	1·2823	1·4143	1·4834	1·7031	2·0261	2·3848	2·5780	2·7806	3·6892	5·0584
0·0000500	0·211	0·217	0·220	0·223	0·229	0·234	0·237	0·246	0·257	0·268	0·273	0·279	0·300	0·325
1/ 20000	0·9524	1·0634	1·1218	1·1821	1·3087	1·4435	1·5140	1·7383	2·0679	2·4340	2·6312	2·8380	3·7653	5·1628
0·0000525	0·216	0·222	0·225	0·228	0·234	0·240	0·243	0·252	0·263	0·275	0·280	0·286	0·307	0·333
1/ 19048	0·9760	1·0897	1·1495	1·2113	1·3411	1·4792	1·5514	1·7812	2·1190	2·4942	2·6962	2·9081	3·8584	5·2903
0·0000550	0·221	0·227	0·230	0·234	0·240	0·246	0·249	0·258	0·270	0·281	0·287	0·292	0·314	0·340
1/ 18182	0·9990	1·1154	1·1766	1·2398	1·3727	1·5140	1·5880	1·8232	2·1689	2·5529	2·7597	2·9766	3·9492	5·4149
0·0000575	0·226	0·232	0·236	0·239	0·245	0·251	0·255	0·264	0·276	0·288	0·293	0·299	0·321	0·348
1/ 17391	1·0214	1·1404	1·2030	1·2677	1·4035	1·5481	1·6237	1·8642	2·2177	2·6103	2·8217	3·0435	4·0380	5·5367
0·0000600	0·231	0·237	0·241	0·244	0·250	0·257	0·260	0·269	0·282	0·294	0·300	0·305	0·328	0·356
1/ 16667	1·0434	1·1650	1·2289	1·2950	1·4337	1·5814	1·6586	1·9043	2·2654	2·6665	2·8824	3·1090	4·1249	5·6558
0·0000625	0·235	0·242	0·246	0·249	0·256	0·262	0·265	0·275	0·287	0·300	0·306	0·312	0·335	0·363
1/ 16000	1·0649	1·1890	1·2543	1·3217	1·4633	1·6140	1·6929	1·9436	2·3122	2·7215	2·9419	3·1731	4·2100	5·7725
0·0000650	0·240	0·247	0·250	0·254	0·261	0·267	0·271	0·280	0·293	0·306	0·312	0·318	0·342	0·370
1/ 15385	1·0860	1·2126	1·2791	1·3479	1·4923	1·6460	1·7264	1·9821	2·3580	2·7754	3·0002	3·2360	4·2934	5·8868
0·0000675	0·245	0·252	0·255	0·259	0·266	0·272	0·276	0·286	0·299	0·312	0·318	0·324	0·348	0·377
1/ 14815	1·1067	1·2357	1·3035	1·3736	1·5208	1·6774	1·7593	2·0199	2·4029	2·8283	3·0574	3·2977	4·3752	5·9990
0·0000700	0·249	0·256	0·260	0·263	0·270	0·277	0·281	0·291	0·304	0·317	0·324	0·330	0·355	0·384
1/ 14286	1·1271	1·2584	1·3275	1·3988	1·5487	1·7082	1·7916	2·0570	2·4470	2·8802	3·1135	3·3582	4·4555	6·1091
0·0000725	0·254	0·261	0·265	0·268	0·275	0·282	0·286	0·296	0·310	0·323	0·329	0·336	0·361	0·391
1/ 13793	1·1470	1·2807	1·3510	1·4236	1·5761	1·7384	1·8233	2·0934	2·4904	2·9312	3·1686	3·4177	4·5344	6·2173
0·0000750	0·258	0·265	0·269	0·273	0·280	0·287	0·291	0·301	0·315	0·328	0·335	0·342	0·367	0·398
1/ 13333	1·1667	1·3026	1·3741	1·4480	1·6031	1·7682	1·8545	2·1292	2·5330	2·9814	3·2228	3·4761	4·6120	6·3237
0·0000800	0·266	0·274	0·278	0·282	0·289	0·297	0·300	0·311	0·325	0·339	0·346	0·353	0·379	0·411
1/ 12500	1·2049	1·3453	1·4192	1·4955	1·6557	1·8262	1·9154	2·1991	2·6161	3·0792	3·3286	3·5902	4·7633	6·5311
0·0000850	0·275	0·283	0·286	0·290	0·298	0·306	0·309	0·321	0·335	0·350	0·357	0·364	0·391	0·423
1/ 11765	1·2420	1·3867	1·4629	1·5415	1·7067	1·8824	1·9743	2·2668	2·6966	3·1740	3·4311	3·7007	4·9100	6·7322
0·0000900	0·283	0·291	0·295	0·299	0·307	0·315	0·318	0·330	0·345	0·360	0·367	0·374	0·402	0·436
1/ 11111	1·2781	1·4270	1·5053	1·5862	1·7562	1·9370	2·0316	2·3325	2·7749	3·2660	3·5306	3·8081	5·0524	6·9274
0·0000950	0·290	0·299	0·303	0·307	0·315	0·323	0·327	0·339	0·354	0·370	0·377	0·384	0·413	0·448
1/ 10526	1·3131	1·4661	1·5466	1·6297	1·8043	1·9901	2·0873	2·3965	2·8509	3·3556	3·6274	3·9125	5·1909	7·1173
0·0001000	0·298	0·306	0·311	0·315	0·323	0·332	0·336	0·348	0·364	0·379	0·387	0·394	0·424	0·459
1/ 10000	1·3472	1·5042	1·5868	1·6721	1·8512	2·0418	2·1416	2·4588	2·9250	3·4428	3·7216	4·0141	5·3258	7·3023
	–	–	–	–	–	–	–	–	–	–	–	–	0·74	0·74

$V_{r(0\cdot5)medial}$ for half-full circular pipes.

k_s = 60·0 mm S = 0·00003 to 0·00010

Gradient	**(Equivalent) Pipe diameters in mm**													
	2400	**2500**	**2550**	**2600**	**2700**	**2800**	**2850**	**3000**	**3200**	**3400**	**3500**	**3600**	**4000**	**4500**
0·000100	0·298	0·306	0·311	0·315	0·323	0·332	0·336	0·348	0·364	0·379	0·387	0·394	0·424	0·459
1/ 10000	1·3472	1·5042	1·5868	1·6721	1·8512	2·0418	2·1416	2·4588	2·9250	3·4428	3·7216	4·0141	5·3258	7·3023
0·000105	0·305	0·314	0·318	0·323	0·331	0·340	0·344	0·356	0·373	0·389	0·396	0·404	0·434	0·470
1/ 9524	1·3805	1·5414	1·6260	1·7134	1·8970	2·0923	2·1945	2·5195	2·9973	3·5278	3·8136	4·1133	5·4573	7·4827
0·000110	0·312	0·321	0·326	0·330	0·339	0·348	0·352	0·365	0·381	0·398	0·406	0·414	0·445	0·482
1/ 9091	1·4130	1·5777	1·6642	1·7537	1·9416	2·1416	2·2461	2·5788	3·0678	3·6109	3·9034	4·2101	5·5858	7·6588
0·000115	0·319	0·329	0·333	0·338	0·347	0·356	0·360	0·373	0·390	0·407	0·415	0·423	0·454	0·492
1/ 8696	1·4448	1·6131	1·7017	1·7932	1·9853	2·1897	2·2966	2·6368	3·1368	3·6921	3·9911	4·3048	5·7114	7·8310
0·000120	0·326	0·336	0·340	0·345	0·354	0·363	0·368	0·381	0·398	0·415	0·424	0·432	0·464	0·503
1/ 8333	1·4759	1·6478	1·7383	1·8318	2·0280	2·2368	2·3461	2·6935	3·2043	3·7715	4·0770	4·3974	5·8343	7·9995
0·000125	0·333	0·343	0·347	0·352	0·362	0·371	0·375	0·389	0·407	0·424	0·432	0·441	0·474	0·513
1/ 8000	1·5063	1·6818	1·7741	1·8696	2·0698	2·2830	2·3945	2·7491	3·2704	3·8493	4·1611	4·4881	5·9546	8·1645
0·000130	0·340	0·349	0·354	0·359	0·369	0·378	0·383	0·397	0·415	0·432	0·441	0·450	0·483	0·524
1/ 7692	1·5362	1·7152	1·8093	1·9066	2·1108	2·3282	2·4419	2·8036	3·3352	3·9256	4·2435	4·5770	6·0726	8·3262
0·000135	0·346	0·356	0·361	0·366	0·376	0·385	0·390	0·404	0·423	0·441	0·449	0·458	0·492	0·533
1/ 7407	1·5655	1·7478	1·8438	1·9429	2·1511	2·3726	2·4884	2·8570	3·3988	4·0004	4·3244	4·6643	6·1883	8·4849
0·000140	0·352	0·363	0·368	0·373	0·383	0·392	0·397	0·412	0·430	0·449	0·458	0·467	0·501	0·543
1/ 7143	1·5942	1·7799	1·8776	1·9786	2·1905	2·4161	2·5341	2·9094	3·4612	4·0738	4·4038	4·7499	6·3019	8·6406
0·000145	0·359	0·369	0·374	0·379	0·389	0·399	0·404	0·419	0·438	0·457	0·466	0·475	0·510	0·553
1/ 6897	1·6224	1·8115	1·9109	2·0136	2·2293	2·4589	2·5790	2·9610	3·5224	4·1459	4·4817	4·8340	6·4135	8·7936
0·000150	0·365	0·375	0·381	0·386	0·396	0·406	0·411	0·426	0·445	0·464	0·474	0·483	0·519	0·562
1/ 6667	1·6502	1·8424	1·9436	2·0481	2·2675	2·5010	2·6231	3·0116	3·5827	4·2168	4·5584	4·9166	6·5231	8·9440
0·000160	0·377	0·388	0·393	0·398	0·409	0·419	0·425	0·440	0·460	0·480	0·489	0·499	0·536	0·581
1/ 6250	1·7043	1·9029	2·0073	2·1153	2·3419	2·5830	2·7092	3·1104	3·7002	4·3552	4·7079	5·0779	6·7371	9·2374
0·000170	0·388	0·400	0·405	0·411	0·422	0·432	0·438	0·454	0·474	0·494	0·504	0·514	0·553	0·599
1/ 5882	1·7568	1·9615	2·0691	2·1804	2·4140	2·6625	2·7926	3·2062	3·8141	4·4893	4·8529	5·2343	6·9445	9·5218
0·000180	0·400	0·411	0·417	0·423	0·434	0·445	0·450	0·467	0·488	0·509	0·519	0·529	0·569	0·616
1/ 5556	1·8078	2·0184	2·1291	2·2436	2·4840	2·7397	2·8735	3·2991	3·9248	4·6195	4·9936	5·3861	7·1459	9·7979
0·000190	0·411	0·422	0·428	0·434	0·446	0·457	0·463	0·480	0·501	0·523	0·533	0·544	0·584	0·633
1/ 5263	1·8573	2·0737	2·1875	2·3051	2·5521	2·8148	2·9523	3·3896	4·0323	4·7461	5·1305	5·5337	7·3418	10·066
0·000200	0·421	0·433	0·439	0·445	0·457	0·469	0·475	0·492	0·514	0·536	0·547	0·558	0·599	0·649
1/ 5000	1·9056	2·1276	2·2443	2·3650	2·6184	2·8880	3·0290	3·4777	4·1371	4·8694	5·2638	5·6775	7·5326	10·328
0·000210	0·432	0·444	0·450	0·456	0·469	0·481	0·487	0·504	0·527	0·550	0·561	0·572	0·614	0·665
1/ 4762	1·9527	2·1801	2·2998	2·4235	2·6831	2·9593	3·1039	3·5636	4·2393	4·9897	5·3938	5·8177	7·7186	10·583
0·000220	0·442	0·455	0·461	0·467	0·480	0·492	0·498	0·516	0·540	0·563	0·574	0·585	0·629	0·681
1/ 4545	1·9986	2·2315	2·3539	2·4805	2·7462	3·0290	3·1769	3·6474	4·3391	5·1072	5·5208	5·9547	7·9003	10·832
0·000230	0·452	0·465	0·471	0·478	0·490	0·503	0·509	0·528	0·552	0·575	0·587	0·598	0·643	0·696
1/ 4348	2·0436	2·2816	2·4068	2·5363	2·8080	3·0971	3·2484	3·7294	4·4367	5·2220	5·6449	6·0886	8·0779	11·076
0·000240	0·461	0·475	0·481	0·488	0·501	0·514	0·520	0·539	0·564	0·588	0·599	0·611	0·657	0·711
1/ 4167	2·0875	2·3307	2·4586	2·5908	2·8684	3·1637	3·3182	3·8097	4·5321	5·3343	5·7663	6·2195	8·2517	11·314
0·000250	0·471	0·485	0·491	0·498	0·511	0·524	0·531	0·550	0·575	0·600	0·612	0·624	0·670	0·726
1/ 4000	2·1306	2·3788	2·5093	2·6443	2·9275	3·2290	3·3867	3·8883	4·6256	5·4443	5·8853	6·3478	8·4219	11·547
0·000260	0·480	0·494	0·501	0·508	0·521	0·535	0·541	0·561	0·587	0·612	0·624	0·636	0·683	0·740
1/ 3846	2·1728	2·4259	2·5591	2·6967	2·9855	3·2930	3·4538	3·9653	4·7172	5·5522	6·0019	6·4736	8·5887	11·776
0·000270	0·489	0·504	0·511	0·518	0·531	0·545	0·552	0·572	0·598	0·623	0·636	0·648	0·696	0·755
1/ 3704	2·2142	2·4721	2·6078	2·7481	3·0424	3·3557	3·5196	4·0408	4·8071	5·6580	6·1162	6·5969	8·7524	12·000
0·000280	0·498	0·513	0·520	0·527	0·541	0·555	0·562	0·582	0·609	0·635	0·647	0·660	0·709	0·768
1/ 3571	2·2548	2·5175	2·6557	2·7985	3·0983	3·4173	3·5842	4·1150	4·8953	5·7618	6·2285	6·7180	8·9130	12·221
0·000290	0·507	0·522	0·529	0·536	0·551	0·565	0·572	0·592	0·619	0·646	0·659	0·672	0·722	0·782
1/ 3448	2·2948	2·5621	2·7027	2·8480	3·1531	3·4778	3·6476	4·1879	4·9820	5·8638	6·3388	6·8369	9·0708	12·437
0·000300	0·516	0·531	0·538	0·546	0·560	0·574	0·582	0·603	0·630	0·657	0·670	0·683	0·734	0·795
1/ 3333	2·3340	2·6059	2·7489	2·8967	3·2070	3·5373	3·7100	4·2595	5·0672	5·9641	6·4471	6·9538	9·2259	12·650
	–	–	–	–	–	–	–	–	–	–	–	–	0·74	0·74

$V_{r(0\cdot5)medial}$ for half-full circular pipes.

k_s = **60·0 mm**
S = 0·00030 to 0·00100

ie hydraulic gradient =
1 in 3333 to 1 in 1000

Water (or sewage) at 15°C;
full bore conditions.

velocities in ms^{-1}
discharges in m^3s^{-1}

Gradient **(Equivalent) Pipe diameters in mm**

Gradient	2400	2500	2550	2600	2700	2800	2850	3000	3200	3400	3500	3600	4000	4500
0·000300	0·516	0·531	0·538	0·546	0·560	0·574	0·582	0·603	0·630	0·657	0·670	0·683	0·734	0·795
1/ 3333	2·3340	2·6059	2·7489	2·8967	3·2070	3·5373	3·7100	4·2595	5·0672	5·9641	6·4471	6·9538	9·2259	12·650
0·000320	0·533	0·548	0·556	0·563	0·579	0·593	0·601	0·622	0·651	0·678	0·692	0·706	0·758	0·821
1/ 3125	2·4106	2·6914	2·8391	2·9918	3·3122	3·6533	3·8317	4·3992	5·2334	6·1597	6·6586	7·1819	9·5285	13·065
0·000340	0·549	0·565	0·573	0·581	0·596	0·612	0·619	0·642	0·671	0·699	0·713	0·727	0·782	0·847
1/ 2941	2·4848	2·7742	2·9265	3·0839	3·4142	3·7658	3·9497	4·5346	5·3945	6·3494	6·8636	7·4030	9·8218	13·467
0·000360	0·565	0·582	0·590	0·598	0·614	0·629	0·637	0·660	0·690	0·720	0·734	0·748	0·804	0·871
1/ 2778	2·5568	2·8547	3·0114	3·1733	3·5132	3·8750	4·0642	4·6661	5·5509	6·5335	7·0626	7·6177	10·107	13·857
0·000380	0·581	0·597	0·606	0·614	0·630	0·647	0·655	0·678	0·709	0·739	0·754	0·769	0·826	0·895
1/ 2632	2·6269	2·9329	3·0939	3·2603	3·6095	3·9812	4·1756	4·7940	5·7031	6·7125	7·2562	7·8265	10·384	14·237
0·000400	0·596	0·613	0·622	0·630	0·647	0·663	0·672	0·696	0·728	0·759	0·774	0·789	0·848	0·918
1/ 2500	2·6952	3·0092	3·1743	3·3450	3·7033	4·0846	4·2841	4·9186	5·8513	6·8870	7·4447	8·0298	10·653	14·607
0·000420	0·610	0·628	0·637	0·646	0·663	0·680	0·688	0·713	0·746	0·777	0·793	0·808	0·869	0·941
1/ 2381	2·7618	3·0835	3·2527	3·4276	3·7948	4·1855	4·3899	5·0401	5·9958	7·0571	7·6286	8·2282	10·917	14·968
0·000440	0·625	0·643	0·652	0·661	0·678	0·696	0·704	0·730	0·763	0·796	0·812	0·827	0·889	0·963
1/ 2273	2·8268	3·1561	3·3293	3·5083	3·8841	4·2841	4·4933	5·1587	6·1369	7·2232	7·8082	8·4218	11·174	15·320
0·000460	0·639	0·657	0·667	0·676	0·694	0·711	0·720	0·746	0·780	0·813	0·830	0·846	0·909	0·985
1/ 2174	2·8903	3·2270	3·4041	3·5872	3·9714	4·3804	4·5943	5·2747	6·2749	7·3855	7·9837	8·6112	11·425	15·664
0·000480	0·653	0·672	0·681	0·690	0·709	0·727	0·736	0·762	0·797	0·831	0·848	0·864	0·929	1·006
1/ 2083	2·9525	3·2964	3·4774	3·6643	4·0569	4·4746	4·6931	5·3882	6·4099	7·5444	8·1555	8·7964	11·670	16·001
0·000500	0·666	0·685	0·695	0·704	0·723	0·742	0·751	0·778	0·813	0·848	0·865	0·882	0·948	1·027
1/ 2000	3·0134	3·3644	3·5491	3·7399	4·1405	4·5669	4·7899	5·4993	6·5421	7·7000	8·3237	8·9778	11·911	16·331
0·000525	0·683	0·702	0·712	0·722	0·741	0·760	0·769	0·797	0·834	0·869	0·887	0·904	0·971	1·052
1/ 1905	3·0878	3·4475	3·6367	3·8323	4·2428	4·6797	4·9082	5·6351	6·7037	7·8902	8·5292	9·1996	12·205	16·735
0·000550	0·699	0·719	0·729	0·739	0·758	0·778	0·787	0·816	0·853	0·889	0·907	0·925	0·994	1·077
1/ 1818	3·1605	3·5287	3·7223	3·9225	4·3427	4·7898	5·0237	5·7678	6·8614	8·0759	8·7300	9·4161	12·493	17·129
0·000575	0·714	0·735	0·745	0·755	0·776	0·795	0·805	0·834	0·872	0·909	0·928	0·946	1·016	1·101
1/ 1739	3·2315	3·6080	3·8060	4·0107	4·4403	4·8975	5·1367	5·8974	7·0157	8·2575	8·9262	9·6278	12·773	17·514
0·000600	0·730	0·751	0·761	0·772	0·792	0·812	0·823	0·852	0·891	0·929	0·948	0·966	1·038	1·125
1/ 1667	3·3011	3·6856	3·8879	4·0970	4·5358	5·0029	5·2472	6·0243	7·1666	8·4351	9·1182	9·8349	13·048	17·890
0·000625	0·745	0·766	0·777	0·788	0·809	0·829	0·839	0·870	0·909	0·948	0·967	0·986	1·060	1·148
1/ 1600	3·3691	3·7616	3·9681	4·1815	4·6294	5·1060	5·3554	6·1485	7·3144	8·6091	9·3063	10·038	13·317	18·259
0·000650	0·759	0·781	0·792	0·803	0·825	0·846	0·856	0·887	0·927	0·967	0·986	1·006	1·081	1·171
1/ 1538	3·4359	3·8361	4·0467	4·2643	4·7211	5·2072	5·4615	6·2703	7·4593	8·7796	9·4906	10·237	13·581	18·621
0·000675	0·774	0·796	0·807	0·818	0·840	0·862	0·872	0·904	0·945	0·985	1·005	1·025	1·101	1·193
1/ 1481	3·5014	3·9092	4·1238	4·3455	4·8110	5·3064	5·5655	6·3898	7·6014	8·9468	9·6715	10·432	13·840	18·976
0·000700	0·788	0·811	0·822	0·833	0·856	0·878	0·888	0·921	0·963	1·004	1·024	1·044	1·122	1·215
1/ 1429	3·5656	3·9810	4·1995	4·4253	4·8993	5·4038	5·6677	6·5070	7·7409	9·1110	9·8490	10·623	14·094	19·324
0·000725	0·802	0·825	0·837	0·848	0·871	0·893	0·904	0·937	0·980	1·021	1·042	1·062	1·141	1·237
1/ 1379	3·6287	4·0515	4·2738	4·5036	4·9860	5·4994	5·7680	6·6222	7·8780	9·2723	10·023	10·811	14·343	19·666
0·000750	0·816	0·839	0·851	0·863	0·886	0·908	0·920	0·953	0·996	1·039	1·060	1·080	1·161	1·258
1/ 1333	3·6908	4·1207	4·3469	4·5806	5·0713	5·5935	5·8666	6·7355	8·0126	9·4309	10·195	10·996	14·589	20·002
0·000800	0·843	0·867	0·879	0·891	0·915	0·938	0·950	0·984	1·029	1·073	1·094	1·116	1·199	1·299
1/ 1250	3·8118	4·2559	4·4895	4·7309	5·2376	5·7769	6·0591	6·9564	8·2755	9·7402	10·529	11·357	15·067	20·658
0·000850	0·869	0·894	0·906	0·918	0·943	0·967	0·979	1·014	1·061	1·106	1·128	1·150	1·236	1·339
1/ 1176	3·9292	4·3869	4·6277	4·8765	5·3989	5·9548	6·2456	7·1705	8·5302	10·040	10·853	11·706	15·531	21·294
0·000900	0·894	0·920	0·932	0·945	0·970	0·995	1·007	1·044	1·091	1·138	1·161	1·183	1·272	1·378
1/ 1111	4·0431	4·5141	4·7619	5·0179	5·5554	6·1274	6·4267	7·3784	8·7775	10·331	11·168	12·046	15·981	21·912
0·000950	0·918	0·945	0·958	0·971	0·997	1·022	1·035	1·072	1·121	1·169	1·193	1·216	1·307	1·415
1/ 1053	4·1539	4·6378	4·8924	5·1554	5·7077	6·2954	6·6028	7·5807	9·0181	10·614	11·474	12·376	16·419	22·512
0·001000	0·942	0·969	0·983	0·996	1·023	1·049	1·062	1·100	1·150	1·199	1·224	1·247	1·341	1·452
1/ 1000	4·2619	4·7583	5·0195	5·2894	5·8560	6·4589	6·7743	7·7776	9·2524	10·890	11·772	12·697	16·846	23·097
	–	–	–	–	–	–	–	–	–	–	–	–	0·74	0·74

$V_{r(0·5)medial}$ for half-full circular pipes.

k_s = 60·0 mm S = 0·00030 to 0·00100

Gradient	**(Equivalent) Pipe diameters in mm**													
	2400	**2500**	**2550**	**2600**	**2700**	**2800**	**2850**	**3000**	**3200**	**3400**	**3500**	**3600**	**4000**	**4500**
0·00100	0·942	0·969	0·983	0·996	1·023	1·049	1·062	1·100	1·150	1·199	1·224	1·247	1·341	1·452
1/ 1000	4·2619	4·7583	5·0195	5·2894	5·8560	6·4589	6·7743	7·7776	9·2524	10·890	11·772	12·697	16·846	23·097
0·00105	0·965	0·993	1·007	1·021	1·048	1·075	1·088	1·127	1·179	1·229	1·254	1·278	1·374	1·488
1/ 952	4·3671	4·8759	5·1435	5·4200	6·0006	6·6185	6·9417	7·9697	9·4809	11·159	12·063	13·011	17·262	23·668
0·00110	0·988	1·017	1·031	1·045	1·073	1·100	1·114	1·154	1·207	1·258	1·283	1·308	1·406	1·523
1/ 909	4·4699	4·9906	5·2645	5·5476	6·1418	6·7742	7·1050	8·1573	9·7041	11·422	12·347	13·317	17·668	24·225
0·00115	1·010	1·040	1·054	1·068	1·097	1·125	1·139	1·180	1·234	1·286	1·312	1·338	1·438	1·557
1/ 870	4·5704	5·1028	5·3828	5·6723	6·2799	6·9265	7·2647	8·3406	9·9222	11·678	12·624	13·616	18·065	24·769
0·00120	1·032	1·062	1·077	1·091	1·120	1·149	1·163	1·205	1·260	1·314	1·340	1·366	1·469	1·591
1/ 833	4·6687	5·2126	5·4986	5·7943	6·4150	7·0755	7·4210	8·5201	10·136	11·930	12·896	13·909	18·454	25·302
0·00125	1·053	1·084	1·099	1·114	1·144	1·173	1·187	1·230	1·286	1·341	1·368	1·395	1·499	1·624
1/ 800	4·7650	5·3201	5·6120	5·9138	6·5473	7·2214	7·5741	8·6958	10·345	12·176	13·162	14·196	18·834	25·824
0·00130	1·074	1·105	1·121	1·136	1·166	1·196	1·211	1·255	1·312	1·368	1·395	1·422	1·528	1·656
1/ 769	4·8594	5·4254	5·7232	6·0309	6·6769	7·3644	7·7241	8·8680	10·550	12·417	13·422	14·477	19·207	26·335
0·00135	1·095	1·126	1·142	1·158	1·188	1·219	1·234	1·278	1·337	1·394	1·422	1·449	1·558	1·687
1/ 741	4·9519	5·5288	5·8322	6·1458	6·8042	7·5048	7·8712	9·0370	10·751	12·653	13·678	14·753	19·573	26·837
0·00140	1·115	1·147	1·163	1·179	1·210	1·241	1·256	1·302	1·361	1·419	1·448	1·476	1·586	1·718
1/ 714	5·0428	5·6303	5·9393	6·2586	6·9290	7·6425	8·0157	9·2028	10·948	12·886	13·929	15·024	19·932	27·329
0·00145	1·134	1·167	1·184	1·200	1·232	1·263	1·279	1·325	1·385	1·444	1·473	1·502	1·614	1·749
1/ 690	5·1321	5·7299	6·0444	6·3694	7·0517	7·7778	8·1576	9·3657	11·142	13·114	14·176	15·290	20·285	27·813
0·00150	1·154	1·187	1·204	1·220	1·253	1·285	1·301	1·348	1·409	1·469	1·499	1·528	1·642	1·779
1/ 667	5·2198	5·8279	6·1478	6·4783	7·1723	7·9108	8·2971	9·5258	11·332	13·338	14·418	15·551	20·632	28·289
0·00160	1·192	1·226	1·243	1·260	1·294	1·327	1·343	1·392	1·455	1·517	1·548	1·578	1·696	1·837
1/ 625	5·3911	6·0191	6·3494	6·6908	7·4075	8·1702	8·5692	9·8383	11·704	13·775	14·891	16·061	21·309	29·217
0·00170	1·228	1·264	1·282	1·299	1·334	1·368	1·385	1·435	1·500	1·564	1·595	1·626	1·748	1·894
1/ 588	5·5570	6·2043	6·5448	6·8968	7·6355	8·4217	8·8330	10·141	12·064	14·199	15·349	16·556	21·965	30·116
0·00180	1·264	1·301	1·319	1·337	1·372	1·407	1·425	1·476	1·544	1·609	1·642	1·674	1·799	1·948
1/ 556	5·7181	6·3842	6·7346	7·0967	7·8569	8·6659	9·0891	10·435	12·414	14·611	15·794	17·036	22·602	30·989
0·00190	1·299	1·336	1·355	1·373	1·410	1·446	1·464	1·517	1·586	1·653	1·687	1·720	1·848	2·002
1/ 526	5·8748	6·5592	6·9192	7·2912	8·0722	8·9034	9·3382	10·721	12·754	15·011	16·227	17·502	23·221	31·838
0·00200	1·332	1·371	1·390	1·409	1·446	1·484	1·502	1·556	1·627	1·696	1·730	1·764	1·896	2·054
1/ 500	6·0275	6·7296	7·0989	7·4807	8·2819	9·1347	9·5808	11·000	13·085	15·401	16·649	17·957	23·824	32·665
0·00210	1·365	1·405	1·424	1·444	1·482	1·520	1·539	1·595	1·667	1·738	1·773	1·808	1·943	2·105
1/ 476	6·1763	6·8958	7·2743	7·6654	8·4865	9·3603	9·8174	11·271	13·409	15·782	17·060	18·401	24·413	33·472
0·00220	1·397	1·438	1·458	1·478	1·517	1·556	1·575	1·632	1·706	1·779	1·815	1·850	1·988	2·154
1/ 455	6·3217	7·0581	7·4455	7·8458	8·6862	9·5806	10·048	11·537	13·724	16·153	17·461	18·834	24·987	34·260
0·00230	1·429	1·470	1·491	1·511	1·551	1·591	1·611	1·669	1·745	1·819	1·856	1·892	2·033	2·203
1/ 435	6·4638	7·2168	7·6128	8·0222	8·8815	9·7959	10·274	11·796	14·033	16·516	17·854	19·257	25·549	35·030
0·00240	1·460	1·502	1·523	1·543	1·585	1·625	1·645	1·705	1·782	1·858	1·896	1·933	2·077	2·250
1/ 417	6·6028	7·3720	7·7766	8·1947	9·0725	10·007	10·495	12·050	14·334	16·872	18·238	19·671	26·098	35·783
0·00250	1·490	1·533	1·554	1·575	1·617	1·659	1·679	1·740	1·819	1·897	1·935	1·972	2·120	2·296
1/ 400	6·7390	7·5240	7·9369	8·3637	9·2596	10·213	10·712	12·298	14·630	17·220	18·614	20·077	26·637	36·521
0·00260	1·519	1·563	1·585	1·607	1·649	1·691	1·712	1·774	1·855	1·934	1·973	2·012	2·162	2·342
1/ 385	6·8725	7·6730	8·0941	8·5294	9·4430	10·415	10·924	12·542	14·920	17·561	18·983	20·475	27·164	37·245
0·00270	1·548	1·593	1·615	1·637	1·681	1·724	1·745	1·808	1·890	1·971	2·011	2·050	2·203	2·386
1/ 370	7·0034	7·8192	8·2483	8·6919	9·6229	10·614	11·132	12·781	15·204	17·895	19·344	20·865	27·682	37·954
0·00280	1·577	1·622	1·645	1·667	1·712	1·755	1·777	1·841	1·925	2·007	2·048	2·087	2·243	2·430
1/ 357	7·1319	7·9627	8·3997	8·8514	9·7995	10·808	11·336	13·015	15·483	18·223	19·699	21·248	28·190	38·651
0·00290	1·604	1·651	1·674	1·697	1·742	1·786	1·808	1·874	1·959	2·043	2·084	2·124	2·283	2·473
1/ 345	7·2582	8·1037	8·5484	9·0081	9·9730	11·000	11·537	13·246	15·757	18·546	20·048	21·624	28·689	39·335
0·00300	1·632	1·679	1·702	1·726	1·772	1·817	1·839	1·906	1·993	2·078	2·119	2·161	2·322	2·515
1/ 333	7·3823	8·2422	8·6946	9·1621	10·143	11·188	11·734	13·472	16·027	18·863	20·391	21·993	29·179	40·007
	–	–	–	–	–	–	–	–	–	–	–	–	0·74	0·74

$V_{r(0·5)medial}$ for half-full circular pipes.

A58

(p.5 of 6)

k_s = 60·0 mm
S = 0·00300 to 0·01000

ie hydraulic gradient =
1 in 333 to 1 in 100

Water (or sewage) at 15°C;
full bore conditions.

velocities in ms^{-1}
discharges in m^3s^{-1}

Gradient **(Equivalent) Pipe diameters in mm**

	2400	**2500**	**2550**	**2600**	**2700**	**2800**	**2850**	**3000**	**3200**	**3400**	**3500**	**3600**	**4000**	**4500**
0·00300	1·632	1·679	1·702	1·726	1·772	1·817	1·839	1·906	1·993	2·078	2·119	2·161	2·322	2·515
1/ 333	7·3823	8·2422	8·6946	9·1621	10·143	11·188	11·734	13·472	16·027	18·863	20·391	21·993	29·179	40·007
0·00320	1·685	1·734	1·758	1·782	1·830	1·877	1·900	1·968	2·058	2·146	2·189	2·232	2·398	2·598
1/ 313	7·6244	8·5126	8·9797	9·4626	10·476	11·555	12·119	13·914	16·552	19·482	21·060	22·715	30·136	41·319
0·00340	1·737	1·788	1·812	1·837	1·886	1·934	1·958	2·029	2·121	2·212	2·256	2·300	2·472	2·678
1/ 294	7·8591	8·7746	9·2561	9·7538	10·799	11·910	12·492	14·342	17·062	20·081	21·708	23·414	31·064	42·591
0·00360	1·788	1·839	1·865	1·890	1·941	1·990	2·015	2·088	2·183	2·276	2·322	2·367	2·544	2·756
1/ 278	8·0869	9·0290	9·5245	10·037	11·112	12·256	12·854	14·758	17·556	20·664	22·337	24·093	31·964	43·826
0·00380	1·837	1·890	1·916	1·942	1·994	2·045	2·070	2·145	2·243	2·338	2·385	2·432	2·613	2·831
1/ 263	8·3086	9·2764	9·7855	10·312	11·416	12·592	13·207	15·162	18·037	21·230	22·949	24·753	32·840	45·027
0·00400	1·884	1·939	1·966	1·993	2·046	2·098	2·124	2·201	2·301	2·399	2·447	2·495	2·681	2·905
1/ 250	8·5244	9·5174	10·040	10·580	11·713	12·919	13·550	15·556	18·506	21·782	23·546	25·396	33·693	46·197
0·00420	1·931	1·987	2·014	2·042	2·096	2·150	2·176	2·255	2·358	2·458	2·508	2·557	2·747	2·976
1/ 238	8·7349	9·7525	10·288	10·841	12·002	13·238	13·884	15·941	18·963	22·319	24·127	26·023	34·525	47·338
0·00440	1·976	2·034	2·062	2·090	2·146	2·200	2·228	2·308	2·413	2·516	2·567	2·617	2·812	3·046
1/ 227	8·9405	9·9820	10·530	11·096	12·285	13·549	14·211	16·316	19·409	22·845	24·695	26·636	35·338	48·452
0·00460	2·021	2·079	2·108	2·137	2·194	2·250	2·278	2·360	2·468	2·573	2·624	2·676	2·875	3·115
1/ 217	9·1415	10·206	10·766	11·345	12·561	13·854	14·530	16·682	19·846	23·358	25·250	27·234	36·132	49·541
0·00480	2·064	2·124	2·154	2·183	2·241	2·298	2·327	2·411	2·521	2·628	2·681	2·733	2·937	3·182
1/ 208	9·3381	10·426	10·998	11·589	12·831	14·152	14·843	17·041	20·272	23·861	25·793	27·820	36·909	50·606
0·00500	2·107	2·168	2·198	2·228	2·287	2·346	2·375	2·461	2·573	2·682	2·736	2·790	2·998	3·248
1/ 200	9·5307	10·641	11·225	11·828	13·095	14·444	15·149	17·393	20·691	24·353	26·325	28·394	37·671	51·650
0·00525	2·159	2·221	2·252	2·283	2·344	2·404	2·433	2·521	2·636	2·748	2·804	2·858	3·072	3·328
1/ 190	9·7660	10·904	11·502	12·121	13·419	14·800	15·523	17·822	21·202	24·954	26·975	29·095	38·601	52·925
0·00550	2·210	2·274	2·305	2·337	2·399	2·460	2·491	2·581	2·698	2·813	2·870	2·926	3·144	3·406
1/ 182	9·9959	11·160	11·773	12·406	13·735	15·149	15·889	18·242	21·700	25·541	27·610	29·780	39·509	54·171
0·00575	2·259	2·325	2·357	2·389	2·453	2·516	2·547	2·639	2·759	2·876	2·934	2·991	3·215	3·483
1/ 174	10·221	11·411	12·037	12·685	14·043	15·489	16·246	18·652	22·188	26·115	28·230	30·449	40·397	55·389
0·00600	2·308	2·375	2·408	2·441	2·506	2·570	2·601	2·695	2·818	2·938	2·997	3·056	3·284	3·558
1/ 167	10·440	11·657	12·296	12·957	14·345	15·822	16·595	19·053	22·665	26·677	28·838	31·104	41·266	56·580
0·00625	2·355	2·424	2·457	2·491	2·557	2·623	2·655	2·751	2·876	2·999	3·059	3·119	3·352	3·631
1/ 160	10·656	11·897	12·550	13·225	14·641	16·149	16·937	19·446	23·133	27·227	29·432	31·745	42·117	57·747
0·00650	2·402	2·472	2·506	2·540	2·608	2·675	2·708	2·805	2·933	3·058	3·120	3·181	3·418	3·703
1/ 154	10·867	12·133	12·798	13·487	14·931	16·469	17·273	19·831	23·591	27·766	30·015	32·374	42·951	58·890
0·00675	2·448	2·519	2·554	2·589	2·657	2·725	2·759	2·859	2·989	3·117	3·179	3·241	3·483	3·773
1/ 148	11·074	12·364	13·042	13·744	15·216	16·782	17·602	20·209	24·040	28·295	30·587	32·991	43·770	60·012
0·00700	2·493	2·565	2·601	2·636	2·706	2·776	2·810	2·911	3·044	3·174	3·237	3·301	3·547	3·843
1/ 143	11·277	12·591	13·282	13·996	15·495	17·090	17·925	20·579	24·482	28·815	31·148	33·596	44·573	61·113
0·00725	2·537	2·610	2·647	2·683	2·754	2·825	2·860	2·963	3·098	3·230	3·295	3·359	3·610	3·911
1/ 138	11·477	12·813	13·517	14·244	15·769	17·393	18·242	20·944	24·915	29·325	31·700	34·191	45·362	62·195
0·00750	2·580	2·655	2·692	2·729	2·801	2·873	2·908	3·014	3·151	3·285	3·351	3·416	3·671	3·977
1/ 133	11·673	13·033	13·748	14·487	16·039	17·690	18·554	21·302	25·341	29·826	32·242	34·775	46·137	63·258
0·00800	2·665	2·742	2·780	2·818	2·893	2·967	3·004	3·112	3·254	3·393	3·461	3·529	3·792	4·108
1/ 125	12·056	13·460	14·199	14·962	16·565	18·270	19·163	22·000	26·172	30·804	33·299	35·916	47·650	65·333
0·00850	2·747	2·826	2·866	2·905	2·982	3·058	3·096	3·208	3·354	3·497	3·568	3·637	3·909	4·234
1/ 118	12·427	13·874	14·636	15·423	17·075	18·833	19·752	22·678	26·978	31·752	34·324	37·021	49·117	67·344
0·00900	2·827	2·908	2·949	2·989	3·069	3·147	3·186	3·301	3·452	3·599	3·671	3·743	4·022	4·357
1/ 111	12·787	14·276	15·060	15·870	17·570	19·379	20·325	23·335	27·760	32·673	35·319	38·095	50·541	69·296
0·00950	2·904	2·988	3·030	3·071	3·153	3·233	3·273	3·392	3·546	3·697	3·772	3·845	4·132	4·476
1/ 105	13·137	14·668	15·473	16·305	18·051	19·910	20·882	23·974	28·520	33·568	36·287	39·139	51·926	71·195
0·01000	2·979	3·066	3·108	3·151	3·235	3·317	3·358	3·480	3·638	3·793	3·870	3·945	4·239	4·593
1/ 100	13·479	15·049	15·875	16·728	18·520	20·427	21·424	24·597	29·261	34·440	37·230	40·155	53·275	73·045
	–	–	–	–	–	–	–	–	–	–	–	–	0·74	0·74

$V_{r(0·5)medial}$ for half-full circular pipes.

k_s = 60·0 mm S = 0·00300 to 0·01000

k_s = 60·0 mm
S = 0·01000 to 0·03000

ie hydraulic gradient =
1 in 100 to 1 in 33·3

Water (or sewage) at 15°C;
full bore conditions.

velocities in ms^{-1}
discharges in m^3s^{-1}

Gradient **(Equivalent) Pipe diameters in mm**

Gradient	**2400**	**2500**	**2550**	**2600**	**2700**	**2800**	**2850**	**3000**	**3200**	**3400**	**3500**	**3600**	**4000**	**4500**
0·01000	2·979	3·066	3·108	3·151	3·235	3·317	3·358	3·480	3·638	3·793	3·870	3·945	4·239	4·593
1/ 100	13·479	15·049	15·875	16·728	18·520	20·427	21·424	24·597	29·261	34·440	37·230	40·155	53·275	73·045
0·01050	3·053	3·141	3·185	3·229	3·315	3·399	3·441	3·566	3·728	3·887	3·965	4·042	4·344	4·706
1/ 95	13·812	15·420	16·267	17·141	18·977	20·931	21·954	25·205	29·984	35·291	38·149	41·147	54·591	74·849
0·01100	3·125	3·215	3·260	3·305	3·393	3·479	3·522	3·650	3·816	3·978	4·058	4·138	4·446	4·817
1/ 91	14·137	15·783	16·650	17·545	19·424	21·424	22·470	25·798	30·690	36·121	39·047	42·115	55·875	76·610
0·01150	3·195	3·288	3·333	3·379	3·469	3·558	3·601	3·732	3·902	4·068	4·150	4·231	4·546	4·925
1/ 87	14·454	16·138	17·024	17·939	19·861	21·906	22·975	26·378	31·379	36·933	39·924	43·062	57·131	78·332
0·01200	3·264	3·358	3·405	3·451	3·543	3·634	3·679	3·812	3·986	4·155	4·239	4·322	4·644	5·031
1/ 83	14·765	16·485	17·390	18·325	20·288	22·377	23·469	26·945	32·054	37·728	40·783	43·988	58·360	80·017
0·01250	3·331	3·428	3·475	3·523	3·616	3·709	3·755	3·891	4·068	4·241	4·326	4·411	4·740	5·135
1/ 80	15·070	16·825	17·749	18·703	20·706	22·838	23·953	27·501	32·715	38·506	41·624	44·895	59·563	81·667
0·01300	3·397	3·495	3·544	3·592	3·688	3·782	3·829	3·968	4·148	4·325	4·412	4·498	4·834	5·237
1/ 77	15·368	17·158	18·100	19·073	21·116	23·290	24·428	28·045	33·363	39·268	42·449	45·785	60·743	83·284
0·01350	3·462	3·562	3·612	3·661	3·758	3·854	3·902	4·043	4·227	4·407	4·496	4·584	4·926	5·336
1/ 74	15·661	17·485	18·445	19·437	21·519	23·734	24·893	28·580	33·999	40·016	43·257	46·657	61·900	84·871
0·01400	3·525	3·627	3·678	3·728	3·827	3·925	3·974	4·117	4·305	4·488	4·579	4·668	5·016	5·434
1/ 71	15·948	17·806	18·783	19·793	21·913	24·170	25·350	29·104	34·623	40·751	44·051	47·513	63·036	86·428
0·01450	3·588	3·692	3·743	3·794	3·895	3·995	4·044	4·190	4·381	4·568	4·660	4·750	5·105	5·530
1/ 69	16·231	18·121	19·116	20·144	22·301	24·598	25·799	29·619	35·236	41·472	44·831	48·354	64·152	87·958
0·01500	3·649	3·755	3·807	3·859	3·962	4·063	4·113	4·262	4·456	4·646	4·739	4·832	5·192	5·625
1/ 67	16·508	18·431	19·443	20·488	22·683	25·018	26·240	30·126	35·838	42·181	45·597	49·181	65·249	89·462
0·01600	3·769	3·878	3·932	3·985	4·092	4·196	4·248	4·402	4·602	4·798	4·895	4·990	5·363	5·809
1/ 63	17·050	19·036	20·080	21·160	23·426	25·839	27·100	31·114	37·013	43·564	47·093	50·794	67·389	92·396
0·01700	3·885	3·997	4·053	4·108	4·217	4·325	4·379	4·537	4·744	4·946	5·045	5·144	5·528	5·988
1/ 59	17·574	19·621	20·698	21·811	24·147	26·634	27·934	32·071	38·152	44·905	48·542	52·357	69·463	95·240
0·01800	3·997	4·113	4·170	4·227	4·340	4·451	4·506	4·669	4·881	5·089	5·192	5·293	5·688	6·162
1/ 56	18·084	20·190	21·298	22·444	24·848	27·406	28·744	33·001	39·259	46·207	49·949	53·875	71·477	98·001
0·01900	4·107	4·226	4·285	4·343	4·459	4·573	4·629	4·797	5·015	5·229	5·334	5·438	5·844	6·331
1/ 53	18·579	20·744	21·882	23·059	25·528	28·157	29·532	33·905	40·334	47·473	51·318	55·351	73·435	100·69
0·02000	4·214	4·336	4·396	4·456	4·575	4·692	4·750	4·921	5·145	5·365	5·472	5·579	5·996	6·495
1/ 50	19·062	21·283	22·451	23·658	26·192	28·889	30·299	34·786	41·382	48·707	52·651	56·789	75·343	103·30
0·02100	4·318	4·443	4·505	4·566	4·687	4·807	4·867	5·043	5·273	5·497	5·608	5·717	6·144	6·656
1/ 47·6	19·533	21·808	23·005	24·242	26·839	29·602	31·047	35·645	42·404	49·910	53·952	58·192	77·204	105·85
0·02200	4·419	4·547	4·611	4·673	4·798	4·921	4·981	5·161	5·397	5·626	5·740	5·851	6·288	6·812
1/ 45·5	19·992	22·321	23·546	24·812	27·470	30·299	31·778	36·484	43·402	51·084	55·221	59·561	79·021	108·34
0·02300	4·519	4·649	4·714	4·778	4·906	5·031	5·093	5·277	5·518	5·753	5·869	5·983	6·430	6·965
1/ 43·5	20·442	22·823	24·076	25·370	28·088	30·980	32·492	37·304	44·378	52·232	56·462	60·900	80·797	110·78
0·02400	4·616	4·749	4·816	4·881	5·011	5·139	5·203	5·391	5·637	5·877	5·995	6·112	6·568	7·115
1/ 41·7	20·881	23·314	24·593	25·916	28·692	31·646	33·191	38·107	45·332	53·356	57·677	62·209	82·534	113·16
0·02500	4·711	4·847	4·915	4·982	5·114	5·245	5·310	5·502	5·753	5·998	6·118	6·238	6·703	7·262
1/ 40·0	21·312	23·795	25·101	26·450	29·283	32·298	33·876	38·892	46·267	54·456	58·866	63·492	84·236	115·50
0·02600	4·804	4·943	5·012	5·081	5·216	5·349	5·415	5·611	5·867	6·117	6·240	6·361	6·836	7·406
1/ 38·5	21·734	24·266	25·598	26·974	29·863	32·938	34·547	39·663	47·183	55·534	60·032	64·750	85·905	117·78
0·02700	4·896	5·038	5·108	5·177	5·315	5·451	5·518	5·718	5·979	6·233	6·358	6·482	6·966	7·547
1/ 37·0	22·148	24·728	26·085	27·488	30·432	33·566	35·205	40·418	48·082	56·592	61·176	65·983	87·541	120·03
0·02800	4·986	5·130	5·201	5·272	5·413	5·551	5·620	5·823	6·088	6·348	6·475	6·601	7·094	7·685
1/ 35·7	22·555	25·182	26·564	27·992	30·991	34·182	35·851	41·160	48·964	57·631	62·298	67·194	89·147	122·23
0·02900	5·074	5·221	5·293	5·366	5·508	5·649	5·719	5·926	6·196	6·460	6·590	6·718	7·220	7·821
1/ 34·5	22·954	25·628	27·034	28·488	31·539	34·787	36·485	41·889	49·831	58·651	63·401	68·383	90·725	124·39
0·03000	5·161	5·310	5·384	5·457	5·603	5·746	5·817	6·027	6·302	6·570	6·702	6·833	7·343	7·955
1/ 33·3	23·346	26·066	27·496	28·975	32·078	35·381	37·109	42·605	50·683	59·654	64·485	69·552	92·276	126·52
	–	–	–	–	–	–	–	–	–	–	–	–	0·74	0·74

$V_{r(0·5)medial}$ for half-full circular pipes.

Annexure 1 to Tables A : Multiplying factors on velocity and discharge for variations of temperature from 15°C in turbulent flow (p.1 of 6)

Temperatures in degrees Celsius are shown at the top of the columns of corresponding factors. Where a factor is associated with Reynolds number value less than 2500, the factor value is in italics.

k_s = 0·006mm D = 20mm

Gradient	0	2·5	5	7·5	10	12·5	17·5	20	22·5	25	27·5	30	32·5	35
0·00100	*0·909*	*0·926*	*0·942*	*0·958*	*0·973*	*0·987*	*1·013*	*1·025*	*1·037*	*1·048*	*1·059*	*1·070*	*1·080*	1·090
0·00300	*0·919*	*0·934*	*0·948*	*0·962*	*0·975*	0·988	1·012	1·022	1·033	1·043	1·053	1·063	1·072	1·081
0·01000	0·927	0·941	0·954	0·967	0·978	0·989	1·010	1·020	1·029	1·038	1·047	1·056	1·064	1·071
0·03000	0·936	0·947	0·959	0·970	0·980	0·990	1·009	1·018	1·026	1·034	1·042	1·050	1·057	1·064
0·10000	0·942	0·953	0·963	0·973	0·983	0·992	1·008	1·016	1·023	1·030	1·037	1·043	1·049	1·055
0·30000	0·949	0·959	0·968	0·977	0·985	0·993	1·007	1·014	1·020	1·026	1·032	1·037	1·043	1·048

k_s = 0·006mm D = 30mm

Gradient	0	2·5	5	7·5	10	12·5	17·5	20	22·5	25	27·5	30	32·5	35
0·00030	*9·909*	*0·926*	*0·942*	*0·958*	*0·972*	*0·987*	*1·013*	*1·025*	*1·037*	*1·048*	*1·059*	*1·070*	*1·081*	1·091
0·00100	*0·919*	*0·935*	*0·949*	*0·963*	0·976	0·988	1·011	1·022	1·033	1·043	1·053	1·062	1·071	1·080
0·00300	0·927	0·941	0·954	0·966	0·978	0·989	1·010	1·020	1·030	1·039	1·047	1·056	1·064	1·072
0·01000	0·934	0·947	0·959	0·970	0·980	0·990	1·009	1·018	1·026	1·035	1·042	1·050	1·057	1·064
0·03000	0·941	0·952	0·963	0·973	0·982	0·991	1·008	1·016	1·024	1·031	1·038	1·045	1·051	1·057
0·10000	0·947	0·957	0·967	0·976	0·984	0·992	1·007	1·014	1·021	1·027	1·033	1·039	1·045	1·050
0·30000	0·954	0·963	0·971	0·979	0·986	0·993	1·006	1·012	1·018	1·023	1·028	1·033	1·038	1·042

k_s = 0·006mm D = 50mm

Gradient	0	2·5	5	7·5	10	12·5	17·5	20	22·5	25	27·5	30	32·5	35
0·00010	*0·913*	*0·929*	*0·945*	*0·960*	*0·974*	0·987	1·012	1·024	1·036	1·046	1·057	1·068	1·077	1·087
0·00030	0·921	0·936	0·950	0·964	0·976	0·988	1·011	1·022	1·032	1·042	1·051	1·061	1·070	1·078
0·00100	0·929	0·943	0·955	0·967	0·979	0·990	1·010	1·020	1·029	1·038	1·046	1·055	1·062	1·070
0·00300	0·935	0·948	0·959	0·970	0·981	0·990	1·009	1·018	1·026	1·034	1·042	1·049	1·057	1·064
0·01000	0·942	0·953	0·963	0·973	0·982	0·991	1·008	1·016	1·023	1·031	1·038	1·044	1·051	1·057
0·03000	0·947	0·957	0·967	0·976	0·984	0·992	1·007	1·014	1·021	1·027	1·034	1·040	1·045	1·051
0·10000	0·953	0·962	0·971	0·979	0·986	0·993	1·006	1·012	1·018	1·024	1·029	1·034	1·039	1·044
0·30000	0·959	0·967	0·975	0·982	0·988	0·994	1·005	1·011	1·016	1·020	1·025	1·029	1·033	1·037

k_s = 0·006mm D = 100mm

Gradient	0	2·5	5	7·5	10	12·5	17·5	20	22·5	25	27·5	30	32·5	35
0·00010	0·927	0·941	0·954	0·966	0·978	0·989	1·010	1·020	1·030	1·039	1·047	1·056	1·064	1·072
0·00030	0·934	0·946	0·958	0·969	0·980	0·990	1·009	1·018	1·027	1·036	1·043	1·051	1·059	1·066
0·00100	0·940	0·951	0·962	0·972	0·982	0·991	1·009	1·017	1·025	1·032	1·039	1·046	1·053	1·060
0·00300	0·944	0·955	0·965	0·974	0·983	0·992	1·008	1·015	1·022	1·029	1·036	1·042	1·049	1·054
0·01000	0·950	0·959	0·968	0·977	0·985	0·993	1·007	1·014	1·020	1·026	1·032	1·038	1·043	1·049
0·03000	0·954	0·963	0·971	0·979	0·986	0·993	1·006	1·012	1·018	1·023	1·029	1·034	1·039	1·043
0·10000	0·960	0·968	0·975	0·982	0·988	0·994	1·005	1·010	1·015	1·020	1·024	1·029	1·033	1·036
0·30000	0·966	0·973	0·979	0·985	0·990	0·995	1·005	1·009	1·013	1·016	1·020	1·023	1·027	1·030

k_s = 0·006mm D = 200mm

Gradient	0	2·5	5	7·5	10	12·5	17·5	20	22·5	25	27·5	30	32·5	35
0·00010	0·938	0·950	0·961	0·971	0·981	0·991	1·009	1·017	1·025	1·033	1·040	1·048	1·055	1·062
0·00030	0·943	0·954	0·964	0·974	0·983	0·992	1·008	1·016	1·023	1·030	1·037	1·044	1·050	1·057
0·00100	0·947	0·957	0·967	0·976	0·984	0·992	1·007	1·014	1·021	1·028	1·034	1·040	1·046	1·052
0·00300	0·952	0·961	0·969	0·978	0·985	0·993	1·007	1·013	1·019	1·025	1·031	1·037	1·042	1·047
0·01000	0·956	0·965	0·972	0·980	0·987	0·994	1·006	1·012	1·017	1·023	1·028	1·033	1·037	1·042
0·03000	0·961	0·968	0·975	0·982	0·988	0·994	1·005	1·010	1·015	1·020	1·024	1·029	1·033	1·037
0·10000	0·966	0·973	0·979	0·985	0·990	0·995	1·005	1·009	1·013	1·017	1·020	1·024	1·027	1·030

k_s = 0·006mm D = 500mm

Gradient	0	2·5	5	7·5	10	12·5	17·5	20	22·5	25	27·5	30	32·5	35
0·00010	0·948	0·958	0·967	0·976	0·984	0·992	1·007	1·014	1·021	1·027	1·034	1·040	1·046	1·051
0·00030	0·952	0·961	0·969	0·978	0·985	0·993	1·007	1·013	1·020	1·025	1·031	1·037	1·042	1·048
0·00100	0·955	0·964	0·972	0·980	0·987	0·993	1·006	1·012	1·018	1·023	1·029	1·034	1·039	1·043
0·00300	0·959	0·967	0·974	0·981	0·988	0·994	1·006	1·011	1·016	1·021	1·026	1·031	1·036	1·039
0·01000	0·963	0·970	0·977	0·983	0·989	0·995	1·005	1·010	1·014	1·019	1·023	1·027	1·031	1·034
0·03000	0·968	0·974	0·980	0·985	0·991	0·995	1·004	1·008	1·012	1·016	1·019	1·023	1·026	1·029
0·10000	0·973	0·979	0·983	0·988	0·992	0·996	1·004	1·007	1·010	1·013	1·015	1·018	1·021	1·023

k_s = 0·006mm D = 1000mm

Gradient	0	2·5	5	7·5	10	12·5	17·5	20	22·5	25	27·5	30	32·5	35
0·00010	0·954	0·963	0·971	0·979	0·986	0·993	1·007	1·013	1·019	1·024	1·030	1·035	1·040	1·045
0·00030	0·957	0·965	0·973	0·980	0·987	0·994	1·006	1·012	1·017	1·023	1·028	1·033	1·038	1·042
0·00100	0·960	0·968	0·975	0·982	0·988	0·994	1·006	1·011	1·016	1·021	1·025	1·030	1·034	1·038
0·00300	0·964	0·971	0·977	0·983	0·989	0·995	1·005	1·010	1·014	1·019	1·023	1·027	1·031	1·034
0·01000	0·968	0·974	0·980	0·985	0·991	0·995	1·004	1·008	1·012	1·016	1·020	1·023	1·026	1·029
0·03000	0·972	0·978	0·983	0·988	0·992	0·996	1·004	1·007	1·010	1·013	1·016	1·019	1·022	1·024
0·10000	0·978	0·982	0·986	0·990	0·994	0·997	1·003	1·005	1·008	1·010	1·012	1·014	1·016	1·018

k_s = 0·006mm D = 2000mm

Gradient	0	2·5	5	7·5	10	12·5	17·5	20	22·5	25	27·5	30	32·5	35
0·00010	0·959	0·967	0·974	0·981	0·988	0·994	1·006	1·011	1·017	1·022	1·027	1·032	1·036	1·041
0·00030	0·962	0·969	0·976	0·982	0·988	0·994	1·005	1·010	1·015	1·020	1·025	1·029	1·033	1·037
0·00100	0·965	0·971	0·978	0·984	0·989	0·995	1·005	1·010	1·014	1·018	1·022	1·026	1·030	1·034
0·00300	0·968	0·974	0·980	0·985	0·991	0·995	1·004	1·008	1·012	1·016	1·020	1·023	1·027	1·030
0·01000	0·972	0·978	0·983	0·988	0·992	0·996	1·004	1·007	1·010	1·014	1·017	1·019	1·022	1·025
0·03000	0·977	0·981	0·986	0·990	0·993	0·997	1·003	1·006	1·008	1·011	1·013	1·015	1·018	1·020

k_s = 0·015mm D = 20mm

Gradient	0	2·5	5	7·5	10	12·5	17·5	20	22·5	25	27·5	30	32·5	35
0·00100	*0·910*	*0·927*	*0·943*	*0·958*	*0·973*	*0·987*	*1·013*	*1·025*	*1·036*	*1·048*	*1·058*	*1·069*	*1·079*	1·089
0·00300	*0·920*	*0·935*	*0·949*	*0·963*	*0·976*	0·988	1·011	1·022	1·032	1·042	1·052	1·061	1·070	1·078
0·01000	0·930	0·943	0·956	0·968	0·979	0·990	1·010	1·019	1·028	1·036	1·045	1·053	1·060	1·068
0·03000	0·938	0·950	0·961	0·972	0·982	0·991	1·009	1·017	1·024	1·032	1·039	1·046	1·052	1·058
0·10000	0·947	0·958	0·967	0·976	0·984	0·992	1·007	1·014	1·020	1·026	1·032	1·038	1·043	1·048
0·30000	0·956	0·965	0·973	0·980	0·987	0·994	1·006	1·011	1·016	1·021	1·026	1·030	1·034	1·038

k_s = 0·015mm D = 30mm

Gradient	0	2·5	5	7·5	10	12·5	17·5	20	22·5	25	27·5	30	32·5	35
0·00030	*9·910*	*0·927*	*0·943*	*0·958*	*0·973*	*0·987*	*1·013*	*1·025*	*1·037*	*1·048*	*1·059*	*1·070*	*1·080*	1·090
0·00100	*0·920*	*0·935*	*0·950*	*0·963*	0·976	0·988	1·011	1·022	1·032	1·042	1·052	1·061	1·070	1·078
0·00300	0·929	0·942	0·955	0·967	0·979	0·989	1·010	1·019	1·029	1·037	1·046	1·054	1·062	1·070
0·01000	0·937	0·949	0·960	0·971	0·981	0·991	1·009	1·017	1·025	1·033	1·040	1·047	1·054	1·060
0·03000	0·944	0·955	0·965	0·975	0·984	0·992	1·008	1·015	1·022	1·028	1·034	1·040	1·046	1·052
0·10000	0·953	0·962	0·971	0·979	0·986	0·993	1·006	1·012	1·018	1·023	1·028	1·033	1·037	1·042
0·30000	0·961	0·969	0·976	0·983	0·989	0·995	1·005	1·010	1·014	1·018	1·022	1·026	1·029	1·033

k_s = 0·015mm D = 50mm

Gradient	0	2·5	5	7·5	10	12·5	17·5	20	22·5	25	27·5	30	32·5	35
0·00010	*0·913*	*0·930*	*0·945*	*0·960*	*0·974*	*0·987*	*1·012*	*1·024*	*1·035*	1·046	1·057	1·067	1·077	1·086
0·00030	*0·922*	*0·937*	0·951	0·964	0·976	0·988	1·011	1·021	1·032	1·041	1·051	1·060	1·069	1·077
0·00100	0·930	0·944	0·956	0·968	0·979	0·990	1·010	1·019	1·028	1·037	1·045	1·053	1·061	1·068
0·00300	0·937	0·949	0·960	0·971	0·981	0·991	1·009	1·017	1·025	1·033	1·040	1·047	1·054	1·061
0·01000	0·945	0·955	0·965	0·975	0·984	0·992	1·008	1·015	1·022	1·028	1·035	1·041	1·047	1·052
0·03000	0·951	0·961	0·970	0·978	0·986	0·993	1·007	1·013	1·019	1·024	1·030	1·035	1·040	1·044
0·10000	0·960	0·968	0·975	0·982	0·988	0·994	1·005	1·010	1·015	1·019	1·024	1·028	1·032	1·035
0·30000	0·967	0·974	0·980	0·986	0·991	0·996	1·004	1·008	1·012	1·015	1·018	1·021	1·024	1·027

k_s = 0·015mm D = 100mm

Gradient	0	2·5	5	7·5	10	12·5	17·5	20	22·5	25	27·5	30	32·5	35
0·00010	0·928	0·942	0·954	0·967	0·978	0·989	1·010	1·020	1·029	1·038	1·047	1·055	1·063	1·071
0·00030	0·935	0·947	0·959	0·970	0·980	0·990	1·009	1·018	1·026	1·035	1·042	1·050	1·057	1·064
0·00100	0·941	0·952	0·963	0·973	0·982	0·991	1·008	1·016	1·024	1·031	1·038	1·045	1·051	1·057
0·00300	0·947	0·957	0·966	0·976	0·984	0·992	1·007	1·014	1·021	1·028	1·034	1·040	1·045	1·051
0·01000	0·953	0·962	0·971	0·979	0·986	0·993	1·006	1·012	1·018	1·024	1·029	1·034	1·039	1·043
0·03000	0·960	0·967	0·975	0·982	0·988	0·994	1·005	1·010	1·015	1·020	1·024	1·028	1·032	1·036
0·10000	0·967	0·974	0·980	0·985	0·991	0·995	1·004	1·008	1·012	1·015	1·019	1·022	1·025	1·028
0·30000	0·974	0·980	0·984	0·989	0·993	0·997	1·003	1·006	1·009	1·011	1·014	1·016	1·018	1·020

k_s = 0·015mm D = 200mm

Gradient	0	2·5	5	7·5	10	12·5	17·5	20	22·5	25	27·5	30	32·5	35
0·00010	0·939	0·950	0·961	0·972	0·982	0·991	1·009	1·017	1·025	1·032	1·040	1·047	1·054	1·060
0·00030	0·944	0·955	0·964	0·974	0·983	0·992	1·008	1·015	1·023	1·029	1·036	1·043	1·049	1·055
0·00100	0·949	0·959	0·968	0·977	0·985	0·993	1·007	1·014	1·020	1·026	1·032	1·038	1·044	1·049
0·00300	0·954	0·963	0·971	0·979	0·986	0·993	1·006	1·012	1·018	1·023	1·029	1·034	1·038	1·043
0·01000	0·960	0·968	0·975	0·982	0·988	0·994	1·005	1·010	1·015	1·020	1·024	1·028	1·032	1·036
0·03000	0·966	0·973	0·979	0·985	0·990	0·995	1·004	1·009	1·012	1·016	1·020	1·023	1·026	1·029
0·10000	0·974	0·979	0·984	0·988	0·993	0·996	1·003	1·006	1·009	1·012	1·015	1·017	1·019	1·021

k_s = 0·015mm D = 500mm

Gradient	0	2·5	5	7·5	10	12·5	17·5	20	22·5	25	27·5	30	32·5	35
0·00010	0·949	0·959	0·968	0·977	0·985	0·993	1·007	1·014	1·021	1·027	1·033	1·039	1·044	1·050
0·00030	0·953	0·962	0·970	0·978	0·986	0·993	1·007	1·013	1·019	1·024	1·030	1·035	1·040	1·045
0·00100	0·958	0·966	0·973	0·981	0·987	0·994	1·006	1·011	1·017	1·022	1·026	1·031	1·036	1·040
0·00300	0·963	0·970	0·977	0·983	0·989	0·996	1·005	1·010	1·014	1·019	1·023	1·027	1·031	1·034
0·01000	0·968	0·975	0·980	0·986	0·991	0·996	1·004	1·008	1·012	1·015	1·019	1·022	1·025	1·027
0·03000	0·974	0·979	0·984	0·989	0·993	0·996	1·003	1·006	1·009	1·012	1·014	1·017	1·019	1·021
0·10000	0·981	0·985	0·988	0·992	0·995	0·997	1·002	1·004	1·007	1·008	1·010	1·012	1·013	1·015

k_s = 0·015mm D = 1000mm

Gradient	0	2·5	5	7·5	10	12·5	17·5	20	22·5	25	27·5	30	32·5	35
0·00010	0·955	0·964	0·972	0·979	0·987	0·993	1·006	1·012	1·018	1·023	1·029	1·034	1·039	1·044
0·00030	0·959	0·967	0·974	0·981	0·988	0·994	1·006	1·011	1·016	1·021	1·026	1·031	1·035	1·039
0·00100	0·963	0·970	0·977	0·983	0·989	0·995	1·005	1·010	1·014	1·019	1·023	1·027	1·030	1·034
0·00300	0·968	0·974	0·980	0·986	0·991	0·995	1·004	1·008	1·012	1·016	1·019	1·022	1·026	1·029
0·01000	0·974	0·979	0·984	0·988	0·992	0·996	1·003	1·007	1·010	1·012	1·015	1·017	1·020	1·022
0·03000	0·979	0·984	0·987	0·991	0·994	0·997	1·003	1·005	1·007	1·009	1·011	1·013	1·015	1·016
0·10000	0·985	0·988	0·991	0·994	0·996	0·998	1·002	1·003	1·005	1·006	1·007	1·009	1·010	1·011

k_s = 0·015mm D = 2000mm

Gradient	0	2·5	5	7·5	10	12·5	17·5	20	22·5	25	27·5	30	32·5	35
0·00010	0·960	0·968	0·975	0·982	0·988	0·994	1·006	1·011	1·016	1·021	1·025	1·030	1·034	1·038
0·00030	0·964	0·971	0·977	0·983	0·989	0·995	1·005	1·010	1·014	1·018	1·023	1·027	1·030	1·034
0·00100	0·968	0·974	0·980	0·986	0·991	0·995	1·004	1·008	1·012	1·016	1·019	1·023	1·026	1·029
0·00300	0·973	0·978	0·983	0·988	0·992	0·996	1·004	1·007	1·010	1·013	1·016	1·019	1·021	1·023
0·01000	0·979	0·983	0·987	0·991	0·994	0·997	1·003	1·005	1·008	1·010	1·012	1·014	1·016	1·017
0·03000	0·984	0·987	0·990	0·993	0·996	0·998	1·002	1·004	1·005	1·007	1·008	1·010	1·011	1·012

k_s = 0·015mm D = 4000mm

Gradient	0	2·5	5	7·5	10	12·5	17·5	20	22·5	25	27·5	30	32·5	35
0·00010	0·965	0·971	0·978	0·984	0·989	0·995	1·005	1·009	1·014	1·018	1·022	1·026	1·030	1·034
0·00030	0·968	0·974	0·980	0·986	0·991	0·995	1·004	1·008	1·012	1·016	1·020	1·023	1·026	1·029
0·00100	0·973	0·978	0·983	0·988	0·992	0·996	1·004	1·007	1·010	1·013	1·016	1·019	1·022	1·024
0·00300	0·977	0·982	0·986	0·990	0·994	0·997	1·003	1·006	1·008	1·011	1·013	1·015	1·017	1·019
0·01000	0·983	0·986	0·990	0·993	0·995	0·998	1·002	1·004	1·006	1·008	1·009	1·011	1·012	1·013

Annexure : Multiplying factors for temperature variations (p.3 of 6)

k_s = 0·030mm D = 20mm

Gradient	0	2·5	5	7·5	10	12·5	17·5	20	22·5	25	27·5	30	32·5	35
0·00100	*0·911*	*0·928*	*0·944*	*0·959*	*0·973*	*0·987*	*1·012*	*1·024*	*1·036*	*1·046*	*1·057*	*1·067*	*1·077*	1·086
0·00300	*0·922*	*0·937*	*0·951*	*0·964*	*0·977*	0·989	1·011	1·021	1·031	1·040	1·049	1·058	1·067	1·075
0·01000	0·933	0·946	0·958	0·970	0·980	0·990	1·009	1·018	1·026	1·034	1·041	1·049	1·056	1·062
0·03000	0·943	0·954	0·964	0·974	0·983	0·992	1·008	1·015	1·022	1·028	1·034	1·040	1·046	1·051
0·10000	0·954	0·963	0·972	0·980	0·987	0·994	1·006	1·012	1·017	1·022	1·026	1·031	1·035	1·039
0·30000	0·964	0·972	0·978	0·984	0·990	0·995	1·005	1·009	1·013	1·016	1·020	1·023	1·026	1·029

k_s = 0·030mm D = 30mm

Gradient	0	2·5	5	7·5	10	12·5	17·5	20	22·5	25	27·5	30	32·5	35
0·00030	*0·911*	*0·928*	*0·943*	*0·959*	*0·973*	*0·987*	*1·013*	*1·025*	*1·036*	*1·047*	*1·058*	*1·068*	*1·078*	1·088
0·00100	*0·922*	*0·937*	*0·951*	*0·964*	0·977	0·989	1·011	1·021	1·031	1·041	1·050	1·059	1·068	1·076
0·00300	0·931	0·944	0·957	0·968	0·979	0·990	1·010	1·019	1·027	1·036	1·043	1·051	1·059	1·066
0·01000	0·941	0·952	0·963	0·973	0·983	0·991	1·008	1·016	1·023	1·030	1·036	1·043	1·049	1·055
0·03000	0·950	0·960	0·969	0·977	0·985	0·993	1·007	1·013	1·019	1·024	1·030	1·035	1·040	1·044
0·10000	0·960	0·968	0·975	0·982	0·989	0·994	1·005	1·010	1·014	1·019	1·023	1·026	1·030	1·033
0·30000	0·970	0·976	0·981	0·987	0·991	0·996	1·004	1·007	1·011	1·014	1·016	1·019	1·022	1·024

k_s = 0·030mm D = 50mm

Gradient	0	2·5	5	7·5	10	12·5	17·5	20	22·5	25	27·5	30	32·5	35
0·00010	*0·914*	*0·930*	*0·945*	*0·960*	*0·974*	*0·987*	*1·012*	*1·024*	*1·035*	1·046	1·056	1·066	1·076	1·085
0·00030	*0·923*	*0·938*	0·951	0·965	0·977	0·989	1·011	1·021	1·031	1·040	1·049	1·059	1·067	1·075
0·00100	0·932	0·945	0·957	0·969	0·980	0·990	1·009	1·018	1·027	1·035	1·043	1·051	1·058	1·065
0·00300	0·940	0·952	0·962	0·973	0·982	0·991	1·008	1·016	1·024	1·031	1·037	1·044	1·050	1·056
0·01000	0·949	0·959	0·968	0·977	0·985	0·993	1·007	1·013	1·020	1·025	1·031	1·036	1·041	1·046
0·03000	0·957	0·966	0·974	0·981	0·988	0·994	1·006	1·011	1·016	1·020	1·025	1·029	1·033	1·037
0·10000	0·967	0·974	0·980	0·985	0·991	0·995	1·004	1·008	1·012	1·015	1·018	1·021	1·024	1·027
0·30000	0·975	0·980	0·985	0·989	0·993	0·997	1·003	1·006	1·008	1·011	1·013	1·015	1·017	1·019

k_s = 0·030mm D = 100mm

Gradient	0	2·5	5	7·5	10	12·5	17·5	20	22·5	25	27·5	30	32·5	35
0·00010	0·929	0·942	0·955	0·967	0·979	0·990	1·010	1·019	1·029	1·037	1·046	1·054	1·062	1·070
0·00030	0·936	0·948	0·960	0·971	0·981	0·991	1·009	1·017	1·026	1·033	1·041	1·048	1·055	1·062
0·00100	0·943	0·954	0·964	0·974	0·983	0·992	1·008	1·015	1·022	1·029	1·036	1·042	1·048	1·054
0·00300	0·950	0·960	0·969	0·977	0·985	0·993	1·007	1·013	1·019	1·025	1·031	1·036	1·041	1·046
0·01000	0·958	0·966	0·974	0·981	0·988	0·994	1·006	1·011	1·016	1·020	1·025	1·029	1·033	1·037
0·03000	0·966	0·973	0·979	0·985	0·990	0·995	1·004	1·008	1·012	1·016	1·019	1·023	1·026	1·028
0·10000	0·974	0·980	0·984	0·989	0·993	0·997	1·003	1·006	1·009	1·011	1·014	1·016	1·018	1·020
0·30000	0·982	0·986	0·989	0·992	0·995	0·998	1·002	1·004	1·006	1·008	1·009	1·011	1·012	1·013

k_s = 0·030mm D = 200mm

Gradient	0	2·5	5	7·5	10	12·5	17·5	20	22·5	25	27·5	30	32·5	35
0·00010	0·940	0·951	0·962	0·972	0·982	0·991	1·008	1·016	1·024	1·032	1·039	1·046	1·052	1·059
0·00030	0·946	0·956	0·966	0·975	0·984	0·992	1·008	1·015	1·022	1·028	1·035	1·041	1·047	1·052
0·00100	0·952	0·961	0·970	0·978	0·986	0·993	1·007	1·013	1·019	1·024	1·030	1·035	1·040	1·045
0·00300	0·958	0·966	0·974	0·981	0·988	0·994	1·006	1·011	1·016	1·021	1·025	1·030	1·034	1·038
0·01000	0·966	0·973	0·979	0·985	0·990	0·995	1·004	1·009	1·013	1·016	1·020	1·023	1·026	1·029
0·03000	0·973	0·978	0·983	0·988	0·992	0·996	1·003	1·007	1·009	1·012	1·015	1·017	1·020	1·022
0·10000	0·981	0·985	0·988	0·992	0·995	0·997	1·002	1·004	1·006	1·008	1·010	1·012	1·013	1·015

k_s = 0·030mm D = 500mm

Gradient	0	2·5	5	7·5	10	12·5	17·5	20	22·5	25	27·5	30	32·5	35
0·00010	0·951	0·960	0·969	0·977	0·985	0·993	1·007	1·013	1·020	1·026	1·031	1·037	1·042	1·048
0·00030	0·955	0·964	0·972	0·980	0·987	0·994	1·006	1·012	1·018	1·023	1·028	1·033	1·037	1·042
0·00100	0·961	0·969	0·976	0·982	0·989	0·994	1·005	1·010	1·015	1·019	1·023	1·028	1·031	1·035
0·00300	0·967	0·974	0·980	0·985	0·991	0·995	1·004	1·008	1·012	1·016	1·019	1·022	1·025	1·028
0·01000	0·974	0·980	0·984	0·989	0·993	0·996	1·003	1·006	1·009	1·012	1·014	1·017	1·019	1·021
0·03000	0·981	0·985	0·988	0·992	0·995	0·997	1·002	1·005	1·007	1·008	1·010	1·012	1·013	1·015
0·10000	0·987	0·990	0·992	0·994	0·996	0·998	1·002	1·003	1·004	1·005	1·007	1·008	1·008	1·009

k_s = 0·030mm D = 1000mm

Gradient	0	2·5	5	7·5	10	12·5	17·5	20	22·5	25	27·5	30	32·5	35
0·00010	0·957	0·965	0·973	0·980	0·987	0·994	1·006	1·012	1·017	1·022	1·027	1·032	1·036	1·041
0·00030	0·962	0·969	0·976	0·983	0·989	0·994	1·005	1·010	1·015	1·019	1·024	1·028	1·032	1·035
0·00100	0·967	0·974	0·980	0·985	0·990	0·995	1·004	1·008	1·012	1·016	1·019	1·023	1·026	1·029
0·00300	0·973	0·978	0·983	0·988	0·992	0·996	1·003	1·005	1·007	1·009	1·011	1·013	1·014	1·016
0·01000	0·980	0·984	0·988	0·991	0·994	0·997	1·003	1·005	1·007	1·009	1·011	1·013	1·014	1·016
0·03000	0·985	0·988	0·991	0·994	0·996	0·998	1·002	1·003	1·005	1·006	1·007	1·009	1·010	1·011
0·10000	0·991	0·993	0·994	0·996	0·997	0·999	1·001	1·002	1·003	1·004	1·005	1·005	1·006	1·007

k_s = 0·030mm D = 2000mm

Gradient	0	2·5	5	7·5	10	12·5	17·5	20	22·5	25	27·5	30	32·5	35
0·00010	0·962	0·970	0·976	0·983	0·989	0·995	1·005	1·010	1·015	1·019	1·023	1·027	1·031	1·035
0·00030	0·967	0·973	0·979	0·985	0·990	0·995	1·004	1·009	1·013	1·016	1·020	1·023	1·027	1·030
0·00100	0·973	0·978	0·983	0·988	0·992	0·996	1·004	1·007	1·010	1·013	1·016	1·018	1·021	1·023
0·00300	0·978	0·983	0·987	0·990	0·994	0·997	1·003	1·005	1·008	1·010	1·012	1·014	1·016	1·017
0·01000	0·984	0·988	0·991	0·993	0·996	0·998	1·002	1·004	1·005	1·007	1·008	1·009	1·011	1·012
0·03000	0·989	0·991	0·994	0·995	0·997	0·999	1·010	1·002	1·003	1·004	1·005	1·006	1·007	1·008

k_s = 0·030mm D = 4000mm

Gradient	0	2·5	5	7·5	10	12·5	17·5	20	22·5	25	27·5	30	32·5	35
0·00010	0·967	0·974	0·980	0·985	0·990	0·995	1·004	1·009	1·013	1·016	1·020	1·023	1·027	1·030
0·00030	0·972	0·977	0·983	0·987	0·992	0·996	1·004	1·007	1·010	1·014	1·017	1·019	1·022	1·025
0·00100	0·978	0·982	0·986	0·990	0·994	0·997	1·003	1·006	1·008	1·010	1·013	1·015	1·017	1·018
0·00300	0·983	0·986	0·990	0·993	0·995	0·998	1·002	1·004	1·006	1·008	1·009	1·011	1·012	1·013
0·01000	0·988	0·991	0·993	0·995	0·997	0·998	1·001	1·003	1·004	1·005	1·006	1·007	1·008	1·009

k_s = 0·060mm *D* = 20mm

Gradient	0	2·5	5	7·5	10	12·5	17·5	20	22·5	25	27·5	30	32·5	35
0·00100	*0·914*	*0·931*	*0·946*	*0·961*	*0·974*	*0·987*	*1·012*	*1·023*	*1·034*	*1·044*	*1·054*	*1·064*	*1·073*	*1·082*
0·00300	*0·926*	*0·941*	*0·954*	*0·966*	*0·978*	0·989	1·010	1·020	1·029	1·037	1·046	1·054	1·061	1·068
0·01000	0·939	0·951	0·962	0·973	0·982	0·991	1·008	1·016	1·023	1·030	1·036	1·043	1·048	1·054
0·03000	0·951	0·961	0·970	0·978	0·986	0·993	1·006	1·012	1·018	1·023	1·028	1·033	1·037	1·041
0·10000	0·963	0·971	0·978	0·984	0·990	0·995	1·005	1·009	1·013	1·016	1·020	1·023	1·026	1·029
0·30000	0·974	0·979	0·984	0·989	0·993	0·997	1·003	1·006	1·009	1·011	1·014	1·016	1·018	1·020

k_s = 0·060mm *D* = 30mm

Gradient	0	2·5	5	7·5	10	12·5	17·5	20	22·5	25	27·5	30	32·5	35
0·00030	*0·912*	*0·929*	*0·945*	*0·960*	*0·974*	*0·987*	*1·012*	*1·024*	*1·035*	*1·046*	*1·056*	*1·066*	*1·076*	1·085
0·00100	*0·925*	*0·939*	*0·953*	*0·966*	0·978	0·989	1·010	1·020	1·030	1·039	1·047	1·056	1·063	1·071
0·00300	0·935	0·948	0·960	0·971	0·981	0·991	1·009	1·017	1·025	1·032	1·039	1·046	1·053	1·059
0·01000	0·947	0·958	0·967	0·976	0·985	0·993	1·007	1·014	1·020	1·026	1·031	1·036	1·041	1·046
0·03000	0·958	0·966	0·974	0·981	0·988	0·994	1·005	1·010	1·015	1·020	1·024	1·028	1·031	1·035
0·10000	0·969	0·976	0·981	0·987	0·991	0·996	1·004	1·007	1·011	1·014	1·016	1·019	1·022	1·024
0·30000	0·978	0·983	0·987	0·991	0·994	0·997	1·003	1·005	1·007	1·009	1·011	1·013	1·014	1·016

k_s = 0·060mm *D* = 50mm

Gradient	0	2·5	5	7·5	10	12·5	17·5	20	22·5	25	27·5	30	32·5	35
0·00010	*0·915*	*0·931*	*0·946*	*0·961*	*0·975*	*0·988*	*1·012*	*1·023*	*1·034*	1·044	1·054	1·064	1·074	1·083
0·00030	*0·925*	*0·940*	0·953	0·966	0·978	0·989	1·010	1·020	1·030	1·039	1·047	1·056	1·064	1·072
0·00100	0·936	0·948	0·960	0·971	0·981	0·991	1·009	1·017	1·025	1·033	1·040	1·047	1·054	1·060
0·00300	0·945	0·956	0·966	0·975	0·984	0·992	1·007	1·014	1·021	1·027	1·033	1·039	1·044	1·049
0·01000	0·956	0·964	0·973	0·980	0·987	0·994	1·006	1·011	1·016	1·021	1·026	1·030	1·034	1·038
0·03000	0·965	0·972	0·979	0·985	0·990	0·995	1·004	1·008	1·012	1·016	1·019	1·022	1·025	1·028
0·10000	0·975	0·980	0·985	0·989	0·993	0·997	1·003	1·006	1·008	1·011	1·013	1·015	1·017	1·019
0·30000	0·983	0·987	0·990	0·993	0·996	0·998	1·002	1·004	1·005	1·007	1·008	1·010	1·011	1·012

k_s = 0·060mm *D* = 100mm

Gradient	0	2·5	5	7·5	10	12·5	17·5	20	22·5	25	27·5	30	32·5	35
0·00010	0·931	0·944	0·956	0·968	0·979	0·990	1·010	1·019	1·028	1·036	1·044	1·052	1·060	1·067
0·00030	0·939	0·950	0·961	0·972	0·982	0·991	1·009	1·016	1·024	1·032	1·039	1·045	1·052	1·058
0·00100	0·947	0·958	0·967	0·976	0·984	0·992	1·007	1·014	1·020	1·026	1·032	1·038	1·043	1·048
0·00300	0·956	0·964	0·972	0·980	0·987	0·994	1·006	1·011	1·017	1·021	1·026	1·031	1·035	1·039
0·01000	0·965	0·972	0·979	0·985	0·990	0·995	1·004	1·009	1·012	1·016	1·019	1·023	1·026	1·029
0·03000	0·974	0·979	0·984	0·989	0·993	0·996	1·003	1·006	1·009	1·012	1·014	1·016	1·018	1·020
0·10000	0·982	0·986	0·989	0·992	0·995	0·998	1·002	1·004	1·006	1·008	1·009	1·011	1·012	1·013
0·30000	0·988	0·991	0·993	0·995	0·997	0·998	1·001	1·003	1·004	1·005	1·006	1·007	1·008	1·008

k_s = 0·060mm *D* = 200mm

Gradient	0	2·5	5	7·5	10	12·5	17·5	20	22·5	25	27·5	30	32·5	35
0·00010	0·942	0·953	0·963	0·973	0·983	0·992	1·008	1·016	1·023	1·030	1·037	1·043	1·049	1·055
0·00030	0·949	0·959	0·968	0·977	0·985	0·993	1·007	1·014	1·020	1·026	1·032	1·037	1·043	1·048
0·00100	0·956	0·965	0·973	0·980	0·987	0·994	1·006	1·011	1·016	1·021	1·026	1·030	1·035	1·039
0·00300	0·964	0·971	0·978	0·984	0·990	0·995	1·005	1·009	1·013	1·017	1·020	1·024	1·027	1·030
0·01000	0·973	0·978	0·983	0·988	0·992	0·996	1·003	1·006	1·009	1·012	1·015	1·017	1·019	1·021
0·03000	0·980	0·984	0·988	0·992	0·995	0·997	1·002	1·005	1·007	1·008	1·010	1·012	1·013	1·015
0·10000	0·987	0·990	0·992	0·995	0·997	0·998	1·002	1·003	1·004	1·005	1·006	1·007	1·008	1·009

k_s = 0·060mm *D* = 500mm

Gradient	0	2·5	5	7·5	10	12·5	17·5	20	22·5	25	27·5	30	32·5	35
0·00010	0·953	0·962	0·971	0·979	0·986	0·993	1·006	1·012	1·018	1·024	1·029	1·034	1·039	1·044
0·00030	0·959	0·967	0·975	0·982	0·988	0·994	1·005	1·011	1·016	1·020	1·025	1·029	1·033	1·037
0·00100	0·967	0·973	0·979	0·985	0·990	0·995	1·004	1·008	1·012	1·016	1·019	1·023	1·026	1·028
0·00300	0·974	0·979	0·984	0·988	0·993	0·996	1·003	1·006	1·009	1·012	1·014	1·017	1·019	1·021
0·01000	0·981	0·985	0·989	0·992	0·995	0·998	1·002	1·004	1·006	1·008	1·010	1·011	1·013	1·014
0·03000	0·987	0·990	0·992	0·994	0·996	0·998	1·002	1·003	1·004	1·005	1·006	1·008	1·008	1·009
0·10000	0·992	0·994	0·995	0·997	0·998	0·999	1·001	1·002	1·003	1·003	1·004	1·005	1·005	1·006

k_s = 0·060mm *D* = 1000mm

Gradient	0	2·5	5	7·5	10	12·5	17·5	20	22·5	25	27·5	30	32·5	35
0·00010	0·960	0·968	0·975	0·982	0·988	0·994	1·005	1·010	1·015	1·020	1·024	1·029	1·033	1·036
0·00030	0·966	0·973	0·979	0·985	0·990	0·995	1·005	1·009	1·013	1·016	1·020	1·023	1·027	1·030
0·00100	0·973	0·978	0·983	0·988	0·992	0·996	1·003	1·007	1·010	1·012	1·015	1·018	1·020	1·022
0·00300	0·979	0·984	0·988	0·991	0·994	0·997	1·003	1·005	1·007	1·009	1·011	1·013	1·014	1·016
0·01000	0·986	0·989	0·992	0·994	0·996	0·998	1·002	1·003	1·005	1·006	1·007	1·008	1·009	1·010
0·03000	0·991	0·993	0·994	0·996	0·997	0·999	1·001	1·002	1·003	1·004	1·005	1·005	1·006	1·007
0·10000	0·994	0·996	0·997	0·998	0·999	0·999	1·001	1·001	1·002	1·002	1·003	1·003	1·003	1·004

k_s = 0·060mm *D* = 2000mm

Gradient	0	2·5	5	7·5	10	12·5	17·5	20	22·5	25	27·5	30	32·5	35
0·00010	0·966	0·973	0·979	0·985	0·990	0·995	1·005	1·009	1·013	1·017	1·020	1·024	1·027	1·030
0·00030	0·972	0·977	0·983	0·987	0·992	0·996	1·004	1·007	1·010	1·013	1·016	1·019	1·021	1·024
0·00100	0·978	0·983	0·987	0·991	0·994	0·997	1·003	1·005	1·007	1·010	1·012	1·014	1·015	1·017
0·00300	0·984	0·988	0·991	0·993	0·996	0·998	1·002	1·004	1·005	1·007	1·008	1·009	1·011	1·012
0·01000	0·990	0·992	0·994	0·996	0·997	0·999	1·001	1·002	1·003	1·004	1·005	1·006	1·007	1·007
0·03000	0·993	0·995	0·996	0·997	0·998	0·999	1·001	1·001	1·002	1·003	1·003	1·004	1·004	1·005

k_s = 0·060mm *D* = 4000mm

Gradient	0	2·5	5	7·5	10	12·5	17·5	20	22·5	25	27·5	30	32·5	35
0·00010	0·972	0·977	0·982	0·987	0·992	0·996	1·004	1·007	1·011	1·014	1·017	1·019	1·022	1·024
0·00030	0·977	0·982	0·986	0·990	0·994	0·997	1·003	1·006	1·008	1·010	1·013	1·015	1·017	1·019
0·00100	0·983	0·987	0·990	0·993	0·995	0·998	1·002	1·004	1·006	1·007	1·009	1·010	1·011	1·013
0·00300	0·988	0·991	0·993	0·995	0·997	0·998	1·001	1·003	1·004	1·005	1·006	1·007	1·008	1·008
0·01000	0·993	0·994	0·996	0·997	0·998	0·999	1·001	1·002	1·002	1·003	1·004	1·004	1·005	1·005

Annexure : Multiplying factors for temperature variations (p. 5 of 6)

k_s = 0·150mm D = 20mm

Gradient	0	5	10	20	25	30	35
0·00100	*0·921*	*0·951*	*0·977*	*1·021*	*1·039*	*1·056*	*1·071*
0·00300	*0·936*	*0·960*	*0·982*	1·016	1·031	1·044	1·055
0·01000	0·951	0·970	0·986	1·012	1·022	1·031	1·039
0·03000	0·964	0·978	0·990	1·008	1·015	1·022	1·027
0·10000	0·976	0·986	0·994	1·005	1·010	1·014	1·017
0·30000	0·985	0·991	0·996	1·003	1·006	1·009	1·011

k_s = 0·150mm D = 30mm

Gradient	0	5	10	20	25	30	35
0·00100	*0·932*	*0·958*	0·980	1·017	1·033	1·047	1·060
0·00300	0·946	0·966	0·984	1·014	1·026	1·036	1·046
0·01000	0·959	0·975	0·989	1·010	1·018	1·026	1·032
0·03000	0·971	0·982	0·992	1·007	1·012	1·018	1·022
0·10000	0·981	0·989	0·995	1·004	1·008	1·011	1·013
0·30000	0·988	0·993	0·997	1·003	1·005	1·007	1·008

k_s = 0·150mm D = 50mm

Gradient	0	5	10	20	25	30	35
0·00010	*0·919*	*0·949*	*0·976*	*1·022*	1·041	1·060	1·076
0·00030	*0·931*	0·957	0·980	1·018	1·034	1·050	1·063
0·00100	0·944	0·965	0·984	1·014	1·027	1·039	1·049
0·00300	0·955	0·973	0·987	1·011	1·021	1·029	1·037
0·01000	0·968	0·980	0·991	1·008	1·014	1·020	1·025
0·03000	0·977	0·986	0·994	1·005	1·010	1·013	1·017
0·10000	0·986	0·991	0·996	1·003	1·006	1·008	1·010
0·30000	0·991	0·995	0·998	1·002	1·004	1·005	1·007

k_s = 0·150mm D = 100mm

Gradient	0	5	10	20	25	30	35
0·00010	0·935	0·959	0·981	1·017	1·033	1·047	1·060
0·00030	0·945	0·966	0·984	1·014	1·027	1·039	1·049
0·00100	0·956	0·973	0·987	1·011	1·021	1·029	1·037
0·00300	0·966	0·979	0·991	1·008	1·015	1·021	1·027
0·01000	0·976	0·986	0·994	1·005	1·010	1·014	1·018
0·03000	0·984	0·990	0·996	1·004	1·007	1·009	1·011
0·10000	0·990	0·994	0·997	1·002	1·004	1·005	1·007
0·30000	0·994	0·996	0·998	1·001	1·002	1·003	1·004

k_s = 0·150mm D = 200mm

Gradient	0	5	10	20	25	30	35
0·00010	0·947	0·967	0·984	1·014	1·026	1·038	1·048
0·00030	0·956	0·973	0·987	1·011	1·021	1·030	1·038
0·00100	0·966	0·979	0·990	1·008	1·016	1·022	1·028
0·00300	0·975	0·985	0·993	1·006	1·011	1·015	1·019
0·01000	0·983	0·990	0·995	1·004	1·007	1·010	1·012
0·03000	0·989	0·993	0·997	1·002	1·004	1·006	1·008
0·10000	0·993	0·996	0·998	1·001	1·003	1·004	1·004

k_s = 0·150mm D = 500mm

Gradient	0	5	10	20	25	30	35
0·00010	0·960	0·975	0·988	1·010	1·019	1·028	1·035
0·00030	0·967	0·980	0·991	1·008	1·015	1·021	1·027
0·00100	0·976	0·985	0·994	1·006	1·011	1·015	1·018
0·00300	0·983	0·990	0·996	1·004	1·007	1·010	1·012
0·01000	0·989	0·994	0·997	1·002	1·004	1·006	1·008
0·03000	0·993	0·996	0·998	1·001	1·003	1·004	1·005
0·10000	0·996	0·998	0·999	0·001	1·002	1·002	1·003

k_s = 0·150mm D = 1000mm

Gradient	0	5	10	20	25	30	35
0·00010	0·967	0·980	0·991	1·008	1·015	1·022	1·028
0·00030	0·974	0·984	0·993	1·006	1·011	1·016	1·020
0·00100	0·982	0·989	0·995	1·004	1·008	1·011	1·013
0·00300	0·988	0·993	0·997	1·003	1·005	1·007	1·009
0·01000	0·992	0·996	0·998	1·002	1·003	1·004	1·005
0·03000	0·995	0·997	0·999	1·001	1·002	1·003	1·003

k_s = 0·150mm D = 2000mm

Gradient	0	5	10	20	25	30	35
0·00010	0·974	0·984	0·993	1·006	1·012	1·017	1·021
0·00030	0·980	0·988	0·995	1·005	1·009	1·012	1·015
0·00100	0·987	0·992	0·996	1·003	1·006	1·008	1·010
0·00300	0·991	0·995	0·998	1·002	1·004	1·005	1·006
0·01000	0·995	0·997	0·999	1·001	1·002	1·003	1·004
0·03000	0·997	0·998	0·999	1·001	1·001	1·002	1·002

k_s = 0·30mm D = 20mm

Gradient	0	5	10	20	25	30	35
0·00100	*0·930*	*0·957*	*0·980*	*1·017*	*1·033*	*1·047*	*1·059*
0·00300	*0·947*	*0·968*	*0·985*	1·013	1·024	1·034	1·042
0·01000	0·963	0·978	0·990	1·009	1·016	1·022	1·028
0·03000	0·975	0·985	0·993	1·006	1·010	1·014	1·018
0·10000	0·984	0·991	0·996	1·003	1·006	1·009	1·011
0·30000	0·990	0·994	0·997	1·002	1·004	1·005	1·006

k_s = 0·30mm D = 30mm

Gradient	0	5	10	20	25	30	35
0·00100	*0·942*	*0·964*	*0·983*	1·014	1·027	1·038	1·048
0·00300	0·956	0·973	0·988	1·010	1·019	1·027	1·034
0·01000	0·970	0·982	0·992	1·007	1·013	1·018	1·022
0·03000	0·980	0·988	0·995	1·004	1·008	1·011	1·014
0·10000	0·988	0·993	0·997	1·003	1·005	1·007	1·008
0·30000	0·993	0·996	0·998	1·002	1·003	1·004	1·005

k_s = 0·30mm D = 50mm

Gradient	0	5	10	20	25	30	35
0·00010	*0·925*	*0·953*	*0·978*	*1·020*	*1·037*	1·053	1·068
0·00030	0·939	0·962	0·982	1·016	1·029	1·042	1·053
0·00100	0·953	0·971	0·987	1·011	1·021	1·030	1·038
0·00300	0·966	0·979	0·991	1·008	1·015	1·021	1·026
0·01000	0·977	0·986	0·994	1·005	1·008	1·013	1·016
0·03000	0·985	0·991	0·996	1·003	1·006	1·008	1·010
0·10000	0·991	0·995	0·998	1·002	1·003	1·005	1·006
0·30000	0·995	0·997	0·999	1·001	1·002	1·003	1·004

k_s = 0·30mm D = 100mm

Gradient	0	5	10	20	25	30	35
0·00010	0·941	0·964	0·983	1·015	1·028	1·041	1·051
0·00030	0·953	0·971	0·987	1·012	1·022	1·031	1·039
0·00100	0·965	0·979	0·990	1·008	1·015	1·022	1·027
0·00300	0·975	0·985	0·993	1·006	1·010	1·015	1·018
0·01000	0·984	0·991	0·996	1·003	1·006	1·009	1·011
0·03000	0·990	0·994	0·997	1·002	1·004	1·005	1·007
0·10000	0·994	0·997	0·998	1·001	1·002	1·003	1·004
0·30000	0·996	0·998	0·999	1·001	1·001	1·002	1·002

k_s = 0·30mm D = 200mm

Gradient	0	5	10	20	25	30	35
0·00010	0·954	0·972	0·987	1·012	1·022	1·031	1·039
0·00030	0·964	0·978	0·990	1·009	1·016	1·023	1·029
0·00100	0·975	0·985	0·993	1·006	1·011	1·015	1·019
0·00300	0·983	0·990	0·995	1·004	1·007	1·010	1·012
0·01000	0·989	0·994	0·997	1·002	1·004	1·006	1·007
0·03000	0·993	0·996	0·998	1·001	1·003	1·004	1·004
0·10000	0·996	0·998	0·999	1·001	1·001	1·002	1·003

k_s = 0·30mm D = 500mm

Gradient	0	5	10	20	25	30	35
0·00010	0·967	0·980	0·991	1·008	1·015	1·021	1·027
0·00030	0·975	0·985	0·993	1·006	1·011	1·015	1·019
0·00100	0·983	0·990	0·996	1·004	1·007	1·010	1·012
0·00300	0·989	0·994	0·997	1·002	1·004	1·006	1·008
0·01000	0·993	0·996	0·998	1·001	1·003	1·004	1·004
0·03000	0·996	0·998	0·999	1·001	1·002	1·002	1·003
0·10000	0·998	0·999	0·999	1·000	1·001	1·001	1·001

k_s = 0·30mm D = 1000mm

Gradient	0	5	10	20	25	30	35
0·00010	0·974	0·984	0·993	1·006	1·011	1·016	1·020
0·00030	0·982	0·989	0·995	1·004	1·008	1·011	1·013
0·00100	0·988	0·993	0·997	1·003	1·005	1·007	1·008
0·00300	0·992	0·996	0·998	1·002	1·003	1·004	1·005
0·01000	0·996	0·997	0·999	1·001	1·002	1·002	1·003

k_s = 0·30mm D = 2000mm

Gradient	0	5	10	20	25	30	35
0·00010	0·981	0·988	0·995	1·004	1·008	1·012	1·015
0·00030	0·987	0·992	0·996	1·003	1·005	1·008	1·010
0·00100	0·992	0·995	0·998	1·002	1·003	1·005	1·006
0·00300	0·995	0·997	0·999	1·001	1·002	1·003	1·003
0·01000	0·997	0·998	0·999	1·001	1·001	1·002	1·002

k_s = 0·60mm D = 20mm

Gradient	0	5	10	20	25	30	35
0·00100	*0·943*	*0·965*	*0·984*	*1·014*	*1·025*	*1·036*	*1·045*
0·00300	*0·960*	*0·976*	*0·989*	1·009	1·017	1·024	1·030
0·01000	0·974	0·985	0·993	1·006	1·011	1·015	1·018
0·03000	0·983	0·990	0·996	1·004	1·007	1·009	1·011
0·10000	0·990	0·994	0·997	1·002	1·004	1·005	1·006
0·30000	0·994	0·997	0·998	1·001	1·002	1·003	1·004

k_s = 0·60mm D = 30mm

Gradient	0	5	10	20	25	30	35
0·00100	*0·954*	*0·972*	*0·987*	1·011	1·020	1·028	1·036
0·00300	0·968	0·981	0·991	1·007	1·013	1·019	1·023
0·01000	0·980	0·988	0·995	1·004	1·008	1·011	1·014
0·03000	0·987	0·993	0·997	1·003	1·005	1·007	1·009
0·10000	0·993	0·996	0·998	1·002	1·003	1·004	1·005
0·30000	0·996	0·997	0·999	1·001	1·002	1·002	1·003

Annexure : Multiplying factors for temperature variations (p. 6 of 6)

k_s = 0·60mm D = 50mm

Gradient	0	5	10	20	25	30	35
0·00010	*0·934*	*0·959*	*0·981*	*1·017*	*1·031*	1·045	1·056
0·00030	0·949	0·969	0·986	1·012	1·023	1·032	1·041
0·00100	0·965	0·979	0·990	1·008	1·015	1·021	1·027
0·00300	0·976	0·986	0·994	1·005	1·010	1·014	1·017
0·01000	0·985	0·991	0·996	1·003	1·006	1·008	1·010
0·03000	0·991	0·995	0·998	1·002	1·004	1·005	1·006
0·10000	0·995	0·997	0·999	1·001	1·002	1·003	1·003

k_s = 0·60mm D = 100mm

Gradient	0	5	10	20	25	30	35
0·00010	0·951	0·970	0·986	1·012	1·023	1·032	1·040
0·00030	0·963	0·978	0·990	1·009	1·016	1·023	1·028
0·00100	0·975	0·985	0·993	1·006	1·010	1·014	1·018
0·00300	0·984	0·990	0·996	1·004	1·007	1·009	1·011
0·01000	0·990	0·994	0·997	1·002	1·004	1·005	1·007
0·03000	0·994	0·996	0·998	1·001	1·002	1·003	1·004
0·10000	0·998	0·998	0·999	1·001	1·001	1·002	1·002

k_s = 0·60mm D = 200mm

Gradient	0	5	10	20	25	30	35
0·00010	0·963	0·978	0·990	1·009	1·016	1·023	1·029
0·00030	0·974	0·984	0·993	1·006	1·011	1·016	1·020
0·00100	0·983	0·990	0·995	1·004	1·007	1·010	1·012
0·00300	0·989	0·993	0·997	1·002	1·004	1·006	1·007
0·01000	0·993	0·996	0·998	1·001	1·003	1·003	1·004

k_s = 0·60mm D = 500mm

Gradient	0	5	10	20	25	30	35
0·00010	0·975	0·985	0·993	1·006	1·011	1·015	1·019
0·00030	0·983	0·990	0·995	1·004	1·007	1·010	1·012
0·00100	0·989	0·994	0·997	1·002	1·004	1·006	1·007
0·00300	0·993	0·996	0·998	1·001	1·003	1·004	1·004
0·01000	0·996	0·998	0·999	1·001	1·001	1·002	1·002

k_s = 0·60mm D = 1000mm

Gradient	0	5	10	20	25	30	35
0·00010	0·982	0·989	0·995	1·004	1·008	1·010	1·013
0·00030	0·988	0·993	0·997	1·003	1·005	1·007	1·008
0·00100	0·993	0·996	0·998	1·002	1·003	1·004	1·005

k_s = 0·60mm D = 2000mm

Gradient	0	5	10	20	25	30	35
0·00010	0·987	0·992	0·997	1·003	1·005	1·007	1·009
0·00030	0·992	0·995	0·998	1·002	1·003	1·005	1·006
0·00100	0·995	0·997	0·999	1·001	1·002	1·003	1·003

k_s = 1·50mm D = 20mm

Gradient	0	5	10	20	25	30	35
0·00100	*0·961*	*0·977*	*0·989*	*1·009*	*1·016*	*1·023*	*1·028*
0·00300	*0·975*	*0·985*	*0·993*	*1·005*	*1·010*	1·014	1·017
0·01000	0·985	0·991	0·996	1·003	1·006	1·008	1·010
0·03000	0·991	0·995	0·998	1·002	1·003	1·005	1·006
0·10000	0·995	0·997	0·999	1·001	1·002	1·003	1·003

k_s = 1·50mm D = 30mm

Gradient	0	5	10	20	25	30	35
0·00100	*0·970*	*0·982*	*0·992*	1·007	1·012	1·017	1·021
0·00300	*0·981*	0·989	0·995	1·004	1·008	1·010	1·013
0·01000	0·989	0·993	0·997	1·002	1·004	1·006	1·007
0·03000	0·993	0·996	0·998	1·001	1·003	1·004	1·004
0·10000	0·996	0·998	0·999	1·001	1·001	1·002	1·002

k_s = 1·50mm D = 50mm

Gradient	0	5	10	20	25	30	35
0·00010	*0·950*	*0·970*	*0·986*	*1·012*	*1·022*	*1·031*	1·038
0·00030	*0·966*	*0·979*	0·991	1·008	1·014	1·020	1·025
0·00100	0·978	0·987	0·994	1·005	1·009	1·012	1·015
0·00300	0·986	0·992	0·996	1·003	1·005	1·007	1·009
0·01000	0·992	0·995	0·998	1·002	1·003	1·004	1·005
0·03000	0·995	0·997	0·999	1·001	1·002	1·002	1·003

k_s = 1·50mm D = 100mm

Gradient	0	5	10	20	25	30	35
0·00010	0·966	0·979	0·991	1·008	1·015	1·020	1·025
0·00030	0·977	0·986	0·994	1·005	1·009	1·013	1·016
0·00100	0·986	0·992	0·996	1·003	1·006	1·008	1·010
0·00300	0·991	0·995	0·998	1·002	1·003	1·005	1·006
0·01000	0·995	0·997	0·999	1·001	1·002	1·003	1·003
0·03000	0·997	0·998	0·999	1·001	1·001	1·002	1·002

k_s = 1·50mm D = 200mm

Gradient	0	5	10	20	25	30	35
0·00010	0·976	0·986	0·994	1·005	1·010	1·014	1·017
0·00030	0·985	0·991	0·996	1·003	1·006	1·009	1·011
0·00100	0·991	0·995	0·998	1·002	1·004	1·005	1·006
0·00300	0·994	0·997	0·999	1·001	1·002	1·003	1·004

k_s = 1·50mm D = 500mm

Gradient	0	5	10	20	25	30	35
0·00010	0·986	0·991	0·996	1·003	1·006	1·008	1·010
0·00030	0·991	0·995	0·998	1·002	1·004	1·005	1·006
0·00100	0·995	0·997	0·999	1·001	1·002	1·003	1·003
0·00300	0·997	0·998	0·999	1·001	1·001	1·002	1·002

k_s = 1·50mm D = 1000mm

Gradient	0	5	10	20	25	30	35
0·00010	0·990	0·994	0·997	1·002	1·004	1·005	1·007
0·00030	0·994	0·996	0·998	1·001	1·002	1·003	1·004

k_s = 1·50mm D = 2000mm

Gradient	0	5	10	20	25	30	35
0·00010	0·993	0·996	0·998	1·001	1·003	1·004	1·004
0·00030	0·996	0·998	0·999	1·001	1·002	1·002	1·003

k_s = 3·0mm D = 30mm

Gradient	0	5	10	20	25	30	35
0·00030	*0·966*	*0·980*	*0·991*	*1·007*	*1·014*	*1·019*	*1·024*
0·00100	*0·980*	*0·988*	*0·995*	*1·004*	*1·008*	1·011	1·014
0·00300	0·988	0·993	0·997	1·003	1·005	1·007	1·008
0·01000	0·993	0·996	0·998	1·001	1·003	1·004	1·005

k_s = 3·0mm D = 50mm

Gradient	0	5	10	20	25	30	35
0·00030	*0·976*	*0·986*	*0·994*	1·005	1·009	1·013	1·016
0·00100	0·986	0·992	0·996	1·003	1·005	1·008	1·009
0·00300	0·992	0·995	0·998	1·002	1·003	1·004	1·006
0·01000	0·995	0·997	0·999	1·001	1·002	1·003	1·003

k_s = 3·0mm D = 100mm

Gradient	0	5	10	20	25	30	35
0·00010	0·976	0·986	0·994	1·005	1·010	1·013	1·017
0·00030	0·985	0·991	0·996	1·003	1·006	1·008	1·010
0·00100	0·991	0·995	0·998	1·002	1·003	1·005	1·006
0·00300	0·995	0·997	0·999	1·001	1·002	1·003	1·003

k_s = 3·0mm D = 200mm

Gradient	0	5	10	20	25	30	35
0·00010	0·984	0·991	0·996	1·003	1·006	1·009	1·011
0·00030	0·990	0·994	0·997	1·002	1·004	1·005	1·006
0·00100	0·994	0·997	0·999	1·001	1·002	1·003	1·004
0·00300	0·997	0·998	0·999	1·001	1·001	1·002	1·002

k_s = 3·0mm D = 500mm

Gradient	0	5	10	20	25	30	35
0·00010	0·991	0·995	0·998	1·002	1·003	1·005	1·006
0·00030	0·995	0·997	0·999	1·001	1·002	1·003	1·004

k_s = 3·0mm D = 1000mm

Gradient	0	5	10	20	25	30	35
0·00010	0·994	0·997	0·998	1·001	1·002	1·003	1·004
0·00030	0·996	0·998	0·999	1·001	1·001	1·002	1·002

k_s = 6·0mm D = 50mm

Gradient	0	5	25	30	35
0·00010	0·975	0·985	1·010	1·014	1·017
0·00030	0·984	0·991	1·006	1·008	1·010
0·00100	0·991	0·995	1·003	1·005	1·006
0·00300	0·995	0·997	1·002	1·003	1·003

k_s = 6·0mm D = 100mm

Gradient	0	5	25	30	35
0·00010	0·985	0·991	1·006	1·008	1·010
0·00030	0·991	0·995	1·004	1·005	1·006
0·00100	0·995	0·997	1·002	1·003	1·003
0·00300	0·997	0·998	1·001	1·002	1·002

k_s = 6·0mm D = 200mm

Gradient	0	5	25	30
0·00010	0·990	0·994	1·004	1·005
0·00030	0·994	0·997	1·002	1·003
0·00100	0·997	0·998	1·001	1·002
0·00300	0·998	0·999	1·001	1·001

k_s = 6·0mm D = 500mm

Gradient	0	5	25	30	35
0·00010	0·995	0·997	1·002	1·003	1·003
0·00030	0·997	0·998	1·001	1·002	1·002

k_s = 6·0mm D = 1000mm

Gradient	0	5	25	30	35
0·00010	0·997	0·998	1·001	1·002	1·002
0·00030	0·998	0·999	1·001	1·001	1·001

k_s = 6·0mm D = 2000mm

Gradient	0	5	25	30
0·00010	0·998	0·999	1·001	1·001
0·00030	0·999	0·999	1·000	1·001

k_s = 15mm D = 100mm

Gradient	0	5	25	30	35
0·00010	0·991	0·995	1·003	1·005	1·006
0·00030	0·995	0·997	1·002	1·003	1·003

k_s = 15mm D = 500mm

Gradient	0	5	25	30	35
0·00010	0·997	0·998	1·001	1·001	1·002
0·00030	0·998	0·999	1·001	1·001	1·001

k_s = 15mm D = 2000mm

Gradient	0	5	25	30
0·00010	0·999	0·999	1·000	1·001
0·00030	0·999	1·000	1·000	1·000

B1 Q varies with D to the exponent x as tabulated

Colebrook-White solutions

DV in metres and metres per second

D/k_s both in millimetres

DV	0·005	0·01	0·02	0·05	0·10	0·20	0·50	1	2	5	10	20	50	100	200
D/k_s															
5	2·85	2·845	2·845	2·845	2·845	2·845	2·845	2·84	2·84	2·84	2·84	2·84	2·84	2·84	2·84
10	2·79	2·785	2·78	2·78	2·78	2·775	2·775	2·775	2·775	2·775	2·775	2·775	2·775	2·775	2·775
20	2·755	2·745	2·74	2·735	2·735	2·735	2·735	2·73	2·73	2·73	2·73	2·73	2·73	2·73	2·73
50	2·74	2·72	2·705	2·70	2·695	2·695	2·69	2·69	2·69	2·69	2·69	2·69	2·69	2·69	2·69
100	2·74	2·71	2·695	2·68	2·675	2·67	2·67	2·67	2·67	2·67	2·67	2·67	2·67	2·67	2·67
200	2·745	2·715	2·69	2·67	2·66	2·655	2·655	2·655	2·65	2·65	2·65	2·65	2·65	2·65	2·65
500	2·75	2·72	2·695	2·67	2·655	2·645	2·64	2·635	2·635	2·635	2·635	2·635	2·635	2·635	2·635
10^3	2·755	2·73	2·705	2·675	2·655	2·645	2·63	2·625	2·625	2·625	2·62	2·62	2·62	2·62	2·62
2×10^3	2·755	2·73	2·705	2·68	2·66	2·645	2·63	2·62	2·615	2·615	2·615	2·615	2·61	2·61	2·61
5×10^3	2·755	2·73	2·71	2·685	2·67	2·655	2·635	2·62	2·615	2·61	2·605	2·605	2·60	2·60	2·60
10^4	2·755	2·73	2·71	2·69	2·675	2·66	2·64	2·625	2·615	2·605	2·60	2·60	2·595	2·595	2·595
2×10^4	2·755	2·735	2·71	2·69	2·675	2·66	2·645	2·63	2·62	2·605	2·60	2·595	2·59	2·59	2·59
5×10^4	2·755	2·735	2·715	2·69	2·675	2·665	2·65	2·635	2·625	2·61	2·60	2·595	2·585	2·585	2·585
10^5	2·755	2·735	2·715	2·69	2·675	2·665	2·65	2·64	2·63	2·615	2·605	2·595	2·585	2·585	2·58
2×10^5	2·755	2·735	2·715	2·69	2·675	2·665	2·65	2·64	2·63	2·62	2·61	2·60	2·585	2·585	2·58
5×10^5	2·755	2·735	2·715	2·69	2·675	2·665	2·65	2·64	2·635	2·62	2·615	2·605	2·595	2·585	2·58
10^6	2·755	2·735	2·715	2·69	2·68	2·665	2·65	2·64	2·635	2·625	2·615	2·61	2·60	2·59	2·58
D/k_s															
R	$4·38 \times10^3$	$8·76 \times10^3$	$1·75 \times10^4$	$4·38 \times10^4$	$8·76 \times10^4$	$1·75 \times10^5$	$4·38 \times10^5$	$8·76 \times10^5$	$1·75 \times10^6$	$4·38 \times10^6$	$8·76 \times10^6$	$1·75 \times10^7$	$4·38 \times10^7$	$8·76 \times10^7$	$1·75 \times10^8$

Equivalent Reynolds number for water at 15°C

$$Q_R = Q_T \left[\frac{D_G}{D_T} \right]^x \left[\frac{S_G}{S_T} \right]^{1/y} \left[\frac{(k_s)_T}{(k_s)_G} \right]^u \qquad (8)$$

For V_R, Q_T is replaced by V_T and the exponent x is replaced by $(x–2)$

$$D_R = D_T \left[\frac{Q_G}{Q_T} \right]^{1/x} \left[\frac{S_T}{S_G} \right]^{1/z} \left[\frac{(k_s)_G}{(k_s)_T} \right]^w \qquad (10)$$

S varies with Q to the exponent y as tabulated

Colebrook-White solutions

DV in metres and metres per second

D/k_s both in millimetres

DV	0·005	0·01	0·02	0·05	0·10	0·20	0·50	1	2	5	10	20	50	100	200
D/k_s															
5	1·98	1·99	1·995	2·00	2·00	2·00	2·00	2·00	2·00	2·00	2·00	2·00	2·00	2·00	2·00
10	1·965	1·98	1·99	1·995	2·00	2·00	2·00	2·00	2·00	2·00	2·00	2·00	2·00	2·00	2·00
20	1·935	1·965	1·98	1·995	1·995	2·00	2·00	2·00	2·00	2·00	2·00	2·00	2·00	2·00	2·00
50	1·88	1·93	1·96	1·985	1·99	1·995	2·00	2·00	2·00	2·00	2·00	2·00	2·00	2·00	2·00
100	1·83	1·885	1·93	1·965	1·985	1·99	1·995	2·00	2·00	2·00	2·00	2·00	2·00	2·00	2·00
200	1·785	1·835	1·89	1·94	1·965	1·985	1·995	1·995	2·00	2·00	2·00	2·00	2·00	2·00	2·00
500	1·745	1·785	1·835	1·895	1·935	1·96	1·985	1·99	1·995	2·00	2·00	2·00	2·00	2·00	2·00
10^3	1·725	1·76	1·80	1·855	1·90	1·935	1·97	1·985	1·99	1·995	2·00	2·00	2·00	2·00	2·00
2×10^3	1·715	1·745	1·78	1·82	1·86	1·90	1·945	1·97	1·985	1·995	1·995	2·00	2·00	2·00	2·00
5×10^3	1·71	1·74	1·76	1·795	1·825	1·86	1·905	1·94	1·965	1·985	1·99	1·995	2·00	2·00	2·00
10^4	1·71	1·735	1·755	1·785	1·81	1·835	1·875	1·91	1·94	1·97	1·985	1·99	1·995	2·00	2·00
2×10^4	1·71	1·735	1·755	1·78	1·80	1·82	1·85	1·88	1·91	1·95	1·97	1·985	1·995	1·995	2·00
5×10^4	1·71	1·73	1·75	1·775	1·795	1·81	1·83	1·85	1·875	1·915	1·94	1·965	1·985	1·99	1·995
10^5	1·71	1·73	1·75	1·775	1·79	1·805	1·825	1·84	1·86	1·89	1·915	1·945	1·97	1·985	1·99
2×10^5	1·71	1·73	1·75	1·775	1·79	1·80	1·82	1·835	1·85	1·87	1·895	1·92	1·95	1·97	1·985
5×10^5	1·71	1·73	1·75	1·775	1·79	1·80	1·82	1·83	1·84	1·855	1·875	1·89	1·92	1·945	1·965
10^6	1·71	1·73	1·75	1·775	1·79	1·80	1·815	1·83	1·84	1·85	1·865	1·88	1·90	1·925	1·945
D/k_s															
R	4·38 $\times10^3$	8·76 $\times10^3$	1·75 $\times10^4$	4·38 $\times10^4$	8·76 $\times10^4$	1·75 $\times10^5$	4·38 $\times10^5$	8·76 $\times10^5$	1·75 $\times10^6$	4·38 $\times10^6$	8·76 $\times10^6$	1·75 $\times10^7$	4·38 $\times10^7$	8·76 $\times10^7$	1·75 $\times10^8$

Equivalent Reynolds number for water at 15°C

$$Q_R = Q_T \left[\frac{D_G}{D_T} \right]^x \left[\frac{S_G}{S_T} \right]^{1/y} \left[\frac{(k_s)_T}{(k_s)_G} \right]^u \qquad (8)$$

For V_R, Q_T is replaced by V_T and the exponent x is replaced by $(x-2)$

$$S_R = S_T \left[\frac{Q_G}{Q_T} \right]^y \left[\frac{D_T}{D_G} \right]^z \left[\frac{(k_s)_G}{(k_s)_T} \right]^v \qquad (9)$$

B3

S varies inversely with *D* to the exponent *z* as tabulated

Colebrook-White solutions

DV in metres and metres per second

D/k_s both in millimetres

DV / D/k_s	0·005	0·01	0·02	0·05	0·10	0·20	0·50	1	2	5	10	20	50	100	200
5	5·65	5·665	5·675	5·68	5·685	5·69	5·685	5·685	5·685	5·685	5·685	5·685	5·685	5·685	5·685
10	5·485	5·52	5·535	5·545	5·55	5·55	5·555	5·555	5·555	5·555	5·555	5·555	5·555	5·555	5·555
20	5·34	5·40	5·43	5·45	5·455	5·46	5·465	5·465	5·465	5·465	5·465	5·465	5·465	5·465	5·465
50	5·15	5·24	5·305	5·35	5·365	5·375	5·38	5·38	5·38	5·385	5·385	5·385	5·385	5·385	5·385
100	5·005	5·11	5·20	5·275	5·305	5·32	5·33	5·335	5·335	5·335	5·34	5·34	5·34	5·34	5·34
200	4·89	4·99	5·085	5·19	5·24	5·27	5·29	5·295	5·30	5·30	5·30	5·30	5·305	5·305	5·305
500	4·79	4·86	4·945	5·06	5·135	5·19	5·23	5·25	5·255	5·26	5·265	5·265	5·265	5·265	5·265
10^3	4·75	4·805	4·86	4·96	5·04	5·115	5·18	5·21	5·225	5·235	5·24	5·24	5·245	5·245	5·245
2×10^3	4·73	4·77	4·815	4·885	4·955	5·03	5·115	5·16	5·19	5·21	5·215	5·22	5·225	5·225	5·225
5×10^3	4·715	4·745	4·775	4·825	4·87	4·93	5·015	5·08	5·13	5·17	5·185	5·195	5·20	5·20	5·205
10^4	4·71	4·74	4·765	4·80	4·835	4·875	4·945	5·01	5·065	5·13	5·155	5·175	5·185	5·185	5·19
2×10^4	4·71	4·735	4·76	4·79	4·81	4·84	4·895	4·945	5·005	5·075	5·12	5·145	5·165	5·17	5·175
5×10^4	4·71	4·735	4·755	4·78	4·80	4·82	4·85	4·885	4·925	4·995	5·05	5·095	5·13	5·15	5·155
10^5	4·71	4·73	4·755	4·775	4·795	4·81	4·835	4·855	4·885	4·94	4·99	5·045	5·095	5·125	5·14
2×10^5	4·71	4·73	4·75	4·775	4·79	4·805	4·825	4·845	4·865	4·90	4·94	4·99	5·05	5·09	5·115
5×10^5	4·71	4·73	4·75	4·775	4·79	4·805	4·82	4·835	4·845	4·87	4·895	4·93	4·985	5·03	5·07
10^6	4·71	4·73	4·75	4·775	4·79	4·805	4·815	4·83	4·84	4·86	4·875	4·90	4·94	4·98	5·025
D/k_s															
R	4·38 $\times10^3$	8·76 $\times10^3$	1·75 $\times10^4$	4·38 $\times10^4$	8·76 $\times10^4$	1·75 $\times10^5$	4·38 $\times10^5$	8·76 $\times10^5$	1·75 $\times10^6$	4·38 $\times10^6$	8·76 $\times10^6$	1·75 $\times10^7$	4·38 $\times10^7$	8·76 $\times10^7$	1·75 $\times10^8$

Equivalent Reynolds number for water at 15°C

$$S_R = S_T \left[\frac{Q_G}{Q_T} \right]^y \left[\frac{D_T}{D_G} \right]^z \left[\frac{(k_s)_G}{(k_s)_T} \right]^v \tag{9}$$

$$D_R = D_T \left[\frac{Q_G}{Q_T} \right]^{1/x} \left[\frac{S_T}{S_G} \right]^{1/z} \left[\frac{(k_s)_G}{(k_s)_T} \right]^w \tag{10}$$

Q varies inversely with k_s to the exponent u as tabulated

Colebrook-White solutions

DV in metres and metres per second

D/k_s both in millimetres

DV	0·005	0·01	0·02	0·05	0·10	0·20	0·50	1	2	5	10	20	50	100	200
D/k_s															
5	0·335	0·34	0·34	0·34	0·345	0·345	0·345	0·345	0·345	0·345	0·345	0·345	0·345	0·345	0·345
10	0·265	0·27	0·275	0·275	0·275	0·275	0·275	0·275	0·275	0·275	0·275	0·275	0·275	0·275	0·275
20	0·21	0·22	0·225	0·23	0·23	0·23	0·23	0·23	0·23	0·23	0·23	0·23	0·23	0·23	0·23
50	0·145	0·16	0·175	0·185	0·19	0·19	0·19	0·19	0·19	0·19	0·19	0·19	0·19	0·19	0·19
100	0·095	0·12	0·14	0·155	0·16	0·165	0·17	0·17	0·17	0·17	0·17	0·17	0·17	0·17	0·17
200	0·06	0·08	0·105	0·125	0·14	0·145	0·15	0·15	0·15	0·15	0·15	0·15	0·15	0·15	0·15
500	0·03	0·04	0·06	0·085	0·105	0·115	0·125	0·13	0·13	0·13	0·135	0·135	0·135	0·135	0·135
10^3	0·015	0·025	0·035	0·06	0·075	0·095	0·11	0·115	0·12	0·12	0·12	0·12	0·12	0·12	0·12
2×10^3	0·01	0·01	0·02	0·035	0·05	0·07	0·09	0·10	0·105	0·11	0·11	0·11	0·11	0·11	0·11
5×10^3	0·005	0·005	0·01	0·015	0·025	0·04	0·06	0·075	0·085	0·095	0·10	0·10	0·10	0·10	0·10
10^4	0·00	0·005	0·005	0·01	0·015	0·02	0·04	0·055	0·065	0·08	0·085	0·09	0·095	0·095	0·095
2×10^4	0·00	0·00	0·00	0·005	0·005	0·01	0·025	0·035	0·05	0·065	0·075	0·08	0·085	0·09	0·09
5×10^4	0·00	0·00	0·00	0·00	0·005	0·005	0·01	0·025	0·025	0·045	0·055	0·065	0·075	0·08	0·08
10^5	0·00	0·00	0·00	0·00	0·00	0·005	0·005	0·01	0·015	0·025	0·04	0·05	0·065	0·07	0·075
2×10^5	0·00	0·00	0·00	0·00	0·00	0·00	0·005	0·005	0·01	0·015	0·025	0·035	0·05	0·06	0·065
5×10^5	0·00	0·00	0·00	0·00	0·00	0·00	0·00	0·00	0·005	0·005	0·01	0·02	0·035	0·045	0·055
10^6	0·00	0·00	0·00	0·00	0·00	0·00	0·00	0·00	0·00	0·005	0·005	0·01	0·02	0·03	0·04
D/k_s															
R	4·38 $\times10^3$	8·76 $\times10^3$	1·75 $\times10^4$	4·38 $\times10^4$	8·76 $\times10^4$	1·75 $\times10^5$	4·38 $\times10^5$	8·76 $\times10^5$	1·75 $\times10^6$	4·38 $\times10^6$	8·76 $\times10^6$	1·75 $\times10^7$	4·38 $\times10^7$	8·76 $\times10^7$	1·75 $\times10^8$

Equivalent Reynolds number for water at 15°C

$$Q_R = Q_T \left[\frac{D_G}{D_T} \right]^x \left[\frac{S_G}{S_T} \right]^{1/y} \left[\frac{(k_s)_T}{(k_s)_G} \right]^u \qquad (8)$$

For V_R, Q_T is replaced by V_T and the exponent x is replaced by $(x-2)$

Note that if the change in value of k_s is sufficient to modify the value of D/k_s as leading to the selection of the appropriate line on the table, this shift should influence accordingly the estimate of medial value of exponent.

B5

S varies with k_s to the exponent v as tabulated

Colebrook-White solutions

DV in metres and metres per second

D/k_s both in millimetres

DV	**0·005**	**0·01**	**0·02**	**0·05**	**0·10**	**0·20**	**0·50**	**1**	**2**	**5**	**10**	**20**	**50**	**100**	**200**
D/k_s															
5	0·67	0·675	0·68	0·685	0·685	0·685	0·685	0·685	0·685	0·685	0·685	0·685	0·685	0·685	0·685
10	0·52	0·535	0·545	0·55	0·55	0·555	0·555	0·555	0·555	0·555	0·555	0·555	0·555	0·555	0·555
20	0·405	0·43	0·45	0·46	0·46	0·465	0·465	0·465	0·465	0·465	0·465	0·465	0·465	0·465	0·465
50	0·27	0·31	0·345	0·365	0·375	0·38	0·38	0·385	0·385	0·385	0·385	0·385	0·385	0·385	0·385
100	0·18	0·225	0·27	0·305	0·32	0·33	0·335	0·335	0·335	0·34	0·34	0·34	0·34	0·34	0·34
200	0·105	0·15	0·195	0·245	0·27	0·285	0·30	0·30	0·30	0·30	0·30	0·305	0·305	0·305	0·305
500	0·05	0·075	0·11	0·165	0·20	0·23	0·25	0·255	0·26	0·265	0·265	0·265	0·265	0·265	0·265
10^3	0·025	0·04	0·065	0·105	0·145	0·18	0·21	0·225	0·235	0·24	0·24	0·24	0·245	0·245	0·245
2×10^3	0·015	0·02	0·035	0·065	0·095	0·13	0·17	0·195	0·21	0·215	0·22	0·225	0·225	0·225	0·225
5×10^3	0·005	0·01	0·015	0·03	0·045	0·07	0·11	0·145	0·165	0·19	0·195	0·20	0·20	0·205	0·205
10^4	0·005	0·005	0·005	0·015	0·025	0·04	0·07	0·10	0·13	0·16	0·175	0·18	0·185	0·19	0·19
2×10^4	0·00	0·00	0·005	0·01	0·015	0·02	0·04	0·065	0·09	0·13	0·15	0·16	0·17	0·175	0·175
5×10^4	0·00	0·00	0·00	0·005	0·005	0·01	0·02	0·03	0·05	0·08	0·11	0·13	0·15	0·155	0·16
10^5	0·00	0·00	0·00	0·00	0·005	0·005	0·01	0·015	0·03	0·05	0·075	0·10	0·125	0·14	0·145
2×10^5	0·00	0·00	0·00	0·00	0·00	0·00	0·005	0·01	0·015	0·03	0·045	0·07	0·10	0·12	0·13
5×10^5	0·00	0·00	0·00	0·00	0·00	0·00	0·00	0·005	0·005	0·015	0·02	0·035	0·065	0·085	0·105
10^6	0·00	0·00	0·00	0·00	0·00	0·00	0·00	0·00	0·005	0·005	0·01	0·02	0·04	0·06	0·08
D/k_s															
R	$4{\cdot}38\times10^3$	$8{\cdot}76\times10^3$	$1{\cdot}75\times10^4$	$4{\cdot}38\times10^4$	$8{\cdot}76\times10^4$	$1{\cdot}75\times10^5$	$4{\cdot}38\times10^5$	$8{\cdot}76\times10^5$	$1{\cdot}75\times10^6$	$4{\cdot}38\times10^6$	$8{\cdot}76\times10^6$	$1{\cdot}75\times10^7$	$4{\cdot}38\times10^7$	$8{\cdot}76\times10^7$	$1{\cdot}75\times10^8$

Equivalent Reynolds number for water at 15°C

$$S_R = S_T \left[\frac{Q_G}{Q_T} \right]^y \left[\frac{D_T}{D_G} \right]^z \left[\frac{(k_s)_G}{(k_s)_T} \right]^v \qquad (9)$$

Note that if the change in value of k_s is sufficient to modify the value of D/k_s as leading to the selection of the appropriate line on the table, this shift should influence accordingly the estimate of medial value of exponent.

D varies with k_s to the exponent w as tabulated

Colebrook-White solutions

DV in metres and metres per second

D/k_s both in millimetres

DV	0·005	0·01	0·02	0·05	0·10	0·20	0·50	1	2	5	10	20	50	100	200
D/k_s															
5	0·12	0·12	0·12	0·12	0·12	0·12	0·12	0·12	0·12	0·12	0·12	0·12	0·12	0·12	0·12
10	0·095	0·095	0·10	0·10	0·10	0·10	0·10	0·10	0·10	0·10	0·10	0·10	0·10	0·10	0·10
20	0·075	0·08	0·08	0·085	0·085	0·085	0·085	0·085	0·085	0·085	0·085	0·085	0·085	0·085	0·085
50	0·05	0·06	0·065	0·07	0·07	0·07	0·07	0·07	0·07	0·07	0·07	0·07	0·07	0·07	0·07
100	0·035	0·045	0·05	0·06	0·06	0·06	0·065	0·065	0·065	0·065	0·065	0·065	0·065	0·065	0·065
200	0·02	0·03	0·04	0·045	0·05	0·055	0·055	0·055	0·055	0·055	0·055	0·055	0·055	0·055	0·055
500	0·01	0·015	0·02	0·03	0·04	0·045	0·05	0·05	0·05	0·05	0·05	0·05	0·05	0·05	0·05
10^3	0·005	0·01	0·015	0·02	0·03	0·035	0·04	0·045	0·045	0·045	0·045	0·045	0·045	0·045	0·045
2×10^3	0·005	0·005	0·005	0·015	0·02	0·025	0·035	0·035	0·04	0·04	0·04	0·045	0·045	0·045	0·045
5×10^3	0·00	0·00	0·005	0·005	0·01	0·015	0·02	0·03	0·035	0·04	0·04	0·04	0·04	0·04	0·04
10^4	0·00	0·00	0·00	0·005	0·005	0·01	0·015	0·02	0·025	0·03	0·035	0·035	0·035	0·035	0·035
2×10^4	0·00	0·00	0·00	0·00	0·005	0·005	0·01	0·015	0·02	0·025	0·03	0·03	0·035	0·035	0·035
5×10^4	0·00	0·00	0·00	0·00	0·00	0·00	0·005	0·005	0·01	0·015	0·02	0·025	0·03	0·03	0·03
10^5	0·00	0·00	0·00	0·00	0·00	0·00	0·00	0·005	0·005	0·01	0·015	0·02	0·025	0·025	0·03
2×10^5	0·00	0·00	0·00	0·00	0·00	0·00	0·00	0·00	0·005	0·005	0·01	0·015	0·02	0·025	0·025
5×10^5	0·00	0·00	0·00	0·00	0·00	0·00	0·00	0·00	0·00	0·005	0·005	0·005	0·015	0·015	0·02
10^6	0·00	0·00	0·00	0·00	0·00	0·00	0·00	0·00	0·00	0·00	0·00	0·005	0·01	0·01	0·015
D/k_s															
R	$4·38\times10^3$	$8·76\times10^3$	$1·75\times10^4$	$4·38\times10^4$	$8·76\times10^4$	$1·75\times10^5$	$4·38\times10^5$	$8·76\times10^5$	$1·75\times10^6$	$4·38\times10^6$	$8·76\times10^6$	$1·75\times10^7$	$4·38\times10^7$	$8·76\times10^7$	$1·75\times10^8$

Equivalent Reynolds number for water at 15°C

$$D_R = D_T \left[\frac{Q_G}{Q_T} \right]^{1/x} \left[\frac{S_T}{S_G} \right]^{1/z} \left[\frac{(k_s)_G}{(k_s)_T} \right]^{w} \qquad (10)$$

Note that if the change in value of k_s is sufficient to modify the value of D/k_s as leading to the selection of the appropriate line on the table, this shift should influence accordingly the estimate of medial value of exponent.

C1

Circular pipe

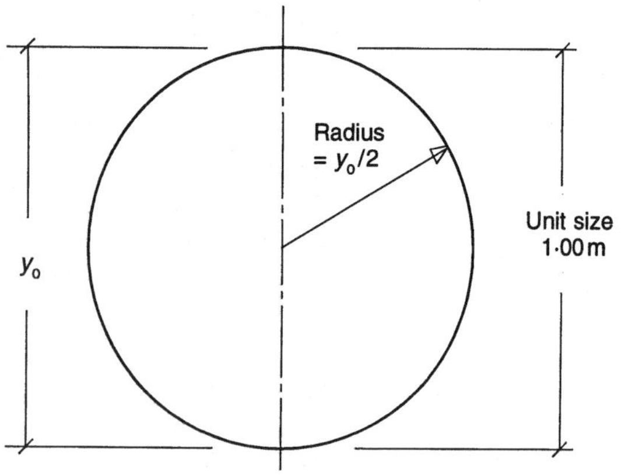

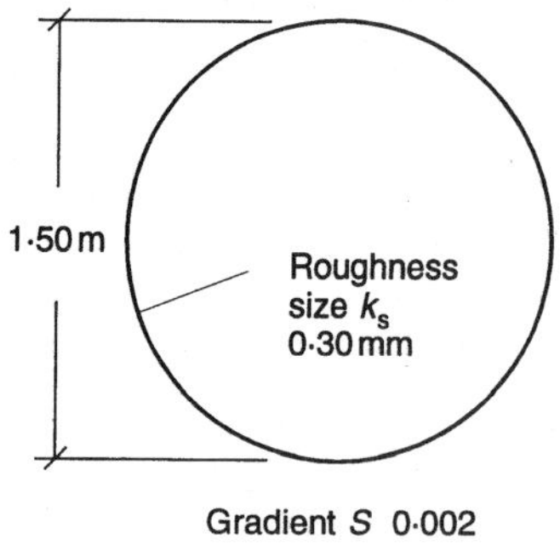

Representative case for discharge ratios

Prop dpth Y	U.equ. dia. $D_{ep(u)}$ (m)	Equiv. disch. factor J	Unit sect. area A_u (m^2)	Unit wetted perim. P_u (m)	Unit surf. brdth $B_{s(u)}$ (m)	Unit mean depth $y_{m(u)}$ (m)	Disch. ratio $Q/Q_{1\cdot00}$ basic	Veloc. ratio V/V_{basic}	Disch. ratio $Q/Q_{1\cdot00}$ adjstd	U.crit. disch. $Q_{c(u)}$ (m^3s^{-1})	Unit depth cntrd $y_{d(u)}$ (m)
0·02	0·0528	0·5849	0·0037	0·2838	0·2800	0·0134	0·001	1·021	0·001	0·0014	0·0080
0·04	0·1047	0·8165	0·0105	0·4027	0·3919	0·0269	0·003	1·006	0·003	0·0054	0·0161
0·06	0·1555	0·9870	0·0192	0·4949	0·4750	0·0405	0·008	0·993	0·008	0·0121	0·0241
0·08	0·2053	1·1246	0·0294	0·5735	0·5426	0·0542	0·014	0·982	0·014	0·0215	0·0322
0·10	0·2541	1·2404	0·0409	0·6435	0·6000	0·0681	0·022	0·971	0·022	0·0334	0·0404
0·11	0·2781	1·2921	0·0470	0·6761	0·6258	0·0751	0·027	0·967	0·026	0·0403	0·0444
0·12	0·3018	1·3403	0·0534	0·7075	0·6499	0·0821	0·032	0·962	0·031	0·0479	0·0485
0·13	0·3253	1·3854	0·0600	0·7377	0·6726	0·0892	0·038	0·958	0·036	0·0561	0·0526
0·14	0·3485	1·4276	0·0668	0·7670	0·6940	0·0963	0·044	0·953	0·042	0·0649	0·0567
0·15	0·3715	1·4674	0·0739	0·7954	0·7141	0·1034	0·051	0·949	0·048	0·0744	0·0608
0·16	0·3942	1·5047	0·0811	0·8230	0·7332	0·1106	0·058	0·946	0·055	0·0845	0·0650
0·17	0·4166	1·5400	0·0885	0·8500	0·7513	0·1178	0·066	0·942	0·062	0·0952	0·0691
0·18	0·4388	1·5732	0·0961	0·8763	0·7684	0·1251	0·074	0·939	0·069	0·1065	0·0732
0·19	0·4607	1·6046	0·1039	0·9021	0·7846	0·1324	0·082	0·936	0·077	0·1184	0·0774
0·20	0·4824	1·6342	0·1118	0·9273	0·8000	0·1398	0·091	0·933	0·085	0·1309	0·0816
0·21	0·5037	1·6622	0·1199	0·9521	0·8146	0·1472	0·100	0·930	0·093	0·1440	0·0857
0·22	0·5248	1·6887	0·1281	0·9764	0·8285	0·1546	0·110	0·927	0·102	0·1578	0·0899
0·23	0·5457	1·7136	0·1365	1·0004	0·8417	0·1621	0·120	0·925	0·111	0·1721	0·0941
0·24	0·5662	1·7372	0·1449	1·0239	0·8542	0·1697	0·130	0·923	0·120	0·1870	0·0983
0·25	0·5865	1·7595	0·1535	1·0472	0·8660	0·1773	0·141	0·921	0·130	0·2025	0·1025
0·26	0·6065	1·7805	0·1623	1·0701	0·8773	0·1850	0·152	0·919	0·140	0·2185	0·1067
0·27	0·6262	1·8003	0·1711	1·0928	0·8879	0·1927	0·163	0·917	0·150	0·2352	0·1110
0·28	0·6457	1·8190	0·1800	1·1152	0·8980	0·2005	0·175	0·916	0·161	0·2524	0·1152
0·29	0·6649	1·8365	0·1890	1·1374	0·9075	0·2083	0·187	0·914	0·171	0·2702	0·1195
0·30	0·6838	1·8530	0·1982	1·1593	0·9165	0·2162	0·200	0·913	0·183	0·2886	0·1237
0·31	0·7024	1·8684	0·2074	1·1810	0·9250	0·2242	0·213	0·912	0·194	0·3075	0·1280
0·32	0·7207	1·8828	0·2167	1·2025	0·9330	0·2322	0·226	0·911	0·206	0·3270	0·1323
0·33	0·7387	1·8963	0·2260	1·2239	0·9404	0·2404	0·239	0·911	0·218	0·3470	0·1366
0·34	0·7565	1·9088	0·2355	1·2451	0·9474	0·2485	0·253	0·910	0·230	0·3676	0·1410
0·35	0·7740	1·9204	0·2450	1·2661	0·9539	0·2568	0·267	0·910	0·243	0·3888	0·1453
0·36	0·7911	1·9312	0·2546	1·2870	0·9600	0·2652	0·281	0·910	0·255	0·4105	0·1496
0·37	0·8080	1·9411	0·2642	1·3078	0·9656	0·2736	0·295	0·910	0·269	0·4327	0·1540
0·38	0·8246	1·9501	0·2739	1·3284	0·9708	0·2821	0·310	0·910	0·282	0·4555	0·1584
0·39	0·8409	1·9584	0·2836	1·3490	0·9755	0·2907	0·325	0·910	0·296	0·4788	0·1628
0·40	0·8569	1·9658	0·2934	1·3694	0·9798	0·2994	0·340	0·910	0·309	0·5027	0·1672
0·41	0·8726	1·9725	0·3032	1·3898	0·9837	0·3082	0·355	0·911	0·323	0·5271	0·1716
0·42	0·8880	1·9784	0·3130	1·4101	0·9871	0·3171	0·371	0·911	0·338	0·5521	0·1760
0·43	0·9031	1·9835	0·3229	1·4303	0·9902	0·3261	0·386	0·912	0·352	0·5775	0·1805
0·44	0·9179	1·9880	0·3328	1·4505	0·9928	0·3353	0·402	0·913	0·367	0·6035	0·1850
0·45	0·9323	1·9917	0·3428	1·4706	0·9950	0·3445	0·418	0·914	0·382	0·6301	0·1895

Circular pipe

C1

continued

Prop dpth Y	U.equ. dia. $D_{ep(u)}$ (m)	Equiv. disch. factor J	Unit sect. area A_u (m^2)	Unit wetted perim. P_u (m)	Unit surf. brdth $B_{s(u)}$ (m)	Unit mean depth $y_{m(u)}$ (m)	Disch. ratio $Q/Q_{1\cdot00}$ basic	Veloc. ratio V/V_{basic}	Disch. ratio $Q/Q_{1\cdot00}$ adjstd	U.crit. disch. $Q_{c(u)}$ (m^3s^{-1})	Unit depth cntrd $y_{d(u)}$ (m)
0·46	0·9465	1·9947	0·3527	1·4907	0·9968	0·3539	0·434	0·914	0·397	0·6571	0·1940
0·47	0·9604	1·9970	0·3627	1·5108	0·9982	0·3634	0·451	0·916	0·413	0·6847	0·1985
0·48	0·9739	1·9987	0·3727	1·5308	0·9992	0·3730	0·467	0·917	0·428	0·7128	0·2031
0·49	0·9871	1·9997	0·3827	1·5508	0·9998	0·3828	0·483	0·918	0·444	0·7415	0·2076
0·50	1·0000	2·0000	0·3927	1·5708	1·0000	0·3927	0·500	0·919	0·460	0·7706	0·2122
0·51	1·0126	1·9997	0·4027	1·5908	0·9998	0·4028	0·517	0·920	0·476	0·8003	0·2168
0·52	1·0248	1·9987	0·4127	1·6108	0·9992	0·4130	0·533	0·922	0·492	0·8306	0·2214
0·53	1·0367	1·9971	0·4227	1·6308	0·9982	0·4234	0·550	0·923	0·508	0·8613	0·2261
0·54	1·0483	1·9949	0·4327	1·6509	0·9968	0·4340	0·567	0·925	0·524	0·8926	0·2308
0·55	1·0595	1·9921	0·4426	1·6710	0·9950	0·4448	0·584	0·926	0·541	0·9245	0·2355
0·56	1·0704	1·9886	0·4526	1·6911	0·9928	0·4558	0·601	0·928	0·557	0·9568	0·2402
0·57	1·0810	1·9846	0·4625	1·7113	0·9902	0·4671	0·617	0·929	0·574	0·9898	0·2449
0·58	1·0912	1·9799	0·4724	1·7315	0·9871	0·4785	0·634	0·931	0·591	1·0232	0·2497
0·59	1·1011	1·9746	0·4822	1·7518	0·9837	0·4902	0·651	0·933	0·607	1·0573	0·2545
0·60	1·1106	1·9688	0·4920	1·7722	0·9798	0·5022	0·668	0·934	0·624	1·0919	0·2593
0·61	1·1197	1·9623	0·5018	1·7926	0·9755	0·5144	0·684	0·936	0·641	1·1271	0·2642
0·62	1·1285	1·9553	0·5115	1·8132	0·9708	0·5269	0·701	0·938	0·657	1·1628	0·2690
0·63	1·1369	1·9477	0·5212	1·8338	0·9656	0·5398	0·718	0·939	0·674	1·1992	0·2739
0·64	1·1449	1·9395	0·5308	1·8546	0·9600	0·5530	0·734	0·941	0·691	1·2362	0·2789
0·65	1·1526	1·9307	0·5404	1·8755	0·9539	0·5665	0·750	0·943	0·707	1·2738	0·2839
0·66	1·1599	1·9213	0·5499	1·8965	0·9474	0·5804	0·766	0·944	0·724	1·3120	0·2889
0·67	1·1667	1·9113	0·5594	1·9177	0·9404	0·5948	0·782	0·946	0·740	1·3510	0·2939
0·68	1·1732	1·9008	0·5687	1·9391	0·9330	0·6096	0·798	0·947	0·756	1·3906	0·2990
0·69	1·1793	1·8896	0·5780	1·9606	0·9250	0·6249	0·814	0·949	0·772	1·4309	0·3041
0·70	1·1849	1·8779	0·5872	1·9823	0·9165	0·6407	0·829	0·950	0·788	1·4720	0·3093
0·71	1·1902	1·8656	0·5964	2·0042	0·9075	0·6571	0·844	0·952	0·804	1·5139	0·3144
0·72	1·1950	1·8526	0·6054	2·0264	0·8980	0·6741	0·859	0·953	0·819	1·5566	0·3197
0·73	1·1994	1·8391	0·6143	2·0488	0·8879	0·6919	0·874	0·955	0·834	1·6001	0·3250
0·74	1·2033	1·8249	0·6231	2·0715	0·8773	0·7103	0·888	0·956	0·849	1·6446	0·3303
0·75	1·2067	1·8101	0·6319	2·0944	0·8660	0·7296	0·902	0·958	0·864	1·6901	0·3357
0·76	1·2097	1·7947	0·6405	2·1176	0·8542	0·7498	0·916	0·959	0·878	1·7367	0·3411
0·77	1·2123	1·7786	0·6489	2·1412	0·8417	0·7710	0·929	0·960	0·892	1·7844	0·3466
0·78	1·2143	1·7619	0·6573	2·1652	0·8285	0·7933	0·942	0·962	0·906	1·8334	0·3521
0·79	1·2158	1·7444	0·6655	2·1895	0·8146	0·8169	0·954	0·963	0·919	1·8837	0·3577
0·80	1·2168	1·7263	0·6736	2·2143	0·8000	0·8420	0·966	0·964	0·932	1·9355	0·3633
0·81	1·2172	1·7075	0·6815	2·2395	0·7846	0·8686	0·978	0·965	0·944	1·9890	0·3691
0·82	1·2171	1·6879	0·6893	2·2653	0·7684	0·8970	0·989	0·967	0·956	2·0443	0·3748
0·83	1·2164	1·6675	0·6969	2·2916	0·7513	0·9276	1·000	0·968	0·967	2·1018	0·3807
0·84	1·2150	1·6464	0·7043	2·3186	0·7332	0·9605	1·010	0·969	0·978	2·1616	0·3866
0·85	1·2131	1·6243	0·7115	2·3462	0·7141	0·9963	1·019	0·970	0·988	2·2241	0·3927
0·86	1·2104	1·6014	0·7186	2·3746	0·6940	1·0354	1·028	0·971	0·998	2·2897	0·3988
0·87	1·2071	1·5775	0·7254	2·4039	0·6726	1·0785	1·036	0·972	1·007	2·3591	0·4050
0·88	1·2029	1·5526	0·7320	2·4341	0·6499	1·1263	1·043	0·974	1·015	2·4328	0·4113
0·89	1·1980	1·5265	0·7384	2·4655	0·6258	1·1800	1·049	0·975	1·023	2·5118	0·4177
0·90	1·1921	1·4992	0·7445	2·4981	0·6000	1·2409	1·055	0·976	1·030	2·5972	0·4242
0·91	1·1853	1·4706	0·7504	2·5322	0·5724	1·3110	1·060	0·977	1·035	2·6906	0·4308
0·92	1·1775	1·4404	0·7560	2·5681	0·5426	1·3933	1·063	0·979	1·040	2·7943	0·4376
0·93	1·1684	1·4085	0·7612	2·6061	0·5103	1·4917	1·066	0·980	1·044	2·9115	0·4445
0·94	1·1579	1·3745	0·7662	2·6467	0·4750	1·6131	1·067	0·981	1·047	3·0472	0·4517
0·95	1·1458	1·3379	0·7707	2·6906	0·4359	1·7681	1·066	0·983	1·048	3·2093	0·4590
0·96	1·1316	1·2980	0·7749	2·7389	0·3919	1·9771	1·064	0·985	1·047	3·4119	0·4665
0·97	1·1148	1·2538	0·7785	2·7934	0·3412	2·2819	1·059	0·987	1·045	3·6829	0·4743
0·98	1·0941	1·2027	0·7816	2·8578	0·2800	2·7916	1·051	0·989	1·040	4·0898	0·4823
0·99	1·0663	1·1389	0·7841	2·9413	0·1990	3·9401	1·038	0·992	1·030	4·8738	0·4908
1·00	1·0000	1·0000	0·7854	3·1416	0·0000	-	1·000	1·000	1·000	-	0·5000

C1(a) Prop. discharges in circular pipes - 'adjusted' values for stated $V_{r(0\cdot5)}$ values
(p.1 of 2)

θ for full pipe is $[\{ k_s/D \} + \{1/(416\cdot81 \times 10^3 D^{3/2} S^{1/2})\}]^{-1}$
(Water at 15°C - k_s and D in m). Values of $V_{r(0\cdot5)}$ are given in Tables A and E

$V_{r(0\cdot5)}$	0·74	0·76	0·78	0·79	0·80	0·81	0·82	0·83	0·84	0·85	0·86	0·87	0·88
θ	64·91	151·3	270·9	345·8	432·8	533·9	651·7	789·5	951·4	1142	1371	1646	1981
P.depth													
0·02	0·0005	0·0006	0·0007	0·0007	0·0007	0·0007	0·0007	0·0007	0·0007	0·0007	0·0007	0·0007	0·0007
0·04	0·0026	0·0028	0·0030	0·0030	0·0030	0·0031	0·0031	0·0031	0·0031	0·0031	0·0031	0·0031	0·0031
0·06	0·0063	0·0067	0·0070	0·0070	0·0071	0·0071	0·0072	0·0072	0·0073	0·0073	0·0073	0·0073	0·0074
0·08	0·0115	0·0122	0·0125	0·0127	0·0128	0·0129	0·0130	0·0130	0·0131	0·0132	0·0132	0·0133	0·0133
0·10	0·0182	0·0191	0·0196	0·0198	0·0200	0·0202	0·0203	0·0204	0·0205	0·0207	0·0208	0·0208	0·0209
0·12	0·0262	0·0274	0·0281	0·0284	0·0287	0·0289	0·0291	0·0293	0·0295	0·0297	0·0298	0·0300	0·0302
0·14	0·0355	0·0370	0·0379	0·0383	0·0386	0·0390	0·0393	0·0396	0·0399	0·0401	0·0404	0·0406	0·0409
0·16	0·0458	0·0477	0·0489	0·0494	0·0499	0·0503	0·0508	0·0512	0·0516	0·0520	0·0524	0·0527	0·0531
0·18	0·0573	0·0595	0·0610	0·0617	0·0623	0·0629	0·0635	0·0640	0·0646	0·0651	0·0656	0·0661	0·0666
0·20	0·0698	0·0723	0·0742	0·0750	0·0758	0·0766	0·0774	0·0781	0·0788	0·0795	0·0802	0·0809	0·0815
0·22	0·0832	0·0861	0·0884	0·0895	0·0905	0·0914	0·0924	0·0933	0·0942	0·0951	0·0960	0·0969	0·0977
0·24	0·0975	0·1009	0·1037	0·1049	0·1061	0·1073	0·1085	0·1096	0·1107	0·1118	0·1130	0·1140	0·1151
0·26	0·1128	0·1166	0·1199	0·1213	0·1228	0·1242	0·1256	0·1270	0·1284	0·1297	0·1311	0·1324	0·1337
0·28	0·1289	0·1333	0·1370	0·1387	0·1405	0·1421	0·1438	0·1454	0·1471	0·1487	0·1503	0·1519	0·1535
0·30	0·1459	0·1509	0·1551	0·1571	0·1591	0·1611	0·1630	0·1649	0·1669	0·1688	0·1707	0·1726	0·1744
0·32	0·1639	0·1694	0·1742	0·1765	0·1788	0·1810	0·1833	0·1855	0·1877	0·1899	0·1921	0·1943	0·1965
0·34	0·1828	0·1888	0·1942	0·1968	0·1994	0·2020	0·2045	0·2070	0·2096	0·2121	0·2146	0·2171	0·2196
0·36	0·2027	0·2092	0·2153	0·2182	0·2211	0·2239	0·2268	0·2296	0·2325	0·2353	0·2381	0·2409	0·2438
0·38	0·2235	0·2306	0·2373	0·2405	0·2437	0·2469	0·2501	0·2532	0·2564	0·2595	0·2627	0·2658	0·2690
0·40	0·2454	0·2530	0·2603	0·2638	0·2673	0·2708	0·2743	0·2778	0·2813	0·2848	0·2883	0·2917	0·2952
0·42	0·2682	0·2764	0·2843	0·2881	0·2920	0·2958	0·2996	0·3034	0·3072	0·3110	0·3148	0·3186	0·3224
0·44	0·2921	0·3008	0·3093	0·3134	0·3176	0·3217	0·3258	0·3299	0·3341	0·3382	0·3423	0·3464	0·3506
0·46	0·3171	0·3262	0·3352	0·3397	0·3441	0·3486	0·3530	0·3574	0·3619	0·3663	0·3707	0·3751	0·3796
0·48	0·3430	0·3526	0·3622	0·3669	0·3716	0·3763	0·3811	0·3858	0·3905	0·3952	0·4000	0·4047	0·4094
0·50	0·3700	0·3800	0·3900	0·3950	0·4000	0·4050	0·4100	0·4150	0·4200	0·4250	0·4300	0·4350	0·4400
0·52	0·3979	0·4083	0·4187	0·4240	0·4292	0·4345	0·4397	0·4450	0·4502	0·4555	0·4608	0·4660	0·4713
0·54	0·4268	0·4374	0·4483	0·4537	0·4592	0·4647	0·4702	0·4757	0·4812	0·4867	0·4921	0·4976	0·5031
0·56	0·4565	0·4673	0·4785	0·4842	0·4899	0·4956	0·5013	0·5070	0·5127	0·5184	0·5241	0·5298	0·5355
0·58	0·4870	0·4980	0·5095	0·5153	0·5212	0·5270	0·5329	0·5388	0·5447	0·5506	0·5564	0·5623	0·5682
0·60	0·5182	0·5293	0·5410	0·5469	0·5529	0·5590	0·5650	0·5710	0·5771	0·5831	0·5891	0·5952	0·6012
0·62	0·5500	0·5611	0·5729	0·5790	0·5851	0·5912	0·5974	0·6035	0·6097	0·6158	0·6220	0·6282	0·6343
0·64	0·5822	0·5932	0·6052	0·6113	0·6175	0·6237	0·6299	0·6362	0·6424	0·6487	0·6549	0·6621	0·6674
0·66	0·6148	0·6257	0·6376	0·6438	0·6500	0·6563	0·6626	0·6688	0·6751	0·6815	0·6877	0·6940	0·7003
0·68	0·6475	0·6582	0·6701	0·6763	0·6825	0·6888	0·6951	0·7014	0·7077	0·7140	0·7203	0·7266	0·7329
0·70	0·6804	0·6907	0·7025	0·7086	0·7148	0·7210	0·7273	0·7336	0·7399	0·7462	0·7525	0·7588	0·7650
0·72	0·7180	0·7230	0·7346	0·7406	0·7467	0·7529	0·7591	0·7653	0·7715	0·7778	0·7840	0·7903	0·7965
0·74	0·7454	0·7549	0·7662	0·7721	0·7781	0·7841	0·7902	0·7964	0·8025	0·8087	0·8148	0·8209	0·8271
0·76	0·7772	0·7862	0·7971	0·8029	0·8087	0·8146	0·8206	0·8266	0·8326	0·8386	0·8446	0·8506	0·8566
0·78	0·8084	0·8168	0·8273	0·8328	0·8385	0·8442	0·8500	0·8558	0·8616	0·8675	0·8733	0·8791	0·8850
0·80	0·8386	0·8464	0·8564	0·8617	0·8671	0·8726	0·8781	0·8837	0·8893	0·8950	0·9006	0·9062	0·9119
0·82	0·8677	0·8748	0·8842	0·8892	0·8944	0·8996	0·9049	0·9102	0·9156	0·9209	0·9263	0·9317	0·9371
0·84	0·8954	0·9019	0·9105	0·9153	0·9201	0·9250	0·9300	0·9350	0·9401	0·9451	0·9502	0·9553	0·9604
0·86	0·9216	0·9272	0·9351	0·9395	0·9440	0·9485	0·9531	0·9578	0·9625	0·9673	0·9720	0·9768	0·9815
0·88	0·9457	0·9506	0·9577	0·9616	0·9657	0·9699	0·9741	0·9784	0·9827	0·9871	0·9914	0·9958	1·0002
0·90	0·9676	0·9716	0·9778	0·9813	0·9850	0·9887	0·9925	0·9963	1·0002	1·0041	1·0080	1·0120	1·0159
0·92	0·9866	0·9898	0·9951	0·9981	1·0012	1·0045	1·0078	1·0111	1·0145	1·0179	1·0214	1·0248	1·0283
0·94	1·0022	1·0045	1·0088	1·0113	1·0139	1·0166	1·0193	1·0221	1·0250	1·0278	1·0307	0·0337	1·0366
0·96	1·0133	1·0148	1·0180	1·0199	1·0219	1·0239	1·0261	1·0282	1·0304	1·0327	1·0349	1·0372	1·0395
0·98	1·0177	1·0184	1·0204	1·0216	1·0229	1·0242	1·0256	1·0270	1·0285	1·0299	1·0314	1·0329	1·0345
1·00	1·0000	1·0000	1·0000	1·0000	1·0000	1·0000	1·0000	1·0000	1·0000	1·0000	1·0000	1·0000	1·0000
$V_{r(0\cdot5)}$	0·74	0·76	0·78	0·79	0·80	0·81	0·82	0·83	0·84	0·85	0·86	0·87	0·88

Prop. discharges in circular pipes - 'adjusted' values for stated $V_{r(0{\cdot}5)}$ values C1(a)

(p.2 of 2)

θ for full pipe is $[\{ k_s/D \} + \{1/(416{\cdot}81\times10^3 D^{3/2} S^{1/2})\}]^{-1}$

(Water at 15°C - k_s and D in m). Values of $V_{r(0{\cdot}5)}$ are given in Tables A and E

$V_{r(0{\cdot}5)}$	0·88	0·89	0·90	0·91	0·92	0·93	0·94	0·95	0·96	0·97	0·98	0·99	0·995
θ	1981	2394	2913	3577	4449	5630	7299	9792	13820	21170	37728	97611	-
P.depth													
0·02	0·0007	0·0007	0·0007	0·0007	0·0007	0·0007	0·0007	0·0007	0·0007	0·0007	0·0007	0·0007	0·0007
0·04	0·0031	0·0031	0·0031	0·0031	0·0031	0·0031	0·0031	0·0031	0·0030	0·0030	0·0031	0·0032	0·0033
0·06	0·0074	0·0074	0·0074	0·0074	0·0074	0·0073	0·0073	0·0073	0·0072	0·0072	0·0073	0·0075	0·0077
0·08	0·0133	0·0133	0·0134	0·0134	0·0134	0·0134	0·0133	0·0134	0·0132	0·0132	0·0134	0·0137	0·0140
0·10	0·0209	0·0210	0·0211	0·0211	0·0212	0·0212	0·0212	0·0213	0·0211	0·0210	0·0214	0·0219	0·0223
0·12	0·0302	0·0303	0·0304	0·0305	0·0306	0·0307	0·0307	0·0309	0·0307	0·0307	0·0312	0·0319	0·0325
0·14	0·0409	0·0411	0·0413	0·0415	0·0417	0·0418	0·0420	0·0422	0·0422	0·0421	0·0428	0·0437	0·0445
0·16	0·0531	0·0534	0·0537	0·0541	0·0543	0·0546	0·0548	0·0552	0·0552	0·0553	0·0562	0·0574	0·0584
0·18	0·0666	0·0671	0·0676	0·0680	0·0685	0·0689	0·0692	0·0698	0·0698	0·0701	0·0712	0·0727	0·0740
0·20	0·0815	0·0822	0·0828	0·0835	0·0841	0·0846	0·0852	0·0859	0·0860	0·0866	0·0879	0·0898	0·0913
0·22	0·0977	0·0986	0·0994	0·1002	0·1010	0·1018	0·1025	0·1035	0·1038	0·1045	0·1062	0·1084	0·1102
0·24	0·1151	0·1162	0·1173	0·1183	0·1193	0·1203	0·1213	0·1225	0·1230	0·1240	0·1260	0·1286	0·1307
0·26	0·1337	0·1351	0·1364	0·1377	0·1390	0·1402	0·1414	0·1429	0·1437	0·1450	0·1473	0·1503	0·1526
0·28	0·1535	0·1551	0·1567	0·1583	0·1598	0·1614	0·1629	0·1646	0·1657	0·1673	0·1699	0·1733	0·1759
0·30	0·1744	0·1763	0·1782	0·1801	0·1819	0·1837	0·1856	0·1876	0·1890	0·1909	0·1939	0·1977	0·2006
0·32	0·1965	0·1986	0·2008	0·2030	0·2052	0·2073	0·2094	0·2118	0·2136	0·2158	0·2191	0·2233	0·2264
0·34	0·2196	0·2221	0·2246	0·2271	0·2296	0·2320	0·2345	0·2372	0·2393	0·2419	0·2455	0·2500	0·2533
0·36	0·2438	0·2466	0·2494	0·2522	0·2551	0·2579	0·2607	0·2637	0·2662	0·2691	0·2731	0·2779	0·2813
0·38	0·2690	0·2721	0·2753	0·2785	0·2816	0·2848	0·2879	0·2912	0·2941	0·2974	0·3016	0·3067	0·3102
0·40	0·2952	0·2987	0·3022	0·3057	0·3092	0·3126	0·3161	0·3198	0·3230	0·3267	0·3311	0·3364	0·3399
0·42	0·3224	0·3262	0·3300	0·3339	0·3377	0·3415	0·3453	0·3493	0·3529	0·3568	0·3615	0·3669	0·3703
0·44	0·3506	0·3547	0·3588	0·3629	0·3671	0·3712	0·3754	0·3796	0·3836	0·3879	0·3927	0·3981	0·4014
0·46	0·3796	0·3840	0·3884	0·3929	0·3973	0·4018	0·4062	0·4107	0·4151	0·4196	0·4246	0·4299	0·4330
0·48	0·4094	0·4141	0·4189	0·4236	0·4283	0·4331	0·4378	0·4426	0·4472	0·4520	0·4570	0·4622	0·4651
0·50	0·4400	0·4450	0·4500	0·4550	0·4600	0·4650	0·4700	0·4750	0·4800	0·4850	0·4900	0·4950	0·4975
0·52	0·4713	0·4765	0·4818	0·4870	0·4923	0·4975	0·5028	0·5080	0·5132	0·5184	0·5234	0·5281	0·5302
0·54	0·5031	0·5086	0·5141	0·5196	0·5250	0·5305	0·5360	0·5413	0·5469	0·5522	0·5570	0·5614	0·5630
0·56	0·5355	0·5412	0·5468	0·5525	0·5582	0·5638	0·5695	0·5750	0·5807	0·5862	0·5909	0·5948	0·5958
0·58	0·5682	0·5741	0·5799	0·5858	0·5916	0·5974	0·6032	0·6088	0·6148	0·6203	0·6248	0·6282	0·6287
0·60	0·6012	0·6072	0·6132	0·6192	0·6251	0·6311	0·6370	0·6427	0·6488	0·6544	0·6587	0·6616	0·6614
0·62	0·6343	0·6404	0·6465	0·6526	0·6587	0·6647	0·6707	0·6765	0·6827	0·6883	0·6923	0·6947	0·6938
0·64	0·6674	0·6736	0·6798	0·6859	0·6921	0·6982	0·7043	0·7100	0·7164	0·7220	0·7257	0·7274	0·7259
0·66	0·7003	0·7066	0·7128	0·7190	0·7252	0·7313	0·7375	0·7432	0·7496	0·7551	0·7586	0·7597	0·7575
0·68	0·7329	0·7392	0·7454	0·7517	0·7578	0·7640	0·7701	0·7758	0·7823	0·7877	0·7908	0·7914	0·7886
0·70	0·7650	0·7713	0·7775	0·7837	0·7899	0·7960	0·8021	0·8077	0·8142	0·8196	0·8223	0·8223	0·8190
0·72	0·7965	0·8027	0·8088	0·8150	0·8211	0·8271	0·8332	0·8387	0·8452	0·8505	0·8529	0·8524	0·8486
0·74	0·8271	0·8332	0·8393	0·8453	0·8513	0·8573	0·8632	0·8686	0·8751	0·8808	0·8824	0·8814	0·8772
0·76	0·8566	0·8626	0·8686	0·8745	0·8804	0·8862	0·8920	0·8973	0·9037	0·9087	0·9106	0·9092	0·9047
0·78	0·8850	0·8908	0·8966	0·9023	0·9081	0·9138	0·9194	0·9245	0·9308	0·9357	0·9374	0·9357	0·9309
0·80	0·9119	0·9175	0·9230	0·9286	0·9342	0·9397	0·9452	0·9500	0·9562	0·9610	0·9624	0·9605	0·9556
0·82	0·9371	0·9424	0·9478	0·9531	0·9584	0·9637	0·9690	0·9737	0·9797	0·9843	0·9855	0·9835	0·9786
0·84	0·9604	0·9655	0·9705	0·9756	0·9807	0·9857	0·9907	0·9952	1·0010	1·0054	1·0065	1·0044	0·9996
0·86	0·9815	0·9863	0·9910	0·9956	1·0005	1·0053	1·0101	1·0142	1·0198	1·0241	1·0250	1·0230	1·0184
0·88	1·0002	1·0046	1·0090	1·0134	1·0178	1·0222	1·0266	1·0305	1·0358	1·0398	1·0406	1·0388	1·0346
0·90	1·0159	1·0199	1·0239	1·0279	1·0319	1·0359	1·0400	1·0435	1·0485	1·0522	1·0530	1·0514	1·0477
0·92	1·0283	1·0318	1·0353	1·0389	1·0424	1·0460	1·0496	1·0528	1·0573	1·0607	1·0615	1·0602	1·0571
0·94	1·0366	1·0396	1·0425	1·0455	1·0485	1·0516	1·0548	1·0575	1·0615	1·0645	1·0652	1·0642	1·0618
0·96	1·0395	1·0419	1·0442	1·0466	1·0490	1·0514	1·0540	1·0562	1·0595	1·0620	1·0626	1·0621	1·0604
0·98	1·0345	1·0360	1·0376	1·0392	1·0408	1·0425	1·0442	1·0458	1·0481	1·0499	1·0505	1·0503	1·0494
1·00	1·0000	1·0000	1·0000	1·0000	1·0000	1·0000	1·0000	1·0000	1·0000	1·0000	1·0000	1·0000	1·0000
$V_{r(0{\cdot}5)}$	0·88	0·89	0·90	0·91	0·92	0·93	0·94	0·95	0·96	0·97	0·98	0·99	0·995

C2

Form 1 egg-shape (3:2 old type)

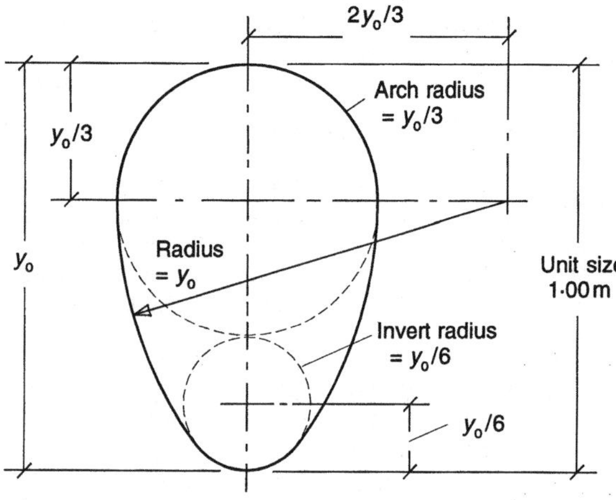

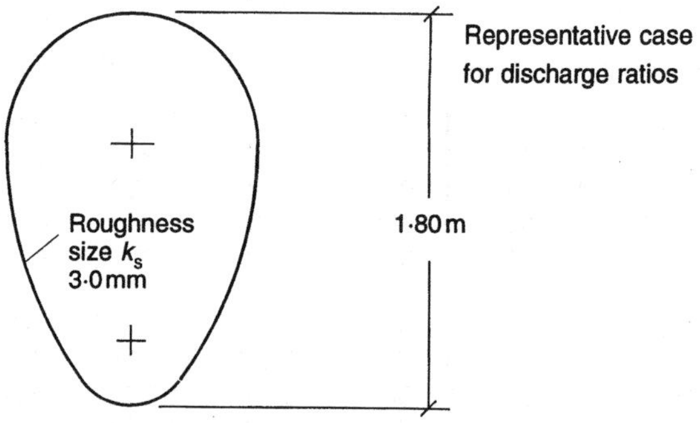

Gradient S 0·0010

Prop dpth Y	U.equ. dia. $D_{ep(u)}$ (m)	Equiv. disch. factor J	Unit sect. area A_u (m^2)	Unit wetted perim. P_u (m)	Unit surf. brdth $B_{s(u)}$ (m)	Unit mean depth $y_{m(u)}$ (m)	Disch. ratio $Q/Q_{1\cdot00}$ basic	Veloc. ratio V/V_{basic}	Disch. ratio $Q/Q_{1\cdot00}$ adjstd	U.crit. disch. $Q_{c(u)}$ (m^{3}s^{-1})	Unit depth cntrd $y_{d(u)}$ (m)
0·02	0·0518	0·9870	0·0021	0·1650	0·1583	0·0135	0·001	0·991	0·001	0·0008	0·0080
0·04	0·1006	1·3403	0·0059	0·2358	0·2166	0·0274	0·003	0·910	0·003	0·0031	0·0162
0·06	0·1463	1·5732	0·0107	0·2921	0·2561	0·0417	0·007	0·849	0·006	0·0068	0·0244
0·08	0·1883	1·7288	0·0161	0·3422	0·2863	0·0563	0·013	0·807	0·010	0·0120	0·0328
0·10	0·2262	1·8167	0·0221	0·3912	0·3146	0·0703	0·020	0·783	0·015	0·0184	0·0411
0·11	0·2440	1·8455	0·0253	0·4154	0·3281	0·0772	0·024	0·776	0·018	0·0220	0·0452
0·12	0·2612	1·8675	0·0287	0·4393	0·3414	0·0840	0·028	0·771	0·021	0·0260	0·0494
0·13	0·2778	1·8843	0·0322	0·4631	0·3543	0·0908	0·033	0·767	0·025	0·0303	0·0535
0·14	0·2939	1·8972	0·0358	0·4868	0·3668	0·0975	0·038	0·766	0·029	0·0350	0·0576
0·15	0·3097	1·9069	0·0395	0·5102	0·3790	0·1042	0·043	0·765	0·033	0·0399	0·0617
0·16	0·3250	1·9141	0·0433	0·5335	0·3910	0·1109	0·049	0·765	0·037	0·0452	0·0658
0·17	0·3400	1·9193	0·0473	0·5566	0·4025	0·1175	0·055	0·766	0·042	0·0508	0·0698
0·18	0·3547	1·9229	0·0514	0·5796	0·4138	0·1242	0·061	0·768	0·047	0·0567	0·0739
0·19	0·3692	1·9252	0·0556	0·6024	0·4248	0·1309	0·068	0·770	0·052	0·0630	0·0779
0·20	0·3833	1·9264	0·0599	0·6251	0·4355	0·1375	0·075	0·772	0·057	0·0696	0·0820
0·21	0·3972	1·9267	0·0643	0·6476	0·4459	0·1442	0·082	0·775	0·063	0·0765	0·0860
0·22	0·4108	1·9262	0·0688	0·6700	0·4561	0·1509	0·090	0·777	0·069	0·0837	0·0900
0·23	0·4242	1·9250	0·0734	0·6923	0·4659	0·1576	0·098	0·780	0·076	0·0913	0·0941
0·24	0·4374	1·9232	0·0781	0·7145	0·4755	0·1643	0·106	0·783	0·083	0·0992	0·0981
0·25	0·4504	1·9210	0·0829	0·7365	0·4848	0·1711	0·115	0·787	0·090	0·1074	0·1021
0·26	0·4632	1·9183	0·0878	0·7585	0·4938	0·1778	0·124	0·790	0·097	0·1160	0·1062
0·27	0·4757	1·9153	0·0928	0·7803	0·5026	0·1847	0·133	0·793	0·105	0·1249	0·1102
0·28	0·4881	1·9119	0·0979	0·8021	0·5111	0·1915	0·143	0·796	0·113	0·1341	0·1142
0·29	0·5003	1·9082	0·1030	0·8237	0·5194	0·1984	0·153	0·800	0·121	0·1437	0·1183
0·30	0·5123	1·9042	0·1083	0·8453	0·5274	0·2053	0·163	0·803	0·130	0·1536	0·1223
0·31	0·5242	1·9000	0·1136	0·8667	0·5351	0·2122	0·174	0·806	0·139	0·1639	0·1263
0·32	0·5358	1·8955	0·1190	0·8881	0·5426	0·2192	0·184	0·810	0·148	0·1744	0·1304
0·33	0·5473	1·8909	0·1244	0·9094	0·5499	0·2263	0·196	0·813	0·158	0·1853	0·1345
0·34	0·5586	1·8860	0·1300	0·9306	0·5569	0·2333	0·207	0·816	0·168	0·1966	0·1385
0·35	0·5698	1·8810	0·1356	0·9517	0·5637	0·2405	0·219	0·820	0·178	0·2082	0·1426
0·36	0·5808	1·8758	0·1412	0·9727	0·5703	0·2477	0·231	0·823	0·189	0·2201	0·1467
0·37	0·5916	1·8704	0·1470	0·9937	0·5766	0·2549	0·243	0·826	0·199	0·2324	0·1507
0·38	0·6023	1·8648	0·1528	1·0146	0·5827	0·2622	0·255	0·829	0·210	0·2449	0·1548
0·39	0·6128	1·8592	0·1586	1·0355	0·5886	0·2695	0·268	0·832	0·222	0·2579	0·1589
0·40	0·6231	1·8533	0·1645	1·0562	0·5942	0·2769	0·281	0·835	0·233	0·2711	0·1630
0·41	0·6333	1·8474	0·1705	1·0770	0·5997	0·2843	0·294	0·838	0·245	0·2847	0·1671
0·42	0·6433	1·8413	0·1765	1·0976	0·6049	0·2919	0·308	0·841	0·257	0·2986	0·1713
0·43	0·6532	1·8351	0·1826	1·1182	0·6098	0·2994	0·321	0·844	0·270	0·3129	0·1754
0·44	0·6629	1·8288	0·1887	1·1388	0·6146	0·3071	0·335	0·847	0·282	0·3275	0·1796
0·45	0·6725	1·8223	0·1949	1·1593	0·6192	0·3148	0·350	0·850	0·295	0·3424	0·1837

Form 1 egg-shape (3:2 old type)

Prop dpth Y	U.equ. dia. $D_{ep(u)}$ (m)	Equiv. disch. factor J	Unit sect. area A_u (m^2)	Unit wetted perim. P_u (m)	Unit surf. brdth $B_{s(u)}$ (m)	Unit mean depth $y_{m(u)}$ (m)	Disch. ratio $Q/Q_{1\cdot00}$ basic	Veloc. ratio V/V_{basic}	Disch. ratio $Q/Q_{1\cdot00}$ adjstd	U.crit. disch. $Q_{c(u)}$ (m^3s^{-1})	Unit depth cntrd $y_{d(u)}$ (m)
0·46	0·6819	1·8157	0·2011	1·1798	0·6235	0·3226	0·364	0·852	0·308	0·3577	0·1879
0·47	0·6911	1·8090	0·2074	1·2002	0·6276	0·3304	0·378	0·855	0·322	0·3733	0·1921
0·48	0·7002	1·8022	0·2137	1·2206	0·6315	0·3383	0·393	0·858	0·335	0·3892	0·1963
0·49	0·7091	1·7953	0·2200	1·2409	0·6352	0·3463	0·408	0·860	0·349	0·4054	0·2005
0·50	0·7179	1·7883	0·2264	1·2612	0·6387	0·3544	0·423	0·863	0·363	0·4220	0·2047
0·51	0·7266	1·7812	0·2328	1·2815	0·6420	0·3626	0·439	0·865	0·377	0·4389	0·2089
0·52	0·7350	1·7740	0·2392	1·3017	0·6450	0·3708	0·454	0·868	0·392	0·4562	0·2132
0·53	0·7434	1·7667	0·2457	1·3219	0·6479	0·3792	0·470	0·870	0·406	0·4737	0·2174
0·54	0·7515	1·7592	0·2522	1·3421	0·6506	0·3876	0·485	0·873	0·421	0·4916	0·2217
0·55	0·7596	1·7517	0·2587	1·3622	0·6530	0·3961	0·501	0·875	0·436	0·5098	0·2260
0·56	0·7674	1·7441	0·2652	1·3824	0·6553	0·4048	0·517	0·877	0·451	0·5284	0·2303
0·57	0·7752	1·7364	0·2718	1·4025	0·6573	0·4135	0·534	0·880	0·467	0·5473	0·2346
0·58	0·7827	1·7286	0·2784	1·4225	0·6591	0·4223	0·550	0·882	0·482	0·5665	0·2389
0·59	0·7901	1·7207	0·2850	1·4426	0·6608	0·4313	0·566	0·884	0·498	0·5860	0·2433
0·60	0·7974	1·7127	0·2916	1·4627	0·6622	0·4403	0·583	0·886	0·514	0·6059	0·2477
0·61	0·8045	1·7046	0·2982	1·4827	0·6635	0·4495	0·599	0·889	0·529	0·6261	0·2520
0·62	0·8115	1·6964	0·3048	1·5027	0·6645	0·4588	0·616	0·891	0·546	0·6466	0·2564
0·63	0·8183	1·6881	0·3115	1·5228	0·6653	0·4682	0·633	0·893	0·562	0·6675	0·2609
0·64	0·8249	1·6798	0·3182	1·5428	0·6660	0·4777	0·650	0·895	0·578	0·6886	0·2653
0·65	0·8314	1·6713	0·3248	1·5628	0·6664	0·4874	0·667	0·897	0·594	0·7102	0·2698
0·66	0·8377	1·6628	0·3315	1·5828	0·6666	0·4973	0·684	0·899	0·611	0·7320	0·2742
0·67	0·8439	1·6542	0·3381	1·6028	0·6666	0·5072	0·701	0·901	0·627	0·7542	0·2787
0·68	0·8499	1·6454	0·3448	1·6228	0·6661	0·5176	0·718	0·903	0·644	0·7769	0·2832
0·69	0·8558	1·6365	0·3515	1·6428	0·6650	0·5285	0·735	0·905	0·661	0·8001	0·2878
0·70	0·8614	1·6274	0·3581	1·6629	0·6633	0·5399	0·752	0·907	0·678	0·8240	0·2923
0·71	0·8669	1·6181	0·3647	1·6830	0·6610	0·5518	0·769	0·909	0·694	0·8484	0·2970
0·72	0·8721	1·6085	0·3713	1·7032	0·6581	0·5643	0·786	0·911	0·711	0·8735	0·3016
0·73	0·8770	1·5986	0·3779	1·7235	0·6545	0·5774	0·802	0·913	0·728	0·8992	0·3063
0·74	0·8817	1·5883	0·3844	1·7440	0·6503	0·5911	0·819	0·915	0·745	0·9255	0·3110
0·75	0·8861	1·5776	0·3909	1·7646	0·6455	0·6056	0·835	0·917	0·762	0·9526	0·3157
0·76	0·8902	1·5665	0·3973	1·7853	0·6400	0·6208	0·852	0·919	0·778	0·9804	0·3206
0·77	0·8940	1·5550	0·4037	1·8062	0·6338	0·6369	0·868	0·921	0·795	1·0089	0·3254
0·78	0·8975	1·5429	0·4100	1·8274	0·6270	0·6540	0·883	0·924	0·811	1·0383	0·3303
0·79	0·9006	1·5303	0·4162	1·8488	0·6194	0·6720	0·899	0·926	0·827	1·0685	0·3353
0·80	0·9033	1·5171	0·4224	1·8704	0·6110	0·6913	0·914	0·928	0·843	1·0998	0·3404
0·81	0·9056	1·5034	0·4284	1·8924	0·6019	0·7118	0·928	0·930	0·859	1·1320	0·3455
0·82	0·9075	1·4890	0·4344	1·9148	0·5919	0·7339	0·943	0·933	0·874	1·1654	0·3507
0·83	0·9090	1·4739	0·4403	1·9375	0·5811	0·7576	0·956	0·935	0·889	1·2001	0·3559
0·84	0·9100	1·4581	0·4460	1·9607	0·5694	0·7833	0·970	0·938	0·904	1·2362	0·3613
0·85	0·9105	1·4414	0·4517	1·9843	0·5568	0·8112	0·982	0·940	0·918	1·2740	0·3667
0·86	0·9104	1·4240	0·4572	2·0086	0·5431	0·8418	0·994	0·943	0·931	1·3136	0·3722
0·87	0·9098	1·4056	0·4625	2·0335	0·5283	0·8756	1·005	0·945	0·945	1·3553	0·3779
0·88	0·9086	1·3863	0·4677	2·0591	0·5122	0·9131	1·016	0·948	0·957	1·3996	0·3836
0·89	0·9067	1·3659	0·4728	2·0856	0·4949	0·9553	1·025	0·951	0·969	1·4470	0·3895
0·90	0·9042	1·3443	0·4776	2·1130	0·4761	1·0032	1·034	0·954	0·980	1·4981	0·3954
0·91	0·9008	1·3213	0·4823	2·1416	0·4556	1·0585	1·042	0·957	0·990	1·5539	0·4016
0·92	0·8965	1·2969	0·4867	2·1716	0·4333	1·1234	1·048	0·960	1·000	1·6155	0·4079
0·93	0·8913	1·2708	0·4909	2·2033	0·4087	1·2011	1·053	0·963	1·008	1·6849	0·4143
0·94	0·8849	1·2427	0·4949	2·2370	0·3816	1·2970	1·057	0·966	1·015	1·7650	0·4210
0·95	0·8772	1·2122	0·4986	2·2734	0·3512	1·4197	1·059	0·970	1·021	1·8603	0·4278
0·96	0·8678	1·1786	0·5019	2·3133	0·3166	1·5851	1·059	0·974	1·025	1·9788	0·4350
0·97	0·8563	1·1408	0·5049	2·3583	0·2764	1·8266	1·056	0·979	1·027	2·1368	0·4424
0·98	0·8418	1·0967	0·5074	2·4112	0·2274	2·2309	1·050	0·984	1·027	2·3733	0·4501
0·99	0·8217	1·0411	0·5094	2·4796	0·1621	3·1429	1·038	0·991	1·022	2·8279	0·4584
1·00	0·7725	0·9181	0·5105	2·6433	0·0000	-	1·000	1·006	1·000	-	0·4674

Rectangular (free surface)

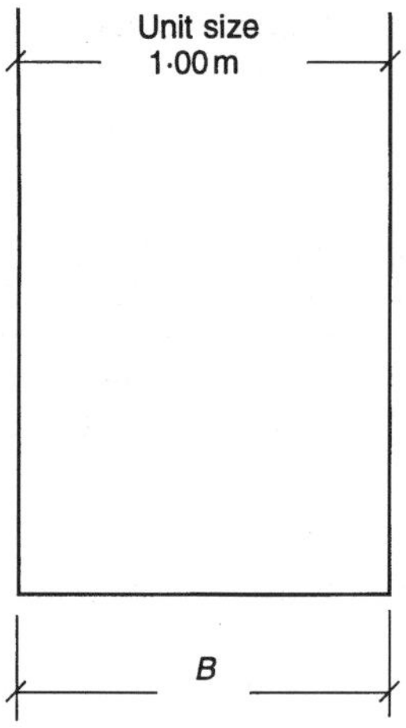

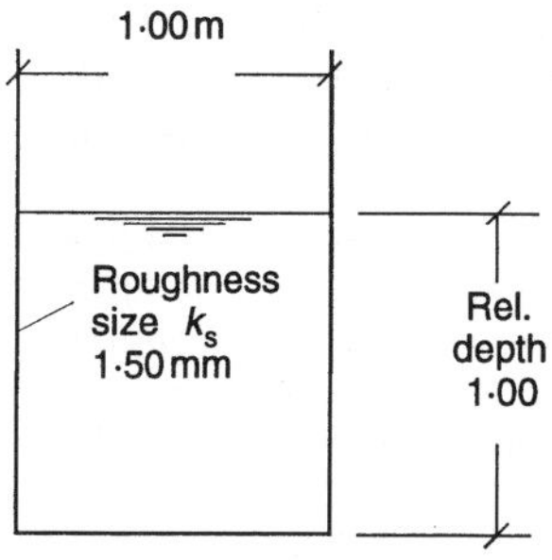

Gradient S 0·002

Representative case for discharge ratios

Rel. dpth Y	U.equ. dia. $D_{ep(u)}$ (m)	Equiv. disch. factor J	Unit sect. area A_u (m^2)	Unit wetted perim. P_u (m)	Unit surf. brdth $B_{s(u)}$ (m)	Unit mean depth $y_{m(u)}$ (m)	Disch. ratio $Q/Q_{1·00}$ basic	Veloc. ratio V/V_{basic}	Disch. ratio $Q/Q_{1·00}$ adjstd	U.crit. disch. $Q_{c(u)}$ (m^3s^{-1})	Unit depth cntrd $y_{d(u)}$ (m)
0·02	0·0769	0·2324	0·0200	1·0400	1·0000	0·0200	0·003	0·942	0·003	0·0089	0·0100
0·04	0·1481	0·4309	0·0400	1·0800	1·0000	0·0400	0·010	0·943	0·010	0·0251	0·0200
0·06	0·2143	0·6011	0·0600	1·1200	1·0000	0·0600	0·019	0·945	0·019	0·0460	0·0300
0·08	0·2759	0·7471	0·0800	1·1600	1·0000	0·0800	0·030	0·946	0·029	0·0709	0·0400
0·10	0·3333	0·8727	0·1000	1·2000	1·0000	0·1000	0·042	0·947	0·041	0·0990	0·0500
0·12	0·3871	0·9807	0·1200	1·2400	1·0000	0·1200	0·056	0·948	0·055	0·1302	0·0600
0·14	0·4375	1·0738	0·1400	1·2800	1·0000	0·1400	0·070	0·949	0·069	0·1640	0·0700
0·16	0·4848	1·1539	0·1600	1·3200	1·0000	0·1600	0·085	0·950	0·084	0·2004	0·0800
0·18	0·5294	1·2229	0·1800	1·3600	1·0000	0·1800	0·102	0·951	0·100	0·2391	0·0900
0·20	0·5714	1·2823	0·2000	1·4000	1·0000	0·2000	0·118	0·951	0·116	0·2801	0·1000
0·22	0·6111	1·3332	0·2200	1·4400	1·0000	0·2200	0·136	0·952	0·134	0·3231	0·1100
0·24	0·6486	1·3769	0·2400	1·4800	1·0000	0·2400	0·154	0·953	0·152	0·3682	0·1200
0·26	0·6842	1·4142	0·2600	1·5200	1·0000	0·2600	0·172	0·954	0·170	0·4152	0·1300
0·28	0·7179	1·4458	0·2800	1·5600	1·0000	0·2800	0·191	0·955	0·189	0·4640	0·1400
0·30	0·7500	1·4726	0·3000	1·6000	1·0000	0·3000	0·210	0·957	0·208	0·5146	0·1500
0·32	0·7805	1·4951	0·3200	1·6400	1·0000	0·3200	0·230	0·959	0·228	0·5669	0·1600
0·34	0·8095	1·5138	0·3400	1·6800	1·0000	0·3400	0·250	0·963	0·249	0·6208	0·1700
0·36	0·8372	1·5292	0·3600	1·7200	1·0000	0·3600	0·270	0·962	0·269	0·6764	0·1800
0·38	0·8636	1·5416	0·3800	1·7600	1·0000	0·3800	0·291	0·962	0·289	0·7336	0·1900
0·40	0·8889	1·5514	0·4000	1·8000	1·0000	0·4000	0·312	0·962	0·310	0·7922	0·2000
0·42	0·9130	1·5589	0·4200	1·8400	1·0000	0·4200	0·333	0·961	0·331	0·8524	0·2100
0·44	0·9362	1·5644	0·4400	1·8800	1·0000	0·4400	0·354	0·961	0·352	0·9140	0·2200
0·46	0·9583	1·5681	0·4600	1·9200	1·0000	0·4600	0·375	0·961	0·373	0·9770	0·2300
0·48	0·9796	1·5701	0·4800	1·9600	1·0000	0·4800	0·397	0·961	0·395	1·0414	0·2400
0·50	1·0000	1·5708	0·5000	2·0000	1·0000	0·5000	0·419	0·961	0·416	1·1072	0·2500
0·52	1·0196	1·5702	0·5200	2·0400	1·0000	0·5200	0·441	0·961	0·438	1·1743	0·2600
0·54	1·0385	1·5685	0·5400	2·0800	1·0000	0·5400	0·463	0·961	0·460	1·2427	0·2700
0·56	1·0566	1·5658	0·5600	2·1200	1·0000	0·5600	0·485	0·961	0·482	1·3123	0·2800
0·58	1·0741	1·5622	0·5800	2·1600	1·0000	0·5800	0·508	0·961	0·505	1·3833	0·2900
0·60	1·0909	1·5578	0·6000	2·2000	1·0000	0·6000	0·530	0·961	0·527	1·4554	0·3000
0·62	1·1071	1·5528	0·6200	2·2400	1·0000	0·6200	0·553	0·962	0·550	1·5288	0·3100
0·64	1·1228	1·5471	0·6400	2·2800	1·0000	0·6400	0·576	0·962	0·573	1·6034	0·3200
0·66	1·1379	1·5409	0·6600	2·3200	1·0000	0·6600	0·599	0·962	0·596	1·6791	0·3300
0·68	1·1525	1·5342	0·6800	2·3600	1·0000	0·6800	0·622	0·962	0·619	1·7560	0·3400
0·70	1·1667	1·5272	0·7000	2·4000	1·0000	0·7000	0·645	0·963	0·642	1·8340	0·3500
0·72	1·1803	1·5197	0·7200	2·4400	1·0000	0·7200	0·668	0·963	0·665	1·9132	0·3600
0·74	1·1935	1·5120	0·7400	2·4800	1·0000	0·7400	0·691	0·963	0·688	1·9935	0·3700
0·76	1·2063	1·5039	0·7600	2·5200	1·0000	0·7600	0·715	0·963	0·712	2·0748	0·3800
0·78	1·2187	1·4956	0·7800	2.5600	1·0000	0·7800	0·738	0·964	0·735	2·1573	0·3900
0·80	1·2308	1·4871	0·8000	2·6000	1·0000	0·8000	0·762	0·964	0·759	2·2408	0·4000

Rel. dpth Y	U.equ. dia. $D_{ep(u)}$ (m)	Equiv. disch. factor J	Unit sect. area A_u (m^2)	Unit wetted perim. P_u (m)	Unit surf. brdth $B_{s(u)}$ (m)	Unit mean depth $y_{m(u)}$ (m)	Disch. ratio $Q/Q_{1\cdot00}$ basic	Veloc. ratio V/V_{basic}	Disch. ratio $Q/Q_{1\cdot00}$ adjstd	U.crit. disch. $Q_{c(u)}$ (m^3s^{-1})	Unit depth cntrd $y_{d(u)}$ (m)
0·82	1·2424	1·4785	0·8200	2·6400	1·0000	0·8200	0·785	0·964	0·783	2·3253	0·4100
0·84	1·2537	1·4697	0·8400	2·6800	1·0000	0·8400	0·809	0·965	0·807	2·4109	0·4200
0·86	1·2647	1·4607	0·8600	2·7200	1·0000	0·8600	0·833	0·965	0·831	2·4975	0·4300
0·88	1·2754	1·4517	0·8800	2·7600	1·0000	0·8800	0·856	0·965	0·855	2·5851	0·4400
0·90	1·2857	1·4426	0·9000	2·8000	1·0000	0·9000	0·880	0·966	0·879	2·6738	0·4500
0·92	1·2958	1·4334	0·9200	2·8400	1·0000	0·9200	0·904	0·966	0·903	2·7634	0·4600
0·94	1·3056	1·4241	0·9400	2·8800	1·0000	0·9400	0·928	0·966	0·927	2·8540	0·4700
0·96	1·3151	1·4149	0·9600	2·9200	1·0000	0·9600	0·952	0·967	0·951	2·9456	0·4800
0·98	1·3243	1·4056	0·9800	2·9600	1·0000	0·9800	0·976	0·967	0·976	3·0381	0·4900
1·00	1·3333	1·3963	1·0000	3·0000	1·0000	1·0000	1·000	0·967	1·000	3·1316	0·5000
1·02	1·3421	1·3870	1·0200	3·0400	1·0000	1·0200	1·024	0·968	1·024	3·2260	0·5100
1·04	1·3506	1·3777	1·0400	3·0800	1·0000	1·0400	1·048	0·968	1·049	3·3213	0·5200
1·06	1·3590	1·3684	1·0600	3·1200	1·0000	1·0600	1·072	0·968	1·074	3·4176	0·5300
1·08	1·3671	1·3591	1·0800	3·1600	1·0000	1·0800	1·097	0·969	1·098	3·5148	0·5400
1·10	1·3750	1·3499	1·1000	3·2000	1·0000	1·1000	1·121	0·969	1·123	3·6128	0·5500
1·12	1·3827	1·3407	1·1200	3·2400	1·0000	1·1200	1·145	0·969	1·147	3·7118	0·5600
1·14	1·3902	1·3316	1·1400	3·2800	1·0000	1·1400	1·170	0·970	1·172	3·8117	0·5700
1·16	1·3976	1·3225	1·1600	3·3200	1·0000	1·1600	1·194	0·970	1·197	3·9124	0·5800
1·18	1·4048	1·3134	1·1800	3·3600	1·0000	1·1800	1·218	0·970	1·222	4·0141	0·5900
1·20	1·4118	1·3045	1·2000	3·4000	1·0000	1·2000	1·243	0·970	1·247	4·1165	0·6000
1·24	1·4253	1·2867	1·2400	3·4800	1·0000	1·2400	1·292	0·971	1·297	4·3241	0·6200
1·28	1·4382	1·2692	1·2800	3·5600	1·0000	1·2800	1·341	0·972	1·347	4·5350	0·6400
1·32	1·4505	1·2519	1·3200	3·6400	1·0000	1·3200	1·390	0·972	1·397	4·7492	0·6600
1·36	1·4624	1·2350	1·3600	3·7200	1·0000	1·3600	1·439	0·973	1·447	4·9667	0·6800
1·40	1·4737	1·2183	1·4000	3·8000	1·0000	1·4000	1·488	0·973	1·498	5·1874	0·7000
1·44	1·4845	1·2020	1·4400	3·8800	1·0000	1·4400	1·538	0·974	1·548	5·4113	0·7200
1·48	1·4949	1·1860	1·4800	3·9600	1·0000	1·4800	1·587	0·974	1·599	5·6384	0·7400
1·52	1·5050	1·1703	1·5200	4·0400	1·0000	1·5200	1·637	0·975	1·650	5·8685	0·7600
1·56	1·5146	1·1549	1·5600	4·1200	1·0000	1·5600	1·687	0·975	1·700	6·1016	0·7800
1·60	1·5238	1·1398	1·6000	4·2000	1·0000	1·6000	1·736	0·976	1·751	6·3378	0·8000
1·64	1·5327	1·1250	1·6400	4·2800	1·0000	1·6400	1·786	0·976	1·802	6·5770	0·8200
1·68	1·5413	1·1106	1·6800	4·3600	1·0000	1·6800	1·836	0·977	1·853	6·8191	0·8400
1·72	1·5495	1·0964	1·7200	4·4400	1·0000	1·7200	1·886	0·977	1·905	7·0640	0·8600
1·76	1·5575	1·0825	1·7600	4·5200	1·0000	1·7600	1·936	0·977	1·956	7·3119	0·8800
1·80	1·5652	1·0690	1·8000	4·6000	1·0000	1·8000	1·985	0·978	2·007	7·5626	0·9000
1·84	1·5726	1·0557	1·8400	4·6800	1·0000	1·8400	2·035	0·978	2·059	7·8160	0·9200
1·88	1·5798	1·0427	1·8800	4·7600	1·0000	1·8800	2·086	0·979	2·110	8·0723	0·9400
1·92	1·5868	1·0300	1·9200	4·8400	1·0000	1·9200	2·136	0·979	2·161	8·3313	0·9600
1·96	1·5935	1·0175	1·9600	4·9200	1·0000	1·9600	2·186	0·979	2·213	8·5930	0·9800
2·00	1·6000	1·0053	2·0000	5·0000	1·0000	2·0000	2·236	0·980	2·265	8·8574	1·0000
2·05	1·6078	0·9904	2·0500	5·1000	1·0000	2·0500	2·299	0·980	2·329	9·1916	1·0250
2·10	1·6154	0·9759	2·1000	5·2000	1·0000	2·1000	2·361	0·981	2·394	9·5299	1·0500
2·15	1·6226	0·9618	2·1500	5·3000	1·0000	2·1500	2·424	0·981	2·458	9·8723	1·0750
2·20	1·6296	0·9481	2·2000	5·4000	1·0000	2·2000	2·487	0·981	2·523	10·219	1·1000
2·25	1·6364	0·9347	2·2500	5·5000	1·0000	2·2500	2·550	0·982	2·588	10·569	1·1250
2·30	1·6429	0·9216	2·3000	5·6000	1·0000	2·3000	2·613	0·982	2·653	10·923	1·1500
2·35	1·6491	0·9089	2·3500	5·7000	1·0000	2·3500	2·676	0·982	2·718	11·281	1·1750
2·40	1·6552	0·8965	2·4000	5·8000	1·0000	2·4000	2·739	0·983	2·783	11·643	1·2000
2·45	1·6610	0·8844	2·4500	5·9000	1·0000	2·4500	2·802	0·983	2·848	12·009	1·2250
2·50	1·6667	0·8727	2·5000	6·0000	1·0000	2·5000	2·865	0·983	2·913	12·379	1·2500
2·55	1·6721	0·8612	2·5500	6·1000	1·0000	2·5500	2·929	0·983	2·975	12·752	1·2750
2·60	1·6774	0·8500	2·6000	6·2000	1·0000	2·6000	2·992	0·982	3·036	13·129	1·3000
2·65	1·6825	0·8390	2·6500	6·3000	1·0000	2·6500	3·055	0·981	3·098	13·509	1·3250
2·70	1·6875	0·8283	2·7000	6·4000	1·0000	2·7000	3·118	0·980	3·159	13·893	1·3500
2·75	1·6923	0·8179	2·7500	6·5000	1·0000	2·7500	3·181	0·979	3·220	14·281	1·3750

Annexure 2 - Change of grouping for correlation of part-full flow

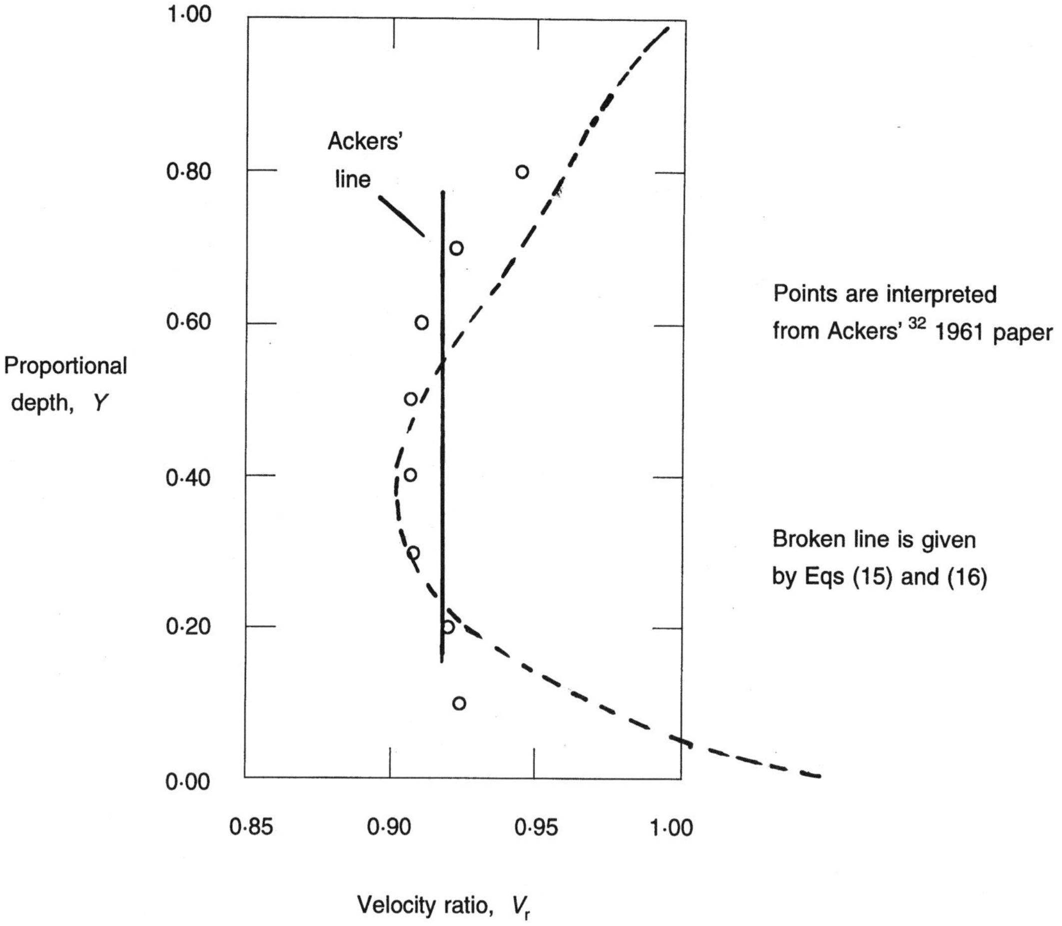

The Colebrook-White equation can be written

$$\frac{V}{\sqrt{(2SgD)}} = 2 \log \left[3{\cdot}71 \left\{ \frac{k_s}{D} + \frac{1}{416814\sqrt{S}D^{1{\cdot}5}} \right\}^{-1} \right]$$

where the RHS is fully adjusted to SI and standard temperature. The large number, 416814, is the consequence of this step. Then the bracketed expression on the RHS can be taken as θ.

Previously the following has been used for θ, although for a more restricted purpose.

$$\theta \approx \left\{ \frac{k_s}{D} + \frac{1}{3600\, D\, S^{1/3}} \right\}^{-1}$$

As with the first evaluation, the expression results from SI and standard water temperature.

In the case of the first evaluation, values of θ for the two Sterling and Knight[31] cases fall into a sequence with the value from Camp[28] and the corresponding values of V_r. This does not happen with the second evaluation. It is possible to correct, say to the 0·009 gradient case, by a constant of 20·1. But this does not give the same value for the 0·001 case as found satisfactory.

Using the first evaluation, the figure shows the extent of agreement obtained with data not used in calibration of the relationships.